List of the Elements with Their Atomic Symbols and Atomic Weights

Name	Symbol	Atomic Number	Atomic Weight	Name	Symbol	Atomic Number	Atomic Weight
Actinium	Ac	89	(227)*	Mendelevium	Md	101	(258)
Aluminum	Al	13	26.981538	Mercury	Hg	80	200.59
Americium	Am	95	(243)	Molybdenum	Mo	42	95.96
Antimony	Sb	51	121.760	Moscovium	Mc	115	(288)
Argon	Ar	18	39.948	Neodymium	Nd	60	144.242
Arsenic	As	33	74.92160	Neon	Ne	10	20.1797
Astatine	At	85	(210)	Neptunium	Np	93	(237)
Barium	Ba	56	137.327	Nickel	Ni	28	58.6934
Berkelium	Bk	97	(247)	Nihonium	Nh	113	(284)
Beryllium	Be	4	9.012182	Niobium	Nb	41	92.90638
Bismuth	Bi	83	208.98040	Nitrogen	N	7	14.0067
Bohrium	Bh	107	(272)	Nobelium	No	102	(259)
Boron	B	5	10.811	Oganesson	Og	118	(294)
Bromine	Br	35	79.904	Osmium	Os	76	190.23
Cadmium	Cd	48	112.411	Oxygen	O	8	15.9994
Calcium	Ca	20	40.078	Palladium	Pd	46	106.42
Californium	Cf	98	(251)	Phosphorus	P	15	30.973762
Carbon	C	6	12.0107	Platinum	Pt	78	195.094
Cerium	Ce	58	140.116	Plutonium	Pu	94	(244)
Cesium	Cs	55	132.90545	Polonium	Po	84	(209)
Chlorine	Cl	17	35.453	Potassium	K	19	39.0983
Chromium	Cr	24	51.9961	Praseodymium	Pr	59	140.90765
Cobalt	Co	27	58.933195	Promethium	Pm	61	(145)
Copernicium	Cn	112	(285)	Protactinium	Pa	91	231.03588
Copper	Cu	29	63.546	Radium	Ra	88	(226)
Curium	Cm	96	(247)	Radon	Rn	86	(222)
Darmstadtium	Ds	110	(281)	Rhenium	Re	75	186.207
Dubnium	Db	105	(268)	Rhodium	Rh	45	102.90550
Dysprosium	Dy	66	162.500	Roentgenium	Rg	111	(280)
Einsteinium	Es	99	(252)	Rubidium	Rb	37	85.4678
Erbium	Er	68	167.259	Ruthenium	Ru	44	101.07
Europium	Eu	63	151.964	Rutherfordium	Rf	104	(265)
Fermium	Fm	100	(257)	Samarium	Sm	62	150.36
Flerovium	Fl	114	(289)	Scandium	Sc	21	44.955912
Fluorine	F	9	18.998403	Seaborgium	Sg	106	(271)
Francium	Fr	87	(223)	Selenium	Se	34	78.96
Gadolinium	Gd	64	157.25	Silicon	Si	14	28.0855
Gallium	Ga	31	69.723	Silver	Ag	47	107.8682
Germanium	Ge	32	72.64	Sodium	Na	11	22.989769
Gold	Au	79	196.96657	Strontium	Sr	38	87.62
Hafnium	Hf	72	178.49	Sulfur	S	16	32.065
Hassium	Hs	108	(270)	Tantalum	Ta	73	180.9479
Helium	He	2	4.002602	Technetium	Tc	43	(98)
Holmium	Ho	67	164.93032	Tellurium	Te	52	127.60
Hydrogen	H	1	1.00794	Tennessine	Ts	117	(292)
Indium	In	49	114.818	Terbium	Tb	65	158.92535
Iodine	I	53	126.90447	Thallium	Tl	81	204.3833
Iridium	Ir	77	192.217	Thorium	Th	90	232.0381
Iron	Fe	26	55.845	Thulium	Tm	69	168.93421
Krypton	Kr	36	83.798	Tin	Sn	50	118.710
Lanthanum	La	57	138.9055	Titanium	Ti	22	47.867
Lawrencium	Lr	103	(262)	Tungsten	W	74	183.84
Lead	Pb	82	207.2	Uranium	U	92	238.02891
Lithium	Li	3	6.941	Vanadium	V	23	50.9415
Livermorium	Lv	116	(293)	Xenon	Xe	54	131.293
Lutetium	Lu	71	174.9668	Ytterbium	Yb	70	173.054
Magnesium	Mg	12	24.3050	Yttrium	Y	39	88.90585
Manganese	Mn	25	54.938045	Zinc	Zn	30	65.38
Meitnerium	Mt	109	(276)	Zirconium	Zr	40	91.224

*Values in parentheses are the mass numbers of the most common or longest lived isotopes of radioactive elements.

Periodic Table of the Elements

Main groups

1 1A	2 2A												13 3A	14 4A	15 5A	16 6A	17 7A	18 8A
1 **H** 1.00794																		2 **He** 4.002602
3 **Li** 6.941	4 **Be** 9.012182												5 **B** 10.811	6 **C** 12.0107	7 **N** 14.0067	8 **O** 15.9994	9 **F** 18.998403	10 **Ne** 20.1797
11 **Na** 22.989769	12 **Mg** 24.3050	3 3B	4 4B	5 5B	6 6B	7 7B	8	9 8B	10	11 1B	12 2B		13 **Al** 26.981538	14 **Si** 28.0855	15 **P** 30.973762	16 **S** 32.065	17 **Cl** 35.453	18 **Ar** 39.948
19 **K** 39.0983	20 **Ca** 40.078	21 **Sc** 44.955912	22 **Ti** 47.867	23 **V** 50.9415	24 **Cr** 51.9961	25 **Mn** 54.938045	26 **Fe** 55.845	27 **Co** 58.933195	28 **Ni** 58.6934	29 **Cu** 63.546	30 **Zn** 65.38		31 **Ga** 69.723	32 **Ge** 72.64	33 **As** 74.92160	34 **Se** 78.96	35 **Br** 79.904	36 **Kr** 83.798
37 **Rb** 85.4678	38 **Sr** 87.62	39 **Y** 88.90585	40 **Zr** 91.224	41 **Nb** 92.90638	42 **Mo** 95.96	43 **Tc** (98)	44 **Ru** 101.07	45 **Rh** 102.90550	46 **Pd** 106.42	47 **Ag** 107.8682	48 **Cd** 112.411		49 **In** 114.818	50 **Sn** 118.710	51 **Sb** 121.760	52 **Te** 127.60	53 **I** 126.90447	54 **Xe** 131.293
55 **Cs** 132.90545	56 **Ba** 137.327	71 **Lu** 174.9668	72 **Hf** 178.49	73 **Ta** 180.9479	74 **W** 183.84	75 **Re** 186.207	76 **Os** 190.23	77 **Ir** 192.217	78 **Pt** 195.094	79 **Au** 196.96657	80 **Hg** 200.59		81 **Tl** 204.3833	82 **Pb** 207.2	83 **Bi** 208.98040	84 **Po** (209)	85 **At** (210)	86 **Rn** (222)
87 **Fr** (223)	88 **Ra** (226)	103 **Lr** (262)	104 **Rf** (265)	105 **Db** (268)	106 **Sg** (271)	107 **Bh** (272)	108 **Hs** (270)	109 **Mt** (276)	110 **Ds** (281)	111 **Rg** (280)	112 **Cn** (285)		113 **Nh** (284)	114 **Fl** (289)	115 **Mc** (288)	116 **Lv** (293)	117 **Ts** (292)	118 **Og** (294)

Transition metals

Main groups

* Lanthanide series

57 *La 138.9055	58 **Ce** 140.116	59 **Pr** 140.90765	60 **Nd** 144.242	61 **Pm** (145)	62 **Sm** 150.36	63 **Eu** 151.964	64 **Gd** 157.25	65 **Tb** 158.92535	66 **Dy** 162.500	67 **Ho** 164.93032	68 **Er** 167.259	69 **Tm** 168.93421	70 **Yb** 173.054

† Actinide series

89 †Ac (227)	90 **Th** 232.0381	91 **Pa** 231.03588	92 **U** 238.02891	93 **Np** (237)	94 **Pu** (244)	95 **Am** (243)	96 **Cm** (247)	97 **Bk** (247)	98 **Cf** (251)	99 **Es** (252)	100 **Fm** (257)	101 **Md** (258)	102 **No** (259)

CHEMISTRY

EIGHTH EDITION

JILL K. ROBINSON

Indiana University

JOHN E. MCMURRY

Cornell University

ROBERT C. FAY

Cornell University

 Pearson

Director of Portfolio Management: Jeanne Zalesky
Executive Courseware Portfolio Manager: Terry Haugen
Content Producer: Shercian Kinosian
Managing Producer: Kristen Flathman
Courseware Director, Content Development: Barbara Yien
Courseware Analysts: Cathy Murphy, Coleen Morrison, Jay McElroy
Courseware Editorial Assistant: Harry Misthos
Rich Media Content Producers: Jenny Moryan, Ziki Dekel
Director MasteringChemistry Content Development: Amir Said
MasteringChemistry Senior Content Producer: Margaret Trombley
MasteringChemistry Content Producers: Meaghan Fallano, Kaitlin Smith
Full-Service Vendor, Project Manager: Pearson CSC, Kelly Murphy
Copyeditor: Pearson CSC
Compositor: Pearson CSC
Art House, Coordinator: Lachina, Rebecca Marshall
Design Manager: Maria Guglielmo Walsh
Interior & Cover Designer: Gary Hespeneide
Rights & Permissions Manager: Ben Ferrini
Rights & Permissions Project Manager: Pearson CSC, Eric Schrader
Rights & Permissions Specialist/Photo Researcher: Pearson CSC, Angelica Aranas
Manufacturing Buyer: Stacey Weinberger
VP, Director of Field Marketing: Tim Galligan
Director of Product Marketing: Allison Rona
Executive Field Marketing Manager: Christopher Barker
Senior Product Marketing Manager: Elizabeth Bell
Cover Photo Credit: Beauty of Science/Science Source

Library of Congress Cataloging-in-Publication Data

Names: Robinson, Jill K. | McMurry, John. | Fay, Robert C., 1936-
Title: Chemistry / Jill K. Robinson (Indiana University), John E. McMurry
  (Cornell University), Robert C. Fay (Cornell University).
Description: Eighth edition. | Hoboken, NJ : Pearson Education, Inc., [2020]
Identifiers: LCCN 2018053050 | ISBN 9780134856230 (casebound)
Subjects: LCSH: Chemistry--Textbooks.
Classification: LCC QD33.2 .M36 2020 | DDC 540--dc23
LC record available at https://lccn.loc.gov/2018053050

5 2019

www.pearson.com

ISBN 10:   0-134-85623-6
ISBN 13:   978-0-134-85623-0 (Student edition)
ISBN 10:   0-135-21012-7
ISBN 13: 978-0-135-21012-3 (Looseleaf Edition)

Brief Contents

Contents

19   Electrochemistry   813

20   Nuclear Chemistry   870

21   Transition Elements and Coordination Chemistry   904

22   The Main-Group Elements   954

List of Interactive Videos ▤

About the Authors

Jill K. Robinson received her Ph.D. in analytical and atmospheric chemistry from the University of Colorado at Boulder. She is a senior lecturer at Indiana University and teaches general, analytical, and environmental chemistry courses. Her clear and relatable teaching style has been honored with several awards including the President's Award for Distinguished Teaching at Indiana University and the J. Calvin Giddings Award for Excellence in Education from the American Chemical Society Division of Analytical Chemistry. She leads workshops to help faculty transition from lecture-based instruction to student-centered pedagogies.

John McMurry, educated at Harvard and Columbia, has taught more than 20,000 students in general and organic chemistry over a 40-year period. An emeritus professor of chemistry at Cornell University, Dr. McMurry previously spent 13 years on the faculty at the University of California at Santa Cruz. He has received numerous awards, including the Alfred P. Sloan Fellowship (1969–71), the National Institute of Health Career Development Award (1975–80), the Alexander von Humboldt Senior Scientist Award (1986–87), and the Max Planck Research Award (1991).

Robert C. Fay, professor emeritus at Cornell University, taught general and inorganic chemistry at Cornell for 45 years beginning in 1962. Known for his clear, well-organized lectures, Dr. Fay was the 1980 recipient of the Clark Distinguished Teaching Award. He has also taught as a visiting professor at Harvard University and the University of Bologna (Italy). A Phi Beta Kappa graduate of Oberlin College, Dr. Fay received his Ph.D. from the University of Illinois. He has been an NSF Science Faculty Fellow at the University of East Anglia and the University of Sussex (England) and a NATO/Heineman Senior Fellow at Oxford University.

Preface

FOR THE STUDENT

Francie came away from her first chemistry lecture in a glow. In one hour she found out that everything was made up of atoms which were in continual motion. She grasped the idea that nothing was ever lost or destroyed. Even if something was burned up or rotted away, it did not disappear from the face of the earth; it changed into something else—gases, liquids, and powders. Everything, decided Francie after that first lecture, was vibrant with life and there was no death in chemistry. She was puzzled as to why learned people didn't adopt chemistry as a religion.

—**Betty Smith,** *A Tree Grows in Brooklyn*

We know that not everyone has such a breathless response to their chemistry lectures, and few would mistake chemistry as a religion, yet chemistry *is* a subject with great logical beauty. We love chemistry because it explains the "why" behind many observations of the world around us and we use it every day to help us make informed choices about our health, lifestyle, and politics. Moreover, chemistry is the fundamental, enabling science that underlies many of the great advances of the last century that have so lengthened and enriched our lives. Chemistry provides a strong understanding of the physical world and will give you the foundation you need to go on and make important contributions to science and humanity.

HOW TO USE THIS BOOK

You no doubt have experience using textbooks and know they are not meant to read like a novel. We have written this book to provide you with a clear, cohesive introduction to chemistry in a way that will help you, as a new student of chemistry, understand and relate to the subject. While you *could* curl up with this book, you will greatly benefit from continually formulating questions and checking your understanding as you *work* through each section. The way this book is designed and written will help you keep your mind active, thus allowing you to digest important concepts as you learn some of the many principles of chemistry.

The 8th edition was revised to create an **interactive** study cycle based on research of effective learning methods. Many common study habits such as highlighting, rereading, and long study sessions create the illusion of fast progress, but these gains fade quickly. More deep and durable learning occurs from self-testing, difficulty in practice, and spaced practice of different skills. Let's see how specific steps in the study cycle use proven strategies to maximize your learning.

Step 1. Learning New Material

The 8th edition eText contains many new interactive features (Big Idea Questions, Interactive Worked Examples, Practice problems, and Figure It Out Questions) that should be used to quiz yourself and receive feedback as you work through the material in each chapter.

- **Narrative:** As you read through the text, always challenge yourself to understand the "why" behind the concept. For example, you will learn that carbon forms four bonds, and the narrative will give the reason why. By gaining a conceptual understanding, you will *not need to memorize* a large collection of facts, making learning and retaining important principles much easier! *Big Idea Questions* were written to help you digest and apply the most important concepts. In the printed book, these questions appear in the margins, and in the eText, they are multiple choice questions with feedback to help you identify common mistakes.

- **Figures:** Figures are not optional! Most summarize and convey important points. *Figure It Out* **Questions** draw your attention to a key principle and provide guidance in interpreting graphs. Answer the question by examining the figure and perhaps rereading the related narrative. We've provided answers to Figure It Out Questions near the figure in the printed book and use an interactive hide-and-reveal feature in the eText.

- **Worked Examples:** Numerous worked examples throughout the text show the approach for solving a certain type of problem. Each worked example uses a step-by-step procedure.

  - **Identify**—The first step in problem solving is to identify key information and classify it as a known or unknown quantity. This step also involves translating between words and chemical symbols. Listing knowns on one side and unknowns on the other organizes the information and makes the process of identifying the correct strategy more visual. The *Identify* step is used in numerical problems.

  - **Strategy**—The strategy describes how to solve the problem without actually solving it. Failing to articulate the needed strategy is a common pitfall; too often students start manipulating numbers and variables without first identifying key equations or making a plan. Articulating a strategy will develop conceptual understanding and is highly preferable to simply memorizing the steps involved in solving a certain type of problem.

  - **Solution**—Once the plan is outlined, the key information is used to answer the question.

  - **Check**—A problem is not completed until you have thought about whether the answer makes sense. Use both your practical knowledge of the world and knowledge of chemistry to evaluate your answer. For example, if heat is added to a sample of liquid water and you are asked to calculate the final temperature, you should critically consider your answer: Is the final temperature lower than the original? Shouldn't adding heat raise the temperature? Is the new temperature above 100 °C, the boiling point of water? The *Check* step is used in problems when the magnitude and sign of a number can be estimated or the physical meaning of the answer verified based on familiar observations.

To test your mastery of the concept explored in Worked Examples, two problems will follow. **PRACTICE** problems are similar in style and complexity to the Worked Example and will test your basic understanding. Interactive Practice Problems are available in the eText and have answer-specific feedback to help you identify common mistakes.

Once you have correctly completed this problem, tackle the **APPLY** problem, in which the concept is used in a new situation to assess a deeper understanding of the topic. Answers to Apply Problems can be found at the end of the book or by using the hide-and-reveal feature in the eText.

- **Interactive Worked Examples:** Each chapter has two video tutorials for challenging problems that model the process of expert thinking. The videos are interactive and ask you to make predictions before moving forward to the complete solution.

- **Conceptual Problems:** Conceptual understanding is a primary focus of this book. Conceptual problems are intended to help you with the critical skill of visualizing the structure and interactions of atoms and molecules while probing your understanding of key principles rather than your ability to correctly use numbers in an equation. The time you spend mastering these problems will provide high long-term returns by solidifying main ideas.

Step 2. Problem-Solving Practice

We achieve more complex and long-lasting learning by practicing problems that require more effort and slow down the pace of learning.

- **End-of-Chapter Problem Sets:** Working problems is essential for success in chemistry! The number and variety of problems at the end of chapter will give you the practice needed to gain mastery of specific concepts. Answers to every other problem are given in the "Answers" section at the back of the book so that you can assess your understanding. Your instructor may assign problems in an online format using the Mastering™ Chemistry platform, which comes with the added benefit of tutorials, feedback, and links to relevant content in the eText.

Step 3. Mastery

Once you have read the chapter and completed the end-of-chapter problems, you will need to review for the exam and assess which topics you have mastered and which still need to be solidified. Inquiry sections and practice tests are chapter capstones that strengthen mental representations by replaying learning and giving it meaning.

- **Inquiries:** Inquiry sections connect chemistry to the world around you by highlighting useful links in the future careers of many science students. Typical themes are materials, medicine, and the environment. The goal of these sections is to deepen your understanding and aid in retention by tying concepts to memorable applications. These sections can be considered as a capstone for each chapter because *Inquiry* problems review several main concepts and calculations. These sections will also help you prepare for professional exams because they were written in the same style as new versions of these exams: a passage of text describing an application followed by a set of questions probing your ability to apply basic scientific concepts to the situation.

- **End-of-Chapter Practice Test and Study Guide:** The end-of-chapter practice test and study guide are useful tools for exam preparation. Each practice test question is linked to a learning objective in the study guide. If you answer a question incorrectly or want more practice on that skill, refer to the study guide, which matches the learning objective to a concept summary, key skills for solving the problem, Worked Examples for assistance, and end-of-chapter problems so that you can practice your mastery of that skill.

For Instructors

NEW TO THIS EDITION

A primary change in the 8th edition is the development of an interactive learning environment. We designed interactive features for the text and classroom based on educational research and strategies proven to help students succeed. Features that help students read a science text and prepare for exams are available for self-assessment in the printed text but are most effectively implemented in the eText. Big Idea Questions, Interactive Worked Examples, and Practice problems have multiple-choice options with answer-specific feedback targeting common mistakes and misconceptions. The eText includes more than 1,000 new interactive features, and students can assess their understanding by answering a question with feedback every one to two pages. In addition to an interactive eText, questions have been developed to help instructors engage students during class using Learning Catalytics, a personal response system used with smart devices. A large body of educational literature has clearly demonstrated increased learning gains, higher attendance, and lower failure rates in classrooms that employ active learning. New interactive features include:

1. **Interactive Big Idea Questions:** Efficient and skilled reading requires students to parse out main ideas and important details and relate new information to prior knowledge. Big Idea Questions probe understanding of important concepts from a text passage. These questions teach students how to actively read a science text by modeling the kinds of questions they should ask themselves and stimulate them to make connections between concepts and mathematical problems. These questions can be found in the margin in the printed text and are multiple-choice questions with specific wrong-answer feedback in the eText.

2. **Interactive Figure It Out Questions:** These questions test knowledge of key principles shown in a figure and the ability to read and interpret graphs. Answers to Figure It Out Questions are provided near the figure in the printed book and use a hide-and-reveal feature in the eText with answer-specific feedback.

3. **Interactive Worked Examples:** Each chapter has two video tutorials featuring lead author Jill Robinson as she models the process of expert problem solving. The videos require students to pause and digest information and then predict how to proceed at key points before moving forward to the complete solution.

4. **Interactive PRACTICE Problems:** These problems follow a Worked Example and test basic understanding. Answers to Practice Problems are provided at the end of the printed book and are multiple-choice questions with specific wrong-answer feedback in the eText. For example, the feedback for Practice problems in the eText provides an opportunity to give remediation in the mathematical operations including the quadratic equation. All steps in solving the algebraic expressions are shown to help students who may need a review.

5. **Interactive APPLY Problems:** These problems follow the Practice Problems and discourage a plug-and-chug approach to problem solving by providing an example of how the same principle can be used in different types of problems with different levels of complexity. Answers to Apply problems are provided at the end of the printed book and use a hide-and-reveal feature in the eText.

6. **Interactive Practice Test Linked to Study Guide:** A useful way for students to review each chapter is by taking the Practice Test, which assesses mastery of chapter learning objectives. The Study Guide provides a targeted follow-up to the Practice Test through the linking of learning objectives to the main lessons in each chapter, associated worked examples, and end-of-chapter problems for more practice. When a

student answers incorrectly in Mastering Chemistry or the eText, the Practice Test automatically links to worked examples and additional practice problems.

7. **Interactive Learning Catalytics Questions:** The Learning Catalytics questions developed for each chapter promote strong conceptual understanding and advanced problem-solving skills. Learning Catalytics includes prebuilt questions for every key topic in chemistry written by lead author Jill Robinson.

Inquiry Sections have been updated and integrated conceptually into each chapter.

Inquiry sections highlight the importance of chemistry, promote student interest, and deepen students' understanding of the content. The Inquiry sections include problems that revisit several chapter concepts and can be covered in class or recitation sections or assigned as homework in Mastering Chemistry. In the 8th edition, the delivery of Inquiry problems in Mastering Chemistry has been improved and new topics have been developed. New Inquiries for the 8th edition are:

- Chapter 2: How can measurements of oxygen and hydrogen isotopes determine past climates?
- Chapter 3: How is the principle of atom economy used to minimize waste in a chemical synthesis?
- Chapter 8: Which is better for human health, natural or synthetic vitamins?
- Chapter 10: How do inhaled anesthetics work?
- Chapter 12: What are quantum dots, and what controls their color?
- Chapter 14: How do enzymes work?
- Chapter 15: How does high altitude affect oxygen transport in the blood?
- Chapter 20: How are radioisotopes used in medicine?

NEW! End-of-chapter problems continually build on concepts and skills from earlier in the chapter.

Educational research shows that interleaved and varied practice with different concepts and skills produces higher learning gains than drilling on a single topic. Section Problems at the end of the chapter now include questions that build on concepts taught earlier in the chapter. In previous editions, Section Problems focused only on learning objectives from that specific section in the text. New questions and questions from the Chapter Problems sections in previous editions that integrate multiple chapter concepts have been incorporated into Section Problems to revisit key ideas on a regular basis and apply them in different situations.

Here is a list of some of the key chemistry content changes made in each chapter:

Chapter 1 Chemical Tools: Experimentation and Measurement

- The scientific method is described in the context of a new case study in the field of nanoscience to help students see the utility of chemistry in solving important world problems.
- Nanotechnology Inquiry problems were updated to promote better understanding of the unique properties of matter on the nanoscale and the size of nanoparticles.
- Figure 1.8 was updated to show the most commonly used laboratory glassware.

Chapter 2 Atoms, Molecules, and Ions

- Several updates to terminology and the periodic table were made. The names of recently discovered elements 113, 115, 117, and 118 were officially assigned in 2016 and listed in Section 2.1 Chemistry and the Elements. A clarification about the definition and common use of the term *atomic mass unit* was added. The atomic mass unit (*amu*) is an obsolete unit, but it is commonly used interchangeably with the correct unit, unified atomic mass unit (*u*). Since 2011, the Union of Pure and Applied Chemistry gives the atomic weights for some elements as a range of values instead of a single value due to isotopic abundances that vary with the source of the sample.
- Section 2.10 Measuring Atomic Weight: Mass Spectrometry was added to describe how atomic weights are experimentally measured. The process of using a mass spectrum to calculate an atomic weight is described in a Worked Example, and follow-up problems and new end-of-chapter problems were written. The description of a mass spectrometer from Chapter 3 was moved into Chapter 2 because it is the instrument used to measure atomic weight.

- In Section 2.12 Ions and Ionic Bonds, additional details on writing formulas for ionic compounds were added for clarification.

- A new Inquiry on isotopes and the climate record provides a strong connection with the Chapter 2 topics of isotopes, atomic weight, and the mole concept.

Chapter 3 Mass Relationships in Chemical Reactions

- Chemical Arithmetic: Stoichiometry was a very long section and contained many concepts. It has been divided into two sections: Section 3.3 Molecular Weight and Molar Mass and Section 3.4 Stoichiometry: Relating Amounts of Reactants and Products.

- A new Inquiry on atom economy concisely summarizes the important concept of relating amounts of reactants and products and introduces green chemistry.

- The section on measuring molecular weight was revised because the mass spectrometer was previously described in Chapter 2 in the section on atomic weight.

Chapter 4 Reactions in Aqueous Solution

- Added a Remember note in the margin at the beginning of Section 4.3 Electrolytes in Aqueous Solution to remind students about the differences between molecules and ions.

- Section 4.7 Acids, Bases, and Neutralization Reactions: Added a Looking Ahead note regarding acids/bases coverage in Chapter 16. Also, added the dissociation reaction for sodium hydroxide and barium hydroxide when discussing strong and weak bases.

- More explanation added to Worked Example 4.12 to help students assign oxidation numbers.

- Section 4.11 Identifying Redox Reactions: New figure shows that silver-colored powdered iron is oxidized by oxygen to produce iron(III)oxide, which is red in color.

Chapter 5 Periodicity and Electronic Structure of Atoms

Chapter 5 contains abstract ideas such as particles behaving as waves and the notion of wave functions of electrons. Eight new figures and descriptive text were added to help students grasp these difficult concepts.

- In Section 5.1 Wave Properties of Radiant Energy and the Electromagnetic Spectrum, the double-slit experiment was described to show that both light and matter have wave properties. New Figure 5.4: Diffraction and interference are phenomena exhibited by waves. New Figure 5.5: Radiant energy exhibits wave properties in a double-slit experiment.

- Section 5.3 Atomic Line Spectra and Quantized Energy: The connection between quantized energy and atomic line spectra was strengthened by condensing content and placing both concepts into the same section. Also, radial distribution plots were added to help visualize the meaning of an orbital and explain electron shielding and the ordering of orbital energies.

- Section 5.4 Wavelike Properties of Matter: de Broglie's Hypothesis: New Figure 5.11: Wave properties of electrons illustrate the different behaviors of particles and waves in a double-slit experiment. Figure also shows that electrons have wave properties, which is a key idea for understanding orbitals.

- Added an electron microscope image that shows individual DNA molecules to illustrate the utility of the wave properties of an electron in Worked Example 5.5 and Apply problem 5.10.

- Section 5.7 The Shapes of Orbitals: New Figure 5.14: Representations of a 1s orbital. New Figure 5.15: Concert hall analogy for radial probability. A figure was added to help explain the concept of radial probability in a familiar way.

- New Figure 5.18: Radial probability plots for the 1s, 2s, and 3s orbitals in a hydrogen atom. Radial probability plots are a useful way to explain the differences in size, energy, and number of nodes for the different s orbitals.

- Section 5.9 Orbital Energy Levels in Multielectron Atoms: New Figure 5.23: Radial distribution plots for 3s, 3p, and 3d orbitals. The penetration of the different orbitals determines the ordering of orbital energies ($3s < 3p < 3d$).

- Section 5.10 Electron Configurations of Multielectron Atoms: New Figure 5.24: Energy levels of orbitals in multielectron atoms was placed in the margin for easy reference when writing electron configurations.

Chapter 6 Ionic Compounds: Periodic Trends and Bonding Theory

- Section 6.1 Electron Configurations of Ions: Added text and a figure to make it more clear why ns electrons are lost before $(n - 1)d$ electrons when forming transition metal ions. A relatively recent article in the *Journal of Chemical Education* describes how many textbooks contain incomplete or inaccurate discussions of this topic. The d orbital collapse for transition metals was described as concisely has possible. (Reference: The Full Story of the Electron Configurations of the Transition Elements, J. Chem Ed., Vol 87, No. 4, April 2010)

- Modified Figure 6.6 so negative electron affinities appear below zero on the graph.

- In the reactions in the Born-Haber cycle, the energy of the reaction is written in units of kJ, not kJ/mol. Figures 6.7 and 6.8 were updated to reflect the change.

- Updated Inquiry questions on ionic liquids.

Chapter 7 Covalent Bonding and Electron-Dot Structures

- Electronegativity was defined earlier in the section to more clearly explain the existence of polar covalent bonds. Electrostatic potential maps of Cl_2, HCl, and $NaCl$ were combined into one figure for comparison and to relate the extent of electron transfer to differences in electronegativity between the elements in the bond.

- Added the topics of dipole moment and percent ionic character to illustrate the extent of electron transfer as a continuum instead of as a sharp cutoff between a polar covalent bond and an ionic bond. A new Worked Example and new Practice and Apply problems were added. End-of-chapter problems were added as well. The content on percent ionic character was moved from Chapter 8 to Chapter 7 because it is much more relevant in this section.

- New Looking Ahead note about intermolecular forces in Section 7.4 A Comparison of Ionic and Covalent Compounds.

- Revised Inquiry Questions.

Chapter 8 Covalent Compounds: Bonding Theories and Molecular Structure

- Developed a new style for representing orbitals in all figures to more clearly show orbital overlap to form chemical bonds in valence bond theory.

- Clarified answer key for orbital overlap diagrams. Terminal atoms that have multiple bonds use the hybrid orbital model.

- References added to help students/instructors learn more about the vague statement "main-group compounds with five and six charge clouds use a more complex bonding pattern that is not easily explained by valence bond theory." The reference appears as a footnote. Some books report that main-group atoms that expand their octets use sp^3d or sp^3d^2 hybrid orbitals, which is not considered an accurate representation based on density functional theory calculations.

- The quantitative aspects of dipole moments were moved to Chapter 7 to help students better understand the differences between a nonpolar covalent bond, polar covalent bond, and ionic bond. A qualitative discussion of dipole moments of molecules is sufficient for Chapter 8 and is aligned with how instructors cover this topic.

- Changed the order of presentation of the different types of intermolecular forces. We now start with London dispersion forces because all molecules have these types of forces. We then get more restrictive and describe polar molecules with dipole-dipole forces, followed by hydrogen bonding, which is more restrictive and a special case of dipole-dipole forces. Finally, ion-dipole is described. The ordering of presentation of forces is from weakest to strongest.

- New Inquiry topic on the difference between natural and synthetic compounds such as vitamins.

Chapter 9 Thermochemistry: Chemical Energy

- A new chapter introduction was written to better connect chapter topics to examples familiar to students.

- Improved the strategy for solving constant-pressure calorimetry problems in Worked Example 9.6.

- Changed the way constant-volume calorimetry was presented to more accurately reflect the way this type of experiment was carried out in the laboratory. A new Worked Example (9.7)

and follow-up problems were written. End-of-chapter problems were revised to fit with this pedagogy.

- Section 9.11 on fossil fuels was removed. This section did not teach any new chemistry content, and the Inquiry on biofuels serves to connect thermochemistry concepts to fuels.

Chapter 10 Gases: Their Properties and Behavior

- Changed formulas for Graham's Law in Section 10.7 Gas Diffusion and Effusion: Graham's Law to replace mass (m) with molar mass (M).

- Removed the section on pollution to shorten the chapter. Most instructors do want to cover some relevant topic about the atmosphere, and the climate change section was improved. Figures on greenhouse gases and climate change were updated to include data from years since the last revision.

- New Inquiry on inhaled anesthetics.

Chapter 11 Liquids and Phase Changes

- The focus of Chapter 11 is on liquids, their properties, and phase changes. The topics of solids and unit cells have been moved to Chapter 12 on solids and solid-state materials.

- A new section on liquid crystals and end-of-chapter problems have been added.

Chapter 12 Solids and Solid-State Materials

- The topics of unit cells of solids and solid-state materials are closely related and are now contained in one chapter. (Chapters 11 and 21 content from the 7th edition is combined to make one coherent unit on solids.)

- Revised Inquiry on quantum dots.

Chapter 13 Solutions and Their Properties

- Added a new figure to show the difference between a solution and colloid using light-scattering properties.

- Divided Section 12.2 from the 7th edition into two new sections to improve the description of the solution-making process.

- Section 13.2 Enthalpy Changes and the Solution Process focuses on describing the intermolecular forces involved in solution formation and the overall effect on the heat of solution.

- New Figure 13.1: A molecular view of the solution making process.

- Section 13.3 Predicting Solubility relates the thermodynamic value of ΔG to the simple rule for solubility "like dissolves like."

- Added a paragraph to Section 13.5 Some Factors That Affect Solubility to explain why increasing temperature increases the solubility of solids but decreases the solubility of gases. A new Big Idea Question highlights this concept.

- Added a figure and description in Section 13.7 Vapor-Pressure Lowering of Solutions: Raoult's Law to illustrate ion pairing and explain why the dissociation of ionic compounds is not complete.

- Section 12.9 from the 7th edition on the fractional distillation of mixtures was deleted. There is already a lot of difficult material in this chapter, and this topic is not covered in most general chemistry courses.

Chapter 14 Chemical Kinetics

- Revised Figure 14.2 and text description to more clearly show how the instantaneous rate is determined from experimental data.

- Worked Example 14.8 (to replace 13.8) was revised to focus on the main idea of calculating half-life and not have students get lost in the details by referring to previous graphs.

- New analogy for rate-limiting step in Section 14.11 Rate Laws for Overall Reactions.

- New Inquiry on enzyme kinetics.

- Data in numerous end-of-chapter problems involving graphing were revised.

Chapter 15 Chemical Equilibrium

- Figure 15.1 was revised to show a macroscale and molecular scale representation of the N_2O_4/NO_2 equilibrium. This figure provides a picture of the data in the concentration versus time graphs in Figures 15.2 and 15.3.

- The feedback for practice problems in the eText provides an opportunity to give remediation in the mathematical operations including the quadratic equation. All steps in solving the algebraic expressions are shown to help students who may need a review.

- Inquiry focus was changed from the general concept of the equilibrium reaction of oxygen and hemoglobin to the more specific focus of the effect of altitude on oxygen supply in muscles.

Chapter 16 Aqueous Equilibria: Acids and Bases

- The procedure for solving acid-base equilibrium problems was reduced from eight steps to five steps, which are simpler to understand. All subsequent worked examples in Chapters 16 and 17 were modified using the new procedure. Figure 16.7 and the description of solving acid–base problems were revised to eliminate wording that was unusual and confusing. Examples are "big" concentrations and "small" concentrations.

- A photo sequence showing the pH change when CO_2 dissolves to produce carbonic acid was added to Worked Example 16.11.

- The Inquiry section was updated to discuss current problems related to acid rain.

Chapter 17 Applications of Aqueous Equilibria

- Section 17.2 The Common-Ion Effect was revised in three ways. The concept of the common-ion effect was presented before mathematical calculations to give students an understanding of the main idea first. Calculating the pH of a weak acid and conjugate base mixture was modified to follow the new simplified approach to solving equilibrium problems given in Figure 16.7. Two example calculations that were repetitive were combined into one example in Worked Example 17.2.

- Section 17.3 Buffer Solutions was rearranged to present the concept of a buffer before showing the calculation of pH change of a buffer upon addition of a strong acid or base. Figure 17.3 describes a buffer by showing pH change after adding a strong base to two different solutions: a strong acid and a buffer. The color change of an acid–base indicator shows that the buffer resists changes in pH. A conceptual Big Idea Question was created on the definition of a buffer.

- The Inquiry section on ocean acidification was updated with recent CO_2 and pH measurements. The problems were revised to promote understanding of the problem and for clarity.

Chapter 18 Thermodynamics: Entropy, Free Energy, and Spontaneity

- The introductory paragraph was revised to include familiar examples to students and review the concepts of reaction direction and extent of reaction.

- Two new figures were created to clarify the question in Worked Example 18.2 on calculating entropy.

- A more realistic example of a process that represents the standard free-energy change was described in Section 18.8 Standard Free-Energy Changes for Reactions.

Chapter 19 Electrochemistry

- In Section 19.1 Balancing Redox Reactions by the Half-Reaction Method, a brief review of oxidation numbers was added that includes a Remember note, a new figure showing oxidation numbers in redox reaction, and a Big Idea Question for students to assess themselves on this important concept from Chapter 4.

- Figure 19.1 showing the steps needed for balancing redox reactions by the half-reaction method was revised to make the individual steps clearer.

- New Worked Example 19.1 (Balancing a Redox Reaction in Acidic Solution): From the previous edition more detail was included so students can more easily follow the steps and canceling process when adding half-reactions.

- Revised Worked Example 19.2 (Balancing a Redox Reaction in Basic Solution): Added more detail so students can more easily follow the steps and canceling process when adding half-reactions.

- It is a convention in electrochemistry to put the anode half-cell on the left and cathode half-cell on the right. Several figures were changed to reflect this common convention.

- Worked Example 19.6 was revised to more clearly show the thought process for determining strengths of reducing agents.
- New Worked Example 19.8 was added on the very important concept of calculating voltage of a galvanic cell (a battery).
- The Inquiry was updated with recent status of commercialization of fuel-cell vehicles.

Chapter 20 Nuclear Chemistry

- In Section 20.3 Nuclear Stability, superheavy elements 113, 115, 117, and 118 were added to the periodic table. The discovery of these elements was connected to nuclear theory and the island of stability.
- In Section 20.3 Nuclear Stability, real examples of nuclear equations were provided instead of general equations to more clearly show how radioactive decay processes affect the neutron to proton ratio.
- Section 20.5 Dating with Radioisotopes was given its own section. The age of artifacts such as the Dead Sea Scrolls were updated based on improved methods of radiocarbon dating. The method of reporting artifact age using the term "Before Present (BP)" with the reference year 1950 was removed because it adds an extra step and is potentially confusing. The age of the object is now reported in the more conventional method of the time frame when the artifact was living. End-of-chapter problems were revised to match this change.
- In Section 20.7 Nuclear Fission and Fusion, Figure 20.9, which provides information on the number of nuclear reactors and nuclear power output worldwide, was updated.
- In Section 20.8 Nuclear Transmutation, information about the nuclear transformation reactions used in the synthesis of new elements $Z = 113–118$ was added, and new problems were written on this topic.
- New Inquiry topic: How are radioisotopes used in medicine? The previous text section was updated and expanded with some recent advances in nuclear medicine such as boron neutron capture therapy.

Chapter 21 Transition Elements and Coordination Chemistry

- Section 20.4 Chemistry of Selected Transition Elements was removed because it did not cover any new chemistry concepts and involved memorization of specific reactions that would not be retained easily. This content is this section is not needed to understand the main concepts of transition metal chemistry such as the color and magnetic properties of complexes.
- Modified Figure 21.9 to label the chelate ring discussed in the text description and added a Figure It Out Question in order to identify a chelate ring.

- Figure 21.24 showing colors of nickel complexes was moved next to text describing the accompanying crystal field diagrams. A description of the connection between the crystal field energy diagrams and the observed color of the complexes was added.
- The section Valence Bond Theory of Coordination Complexes is now placed at the end of the chapter to strengthen the connection between the color of coordination compounds and crystal field theory. The key terms *high-spin* and *low-spin complex* are now defined based on crystal field theory instead of valence bond theory.
- Also, crystal field theory was developed before valence bond theory. The text was modified to reiterate how crystal field theory is different from bonding theories based on quantum mechanics. (Also, many books do not cover valence bond theory of coordination complexes, so placing it last gives instructors the option to omit it.)

Chapter 22 The Main-Group Elements

- The chemistry of each main group was merged into its own section and the content trimmed to avoid excessive memorization.
- Continued emphasis on relating main-group chemistry to previous topics in the book such as periodic trends, bonding, structure, equilibrium, and acid-base chemistry. New end-of-chapter problems were written with emphasis on reviewing important chemical principles.

Chapter 23 Organic and Biological Chemistry

- Section 23.3 Naming Organic Compounds, was removed because the focus of the chapter is on bonding and structure, and naming is not needed to address these topics.
- In Section 23.1 Organic Molecules and Their Structures: Constitutional Isomers on organic molecules and their structures, the concept of constitutional isomers (instead of simply isomers) was stressed. This allows other important types of isomers such as enantiomers and cis-trans isomers to be distinguished and addressed in later sections.
- Unnumbered figure of 2-methylbutane was revised to more clearly show the zigzag structure of the carbon chain, which serves as the basis for organic line drawings.
- New Section 23.2 Stereoisomers: Chiral Molecules. Chirality is an extremely important concept with organic molecules, and the topic warrants its own section. Worked Examples and a set of end-of-chapter problems were developed.
- New Worked Example 23.4: Interpreting Line Drawings for Molecules with Functional Groups.
- New Inquiry on chiral molecules and their biological response to connect with new Section 23.2 Stereoisomers: Chiral Molecules on chiral molecules.

ACKNOWLEDGMENTS

Our thanks go to our families and to the many talented people who helped bring this new edition into being. We are grateful to Terry Haugen, Executive Courseware Portfolio Manager, for his insight and suggestions that improved the book; to Cathy Murphy, Coleen Morrison, and Jay McElroy for their critical review that made the manuscript and art program more understandable for students; to Elizabeth Bell, Senior Product Marketing Manager, who brought new energy to describing features of the 8th edition; and to Shercian Kinosian and Kelly Murphy for their production and editorial efforts. Thank you to Rebecca Marshall for coordinating art production and to Angelica D. Aranas for her photo research efforts.

   We are particularly pleased to acknowledge the outstanding contributions of several colleagues who created the many important supplements that turn a textbook into a complete package:

- The author who updated the accompanying Test Bank.
- Joseph Topich, *Virginia Commonwealth University*, who prepared both the full and partial solutions manuals.
- Mark Benvenuto, *University of Detroit Mercy,* who contributed valuable content for the Instructor Resources.
- Kristi Mock, *University of Toledo,* who prepared the Student Study Guide to accompany this 8th edition.
- Dennis Taylor, *Clemson University*, who prepared the Instructor Resource Manual.
- Sandra Chimon-Rogers, *Calumet College of St. Joseph,* who updated the Laboratory Manual.

   Finally, we want to thank all accuracy reviewers, text reviewers, and our colleagues at so many other institutions who read, criticized, and improved our work.

Jill K. Robinson
John McMurry
Robert C. Fay

REVIEWERS FOR THE EIGHTH EDITION

Stanley Bajue, Medger Evers College
Joe Casalnuovo, Cal Poly, Pomona
Kathryn Davis, Manchester University
Sarah Edwards, Western Kentucky University
Stacy O'Reilly, Butler University

Gabriela Smeureanu, Hunter College
Lucinda Spryn, Thomas Nelson Community College
Joe Topich, Virginia Commonwealth University
Ken Tyrrell, Connors State College
Zachary Varpness, Chadron State College

REVIEWERS OF THE PREVIOUS EDITIONS OF *CHEMISTRY*

James Almy, Golden West College
Laura Andersson, Big Bend Community College
David Atwood, University of Kentucky
James Ayers, Colorado Mesa University
Mufeed Basti, North Carolina A&T State University
David S. Ballantine, Northern Illinois University
Debbie Beard, Mississippi State University
Robert Blake, Glendale Community College
Ronald Bost, North Central Texas University
Danielle Brabazon, Loyola College
Gary Buckley, Cameron University
Robert Burk, Carleton University
Ken Capps, Central FL Community College
Joe Casalnuovo, Cal Poly Pomona
Myron Cherry, Northeastern State University
Sandra Chimon-Rogers, Calumet College of St. Joseph
Allen Clabo, Francis Marion University
Claire Cohen, University of Toledo
Paul Cohen, University of New Jersey
Katherine Covert, West Virginia University
David De Haan, University of San Diego
Nordulf W. G. Debye, Towson University
Dean Dickerhoof, Colorado School of Mines
David Dobberpuhl, Creighton University
Kenneth Dorris, Lamar University
Jon A. Draeger, University of Pittsburgh at Bradford
Brian Earle, Cedar Valley College
Amina El- Ashmawy, Collin County Community College
Joseph W. Ellison, United States Military Academy at West Point
Erik Eriksson, College of the Canyons
Peter M. Fichte, Coker College
Kathy Flynn, College of the Canyons
Joanne Follweiler, Lafayette College
Ted Foster, Folsom Lake College
Cheryl Frech, University of Central Oklahoma
Mark Freilich, University of Memphis
Mark Freitag, Creighton University
Travis Fridgen, Memorial University of Newfoundland
Chammi Gamage-Miller, Blinn College–Bryan Campus
Rachel Garcia, San Jacinto College
Katherine Geiser-Bush, Durham Technical Community College
Jack Goldsmith, University of South Carolina Aiken
Carolyn Griffin, Grand Canyon University
Nathanial Grove, UNC Wilmington
Thomas Grow, Pensacola Junior College
Mildred Hall, Clark State University

Tracy A. Halmi, Pennsylvania State University Erie
Keith Hansen, Lamar University
Lois Hansen-Polcar, Cuyahoga Community College
Wesley Hanson, John Brown University
Alton Hassell, Baylor University
Michael Hauser, St. Louis Community College–Meramec
M. Dale Hawley, Kansas State University
Patricia Heiden, Michigan Tech University
Sherman Henzel, Monroe Community College
Thomas Hermann, University of California–San Diego
Thomas Herrington, University of San Diego
Margaret E. Holzer, California State University–Northridge
Geoff Hoops, Butler University
Todd Hopkins, Baylor University
Narayan S. Hosmane, Northern Illinois University
Jeff Joens, Florida International University
Andy Jorgensen, University of Toledo
Jerry Keister, University of Buffalo
Chulsung Kim, University of Dubuque
Angela King, Wake Forest University
Regis Komperda, Wright State University
Ranjit Koodali, University of South Dakota
Peter Kuhlman, Denison University
Valerie Land, University of Arkansas Community College
John Landrum, Florida International University
Leroy Laverman, University of California–Santa Barbara
Celestia Lau, Lorain County Community College
Stephen S. Lawrence, Saginaw Valley State University
David Leddy, Michigan Technological University
Shannon Lieb, Butler University
Don Linn, IUPU Fort Wayne
Karen Linscott, Tri-County Technical College
Irving Lipschitz, University of Massachusetts–Lowell
Rosemary Loza, Ohio State University
Rudy Luck, Michigan Technological University
Rod Macrae, Marian University
Riham Mahfouz, Thomas Nelson Community College
Ashley Mahoney, Bethel College
Jack F. McKenna, St. Cloud State University
Craig McLauchlan, Illinois State University
Iain McNab, University of Toronto
Christina Mewhinney, Eastfield College
David Miller, California State University–Northridge
Rebecca S. Miller, Texas Tech University
Abdul Mohammed, North Carolina A&T State University
Linda Mona, United States Naval Academy

Give students a robust conceptual foundation while building critical problem-solving skills

Robinson/McMurry/Fay's *Chemistry*, known for a conceptual focus, extensive worked examples, and thoroughly constructed connections between organic, biological, and general chemistry, illustrates the application of chemistry to students' lives and careers. With the **8th edition**, lead author Jill Robinson draws upon her exceptional teaching skills to create more engaging, active learning opportunities for students and faculty, including new interactive experiences that help identify and address students' preconceptions. In **Mastering Chemistry and the Pearson eText,** a new media program increases students' awareness of their learning process and allows instructors to choose the level of interactivity appropriate for their classroom.

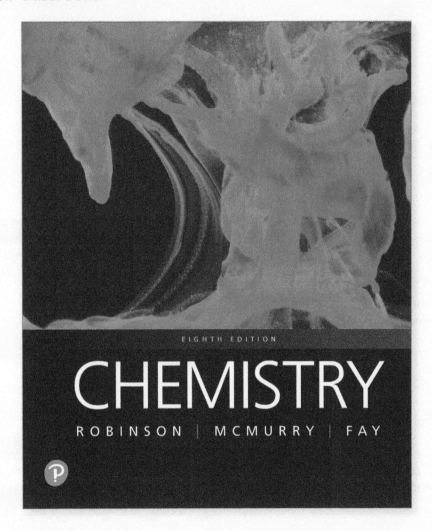

EIGHTH EDITION

CHEMISTRY

ROBINSON | MCMURRY | FAY

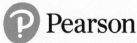

Pearson

Facilitate active learning with new eText interactives

NEWLY INTERACTIVE! Big Idea questions provide new interactivity within the eText and teach students how to actively read a science text by modeling the kinds of questions they should ask themselves, prompting them to summarize main points, and stimulating them to make connections between concepts and mathematical problems. These questions help ensure students are familiar with main concepts and terms before coming to class. Activities like this also improve retention and comprehension by asking students to answer questions at the end of reading a section for retrieval practice.

BIG IDEA Question 4

For which object would be it impossible to accurately know both the position and velocity: a thrown baseball, an electron in a cathode-ray tube, both, or neither?

P. 176

NEWLY INTERACTIVE! Figure It Out questions provide new interactivity within the eText so students can test themselves at the point of learning and receive instant, answer-specific feedback written by author Jill Robinson. Questions appear with select figures and encourage students to look at each illustration more carefully and recognize general lessons offered in the figure.

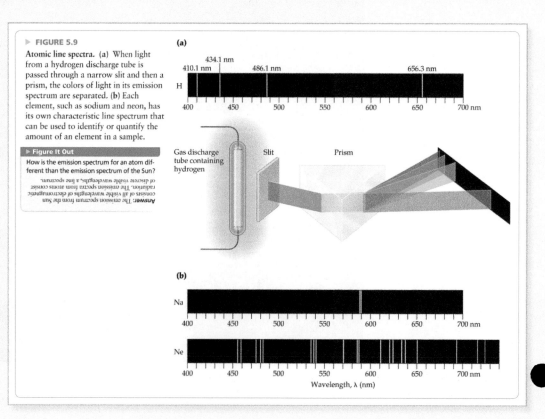

▶ **FIGURE 5.9**

Atomic line spectra. (a) When light from a hydrogen discharge tube is passed through a narrow slit and then a prism, the colors of light in its emission spectrum are separated. **(b)** Each element, such as sodium and neon, has its own characteristic line spectrum that can be used to identify or quantify the amount of an element in a sample.

▶ **Figure It Out**

How is the emission spectrum for an atom different than the emission spectrum of the Sun?

Answer: The emission spectrum from the Sun consists of all visible wavelengths of electromagnetic radiation. The emission spectra from atoms consist of discrete visible wavelengths, a line spectrum.

P. 170

Build students' problem-solving skills

Go to eText

WORKED EXAMPLE 6.4

Higher Ionization Energies

Which has the larger fifth ionization energy, Ge or As?

STRATEGY

Look at their positions in the periodic table, and write the electron configuration. If the fifth electron must be removed from an inner shell, then that element will have the larger ionization energy.

SOLUTION

Ge: $[Ar]\,4s^2\,3d^{10}\,4p^2$ As: $[Ar]\,4s^2\,3d^{10}\,4p^3$

The group 4A element germanium has four valence-shell electrons and thus has four relatively low ionization energies, whereas the group 5A element arsenic has five valence-shell electrons and has five low ionization energies. Germanium has a larger E_{i5} than arsenic because the fifth electron to be removed in Ge occupies a lower electron shell ($n = 3$).

▶ PRACTICE 6.7 Which has the largest third ionization energy: Be, C, or N?

▶ CONCEPTUAL APPLY 6.8 The figure on the left represents the successive ionization energies of an atom in the third period of the periodic table. Which atom is this most likely to be?

All WORKED EXAMPLES with this icon have an interactive video in the eText.

P. 218

NEW! Problem-Solving Videos feature Jill Robinson working out problems in the text to help students see every step of the process. These dynamic videos are provided in Mastering Chemistry, referenced in the print text, and also embedded in the eText. These are also assignable in Mastering Chemistry along with follow-up assessment questions.

NEW! Practice Tests provide assessment for chapter learning objectives and additional instruction in concepts and problem-solving skills in the printed text as well as in the eText and Mastering Chemistry. When students miss a problem, they are instructed to consult the study guide for additional instruction and practice problems. When a student answers incorrectly in Mastering Chemistry or the eText, their practice exam automatically links to worked examples and additional practice problems.

PRACTICE TEST

After studying this chapter, you can assess your understanding with these practice test questions, which are correlated with chapter learning objectives. If you answer a question incorrectly, refer to the learning objectives in the end-of-chapter Study Guide for assistance. The Study Guide provides a conceptual summary, references a Worked Example to model how to solve the problem, and gives additional problems for more practice.

1. Refer to a periodic table. Which pair of elements do you expect to be most similar in their chemical properties? (LO 2.3)
   (a) K and Cu   (b) O and Se
   (c) Be and B   (d) Rb and Sr

2. Identify the location of the element in period 4, group 6A and classify it as a metal, nonmetal, or semimetal. (LO 2.2)

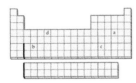

   (a) Element in position a; nonmetal
   (b) Element in position b; metal
   (c) Element in position c; semimetal
   (d) Element in position d; metal

3. Which description of an element is incorrectly matched with its location in the periodic table? (LO 2.5–2.6)

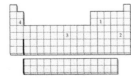

   (a) Element 3—An element in the transition metal group that is a good conductor of electricity.
   (b) Element 2—An element that is in the halogen group and does not conduct electricity.
   (c) Element 4—An element in alkali metal group that is found in its pure form in nature.
   (d) Element 1—An element that is a solid at room temperature, brittle, and a poor conductor of electricity.

4. A compound containing sulfur and fluorine contains 8.00 g of S and 9.50 g of F. Which combination of S and F masses represents a *different* compound that obeys the Law of Multiple Proportions? (LO 2.8)
   (a) 32.0 g of S and 38.0 g of F
   (b) 4.00 g of S and 4.75 g of F
   (c) 8.00 g of S and 10.5 g of F
   (d) 16.0 g of S and 57.0 g of F

5. Which experiment and subsequent observation led to the discovery that atoms contain negatively charged particles, now known as electrons? (LO 2.10–2.12)
   (a) Oil is sprayed into a chamber and the speed at which the oil droplets fall is measured with and without an applied voltage. X rays in the chamber knock electrons out of air molecules. The electrons stick to the oil producing an overall negative charge on the drops. Adjusting the voltage changes the speed at which the negatively charged oil droplets fall.
   (b) When a high voltage is applied across metal electrodes at opposite ends of a sealed glass tube, a cathode ray is produced. The cathode ray is repelled by a negatively charged plate.
   (c) A radioactive substance emits alpha particles, which are directed at a thin gold foil. Most of the alpha particles pass through the foil, but a few alpha particles are slightly deflected and some even bounce back toward the radioactive source.
   (d) The mass of different elements in a pure chemical compound are measured. Different samples of the compound always contains the same proportion of elements by mass.

6. How many protons, neutrons, and electrons are present in an atom of $^{206}_{82}Pb$? (LO 2.14)
   (a) 82 protons, 206 neutrons, 82 electrons
   (b) 124 protons, 82 neutrons, 124 electrons
   (c) 82 protons, 124 neutrons, 82 electrons
   (d) 82 protons, 82 neutrons, 124 electrons

Engage students

Click on the different series of emission lines of atomic hydrogen to observe the nature of the representative transitions from excited to ground states.

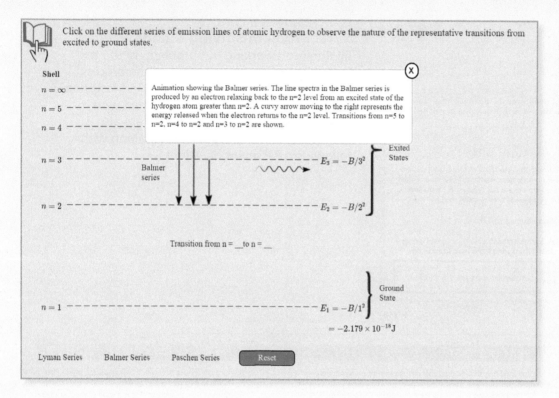

Animation showing the Balmer series. The line spectra in the Balmer series is produced by an electron relaxing back to the n=2 level from an excited state of the hydrogen atom greater than n=2. A curvy arrow moving to the right represents the energy released when the electron returns to the n=2 level. Transitions from n=5 to n=2, n=4 to n=2 and n=3 to n=2 are shown.

Shell

$n = \infty$

$n = 5$

$n = 4$

$n = 3$ $E_3 = -B/3^2$ Exited States

Balmer series

$n = 2$ $E_2 = -B/2^2$

Transition from n = ___ to n = ___

Ground State

$n = 1$ $E_1 = -B/1^2$

$= -2.179 \times 10^{-18} J$

Lyman Series   Balmer Series   Paschen Series   Reset

Interactive Simulations

cover some of the most difficult chemistry concepts. Written by leading authors in simulation development, these increase students' understanding of chemistry and clearly illustrate cause-and-effect relationships.

Interactive solutions are assignable in Mastering Chemistry

and include hints and wrong answer feedback to help students right when they need it.

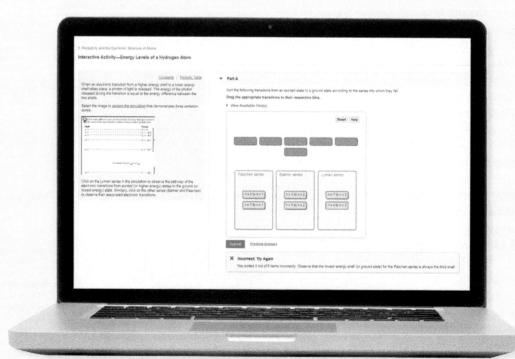

in learning chemistry

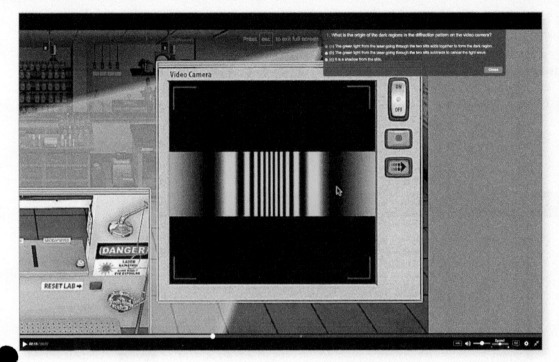

Pause and Predict Video Quizzes bring chemistry to life with lab demonstrations illustrating key topics in general chemistry.

Students are asked to predict the outcome of experiments as they watch the videos. A set of multiple-choice questions, with hints and wrong answer feedback, challenge students to apply the concepts from the video to related scenarios.

Give students anytime, anywhere access with Pearson eText

Pearson eText is a simple-to-use, mobile-optimized, personalized reading experience available within Mastering. It allows students to easily highlight, take notes, and review key vocabulary all in one place—even when offline. Seamlessly integrated videos, rich media, and interactive self-assessment questions engage students and give them access to the help they need, when they need it. Pearson eText is available within Mastering when packaged with a new book; students can also purchase Mastering with Pearson eText online.

Reach every student with Mastering Chemistry

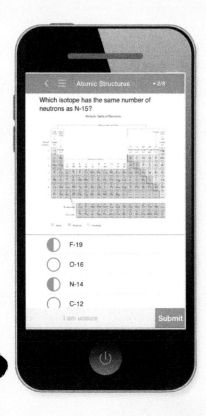

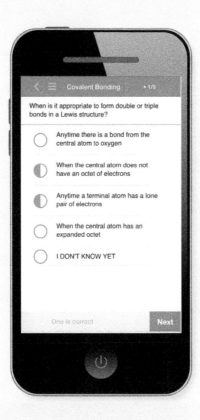

Dynamic Study Modules in Mastering Chemistry help students study effectively—and at their own pace—by keeping them motivated and engaged. The assignable modules rely on the latest research in cognitive science using methods—such as adaptivity, gamification, and intermittent rewards—to stimulate learning and improve retention.

The Chemistry Primer relies on videos, hints, and feedback to refresh students' math skills in the context of chemistry and prepares them for success in the course. These tutorials can be assigned before the course begins or throughout the course as just-in-time remediation. They ensure students practice and maintain their math skills while building their chemical literacy.

Instructor support you can rely on

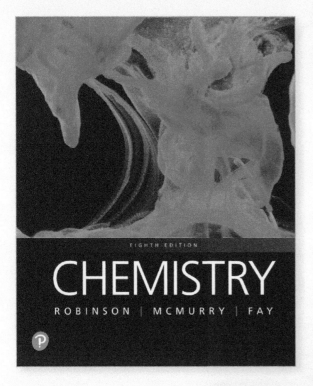

Chemistry includes a full suite of instructor support materials in the Instructor Resources area in Mastering Chemistry. Resources include new Ready-to-Go Teaching Modules, PowerPoint lecture outlines, all images and worked examples from the text, and a testbank.

NEW! Ready-to-Go Teaching Modules provide organized material for every tough topic in General Chemistry. Created for and by instructors, the modules provide a guide for easy-to-use assignments for before and after class plus in-class activities with clicker questions and questions in Learning Catalytics™. Modules can be easily accessed via Mastering Chemistry.

chapter 1

Chemical Tools: Experimentation and Measurement

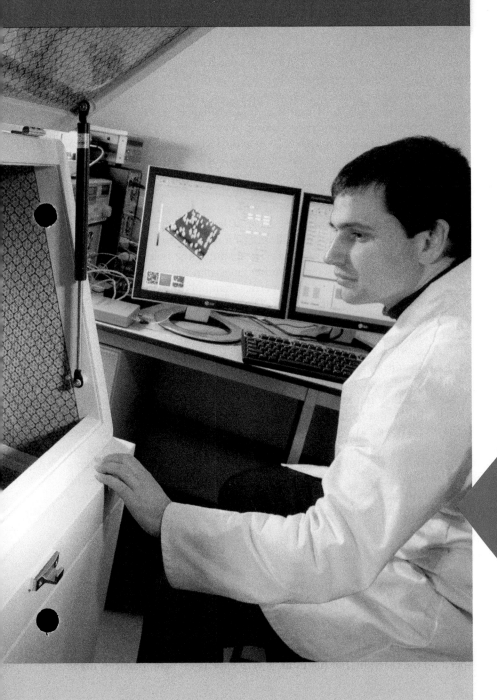

Contents

Instruments for scientific measurements have changed greatly over the centuries. Modern technology has enabled scientists to make images of extremely tiny particles, even individual atoms, using instruments like this atomic force microscope.

What are the unique properties of nanoscale (1 nm = 10^{-9} m) materials?

The answer to this question can be found on page 23 in the INQUIRY ?

1

▲ The sequence of the approximately 5.8 billion nucleic acid units, or *nucleotides*, present in the human genome has been determined using instruments like this automated DNA sequencer.

L ife has changed more in the past two centuries than in all the previously recorded span of human history. The Earth's population has increased sevenfold since 1800, and life expectancy has nearly doubled because of our ability to synthesize medicines, control diseases, and increase crop yields. Methods of transportation have changed from horses and buggies to automobiles and airplanes because of our ability to harness the energy in petroleum. Many goods are now made of polymers and ceramics instead of wood and metal because of our ability to manufacture materials with properties unlike any found in nature.

In one way or another, all these changes involve **chemistry**, the study of the composition, properties, and transformations of matter. Chemistry is deeply involved in both the changes that take place in nature and the profound social changes of the past two centuries. In addition, chemistry is central to the current revolution in molecular biology that is revealing the details of how life is genetically regulated. No educated person today can understand the modern world without a basic knowledge of chemistry.

1.1 THE SCIENTIFIC METHOD: NANOPARTICLE CATALYSTS FOR FUEL CELLS

By opening this book, you have already decided that you need to know more about chemistry to pursue your future goals. Perhaps you want to learn how living organisms function, how medicines are made, how human activities change the environment, or how alternative fuels produce clean energy. A good place to start is by learning the experimental approach used by scientists to make new discoveries. Do not worry if you do not understand all the details of the chemistry yet, as our focus is on the process of modern interdisciplinary research.

Let's examine a nanoscience application to illustrate the scientific method and how chemical principles are applied to make materials with novel properties. **Nanoscience** is the production and study of structures that have at least one dimension between 1 and 100 nm, where one nanometer is one billionth of a meter. Research on nanomaterials is a fast-growing, multidisciplinary enterprise spanning the fields of chemistry, physics, biology, medicine, materials science, and engineering. Inorganic crystals that have nanoscale dimensions exhibit different properties than bulk material as described in more detail the Inquiry section of this chapter. The properties depend on the size of the particle and can be tuned for applications such as tools for diagnosing and treating disease or platforms for sustainable energy.

One research area is the use of nanoparticle catalysts for reactions occurring in fuel cells. A **catalyst** is a substance that speeds up the rate of a chemical reaction. A **fuel cell** is a device that uses a fuel such as hydrogen to produce electricity. Fuel cells operate much like a battery, but they require a continuous input of fuel. Two reactions occur at two different electrodes in a hydrogen fuel cell. At one electrode, hydrogen (H_2) is converted to protons (H^+), and at the other electrode, oxygen (O_2) reacts with protons to produce water. The reactions in the fuel cell involve a transfer of electrons and are called **redox reactions**. The electrons produced by reaction 1 (below) travel through a wire and are used in reaction 2. The movement of electrons through a wire generates electricity. A fuel cell is considered to be *zero emission* because the overall reaction of hydrogen with oxygen produces electricity but pure water is the only product.

Reaction 1: $2\,H_2(g) \longrightarrow 4\,H^+(aq) + 4\,e^-$

Reaction 2: $O_2(g) + 4\,H^+(aq) + 4\,e^- \longrightarrow 2\,H_2O(l)$

Overall reaction: $2\,H_2(g) + O_2(g) \longrightarrow 2\,H_2O(l)$

Fuel cells are a promising technology in the quest for a carbon-neutral energy economy, but one obstacle to their use is the slow rate of conversion of oxygen to water in reaction 2. Platinum particles coated on the surface of the electrode have been used as a catalyst to speed up the reaction, but platinum is very expensive. Nanoparticles made

LOOKING AHEAD . . .
The rates of chemical reactions and how they are increased by catalysts are described in Chapter 14.

LOOKING AHEAD . . .
Chapter 4 describes different types of reactions including **redox reactions** that involve a transfer of electrons. We'll see in Chapter 19 how redox reactions can be used to generate electricity in a **fuel cell**.

from palladium alloys have shown promise as a cost-effective alternative catalyst. An **alloy** is a mixture of metals, and therefore a palladium alloy is a mixture of palladium (Pd) and some other metal such as copper (Cu).

In order to develop a useful catalyst for hydrogen fuel cells, chemists apply the scientific method to carefully control different characteristics of PdCu nanoparticles and measure their effect on the rate of the oxygen reaction. Some characteristics of nanoparticles that can be varied are relative amounts of palladium and copper, the size of the particles, and the shape of the particles. Amazingly nanoparticles exist in a variety of shapes including spheres, cubes, and octopods (**FIGURE 1.1**)!

BIG IDEA Question 1   Go to eText

What is an obstacle to the widespread use of hydrogen fuel cells, and how can nanoparticles be used to overcome the problem?

◀ **FIGURE 1.1**

From left to right, scanning electron microscopy images of octahedral gold nanoparticles, cubic palladium nanoparticles, and eight-branched gold-palladium nanoparticles called "octopods." Note that the orientation of the octopodal nanoparticles allows only four of the branches to be viewed at a time.

Images courtesy of the Skrabalak research group at Indiana University.

The Scientific Method

A general approach to research is called the **scientific method.** The scientific method is an iterative process involving the formulation of questions arising from observations, careful design of experiments, and thoughtful analysis of results. The scientific method involves identifying ways to test the validity of new ideas, and seldom is there only one way to go about it. The main elements of the scientific method, outlined in **FIGURE 1.2,** are the following:

- *Observation.* **Observations** are a systematic recording of natural phenomena and may be **qualitative,** descriptive in nature, or **quantitative,** involving measurements.
- *Hypothesis.* A **hypothesis** is a possible explanation for the observation developed based upon facts collected from previous experiments as well as scientific knowledge

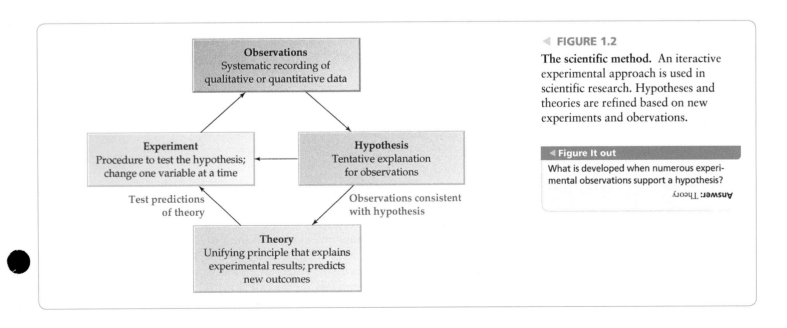

◀ **FIGURE 1.2**

The scientific method. An iterative experimental approach is used in scientific research. Hypotheses and theories are refined based on new experiments and obervations.

◀ **Figure It out**

What is developed when numerous experimental observations support a hypothesis?

Answer: Theory

and intuition. The hypothesis may not be correct, but it must be testable with an experiment.

- *Experiment.* An **experiment** is a procedure for testing the hypothesis. Experiments are most useful when they are performed in a *controlled* manner, meaning that only one variable is changed at a time while all others remain constant.

- *Theory.* A **theory** is developed from a hypothesis consistent with experimental data and is a unifying principle that explains experimental results. It also makes predictions about related systems, and new experiments are carried out to verify the theory.

Keep in mind as you study chemistry or any other science that theories can never be absolutely proven. There's always the chance that a new experiment might give results that can't be explained by present theory. All a theory can do is provide the best explanation that we can come up with at the present time. Science is an ever-changing field where new observations are made with increasingly sophisticated equipment; it is always possible that existing theories may be modified in the future.

Scientific research begins with a driving question that is frequently based on experimental observations or a desire to learn about the unknown. In the case of PdCu nanoparticles, an observed increase in the fuel cell reaction rate led to the question "How can variables related to size, shape, and composition of nanoparticles be controlled to optimize catalytic activity?" Professor Sara Skrabalak at Indiana University leads a team of scientists researching methods for synthesizing high-quality nanomaterials, where these variables are precisely controlled. Although previous research projects involving numerous techniques attempted to control the size, shape, and composition of the nanoparticles, the distributions of palladium and copper atoms in the crystals were found to be statistically random. **FIGURE 1.3a** illustrates a random or disordered arrangement of Pd and Cu atoms in a crystal. **FIGURE 1.3b** illustrates an ordered arrangement with a pattern of alternating Pd and Cu atoms. Without fixed arrangements of atoms, it is impossible to correlate chemical structure with properties such as catalytic activity.

▶ **FIGURE 1.3**

Simple schematic of the arrangement of Pd and Cu atoms in a nanocrystal. (a) A disordered arrangement of Pd and Cu atoms does not have a repeating pattern. (b) An ordered arrangement of atoms has the repeating pattern of alternating Pd and Cu atoms.

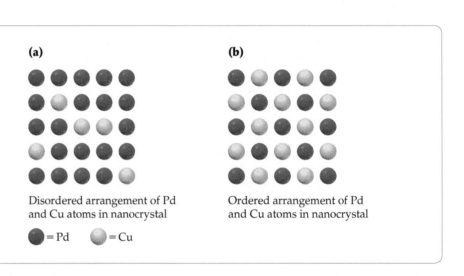

(a)

Disordered arrangement of Pd and Cu atoms in nanocrystal

(b)

Ordered arrangement of Pd and Cu atoms in nanocrystal

● = Pd   ● = Cu

The general hypothesis that the Skrabalak group tested was that larger PdCu nanoparticles with lower surface energies would facilitate the transition from disordered to ordered structures. Student researchers carefully controlled the rate of particle growth by depositing palladium and copper on the surface of a smaller particle. Then various imaging techniques were used to elucidate the atomic-level structure of the nanoparticles and measure their size distribution. Electron microscopy data revealed a thin shell of Pd over an ordered PdCu core called the B2 phase. **FIGURE 1.4a** shows a

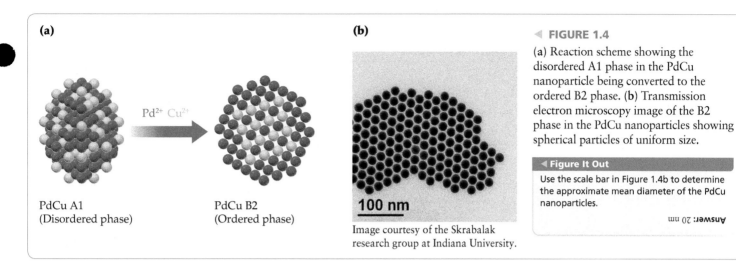

(a)

Pd²⁺ Cu²⁺

PdCu A1
(Disordered phase)

PdCu B2
(Ordered phase)

(b)

100 nm

Image courtesy of the Skrabalak
research group at Indiana University.

◀ **FIGURE 1.4**

(a) Reaction scheme showing the disordered A1 phase in the PdCu nanoparticle being converted to the ordered B2 phase. (b) Transmission electron microscopy image of the B2 phase in the PdCu nanoparticles showing spherical particles of uniform size.

◀ **Figure It Out**

Use the scale bar in Figure 1.4b to determine the approximate mean diameter of the PdCu nanoparticles.

Answer: 20 nm

random distribution of Pd and Cu atoms in the A1 phase that was converted by new synthesis methods to the ordered B2 phase. **FIGURE 1.4b** shows a transmission electron microscope image of PdCu nanoparticles in the B2 phase. The spherical particles have a uniform size distribution with a mean diameter of 18.9 nm.

Many iterations of the scientific method were used by the researchers to devise controlled synthesis techniques for PdCu nanoparticles. Observations from experiments led to new hypotheses and additional experiments to test them. Once studies of the growth mechanism enabled reproducible synthesis of ordered PdCu nanoparticles, they were tested for catalytic activity. The ordered nanoparticles exhibited superior catalytic activity in increasing the rate of oxygen reaction in the fuel cell when compared with PdCu nanoparticles with disordered structures. In summary, Professor Skrabalak's research on nanomaterial synthesis leads to the design of better nanoparticle catalysts for fuel cells and other applications.

Many different chemical principles that you will learn about in this book are central to the design of nanomaterials. In Chapter 8, Bonding Theories and Molecular Structure, you will learn about bonds and forces that cause atoms to aggregate into nanoparticles. Chapter 4, Reactions in Aqueous Solutions, describes how to calculate solution concentrations important in synthetic techniques. Rates of reactions and factors that influence them are explored in Chapter 14, Kinetics. In Chapter 19, on electrochemistry, redox reactions central to forming nanoparticles are described.

At universities around the world, students participate in research projects like the one on nanoparticle synthesis and characterization just described. It is the authors' sincere hope that by reading this book you can gain an appreciation for how chemistry is used in solving many of the world's problems and you become competent with the essential chemical principles needed to contribute to important research projects.

▲ Professor Sara Skrabalak in the lab with undergraduate student researchers working on the synthesis of nanoparticles.
Photo courtesy of Indiana University.

1.2 MEASUREMENTS: SI UNITS AND SCIENTIFIC NOTATION

Chemistry is an experimental science. But if our experiments are to be reproducible, we must be able to fully describe the substances we're working with—their amounts, volumes, temperatures, and so forth. Thus, one of the most important requirements in chemistry is that we have a way to measure things.

Under an international agreement concluded in 1960, scientists throughout the world now use the International System of Units for measurement, abbreviated **SI unit** for the French *Système Internationale d'Unités*. Based on the metric system, which is used in all industrialized countries of the world except the United States, the SI system has seven fundamental units (**TABLE 1.1**). These seven fundamental units, along with others derived from them, suffice for all scientific measurements. We'll look at three of

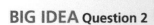

Go to
eText

BIG IDEA Question 2

What are the fundamental SI units of measure for mass, length, and temperature?

TABLE 1.1   The Seven Fundamental SI Units of Measure

Physical Quantity	Name of Unit	Abbreviation
Mass	kilogram	kg
Length	meter	m
Temperature	kelvin	K
Amount of substance	mole	mol
Time	second	s
Electric current	ampere	A
Luminous intensity	candela	cd

the most common units in this chapter—those for mass, length, and temperature—and will discuss others as the need arises in later chapters.

One problem with any system of measurement is that the sizes of the units often turn out to be inconveniently large or small. For example, a chemist describing the diameter of a sodium atom (0.000 000 000 372 m) would find the meter (m) to be inconveniently large, but an astronomer describing the average distance from the Earth to the Sun (150,000,000,000 m) would find the meter to be inconveniently small. For this reason, SI units are modified through the use of prefixes when they refer to either smaller or larger quantities. Thus, the prefix *milli-* means one-thousandth, and a *milli*meter (mm) is 1/1000 of 1 meter. Similarly, the prefix *kilo-* means one thousand, and a *kilo*meter (km) is 1000 meters. (Note that the SI unit for mass [kilogram] already contains the *kilo-* prefix.) A list of prefixes is shown in **TABLE 1.2**, with the most commonly used ones in red.

Notice how numbers that are either very large or very small are indicated in Table 1.2 using an exponential format called **scientific notation**. For example, the number 55,000 is written in scientific notation as 5.5×10^4 and the number 0.003 20 as 3.20×10^{-3}.

TABLE 1.2   Some Prefixes for Multiples of SI Units. Common prefixes and symbols in the chemical sciences are shown in red

Factor	Prefix	Symbol	Example
$1,000,000,000,000 = 10^{12}$	tera	T	1 teragram (Tg) = 10^{12} g
$1,000,000,000 = 10^9$	giga	G	1 gigameter (Gm) = 10^9 m
$1,000,000 = 10^6$	mega	M	1 megameter (Mm) = 10^6 m
$1000 = 10^3$	kilo	k	1 kilogram (kg) = 10^3 g
$100 = 10^2$	hecto	h	1 hectogram (hg) = 100 g
$10 = 10^1$	deka	da	1 dekagram (dag) = 10 g
$0.1 = 10^{-1}$	deci	d	1 decimeter (dm) = 0.1 m
$0.01 = 10^{-2}$	centi	c	1 centimeter (cm) = 0.01 m
$0.001 = 10^{-3}$	milli	m	1 milligram (mg) = 0.001 g
*$0.000\ 001 = 10^{-6}$	micro	μ	1 micrometer (μm) = 10^{-6} m
*$0.000\ 000\ 001 = 10^{-9}$	nano	n	1 nanosecond (ns) = 10^{-9} s
*$0.000\ 000\ 000\ 001 = 10^{-12}$	pico	p	1 picosecond (ps) = 10^{-12} s
*$0.000\ 000\ 000\ 000\ 001 = 10^{-15}$	femto	f	1 femtomole (fmol) = 10^{-15} mol

*For very small numbers, it is becoming common in scientific work to leave a thin space every three digits to the right of the decimal point, analogous to the comma placed every three digits to the left of the decimal point in large numbers.

Review Appendix A if you are uncomfortable with scientific notation or if you need to brush up on how to do mathematical manipulations on numbers with exponents.

Notice also that all measurements contain both a number and a unit label. A number alone is not much good without a unit to define it. If you asked a friend how far it was to the nearest tennis court, the answer "3" alone wouldn't tell you much: 3 blocks? 3 kilometers? 3 miles? Worked Example 1.1 explains how to write a number in scientific notation and represent the unit in prefix notation.

WORKED EXAMPLE 1.1

Expressing Measurements Using Scientific Notation and SI Units

Express the following quantities in scientific notation and then express the number and unit with the most appropriate prefix.

(a) The diameter of a sodium atom, 0.000 000 000 372 m
(b) The distance from the Earth to the Sun, 150,000,000,000 m

---

STRATEGY

To write a number in scientific notation, shift the decimal point to the right or left by n places until you obtain a number between 1 and 10. If the decimal is shifted to the right, n is negative, and if the decimal is shifted to the left, n is positive. Then multiply the result by 10^n. Choose a prefix for the unit that is close to the exponent of the number written in scientific notation.

SOLUTION

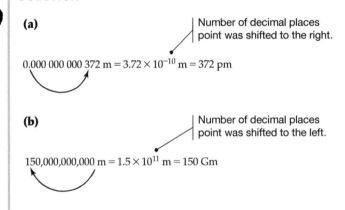

(a)

Number of decimal places point was shifted to the right.

$$0.000\ 000\ 000\ 372\ \text{m} = 3.72 \times 10^{-10}\ \text{m} = 372\ \text{pm}$$

(b)

Number of decimal places point was shifted to the left.

$$150{,}000{,}000{,}000\ \text{m} = 1.5 \times 10^{11}\ \text{m} = 150\ \text{Gm}$$

▶ **PRACTICE 1.1** Express the diameter of a nanoparticle (0.000 000 050 m) in scientific notation, and then express the number and unit with the most appropriate prefix.

▶ **APPLY 1.2** Express the following quantities in scientific notation using fundamental SI units of mass and length given in Table 1.1.

(a) The diameter of a human hair, 70 μm
(b) The mass of carbon dioxide emitted from a large power plant each year, 20 Tg

All **PRACTICE** and **APPLY** problems are interactive in the eText.

1.3 MASS AND ITS MEASUREMENT

Mass is defined as the amount of *matter* in an object. **Matter,** in turn, is a catch-all term used to describe anything with a physical presence—anything you can touch, taste, or smell. (Stated more scientifically, matter is anything that has mass.) Mass is measured in SI units by the **kilogram** (**kg**; 1 kg = 2.205 U.S. lb). Because the kilogram is too large for many purposes in chemistry, the metric **gram** (**g**; 1 g = 0.001 kg), the **milligram** (**mg**; 1 mg = 0.001 g = 10^{-6} kg), and the **microgram** (**μg**; 1 μg = 0.001 mg = 10^{-6} g = 10^{-9} kg) are more commonly used. (The symbol **μ** is the lowercase Greek letter mu.) One gram is a bit less than half the mass of a new U.S. dime.

▲ The mass of a U.S. dime is approximately 2.27 g.

$$1 \text{ kg} = 1000 \text{ g} = 1{,}000{,}000 \text{ mg} = 1{,}000{,}000{,}000 \ \mu\text{g} \qquad (2.205 \text{ lb})$$
$$1 \text{ g} = 1000 \text{ mg} = 1{,}000{,}000 \ \mu\text{g} \qquad (0.035\ 27 \text{ oz})$$
$$1 \text{ mg} = 1000 \ \mu\text{g}$$

The standard kilogram is set as the mass of a cylindrical bar of platinum–iridium alloy stored in a vault in a suburb of Paris, France. There are 40 copies of this bar distributed throughout the world, with two (Numbers 4 and 20) stored at the U.S. National Institute of Standards and Technology near Washington, D.C.

The terms *mass* and *weight*, although often used interchangeably, have quite different meanings. *Mass* is a physical property that measures the amount of matter in an object, whereas *weight* measures the force with which gravity pulls on an object. Mass is independent of an object's location: your body has the same amount of matter whether you're on Earth or on the moon. Weight, however, *does* depend on an object's location. If you weigh 140 lb on Earth, you would weigh only about 23 lb on the moon, which has a lower gravity than the Earth.

At the same location on Earth, two objects with identical masses experience an identical pull of the Earth's gravity and have identical weights. Thus, the mass of an object can be measured by comparing its weight to the weight of a reference standard of known mass. Much of the confusion between mass and weight is simply due to a language problem. We speak of "weighing" when we really mean that we are measuring mass by comparing two weights. **FIGURE 1.5** shows balances typically used for measuring mass in the laboratory.

BIG IDEA Question 3 Go to eText

Which prefix for the unit of grams is most appropriate for reporting the mass of a grain of sand?

▶ **FIGURE 1.5**

Some balances used for measuring mass in the laboratory.

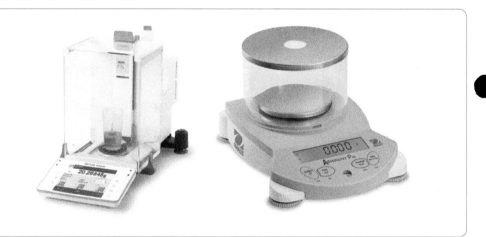

BIG IDEA Question 4 Go to eText

Which prefix for the unit of meter is most appropriate for reporting the diameter of a molecule?

1.4 LENGTH AND ITS MEASUREMENT

The **meter (m)** is the standard unit of length in the SI system. Although originally defined in 1790 as being 1 ten-millionth of the distance from the equator to the North Pole, the meter was redefined in 1889 as the distance between two thin lines on a bar of platinum–iridium alloy stored near Paris, France. To accommodate an increasing need for precision, the meter was redefined again in 1983 as equal to the distance traveled by light through a vacuum in 1/299,792,458 second. Although this new definition isn't as easy to grasp as the distance between two scratches on a bar, it has the great advantage that it can't be lost or damaged.

One meter is 39.37 inches, about 10% longer than an English yard and much too large for most measurements in chemistry. Other more commonly used measures of length are the **centimeter (cm;** 1 cm = 0.01 m, a bit less than half an inch), the **millimeter (mm;** 1 mm = 0.001 m, about the thickness of a U.S. dime), the **micrometer** (μ**m;** 1 μm = 10^{-6} m), the **nanometer (nm;** 1 nm = 10^{-9} m), and the **picometer** (**pm;** 1 pm = 10^{-12} m). Thus, a chemist might refer to the diameter of a sodium atom as 372 pm (3.72×10^{-10} m).

$$1 \text{ m} = 100 \text{ cm} = 1000 \text{ mm} = 1,000,000 \text{ } \mu m = 1,000,000,000 \text{ nm} \quad (1.0936 \text{ yd})$$
$$1 \text{ cm} = 10 \text{ mm} = 10,000 \text{ } \mu m = 10,000,000 \text{ nm} \quad (0.3937 \text{ in.})$$
$$1 \text{ mm} = 1000 \text{ } \mu m = 1,000,000 \text{ nm}$$

1.5 TEMPERATURE AND ITS MEASUREMENT

Just as the kilogram and the meter are slowly replacing the pound and the yard as common units for mass and length measurement in the United States, the **Celsius degree** (°C) is slowly replacing the degree **Fahrenheit** (°F) as the common unit for temperature measurement. In scientific work, however, the **kelvin (K)** has replaced both. (Note that we say only "kelvin," not "degree kelvin.")

For all practical purposes, the kelvin and the degree Celsius are the same—both are one-hundredth of the interval between the freezing point of water and the boiling point of water at standard atmospheric pressure. The only real difference between the two units is that the numbers assigned to various points on the scales differ. Whereas the Celsius scale assigns a value of 0 °C to the freezing point of water and 100 °C to the boiling point of water, the Kelvin scale assigns a value of 0 K to the coldest possible temperature, -273.15 °C, sometimes called *absolute zero*. Thus, 0 K = -273.15 °C and 273.15 K = 0 °C. For example, a warm spring day with a Celsius temperature of 25 °C has a Kelvin temperature of $25 + 273.15 = 298$ K.

▲ The length of the bacteria on the tip of this pin is about 5×10^{-7} m or 500 nm.

> **Relationship between the Kelvin and Celsius scales**
> $$\text{Temperature in K} = \text{Temperature in °C} + 273.15$$
> $$\text{Temperature in °C} = \text{Temperature in K} - 273.15$$

In contrast to the Kelvin and Celsius scales, the common Fahrenheit scale specifies an interval of 180° between the freezing point (32 °F) and the boiling point (212 °F) of water. Thus, it takes 180 degrees Fahrenheit to cover the same range as 100 degrees Celsius (or kelvins), and a degree Fahrenheit is therefore only $100/180 = 5/9$ as large as a degree Celsius. **FIGURE 1.6** compares the Fahrenheit, Celsius, and Kelvin scales.

Two adjustments are needed to convert between Fahrenheit and Celsius scales—one to adjust for the difference in degree size and one to adjust for the difference in zero points. The size adjustment is made using the relationships 1 °C = (9/5) °F

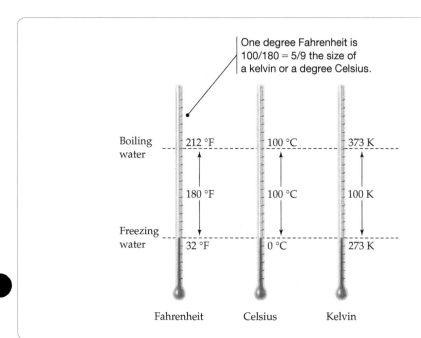

One degree Fahrenheit is $100/180 = 5/9$ the size of a kelvin or a degree Celsius.

Boiling water — 212 °F / 100 °C / 373 K

180 °F / 100 °C / 100 K

Freezing water — 32 °F / 0 °C / 273 K

Fahrenheit   Celsius   Kelvin

◀ **FIGURE 1.6**

A comparison of the Fahrenheit, Celsius, and Kelvin temperature scales.

◀ Figure It Out

Which represents the largest increase in temperature: +10 °F, +10 °C, or +10 K?

Answer: Temperature changes of +10 °C or +10 K are equal and larger than +10 °F.

4096

and $1\,°F = (5/9)\,°C$. The zero-point adjustment is made by remembering that the freezing point of water is higher by 32 on the Fahrenheit scale than on the Celsius scale. Thus, if you want to convert from Celsius to Fahrenheit, you do a size adjustment (multiply °C by 9/5) and then a zero-point adjustment (add 32). If you want to convert from Fahrenheit to Celsius, you find out how many Fahrenheit degrees there are above freezing (by subtracting 32) and then do a size adjustment (multiply by 5/9). The following formulas describe the conversions:

Celsius to Fahrenheit	Fahrenheit to Celsius
$°F = \left(\dfrac{9\,°F}{5\,°C} \times °C\right) + 32\,°F$	$°C = \dfrac{5\,°C}{9\,°F} \times (°F - 32\,°F)$

Worked Example 1.2 shows how to convert between temperature scales and estimate the answer. Before tackling Worked Example 1.2, we'd like to point out that the Worked Examples in this book suggest a series of steps useful in organizing and analyzing information.

Problem-Solving Steps In Worked Examples

IDENTIFY

Classify pertinent information as known or unknown. (The quantity needed in the answer will, of course, be unknown.) Specify units and symbols to help identify necessary equations and procedures.

STRATEGY

Find a relationship between the known information and unknown answer, and plan a strategy for getting from one to the other.

SOLUTION

Solve the problem.

CHECK

If possible, make a rough estimate to be sure your calculated answer is reasonable, and think about the number and sign to make sure it makes sense.

WORKED EXAMPLE 1.2

Converting between Temperature Scales

The normal body temperature of a healthy adult is 98.6 °F. What is this value on both Celsius and Kelvin scales?

IDENTIFY

Known	Unknown
Temperature, 98.6 °F	Temperature in units of °C and K

STRATEGY

Use the formulas for converting Fahrenheit to Celsius and Celsius to Kelvin.

SOLUTION

Set up an equation using the temperature conversion formula for changing from Fahrenheit to Celsius:

$$°C = \left(\frac{5\,°C}{9\,°F}\right)(98.6\,°F - 32\,°F) = 37.0\,°C$$

Converting to kelvin gives a temperature of $37.0 + 273.15 = 310.2$ K.

CHECK

A useful way to double-check a calculation is to estimate the answer. Body temperature in °F is first rounded to the nearest whole number, 99. To account for the difference in zero points of the two scales, 32 is subtracted: 99 − 32 = 67. Because a degree Fahrenheit is only 5/9 as large as a degree Celsius, the next step is to multiply by 5/9, which can be approximated by dividing by two: 67/2 = 33.5. The estimate is only slightly lower than the calculated answer (37.0), indicating the mathematical operations have most likely been performed correctly. Estimating Fahrenheit to Celsius conversions is useful as daily temperatures are reported on these two different scales throughout the world.

▶ **PRACTICE 1.3**   The melting point of table salt is 1474 °F. What temperature is this on the Celsius and Kelvin scales?

▶ **APPLY 1.4**   The metal gallium has a relatively low melting point for a metal, 302.91 K. If the temperature in the cargo compartment carrying a shipment of gallium has a temperature of 88 °F, is the gallium in the solid or liquid state?

▲ The melting point of sodium chloride is 1474 °F.

1.6 DERIVED UNITS: VOLUME AND ITS MEASUREMENT

Look back at the seven fundamental SI units given in Table 1.1, and you'll find that measures for such familiar quantities as area, volume, density, speed, and pressure are missing. All are examples of *derived* quantities rather than fundamental quantities because they can be expressed using one or more of the seven base units (**TABLE 1.3**).

Volume, the amount of space occupied by an object, is measured in SI units by the **cubic meter** (m^3), defined as the amount of space occupied by a cube 1 meter on edge (**FIGURE 1.7**).

A cubic meter equals 264.2 U.S. gallons, much too large a quantity for normal use in chemistry. As a result, smaller, more convenient measures are commonly employed. Both the **cubic decimeter** (dm^3) ($1 \, dm^3 = 0.001 \, m^3$), equal in size to the more familiar metric **liter** (**L**), and the **cubic centimeter** (cm^3) ($1 \, cm^3 = 0.001 \, dm^3 = 10^{-6} \, m^3$), equal in size to the metric **milliliter** (**mL**), are particularly convenient. Slightly larger than 1 U.S. quart, a liter has the volume of a cube 1 dm on edge. Similarly, a milliliter has the volume of a cube 1 cm on edge (Figure 1.7).

$$1 \, m^3 = 1000 \, dm^3 = 1{,}000{,}000 \, cm^3 \qquad (264.2 \, gal)$$

$$1 \, dm^3 = 1L = 1000 \, mL \qquad (1.057 \, qt)$$

FIGURE 1.8 shows some of the equipment frequently used in the laboratory for measuring liquid volume.

 Go to eText

BIG IDEA Question 5

What is the edge length of a cube with a volume of 1 L?

TABLE 1.3 Some Derived Quantities		
Quantity	Definition	Derived Unit (Name)
Area	Length times length	m^2
Volume	Area times length	m^3
Density	Mass per unit volume	kg/m^3
Speed	Distance per unit time	m/s
Acceleration	Change in speed per unit time	m/s^2
Force	Mass times acceleration	$(kg{\cdot}m)/s^2$ (newton, N)
Pressure	Force per unit area	$kg/(m{\cdot}s^2)$ (pascal, Pa)
Energy	Force times distance	$(kg{\cdot}m^2)/s^2$ (joule, J)

▶ **FIGURE 1.7**

Units for measuring volume. A cubic meter is the volume of a cube 1 meter along each edge.

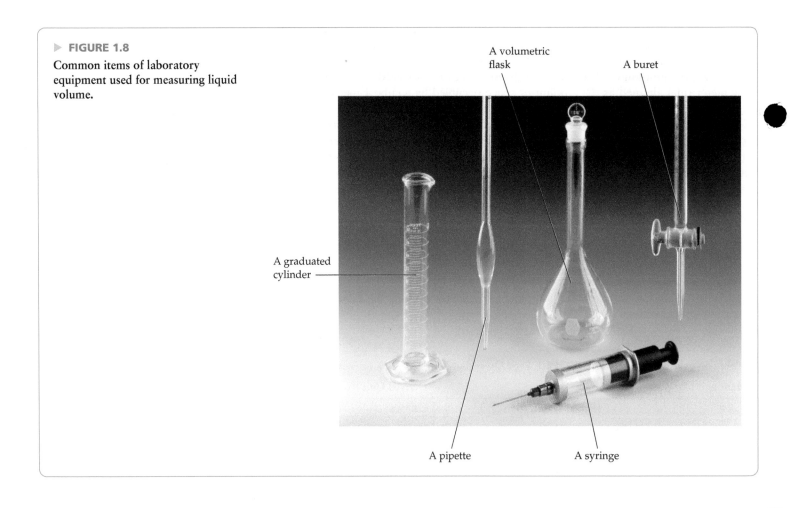

$$1 m^3 = 1000 dm^3$$
$$1 dm^3 = 1 L$$
$$= 1000 cm^3$$
$$1 cm^3 = 1 mL$$

Each cubic meter contains 1000 cubic decimeters (liters).

Each cubic decimeter contains 1000 cubic centimeters (milliliters).

▶ **FIGURE 1.8**

Common items of laboratory equipment used for measuring liquid volume.

A volumetric flask

A buret

A graduated cylinder

A pipette

A syringe

1.7 DERIVED UNITS: DENSITY AND ITS MEASUREMENT

The relationship between the mass of an object and its volume is called *density*. **Density** is calculated as the mass of an object divided by its volume and is expressed in the SI derived unit g/mL for a liquid or g/cm³ for a solid. The densities of some common materials are given in **TABLE 1.4.**

TABLE 1.4 Densities of Some Common Materials	
Substance	Density (g /cm³)
Ice (0 °C)	0.917
Water (3.98 °C)	1.0000
Gold	19.31
Helium (25 °C)	0.000 164
Air (25 °C)	0.001 185
Human fat	0.94
Human muscle	1.06
Cork	0.22–0.26
Balsa wood	0.12
Earth	5.54

$$Density = \frac{Mass\ (g)}{Volume\ (mL\ or\ cm^3)}$$

Because most substances change in volume when heated or cooled, densities are temperature dependent. At 3.98 °C for example, a 1.0000 mL container holds exactly 1.0000 g of water (density = 1.0000 g/mL). As the temperature is raised, however, the volume occupied by the water expands so that only 0.9584 g fits in the 1.0000 mL container at 100 °C (density = 0.9584 g/mL). *When reporting a density, the temperature must also be specified.*

Although most substances expand when heated and contract when cooled, water behaves differently. Water contracts when cooled from 100 °C to 3.98 °C, but below this temperature it begins to expand again. Thus, the density of liquid water is at its maximum of 1.0000 g/mL at 3.98 °C but decreases to 0.999 87 g/mL at 0 °C (**FIGURE 1.9**). When freezing occurs, the density drops still further to a value of 0.917 g/cm³ for ice at 0 °C. Ice and any other substance with a density less than that of water will float, but any substance with a density greater than that of water will sink.

Knowing the density of a substance, particularly a liquid, can be very useful because it's often easier to measure a liquid by volume than by mass. Suppose, for example, that you needed 1.55 g of ethyl alcohol. Rather than trying to weigh exactly the right amount, it would be much easier to look up the density of ethyl alcohol (0.7893 g/mL at 20 °C) and measure the correct volume with a syringe as shown in Figure 1.8.

$$Density = \frac{Mass}{Volume} \quad so \quad Volume = \frac{Mass}{Density}$$

$$Volume = \frac{1.55\ g\ ethyl\ alcohol}{0.7893\ \frac{g}{mL}} = 1.96\ mL\ ethyl\ alcohol$$

▲ Which weighs more, the brass weight or the pillow? Actually, both have identical masses and weights, but the brass has a higher density because its volume is smaller.

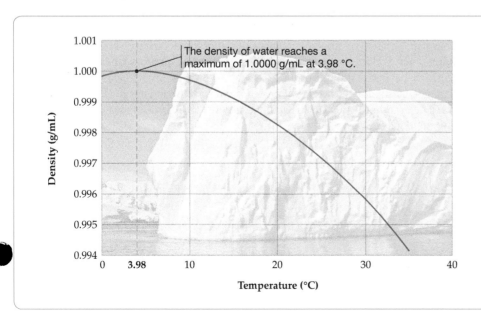

The density of water reaches a maximum of 1.0000 g/mL at 3.98 °C.

Density (g/mL) vs Temperature (°C)

◀ **FIGURE 1.9**

The density of water at different temperatures.

◀ **Figure It Out**

Would 10.0 mL of water at 10 °C have the same mass as 10.0 mL of water at 25 °C?

Answer: No, the density of water changes with temperature. 10.0 mL of water at 10 °C would have a higher mass than 10.0 mL of water at 25 °C because the density is higher at 10 °C.

— **WORKED EXAMPLE 1.3**

Relating Mass and Volume Using Density

What is the volume in cm^3 of 201 g of gold?

IDENTIFY

Known	Unknown
Mass of gold (201 g)	Volume of gold (cm^3)

STRATEGY

Rearrange the equation for density to solve for volume. Use the known value for the density of gold in Table 1.4 and the mass of gold in the equation.

SOLUTION

$$\text{Volume} = \frac{201 \text{ g gold}}{19.31 \text{ g}/cm^3} = 10.4 \text{ } cm^3 \text{ gold}$$

CHECK

The mass of gold (201 g) is roughly 10 times larger than the density (19.31 g), so the estimate for volume is 10 cm^3, which agrees with the calculated answer.

▶ **PRACTICE 1.5**   Chloroform, a substance once used as an anesthetic, has a density of 1.483 g/mL at 20 °C. How many milliliters would you use if you needed 9.37 g? (1 mL = 1 cm^3)

▶ **APPLY 1.6**   You are beachcombing on summer vacation and find a silver bracelet. You take it to the jeweler, and he tells you that it is silver plated and will give you $10 for it. You do not want to be swindled, so you take the bracelet to your chemistry lab and find its mass on a balance (80.0 g). To measure the volume, you place the bracelet in a graduated cylinder (Figure 1.8) containing 10.0 mL of water at 20 °C. The final volume in the graduated cylinder after the bracelet has been added is 17.61 mL. The density of silver at 20 °C is 10.5 g/cm^3 and 1 cm^3 = 1 mL. What can you conclude about the identity of the metal in the bracelet?

▲ How might you determine if this bracelet is pure silver?

1.8 DERIVED UNITS: ENERGY AND ITS MEASUREMENT

The word *energy* is familiar to everyone but is surprisingly hard to define in simple, nontechnical terms. A good working definition, however, is to say that **energy** is the capacity to supply heat or do work. The water falling over a dam, for instance, contains energy that can be used to turn a turbine and generate electricity. A tank of propane gas contains energy that, when released in the chemical process of combustion, can heat a house or barbecue a hamburger.

Energy is classified as either *kinetic* or *potential*. **Kinetic energy** (E_K) is the energy of motion. The amount of kinetic energy in a moving object with mass m and velocity v is given by the equation

$$E_K = \frac{1}{2}mv^2$$

The larger the mass of an object and the larger its velocity, the larger the amount of kinetic energy. Thus, water that has fallen over a dam from a great height has a greater velocity and more kinetic energy than the same amount of water that has fallen only a short distance.

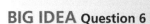

Go to eText

BIG IDEA Question 6

Which of the following statements describes potential energy?
(a) Atoms in a crystal vibrate at temperatures above 0 K.
(b) A negatively charged electron is attracted to a positively charged nucleus.
(c) At room temperature, the average speed of an oxygen molecule is approximately 500 m/s.

Potential energy (E_p), by contrast, is stored energy—perhaps stored in an object because of its height or in a molecule because of chemical reactions it can undergo. The water sitting in a reservoir behind the dam contains potential energy because of its height above the stream at the bottom of the dam. When the water is allowed to fall, its potential energy is converted into kinetic energy. Propane and other substances used as fuels contain potential energy because they can undergo a combustion reaction with oxygen that releases **energy** as heat and work.

The units for energy, $(kg·m^2)/s^2$, follow from the expression for kinetic energy, $E_K = 1/2mv^2$. If, for instance, your body has a mass of 50.0 kg (about 110 lb) and you are riding a bicycle at a velocity of 10.0 m/s (about 22 mi/h), your kinetic energy is 2500 $(kg·m^2)/s^2$.

LOOKING AHEAD . . .
In Chapter 9, we'll calculate the amount of **energy** released or absorbed during a chemical reaction.

$$E_K = \frac{1}{2}mv^2 = \frac{1}{2}(50.0 \text{ kg})\left(10.0\frac{m}{s}\right)^2 = 2500\frac{kg·m^2}{s^2} = 2500 \text{ J}$$

The SI derived unit for energy $(kg·m^2)/s^2$ is given the name **joule (J)** after the English physicist James Prescott Joule (1818–1889). The joule is a fairly small amount of energy—it takes roughly 100,000 J to heat a coffee cup full of water from room temperature to boiling—so kilojoules (kJ) are more frequently used in chemistry.

In addition to the SI energy unit joule, some chemists and biochemists still use the unit calorie (cal, with a lowercase c). Originally defined as the amount of energy necessary to raise the temperature of 1 g of water by 1 °C (specifically, from 14.5 °C to 15.5 °C), one calorie is now defined as exactly 4.184 J.

$$1 \text{ cal} = 4.184 \text{ J (exactly)}$$

Nutritionists use the somewhat confusing unit Calorie (Cal, with a capital C), which is equal to 1000 calories, or 1 kilocalorie (kcal).

$$1 \text{ Cal} = 1000 \text{ cal} = 1 \text{ kcal} = 4.184 \text{ kJ}$$

The energy value, or caloric content, of food is measured in Calories. Thus, the statement that a banana contains 70 Calories means that 70 Cal (70 kcal, or 290 kJ) of energy is released when the banana is used by the body for fuel.

(a)   **(b)**

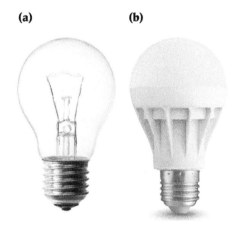

▲ **(a)** A 75-watt incandescent bulb uses energy at the rate of 75 J/s. Only about 5% of that energy appears as light, however; the remaining 95% is given off as heat. **(b)** Energy-efficient light-emitting diode (LED) bulbs are replacing incandescent lights.

WORKED EXAMPLE 1.4

Calculating Kinetic Energy

(a) What is the kinetic energy in joules of a 2360 lb (1070 kg) car moving at 63.3 mi/h (28.3 m/s)? Express the number in scientific notation.
(b) Express the number and unit using an appropriate prefix.

IDENTIFY

Known	Unknown
Mass (1070 kg), velocity (28.3 m/s)	Kinetic energy (E_K) in units of joules (J)

STRATEGY

Use the formula to calculate kinetic energy. If mass is in units of kg and velocity is in units of m/s, then energy will be calculated in units of joules because 1 joule $= \dfrac{1 \text{ kg·m}^2}{s^2}$.

SOLUTION

(a) $E_K = \dfrac{1}{2}mv^2 = \dfrac{1}{2}(1070 \text{ kg})\left(28.3\dfrac{m}{s}\right)^2 = 428,476\dfrac{kg·m^2}{s^2}$

$= 4.28476 \times 10^5 \text{ J}$

(b) Kilo (10^3) and mega (10^6) are both prefixes with exponents similar to the answer in part **(a)**. Therefore, both 428.476 kJ and 0.428 476 MJ are reasonable answers.

CHECK

An answer with a large magnitude should be expected as a car has a very large mass. To estimate the magnitude of kinetic energy, express mass and velocity in the equation using exponents only.

$$E_K \approx (10^3 \text{ kg})\left(\frac{10^1 m}{s}\right)^2 = 10^5\frac{kg·m^2}{s^2}$$

The estimated power of 10 agrees with the detailed calculation.

continued on next page

▶ **PRACTICE 1.7** Some radioactive materials emit a type of radiation called alpha particles at high velocity.

What is the kinetic energy in joules of an alpha particle with a mass of 6.6×10^{-24} g and a speed of 1.5×10^7 m/s? Express the number in scientific notation.

▶ **APPLY 1.8** A baseball with a mass of 450 g has a kinetic energy of 406 J. Calculate the velocity of the baseball in units of m/s.

1.9 ACCURACY, PRECISION, AND SIGNIFICANT FIGURES IN MEASUREMENT

Measuring things, whether in cooking, construction, or chemistry, is something that most of us do every day. But how good are those measurements? Any measurement is only as good as the skill of the person doing the work and the reliability of the equipment being used. You've probably noticed, for instance, that you often get slightly different readings when you weigh yourself on a bathroom scale and on a scale at the doctor's office, so there's always some uncertainty about your real weight. The same is true in chemistry—there is always some uncertainty in the value of a measurement.

In talking about the degree of uncertainty in a measurement, we use the words *accuracy* and *precision*. Although most of us use the words interchangeably in daily life, there's actually an important distinction between them. **Accuracy** refers to how close to the true value a given measurement is, whereas **precision** refers to how well a number of independent measurements agree with one another. To see the difference, imagine that you weigh a tennis ball whose true mass is 54.441 778 g. Assume that you take three independent measurements on each of three different types of balance to obtain the data shown in the following table.

▲ This tennis ball has a mass of about 54 g.

Measurement #	Bathroom Scale	Lab Balance	Analytical Balance
1	0.1 kg	54.4 g	54.4418 g
2	0.0 kg	54.5 g	54.4417 g
3	0.1 kg	54.3 g	54.4418 g
(average)	(0.07 kg)	(54.4 g)	(54.4418 g)

If you use a bathroom scale, your measurement (average = 0.07 kg) is neither accurate nor precise. Its accuracy is poor because it measures to only one digit that is far from the true value, and its precision is poor because any two measurements may differ substantially. If you now weigh the ball on an inexpensive laboratory balance, the value you get (average = 54.4 g) has three digits and is fairly accurate, but it is still not very precise because the three readings vary from 54.3 g to 54.5 g, perhaps due to air movements in the room or a sticky mechanism. Finally, if you weigh the ball on an expensive analytical balance like those found in research laboratories, your measurement (average = 54.4418 g) is both precise and accurate. It's accurate because the measurement is very close to the true value, and it's precise because it has six digits that vary little from one reading to another.

To indicate the uncertainty in a measurement, *the value you record should use all the digits you are sure of plus one additional digit that you estimate.* In reading a thermometer that has a mark for each degree, for example, you could be certain about the digits of the nearest mark—say, 25 °C—but you would have to estimate between two marks—say, between 25 °C and 26 °C—to obtain a value of 25.3 °C.

The total number of digits recorded for a measurement is called the measurement's number of **significant figures.** For example, the mass of the tennis ball as determined on the single-pan balance (54.4 g) has three significant figures, whereas the mass determined on the analytical balance (54.4418 g) has six significant figures. All digits but the last are certain; the final digit is an estimate, which we generally assume to have an error of plus or minus one (± 1).

Finding the number of significant figures in a measurement is usually easy but can be troublesome if zeros are present. Look at the following four quantities:

4.803 cm	Four significant figures: 4, 8, 0, 3
0.006 61 g	Three significant figures: 6, 6, 1
55.220 K	Five significant figures: 5, 5, 2, 2, 0
34,200 m	Anywhere from three (3, 4, 2) to five (3, 4, 2, 0, 0) significant figures

The following rules cover the different situations that arise:

1. **Zeros in the middle of a number are like any other digit; they are always significant.** Thus, 4.803 cm has four significant figures.
2. **Zeros at the beginning of a number are not significant; they act only to locate the decimal point.** Thus, 0.006 61 g has three significant figures. (Note that 0.006 61 g can be rewritten as 6.61×10^{-3} g or as 6.61 mg.)
3. **Zeros at the end of a number and after the decimal point are always significant.** The assumption is that these zeros would not be shown unless they were significant. Thus, 55.220 K has five significant figures. (If the value were known to only four significant figures, we would write 55.22 K.)
4. **Zeros at the end of a number and before the decimal point may or may not be significant.** We can't tell whether they are part of the measurement or whether they just locate the decimal point. Thus, 34,200 m may have three, four, or five significant figures. Often, however, a little common sense is helpful. A temperature reading of 20 °C probably has two significant figures rather than one, since one significant figure would imply a temperature anywhere from 10 °C to 30 °C and would be of little use. Similarly, a volume given as 300 mL probably has three significant figures. On the other hand, a figure of 93,000,000 mi for the distance between the Earth and the Sun probably has only two or three significant figures.

The fourth rule shows why it's helpful to write numbers in scientific notation rather than in ordinary notation. Doing so makes it possible to indicate the number of significant figures. Thus, writing the number 34,200 as 3.42×10^4 indicates three significant figures. but writing it as 3.4200×10^4 indicates five significant figures.

One further point about significant figures: certain numbers, such as those obtained when counting objects, are exact and have an effectively infinite number of significant figures. A week has exactly 7 days, for instance, not 6.9 or 7.0 or 7.1, and a foot has exactly 12 inches, not 11.9 or 12.0 or 12.1. In addition, the power of 10 used in scientific notation is an exact number. That is, the number 10^3 is exactly 1000, but the number 1×10^3 has one significant figure.

WORKED EXAMPLE 1.5

Significant Figures

How many significant figures does each of the following measurements have?

(a) 0.036 653 m   (b) 7.2100×10^{-3} g   (c) 72,100 km   (d) $25.03

SOLUTION

(a) 5 (by rule 2)   (b) 5 (by rule 3)
(c) 3, 4, or 5 (by rule 4)   (d) $25.03 is an exact number

continued on next page

▲ What is the volume in this buret (APPLY 1.10)?

▲ Calculators often display more figures than are justified by the precision of the data.

▶ **PRACTICE 1.9**   How many significant figures does each of the following quantities have?

(a) 0.003 00 mL   (b) 2070 mi   (c) 47.60 mL

▶ **APPLY 1.10**   Read the volume of the buret and report your answer to the correct number of significant figures. The volume is indicated by the bottom of the meniscus. (Remember that the value you record should include all the digits that can be determined from the gradation marks plus one additional digit that is estimated.)

CONCEPTUAL WORKED EXAMPLE 1.6

Determining Precision and Accuracy in a Set of Measurements

Which dartboard represents low accuracy but high precision?

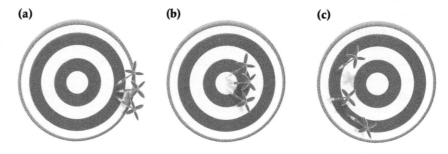

(a)   (b)   (c)

SOLUTION

Precision refers to how close the darts are to one another, and accuracy refers to how close they are to the center of the target. Dartboard (a) has low accuracy because the darts are far from the center but high precision because the three darts are all in the same location.

▶ **PRACTICE 1.11**   Examine the figure in Worked Example 1.6. Which dartboard has low accuracy and precision?

▶ **APPLY 1.12**   A 1.000 mL sample of acetone, a common solvent used as a paint remover, was placed in a small vial whose mass was known to be 4.002 g. The following values were obtained when the acetone-filled vial was weighed: 4.531 g, 4.525 g, and 4.537 g. How would you characterize the precision and accuracy of these measurements if the true mass of the acetone was 0.7795 g?

1.10 SIGNIFICANT FIGURES IN CALCULATIONS

It often happens, particularly when doing arithmetic on a calculator, that a quantity appears to have more significant figures than are really justified. You might calculate the gas mileage of your car, for instance, by finding that it takes 11.70 gallons of gasoline to drive 278 miles:

$$\text{Mileage} = \frac{\text{Miles}}{\text{Gallons}} = \frac{278 \text{ mi}}{11.70 \text{ gal}} = 23.760\ 684 \text{ mi/gal (mpg)}$$

Although the answer on the calculator has eight digits, your measurement is really not as precise as it appears. In fact, your answer is precise to only three significant figures and should be **rounded off** to 23.8 mi/gal by removing all nonsignificant figures.

How do you decide how many figures to keep and how many to ignore? For most purposes, a simple procedure using just two rules is sufficient.

1. **In carrying out a multiplication or division, the answer can't have more significant figures than either of the original numbers.** If you think about it, this rule is just common sense. If you don't know the number of miles you drove to better than three significant figures (278 could mean 277, 278, or 279), you certainly can't calculate your mileage to more than the same number of significant figures.

$$\underset{\substack{\text{Four significant} \\ \text{figures}}}{\overset{\substack{\text{Three significant} \\ \text{figures}}}{\frac{278 \text{ mi}}{11.70 \text{ gal}}}} = \underset{\substack{\text{Three significant} \\ \text{figures}}}{23.8 \text{ mi/gal}}$$

2. **In carrying out an addition or subtraction, the answer can't have more digits to the right of the decimal point than either of the original numbers.** For example, if you have 3.18 L of water and you add 0.013 15 L more, you now have 3.19 L. Again, this rule is just common sense. If you don't know the volume you started with past the second decimal place, you can't know the total of the combined volumes past the same decimal place.

$$
\begin{array}{l}
3.18?\ ?? \quad \longleftarrow \text{Ends two places past decimal point}\\
+\ 0.013\ 15 \quad \longleftarrow \text{Ends five places past decimal point}\\
\hline
3.19?\ ?? \quad \longleftarrow \text{Ends two places past decimal point}
\end{array}
$$

Once you decide how many digits to retain for your answer, the rules for rounding off numbers are as follows:

1. **If the first digit you remove is less than 5, round down by dropping it and all following digits.** Thus, 5.664 becomes 5.66 when rounded to three significant figures because the first of the dropped digits (4) is less than 5.
2. **If the first digit you remove is 5 or greater, round up by adding 1 to the digit on the left.** Thus, 5.664 becomes 5.7 when rounded to two significant figures because the first of the dropped digits (6) is greater than 5.

WORKED EXAMPLE 1.7

Significant Figures in Calculations

It takes 9.25 hours to fly from London, England, to Chicago, Illinois, a distance of 3952 miles. What is the average speed of the airplane in miles per hour?

IDENTIFY

Known	Unknown
Time (9.25 h), distance (3952 mi)	Speed (mi/h)

STRATEGY

Set up a mathematical expression and solve for the answer. Use rules for significant figures in mathematical operations to determine the number of significant figures in the answer.

SOLUTION

First, set up an equation dividing the number of miles flown by the number of hours:

$$\text{Average speed} = \frac{3952 \text{ mi}}{9.25 \text{ h}} = 427.243\ 24 \text{ mi/h}$$

Next, decide how many significant figures should be in your answer. Because the problem involves division and because one of the quantities you started with (9.25 h) has

continued on next page

only three significant figures, the answer must also have three significant figures. Finally, round off your answer. The first digit to be dropped (2) is less than 5, so the answer 427.243 24 must be rounded off to 427 mi/h.

In doing this or any other problem, use all figures, significant or not, for the calculation and then round off the final answer. Don't round off at any intermediate step.

▶ **PRACTICE 1.13**   Carry out the following calculations, expressing each result with the correct number of significant figures:

(a) 24.567 g + 0.044 78 g = ? g
(b) 4.6742 g ÷ 0.003 71 L = ? g/L

▶ **APPLY 1.14**   A sodium chloride solution was prepared in the following manner:

- A 25.0 mL volumetric flask (Figure 1.8) was placed on an analytical balance and found to have a mass of 35.6783 g.
- Sodium chloride was added to flask, and the mass of the solid + flask was 36.2365 g.
- The flask was filled to the mark with water and mixed well.

Calculate the concentration of the sodium chloride solution in units of g/mL, and give the answer in scientific notation with the correct number of significant figures.

1.11 CONVERTING FROM ONE UNIT TO ANOTHER

Because so many scientific activities involve numerical calculations—measuring, weighing, preparing solutions, and so forth—it's often necessary to convert a quantity from one unit to another. Converting between units isn't difficult; we all do it every day. If you run 7.5 laps around a 400-meter track, for instance, you have to convert between the distance unit *lap* and the distance unit *meter* to find that you have run 3000 m (7.5 laps times 400 meters/lap). Converting from one scientific unit to another is just as easy.

$$7.5 \text{ laps} \times \frac{400 \text{ meters}}{1 \text{ lap}} = 3000 \text{ meters}$$

The simplest way to carry out calculations that involve different units is to use the **dimensional-analysis method.** In this method, a quantity described in one unit is converted into an equivalent quantity with a different unit by multiplying with a **conversion factor** that expresses the relationship between units.

Original quantity × Conversion factor = Equivalent quantity

As an example, we know from Section 1.4 that 1 meter equals 39.37 inches. Writing this relationship as a ratio restates it in the form of a conversion factor, either meters per inch or inches per meter.

Conversion factors between meters and inches

$$\frac{1 \text{ m}}{39.37 \text{ in.}} \quad \text{or} \quad \frac{39.37 \text{ in.}}{1 \text{ m}}$$

The key to the dimensional-analysis method of problem solving is that units are treated like numbers and can thus be multiplied and divided just as numbers can. The idea when solving a problem is to set up an equation so that unwanted units cancel, leaving only the desired units. Usually it's best to start by writing what you know and then manipulating that known quantity. For example, say you know your height is 69.5 inches and you want to find it in meters. Begin by writing your height in inches, and then set up an equation multiplying your height by the conversion factor meters per inch:

$$69.5 \text{ in.} \times \frac{1 \text{ m}}{39.37 \text{ in.}} = 1.77 \text{ m}$$

Starting quantity   Conversion factor   Equivalent quantity

The unit "in." cancels because it appears both above and below the division line, so the only unit that remains is "m."

▲ Speed skaters have to convert from laps to meters to find out how far they have gone.

The dimensional-analysis method gives the right answer only if the conversion factor is arranged so that the unwanted units cancel. If the equation is set up in any other way, the units won't cancel properly and you won't get the right answer. Thus, if you were to multiply your height in inches by an inverted conversion factor of inches per meter rather than meters per inch, you would end up with an incorrect answer expressed in meaningless units.

$$\text{Wrong!} \quad 69.5 \text{ in} \times \frac{39.37 \text{ in.}}{1 \text{ m}} = 2740 \text{ in.}^2/\text{m} \quad ??$$

The main drawback to using the dimensional-analysis method is that it's easy to get the right answer without really understanding what you're doing. It's therefore best after solving a problem to think through a rough estimate to check your work. If your estimate isn't close to the answer you get from the detailed solution, there's a misunderstanding somewhere and you should think through the problem again.

Even if you don't make an estimate, it's important to be sure that your calculated answer makes sense. If, for example, you were trying to calculate the volume of a human cell and you came up with the answer 5.3 cm³, you should realize that such an answer couldn't possibly be right. Cells are too tiny to be distinguished with the naked eye, but a volume of 5.3 cm³ is about the size of a walnut. Worked Examples 1.8 and 1.9 show how to devise strategies and estimate answers when converting units using dimensional analysis.

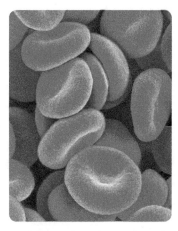

▲ What is the volume of a red blood cell?

Go to eText

WORKED EXAMPLE 1.8

Unit Conversions Using Significant Figures

The Bugatti Veyron Super Sport is the fastest production sports car in the world, with a top speed of 267 miles per hour. What is this speed (reported to the correct number of significant figures) in units of

(a) kilometers per hour?   (b) meters per second?

IDENTIFY

Known	Unknown
Speed (267 mi/h)	Speed (km/h) and speed (m/s)

STRATEGY

(a) Find the conversion factor between km and mi on the inside back cover of this book, and use the dimensional-analysis method to set up an equation so the "mi" units cancel.

(b) Let's begin with our answer from part (a) with the speed in km/h; the unknown is the speed in m/s. Set up a series of conversion factors so that units of "km" and "h" cancel and you are left with units of "m" in the numerator and "s" in the denominator.

SOLUTION

(a) $\dfrac{267 \text{ mi}}{1 \text{ h}} \times \dfrac{1.609 \text{ km}}{1 \text{ mi}} = 429.603 \dfrac{\text{km}}{\text{h}} = 430. \dfrac{\text{km}}{\text{h}}$

(b) $\dfrac{430. \text{ km}}{1 \text{ h}} \times \dfrac{1000 \text{ m}}{1 \text{ km}} \dfrac{1 \text{ h}}{60 \text{ min.}} \times \dfrac{1 \text{ min.}}{60 \text{ s}} = 119.444 \dfrac{\text{m}}{\text{s}} = 119 \dfrac{\text{m}}{\text{s}}$

A very fast car!

CHECK

(a) The answer is certainly large, perhaps several hundred kilometers per hour (km/h). A better estimate is to realize that, because 1 mi = 1.609 km, it takes about $1\frac{1}{2}$ times as many kilometers as miles to measure the same distance. Thus, 267 mi is about 400 km, and 267 mi/h is about 400 km/h. The estimate agrees with the detailed solution.

(b) At first glance, the answer makes sense as the speed is very high. A top sprinter can run 100 m in about 10 seconds so the car is roughly ten times faster. This seems reasonable. This is a difficult problem to estimate, however, because it requires several different conversions. It's therefore best to think the problem through one step at a time, writing down the intermediate estimates:

- Because 1 km = 1000 m then the speed is 430,000 m/h or 4.3×10^5 m/h.
- Changing units of time from hours to seconds should decrease the number significantly because the car will travel a shorter distance in 1 second than in 1 hour. Because there are 3600 (3.6×10^3) seconds in 1 hour, we can estimate by dividing the speed 4.3×10^5 m/h by 3.6×10^3 s/h

$$\frac{10^5 \text{ m/h}}{10^3 \text{ s/h}} = 10^2 \text{ m/s}$$

Making estimates using powers of 10 is very useful in checking to make sure that your answer is of the correct magnitude. This estimate agrees with the detailed solution.

continued on next page

▶ **PRACTICE 1.15** Gemstones are weighed in *carats*, with 1 carat = 200 mg (exactly). What is the mass in grams of the Hope Diamond, the world's largest blue diamond at 44.4 carats? What is this mass in ounces? (1 oz = 28.35 g)

▶ **APPLY 1.16** A pure diamond has a density of 3.52 g/cm^3. Set up a dimensional-analysis equation to find the volume (cm^3) of the Hope Diamond (PRACTICE 1.15).

All **WORKED EXAMPLES** with this icon have an interactive video in the eText.

▲ Volcano on Krakatau in Indonesia

WORKED EXAMPLE 1.9

Unit Conversions with Squared and Cubed Units

The volcanic explosion that destroyed the Indonesian island of Krakatau on August 27, 1883, released an estimated 4.3 cubic miles (mi^3) of debris into the atmosphere and affected global weather for years. In SI units, how many cubic meters (m^3) of debris were released?

IDENTIFY

Known	Unknown
Volume (4.3 mi^3)	Volume (m^3)

STRATEGY

It's probably simplest to convert first from mi^3 to km^3 and then convert km^3 to m^3. *Notice that the entire conversion factor is cubed.*

SOLUTION

$$4.3 \ \text{mi}^3 \times \left(\frac{1 \ \text{km}}{0.6214 \ \text{mi}} \right)^3 = 17.92 \ \text{km}^3$$

$$17.92 \ \text{km}^3 \times \left(\frac{1000 \ \text{m}}{1 \ \text{km}} \right)^3 = 1.792 \times 10^{10} \ \text{m}^3$$

$$= 1.8 \times 10^{10} \ \text{m}^3 \quad \text{Rounded off}$$

CHECK

One meter is much less than 1 mile, so it takes a large number of cubic meters to equal 1 mi^3, and the answer is going to be very large. Because 1 km is about 0.6 mi, 1 km^3 is about $(0.6)^3 = 0.2$ times as large as 1 mi^3. Thus, each mi^3 contains about 5 km^3, and 4.3 mi^3 contains about 20 km^3. Each km^3, in turn, contains $(1000 \ \text{m})^3 = 10^9 \ \text{m}^3$. Thus, the volume of debris from the Krakatau explosion was about $20 \times 10^9 \ \text{m}^3$, or $2 \times 10^{10} \ \text{m}^3$. The estimate agrees with the detailed solution.

▶ **PRACTICE 1.17** The maximum dimensions of a soccer field are 90.0 m wide and 120.0 m long, giving an area of $1.08 \times 10^4 \ \text{m}^2$. What is the area of the soccer field in square feet? (1 m = 3.28 ft)

▶ **APPLY 1.18** How large, in cubic centimeters, is the volume of a red blood cell (in cm^3) if the cell has a cylindrical shape with a diameter of 6×10^{-6} m and a height of 2×10^{-6} m? What is the volume in pL?

INQUIRY ❓ What are the unique properties of nanoscale materials?

Imagine a world of new lightweight replacements for metals, synthetic scaffolds on which bones can be regrown, drugs that target and kill cancer cells with a minimum of side effects, and faster, smaller computers. Those are but a few of the developments that might emerge from nanoscience, one of the hottest research areas in science today.

The nanoscale is generally defined as including any material of which at least one dimension is 1 to 100 nanometers in length. New tools enable scientists to explore and understand matter in this extremely small realm is ways that were not possible a only a few years ago.

To appreciate the extremely small size of nanomaterials, it is useful to compare the sizes of objects on different scales.

Macroscale items are large enough to be observed with the human eye and are measured with instruments such as rulers and calipers. The **microscale** is a smaller size regime and is so named because dimensions of materials are in the micrometer range ($1 \ \mu m = 1 \times 10^{-6}$ m). Microscale objects, such as cells, cannot be seen with the human eye and therefore must be imaged with an optical microscope. One thousand times smaller than the microscale is the **nanoscale**, representing particles with nanometer-sized ($1 \ nm = 1 \times 10^{-9}$ m) dimensions. Atoms and molecules are nanoscale entities that use specialized instruments such as electron microscopes and atomic force microscopes for imaging. **FIGURE 1.10** depicts the scale regimes and representative objects.

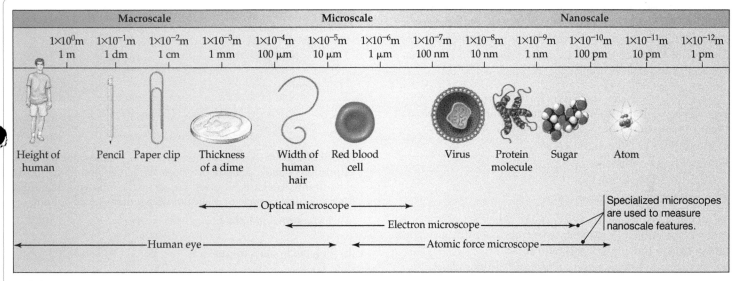

▲ **FIGURE 1.10**

Size scale of macroscopic and microscopic objects.

> ▲ **Figure It Out**
>
> What objects can be seen with a microscope but not the human eye? What particles can be imaged with an atomic force microscope but not an optical microscope?
>
> **Answer:** Cells can be seen with a microscope but not the eye. Molecules and viruses can be seen with an atomic force microscope but not an optical microscope. Some very high resolution images of individual atoms have been recorded with an atomic force microscope.

▶ A bar of gold (left) and solutions of various-sized gold nanoparticles (right). Gold does not retain its characteristic color at the nanoscale. The red solution in the vial on the left contains gold nanoparticles with diameters of 3–30 nm, and particle size increases going to the right. The violet color on the right is characteristic of gold particles hundreds of nanometers in diameter.

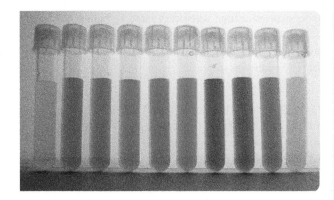

Nanotechnology is an exciting research frontier because reducing the size of an object to nanometer proportions alters its properties. You may have learned in previous science classes that a substance has the same properties regardless of how much is present. For instance, gold is a yellow, shiny material that has a high melting point and conducts electricity. It is relatively *inert*, or unreactive, which is why it is useful for making jewelry or money. Gold has exactly the same properties in a large bar as it does in a tiny flake. But these properties do not extend to gold nanoparticles. Gold nanoparticles are chemically reactive and range in color from red to purple. In general, *nanoparticles have unique properties that vary with size and composition.* They tend to have lower melting points, different colors, and greater reactivity than the material in bulk.

The *surface area-to-volume ratio* of a particle is a measure that can explain why some properties change with particle size. In small particles, a substantial portion of atoms are on the surface, which makes them more reactive. **FIGURE 1.11** illustrates how the size of a cubic object influences this quantity. The surface area (*SA*) is calculated by multiplying the number of sides by the area of each side, $A = (l \times w)$, and the volume (*V*) is calculated using the formula $V = l \times w \times h$. Both the surface area and volume on the cube with the side length of 2 cm are larger than the cube with a 1 cm side. However, the surface area to volume ratio is smaller for the larger cube.

PROBLEM 1.19 Refer to Figure 1.10. What object(s) can be seen with an optical microscope but not the human eye? Select all the correct answers.
(a) An tiny ant (1 mm long)
(b) A cell (5 μm radius)
(c) A virus (50 nm radius)
(d) A molecule (1 nm radius)

PROBLEM 1.20 Refer to Figure 1.10. What object(s) can be seen with an atomic force microscope but not with an optical microscope? Select all the correct answers.
(a) An tiny ant (1 mm long)   (b) A cell (5 μm radius)
(c) A virus (50 nm radius)   (d) A molecule (1 nm radius)

PROBLEM 1.21 Use Figure 1.10 to estimate in powers of 10
(a) how many times larger the diameter of a human hair is than a 10 nm gold nanoparticle.
(b) how many times larger a red blood cell is than a glucose molecule.

PROBLEM 1.22 On the nanoscale, materials often exhibit novel properties not observed at other scales. Which properties of gold nanoparticles change when compared to bulk gold? Select all the correct answers.
(a) Color changes from shiny yellow to reddish-purple
(b) The reactivity increases
(c) The melting point decreases

PROBLEM 1.23 Refer to Figure 1.11. Which cube has a greater surface area to volume ratio?
(a) A cube with an edge length of 1 cm
(b) A cube with an edge length of 2 cm

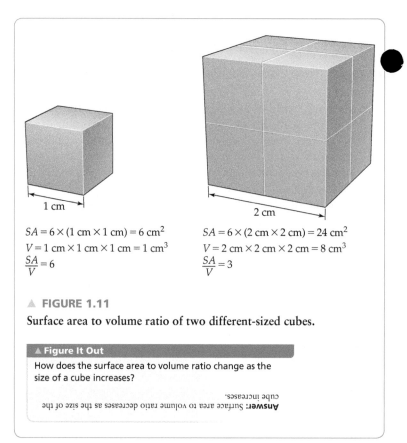

$SA = 6 \times (1\ \text{cm} \times 1\ \text{cm}) = 6\ \text{cm}^2$
$V = 1\ \text{cm} \times 1\ \text{cm} \times 1\ \text{cm} = 1\ \text{cm}^3$
$\frac{SA}{V} = 6$

$SA = 6 \times (2\ \text{cm} \times 2\ \text{cm}) = 24\ \text{cm}^2$
$V = 2\ \text{cm} \times 2\ \text{cm} \times 2\ \text{cm} = 8\ \text{cm}^3$
$\frac{SA}{V} = 3$

▲ **FIGURE 1.11**

Surface area to volume ratio of two different-sized cubes.

▲ **Figure It Out**

How does the surface area to volume ratio change as the size of a cube increases?

Answer: Surface area to volume ratio decreases as the size of the cube increases.

PROBLEM 1.24 Catalytic converters use nanoscale particles of precious metals such as platinum to change pollutants in car exhaust into less harmful gases. Calculate the following quantities for a spherical particle of platinum with a diameter of 5.0 nm.
(a) Surface area in units of μm^2 ($SA = 4\pi r^2$)
(b) Volume in units of μm^3 $\left(V = \frac{4}{3}\pi r^3 \right)$
(c) Surface area to volume ratio in units of μm^{-1}
(d) A 5.0 μm diameter particle has a surface area to volume ratio of 0.6 μm^{-1}. How many times larger is the surface area to volume ratio of the 5.0 nm particle than the 5.0 μm particle?

PROBLEM 1.25 Platinum is an expensive and rare metal used in catalytic converters. Much research has been devoted to maximizing reactive properties of other less expensive metals by reducing their size to the nanoscale. Smaller particles are more reactive because they have a larger surface area to volume ratio. Which type of an atom in a nanoparticle is more reactive?
(a) An atom on the surface of a particle
(b) An atom in the interior of a particle
(c) The location of an atom in the nanoparticle does not influence its reactivity.

PROBLEM 1.26 Calculate the percentage of atoms on the surface of a cubic nanoparticle if the diameter of the atoms is 250 pm and the edge length of the particle is
(a) 5.0 nm   (b) 10.0 nm

STUDY GUIDE

Section	Concept Summary	Learning Objectives	Test Your Understanding
1.1 The Scientific Method: Nanoparticle Catalysts for Fuel Cells	The **scientific method** is an iterative process used to perform research. A driving question, often based upon observations, is the first step. Next a **hypothesis** is developed to explain the observation. **Experiments** are designed to test the hypothesis, and the results are used to verify or modify the original hypothesis. **Theories** arise when numerous experiments validate a hypothesis and are used to make new predictions. **Models** are simplified representations of complex systems that help make theories more concrete.	**1.1** Identify the steps in the scientific method.	Problems 1.32, 1.34
		1.2 Differentiate between a qualitative and quantitative measurement.	Problems 1.34–1.35
1.2 Measurements: SI Units and Scientific Notation	Accurate measurement is crucial to scientific experimentation. Scientists use units of measure established by the *Système Internationale* (**SI units**). There are seven fundamental SI units, together with other derived units (Table 1.1).	**1.3** Write numbers in scientific notation and use prefixes for multiples of SI units.	Worked Example 1.1; Problems 1.40, 1.44, 1.46, 1.48
1.3 Mass and Its Measurement	**Mass,** the amount of matter in an object, is measured in the SI unit of **kilograms (kg).**	**1.4** Convert between different prefixes used in mass measurements.	Problem 1.50
1.4 Length and Its Measurement	**Length** is measured in the SI unit of **meters (m).**	**1.5** Convert between different prefixes used in length measurements.	Problems 1.49, 1.51
1.5 Temperature and Its Measurement	**Fahrenheit (°F)** is the most common unit for measuring temperature in the United States, whereas **Celsius (°C)** is more common in other parts of the world. **Kelvin (K)** is the standard temperature unit in scientific work.	**1.6** Convert between common units of temperature measurements.	Worked Example 1.2; Problems 1.54, 1.56, 1.60
1.6 Derived Units: Volume and Its Measurement	Volume, the amount of space occupied by an object, is measured in SI units by the **cubic meter (m^3).**	**1.7** Convert between SI and metric units of volume.	Problems 1.64–1.65
		1.8 Convert between different prefixes used in volume measurements.	Problem 1.63
1.7 Derived Units: Density and Its Measurement	**Density** is a property that relates mass to volume and is measured in the derived SI unit g/cm^3 or g/mL.	**1.9** Calculate mass, volume, or density using the formula for density.	Worked Example 1.3; Problems 1.66, 1.68, 1.74, 1.76
		1.10 Predict whether a substance will float or sink in another substance based on density.	Problem 1.31
1.8 Derived Units: Energy and Its Measurement	**Energy** is the capacity to supply heat or do work and is measured in the derived SI unit ($kg \cdot m^2/s^2$), or **joule (J).** Energy is of two kinds, potential and kinetic. **Kinetic energy (E_K)** is the energy of motion, and **potential energy (E_P)** is stored energy.	**1.11** Calculate kinetic energy of a moving object.	Worked Example 1.4; Problems 1.78–1.79
		1.12 Convert between common energy units.	Problems 1.82–1.83
1.9 Accuracy, Precision, and Significant Figures in Measurement	If measurements are **accurate,** they are close to the true value, and if measurements are **precise,** they are reproducible or close to one another.	**1.13** Specify the number of significant figures in a measurement.	Worked Example 1.5; Problems 1.84, 1.86
		1.14 Evaluate the level of accuracy and precision in a data set.	Worked Example 1.6; Problem 1.12
		1.15 Report a measurement to the appropriate number of significant figures.	Problems 1.28–1.29

Section	Concept Summary	Learning Objectives	Test Your Understanding
1.10 Significant Figures in Calculations	It's important when measuring physical quantities or carrying out calculations to indicate the precision of the measurement by **rounding off** the result to the correct number of **significant figures.**	**1.16** Report the result of mathematical calculation to the correct number of significant figures.	Worked Example 1.7; Problems 1.92–1.93
1.11 Converting from One Unit to Another	Because many experiments involve numerical calculations, it's often necessary to manipulate and convert different units of measure. The simplest way to carry out such conversions is to use the **dimensional-analysis method,** in which an equation is set up so that unwanted units cancel and only the desired units remain.	**1.17** Convert from one unit to another using conversion factors.	Worked Examples 1.8–1.9; Problems 1.94, 1.98, 1.102, 1.106

KEY TERMS

accuracy 16
alloy 3
Celsius degree (°C) 9
centimeter (cm) 8
chemistry 2
conversion factor 20
cubic centimeter (cm^3) 11
cubic decimeter (dm^3) 11
cubic meter (m^3) 11
density 13
dimensional-analysis method 20
energy 14

experiment 4
Fahrenheit (°F) 9
gram (g) 7
hypothesis 3
joule (J) 15
kelvin (K) 9
kilogram (kg) 7
kinetic energy (E_K) 14
liter (L) 11
macroscale 23
mass 7
matter 7

meter (m) 8
microgram (μg) 7
micrometer (μm) 8
microscale 23
milligram (mg) 7
milliliter (mL) 11
millimeter (mm) 8
nanometer (nm) 8
nanoscale 23
nanoscience 2
observation 3
picometer (pm) 8

potential energy (E_P) 15
precision 16
qualitative 3
quantitative 3
rounded off 18
scientific method 3
scientific notation 6
SI unit 5
significant figure 16
theory 4

KEY EQUATIONS

- **Relationship between the Kelvin and Celsius Scales (Section 1.5)**
  Temperature in K = Temperature in °C + 273.15
  Temperature in °C = Temperature in K − 273.15
- **Converting between Celsius and Fahrenheit Temperatures (Section 1.5)**

$$°F = \left(\frac{9 \, °F}{5 \, °C} \times °C \right) + 32 \, °F \quad °C = \frac{5 \, °C}{9 \, °F} \times (°F - 32 \, °F)$$

- **Calculating Density (Section 1.7)**

$$\text{Density} = \frac{\text{Mass (g)}}{\text{Volume (mL or } cm^3)}$$

- **Calculating Kinetic Energy (Section 1.8)**

$$E_K = \frac{1}{2} m v^2$$

PRACTICE TEST

After studying this chapter, you can assess your understanding with these practice test questions, which are correlated with chapter learning objectives. If you answer a question incorrectly, refer to the learning objectives in the end-of-chapter Study Guide for assistance. The Study Guide provides a conceptual summary, references a Worked Example to model how to solve the problem, and gives additional problems for more practice.

1. Which of the following statements is a hypothesis about the synthesis of gold nanoparticles? **(LO 1.1)**
   (a) Adding a salt solution to gold nanoparticles causes the color to change from red to blue.
   (b) To examine the effect of salt on gold nanoparticles, variable concentrations of salt are added to the nanoparticles and the results are measured.

(c) A solution of gold nanoparticles with an average diameter of 30 nm has a wavelength of maximum absorption of 450 nm and is a reddish-orange color.
(d) Adding a substance with a negative charge to the surface of the nanoparticles creates repulsive forces that stabilize small particle sizes.

2. Convert 0.055 milliseconds to seconds, and write the answer in scientific notation. **(LO 1.3)**
   (a) 5.5×10^{-3} s
   (b) 5.5×10^{-4} s
   (c) 5.5×10^{-5} s
   (d) 5.5×10^{-7} s

3. Which quantity represents the largest mass? **(LO 1.4)**
   (a) $2.5 \times 10^7 \, \mu g$
   (b) 2.5×10^2 mg
   (c) 2.5×10^8 ng
   (d) 2.5×10^{-3} kg

4. A mammalian HELA cell has a diameter of 2×10^{-5} m. Report the diameter of the cell using the most appropriate prefix on the base unit of meter. **(LO 1.5)**
   (a) $2 \mu m$ (b) $20 \mu m$
   (c) 200 nm   (d) 0.2 mm

5. The temperature on the surface of the Sun is 5778 K. What is the temperature in degrees Fahrenheit? **(LO 1.6)**
   (a) $3344 \,°F$ (b) $3040 \,°F$
   (c) $10{,}920 \,°F$ (d) $9941 \,°F$

6. Calculate the volume in liters of a rectangular object with dimensions 13.0 cm $\times$ 11.0 cm $\times$ 12.0 cm. **(LO 1.7)**
   (a) 1720 L   (b) 1.72 L
   (c) 14.3 L   (d) 2.41 L

7. A 25.5 g sample of a metal was placed into water in a graduated cylinder. The metal sank to the bottom, and the water level rose from 15.7 mL to 25.3 mL. What is the identity of the metal? **(LO 1.9)**
   (a) Tin (density $= 7.31$ g/cm^3)
   (b) Lead (density $= 11.34$ g/cm^3)
   (c) Silver (density $= 10.49$ g/cm^3)
   (d) Aluminum (density $= 2.64$ g/cm^3)

8. Consider 20 mL samples of the following liquids. Which sample has the largest mass? **(LO 1.9)**
   (a) Water (density $= 1.0$ g/mL)
   (b) Glycerol (density $= 1.26$ g/mL)
   (c) Ethanol (density $= 0.79$ g/mL)
   (d) Acetic acid (density $= 1.05$ g/mL)

9. The cylinder contains two liquids that do not mix with one another: water (density $= 1.0$ g/mL) and vegetable oil (density $= 0.93$ g/mL). Four different pieces of plastic are added to the cylinder. Which type of plastic is at the position indicated by the square object in the figure? **(LO 1.10)**
   (a) Polyvinyl chloride
       (density $= 1.26$ g/mL)
   (b) Polypropylene
       (density $= 0.90$ g/mL)
   (c) High-density polyethylene
       (density $= 0.96$ g/mL)
   (d) Polyethylene terephthalate
       (density $= 1.38$ g/mL)

10. An electron with a mass of 9.1×10^{-28} g is traveling at 1.8×10^7 m/s in an electron microscope. Calculate the kinetic energy of electron in units of joules, and report your answer in scientific notation. **(LO 1.11)**
    (a) 1.5×10^{-16} J   (b) 1.6×10^{-20} J
    (c) 2.9×10^{-13} J   (d) 2.9×10^{-10} J

11. Report the reading on the buret to the correct number of significant figures. **(LO 1.15)**
    (a) 1 mL
    (b) 1.4 mL
    (c) 1.40 mL
    (d) 1.400 mL

12. A scientist uses an uncalibrated pH meter and measures the pH of a rainwater sample four times. A different pH meter was calibrated using several solutions with known pH. The true pH of the rain was found by the calibrated pH meter to be 5.12. What can be said about the level of accuracy and precision of the uncalibrated pH meter? **(LO 1.14)**

pH meter	pH
1	5.68
2	5.61
3	5.71
4	5.63

    (a) The uncalibrated pH meter is accurate and precise.
    (b) The uncalibrated pH meter is neither accurate nor precise.
    (c) The uncalibrated pH meter is accurate but not precise.
    (d) The uncalibrated pH meter is precise but not accurate.

13. Perform the calculation, and report the answer to the correct number of significant figures. **(LO 1.16)**

$$\frac{\left(\dfrac{0.368}{1.001 \times 10^2}\right)}{(25.26 - 1.50)} = ?$$

    (a) 1.5×10^{-4} (b) 1.55×10^{-4}
    (c) 1.547×10^{-4} (d) 1.5473×10^{-4}

14. A person runs at a pace of 6.52 mi/hr. How long does it take the person to run a 15.0 km race? (1 mi $= 1.61$ km) **(LO 1.17)**
    (a) 85.7 min   (b) 222 min
    (c) 50.0 min   (d) 93.4 min

15. Aerogels are transparent, low-density materials that are nearly 99.8% empty space and excellent insulators against hot and cold. The density of a silica-based aerogel is 3.0 mg/cm^3. What is the density in units of g/m^3? **(LO 1.17)**
    (a) 3.0×10^{-3} g/m^3 (b) 3.0×10^1 g/m^3
    (c) 3.0 g/m^3 (d) 3.0×10^3 g/m^3

Answers:

1. d, 2. c, 3. a, 4. b, 5. d, 6. b, 7. d, 8. b, 9. c, 10. a, 11. c, 12. d, 13. b, 14. a, 15. d

| Mastering **Chemistry** | provides end-of-chapter exercises, feedback-enriched tutorial problems, animations, and interactive activities to encourage problem-solving practice and deeper understanding of key concepts and topics. |

RAN *Randomized in Mastering Chemistry*

CONCEPTUAL PROBLEMS

Problems at the end of each chapter begin with a section called "Conceptual Problems." The problems in this section are visual or abstract rather than numerical and are intended to probe your understanding rather than your facility with numbers and formulas. Answers to even-numbered problems (in color) can be found at the end of the book following the appendices. Problems 1.1–1.26 appear within the chapter.

1.27 Which block in each of the following drawings of a balance is more dense, red or green? Explain.

(a)   **(b)**

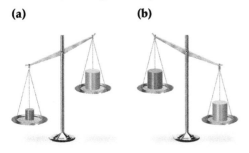

1.28 What is the temperature reading on the following Celsius thermometer? How many significant figures do you have in your answer?

1.29 How many milliliters of water does the graduated cylinder in (a) contain, and how tall in centimeters is the paper clip in (b)? How many significant figures do you have in each answer?

(a)   **(b)**

1.30 Assume that you have two graduated cylinders, one with a capacity of 5 mL (a) and the other with a capacity of 50 mL (b). Draw a line in each, showing how much liquid you would add if you needed to measure 2.64 mL of water. Which cylinder will give the more accurate measurement? Explain.

(a)   **(b)**

1.31 The following cylinder contains three liquids that do not mix with one another: water (density = 1.0 g/mL), vegetable oil (density = 0.93 g/mL), and mercury (density = 13.5 g/mL). Which liquid is which?

SECTION PROBLEMS

The Section Problems at the end of each chapter cover specific topics from the various sections of the chapter. These problems are presented in pairs, with each even-numbered problem followed by an odd-numbered one requiring similar skills. Even-numbered problems (in color) are answered at the end of the book following the appendixes.

Scientific Method (Section 1.1)

1.32 The following statements pertain to the development of the theory of combustion by the French chemist Lavoisier in the eighteenth century. Match the statement with the appropriate step (observation, hypothesis, experiment designed to test hypothesis) in the scientific method.

(a) A metal is burned in a closed container, and the change in mass of the solid and volume of the gas is measured.

(b) Oxygen gas combines with a substance during its combustion.

(c) Combustion of a metal in a closed container ceases after a length of time.

1.33 The following statements pertain to the development of the theory of the structure of DNA. Match the statement with the appropriate step (observation, hypothesis, experiment designed to test hypothesis) in the scientific method.

(a) Two strands of DNA wind around one another in a helical structure.

(b) In a sample of DNA, there are equal amounts of the bases A and T and equal amounts of the bases C and G.

(c) Direct X rays at a sample of crystallized DNA, and interpret the diffraction pattern for structural information.

1.34 Label the following statements about the world's largest gold bar as quantitative or qualitative observations. (This gold bar was worth approximately $10.25 million in 2013.)

(a) The melting point of gold is 1064.2 °C.

(b) The volume of the gold bar is 15,730 cm^3.

(c) Gold metal is a conductor of electricity.

(d) The mass of the gold bar is 250 kg.

(e) The gold bar is yellow and shiny.

1.35 Label the following statements as quantitative or qualitative observations.

(a) An object weighs less on the moon than on Earth.

(b) An object that weighs 50 pounds on Earth only weighs 8.3 pounds on the moon.

(c) The freezing point of water is cold.

(d) The freezing point of water is 0 °C

1.36 Refer to Figure 1.2. What is developed when numerous observations support a hypothesis?

1.37 What is the difference between a hypothesis and theory?

(a) A hypothesis provides an explanation for a phenomenon, but a theory does not.

(b) A theory provides an explanation for a phenomenon, but a hypothesis does not.

(c) Both a theory and a hypothesis provide an explanation for a phenomenon, but a theory has been upheld by experimental observations.

SI Units and Scientific Notation (Section 1.2)

1.38 What SI units are used for measuring the following quantities? For derived units, express your answers in terms of the six fundamental units.

(a) Mass   (b) Length

(c) Temperature   (d) Volume

(e) Energy   (f) Density

1.39 Prefixes for multiples of SI units are used to express large and small quantities. Complete the following table. The first row is completed as a model.

Prefix	Abbreviation	Exponential Factor
mega	M	10^6
	μ	
kilo		
		10^{-3}

Prefix	Abbreviation	Exponential Factor
nano		
	G	
		10^{-12}

1.40 Complete the following equivalent expressions by filling in the blanks.

(a) 1 km = _____ m

(b) 1 m = _____ km

(c) 1 mmol = _____ mol

(d) 1 mol = _____ mmol

1.41 Complete the following equivalent expressions by filling in the blanks.

(a) 1 L = _____ μL   (b) 1 μL = _____ L

(c) 1 L = _____ nL   (d) 1 nL = _____ L

1.42 Bottles of wine sometimes carry the notation "Volume = 75 cL." What does the unit cL mean?
RAN

1.43 Which quantity in each of the following pairs is larger?
RAN
(a) 5.63×10^6 cm or 6.02×10^1 km

(b) 46 μs or 3.2×10^{-2} ms

(c) 200,098 g or 17×10^1 kg

1.44 Which quantity in each of the following pairs is smaller?
RAN
(a) 154 pm or 7.7×10^{-9} cm

(b) 1.86×10^{11} μm or 2.02×10^2 km

(c) 2.9 GA or 3.1×10^{15} μA

1.45 How many picograms are in 1 mg? In 35 ng?
RAN

1.46 How many microliters are in 1 L? In 20 mL?
RAN

1.47 Carry out the following conversions.
RAN
(a) 5 pm = _____ cm = _____ nm

(b) 8.5 cm^3 = _____ m^3 = _____ mm^3

(c) 65.2 mg = _____ g = _____ pg

1.48 Express the following measurements in scientific notation.

(a) 453.32 mg   (b) 0.000 042 1 mL

(c) 667,000 g

1.49 Convert the following measurements from scientific notation to standard notation.
RAN
(a) 3.221×10^{-3} mm   (b) 8.940×10^5 m

(c) $1.350\ 82 \times 10^{-12}$ m^3 (d) 6.4100×10^2 km

Measurement of Mass, Length, and Temperature (Sections 1.3–1.5)

1.50 An experimental procedure call for 250 mg of calcium carbonate. The balance in the laboratory measures mass in grams and reads to four decimal places. Which reading on the balance corresponds to 250 mg?

(a) 0.0250   (b) 0.2500   (c) 0.0025

1.51 A virus has a diameter of 5.2×10^{-8} m. What is the most appropriate prefix for reporting the diameter of the virus?

1.52 Which is larger, a Fahrenheit degree or a Celsius degree? By how much?

1.53 What is the difference between a kelvin and a Celsius degree?

1.54 The normal body temperature of a goat is 39.9 °C, and that of an Australian spiny anteater is 22.2 °C. Express these temperatures in degrees Fahrenheit.

1.55 Of the 90 or so naturally occurring elements, only four are liquid near room temperature: mercury (melting point = −38.87 °C), bromine (melting point = −7.2 °C), cesium (melting point = 28.40 °C), and gallium (melting point = 29.78 °C). Convert these melting points to degrees Fahrenheit.

1.56 Suppose that your oven is calibrated in degrees Fahrenheit RAN but a recipe calls for you to bake at 175 °C. What oven setting should you use?

1.57 Tungsten, the element used to make filaments in light bulbs, has a melting point of 6192 °F. Convert this temperature to degrees Celsius and to kelvin.

1.58 Suppose you were dissatisfied with both Celsius and Fahrenheit RAN units and wanted to design your own temperature scale based on ethyl alcohol (ethanol). On the Celsius scale, ethanol has a melting point of −117.3 °C and a boiling point of 78.5 °C, but on your new scale calibrated in units of degrees ethanol, °E, you define ethanol to melt at 0 °E and boil at 200 °E.

   (a) How does your ethanol degree compare in size with a Celsius degree?

   (b) How does an ethanol degree compare in size with a Fahrenheit degree?

   (c) What are the melting and boiling points of water on the ethanol scale?

   (d) What is normal human body temperature (98.6 °F) on the ethanol scale?

   (e) If the outside thermometer reads 130 °E, how would you dress to go out?

1.59 Answer parts (a)–(d) of Problem 1.58 assuming that your RAN new temperature scale is based on ammonia, NH_3. On the Celsius scale, ammonia has a melting point of −77.7 °C and a boiling point of −33.4 °C, but on your new scale calibrated in units of degrees ammonia, °A, you define ammonia to melt at 0 °A and boil at 100 °A.

1.60 Sodium chloride has a melting point of 1074 K and a boil RAN ing point of 1686 K. Convert these temperatures to degrees Celsius and to degrees Fahrenheit.

1.61 A 125 mL sample of water at 293.2 K was heated for 8 min, RAN 25 s so as to give a constant temperature increase of 3.0 °F/min. What is the final temperature of the water in degrees Celsius?

Derived Units: Volume and Density (Sections 1.6–1.7)

1.62 What is the difference between a derived SI unit and a fundamental SI unit? Give an example of each.

1.63 Which volume in each pair is larger, and by approximately how much?

   (a) 1000 mL or 10 L   (b) 1 m³ or 1 dL

1.64 What is the volume in L of a cube with an edge length of 7.0 dm?

1.65 What is the volume in mL of a cube with an edge length of 2.5 cm?

1.66 What is the density of glass in g/cm³ if a sample weighing RAN 27.43 g has a volume of 12.40 cm³?

1.67 What is the density of lead in g/cm³ if a sample weighing RAN 206.77 g has a volume of 15.50 cm³?

1.68 A vessel contains 4.67 L of bromine whose density is 3.10 g/cm³. RAN What is the mass of the bromine in the vessel (in kilograms)?

1.69 Aspirin has a density of 1.40 g/cm³. What is the volume in RAN cubic centimeters of an aspirin tablet weighing 250 mg? Of a tablet weighing 500 lb?

1.70 Gaseous hydrogen has a density of 0.0899 g/L at 0 °C, and RAN gaseous chlorine has a density of 3.214 g/L at the same temperature. How many liters of each would you need if you wanted 1.0078 g of hydrogen and 35.45 g of chlorine?

1.71 The density of silver is 10.5 g/cm³. What is the mass (in kilo RAN grams) of a cube of silver that measures 0.62 m on each side?

1.72 What is the density of lead in g/cm³ if a rectangular bar mea RAN suring 0.50 cm in height, 1.55 cm in width, and 25.00 cm in length has a mass of 220.9 g?

1.73 What is the density of lithium metal in g/cm³ if a cylindrical wire with a diameter of 2.40 mm and a length of 15.0 cm has a mass of 0.3624 g?

1.74 You would like to determine if a set of antique silverware RAN is pure silver. The mass of a small fork was measured on a balance and found to be 80.56 g. The volume was found by dropping the fork into a graduated cylinder initially containing 10.0 mL of water. The volume after the fork was added was 15.90 mL. Calculate the density of the fork. If the density of pure silver at the same temperature is 10.5 g/cm³, is the fork pure silver?

1.75 An experiment is performed to determine if pennies are made RAN of pure copper. The mass of 10 pennies was measured on a balance and found to be 24.656 g. The volume was found by dropping the 10 pennies into a graduated cylinder initially containing 10.0 mL of water. The volume after the pennies were added was 12.90 mL. Calculate the density of the pennies. If the density of pure copper at the same temperature is 8.96 g/cm³, are the pennies made of pure copper?

1.76 The density of chloroform, a widely used organic solvent, is RAN 1.4832 g/mL at 20 °C. How many milliliters would you use if you wanted 112.5 g of chloroform?

1.77 More sulfuric acid (density = 1.8302 g/cm³) is produced than RAN any other chemical—approximately 3.6×10^{11} lb/yr worldwide. What is the volume of this amount in liters?

Energy (Section 1.8)

1.78 Which has more kinetic energy, a 1400 kg car moving at 115 km/h or a 12,000 kg truck moving at 38 km/h?

1.79 Assume that the kinetic energy of a 1400 kg car moving at RAN 115 km/h (Problem 1.78) is converted entirely into heat. How many calories of heat are released, and what amount of water in liters could be heated from 20.0 °C to 50.0 °C by the car's energy? (One calorie raises the temperature of 1 mL of water by 1 °C)

1.80 The combustion of 45.0 g of methane (natural gas) releases RAN 2498 kJ of heat energy. How much energy in kilocalories (kcal) would combustion of 0.450 ounces of methane release?

1.81 Sodium (Na) metal undergoes a chemical reaction with chlo RAN rine (Cl) gas to yield sodium chloride, or common table salt. If 1.00 g of sodium reacts with 1.54 g of chlorine, 2.54 g of sodium chloride is formed and 17.9 kJ of heat is released. How much sodium and how much chlorine in grams would have to react to release 171 kcal of heat?

1.82 A Big Mac hamburger from McDonald's contains 540 Calories.

   (a) How many kilojoules does a Big Mac contain?

   (b) For how many hours could the amount of energy in a Big Mac light a 100-watt light bulb? (1 watt = 1 J/s)

1.83 A 20 fluid oz. soda contains 238 Calories.
RAN
   (a) How many kilojoules does the soda contain?

   (b) For how many hours could the amount of energy in the soda light a 75-watt light bulb? (1 watt = 1 J/s)

Accuracy, Precision, and Significant Figures (Sections 1.9–1.10)

1.84 Which of the following statements uses exact numbers?

(a) 1 ft = 12 in.

(b) 1 cal = 4.184 J

(c) The height of Mt. Everest is 29,035 ft.

(d) The world record for the 1-mile run, set by Morocco's Hicham el Guerrouj in July 1999, is 3 minutes, 43.13 seconds.

1.85 What is the difference in mass between a nickel that weighs
RAN 4.8 g and a nickel that weighs 4.8673 g?

1.86 How many significant figures are in each of the following measurements?

(a) 35.0445 g (b) 59.0001 cm (c) 0.030 03 kg

(d) 0.004 50 m (e) 67,000 m² (f) 3.8200 × 10³ L

1.87 How many significant figures are in each of the following
RAN measurements?

(a) $130.95 (b) 2000.003 g (c) 5 ft 3 in.

(d) 510 J (e) 5.10×10^2 J (f) 10 students

1.88 The Vehicle Assembly Building at the John F. Kennedy Space Center in Cape Canaveral, Florida, is the largest building in the world, with a volume of 3,666,500 m³.

(a) Round off this quantity to four significant figures and then to two significant figures.

(b) Express the answers in scientific notation.

1.89 The diameter of the Earth at the equator is 7926.381 mi.
RAN
(a) Round off this quantity to four significant figures and then to two significant figures.

(b) Express the answers in scientific notation.

1.90 Round off the following quantities to the number of signifi-
RAN cant figures indicated in parentheses.

(a) 35,670.06 m (4, 6) (b) 68.507 g (2, 3)

(c) 4.995×10^3 cm (3) (d) $2.309\,85 \times 10^{-4}$ kg (5)

1.91 Round off the following quantities to the number of signifi-
RAN cant figures indicated in parentheses.

(a) 7.0001 kg (4) (b) 1.605 km (3)

(c) 13.2151 g/cm³ (3) (d) 2,300,000.1 (7)

1.92 Express the results of the following calculations with the
RAN correct number of significant figures.

(a) 4.884 × 2.05 (b) 94.61 ÷ 3.7

(c) 3.7 ÷ 94.61 (d) 5502.3 + 24 + 0.01

(e) 86.3 + 1.42 − 0.09 (f) 5.7 × 2.31

1.93 Express the results of the following calculations with the
RAN correct number of significant figures.

(a) $\dfrac{3.41 - 0.23}{5.233} \times 0.205$ (b) $\dfrac{5.556 \times 2.3}{4.223 - 0.08}$

Unit Conversions (Section 1.11)

1.94 Carry out the following conversions.

(a) How many grams of meat are in a quarter-pound hamburger (0.25 lb)?

(b) How tall in meters is the Willis Tower, formerly called the Sears Tower, in Chicago (1454 ft)?

(c) How large in square meters is the land area of Australia (2,941,526 mi²)?

1.95 Convert the following quantities into SI units with the correct
RAN number of significant figures.

(a) 5.4 in. (b) 66.31 lb (c) 0.5521 gal

(d) 65 mi/h (e) 978.3 yd³ (f) 2.380 mi²

1.96 The world record for the women's outdoor 20,000-meter run, set
RAN in 2000 by Tegla Loroupe, is 1:05:26.6 (seconds are given to the nearest tenth). What was her average speed, expressed in miles per hour with the correct number of significant figures? (Assume that the race distance is accurate to 5 significant figures.)

1.97 In the United States, the emissions limit for carbon monoxide in motorcycle engine exhaust is 12.0 g of carbon monoxide per kilometer driven. What is this limit expressed in mg per mile with the correct number of significant figures?

1.98 The volume of water used for crop irrigation is measured in
RAN acre-feet, where 1 acre-foot is the amount of water needed to cover 1 acre of land to a depth of 1 ft.

(a) If there are 640 acres per square mile, how many cubic feet of water are in 1 acre-foot?

(b) How many acre-feet are in Lake Erie (total volume = 116 mi³)?

1.99 The height of a horse is usually measured in *hands* instead of
RAN in feet, where 1 hand equals 1/3 ft (exactly).

(a) How tall in centimeters is a horse of 18.6 hands?

(b) What is the volume in cubic meters of a box measuring 6 × 2.5 × 15 hands?

1.100 Weights in England are commonly measured in *stones*,
RAN where 1 stone = 14 lb. What is the weight in pounds of a person who weighs 8.65 stones?

1.101 Concentrations of substances dissolved in solution are often
RAN expressed as mass per unit volume. For example, normal human blood has a cholesterol concentration of about 200 mg/100 mL. Express this concentration in the following units.

(a) mg/L (b) μg/mL

(c) g/L (d) ng/μL

(e) How much total blood cholesterol in grams does a person have if the normal blood volume in the body is 5 L?

1.102 Administration of digitalis, a drug used to control atrial fibrilla-
RAN tion in heart patients, must be carefully controlled because even a modest overdose can be fatal. To take differences between patients into account, drug dosages are prescribed in terms of mg/kg body weight. Thus, a child and an adult differ greatly in weight, but both receive the same dosage per kilogram of body weight. At a dosage of 20 μg/kg body weight, how many milligrams of digitalis should a 160 lb patient receive?

1.103 Among many alternative units that might be considered as a
RAN measure of time is the *shake* rather than the second. Based on the expression "faster than a shake of a lamb's tail," we'll define 1 shake as equal to 2.5×10^{-4} s. If a car is traveling at 55 mi/h, what is its speed in cm/shake?

1.104 Which is larger in each pair, and by approximately how much?

(a) A liter or a quart (b) A mile or a kilometer

(c) A gram or an ounce (d) A centimeter or an inch

1.105 The density of polystyrene, a plastic commonly used to make
RAN CD cases and transparent cups, is 0.037 lbs/in³. Calculate the density in units of g/cm³.

1.106 The density of polypropylene, a plastic commonly used to make
RAN bottle caps, yogurt containers, and carpeting, is 0.55 oz/in³. Calculate the density in units of g/cm³.

MULTICONCEPT PROBLEMS

1.107 A large tanker truck for carrying gasoline has a capacity of
RAN 3.4×10^4 L.

(a) What is the tanker's capacity in gallons?

(b) If the retail price of gasoline is $3.00 per gallon, what is the value of the truck's full load of gasoline?

1.108 A 1.0-ounce piece of chocolate contains 15 mg of caffeine, and
RAN a 6.0-ounce cup of regular coffee contains 105 mg of caffeine. How much chocolate would you have to consume to get as much caffeine as you would from 2.0 cups of coffee?

1.109 When an irregularly shaped chunk of silicon weighing 8.763 g was placed in a graduated cylinder containing 25.00 mL of water, the water level in the cylinder rose to 28.76 mL. What is the density of silicon in g/cm^3?

1.110 Lignum vitae is a hard, durable, and extremely dense wood
RAN used to make ship bearings. A sphere of this wood with a diameter of 7.60 cm has a mass of 313 g.

(a) What is the density of the lignum vitae sphere?

(b) Will the sphere float or sink in water?

(c) Will the sphere float or sink in chloroform? (The density of chloroform is 1.48 g/mL.)

1.111 Answer the following questions.
RAN
(a) An old rule of thumb in cooking says: "A pint's a pound the world around." What is the density in g/mL of a substance for which 1 pt = 1 lb exactly?

(b) There are exactly 640 acres in 1 square mile. How many square meters are in 1 acre?

(c) A certain type of wood has a density of $0.40 \ g/cm^3$. What is the mass of 1.0 cord of this wood in kg, where 1 cord is 128 cubic feet of wood?

(d) A particular sample of crude oil has a density of 0.85 g/mL. What is the mass of 1.00 barrel of this crude oil in kg, where a barrel of oil is exactly 42 gallons?

(e) A gallon of ice cream contains exactly 32 servings, and each serving has 165 Calories, of which 30.0% are derived from fat. How many Calories derived from fat would you consume if you ate one half of a gallon of ice cream?

1.112 A bag of Hershey's Kisses contains the following information:
RAN
Serving size: 9 pieces = 41 g

Calories per serving: 230

Total fat per serving: 13 g

(a) The bag contains 2.0 lbs of Hershey's Kisses. How many Kisses are in the bag?

(b) The density of a Hershey's Kiss is 1.4 g/mL. What is the volume of a single Hershey's Kiss?

(c) How many Calories are in one Hershey's Kiss?

(d) Each gram of fat yields 9 Calories when metabolized. What percent of the calories in Hershey's Kisses are derived from fat?

1.113 Vinaigrette salad dressing consists mainly of oil and vinegar.
RAN The density of olive oil is $0.918 \ g/cm^3$, the density of vinegar is $1.006 \ g/cm^3$, and the two do not mix. If a certain mixture of olive oil and vinegar has a total mass of 397.8 g and a total volume of $422.8 \ cm^3$, what is the volume of oil and what is the volume of vinegar in the mixture?

1.114 At a certain point, the Celsius and Fahrenheit scales "cross," giving the same numerical value on both. At what temperature does this crossover occur?

1.115 Imagine that you place a cork measuring 1.30 cm $\times$ 5.50 cm $\times$ 3.00 cm in water and that on top of the cork you place a small cube of lead measuring 1.15 cm on each edge. The density of cork is $0.235 \ g/cm^3$, and the density of lead is $11.35 \ g/cm^3$. Will the combination of cork plus lead float or sink?

1.116 A calibrated flask was filled to the 25.00 mL mark with ethyl alcohol. By weighing the flask before and after adding the alcohol, it was determined that the flask contained 19.7325 g of alcohol. In a second experiment, 25.0920 g of metal beads were added to the flask, and the flask was again filled to the 25.00 mL mark with ethyl alcohol. The total mass of the metal plus alcohol in the flask was determined to be 38.4704 g. What is the density of the metal in g/mL?

1.117 Brass is a copper–zinc alloy. What is the mass in grams of a
RAN brass cylinder having a length of 1.62 in. and a diameter of 0.514 in. if the composition of the brass is 67.0% copper and 33.0% zinc by mass? The density of copper is $8.92 \ g/cm^3$, and the density of zinc is $7.14 \ g/cm^3$. Assume that the density of the brass varies linearly with composition.

1.118 Ocean currents are measured in *Sverdrups* (sv) where 1 sv $= 10^9 \ m^3/s$. The Gulf Stream off the tip of Florida, for instance, has a flow of 35 sv.

(a) What is the flow of the Gulf Stream in milliliters per minute?

(b) What mass of water in the Gulf Stream flows past a given point in 24 hours? The density of seawater is 1.025 g/mL.

(c) How much time is required for 1 petaliter (PL; 1 PL = 10^{15} L) of seawater to flow past a given point?

1.119 The element gallium (Ga) has the second-largest liquid range
RAN of any element, melting at 29.78 °C and boiling at 2204 °C at atmospheric pressure.

(a) What is the density of gallium in g/cm^3 at 25 °C if a 1 in. cube has a mass of 0.2133 lb?

(b) Assume that you construct a thermometer using gallium as the fluid instead of mercury and that you define the melting point of gallium as 0 °G and the boiling point of gallium as 1000 °G. What is the melting point of sodium chloride (801 °C) on the gallium scale?

chapter 2

Atoms, Molecules, and Ions

Contents

If you could take a large piece of a pure element such as carbon and cut it into ever smaller and smaller pieces, you would find that it is made of a vast number of tiny fundamental units that we call *atoms*. Scientists study the composition of matter on the atomic level to learn about the environment and how it has changed over time.

How can isotopes of hydrogen and oxygen in ice core samples be used to determine past climates?

The answer to this question can be found on page 69 in the INQUIRY ?

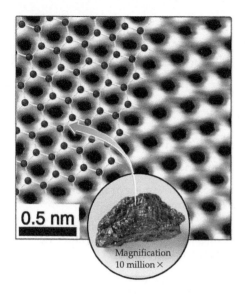

0.5 nm

Magnification
10 million ×

▲ A scanning tunelling microscope generated this image of individual carbon atoms in the small flake of highly ordered pyrolytic graphite. (Image courtesy of Tait research lab at Indiana University.)

▲ Samples of silver, mercury, and sulfur (from top to bottom).

People have always been fascinated by changes, particularly by those that are dramatic or useful. In the ancient world, the change that occurred when a stick of wood burned, gave off heat, and turned into a small pile of ash was especially important. Similarly, the change that occurred when a reddish lump of rock (iron ore) was heated with charcoal and produced a gray metal (iron) useful for making weapons, tools, and other implements was of enormous value. Observing such changes eventually caused philosophers to think about what different materials might be composed of and led to the idea of fundamental substances that we today call *elements*.

At the same time philosophers were pondering the question of elements, they were also thinking about related matters: What makes up an element? Is matter continuously divisible into ever smaller and smaller pieces, or is there an ultimate limit? Can you cut a piece of gold in two, take one of the pieces and cut *it* in two, and so on infinitely, or is there a point at which you must stop? Most thinkers, including Plato and Aristotle, believed that matter is continuously divisible, but the Greek philosopher Democritus (460–370 B.C.) disagreed. Democritus proposed that matter is composed of tiny, discrete particles, which we now call *atoms,* from the Greek word *atomos,* meaning "indivisible." Little else was learned about elements and atoms until the birth of modern experimental science some 2000 years later.

Although atomic theory existed for many years, atoms were never directly "seen" until the invention of specialized instruments such as the scanning tunneling microscope (STM) and the atomic force microscope (AFM) in the 1980s. An STM was used to create the image of individual carbon atoms in the margin photo. Now, for the first time, matter can be directly manipulated and studied at the atomic scale. Individual atoms can even be moved in an attempt to build materials in novel ways. The control and study of matter at the atomic level has important implications in computing, biochemistry, and industrial products such as batteries, catalysts, and plastics.

2.1 CHEMISTRY AND THE ELEMENTS

Everything you see around you is formed from one or more of 118 presently known *elements*. An **element** is a fundamental substance that can't be chemically changed or broken down into anything simpler while still retaining the properties of that element. Mercury, silver, and sulfur are common examples, as listed in **TABLE 2.1.**

TABLE 2.1 Names and Symbols of Some Common Elements. Latin names from which the symbols of some elements are derived are shown in parentheses.

Aluminum	**Al**	Chlorine	**Cl**	Manganese	**Mn**	Copper (*cuprum*)	**Cu**
Argon	**Ar**	Fluorine	**F**	Nitrogen	**N**	Iron (*ferrum*)	**Fe**
Barium	**Ba**	Helium	**He**	Oxygen	**O**	Lead (*plumbum*)	**Pb**
Boron	**B**	Hydrogen	**H**	Phosphorus	**P**	Mercury (*hydrargyrum*)	**Hg**
Bromine	**Br**	Iodine	**I**	Silicon	**Si**	Potassium (*kalium*)	**K**
Calcium	**Ca**	Lithium	**Li**	Sulfur	**S**	Silver (*argentum*)	**Ag**
Carbon	**C**	Magnesium	**Mg**	Zinc	**Zn**	Sodium (*natrium*)	**Na**

Actually, the previous statement about everything being made of one or more of 118 elements is an exaggeration because only about 90 of the 118 occur naturally. The remaining 28 have been produced artificially by nuclear chemists using high-energy particle accelerators.

Furthermore, only 83 of the 90 or so naturally occurring elements are found in any appreciable abundance. Hydrogen is thought to account for approximately 75% of the observed mass in the universe; oxygen and silicon together account for 75% of the mass of the Earth's crust; and oxygen, carbon, hydrogen, and nitrogen make up more than 95% of the mass of the human body (**FIGURE 2.1**). By contrast, there is probably less than 20 grams of the element francium (Fr) dispersed over the entire Earth at any one time. Francium is an unstable radioactive element, atoms of which are continually being formed and destroyed. We'll discuss **radioactivity** in Chapter 20.

LOOKING AHEAD . . .
As we'll see in Chapter 20, **radioactivity** is the spontaneous decay of certain unstable atoms with emission of some form of radiation.

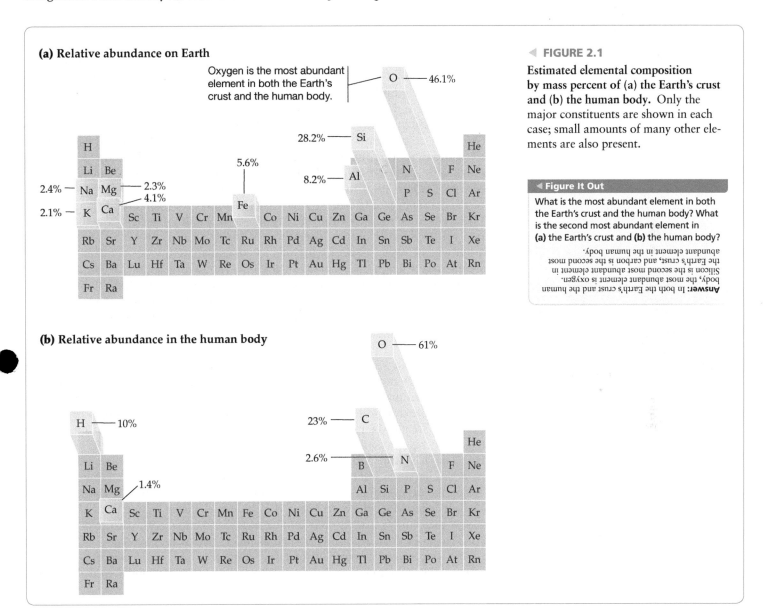

(a) Relative abundance on Earth

Oxygen is the most abundant element in both the Earth's crust and the human body.

O — 46.1%

28.2% — Si

5.6%

8.2% — Al

2.4% — Na Mg — 2.3%
4.1%

2.1% — K Ca

Fe

(b) Relative abundance in the human body

O — 61%

23% — C

H — 10%

2.6% — B

1.4%

◀ **FIGURE 2.1**

Estimated elemental composition by mass percent of (a) the Earth's crust and (b) the human body. Only the major constituents are shown in each case; small amounts of many other elements are also present.

◀ **Figure It Out**

What is the most abundant element in both the Earth's crust and the human body? What is the second most abundant element in **(a)** the Earth's crust and **(b)** the human body?

Answer: In both the Earth's crust and the human body, the most abundant element is oxygen. Silicon is the second most abundant element in the Earth's crust, and carbon is the second most abundant element in the human body.

For simplicity, chemists refer to specific elements using one- or two-letter symbols. As shown by the examples in Table 2.1, the first letter of an element's symbol is always capitalized and the second letter, if any, is lowercase. Many of the symbols are just the first one or two letters of the element's English name: H = hydrogen, C = carbon, Al = aluminum, and so forth. Other symbols derive from Latin or other languages: Na = sodium (Latin, *natrium*), Pb = lead (Latin, *plumbum*), and W = tungsten (German, *wolfram*). The names, symbols, and other information about all 118 known elements are given inside the front cover of this book, organized in a format you've undoubtedly seen before called the **periodic table**.

In 2016, the International Union of Pure and Applied Chemistry (IUPAC) approved the names and symbols of four recently discovered man-made elements. Element 113

Go to
eText

BIG IDEA Question 1

Which of the following statements about elements is *false*?

(a) Elements cannot be chemically changed or broken down into anything simpler and still be the same element.

(b) All naturally occurring elements are stable and nonradioactive.

(c) Elements exist in widely varying amounts on Earth.

(d) Elements are represented using one- or two-letter symbols that are often derived from the element's Latin name.

▲ Left to right, samples of chlorine, bromine, and iodine, one of Döbereiner's triads of elements with similar chemical properties.

was named nihonium (Nh) by the Japanese scientists who created it because *nihon* is one way to say "Land of the Rising Sun" in Japanese. Element 115 was named moscovium (Mc), and element 117 was named tennesine (Ts) by their joint discoverers at the Institute for Nuclear Research in Dubna, Russia, and Oak Ridge National Laboratories in Tennessee, United States. Element 118, named oganesson (Og), recognizes Yuri Oganessian for his pioneering achievements in the discovery of superheavy elements.

2.2 ELEMENTS AND THE PERIODIC TABLE

Ten elements have been known since the beginning of recorded history: antimony (Sb), carbon (C), copper (Cu), gold (Au), iron (Fe), lead (Pb), mercury (Hg), silver (Ag), sulfur (S), and tin (Sn). The first "new" element to be found in several thousand years was arsenic (As), discovered in about 1250. In fact, only 24 elements were known when the United States was founded in 1776.

As the pace of scientific discovery quickened in the late 1700s and early 1800s, chemists began to look for similarities among elements that might allow general conclusions to be drawn. Particularly important among the early successes was Johann Döbereiner's observation in 1829 that there were several *triads*, or groups of three elements, that appeared to behave similarly. Calcium (Ca), strontium (Sr), and barium (Ba) form one such triad; chlorine (Cl), bromine (Br), and iodine (I) form another; and lithium (Li), sodium (Na), and potassium (K) form a third. By 1843, 16 such triads were known and chemists were searching for an explanation.

Numerous attempts were made in the mid-1800s to account for the similarities among groups of elements, but the breakthrough came in 1869 when the Russian chemist Dmitri Mendeleev created the forerunner of the modern periodic table. Mendeleev's creation is an ideal example of how a scientific theory develops. At first, there is only disconnected information—a large number of elements and many observations about their properties and behavior. As more and more facts become known, people try to organize the data in ways that make sense until ultimately a consistent theory emerges.

A good theory must do two things: It must explain known facts, and it must make predictions about phenomena yet unknown. If the predictions are tested and found to be true, then the theory is a good one and will stand until additional facts are discovered that require it to be modified or discarded. Mendeleev's theory about how known chemical information could be organized passed all tests. Not only did the periodic table arrange data in a useful and consistent way to explain known facts about chemical reactivity, it also led to several remarkable predictions that were later found to be accurate.

Using the experimentally observed chemistry of the elements as his primary organizing principle, Mendeleev arranged the known elements in order of the relative masses of their atoms (called their *atomic weights*, Section 2.9), with hydrogen = 1, and then grouped them according to their chemical reactivity. On so doing, he realized that there were several "holes" in the table, some of which are shown in **FIGURE 2.2**. The chemical behavior of aluminum (relative mass ≈ 27.3) is similar to that of boron

▶ **FIGURE 2.2**

A portion of Mendeleev's periodic table. The table shows the relative masses of atoms as known at the time and some of the holes representing unknown elements.

H = 1								
Li = 7	Be = 9.4			B = 11	C = 12	N = 14	O = 16	F = 19
Na = 23	Mg = 24			Al = 27.3	Si = 28	P = 31	S = 32	Cl = 35.5
K = 39	Ca = 40	?, Ti, V, Cr, Mn, Fe, Co, Ni, Cu, Zn	? = 68	? = 72	As = 75	Se = 78	Br = 80	

There was an unknown element beneath aluminum (Al), which was later discovered and named gallium (Ga).

There was another unknown element beneath silicon (Si), which was later discovered and named germanium (Ge).

▶ **Figure It Out**

In the 1800s, Li, Na, and K were grouped as a triad due to their similar reactivity. How is this triad represented in Mendeleev's periodic table?

Answer: Elements with similar reactivity were placed in vertical groups.

(relative mass ≈ 11), but there was no element known at the time that fit into the slot below aluminum. In the same way, silicon (relative mass ≈ 28) is similar in many respects to carbon (relative mass ≈ 12), but there was no element known that fit below silicon.

Looking at the holes in the table, Mendeleev predicted that two then-unknown elements existed and might be found at some future time. Furthermore, he predicted with remarkable accuracy what the properties of these unknown elements would be. The element immediately below aluminum, which he called *eka*-aluminum from a Sanskrit word meaning "first," should have a relative mass near 68 and should have a low melting point. Gallium, discovered in 1875, has exactly these properties. The element below silicon, which Mendeleev called *eka*-silicon, should have a relative mass near 72 and should be dark gray in color. Germanium, discovered in 1886, fits the description perfectly (**TABLE 2.2**).

TABLE 2.2   A Comparison of Predicted and Observed Properties for Gallium (*eka*-Aluminum) and Germanium (*eka*-Silicon)

Element	Property	Mendeleev's Prediction	Observed Property
Gallium	Relative mass	68	69.7
	Density	5.9 g/cm^3	5.91 g/cm^3
	Melting point	Low	29.8 °C
Germanium	Relative mass	72	72.6
	Density	5.5 g/cm^3	5.35 g/cm^3
	Color	Dark gray	Light gray

▲ Gallium is a shiny, low-melting metal.

▲ Germanium is a hard, gray semimetal.

In the modern periodic table, shown in **FIGURE 2.3**, elements are placed on a grid with seven horizontal rows, called **periods,** and 18 vertical columns, called **groups.** When organized in this way, *the elements in a given group have similar chemical properties.* Lithium, sodium, potassium, and the other metallic elements in group 1A behave similarly. Beryllium, magnesium, calcium, and the other elements in group 2A behave similarly. Fluorine, chlorine, bromine, and the other elements in group 7A behave similarly, and so on throughout the table. (Mendeleev, by the way, was completely unaware of the existence of the group 8A elements—He, Ne, Ar, Kr, Xe, and Rn—because none were known when he constructed his table. All are colorless, odorless gases with little or no chemical reactivity, and none were discovered until 1894, when argon was first isolated.)

The overall form of the periodic table is well accepted, but chemists in different countries have historically used different conventions for labeling the groups. To resolve these difficulties, an international standard calls for numbering the groups from 1 to 18 going left to right. This standard has not yet found complete acceptance, however, and we'll continue to use the U.S. system of numbers and capital letters—group 3B instead of group 3 and group 7A instead of group 17, for example. Labels for the newer system are also shown in Figure 2.3.

One further note: There are actually 32 groups in the periodic table rather than 18, but to make the table fit manageably on a page, the 14 elements beginning with lanthanum (the *lanthanides*) and the 14 beginning with actinium (the *actinides*) are pulled out and shown below the others. These groups are not numbered.

We'll see repeatedly throughout this book that the periodic table of the elements is the most important organizing principle in chemistry. The time you take now to familiarize yourself with the layout and organization of the periodic table will pay off later on. Notice in Figure 2.3, for instance, that there is a regular progression in the size of the seven periods (rows). The first period has only 2 elements, hydrogen (H) and helium (He); the second and third periods have 8 elements each; the fourth and fifth periods have 18 elements each; and the sixth and seventh periods, which include the lanthanides and actinides, have 32 elements each. We'll see in Chapter 5 that this regular progression in the periodic table reflects a similar regularity in the structure of atoms.

Notice also that not all groups in the periodic table have the same number of elements. Groups 1, 2, and 13–18 are called the **main groups.** Most of the elements

LOOKING AHEAD . . .
We'll see in Chapter 5 that the detailed structure of a given atom is related to its position in the periodic table.

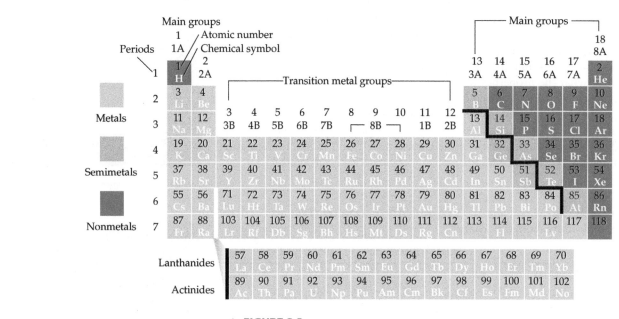

▲ **FIGURE 2.3**

The periodic table. Each element is identified by a one- or two-letter symbol and is characterized by an atomic number (Section 2.8). Elements are organized into 18 vertical columns, or *groups*, and seven horizontal rows, or *periods*. Groups 1, 2, and 13–18, are the *main groups*, and groups 3–12 are the *transition metal groups*. The 14 elements beginning with lanthanum are the *lanthanides*, and the 14 elements beginning with actinium are the *actinides*. Together, the lanthanides and actinides are known as the *inner transition metal groups*.

Those elements (except hydrogen) on the left side of the zigzag line are metals; those elements (plus hydrogen) to the right of the line are nonmetals; and seven of the nine elements abutting the line are metalloids, or semimetals.

▲ **Figure It Out**

What is the symbol for the element located in period 5, group 16? Classify the element as a metal, semimetal, or nonmetal.

Answer: Te, semimetal.

BIG IDEA Question 2

Go to eText

How are elements with similar chemical properties positioned in the periodic table?

on which life is based—carbon, hydrogen, nitrogen, oxygen, and phosphorus, for instance—are main-group elements. Group 3–12 in the middle of the table are called the **transition metal groups**. Most of the metals you're probably familiar with—iron, copper, zinc, and gold, for instance—are transition metals. And the 14 groups shown separately at the bottom of the table are called the **inner transition metal groups**.

2.3 SOME COMMON GROUPS OF ELEMENTS AND THEIR PROPERTIES

Any characteristic that can be used to describe or identify matter is called a **property**. Examples include volume, amount, odor, color, and temperature. Still other properties include such characteristics as melting point, solubility, and chemical behavior. For example, we might list some properties of sodium chloride (table salt) by saying that it melts at 1474 °F (or 801 °C), dissolves in water, and undergoes a chemical reaction when it comes into contact with a silver nitrate solution.

Properties can be classified as either *intensive* or *extensive,* depending on whether the value of the property changes with the amount of the sample. **Intensive properties,** like temperature and melting point, have values that do not depend on the amount of sample: A small ice cube might have the same temperature as a massive iceberg. **Extensive properties,** like length and volume, have values that *do* depend on the sample size: An ice cube is much smaller than an iceberg.

Properties can also be classified as either *physical* or *chemical,* depending on whether the property involves a change in the chemical makeup of a substance. **Physical properties** are characteristics that do not involve a change in a sample's chemical makeup, whereas **chemical properties** are characteristics that *do* involve a change in chemical makeup. The melting point of ice, for instance, is a physical property because melting causes the

▲ Addition of a solution of silver nitrate to a solution of sodium chloride yields a white precipitate of solid silver chloride.

water to change only in form, from solid to liquid, but not in chemical makeup. Water has the exact same chemical composition (H_2O) in both states of matter. The rusting of an iron (Fe) bicycle left in the rain is a chemical property, however, because iron combines with oxygen and moisture from the air to give the new substance, rust (which has the formula Fe_2O_3). **TABLE 2.3** lists other examples of physical and chemical properties.

TABLE 2.3　Some Examples of Physical and Chemical Properties

Physical Properties		Chemical Properties
Temperature	Amount	Rusting (of iron)
Color	Odor	Combustion (of gasoline)
Melting point	Solubility	Tarnishing (of silver)
Electrical conductivity	Hardness	Cooking (of an egg)

As noted previously, the elements in a group of the periodic table often show remarkable similarities in their chemical properties. Look at the following groups, for instance, to see some examples:

- **Group 1A—Alkali metals**　Lithium (Li), sodium (Na), potassium (K), rubidium (Rb), and cesium (Cs) are soft, silvery metals. All react rapidly, often violently, with water to form products that are highly alkaline, or *basic*—hence the name *alkali metals*. Because of their high reactivity, the alkali metals are never found in nature in the pure state but only in combination with other elements. Francium (Fr) is also an alkali metal, but, as noted previously, it is so rare that little is known about it.

  Note that group 1A also contains hydrogen (H) even though, as a colorless gas, it is completely different in appearance and behavior from the alkali metals. We'll see the reason for this classification in Section 5.13.

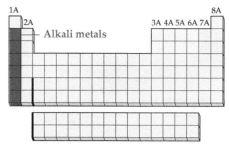

- **Group 2A—Alkaline earth metals**　Beryllium (Be), magnesium (Mg), calcium (Ca), strontium (Sr), barium (Ba), and radium (Ra) are also lustrous, silvery metals but are less reactive than their neighbors in group 1A. Like the alkali metals, the alkaline earths are never found in nature in the pure state.

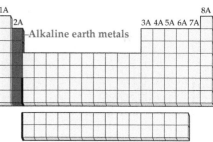

▲ Sodium, one of the alkali metals, reacts violently with water to yield hydrogen gas and an alkaline (basic) solution.

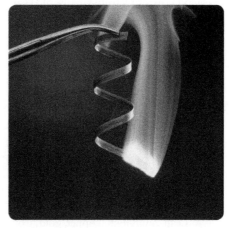

▲ Magnesium, one of the alkaline earth metals, burns in air.

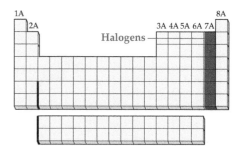

- **Group 7A—Halogens**　Fluorine (F), chlorine (Cl), bromine (Br), and iodine (I) are colorful, corrosive nonmetals. They are found in nature only in combination with other elements, such as with sodium in table salt (sodium chloride, NaCl). In fact, the group name *halogen* is taken from the Greek word *hals,* meaning "salt." Astatine (At) is also a halogen, but it exists in such tiny amounts that little is known about it.

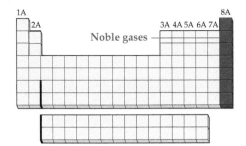

- **Group 8A—Noble gases**   Helium (He), neon (Ne), argon (Ar), krypton (Kr), xenon (Xe), and radon (Rn) are colorless gases with very low chemical reactivity. Helium and neon don't combine with any other element; argon, krypton, and xenon combine with very few.

▲ Bromine, a halogen, is a corrosive dark red liquid at room temperature.

▲ Neon, one of the noble gases, is used in orange-colored neon lights and signs.

As indicated in Figure 2.3, the elements of the periodic table are often divided into three major categories: metals, nonmetals, and semimetals.

- **Metals**   Metals, the largest category of elements, are found on the left side of the periodic table, bounded on the right by a zigzag line running from boron (B) at the top to astatine (At) at the bottom. The metals are easy to characterize by their appearance. All except mercury are solid at room temperature, and most have the silvery shine we normally associate with metals. In addition, metals are generally malleable rather than brittle, can be twisted and drawn into wires without breaking, and are good conductors of heat and electricity.

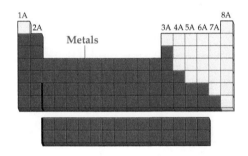

- **Nonmetals**   Except for hydrogen, nonmetals are found on the right side of the periodic table and, like metals, are easy to characterize by their appearance. Eleven of the seventeen nonmetals are gases, one is a liquid (bromine), and only five are solids at room temperature (carbon, phosphorus, sulfur, selenium, and iodine). None are silvery in appearance, and several are brightly colored. The solid nonmetals are brittle rather than malleable and are poor conductors of heat and electricity.

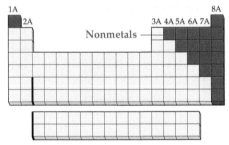

▲ Lead, aluminum, copper, gold, iron, and silver (clockwise from left) are typical metals. All conduct electricity and can be drawn into wires.

▲ Bromine, carbon, phosphorus, and sulfur (clockwise from top left) are typical nonmetals. None conduct electricity or can be made into wires.

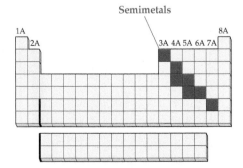

- **Semimetals**   Seven of the nine elements adjacent to the zigzag boundary between metals and nonmetals—boron, silicon, germanium, arsenic, antimony, tellurium, and astatine—are called semimetals because their properties are intermediate between those of their metallic and nonmetallic neighbors. Although most are silvery in appearance and all are solid at room temperature, semimetals are brittle rather than malleable and tend to be poor conductors of heat and electricity. Silicon, for example, is a widely used *semiconductor,* a substance whose electrical conductivity is intermediate between that of a metal and an insulator.

CONCEPTUAL WORKED EXAMPLE 2.1

Classifying Elements Based upon Properties and Location in the Periodic Table

An element is a colorless gas at room temperature and pressure. It is highly reactive and therefore only found in nature combined with other elements. Classify the element as a metal, nonmetal, or semimetal, and specify its general location in the periodic table.

STRATEGY

Match the description of the element to the properties of the main categories of elements and the four specific groups: alkali metals, alkaline earth metals, halogens, and noble gases.

SOLUTION

The element is a nonmetal because it is a gas. All metals (with the exception of mercury) and semimetals are solids at room temperature and pressure. Its high reactivity indicates it is a halogen, but it cannot be bromine or iodine since these elements are not gases at room temperature.

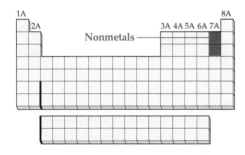

▶ **CONCEPTUAL PRACTICE 2.1** An element is a shiny, silver-colored solid at room temperature and pressure. It conducts electricity and can be found in nature in its pure form. Which element on the periodic table (a–d) would have these properties?

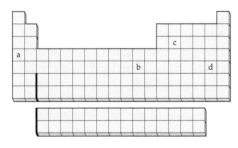

▶ **CONCEPTUAL APPLY 2.2** An element is indicated in the periodic table.

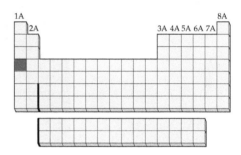

(a) Give the symbol and name of the element, and classify it as a metal, nonmetal, or semimetal.
(b) Predict the appearance of the element at room temperature and pressure.
(c) What would happen if you were to hit a pure sample of this element with a hammer?
(d) Could this element be used to make a wire to conduct electricity in your home?

All **PRACTICE** and **APPLY** problems are interactive in the eText.

2.4 OBSERVATIONS SUPPORTING ATOMIC THEORY: THE CONSERVATION OF MASS AND THE LAW OF DEFINITE PROPORTIONS

The notion that matter was composed of atoms was an accepted scientific theory long before the technology to "see" individual atoms was invented. Hundreds of years ago, scientists designed clever experiments and made careful measurements to obtain results that supported atomic theory. The Englishman Robert Boyle (1627–1691) is generally credited with being the first to study chemistry as a separate intellectual discipline and the first to carry out rigorous chemical experiments. Through a careful series of experiments into the nature and behavior of gases, Boyle provided clear evidence for the atomic makeup of matter. In addition, Boyle was the first to clearly define an element as a substance that cannot be chemically broken down further and to suggest that a substantial number of different elements might exist. Atoms of these different elements, in turn, can join together in different ways to yield a vast number of different substances we call **chemical compounds.**

Progress in chemistry was slow in the decades following Boyle, and it was not until the work of Joseph Priestley (1733–1804) that the next great leap was made. Priestley prepared and isolated the gas oxygen in 1774 by heating the compound mercury oxide (HgO) according to the chemical equation we would now write as $2 \, HgO \rightarrow 2 \, Hg + O_2$.

▶ Heating the red powder HgO causes it to decompose into the silvery liquid mercury and the colorless gas oxygen.

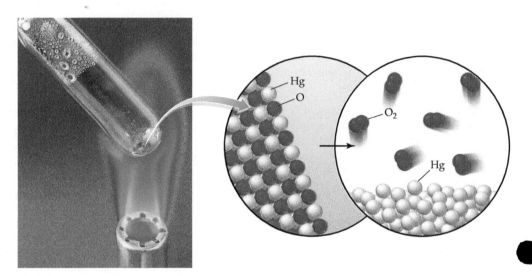

In this standard format for writing chemical transformations, each compound is described by its **chemical formula,** which lists the symbols of its constituent elements and uses subscripts to indicate the number of atoms of each. If no subscript is given, the number 1 is understood. Thus, sodium chloride (table salt) is written as NaCl, water as H_2O, and sucrose (table sugar) as $C_{12}H_{22}O_{11}$. A chemical reaction is written in a standard format called a **chemical equation,** in which the reactant substances undergoing change are written on the left, the product substances being formed are written on the right, and an arrow is drawn between them to indicate the direction of the chemical transformation.

Soon after Priestley's discovery, Antoine Lavoisier (1743–1794) showed that oxygen is the key substance involved in combustion. Furthermore, Lavoisier demonstrated with careful measurements that when combustion is carried out in a closed container, the mass of the combustion products exactly equals the mass of the starting reactants. When hydrogen gas burns and combines with oxygen to yield water (H_2O), for instance, the mass of the water formed is equal to the mass of the hydrogen and oxygen consumed. Called the **law of mass conservation,** this principle is a cornerstone of chemical science.

Law of mass conservation   Mass is neither created nor destroyed in chemical reactions.

It's easy to demonstrate the law of mass conservation by carrying out an experiment like that shown in **FIGURE 2.4.** If 3.25 g of mercury nitrate [$Hg(NO_3)_2$] and 3.32 g of potassium iodide (KI) are each dissolved in water and the solutions are mixed, an immediate chemical reaction occurs leading to the formation of the insoluble orange solid mercury iodide (HgI_2). Filtering the reaction mixture gives 4.55 g of mercury iodide, and evaporation of the water from the remaining solution leaves 2.02 g of potassium nitrate (KNO_3). Thus, the combined mass of the reactants ($3.25\ g + 3.32\ g = 6.57\ g$) is exactly equal to the combined mass of the products ($4.55\ g + 2.02\ g = 6.57\ g$).

The combined masses of these two reactants . . .	. . . equal the combined masses of these two products.

$$Hg(NO_3)_2 + 2\,KI \longrightarrow HgI_2 + 2\,KNO_3$$

| Mercury nitrate | Potassium iodide | Mercury iodide | Potassium nitrate |

| Known amounts of solid KI and solid $Hg(NO_3)_2$ are weighed and then dissolved in water. | The solutions are mixed to give solid HgI_2, which is removed by filtration. | The solution that remains is evaporated to give solid KNO_3. On weighing, the combined masses of the products equal the combined masses of the reactants. |

▲ **FIGURE 2.4**

An illustration of the law of mass conservation. In any chemical reaction, the combined mass of the final products equals the combined mass of the starting reactants.

Further investigations in the decades following Lavoisier led the French chemist Joseph Proust (1754–1826) to formulate a second fundamental chemical principle that we now call the **law of definite proportions:**

Law of definite proportions Different samples of a pure chemical compound always contain the same proportion of elements by mass.

Every sample of water (H_2O) contains 1 part hydrogen and 8 parts oxygen by mass, every sample of carbon dioxide (CO_2) contains 3 parts carbon and 8 parts oxygen by mass, and so on. *Elements combine in specific proportions, not in random proportions.*

2.5 THE LAW OF MULTIPLE PROPORTIONS AND DALTON'S ATOMIC THEORY

At the same time that Proust was formulating the law of definite proportions, the English schoolteacher John Dalton (1766–1844) was exploring along similar lines. His work led him to propose what has come to be called the **law of multiple proportions:**

Law of multiple proportions Elements can combine in different ways to form different chemical compounds whose mass ratios are simple whole-number multiples of each other.

 Go to eText

BIG IDEA Question 3

Propane burns in oxygen according to the following chemical equation:

$$C_3H_8 + 5\,O_2 \longrightarrow 3\,CO_2 + 4\,H_2O$$

If 10.0 g of propane reacts completely with 36.4 g of oxygen and 30.0 g of carbon dioxide is produced, what mass of water is generated in the reaction?

▲ The brown gas nitrogen dioxide, NO_2, is an important component of urban pollution.

The key to Dalton's proposition was his realization that the *same* elements sometimes combine in different ratios to give *different* chemical compounds. For example, oxygen and nitrogen can combine either in a 7:8 mass ratio to make the compound we know today as nitric oxide (NO) or in a 7:16 mass ratio to make the compound we know as nitrogen dioxide (NO_2). The second compound contains exactly twice as much oxygen as the first.

NO: 7 g nitrogen per 8 g oxygen   N:O mass ratio = 7:8

NO_2:   7 g nitrogen per 16 g oxygen   N:O mass ratio = 7:16

Comparison of N:O ratios in NO and NO_2 $\dfrac{\text{N:O mass ratio in NO}}{\text{N:O mass ratio in NO}_2} = \dfrac{(7 \text{ g N})/(8 \text{ g O})}{(7 \text{ g N})/(16 \text{ g O})} = 2$

This result makes sense only if we assume that matter is composed of discrete atoms that have characteristic masses and combine with one another in specific and well-defined ways (**FIGURE 2.5**).

▶ **FIGURE 2.5**

An illustration of Dalton's law of multiple proportions.

▶ **Figure It Out**

Does a compound with the formula N_2O_5 follow the law of multiple proportions?

Answer: Yes, any chemical formula with whole-number subscripts follows the law of multiple proportions.

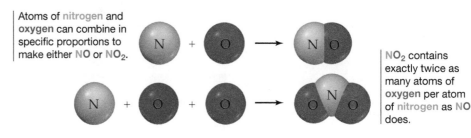

Atoms of nitrogen and oxygen can combine in specific proportions to make either NO or NO_2.

NO_2 contains exactly twice as many atoms of oxygen per atom of nitrogen as NO does.

▲ These samples of sulfur and carbon have different masses but contain the same number of atoms.

Taking all three laws together—the law of mass conservation, the law of definite proportions, and the law of multiple proportions—ultimately led Dalton to propose a new theory of matter. He reasoned as follows:

● *Elements are made up of tiny particles called* **atoms.** Although Dalton didn't know what atoms were like, he nevertheless felt they were necessary to explain why there were so many different elements.

● *Each element is characterized by the mass of its atoms. Atoms of the same element have the same mass, but atoms of different elements have different masses.* Dalton realized that there must be some feature that distinguishes the atoms of one element from those of another. Because Proust's law of definite proportions showed that elements always combine in specific mass ratios, Dalton reasoned that the distinguishing feature among atoms of different elements must be mass.

● *The chemical combination of elements to make different chemical compounds occurs when whole numbers of atoms join in fixed proportions.* Only if whole numbers of atoms combine will different samples of a pure chemical compound always contain the same proportion of elements by mass (the law of definite proportions and the law of multiple proportions). Fractional parts of atoms are never involved in chemical reactions.

● *Chemical reactions only rearrange how atoms are combined in chemical compounds; the atoms themselves don't change.* Dalton realized that atoms must be chemically indestructible for the law of mass conservation to be valid. If the same numbers and kinds of atoms are present in both reactants and products, then the masses of reactants and products must also be the same.

Not everything that Dalton proposed was correct. He thought, for instance, that water had the formula HO rather than H_2O. Nevertheless, his atomic theory of matter was ultimately accepted and came to form a cornerstone of modern chemical science.

— WORKED EXAMPLE 2.2

Using the law of Multiple Proportions

Methane and propane are both constituents of natural gas. A sample of methane contains 5.70 g of carbon atoms and 1.90 g of hydrogen atoms combined in a certain way, whereas a sample of propane contains 4.47 g of carbon atoms and 0.993 g of hydrogen atoms combined in a different way. Show that the two compounds obey the law of multiple proportions.

STRATEGY

Find the C:H mass ratio in each compound, and then compare the ratios to see whether they are simple multiples of each other.

SOLUTION

$$\text{Methane:}\quad \text{C:H mass ratio} = \frac{5.70 \text{ g C}}{1.90 \text{ g H}} = 3.00$$

$$\text{Propane:}\quad \text{C:H mass ratio} = \frac{4.47 \text{ g C}}{0.993 \text{ g H}} = 4.50$$

$$\frac{\text{C:H mass ratio in methane}}{\text{C:H mass ratio in propane}} = \frac{3.00}{4.50} = \frac{2}{3}$$

▲ Sulfur burns with a bluish flame to yield colorless SO_2 gas.

▶ **PRACTICE 2.3**   Compounds A and B are colorless gases obtained by combining sulfur with oxygen. Compound A results from combining 6.00 g of sulfur with 5.99 g of oxygen, and compound B results from combining 8.60 g of sulfur with 12.88 g of oxygen. What is the mass ratio sulfur to oxygen in compound A and in compound B? Are the mass ratios related by a simple multiple?

▶ **APPLY 2.4**   If the chemical formula of compound A in Practice 2.3 is SO_2, what is the chemical formula of compound B?

2.6 ATOMIC STRUCTURE: ELECTRONS

Dalton's atomic theory leaves unanswered the obvious question: What makes up an atom? Dalton himself had no way of answering this question, and it was not until nearly a century later that experiments by the English physicist J. J. Thomson (1856–1940) provided some clues. Thomson's experiments involved the use of *cathode-ray tubes* (CRTs), early predecessors of the tubes found in older televisions and computer displays.

As shown in **FIGURE 2.6a**, a cathode-ray tube is a sealed glass vessel from which the air has been removed and in which two thin pieces of metal, called *electrodes,* have been sealed. When a sufficiently high voltage is applied across the electrodes, an electric current flows through the tube from the negatively charged electrode (the *cathode*) to the positively charged electrode (the *anode*). If the tube is not fully evacuated but still contains a small amount of air or other gas, the flowing current is visible as a glow called a *cathode ray.*

Experiments by a number of physicists in the 1890s had shown that cathode rays can be deflected by bringing either a magnet or an electrically charged plate near the tube (**FIGURE 2.6b**). Because the beam is produced at a negative electrode and is deflected toward a positive plate, Thomson proposed that cathode rays must consist of tiny, negatively charged particles, which we now call **electrons.** Furthermore, because electrons are emitted from electrodes made of many different metals, all these different metals must contain electrons.

Thomson reasoned that the amount of deflection of the electron beam in a cathode-ray tube due to a nearby magnetic or electric field should depend on three factors:

1. *The strength of the deflecting magnetic or electric field.* The stronger the magnet or the higher the voltage on the charged plate, the greater the deflection.
2. *The magnitude of the negative charge on the electron.* The larger the charge on the particle, the greater its interaction with the magnetic or electric field and the greater the deflection.
3. *The mass of the electron.* The lighter the particle, the greater its deflection (just as a Ping-Pong ball is more easily deflected than a bowling ball).

(a) The electron beam ordinarily travels in a straight line.

(b) The beam is deflected by either a magnetic field or an electric field.

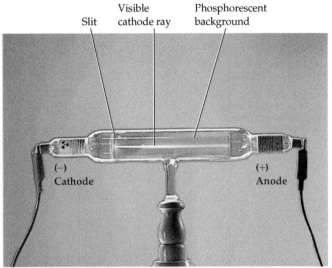

Slit

Visible
cathode ray

Phosphorescent
background

(−)
Cathode

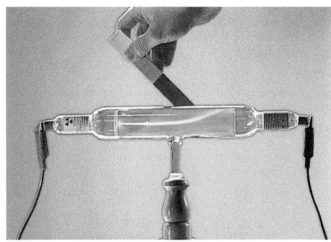

(+)
Anode

▲ **Figure It Out**

What can be deduced about the charge
on an electron from the cathode-ray tube
experiment?

Answer: Electrons are negatively charged because
they are deflected toward a positive plate.

▲ **FIGURE 2.6**

A cathode-ray tube. In a cathode-ray tube, a stream of electrons emitted from the negatively
charged cathode passes through a slit, moves toward the positively charged anode, and is
detected by a phosphorescent strip.

By carefully measuring the amount of deflection caused by electric and magnetic
fields of known strength, Thomson was able to calculate the ratio of the electron's
electric charge to its mass: its *charge-to-mass ratio, e/m.* The modern value is

$$\frac{e}{m} = 1.758\ 820 \times 10^8 \ \text{C/g}$$

where e is the magnitude of the charge on the electron in coulombs (C) and m is the
mass of the electron in grams. (We'll say more about **coulombs** and **electrical charge** in
Chapter 19.) Note that because e is defined as a positive quantity, the actual (negative)
charge on the electron is $-e$.

Thomson was able to measure only the ratio of charge to mass, not charge or mass
itself, and it was left to the American R. A. Millikan (1868–1953) to devise a method for
measuring the mass of an electron (**FIGURE 2.7**). In Millikan's experiment, a fine mist of
oil was sprayed into a chamber, and the tiny droplets were allowed to fall between two
horizontal plates. Observing the droplets through a telescopic eyepiece made it possible to
determine how rapidly they fell through the air, which in turn allowed their masses to be
calculated. The droplets were then given a negative charge by irradiating them with X rays.
X rays knocked electrons from gas molecules in the surrounding air, and the electrons stuck
to the oil droplet. By applying a voltage to the plates, with the upper plate positive, it was
possible to counteract the downward fall of the charged droplets and keep them suspended.

With the voltage on the plates and the mass of the droplets known, Millikan was able
to show that the charge on a given droplet was always a small whole-number multiple of e,
whose modern value is $1.602\ 177 \times 10^{-19}$ C. Substituting the value of e into Thomson's
charge-to-mass ratio then gives the mass m of the electron as $9.109\ 382 \times 10^{-28}$ g:

$$\text{Because} \quad \frac{e}{m} = 1.758\ 820 \times 10^8 \ \text{C/g}$$

$$\text{then} \quad m = \frac{e}{1.758\ 820 \times 10^8 \ \text{C/g}} = \frac{1.602\ 177 \times 10^{-19} \ \cancel{C}}{1.758\ 820 \times 10^8 \ \cancel{C}/\text{g}}$$

$$= 9.109\ 386 \times 10^{-28} \ \text{g}$$

LOOKING AHEAD . . .

We'll see in Chapter 19 that **coloumbs**
and **electrical charge** are fundamental to
electrochemistry—the area of chemistry
involving batteries, fuel cells, and
electroplating of metals.

Go to
eText

BIG IDEA Question 4

What properties of the electron were
measured in Thomson's cathode-ray
experiment and Millikan's oil drop
experiment?

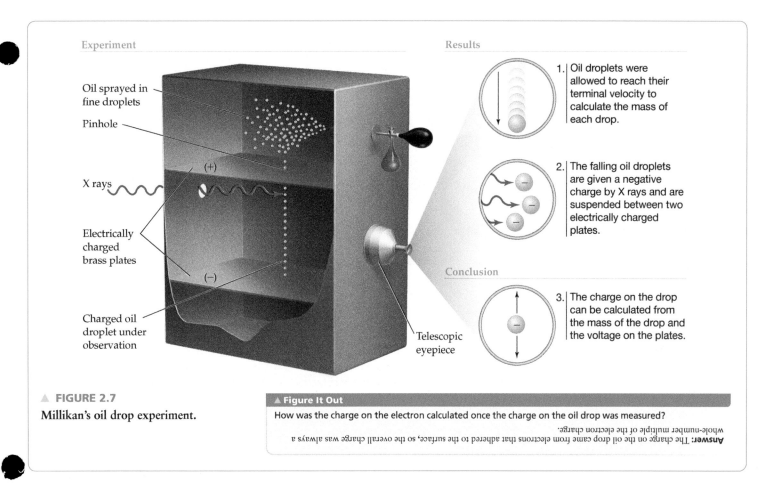

Experiment

Results

1. Oil droplets were allowed to reach their terminal velocity to calculate the mass of each drop.

Oil sprayed in fine droplets

Pinhole

(+)

X rays

Electrically charged brass plates

(−)

Charged oil droplet under observation

Telescopic eyepiece

2. The falling oil droplets are given a negative charge by X rays and are suspended between two electrically charged plates.

Conclusion

3. The charge on the drop can be calculated from the mass of the drop and the voltage on the plates.

▲ **FIGURE 2.7**
Millikan's oil drop experiment.

▲ **Figure It Out**

How was the charge on the electron calculated once the charge on the oil drop was measured?

Answer: The charge on the oil drop came from electrons that adhered to the surface, so the overall charge was always a whole-number multiple of the electron charge.

2.7 ATOMIC STRUCTURE: PROTONS AND NEUTRONS

Think about the consequences of Thomson's cathode-ray experiments. Because matter is electrically neutral overall, the fact that the atoms in an electrode can give off negatively charged particles (electrons) must mean that those same atoms also contain positively charged particles for electrical balance. The search for those positively charged particles and for an overall picture of atomic structure led to a landmark experiment published in 1911 by the New Zealand physicist Ernest Rutherford (1871–1937).

Rutherford's work involved the use of *alpha* (α) *particles,* a type of emission previously found to be given off by a number of naturally occurring radioactive elements, including radium, polonium, and radon. Rutherford knew that alpha particles are about 7000 times more massive than electrons and that they have a positive charge that is twice the magnitude of the charge on an electron but opposite in sign.

When Rutherford directed a beam of alpha particles at a thin gold foil, he found that almost all the particles passed through the foil undeflected. A very small number, however (about 1 of every 20,000), were deflected at an angle, and a few actually bounced back toward the particle source (**FIGURE 2.8**). Rutherford described his reaction to the results as follows:

> *It was quite the most incredible event that has ever happened to me in my life. It was almost as incredible as if you fired a 15-inch shell at a piece of tissue paper and it came back and hit you.*

Rutherford explained his results by proposing that a metal atom must be almost entirely empty space and have its mass concentrated in a tiny central core that he called the **nucleus.** If the nucleus contains the atom's positive charges and most of its mass, and if the electrons are a relatively large distance away, then it is clear why the observed scattering results are obtained: most alpha particles encounter empty space as they fly

An alpha particle
(relative mass = 7000;
charge = $+2e$)

An electron
(relative mass = 1;
charge = $-1e$)

Experiment

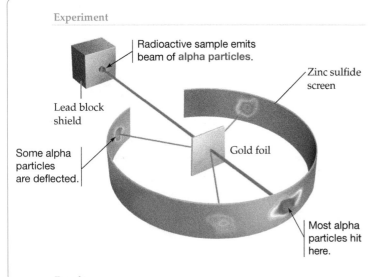

Radioactive sample emits beam of **alpha particles**.

Lead block shield

Some alpha particles are deflected.

Zinc sulfide screen

Gold foil

Most alpha particles hit here.

Conclusion

Because the majority of particles are not deflected, the gold atoms must be almost entirely empty space. The atom's mass is concentrated in a tiny dense core, which deflects the occasional alpha particle.

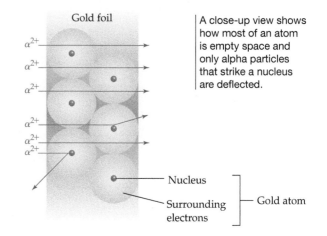

Gold foil

α^{2+}

A close-up view shows how most of an atom is empty space and only alpha particles that strike a nucleus are deflected.

Nucleus

Surrounding electrons

Gold atom

Results

When a beam of **alpha particles** is directed at a thin gold foil, most particles pass through undeflected, but some are deflected at large angles and a few bounce back toward the particle source.

▲ **FIGURE 2.8**

Rutherford's scattering experiment.

▲ **Figure It Out**

What was the surprising result in the gold foil experiment? What conclusions about atomic structure were drawn from this observation?

Answer: Some alpha particles bounced backward. This observation could only be explained by the alpha particle interacting with a dense, positively charged region in a gold atom (the nucleus).

▲ The relative size of the nucleus in an atom is roughly the same as that of a pea in the middle of this huge stadium.

through the foil. Only when a positive alpha particle chances to come near a small but massive positive nucleus is it repelled strongly enough to make it bounce backward.

Modern measurements show that an atom has a diameter of roughly 10^{-10} m and that a nucleus has a diameter of about 10^{-15} m. It's difficult to imagine from these numbers alone, though, just how small a nucleus really is. For comparison purposes, if an atom were the size of a large domed stadium, the nucleus would be approximately the size of a small pea in the center of the playing field.

Further experiments by Rutherford and others between 1910 and 1930 showed that a nucleus is composed of two kinds of particles, called *protons* and *neutrons*. **Protons** have a mass of $1.672\ 622 \times 10^{-24}$ g (about 1836 times that of an electron) and are positively charged. Because the charge on a proton is opposite in sign but equal in size to that on an electron, the numbers of protons and electrons in a neutral atom are equal. **Neutrons** ($1.674\ 927 \times 10^{-24}$ g) are almost identical in mass to protons but carry no charge, and the number of neutrons in a nucleus is not directly related to the numbers of protons and electrons. **TABLE 2.4** compares the three fundamental subatomic particles, and **FIGURE 2.9** gives an overall view of the atom.

TABLE 2.4   A Comparison of Subatomic Particles

Particle	Mass		Charge	
	Grams	u*	Coulombs	e
Electron	$9.109\ 382 \times 10^{-28}$	$5.485\ 799 \times 10^{-4}$	$-1.602\ 176 \times 10^{-19}$	-1
Proton	$1.672\ 622 \times 10^{-24}$	$1.007\ 276$	$+1.602\ 176 \times 10^{-19}$	$+1$
Neutron	$1.674\ 927 \times 10^{-24}$	$1.008\ 665$	0	0

*The unified atomic mass unit (u) is defined in Section 2.9.

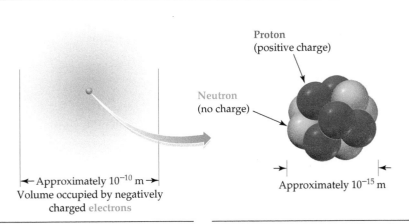

◀ **FIGURE 2.9**
Size scale of the atom.

◀ **Figure It Out**
How many times larger is the diameter of the atom in comparison to the diameter of the nucleus?

Answer: 10^5 or 100,000.

Proton
(positive charge)

Neutron
(no charge)

←Approximately 10^{-10} m→
Volume occupied by negatively charged electrons

Approximately 10^{-15} m

A number of electrons equal to the number of protons move about the nucleus and account for most of the atom's volume.

The protons and neutrons in the nucleus take up very little volume but contain essentially all the atom's mass.

WORKED EXAMPLE 2.3

Calculations Using Atomic Size

Ordinary "lead" pencils actually are made of a form of carbon called *graphite*. If a pencil line is 0.35 mm wide and the diameter of a carbon atom is 1.5×10^{-10} m, how many atoms wide is the line?

IDENTIFY

Known	Unknown
Diameter of a C atom $\left(\dfrac{1.5 \times 10^{-10} \text{ m}}{1 \text{ atom}}\right)$	Number of C atoms in width of line
Pencil line = 0.35 mm	

STRATEGY

Begin with the known information that is not a conversion factor (width of the pencil line). Next set up an equation using appropriate conversion factors so that the unwanted units cancel. Conversion factors between prefixes are also needed in this problem.

SOLUTION

$$\text{Atoms} = 0.35 \text{ mm} \times \frac{1 \text{ m}}{1000 \text{ mm}} \times \frac{1 \text{ atom}}{1.5 \times 10^{-10} \text{ m}}$$

$$= 2.3 \times 10^6 \text{ atoms}$$

CHECK

A single carbon atom is about 10^{-10} m across, so it takes 10^{10} carbon atoms placed side by side to stretch 1 m, 10^7 carbon atoms to stretch 1 mm, and about 0.3×10^7 (or 3×10^6; 3 *million*) carbon atoms to stretch 0.35 mm. The estimate agrees with the solution.

▶ **PRACTICE 2.5** The gold foil that Rutherford used in his scattering experiment had a thickness of approximately 0.005 mm. If a single gold atom has a diameter of 2.9×10^{-10} m, how many atoms thick was Rutherford's foil?

▶ **APPLY 2.6** A small speck of carbon, the size of a pinhead, contains about 10^{19} atoms, the diameter of a carbon atom is 1.5×10^{-10} m, and the circumference of the Earth at the equator is 40,075 km. How many times around the Earth would the atoms from this speck of carbon extend if they were laid side by side?

2.8 ATOMIC NUMBERS

Thus far, we've described atoms only in general terms and have not yet answered the most important question: What is it that makes one atom different from another? How, for example, does an atom of gold differ from an atom of carbon? The answer turns out to be quite simple. *Elements differ from one another according to the number of protons in the nucleus,* a value called the element's **atomic number** (Z). That is, all atoms of a given element contain the same number of protons in their nuclei. All hydrogen atoms,

REMEMBER . . .
The periodic table is arranged by increasing atomic number of the elements (Figure 2.3). The atomic number is defined as the number of protons in the nucleus and gives the identity of the element.

atomic number 1, have 1 proton; all helium atoms, atomic number 2, have 2 protons; all carbon atoms, atomic number 6, have 6 protons; and so on. In addition, every neutral atom contains a number of electrons equal to its number of protons.

Atomic number (Z)

     = Number of protons in an atom's nucleus
     = Number of electrons around the nucleus in a neutral atom

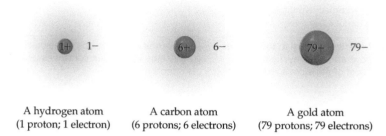

A hydrogen atom	A carbon atom	A gold atom
(1 proton; 1 electron)	(6 protons; 6 electrons)	(79 protons; 79 electrons)

In addition to protons, the nuclei of all atoms (other than hydrogen) also contain neutrons. The sum of the numbers of protons (Z) and neutrons (N) in an atom is called the atom's **mass number** (A). That is, $A = Z + N$.

Mass number (A) = Number of protons (Z) + number of neutrons (N)

Most hydrogen atoms have 1 proton and no neutrons, so their mass number is $A = 1 + 0 = 1$. Most helium atoms have 2 protons and 2 neutrons, so their mass number is $A = 2 + 2 = 4$. Most carbon atoms have 6 protons and 6 neutrons, so their mass number is $A = 6 + 6 = 12$. Except for hydrogen, stable atoms always contain at least as many neutrons as protons, although there is no simple way to predict how many neutrons a given atom will have.

Notice that we said *most* hydrogen atoms have mass number 1, *most* helium atoms have mass number 4, and *most* carbon atoms have mass number 12. In fact, different atoms of the same element can have different mass numbers depending on how many neutrons they have. Atoms with identical atomic numbers but different mass numbers are called **isotopes.** Hydrogen, for example, has three isotopes.

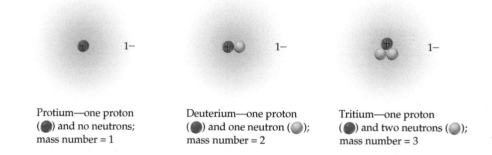

Protium—one proton (●) and no neutrons; mass number = 1

Deuterium—one proton (●) and one neutron (○); mass number = 2

Tritium—one proton (●) and two neutrons (○); mass number = 3

All hydrogen atoms have 1 proton in their nucleus (otherwise they wouldn't be hydrogen), but 99.985% of them have no neutrons. These hydrogen atoms, called *protium,* have mass number 1. In addition, 0.015% of hydrogen atoms, called *deuterium,* have 1 neutron and mass number 2. Still other hydrogen atoms, called *tritium,* have 2 neutrons and mass number 3. An unstable, radioactive isotope, tritium occurs only in trace amounts on Earth but is made artificially in nuclear reactors. As other examples, there are 15 known isotopes of nitrogen, only 2 of which occur naturally on Earth, and 25 known isotopes of uranium, only 3 of which occur naturally. In total, more than 3600 isotopes of the 118 known elements have been identified.

A specific isotope is represented by its element symbol accompanied by its mass number as a left superscript and its atomic number as a left subscript. Thus, protium is represented as ^1_1H, deuterium as ^2_1H, and tritium as ^3_1H. Similarly, the two naturally occurring isotopes of nitrogen are represented as $^{14}_7N$ (spoken as "nitrogen-14") and $^{15}_7N$ (nitrogen-15). The number of neutrons in an isotope is not given explicitly but can be calculated by subtracting the atomic number (subscript) from the mass number (superscript). For example, subtracting the atomic number 7 from the mass number 14 indicates that a $^{14}_7N$ atom has 7 neutrons.

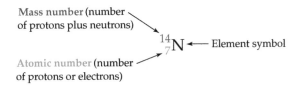

The number of neutrons in an atom has relatively little effect on the atom's chemical properties. The chemical behavior of an element is determined almost entirely by the number of electrons it has, which in turn is determined by the number of protons in its nucleus. All three isotopes of hydrogen therefore behave similarly (although not identically) in their chemical reactions.

BIG IDEA Question 5

The number of which subatomic particles determines the identity of an atom and the mass number (*A*) of an atom?

WORKED EXAMPLE 2.4

Interpreting Isotope Symbols

The isotope of uranium used to generate nuclear power is $^{235}_{92}U$. How many protons, neutrons, and electrons does an atom of $^{235}_{92}U$ have?

STRATEGY

The atomic number (subscript 92) in the symbol $^{235}_{92}U$ indicates the number of protons and electrons in the atom. The number of neutrons is the difference between the mass number (superscript 235) and the atomic number (92).

SOLUTION

An atom of $^{235}_{92}U$ has 92 protons, 92 electrons, and $235 - 92 = 143$ neutrons.

▶ **PRACTICE 2.7** The isotope $^{75}_{34}Se$ is used medically for the diagnosis of pancreatic disorders. How many protons, neutrons, and electrons does an atom of $^{75}_{34}Se$ have?

▶ **APPLY 2.8** Element X is toxic to humans in high concentration but is essential to life in low concentrations. Identify element X, whose atoms contain 24 protons, and write the symbol for the isotope of X that has 28 neutrons. (Refer to Figure 2.3, the periodic table.)

▲ Uranium-235 is used as fuel in this nuclear-powered icebreaker.

2.9 ATOMIC WEIGHTS AND THE MOLE

Pick up a pencil, and look at the small amount of tip visible. How many carbon atoms in the graphite "pencil lead" do you think are in the tip? One thing is certain: Atoms are so tiny that the number needed to make a visible sample is enormous. In fact, even the smallest speck of dust visible to the naked eye contains at least 10^{17} atoms. Thus, the mass in grams of a single atom is much too small a number for convenience, so chemists use a unit called the **unified atomic mass unit (u)** (also known as a *dalton* [Da] in biological work). One unified atomic mass unit is defined as exactly 1/12 the mass of an atom of $^{12}_6$C and is equal to $1.660\ 539 \times 10^{-24}$ g.

$$1\ \text{u} = \frac{\text{Mass of one } ^{12}_6\text{C atom}}{12} = 1.660\ 539 \times 10^{-24}\ \text{g}$$

Because the mass of an atom's electrons is negligible compared to the mass of its protons and neutrons, defining 1 u as 1/12 the mass of a $^{12}_6$C atom means that protons and neutrons both have a mass of almost exactly 1 u (Table 2.4). Thus, the mass of a specific atom in unified atomic mass units—called the atom's **atomic mass**—is numerically close to the atom's mass number. A $^1_1$H atom, for instance, has a mass of 1.007 825; a $^{235}_{92}$U atom has an atomic mass of 235.043 930; and so forth. In practice, atomic masses are taken to be dimensionless, and the unit *u* is understood rather than specified.

Most elements occur naturally as a mixture of different isotopes. Thus, if you look at the periodic table inside the front cover, you'll see listed below the symbol for each element a value called the element's *atomic weight*. Again, the unit *u* is understood but not specified.

6 ← Atomic number
C ← Symbol
12.011 ← Atomic weight

An element's **atomic weight** is the weighted average of the isotopic masses of the element's naturally occurring isotopes. Carbon, for example, occurs on Earth as a mixture of two major isotopes, $^{12}_6$C (98.89% natural abundance) and $^{13}_6$C (1.11% natural abundance). Although the isotopic mass of any individual carbon atom is either exactly 12 (a carbon-12 atom) or 13.0034 (a carbon-13 atom), the *average* atomic mass—that is, the atomic weight—of a large collection of carbon atoms is 12.011. A third carbon isotope, $^{14}_6$C, also exists, but its natural abundance is so small that it can be ignored when calculating atomic weight.

Atomic weight of C = (Mass of $^{12}_6$C)(Abundance of $^{12}_6$C) + (Mass of $^{13}_6$C)(Abundance of $^{13}_6$C)

= (12)(0.9889) + (13.0034)(0.0111)

= 11.867 + 0.144 = 12.011

WORKED EXAMPLE 2.5

Calculating an Atomic Weight

Chlorine has two naturally occurring isotopes: $^{35}_{17}$Cl, with a natural abundance of 75.76% and an isotopic mass of 34.969; and $^{37}_{17}$Cl, with a natural abundance of 24.24% and an isotopic mass of 36.966. What is the atomic weight of chlorine?

IDENTIFY

Known	Unknown
Isotope masses and abundances	Atomic weight of chlorine

(75.76% of isotope with mass 34.969 and 24.24% of isotope with mass 36.966)

STRATEGY

The atomic weight of an element is the weighted average of the isotopic masses, which equals the sum of the masses of each isotope times the natural fractional abundance of that isotope:

Atomic weight = (Mass of $^{35}_{17}$Cl)(Abundance of $^{35}_{17}$Cl)

+ (Mass of $^{37}_{17}$Cl)(Abundance of $^{37}_{17}$Cl)

SOLUTION

Atomic weight = $(34.969)(0.7576) + (36.966)(0.2424) = 35.45$

CHECK

The atomic weight is somewhere between 35 and 37, the masses of the two individual isotopes. It is closer to 35 since this is the mass of the more abundant isotope.

▶ **PRACTICE 2.9** Copper metal has two naturally occurring isotopes: copper-63 (69.15%; isotopic mass = 62.93) and copper-65 (30.85%; isotopic mass = 64.93). Calculate the atomic weight of copper, and check your answer in a periodic table.

▶ **APPLY 2.10** Gallium has two naturally occurring isotopes: gallium-69 (isotopic mass = 68.9256) and gallium-71 (isotopic mass = 70.9247). The atomic weight of gallium is 69.7231.

(a) Without doing any calculations, state which isotope has the greater abundance.

(b) Calculate the percent abundance of each isotope.

All **WORKED EXAMPLES** with this icon have an interactive video in the eText.

Atomic weights of some elements are not constants of nature because natural isotopic abundances can vary depending on the source of a sample. In 2009 and 2011, the Commission on Isotopic Abundances and Atomic Weights replaced single-value standard atomic weight values with atomic weight intervals for 12 elements (H, Li, B, C, N, O, Mg, Si, S, Cl, Br, and Tl). For example, the standard atomic weight of nitrogen became the interval [14.00643,14.00728]. For practical purposes, a single representative value for nitrogen (14.0067) and other elements is given in the periodic table on the inside front cover of this book.

Atomic weights are extremely useful because they are conversion factors between numbers of atoms and masses; that is, they allow us to *count* a large number of atoms by *weighing* a sample of the substance. For instance, knowing that carbon has an atomic weight of 12.011 lets us calculate that a small pencil tip made of carbon and weighing 15 mg (1.5×10^{-2} g) contains 7.5×10^{20} atoms:

$$(1.5 \times 10^{-2} \text{ g})\left(\frac{1 \text{ u}}{1.6605 \times 10^{-24} \text{ g}}\right)\left(\frac{1 \text{ C atom}}{12.011 \text{ u}}\right) = 7.5 \times 10^{20} \text{ C atoms}$$

When referring to the enormous numbers of atoms that make up the visible amounts we typically deal with, chemists use the fundamental SI unit for amount called a *mole*, abbreviated *mol*. One **mole** of any element is the amount whose mass in grams, called its **molar mass**, is numerically equal to its atomic weight. One mole of carbon atoms has a mass of 12.011 g, and one mole of silver atoms has a mass of 107.868 g. Molar mass thus acts as a conversion factor that lets you convert between mass in grams and number of atoms. *Whenever you have the same number of moles of different elements, you also have the same number of atoms.*

How many atoms are there in a mole? Experiments show that one mole of any element contains 6.022 141 $\times 10^{23}$ atoms, a value called **Avogadro's number,** abbreviated N_A, after the Italian scientist who first recognized the importance of the mass/number relationship. Avogadro's number of atoms of any element—that is, one mole—has a mass in grams equal to the element's atomic weight.

It's hard to grasp the magnitude of a quantity as large as Avogadro's number, but some comparisons might give you a sense of scale: The age of the universe in seconds (13.7 billion years, or 4.32×10^{17} s) is less than a millionth the size of Avogadro's number. The number of liters of water in the world's oceans (1.3×10^{21} L) is less than one-hundredth the size of Avogadro's number. The mass

▲ These samples of helium, sulfur, copper, and mercury each contain 1 mol. Do they have the same mass?

of the Earth in kilograms (5.98×10^{24} kg) is only 10 times Avogadro's number. We'll return to the mole throughout the book and see some of its many uses in Chapter 3.

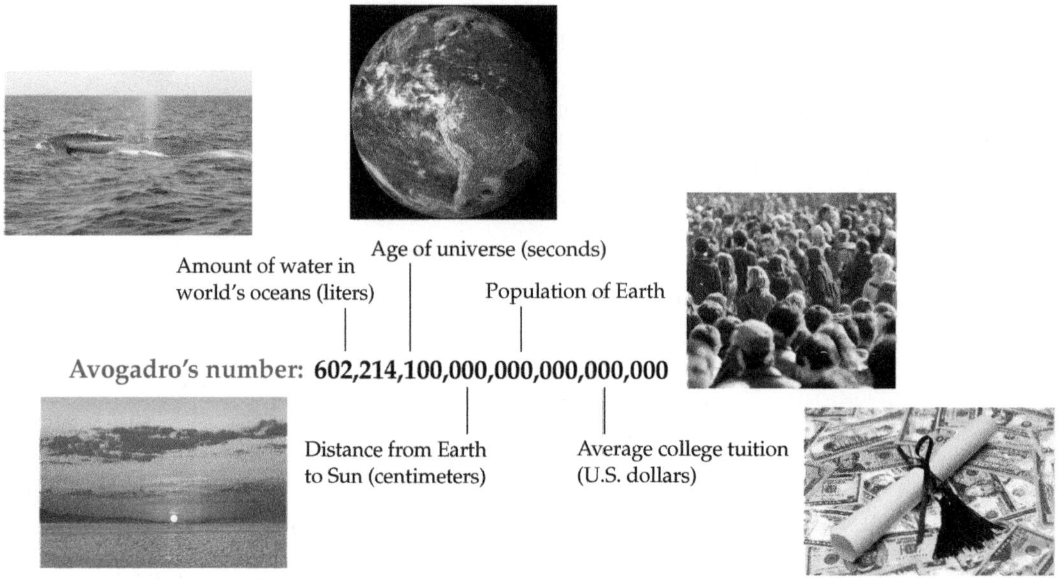

Amount of water in world's oceans (liters)

Age of universe (seconds)

Population of Earth

Avogadro's number: **602,214,100,000,000,000,000,000**

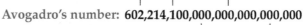

Distance from Earth to Sun (centimeters)

Average college tuition (U.S. dollars)

WORKED EXAMPLE 2.6

Go to eText

Converting Between Mass and Numbers of Moles and Atoms

How many moles and how many atoms of silicon are in a sample weighing 10.53 g? The atomic weight of silicon is 28.0855.

IDENTIFY

Known	Unknown
Mass of sample 10.53 g	Moles and number of atoms of silicon
Atomic weight of silicon (28.0855)	

STRATEGY

The molar mass (28.0855 g/mol) is numerically equivalent to the atomic weight. Use molar mass to convert between mass and number of moles, and then use Avogadro's number to convert between moles and number of atoms.

SOLUTION

$$(10.53 \text{ g Si})\left(\frac{1 \text{ mol Si}}{28.0855 \text{ g Si}}\right) = 0.37492 \text{ mol Si}$$

$$(0.37492 \text{ mol Si})\left(\frac{6.022 \times 10^{23} \text{ atoms Si}}{1 \text{ mol Si}}\right) = 2.258 \times 10^{23} \text{ atoms Si}$$

CHECK

A mass of 10.53 g of silicon is a bit more than one-third the molar mass of silicon (28.0855 g/mol), so the sample contains a bit more than 0.33 mol. This number of moles, in turn, contains a bit more than one-third of Avogadro's number of atoms, or about 2×10^{23} atoms.

▶ **PRACTICE 2.11** How many moles and how many atoms of platinum are in a ring with a mass of 9.50 g? Use the periodic table on the inside front cover of the book to look up the atomic weight of platinum.

▶ **APPLY 2.12** If 2.26×10^{22} atoms of element Y have a mass of 1.50 g, what is the identity of Y?

2.10 MEASURING ATOMIC WEIGHT: MASS SPECTROMETRY

The most common method of measuring an element's atomic weight is with an instrument called a **mass spectrometer.** More than 20 different kinds of mass spectrometers are commercially available, depending on the intended application, but the electron-impact, magnetic-sector instrument shown in **FIGURE 2.10** is particularly common. In this instrument, the sample is vaporized and injected as a dilute gas into an evacuated chamber, where it is bombarded with a beam of high-energy electrons. The electron beam knocks other electrons from atoms in the sample, which become positively charged ions. An ion is a charged particle and will be discussed further in Section 2.12. A reaction for ionization of an atom of boron (B) by a high-energy electron into a positively charged boron ion is:

$$B\ (g) + e^-\ (\text{high-energy}) \rightarrow B^+\ (g) + 2\ e^-$$

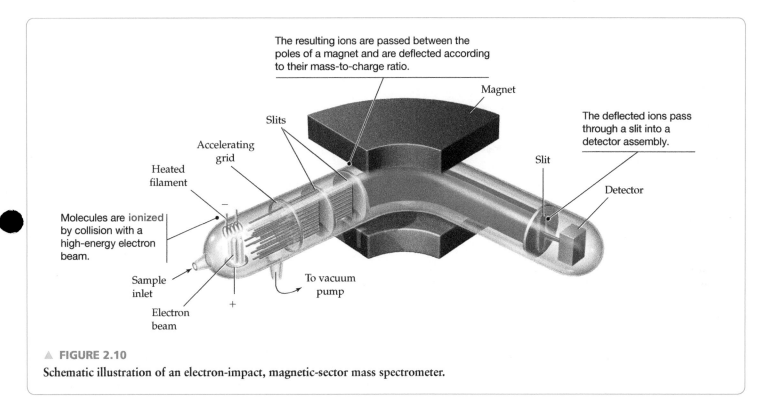

The resulting ions are passed between the poles of a magnet and are deflected according to their mass-to-charge ratio.

Magnet

Slits

Accelerating grid

Heated filament

Molecules are ionized by collision with a high-energy electron beam.

Sample inlet

Electron beam

To vacuum pump

Slit

The deflected ions pass through a slit into a detector assembly.

Detector

▲ **FIGURE 2.10**

Schematic illustration of an electron-impact, magnetic-sector mass spectrometer.

Ionization is necessary because electric and magnetic fields will exert a force on only a charged species, not a neutral molecule. The various ions of different masses are then accelerated by an electric field and passed between the poles of a strong magnet, which deflects them through a curved, evacuated pipe. The radius of deflection of a charged ion such as B^+, as it passes between the magnet poles, depends on its mass-to-charge ratio. If we assume that all ions have the same charge (+1), then the radius of deflection depends only on mass. Lighter ions are deflected to a greater degree than heavier ones in a similar manner to vehicles rounding a turn at high speed. A small, light race car will be able to make a tighter turn than a heavy semi truck.

By varying the strength of the magnetic field, it's possible to focus ions of different mass through a slit at the end of the curved pipe and onto a detector assembly. The **mass spectrum** that results is plotted as a graph of ion intensity versus mass-to-charge ratio. The intensity (height) of the peaks gives a direct measure of the relative abundance of the isotopes. **FIGURE 2.11** shows the mass spectrum of a sample of boron with two isotopes. The most abundant isotope has the tallest peak and is often given an arbitrary intensity value of 100. An intensity of 100 is not to be confused with the percent isotope abundance, which totals 100% for the naturally occurring isotopes. Worked Example 2.7 shows how to calculate the atomic weight of boron from the mass spectrum.

▶ **FIGURE 2.11**

Mass spectrum of a sample of boron. The ^{11}B isotope with a mass of 11.0093 is more abundant than the ^{10}B isotope with a mass of 10.0129.

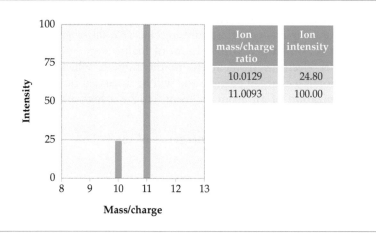

Ion mass/charge ratio	Ion intensity
10.0129	24.80
11.0093	100.00

WORKED EXAMPLE 2.7

Determination of Atomic Weight from a Mass Spectrum

Use data from the mass spectrum of a sample of boron in Figure 2.11 to calculate the atomic weight of boron.

IDENTIFY

Known	Unknown
^{10}B (Mass/charge ratio = 10.0129 and Intensity = 24.80)	Atomic weight of boron
^{11}B (Mass/charge ratio = 11.0093 and Intensity = 100.00)	

STRATEGY

Calculate the fractional abundance of each isotope from the intensity of the isotope peaks. Then use the formula for calculating atomic weight using isotopic masses and fractional abundances (Worked Example 2.5).

SOLUTION

To find the percent abundance of each isotope, we must first find the total intensity by adding the intensity of each isotope. ^{10}B has an intensity of 24.80, and ^{11}B has an intensity of 100.00, giving a total intensity of 124.80. The fractional abundance for each isotope is the intensity of isotope divided by the total intensity.

Fractional abundance of ^{10}B: $\dfrac{24.80}{124.80} = 0.1987$

Fractional abundance of ^{11}B: $\dfrac{100.0}{124.80} = 0.8013$

The formula for calculating atomic weight can now be applied. The charge on each ion is +1; therefore, the mass/charge ratio is equal to the mass of the isotope.

Atomic weight = (Mass of ^{10}B)(Abundance of ^{10}B)
                      + (Mass of ^{11}B)(Abundance of ^{11}B)

Atomic weight = (10.0129)(0.1987) + (11.0093)(0.8013)
              = 10.81

CHECK

The isotope of boron with a mass close to 11 has an abundance that is four times greater than the isotope of boron with a mass near 10. Therefore, the atomic weight of boron should be closer to 11 than 10. The calculated answer of 10.81 agrees with the prediction.

▶ **PRACTICE 2.13**   Use the data from the mass spectrum of a sample of an element to calculate the element's atomic weight. Identify the element.

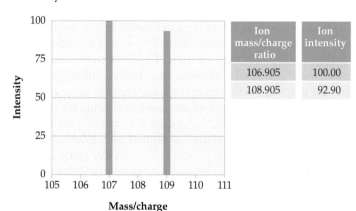

Ion mass/charge ratio	Ion intensity
106.905	100.00
108.905	92.90

▶ **APPLY 2.14**   Use the data from the mass spectrum of a sample of an element to calculate the element's atomic weight. Identify the element.

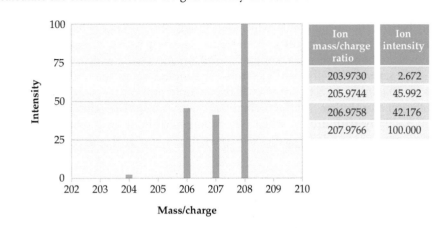

Ion mass/charge ratio	Ion intensity
203.9730	2.672
205.9744	45.992
206.9758	42.176
207.9766	100.000

2.11 MIXTURES AND CHEMICAL COMPOUNDS; MOLECULES AND COVALENT BONDS

Although only 90 elements occur naturally, there are far more than 90 different substances on Earth. Water, sugar, protein in food, and cotton or rayon in clothing are familiar substances that are not pure elements. Matter, anything that has mass and occupies volume, can be classified as either mixtures or pure substances (**FIGURE 2.12**). Pure substances, in turn, can be either elements or chemical compounds.

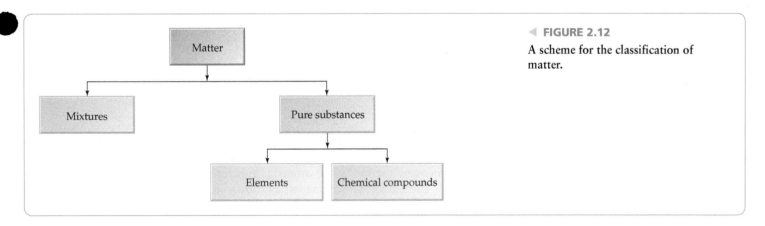

◀ **FIGURE 2.12**
A scheme for the classification of matter.

A **mixture** is simply a blend of two or more substances added together in some arbitrary proportion without chemically changing the individual substances themselves. Thus, the constituent units in the mixture are not all the same, and the proportion of the units is variable. Hydrogen gas and oxygen gas, for instance, can be mixed in any ratio without changing them (as long as there is no flame nearby to initiate reaction), just as a spoonful of sugar and a spoonful of salt can be mixed.

A chemical compound, in contrast to a mixture, is a pure substance that is formed when atoms of different elements combine in a specific way to create a new material with properties completely unlike those of its constituent elements. A chemical compound has a constant composition throughout, and its constituent units are all identical. For example, when atoms of sodium (a soft, silvery metal) combine with atoms of chlorine (a toxic, yellow-green gas), the familiar white solid called *sodium chloride* (table salt) is formed. Similarly, when two atoms of hydrogen combine with one atom of oxygen, water is formed.

To see how a chemical compound is formed, imagine what must happen when two atoms approach each other at the beginning of a chemical reaction. Because the electrons of an atom occupy a much greater volume than the nucleus, it's the electrons that actually

▲ The crystalline quartz sand on this beach is a pure compound (SiO_2), but the seawater is a liquid mixture of many compounds dissolved in water.

make the contact when atoms collide. Thus, it's the electrons that form the connections, or **chemical bonds,** that join atoms together in compounds. Chemical bonds between atoms are usually classified as either *covalent* or *ionic*. As a general rule, covalent bonds occur primarily between nonmetal atoms, while ionic bonds occur primarily between metal and nonmetal atoms. Let's look briefly at both kinds, beginning with covalent bonds.

A **covalent bond,** the most common kind of chemical bond, results when two atoms *share* several (usually two) electrons. A simple way to think about a covalent bond is to imagine it as a tug-of-war. If two people pull on the same rope, they are effectively joined together. Neither person can escape from the other as long as both hold on. Similarly with atoms, when two atoms both hold on to some shared electrons, the atoms are bonded together (**FIGURE 2.13**).

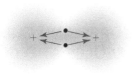

Similarly, two atoms are joined together when both nuclei (+) tug on the same electrons (dots).

The two teams are joined together because both are tugging on the same rope.

▲ **FIGURE 2.13**

A covalent bond between atoms is analogous to a tug-of-war.

The unit of matter that results when two or more atoms are joined by covalent bonds is called a **molecule.** A hydrogen chloride (HCl) molecule results when a hydrogen atom and a chlorine atom share two electrons. A water (H_2O) molecule results when each of two hydrogen atoms shares two electrons with a single oxygen atom. An ammonia (NH_3) molecule results when each of three hydrogen atoms shares two electrons with a nitrogen atom, and so on. To visualize these and other molecules, it helps to imagine the individual atoms as spheres joined together to form molecules with specific three-dimensional shapes, as shown in **FIGURE 2.14**. *Ball-and-stick* models specifically indicate the covalent bonds between atoms, while *space-filling* models portray overall molecular shape but don't explicitly show covalent bonds.

▶ **FIGURE 2.14**

Molecular models. Drawings such as these help in visualizing molecules.

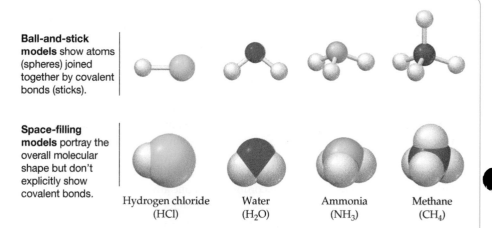

Ball-and-stick models show atoms (spheres) joined together by covalent bonds (sticks).

Space-filling models portray the overall molecular shape but don't explicitly show covalent bonds.

Hydrogen chloride (HCl)   Water (H_2O)   Ammonia (NH_3)   Methane (CH_4)

CONCEPTUAL WORKED EXAMPLE 2.8

Visual Representations of Mixtures and Compounds

Which of the following drawings represents a mixture, which a pure compound, and which an element?

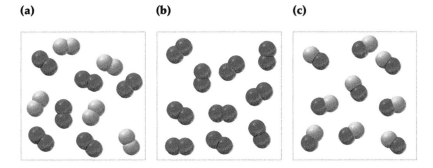

(a)   **(b)**   **(c)**

STRATEGY

Most people (professional chemists included) find chemistry easier to grasp when they can visualize the behavior of atoms, thereby turning symbols into pictures. The Conceptual Problems in this text are intended to help you do that, frequently representing atoms and molecules as collections of spheres. Don't take the pictures literally; focus instead on interpreting what they represent. An element contains only one kind of atom, while a compound contains two or more different elements bonded together. A pure substance contains only one type of element or compound, while a mixture contains two or more substances.

SOLUTION

Drawing (a) represents a mixture of two diatomic elements, one composed of two red atoms and one composed of two blue atoms. Drawing (b) represents molecules of a pure diatomic element because all atoms are identical. Drawing (c) represents molecules of a pure compound composed of one red and one blue atom.

▶ **CONCEPTUAL PRACTICE 2.15** Which of the following drawings represents a pure sample of hydrogen peroxide (H_2O_2) molecules? The red spheres represent oxygen atoms, and the ivory spheres represent hydrogen.

(a)   **(b)**   **(c)**   **(d)**

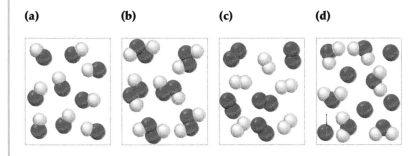

▶ **CONCEPTUAL APPLY 2.16**   Red and blue spheres represent atoms of different elements.

(a) Which drawing(s) illustrate a pure substance?
(b) Which drawing(s) illustrate a mixture?
(c) Which two drawings illustrate the law of multiple proportions?

(a)   **(b)**

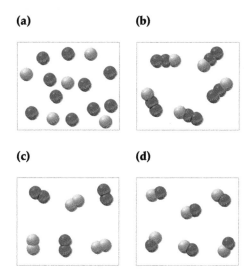

(c)   **(d)**

Chemists normally represent a molecule by giving its **structural formula,** which shows the specific connections between atoms and therefore gives much more information than the chemical formula alone. Ethyl alcohol, for example, has the chemical formula C_2H_6O and the following structural formula:

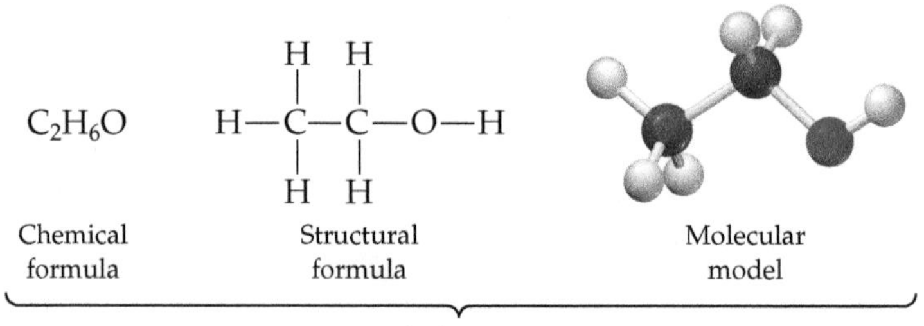

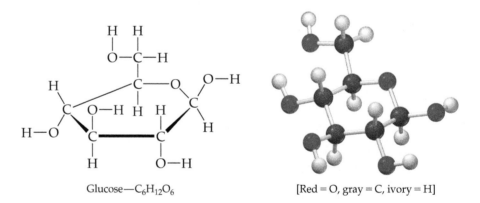

C_2H_6O

Chemical
formula

Structural
formula

Molecular
model

Ethyl alcohol

A structural formula uses lines between atoms to indicate the covalent bonds. Thus, the 2 carbon atoms in ethyl alcohol are covalently bonded to each other, the oxygen atom is bonded to one of the carbon atoms, and the 6 hydrogen atoms are distributed 3 to one carbon, 2 to the other carbon, and 1 to the oxygen.

Structural formulas are particularly important in *organic chemistry*—the chemistry of carbon compounds—where the behavior of large, complex molecules is almost entirely governed by their structure. Take even a relatively simple substance like glucose, for instance. The molecular formula of glucose, $C_6H_{12}O_6$, tells nothing about how the atoms are connected. In fact, you could probably imagine a great many different ways in which the 24 atoms might be connected. The structural formula for glucose, however, shows that 5 carbons and 1 oxygen form a ring of atoms, with the remaining 5 oxygens each bonded to 1 hydrogen and bonded to different carbons.

Glucose—$C_6H_{12}O_6$

[Red = O, gray = C, ivory = H]

Some elements even exist as molecules rather than as individual atoms. Hydrogen, nitrogen, oxygen, fluorine, chlorine, bromine, and iodine all exist as *diatomic* (two-atom) molecules whose two atoms are held together by covalent bonds. We therefore have to write them as such—H_2, N_2, O_2, F_2, Cl_2, Br_2, and I_2—when using any of these elements in a chemical equation. Notice that all these diatomic elements except hydrogen cluster toward the far right side of the periodic table.

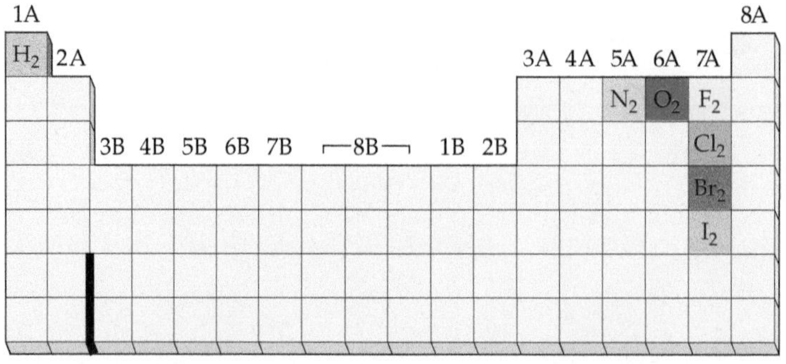

— CONCEPTUAL WORKED EXAMPLE 2.9

Converting between Structural and Molecular Formulas

Propane, C_3H_8, has a structure in which the three carbon atoms are bonded in a row, each end carbon is bonded to three hydrogens, and the middle carbon is bonded to two hydrogens. Draw the structural formula, using lines between atoms to represent covalent bonds.

SOLUTION

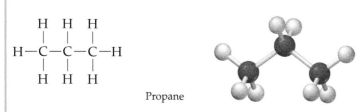

Propane

▶CONCEPTUAL APPLY 2.18   Adrenaline, the so-called flight-or-fight hormone, can be represented by the following ball-and-stick model. What is the chemical formula of adrenaline? (Gray = C, ivory = H, red = O, blue = N)

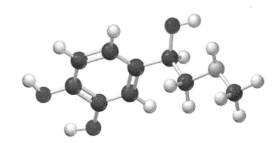

▶CONCEPTUAL PRACTICE 2.17   Thymine, one of the four bases in deoxyribonucleic acid (DNA), has the following structure. What is the chemical formula of thymine? In writing the formula, list the element symbols in alphabetical order, and give the number of each element as a subscript.

2.12 IONS AND IONIC BONDS

In contrast to a covalent bond, an **ionic bond** results not from a sharing of electrons but from a transfer of one or more electrons from one atom to another. As noted previously, ionic bonds generally form between a metal and a nonmetal. Metals, such as sodium, magnesium, and zinc, tend to give up electrons, whereas nonmetals, such as oxygen, nitrogen, and chlorine, tend to accept electrons.

For example, when sodium metal comes in contact with chlorine gas, a sodium atom gives an electron to a chlorine atom, resulting in the formation of two charged particles, called **ions.** Because a sodium atom loses one electron, it loses one negative charge and becomes an Na^+ ion with a charge of +1. Such positive ions are called **cations** (pronounced **cat**-ions). Conversely, because a chlorine atom gains an electron, it gains a negative charge and becomes a Cl^- ion with a charge of −1. Such negative ions are called **anions** (**an**-ions).

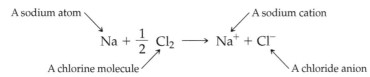

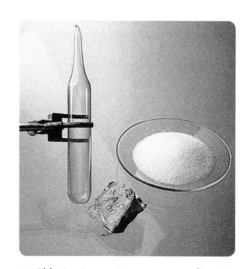

▲ Chlorine is a toxic green gas, sodium is a reactive metal, and sodium chloride is a harmless white solid.

Showing the ions in the preceding chemical equation is useful for keeping track of charged species, but according to convention, ions are not shown in the chemical equation. The reaction of sodium and chlorine is written as:

$$Na + \frac{1}{2} Cl_2 \longrightarrow NaCl$$

A similar reaction takes place when magnesium and chlorine molecules (Cl_2) come in contact to form $MgCl_2$. A magnesium atom transfers an electron to each of two chlorine atoms, yielding the doubly charged Mg^{2+} cation and two Cl^- anions.

$$Mg + Cl_2 \longrightarrow Mg^{2+} + Cl^- + Cl^-$$

The chemical equation for reaction of magnesium with chlorine is:

$$Mg + Cl_2 \longrightarrow MgCl_2$$

Because opposite charges attract, positively charged cations such as Na^+ and Mg^{2+} experience a strong electrical attraction to negatively charged anions like Cl^-, an attraction that we call an ionic bond. Unlike what happens when covalent bonds are formed, though, we can't really talk about discrete Na^+Cl^- *molecules* under normal conditions. We can speak only of an **ionic solid,** in which equal numbers of Na^+ and Cl^- ions are packed together in a regular way (**FIGURE 2.15**). In a crystal of table salt, for instance, each Na^+ ion is surrounded by six nearby Cl^- ions, and each Cl^- ion is surrounded by six nearby Na^+ ions, but we can't specify what pairs of ions "belong" to each other as we can with atoms in covalent molecules.

▶ **FIGURE 2.15**

The arrangement of Na^+ and Cl^- ions in a crystal of sodium chloride. There is no discrete "molecule" of NaCl. Instead, the entire crystal is an ionic solid.

▶ **Figure It Out**

Which element loses electrons and which element gains electrons when a metal and nonmetal form an ionic compound?

Answer: The metal loses electrons to form a positively charged cation. The nonmetal gains electrons to form a negatively charged anion.

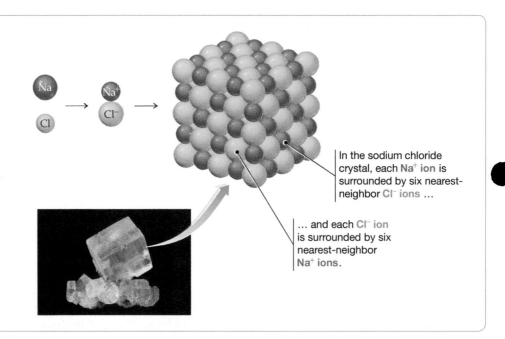

In the sodium chloride crystal, each Na^+ ion is surrounded by six nearest-neighbor Cl^- ions …

… and each Cl^- ion is surrounded by six nearest-neighbor Na^+ ions.

Charged, covalently bonded groups of atoms, called **polyatomic ions,** are also common—ammonium ion (NH_4^+), hydroxide ion (OH^-), nitrate ion (NO_3^-), and the doubly charged sulfate ion (SO_4^{2-}) are examples (**FIGURE 2.16**). You can think of these polyatomic ions as charged molecules because they consist of specific numbers and kinds of atoms joined together by covalent bonds, with the overall unit having a positive or negative charge. When writing the formulas of substances that contain more than one of these ions, parentheses are placed around the entire polyatomic unit. The formula $Ba(NO_3)_2$, for instance, indicates a substance made of Ba^{2+} cations and NO_3^- polyatomic anions in a 1:2 ratio. We'll learn how to name compounds with these ions in Section 2.13.

There is no net charge on any ionic compound so the total number of positive charges must equal the total number of negative charges. Creating an overall neutral charge serves as a guide for writing the formulas of ionic compounds. For example, the formula for the ionic compound formed from sodium (Na^+) and bromide (Br^-) ions is NaBr because one positive charge from the sodium ion balances one negative charge from the bromide ion. The formula for the ionic compound formed from sodium (Na^+) and sulfate (SO_4^{2-}) ions is Na_2SO_4 because two sodium ions each with a +1 charge are needed to balance the −2 charge on sulfate. The formula for the ionic compound formed from sodium (Na^+) and phosphate (PO_4^{3-}) ions is Na_3PO_4 because three sodium ions each with a +1 charge are needed to balance the −3 charge on phosphate.

WORKED EXAMPLE 2.10

Identifying Ionic and Molecular Compounds

Which of the following compounds would you expect to be ionic and which molecular?
A molecular compound has covalent bonds.

(a) BaF_2 (b) SF_4 (c) PH_3 (d) CH_3OH

STRATEGY

Remember that covalent bonds generally form between nonmetal atoms, while ionic
bonds form between metal and nonmetal atoms.

SOLUTION

Compound (a) is composed of a metal (barium) and a nonmetal (fluorine) and is likely
to be ionic. Compounds (b)–(d) are composed entirely of nonmetals and therefore are
probably molecular.

▶ **PRACTICE 2.19** Which of the following is an ionic compound?

(a) LiBr   (b) $SiCl_4$ (c) NF_3

▶ **CONCEPTUAL APPLY 2.20** Which of the following drawings most likely represents an
ionic compound and which a molecular compound? Explain.

(a)   **(b)**

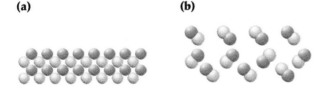

2.13 NAMING CHEMICAL COMPOUNDS

In the early days of chemistry, when few pure substances were known, newly discovered
compounds were often given fanciful names—morphine, quicklime, potash, and barbituric acid (said to be named by its discoverer in honor of his friend Barbara) to cite
a few. Today, with more than 40 million pure compounds known, there would be
chaos without a systematic method for naming compounds. Every chemical compound
must be given a name that not only defines it uniquely but also allows chemists (and
computers) to know its chemical structure.

Different kinds of compounds are named by different rules. Ordinary table salt,
for instance, is named *sodium chloride* because of its formula NaCl, but common table
sugar ($C_{12}H_{22}O_{11}$) is named β-D-*fructofuranosyl-α-D-glucopyranoside* because of special rules for carbohydrates. (Organic compounds often have quite complex structures
and correspondingly complex names, though we'll not discuss them in this text.) We'll
begin by seeing how to name simple ionic compounds and then introduce additional
rules in later chapters as the need arises.

Naming Binary Ionic Compounds

Binary ionic compounds—those made of only two elements—are named by identifying
first the positive ion and then the negative ion. The positive ion takes the same name as
the element, while the negative ion takes the first part of its name from the element and
then adds the ending *-ide*. For example, KBr is named potassium bromide: *potassium*
for the K⁺ ion and *bromide* for the negative Br⁻ ion derived from the element *brom*ine.

       LiF $CaBr_2$ $AlCl_3$
   Lithium fluoride   Calcium bromide   Aluminum chloride

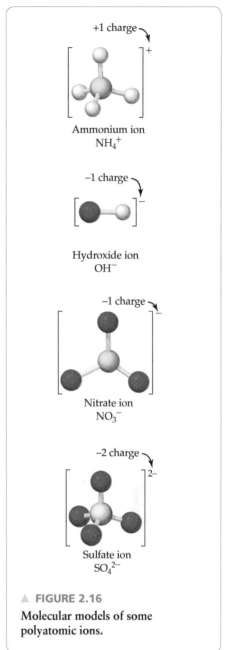

▲ **FIGURE 2.16**

Molecular models of some
polyatomic ions.

▲ Morphine, a pain-killing agent found
in the opium poppy, was named after
Morpheus, the Greek god of dreams.

FIGURE 2.17 shows some common main-group ions, and **FIGURE 2.18** shows some common transition metal ions. There are several interesting points about Figure 2.17. Notice, for instance, that metals tend to form cations and nonmetals tend to form anions. Also note that elements within a given group of the periodic table form ions with the same charge and that the charge is related to the group number. Main-group metals usually form cations whose charge is equal to the group number. For example, group 1A elements form singly positive ions (M^+, where M is a metal), group 2A elements form doubly positive ions (M^{2+}), and group 3A elements form triply positive ions (M^{3+}). Main-group nonmetals usually form anions whose charge is equal to the group number in the U.S. system minus eight. Thus, group 6A elements form doubly negative ions ($6 - 8 = -2$), group 7A elements form singly negative ions ($7 - 8 = -1$), and group 8A elements form no ions at all ($8 - 8 = 0$). We'll see the reason for this behavior in Chapter 6.

Notice also, in both Figures 2.17 and 2.18, that some metals form more than one kind of cation. Iron, for instance, forms both the doubly charged Fe^{2+} ion and the triply charged Fe^{3+} ion. In naming these ions, we distinguish between them by using a Roman numeral in parentheses to indicate the number of charges. Thus, $FeCl_2$ is named iron(II) chloride and $FeCl_3$ is iron(III) chloride. Alternatively, an older method distinguishes between the ions by using the Latin name of the element (*ferrum* in the case of iron) together with the ending *-ous* for the ion with lower charge and *-ic* for the ion with higher charge. Thus, $FeCl_2$ is sometimes called ferrous chloride and $FeCl_3$ is called ferric chloride. Although still in use, this older naming system is being phased out, and we'll rarely use it in this book.

▲ Crystals of iron(II) chloride tetrahydrate are greenish, and crystals of iron(III) chloride hexahydrate are brownish yellow.

▶ **FIGURE 2.17**

Main-group cations (blue) and anions (red). A *cation* bears the same name as the element it is derived from; an *anion* name has an *-ide* ending.

▶ **Figure It Out**

What charge is formed on group 2A and 6A elements?

Answer: Group 2A = +2 and group 6A = −2.

1 1A	2 2A		13 3A	14 4A	15 5A	16 6A	17 7A	18 8A
H^+ H^- Hydride								
Li^+	Be^{2+}				N^{3-} Nitride	O^{2-} Oxide	F^- Fluoride	
Na^+	Mg^{2+}		Al^{3+}			S^{2-} Sulfide	Cl^- Chloride	
K^+	Ca^{2+}		Ga^{3+}			Se^{2-} Selenide	Br^- Bromide	
Rb^+	Sr^{2+}		In^{3+}	Sn^{2+} Sn^{4+}		Te^{2-} Telluride	I^- Iodide	
Cs^+	Ba^{2+}		Tl^+ Tl^{3+}	Pb^{2+} Pb^{4+}				

▶ **FIGURE 2.18**

Common transition metal ions. Only ions that exist in aqueous solution are shown.

▶ **Figure It Out**

Can the charge on a transition metal ion be predicted from its group number?

Answer: No, many transition metals form different charge states. The charge on the transition metal ion is specified in the formula or name.

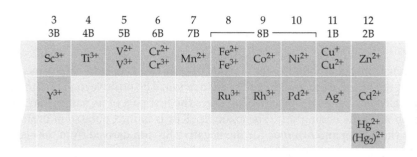

3 3B	4 4B	5 5B	6 6B	7 7B	8	9 8B	10	11 1B	12 2B
Sc^{3+}	Ti^{3+}	V^{2+} V^{3+}	Cr^{2+} Cr^{3+}	Mn^{2+}	Fe^{2+} Fe^{3+}	Co^{2+}	Ni^{2+}	Cu^+ Cu^{2+}	Zn^{2+}
Y^{3+}					Ru^{3+}	Rh^{3+}	Pd^{2+}	Ag^+	Cd^{2+}
									Hg^{2+} $(Hg_2)^{2+}$

Fe^{2+}	Fe^{3+}	Sn^{2+}	Sn^{4+}
Iron(II) ion	Iron(III) ion	Tin(II) ion	Tin(IV) ion
Ferrous ion	Ferric ion	Stannous ion	Stannic ion
(From the Latin *ferrum* = iron)		(From the Latin *stannum* = tin)	

In any neutral compound, the total number of positive charges must equal the total number of negative charges. Thus, you can always figure out the number of positive charges on a metal cation by counting the number of negative charges on the associated anion(s). In $FeCl_2$, for example, the iron ion must be Fe(II) because there are two Cl^- ions associated with it. Similarly, in $TiCl_3$, the titanium ion is Ti(III) because there are three Cl^- anions associated with it. As a general rule, a Roman numeral is needed for transition-metal compounds to avoid ambiguity. In addition, the main-group metals tin (Sn), thallium (Tl), and lead (Pb) can form more than one kind of ion and need Roman numerals for naming their compounds. Metals in group 1A and group 2A form only one cation, however, so Roman numerals are not needed.

WORKED EXAMPLE 2.11

Converting between Names and Formulas for Binary Ionic Compounds

Give systematic names for the following compounds:

(a) $BaCl_2$ (b) $CrCl_3$ (c) PbS   (d) Fe_2O_3

STRATEGY

Name the cation with the name of the element and the anion using the first part of the element name + "ide." If the cation is a transition metal, then the charge is specified with Roman numerals. Figure out the number of positive charges on each transition metal cation by counting the number of negative charges on the associated anion(s). Refer to Figures 2.17 and 2.18 as necessary.

SOLUTION

(a) Barium chloride — No Roman numeral is necessary because barium, a group 2A element, forms only Ba^{2+}.

(b) Chromium(III) chloride — The Roman numeral III is necessary to specify the +3 charge on chromium (a transition metal).

(c) Lead(II) sulfide — The sulfide anion (S^{2-}) has a double negative charge, so the lead cation must be doubly positive.

(d) Iron(III) oxide — The three oxide anions (O^{2-}) have a total negative charge of −6, so the two iron cations must have a total charge of +6. Thus, each is Fe(III).

▶ **PRACTICE 2.21**   Write formulas for the following compounds:

(a) Magnesium fluoride
(b) Tin(IV) oxide

▶ **CONCEPTUAL APPLY 2.22**   Three binary ionic compounds are represented on the following periodic table: red with red, green with green, and blue with blue. Name each, and write its likely formula.

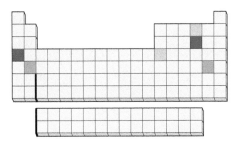

Naming Compounds with Polyatomic Ions

Ionic compounds that contain polyatomic ions are named in the same way as binary ionic compounds: First the cation is identified and then the anion. For example, $Ba(NO_3)_2$ is called *barium nitrate* because Ba^{2+} is the cation and the NO_3^- polyatomic anion has the name *nitrate*. Unfortunately, there is no simple systematic way of naming the polyatomic ions themselves, so it's necessary to memorize the names, formulas, and charges of the most common ones, listed in **TABLE 2.5**. The ammonium ion (NH_4^+) is the only cation on the list; all the others are anions.

TABLE 2.5   Some Common Polyatomic Ions

Formula	Name	Formula	Name
Cation		**Singly charged anions (continued)**	
NH_4^+	Ammonium	NO_2^-	Nitrite
		NO_3^-	Nitrate
Singly charged anions		**Doubly charged anions**	
$CH_3CO_2^-$	Acetate	CO_3^{2-}	Carbonate
CN^-	Cyanide	CrO_4^{2-}	Chromate
ClO^-	Hypochlorite	$Cr_2O_7^{2-}$	Dichromate
ClO_2^-	Chlorite	O_2^{2-}	Peroxide
ClO_3^-	Chlorate	HPO_4^{2-}	Hydrogen phosphate
ClO_4^-	Perchlorate	SO_3^{2-}	Sulfite
$H_2PO_4^-$	Dihydrogen phosphate	SO_4^{2-}	Sulfate
HCO_3^-	Hydrogen carbonate (or bicarbonate)	$S_2O_3^{2-}$	Thiosulfate
HSO_4^-	Hydrogen sulfate (or bisulfate)	**Triply charged anion**	
OH^-	Hydroxide	PO_4^{3-}	Phosphate
MnO_4^-	Permanganate		

Several points about the ions in Table 2.5 need special mention. First, note that the names of most polyatomic anions end in *-ite* or *-ate*. Only hydroxide (OH^-), cyanide (CN^-), and peroxide (O_2^{2-}) have the *-ide* ending. Second, note that several of the ions form a series of **oxoanions,** binary polyatomic anions in which an atom of a given element is combined with different numbers of oxygen atoms—hypochlorite (ClO^-), chlorite (ClO_2^-), chlorate (ClO_3^-), and perchlorate (ClO_4^-), for example. When there are only two oxoanions in a series, as with sulfite (SO_3^{2-}) and sulfate (SO_4^{2-}), the ion with fewer oxygens takes the *-ite* ending and the ion with more oxygens takes the *-ate* ending.

SO_3^{2-}	Sul*fite* ion (fewer oxygens)	SO_4^{2-}	Sul*fate* ion (more oxygens)
NO_2^-	Nit*rite* ion (fewer oxygens)	NO_3^-	Nit*rate* ion (more oxygens)

When there are more than two oxoanions in a series, the prefix *hypo-* (meaning "less than") is used for the ion with the fewest oxygens, and the prefix *per-* (meaning "more than") is used for the ion with the most oxygens.

ClO^-	*Hypo*chlorite ion (less oxygen than chlorite)
ClO_2^-	Chlorite ion
ClO_3^-	Chlorate ion
ClO_4^-	*Per*chlorate iron (more oxygen than chlorate)

Third, note that several pairs of ions are related by the presence or absence of a hydrogen ion. The hydrogen carbonate anion (HCO_3^-) differs from the carbonate anion (CO_3^{2-}) by the presence of H^+, and the hydrogen sulfate anion (HSO_4^-) differs from the sulfate anion (SO_4^{2-}) by the presence of H^+. The ion that has the additional hydrogen is sometimes referred to using the prefix *bi-*, although this usage is now discouraged; for example, $NaHCO_3$ is sometimes called sodium bicarbonate.

HCO_3^-	Hydrogen carbonate (*bi*carbonate) ion	CO_3^{2-}	Carbonate ion
HSO_4^-	Hydrogen sulfate (*bi*sulfate) ion	SO_4^{2-}	Sulfate ion

— WORKED EXAMPLE 2.12

Converting between Names and Formulas for Compounds with Polyatomic Ions

Give systematic names for the following compounds:

(a) $LiNO_3$ (b) $KHSO_4$ (c) $CuCO_3$ (d) $Fe(ClO_4)_3$

STRATEGY

Name the cation first and the anion second. Unfortunately, there is no alternative: The names and charges of the common polyatomic ions must be memorized. Refer to Table 2.5 if you need help.

SOLUTION

(a) Lithium nitrate — Lithium (group 1A) forms only the Li^+ ion and does not need a Roman numeral.

(b) Potassium hydrogen sulfate — Potassium (group 1A) forms only the K^+ ion.

(c) Copper(II) carbonate — The carbonate ion has a -2 charge, so copper must be $+2$. A Roman numeral is needed because copper, a transition metal, can form more than one ion.

(d) Iron(III) perchlorate — There are three perchlorate ions, each with a -1 charge, so the iron must have a $+3$ charge.

▶ **PRACTICE 2.23** Write the formula for iron(III) carbonate.

▶ **CONCEPTUAL APPLY 2.24** The following drawings are those of solid ionic compounds, with red spheres representing the cations and blue spheres representing the anions in each.

(1)   **(2)**

Which of the following formulas are consistent with each drawing?

(a) LiBr   (b) $NaNO_2$ (c) $CaCl_2$

Naming Binary Molecular Compounds

Binary molecular compounds—those made of only two covalently bonded elements—are named in much the same way as binary ionic compounds. One of the elements in the compound is more electron-poor, or *cationlike*, and the other element is more electron-rich, or *anionlike*. As with ionic compounds, the cationlike element takes the name of the element itself, and the anionlike element takes an *-ide* ending. The compound HF, for example, is called *hydrogen fluoride*.

HF   Hydrogen is more cationlike because it is farther left in the periodic table, and fluoride is more anionlike because it is farther right. The compound is therefore named *hydrogen fluoride*.

We'll see a quantitative way to decide which element is more cationlike and which is more anionlike in Section 7.3 but you might note for now that it's usually possible to decide by looking at the relative positions of the elements in the periodic table. The farther left and toward the bottom of the periodic table an element occurs, the more likely it is to be cationlike; the farther right and toward the top an element occurs (except for the noble gases), the more likely it is to be anionlike.

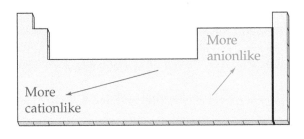

TABLE 2.6 Numerical Prefixes for Naming Compounds

Prefix	Meaning
mono-	1
di-	2
tri-	3
tetra-	4
penta-	5
hexa-	6
hepta-	7
octa-	8
nona-	9
deca-	10

The following examples show how this generalization applies:

CO Carbon monoxide (C is in group 4A; O is in group 6A)
CO_2 Carbon dioxide
PCl_3 Phosphorus trichloride (P is in group 5A; Cl is in group 7A)
SF_4 Sulfur tetrafluoride (S is in group 6A; F is in group 7A)
N_2O_4 Dinitrogen tetroxide (N is in group 5A; O is in group 6A)

Because nonmetals often combine with one another in different proportions to form different compounds, numerical prefixes are usually included in the names of binary molecular compounds to specify the numbers of each kind of atom present. The compound CO, for example, is called carbon *mon*oxide, and CO_2 is called carbon *di*oxide. **TABLE 2.6** lists the most common numerical prefixes. Note that when the prefix ends in *a* or *o* (but not *i*) and the anion name begins with a vowel (*oxide*, for instance), the *a* or *o* on the prefix is dropped to avoid having two vowels together in the name. Thus, we write carbon *mon*oxide rather than carbon *mono*oxide for CO and dinitrogen *tetr*oxide rather than dinitrogen *tetra*oxide for N_2O_4. A *mono*- prefix is not used for the atom named first: CO_2 is called carbon dioxide rather than monocarbon dioxide.

WORKED EXAMPLE 2.13

Converting between Names and Formulas for Binary Molecular Compounds

Give systematic names for the following compounds:

(a) PCl_3 (b) N_2O_3 (c) P_4O_7 (d) BrF_3

STRATEGY

Look at a periodic table to see which element in each compound is more cationlike (located farther to the left or lower) and which is more anionlike (located farther to the right or higher). Then name the compound using the appropriate numerical prefix to specify the number of atoms.

SOLUTION

(a) Phosphorus trichloride   (b) Dinitrogen trioxide
(c) Tetraphosphorus heptoxide   (d) Bromine trifluoride

PRACTICE 2.25   Write the formula for dinitrogen pentoxide.

CONCEPTUAL APPLY 2.26   Give systematic names for the following compounds:

(a)   **(b)**

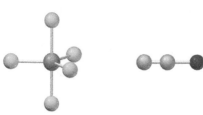

Purple = P, green = Cl   Blue = N, red = O

INQUIRY ? How can measurements of oxygen and hydrogen isotopes in ice cores determine past climates?

Climate change refers to variations in average weather conditions, temperatures, and rainfall over an extended period of time. Climate research also includes tracking the number and severity of extreme weather events such as heat waves, tornadoes, and hurricanes. The Earth's climate has varied over geological time due to a number of natural causes including variations in the Earth's orbit, the sun's intensity, particulates from volcanic eruptions, and levels of greenhouse gases.

The term *climate change* or *global warming* is most often associated with the pronounced warming of the climate from the mid- to late twentieth century, which is largely attributed to increased levels of carbon dioxide in the atmosphere from burning fossil fuels. Examine **FIGURE 2.19,** which shows the change in global surface temperature (°C) from 1880 to the present relative to the long-term average temperature from 1901–2000. Temperatures measured on land and at sea show that Earth's globally averaged surface temperature is rising. Though warming has not been uniform across the planet, the upward trend in the globally averaged temperature shows that more areas are warming

than cooling. For the past 45 years, global surface temperature rose at an average rate of about 0.17 °C (about 0.3 °Fahrenheit) per decade—more than twice as fast as the 0.07 °C per decade increase observed for the entire period of recorded observations (1880–2015). Remarkably, all 16 years of the twenty-first century rank among the 17 warmest years on record.

In recent history, scientists have been recording the Earth's average temperature using satellite measurements and data from numerous weather and research stations. Surrounding temperatures and other information are used to fill in data from areas that have few measurements. This process provides a consistent, reliable method for monitoring changes in Earth's surface temperature over time. However, a long-term record of past climate is needed to put the recent warming trends in context. Analyzing a historical temperature record enables scientists to evaluate previous rates of temperature change and the magnitude of temperature changes that caused drastic differences in climate such as ice ages.

Clues to past climates are etched on our planet in polar ice caps, cave rocks, coral reefs, and tree rings. Measuring oxygen and hydrogen isotope ratios in polar and glacial ice can be used to reconstruct past global temperatures. Sensitive mass spectrometers (Section 2.10) are used to measure isotope ratios in water from ice core samples. **FIGURE 2.20** shows a near linear relationship between the difference in the $^{18}O/^{16}O$ ratio in snowfall and the mean annual temperature for that site. Isotopic ratios are a measure of temperature because more energy

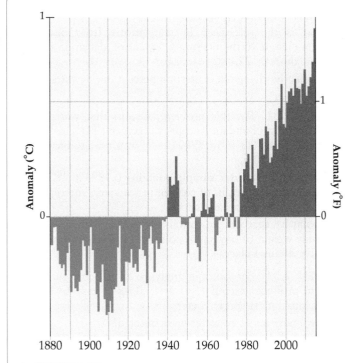

▲ **FIGURE 2.19**

Global land and ocean temperature anomalies.
Annual global temperatures since 1880 compared to the long-term average temperature from 1901–2000. The zero line represents the long-term average global temperature, and the blue and red bars show the difference above or below average for each year. Data come from a combined set of land-based weather stations and sea-surface temperature measurements.

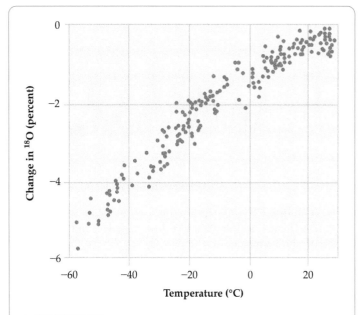

▲ **FIGURE 2.20**

Difference in the $^{18}O/^{16}O$ ratio in ice core samples related to the average temperature when the snowfall occurred.

is required to evaporate water molecules containing a heavy isotope from the surface of the ocean than water molecules with lighter isotopes. As warm air is transported to cold, polar regions, water molecules containing heavier isotopes preferentially precipitate. Therefore, the ratio of heavier isotopes to lighter isotopes ($^{18}O/^{16}O$ and $^{2}H/^{1}H$) in precipitation increases with warmer temperatures. Cold locations such as Antarctica have about 5% less ^{18}O than warm ocean water.

Plotting either $^{18}O/^{16}O$ or $^{2}H/^{1}H$ with ice core depth reveals oscillations in temperature as a function of time. Ice core samples are dated by number of layers and depth. The data in **FIGURE 2.21** was generated from an Antarctic ice core which extends 3 km in length and dates back 800,000 years. The top graph shows variations in the amount of the heavy isotope of water (^{2}H, deuterium), and the bottom graph shows the correlation with temperature. The data reveals cold glacial periods interspersed by warm periods roughly every 100,000 years. Historical climate records show that the Earth has experienced warm and cool periods, but the rate of warming in the past 50 years is unprecedented. Past global temperatures can also be correlated to levels of greenhouse gases such as carbon dioxide trapped in bubbles in the ice. Section 10.11 discusses the effect of greenhouse gases on climate, and Figure 10.25 shows the correlation between carbon dioxide concentration and global temperature.

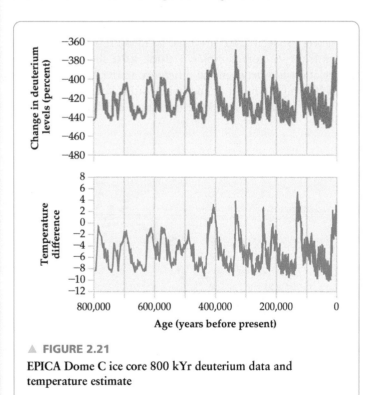

▲ **FIGURE 2.21**

EPICA Dome C ice core 800 kYr deuterium data and temperature estimate

PROBLEM 2.27   Global climate is affected by variations in
(a) the Earth's orbit around the Sun.
(b) particulates in the atmosphere from pollution or volcanoes.
(c) the Sun's intensity.
(d) greenhouse gas levels.
(e) all of the above.

PROBLEM 2.28   Use Figure 2.19 to determine if the following statements are true or false.
(a) Each year after 1950 had an average global temperature higher than the 1901–2000 average temperature.
(b) Prior to 1940, average global temperatures were lower than the 1901–2000 average temperature.
(c) Warming in the time period 1970–2015 occurred at a faster rate than warming in the time period 1900–1950.
(d) Global average temperatures exhibited a cooling trend from 1880–1910.

PROBLEM 2.29   How many protons, neutrons, and electrons are in ^{18}O and ^{2}H?

PROBLEM 2.30   Which sample of H_2O has a higher ratio of $^{18}O/^{16}O$: warm seawater near the equator or snow falling in Antarctica?

PROBLEM 2.31   The last ice age occurred from 110,000 to 11,700 years ago. Use Figure 2.21 to answer questions about variations in temperature and change in deuterium percent between a warm climate and an ice age.
(a) What is the difference in temperature from our current warm climate to the maximum extent of glaciation occurring approximately 22,000 years ago?
(b) What is the change in deuterium percent during the same time period?

PROBLEM 2.32   For this problem, assume that water consists only of the most abundant isotopes of oxygen (^{16}O and ^{18}O). The atomic mass for ^{16}O is 15.9949146 u, and the atomic mass for ^{18}O is 17.9991610 u.
(a) A standard seawater sample contains 0.1995% ^{18}O. Calculate the atomic weight of oxygen in seawater, and report your answer to five decimal places.
(b) A polar ice core sample contains 0.1971% ^{18}O. Calculate the atomic weight of oxygen in polar ice, and report your answer to five decimal places.

PROBLEM 2.33   For isotopic analysis, an ice core sample was heated to produce gaseous H_2O. If 1.00 μg of gaseous H_2O was injected into a mass spectrometer:
(a) How many moles of water were injected?
(b) If the sample contains 0.0156% deuterium, how many deuterium atoms were injected?

STUDY GUIDE

Section	Concept Summary	Learning Objectives	Test Your Understanding
2.1 Chemistry and the Elements	All matter is formed from one or more of the 118 presently known **elements**—fundamental substances that can't be chemically broken down. Elements are symbolized by one- or two-letter abbreviations.	**2.1** Write symbols to represent element names.	Problems 2.48, 2.50, 2.52
2.2 Elements and the Periodic Table	Elements are organized into a **periodic table** with **groups** (columns) and **periods** (rows). Elements in the same groups show similar chemical behavior. Elements are classified as **metals, nonmetals,** or **semimetals.**	**2.2** Identify the location of metals, nonmetals, and semimetals on the periodic table.	Problems 2.37, 2.62, 2.63
		2.3 Indicate the atomic number, group number, and period number for an element whose position in the periodic is given.	Problem 2.36
		2.4 Identify groups as main group, transition, metal group, or inner transition metal group.	Problems 2.59, 2.61
2.3 Some Common Groups of Elements and Their Properties	The characteristics, or **properties,** that are used to describe matter can be classified in several ways. **Physical properties** are those that can be determined without changing the chemical composition of the sample, whereas **chemical properties** are those that do involve a chemical change. Intensive properties are those whose values do not depend on the size of the sample, whereas **extensive properties** are those that do depend on sample amount.	**2.5** Specify the location and give examples of elements in the alkali metal, alkaline earth metal, halogen, noble gas groups.	Problem 2.35, 2.64–2.67
		2.6 Classify an element as a metal, nonmetal, or semimetal using its properties.	Worked Example 2.1; Problems 2.1–2.2, 2.68–2.71
2.4 Observations Supporting Atomic Theory: The Conservation of Mass and the Law of Definite Proportions	Elements join together in different ways to make **chemical compounds,** and a pure compound always has the same proportion of elements by mass. During a chemical reaction, the **law of mass conservation** applies, and the mass of reactants is the same as the mass of products.	**2.7** Determine the mass of the products in a reaction using the law of mass conservation.	Problems 2.78–2.79
2.5 The Law of Multiple Proportions and Dalton's Atomic Theory	**Elements** are made of tiny particles called **atoms,** which can combine in simple numerical ratios according to the **law of multiple proportions.**	**2.8** Demonstrate the law of multiple proportions using mass composition of two compounds of the same elements.	Worked Example 2.2; Problems 2.80, 2.82
		2.9 Determine the formula of a compound given mass composition data for two compounds and the formula of one compound.	Problems 2.83–2.84
2.6 Atomic Structure: Electrons	Atoms are composed of three fundamental particles: **protons** are positively charged, **electrons** are negatively charged, and **neutrons** are neutral.	**2.10** Describe Thomson's cathode-ray experiment and what it contributed to the current model of atomic structure.	Problems 2.86–2.88
		2.11 Describe Millikan's oil drop experiment and what it contributed to the current model of atomic structure.	Problems 2.89–2.90
2.7 Atomic Structure: Proton and Neutrons	According to the nuclear model of an atom proposed by Ernest Rutherford, protons and neutrons are clustered into a dense core called the **nucleus,** while electrons move around the nucleus at a relatively great distance.	**2.12** Describe Rutherford's gold foil experiment and what it contributed to the current model of atomic structure.	Problems 2.91–2.92
		2.13 Calculate the number of atoms in sample given the size of the atom.	Problems 2.93–2.94

Section	Concept Summary	Learning Objectives	Test Your Understanding
2.8 Atomic Numbers	Elements differ from one another according to how many protons their atoms contain, a value called the **atomic number (Z)** of the element. The sum of an atom's protons and neutrons is its **mass number (A)**. Although all atoms of a specific element have the same atomic number, different atoms of an element can have different mass numbers depending on how many neutrons they have. Atoms with identical atomic numbers but different mass numbers are called **isotopes.**	**2.14** Determine the mass number, atomic number, and number of protons neutrons and electrons from an isotope symbol.	Worked Example 2.4; Problems 2.100–2.102, 2.102, 2.104, 2.106
2.9 Atomic Weights and the Mole	Atomic weights are measured using the **unified atomic mass unit (u)**, defined as 1/12 the mass of a ^{12}C atom. Because both protons and neutrons have a mass of approximately 1, the mass of an atom in unified atomic mass units is numerically close to the atom's mass number. The element's **atomic weight** is a weighted average of the isotopic masses of its naturally occurring isotopes. When referring to the enormous numbers of atoms that make up visible amounts of matter, the fundamental SI unit called a *mole* is used. One **mole** is the amount whose mass in grams, called its **molar mass,** is numerically equal to the atomic weight. Numerically, one mole of any element contains 6.022×10^{23} atoms, a value called **Avogadro's number (N_A).**	**2.15** Calculate atomic weight given the fractional abundance and mass of each isotope.	Worked Example 2.5; Problems 2.116, 2.118, 2.120
		2.16 Convert between grams and numbers of moles or atoms using molar mass and Avogadro's number.	Worked Example 2.6; Problems 2.124–2.125
2.10 Measuring Atomic Weight: Mass Spectrometry	A **mass spectrometer** separates gaseous ions based on their mass-to-charge ratio. The mass spectrum records the intensity (the number) of ions on the y-axis and the mass-to-charge ratio of the ions on the x-axis. Mass spectral data can be used to calculate the atomic weight of an element.	**2.17** Use data from a mass spectrum to calculate the atomic weight of an element.	Worked Example 2.7; Problems 2.13–2.14, 2.132–2.133
2.11 Mixtures and Chemical Compounds; Molecules and Covalent Bonds	Most substances are **chemical compounds,** formed when atoms of two or more elements combine in a **chemical reaction.** The atoms in a compound are held together by one of two kinds of **chemical bonds. Covalent bonds** form when two atoms share electrons to give a new unit of matter called a **molecule.**	**2.18** Classify matter as a mixture, pure substance, element, or compound.	Worked Example 2.8; Problems 2.15–2.16 2.42
		2.19 Convert between structural formulas, ball-and-stick models, and chemical formulas.	Worked Example 2.9; Problems 2.43–2.44
2.12 Ions and Ionic Bonds	**Ionic bonds** form when one atom completely transfers one or more electrons to another atom, resulting in the formation of **ions.** Positively charged ions (**cations**) are strongly attracted to negatively charged ions (**anions**) by electrical forces.	**2.20** Classify bonds as ionic or covalent.	Worked Example 2.10; Problems 2.134–2.135
		2.21 Determine the number of electrons and protons from chemical symbol and charge.	Problems 2.41, 2.138–2.139
		2.22 Match the molecular representation of an ionic compound with its chemical formula.	Problem 2.45
2.13 Naming Chemical Compounds	Chemical compounds are named systematically by following a series of rules. Binary ionic compounds are named by identifying first the positive ion and then the negative ion. Binary molecular compounds are similarly named by identifying the cationlike and anionlike elements. Naming compounds with **polyatomic ions** involves memorizing the names and formulas of the most common ones.	**2.23** Convert between name and formula for ionic compounds.	Worked Examples 2.11–2.12; Problems 2.146, 2.148, 2.150, 2.152
		2.24 Convert between name and formula for binary molecular compounds.	Worked Example 2.13; Problems 2.161–2.162

KEY TERMS

anion *61*	covalent bond *58*	law of definite proportions *43*	nucleus *47*
atom *44*	electron *45*	law of mass conservation *42*	oxoanion *66*
atomic mass *52*	element *34*	law of multiple proportions *43*	period *37*
atomic number (Z) *49*	extensive property *38*	main group *37*	periodic table *35*
atomic weight *52*	group *37*	mass number *50*	physical property *38*
Avogadro's number *53*	inner transition metal	mass spectrometer *55*	polyatomic ion *62*
cation *61*	group *38*	mass spectrum *55*	property *38*
chemical bond *58*	intensive property *38*	mixture *57*	proton *48*
chemical compound *42*	ion *61*	molar mass *53*	structural formula *60*
chemical equation *42*	ionic bond *61*	mole *53*	transition metal group *38*
chemical formula *42*	ionic solid *62*	molecule *58*	unified atomic mass
chemical property *38*	isotope *50*	neutron *48*	unit (u) *52*

PRACTICE TEST

After studying this chapter, you can assess your understanding with these practice test questions, which are correlated with chapter learning objectives. If you answer a question incorrectly, refer to the learning objectives in the end-of-chapter Study Guide for assistance. The Study Guide provides a conceptual summary, references a Worked Example to model how to solve the problem, and gives additional problems for more practice.

1. Refer to a periodic table. Which pair of elements do you expect to be most similar in their chemical properties? **(LO 2.3)**

(a) K and Cu   (b) O and Se

(c) Be and B   (d) Rb and Sr

2. Identify the location of the element in period 4, group 6A and classify it as a metal, nonmetal, or semimetal. **(LO 2.2)**

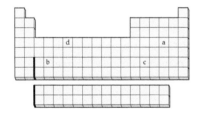

(a) Element in position a; nonmetal

(b) Element in position b; metal

(c) Element in position c; semimetal

(d) Element in position d; metal

3. Which description of an element is incorrectly matched with its location in the periodic table? **(LO 2.5–2.6)**

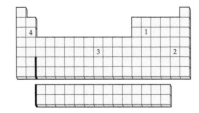

(a) Element 3—An element in the transition metal group that is a good conductor of electricity.

(b) Element 2—An element that is in the halogen group and does not conduct electricity.

(c) Element 4—An element in alkali metal group that is found in its pure form in nature.

(d) Element 1—An element that is a solid at room temperature, brittle, and a poor conductor of electricity.

4. A compound containing sulfur and fluorine contains 8.00 g of S and 9.50 g of F. Which combination of S and F masses represents a *different* compound that obeys the Law of Multiple Proportions? **(LO 2.8)**

(a) 32.0 g of S and 38.0 g of F

(b) 4.00 g of S and 4.75 g of F

(c) 8.00 g of S and 10.5 g of F

(d) 16.0 g of S and 57.0 g of F

5. Which experiment and subsequent observation led to the discovery that atoms contain negatively charged particles, now known as electrons? **(LO 2.10–2.12)**

(a) Oil is sprayed into a chamber and the speed at which the oil droplets fall is measured with and without an applied voltage. X rays in the chamber knock electrons out of air molecules. The electrons stick to the oil producing an overall negative charge on the drops. Adjusting the voltage changes the speed at which the negatively charged oil droplets fall.

(b) When a high voltage is applied across metal electrodes at opposite ends of a sealed glass tube, a cathode ray is produced. The cathode ray is repelled by a negatively charged plate.

(c) A radioactive substance emits alpha particles, which are directed at a thin gold foil. Most of the alpha particles pass through the foil, but a few alpha particles are slightly deflected and some even bounce back toward the radioactive source.

(d) The mass of different elements in a pure chemical compound are measured. Different samples of the compound always contains the same proportion of elements by mass.

6. How many protons, neutrons, and electrons are present in an atom of $^{206}_{82}Pb$? **(LO 2.14)**

(a) 82 protons, 206 neutrons, 82 electrons

(b) 124 protons, 82 neutrons, 124 electrons

(c) 82 protons, 124 neutrons, 82 electrons

(d) 82 protons, 82 neutrons, 124 electrons

7. What is the atomic weight of an element that consists of two naturally occurring isotopes? The first isotope has a mass of 84.911 and an abundance of 72.17% and the second isotope has a mass of 86.909 and an abundance of 27.83%. **(LO 2.15)**
   (a) 85.47
   (b) 86.35
   (c) 85.91
   (d) 85.17

8. Which sample has the greatest mass? **(LO 2.16)**
   (a) 5.5 mol of C
   (b) 2.1 mol of S
   (c) 4.2 mol of Be
   (d) 0.52 mol of Ag

9. How many atoms are present in 1.2 g of gold? **(LO 2.16)**
   (a) 2.5×10^{21}
   (b) 1.4×10^{26}
   (c) 7.2×10^{23}
   (d) 3.7×10^{21}

10. Bromine has two naturally occurring isotopes; ^{79}Br (mass of 78.918) and ^{81}Br (mass of 80.916). If the atomic weight of bromine is 79.904, predict the mass spectrum of a sample of bromine atoms. **(LO 2.17)**

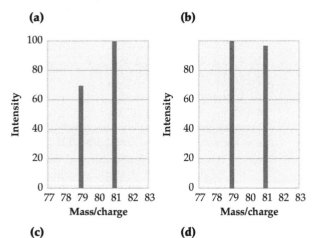

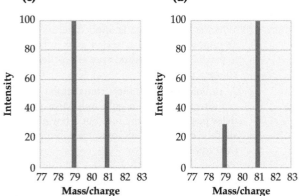

11. The molecular illustration represents **(LO 2.18)**

   (a) a pure element
   (b) a mixture of two elements
   (c) a pure compound
   (d) a mixture of an element and a compound

12. Which of the following compounds would you expect to have covalent bonds? **(LO 2.20)**
   (a) Na_2O
   (b) PBr_3
   (c) $CaBr_2$
   (d) MgS

13. How many protons, neutrons, and electrons are in $^{107}_{47}Ag^+$? **(LO 2.21)**
   (a) protons = 47, neutrons = 60, electrons = 46
   (b) protons = 47, neutrons = 107, electrons = 48
   (c) protons = 60, neutrons = 47, electrons = 47
   (d) protons = 47, neutrons = 107, electrons = 46

14. What is the correct formula for sodium phosphate? **(LO 2.24)**
   (a) Na_3PO_4
   (b) Na_3P
   (c) $NaPO_4$
   (d) $Na(PO_4)_2$

15. Which of the following compounds is *incorrectly* named? **(LO 2.23–2.25)**
   (a) CaO; calcium oxide
   (b) $FeBr_2$; iron dibromide
   (c) N_2O_5; dinitrogen pentoxide
   (d) CrO_3; chromium(VI) oxide

Answers:
1. b, 2. a, 3. c, 4. d, 5. b, 6. c, 7. a, 8. b, 9. d, 10. b, 11. c, 12. b, 13. a, 14. a, 15. b

| **Mastering Chemistry** | provides end-of-chapter exercises, feedback-enriched tutorial problems, animations, and interactive activities to encourage problem-solving practice and deeper understanding of key concepts and topics. |

RAN   *Randomized in Mastering Chemistry*

CONCEPTUAL PROBLEMS

Problems 2.1–2.33 appear within the chapter.

2.34 Where on the following outline of a periodic table are the indicated elements or groups of elements?

(a) Alkali metals   (d) Transition metals

(b) Halogens   (e) Hydrogen

(c) Alkaline earth metals   (f) Helium

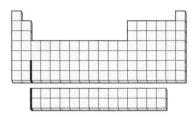

2.35 Is the red element on the following periodic table likely to be a gas, a liquid, or a solid? What is the atomic number of the blue element? What is the group number of the green, blue, and red elements? Name at least one other element that is chemically similar to the green element.

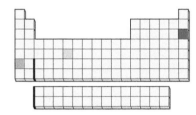

2.36 The element indicated on the following periodic table is used in smoke detectors. Identify it, give its atomic number, and tell what kind of group it's in.

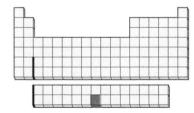

2.37 Identify the three elements indicated on the periodic table,
RAN   and give the group that they are in. Classify these elements as metals, nonmetals, or semimetals. Would you expect these elements to have similar or different chemical reactivity?

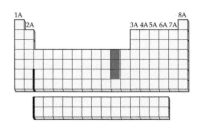

2.38 If yellow spheres represent sulfur atoms and red spheres represent oxygen atoms, which of the following drawings shows a collection of sulfur dioxide (SO_2) units?

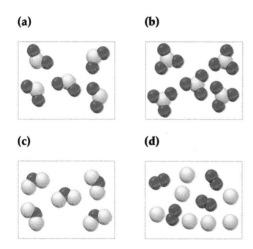

2.39 Assume that the mixture of substances in drawing (a) undergoes a reaction. Which of the drawings (b)–(d) represents a product mixture consistent with the law of mass conservation?

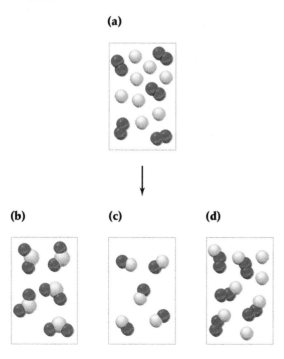

2.40 In the following drawings, red spheres represent protons, and blue spheres represent neutrons. Which of the drawings represent different isotopes of the same element, and which represents a different element altogether?

RAN

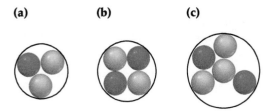

(a)   (b)   (c)

2.41 Which of the following three drawings represents a neutral Na atom, which represents a Ca atom with two positive electrical charges (Ca^{2+}), and which represents an F atom with one minus charge (F^-)?

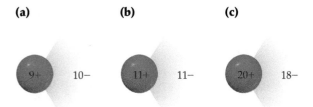

(a)   (b)   (c)

9+   10–   11+   11–   20+   18–

2.42 In the following drawings, red and blue spheres represent atoms of different elements. Match the molecular pictures (a)–(c) with the following descriptions:
(i) a pure substance consisting of a compound
(ii) a pure substance consisting of an element
(iii) a mixture of elements

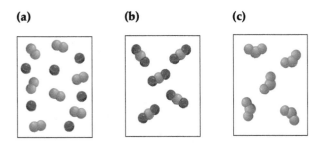

(a)   (b)   (c)

2.43 Methionine, one of the 20 amino acid building blocks from which proteins are made, has the following structure. What is the chemical formula of methionine? In writing the formula, list the element symbols in alphabetical order, and give the number of each element as a subscript.

RAN

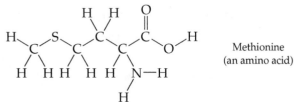

Methionine
(an amino acid)

2.44 Give molecular formulas corresponding to each of the following ball-and-stick molecular representations (red = O, gray = C, blue = N, ivory = H). In writing the formula, list the elements in alphabetical order.

(a) Alanine
(an amino acid)

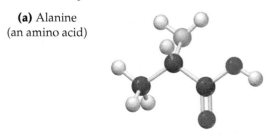

(b) Ethylene glycol
(automobile antifreeze)

(c) Acetic acid
(vinegar)

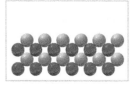

2.45 In the following drawings, red spheres represent cations, and blue spheres represent anions. Match the drawings (a)–(d) with the following ionic compounds:
(i) $Ca_3(PO_4)_2$ (ii) Li_2CO_3
(iii) $FeCl_2$ (iv) $MgSO_4$

(a)   (b)

(c)   (d)

SECTION PROBLEMS

Elements and the Periodic Table (Sections 2.1–2.3)

2.46 How many elements are presently known? About how many occur naturally?

2.47 Which element accounts for roughly 75% of the observed mass of the universe? Which four elements make up 95% of the mass of the human body?

2.48 Look at the alphabetical list of elements inside the front
RAN cover. What are the symbols for the following elements?
  (a) Gadolinium (used in color TV screens)
  (b) Germanium (used in semiconductors)
  (c) Technetium (used in biomedical imaging)
  (d) Arsenic (used in pesticides)

2.49 Look at the alphabetical list of elements inside the front
RAN cover. What are the symbols for the following elements?
  (a) Cadmium (used in rechargeable Ni-cad batteries)
  (b) Iridium (used for hardening alloys)
  (c) Beryllium (used in the space shuttle)
  (d) Tungsten (used in light bulbs)

2.50 Look at the alphabetical list of elements inside the front cover.
RAN Give the names corresponding to the following symbols:
  (a) Te   (b) Re   (c) Be
  (d) Ar   (e) Pu

2.51 Look at the alphabetical list of elements inside the front cover.
RAN Give the names corresponding to the following symbols:
  (a) B   (b) Rh   (c) Cf
  (d) Os   (e) Ga

2.52 What is wrong with each of the following statements?
  (a) The symbol for tin is Ti.
  (b) The symbol for manganese is Mg.
  (c) The symbol for potassium is Po.
  (d) The symbol for helium is HE.

2.53 What is wrong with each of the following statements?
  (a) The symbol for carbon is ca.
  (b) The symbol for sodium is So.
  (c) The symbol for nitrogen is Ni.
  (d) The symbol for chlorine is Cr.

2.54 Examine Figure 2.2, A portion of Mendeleev's periodic table.
  (a) Which characteristic was used to organize the elements in the table?
  (b) How was Mendeleev able to predict with a high level of accuracy the properties of the undiscovered elements beneath Al and Si in the periodic table?

2.55 Examine Figure 2.3, The periodic table.
  (a) Which characteristic was used to organize the elements in the table?
  (b) A horizontal row is called a _____.
  (c) A vertical column is called a _____.
  (d) Groups 1A–8A (also known as 1, 2, and 13–18) are called _____ groups.
  (e) Groups 1B–8B (also known as 3–12) are called _____ groups.
  (f) Classify the following three elements as metals, nonmetals, or semimetals: Mo, Br, Si.

  (g) What is the symbol for the element located in period 3, group 4A?

2.56 What are the rows called and what are the columns called in the periodic table?

2.57 How many groups are there in the periodic table? How are they labeled?

2.58 What common characteristics do elements within a group of the periodic table have?

2.59 Where in the periodic table are the main-group elements found? Where are the transition metal groups found?

2.60 Where in the periodic table are the metallic elements found? Where are the nonmetallic elements found?

2.61 What is a semimetal, and where in the periodic table are semimetals found?

2.62 Classify the following elements as metals, nonmetals, or
RAN semimetals:
  (a) Ti   (b) Te   (c) Se
  (d) Sc   (e) Si

2.63 Classify the following elements as metals, nonmetals, or
RAN semimetals:
  (a) Ar   (b) Sb   (c) Mo
  (d) Cl   (e) N   (f) Mg

2.64 List several general properties of the following groups:
  (a) Alkali metals   (b) Noble gases   (c) Halogens

2.65 (a) Without looking at a periodic table, list as many alkali
RAN   metals as you can. (There are five common ones.)
  (b) Without looking at a periodic table, list as many alkaline earth metals as you can. (There are five common ones.)

2.66 Without looking at a periodic table, list as many halogens as you can. (There are four common ones.)

2.67 Without looking at a periodic table, list as many noble gases as you can. (There are six common ones.)

2.68 At room temperature, a certain element is found to be a soft, silver-colored solid that reacts violently with water and is a good conductor of electricity. Is the element likely to be a metal, a nonmetal, or a semimetal?

2.69 At room temperature, a certain element is found to be a shiny, silver-colored solid that is a poor conductor of electricity. When a sample of the element is hit with a hammer, it shatters. Is the element likely to be a metal, a nonmetal, or a semimetal?

2.70 At room temperature, a certain element is yellow crystalline
RAN solid. It does not conduct electricity and when hit with a hammer, it shatters. Is the element likely to be a metal, a nonmetal, or a semimetal?

2.71 At room temperature, a certain element is a colorless, unre-
RAN active gas. Is the element likely to be a metal, a nonmetal, or a semimetal?

2.72 In which of the periodic groups 1A, 2A, 5A, and 7A is the first letter of all elements' symbol the same as the first letter of their name?

2.73 For which elements in groups 1A, 2A, 5A, and 7A of the periodic table does the first letter of their symbol differ from the first letter of their name?

2.74 Which type of property—intensive or extensive—does not depend on the amount of substance present?

2.75 Label the following properties as intensive or extensive: density, volume, mass, electrical conductivity.

Atomic Theory (Sections 2.4–2.5)

2.76 How does Dalton's atomic theory account for the law of mass conservation and the law of definite proportions?

2.77 What is the law of multiple proportions, and how does Dalton's atomic theory account for it?

2.78 A sample of mercury with a mass of 114.0 g was combined
RAN with 12.8 g of oxygen gas, and the resulting reaction gave 123.1 g of mercury(II) oxide. How much oxygen was left over after the reaction was complete?

2.79 A sample of $CaCO_3$ was heated, causing it to form CaO
RAN and CO_2 gas. Solid CaO remained behind, while the CO_2 escaped to the atmosphere. If the $CaCO_3$ weighed 612 g and the CaO weighed 343 g, how many grams of CO_2 were formed in the reaction?

2.80 In methane, one part hydrogen combines with three parts carbon by mass. If a sample of a compound containing only carbon and hydrogen contains 32.0 g of carbon and 8.0 g of hydrogen, could the sample be methane? If the sample is not methane, show that the law of multiple proportions is followed for methane and this other substance.

2.81 In borane, one part hydrogen combines with 3.6 parts boron by mass. A compound containing only hydrogen and boron contains 6.0 g of hydrogen and 43.2 g of boron. Could this compound be borane? If it is not borane, show that the law of multiple proportions is followed for borane and this other substance.

2.82 Benzene, ethane, and ethylene are just three of a large number of *hydrocarbons*—compounds that contain only carbon and hydrogen. Show how the following data are consistent with the law of multiple proportions.

Compound	Mass of Carbon in 5.00 g Sample	Mass of Hydrogen in 5.00 g Sample
Benzene	4.61 g	0.39 g
Ethane	4.00 g	1.00 g
Ethylene	4.29 g	0.71 g

2.83 The atomic weight of carbon (12.011) is approximately 12 times that of hydrogen (1.008).
  (a) Show how you can use this knowledge to calculate possible formulas for benzene, ethane, and ethylene (Problem 2.82).
  (b) Show how your answer to part (a) is consistent with the actual formulas for benzene (C_6H_6), ethane (C_2H_6), and ethylene (C_2H_4).

2.84 Two compounds containing carbon and oxygen have the
RAN following percent composition by mass.

  Compound 1: 42.9% carbon and 57.1% oxygen
  Compound 2: 27.3% carbon and 72.7% oxygen

  Show that the law of multiple proportions is followed. If the formula of the first compound is CO, what is the formula of the second compound?

2.85 In addition to carbon monoxide (CO) and carbon dioxide
RAN (CO_2), there is a third compound of carbon and oxygen called *carbon suboxide*. If a 2.500 g sample of carbon suboxide contains 1.32 g of C and 1.18 g of O, show that the law of multiple proportions is followed. What is a possible formula for carbon suboxide?

Elements and Atoms (Sections 2.6–2.8)

2.86 The results from Thomson's cathode-ray tube experiment
RAN led to the discovery of which subatomic particle?

2.87 What affects the magnitude of the deflection of the cathode ray in Thomson's experiment?

2.88 Label the following statements about J. J. Thomson's cathode-ray tube experiments shown in Figure 2.6 as true or false.
  (a) When a high voltage is applied to metal electrodes in a sealed, evacuated glass tube, an electric current flows.
  (b) A cathode ray is a stream of charged particles.
  (c) The cathode ray is deflected away from a positively charged plate.
  (d) Many different types of metal electrodes are capable of producing a cathode ray.
  (e) A cathode ray is made up of protons.
  (f) By measuring the deflection of the cathode ray beam caused by electric fields of known strength, the charge-to-mass ratio of the electron was calculated.

2.89 Fill in the blanks in the description of Millikan's oil drop experiment, shown in Figure 2.7.

  A fine mist of oil was sprayed into a chamber, and the velocity of the oil droplets was measured using a telescopic eyepiece. Knowing the terminal velocity of a falling oil droplet allows the _____ (mass/charge/volume) of the oil droplet to be calculated.

  Energetic X rays remove _____ (electrons/protons) from air molecules, which adhere to the surface of the oil droplet, giving it a _____ (negative/positive) charge. The charged oil droplets fall between two horizontal plates with an applied voltage. The top plate is given a _____ (negative/positive) charge, which slows the fall of the drop and suspends it between the two plates. The overall charge on the drop can be calculated from the mass of the drop and the voltage on the plates. The charge on the oil drop was found to be _____ (a fraction/a whole number multiple) of the charge of an electron. The oil drop experiment measured the _____ (mass/charge) of the electron, allowing the _____ (mass/charge) of the electron to be calculated from J. J. Thomson's charge-to-mass ratio measurements.

2.90 Which of the following charges is *not* possible for the over-
RAN all charge on an oil droplet in Millikan's experiment? For this problem, we'll round the currently accepted charge of an electron to 1.602×10^{-19} C.
  (a) -1.010×10^{-18} C   (b) -8.010×10^{-19} C
  (c) -2.403×10^{-18} C

2.91 What discovery about atomic structure was made from the results of Rutherford's gold foil experiment?

2.92 Prior to Rutherford's gold foil experiment, the "plum pudding" model of the atom represented atomic structure. In this model, the atom is composed of electrons interspersed within a positive cloud of charge. If this were the correct model of the atom, predict how the results of Rutherford's experiment would have been different.
  (a) The alpha particles would pass right through the gold foil with little to no deflection.
  (b) Most of the alpha particles would be deflected back toward the source.

(c) Most of the alpha particles would be absorbed by the atom and neither pass through nor be deflected from the gold foil.

2.93 A period at the end of sentence written with a graphite pen-
RAN cil has a diameter of 1 mm. If the period represented the nucleus, approximately how large is the diameter of the entire atom in units of m?

2.94 A 1/4-inch-thick lead sheet is used for protection from medi-
RAN cal X rays. If a single lead atom has a diameter 350 pm, how many atoms thick is the lead sheet?

2.95 A period at the end of sentence written with a graphite pen-
RAN cil has a diameter of 1 mm. How many carbon atoms would it take to line up across the period if a single carbon atom has a diameter of 150 pm?

2.96 What is the difference between an atom's atomic number and its mass number?

2.97 What is the difference between an element's atomic number and its atomic weight?

2.98 The subscript giving the atomic number of an atom is often left off when writing an isotope symbol. For example, $^{13}_{6}C$ is often written simply as ^{13}C. Why is this allowed?

2.99 Iodine has a *lower* atomic mass than tellurium
RAN (126.90 for iodine, 127.60 for tellurium) even though it has a *higher* atomic number (53 for iodine, 52 for tellurium). Explain.

2.100 Give the names and symbols for the following elements:
(a) An element with atomic number 6
(b) An element with 18 protons in its nucleus
(c) An element with 23 electrons

2.101 The radioactive isotope cesium-137 was produced in large amounts in fallout from the 1985 nuclear power plant disaster at Chernobyl, Ukraine. Write the symbol for this isotope in standard format.

2.102 Write symbols for the following isotopes:
RAN (a) Radon-220 (b) Polonium-210 (c) Gold-197

2.103 Write symbols for the following isotopes:
(a) $Z = 58$ and $A = 140$ (b) $Z = 27$ and $A = 60$

2.104 How many protons, neutrons, and electrons are in each of
RAN the following atoms?
(a) $^{15}_{7}N$ (b) $^{60}_{27}Co$ (c) $^{131}_{53}I$ (d) $^{148}_{58}Ce$

2.105 How many protons and neutrons are in the nucleus of the following atoms?
(a) ^{27}Al (b) ^{32}S (c) ^{64}Zn (d) ^{207}Pb

2.106 Identify the following elements:
RAN (a) $^{24}_{12}X$ (b) $^{58}_{28}X$ (c) $^{104}_{46}X$ (d) $^{183}_{74}X$

2.107 Identify the following elements:
(a) $^{202}_{80}X$ (b) $^{195}_{78}X$ (c) $^{184}_{76}X$ (d) $^{209}_{83}X$

2.108 Which of the following isotope symbols can't be correct?
$^{18}_{9}F$ $^{12}_{5}C$ $^{33}_{35}Br$ $^{18}_{8}O$ $^{11}_{5}Bo$

2.109 Which of the following isotope symbols can't be correct?
RAN $^{14}_{7}Ni$ $^{131}_{54}Xe$ $^{54}_{26}Fe$ $^{73}_{23}Ge$ $^{1}_{2}He$

2.110 Fluorine occurs naturally as a single isotope. How many protons, neutrons, and electrons are present in deuterium fluoride (^{2}HF)? (Deuterium is ^{2}H.)

2.111 Hydrogen has three isotopes (^{1}H, ^{2}H, and ^{3}H), and chlorine has two isotopes (^{35}Cl and ^{37}Cl). How many isotopic kinds of HCl are there? Write the formula for each, and tell how many protons, neutrons, and electrons each contains.

Atomic Weight, Moles, and Mass Spectrometry (Sections 2.9–2.10)

2.112 The unified atomic mass unit (u) is defined as exactly 1/12 the mass of a neutral atom of:
(a) ^{1}H (b) ^{12}C (c) ^{14}C (d) ^{16}O

2.113 (a) The unified atomic mass unit (u) is used to represent the extremely small mass of atoms. How many grams are equivalent to 1 u?
(b) The mole is a unit used to represent a very large number of atoms. How many atoms are equivalent to 1 mol of atoms?

2.114 Match the descriptions (a)–(e) with the following terms: atomic weight, atomic mass, mass number, atomic number, molar mass.
(a) The mass of a specific atom such as one atom of ^{13}C
(b) The quantity determined by the number of protons in an element.
(c) The number of grams in 1 mol of an element
(d) The number of protons and neutrons in an element
(e) The weighted average of the isotopic masses of an element's naturally occurring isotopes

2.115 Label the following statements as true or false.
(a) The atomic weight and the atomic number of element have the same numerical value.
(b) The molar mass in grams for an element and the atomic weight have the same numerical value.

2.116 Copper has two naturally occurring isotopes, including ^{65}Cu. Look at the periodic table, and tell whether the second isotope is ^{63}Cu or ^{66}Cu.

2.117 Sulfur has four naturally occurring isotopes, including ^{33}S,
RAN ^{34}S, and ^{36}S. Look at the periodic table, and tell whether the fourth isotope is ^{32}S or ^{35}S.

2.118 Naturally occurring boron consists of two isotopes: ^{10}B (19.9%) with an isotopic mass of 10.0129 and ^{11}B (80.1%) with an isotopic mass of 11.009 31. What is the atomic weight of boron? Check your answer by looking at a periodic table.

2.119 Naturally occurring silver consists of two isotopes: ^{107}Ag (51.84%) with an isotopic mass of 106.9051 and ^{109}Ag (48.16%) with an isotopic mass of 108.9048. What is the atomic weight of silver? Check your answer in a periodic table.

2.120 Magnesium has three naturally occurring isotopes: ^{24}Mg (23.985) with 78.99% abundance, ^{25}Mg (24.986) with 10.00% abundance, and a third with 11.01% abundance. Look up the atomic weight of magnesium, and then calculate the mass of the third isotope.

2.121 A sample of naturally occurring silicon consists of ^{28}Si (27.9769), ^{29}Si (28.9765), and ^{30}Si (29.9738). If the atomic weight of silicon is 28.0855 and the natural abundance of ^{29}Si is 4.68%, what are the natural abundances of ^{28}Si and ^{30}Si?

2.122 Copper metal has two naturally occurring isotopes: copper-63
RAN (69.15%; isotopic mass = 62.93) and copper-65 (30.85%; isotopic mass 64.93). Calculate the atomic weight of copper, and check your answer in the periodic table.

2.123 Germanium has five naturally occurring isotopes: ^{70}Ge,
RAN 20.5%, 69.924; ^{72}Ge, 27.4%, 71.922; ^{73}Ge, 7.8%, 72.923; ^{74}Ge, 36.5%, 73.921; and ^{76}Ge, 7.8%, 75.921. What is the atomic weight of germanium?

2.124 What is the mass in grams of each of the following samples?
RAN (a) 1.505 mol of Ti (b) 0.337 mol of Na
(c) 2.583 mol of U

2.125 How many moles are in each of the following samples?

(a) 11.51 g of Ti (b) 29.127 g of Na

(c) 1.477 kg of U

2.126 If the atomic weight of an element is x, what is the mass in grams
RAN of 6.02×10^{23} atoms of the element? How does your answer
compare numerically with the atomic weight of element x?

2.127 If the atomic weight of an element is x, what is the mass in
RAN grams of 3.17×10^{20} atoms of the element?

2.128 If 6.02×10^{23} atoms of element Y have a mass of 83.80 g,
what is the identity of Y?

2.129 If 4.61×10^{21} atoms of element Z have a mass of 0.815 g,
what is the identity of Z?

2.130 Refer to Figure 2.10 showing a schematic illustration of a
mass spectrometer.

(a) What is the purpose of bombarding the gaseous atoms
with an electron beam?

(b) Compare two ions with a +1 charge traveling through
the curved, evacuated tube in the mass spectrometer.
Will a heavier ion or lighter ion be deflected to a greater
degree by the magnetic field?

(c) Under a given set of experimental conditions ions of
a certain mass-to-charge ratio pass through a slit and
strike the detector. What experimental variable in the
mass spectrometer is altered so that a different mass-to-
charge ratio ion strikes the detector?

2.131 Copper has two naturally occurring isotopes, ^{63}Cu (relative abundance = 69.17%) and ^{65}Cu (relative abundance = 30.83%).
Select which mass spectrum represents a sample of copper.

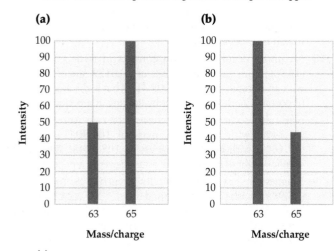

2.132 Use the data from the mass spectrum of a sample of an element
to calculate the element's atomic weight. Identify the element.

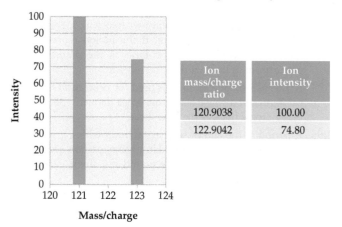

Ion mass/charge ratio	Ion intensity
120.9038	100.00
122.9042	74.80

2.133 Use the data from the mass spectrum of a sample of an element
to calculate the element's atomic weight. Identify the element.

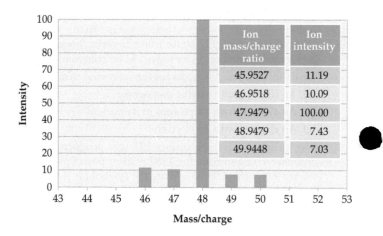

Ion mass/charge ratio	Ion intensity
45.9527	11.19
46.9518	10.09
47.9479	100.00
48.9479	7.43
49.9448	7.03

Chemical Compounds (Sections 2.11–2.12)

2.134 What is the difference between a covalent bond and an ionic
bond?

2.135 Which of the following bonds are likely to be covalent and
which ionic? Explain.

(a) B—Br (b) Na—Br

(c) Br—Cl (d) O—Br

2.136 The symbol CO stands for carbon monoxide, but the sym-
RAN bol Co stands for the element cobalt. Explain.

2.137 Correct the error in each of the following statements:

(a) The formula of ammonia is NH3.

(b) Molecules of potassium chloride have the formula KCl.

(c) Cl^- is a cation.

(d) CH_4 is a polyatomic ion.

2.138 How many protons and electrons are in each of the following ions?

(a) Be^{2+} (b) Rb^+

(c) Se^{2-} (d) Au^{3+}

2.139 What is the identity of the element X in the following ions?

(a) X^{2+}, a cation that has 36 electrons

(b) X^-, an anion that has 36 electrons

2.140 The structural formula of isopropyl alcohol, better known as "rubbing alcohol," is shown. What is the chemical formula of isopropyl alcohol?

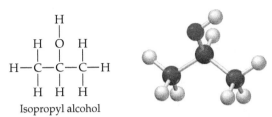

Isopropyl alcohol

2.141 Lactic acid, a compound found both in sour milk and in tired muscles, has the structure shown. What is its chemical formula?

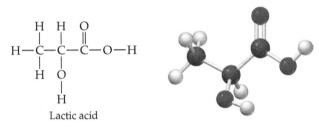

Lactic acid

2.142 Butane, the fuel used in disposable lighters, has the formula C_4H_{10}. The carbon atoms are connected in the sequence C—C—C—C, and each carbon has four covalent bonds. Draw the structural formula of butane.

2.143 Cyclohexane, C_6H_{12}, is an important starting material used in the industrial synthesis of nylon. Each carbon has four covalent bonds, two to hydrogen and two to other carbons. Draw the structural formula of cyclohexane.

2.144 Isooctane, the substance in gasoline from which the term *octane rating* derives, has the formula C_8H_{18}. Each carbon has four covalent bonds, and the atoms are connected in the sequence shown. Draw the complete structural formula of isooctane.

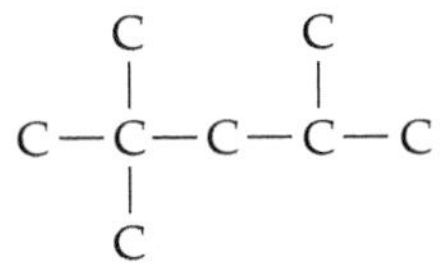

$$
\begin{array}{ccc}
& C & C \\
& | & | \\
C-&C-C-C&-C \\
& | & \\
& C &
\end{array}
$$

2.145 Fructose, $C_6H_{12}O_6$, is the sweetest naturally occurring sugar and is found in many fruits and berries. Each carbon has four covalent bonds, each oxygen has two covalent bonds, each hydrogen has one covalent bond, and the atoms are connected in the sequence shown. Draw the complete structural formula of fructose.

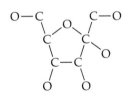

Naming Compounds (Section 2.13)

2.146 Give systematic names for the following binary compounds:
(a) CsF   (b) K_2O (c) CuO

2.147 Give systematic names for the following binary compounds:
(a) BaS   (b) $BeBr_2$ (c) $FeCl_3$

2.148 Write formulas for the following binary compounds:
(a) Potassium chloride   (b) Tin(II) bromide
(c) Calcium oxide   (d) Barium chloride
(e) Aluminum hydride

2.149 Write formulas for the following binary compounds:
RAN
(a) Vanadium(III) chloride
(b) Manganese(IV) oxide
(c) Copper(II) sulfide
(d) Aluminum oxide

2.150 Write formulas for the following compounds:
RAN
(a) Calcium acetate
(b) Iron(II) cyanide
(c) Sodium dichromate
(d) Chromium(III) sulfate
(e) Mercury(II) perchlorate

2.151 Write formulas for the following compounds:
RAN
(a) Lithium phosphate
(b) Magnesium hydrogen sulfate
(c) Manganese(II) nitrate
(d) Chromium(III) sulfate

2.152 Give systematic names for the following compounds:
RAN
(a) $Ca(ClO)_2$ (b) $Ag_2S_2O_3$
(c) NaH_2PO_4 (d) $Sn(NO_3)_2$
(e) $Pb(CH_3CO_2)_4$ (f) $(NH_4)_2SO_4$

2.153 Name the following ions:
RAN
(a) Ba^{2+} (b) Cs^+
(c) V^{3+} (d) HCO_3^-
(e) NH_4^+ (f) Ni^{2+}
(g) NO_2^- (h) ClO_2^-
(i) Mn^{2+} (j) ClO_4^-

2.154 What are the formulas of the compounds formed from the following ions?
RAN
(a) Ca^{2+} and Br^- (b) Ca^{2+} and SO_4^{2-}
(c) Al^{3+} and SO_4^{2-}

2.155 What are the formulas of the compounds formed from the following ions?
RAN
(a) Na^+ and NO_3^- (b) K^+ and SO_4^{2-}
(c) Sr^{2+} and Cl^-

2.156 Write formulas for compounds of calcium with each of the following:
RAN
(a) Chlorine   (b) Oxygen   (c) Sulfur

2.157 Write formulas for compounds of rubidium with each of the following:
RAN
(a) Bromine   (b) Nitrogen   (c) Selenium

2.158 Give the formulas and charges of the following ions:
RAN
(a) Sulfite ion   (b) Phosphate ion
(c) Zirconium(IV) ion   (d) Chromate ion
(e) Acetate ion   (f) Thiosulfate ion

2.159 What are the charges on the positive ions in the following compounds?
(a) $Zn(CN)_2$ (b) $Fe(NO_2)_3$
(c) $Ti(SO_4)_2$ (d) $Sn_3(PO_4)_2$
(e) Hg_2S (f) MnO_2
(g) KIO_4 (h) $Cu(CH_3CO_2)_2$

2.160 Name the following binary molecular compounds:
RAN **(a)** CCl_4 **(b)** ClO_2
     (c) N_2O **(d)** N_2O_3

2.161 Give systematic names for the following compounds:
RAN **(a)** NCl_3 **(b)** P_4O_6 **(c)** S_2F_2

2.162 Name the following binary compounds of nitrogen and
RAN oxygen:
     (a) NO   **(b)** N_2O **(c)** NO_2
     (d) N_2O_4 **(e)** N_2O_5

2.163 Name the following binary compounds of sulfur and oxygen:
RAN **(a)** SO   **(b)** S_2O_2 **(c)** S_5O
     (d) S_7O_2 **(e)** SO_3

2.164 Fill in the missing information to give formulas for the following compounds:
     (a) $Na_?SO_4$ **(b)** $Ba_?(PO_4)_?$ **(c)** $Ga_?(SO_4)_?$

2.165 Write formulas for each of the following compounds:
     (a) Sodium peroxide   **(b)** Aluminum bromide
     (c) Chromium(III) sulfate

MULTICONCEPT PROBLEMS

2.166 Ammonia (NH_3) and hydrazine (N_2H_4) are both compounds
RAN of nitrogen and hydrogen. Based on the law of multiple proportions, how many grams of hydrogen would you expect 2.34 g of nitrogen to combine with to yield ammonia? To yield hydrazine?

2.167 If 3.670 g of nitrogen combines with 0.5275 g of hydrogen
RAN to yield compound X, how many grams of nitrogen would combine with 1.575 g of hydrogen to make the same compound? Is X ammonia (NH_3) or hydrazine (N_2H_4)?

2.168 Prior to 1961, the atomic mass unit (*amu*) was defined as
RAN 1/16 the mass of the atomic weight of oxygen; that is, the atomic weight of oxygen was defined as exactly 16. What was the mass of a ^{12}C atom prior to 1961 if the atomic weight of oxygen on today's scale is 15.9994?

2.169 What was the mass in atomic mass units of a ^{40}Ca atom
RAN prior to 1961 if its mass on today's scale is 39.9626? (See Problem 2.168.)

2.170 The *molecular weight* of a compound is the sum of the
RAN atomic masses of all atoms in the molecule. What is the molecular mass of acetaminophen ($C_8H_9NO_2$), the active ingredient in Tylenol?

2.171 The *mass percent* of an element in a compound is the mass
RAN of the element (total mass of the element's atoms in the compound) divided by the mass of the compound (total mass of all atoms in the compound) times 100%. What is the mass percent of each element in acetaminophen? (See Problem 2.170.)

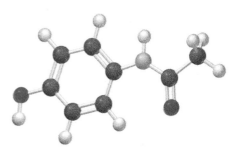

Acetaminophen ($C_8H_9NO_2$)

2.172 In an alternate universe, the smallest negatively charged
RAN particle, analogous to our electron, is called a blorvek. To determine the charge on a single blorvek, an experiment like Millikan's with charged oil droplets was carried out, and the following results were recorded:

Droplet Number	Charge (C)
1	7.74×10^{-16}
2	4.42×10^{-16}
3	2.21×10^{-16}
4	4.98×10^{-16}
5	6.64×10^{-16}

(a) Based on these observations, what is the largest possible value for the charge on a blorvek?

(b) Further experiments found a droplet with a charge of 5.81×10^{-16} C. Does this new result change your answer to part (a)? If so, what is the new largest value for the blorvek's charge?

chapter 3

Mass Relationships in Chemical Reactions

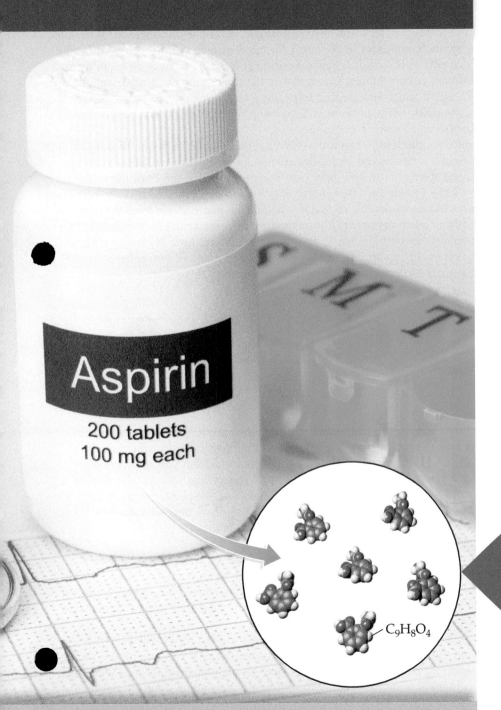

Aspirin

200 tablets
100 mg each

$C_9H_8O_4$

Contents

Mass relationships between reactants and products can be used to calculate the amount of acetylsalicylic acid formed by the reaction used in the pharmaceutical industry.

How Is the Principle of Atom Economy Used to Minimize Waste in a Chemical Synthesis?

The answer to this question can
be found on page 105 in the INQUIRY ?

It's important to realize that chemical *reactions*—the change of one substance into another—are at the heart of the science. Nearly every process that occurs in your body, including vision, the sensation of pain, and the conversion of food to energy, is in essence a series of chemical reactions. Likewise, products we use every day such as dyes, plastics, computer chips, and metals are manufactured by chemical reactions.

In this chapter, we'll begin learning how to describe chemical reactions by first looking at conventions for writing chemical equations. Next, we'll examine mass relationships between reactants and products in chemical reactions, which allow us to calculate how much product can be made from a given amount of starting material. Finally, we'll see how chemical formulas are determined and molecular weights are measured.

3.1 REPRESENTING CHEMISTRY ON DIFFERENT LEVELS

Before starting this chapter, let's first answer a simple yet important question: What do numbers and symbols represent in chemical formulas and equations? Answering this question isn't as easy as it sounds because a chemical symbol can have different meanings under different circumstances. Chemists use the same symbols to represent chemistry on both a small-scale, microscopic level and a large-scale, macroscopic level and tend to not distinguish between what is happening on each of the two levels, which can be very confusing to newcomers to the field.

On the microscopic level, chemical symbols represent the behavior of individual atoms and molecules. Atoms and molecules are much too small to see, but we can nevertheless describe their microscopic behavior. For example, we can read the equation $2 H_2 + O_2 \rightarrow 2 H_2O$ to mean "Two molecules of hydrogen react with one molecule of oxygen to yield two molecules of water." It's on the microscopic level that we try to understand how reactions occur. Although simplistic, we visualize a molecule as a collection of spheres stuck together. In trying to understand how H_2 reacts with O_2, for example, you might picture H_2 and O_2 molecules as made of two spheres pressed together and a water molecule as made of three spheres.

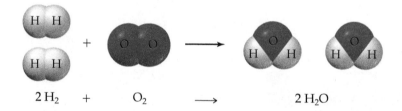

$$2 H_2 \quad + \quad O_2 \quad \longrightarrow \quad 2 H_2O$$

On the macroscopic level, formulas and equations represent the large-scale behaviors of atoms and molecules that give rise to visible properties. In other words, the symbols H_2, O_2, and H_2O represent not just single molecules but vast numbers of molecules that together have a set of measurable physical properties. A collection of a large number of H_2O molecules appears to us as a colorless liquid that freezes at 0 °C and boils at 100 °C. Clearly, it's this macroscopic behavior we deal with in the laboratory when we weigh specific amounts of reactants, place them in a flask, and observe visible changes. We can also read the equation $2 H_2 + O_2 \rightarrow 2 H_2O$ to mean "Two moles of hydrogen react with one mole of oxygen to yield two moles of water."

In the same way, a single atom of copper does not conduct electricity and has no color on a microscopic level. On a macroscopic level, however, a large collection of copper atoms appears to us as a shiny, reddish-brown solid that can be drawn into electrical wires or made into coins.

A chemical formula or equation can be read either on the macroscopic level or on the microscopic level. The symbol H_2O can be interpreted either as one tiny, invisible molecule or as a vast collection of molecules large enough to swim in. You will learn to interpret a chemical formula or equation differently depending upon the context in which it is presented.

BIG IDEA Question 1 Go to eText

Write a sentence to describe the reaction below. (Gray = C, red = O, ivory = H.) (Hint: CH_4 is methane.)

3.2 BALANCING CHEMICAL EQUATIONS

Preceding chapters provided several examples of reactions: hydrogen reacting with oxygen to yield water, sodium reacting with chlorine to yield sodium chloride, and mercury(II) nitrate reacting with potassium iodide to yield mercury(II) iodide:

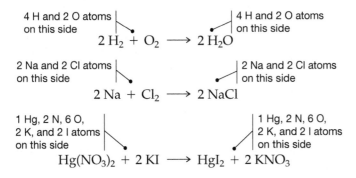

4 H and 2 O atoms on this side
4 H and 2 O atoms on this side
$$2\,H_2 + O_2 \longrightarrow 2\,H_2O$$

2 Na and 2 Cl atoms on this side
2 Na and 2 Cl atoms on this side
$$2\,Na + Cl_2 \longrightarrow 2\,NaCl$$

1 Hg, 2 N, 6 O, 2 K, and 2 I atoms on this side
1 Hg, 2 N, 6 O, 2 K, and 2 I atoms on this side
$$Hg(NO_3)_2 + 2\,KI \longrightarrow HgI_2 + 2\,KNO_3$$

Look carefully at how these equations are written. Because hydrogen, oxygen, and chlorine exist as covalent diatomic molecules, we must write them as H_2, O_2, and Cl_2 rather than as isolated atoms (Section 2.11). Now, look at the atoms on each side of the reaction arrow. Although we haven't explicitly stated it yet, chemical equations are always written so that they are **balanced;** that is, the numbers and kinds of atoms on both sides of the reaction arrow are the same. This requirement is a consequence of the law of mass conservation (Section 2.4). Because atoms are neither created nor destroyed in chemical reactions, the kinds of atoms and numbers of atoms must remain the same in both products and reactants.

Balancing a chemical equation involves finding out how many *formula units* of each different substance take part in the reaction. A **formula unit,** as its name implies, is one unit—whether atom, ion, or molecule—corresponding to a given formula. One formula unit of NaCl is one Na^+ ion and one Cl^- ion, one formula unit of $MgBr_2$ is one Mg^{2+} ion and two Br^- ions, and one formula unit of H_2O is one H_2O molecule.

Complicated equations generally need to be balanced using a systematic method, as will be shown in later chapters, whereas simpler equations can often be balanced using a mixture of common sense and trial and error. Four steps in balancing a chemical equation are:

1. *Write an unbalanced equation using the correct chemical formula unit for each reactant and product.* For the reaction of ammonia (NH_3) with oxygen to form nitrogen monoxide and water, we begin by writing:

$$NH_3 + O_2 \longrightarrow NO + H_2O \qquad \text{Unbalanced}$$

2. *Find suitable coefficients—the numbers placed before formulas to indicate how many formula units of each substance are required to balance the equation.* Only these coefficients can be changed when balancing an equation; the formulas themselves can't be changed. It is best to begin by balancing the elements that appear in only two species in the equation. In the reaction of ammonia with oxygen, the element H appears in NH_3 and H_2O and the element N appears in NH_3 and NO. To balance H, a coefficient of 2 is placed in front of NH_3 and a coefficient of 3 is placed in front of H_2O. There are now 6 H atoms in both reactants and products.

$$2\,NH_3 + O_2 \longrightarrow NO + 3\,H_2O \qquad \text{Balanced for H}$$

To balance N, a coefficient of 2 is needed in front of NO because there are 2 N atoms in the reactants.

$$2\,NH_3 + O_2 \longrightarrow 2\,NO + 3\,H_2O \qquad \text{Balanced for H and N}$$

REMEMBER . . .
According to the law of mass conservation, mass is neither created nor destroyed in chemical reactions (Section 2.4).

Elements that are not combined with other elements should be balanced last as changing the coefficient will not impact other species in the equation. In this case, adding a coefficient of 5/2 in front of O_2 in the reactants will balance O_2 because there are 5 O_2 atoms in the products.

$$2\,NH_3 + \frac{5}{2}\,O_2 \longrightarrow 2\,NO + 3\,H_2O \qquad \text{Balanced for H and N and O}$$

3. *Report coefficients to their smallest whole-number values.* The equation for the reaction of ammonia with oxygen is now balanced, but it is common to use whole-number coefficients in balanced equations. Therefore, multiply all the coefficients by 2 to get the final balanced equation.

$$4\,NH_3 + 5\,O_2 \longrightarrow 4\,NO + 6\,H_2O \qquad \text{Balanced}$$

During the trial-and-error process for balancing, you may arrive at an equation that has coefficients that need to be reduced. If you had arrived at the balanced equation

$$8\,NH_3 + 10\,O_2 \longrightarrow 8\,NO + 12\,H_2O$$

it would be necessary to divide by a common divisor to give the smallest whole-number values. Dividing all the coefficients by 2 would result in the correct balanced equation.

4. *Check your answer by making sure that the numbers and kinds of atoms are the same on both sides of the equation.*

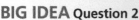
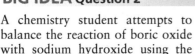

> **BIG IDEA** Question 2   Go to eText
>
> A chemistry student attempts to balance the reaction of boric oxide with sodium hydroxide using the following coefficients. Is the reaction balanced?
>
> $$B_2O_3 + 6\,NaOH \longrightarrow 2\,Na_3BO_3 + 2\,H_2O$$

4 N, 12 H, and 10 O atoms on this side   4 N, 12 H, and 10 O atoms on this side

$$4\,NH_3 + 5\,O_2 \longrightarrow 4\,NO + 6\,H_2O$$

Let's work through some additional examples.

CONCEPTUAL WORKED EXAMPLE 3.1

Visualizing Atoms and Molecules in a Chemical Reaction

Write a balanced equation for the reaction of element A (red spheres) with element B (blue spheres) as represented below:

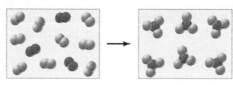

STRATEGY

Balancing the reactions shown in this molecular representation is just a matter of counting the numbers of reactant and product formula units. In this example, the reactant box contains three red A_2 molecules and nine blue B_2 molecules, whereas the product box contains six AB_3 molecules with no reactant left over.

SOLUTION

$$3\,A_2 + 9\,B_2 \longrightarrow 6\,AB_3 \quad \text{dividing by 3 reduces the equation to} \quad A_2 + 3\,B_2 \longrightarrow 2\,AB_3$$

CHECK

In any balanced equation, the numbers and kinds of atoms must be the same on both sides.

2 A and 6 B atoms on this side   2 A and 6 B atoms on this side

$$A_2 + 3\,B_2 \longrightarrow 2\,AB_3$$

▶ **CONCEPTUAL PRACTICE 3.1** Write a balanced equation for the reaction of element A (red spheres) with element B (green spheres) as represented below:

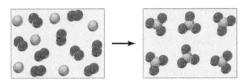

▶ **CONCEPTUAL APPLY 3.2** If blue spheres represent nitrogen atoms and red spheres represent oxygen atoms, which box represents reactants and which box represents products for the reaction $2NO(g) + O_2(g) \longrightarrow 2NO_2(g)$?

(a)   **(b)**   **(c)**   **(d)**

All **PRACTICE** and **APPLY** problems are interactive in the eText.

WORKED EXAMPLE 3.2

Balancing a Chemical Equation

Propane, C_3H_8, is a colorless, odorless gas often used as a heating and cooking fuel in campers and rural homes. Write a balanced equation for the combustion reaction of propane with oxygen to yield carbon dioxide and water.

STRATEGY AND SOLUTION

Follow the four steps described in the text:

Step 1. Write the unbalanced equation using correct chemical formulas for all substances:

$$C_3H_8 + O_2 \longrightarrow CO_2 + H_2O \qquad \text{Unbalanced}$$

Step 2. Find coefficients to balance the equation. Begin by balancing the elements that appear in two species; in this reaction, these are C and H. Look at the unbalanced equation, and note that there are 3 carbon atoms on the left side of the equation but only 1 on the right side. If we add a coefficient of 3 to CO_2 on the right, the carbons balance:

$$C_3H_8 + O_2 \longrightarrow 3\,CO_2 + H_2O \qquad \text{Balanced for C}$$

Next, look at the number of hydrogen atoms. There are 8 hydrogens on the left but only 2 (in H_2O) on the right. By adding a coefficient of 4 to the H_2O on the right, the hydrogens balance:

$$C_3H_8 + O_2 \longrightarrow 3\,CO_2 + 4\,H_2O \qquad \text{Balanced for C and H}$$

Find the coefficient for O_2 last as oxygen is not combined with other elements. Look at the number of oxygen atoms. There are 2 on the left but 10 on the right. By adding a coefficient of 5 to the O_2 on the left, the oxygens balance:

$$C_3H_8 + 5\,O_2 \longrightarrow 3\,CO_2 + 4\,H_2O \qquad \text{Balanced for C, H, and O}$$

Step 3. Make sure that the coefficients are reduced to their smallest whole-number values. In fact, our answer is already correct, but we might have arrived at a different answer through trial and error:

$$2\,C_3H_8 + 10\,O_2 \longrightarrow 6\,CO_2 + 8\,H_2O$$

Although the preceding equation is balanced, the coefficients are not the smallest whole numbers. It would be necessary to divide all coefficients by 2 to reach the final equation. A coefficient of 1 is never written but is implied if no other coefficient is given.

$$C_3H_8 + 5\,O_2 \longrightarrow 3\,CO_2 + 4\,H_2O$$

continued on next page

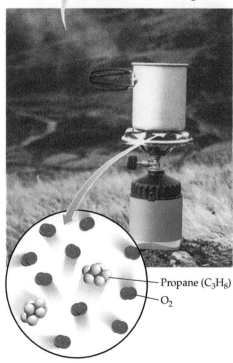
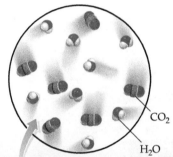

▲ Propane is used as a fuel in camp stoves and rural homes.

▲ Violent reaction of potassium chlorate with table sugar.

Step 4. Check the answer by counting the numbers and kinds of atoms on both sides of the equation to make sure that they're the same:

3 C, 8 H, and 10 O atoms on this side   3 C, 8 H, and 10 O atoms on this side

$$C_3H_8 + 5 O_2 \longrightarrow 3 CO_2 + 4 H_2O$$

▶ **PRACTICE 3.3**   Balance the equation for synthesis of hydrazine for rocket fuel.

$$NH_3 + Cl_2 \longrightarrow N_2H_4 + NH_4Cl$$

▶ **APPLY 3.4**   The major ingredient in ordinary safety matches is potassium chlorate, $KClO_3$, a substance that can act as a source of oxygen in combustion reactions. Its reaction with ordinary table sugar (sucrose, $C_{12}H_{22}O_{11}$), for example, occurs violently to yield potassium chloride, carbon dioxide, and water. Write a balanced equation for the reaction.

3.3 MOLECULAR WEIGHT AND MOLAR MASS

Imagine a laboratory experiment—perhaps the reaction of ethylene, C_2H_4, with hydrogen chloride, HCl, to prepare ethyl chloride, C_2H_5Cl, a colorless, low-boiling liquid that doctors and athletic trainers use as a spray-on anesthetic for minor injuries. You might note that in writing this and other equations, the designations (g) for gas, (l) for liquid, (s) for solid, and (aq) for aqueous solutions are often appended to the symbols of reactants and products to show their physical state. We'll do this frequently from now on.

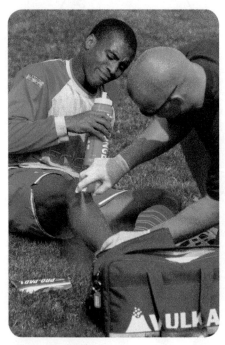

▲ Ethyl chloride is often used as a spray-on anesthetic for athletic injuries.

| $C_2H_4(g)$ | + | HCl(g) | | $C_2H_5Cl(l)$ |
| Ethylene | | Hydrogen chloride | | Ethyl chloride (an anesthetic) |

How much ethylene and how much hydrogen chloride should you use for your experiment? According to the coefficients of the balanced equation, you need a 1:1 numerical ratio of the two reactants. In the laboratory, you can't count the reactant molecules; you have to weigh them. That is, you must convert a *number* ratio of reactant molecules, as given by coefficients in the balanced equation, into a *mass* ratio to be sure that you are using the right amounts.

Mass ratios are determined by using the *molecular weights* of the substances involved in a reaction. Just as the **atomic weight** of an element is the average mass of the element's *atoms* (Section 2.9), the **molecular weight** of a substance is the average mass of the substance's *molecules*. Numerically, molecular weight (or, more generally, **formula weight** to include both ionic and molecular substances) equals the sum of the atomic weights of all atoms in the molecule.

Molecular weight   Sum of atomic weights of all atoms in a molecule.
Formula weight   Sum of atomic weights of all atoms in a formula unit of any compound, molecular or ionic.

As examples, the molecular weight of ethylene is 28.0, the molecular weight of hydrogen chloride is 36.5, and the molecular weight of ethyl chloride is 64.5. (These numbers are rounded off to one decimal place for convenience; the actual values are known more precisely.)

For ethylene, C_2H_4:	For hydrogen chloride, HCl:	For ethyl chloride, C_2H_5Cl:
Atomic weight of 2 C = (2)(12.0) = 24.0	Atomic weight of H $\quad$ = 1.0	Atomic weight of 2 C = (2)(12.0) = 24.0
Atomic weight of 4 H = (4)(1.0) $\;$ = $\;$ 4.0	Atomic weight of Cl = 35.5	Atomic weight of 5 H = (5)(1.0) $\;$ = $\;$ 5.0
Molecular weight of C_2H_4 $\quad$ = 28.0	Molecular weight of HCl = 36.5	Atomic weight of Cl $\quad$ = 35.5
		Molecular weight of C_2H_5Cl $\quad$ = 64.5

How do we use molecular weights? We saw in Section 2.9 that one **mole** of any element is the amount whose mass in grams, or *molar mass*, is numerically equal to the element's atomic weight. In the same way, one mole of any chemical compound is the amount whose mass in grams is numerically equal to the compound's molecular weight (or formula weight) and contains **Avogadro's number** of formula units (6.022×10^{23}). Thus, 1 mol of ethylene has a mass of 28.0 g, 1 mol of HCl has a mass of 36.5 g, and 1 mol of C_2H_5Cl has a mass of 64.5 g.

Mol. wt. of HCl = 36.5	Molar mass of HCl = 36.5 g/mol	1 mol of HCl = 6.022×10^{23} HCl molecules
Mol. wt. of C_2H_4 = 28.0	Molar mass of C_2H_4 = 28.0 g/mol	1 mol of C_2H_4 = 6.022×10^{23} C_2H_4 molecules
Mol. wt. of C_2H_5Cl = 64.5	Molar mass of C_2H_5Cl = 64.5 g/mol	1 mol of C_2H_5Cl = 6.022×10^{23} C_2H_5Cl molecules

WORKED EXAMPLE 3.3

Calculating a Molecular Weight

What is the molecular weight of glucose $(C_6H_{12}O_6)$, and what is its molar mass in grams per mole?

IDENTIFY

Known	Unknown
Chemical formula of glucose $(C_6H_{12}O_6)$	Molecular weight and molar mass of glucose

STRATEGY

The molecular weight of a substance is the sum of the atomic weights of the constituent atoms. List the elements present in the molecule, and look up the atomic weight of each (we'll round off to one decimal place for convenience):

$$C\,(12.0) \qquad H\,(1.0) \qquad O\,(16.0)$$

Then multiply the atomic weight of each element by the number of times that element appears in the chemical formula, and total the results.

SOLUTION

$$C_6\,(6 \times 12.0) = 72.0$$
$$H_{12}\,(12 \times 1.0) = 12.0$$
$$O_6\,(6 \times 16.0) = 96.0$$
$$\text{Mol. wt. of } C_6H_{12}O_6 = 180.0$$

Because one *molecule* of glucose has a mass of 180.0 u, 1 *mol* of glucose has a mass of 180.0 g. Thus, the molar mass of glucose is 180.0 g/mol.

▶ **PRACTICE 3.5** Calculate the molecular weight of sulfuric acid (H_2SO_4).

▶ **CONCEPTUAL APPLY 3.6** Use the structural formula of sucrose to determine its molecular weight and molar mass in grams per mole. (Gray = C, red = O, ivory = H.)

REMEMBER . . .
The **mole** is the fundamental SI unit for measuring the amount of matter. One mole of any substance—atom, ion, or molecule—is the amount whose mass in grams is numerically equal to the substance's atomic or formula weight. One mole contains **Avogadro's number** (6.022×10^{23}) of formula units (Section 2.9).

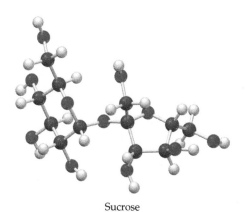

Sucrose

WORKED EXAMPLE 3.4

Interconverting Mass and Moles

How many moles of glucose, which is used to treat low blood sugar, are in a tablet containing 2.00 g? (The molar mass of glucose, $C_6H_{12}O_6$, was calculated in Worked Example 3.3.)

IDENTIFY

Known	Unknown
Mass of glucose (2.00 g)	Moles of glucose

STRATEGY

The known quantity (grams) can be converted to the unknown quantity (moles) using the molar mass of glucose as a conversion factor. Set up an equation so that the unwanted unit cancels.

SOLUTION

$$2.00 \text{ g glucose} \times \frac{1 \text{ mol glucose}}{180.0 \text{ g glucose}} = 0.0111 \text{ mol glucose}$$
$$= 1.11 \times 10^{-2} \text{ mol glucose}$$

CHECK

Because the molar mass of glucose is 180.0 g/mol, 1 mol of glucose has a mass of 180.0 g. Thus, 2.00 g of glucose is a bit more than one-hundredth of a mole, or 0.01 mol. The estimate agrees with the detailed solution.

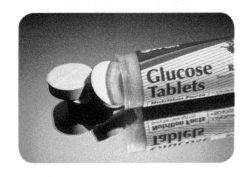

▶ **PRACTICE 3.7** How many moles are in 5.26 g of $NaHCO_3$, the main ingredient in Alka-Seltzer tablets?

▶ **APPLY 3.8** When a diabetic experiences low blood glucose, possibly due to an excess of insulin or increased levels of exercise, the treatment is consumption of glucose tablets.

(a) How many grams of glucose are in the recommended amount for treatment of an adult, 0.0833 mol glucose?
(b) A typical tablet contains 3.75 g of glucose. How many tablets should be eaten?
(c) How many molecules of glucose are in 0.0833 mol?

3.4 STOICHIOMETRY: RELATING AMOUNTS OF REACTANTS AND PRODUCTS

Stoichiometry (stoy-key-*ahm*-uh-tree; from the Greek *stoichion,* "element," and *metron,* "measure") refers to the chemical arithmetic needed to relate amounts of reactants and products in a chemical reaction. Questions such as "What mass of product can be made from a given amount of reactant?" or "What mass of one reactant is needed to completely react with the other reactant?" can be answered using stoichiometry.

In any balanced chemical equation, the coefficients tell the number of formula units, and thus the number of moles, of each substance in the reaction. You can then use molar masses as conversion factors to calculate reactant and product masses. If you saw the following balanced equation for the industrial synthesis of ammonia, for instance, you would know that 3 mol of $H_2(g)$ (3 mol $\times$ 2.0 g/mol = 6.0 g) is needed for reaction with 1 mol of $N_2(g)$ (28.0 g) to yield 2 mol of $NH_3(g)$ (2 mol $\times$ 17.0 g/mol = 34.0 g).

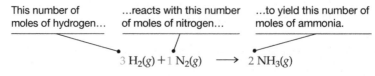

This number of moles of hydrogen...   ...reacts with this number of moles of nitrogen...   ...to yield this number of moles of ammonia.

$$3 \text{ H}_2(g) + 1 \text{ N}_2(g) \longrightarrow 2 \text{ NH}_3(g)$$

Let's look again at the reaction of ethylene with HCl (given at the start of Section 3.3) and assume that we have 15.0 g of ethylene and need to know how many grams of HCl to use in the reaction.

$$C_2H_4(g) + HCl(g) \longrightarrow C_2H_5Cl(l)$$

According to the coefficients in the balanced equation, 1 molecule of HCl reacts with 1 molecule of ethylene, so 1 mol of HCl is needed for reaction with each mole of ethylene. To find out how many grams of HCl are needed to react with 15.0 g of ethylene, we first have to find out how many moles of ethylene are in 15.0 g. We do this gram-to-mole conversion by calculating the molar mass of ethylene and using that value as a conversion factor:

Mol. wt. of $C_2H_4 = (2 \times 12.0) + (4 \times 1.0) = 28.0$

Molar mass of $C_2H_4 = 28.0$ g/mol

Moles of $C_2H_4 = 15.0 \text{ g ethylene} \times \dfrac{1 \text{ mol ethylene}}{28.0 \text{ g ethylene}} = 0.536$ mol ethylene

Now that we know how many moles of ethylene we have (0.536 mol), we also know from the balanced equation how many moles of HCl we need (0.536 mol), and we have to do a mole-to-gram conversion to find the mass of HCl required. Once again, the conversion is done by calculating the molar mass of HCl and using that value as a conversion factor:

Mol. wt. of HCl $= 1.0 + 35.5 = 36.5$

Molar mass of HCl $= 36.5$ g/mol

Grams of HCl $= 0.536 \text{ mol } C_2H_4 \times \dfrac{1 \text{ mol HCl}}{1 \text{ mol } C_2H_4} \times \dfrac{36.5 \text{ g HCl}}{1 \text{ mol HCl}} = 19.6$ g HCl

Thus, 19.6 g of HCl is needed to react with 15.0 g of ethylene.

Look carefully at the sequence of steps in the calculation just completed. *Moles* (numbers of molecules) are given by the coefficients in the balanced equation, but *grams* are used to weigh reactants in the laboratory. Moles tell us *how many molecules* of each reactant are needed, whereas grams tell us *how much mass* of each reactant is needed.

Moles $\longrightarrow$ Numbers of molecules or formula units

Grams $\longrightarrow$ Mass

The flow diagram in **FIGURE 3.1** illustrates the necessary conversions. Note again that you can't go directly from the number of grams of one reactant to the number of grams of another reactant. You *must* first convert to moles.

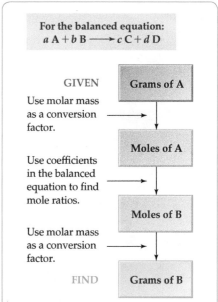

For the balanced equation:
$$a\,A + b\,B \longrightarrow c\,C + d\,D$$

GIVEN — **Grams of A**

Use molar mass as a conversion factor.

Moles of A

Use coefficients in the balanced equation to find mole ratios.

Moles of B

Use molar mass as a conversion factor.

FIND — **Grams of B**

▲ **FIGURE 3.1**

Stoichiometry: Conversions between moles and grams for a chemical reaction. The numbers of moles tell how many molecules of each reactant are needed, as given by the coefficients of the balanced equation; the numbers of grams tell what mass of each reactant is needed.

▲ **Figure It Out**

Why does a gram-to-mole conversion need to be done when relating the mass of one reactant to another?

Answer: Relative amounts of reactants and products for a reaction can be found from coefficients in the balanced equation. These coefficients represent moles or number of atoms/molecules, not mass.

Go to
eText

WORKED EXAMPLE 3.5

Relating the Masses of Reactants and Products

Aqueous solutions of sodium hypochlorite (NaOCl), best known as household bleach, are prepared by reaction of sodium hydroxide with chlorine. How many grams of NaOH are needed to react with 25.0 g of Cl_2?

$$2 \text{ NaOH}(aq) + Cl_2(g) \longrightarrow \text{NaOCl}(aq) + \text{NaCl}(aq) + H_2O(l)$$

IDENTIFY

Known	Unknown
Mass of Cl_2 (25.0 g)	Mass of NaOH (g)
Balanced reaction	

STRATEGY

The goal is to relate the known amount of one reactant (Cl_2) with the other reactant (NaOH). Finding the relationships between quantities of reactants requires working in moles and using the balanced equation to relate amounts. Molar masses are used to interconvert between moles and grams. Use the general strategy outlined in Figure 3.1.

continued on next page

▲ Household bleach is an aqueous solution of NaOCl, made by reaction of NaOH with Cl_2.

SOLUTION

Step 1. Convert grams of Cl_2 to moles of Cl_2. This gram-to-mole conversion is done in the usual way, using the molar mass of Cl_2 (70.9 g/mol) as the conversion factor:

$$25.0 \text{ g } Cl_2 \times \frac{1 \text{ mol } Cl_2}{70.9 \text{ g } Cl_2} = 0.353 \text{ mol } Cl_2$$

Step 2. Convert moles of Cl_2 to moles of NaOH. The coefficients in the balanced equation show that each mole of Cl_2 reacts with 2 mol of NaOH.

$$0.353 \text{ mol } Cl_2 \times \frac{2 \text{ mol NaOH}}{1 \text{ mol } Cl_2} = 0.706 \text{ mol NaOH}$$

Step 3. Convert moles of NaOH to grams of NaOH. Carry out a mole-to-gram conversion using the molar mass of NaOH (40.0 g/ mol) as a conversion factor to find that 28.2 g of NaOH is required for the reaction:

$$0.706 \text{ mol NaOH} \times \frac{40.0 \text{ g NaOH}}{1 \text{ mol NaOH}} = 28.2 \text{ g NaOH}$$

The problem can also be worked by combining the steps and setting up one large equation:

$$\text{Grams of NaOH} = 25.0 \text{ g } Cl_2 \times \frac{1 \text{ mol } Cl_2}{70.9 \text{ g } Cl_2} \times \frac{2 \text{ mol NaOH}}{1 \text{ mol } Cl_2} \times \frac{40.0 \text{ g NaOH}}{1 \text{ mol NaOH}}$$
$$= 28.2 \text{ g NaOH}$$

CHECK

The molar mass of NaOH is about half that of Cl_2, and 2 mol of NaOH is needed per 1 mol of Cl_2. Thus, the needed mass of NaOH will be similar to that of Cl_2, or about 25 g.

▸ **PRACTICE 3.9**   Aspirin is prepared by reaction of salicylic acid ($C_7H_6O_3$) with acetic anhydride ($C_4H_6O_3$) according to the following equation:

$$C_7H_6O_3(s) + C_4H_6O_3(l) \longrightarrow C_9H_8O_4(s) + CH_3CO_2H(l)$$
Salicylic   Acetic   Aspirin   Acetic acid
acid   anhydride

How many grams of acetic anhydride are needed to react with 4.50 g of salicylic acid?

▸ **APPLY 3.10**   Refer to the balanced reaction for the synthesis of aspirin in Problem 3.9.

(a) How many grams of salicylic acid are needed to make 10.0 g of aspirin?
(b) How many grams of acetic acid are formed as a by-product when 10.0 g of aspirin are synthesized?

All **WORKED EXAMPLES** with this icon [Go to eText] have an interactive video in the eText.

[Go to eText]

BIG IDEA Question 3

The following diagrams represent the reaction of A_2 (red spheres) with B_2 (blue spheres) to form AB_3. Which diagram represents a reaction carried out with a percent yield less than 100%?

(a)

(b)

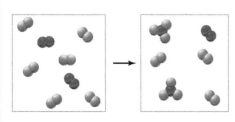

3.5 YIELDS OF CHEMICAL REACTIONS

In the stoichiometry examples worked out in the preceding section, we made the unstated assumption that all reactions "go to completion." That is, we assumed that all reactant molecules are converted to products. In fact, few reactions behave so nicely. More often, a large majority of molecules react as expected, but other processes, or *side reactions*, also occur. Thus, the amount of product actually formed, called the **yield** of the reaction, is usually less than the amount predicted by calculations.

The amount of product actually formed in a reaction divided by the amount theoretically possible and multiplied by 100% is the reaction's **percent yield.** For example, if a given reaction *could* provide 6.9 g of a product according to its stoichiometry but actually provides only 4.7 g, then its percent yield is $4.7/6.9 \times 100, = 68\%$.

$$\text{Percent yield} = \frac{\text{Actual yield of product}}{\text{Theoretical yield of product}} \times 100\%$$

Worked Example 3.6 shows how to calculate and use percent yield.

WORKED EXAMPLE 3.6

Calculating Percent Yield

Methyl *tert*-butyl ether (MTBE, $C_5H_{12}O$), a gasoline additive now being phased out in many places because of health concerns, can be made by reaction of isobutylene (C_4H_8) with methanol (CH_4O). What is the percent yield of the reaction if 32.8 g of MTBE is obtained from reaction of 26.3 g of isobutylene with sufficient methanol?

$$C_4H_8(g) + CH_4O(l) \longrightarrow C_5H_{12}O(l)$$

Isobutylene   Methyl *tert*-butyl ether (MTBE)

Methyl *tert*-butyl ether

IDENTIFY

Known	Unknown
Balanced reaction	Percent yield of reaction
Mass of isobutylene (26.3 g)	
Methanol (sufficient amount to react with isobutylene)	
Mass of MTBE (32.8 g) (actual yield of product)	

STRATEGY

To calculate the unknown quantity (percent yield), the actual yield and the theoretical yield must be known. Since the actual yield (32.8 g MTBE) is given in the problem, we need to calculate the theoretical yield, which is the amount of MTBE that could be produced from the complete reaction of 26.3 g of isobutylene.

Step 1. Calculate the molar masses of reactants and products to use as conversion factors.

Step 2. Find the theoretical amount of product, MTBE, using the coefficients from the balanced equation and molar masses as conversion factors.

Step 3. Use the equation for percent yield.

SOLUTION

Step 1. Calculation of Molar Masses

Isobutylene, C_4H_8:
Mol. wt. = $(4 \times 12.0) + (8 \times 1.0) = 56.0$
Molar mass of isobutylene = 56.0 g/mol

MTBE, $C_5H_{12}O$:
Mol. wt. = $(5 \times 12.0) + (12 \times 1.0) + 16.0 = 88.0$
Molar mass of MTBE = 88.0 g/mol

Step 2. Theoretical Amount of Product
To calculate the amount of MTBE that could theoretically be produced from 26.3 g of isobutylene, we first have to find the number of moles of reactant, using molar mass as the conversion factor:

$$26.3 \text{ g isobutylene} \times \frac{1 \text{ mol isobutylene}}{56.0 \text{ g isobutylene}} = 0.470 \text{ mol isobutylene}$$

According to the balanced equation, 1 mol of product is produced per mole of reactant, so we know that 0.470 mol of isobutylene can theoretically yield 0.470 mol of MTBE. Finding the mass of this MTBE requires a mole-to-mass conversion:

$$0.470 \text{ mol isobutylene} \times \frac{1 \text{ mol MTBE}}{1 \text{ mol isobutylene}} \times \frac{88.0 \text{ g MTBE}}{1 \text{ mol MTBE}}$$
$$= 41.4 \text{ g MTBE}$$

Step 3. Percent Yield Equation

Dividing the actual amount by the theoretical amount and multiplying by 100% gives the percent yield:

$$\frac{32.8 \text{ g MTBE}}{41.4 \text{ g MTBE}} \times 100\% = 79.2\%$$

CHECK

It is difficult to estimate a magnitude of the number for percent yield as the calculation involves multiple steps. However, the value for percent yield must be between 0% and 100%; therefore, the answer of 79.2% is in the correct range.

▶ **PRACTICE 3.11** Ethyl alcohol is prepared industrially by the reaction of ethylene, C_2H_4, with water. What is the percent yield of the reaction if 4.6 g of ethylene gives 4.7 g of ethyl alcohol?

$$C_2H_4(g) + H_2O(l) \longrightarrow C_2H_6O(l)$$

Ethylene   Ethyl alcohol

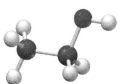

Ethyl alcohol

▶ **APPLY 3.12**

(a) Diethyl ether ($C_4H_{10}O$), the "ether" used medically as an anesthetic, is prepared commercially by treatment of ethyl alcohol (C_2H_6O) with an acid. How many grams of diethyl ether would you obtain from 40.0 g of ethyl alcohol if the percent yield of the reaction is 87.0%?

$$2 \; C_2H_6O(l) \xrightarrow{\text{Acid}} C_4H_{10}O(l) + H_2O(l)$$

Ethyl alcohol   Diethyl ether

(b) How many grams of ethyl alcohol would be needed to produce 100.0 g of diethyl ether if the percent yield of reaction is 87.0%?

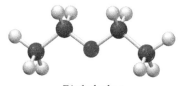

Diethyl ether

Go to
eText

BIG IDEA Question 4

A sandwich consists of one piece of cheese and two slices of bread. An equation for making a sandwich can be written as:

$$C + 2B \longrightarrow S$$

where C represents a piece of cheese, B represents a slice of bread, and S represents a sandwich.

What is the limiting reactant, and how many sandwiches can be made if there are 7 pieces of cheese and 10 slices of bread?

3.6 REACTIONS WITH LIMITING AMOUNTS OF REACTANTS

Because chemists usually write balanced equations, it's easy to get the impression that reactions are always carried out using exactly the right proportions of reactants. In fact, this is often not the case. Many reactions are carried out using an excess amount of one reactant—more than is actually needed according to stoichiometry. Look, for instance, at the industrial synthesis of ethylene glycol, $C_2H_6O_2$, a substance used both as automobile antifreeze and as a starting material for the preparation of polyester polymers. Approximately 18 million metric tons of ethylene glycol are prepared each year worldwide by reaction of ethylene oxide, C_2H_4O, with water at high temperature (1 metric ton = 1000 kg = 2205 lb).

$$
\begin{array}{ccccc}
C_2H_4O\ (g) & + & H_2O\ (l) & \xrightarrow{\text{Heat}} & C_2H_6O_2\ (l) \\
\text{Ethylene oxide} & & \text{Water} & & \text{Ethylene glycol}
\end{array}
$$

Because water is so cheap and so abundant, it doesn't make sense to worry about using exactly 1 mol of water for each mole of ethylene oxide. Rather, it's much easier to use an excess of water to be certain that enough is present to entirely consume the more valuable ethylene oxide reactant. Of course, when an excess of water is present, only the amount required by stoichiometry undergoes reaction. The excess water does not react and remains unchanged.

Whenever the ratios of reactant molecules used in an experiment are different from those given by the coefficients of the balanced equation, a surplus of one reactant is left over after the reaction is finished. Thus, the extent to which a chemical reaction takes place depends on the reactant that is present in limiting amount—the **limiting reactant**. The other reactant is said to be the *excess reactant*.

The situation with excess reactants and limiting reactants is analogous to what sometimes happens with people and chairs. If there are five people in a room but only three chairs, then only three people can sit while the other two stand because the number of people sitting is limited by the number of available chairs. The chairs are analogous to the limiting reactant, whereas the people are the excess reactant. Worked Example 3.7 shows how to visualize a limiting reactant problem.

CONCEPTUAL WORKED EXAMPLE 3.7

Identifying a Limiting Reactant from a Molecular Representation

Examine the balanced reaction for the production of ethylene glycol from ethylene oxide and water and the graphical molecular representation shown below.

$$
\begin{array}{ccccc}
C_2H_4O & + & H_2O & \xrightarrow{\text{Heat}} & C_2H_6O_2 \\
\text{Ethylene oxide} & & \text{Water} & & \text{Ethylene glycol}
\end{array}
$$

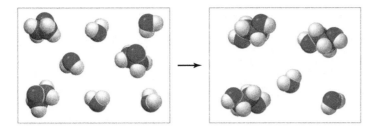

(a) Identify the limiting and the excess reactant.
(b) How many molecules of excess reactant are left over after the reaction occurs?
(c) How many molecules of product can be made?

STRATEGY

Count the numbers of reactant and product molecules and use coefficients from the balanced equation to relate them to one another.

SOLUTION

(a) Count the number of each type of molecule in the box on the reactant side of the equation. There are 3 ethylene oxide molecules and 5 water molecules. According to the balanced equation the stoichiometry between the reactants is 1:1. Therefore, 5 ethylene oxide molecules would be needed to react with 5 water molecules. Since there are only 3 ethylene oxide molecules, it is the limiting reactant, and water is in excess.
(b) Count the number of water molecules on the product side of the equation. There are 2 water molecules that have not reacted, and water is called the excess reactant.
(c) Count the number of ethylene glycol molecules on the product side of the equation. There are 3 ethylene glycol molecules present.

   Therefore, the reaction of 3 ethylene oxide molecules with 5 water molecules results in 3 ethylene glycol molecules with 2 water molecules left over.

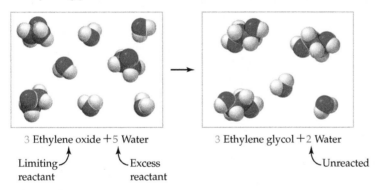

CONCEPTUAL PRACTICE 3.13   The following diagram represents the reaction of A (red spheres) with B_2 (blue spheres):

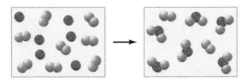

(a) Write a balanced equation for the reaction.
(b) Identify the limiting and excess reactant.
(c) How many molecules of product are made?

CONCEPTUAL APPLY 3.14   Draw a diagram similar to the one shown in Problem 3.13 for the following reaction, when 8 molecules of AB react with 6 molecules of B_2. Represent each atom as a sphere labeled with the symbol A or B. Specify the limiting and excess reactant.

$$2\ AB + B_2 \longrightarrow 2\ AB_2$$

FIGURE 3.2 is a flow chart that summarizes the steps needed for identifying a limiting reactant and calculating the theoretical mass of product formed when specified amounts of reactants are mixed. The process involves calculating the amount of product that can be made if each reactant is completely used up. The limiting reactant can be determined by comparing the amount of product formed from each reactant. The limiting reactant will form the *lowest* amount of product, which is the *theoretical yield*. Just as in previous stoichiometry examples, amounts of reactants and products are related from the balanced equation and number of moles. Worked Example 3.8 shows how to determine the limiting reactant and how to calculate the amount of product and excess reactant.

▶ **FIGURE 3.2**

Steps for identifying a limiting reactant and calculating theoretical yield.

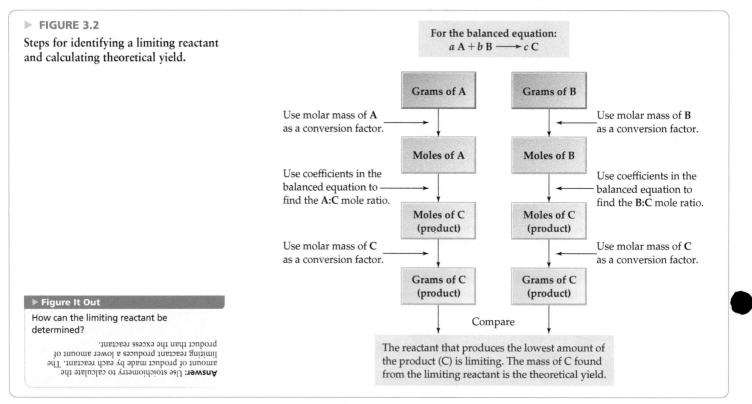

▶ **Figure It Out**

How can the limiting reactant be determined?

Answer: Use stoichiometry to calculate the amount of product made by each reactant. The limiting reactant produces a lower amount of product than the excess reactant.

Go to
eText

WORKED EXAMPLE 3.8

Calculating the Amount of Product or Excess Reactant When One Reactant Is Limiting

Cisplatin, an anticancer agent used for the treatment of solid tumors, is prepared by the reaction of ammonia with potassium tetrachloroplatinate. Assume that 10.0 g of K_2PtCl_4 and 10.0 g of NH_3 are allowed to react.

$$K_2PtCl_4(aq) \; + \; 2\,NH_3(aq) \; \longrightarrow \; Pt(NH_3)_2Cl_2(s) \; + \; 2\,KCl(aq)$$
Potassium tetrachloroplatinate   Cisplatin

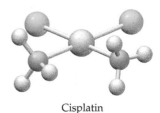

Cisplatin

(a) Which reactant is limiting, and which is in excess? How many grams of cisplatin are formed?
(b) How many grams of the excess reactant are consumed, and how many grams remain?

IDENTIFY

Known	Unknown
Balanced reaction	Limiting reactant
Mass of NH_3 (10.0 g)	Mass of $Pt(NH_3)_2Cl_2$ (g)
Mass of K_2PtCl_4 (10.0 g)	Mass of excess reactant consumed (g)
	Mass of excess reactant left over (g)

STRATEGY

(a) Because the amount of each reactant is known, the procedure outlined in Figure 3.2 can be used to determine which reactant is limiting. Use stoichiometry to calculate the amount of product that can be made from the amount of each reactant given. The reactant that results in the least amount of product is the limiting reactant and determines the maximum amount of product that can be made.

(b) Stoichiometry can also be used to relate the amount of limiting reactant to the amount of excess reactant.

SOLUTION

(a) **Step 1.** Calculation of Molar Masses
Finding the molar amounts of reactants always begins by calculating formula weight and using molar masses as conversion factors:

Form. weight of K_2PtCl_4 = $(2 \times 39.1) + 195.1$
$+ (4 \times 35.5) = 415.3$
Molar mass of K_2PtCl_4 = 415.3 g/mol
Mol. wt. of NH_3 = $14.0 + (3 \times 1.0) = 17.0$
Molar mass of NH_3 = 17.0 g/mol
Mol. wt. of $Pt(NH_3)_2Cl_2$ = $195.1 + (2 \times 17.0)$
$+ (2 \times 35.5) = 300.1$
Molar mass of $Pt(NH_3)_2Cl_2$ = 300.1 g/mol

Step 2. Amount of Product Made from Each Reactant
Taking the given amount of *each* reactant, use stoichiometry to find the amount of product that can be made if the entire amount reacted.

$$10.0 \text{ g } K_2PtCl_4 \times \frac{1 \text{ mol } K_2PtCl_4}{415.3 \text{ g } K_2PtCl_4} \times \frac{1 \text{ mol } Pt(NH_3)_2Cl_2}{1 \text{ mol } K_2PtCl_4}$$

$$\times \frac{300.1 \text{ g } Pt(NH_3)_2Cl_2}{1 \text{ mol } Pt(NH_3)_2Cl_2} = 7.23 \text{ g } Pt(NH_3)_2Cl_2$$

$$10.0 \text{ g } NH_3 \times \frac{1 \text{ mol } NH_3}{17.0 \text{ g } NH_3} \times \frac{1 \text{ mol } Pt(NH_3)_2Cl_2}{2 \text{ mol } NH_3}$$

$$\times \frac{300.1 \text{ g } Pt(NH_3)_2Cl_2}{1 \text{ mol } Pt(NH_3)_2Cl_2} = 88.3 \text{ g } Pt(NH_3)_2Cl_2$$

These calculations tell us that K_2PtCl_4 is the limiting reactant because it produces the fewest grams of product. The excess reactant is NH_3 because its consumption would produce a larger amount of product. Only 7.23 g of $Pt(NH_3)_2Cl_2$ can be produced, given the initial amount of K_2PtCl_4.

(b) With the identities of the excess reactant and limiting reactant known, we can use stoichiometry to find out how much NH_3 reacts and how much is left over. We know that all the limiting reactant (K_2PtCl_4) is used up; stoichiometry allows us to calculate the amount of NH_3 that reacts as follows:

$$10.0 \text{ g } K_2PtCl_4 \times \frac{1 \text{ mol } K_2PtCl_4}{415.3 \text{ g } K_2PtCl_4} \times \frac{2 \text{ mol } NH_3}{1 \text{ mol } K_2PtCl_4}$$

$$\times \frac{17.0 \text{ g } NH_3}{1 \text{ mol } NH_3} = 0.819 \text{ g } NH_3$$

Grams of unreacted NH_3 = $(10.0 \text{ g} - 0.819 \text{ g}) = 9.2 \text{ g } NH_3$

CHECK

It is reasonable that K_2PtCl_4 (415.3 g/mol) is the limiting reactant because it has a much higher molar mass than NH_3 (17.0 g/mol) and the reaction started with an equal number of grams of each. The amount of product, $Pt(NH_3)_2Cl_2$, can be estimated from reaction stoichiometry with the limiting reactant and relative molar masses. Because 1 mol of K_2PtCl_4 will produce 1 mol of $Pt(NH_3)_2Cl_2$ and the molar mass of $Pt(NH_3)_2Cl_2$ (300.1 g/mol) is about $\frac{3}{4}$ of the molar mass of K_2PtCl_4 (415.3 g/mol), then the amount produced (7.23 g) should be about $\frac{3}{4}$ the mass of the initial amount (10.0 g).

▶ **PRACTICE 3.15** Lithium oxide is used aboard the space shuttle to remove water from the air supply.

(a) If 80.0 kg of water is to be removed and 65 kg of Li_2O is available, which reactant is limiting?
(b) How many kilograms of the excess reactant remain?

$$Li_2O(s) + H_2O(g) \longrightarrow 2 \text{ LiOH}(s)$$

▶ **APPLY 3.16** After lithium hydroxide is produced aboard the space shuttle by reaction of Li_2O with H_2O (Problem 3.15), it is used to remove exhaled carbon dioxide from the air supply. Initially 400.0 g of LiOH were present and 500.0 g of $LiHCO_3$ have been produced. Can the reaction remove any additional CO_2 from the air? If so, how much?

$$\text{LiOH}(s) + CO_2(g) \longrightarrow LiHCO_3(s)$$

3.7 PERCENT COMPOSITION AND EMPIRICAL FORMULAS

All the substances we've dealt with thus far have had known formulas. When a new compound is made in the laboratory or found in nature, however, its formula must be experimentally determined.

Determining the formula of a new compound begins with analyzing the substance to discover what elements it contains and how much of each element is present—that is, to find its *composition*. The **percent composition** of a compound is expressed by identifying the elements present and giving the mass percent of each. For example, we express the percent composition of a certain colorless liquid found in gasoline by saying that it contains 84.1% carbon and 15.9% hydrogen by mass. In other words, a 100.0 g sample of the compound contains 84.1 g of carbon atoms and 15.9 g of hydrogen atoms.

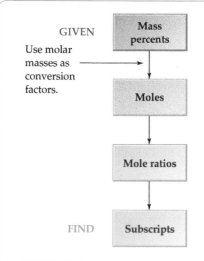

GIVEN

Use molar masses as conversion factors.

Mass percents

↓

Moles

↓

Mole ratios

↓

FIND   Subscripts

▲ **FIGURE 3.3**

Calculating the formula of a compound from its percent composition.

▲ **Figure It Out**

A mole ratio is the number of moles of one element divided by the number of moles of another element. How do mole ratios relate to subscripts in a chemical formula?

Answer: The subscripts in a chemical formula represent moles. For example, in 1 mole of C_2H_4, there are two moles of C atoms and four moles of hydrogen atoms. The mole ratio is the same as the ratio of subscripts.

Knowing a compound's percent composition makes it possible to calculate the compound's chemical formula. As shown in **FIGURE 3.3,** the strategy is to find the relative number of moles of each element in the compound and then use those numbers to establish the mole ratios of the elements. The mole ratios, in turn, correspond to the subscripts in the chemical formula.

Let's use for our example the colorless liquid whose composition is 84.1% carbon and 15.9% hydrogen by mass. Arbitrarily taking 100 g of the substance to make the calculation easier, we find by using molar masses as conversion factors that the 100 g contains:

$$84.1 \text{ g C} \times \frac{1 \text{ mol C}}{12.01 \text{ g C}} = 7.00 \text{ mol C}$$

$$15.9 \text{ g H} \times \frac{1 \text{ mol H}}{1.008 \text{ g H}} = 15.8 \text{ mol H}$$

With the relative numbers of moles of C and H known, we find the mole ratio by dividing both by the smaller number (7.00):

$$C_{\left(\frac{7.00}{7.00}\right)} H_{\left(\frac{15.8}{7.00}\right)} = C_1 H_{2.26}$$

The C:H mole ratio of 1:2.26 means that we can write $C_1H_{2.26}$ as a temporary formula for the liquid. Multiplying the subscripts by small integers in a trial-and-error procedure until whole numbers are found then gives the **empirical formula,** which gives the smallest whole-number ratios of atoms in the compound. In the present instance, we need to multiply the subscripts by 4 to obtain the empirical formula C_4H_9. (The subscripts may not always be *exact* integers because of small errors in the data, but the discrepancies should be small.)

$$C_{(1 \times 4)} H_{(2.26 \times 4)} = C_4 H_{9.04} = C_4 H_9$$

An empirical formula determined from percent composition tells only the ratios of atoms in a compound. The **molecular formula,** which tells the *actual numbers* of atoms in a molecule, can be either the same as the empirical formula or a multiple of it. To determine the molecular formula, it's necessary to know the molecular weight of the substance. In the present instance, the molecular weight of our compound (octane) is 114.2, which is a simple multiple of the empirical molecular weight for C_4H_9 (57.1).

To find the multiple, divide the molecular weight by the empirical formula weight:

$$\text{Multiple} = \frac{\text{Molecular weight}}{\text{Empirical formula weight}} = \frac{114.2}{57.1} = 2.00$$

Then multiply the subscripts in the empirical formula by this multiple to obtain the molecular formula. In our example, the molecular formula of octane is $C_{(4 \times 2)}H_{(9 \times 2)}$, or C_8H_{18}.

Just as we can find the empirical formula of a substance from its percent composition, we can also find the percent composition of a substance from its empirical (or molecular) formula. The strategies for the two kinds of calculations are exactly opposite. Aspirin, for example, has the molecular formula $C_9H_8O_4$ and thus has a C:H:O mole ratio of 9:8:4. We can convert this mole ratio into a mass ratio, and thus into percent composition, by carrying out mole-to-gram conversions.

Let's assume we start with 1 mol of compound to simplify the calculation:

$$1 \text{ mol aspirin} \times \frac{9 \text{ mol C}}{1 \text{ mol aspirin}} \times \frac{12.0 \text{ g C}}{1 \text{ mol C}} = 108 \text{ g C}$$

$$1 \text{ mol aspirin} \times \frac{8 \text{ mol H}}{1 \text{ mol aspirin}} \times \frac{1.01 \text{ g H}}{1 \text{ mol H}} = 8.08 \text{ g H}$$

$$1 \text{ mol aspirin} \times \frac{4 \text{ mol O}}{1 \text{ mol aspirin}} \times \frac{16.0 \text{ g O}}{1 \text{ mol O}} = 64.0 \text{ g O}$$

Dividing the mass of each element by the total mass and multiplying by 100% then gives the percent composition:

$$\text{Total mass of 1 mol aspirin} = 108 \text{ g} + 8.08 \text{ g} + 64.0 \text{ g} = 180 \text{ g}$$

$$\%C = \frac{108 \text{ g C}}{180 \text{ g}} \times 100\% = 60.0\%$$

$$\%H = \frac{8.08 \text{ g H}}{180 \text{ g}} \times 100\% = 4.49\%$$

$$\%O = \frac{64.0 \text{ g O}}{180 \text{ g}} \times 100\% = 35.6\%$$

The answer can be checked by confirming that the sum of the mass percentages is within a rounding error of 100%: 60.0% + 4.49% + 35.6, = 100.1%.

Worked Example 3.10 further illustrates conversions between percent composition and empirical formulas.

WORKED EXAMPLE 3.9

Calculating Empirical Formulas, Molecular Formulas, and Percent Composition

(a) Vitamin C (ascorbic acid) contains 40.92% C, 4.58% H, and 54.50% O by mass. What is the empirical formula of ascorbic acid?

(b) Use the structure of ascorbic acid to determine the molecular formula for ascorbic acid. (Gray = C, red = O, ivory = H.) By what integer should the empirical formula be multiplied to convert it to the molecular formula?

(a) IDENTIFY

Known	Unknown
Percent composition of ascorbic acid	Empirical formula of ascorbic acid
(40.92% C, 4.58% H, and 54.50% O by mass)	

Ascorbic acid

STRATEGY

Figure 3.3 outlines the procedure for converting from percent composition to empirical formula. Assume 100.0 g of ascorbic acid, convert grams of each element to moles, and divide by the smallest number of moles to obtain mole ratios for subscripts.

SOLUTION

(a) **Step 1.** Per Figure 3.3, our first step is to convert the mass of each element in the sample to moles.

$$40.92 \text{ g C} \times \frac{1 \text{ mol C}}{12.0 \text{ g C}} = 3.41 \text{ mol C}$$

$$4.58 \text{ g H} \times \frac{1 \text{ mol H}}{1.01 \text{ g H}} = 4.53 \text{ mol H}$$

$$54.50 \text{ g O} \times \frac{1 \text{ mol O}}{16.0 \text{ g O}} = 3.41 \text{ mol H}$$

Step 2. Find mole ratios. Dividing each of the three numbers by the smallest (3.41 mol) gives a C:H:O mole ratio of 1:1.33:1 and a temporary formula of $C_1H_{1.33}O_1$.

Step 3. Determine what whole-number subscripts belong in the empirical formula. Multiplying the subscripts by small integers, such as 1, 2, 3, or 4, in a trial-and-error procedure until whole numbers are found gives the empirical formula: $C_{(3 \times 1)}H_{(3 \times 1.33)}O_{(3 \times 1)} = C_3H_4O_3$.

(b) IDENTIFY

Known	Unknown
Empirical formula from part (a)	Molecular formula
Ball-and-stick model for ascorbic acid	

continued on next page

STRATEGY

Find the molecular formula by counting atoms in the structural formula provided. Determine the multiplication factor needed to turn subscripts in the empirical formula into the molecular formula.

SOLUTION

Counting the atoms in the structural formula for ascorbic acid gives a molecular formula of $C_6H_8O_6$. If the subscripts in the empirical formula weight ($C_3H_4O_3$) are multiplied by 2, then the molecular formula is obtained.

CHECK

The empirical formula can be multiplied by a whole number to obtain the molecular formula. This is strong evidence that the empirical formula has been determined correctly.

▶ **PRACTICE 3.17** What is the empirical formula of the ingredient in Bufferin tablets that has the percent composition 14.25% C, 56.93% O, and 28.83% Mg by mass?

▶ **CONCEPTUAL APPLY 3.18** Use the structural formula for glucose to determine the molecular formula. What is the empirical formula, and what is the percent composition of each atom in glucose?

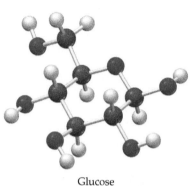

Glucose

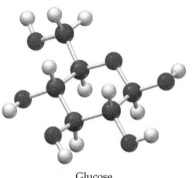

GIVEN

Burn sample in excess oxygen

Weigh sample

↓

Weigh CO_2 and H_2O

↓

Calculate moles of C and H

↓

Calculate C:H mole ratio

↓

FIND

Calculate empirical formula

▲ **FIGURE 3.4**

Determining an empirical formula from combustion analysis of a compound containing C and H.

> ▲ **Figure It Out**
>
> Generally explain how the mass of CO_2 and H_2O produced during combustion analysis is useful in determining the empirical formula of the hydrocarbon.
>
> **Answer:** All of the CO_2 comes from carbon in the compound. Similarly, all of the H_2O comes from hydrogen in the original compound. Therefore, the mass of these compounds can be used to find the number of moles of C and H in the original compound.

3.8 DETERMINING EMPIRICAL FORMULAS: ELEMENTAL ANALYSIS

One of the most common methods used to determine percent composition and empirical formulas, particularly for organic compounds containing carbon and hydrogen, is *combustion analysis*. In this method, a compound of unknown composition is burned with oxygen to produce the volatile combustion products CO_2 and H_2O, which are separated and have their amounts determined by an automated instrument. Methane (CH_4), for instance, burns according to the balanced equation:

$$CH_4(g) + 2\ O_2(g) \longrightarrow CO_2(g) + 2\ H_2O(g)$$

With the amounts of the carbon-containing product (CO_2) and hydrogen-containing product (H_2O) established, the strategy is to calculate the number of moles of carbon and hydrogen in the products, from which we can find the C:H mole ratio of the starting compound. This information, in turn, provides the chemical formula, as outlined by the flow diagram in **FIGURE 3.4**.

As an example of how combustion analysis works, imagine that we have a sample of a pure substance— naphthalene, which is often used for household moth balls. We weigh a known amount of the sample, burn it in pure oxygen, and then analyze the products. A sample of 0.330 g of naphthalene reacts with O_2 and 1.133 g of CO_2 and 0.185 g of H_2O are formed. The first thing we need to find is the number of moles of carbon and hydrogen in the CO_2 and H_2O products so that we can calculate the number of moles of each element originally present in the naphthalene sample.

$$\text{Moles of C in 1.133 g } CO_2 = 1.133 \text{ g } CO_2 \times \frac{1 \text{ mol } CO_2}{44.01 \text{ g } CO_2} \times \frac{1 \text{ mol C}}{1 \text{ mol } CO_2}$$

$$= 0.02574 \text{ mol C}$$

$$\text{Moles of H in 0.185 g } H_2O = 0.185 \text{ g } H_2O \times \frac{1 \text{ mol } H_2O}{18.02 \text{ g } H_2O} \times \frac{2 \text{ mol H}}{1 \text{ mol } H_2O}$$

$$= 0.0205 \text{ mol H}$$

Although it's not necessary in this instance because naphthalene contains only carbon and hydrogen, we can make sure that all the mass is accounted for and that no other elements are present. To do so, we carry out mole-to-gram conversions to find the number of grams of C and H in the starting sample:

$$\text{Mass of C} = 0.02574 \ \cancel{\text{mol C}} \times \frac{12.01 \text{ g C}}{1 \cancel{\text{mol C}}} = 0.3091 \text{ g C}$$

$$\text{Mass of H} = 0.0205 \ \cancel{\text{mol H}} \times \frac{1.01 \text{ g H}}{1 \cancel{\text{mol H}}} = 0.0207 \text{ g H}$$

$$\text{Total mass of C and H} = 0.3091 \text{ g} + 0.0207 \text{ g} = 0.3298 \text{ g}$$

Because the total mass of the C and H in the products (0.3298 g) is the same as the mass of the starting sample (0.330 g), we know that no other elements are present in naphthalene.

With the relative number of moles of C and H in naphthalene known, divide the larger number of moles by the smaller number to get the formula $C_{1.26}H_1$:

$$C_{\left(\frac{0.02574}{0.0205}\right)}H_{\left(\frac{0.0205}{0.0205}\right)} = C_{1.26}H_1$$

Then multiply the subscripts by small integers in a trial-and-error procedure until whole numbers are found to obtain the whole-number formula C_5H_4:

Multiply subscripts by 2: $C_{(1.26 \times 2)}H_{(1 \times 2)} = C_{2.52}H_2$

Multiply subscripts by 3: $C_{(1.26 \times 3)}H_{(1 \times 3)} = C_{3.78}H_3$

Multiply subscripts by 4: $C_{(1.26 \times 4)}H_{(1 \times 4)} = C_{5.04}H_4 = C_5H_4$ (Both subscripts are integers.)

Elemental analysis provides only an empirical formula. To determine the molecular formula, it's also necessary to know the substance's molecular weight. The next section will show mass spectrometry data for determining the molecular weight of compounds. In the present problem, the molecular weight of naphthalene is 128.2, or twice the empirical formula weight of C_5H_4 (64.1). Thus, the molecular formula of naphthalene is $C_{(2 \times 5)}H_{(2 \times 4)} = C_{10}H_8$.

Worked Example 3.10 shows an example of combustion analysis when the sample contains oxygen in addition to carbon and hydrogen.

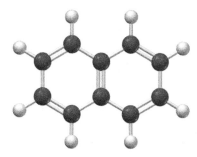

Naphthalene

WORKED EXAMPLE 3.10

Calculating an Empirical Formula and a Molecular Formula from a Combustion Analysis

Caproic acid, the substance responsible for the aroma of goats, dirty socks, and old shoes, contains carbon, hydrogen, and oxygen. On combustion analysis, a 0.450 g sample of caproic acid gives 0.418 g of H_2O and 1.023 g of CO_2. What is the empirical formula of caproic acid? If the molecular weight of caproic acid is 116.2, what is the molecular formula?

IDENTIFY

Known	Unknown
Mass of caproic acid (0.450 g)	Empirical formula
Mass of H_2O (0.418 g) and CO_2 (1.023 g)	Molecular formula
Mol. wt. of caproic acid (116.2)	

STRATEGY

Use the procedure outlined in Figure 3.4 to turn combustion analysis data into an empirical formula. This molecule also contains oxygen, and because oxygen yields no combustion products, its presence in a molecule can't be directly detected by

continued on next page

Caproic acid

combustion analysis. Rather, the presence of oxygen must be inferred by subtracting the calculated masses of C and H from the total mass of the sample. (See Steps 2 and 3 illustrating how the amount of oxygen can be determined.)

SOLUTION

Step 1. Find the molar amounts of C and H in the sample:

$$\text{Moles of C} = 1.023 \text{ g CO}_2 \times \frac{1 \text{ mol CO}_2}{44.01 \text{ g CO}_2} \times \frac{1 \text{ mol C}}{1 \text{ mol CO}_2} = 0.023\ 24 \text{ mol C}$$

$$\text{Moles of H} = 0.418 \text{ g H}_2\text{O} \times \frac{1 \text{ mol H}_2\text{O}}{18.02 \text{ g H}_2\text{O}} \times \frac{2 \text{ mol H}}{1 \text{ mol H}_2\text{O}} = 0.0464 \text{ mol H}$$

Step 2. Find the number of grams of each element in the sample:
Start with C and H as these can be determined from the number of moles found in Step 1.

$$\text{Mass of C} = 0.023\ 24 \text{ mol C} \times \frac{12.01 \text{ g C}}{1 \text{ mol C}} = 0.2791 \text{ g C}$$

$$\text{Mass of H} = 0.0464 \text{ mol H} \times \frac{1.01 \text{ g H}}{1 \text{ mol H}} = 0.0469 \text{ g H}$$

Subtracting the masses of C and H from the mass of the starting sample indicates that 0.124 g is unaccounted for:

$$0.450 \text{ g} - (0.2791 \text{ g} + 0.0469 \text{ g}) = 0.124 \text{ g}$$

Step 3. Find the moles of oxygen:
Because we are told that oxygen is also present in the sample, the "missing" mass must be due to oxygen, which can't be detected by combustion. We therefore need to find the number of moles of oxygen in the sample:

$$\text{Moles of O} = 0.124 \text{ g O} \times \frac{1 \text{ mol O}}{16.00 \text{ g O}} = 0.007\ 75 \text{ mol O}$$

Step 4. Find the mole ratios of the elements:
Knowing the relative numbers of moles of all three elements, C, H, and O, we divide the three numbers of moles by the smallest number (0.007 75 mol of oxygen) to arrive at a C:H:O ratio of 3:6:1.

$$C_{\left(\frac{0.02324}{0.00775}\right)} H_{\left(\frac{0.0464}{0.00775}\right)} O_{\left(\frac{0.00775}{0.00775}\right)} = C_3H_6O$$

Step 5. Find the molecular formula:
The empirical formula of caproic acid is, therefore, C_3H_6O, and the empirical formula weight is 58.1. Because the molecular weight of caproic acid is 116.2, or twice the empirical formula weight, the molecular formula of caproic acid must be $C_{(2 \times 3)}H_{(2 \times 6)}O_{(2 \times 1)} = C_6H_{12}O_2$.

CHECK

If a simple empirical formula is obtained and the molecular weight is a whole-number multiple of the empirical formula weight, then the formulas have most likely been correctly determined.

Menthol

▶ **PRACTICE 3.19**   Menthol, a flavoring agent obtained from peppermint oil, contains carbon, hydrogen, and oxygen. On combustion analysis, 1.00 g of menthol yields 1.161 g of H_2O and 2.818 g of CO_2. What is the empirical formula of menthol? Check your answer with the structural formula provided.

▶ **APPLY 3.20**   Combustion analysis is performed on 0.50 g of a hydrocarbon, and 1.55 g of CO_2 and 0.697 g of H_2O are produced. What is the empirical formula of the hydrocarbon? If the molar mass is 142.0 g/mol, what is the molecular formula?

3.9 DETERMINING MOLECULAR WEIGHTS: MASS SPECTROMETRY

As we saw in the previous section, determining a compound's molecular formula requires knowledge of its molecular weight. But how is molecular weight determined?

The most common method of determining both atomic and molecular weights is with an instrument called a *mass spectrometer*. Molecules must be vaporized and ionized for analysis in a mass spectrometer. Ionization occurs by bombarding molecules with a beam of high-energy electrons that knock an electron out of each molecule, giving it a +1 charge. A reaction for ionization of a molecule (M) is shown below:

$$M(g) + e^-_{\text{high energy}} \longrightarrow M^+(g) + 2\,e^-$$

Ionization is necessary as electric and magnetic fields will only exert a force on a charged species, not a neutral molecule. Some of these ionized molecules survive, and others fragment into smaller ions. Collisions with the electron beam have sufficient energy to not only ionize the molecule but also break bonds. Fragment ions from the molecule with various mass-to-charge ratios are separated, quantified, and displayed in a mass spectrum.

Although a typical mass spectrum contains ions of many different mass-to-charge ratios, the heaviest ion is generally due to the ionized molecule itself, the molecular ion (M^+). By measuring the mass of this molecular ion, the molecular weight of the molecule can be determined. Naphthalene, for example, gives rise to an intense peak at a mass-to-charge ratio of 128 in its spectrum, consistent with a molecular formula of $C_{10}H_8$ (**FIGURE 3.5**). There is a small peak at mass/charge 129 in the spectrum that arises from the presence of the carbon-13 isotope in a naturally occurring sample of naphthalene, $^{13}C_{10}{}^1H_8$. The intensity is lower because the abundance of carbon-13 is low, approximately 1%. Most naturally occurring carbon is the isotope carbon-12. Modern mass spectrometers are so precise that molecular weights can often be measured to seven significant figures. A $^{12}C_{10}{}^1H_8$ molecule of naphthalene has a molecular weight of 128.0626 as measured by mass spectrometry. High mass accuracy is often needed to make an identification of a compound. For example, two compounds with different molecular formulas can have very similar masses, $C_5H_8O = 84.0570$ and $C_6H_{12} = 84.0934$. Highly accurate mass measurements are frequently used to confirm the identity of molecules synthesized in laboratories.

> **REMEMBER . . .**
> A mass spectrometer is an instrument that uses a magnetic field to separate ions of different mass-to-charge ratios (Section 2.10, Figure 2.10).

> **REMEMBER . . .**
> Atoms with identical atomic numbers but with different mass numbers are called isotopes. Carbon has several isotopes, of which only ^{12}C and ^{13}C are stable (Section 2.8).

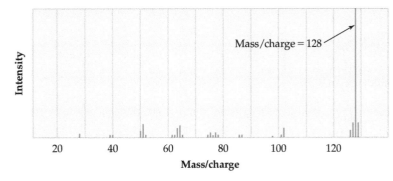

A mass spectrum for naphthalene

Mass/charge = 128

Intensity

20   40   60   80   100   120

Mass/charge

◄ **Figure It Out**

Peaks for ions with mass/charge ratios of 102, 128, and 129 are present in the mass spectrum. Match each peak with its correct description.
(a) A fragment ion with fewer atoms than the naphthalene molecule
(b) A naphthalene ion with a +1 charge that contains one carbon-13 atom
(c) A naphthalene ion with a +1 charge (also called the molecular ion)

Answer:
(a) mass/charge 102
(b) mass/charge 129
(c) mass/charge 128

▲ **FIGURE 3.5**

A mass spectrum of naphthalene, molecular weight = 128, showing peaks of different mass to charge ratios on the horizontal axis.

WORKED EXAMPLE 3.11

Determination of Molecular Formula from Combustion Analysis and a Mass Spectrum

A compound has an empirical formula of CH as determined from combustion analysis. The mass spectrum for the compound is shown in **FIGURE 3.6.** What is the molecular formula?

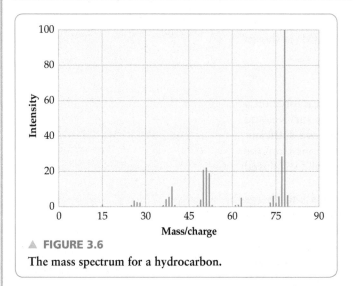

▲ **FIGURE 3.6**

The mass spectrum for a hydrocarbon.

IDENTIFY

Known	Unknown
Empirical formula (CH)	Molecular formula
Mass spectrum (used to find molecular weight)	

STRATEGY

Step 1. The molecular weight can be found by interpreting the mass spectrum.

Step 2. Compute the empirical formula weight, and use the equation to find the integer multiple needed to convert the empirical formula into the molecular formula.

$$\text{Multiple} = \frac{\text{Molecular weight}}{\text{Empirical formula weight}}$$

SOLUTION

Step 1. The molecular weight is 78 as determined from the most intense peak with the highest mass/charge ratio. The peak at 79 has the greatest mass/charge ratio but a very low intensity. This peak is most likely due to the abundance of carbon-13 isotope in the natural sample.

Step 2. The empirical formula weight of CH is 13. The whole-number multiple for converting the empirical formula into the molecular formula can be found by substituting values into the equation.

$$\text{Multiple} = \frac{\text{Molecular weight}}{\text{Empirical formula weight}} = \frac{78}{13} = 6$$

The molecular formula can be found by multiplying subscripts of the empirical formula by 6:

$$C_{(1\times6)}H_{(1\times6)} = C_6H_6$$

▶ **PRACTICE 3.21**   A compound has an empirical formula of C_6H_5 as determined from combustion analysis. The mass spectrum for the compound is shown in **FIGURE 3.7.** What is the molecular formula?

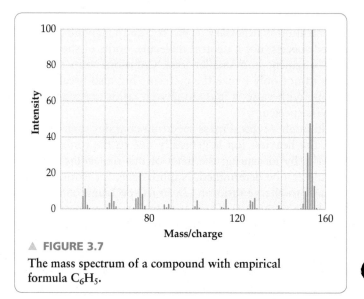

▲ **FIGURE 3.7**

The mass spectrum of a compound with empirical formula C_6H_5.

▶ **APPLY 3.22**   Combustion analysis was performed on 1.00 g of a compound containing C, H, and N, and 2.79 g of CO_2 and 0.57 g of H_2O were produced. Given the mass spectrum for the compound in **FIGURE 3.8,** what is the empirical formula? What is the molecular formula?

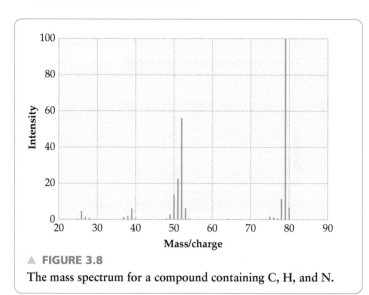

▲ **FIGURE 3.8**

The mass spectrum for a compound containing C, H, and N.

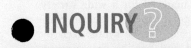

INQUIRY ❓ How is the principle of atom economy used to minimize waste in a chemical synthesis?

Chemical synthesis, combining atoms of different elements to make new compounds, is central to the global economy and a source of many products that enhance our lives. Dyes, fertilizers, plastics, synthetic fabrics, medicines, and electronic components are familiar examples of substances produced by chemical reactions. In the past, rapid and economic production methods have taken precedence over environmental considerations. Many chemical processes use large amounts of energy, non-renewable, petroleum-based feedstocks and hazardous materials that pollute the environment. However, as dangers of commonly used chemicals have been discovered, scientists have begun to change their approach to chemical synthesis.

Green chemistry is the design of chemical products and processes that reduce or eliminate hazardous substances. It is different than remediation in that it aims to eliminate pollution by *preventing* it from happening in the first place. Green chemistry focuses on developing reactions that minimize energy, use benign or renewable starting materials, and generate waste materials that can be reused, recycled, or biodegraded. Adoption of green chemistry technologies provides economic benefits, improved safety, and the promise of a sustainable future.

Chemists use green chemistry principles to design processes at the atomic level to prevent the formation of pollutants and waste. If a large proportion of atoms in a reaction ends up in waste products, production costs can be high, and resources are used ineffectively. One way to evaluate the efficiency of a reaction is percent yield (Section 3.4), which measures the extent to which reactants undergo a reaction to form products. While percent yield describes the completeness of a reaction or the presence of side reactions, it does not take into account the fraction of reactant atoms that end up in the desired product. Let's examine a reaction for the synthesis of aspirin from salicylic acid and acetic anhydride. The aspirin molecule is the desired product, and the other product, acetic anhydride, consists of atoms that are "wasted" in the reaction. Even if the synthesis occurs with 100% yield, some atoms will be left over and not incorporated into the desired pharmaceutical product, aspirin.

Atom economy is a concept conceived by Stanford chemistry professor Barry Trost, which states that it is *best to have all or most starting atoms end up in the desired product rather than in waste by-products.* We can think of it as the efficiency of the reaction in terms of number of atoms and can calculate it as follows:

$$\text{Percent atom economy} = \frac{\text{Molecular weight (desired product)}}{\Sigma \text{Molecular weight (reactants)}} \times 100\%$$

where Σ (sigma) means "sum." The numerator is the molecular weight of the desired product, and the denominator is the sum of the molecular weights of the reactants.

Let's calculate the percent **atom economy** in the reaction between salicylic acid and acetic anhydride to make aspirin. First, we must calculate the molecular weights of each reactant (salicylic acid and acetic anhydride) and the desired product (aspirin). Examine the structural formula of each substance, and count the number of each type of atom to determine the molecular formula. Counting the number of carbon, oxygen, and hydrogen atoms in the structural formulas gives the following molecular formulas: salicylic acid ($C_7H_6O_3$), acetic anhydride ($C_4H_6O_3$), and aspirin ($C_9H_8O_4$). We calculate the molecular weight (Section 3.3) by multiplying the atomic weight of each element by the number of times that element appears in the molecular formula and then sum the results.

Salicylic acid ($C_7H_6O_3$)	Acetic anhydride ($C_4H_6O_3$)	Aspirin ($C_9H_8O_4$)
$C_7 = (7)(12.0)$ $= 84.0$	$C_4 = (4)(12.0)$ $= 48.0$	$C_9 = (9)(12.0)$ $= 108.0$
$H_6 = (6)(1.0)$ $= 6.0$	$H_6 = (6)(1.0)$ $= 6.0$	$H_8 = (8)(1.0)$ $= 8.0$
$O_3 = (3)(16.0)$ $= 48.0$	$O_3 = (3)(16.0)$ $= 48.0$	$O_4 = (4)(16)$ $= 64.0$
Mol. wt. = 138.0	Mol. wt. = 102.0	Mol. wt. = 180.0

We can apply the formula for percent atom economy using the molecular weight of aspirin in the numerator and the sum

Salicylic acid

Acetic anhydride

Aspirin

Acetic acid

of the molecular weights of salicylic acid and acetic anhydride in the denominator.

$$\text{Percent atom economy} = \frac{(180.0)}{(138.0 + 102.0)} \times 100\% = 75.0\%$$

The percent atom economy calculation tells us that the reaction is 75.0% efficient in its utilization of matter. The molecule acetic acid (CH_3COOH) is a "by-product" because it is not desired in the synthesis. Thus, 2 carbon atoms, 4 hydrogen atoms, and 2 oxygen atoms are considered to be waste in the production of one aspirin molecule. The law of mass conservation (Section 2.4) states that "mass is neither created nor destroyed in chemical reactions." Atom economy illustrates this law because chemical reactions involve breaking and forming bonds between atoms, but the kind and total number of atoms remain the same. Green chemistry involves designing reactions that maximize the number of reactant atoms that end up in in the desired product and not in wasteful by-products.

PROBLEM 3.23   What is the goal of green chemistry?

(a) Design chemical products and processes with the lowest cost of raw materials

(b) Design safer chemical products and processes that reduce or eliminate the generation of hazardous substances

(c) Design chemical products and processes to remediate hazardous waste sites

(d) All of the above

PROBLEM 3.24   Match the terms *percent yield* and *percent atom economy* with their descriptions.

(a) Efficiency of a reaction in converting reactants to products

(b) Efficiency of a reaction in terms of number of reactant atoms incorporated into the desired product

PROBLEM 3.25   Examine two reactions important in chemical synthesis of organic compounds.

Reaction 1: An addition reaction where two molecules are combined to form a larger molecule.

Desired product

Reaction 2: A substitution reaction where an atom or group of atoms is replaced by a different atom.

Desired product

(a) Without performing any calculations, predict which reaction has a higher percent atom economy.

(b) Calculate the percent atom economy for both reactions.

PROBLEM 3.26   Propene is a raw material for a wide variety of products including the polymer polypropylene used in plastic wrap and Styrofoam cups. Propene can be synthesized by mixing propanol with sulfuric acid and heating the mixture. (Note: Sulfuric acid, H_2SO_4, is a catalyst that can be recovered, so it is not considered in atom economy calculations.)

Propanol

Propene   Water

(a) Calculate the percent yield if 23.50 grams of propanol reacted to produce 10.15 grams of propene.

(c) Calculate the atom economy for the synthesis of propene from propanol.

PROBLEM 3.27   Ibuprofen (the active ingredient in the over-the-counter drugs Advil and Motrin) is a molecule that alleviates pain and reduces fever and swelling. The ball-and-stick model of ibuprofen is shown. (Gray = C, ivory = H, red = O.)

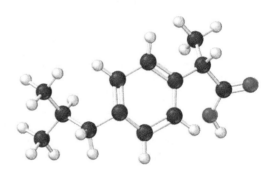

(a) What is the molecular formula of ibuprofen?

(b) What is the molecular weight of ibuprofen?

(c) What is the percent composition by mass of each element in ibuprofen?

PROBLEM 3.28   The original synthesis for ibuprofen, developed in the 1960s, had a percent atom economy of 40.0%. In the 1990s, BHC Co. developed a "greener" three-step synthesis for ibuprofen with a percent atom economy of 77.5%. In the three-step synthesis, 4 moles of H, 2 moles of C, and 2 moles of O are wasted for every mole of ibuprofen produced.

(a) Calculate the total mass (in g) wasted for every one mole of ibuprofen produced.

(b) Yearly production of ibuprofen is approximately 30 million pounds. Calculate the number of moles of ibuprofen produced each year. (1 kg = 2.20 lbs.)

(c) Calculate the total mass (in kg) wasted in the annual production ibuprofen by BHC Co.'s three-step synthesis.

STUDY GUIDE

Section	Concept Summary	Learning Objectives	Test Your Understanding
3.1–3.2 Balancing Chemical Equations	Because mass is neither created nor destroyed in chemical reactions, all chemical equations must be balanced—that is, the numbers and kinds of atoms on both sides of the reaction arrow must be the same. A balanced equation tells the number ratio of reactant and product formula units in a reaction. The coefficients represent the number of moles or number of molecules/atoms of a reactant or product.	**3.1** Visualize bonds broken and formed in a chemical reaction and relate numbers of molecules or atoms to the balanced reaction.	Worked Example 3.1; Problems 3.29–3.30
		3.2 Balance a chemical reaction given the formulas of reactants and products.	Worked Example 3.2; Problems 3.38, 3.40
3.3–3.4 Molar Mass and Stoichiometry	Just as atomic weight is the mass of an atom, molecular weight is the mass of a molecule. The analogous term formula weight is used for ionic and other nonmolecular substances. Molecular weight is the sum of the atomic masses of all atoms in the molecule. One mole of a substance is the amount whose mass in grams is numerically equal to the substance's molecular or formula mass. Carrying out chemical calculations using mass–mole relationships is called **stoichiometry** and is done using molar masses and ratios of coefficients in the balanced equation as conversion factors.	**3.3** Calculate formula weight, molecular weight, and molar mass given a chemical formula or structure.	Worked Example 3.3; Problems 3.31, 3.44, 3.46
		3.4 Convert between mass, moles, and molecules or atoms of a substance.	Worked Example 3.4; Problems 3.48, 3.50, 3.56
		3.5 Relate the amount (moles or mass) of reactants and products in a balanced equation using stoichiometry.	Worked Example 3.5; Problems 3.62, 3.64
3.5 Yields of Chemical Reactions	The amount of product actually formed in a reaction—the reaction's yield—is often less than the amount theoretically possible. Dividing the actual amount by the theoretical amount and multiplying by 100% gives the reaction's percent yield.	**3.6** Calculate the percent yield of a reaction.	Worked Example 3.6; Problems 3.11–3.12, 3.72
3.6 Reactions with Limiting Amounts of Reactants	Often, reactions are carried out with an excess of one reactant beyond that called for by the balanced equation. In such cases, the extent to which the reaction takes place depends on the reactant present in limiting amount, the limiting reactant.	**3.7** Determine the relative amounts of atoms or molecules in the reactants and products of a balanced reaction given a molecular representation.	Worked Example 3.7; Problems 3.13–3.14, 3.32
		3.8 Determine which reactant is limiting and calculate the theoretical yield of the product and the amount of excess reactant.	Worked Example 3.8; Problems 3.74, 3.77, 3.84
		3.9 Calculate percent yield when one reactant is limiting.	Problems 3.80, 3.82
3.7–3.8 Percent Composition, Empirical Formulas, and Combustion Analysis	The chemical makeup of a substance is described by its percent composition—the percentage of the substance's mass due to each of its constituent elements. Elemental analysis is used to calculate a substance's empirical formula, which gives the smallest whole-number ratio of atoms of the elements in the compound. To determine the molecular formula, which may be a simple multiple of the empirical formula, it's also necessary to know the substance's molecular weight.	**3.10** Calculate the percent composition, given a chemical formula or structure.	Problems 3.86–3.87
		3.11 Determine the empirical and molecular formula, given the mass percent composition and molecular weight of a compound.	Worked Example 3.9; Problems 3.88, 3.90
		3.12 Determine the empirical and molecular formula, given combustion analysis data and molecular weight.	Worked Example 3.10; Problems 3.92, 3.94, 3.100
3.9 Determining Molecular Weights: Mass Spectrometry	Molecular weights are experimentally measured by mass spectrometry. A mass spectrometer separates ions of different masses by altering the direction they travel using a magnetic field.	**3.13** Determine empirical and molecular formulas using data from a mass spectrum and combustion analysis.	Worked Example 3.11; Problems 3.106–3.107

KEY TERMS

atom economy *105*	formula unit *85*	molecular formula *98*	stoichiometry *90*
balanced equation *85*	formula weight *88*	molecular weight *88*	yield *92*
coefficient *85*	green chemistry *105*	percent composition *97*	
empirical formula *98*	limiting reactant *94*	percent yield *92*	

KEY EQUATIONS

• **Molecular and Formula Weight (Section 3.3)**

Molecular Weight Sum of atomic weights of all atoms in a molecule.
Formula Weight Sum of atomic weights of all atoms in a formula unit of any compound, molecular or ionic.

• **Percent Yield (Section 3.4)**

$$\text{Percent yield} = \frac{\text{Actual yield of product}}{\text{Theoretical yield of product}} \times 100\%$$

PRACTICE TEST

After studying this chapter, you can assess your understanding with these practice test questions, which are correlated with chapter learning objectives. If you answer a question incorrectly, refer to the learning objectives in the end-of-chapter Study Guide for assistance. The Study Guide provides a conceptual summary, references a Worked Example to model how to solve the problem, and gives additional problems for more practice.

1. The reaction of A_2 (red spheres) with B_2 (blue spheres) is shown in the diagram. What is the balanced chemical equation? **(LO 3.1)**

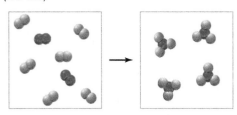

  (a) $2\,A_2 + 6\,B_2 \longrightarrow 4\,AB_3$
  (b) $4\,A + 12\,B \longrightarrow 4\,AB_3$
  (c) $4\,A + 12\,B \longrightarrow A_4 + B_{12}$
  (d) $A_2 + 3\,B_2 \longrightarrow 2\,AB_3$

2. What are the coefficients in the balanced equation for the combustion of ethanol? **(LO 3.2)**

$$\underline{\quad}\ C_2H_6O(l) + \underline{\quad}\ O_2(g) \longrightarrow \underline{\quad}\ CO_2(g) + \underline{\quad}\ H_2O(l)$$

  (a) 1, 3, 2, 3   (b) 2, 3, 4, 3
  (c) 2, 7, 4, 6   (d) 1, 4, 2, 3

3. The ball-and-stick molecular model is a representation of caffeine. Calculate the molecular weight of caffeine. (Gray = C, red = O, blue = N, ivory = H.) **(LO 3.3)**

  (a) 194.2   (b) 182.2
  (c) 192.2   (d) 180.2

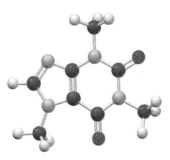

4. A can of diet soda contains 180 mg of the low-calorie sugar substitute aspartame ($C_{14}H_{18}N_2O_5$). How many molecules of aspartame are in the can of soda? **(LO 3.4)**

  (a) 3.7×10^{23} (b) 3.7×10^{20}
  (c) 3.2×10^{25} (d) 1.2×10^{22}

5. How many moles of chloride ions are in 7.75 g of $FeCl_3$? **(LO 3.4)**

  (a) 4.78×10^{-2} (b) 1.59×10^{-2}
  (c) 1.43×10^{-1} (d) 1.91×10^{-1}

6. One way to make coal burning better for the environment is to remove carbon dioxide from the exhaust gases released from power plants using a compound containing an amine ($-NH2$) group. The reaction between carbon dioxide and monoethanolamine is:

$$CO_2(g) + 2\ HOCH_2CH_2NH_2(aq) \longrightarrow$$
$$HOCH_2CH_2NH_3{}^+(aq) + HOCH_2CH_2NHCO_2{}^-(aq)$$

What mass of monoethanoloamine is required to react with 1.0 kg of carbon dioxide? **(LO 3.5)**

  (a) 2.8 kg   (b) 1.1 kg
  (c) 0.93 kg   (d) 0.53 kg

7. If 42.85 grams of salicylic acid reacts with excess acetic anhydride and produces 48.47 grams of aspirin, what is the percent yield of the reaction? (LO 3.6)

$$\underset{\substack{\text{Salicylic}\\\text{acid}}}{C_7H_6O_3(s)} + \underset{\substack{\text{Acetic}\\\text{anhydride}}}{C_4H_6O_3(l)} \longrightarrow \underset{\text{Aspirin}}{C_9H_8O_4(s)} + \underset{\text{Acetic acid}}{CH_3CO_2H(l)}$$

  (a) 88.40%   (b) 64.69%
  (c) 86.72%   (d) 78.74%

8. The diagram represents a mixture of AB_2 and B_2 before it reacts to form AB_3. (Red spheres = A, blue spheres = B.) Which reactant is limiting, and how many AB_3 molecules are formed? (**LO 3.7**)

   (a) B_2 is limiting, and 10 molecules of AB_3 are formed.
   (b) B_2 is limiting, and 4 molecules of AB_3 are formed.
   (c) AB_2 is limiting, and 6 molecules of AB_3 are formed.
   (d) AB_2 is limiting, and 4 molecules of AB_3 are formed.

9. If 2.00 moles of nitrogen and 5.50 moles of hydrogen are placed in a reaction vessel and react to form ammonia, what is the theoretical yield of ammonia (NH_3)? (**LO 3.8**)

$$N_2(g) + 3 H_2(g) \longrightarrow 2 NH_3(g)$$

   (a) 31.2 g   (b) 62.3 g
   (c) 93.7 g   (d) 34.1 g

10. Silver sulfide, the tarnish on silverware, comes from the reaction of silver metal with hydrogen sulfide (H_2S). The unbalanced equation is:

$$Ag + H_2S + O_2 \longrightarrow Ag_2S + H_2O \quad \text{Unbalanced}$$

   If the reaction was used intentionally to prepare Ag_2S, how many grams would be formed from 496 g of Ag, 80.0 g of H_2S, and excess O_2 if the reaction takes place in 90% yield? (**LO 3.9**)

   (a) 525 g   (b) 1139 g
   (c) 583 g   (d) 1025 g

11. What is the percent composition by mass of Mn in potassium permanganate, $KMnO_4$? (**LO 3.10**)

   (a) 22.6%   (b) 34.8%
   (c) 49.9%   (d) 54.9%

12. Dimethylhydrazine, a colorless liquid used as a rocket fuel, is 40.0% C, 13.3% H, and 46.7% N. What is the empirical formula? (**LO 3.11**)

   (a) CH_4N (b) CH_2N
   (c) C_2H_4N (d) $C_2H_5N_2$

13. Lactic acid forms in muscle tissue after strenuous exercise. Elemental analysis shows that lactic acid is 40.0% carbon, 6.71% hydrogen, and 53.3% oxygen by mass. If the molecular weight of lactic acid is 90.08, what is the molecular formula? (**LO 3.11**)

   (a) CH_2O (b) $C_3H_6O_3$
   (c) $C_4H_8O_4$ (d) $C_4H_{10}O_2$

14. Combustion analysis is performed on 0.50 g of a hydrocarbon and 1.55 g of CO_2, and 0.697 g of H_2O are produced. The mass spectrum for the hydrocarbon is provided below. What is the molecular formula? (**LO 3.12 and 3.13**)

   (a) C_5H_{11} (b) C_8H_{18}
   (c) $C_{11}H_{10}$ (d) $C_{10}H_{22}$

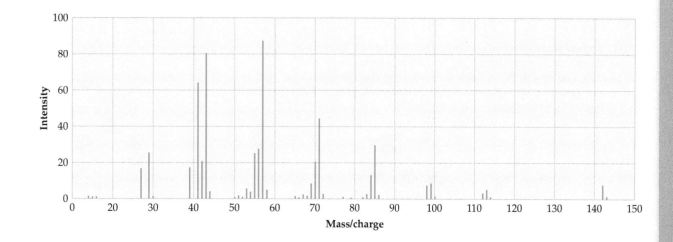

Answers:

1. d, 2. a, 3. a, 4. b, 5. c, 6. a, 7. c, 8. d, 9. b, 10. a, 11. b, 12. a, 13. b, 14. d

Mastering Chemistry provides end-of-chapter exercises, feedback-enriched tutorial problems, animations, and interactive activities to encourage problem-solving practice and deeper understanding of key concepts and topics.

RAN *Randomized in Mastering Chemistry*

CONCEPTUAL PROBLEMS

Problems 3.1–3.28 appear within the chapter.

3.29 The reaction of A (red spheres) with B (blue spheres) is shown in the following diagram:

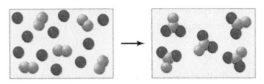

Which equation best describes the stoichiometry of the reaction?

(a) $A_2 + 2\,B \longrightarrow A_2B_2$

(b) $10\,A + 5\,B_2 \longrightarrow 5\,A_2B_2$

(c) $2\,A + B_2 \longrightarrow A_2B_2$

(d) $5\,A + 5\,B_2 \longrightarrow 5\,A_2B_2$

3.30 The diagrams represent a reaction on the molecular level. Atoms of A are represented with red spheres, and atoms of B are represented with blue spheres. Write a balanced chemical equation.

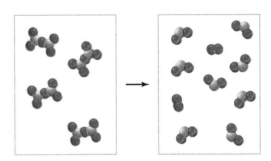

3.31 Fluoxetine, marketed as an antidepressant under the name *Prozac*, can be represented by the following ball-and-stick molecular model. Write the molecular formula for fluoxetine, and calculate its molecular weight. (Red = O, gray = C, blue = N, yellow-green = F, ivory = H.)

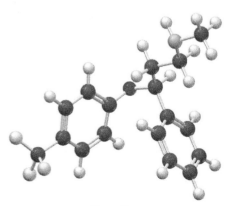

Fluoxetine

3.32 The following diagram represents the reaction of A_2 (red spheres) with B_2 (blue spheres):

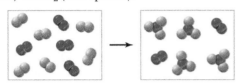

(a) Write a balanced equation for the reaction, and identify the limiting reactant.

(b) How many moles of product can be made from 1.0 mol of A_2 and 1.0 mol of B_2?

3.33 What is the percent composition of cysteine, one of the 20 amino acids commonly found in proteins? (Gray = C, red = O, blue = N, yellow = S, ivory = H.)

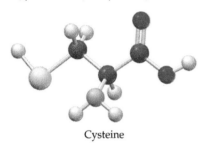

Cysteine

3.34 Cytosine, a constituent of deoxyribonucleic acid (DNA), can be represented by the following molecular model. If 0.001 mol of cytosine is submitted to combustion analysis, how many moles of CO_2 and how many moles of H_2O would be formed? (Gray = C, red = O, blue = N, ivory = H.)

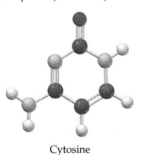

Cytosine

3.35 A hydrocarbon of unknown formula C_xH_y was submitted to combustion analysis with the following results. What is the empirical formula of the hydrocarbon?

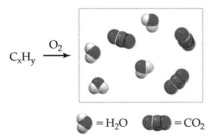

$$C_xH_y \xrightarrow{O_2}$$

= H_2O = CO_2

SECTION PROBLEMS

Balancing Equations (Section 3.2)

3.36 Which of the following equations is balanced?

(a) The development reaction in silver-halide photography:

$$2\ AgBr + 2\ NaOH + C_6H_6O_2 \longrightarrow$$
$$2\ Ag + H_2O + 2\ NaBr + C_6H_4O_2$$

(b) The preparation of household bleach:

$$2\ NaOH + Cl_2 \longrightarrow NaOCl + NaCl + H_2O$$

3.37 Which of the following equations is balanced? Balance any that need it.

(a) The thermite reaction, used in welding:

$$Al + Fe_2O_3 \longrightarrow Al_2O_3 + Fe$$

(b) The photosynthesis of glucose from CO_2:

$$6\ CO_2 + 6\ H_2O \longrightarrow C_6H_{12}O_6 + 6\ O_2$$

(c) The separation of gold from its ore:

$$Au + 2\ NaCN + O_2 + H_2O \longrightarrow$$
$$NaAu(CN)_2 + 3\ NaOH$$

3.38 Balance the following equations.

(a) $Mg + HNO_3 \longrightarrow H_2 + Mg(NO_3)_2$

(b) $CaC_2 + H_2O \longrightarrow Ca(OH)_2 + C_2H_2$

(c) $S + O_2 \longrightarrow SO_3$

(d) $UO_2 + HF \longrightarrow UF_4 + H_2O$

3.39 Balance the following equations.

(a) The explosion of ammonium nitrate:
$$NH_4NO_3 \longrightarrow N_2 + O_2 + H_2O$$

(b) The spoilage of wine into vinegar:
$$C_2H_6O + O_2 \longrightarrow C_2H_4O_2 + H_2O$$

(c) The burning of rocket fuel:
$$C_2H_8N_2 + N_2O_4 \longrightarrow N_2 + CO_2 + H_2O$$

3.40 Balance the following equations.

RAN (a) $SiCl_4 + H_2O \longrightarrow SiO_2 + HCl$

(b) $P_4O_{10} + H_2O \longrightarrow H_3PO_4$

(c) $CaCN_2 + H_2O \longrightarrow CaCO_3 + NH_3$

(d) $NO_2 + H_2O \longrightarrow HNO_3 + NO$

3.41 Balance the following equations.

(a) $VCl_3 + Na + CO \longrightarrow V(CO)_6 + NaCl$

(b) $RuI_3 + CO + Ag \longrightarrow Ru(CO)_5 + AgI$

(c) $CoS + CO + Cu \longrightarrow Co_2(CO)_8 + Cu_2S$

3.42 Balance the following equations.

(a) $C_6H_5NO_2 + O_2 \longrightarrow CO_2 + H_2O + NO_2$

(b) $Au + H_2SeO_4 \longrightarrow Au_2(SeO_4)_3 + H_2SeO_3 + H_2O$

(c) $NH_4ClO_4 + Al \longrightarrow Al_2O_3 + N_2 + Cl_2 + H_2O$

3.43 Balance the following equations.

RAN (a) $CO(NH_2)_2(aq) + HOCl(aq) \longrightarrow$
$$NCl_3(aq) + CO_2(aq) + H_2O(l)$$

(b) $Ca_3(PO_4)_2(s) + SiO_2(s) + C(s) \longrightarrow$
$$P_4(g) + CaSiO_3(l) + CO(g)$$

Molecular Weights and Molar Mass (Section 3.3)

3.44 What are the molecular (formula) weights of the following
RAN substances?

(a) Hg_2Cl_2 (calomel, used at one time as a bowel purgative)

(b) $C_4H_8O_2$ (butyric acid, responsible for the odor of rancid butter)

(c) CF_2Cl_2 (a chlorofluorocarbon that destroys the stratospheric ozone layer)

3.45 What are the formulas of the following substances?

(a) $PCl_?$; mol. weight = 137.3

(b) Nicotine, $C_{10}H_{14}N_?$; mol. weight = 162.2

3.46 What are the molecular weights of the following pharmaceuticals?

(a) $C_{33}H_{35}FN_2O_5$ (atorvastatin, lowers blood cholesterol)

(b) $C_{22}H_{27}F_3O_4S$ (fluticasone, anti-inflammatory)

(c) $C_{16}H_{16}ClNO_2S$ (clopidogrel, inhibits blood clots)

3.47 What are the molecular weights of the following herbicides?

(a) $C_6H_6Cl_2O_3$ (2, 4-dichlorophenoxyacetic acid, effective on broadleaf plants)

(b) $C_{15}H_{22}ClNO_2$ (metolachlor, pre-emergent herbicide)

(c) $C_8H_6Cl_2O_3$ (dicamba, effective on broadleaf plants)

3.48 How many grams are in a mole of each of the following
RAN substances?

(a) Ti   (b) Br_2 (c) Hg   (d) H_2O

3.49 How many moles are in a gram of each of the following
RAN substances?

(a) Cr   (b) Cl_2 (c) Au   (d) NH_3

3.50 How many moles of ions are in 27.5 g of $MgCl_2$?
RAN

3.51 How many moles of anions are in 35.6 g of AlF_3?
RAN

3.52 What is the molecular weight of chloroform if 0.0275 mol
RAN weighs 3.28 g?

3.53 What is the molecular weight of cholesterol if 0.5731 mol
RAN weighs 221.6 g?

3.54 Iron(II) sulfate, $FeSO_4$, is prescribed for the treatment of
RAN anemia. How many moles of $FeSO_4$ are present in a standard 300 mg tablet? How many iron(II) ions?

3.55 The "lead" in lead pencils is actually almost pure carbon, and the mass of a period mark made by a lead pencil is about 0.0001 g. How many carbon atoms are in the period?

3.56 An average cup of coffee contains about 125 mg of caffeine,
RAN $C_8H_{10}N_4O_2$. How many moles of caffeine are in a cup? How many molecules of caffeine?

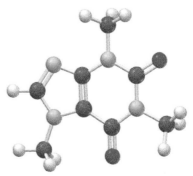

Caffeine

3.57 What is the mass in grams of each of the following samples?
RAN
(a) 0.0015 mol of sodium

(b) 0.0015 mol of lead

(c) 0.0015 mol of diazepam (Valium), $C_{16}H_{13}ClN_2O$

3.58 A sample that weighs 25.12 g contains 6.022×10^{23} particles.
RAN If 25.00% of the total number of particles are argon atoms and 75.00% are another element, what is the chemical identity of the other constituent?

3.59 A sample that weighs 107.75 g is a mixture of 30% helium
RAN atoms and 70% krypton atoms. How many particles are present in the sample?

3.60 Titanium metal is obtained from the mineral rutile, TiO_2. How
RAN many kilograms of rutile are needed to produce 100.0 kg of Ti?

3.61 Iron metal can be produced from the mineral hematite,
RAN Fe_2O_3, by reaction with carbon. How many kilograms of iron are present in 105 kg of hematite?

Stoichiometry (Section 3.4)

3.62 In the preparation of iron from hematite, Fe_2O_3 reacts with
RAN carbon:

$$Fe_2O_3 + C \longrightarrow Fe + CO_2 \qquad \text{Unbalanced}$$

(a) Balance the equation.

(b) How many moles of carbon are needed to react with 525 g of hematite?

(c) How many grams of carbon are needed to react with 525 g of hematite?

3.63 An alternative method for preparing pure iron from Fe_2O_3 is
RAN by reaction with carbon monoxide:

$$Fe_2O_3 + CO \longrightarrow Fe + CO_2 \qquad \text{Unbalanced}$$

(a) Balance the equation.

(b) How many grams of CO are needed to react with 3.02 g of Fe_2O_3?

(c) How many grams of CO are needed to react with 1.68 mol of Fe_2O_3?

3.64 Magnesium metal burns in oxygen to form magnesium
RAN oxide, MgO.

(a) Write a balanced equation for the reaction.

(b) How many grams of oxygen are needed to react with 25.0 g of Mg? How many grams of MgO will result?

(c) How many grams of Mg are needed to react with 25.0 g of O_2? How many grams of MgO will result?

3.65 Ethylene gas, C_2H_4, reacts with water at high temperature to
RAN yield ethyl alcohol, C_2H_6O.

(a) How many grams of ethylene are needed to react with 0.133 mol of H_2O? How many grams of ethyl alcohol will result?

(b) How many grams of water are needed to react with 0.371 mol of ethylene? How many grams of ethyl alcohol will result?

3.66 Pure oxygen was first made by heating mercury(II) oxide:
RAN
$$HgO \xrightarrow{\text{Heat}} Hg + O_2 \qquad \text{Unbalanced}$$

(a) Balance the equation.

(b) How many grams of mercury and how many grams of oxygen are formed from 45.5 g of HgO?

(c) How many grams of HgO would you need to obtain 33.3 g of O_2?

3.67 Titanium dioxide (TiO_2), the substance used as the pigment
RAN in white paint, is prepared industrially by reaction of $TiCl_4$ with O_2 at high temperature.

$$TiCl_4 + O_2 \xrightarrow{\text{Heat}} TiO_2 + 2 Cl_2$$

How many kilograms of TiO_2 can be prepared from 5.60 kg of $TiCl_4$?

3.68 Silver metal reacts with chlorine (Cl_2) to yield silver chlo-
RAN ride. If 2.00 g of Ag reacts with 0.657 g of Cl_2, what is the empirical formula of silver chloride?

3.69 Aluminum reacts with oxygen to yield aluminum oxide. If
5.0 g of Al reacts with 4.45 g of O_2, what is the empirical formula of aluminum oxide?

3.70 The industrial production of hydriodic acid takes place by
RAN treatment of iodine with hydrazine (N_2H_4):

$$2 I_2 + N_2H_4 \longrightarrow 4 HI + N_2$$

(a) How many grams of I_2 are needed to react with 36.7 g of (N_2H_4)?

(b) How many grams of HI are produced from the reaction of 115.7 g of N_2H_4 with excess iodine?

3.71 An alternative method for producing hydriodic acid is the
RAN reaction of iodine with hydrogen sulfide:

$$H_2S + I_2 \longrightarrow 2 HI + S$$

(a) How many grams of I_2 are needed to react with 49.2 g of H_2S?

(b) How many grams of HI are produced from the reaction of 95.4 g of H_2S with excess I_2?

Reaction Yield and Limiting Reactants (Sections 3.5–3.6)

3.72 Nickel(II) sulfate, used for nickel plating, is prepared by treat-
RAN ment of nickel(II) carbonate with sulfuric acid:

$$NiCO_3 + H_2SO_4 \longrightarrow NiSO_4 + CO_2 + H_2O$$

(a) How many grams of H_2SO_4 are needed to react with 14.5 g of $NiCO_3$?

(b) How many grams of $NiSO_4$ are obtained if the yield is 78.9%?

3.73 Hydrazine, N_2H_4, once used as a rocket propellant, reacts
RAN with oxygen:

$$N_2H_4 + O_2 \longrightarrow N_2 + 2 H_2O$$

(a) How many grams of O_2 are needed to react with 50.0 g of N_2H_4?

(b) How many grams of N_2 are obtained if the yield is 85.5%?

3.74 Assume that you have 1.39 mol of H_2 and 3.44 mol of N_2.
RAN How many grams of ammonia (NH_3) can you make, and how many grams of which reactant will be left over?

$$3 H_2 + N_2 \longrightarrow NH_3$$

3.75 Hydrogen and chlorine react to yield hydrogen chloride:
RAN $H_2 + Cl_2 \longrightarrow 2 HCl$. How many grams of HCl are formed from reaction of 3.56 g of H_2 with 8.94 g of Cl_2? Which reactant is limiting?

3.76 How many grams of the dry-cleaning solvent 1,2-dichloroethane (also called ethylene chloride), $C_2H_4Cl_2$, can be prepared by reaction of 15.4 g of ethylene, C_2H_4, with 3.74 g of Cl_2?

$$C_2H_4 + Cl_2 \longrightarrow C_2H_4Cl_2$$

1,2-Dichloroethane
(ethylene chloride)

3.77 How many grams of each product result from the following reactions, and how many grams of which reactant is left over?
(a) $(1.3 \text{ g NaCl}) + (3.5 \text{ g AgNO}_3) \longrightarrow$
$$(x \text{ g AgCl}) + (y \text{ g NaNO}_3)$$
(b) $(2.65 \text{ g BaCl}_2) + (6.78 \text{ g H}_2\text{SO}_4) \longrightarrow$
$$(x \text{ g BaSO}_4) + (y \text{ g HCl})$$

3.78 Limestone ($CaCO_3$) reacts with hydrochloric acid according to the equation $CaCO_3 + 2 HCl \longrightarrow CaCl_2 + H_2O + CO_2$. If 1.00 mol of CO_2 has a volume of 22.4 L under the reaction conditions, how many liters of gas can be formed by reaction of 2.35 g of $CaCO_3$ with 2.35 g of HCl? Which reactant is limiting?

3.79 Sodium azide (NaN_3) yields N_2 gas when heated to 300 °C, a reaction used in automobile air bags. If 1.00 mol of N_2 has a volume of 47.0 L under the reaction conditions, how many liters of gas can be formed by heating 38.5 g of NaN_3? The reaction is
$$2 NaN_3 \longrightarrow 3 N_2(g) + 2 Na$$

3.80 Acetic acid (CH_3CO_2H) reacts with isopentyl alcohol ($C_5H_{12}O$) to yield isopentyl acetate ($C_7H_{14}O_2$), a fragrant substance with the odor of bananas. If the yield from the reaction of acetic acid with isopentyl alcohol is 45%, how many grams of isopentyl acetate are formed from 3.58 g of acetic acid and 4.75 g of isopentyl alcohol? The reaction is
$$CH_3CO_2H + C_5H_{12}O \longrightarrow C_7H_{14}O_2 + H_2O$$

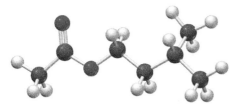

Isopentyl acetate

3.81 Cisplatin [$Pt(NH_3)_2Cl_2$], a compound used in cancer treatment, is prepared by reaction of ammonia with potassium tetrachloroplatinate:
$$K_2PtCl_4 + 2 NH_3 \longrightarrow 2 KCl + Pt(NH_3)_2Cl_2$$
How many grams of cisplatin are formed from 55.8 g of K_2PtCl_4 and 35.6 g of NH_3 if the reaction takes place in 95% yield based on the limiting reactant?

3.82 If 1.87 g of acetic acid (CH_3COOH) reacts with 2.31 g of isopentyl alcohol ($C_5H_{12}O$) to give 2.96 g of isopentyl acetate ($C_7H_{14}O_2$), what is the percent yield of the reaction?

3.83 If 3.42 g of K_2PtCl_4 and 1.61 g of NH_3 give 2.08 g of cisplatin (Problem 3.81), what is the percent yield of the reaction?

3.84 The reaction of tungsten hexachloride (WCl_6) with bismuth gives hexatungsten dodecachloride (W_6Cl_{12}).
$$WCl_6 + Bi \longrightarrow W_6Cl_{12} + BiCl_3 \quad \text{Unbalanced}$$
(a) Balance the equation.
(b) How many grams of bismuth react with 150.0 g of WCl_6?
(c) When 228 g of WCl_6 react with 175 g of Bi, how much W_6Cl_{12} is formed based on the limiting reactant?

3.85 Sodium borohydride, $NaBH_4$, a substance used in the synthesis of many pharmaceutical agents, can be prepared by reaction of NaH with B_2H_6 according to the equation
$$2 NaH + B_2H_6 \longrightarrow 2 NaBH_4$$
(a) How many grams of $NaBH_4$ can be prepared by reaction between 8.55 g of NaH and 6.75 g of B_2H_6?
(b) Which reactant is limiting, and how many grams of the excess reactant will be left over?

Percent Composition and Empirical Formulas (Section 3.7)

3.86 Urea, a substance commonly used as a fertilizer, has the formula CH_4N_2O. What is its percent composition by mass?

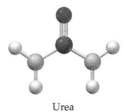

Urea

3.87 Calculate the mass percent composition of each of the following substances.
(a) Malachite, a copper-containing mineral: $Cu_2(OH)_2CO_3$
(b) Acetaminophen, a headache remedy: $C_8H_9NO_2$
(c) Prussian blue, an ink pigment: $Fe_4[Fe(CN)_6]_3$

3.88 What are the empirical formulas of substances with the following mass percent compositions?
(a) Aspirin: 4.48% H, 60.00% C, 35.52% O
(b) Ilmenite (a titanium-containing ore): 31.63% O, 31.56% Ti, 36.81% Fe
(c) Sodium thiosulfate (photographic "fixer"): 30.36% O, 29.08% Na, 40.56% S

3.89 Ferrocene, a substance proposed for use as a gasoline additive, has the percent composition 5.42% H, 64.56% C, and 30.02% Fe. What is the empirical formula of ferrocene?

3.90 What is the empirical formula of stannous fluoride, the first fluoride compound added to toothpaste to protect teeth against decay? Its mass percent composition is 24.25% F, 75.75% Sn.

3.91 What are the empirical formulas of each of the following substances?
(a) Ibuprofen, a headache remedy: 75.69% C, 15.51% O, 8.80% H
(b) Magnetite, a naturally occurring magnetic mineral: 72.36% Fe, 27.64% O
(c) Zircon, a mineral from which cubic zirconia is made: 34.91% O, 15.32% Si, 49.77% Zr

Formulas and Elemental Analysis (Section 3.8)

3.92 An unknown liquid is composed of 5.57% H, 28.01% Cl, and 66.42% C. The molecular weight found by mass spectrometry is 126.58. What is the molecular formula of the compound?

3.93 An unknown liquid is composed of 34.31% C, 5.28% H, and 60.41% I. The molecular weight found by mass spectrometry is 210.06. What is the molecular formula of the compound?

3.94 Combustion analysis of 45.62 mg of toluene, a commonly used solvent, gives 35.67 mg of H_2O and 152.5 mg of CO_2. What is the empirical formula of toluene?

3.95 Coniine, a toxic substance isolated from poison hemlock, contains only carbon, hydrogen, and nitrogen. Combustion analysis of a 5.024 mg sample yields 13.90 mg of CO_2 and 6.048 mg of H_2O. What is the empirical formula of coniine?

3.96 Cytochrome *c* is an iron-containing enzyme found in the cells of all aerobic organisms. If cytochrome *c* is 0.43% Fe by mass, what is its minimum molecular weight?

3.97 Nitrogen fixation in the root nodules of peas and other leguminous plants is carried out by the molybdenum-containing enzyme *nitrogenase*. What is the molecular mass of nitrogenase if the enzyme contains two molybdenum atoms and is 0.0872% Mo by mass?

3.98 Disilane, Si_2H_x, is analyzed and found to contain 90.28%
RAN silicon by mass. What is the value of *x*?

3.99 A certain metal sulfide, MS_2, is used extensively as a high-temperature lubricant. If MS_2 is 40.06% sulfur by mass, what is the identity of the metal M?

3.100 Combustion analysis of a 31.472 mg sample of the widely
RAN used flame retardant Decabrom gave 1.444 mg of CO_2. Is the molecular formula of Decabrom $C_{12}Br_{10}$ or $C_{12}Br_{10}O$?

3.101 The stimulant amphetamine contains only carbon, hydrogen, and nitrogen. Combustion analysis of a 42.92 mg sample of amphetamine gives 37.187 mg of H_2O and 125.75 mg of CO_2. If the molar mass of amphetamine is less than 160 g/mol, what is its molecular formula?

Amphetamine

Mass Spectrometry (Section 3.9)

3.102 Describe the path of a neutral molecule in the mass spectrometer. Why is ionization a necessary first step?

3.103 Why is high precision and accuracy in molecular mass measurement needed in identifying the formula of molecules made in the lab?

3.104 The molecular weight of an organic compound was found
RAN by mass spectrometry to be 70.042 11. Is the sample C_5H_{10}, C_4H_6O, or $C_3H_6N_2$? Exact masses of elements are: 1.007 825 (^1H); 12.000 00 (^{12}C); 14.003 074 (^{14}N); 15.994 915 (^{16}O).

3.105 The mass of an organic compound was found by mass spec-
RAN trometry to be 58.077 46. Is the sample C_4H_{10}, C_3H_6O, or $C_2H_6N_2$? Exact masses of elements are: 1.007 825 (^1H); 12.000 00 (^{12}C); 14.003 074 (^{14}N); 15.994 915 (^{16}O).

3.106 (a) Combustion analysis of 50.0 mg of benzene, a commonly
RAN used solvent composed of carbon and hydrogen, gives 34.6 mg of H_2O and 169.2 mg of CO_2. What is the empirical formula of benzene?

   (b) Given the mass spectrum of benzene, identify the molecular weight and give the molecular formula.

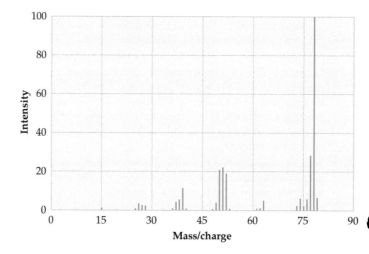

3.107 (a) Combustion analysis of 150.0 mg of 1,2,3,benzenetriol,
RAN a compound composed of carbon, hydrogen, and oxygen, gives 64.3 mg of H_2O and 314.2 mg of CO_2. What is the empirical formula of 1,2,3,benzenetriol?

   (b) Given the mass spectrum of 1,2,3,benzenetriol, identify the molecular weight and give the molecular formula.

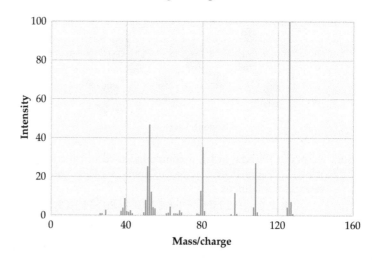

MULTICONCEPT PROBLEMS

3.108 The molecular weight of ethylene glycol is 62.0689 when calculated using the atomic weights found in a standard periodic table, yet the molecular weight determined experimentally by high-resolution mass spectrometry is 62.0368. Explain the discrepancy.

3.109 The molar mass of HCl is 36.5 g/mol, and the average mass per HCl molecule is 36.5 u. Use the fact that $1 \text{ u} = 1.6605 \times 10^{2\,24}\text{g}$ to calculate Avogadro's number.

3.110 *RAN* Assume that gasoline has the formula C_8H_{18} and has a density of 0.703 g/mL. How many pounds of CO_2 are produced from the complete combustion of 1.00 gal of gasoline?

3.111 Compound X contains only carbon, hydrogen, nitrogen, and chlorine. When 1.00 g of X is dissolved in water and allowed to react with excess silver nitrate, $AgNO_3$, all the chlorine in X reacts and 1.95 g of solid AgCl is formed. When 1.00 g of X undergoes complete combustion, 0.900 g of CO_2 and 0.735 g of H_2O are formed. What is the empirical formula of X?

3.112 A pulverized rock sample believed to be pure calcium carbonate, $CaCO_3$, is subjected to chemical analysis and found to contain 51.3% Ca, 7.7% C, and 41.0% O by mass. Why can't this rock sample be pure $CaCO_3$?

3.113 *RAN* A certain alcoholic beverage contains only ethanol (C_2H_6O) and water. When a sample of this beverage undergoes combustion, the ethanol burns but the water simply evaporates and is collected along with the water produced by combustion. The combustion reaction is

$$C_2H_6O(l) + 3\ O_2(g) \longrightarrow 2\ CO_2(g) + 3\ H_2O(g)$$

When a 10.00 g sample of this beverage is burned, 11.27 g of water is collected. What is the mass in grams of ethanol, and what is the mass of water in the original sample?

3.114 *RAN* A mixture of FeO and Fe_2O_3 with a mass of 10.0 g is converted to 7.43 g of pure Fe metal. What are the amounts in grams of FeO and Fe_2O_3 in the original sample?

3.115 *RAN* A compound of formula XCl_3 reacts with aqueous $AgNO_3$ to yield solid AgCl according to the following equation:

$$XCl_3(aq) + 3\ AgNO_3(aq) \longrightarrow X(NO_3)_3(aq) + 3\ AgCl(s)$$

When a solution containing 0.634 g of XCl_3 was allowed to react with an excess of aqueous $AgNO_3$, 1.68 g of solid AgCl was formed. What is the identity of the atom X?

3.116 *RAN* When eaten, dietary carbohydrates are digested to yield glucose ($C_6H_{12}O_6$), which is then metabolized to yield carbon dioxide and water:

$$C_6H_{12}O_6 + O_2 \longrightarrow CO_2 + H_2O \qquad \text{Unbalanced}$$

Balance the equation, and calculate both the mass in grams and the volume in liters of the CO_2 produced from 66.3 g of glucose, assuming that 1 mol of CO_2 has a volume of 25.4 L at normal body temperature.

3.117 *RAN* A copper wire having a mass of 2.196 g was allowed to react with an excess of sulfur. The excess sulfur was then burned, yielding SO_2 gas. The mass of the copper sulfide produced was 2.748 g.
  (a) What is the percent composition of copper sulfide?
  (b) What is its empirical formula?
  (c) Calculate the number of copper ions per cubic centimeter if the density of the copper sulfide is 5.6 g/cm^3.

3.118 *RAN* Element X, a member of group 5A, forms two chlorides, XCl_3 and XCl_5. Reaction of an excess of Cl_2 with 8.729 g of XCl_3 yields 13.233 g of XCl_5. What is the atomic weight and the identity of the element X?

3.119 *RAN* A mixture of XCl_3 and XCl_5 weighing 10.00 g contains 81.04% Cl by mass. How many grams of XCl_3 and how many grams of XCl_5 are present in the mixture?

3.120 *RAN* Ammonium nitrate, a potential ingredient of terrorist bombs, can be made nonexplosive by addition of diammonium hydrogen phosphate, $(NH_4)_2HPO_4$. Analysis of such a $NH_4NO_3 - (NH_4)_2HPO_4$ mixture showed the mass percent of nitrogen to be 30.43%. What is the mass ratio of the two components in the mixture?

3.121 *RAN* Window glass is typically made by mixing soda ash (Na_2CO_3), limestone ($CaCO_3$), and silica sand (SiO_2) and then heating to 1500 °C to drive off CO_2 from the (Na_2CO_3) and $CaCO_3$. The resultant glass consists of about 12% Na_2O by mass, 13% CaO by mass, and 75% SiO_2 by mass. How much of each reactant would you start with to prepare 0.35 kg of glass?

3.122 An unidentified metal M reacts with an unidentified halogen X to form a compound MX_2. When heated, the compound decomposes by the reaction:

$$2\ MX_2(s) \longrightarrow 2\ MX(s) + X_2(g)$$

When 1.12 g of MX_2 is heated, 0.720 g of MX is obtained, along with 56.0 mL of X_2 gas. Under the conditions used, 1.00 mol of the gas has a volume of 22.41 L.
  (a) What is the atomic weight and identity of the halogen X?
  (b) What is the atomic weight and identity of the metal M?

3.123 Ethylene glycol, commonly used as automobile antifreeze, contains only carbon, hydrogen, and oxygen. Combustion analysis of a 23.46 mg sample yields 20.42 mg of H_2O and 33.27 mg of CO_2. What is the empirical formula of ethylene glycol? What is its molecular formula if it has a molecular weight of 62.0?

3.124 *RAN* (a) Polychlorinated biphenyls (PCBs) were compounds used as coolants in transformers and capacitors, but their production was banned by the U.S. Congress in 1979 because they are highly toxic and persist in the environment. When 1.0 g of a PCB containing carbon, hydrogen, and chlorine was subjected to combustion analysis, 1.617 g of CO_2. and 0.138 g of H_2O were produced. What is the empirical formula?
  (b) If the molecular weight is 326.26, what is the molecular formula?
  (c) Can combustion analysis be used to determine the empirical formula of a compound containing carbon, hydrogen, oxygen, and chlorine?

chapter 4

Reactions in Aqueous Solution

Contents

Many chemical reactions in your body occur in aqueous solution and fall into the general categories of precipitation, acid -base, and redox reactions. Vigorous exercise depletes fluids and electrolytes, upsetting the body's delicate chemical balance.

How do sports drinks replenish the substances lost in sweat?

The answer to this question can be found on page 148 in the INQUIRY

ur world is based on water. Approximately 71% of the Earth's surface is covered by water, and another 3% is covered by ice; 66% of the mass of an adult human body is water, and water is needed to sustain all living organisms. It's therefore not surprising that a large amount of important chemistry, including all those reactions that happen in our bodies, takes place in water—that is, in *aqueous solution.*

We saw in the previous chapter how chemical reactions are described and how the specific mass relationships among reactants and products in reactions can be calculated. In this chapter, we'll continue the study of chemical reactions by describing some ways that chemical reactions can be classified and by exploring some general ways in which reactions take place.

4.1 SOLUTION CONCENTRATION: MOLARITY

For a chemical reaction to occur, the reacting molecules or ions must come into contact. This means that the reactants must be mobile, which in turn means that many chemical reactions are carried out in the liquid state or in solution rather than in the solid state. It's therefore necessary to have a standard means to describe exact quantities of substances in solution.

As we've seen, stoichiometry calculations for chemical reactions always require working in moles. Thus, the most useful means of expressing a solution's concentration is **molarity (M)**, the number of moles of a substance, or **solute,** dissolved in enough solvent to make 1 liter of solution. For example, a solution made by dissolving 1.00 mol (58.5 g) of NaCl in enough water to give 1.00 L of solution has a concentration of 1.00 mol/L, or 1.00 M. The molarity of any solution is found by dividing the number of moles of solute by the number of liters of solution:

$$\text{Molarity (M)} = \frac{\text{Moles of solute}}{\text{Liters of solution}}$$

Note that it's the final volume of the *solution* that's important, not the starting volume of the *solvent* used. The final volume of the solution might be a bit larger than the volume of the solvent because of the additional volume of the solute. In practice, a solution of known molarity is prepared by weighing an appropriate amount of solute and placing it in a container called a *volumetric flask,* as shown in **FIGURE 4.1.** Enough solvent is added to dissolve the solute, and further solvent is added until an accurately calibrated final volume is reached. The solution is then gently mixed to reach a uniform concentration.

Molarity can be used as a conversion factor to relate a solution's volume to the number of moles of solute. If we know the molarity and volume of a solution, we can calculate the number of moles of solute. If we know the number of moles of solute and the molarity of the solution, we can find the solution's volume. We walk through these calculations in Worked Examples 4.1 and 4.2.

$$\text{Molarity} = \frac{\text{Moles of solute}}{\text{Volume of solution(L)}}$$

$$\text{Moles of solute} = \text{Molarity} \times \text{Volume of solution} \qquad \text{Volume of solution} = \frac{\text{Moles of solute}}{\text{Molarity}}$$

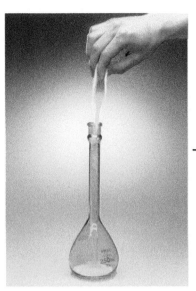

| A measured number of moles of solute is placed in a volumetric flask. | Enough solvent is added to dissolve the solute by swirling. | Further solvent is added to reach the calibration mark on the neck of the flask, and the solution is mixed until uniform. |

▲ **FIGURE 4.1**

Preparing A Solution Of Known Molarity.

▲ **Figure It Out**

Why is it incorrect to add 250 mL of solvent to the flask before dissolving the measured number of moles of solute?

Answer: The final volume of the solution would be greater than 250 mL, and the actual molarity would be lower than expected.

WORKED EXAMPLE 4.1

Calculating the Molarity of a Solution

What is the molarity of a solution made by dissolving 2.355 g of sulfuric acid (H_2SO_4) in water and diluting to a final volume of 50.0 mL?

IDENTIFY

Known	Unknown
Mass of solute (2.355 g of sulfuric acid)	Molarity (M = mol/L)
Volume of solution (50.0 mL)	

STRATEGY

Molarity is the number of moles of solute per liter of solution. Thus, it's necessary to find the number of moles of sulfuric acid in 2.355 g and then divide by the volume of the solution in liters.

SOLUTION

Mol. wt. $H_2SO_4 = (2 \times 1.0) + (32.1) + (4 \times 16.0) = 98.1$

Molar mass of $H_2SO_4 = 98.1$ g/mol

$$2.355 \text{ g } H_2SO_4 \times \frac{1 \text{ mol } H_2SO_4}{98.1 \text{ g } H_2SO_4} = 0.0240 \text{ mol } H_2SO_4$$

$$\frac{0.0240 \text{ mol } H_2SO_4}{0.0500 \text{ L}} = 0.480 \text{ M}$$

The solution has a sulfuric acid concentration of 0.480 M.

▶ **PRACTICE 4.1**   A sweetened iced tea beverage contains 43.0 g of sucrose ($C_{12}H_{22}O_{11}$) in a total volume of 355 mL. Calculate the molarity of sucrose in the drink.

▶ **APPLY 4.2**   An intravenous solution was prepared by adding 13.252 g of dextrose ($C_6H_{12}O_6$) and 0.686 g of sodium chloride to a 250.0 mL volumetric flask and diluting to the calibration mark with water. What is the molarity of each component of the solution?

All **PRACTICE** and **APPLY** problems are interactive in the eText.

— WORKED EXAMPLE 4.2

Calculating the Number of Moles of Solute in a Solution

Hydrochloric acid (HCl) is sold commercially as a 12.0 M aqueous solution. How many moles of HCl are in 300.0 mL of 12.0 M solution?

IDENTIFY

Known	Unknown
Volume of solution (300.0 mL)	Moles of solute (mol HCl)
Molarity of solution (12.0 M)	

STRATEGY

The number of moles of solute is calculated by multiplying the molarity of the solution by its volume.

SOLUTION

Moles of HCl = (Molarity of solution) × (Volume of solution)

$$= \frac{12.0 \text{ mol HCl}}{1 \text{ } L} \times 0.3000 \text{ } L = 3.60 \text{ mol HCl}$$

There are 3.60 mol of HCl in 300.0 mL of 12.0 M solution.

CHECK

One liter of 12.0 M HCl solution contains 12 mol of HCl, so 300 mL (0.3 L) of solution contains 0.3 × 12 = 3.6 mol.

▶ **PRACTICE 4.3** How many moles of solute are present in 125 mL of 0.20 M $NaHCO_3$?

▶ **APPLY 4.4** The concentration of cholesterol ($C_{27}H_{46}O$) in normal blood is approximately 0.0050 M.

(a) How many grams of cholesterol are in 750 mL of blood?
(b) A medical test for cholesterol requires 25 mg. How many milliliters of blood are needed to perform the test?

Cholesterol

All **WORKED EXAMPLES** with this icon have an interactive video in the eText.

4.2 DILUTING CONCENTRATED SOLUTIONS

For convenience, chemicals are sometimes bought and stored as concentrated solutions, which are then diluted before use. Aqueous hydrochloric acid, for example, is sold commercially as a 12.0 M solution, yet it is most commonly used in the laboratory after dilution with water to a final concentration of either 6.0 M or 1.0 M.

Concentrated solution + Solvent ⟶ Dilute solution

The main thing to remember when diluting a concentrated solution is that the number of moles of solute is constant; only the volume of the solution is changed by adding more solvent. Because the number of moles of solute can be calculated by multiplying molarity times volume, we can set up the following equation:

> **Moles of solute (constant)** = Molarity × Volume
> $$= M_i \times V_i = M_f \times V_f$$

where M_i is the initial molarity, V_i is the initial volume, M_f is the final molarity, and V_f is the final volume after dilution. Rearranging this equation into a more useful form shows that the molar concentration after dilution (M_f) can be found by multiplying the initial concentration (M_i) by the ratio of initial and final volumes (V_i/V_f):

$$M_f = M_i \times \frac{V_i}{V_f}$$

Suppose, for example, that we dilute 50.0 mL of a solution of 2.00 M H_2SO_4 to a volume of 200.0 mL. The solution volume *increases* by a factor of 4 (from 50 mL

▲ Just as the frozen orange juice concentrate must be diluted before use by adding water, many chemical solutions must also be diluted.

Go to eText

BIG IDEA Question 1

10.00 mL of an aqueous 1.0 M NaCl solution is added to a 100.0 mL volumetric flask. Water is added to fill the flask to the calibration mark. Which quantity does not change when the solution is diluted? (the number of moles of NaCl, the number of moles of water, or the molarity of the solution)

to 200 mL), so the concentration of the solution must *decrease* by a factor of 4 (from 2.00 M to 0.500 M):

$$M_f = 2.00 \text{ M} \times \frac{50.0 \text{ mL}}{200.0 \text{ mL}} = 0.500 \text{ M}$$

In practice, dilutions are usually carried out as shown in **FIGURE 4.2**. The volume to be diluted is withdrawn using a calibrated tube called a *pipet*, placed in an empty volumetric flask of the chosen volume, and diluted to the calibration mark on the flask. The one common exception to this order of steps is when diluting a strong acid such as H_2SO_4, where a large amount of heat is released. In such instances, it is much safer to add the acid slowly to the water rather than adding water to the acid.

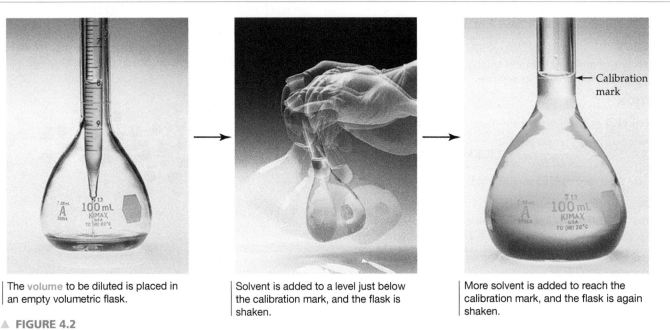

| The volume to be diluted is placed in an empty volumetric flask. | Solvent is added to a level just below the calibration mark, and the flask is shaken. | More solvent is added to reach the calibration mark, and the flask is again shaken. |

▲ **FIGURE 4.2**

The procedure for diluting a concentrated solution.

WORKED EXAMPLE 4.3

Diluting a Solution

What volume of 1.000 M NaOH solution is required to prepare 500.0 mL of 0.2500 M NaOH?

IDENTIFY

Known	Unknown
Final volume (V_f = 500.0 mL)	Initial volume (V_i)
Final concentration (M_f = 0.2500 M)	
Initial concentration (M_i = 1.000 M)	

STRATEGY

The initial volume (V_i) can be found by rearranging the equation $M_i \times V_i = M_f \times V_f$ to solve for the unknown, $V_i = (M_f/M_i) \times V_f$.

SOLUTION

$$V_i = \frac{M_f}{M_i} \times V_f = \frac{0.2500 \text{ M}}{1.000 \text{ M}} \times 500.0 \text{ mL} = 125.0 \text{ mL}$$

CHECK

Because the concentration decreases by a factor of 4 after dilution (from 1.000 M to 0.2500 M), the volume must increase by a factor of 4. Thus, to prepare 500.0 mL of solution, we should start with 500.0/4 = 125.0 mL.

▶ **PRACTICE 4.5** What is the final concentration if 75.0 mL of a 3.50 M glucose solution is diluted to a volume of 400.0 mL?

▶ **APPLY 4.6** Sulfuric acid is normally purchased at a concentration of 18.0 M. How would you prepare 250.0 mL of 0.500 M aqueous H_2SO_4? (Remember to add acid to water rather than water to acid.)

4.3 ELECTROLYTES IN AQUEOUS SOLUTION

Before beginning a study of reactions in water, let's look at some general properties of aqueous solutions. We all know that both sugar (sucrose) and table salt (NaCl) dissolve in water. The solutions that result, though, are quite different. When sucrose, a molecular substance, dissolves in water, the resulting solution contains neutral sucrose **molecules** surrounded by water. When NaCl, an ionic substance, dissolves in water, the solution contains separate Na^+ and Cl^- **ions** surrounded by water. Because of the presence of the charged ions, the NaCl solution conducts an electric current, but the sucrose solution does not.

$$C_{12}H_{22}O_{11}(s) \xrightarrow{H_2O} C_{12}H_{22}O_{11}(aq)$$
$$\text{Sucrose}$$
$$NaCl(s) \xrightarrow{H_2O} Na^+(aq) + Cl^-(aq)$$

The electrical conductivity of an aqueous NaCl solution is easy to demonstrate using a battery, a light bulb, and several pieces of wire, connected as shown in **FIGURE 4.3**. When the wires are dipped into an aqueous NaCl solution, the positively charged Na^+ ions move through the solution toward the wire connected to the negatively charged terminal of the battery, and the negatively charged Cl^- ions move toward the wire connected to the positively charged terminal of the battery. The resulting movement of electrical charges allows a current to flow, so the bulb lights. When the wires are dipped into an aqueous sucrose solution, however, there are no ions to carry the current, so the bulb remains dark.

> **REMEMBER ...**
> A **molecule** is a unit of matter that results when two or more nonmetal atoms are joined by covalent bonds in which electrons are shared. An ionic substance is formed when a metal and nonmetal atom form an ionic bond in which electrons are transferred from the metal to the nonmetal to form **ions** (Section 2.11).

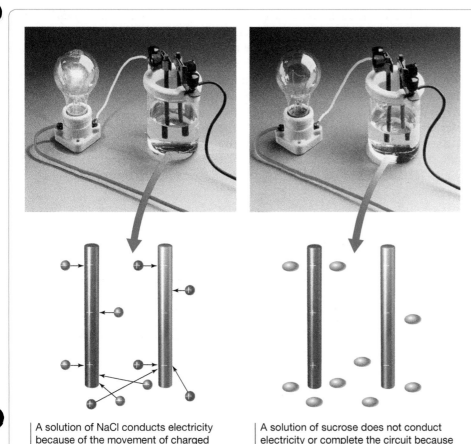

◀ **FIGURE 4.3**

Testing the conductivity of aqueous solutions.

◀ Figure It Out

Describe the appearance of the light bulb if it were placed in 1.0 M solutions of (a) HNO_3, (b) HF, and (c) C_2H_5OH. Refer to Table 4.1 for classification of these compounds.

Answer: (a) The bulb would be bright since HNO_3 is a strong electrolyte and dissociates nearly completely into ions. (b) The bulb would light dimly because HF is a weak electrolyte and dissociates only to a small extent. (c) The bulb would not light because ethanol is a nonelectrolyte.

A solution of NaCl conducts electricity because of the movement of charged particles (ions), thereby completing the circuit and allowing the bulb to light.

A solution of sucrose does not conduct electricity or complete the circuit because it contains no mobile charged particles. The bulb therefore remains dark.

Substances such as NaCl or KBr, which dissolve in water to produce conducting solutions of ions, are called **electrolytes.** Substances such as sucrose or ethyl alcohol, which do not produce ions in aqueous solution, are **nonelectrolytes.** Most electrolytes are ionic compounds, but some are molecular. Hydrogen chloride, for instance, is a gaseous molecular compound when pure but **dissociates,** or splits apart, to give H^+ and Cl^- ions when it dissolves in water.

$$HCl(g) \xrightarrow{\text{H}_2\text{O}} H^+(aq) + Cl^-(aq)$$

Compounds that dissociate to a large extent (70–100%) into ions when dissolved in water are said to be **strong electrolytes,** while compounds that dissociate to only a small extent are **weak electrolytes.** Potassium chloride and most other ionic compounds, for instance, are largely dissociated in dilute solution and are thus strong electrolytes. Acetic acid (CH_3CO_2H), by contrast, dissociates only to the extent of about 1.3% in a 0.10 M solution and is a weak electrolyte. As a result, a 0.10 M solution of acetic acid is only weakly conducting, and the bulb in Figure 4.3 would only light dimly.

For 0.10 M solutions:
$$KCl(aq) \rightleftharpoons K^+(aq) + Cl^-(aq) \quad \text{•} \dashv \text{ Strong electrolyte}$$
$$\quad\quad (2\%) \quad\quad\quad\quad (98\%)$$

$$CH_3CO_2H(aq) \rightleftharpoons H^+(aq) + CH_3CO_2^-(aq) \quad \text{•} \dashv \text{ Weak electrolyte}$$
$$\quad\quad (99\%) \quad\quad\quad\quad (1\%)$$

Note that a forward-and-backward double arrow ($\rightleftharpoons$) is used in the dissociation equation to indicate that the reaction takes place simultaneously in both directions. That is, dissociation is a dynamic process in which an *equilibrium* is established between the forward and reverse reactions. Dissociation of acetic acid takes place in the forward direction, while recombination of H^+ and $CH_3CO_2^-$ ions takes place in the reverse direction. The size of the equilibrium arrow indicates whether the equilibrium reaction forms mostly products or mostly reactants. Ultimately, the concentrations of the reactants and products reach constant values and no longer change with time. We'll learn much more about **chemical equilibria** in Chapter 15.

LOOKING AHEAD . . .
A **chemical equilibrium,** as we'll see in Chapter 15, is the state in which a reaction takes place in both forward and backward directions so that the concentrations of products and reactants remain constant over time.

TABLE 4.1 lists some common substances classified according to their electrolyte strength. Note that pure water is a nonelectrolyte because it does not dissociate appreciably into H^+ and OH^- ions. We'll explore the dissociation of water in more detail in Section 16.4.

BIG IDEA Question 2   Go to eText

The light bulb conductivity apparatus shown in Figure 4.3 is dipped in a 1.0 M aqueous solution of rubidium bromide. Predict whether the light from the bulb will be bright, dim, or absent.

TABLE 4.1   Electrolyte Classification of Some Common Substances

Strong Electrolytes	Weak Electrolytes	Nonelectrolytes
HCl, HBr, HI	CH_3CO_2H	H_2O
$HClO_4$	HF	CH_3OH (methyl alcohol)
HNO_3	HCN	C_2H_5OH (ethyl alcohol)
H_2SO_4		$C_{12}H_{22}O_{11}$ (sucrose)
KBr		Most compounds of carbon (organic compounds)
NaCl		
NaOH, KOH		
Other soluble ionic compounds		

— **WORKED EXAMPLE 4.4**

Calculating the Concentration of Ions in a Solution

What is the total molar concentration of ions in a 0.350 M solution of the strong electrolyte Na_2SO_4, assuming complete dissociation?

IDENTIFY

Known	Unknown
Molar concentration of compound (0.350 M Na_2SO_4)	Total molar concentration of ions

STRATEGY

The ions come from the dissociation of Na_2SO_4, and therefore a reaction for dissolving Na_2SO_4 in water is written. The balanced reaction shows that 3 mol of ions are formed from 1 mol of the compound: 2 mol of Na^+ and 1 mol of SO_4^{2-}.

$$Na_2SO_4(s) \xrightarrow{H_2O} 2\,Na^+(aq) + SO_4^{2-}(aq)$$

SOLUTION

Assuming complete dissociation, the total molar concentration of ions is three times the molarity of Na_2SO_4, or 1.05 M:

$$\frac{0.350\ \text{mol } Na_2SO_4}{1\ L} \times \frac{3\ \text{mol ions}}{1\ \text{mol } Na_2SO_4} = 1.05\ \text{M}$$

▶ **PRACTICE 4.7**   What is the molar concentration of Br^- ions in a 0.225 M aqueous solution of $FeBr_3$, assuming complete dissociation?

▶ **CONCEPTUAL APPLY 4.8**   Three different substances, A_2X, A_2Y, and A_2Z, are dissolved in water, with the following results. (Water molecules are omitted for clarity.)

(a) Which of the substances is the strongest electrolyte, and which is the weakest? Explain.
(b) What is the molar concentration of A ions and Y ions in a 0.350 M solution of A_2Y?
(c) What is the percent ionization of A_2X indicated in the graphical representation?

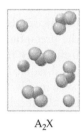  

    A_2X A_2Y A_2Z

4.4 TYPES OF CHEMICAL REACTIONS IN AQUEOUS SOLUTION

Many common chemical reactions that take place in aqueous solution fall into one of three general categories: *precipitation reactions, acid–base neutralization reactions,* and *oxidation–reduction reactions.* Let's look briefly at an example of each before looking at them in more detail in subsequent sections.

- In **precipitation reactions,** soluble ionic reactants (strong electrolytes) yield an insoluble solid product called a *precipitate,* which falls out of the solution, thereby removing some of the dissolved ions. Most precipitations take place when the anions and cations of two ionic compounds change partners. For example, an aqueous solution of lead(II) nitrate reacts with an aqueous solution of potassium iodide to yield an aqueous solution of potassium nitrate plus an insoluble yellow precipitate of lead(II) iodide:

$$Pb(NO_3)_2(aq) + 2\,KI(aq) \rightarrow 2\,KNO_3(aq) + PbI_2(s)$$

- In **acid–base neutralization reactions,** an acid reacts with a base to yield water plus an ionic compound called a *salt.* Acids are compounds that produce H^+ ions when dissolved in water, and bases are compounds that produce OH^- ions when dissolved in water. Thus, a neutralization reaction removes H^+ and OH^- ions from solution, just as a precipitation reaction removes metal and nonmetal ions. The

▲ Reaction of aqueous lead(II) nitrate with aqueous potassium iodide gives a yellow precipitate of lead(II) iodide.
$Pb(NO_3)_2(aq) + 2\,KI(aq) \rightarrow$
$PbI_2(s) + 2\,KNO_3(aq)$

Go to
eText

BIG IDEA Question 3

Classify the reaction as a precipitation, acid-base neutralization, or redox reaction.

$$Cu^{2+}(aq) + Zn(s) \longrightarrow$$
$$Cu(s) + Zn^{2+}(aq)$$

reaction between hydrochloric acid and aqueous sodium hydroxide to yield water plus aqueous sodium chloride is a typical example:

$$HCl(aq) + NaOH(aq) \longrightarrow H_2O(l) + NaCl(aq)$$

• In **oxidation–reduction reactions,** or **redox reactions,** one or more electrons are transferred between reaction partners (atoms, molecules, or ions). As a result of this electron transfer, the charges on atoms in the various reactants change. When metallic magnesium reacts with aqueous hydrochloric acid, for instance, a magnesium atom gives an electron to each of two H^+ ions, forming an Mg^{2+} ion and a H_2 molecule. The charge on the magnesium changes from 0 to +2, and the charge on each hydrogen changes from +1 to 0:

$$Mg(s) + 2\,HCl(aq) \longrightarrow MgCl_2(aq) + H_2(g)$$

4.5 AQUEOUS REACTIONS AND NET IONIC EQUATIONS

The equations we've been writing up to this point have all been **molecular equations.** That is, all the substances involved in the reactions have been written using their complete formulas as if they were *molecules*. In the previous section, for instance, we wrote the precipitation reaction of lead(II) nitrate with potassium iodide to yield solid PbI_2 using only the parenthetical **state abbreviation** (*aq*) to indicate that the substances are dissolved in aqueous solution. Nowhere in the equation was it indicated that ions were involved. There are, however, more accurate ways of writing equations to reflect the chemical changes that occur during reactions, as the following series of examples illustrates.

REMEMBER . . .

The physical state of a substance in a chemical reaction is often indicated with a parenthetical **state abbreviation** (*s*) for solid, (*l*) for liquid, (*g*) for gas, and (*aq*) for aqueous solution (Section 3.3).

A Molecular Equation

$$Pb(NO_3)_2(aq) + 2\,KI(aq) \longrightarrow 2\,KNO_3(aq) + PbI_2(s)$$

This equation implies that molecules are interacting. It is the case, however, that lead nitrate, potassium iodide, and potassium nitrate are strong electrolytes that dissolve in water to yield solutions of ions. Thus, it's more accurate to write the precipitation reaction as an **ionic equation,** in which all the ions are explicitly shown.

An Ionic Equation

$$Pb^{2+}(aq) + 2\,NO_3^-(aq) + 2\,K^+(aq) + 2\,I^-(aq) \longrightarrow 2\,K^+(aq) + 2\,NO_3^-(aq) + PbI_2(s)$$

This ionic equation shows that the NO_3^- and K^+ ions undergo no change during the reaction. Instead, they appear on both sides of the reaction arrow and act merely as **spectator ions,** whose only role is to balance the charge. A **net ionic equation** gives only the species that react (the Pb^{2+} and I^- ions in this instance) because spectator ions are canceled from both sides of the equation.

An Ionic Equation

$$Pb^{2+}(aq) + 2\,\cancel{NO_3^-}(aq) + 2\,\cancel{K^+}(aq) + 2\,I^-(aq) \longrightarrow 2\,\cancel{K^+}(aq) + 2\,\cancel{NO_3^-}(aq) + PbI_2(s)$$

A Net Ionic Equation

$$Pb^{2+}(aq) + 2\,I^-(aq) \longrightarrow PbI_2(s)$$

Go to
eText

BIG IDEA Question 4

What is a spectator ion?

Leaving the spectator ions out of a net ionic equation doesn't mean that their presence is irrelevant. If a reaction occurs by mixing a solution of Pb^{2+} ions with a solution of I^- ions, then those solutions must also contain additional ions to balance the charge in each. That is, the Pb^{2+} solution must also contain an anion, and the I^- solution must also contain a cation. Leaving these other ions out of the net ionic equation only implies that these ions do not undergo a chemical reaction. Any nonreactive spectator ion could serve to balance charge.

WORKED EXAMPLE 4.5

Writing a Net Ionic Equation

Aqueous hydrochloric acid reacts with zinc metal to yield hydrogen gas and aqueous zinc chloride. Write a net ionic equation for the process, and identify the spectator ions.

$$2\,HCl(aq) + Zn(s) \longrightarrow H_2(g) + ZnCl_2(aq)$$

STRATEGY

First, write the ionic equation, listing all the species present in solution. Table 4.1 tells us that both HCl (a molecular compound) and $ZnCl_2$ (a soluble ionic compound) are strong electrolytes that exist as ions in solution. Then find the ions that are present on both sides of the reaction arrow—the spectator ions—and cancel them to leave the net ionic equation.

SOLUTION

Ionic Equation

$$2\,H^+(aq) + 2\,\cancel{Cl^-}(aq) + Zn(s) \longrightarrow H_2(g) + Zn^{2+}(aq) + 2\,\cancel{Cl^-}(aq)$$

Net Ionic Equation

$$2\,H^+(aq) + Zn(s) \longrightarrow H_2(g) + Zn^{2+}(aq)$$

Chloride ion is the only spectator ion in this reaction.

▲ Zinc metal reacts with aqueous hydrochloric acid to give hydrogen gas and aqueous Zn^{2+} ions.

▶ **PRACTICE 4.9** Write net ionic equations for the following reaction:

$$2\,AgNO_3(aq) + Na_2CrO_4(aq) \longrightarrow Ag_2CrO_4(s) + 2\,NaNO_3(aq)$$

▶ **CONCEPTUAL APPLY 4.10** Two clear, colorless solutions are mixed, and a white precipitate forms and settles to the bottom of the container. The diagram on the right represents the ions in solution and the solid after the reaction has occurred. Identify the spectator ions, and give the balanced net ionic equation and molecular equation for the reaction.

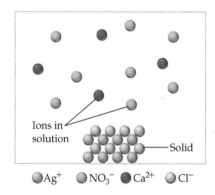

Ions in solution

Solid

Ag^+ NO_3^- Ca^{2+} Cl^-

4.6 PRECIPITATION REACTIONS AND SOLUBILITY GUIDELINES

How can you predict whether a precipitation reaction will occur when mixing aqueous solutions of two substances? To do that, you must know the **solubility** of each potential product—how much of each compound will dissolve in a given amount of solvent at a given temperature. If a substance has a high solubility in water, no precipitate will form. If a substance has a low solubility in water, it's likely to precipitate from an aqueous solution.

Solubility is a complex matter, and it's not always possible to make correct predictions. Furthermore, solubility depends on the concentrations of the reactant ions, and the very words *soluble* and *insoluble* are imprecise. Solubility is a continuum with some substances having a high solubility and others having a low solubility. In this section, we will define a substance as soluble if it dissolves to give a concentration of 0.01 M or greater. Solubility can be predicted by looking at the cations and anions that make up the compound. A compound is soluble if it meets either (or both) of the following criteria:

1. **A compound is soluble if it contains one of the following** *cations*:

   - Li^+, Na^+, K^+, Rb^+, Cs^+ (group 1A cations)
   - NH_4^+ (ammonium ion)

That is, essentially all ionic compounds containing an alkali metal cation or ammonium cation are soluble in water and will not precipitate, regardless of the anions present.

2. **A compound is soluble if it contains one of the following** *anions:*

- Cl^-, Br^-, I^- (halide)
  except: Ag^+, Hg_2^{2+}, and Pb^{2+} halides

- NO_3^- (nitrate), ClO_4^- (perchlorate), $CH_3CO_2^-$ (acetate), and SO_4^{2-} (sulfate)
  except: Sr^{2+}, Ba^{2+}, Hg_2^{2+}, and Pb^{2+} sulfates

That is, most ionic compounds containing a halide, nitrate, perchlorate, acetate, or sulfate anion are soluble in water and will not precipitate regardless of the cations present. The exceptions that *will* precipitate are silver(I), mercury(I) and lead(II) halides and strontium, barium, mercury(I), and lead(II) sulfates.

On the other hand, a compound that does *not* contain one of the cations or anions listed above is *not* soluble. Thus, carbonates (CO_3^{2-}), sulfides (S^{2-}), phosphates (PO_4^{3-}), and hydroxides (OH^-) are generally not soluble unless they contain an alkali metal or ammonium cation. The main exceptions are the sulfides and hydroxides of Ca^{2+}, Sr^{2+}, and Ba^{2+}. Solubility guidelines are summarized in **TABLE 4.2.**

Go to eText

BIG IDEA Question 5

Use the solubility rules in Table 4.2 to classify each of the ionic compounds Na_2CO_3 and $Fe(OH)_3$ as soluble or insoluble.

TABLE 4.2   Solubility Guidelines for Ionic Compounds in Water

Soluble Compounds	Common Exceptions
Li^+, Na^+, K^+, Rb^+, Cs^+ (group 1A cations)	None
NH_4^+ (ammonium ion)	None
Cl^-, Br^-, I^- (halide)	Halides of Ag^+, Hg_2^{2+}, Pb^{2+}
NO_3^- (nitrate)	None
ClO_4^- (perchlorate)	None
$CH_3CO_2^-$ (acetate)	None
SO_4^{2-} (sulfate)	Sulfates of Sr^{2+}, Ba^{2+}, Hg_2^{2+}, Pb^{2+}
Insoluble Compounds	**Common Exceptions**
CO_3^{2-} (carbonate)	Carbonates of group 1A cations, NH_4^+
S^{2-} (sulfide)	Sulfides of group 1A cations, NH_4^+, Ca^{2+}, Sr^{2+}, and Ba^{2+}
PO_4^{3-} (phosphate)	Phosphates of group 1A cations, NH_4^+
OH^- (hydroxide)	Hydroxides of group 1A cations, NH_4^+, Ca^{2+}, Sr^{2+}, and Ba^{2+}

You might notice that most of the ions that impart solubility to compounds are singly charged—either singly positive (Li^+, Na^+, K^+, Rb^+, Cs^+, NH_4^+) or singly negative (Cl^-, Br^-, I^-, NO_3^-, ClO_4^-, $CH_3CO_2^-$). Very few doubly charged ions or triply charged ions form soluble compounds. This solubility behavior arises because of the relatively strong ionic bonds in compounds containing ions with multiple charges. The greater the strength of the ionic bonds holding ions together in a crystal, the more difficult it is to break those bonds apart during the solution process. We'll return to this topic in Section 6.8.

Using the solubility guidelines makes it possible not only to predict whether a precipitate will form when solutions of two ionic compounds are mixed but also to prepare a specific compound by purposefully carrying out a precipitation. If, for example, you wanted to prepare a sample of solid silver carbonate, Ag_2CO_3, you could mix a solution of $AgNO_3$ with a solution of Na_2CO_3. Both starting compounds are soluble in water, as is $NaNO_3$. Silver carbonate is the only insoluble combination of ions and will therefore precipitate from solution.

▲ Reaction of aqueous $AgNO_3$ with aqueous Na_2CO_3 gives a white precipitate of Ag_2CO_3.

$$2\,AgNO_3(aq) + Na_2CO_3(aq) \longrightarrow Ag_2CO_3(s) + 2\,NaNO_3(aq)$$

WORKED EXAMPLE 4.6

Predicting the Product of a Precipitation Reaction

Will a precipitation reaction occur when aqueous solutions of $CdCl_2$ and $(NH_4)_2S$ are mixed? If so, write the net ionic equation.

STRATEGY

Determine the possible products of the reaction by combining the cation from one reactant with the anion from the other reactant.

$$CdCl_2(aq) + (NH_4)_2S(aq) \longrightarrow CdS(?) + 2\,NH_4Cl(?)$$

Next predict the solubility of each product using the guidelines in Table 4.2.

SOLUTION

Of the two possible products, the solubility guidelines predict that CdS, a sulfide, is insoluble and that NH_4Cl, an ammonium

compound and a halide, is soluble. Thus, a precipitation reaction will likely occur:

$$Cd^{2+}(aq) + S^{2-}(aq) \longrightarrow CdS(s)$$

▶ **PRACTICE 4.11**   Will a precipitation reaction occur when aqueous solutions of $AgClO_4$ and $CaBr_2$ are mixed? If so, write the net ionic equation.

▶ **APPLY 4.12**   How might you use a precipitation reaction to prepare a sample of $Ca_3(PO_4)_2$? Write the net ionic equation.

CONCEPTUAL WORKED EXAMPLE 4.7

Visualizing Stoichiometry in Precipitation Reactions

When aqueous solutions of two ionic compounds are mixed, the following results are obtained:

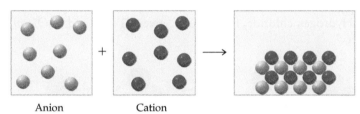

(Only the anion of the first compound, represented by blue spheres, and the cation of the second compound, represented by red spheres, are shown.) Which cation and anion combinations are compatible with the observed results?
   Anions: NO_3^-, Cl^-, CO_3^{2-}, PO_4^{3-}
   Cations: Ca^{2+}, Ag^+, K^+, Cd^{2+}

STRATEGY

The drawing represents a precipitation reaction because it shows that ions in solution fall to the bottom of the container in an ordered arrangement. Counting the spheres shows that the cation and anion react in equal numbers (8 of each), so they must have the same number of charges—either both singly charged or both doubly charged. (There is no triply charged cation in the list.) Look at all the possible combinations, and decide which would precipitate.

SOLUTION

   Possible combinations of singly charged ions: $AgNO_3$, KNO_3, AgCl, KCl
   Possible combinations of doubly charged ions: $CaCO_3$, $CdCO_3$

Of the possible combinations, AgCl, $CaCO_3$, and $CdCO_3$ are insoluble. Therefore, the precipitate could arise from three possible combinations of anions and cations (1) Ag^+ and Cl^-, (2) Ca^{2+} and CO_3^{2-}, or (3) Cd^{2+} and CO_3^{2-}.

▶ **CONCEPTUAL PRACTICE 4.13**   An aqueous solution containing an anion, represented by blue spheres, is added to another

solution containing a cation, represented by red spheres, and the following result is obtained.

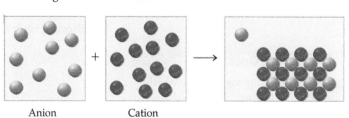

Which cation and anion combinations are compatible with the observed results?
   Anions: S^{2-}, PO_4^{3-}
   Cations: Mg^{2+}, Fe^{3+}, NH_4^+

▶ **CONCEPTUAL APPLY 4.14**   A solution containing the compounds $Pb(NO_3)_2$, KBr, and $Ba(CH_3CO_2)_2$ is prepared, and a white precipitate forms. Identify the precipitate, and draw a diagram similar to the one in Worked Example 4.7 to represent the precipitation reaction.

4.7 ACIDS, BASES, AND NEUTRALIZATION REACTIONS

In 1777, the French chemist Antoine Lavoisier (1743–1794) proposed that all acids contain a common element: oxygen. In fact, the word *oxygen* is derived from a Greek phrase meaning "acid former." Lavoisier's idea had to be modified, however, when the English chemist Sir Humphrey Davy (1778–1829) showed in 1810 that muriatic acid (now called hydrochloric acid) contains only hydrogen and chlorine but no oxygen. Davy's studies thus suggested that the common element in acids is *hydrogen,* not oxygen.

LOOKING AHEAD . . .
Acids and bases are discussed in detail in Chapter 16.

Swedish chemist Svante Arrhenius (1859–1927) clarified the relationship between acidic behavior and the presence of hydrogen in a compound in 1887. Arrhenius proposed that an **acid** is a substance that dissociates in water to give hydrogen ions (H^+) and a **base** is a substance that dissociates in water to give hydroxide ions (OH^-):

$$\textbf{An } \text{acid}\quad HA(aq) \longrightarrow H^+(aq) + A^-(aq)$$
$$\textbf{A } \text{base}\quad MOH(aq) \longrightarrow M^+(aq) + OH^-(aq)$$

In these equations, HA is a general formula for an acid—for example, HCl or HNO_3—and MOH is a general formula for a metal hydroxide—for example, NaOH or KOH.

Although convenient to use in equations, the symbol $H^+(aq)$ does not really represent the structure of the ion present in aqueous solution. As a bare hydrogen nucleus—a proton—with no electron nearby, H^+ is much too reactive to exist by itself. Rather, the H^+ bonds to the oxygen atom of a water molecule and forms the more stable **hydronium ion, H_3O^+.** We'll sometimes write $H^+(aq)$ for convenience, particularly when balancing equations, but will more often write $H_3O^+(aq)$ to represent an aqueous acid solution. Hydrogen chloride, for instance, gives $Cl^-(aq)$ and $H_3O^+(aq)$ when it dissolves in water.

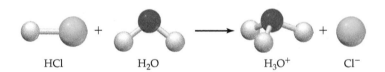

| HCl | | H_2O | | H_3O^+ | | Cl^- |

Different acids dissociate to different extents in aqueous solution. Acids that dissociate to a large extent are strong electrolytes and **strong acids,** whereas acids that dissociate to only a small extent are weak electrolytes and **weak acids.** We've already seen in Table 4.1, for instance, that HCl, $HClO_4$, HNO_3, and H_2SO_4 are strong electrolytes and therefore strong acids, while CH_3CO_2H and HF are weak electrolytes and therefore weak acids. You might note that acetic acid actually contains four hydrogens, but only the one bonded to the oxygen atom dissociates. We will explain the effect of molecular structure on acid dissociation in Chapter 16.

BIG IDEA Question 6 Go to eText

The equation for the reaction of ethylamine with water is as follows:

$$CH_3NH_2(aq) + H_2O(l) \rightleftharpoons$$
$$CH_3NH_3^+(aq) + OH^-(aq)$$

Classify methylamine as a strong acid, a weak acid, a strong base, or a weak base.

Acetic acid

Nonacidic

Acidic

Different acids can have different numbers of acidic hydrogens and yield different numbers of H_3O^+ ions in solution. Hydrochloric acid (HCl) is said to be a **monoprotic acid** because it provides only one H^+ ion, but sulfuric acid (H_2SO_4) is a **diprotic acid** because it can provide two H^+ ions. Phosphoric acid (H_3PO_4) is a **triprotic acid** and can provide three H^+ ions. With sulfuric acid, the first dissociation of an H^+ is complete—all H_2SO_4 molecules lose one H^+—but the second dissociation is incomplete, as indicated by the double arrow in the following equation:

Sulfuric acid: $H_2SO_4(aq) + H_2O(l) \longrightarrow HSO_4^-(aq) + H_3O^+(aq)$
$HSO_4^-(aq) + H_2O(l) \rightleftharpoons SO_4^{2-}(aq) + H_3O^+(aq)$

With phosphoric acid, none of the three dissociations is complete:

Phosphoric acid: $H_3PO_4(aq) + H_2O(l) \rightleftharpoons H_2PO_4^-(aq) + H_3O^+(aq)$
$H_2PO_4^-(aq) + H_2O(l) \rightleftharpoons HPO_4^{2-}(aq) + H_3O^+(aq)$
$HPO_4^{2-}(aq) + H_2O(l) \rightleftharpoons PO_4^{3-}(aq) + H_3O^+(aq)$

Bases, like acids, can also be either strong or weak, depending on the extent to which they produce OH^- ions in aqueous solution. Most metal hydroxides, such as NaOH and $Ba(OH)_2$, are strong electrolytes and **strong bases.** The dissociation reactions for sodium hydroxide and barium hydroxide are:

$NaOH(aq) \longrightarrow Na^+(aq) + OH^-(aq)$
$Ba(OH)_2(aq) \longrightarrow Ba^{2+}(aq) + 2\ OH^-(aq)$

Ammonia (NH_3) is a weak electrolyte and a **weak base.** Ammonia is a weak base because it reacts to a small extent with water to yield NH_4^+ and OH^- ions. In fact, aqueous solutions of ammonia are often called *ammonium hydroxide,* although this is really a misnomer because the concentrations of NH_4^+ and OH^- ions are low.

$NH_3(g) + H_2O(l) \rightleftharpoons NH_4^+(aq) + OH^-(aq)$

As with the dissociation of acetic acid, discussed in Section 4.3, the reaction of ammonia with water takes place only to a small extent (about 1%). Most of the ammonia remains unreacted, and we therefore write the reaction with a double arrow to show that a dynamic equilibrium exists between the forward and reverse reactions.

TABLE 4.3 summarizes the names, formulas, and classification of some common acids and bases.

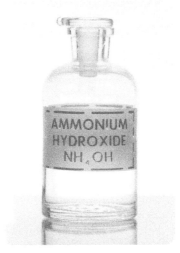

⬛ Shouldn't this bottle be labeled "Aqueous Ammonia" rather than "Ammonium Hydroxide"?

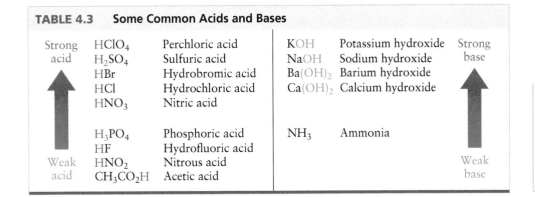

TABLE 4.3	Some Common Acids and Bases				
Strong acid	$HClO_4$	Perchloric acid	KOH	Potassium hydroxide	Strong base
	H_2SO_4	Sulfuric acid	NaOH	Sodium hydroxide	
	HBr	Hydrobromic acid	$Ba(OH)_2$	Barium hydroxide	
	HCl	Hydrochloric acid	$Ca(OH)_2$	Calcium hydroxide	
	HNO_3	Nitric acid			
	H_3PO_4	Phosphoric acid	NH_3	Ammonia	
	HF	Hydrofluoric acid			
Weak acid	HNO_2	Nitrous acid			Weak base
	CH_3CO_2H	Acetic acid			

BIG IDEA Question 7

Which solution will have the highest concentration of H_3O^+ ions: 1.0 M HNO_3, 1.0 M HF, 1.0 M NaOH, or 1.0 M NH_3? Refer to Table 4.3, which lists some common acids and bases.

Naming Acids

Most acids are **oxoacids;** that is, they contain oxygen in addition to hydrogen and other elements. When dissolved in water, an oxoacid yields one or more H^+ ions and an **oxoanion** like one of those listed in **TABLE 4.4** and discussed previously in Section 2.13.

The names of oxoacids are related to the names of the corresponding oxoanions, with the *-ite* or *-ate* ending of the anion name replaced by *-ous acid* or *-ic acid,* respectively. In other words, the acid with fewer oxygens has an *-ous* ending, and the acid with more oxygens has an *-ic* ending. The compound HNO_2, for example, is called *nitrous acid* because it has fewer oxygens and yields the nit*rite* ion (NO_2^-) when dissolved in water, while HNO_3 is called *nitric acid* because it has more oxygen atoms and yields the nit*rate* ion (NO_3^-) when dissolved in water.

REMEMBER . . .
Oxoanions are polyatomic anions in which an atom of a given element is combined with different numbers of oxygen atoms (Section 2.12).

TABLE 4.4 Common Oxoacids and Their Anions

Oxoacid		Oxoanion	
HNO_2	Nitrous acid	NO_2^-	Nitrite ion
HNO_3	Nitric acid	NO_3^-	Nitrate ion
H_3PO_4	Phosphoric acid	PO_4^{3-}	Phosphate ion
H_2SO_3	Sulfurous acid	SO_3^{2-}	Sulfite ion
H_2SO_4	Sulfuric acid	SO_4^{2-}	Sulfate ion
$HClO$	Hypochlorous acid	ClO^-	Hypochlorite ion
$HClO_2$	Chlorous acid	ClO_2^-	Chlorite ion
$HClO_3$	Chloric acid	ClO_3^-	Chlorate ion
$HClO_4$	Perchloric acid	ClO_4^-	Perchlorate ion

Nitrous Acid Gives Nitrite Ion

$$HNO_2(aq) + H_2O(l) \xrightleftharpoons{\text{Dissolve in water}} H_3O^+(aq) + NO_2^-(aq)$$

Nitric Acid Gives Nitrate Ion

$$HNO_3(aq) + H_2O(l) \xrightarrow{\text{Dissolve in water}} H_3O^+(aq) + NO_3^-(aq)$$

In a similar way, hypochlorous acid yields the hypochlorite ion, chlorous acid yields the chlorite ion, chloric acid yields the chlorate ion, and perchloric acid yields the perchlorate ion (Table 4.4).

In addition to the oxoacids, there are a small number of other common acids, such as HCl, that do not contain oxygen. For such compounds, the prefix *hydro-* and the suffix *-ic acid* are used for the aqueous solution.

Hydrogen Chloride Gives *Hydrochloric Acid*

$$HCl(g) + H_2O(l) \xrightarrow{\text{Dissolve in water}} H_3O^+(aq) + Cl^-(aq)$$

Hydrogen Cyanide Gives *Hydrocyanic Acid*

$$HCN(g) + H_2O(l) \xrightleftharpoons{\text{Dissolve in water}} H_3O^+(aq) + CN^-(aq)$$

WORKED EXAMPLE 4.8

Naming Acids

Name the following acids:
(a) HBrO(aq)
(b) H₂S(aq)

STRATEGY

To name an acid, look at its formula, and decide whether the compound is an oxoacid. If so, the name must reflect the number of oxygen atoms, according to Table 4.4. If the compound is not an oxoacid, it is named using the prefix *hydro-* and the suffix *-ic acid*.

SOLUTION

(a) This compound is an oxoacid that yields hypobromite ion (BrO⁻) when dissolved in water. Its name is *hypobromous acid*.

(b) This compound is not an oxoacid but yields sulfide ion when dissolved in water. As a pure gas, H₂S is named hydrogen sulfide. In water solution, it is called *hydrosulfuric acid*.

▶ **PRACTICE 4.15** Name the acids HI and HBrO₂.

▶ **APPLY 4.16** Give likely chemical formulas corresponding to the following names:

(a) Phosphorous acid   (b) Hydroselenic acid

Neutralization Reactions

When an acid and a base are mixed in the right stoichiometric proportions, both acidic and basic properties disappear because of a neutralization reaction that produces water and an ionic **salt**. The anion of the salt (A^-) comes from the acid, and the cation of the salt (M^+) comes from the base:

> **A neutralization reaction**
>
> $$\underset{\text{Acid}}{HA(aq)} + \underset{\text{Base}}{MOH(aq)} \longrightarrow \underset{\text{Water}}{H_2O(l)} + \underset{\text{A salt}}{MA(aq)}$$

Because salts are generally strong electrolytes in aqueous solution, we can write the neutralization reaction of a strong acid with a strong base as an ionic equation:

$$H^+(aq) + A^-(aq) + M^+(aq) + OH^-(aq) \longrightarrow H_2O(l) + M^+(aq) + A^-(aq)$$

Canceling the ions that appear on both sides of the ionic equation, A^- and M^+, gives the net ionic equation, which describes the reaction of any strong acid with any strong base in water.

Net Ionic Equation

$$H^+(aq) + OH^-(aq) \longrightarrow H_2O(l)$$
$$\text{or} \quad H_3O^+(aq) + OH^-(aq) \longrightarrow 2\, H_2O(l)$$

For the reaction of a weak acid with a strong base, a similar neutralization occurs, but we must write the molecular formula of the acid rather than simply H^+ (aq) because the dissociation of the acid in water is incomplete. Instead, the acid exists primarily as the neutral molecule. In the reaction of the weak acid HF with the strong base KOH, for example, we write the net ionic equation as

$$HF(aq) + OH^-(aq) \longrightarrow H_2O(l) + F^-(aq)$$

WORKED EXAMPLE 4.9

Writing Ionic and Net Ionic Equations for an Acid–Base Reaction

Write both an ionic equation and a net ionic equation for the neutralization reaction of aqueous HBr and aqueous $Ba(OH)_2$.

STRATEGY

Hydrogen bromide is a strong acid whose aqueous solution contains H^+ ions and Br^- ions. Barium hydroxide is a strong base whose aqueous solution contains Ba^{2+} and OH^- ions. Thus, we have a mixture of four different ions on the reactant side. Two HBr molecules are needed to react with one formula unit of $Ba(OH)_2$ because there are two hydroxide ions in barium hydroxide. Write the neutralization reaction as an ionic equation, and then cancel spectator ions to give the net ionic equation.

SOLUTION

Ionic Equation

$$2\, H^+(aq) + 2\, Br^-(aq) + Ba^{2+}(aq) + 2\, OH^-(aq) \longrightarrow 2\, H_2O(l) + 2\, Br^-(aq) + Ba^{2+}(aq)$$

Net Ionic Equation

$$2H^+(aq) + 2OH^-(aq) \longrightarrow 2H_2O(l)$$
$$\text{or} \quad H^+(aq) + OH^-(aq) \longrightarrow H_2O(l)$$

The reaction of HBr with $Ba(OH)_2$ involves the combination of a proton (H^+) from the acid with OH^- from the base to yield water and an aqueous salt ($BaBr_2$).

continued on next page

Milk of
Magnesia
LAXATIVE/ANTACID

▶ **PRACTICE 4.17** Write a balanced net ionic equation for the following acid-base reaction:

$$Ca(OH)_2(aq) + 2\ CH_3CO_2H(aq) \rightarrow ?$$

▶ **APPLY 4.18** Milk of magnesia (active ingredient: magnesium hydroxide) is used as an antacid to treat indigestion and heartburn. Write a balanced ionic equation and net ionic equation for the reaction of magnesium hydroxide with stomach acid, hydrochloric acid.

4.8 SOLUTION STOICHIOMETRY

We learned in Section 4.1 that molarity is a conversion factor between numbers of moles of solute and the volume of a solution. Thus, if we know the volume and molarity of a solution, we can calculate the number of moles of solute. If we know the number of moles of solute and molarity, we can find the volume.

As indicated by the flow diagram in **FIGURE 4.4,** using molarity is critical for carrying out stoichiometry calculations on substances in solution. Molarity makes it possible to calculate the volume of one solution needed to react with a given volume of another solution. This sort of calculation is particularly important in the chemistry of acids and bases, as shown in Worked Example 4.10.

For the balanced equation:
$$a\,A + b\,B \longrightarrow c\,C + d\,D$$

GIVEN

Use molarity as a conversion factor.

→ Volume of solution of A

→ Moles of A

Use coefficients in the balanced equation to find A:B mole ratio.

→ Moles of B needed

Use molarity as a conversion factor.

FIND → Volume of solution of B

▲ **FIGURE 4.4**

Using molarity as a conversion factor between moles and volume in stoichiometry calculations.

▲ **Figure It Out**

Show the conversion factors needed to convert from liters of solution A to liters of solution B. Assume the molarity of solution A = X, the stoichiometric coefficients for A and B are a and b, respectively, and the molarity of solution B = Y.

Answer:

$$L_A \times \frac{X\ mol\ A}{1\ L} \times \frac{b\ mol\ B}{a\ mol\ A} \times \frac{1\ L}{Y\ mol\ B} = L_B$$

— **WORKED EXAMPLE 4.10**

Reaction Stoichiometry in Solution

Stomach acid, a dilute solution of HCl in water, can be neutralized by reaction with sodium hydrogen carbonate, $NaHCO_3$, according to the equation

$$HCl(aq) + NaHCO_3(aq) \longrightarrow NaCl(aq) + H_2O(l) + CO_2(g)$$

How many milliliters of 0.125 M $NaHCO_3$ solution are needed to neutralize 18.0 mL of 0.100 M HCl?

IDENTIFY

Known	Unknown
Volume of HCl (18.0 mL)	Volume of $NaHCO_3$ (mL)
Concentration of HCl (0.100 M)	
Concentration of $NaHCO_3$ (0.125 M)	
Balanced reaction	

STRATEGY

Solving stoichiometry problems always requires finding the number of moles of one reactant and using the coefficients of the balanced equation to find the number of moles of the other reactant. Molarity is used as a conversion factor between volume and moles. Figure 4.4 outlines the steps for relating the volumes of reactants or products.

SOLUTION

We first have to find how many moles of HCl are in 18.0 mL of a 0.100 M solution by multiplying volume times molarity:

$$\text{Moles of HCl} = 18.0\ mL \times \frac{1\ L}{1000\ mL} \times \frac{0.100\ mol}{1\ L} = 1.80 \times 10^{-3}\ mol\ HCl$$

Next, find the moles of $NaHCO_3$ using the coefficients from the balanced equation.

$$1.80 \times 10^{-3} \text{ mol HCl} \times \frac{1 \text{ mol } NaHCO_3}{1 \text{ mol HCl}} = 1.80 \times 10^{-3} \text{ mol } NaHCO_3$$

Find the volume of $NaHCO_3$ by using the molarity to convert between moles and volume.

$$1.80 \times 10^{-3} \text{ mol } NaHCO_3^- \times \frac{1 \text{ L solution}}{0.125 \text{ mol } NaHCO_3} \times \frac{1000 \text{ mL}}{1 \text{ L solution}} = 14.4 \text{ mL solution}$$

Thus, 14.4 mL of the 0.125 M $NaHCO_3$ solution is needed to neutralize 18.0 mL of the 0.100 M HCl solution.

CHECK

The balanced equation shows that HCl and $NaHCO_3$ react in a 1:1 molar ratio, and we are told that the concentrations of the two solutions are about the same. Thus, the volume of the $NaHCO_3$ solution must be about the same as that of the HCl solution.

▶ **PRACTICE 4.19**   What volume of 0.250 M H_2SO_4 is needed to react with 50.0 mL of 0.100 M NaOH? The equation is

$$H_2SO_4(aq) + 2 \text{ NaOH}(aq) \longrightarrow Na_2SO_4(aq) + 2 H_2O(l)$$

▶ **APPLY 4.20**   What is the molarity of a HNO_3 solution if 68.5 mL is needed to react with 25.0 mL of 0.150 M KOH solution? The equation is

$$HNO_3(aq) + KOH(aq) \longrightarrow KNO_3(aq) + H_2O(l)$$

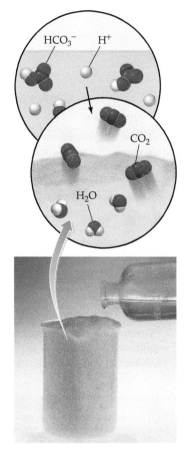

▲ Neutralization of sodium hydrogen carbonate with acid leads to release of CO_2 gas, visible in this fizzing solution.

4.9 MEASURING THE CONCENTRATION OF A SOLUTION: TITRATION

Measuring the concentration of solutions is important for a variety of reasons. For example, the concentration of ions such as H^+, OH^-, Fe^{3+}, Mg^{2+}, Ca^{2+}, and Cl^- is important in the quality of water used for drinking, irrigation, and industrial processes. Furthermore, solutions prepared by the methods described in Section 4.1 may not result in the exact molarity calculated by dividing moles of solute by volume of solution. This discrepancy arises because chemicals often cannot be purchased in their pure form or they react with other chemicals in the solvent or in the air. A technique frequently used for determining a solution's exact molarity is called a *titration*.

Titration is a procedure for determining the concentration of a solution by allowing a measured volume of that solution to react with a second solution of another substance (the *standard solution*) whose concentration is known. By finding the volume of the standard solution that reacts with the measured volume of the first solution, the concentration of the first solution can be calculated. (It's necessary, though, that the reaction go to completion and have a yield of 100%.)

To see how titration works, let's imagine that we have an HCl solution (an acid) whose concentration we want to find by allowing it to react with NaOH (a base) in an acid–base neutralization reaction. The balanced equation is

$$\text{NaOH}(aq) + \text{HCl}(aq) \longrightarrow \text{NaCl}(aq) + H_2O(l)$$

We'll begin the titration by measuring out a known volume of the HCl solution and adding a small amount of an *indicator,* a compound that undergoes a color change during the course of the reaction. The compound phenolphthalein, for instance, is colorless in acid solution but turns red in base solution. Next, we fill a calibrated glass tube called a *buret* with an NaOH standard solution of known concentration and slowly add the NaOH to the HCl. When the phenolphthalein just begins to turn pink, all the HCl has completely reacted, and the solution now has a tiny amount of excess NaOH. By then reading from the buret to find the volume of the NaOH standard solution that has been added to react with the known volume of HCl solution, we can calculate the concentration of the HCl. The strategy is summarized in **FIGURE 4.5**, and the procedure is shown in **FIGURE 4.6**.

▶ **FIGURE 4.5**

A flow diagram for an acid–base titration. The calculations needed to determine the concentration of an HCl solution by titration with an NaOH standard solution are summarized.

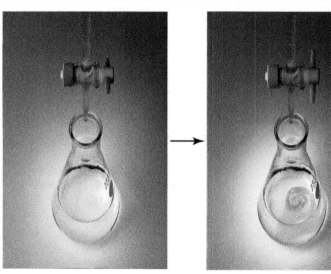

For the balanced equation:
$$NaOH + HCl \longrightarrow NaCl + H_2O$$

GIVEN

| Volume of NaOH | → | Moles of NaOH | → | Moles of HCl | → | Molarity of HCl |

FIND

Use molarity of NaOH as a conversion factor.

Use coefficients in the balanced equation to find mole ratios.

Divide by volume of HCl.

▶ **FIGURE 4.6**

Titration of an acid solution of unknown concentration with a base solution of known concentration.

▶ **Figure It Out**

What does the color change indicate in the chemical reaction between HCl and NaOH?

Answer: The HCl is completely reacted, and there is excess base in solution. The first faint, permanent color change indicates the end of the titration.

A measured volume of acid solution is placed in a flask, and phenolphthalein indicator is added.

Base solution of known concentration is added from a buret until the indicator changes color. Reading the volume of base from the buret allows calculation of the acid concentration.

Go to
eText

─ **WORKED EXAMPLE 4.11**

Determining the Concentration of a Solution Using a Titration Procedure

A 20.0 mL sample of hydrochloric acid (HCl) is titrated and found to react with 42.6 mL of 0.100 M NaOH. What is the molarity of the hydrochloric acid solution?

IDENTIFY

Known	Unknown
Volume of HCl solution (20.0 mL)	Concentration of HCl solution (M)
Volume of NaOH solution (42.6 mL)	
Concentration of NaOH solution (0.100 M)	
Balanced reaction	

STRATEGY

Solving stoichiometry problems always requires finding the number of moles of one reactant and using the coefficients of the balanced equation to find the number of moles of the other reactant. Molarity is used as a conversion factor between volume and moles. Figure 4.5 outlines the steps for finding concentration in a titration experiment.

SOLUTION

Using the molarity of the NaOH standard solution as a conversion factor, we can calculate the number of moles of NaOH undergoing reaction:

$$\text{Moles of NaOH} = 0.0426 \text{ L NaOH} \times \frac{0.100 \text{ mol NaOH}}{1 \text{ L NaOH}}$$

$$= 0.004\ 26 \text{ mol NaOH}$$

According to the balanced equation, the number of moles of HCl is the same as that of NaOH:

$$\text{Moles of HCl} = 0.004\ 26 \text{ mol NaOH} \times \frac{1 \text{ mol HCl}}{1 \text{ mol NaOH}}$$

$$= 0.004\ 26 \text{ mol HCl}$$

Dividing the number of moles of HCl by the volume then gives the molarity of the HCl:

$$\text{HCl molarity} = \frac{0.004\ 26 \text{ mol HCl}}{0.0200 \text{ L HCl}} = 0.213 \text{ M HCl}$$

CHECK

The volume of NaOH used in the titration is just over twice the volume of the HCl solution. Therefore, the concentration of HCl solution must be just over twice the concentration of the NaOH solution.

▶ **PRACTICE 4.21** A 25.0 mL sample of vinegar (dilute acetic acid, CH_3CO_2H is titrated and found to react with 94.7 mL of 0.200 M NaOH. What is the molarity of the acetic acid solution? The reaction is

$$NaOH(aq) + CH_3CO_2H(aq) \longrightarrow CH_3CO_2Na(aq) + H_2O(l)$$

▶ **CONCEPTUAL APPLY 4.22** Assume that the buret contains H^+ ions, the flask contains OH^- ions, and each has a volume of 100 mL. How many milliliters would you need to add from the buret to the flask to neutralize all the OH^- ions in a titration procedure? The equation is $H^+(aq) + OH^-(aq) \longrightarrow H_2O(l)$.

Assume the number of particles in the containers is directly proportional to the concentration.

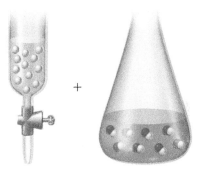

4.10 OXIDATION–REDUCTION (REDOX) REACTIONS

Purple aqueous permanganate ion, MnO_4^-, reacts with aqueous Fe^{2+} ion to yield Fe^{3+} and pale pink Mn^{2+}:

$$MnO_4^-(aq) + 5 \text{ Fe}^{2+}(aq) + 8 \text{ H}^+(aq) \longrightarrow Mn^{2+}(aq) + 5 \text{ Fe}^{3+}(aq) + 4 \text{ H}_2O(l)$$

Magnesium metal burns in air with an intense white light to form solid magnesium oxide:

$$2 \text{ Mg}(s) + O_2(g) \longrightarrow 2 \text{ MgO}(s)$$

Red phosphorus reacts with liquid bromine to form liquid phosphorus tribromide:

$$2 \text{ P}(s) + 3 \text{ Br}_2(l) \longrightarrow 2 \text{ PBr}_3(l)$$

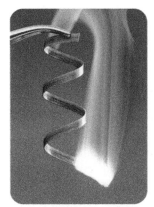

▲ Magnesium metal burns in air to give MgO. Elemental phosphorus reacts spectacularly with bromine to give PBr_3.

Although these and many thousands of other reactions appear unrelated, and many don't even take place in aqueous solution, all are oxidation–reduction (redox) reactions.

Historically, the word *oxidation* referred to the combination of an element with oxygen to yield an oxide, and the word *reduction* referred to the removal of oxygen from an oxide to yield the element. Such oxidation–reduction processes have been crucial to the development of human civilization and still have enormous commercial value. The oxidation (rusting) of iron metal by reaction with moist air has been known for millennia and is still a serious problem that causes enormous structural damage to buildings, boats, and bridges. The reduction of iron ore (Fe_2O_3) with charcoal (C) to make iron metal has been carried out since prehistoric times and is still used today in the initial stages of steelmaking.

$$4 \, Fe(s) + 3 \, O_2(g) \longrightarrow 2 \, Fe_2O_3(s)$$ Rusting of iron: an oxidation of Fe
$$2 \, Fe_2O_3(s) + 3 \, C(s) \longrightarrow 4 \, Fe(s) + 3 \, CO_2(g)$$ Manufacture of iron: a reduction of Fe_2O_3

Today, the words *oxidation* and *reduction* have taken on a much broader meaning. **Oxidation** is now defined as the loss of one or more electrons by a substance, whether element, compound, or ion, and **reduction** is the gain of one or more electrons by a substance. Thus, an oxidation–reduction, or redox, reaction is any process in which electrons are transferred from one substance to another.

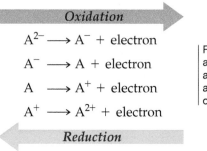

How can you tell when a redox reaction takes place? The answer is that you assign to each atom in a compound a value called an **oxidation number** (or *oxidation state*), which indicates whether the atom is neutral, electron-rich, or electron-poor. By comparing the oxidation number of an atom before and after reaction, you can tell whether the atom has gained or lost electrons. Note that oxidation numbers don't necessarily imply ionic charges; they are just a convenient device to help keep track of electrons during redox reactions.

The rules for assigning oxidation numbers are as follows:

1. **An atom in its elemental state has an oxidation number of 0. For example:**

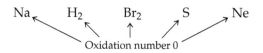

2. **An atom in a monatomic ion has an oxidation number identical to its charge.** Review Section 2.13 to see the charges on some common ions. For example:

$$\begin{array}{ccccc} Na^+ & Ca^{2+} & Al^{3+} & Cl^- & O^{2-} \\ \uparrow & \uparrow & \uparrow & \uparrow & \uparrow \\ +1 & +2 & +3 & -1 & -2 \end{array}$$

3. **An atom in a polyatomic ion or in a molecular compound usually has the same oxidation number it would have if it were a monatomic ion.** In the hydroxide ion (OH^-), for instance, the hydrogen atom has an oxidation number of $+1$, as if it were H^+, and the oxygen atom has an oxidation number of -2, as if it were a monatomic O^{2-} ion.

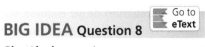

BIG IDEA Question 8   Go to eText

Classify the reaction
$Cu^{2+} + e^- \longrightarrow Cu^+$ as an oxidation or reduction reaction, and explain your reasoning.

$$\underset{\substack{\uparrow \\ +1}}{H} - \underset{\substack{\uparrow \\ -2}}{O} - \underset{\substack{\uparrow \\ +1}}{H} \qquad [\underset{\substack{\uparrow \\ -2}}{O} - \underset{\substack{\uparrow \\ +1}}{H}]^- \qquad \underset{\substack{\uparrow \\ +1}}{H} - \underset{\substack{\uparrow \\ -3}}{N} - \underset{\substack{\uparrow \\ +1}}{H}$$
$$\overset{H \leftarrow +1}{\underset{|}{}}$$

In general, the farther left an element is in the periodic table, the more probable that it will be *cationlike*. Metals, therefore, usually have positive oxidation numbers. The farther right an element is in the periodic table, the more probable that it will be *anionlike*. Nonmetals, such as O, N, and the halogens, usually have negative oxidation numbers. We'll see the reasons for these trends in Sections 6.3–6.5.

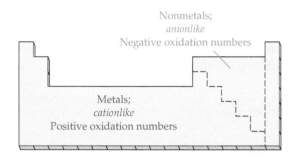

(a) **Hydrogen can be either +1 or −1.** When bonded to a metal, such as Na or Ca, hydrogen has an oxidation number of −1. When bonded to a nonmetal, such as C, N, O, or Cl, hydrogen has an oxidation number of +1.

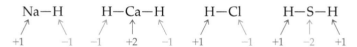

(b) **Oxygen usually has an oxidation number of −2.** The major exception is in compounds called *peroxides*, which contain either the O_2^{2-} ion or an O—O covalent bond in a molecule. Both oxygen atoms in a peroxide have an oxidation number of −1.

$$\begin{array}{ccc} \text{H—O—H} & \text{H—O—O—H} & [\text{O—O}]^{2-} \\ {}^{+1}\ {}_{-2}\ {}^{+1} & {}^{+1}\ {}_{-1}\ {}_{-1}\ {}^{+1} & {}^{-1}\quad {}^{-1} \end{array}$$

(c) **Halogens usually have an oxidation number of −1.** The major exception is in compounds of chlorine, bromine, or iodine in which the halogen atom is bonded to oxygen. In such cases, the oxygen has an oxidation number of −2, and the halogen has a positive oxidation number. In Cl_2O, for instance, the O atom has an oxidation number of −2, and each Cl atom has an oxidation number of +1.

$$\begin{array}{cc} \text{Cl—O—Cl} & \text{H—O—Br} \\ {}^{+1}\ {}_{-2}\ {}^{+1} & {}^{+1}\ {}_{-2}\ {}^{+1} \end{array}$$

4. **The sum of the oxidation numbers is 0 for a neutral compound and is equal to the net charge for a polyatomic ion.** This rule is particularly useful for finding the oxidation number of an atom in difficult cases. The general idea is to assign oxidation numbers to the "easy" atoms first and then find the oxidation number of the "difficult" atom by subtraction. For example, suppose we want to know the oxidation number of the sulfur atom in sulfuric acid, H_2SO_4. Since each H atom is +1 and each O atom is −2, the S atom must have an oxidation number of +6 for the compound to have no net charge:

$$\begin{array}{l} H_2SO_4 \qquad 2(+1) + (?) + 4(-2) = 0 \text{ net charge} \\ {}^{+1}\ {}_{?}\ {}_{-2} \qquad ? = 0 - 2(+1) - 4(-2) = +6 \end{array}$$

To find the oxidation number of the chlorine atom in the perchlorate anion (ClO_4^-), we know that each oxygen is −2, so the Cl atom must have an oxidation number of +7 for there to be a net charge of −1 on the ion:

$$ClO_4^- \qquad ? + 4(-2) = -1 \text{ net charge}$$

$$? \qquad -2 \qquad ? = -1 - 4(-2) = +7$$

To find the oxidation number of the nitrogen atom in the ammonium cation (NH_4^+), we know that each H atom is +1, so the N atom must have an oxidation number of −3 for the ion to have a net charge of +1:

$$NH_4^+ \qquad ? + 4(+1) = +1 \text{ net charge}$$

$$? \qquad +1 \qquad ? = +1 - 4(+1) = -3$$

WORKED EXAMPLE 4.12

Assigning Oxidation Numbers

Assign oxidation numbers to each atom in the following substances:

(a) CdS (b) AlH_3 (c) $S_2O_3^{2-}$ (d) $Na_2Cr_2O_7$

STRATEGY

First assign oxidation numbers to elements that usually form oxidation states equal to the charge on the monoatomic ion, such as H, O, and other main group elements. Then find the unknown oxidation number of an element by summing oxidation numbers of all elements to give the overall charge on the species.

(a) The sulfur atom in S^{2-} has an oxidation number of −2, so Cd must be +2.
(b) H bonded to a metal has the oxidation number −1, so Al must be +3.
(c) O usually has the oxidation number −2, so S must be +2 for the anion to have a net charge of −2: for ($2\ S^{+2}$) ($3\ O^{-2}$), $2(+2) + 3(-2) = -2$ net charge.
(d) Na is always +1, and oxygen is −2, so Cr must be +6 for the compound to be neutral: for ($2\ Na^+$) ($2\ Cr^{+6}$) ($7\ O^{-2}$), $2(+1) + 2(+6) + 7(-2) = 0$ net charge.

SOLUTION

(a) CdS (b) AlH_3 (c) $S_2O_3^{2-}$ (d) $Na_2Cr_2O_7$
   ↑ ↑   ↑ ↑   ↑ ↑   ↑ ↑ ↑
  +2 −2   +3 −1   +2 −2   +1 +6 −2

▶ **PRACTICE 4.23** Assign an oxidation number to each atom in the following compounds:

(a) $SnCl_4$ (b) CrO_3 (c) $VOCl_3$
(d) V_2O_3 (e) HNO_3 (f) $FeSO_4$

▶ **APPLY 4.24** Chlorine can have several different oxidation numbers ranging in value from −1 to +7.

(a) Write the formula and give the name of the chlorine oxide compound in which chlorine has an oxidation number of +2, +3, +6, and +7.
(b) Based on oxidation numbers, which chlorine oxide from part (a) cannot react with molecular oxygen?

4.11 IDENTIFYING REDOX REACTIONS

Once oxidation numbers are assigned, it's clear why all the reactions mentioned in the previous section are redox processes. Take the rusting of iron, for example. Two of the reactants, Fe and O_2, are neutral elements and have oxidation numbers of 0. In the product, however, the oxygen atoms have an oxidation number of −2 and the iron atoms have an oxidation number of +3. Thus, Fe has undergone a change from 0 to +3 (a loss of electrons, or oxidation), and O has undergone a change from 0 to −2 (a gain of electrons, or reduction). Note that the total number of electrons given up by the atoms being oxidized (4 Fe × 3 electrons/Fe = 12 electrons) is the same as the number gained by the atoms being reduced (6 O × 2 electrons/O = 12 electrons).

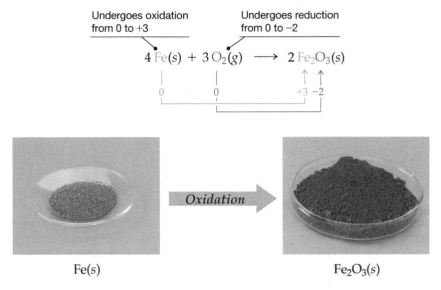

Fe(s) $Fe_2O_3(s)$

▲ Silver-colored powdered iron is oxidized by oxygen to produce iron(III)oxide, which is red in color.

A similar analysis can be carried out for the production of iron metal from its ore. The iron atom is reduced because it goes from an oxidation number of +3 in the reactant (Fe_2O_3) to 0 in the product (Fe). At the same time, the carbon atom is oxidized because it goes from an oxidation number of 0 in the reactant (C) to +4 in the product (CO_2). The oxygen atoms undergo no change because they have an oxidation number of −2 in both reactant and product. The total number of electrons given up by the atoms being oxidized (3 C × 4 electrons/C = 12 electrons) is the same as the number gained by the atoms being reduced (4 Fe × 3 electrons/Fe = 12 electrons).

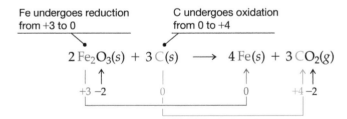

BIG IDEA Question 9   Go to eText

What is the oxidizing agent in the reaction between $Fe_2O_3(s)$ and $C(s)$?

As these examples show, oxidation and reduction reactions, called **half-reactions**, always occur together. A redox reaction consists of two half-reactions; one oxidation half-reaction and one reduction half-reaction. Whenever one atom loses one or more electrons, another atom must gain those electrons. The substance that *causes* a reduction by giving up electrons—the iron atom in the reaction of Fe with O_2 and the carbon atom in the reaction of C with Fe_2O_3—is called a **reducing agent.** The substance that causes an oxidation by accepting electrons—the oxygen atom in the reaction of Fe with O_2 and the iron atom in the reaction of C with Fe_2O_3—is called an **oxidizing agent.** The reducing agent is itself oxidized when it gives up electrons, and the oxidizing agent is itself reduced when it accepts electrons.

Reducing agent	Oxidizing agent
• Causes reduction	• Causes oxidation
• Loses one or more electrons	• Gains one or more electrons
• Undergoes oxidation	• Undergoes reduction
• Oxidation number of atom increases	• Oxidation number of atom decreases

LOOKING AHEAD . . .
Applications of **redox reactions** such as batteries are discussed in Chapter 19 on electrochemistry.

We'll see in Chapter 19 that **redox reactions** are common for almost every element in the periodic table except for the noble-gas elements of group 8A. In general, metals give up electrons and act as reducing agents, while reactive nonmetals such as O_2 and the halogens accept electrons and act as oxidizing agents.

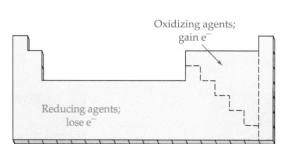

Different metals can give up different numbers of electrons in redox reactions. Lithium, sodium, and the other group 1A elements give up only one electron and become monopositive ions with oxidation numbers of +1. Beryllium, magnesium, and the other group 2A elements, however, typically give up two electrons and become dipositive ions. The transition metals in the middle of the periodic table can give up a variable number of electrons to yield more than one kind of ion depending on the exact reaction. Titanium, for example, can react with chlorine to yield either $TiCl_3$ or $TiCl_4$. Because a chloride ion has a -1 oxidation number, the titanium atom in $TiCl_3$ must have a +3 oxidation number, and the titanium atom in $TiCl_4$ must be +4.

WORKED EXAMPLE 4.13

Identifying Oxidizing and Reducing Agents

Assign oxidation numbers to all atoms, tell in each case which substance is undergoing oxidation and which reduction, and identify the oxidizing and reducing agents.

(a) $Ca(s) + 2\ H^+(aq) \longrightarrow Ca^{2+}(aq) + H_2(g)$
(b) $2\ Fe^{2+}(aq) + Cl_2(aq) \longrightarrow 2\ Fe^{3+}(aq) + 2\ Cl^-(aq)$

STRATEGY AND SOLUTION

(a) The elements Ca and H_2 have oxidation numbers of 0; Ca^{2+} is +2, and H^+ is +1. Ca is oxidized because its oxidation number increases from 0 to +2, and H^+ is reduced because its oxidation number decreases from +1 to 0. The reducing agent is the substance that gives away electrons, thereby going to a higher oxidation number, and the oxidizing agent is the substance that accepts electrons, thereby going to a lower oxidation number. In the present case, calcium is the reducing agent and H^+ is the oxidizing agent.

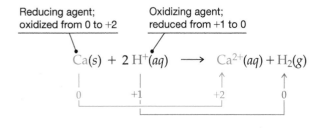

(b) Atoms of the neutral element Cl_2 have an oxidation number of 0; the monatomic ions have oxidation numbers equal to their charge:

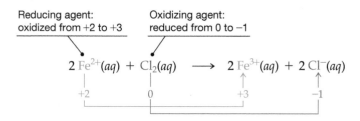

Fe^{2+} is oxidized because its oxidation number increases from +2 to +3, and Cl_2 is reduced because its oxidation number decreases from 0 to -1. Fe^{2+} is the reducing agent, and Cl_2 is the oxidizing agent.

▶ **PRACTICE 4.25** Identify the oxidizing agent and reducing agent in the reaction:

$$4\ NH_3(g) + 5\ O_2(g) \longrightarrow 4\ NO(g) + 6\ H_2O(l)$$

▶ **APPLY 4.26** Police often use a Breathalyzer test to determine the ethanol (C_2H_5OH) content in a person's blood. The test involves a redox reaction that produces a color change. Potassium dichromate is reddish orange, and chromium(III) sulfate is green. The balanced reaction is:

$$2\ K_2Cr_2O_7(aq) + 3\ C_2H_5OH(g) + 8\ H_2SO_4(aq) \longrightarrow$$
$$2\ Cr_2(SO_4)_3(aq) + 2\ K_2SO_4(aq)$$
$$+ 3\ CH_3COOH(aq) + 11\ H_2O(l)$$

(a) Identify the element that gets oxidized and the element that gets reduced.

(b) Give the oxidizing agent and the reducing agent.

▲ A Breathalyzer test measures alcohol concentration in exhaled breath using a redox reaction.

4.12 THE ACTIVITY SERIES OF THE ELEMENTS

The reaction of an aqueous cation, usually a metal ion, with a free element to give a different cation and a different element is among the simplest of all redox processes. Aqueous copper(II) ion reacts with iron metal, for example, to give iron(II) ion and copper metal (**FIGURE 4.7**):

$$Fe(s) + Cu^{2+}(aq) \longrightarrow Fe^{2+}(aq) + Cu(s)$$

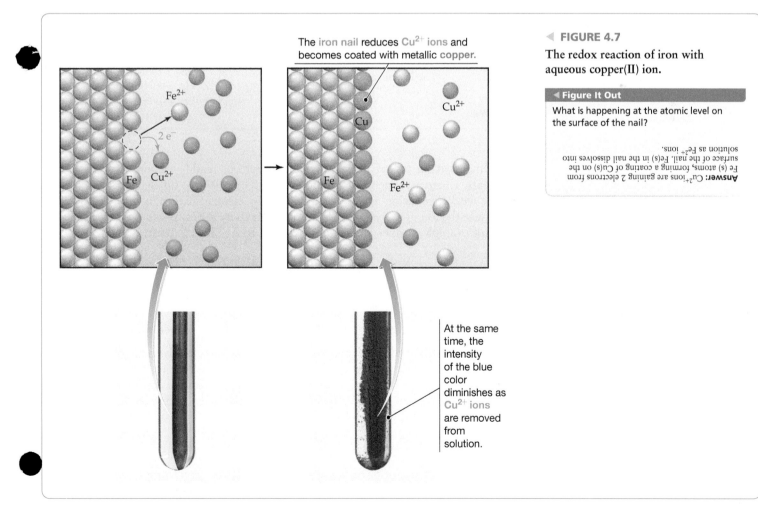

The iron nail reduces Cu^{2+} ions and becomes coated with metallic copper.

◀ **FIGURE 4.7**

The redox reaction of iron with aqueous copper(II) ion.

◀ **Figure It Out**

What is happening at the atomic level on the surface of the nail?

Answer: Cu^{2+} ions are gaining 2 electrons from Fe (s) atoms, forming a coating of Cu(s) on the surface of the nail. Fe(s) in the nail dissolves into solution as Fe^{2+} ions.

At the same time, the intensity of the blue color diminishes as Cu^{2+} ions are removed from solution.

Whether a reaction occurs between a given ion and a given element depends on the relative ease with which the various substances gain or lose electrons—that is, on how easily each substance is reduced or oxidized. By noting the results from a succession of different reactions, it's possible to construct an **activity series,** which ranks the elements in order of their reducing ability in aqueous solution (**TABLE 4.5**).

BIG IDEA Question 10   Go to eText

Which metal will react with an acidic solution but not liquid water to produce H_2 gas: K, Ba, Ag, or Zn? (Refer to Table 4.5.)

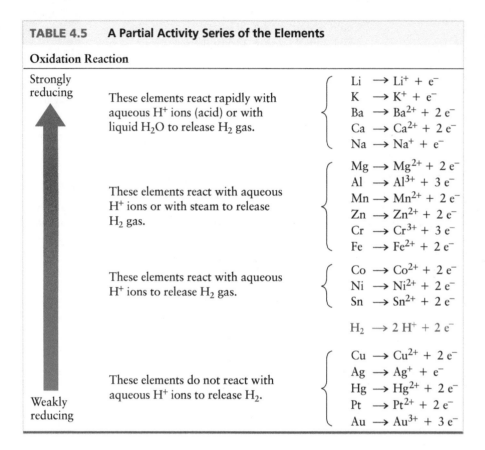

TABLE 4.5   A Partial Activity Series of the Elements

Oxidation Reaction

Strongly reducing → Weakly reducing

These elements react rapidly with aqueous H^+ ions (acid) or with liquid H_2O to release H_2 gas.

$$Li \rightarrow Li^+ + e^-$$
$$K \rightarrow K^+ + e^-$$
$$Ba \rightarrow Ba^{2+} + 2\,e^-$$
$$Ca \rightarrow Ca^{2+} + 2\,e^-$$
$$Na \rightarrow Na^+ + e^-$$

These elements react with aqueous H^+ ions or with steam to release H_2 gas.

$$Mg \rightarrow Mg^{2+} + 2\,e^-$$
$$Al \rightarrow Al^{3+} + 3\,e^-$$
$$Mn \rightarrow Mn^{2+} + 2\,e^-$$
$$Zn \rightarrow Zn^{2+} + 2\,e^-$$
$$Cr \rightarrow Cr^{3+} + 3\,e^-$$
$$Fe \rightarrow Fe^{2+} + 2\,e^-$$

These elements react with aqueous H^+ ions to release H_2 gas.

$$Co \rightarrow Co^{2+} + 2\,e^-$$
$$Ni \rightarrow Ni^{2+} + 2\,e^-$$
$$Sn \rightarrow Sn^{2+} + 2\,e^-$$

$$H_2 \rightarrow 2\,H^+ + 2\,e^-$$

These elements do not react with aqueous H^+ ions to release H_2.

$$Cu \rightarrow Cu^{2+} + 2\,e^-$$
$$Ag \rightarrow Ag^+ + e^-$$
$$Hg \rightarrow Hg^{2+} + 2\,e^-$$
$$Pt \rightarrow Pt^{2+} + 2\,e^-$$
$$Au \rightarrow Au^{3+} + 3\,e^-$$

Elements at the top of Table 4.5 give up electrons readily and are stronger reducing agents, whereas elements at the bottom of the table give up electrons less readily and are weaker reducing agents. As a result, any element higher in the activity series will reduce the ion of any element lower in the activity series. Because copper is above silver, for example, copper metal gives electrons to Ag^+ ions (**FIGURE 4.8**).

$$Cu(s) + 2\,Ag^+(aq) \longrightarrow Cu^{2+}(aq) + 2\,Ag(s)$$

Conversely, because gold is below silver in the activity series, gold metal does not give electrons to Ag^+ ions.

$$Au(s) + 3\,Ag^+(aq) \nrightarrow Au^{3+}(aq) + 3\,Ag(s) \qquad \textit{Does not occur}$$

The position of hydrogen in the activity series is particularly important because it indicates which metals react with aqueous acid (H^+) to release H_2 gas. The metals at the top of the series—the alkali metals of group 1A and alkaline earth metals of group 2A—are such powerful reducing agents that they react even with pure water, in which the concentration of H^+ is very low:

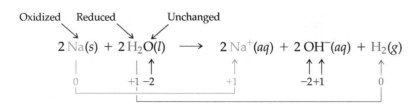

In contrast, the metals in the middle of the series react with aqueous acid but not with water, and the metals at the bottom of the series react with neither aqueous acid nor water:

$$Fe(s) + 2\,H^+(aq) \longrightarrow Fe^{2+}(aq) + H_2(g)$$
$$Ag(s) + H^+(aq) \longrightarrow \text{No reaction}$$

Notice that the most easily oxidized metals—those at the top of the activity series—are on the left of the periodic table. Conversely, the least easily oxidized metals—those at the bottom of the activity series—are in the transition metal groups closer to the right side of the table.

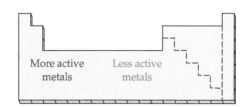

More active metals   Less active metals

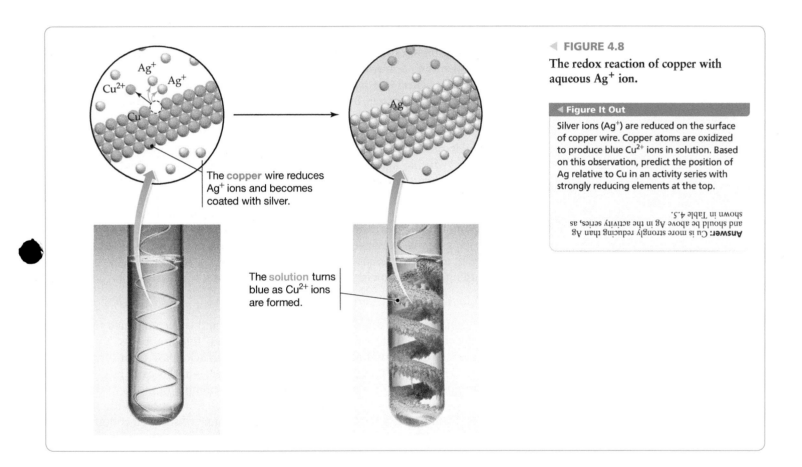

◀ **FIGURE 4.8**

The redox reaction of copper with aqueous Ag^+ ion.

◀ **Figure It Out**

Silver ions (Ag^+) are reduced on the surface of copper wire. Copper atoms are oxidized to produce blue Cu^{2+} ions in solution. Based on this observation, predict the position of Ag relative to Cu in an activity series with strongly reducing elements at the top.

Answer: Cu is more strongly reducing than Ag and should be above Ag in the activity series, as shown in Table 4.5.

The copper wire reduces Ag^+ ions and becomes coated with silver.

The solution turns blue as Cu^{2+} ions are formed.

WORKED EXAMPLE 4.14

Predicting the Products of a Redox Reaction

Predict whether the following redox reactions will occur:
(a) $Hg^{2+}(aq) + Zn(s) \longrightarrow Hg(l) + Zn^{2+}(aq)$
(b) $2\,H^+(aq) + Cu(s) \longrightarrow H_2(g) + Cu^{2+}(aq)$

STRATEGY

Look at Table 4.5 to find the relative reactivities of the elements.

SOLUTION

(a) Zinc is above mercury in the activity series, so this reaction will occur.
(b) Copper is below hydrogen in the activity series, so this reaction will not occur.

continued on next page

▶ **PRACTICE 4.27** Predict whether the following reaction will occur:

$$Ca^{2+}(aq) + Mg(s) \longrightarrow Ca(s) + Mg^{2+}(aq)$$

▶ **APPLY 4.28** Use the following reactions to arrange the elements **A, B, C,** and **D** in order of their strength as a reducing agent from strongest to weakest.

$$A + D^+ \longrightarrow A^+ + D \qquad C^+ + D \longrightarrow C + D^+$$
$$B^+ + D \longrightarrow B + D^+ \qquad B + C^+ \longrightarrow B^+ + C$$

4.13 REDOX TITRATIONS

REMEMBER . . .
The reaction used for a **titration** must go to
completion and have a yield of 100%
(Section 4.9).

We saw in Section 4.9 that the concentration of an acid or base solution can be determined by **titration**. A measured volume of the acid or base solution of unknown concentration is placed in a flask, and a base or acid solution of known concentration is slowly added from a buret. By measuring the volume of the added solution necessary for a complete reaction, as signaled by the color change of an indicator, the unknown concentration can be calculated.

A similar procedure can be carried out to determine the concentration of many oxidizing or reducing agents using a *redox titration*. All that's necessary is that the substance whose concentration you want to determine undergo an oxidation or reduction reaction in 100% yield and that there be some means, such as a color change, to indicate when the reaction is complete. The color change might be due to one of the substances undergoing reaction or to some added indicator.

Let's imagine that we have a potassium permanganate solution whose concentration we want to find. Aqueous $KMnO_4$ reacts with oxalic acid, $H_2C_2O_4$, in acidic solution according to the following net ionic equation (K^+ is a spectator ion):

$$5\ H_2C_2O_4(aq) + 2\ MnO_4^-(aq) + 6\ H^+(aq) \longrightarrow 10\ CO_2(g) + 2\ Mn^{2+}(aq) + 8\ H_2O(l)$$

The reaction goes to completion with 100% yield and is accompanied by a sharp color change when the intense purple color of the MnO_4^- ion disappears.

The strategy used is outlined in **FIGURE 4.9.** As with acid–base titrations, the general idea is to measure a known amount of one substance—in this case, $H_2C_2O_4$—and use mole ratios from the balanced equation to find the number of moles of the second substance—in this case, $KMnO_4$—necessary for complete reaction. With the molar amount of $KMnO_4$ thus known, titration gives the volume of solution containing that amount. Dividing the number of moles by the volume gives the concentration.

▶ **FIGURE 4.9**

A summary of calculations for determining the concentration of a $KMnO_4$ solution by redox titration of $H_2C_2O_4$.

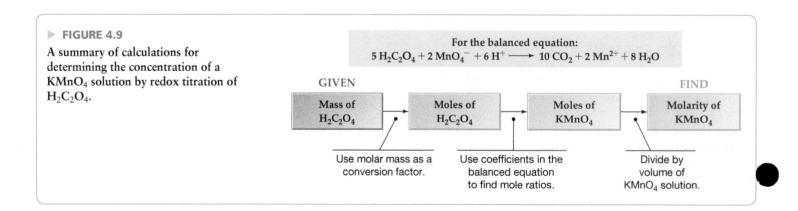

For the balanced equation:
$$5\ H_2C_2O_4 + 2\ MnO_4^- + 6\ H^+ \longrightarrow 10\ CO_2 + 2\ Mn^{2+} + 8\ H_2O$$

GIVEN   FIND

| Mass of $H_2C_2O_4$ | Moles of $H_2C_2O_4$ | Moles of $KMnO_4$ | Molarity of $KMnO_4$ |

Use molar mass as a conversion factor.   Use coefficients in the balanced equation to find mole ratios.   Divide by volume of $KMnO_4$ solution.

As an example of how the procedure works, let's carefully weigh some amount of $H_2C_2O_4$—, 0.2585 g—and dissolve it in approximately 100 mL of 0.5 M H_2SO_4. The exact volume isn't important because we're concerned only with the amount of dissolved $H_2C_2O_4$, not with its concentration. Next, we place an aqueous $KMnO_4$ solution of unknown concentration in a buret and slowly add it to the $H_2C_2O_4$ solution. The purple color of the added MnO_4^- initially disappears as reaction occurs, but we continue the addition until a faint color persists, indicating that all the $H_2C_2O_4$ has reacted and that MnO_4^- ion is no longer being reduced. At this equivalence point, or *end point,* of the titration, we might find that 22.35 mL of the $KMnO_4$ solution has been added (**FIGURE 4.10**).

To calculate the molarity of the $KMnO_4$ solution, we need to find the number of moles of $KMnO_4$ present in 22.35 mL of solution used for titration. We do this by following the procedure outlined in Figure 4.9 and first calculating the number of moles of oxalic acid that react with the permanganate ion. A gram-to-mole conversion is done using the molar mass of $H_2C_2O_4$ as the conversion factor:

$$\text{Moles of } H_2C_2O_4 = 0.2585 \text{ g } H_2C_2O_4 \times \frac{1 \text{ mol } H_2C_2O_4}{90.03 \text{ g } H_2C_2O_4}$$

$$= 2.871 \times 10^{-3} \text{ mol } H_2C_2O_4$$

According to the balanced equation, 5 mol of oxalic acid react with 2 mol of permanganate ion. Thus, we can calculate the number of moles of $KMnO_4$ that react with 2.871×10^{-3} mol of $H_2C_2O_4$:

$$\text{Moles of } KMnO_4 = 2.871 \times 10^{-3} \text{ mol } H_2C_2O_4 \times \frac{2 \text{ mol } KMnO_4}{5 \text{ mol } H_2C_2O_4}$$

$$= 1.148 \times 10^{-3} \text{ mol } KMnO_4$$

Knowing both the number of moles of $KMnO_4$ that react (1.148×10^{-3} mol) and the volume of the $KMnO_4$ solution (22.35 mL), we can calculate the molarity:

$$\text{Molarity} = \frac{1.148 \times 10^{-3} \text{ mol } KMnO_4}{22.35 \text{ mL}} \times \frac{1000 \text{ mL}}{1 \text{ L}} = 0.051\,36 \text{ M}$$

The molarity of the $KMnO_4$ solution is 0.051 36 M.

WORKED EXAMPLE 4.15

Using a Redox Reaction to Determine a Solution's Concentration

The concentration of an aqueous I_3^- solution can be determined by titration with aqueous sodium thiosulfate, $Na_2S_2O_3$, in the presence of a starch indicator. The starch turns from deep blue to colorless when all the I_3^- has reacted. What is the molar concentration of I_3^- in an aqueous solution if 24.55 mL of 0.102 M $Na_2S_2O_3$ is needed for complete reaction with 10.00 mL of the I_3^- solution? The net ionic equation is

$$2\,S_2O_3^{2-}\,(aq) + I_3^-(aq) \rightarrow S_4O_6^{2-}\,(aq) + 3\,I^-(aq)$$

IDENTIFY

Known	Unknown
Volume of I_3^- solution (10.00 mL)	Molarity of I_3^- solution
Volume of $Na_2S_2O_3$ solution (24.55 mL)	
Molarity of $Na_2S_2O_3$ solution (0.102 M)	

continued on next page

A precise amount of oxalic acid is weighed and dissolved in aqueous H_2SO_4.

Aqueous $KMnO_4$ of unknown concentration is added from a buret until …

… the purple color persists, indicating that all of the oxalic acid has reacted.

▲ **FIGURE 4.10**

The redox titration of oxalic acid, $H_2C_2O_4$, with $KMnO_4$.

▲ **Figure It Out**

Why does the solution in the flask remain clear until the end point of the titration is reached?

Answer: The purple titrant MnO_4^- is reduced to colorless Mn^{2+} by $H_2C_2O_4$. As soon as the $H_2C_2O_4$ is used up, additional MnO_4^- added to the flask generates the purple end point.

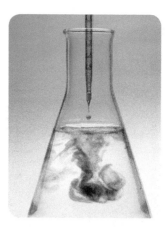

▲ The reddish I_3^- *solution* turns a deep blue color when it is added to a solution containing a small amount of starch.

STRATEGY

The procedure is similar to that outlined in Figure 4.9 except that volume of the $Na_2S_2O_3$ solution can be used to find moles of $S_2O_3^{2-}$ instead of a gram-to-mole conversion.

SOLUTION

We first need to find the number of moles of thiosulfate ion used for the titration:

$$24.55 \text{ mL} \times \frac{1 \text{ L}}{1000 \text{ mL}} \times \frac{0.102 \text{ mol } S_2O_3^{2-}}{1 \text{ L}} = 2.50 \times 10^{-3} \text{ mol } S_2O_3^{2-}$$

According to the balanced equation, 2 mol of $S_2O_3^{2-}$ ion react with 1 mol of I_3^- ion. Thus, we can find the number of moles of I_3^- ion:

$$2.50 \times 10^{-3} \text{ mol } S_2O_3^{2-} \times \frac{1 \text{ mol } I_3^-}{2 \text{ mol } S_2O_3^{2-}} = 1.25 \times 10^{-3} \text{ mol } I_3^-$$

Knowing both the number of moles of I_3^- (1.25×10^{-3} mol) and the volume of the I_3^- solution (10.00 mL), let us calculate molarity:

$$\frac{1.25 \times 10^{-3} \text{ mol } I_3^-}{10.00 \text{ mL}} \times \frac{10^3 \text{ mL}}{1 \text{ L}} = 0.125 \text{ M}$$

The molarity of the I_3^- solution is 0.125 M.

CHECK

According to the balanced equation, the amount of $S_2O_3^{2-}$ needed for the reaction (2 mol) is twice the amount of I_3^- (1 mol). The titration results indicate that the volume of the $S_2O_3^{2-}$ solution (24.55 mL) is a little over twice the volume of the I_3^- solution (10.00 mL). Thus, the concentrations of the two solutions must be about the same—approximately 0.1 M.

▶**PRACTICE 4.29** What is the molar concentration of Fe^{2+} ion in an aqueous solution if 31.50 mL of 0.105 M $KBrO_3$ is required for complete reaction with 10.00 mL of the Fe^{2+} solution? The net ionic equation is:

$$6 \text{ Fe}^{2+}(aq) + \text{BrO}_3^-(aq) + 6 \text{ H}^+(aq) \longrightarrow 6 \text{ Fe}^{3+}(aq) + \text{Br}^-(aq) + 3 \text{ H}_2O(l)$$

▶**APPLY 4.30** Iron(II) sulfate is a soluble ionic compound added as a source of iron in vitamin tablets. Determine the mass of iron (mg) in one tablet that has been dissolved in 10.0 mL of water and titrated with 14.92 mL of 0.0100 M $K_2Cr_2O_7$ solution. The net ionic equation is:

$$\text{Cr}_2O_7^{2-}(aq) + 6 \text{ Fe}^{2+}(aq) + 14 \text{ H}^+(aq) \longrightarrow 2 \text{ Cr}^{3+}(aq) + 6 \text{ Fe}^{3+}(aq) + 7 \text{ H}_2O(l)$$

4.14 SOME APPLICATIONS OF REDOX REACTIONS

Redox reactions take place with every element in the periodic table except helium and neon and occur in a vast number of processes throughout nature, biology, and industry. Here are just a few examples:

● **Combustion.** *Combustion* is the burning of a fuel by oxidation with oxygen in air. Gasoline, fuel oil, natural gas, wood, paper, and other organic substances of carbon and hydrogen are the most common fuels. Even some metals, such as magnesium and calcium, will burn in air.

$$\text{CH}_4(g) + 2 \text{ O}_2(g) \longrightarrow \text{CO}_2(g) + 2 \text{ H}_2O(l)$$
Methane
(Natural gas)

● **Bleaching.** *Bleaching* uses redox reactions to decolorize or lighten colored materials. Dark hair is bleached to turn it blond, clothes are bleached to remove stains, wood pulp is bleached to make white paper, and so on. The exact oxidizing agent used depends on the situation—hydrogen peroxide (H_2O_2) is used for hair, sodium hypochlorite (NaOCl) is used for clothes, and ozone or chlorine dioxide is used for wood pulp—but the principle is always the same. In all cases, colored impurities are destroyed by reaction with a strong oxidizing agent.

- **Batteries.** Although they *come* in many types and sizes, all types of batteries are powered by redox reactions. In a typical redox reaction carried out in the laboratory—say, the reaction of zinc metal with Ag^+ to yield Zn^{2+} and silver metal—the reactants are simply mixed in a flask and electrons are transferred by direct contact between them. In a battery, however, the two reactants are kept in separate compartments and the electrons are transferred through a wire running between them.

▲ Dark hair can be bleached by a redox reaction with hydrogen peroxide.

    The inexpensive alkaline battery commonly used in flashlights and other small household items uses a thin steel can containing zinc powder and a paste of potassium hydroxide as one reactant, separated by paper from a paste of powdered carbon and manganese dioxide as the other reactant. A graphite rod with a metal cap sticks into the MnO_2 to provide electrical contact. When the can and the graphite rod are connected by a wire, zinc sends electrons flowing through the wire toward the MnO_2 in a redox reaction. The resultant electrical current can be used to light a bulb or power a small electronic device. The reaction is

$$Zn(s) + 2\ MnO_2(s) \longrightarrow ZnO(s) + Mn_2O_3(s)$$

We'll look at the chemistry of batteries in more detail in Section 19.10.

- **Metallurgy.** *Metallurgy,* the extraction and purification of metals from their ores, makes use of numerous redox processes. Metallic zinc is prepared by reduction of ZnO with *coke,* a form of carbon:

$$ZnO(s) + C(s) \longrightarrow Zn(s) + CO(g)$$

- **Corrosion.** *Corrosion* is the *deterioration* of a metal by oxidation, such as the rusting of iron in moist air. The economic consequences of rusting are enormous: It has been estimated that up to one-fourth of the iron produced in the United States is used to replace bridges, buildings, and other structures that have been destroyed by corrosion. (The raised dot in the formula $Fe_2O_3 \cdot H_2O$ for rust indicates that one water molecule is associated with each Fe_2O_3 in an unspecified way.)

$$4\ Fe(s) + 3\ O_2(g) \xrightarrow{H_2O} 2\ Fe_2O_3 \cdot H_2O(s)$$
$$\text{Rust}$$

- **Respiration.** The term *respiration* refers to the processes of *breathing* and using oxygen for the many biological redox reactions that provide the energy needed by living organisms. The energy is released from food molecules slowly and in complex, multi-step pathways, but the overall result of respiration is similar to that of a combustion reaction. For example, the simple sugar glucose ($C_6H_{12}O_6$) reacts with O_2 to give CO_2 and H_2O according to the following equation:

$$\underset{\substack{\text{Glucose} \\ \text{(a carbohydrate)}}}{C_6H_{12}O_6} + 6\ O_2 \rightarrow 6\ CO_2 + 6\ H_2O + \text{energy}$$

INQUIRY ?   **How do sports drinks replenish the substances lost in sweat?**

Athletes consume sports drinks, such as Gatorade and Powerade, during exercise. How do these drinks help them perform better and recover more quickly? Sports drinks were first developed in 1965 when University of Florida football coaches noted that players became extremely fatigued, lost significant amounts of weight, and seldom needed to urinate after exercising in the heat. The team consulted with Robert Cade (1927–2007), a kidney specialist at the University of Florida's College of Medicine, who speculated that electrolytes lost in sweat were upsetting the body's delicate chemical balance. Sodium and potassium ions were of primary concern due to their importance in nerve and muscle function, regulation of body heat, distribution of water, and transport of solutes such as glucose for energy.

To test his hypothesis, Cade and a team of researchers studied the fluids of freshman players before and after exercising vigorously in the heat. The results were staggering; after exercise the players had an electrolyte imbalance, low blood sugar, and decreased total blood volume; leading to diminished physical performance and in some cases extreme heat exhaustion. Cade's team created a drink to replace the fluids and electrolytes lost through sweat and the carbohydrates burned for energy. The first batch contained water, salt, sugar, and lemon juice.

By 1966, the drink known as *Gatorade* became a staple for the team, and hospitalizations of players due to heat exhaustion became almost nonexistent. The Gators advanced to the Orange Bowl for the first time in the school's history. The university released an official statement about Gatorade in late December 1966 that the *Florida Times-Union* summed up with this headline: "One Lil' Swig of That Kickapoo Juice and Biff, Bam, Sock—It's Gators, 8-2."

▲ Gatorade and other sports drinks contain electrolytes and conduct electricity.

PROBLEM 4.31   A vitamin-fortified brand of a sports beverage contains sodium chloride (NaCl), sodium citrate ($NaC_6H_7O_7$), and potassium dihydrogen phosphate (KH_2PO_4) as well as the substances whose structures are given in the figure.

(a) Use Table 4.1 (Electrolyte Classification of Some Common Substances) to classify the components of the sports drink as a strong electrolyte, weak electrolyte, or nonelectrolyte.

(b) Identify which substances replenish electrolytes with important biological functions.

Citric acid ($C_6H_8O_7$)

Vitamin B3 ($C_6NH_5O_2$)

Fructose ($C_6H_{12}O_6$)

PROBLEM 4.32   The nutritional label on Powerade specifies that there are 150 mg of sodium and 35 mg of potassium in 360 mL of the beverage. Calculate the concentration of sodium and potassium ions in units of molarity.

PROBLEM 4.33   The concentration of sodium ions in Powerade is 0.416 mg/mL. Imagine that you want to prepare your own drink with the same concentration of sodium ions. How many grams of sodium chloride are needed to prepare 0.500 L of solution?

PROBLEM 4.34   One way to analyze a sports drink for the concentration of chloride ions is to add silver ions and weigh the resulting AgCl precipitate. One problem with the analysis is that many sports drinks contain phosphate ion (PO_4^{3-}), which will also precipitate with silver, thus interfering with the chloride measurement. The phosphate ion can be removed by precipitation prior to the analysis of chloride.

(a) Use the solubility guidelines (Table 4.2) to choose a cation from the list below that would form a precipitate with phosphate but not with chloride.

$$K^+, Ba^{2+}, Pb^{2+}, NH_4^+$$

(b) Write the net ionic reaction for the precipitation reaction from part (a).

PROBLEM 4.35   To measure the concentration of chloride ions in a sports beverage, an excess of silver ions were added to 100.0 mL of the drink. A white precipitate of silver chloride was isolated by filtration, dried, and found to have a mass of 172 mg. Calculate the concentration of chloride ion in the drink in units of molarity.

PROBLEM 4.36   The flavor of the first batch of Gatorade was improved by adding lemon juice, which contains citric acid ($H_3C_6H_5O_7$). Citric acid is still added as flavoring to sports drinks today. The concentration of citric acid in a beverage was determined by titration with sodium hydroxide according to the reaction:

$$H_3C_6H_5O_7(aq) + 3\,NaOH(aq) \longrightarrow Na_3C_6H_5O_7(aq) + 3\,H_2O(l)$$

If 25.0 mL of the beverage required 35.6 mL of 0.0400 M NaOH for a complete reaction, calculate the molarity of citric acid.

STUDY GUIDE

Section	Concept Summary	Learning Objectives	Test Your Understanding
4.1 Solution Concentration: Molarity	The concentration of a substance in solution is usually expressed as **molarity (M)**, defined as the number of moles of a substance (**solute**) dissolved per liter of solution. A solution's molarity acts as a conversion factor between solution volume and number of moles of solute, making it possible to carry out stoichiometry calculations on solutions.	**4.1** Calculate the molarity of a solution given the mass of solute and total volume.	Worked Example 4.1; Problems 4.1–4.2, 4.50–4.51
		4.2 Calculate the a amount of solute in a given volume of solution with a known molarity.	Worked Example 4.2; Problems 4.3–4.4, 4.46, 4.53
		4.3 Describe the proper technique for preparing solutions of known molarity.	Problems 4.55–4.56
4.2 Diluting Concentrated Solutions	When carrying out a dilution, only the volume is changed by adding solvent; the amount of solute is unchanged.	**4.4** Calculate the concentration of a solution that has been diluted.	Worked Example 4.3; Problems 4.5–4.6, 4.59–4.60
		4.5 Describe the proper technique for diluting solutions.	Problems 4.60–4.61
4.3 Electrolytes in Aqueous Solution	Many reactions take place in aqueous solution. Substances whose aqueous solutions contain ions conduct electricity and are called **electrolytes.** Ionic compounds, such as NaCl, and molecular compounds that **dissociate** substantially into ions when dissolved in water are **strong electrolytes.** Substances that dissociate to only a small extent are **weak electrolytes,** and substances that do not produce ions in aqueous solution are **nonelectrolytes.**	**4.6** Classify a substance as a strong, weak, or nonelectrolyte. (Table 4.1)	Problems 4.62–4.63
		4.7 Calculate the concentration of ions in a strong electrolyte solution.	Worked Example 4.4; Problems 4.7–4.8, 4.68, 4.70
4.4 Types of Chemical Reactions in Aqueous Solution	Aqueous reactions can be classified into three major groups. **Precipitation reactions** occur when solutions of two ionic substances are mixed and a precipitate settles out of the solution. **Acid–base neutralization reactions** occur when an acid is mixed with a base, yielding water and an ionic **salt. Oxidation–reduction reactions,** or **redox reactions,** occur when one or more electrons are transferred between reaction partners.	**4.8** Classify a reaction as a precipitation, acid–base neutralization, or oxidation–reduction (redox) reaction.	Problems 4.72–4.73
4.5 Aqueous Reactions and Net Ionic Equations	Aqueous ionic compounds exist as cations and anions in solution. An **ionic equation** shows all the ions in a reaction, and a **net ionic equation** shows only the ions that take part in a reaction. **Spectator ions** are present to balance charge but do not take part in the chemical reaction.	**4.9** Write a net ionic equation and identify spectator ions given the molecular equation.	Worked Example 4.5; Problems 4.9–4.10, 4.74
4.6 Precipitation Reactions and Solubility Guidelines	Solubility guidelines (Table 4.2) are used to predict which combinations of anions and cations in ionic compounds will be soluble and insoluble. To predict whether a precipitate will form in a reaction, write the formula of possible products and determine **solubility.**	**4.10** Use the solubility guidelines to predict the solubility of an ionic compound in water.	Problems 4.76–4.77
		4.11 Predict whether a precipitation reaction will occur and write the ionic and net ionic equations.	Worked Examples 4.6–4.7; Problems 4.11–4.14, 4.78, 4.80
4.7 Acids, Bases, and Neutralization Reactions	An acid is a substance that dissociates in water to give hydrogen (H^+) ions, and a base is a substance that dissociates to give hydroxide ions (OH^-). The neutralization of a strong acid with a strong base can be written as a net ionic equation, in which nonparticipating, spectator ions are not specified: $$H^+(aq) + OH^-(aq) \longrightarrow H_2O(l)$$	**4.12** Convert between name and formula for an acid.	Worked Example 4.8; Problems 4.15–4.16
		4.13 Classify acids as strong or weak based on the molecular picture of dissociation.	Problem 4.40
		4.14 Write the ionic equation and net ionic equation for an acid–base neutralization reaction.	Worked Example 4.9; Problems 4.17–4.18, 4.96, 4.98

Section	Concept Summary	Learning Objectives	Test Your Understanding
4.8 Solution Stoichiometry	Stoichiometry calculations are performed by relating amounts of reactants and products in a balanced equation in units of moles since stoichiometric coefficients refer to moles. Molarity is a conversion factor between numbers of moles of solute and the volume of a solution.	**4.15** Convert between moles and volume using molarity in stoichiometry calculations.	Worked Example 4.10; Problems 4.19–4.20, 4.100–4.101
4.9 Measuring the Concentration of a Solution: Titration	**Titration** is a technique used to find the exact concentration of a solution. A fixed volume of solution with unknown concentration is added to a flask. A solution with a known concentration (titrant) is added from a buret until the reaction is complete. The measured volume of titrant and reaction stoichiometry are used to calculate the concentration of the solution in the flask.	**4.16** Determine the concentration of a solution using titration data.	Worked Example 4.11; Problems 4.21, 4.102–4.103
		4.17 Interpret molecular representations of substances in solution in a titration procedure.	Problems 4.22, 4.42–4.43
4.10 Oxidation–Reduction (Redox) Reactions	**Oxidation** is the loss of one or more electrons; a **reduction** is the gain of one or more electrons. Redox reactions can be identified by assigning to each atom in a substance an **oxidation number,** which provides a measure of whether the atom is neutral, electron-rich, or electron-poor. Comparing the oxidation numbers of an atom before and after reaction shows whether the atom has gained or lost electrons.	**4.18** Assign oxidation numbers to atoms in a compound.	Worked Example 4.12; Problems 4.23–4.24, 4.108, 4.110
4.11 Identifying Redox Reactions	Oxidations and reductions must occur together. Whenever one substance loses one or more electrons (is oxidized), another substance gains the electrons (is reduced). The substance that causes a reduction by giving up electrons is called a **reducing agent.** The substance that causes an oxidation by accepting electrons is called an **oxidizing agent.** The reducing agent is itself oxidized when it gives up electrons, and the oxidizing agent is itself reduced when it accepts electrons.	**4.19** Identify redox reactions, oxidizing agents, and reducing agents.	Worked Example 4.13; Problems 4.25–4.26, 4.118–4.119
4.12 The Activity Series of the Elements	Among the simplest of redox processes is the reaction of an aqueous cation, usually a metal ion, with a free element to give a different ion and a different element. Noting the results from a succession of different reactions makes it possible to organize an **activity series,** which ranks the elements in order of their reducing ability in aqueous solution.	**4.20** Use the location of elements in the periodic table and activity series to predict if a redox reaction will occur.	Worked Example 4.14; Problems 4.27–4.28, 4.45, 4.120
		4.21 Develop an activity series and predict if a redox reaction will occur based on experimental data provided.	Problems 4.122–4.123
4.13–4.14 Redox Titration and Applications of Redox Reactions	The concentration of an oxidizing agent or a reducing agent in solution can be determined by a redox titration.	**4.22** Use a redox titration to determine the concentration of an oxidizing or reducing agent in solution.	Worked Example 4.15; Problems 4.29, 4.124, 4.126, 4.132

KEY TERMS

acid *128*
acid–base neutralization
  reaction *123*
activity series *142*
base *128*
diprotic acid *128*
dissociate *122*
electrolyte *122*
half-reaction *139*
hydronium ion, H_3O^+ *128*

ionic equation *124*
molarity (M) *117*
molecular equation *124*
monoprotic acid *128*
net ionic equation *124*
nonelectrolyte *122*
oxidation *136*
oxidation number *136*
oxidation–reduction
  reaction *124*

oxidizing agent *139*
oxoacid *129*
precipitation reaction *123*
redox reaction *124*
reducing agent *150*
reduction *136*
salt *131*
solubility *125*
solute *117*
spectator ion *124*

strong acid *128*
strong base *129*
strong electrolyte *122*
titration *133*
triprotic acid *128*
weak acid *128*
weak base *129*
weak electrolyte *122*

KEY EQUATIONS

• **Molarity (Section 4.1)**

$$\text{Molarity (M)} = \frac{\text{Moles of solute}}{\text{Liters of solution}}$$

• **Dilution (Section 4.2)**

$$M_i \times V_i = M_f \times V_f$$

PRACTICE TEST

After studying this chapter, you can assess your understanding with these practice test questions, which are correlated with chapter learning objectives. If you answer a question incorrectly, refer to the learning objectives in the end-of-chapter Study Guide for assistance. The Study Guide provides a conceptual summary, references a Worked Example to model how to solve the problem, and gives additional problems for more practice.

1. What is the molarity of a solution prepared by dissolving 10.19 g of ethanol (CH_3CH_2OH) in enough water to produce 250.0 mL of solution? **(LO 4.1)**
   (a) 0.8848 M   (b) 18.08 M
   (c) 1.130 M   (d) 0.01808 M

2. What is the mass of chloride ions in 375.0 mL of solution with a magnesium chloride concentration of 0.250 M? **(LO 4.2)**
   (a) 3.32 g   (b) 47.3 g
   (c) 23.6 g   (d) 6.65 g

3. What volume of a 2.00 M stock solution of NaOH is required to prepare 50.0 mL of 0.400 M NaOH? **(LO 4.4)**
   (a) 15.0 mL   (b) 1.00 mL
   (c) 10.0 mL   (d) 4.00 mL

Refer to the figure to answer questions 4 and 5. The images are a molecular representation of three different substances, AX_3, AY_3, and AZ_3, dissolved in water. (Water molecules are omitted for clarity.)

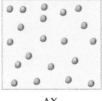
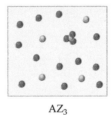

AX₃   AY₃   AZ₃

4. Which of the substances is the weakest electrolyte? **(LO 4.6)**
   (a) AX_3
   (b) AY_3
   (c) AZ_3
   (d) All of the substances are strong electrolytes.

5. What are the molar concentrations of A ions and X ions in a 0.500 M solution of AX_3? **(LO 4.7)**
   (a) 0.500 M A and 0.500 M X
   (b) 0.500 M A and 0.167 M X
   (c) 1.500 M A and 0.500 M X
   (d) 0.500 M A and 1.500 M X

6. Which of the following substances will produce a solution that *does not* conduct electricity when it dissolves in water? **(LO 4.6)**
   (a) NaOH   (b) HNO_3
   (c) Na_2SO_4 (d) CH_3OH

7. Which of the following solutions *will not* form a precipitate when added to 10 mL of 0.10 M KOH? **(LO 4.10, 4.11)**
   (a) 10 mL of 0.10 NH_4Cl
   (b) 10 mL of 0.10 M $PbSO_4$
   (c) 10 mL of 0.10 M $Fe(NO_3)_3$
   (d) 10 mL of 0.10 M $AgCH_3CO_2$

8. Write a net ionic equation for the reaction that occurs when 10 mL of 0.5 M ammonium carbonate is mixed with 10 mL of 0.5 M silver nitrate. **(LO 4.9, 4.11)**
   (a) $NH_4^+(aq) + NO_3^-(aq) \longrightarrow NH_4NO_3(s)$
   (b) $Ag^+(aq) + CO_3^{2-}(aq) \longrightarrow AgCO_3^-(s)$
   (c) $2\,Ag^+(aq) + CO_3^{2-}(aq) \longrightarrow Ag_2CO_3(s)$
   (d) A net ionic reaction cannot be written because a reaction does not take place.

9. When 75.0 mL of a 0.100 M lead(II) nitrate solution is mixed with 100.0 mL of a 0.190 M potassium iodide solution, a yellow-orange precipitate of lead(II) iodide is formed. What is the mass in grams of lead(II) iodide formed? Assume the reaction goes to completion. **(LO 4.11, 4.15)**
   (a) 1.729 g   (b) 3.458 g
   (c) 4.380 g   (d) 8.760 g

10. What volume of 0.250 M HCl is needed to react completely with 25.00 mL of 0.375 M Na_2CO_3? **(LO 4.15)**

    $2\,HCl(aq) + Na_2CO_3(aq) \longrightarrow 2\,NaCl(aq) + H_2O(l) + CO_2(g)$

    (a) 75.0 mL   (b) 18.8 mL
    (c) 37.5 mL   (d) 33.3 mL

11. Succinic acid, an intermediate in the metabolism of food molecules, has a molecular weight of 118.1. When 1.926 g of succinic acid was dissolved in water and titrated, 65.20 mL of 0.5000 M NaOH solution was required to neutralize the acid. How many acidic hydrogens are there in a molecule of succinic acid? **(LO 4.16)**
    (a) 1   (b) 2   (c) 3   (d) 4

12. Assign oxidation numbers to each atom in Borax, $Na_2B_4O_7$, a mineral used in laundry detergent. **(LO 4.18)**

	Na	B	O
(a)	+2	+3	−2
(b)	−1	−3	+2
(c)	+1	+3	−2
(d)	+1	+2	−1

13. Identify the element that gets oxidized and the oxidizing agent in the reaction. (**LO 4.19**)

$$HCrO_4^-(aq) + H_2S(aq) \longrightarrow Cr_2O_3(s) + SO_4^{2-}$$

	Element oxidized	Oxidizing agent
(a)	O	$HCrO_4^-$
(b)	S	$HCrO_4^-$
(c)	Cr	H_2S
(d)	Cr	$HCrO_4^-$

14. The most strongly reducing elements are listed at the top of the partial activity series table provided. Use the activity series to predict which reaction will occur. (**LO 4.20**)

$$K(s) \longrightarrow K^+(aq) + e^-$$
$$Mg(s) \longrightarrow Mg^{2+}(aq) + 2e^-$$
$$H_2(g) \longrightarrow 2 H^+(aq) + 2e^-$$
$$Ag(s) \longrightarrow Ag^+(aq) + e^-$$

(a) $Mg(s) + 2 K^+(aq) \longrightarrow Mg^{2+}(aq) + 2 K(s)$
(b) $2 Ag(s) + 2 H^+(aq) \longrightarrow 2 Ag^+(aq) + H_2(g)$
(c) $H_2(g) + Mg^{2+}(aq) \longrightarrow 2 H^+(aq) + Mg(s)$
(d) $Mg(s) + 2 Ag^+(aq) \longrightarrow Mg^{2+}(aq) + 2 Ag(s)$

15. The concentration of a solution of potassium permanganate, $KMnO_4$, can be determined by titration with a known amount of oxalic acid, $H_2C_2O_4$, according to the following equation:

$$5 H_2C_2O_4(aq) + 2 KMnO_4(aq) + 3 H_2SO_4(aq) \longrightarrow$$
$$10 CO_2(g) + 2 MnSO_4(aq) + K_2SO_4(aq) + 8 H_2O(l)$$

What is the concentration of a $KMnO_4$ solution if 22.35 mL reacts with 0.5170 g of oxalic acid? (**LO 4.22**)

(a) 0.6423 M   (b) 0.1028 M
(c) 0.4161 M   (d) 0.2569 M

Answers:

1. a, 2. d, 3. c, 4. b, 5. d, 6. d, 7. a, 8. c, 9. b, 10. a, 11. b, 12. c, 13. b, 14. d, 15. b

---

Mastering Chemistry   provides end-of-chapter exercises, feedback-enriched tutorial problems, animations, and interactive activities to encourage problem-solving practice and deeper understanding of key concepts and topics.

RAN   *Randomized in Mastering Chemistry*

CONCEPTUAL PROBLEMS

Problems 4.1–4.36 appear within the chapter.

4.37   Box (a) represents 1.0 mL of a solution of particles at a
RAN   given concentration. Which of the boxes (b)–(d) represents 1.0 mL of the solution that results after (a) has been diluted by doubling the volume of its solvent?

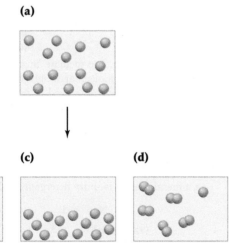

(a)

(b)   (c)   (d)

4.38   Assume that an aqueous solution of a cation, represented as a red sphere, is allowed to mix with a solution of an anion, represented as a yellow sphere. Three possible outcomes are represented by boxes (1)–(3):

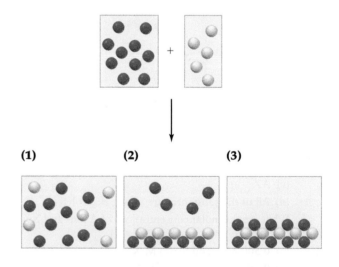

(1)   (2)   (3)

Which outcome corresponds to each of the following reactions?
(a) $2 Na^+(aq) + CO_3^{2-}(aq) \rightarrow$
(b) $Ba^{2+}(aq) + CrO_4^{2-}(aq) \rightarrow$
(c) $2 Ag^+(aq) + SO_3^{2-}(aq) \rightarrow$

4.39 Assume that an aqueous solution of a cation, represented as a blue sphere, is allowed to mix with a solution of an anion, represented as a red sphere, and that the following result is obtained:

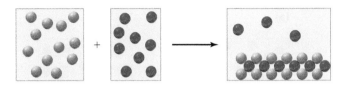

Which combinations of cation and anion, chosen from the following lists, are compatible with the observed results? Explain.

Cations: Na^+, Ca^{2+}, Ag^+, Ni^{2+}

Anions: Cl^-, CO_3^{2-}, CrO_4^{2-}, NO_3^-

4.40 The following pictures represent aqueous solutions of three acids HA (A = X, Y, or Z), with surrounding water molecules omitted for clarity. Which of the three is the strongest acid, and which is the weakest?

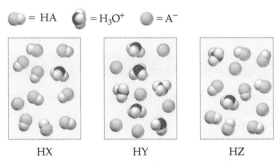

= HA = H_3O^+ = A^-

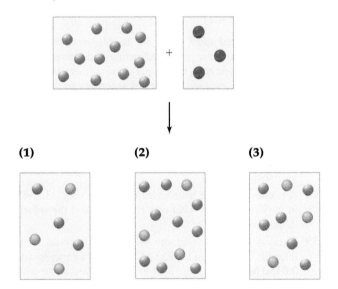

HX   HY   HZ

4.41 Assume that an aqueous solution of OH^-, represented as a blue sphere, is allowed to mix with a solution of an acid H_nA, represented as a red sphere. Three possible outcomes are depicted by boxes (1)–(3), where the green spheres represent A^{n-}, the anion of the acid:

+

↓

(1)   (2)   (3)

Which outcome corresponds to each of the following reactions?

(a) $HF + OH^- \longrightarrow H_2O + F^-$

(b) $H_2SO_3 + 2\,OH^- \longrightarrow 2\,H_2O + SO_3^{2-}$

(c) $H_3PO_4 + 3\,OH^- \rightarrow 3\,H_2O + PO_4^{3-}$

4.42 The concentration of an aqueous solution of NaOCl (sodium hypochlorite; the active ingredient in household bleach) can be determined by a redox titration with iodide ion in acidic solution:

$$OCl^-(aq) + 2I^-(aq) + 2H^+(aq) \rightarrow Cl^-(aq) + I_2(aq) + H_2O(l)$$

Assume that the blue spheres in the buret represent I^- ions, the red spheres in the flask represent OCl^- ions, the concentration of the I^- ions in the buret is 0.120 M, and the volumes in the buret and the flask are identical. What is the concentration of NaOCl in the flask? What percentage of the I^- solution in the buret must be added to the flask to react with all the OCl^- ions?

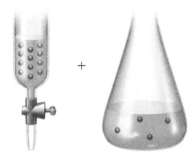

+

4.43 Assume that the electrical conductivity of a solution depends on the total concentration of dissolved ions and that you measure the conductivity of three different solutions while carrying out titration procedures:

(a) Begin with 1.00 L of 0.100 M KCl, and titrate by adding 0.100 M $AgNO_3$.

(b) Begin with 1.00 L of 0.100 M HF, and titrate by adding 0.100 M KOH.

(c) Begin with 1.00 L of 0.100 M $BaCl_2$, and titrate by adding 0.100 M Na_2SO_4.

Which of the following graphs corresponds to which titration?

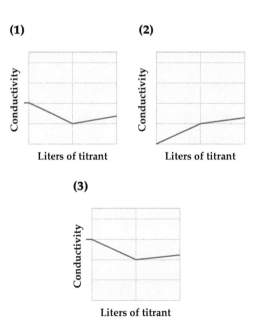

(1)   **(2)**

(3)

4.44 Based on the positions in the periodic table, which of the following reactions would you expect to occur?

(a) $Red^+ + Green \rightarrow Red + Green^+$

(b) $Blue + Green^+ \rightarrow Blue^+ + Green$

(c) $Red + Blue^+ \rightarrow Red^+ + Blue$

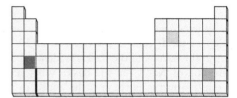

4.45 The following two redox reactions occur between aqueous cations and solid metals. Will a solution of green cations react with solid blue metal? Explain.

(a)

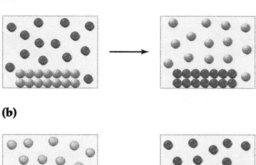

(b)

SECTION PROBLEMS

Molarity (Section 4.1)

4.46 How many moles of solute are present in each of the following solutions?
RAN

(a) 35.0 mL of 1.200 M HNO_3

(b) 175 mL of 0.67 M glucose ($C_6H_{12}O_6$)

4.47 How many grams of solute would you use to prepare each of the following solutions?
RAN

(a) 250.0 mL of 0.600 M ethyl alcohol (C_2H_6O)

(b) 167 mL of 0.200 M boric acid (H_3BO_3)

4.48 How many milliliters of a 0.45 M $BaCl_2$ solution contain 15.0 g of $BaCl_2$?
RAN

4.49 How many milliliters of a 0.350 M KOH solution contain 0.0171 mol of KOH?
RAN

4.50 The sterile saline solution used to rinse contact lenses can be made by dissolving 400 mg of NaCl in sterile water and diluting to 100 mL. What is the molarity of the solution?
RAN

4.51 The concentration of glucose ($C_6H_{12}O_6$) in normal blood is approximately 90 mg per 100 mL. What is the molarity of the glucose?

4.52 Copper reacts with dilute nitric acid according to the following equation:
RAN

$$3\ Cu(s) + 8\ HNO_3(aq) \longrightarrow 3\ Cu(NO_3)_2(aq) + 2\ NO(g) + 4\ H_2O(l)$$

If a copper penny weighing 3.045 g is dissolved in a small amount of nitric acid and the resultant solution is diluted to 50.0 mL with water, what is the molarity of the $Cu(NO_3)_2$?

4.53 The estimated concentration of gold in the oceans is 1.0×10^{-11} g/mL.
RAN

(a) Express the concentration in mol/L.

(b) Assuming that the volume of the oceans is 1.3×10^{21} L, estimate the amount of dissolved gold in grams in the oceans.

4.54 How many grams of solute would you use to prepare the following solutions?
RAN

(a) 500.0 mL of 1.25 M NaOH

(b) 1.50 L of 0.250 M glucose ($C_6H_{12}O_6$)

4.55 How would you prepare 500 mL of a 0.330 M solution of $CaCl_2$ from solid $CaCl_2$? Specify the glassware that should be used.
RAN

4.56 How would you prepare 250 mL of a 0.100 M solution of fluoride ions from solid CaF_2? Specify the glassware that should be used.

Dilutions (Section 4.2)

4.57 Pennies minted after 1982 are mostly zinc (97.5%) with a copper cover. If a post-1982 penny is dissolved in a small amount of nitric acid, the copper coating reacts as in Problem 4.52, and the exposed zinc reacts according to the following equation:
RAN

$$Zn(s) + 2\ HNO_3(aq) \longrightarrow Zn(NO_3)_2(aq) + H_2(g)$$

For a penny that weighs 2.482 g, what is the molarity of the $Zn(NO_3)_2$ if the resultant solution is diluted to 250.0 mL with water?

4.58 A bottle of 12.0 M hydrochloric acid has only 35.7 mL left in it. What will the HCl concentration be if the solution is diluted to 250.0 mL?
RAN

4.59 What is the volume of the solution that would result by diluting 70.00 mL of 0.0913 M NaOH to a concentration of 0.0150 M?
RAN

4.60 How would you prepare 250 mL of a 0.100 M solution of chloride ions from a 3.00 M stock solution of $CaCl_2$? Specify the glassware that should be used.

4.61 How would you prepare 250 mL of 0.150 M solution of $CaCl_2$ from a 3.00 M stock solution? Specify the glassware that should be used.
RAN

Electrolytes (Section 4.3)

4.62 The following aqueous solutions were tested with a light
RAN bulb conductivity apparatus, as shown in Figure 4.3. What
result—dark, dim, or bright—do you expect from each?

(a) 0.10 M potassium chloride

(b) 0.10 M methanol

(c) 0.10 M acetic acid

4.63 The following aqueous solutions were tested with a light
bulb conductivity apparatus, as shown in Figure 4.3. What
result—dark, dim, or bright—do you expect from each?

(a) 0.10 M hydrofluoric acid

(b) 0.10 M sodium chloride

(c) 0.10 M glucose ($C_6H_{12}O_6$)

4.64 Individual solutions of $Ba(OH)_2$ and H_2SO_4 both conduct
electricity, but the conductivity disappears when equal molar
amounts of the solutions are mixed. Explain.

4.65 A solution of HCl in water conducts electricity, but a solution of HCl in chloroform, $CHCl_3$, does not. What does this
observation tell you about how HCl exists in water and how
it exists in chloroform?

4.66 Classify each of the following substances as a strong electrolyte, weak electrolyte, or nonelectrolyte.

(a) HBr   (b) HF

(c) $NaClO_4$ (d) $(NH_4)_2CO_3$

(e) NH_3 (f) Ethyl alcohol

4.67 Is it possible for a molecular substance to be a strong electrolyte? Explain.

4.68 What is the total molar concentration of ions in each of the
RAN following solutions, assuming complete dissociation?

(a) A 0.750 M solution of K_2CO_3

(b) A 0.355 M solution of $AlCl_3$

4.69 What is the total molar concentration of ions in each of the
RAN following solutions?

(a) A 1.250 M solution of CH_3OH

(b) A 0.225 M solution of $HClO_4$

4.70 *Ringer's solution,* used in the treatment of burns and wounds,
RAN is prepared by dissolving 4.30 g of NaCl, 0.150 g of KCl, and
0.165 g of $CaCl_2$ in water and diluting to a volume of 500.0 mL.
What is the molarity of each of the component ions in the
solution?

4.71 What is the molarity of each ion in a solution prepared by dis-
RAN solving 0.550 g of Na_2SO_4, 1.188 g of Na_3PO_4, and 0.223 g
of Li_2SO_4 in water and diluting to a volume of 100.00 mL?

Net Ionic Equations and Aqueous Reactions (Sections 4.4–4.5)

4.72 Classify each of the following reactions as a precipitation,
acid–base neutralization, or oxidation–reduction.

(a) $Hg(NO_3)_2(aq) + 2\,NaI(aq) \rightarrow 2\,NaNO_3(aq) + HgI_2(s)$

(b) $2\,HgO(s) \xrightarrow{\text{heat}} 2\,Hg(l) + O_2(g)$

(c) $H_3PO_4(aq) + 3\,KOH(aq) \rightarrow K_3PO_4(aq) + 3\,H_2O(l)$

4.73 Classify each of the following reactions as a precipitation,
acid–base neutralization, or oxidation–reduction.

(a) $S_8(s) + 8\,O_2(g) \rightarrow 8\,SO_2(g)$

(b) $NiCl_2(aq) + Na_2S(aq) \rightarrow NiS(s) + 2\,NaCl(aq)$

(c) $2\,CH_3CO_2H(aq)\ 1\ Ba(OH)_2(aq) \rightarrow$
$(CH_3CO_2)_2Ba(aq) + 2\,H_2O(l)$

4.74 Write net ionic equations for the reactions listed in
RAN Problem 4.72.

4.75 Write net ionic equations for the reactions listed in
RAN Problem 4.73.

Precipitation Reactions and Solubility Guidelines (Section 4.6)

4.76 Which of the following substances are likely to be soluble in
RAN water?

(a) $PbSO_4$ (b) $Ba(NO_3)_2$

(c) $SnCO_3$ (d) $(NH_4)_3PO_4$

4.77 Which of the following substances are likely to be soluble in
RAN water?

(a) ZnS   (b) $AU_2(CO_3)_3$

(c) $PbCl_2$ (d) Na_2S

4.78 Predict whether a precipitation reaction will occur when
RAN aqueous solutions of the following substances are mixed. For
those that form a precipitate, write the net ionic reaction.

(a) $NaOH + HClO_4$ (b) $FeCl_2 + KOH$

(c) $(NH_4)_2SO_4 + NiCl_2$ (d) $CH_3CO_2Na + HCl$

4.79 Predict whether a precipitation reaction will occur when
RAN aqueous solutions of the following substances are mixed. For
those that form a precipitate, write the net ionic reaction.

(a) $MnCl_2 + Na_2S$

(b) $HNO_3 + CuSO_4$

(c) $Hg(NO_3)_2 + Na_3PO_4$

(d) $Ba(NO_3)_2 + KOH$

4.80 Which of the following solutions will not form a precipitate
RAN when added to 0.10 M $BaCl_2$?

(a) 0.10 M $LiNO_3$ (b) 0.10 M K_2SO_4

(c) 0.10 M $AgNO_3$

4.81 Which of the following solutions will not form a precipitate
RAN when added to 0.10 M NaOH?

(a) 0.10 M $MgBr_2$ (b) 0.10 M NH_4Br

(c) 0.10 M $FeCl_2$

4.82 How would you prepare the following substances by a pre-
RAN cipitation reaction?

(a) $PbSO_4$ (b) $Mg_3(PO_4)_2$

(c) $ZnCrO_4$

4.83 How would you prepare the following substances by a precipitation reaction?

(a) $Al(OH)_3$ (b) FeS

(c) $CoCO_3$

4.84 What are the mass and the identity of the precipitate that
RAN forms when 30.0 mL of 0.150 M HCl reacts with 25.0 mL
of 0.200 M $AgNO_3$?

4.85 What are the mass and the identity of the precipitate that
RAN forms when 55.0 mL of 0.100 M $BaCl_2$ reacts with 40.0 mL
of 0.150 M Na_2CO_3?

4.86 Assume that you have an aqueous mixture of $NaNO_3$ and
$AgNO_3$. How could you use a precipitation reaction to separate the two metal ions?

4.87 Assume that you have an aqueous mixture of $BaCl_2$ and
$CuCl_2$. How could you use a precipitation reaction to separate the two metal ions?

4.88 Assume that you have an aqueous solution of an unknown
salt. Treatment of the solution with dilute NaOH, Na_2SO_4,

and KCl produces no precipitate. Which of the following cations might the solution contain?

(a) Ag^+ (b) Cs^+

(c) Ba^{2+} (d) NH_4^+

4.89 Assume that you have an aqueous solution of an unknown salt. Treatment of the solution with dilute $BaCl_2$, $AgNO_3$, and $Cu(NO_3)_2$ produces no precipitate. Which of the following anions might the solution contain?

(a) Cl^- (b) NO_3^-

(c) OH^- (d) SO_4^{2-}

4.90 How could you use a precipitation reaction to separate each of the following pairs of cations? Write the formula for each reactant you would add, and write a balanced net ionic equation for each reaction.

(a) K^+ and Hg_2^{2+} (b) Pb^{2+} and Ni^{2+}

(c) Ca^{2+} and NH_4^+ (d) Fe^{2+} and Ba^{2+}

4.91 How could you use a precipitation reaction to separate each of the following pairs of anions? Write the formula for each reactant you would add, and write a balanced net ionic equation for each reaction.

(a) Cl^- and NO_3^- (b) S^{2-} and SO_4^{2-}

(c) SO_4^{2-} and CO_3^{2-} (d) OH^- and ClO_4^-

4.92 The following three solutions are mixed: 100.0 mL of 0.100 M Na_2SO_4, 50.0 mL of 0.300 M $ZnCl_2$, and 100.0 mL of 0.200 M $Ba(CN)_2$.
RAN

(a) What ionic compounds will precipitate out of solution?

(b) What is the molarity of each ion remaining in the solution assuming complete precipitation of all insoluble compounds?

4.93 A 250.0 g sample of a white solid is known to be a mixture of
RAN KNO_3, $BaCl_2$, and NaCl. When 100.0 g of this mixture is dissolved in water and allowed to react with excess H_2SO_4, 67.3 g of a white precipitate is collected. When the remaining 150.0 g of the mixture is dissolved in water and allowed to react with excess $AgNO_3$, 197.6 g of a second precipitate is collected.

(a) What are the formulas of the two precipitates?

(b) What is the mass of each substance in the original 250 g mixture?

Acids, Bases, and Neutralization Reactions (Section 4.7)

4.94 Assume that you are given a solution of an unknown acid or base. How can you tell whether the unknown substance is acidic or basic?

4.95 Why do we use a double arrow ($\rightleftharpoons$) to show the dissociation of a weak acid or weak base in aqueous solution?

4.96 Write balanced ionic equations for the following reactions.

(a) Aqueous perchloric acid is neutralized by aqueous calcium hydroxide.

(b) Aqueous sodium hydroxide is neutralized by aqueous acetic acid.

4.97 Write balanced ionic equations for the following reactions.
RAN
(a) Aqueous hydrobromic acid is neutralized by aqueous calcium hydroxide.

(b) Aqueous barium hydroxide is neutralized by aqueous nitric acid.

4.98 Write balanced net ionic equations for the following reactions.

(a) $LiOH(aq) + HI(aq) \longrightarrow ?$

(b) $HBr(aq) + Ca(OH)_2(aq) \rightarrow ?$

4.99 Write balanced net ionic equations for the following reactions. Note that $HClO_3$ is a strong acid.

(a) $Fe(OH)_3(s) + H_2SO_4(aq) \rightarrow ?$

(b) $HClO_3(aq) + NaOH(aq) \longrightarrow ?$

Solution Stoichiometry and Titration (Sections 4.8–4.9)

4.100 A flask containing 450 mL of 0.500 M HBr was accidentally
RAN knocked to the floor. How many grams of K_2CO_3 would you need to put on the spill to neutralize the acid according to the following equation?

$$2\ HBr(aq) + K_2CO_3(aq) \longrightarrow 2\ KBr(aq) + CO_2(g) + H_2O(l)$$

4.101 The odor of skunks is caused by chemical compounds called
RAN *thiols*. These compounds, of which butanethiol ($C_4H_{10}S$) is a representative example, can be deodorized by reaction with household bleach (NaOCl) according to the following equation:

$$2\ C_4H_{10}S + NaOCl(aq) \longrightarrow C_8H_{18}S_2 + NaCl + H_2O(aq)$$

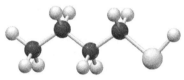

Butanethiol

How many grams of butanethiol can be deodorized by reaction with 5.00 mL of 0.0985 M NaOCl?

4.102 Potassium permanganate ($KMnO_4$) reacts with oxalic acid
RAN ($H_2C_2O_4$) in aqueous sulfuric acid according to the following equation:

$$2\ KMnO_4 + 5\ H_2C_2O_4 + 3\ H_2SO_4 \longrightarrow$$
$$2\ MnSO_4 + 10\ CO_2 + 8\ H_2O + K_2SO_4$$

How many milliliters of a 0.250 M $KMnO_4$ solution are needed to react completely with 3.225 g of oxalic acid?

4.103 Oxalic acid, $H_2C_2O_4$, is a toxic substance found in spinach leaves. What is the molarity of a solution made by dissolving 12.0 g of oxalic acid in enough water to give 400.0 mL of solution? How many milliliters of 0.100 M KOH would you need to titrate 25.0 mL of the oxalic acid solution according to the following equation?

$$H_2C_2O_4(aq) + 2\ KOH(aq) \longrightarrow K_2C_2O_4(aq) + 2\ H_2O(l)$$

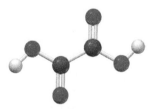

Oxalic acid

4.104 How many milliliters of 1.00 M KOH must be added to
RAN neutralize the following solutions?

(a) A mixture of 0.240 M LiOH (25.0 mL) and 0.200 M HBr (75.0 mL)

(b) A mixture of 0.300 M HCl (45.0 mL) and 0.250 M NaOH (10.0 mL)

4.105 How many milliliters of 2.00 M HCl must be added to neu-
RAN tralize the following solutions?

(a) A mixture of 0.160 M HNO_3 (100.0 mL) and 0.100 M KOH (400.0 mL)

(b) A mixture of 0.120 M NaOH (350.0 mL) and 0.190 M HBr (150.0 mL)

4.106 If the following solutions are mixed, is the resulting solution
RAN acidic, basic, or neutral?

(a) 50.0 mL of 0.100 M HBr and 30.0 mL of 0.200 M KOH

(b) 100.0 mL of 0.0750 M HCl and 75.0 mL of 0.100 M $Ba(OH)_2$

4.107 If the following solutions are mixed, is the resulting solution
RAN acidic, basic, or neutral?

(a) 65.0 mL of 0.0500 M $HClO_4$ and 40.0 mL of 0.0750 M NaOH

(b) 125.0 mL of 0.100 M HNO_3 and 90.0 mL of 0.0750 M $Ca(OH)_2$

Oxidation Numbers (Section 4.10)

4.108 Assign oxidation numbers to each element in the following compounds.

(a) NO_2 (b) SO_3 (c) $COCl_2$

(d) CH_2Cl_2 (e) $KClO_3$ (f) HNO_3

4.109 Assign oxidation numbers to each element in the following compounds.

(a) $VOCl_3$ (b) $CuSO_4$ (c) CH_2O

(d) Mn_2O_7 (e) OsO_4 (f) H_2PtCl_6

4.110 Assign oxidation numbers to each element in the following ions.

(a) ClO_3^- (b) SO_3^{2-} (c) $C_2O_4^{2-}$

(d) NO_2^- (e) BrO^- (f) AsO_4^{3-}

4.111 Assign oxidation numbers to each element in the following ions.

(a) $Cr(OH)_4^-$ (b) $S_2O_3^{2-}$ (c) NO_3^-

(d) MnO_4^{2-} (e) HPO_4^{2-} (f) $V_2O_7^{4-}$

4.112 Nitrogen can have several different oxidation numbers rang-
RAN ing in value from -3 to $+5$.

(a) Write the formula and give the name of the nitrogen oxide compound in which nitrogen has an oxidation number of $+1$, $+2$, $+4$, and $+5$.

(b) Based on oxidation numbers, which nitrogen oxide from part (a) cannot react with molecular oxygen?

4.113 Phosphorus can have several different oxidation numbers ranging in value from -3 to $+5$.

(a) When phosphorus burns in air or oxygen, it yields either tetraphosphorus hexoxide or tetraphosphorus decoxide. Write the formula and give the oxidation number for each compound.

(b) Based on oxidation numbers, which phosphorus oxide compound from part (a) was formed by combustion with a limited supply of oxygen?

Redox Reactions (Section 4.11)

4.114 Where in the periodic table are the best reducing agents found? The best oxidizing agents?

4.115 Where in the periodic table are the most easily reduced elements found? The most easily oxidized?

4.116 In each of the following instances, tell whether the substance gains electrons or loses electrons in a redox reaction.

(a) An oxidizing agent

(b) A reducing agent

(c) A substance undergoing oxidation

(d) A substance undergoing reduction

4.117 Tell for each of the following substances whether the oxidation number increases or decreases in a redox reaction.

(a) An oxidizing agent

(b) A reducing agent

(c) A substance undergoing oxidation

(d) A substance undergoing reduction

4.118 Which element is oxidized and which is reduced in each of
RAN the following reactions?

(a) $Ca(s) + Sn^{2+}(aq) \rightarrow Ca^{2+}(aq) + Sn(s)$

(b) $ICl(s) + H_2O(l) \rightarrow HCl(aq) + HOI(aq)$

4.119 Which element is oxidized and which is reduced in each of the following reactions?

(a) $Si(s) + 2\,Cl_2(g) \rightarrow SiCl_4(l)$

(b) $Cl_2(g) + 2\,NaBr(aq) \rightarrow Br_2(aq) + 2\,NaCl(aq)$

Activity Series (Section 4.12)

4.120 Use the activity series of metals (Table 4.5) to predict the
RAN outcome of each of the following reactions. If no reaction occurs, write *NR*.

(a) $Na^+(aq) + Zn(s) \rightarrow ?$ (b) $HCl(aq) + Pt(s) \rightarrow ?$

(c) $Ag^+(aq) + Au(s) \rightarrow ?$ (d) $Au^{3+}(aq) + Ag(s) \rightarrow ?$

4.121 Neither strontium (Sr) nor antimony (Sb) is shown in the activity series of Table 4.5. Based on their positions in the periodic table, which would you expect to be the better reducing agent? Will the following reaction occur? Explain.

$$2\,Sb^{3+}(aq) + 3\,Sr(s) \longrightarrow 2\,Sb(s) + 3\,Sr^{2+}(aq)$$

4.122 (a) Use the following reactions to arrange the elements **A**, **B**, **C**, and **D** in order of their decreasing ability as reducing agents:

$$A + B^+ \longrightarrow A^+ + B \quad C^+ + D \longrightarrow \text{no reaction}$$
$$B + D^+ \longrightarrow B^+ + D \quad B + C^+ \longrightarrow B^+ + C$$

(b) Which of the following reactions would you expect to occur according to the activity series you established in part (a)?

(1) $A^+ + C \longrightarrow A + C^+$

(2) $A^+ + D \longrightarrow A + D^+$

4.123 (a) Use the following reactions to arrange the elements **A**, **B**, **C**, and **D** in order of their decreasing ability as reducing agents:

$$2\,A + B^{2+} \longrightarrow 2\,A^+ + B \quad B + D^{2+} \longrightarrow B^{2+} + D$$
$$A^+ + C \longrightarrow \text{no reaction} \quad 2C + B^{2+} \longrightarrow 2\,C^+ + B$$

(b) Which of the following reactions would you expect to occur according to the activity series you established in part (a)?

(1) $2\,A^+ + D \longrightarrow 2\,A + D^{2+}$

(2) $D^{2+} + 2\,C \longrightarrow D + 2\,C^+$

Redox Titrations (Section 4.13)

4.124 Iodine, I_2, reacts with aqueous thiosulfate ion in neutral solu-
RAN tion according to the balanced equation

$$I_2(aq) + 2\,S_2O_3^{2-}(aq) \longrightarrow S_4O_6^{2-}(aq) + 2\,I^-(aq)$$

How many grams of I_2 are present in a solution if 35.20 mL of 0.150 M $Na_2S_2O_3$ solution is needed to titrate the I_2 solution?

4.125 How many milliliters of 0.250 M $Na_2S_2O_3$ solution is
RAN needed for complete reaction with 2.486 g of I_2 according to the equation in Problem 4.124?

4.126 Dichromate ion, $Cr_2O_7^{2-}$, reacts with aqueous iron(II) ion in
RAN acidic solution according to the balanced equation

$$Cr_2O_7^{2-}(aq) + 6\ Fe^{2+}(aq) + 14\ H^+(aq) \longrightarrow$$
$$2\ Cr^{3+}(aq) + 6\ Fe^{3+}(aq) + 7\ H_2O(l)$$

What is the concentration of Fe^{2+} if 46.99 mL of 0.2004 M $K_2Cr_2O_7$ is needed to titrate 50.00 mL of the Fe^{2+} solution?

4.127 A volume of 18.72 mL of 0.1500 M $K_2Cr_2O_7$ solution was
RAN required to titrate a sample of $FeSO_4$ according to the equation in Problem 4.126. What is the mass of the sample?

4.128 What is the molar concentration of As(III) in a solution if
RAN 22.35 mL of 0.100 M $KBrO_3$ is needed for complete reaction with 50.00 mL of the As(III) solution? The balanced equation is:

$$3\ H_3AsO_3(aq) + BrO_3^-(aq) \longrightarrow Br^-(aq) + 3\ H_3AsO_4(aq)$$

4.129 Standardized solutions of $KBrO_3$ are frequently used in
RAN redox titrations. The necessary solution can be made by dissolving $KBrO_3$ in water and then titrating it with an As(III) solution. What is the molar concentration of a $KBrO_3$ solution if 28.55 mL of the solution is needed to titrate 1.550 g of As_2O_3? See Problem 4.128 for the balanced equation. (As_2O_3 dissolves in aqueous acid solution to yield H_3AsO_3: $As_2O_3 + 3\ H_2O \rightarrow 2\ H_3AsO_3$.)

4.130 The metal content of iron in ores can be determined by a
RAN redox procedure in which the sample is first oxidized with Br_2 to convert all the iron to Fe^{3+} and then titrated with Sn^{2+} to reduce the Fe^{3+} to Fe^{2+}. The balanced equation is:

$$2\ Fe^{3+}(aq) + Sn^{2+}(aq) \longrightarrow 2\ Fe^{2+}(aq) + Sn^{4+}(aq)$$

What is the mass percent Fe in a 0.1875 g sample of ore if 13.28 mL of a 0.1015 M Sn^{2+} solution is needed to titrate the Fe^{3+}?

4.131 The concentration of the Sn^{2+} solution used in Problem 4.130
RAN can be found by letting it react with a known amount of Fe^{2+}. What is the molar concentration of an Sn^{2+} solution if 23.84 mL is required for complete reaction with 1.4855 g of Fe_2O_3?

4.132 Alcohol levels in blood can be determined by a redox reaction
RAN with potassium dichromate according to the balanced equation

$$C_2H_5OH(aq) + 2\ Cr_2O_7^{2-}(aq) + 16\ H^+(aq) \longrightarrow$$
$$2\ CO_2(g) + 4\ Cr^{3+}(aq) + 11\ H_2O(l)$$

What is the blood alcohol level in mass percent if 8.76 mL of 0.049 88 M $K_2Cr_2O_7$ is required for complete reaction with a 10.002 g sample of blood?

4.133 Calcium levels in blood can be determined by adding oxa-
RAN late ion to precipitate calcium oxalate, CaC_2O_4, followed by dissolving the precipitate in aqueous acid and titrating the resulting oxalic acid ($H_2C_2O_4$) with $KMnO_4$:

$$5\ H_2C_2O_4(aq) + 2\ MnO_4^-(aq) + 6\ H^+(aq) \longrightarrow$$
$$10\ CO_2(g) + 2\ Mn^{2+}(aq) + 8\ H_2O(l)$$

How many milligrams of Ca^{2+} are present in 10.0 mL of blood if 21.08 mL of 0.000 988 M $KMnO_4$ solution is needed for the titration?

MULTICONCEPT PROBLEMS

4.134 Assume that you have 1.00 g of a mixture of benzoic acid
RAN (Mol. wt. = 122) and gallic acid (Mol. wt. = 170)), both of which contain one acidic hydrogen that reacts with NaOH. On titrating the mixture with 0.500 M NaOH, 14.7 mL of base is needed to completely react with both acids. What mass in grams of each acid is present in the original mixture?

4.135 A compound with the formula $XOCl_2$ reacts with water,
yielding HCl and another acid H_2XO_3, which has two acidic hydrogens that react with NaOH. When 0.350 g of $XOCl_2$ was added to 50.0 mL of water and the resultant solution was titrated, 96.1 mL of 0.1225 M NaOH was required to react with all the acid.

(a) Write a balanced equation for the reaction of $XOCl_2$ with H_2O.

(b) What are the atomic mass and identity of element X?

4.136 A procedure for determining the amount of iron in a sample
RAN is to convert the iron to Fe^{2+} and then titrate it with a solution of $Ce(NH_4)_2(NO_3)_6$:

$$Fe^{2+}(aq) + Ce^{4+}(aq) \longrightarrow Fe^{3+}(aq) + Ce^{3+}(aq)$$

What is the mass percent of iron in a sample if 1.2284 g of the sample requires 54.91 mL of 0.1018 M $Ce(NH_4)_2(NO_3)_6$ for complete reaction?

4.137 Some metals occur naturally in their elemental state while
others occur as compounds in ores. Gold, for instance, is found as the free metal; mercury is obtained by heating mercury(II) sulfide ore in oxygen; and zinc is obtained by heating zinc(II) oxide ore with coke (carbon). Judging from their positions in the activity series, which of the metals silver, platinum, and chromium would probably be obtained by

(a) finding it in its elemental state?

(b) heating its sulfide with oxygen?

(c) heating its oxide with coke?

4.138 A sample weighing 14.98 g and containing a small amount
RAN of copper was treated to give a solution containing aqueous Cu^{2+} ions. Sodium iodide was then added to yield solid copper(I) iodide plus I_3^- ion, and the I_3^- was titrated with thiosulfate, $S_2O_3^{2-}$. The titration required 10.49 mL of 0.100 M $Na_2S_2O_3$ for complete reaction. What is the mass percent copper in the sample? The balanced equations are

$$2\ Cu^{2+}(aq) + 5\ I^-(aq) \longrightarrow 2\ CuI(s) + I_3^-(aq)$$
$$I_3^-(aq) + 2\ S_2O_3^{2-}(aq) \longrightarrow 3\ I^-(aq) + S_4O_6^{2-}(aq)$$

4.139 The solubility of an ionic compound can be described quan-
RAN titatively by a value called the *solubility product constant,* K_{sp}. For the general solubility process

$$A_aB_b \rightleftharpoons a\ A^{n+} + b\ B^{m-}, K_{sp} = [A^{n+}]^a\ [B^{m-}]^b.$$

The brackets refer to concentrations in moles per liter.

(a) Write the expression for the solubility product constant of Ag_2CrO_4.

(b) If $K_{sp} = 1.1 \times 10^{-12}$ for Ag_2CrO_4, what are the molar concentrations of Ag^+ and CrO_4^{2-} in solution?

4.140 Write the expression for the solubility product constant of
RAN MgF_2 (see Problem 4.139). If $[Mg^{2+}] = 2.6 \times 10^{-4}$ mol/L
in a solution, what is the value of K_{sp}?

4.141 A 100.0 mL solution containing aqueous HCl and HBr was
RAN   titrated with 0.1235 M NaOH. The volume of base required
to neutralize the acid was 47.14 mL. Aqueous $AgNO_3$ was
then added to precipitate the Cl^- and Br^- ions as AgCl and
AgBr. The mass of the silver halides obtained was 0.9974 g.
What are the molarities of the HCl and HBr in the original
solution?

4.142 A mixture of CuO and Cu_2O with a mass of 10.50 g is reduced
RAN   to give 8.66 g of pure Cu metal. What are the amounts in
grams of CuO and Cu_2O in the original mixture?

4.143 A sample of metal (**M**) reacted with both steam and aqueous
RAN   HCl to release H_2 but did not react with water at room temperature. When 1.000 g of the metal was burned in oxygen,
it formed 1.890 g of a metal oxide, M_2O_3. What is the identity of the metal?

4.144 An unknown metal (**M**) was found not to react with either
RAN   water or steam, but its reactivity with aqueous acid was
not investigated. When a 1.000 g sample of the metal was
burned in oxygen and the resulting metal oxide converted to
a metal sulfide, 1.504 g of sulfide was obtained. What is the
identity of the metal?

4.145 A mixture of acetic acid (CH_3CO_2H; monoprotic) and oxalic
RAN   acid ($H_2C_2O_4$; diprotic) requires 27.15 mL of 0.100 M
NaOH to neutralize it. When an identical amount of the
mixture is titrated, 15.05 mL of 0.0247 M $KMnO_4$ is needed
for complete reaction. What is the mass percent of each acid
in the mixture? (Acetic acid does not react with MnO_4^-. The
equation for the reaction of oxalic acid with MnO_4^- was
given in Problem 4.133.)

4.146 Iron content in ores can be determined by a redox procedure
RAN   in which the sample is first reduced with Sn^{2+}, as in Problem
4.130, and then titrated with $KMnO_4$ to oxidize the Fe^{2+} to
Fe^{3+}. The balanced equation is

$$MnO_4^-(aq) + 5\ Fe^{2+}(aq) + 8\ H^+(aq) \longrightarrow$$
$$Mn^{2+}(aq) + 5\ Fe^{3+}(aq) + 4\ H_2O(l)$$

What is the mass percent Fe in a 2.368 g sample if 48.39 mL
of a 0.1116 M $KMnO_4$ solution is needed to titrate the Fe^{3+}?

4.147 A mixture of $FeCl_2$ and NaCl is dissolved in water, and addi-
RAN   tion of aqueous silver nitrate then yields 7.0149 g of a precipitate. When an identical amount of the mixture is titrated
with MnO_4^-, 14.28 mL of 0.198 M $KMnO_4$ is needed for
complete reaction. What are the mass percents of the two
compounds in the mixture? (Na^+ and Cl^- do not react with
MnO_4^-. The equation for the reaction of Fe^{2+} with MnO_4^-
was given in Problem 4.146.)

4.148 Salicylic acid, used in the manufacture of aspirin, contains
only the elements C, H, and O and has only one acidic
hydrogen that reacts with NaOH. When 1.00 g of salicylic
acid undergoes complete combustion, 2.23 g CO_2 and 0.39 g
H_2O are obtained. When 1.00 g of salicylic acid is titrated
with 0.100 M NaOH, 72.4 mL of base is needed for complete reaction. What are the empirical and molecular formulas of salicylic acid?

4.149 Compound X contains only the elements C, H, O, and S. A
5.00 g sample undergoes complete combustion to give 4.83 g
of CO_2, 1.48 g of H_2O, and a certain amount of SO_2 that

is further oxidized to SO_3 and dissolved in water to form
sulfuric acid, H_2SO_4. On titration of the H_2SO_4, 109.8 mL
of 1.00 M NaOH is needed for complete reaction. (Both H
atoms in sulfuric acid are acidic and react with NaOH.)

(a) What is the empirical formula of X?

(b) When 5.00 g of X is titrated with NaOH, it is found that X
has two acidic hydrogens that react with NaOH and that
54.9 mL of 1.00 M NaOH is required to completely neutralize the sample. What is the molecular formula of X?

4.150 A 1.268 g sample of a metal carbonate (MCO_3) was treated
RAN   with 100.00 mL of 0.1083 M sulfuric acid (H_2SO_4), yielding
CO_2 gas and an aqueous solution of the metal sulfate (MSO_4).
The solution was boiled to remove all the dissolved CO_2 and
was then titrated with 0.1241 M NaOH. A 71.02 mL volume
of NaOH was required to neutralize the excess H_2SO_4.

(a) What is the identity of the metal M?

(b) How many liters of CO_2 gas were produced if the density of CO_2 is 1.799 g/L?

4.151 Element M is prepared industrially by a two-step procedure
RAN   according to the following (unbalanced) equations:

(1) $M_2O_3(s) + C(s) + Cl_2(g) \longrightarrow MCl_3(l) + CO(g)$
(2) $MCl_3(l) + H_2(g) \longrightarrow M(s) + HCl(g)$

Assume that 0.855 g of M_2O_3 is submitted to the reaction
sequence. When the HCl produced in step (2) is dissolved in
water and titrated with 0.511 M NaOH, 144.2 mL of the
NaOH solution is required to neutralize the HCl.

(a) Balance both equations.

(b) What is the atomic mass of element M, and what is its
identity?

(c) What mass of M in grams is produced in the reaction?

4.152 Assume that you dissolve 10.0 g of a mixture of NaOH
RAN   and $Ba(OH)_2$ in 250.0 mL of water and titrate with 1.50 M
hydrochloric acid. The titration is complete after 108.9 mL
of the acid has been added. What is the mass in grams of
each substance in the mixture?

4.153 Four solutions are prepared and mixed in the following order:

(a) Start with 100.0 mL of 0.100 M $BaCl_2$

(b) Add 50.0 mL of 0.100 M $AgNO_3$

(c) Add 50.0 mL of 0.100 M H_2SO_4

(d) Add 250.0 mL of 0.100 M NH_3

Write an equation for any reaction that occurs after each step,
and calculate the concentrations of Ba^{2+}, Cl^-, NO_3^-, NH_3,
and NH_4^+ in the final solution, assuming that all reactions
go to completion.

4.154 To 100.0 mL of a solution that contains 0.120 M $Cr(NO_3)_2$
RAN   and 0.500 M HNO_3 is added to 20.0 mL of 0.250 M
$K_2Cr_2O_7$. The dichromate and chromium(II) ions react to
give chromium(III) ions.

(a) Write a balanced net ionic equation for the reaction.

(b) Calculate the concentrations of all ions in the solution
after reaction. Check your concentrations to make sure
that the solution is electrically neutral.

4.155 Sodium nitrite, $NaNO_2$, is frequently added to processed
RAN   meats as a preservative. The amount of nitrite ion in a sample
can be determined by acidifying to form nitrous acid (HNO_2),
letting the nitrous acid react with an excess of iodide ion, and
then titrating the I_3^- ion that results with thiosulfate solution

in the presence of a starch indicator. The unbalanced equations are

(1) $HNO_2 + I^- \longrightarrow NO + I_3^-$ (in acidic solution)

(2) $I_3^- + S_2O_3^{2-} \longrightarrow I^- + S_4O_6^{2-}$

(a) Balance the two redox equations.

(b) When a nitrite-containing sample with a mass of 2.935 g was analyzed, 18.77 mL of 0.1500 M $Na_2S_2O_3$ solution was needed for the reaction. What is the mass percent of NO_2^- ion in the sample?

4.156 Brass is an approximately 4:1 alloy of copper and zinc, along
RAN with small amounts of tin, lead, and iron. The mass percents of copper and zinc can be determined by a procedure that begins with dissolving the brass in hot nitric acid. The resulting solution of Cu^{2+} and Zn^{2+} ions is then treated with aqueous ammonia to lower its acidity, followed by addition of sodium thiocyanate (NaSCN) and sulfurous acid (H_2SO_3) to precipitate copper(I) thiocyanate (CuSCN). The solid CuSCN is collected, dissolved in aqueous acid, and treated with potassium iodate (KIO_3) to give iodine, which is then titrated with aqueous sodium thiosulfate ($Na_2S_2O_3$). The filtrate remaining after CuSCN has been removed is neutralized by addition of aqueous ammonia, and a solution of diammonium hydrogen phosphate (($NH_4)_2HPO_4$) is added to yield a precipitate of zinc ammonium phosphate ($ZnNH_4PO_4$). Heating the precipitate to 900 °C converts it to zinc pyrophosphate ($Zn_2P_2O_7$), which is weighed. The equations are

(1) $Cu(s) + NO_3^-(aq) \longrightarrow Cu^{2+}(aq) + NO(g)$ (in acid)

(2) $Cu^{2+}(aq) + SCN^-(aq) + HSO_3^-(aq) \longrightarrow$
$\qquad CuSCN(s) + HSO_4^-(aq)$ (in acid)

(3) $Cu^+(aq) + IO_3^-(aq) \longrightarrow Cu^{2+}(aq) + I_2(aq)$ (in acid)

(4) $I_2(aq) + S_2O_3^{2-}(aq) \longrightarrow I^-(aq) + S_4O_6^{2-}(aq)$ (in acid)

(5) $ZnNH_4PO_4(s) \longrightarrow Zn_2P_2O_7(s) + H_2O(g) + NH_3(g)$

(a) Balance all equations.

(b) When a brass sample with a mass of 0.544 g was subjected to the preceding analysis, 10.82 mL of 0.1220 M sodium thiosulfate was required for the reaction with iodine. What is the mass percent copper in the brass?

(c) The brass sample in part (b) yielded 0.246 g of $Zn_2P_2O_7$. What is the mass percent zinc in the brass?

4.157 A certain metal sulfide, MS_n (where n is a small integer), is
RAN widely used as a high-temperature lubricant. The substance is prepared by reaction of the metal pentachloride (MCl_5) with sodium sulfide (Na_2S). Heating the metal sulfide to 700 °C in air gives the metal trioxide (MO_3) and sulfur dioxide (SO_2), which reacts with Fe^{3+} ion under aqueous acidic conditions to give sulfate ion (SO_4^{2-}). Addition of aqueous $BaCl_2$ then forms a precipitate of $BaSO_4$. The unbalanced equations are:

(1) $MCl_5(s) + Na_2S(s) \longrightarrow MS_n(s) + S(l) + NaCl(s)$

(2) $MS_n(s) + O_2(g) \longrightarrow MO_3(s) + SO_2(g)$

(3) $SO_2(g) + Fe^{3+}(aq) \longrightarrow Fe^{2+}(aq) + SO_4^{2-}(aq)$ (in acid)

(4) $SO_4^{2-}(aq) + Ba^{2+}(aq) \longrightarrow BaSO_4(s)$

Assume that you begin with 4.61 g of MCl_5 and that reaction (1) proceeds in 91.3% yield. After oxidation of the MS_n product, oxidation of SO_2, and precipitation of sulfate ion, 7.19 g of $BaSO_4(s)$ is obtained.

(a) How many moles of sulfur are present in the MS_n sample?

(b) Assuming several possible values for n ($n = 1, 2, 3 \ldots$), what is the atomic weight of M in each case?

(c) What is the likely identity of the metal M, and what is the formula of the metal sulfide MS_n?

(d) Balance all equations.

4.158 On heating a 0.200 g sample of a certain semimetal M in air, the corresponding oxide M_2O_3 was obtained. When the oxide was dissolved in aqueous acid and titrated with $KMnO_4$, 10.7 mL of 0.100 M MnO_4^- was required for complete reaction. The unbalanced equation is

$H_3MO_3(aq) + MnO_4^-(aq) \longrightarrow H_3MO_4(aq)$
$\qquad\qquad\qquad\qquad + Mn^{2+}(aq)$ (in acid)

(a) Balance the equation.

(b) How many moles of oxide were formed, and how many moles of semimetal were in the initial 0.200 g sample?

(c) What is the identity of the semimetal M?

chapter 5

Periodicity and the Electronic Structure of Atoms

Contents

The color emitted by elements heated in a flame arises from electrons moving between different energy levels in an atom. Atoms emit characteristic colors because each has its own unique energy levels, referred to as electronic structure.

How does knowledge of atomic emission spectra help us build more efficient light bulbs?

The answer to this question can be found on page 195 in the INQUIRY ?

The periodic table, introduced in Section 2.2, is the most important organizing principle in chemistry. If you know the properties of any one element in a group, or column, of the periodic table, you can make a good guess at the properties of every other element in the same group and even of the elements in neighboring groups. Although the periodic table was originally constructed from empirical observations, its scientific underpinnings have long been established and are well understood.

To see why it's called the *periodic* table, look at the graph of atomic radius versus atomic number in **FIGURE 5.1,** which shows a periodic rise-and-fall pattern. Beginning on the left with atomic number 1 (hydrogen), the size of the atoms increases to a maximum at atomic number 3 (lithium), then decreases to a minimum, then increases again to a maximum at atomic number 11 (sodium), and then decreases in a repeating pattern. It turns out that all the maxima occur for atoms of group 1A elements—Li, Na, K, Rb, Cs, and Fr—and that the minima occur for atoms of the group 7A elements—F, Cl, Br, and I.

There's nothing unique about the periodicity of atomic radii shown in Figure 5.1. Any of several dozen other physical or chemical properties could be plotted in a similar way with similar results. We'll look at several examples of such periodicity in Chapters 5 and 6 and explain why it occurs.

▶ **FIGURE 5.1**

A graph of atomic radius in picometers (pm) versus atomic number. A clear rise-and-fall pattern of periodicity is evident. (Accurate data are not available for the group 8A elements.)

▶ **Figure It Out**

What is the periodic trend in atomic radius across a period? Down a group?

Answer: In general, atomic radius decreases across a period and increases down a group.

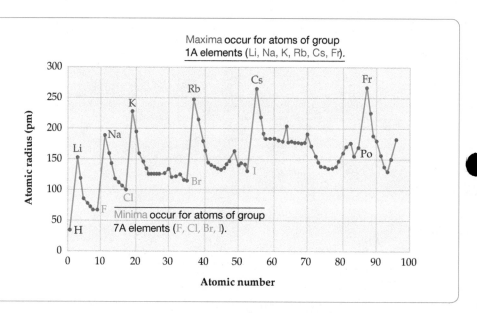

5.1 WAVE PROPERTIES OF RADIANT ENERGY AND THE ELECTROMAGNETIC SPECTRUM

What fundamental property of atoms is responsible for the periodic variations we observe in atomic radii and in so many other characteristics of the elements? This question occupied the thoughts of chemists for more than 50 years after Mendeleev, and it was not until well into the 1920s that the answer was established. To understand how the answer slowly emerged, it's necessary to look first at the nature of visible light and other forms of *radiant energy. Spectroscopy,* the study of the interaction of radiant energy with matter, has provided immense insight into atomic structure.

Although they appear quite different to our senses, visible light, infrared radiation, microwaves, radio waves, and X rays are all different forms of *electromagnetic radiation* or **radiant energy.** Collectively, they make up the **electromagnetic spectrum,** shown in **FIGURE 5.2.**

Electromagnetic energy traveling through a vacuum behaves in some ways like ocean waves traveling through water. Like ocean waves, electromagnetic energy is characterized by a *frequency,* a *wavelength,* and an *amplitude.* If you could stand in one place and

▲ Ocean waves, like electromagnetic waves, are characterized by a wavelength, a frequency, and an amplitude.

look at a sideways, cutaway view of an ocean wave moving through the water, you would see a regular rise-and-fall pattern like that in **FIGURE 5.3.**

The **frequency** (ν, Greek nu) of a wave is simply the number of wave peaks that pass by a given point per unit time, usually expressed in units of reciprocal seconds,

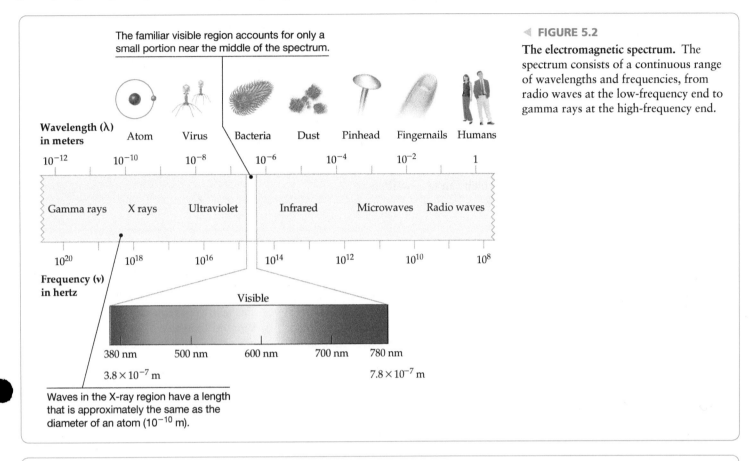

The familiar visible region accounts for only a small portion near the middle of the spectrum.

Wavelength (λ) in meters

Atom   Virus   Bacteria   Dust   Pinhead   Fingernails   Humans

10^{-12} 10^{-10} 10^{-8} 10^{-6} 10^{-4} 10^{-2} 1

Gamma rays   X rays   Ultraviolet   Infrared   Microwaves   Radio waves

10^{20} 10^{18} 10^{16} 10^{14} 10^{12} 10^{10} 10^{8}

Frequency (ν) in hertz

Visible

380 nm   500 nm   600 nm   700 nm   780 nm

3.8×10^{-7} m 7.8×10^{-7} m

Waves in the X-ray region have a length that is approximately the same as the diameter of an atom (10^{-10} m).

◀ **FIGURE 5.2**

The electromagnetic spectrum. The spectrum consists of a continuous range of wavelengths and frequencies, from radio waves at the low-frequency end to gamma rays at the high-frequency end.

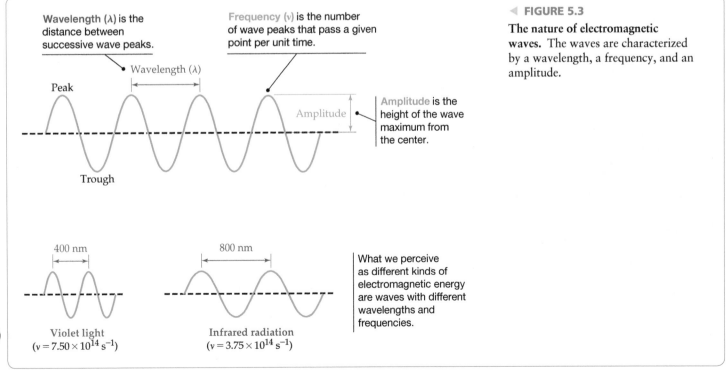

Wavelength (λ) is the distance between successive wave peaks.

Frequency (ν) is the number of wave peaks that pass a given point per unit time.

Peak

Wavelength (λ)

Amplitude

Trough

Amplitude is the height of the wave maximum from the center.

400 nm

800 nm

Violet light ($\nu = 7.50 \times 10^{14}$ s^{-1})

Infrared radiation ($\nu = 3.75 \times 10^{14}$ s^{-1})

What we perceive as different kinds of electromagnetic energy are waves with different wavelengths and frequencies.

◀ **FIGURE 5.3**

The nature of electromagnetic waves. The waves are characterized by a wavelength, a frequency, and an amplitude.

or **hertz** (**Hz;** 1 Hz = 1 s^{-1}). The **wavelength** ($\boldsymbol{\lambda}$, Greek lambda) of the wave is the distance from one wave peak to the next, and the **amplitude** of the wave is the height of the wave, measured from the center line between peak and trough. Physically, what we perceive as the intensity of electromagnetic energy is proportional to the square of the wave amplitude. A faint beam and a blinding glare of light may have the same wavelength and frequency, but they differ greatly in amplitude.

Multiplying the wavelength of a wave in meters (m) by its frequency in reciprocal seconds (s^{-1}) gives the speed of the wave in meters per second (m/s). The rate of travel of all electromagnetic energy in a vacuum is a constant value, commonly called the speed of light and abbreviated *c*. Its numerical value is defined as exactly 2.997 924 58 $\times$ 10^8 m/s, usually rounded off to 3.00 $\times$ 10^8 m/s:

$$\text{Wavelength} \times \text{Frequency} = \text{Speed}$$

$$\lambda(\text{m}) \times \nu(\text{s}^{-1}) = c(\text{m/s})$$

which can be rewritten as

$$\lambda = \frac{c}{\nu} \quad \text{or} \quad \nu = \frac{c}{\lambda}$$

This equation says that frequency and wavelength are inversely related: Electromagnetic energy with a longer wavelength has a lower frequency, and energy with a shorter wavelength has a higher frequency. Worked Example 5.1 demonstrates how to convert between the wavelength and frequency of electromagnetic radiation.

WORKED EXAMPLE 5.1

Calculating Frequency from Wavelength

The light blue glow given off by mercury streetlamps has a wavelength of 436 nm. What is its frequency in hertz?

IDENTIFY

Known	Unknown
Wavelength (λ)	Frequency (ν)

STRATEGY

Use the equation that relates wavelength and frequency. Don't forget to convert from nanometers to meters as units on the speed of light (*c*) are m/s.

SOLUTION

$$\nu = \frac{c}{\lambda} = \frac{\left(3.00 \times 10^8 \frac{\text{m}}{\text{s}}\right)}{(436 \text{ nm})\left(\frac{1 \text{ m}}{10^9 \text{ nm}}\right)}$$

$$= 6.88 \times 10^{14} \text{ s}^{-1} = 6.88 \times 10^{14} \text{ Hz}$$

The frequency of the light is 6.88 $\times$ 10^{14} s^{-1}, or 6.88 $\times$ 10^{14} Hz.

PRACTICE 5.1   What is the wavelength in meters of an FM radio wave with frequency ν = 102.5 MHz?

▲ Does the blue glow from this mercury lamp correspond to a longer or shorter wavelength than the yellow glow from a sodium lamp?

▶ **CONCEPTUAL APPLY 5.2**   Two electromagnetic waves are represented below.

(a)   **(b)**

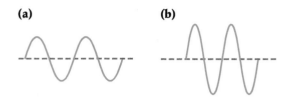

(a) Which wave has the higher frequency?
(b) Which wave represents a more intense (brighter) beam of light?
(c) Which wave represents blue light, and which represents red light?

All **PRACTICE** and **APPLY** problems are interactive in the eText.

Since we cannot observe visible light traveling as a wave with our eyes, you may be wondering how we know that radiant energy has wave properties. Scientists have performed experiments testing for the presence of physical phenomena exhibited by waves, such as diffraction and interference. Diffraction is the bending of a light wave around an object, as shown in **FIGURE 5.4a.** Interference occurs when two or more waves superpose to form a new wave, as shown in **FIGURE 5.4b.** *Constructive interference* occurs when two waves of the same frequency are in phase with the crests and troughs aligned. The sum of the amplitudes of both waves at each point gives a new wave with the same frequency but larger amplitude. *Destructive interference* occurs when two waves of the same frequency are out of phase. The crest from one wave is aligned with the trough from the other wave. The sum of the amplitude of each wave at each point gives a new wave with zero amplitude.

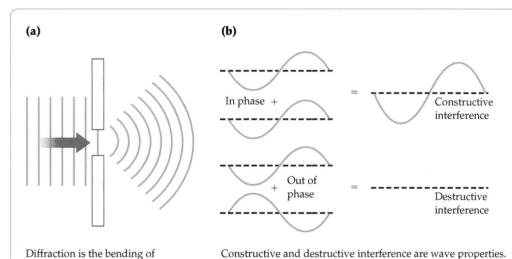

(a)

Diffraction is the bending of waves around an object.

(b)

In phase +

=

Constructive interference

Out of phase

+

=

Destructive interference

Constructive and destructive interference are wave properties.

◀ **FIGURE 5.4**

Diffraction and interference are phenomena exhibited by waves.
(a) Diffraction is the bending of a wave as it travels through a slit.
(b) Constructive interference occurs when two waves of the same frequency are in phase, giving a new wave of the same frequency but larger amplitude. Destructive interference occurs when the waves are out of phase, giving a new wave of zero amplitude.

In the early 1800s, Thomas Young (1773–1829) demonstrated that light consists of waves in an experiment known as the "double-slit" experiment (**FIGURE 5.5**). If light waves of a single wavelength are in phase and directed at a plate with two parallel slits, an interference pattern is observed on a detector on the other side. When light

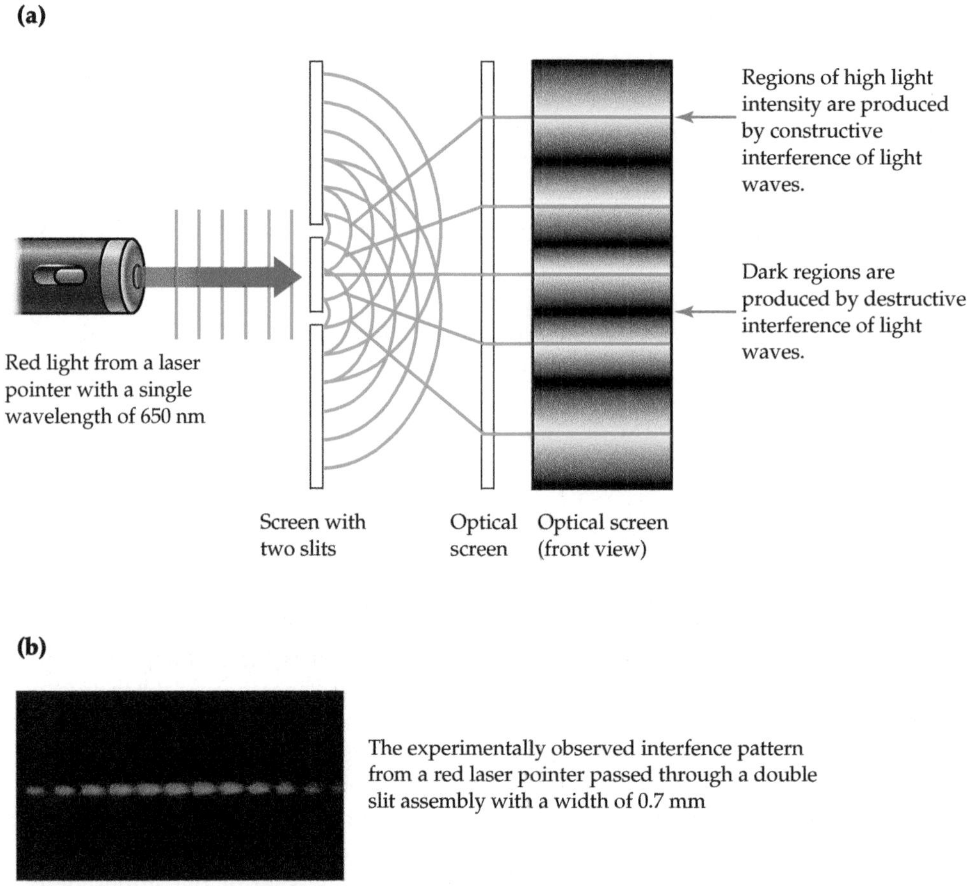

▶ **FIGURE 5.5**

Radiant energy exhibits wave properties in a double-slit experiment. (a) Light waves are directed at a barrier with two narrow slits. The waves are diffracted as they travel through the slits and then interfere constructively and destructively to produce an interference pattern of light and dark bands. (b) The interference pattern that results from red light traveling through two slits.

▶ **Figure It Out**

What wave properties are observed in the double-slit experiment?

Answer: Diffraction and interference

(a)

Red light from a laser pointer with a single wavelength of 650 nm

Screen with two slits

Optical screen

Optical screen (front view)

Regions of high light intensity are produced by constructive interference of light waves.

Dark regions are produced by destructive interference of light waves.

(b)

The experimentally observed interference pattern from a red laser pointer passed through a double slit assembly with a width of 0.7 mm

waves pass through the two slits, they are diffracted; the diffracted waves interfere constructively to produce bright bands and destructively to produce dark bands at the detector. The wave properties of light cause the observed interference pattern of bright and dark bands. We will see in Section 5.4 that particles such as high-speed electrons also exhibit wave properties in a similar double-slit experiment.

5.2 PARTICLELIKE PROPERTIES OF RADIANT ENERGY: THE PHOTOELECTRIC EFFECT AND PLANCK'S POSTULATE

One important step toward developing a model of atomic structure came in 1905, when Albert Einstein (1879–1955) proposed an explanation of the **photoelectric effect.** Scientists had known since the late 1800s that irradiating a clean metal surface with light causes electrons to be ejected from the metal (**FIGURE 5.6**). Furthermore, the frequency of the light used for the irradiation must be above some threshold value, which is different for every metal. Blue light ($\nu \approx 6.5 \times 10^{14}$ Hz) causes metallic sodium to emit electrons, for example, but red light ($\nu \approx 4.5 \times 10^{14}$ Hz) has no effect on sodium.

Einstein explained the photoelectric effect by assuming that a beam of light behaves as if it were a stream of small particles, called **photons,** whose energy (E) is related to their frequency, ν (or wavelength, λ) by an equation called **Planck's postulate** after the German physicist Max Planck (1858–1947).

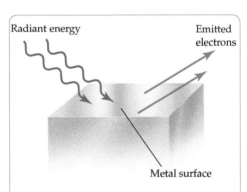

Radiant energy

Emitted electrons

Metal surface

▲ **FIGURE 5.6**

The photoelectric effect. Irradiation of a clean metal surface with radiant energy above a threshold frequency causes electrons to be emitted.

Planck's postulate $E = h\nu = \dfrac{hc}{\lambda}$

The proportionality constant h represents a fundamental physical constant that we now call Planck's constant and that has the value $h = 6.626 \times 10^{-34}$ J·s. For example, one photon of red light with a frequency $\nu = 4.62 \times 10^{14}$ s^{-1} (wavelength $\lambda = 649$ nm) has an energy of 3.06×10^{-19} J. [Recall from Section 1.8 that the SI unit for energy is the joule (J), where 1 J = 1 (kg·m^2)/s^2.]

$$E = h\nu = (6.626 \times 10^{-34} \text{ J·s})(4.62 \times 10^{14} \text{ s}^{-1}) = 3.06 \times 10^{-19} \text{ J}$$

You might also recall from Section 2.9 that 1 mole (mol) of anything is the amount that contains Avogadro's number (6.022×10^{23}) of entities. Thus, it's often convenient to express electromagnetic energy on a per-mole basis rather than a per-photon basis. Multiplying the per-photon energy of 3.06×10^{-19} J by Avogadro's number gives an energy of 184 kJ/mol.

$$\left(3.06 \times 10^{-19} \frac{\text{J}}{\text{photon}}\right)\left(6.022 \times 10^{23} \frac{\text{photon}}{\text{mol}}\right) = 1.84 \times 10^5 \frac{\text{J}}{\text{mol}}$$

$$\left(1.84 \times 10^5 \frac{\text{J}}{\text{mol}}\right)\left(\frac{1 \text{ kJ}}{1000 \text{ J}}\right) = 184 \text{ kJ/mol}$$

Higher frequencies and shorter wavelengths correspond to higher-energy radiation, while lower frequencies and longer wavelengths correspond to lower energy. Blue light ($\lambda \approx 450$ nm), for instance, has a shorter wavelength and is more energetic than red light ($\lambda \approx 650$ nm). Similarly, an X ray ($\lambda \approx 1$ nm) has a shorter wavelength and is more energetic than an FM radio wave ($\lambda \approx 10^{10}$ nm, or 10 m).

If the frequency (or energy) of the photon striking a metal is below a minimum value, no electron is ejected. Above the threshold level, however, sufficient energy is transferred from the photon to an electron to overcome the attractive forces holding the electron to the metal (**FIGURE 5.7**). The amount of energy necessary to eject an electron is called the work function (Φ) of the metal and is lowest for the group 1A and group 2A elements. That is, elements on the left side of the periodic table hold their electrons less tightly than other metals and lose them more readily (**TABLE 5.1**).

Note again that the energy of an individual photon depends only on its frequency (or wavelength), not on the intensity of the light beam. The intensity of a light beam is a measure of the *number* of photons in the beam, whereas frequency is a measure of the *energies* of those photons. A low-intensity beam of high-energy photons might easily knock a few electrons loose from a metal, but a high-intensity beam of low-energy photons might not be able to knock loose a single electron. As a rough analogy, think of throwing balls of different masses at a glass window. A thousand Ping-pong balls (lower energy) would only bounce off the window, but a single baseball (higher energy) would break the glass. In the same way, low-energy photons have no effect on the the metal surface, but a single photon at or above the threshold energy can dislodge an electron.

▲ Night-vision goggles are based on the photoelectric effect of metal alloys that emit electrons when irradiated with infrared light.

TABLE 5.1 Work Functions of Some Common Metals

Element	Work Function (Φ) (kJ/mol)
Cs	188
K	221
Na	228
Ca	277
Mg	353
Cu	437
Fe	451

Go to eText

BIG IDEA Question 1

When light with an energy of 277 kJ/mol (432 nm) shines on a piece of calcium, electrons are ejected in a phenomenon called the *photoelectric effect*. What is a different wavelength of light that would also cause the photoelectric effect in calcium?

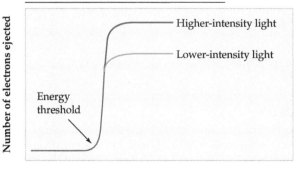

A plot of the number of electrons ejected from a metal surface versus light frequency shows a threshold value.

Number of electrons ejected

Higher-intensity light

Lower-intensity light

Energy threshold

Frequency →

Increasing the intensity of the light while keeping the frequency constant increases the number of ejected electrons but does not change the threshold value.

◄ **FIGURE 5.7**

Dependence of the photoelectric effect on frequency.

◄ Figure It Out

Would electrons be ejected from a metal surface if (a) very bright light with a frequency below the threshold is used or (b) very dim light with a frequency above the threshold is used? How does this observation support the theory that light has particlelike properties?

Answer: (a) No electrons ejected. (b) Low number of electrons ejected. Only light above a certain frequency has enough energy to eject an electron. This is analogous to a "particle" such as a baseball being able to break a window, whereas a Ping-pong ball cannot do so.

▲ A glass window can be broken by a single baseball, but a thousand Ping-pong balls would only bounce off.

The main conclusion from Einstein's work was that the behavior of light and other forms of electromagnetic energy is more complex than had been formerly believed. In addition to behaving as waves, *light energy can also behave as small particles*. The idea might seem strange at first but becomes less so if you think of light as analogous to matter. Both are said to be **quantized,** meaning that both matter and electromagnetic energy occur only in discrete amounts. Just as there can be either 1 or 2 hydrogen atoms but not 1.5 or 1.8, there can be 1 or 2 photons of light but not 1.5 or 1.8. A **quantum** is the smallest possible unit of a quantity, just like an atom is the smallest possible quantity of an element. The quantum of energy corresponding to one photon of light is almost inconceivably small, just as the amount of matter in one atom is inconceivably small, but the idea is the same. Planck's postulate and Einstein's photons with particlelike properties were a dramatic departure from the laws of classical physics at the time and ultimately led to a revolution in the way that scientists thought about the structure of the atom.

WORKED EXAMPLE 5.2

Calculating the Energy of a Photon

What is the energy in kilojoules per mole of radar waves with $\nu = 3.35 \times 10^8$ Hz?

IDENTIFY

Known	Unknown
Frequency (ν)	Energy (E)

STRATEGY

The energy of a photon with frequency ν can be calculated with the equation $E_{photon} = h\nu$. To find the energy per mole of photons, the energy of one photon must be multiplied by Avogadro's number (Section 2.9).

SOLUTION

$$E = h\nu = (6.626 \times 10^{-34} \text{ J} \cdot \text{s})(3.35 \times 10^8 \text{ s}^{-1}) = 2.22 \times 10^{-25} \text{ J}$$

$$\left(2.22 \times 10^{-25} \frac{\text{J}}{\text{photon}}\right)\left(6.022 \times 10^{23} \frac{\text{photon}}{\text{mol}}\right) = 0.134 \text{ J/mol}$$

$$= 1.34 \times 10^{-4} \text{ kJ/mol}$$

▶ **PRACTICE 5.3**   The biological effects of a given dose of electromagnetic energy generally become more serious as the energy of the radiation increases: Infrared radiation has a pleasant warming effect, ultraviolet radiation causes tanning and burning, and X rays can cause considerable tissue damage. What energies in kilojoules per mole are associated with the following wavelengths: infrared radiation with $\lambda = 1.55 \times 10^{-6}$ m, ultraviolet light with $\lambda = 250$ nm, and X rays with $\lambda = 5.49$ nm?

▶ **APPLY 5.4**   It requires 74 kJ to heat a cup of water from room temperature to boiling in the microwave oven. If the wavelength of microwave radiation is 2.3×10^{-3} m, how many moles of photons are required to heat the water?

WORKED EXAMPLE 5.3

Calculating the Frequency for a Photoelectric Effect Given a Work Function

The work function of lithium metal is $\Phi = 283$ kJ/mol. What is the minimum frequency of light needed to eject electrons from lithium?

IDENTIFY

Known	Unknown
Work Function (Φ) = Energy (E)	Frequency (ν)

STRATEGY

The work function of a metal is the minimum energy needed to eject electrons: $E = 283$ kJ/mol for lithium. Convert the work function from kJ/mol to J/photon to find the energy of a single photon, and rearrange Planck's equation $E = h\nu$ to solve for frequency, $\nu = E/h$.

SOLUTION

$$E = \left(\frac{283 \text{ kJ}}{\text{mol}}\right)\left(\frac{1000 \text{ J}}{1 \text{ kJ}}\right)\left(\frac{1 \text{ mol}}{6.022 \times 10^{23} \text{ photons}}\right) = 4.70 \times 10^{-19} \frac{\text{J}}{\text{photon}}$$

$$\nu = \frac{E}{h} = \frac{4.70 \times 10^{-19} \text{ J}}{(6.626 \times 10^{-34} \text{ J} \cdot \text{s})} = 7.90 \times 10^{14} \text{ s}^{-1} \text{ or } 7.09 \times 10^{14} \text{ Hz}$$

▶ **PRACTICE 5.5**   The work function of zinc metal is 350 kJ/mol. Will photons of violet light with $\lambda = 390$ nm cause electrons to be ejected from a sample of zinc?

▶ **CONCEPTUAL APPLY 5.6**   Compare the two elements Rb and Ag.

(a) Which element do you predict to have a higher work function?
(b) Which element exhibits the photoelectric effect at a longer wavelength of light?

5.3 ATOMIC LINE SPECTRA AND QUANTIZED ENERGY

Now that we understand that electromagnetic radiation has both wavelike and particlelike properties, we can examine how its interaction with matter provides clues about atomic structure. Let's compare the emission of light from an atom with the emission of light from a light bulb. The light that we see from the Sun or from a typical light bulb is "white" light, meaning that it consists of an essentially continuous distribution of wavelengths spanning the entire visible region of the electromagnetic spectrum. That white light actually consists of a spectrum of many colors of light is made evident when a narrow beam of white light is passed through a glass prism to produce a "rainbow" of colors (**FIGURE 5.8a**). This happens because the different wavelengths contained in white light travel through the glass at different speeds. The prism separates the white light into its component colors, ranging from red at the long-wavelength end of the spectrum (780 nm) to violet at the short-wavelength end (380 nm). This separation into colors also occurs when light travels through water droplets in the air, forming a rainbow, or through oriented ice crystals in clouds, causing a *parhelion*, or sundog (**FIGURE 5.8b**).

(a)   **(b)**

◀ **FIGURE 5.8**

Separation of white light into its constituent colors. (a) When a narrow beam of ordinary white light is passed through a glass prism, different wavelengths travel through the glass at different rates and appear as different colors. A similar effect occurs when light passes through water droplets in the air, forming a rainbow, or (b) through ice crystals in clouds, causing an unusual weather phenomenon called a *parhelion*, or sundog.

What do visible light and other kinds of electromagnetic energy have to do with atomic structure? It turns out that atoms give off light when heated or otherwise energetically excited, thereby providing a clue to their atomic makeup. Unlike the white light from the Sun, though, an energetically excited atom emits light not in a continuous distribution of wavelengths but only at certain specific wavelengths. When passed first through a narrow slit and then through a prism, the light emitted by an excited atom is found to consist of only a few wavelengths rather than a full rainbow of colors, giving a series of discrete lines on an otherwise dark background—a **line spectrum** that is unique for each element. If hydrogen atoms are electrically excited in a discharge tube, they give off a pinkish light made of several different colors (**FIGURE 5.9a**). Other elements also produce line spectra upon excitation such as neon in signs or sodium salts in a flame (**FIGURE 5.9b**). In fact, the brilliant colors of fireworks are produced by mixtures of metal atoms that have been heated by explosive powder.

▶ **FIGURE 5.9**

Atomic line spectra. (a) When light from a hydrogen discharge tube is passed through a narrow slit and then a prism, the colors of light in its emission spectrum are separated. (b) Each element, such as sodium and neon, has its own characteristic line spectrum that can be used to identify or quantify the amount of an element in a sample.

▶ **Figure It Out**

How is the emission spectrum for an atom different than the emission spectrum of the Sun?

Answer: The emission spectrum from the Sun consists of all visible wavelengths of electromagnetic radiation. The emission spectra from atoms consist of discrete visible wavelengths, a line spectrum.

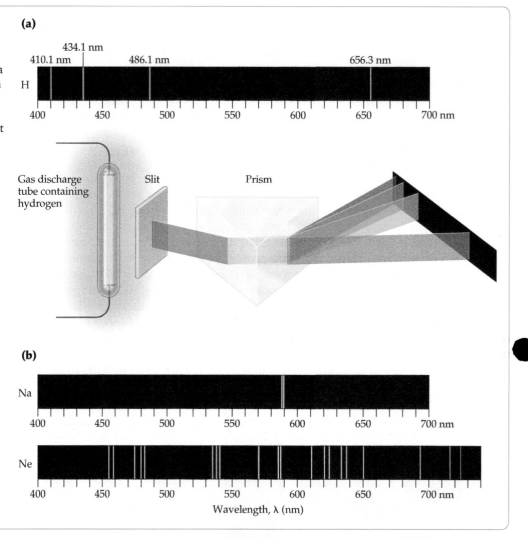

Soon after the discovery that energetic atoms emit light of specific wavelengths, chemists began cataloging the line spectra of various elements. They rapidly found that each element has its own unique spectral "signature," and they began using the results to identify the elements present in minerals and other substances.

▶ Electronically excited hydrogen atoms give off pink light and neon atoms emit orange light in discharge tubes. Thermally excited sodium atoms emit yellow light in a flame.

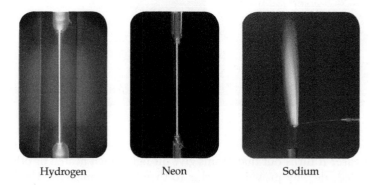

Hydrogen   Neon   Sodium

As often happens in science, experimental results are obtained before a theory to explain them is developed. The discovery of atomic line spectra was made decades before a theory of atomic structure to explain the spectra was developed. During the time period 1900–1911, scientists conducted key experiments and developed new theories that helped to solve the puzzle of line spectra. Among them were Planck's postulate of quantized energy (1900), Einstein's concept of photons and the photo-electric effect (1905), and **Rutherford's nuclear model** of the atom (1911). Building upon these seminal discoveries, Niels Bohr (1885–1962), a Danish physicist working in Rutherford's lab, proposed an atomic model that predicted the existence of line spectra. His model of the hydrogen atom described a small, positively charged nucleus with an electron circling around it, much as a planet orbits the Sun. Bohr postulated that the energy levels of the orbits are *quantized* so that only certain specific orbits corresponding to certain specific energies for the electron are available. You might think of the quantized nature of orbits in terms of an analogy: climbing stairs versus a ramp. The height of a ramp changes continuously, but stairs change height only in discrete amounts; the height reached by climbing each stair is thus quantized.

The success of the Bohr model was that it was able to explain the line spectrum observed for hydrogen. Each orbit has its own radius, referred to as *n*, which is directly related to energy. As the radius increases, the energy also increases. Thus, $n = 2$ has greater energy than $n = 1$ and $n = 3$ has greater energy than $n = 2$, and so on. In Bohr's model, there is no change in energy when an electron moves within its orbit, but when an electron falls into a lower orbit, it emits a photon whose energy equals the difference in the energies of the two orbits (**FIGURE 5.10**).

$$\Delta E = E_{final} - E_{initial} = h\nu$$

Quantization of energy in the atom arises from the fact that an electron cannot reside between orbits in the Bohr model, just like you cannot stand between steps on the stairs.

▲ A ramp changes height continuously, but stairs are quantized, changing height only in discrete amounts. In the same way, electromagnetic energy is not continuous but is emitted only in discrete amounts.

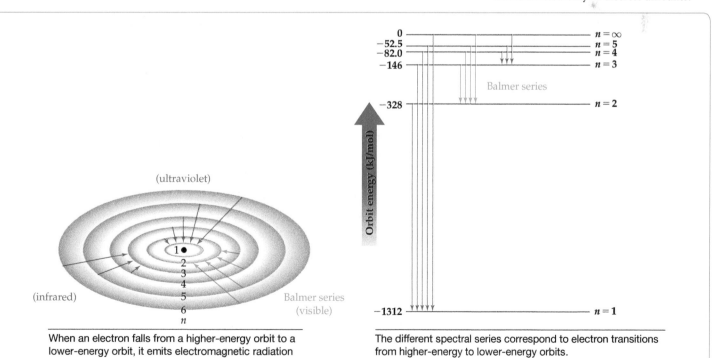

When an electron falls from a higher-energy orbit to a lower-energy orbit, it emits electromagnetic radiation whose frequency corresponds to the energy difference between the orbits.

The different spectral series correspond to electron transitions from higher-energy to lower-energy orbits.

▲ **FIGURE 5.10**

The Bohr model for hydrogen predicts atomic line spectra.

▲ **Figure It Out**

Which electron transition corresponds to the emission of light with the most energy: $n = 5$ to $n = 3$, $n = 5$ to $n = 2$, $n = 5$, to $n = 1$?

Answer: $n = 5$ to $n = 1$ as shown by the difference in energy between the orbits.

When energy is absorbed by an atom, an electron moves from a lower- to higher-energy orbit ($n_{initial} < n_{final}$). Conversely, when an electron falls from a higher- to lower-energy orbit, energy is released ($n_{inital} > n_{final}$). In the hydrogen atom, electrons moving from $n = 6, 5, 4, 3$ to $n = 2$ result in emission of wavelengths in the visible region of the electromagnetic spectrum (Balmer series). As seen in Figure 5.10, electrons moving from $n = 6, 5, 4, 3$ to $n = 1$ correspond to greater energy than the Balmer series and result in emission of photons in the ultraviolet region of the spectrum. Similarly, electronic transitions from $n = 6, 5, 4$ to $n = 3$ are lower in energy than the Balmer series and correlate to spectral lines in the infrared.

Prior to Bohr's model of the atom, the combined work of Johann Balmer (1825–1898) and Johannes Rydberg (1854–1919) led to a formula describing the relation between spectral lines in hydrogen. The wavelengths of the lines in the hydrogen spectrum can be expressed by the **Balmer–Rydberg equation.**

Balmer–Rydberg equation

$$\frac{1}{\lambda} = R_\infty \left[\frac{1}{m^2} - \frac{1}{n^2} \right] \qquad \text{where } R_\infty = 1.097 \times 10^{-2} \text{ nm}^{-1}$$

m — *n* level of lower-energy orbit

n — *n* level of higher-energy orbit

The Rydberg constant, R_∞, has a value of 1.097×10^{-2} nm^{-1}, and the variables m and n in the Balmer–Rydberg equation for hydrogen represent the energy levels of the orbits in the Bohr model. The variable n corresponds to the *n value* of the higher-energy orbit, and the variable m corresponds to the *n value* of the lower-energy orbit closer to the nucleus.

Notice in Figure 5.10 that as n becomes larger and approaches infinity, the energy difference between $n = \infty$ and $n = 1$ converges to a value of 1312 kJ/mol. That is, 1312 kJ is released when electrons come from a great distance (the "infinite" level) and add to H$^+$ to give a mole of hydrogen atoms, each with an electron in its lowest energy state:

$$\text{H}^+ + \text{e}^- \longrightarrow \text{H} + \text{Energy} \qquad (1312 \text{ kJ/mol})$$

Because the energy released upon adding an electron to H$^+$ is equal to the energy absorbed on removing an electron from a hydrogen atom, we can also say that 1312 kJ/mol is required to remove the electron from a hydrogen atom. We'll see in the next chapter that the amount of energy necessary to remove an electron from a given atom provides an important clue about that element's chemical reactivity.

Although the Bohr model was very successful in accounting for the line spectrum of hydrogen, it suffered from several limitations.

- It failed to predict the spectrum of any atom other than hydrogen and only works for *one-electron species* such as H, He$^+$, or Li^{2+}. If more than one electron exists, interactions such as electron–electron repulsions must be accounted for in a more complex model.

- It does not give an accurate depiction of electron location. *Electrons do not move in fixed, defined orbits.* In fact, we can never know the precise location of an electron in the atom and can only define probabilities of an electron existing within a given volume of space.

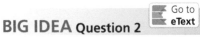

BIG IDEA Question 2   Go to eText

According to modern atomic theory, which part of Bohr's atomic model is no longer accepted as correct?

However, the fundamental idea of quantized energy levels for the electron was an important theory for which Bohr was awarded the Nobel Prize in Physics in 1922. The next several sections will expand on Bohr's ideas and further describe the modern model for electrons in an atom.

Go to eText

WORKED EXAMPLE 5.4

Relating the Bohr Model and the Balmer–Rydberg Equation

(a) Use the Balmer–Rydberg equation to calculate the wavelength of the photon emitted when an electron falls from the $n = 4$ level to the $n = 1$ level in the hydrogen atom.

(b) Calculate the energy of the photon in units of kJ/mol. Check that the energy of the photon corresponds to the difference in energy between the $n = 4$ and $n = 1$ level in Figure 5.10.

(a) STRATEGY

Use the Balmer–Rydberg equation and identify the values of m and n. n represents the highest energy level, $n = 4$, and m represents the lowest energy level, $n = 1$. Therefore, in the Rydberg equation $n = 4$ and $m = 1$.

SOLUTION

Solving the equation for wavelength gives

$$\frac{1}{\lambda} = R_{\infty}\left[\frac{1}{m^2} - \frac{1}{n^2}\right] = (1.097 \times 10^{-2}\ nm^{-1})\left[\frac{1}{1^2} - \frac{1}{4^2}\right]$$

$$= 1.028 \times 10^{-2}\ nm^{-1}$$

$$\text{or}\quad \lambda = \frac{1}{1.028 \times 10^{-2}\ nm^{-1}} = 97.3\ nm$$

(b) STRATEGY

Use Planck's postulate to convert between wavelength and energy. To use this equation, first convert the units on wavelength to meters because the speed of light is expressed in units of (m/s). Planck's equation gives the energy in units of J/photon, which must be converted to kJ/mol.

SOLUTION

$$E = \frac{hc}{\lambda} = \frac{(6.626 \times 10^{-34}\ J \cdot s)\left(3.00 \times 10^8\ \frac{m}{s}\right)}{(97.2\ nm)\left(\frac{1\ m}{10^9\ nm}\right)} = 2.05 \times 10^{-18}\ \frac{J}{photon}$$

$$\left(2.05 \times 10^{-18}\ \frac{J}{photon}\right)\left(6.022 \times 10^{23}\ \frac{photon}{mol}\right) = 1.23 \times 10^6\ J/mol$$

$$= 1.23 \times 10^3\ kJ/mol$$

CHECK

The wavelength of the photon released in the $n = 4$ to $n = 1$ transition is in the ultraviolet region of the spectrum which is consistent with theory. Figure 5.10 gives the energy in kJ/mol for each n level and thus $\Delta E = E_{final} - E_{initial} = [-1312\ kJ/mol - (-82\ kJ/mol)] = -1.23 \times 10^3\ kJ/mol$. The magnitude is consistent with the photon energy, and the negative sign indicates energy was emitted from the atom because the electron fell from a higher energy level to a lower energy level. A positive sign of ΔE would mean that energy was absorbed and represents an electron moving from lower to higher energy.

▶ **PRACTICE 5.7** The Balmer equation can be extended beyond the visible portion of the electromagnetic spectrum to include lines in the ultraviolet. What is the wavelength in nanometers and energy in kJ/mol of ultraviolet light in the Balmer series corresponding to a value of $n = 7$? (Recall that $m = 2$ in the Balmer series.)

▶ **APPLY 5.8**

(a) What is the longest-wavelength line in nanometers in the infrared series for hydrogen where $m = 3$?

(b) What is the shortest-wavelength line in nanometers in the infrared series for hydrogen where $m = 3$? (Hint: $n = \infty$)

All **WORKED EXAMPLES** with this icon Go to eText have an interactive video in the eText.

5.4 WAVELIKE PROPERTIES OF MATTER: DE BROGLIE'S HYPOTHESIS

The analogy between matter and radiant energy developed in the early 1900s was further extended in 1924 by the French physicist Louis de Broglie (1892–1987). De Broglie suggested that if *light* can behave in some respects like *matter*, then perhaps *matter* can behave in some respects like *light*. That is, perhaps matter is wavelike as well as particlelike.

In developing his theory about the wavelike behavior of matter, de Broglie focused on the inverse relationship between energy and wavelength for photons:

$$\text{Since}\quad E = \frac{hc}{\lambda}\quad \text{then}\quad \lambda = \frac{hc}{E}$$

Using the famous equation $E = mc^2$ proposed in 1905 by Einstein as part of his special theory of relativity, and substituting for E, then gives

$$\lambda = \frac{hc}{E} = \frac{hc}{mc^2} = \frac{h}{mc}$$

De Broglie suggested that a similar equation might be applied to moving particles like electrons by replacing the speed of light, c, by the speed of the particle, v. The resultant **de Broglie equation** allows calculation of a "wavelength" of an electron or of any other particle or object of mass m moving at velocity v:

> **de Broglie equation** $\quad \lambda = \dfrac{h}{mv}$

Worked Example 5.5 demonstrates how to use the de Broglie equation to calculate the wavelength of a particle in motion.

WORKED EXAMPLE 5.5

Calculating the de Broglie Wavelength of Moving Particles

Calculate the wavelength of an electron in a hydrogen atom with a mass of 9.11×10^{-31} kg and a velocity v of 2.2×10^6 m/s (about 1% of the speed of light).

STRATEGY

Since mass and velocity are known, the de Broglie equation can be used to calculate wavelength. Note that mass and velocity must be in SI units of kg and m/s.

SOLUTION

Planck's constant, which is usually expressed in units of joule seconds (J·s), is expressed for the present purposes in units of $(kg \cdot m^2)/s$ [1 J = 1 $(kg \cdot m^2)/s^2$].

$$\lambda = \frac{h}{mv} = \frac{6.626 \times 10^{-34} \dfrac{kg \cdot m^2}{s}}{(9.11 \times 10^{-31} \, kg)\left(2.2 \times 10^6 \dfrac{m}{s}\right)} = 3.3 \times 10^{-10} \text{ m}$$

▶ **PRACTICE 5.9** What is the de Broglie wavelength in meters of a small car with a mass of 1150 kg traveling at a velocity of 55.0 mi/h? (1 mi = 1609.3 m)?

▶ **APPLY 5.10** Electron microscopes make use of the de Broglie wavelength of high-speed electrons to image tiny objects. The magnification of an electron microscope ($\approx 10^6$ times) is much greater than an optical microscope using visible light ($\approx 10^3$ times) because the wavelengths of high-speed electrons are much smaller than the wavelengths of visible light. To produce a high-quality image, the wavelength of the electron must be about 10 times less than the diameter of the particle. What electron velocity is required if a 1 nm particle is to be imaged?

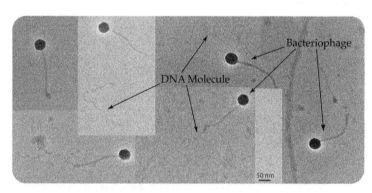

▲ An image of a bacteriophage releasing single molecules of DNA. The width of the DNA molecules is about 2 nm. Electron microscopes are able to image much smaller objects than optical microscopes by using a beam of high-speed electrons with short de Broglie wavelengths.

What does it mean to say that light and matter act both as waves and as particles? On the everyday human scale, the answer is "not much." We have never directly observed the wave nature of familiar objects such as a baseball with our eyes because the de Broglie wavelengths of macroscopic objects are much too small. The problem in trying to understand the dual wave/particle description of light and matter is that our common sense isn't up to the task. Our intuition has been developed from personal experiences, using our eyes and other senses to tell us how light and matter are "supposed" to behave. It is difficult to imagine a particle simultaneously behaving as a wave as we have never observed such a phenomenon.

In contrast, on the atomic scale, where distances and masses are so tiny, light and matter behave in a manner different from what we're used to. Experiments have been performed that support the theory that subatomic particles such as electrons exhibit wave properties.

Let's take a look at what happens when a beam of electrons is directed at a plate with two parallel slits. To start off, imagine throwing tennis balls at a wall with two slits in it. Some balls will bounce off the wall, but some will travel through the slits. If there's another wall behind the first, the balls that have traveled through the slits will hit it. If you mark all the spots where a ball has hit the second wall, you would expect to see two strips of marks roughly the same shape as the slits. We know that electrons are tiny particles with a mass of 9.11×10^{-31} kg. If electrons behave as particles, they would also produce a pattern of two strips after passing through a barrier with two slits (**FIGURE 5.11a**). However, when a beam of electrons is directed at two slits, an interference pattern of bright and dark bands appears on the detector and resembles the interference pattern obtained with light waves (**FIGURE 5.11b**). Recall from Section 5.1 that when a light ray was directed at a double slit, an interference pattern was observed as a result of constructive and destructive interference of diffracted light waves. The image produced from directing a beam of electrons at a double slit is shown in **FIGURE 5.11c**. Observations from the double-slit experiment support the theory that subatomic particles such as electrons exhibit wave properties.

BIG IDEA Question 3   ⬛ Go to eText

Figure 5.11 shows the behavior of particles and electrons directed at a barrier with two slits. What can be concluded about the behavior of electrons from the results of the experiment?

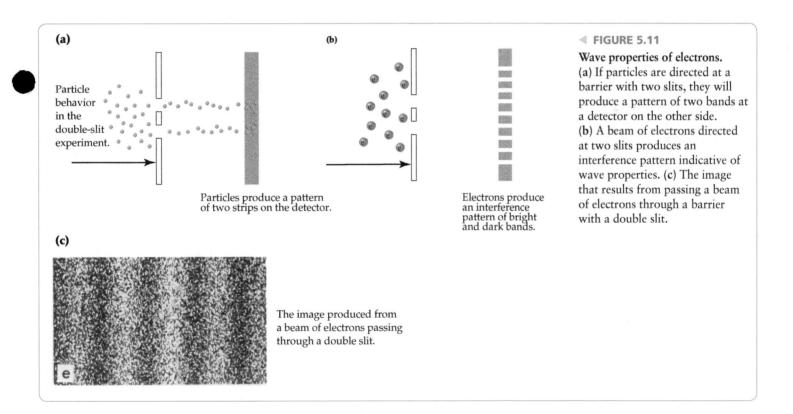

◀ FIGURE 5.11

Wave properties of electrons.
(a) If particles are directed at a barrier with two slits, they will produce a pattern of two bands at a detector on the other side.
(b) A beam of electrons directed at two slits produces an interference pattern indicative of wave properties. **(c)** The image that results from passing a beam of electrons through a barrier with a double slit.

(a) Particle behavior in the double-slit experiment.

Particles produce a pattern of two strips on the detector.

(b) Electrons produce an interference pattern of bright and dark bands.

(c) The image produced from a beam of electrons passing through a double slit.

5.5 THE QUANTUM MECHANICAL MODEL OF THE ATOM: HEISENBERG'S UNCERTAINTY PRINCIPLE

With the particlelike nature of energy and the wavelike nature of matter established, let's return to the problem of atomic structure. Several models of atomic structure were proposed in the late nineteenth and early twentieth centuries, such as the Bohr model described in Section 5.3. The Bohr model was important historically because of its conclusion that electrons have only specific energy levels available to them, but as we noted the model fails for atoms with more than one electron.

The breakthrough in understanding atomic structure came in 1926, when the Austrian physicist Erwin Schrödinger (1887–1961) proposed what has come to be called the **quantum mechanical model** of the atom. The fundamental idea behind the model is that it's best to abandon the notion of an electron as a small particle moving around the nucleus in a defined path and to concentrate instead on the electron's wavelike properties. In fact, it was shown in 1927 by Werner Heisenberg (1901–1976) that it is *impossible* to know precisely where an electron is and what path it follows—a statement called the **Heisenberg uncertainty principle.**

The Heisenberg uncertainty principle can be understood by imagining what would happen if we tried to determine the position of an electron at a given moment. For us to "see" the electron, light photons of an appropriate frequency would have to interact with and bounce off the electron. But such an interaction would transfer energy from the photon to the electron, thereby increasing the energy of the electron and making it move faster. Thus, the very act of determining the electron's position would make that position change.

In mathematical terms, Heisenberg's principle states that the uncertainty in the electron's position, Δx, times the uncertainty in its momentum, Δmv, is equal to or greater than the quantity $h/4\pi$:

$$\text{Heisenberg uncertainty principle}\quad (\Delta x)(\Delta mv) \geq \frac{h}{4\pi}$$

According to this equation, we can never know both the position and the velocity of an electron (or of any other object) beyond a certain level of precision. If we know the *velocity* with a high degree of certainty (Δmv is small), then the *position* of the electron must be uncertain (Δx must be large). Conversely, if we know the position of the electron exactly (Δx is small), then we can't know its velocity (Δmv must be large). As a result, an electron will always appear as something of a blur whenever we attempt to make any physical measurements of its position and velocity.

A brief calculation can help make the conclusions of the uncertainty principle clearer. As mentioned in the previous section, the mass m of an electron is 9.11×10^{-31} kg and the velocity v of an electron in a hydrogen atom is 2.2×10^6 m/s. If we assume that the velocity is known to within 10%, or 0.2×10^6 m/s, then the uncertainty in the electron's position in a hydrogen atom is greater than 3×10^{-10} m, or 300 pm. But since the diameter of a hydrogen atom is only 240 pm, *the uncertainty in the electron's position is similar in size to the atom itself!*

▲ Even the motion of very fast objects such as bullets can be captured in daily life. On the atomic scale, however, velocity and position can't both be known precisely.

$$\text{If}(\Delta x)(\Delta mv) \geq \frac{h}{4\pi} \quad \text{then } (\Delta x) \geq \frac{h}{(4\pi)(\Delta mv)}$$

$$\Delta x \geq \frac{6.626 \times 10^{-34}\ \frac{\text{kg} \cdot \text{m}^2}{\text{s}}}{(4)(3.1416)(9.11 \times 10^{-31}\ \text{kg})\left(0.2 \times 10^6\ \frac{\text{m}}{\text{s}}\right)}$$

$$\Delta x \geq 3 \times 10^{-10}\ \text{m} \quad \text{or} \quad 300\ \text{pm}$$

When the mass m of an object is relatively large, as in daily life, then both Δx and Δv in the Heisenberg relationship are very small, so we have no problem in measuring both position and velocity for visible objects. The problem arises only on the atomic scale.

 Go to eText

BIG IDEA Question 4

For which object would be it impossible to accurately know both the position and velocity: a thrown baseball, an electron in a cathode-ray tube, both, or neither?

5.6 THE QUANTUM MECHANICAL MODEL OF THE ATOM: ORBITALS AND QUANTUM NUMBERS

Schrödinger's quantum mechanical model of atomic structure is framed in the form of a mathematical expression called a *wave equation* because it is similar in form to the equation used to describe the motion of ordinary waves in fluids. The solutions to the wave equation are called **wave functions,** or **orbitals,** and are represented by the symbol ψ (Greek psi). The best way to think about an electron's wave function is to regard it as an expression whose square, ψ^2, defines the probability of finding the electron within a given volume of space around the nucleus. It is important to distinguish the difference between "orbits" in the Bohr model of the atom and "orbitals" in the quantum mechanical model. An orbit defines a specific location for the electron while the orbital is a mathematical equation. As Heisenberg showed, we can never be completely certain about an electron's position. A wave function, however, tells where the electron will most probably be found.

$$\text{Wave equation} \xrightarrow{\text{Solve}} \text{Wave function or orbital } (\psi) \longrightarrow \text{Probability of finding electron in a region of space } (\psi^2)$$

A wave function is characterized by three parameters called **quantum numbers,** represented as n, l, and m_l, which describe the energy level of the orbital and the three-dimensional shape of the region in space occupied by a given electron.

- The **principal quantum number** (n) is a positive integer ($n = 1, 2, 3, 4, \ldots$) on which the size and energy level of the orbital primarily depend. For hydrogen and other one-electron atoms, such as He^+, the energy of an orbital depends only on n. For atoms with more than one electron, the energy level of an orbital depends both on n and on the l quantum number.

  As the value of n increases, the number of allowed orbitals increases and the size of those orbitals becomes larger, thus allowing an electron to be farther from the nucleus. Because it takes energy to separate a negative charge from a positive charge, this increased distance between the electron and the nucleus means that the energy of the electron in the orbital increases as the quantum number n increases.

  We often speak of orbitals as being grouped according to the principal quantum number n into successive layers, or **shells,** around the nucleus. Those orbitals with $n = 3$, for example, are said to be in the third shell.

- The **angular-momentum quantum number** (l) defines the three-dimensional shape of the orbital. For an orbital whose principal quantum number is n, the angular-momentum quantum number l can have any integral value from 0 to $n - 1$. Thus, within each shell, there are n different shapes for orbitals.

  If $n = 1$, then $l = 0$
  If $n = 2$, then $l = 0$ or 1
  If $n = 3$, then $l = 0, 1,$ or 2
  $\ldots$ and so on

  Just as it's convenient to think of orbitals as being grouped into shells according to the principal quantum number n, we often speak of orbitals within a shell as being further grouped into **subshells** according to the angular-momentum quantum number l. Different subshells are usually designated by letters rather than by numbers, following the order s, p, d, f, g. (Historically, the letters s, p, d, and f arose from the use of the words *sharp, principal, diffuse,* and *fundamental* to describe various lines in atomic spectra.) After f, successive subshells are designated alphabetically: g, h, and so on.

Quantum number l:	0	1	2	3	4	$\ldots$
Subshell notation:	s	p	d	f	g	$\ldots$

  As an example, an orbital with $n = 3$ and $l = 2$ is a $3d$ orbital: 3 to represent the third shell and d to represent the $l = 2$ subshell.

- The **magnetic quantum number** (m_l) defines the spatial orientation of the orbital with respect to a standard set of coordinate axes. For an orbital whose angular-momentum quantum number is l, the magnetic quantum number m_l can have any integral value from $-l$ to $+l$. Thus, within each subshell—orbitals with the same shape, or value of l —there are $2l + 1$ different spatial orientations for those orbitals. We'll explore this point further in the next section.

If $l = 0$, then $m_l = 0$
If $l = 1$, then $m_l = -1, 0,$ or $+1$
If $l = 2$, then $m_l = -2, -1, 0, +1,$ or $+2$
... and so forth

A summary of the allowed combinations of quantum numbers for the first four shells is given in **TABLE 5.2**.

BIG IDEA Question 5

What are the values of n, l, and m_l for an electron in a $4d$ orbital?

TABLE 5.2   Allowed Combinations of Quantum Numbers n, l, and m_l for the First Four Shells

n	l	m_l	Orbital Notation	Number of Orbitals in Subshell	Number of Orbitals in Shell
1	0	0	$1s$	1	1
2	0	0	$2s$	1	4
	1	$-1, 0, +1$	$2p$	3	
3	0	0	$3s$	1	9
	1	$-1, 0, +1$	$3p$	3	
	2	$-2, -1, 0, +1, +2$	$3d$	5	
4	0	0	$4s$	1	16
	1	$-1, 0, +1$	$4p$	3	
	2	$-2, -1, 0, +1, +2$	$4d$	5	
	3	$-3, -2, -1, 0, +1, +2, +3$	$4f$	7	

The energy levels of various orbitals are shown in **FIGURE 5.12**. As noted earlier in this section, the energy levels of different orbitals in a hydrogen atom depend only on the principal quantum number n, but the energy levels of orbitals in multielectron atoms depend on both n and l. In other words, the orbitals in a given shell all have the same energy for hydrogen but have slightly different energies for other atoms,

▶ **FIGURE 5.12**

Energy levels of atomic orbitals: (a) hydrogen and (b) a typical multielectron atom. The differences between energies of various subshells in (b) are exaggerated for clarity.

▶ **Figure It Out**

Rank the orbitals ($4s$, $4p$, $4d$, and $4f$) from lowest to highest energy in a hydrogen atom and a multielectron atom.

Answer: In hydrogen the energies of $4s$, $4p$, $4d$, and $4f$ are all equivalent. In a multielectron atom the order of energy from lowest to highest is $4s > 4p > 4d > 4f$.

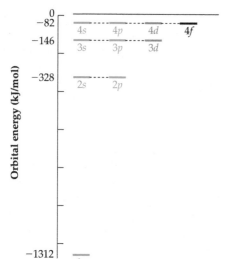

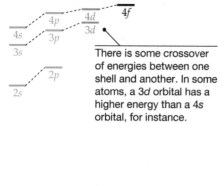

There is some crossover of energies between one shell and another. In some atoms, a $3d$ orbital has a higher energy than a $4s$ orbital, for instance.

depending on their subshell. In fact, there is even some crossover of energies between one shell and another. A 3*d* orbital in some multielectron atoms has a higher energy than a 4*s* orbital, for instance.

Go to
eText

WORKED EXAMPLE 5.6

Assigning Quantum Numbers to an Orbital

Give the possible combinations of quantum numbers for a 4*p* orbital.

STRATEGY

The principal quantum number n gives the shell number, which is 4 in this case. The angular-momentum quantum number l gives the subshell designation. For a *p* orbital $l = 1$. The magnetic quantum number m_l is related to the spatial orientation of the orbital and can have any of the three values from $-l$ to $+l$. For $l = 1$, values for $m_l = -1, 0$, or $+1$.

SOLUTION

The allowable combinations are

$$n = 4, l = 1, m_l = -1 \qquad n = 4, l = 1, m_l = 0 \qquad n = 4, l = 1, m_l = +1$$

▶ **PRACTICE 5.11** Give the orbital notation for an electron in an orbital with the quantum numbers: $n = 4, l = 3, m_l = -2$

▶ **APPLY 5.10** Extend Table 5.2 to show allowed combinations of quantum numbers when $n = 5$. How many orbitals are in the fifth shell?

5.7 THE SHAPES OF ORBITALS

We said in the previous section that the square of a wave function, or orbital, describes the probability of finding the electron within a specific region of space. The shape of that spatial region is defined by the angular-momentum quantum number l, with $l = 0$ called an *s* orbital, $l = 1$ a *p* orbital, $l = 2$ a *d* orbital, and $l = 3$ an *f* orbital. Of the various possibilities, *s*, *p*, *d*, and *f* orbitals are the most important because these are the only ones actually occupied in known elements. Let's look at each of the four individually.

s Orbitals

As described in Section 5.7, wave functions (ψ), or orbitals, are solutions to the wave equation and predict the allowed energy states of the electron. The square of the wave function (ψ^2) is the probability of finding an electron in a given region of space. **FIGURE 5.13a** represents ψ^2 using dots in an electron density plot for a 1*s* orbital. The higher concentration of dots near the nucleus corresponds to a greater probability of finding the electron. Note that all *s* orbitals are spherical, meaning the probability of finding an electron in an *s* orbital depends only on the distance from the nucleus and not on the direction. The value of electron density (ψ^2) for a 1*s* orbital is greatest near the nucleus and drops off as the distance (r) from the nucleus increases (**FIGURE 5.13b**). Although the electron density is greatest close to the nucleus, the **radial distribution plot** (**FIGURE 5.13c**) more accurately represents the probability of finding the electron in a thin shell at any given distance (r) from the nucleus. The radial probability function is the electron density (ψ^2) times the surface area of a sphere with radius (r). At the nucleus, the surface area of the sphere is zero because $r = 0$ and surface area $= 4\pi r^2$. As r increases, the radial probability function increases because the increase in surface area of the sphere outweighs the decrease in electron density. At even larger values of r, the radial probability function decreases due to the exponential decrease of electron density, which now outweighs the increase in surface area.

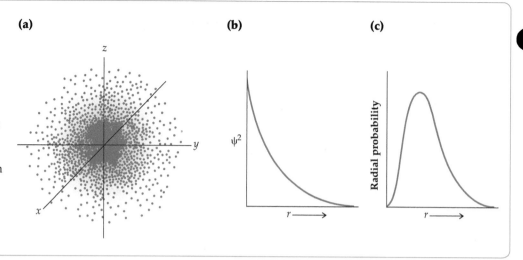

▶ **FIGURE 5.13**

Representations of a 1s orbital. **(a)** Electron density distribution for a 1s orbital. A higher concentration of dots represents a greater probability of finding the electron in a given volume. **(b)** Probability density (ψ^2) as a function of distance from the nucleus (r) for a 1s orbital. **(c)** The radial distribution plot for a 1s orbital shows the probability of finding an electron in a thin shell at a given distance (r) from the nucleus.

A simple way to visualize radial probability function is to consider the seats in a concert hall (**FIGURE 5.14**). Notice that row 1 has the highest percentage of seats occupied, with three out of three seats filled (100%). The percentage of seats filled in subsequent rows is lower. Row 2 has four out five seats filled (80%), row 3 has five out of seven seats filled (71%), and row 4 has four out of nine seats filled (44%). The percentage of seats filled can be thought of as the electron density (ψ^2) in an orbital. Notice that the percentage of seats filled decreases with distance from the stage, just like the electron density decreases with distance from the nucleus. The radial probability, or the chance of finding the electron at given distance from the nucleus, can be thought of as the total number of people in each row. The numbers of people in each row are: row 1 = 3 people, row 2 = 4 people, row 3 = 5 people, and row 4 = 4 people. Let's compare row 1 to row 3. Row 1 has the highest percentage of filled seats but only 3 seats possible, resulting in 3 people. Row 3 has a lower percentage of filled seats but more total seats, resulting in 5 people. If we were to think of the rows in the concert hall as a radial distribution plot, row 3 would have the highest radial probability because it has the greatest number of people.

▶ **FIGURE 5.14**

Concert hall analogy for radial probability. The percentage of seats filled in each row can be thought of as the probability of finding the electron as a distance from the nucleus increases. The percentage of seats filled decreases as distance from the stage increases. The radial probability can be thought of as the total number of people in each row. Row 3 has the highest radial probability because it has the largest number of seats occupied (5).

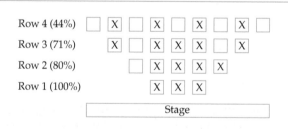

As shown in Figure 5.13b, the value of ψ^2 for an s orbital is greatest near the nucleus and then drops off rapidly as the distance from the nucleus increases, although it never goes all the way to zero, even at a large distance. As a result, there is no definite boundary to the atom and no definite size. For purposes like that of **FIGURE 5.15,** however, we usually imagine a boundary surface enclosing the volume where an electron has a 90% chance of being found.

Although all s orbitals are spherical, there are significant differences among the s orbitals in different shells. For one thing, the size of the s orbital increases in successively higher shells, implying that an electron in an outer-shell s orbital is farther from the nucleus on average than an electron in an inner-shell s orbital. For another thing, the electron distribution in an outer-shell s orbital has more than one region of high probability. As shown in Figure 5.15, a 2s orbital is essentially a sphere within a

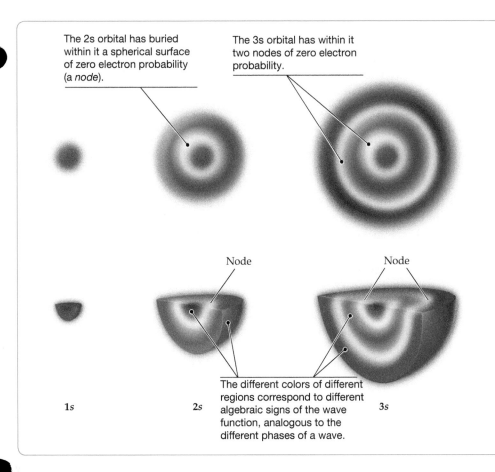

The 2s orbital has buried within it a spherical surface of zero electron probability (a *node*).

The 3s orbital has within it two nodes of zero electron probability.

Node

Node

The different colors of different regions correspond to different algebraic signs of the wave function, analogous to the different phases of a wave.

1s   2s   3s

◀ **FIGURE 5.15**

Representations of 1s, 2s, and 3s orbitals. Slices through these spherical orbitals are shown on the top and cutaway views on the bottom, with the probability of finding an electron represented by the density of the shading.

◀ **Figure It Out**

What are the differences between 1s, 2s, and 3s orbitals?

Answer: The size increases from 1s, to 2s, to 3s. Also, the 2s orbital has one node and the 3s orbital has two nodes, or points where the probability of finding an electron is zero.

sphere and has two regions of high probability, separated by a surface of zero probability called a **node.** Similarly, a 3s orbital has three regions of high probability and two spherical nodes.

The concept of an orbital node—a surface of zero electron probability separating regions of nonzero probability—is difficult to grasp because it raises the question "How does an electron get from one region of the orbital to another if it's not allowed to be at the node?" The question is misleading, though, because it assumes particlelike behavior for the electron rather than wavelike behavior.

In fact, nodes are an intrinsic property of waves, from moving waves of water in the ocean to the stationary, or standing, wave generated by vibrating a rope or guitar string (**FIGURE 5.16**). A node simply corresponds to the zero-amplitude part of the

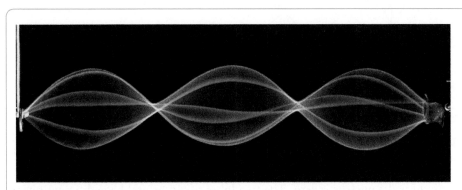

When a rope is fixed at one end and vibrated rapidly at the other, a standing wave is generated.

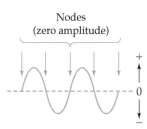

Nodes (zero amplitude)

◀ **FIGURE 5.16**

A standing wave in a vibrating rope.

The wave has two phases with different algebraic signs, + and −, separated by zero-amplitude regions, called *nodes*.

wave. On either side of the node is a nonzero wave amplitude. Note that a wave has two **phases**—peaks above the zero line and troughs below—corresponding to different algebraic signs, + and −. Similarly, the different regions of 2s and 3s orbitals have different phases, + and −, as indicated in Figure 5.15 by different colors.

Another way to visualize differences in the 1s, 2s, and 3s orbitals in the hydrogen atom is to examine the radial probability plots shown in **FIGURE 5.17.** The 1s orbital has a maximum in its radial probability function at a smaller r value than the maximum in the 2s orbital. This means that, on average, an electron in a 1s orbital spends more time closer to the nucleus than an electron in a 2s orbital. Similarly, the maximum in the radial probability for the 2s orbital is at a smaller r value than the maximum for the 3s orbital, meaning the electron in the 2s orbital has a greater probability of being closer to the nucleus than an electron is a 3s orbital. The radial probability plots for the s orbitals show that increasing the n level increases the size of the orbital. Larger orbitals are higher in energy than smaller orbitals because the electron has a greater probability of being further from the nucleus and it requires energy to separate a positive and negative charge.

Orbital size and energy: 1s < 2s < 3s

The presence of nodes in the 2s and 3s orbitals shown in Figure 5.15 is also evident in the radial probability plots. The 2s orbital has radial probability of zero at a distance of just over 100 pm from the nucleus corresponding to a node. The 3s orbital has two nodes shown by a radial probability of zero near 100 pm and 375 pm.

▶ **FIGURE 5.17**

Radial probability plots for the 1s, 2s, and 3s orbitals in a hydrogen atom.

▶ **Figure It Out**

How many nodes are present in a 1s, 2s, and 3s orbital?

Answer: Nodes have a radial probability value of zero. There are no nodes in a 1s orbital. In a 2s orbital, there is one node near 100 pm. In a 3s orbital, there are two nodes near 100 pm and 330 pm.

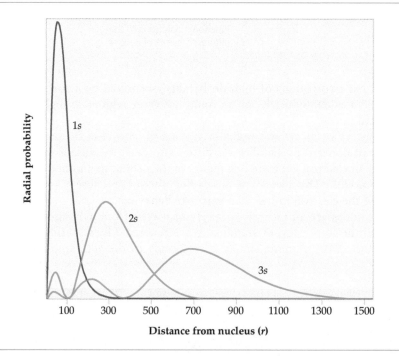

Distance from nucleus (r)

p Orbitals

The p orbitals are dumbbell-shaped rather than spherical, with their electron distribution concentrated in identical lobes on either side of the nucleus and separated by a planar node cutting through the nucleus. As a result, the probability of finding a p electron near the nucleus is zero. The two lobes of a p orbital have different phases, as indicated in **FIGURE 5.18** by different colors. We'll see in Chapter 7 that these phases are crucial for bonding because only lobes of the same phase can interact in forming covalent chemical bonds.

There are three allowable values of m_l when $l = 1$, so each shell beginning with the second has three p orbitals, which are oriented in space at 90° angles to one another along the three coordinate axes x, y, and z. The three p orbitals in the second shell, for

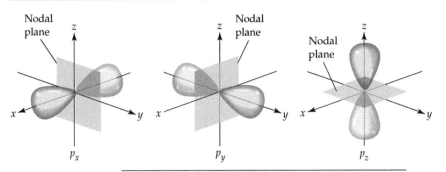

Each *p* orbital has two lobes of high electron probability separated by a nodal plane passing through the nucleus.

p_x p_y p_z

The different colors of the lobes represent different algebraic signs, analogous to the different phases of a wave.

◀ **FIGURE 5.18**

Representations of the three 2*p* orbitals. Each orbital is dumbbell-shaped and oriented in space along one of the three coordinate axes *x*, *y*, or *z*.

example, are designated $2p_x$, $2p_y$, and $2p_z$. As you might expect, *p* orbitals in the third and higher shells are larger than those in the second shell and extend farther from the nucleus. Their shape is roughly the same, however.

d and *f* Orbitals

The third and higher shells each contain five *d* orbitals, which differ from their *s* and *p* counterparts because they have two different shapes. Four of the five *d* orbitals are cloverleaf-shaped and have four lobes of maximum electron probability separated by two nodal planes through the nucleus (**FIGURE 5.19a–d**). The fifth *d* orbital is similar in shape to a p_z orbital but has an additional donut-shaped region of electron probability centered in the *xy* plane (**FIGURE 5.19e**). In spite of their different shapes, all five *d* orbitals in a given shell have the same energy. As with *p* orbitals, alternating lobes of the *d* orbitals have different phases.

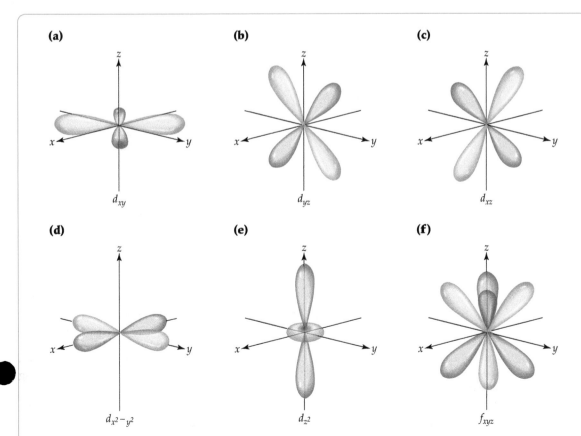

(a) d_{xy} **(b)** d_{yz} **(c)** d_{xz} **(d)** $d_{x^2-y^2}$ **(e)** d_{z^2} **(f)** f_{xyz}

◀ **FIGURE 5.19**

Representations of the five 3*d* orbitals. Four of the orbitals are shaped like a cloverleaf (a–d), and the fifth is shaped like an elongated dumbbell inside a donut (e). Also shown is one of the seven 4*f* orbitals (f). As with *p* orbitals in Figure 5.18, the different colors of the lobes reflect different phases.

You've probably noticed that both the number of nodal planes through the nucleus and the overall geometric complexity of the orbitals increases with the l quantum number of the subshell: An s orbital has one lobe and no nodal plane through the nucleus; a p orbital has two lobes and one nodal plane; and a d orbital has four lobes and two nodal planes. The seven f orbitals are more complex still, having eight lobes of maximum electron probability separated by three nodal planes through the nucleus. (**FIGURE 5.19f** shows one of the seven $4f$ orbitals.) Most of the elements we'll deal with in the following chapters don't use f orbitals in bonding, however, so we won't spend time on them.

CONCEPTUAL WORKED EXAMPLE 5.7

Assigning Quantum Numbers to an Orbital

Give a possible combination of n and l quantum numbers for the following fourth-shell orbital:

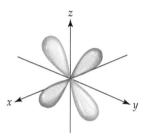

STRATEGY

Since the orbital is in the fourth shell, $n = 4$. Determine the angular momentum quantum number l from the orbital shape. The clover leaf shape represents a d orbital and therefore $l = 2$.

SOLUTION

$n = 4, l = 2$

▸ **CONCEPTUAL PRACTICE 5.13**   Give a possible combination of n and l quantum numbers for the following orbital.

▸ **CONCEPTUAL APPLY 5.14**   How many nodal planes through the nucleus do you think a g orbital has?

5.8 ELECTRON SPIN AND THE PAULI EXCLUSION PRINCIPLE

The three quantum numbers n, l, and m_l discussed in Section 5.6 define the energy, shape, and spatial orientation of orbitals, but they don't quite tell the whole story. When the line spectra of many multielectron atoms are studied in detail, it turns out that some lines actually occur as very closely spaced pairs. (You can see this pairing if you look closely at the visible spectrum of sodium in Figure 5.9.) Thus, there are more energy levels than simple quantum mechanics predicts, and a fourth quantum number is required. Denoted m_s, this fourth quantum number is related to a property called *electron spin*.

In some ways, electrons behave as if they were spinning around an axis, somewhat as the Earth spins daily. This spinning charge gives rise to a tiny magnetic field and to a **spin quantum number** (m_s), which can have either of two values, $+1/2$ or $-1/2$ (**FIGURE 5.20**). A spin of $+1/2$ is usually represented by an up arrow ($\uparrow$), and a spin of $-1/2$ by a down arrow ($\downarrow$). Note that the value of m_s is independent of the other three quantum numbers, unlike the values of n, l, and m_l, which are interrelated.

The importance of the spin quantum number comes when electrons occupy specific orbitals in multielectron atoms. According to the **Pauli exclusion principle**, proposed in 1925 by the Austrian physicist Wolfgang Pauli (1900–1958), no two electrons in an atom can have the same four quantum numbers. In other words, the set of four quantum numbers associated with an electron acts as a unique "address" for that electron in an atom, and no two electrons can have the same address.

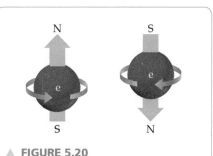

▲ **FIGURE 5.20**

Electron spin. Electrons behave in some ways as if they were tiny charged spheres spinning around an axis. This **spin** (blue arrow) gives rise to a tiny **magnetic field** (green arrow) and to a fourth quantum number, m_s, which can have a value of either $+1/2$ or $-1/2$.

> **Pauli exclusion principle**   No two electrons in an atom can have the same four quantum numbers.

What are the consequences of the Pauli exclusion principle? Electrons that occupy the same orbital have the same three quantum numbers, n, l, and m_l. But if they have the same values for n, l, and m_l, they must have different values for the fourth quantum number, m_s: either $m_s = +1/2$ or $m_s = -1/2$. Thus, an orbital can hold only two electrons, which must have opposite spins.

5.9 ORBITAL ENERGY LEVELS IN MULTIELECTRON ATOMS

BIG IDEA Question 6

A set of quantum numbers for one of the $4s$ electrons in calcium is

$$n = 4, l = 0, m_l = 0, m_s = +1/2$$

What is the set of four quantum numbers for the other $4s$ electron?

As we said in Section 5.6, the energy level of an orbital in a hydrogen atom, which has only one electron, is determined by its principal quantum number n. Within a shell, all hydrogen orbitals have the same energy, independent of their other quantum numbers. For example, in hydrogen the $2s$ and $2p$ orbitals have the same energy, and the $3s$, $3d$, and $3p$ orbitals have the same energy. The situation is different in multielectron atoms, however, where the energy level of a given orbital depends not only on the shell but also on the subshell. The s, p, d, and f orbitals within a given shell have slightly different energies in a multielectron atom, as shown previously in Figure 5.12. Upon inspection of Figure 5.12, we find that in $n = 3$ for multielectron atoms the ordering of orbital energies is $3s < 3p < 3d$.

The difference in energy between subshells in multielectron atoms results from electron–electron repulsions. In hydrogen, the only electrical interaction is the attraction of the positive nucleus for the negative electron, but in multielectron atoms there are many different interactions. Not only are there the attractions of the nucleus for each electron, there are also the repulsions between every electron and each of its neighbors.

The repulsion of outer-shell electrons by inner-shell electrons is particularly important because the outer-shell electrons are pushed farther away from the nucleus and are thus held less tightly. Part of the attraction of the nucleus for an outer electron is thereby canceled, an effect we describe by saying that the outer electrons are *shielded* from the nucleus by the inner electrons (**FIGURE 5.21**). The nuclear charge actually felt by an electron, called the **effective nuclear charge**, Z_{eff}, is often substantially lower than the actual nuclear charge Z.

> **Effective nuclear charge** $Z_{eff} = Z_{actual} -$ Electron shielding

▶ **FIGURE 5.21**

The origin of electron shielding and Z_{eff}. The outer electrons feel a diminished nuclear attraction because inner electrons shield them from the full charge of the nucleus.

▶ **Figure It Out**

Which electron has a lower value of Z_{eff}: an electron in a 2s orbital or an electron in a 3s orbital?

Answer: An electron in a 3s orbital has a lower Z_{eff} value because it is shielded by 1s, 2s, and 2p electrons. An electron in a 2s orbital is only shielded by 1s electrons.

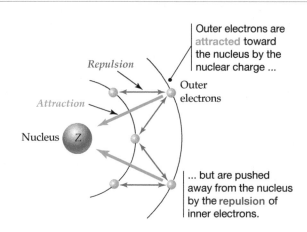

Repulsion

Attraction

Nucleus   Z

Outer electrons are attracted toward the nucleus by the nuclear charge ...

Outer electrons

... but are pushed away from the nucleus by the repulsion of inner electrons.

How does electron shielding lead to energy differences among orbitals within a shell? The answer is a consequence of the difference in orbital shapes. Compare the shapes of *s*, *p*, and *d* orbitals. The 3s orbital is spherical in shape and has a high electron density near the nucleus, the 3p orbitals are dumbbell shaped with a node at the nucleus, and four of the 3d orbitals are shaped like a cloverleaf with a node at the nucleus. **FIGURE 5.22** shows the radial probability plot for the 3s, 3p, and 3d orbitals. The 3s orbital has a maximum radial probability at a distance further from the nucleus than the maximum in either the 3p or 3d orbital. If on average an electron in a 3s orbital is farther from the nucleus than an electron in a 3p or 3d orbital, we might expect the 3s orbital to be highest in energy. However, the 3s orbital is actually the lowest in energy due to its ability to penetrate closer to the nucleus and into the region occupied by inner electrons. The small bump in the curve for the 3s orbital represents a significant probability of finding the electron closer to the nucleus than in the 3p or 3d orbitals. The effect is that electrons in a 3s orbital are less efficiently shielded by inner electrons than 3p or 3d electrons and experience the highest effective nuclear charge (Z_{eff}). A higher effective nuclear charge corresponds to a lower energy for the electron. Similarly, an electron in a 3p orbital is lower in energy than an electron in a 3d orbital due to its ability to penetrate closer to the nucleus.

▶ **FIGURE 5.22**

Radial distribution plots for 3s, 3p, and 3d orbitals. The penetration of the different orbitals determines the ordering of orbital energies ($3s < 3p < 3d$).

▶ **Figure It Out**

Electrons in which orbital are most effectively shielded by inner electrons: 3s, 3p, 3d?

Answer: The 3s and 3p orbitals penetrate closer to the nucleus than the 3d orbital, and thus inner electrons most effectively shield 3d electrons from the nucleus.

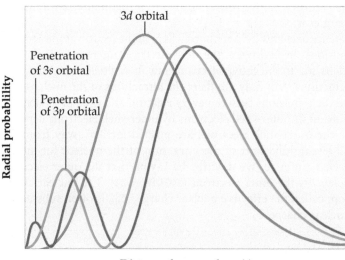

3d orbital

Penetration of 3s orbital

Penetration of 3p orbital

Radial probability

Distance from nucleus (r)

In summary, within any given shell, a lower value of the angular-momentum quantum number l corresponds to a higher Z_{eff} and to a lower energy for the electron.

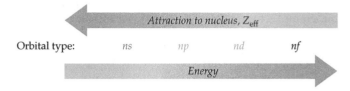

The idea that electrons in different orbitals are shielded differently and feel different values of Z_{eff} is a very useful one to which we'll return on several occasions to explain various chemical phenomena.

5.10 ELECTRON CONFIGURATIONS OF MULTIELECTRON ATOMS

All the parts are now in place to provide an electronic description for every element. Knowing the relative energies of the various orbitals, we can predict for each element which orbitals are occupied by electrons—the element's **electron configuration.**

A set of three rules called the **aufbau principle,** from the German word for "building up," guides the filling order of orbitals. In general, each successive electron added to an atom occupies the lowest-energy orbital available. The resultant lowest-energy configuration is called the **ground-state electron configuration** of the atom. Often, several orbitals will have the same energy level—for example, the three p orbitals or the five d orbitals in a given subshell. Orbitals that have the same energy level are said to be **degenerate.**

Rules of the Aufbau Principle:

1. **Lower-energy orbitals fill before higher-energy orbitals. FIGURE 5.23** shows the ordering of energy levels for orbitals in multielectron atoms. For elements through atomic number 20 (Ca), the energy of the atomic orbitals is as shown. For element 21 (Sc) at the start of the transition metals, the $3d$ orbitals become lower in energy than the $4s$ orbital. The switching of orbital energies is also true for the $4d$ and $5s$ orbitals in the second row of transition metal elements. However, when writing electron configurations, always follow the ordering of energies given in Figure 5.23 because it reliably predicts the correct electron configurations of elements.
2. **An orbital can hold only two electrons, which must have opposite spins.** This is just a restatement of the Pauli exclusion principle (Section 5.8), emphasizing that no two electrons in an atom can have the same four quantum numbers.
3. **If two or more degenerate orbitals are available, one electron goes into each until all are half-full,** a statement called **Hund's rule.** Only then does a second electron fill one of the orbitals. Furthermore, the electrons in each of the singly occupied orbitals must have the same value for their spin quantum number.

> **Hund's rule** If two or more orbitals with the same energy are available, one electron goes in each until all are half full. The value of the spin quantum number of electrons in the half-filled orbitals will be the same.

Hund's rule applies because electrons repel one another and therefore remain as far apart as possible. Electrons will be farther apart and lower in energy if they are in different orbitals describing different spatial regions than if they are in the same orbital occupying the same region. It also turns out that electrons in half-filled orbitals stay farther apart on average if they have the same spin rather than opposite spins.

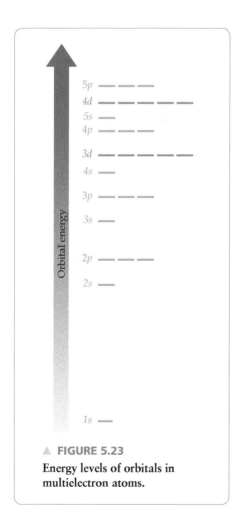

▲ **FIGURE 5.23**
Energy levels of orbitals in multielectron atoms.

Go to
eText

BIG IDEA Question 7

Which of the orbital-filling diagrams violates the rules of the aufbau principle?

(a)

$$\frac{\uparrow\downarrow}{3s} \quad \frac{\quad}{} \; \frac{\quad}{3p} \; \frac{\quad}{}$$

(b)

$$\frac{\uparrow\downarrow}{3s} \quad \frac{\uparrow\downarrow}{} \; \frac{\uparrow}{3p} \; \frac{\uparrow}{}$$

(c)

$$\frac{\uparrow\downarrow}{3s} \quad \frac{\uparrow\downarrow}{} \; \frac{\quad}{3p} \; \frac{\quad}{}$$

Electron configurations are normally represented by listing the n quantum number and the s, p, d, or f designation of the occupied orbitals, beginning with the lowest energy one, and with the number of electrons occupying each orbital indicated as a superscript. Let's look at some examples to see how the rules of the aufbau principle are applied.

- **Hydrogen:** Hydrogen has only one electron, which must go into the lowest-energy, $1s$ orbital. Thus, the ground-state electron configuration of hydrogen is $1s^1$.

$$\text{H: } 1s^1$$

- **Helium:** Helium has two electrons, both of which fit into the lowest-energy $1s$ orbital. The two electrons have opposite spins.

$$\text{He: } 1s^2$$

- **Lithium and beryllium:** With the $1s$ orbital full, the third and fourth electrons go into the next available orbital, $2s$.

$$\text{Li: } 1s^2\, 2s^1 \qquad \text{Be: } 1s^2\, 2s^2$$

- **Boron through neon:** In the six elements from boron through neon, electrons fill the three $2p$ orbitals successively. Because these three $2p$ orbitals have the same energy, they are degenerate and are thus filled according to Hund's rule. In carbon, for instance, the two $2p$ electrons occupy different orbitals, which can be arbitrarily specified as $2p_x$, $2p_y$, or $2p_z$ when writing the electron configuration. The same is true of nitrogen, whose three $2p$ electrons must be in three different orbitals. Per Hund's rule, the electrons in each of the singly occupied carbon and nitrogen $2p$ orbitals must have the same value of the spin quantum number—either $+1/2$ or $-1/2$—but this is not usually noted in the written electron configuration.

For clarity, we sometimes specify electron configurations using *orbital-filling diagrams,* in which electrons are represented by arrows. The two values of the spin quantum numbers are indicated by having the arrow point either up or down. An up–down pair indicates that an orbital is filled, while a single up (or down) arrow indicates that an orbital is half filled. Note in the diagrams for carbon and nitrogen that the degenerate $2p$ orbitals are half filled rather than filled, according to Hund's rule, and that the electron spin is the same in each.

Electron configuration		Orbital-filling diagram
B: $1s^2\, 2s^2\, 2p^1$	or	$\dfrac{\downarrow\uparrow}{1s} \quad \dfrac{\downarrow\uparrow}{2s} \quad \dfrac{\uparrow}{} \; \dfrac{\quad}{2p} \; \dfrac{\quad}{}$
C: $1s^2\, 2s^2\, 2p_x^{\,1}\, 2p_y^{\,1}$	or	$\dfrac{\downarrow\uparrow}{1s} \quad \dfrac{\downarrow\uparrow}{2s} \quad \dfrac{\uparrow}{} \; \dfrac{\uparrow}{2p} \; \dfrac{\quad}{}$
N: $1s^2\, 2s^2\, 2p_x^{\,1}\, 2p_y^{\,1}\, 2p_z^{\,1}$	or	$\dfrac{\downarrow\uparrow}{1s} \quad \dfrac{\downarrow\uparrow}{2s} \quad \dfrac{\uparrow}{} \; \dfrac{\uparrow}{2p} \; \dfrac{\uparrow}{}$

From oxygen through neon, the three $2p$ orbitals are successively filled. For fluorine and neon, it's no longer necessary to distinguish among the different $2p$ orbitals, so we can simply write $2p^5$ and $2p^6$.

O: $1s^2\, 2s^2\, 2p_x^{\,2}\, 2p_y^{\,1}\, 2p_z^{\,1}$	or	$\dfrac{\downarrow\uparrow}{1s} \quad \dfrac{\downarrow\uparrow}{2s} \quad \dfrac{\downarrow\uparrow}{} \; \dfrac{\uparrow}{2p} \; \dfrac{\uparrow}{}$
F: $1s^2\, 2s^2\, 2p^5$	or	$\dfrac{\downarrow\uparrow}{1s} \quad \dfrac{\downarrow\uparrow}{2s} \quad \dfrac{\downarrow\uparrow}{} \; \dfrac{\downarrow\uparrow}{2p} \; \dfrac{\uparrow}{}$
Ne: $1s^2\, 2s^2\, 2p^6$	or	$\dfrac{\downarrow\uparrow}{1s} \quad \dfrac{\downarrow\uparrow}{2s} \quad \dfrac{\downarrow\uparrow}{} \; \dfrac{\downarrow\uparrow}{2p} \; \dfrac{\downarrow\uparrow}{}$

- **Sodium and magnesium:** The $3s$ orbital is filled next, giving sodium and magnesium the ground-state electron configurations shown. Note that we often write the configurations in a shorthand version by giving the symbol of the noble gas in the previous row to indicate electrons in filled shells and then specifying only those electrons in partially filled shells.

Neon configuration

Na: $1s^2\,2s^2\,2p^6\,3s^1$ or $[Ne]\,3s^1$

Mg: $1s^2\,2s^2\,2p^6\,3s^2$ or $[Ne]\,3s^2$

- **Aluminum through argon:** The $3p$ orbitals are filled according to the same rules used previously for filling the $2p$ orbitals of boron through neon. Rather than explicitly identify which of the degenerate $3p$ orbitals are occupied in Si, P, and S, we'll simplify the writing by giving just the total number of electrons in the subshell. For example, we'll write $3p^2$ for silicon rather than $3p_x^1\,3p_y^1$.

Al: $[Ne]\,3s^2\,3p^1$ **Si:** $[Ne]\,3s^2\,3p^2$ **P:** $[Ne]\,3s^2\,3p^3$

S: $[Ne]\,3s^2\,3p^4$ **Cl:** $[Ne]\,3s^2\,3p^5$ **Ar:** $[Ne]\,3s^2\,3p^6$

- **Elements past argon:** Following the filling of the $3p$ subshell in argon, the first crossover in the orbital filling order is encountered. Rather than continue filling the third shell by populating the $3d$ orbitals, the next two electrons in potassium and calcium go into the $4s$ subshell. Only then does filling of the $3d$ subshell occur to give the first transition metal series from scandium through zinc.

K: $[Ar]\,4s^1$ **Ca:** $[Ar]\,4s^2$ **Sc:** $[Ar]\,4s^2\,3d^1 \longrightarrow$ **Zn:** $[Ar]\,4s^2\,3d^{10}$

The experimentally determined ground-state electron configurations of the elements are shown in **FIGURE 5.24.**

5.11 ANOMALOUS ELECTRON CONFIGURATIONS

The guidelines discussed in the previous section for determining ground-state electron configurations work well but are not completely accurate. A careful look at Figure 5.24 shows that 90 electron configurations are correctly accounted for by the rules but that 21 of the predicted configurations are incorrect.

The reasons for the anomalies often have to do with the unusual stability of both half-filled and fully filled subshells. Chromium, for example, which we would predict to have the configuration $[Ar]\,4s^2\,3d^4$, actually has the configuration $[Ar]\,4s^1\,3d^5$. By moving an electron from the $4s$ orbital to an energetically similar $3d$ orbital, chromium trades one filled subshell $(4s^2)$ for two half-filled subshells $(4s^1\,3d^5)$, thereby allowing the two electrons to be farther apart. In the same way, copper, which we would predict to have the configuration $[Ar]\,4s^2\,3d^9$, actually has the configuration $[Ar]\,4s^1\,3d^{10}$. By transferring an electron from the $4s$ orbital to a $3d$ orbital, copper trades one filled subshell $(4s^2)$ for a different filled subshell $(3d^{10})$ and gains a half-filled subshell $(4s^1)$.

Most of the anomalous electron configurations shown in Figure 5.24 occur in elements with atomic numbers greater than $Z = 40$, where the energy differences between subshells are small. In all cases, the transfer of an electron from one subshell to another lowers the total energy of the atom because of a decrease in electron–electron repulsions.

5.12 ELECTRON CONFIGURATIONS AND THE PERIODIC TABLE

Why are electron configurations so important, and what do they have to do with the periodic table? The answers emerge when you look closely at Figure 5.24. Focusing only on the electrons in the outermost shell, called the **valence shell**, *all the elements in a given group of the periodic table have similar valence-shell electron configurations* (**TABLE 5.3**).

TABLE 5.3 Valence-Shell Electron Configurations of Main-Group Elements

Group	Valence-Shell Electron Configuration	
1A	ns^1	(1 total)
2A	ns^2	(2 total)
3A	ns^2np^1	(3 total)
4A	ns^2np^2	(4 total)
5A	ns^2np^3	(5 total)
6A	ns^2np^4	(6 total)
7A	ns^2np^5	(7 total)
8A	ns^2np^6	(8 total)

1 1A	2 2A	3 3B	4 4B	5 5B	6 6B	7 7B	8	9 8B	10	11 1B	12 2B	13 3A	14 4A	15 5A	16 6A	17 7A	18 8A
1 H $1s^1$																	2 He $1s^2$
3 Li $2s^1$	4 Be $2s^2$											5 B $2s^22p^1$	6 C $2s^22p^2$	7 N $2s^22p^3$	8 O $2s^22p^4$	9 F $2s^22p^5$	10 Ne $2s^22p^6$
11 Na $3s^1$	12 Mg $3s^2$											13 Al $3s^23p^1$	14 Si $3s^23p^2$	15 P $3s^23p^3$	16 S $3s^23p^4$	17 Cl $3s^23p^5$	18 Ar $3s^23p^6$
19 K $4s^1$	20 Ca $4s^2$	21 Sc $4s^23d^1$	22 Ti $4s^23d^2$	23 V $4s^23d^3$	24 Cr $4s^13d^5$	25 Mn $4s^23d^5$	26 Fe $4s^23d^6$	27 Co $4s^23d^7$	28 Ni $4s^23d^8$	29 Cu $4s^13d^{10}$	30 Zn $4s^23d^{10}$	31 Ga $4s^23d^{10}4p^1$	32 Ge $4s^23d^{10}4p^2$	33 As $4s^23d^{10}4p^3$	34 Se $4s^23d^{10}4p^4$	35 Br $4s^23d^{10}4p^5$	36 Kr $4s^23d^{10}4p^6$
37 Rb $5s^1$	38 Sr $5s^2$	39 Y $5s^24d^1$	40 Zr $5s^24d^2$	41 Nb $5s^14d^4$	42 Mo $5s^14d^5$	43 Tc $5s^24d^5$	44 Ru $5s^14d^7$	45 Rh $5s^14d^8$	46 Pd $4d^{10}$	47 Ag $5s^14d^{10}$	48 Cd $5s^24d^{10}$	49 In $5s^24d^{10}5p^1$	50 Sn $5s^24d^{10}5p^2$	51 Sb $5s^24d^{10}5p^3$	52 Te $5s^24d^{10}5p^4$	53 I $5s^24d^{10}5p^5$	54 Xe $5s^24d^{10}5p^6$
55 Cs $6s^1$	56 Ba $6s^2$	71 Lu $6s^24f^{14}5d^1$	72 Hf $6s^24f^{14}5d^2$	73 Ta $6s^24f^{14}5d^3$	74 W $6s^24f^{14}5d^4$	75 Re $6s^24f^{14}5d^5$	76 Os $6s^24f^{14}5d^6$	77 Ir $6s^24f^{14}5d^7$	78 Pt $6s^14f^{14}5d^9$	79 Au $6s^14f^{14}5d^{10}$	80 Hg $6s^24f^{14}5d^{10}$	81 Tl $6s^24f^{14}5d^{10}6p^1$	82 Pb $6s^24f^{14}5d^{10}6p^2$	83 Bi $6s^24f^{14}5d^{10}6p^3$	84 Po $6s^24f^{14}5d^{10}6p^4$	85 At $6s^24f^{14}5d^{10}6p^5$	86 Rn $6s^24f^{14}5d^{10}6p^6$
87 Fr $7s^1$	88 Ra $7s^2$	103 Lr $7s^25f^{14}6d^1$	104 Rf $7s^25f^{14}6d^2$	105 Db $7s^25f^{14}6d^3$	106 Sg $7s^25f^{14}6d^4$	107 Bh $7s^25f^{14}6d^5$	108 Hs $7s^25f^{14}6d^6$	109 Mt $7s^25f^{14}6d^7$	110 Ds $7s^15f^{14}6d^9$	111 Rg $7s^15f^{14}6d^{10}$	112 Cn $7s^25f^{14}6d^{10}$	113 Nh	114 Fl	115 Mc	116 Lv	117 Ts	118 Og

57 La $6s^25d^1$	58 Ce $6s^24f^15d^1$	59 Pr $6s^24f^3$	60 Nd $6s^24f^4$	61 Pm $6s^24f^5$	62 Sm $6s^24f^6$	63 Eu $6s^24f^7$	64 Gd $6s^24f^75d^1$	65 Tb $6s^24f^9$	66 Dy $6s^24f^{10}$	67 Ho $6s^24f^{11}$	68 Er $6s^24f^{12}$	69 Tm $6s^24f^{13}$	70 Yb $6s^24f^{14}$
89 Ac $7s^26d^1$	90 Th $7s^26d^2$	91 Pa $7s^25f^26d^1$	92 U $7s^25f^36d^1$	93 Np $7s^25f^46d^1$	94 Pu $7s^25f^6$	95 Am $7s^25f^7$	96 Cm $7s^25f^76d^1$	97 Bk $7s^25f^9$	98 Cf $7s^25f^{10}$	99 Es $7s^25f^{11}$	100 Fm $7s^25f^{12}$	101 Md $7s^25f^{13}$	102 No $7s^25f^{14}$

▲ **FIGURE 5.24**

Outer-shell, ground-state electron configurations of the elements.

▲ **Figure It Out**

Identify the elements in period 5 with an anomalous electron configuration.

Answer: Nb, Mo, Ru, Rh, Pd, Ag

The group 1A elements, for example, all have an s^1 valence-shell configuration; the group 2A elements have an s^2 valence-shell configuration; the group 3A elements have an $s^2 p^1$ valence-shell configuration; and so on across every group of the periodic table (except for the small number of anomalies). Furthermore, because the valence-shell electrons are outermost and least tightly held, they are the most important for determining an element's properties, thus explaining why the elements in a given group of the periodic table have similar chemical behavior.

The periodic table can be divided into four regions, or blocks, of elements according to the orbitals being filled (**FIGURE 5.25**). The group 1A and group 2A elements on the left side of the table are called the *s*-**block elements** because they result from the filling of an *s* orbital, the group 3A–8A elements on the right side of the table are the *p*-**block elements** because they result from the filling of *p* orbitals, the transition metal *d*-**block elements** in the middle of the table result from the filling of *d* orbitals, and the lanthanide/actinide *f*-**block elements** detached at the bottom of the table result from the filling of *f* orbitals.

Thinking of the periodic table as outlined in Figure 5.25 provides a useful way to remember the order of orbital filling. Beginning at the top left corner of the periodic table and going across successive rows gives the correct orbital-filling order. The first row of the periodic table, for instance, contains only the two *s*-block elements H and He, so the first available *s* orbital (1*s*) is filled first. The second row begins with two *s*-block elements (Li and Be) and continues with six *p*-block elements (B through Ne), so the next available *s* orbital (2*s*) and then the first available *p* orbitals (2*p*) are filled. Moving similarly across the third row, the 3*s* and 3*p* orbitals are filled. The fourth row again starts with two *s*-block elements (K and Ca) but is then followed by 10 *d*-block elements (Sc through Zn) and six *p*-block elements (Ga through Kr). Thus, the order of orbital filling is 4*s* followed by the first available *d* orbitals (3*d*) followed by 4*p*. Continuing through successive rows of the periodic table gives the entire filling order:

$$1s \rightarrow 2s \rightarrow 2p \rightarrow 3s \rightarrow 3p \rightarrow 4s \rightarrow 3d \rightarrow 4p \rightarrow 5s \rightarrow 4d \rightarrow$$
$$5p \rightarrow 6s \rightarrow 4f \rightarrow 5d \rightarrow 6p \rightarrow 7s \rightarrow 5f \rightarrow 6d \rightarrow 7p$$

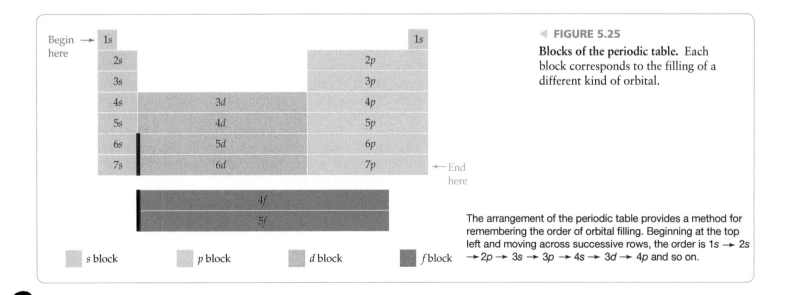

◄ **FIGURE 5.25**

Blocks of the periodic table. Each block corresponds to the filling of a different kind of orbital.

The arrangement of the periodic table provides a method for remembering the order of orbital filling. Beginning at the top left and moving across successive rows, the order is 1s → 2s → 2p → 3s → 3p → 4s → 3d → 4p and so on.

CONCEPTUAL WORKED EXAMPLE 5.8

Assigning a Ground-State Electron Configuration to an Atom

Give the ground-state electron configuration of arsenic, $Z = 33$, and draw an orbital-filling diagram, indicating the electrons as up or down arrows.

STRATEGY

Think of the periodic table as having s, p, d, and f blocks of elements, as shown in Figure 5.26. Start with hydrogen at the upper left, and fill orbitals until 33 electrons have been added. Remember that only two electrons can go into an orbital and that each one of a set of degenerate orbitals must be half filled before any one can be completely filled.

SOLUTION

As: $1s^2\,2s^2\,2p^6\,3s^2\,3p^6\,4s^2\,3d^{10}\,4p^3$ or $[Ar]\,4s^2\,3d^{10}\,4p^3$

An orbital-filling diagram indicates the electrons in each orbital as arrows. Note that the three $4p$ electrons all have the same spin as the orbitals fill singly before electrons pair according to Hund's rule.

As: $[Ar]$ $\underset{4s}{\underline{\text{↓↑}}}$ $\underset{3d}{\underline{\text{↓↑}}\ \underline{\text{↓↑}}\ \underline{\text{↓↑}}\ \underline{\text{↓↑}}\ \underline{\text{↓↑}}}$ $\underset{4p}{\underline{\text{↑}}\ \underline{\text{↑}}\ \underline{\text{↑}}}$

▶ **CONCEPTUAL PRACTICE 5.15**   Give the expected ground state electron configuration for Ti ($Z = 22$), and draw an orbital-filling diagram.

▶ **CONCEPTUAL APPLY 5.16**   Identify the atoms with the following ground-state electron configuration:

(a) $[Kr]$ $\underset{5s}{\underline{\text{↓↑}}}$ $\underset{4d}{\underline{\text{↑}}\ \underline{\text{↑}}\ \underline{\text{↑}}\ \underline{\text{↑}}\ \underline{\text{↑}}}$ $\underset{5p}{\underline{\quad}\ \underline{\quad}\ \underline{\quad}}$

(b) $[Ar]$ $\underset{4s}{\underline{\text{↓↑}}}$ $\underset{3d}{\underline{\text{↓↑}}\ \underline{\text{↓↑}}\ \underline{\text{↓↑}}\ \underline{\text{↑}}\ \underline{\text{↑}}}$ $\underset{4p}{\underline{\quad}\ \underline{\quad}\ \underline{\quad}}$

5.13 ELECTRON CONFIGURATIONS AND PERIODIC PROPERTIES: ATOMIC RADII

We began this chapter by saying that atomic radius is one of many elemental properties to show periodic behavior. You might wonder, though, how we can talk about a definite size for an atom, having said in Section 5.6 that the electron clouds around atoms have no specific boundaries. What's usually done is to define an atom's radius as being half the distance between the nuclei of two identical atoms when they are bonded together. In Cl_2, for example, the distance between the two chlorine nuclei is 198 pm; in diamond (elemental carbon), the distance between two carbon nuclei is 154 pm. Thus, we say that the atomic radius of chlorine is half the Cl—Cl distance, or 99 pm, and the atomic radius of carbon is half the C—C distance, or 77 pm.

It's possible to check the accuracy of atomic radii by making sure that the assigned values are additive. For instance, since the atomic radius of Cl is 99 pm and the atomic radius of C is 77 pm, the distance between Cl and C nuclei when those two atoms are bonded together ought to be roughly 99 pm + 77 pm, or 176 pm. In fact, the measured distance between chlorine and carbon in chloromethane (CH_3Cl) is 178 pm, remarkably close to the expected value.

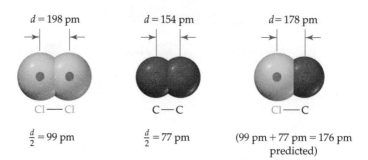

$d = 198$ pm $d = 154$ pm $d = 178$ pm

Cl—Cl   C—C   Cl—C

$\frac{d}{2} = 99$ pm $\frac{d}{2} = 77$ pm   (99 pm + 77 pm = 176 pm predicted)

As shown pictorially in **FIGURE 5.26** and graphically in Figure 5.1 at the beginning of this chapter, a comparison of atomic radius versus atomic number shows a periodic rise-and-fall pattern. Atomic radii increase going down a group of the periodic table (Li < Na < K < Rb < Cs, for instance) but decrease going across a row from left to right (Na > Mg > Al > Si > P > S > Cl, for instance). How can this behavior be explained?

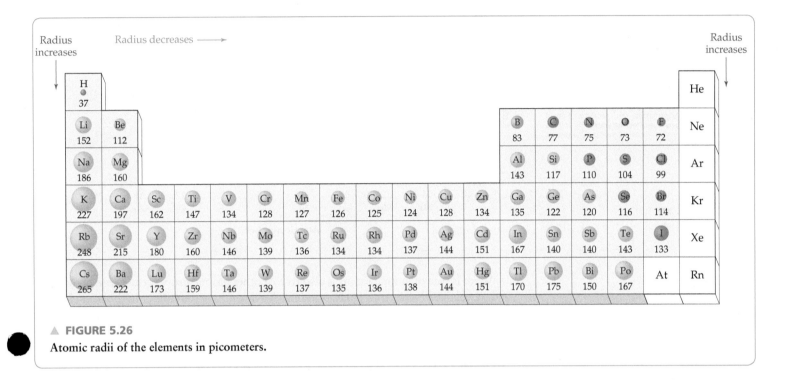

▲ **FIGURE 5.26**

Atomic radii of the elements in picometers.

The increase in radius going down a group of the periodic table occurs because successively larger valence-shell orbitals are occupied. In Li, for example, the outermost occupied shell is the second one ($2s^1$); in Na it's the third one ($3s^1$); in K it's the fourth one ($4s^1$); and so on through Rb ($5s^1$), Cs ($6s^1$), and Fr ($7s^1$). Because larger shells are occupied, the atomic radii are also larger.

The decrease in radius from left to right across the periodic table occurs because of an increase in effective nuclear charge caused by the increasing number of protons in the nucleus. As we saw in Section 5.9, Z_{eff}, the effective nuclear charge actually felt by an electron, is lower than the true nuclear charge Z because of shielding by other electrons in the atom. The amount of shielding felt by an electron depends on both the shell and subshell of the other electrons with which it is interacting. As a general rule, a valence-shell electron is:

• Strongly shielded by electrons in inner shells, which are closer to the nucleus.

• Less strongly shielded by other electrons in the same shell, according to the order $s > p > d > f$.

• Only weakly shielded by other electrons in the same subshell, which are at the same distance from the nucleus.

Going across the third period from Na to Cl, for example, each additional electron adds to the same shell (from $3s^1$ for Na to $3s^2\,3p^5$ for Cl). Because electrons in the same shell are at approximately the same distance from the nucleus, they are relatively ineffective at shielding one another. At the same time, though, the nuclear charge Z increases from +11 for Na to +17 for Cl.

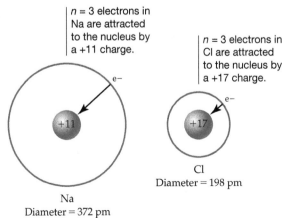

$n = 3$ electrons in Na are attracted to the nucleus by a +11 charge.

$n = 3$ electrons in Cl are attracted to the nucleus by a +17 charge.

Na
Diameter = 372 pm

Cl
Diameter = 198 pm

Thus, the *effective* nuclear charge for the valence-shell electrons increases across the period, drawing all the valence-shell electrons closer to the nucleus and progressively shrinking the atomic radii (**FIGURE 5.27**).

▶ **FIGURE 5.27**

Plots of atomic radius and calculated Z_{eff} for the highest-energy electron versus atomic number.

▶ **Figure It Out**

What is the relationship between Z_{eff} and atomic radius? Explain how Z_{eff} influences atomic radius.

Answer: As Z_{eff} increases atomic radius decreases. Higher Z_{eff} means the electrons experience a greater attraction to the nucleus, resulting in a smaller atomic radius.

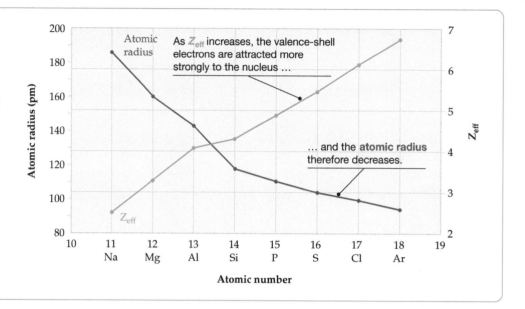

What is true of atomic radius is also true of other atomic properties, whose periodicity can be explained by electron configurations. We'll continue the subject in the next chapter.

WORKED EXAMPLE 5.9

Predicting the Relative Size of Atomic Radii

Which atom in each of the following pairs would you expect to be larger? Explain.

(a) Mg or Ba
(b) W or Hf

STRATEGY

Atomic radius decreases across a period as the effective nuclear charge increases (Z_{eff}), thus pulling electrons to the nucleus with stronger force. Atomic radius increases down a period as a higher-energy valence shell is occupied.

SOLUTION

(a) Ba (valence shell of $n = 6$) is larger than Mg (valence shell of $n = 3$).
(b) Hf (nuclear charge = +72) is larger than W (nuclear charge = +74).

▶ **PRACTICE 5.17**   Order the atoms from smallest to largest: Cs, Si, Sn.

▶ **APPLY 5.18**   Predict which bond length will be the longest: C—F, C—I, C—Br.

INQUIRY ? How does knowledge of atomic emission spectra help us build more efficient light bulbs?

In the standard incandescent light bulb that has been used for more than a century, an electrical current passes through a thin tungsten filament, which is thereby heated and begins to glow. The wavelengths and intensity of the light emitted depend on the temperature of the glowing filament—typically about 2500 °C—and cover the range from ultraviolet (200–400 nm), through the visible (400–800 nm), to the infrared (800–2000 nm). The ultraviolet frequencies are blocked by the glass of the bulb, the visible frequencies pass through (the whole point of the light bulb, after all), and the infrared frequencies warm the bulb and its surroundings.

Despite its long history, an incandescent bulb is actually an extremely inefficient device. In fact, only about 5% of the electrical energy consumed by the bulb is converted into visible light, with most of the remaining 95% converted into heat. Thus, many households and businesses are replacing their incandescent light bulbs with modern compact fluorescent bulbs in an effort to use less energy. (Not that fluorescent bulbs are terribly efficient themselves—only about 20% of the energy they consume is converted into light—but that still makes them about four times better than incandescents.)

A fluorescent bulb is, in essence, a variation of the cathode-ray tube described in Section 2.6. The bulb has two main parts, an argon-filled glass tube (either straight or coiled) containing a small amount of mercury vapor, and electronic circuitry that provides a controlled high-voltage current. The current passes through a filament, which heats up and emits a flow of electrons through the tube. In the tube, some of the flowing electrons collide with mercury atoms, transferring their kinetic energy and exciting mercury electrons to higher-energy orbitals. Photons are then released when the excited mercury electrons fall back to the ground state, generating an atomic line spectrum as shown in **FIGURE 5.28**.

Some photons emitted by the excited mercury atoms are in the visible range and contribute to the light we observe, but most are in the ultraviolet range at 185 nm and 254 nm and are invisible to our eyes. To capture this ultraviolet energy, fluorescent bulbs are coated on the inside with a *phosphor*, a substance that absorbs the ultraviolet light and re-emits the energy as visible light. As a result, fluorescent lights waste much less energy than incandescent bulbs.

Many different phosphors are used in fluorescent lights, each emitting its own line spectrum with visible light of various colors. Typically, a so-called triphosphor mixture is used, consisting of several complex metal oxides and rare-earth ions: Y_2O_3:Eu^{3+} (red emitting), $CeMgAl_{11}O_{19}$:Tb^{3+} (green emitting), and $BaMgAl_{10}O_{17}$:Eu^{2+} (blue emitting). The final color that results can be tuned as desired by the manufacturer, but typically the three emissions together are distributed fairly evenly over the visible spectrum to provide a color reproduction that our eyes perceive as natural white light (**FIGURE 5.29**).

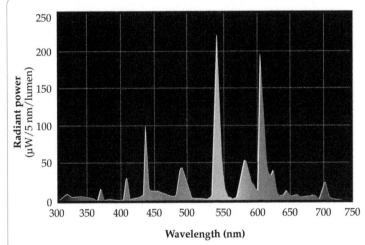

▲ **FIGURE 5.29**

The triphosphor spectrum emitted from a typical fluorescent bulb. The triphosphor spectrum is distributed over the visible spectrum and is perceived by our eyes as white light.

▲ Compact fluorescent bulbs like those shown here are a much more energy-efficient way to light a home than typical incandescent light bulbs.

▶ **FIGURE 5.28**

Line emission spectrum for mercury.

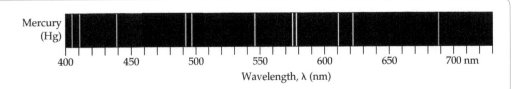

PROBLEM 5.19   What percentage of electrical energy is converted to light energy in
(a) an incandescent bulb?
(b) a fluorescent bulb?

PROBLEM 5.20   What energy source is used to excite mercury atoms in the fluorescent bulb so they emit a line spectrum?
(a) Thermal energy from a flame
(b) Kinetic energy of high speed electrons
(c) Light emitted from the phosphor coated on the inside of the bulb

PROBLEM 5.21
(a) How is the emission spectrum from the triphosphor in the fluorescent bulb different than the emission spectrum from a "white light" source such as the sun or an incandescent bulb?
(b) Why does the fluorescent bulb appear as "white light" to our eyes?

PROBLEM 5.22   Mercury vapor is contained inside the fluorescent bulb.
(a) Use the periodic table on the inside front cover of the book to write the electron configuration for the ground state of mercury.
(b) Sketch the orbital filling diagram for the ground state of mercury.
(c) How many unpaired electrons are in the ground state of mercury?

PROBLEM 5.23   When electricity is used to add energy to the mercury atoms, an electron in a 6s orbital can be "excited" into a 6p orbital.
(a) Write the electron configuration for a mercury atom in the excited state.
(b) How many unpaired electrons are there in the excited state of mercury?

PROBLEM 5.24   Atomic emission spectra arise from electron transitions from higher-energy orbitals to lower-energy orbitals. The blue line at 434.7 nm in the emission spectrum for mercury (Figure 5.28) arises from an electron moving from a 7d to a 6p orbital.
(a) Give the n and l quantum numbers and give the range of possible m_l quantum numbers for the higher-energy 7d orbital.
(b) Give the n and l quantum numbers and give the range of possible m_l quantum numbers for the lower energy 6p orbital.
(c) Calculate the energy difference between the 7d and 6p orbital in units of kJ/mol.

PROBLEM 5.25   Three different wavelengths in the line spectrum of mercury and three electronic transitions in mercury are given. Match the three wavelengths with the correct electronic transition between orbitals.

Emission Lines: 185.0 nm, 140.2 nm, 126.8 nm
Electronic Transitions: $8p \rightarrow 6s$, $7p \rightarrow 6s$, $6p \rightarrow 6s$

STUDY GUIDE

Section	Concept Summary	Learning Objectives	Test Your Understanding
5.1 Wave Properties of Radiant Energy and the Electromagnetic Spectrum	Understanding the nature of atoms and molecules begins with an understanding of light and other kinds of **radiant energy** that make up the **electromagnetic spectrum**. An electromagnetic wave travels through a vacuum at the speed of light (c) and is characterized by its **frequency (ν)**, **wavelength (λ)**, and **amplitude**.	5.1 Label the wavelength, frequency, and amplitude in an electromagnetic wave.	Problems 5.2, 5.26
		5.2 Convert between wavelength and frequency of electromagnetic radiation.	Worked Example 5.1; Problems 5.38, 5.40
5.2 Particle-Like Properties of Radiant Energy: The Photoelectric Effect and Planck's Postulate	Electromagnetic energy exists only in discrete amounts, referred to as a **quantum**. The energy of one quantum is related to the frequency of an electromagnetic wave and is given by **Planck's postulate**. In the **photoelectric effect**, light above a threshold energy is able to remove electrons from a substance. Einstein proposed that light behaves like particles (photons) in his explanation of the photoelectric effect.	5.3 Calculate the energy of electromagnetic radiation in units of J/photon or kJ/mol, when given the frequency or wavelength.	Worked Example 5.2; Problems 5.42, 5.44, 5.46
		5.4 Describe the photoelectric effect and explain how it supports the theory of particle-like properties of light.	Problems 5.52–5.53
		5.5 Calculate the frequency or wavelength of radiation needed to produce the photoelectric effect given the work function of a metal.	Worked Example 5.3; Problems 5.54–5.55
5.3 Atomic Line Spectra and Quantized Energy	Unlike the white light of the Sun, which consists of a nearly continuous distribution of wavelengths, the light emitted by an excited atom consists of only a few discrete wavelengths, a **line spectrum**. The **Balmer–Rydberg equation** is a mathematical model used to calculate the wavelengths in the emission spectrum for hydrogen. Bohr's model of the atom depicts a dense, positively charged nucleus circled by electrons in orbits of different radii. The model is successful in predicting line spectra because it presumes the energy of the orbits is **quantized**. The Bohr model is incorrect, however, in describing fixed electron orbits. Another limitation of the Bohr model is that it is only accurately describes one-electron species.	5.6 Describe the difference between a continuous spectrum and a line spectrum.	Problems 5.56–5.57
		5.7 Compare the wavelength and frequency of different electron transitions in the Bohr model of the atom.	Problem 5.27
		5.8 Use the Balmer–Rydberg equation to calculate the wavelength and energies of electron transitions in the hydrogen atom.	Worked Example 5.4; Problems 5.58, 5.60, 5.62, 5.64
5.4 Wave-Like Properties of Matter: de Broglie's Hypothesis	Just as light behaves in some respects like a stream of small particles (**photons**), electrons and other tiny units of matter behave in some respects like waves. The wavelength of a particle of mass m traveling at a velocity v is given by the **de Broglie equation**.	5.9 Calculate the wavelength of a moving object using the de Broglie equation.	Worked Example 5.5; Problems 5.66–5.67, 5.71
		5.10 Explain why the wavelength of macroscopic objects cannot be observed.	Problem 5.68–5.69
5.5 The Quantum Mechanical Model of the Atom: Heisenberg's Uncertainty Principle	According to **Heisenberg's uncertainty principle**, we can never know both the position and the velocity of an electron (or of any other object) beyond a certain level of precision.	5.11 Calculate the uncertainty in the position of the moving object if the velocity is known.	Problems 5.72–5.73

Section	Concept Summary	Learning Objectives	Test Your Understanding
5.6 The Quantum Mechanical Model of the Atom: Orbitals and Quantum Numbers	The **quantum mechanical model** proposed in 1926 by Erwin Schrödinger describes an atom by a mathematical equation similar to that used to describe wave motion. The behavior of each electron in an atom is characterized by a **wave function,** or **orbital,** whose square defines the probability of finding the electron in a given volume of space. Each wave function has a set of three parameters called **quantum numbers.** The **principal quantum number** n defines the size of the orbital; the **angular-momentum quantum number** l defines the shape of the orbital; and the **magnetic quantum number** m_l defines the spatial orientation of the orbital.	**5.12** Identify and write sets of quantum numbers that describe electrons in different types of orbitals.	Worked Example 5.6; Problems 5.76, 5.78, 5.80, 5.86
5.7 The Shapes of Orbitals	The square of a wave function ψ^2, or orbital, describes the probability of finding the electron within a specific region of space. The shape of that spatial region is defined by the angular-momentum quantum number l, with $l = 0$ called an s orbital, $l = 1$ a p orbital, $l = 2$ a d orbital, and $l = 3$ an f orbital.	**5.13** Identify an orbital based on its shape and describe it using a set of quantum numbers.	Worked Example 5.7; Problems 5.13, 5.28, 5.93
		5.14 Locate the nodal planes in different types of orbitals and different shells.	Problems 5.90–5.92
5.8 Electron Spin and the Pauli Exclusion Principle	A fourth quantum number m_s, describing electron spin, differentiates the two electrons that may occupy a given orbital. The **Pauli exclusion principle** states that no two electrons in an atom can have the same four quantum numbers. The **spin quantum number** m_s specifies the electron spin as either $+1/2$ or $-1/2$. Electrons that occupy the same orbital must have opposing spins.	**5.15** Assign a set of four quantum numbers for electrons in an atom.	Problems 5.80–5.81 5.86–5.89
5.9 Orbital Energy Levels in Multielectron Atoms	Within a shell, all hydrogen orbitals have the same energy. In multielectron atoms, the s, p, d, and f orbitals within a given shell have slightly different energies. The difference in energy between subshells in multielectron atoms results from the different shapes of the orbitals that lead to differing degrees of electron shielding.	**5.16** Explain how electron shielding gives the order of subshells from lowest to highest in energy.	Problems 5.94–5.97
		5.17 Predict the order of filling of subshells based upon energy.	Problems 5.100–5.103
5.10–5.11 Electron Configurations of Multielectron Atoms	The **ground-state electron configuration** of a multielectron atom is arrived at by following a series of rules called the **aufbau principle.** 1. The lowest-energy orbitals fill first. 2. Only two electrons go into any one **orbital (Pauli exclusion principle),** and these must be of opposite spins. 3. If two or more orbitals are equal in energy (degenerate), each is half filled before any one is completely filled (**Hund's rule**). The aufbau principle usually correctly predicts the electron configuration of elements; however, 21 elements have anomalous configurations due to the unusual stability of both half-filled and fully filled subshells (Figure 5.17).	**5.18** Assign electron configurations to atoms in their ground state.	Worked Example 5.8; Problems 5.104–5.105, 5.109–5.110
5.12 Electron Configurations and the Periodic Table	The periodic table is the most important organizing principle of chemistry. It is successful because elements in each group of the periodic table have similar **valence-shell** electron configurations and therefore have similar properties. The periodic table can be used to predict the order that electrons fill orbitals (Figure 5.26).	**5.19** Draw orbital filling diagrams for the ground state of an atom and determine the number of unpaired electrons.	Worked Example 5.8; Problems 5.15–5.16, 5.106–5.108
		5.20 Identify atoms from orbital filling diagrams or electron configurations.	Problems 5.16, 5.31–5.32
5.13 Electron Configurations and Periodic Properties: Atomic Radii	Atomic radii of elements show a periodic rise-and-fall pattern according to the positions of the elements in the table. Atomic radii increase going down a group because n increases, and they decrease from left to right across a period because the **effective nuclear charge (Z_{eff})** increases.	**5.21** Explain the periodic trend in atomic radii.	Problems 5.118–5.119
		5.22 Predict the relative size of atoms based upon their position in the periodic table.	Worked Example 5.9; Problems 5.120–5.123

KEY TERMS

amplitude 164
angular-momentum quantum
   number (*l*) 177
aufbau principle 187
Balmer–Rydberg equation 172
d-block element 191
de Broglie equation 174
degenerate 187
effective nuclear charge
   (Z_{eff}) 185
electromagnetic spectrum 162
electron configuration 187
f-block element 191

frequency (*ν*) 163
ground-state electron
   configuration 187
Heisenberg uncertainty
   principle 176
hertz (Hz) 164
Hund's rule 187
line spectrum 169
magnetic quantum number
   (m_l) 177
node 181
orbital 177
p-block element 191

Pauli exclusion principle 185
phase 182
photon 166
photoelectric effect 166
Planck's postulate 166
principal quantum
   number (*n*) 177
quantum 168
quantized 168
quantum mechanical
   model 176
quantum number 177
radial distribution plot 179

radiant energy 162
s-block element 191
shell 177
spin quantum number
   (m_s) 185
subshell 177
valence shell 189
wave function 177
wavelength (λ) 164

KEY EQUATIONS

- **The Relationship between Frequency (*ν*) and Wavelength (λ) (Section 5.1)**

$$\lambda = \frac{c}{\nu} \quad \text{or} \quad \nu = \frac{c}{\lambda}$$

where *c* is the speed of light 3.00×10^8 m/s.

- **Planck's Postulate Relating Energy (*E*) and Frequency (*ν*) or Wavelength (λ) (Section 5.2)**

$$E = h\nu = \frac{hc}{\lambda}$$

where *h* is Planck's constant 6.626×10^{-34} J·s.

- **The Balmer–Rydberg Equation to Account for the Line Spectrum of Hydrogen (Section 5.3)**

$$\frac{1}{\lambda} = R_\infty \left[\frac{1}{m^2} - \frac{1}{n^2} \right] \quad \text{or} \quad \nu = R_\infty \cdot c \left[\frac{1}{m^2} - \frac{1}{n^2} \right]$$

where *m* and *n* are integers with $n > m$ and R_∞ is the Rydberg constant.

- **De Broglie's Hypothesis Relating Wavelength (λ), Mass (*m*), and Velocity (*v*) (Section 5.5)**

$$\lambda = \frac{h}{mv}$$

- **Heisenberg's Uncertainty Principle (Section 5.6)**

$$(\Delta x)(\Delta mv) \geq \frac{h}{4\pi}$$

PRACTICE TEST

After studying this chapter, you can assess your understanding with these practice test questions, which are correlated with chapter learning objectives. If you answer a question incorrectly, refer to the learning objectives in the end-of-chapter Study Guide for assistance. The Study Guide provides a conceptual summary, references a Worked Example to model how to solve the problem, and gives additional problems for more practice.

1. Which wave corresponds to higher energy radiation? **(LO 5.1)**

(a)

(b)

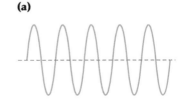

2. What is the frequency of the light wave in Hz? **(LO 5.2)**

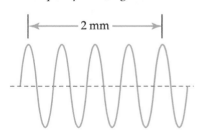

(a) 6.0×10^{11} Hz   (b) 6.0×10^8 Hz
(c) 1.5×10^{11} Hz   (d) 1.5×10^8 Hz

3. What is the energy (in kJ) of one mole of photons of ultraviolet light with a wavelength of 85 nm? **(LO 5.3)**

(a) 1.4×10^{-6} kJ   (b) 1.4×10^3 kJ
(c) 2.4×10^{14} kJ   (d) 2.4×10^{-15} kJ

4. Which type of electromagnetic radiation will cause the greatest number of electrons to be ejected from zinc metal with a work function of 350 kJ/mol? **(LO 5.4, 5.5)**
   (a) Dim light with a wavelength of 320 nm
   (b) Dim light with a wavelength of 360 nm
   (c) Bright light with a wavelength of 360 nm
   (d) Bright light with a wavelength of 375 nm

5. The atomic emission spectrum for hydrogen in the visible region is shown. **(LO 5.6)**

   410 nm 434 nm   486 nm   656 nm

   What can be concluded from observing the four lines?
   (a) The hydrogen molecules have the formula H_4.
   (b) Since there are only four lines in the visible region of the spectrum, there must be more lines in the ultraviolet region of the spectrum.
   (c) There are four electrons in an excited hydrogen atom.
   (d) Only certain energies are allowed for the electron in a hydrogen atom.

6. When a copper salt such as $Cu(NO_3)_2$ is burned in a flame, a blue-green color is emitted. Which figure represents the emission spectrum for the element copper? **(LO 5.6)?**

   (a)
   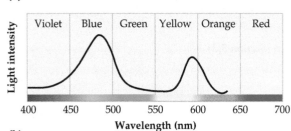

   (b)
   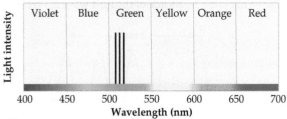

   (c)
   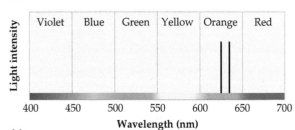

   (d)
   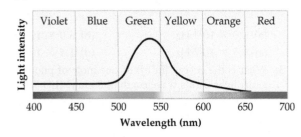

7. Which arrow in the energy diagram for an atom represents the *absorption* of light with the *shortest wavelength*? **(LO 5.7)**

   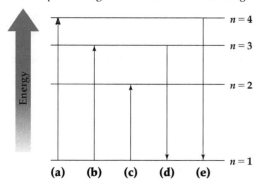
   (a)   (b)   (c)   (d)   (e)

8. Calculate the wavelength in nm of the light emitted when an electron makes a transition from an orbital in $n = 5$ to an orbital in $n = 2$ in the hydrogen atom. **(LO 5.8)**
   (a) 2.31×10^{-3} nm
   (b) 4.34×10^{-2} nm
   (c) 231 nm
   (d) 434 nm

9. What is the wavelength in meters of an electron (mass $\times$ 9.11×10^{-28} g) that has been accelerated to a speed of 2.1×10^7 m/s? **(LO 5.9)**
   (a) 3.5×10^{-14} m
   (b) 2.9×10^{-7} m
   (c) 3.5×10^{-11} m
   (d) 2.9×10^{10} m

10. What are the possible values of n, l, and m_l for an electron in a $5p$ orbital? **(LO 5.12)**

	n	l	m_l
(a)	5	0	0
(b)	1, 2, 3, 4, 5	1, 2	−2, −1, 0, +1, or +2
(c)	5	2	−2, −1, 0, +1, or +2
(d)	5	1	−1, 0, +1

11. What are the possible values of n, l, and m_l for the orbital shown? **(LO 5.13)**

	n	l	m_l
(a)	2	0	0
(b)	3	0	0
(c)	3	1	−1, 0, +1
(d)	4	2	−2, −1, 0, +1, or +2

12. For a multielectron atom, a $3s$ orbital lies lower in energy than a $3p$ orbital because **(LO 5.16)**
    (a) a $3p$ orbital has more nodal surfaces than a $3s$ orbital.
    (b) an electron in a $3p$ orbital has a higher probability of being closer to the nucleus than an electron in a $3s$ orbital.

(c) inner electrons shield electrons in a $3p$ orbital more effectively than electrons in a $3s$ orbital.

(d) the energy of the electron can be spread between three $3p$ orbitals instead of only one $3s$ orbital.

13. Which element has the ground-state electron configuration of $[Ne]3s^2 3p^4$? **(LO 5.18)**

(a) S   (b) Si

(c) Cr   (d) Te

14. Which is the correct orbital filling diagram for the ground-state of Fe? **(LO 5.19)**

(a)

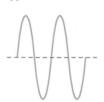

(b)

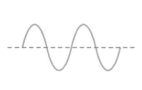

(c)

[Ar]   ↑↓   ↑↓ ↑↓ ↑↓ __ __   __ __ __
         $4s$ $3d$ $4p$

(d)

[Ar]   __   ↑↓ ↑↓ ↑↓ ↑ ↑   __ __ __
         $4s$ $3d$ $4p$

15. Which element has the largest atomic radius? **(LO 5.20)**

(a) Rb   (b) Co

(c) Mg   (d) As

Answers:

1. b, 2. a, 3. b, 4. a, 5. d, 6. c, 7. a, 8. d, 9. c, 10. d, 11. b, 12. c, 13. a, 14. a, 15, a

---

Mastering Chemistry   provides end-of-chapter exercises, feedback-enriched tutorial problems, animations, and interactive activities to encourage problem-solving practice and deeper understanding of key concepts and topics.

RAN   *Randomized in Mastering Chemistry*

CONCEPTUAL PROBLEMS

Problems 5.1–5.25 appear within the chapter.

5.26 Two electromagnetic waves are represented below.

(a) Which wave has the greater intensity?

(b) Which wave corresponds to higher-energy radiation?

(c) Which wave represents yellow light, and which represents infrared radiation?

(i)   **(ii)**

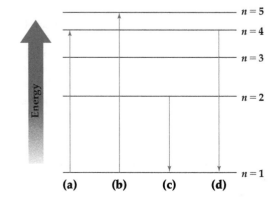

5.27 The following diagram shows the energy levels of the differ-
RAN ent shells in the hydrogen atom.

(a) Which transition corresponds to absorption of light with the longest wavelength?

(b) Which transition corresponds to emission of light with the shortest wavelength?

5.28 Identify each of the following orbitals, and give n and l quantum numbers for each.

(a)   **(b)**

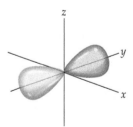   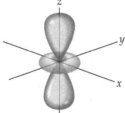

(in third shell)   (in fourth shell)

5.29 Where on the blank outline of the periodic table do elements that meet the following descriptions appear?

(a) Elements with the valence-shell ground-state electron configuration $ns^2\,np^5$

(b) An element with the ground-state electron configuration $[Ar]\,4s^2\,3d^{10}\,4p^5$

(c) Elements with electrons whose largest principal quantum number is $n = 4$

(d) Elements that have only one unpaired *p* electron

(e) The *d*-block elements

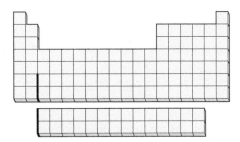

5.30 One of the elements shown on the following periodic table has an anomalous ground-state electron configuration. Which is it—red, blue, or green—and why?

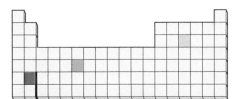

5.31 What atom has the following orbital-filling diagram?

5.32 The following orbital-filling diagram represents an excited state rather than the ground state of an atom. Identify the atom, and give its ground-state electron configuration.

$$[\text{Ar}] \quad \underset{4s}{\uparrow\downarrow} \quad \underset{3d}{\uparrow\downarrow \ \uparrow\downarrow \ \uparrow\downarrow \ \uparrow\downarrow \ \uparrow\downarrow} \quad \underset{4p}{\uparrow\downarrow \ \uparrow\downarrow \ \_\_\_}$$

5.33 Which of the following three spheres represents a Ca atom, which an Sr atom, and which a Br atom?

$r = 215$ pm $r = 114$ pm $r = 197$ pm

SECTION PROBLEMS

Wave Properties of Radiant Energy (Section 5.1)

5.34 Which has the higher frequency, red light or violet light? Which has the longer wavelength? Which has the greater energy?

5.35 Which has the higher frequency, infrared light or ultraviolet light? Which has the longer wavelength? Which has the greater energy?

5.36 The Hubble Space Telescope detects electromagnetic energy in the wavelength range 1.15×10^{-7} m to 2.0×10^{-6} m. What region of the electromagnetic spectrum is found completely within this range? What regions fall partially in this range?

5.37 The Green Bank Telescope in West Virginia—the world's largest steerable radio telescope—detects frequencies from 290 MHz to 90 GHz. What region or regions of the electromagnetic spectrum are found completely or partially within its detection range?

5.38 What is the wavelength in meters of ultraviolet light with RAN $\nu = 5.5 \times 10^{15}$ s^{-1}?

5.39 What is the frequency of a microwave with $\lambda = 4.33 \times$ RAN 10^{-3} m?

5.40 A certain cellular telephone transmits at a frequency of 825 RAN MHz and receives at a frequency of 875 MHz.

(a) What is the wavelength of the transmitted signal in cm?

(b) What is the wavelength of the received signal in cm?

5.41 Optical fibers allow the fast transmission of vast amounts of RAN data. In one type of fiber, the wavelength of transmitted light is 1.3×10^3 nm.

(a) What is the frequency of the light?

(b) Fiber optic cable is available in 12 km lengths. How long will it take for a signal to travel that distance assuming that the speed of light in the cable is the same as in a vacuum?

Particlelike Properties of Radiant Energy (Section 5.2)

5.42 Calculate the energies of the following waves in kilojoules per RAN mole, and tell which member of each pair has the higher value.

(a) An FM radio wave at 99.5 MHz and an AM radio wave at 1150 kHz

(b) An X ray with $\lambda = 3.44 \times 10^{-9}$ m and a microwave with $\lambda = 6.71 \times 10^{-2}$ m

5.43 The magnetic resonance imaging (MRI) body scanners used in hospitals operate with 400 MHz radio frequency energy. How much energy does this correspond to in kilojoules per mole?

5.44 What is the wavelength in meters of photons with the fol- RAN lowing energies? In what region of the electromagnetic spectrum does each appear?

(a) 90.5 kJ/mol   (b) 8.05×10^{-4} kJ/mol

(c) 1.83×10^3 kJ/mol

5.45 What is the energy of each of the following photons in kilojoules per mole?

(a) $\nu = 5.97 \times 10^{19}$ s^{-1} (b) $\nu = 1.26 \times 10^6$ s^{-1}

(c) $\lambda = 2.57 \times 10^2$ m

5.46 You're probably familiar with using Scotch tape for wrapping presents but may not know that it can also generate electromagnetic radiation. When Scotch tape is unrolled in a vacuum (but not in air), photons with a range of frequencies around $\nu = 2.9 \times 10^{18}$ s^{-1} are emitted in nanosecond bursts.

(a) What is the wavelength in meters of photons with $\nu = 2.9 \times 10^{18}$ s^{-1}?

(b) What is the energy in kJ/mol of photons with $\nu = 2.9 \times 10^{18}$ s^{-1}?

(c) What type of electromagnetic radiation are these photons?

5.47 Hard wintergreen-flavored candies are *triboluminescent,* meaning that they emit flashes of light when crushed. (You can see it for yourself if you look in a mirror while crunching a wintergreen Life Saver in your mouth in a dark room.) The strongest emission is around $\lambda = 450$ nm.

(a) What is the frequency in s^{-1} of photons with $\lambda = 450$ nm?

(b) What is the energy in kJ/mol of photons with $\lambda = 450$ nm?

(c) What is the color of the light with $\lambda = 450$ nm?

5.48 The *second* in the SI system is defined as the duration of 9,192,631,770 periods of radiation corresponding to the transition between two energy levels of a cesium-133 atom. What is the energy difference between the two levels in kilojoules per mole?

5.49 Photochromic sunglasses, which darken when exposed to light, contain a small amount of colorless AgCl embedded in the glass. When irradiated with light, metallic silver atoms are produced and the glass darkens: $AgCl \longrightarrow Ag + Cl$. Escape of the chlorine atoms is prevented by the rigid structure of the glass, and the reaction therefore reverses as soon as the light is removed. If 310 kJ/mol of energy is required to make the reaction proceed, what wavelength of light is necessary?

5.50 The data encoded on CDs, DVDs, and Blu-ray discs is read by lasers. What is the wavelength in nanometers and the energy in joules of the following lasers?

(a) CD laser, $\nu = 3.85 \times 10^{14}$ s^{-1}

(b) DVD laser, $\nu = 4.62 \times 10^{14}$ s^{-1}

(c) Blu-ray laser, $\nu = 7.41 \times 10^{14}$ s^{-1}

5.51 The semimetal germanium is used as a component in photodetectors, which generate electric current when exposed to light. If a germanium photodetector responds to photons in the range $\lambda = 400 - 1700$ nm, will the following light sources be detected?

(a) A laser with $\nu = 4.35 \times 10^{14}$ s^{-1}

(b) Photons with $E = 43$ kJ/mol

(c) Electromagnetic radiation with $\nu = 706$ THz

5.52 The work function of cesium metal is 188 kJ/mol, which
RAN corresponds to light with a wavelength of 637 nm. Which of the following will cause the smallest number of electrons to be ejected from cesium?

(a) High-amplitude wave with a wavelength of 500 nm

(b) Low-amplitude wave with a wavelength of 500 nm

(c) High-amplitude wave with a wavelength of 650 nm

(d) Low-amplitude wave with a wavelength of 650 nm

5.53 The work function of calcium metal is kJ/mol, which corre-
RAN sponds to light with a wavelength of 432 nm. Which of the following will cause the largest number of electrons to be ejected from cesium?

(a) High-amplitude wave with a wavelength of 400 nm

(b) Low-amplitude wave with a wavelength of 400 nm

(c) High-amplitude wave with a wavelength of 450 nm

(d) Low-amplitude wave with a wavelength of 450 nm

5.54 The work function of silver metal is 436 kJ/mol. What frequency
RAN of light is needed to eject electrons from a sample of silver?

5.55 Cesium metal is frequently used in photoelectric cells because the amount of energy necessary to eject electrons from a cesium surface is relatively small—only 206.5 kJ/mol. What wavelength of light in nanometers does this correspond to?

Atomic Line Spectra and Quantized Energy (Section 5.3)

5.56 Spectroscopy is a technique that uses the interaction of radiant energy with matter to identify or quantify a substance in a sample. A deuterium lamp is often used a light source in the ultraviolet region of the spectrum and the emission spectrum is shown. Is this a continuous or line emission spectrum?

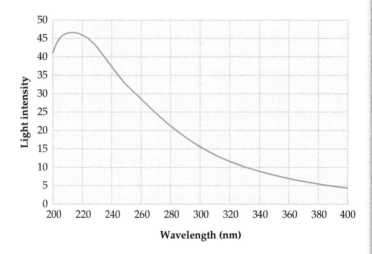

5.57 Sodium-vapor lamps are a common source of lighting. The emission spectrum from this type of lamp is shown. Is this a continuous or line emission spectrum?

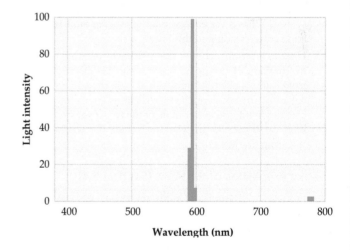

5.58 According to the equation for the Balmer line spectrum of hydrogen, a value of $n = 3$ gives a red spectral line at 656.3 nm, a value of $n = 4$ gives a green line at 486.1 nm, and a value of $n = 5$ gives a blue line at 434.0 nm. Calculate the energy in kilojoules per mole of the radiation corresponding to each of these spectral lines.

5.59 According to the equation for the Balmer line spectrum of
RAN hydrogen, a value of $n = 4$ gives a red spectral line at 486.1 nm. Calculate the energy in kilojoules per mole of the radiation corresponding to this spectral line.

5.60 Calculate the wavelength and energy in kilojoules necessary
RAN to completely remove an electron from the first shell ($m = 1$) of a hydrogen atom ($R_\infty = 1.097 \times 10^{-2}$ nm^{-1}).

5.61 Calculate the wavelength and energy in kilojoules necessary to
RAN completely remove an electron from the second shell ($m = 2$)
of a hydrogen atom ($R_\infty = 1.097 \times 10^{-2}$ nm^{-1}).

5.62 One series of lines of the hydrogen spectrum is caused by emission of energy accompanying the fall of an electron from outer shells to the fourth shell. The lines can be calculated using the Balmer–Rydberg equation:

$$\frac{1}{\lambda} = R_\infty \left[\frac{1}{m^2} - \frac{1}{n^2} \right]$$

where $m = 4$, $R_\infty = 1.097 \times 10^{-2}$ nm^{-1}, and n is an integer greater than 4. Calculate the wavelengths in nanometers and energies in kilojoules per mole of the first two lines in the series. In what region of the electromagnetic spectrum do they fall?

5.63 One series of lines of the hydrogen spectrum is caused by
RAN emission of energy accompanying the fall of an electron from outer shells to the third shell. The lines can be calculated using the Balmer–Rydberg equation:

$$\frac{1}{\lambda} = R_\infty \left[\frac{1}{m^2} - \frac{1}{n^2} \right]$$

where $m = 3$, $R_\infty = 1.097 \times 10^{-2}$ nm^{-1}, and n is an integer greater than 3. Calculate the wavelengths in nanometers and energies in kilojoules per mole of the first two lines in the series. In what region of the electromagnetic spectrum do they fall?

5.64 Use the Balmer equation to calculate the wavelength in nano-
RAN meters of the spectral line for hydrogen when $n = 6$ and $m = 2$. What is the energy in kilojoules per mole of the radiation corresponding to this line?

5.65 Lines in a certain series of the hydrogen spectrum are caused by emission of energy accompanying the fall of an electron from outer shells to the fifth shell. Use the Balmer–Rydberg equation to calculate the wavelengths in nanometers and energies in kilojoules per mole of the two longest-wavelength lines in the series. In what region of the electromagnetic spectrum do they fall?

Wavelike Properties of Particles (Section 5.4)

5.66 Protons and electrons can be given very high energies in particle accelerators. What is the wavelength in meters of an electron (mass $= 9.11 \times 10^{-31}$ kg) that has been accelerated to 5% of the speed of light? In what region of the electromagnetic spectrum is this wavelength?

5.67 What is the wavelength in meters of a proton (mass $= 1.673 \times$
RAN 10^{-24} g) that has been accelerated to 1% of the speed of light? In what region of the electromagnetic spectrum is this wavelength?

5.68 What is the de Broglie wavelength in meters of a baseball
RAN weighing 145 g and traveling at 156km/h? Why do we not observe this wavelength?

5.69 What is the de Broglie wavelength in meters of a mosquito
RAN weighing 1.55 mg and flying at 1.38 m/s? Why do we not observe this wavelength?

5.70 At what speed in meters per second must a 145 g baseball be
RAN traveling to have a de Broglie wavelength of 0.500 nm?

5.71 What velocity would an electron (mass $= 9.11 \times 10^{-31}$ kg)
RAN need for its de Broglie wavelength to be that of red light (750 nm)?

Orbitals and Quantum Mechanics (Sections 5.5–5.8)

5.72 Use the Heisenberg uncertainty principle to calculate the
RAN uncertainty in meters in the position of a honeybee weighing 0.68 g and traveling at a velocity of 0.85 m/s. Assume that the uncertainty in the velocity is 0.1 m/s.

5.73 The mass of a helium atom is 4.0026 amu, and its average
RAN velocity at 25 °C is 1.36×10^3 m/s. What is the uncertainty in meters in the position of a helium atom if the uncertainty in its velocity is 1%?

5.74 What is the Heisenberg uncertainty principle, and how does it affect our description of atomic structure?

5.75 Why do we have to use an arbitrary value such as 90% to determine the spatial limitations of an orbital?

5.76 What are the four quantum numbers, and what does each specify?

5.77 Tell which of the following combinations of quantum numbers are not allowed. Explain your answers.
(a) $n = 3, l = 0, m_l = -1$ (b) $n = 3, l = 1, m_l = 1$
(c) $n = 4, l = 4, m_l = 0$

5.78 Give the allowable combinations of quantum numbers for
RAN each of the following electrons.
(a) A 4s electron   (b) A 3p electron
(c) A 5f electron   (d) A 5d electron

5.79 Give the orbital designations of electrons with the following quantum numbers.
(a) $n = 3, l = 0, m_l = 0$ (b) $n = 2, l = 1, m_l = -1$
(c) $n = 4, l = 3, m_l = -2$ (d) $n = 4, l = 2, m_l = 0$

5.80 Which of the following combinations of quantum numbers can refer to an electron in a ground-state cobalt atom ($Z = 27$)?
(a) $n = 3, l = 0, m_l = 2$ (b) $n = 4, l = 2, m_l = -2$
(c) $n = 3, l = 1, m_l = 0$

5.81 Which of the following combinations of quantum numbers
RAN can refer to an electron in a ground-state selenium atom ($Z = 34$)?
(a) $n = 3, l = 3, m_l = 2$ (b) $n = 4, l = 2, m_l = -2$
(c) $n = 4, l = 1, m_l = 0$

5.82 What is the maximum number of electrons in an atom whose highest-energy electrons have the principal quantum number $n = 5$?

5.83 What is the maximum number of electrons in an atom whose highest-energy electrons have the principal quantum number $n = 4$ and the angular-momentum quantum number $l = 0$?

5.84 Sodium atoms emit light with a wavelength of 330 nm when an electron moves from a 4p orbital to a 3s orbital. What is the energy difference between the orbitals in kilojoules per mole?

5.85 Excited rubidium atoms emit red light with $\lambda = 795$ nm. What is the energy difference in kilojoules per mole between orbitals that give rise to this emission?

5.86 Assign a set of four quantum numbers to each electron in carbon.

5.87 Assign a set of four quantum numbers to each electron in oxygen.

5.88 Assign a set of four quantum numbers for the outermost
RAN two electrons in Sr.

5.89 Which of the following is a valid set of four quantum num-
RAN bers for a *d* electron in Mo?
  (a) $n = 4, l = 1, m_l = 0, m_s = +1/2$
  (b) $n = 5, l = 2, m_l = 0, m_s = -1/2$
  (c) $n = 4, l = 2, m_l = -1, m_s = +1/2$
  (d) $n = 5, l = 2, m_l = -3, m_s = +1/2$

5.90 How many nodal surfaces does a 4*s* orbital have? Draw a
RAN cutaway representation of a 4*s* orbital showing the nodes
  and the regions of maximum electron probability.

5.91 How does the number of nodal planes change as the *l* quantum number of the subshell changes?

5.92 What do the different colors in the two lobes of the *p* orbitals specify?

5.93 What is the *l* quantum of an orbital that has four lobes that
  resemble a cloverleaf?

Orbital Energy Levels in Multielectron Atoms (Section 5.9)

5.94 What is meant by the term *effective nuclear charge, Z_{eff},* and
  what causes it?

5.95 How does electron shielding in multielectron atoms give rise
  to energy differences among 3*s*, 3*p*, and 3*d* orbitals?

5.96 Order the electrons in the following orbitals according to
  their shielding ability: 4*s*, 4*d*, 4*f*,

5.97 Order the following elements according to increasing Z_{eff}:
  Ca, Se, Kr, K.

5.98 Why does the number of elements in successive periods of
  the periodic table increase by the progression 2, 8, 18, 32?

5.99 Which two of the four quantum numbers determine the energy
  level of an orbital in a multielectron atom?

5.100 Which orbital in each of the following pairs is higher in
  energy?
  (a) 5*p* or 5*d*   (b) 4*s* or 3*p*
  (c) 6*s* or 4*d*

5.101 Order the orbitals for a multielectron atom in each of the
  following lists according to increasing energy.
  (a) 4*d*, 3*p*, 2*p*, 5*s*   (b) 2*s*, 4*s*, 3*d*, 4*p*
  (c) 6*s*, 5*p*, 3*d*, 4*p*

5.102 According to the aufbau principle, which orbital is filled
  immediately *after* each of the following in a multielectron
  atom?
  (a) 4*s*   (b) 3*d*   (c) 5*f*   (d) 5*p*

5.103 According to the aufbau principle, which orbital is filled
  immediately *before* each of the following?
  (a) 3*p*   (b) 4*p*   (c) 4*f*   (d) 5*d*

Electron Configurations (Sections 5.10–5.12)

5.104 Give the expected ground-state electron configurations for
  the following elements.
  (a) Ti   (b) Ru   (c) Sn
  (d) Sr   (e) Se

5.105 Give the expected ground-state electron configurations for
  atoms with the following atomic numbers.
  (a) Z = 55   (b) Z = 40
  (c) Z = 80   (d) Z = 62

5.106 Draw orbital-filling diagrams for the following atoms. Show
  each electron as an up or down arrow, and use the abbreviation of the preceding noble gas to represent inner-shell
  electrons.
  (a) Rb   (b) W   (c) Ge   (d) Zr

5.107 Draw orbital-filling diagrams for atoms with the following atomic numbers. Show each electron as an up or down
  arrow, and use the abbreviation of the preceding noble gas
  to represent inner-shell electrons.
  (a) Z = 25   (b) Z = 56
  (c) Z = 28   (d) Z = 47

5.108 How many unpaired electrons are present in each of the following ground-state atoms?
  (a) O   (b) Si   (c) K   (d) As

5.109 Identify the following atoms.
  (a) It has the ground-state electron configuration [Ar] $4s^2$
    $3d^{10} 4p^1$.
  (b) It has the ground-state electron configuration [Kr] $4d^{10}$.

5.110 Write the symbol, give the ground-state electron configuration, and draw an orbital-filling diagram for each of the following atoms. Use the abbreviation of the preceding noble
  gas to represent the inner-shell electrons.
  (a) The heaviest alkaline earth metal
  (b) The lightest transition metal
  (c) The heaviest actinide metal
  (d) The lightest semimetal
  (e) The group 6A element in the fifth period

5.111 Given the subshells 1*s*, 2*s*, 2*p*, 3*s*, 3*p* and 3*d*, identify those
  that meet the following descriptions.
  (a) Has $l = 2$
  (b) Can have $m_l = -1$
  (c) Is empty in a nitrogen atom
  (d) Is full in a carbon atom
  (e) Contains the outermost electrons in a beryllium atom
  (f) Can contain two electrons, both with spin $m_s = +1/2$

5.112 At what atomic number is the filling of a *g* orbital likely to
  begin?

5.113 Assuming that *g* orbitals fill according to Hund's rule, what
  is the atomic number of the first element to have a filled *g*
  orbital?

5.114 Take a guess. What do you think is a likely ground-state
RAN electron configuration for the sodium *ion*, Na^+, formed by
  loss of an electron from a neutral sodium atom?

5.115 Take a guess. What is a likely ground-state electron configu-
RAN ration for the chloride ion, Cl^-, formed by adding an electron to a neutral chlorine atom?

5.116 What is the expected ground-state electron configuration of
  the recently discovered element with $Z = 116$?

5.117 What is the atomic number and expected ground-state elec-
RAN tron configuration of the yet undiscovered element directly
  below Fr in the periodic table?

Electron Configurations and Periodic Properties (Section 5.13)

5.118 Why do atomic radii increase going down a group of the periodic table?

5.119 Why do atomic radii decrease from left to right across a period of the periodic table?

5.120 Order the following atoms according to increasing atomic radius: S, F, O.

5.121 Order the following atoms according to increasing atomic
RAN   radius: Rb, Cl, As, K.

5.122 Which atom in each of the following pairs has a larger radius?

(a) Na or K   (b) V or Ta
(c) V or Zn   (d) Li or Ba

5.123 Which atom in each of the following pairs has a larger
RAN   radius?

(a) C or Ge   (b) Ni or Pt
(c) Sn or I   (d) Na or Rb

MULTICONCEPT PROBLEMS

5.124 Orbital energies in single-electron atoms or ions, such as He$^+$,
RAN   can be described with an equation similar to the Balmer–Rydberg equation:

$$\frac{1}{\lambda} = Z^2 R \left[\frac{1}{m^2} - \frac{1}{n^2} \right]$$

where Z is the atomic number. What wavelength of light in nanometers is emitted when the electron in He$^+$ falls from $n = 3$ to $n = 2$?

5.125 Like He$^+$, the Li^{2+} ion is a single-electron system (Problem
RAN   5.124). What wavelength of light in nanometers must be absorbed to promote the electron in Li^{2+} from $n = 1$ to $n = 4$?

5.126 Imagine a universe in which the four quantum numbers can
RAN   have the same possible values as in our universe except that the angular-momentum quantum number l can have integral values of $0, 1, 2, \ldots, n + 1$ (instead of $0, 1, 2, \ldots, n - 1$).

(a) How many elements would be in the first two rows of the periodic table in this universe?

(b) What would be the atomic number of the element in the second row and fifth column?

(c) Draw an orbital-filling diagram for the element with atomic number 12.

5.127 Draw orbital-filling diagrams for the following atoms. Show each electron as an up or down arrow, and use the abbreviation of the preceding noble gas to represent inner-shell electrons.

(a) Sr   (b) Cd
(c) Has $Z = 22$ (d) Has $Z = 34$

5.128 The atomic radii of Y (180 pm) and La (187 pm) are significantly different, but the radii of Zr (160 pm) and Hf (159 pm) are essentially identical. Explain.

5.129 One method for calculating Z_{eff} is to use the equation

$$Z_{eff} = \sqrt{\frac{(E)(n^2)}{1312 \text{ kJ/mol}}}$$

where E is the energy necessary to remove an electron from an atom and n is the principal quantum number of the electron. Use this equation to calculate Z_{eff} values for the highest-energy electrons in potassium ($E = 418.8$ kJ/mol) and krypton ($E = 1350.7$ kJ/mol).

5.130 One watt (W) is equal to 1 J/s. Assuming that 5.0% of the
RAN   energy output of a 75 W light bulb is visible light and that the average wavelength of the light is 550 nm, how many photons are emitted by the light bulb each second?

5.131 Microwave ovens work by irradiating food with microwave
RAN   radiation, which is absorbed and converted into heat. Assuming that radiation with $\lambda = 15.0$ cm is used, that all the energy is converted to heat, and that 4.184 J is needed to raise the temperature of 1.00 g of water by 1.00 °C, how many photons are necessary to raise the temperature of a 350 mL cup of water from 20 °C to 95 °C?

5.132 The amount of energy necessary to remove an electron from an atom is a quantity called the *ionization energy, E_i.* This energy can be measured by a technique called *photoelectron spectroscopy,* in which light of wavelength λ is directed at an atom, causing an electron to be ejected. The kinetic energy of the ejected electron (E_k) is measured by determining its velocity, v ($E_k = mv^2/2$), and E_i is then calculated using the conservation of energy principle. That is, the energy of the incident light equals E_i plus E_k. What is the ionization energy of selenium atoms in kilojoules per mole if light with $\lambda = 48.2$ nm produces electrons with a velocity of 2.371×10^6 m/s? The mass, m, of an electron is 9.109×10^{-31} kg.

5.133 X rays with a wavelength of 1.54×10^{-10} m are produced when a copper metal target is bombarded with high-energy electrons that have been accelerated by a voltage difference of 30,000 V. The kinetic energy of the electrons equals the product of the voltage difference and the electronic charge in coulombs, where 1 volt-coulomb = 1 J.

(a) What is the kinetic energy in joules and the de Broglie wavelength in meters of an electron that has been accelerated by a voltage difference of 30,000 V?

(b) What is the energy in joules of the X rays emitted by the copper target?

5.134 In the Bohr model of atomic structure, electrons are constrained to orbit a nucleus at specific distances, given by the equation

$$r = \frac{n^2 a_0}{Z}$$

where r is the radius of the orbit, Z is the charge on the nucleus, a_0 is the *Bohr radius* and has a value of 5.292×10^{-11} m, and n is a positive integer ($n = 1, 2, 3, \ldots$) like a principal quantum number. Furthermore, Bohr concluded that the energy level E of an electron in a given orbit is

$$E = \frac{-Ze^2}{2r}$$

where e is the charge on an electron. Derive an equation that will let you calculate the difference ΔE between any two energy levels. What relation does your equation have to the Balmer–Rydberg equation?

5.135 Assume that the rules for quantum numbers are different and
RAN   that the spin quantum number m_s can have any of three values,
$m_s = -1/2, 0, +1/2$, while all other rules remain the same.

(a) Draw an orbital-filling diagram for the element with
$Z = 25$, showing the individual electrons in the outermost subshell as up arrows, down arrows, or 0. How
many partially filled orbitals does the element have?

(b) What is the atomic number of the element in the third
column of the fourth row under these new rules? What
block does it belong to (s, p, d, or f)?

5.136 A minimum energy of 7.21×10^{-19} J is required to produce
the photoelectric effect in chromium metal.

(a) What is the minimum frequency of light needed to remove
an electron from chromium?

(b) Light with a wavelength of 2.50×10^{-7} m falls on a
piece of chromium in an evacuated glass tube. What is
the minimum de Broglie wavelength of the emitted electrons? (Note that the energy of the incident light must be
conserved; that is, the photon's energy must equal the sum
of the energy needed to eject the electron plus the kinetic
energy of the electron.)

5.137 A photon produced by an X-ray machine has an energy of
RAN 4.70×10^{-16} J.

(a) What is the frequency of the photon?

(b) What is the wavelength of radiation of frequency (a)?

(c) What is the velocity of an electron with a de Broglie
wavelength equal to (b)?

(d) What is the kinetic energy of an electron traveling at
velocity (c)?

5.138 An energetically excited hydrogen atom has its electron in a
RAN $5f$ subshell. The electron drops down to the $3d$ subshell,
releasing a photon in the process.

(a) Give the n and l quantum numbers for both subshells,
and give the range of possible m_l quantum numbers.

(b) What wavelength of light is emitted by the process?

(c) The hydrogen atom now has a single electron in the
$3d$ subshell. What is the energy in kJ/mol required to
remove this electron?

5.139 Consider the noble gas xenon.

(a) Write the electron configuration of xenon using the abbreviation of the previous noble gas.

(b) When xenon absorbs 801 kJ/mol of energy, it is excited
into a higher-energy state in which the outermost electron has been promoted to the next available subshell.
Write the electron configuration for this excited xenon.

(c) The energy required to completely remove the outermost
electron from the excited xenon atom is 369 kJ/mol,
almost identical to that of cesium (376 kJ/mol). Explain.

chapter 6

Ionic Compounds: Periodic Trends and Bonding Theory

Contents

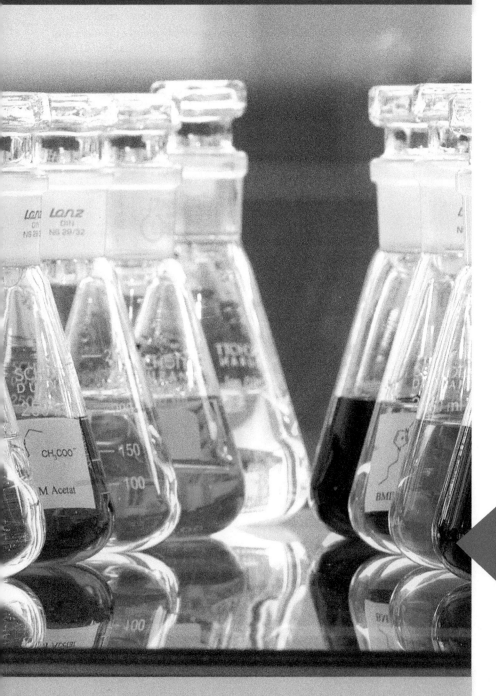

Most ionic compounds, such as table salt, are crystalline solids. Surprisingly, the ionic compound in this flask is a liquid used as a solvent in chemical processes.

How do ionic liquids lead to more environmentally friendly processes?

The answer to this question can be found on page 228 in the INQUIRY ?

Having described the electronic structure of isolated atoms in Chapter 5, let's now extend that description to atoms in chemical compounds. What is the force that holds atoms together in chemical compounds? Certainly there must be *some* force holding atoms together; otherwise, they would simply fly apart and no chemical compounds could exist. As we saw in Section 2.11, the forces that hold atoms together are called *chemical bonds* and are of two types: covalent bonds and ionic bonds. As a general rule, ionic bonds form primarily between a metal atom and a nonmetal atom, while covalent bonds form primarily between two nonmetal atoms. Electrostatic forces, the attraction of positive and negative charges, are the fundamental reason that atoms come together to form a bond, but the types of charges in ionic and covalent bonds are different. In Chapters 6, 7, and 8, we'll look at the nature of chemical bonds and at the energy changes that accompany their formation and breakage. We'll begin in the present chapter with a look at ions and the formation of ionic bonds.

6.1 ELECTRON CONFIGURATIONS OF IONS

We learned in Sections 4.10 and 4.11 during the discussion of redox reactions, that metals (left side of the periodic table) tend to give up electrons in their chemical reactions and form cations. Conversely, halogens and some other nonmetals (right side of the table) tend to accept electrons in their chemical reactions and form anions. What are the **ground-state electron configurations** of the resultant ions?

Ions of main-group elements (groups 1A–7A) lose or gain electrons to achieve a noble-gas configuration. Atoms or ions with the same electron configuration are **isoelectronic;** thus main group ions are isoelectronic to noble gases. The **aufbau principle** (Section 5.10) applies to the formation of ionic compounds: Electrons given up by a metal in forming a cation come from the highest-energy occupied orbital, while the electrons that are accepted by a nonmetal in forming an anion go into the lowest-energy unoccupied orbital. When a sodium atom $(1s^2\ 2s^2\ 2p^6\ 3s^1)$ reacts with a chlorine atom and gives up an electron, for example, the valence-shell $3s$ electron of sodium is lost, giving an Na^+ ion with the noble-gas electron configuration of neon $(1s^2\ 2s^2\ 2p^6)$. At the same time, when the chlorine atom $(1s^2\ 2s^2\ 2p^6\ 3s^2\ 3p^5)$ accepts an electron from sodium, the electron fills the remaining vacancy in the $3p$ subshell to give a Cl^- ion with the noble-gas electron configuration of argon $(1s^2\ 2s^2\ 2p^6\ 3s^2\ 3p^6)$.

$$Na: 1s^2\ 2s^2\ 2p^6 3s^1 \xrightarrow{-e^-} Na^+: 1s^2 2s^2 2p^6 \text{ or } [Ne]$$

$$Cl: 1s^2\ 2s^2\ 2p^6\ 3s^2\ 3p^5 \xrightarrow{+e^-} Cl^-: 1s^2\ 2s^2\ 2p^6\ 3s^2\ 3p^6 \text{ or } [Ar]$$

What is true for sodium is also true for the other elements in group 1A: All form positive ions by losing their valence-shell s electron when they undergo reaction, and all the resultant ions have noble-gas electron configurations. Similarly for the elements in group 2A: All form a doubly positive ion when they react, losing both their valence-shell s electrons. An Mg atom $(1s^2\ 2s^2\ 2p^6\ 3s^2)$, for example, goes to an Mg^{2+} ion with the neon configuration $1s^2\ 2s^2\ 2p^6$ by loss of its two $3s$ electrons.

$$\text{Group 1A atom: } [\text{Noble gas}]\ ns^1 \xrightarrow{-e^-} \text{Group 1A ion}^+: [\text{Noble gas}]$$

$$\text{Group 2A atom: } [\text{Noble gas}]\ ns^2 \xrightarrow{-2e^-} \text{Group 2A ion}^{2+}: [\text{Noble gas}]$$

Just as the group 1A and group 2A metals *lose* the appropriate number of electrons to yield ions with noble-gas configurations, the group 6A and group 7A nonmetals *gain* the appropriate number of electrons when they react with metals. The halogens in group 7A gain one electron to form singly charged anions with noble-gas configurations, and the elements in group 6A gain two electrons to form doubly charged anions with noble-gas configurations. Oxygen $(1s^2\ 2s^2\ 2p^4)$, for example, becomes the O^{2-} ion with the neon configuration $(1s^2\ 2s^2\ 2p^6)$ when it reacts with a metal:

$$\text{Group 6A atom: } [\text{Noble gas}]\ ns^2\ np^4 \xrightarrow{+2e^-} \text{Group 6A ion}^{2-}: [\text{Noble gas}]\ ns^2\ np^6$$

$$\text{Group 7A atom: } [\text{Noble gas}]\ ns^2\ np^5 \xrightarrow{+e^-} \text{Group 7A ion}^-: [\text{Noble gas}]\ ns^2\ np^6$$

REMEMBER . . .
The **ground-state electron configuration** of an atom or ion is a description of the atomic orbitals that are occupied in the lowest-energy state of the atom or ion (Section 5.10).

REMEMBER . . .
According to the **aufbau principle,** lower-energy orbitals fill before higher-energy ones and an orbital can hold only two electrons, which have opposite spins (Section 5.10).

BIG IDEA Question 1

How many electrons is nitrogen likely to gain or lose when it forms an ion? Hint: Write the electron configuration for nitrogen $(Z = 7)$.

TABLE 6.1   Some Common Main-Group Ions and Their Noble-Gas Electron Configurations

Group 1A	Group 2A	Group 3A	Group 6A	Group 7A	Electron Configuration
H^+					[None]
H^-					[He]
Li^+	Be^{2+}				[He]
Na^+	Mg^{2+}	Al^{3+}	O^{2-}	F^-	[Ne]
K^+	Ca^{2+}	*Ga^{3+}	S^{2-}	Cl^-	[Ar]
Rb^+	Sr^{2+}	*In^{3+}	Se^{2-}	Br^-	[Kr]
Cs^+	Ba^{2+}	*Tl^{3+}	Te^{2-}	I^-	[Xe]

*These ions don't have a true noble-gas electron configuration because they have an additional filled *d* subshell.

The formulas and electron configurations of the most common main-group ions are listed in **TABLE 6.1**.

The situation is a bit different for ion formation from the transition-metal elements than it is for the main-group elements. Transition metals react with nonmetals to form cations by first losing their valence-shell *s* electrons and then losing one or more *d* electrons. As a result, all the remaining valence electrons in transition-metal cations occupy *d* orbitals. Iron, for instance, forms the Fe^{2+} ion by losing its two 4*s* electrons and forms the Fe^{3+} ion by losing two 4*s* electrons and one 3*d* electron:

$$\textbf{Fe: } [Ar]\, 4s^2\, 3d^6 \xrightarrow{-2e^-} Fe^{2+} : [Ar]\, 3d^6$$

$$\textbf{Fe: } [Ar]\, 4s^2\, 3d^6 \xrightarrow{-3e^-} Fe^{3+} : [Ar]\, 3d^5$$

It may seem strange that in writing electron configurations, the 3*d* electrons are added *after* the 4*s* electrons, whereas ion formation from a transition metal removes the 4*s* electrons *before* the 3*d* electrons. According to the aufbau principle, lower-energy orbitals fill before higher-energy orbitals. The process for writing electron configurations outlined in Section 5.12 fills the 4*s* orbital before the 3*d* orbital, implying the 4*s* orbital is lower in energy. When forming ions, electrons are generally removed from the higher-energy orbital first, which implies the 4*s* orbital is higher in energy than the 3*d* orbital for the transition elements. Let's examine the energies of the 4*s* and 3*d* orbitals as a function of atomic number (**FIGURE 6.1**) to explain the apparent contradiction in orbital energies.

Figure 6.1 shows that for the fourth-period elements potassium and calcium ($Z = 19$ and 20), the 4*s* orbitals are lower in energy than the 3*d* orbitals and fill first when writing electron configurations.

Orbital energies for K and Ca: $4s < 3d < 4p$

K: $[Ar]4s^1$ **Ca:** $[Ar]4s^2$

However, starting with the transition metal scandium ($Z = 21$), the 3*d* orbitals are lower in energy than the 4*s* orbitals due to increased nuclear attraction and imperfect shielding by the inner-core electrons.

Orbital energies for transition elements ($Z \geq 21$): $3d < 4s < 4p$

If electrons fill the lowest-energy orbital first, we predict that the 3*d* orbitals fill before the 4*s* orbital. Let's examine the process of adding electrons to the Sc^{3+} ion to form the neutral atom to show this is the case. Sc^{3+} is isoelectronic to Ar, and when an electron is added to form Sc^{2+}, it goes into a 3*d* orbital, giving the electron configuration $[Ar]3d^1$. Surprisingly, the next two electrons added to form Sc^+ and the neutral Sc atom go into the 4*s* orbital. The reason is that the 3*d* orbitals are compact compared

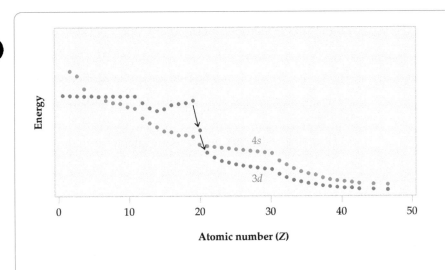

◀ **FIGURE 6.1**

The energy of 4s and 3d orbitals in multielectron atoms. The 4s orbital is lower in energy than the 3d orbital from $Z = 4$ to $Z = 20$. Starting at the transition metal elements ($Z \geq 21$), the 4s orbital is higher in energy than the 3d orbital. When forming transition metal ions, electrons are removed from the 4s orbital before the 3d orbital because it is higher in energy.

◀ **Figure It out**

What is the atomic number at which the 4s orbitals become higher in energy than the 3d orbitals?

Answer: Atomic number 21, the start of the transition metal series.

to the 4s orbital and electron–electron repulsions caused by adding another electron to a 3d orbital outweigh the increase in energy of the 4s orbital.

$$Sc^{3+}: [Ar] \xrightarrow{+e^-} Sc^{2+}: [Ar]3d^1$$

$$Sc^{2+}: [Ar]3d^1 \xrightarrow{+e^-} Sc^+: [Ar]4s^13d^1$$

$$Sc^+: [Ar]4s^13d^1 \xrightarrow{+e^-} Sc: [Ar]4s^23d^1$$

Similarly, electrons add to two of the 3d orbitals before the 4s orbital when added to the Ti^{4+} ion.

$$Ti^{4+}: [Ar] \xrightarrow{+e^-} Ti^{3+}: [Ar]3d^1$$

$$Ti^{3+}: [Ar]3d^1 \xrightarrow{+e^-} Ti^{2+}: [Ar]3d^2$$

$$Ti^{2+}: [Ar]3d^2 \xrightarrow{+e^-} Ti^+: [Ar]4s^13d^2$$

$$Ti^+: [Ar]4s^13d^2 \xrightarrow{+e^-} Ti: [Ar]4s^23d^2$$

Notice that the electron configurations for neutral Ti and Sc atoms are the same as those predicted by filling the 4s orbital first. Therefore, using the periodic table to write electron configurations as described in Section 5.12 is a method that gives the correct electron configuration. However, the 4s orbital is higher in energy than the 3d orbitals for the transition metal elements, which is why electrons are removed from the 4s orbital before the 3d orbital when writing electron configurations for transition metal ions. A similar shift in energies for the 5s and 4d orbitals occurs at the element yttrium (Y) ($Z = 39$). For fifth-period transition metal elements, the 5s orbital is higher in energy than the 4d orbital, and 5s electrons are removed first when forming cations.

WORKED EXAMPLE 6.1

Writing Ground-State Electron Configurations for Ions

Predict the ground-state electron configuration for each of the following ions.

(a) Se^{2-} (b) Cs^+ (c) Cr^{3+}

STRATEGY

First write the electron configuration for the atom. To account for charge of the ion, add electrons for anions and remove electrons for cations. Remember transition metal cations lose outer shell s electrons before d electrons.

SOLUTION

(a) The valence electron configuration for Se is $[Ar] 4s^2 4p^4$. Add two electrons to form the Se^{2-} ion:

$$Se: [Ar] 4s^2 4p^4 \xrightarrow{+2e^-} Se^{2-}: [Ar] 4s^2 4p^6 \text{ or } [Kr]$$

continued on next page

(b) The electron configuration for Cs is $[Xe]\,6s^1$. Remove one electron to form the Cs^+ ion:

$$Cs: [Xe]\,6s^1 \xrightarrow{-e^-} Cs^+: [Kr]\,5s^2\,5p^6 \text{ or } [Xe]$$

(c) The electron configuration for Cr is $[Ar]\,4s^1\,3d^5$. Three electrons must be removed to form Cr^{3+}. First remove one electron from the 4s orbital and then two electrons from the 3d orbitals to form the Cr^{3+} ion:

$$Cr: [Ar]\,4s^1\,3d^5 \xrightarrow{-3e^-} Cr^{3+}: [Ar]\,3d^3$$

▶ **PRACTICE 6.1**   Predict the ground-state electron configuration for each of the following ions:

(a) Ra^{2+} **(b)** Ni^{2+} **(c)** N^{3-}

▶ **APPLY 6.2**   Which of the following sets of ions are isoelectronic?

(a) F^-, Cl^-, Br^- **(b)** Ti^{4+}, Ca^{2+}, Cl^-
(c) Na^+, Mg^{2+}, Al^{3+}

All **PRACTICE** and **APPLY** problems are interactive in the eText.

6.2 IONIC RADII

Just as there are systematic differences in **atomic radii** (Section 5.13), there are also systematic differences in the radii of ions. As shown in **FIGURE 6.2** for the elements of groups 1A and 2A, atoms shrink dramatically when an electron is removed to form a cation. The radius of an Na atom, for example, is 186 pm, but that of an Na^+ cation is 102 pm. Similarly, the radius of an Mg atom is 160 pm and that of an Mg^{2+} cation is 72 pm.

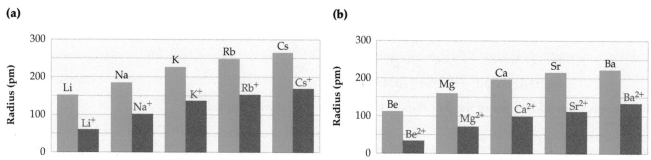

(a)   **(b)**

Cations are smaller than the corresponding neutral atoms, both because the principal quantum number of the valence-shell electrons is smaller for the cations than it is for the neutral atoms and because Z_{eff} is larger.

▲ **FIGURE 6.2**

Radii of (a) group 1A atoms and their cations;
(b) group 2A atoms and their cations.

▲ **Figure It out**

What is the magnitude of the change in radius that occurs when group 1A and group 2A elements form cations?

Answer: Formation of the cations of group 1A and group 2A elements decreases the atomic radius by approximately a factor of 2, a large change.

The cation that results when an electron is removed from a neutral atom is smaller than the original atom both because the electron is removed from a large, valence-shell orbital and because there is an increase in the **effective nuclear charge, Z_{eff},** for the remaining electrons (Section 5.9). On going from a neutral Na atom to a charged Na^+ cation, for example, the electron configuration changes from $1s^2\,2s^2\,2p^6\,3s^1$ to $1s^2\,2s^2\,2p^6$.

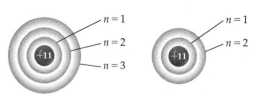

$$\textbf{Na: } 1s^2\,2s^2\,2p^6\,3s^1 \xrightarrow{-e^-} \textbf{Na}^+\textbf{: } 1s^2\,2s^2\,2p^6 \text{ or } [Ne]$$

The valence shell of the Na *atom* is the *third* shell, but the valence shell of the Na^+ *cation* is the *second* shell. Thus, the Na^+ ion has a smaller valence shell than the Na atom and therefore a smaller size. In addition, the effective nuclear charge felt by the valence-shell electrons is greater in the Na⁺ cation than in the neutral atom. The Na atom has 11 protons and 11 electrons, but the Na⁺ cation has 11 protons and only 10 electrons. The smaller number of electrons in the cation means that they shield one another to a lesser extent and therefore are pulled in more strongly toward the nucleus.

The same effects felt by the group 1A elements when a single electron is lost are felt by the group 2A elements when two electrons are lost. For example, loss of two valence-shell electrons from an Mg atom ($1s^2\,2s^2\,2p^6\,3s^2$) gives the Mg^{2+} cation ($1s^2\,2s^2\,2p^6$). The smaller valence shell of the Mg^{2+} cation and the increase in effective nuclear charge combine to cause a dramatic shrinkage. A similar shrinkage occurs whenever any of the metal atoms on the left-hand two-thirds of the periodic table is converted into a cation.

$$\text{Cl: [Ne] } 3s^2\,3p^5 \xrightarrow{+e^-} \text{Cl}^-\text{: [Ne] } 3s^2\,3p^6$$

Just as neutral atoms shrink when converted to cations by loss of one or more electrons, they expand when converted to anions by gain of one or more electrons. As shown in **FIGURE 6.3** for the group 7A elements (halogens), the expansion is dramatic. Chlorine, for example, nearly doubles in radius, from 99 pm for the neutral atom to 184 pm for the chloride anion.

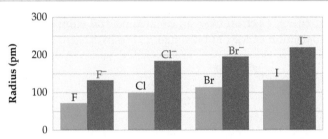

Anions are larger than their neutral atoms because of additional electron–electron repulsions and a decrease in Z_{eff}.

◀ **FIGURE 6.3**
Radii of the group 7A atoms (halogens) and their anions.

The expansion that occurs when a group 7A atom gains an electron to yield an anion can't be accounted for by a change in the quantum number of the valence shell because the added electron simply completes an already occupied p subshell. For instance, [Ne] $3s^2\,3p^5$ for a Cl atom becomes [Ne] $3s^2\,3p^6$ for a Cl⁻ anion. Thus, the expansion is due entirely to the decrease in effective nuclear charge and the increase in electron–electron repulsions that occur when an extra electron is added.

BIG IDEA Question 2
Go to eText

How does the radius of an anion compare to the radius of the neutral atom? What is the reason for the difference?

WORKED EXAMPLE 6.2

Predicting Relative Sizes of Ions

Which atom or ion in each of the following pairs would you expect to be larger?

(a) S and S^{2-} (b) Cl⁻ and I⁻   (c) Ni and Ni^{2+}

STRATEGY

For each set, compare the subshell of the valence electrons and the effective nuclear charge (Z_{eff}). Larger size correlates with higher n levels of the valence electrons and smaller Z_{eff}.

SOLUTION

(a) S^{2-} is larger than S because Z_{eff} decreases as the proton-to-electron ratio decreases. The proton-to-electron ratio in S^{2-} (16p/18e⁻) is smaller than in S (16p/16e⁻). Also, two additional electrons in the $3p$ subshell increase electron–electron repulsions, causing an increase in size.

(b) I⁻ is larger than Cl⁻. The valence electrons in I⁻ are in the $5p$ subshell while the valence electrons in Cl⁻ are in the $3p$ subshell.

(c) Ni is larger than Ni^{2+} because two electrons are removed from the $4s$ orbital, which is larger than the $3d$ orbital. Z_{eff} also increases for Ni^{2+} because the ratio of protons to electrons increases; Ni (28p/28e⁻) and Ni^{2+} (28p/26e⁻).

continued on next page

▶ **PRACTICE 6.3**   Which atom or ion has the largest radius: Fe, Fe^{2+}, or Fe^{3+}?

▶ **CONCEPTUAL APPLY 6.4**   Which of the following spheres represents a K^+ ion, which a Ca^{2+} ion, and which a Cl^- ion?

$r = 184$ pm $r = 133$ pm $r = 100$ pm

6.3 IONIZATION ENERGY

We saw in the previous chapter that the absorption of electromagnetic energy by an atom leads to a change in electron configuration. When energy is added, a valence-shell electron is promoted from a lower-energy orbital to a higher-energy one with a larger principal quantum number n. If enough energy is absorbed, the electron can even be removed completely from the atom, leaving behind a cation. The amount of energy necessary to remove the highest-energy electron from an isolated neutral atom in the gaseous state is called the atom's **ionization energy**, abbreviated E_i. For hydrogen, $E_i = 1312.0$ kJ/mol.

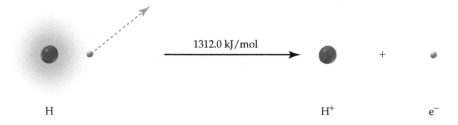

1312.0 kJ/mol

H H^+ e^-

As shown by the plot in **FIGURE 6.4,** ionization energies differ widely, from a low of 375.7 kJ/mol for cesium to a high of 2372.3 kJ/mol for helium. Furthermore, the data show a clear periodicity. The minimum E_i values correspond to the group 1A elements (alkali metals), the maximum E_i values correspond to the group 8A elements (noble gases), and a gradual increase in E_i occurs from left to right across a row of the periodic table—from Na to Ar, for example. Note that all the values are positive, meaning that energy must always be added to remove an electron from an atom.

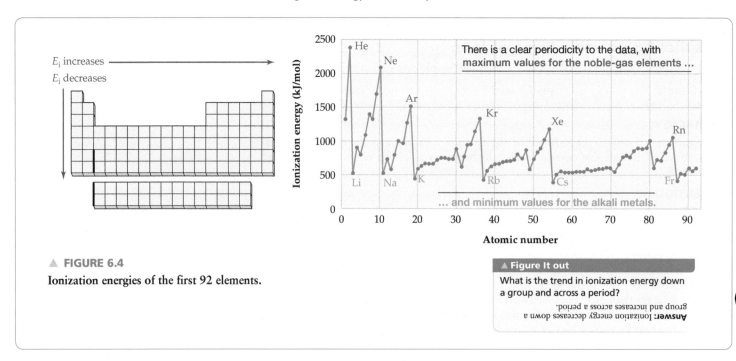

E_i increases ⟶

E_i decreases

There is a clear periodicity to the data, with maximum values for the noble-gas elements ...

... and minimum values for the alkali metals.

Ionization energy (kJ/mol)

Atomic number

▲ **FIGURE 6.4**

Ionization energies of the first 92 elements.

▲ **Figure It out**

What is the trend in ionization energy down a group and across a period?

Answer: Ionization energy decreases down a group and increases across a period.

The periodicity evident in Figure 6.4 can be explained by electron configurations. Atoms of the group 8A elements have filled valence subshells, either s for helium or both s and p for the other noble gases. As described in Section 5.13, an electron in a filled valence subshell feels a relatively high Z_{eff} because electrons in the same subshell don't shield one another very strongly. As a result, the electrons are held tightly to the nucleus, the radius of the atom is small, and the energy necessary to remove an electron is relatively large. Atoms of group 1A elements, by contrast, have only a single s electron in their valence shell. This single valence electron is shielded from the nucleus by all the inner-shell electrons, called the **core electrons**, resulting in a low Z_{eff}. The valence electron is thus held loosely, and the energy necessary to remove it is relatively small.

The plot of ionization energies in Figure 6.4 shows other trends in the data beyond the obvious periodicity. One such trend is that ionization energies gradually decrease going down a group in the periodic table, from He to Rn and from Li to Fr, for instance. As the atomic number increases going down a group, both the principal quantum number of the valence-shell electrons and their average distance from the nucleus also increase. As a result, the valence-shell electrons are less tightly held and E_i is smaller.

Yet another point about the E_i data is that minor irregularities occur across a row of the periodic table. A close look at E_i values of the first 20 elements (**FIGURE 6.5**) shows that the E_i of beryllium is larger than that of its neighbor boron and the E_i of nitrogen is larger than that of its neighbor oxygen. Similarly, magnesium has a larger E_i than aluminum and phosphorus has a slightly larger E_i than sulfur.

> **REMEMBER . . .**
> Valence electrons are strongly shielded by inner-shell (core) electrons. They are less strongly shielded by other electrons in the same shell, and the magnitude of shielding follows the order s > p > d > f.

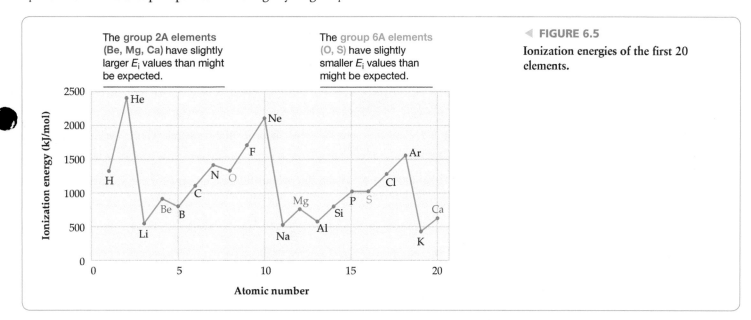

The group 2A elements (Be, Mg, Ca) have slightly larger E_i values than might be expected.

The group 6A elements (O, S) have slightly smaller E_i values than might be expected.

◀ **FIGURE 6.5**

Ionization energies of the first 20 elements.

The slightly enlarged E_i values for the group 2A elements Be, Mg, and others can be explained by their electron configurations. Compare beryllium with boron, for instance. A $2s$ electron is removed on ionization of beryllium, but a $2p$ electron is removed on ionization of boron:

Go to eText

BIG IDEA Question 3

How does the first ionization energy of elements change across a period? What is the reason for the change?

s electron removed

$$\text{Be } (1s^2\, 2s^2) \longrightarrow \text{Be}^+ (1s^2\, 2s^1) + e^- \qquad E_i = 899.4 \text{ kJ/mol}$$

p electron removed

$$\text{B } (1s^2\, 2s^2\, 2p^1) \longrightarrow \text{B}^+ (1s^2\, 2s^2) + e^- \qquad E_i = 800.6 \text{ kJ/mol}$$

Because a $2s$ electron spends more time closer to the nucleus than a $2p$ electron, it is held more tightly and is harder to remove. Thus, the E_i of beryllium is larger than that of boron. Put another way, the $2p$ electron of boron is shielded somewhat by the $2s$ electrons and is thus more easily removed than a $2s$ electron of beryllium.

The lowered E_i values for atoms of group 6A elements can be explained by their electron configurations as well. Comparing nitrogen with oxygen, for instance, the nitrogen electron is removed from a half-filled orbital, whereas the oxygen electron is removed from a filled orbital:

Half-filled orbital

$$N\ (1s^2\ 2s^2\ 2p_x{}^1\ 2p_y{}^1\ 2p_z{}^1) \longrightarrow N^+\ (1s^2\ 2s^2\ 2p_x{}^1\ 2p_y{}^1) + e^- \qquad E_i = 1402.3\ kJ/mol$$

Filled orbital

$$O\ (1s^2\ 2s^2\ 2p_x{}^2\ 2p_y{}^1\ 2p_z{}^1) \longrightarrow O^+\ (1s^2\ 2s^2\ 2p_x{}^1\ 2p_y{}^1\ 2p_z{}^1) + e^- \quad E_i = 1313.9\ kJ/mol$$

Because electrons repel one another and tend to stay as far apart as possible, electrons that are forced together in a filled orbital are slightly higher in energy than those in a half-filled orbital, so removing one is slightly easier. Thus, oxygen has a smaller E_i than nitrogen.

WORKED EXAMPLE 6.3

Predicting Ionization Energies

Arrange the elements Se, Cl, and S in order of increasing ionization energy.

STRATEGY

Ionization energy generally increases from left to right across a row of the periodic table and decreases from top to bottom down a group.

SOLUTION

The order is Se < S < Cl.

▶ **PRACTICE 6.5** Use the periodic table to order the elements from lowest to highest ionization energy: Se, O, Rb.

▶ **CONCEPTUAL APPLY 6.6** Given the orbital filling diagrams for the valence electrons of elements, rank them from lowest to highest ionization energy.

(a) ↓↑ ↑ ↑ ↑
     3s   3p

(b) ↓↑ ↓↑ ↓↑ ↓↑
     3s   3p

(c) ↓↑ — — —
     5s   5p

(d) ↓↑ ↓↑ ↓↑ ↑
     2s   2p

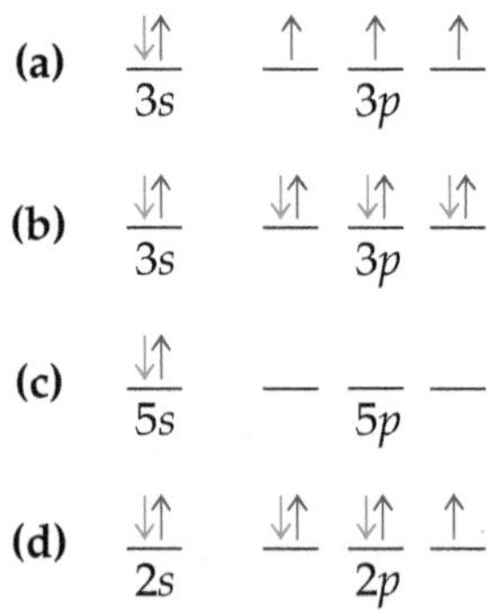

6.4 HIGHER IONIZATION ENERGIES

Ionization is not limited to the loss of a single electron from an atom. Two, three, or even more electrons can be lost sequentially from an atom, and the amount of energy associated with each step can be measured.

$$M + Energy \longrightarrow M^+ + e^- \qquad \text{First ionization energy } (E_{i1})$$

$$M^+ + Energy \longrightarrow M^{2+} + e^- \qquad \text{Second ionization energy}(E_{i2})$$

$$M^{2+} + Energy \longrightarrow M^{3+} + e^- \qquad \text{Third ionization energy } (E_{i3})$$

... and so forth

Successively larger amounts of energy are required for each ionization step because it is much harder to pull a negatively charged electron away from a positively charged ion than from a neutral atom. Interestingly, though, the energy differences between successive steps vary dramatically from one element to another. Removing the second electron from sodium takes nearly 10 times as much energy as removing the first one (4562 versus 496 kJ/mol), but removing the second electron from magnesium takes only twice as much energy as removing the first one (1451 versus 738 kJ/mol).

Large jumps in successive ionization energies are also found for other elements, as is indicated by the zigzag line in **TABLE 6.2**. Magnesium has a large jump between its second and third ionization energies, aluminum has a large jump between its third and fourth ionization energies, silicon has a large jump between its fourth and fifth ionization energies, and so on.

TABLE 6.2   Higher Ionization Energies (kJ/mol) for Main-Group Third-Row Elements

Group	1A	2A	3A	4A	5A	6A	7A	8A
E_i Number	Na	Mg	Al	Si	P	S	Cl	Ar
E_{i1}	496	738	578	787	1,012	1,000	1,251	1,520
E_{i2}	4,562	1,451	1,817	1,577	1,903	2,251	2,297	2,665
E_{i3}	6,912	7,733	2,745	3,231	2,912	3,361	3,822	3,931
E_{i4}	9,543	10,540	11,575	4,356	4,956	4,564	5,158	5,770
E_{i5}	13,353	13,630	14,830	16,091	6,273	7,013	6,540	7,238
E_{i6}	16,610	17,995	18,376	19,784	22,233	8,495	9,458	8,781
E_{i7}	20,114	21,703	23,293	23,783	25,397	27,106	11,020	11,995

The zigzag line marks the large jumps in ionization energies.

The large increases in ionization energies highlighted by the zigzag line in Table 6.2 can be understood by examining electron configurations.

Let's first examine sodium. The equations representing the first and second ionization energies in sodium are:

$$\text{Na}\ (1s^2 2s^2 2p^6 3s^1) + \text{Energy} \longrightarrow \text{Na}^+(1s^2 2s^2 2p^6) + e^- \quad \text{First ionization energy } (E_{i1})$$
$$\text{Na}^+(1s^2 2s^2 2p^6) + \text{Energy} \longrightarrow \text{Na}^{2+}(1s^2 2s^2 2p^5) + e^- \quad \text{Second ionization energy } (E_{i2})$$

In sodium, the large jump in ionization energy between E_{i1} and E_{i2} can be attributed to the difference in energy required to remove an electron from the $3s$ subshell and the $2p$ subshell. The $3s$ subshell is further from the nucleus and more shielded than the $2p$ subshell; thus the first ionization requires less energy. In other words, Z_{eff} is lower for an electron in a $3s$ orbital compared to a $2p$ orbital.

Now let's examine magnesium, which has a large jump between E_{i2} and E_{i3}. The equations representing the first three ionization energies of magnesium are:

$$\text{Mg}\ (1s^2 2s^2 2p^6 3s^2) + \text{Energy} \longrightarrow \text{Mg}^+(1s^2 2s^2 2p^6 3s^1) + e^- \quad \text{First ionization energy } (E_{i1})$$
$$\text{Mg}^+(1s^2 2s^2 2p^6 3s^1) + \text{Energy} \longrightarrow \text{Mg}^{2+}(1s^2 2s^2 2p^6) + e^- \quad \text{Second ionization energy } (E_{i2})$$
$$\text{Mg}^{2+}(1s^2 2s^2 2p^6) + \text{Energy} \longrightarrow \text{Mg}^{3+}(1s^2 2s^2 2p^5) + e^- \quad \text{Third ionization energy } (E_{i3})$$

In a similar manner to sodium, the large increase in ionization energy arises when the electron must be removed from inner core electrons ($2p$) in the third ionization. It's relatively easier to remove an electron from a *partially* filled valence shell because Z_{eff} is lower, but it's relatively harder to remove an electron from a *filled* valence shell because Z_{eff} is higher. In other words, ions formed by reaction of main-group elements usually have filled s and p subshells (a noble-gas electron configuration), which corresponds to having eight electrons (an *octet*) in the valence shell of an atom or ion. Sodium ($[Ne]\,3s^1$) loses only one electron easily, magnesium ($[Ne]\,3s^2$) loses only two electrons easily, aluminum ($[Ne]\,3s^2\,3p^1$) loses only three electrons easily, and so on across the row.

$$Na\ (1s^2\,2s^2\,2p^6\,3s^1) \longrightarrow Na^+\ (1s^2\,2s^2\,2p^6) + e^-$$

8 electrons in outer (2nd) shell

$$Mg\ (1s^2\,2s^2\,2p^6\,3s^2) \longrightarrow Mg^{2+}\ (1s^2\,2s^2\,2p^6) + 2\,e^-$$

$$Al\ (1s^2\,2s^2\,2p^6\,3s^2\,3p^1) \longrightarrow Al^{3+}\ (1s^2\,2s^2\,2p^6) + 3\,e^-$$

$$\vdots \qquad\qquad\qquad \vdots$$

$$Cl\ (1s^2\,2s^2\,2p^6\,3s^2\,3p^5) \longrightarrow Cl^{7+}\ (1s^2\,2s^2\,2p^6) + 7\,e^-$$

WORKED EXAMPLE 6.4

Higher Ionization Energies

Which has the larger fifth ionization energy, Ge or As?

STRATEGY

Look at their positions in the periodic table, and write the electron configuration. If the fifth electron must be removed from an inner shell, then that element will have the larger ionization energy.

SOLUTION

Ge: $[Ar]\,4s^2\,3d^{10}\,4p^2$ \hspace{2cm} As: $[Ar]\,4s^2\,3d^{10}\,4p^3$

The group 4A element germanium has four valence-shell electrons and thus has four relatively low ionization energies, whereas the group 5A element arsenic has five valence-shell electrons and has five low ionization energies. Germanium has a larger E_{i5} than arsenic because the fifth electron to be removed in Ge occupies a lower electron shell ($n = 3$).

▶ **PRACTICE 6.7**   Which has the largest third ionization energy: Be, C, or N?

▶ **CONCEPTUAL APPLY 6.8**   The figure on the left represents the successive ionization energies of an atom in the third period of the periodic table. Which atom is this most likely to be?

All **WORKED EXAMPLES** with this icon [Go to eText] have an interactive video in the eText.

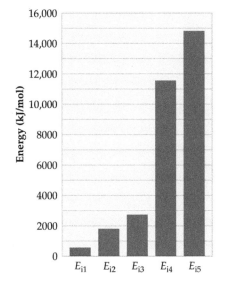

6.5 ELECTRON AFFINITY

Just as it's possible to measure the energy change on *removing* an electron from an atom to form a *cation,* it's also possible to measure the energy change on *adding* an electron to an atom to form an *anion.* An element's **electron affinity** (E_{ea}) is the energy change that occurs when an electron is added to an isolated atom in the gaseous state.

Ionization energies (Section 6.3) are always positive because energy must always be added to separate a negatively charged electron from the resultant positively charged cation. Electron affinities, however, are generally negative because energy is usually

---

*We have defined E_{ea} as the energy *released* when a neutral atom *gains* an electron to form an anion and have given it a negative sign. Some books and reference sources adopt the opposite point of view, defining E_{ea} as the energy *gained* when an anion *loses* an electron to form a neutral atom and giving it a positive value. The two definitions are simply the reverse of one another, so the sign of the energy change is also reversed.

released when a neutral atom adds an additional electron.* We'll see in Chapter 9 that this same convention is used throughout chemistry: A positive energy change means that energy is added, and a negative energy change means that energy is released.

The more negative the E_{ea}, the greater the tendency of the atom to accept an electron and the more stable the anion that results. In contrast, an atom that forms an unstable anion by addition of an electron has, in principle, a positive value of E_{ea}, but no experimental measurement can be made because the process does not take place. All we can say is that the E_{ea} for such an atom is greater than zero. The E_{ea} of hydrogen, for instance, is -72.8 kJ/mol, meaning that energy is released and the H^- anion is stable. The E_{ea} of neon, however, is greater than 0 kJ/mol, meaning that Ne does not add an electron and the Ne^- anion is not stable.

$$H\,(1\,s^1) + e^- \longrightarrow H^-(1\,s^2) + 72.8 \text{ kJ/mol} \qquad\qquad E_{ea} = -72.8 \text{ kJ}/mol$$
$$Ne\,(1s^2\,2s^2\,2p^6) + e^- + \text{Energy} \longrightarrow Ne^-(1s^2\,2s^2\,2p^6\,3s^1) \quad E_{ea} > 0 \text{ kJ/mol}$$

As with ionization energies, electron affinities show a periodicity that is related to the electron configurations of the elements. The data in **FIGURE 6.6** indicate that group 7A elements have the most negative electron affinities, corresponding to the largest release of energy, while group 2A and group 8A elements have near-zero or positive electron affinities, corresponding to a small release or even an absorption of energy.

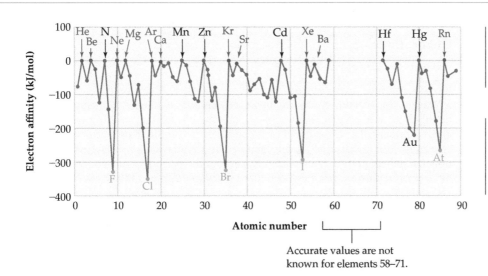

A negative value for E_{ea}, such as those for the group 7A elements (halogens), means that energy is released when an electron adds to an atom.

A value of zero, such as those for the group 2A elements (alkaline earths) and group 8A elements (noble gases), means that energy is absorbed but the exact amount can't be measured.

▲ **FIGURE 6.6**

Electron affinities for elements 1–57 and 72–86.

▲ **Figure It out**

Which group in the periodic table has the most negative electron affinities? What does a negative electron affinity mean?

Answer: The large negative E_{ea} for the halogens indicates that they form stable anions with a large release of energy.

The value of an element's electron affinity is due to an interplay of several offsetting factors. Attraction between the additional electron and the nucleus favors a negative E_{ea}, but the increase in electron–electron repulsions that results from addition of the extra electron favors a positive E_{ea}.

Large negative E_{ea}'s are found for the halogens (F, Cl, Br, I) because each of these elements has both a high Z_{eff} and room in its valence shell for an additional electron. *Halide ions*, halogens with a negative charge such as fluoride (F^-), have a noble-gas electron configuration with filled s and p sublevels, and the attraction between the additional electron and the atomic nucleus is high. Positive E_{ea}'s are found for the noble-gas elements (He, Ne, Ar, Kr, Xe), however, because the s and p sublevels in these elements are already full, so the additional electron must go into the next higher shell, where it is shielded from the nucleus and feels a relatively low Z_{eff}. The attraction

Go to eText

BIG IDEA Question 4

Does an element with a large negative electron affinity have a low or high effective nuclear charge (Z_{eff})?

of the nucleus for the added electron is therefore small and is outweighed by the additional electron–electron repulsions.

A halogen: Cl $(\ldots 3s^2\, 3p^5) + e^- \longrightarrow$ Cl$^-(\ldots 3s^2\, 3p^6)$ $E_{ea} = -348.6$ kJ/mol
A noble gas: Ar$(\ldots 3s^2\, 3p^6) + e^- \longrightarrow$ Ar$^-$ $(\ldots 3s^2\, 3p^6 4s^1)$ $E_{ea} > 0$ kJ/mo

In looking for other trends in the data of Figure 6.6, the near-zero E_{ea}'s of the alkaline earth metals (Be, Mg, Ca, Sr, Ba) are particularly striking. Atoms of these elements have filled s subshells, which means that the additional electron must go into a p subshell. The higher energy of the p subshell, together with a relatively low Z_{eff} for elements on the left side of the periodic table, means that alkaline earth atoms accept an electron reluctantly and have E_{ea} values near zero.

An alkaline earth metal: Mg$(\ldots 3s^2) + e^- \longrightarrow$ Mg$^-(\ldots 3s^2 3p^1)$ $E_{ea} \approx 0$ kJ/mol

---

WORKED EXAMPLE 6.5

Comparing Electron Affinities of Different Elements

Why does nitrogen have a less favorable (more positive) E_{ea} than its neighbors on either side, C and O?

STRATEGY AND SOLUTION

The magnitude of an element's E_{ea} depends on the element's valence-shell electron configuration. The electron configurations of C, N, and O are

Carbon: $1s^2\, 2s^2\, 2p_x^1\, 2p_y^1$ Nitrogen: $1s^2\, 2s^2\, 2p_x^1\, 2p_y^1\, 2p_z^1$

Oxygen: $1s^2\, 2s^2\, 2p_x^2\, 2p_y^1\, 2p_z^1$

Carbon has only two electrons in its $2p$ subshell and can readily accept another in its vacant $2p_z$ orbital. Nitrogen, however, has a half-filled $2p$ subshell, so the additional electron must pair up in a $2p$ orbital where it feels a repulsion from the electron already present. Thus, the E_{ea} of nitrogen is less favorable than that of carbon. Oxygen also must add an electron to an orbital that already has one electron, but the additional stabilizing effect of increased Z_{eff} across the periodic table counteracts the effect of electron repulsion, resulting in a more favorable E_{ea} for O than for N.

▶ **PRACTICE 6.9** Order the following elements from least to most favorable electron affinity: Ge, As, Br. (The most favorable electron affinity has the largest negative value.)

▶ **CONCEPTUAL APPLY 6.10** Which of the indicated three elements has the least favorable E_{ea}, and which has the most favorable E_{ea}?

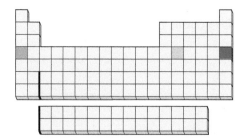

6.6 THE OCTET RULE

Let's list the important points discussed in the previous four sections and see if we can draw some general conclusions:

- **Group 1A elements have a relatively low E_{i1},** so they tend to lose their ns^1 valence-shell electron easily when they react, thereby adopting the electron configuration of the noble gas in the previous row of the periodic table.

- **Group 2A elements have relatively low E_{i1} and E_{i2},** so they tend to lose both their ns^2 valence-shell electrons easily when they react and adopt a noble-gas electron configuration.
- **Group 7A elements have a relatively large negative E_{ea},** so they tend to gain one electron easily when they react, changing from $ns^2\, np^5$ to $ns^2\, np^6$ and thereby adopting the configuration of the neighboring noble gas in the same row.
- **Group 8A (noble gas) elements are essentially inert** and undergo very few reactions. They neither gain nor lose electrons easily.

All these observations can be gathered into a single statement called the **octet rule**.

> **Octet rule** Main-group elements tend to undergo reactions that leave them with eight outer-shell electrons. That is, main-group elements react so that they attain a noble-gas electron configuration with filled s and p sublevels in their valence electron shell.

As we'll see in the next chapter, there are exceptions to the octet rule, particularly for elements in the third and lower rows of the periodic table. Nevertheless, the rule is useful for making predictions and for providing insights about chemical bonding.

Why does the octet rule work? What factors determine how many electrons an atom is likely to gain or lose? Clearly, electrons are most likely to be lost if they are held loosely in the first place—that is, if they feel a relatively low effective nuclear charge, Z_{eff}, and have lower ionization energies. Valence-shell electrons in the group 1A, 2A, and 3A metals, for instance, are shielded from the nucleus by core electrons, feel a low Z_{eff}, and are therefore lost relatively easily. Once the next lower noble-gas configuration is reached, though, loss of an additional electron suddenly becomes much more difficult because it must come from an inner shell, where it feels a much higher Z_{eff}.

$$\frac{\downarrow\uparrow}{ns^2} \qquad \frac{\downarrow\uparrow}{} \frac{\downarrow\uparrow}{} \frac{\downarrow\uparrow}{np^6} \qquad \frac{\uparrow}{(n+1)s^1} \qquad \left|\begin{array}{l}\text{Strongly shielded; low } Z_{eff}; \\ \text{easy to remove}\end{array}\right.$$

$$\frac{\downarrow\uparrow}{ns^2} \qquad \frac{\downarrow\uparrow}{} \frac{\downarrow\uparrow}{} \frac{\downarrow\uparrow}{np^6} \qquad \frac{}{(n+1)s} \qquad \left|\begin{array}{l}\text{Poorly shielded; high } Z_{eff}; \\ \text{hard to remove}\end{array}\right.$$

Conversely, electrons are most likely to be gained if they can be held tightly by a high Z_{eff}. Valence-shell electrons in the group 6A and 7A elements, for example, are poorly shielded, feel high values of Z_{eff}, and aren't lost easily. The high Z_{eff} thus makes possible the gain of one or more additional electrons into vacant valence-shell orbitals. Once the noble-gas configuration is reached, though, lower-energy orbitals are no longer available. An additional electron would have to be placed in a higher-energy orbital, where it would feel only a low Z_{eff}.

$$\frac{\downarrow\uparrow}{ns^2} \qquad \frac{\downarrow\uparrow}{} \frac{\downarrow\uparrow}{} \frac{\downarrow\uparrow}{np^6} \qquad \frac{}{(n+1)s} \qquad \left|\begin{array}{l}\text{Strongly shielded; low } Z_{eff}; \\ \text{hard to add}\end{array}\right.$$

$$\frac{\downarrow\uparrow}{ns^2} \qquad \frac{\downarrow\uparrow}{} \frac{\downarrow\uparrow}{} \frac{\uparrow}{np^5} \qquad \frac{}{(n+1)s} \qquad \left|\begin{array}{l}\text{Poorly shielded; high } Z_{eff}; \\ \text{easy to add}\end{array}\right.$$

Eight is therefore the magic number for valence-shell electrons. Taking electrons *from* a filled octet is difficult because they are tightly held by a high Z_{eff}; adding more electrons *to* a filled octet is difficult because, with s and p sublevels full, no low-energy orbital is available.

BIG IDEA Question 5

Why do groups 1A and 2A elements lose electrons when forming ions?

▲ The reaction of sodium chloride with chlorine gas to produce solid sodium chloride releases energy as shown by the presence of a flame.

6.7 IONIC BONDS AND THE FORMATION OF IONIC SOLIDS

An **ionic bond** occurs when one atom transfers an electron to another, creating an electrostatic attraction between positively charged cations and negatively charged anions. A familiar example of an ionic compound is table salt, NaCl, consisting of Na^+ and Cl^- ions. Sodium metal directly reacts with chlorine gas to form ions because sodium gives up an electron relatively easily (that is, has a small positive ionization energy) and chlorine accepts an electron easily (that is, has a large negative electron affinity). In general, an element with a small E_i can transfer an electron to an element with the negative E_{ea}, yielding a cation and an anion. The electron transfer from sodium to chlorine to form ions is:

$$Na + Cl \longrightarrow Na^+ Cl^-$$

$$\underset{1s^2\,2s^2\,2p^6\,3s^1}{} \quad \underset{1s^2\,2s^2\,2p^6\,3s^2\,3p^5}{} \qquad \underset{1s^2\,2s^2\,2p^6}{} \quad \underset{1s^2\,2s^2\,2p^6\,3s^2\,3p^6}{}$$

What about the overall energy change, ΔE, for the reaction of sodium with chlorine to yield Na^+ and Cl^- ions in the ionic solid sodium chloride? As shown in the margin photo, when molten sodium chloride is placed into contact with chlorine gas, a fiery reaction takes place, and the white solid NaCl is produced. The reaction of sodium with chlorine releases energy as heat. (The Greek capital letter delta, Δ, is used to represent a change in the value of the indicated quantity, in this case an energy change ΔE.) ΔE is negative because the formation of sodium chloride from the elements releases heat. Let's examine the factors that result in the release of energy. It's apparent from E_i and E_{ea} values that the amount of energy released when a chlorine atom accepts an electron [$E_{ea} = -348.6$ kJ per mol of $Cl(g)$]. is insufficient to offset the amount absorbed when a sodium atom loses an electron [$E_i = +495.8$ kJ per mol of $Na(g)$].

$$
\begin{array}{lll}
E_i \text{ for Na} & = +495.8 \text{ kJ} & \text{(Unfavorable)} \\
\underline{E_{ea} \text{ for Cl}} & \underline{= -348.6 \text{ kJ}} & \underline{\text{(Favorable)}} \\
\Delta E & = +147.2 \text{ kJ} & \text{(Unfavorable)}
\end{array}
$$

The net ΔE for the reaction of sodium and chlorine atoms would be unfavorable by $+147.2$ kJ and no reaction would occur, unless some other factors were involved.

This additional factor, which is more than enough to overcome the unfavorable energy change of electron transfer, is the large gain in stability due to the electrostatic attractions between product anions and cations in the formation of an ionic solid.

The actual reaction of solid sodium metal with gaseous chlorine molecules to form solid sodium chloride occurs all at once rather than in a stepwise manner, but it's easier to make an energy calculation if we imagine a series of hypothetical steps for which exact energy changes can be measured experimentally. There are five steps to take into account to calculate the overall energy change.

REMEMBER . . .
In an ionic solid, oppositely charged ions are attracted to one another by ionic bonds and are packed together in a regular way (Section 2.11).

Step ① Solid Na metal is first converted into isolated, gaseous Na atoms, a process called *sublimation*. Because energy must be added to disrupt the forces holding atoms together in a solid, the heat of sublimation has a positive value: +107.3 kJ per mol of $Na(s)$.

$$Na(s) \longrightarrow Na(g)$$
$$\text{+107.3 kJ per mol of } Na(s)$$

Step ② Gaseous Cl_2 molecules are split into individual Cl atoms. Energy must be added to break molecules apart, and the energy required for bond breaking therefore has a positive value: +243 kJ per mol of Cl_2 (or 122 kJ for 1/2 mol of Cl_2). We'll look further into bond dissociation energies in Section 7.2.

$$1/2 \, Cl_2(g) \longrightarrow Cl(g)$$
$$\text{+122 kJ per 1/2 mol of } Cl_2(g)$$

Step ③ Isolated Na atoms are ionized into Na^+ ions plus electrons. The energy required is the first ionization energy of sodium (E_{i1}) and has a positive value: +495.8 kJ per mol of Na(g).

$$Na(g) \longrightarrow Na^+(g) + e^-$$
$$\text{+495.8 kJ per mol of } Na(g)$$

Step ④ Cl^- ions are formed from Cl atoms by addition of an electron. The energy released is the electron affinity of chlorine (E_{ea}) and has a negative value: −348.6 kJ per mol of $Cl(g)$.

$$Cl(g) + e^- \longrightarrow Cl^-(g)$$
$$\text{−348.6 kJ per mol of } Cl(g)$$

Step ⑤ Lastly, solid NaCl is formed from isolated gaseous Na^+ and Cl^- ions. The energy change is a measure of the overall electrostatic interactions between ions in the solid. It is the amount of energy released when isolated ions condense to form a solid, and it has a negative value: −787 kJ per mol of $Na^+(g)$ and $Cl^-(g)$ ions that react to form one mol of NaCl(s).

$$Na^+(g) + Cl^-(g) \longrightarrow NaCl(s)$$
$$\text{−787 kJ per mol of } Na^+(g)$$
$$\text{and } Cl^-(g) \text{ ions}$$

Net reaction:

$$Na(s) + 1/2Cl_2(g) \longrightarrow NaCl(s)$$
$$\text{−411 kJ per mol of } NaCl(s)$$

Net energy change (ΔE):

The five hypothetical steps in the reaction between sodium metal and gaseous chlorine are depicted in **FIGURE 6.7** in a pictorial format called a **Born–Haber cycle**, which shows how each step contributes to the overall energy change and how the net process is the sum of the individual steps. As indicated in the diagram, steps 1, 2, and 3 have positive values and absorb energy, while steps 4 and 5 have negative values and release energy. The largest contribution is step 5, which measures the electrostatic forces between ions in the solid product—that is, the strength of the ionic bonding. Were it not for this large amount of stabilization of the solid due to ionic bonding, no reaction would take place.

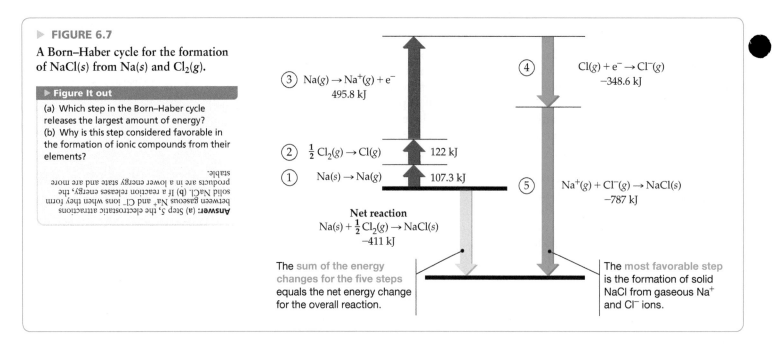

▶ **FIGURE 6.7**

A Born–Haber cycle for the formation of NaCl(s) from Na(s) and Cl$_2$(g).

▶ **Figure It out**

(a) Which step in the Born–Haber cycle releases the largest amount of energy?
(b) Why is this step considered favorable in the formation of ionic compounds from their elements?

Answer: (a) Step 5, the electrostatic attractions between gaseous Na$^+$ and Cl$^-$ ions when they form solid NaCl. (b) If a reaction releases energy, the products are in a lower energy state and are more stable.

③ Na(g) → Na$^+$(g) + e$^-$
495.8 kJ

② $\frac{1}{2}$ Cl$_2$(g) → Cl(g)   122 kJ

① Na(s) → Na(g)   107.3 kJ

④ Cl(g) + e$^-$ → Cl$^-$(g)
−348.6 kJ

⑤ Na$^+$(g) + Cl$^-$(g) → NaCl(s)
−787 kJ

Net reaction
Na(s) + $\frac{1}{2}$ Cl$_2$(g) → NaCl(s)
−411 kJ

The sum of the energy changes for the five steps equals the net energy change for the overall reaction.

The most favorable step is the formation of solid NaCl from gaseous Na$^+$ and Cl$^-$ ions.

A similar Born–Haber cycle for the reaction of magnesium with chlorine shows the energy changes involved in the reaction of an alkaline earth element (**FIGURE 6.8**). As in the reaction of sodium and chlorine to form NaCl, there are five contributions to the overall energy change. First, solid magnesium metal must be converted into isolated gaseous magnesium atoms (sublimation). Second, the bond in Cl$_2$ molecules must be broken to yield chlorine atoms. Third, the magnesium atoms must lose two electrons to form Mg^{2+} ions. Fourth, the chlorine atoms formed in step 2 must accept electrons to form Cl$^-$ ions. Fifth, the gaseous ions must combine to form the ionic solid, MgCl$_2$. As the Born–Haber cycle indicates, it is the large contribution from ionic bonding that releases enough energy to drive the entire process.

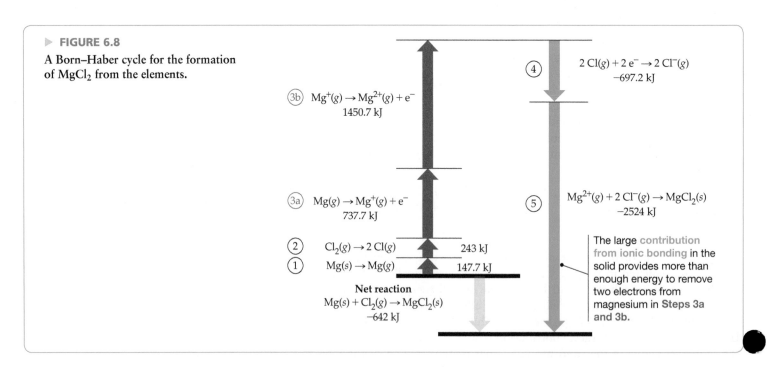

▶ **FIGURE 6.8**

A Born–Haber cycle for the formation of MgCl$_2$ from the elements.

③b Mg$^+$(g) → Mg^{2+}(g) + e$^-$
1450.7 kJ

③a Mg(g) → Mg$^+$(g) + e$^-$
737.7 kJ

② Cl$_2$(g) → 2 Cl(g)   243 kJ
① Mg(s) → Mg(g)   147.7 kJ

④ 2 Cl(g) + 2 e$^-$ → 2 Cl$^-$(g)
−697.2 kJ

⑤ Mg^{2+}(g) + 2 Cl$^-$(g) → MgCl$_2$(s)
−2524 kJ

Net reaction
Mg(s) + Cl$_2$(g) → MgCl$_2$(s)
−642 kJ

The large contribution from ionic bonding in the solid provides more than enough energy to remove two electrons from magnesium in **Steps 3a** and **3b**.

Go to
eText

WORKED EXAMPLE 6.7

Calculating the Energy Change in the Formation of an Ionic Compound

Calculate the net energy change in kilojoules per mole of KF(s) that takes place on formation of KF(s) from the elements: $K(s) + 1/2 \, F_2(g) \longrightarrow KF(s)$. The following information is needed:

Heat of sublimation for K(s) = 89.2 kJ/mol E_{ea} for F(g) = −328 kJ/mol

Bond dissociation energy for $F_2(g)$ = 158 kJ/mol E_i for K(g) = 418.8 kJ/mol

Electrostatic interactions in KF(s) = −821 kJ/mol

STRATEGY

Write a chemical reaction for each step in the process of forming the ionic solid KF and draw a Born–Haber cycle similar to Figure 6.7. The sum of the energy changes for each step should give the energy change for the overall reaction.

SOLUTION

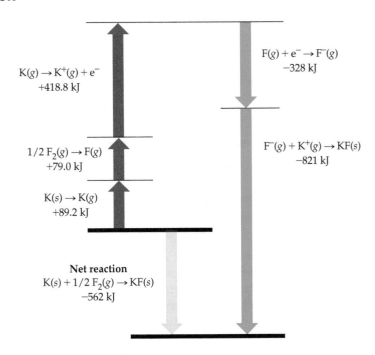

$F(g) + e^- \rightarrow F^-(g)$
−328 kJ

$K(g) \rightarrow K^+(g) + e^-$
+418.8 kJ

$F^-(g) + K^+(g) \rightarrow KF(s)$
−821 kJ

$1/2 \, F_2(g) \rightarrow F(g)$
+79.0 kJ

$K(s) \rightarrow K(g)$
+89.2 kJ

Net reaction
$K(s) + 1/2 \, F_2(g) \rightarrow KF(s)$
−562 kJ

▶ **PRACTICE 6.13** Calculate the net energy change in kilojoules per mole that takes place on formation of $MgF_2(s)$ from the elements: $Mg(s) + F_2(g) \longrightarrow MgF_2(s)$. The following information is needed:

Heat of sublimation for Mg(s) = 147.7 kJ/mol E_{ea} for F(g) = −328 kJ/mol

Bond dissociation energy for $F_2(g)$ = 158 kJ/mol E_{i1} for Mg(g) = 737.7 kJ/mol

Electrostatic interactions in $MgF_2(s)$ = −2957 kJ/mol E_{i2} for Mg(g) = 1450.7 kJ/mol

▶ **APPLY 6.14** Calculate the energy of electrostatic attractions of LiCl(s). The following information is needed.

Heat of sublimation for Li(s) = 161 kJ/mol E_{ea} for Cl(g) = −349 kJ/mol

Bond dissociation energy for $Cl_2(g)$ = 243 kJ/mol E_i for Li(g) = 520 kJ/mol

$Li(s) + 1/2 \, Cl_2(g) \longrightarrow LiCl(s)$ = −409 kJ/mol

6.8 LATTICE ENERGIES IN IONIC SOLIDS

The measure of the electrostatic interaction energies between ions in a solid—and thus the measure of the strength of the solid's ionic bonds—is called the **lattice energy** (U). By convention, lattice energy is defined as the amount of energy that must be added to break up an ionic solid into its individual gaseous ions, so it has a positive value. The formation of a solid from ions is the reverse of the breakup, and thus has a negative value; for example, in step 5 in the Born–Haber cycle illustrated in Figure 6.7, the formation of NaCl has a negative value, $-U$:

$$NaCl(s) \longrightarrow Na^+(g) + Cl^-(g) \qquad U = +787 \text{ kJ/mol} \qquad \text{(Energy absorbed)}$$
$$Na^+(g) + Cl^-(g) \longrightarrow NaCl(s) \qquad -U = -787 \text{ kJ/mol} \qquad \text{(Energy released)}$$

What factors affect the magnitude of the lattice energy in an ionic compound? The lattice energy depends on the strength of the ionic bond between the cations and anions in the ionic compound. **Coulomb's law** describes the force (F) that results from the interaction of electric charges and is equal to a constant k times the product of the charges on the ions, z_1 and z_2, divided by the square of the distance d between their centers (nuclei):

$$\text{Coulomb's law} \quad F = k \times \frac{z_1 z_2}{d^2}$$

Because energy is equal to force times distance, the lattice energy is

 Go to
eText

BIG IDEA Question 6

How does the magnitude of the charge on the ions affect lattice energy? How does the radius of the ions influence lattice energy?

$$\text{Lattice energy} \quad U = F \times d = k \times \frac{z_1 z_2}{d}$$

The value of the constant k depends on the arrangement of the ions in the specific compound and is different for different substances.

Lattice energies are large when the distance d between ions is small and when the charges z_1 and z_2 are large. A small distance d means that the ions are close together, which implies that they have small ionic radii. Thus, if z_1 and z_2 are held constant, the largest lattice energies belong to compounds formed from the smallest ions, as listed in **TABLE 6.3.**

TABLE 6.3   Lattice Energies of Some Ionic Solids (kJ/mol)

Cation	Anion				
	F^-	Cl^-	Br^-	I^-	O^{2-}
Li^+	1036	853	807	757	2925
Na^+	923	787	747	704	2695
K^+	821	715	682	649	2360
Be^{2+}	3505	3020	2914	2800	4443
Mg^{2+}	2957	2524	2440	2327	3791
Ca^{2+}	2630	2258	2176	2074	3401

Within a series of compounds that have the same anion but different cations, lattice energy increases as the cation becomes smaller. Comparing LiF, NaF, and KF, for example, cation size follows the order $K^+ > Na^+ > Li^+$, so lattice energies follow the order LiF > NaF > KF. Similarly, within a series of compounds that have the same cation but different anions, lattice energy increases as anion size decreases. Comparing LiF, LiCl, LiBr, and LiI, for example, anion size follows the order $I^- > Br^- > Cl^- > F^-$, so lattice energies follow the reverse order LiF > LiCl > LiBr > LiI.

Table 6.3 also shows that compounds of ions with higher charges have larger lattice energies than compounds of ions with lower charges. In comparing NaI, MgI_2, and AlI_3, for example, the order of charges on the cations is $Al^{3+} > Mg^{2+} > Na^+$, and the order of lattice energies is $AlI_3 > MgI_2 > NaI$.

Go to
eText

WORKED EXAMPLE 6.8

Lattice Energies

Which has the larger lattice energy, NaCl or CsI?

STRATEGY

The magnitude of a substance's lattice energy is affected both by the charges on its constituent ions and by the sizes of those ions. The higher the charges on the ions and the smaller the sizes of the ions, the larger the lattice energy. In this case, all four ions—Na^+, Cs^+, Cl^-, and I^-—are singly charged, so they differ only in size.

SOLUTION

Because Na^+ is smaller than Cs^+ and Cl^- is smaller than I^-, the distance between ions is smaller in NaCl than in CsI. Thus, NaCl has the larger lattice energy.

▶ **PRACTICE 6.15** Which substance has the largest lattice energy: SrF_2, CuI_2, CsI, or MgO?

▶ **CONCEPTUAL APPLY 6.16** One of the following pictures represents NaCl and one represents MgO. Which is which, and which has the larger lattice energy?

(a)   **(b)**

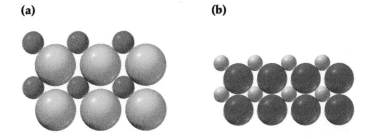

INQUIRY ? How do ionic liquids lead to more environmentally friendly processes?

When you think of ionic compounds, you probably think of crystalline, high-melting solids: sodium chloride (mp = 801 °C), magnesium oxide (mp = 2825 °C), lithium carbonate (mp = 732 °C), and so on. It's certainly true that many ionic compounds fit that description, but not all. Some ionic compounds are actually liquid at room temperature. Ionic liquids, in fact, have been known for nearly a century—the first such compound to be discovered was ethylammonium nitrate, $CH_3CH_2NH_3^+ NO_3^-$, with a melting point of just 12 °C (54 °F).

One defining property of a liquid is the ability to flow. To do this, ions in a lattice must move positions relative to one another and overcome the electrostatic forces between them. Therefore, the attractive forces between ions in an ionic liquid are lower than those in an ionic solid, where the position of ions is rigid and fixed. Generally speaking, the ionic liquids used today are salts in which the cation has an irregular shape and in which one or both of the ions are large and bulky so that the charges are dispersed over a large volume. Both factors minimize the crystal lattice energy, thereby making the solid less stable and favoring the liquid. Typical cations are derived from nitrogen-containing organic compounds called *amines,* either tetrabutylammonium ions or N-alkylpyridinium ions.

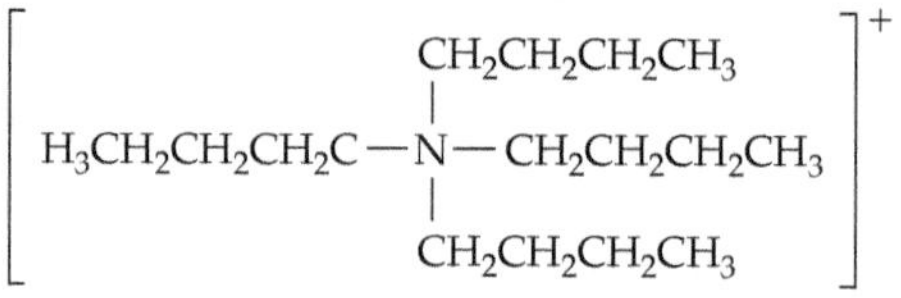

Tetrabutylammonium ion

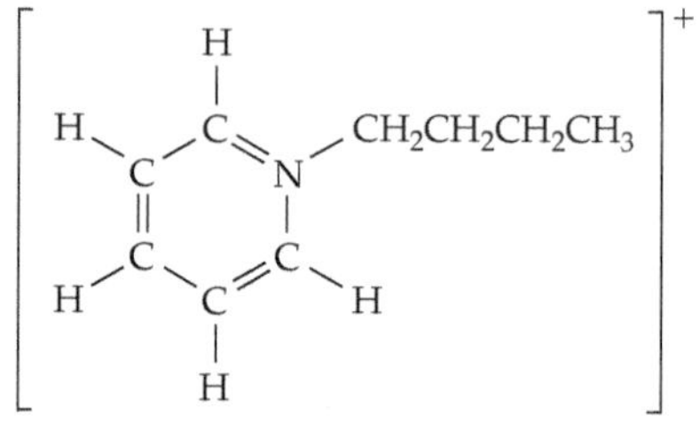

N-alkylpyridinium ion

Anions are just as varied as the cations, and more than 500 different ionic liquids with different anion/cation combinations are commercially available. Hexafluorophosphate, tetrafluoroborate, alkyl sulfates, trifluoromethanesulfonate, and halides are typical anions.

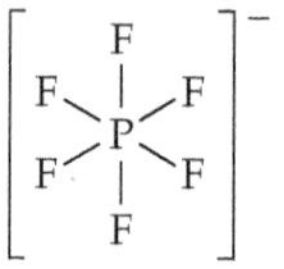

Hexafluorophosphate 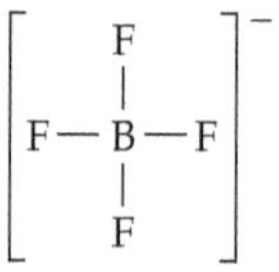   Tetrafluoroborate

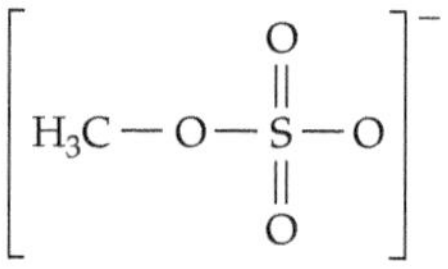

Methyl sulfate 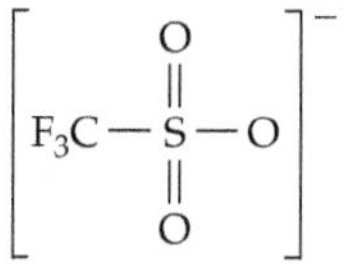   Trifluoromethanesulfonate   Halide

▲ Several hundred different ionic liquids, such as these from the German company BASF, are manufactured and sold commercially for use in a wide variety of chemical processes.

For many years, ionic liquids were just laboratory curiosities. More recently, though, they have been found to be excellent solvents, particularly for use in green chemistry processes like those described in the Chapter 3 Inquiry. Ionic liquids have many useful properties:

• They dissolve many different types of compounds, giving highly concentrated solutions and thereby minimizing the amount of liquid needed.

- They can be fine-tuned for use in specific reactions by varying cation and anion structures.
- They are nonflammable.
- They are stable at high temperatures.
- They do not evaporate readily.
- They are generally recoverable and can be reused many times.

Among their potential applications, ionic liquids are now being explored as replacements for toxic or flammable organic solvents in many industrial processes, for use as electrolytes in high-temperature batteries, and as solvents for the extraction of heavy organic materials from oil shale. Since most pharmaceuticals are salts, ionic liquid drugs are a novel way to produce a dual-action substance. Combining a pharmaceutically active cation with a pharmaceutically active anion leads to an ionic liquid in which the actions of two drugs are combined. Ionic liquids are also being used to store the Sun's thermal energy. During the day, heat from the Sun melts the ionic substance into a liquid at a moderate temperature. As the temperature drops at night, energy is released when the ionic liquid freezes into a solid. We'll be hearing much more about ionic liquids in the coming years.

PROBLEM 6.17   What are the benefits of replacing organic solvents with ionic liquids in industrial processes? Select all the correct answers.

(a) Ionic liquids are nonflammable.
(b) Ionic liquids are obtained from renewable resources.
(c) Ionic liquids can be recycled and used again.
(d) Ionic liquids can be used at high temperatures.

PROBLEM 6.18   What structural features do ionic liquids have that prevent them from forming solids easily?

PROBLEM 6.19   Compare the following two ionic liquids: tetraheptylammonium bromide and tetraheptylammonium iodide. The structure of the tetraheptylammonium ion is:

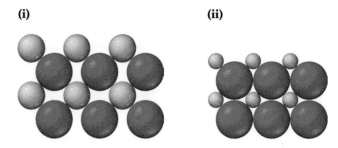

(a) Which picture corresponds to tetraheptylammonium bromide and which to tetraheptylammonium iodide?

(i)   **(ii)**

(b) Which ionic liquid has a larger lattice energy?
(c) One ionic liquid has a melting point of 88 °C and the other has a melting point of 39 °C. Match the ionic liquid to its melting point.

PROBLEM 6.20   An ionic liquid consisting of a bulky amine cation and chloride anion has a melting point of 41 °C. For a green chemistry solvent application, it is desirable to have a melting point below room temperature (25 °C) so energy is not required to heat the compound. The melting point of the ionic liquid can be altered by replacing chloride with a different anion. The anion possibilities are: F^-, Se^{2-}, O^{2-}, and Br^-.

(a) Write the electron configuration for each of these ions.
(b) Which ions are isoelectronic?
(c) Which ion is the best choice for replacing the chloride ion to make an ionic liquid with a lower melting temperature?

STUDY GUIDE

Section	Concept Summary	Learning Objectives	Test Your Understanding
6.1 Electron Configurations of Ions	Metallic elements, on the left side of the periodic table, tend to give up electrons to form cations, while the halogens and a few other nonmetallic elements, on the right side of the table, tend to accept electrons to form anions. The electrons given up by a main-group metal in forming a cation come from the highest-energy occupied orbital, while the electrons that are accepted by a nonmetal in forming an anion go into the lowest-energy unoccupied orbital.	**6.1** Write ground-state electron configurations for main group and transition metal ions.	Worked Example 6.1; Problems 6.21, 6.40, 6.42, 6.44
		6.2 Determine the number of unpaired electrons in main group and transition metal ions.	Problems 6.46–6.47
6.2 Ionic Radii	Cations have smaller atomic radii than corresponding atoms because removing electrons increases the **effective nuclear charge** (Z_{eff}) and in some cases decreases the valence shell. Anions have larger atomic radii than corresponding atoms because gaining electron decreases Z_{eff} and creates additional electron–electron repulsive forces.	**6.3** Predict the relative size of anions, cations, and neutral atoms.	Worked Example 6.2; Problems 6.22, 6.48–6.49, 6.51
		6.4 Predict the relative size of isoelectronic ions.	Problems 6.52, 6.54
6.3 Ionization Energy	The amount of energy necessary to remove a valence electron from an isolated neutral atom is called the atom's **ionization energy** (E_i). Ionization energies are smallest for metallic elements on the left side of the periodic table and largest for nonmetallic elements on the right side. As a result, metals usually give up electrons and act as reducing agents in chemical reactions.	**6.5** Order elements from lowest to highest ionization energy.	Worked Example 6.3; Problems 6.6, 6.58–6.59
		6.6 Explain the periodic trend in ionization energy.	Problem 6.56
6.4 Higher Ionization Energies	Ionization is not limited to the removal of a single electron from an atom. Two, three, or even more electrons can be removed sequentially from an atom, although larger amounts of energy are required for each successive ionization step. In general, valence-shell electrons are much more easily removed than **core electrons.**	**6.7** Compare successive ionization energies for different elements.	Worked Example 6.4; Problems 6.24, 6.60, 6.62
		6.8 Identify elements based on values of successive ionization energies.	Problems 6.8, 6.25, 6.64–6.65
6.5 Electron Affinity	The amount of energy released or absorbed when an electron adds to an isolated neutral atom is called the atom's **electron affinity** (E_{ea}). By convention, a negative E_{ea} corresponds to a release of energy and a positive E_{ea} corresponds to an absorption of energy. Electron affinities are most negative for group 7A elements and most positive for group 2A and 8A elements. As a result, the group 7A elements usually accept electrons and in chemical reactions.	**6.9** Compare the value of electron affinity for different elements.	Worked Example 6.5; Problems 6.68–6.69
		6.10 Explain the periodic trend in electron affinity.	Problems 6.70–6.73
6.6 The Octet Rule	In general, reactions of main-group elements can be described by the **octet rule,** which states that these elements tend to undergo reactions so as to attain a noble-gas electron configuration with filled s and p subshells within their valence shell. Elements on the left side of the periodic table tend to give up electrons until a noble-gas configuration is reached, elements on the right side of the table tend to accept electrons until a noble-gas configuration is reached, and the noble gases themselves are essentially unreactive.	**6.11** Use the octet rule to predict charges on main group ions, electron configurations of main group ions, and formulas for ionic compounds.	Worked Example 6.6; Problems 6.31, 6.74, 6.76, 6.78

Section	Concept Summary	Learning Objectives	Test Your Understanding
6.7 Ionic Bonds and the Formation of Ionic Solids	Main-group metals in groups 1A and 2A react with nonmetals in groups 5A–7A, during which the metal loses one or more electrons to the nonmetal. The product, such as NaCl, is an ionic solid that consists of metal cations and halide anions electrostatically attracted to one another by **ionic bonds**.	**6.12** Match the formula of an ionic compound with its molecular image.	Problems 6.26–6.28
		6.13 Draw a Born–Haber cycle and calculate the energy change that occurs when an ionic compound is formed from its elements.	Worked Example 6.7; Problems 6.32–6.33, 6.82, 6.84, 6.86
6.8 Lattice Energies in Ionic Solids	The sum of the interaction energies among all ions in a crystal is called the crystal's **lattice energy** (U). The higher the charges on the ions and the smaller the sizes of the ions, the larger the lattice energy.	**6.14** Predict the relative magnitude of lattice energy given the formula or molecular representation of an ionic compound.	Worked Example 6.8; Problems 6.29–6.30, 6.98–6.99

KEY TERMS

Born–Haber cycle 223
core electron 215
Coulomb's law 226

electron affinity (E_{ea}) 218
ionic bond 222
ionization energy (E_i) 214

isoelectronic 209
lattice energy (U) 226
octet rule 221

KEY EQUATIONS

- **Coulomb's Law (Section 6.8)**

$$F = k \times \frac{z_1 z_2}{d^2}$$

- **Lattice Energy (U) (Section 6.8)**

$$U = F \times d = k \times \frac{z_1 z_2}{d}$$

PRACTICE TEST

After studying this chapter, you can assess your understanding with these practice test questions, which are correlated with chapter learning objectives. If you answer a question incorrectly, refer to the learning objectives in the end-of-chapter Study Guide for assistance. The Study Guide provides a conceptual summary, references a Worked Example to model how to solve the problem, and gives additional problems for more practice.

1. What is the ground-state electron configuration for the Mg^{2+} ion? (LO 6.1)
   (a) $1s^2 2s^2 2p^6$
   (b) $1s^2 2s^2 2p^6 3s^2$
   (c) $1s^2 2s^2 2p^6 3s^2 3p^2$
   (d) $1s^2 2s^2 2p^6 3s^2 3p^6$

2. Give the ground state electron configuration and number of unpaired electrons in a Ru^{2+} ion. (LO 6.1, 6.2)
   (a) $[Kr]5s^2 4d^4$ 0 unpaired electrons
   (b) $[Kr]5s^2 4d^6$ 0 unpaired electrons
   (c) $[Kr]4d^6$ 4 unpaired electrons
   (d) $[Kr]5s^2 4d^4$ 4 unpaired electrons

3. Identify the *false* statement about atomic and ionic radii. (LO 6.3)
   (a) I^- has a larger radius than Br^-
   (b) Ba^{2+} has a smaller radius than Ba
   (c) Te has a larger radius than Te^{2-}
   (d) Sr^{2+} has a smaller radius than Se^{2-}

4. Arrange the ions Rb^+, Br^-, and Sr^{2+} from the smallest to the largest. (LO 6.4)
   (a) $Br^- < Rb^+ < Sr^{2+}$
   (b) $Sr^{2+} < Br^- < Rb^+$
   (c) $Rb^+ < Sr^{2+} < Br^-$
   (d) $Sr^{2+} < Rb^+ < Br^-$

5. Which of the following processes requires the largest input of energy? (LO 6.5)
   (a) $K(g) \longrightarrow K^+(g) + e^-$
   (b) $Na(g) \longrightarrow Na^+(g) + e^-$
   (c) $Ba(g) \longrightarrow Ba^+(g) + e^-$
   (d) $Na^+(g) \longrightarrow Na^{2+}(g) + e^-$

6. Phosphorus has a _____ ionization energy than magnesium because _____. (LO 6.5, 6.6)
   (a) larger; the electron in phosphorus is in a higher n level than the electron in magnesium
   (b) larger; the electron in phosphorus has a higher Z_{eff} than the electron in magnesium
   (c) smaller; the electron in phosphorus is in a lower n level than the electron in magnesium
   (d) smaller; the electron in phosphorus has a lower Z_{eff} than the electron in magnesium

7. The successive ionization energies for a second-period element are given. What is the identity of the element? (LO 6.8)
   $E_{i1} = 1402$ kJ/mol
   $E_{i2} = 2856$ kJ/mol
   $E_{i3} = 4578$ kJ/mol
   $E_{i4} = 7475$ kJ/mol
   $E_{i5} = 9445$ kJ/mol
   $E_{i6} = 53,266$ kJ/mol
   $E_{i7} = 64,630$ kJ/mol
   (a) Be
   (b) C
   (c) N
   (d) F

8. Which of the following processes will release the most energy? (**LO 6.9**)

  (a) $Cl(g) + e^- \longrightarrow Cl^-(g)$
  (b) $Kr(g) + e^- \longrightarrow Kr^-(g)$
  (c) $Rb(g) + e^- \longrightarrow Rb^-(g)$
  (d) $Cl^-(g) + e^- \longrightarrow Cl^{2-}(g)$

9. Elements that have large negative electron affinities generally have (**LO 6.10**)

  (a) high values for Z_{eff} and a vacancy in a valence orbital.
  (b) low values for Z_{eff} and a vacancy in a valence orbital.
  (c) high values for Z_{eff} and filled valence orbitals.
  (d) low values for Z_{eff} and filled valence orbitals.

10. Predict the formula of the ionic compound that forms between potassium and sulfur. (**LO 6.11**)

  (a) KS
  (b) KS_2
  (c) K_2S_2
  (d) K_2S

11. Which molecular scale image best represents the ionic compound that forms between cesium and chlorine? (Cesium is represented by red circles, and chlorine is represented by blue circles.) (**LO 6.12**)

(a)   **(b)**

(c)   **(d)**

12. Given the following information, construct a Born–Haber cycle to calculate the lattice energy of $CaCl_2$ (s). (**LO 6.13**)

  Net energy change for the formation of $CaCl_2$ (s) from Ca(s) and Cl_2 (g) = -795.4 kJ/mol
  Heat of sublimation for Ca(s) = $+178$ kJ/mol
  E_{i1} for Ca(g) = $+590$ kJ/mol
  E_{i2} for Ca(g) = $+1145$ kJ/mol
  Bond dissociation energy for $Cl_2(g)$ = $+243$ kJ/mol
  E_{ea1} for Cl(g) = -348.6 kJ/mol

  (a) 2603 kJ/mol   (b) 2254 kJ/mol
  (c) 2481 kJ/mol   (d) 1663 kJ/mol

13. Select the compound with the highest lattice energy. (**LO 6.14**)

  (a) CaS   (b) $SrCl_2$
  (c) NaI   (d) $CuBr_2$

Answers:

1. a, 2. c, 3. c, 4. d, 5. d, 6. b, 7. c, 8. a, 9. a, 10. d, 11. c, 12. b, 13. a

---

Mastering Chemistry   provides end-of-chapter exercises, feedback-enriched tutorial problems, animations, and interactive activities to encourage problem-solving practice and deeper understanding of key concepts and topics.

RAN   *Randomized in Mastering Chemistry*

CONCEPTUAL PROBLEMS

Problems 6.1–6.20 appear within the chapter.

6.21 Where on the periodic table would you find the element that has an ion with each of the following electron configurations? Identify each ion.

  (a) 3+ion: $1s^2 2s^2 2p^6$
  (b) 3+ ion: $[Ar] 3d^3$
  (c) 2+ ion: $[Kr] 5s^2 4d^{10}$
  (d) 1+ ion: $[Kr] 4d^{10}$

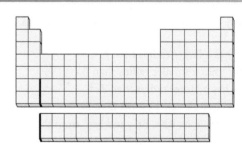

6.22 Which of the following spheres is likely to represent a metal atom and which a nonmetal atom? Which sphere in the products represents a cation and which an anion?

6.23 Circle the approximate part or parts of the periodic table where the following elements appear:
(a) Elements with the smallest values of E_{i1}
(b) Elements with the largest atomic radii
(c) Elements with the most negative values of E_{ea}

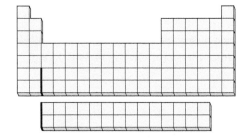

6.24 Order the indicated three elements according to the ease with which each is likely to lose its third electron.

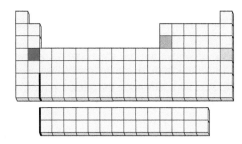

6.25 This figure represents the successive ionization energy of an atom
RAN in the third period of the periodic table. Which atom is this most likely to be?

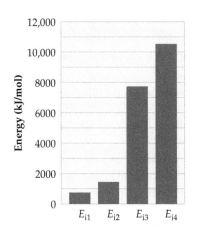

6.26 In the following drawings, red spheres represent cations and blue spheres represent anions. Match each of the drawings (a)–(d) with the following ionic compounds.
(i) $Ca_3(PO_4)_2$ (ii) Li_2CO_3 (iii) $FeCl_2$ (iv) $MgSO_4$

(a)   (b)

(c)   (d)

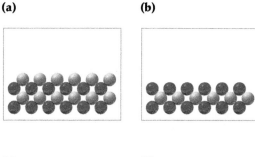

6.27 Which of the following drawings is more likely to represent an ionic compound and which a covalent compound?

(a)   (b)

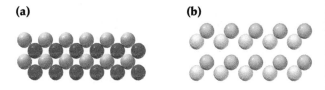

6.28 Each of the pictures (a)–(d) represents one of the following substances at 25 °C: sodium, chlorine, iodine, sodium chloride. Which picture corresponds to which substance?

(a)   (b)   (c)   (d)

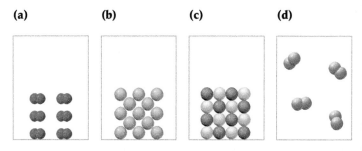

6.29 Which of the following alkali metal halides has the largest lattice energy, and which has the smallest lattice energy? Explain.

(a)   (b)   (c)

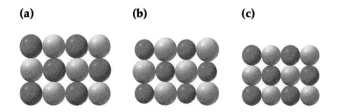

6.30 Which of the following alkali metal halides has the larger lattice energy, and which has the smaller lattice energy? Explain.

(a)   **(b)**

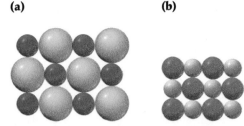

6.31 Three binary compounds are represented on the following drawing: red with red, blue with blue, and green with green. Give a likely formula for each compound.

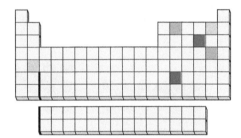

6.32 Given the following values for steps in the formation of LiCl(s) from its elements, draw a Born–Haber cycle similar to that shown in Figure 6.7.

E_{ea} for Cl(g) = −348.6 kJ/mol
Heat of sublimation for Li(s) = +159.4 kJ/mol
E_{i1} for Li(g) = +520 kJ/mol
Bond dissociation energy for $Cl_2(g)$ = +243 kJ/mol
Lattice energy for LiCl(s) = +853 kJ/mol

6.33 Given the following values for steps in the formation of CaO(s) from its elements, draw a Born–Haber cycle similar to that shown in Figure 6.7.

E_{ea1} for O(g) = −141 kJ/mol
E_{ea2} for O(g) = 745.1 kJ/mol
Heat of sublimation for Ca(s) = 178 kJ/mol
E_{i1} for Ca(g) = 590 kJ/mol
E_{i1} for Ca(g) = 1145 kJ/mol
Bond dissociation energy for $O_2(g)$ = 498 kJ/mol
Lattice energy for CaO(s) = 3401 kJ/mol

SECTION PROBLEMS

Electron Configuration of Ions (Section 6.1)

6.34 What is the difference between a covalent bond and an ionic bond?

6.35 Characterize bonds between the two atoms as covalent or ionic.
RAN
(a) Cl and F   (b) Rb and F
(c) Na and S   (d) N and S

6.36 What is the difference between a molecule and an ion?

6.37 Label the following species as molecules or ions.
RAN
(a) NO_3^- (b) NH_4^+
(c) NO_2 (d) $CH_3CO_2^-$

6.38 How many protons and electrons are in each of the following ions?
(a) Be^{2+} (b) Rb^+
(c) Se^{2-} (d) Au^{3+}

6.39 What is the identity of the element X in the following ions?
(a) X^{2+}, a cation that has 36 electrons
(b) X^-, an anion that has 36 electrons

6.40 What are the likely ground-state electron configurations of the following cations?
(a) La^{3+} (b) Ag^+ (c) Sn^{2+}

6.41 What are the likely ground-state electron configurations of the following anions?
(a) Se^{2-} (b) N^{3-}

6.42 What is the electron configuration of Ca^{2+}? What is the electron configuration of Ti^{2+}?
RAN

6.43 Identify the element whose 2+ ion has the ground-state electron configuration $[Ar]\, 3d^{10}$.

6.44 What doubly positive ion has the ground-state electron configuration $1s^2\, 2s^2\, 2p^6$?
RAN

6.45 What tripositive ion has the electron configuration $[Kr]\, 4d^3$? What neutral atom has the electron configuration $[Kr]\, 5s^2\, 4d^2$?
RAN

6.46 There are two elements in the transition-metal series Sc through Zn that have four unpaired electrons in their 2+ ions. Identify them.

6.47 Which element in the transition-metal series Sc through Zn has five unpaired electrons in its +3 ion?
RAN

Ionic Radii (Section 6.2)

6.48 Which atom or ion in the following pairs would you expect to be larger?
RAN
(a) S or S^{2-}
(b) Ca or Ca^{2+}
(c) O^- or O^{2-}

6.49 Which atom or ion in the following pairs would you expect to be larger?
RAN
(a) Rb or Rb^+ (b) N or N^{3-}
(c) Cr^{3+} or Cr^{6+}

6.50 Order the following ions from smallest to largest: Sr^{2+}, Se^{2-}, Br^-, Rb^+.
RAN

6.51 Order the following ions from smallest to largest: Mg^{2+}, O^{2-}, F^-, Na^+.
RAN

6.52 Which ion has a larger atomic radius, Cu^+ or Cu^{2+}? Explain your reasoning.

6.53 Which ion has a larger atomic radius, Fe^{2+} or Fe^{3+}? Explain your reasoning.

6.54 The following ions all have the same number of electrons: Ti^{4+}, Sc^{3+}, Ca^{2+}, S^{2-}. Order them according to their expected sizes, and explain your answer.

6.55 Which of the ions Se^{2-}, F^-, O^{2-}, and Rb^+ has the largest radius? Explain.

Ionization Energy (Section 6.3)

6.56 Which group of elements in the periodic table has the largest E_{i1}, and which group has the smallest? Explain.

6.57 Which element in the periodic table has the smallest ionization energy? Which has the largest?

6.58 Which element in each of the following sets has the smallest first ionization energy, and which has the largest?

(a) Li, Ba, K   (b) B, Be, Cl

(c) Ca, C, Cl

6.59 Order the elements in each set from the smallest to largest first ionization energy.

(a) Na, I, P   (b) P, Sr, Mg

(c) Ca, Cs, Se

Higher Ionization Energies (Section 6.4)

6.60 (a) Which has the smaller second ionization energy, K or Ca?

RAN (b) Which has the larger third ionization energy, Ga or Ca?

6.61 (a) Which has the smaller fourth ionization energy, Sn or Sb?

(b) Which has the larger sixth ionization energy, Se or Br?

6.62 Three atoms have the following electron configurations:

(a) $1s^2\,2s^2\,2p^6\,3s^2\,3p^3$

(b) $1s^2\,2s^2\,2p^6\,3s^2\,3p^6$

(c) $1s^2\,2s^2\,2p^6\,3s^2\,3p^6\,4s^2$

Which of the three has the largest E_{i2}? Which has the smallest E_{i7}?

6.63 Three atoms have the following electron configurations:

(a) $1s^2\,2s^2\,2p^6\,3s^2\,3p^1$

(b) $1s^2\,2s^2\,2p^6\,3s^2\,3p^5$

(c) $1s^2\,2s^2\,2p^6\,3s^2\,3p^6\,4s^1$

Which of the three has the largest E_{i1}? Which has the smallest E_{i4}?

6.64 The first four ionization energies in kJ/mol of a certain second-row element are 801, 2427, 3660, and 25,025. What is the likely identity of the element?

6.65 The first four ionization energies in kJ/mol of a certain second-row element are 900, 1757, 14,849, and 21,007. What is the likely identity of the element?

Electron Affinity (Section 6.5)

6.66 What is the relationship between the electron affinity of a singly charged cation such as Na^+ and the ionization energy of the neutral atom?

6.67 What is the relationship between the ionization energy of a singly charged anion such as Cl^- and the electron affinity of the neutral atom?

6.68 Which has the more negative electron affinity, Na^+ or Na? Na^+ or Cl?

6.69 Which has the more negative electron affinity, Br or Br^-?

6.70 Why is energy usually released when an electron is added to a neutral atom but absorbed when an electron is removed from a neutral atom?

6.71 Why does ionization energy increase regularly across the periodic table from group 1A to group 8A, whereas electron affinity increases irregularly from group 1A to group 7A and then falls dramatically for group 8A?

6.72 No element has a negative second electron affinity. That is, the process $A^-(g) + e^- \rightarrow A^{2-}(g)$ is unfavorable for every element. Suggest a reason.

6.73 Why does phosphorus have a less negative electron affinity than its neighbors silicon and sulfur?

Octet Rule (Section 6.6)

6.74 What noble-gas configurations and charge are the following

RAN elements likely to attain in reactions in which they form ions?

(a) N   (b) Ca   (c) S   (d) Br

6.75 What noble-gas configurations and charge are the following elements likely to attain in reactions in which they form ions?

(a) I   (b) O   (c) Al   (d) Ca

6.76 Each of the following pairs of elements will react to form

RAN a binary ionic compound. Write the formula of each compound formed, and give its name.

(a) Magnesium and chlorine

(b) Calcium and oxygen

(c) Lithium and nitrogen

(d) Aluminum and oxygen

6.77 Each of the following pairs of elements will react to form a binary ionic compound. Write the formula of each compound formed, and give its name.

(a) Sodium and iodine   (b) Potassium and sulfur

(c) Lithium and nitrogen   (d) Barium and flourine

6.78 Element X reacts with element Y to give a product containing X^{3+} ions and Y^{2-} ions.

(a) Is element X likely to be a metal or a nonmetal? Explain.

(b) Is element Y likely to be a metal or a nonmetal? Explain.

(c) What is the formula of the product?

(d) In what groups of the periodic table are elements X and Y likely to be found?

6.79 Element X reacts with element Y to give a product containing X^{2+} ions and Y^- ions.

(a) Is element X likely to be a metal or a nonmetal? Explain.

(b) Is element Y likely to be a metal or a nonmetal? Explain.

(c) What is the formula of the product?

(d) In what groups of the periodic table are elements X and Y likely to be found?

Formation of Ionic Compounds (Section 6.7)

6.80 Calculate the energy change in kilojoules per mole when lithium atoms lose an electron to bromine atoms to form isolated Li^+ and Br^- ions. [The E_i for $Li(g)$ is 520 kJ/mol; the E_{ea} for $Br(g)$ is -325 kJ/mol.] Will a lithium atom transfer an electron to a bromine atom to form isolated $Li^+(g)$ and $Br^-(g)$ ions? Explain.

6.81 Cesium has the smallest ionization energy of all elements (376 kJ/mol), and chlorine has the most negative electron affinity (-349 kJ/mol). Will a cesium atom transfer an electron to a chlorine atom to form isolated $Cs^+(g)$ and $Cl^-(g)$ ions? Explain.

6.82 Find the lattice energy of LiBr(s) in Table 6.3, and calculate the energy change in kilojoules for the formation of one mole of solid LiBr from the elements. [The sublimation energy for Li(s) is +159.4 kJ/mol, the bond dissociation energy of $Br_2(g)$ is +224 kJ/mol, and the energy necessary to convert $Br_2(l)$ to $Br_2(g)$ is 30.9 kJ/mol.]

6.83 Look up the lattice energies in Table 6.3, and calculate the energy change in kilojoules per mole for the formation of the following substances from their elements.

(a) LiF(s) [The sublimation energy for Li(s) is +159.4 kJ/mol, the E_i for Li(g) is 520 kJ/mol, the E_{ea} for F(g) is −328 kJ/mol, and the bond dissociation energy of $F_2(g)$ is +158 kJ/mol.]

(b) $CaF_2(s)$ [The sublimation energy for Ca(s) is +178.2 kJ/mol, E_{i1} = +589.8 kJ/mol, and E_{i2} = +1145 kJ/mol.]

6.84 Born–Haber cycles, such as those shown in Figures 6.7 and 6.8, are called *cycles* because they form closed loops. If any five of the six energy changes in the cycle are known, the value of the sixth can be calculated. Use the following five values to calculate the lattice energy in kilojoules per mole for sodium hydride, NaH(s):

E_{ea} for H(g) = −72.8 kJ/mol

E_{i1} for Na(g) = +495.8 kJ/mol

Heat of sublimation for Na(s) = +107.3 kJ/mol

Bond dissociation energy for $H_2(g)$ = +435.9 kJ/mol

Net energy change for the formation of $NaH_2(s)$ from its elements = −60 kJ/mol

6.85 Calculate a lattice energy for $CaH_2(s)$ in kilojoules per mole using the following information:

E_{ea} for H(g) = −72.8 kJ/mol

E_{i1} for Ca(g) = +589.8 kJ/mol

E_{i2} for Ca(g) = +1145 kJ/mol

Heat of sublimation for Ca(s) = +178.2 kJ/mol

Bond dissociation energy for $H_2(g)$ = +435.9 kJ/mol

Net energy change for the formation of $CaH_2(s)$ from its elements = −186.2 kJ/mol

6.86 Calculate the overall energy change in kilojoules per mole for the formation of CsF(s) from its elements using the following data:

E_{ea} for F(g) = −328 kJ/mol

E_{i1} for Cs(g) = +375.7 kJ/mol

E_{i2} for Cs(g) = +2422 kJ/mol

Heat of sublimation for Cs(s) = +76.1 kJ/mol

Bond dissociation energy for $F_2(g)$ = +158 kJ/mol

Lattice energy for CsF(s) = +740 kJ/mol

6.87 The estimated lattice energy for $CsF_2(s)$ is +2347 kJ/mol. Use the data given in Problem 6.86 to calculate an overall energy change in kilojoules per mole for the formation of $CsF_2(s)$ from its elements. Does the overall reaction absorb energy or release it? In light of your answer to Problem 6.86, which compound is more likely to form in the reaction of cesium with fluorine, CsF or CsF_2?

6.88 Calculate the overall energy change in kilojoules per mole for the formation of CaCl(s) from the elements. The following data are needed:

E_{ea} for Cl(g) = −348.6 kJ/mol

E_{i1} for Ca(g) = +589.8 kJ/mol

E_{i2} for Ca(g) = +1145 kJ/mol

Heat of sublimation for Ca(s) = +178.2 kJ/mol

Bond dissociation energy for $Cl_2(g)$ = +243 kJ per mol of Cl_2 (g)

Lattice energy for $CaCl_2(s)$ = +2258 kJ/mol

Lattice energy for CaCl(s) = +717 kJ/mol (estimated)

6.89 Use the data in Problem 6.88 to calculate an overall energy change for the formation of $CaCl_2(s)$ from the elements. Which is more likely to form, CaCl or $CaCl_2$?

6.90 Use the data and the result in Problem 6.84 to draw a Born–Haber cycle for the formation of NaH(s) from its elements.

6.91 Use the data and the result in Problem 6.83(a) to draw a Born–Haber cycle for the formation of LiF(s) from its elements.

6.92 Calculate overall energy changes in kilojoules per mole for the formation of MgF(s) and $MgF_2(s)$ from their elements. In light of your answers, which compound is more likely to form in the reaction of magnesium with fluorine, MgF or MgF_2? The following data are needed:

E_{ea} for F(g) = −328 kJ/mol

E_{i1} for Mg(g) = +737.7 kJ/mol

E_{i2} for Mg(g) = +1450.7 kJ/mol

Heat of sublimation for Mg(s) = +147.7 kJ/mol

Bond dissociation energy for $F_2(g)$ = +158 kJ/mol

Lattice energy for $MgF_2(s)$ = +2952 kJ/mol

Lattice energy for MgF(s) = 930 kJ/mol (estimated)

6.93 Draw Born–Haber cycles for the formation of both MgF and MgF_2 (Problem 6.92).

6.94 We saw in Section 6.7 that the reaction of solid sodium with gaseous chlorine to yield solid sodium chloride (Na^+Cl^-) is favorable by 411 kJ/mol. Calculate the energy change for the alternative reaction that yields chlorine sodide (Cl^+Na^-), and then explain why sodium chloride formation is preferred.

$$2\,Na(s) + Cl_2(g) \rightarrow 2\,Cl^+Na^-(s)$$

Assume that the lattice energy for Cl^+Na^- is the same as that for Na^+Cl^-. The following data are needed in addition to that found in Section 6.7:

E_{ea} for Na(g) = −52.9 kJ/mol

E_{i1} for Cl(g) = +1251 kJ/mol

6.95 Draw a Born–Haber cycle for the reaction of sodium with chlorine to yield chlorine sodide (Problem 6.94).

6.96 Use the following information plus the data given in Tables 6.2 and 6.3 to calculate the second electron affinity, E_{ea2}, of oxygen. Is the O^{2-} ion stable in the gas phase? Why is it stable in solid MgO?

Heat of sublimation for Mg(s) = +147.7 kJ/mol

Bond dissociation energy for $O_2(g)$ = +498.4 kJ/mol

E_{ea1} for O(g) = −141.0 kJ/mol

Net energy change for formation of MgO(s) from its elements = −601.7 kJ/mol

6.97 Given the following information, construct a Born–Haber cycle to calculate the lattice energy of $CaC_2(s)$.

Net energy change for the formation of $CaC_2(s)$ = −60 kJ/mol

Heat of sublimation for Ca(s) = +178 kJ/mol

E_{i1} for Ca(g) = +590 kJ/mol

E_{i2} for Ca(g) = +1145 kJ/mol

Heat of sublimation for $C(s) = +717$ kJ/mol

Bond dissociation energy for $C_2(g) = +614$ kJ/mol

E_{ea1} for $C_2(g) = -315$ kJ/mol

E_{ea2} for $C_2(g) = +410$ kJ/mol

Lattice Energy (Section 6.8)

6.98 Order the following compounds according to their expected lattice energies: LiCl, KCl, KBr, $MgCl_2$.

6.99 Order the following compounds according to their expected lattice energies: $AlBr_3$, $MgBr_2$, LiBr, CaO.

MULTICONCEPT PROBLEMS

6.100 Many early chemists noted a diagonal relationship among elements in the periodic table, whereby a given element is sometimes more similar to the element below and to the right than it is to the element directly below. Lithium is more similar to magnesium than to sodium, for example, and boron is more similar to silicon than to aluminum. Use your knowledge about the periodic trends of such properties as atomic radii and Z_{eff} to explain the existence of diagonal relationships.

6.101 Heating elemental cesium and platinum together for two days at 973 K gives a dark red ionic compound that is 57.67% Cs and 42.33% Pt.

(a) What is the empirical formula of the compound?

(b) What are the charge and electron configuration of the cesium ion?

(c) What are the charge and electron configuration of the platinum ion?

6.102 Given the following information, construct a Born–Haber cycle to calculate the lattice energy of $CrCl_2I(s)$:

Net energy change for the formation of $CrCl_2I(s) = -420$ kJ/mol

Bond dissociation energy for $Cl_2(g) = +243$ kJ/mol

Bond dissociation energy for $I_2(g) = +151$ kJ/mol

Heat of sublimation for $I_2(s) = +62$ kJ/mol

Heat of sublimation for $Cr(s) = +397$ kJ/mol

E_{i1} for $Cr(g) = 652$ kJ/mol

E_{i2} for $Cr(g) = 1588$ kJ/mol

E_{i3} for $Cr(g) = 2882$ kJ/mol

E_{ea} for $Cl(g) = -349$ kJ/mol

E_{ea} for $I(g) = -295$ kJ/mol

6.103 Consider the electronic structure of the element bismuth.

(a) The first ionization energy of bismuth is $E_{i1} = +703$ kJ/mol. What is the longest possible wavelength of light that could ionize an atom of bismuth?

(b) Write the electron configurations of neutral Bi and the Bi^+ cation.

(c) What are the n and l quantum numbers of the electron removed when Bi is ionized to Bi^+?

(d) Would you expect element 115 to have an ionization energy greater than, equal to, or less than that of bismuth? Explain.

6.104 Iron is commonly found as Fe, Fe^{2+}, and Fe^{3+}.

(a) Write electron configurations for each of the three.

(b) What are the n and l quantum numbers of the electron removed on going from Fe^{2+} to Fe^{3+}?

(c) The third ionization energy of Fe is $E_{i3} = +2952$ kJ/mol. What is the longest wavelength of light that could ionize $Fe^{2+}(g)$ to $Fe^{3+}(g)$?

(d) The third ionization energy of Ru is less than the third ionization energy of Fe. Explain.

6.105 The ionization energy of an atom can be measured by photoelectron spectroscopy, in which light of wavelength λ is directed at an atom, causing an electron to be ejected. The kinetic energy of the ejected electron (E_K) is measured by determining its velocity, v since $E_K = 1/2\ mv^2$. The E_i is then calculated using the relationship that the energy of the incident light equals the sum of E_i plus E_K.

(a) What is the ionization energy of rubidium atoms in kilojoules per mole if light with $\lambda = 58.4$ nm produces electrons with a velocity of 2.450×10^6 m/s? (The mass of an electron is 9.109×10^{-31} kg.)

(b) What is the ionization energy of potassium in kilojoules per mole if light with $\lambda = 142$ nm produces electrons with a velocity of 1.240×10^6 m/s?

chapter 7

Covalent Bonding and Electron-Dot Structures

Contents

Parathion is an organophosphate insecticide used to control pests in many crops such as corn, wheat, and cotton. The toxicity of organophosphate compounds is influenced by the properties of chemical bonds discussed in this chapter.

How does bond polarity affect the toxicity of organophosphate insecticides?

The answer to this question can be found on page 265 in the INQUIRY ?

The chapter-opening image shows the chemical structure of the insecticide parathion. In this *compound,* all the bonds are between two nonmetal elements such as C—C, C—H, C—O, and P—S. We saw in the previous chapter that an ionic bond between a metal and a reactive nonmetal is typically formed by the transfer of electrons between atoms. The metal loses one or more electrons and becomes a cation, while the nonmetal atom gains one or more electrons and becomes an anion. The oppositely charged ions are held together by electrostatic attractions called *ionic bonds.*

How, though, do bonds form between atoms of the same or similar elements? Simply put, the answer is that the bonds in such compounds are formed by the *sharing* of electrons between atoms rather than by the transfer of electrons from one atom to another. As we saw in Section 2.10, a bond formed by the sharing of electrons is called a *covalent bond,* and the unit of matter held together by one or more covalent bonds is called a *molecule.* We'll explore the nature of covalent bonding in this chapter.

7.1 COVALENT BONDING IN MOLECULES

The covalent bonding model involves a *sharing* of electrons between two atoms, in contrast to the ionic bond, in which a *transfer* of electrons occurs. To see how the formation of a covalent bond between atoms can be described, let's look at the H—H bond in the H_2 molecule as the simplest example. When two hydrogen atoms come close together, electrostatic interactions develop between them. The two positively charged nuclei repel each other, and the two negatively charged electrons repel each other, but each nucleus attracts both electrons (**FIGURE 7.1**). If the attractive forces are stronger than the repulsive forces, a covalent bond forms, holding the two atoms together. The two shared electrons occupy the region between the nuclei.

In essence, the shared electrons act as a kind of "glue" to bind the two atoms into an H_2 molecule. Both nuclei are simultaneously attracted to the same electrons and are therefore held together, much as two tug-of-war teams pulling on the same rope are held together.

The magnitudes of the various attractive and repulsive forces between nuclei and electrons in a covalent bond depend on how close the atoms are. If the hydrogen atoms are too far apart, the attractive forces are small and no bond exists. If the hydrogen atoms are too close together, the repulsive interaction between the nuclei becomes so strong that it pushes the atoms apart. Thus, there is an optimum distance between nuclei called the *bond length* where net attractive forces are maximized and the H—H molecule is most stable. In the H_2 molecule, the bond length is 74 pm. On a graph of energy versus internuclear distance, the bond length is the H—H distance in the minimum energy, most stable arrangement (**FIGURE 7.2**).

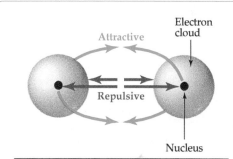

The nucleus–electron attractions are greater than the nucleus–nucleus and electron–electron repulsions, resulting in a net attractive force that binds the atoms together.

▲ **FIGURE 7.1**

A covalent H—H bond. The bond is the net result of attractive and repulsive electrostatic forces.

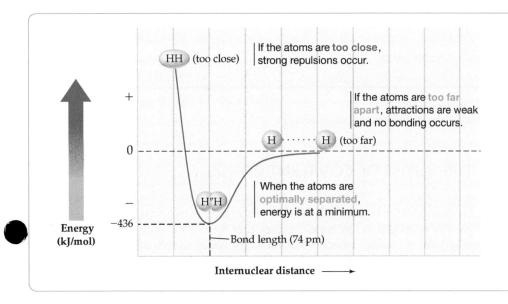

◀ **FIGURE 7.2**

A graph of energy versus internuclear distance for the H_2 molecule.

◀ **Figure It out**

Why is 74 pm the bond length of H_2?

Answer: At an internuclear distance of 74 pm, the energy is lowest because attractive forces are maximized. The lowest energy state is the most stable.

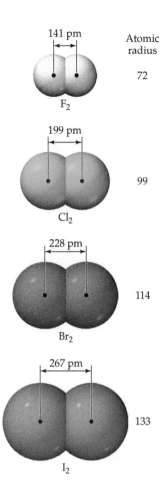

141 pm
F₂
Atomic radius
72

199 pm
Cl₂
99

228 pm
Br₂
114

267 pm
I₂
133

Every bond in every molecule has its own specific **bond length**, the distance between nuclei of two bonded atoms. Not surprisingly, though, bonds between the same pairs of atoms usually have similar lengths. For example, carbon–carbon single bonds usually have lengths in the range 152.0–153.5 pm regardless of the exact structure of the molecule. Note in the following examples that covalent bonds are indicated by **lines between atoms,** as described in Section 2.10.

Lines between the atoms indicate a shared electron pair in covalent bonds.

Ethane
(C—C) Bond length: 153.5 pm

Propane
153.2 pm

Butane
153.1 pm

Because similar bonds have similar lengths, it's possible to construct a table of average values to compare different kinds of bonds (**TABLE 7.1**). Keep in mind, though, that the actual value in a specific molecule might vary by $\pm 10\%$ from the average.

Every covalent bond has its own characteristic length that leads to maximum stability and that is roughly predictable from the knowledge of atomic radii (Section 5.13). For example, because the atomic radius of hydrogen is 37 pm and the atomic radius of chlorine is 99 pm, the H—Cl bond length in a hydrogen chloride molecule should be approximately 37 pm + 99 pm = 136 pm. The actual value is 127 pm. Note that in the series of diatomic halogen molecules that bond length increases down the group as the atomic radius of the element increases.

TABLE 7.1   Average Bond Lengths (pm)

H—H	74[a]	C—H	110	N—H	98	O—F	130	I—I	267[a]
H—C	110	C—C	154	N—C	147	O—Cl	165	S—F	168
H—F	92[a]	C—F	141	N—F	134	O—Br	180	S—Cl	203
H—Cl	127[a]	C—Cl	176	N—Cl	169	O—I	199	S—Br	218
H—Br	142[a]	C—Br	191	N—Br	184	O—N	136	S—S	208
H—I	161[a]	C—I	176	N—N	140	O—O	132		
H—N	98	C—N	147	N—O	136	F—F	141[a]		
H—O	94	C—O	143	O—H	94	Cl—Cl	199[a]		
H—S	132	C—S	181	O—C	143	Br—Br	228[a]		

Multiple covalent bonds[b]

C=C	134	C≡C	120	C=O	121	O=O	121[a]	N≡N	113[a]

[a]Exact value.

[b]We'll discuss multiple covalent bonds in Section 7.5.

7.2 STRENGTHS OF COVALENT BONDS

Look again at Figure 7.2, the graph of energy versus internuclear distance for the H_2 molecule, and note how the H_2 molecule is lower in energy than two separate hydrogen atoms. When pairs of hydrogen atoms bond together, they form lower-energy H_2 molecules and release 436 kJ/mol. In other words, 436 kJ must be *added* to split 1 mol of H_2 molecules apart into 2 mol of hydrogen atoms.

BIG IDEA Question 1

Select the *false* statement about chemical bonds.
(a) The bond length corresponds to the distance between nuclei that is the highest energy state.
(b) The same type of bond, such as a carbon–carbon single bond, has approximately the same length in many different molecules.
(c) A carbon–carbon double bond has a shorter bond length than a carbon–carbon single bond.

The amount of energy that must be supplied to break a chemical bond in an isolated molecule in the gaseous state—and thus the amount of energy released when the bond forms—is called the **bond dissociation energy (D).** Bond dissociation energies are always positive because energy must always be supplied to break a bond. Conversely, the amount of energy released on forming a bond always has a negative value.

Similar to bond length, every bond has its own specific bond dissociation energy and bonds between the same pairs of atoms usually have similar D values. For example, carbon–carbon single bonds usually have D values of approximately 350–380 kJ/mol regardless of the exact structure of the molecule.

$$
\begin{array}{ccc}
\text{Ethane} & \text{Propane} & \text{Butane} \\
D = 377\ \text{kJ/mol} & D = 370\ \text{kJ/mol} & D = 372\ \text{kJ/mol}
\end{array}
$$

The average bond dissociation energies for specific types of bonds are listed in **TABLE 7.2.** These also vary by $\pm 10\%$ depending on the actual structure of molecule. Bond dissociation energies cover a wide range, from a low of 151 kJ/mol for the I—I bond to a high of 570 kJ/mol for the H—F bond. As a rule of thumb, though, most of the bonds commonly encountered in naturally occurring molecules (C—H, C—C, C—O) have values in the range of 350–400 kJ/mol.

TABLE 7.2   **Average Bond Dissociation Energies, *D* (kJ/mol)**

H—H	436[a]	C—H	410	N—H	390	O—F	180	I—I	151[a]
H—C	410	C—C	350	N—C	300	O—Cl	200	S—F	310
H—F	570[a]	C—F	450	N—F	270	O—Br	210	S—Cl	250
H—Cl	432[a]	C—Cl	330	N—Cl	200	O—I	220	S—Br	210
H—Br	366[a]	C—Br	270	N—Br	240	O—N	200	S—S	225
H—I	298[a]	C—I	240	N—N	240	O—O	180		
H—N	390	C—N	300	N—O	200	F—F	159[a]		
H—O	460	C—O	350	O—H	460	Cl—Cl	243[a]		
H—S	340	C—S	260	O—C	350	Br—Br	193[a]		
Multiple covalent bonds[b]									
C=C	728[c]	C≡C	965[d]	C=O	732	O=O	498[a]	N≡N	945[a]

[a]Exact value.

[b]We'll discuss multiple covalent bonds in Section 7.5.

[c]Value in ethene.

[d]Value in ethyne.

Periodic properties and bond lengths can be used to explain why some bonds are stronger than others. Let's examine the bond lengths and values of D for a series of hydrogen halide bonds:

Bond	Bond Dissociation Energy (D) (kJ/mol)	Bond Length (pm)
H—F	570	92
H—Cl	432	127
H—Br	366	142
H—I	298	161

Go to
eText

BIG IDEA Question 2

Which molecule has the largest bond dissociation energy (D) for the carbon–oxygen bond?

(a)

:C≡O:

(b)

:O=C=O:

(c)

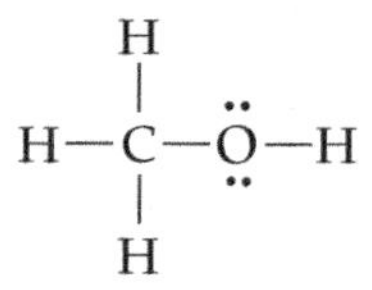

$$\begin{array}{c} H \\ | \\ H-C-\overset{..}{\underset{..}{O}}-H \\ | \\ H \end{array}$$

As we proceed down through the period of halogens, the value of D becomes smaller, meaning the bond is weaker. As the atomic radius of the halogen increases, the shared electrons are farther away and more shielded from the positively charged nucleus, leading to a longer and weaker bond. Although there are exceptions, *shorter bonds are typically stronger*. For example, we predict the F—F bond to be stronger than the Cl—Cl bond because fluorine atoms are smaller, but in fact the bond dissociation energy for F—F is 159 kJ/mol compared to 243 kJ/mol for Cl—Cl. Chapter 8 will describe two theories for covalent bonding, molecular orbital theory and valence bond theory, used to explain observed bond properties, including length and strength. We will see that certain models work well in explaining and predicting some properties while other models are better suited for different properties.

The correlation between bond length and strength also holds true for multiple bonds. In speaking of molecules with multiple bonds, we often use the term **bond order** to refer to the number of electron pairs shared between atoms. Thus, the F—F bond in the F_2 molecule has a bond order of 1, the O=O bond in the O_2 molecule has a bond order of 2, and the N≡N bond in the N_2 molecule has a bond order of 3.

Multiple bonds are both shorter and stronger than their corresponding single-bond counterparts because there are more shared electrons holding the atoms together. Compare, for example, the O=O double bond in O_2 with the O—O single bond in H_2O_2 (hydrogen peroxide), and compare the N≡N triple bond in N_2 with the N—N single bond in N_2H_4 (hydrazine):

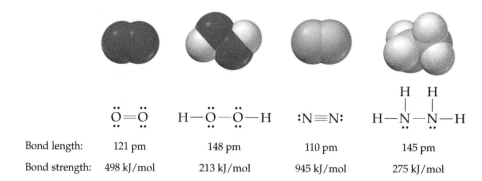

$\overset{..}{\underset{..}{O}}=\overset{..}{\underset{..}{O}}$	$H-\overset{..}{\underset{..}{O}}-\overset{..}{\underset{..}{O}}-H$	:N≡N:	$\begin{array}{c} H \quad H \\ \| \quad \| \\ H-\overset{}{\underset{..}{N}}-\overset{}{\underset{..}{N}}-H \end{array}$
Bond length: 121 pm	148 pm	110 pm	145 pm
Bond strength: 498 kJ/mol	213 kJ/mol	945 kJ/mol	275 kJ/mol

7.3 POLAR COVALENT BONDS: ELECTRONEGATIVITY

Up to this point, a given bond has been described as either purely ionic, with electrons completely transferred, or purely covalent, with electrons shared equally. In fact, though, ionic and covalent bonds represent only the two extremes of a continuous range of possibilities. Between these two extremes are the large majority of bonds in which the bonding electrons are shared unequally between two atoms but are not completely transferred, called **polar covalent bonds** (**FIGURE 7.3**). The lowercase Greek letter delta (δ) is used to denote the resultant partial charges on the atoms, either partial positive ($\delta+$) for the atom that has a smaller share of the bonding electrons or partial negative ($\delta-$) for the atom that has a larger share.

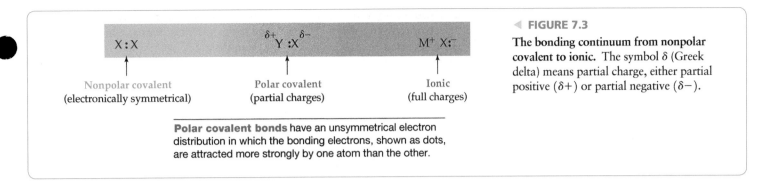

\delta+

The bonding continuum from nonpolar covalent to ionic. The symbol δ (Greek delta) means partial charge, either partial positive (δ+) or partial negative (δ−).

X:X $\overset{\delta+}{Y}:\overset{\delta-}{X}$ $M^+ X:^-$

Nonpolar covalent   Polar covalent   Ionic
(electronically symmetrical)   (partial charges)   (full charges)

Polar covalent bonds have an unsymmetrical electron distribution in which the bonding electrons, shown as dots, are attracted more strongly by one atom than the other.

Bond polarity results from differences in **electronegativity (EN)**, the ability of an atom in a molecule to attract the shared electrons in a covalent bond. As shown in **FIGURE 7.4,** electronegativities are expressed on a unitless scale, with fluorine, the most highly electronegative element, assigned a value of 4.0. Metallic elements on the left of the periodic table attract electrons only weakly and are the least electronegative elements. Halogens and other reactive nonmetals in the upper right of the table attract electrons strongly and are the most electronegative. Figure 7.4 shows that electronegativity generally increases across a period and decreases down a group.

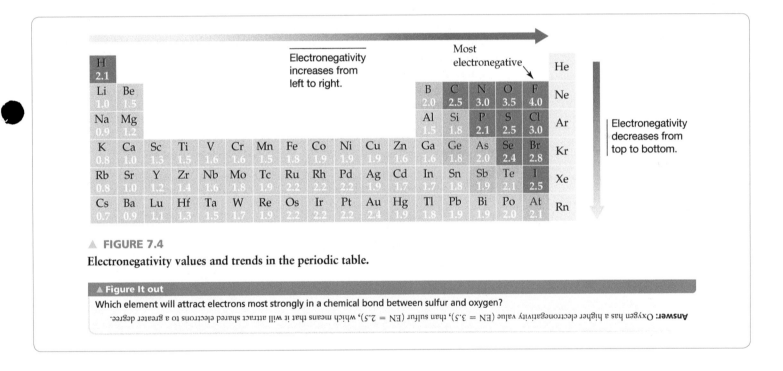

▲ **FIGURE 7.4**

Electronegativity values and trends in the periodic table.

▲ **Figure It out**

Which element will attract electrons most strongly in a chemical bond between sulfur and oxygen?

Answer: Oxygen has a higher electronegativity value (EN = 3.5), than sulfur (EN = 2.5), which means that it will attract shared electrons to a greater degree.

The polarity of a bond is another important factor that influences bond dissociation energy. The American chemist Linus Pauling (1901–1994) developed the electronegativity scale shown in Figure 7.4 by examining extra bond strength due to bond polarity. For example, we would expect the HF bond dissociation energy to be approximately the average of the bond dissociation energies of H_2 (436 kJ/mol) and F_2 (159 kJ/mol), or about 297 kJ/mol. However, the HF bond strength is much higher (570 kJ/mol). Pauling explained this discrepancy through bond polarity. The H_2 bond is nonpolar while the HF bond is highly polar, resulting in a partial negative charge on fluorine and a partial positive charge on hydrogen. The attraction between these partial charges *increases* the energy required to break the bond. In general, *increased bond polarity leads to increased bond strength.* Using the bond dissociation energies of many compounds, Pauling arrived at a scale of relative electronegativity values.

We can most easily visualize the extent of electron transfer between elements with different electronegativity values using *electrostatic potential maps,* which use color to portray the calculated electron distribution in an isolated, gas-phase molecule. Yellow-green represents a neutral, nonpolar atom; blue represents a deficiency of electrons on an atom (partial positive charge); and red represents a surplus of electrons on an atom (partial negative charge). As examples of different points along the bonding spectrum, let's look at the three substances Cl_2, HCl, and NaCl.

Nonpolar covalent. The bond in a chlorine molecule is nonpolar covalent because the Cl atoms have the same electronegativity value. The bonding electrons are attracted equally to the two identical chlorine atoms. A similar situation exists in all such molecules that contain a covalent bond between two identical atoms. Neither chlorine atom has a partial positive or partial negative charge as shown by their identical yellow-green coloration in the electrostatic potential map shown in **FIGURE 7.5.**

HCl: Polar covalent. The bond in a hydrogen chloride molecule is polar covalent. The chlorine atom with an electronegativity value of 3.0 attracts the bonding electron pair more strongly than hydrogen with an electronegativity value of 2.1. The electrostatic potential map in Figure 7.5 shows an unsymmetrical distribution of electrons. Chlorine thus has a partial negative charge (orange-red in the electrostatic potential map), and hydrogen has a partial positive charge (blue in the electrostatic potential map).

NaCl: Ionic. The bond in solid sodium chloride is largely ionic due to the large electronegativity difference between chlorine (EN = 3.0) and sodium (EN = 0.9). We can describe the bond as electrostatic attraction between Na^+ and Cl^-. However, experiments show that the NaCl bond is only about 80% ionic and that the electron transferred from Na to Cl still spends some of its time near sodium. Thus, the electron-poor sodium atom is blue in an electrostatic potential map, while the electron-rich chlorine is red (Figure 7.5).

Cl_2: Nonpolar covalent bond (ΔEN= 0)

Yellow-green represents a neutral atom.

Cl:Cl

The two bonding electrons, shown here as dots, are symmetrically distributed between the two Cl atoms.

HCl: Polar covalent bond (ΔEN= 0.9)

Red indicates a partial negative charge, and blue indicates a partial positive charge.

$^{\delta+}H — Cl^{\delta-}$

$[H :Cl]$

The two bonding electrons are attracted more strongly by Cl than by H.

NaCl: Ionic bond (ΔEN= 2.1)

Red indicates a partial negative charge, and blue indicates a partial positive charge.

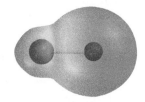

$Na^+ Cl^-$

An ionic bond represents a transfer of an electron from sodium to chlorine.

▲ **FIGURE 7.5**

Electrostatic potential maps show the extent of electron transfer between elements with different electronegativity values in a chemical bond.

WORKED EXAMPLE 7.1

Classifying Bond Type

Classify the carbon–chlorine bond in chloroform, $CHCl_3$, and the rubidium–bromine bond in rubidium bromide as nonpolar covalent, polar covalent, or ionic.

STRATEGY

Calculate the electronegativity difference between elements using values in Figure 7.4, and follow the classification guidelines.

- Bonds between atoms with the same or similar electronegativity are usually nonpolar covalent.
- Bonds between atoms whose electronegativities differ by more than two units are largely ionic.
- Bonds between atoms whose electronegativities differ by less than two units are usually polar covalent.

SOLUTION

We can be reasonably sure that a C—Cl, bond in chloroform, $CHCl_3$, is polar covalent, while an Rb^+Br^- bond in rubidium bromide is largely ionic.

$$H—\overset{\displaystyle Cl}{\underset{\displaystyle Cl}{C}}—Cl \qquad \text{(Polar covalent)} \qquad\qquad Rb^+\overset{\curvearrowleft}{}Br^- \quad \text{(Ionic)}$$

Chlorine:	EN = 3.0		Bromine:	EN = 2.8
Carbon:	EN = 2.5		Rubidium:	EN = 0.8

Difference = 0.5   Difference = 2.0

▶ **PRACTICE 7.1** Use the electronegativity values in Figure 7.4 to predict whether the bonds in the following compounds are nonpolar covalent, polar covalent, or ionic. Which is incorrectly labeled?

(a) $SiCl_4$ (nonpolar covalent)
(b) CsBr (ionic)
(c) F_2 (nonpolar covalent)
(d) CF_4 (polar covalent)

▶ **APPLY 7.2** An electrostatic potential map of water is shown below. Which atom, H or O, is positively polarized (electron-poor), and which is negatively polarized (electron-rich)? Is this polarity pattern consistent with the electronegativity values of O and H given in Figure 7.4?

Water

All **PRACTICE** and **APPLY** problems are interactive in the eText.

Worked Example 7.1 shows that differences in electronegativity values between elements can be used as a general guideline for classifying a bond as nonpolar covalent, polar covalent, or ionic. A general rule of thumb is that an ionic bond occurs when the difference in electronegativity between two elements is 2.0 or greater. In reality, there is little difference in bond polarity between a bond with $\Delta EN = 2.0$ and $\Delta EN = 1.9$. The measure of bond polarity in a diatomic molecule is a quantity called the **dipole moment**, μ (Greek mu), which is defined as the magnitude of the charge Q at either end of the molecular dipole times the distance r between the charges: $\mu = Q \times r$. Dipole moments are expressed in *debyes* (D), where $1\ D = 3.336 \times 10^{-30}$ coulomb meters (C·m) in SI units. To calibrate your thinking, the charge on an electron is 1.60×10^{-19} C. Thus, if a proton and an electron were separated by 100 pm (a bit less than the length of a typical covalent bond), then the dipole moment would be 1.60×10^{-29} C·m, or 4.80 D:

$$\mu = Q \times r$$
$$\mu = (1.60 \times 10^{-19}\ C)(100 \times 10^{-12}\ m)\left(\frac{1\ D}{3.336 \times 10^{-30}\ C \cdot m}\right) = 4.80\ D$$

Calculating the **percent ionic character** of a bond is a useful way to illustrate the extent of electron transfer as a continuum instead of as a sharp cutoff between a polar covalent bond and an ionic bond. A covalent bond with equal sharing electrons has 0% ionic character, and a perfect ionic bond has 100% ionic character. One method of estimating the percent ionic character of a bond is to compute the ratio of the observed dipole moment to the calculated dipole moment when charge separation is complete and then multiply the result by 100. Worked Example 7.2 shows how to calculate percent ionic character.

BIG IDEA Question 3

Two diatomic molecules A—B and C—D have identical bond lengths. The difference in electronegativity between A and B is 1.0, and the difference in electronegativity between C and D is 0.5. Which diatomic molecule will have the largest dipole moment?

—— **WORKED EXAMPLE 7.2**

Calculating Percent Ionic Character from a Dipole Moment

The dipole moment of NaCl in the gas phase is 9.00 D, and the distance between atoms is 236 pm. What is the percent ionic character of the NaCl bond?

continued on next page

STRATEGY

If NaCl were 100% ionic, a negative charge (Cl⁻) would be separated from a positive charge (Na⁺) by 236 pm. Calculate the dipole moment if charge separation were complete, and compare the calculated value to the actual value.

SOLUTION

The calculated dipole moment of complete charge separation is:

$$\mu = Q \times r$$

$$\mu = (1.60 \times 10^{-19}\ C)(236 \times 10^{-12}\ m) = 3.776 \times 10^{-29}\ C \cdot m$$

Converting to units of debye:

$$(3.776 \times 10^{-29}\ C \cdot m)\left(\frac{1\ D}{3.336 \times 10^{-30}\ C \cdot m}\right) = 11.3\ D$$

The observed dipole moment of 9.00 D for NaCl implies that the Na⁺Cl⁻ bond is 79.6% ionic:

$$\frac{9.00\ D}{11.3\ D} \times 100\% = 79.6\%$$

$$\text{Percent ionic character} = \frac{\text{experimental dipole moment } (\mu)}{\text{theoretical dipole moment } (\mu)} \times 100\%$$

▶ **PRACTICE 7.3** The dipole moment of AgCl in the gas phase is μ = 6.08 D, and the bond length is 228 pm. Calculate the percent ionic character of the Ag—Cl bond.

▶ **APPLY 7.4** Predict which bond has greater percent ionic character, Li—H or H—F, using electronegativity values in Figure 7.4. Check your prediction using the following data: the dipole moment of LiH is μ = 6.00 D, and the bond length is 160 pm; the dipole moment of HF is μ = 1.83 D, and the bond length is 92 pm.

7.4 A COMPARISON OF IONIC AND COVALENT COMPOUNDS

Look at the comparison between NaCl and HCl in **TABLE 7.3** to get an idea of the difference between ionic and covalent compounds. Sodium chloride, an ionic compound, is a white solid with a melting point of 801 °C and a boiling point of 1465 °C. Hydrogen chloride, a covalent compound, is a colorless gas with a melting point of −115 °C and a boiling point of −84.9 °C. What accounts for such large differences in properties between ionic compounds and covalent compounds?

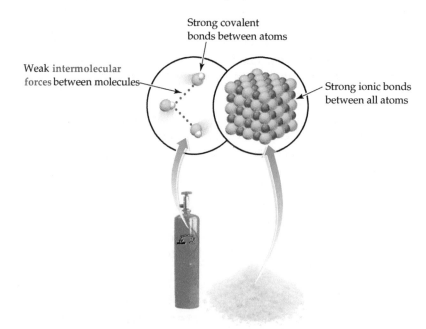

BIG IDEA Question 4

Go to eText

Do ionic or covalent compounds tend to have higher melting and boiling points? Why?

▲ Sodium chloride, an ionic compound, is a white, crystalline solid that melts at 801 °C. Hydrogen chloride, a molecular compound, is a gas at room temperature.

TABLE 7.3 Some Physical Properties of NaCl and HCl		
Property	NaCl	HCl
Formula mass	58.44 amu	36.46 amu
Physical appearance	White solid	Colorless gas
Type of bond	Ionic	Covalent
Melting point	801°C	−115 °C
Boiling point	1465°C	−84.9 °C

Ionic compounds are high-melting solids because there are strong ionic bonds between each atom in a lattice structure. As discussed previously in Section 2.11, a visible sample of sodium chloride consists of not NaCl molecules but a vast three-dimensional network of ions in which each Na^+ cation is attracted to many surrounding Cl^- anions and each Cl^- ion is attracted to many surrounding Na^+ ions. To melt or boil sodium chloride, all the ions must be separated from one another. This means that every ionic attraction in the entire crystal—the lattice energy (*U*)—must be overcome, a process that requires a large amount of energy.

Covalent compounds, by contrast, are low-melting solids, liquids, or even gases. A sample of a covalent compound, such as hydrogen chloride, consists of discrete HCl molecules. In order to melt or boil a covalent compound, forces between molecules must be broken. In covalent compounds, bonds exist *within* molecules but not *between* molecules. Attractive forces between molecules do exist and are called intermolecular forces. *Intermolecular forces are weaker than ionic or covalent bonds,* and therefore relatively little energy is required to overcome them to melt or boil the substance. We'll look at the nature of intermolecular forces in Chapter 8.

REMEMBER . . .
Lattice energy (*U*) is the amount of energy that must be supplied to break an ionic solid into its individual gaseous ions and is thus a measure of the strength of the crystal's ionic bonds (Section 6.8.)

LOOKING AHEAD . . .
The strength of intermolecular forces for molecular substances influence many properties such as melting and boiling points. Types of intermolecular forces are described in Section 8.6.

7.5 ELECTRON-DOT STRUCTURES: THE OCTET RULE

One way to picture the sharing of electrons between atoms in covalent or polar covalent bonds is to use electron-dot structures, or *Lewis structures,* named after G. N. Lewis (1875–1946) of the University of California at Berkeley. An electron-dot structure represents an atom's valence electrons by dots, and the placement of those dots indicates how the valence electrons are distributed in a molecule. The central idea is that when atoms form a bond they share electrons to achieve a *complete valence shell* or *noble-gas electron configuration.*

The electron-dot model, like most models, is a simplified picture of bonding that has been developed because it is useful in making predictions. Just as a blueprint serves as a model for a building, the Lewis structures can be thought of as a "blueprint" containing a set of rules for combining atoms to make molecules. Some key uses of the electron-dot model for bonding are in predicting molecular formulas, reactivity, and shape.

We can write an electron-dot structure for an H_2 molecule by drawing a pair of dots between the hydrogen atoms, indicating that the hydrogens share the pair of electrons in a covalent bond:

An electron-pair bond

$$H\cdot \quad \cdot H \quad \longrightarrow \quad H\!:\!H$$

Two hydrogen atoms   A hydrogen molecule

By sharing two electrons in a covalent bond, each hydrogen effectively has one electron pair and the stable, $1s^2$ electron configuration of the noble gas helium.

This hydrogen shares an electron pair ...   ... and this hydrogen shares an electron pair.

(a)

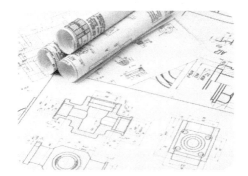

(b)

▲ Lewis structures can be thought of as a "blueprint" that provides a set of guidelines for combining atoms to make molecules and predict their three-dimensional structure.

Atoms other than hydrogen also form covalent bonds by sharing electron pairs, and we can draw the electron-dot structures of the resultant molecules by assigning the correct number of valence electrons to each atom. Group 3A atoms, such as boron, have three valence electrons; group 4A atoms, such as carbon, have four valence electrons; and so on across the periodic table. The group 7A element fluorine has seven valence electrons, and an electron-dot structure for the F_2 molecule shows how a covalent bond can form:

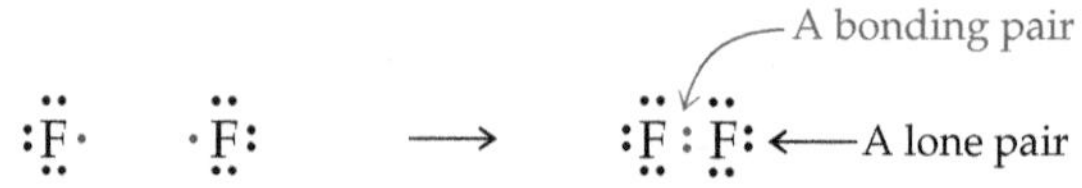

Two F atoms (seven valence electrons per atom) — An F_2 molecule (each F is surrounded by eight valence electrons)

Six of the seven valence electrons in a fluorine atom are already paired in three filled atomic orbitals and thus are not shared in bonding. The seventh fluorine valence electron, however, is unpaired and can be shared in a covalent bond with another fluorine. Each atom in the resultant F_2 molecule thereby gains a noble-gas configuration with eight valence-shell electrons and thus obeys the **octet rule,** discussed in Section 6.6. The three pairs of nonbonding electrons on each fluorine atom are called **lone pairs,** or *nonbonding pairs,* and the shared electrons are called a **bonding pair.**

REMEMBER . . .
According to the **octet rule,** main-group elements tend to undergo reactions that give them a noble-gas electron configuration with filled *s* and *p* subshells in their valence electron shell (Section 6.6.)

The tendency of main-group atoms to fill their *s* and *p* subshells and thereby achieve a noble-gas configuration when they form bonds is an important guiding principle that makes it possible to predict the formulas and electron-dot structures of a great many molecules. As a general rule, a main-group atom shares as many of its valence-shell electrons as possible, either until it has no more to share or until it reaches an octet configuration. The following guidelines apply:

- **Group 3A elements,** such as boron, have three valence electrons and can therefore form three electron-pair bonds in neutral molecules such as borane, BH_3. The boron atom in the resultant molecule has only three bonding pairs of electrons, however, and can't reach an electron octet. (The bonding situation in BH_3 is actually more complicated than suggested here; we'll deal with it in Section 22.5.)

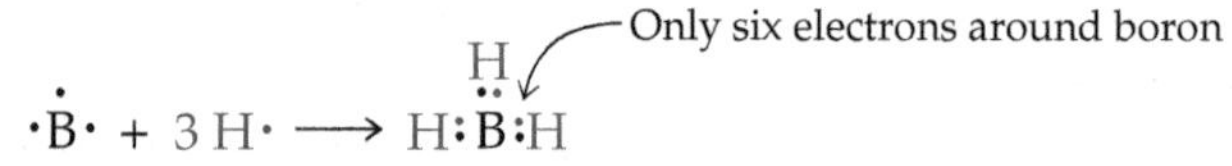

Borane

- **Group 4A elements,** such as carbon, have four valence electrons and form four bonds, as in methane, CH_4. The carbon atom in the resultant molecule has four bonding pairs of electrons.

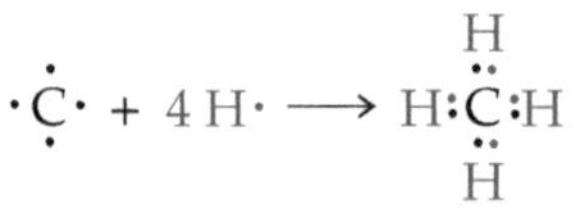

Methane

- **Group 5A elements,** such as nitrogen, have five valence electrons and form three bonds, as in ammonia, NH_3. The nitrogen atom in the resultant molecule has three bonding pairs of electrons and one lone pair.

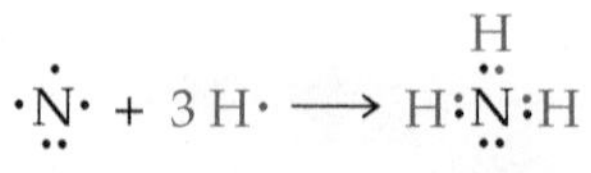

Ammonia

- **Group 6A elements,** such as oxygen, have six valence electrons and form two bonds, as in water, H_2O. The oxygen atom in the resultant molecule has two bonding pairs of electrons and two lone pairs.

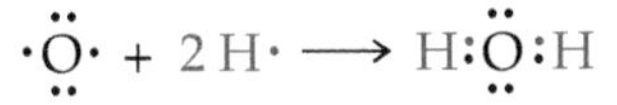

$$\cdot \ddot{O} \cdot + 2\,H \cdot \longrightarrow H\!:\!\ddot{O}\!:\!H$$
Water

- **Group 7A elements** (halogens), such as fluorine, have seven valence electrons and form one bond, as in hydrogen fluoride, HF. The fluorine atom in the resultant molecule has one bonding pair of electrons and three lone pairs.

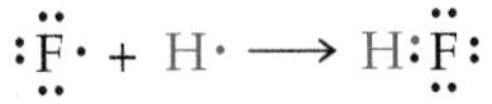

$$:\!\ddot{F} \cdot + H \cdot \longrightarrow H\!:\!\ddot{F}\!:$$
Hydrogen
fluoride

- **Group 8A elements** (noble gases), such as neon, rarely form covalent bonds because they already have valence-shell octets.

$$:\!\ddot{Ne}\!:$$ Does not form covalent bonds

TABLE 7.4 summarizes these conclusions.

TABLE 7.4	Covalent Bonding for Second-Row Elements		
Group	Number of Valence Electrons	Number of Bonds	Example
3A	3	3	BH_3
4A	4	4	CH_4
5A	5	3	NH_3
6A	6	2	H_2O
7A	7	1	HF
8A	8	0	Ne

Not all covalent bonds contain just one shared electron pair, or **single bond,** like those just discussed. In molecules such as O_2, N_2, and many others, the atoms share more than one pair of electrons, leading to the formation of *multiple* covalent bonds. The oxygen atoms in the O_2 molecule, for example, reach valence-shell octets by sharing two pairs, or four electrons, in a **double bond.** Similarly, the nitrogen atoms in the N_2 molecule share three pairs, or six electrons, in a **triple bond.**

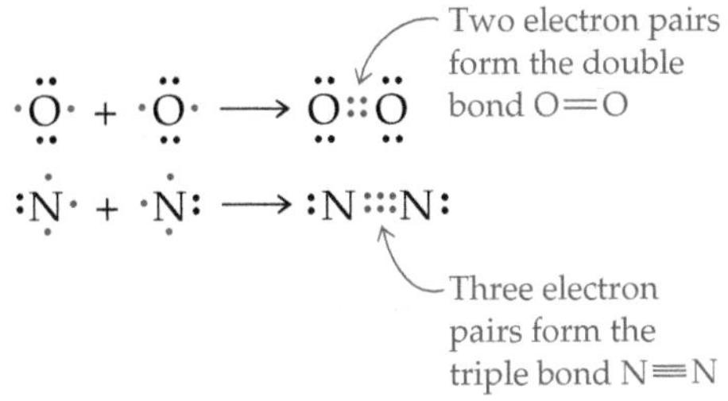

$$\cdot \ddot{O} \cdot + \cdot \ddot{O} \cdot \longrightarrow \ddot{O}\!::\!\ddot{O}$$
Two electron pairs form the double bond O=O

$$:\!\dot{N} \cdot + \cdot \dot{N}\!: \longrightarrow :\!N\!:::\!N\!:$$
Three electron pairs form the triple bond N≡N

One final point about covalent bonds involves the origin of the bonding electrons. Although most covalent bonds form when two atoms each contribute one electron, bonds can also form when one atom donates both electrons (a lone pair) to another atom that has a vacant valence orbital. The ammonium ion (NH_4^+), for instance, forms when the two lone-pair electrons from the nitrogen atom of ammonia, $:NH_3$, bond to H^+. Such bonds are sometimes called **coordinate covalent bonds,** which form when one atom donates an electron pair to an empty orbital on another atom.

An ordinary covalent bond—each atom donates one electron.

$$H\cdot + \cdot H \longrightarrow H \colon H$$

A coordinate covalent bond—the nitrogen atom donates both electrons.

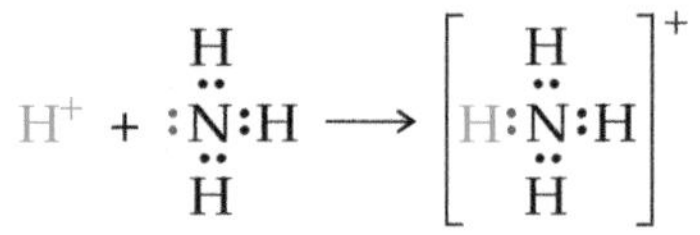

$$H^+ + \ddot{\underset{\displaystyle H}{\overset{\displaystyle H}{N}}}\colon H \longrightarrow \left[H \colon \ddot{\underset{\displaystyle H}{\overset{\displaystyle H}{N}}}\colon H \right]^+$$

Note that the nitrogen atom in the ammonium ion (NH_4^+) has more than the usual number of bonds—four instead of three—but that it still has an octet of valence electrons. Nitrogen, oxygen, phosphorus, and sulfur form coordinate covalent bonds frequently.

WORKED EXAMPLE 7.3

Drawing an Electron-Dot Structure

Draw an electron-dot structure for phosphine, PH_3.

STRATEGY

The number of covalent bonds formed by a main-group element depends on the element's group number. Phosphorus, a group 5A element, has five valence electrons and can achieve a valence-shell octet by forming three bonds and leaving one lone pair. Each hydrogen supplies one electron.

SOLUTION

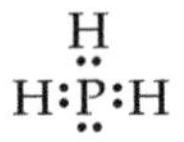

$$H \colon \ddot{\underset{\displaystyle \cdot \cdot}{P}}\colon H \qquad \text{Phosphine}$$
with H above P

▶ **PRACTICE 7.5**   Select the correct electron-dot structure for H_2S, hydrogen sulfide, a poisonous gas produced by rotten eggs.

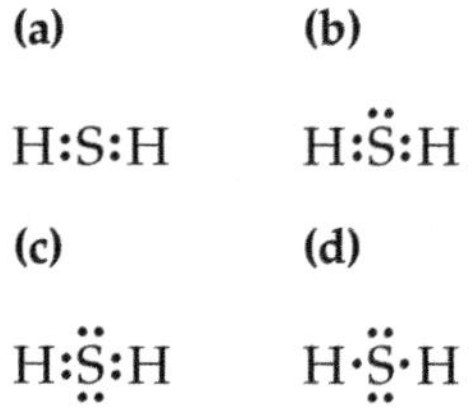

(a)   **(b)**

$H\colon S\colon H$ $H\colon \ddot{S}\colon H$

(c)   **(d)**

$H\colon \ddot{\underset{\displaystyle \cdot\cdot}{S}}\colon H$ $H\cdot \ddot{\underset{\displaystyle \cdot\cdot}{S}}\cdot H$

▶ **APPLY 7.6**   Use the octet rule to predict the molecular formula of compounds that form between the elements

(a) oxygen and fluorine.
(b) silicon and chlorine.

7.6 PROCEDURE FOR DRAWING ELECTRON-DOT STRUCTURES

In drawing structures, we indicate a two-electron covalent bond by a line. Similarly, we'll use two lines between atoms to represent four shared electrons (two pairs) in a double bond and three lines to represent six shared electrons (three pairs) in a triple bond. Arranging individual electrons when drawing electron-dot structures can be tedious, especially in large molecules or in those with multiple bonds. Instead, a general method of drawing electron-dot structures is outlined below.

Drawing Electron-Dot Structures

Step 1. Find the total number of valence electrons in the molecule or ion. Find the sum of the valence electrons for all atoms, add one additional electron for each negative charge in an anion, and subtract one electron for each positive charge in a cation. In SF_4, for example, the total is 34 (6 from sulfur and 7 from each of 4 fluorines). In OH^-, the total is 8 (6 from oxygen, 1 from hydrogen, and 1 for the negative charge). In NH_4^+, the total is 8 (5 from nitrogen and 1 from each of 4 hydrogens minus 1 for the positive charge).

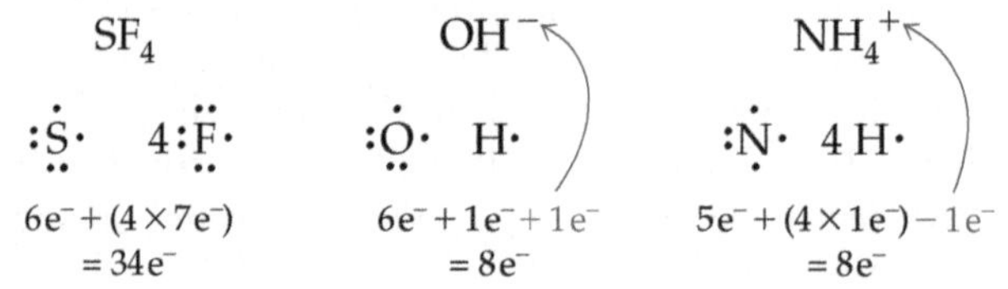

$$SF_4 \qquad\qquad OH^- \qquad\qquad NH_4^+$$

$$:\!\dot{\underset{\cdot\cdot}{S}}\!\cdot \quad 4:\!\ddot{\underset{\cdot\cdot}{F}}\!\cdot \qquad :\!\dot{\underset{\cdot\cdot}{O}}\!\cdot \;\; H\cdot \qquad :\!\dot{N}\!\cdot \quad 4\,H\cdot$$

$$6e^- + (4\times 7e^-) \qquad 6e^- + 1e^- + 1e^- \qquad 5e^- + (4\times 1e^-) - 1e^-$$
$$= 34e^- \qquad\qquad = 8e^- \qquad\qquad = 8e^-$$

Step 2. Decide what the connections are between atoms, and draw lines to represent the bonds. Often, you'll be told the connections; other times you'll have to guess. Guidelines for connecting atoms in an electron-dot structure are:

• The central atom is typically the one with the lowest electronegativity (except H).

• Hydrogen and the halogens usually form only one bond.

• Elements in the second row usually form the number of bonds given in Table 7.4.

• *(Octet Rule Exception)* Elements in Group 3A, such as B and Al, are frequently **electron deficient,** meaning they are surrounded by less than eight electrons. The electron-dot structure for BH_3 in Section 7.5 is a representative example; B has only three bonds, or a total of six electrons.

• *(Octet Rule Exception)* Elements in the third row and lower often have an **expanded octet,** as they form more bonds than predicted by the octet rule. Atoms of these elements are larger than their second row counterparts, can accommodate more than four atoms around them, and therefore form more than four bonds (**FIGURE 7.6**). The second-row element nitrogen, for instance, bonds to only three chlorine atoms in forming NCl_3 and thus obeys the octet rule, while the third-row element phosphorus bonds to five chlorine atoms in forming PCl_5 and thus does not follow the octet rule.

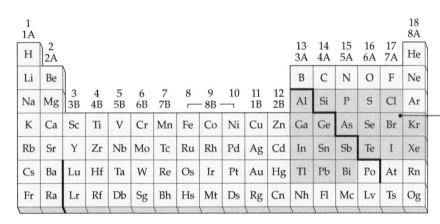

Atoms of these elements, all of which are in the third row or lower, are larger than their second-row counterparts and can therefore accommodate more bonded atoms.

▲ **FIGURE 7.6**

The octet rule occasionally fails for the main-group elements shown in blue.

▲ **Figure It out**

Which of the following elements can have an expanded octet when they are the central atom in an electron-dot structure: Cl, O, S, I, C?

Answer: Cl, S, I

If, for example, you were asked to predict the connections in SF_4, a good guess would be that each fluorine forms one bond to sulfur, which occurs as the central atom.

Sulfur tetrafluoride, SF_4

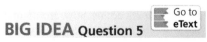
Go to
eText

BIG IDEA Question 5

What criteria must be met in a valid electron-dot structure?

Step 3. Subtract the number of valence electrons used for bonding from the total number calculated in Step 1 to find the number that remain. Assign as many of these remaining electrons as necessary to the terminal atoms (other than hydrogen) so that each has an octet. In SF_4, 8 of the 34 total valence electrons are used in covalent bonding, leaving $34 - 8 = 26$. Twenty-four of the 26 are assigned to the four terminal fluorine atoms to reach an octet configuration for each:

$8 + 24 = 32$ electrons distributed

Step 4. If unassigned electrons remain after Step 3, place them on the central atom. In SF_4, 32 of the 34 electrons have been assigned, leaving the final 2 to be placed on the central S atom. Sulfur is in the third row and is large enough to expand its octet.

34 electrons distributed

Step 5. Form multiple bonds to fulfill an incomplete octet on the central atom. If no unassigned electrons remain after Step 3 but the central atom does not yet have an octet, use one or more lone pairs of electrons from a neighboring atom to form a multiple bond (either double or triple). Oxygen, carbon, nitrogen, and sulfur often form multiple bonds. Worked Example 7.5 shows how to deal with such a case.

Drawing Electron-Dot Structures for Molecules with One Central Atom

Worked Examples 7.4 and 7.5 demonstrate how to draw electron-dot structures for molecules with one central atom.

WORKED EXAMPLE 7.4

Drawing Electron-Dot Structures for Molecules with One Central Atom and Single Bonds

Draw an electron-dot structure for OF_2.

STRATEGY

Follow the five steps outlined at the start of this section under "Drawing Electron-Dot Structures."

SOLUTION

Step 1. First count the total number of valence electrons. Oxygen has 6 and each fluorine has 7 for a total of 20.

Step 2. Next, decide how atoms may be connected and draw lines to indicate bonds. Since oxygen is less electronegative than fluorine and halogens typically only form one bond, place oxygen in the center.

$$F-O-F$$

Step 3. Subtract the number of valence electrons used for bonding (4 total, 2 from each bond) from the total number calculated in Step 1 (20) to find the number that remain ($20 - 4 = 16$). Assign as many of these remaining electrons as necessary to the terminal atoms so that each has an octet. Each fluorine atom

requires three lone pairs to satisfy the octet rule for a total of 12 nonbonding electrons.

$$:\overset{..}{\underset{..}{F}}-O-\overset{..}{\underset{..}{F}}:$$ $4 + 12 = 16$ electrons distributed

Step 4. If unassigned electrons remain after Step 3, place them on the central atom.

$$:\overset{..}{\underset{..}{F}}-\overset{..}{\underset{..}{O}}-\overset{..}{\underset{..}{F}}:$$ 20 electrons distributed

Step 5. If the central atom does not have an octet, make multiple bonds. This step is not necessary because the central atom oxygen has a complete octet.

CHECK

Electron-dot structures can always be checked by evaluating two criteria. (1) Does the final structure have the same number of electrons as the total number of valence electrons in molecule? (2) Is the octet rule satisfied for each atom? The structure for OF_2 contains 20 electrons, and each atom has an octet; therefore, a valid electron-dot structure has been drawn.

▶ **PRACTICE 7.7**   Identify the correct electron-dot structure for $POCl_3$. Explain what is wrong with the incorrect structures.

(a)   (b)   (c)   (d)

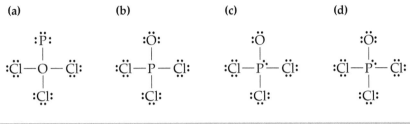

▶ **APPLY 7.8**   Draw an electron-dot structure for each of the following molecules or ions:

(a) CH_2F_2 (b) $AlCl_3$ (c) ClO_4^-
(d) PCl_5 (e) $XeOF_4$ (f) NH_4^+

Go to
eText

— **WORKED EXAMPLE 7.5**

Drawing Electron-Dot Structures for Molecules with One Central Atom and Multiple Bonds

Draw an electron-dot structure for formaldehyde, CH_2O, a compound used in manufacturing the adhesives for making plywood and particle board.

STRATEGY

Follow the five steps outlined in the list titled "Drawing Electron-Dot Structures."

SOLUTION

Step 1. Count the total number of valence electrons. Carbon has 4, each hydrogen has 1, and the oxygen has 6 for a total of 12.

Step 2. Decide on the probable connections between atoms, and draw a line to indicate each bond. In the case of formaldehyde, the less electronegative atom (carbon) is the central atom, and both hydrogens and the oxygen are bonded to carbon:

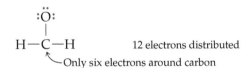

Steps 3 and 4. Six of the 12 valence electrons are used for bonds, leaving 6 for assignment to the terminal oxygen atom. There are no more electrons to add to the central carbon atom.

$\ddot{\text{O}}:$
|
H—C—H   12 electrons distributed
↳ Only six electrons around carbon

Step 5. At this point, all the valence electrons are assigned, but the central carbon atom still does not have an octet. To achieve an octet, we move two of the oxygen electrons from a lone pair into a bonding pair, generating a carbon–oxygen double bond that satisfies the octet rule for both oxygen and carbon.

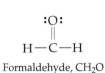

Formaldehyde, CH_2O

CHECK

The structure for CH_2O contains the correct number of valence electrons (12), and each atom has an octet.

▶ **PRACTICE 7.9**   Select the correct electron-dot structure for carbon monoxide, CO.

(a)   (b)

$:\ddot{\text{C}}—\ddot{\text{O}}:$ $:\text{C}{=}\text{O}:$

(c)   (d)

$:\text{C}{=}\text{O}:$ $:\text{C}{\equiv}\text{O}:$

▶ **APPLY 7.10**   Identify the correct electron-dot structure(s) for SCN^-.

(a)   (b)

$\left[:\ddot{\text{S}}—\text{C}{=}\ddot{\text{N}}:\right]^-$ $\left[:\ddot{\text{S}}{=}\text{C}{=}\ddot{\text{N}}:\right]^-$

(c)   (d)

$\left[:\ddot{\text{S}}—\text{C}{\equiv}\text{N}:\right]^-$ $\left[:\ddot{\text{S}}—\ddot{\text{C}}—\ddot{\text{N}}:\right]^-$

(e) Both (b) and (c) are correct.
(f) All structures shown in (a)–(d) are correct.

All **WORKED EXAMPLES** with this icon ⎯ Go to eText ⎯ have an interactive video in the eText.

7.7 DRAWING ELECTRON-DOT STRUCTURES FOR RADICALS

Except for hydrogen, nearly all atoms in all the 60 million known chemical compounds follow the octet rule. As a result, most compounds have an even number of electrons, which are paired up in filling orbitals. A very few substances, called **radicals** (or sometimes *free radicals*), have an odd number of electrons, meaning that at least one of their electrons must be unpaired in a half-filled orbital.

Because of their unpaired electron, radicals are usually very reactive, rapidly undergoing some kind of reaction that allows them to pair their electrons and form a more stable product. This pairing of electrons is why isolated halogen atoms dimerize to form halogen molecules, and nitrogen dioxide (NO_2) dimerizes to form dinitrogen tetroxide (N_2O_4).

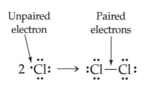

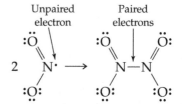

Despite their high reactivity, small amounts of radicals are involved in many common processes, including the combustion of fuels, the industrial preparation of polyethylene and other polymers, and the ongoing destruction of the Earth's ozone layer in the upper atmosphere by chlorofluorocarbons. Perhaps more surprisingly, radicals are also present in the human body, where they both mediate important life processes and act as causal factors for many diseases, including cancer, emphysema, stroke, and diabetes. Even the changes that take place as a result of normal aging are thought to involve radicals.

Nitric oxide (NO) serves several useful purposes in the body, and in 1998 the Nobel Prize in Physiology and Medicine was awarded to scientists who discovered that NO is an important signaling molecule in the cardiovascular system. Nitric oxide, also known as the endothelium-derived relaxing factor (EDRF), is released by cells lining the interior surface of blood vessels. The release of NO causes the muscles surrounding vessels to relax, thereby dilating blood vessels and increasing blood flow. Hair growth, pulmonary hypertension, and angina are all treated by drugs that affect NO release.

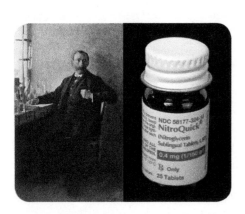

▲ Alfred Nobel used his fortune from the invention of dynamite, a formulation with the nitroglycerin molecule, to institute the highly prestigious Nobel prizes. He found it ironic that later in life his doctor prescribed nitroglycerin, which releases nitric oxide, to treat his heart condition.

WORKED EXAMPLE 7.6

Drawing an Electron-Dot Structure for a Radical

Draw an electron-dot structure for nitric oxide, NO.

STRATEGY

Follow the five steps outlined in the list titled "Drawing Electron-Dot Structures" (Section 7.6). In a radical, the *least electronegative* atom will have an odd number of electrons and an incomplete octet.

SOLUTION

Step 1. Count the total number of valence electrons. Nitrogen has 5 and oxygen has 6 for a total of 11. An odd number of electrons indicates the presence of a radical.

Step 2. Determine the connectivity of atoms. Nitrogen and oxygen are bonded together as a simple diatomic molecule.

Steps 3 and **4.** Subtract the number of valence electrons used for bonding (2) from the total number calculated in Step 1 (11) to find the number that remain (9). Assign as many of these remaining electrons as necessary to the terminal atoms. Give the more electronegative atom, oxygen, an octet, and place any remaining electrons on nitrogen.

$$\cdot \ddot{N}{-}\ddot{O}: \qquad \text{11 electrons distributed}$$

Step 5. All the valence electrons are assigned, but nitrogen is not close to satisfying its octet with only five electrons in the structure. We therefore move two of the oxygen electrons from a lone pair into a bonding pair, generating a double bond. The octet rule is satisfied for oxygen as it is the more electronegative atom and nitrogen is now closer to an octet. The least electronegative atom will never have an octet since a radical has an odd number of electrons.

$$\cdot\ddot{\text{N}}=\ddot{\text{O}}$$ 11 electrons distributed

CHECK

The structure for NO contains 11 electrons, and the most electronegative atom (O) has a complete octet.

▶ **PRACTICE 7.11** Select the best electron-dot structure for the ClO_2 radical, which is involved in the depletion of stratospheric ozone.

(a)   **(c)**

$:\ddot{\text{O}}-\ddot{\text{Cl}}-\ddot{\text{O}}:$ $:\ddot{\text{O}}-\ddot{\text{Cl}}-\ddot{\text{O}}:$

(b)   **(d)**

$:\ddot{\text{O}}-\ddot{\text{Cl}}=\text{O}:$ $:\ddot{\text{O}}-\ddot{\text{Cl}}-\dot{\text{O}}:$

▶ **APPLY 7.12** Which oxygen species do you predict to be most reactive: O_3, O_2, or O_2^-?

7.8 ELECTRON-DOT STRUCTURES OF COMPOUNDS CONTAINING ONLY HYDROGEN AND SECOND-ROW ELEMENTS

Many of the naturally occurring compounds on which life is based—proteins, fats, carbohydrates, and numerous others—contain only hydrogen and one or more of the second-row elements carbon, nitrogen, and oxygen. Molecules that contain carbon atoms bonded together in a chain are called **organic compounds.** Electron-dot structures are relatively easy to draw for such compounds because the octet rule is almost always followed and the number of bonds formed by each element is predictable (Table 7.4).

For molecules that contain second-row atoms in addition to hydrogen, the second-row atoms are bonded to one another in a central core, with hydrogens on the periphery. In ethane (C_2H_6), for instance, two carbon atoms, each of which forms four bonds, combine with six hydrogens, each of which forms one bond. Joining the two carbon atoms and adding the appropriate number of hydrogens to each yields only one possible structure:

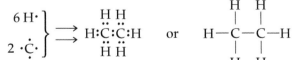

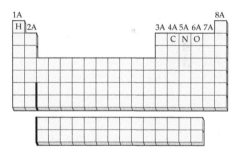

Ethane, C_2H_6

For larger molecules that contain numerous second-row atoms, there is usually more than one possible electron-dot structure. In such cases, some additional knowledge about the order of connections among atoms is necessary before a structure can be drawn.

BIG IDEA Question 6   Go to eText

In an electron-dot structure, which second-row element typically forms four bonds, and which second-row element typically forms two bonds and has two lone pairs of electrons?

WORKED EXAMPLE 7.7

Drawing an Electron-Dot Structure for Molecules with More Than One Central Atom

Draw an electron-dot structure for acetylene, C_2H_2, commonly used as a fuel for torches and as a starting material in chemical synthesis.

STRATEGY

Follow the five steps for drawing electron-dot structures outlined in Section 7.6. For molecules with more than one central atom, the second-row atoms are bonded to one another in a central core.

SOLUTION

Step 1. Count the total number of valence electrons in the molecule. Each carbon has 4 and each hydrogen has 1 for a total of 10.

Step 2. Determine the connectivity of atoms. Connect the second row atoms (C) in a chain.

$$H-C-C-H$$

Steps 3 and **4.** Subtract the number of valence electrons used for bonding (6) from the total number calculated in step 1 (10) to find the number that remain (4). Assign as many of these remaining electrons as necessary to the terminal atoms, except hydrogen. Since four electrons remain and all terminal atoms are hydrogen, place the electrons on the central carbon atoms.

$$H-\ddot{C}-\ddot{C}-H$$

Step 5. All the valence electrons are assigned, but the two central carbon atoms do not have an octet. We can move electron pairs on the two carbon atoms to form a triple bond between them, thus satisfying the octet rule for carbon. The valence shell for hydrogen is also filled with two electrons.

$$H-C\equiv C-H$$

CHECK

The structure for C_2H_2 contains the correct number of valence electrons (10), and each atom has a complete valence shell: carbon with 8 and hydrogen with 2.

▶ **PRACTICE 7.13**   Draw an electron-dot structure for the following molecules.

(a) Methylamine, CH_5N (b) Ethylene, C_2H_4
(c) Hydrogen peroxide, H_2O_2 (d) Hydrazine, N_2H_4

▶ **APPLY 7.14**   There are two molecules with the formula C_2H_6O. Draw electron-dot structures for both. (Hint: The connection of atoms is different in the two structures.)

CONCEPTUAL WORKED EXAMPLE 7.8

Identifying Multiple Bonds and Lone Pairs in Organic Molecules

The molecular model a representation of histidine, an amino acid constituent of proteins. Only the connections between atoms are shown; multiple bonds are not indicated. Give the chemical formula of histidine, and complete the structure by showing where the multiple bonds and lone pairs are located. (Red = O, gray = C, blue = N, ivory = H.)

Histidine

STRATEGY

Count the atoms of each element to find the formula. Then look at each atom in the structure to find what is needed for completing its octet. Table 7.4 indicates the number of bonds formed by elements in different groups of the periodic table. Each carbon (gray) should have four bonds, each oxygen (red) should have two bonds, and each nitrogen (blue) should have three bonds. Complete the octet for each atom by adding lone pairs; carbon already has a complete octet with four bonds, nitrogen needs one lone pair of electrons, and oxygen needs two lone pairs of electrons.

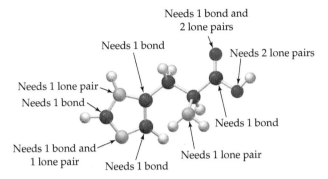

Needs 1 bond and
2 lone pairs

Needs 2 lone pairs

Needs 1 bond

Needs 1 lone pair

Needs 1 bond

Needs 1 bond

Needs 1 bond and
1 lone pair

Needs 1 lone pair

Needs 1 bond

SOLUTION

Histidine has the formula $C_6H_9N_3O_2$.

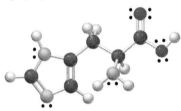

CHECK

It would be time-consuming to count all electrons in this structure and compare the number to total valence electrons. The best way to check a structure for a large molecule is to ensure that each atom (except H) has an octet and the bonding patterns outlined in Table 7.4 are followed.

▶ **PRACTICE 7.15** The following structure is a representation of cytosine, a constituent of the DNA found in all living cells. Only the connections between atoms are shown; multiple bonds are not indicated. Give the formula of cytosine, and complete the structure by showing where the multiple bonds and lone pairs are located. (Red = O, gray = C, blue = N, ivory = H.)

Cytosine

▶ **APPLY 7.16** Draw two possible electron-dot structures for the molecules with the following formulas. You may connect the atoms in different ways to draw the structures. (Hint: Look at the structure in Worked Example 7.8 for some common ways that second row atoms are connected.)

(a) C_6H_7N (b) $C_2H_5O_2N$

7.9 ELECTRON-DOT STRUCTURES AND RESONANCE

The stepwise procedure given in Section 7.6 for drawing electron-dot structures sometimes leads to an interesting problem. Look at ozone, O_3, for instance. Step 1 indicates the molecule has 18 valence electrons, and following Steps 2–4 we draw the following structure:

$$:\ddot{O}-\ddot{O}-\ddot{O}:$$

We find at this point that the central atom does not yet have an octet, and we therefore have to move one of the lone pairs of electrons from a terminal oxygen to become a bonding pair, giving the central oxygen an octet. But from which of the terminal oxygens should we take a lone pair? Moving a lone pair from either the "right-hand" or "left-hand" oxygen produces acceptable structures:

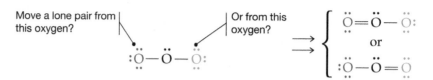

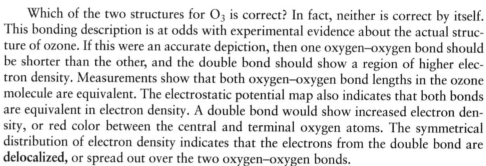

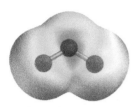

▲ Electrostatic potential map for ozone showing two equivalent oxygen–oxygen bonds.

Which of the two structures for O_3 is correct? In fact, neither is correct by itself. This bonding description is at odds with experimental evidence about the actual structure of ozone. If this were an accurate depiction, then one oxygen–oxygen bond should be shorter than the other, and the double bond should show a region of higher electron density. Measurements show that both oxygen–oxygen bond lengths in the ozone molecule are equivalent. The electrostatic potential map also indicates that both bonds are equivalent in electron density. A double bond would show increased electron density, or red color between the central and terminal oxygen atoms. The symmetrical distribution of electron density indicates that the electrons from the double bond are **delocalized,** or spread out over the two oxygen–oxygen bonds.

Bond lengths and bond strengths for ozone can be measured experimentally and are compared with average single (O—O) and double (O=O) bonds from other molecules. The characteristics of the oxygen–oxygen bond in ozone are between those of a single and double bond. The two bonds in ozone have an identical length of 128 pm, and we describe ozone as having a bond order of 1.5, midway between pure single bonds and pure double bonds.

	(O=O) in O_2	Average (O—O)	Oxygen–Oxygen Bond in Ozone
Bond Length (pm)	121	132	128
Bond Strength (kJ/mol)	498	146	374

Noting that the electron-dot model is at odds with the actual structure of ozone, we underscore that Lewis theory is just a model for bonding that chemists use to make predictions about how atoms combine. Lewis theory is extremely useful, but keep in mind that all models have limitations and may not be a true representation of reality.

How does the "hybrid" nature of the bonds in ozone arise? In 1930, Linus Pauling developed the concept of **resonance theory,** which describes molecules as a composite of two or more electron-dot structures. Curved arrows are a bookkeeping method showing how electrons can be rearranged to create a different electron-dot structure. Note that electrons are not really flowing through the molecule; the curved arrows are used to keep track of electron changes in dot structures. The tail indicates which electrons will move, the arrowhead indicates where they will now reside. Electrons can move into a new bond or onto an atom as a lone pair. When drawing resonance structures, do not move atoms, only electrons! Curved arrows show how the two resonance structures for the ozone molecule are interconverted.

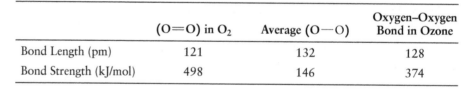

Resonance structures are indicated when two (or more) alternative electron-dot structures are connected by a double-headed *resonance arrow* (↔). A straight, double-headed arrow always indicates resonance; it is never used for any other purpose. Two resonance structures can be drawn for ozone and the *average* of these two structures is called a **resonance hybrid.** Note that the two resonance forms differ only in the placement of valence shell electrons (both bonding and nonbonding). The total number of valence electrons is the same in both structures, the connections between atoms remain the same, and the relative positions of the atoms remain the same.

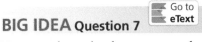

BIG IDEA Question 7

How can the molecular structure of ozone be described?

This double-headed arrow means that structures on either side are contributors to the resonance hybrid.

A solid and dashed line indicates a bond with properties between those of a single and double bond.

$$\left[:\ddot{O}-\ddot{O}=\ddot{O} \longleftrightarrow \ddot{O}=\ddot{O}-\ddot{O}:\right] \qquad O\overset{\cdots}{=}\ddot{O}\overset{\cdots}{=}O$$

Resonance hybrid

Recognizing that single electron-dot structures can't be written for all molecules tells us that the electron-dot model oversimplifies how bonds form and how electrons are distributed within a molecule. We will present a more accurate way of describing electron distributions, called *molecular orbital theory,* in Chapter 8. Chemists nevertheless make routine use of electron-dot and resonance structures because they are a simple and powerful tool for predicting molecular structure and reactive sites in molecules.

WORKED EXAMPLE 7.9

Drawing Resonance Structures

The nitrate ion, NO_3^-, has three equivalent nitrogen–oxygen bonds, and its electronic structure is a resonance hybrid of three electron-dot structures.

(a) Draw an electron-dot structure for nitrate.
(b) Use curved arrows to convert the structure drawn in part (a) into two other resonance structures.
(c) Draw all three resonance structures and a resonance hybrid.

(a) STRATEGY AND SOLUTION

Begin as you would for drawing any electron-dot structure. There are 24 valence electrons in the nitrate ion: 5 from nitrogen, 6 from each of 3 oxygens, and 1 for the negative charge. The three equivalent oxygens are all bonded to nitrogen, the less electronegative central atom:

O
|
N
O   O   6 of 24 valence electrons assigned

Distributing the remaining 18 valence electrons among the three terminal oxygen atoms completes the octet of each oxygen but leaves nitrogen with only 6 electrons.

$$\left[\begin{array}{c} :\ddot{O}: \\ | \\ N \\ :\ddot{O} \qquad \ddot{O}: \end{array}\right]^-$$

To give nitrogen an octet, one of the oxygen atoms must use a lone pair to form a nitrogen–oxygen double bond.

$$\left[\begin{array}{c} \ddot{O} \\ \| \\ N \\ :\ddot{O} \qquad \ddot{O}: \end{array}\right]^-$$

(b) STRATEGY AND SOLUTION

Use curved arrows to change the location of electrons, creating a new electron-dot structure. Do not move atoms or break single bonds when drawing resonance structures!

$$\left[\begin{array}{c} \dot{\ddot{O}} \\ \| \\ N \\ :\ddot{O} \qquad \ddot{O}: \end{array}\right]^- \longleftrightarrow \left[\begin{array}{c} :\ddot{O}: \\ | \\ N \\ :\ddot{O} \qquad \ddot{O}: \end{array}\right]^-$$

$$\left[\begin{array}{c} \dot{\ddot{O}} \\ \| \\ N \\ :\ddot{O} \qquad \ddot{O}: \end{array}\right]^- \longleftrightarrow \left[\begin{array}{c} :\ddot{O}: \\ | \\ N \\ :O \qquad \ddot{O}: \end{array}\right]^-$$

(c) STRATEGY AND SOLUTION

Show all the valid electron-dot structures connected by the resonance arrow(s). The resonance hybrid is considered to be an average of all three structures and is drawn with three equivalent nitrogen–oxygen bonds. The bonds are represented with a solid line and a dashed line to indicate that they are between a single and a double bond.

continued on next page

SOLUTION

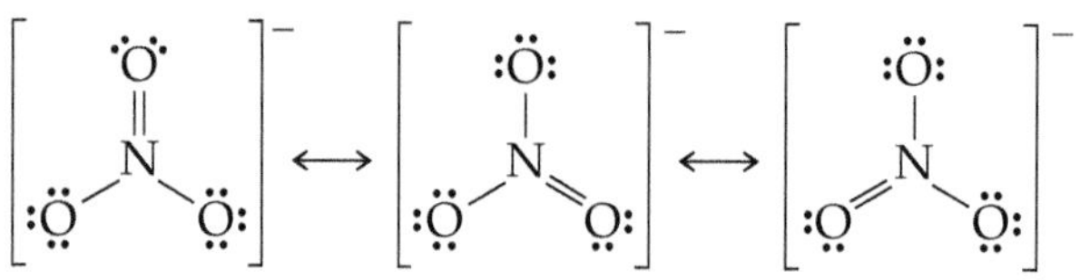

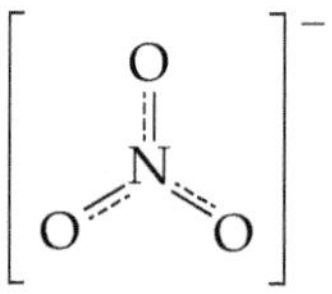

Resonance hybrid

CHECK

All structures have the correct number of valence electrons and satisfy the octet rule.

▶ **PRACTICE 7.17**   Called "laughing gas," nitrous oxide (N_2O) is sometimes used by dentists as an anesthetic.

(a) Given the connections N—N—O draw an electron-dot structure.

(b) Use curved arrows to show how the structure in part (a) can be converted into a resonance structure.

▶ **APPLY 7.18**   Draw as many resonance structures as possible for each of the following molecules or ions, giving all atoms (except H) octets. Use curved arrows to show how one structure can be converted into another.

(a) SO_2 (b) CO_3^{2-} (c) BF_3

Go to
eText

CONCEPTUAL WORKED EXAMPLE 7.10

Drawing Resonance Structures in Organic Compounds

A structural formula for histidine (Worked Example 7.8) is shown. Evaluate whether the following arrows result in a valid electron-dot structure for histidine. If so, draw the resonance structure.

(a)

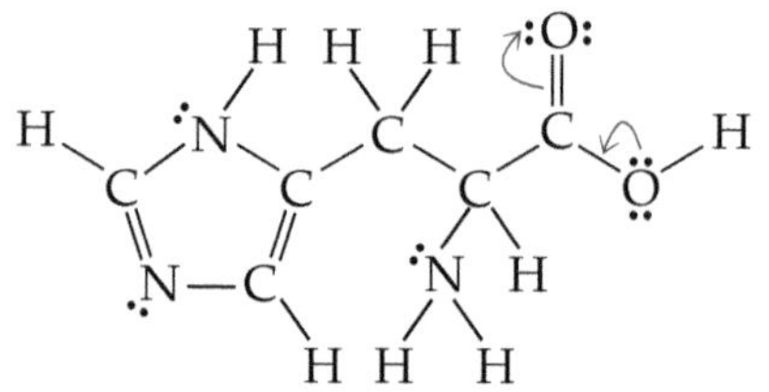

(b)

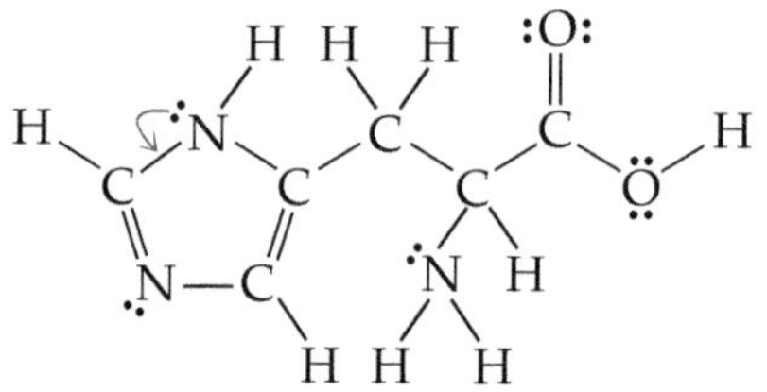

STRATEGY

Rearrange electrons as indicated by the curved arrows, check to make sure that each atom satisfies the octet rule. An invalid electron-dot structure results when the octet rule is exceeded.

SOLUTION

(a) The octet rule is satisfied for each atom. The resonance structure is seen below.

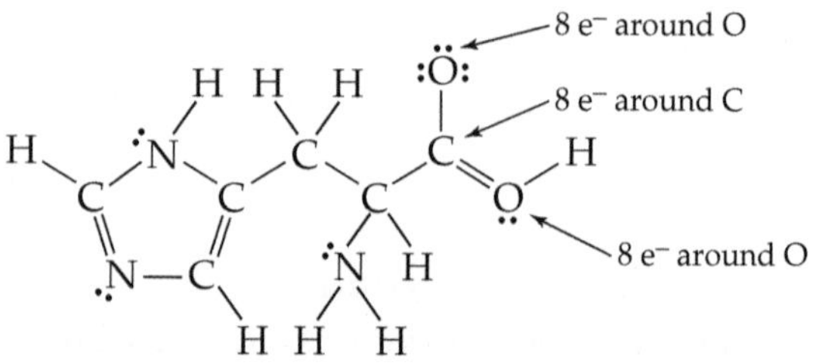

8 e⁻ around O
8 e⁻ around C
8 e⁻ around O

(b) The arrow results in an invalid electron-dot structure because C has five bonds (10 e⁻) and therefore exceeds an octet. A resonance structure cannot be drawn from the arrow indicated.

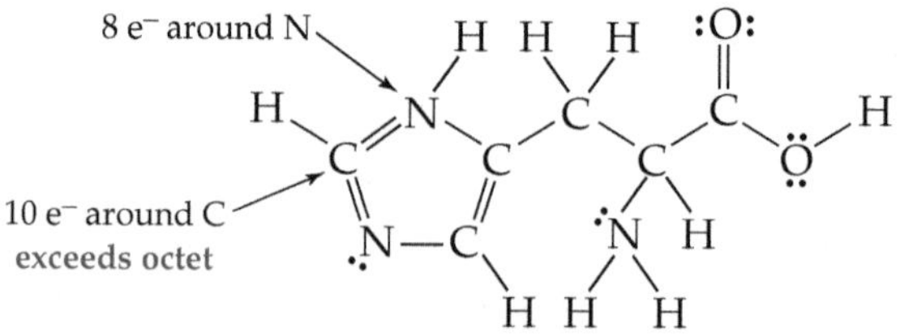

8 e⁻ around N
10 e⁻ around C exceeds octet

▶ **PRACTICE 7.19**   Evaluate whether the following arrows result in a valid electron-dot structure for histidine. If so, draw the resonance structure.

(a)

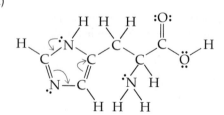

(b)

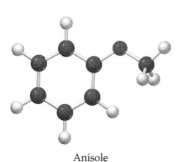

Anisole

▶ **APPLY 7.20** The following structure shows the connections between atoms in anisole, a compound used in perfumery. (Red = O, gray = C, ivory = H.)

(a) Draw a structural formula for anisole showing positions of multiple bonds and lone pairs.

(b) Use curved arrows to show how to convert between the original structure and two additional resonance structures.

7.10 FORMAL CHARGES

Closely related to the ideas of electronegativity and polar covalent bonds discussed in Section 7.3 is the concept of **formal charge** on specific atoms in electron-dot structures. Formal charges result from a kind of electron "bookkeeping" and are calculated as the difference between the number of valence electrons around an atom in a given electron-dot structure compared to the number of valence electrons in the isolated atom. If an atom in an electron-dot structure has a different number of electrons than its number of valence electrons, then the atom has either gained or lost electrons and thus has a formal charge. If the atom in a molecule has more electrons than the isolated atom, it has a negative formal charge; if it has fewer electrons, it has a positive formal charge.

$$\text{Formal charge} = \begin{pmatrix} \text{Number of} \\ \text{valence electrons} \\ \text{in free atom} \end{pmatrix} - \begin{pmatrix} \text{Number of} \\ \text{valence electrons} \\ \text{in bonded atom} \end{pmatrix}$$

In counting the number of valence electrons in a bonded atom, we need to distinguish between unshared, nonbonding electrons and shared, bonding electrons. For bookkeeping purposes, we can think of an atom as "owning" all of its nonbonding electrons but only half of its bonding electrons because the bonding electrons are shared with another atom. Thus, the definition of formal charge can be rewritten as:

$$\text{Formal charge} = \begin{pmatrix} \text{Number of} \\ \text{valence electrons} \\ \text{in free atom} \end{pmatrix} - \frac{1}{2}\begin{pmatrix} \text{Number of} \\ \text{bonding} \\ \text{electrons} \end{pmatrix} - \begin{pmatrix} \text{Number of} \\ \text{nonbonding} \\ \text{electrons} \end{pmatrix}$$

In the ammonium ion (NH_4^+), for instance, each of the four equivalent hydrogen atoms has 2 valence electrons in its covalent bond to nitrogen and the nitrogen atom has 8 valence electrons, 2 from each of its four $N—H$ bonds:

$$\left[\begin{array}{c} H \\ H : \ddot{N} : H \\ \ddot{H} \end{array} \right]^{+}$$

Ammonium ion
8 valence electrons around nitrogen
2 valence electrons around each hydrogen

For bookkeeping purposes, each hydrogen atom owns half of its 2 shared bonding electrons, or 1, while the nitrogen atom owns half of its 8 shared bonding electrons, or 4. Because an isolated hydrogen atom has 1 electron and the hydrogens in the ammonium ion each still own 1 electron, they have neither gained nor lost electrons and thus have no formal charge. An isolated nitrogen atom, however, has 5 valence electrons, while the nitrogen atom in NH_4^+ owns only 4 and thus has a formal charge of +1. The sum of the formal charges on all the atoms (+1 in this example) must equal the overall charge on the ion.

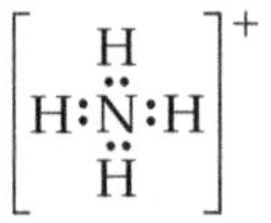

For hydrogen:	Isolated hydrogen valence electrons	1
	Bound hydrogen bonding electrons	2
	Bound hydrogen nonbonding electrons	0

$$\text{Formal charge} = 1 - \frac{1}{2}(2) - 0 = 0$$

For nitrogen:	Isolated nitrogen valence electrons	5
	Bound nitrogen bonding electrons	8
	Bound nitrogen nonbonding electrons	0

$$\text{Formal charge} = 5 - \frac{1}{2}(8) - 0 = +1$$

The value of formal charge calculations comes from their application to the resonance structures described in the previous section. It often happens that the resonance structures of a given substance are not equivalent. One of the structures may be "better" than the others, meaning that it approximates the actual electronic structure of the substance more closely. The resonance hybrid in such cases is thus weighted more strongly toward the more favorable structure.

When evaluating the relative importance of different resonance structures, three criteria are frequently employed.

- Smaller formal charges (either positive or negative) are preferable to larger ones. Zero is preferred over −1, but −1 is preferred over −2.
- Negative formal charges should reside on more electronegative atoms.
- Like charges should not be on adjacent atoms.

We can use these criteria to evaluate the resonance structures for N_2O. Formal charges have been assigned to each atom in the structure:

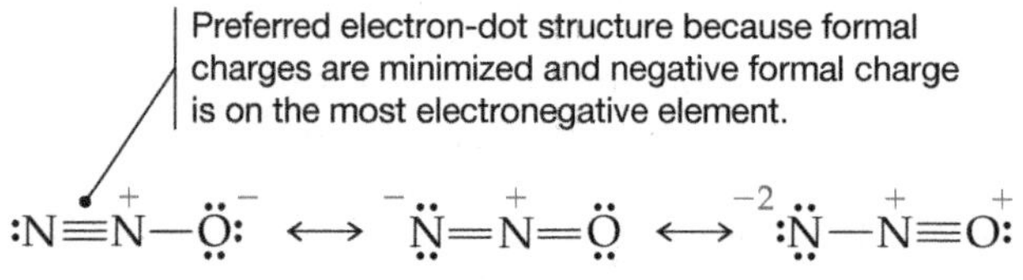

Preferred electron-dot structure because formal charges are minimized and negative formal charge is on the most electronegative element.

Notice that when the formal charges assigned in each resonance structure are added together they give the overall charge on N_2O, which is zero. The first and second structures are preferred over the third because the magnitude of the formal charges is lower. The first structure is "best" because the negative charge resides on the more electronegative atom oxygen. The best structure based on formal charge will make the largest contribution to the resonance hybrid. In this case, in N_2O the nitrogen–oxygen bond will have more single-bond character while the nitrogen–nitrogen bond will have more triple bond character.

Go to
eText

── **WORKED EXAMPLE 7.11**

Calculating Formal Charges

Calculate the formal charge on each atom in the following electron-dot structure for SO_2:

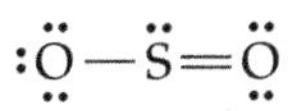

$$:\ddot{O}-\ddot{S}=\ddot{O}$$

STRATEGY

Use the formula for calculating formal charge. Find the number of valence electrons on each atom (its periodic group number). Then subtract half the number of the atom's bonding electrons and all of its nonbonding electrons.

SOLUTION

For sulfur:	Isolated sulfur valence electrons	6
	Sulfur bonding electrons	6
	Sulfur nonbonding electrons	2
	Formal charge = $6 - \frac{1}{2}(6) - 2 = +1$	
For singly bonded oxygen:	Isolated oxygen valence electrons	6
	Oxygen bonding electrons	2
	Oxygen nonbonding electrons	6
	Formal charge = $6 - \frac{1}{2}(2) - 6 = -1$	
For doubly bonded oxygen:	Isolated oxygen valence electrons	6
	Oxygen bonding electrons	4
	Oxygen nonbonding electrons	4
	Formal charge = $6 - \frac{1}{2}(4) - 4 = 0$	

The sulfur atom of SO_2 has a formal charge of +1, and the singly bonded oxygen atom has a formal charge of −1. We might therefore write the structure for SO_2 as

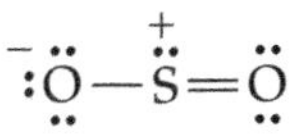

$$:\overset{-}{\ddot{O}}-\overset{+}{\ddot{S}}=\ddot{O}$$

CHECK

The sum of the formal charges must equal the overall charge on the molecule or ion. In this case, the sum of formal charges is zero, which equals the charge on the neutral SO_2 molecule.

▶ **PRACTICE 7.21** Calculate the formal charge on each atom in the following structure.

Cyanate ion: $\left[\ddot{N}=C=\ddot{O} \right]^{-}$

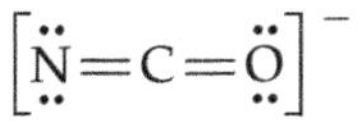

▶ **APPLY 7.22** Start with the electron-dot structure for the cyanate ion shown in Problem 7.21.

(a) Use curved arrows to convert the original structure into two additional electron-dot structures.
(b) Calculate the formal charge on each atom in all three resonance structures, and decide which makes the largest contribution to the resonance hybrid.
(c) Which bond do you predict to be the shortest (carbon–nitrogen or carbon–oxygen)?

── **WORKED EXAMPLE 7.12**

Calculating Formal Charges in Organic Compounds

Calculate formal charges on the C, O, and N atoms in the two resonance structures for acetamide, a compound related to proteins. Decide which structure makes a larger contribution to the resonance hybrid.

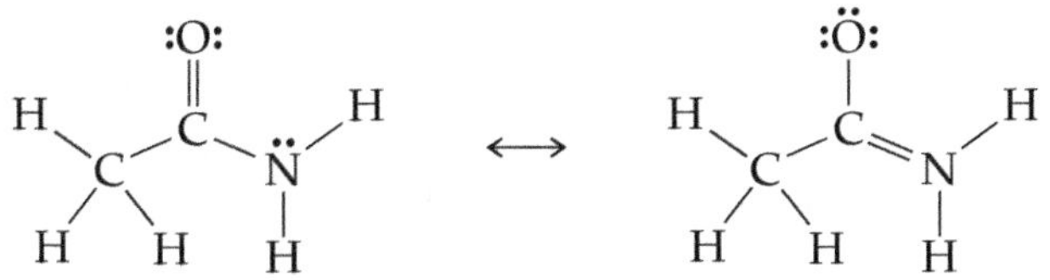

continued on next page

STRATEGY

Use the formula for calculating formal charge. Evaluate the importance of the resonance structures using the three criteria listed in this section: lowest formal charge, negative formal charge on electronegative atoms, and separation of like charges.

SOLUTION

The structure without formal charges makes a larger contribution to the resonance hybrid because energy is required to separate + and − charges. Thus, the actual electronic structure of acetamide is closer to that of the more favorable, lower energy structure.

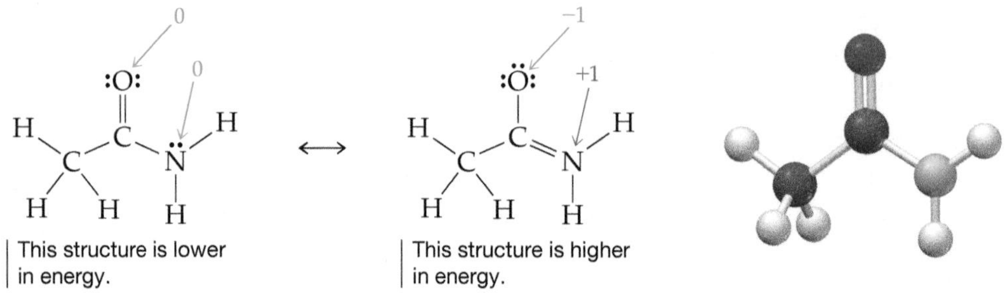

This structure is lower in energy.

This structure is higher in energy.

Acetamide

▶ **PRACTICE 7.23**   Calculate formal charges on the C and O atoms in two resonance structures for acetic acid, the main component of vinegar. Decide which structure makes a larger contribution to the resonance hybrid.

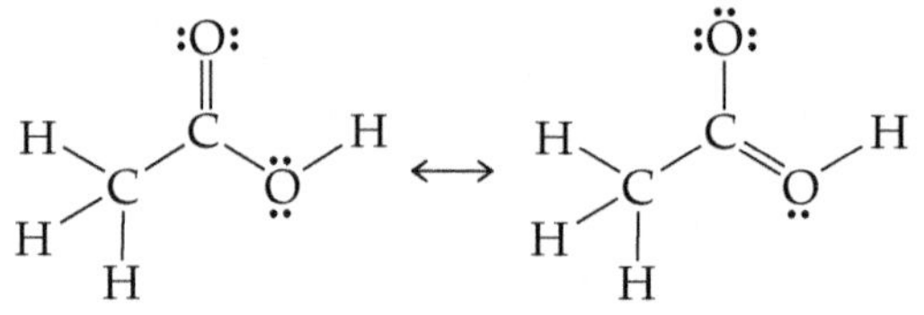

▶ **APPLY 7.24**   Three resonance structures for anisole (Problem 7.20) are shown. Calculate formal charges on C and O atoms and decide which structure makes the largest contribution to the resonance hybrid.

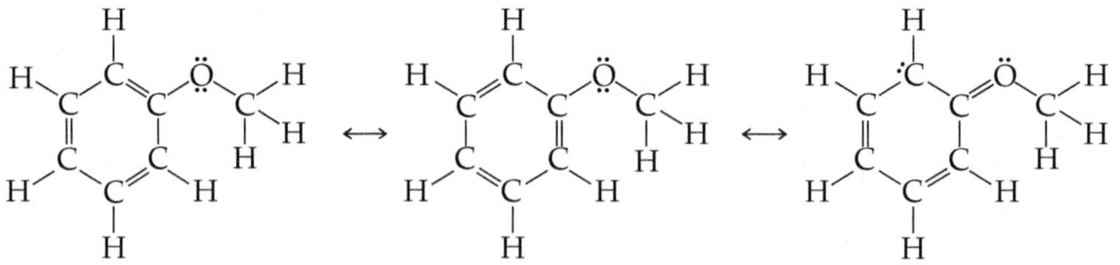

▲ Organophosphate insecticides are used on many crops throughout the world.

$$O_2N-C \quad \text{(ring)} \quad C-O-\overset{\overset{S}{\parallel}}{\underset{OCH_2CH_3}{P}}-OCH_2CH_3$$

Parathion

$$H_3CH_2CSH_2CH_2C-S-\overset{\overset{S}{\parallel}}{\underset{OCH_2CH_3}{P}}-OCH_2CH_3$$

Disulfoton

Organophosphates are a class of chemical compounds that have been widely used as insecticides. They are beneficial from an environmental perspective because they are nonpersistent, meaning they are degraded rapidly into harmless, water-soluble products. A general chemical structure for an organophosphate is

$$X-\overset{\overset{O}{\parallel}}{\underset{OR_2}{P}}-OR_1$$

R is a general designation for an organic (carbon-containing) group.

R_1 and R_2 could be the same or different.

X represents a halogen, SR, or OR group.

This class of insecticides must be highly potent due to their short lifetime, and changes in their chemical structure can be used to "tune" toxicity. In fact, nerve agents are highly toxic organophosphate compounds in which substituent groups X and R have been changed. In this Inquiry, we will examine how bond polarity influences the potency of these compounds and how substituting atoms of different electronegativity can make these compounds less toxic.

Examine the chemical structure of parathion and disulfoton, two organophosphate insecticides used to control a variety of pests that attack many field and vegetable crops. The double bond between phosphorus and oxygen in the general structure of an organophosphate has been replaced with a double bond to sulfur.

The chemical change has been made to prevent massive poisoning in humans when these compounds are applied in commercial farming or residential environments. Organophosphates are toxic because they interfere with the normal nerve transduction process in organisms. The organophosphate molecule reacts with an enzyme called cholinesterase, inhibiting its function and causing nerves to fire continuously, leading to paralysis and death. For the reaction between the organophosphate insecticide and the enzyme to occur, the phosphorus atom must bear a positive charge. Greater positive charge leads to increased rate of reaction and increased toxicity of the insecticide.

FIGURE 7.7 shows electrostatic potential maps and atomic charges of an organophosphate with a phosphorus–oxygen double bond (P=O) and a phosphorus–sulfur (P=S) double bond. Since oxygen is more electronegative than sulfur it has greater ability to pull shared electrons toward itself. Therefore, the molecule with the (P=O) bond is more toxic due to the greater positive charge on phosphorus (+1.80). In the less toxic form (P=S), the positive charge on phosphorus is lower (+1.30). The electrostatic potential map reflects the higher negative charge on oxygen as indicated by the strong red color. The sulfur atom is larger and less electronegative, and the lower negative charge is depicted with a combination of yellow and red.

Organophosphates with a P=S bond are still highly potent insecticides due to differences in biochemistry between insects and mammals. Once inside the insect, the sulfur atom is rapidly converted back into oxygen by oxidative enzymes, thus creating a potent neurotoxin. Animals have much lower levels of these enzymes, so the conversion of sulfur to oxygen does not occur to a significant extent. Therefore, when these compounds enter mammals they remain in a less toxic form but are converted to the highly toxic form in insects. Although the P=S bond makes the organophosphate less toxic, care must still be taken in applying these insecticides; several cases of human poisoning and even death have occurred.

PROBLEM 7.25 The toxicity of the organophosphate insecticides can be changed by substituting a sulfur for an oxygen atom in the chemical formula.

(a) Use Figure 7.4 to determine which bond is more polar, phosphorus-sulfur or phosphorus-oxygen.

(b) Which bond, phosphorus-sulfur or phosphorus-oxygen, leads to a more toxic insecticide? Explain.

▶ **FIGURE 7.7**

Molecular modeling of organophosphate compounds. In organophosphate insecticides, the positive charge on the phosphorus atom can be altered by binding it to elements with varying electronegativity. Oxygen is more electronegative than sulfur; thus the P=O is more polar than the P=S bond, and the positive charge on phosphorus is larger.

$$H_3CO-\overset{\overset{\displaystyle S}{\|}}{\underset{\underset{\displaystyle OCH_3}{|}}{P}}-OCH_3$$

Less toxic insecticide

←S = −0.6

P = +1.3

$$H_3CO-\overset{\overset{\displaystyle O}{\|}}{\underset{\underset{\displaystyle OCH_3}{|}}{P}}-OCH_3$$

More toxic insecticide

←O = −0.8

P = +1.8

PROBLEM 7.26 Consider the substituted phenyldiethylphosphate insecticide with the following structure:

Which chemical group at position X would result in greatest toxicity?
(a) I, Br, Cl   (b) CF_3, CHF_2, CH_2F, CH_3

PROBLEM 7.27 In organophosphate compounds, phosphorus has an expanded octet. Why can phosphorus accommodate more than eight electrons in its electron-dot structure?

PROBLEM 7.28 The following structure is a representation of the organophosphate insecticide diazinon. Only the connections between atoms are shown; multiple bonds are not indicated. Give the formula of diazinon, and complete the structure by showing where the multiple bonds and lone pairs are located. (Red = O, gray = C, yellow = S, blue = N, ivory = H, purple = P.) (Hint: There is a double bond between phosphorus and sulfur.)

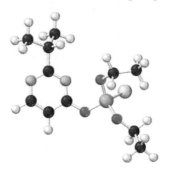

PROBLEM 7.29 The electron-dot structure for the nerve agent sarin is shown. Calculate the formal charges on P and the atoms bonded to it: F, C, and two O's.

PROBLEM 7.30 Draw the new electron-dot structures indicated by the curved arrows in the sarin molecule. Compare formal charges to those calculated in Problem 7.29. Use formal charge to evaluate which structure makes the largest contribution to the resonance hybrid.

(a)   (b)

STUDY GUIDE

Section	Concept Summary	Learning Objectives	Test Your Understanding
7.1 Covalent Bonding in Molecules	A **covalent bond** results from the sharing of electrons between atoms. The force joining atoms is the electrostatic attraction of the negatively charged electrons in each atom to the positively charged nuclei of the two atoms.	**7.1** Describe the difference between an ionic and covalent bond.	Problem 7.40
		7.2 Describe the changes in energy that occur as two nuclei approach to form a covalent bond.	Problems 7.33–7.34, 7.47
7.2 Strengths of Covalent Bonds	Every covalent bond has a specific **bond length** that leads to optimum stability and a specific **bond dissociation energy** (*D*) that describes the strength of the bond. Energy is released when a bond is formed; energy is absorbed when a bond is broken.	**7.3** Predict trends in bond length and bond dissociation energy based on bond order and atomic size.	Problems 7.42, 7.44
7.3 Polar Covalent Bonds: Electronegativity	In a bond between dissimilar atoms, such as that in HCl, one atom often attracts the bonding electrons more strongly than the other, giving rise to a **polar covalent bond.** Bond polarity is due to differences in **electronegativity (EN)**, the ability of an atom in a molecule to attract shared electrons. Electronegativity increases from left to right across a row and decreases from top to bottom in a group of the periodic table.	**7.4** Rank elements by increasing value of electronegativity.	Problem 7.48
		7.5 Classify bonds as nonpolar covalent, polar covalent, or ionic.	Worked Example 7.1; Problems 7.50, 7.54
		7.6 Interpret electrostatic potential maps to determine regions of high and low electron density.	Problems 7.33–7.34
		7.7 Calculate the percent ionic character of a bond.	Worked Example 7.2; Problems 7.60–7.61
7.4 A Comparison of Ionic and Covalent Compounds	Ionic compounds are solids with high melting points because strong ionic bonds exist between all atoms in a crystal. In general, covalent compounds have lower melting points because they are molecules held together by inter-molecular forces that are much weaker than actual bonds.	**7.8** Explain the different physical properties of ionic and covalent compounds.	Problems 7.35, 7.62–7.63
7.5 Electron-Dot Structures: The Octet Rule	As a general rule, a main-group atom shares as many of its valence-shell electrons as possible, either until it has no more to share or until it reaches an octet. Atoms in the third and lower rows of the periodic table can accommodate more than the number of bonds predicted by the octet rule.	**7.9** Draw an electron-dot structure that satisfies the octet rule.	Worked Example 7.3; Problem 7.3
7.6 Procedure for Drawing Electron-Dot Structures	An **electron-dot structure** represents an atom's valence electrons by dots and shows the two electrons in a **single bond** as a line between atoms. Similarly, a **double bond** is represented as four dots or two lines between atoms, and a **triple bond** is represented as six dots or three lines between atoms.	**7.10** Use the five-step procedure to draw electron-dot structures for molecules, including those with expanded octets and those containing multiple bonds.	Worked Examples 7.4–7.5; Problems 7.66, 7.68, 7.70, 7.74
7.7 Drawing Electron-Dot Structures for Radicals	**Radicals** are highly reactive substances that contain an odd number of electrons. The least electronegative atom does not have a complete octet in the electron-dot structure for a radical.	**7.11** Draw electron-dot structures for radicals.	Worked Example 7.6; Problems 7.9–7.10
7.8 Electron-Dot Structures of Compounds Containing Only Hydrogen and Second-Row Elements	Many biological and organic compounds contain only second row elements and hydrogen. Drawing electron-dot structures is straightforward because they follow the pattern of bonding outlined in Table 7.4.	**7.12** Draw electron-dot structures for molecules with more than one central atom.	Worked Examples 7.7–7.8; Problems 7.38, 7.80, 7.82

Section	Concept Summary	Learning Objectives	Test Your Understanding
7.9 Electron-Dot Structures and Resonance	Some molecules with multiple bonds can be represented by more than one electron-dot structure. In such cases, no single structure is adequate by itself. The actual electronic structure of the molecule is a **resonance hybrid** of the different individual structures.	**7.13** Draw resonance structures and use curved arrows to depict how one structure can be converted to another.	Worked Examples 7.9–7.10; Problems 7.84, 7.86, 7.88, 7.91
		7.14 Describe the bonding in a molecule using the resonance hybrid model.	Problems 7.90, 7.93
7.10 Formal Charges	**Formal charge** denotes whether an atom has more or fewer electrons in an electron-dot structure compared to its number of valence electrons. Formal charges are used to assess the relative contribution of individual resonance structures to the resonance hybrid.	**7.15** Calculate the formal charge on atoms in an electron-dot structure.	Worked Example 7.11; Problems 7.94, 7.96, 7.100
		7.16 Use the formal charge to evaluate the contribution of different resonance structures to the resonance hybrid.	Worked Example 7.12; Problems 7.98, 7.102, 7.105

KEY TERMS

bond dissociation
   energy (D) *241*
bond length *240*
bond order *242*
bonding pair *248*
coordinate covalent bond *249*

delocalized *258*
dipole moment *245*
double bond *249*
electron deficient *251*
electron-dot structure *247*
electronegativity (EN) *243*

expanded octet *251*
formal charge   *261*
lines between atoms *240*
lone pair *248*
organic compounds *255*
percent ionic character *245*

polar covalent bond *242*
radicals *254*
resonance hybrid *258*
resonance theory *258*
single bond *249*
triple bond *249*

KEY EQUATIONS

- **Formal Charge (Section 7.10)**

$$\text{Formal charge} = \left(\begin{array}{c}\text{Number of}\\\text{valence electrons}\\\text{in free atom}\end{array}\right) - \frac{1}{2}\left(\begin{array}{c}\text{Number of}\\\text{bonding}\\\text{electrons}\end{array}\right) - \left(\begin{array}{c}\text{Number of}\\\text{nonbonding}\\\text{electrons}\end{array}\right)$$

PRACTICE TEST

After studying this chapter, you can assess your understanding with these practice test questions, which are correlated with chapter learning objectives. If you answer a question incorrectly, refer to the learning objectives in the end-of-chapter Study Guide for assistance. The Study Guide provides a conceptual summary, references a Worked Example to model how to solve the problem, and gives additional problems for more practice.

1. The graph shows how potential energy changes as a function of the distance between two atoms. (**LO 7.2**)

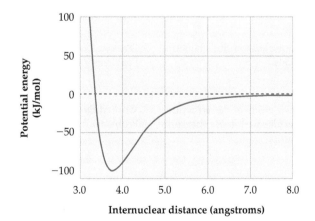

What is the length of the bond between the two atoms?
   (a) 3.4 angstroms   (b) 3.8 angstroms
   (c) 6.0 angstroms   (d) 8.0 angstroms

2. Which molecule has the shortest carbon–oxygen bond? (**LO 7.3, 7.9**)
   (a) CH_3OH (b) CO
   (c) CO_2 (d) HCOOH

3. The compounds below are paired with a type of bonding. Which type of bonding is *incorrectly* classified? (**LO 7.5**)
   (a) Rb_2O (polar covalent)
   (b) SO_2 (polar covalent)
   (c) O_3 (nonpolar covalent)
   (d) KBr   (ionic)
   (e) HF   (polar covalent)

4. Which bond in the amino acid cysteine is most polar? (**LO 7.6**)

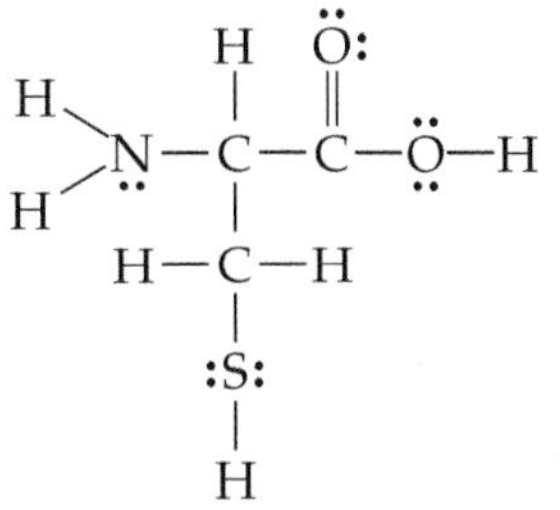

(a) N—H   (b) S—H   (c) C—O

(d) O—H   (e) N—C

5. Which compound is a solid at room temperature? (**LO 7.8**)

(a) H_2O (b) Na_2S

(c) SO_3 (d) Cl_2

6. Select the correct electron-dot structure for the sulfite ion (SO_3^{2-}). (**LO 7.10**)

(a)   (b)

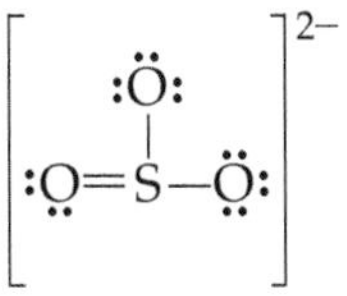

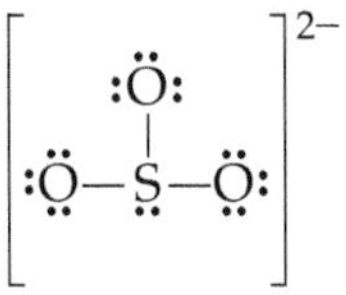

(c) 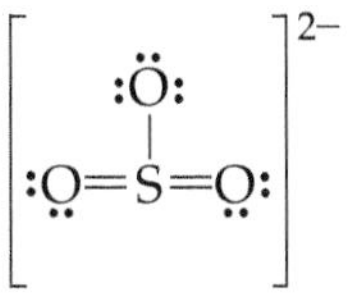   (d)

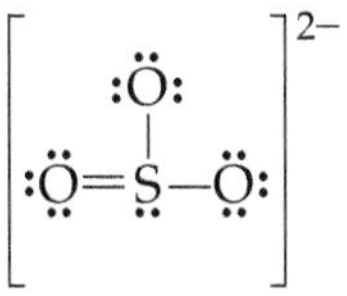

7. Select the correct electron-dot structure for $SeCl_2$. (**LO 7.10**)

(a)   (b)

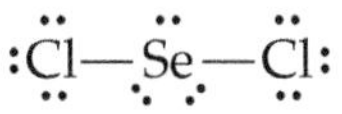

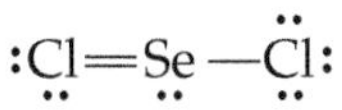

(c) 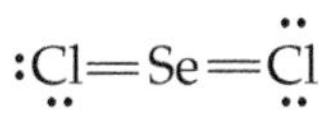   (d)

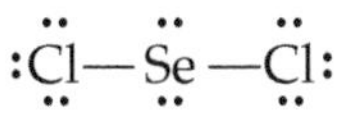

8. Select the correct electron-dot structure for ClF_3. (**LO 7.10**)

(a)   (b)

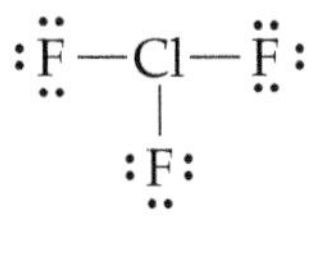

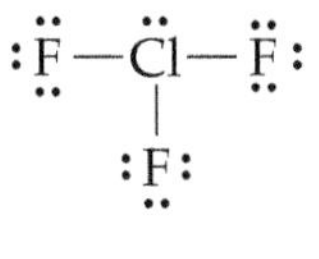

(c) 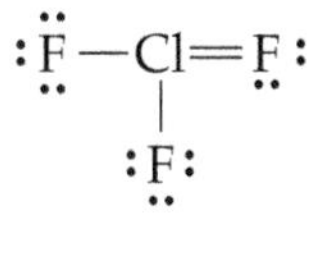   (d)

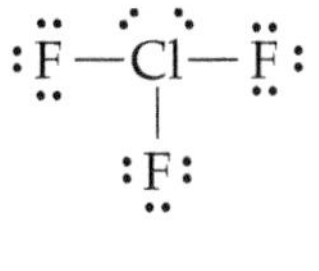

9. Which figure shows the correct electron-dot structure for a molecule with the following connections of atoms? (**LO 7.12**)

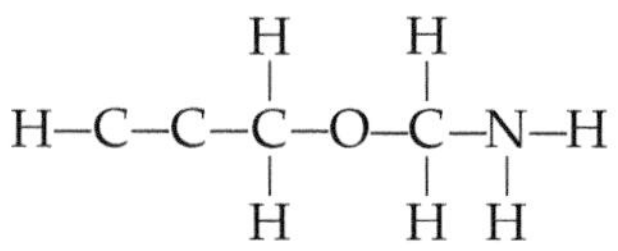

(a) 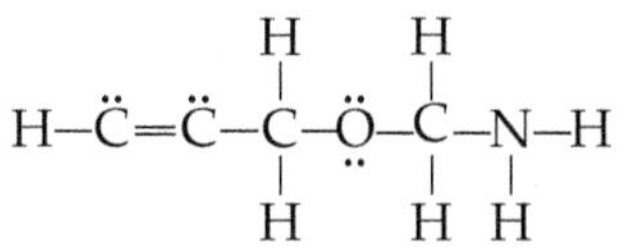   (b)

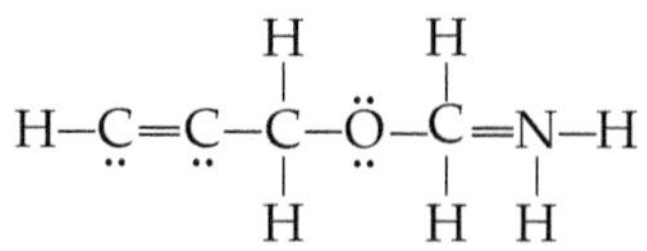

(c) 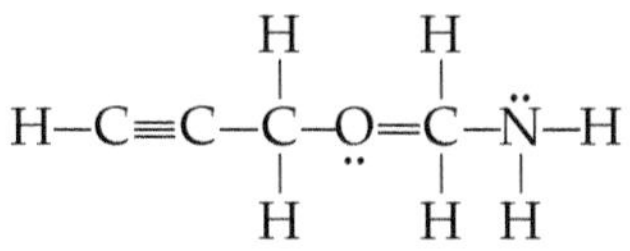   (d)

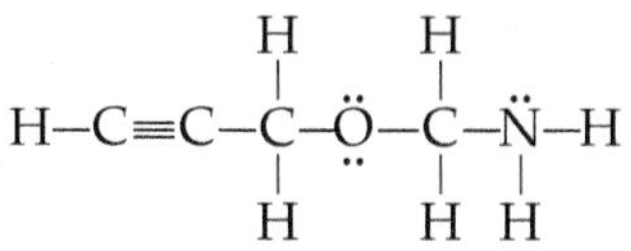

10. Which of the following pairs represent resonance structures? (**LO 7.13**)

(a)

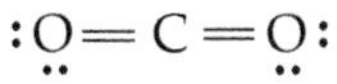

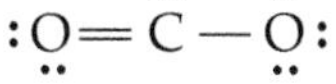

(b)

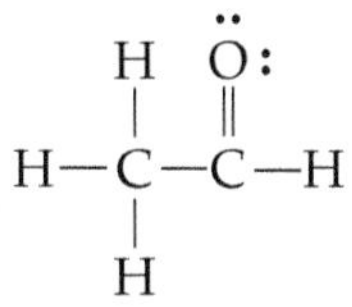

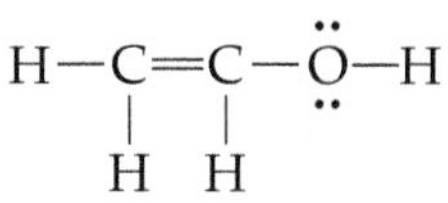

(c)

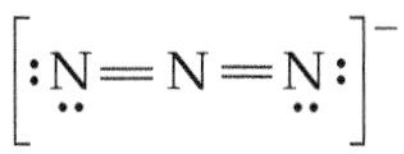

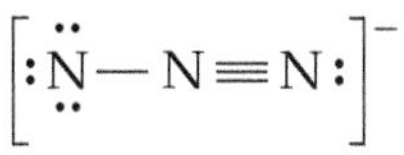

11. The structure for the DNA base cytosine is shown.

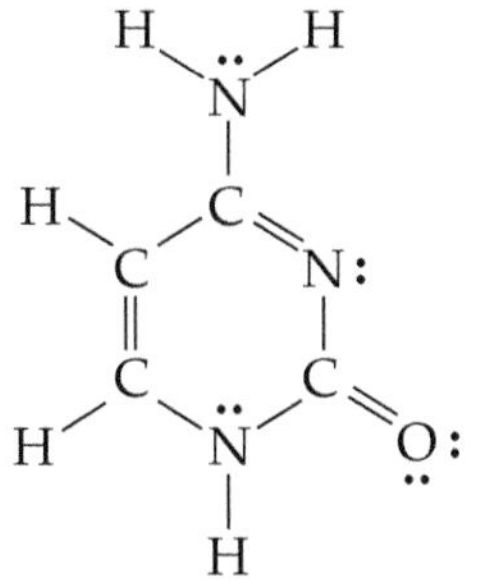

Which of the following is *not* a resonance structure of cytosine? (**LO 7.13**)

(a)

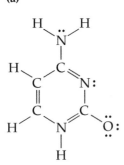

(b)

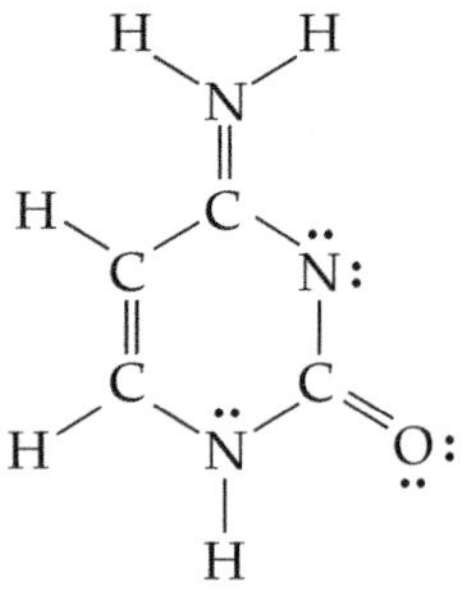

(c)

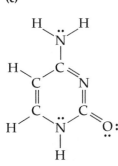

(d)

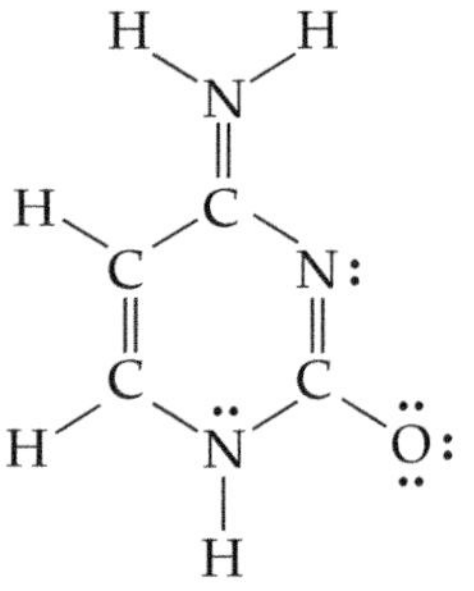

12. What is the best prediction for the carbon–oxygen bond length in the carbonate anion, CO_3^{2-}? (**LO 7.14**)

Data for average carbon-oxygen bond lengths:

C—O (143 pm)   C=O (121 pm)   C≡O (113 pm)

(a) 143 pm   (b) 132 pm
(c) 121 pm   (d) 118 pm

13. What is the formal charge on bromine in the bromate ion? (**LO 7.15**)

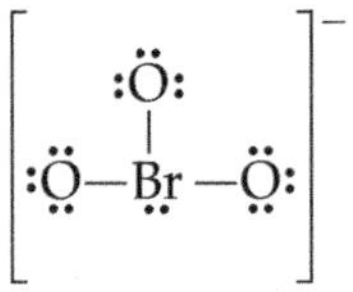

(a) +1   (b) +2
(c) +3   (d) −2

14. Which is the best electron-dot structure for the thiosulfate ion ($S_2O_3^{2-}$) based on the rules of formal charge? (**LO 7.16**)

(a)   **(b)**

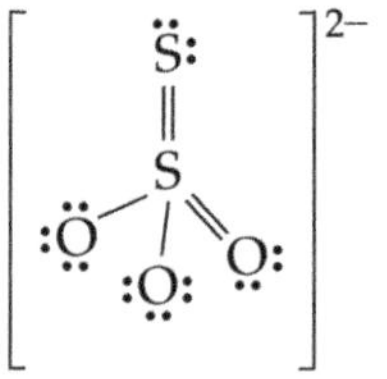

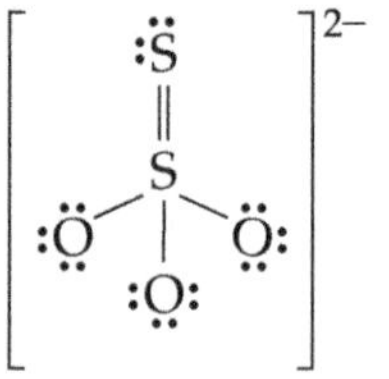

(c) 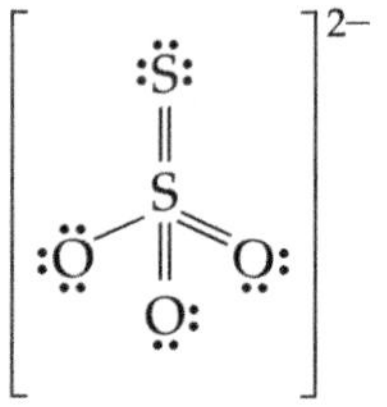   **(d)**

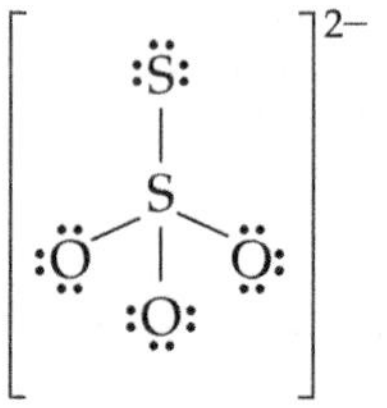

15. Use formal charge to select which resonance structure makes the largest contribution to the resonance hybrid. (**LO 7.16**)

(a) Structure I
(b) Structure II
(c) Structure III
(d) All structures are equivalent and make the same contribution to the resonance hybrid.

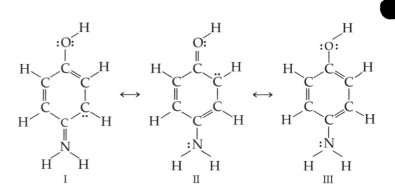

Answers:
1. b, 2. b, 3. a, 4. d, 5. b, 6. b, 7. d, 8. d, 9. d, 10. c, 11. c, 12. b, 13. b, 14. a, 15. c

Mastering Chemistry provides end-of-chapter exercises, feedback-enriched tutorial problems, animations, and interactive activities to encourage problem-solving practice and deeper understanding of key concepts and topics.

RAN *Randomized in Mastering Chemistry*

CONCEPTUAL PROBLEMS

Problems 7.1–7.30 appear within the chapter.

7.31 The following diagram shows the potential energy of two atoms as a function of internuclear distance. Match the descriptions with the indicated letter on the plot.

(a) Repulsive forces are high between the two atoms.

(b) The two atoms neither exert attractive nor repulsive forces on one another.

(c) The attractive forces between atoms are maximized, resulting in the lowest energy state.

(d) Attractive forces between atoms are present but are not at maximum strength.

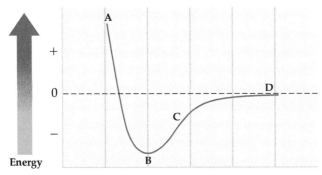

7.32 The following diagram shows the potential energy of two atoms as a function of internuclear distance. Which bond is the strongest? Which bond is the longest?

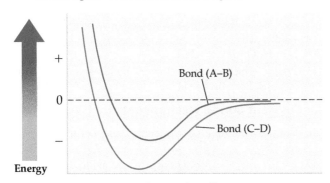

7.33 Two electrostatic potential maps are shown, one of methyllithium (CH_3Li) and the other of chloromethane (CH_3Cl). Based on their polarity patterns, which do you think is which?

(a)　　　　　　**(b)**

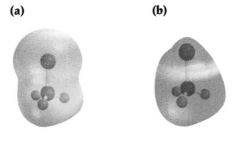

7.34 Electrostatic potential maps of acetaldehyde (C_2H_4O), ethane (C_2H_6), ethanol (C_2H_6O), and fluorethane (C_2H_5F) are shown. Which do you think is which?

(a)　　　　　　　　　　**(b)**

(c)　　　　　　　　　　**(d)**

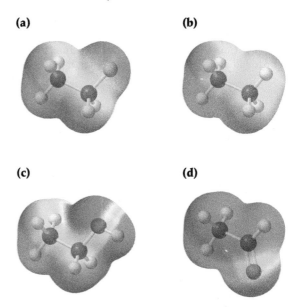

7.35 Which of the following drawings is most likely to represent an ionic compound and which a covalent compound?

(a)　　　　　　**(b)**　　　　　　**(c)**

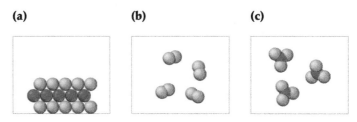

7.36 The following ball-and-stick molecular model is a representation of acetaminophen, the active ingredient in such over-the-counter headache remedies as Tylenol. (Red = O, gray = C, blue = N, ivory = H.) Give the formula of acetaminophen, and indicate the positions of the double bonds and lone pairs.

Acetaminophen

7.37 The following ball-and-stick molecular model is a representation of thalidomide, a drug that causes birth defects when taken by expectant mothers but is valuable for its use against leprosy. The lines indicate only the connections between atoms, not whether the bonds are single, double, or triple.

(Red = O, gray = C, blue = N, ivory = H.) Give the formula of thalidomide, and indicate the positions of multiple bonds and lone pairs.

Thalidomide

7.38 Sinapaldehyde, a compound present in the toasted wood used for aging wine, has the following connections among atoms. Complete the electron-dot structure for sinapaldehyde showing lone pairs and identifying multiple bonds.

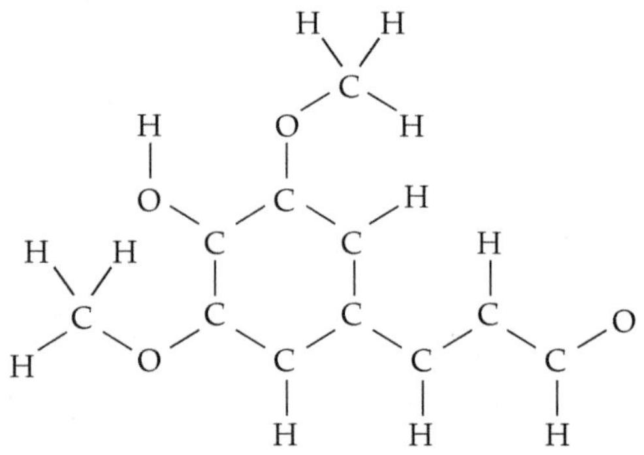

Sinapaldehyde

7.39 Vitamin C (ascorbic acid) has the following connections among atoms. Complete the electron-dot structure for vitamin C, showing lone pairs and identifying multiple bonds.

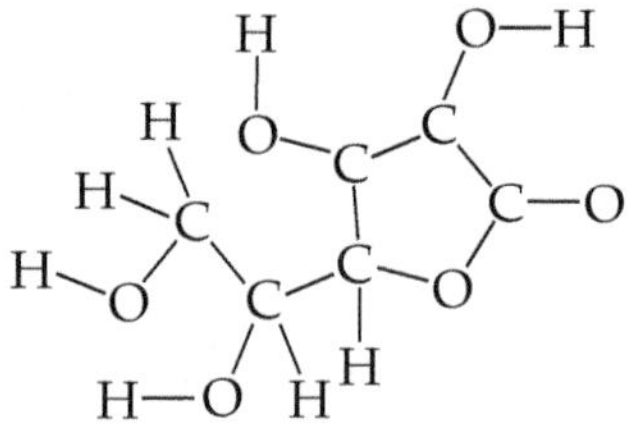

Vitamin C

SECTION PROBLEMS

Covalent Bonding (Section 7.1)

7.40 Match the following descriptions with the type of bond (ionic, nonpolar covalent, covalent).

(a) One or more electrons are transferred from a metal to a nonmetal atom.

(b) Electrons are shared equally between two atoms.

(c) Electrons are shared unequally by two atoms.

7.41 Why do two atoms come together to form a covalent bond?

RAN (a) Attractive forces between the positively charged nuclei and the electrons in both atoms occur when the atoms are close together.

(b) Repulsive forces between protons in the nuclei of the two atoms are minimized when the atoms are close together.

(c) Repulsive forces between electrons in the two atoms are minimized when the atoms are close together.

Strengths of Covalent Bonds (Section 7.2)

7.42 Explain the difference in the bond dissociation energies for the following bonds: (C—F, 450 kJ/mol), (N—F, 270 kJ/mol), (O—F, 180 kJ/mol), (F—F, 159 kJ/mol).

7.43 Explain the difference in the bond dissociation energies for the following bonds: (C—F, 450 kJ/mol), (C—Cl, 330 kJ/mol), (C—Br, 270 kJ/mol), (C—I, 240 kJ/mol).

7.44 Predict which of the following bonds should be strongest:
RAN N—H, O—H, S—H.

7.45 Predict which of the following bonds should be weakest:
RAN Cl—Cl, Br—Br, I—I.

Polar Covalent Bonds: Electronegativity (Section 7.3)

7.46 What general trends in electronegativity occur in the periodic table?

7.47 Predict the electronegativity of the undiscovered element with Z = 119.

7.48 Order the following elements according to increasing electro-
RAN negativity: Li, Br, Pb, K, Mg, C.

7.49 Order the following elements according to decreasing electronegativity: C, Ca, Cs, Cl, Cu.

7.50 Which of the following substances contain bonds that are largely ionic, and which contain bonds that are covalent?

(a) HF   (b) HI   (c) $PdCl_2$

(d) BBr_3 (e) NaOH   (f) CH_3Li

7.51 Use the electronegativity data in Figure 7.4 to predict which bond in each of the following pairs is more polar.

    (a) C—H or C—Cl

    (b) Si—Li or Si—Cl

    (c) N—Cl or N—Mg

7.52 Show the direction of polarity for each of the bonds in
RAN Problem 7.51 using the $\delta+/\delta-$ notation.

7.53 Show the direction of polarity for each of the covalent bonds
RAN in Problem 7.50 using the $\delta+/\delta-$ notation.

7.54 Which of the substances $CdBr_2$, P_4, BrF_3, MgO, NF_3, $BaCl_2$,
RAN $POCl_3$, and LiBr contain bonds that are:

    (a) largely ionic?

    (b) nonpolar covalent?

    (c) polar covalent?

7.55 Which of the substances S_8, $CaCl_2$, $SOCl_2$, NaF, CBr_4, BrCl,
RAN LiF, and AsH_3 contain bonds that are:

    (a) largely ionic?

    (b) nonpolar covalent?

    (c) polar covalent?

7.56 Order the following compounds according to the increasing ionic character of their bonds: CCl_4, $BaCl_2$, $TiCl_3$, ClO_2.

7.57 Order the following compounds according to the increasing ionic character of their bonds: NH_3, NCl_3, Na_3N, NO_2.

7.58 Using only the elements P, Br, and Mg, give formulas for the following.

    (a) An ionic compound

    (b) A molecular compound with polar covalent bonds that obeys the octet rule and has no formal charges

7.59 Using only the elements Ca, Cl, and Si, give formulas for the following.

    (a) An ionic compound

    (b) A molecular compound with polar covalent bonds that obeys the octet rule and has no formal charges

7.60 The dipole moment of BrCl is 0.518 D, and the distance between atoms is 213.9 pm. What is the percent ionic character of the BrCl bond?

7.61 The dipole moment of ClF is 0.887 D, and the distance between atoms is 162.8 pm. What is the percent ionic character of the ClF bond?

A Comparison of Ionic and Covalent Compounds (Section 7.4)

7.62 Which of the following is most likely to be a gas at room
RAN temperature (25°C)?

    (a) NH_3 (b) K_3N (c) Ca_3N_2

7.63 Which of the following is most likely to be a solid at room
RAN temperature (25°C)?

    (a) H_2S (b) SO_2 (c) Na_2S

Electron-Dot Structures and Resonance (Sections 7.5–7.7)

7.64 Why does the octet rule apply primarily to main-group elements, not to transition metals?

7.65 Which of the following substances contains an atom that does not follow the octet rule?

    (a) $AlCl_3$ (b) PCl_3

    (c) PCl_5 (d) $SiCl_4$

7.66 Draw electron-dot structures for the following molecules or ions.

    (a) CBr_4 (b) NCl_3 (c) C_2H_5Cl

    (d) $BF_4{}^2$ (e) $O_2{}^{2-}$ (f) NO^+

7.67 Draw electron-dot structures for the following molecules, which contain atoms from the third row or lower.

    (a) $SbCl_3$ (b) KrF_2 (c) ClO_2

    (d) PF_5 (e) H_3PO_4 (f) $SeCl_2$

7.68 Identify the correct electron-dot structure for $XeF_5{}^+$.
RAN

    (a)   **(b)**   **(c)**

7.69 Draw an electron-dot structure for the hydronium ion, H_3O^+, and show how a coordinate covalent bond is formed by the reaction of H_2O with H^+.

7.70 Oxalic acid, $H_2C_2O_4$, is a mildly poisonous substance found in the leaves of rhubarb, spinach, and many other plants. (You'd have to eat about 15 pounds or so of spinach leaves to ingest a lethal amount.) If oxalic acid has a C—C single bond and no C—H bond, draw its electron-dot structure showing lone pairs and identifying any multiple bonds.

7.71 Draw an electron-dot structure for carbon disulfide, CS_2, showing lone pairs and identifying any multiple bonds.

7.72 Identify the third-row elements, X, that form the following ions.

    (a)   **(b)**

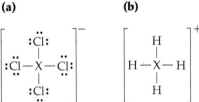

7.73 Identify the fourth-row elements, X, that form the following compounds.

    (a)   **(b)**

7.74 Write electron-dot structures for molecules with the following connections, showing lone pairs and identifying any multiple bonds.

(a)

$$Cl—\overset{\displaystyle O}{\underset{\displaystyle H}{C}}—O—\overset{\displaystyle H}{\underset{\displaystyle H}{C}}—H$$

(b)

$$H—\overset{\displaystyle H}{\underset{\displaystyle H}{C}}—C—C—H$$

7.75 Write electron-dot structures for molecules with the following connections, showing lone pairs and identifying any multiple bonds.

(a)

$$H—\overset{\displaystyle O}{C}—\overset{\displaystyle H}{N}—H$$

(b)

$$H—\overset{\displaystyle H}{\underset{\displaystyle H}{C}}—C—N—O$$

7.76 Which compound do you expect to have the stronger N—N bond, N_2H_2 or N_2H_4? Explain.

7.77 Which compound do you expect to have the stronger N—O bond, NO or NO_2? Explain.

7.78 Draw an electron-dot structure for each of the following substances.
(a) $F_3S—S—F$
(b) $CH_3—C = C—CO_2^-$

7.79 Write an electron-dot structure for chloral hydrate, also known in old detective novels as "knockout drops."

$$Cl—\overset{\displaystyle Cl}{\underset{\displaystyle Cl}{C}}—\overset{\displaystyle O—H}{\underset{\displaystyle H}{C}}—O—H \quad \text{Chloral hydrate}$$

Electron-Dot Structures for Molecules with Second-Row Elements (Section 7.8)

7.80 Methylphenidate ($C_{14}H_{19}NO_2$), marketed as Ritalin, is often prescribed for attention deficit disorder. Complete the following electron-dot structure, showing the position of any multiple bonds and lone pairs.

Methylphenidate
(Ritalin)

7.81 Pregabalin ($C_8H_{17}NO_2$), marketed as Lyrica, is an anticonvulsant drug prescribed for treatment of seizures. Complete the following electron-dot structure, showing the position of any multiple bonds and lone pairs.

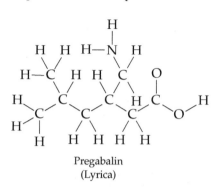

Pregabalin
(Lyrica)

7.82 The following molecular model is that of aspartame, $C_{14}H_{18}N_2O_5$, known commercially as NutraSweet. Only the connections between atoms are shown; multiple bonds are not indicated. Complete the structure by indicating the positions of the multiple bonds and lone pairs.

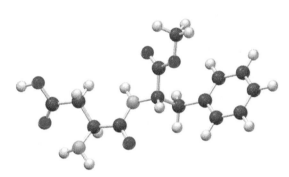

Aspartame

7.83 Ibuprofen ($C_{13}H_{18}O_2$), marketed under such brand names as Advil and Motrin, is a drug sold over the counter for treatment of pain and inflammation. Complete the structure of ibuprofen by adding hydrogen atoms and lone pairs where needed.

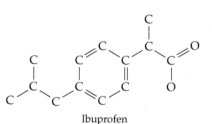

Ibuprofen

Resonance (Section 7.9)

7.84 Draw as many resonance structures as you can that obey the octet rule for each of the following molecules or ions. Use curved arrows to depict the conversion of one structure into another.
(a) HN_3 (b) SO_3 (c) SCN^-

7.85 Draw as many resonance structures as you can for the following nitrogen-containing compounds. Not all will obey the octet rule. Use curved arrows to depict the conversion of one structure into another.
(a) N_2O (b) NO
(c) NO_2 (d) $N_2O_3(ONNO_2)$

7.86 Which of the following pairs of structures represent resonance forms, and which do not?

(a)

$$H-C\equiv N-\ddot{O}: \quad \text{and} \quad H-C=\ddot{N}-\ddot{O}:$$

(b)

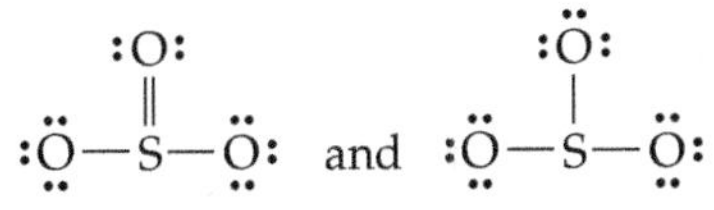

(c)

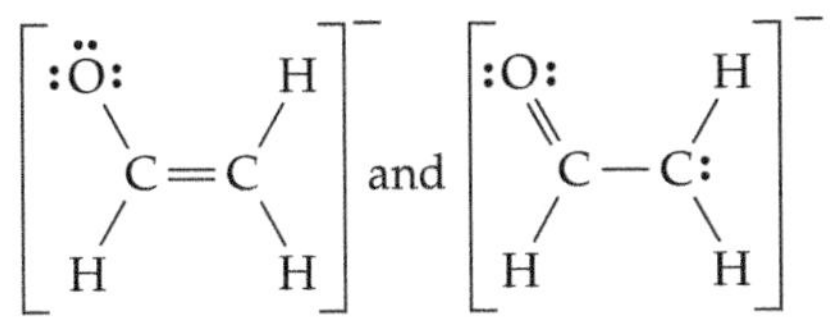

(d)

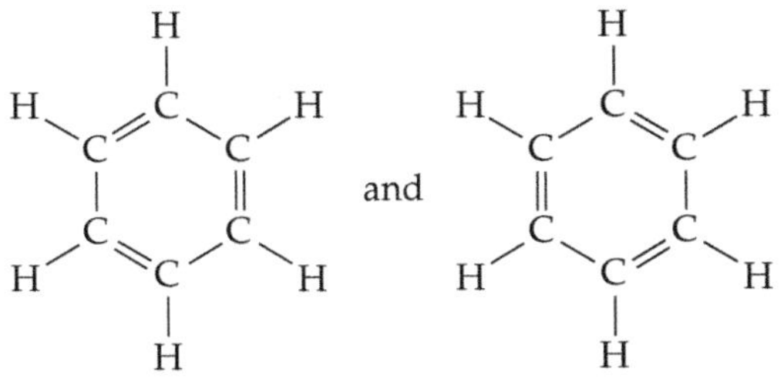

7.87 Which of the following pairs of structures represent resonance forms, and which do not?

(a)

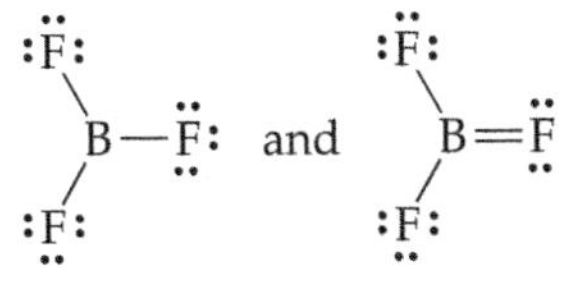

(b)

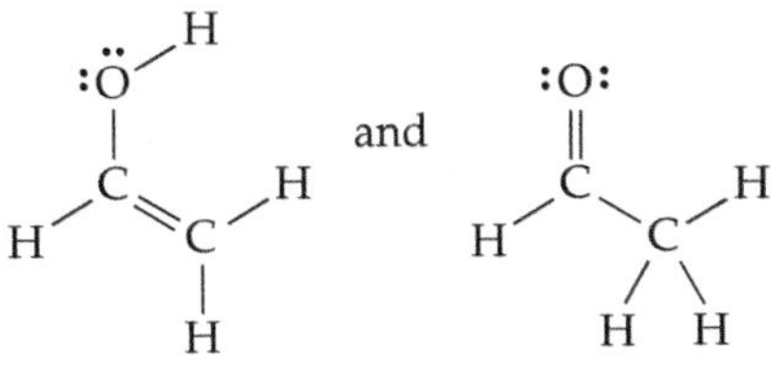

(c)

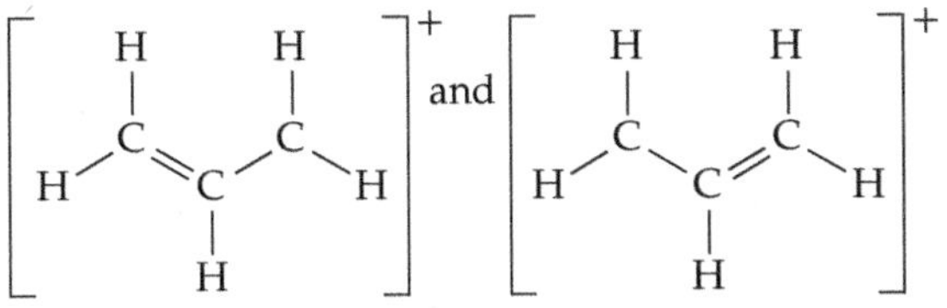

7.88 Draw as many resonance structures as you can that obey the octet rule for phenol (C_6H_6O). Use curved arrows to depict the conversion of one structure into another.

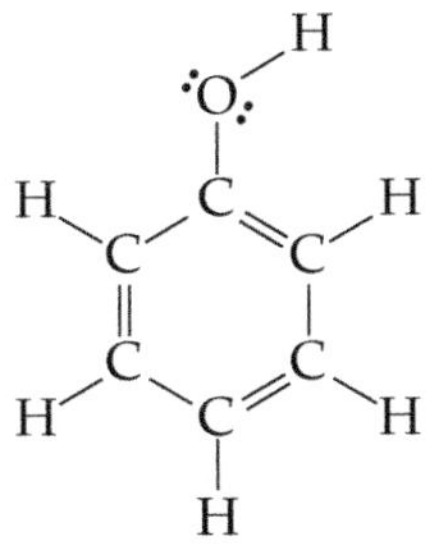

7.89 Draw as many resonance structures as you can that obey the octet rule for C_2H_5N. Use curved arrows to depict the conversion of one structure into another.

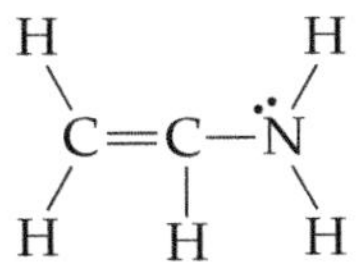

7.90 Benzene has the following structural formula.

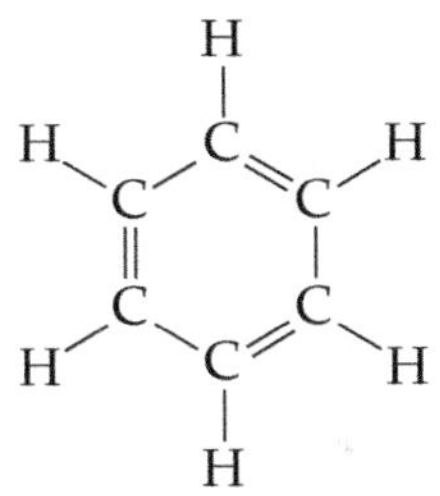

(a) Use curved arrows to show how to convert the original structure into a resonance structure.

(b) Which statement best describes the carbon–carbon bonds in benzene?

   (i) Three carbon–carbon bonds are longer and weaker than the other three carbon–carbon bonds.

   (ii) All six carbon–carbon bonds are identical, and their length and strength are between a double and single bond.

   (iii) The length of carbon–carbon double bond switches back and forth between the length of a double and a single bond.

7.91 Draw three resonance structures for sulfur tetroxide, SO_4, whose connections are shown below. (This is a neutral molecule; it is not a sulfate ion.) Assign formal charges to the atoms in each structure.

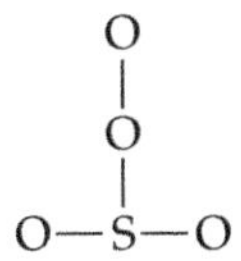

Sulfur tetroxide

7.92 Some mothballs used when storing clothes are made of naph-thalene ($C_{10}H_8$), which has the following incomplete structure.

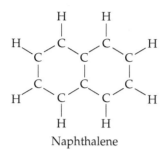

Naphthalene

(a) Add double bonds where needed to draw a complete electron-dot structure.

(b) Starting from this structure, use curved arrows to indicate how a new resonance structure can be drawn.

7.93 Four different structures (a), (b), (c), and (d) can be drawn for compounds named dibromobenzene, but only three different compounds actually exist. Explain.

(a)   (b)

(c)   (d)

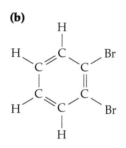

Formal Charges (Section 7.10)

7.94 Draw an electron-dot structure for carbon monoxide, CO, and assign formal charges to both atoms.

7.95 Assign formal charges to the atoms in the following structures.

(a)   (b)

$$H-N-\ddot{O}-H$$

$$\left[H-\ddot{N}-\underset{\underset{H}{|}}{\overset{\overset{H}{|}}{C}}-H \right]^{-}$$

(c)

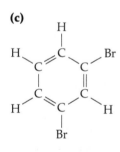

7.96 Assign formal charges to the atoms in the following resonance forms of ClO_2^-.

$$\left[:\ddot{O}-\ddot{C}l-\ddot{O}: \right]^{-} \longleftrightarrow \left[:\ddot{O}-\ddot{C}l=\ddot{O} \right]^{-}$$

7.97 Assign formal charges to the atoms in the following resonance forms of H_2SO_3.

$$H\ddot{O}-\overset{\overset{:O:}{||}}{S}-\ddot{O}H \longleftrightarrow H\ddot{O}-\overset{\overset{:\ddot{O}:}{|}}{S}-\ddot{O}H$$

7.98 Assign formal charges to the atoms in the following structures. Which of the two do you think is the more important contributor to the resonance hybrid?

(a)   (b)

7.99 Calculate formal charges for the C and O atoms in the following two resonance structures. Which structure do you think is the more important contributor to the resonance hybrid? Explain.

7.100 Draw two electron-dot resonance structures that obey the octet rule for trichloronitromethane, CCl_3NO_2, and show the formal charges on N and O in both structures. (Carbon is connected to the chlorines and to nitrogen; nitrogen is also connected to both oxygens.)

7.101 Draw two electron-dot resonance structures that obey the octet rule for nitrosyl chloride, NOCl (nitrogen is the central atom). Show formal charges, if present, and predict which of the two structures is a larger contributor to the resonance hybrid.

7.102 Draw the resonance structure indicated by the curved arrows. Assign formal charges, and evaluate which of the two structures is a larger contributor to the resonance hybrid.

7.103 Draw the resonance structure indicated by the curved arrows. Assign formal charges, and evaluate which of the two structures is a larger contributor to the resonance hybrid.

7.104 Boron trifluoride reacts with dimethyl ether to form a compound with a coordinate covalent bond. Assign formal charges to the B and O atoms in both the reactants and products.

Boron   Dimethyl
trifluoride   ether

7.105 Thiofulminic acid, $H-C\equiv N-S$, has recently been detected at very low temperatures.

    (a) Draw an electron-dot structure for thiofulminic acid, and assign formal charges.

    (b) A related compound with the same formula and the connection $H-N-C-S$ is also known. Draw an electron-dot structure for this related compound, and assign formal charges.

    (c) Which of the two molecules is likely to be more stable? Explain.

7.106 Draw two resonance structures for methyl isocyanate, CH_3NCO, a toxic gas that was responsible for the deaths of at least 3000 people when it was accidentally released into the atmosphere in December 1984 in Bhopal, India. Assign formal charges to the atoms in each resonance structure.

7.107 In the cyanate ion, OCN^-, carbon is the central atom.

    (a) Draw as many resonance structures as you can for OCN^-, and assign formal charges to the atoms in each.

    (b) Which resonance structure makes the greatest contribution to the resonance hybrid? Which makes the least contribution? Explain.

MULTICONCEPT PROBLEMS

7.108 The N_2O_5 molecule has nitrogen–oxygen bonds but no nitrogen–nitrogen bonds nor oxygen–oxygen bonds. Draw eight resonance structures for N_2O_5, and assign formal charges to the atoms in each. Which resonance structures make the more important contributions to the resonance hybrid?

7.109 Sulfur reacts with chlorine to give a product that contains 47.5% by mass sulfur and 52.5% by mass chlorine and has no formal charges on any of its atoms. Draw the electron-dot structure of the product.

7.110 Sulfur reacts with ammonia to give a product A that contains 69.6% by mass sulfur and 30.4% by mass nitrogen and has a molar mass of 184.3 g.

    (a) What is the formula of product A

    (b) The S and N atoms in the product A alternate around a ring, with half of the atoms having formal charges. Draw two possible electron-dot structures for A.

    (c) When compound A is heated with metallic silver at 250 °C a new product **B** is formed. Product **B** has the same percent composition as A but has a molar mass of 92.2 g. Draw two possible electron-dot structures for **B**, which, like A, also has a ring structure.

7.111 The neutral OH molecule has been implicated in certain ozone-destroying processes that take place in the upper atmosphere.

    (a) Draw electron-dot structures for the OH molecule and the OH^- ion.

    (b) Electron affinity can be defined for molecules just as it is defined for single atoms. Assuming that the electron added to OH is localized in a single atomic orbital on one atom, identify which atom is accepting the electron, and give the n and l quantum numbers of the atomic orbital.

    (c) The electron affinity of OH is similar to but slightly more negative than that of O atoms. Explain.

7.112 Suppose that the Pauli exclusion principle were somehow changed to allow three electrons per orbital rather than two.

    (a) Instead of an octet, how many outer-shell electrons would be needed for a noble-gas electron configuration?

    (b) How many electrons would be shared in a covalent bond?

    (c) Give the electron configuration, and draw an electron-dot structure for element X with $Z = 12$.

    (d) Draw an electron-dot structure for the molecule X_2.

7.113 The dichromate ion, $Cr_2O_7^{2-}$, has neither $Cr-Cr$ nor $O-O$ bonds.
RAN

    (a) Taking both $4s$ and $3d$ electrons into account, draw an electron-dot structure that minimizes the formal charges on the atoms.

    (b) How many outer-shell electrons does each Cr atom have in your electron-dot structure?

chapter 8

Covalent Compounds: Bonding Theories and Molecular Structure

Contents

We obtain ascorbic acid, also known as vitamin C, by eating citrus fruits or taking supplements. Vitamin C repairs tissues, produces neurotransmitters, and activates enzymes. The biological function of molecules depends on molecular shape and intermolecular forces, two properties of molecules we'll discuss in this chapter.

Which is better for human health: natural or synthetic vitamins?

The answer to this question can be found on page 314 in the INQUIRY ?

hy is molecular structure important? We can predict properties and behavior by examining the three-dimensional shape and charge distribution in a compound. Melting and boiling points, solubility, and reactivity all depend on molecular structure. Molecular structure also plays an important role in biological processes like smell, taste, and nerve impulses. The electron-dot structures described in Chapter 7 provide a straightforward way to indicate covalent bonds within a molecule. However, electron dot structures are two-dimensional, while molecules are three-dimensional. We need more information in order to form a detailed picture of molecular structure.

In this chapter, we will build upon the concepts of electron-dot structures and polar covalent bonds to describe shapes of molecules and interactions between them. We will also develop more complex models of chemical bonding including valence bond theory and molecular orbital theory that can be used to predict structure and reactivity.

8.1 MOLECULAR SHAPES: THE VSEPR MODEL

Look at the following ball-and-stick models of water, ammonia, and methane. Each of these molecules—and every other molecule as well—has a specific three-dimensional shape. Below the ball-and-stick model is the structural model that uses solid wedges and dashed lines to represent three-dimensional shape. Solid lines are assumed to be in the plane of the paper, dashed lines recede behind the plane of the paper away from the viewer, and heavy, wedged lines protrude out of the paper toward the viewer. It is a useful skill to be able to draw molecules using this type of three-dimensional representation.

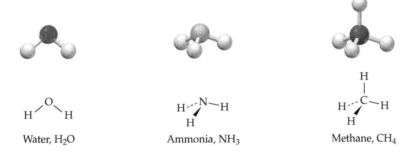

Water, H_2O Ammonia, NH_3 Methane, CH_4

The shape of a molecule depends on the electronic structure of its atoms. That shape can often be predicted using what is called the **valence-shell electron-pair repulsion (VSEPR) model**. Electrons in bonds and in lone pairs can be thought of as "charge clouds" that repel one another and stay as far apart as possible, thus causing molecules to assume specific shapes. There are only two steps to remember in applying the VSEPR model:

Applying the VSEPR Model

Step 1. Write an electron-dot structure for the molecule, as described in Section 7.6, and count the number of electron charge clouds surrounding the atom of interest. A charge cloud is simply a group of electrons, either in a bond or in a lone pair, that occupy a region of space around an atom. The following bonds and electron groups represent one charge cloud: a single electron (in a radical species), a lone pair of electrons, a single bond, a double bond, and a triple bond. The total number of charge clouds around an atom is the number of regions of space that contain electrons. For example, an atom with four single bonds has four charge clouds. An atom with a lone pair, a single bond, and a double bond has three charge clouds.

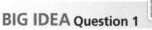
Go to
eText

BIG IDEA Question 1

According to the VSEPR model, what determines the three-dimensional shape of a molecule?

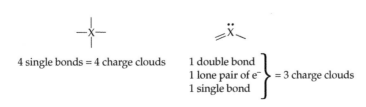

4 single bonds = 4 charge clouds   1 double bond
1 lone pair of e^- } = 3 charge clouds
1 single bond

Step 2. Predict the geometric arrangement of charge clouds around each atom by assuming that the clouds are oriented in space as far away from one another as possible. How they achieve this orientation depends on their number. Let's look at the possibilities.

Two Charge Clouds

When there are only two charge clouds on an atom, as occurs on the carbon atoms of CO_2 (two double bonds) and HCN (one single bond and one triple bond), the clouds are farthest apart when they point in opposite directions. Thus, CO_2 and HCN are linear molecules with **bond angles** of 180°.

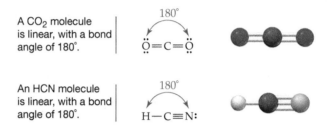

A CO_2 molecule is linear, with a bond angle of 180°.

An HCN molecule is linear, with a bond angle of 180°.

Three Charge Clouds

When there are three charge clouds on an atom, as occurs on the carbon atom of formaldehyde (two single bonds and one double bond) and the sulfur atom of SO_2 (one single bond, one double bond, and one lone pair), the clouds are farthest apart when they lie in the same plane and point to the corners of an equilateral triangle. Thus, a formaldehyde molecule has a *trigonal planar shape*, with H—C—H and H—C=O bond angles near 120°. Similarly, an SO_2 molecule has a trigonal planar arrangement of its three charge clouds on sulfur, but one corner of the triangle is occupied by a lone pair and two corners by oxygen atoms. The molecule therefore has a bent shape, with an O—S=O bond angle of approximately 120°. Note that the number of charge clouds determines the bond angles, but the actual shape of the molecule is given by positions of atoms only and lone pairs of electrons are not "seen." For consistency, we'll use the word *shape* to refer to the overall arrangement of atoms in a molecule, not to the geometric arrangement of charge clouds around a specific atom.

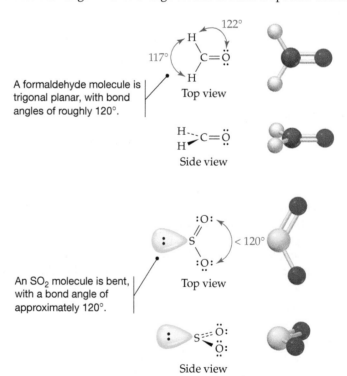

A formaldehyde molecule is trigonal planar, with bond angles of roughly 120°.

Top view

Side view

An SO_2 molecule is bent, with a bond angle of approximately 120°.

Top view

Side view

Four Charge Clouds

When there are four charge clouds on an atom, as occurs on the central atoms in CH_4 (four single bonds), NH_3 (three single bonds and one lone pair), and H_2O (two single bonds and two lone pairs), the clouds are farthest apart if they extend toward the corners of a regular tetrahedron. As illustrated in **FIGURE 8.1,** a regular tetrahedron is a geometric solid whose four identical faces are equilateral triangles. The central atom lies in the center of the tetrahedron, the charge clouds point toward the four corners, and the angle between two lines drawn from the center to any two corners is 109.5°.

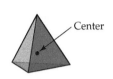 

A regular tetrahedron

The atom is located in the **center** of a regular tetrahedron.

The four charge clouds point to the **four corners** of the tetrahedron.

A tetrahedral molecule

109.5°

The angle between any two bonds is 109.5°.

◀ **FIGURE 8.1**

The tetrahedral geometry of an atom with four charge clouds.

◀ **Figure It out**

Why do four charge clouds orient in a tetrahedral geometry?

Answer: The electron charge clouds repel one another and a tetrahedral geometry maximizes the angles between clouds.

Because valence electron octets are so common, particularly for second-row elements, the atoms in a great many molecules have geometries based on the tetrahedron. Methane (CH_4), for example, has a tetrahedral shape, with H—C—H bond angles of 109.5°. In ammonia (NH_3), the nitrogen atom has a tetrahedral arrangement of its four charge clouds, but one corner of the tetrahedron is occupied by a lone pair, resulting in a *trigonal pyramidal* shape for the molecule. Similarly, H_2O has two corners of the tetrahedron occupied by lone pairs and thus has a *bent* shape.

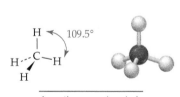 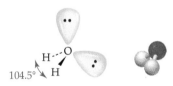

A methane molecule is tetrahedral, with bond angles of 109.5°.

CH_4

An ammonia molecule is trigonal pyramidal, with bond angles of 107°.

NH_3

A water molecule is bent, with a bond angle of 104.5°.

H_2O

Note that the H—N—H bond angles in ammonia (107°) and the H—O—H bond angle in water (104.5°) are less than the ideal 109.5° tetrahedral value. The angles are diminished somewhat from the tetrahedral value because of the presence of lone pairs. Charge clouds of lone-pair electrons spread out more than charge clouds of bonding electrons because they aren't confined to the space between two nuclei. As a result, the somewhat enlarged lone-pair charge clouds tend to compress the bond angles in the rest of the molecule.

Five Charge Clouds

Five charge clouds, such as are found on the central atoms in PCl_5, SF_4, ClF_3, and I_3^-, are oriented toward the corners of a geometric figure called a *trigonal bipyramid.* Three clouds lie in a plane and point toward the corners of an equilateral triangle, the fourth cloud points directly up, and the fifth cloud points down:

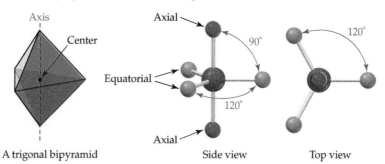

A trigonal bipyramid   Side view   Top view

Trigonal bipyramidal geometry differs from the linear, trigonal planar, and tetrahedral geometries discussed previously because it has two kinds of positions—three *equatorial* positions (around the "equator" of the bipyramid) and two *axial* positions (along the "axis" of the bipyramid). The three equatorial positions are at angles of 120° to one another and at an angle of 90° to the axial positions. The two axial positions are at angles of 180° to each other and at an angle of 90° to the equatorial positions.

Different substances containing a trigonal bipyramidal arrangement of charge clouds on an atom adopt different shapes, depending on whether the five charge clouds contain bonding or nonbonding electrons. Phosphorus pentachloride, for instance, has all five positions around phosphorus occupied by chlorine atoms and thus has a trigonal bipyramidal shape:

A PCl_5 molecule is trigonal bipyramidal.

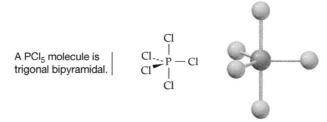

Go to
eText

BIG IDEA Question 2

In a molecule with four bonds and one lone pair of electrons (five charge clouds), what is the position of the lone pair?

The sulfur atom in SF_4 is bonded to four other atoms and has one nonbonding electron lone pair. Because an electron lone pair spreads out and occupies more space than a bonding pair, the nonbonding electrons in SF_4 occupy an equatorial position where they are close to (90° away from) only two charge clouds. Were they instead to occupy an axial position, they would be close to three charge clouds. As a result, SF_4 has a shape often described as that of a seesaw. The two axial bonds form the board, and the two equatorial bonds form the legs of the seesaw.

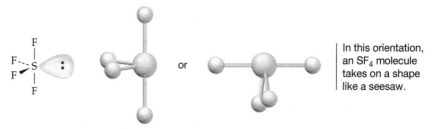

In this orientation, an SF_4 molecule takes on a shape like a seesaw.

The chlorine atom in ClF_3 is bonded to three other atoms and has two nonbonding electron lone pairs. Both lone pairs occupy equatorial positions, resulting in a T shape for the ClF_3 molecule.

A ClF_3 molecule is T-shaped.

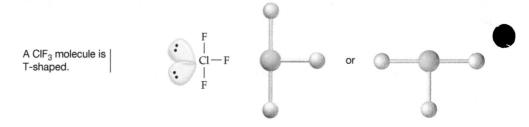

The central iodine atom in the I_3^- ion is bonded to two other atoms and has three lone pairs. All three lone pairs occupy equatorial positions, resulting in a linear shape for I_3^-.

An I_3^- ion is linear.

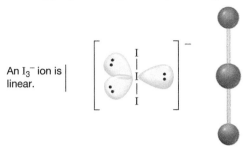

Six Charge Clouds

Six charge clouds around an atom orient toward the six corners of a regular octahedron, a geometric solid whose eight faces are equilateral triangles. All six positions are equivalent, and the angle between any two adjacent positions is 90°.

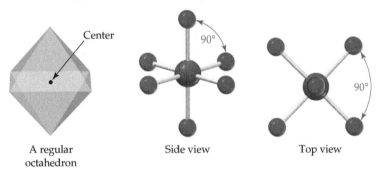

Center

90°

90°

A regular octahedron

Side view

Top view

As was true in the case of five charge clouds, different shapes are possible for molecules having atoms with six charge clouds, depending on whether the clouds are of bonding or nonbonding electrons. In sulfur hexafluoride, for instance, all six positions around sulfur are occupied by fluorine atoms:

An SF_6 molecule is octahedral.

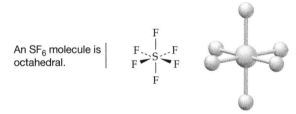

The antimony atom in the $SbCl_5^{2-}$ ion also has six charge clouds but is bonded to only five atoms and has one nonbonding electron lone pair. As a result, the ion has a *square pyramidal* shape—a pyramid with a square base:

An $SbCl_5^{2-}$ ion has a square pyramidal shape.

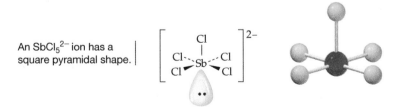

The xenon atom in XeF_4 is bonded to four atoms and has two lone pairs. The lone pairs orient as far away from each other as possible to minimize electronic repulsions, giving the molecule a *square planar* shape:

An XeF_4 molecule has a square planar shape.

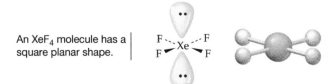

All the geometries for two to six charge clouds are summarized in **TABLE 8.1**.

WORKED EXAMPLE 8.1

Using the VSEPR Model to Predict Molecular Shape

Predict the shape of BrF_5.

STRATEGY

First, draw an electron-dot structure for BrF_5 to determine that the central bromine atom has six charge clouds (five bonds and one lone pair). Then predict how the six charge clouds are arranged.

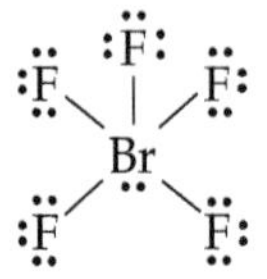

Bromine pentafluoride

SOLUTION

Six charge clouds imply an octahedral arrangement. The five attached atoms and one lone pair give BrF_5 a square pyramidal shape:

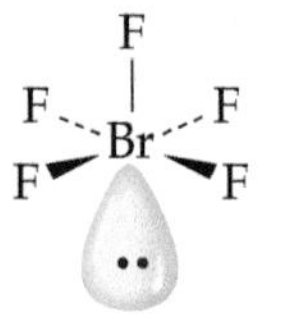

▶ **PRACTICE 8.1** What is the shape of ICl_4^-?

▶ **CONCEPTUAL APPLY 8.2** What is the number and geometric arrangement of charge clouds around the central atom in each of the following molecular models? What is the shape of the molecule?

(a) **(b)**

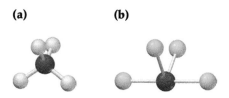

All **PRACTICE** and **APPLY** problems are interactive in the eText.

Shapes of Larger Molecules

We can also predict the geometries around individual atoms in larger molecules from the rules summarized in Table 8.1. For example, each of the two carbon atoms in ethylene ($H_2C=CH_2$) has three charge clouds, giving rise to trigonal planar geometry for each carbon. The molecule as a whole has a planar shape, with $H-C-C$ and $H-C-H$ bond angles of approximately 120°.

Each carbon atom in ethylene has trigonal planar geometry. As a result, the entire molecule is planar, with bond angles of 120°.

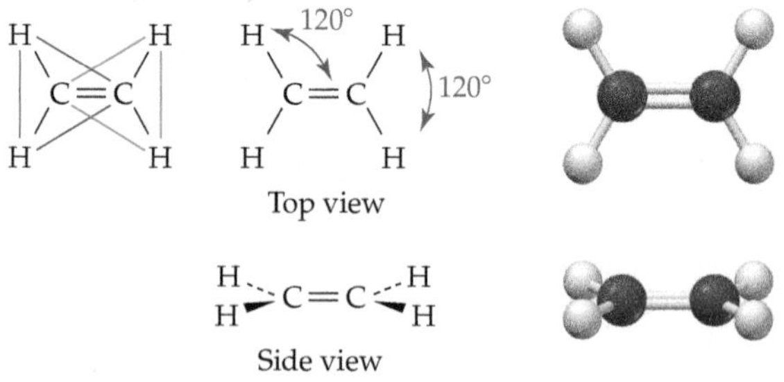

Top view

Side view

Carbon atoms bonded to four other atoms are each at the center of a tetrahedron. As shown below for ethane, H_3C-CH_3, the two tetrahedrons are joined so that the central carbon atom of one is a corner atom of the other.

Each carbon atom in ethane has tetrahedral geometry, with bond angles of 109.5°.

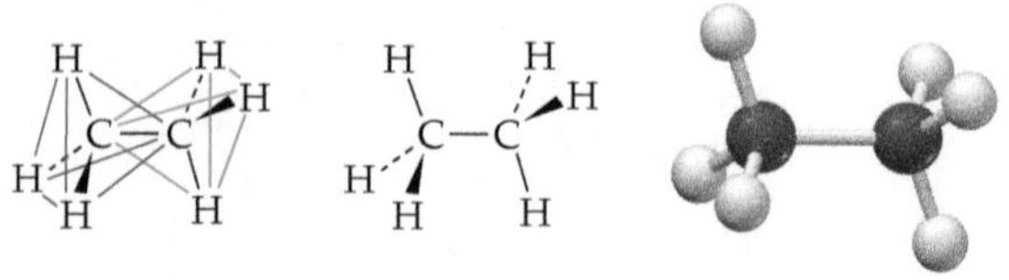

TABLE 8.1   Geometry Around Atoms with 2, 3, 4, 5, and 6 Charge Clouds

Number of Bonds	Number of Lone Pairs	Number of Charge Clouds	Molecular Shape (Geometry)	Example
2	0	2	Linear	$O{=}C{=}O$
3	0	3	 Trigonal planar	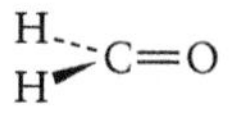
2	1		 Bent	
4	0	4	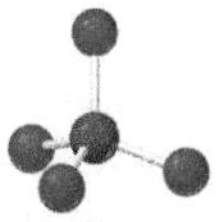 Tetrahedral	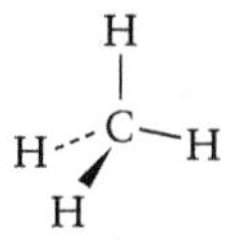
3	1		 Trigonal pyramidal	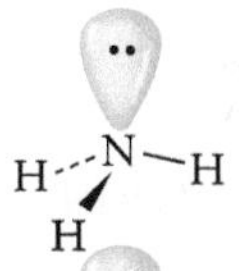
2	2		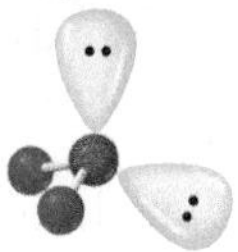 Bent	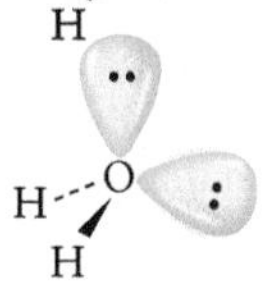
5	0	5	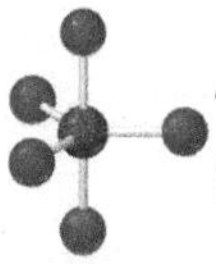 Trigonal bipyramidal	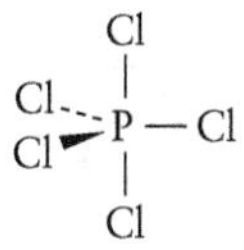
4	1		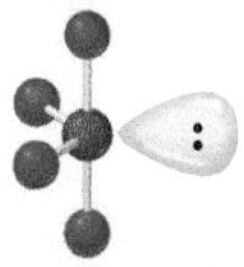 Seesaw	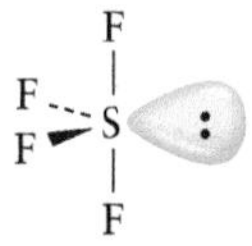
3	2		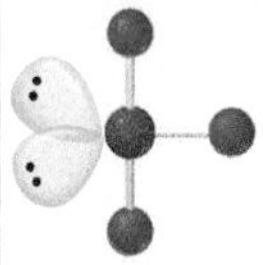 T-shaped	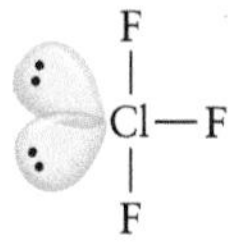
2	3		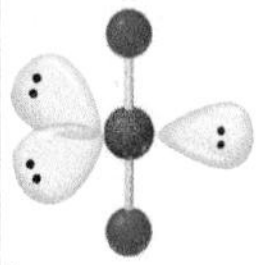 Linear	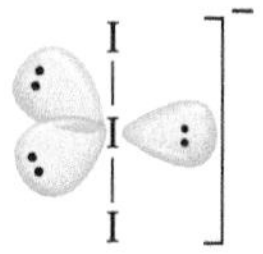
6	0	6	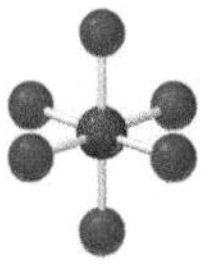 Octahedral	
5	1		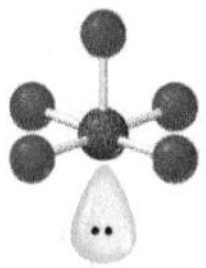 Square pyramidal	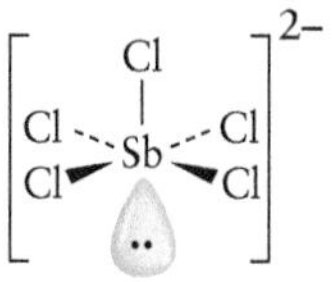
4	2		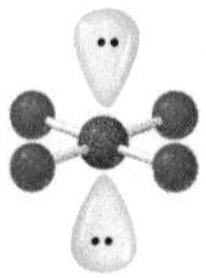 Square planar	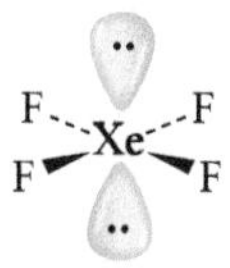

CONCEPTUAL WORKED EXAMPLE 8.2

Predicting Shape in Molecules with More Than One Central Atom

Pyruvic acid is a key substance in the metabolism of both carbohydrates and several amino acids. Describe the geometry around each of the central atoms in pyruvic acid, and draw the overall shape of the molecule.

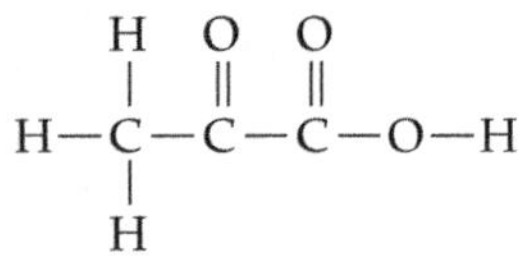

$$\text{H} \quad \text{O} \quad \text{O}$$
$$| \quad \| \quad \|$$
$$\text{H}-\text{C}-\text{C}-\text{C}-\text{O}-\text{H}$$
$$|$$
$$\text{H}$$

Pyruvic acid

STRATEGY

Use the VSEPR model to predict the geometric arrangement of charge clouds around each of the three carbon atoms. Draw the overall molecule showing the bond angles predicted by VSEPR and using dashed lines and wedges to represent three-dimensional shape.

SOLUTION

The carbon on the left has four charge clouds (four single bonds) and tetrahedral geometry with 109.5° bond angles. Both the carbon in the middle and on the right have three charge clouds (two single bonds and one double bond) and trigonal planar geometry with 120° bond angles.

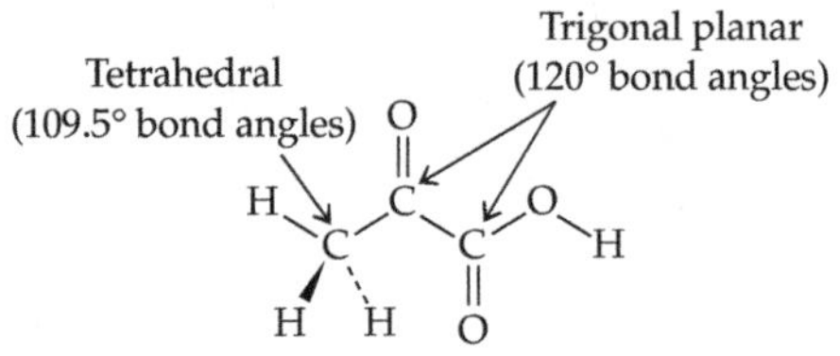

Tetrahedral
(109.5° bond angles)

Trigonal planar
(120° bond angles)

▶ **PRACTICE 8.3** Acetic acid, CH_3CO_2H, is the main organic constituent of vinegar. Draw an electron-dot structure for acetic acid and give the bond angles around each carbon atom. (The two carbons are connected by a single bond, and both oxygens are connected to the same carbon.)

▶ **APPLY 8.4** Benzene, C_6H_6, is a cyclic molecule in which all six carbon atoms are joined in a ring, with each carbon also bonded to one hydrogen. Draw an electron-dot structure for benzene and give the bond angles around each carbon. Draw a three-dimensional representation of the molecule.

8.2 VALENCE BOND THEORY

The VSEPR model discussed in the previous section provides a simple way to predict molecular shapes, but it says nothing about the electronic nature of covalent bonds. To describe bonding, two models, called *valence bond theory* and *molecular orbital theory*, have been developed. We'll look first at valence bond theory, followed by molecular orbital theory in Sections 8.7 and 8.8.

Valence bond theory provides an easily visualized orbital picture of how electron pairs are shared in a covalent bond. In essence, a covalent bond results when two atoms approach each other closely enough so that a singly occupied valence orbital on one atom spatially *overlaps* a singly occupied valence orbital on the other atom. The now-paired electrons in the overlapping orbitals are attracted to the nuclei of both atoms and thus bond the two atoms together. In the H_2 molecule, for instance, the H—H bond results from the overlap of two singly occupied hydrogen 1s orbitals.

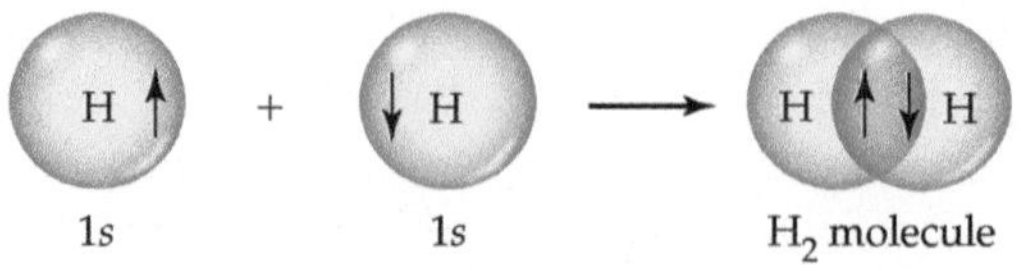

H ↑   +   ↓ H   ⟶   H ↑↓ H

1s   1s H_2 molecule

Recall that atomic orbitals arise from the Schrödinger wave equation and that the two lobes of a p atomic orbital have different phases, as represented by different colors. In the valence bond model, the two overlapping orbital lobes must be of the same phase. The phases of orbitals are analogous to the phase of wave, either positive (up) or negative (down). Overlapping lobes of the same phase are similar to adding two waves that are in phase. The amplitude of the resultant wave increases, and in valence bond theory, the probability of finding the electron increases. Also, the strength of the covalent bond that forms depends on the amount of orbital overlap: the greater the overlap, the stronger the bond. This, in turn, means that bonds formed by overlap of other than s orbitals have a directionality to them. In the F_2 molecule, for instance, each fluorine atom has the ground state electron configuration $[\text{He}]\,2s^2\,2p_x^2\,2p_y^2\,2p_z^1$ and the F—F bond results from the overlap of two singly occupied $2p$ orbitals. The two $2p$ orbitals must point directly at each other for optimum overlap to occur, and the F—F bond forms along the orbital axis. Such bonds that result from head-on orbital overlap are called **sigma (σ) bonds.**

REMEMBER . . .

The Schrödinger wave equation focuses on the wavelike properties of atoms to describe the quantum mechanical model of atomic structure. The solutions to the wave equation are called wave functions, or orbitals (Section 5.6).
The two lobes of a p orbital have different mathematical signs in the wave function, corresponding to the different phases of a wave (Section 5.7).

Bonds form between two lobes of the same phase.

A **sigma (σ) bond** forms from head-on orbital overlap.

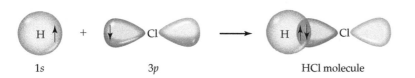

$2p$ $2p$ F_2 molecule

In HCl, the covalent bond involves overlap of a hydrogen $1s$ orbital with a chlorine $3p$ orbital and forms along the p-orbital axis:

$1s$ $3p$ HCl molecule

BIG IDEA Question 3

What type of orbital overlap results in a chemical bond in valence bond theory?

We can summarize the key ideas of valence bond theory as follows.

Principles of Valence Bond Theory

- Covalent bonds are formed by overlap of atomic orbitals, each of which contains one electron of opposite spin. The two overlapping lobes must be of the same phase.

- Each of the bonded atoms maintains its own atomic orbitals, but the electron pair in the overlapping orbitals is shared by both atoms.

- The greater the amount of orbital overlap, the stronger the bond. This leads to a directional character for the bond when other than s orbitals are involved.

8.3 HYBRIDIZATION AND sp^3 HYBRID ORBITALS

How does valence bond theory describe the electronic structure of complex polyatomic molecules, and how does it account for the observed geometries around atoms in molecules? Let's look first at a simple tetrahedral molecule such as methane, CH_4. There are several problems to be dealt with.

Carbon has the ground-state electron configuration $[\text{He}]2s^2\,2p_x^1\,2p_y^1$. It thus has four valence electrons, two of which are paired in a $2s$ orbital and two of which are unpaired in different $2p$ orbitals that we'll arbitrarily designate as $2p_x$ and $2p_y$. But how can carbon form four bonds if two of its valence electrons are already paired and only two unpaired electrons remain for sharing? The answer is that an electron must be promoted from the lower-energy $2s$ orbital to the vacant, higher-energy $2p_z$ orbital, giving an *excited-state configuration* $[He] 2s^1 2p_x^1 2p_y^1 2p_z^1$ that has *four* unpaired electrons and can thus form four bonds.

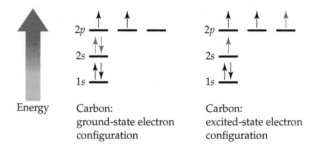

A second problem is more difficult to resolve: If excited-state carbon uses two kinds of orbitals for bonding, $2s$ and $2p$, how can it form four *equivalent* bonds? Furthermore, if the three $2p$ orbitals in carbon are at angles of 90° to one another, and if the $2s$ orbital has no directionality, how can carbon form bonds with angles of 109.5° directed to the corners of a regular tetrahedron? The answers to these questions were provided in 1931 by Linus Pauling, who introduced the idea of *hybrid orbitals*.

Pauling showed how the quantum mechanical wave functions for s and p atomic orbitals derived from the Schrödinger wave equation can be mathematically combined to form a new set of equivalent wave functions called **hybrid atomic orbitals.** When one s orbital combines with three p orbitals, as occurs in an excited-state carbon atom, four equivalent hybrid orbitals, called sp^3 **hybrid orbitals,** result. (The superscript 3 in the name sp^3 tells how many p atomic orbitals are combined to construct the hybrid orbitals, not how many electrons occupy the orbital.)

 Go to eText

BIG IDEA Question 4

Why is it necessary to use hybrid orbitals rather than atomic orbitals in valence bond theory?

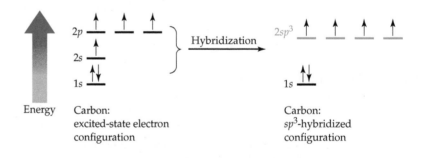

Each of the four equivalent sp^3 hybrid orbitals has two lobes of different phase like an atomic p orbital (Section 5.7), but one of the lobes is larger than the other, giving the orbital a directionality. The four large lobes are oriented toward the four corners of a tetrahedron at angles of 109.5°, as shown in **FIGURE 8.2.** For consistency in the use of colors, we'll routinely show the different phases of orbitals in red and blue and will show the large lobes of the resultant hybrid orbitals in green.

The shared electrons in a covalent bond made with a spatially directed hybrid orbital spend most of their time in the region between the two bonded nuclei. As a result, covalent bonds made with sp^3 hybrid orbitals are often strong ones. In fact, the energy released on forming the four strong C—H bonds in CH_4 more than compensates for the energy required to form the excited state of carbon. **FIGURE 8.3** shows how the four C—H sigma bonds in methane can form by head-on overlap of carbon sp^3 hybrid orbitals with hydrogen $1s$ orbitals.

The same kind of sp^3 hybridization that describes the bonds to carbon in the tetrahedral methane molecule can also be used to describe bonds to nitrogen in the trigonal

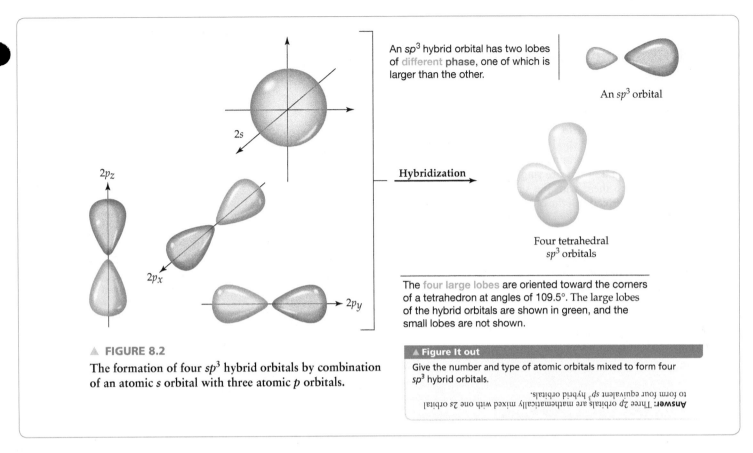

An sp^3 hybrid orbital has two lobes of different phase, one of which is larger than the other.

An sp^3 orbital

Hybridization

Four tetrahedral sp^3 orbitals

The four large lobes are oriented toward the corners of a tetrahedron at angles of 109.5°. The large lobes of the hybrid orbitals are shown in green, and the small lobes are not shown.

▲ **FIGURE 8.2**

The formation of four sp^3 hybrid orbitals by combination of an atomic *s* orbital with three atomic *p* orbitals.

▲ **Figure It out**

Give the number and type of atomic orbitals mixed to form four sp^3 hybrid orbitals.

Answer: Three $2p$ orbitals are mathematically mixed with one $2s$ orbital to form four equivalent sp^3 hybrid orbitals.

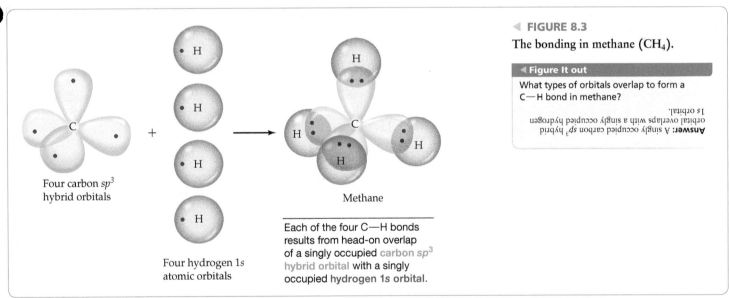

Four carbon sp^3 hybrid orbitals

Four hydrogen $1s$ atomic orbitals

Methane

Each of the four C—H bonds results from head-on overlap of a singly occupied carbon sp^3 hybrid orbital with a singly occupied hydrogen 1s orbital.

◀ **FIGURE 8.3**

The bonding in methane (CH_4).

◀ **Figure It out**

What types of orbitals overlap to form a C—H bond in methane?

Answer: A singly occupied carbon sp^3 hybrid orbital overlaps with a singly occupied hydrogen 1s orbital.

pyramidal ammonia molecule, to oxygen in the bent water molecule, and to all other atoms that the VSEPR model predicts to have a tetrahedral arrangement of four charge clouds.

Methane, CH_4

Ammonia, NH_3

Water, H_2O

WORKED EXAMPLE 8.3

Identifying Orbital Overlap in Tetrahedral Geometry

Identify the orbitals that overlap to form the P—Cl bond in PCl_3.

STRATEGY

Write the electron-dot structure for PCl_3 and determine the geometry and hybrid orbitals around the central atom. Determine which orbitals on the terminal atoms are singly occupied and overlap them with the singly occupied hybrid orbitals on the central atom to form a bond.

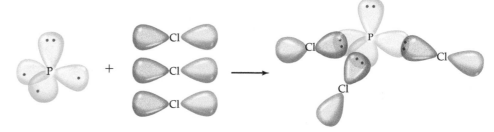

Four phosphorus sp^3 hybrid orbitals

Three chlorine $3p$ atomic orbitals

Each of the three P—Cl bonds results from head-on overlap of a singly occupied phosphorus sp^3 hybrid orbital with a singly occupied chlorine $3p$ orbital.

SOLUTION

The electron-dot structure for PCl_3 has three single bonds and one lone pair of electrons. There are four charge clouds in a tetrahedral arrangement that corresponds to sp^3 hybrid orbitals. The shape of PCl_3 is trigonal pyramidal because there are three bonds and one lone pair.

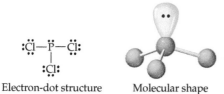

Electron-dot structure

Molecular shape
Trigonal pyramidal

▶ **PRACTICE 8.5** Identify the orbitals that overlap to form the C—Cl bonds and C—H bonds in CH_2Cl_2.

▶ **APPLY 8.6** Describe the bonding in propane, C_3H_8, a fuel often used to heat rural homes and campers. Give the kinds of orbitals on each atom that overlap to form the C—C and C—H bonds.

8.4 OTHER KINDS OF HYBRID ORBITALS

Other geometries shown in Table 8.1 can also be accounted for by specific kinds of orbital hybridization, although hybrid orbitals are not a useful model with atoms with five and six charge clouds. Let's look at each.

sp^2 Hybridization

Atoms with three charge clouds undergo hybridization by combination of one atomic s orbital with two p orbitals, resulting in three sp^2 **hybrid orbitals**. These three sp^2 hybrids lie in a plane and are oriented toward the corners of an equilateral triangle at angles of 120° to one another. One p orbital remains unchanged and is oriented at a 90° angle to the plane of the sp^2 hybrids, as shown in **FIGURE 8.4**.

The presence of the unhybridized p orbital on an sp^2-hybridized atom has some interesting consequences. Look, for example, at ethylene, H_2C=CH_2, a colorless gas used as starting material for the industrial preparation of polyethylene. Each carbon atom in ethylene has three charge clouds and is sp^2-hybridized. The two sp^2-hybridized carbon atoms approach each other with sp^2 orbitals aligned head-on to form a sigma (σ) bond. The unhybridized p orbitals on the carbons likewise approach each other and form a bond, but in a parallel, sideways manner rather than head-on. Such a sideways bond, in which the shared electrons occupy regions above and below a line connecting the nuclei rather than directly between the nuclei, is called a **pi** (π) **bond** (**FIGURE 8.5**). A double bond, such as the C=C bond in ethylene, consists of one sigma bond and one pi bond. In addition, four C—H bonds form in ethylene by overlap of the remaining four sp^2 orbitals with hydrogen $1s$ orbitals.

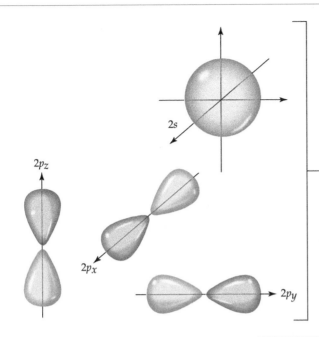

An sp^2 hybrid orbital has two lobes of different phase, one of which is larger than the other.

An sp^2 orbital

Hybridization →

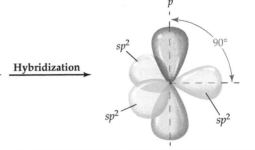

The hybrid orbitals lie in a plane at angles of 120° to one another, and one unhybridized p orbital remains, oriented at a 90° angle to the sp^2 hybrids. The large lobes of the hybrid orbitals are shown in green, and the small lobes are not shown.

▲ **FIGURE 8.4**

The formation of sp^2 hybrid orbitals by combination of one s orbital and two p orbitals.

▲ **Figure It out**

Give the number and type of atomic orbitals mixed to form three sp^2 hybrid orbitals. Which atomic orbital is not mixed, and what is its orientation relative to the sp^2 hybrid orbitals?

Answer: One 2s orbital and two 2p orbitals are mathematically mixed to form three sp^2 hybrid orbitals. One 2p atomic orbital remains, which is oriented 90° to the sp^2 hybrid orbitals.

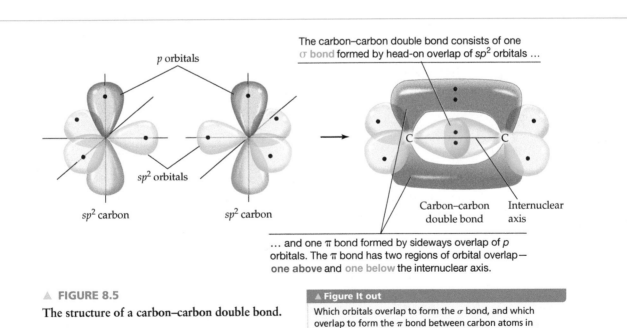

The carbon–carbon double bond consists of one σ bond formed by head-on overlap of sp^2 orbitals ...

p orbitals

sp^2 orbitals

sp^2 carbon sp^2 carbon

Carbon–carbon double bond   Internuclear axis

... and one π bond formed by sideways overlap of p orbitals. The π bond has two regions of orbital overlap— one above and one below the internuclear axis.

▲ **FIGURE 8.5**

The structure of a carbon–carbon double bond.

▲ **Figure It out**

Which orbitals overlap to form the σ bond, and which overlap to form the π bond between carbon atoms in ethylene?

Answer: The σ bond is formed by overlap of C(sp^2) with C(sp^2), and the π bond is formed by overlap of C(2p) with C(2p).

The π bond has two regions of orbital overlap, one above and one below a line drawn between the nuclei, known as the *internuclear axis*. Both regions are part of the same bond, and the two shared electrons are spread over both regions. As always, the p lobes must be of the same phase for overlap leading to bond formation. The net result of both σ and π overlap is the sharing of four electrons and the formation of a carbon–carbon double bond.

WORKED EXAMPLE 8.4

Identifying Hybridization and Orbital Overlap in Single and Double Bonds

Propylene, C_3H_6, a starting material used to manufacture polypropylene polymers, has its three carbon atoms connected in a row. Draw the overall shape of the molecule, and indicate what kinds of orbitals on each atom overlap to form the carbon–carbon and carbon–hydrogen bonds.

STRATEGY

Draw an electron-dot structure of the molecule and count the number of charge clouds around each carbon atom. Use the number of charge clouds and the VSEPR model to predict geometry and hybridization.

SOLUTION

A molecule with the formula C_3H_6 and its carbons connected in a row does not have enough hydrogen atoms to give each carbon four single bonds. Thus, propylene must contain at least one carbon–carbon double bond. In the structure of polypropylene, the carbon atom on the left has four charge clouds (four single bonds) and tetrahedral geometry with sp^3 hybrid orbitals. The other carbon atoms have three charge clouds (two single bonds and one double bond) and trigonal planar geometry with sp^2 hybrid orbitals. The carbon on the left has bond angles of 109.5° and the other carbon atoms have bond angles of 120°.

The carbon–carbon single bond is formed by an overlap of an sp^3 orbital and an sp^2 orbital. The carbon–carbon double bond consists of a σ bond formed from head-on overlap of two sp^2 hybrid orbitals and a π bond formed from sideways overlap of two p orbitals. Each carbon–hydrogen bond is formed from overlap of a hydrogen $1s$ orbital with the hybrid orbital on the carbon to which it is attached.

▶ **PRACTICE 8.7**   Describe the hybridization of the carbon atom in formaldehyde, $H_2C{=}O$, and specify what kinds of orbitals on each atom overlap to form the bonds in the molecule.

▶ **APPLY 8.8**   Describe the hybridization of each carbon atom in cyclopentene (C_5H_8). Tell what kinds of orbitals on each atom overlap to form the carbon–carbon and carbon–hydrogen bonds.

Trigonal planar [sp^2]

Tetrahedral [sp^3]

Propylene

Cyclopentene

$C(sp^3)-C(sp^2)$

$C(sp^2)-H(1s)$

σ bond: $C(sp^2)-C(sp^2)$

π bond: $C(2p)-C(2p)$

$C(sp^3)-H(1s)$

$C(sp^2)-H(1s)$

All **WORKED EXAMPLES** with this icon have an interactive video in the eText.

sp Hybridization

Atoms with two charge clouds undergo hybridization by combination of one atomic *s* orbital with one *p* orbital, resulting in two **sp hybrid orbitals** that are oriented 180°

from each other. Since only one *p* orbital is involved when an atom undergoes *sp* hybridization, the other two *p* orbitals are unchanged and are oriented at 90° angles to the *sp* hybrids, as shown in **FIGURE 8.6**.

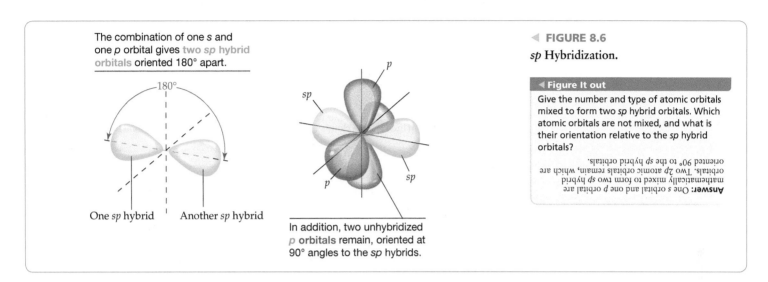

The combination of one *s* and one *p* orbital gives two *sp* hybrid orbitals oriented 180° apart.

180°

One *sp* hybrid   Another *sp* hybrid

In addition, two unhybridized *p* orbitals remain, oriented at 90° angles to the *sp* hybrids.

◀ **FIGURE 8.6**

sp **Hybridization.**

◀ **Figure It out**

Give the number and type of atomic orbitals mixed to form two *sp* hybrid orbitals. Which atomic orbitals are not mixed, and what is their orientation relative to the *sp* hybrid orbitals?

Answer: One *s* orbital and one *p* orbital are mathematically mixed to form two *sp* hybrid orbitals. Two 2*p* atomic orbitals remain, which are oriented 90° to the *sp* hybrid orbitals.

One of the simplest examples of *sp* hybridization occurs in acetylene, H—C≡C—H, a colorless gas used in welding. Both carbon atoms in the acetylene molecule have linear geometry and are *sp*-hybridized. When the two *sp*-hybridized carbon atoms approach each other with their *sp* orbitals aligned head-on for σ bonding, the unhybridized *p* orbitals on each carbon are aligned for π bonding. Two *p* orbitals are aligned in an up/down position, and two are aligned in an in/out position. Thus, there are two mutually perpendicular π bonds that form by sideways overlap of *p* orbitals, along with one σ bond that forms by head-on overlap of the *sp* orbitals. The net result is the sharing of six electrons and formation of a triple bond (**FIGURE 8.7**). In addition, two C—H bonds form in acetylene by overlap of the remaining two *sp* orbitals with hydrogen 1*s* orbitals.

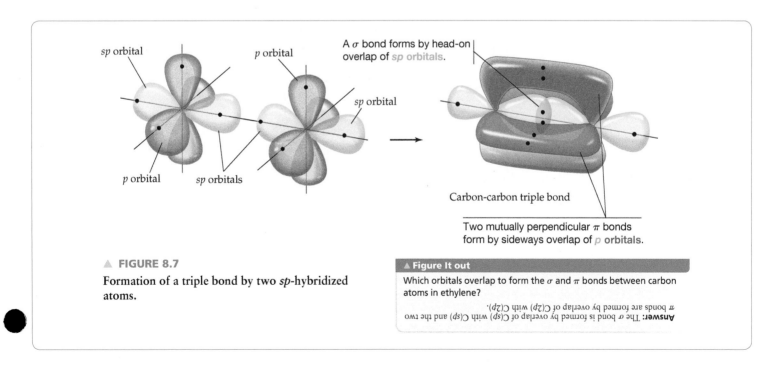

sp orbital   *p* orbital

A σ bond forms by head-on overlap of *sp* orbitals.

sp orbital

p orbital   *sp* orbitals

Carbon-carbon triple bond

Two mutually perpendicular π bonds form by sideways overlap of *p* orbitals.

▲ **FIGURE 8.7**

Formation of a triple bond by two *sp*-hybridized atoms.

▲ **Figure It out**

Which orbitals overlap to form the σ and π bonds between carbon atoms in ethylene?

Answer: The σ bond is formed by overlap of C(*sp*) with C(*sp*) and the two π bonds are formed by overlap of C(2*p*) with C(2*p*).

WORKED EXAMPLE 8.5

Identifying Hybridization and Orbital Overlap in Single and Double Bonds

Describe the hybridization of the carbon atoms in allene, $H_2C=C=CH_2$, and make a rough sketch of the molecule showing orbital overlap in bond formation.

STRATEGY

Draw an electron-dot structure to find the number of charge clouds on each atom.

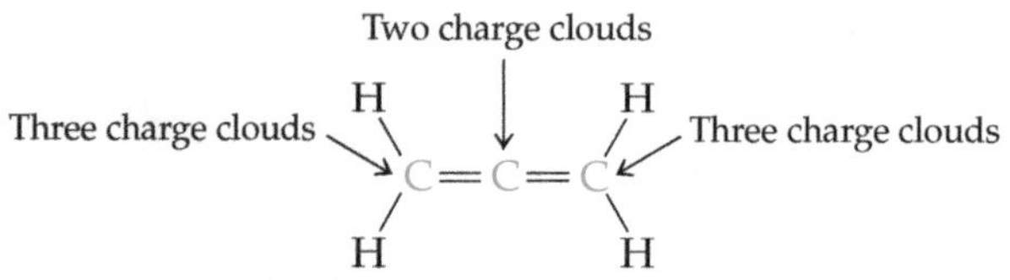

Two charge clouds

Three charge clouds → C=C=C ← Three charge clouds

(with H atoms as shown)

Then predict the geometry around each atom using the VSEPR model (Table 8.1).

SOLUTION

Because the central carbon atom in allene has two charge clouds (two double bonds), it has a linear geometry and is sp-hybridized. Because the two terminal carbon atoms have three charge clouds each (one double bond and two single bonds), they have trigonal planar geometry and are sp^2-hybridized. The central carbon uses its sp orbitals to form two σ bonds at 180° angles and uses its two unhybridized p orbitals to form π bonds, one to each of the terminal carbons. The unhybridized p orbitals are oriented at 90° from each other and are designated at $2p_x$ and $2p_y$ to show this spatial arrangement. Each terminal carbon atom uses an sp^2 orbital for σ bonding to carbon, a p orbital for π bonding, and its two remaining sp^2 orbitals for C—H bonds.

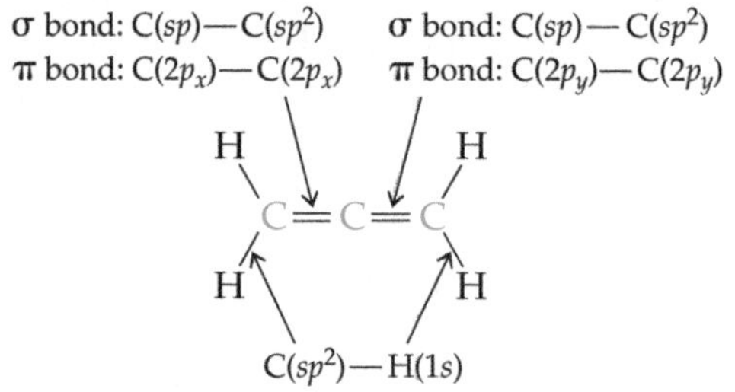

σ bond: $C(sp)$—$C(sp^2)$ σ bond: $C(sp)$—$C(sp^2)$
π bond: $C(2p_x)$—$C(2p_x)$ π bond: $C(2p_y)$—$C(2p_y)$

H   H
  \   ↓   /
   C=C=C
  /   \
H   H

$C(sp^2)$—$H(1s)$

Note that the mutually perpendicular arrangement of the two π bonds results in a similar perpendicular arrangement of the two CH_2 groups.

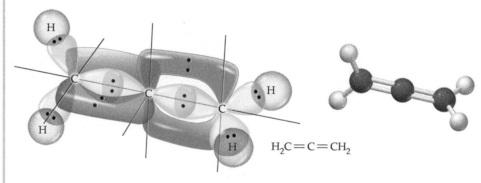

$H_2C=C=CH_2$

▶ **PRACTICE 8.9**   Which orbitals overlap to form the sigma and pi bonds in the following structure of carbon dioxide?

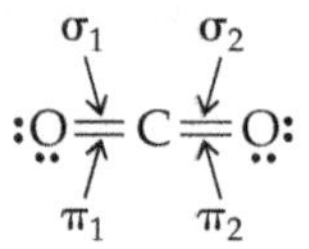

σ_1 σ_2

$:O \lessgtr C \lessgtr O:$

π_1 π_2

▶ **APPLY 8.10**   Describe the hybridization of the carbon atom in the hydrogen cyanide molecule, $H—C\equiv N$, and make a rough sketch to show the hybrid orbitals it uses for bonding.

Atoms with Five and Six Charge Clouds

Main-group atoms with five or six charge clouds, such as the phosphorus in PCl_5 and the sulfur in SF_6, were at one time thought to undergo hybridization by combination of five and six atomic orbitals, respectively. Because a given shell has a total of only four s and p orbitals, however, the need to use five or six orbitals implies that d orbitals are involved. As we'll see in Section 21.11, hybridization involving d orbitals is indeed involved for many compounds of transition metals. Recent quantum mechanical calculations indicate, however, that main-group compounds do not use d orbitals in hybridization but instead use a more complex bonding pattern that is not easily explained by valence bond theory.*

A summary of the three common kinds of hybridization for main-group elements and the geometry that each corresponds to is given in **TABLE 8.2**.

TABLE 8.2 Hybrid Orbitals and Their Geometry

Number of Charge Clouds	Arrangement of Charge Clouds	Hybridization
2	Linear	sp
3	Trigonal planar	sp^2
4	Tetrahedral	sp^3

8.5 POLAR COVALENT BONDS AND DIPOLE MOMENTS

Molecular polarity is an important characteristic that influences both chemical and physical properties of molecules. To understand the notion of molecular polarity, it's first necessary to develop the ideas of *bond dipoles* and *dipole moments*. We saw in Section 7.3 that polar covalent bonds form between atoms of different electronegativity. In chloromethane (CH_3Cl), chlorine is more electronegative than carbon so the chlorine atom attracts the electrons in the C—Cl bond more strongly than carbon does. The C—Cl bond is therefore polarized so that the chlorine atom is slightly electron-rich ($\delta-$) and the carbon atom is slightly electron-poor ($\delta+$).

Because the polar C—Cl bond in chloromethane has a positive end and a negative end, we describe it as being a bond **dipole**, and we often represent the dipole using an arrow with a cross at one end (↦) to indicate the direction of electron displacement. The point of the arrow represents the negative end of the dipole ($\delta-$), and the crossed end (which looks like a plus sign) represents the positive end ($\delta+$). The C—H bond in chloromethane also has a small bond dipole because carbon is slightly more electronegative than H. In theory, we can draw a bond dipole with the arrow pointing toward the partially negative carbon atom, but in practice the C—H bonds are considered to be nearly nonpolar. Due to the small difference in electronegativity between carbon and hydrogen, we typically do not draw bond dipoles for C—H bonds.

Just as individual bonds in molecules are often polar, molecules as a whole are also often polar because of the net sum of individual bond polarities and lone-pair contributions. The overall molecular polarity of CH_3Cl is clearly visible in an electrostatic potential map, which shows the electron-rich chlorine atom as red and the electron-poor remainder of the molecule as blue-green. The resultant *molecular dipole* can be

REMEMBER . . .
Polar covalent bonds are those in which the bonding electrons are shared unequally between two atoms but are not completely transferred. Thus, they are intermediate between nonpolar covalent bonds and ionic bonds (Section 7.3).

REMEMBER . . .
Electronegativity is the ability of an atom in a molecule to attract the shared electrons in a covalent bond. Metals on the left of the periodic table are the least electronegative, while halogens and other reactive nonmetals in the upper right of the table are the most electronegative (Section 7.3).

REMEMBER . . .
An electrostatic potential map uses color to portray the calculated electron distribution in a molecule. Electron-rich regions are red, and electron-poor regions are blue (Section 7.3).

---

*For more information on the theory of bonding in molecules with an expanded octet, see the following references.

Noury, S., Silvi, B., & Gillespie, R. Chemical Bonding in Hypervalent Molecules: Is the Octet Rule Relevant?, *Inorg. Chem.* Vol. 41, 2002, p. 2164.

T. A. Mitchell et al. Predicting the Stability of Hypervalent Molecules, *Journal of Chemical Education*, Vol. 84, 2007, p. 629.

Galbraith, J. On the Role of d Orbital Hybridization in the Chemistry Curriculum, *Journal of Chemical Education*, Vol. 84, 2007, p. 783.

TABLE 8.3 Dipole Moments of Some Common Compounds	
Compound	Dipole Moment (D)
NaCl[a]	9.0
CH$_3$Cl	1.90
H$_2$O	1.85
NH$_3$	1.47
HCl	1.11
CO$_2$	0
CCl$_4$	0

[a]Measured in the gas phase.

looked at in the following way: Consider the symmetry in the arrangement of bonds and lone pairs of electrons around the central atom. If the arrangement is asymmetrical, then the molecule has net polarity. In chloromethane, there are four bonds in a tetrahedral arrangement. The C—Cl is polar while the three C—H bonds are nonpolar, an asymmetrical arrangement of bonds around the central atom.

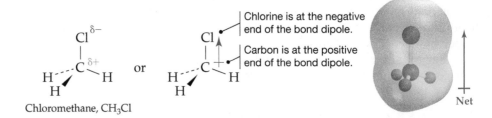

Chloromethane, CH$_3$Cl

REMEMBER . . .
The **dipole moment** (μ) is defined as the magnitude of the charge (Q) at either end of the molecular dipole times the distance (r) between the charges: Dipole moments are expressed in *debyes* (D), where 1 D = 3.336 × 10^{-30} coulomb meters in SI units (Section 7.3).

It's relatively easy to measure dipole moments experimentally, and values for some common substances are given in **TABLE 8.3**. The measure of net molecular polarity is a quantity called the **dipole moment** (μ), (Greek mu).

Perhaps it's not surprising that by far the largest dipole moment listed in Table 8.3 belongs to the ionic compound NaCl, which exists in the gas phase as a pair of Na$^+$ and Cl$^-$ ions held together by a strong ionic bond. The molecular compounds water and ammonia also have substantial dipole moments because both oxygen and nitrogen are electronegative relative to hydrogen and because both O and N have lone pairs of electrons that make substantial contributions to net molecular polarity:

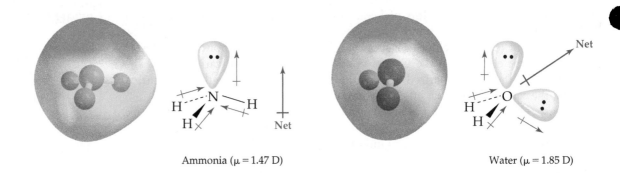

Ammonia ($\mu = 1.47$ D)   Water ($\mu = 1.85$ D)

In contrast with water and ammonia, carbon dioxide and tetrachloromethane (CCl$_4$) have zero dipole moments. Molecules of both substances contain *individual* polar covalent bonds, but because of the symmetry of their structures, the individual bond polarities exactly cancel.

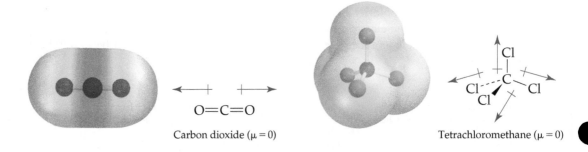

Carbon dioxide ($\mu = 0$)   Tetrachloromethane ($\mu = 0$)

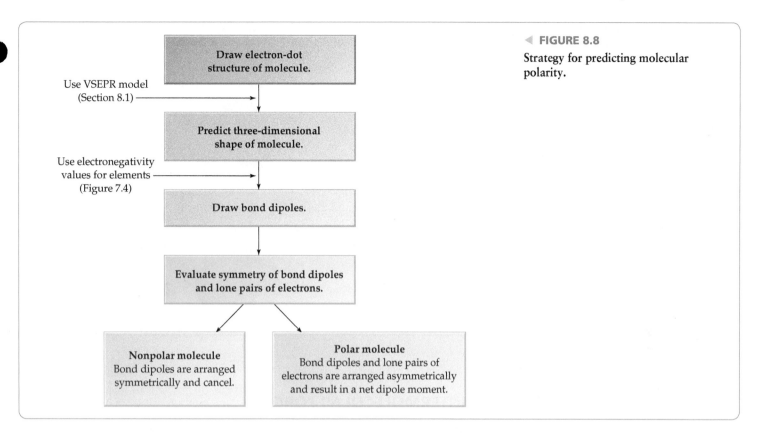

◀ **FIGURE 8.8**
Strategy for predicting molecular polarity.

FIGURE 8.8 outlines the strategy for predicting molecular polarity. The first step is to draw an electron-dot structure and use the VSEPR model (Section 8.1) to determine three-dimensional molecular shape. Next draw bond dipoles using electronegativity values (Figure 7.4) and examine symmetry in bond dipoles and lone pairs to determine if a molecule has a dipole moment.

Worked Example 8.6 illustrates the process for predicting molecular polarity.

CONCEPTUAL WORKED EXAMPLE 8.6

Predicting the Presence of a Dipole Moment

Would you expect vinyl chloride ($H_2C{=}CHCl$), the starting material used for preparation of poly(vinyl chloride) polymer, to have a dipole moment? If so, indicate the direction.

STRATEGY AND SOLUTION

Follow the procedure outlined in Figure 8.8 for determining molecular polarity.

The electron-dot structure for vinyl chloride is:

Then use the VSEPR model described in Section 8.1 to predict the molecular shape of vinyl chloride. Each carbon has trigonal planar geometry because both carbon atoms have three

charge clouds. The molecule as a whole is planar because the unhybridized carbon $2p$ orbitals in the pi part of the $C{=}C$ double bond overlap in a sideways manner:

Then assign polarities to the individual bonds according to the differences in electronegativity of the bonded atoms, and make a reasonable guess about the overall polarity that would result

continued on next page

by evaluating symmetry in bond dipoles and summing the individual contributions. Only the C—Cl bond has a substantial polarity, giving the molecule a net polarity in the direction of the chlorine atom:

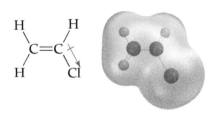

▶ **CONCEPTUAL PRACTICE 8.11**   Which of the following compounds has a dipole moment? Draw the direction of the dipole moment of this compound.

(a) SF_6 (b) SO_3 (c) BrF_3

▶ **CONCEPTUAL APPLY 8.12**   Match the following electrostatic potential maps with their corresponding molecule: CH_3F, CHF_3, CF_4, CH_2F_2.

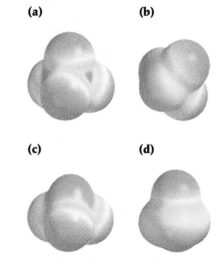

(a)   (b)

(c)   (d)

8.6 INTERMOLECULAR FORCES

Now that we know a bit about molecular polarities, let's see how they give rise to some of the forces that occur between molecules, called **intermolecular forces.** These forces are different than covalent bonds between atoms *within* a molecule. For example, in H_2O there are two covalent bonds between hydrogen and oxygen, but there are also attractive forces between two H_2O molecules that we refer to as intermolecular forces. (Strictly speaking, the term *intermolecular* refers only to molecular substances, but we'll use it generally to refer to interactions among all kinds of particles, including molecules, ions, and atoms.)

Intermolecular forces influence many important macroscopic properties of matter such as solubility, melting point, and boiling point. They also play a key role in stabilizing the shapes and interactions of biomolecules, as we'll see at the end of this section. To understand how the strength of a substance's intermolecular forces influences its melting and boiling point we must first visualize the difference between a gas, liquid, and solid at the molecular level. **FIGURE 8.9** shows that the particles in a gas are free to move around at random and fill the volume of a container because the attractive forces (intermolecular forces) between particles are weak. Liquids and solids are distinguished from gases by the presence of substantial attractive forces between particles. In liquids, these attractive forces are strong enough to hold the particles in close contact while still allowing them to move freely past one another. In solids, the forces are so strong that they hold the particles rigidly in place and prevent their movement.

Temperature influences the phase of matter because at higher temperatures particles have greater kinetic energy and can overcome the intermolecular forces holding them together. We are familiar with a sample of H_2O that exists either as solid ice,

◀ **FIGURE 8.9**

A molecular comparison of gases, liquids, and solids.

◀ **Figure It out**

Which phase of matter has the strongest intermolecular forces?

Answer: Solid

In gases, the particles feel little attraction for one another and are free to move about randomly.

In liquids, the particles are held close together by attractive forces but are free to move around one another.

In solids, the particles are held in an ordered arrangement.

liquid water, or gaseous vapor, depending on its temperature. Nitrogen (N_2) offers another example; nitrogen is a gas at higher temperatures but becomes a liquid at low temperature (**FIGURE 8.10**).

Intermolecular forces as a whole are usually called **van der Waals forces** after the Dutch scientist Johannes van der Waals (1837–1923). These forces are of several different types, including *London dispersion forces, dipole-dipole forces,* and *hydrogen bonds.* In addition, *ion–dipole forces* exist between ions and molecules. All these intermolecular forces are electrostatic in origin and result from the mutual attraction of unlike charges or the mutual repulsion of like charges. If the particles are ions, then full charges are present and the ion–ion attractions are so strong (energies on the order of 500–1000 kJ/mol) that they give rise to what we call **ionic bonds**. If the particles are neutral, then only partial charges are present, but even so, the attractive forces can be substantial.

REMEMBER . . .

Ionic bonds generally form between the cation of a metal and the anion of a reactive nonmetal (Section 6.7).

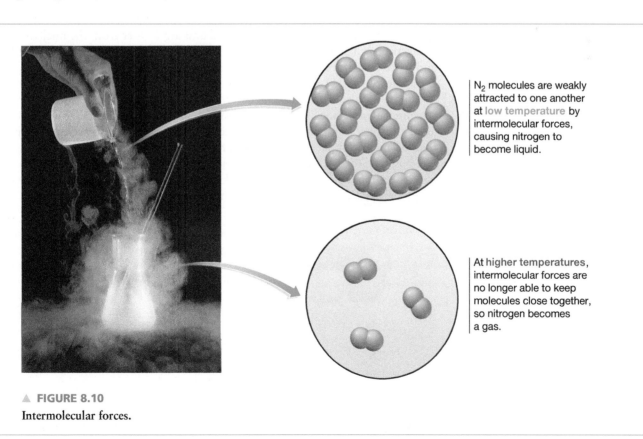

N_2 molecules are weakly attracted to one another at low temperature by intermolecular forces, causing nitrogen to become liquid.

At higher temperatures, intermolecular forces are no longer able to keep molecules close together, so nitrogen becomes a gas.

▲ **FIGURE 8.10**
Intermolecular forces.

London Dispersion Forces

All atoms and molecules, regardless of structure, experience **London dispersion forces,** which result from the motion of electrons, which in turn creates temporary dipole moments in molecules. Consider a simple nonpolar molecule like Br_2. Averaged over time, the distribution of electrons throughout the molecule is symmetrical, but at any given instant there may be more electrons at one end of the molecule than at the other, giving the molecule a short-lived dipole moment. This instantaneous dipole on one molecule can affect the electron distributions in neighboring molecules and *induce* temporary dipoles in those neighbors (**FIGURE 8.11**). As a result, weak attractive forces develop and Br_2 is a liquid at room temperature rather than a gas.

▶ **FIGURE 8.11**

London dispersion forces.

 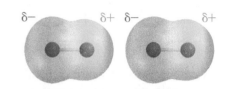

Averaged over time, the electron distribution in a Br_2 molecule is symmetrical.

At any given instant, the electron distribution in a molecule may be unsymmetrical, resulting in a temporary dipole and inducing a complementary attractive dipole in neighboring molecules.

TABLE 8.4 Melting Points and Boiling Points of the Halogens

Halogen	mp (K)	bp (K)
F_2	53.5	85.0
Cl_2	171.6	239.1
Br_2	265.9	331.9
I_2	386.8	457.5

Dispersion forces are generally small, with energies in the range 1–10 kJ/mol, and their exact magnitude depends on the ease with which a molecule's electron cloud can be distorted by a nearby electric field, a property referred to as **polarizability.** A smaller molecule or lighter atom is less polarizable and has smaller dispersion forces because it has only a few, tightly held electrons. A larger molecule or heavier atom, however, is more polarizable and has larger dispersion forces because it has many electrons, some of which are less tightly held and are farther from the nucleus. Among the halogens, for instance, the F_2 molecule is small and less polarizable, while I_2 is larger and more polarizable. As a result, F_2 has smaller dispersion forces and is a gas at room temperature, while I_2 has larger dispersion forces and is a solid (**TABLE 8.4**).

Shape is also important in determining the magnitude of the dispersion forces affecting a molecule. More spread-out shapes, which maximize molecular surface area, allow greater contact between molecules and give rise to higher dispersion forces than do more compact shapes, which minimize molecular contact. Pentane, for example, boils at 309.2 K, whereas 2,2-dimethylpropane boils at 282.6 K. Both substances have the same molecular formula, C_5H_{12}, but pentane is longer and somewhat spread out, whereas 2,2-dimethylpropane is more spherical and compact (**FIGURE 8.12**).

▶ **FIGURE 8.12**

The effect of molecular shape on London dispersion forces.

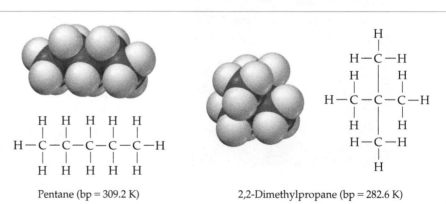

Pentane (bp = 309.2 K)

2,2-Dimethylpropane (bp = 282.6 K)

Longer, less compact molecules like pentane feel stronger dispersion forces and consequently have higher boiling points.

More compact molecules like 2,2-dimethylpropane feel weaker dispersion forces and have lower boiling points.

Dipole–Dipole Forces

Neutral but polar molecules experience **dipole–dipole forces** as the result of electrical inter-actions among dipoles on neighboring molecules. The forces can be either attractive or repulsive, depending on the orientation of the molecules (**FIGURE 8.13**), and the net force in a large collection of molecules is a summation of many individual interactions of both types. In molecules with similar molecular weights, dipole–dipole forces are stronger than London dispersion forces. Dipole–dipole forces have energies on the order of 3–4 kJ/mol.

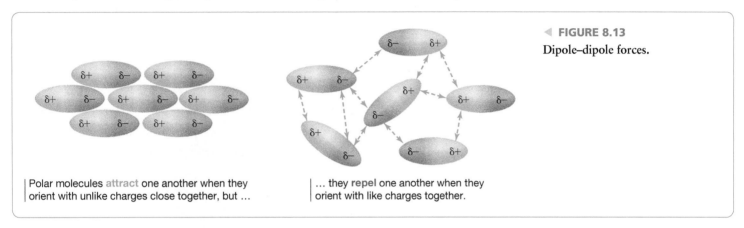

◀ **FIGURE 8.13**
Dipole–dipole forces.

Polar molecules **attract** one another when they orient with unlike charges close together, but …

… they **repel** one another when they orient with like charges together.

Butane, for instance, is a nonpolar molecule with a molecular weight of 58 and a boiling point of −0.5 °C, while acetone has the same molecular weight yet boils 57 °C higher because it is polar.

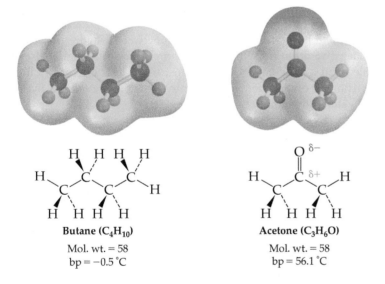

Butane (C_4H_{10})
Mol. wt. = 58
bp = −0.5 °C

Acetone (C_3H_6O)
Mol. wt. = 58
bp = 56.1 °C

The strength of a given dipole–dipole interaction depends on the sizes of the dipole moments involved. The more polar the substance, the greater the strength of its dipole–dipole interactions. **TABLE 8.5** lists several substances with similar molecular weights but different dipole moments. Note the rough correlation between dipole moment and boiling point. The larger the dipole moment, the stronger the intermolecular forces and the greater the amount of heat that must be added to overcome those forces. Thus, substances with higher dipole moments generally have higher boiling points.

TABLE 8.5 Comparison of Molecular Weights, Dipole Moments, and Boiling Points

Substance	Mol. Wt.	Dipole Moment (D)	bp (K)
$CH_3CH_2CH_3$	44.10	0.08	231
CH_3OCH_3	46.07	1.30	248
CH_3CN	41.05	3.93	355

BIG IDEA Question 5

Consider three molecules with similar molecular weights. Would the molecule with the highest or lowest dipole moment have the highest boiling point?

Hydrogen Bonds

LOOKING AHEAD . . .
In Section 23.11, we'll examine the molecular composition of **DNA** and the role of hydrogen bonding in structure and replication.

In many ways, *hydrogen bonds* are responsible for life on Earth. They cause water to be a liquid rather than a gas at ordinary temperatures, and they are the primary intermolecular force that holds huge biomolecules in the shapes needed to play their essential roles in biochemistry. **Deoxyribonucleic acid (DNA)**, a molecule that stores our genetic code, contains two enormously long molecular strands that are coiled around each other and held together by hydrogen bonds.

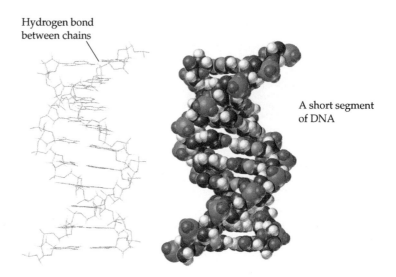

Hydrogen bond between chains

A short segment of DNA

A **hydrogen bond** is an attractive interaction between a hydrogen atom bonded to a very electronegative atom (O, N, or F) and an electron-rich region elsewhere in the same molecule or in a different molecule. Most often, this electron-rich region is an unshared electron pair on another electronegative atom. For example, hydrogen bonds occur in both water and ammonia:

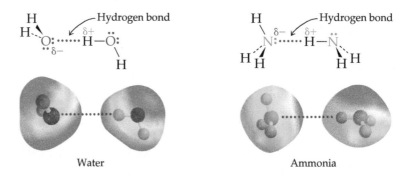

Water   Ammonia

Hydrogen bonds arise because O—H, N—H, and F—H bonds are highly polar, with a partial positive charge on the hydrogen and a partial negative charge on the electronegative atom. In addition, the hydrogen atom has no core electrons to shield its nucleus and it has a small size so it can be approached closely by other molecules. As a result, the dipole–dipole attraction between the hydrogen and an unshared electron pair on a nearby atom is unusually strong, giving rise to a hydrogen bond. Water, in particular, is able to form a vast three-dimensional network of hydrogen bonds because each H_2O molecule has two hydrogens and two electron pairs (**FIGURE 8.14**).

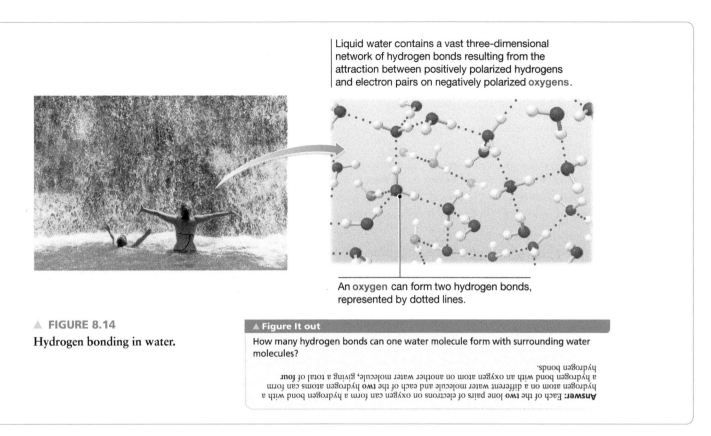

Liquid water contains a vast three-dimensional network of hydrogen bonds resulting from the attraction between positively polarized hydrogens and electron pairs on negatively polarized oxygens.

An oxygen can form two hydrogen bonds, represented by dotted lines.

▲ **FIGURE 8.14**

Hydrogen bonding in water.

▲ **Figure It out**

How many hydrogen bonds can one water molecule form with surrounding water molecules?

Answer: Each of the two lone pairs of electrons on oxygen can form a hydrogen bond with a hydrogen atom on a different water molecule and each of the two hydrogen atoms can form a hydrogen bond with an oxygen atom on another water molecule, giving a total of four hydrogen bonds.

Hydrogen bonds can be quite strong, with energies up to 40 kJ/mol. To see one effect of hydrogen bonding, look at **TABLE 8.6,** which plots the boiling points of the covalent binary hydrides for the group 4A–7A elements. As you might expect, the boiling points generally increase with molecular weight down a group of the periodic table as a result of increased London dispersion forces—for example, $CH_4 < SiH_4 < GeH_4 < SnH_4$. Three substances, however, are clearly anomalous: NH_3, H_2O, and HF. All three have higher boiling points than might be expected because of the hydrogen bonds they contain.

TABLE 8.6 Boiling Points of the Covalent Binary Hydrides of Groups 4A, 5A, 6A, and 7A.

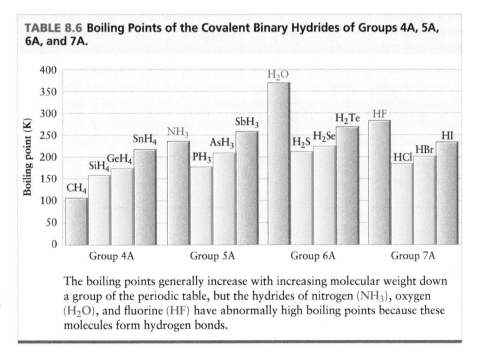

The boiling points generally increase with increasing molecular weight down a group of the periodic table, but the hydrides of nitrogen (NH_3), oxygen (H_2O), and fluorine (HF) have abnormally high boiling points because these molecules form hydrogen bonds.

CONCEPTUAL WORKED EXAMPLE 8.7

Drawing Hydrogen Bonds

Methanol (CH_3OH) is a liquid that can be used in fuel cells. Draw all the hydrogen bonds that can form when one methanol molecule is surrounded by other methanol molecules in a liquid sample.

STRATEGY

Hydrogen bonding occurs as a result of a strong dipole–dipole interaction and follows the pattern shown below. X represents one of the highly electronegative atoms F, O, or N. Align two CH_3OH molecules to fit the pattern and draw the resulting hydrogen bonds.

$$\overset{\delta+}{X-H}\cdots\cdots\overset{\delta-}{:X}$$

SOLUTION

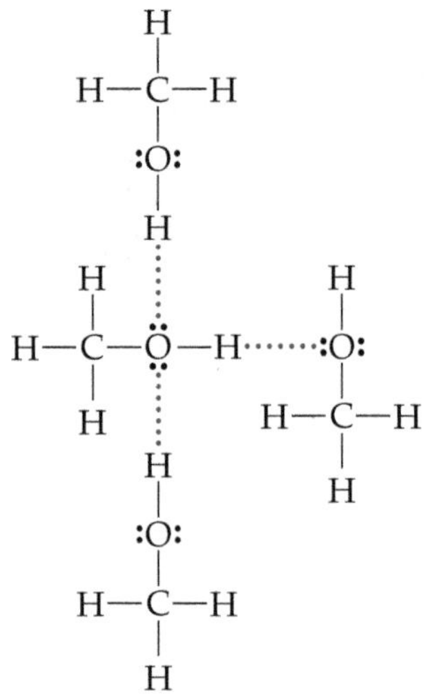

▶ **CONCEPTUAL PRACTICE 8.13** A liquid sample contains formaldehyde (CH_2O) dissolved in water. Which of the following illustrations is an *incorrect* depiction of hydrogen bonding that can occur in this solution?

(a)   **(b)** 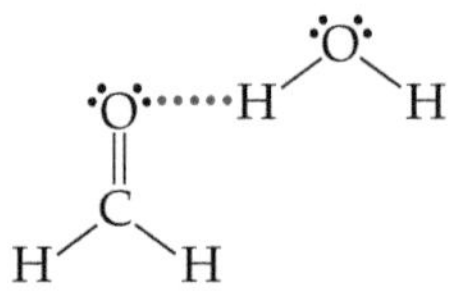   **(c)**

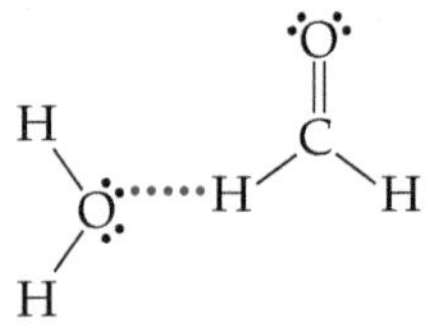

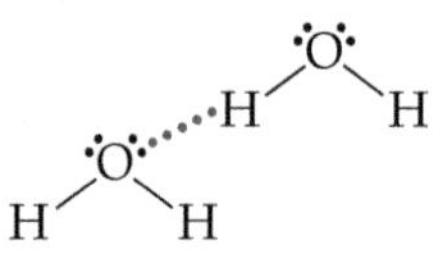

▶ **CONCEPTUAL APPLY 8.14** Hydrogen bonding between pairs of guanine (G)–cytosine (C) molecules and adenine (A)–thymine (T) molecules holds two DNA strands together. In the diagram below, these pairs of molecules are aligned as they exist in two DNA strands. The red squiggly line represents where the molecule is attached to the sugar-phosphate backbone of DNA.

(a) Draw the hydrogen bonds that occur in the G–C pair and A–T pair.
(b) Which region of DNA would have the higher melting point; regions high in A–T pairs or regions high in G–C pairs?

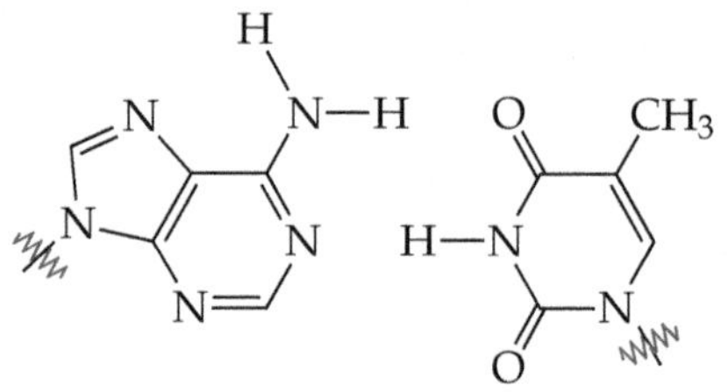

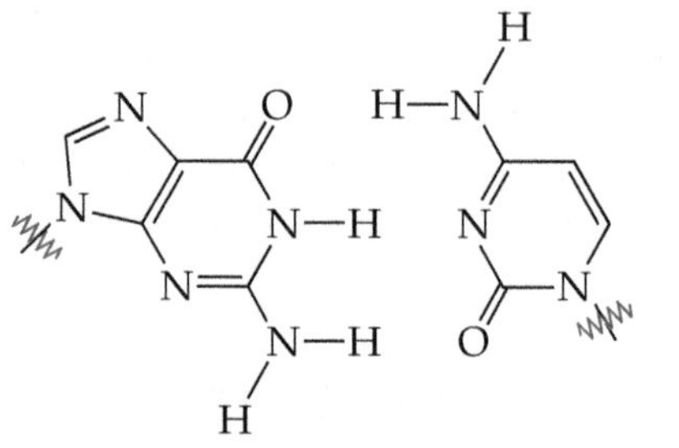

Ion–Dipole Forces

We said in the previous section that a molecule has a net polarity and an overall dipole moment if the sum of its individual bond dipoles is nonzero. One side of the molecule has a net excess of electrons and a partial negative charge ($\delta-$), while the other side has a net deficiency of electrons and a partial positive charge ($\delta+$). An **ion–dipole force** is the result of electrical interactions between an ion and the partial charges on a polar molecule (**FIGURE 8.15**).

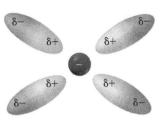 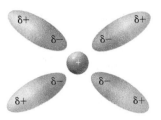

Polar molecules orient toward ions so that the positive end of the dipole is near an anion and ...

... the negative end of the dipole is near a cation.

◀ **FIGURE 8.15**
Ion–dipole forces.

The favored orientation of a polar molecule in the presence of ions is one where the positive end of the molecular dipole is near an anion and the negative end of the dipole is near a cation. The magnitude of the interaction energy E depends on the charge on the ion z, the strength of the dipole as measured by its dipole moment μ, and the inverse square of the distance r from the ion to the dipole: $E \propto z\mu/r^2$. Ion–dipole forces are highly variable in strength and are particularly important in aqueous solutions of ionic substances such as NaCl, in which polar water molecules surround the ions. We'll explore the **formation of solutions** in more detail in Chapter 13.

TABLE 8.7 compares the various kinds of intermolecular forces.

LOOKING AHEAD . . .
In Section 13.2, we'll look at the energy changes accompanying the **formation of solutions** and at how those energy changes are affected by different kinds of intermolecular forces.

TABLE 8.7   A Comparison of Intermolecular Forces

Force	Strength	Characteristics
Ion–dipole	Highly variable (10–70 kJ/mol)	Occurs between ions and polar molecules
Dipole–dipole	Weak (3–4 kJ/mol)	Occurs between polar molecules
London dispersion	Weak (1–10 kJ/mol)	Occurs between all molecules; strength depends on size, polarizability
Hydrogen bond	Moderate (10–40 kJ/mol)	Occurs between molecules with O—H, N—H, and F—H bonds

WORKED EXAMPLE 8.8

Identifying Intermolecular Forces

Identify the kinds of intermolecular forces in the following substances:

(a) HCl
(b) CH_3CH_3 (ethane)
(c) CH_3NH_2 (methylamine)
(d) Kr

STRATEGY

Determine the structure of each substance, and decide what intermolecular forces are present. Remember: All molecules have dispersion forces; polar molecules have dipole–dipole forces; and molecules with O—H, N—H, or F—H bonds have hydrogen bonds.

SOLUTION

(a) HCl is a polar molecule but can't form hydrogen bonds. It has dipole–dipole forces and dispersion forces.
(b) CH_3CH_3 is a nonpolar molecule and has only dispersion forces.

Ethane

(c) CH_3NH_2 is a polar molecule that can form hydrogen bonds. In addition, it has dipole–dipole forces and dispersion forces.

Methylamine

(d) Kr is nonpolar and has only dispersion forces.

continued on next page

▶ **PRACTICE 8.15** Determine the types of intermolecular forces in the following substances: Cl_2, CCl_4, CH_3F, and HF. Which statement about intermolecular forces is *false*?

(a) CH_3F has dispersion forces, dipole–dipole forces, and hydrogen bonding.
(b) CCl_4 has stronger dispersion forces than Cl_2.
(c) HF has dispersion forces, dipole–dipole forces, and hydrogen bonding.
(d) CCl_4 only has dispersion forces.

▶ **APPLY 8.16** Consider the kinds of intermolecular forces present in the following compounds, and rank the substances in likely order of increasing boiling point: H_2S(mol. wt. = 34), CH_3OH(mol. wt. = 32), C_2H_6 (mol. wt. = 30), Ar (mol. wt. = 40).

8.7 MOLECULAR ORBITAL THEORY: THE HYDROGEN MOLECULE

The valence bond model that describes covalent bonding through orbital overlap is easy to visualize and leads to a satisfactory description for most molecules. It does, however, have some problems. Perhaps the most serious flaw in the valence bond model is that it sometimes leads to an incorrect electronic description. For this reason, another bonding description called **molecular orbital (MO) theory** is often used. The molecular orbital model is more complex and less easily visualized than the valence bond model, particularly for larger molecules, but it sometimes gives a more satisfactory accounting of chemical and physical properties.

To introduce some of the basic ideas of molecular orbital theory, let's look again at orbitals. The concept of an orbital derives from the quantum mechanical wave equation, in which the square of the wave function gives the probability of finding an electron within a given region of space. The kinds of orbitals that we've been concerned with up to this point are called *atomic orbitals* because they are characteristic of individual atoms. Atomic orbitals on the same atom can combine to form hybrids, and atomic orbitals on different atoms can overlap to form covalent bonds, but the orbitals and the electrons in them remain localized on specific atoms.

> **Atomic orbital** A wave function whose square gives the probability of finding an electron within a given region of space *in an atom.*

Molecular orbital theory takes a different approach to bonding by considering the molecule as a whole rather than concentrating on individual atoms. A **molecular orbital** is to a molecule what an atomic orbital is to an atom.

> **Molecular orbital** A wave function whose square gives the probability of finding an electron within a given region of space in a molecule.

Like atomic orbitals, molecular orbitals have specific energy levels and specific shapes, and they can be occupied by a maximum of two electrons with opposite spins. The energy and shape of a molecular orbital depend on the size and complexity of the molecule and can thus be fairly complicated, but the fundamental analogy between atomic and molecular orbitals remains. Let's look at the molecular orbital description of the simple diatomic molecule H_2 to see some general features of MO theory.

Imagine what might happen when two isolated hydrogen atoms approach each other and begin to interact. The 1s orbitals begin to blend together, and the electrons spread out over both atoms. Molecular orbital theory says that there are two ways for the orbital interaction to occur—an additive way and a subtractive way. The additive interaction leads to formation of a molecular orbital that is roughly egg-shaped, whereas the subtractive interaction leads to formation of a molecular orbital that contains a node between atoms (**FIGURE 8.16**).

REMEMBER . . .
A node is a surface of zero electron probability separating regions of nonzero probability within an orbital (Section 5.7).

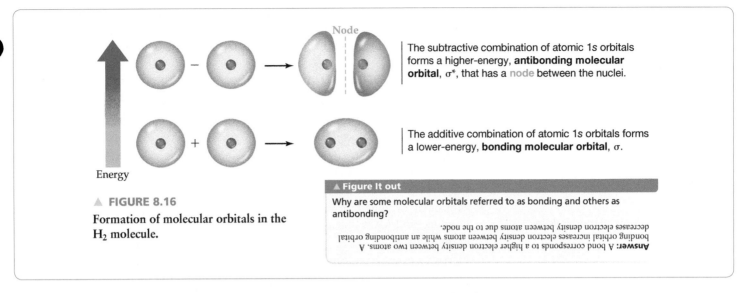

The subtractive combination of atomic 1s orbitals forms a higher-energy, **antibonding molecular orbital**, σ^*, that has a node between the nuclei.

The additive combination of atomic 1s orbitals forms a lower-energy, **bonding molecular orbital**, σ.

▲ **FIGURE 8.16**

Formation of molecular orbitals in the H_2 molecule.

▲ Figure It out

Why are some molecular orbitals referred to as bonding and others as antibonding?

Answer: A bond corresponds to a higher electron density between two atoms. A bonding orbital increases electron density between atoms while an antibonding orbital decreases electron density between atoms due to the node.

The additive combination, denoted σ, is lower in energy than the two isolated 1s orbitals and is called a **bonding molecular orbital** because any electrons it contains spend most of their time in the region between the two nuclei, bonding the atoms together. The subtractive combination, denoted σ^* and spoken as "sigma star," is higher in energy than the two isolated 1s orbitals and is called an **antibonding molecular orbital.** Any electrons it contains can't occupy the central region between the nuclei and can't contribute to bonding.

Diagrams such as the one shown in **FIGURE 8.17** are used to show the energy relationships of the various orbitals. The two isolated H atomic orbitals are shown on either side, and the two H_2 molecular orbitals are shown in the middle. Each of the starting hydrogen atomic orbitals has one electron, which pair up and occupy the lower-energy bonding MO after covalent bond formation.

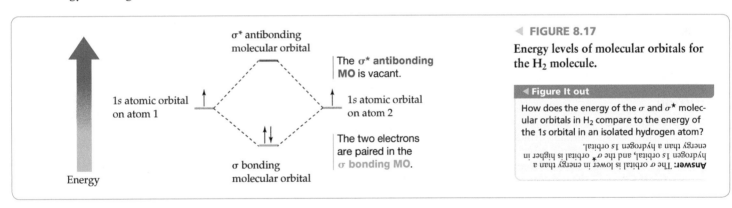

σ^* antibonding molecular orbital

1s atomic orbital on atom 1

The σ^* antibonding **MO** is vacant.

1s atomic orbital on atom 2

The two electrons are paired in the σ bonding MO.

σ bonding molecular orbital

Energy

◄ **FIGURE 8.17**

Energy levels of molecular orbitals for the H_2 molecule.

◄ Figure It out

How does the energy of the σ and σ^* molecular orbitals in H_2 compare to the energy of the 1s orbital in an isolated hydrogen atom?

Answer: The σ orbital is lower in energy than a hydrogen 1s orbital, and the σ^* orbital is higher in energy than a hydrogen 1s orbital.

Similar MO diagrams can be drawn and predictions about stability can be made for diatomic species such as H_2^-. For example, we might imagine constructing the H_2^- ion by bringing together a neutral H atom with one electron and an H^- anion with two electrons. Since the resultant H_2^- ion has three electrons, two of them will occupy the lower-energy bonding σ MO and one will occupy the higher-energy antibonding σ^* MO as shown in **FIGURE 8.18.** Two electrons are lowered in energy while only one electron is raised in energy, so a net gain in stability results. We therefore predict (and find experimentally) that the H_2^- ion is stable.

Bond orders—the number of electron pairs shared between atoms—can be calculated from MO diagrams by subtracting the number of antibonding electrons from the number of bonding electrons and dividing the difference by 2:

$$\text{Bond order} = \frac{\left(\begin{array}{c}\text{Number of}\\\text{bonding electrons}\end{array}\right) - \left(\begin{array}{c}\text{Number of}\\\text{antibonding electrons}\end{array}\right)}{2}$$

REMEMBER . . .

Bond order is the number of electron pairs shared between atoms. Higher bond orders correspond to shorter and stronger bonds (Section 7.2).

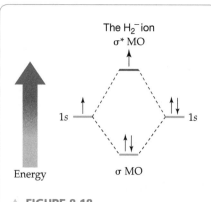

▲ **FIGURE 8.18**
Energy levels of molecular orbitals for the stable H_2^- ion.

The H_2 molecule, for instance, has a bond order of 1 because it has two bonding electrons and no antibonding electrons. In the same way, the H_2^- ion has a bond order of ½.

The key ideas of the molecular orbital theory of bonding can be summarized as follows.

Key Ideas of Molecular Orbital Theory

- Molecular orbitals are to molecules what atomic orbitals are to atoms. A molecular orbital describes a region of space in a molecule where electrons are most likely to be found, and it has a specific size, shape, and energy level.
- Molecular orbitals are formed by combining atomic orbitals on different atoms. The number of molecular orbitals formed is the same as the number of atomic orbitals combined.
- Molecular orbitals that are lower in energy than the starting atomic orbitals are bonding, and MOs that are higher in energy than the starting atomic orbitals are antibonding.
- Electrons occupy molecular orbitals beginning with the MO of lowest energy. A maximum of two electrons can occupy each orbital, and their spins are paired.
- Bond order can be calculated by subtracting the number of electrons in antibonding MOs from the number in bonding MOs and dividing the difference by 2.

WORKED EXAMPLE 8.9

Constructing an MO Diagram for a First Row Diatomic Molecule

Construct an MO diagram for the He_2 molecule. What is its bond order? Is this molecule likely to be stable?

STRATEGY

Count the total number of valence electrons in the He_2 molecule and fill orbitals in the MO diagram starting with the lowest energy. Calculate the bond order according to the equation in this section. Bond orders greater than zero represent stable species; the higher the bond order, the stronger the bond.

SOLUTION

The hypothetical He_2 molecule has 4 electrons, 2 of which occupy the lower-energy bonding orbital, and 2 of which occupy the higher-energy antibonding orbital. Since the decrease in energy for the 2 bonding electrons is counteracted by the increase in energy for the 2 antibonding electrons, the He_2 molecule has no net bonding energy and is not stable. The hypothetical He_2 molecule has 2 bonding and 2 antibonding electrons and a bond order of $(2 - 2)/2 = 0$, which accounts for the instability of He_2.

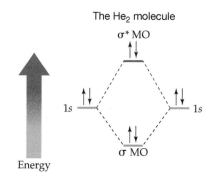

▶ **PRACTICE 8.17** Construct an MO diagram for the He_2^+ ion. What is its bond order? Is this ion likely to be stable?

▶ **APPLY 8.18** According to MO theory, is He_2^{2+} predicted to have a stronger or weaker bond than He_2^+?

8.8 MOLECULAR ORBITAL THEORY: OTHER DIATOMIC MOLECULES

Having looked at bonding in the H_2 molecule, let's move up a level in complexity by looking at the bonding in several second-row diatomic molecules—N_2, O_2, and F_2. The valence bond model developed in Section 8.2 predicts that the nitrogen atoms in N_2 are triply bonded and have one lone pair each, that the oxygen atoms in O_2 are doubly bonded and have two lone pairs each, and that the fluorine atoms in F_2 are singly bonded and have three lone pairs each:

Valence bond
theory predicts: $:N{\equiv}N:$ $\ddot{O}{=}\ddot{O}$ $:\ddot{F}{-}\ddot{F}:$

                      1 σ bond   1 σ bond   1 σ bond
                      and 2 π bonds   and 1 π bond

Unfortunately, this simple valence bond picture can't be right because it predicts that the electrons in all three molecules are *spin-paired*. In other words, the electron-dot structures indicate that the occupied atomic orbitals in all three molecules contain two electrons each. It can be demonstrated experimentally, however, that the O_2 molecule has two electrons that are not spin-paired and that these electrons therefore must be in different, singly occupied orbitals.

Experimental evidence for the electronic structure of O_2 rests on the observation that substances with unpaired electrons are attracted by magnetic fields and are thus said to be **paramagnetic**. The more unpaired electrons a substance has, the stronger the paramagnetic attraction. Substances whose electrons are all spin-paired, by contrast, are weakly repelled by magnetic fields and are said to be **diamagnetic**. Both N_2 and F_2 are diamagnetic, just as predicted by their electron-dot structures, but O_2 is paramagnetic. When liquid O_2 is poured over the poles of a strong magnet, the O_2 sticks to the poles, as shown in **FIGURE 8.19**.

Why is O_2 paramagnetic? Electron-dot structures and valence bond theory fail to answer this question, but MO theory explains the experimental results nicely. In a molecular orbital description of N_2, O_2, and F_2, two atoms come together and their valence-shell atomic orbitals interact to form molecular orbitals. Four orbital interactions occur, leading to the formation of four bonding MOs and four antibonding MOs, whose relative energies are shown in **FIGURE 8.20**. (Note that the relative energies of the σ_{2p} and π_{2p} orbitals in N_2 are different from those in O_2 and F_2.)

▲ **FIGURE 8.19**
Paramagnetism.

▲ **Figure It out**
Why does liquid O_2 stick to the poles of a magnet?

Answer: Molecules with unpaired electrons are paramagnetic and are attracted to a magnetic field.

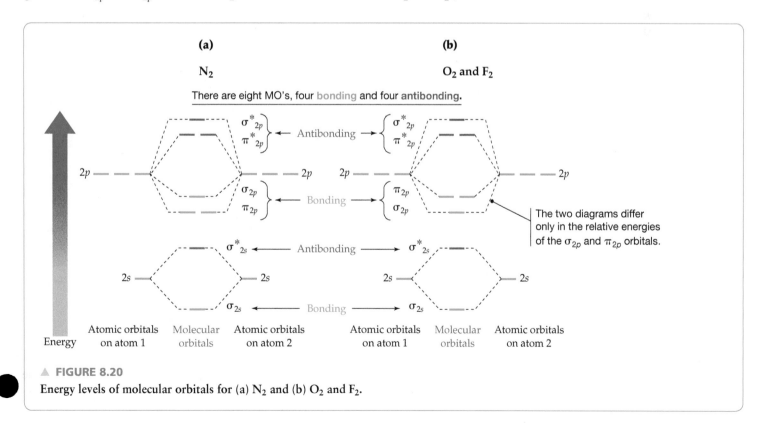

(a)   **(b)**
N_2 O_2 and F_2

There are eight MO's, four bonding and four antibonding.

| Atomic orbitals on atom 1 | Molecular orbitals | Atomic orbitals on atom 2 | Atomic orbitals on atom 1 | Molecular orbitals | Atomic orbitals on atom 2 |

Energy

The two diagrams differ only in the relative energies of the σ_{2p} and π_{2p} orbitals.

▲ **FIGURE 8.20**
Energy levels of molecular orbitals for (a) N_2 and (b) O_2 and F_2.

Go to
eText

BIG IDEA Question 6

What property of oxygen gas can be explained by molecular orbital theory but not valence bond theory?

REMEMBER . . .

Degenerate orbitals are those that have the same energy (Section 5.10).

The diagrams in Figure 8.20 show the following orbital interactions:

● The 2s orbitals interact to give σ_{2s} and σ^*_{2s} MOs.

● The two 2p orbitals that lie on the internuclear axis interact head-on to give σ_{2p} and σ^*_{2p} MOs.

● The two remaining pairs of 2p orbitals that are perpendicular to the internuclear axis interact in a sideways manner to give two degenerate π_{2p} and two degenerate π^*_{2p} MOs oriented 90° apart.

We should also point out that MO diagrams like those in Figure 8.20 are usually obtained from mathematical calculations and can't necessarily be predicted. MO theory is therefore less easy to visualize and understand on an intuitive level than valence bond theory. The shapes of the σ_{2p}, σ^*_{2p}, π_{2p}, and π^*_{2p} MOs are shown in **FIGURE 8.21**.

▶ **FIGURE 8.21**

Formation of (a) σ_{2p} and σ^*_{2p} MOs by head-on interaction of two p atomic orbitals, and (b) π_{2p} and π^*_{2p} MOs by sideways interaction.

▶ **Figure It out**

Which types of molecular orbitals are formed by interacting lobes of orbitals with same phase? Which type of molecular orbitals have a node between the nuclei?

Answer: Bonding molecular orbitals; antibonding molecular orbitals.

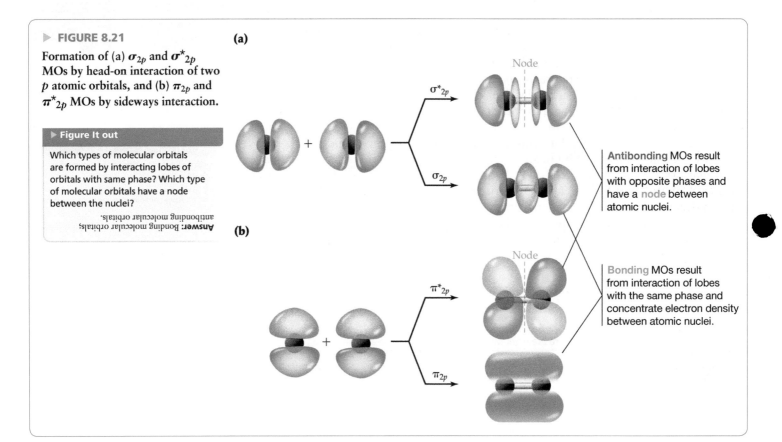

(a)

Node

σ^*_{2p}

σ_{2p}

(b)

Node

π^*_{2p}

π_{2p}

Antibonding MOs result from interaction of lobes with opposite phases and have a node between atomic nuclei.

Bonding MOs result from interaction of lobes with the same phase and concentrate electron density between atomic nuclei.

When appropriate numbers of valence electrons are added to occupy the molecular orbitals, the results shown in **FIGURE 8.22** are obtained. Both N_2 and F_2 have all their electrons spin-paired, but O_2 has two unpaired electrons in the degenerate π^*_{2p} orbitals. Both N_2 and F_2 are therefore diamagnetic, whereas O_2 is paramagnetic.

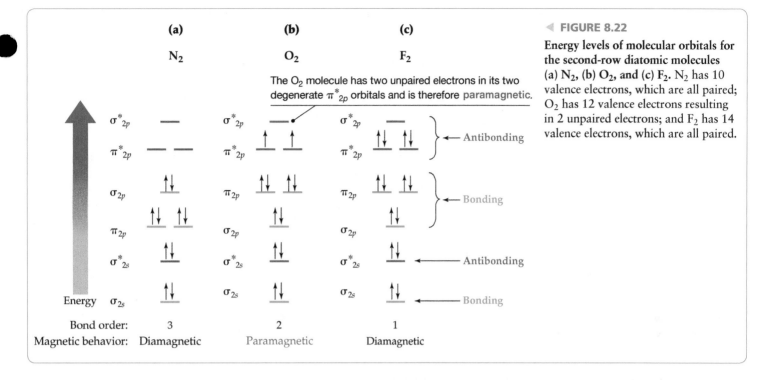

◀ **FIGURE 8.22**

Energy levels of molecular orbitals for the second-row diatomic molecules (a) N$_2$, (b) O$_2$, and (c) F$_2$. N$_2$ has 10 valence electrons, which are all paired; O$_2$ has 12 valence electrons resulting in 2 unpaired electrons; and F$_2$ has 14 valence electrons, which are all paired.

WORKED EXAMPLE 8.10

Constructing an MO Diagram for a Second-Row Diatomic Molecule

Nitrogen monoxide (NO), typically called nitric oxide, is a diatomic molecule with a molecular orbital diagram similar to that of N$_2$ (Figure 8.22a). Show the MO diagram for nitric oxide, and predict its bond order. Is NO diamagnetic or paramagnetic?

STRATEGY

Count the number of valence electrons in NO, and assign them to the available MOs shown in Figure 8.22a, beginning with the lowest energy orbital.

SOLUTION

Nitric oxide has 5 valence electrons from N and 6 from O for a total of 11. The 11 electrons are assigned to MOs in Figure 8.22a. NO has 8 bonding and 3 antibonding electrons and a bond order of $(8 - 3)/2 = 2.5$. Because it has one unpaired electron, it is paramagnetic.

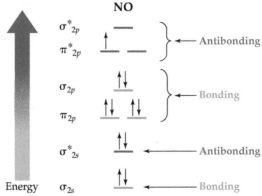

▶ **PRACTICE 8.19**   The B$_2$ molecule has a MO diagram similar to that of N$_2$ in Figure 8.22a. What is the bond order of B$_2$ and is it diamagnetic or paramagnetic?

▶ **APPLY 8.20**   Determine the bond order for the species: O$_2{}^{2-}$, O$_2{}^{-}$, O$_2$, O$_2{}^{+}$, O$_2{}^{2+}$. List them in order of increasing bond energy and increasing bond length.

8.9 COMBINING VALENCE BOND THEORY AND MOLECULAR ORBITAL THEORY

Whenever two different theories explain the same concept, the question comes up: Which theory is better? The question isn't easy to answer, though, because it depends on what is meant by "better." Valence bond theory is better because of its simplicity and ease of visualization, but MO theory is better because of its accuracy. Best of all, though, is a joint use of the two theories that combines the strengths of both.

Valence bond theory has two main problems: (1) It incorrectly predicts aspects of electronic structure such as magnetic properties and energy levels. In the O_2 molecule, all electrons are paired in the electron-dot structure even though experimental data indicates that it is paramagnetic (has unpaired electrons). (2) Valence bond theory cannot adequately represent bonding in molecules with a single electron-dot structure (such as O_3). No single structure is adequate; the concept of a resonance hybrid, a blend of multiple electron-dot structures (Section 7.9) is not well described by valence bond theory. The first problem occurs infrequently, but the second is much more common. To better deal with resonance, chemists often use a combination of bonding theories in which the σ bonds in a given molecule are described by valence bond theory and π bonds in the same molecule are described by MO theory.

Take O_3, for instance. Valence bond theory says that ozone is a resonance hybrid of two equivalent structures, both of which have two $O-O$ σ bonds and one $O=O$ π bond (Section 7.9). One structure has a lone pair of electrons in the p orbital on the left-hand oxygen atom and a π bond to the right-hand oxygen. The other structure has a lone pair of electrons in the p orbital on the right-hand oxygen and a π bond to the left-hand oxygen. The actual structure of O_3 is an average of the two resonance forms in which four electrons occupy the entire region encompassed by the overlapping set of three p orbitals. The only difference between the resonance structures is in the placement of p electrons. The atoms themselves are in the same positions in both, and the geometries are the same in both (**FIGURE 8.23**).

▶ **FIGURE 8.23**

The structure of ozone.

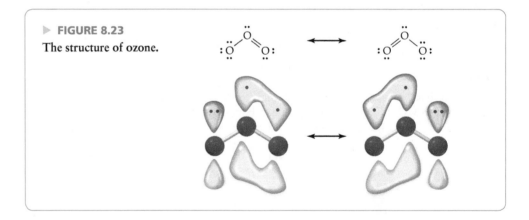

Valence bond theory thus gives a good description of the ozone $O-O$ σ bonds, whose electrons are localized between specific pairs of atoms, but a poor description of the π bonds among p atomic orbitals, whose four electrons are spread out, or *delocalized,* over the molecule. Yet this is exactly what MO theory does best—describe bonds in which electrons are delocalized throughout a molecule. Thus, a combination of valence bond theory and MO theory is used. The σ bonds are best described in valence bond terminology as being localized between pairs of atoms, and the π electrons are best described by MO theory as being delocalized over the entire molecule. The lowest energy π molecular orbital best represents the resonance hybrid, which is a blend of the two resonance structures of ozone.

Resonance hybrid

Lowest-energy
π molecular orbital

WORKED EXAMPLE 8.11

Combining Valence Bond and MO Theory

Acetamide (C_2H_5NO), an organic compound that has been detected in interstellar space, has the following connections among atoms and can be drawn using two nonequivalent electron-dot resonance structures. Draw both, and sketch a π molecular orbital showing how the π electrons are delocalized over the oxygen and nitrogen atoms.

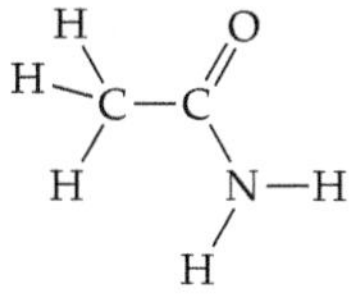

Acetamide

STRATEGY

Draw the two electron-dot structures in the usual way (Section 7.9), including all nonbonding electrons. The two structures differ in that one has a carbon-oxygen double bond and a lone pair on nitrogen, while the other has a carbon–nitrogen double bond and an additional lone pair on oxygen. Thus, a π molecular orbital corresponding to the O—C—N resonance includes all three atoms.

SOLUTION

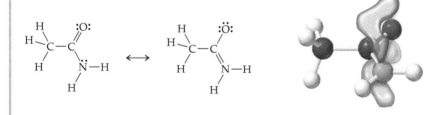

▶ **PRACTICE 8.21** Draw two resonance structures for the formate ion, HCO_2^-, and sketch a π molecular orbital showing how the π electrons are delocalized over both oxygen atoms.

▶ **APPLY 8.22** Draw two resonance structures for the benzene molecule, C_6H_6, and sketch a π molecular orbital showing how the π electrons are delocalized over the carbon atoms.

INQUIRY ? Which is better for human health, natural or synthetic vitamins?

Prior to the development of the chemical industry in the late nineteenth and early twentieth centuries, only substances from natural sources were available for treating our diseases, dying our clothes, and cleansing and perfuming our bodies. Extracts of the opium poppy and willow bark, for instance, have been used since the seventeenth century for the relief of pain. The prized purple dye called *Tyrian purple*, obtained from a Middle Eastern mollusk, has been known since antiquity. Oils distilled from bergamot, sweet bay, rose, and lavender have been employed for centuries in making perfume.

▲ Whether from the laboratory or from food, the vitamin C is the same.

Many of these *natural products* were first used without any knowledge of their chemical composition. As chemistry developed, though, scientists learned how to work out the structures of the compounds in natural products. The disease-curing properties of limes and other citrus fruits, for example, were known for centuries, but the chemical structure of vitamin C, the active ingredient, was not determined until 1933. Today there is a revival of interest in folk remedies, and a large effort is being made to identify medicinally important chemical compounds found in plants.

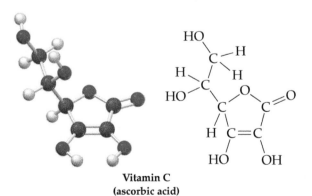

Vitamin C
(ascorbic acid)

Once a structure is known, organic chemists try to synthesize the compound in the laboratory. If the starting materials are inexpensive and the synthesis process is simple enough, it may become more economical to manufacture a compound than to isolate it from a plant or bacterium. In the case of vitamin C, a complete synthesis was achieved in 1933, and it is now much cheaper to synthesize it starting from glucose than to extract it from citrus or other natural sources. Worldwide, more than 110,000 metric tons of vitamin C are synthesized each year.

But is the "synthetic" vitamin C as good as the "natural" one? Some people still demand vitamins only from natural sources, assuming that natural is somehow better. Although eating an orange is probably better than taking a tablet, the difference lies in the many other substances present in the orange. The vitamin C itself is exactly the same, just as the NaCl produced by reacting sodium and chlorine in the laboratory is exactly the same as the NaCl found in the ocean. Natural and synthetic compounds are identical in all ways; neither is better than the other.

PROBLEM 8.23 Refer to the structure of vitamin C below to answer the following questions.

(a) What are the bond angles around the carbon atoms labeled as C_a, C_b, and C_c in the structure?

(b) What is the hybridization of the carbon atoms labeled as C_a, C_b, and C_c in the structure?

(c) What orbitals overlap to form the carbon-carbon double bond in the five-membered ring?

PROBLEM 8.24 Vitamin C is a water-soluble substance because it forms hydrogen bonds with surrounding water molecules. Which of the interactions in the figure (labeled *1–4*) are hydrogen bonds that can form between vitamin C and surrounding water molecules?

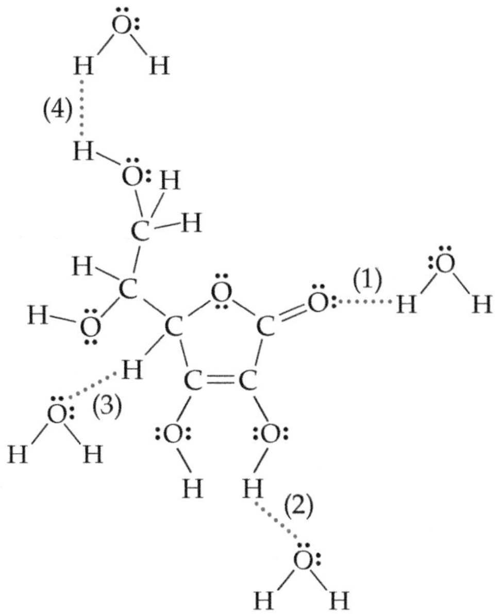

PROBLEM 8.25 Caffeine is the most widely used stimulant psychoactive drug in the United States. It is found naturally in the seeds and leaves of some plants such as tea and coffee and can be synthesized in the laboratory. The following structure shows the connections of atoms in caffeine. Add multiple bonds and lone pairs of electrons to draw an electron-dot structure in which the formal charge on all atoms is zero.

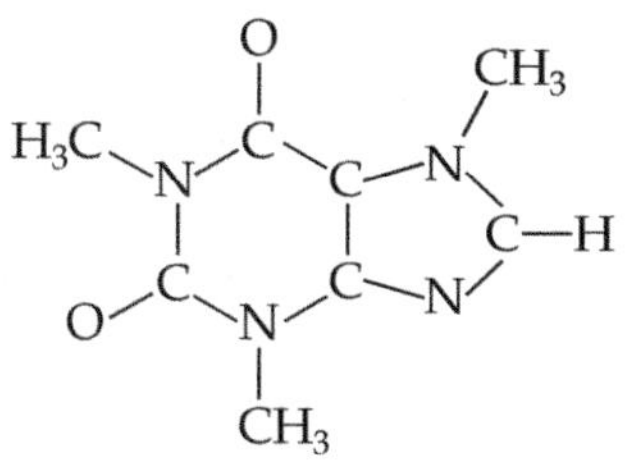

(a) How many lone pairs of electrons are in the structure? How many π bonds are in the structure?

(b) What types of intermolecular forces are present in a sample of caffeine?

(c) Would you expect caffeine to be a solid, liquid, or gas at room temperature? Explain your reasoning.

PROBLEM 8.26 Salicin is a molecule extracted from willow bark to treat pain. Which six-membered ring in salicin has a π molecular orbital that delocalizes electron density?

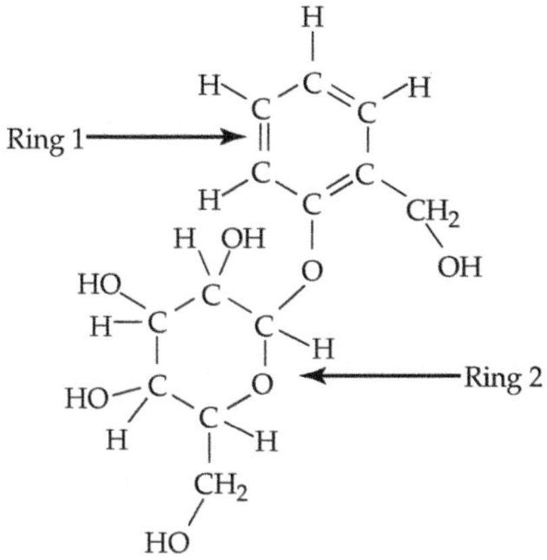

STUDY GUIDE

Section	Concept Summary	Learning Objectives	Test Your Understanding
8.1 Molecular Shapes: The VSEPR Model	Molecular shape can often be predicted using the **valence-shell electron-pair repulsion (VSEPR) model,** which treats the electrons around atoms as charge clouds that repel one another and therefore orient themselves as far away from one another as possible. Atoms with two charge clouds adopt a linear arrangement of the clouds, atoms with three charge clouds adopt a trigonal planar arrangement, and atoms with four charge clouds adopt a tetrahedral arrangement. Similarly, atoms with five charge clouds are trigonal bipyramidal and atoms with six charge clouds are octahedral.	**8.1** Use the VSEPR model to predict shape and bond angles in a molecule or ion with one central atom.	Worked Example 8.1; Problems 8.42–8.49
		8.2 Use the VSEPR model to predict bond angles in a molecule with more than one central atom.	Worked Example 8.2; Problems 8.50–8.55
8.2 Valence Bond Theory	According to **valence bond theory,** covalent bond formation occurs by the overlap of two singly occupied atomic orbital lobes of the same phase, either head-on along the internuclear axis to form a σ **bond** or sideways above and below the internuclear axis to form a π **bond.**	**8.3** Describe the difference between a sigma and a pi bond in valence bond theory.	Problem 8.58
8.3–8.4 Hybrid Orbitals	The observed geometry of covalent bonding in main-group compounds is described by assuming that s and p atomic orbitals combine to generate hybrid orbitals, which are strongly oriented in specific directions: sp **hybrid orbitals** have linear geometry, sp^2 **hybrid orbitals** have trigonal planar geometry, and sp^3 **hybrid orbitals** have tetrahedral geometry. The bonding of main-group atoms with five and six charge clouds is more complex.	**8.4** Write an electron-dot structure for a molecule and determine hybridization and bond angles.	Problems 8.32–8.33, 8.62–8.65
		8.5 Identify which orbitals overlap to form σ and π bonds in molecules.	Worked Examples 8.3–8.5; Problems 8.66–8.72
8.5 Polar Covalent Bonds and Dipole Moments	The difference in electronegativity between elements in a bond can result in a polar covalent bond. If the bond dipoles and lone pairs of electrons occur in an asymmetric arrangement around the center of the molecule, the molecule has a net polarity, a property measured by the **dipole moment.**	**8.6** Predict whether a given molecule has a dipole moment and draw its direction.	Worked Example 8.6; Problems 8.35, 8.78, 8.80
		8.7 Interpret electrostatic potential maps of molecules.	Problems 8.34, 8.36–8.37
8.6 Intermolecular Forces	**Intermolecular forces,** collectively known as van der Waals forces, are the attractions responsible for holding particles together in the solid and liquid phases. There are several kinds of intermolecular forces, all which arise from electrostatic attractions. **London dispersion forces** are characteristic of all molecules and result from the presence of temporary dipole moments caused by momentarily unsymmetrical electron distributions. **Dipole–dipole forces** occur between two polar molecules. A **hydrogen bond** is the attraction between a positively polarized hydrogen bonded to O, N, or F and a lone pair of electrons on an O, N, or F atom. In addition, **ion–dipole forces** occur between an ion and a polar molecule.	**8.8** Identify the types of intermolecular forces experienced by a molecule.	Worked Example 8.8; Problems 8.15, 8.88–8.89
		8.9 Relate the strength of intermolecular forces to physical properties such as melting point and boiling point.	Problems 8.16, 8.90–8.91
		8.10 Sketch the hydrogen bonding that occurs between two molecules.	Worked Example 8.7; Problems 8.14, 8.94–8.95

Section	Concept Summary	Learning Objectives	Test Your Understanding
8.7– 8.8 Molecular Orbital Theory	**Molecular orbital theory** sometimes gives a more accurate picture of electronic structure than the valence bond model. A **molecular orbital** is a wave function whose square gives the probability of finding an electron in a given region of space in a molecule. Combination of two atomic orbitals gives two molecular orbitals, a **bonding MO** that is lower in energy than the starting atomic orbitals and an **antibonding MO** that is higher in energy than the starting atomic orbitals. Molecular orbital theory is particularly useful for describing delocalized π bonding in molecules.	**8.11** Interpret the molecular orbital diagram for a first row diatomic molecule or ion.	Worked Example 8.9; Problems 8.17–8.18
		8.12 Interpret the molecular orbital diagram for a second-row diatomic molecule or ion. Calculate the bond order and predict magnetic properties.	Worked Example 8.10; Problems 8.98, 8.100, 8.102
8.9 Combining Valence Bond Theory and Molecular Orbital Theory	Molecules that exhibit resonance are often described by a combination of valence bond theory and molecular orbital theory. Valence bond theory is a simple model that works well for describing σ bonds but often inaccurately predicts properties such as magnetism. Molecular orbital theory is a more sophisticated model for describing electron delocalization in π bonds and more accurately predicts properties.	**8.13** Draw orbital overlap diagrams for molecules and describe the use of both valence bond theory and molecular orbital theory.	Worked Example 8.11; Problems 8.21–8.22, 8.108–8.109

KEY TERMS

antibonding molecular orbital *307*
bond angle *280*
bonding molecular orbital *307*
deoxyribonuclease (DNA) *302*
diamagnetic *309*
dipole *295*

dipole–dipole force *301*
hybrid atomic orbital *288*
hydrogen bond *302*
intermolecular force *298*
ion–dipole force *304*
London dispersion force *300*
molecular orbital *306*

molecular orbital (MO) theory *306*
paramagnetic *309*
pi (π) bond *290*
polarizability *300*
sigma (σ) bond *287*
sp hybrid orbital *292*

sp^2 hybrid orbital *290*
sp^3 hybrid orbital *288*
valence bond theory *286*
valence-shell electron-pair repulsion (VSEPR) model *279*
van der Waals forces *299*

PRACTICE TEST

After studying this chapter, you can assess your understanding with these practice test questions, which are correlated with chapter learning objectives. If you answer a question incorrectly, refer to the learning objectives in the end-of-chapter Study Guide for assistance. The Study Guide provides a conceptual summary, references a Worked Example to model how to solve the problem, and gives additional problems for more practice.

1. Which image shows the correct three-dimensional shape of SeF_4? **(LO 8.1)**

(a)

F—Se—F with F above and F below

(b)

Se with F atoms (wedge structure)

(c)

Se with F atoms (wedge structure)

(d)

F—Se—F with F atoms (wedge structure)

2. Give the geometry and approximate bond angles around the central atom in CCl_3^-. **(LO 8.1)**
   (a) Trigonal planar, 120°
   (b) Trigonal pyramidal, 109.5°
   (c) Trigonal pyramidal, 120°
   (d) Bent, 109.5°

3. Give the geometry and approximate bond angles around the central carbon atom in OCN^-. **(LO 8.1)**
   (a) Linear, 180° (b) Bent, 120°
   (c) Bent, 109.5° (d) Trigonal planar, 120°

4. Lactic acid is a waste product of glucose metabolism. The connections between atoms in lactic acid are shown. Complete the electron dot structure by adding double bonds and lone pairs of electrons. Give the approximate bond angles around the numbered carbon and oxygen atoms. **(LO 8.2)**

$$H-C_1-C-C_2-O-H$$

(a) $C_1 = 90°$, $O_1 = 180°$, $C_2 = 90°$
(b) $C_1 = 109.5°$, $O_1 = 180°$, $C_2 = 120°$
(c) $C_1 = 109.5°$, $O_1 = 109.5°$, $C_2 = 109.5°$
(d) $C_1 = 109.5°$, $O_1 = 109.5°$, $C_2 = 120°$

5. The following diagrams illustrate p-p orbital overlap or s-p orbital overlap. Which diagram represents a π bond in valence bond theory? (**LO 8.3**)

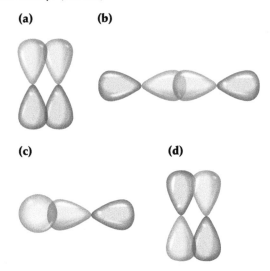

(a)   **(b)**

(c)   **(d)**

6. Which of the following molecules has a central atom with sp^3 hybridization? (**LO 8.4**)
   (a) PCl_5 (b) OF_2
   (c) CO_2 (d) SF_4

7. The connections between atoms in the amino acid histidine are shown. Complete the electron-dot structure by adding multiple bonds and lone pairs of electrons. Give the hybridization on the numbered carbon and nitrogen atoms. (**LO 8.4**)

(a) $N_1 = sp^2$, $N_2 = sp$, $C_1 = sp^2$, $C_2 = sp^3$
(b) $N_1 = sp^2$, $N_2 = sp^2$, $C_1 = sp^3$, $C_2 = sp^2$
(c) $N_1 = sp^3$, $N_2 = sp$, $C_1 = sp^2$, $C_2 = sp^3$
(d) $N_1 = sp^3$, $N_2 = sp^2$, $C_1 = sp^2$, $C_2 = sp^3$

Use the chemical structure for Tagamet, a drug used to treat peptic ulcers and heartburn, to answer questions 8 and 9.

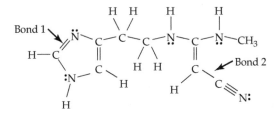

8. Which orbitals overlap to form bond 1? (**LO 8.5**)
   (a) σ bond: N (sp^2) with C (sp^2); π bond: N (p) with C (p)
   (b) N (sp^2) with C (sp^2)
   (c) σ bond: N (sp^2) with C (sp^2); π bond: N (sp^2) with C (sp^2)
   (d) σ bond: N (sp) with C (sp^2); π bond: N (p) with C (p)

9. Which orbitals overlap to form bond 2? (**LO 8.5**)
   (a) C (sp) with C (sp)   (b) C (sp) with C (sp^2)
   (c) C (sp^2) with C (sp^2)   (d) C (sp^3) with C (sp^2)

10. Which molecule has polar bonds but is nonpolar? (**LO 8.6**)
    (a) SF_6 (b) SF_2
    (c) F_2 (d) NF_3

11. What types of intermolecular forces exist in a sample of acetone? (**LO 8.8**)

$$\text{H}\quad\ddot{\text{O}}\text{:}\quad\text{H}$$
$$\text{H}-\text{C}-\text{C}-\text{C}-\text{H}$$
$$\text{H}\qquad\text{H}$$

   (a) Dispersion forces
   (b) Dispersion forces and dipole–dipole forces
   (c) Dipole–dipole forces
   (d) Dispersion forces, dipole–dipole forces, and hydrogen bonding

12. Arrange the following substances from *lowest* to *highest* boiling point. (**LO 8.9**)

$$CaF_2,\ CF_4,\ CH_4,\ CHF_3$$

   (a) $CH_4 < CF_4 < CHF_3 < CaF_2$
   (b) $CF_4 < CH_4 < CHF_3 < CaF_2$
   (c) $CaF_2 < CH_4 < CHF_3 < CF_4$
   (d) $CH_4 < CHF_3 < CaF_2 < CF_4$

13. Arrange the following molecules from *lowest* to *highest* boiling point. (**LO 8.9**)

(I)   (II)

(III)   (IV)

   (a) I < II < III < IV
   (b) II < III < I < IV
   (c) IV < II < III < I
   (d) IV < I < II < III

14. The DNA base thymine dissolves in water due to hydrogen bonding. Which of the following hydrogen bonds drawn between thymine and surrounding water molecules are valid? (**LO 8.10**)

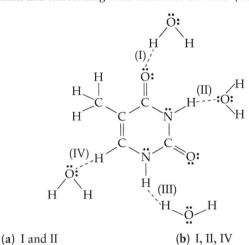

(a) I and II   (b) I, II, IV
(c) I, II, III   (d) III and IV

15. The C_2 molecule has a MO diagram similar to N_2 (Figure 8.22a). What is the bond order of C_2 and is it paramagnetic or diamagnetic? (**LO 8.12**)

(a) Bond order = 2, diamagnetic
(b) Bond order = 2, paramagnetic
(c) Bond order = 0, paramagnetic
(d) Bond order = 3/2, diamagnetic

Answers:
1. c, 2. b, 3. a, 4. d, 5. a, 6. b, 7. d, 8. a, 9. b, 10. a, 11. b, 12. a, 13. d, 14. a, 15. b

---

Mastering Chemistry provides end-of-chapter exercises feedback-enriched tutorial problems, animations, and interactive activities to encourage problem-solving practice and deeper understanding of key concepts and topics.

RAN *Randomized in Mastering Chemistry*

CONCEPTUAL PROBLEMS

Problems 8.1–8.26 appear within the chapter.

8.27 What is the geometry around the central atom in each of the following molecular models?

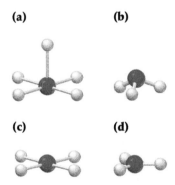

8.28 What is the geometry around the central atom in each of the following molecular models? (There may be a "hidden" atom directly behind a visible atom in some cases.)

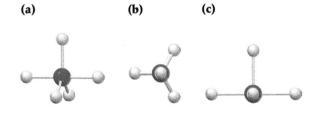

8.29 Three of the following molecular models have a tetrahedral central atom, and one does not. Which is the odd one? (There may be a "hidden" atom directly behind a visible atom in some cases.)

(a)   (b)   (c)   (d)

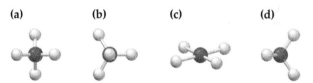

8.30 Identify each of the following sets of hybrid orbitals.

(a)   (b)   (c)

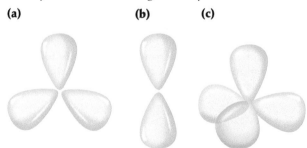

8.31 The VSEPR model is a simple predictive tool that is usually, but not always, correct. Take urea, for instance, a waste product excreted in animal urine:

Urea

What hybridization would you expect for the C and N atoms in urea according to the VSEPR model, and what approximate values would you expect for the various bond angles? What are the actual hybridizations and bond angles based on the molecular model shown? (Red = O, gray = C, blue = N, ivory = H.)

8.32 The following ball-and-stick molecular model is a representation of acetaminophen, the active ingredient in such over-the-counter headache remedies as Tylenol. (Red = O, gray = C, blue = N, ivory = H.)

(a) What is the formula of acetaminophen?

(b) Indicate the positions of the multiple bonds in acetaminophen.

(c) What is the geometry around each carbon?

(d) What is the hybridization of each carbon?

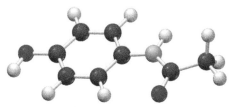

Acetaminophen

8.33 The following ball-and-stick molecular model is a representation of thalidomide, a drug that causes birth defects when taken by expectant mothers but is valuable for its use against leprosy. The lines indicate only the connections between atoms, not whether the bonds are single, double, or triple. (Red = O, gray = C, blue = N, ivory = H.)

(a) What is the formula of thalidomide?

(b) Indicate the positions of the multiple bonds in thalidomide.

(c) What is the geometry around each carbon?

(d) What is the hybridization of each carbon?

Thalidomide

8.34 Ethyl acetate, $CH_3CO_2CH_2CH_3$, is commonly used as a solvent and nail-polish remover. Look at the following electrostatic potential map of ethyl acetate, and explain the observed polarity.

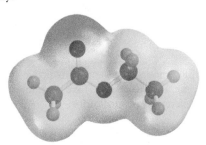

8.35 The dipole moment of methanol is $\mu = 1.70$ D. Use arrows to indicate the direction in which electrons are displaced.

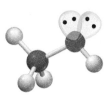

8.36 Methylamine, CH_3NH_2, is responsible for the odor of rotting fish. Look at the following electrostatic potential map of methylamine, and explain the observed polarity.

8.37 Two dichloroethylene molecules with the same chemical formula $(C_2H_2Cl_2)$, but different arrangements of atoms are shown.

cis 1,2 dichloroethylene   *trans* 1,2 dichloroethylene

(a) Match each form of dichloroethylene to its electrostatic potential map.

(i)   **(ii)**

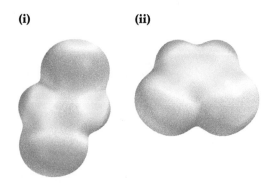

(b) Which form of dichloroethylene has a dipole moment of 2.39 D, and which has a dipole moment of 0.00 D?

(c) Which form of dichloroethylene has the highest boiling point?

SECTION PROBLEMS

The VSEPR Model (Section 8.1)

8.38 What geometric arrangement of charge clouds do you expect for atoms that have the following number of charge clouds?
  (a) 3   (b) 5   (c) 2   (d) 6

8.39 What shape do you expect for molecules that meet the following descriptions?
  (a) A central atom with two lone pairs and three bonds to other atoms
  (b) A central atom with two lone pairs and two bonds to other atoms
  (c) A central atom with two lone pairs and four bonds to other atoms

8.40 How many charge clouds are there around the central atom in molecules that have the following geometry?
  (a) Tetrahedral   (b) Octahedral
  (c) Bent   (d) Linear
  (e) Square pyramidal   (f) Trigonal pyramidal

8.41 How many charge clouds are there around the central atom in molecules that have the following geometry?
  (a) Seesaw   (b) Square planar
  (c) Trigonal bipyramidal   (d) T-shaped
  (e) Trigonal planar   (f) Linear

8.42 What shape do you expect for each of the following molecules?
RAN (a) H_2Se (b) $TiCl_4$
  (c) O_3 (d) GaH_3

8.43 What shape do you expect for each of the following molecules?
  (a) XeO_4 (b) SO_2Cl_2
  (c) OsO_4 (d) SeO_2

8.44 What shape do you expect for each of the following molecules or ions?
  (a) SbF_5 (b) IF_4^+
  (c) SeO_3^{2-} (d) CrO_4^{2-}

8.45 Predict the shape of each of the following ions.
  (a) NO_3^- (b) NO_2^+ (c) NO_2^-

8.46 What shape do you expect for each of the following anions?
RAN (a) PO_4^{3-} (b) MnO_4^-
  (c) SO_4^{2-} (d) SO_3^{2-}
  (e) ClO_4^- (f) SCN^-

8.47 What shape do you expect for each of the following cations?
  (a) XeF_3^+ (b) SF_3^+
  (c) ClF_2^+ (d) CH_3^+

8.48 What bond angles do you expect for each of the following?
  (a) The F—S—F angle in SF_2
  (b) The H—N—N angle in N_2H_2
  (c) The F—Kr—F angle in KrF_4
  (d) The Cl—N—O angle in NOCl

8.49 What bond angles do you expect for each of the following?
  (a) The Cl—P—Cl angle in PCl_6^-
  (b) The Cl—I—Cl angle in ICl_2^-
  (c) The O—S—O angle in SO_4^{2-}
  (d) The O—B—O angle in BO_3^{3-}

8.50 Acrylonitrile is used as the starting material for manufacturing acrylic fibers. Predict values for all bond angles in acrylonitrile.

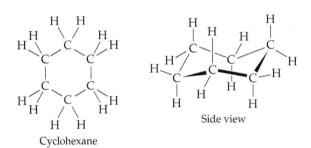

Acrylonitrile

8.51 Predict values for all bond angles in dimethyl sulfoxide, a powerful solvent used in veterinary medicine to treat inflammation.

$$H_3C—\overset{\displaystyle :\ddot{O}:}{\underset{\displaystyle \cdot\cdot}{S}}—CH_3 \quad \text{Dimethyl sulfoxide}$$

8.52 Oceanographers study the mixing of water masses by releasing tracer molecules at a site and then detecting their presence at other places. The molecule trifluoromethylsulfur pentafluoride is one such tracer. Draw an electron-dot structure for CF_3SF_5, and predict the bond angles around both carbon and sulfur.

8.53 A potential replacement for the chlorofluorocarbon refrigerants that harm the Earth's protective ozone layer is a compound called E143a, or trifluoromethyl methyl ether, F_3COCH_3. Draw an electron-dot structure for F_3COCH_3, and predict the geometry around both the carbons and the oxygen.

8.54 Explain why cyclohexane, a substance that contains a six-membered ring of carbon atoms, is not flat but instead has a puckered, nonplanar shape. Predict the values of the C—C—C bond angles.

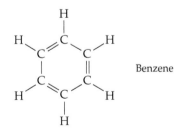

Cyclohexane   Side view

8.55 Like cyclohexane (Problem 8.54), benzene also contains a six-membered ring of carbon atoms, but it is flat rather than puckered. Explain, and predict the values of the C—C—C bond angles.

Benzene

8.56 Use VSEPR theory to answer the following questions:
  (a) Which molecule, BF_3 or PF_3, has the smaller F—X—F angles?
  (b) Which ion, PCl_4^+ or ICl_2^-, has the smaller Cl—X—Cl angles?
  (c) Which ion, CCl_3^- or PCl_6^-, has the smaller Cl—X—Cl angles?

8.57 Draw an electron-dot structure for each of the following substances, and predict the molecular geometry of every nonterminal atom.

(a) $F_3S—S—F$

(b) $CH_3—C≡C—CO_2^-$

Valence Bond Theory and Hybridization
(Sections 8.2–8.4)

8.58 What is the difference in spatial distribution between electrons in a π bond and electrons in a σ bond?

8.59 The average C—C bond dissociation energy (D) is 350 kJ/mol, and the average C=C bond dissociation energy is 728 kJ/mol. Based on these values, which is stronger: a σ or a π bond?

8.60 What hybridization do you expect for atoms that have the following numbers of charge clouds?

(a) 2   (b) 3   (c) 4

8.61 What spatial arrangement of charge clouds corresponds to each of the following kinds of hybridization?

(a) sp^3 (b) sp^2 (c) sp

8.62 What hybridization would you expect for the indicated atom in each of the following molecules?

(a) $H_2C=O$ (b) BH_3

(c) CH_3SH (d) $H_2C=NH$

8.63 What hybridization would you expect for the indicated atom in each of the following ions?

(a) BH_4^- (b) HCO_2^-

(c) CH_3^+ (d) CH_3^-

8.64 Oxaloacetic acid is an intermediate involved in the citric acid cycle of food metabolism. What is the hybridization of the various carbon atoms in oxaloacetic acid, and what are the approximate values of the various bond angles?

$$H—O—\overset{\overset{O}{\|}}{C}—\overset{\overset{O}{\|}}{C}—\overset{\overset{H}{|}}{\underset{\underset{H}{|}}{C}}—\overset{\overset{O}{\|}}{C}—O—H \quad \text{Oxaloacetic acid}$$

8.65 The atoms in the amino acid glycine are connected as shown:

$$H—\overset{\overset{H}{|}}{\underset{\underset{H}{|}}{N}}—\overset{\overset{H}{|}}{\underset{\underset{H}{|}}{C}}—\overset{\overset{O}{\|}}{C}—O—H \quad \text{Glycine}$$

(a) Draw an electron-dot structure for glycine, showing lone pairs and identifying any multiple bonds.

(b) Predict approximate values for the H—C—H, O—C—O, and H—N—H bond angles.

(c) Which hybrid orbitals are used by the C and N atoms?

8.66 Describe the hybridization of the carbon atom in the poisonous gas phosgene, Cl_2CO, and make a rough sketch of the molecule showing its hybrid orbitals and π bonds.

8.67 Describe the hybridization of each carbon atom in propyne (C_3H_4), and make a rough sketch of the molecule showing its hybrid orbitals and π bonds.

$$H—C≡C—\overset{\overset{H}{|}}{\underset{\underset{H}{|}}{C}}—H$$

8.68 Bupropion, marketed as Wellbutrin, is a heavily prescribed medication used in the treatment of depression. Complete the following electron-dot structure for Bupropion by adding lone pairs of electrons.

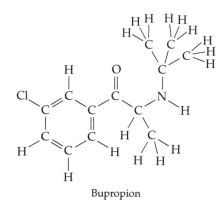

Bupropion

(a) How many σ and how many π bonds are in the molecule?

(b) Give the hybridization of each carbon atom in the molecule.

(c) Give the bond angles of each carbon atom.

(d) Give the hybridization of the N atom in the molecule.

8.69 Efavirenz, marketed as Sustiva, is a medication used in the treatment of human immunodeficiency virus (HIV). Complete the following electron-dot structure for Efavirenz by adding lone pairs of electrons.

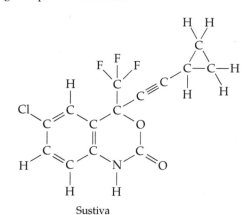

Sustiva

(a) How many σ and how many π bonds are in the molecule?

(b) Give the hybridization of each carbon atom in the molecule.

(c) Give the hybridization of the N atom in the molecule.

(d) Give the hybridization of the oxygen atoms in the molecule.

8.70 What is the hybridization of the B and N atoms in borazine, what are the values of the B—N—B and N—B—N bond angles, and what is the overall shape of the molecule?

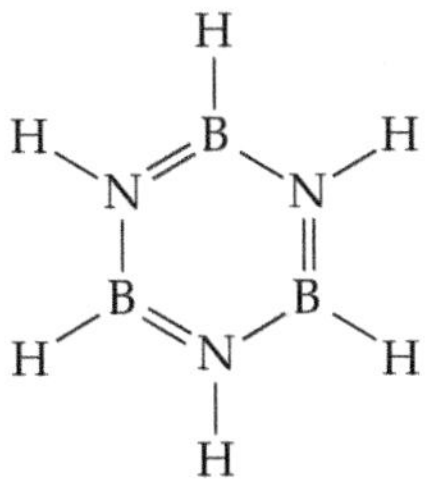

Borazine

8.71 Benzyne, C_6H_4, is a highly energetic and reactive molecule. What hybridization do you expect for the two triply bonded carbon atoms? What are the "theoretical" values for the C—C≡C bond angles? Why do you suppose benzyne is so reactive?

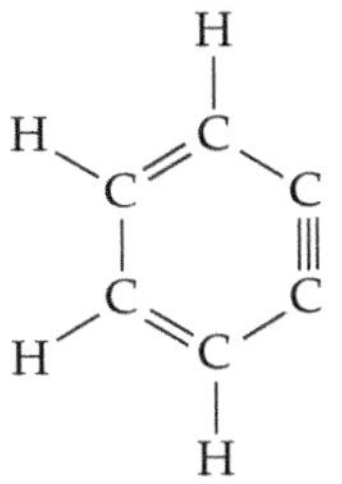

Benzyne

8.72 Aspirin has the following connections among atoms. Complete the electron-dot structure for aspirin, tell how many σ bonds and how many π bonds the molecule contains, and tell the hybridization of each carbon atom.

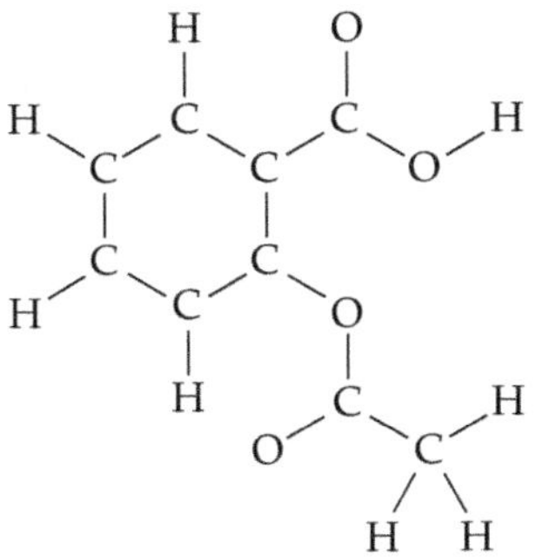

Aspirin

8.73 The cation $[H—C—N—Xe—F]^+$ is entirely linear. Draw an electron-dot structure consistent with that geometry, and tell the hybridization of the C and N atoms.

8.74 Acrylonitrile (C_3H_3N) is a molecule that is polymerized to make carpets and fabrics. The connections between atoms are shown.

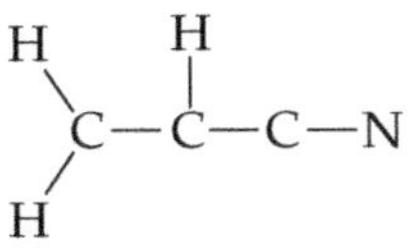

(a) Complete the electron-dot structure for acrylonitrile by adding multiple bonds and lone pairs of electrons.

(b) Tell how many σ bonds and how many π bonds the molecule contains.

(c) Tell the hybridization of each carbon atom.

(d) Identify the shortest bond in the molecule.

8.75 The odor of cinnamon oil is due to cinnamaldehyde, C_9H_8O. What is the hybridization of each carbon atom in cinnamaldehyde? How many σ and how many π bonds does cinnamaldehyde have?

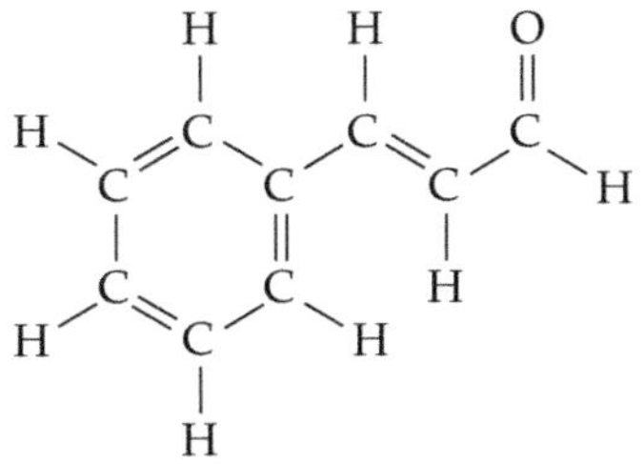

Cinnamaldehyde

8.76 The following molecular model is a representation of caffeine. Identify the position(s) of multiple bonds in caffeine, and tell the hybridization of each carbon atom. (Red = O, blue = N, gray = C, ivory = H.)

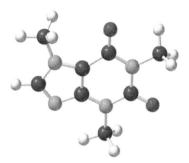

Caffeine

8.77 The following molecular model is a representation of ascorbic acid, or vitamin C. Identify the position(s) of multiple bonds in ascorbic acid, and tell the hybridization of each carbon atom. (Red = O, gray = C, ivory = H.)

Ascorbic acid

Dipole Moments (Sections 8.5)

8.78 Which of the following substances would you expect to have a nonzero dipole moment? Explain, and show the direction of each.

(a) Cl_2O

(b) XeF_4

(c) Chloroethane, CH_3CH_2Cl

(d) BF_3

8.79 Which of the following substances would you expect to have a nonzero dipole moment? Explain, and show the direction of each.

(a) NF_3

(b) CH_3NH_2

(c) XeF_2

(d) PCl_5

8.80 Why is the dipole moment of SO_2 1.63 D but that of CO_2 is zero?

8.81 Draw three-dimensional structures of PCl_3 and PCl_5, and then explain why one of the molecules has a dipole moment and one does not.

8.82 The class of ions $PtX_4{}^{2-}$, where X is a halogen, has a square planar geometry.

(a) Draw a structure for a $PtBr_2Cl_2{}^{2-}$ ion that has no dipole moment.

(b) Draw a structure for a $PtBr_2Cl_2{}^{2-}$ ion that has a dipole moment.

8.83 Of the two compounds SiF_4 and SF_4, which is polar and which is nonpolar?

8.84 Why don't all molecules with polar covalent bonds have dipole moments?

8.85 Fluorine is more electronegative than chlorine, yet fluoromethane (CH_3F; $\mu = 1.86$ D) has a smaller dipole moment than chloromethane (CH_3Cl; $\mu = 1.90$ D). Explain.

Intermolecular Forces (Section 8.6)

8.86 What is the difference between London dispersion forces and dipole–dipole forces?

8.87 Which substance in each of the following pairs would you
RAN expect to have larger dispersion forces?

(a) Ethane, C_2H_6, or octane, C_8H_{18}

(b) HCl or HI

(c) H_2O or H_2Se

8.88 What are the most important kinds of intermolecular forces present in each of the following substances?

(a) Chloroform, $CHCl_3$

(b) Oxygen, O_2

(c) Polyethylene, C_nH_{2n+2}

(d) Methanol, CH_3OH

8.89 Of the substances Xe, CH_3Cl, and HF, which has:

(a) The smallest dipole–dipole forces?

(b) The largest hydrogen bond forces?

(c) The largest dispersion forces?

8.90 Methanol (CH_3OH; bp = 65 °C) boils nearly 230 °C higher than methane (CH_4; bp = −164 °C), but 1-decanol ($C_{10}H_{21}OH$; bp = 231 °C) boils only 57 °C higher than decane ($C_{10}H_{22}$; bp = 174 °C). Explain.

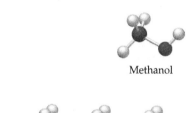

Methanol

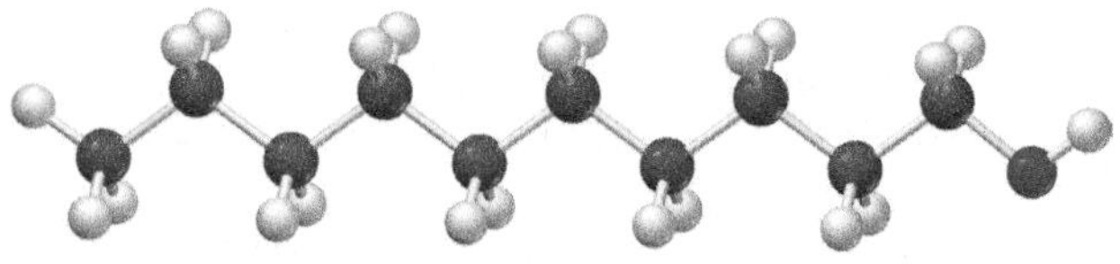

1-Decanol

8.91 For each of the following substances, identify the intermolecular force or forces that predominate. Using your knowledge of the relative strengths of the various forces, rank the substances in order of their normal boiling points: Al_2O_3, F_2, H_2O, Br_2, ICl, NaCl.

8.92 Draw a picture showing how hydrogen bonding takes place between two ammonia molecules.

8.93 1,3-Propanediol can form *intra*molecular as well as *inter*molecular hydrogen bonds. Draw a structure of 1,3-propanediol showing an intramolecular hydrogen bond.

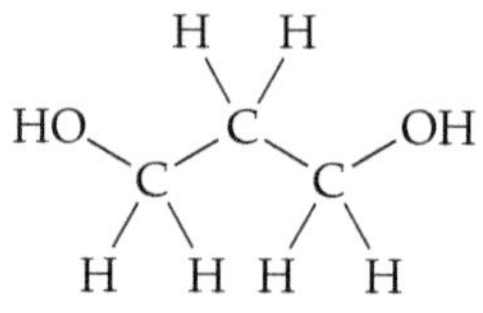

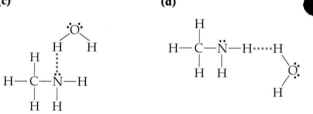

1,3-Propanediol

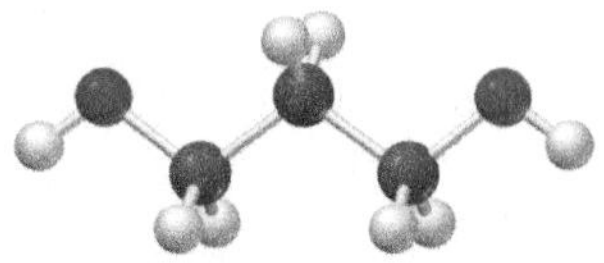

8.94 A liquid sample contains methylamine (CH_3NH_2) dissolved in water. Which of the following illustrations depicts the hydrogen bonding that occurs between methylamine and water?

(a)

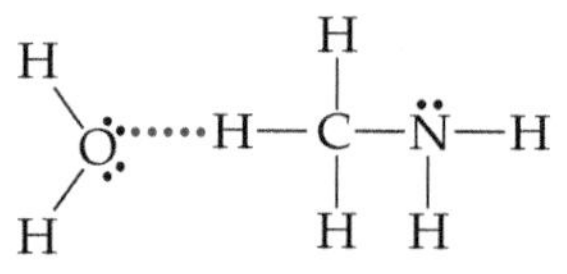

(b)

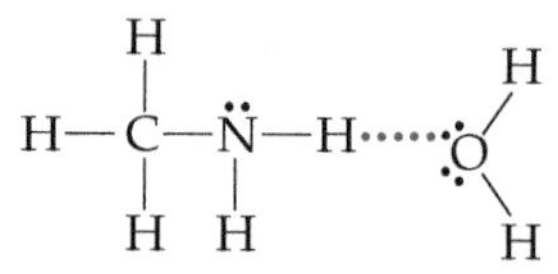

(c)

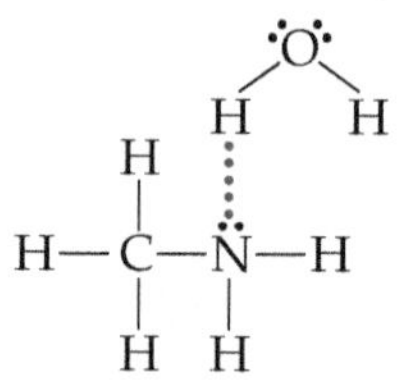

(d)

8.95 Dimethyl ether has the following structure.

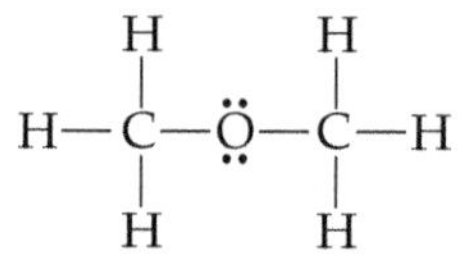

(a) Does hydrogen bonding occur in a pure sample of dimethyl ether? If so, sketch the hydrogen bonds.

(b) Which of the following illustrations depicts the hydrogen bonding that occurs between dimethyl ether and water?

(i)

(ii)

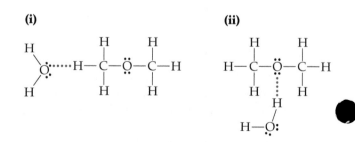

Molecular Orbital Theory (Sections 8.7–8.9)

8.96 What is the difference in spatial distribution between electrons in a bonding MO and electrons in an antibonding MO?

8.97 For a given type of MO, use a σ_{2s} as an example, is the bonding or antibonding orbital higher in energy? Explain.

8.98 Use the MO energy diagram in Figure 8.22b to describe the
RAN bonding in O_2^+, O_2, and O_2^-. Which of the three is likely to be stable? What is the bond order of each? Which contain unpaired electrons?

8.99 Use the MO energy diagram in Figure 8.22a to describe the
RAN bonding in N_2^+, N_2, and N_2^-. Which of the three is likely to be stable? What is the bond order of each? Which contain unpaired electrons?

8.100 The C_2 molecule can be represented by an MO diagram sim-
RAN ilar to that in Figure 8.22a.

(a) What is the bond order of C_2?

(b) To increase the bond order of C_2, should you add or remove an electron?

(c) Give the charge and the bond order of the new species made in part (b).

8.101 Look at the molecular orbital diagram for O_2 in Figure 8.22b,
RAN and answer the following questions.

(a) What is the bond order of O_2?

(b) To increase the bond order of O_2, should you add or remove an electron?

(c) Give the charge and the bond order of the new species made in part (b).

8.102 Look at the MO diagrams of corresponding neutral diatomic species in Figure 8.22, and predict whether each of the following ions is diamagnetic or paramagnetic. Diagrams for Li_2 and C_2 are similar to N_2; Cl_2 is similar to F_2.

(a) C_2^{2-} (b) C_2^{2+} (c) F_2^-

(d) Cl_2 (e) Li_2^+

8.103 Look at the MO diagrams of corresponding neutral diatomic species in Figure 8.22, and predict whether each of the following ions is diamagnetic or paramagnetic. MO diagrams for Li_2 and C_2 are similar to N_2; Cl_2 is similar to F_2.

(a) O_2^{2+} (b) N_2^{2+} (c) C_2^+

(d) F_2^{2+} (e) Cl_2^+

8.104 Draw a molecular orbital energy diagram for Li_2. What is the bond order? Is the molecule likely to be stable? Explain.

8.105 Calcium carbide, CaC_2, reacts with water to produce acetylene, C_2H_2, and is sometimes used as a convenient source of that substance. Use the MO energy diagram in Figure 8.22a to describe the bonding in the carbide anion, C_2^{2-}. What is its bond order?

8.106 At high temperatures, sulfur vapor is predominantly in the form of $S_2(g)$ molecules.

(a) Assuming that the molecular orbitals for third-row diatomic molecules are analogous to those for second-row molecules, construct an MO diagram for the valence orbitals of $S_2(g)$.

(b) Is S_2 likely to be paramagnetic or diamagnetic?

(c) What is the bond order of $S_2(g)$?

(d) When two electrons are added to S_2, the disulfide ion S_2^{2-} is formed. Is the bond length in S_2^{2-} likely to be shorter or longer than the bond length in S_2? Explain.

8.107 Carbon monoxide is produced by incomplete combustion of fossil fuels.

(a) Give the electron configuration for the valence molecular orbitals of CO. The orbitals have the same energy order as those of the N_2 molecule.

(b) Do you expect CO to be paramagnetic or diamagnetic?

(c) What is the bond order of CO? Does this match the bond order predicted by the electron-dot structure?

(d) CO can react with OH^- to form the formate ion, HCO_2^-. Draw an electron-dot structure for the formate ion, and give any resonance structures if appropriate.

8.108 Make a sketch showing the location and geometry of the p orbitals in the nitrite ion, NO_2^-. Describe the bonding in this ion using a localized valence bond model for σ bonding and a delocalized MO model for π bonding.

8.109 Make a sketch showing the location and geometry of the p orbitals in the allyl cation. Describe the bonding in this cation using a localized valence bond model for σ bonding and a delocalized MO model for π bonding.

$$\begin{array}{c} H \\ | \\ H_2C{=}C{-}CH_2{}^+ \end{array} \quad \text{Allyl cation}$$

MULTICONCEPT PROBLEMS

8.110 Propose structures for molecules that meet the following
RAN descriptions.

(a) Contains a C atom that has two π bonds and two σ bonds

(b) Contains an N atom that has one π bond and two σ bonds

(c) Contains an S atom that has a coordinate covalent bond

8.111 In the cyanate ion, OCN^-, carbon is the central atom.

(a) Draw as many resonance structures as you can for OCN^-, and assign formal charges to the atoms in each.

(b) Which resonance structure makes the greatest contribution to the resonance hybrid? Which makes the least contribution? Explain.

(c) Is OCN^- linear or bent? Explain.

(d) Which hybrid orbitals are used by the C atom, and how many π bonds does the C atom form?

8.112 The ion I_5^- is shaped like a big "V." Draw an electron-dot structure consistent with this overall geometry.

8.113 The dichromate ion, $Cr_2O_7^{2-}$, has neither Cr—Cr nor O—O bonds.

(a) Taking both $4s$ and $3d$ electrons into account, draw an electron-dot structure that minimizes the formal charges on the atoms.

(b) How many outer-shell electrons does each Cr atom have in your electron-dot structure? What is the likely geometry around the Cr atoms?

8.114 Just as individual bonds in a molecule are often polar, molecules as a whole are also often polar because of the net sum of individual bond polarities. There are three possible structures for substances with the formula $C_2H_2Cl_2$, two of which are polar overall and one of which is not.

(a) Draw the three possible structures for $C_2H_2Cl_2$, predict an overall shape for each, and explain how they differ.

(b) Which of the three structures is nonpolar, and which two are polar? Explain.

(c) Two of the three structures can be interconverted by a process called cis–trans isomerization, in which rotation around the central carbon–carbon bond takes place when the molecules are irradiated with ultraviolet light. If light with a wavelength of approximately 200 nm is required for isomerization, how much energy in kJ/mol is involved?

(d) Sketch the orbitals involved in the central carbon–carbon bond, and explain why so much energy is necessary for bond rotation to occur.

8.115 Cyclooctatetraene dianion, $C_8H_8^{2-}$, is an organic ion with the structure shown. Considering only the π bonds and not the σ bonds, cyclooctatetraene dianion can be described by the following energy diagrams of its π molecular orbitals:

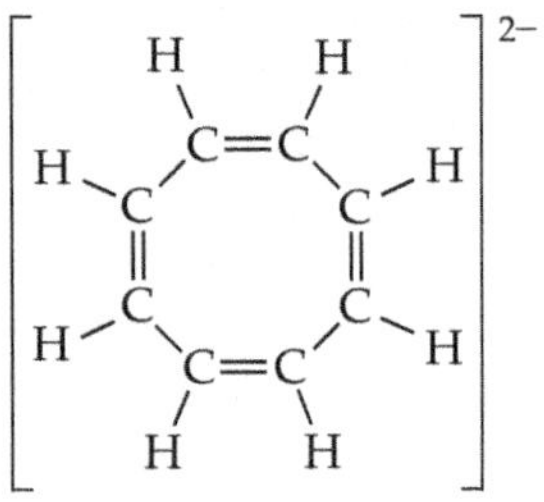

(a) What is the hybridization of the 8 carbon atoms?

(b) Three of the π molecular orbitals are bonding, three are antibonding, and two are *nonbonding*, meaning that they have the same energy level as isolated p orbitals. Which is which?

(c) Complete the MO energy diagram by assigning the appropriate numbers of p electrons to the various molecular orbitals, indicating the electrons using up/down arrows ($\uparrow\downarrow$).

(d) Based on your MO energy diagram, is the dianion paramagnetic or diamagnetic?

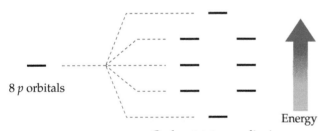

Cyclooctatetraene dianion

8 p orbitals

Energy

Cyclooctatetraene dianion
pi (π) molecular orbitals

chapter 9

Thermochemistry: Chemical Energy

Contents

Biofuels made in a photobioreactor from algae are a renewable alternative to fossil fuels.

How do we determine the energy content of biofuels?

The answer to this question can be found on page 359 in the INQUIRY ?

Energy is important is every aspect of our lives. The food we eat provides energy to move, grow, think, and maintain a body temperature of 98.6 °F. Fossil fuels like coal, oil, and gas provide energy for electricity generation, manufacturing processes, and transportation. Projected petroleum shortages and the undesirable consequences of burning fossil fuels, including climate change and pollution, have prompted an increase in the use of renewable sources of energy such as solar cells, wind farms, and biofuels. The Inquiry at the end of the chapter describes how biofuels are made and explores their energy content.

We'll examine different forms of energy in this chapter and describe the subject of **thermochemistry**, the absorption or release of heat energy that accompanies chemical reactions. We'll also examine how heat transfer influences the very important question "Why do chemical reactions occur?" Stated simply, less stable substances with higher energy are generally converted to more stable substances with lower energy.

9.1 ENERGY AND ITS CONSERVATION

REMEMBER . . .
Energy is the capacity to supply heat or do work and is classified as either **kinetic energy**—the energy of motion—or **potential energy**—stored energy not yet released (Section 1.8).

The word *energy*, though familiar to everyone, is surprisingly hard to define in simple, nontechnical terms. A good working definition, however, is to say that **energy** is the capacity to do work or supply heat. The water falling over a dam, for instance, contains energy that can be used to turn a turbine and generate electricity. A tank of propane gas contains energy that, when released in the chemical process of combustion, can heat a house or barbecue a hamburger.

Energy is classified as either *kinetic* or *potential*. **Kinetic energy (E_K)** is the energy of motion.

Kinetic energy $E_K = \frac{1}{2}mv^2$ where m = mass and v = velocity

The derived SI unit for energy ($kg \cdot m^2/s^2$) follows from the expression for kinetic energy, $E_K = (1/2)mv^2$, and is given the name **joule (J)**.

$$1\ (kg \cdot m^2/s^2) = 1\ J$$

Potential energy (E_P), by contrast, is stored energy—perhaps stored in an object because of its height or in a molecule because of reactions it can undergo. The water sitting in a reservoir behind a dam contains potential energy because of its height above the stream at the bottom of the dam. When the water is allowed to fall, its potential energy is converted to kinetic energy. Propane and other substances used as fuels contain potential energy because they can undergo a reaction with oxygen—a *combustion reaction*—that releases heat.

Let's pursue the relationship between potential energy and kinetic energy a bit further. According to the **conservation of energy law,** energy can be neither created nor destroyed; it can only be converted from one form into another.

Conservation of energy law Energy cannot be created or destroyed; it can only be converted from one form into another.

To take an example, think about a hydroelectric dam. The water sitting motionless in the reservoir behind the dam has potential energy because of its height above the outlet stream, but it has no kinetic energy because it isn't moving ($v = 0$). When the water falls through the penstocks of the dam, however, its height and potential energy decrease while its velocity and kinetic energy increase. The moving water then spins the turbine of a generator, converting its kinetic energy into electrical energy (**FIGURE 9.1**).

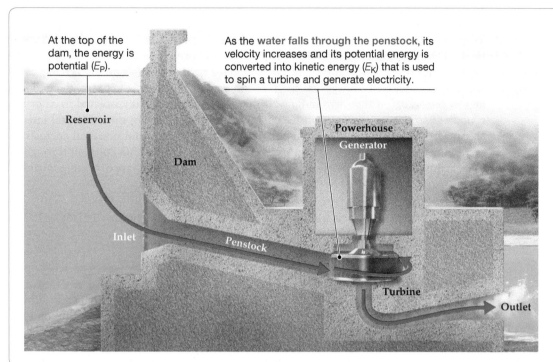

At the top of the dam, the energy is potential (E_P).

As the water falls through the penstock, its velocity increases and its potential energy is converted into kinetic energy (E_K) that is used to spin a turbine and generate electricity.

Reservoir

Powerhouse

Generator

Dam

Inlet

Penstock

Turbine

Outlet

◀ **FIGURE 9.1**

Conservation of energy. The total amount of energy contained by the water in the reservoir is constant.

The conversion of the kinetic energy in falling water into electricity illustrates several other important points about energy. One is that energy has many forms. Thermal energy, for example, seems different from the kinetic energy of falling water, yet is really quite similar. *Thermal energy* is just the kinetic energy of molecular motion, which we measure by finding the **temperature** of an object. An object has a low temperature and we perceive it as cold if its atoms or molecules are moving slowly. Conversely, an object has a high temperature and we perceive it as hot if its atoms or molecules are moving rapidly and are colliding forcefully with a thermometer or other measuring device.

Heat, in turn, is the amount of thermal energy transferred from one object to another as the result of a temperature difference between the two. Rapidly moving molecules in a hotter object collide with more slowly moving molecules in a colder object, transferring kinetic energy and causing the slower moving molecules to speed up.

Chemical energy is another kind of energy that seems different from that of the water in a reservoir, yet again is really quite similar. Chemical energy is a kind of potential energy in which chemical bonds act as the storage medium. Just as water releases its potential energy when it falls to a lower height, chemicals can release their potential energy in the form of heat or light when they undergo reactions and form lower-energy products. We'll explore this topic shortly.

A second point illustrated by falling water involves the conservation of energy law. To keep track of all the energy involved, it's necessary to take into account the entire chain of events that ensue from the falling water: the sound of the crashing water, the heating of the rocks at the bottom of the dam, the driving of turbines and electrical generators, the transmission of electrical power, the appliances powered by the electricity, and so on. Carrying the process to its logical extreme, it's necessary to take the entire universe into account when keeping track of all the energy in the water because the energy lost in one form *always* shows up elsewhere in another form. So important is the conservation of energy law that it's also known as the **first law of thermodynamics.**

Go to eText

BIG IDEA Question 1

When natural gas is burned to heat a home in the winter, _____ energy is converted to _____ energy. (Fill in the blanks with either *kinetic* or *potential*.)

First law of thermodynamics   Energy cannot be created or destroyed; it can only be converted from one form into another.

9.2 INTERNAL ENERGY AND STATE FUNCTIONS

When keeping track of the energy changes in a chemical reaction, it's often helpful to think of the reaction as being isolated from the world around it. The substances we focus on in an experiment—the starting reactants and the final products—are collectively called the **system,** while everything else—the reaction flask, the solvent, the room, the building, and so on—is called the **surroundings.** If the system could be truly isolated from its surroundings so that no energy transfer could occur between them, then the total **internal energy** (E) of the system, defined as the sum of all the kinetic and potential energies for every molecule or ion in the system, would be conserved and remain constant throughout the reaction. In fact, this assertion is just a restatement of the first law of thermodynamics.

> **First law of thermodynamics (restated)**   The total internal energy E of an isolated system is constant.

In practice, of course, it's not possible to truly isolate a chemical reaction from its surroundings. In any real situation, the chemicals are in physical contact with the walls of a flask or container, and the container itself is in contact with the surrounding air or laboratory bench. What's important, however, is not that the system be isolated but that we be able to measure accurately any energy that enters the system from the surroundings or leaves the system and flows to the surroundings (**FIGURE 9.2**). That is, we must be able to measure any *change* in the internal energy of the system, ΔE. The energy change ΔE represents the difference in internal energy between the final state of the system after reaction and the initial state of the system before reaction:

$$\Delta E = E_{final} - E_{initial}$$

By convention, energy changes are measured from the point of view of the system. Any energy that flows *from* the system *to* the surroundings has a negative sign because the system has lost it (that is, E_{final} is smaller than $E_{initial}$). Any energy that flows *to* the system *from* the surroundings has a positive sign because the system has gained it E_{final} is larger than $E_{initial}$).

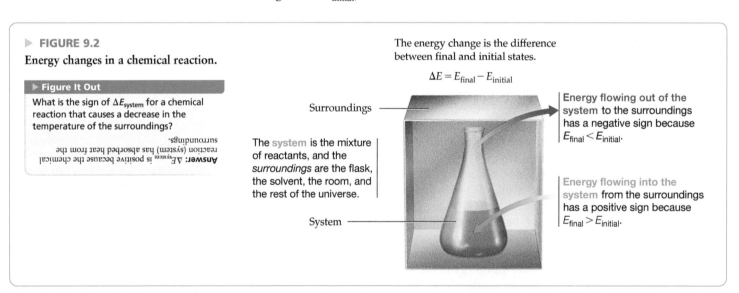

▶ **FIGURE 9.2**

Energy changes in a chemical reaction.

▶ **Figure It Out**

What is the sign of ΔE_{system} for a chemical reaction that causes a decrease in the temperature of the surroundings?

Answer: ΔE_{system} is positive because the chemical reaction (system) has absorbed heat from the surroundings.

The system is the mixture of reactants, and the *surroundings* are the flask, the solvent, the room, and the rest of the universe.

Surroundings

System

The energy change is the difference between final and initial states.

$\Delta E = E_{final} - E_{initial}$

Energy flowing out of the system to the surroundings has a negative sign because $E_{final} < E_{initial}$.

Energy flowing into the system from the surroundings has a positive sign because $E_{final} > E_{initial}$.

If, for instance, we were to burn 1.00 mol of methane in the presence of 2.00 mol of oxygen, 802 kJ would be released as heat and transferred from the system to the surroundings. The system has 802 kJ less energy, so $\Delta E = -802$ kJ. This energy flow can be detected and measured by placing the reaction vessel in a water bath and noting the temperature rise of the bath during the reaction.

$$CH_4(g) + 2\,O_2(g) \longrightarrow CO_2(g) + 2\,H_2O(g) + 802\text{ kJ energy} \qquad \Delta E = -802\text{ kJ}$$

The methane combustion experiment tells us that the products of the reaction, $CO_2(g)$ and $2 H_2O(g)$, have 802 kJ less internal energy than the reactants, $CH_4(g)$ and $2 O_2(g)$, even though we don't know the exact values at the beginning ($E_{initial}$) and end (E_{final}) of the reaction. Note that the value $\Delta E = -802$ kJ for the reaction refers to the energy released when reactants are converted to products *in the molar amounts represented by coefficients in the balanced equation*. That is, 802 kJ is released when 1 mol of gaseous methane reacts with 2 mol of gaseous oxygen to give 1 mol of gaseous carbon dioxide and 2 mol of gaseous water vapor (**FIGURE 9.3**).

The internal energy of a system depends on many things: chemical identity, sample size, temperature, pressure, physical state (gas, liquid, or solid), and so forth. What the internal energy does not depend on is the system's past history. It doesn't matter what the system's temperature or physical state was an hour ago, and it doesn't matter how the chemicals were obtained. All that matters is the present condition of the system. Thus, internal energy is said to be a **state function**, one whose value depends only on the present state of the system. Pressure, volume, and temperature are other examples of state functions, but work and heat are not.

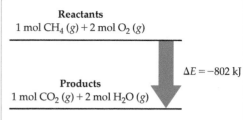

Reactants
1 mol CH_4 (g) + 2 mol O_2 (g)

$\Delta E = -802$ kJ

Products
1 mol CO_2 (g) + 2 mol H_2O (g)

▲ **FIGURE 9.3**

Energy change in the combustion of one mole of methane (CH_4).

State function   A function or property whose value depends only on the present state, or condition, of the system, not on the path used to arrive at that state.

We can illustrate the idea of a state function by imagining a cross-country trip, say from the Artichoke Capitol of the World (Castroville, California), to the Hub of the Universe (Boston, Massachusetts). You are the system, and your position is a state function because how you got to wherever you are is irrelevant. Because your position is a state function, the *change* in your position after you complete your travel (Castroville and Boston are about 2720 miles apart) is independent of the path you take, whether through North Dakota or Louisiana (**FIGURE 9.4**).

The cross-country trip shown in Figure 9.4 illustrates an important point about state functions: their reversibility. Imagine that, after traveling from Castroville to Boston, you turn around and go back. Because your final position is now identical to your initial position, the change in your position is zero. The overall change in any state function is zero when the system returns to its original condition. For a nonstate function, however, the overall change is not zero when the system returns to its original condition. Any work you do in making the trip is not recovered when you return to your initial position, and any money or time you spend does not reappear.

Measurements of the caloric content of food are based on the principle that energy changes in chemical reactions are state functions. The energy content of biological molecules such as carbohydrates and fats can be measured by burning them in oxygen

▲ Natural gas (methane, CH_4) burns and releases energy as heat to its surroundings.

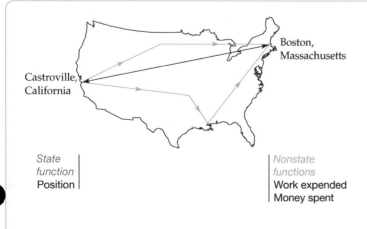

Castroville, California

Boston, Massachusetts

State function	*Nonstate functions*
Position	Work expended
	Money spent

◀ **FIGURE 9.4**

State functions. Because your position is a state function, the change in your position on going from Castroville, California, to Boston, Massachusetts, is independent of the path you take.

◀ **Figure It Out**

Which of the following is a state function: the speed of the car on the highway or the distance of the route from Castroville to Boston?

Answer: The speed of the car is a state function because it does not depend on the pathway taken to achieve it. The car is traveling at 60 mph whether it has sped up from 30 mph or slowed down from 75 mph. The distance of the trip is a nonstate function because it depends on the pathway taken.

in a calorimetry experiment as described in Section 9.7. The energy change is the same as when these compounds are "burned" for energy in your body because both processes have an identical overall chemical reaction.

$$C_6H_{12}O_6(s) + 6\,O_2(g) \xrightarrow{\text{combustion or metabolism}} 6\,CO_2(g) + 6\,H_2O(l) \quad \Delta E = -2810\text{ kJ}$$

REMEMBER . . .
Energy is frequently reported in different units and common conversion factors are 4.184 J = 1 cal and 1000 cal = 1 Cal (Section 1.8).

The same amount of energy is released whether glucose is directly burned or whether it is metabolized to carbon dioxide and water in a series of reactions in your body because state functions do not depend on the pathway taken from the initial to final state. Energy content of foods is typically reported in units of **Calories (Cal)** per gram, and the energy change for the combustion of glucose can be converted to the unit given on food labels as follows:

$$\frac{2810\text{ kJ}}{1\text{ mol }C_6H_{12}O_6} \times \frac{1000\text{ J}}{1\text{ kJ}} \times \frac{1\text{ cal}}{4.184\text{ J}} \times \frac{1\text{ Cal}}{1000\text{ cal}} \times \frac{1\text{ mol }C_6H_{12}O_6}{180.0\text{ g }C_6H_{12}O_6} = 3.73\text{ Cal/g}$$

9.3 EXPANSION WORK

Work is a type of energy transfer that comes in many forms. In physics, **work (w)** is defined as the force (F) that produces the movement of an object times the distance moved (d):

$$\text{Work} = \text{Force} \times \text{Distance}$$

$$w = F \times d$$

When you run up stairs, for instance, your leg muscles provide a force sufficient to overcome gravity and lift you higher. When you swim, you provide a force sufficient to push water out of the way and pull yourself forward.

▲ The runner going up the steps is doing a lot of work to overcome gravity.

The most common type of work encountered in chemical systems is the *expansion work* (also called *pressure–volume*, or *PV, work*) done as the result of a volume change in the system. In the combustion reaction of propane (C_3H_8) with oxygen, for instance, the balanced equation says that 7 mol of products come from 6 mol of reactants:

$$\underbrace{C_3H_8(g) + 5\,O_2(g)}_{\text{6 mol of gas}} \longrightarrow \underbrace{3\,CO_2(g) + 4\,H_2O(g)}_{\text{7 mol of gas}}$$

If the reaction takes place inside a container outfitted with a movable piston, the greater volume of gas in the product will force the piston outward against the pressure of the atmosphere (P), moving air molecules aside and thereby doing work (**FIGURE 9.5**).

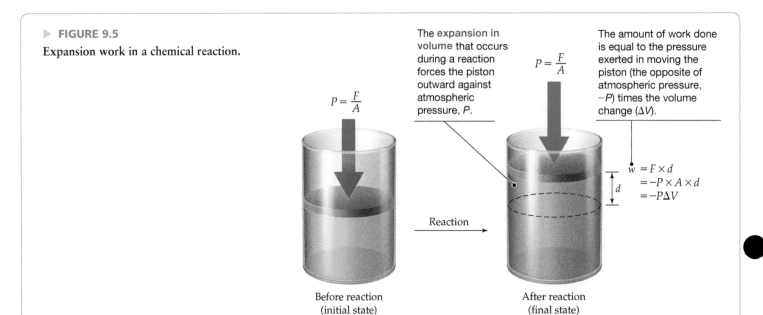

▶ **FIGURE 9.5**

Expansion work in a chemical reaction.

The expansion in volume that occurs during a reaction forces the piston outward against atmospheric pressure, P.

$$P = \frac{F}{A}$$

The amount of work done is equal to the pressure exerted in moving the piston (the opposite of atmospheric pressure, $-P$) times the volume change (ΔV).

$$P = \frac{F}{A}$$

$$\begin{aligned} w &= F \times d \\ &= -P \times A \times d \\ &= -P\Delta V \end{aligned}$$

Before reaction (initial state)

After reaction (final state)

Reaction

A short calculation gives the exact amount of work done during the expansion. We know from physics that force (F) is area (A) times pressure (P). In Figure 9.5, the force that the expanding gas exerts is the area of the piston times the pressure that the gas exerts against the piston. This pressure is equal in magnitude but opposite in sign to the external atmospheric pressure that opposes the movement, so it has the value $-P$.

$$F = -P \times A \quad \text{where } P \text{ is the external atmospheric pressure}$$

If the piston is pushed out a distance d, then the amount of work done is equal to force times distance, or pressure times area times distance:

$$w = F \times d = -P \times A \times d$$

This equation can be simplified by noticing that the area of the piston times the distance the piston moves is just the volume change in the system: $\Delta V = A \times d$. Thus, the amount of work done is equal to the pressure the gas exerts against the piston times the volume change, hence the name PV work:

A negative value ↘ ↙ A positive value

$$w = -P\Delta V \qquad \text{Work done during expansion}$$

What about the sign of the work done during the expansion? Because the work is done by the system to move air molecules aside as the piston rises, work energy must be leaving the system. Thus, the negative sign of the work in the preceding equation is consistent with the convention previously established for ΔE (Section 9.2), whereby we always adopt the point of view of the system. Any energy that flows out of the system has a negative sign because the system has lost it ($E_{final} < E_{initial}$).

If the pressure is given in the unit atmospheres (atm) and the volume change is given in liters, then the amount of work done has the derived unit liter atmosphere (L·atm), where 1 atm $= 101 \times 10^3$ kg/(m·s^2). Thus, 1 L·atm $= 101$ J:

$$1\,\text{L·atm} = (1\,\text{L})\left(\frac{10^{-3}\,\text{m}^3}{1\,\text{L}}\right)\left(101 \times 10^3\,\frac{\text{kg}}{\text{m·s}^2}\right) = 101\,\frac{\text{kg·m}^2}{\text{s}^2} = 101\,\text{J}$$

When a reaction takes place with a contraction in volume rather than an expansion, the ΔV term has a negative sign and the work has a positive sign. This is again consistent with adopting the point of view of the system because the system has now gained work energy ($E_{final} > E_{initial}$). An example is the industrial synthesis of ammonia by reaction of hydrogen with nitrogen. Four moles of gaseous reactants yield only 2 mol of gaseous products, so the volume of the system contracts and work is gained by the system.

$$\underbrace{3\,H_2(g) + N_2(g)}_{\text{4 mol of gas}} \longrightarrow \underbrace{2\,NH_3(g)}_{\text{2 mol of gas}}$$

A positive value ↘ ↙ A negative value

$$w = -P\Delta V \qquad \text{Work gained during contraction}$$

If there is no volume change, then $\Delta V = 0$ and there is no work. Such is the case for the combustion of methane, where 3 mol of gaseous reactants give 3 mol of gaseous products: $CH_4(g) + 2\,O_2(g) \longrightarrow CO_2(g) + 2\,H_2O(g)$

WORKED EXAMPLE 9.1

Calculating the Amount of *PV* Work

Calculate the work in kilojoules done during a reaction in which the volume expands from 12.0 L to 14.5 L against an external pressure of 5.0 atm.

IDENTIFY

Known	Unknown
Pressure (P) = 5.0 atm	Work (w)
Initial and Final Volume (V)	

STRATEGY

Expansion work done during a chemical reaction is calculated with the formula $w = -P\Delta V$, where P is the external pressure opposing the change in volume. In this instance, $P = 5.0$ atm and $\Delta V = (14.5 - 12.0)\,\text{L} = 2.5\,\text{L}$.

continued on next page

SOLUTION

$$w = -(5.0 \text{ atm})(2.5 \text{ L}) = -12.5 \text{ L} \cdot \text{atm}$$

$$(-12.5 \text{ L} \cdot \text{atm})\left(101 \frac{\text{J}}{\text{L} \cdot \text{atm}}\right) = -1.3 \times 10^3 \text{ J} = -1.3 \text{ kJ}$$

CHECK

It is always useful to check the sign of energy change to ensure it makes sense. Changes in energy and work are expressed from the perspective of the system. In this case, the system has done work on the surroundings by expanding the volume, and thus the sign of w should be negative.

▶ **PRACTICE 9.1**   Calculate the change in energy as work (kJ) that occurs during a synthesis of ammonia in which the volume contracts from 8.60 L to 4.30 L at a constant external pressure of 44.0 atm. In which direction does the work energy flow?

▶ **CONCEPTUAL APPLY 9.2**   How much work is done in kilojoules, and in which direction, as a result of the following reaction?

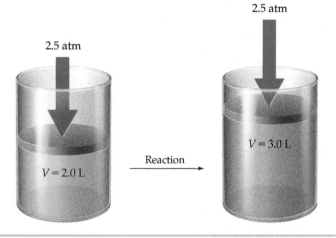

All **PRACTICE** and **APPLY** problems are interactive in the eText.

9.4 ENERGY AND ENTHALPY

BIG IDEA Question 2   Go to eText

Consider the freezing of liquid water to ice at temperatures below 0 °C.

$$H_2O(l) \longrightarrow H_2O(s)$$

Does the energy transfer occur primarily as heat or work, and what is the sign of the change?

We've seen up to this point that a system can exchange energy with its surroundings either by transferring heat or by doing work. The burning of gasoline (C_8H_{18}) in your car's engine heats up the surroundings causing the product gases CO_2 and H_2O to expand and do work by pushing the piston (**FIGURE 9.6**). Using the symbol q to represent transferred heat and the formula $w = -P\Delta V$, we can represent the total change in internal energy of a system, ΔE, as follows:

Change in the internal energy of a system (ΔE) $\Delta E = q + w = q - P\Delta V$

where q has a positive sign if the system gains heat and a negative sign if the system loses heat.

Rearranging the equation to solve for q gives the amount of heat transferred:

$$q = \Delta E + P\Delta V$$

▶ **FIGURE 9.6**

Energy is transferred as heat and work when gasoline (C_8H_{18}) burns in a car's engine.

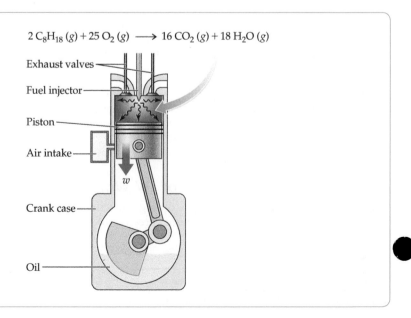

$2 C_8H_{18}(g) + 25 O_2(g) \longrightarrow 16 CO_2(g) + 18 H_2O(g)$

Let's look at two ways in which a chemical reaction might be carried out. On the one hand, a reaction might be carried out in a closed container with a constant volume, so that $\Delta V = 0$. In such a case, no PV work can be done so the energy change in the system is due entirely to heat transfer. We indicate this heat transfer at constant volume by the symbol q_v.

$$q_v = \Delta E \quad \text{At constant volume; } \Delta V = 0$$

Alternatively, a reaction might be carried out in an open flask or other apparatus that keeps the pressure constant and allows the volume of the system to change freely. In such a case, $\Delta V \neq 0$ and the energy change in the system is due to both heat transfer and PV work. We indicate the heat transfer at constant pressure by the symbol q_p:

$$q_p = \Delta E + P\Delta V \quad \text{At constant pressure}$$

▲ Chemical reactions are often carried out in open vessels at constant atmospheric pressure.

Because reactions carried out at constant pressure in open containers are so common in chemistry, the heat change q_p for such a process is given a special symbol and is called the **heat of reaction**, or **enthalpy change** (ΔH). The **enthalpy** (H) of a system is the name given to the quantity $E + PV$.

Enthalpy change (ΔH) $q_p = \Delta E + P\Delta V = \overset{\displaystyle\frown\ \text{Enthalpy change}}{\Delta H}$

Note that only the enthalpy *change* during a reaction is important. As with internal energy, enthalpy is a state function whose value depends only on the current state of the system, not on the path taken to arrive at that state. Thus, we don't need to know the exact value of the system's enthalpy before and after a reaction. We need to know only the difference between final and initial states:

$$\Delta H = H_{\text{final}} - H_{\text{initial}}$$
$$= H_{\text{products}} - H_{\text{reactants}}$$

How large is the difference between $q_v = \Delta E$, the heat flow at constant volume, and $q_p = \Delta H$, the heat flow at constant pressure? Let's look again at the combustion reaction of propane, C_3H_8, with oxygen as an example. When the reaction is carried out in a closed container at constant volume, no PV work is possible so all the energy released is released as heat: $\Delta E = -2046$ kJ. When the same reaction is carried out in an open container at constant pressure, however, only 2044 kJ of heat is released ($\Delta H = -2044$ kJ). The difference, 2 kJ, is due to the small amount of expansion work done against the atmosphere as 6 mol of gaseous reactants are converted into 7 mol of gaseous products.

$$C_3H_8(g) + 5\, O_2(g) \longrightarrow 3\, CO_2(g) + 4\, H_2O(g) \qquad \Delta E = -2046 \text{ kJ}$$
$$\text{Propane}$$
$$\Delta H = -2044 \text{ kJ}$$
$$P\Delta V = +2 \text{ kJ}$$

That is:

$$\begin{bmatrix} q_v = q_p + w \\ \Delta E = \Delta H - P\Delta V \\ -2046 \text{ kJ} = -2044 \text{ kJ} - (+2 \text{ kJ}) \end{bmatrix}$$

Go to eText

BIG IDEA Question 3

For which type of chemical reaction does the change in enthalpy equal the change in energy ($\Delta H = \Delta E$)?

What is true of the reaction of propane and oxygen is also true of most other reactions: The difference between ΔH and ΔE is usually small, so the two quantities are nearly equal. Of course, if no volume change occurs and no work is done, such as in the combustion of methane in which 3 mol of gaseous reactants give 3 mol of gaseous products, then ΔH and ΔE are the same:

$$CH_4(g) + 2\ O_2(g) \longrightarrow CO_2(g) + 2\ H_2O(g) \quad \Delta E = \Delta H = -802\ kJ$$

Although the amount of work is small compared to heat in most chemical reactions such as the combustion of propane, a significant amount of work can be obtained by engineering systems that convert heat into work. In the example of a car's engine, most of the work done on the pistons comes from the expansion of the product gases as a result of their temperature increase from the heat transfer of the reaction.

WORKED EXAMPLE 9.2

Calculating Internal Energy Change (ΔE) for a Reaction

The reaction of nitrogen with hydrogen to make ammonia has $\Delta H = -92.2\ kJ$. What is the value of ΔE in kilojoules if the reaction is carried out at a constant pressure of 40.0 atm and the volume change is $-1.12\ L$?

$$N_2(g) + 3\ H_2(g) \longrightarrow 2\ NH_3(g) \quad \Delta H = -92.2\ kJ$$

IDENTIFY

Known	Unknown
Change in enthalpy ($\Delta H = -92.2\ kJ$)	Change in internal energy (ΔE)
Pressure ($P = 40.0\ atm$)	
Volume Change ($\Delta V = -1.12\ L$)	

STRATEGY

We are given an enthalpy change ΔH, a volume change ΔV, and a pressure P and asked to find an energy change ΔE. Rearrange the equation $\Delta H = \Delta E + P\Delta V$ to the form $\Delta E = \Delta H - P\Delta V$ and substitute the appropriate values for ΔH, P, and ΔV.

SOLUTION

$\Delta E = \Delta H - P\Delta V$

where $\quad \Delta H = -92.2\ kJ$

$\quad P\Delta V = (40.0\ atm)(-1.12\ L) = -44.8\ L \cdot atm$

$$= (-44.8\ L \cdot atm)\left(101\ \frac{J}{L \cdot atm}\right) = -4520\ J = -4.52\ kJ$$

$\Delta E = (-92.2\ kJ) - (-4.52\ kJ) = -87.7\ kJ$

CHECK

The sign of ΔE is similar in size and magnitude ΔH, which is to be expected because energy transfer as work is usually small compared to heat.

▶ **PRACTICE 9.3**   The reaction between hydrogen and oxygen to yield water vapor has $\Delta H = -484\ kJ$. What is the value of ΔE in kilojoules for the reaction of 2.00 mol of H_2 with 1.00 mol of O_2 at atmospheric pressure if the volume change is $-24.4\ L$?

$$2\ H_2(g) + O_2(g) \longrightarrow 2\ H_2O(g) \quad \Delta H° = -484\ kJ$$

▶ **CONCEPTUAL APPLY 9.4**   The following reaction has $\Delta E = -186\ kJ/mol$.

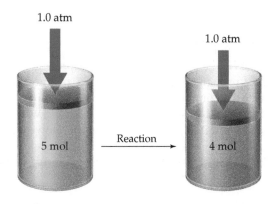

(a) Is the sign of $P\Delta V$ positive or negative? Explain.
(b) What is the sign and approximate magnitude of ΔH? Explain.

9.5 THERMOCHEMICAL EQUATIONS AND THE THERMODYNAMIC STANDARD STATE

A **thermochemical equation** gives a balanced chemical equation along with the value of the enthalpy change (ΔH), the amount of heat released or absorbed when reactants are converted to products. In the combustion of propane the thermochemical equation is:

$$C_3H_8(g) + 5\ O_2(g) \longrightarrow 3\ CO_2(g) + 4\ H_2O(g) \quad \Delta H = -2044\ kJ$$

To ensure that all measurements are reported in the same way so that different reactions can be compared, a set of conditions called the **thermodynamic standard state** has been defined.

> **Thermodynamic standard state**   Most stable form of a substance at 1 atm pressure*
> and at a specified temperature, usually 25 °C; 1 M concentration for all substances
> in solution.

Measurements made under these standard conditions are indicated by addition of the superscript ° to the symbol of the quantity reported. Thus, an enthalpy change measured under standard conditions is called a **standard enthalpy of reaction** and is indicated by the symbol $\Delta H°$. The reaction of propane with oxygen in the thermodynamic standard state is written as:

$$C_3H_8(g) + 5\,O_2(g) \longrightarrow 3\,CO_2(g) + 4\,H_2O(g) \quad \Delta H° = -2044 \text{ kJ } (25 \text{ °C, 1 atm})$$

A thermochemical equation specifies the amount of each substance, and therefore the equation for combustion of propane above means that the reaction of 1 mol of propane gas with 5 mol of oxygen gas to give 3 mol of CO_2 gas and 4 mol of water vapor releases 2044 kJ. The amount of heat released in a specific reaction, however, depends on the amounts of reactants. Thus, reaction of 2.000 mol of propane with 10.00 mol of O_2 releases 2.000×2044 kJ $= 4088$ kJ.

$$2\,C_3H_8(g) + 10\,O_2(g) \longrightarrow 6\,CO_2(g) + 8\,H_2O(g) \quad \Delta H° = -4088 \text{ kJ}$$

It should also be emphasized that $\Delta H°$ values refer to the reaction going in the direction written. For the *reverse reaction, the sign of $\Delta H°$ must be changed.* Because of the reversibility of state functions (Section 9.2), the enthalpy change for a reverse reaction is equal in magnitude but opposite in sign to that for the corresponding forward reaction. Now consider the reaction where gaseous carbon dioxide and water vapor react to form propane and oxygen (the reverse of the combustion of propane). Heat must be absorbed for the reaction to occur and for the formation of one mole of propane the value of $\Delta H° = +2044$ kJ:

$$3\,CO_2(g) + 4\,H_2O(g) \longrightarrow C_3H_8(g) + 5\,O_2(g) \quad \Delta H° = +2044 \text{ kJ}$$
$$C_3H_8(g) + 5\,O_2(g) \longrightarrow 3\,CO_2(g) + 4\,H_2O(g) \quad \Delta H° = -2044 \text{ kJ}$$

Note that the physical states of reactants and products must be specified as solid (s), liquid (l), gaseous (g), or aqueous (aq) when enthalpy changes are reported. The enthalpy change for the reaction of propane with oxygen is $\Delta H° = -2044$ kJ if water is produced as a gas but $\Delta H° = -2220$ kJ if water is produced as a liquid.

$$C_3H_8(g) + 5\,O_2(g) \longrightarrow 3\,CO_2(g) + 4\,H_2O(g) \quad \Delta H° = -2044 \text{ kJ}$$
$$C_3H_8(g) + 5\,O_2(g) \longrightarrow 3\,CO_2(g) + 4\,H_2O(l) \quad \Delta H° = -2220 \text{ kJ}$$

The difference of 176 kJ between the values of $\Delta H°$ for the two reactions arises because the conversion of liquid water to gaseous water requires energy. If liquid water is produced, $\Delta H°$ is larger (more negative), but if gaseous water is produced, $\Delta H°$ is smaller (less negative) because 44.0 kJ/mol is needed for the vaporization.

$$H_2O(l) \longrightarrow H_2O(g) \quad \Delta H° = 44.0 \text{ kJ}$$

or $\quad 4\,H_2O(l) \longrightarrow 4\,H_2O(g) \quad \Delta H° = 176 \text{ kJ}$

Worked Example 9.3 illustrates how to calculate the amount of heat transfer for a given reaction given the amounts of reactants and direction of reaction.

BIG IDEA Question 4

Given the enthalpy change for the reaction:

$$H_2(g) + \frac{1}{2}O_2(g) \longrightarrow H_2O(g)$$

$$\Delta H = -241.8 \text{ kJ}$$

what is the enthalpy change for the reaction below?

$$2\,H_2O(g) \longrightarrow 2\,H_2(g) + O_2(g)$$

$$\Delta H = ? \text{ kJ}$$

---

*The standard pressure, listed here and in most other books as 1 atmosphere (atm), has been redefined to be 1 bar, which is equal to 0.986 923 atm. The difference is small, however.

WORKED EXAMPLE 9.3

Calculating the Amount of Heat Released in a Reaction

How much heat in kilojoules is evolved when 5.00 g of propane reacts with excess O_2?

$$C_3H_8(g) + 5\ O_2(g) \longrightarrow 3\ CO_2(g) + 4\ H_2O(g) \quad \Delta H° = -2044 \text{ kJ}$$

IDENTIFY

Known	Unknown
Amount of propane (5.00 g)	Heat transfer (kJ)
Balanced thermochemical equation	

STRATEGY

According to the balanced equation, 2044 kJ of heat is evolved from the reaction of 1 mol of propane. To find out how much heat is evolved from the reaction of 5.00 g of propane, we have to convert from grams to moles and then use the thermochemical equation to relate quantity to heat.

SOLUTION

The molar mass of propane (C_3H_8) is 44.09 g/mol, so 5.00 g of C_3H_8 equals 0.113 mol:

$$5.00 \text{ g } C_3H_8 \times \frac{1 \text{ mol } C_3H_8}{44.09 \text{ g } C_3H_8} = 0.1134 \text{ mol } C_3H_8$$

Because 1 mol of C_3H_8 releases 2044 kJ of heat, 0.1134 mol of C_3H_8 releases 231 kJ of heat:

$$0.1134 \text{ mol } C_3H_8 \times \frac{2044 \text{ kJ}}{1 \text{ mol } C_3H_8} = 232 \text{ kJ}$$

CHECK

Since the molar mass of C_3H_8 is about 44 g/mol, 5 g of C_3H_8 is roughly 0.1 mol. Therefore, the heat evolved is roughly 1/10 of the amount released for 1 mol of C_3H_8, and the estimate yields an answer of 204 kJ, which is reasonably close to 232 kJ.

▶ **PRACTICE 9.5** Use the following thermochemical equation to calculate how much heat in kilojoules is evolved or absorbed when 10.0 g of liquid water are converted to hydrogen and oxygen gas.

$$2\ H_2(g) + O_2(g) \longrightarrow 2\ H_2O(l) \quad \Delta H° = -571.6 \text{ kJ}$$

▶ **APPLY 9.6** Approximately, 1.8×10^6 kJ of energy is required to heat an average home in the Midwest during the month of January. How much propane (kg) must be burned to provide 1.8×10^6 kJ of energy?

$$C_3H_8(g) + 5\ O_2(g) \longrightarrow 3\ CO_2(g) + 4\ H_2O(g) \quad \Delta H° = -2044 \text{ kJ}$$

9.6 ENTHALPIES OF CHEMICAL AND PHYSICAL CHANGES

Almost every change in a system involves either a gain or a loss of enthalpy. The change can be either physical, such as the melting of a solid to a liquid, or chemical, such as the burning of propane. Let's look at examples of both kinds.

Enthalpies of Chemical Change

Enthalpy change is often called a *heat of reaction* because it is a measure of the heat flow into or out of a system at constant pressure. If the products of a reaction have more enthalpy than the reactants, then heat has flowed into the system from the surroundings and ΔH has a positive sign. Reactions in which the system gains heat are said to be **endothermic** (*endo* means "within," so heat flows in). The reaction of 1 mol

▲ A balloon filled with hydrogen burns in air to release energy to the surroundings.

of barium hydroxide octahydrate* with ammonium chloride, for example, absorbs 80.3 kJ from the surroundings ($\Delta H° = +80.3$ kJ). The surroundings, having lost heat, become cold—so cold, in fact, that the temperature drops below freezing (**FIGURE 9.7**).

$$Ba(OH)_2 \cdot 8\, H_2O(s) + 2\, NH_4Cl(s) \longrightarrow$$
$$BaCl_2(aq) + 2\, NH_3(aq) + 10\, H_2O(l) \quad \Delta H° = +80.3 \text{ kJ}$$

If the products of a reaction have less enthalpy than the reactants, then heat has flowed out of the system to the surroundings and ΔH has a negative sign. Reactions that lose heat to the surroundings are said to be **exothermic** (*exo* means "out," so heat flows out). The thermite reaction of aluminum with iron(III) oxide, for instance, releases so much heat ($\Delta H° = -852$ kJ), and the surroundings get so hot, that the reaction is used in construction work to weld iron.

$$2\, Al(s) + Fe_2O_3(s) \longrightarrow 2\, Fe(s) + Al_2O_3(s) \quad \Delta H° = -852 \text{ kJ}$$

As noted previously, the value of $\Delta H°$ given for an equation assumes that the equation is balanced to represent the numbers of moles of reactants and products, that all substances are in their standard states, and that the physical state of each substance is as specified.

Enthalpies of Physical Change

What would happen if you started with a block of ice at a low temperature, say -10 °C, and slowly increased its enthalpy by adding heat? The initial input of heat would cause the temperature of the ice to rise until it reached 0 °C. Additional heat would then cause the ice to melt without raising its temperature as the added energy is expended in overcoming the intermolecular forces that hold H_2O molecules together in the ice crystal. The amount of heat necessary to melt a substance without changing its temperature is called the *enthalpy of fusion,* or *heat of fusion* (ΔH_{fusion}). For H_2O, $\Delta H_{fusion} = 6.01$ kJ/mol at 0 °C and the process is endothermic because heat is absorbed. The thermochemical equation representing the heat of fusion of water is:

$$H_2O(s) \longrightarrow H_2O(l) \quad \Delta H_{fusion} = +6.01 \text{ kJ/mol}$$

Once the ice has melted, further input of heat raises the temperature of the liquid water until it reaches 100 °C, and adding still more heat then causes the water to boil. Once again, energy is necessary to overcome the intermolecular forces holding molecules together in the liquid, so the temperature does not rise again until all the liquid has been converted into vapor. The amount of heat required to vaporize a substance without changing its temperature is called the *enthalpy of vaporization,* or *heat of vaporization* (ΔH_{vap}). For H_2O, $\Delta H_{vap} = 40.7$ kJ/mol at 100 °C, and the process is endothermic. The thermochemical equation representing the heat of vaporization of water is:

$$H_2O(l) \longrightarrow H_2O(g) \quad \Delta H_{vap} = +40.7 \text{ kJ/mol}$$

Another kind of physical change in addition to melting and boiling is **sublimation**, the direct conversion of a solid to a vapor without going through a liquid state. Solid CO_2 (dry ice), for example, changes directly from solid to vapor at atmospheric pressure without first melting to a liquid. Since enthalpy is a state function, the enthalpy change on going from solid to vapor must be constant regardless of the path taken. Thus, at a given temperature, a substance's *enthalpy of sublimation,* or *heat of sublimation* (ΔH_{subl}), equals the sum of the heat of fusion and the heat of vaporization (**FIGURE 9.8**).

▲ **FIGURE 9.7**
The endothermic reaction of barium hydroxide octahydrate with ammonium chloride. The reaction draws so much heat from the surroundings that the temperature falls below 0 °C.

▲ **Figure It Out**
Identify the system and the surroundings in the photo of the endothermic reaction.

The surroundings are the solution, beaker, tabletop, temperature probe, and air in the room.

$$BaCl_2(aq) + 2\, NH_3(aq) + 10\, H_2O(l)$$
$$Ba(OH)_2 \cdot 8\, H_2O(s) + 2\, NH_4Cl(s) \longrightarrow$$

Answer: The system is the reaction.

*Barium hydroxide octahydrate, $Ba(OH)_2 \cdot 8\, H_2O$, is a crystalline compound that contains eight water molecules clustered around the barium ion.

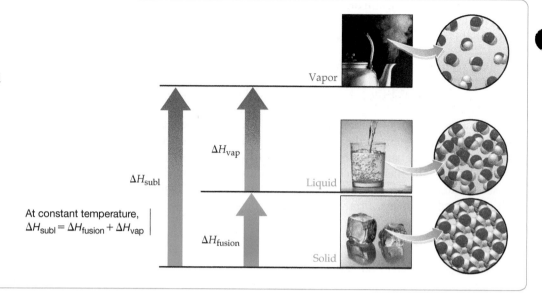

▶ **FIGURE 9.8**

Enthalpy as a state function.
Because enthalpy is a state function, the value of the enthalpy change from solid to vapor does not depend on the path taken between the two states.

Vapor

ΔH_{vap}

ΔH_{subl}

Liquid

At constant temperature,
$\Delta H_{subl} = \Delta H_{fusion} + \Delta H_{vap}$

ΔH_{fusion}

Solid

WORKED EXAMPLE 9.4

Classifying Endo- and Exothermic Processes

Indicate the direction of heat transfer between the system and the surroundings, classify the following processes as endo- or exothermic, and give the sign of $\Delta H°$.
(a) Water freezing on the surface of a lake in the winter.
(b) Sweat evaporating from your skin.
(c) Dissolving the salt magnesium sulfate in water to produce a hand warmer.

STRATEGY

Write an equation for each process and evaluate whether heat is absorbed or released based on observed temperature changes.

SOLUTION

(a) It is difficult to observe the temperature change in the surroundings as water freezes. However, it is intuitive that heat must be absorbed from the surroundings to melt solid ice into liquid water. In the reverse process of freezing, $H_2O(l) \longrightarrow H_2O(s)$, the system releases heat to the surroundings. Therefore, the process is exothermic and $\Delta H°$ is negative.

(b) When sweat evaporates, liquid water is turned into gaseous water, $H_2O(l) \longrightarrow H_2O(g)$, and the system absorbs heat from the surroundings. Sweating cools you down because the phase change absorbs heat from your body (part of the surroundings). Therefore, the sign $\Delta H°$ is positive and the process is endothermic.

(c) In a hand warmer, the temperature of the solution increases because the system, $MgSO_4(s) \longrightarrow Mg^{2+}(aq) + SO_4^{2-}(aq)$, transfers heat to the surroundings. Therefore, the process is exothermic and $\Delta H°$ is negative.

▶ **PRACTICE 9.7** Classify the following reaction as endo- or exothermic, and give the sign of $\Delta H°$.

250 mL of 1.0 M NaOH was added to 250 mL of 1.0 M HCl, and the temperature of the solution increased from 23.4 °C to 30.4 °C.

▶ **APPLY 9.8** Instant hot packs and cold packs contain a solid salt in a capsule that is broken and dissolved in water to produce a temperature change. Which of the following salts would result in greatest decrease in temperature when dissolved in 100.0 mL of water in a cold pack? Use the following thermochemical data to answer the question.

$LiF(s) \longrightarrow Li^+(aq) + F^-(aq)$ $\Delta H° = +5.5$ kJ

$LiCl(s) \longrightarrow Li^+(aq) + Cl^-(aq)$ $\Delta H° = -37.1$ kJ

$NH_4NO_3(s) \longrightarrow NH_4^+(aq) + NO_3^-(aq)$ $\Delta H° = +25.7$ kJ

$CaCl_2(s) \longrightarrow Ca^{2+}(aq) + 2Cl^-(aq)$ $\Delta H° = -81.8$ kJ

(a) 10.0 g of LiF   (b) 10.0 g of LiCl
(c) 10.0 g of NH₄NO₃   (d) 10.0 g of CaCl₂

▲ Instant cold packs are used to treat athletic injuries.

9.7 CALORIMETRY AND HEAT CAPACITY

The amount of heat transferred during a reaction can be measured with a device called a *calorimeter*, shown schematically in **FIGURE 9.9**. At its simplest, a calorimeter is just an insulated vessel with a stirrer, a thermometer, and a loose-fitting lid to keep the contents at atmospheric pressure. The reaction is carried out inside the vessel, and the heat evolved or absorbed is calculated from the temperature change. Because the pressure inside the calorimeter is constant (atmospheric pressure), the temperature measurement makes it possible to calculate the enthalpy change ΔH during a reaction. This type of calorimeter is used to measure the enthalpy change for the reaction of a salt dissolving in water such as the reaction that occurs in hot packs and cold packs.

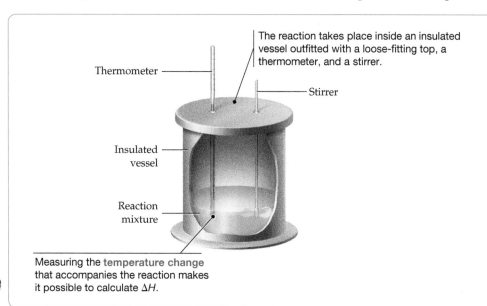

The reaction takes place inside an insulated vessel outfitted with a loose-fitting top, a thermometer, and a stirrer.

Thermometer

Stirrer

Insulated vessel

Reaction mixture

Measuring the **temperature change** that accompanies the reaction makes it possible to calculate ΔH.

◀ **FIGURE 9.9**

A calorimeter for measuring the heat flow in a reaction at constant pressure (ΔH).

 Go to eText

BIG IDEA Question 5

A reaction in aqueous solution is carried out in the calorimeter shown in Figure 9.9, and the temperature of the solution decreases. What is the sign of ΔH for the reaction?

A device called a *bomb calorimeter* is used to measure the heat released during a combustion reaction, or burning of a flammable substance. (More generally, a *combustion* reaction is any reaction that produces a flame.) The sample is placed in a small cup and sealed in an oxygen atmosphere inside a steel bomb that is itself placed in an insulated, water-filled container (**FIGURE 9.10**). The reactants are ignited electrically, and the evolved heat is calculated from the temperature change of the surrounding water. Since the reaction takes place at constant volume rather than constant pressure, the measurement provides a value for ΔE rather than ΔH. The caloric content of food, such as a potato chip, can be measured by burning it in a bomb calorimeter.

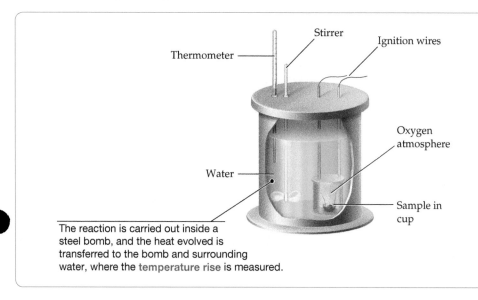

Stirrer

Ignition wires

Thermometer

Oxygen atmosphere

Water

Sample in cup

The reaction is carried out inside a steel bomb, and the heat evolved is transferred to the bomb and surrounding water, where the **temperature rise** is measured.

◀ **FIGURE 9.10**

A bomb calorimeter for measuring the heat evolved at constant volume (ΔE) in a combustion reaction.

How can the temperature change inside a calorimeter be used to calculate ΔH (or ΔE) for a reaction? When a calorimeter and its contents absorb a given amount of heat, the temperature rise that results depends on the calorimeter's *heat capacity*. **Heat capacity (C)** is the amount of heat required to raise the temperature of an object or substance by a given amount, a relationship that can be expressed by the equation

$$C = \frac{q}{\Delta T}$$

where q is the quantity of heat transferred and ΔT is the temperature change that results ($\Delta T = T_{\text{final}} - T_{\text{initial}}$).

The greater the heat capacity, the greater the amount of heat needed to produce a given temperature change. A bathtub full of water, for instance, has a greater heat capacity than a cup full of water, and it therefore takes far more heat to warm the tubful than the cupful. The exact amount of heat absorbed is equal to the heat capacity times the temperature rise:

$$q = C \times \Delta T$$

REMEMBER . . .
Extensive properties, such as length and volume, have values that depend on the sample size. Intensive properties, such as temperature and melting point, have values that do not depend on the amount of the sample (Section 2.3).

Heat capacity is an **extensive property**, so its value depends on both the size of an object and its composition. To compare different substances, it's useful to define a quantity called the *specific heat capacity*, or simply **specific heat (c)**, the amount of heat necessary to raise the temperature of 1 g of a substance by 1 °C. The amount of heat necessary to raise the temperature of a given object is the specific heat times the mass of the object times the rise in temperature:

$$q = \text{Specific heat } (c) \times \text{Mass of substance } (m) \times \Delta T$$

Worked Example 9.5 shows how specific heats (c) and heat capacities (C) are determined.

Closely related to specific heat is the **molar heat capacity (C_m)**, defined as the amount of heat necessary to raise the temperature of 1 mol of a substance by 1 °C. The amount of heat necessary to raise the temperature of a given number of moles of a substance is thus

$$q = C_m \times \text{Moles of a substance } (n) \times \Delta T$$

Values of specific heats and molar heat capacities for some common substances are given in **TABLE 9.1**. The values are temperature dependent, so the temperatures at which the measurements are taken must be specified.

As indicated in Table 9.1, the specific heat of liquid water is considerably higher than that of most other substances, so a large transfer of heat is necessary to either cool or warm a given amount of water. One consequence is that large lakes or other bodies of water tend to moderate the air temperature in surrounding areas. Another consequence is that the human body, which is about 60% water, is able to maintain a relatively steady internal temperature under changing outside conditions.

▲ Lake Chelan in the North Cascades of Washington State is the third deepest freshwater lake in the United States at 1486 ft. Such large masses of water moderate the temperature of the surroundings because of their high heat capacity.

TABLE 9.1   Specific Heats and Molar Heat Capacities for Some Common Substances at 25 °C

Substance	Specific Heat (c) J/(g · °C)	Molar Heat Capacity (C_m) J/(mol · °C)
Air (dry)	1.01	29.1
Aluminum	0.897	24.2
Copper	0.385	24.4
Gold	0.129	25.4
Iron	0.449	25.1
Mercury	0.140	28.0
NaCl	0.859	50.2
Water(s)[a]	2.03	36.6
Water(l)	4.179	75.3

[a]At −11 °C

── WORKED EXAMPLE 9.5

Calculating Specific Heat (*C*) and Molar Heat Capacity (*C_M*)

What is the specific heat of silicon in $J/(g \cdot °C)$ if it takes 192 J to raise the temperature of a 45.0 g block by 6.0 °C? What is the molar heat capacity of silicon in $J/(mol \cdot °C)$?

IDENTIFY

Known	Unknown
Heat transfer ($q = 192$ J)	Specific heat (c)
Mass of Si ($m = 45.0$ g)	Molar heat capacity (C_m)
Change in temperature ($\Delta T = 6.0$ °C)	

STRATEGY

To find a specific heat of a substance, calculate the amount of energy necessary to raise the temperature of 1 g of the substance by 1 °C. To calculate molar heat capacity, use the molar mass of silicon to convert from grams to moles in specific heat capacity.

SOLUTION

$$\text{Specific heat } (c) \text{ of Si} = \frac{192 \text{ J}}{(45.0 \text{ g})(6.0 °C)} = 0.71 \text{ J}/(g \cdot °C)$$

$$\text{Molar heat capacity}(C_m): \frac{0.71 \text{ J}}{(g \cdot °C)} \times \frac{28.1 \text{ g Si}}{1 \text{ mol Si}} = \frac{20.0 \text{ J}}{(\text{mol} \cdot °C)}$$

CHECK

Table 9.1 lists specific and molar heat capacities for several substances. The values calculated are within the range given in the table and are therefore reasonable numbers.

▶ **PRACTICE 9.9**   What is the specific heat of lead in $J/(g \cdot °C)$ if it takes 97.2 J to raise the temperature of a 75.0 g block by 10.0 °C? What is the molar heat capacity of lead in $J/(mol \cdot °C)$?

▶ **APPLY 9.10**   Calculate the heat capacity (*C*) of a bomb calorimeter if 1650 J raised the temperature of the entire calorimeter by 2.00 °C.

Go to
eText

── WORKED EXAMPLE 9.6

Calculating Δ*H* in a Constant-Pressure Calorimetry Experiment

Aqueous silver ion reacts with aqueous chloride ion to yield a white precipitate of solid silver chloride:

$$Ag^+(aq) + Cl^-(aq) \longrightarrow AgCl(s)$$

When 10.0 mL of 1.00 *M* $AgNO_3$ solution is added to 10.0 mL of 1.00 M NaCl solution at 25.0 °C in a calorimeter, a white precipitate of AgCl forms and the temperature of the aqueous mixture increases to 32.6 °C. Assuming that the specific heat of the aqueous mixture is 4.18 $J/(g \cdot °C)$, that the density of the mixture is 1.00 g/mL, and that the calorimeter itself absorbs a negligible amount of heat, calculate Δ*H* in kilojoules/mol AgCl for the reaction.

IDENTIFY

Known	Unknown
Volume of $AgNO_3$ solution (10.0 mL)	Enthalpy change for reaction (Δ*H*)
Concentration of $AgNO_3$ solution (1.00 M)	
Volume of NaCl solution (10.0 mL)	
Concentration of NaCl solution (1.00 M)	
Initial temperature (25.0 °C)	
Final temperature (32.6 °C)	
Specific heat of mixture ($c = 4.18$ J/(g·°C))	
Density of mixture ($d = 1.00$ g/mL)	

continued on next page

STRATEGY

In a calorimetry experiment, the heat transferred to the surroundings is equal in magnitude but opposite in sign to the heat transferred by the reaction: $(q_{\text{reaction}} = -q_{\text{surroundings}})$. The solution is the surroundings and the specific heat, mass, and temperature change for the solution are given. Dividing the heat transfer of the reaction (q_{reaction}) by the number of moles gives the enthalpy change (ΔH) for the reaction.

SOLUTION

Step 1. Calculate the heat transfer for the reaction (q_{reaction}).

$$q_{\text{reaction}} = -q_{\text{surroundings}}$$
$$q_{\text{reaction}} = -q_{\text{solution}}$$

The specific heat, mass, and temperature change for the solution are given and can be used to calculate q_{reaction}.

$$q_{\text{reaction}} = -(c \times m \times \Delta T)_{\text{solution}}$$

$$q_{\text{reaction}} = -\left[\left(\frac{4.18 \text{ J}}{g \cdot {}^\circ C}\right)(20.0 \text{ g})(7.6 \,{}^\circ C)\right] = -635.4 \text{ J}$$

Step 2. Dividing the heat transfer of the reaction by the number of moles gives the enthalpy change (ΔH) for the reaction. According to the balanced equation, the number of moles of AgCl produced equals the number of moles of Ag^+ (or Cl^-) reacted:

$$\text{Moles of } Ag^+ = (10.0 \text{ mL})\left(\frac{1 \text{ L}}{1000 \text{ mL}}\right)\left(\frac{1.00 \text{ mol } Ag^+}{1 \text{ mL}}\right) = 1.00 \times 10^{-2} \text{ mol } Ag^+$$

$$\text{Moles of AgCl} = 1.00 \times 10^{-2} \text{ mol AgCl}$$

$$\Delta H = \frac{-635.4 \text{ J}}{1.00 \times 10^{-2} \text{ mol AgCl}} \times \frac{1 \text{ kJ}}{1000 \text{ J}} = -63.5 \text{ kJ/mol AgCl}$$

Therefore, $\Delta H = -63.5$ kJ (negative because heat is released by the reaction to warm the solution).

▶ **PRACTICE 9.11**   When 25.0 mL of 1.00 M H_2SO_4 is added to 50.0 mL of 1.00 M NaOH at 25.0 °C in a calorimeter, the temperature of the aqueous solution increases to 33.9 °C. Assuming that the specific heat of the solution is 4.18 J/(g • °C), that its density is 1.00 g/mL, and that the calorimeter itself absorbs a negligible amount of heat, calculate ΔH in kilojoules/mol H_2SO_4 for the reaction.

$$H_2SO_4(aq) + 2 \text{ NaOH}(aq) \longrightarrow 2 \text{ } H_2O(l) + Na_2SO_4(aq)$$

▶ **APPLY 9.12**   When 15.0 g of NH_4NO_3 is dissolved in 150.0 mL of water, the temperature of the liquid decreases from 24.5 °C to 17.5 °C. (Assume the specific heat capacity of the solution is 4.18 J/g • °C and no heat is transferred to the calorimeter.) Calculate ΔH in kJ/mol NH_4NO_3 for the reaction.

$$NH_4NO_3(s) \longrightarrow NH_4^+(aq) + NO_3^-(aq) \quad \Delta H = ?$$

All **WORKED EXAMPLES** with this icon [Go to eText] have an interactive video in the eText.

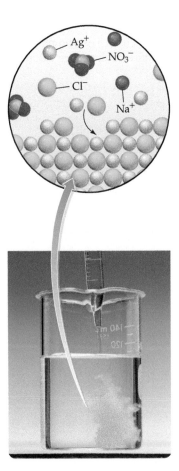

▲ The reaction of aqueous $AgNO_3$ with aqueous NaCl to yield solid AgCl is an exothermic process.

— **WORKED EXAMPLE 9.7**

Calculating ΔE in Constant-Volume Calorimetry

When 2.00 g of glucose, $C_6H_{12}O_6$, is burned in a bomb calorimeter, the temperature rises by 18.5 °C. If the heat capacity for the calorimeter is 1675 J/°C, calculate the combustion energy (ΔE) in kJ per mole of glucose.

$$C_6H_{12}O_6(s) + 6 \text{ } O_2(g) \longrightarrow 6 \text{ } CO_2(g) + 6 \text{ } H_2O(l) \quad \Delta E = ?$$

IDENTIFY

Known	Unknown
2.00 g of $C_6H_{12}O_6$	ΔE (kJ)
$\Delta T = 18.5\ °C$	
Heat capacity (1675 J/°C)	

STRATEGY

In a calorimetry experiment, the heat transferred to the surroundings is equal in magnitude but opposite in sign to the heat transferred by the reaction: $(q_{reaction} = -q_{surroundings})$. The calorimeter (steel bomb and water) is the surroundings, and the amount of heat transfer is calculated using the equation $q = C \times \Delta T$. Dividing the heat transfer of the reaction $(q_{reaction})$ by the number of moles gives the energy change (ΔE) for the reaction.

SOLUTION

Step 1. Calculate the heat transfer for the reaction $(q_{reaction})$.

$$q_{reaction} = -q_{surroundings}$$

$$q_{reaction} = -q_{calorimeter}$$

$$q_{reaction} = -(C \times \Delta T)$$

$$q_{reaction} = -\left[\left(\frac{1675\ J}{°C}\right)(18.6\ °C)\right] = -3.116 \times 10^4\ J$$

Step 2. Dividing the heat transfer of the reaction by the number of moles gives the energy change (ΔE) for the reaction.

$$\frac{-3.116 \times 10^4\ J}{2.00\ g\ C_6H_{12}O_6} \times \frac{180.1\ g\ C_6H_{12}O_6}{1\ mol\ C_6H_{12}O_6} \times \frac{1\ kJ}{1000\ J} = \frac{-2.81 \times 10^3\ kJ}{mol\ C_6H_{12}O_6}$$

Since the combustion reaction is written for 1 mol of glucose, $\Delta E = -2.81 \times 10^3$ kJ.

CHECK

The combustion reaction releases energy and warms the surrounding, so the sign of ΔE is negative. Many people who keep track of calories for a healthy diet know that carbohydrates contain about 4 Cal/g. Glucose is a carbohydrate, so we can convert from kJ/mol to Cal/g to check the validity of the answer.

$$\frac{2.81 \times 10^3\ kJ}{mol\ C_6H_{12}O_6} \times \frac{1\ mol\ C_6H_{12}O_6}{180.1\ g\ C_6H_{12}O_6} \times \frac{1\ Cal}{4.18\ kJ} = 3.73\ Cal/g \approx 4\ Cal/g$$

▶ **PRACTICE 9.13** When 1.00 g of toluene, C_7H_8, is burned in a bomb calorimeter, the temperature rises by 13.0 °C. If the heat capacity of the calorimeter is 3.21 kJ/°C, calculate the combustion energy (ΔE) in kJ per mole of toluene.

▶ **APPLY 9.14** The combustion of octane has $\Delta E = -5468$ kJ/mol C_8H_{18}. When 1.00 g of octane, C_8H_{18}, is burned in a bomb calorimeter, the temperature rises from 25.67 °C to 37.15 °C. Calculate the heat capacity of the bomb calorimeter in kJ/°C.

9.8 HESS'S LAW

Now that we've discussed in general terms the energy changes that occur during chemical reactions, let's look at a specific example in detail. In particular, let's look at the *Haber process*, the industrial method by which more than 120 million metric tons of ammonia is produced each year worldwide, primarily for use as fertilizer (1 metric ton = 1000 kg). The reaction of hydrogen with nitrogen to make ammonia is exothermic, with $\Delta H° = -92.2$ kJ.

$$3\ H_2(g) + N_2(g) \longrightarrow 2\ NH_3(g) \quad \Delta H° = -92.2\ kJ$$

If we dig into the details of the reaction, we find that it's not as simple as it looks. In fact, the overall reaction occurs in a series of steps, with hydrazine (N_2H_4) produced at an intermediate stage:

$$2 H_2(g) + N_2(g) \longrightarrow N_2H_4(g) \xrightarrow{H_2} 2 NH_3(g)$$
$$\text{Hydrazine} \qquad \text{Ammonia}$$

The enthalpy change for the conversion of hydrazine to ammonia can be measured as $\Delta H° = -187.6$ kJ, but if we wanted to measure $\Delta H°$ for the formation of hydrazine from hydrogen and nitrogen, we would have difficulty because the reaction doesn't go to completion. Some of the hydrazine is converted into ammonia and some of the starting nitrogen remains.

Fortunately, there's a way around the difficulty—a way that makes it possible to measure an enthalpy change indirectly when a direct measurement can't be made. The trick is to realize that, because enthalpy is a state function, ΔH is the same no matter what path is taken between two states. Thus, the sum of the enthalpy changes for the individual steps in a sequence must equal the enthalpy change for the overall reaction, a statement known as **Hess's law.**

Hess's law   The overall enthalpy change for a reaction is equal to the sum of the enthalpy changes for the individual steps in the reaction.

Reactants and products in the individual steps can be added and subtracted like algebraic quantities in determining the overall equation. In the synthesis of ammonia, for example, the sum of steps 1 and 2 is equal to the overall reaction. Thus, the sum of the enthalpy changes for steps 1 and 2 is equal to the enthalpy change for the overall reaction. With this knowledge, we can calculate the enthalpy change for step 1. **FIGURE 9.11** shows the situation pictorially, and Worked Example 9.8 gives an example of Hess's law calculations.

Step 1. $2 H_2(g) + N_2(g) \longrightarrow N_2H_4(g)$ $\Delta H°_1 = ?$

Step 2. $N_2H_4(g) + H_2(g) \longrightarrow 2 NH_3(g)$ $\Delta H°_2 = -187.6$ kJ

Overall reaction $3 H_2(g) + N_2(g) \longrightarrow 2 NH_3(g)$ $\Delta H°_{reaction} = -92.2$ kJ

Since $\Delta H°_1 + \Delta H°_2 = \Delta H°_{reaction}$

then $\Delta H°_1 = \Delta H°_{reaction} - \Delta H°_2$

$$= (-92.2 \text{ kJ}) - (-187.6 \text{ kJ}) = +95.4 \text{ kJ}$$

▶ **FIGURE 9.11**

Enthalpy changes for steps in the synthesis of ammonia from nitrogen and hydrogen. If $\Delta H°$ values for step 2 and for the overall reaction are known, then ΔH for step 1 can be calculated.

▶ **Figure It Out**

In a different reaction that occurs by two steps, what is $\Delta H°$ for step 1 if $\Delta H°$ for step 2 is +25.0 kJ and $\Delta H°$ for the overall reaction is −25.0 kJ?

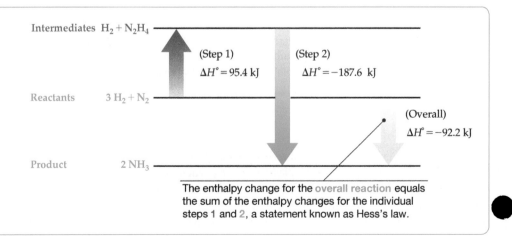

Intermediates $H_2 + N_2H_4$

(Step 1)   (Step 2)
$\Delta H° = 95.4$ kJ $\Delta H° = -187.6$ kJ

Reactants $3 H_2 + N_2$

(Overall)
$\Delta H° = -92.2$ kJ

Product $2 NH_3$

The enthalpy change for the overall reaction equals the sum of the enthalpy changes for the individual steps 1 and 2, a statement known as Hess's law.

WORKED EXAMPLE 9.8

Using Hess's Law to Calculate $\Delta H°$

Methane, the main constituent of natural gas, burns in oxygen to yield carbon dioxide and water:

$$CH_4(g) + 2\,O_2(g) \longrightarrow CO_2(g) + 2\,H_2O(l)$$

Use the following information to calculate $\Delta H°$ in kilojoules for the combustion of methane:

$CH_4(g) + O_2(g) \longrightarrow CH_2O(g) + H_2O(g)$	$\Delta H° = -275.6$ kJ
$CH_2O(g) + O_2(g) \longrightarrow CO_2(g) + H_2O(g)$	$\Delta H° = -526.7$ kJ
$H_2O(l) \longrightarrow H_2O(g)$	$\Delta H° = 44.0$ kJ

STRATEGY

It often takes some trial and error, but the idea is to combine the individual reactions so that their sum is the desired reaction. The important points are that:

- All the reactants $[CH_4(g)$ and $O_2(g)]$ must appear on the left.
- All the products $[CO_2(g)$ and $H_2O(l)]$ must appear on the right.
- All intermediate products $[CO_2(g)$ and $H_2O(g)]$ must occur on *both* the left and the right so that they cancel.
- A reaction written in the reverse of the direction given $[H_2O(g) \longrightarrow H_2O(l)]$ must have the sign of its $\Delta H°$ reversed (Section 9.5).
- If a reaction is multiplied by a coefficient $[H_2O(g) \longrightarrow H_2O(l)$ is multiplied by 2], then $\Delta H°$ for the reaction must be multiplied by that same coefficient (Section 9.5).

SOLUTION

$CH_4(g) + O_2(g) \longrightarrow \cancel{CH_2O(g)} + \cancel{H_2O(g)}$	$\Delta H° = -275.6$ kJ
$\cancel{CH_2O(g)} + O_2(g) \longrightarrow CO_2(g) + \cancel{H_2O(g)}$	$\Delta H° = -526.7$ kJ
$2\,[\cancel{H_2O(g)} \longrightarrow H_2O(l)]\ \ 2\,[\Delta H° = -44.0\text{ kJ}] =$	-88.0 kJ
$CH_4(g) + 2\,O_2(g) \longrightarrow CO_2(g) + 2\,H_2O(l)$	$\Delta H° = -890.3$ kJ

▶ **PRACTICE 9.15**   *Water gas* is the name for the mixture of CO and H_2 prepared by reaction of steam with carbon at 1000 °C:

$$C(s) + H_2O(g) \longrightarrow \underset{\text{"Water gas"}}{CO(g) + H_2(g)}$$

The hydrogen is then purified and used as a starting material for preparing ammonia. Use the following information to calculate $\Delta H°$ in kilojoules for the water–gas reaction:

$C(s) + O_2(g) \longrightarrow$	$CO_2(g)$	$\Delta H° = -393.5$ kJ
$2\,CO(g) + O_2(g) \longrightarrow$	$2\,CO_2(g)$	$\Delta H° = -566.0$ kJ
$2\,H_2(g) + O_2(g) \longrightarrow$	$2\,H_2O(g)$	$\Delta H° = -483.6$ kJ

▶ **CONCEPTUAL APPLY 9.16**   The reaction of A with B to give D proceeds in two steps and can be represented by the following Hess's law diagram.

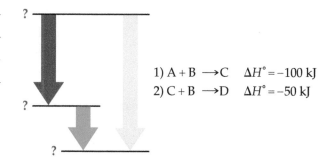

1) $A + B \longrightarrow C$ $\Delta H° = -100$ kJ
2) $C + B \longrightarrow D$ $\Delta H° = -50$ kJ

(a) What is the equation and $\Delta H°$ for the net reaction?
(b) Which arrow on the diagram corresponds to which step, and which arrow corresponds to the net reaction?
(c) The diagram shows three energy levels. Which substances are represented by each energy level?

9.9 STANDARD HEATS OF FORMATION

Where do the $\Delta H°$ values we've been using in previous sections come from? There are so many chemical reactions—hundreds of millions are known—that it's impossible to measure $\Delta H°$ for all of them. A more efficient method is to measure **standard heats of formation** $(\Delta H°_f)$ for different substances and use those values to calculate $\Delta H°$ for any reaction that involves those substances.

> **Standard heat of formation** The enthalpy change $(\Delta H°_f)$ for the formation of 1 mol of a substance in its standard state from its constituent elements in their standard states.

Note several points about this definition. First, the "reaction" to form a substance from its constituent elements can be (and often is) hypothetical. We can't combine carbon and hydrogen in the laboratory to make methane, for instance, yet the heat of formation for methane is $\Delta H°_f = -74.8$ kJ/mol, which corresponds to the standard enthalpy change for the hypothetical reaction

$$C(s) + 2 H_2(g) \longrightarrow CH_4(g) \quad \Delta H° = -74.8 \text{ kJ}$$

Second, each substance in the reaction must be in its most stable, standard form at 1 atm pressure and the specified temperature (usually 25 °C). Carbon, for instance, is most stable as solid graphite rather than as diamond under these conditions, and hydrogen is most stable as gaseous H_2 molecules rather than as H atoms. **TABLE 9.2** gives standard heats of formation for some common substances, and Appendix B gives a more detailed list.

No elements are listed in Table 9.2 because, by definition, the most stable form of any element, such as Ar(g) or Au(s), in its standard state has $\Delta H°_f = 0$ kJ. The enthalpy change for the formation of an element from itself is zero. Defining $\Delta H°_f$ as zero for all elements thus establishes a thermochemical "sea level," or reference point, from which other enthalpy changes are measured.

TABLE 9.2 Standard Heats of Formation for Some Common Substances at 25 °C

Substance	Formula	$\Delta H°_f$ (kJ/mol)	Substance	Formula	$\Delta H°_f$ (kJ/mol)
Acetylene	$C_2H_2(g)$	227.4	Hydrogen chloride	$HCl(g)$	-92.3
Ammonia	$NH_3(g)$	-46.1	Iron(III) oxide	$Fe_2O_3(s)$	-824.2
Carbon dioxide	$CO_2(g)$	-393.5	Magnesium carbonate	$MgCO_3(s)$	-1095.8
Carbon monoxide	$CO(g)$	-110.5	Methane	$CH_4(g)$	-74.8
Ethanol	$C_2H_5OH(l)$	-277.7	Nitric oxide	$NO(g)$	91.3
Ethylene	$C_2H_4(g)$	52.3	Water (g)	$H_2O(g)$	-241.8
Glucose	$C_6H_{12}O_6(s)$	-1273.3	Water (l)	$H_2O(l)$	-285.8

How can standard heats of formation be used for thermochemical calculations? The standard enthalpy change for any chemical reaction is found by subtracting the sum of the heats of formation of all reactants from the sum of the heats of formation of all products, with each heat of formation multiplied by the coefficient of that substance in the balanced equation.

$$\Delta H°_{\text{reaction}} = \Delta H°_f(\text{Products}) - \Delta H°_f(\text{Reactants})$$

To find $\Delta H°$ for the reaction

$$a\,A + b\,B + \cdots \longrightarrow c\,C + d\,D + \cdots$$

| Subtract the sum of the heats of formation for these reactants . . . | . . . from the sum of the heats of formation for these products. |

$$\Delta H°_{reaction} = [c\,\Delta H°_f(C) + d\,\Delta H°_f(D) + \cdots] - [a\,\Delta H°_f(A) + b\,\Delta H°_f(B) + \cdots]$$

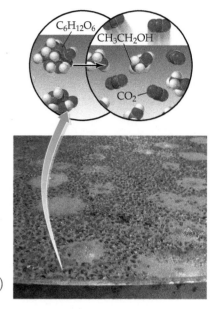

As an example, let's calculate $\Delta H°$ for the fermentation of glucose to make ethyl alcohol (ethanol), the reaction that occurs during the production of alcoholic beverages:

$$C_6H_{12}O_6(s) \longrightarrow 2\,C_2H_5OH(l) + 2\,CO_2(g) \qquad \Delta H° = ?$$

Using the data in Table 9.2 gives the following answer:

$$\begin{aligned}\Delta H° &= [2\,\Delta H°_f(\text{Ethanol}) + 2\,\Delta H°_f(CO_2)] - [\Delta H°_f(\text{Glucose})]\\ &= (2\text{ mol})(-277.7\text{ kJ/mol}) + (2\text{ mol})(-393.5\text{ kJ/mol}) - (1\text{ mol})(-1273.3\text{ kJ/mol})\\ &= -69.1\text{ kJ}\end{aligned}$$

▲ Fermentation of the sugar from grapes yields the ethyl alcohol in wine.

The fermentation reaction is exothermic by 69.1 kJ.

Why does this calculation "work"? It works because enthalpy is a state function and the calculation is really just an application of Hess's law. That is, the sum of the individual equations corresponding to the heat of formation for each substance in the reaction equals the enthalpy change for the overall reaction:

(1)	$C_6H_{12}O_6(s) \longrightarrow 6\,C(s) + 6\,H_2(g) + 3\,O_2(g)$	$-\Delta H°_f =$	$+1273.3$ kJ
(2)	$2\,[2\,C(s) + 3\,H_2(g) + 1/2\,O_2(g) \longrightarrow C_2H_5OH(l)]$	$2\,[\Delta H°_f = -277.7\text{ kJ}] =$	-555.4 kJ
(3)	$2\,[C(s) + O_2(g) \longrightarrow CO_2(g)]$	$2\,[\Delta H°_f = -393.5\text{ kJ}] =$	-787.0 kJ
(Net)	$C_6H_{12}O_6(s) \longrightarrow 2\,C_2H_5OH(l) + 2\,CO_2(g)$	$\Delta H° =$	-69.1 kJ

Note that reaction (1) represents the formation of glucose from its elements written in reverse, so the sign of $\Delta H°_f$ is reversed. Note also that reactions (2) and (3), which represent the formation of ethyl alcohol and carbon dioxide, respectively, are multiplied by 2 to arrive at the balanced equation for the overall reaction.

WORKED EXAMPLE 9.9

Using Standard Heats of Formation to Calculate $\Delta H°$

Oxyacetylene welding torches burn acetylene gas, $C_2H_2(g)$. Use the information in Table 9.2 to calculate $\Delta H°$ in kilojoules for the combustion reaction of acetylene to yield $CO_2(g)$ and $H_2O(g)$.

STRATEGY

Write the balanced equation and look up the appropriate heats of formation for each reactant and product in Table 9.2. Carry out the calculation using the general formula:

$$\Delta H°_{reaction} = \Delta H°_f\,(\text{Products}) - \Delta H°_f\,(\text{Reactants})$$

Multiply each $\Delta H°_f$ by the coefficient given in the balanced equation and remember that $\Delta H°_f(O_2) = 0$ kJ/mol because O_2 is in its elemental form.

continued on next page

SOLUTION

The balanced equation is

$$2\ C_2H_2(g)\ +\ 5\ O_2(g)\ \longrightarrow\ 4\ CO_2(g)\ +\ 2\ H_2O(g)$$

The necessary heats of formation are

$$\Delta H^\circ_f\,[C_2H_2(g)]\ =\ 227.4\ \text{kJ/mol} \qquad \Delta H^\circ_f\,[H_2O(g)]\ =\ -241.8\ \text{kJ/mol}$$
$$\Delta H^\circ_f\,[CO_2(g)]\ =\ -393.5\ \text{kJ/mol}$$

The standard enthalpy change for the reaction is

$$\Delta H^\circ\ =\ [4\ \Delta H^\circ_f\,(CO_2)]\ +\ 2\ \Delta H^\circ_f\,(H_2O)]\ -\ [2\ \Delta H^\circ_f\,(C_2H_2)]$$
$$=\ (4\ \text{mol})(-393.5\ \text{kJ/mol})\ +\ (2\ \text{mol})(-241.8\ \text{kJ/mol})\ -\ (2\ \text{mol})(227.4\ \text{kJ/mol})$$
$$=\ -2512.4\ \text{kJ}$$

▶ **PRACTICE 9.17**   Use the information in Table 9.2 to calculate ΔH° in kilojoules for the reaction of ammonia (NH_3) with O_2 to yield nitric oxide (NO) and $H_2O(g)$, a step in the Ostwald process for the commercial production of nitric acid.

▶ **APPLY 9.18**   The thermochemical equation for the combustion of octane (C_8H_{18}) in your car's engine is: $2\ C_8H_{18}(l)\ +\ 25\ O_2(g)\ \longrightarrow\ 8\ CO_2(g)\ +\ 9H_2O(l)$ $\Delta H^\circ\ =\ -5220\ \text{kJ}$

Use the information in Table 9.2 to calculate $\Delta H^\circ_f\,[C_8H_{18}(l)]$.

9.10 BOND DISSOCIATION ENERGIES

The procedure described in the previous section for determining heats of reaction from heats of formation is extremely useful, but it still presents a problem. To use the method, it's necessary to know ΔH°_f for every substance in a reaction. This implies, in turn, that vast numbers of measurements are needed because there are over 40 million known chemical compounds. In practice, though, only a few thousand ΔH°_f values have been determined.

For those reactions where insufficient ΔH°_f data are available to allow an exact calculation of ΔH°, it's often possible to estimate ΔH° by using the average **bond dissociation energies (D)** discussed previously in Section 7.2. Although we didn't identify it as such at the time, a bond dissociation energy is really just a standard enthalpy change for the corresponding bond-breaking reaction.

For the reaction $X - Y \longrightarrow X + Y$ $\Delta H^\circ = D =$ Bond dissociation energy

When we say, for example, that the bond dissociation energy of Cl_2 is 243 kJ/mol, we mean that the standard enthalpy change for the reaction $Cl_2(g) \longrightarrow 2\ Cl(g)$ is $\Delta H^\circ = 243$ kJ. Bond dissociation energies are always positive because energy must always be put into bonds to break them. Average bond dissociation energies were given in Table 9.2 and are reproduced for easy reference in **TABLE 9.3.**

Applying Hess's law, we can calculate an approximate enthalpy change for any reaction by subtracting the sum of the bond dissociation energies in the products from the sum of the bond dissociation energies in the reactants:

$$\Delta H^\circ\ =\ D(\text{Reactant bonds})\ -\ D(\text{Product bonds})$$

In the reaction of H_2 with Cl_2 to yield HCl, for example, the reactants have one Cl—Cl bond and one H—H bond, while the product has two H—Cl bonds.

$$H_2(g)\ +\ Cl_2(g)\ \longrightarrow\ 2\ HCl(g)$$

REMEMBER . . .
Bond dissociation energy (*D*) is the amount of energy that must be supplied to break a chemical bond in an isolated molecule in the gaseous state and is thus the amount of energy released when the bond forms (Section 7.2).

TABLE 9.3   Average Bond Dissociation Energies, *D* (kJ/mol)

H—H	436[a]	C—H	410	N—H	390	O—F	180	I—I	151[a]
H—C	410	C—C	350	N—C	300	O—Cl	200	S—F	310
H—F	570[a]	C—F	450	N—F	270	O—Br	210	S—Cl	250
H—Cl	432[a]	C—Cl	330	N—Cl	200	O—I	220	S—Br	210
H—Br	366[a]	C—Br	270	N—Br	240	O—N	200	S—S	225
H—I	298[a]	C—I	240	N—N	240	O—O	180		
H—N	390	C—N	300	N—O	200	F—F	159[a]		
H—O	460	C—O	350	O—H	460	Cl—Cl	2[a]		
H—S	340	C—S	260	O—C	350	Br—Br	193[a]		

Multiple covalent bonds

C=C	728	C≡C	965	C=O	732	O=O	498[a]	N≡N	945[a]

[a]Exact value.

According to the data in Table 9.3, the bond dissociation energy of Cl_2 is 243 kJ/mol, that of H_2 is 436 kJ/mol, and that of HCl is 432 kJ/mol. We can thus calculate an approximate standard enthalpy change for the reaction.

$$\Delta H° = D(\text{Reactant bonds}) - D(\text{Product bonds})$$
$$= (D_{Cl-Cl} + D_{H-H}) - (2\,D_{H-Cl})$$
$$= [(1\text{ mol})(243\text{ kJ/mol}) + (1\text{ mol})(436\text{ kJ/mol})] - (2\text{ mol})(432\text{ kJ/mol})$$
$$= -185\text{ kJ}$$

The reaction is exothermic by approximately 185 kJ.

WORKED EXAMPLE 9.10

Using Bond Dissociation Energies to Calculate Δ*H*°

Use the data in Table 9.3 to find an approximate $\Delta H°$ in kilojoules for the industrial synthesis of chloroform by reaction of methane with Cl_2.

$$CH_4(g) + 3\,Cl_2(g) \longrightarrow CHCl_3(g) + 3\,HCl(g)$$

STRATEGY

Identify all the bonds in the reactants and products, and look up the appropriate bond dissociation energies in Table 9.3. Then subtract the sum of the bond dissociation energies in the products from the sum of the bond dissociation energies in the reactants to find the enthalpy change for the reaction.

SOLUTION

The reactants have four C—H bonds and three Cl—Cl bonds; the products have one C—H bond, three C—Cl bonds, and three H—Cl bonds. The bond dissociation energies from Table 9.3 are:

$$C-H \quad D = 410\text{ kJ/mol} \quad Cl-Cl \quad D = 243\text{ kJ/mol}$$
$$C-Cl \quad D = 330\text{ kJ/mol} \quad H-Cl \quad D = 432\text{ kJ/mol}$$

continued on next page

Ethyl alcohol

Subtracting the product bond dissociation energies from the reactant bond dissociation energies gives the enthalpy change for the reaction:

$$\Delta H° = [3 \, D_{Cl-Cl} + 4 \, D_{C-H}] - [D_{C-H} + 3 \, D_{H-Cl} + 3 \, D_{C-Cl}]$$

$$= [(3 \text{ mol})(243 \text{ kJ/mol}) + (4 \text{ mol})(410 \text{ kJ/mol})] - [(1 \text{ mol})(410 \text{ kJ/mol})$$

$$+ (3 \text{ mol})(432 \text{ kJ/mol}) + (3 \text{ mol})(330 \text{ kJ/mol})]$$

$$= -327 \text{ kJ}$$

The reaction is exothermic by approximately 330 kJ.

▶ **PRACTICE 9.19** Use the data in Table 9.3 to calculate an approximate $\Delta H°$ in kilojoules for the industrial synthesis of ethyl alcohol from ethylene:

$$C_2H_4(g) + H_2O(g) \longrightarrow C_2H_5OH(g).$$

▶ **APPLY 9.20** Benzene (C_6H_6) has two resonance structures, meaning each carbon–carbon bond is equivalent, with a bond strength between a single C—C bond and double C=C.

Estimate the carbon–carbon bond strength in benzene given:

$$2 \, C_6H_6(g) + 15 \, O_2(g) \longrightarrow 12 \, CO_2(g) + 6 \, H_2O(g) \quad \Delta H° = -6339 \text{ kJ}$$

Bond dissociation data can be found in Table 9.3. (The strength of the O=O bond in O_2 is 498 kJ/mol, and that of a C=O bond in CO_2 is 804 kJ/mol.)

9.11 AN INTRODUCTION TO ENTROPY

We said in the introduction to this chapter that chemical reactions (and physical processes) occur when the final state is more stable than the initial state. Because less stable substances generally have higher internal energy and are converted into more stable substances with lower internal energy, energy is generally released in chemical reactions. At the same time, though, we've seen that some reactions and processes occur even though they absorb rather than release energy. The endothermic reaction of barium hydroxide octahydrate with ammonium chloride shown previously in Figure 9.7, for example, absorbs 80.3 kJ of heat ($\Delta H° = +80.3$ kJ) and leaves the surroundings so cold that the temperature drops below 0 °C.

$$Ba(OH)_2 \cdot 8 \, H_2O(s) + 2 \, NH_4Cl(s) \longrightarrow BaCl_2(aq) + 2 \, NH_3(aq) + 10 \, H_2O(l)$$
$$\Delta H° = +80.3 \text{ kJ}$$

An example of a physical process that takes place spontaneously yet absorbs energy takes place every time an ice cube melts. At a temperature of 0 °C, ice spontaneously absorbs heat from the surroundings to turn from solid into liquid water.

Before exploring the situation further, it's important to understand what the word *spontaneous* means in chemistry, for it's not quite the same as in everyday language. In chemistry, a **spontaneous process** is one that, once started, proceeds on its own without a continuous external influence. The change need not happen quickly, like a spring uncoiling or a sled going downhill. It can also happen slowly, like the gradual rusting away of an iron bridge or abandoned car. A *nonspontaneous* process, by contrast, takes place only in the presence of a continuous external influence. Energy must be

▲ Sledding downhill is a spontaneous process that, once started, continues on its own. Dragging the sled back uphill is a nonspontaneous process that requires a continuous input of energy.

continuously expended to recoil a spring or to push a sled uphill. When the external influence stops, the process also stops.

Note that the reaction of barium hydroxide octahydrate with ammonium chloride and the melting ice cube absorb heat yet still take place spontaneously. What's going on? There must be some other factor in addition to energy that determines whether a reaction or process will occur. We'll take only a brief look at this additional factor now and return for a more in-depth study in Chapter 18 on **thermodynamics.**

What do the reaction of barium hydroxide octahydrate and the melting of an ice cube have in common that allows the two processes to take place spontaneously even though they absorb heat? The common feature of these and all other processes that absorb heat yet occur spontaneously is an increase in the amount of molecular randomness of the system. The eight water molecules rigidly held in the $Ba(OH)_2 \cdot 8\,H_2O$ crystal break loose and become free to move about randomly in the aqueous liquid product. Similarly, the rigidly held H_2O molecules in the ice lose their crystalline ordering and move around freely in liquid water.

The amount of molecular randomness in a system is called the system's **entropy** (S). Entropy has the units J/K (not kJ/K) and is a quantity that can be determined for pure substances. The larger the value of S, the greater the molecular randomness of the particles in the system. Gases, for example, have more randomness and higher entropy than liquids, and liquids have more randomness and higher entropy than solids (**FIGURE 9.12**).

LOOKING AHEAD . . .
In Chapter 18, we provide a consolidated study of **thermodynamics,** including a more detailed discussion of entropy and free energy.

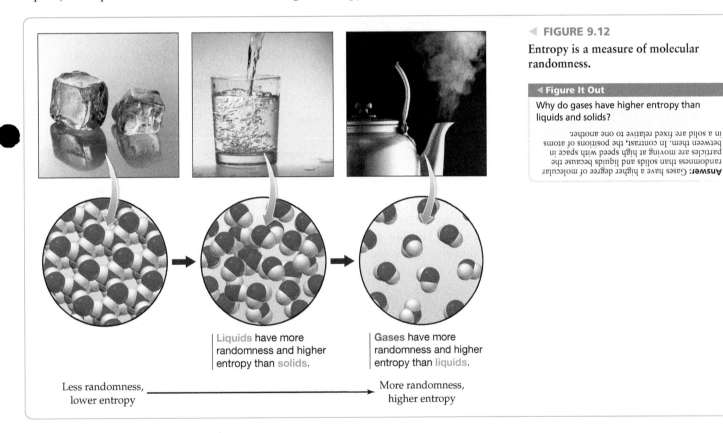

◄ **FIGURE 9.12**

Entropy is a measure of molecular randomness.

◄ Figure It Out

Why do gases have higher entropy than liquids and solids?

Answer: Gases have a higher degree of molecular randomness than solids and liquids because the particles are moving at high speed with space in between them. In contrast, the positions of atoms in a solid are fixed relative to one another.

Liquids have more randomness and higher entropy than solids.

Gases have more randomness and higher entropy than liquids.

Less randomness, lower entropy ————————————→ More randomness, higher entropy

A change in entropy is represented as $\Delta S = S_{final} - S_{initial}$. When randomness increases, as it does when barium hydroxide octahydrate reacts or ice melts, ΔS has a positive value because $S_{final} > S_{initial}$. The reaction of $Ba(OH)_2 \cdot 8\,H_2O(s)$ with $NH_4Cl(s)$ has $\Delta S° = +428$ J/K, and the melting of ice has $\Delta S° = +22.0$ J/(K·mol). When randomness decreases, ΔS is negative because $S_{final} < S_{initial}$. The freezing of water, for example, has $\Delta S° = -22.0$ J/(K·mol). (As with $\Delta H°$, the superscript ° is used in $\Delta S°$ to refer to the standard entropy change in a reaction where products and reactants are in their standard states.)

Thus, two factors determine the spontaneity of a chemical or physical change in a system: a release or absorption of heat (ΔH) and an increase or decrease in molecular randomness (ΔS). To decide whether a process is spontaneous, both enthalpy and entropy changes must be taken into account.

Spontaneous process:

Favored by decrease in H (negative ΔH)

Favored by increase in S (positive ΔS)

Nonspontaneous process:

Favored by increase in H (positive ΔH)

Favored by decrease in S (negative ΔS)

Note that the two factors don't have to operate in the same direction. Thus, it's possible for a process to be *disfavored* by enthalpy (endothermic, positive ΔH) yet still be spontaneous because it is strongly *favored* by entropy (positive ΔS). The melting of ice $[\Delta H° = +6.01 \text{ kJ/mol}; \Delta S° = +22.0 \text{ J/(K·mol)}]$ is just such a process, as is the reaction of barium hydroxide octahydrate with ammonium chloride ($\Delta H° = +80.3 \text{ kJ}$; $\Delta S° = +428 \text{ J/K}$). In the latter case, 3 mol of solid reactants produce 10 mol of liquid water, 2 mol of dissolved ammonia, and 3 mol of dissolved ions (1 mol of Ba^{2+} and 2 mol of Cl^-), with a consequent large increase in molecular randomness:

$$\underbrace{Ba(OH)_2 \cdot 8\,H_2O(s) + 2\,NH_4Cl(s)}_{\text{3 mol solid reactants}} \longrightarrow \underset{\substack{\uparrow \\ \text{3 mol} \\ \text{dissolved ions}}}{BaCl_2(aq)} + \underset{\substack{\uparrow \\ \text{2 mol dissolved} \\ \text{molecules}}}{2\,NH_3(aq)} + \underset{\substack{\uparrow \\ \text{10 mol} \\ \text{liquid} \\ \text{water molecules}}}{10\,H_2O(l)}$$

$$\Delta H° = +80.3 \text{ kJ} \longleftarrow \text{Unfavorable}$$

$$\Delta S° = +428 \text{ J/K} \longleftarrow \text{Favorable}$$

Conversely, it's also possible for a process to be favored by enthalpy (exothermic, negative ΔH) yet be nonspontaneous because it is strongly disfavored by entropy (negative ΔS). The conversion of liquid water to ice is nonspontaneous above 0 °C, for example, because the process is disfavored by entropy $[\Delta S° = -22.0 \text{ J/(K·mol)}]$ even though it is favored by enthalpy ($\Delta H° = -6.01 \text{ kJ/mol}$).

WORKED EXAMPLE 9.11

Predicting the Sign of ΔS for a Reaction

Predict whether $\Delta S°$ is likely to be positive or negative for each of the following reactions:

(a) $H_2C{=}CH_2(g) + Br_2(g) \longrightarrow BrCH_2CH_2Br(l)$
(b) $2\,C_2H_6(g) + 7\,O_2(g) \longrightarrow 4\,CO_2(g) + 6\,H_2O(g)$

STRATEGY

Look at each reaction, and try to decide whether molecular randomness increases or decreases. Reactions that increase the number of gaseous molecules generally have a positive ΔS, while reactions that decrease the number of gaseous molecules have a negative ΔS.

SOLUTION

(a) The amount of molecular randomness in the system decreases when 2 mol of gaseous reactants combine to give 1 mol of liquid product, so the reaction has a negative $\Delta S°$.
(b) The amount of molecular randomness in the system increases when 9 mol of gaseous reactants give 10 mol of gaseous products, so the reaction has a positive $\Delta S°$.

▶ **PRACTICE 9.21**   Ethane, C_2H_6, can be prepared by the reaction of acetylene, C_2H_2, with hydrogen. Is $\Delta S°$ for the reaction likely to be positive or negative? Explain.

$$C_2H_2(g) + 2\,H_2(g) \longrightarrow C_2H_6(g)$$

▶ **CONCEPTUAL APPLY 9.22**   Is the reaction represented in the following drawing likely to have a positive or a negative value of $\Delta S°$? Explain.

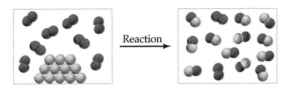

9.12 AN INTRODUCTION TO FREE ENERGY

How do we weigh the relative contributions of enthalpy changes (ΔH) and entropy changes (ΔS) to the overall spontaneity of a process? To take both factors into account when deciding the spontaneity of a chemical reaction or other process, we define a quantity called the **Gibbs free-energy change (ΔG)** which is related to ΔH and ΔS by the equation $\Delta G = \Delta H - T\Delta S$.

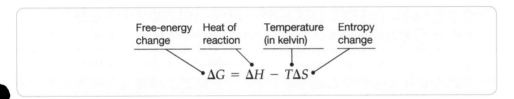

The value of the free-energy change ΔG determines whether a chemical or physical process will occur spontaneously. If ΔG has a negative value, free energy is released and the process is spontaneous. If ΔG has a value of 0, the process is neither spontaneous nor nonspontaneous but instead is at an equilibrium. And if ΔG has a positive value, free energy is absorbed and the process is nonspontaneous.

$\Delta G < 0$ Process is spontaneous

$\Delta G = 0$ Process is at equilibrium—neither spontaneous nor nonspontaneous

$\Delta G > 0$ Process is nonspontaneous

Because the $T\Delta S$ term in the free-energy equation is temperature dependent, we deduce that some processes might be either spontaneous or nonspontaneous depending on the temperature. At low temperature, for instance, an unfavorable (positive) ΔH term might be larger than a favorable (positive) $T\Delta S$ term, but at higher temperature, the $T\Delta S$ term might be larger. Thus, an endothermic process that is nonspontaneous at low temperature can become spontaneous at higher temperature. This, in fact, is exactly what happens in the ice–water transition. At a temperature below 0 °C, the melting of ice is nonspontaneous because the unfavorable ΔH term outweighs the favorable $T\Delta S$ term. At a temperature above 0 °C, however, the melting of ice is spontaneous because the favorable $T\Delta S$ term outweighs the unfavorable ΔH term (**FIGURE 9.13**). At exactly 0 °C, the two terms are balanced.

$$\Delta G° = \Delta H° - T\Delta S°$$

At -10 °C (263 K): $\Delta G° = 6.01\,\dfrac{kJ}{mol} - (263\ K)\left(0.0220\,\dfrac{kJ}{K \cdot mol}\right) = +0.22\ kJ/mol$ (nonspontaneous)

At 0 °C (273 K): $\Delta G° = 6.01\,\dfrac{kJ}{mol} - (273\ K)\left(0.0220\,\dfrac{kJ}{K \cdot mol}\right) = 0.00\ kJ/mol$ (equilibrium)

At $+10$ °C (283 K): $\Delta G° = 6.01\,\dfrac{kJ}{mol} - (283\ K)\left(0.0220\,\dfrac{kJ}{K \cdot mol}\right) = -0.22\ kJ/mol$ (spontaneous)

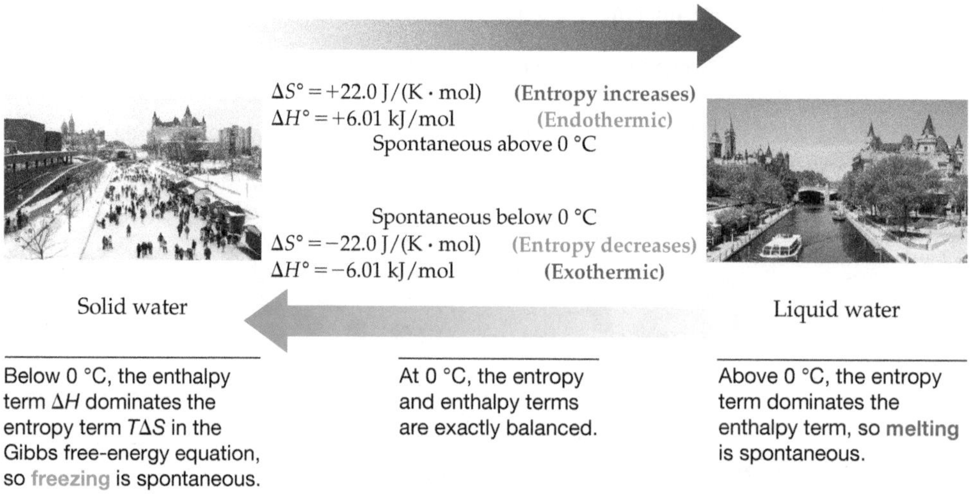

> ▶ **FIGURE 9.13**
>
> **Melting and freezing.** The melting of ice is disfavored by enthalpy ($\Delta H > 0$) but favored by entropy ($\Delta S > 0$). The freezing of water is favored by enthalpy ($\Delta H < 0$) but disfavored by entropy ($\Delta S < 0$).

$\Delta S° = +22.0$ J/(K · mol)   (Entropy increases)
$\Delta H° = +6.01$ kJ/mol   (Endothermic)
      Spontaneous above 0 °C

      Spontaneous below 0 °C
$\Delta S° = -22.0$ J/(K · mol)   (Entropy decreases)
$\Delta H° = -6.01$ kJ/mol   (Exothermic)

Solid water   Liquid water

| Below 0 °C, the enthalpy term ΔH dominates the entropy term $T\Delta S$ in the Gibbs free-energy equation, so freezing is spontaneous. | At 0 °C, the entropy and enthalpy terms are exactly balanced. | Above 0 °C, the entropy term dominates the enthalpy term, so melting is spontaneous. |

An example of a chemical reaction in which temperature controls spontaneity is that of carbon with water to yield carbon monoxide and hydrogen. The reaction has an unfavorable ΔH term (positive) but a favorable $T\Delta S$ term (positive) because randomness increases when a solid and 1 mol of gas are converted into 2 mol of gas:

$$C(s) + H_2O(g) \longrightarrow CO(g) + H_2(g) \quad \Delta H° = +131 \text{ kJ} \quad \text{Unfavorable}$$

$$\Delta S° = +134 \text{ J/K} \quad \text{Favorable}$$

No reaction occurs if carbon and water are mixed at room temperature because the unfavorable ΔH term outweighs the favorable $T\Delta S$ term. At approximately 978 K (705 °C), however, the reaction becomes spontaneous because the favorable $T\Delta S$ term becomes larger than the unfavorable ΔH term. Below 978 K, ΔG has a positive value; at 978 K, $\Delta G = 0$; and above 978 K, ΔG has a negative value. (The calculation is not exact because values of ΔH and ΔS themselves vary somewhat with temperature.)

$$\Delta G° = \Delta H° - T\Delta S°$$

At 695 °C (968 K): $\Delta G° = 131 \text{ kJ} - (968 \text{ K})\left(0.134 \dfrac{\text{kJ}}{\text{K}}\right) = +1 \text{ kJ (nonspontaneous)}$

At 705 °C (978 K): $\Delta G° = 131 \text{ kJ} - (978 \text{ K})\left(0.134 \dfrac{\text{kJ}}{\text{K}}\right) = 0 \text{ kJ (equilibrium)}$

At 715 °C (988 K): $\Delta G° = 131 \text{ kJ} - (988 \text{ K})\left(0.134 \dfrac{\text{kJ}}{\text{K}}\right) = -1 \text{ kJ (spontaneous)}$

The reaction of carbon with water is, in fact, the first step of an industrial process for manufacturing methanol (CH_3OH). As supplies of natural gas and oil diminish, this reaction may become important for the manufacture of synthetic fuels.

A process is at equilibrium when it is balanced between spontaneous and nonspontaneous—that is, when $\Delta G = 0$ and it is energetically unfavorable to go either from reactants to products or from products to reactants. Thus, at the equilibrium point, we can set up the equation

$$\Delta G = \Delta H - T\Delta S = 0 \quad \text{At equilibrium}$$

Solving this equation for T gives

$$T = \frac{\Delta H}{\Delta S}$$

which makes it possible to calculate the temperature at which a changeover in behavior between spontaneous and nonspontaneous occurs. Using the known values of $\Delta H°$

and $\Delta S°$ for the melting of ice, for instance, we find that the point at which liquid water and solid ice are in equilibrium is

$$T = \frac{\Delta H°}{\Delta S°} = \frac{6.01 \text{ kJ}}{0.0220 \dfrac{\text{kJ}}{\text{K}}} = 273 \text{ K} = 0 \text{ °C}$$

Not surprisingly, the ice–water equilibrium point is 273 K, or 0 °C, the melting point of ice.

In the same way, the temperature at which the reaction of carbon with water changes between spontaneous and nonspontaneous is 978 K, or 705 °C:

$$T = \frac{\Delta H°}{\Delta S°} = \frac{131 \text{ kJ}}{0.134 \dfrac{\text{kJ}}{\text{K}}} = 978 \text{ K}$$

This section and the preceding one serve only as an introduction to entropy and free energy. In Chapter 18, we will take a more in-depth look at these two important topics.

CONCEPTUAL WORKED EXAMPLE 9.12

Predicting the Signs of ΔH, ΔS, and ΔG for a Reaction

What are the signs of ΔH, ΔS, and ΔG for the following nonspontaneous transformation?

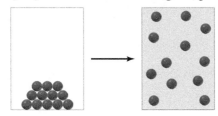

STRATEGY

First, decide what kind of process is represented in the drawing. Then decide whether the process increases or decreases the entropy of the system and whether it is exothermic or endothermic.

SOLUTION

The drawing shows ordered particles in a solid subliming to give a gas. Formation of a gas from a solid increases molecular randomness, so ΔS is positive. Furthermore, because we're told that the process is nonspontaneous, ΔG is also positive. Because the process is favored by ΔS (positive) yet still nonspontaneous, ΔH must be unfavorable (positive). This makes sense, because conversion of a solid to a liquid or gas requires energy and is always endothermic.

▶ **CONCEPTUAL PRACTICE 9.23**   What are the signs of ΔH, ΔS, and ΔG for the following spontaneous reaction?

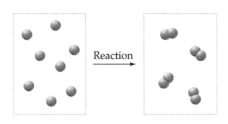

▶ **CONCEPTUAL APPLY 9.24**   The following reaction is exothermic:

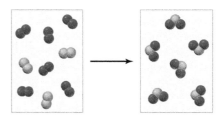

(a) Write a balanced equation for the reaction. (Red spheres = A; Green spheres = B.)
(b) What are the signs of ΔH and ΔS for the reaction?
(c) Is the reaction likely to be spontaneous at low temperatures only, at high temperatures only, or at all temperatures? Explain.

WORKED EXAMPLE 9.13

Using the Free-Energy Equation to Calculate Equilibrium Temperature

Lime (CaO) is produced by heating limestone ($CaCO_3$) to drive off CO_2 gas, a reaction used to make Portland cement. Is the reaction spontaneous under standard conditions at 25 °C? Calculate the temperature at which the reaction becomes spontaneous.

$$CaCO_3(s) \longrightarrow CaO(s) + CO_2(g) \quad \Delta H° = 179.2 \text{ kJ}; \quad \Delta S° = 160.0 \text{ J/K}$$

STRATEGY

The spontaneity of the reaction at a given temperature can be found by determining whether ΔG is positive or negative at that temperature. The changeover point between spontaneous and nonspontaneous can be found by setting $\Delta G = 0$ and solving for T.

SOLUTION

At 25 °C (298 K), we have

$$\Delta G = \Delta H - T\Delta S = 179.2 \text{ kJ} - (298 \text{ K})\left(0.1600 \frac{\text{kJ}}{\text{K}}\right) = +131.5 \text{ kJ}$$

Because ΔG is positive at this temperature, the reaction is nonspontaneous. The changeover point between spontaneous and nonspontaneous is approximately

$$T = \frac{\Delta H}{\Delta S} = \frac{179.2 \text{ kJ}}{0.1600 \frac{\text{kJ}}{\text{K}}} = 1120 \text{ K}$$

The reaction becomes spontaneous above approximately 1120 K (847 °C).

▶ **PRACTICE 9.25**   Is the Haber process for the industrial synthesis of ammonia spontaneous or nonspontaneous under standard conditions at 25 °C? At what temperature (°C) does the changeover occur?

$$N_2(g) + 3 H_2(g) \longrightarrow 2 NH_3(g) \quad \Delta H° = -92.2 \text{ kJ}; \Delta S° = -199 \text{ J/K}$$

▶ **APPLY 9.26**   A certain reaction has $\Delta H° = +75.0$ kJ; $\Delta S° = +231$ J/K. At what temperature is the reaction at equilibrium? Should the temperature be increased or decreased to make the reaction spontaneous?

INQUIRY ? How do we determine the energy content of biofuels?

The petroleum era began in August 1859, when the world's first oil well was drilled near Titusville, Pennsylvania. Since that time, approximately 135 billion barrels of petroleum have been used throughout the world, primarily as fuel for automobiles. Current world consumption is approximately 3.4×10^{10} barrels per year, and currently known recoverable reserves are estimated at 1.7×10^{12} barrels. Thus, the world's known petroleum reserves will be exhausted in approximately 50 years at the current rate of consumption. Thus, alternative energy sources are needed.

▲ Algae grown in farms contains up to 40% oil, which can be used to produce biodiesel fuel.

▲ Vegetable oil from the bright yellow rapeseed plant is a leading candidate for large-scale production of biodiesel fuel.

Of the various alternative energy sources now being explored, biofuels—fuels derived from recently living organisms such as trees, corn, sugar cane, and rapeseed—look promising. Biofuels are renewable and they are more carbon neutral than fossil fuels, meaning that the amount of CO_2 released to the environment during the manufacture and burning of a biofuel is similar to the amount of CO_2 removed from the environment by photosynthesis during the plant's growth.

The two biofuels receiving the most attention at present are ethanol and biodiesel. Ethanol is simply ethyl alcohol, the same substance found in alcoholic drinks and produced in the same way: by yeast-catalyzed fermentation of carbohydrate.

Glucose
($C_6H_{12}O_6$)

Yeast enzymes →

2 Ethanol + 2 CO$_2$

Ethanol
(C_2H_5OH)

The only difference between beverage ethanol and fuel ethanol is the source of the sugar. Beverage ethanol comes primarily from fermentation of sugar in grapes (for wine) or grains (for distilled liquors), while fuel ethanol comes primarily from fermentation of cane sugar or corn. Researchers are working on developing economical methods of converting cheap cellulose-based agricultural and logging wastes into fermentable sugars.

Biodiesel consists primarily of organic compounds called long-chain methyl esters, which are produced by reaction of common vegetable oils with methyl alcohol in the presence of an acid catalyst.

Any vegetable oil can be used to produce biodiesel, but rapeseed oil and soybean oil are the most common. Algae is another renewable source of oil that can be converted to biodiesel and fast-growing algae can produce up to 60 times more oil per acre than land-based plants. Like plants, algae need CO_2 to grow, and it is possible to place large algae farms next to power plants and use the waste CO_2 exhaust gas to feed the algae ponds.

How is the energy content of biofuels determined?

The amount of energy released when a substance is burned is called its **heat of combustion** ($\Delta H°_c$), or *combustion enthalpy*, and is the standard enthalpy change for the reaction of 1 mol of the substance with oxygen. Liquid octane (C_8H_{18}) has $\Delta H°_c = -5470$ kJ/mol.

$$C_8H_{18}(l) + \frac{25}{2} O_2(g) \longrightarrow 8 CO_2(g) + 9 H_2O(l)$$
$$\Delta H°_c = -5470 \text{ kJ/mol}$$

We can determine heats of combustion using several methods described in this chapter. We can burn fuels in a bomb calorimeter to measure the amount of energy transfer, use bond dissociation energies to estimate heats of combustion for gas-phase fuels, and use standard enthalpy of formation data ($\Delta H°_f$) to calculate the enthalpy change for a combustion reaction.

To compare the efficiency of different fuels, it's more useful to calculate combustion enthalpies per gram or per milliliter of substance rather than per mole (**TABLE 9.4**). For applications in which weight is important, as in rocket engines, hydrogen is ideal because its combustion enthalpy per gram is the highest of any known fuel. For applications in which volume is important, as in automobiles, a mixture of hydrocarbons—compounds containing carbon and hydrogen, such as those in gasoline—are most efficient because hydrocarbon combustion enthalpies on a per-volume basis are relatively high.

TABLE 9.4   Combustion Enthalpies of Some Fuels

Fuel	$\Delta H°_c(kJ/mol)$	$\Delta H°_c(kJ/g)$	$\Delta H°_c(kJ/ml)$
Hydrogen, $H_2(g)$	−285.8	−141.8	−9.9[a]
Ethanol, $C_2H_5OH(l)$	−1366.8	−29.7	−23.4
Octane, $C_8H_{18}(l)$	−5470	−47.9	−33.6
Biodiesel, $C_{17}H_{35}CO_2CH_3(l)$ from stearic acid	−11,175	−37.8	−34.5

[a]Calculated for compressed liquid at 0 °C.

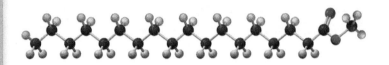

A typical long-chain methyl ester in biodiesel

How does the energy content of biofuels compare to gasoline?

As shown in Table 9.4, ethanol has a lower combustion enthalpy per volume than octane, a representative hydrocarbon in gasoline. Therefore, ethanol is frequently blended with gasoline to increase the amount of energy in a tank of gas. E10 is gasoline with 10% ethanol content, and E85 is gasoline with 85% ethanol content. Biodiesel has a similar heat of combustion on a per-volume basis as octane and is typically mixed with petroleum-based diesel (up to 30% biodiesel) for use in trucks and automobiles.

PROBLEM 9.27   Which biofuel is produced by fermenting corn? Which biofuel is produced by a reaction of vegetable oil and methyl alcohol?

PROBLEM 9.28   Why are biofuels better for the environment than gasoline?

PROBLEM 9.29
(a) Balance the equation for the combustion of liquid ethanol (C_2H_6O).

$$C_2H_6O(l) + O_2(g) \longrightarrow CO_2(g) + H_2O(l)$$

(b) Use the information in Table 9.2 and [$\Delta H°_f$ for $C_2H_6O(l) = -277.7$ kJ/mol] to calculate the combustion enthalpy ($\Delta H°_c$) in kilojoules per mole of ethanol.

PROBLEM 9.30   Bond dissociation energies can be used to estimate the enthalpy change for reaction if the reactants and products are in the gas phase. Calculate an approximate heat of combustion for gaseous ethanol in kilojoules per mole using the bond dissociation energies in Table 9.3. (The bond dissociation energy of the C=O bond in CO_2 is 804 kJ/mol.) The structural formula for ethanol is:

$$
\begin{array}{c}
\text{H} \quad \text{H} \\
| \qquad | \\
\text{H}-\text{C}-\text{C}-\text{O}-\text{H} \\
| \qquad | \\
\text{H} \quad \text{H}
\end{array}
$$

PROBLEM 9.31
(a) Calculate the PV work in kilojoules done during the combustion of 1 mol of ethanol in which the volume contracts from 73.5 L to 49.0 L at a constant external pressure of 1 atm. In which direction does the work energy flow? (1 L·atm = 101.325 J.)

(b) If $\Delta H = -1367$ kJ/mol, calculate the change in internal energy (ΔE) for the combustion of 1 mole of ethanol.

PROBLEM 9.32   When 0.350 g of biodiesel ($C_{19}H_{38}O_2$) is burned in a bomb calorimeter, the temperature rises by 5.29 °C. If the heat capacity of the bomb calorimeter is 2.50 kJ/°C, calculate the combustion energy in units of kJ/g.

PROBLEM 9.33   A 12.0 gallon fuel tank can hold 39.9 kg of biodiesel. Use the following thermochemical equation to calculate the total amount of heat (MJ) released by burning one tank of biodiesel. Compare your answer to a 12.0-gallon tank of octane (main component of gasoline) that burns to release 1530 MJ.

$$C_{19}H_{38}O_2(l) + 27.5\ O_2(g) \longrightarrow 19\ CO_2(g) + 19\ H_2O(l)$$
$$\Delta H° = -11,236\ kJ$$

STUDY GUIDE

Section	Concept Summary	Learning Objectives	Test Your Understanding
9.1 Energy and Its Conservation	Energy is either *kinetic* or *potential*. **Kinetic energy** (E_K) is the energy of motion. Its value depends on both the mass m and velocity v of an object according to the equation $E_K = (1/2)mv^2$. **Potential energy** (E_P) is the energy stored in an object because of its position or in a chemical substance because of its composition.	**9.1** Calculate the kinetic energy of an object in motion.	Problems 9.46–9.47
		9.2 Convert between different units of energy.	Problems 9.52–9.53
9.2 Internal Energy and State Functions	According to the **conservation of energy law,** also known as the **first law of thermodynamics,** energy can be neither created nor destroyed. Thus, the total energy of an isolated system is constant. The total **internal energy** (E) of a system—the sum of all kinetic and potential energies for each particle in the system—is a **state function** because its value depends only on the present condition of the system, not on how that condition was reached.	**9.3** Identify state functions.	Problems 9.48–9.49
		9.4 Identify the sign of heat and work.	Problems 9.34–9.36
9.3 Expansion Work	**Work** (w) is defined as the distance moved times the force that produces the motion. In chemistry, most work is expansion work (*PV* work) done as the result of a volume change during a reaction when air molecules are pushed aside. The amount of work done by an expanding gas is given by the equation $w = -P\Delta V$, where P is the pressure against which the system must push and ΔV is the change in volume of the system.	**9.5** Calculate *PV* work.	Worked Example 9.1; Problems 9.37, 9.54, 9.56, 9.60
9.4 Energy and Enthalpy	The total internal energy change that takes place during a reaction is the sum of the heat transferred (q) and the work done ($-P\Delta V$). The equation $\Delta H = \Delta E + P\Delta V$, where ΔH is the **enthalpy change** of the system, is a fundamental equation of thermochemistry. In general, the $P\Delta V$ term is much smaller than the ΔE term, so that the total internal energy change of a reacting system is approximately equal to ΔH, also called the **heat of reaction.**	**9.6** Calculate the internal energy change (ΔE) for a reaction.	Worked Example 9.2; Problems 9.62, 9.64–9.65
9.5–9.6 Thermochemical Equations and Enthalpies of Chemical and Physical Changes	**Thermochemical equations** show a balanced chemical equation along with the value of the enthalpy change (ΔH) for the reaction. The amount of heat transferred depends on the quantity of reactants and products, and the thermochemical equation specifies this relationship for chemical and physical changes. Reactions that have a negative ΔH are **exothermic** because heat is lost by the system, and reactions that have a positive ΔH are **endothermic** because heat is absorbed by the system.	**9.7** Calculate the amount of heat transferred given a thermochemical equation and the amount of reactant or product.	Worked Example 9.3; Problems 9.68–9.72
		9.8 Classify reaction as endothermic or exothermic and give the sign of ΔH.	Worked Example 9.4; Problems 9.66–9.67, 9.74–9.75
9.7 Calorimetry and Heat Capacity	**Heat capacity** (C) is a value that specifies how much heat is required to raise the temperature of an object by 1 °C. Similarly, **specific heat** (c) and **molar heat capacity** (C_m) are values that specify the amount of heat required to raise the temperature of a given amount of a substance by 1 °C. A calorimeter is a device used to determine ΔH of a reaction by surrounding the reaction with water and measuring the temperature change of the water.	**9.9** Calculate heat capacities, temperature changes, or heat transfer using the equation for heat capacity (C), specific heat (c), or molar heat capacity (C_m).	Worked Example 9.5; Problems 9.78–9.82
		9.10 Calculate enthalpy changes in a calorimetry experiment.	Worked Examples 9.6–9.7 Problems 9.82, 9.84, 9.86, 9.88

Section	Concept Summary	Learning Objectives	Test Your Understanding
9.8 Hess's Law	Because enthalpy is a state function, ΔH is the same regardless of the path taken between reactants and products. Thus, the sum of the enthalpy changes for the individual steps in a reaction is equal to the overall enthalpy change for the entire reaction, a relationship known as **Hess's law.**	**9.11** Use Hess's Law to find ΔH for an overall reaction, given reaction steps and their ΔH values.	Worked Example 9.8; Problems 9.38–9.39, 9.93–9.94, 9.96
9.9 Standard Heats of Formation	Hess's law also makes it possible to calculate the enthalpy change of any reaction if the standard heats of formation ($\Delta H°_f$) are known for the reactants and products. The **standard heat of formation**($\Delta H°_f$) is the enthalpy change for the hypothetical formation of 1 mol of a substance in its **thermodynamic standard state** from the most stable forms of the constituent elements in their standard states (1 atm pressure and a specified temperature, usually 25 °C).	**9.12** Identify standard states of elements.	Problems 9.100–9.101
		9.13 Write standard enthalpy of formation reactions ($\Delta H°_f$) for compounds from their elements.	Problems 9.102–9.103
		9.14 Use values of ($\Delta H°_f$) for elements and compounds to calculate $\Delta H°$ for a reaction.	Worked Example 9.9; Problems 9.106, 9.108, 9.110
9.10 Bond Dissociation Energies	Breaking a bond is an endothermic process and forming a bond is an exothermic process, which is the underlying reason for enthalpy changes of chemical reactions. Bond dissociation energies (D) can be used to estimate $\Delta H°$ for a gas-phase reaction by adding up all the bonds broken in the reactants and subtracting all the bonds formed in the products.	**9.15** Use bond dissociation energies to estimate $\Delta H°$ for a reaction.	Worked Example 9.10; Problems 9.19–9.20, 9.118–9.121
9.11–9.12 An Introduction to Entropy and Free Energy	In addition to enthalpy, **entropy** (S)—a measure of the amount of molecular randomness in a system—is important in determining whether a process will occur spontaneously. Together, changes in enthalpy and entropy define a quantity called the **Gibbs free-energy change** (ΔG) according to the equation $\Delta G = \Delta H - T\Delta S$. If ΔG is negative, the reaction is a **spontaneous process**; if ΔG is positive, the reaction is nonspontaneous.	**9.16** Predict the sign of the entropy change (ΔS) given the chemical equation or a molecular diagram.	Worked Example 9.11; Problems 9.21– 9.22, 9.126–9.127, 9.130
		9.17 Predict the sign of ΔG, ΔH, and ΔS, using the equation for Gibbs free-energy change (ΔG).	Worked Example 9.12; Problems 9.23– 9.24, 9.40, 9.42, 9.129, 9.132
		9.18 Use the Gibbs free-energy equation to calculate the temperature at which a reaction crosses over from spontaneous to nonspontaneous.	Worked Example 9.13; Problems 9.134–9.138

KEY TERMS

conservation of energy law *328*
endothermic *338*
energy *328*
enthalpy (*H*) *335*
enthalpy change (ΔH) *335*
entropy (*S*) *353*
exothermic *339*
first law of
  thermodynamics *329*
Gibbs free-energy change
  (ΔG) *355*

heat *329*
heat capacity (*C*) *342*
heat of combustion
  ($\Delta H°_c$) *359*
heat of reaction (ΔH) *335*
Hess's law *346*
internal energy (*E*) *330*
joule (J) *328*
kinetic energy (E_K) *328*
molar heat capacity
  (C_m) *342*

potential energy (E_P) *328*
specific heat (*c*) *342*
spontaneous process *352*
standard enthalpy of reaction
  ($\Delta H°$) *337*
standard heats of formation
  ($\Delta H°_f$) *348*
state function *331*
sublimation *339*
system *330*
surroundings *330*

temperature *329*
thermochemical equation *336*
thermochemistry *328*
thermodynamic standard
  state *337*
work (*w*) *332*

KEY EQUATIONS

- **Kinetic Energy (Section 9.1)**

$$E_K = \frac{1}{2}mv^2 \quad \text{where } m = \text{mass and } v = \text{velocity}$$

- **Work (Section 9.3)**

$$\text{Work } (w) = \text{Force } (F) \times \text{Distance } (d) = -P\Delta V$$

- **Internal Energy (Section 9.4)**

$$\Delta E = q + w = q - P\Delta V$$

- **Enthalpy (Section 9.4)**

$$\Delta H = q_p = \Delta E + P\Delta V \quad \text{where } q_p = \text{heat at constant pressure, } P = \text{pressure}$$
$$\Delta E = \text{internal energy change, and } \Delta V = \text{volume change}$$

- **Heat Capacity (Section 9.7)**

$$C = \frac{q}{\Delta T}$$

- **Heat Transfer (Section 9.7)**

$$q = C \times \Delta T$$
$$q = \text{Specific heat } (c) \times \text{Mass of substance } (m) \times \Delta T$$
$$q = C_m \times \text{Moles of substance} \times \Delta T$$

- **Heat of Reaction (Section 9.9)**

For the reaction $a\,A + b\,B + \ldots \longrightarrow c\,C + d\,D + \ldots$
$$\Delta H°_{reaction} = [c\,\Delta H°_f(C) + d\,\Delta H°_f(D) + \ldots] - [a\,\Delta H°_f(A) + b\,\Delta H°_f(B) + \ldots]$$

- **Free-Energy Change (Section 9.12)**

$$\Delta G = \Delta H - T\Delta S \text{ where } \Delta H = \text{enthalpy change, } T = \text{temperature, } \Delta S = \text{entropy change}$$

PRACTICE TEST

After studying this chapter, you can assess your understanding with these practice test questions, which are correlated with chapter learning objectives. If you answer a question incorrectly, refer to the learning objectives in the end-of-chapter Study Guide for assistance. The Study Guide provides a conceptual summary, references a Worked Example to model how to solve the problem, and gives additional problems for more practice.

1. When a Cheeto is burned in air to produce carbon dioxide and water, the amount of energy released is _____ the amount of energy released when a person eats a Cheeto and metabolizes the carbohydrates into carbon dioxide and water because energy change is _____. (LO 9.3)
   (a) the same as; a state function
   (b) the same as; a path function
   (c) different than; a state function
   (d) different than; a path function

2. A room-temperature balloon filled with air is placed in the freezer and the balloon contracts. What is the sign of q and w for the air inside the balloon? (LO 9.4)
   (a) $q = +, w = -$ (b) $q = +, w = +$
   (c) $q = -, w = -$ (d) $q = -, w = +$

3. Calculate the energy change as work when a gas expands from 15 L to 35 L against a constant external pressure of 1.5 atm. (LO 9.5)
   (a) -5.3 kJ   (b) $+3.0$ kJ   (c) -3.0 kJ
   (d) $+5.3$ kJ   (e) $+30$ kJ

4. For which of the following reactions are ΔE and ΔH equal? (LO 9.6)
   (a) $CO_2(g) + H_2O(l) \longrightarrow H_2CO_3(aq)$
   (b) $2\,NaHCO_3(s) \longrightarrow Na_2CO_3(s) + H_2O(g) + CO_2(g)$
   (c) $2\,H_2(g) + O_2(g) \longrightarrow 2\,H_2O(g)$
   (d) $CH_4(g) + 2\,O_2(g) \longrightarrow CO_2(g) + 2\,H_2O(g)$

5. Isooctane is the primary component of gasoline and burns in air to produce water and carbon dioxide.

$$2\,C_8H_{18}(l) + 25\,O_2(g) \longrightarrow 16\,CO_2(g) + 18\,H_2O(l)$$
$$\Delta H° = -10,940 \text{ kJ}$$

How much energy is released if 100.0 mL of isooctane (density = 0.690 g/mL) are burned? (LO 9.7)
   (a) 3300 kJ   (b) 6620 kJ
   (c) 6950 kJ   (d) 3.02 kJ

6. Several processes are given in the table and labeled as endothermic or exothermic and given a sign for $\Delta H°$. Which process is labeled with the correct sign of $\Delta H°$ and correct classification as endothermic or exothermic? **(LO 9.8)**

Process	Endo- or Exothermic?	Sign of $\Delta H°$
(a) Ammonium nitrate dissolves in water, and the temperature of the solution decreases.	Endothermic	$\Delta H° = +$
(b) Methane, the main component of natural gas, is burned to produce a flame on a stovetop.	Endothermic	$\Delta H° = -$
(c) Water freezes into ice in the freezer.	Endothermic	$\Delta H° = +$
(d) Rubbing alcohol evaporates from your skin.	Exothermic	$\Delta H° = -$

7. How much heat is required to raise a 50.0 g piece of iron from 25 °C to its melting point of 1538 °C? The specific heat capacity for iron is 0.451 J/g·°C. **(LO 9.9)**
(a) 34.1 kJ   (b) 168 kJ
(c) 12.1 kJ   (d) 6.78 kJ

8. Metal spheres are sold as a way to cool drinks without diluting them. Four metal spheres, each with a mass of 20.0 g, are cooled in the freezer to 0 °C. Which metal would decrease the temperature of a warm drink the most? **(LO 9.9)**
(a) Gold, $c = 0.128$ J/(g ·°C)
(b) Copper, $c = 0.385$ J/(g·°C)
(c) Aluminum, $c = 0.902$ J/(g·°C)
(d) Stainless steel, $c = 0.460$ J/(g·°C)

9. A 25.0 g piece of granite at 100.0 °C was added to 100.0 g of water at 25.0 °C, and the temperature rose to 28.4 °C. What is the specific heat capacity of the granite? (The specific heat capacity for water is 4.18 J/(g·°C).) **(LO 9.10)**
(a) 0.563 J/(g·°C)   (b) 1.53 J/(g·°C)
(c) 0.992 J/(g ·°C)   (d) 0.794 J/(g ·°C)

10. When 12.5 g of NH_4NO_3 is dissolved in 150.0 g of water at 25.0 °C in a coffee cup calorimeter, the final temperature of the solution is 19.7 °C. Assume that the specific heat of the solution is the same as that of water, 4.18 J/(g·°C). What is ΔH per mol of NH_4NO_3? **(LO 9.10)**

$$NH_4NO_3(s) \longrightarrow NH_4^+(aq) + NO_3^-(aq) \quad \Delta H = ?$$

(a) +3.60 kJ   (b) +23.0 kJ
(c) +21.3 kJ   (d) −3.60 kJ

11. Calculate the enthalpy change for the reaction

$$C(s) + 2 H_2(g) \longrightarrow CH_4(g) \quad \Delta H = ?$$

Given the enthalpy values for the following reactions. **(LO 9.11)**

$CH_4(g) + 2 O_2(g) \longrightarrow CO_2(g) + 2 H_2O(l) \quad \Delta H = -890.4$ kJ
$C(s) + O_2(g) \longrightarrow CO_2(g) \qquad\qquad\quad \Delta H = -393.5$ kJ
$H_2(g) + \dfrac{1}{2} O_2(g) \longrightarrow H_2O(l) \qquad\qquad \Delta H = -285.8$ kJ

(a) −1569.7 kJ   (b) +211.1 kJ
(c) −1855.5 kJ   (d) −74.7 kJ

12. A table of standard enthalpies of formation ($\Delta H°_f$) gives a value of −467.9 kJ/mol for $NaNO_3(s)$. Which reaction has a $\Delta H°$ value of −467.9 kJ? **(LO 9.13)**
(a) $Na^+(aq) + NO_3^-(aq) \longrightarrow NaNO_3(s)$
(b) $Na(s) + N(g) + O_3(g) \longrightarrow NaNO_3(s)$
(c) $Na(s) + \dfrac{1}{2} N_2(g) + \dfrac{3}{2} O_2(g) \longrightarrow NaNO_3(s)$
(d) $2 Na(s) + N_2(g) + 3 O_2(g) \longrightarrow 2 NaNO_3(s)$

13. What is ΔH for the explosion of nitroglycerin? **(LO 9.14)**

$$2 C_3H_5(NO_3)_3(l) \longrightarrow 3 N_2(g) + \dfrac{1}{2} O_2(g) + 6 CO_2(g)$$
$$+ 5 H_2O(g) \quad \Delta H = ?$$

Substance	$\Delta H°_f$ (kJ/mol)
$C_3H_5(NO_3)_3$ (l)	−364.0
H_2O (g)	−285.8
CO_2 (g)	−393.5

(a) −315.0 kJ   (b) −4517 kJ
(c) −3425 kJ   (d) −3062 kJ

14. Use bond dissociation energies to calculate the enthalpy change for the reaction: **(LO 9.15)**

$$C_2H_2(g) + HCl(g) \longrightarrow CH_2CHCl(g) \quad \Delta H = ?$$

Type of Bond	Bond Dissociation Energy (kJ/mol)
C—C	350
C=C	728
C≡C	965
H—Cl	432
C—Cl	330
C—H	410

(a) 42 kJ   (b) −71 kJ
(c) −308 kJ   (d) −42 kJ

15. Consider the following *endothermic* reaction of gaseous AB_3 molecules with A_2 molecules. **(LO 9.16, 9.17)**

Identify the *true* statement about the spontaneity of the reaction.
(a) The reaction is likely to be spontaneous at high temperatures.
(b) The reaction is likely to be spontaneous at low temperatures.
(c) The reaction is always spontaneous.
(d) The reaction is never spontaneous.

Answers:
1. a, 2. d, 3. c, 4. d, 5. a, 6. a, 7. a, 8. c, 9. d, 10. b, 11. d, 12. c, 13. d, 14. b, 15. a

Mastering **Chemistry**   provides end-of-chapter exercises feedback-enriched tutorial problems, animations, and interactive activities to encourage problem-solving practice and deeper understanding of key concepts and topics.

RAN   *Randomized in Mastering Chemistry*

CONCEPTUAL PROBLEMS

Problems 9.1–9.33 appear within the chapter.

9.34 A piece of dry ice (solid CO_2) is placed inside a balloon, and
RAN the balloon is tied shut. Over time, the carbon dioxide sublimes, causing the balloon to increase in volume. Give the sign of the enthalpy change and the sign of work for the sublimation of CO_2.

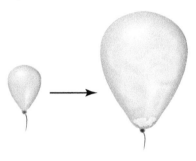

9.35 Imagine a reaction that results in a change in both volume and temperature:

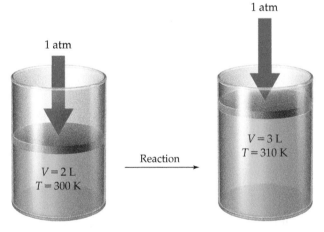

(a) Has any work been done? If so, is its sign positive or negative?

(b) Has there been an enthalpy change? If so, what is the sign of ΔH? Is the reaction exothermic or endothermic?

9.36 Redraw the following diagram to represent the situation (a) when work has been gained by the system and (b) when work has been lost by the system.

9.37 A reaction is carried out in a cylinder fitted with a movable
RAN piston. The starting volume is $V = 5.00$ L, and the apparatus is held at constant temperature and pressure. Assuming that $\Delta H = -35.0$ kJ and $\Delta E = -34.8$ kJ, redraw the piston to show its position after reaction. Does V increase, decrease, or remain the same?

9.38 The reaction of A with B to give D proceeds in two steps:
RAN

$$(1)\ A + B \longrightarrow C \qquad \Delta H° = -20\ kJ$$
$$(2)\ C + B \longrightarrow D \qquad \Delta H° = +50\ kJ$$
$$(3)\ A + 2\,B \longrightarrow D \qquad \Delta H° = ?$$

(a) Which Hess's law diagram represents the reaction steps and the overall reaction?

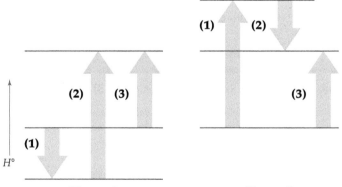

Diagram 1   Diagram 2

(b) What is the value of $\Delta H°$ for the overall reaction

$$A + 2\,B \longrightarrow D \quad \Delta H° = ?$$

9.39 Acetylene, C_2H_2, reacts with H_2 in two steps to yield ethane, CH_3CH_3:

(1) $HC \equiv CH + H_2 \rightarrow H_2C = CH_2 \qquad \Delta H° = -175.1 \text{ kJ}$

(2) $H_2C = CH_2 + H_2 \rightarrow CH_3CH_3 \qquad \Delta H° = -136.3 \text{ kJ}$

Net $HC \equiv CH + 2H_2 \longrightarrow CH_3CH_3 \quad \Delta H° = -311.4 \text{ kJ}$

Which arrow (**a–c**) in the Hess's law diagram corresponds to which step, and which arrow corresponds to the net reaction? Where are the reactants located on the diagram, and where are the products located?

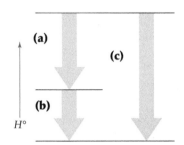

Hess's law diagram

9.40 The following reaction is exothermic:

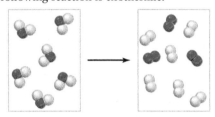

(a) Write a balanced equation for the reaction (red spheres represent A atoms and ivory spheres represent B atoms).

(b) What are the signs (+ or −) of ΔH and ΔS for the reaction?

(c) Is the reaction likely to be spontaneous at lower temperatures only, at higher temperatures only, or at all temperatures?

9.41 The following drawing portrays a reaction of the type $A \rightarrow B + C$, where the different colored spheres represent different molecular structures. Assume that the reaction has $\Delta H° = +55 \text{ kJ}$. Is the reaction likely to be spontaneous at all temperatures, nonspontaneous at all temperatures, or spontaneous at some but nonspontaneous at others? Explain.

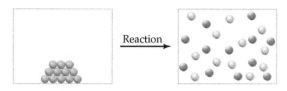

Reaction

9.42 What are the signs of ΔH, ΔS, and ΔG for the following spontaneous change? Explain.

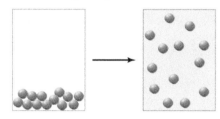

9.43 The following reaction of A_3 molecules is spontaneous.

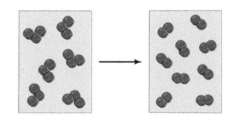

(a) Write a balanced equation for the reaction.

(b) What are the signs of ΔH, ΔS, and ΔG for the reaction? Explain.

SECTION PROBLEMS

Heat, Work, and Energy (Section 9.1–9.3)

9.44 What is the difference between heat and temperature? Between work and energy? Between kinetic energy and potential energy?

9.45 What is internal energy?

9.46 Which has more kinetic energy, a 1400-kg car moving at 115 km/h or a 12,000-kg truck moving at 38 km/h?

9.47 Assume that the kinetic energy of a 1400-kg car moving at 115 km/h (Problem 9.46) could be converted entirely into heat. What amount of water could be heated from 20 °C to 50 °C by the car's energy? 4.184 J are required to heat 1 g of water by 1 °C.

9.48 Which of the following are state functions, and which are not?

(a) The temperature of an ice cube

(b) The volume of an aerosol can

(c) The amount of time required for Paula Radcliffe to run her world-record marathon: 2:15:25

9.49 Which of the following are state functions, and which are not?

(a) The distance from your dorm room to your chemistry class

(b) The temperature in the room of your chemistry class

(c) The balance in your bank account

9.50 Calculate the work done in joules by a chemical reaction if the volume increases from 3.2 L to 3.4 L against a constant external pressure of 3.6 atm. What is the sign of the energy change?

9.51 The addition of H_2 to $C=C$ double bonds is an important reaction used in the preparation of margarine from vegetable oils. If 50.0 mL of H_2 and 50.0 mL of ethylene (C_2H_4) are allowed to react at 1.5 atm, the product ethane (C_2H_6) has a volume of 50.0 mL. Calculate the amount of PV work done, and tell the direction of the energy flow.

$$C_2H_4(g) + H_2(g) \longrightarrow C_2H_6(g)$$

9.52 Assume that the nutritional content of an apple—say, 50 Cal
RAN (1 Cal = 1000 cal)—could be used to light a light bulb. For
how many minutes would there be light from each of the
following?

(a) A 100-watt incandescent bulb (1 W = 1 J/s)

(b) A 23-watt compact fluorescent bulb, which provides a
similar amount of light

9.53 A double cheeseburger has a caloric content of 440 Cal
RAN (1 Cal = 1000 cal). If this energy could be used to operate
a television set, for how many hours would the following
sets run?

(a) A 275-watt 46-in. plasma TV (1 W = 1 J/s)

(b) A 175-watt 46-in. LCD TV

9.54 A reaction inside a cylindrical container with a movable
RAN piston causes the volume to change from 12.0 L to 18.0 L
while the pressure outside the container remains constant at
0.975 atm. (The volume of a cylinder is $V = \pi r^2 h$, where h
is the height; 1 L·atm = 101.325 J.)

(a) What is the value in joules of the work w done during
the reaction?

(b) The diameter of the piston is 17.0 cm. How far does the
piston move?

9.55 At a constant pressure of 0.905 atm, a chemical reaction
RAN takes place in a cylindrical container with a movable piston
having a diameter of 40.0 cm. During the reaction, the height
of the piston drops by 65.0 cm. (The volume of a cylinder is
$V = \pi r^2 h$, where h is the height; 1 L atm = 101.3 J.)

(a) What is the change in volume in liters during the
reaction?

(b) What is the value in joules of the work w done during
the reaction?

9.56 When a sample of a hydrocarbon fuel is ignited and burned in
RAN oxygen, the internal energy decreases by 7.20 kJ. If 5670 J of
heat were transferred to the surroundings, what is the sign and
magnitude of work? If the reaction took place in an environment with a pressure of 1 atm, what was the volume change?

9.57 Used in welding metals, the reaction of acetylene with oxy-
RAN gen has $\Delta H° = -1256.2$ kJ:

$$C_2H_2(g) + 5/2\ O_2(g) \longrightarrow H_2O(g) + 2\ CO_2(g)$$
$$\Delta H° = -1256.2\ \text{kJ}$$

How much PV work is done in kilojoules and what is the
value of ΔE in kilojoules for the reaction of 6.50 g of acetylene at atmospheric pressure if the volume change is -2.80 L?

Energy and Enthalpy (Section 9.4)

9.58 What is the difference between the internal energy change
ΔE and the enthalpy change ΔH? Which of the two is measured at constant pressure and which at constant volume?

9.59 Under what circumstances are ΔE and ΔH essentially equal?

9.60 The explosion of 2.00 mol of solid trinitrotoluene (TNT;
RAN $C_7H_5N_3O_6$) with a volume of approximately 274 mL produces gases with a volume of 448 L at room temperature
and 1.0 atm pressure. How much PV work in kilojoules is
done during the explosion?

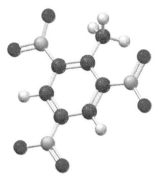

Trinitrotoluene

$$2\ C_7H_5N_3O_6(s) \rightarrow 12\ CO(g) + 5\ H_2(g) + 3\ N_2(g) + 2\ C(s)$$

9.61 The reaction between hydrogen and oxygen to yield water
RAN vapor has $\Delta H° = -484$ kJ. How much PV work is done,
and what is the value of ΔE in kilojoules for the reaction of
0.50 mol of H_2 with 0.25 mol of O_2 at atmospheric pressure
if the volume change is -5.6 L?

$$2\ H_2(g) + O_2(g) \longrightarrow 2\ H_2O(g) \quad \Delta H° = -484\ \text{kJ}$$

9.62 The enthalpy change for the reaction of 50.0 mL of ethylene
RAN with 50.0 mL of H_2 at 1.5 atm pressure (Problem 9.51) is
$\Delta H = -0.31$ kJ. What is the value of ΔE?

9.63 Assume that a particular reaction evolves 244 kJ of heat and
RAN that 35 kJ of PV work is gained by the system. What are the
values of ΔE and ΔH for the system? For the surroundings?

9.64 What is the enthalpy change (ΔH) for a reaction at a con-
RAN stant pressure of 1.00 atm if the internal energy change (ΔE)
is 44.0 kJ and the volume increase is 14.0 L? (1 L·atm =
101.325 J.)

9.65 A reaction takes place at a constant pressure of 1.10 atm with
RAN an internal energy change (ΔE) of 71.5 kJ and a volume
decrease of 13.6 L. What is the enthalpy change (ΔH) for the
reaction? (1 L·atm = 101.325 J.)

Thermochemical Equations for Chemical and Physical Changes (Section 9.5–9.6)

9.66 Indicate the direction of heat transfer between the system
and the surroundings, classify the following processes as
endo- or exothermic, and give the sign of $\Delta H°$.

(a) Evaporating rubbing alcohol from your skin

(b) Solidifying molten gold into a gold bar

(c) Melting solid gallium metal with the heat from your hand

9.67 Indicate the direction of heat transfer between the system
and the surroundings, classify the following processes as
endo- or exothermic, and give the sign of $\Delta H°$.

(a) $N_2(g) + O_2(g) \rightarrow 2\ NO(g) \quad \Delta H° = +182.6$ kJ

(b) $2\ H_2O(g) \rightarrow 2\ H_2(g) + O_2\ (g) \quad \Delta H° = +483.6$ kJ

(c) $H_2(g) + Cl_2(g) \rightarrow 2HCl(g) \quad \Delta H° = -184.6$ kJ

9.68 The familiar "ether" used as an anesthetic agent is diethyl
RAN   ether, $C_4H_{10}O$. Its heat of vaporization is +26.5 kJ/mol at
its boiling point. How much energy in kilojoules is required
to convert 100 mL of diethyl ether at its boiling point from
liquid to vapor if its density is 0.7138 g/mL?

Diethyl ether

9.69 How much energy in kilojoules is required to convert 100 mL
RAN   of water at its boiling point from liquid to vapor, and how
does this compare with the result calculated in Problem 9.68
for diethyl ether? $[\Delta H_{vap}(H_2O) = +40.7 \text{ kJ/mol}]$

9.70 Aluminum metal reacts with chlorine with a spectacular dis-
RAN   play of sparks:

$$2 \text{ Al}(s) + 3 \text{ Cl}_2(g) \rightarrow 2 \text{ AlCl}_3(s) \quad \Delta H° = -1408.4 \text{ kJ}$$

How much heat in kilojoules is released on reaction of 5.00 g
of Al?

9.71 How much heat in kilojoules is evolved or absorbed in the
RAN   reaction of 1.00 g of Na with H_2O? Is the reaction exothermic or endothermic?

$$2 \text{ Na}(s) + 2 \text{ H}_2O(l) \rightarrow 2 \text{ NaOH}(aq) + \text{H}_2(g)$$
$$\Delta H° = -368.4 \text{ kJ}$$

9.72 How much heat in kilojoules is evolved or absorbed in each
of the following reactions?
(a) Burning of 15.5 g of propane:

$$C_3H_8(g) + 5 \text{ O}_2(g) \rightarrow 3 \text{ CO}_2(g) + 4 \text{ H}_2O(l)$$
$$\Delta H° = -2220 \text{ kJ}$$

(b) Reaction of 4.88 g of barium hydroxide octahydrate
with ammonium chloride:

$$\text{Ba(OH)}_2 \cdot 8 \text{ H}_2O(s) + 2 \text{ NH}_4Cl(s) \rightarrow$$
$$\text{BaCl}_2(aq) + 2 \text{ NH}_3(aq) + 10 \text{ H}_2O(l) \quad \Delta H° = +80.3 \text{ kJ}$$

9.73 Nitromethane (CH_3NO_2), sometimes used as a fuel in drag
RAN   racers, burns according to the following equation. How much
heat is released by burning 100.0 g of nitromethane?

$$4 \text{ CH}_3NO_2(l) + 7 \text{ O}_2(g) \longrightarrow 4 \text{ CO}_2(g) + 6 \text{ H}_2O(g) + 4 \text{ NO}_2(g)$$
$$\Delta H° = -2441.6 \text{ kJ}$$

9.74 How much heat in kilojoules is evolved or absorbed in the
RAN   reaction of 2.50 g of Fe_2O_3 with enough carbon monoxide to produce iron metal? Is the process exothermic or
endothermic?

$$Fe_2O_3(s) + 3 \text{ CO}(g) \rightarrow 2 \text{ Fe}(s) + 3 \text{ CO}_2(g)$$
$$\Delta H° = -24.8 \text{ kJ}$$

9.75 How much heat in kilojoules is evolved or absorbed in the reac-
RAN   tion of 233.0 g of calcium oxide with enough carbon to produce
calcium carbide? Is the process exothermic or endothermic?

$$\text{CaO}(s) + 3 \text{ C}(s) \rightarrow \text{CaC}_2(s) + \text{CO}(g)$$
$$\Delta H° = 464.6 \text{ kJ}$$

Calorimetry and Heat Capacity (Section 9.7)

9.76 What is the difference between heat capacity and specific heat?

9.77 Does a measurement carried out in a bomb calorimeter give
a value for ΔH or ΔE? Explain.

9.78 Sodium metal is sometimes used as a cooling agent in heat-
exchange units because of its relatively high molar heat
capacity of 28.2 J/(mol·°C). What is the specific heat and
molar heat capacity of sodium in J/g·°C?

9.79 Titanium metal is used as a structural material in many high-
tech applications, such as in jet engines. What is the specific
heat of titanium in J/(g·°C) if it takes 89.7 J to raise the
temperature of a 33.0 g block by 5.20 °C? What is the molar
heat capacity of titanium in J/(mol·°C)?

9.80 Assuming that Coca-Cola has the same specific heat as water
RAN   [4.18 J/(g·°C)], calculate the amount of heat in kilojoules
transferred when one can (about 350 g) is cooled from 25 °C
to 3 °C.

9.81 Calculate the amount of heat required to raise the tempera-
RAN   ture of 250.0 g (approximately 1 cup) of hot chocolate from
25.0 °C to 80.0 °C. Assume hot chocolate has the same specific heat as water [4.18 J/(g·°C)].

9.82 Instant cold packs used to treat athletic injuries contain solid
RAN NH_4NO_3 and a pouch of water. When the pack is squeezed,
the pouch breaks and the solid dissolves, lowering the temperature because of the endothermic reaction

$$\text{NH}_4NO_3(s) \longrightarrow \text{NH}_4NO_3(aq) \quad \Delta H = +25.7 \text{ kJ}$$

What is the final temperature in a squeezed cold pack that
contains 50.0 g of NH_4NO_3 dissolved in 125 mL of water?
Assume a specific heat of 4.18 J/(g·°C) for the solution, an
initial temperature of 25.0 °C, and no heat transfer between
the cold pack and the environment.

9.83 Instant hot packs contain a solid and a pouch of water. When
RAN   the pack is squeezed, the pouch breaks and the solid dissolves, increasing the temperature because of the exothermic
reaction. The following reaction is used to make a hot pack:

$$\text{LiCl}(s) \xrightarrow{\text{H}_2O} \text{Li}^+(aq) + \text{Cl}^-(aq) \quad \Delta H = -36.9 \text{ kJ}$$

What is the final temperature in a squeezed hot pack that
contains 25.0 g of LiCl dissolved in 125 mL of water?
Assume a specific heat of 4.18 J/(g·°C) for the solution, an
initial temperature of 25.0 °C, and no heat transfer between
the hot pack and the environment.

9.84 When 1.045 g of CaO is added to 50.0 mL of water at 25.0 °C
RAN   in a calorimeter, the temperature of the water increases to 32.3 °C.
Assuming that the specific heat of the solution is 4.18 J/(g·°C)
and that the calorimeter itself absorbs a negligible amount of
heat, calculate ΔH in kilojoules/mol $Ca(OH)_2$ for the reaction

$$\text{CaO}(s) + \text{H}_2O(l) \rightarrow \text{Ca(OH)}_2(aq)$$

9.85 When a solution containing 8.00 g of NaOH in 50.0 g of
water at 25.0 °C is added to a solution of 8.00 g of HCl in
250.0 g of water at 25.0 °C in a calorimeter, the temperature
of the solution increases to 33.5 °C. Assuming that the specific heat of the solution is 4.18 J/(g·°C) and that the calorimeter itself absorbs a negligible amount of heat, calculate
ΔH in kilojoules/mol for the reaction

$$\text{NaOH}(aq) + \text{HCl}(aq) \rightarrow \text{NaCl}(aq) + \text{H}_2O(l)$$

When the experiment is repeated using a solution of 10.00 g of HCl in 248.0 g of water, the same temperature increase is observed. Explain.

9.86 When 0.187 g of benzene, C_6H_6, is burned in a bomb calo-
RAN rimeter the temperature rises by 3.45 °C. If the heat capacity of the calorimeter is 2.46 kJ/°C, calculate the combustion energy (ΔE) for benzene in units of kJ/g and kJ/mol.

Benzene

9.87 When 0.500 g of ethanol, C_2H_6O, is burned in a bomb calo-
RAN rimeter the temperature rises by 9.83 °C. If the heat capacity of the calorimeter is 1.50 kJ/°C, calculate the combustion energy (ΔE) for ethanol in units of kJ/g and kJ/mol.

9.88 When 1.50 g of magnesium metal is allowed to react with 200 mL
RAN of 6.00 M aqueous HCl, the temperature rises from 25.0 °C to 42.9 °C. Calculate ΔH in kilojoules for the reaction, assuming that the heat capacity of the calorimeter is 776 J/°C, that the specific heat of the final solution is the same as that of water [4.18 J/(g·°C)] and that the density of the solution is 1.00 g/mL.

9.89 A 110.0 g piece of molybdenum metal is heated to 100.0 °C and placed in a calorimeter that contains 150.0 g of water at 24.6 °C. The system reaches equilibrium at a final temperature of 28.0 °C. Calculate the specific heat of molybdenum metal in J/g·°C. The specific heat of water is 4.18 J/g·°C.

9.90 Citric acid has three dissociable hydrogens. When 5.00 mL of
RAN 0.64 M citric acid and 45.00 mL of 0.77 M NaOH are mixed at an initial temperature of 26.0 °C, the temperature rises to 27.9 °C as the citric acid is neutralized. The combined mixture has a mass of 51.6 g and a specific heat of 4.0 J/(g·°C). Assuming that no heat is transferred to the surroundings, calculate the enthalpy change for the reaction of 1.00 mol of citric acid in kJ. Is the reaction exothermic or endothermic?

Citric acid

9.91 Assume that 100.0 mL of 0.200 M CsOH and 50.0 mL of 0.400 M HCl are mixed in a calorimeter. The solutions start out at 22.50 °C, and the final temperature after reaction is 24.28 °C. The densities of the solutions are all 1.00 g/mL, and the specific heat of the mixture is 4.2 J/(g·°C). What is the enthalpy change for the neutralization reaction of 1.00 mol of CsOH in kJ?

Hess's Law (Section 9.8)

9.92 What is Hess's law, and why does it "work"?

9.93 The following steps occur in the reaction of ethyl alcohol
RAN (CH_3CH_2OH) with oxygen to yield acetic acid (CH_3CO_2H). Show that equations 1 and 2 sum to give the net equation and calculate $\Delta H°$ for the net equation.

(1) $CH_3CH_2OH(l) + 1/2\,O_2(g) \rightarrow$
$\qquad CH_3CHO(g) + H_2O(l) \quad \Delta H° = -174.2$ kJ

(2) $CH_3CHO(g) + 1/2\,O_2(g) \rightarrow$
$\qquad \underline{CH_3CO_2H(l) \qquad\qquad \Delta H° = -318.4$ kJ}

(Net) $CH_3CH_2OH(l) + O_2(g) \rightarrow$
$\qquad CH_3CO_2H(l) + H_2O(l) \quad \Delta H° = ?$

9.94 The industrial degreasing solvent methylene chloride, CH_2Cl_2, is prepared from methane by reaction with chlorine:

$$CH_4(g) + 2\,Cl_2(g) \rightarrow CH_2Cl_2(g) + 2\,HCl(g)$$

Use the following data to calculate $\Delta H°$ in kilojoules for the reaction:

$CH_4(g) + Cl_2(g) \rightarrow CH_3Cl(g) + HCl(g) \quad \Delta H° = -98.3$ kJ
$CH_3Cl(g) + Cl_2(g) \rightarrow CH_2Cl_2(g) + HCl(g) \quad \Delta H° = -104$ kJ

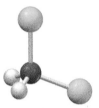

Methylene chloride

9.95 Hess's law can be used to calculate reaction enthalpies for hypothetical processes that can't be carried out in the laboratory. Set up a Hess's law cycle that will let you calculate $\Delta H°$ for the conversion of methane to ethylene:

$$2\,CH_4(g) \rightarrow C_2H_4(g) + 2\,H_2(g)$$

You can use the following information:

$2\,C_2H_6(g) + 7\,O_2(g) \rightarrow 4\,CO_2(g) + 6\,H_2O(l)$
$\qquad\qquad\qquad\qquad\qquad \Delta H° = -3120.8$ kJ

$CH_4(g) + 2\,O_2(g) \rightarrow CO_2(g) + 2\,H_2O(l)$
$\qquad\qquad\qquad\qquad\qquad \Delta H° = -890.3$ kJ

$C_2H_4(g) + H_2(g) \rightarrow C_2H_6(g) \qquad \Delta H° = -136.3$ kJ

$H_2O(l) \quad \Delta H°_f = -285.8$ kJ/mol

9.96 Find $\Delta H°$ in kilojoules for the reaction of nitric oxide with oxygen, $2\,NO(g) + O_2(g) \rightarrow N_2O_4(g)$, given the following data:

$N_2O_4(g) \rightarrow 2\,NO_2(g) \qquad\qquad \Delta H° = 55.3$ kJ
$NO(g) + 1/2\,O_2(g) \rightarrow NO_2(g) \quad \Delta H° = -58.1$ kJ

9.97 Set up a Hess's law cycle, and use the following information to calculate $\Delta H°_f$ for aqueous nitric acid, $HNO_3(aq)$. You will need to use fractional coefficients for some equations.

$$3\ NO_2(g) + H_2O(l) \rightarrow 2\ HNO_3(aq) + NO(g)$$
$$\Delta H° = -137.3\ kJ$$
$$2\ NO(g) + O_2(g) \rightarrow 2\ NO_2(g) \qquad \Delta H° = -116.2\ kJ$$
$$4\ NH_3(g) + 5\ O_2(g) \rightarrow 4\ NO(g) + 6\ H_2O(l)$$
$$\Delta H° = -1165.2\ kJ$$

$$NH_3(g) \quad \Delta H°_f = -46.1\ kJ/mol$$
$$H_2O(l) \quad \Delta H°_f = -285.8\ kJ/mol$$

Heats of Formation (Section 9.9)

9.98 What is a compound's standard heat of formation?

9.99 How is the standard state of an element defined? Why do elements always have $\Delta H°_f = 0$?

9.100 What phase of matter is associated with the standard states
RAN of the following elements and compounds?
(a) Cl_2 (b) Hg   (c) CO_2 (d) Ga

9.101 What is the phase of the standard states of the following ele-
RAN ments and compounds?
(a) NH_3 (b) Fe   (c) N_2 (d) Br_2

9.102 Write balanced equations for the formation of the following compounds from their elements.
(a) Iron(III) oxide
(b) Sucrose (table sugar, $C_{12}H_{22}O_{11}$)
(c) Uranium hexafluoride (a solid at 25 °C)

9.103 Write balanced equations for the formation of the following
RAN compounds from their elements.
(a) Ethanol (C_2H_6O)
(b) Sodium sulfate
(c) Dichloromethane (a liquid, CH_2Cl_2)

9.104 Sulfuric acid (H_2SO_4), the most widely produced chemical in the world, is made by a two-step oxidation of sulfur to sulfur trioxide, SO_3, followed by reaction with water. Calculate $\Delta H°_f$ for SO_3 in kJ/mol, given the following data:

$$S(s) + O_2(g) \rightarrow SO_2(g) \qquad \Delta H° = -296.8\ kJ$$
$$SO_2(g) + 1/2\ O_2(g) \rightarrow SO_3(g) \quad \Delta H° = -98.9\ kJ$$

9.105 Calculate $\Delta H°_f$ in kJ/mol for benzene, C_6H_6, from the following data:

$$2\ C_6H_6(l) + 15\ O_2(g) \rightarrow 12\ CO_2(g) + 6\ H_2O(l)$$
$$\Delta H° = -6534\ kJ$$
$$\Delta H°_f\ (CO_2) = -393.5\ kJ/mol$$
$$\Delta H°_f\ (H_2O) = -285.8\ kJ/mol$$

9.106 The standard enthalpy change for the reaction of $SO_3(g)$ with
RAN $H_2O(l)$ to yield $H_2SO_4(aq)$ is $\Delta H° = -227.8\ kJ$. Use the information in Problem 9.104 to calculate $\Delta H°_f$ for $H_2SO_4(aq)$ in kJ/mol. [For $H_2O(l)$, $\Delta H°_f = -285.8\ kJ/mol$.]

9.107 Acetic acid (CH_3CO_2H), whose aqueous solutions are known as vinegar, is prepared by reaction of ethyl alcohol (CH_3CH_2OH) with oxygen:

$$CH_3CH_2OH(l) + O_2(g) \rightarrow CH_3CO_2H(l) + H_2O(l)$$

Use the following data to calculate $\Delta H°$ in kilojoules for the reaction:

$$\Delta H°_f\ [CH_3CH_2OH(l)] = -277.7\ kJ/mol$$
$$\Delta H°_f[CH_3CO_2H(l)] = -484.5\ kJ/mol$$
$$\Delta H°_f\ [H_2O(l)] = -285.8\ kJ/mol$$

9.108 Styrene (C_8H_8), the precursor of polystyrene polymers, has a standard heat of combustion of $-4395\ kJ/mol$. Write a balanced equation for the combustion reaction, and calculate $\Delta H°_f$ for styrene in kJ/mol.

$$\Delta H°_f\ [CO_2(g)] = -393.5\ kJ/mol;$$
$$\Delta H°_f\ [H_2O(l)] = -285.8\ kJ/mol$$

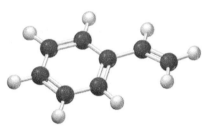

Styrene

9.109 Methyl *tert*-butyl ether (MTBE), $C_5H_{12}O$, a gasoline additive used to boost octane ratings, has $\Delta H°_f = -313.6\ kJ/mol$. Write a balanced equation for its combustion reaction, and calculate its standard heat of combustion in kilojoules.

9.110 Methyl *tert*-butyl ether is prepared by reaction of methanol
RAN $(l)(\Delta H°_f = -239.2\ kJ/mol)$ with 2-methyl-propene (g), according to the equation

$$\begin{array}{c} CH_3 \\ | \\ CH_3-C=CH_2 + CH_3OH \longrightarrow \\ \text{2-Methylpropene} \end{array}$$

$$\begin{array}{c} CH_3 \\ | \\ CH_3-C-O-CH_3 \quad \Delta H° = -57.5\ kJ \\ | \\ CH_3 \\ \text{Methyl } \textit{tert}\text{-butyl ether} \end{array}$$

Calculate $\Delta H°_f$ in kJ/mol for 2-methylpropene.

9.111 One possible use for the cooking fat left over after making french fries is to burn it as fuel. Write a balanced equation, and use the following data to calculate the amount of energy released in kJ/mL from the combustion of cooking fat:

Formula $= C_{51}H_{88}O_6 \quad \Delta H°_f = -1310\ kJ/mol$
Density $= 0.94\ g/ml$

9.112 Given the standard heats of formation shown in Appendix B, what is $\Delta H°$ in kilojoules for the reaction $CaCO_3(s) \rightarrow CaO(s) + CO_2(g)$?

9.113 Given the standard heats of formation shown in Appendix B, what is $\Delta H°$ in kilojoules for the reaction

$$3\ N_2O_4(g) + 2\ H_2O(l) \rightarrow 4\ HNO_3(aq) + 2\ NO(g).$$

9.114 Calculate $\Delta H°$ in kilojoules for the synthesis of lime (CaO)
RAN   from limestone ($CaCO_3$), the key step in the manufacture of
cement.

$$CaCO_3(s) \rightarrow CaO(s) + CO_2(g) \quad \Delta H°_f[CaCO_3(s)] =$$
$$-1207.6 \text{ kJ/mol}$$
$$\Delta H°_f[CaO(s)] = -634.9 \text{ kJ/mol}$$
$$\Delta H°_f[CO_2(g)] = -393.5 \text{ kJ/mol}$$

9.115 Use the information in Table 9.2 to calculate $\Delta H°$ in kilojoules
for the photosynthesis of glucose ($C_6H_{12}O_6$) and O_2 from
CO_2 and liquid H_2O, a reaction carried out by all green plants.

9.116 Use the data in Appendix B to find standard enthalpies of
reaction in kilojoules for the following processes:
(a) $C(s) + CO_2(g) \rightarrow 2 CO(g)$
(b) $2 H_2O_2(aq) \rightarrow 2 H_2O(l) + O_2(g)$
(c) $Fe_2O_3(s) + 3 CO(g) \rightarrow 2 Fe(s) + 3 CO_2(g)$

9.117 Isooctane, C_8H_{18}, is the component of gasoline from which
the term *octane rating* derives.

Isooctane

(a) Write a balanced equation for the combustion of isooctane (l) with O_2 to yield $CO_2(g)$ and $H_2O(l)$.
(b) The standard molar heat of combustion for isooctane (l)
is -5461 kJ/mol. Calculate $\Delta H°_f$ for isooctane (l).

Bond Dissociation Energies (Section 9.10)

9.118 Calculate an approximate heat of combustion for ethane
RAN   (C_2H_6) in kilojoules by using the bond dissociation energies
in Table 9.3. (The strength of the O=O bond is 498 kJ/
mol, and that of a C=O bond in CO_2 is 804 kJ/mol.)

9.119 Use the data in Table 9.3 to calculate an approximate $\Delta H°$
RAN   in kilojoules for the synthesis of hydrazine from ammonia:
$2 NH_3(g) + Cl_2(g) \rightarrow N_2H_4(g) + 2 HCl(g)$.

9.120 Use the average bond dissociation energies in Table 9.3 to
RAN   calculate approximate reaction enthalpies in kilojoules for
the following processes:
(a) $2 CH_4(g) \rightarrow C_2H_6(g) + H_2(g)$
(b) $C_2H_6(g) + F_2(g) \rightarrow C_2H_5F(g) + HF(g)$
(c) $N_2(g) + 3 H_2(g) \rightarrow 2 NH_3(g)$

9.121 Use the bond dissociation energies in Table 9.3 to calculate
RAN   an approximate $\Delta H°$ in kilojoules for the high-temperature
industrial synthesis of isopropyl alcohol (rubbing alcohol)
by reaction of water vapor with propene.

$$\begin{array}{cc} & \text{OH} \\ & | \\ CH_3CH{=}CH_2 + H_2O \longrightarrow & CH_3CHCH_3 \\ \text{Propene} & \text{Isopropyl alcohol} \end{array}$$

Free Energy and Entropy (Section 9.11–9.12)

9.122 What does entropy measure?

9.123 What are the two terms that make up the free-energy change
for a reaction, ΔG, and which of the two is usually more
important?

9.124 How is it possible for a reaction to be spontaneous yet
endothermic?

9.125 Is it possible for a reaction to be nonspontaneous yet exothermic? Explain.

9.126 Tell whether the entropy changes for the following processes
are likely to be positive or negative.
(a) The fizzing of a newly opened can of soda
(b) The growth of a plant from seed

9.127 Tell whether the entropy changes, ΔS, for the following processes are likely to be positive or negative.
(a) The conversion of liquid water to water vapor at 100 °C
(b) The freezing of liquid water to ice at 0 °C
(c) The eroding of a mountain by a glacier

9.128 Tell whether the free-energy changes, ΔG, for the processes
RAN   listed in Problem 9.127 are likely to be positive, negative, or
zero.

9.129 When a bottle of perfume is opened, odorous molecules mix
with air and slowly diffuse throughout the entire room. Is
ΔG for the diffusion process positive, negative, or zero?
What about ΔH and ΔS for the diffusion?

9.130 One of the steps in the cracking of petroleum into gasoline
involves the thermal breakdown of large hydrocarbon molecules into smaller ones. For example, the following reaction
might occur:

$$C_{11}H_{24} \longrightarrow C_4H_{10} + C_4H_8 + C_3H_6$$

Is ΔS for this reaction likely to be positive or negative? Explain.

9.131 The commercial production of 1,2-dichloroethane, a solvent
used in dry cleaning, involves the reaction of ethylene with
chlorine:

$$C_2H_4(g) + Cl_2(g) \longrightarrow C_2H_4Cl_2(l)$$

Is ΔS for this reaction likely to be positive or negative? Explain.

9.132 Tell whether reactions with the following values of ΔH and
RAN ΔS are spontaneous or nonspontaneous and whether they
are exothermic or endothermic.
(a) $\Delta H = -48$ kJ; $\Delta S = +135$ J/K at 400 K
(b) $\Delta H = -48$ kJ; $\Delta S = -135$ J/K at 400 K
(c) $\Delta H = +48$ kJ; $\Delta S = +135$ J/K at 400 K
(d) $\Delta H = +48$ kJ; $\Delta S = -135$ J/K at 400 K

9.133 Tell whether reactions with the following values of ΔH and
ΔS are spontaneous or nonspontaneous and whether they
are exothermic or endothermic.
(a) $\Delta H = -128$ kJ; $\Delta S = 35$ J/K at 500 K
(b) $\Delta H = +67$ kJ; $\Delta S = -140$ J/K at 250 K
(c) $\Delta H = +75$ kJ; $\Delta S = 95$ J/K at 800 K

9.134 Suppose that a reaction has $\Delta H = -33$ kJ and $\Delta S = -58$ J/K.
RAN   At what temperature will it change from spontaneous to
nonspontaneous?

9.135 Suppose that a reaction has $\Delta H = +41$ kJ and $\Delta S = -27$ J/K. At what temperature, if any, will it change between
spontaneous and nonspontaneous?

9.136 Which of the reactions (a)–(d) in Problem 9.132 are sponta-
RAN neous at all temperatures, which are nonspontaneous at all
temperatures, and which have an equilibrium temperature?

9.137 Vinyl chloride ($H_2C = CHCl$), the starting material used
RAN in the industrial preparation of poly(vinyl chloride), is prepared by a two-step process that begins with the reaction of
Cl_2 with ethylene to yield 1,2-dichloroethane:

$$Cl_2(g) + H_2C = CH_2(g) \rightarrow ClCH_2CH_2Cl(l)$$

$$\Delta H° = -217.5 \text{ kJ}$$

$$\Delta S° = -233.9 \text{ J/K}$$

Vinyl chloride

(a) Tell whether the reaction is favored by entropy, by enthalpy,
by both, or by neither, and then calculate $\Delta G°$ at 298 K.

(b) Tell whether the reaction has an equilibrium temperature between spontaneous and nonspontaneous. If yes,
calculate the equilibrium temperature.

9.138 Ethyl alcohol has $\Delta H_{fusion} = 5.02 \text{ kJ/mol}$ and melts at
-114.1 °C. What is the value of ΔS_{fusion} for ethyl alcohol?

9.139 Chloroform has $\Delta H_{vaporization} = 29.2 \text{ kJ/mol}$ and boils at
61.2 °C. What is the value of $\Delta S_{vaporization}$ for chloroform?

9.140 The boiling point of a substance is defined as the temperature
at which liquid and vapor coexist in equilibrium. Use the heat
of vaporization ($\Delta H_{vap} = 30.91 \text{ kJ/mol}$) and the entropy
of vaporization [$\Delta S_{vap} = 93.2 \text{ J/(K·mol)}$] to calculate the
boiling point (°C) of liquid bromine.

9.141 What is the melting point of benzene in kelvin if
$\Delta H_{fusion} = 9.95 \text{ kJ/mol}$ and $\Delta S_{fusion} = 35.7 \text{ J/ (K·mol)}$?

9.142 Metallic mercury is obtained by heating the mineral cinnabar (HgS) in air:

$$HgS(s) + O_2(g) \rightarrow Hg(l) + SO_2(g)$$

(a) Use the data in Appendix B to calculate $\Delta H°$ in kilojoules
for the reaction.

(b) The entropy change for the reaction is $\Delta S° = +36.7 \text{ J/K}$.
Is the reaction spontaneous at 25 °C?

(c) Under what conditions, if any, is the reaction nonspontaneous? Explain.

9.143 Methanol (CH_3OH) is made industrially in two steps from CO
and H_2. It is so cheap to make that it is being considered for use
as a precursor to hydrocarbon fuels, such as methane (CH_4):

$Step\ 1.$ $CO(g) + 2 H_2(g) \rightarrow CH_3OH(l)$

$$\Delta S° = -332 \text{ J/K}$$

$Step\ 2.$ $CH_3OH(l) \rightarrow CH_4(g) + 1/2 O_2(g)$

$$\Delta S° = 162 \text{ J/K}$$

(a) Calculate $\Delta H°$ in kilojoules for step 1.

(b) Calculate $\Delta G°$ in kilojoules for step 1.

(c) Is step 1 spontaneous at 298 K?

(d) Which term is more important, $\Delta H°$ or $\Delta S°$?

(e) In what temperature range is step 1 spontaneous?

(f) Calculate $\Delta H°$ for step 2.

(g) Calculate $\Delta G°$ for step 2.

(h) Is step 2 spontaneous at 298 K?

(i) Which term is more important, $\Delta H°$ or $\Delta S°$?

(j) In what temperature range is step 2 spontaneous?

(k) Calculate an overall $\Delta G°$, $\Delta H°$, and $\Delta S°$ for the formation of CH_4 from CO and H_2.

(l) Is the overall reaction spontaneous at 298 K?

(m) If you were designing a production facility, would you
plan on carrying out the reactions in separate steps or
together? Explain.

MULTICONCEPT PROBLEMS

9.144 Ethyl chloride (C_2H_5Cl), a substance used as a topical anes-
RAN thetic, is prepared by reaction of ethylene with hydrogen
chloride:

$$C_2H_4(g) + HCl(g) \longrightarrow C_2H_5Cl(g) \quad \Delta H° = -72.3 \text{ kJ}$$

Ethyl chloride

How much PV work is done in kilojoules, and what is the
value of ΔE in kilojoules if 89.5 g of ethylene and 125 g of
HCl are allowed to react at atmospheric pressure and the
volume change is -71.5 L?

9.145 We said in Section 9.1 that the potential energy of water at
the top of a dam or waterfall is converted into heat when
the water dashes against rocks at the bottom. The potential
energy of the water at the top is equal to $E_P = mgh$, where
m is the mass of the water, g is the acceleration of the falling

water due to gravity ($g = 9.81 \text{ m/s}^2$), and h is the height of
the water. Assuming that all the energy is converted to heat,
calculate the temperature rise of the water in degrees Celsius
after falling over California's Yosemite Falls, a distance of
739 m. The specific heat of water is 4.18 J/(g·K).

9.146 For a process to be spontaneous, the total entropy of the
RAN system and its surroundings must increase; that is

$$\Delta S_{total} = \Delta S_{system} + \Delta S_{surr} > 0 \quad \text{for a spontaneous process}$$

Furthermore, the entropy change in the surroundings, ΔS_{surr},
is related to the enthalpy change for the process by the equation $\Delta S_{surr} = -\Delta H/T$.

(a) Since both ΔG and ΔS_{total} offer criteria for spontaneity,
they must be related. Derive a relationship between them.

(b) What is the value of ΔS_{surr} for the photosynthesis of glucose from CO_2 at 298 K?

$$6 CO_2(g) + 6 H_2O(l) \rightarrow C_6H_{12}O_6(s) + 6 O_2(g)$$

$$\Delta G° = 2879 \text{ kJ}$$

$$\Delta S° = -262 \text{ J/K}$$

9.147 Given 400.0 g of hot tea at 80.0 °C, what mass of ice at 0 °C
RAN   must be added to obtain iced tea at 10.0 °C? The specific
heat of the tea is 4.18 J/(g·°C) and ΔH_{fusion} for ice is
+6.01 kJ/mol.

9.148 Imagine that you dissolve 10.0 g of a mixture of $NaNO_3$ and
RAN   KF in 100.0 g of water and find that the temperature rises
by 2.22 °C. Using the following data, calculate the mass of
each compound in the original mixture. Assume that the
specific heat of the solution is 4.18 J/(g·°C)

$$NaNO_3(s) \rightarrow NaNO_3(aq) \qquad \Delta H = +20.4 \text{ kJ/mol}$$

$$KF(s) \rightarrow KF(aq) \qquad \Delta H = -17.7 \text{ kJ/mol}$$

9.149 Consider the reaction: $4\,CO(g) + 2\,NO_2(g) \longrightarrow 4\,CO_2(g) +$
$N_2(g)$. Using the following information, determine $\Delta H°$ for
the reaction at 25 °C.

$NO(g)$	$\Delta H°_f = +91.3 \text{ kJ/mol}$
$CO_2(g)$	$\Delta H°_f = -393.5 \text{ kJ/mol}$

$$2\,NO(g) + O_2(g) \rightarrow 2\,NO_2(g) \quad \Delta H° = -116.2 \text{ kJ}$$

$$2\,CO(g) + O_2(g) \rightarrow 2\,CO_2(g) \quad \Delta H° = -566.0 \text{ kJ}$$

9.150 The reaction $S_8(g) \rightarrow 4\,S_2(g)$ has $\Delta H° = +237 \text{ kJ}$

(a) The S_8 molecule has eight sulfur atoms arranged in a
ring. What are the hybridization and geometry around
each sulfur atom in S_8?

(b) The average S—S bond dissociation energy is 225 kJ/mol.
Using the value of $\Delta H°$ given above, what is the S=S
double bond energy in $S_2(g)$?

(c) Assuming that the bonding in S_2 is similar to the bonding
in O_2, give a molecular orbital description of the bonding in S_2. Is S_2 likely to be paramagnetic or diamagnetic?

9.151 Phosgene, $COCl_2(g)$, is a toxic gas used as an agent of war-
RAN   fare in World War I.

(a) Draw an electron-dot structure for phosgene.

(b) Using the table of bond dissociation energies (Table 9.3)
and the value $\Delta H°_f = 716.7 \text{ kJ/mol}$ for $C(g)$, estimate
$\Delta H°_f$ for $COCl_2(g)$ at 25 °C. Compare your answer to
the actual $\Delta H°_f$ given in Appendix B, and explain why
your calculation is only an estimate.

9.152 Acid spills are often neutralized with sodium carbonate or
RAN   sodium hydrogen carbonate. For neutralization of acetic
acid, the unbalanced equations are

(1) $CH_3CO_2H(l) + Na_2CO_3(s) \rightarrow$
$\qquad CH_3CO_2Na(aq) + CO_2(g) + H_2O(l)$
(2) $CH_3CO_2H(l) + NaHCO_3(s)$
$\qquad CH_3CO_2Na(aq) + CO_2(g) + H_2O(l)$

(a) Balance both equations.

(b) How many kilograms of each substance is needed to neutralize a 1.000-gallon spill of pure acetic acid (density =
1.049 g/mL)?

(c) How much heat in kilojoules is absorbed or liberated in
each reaction? See Appendix B for standard heats of formation; $\Delta H°_f = -726.1 \text{ kJ/mol}$ for $CH_3CO_2\,Na(aq)$.

9.153 (a) Write a balanced equation for the reaction of potassium
RAN   metal with water.

(b) Use the data in Appendix B to calculate $\Delta H°$ for the
reaction of potassium metal with water.

(c) Assume that a chunk of potassium weighing 7.55 g is
dropped into 400.0 g of water at 25.0 °C. What is the
final temperature of the water if all the heat released is
used to warm the water?

(d) What is the molarity of the KOH solution prepared in
part (c), and how many milliliters of 0.554 M H_2SO_4 are
required to neutralize it?

9.154 Hydrazine, a component of rocket fuel, undergoes combus-
RAN   tion to yield N_2 and H_2O:

$$N_2H_4(l) + O_2(g) \rightarrow N_2(g) + 2\,H_2O(l)$$

(a) Draw an electron-dot structure for hydrazine, predict
the geometry about each nitrogen atom, and tell the
hybridization of each nitrogen.

(b) Use the following information to set up a Hess's law
cycle, and then calculate $\Delta H°$ for the combustion reaction. You will need to use fractional coefficients for
some equations.

$$2\,NH_3(g) + 3\,N_2O(g) \rightarrow 4\,N_2(g) + 3\,H_2O(l)$$
$$\Delta H° = -1011.2 \text{ kJ}$$
$$N_2O(g) + 3\,H_2(g) \rightarrow N_2H_4(l) + H_2O(l)$$
$$\Delta H° = -317.2 \text{ kJ}$$
$$4\,NH_3(g) + O_2(g) \rightarrow 2\,N_2H_4(l) + 2\,H_2O(l)$$
$$\Delta H° = -286.0 \text{ kJ}$$
$$H_2O(l) \quad \Delta H°_f = -285.8 \text{ kJ/mol}$$

(c) How much heat is released on combustion of 100.0 g of
hydrazine?

9.155 Reaction of gaseous fluorine with compound X yields a single product Y, whose mass percent composition is 61.7% F
and 38.3% Cl.

(a) What is a probable molecular formula for product Y,
and what is a probable formula for X?

(b) Draw an electron-dot structure for Y, and predict the
geometry around the central atom.

(c) Calculate $\Delta H°$ for the synthesis of Y using the following
information:

$$2\,ClF(g) + O_2(g) \rightarrow Cl_2O(g) + OF_2(g)$$
$$\Delta H° = +205.4 \text{ kJ}$$
$$2\,ClF_3(l) + 2\,O_2(g) \rightarrow Cl_2O(g) + 3\,OF_2(g)$$
$$\Delta H° = +532.8 \text{ kJ}$$
$$OF_2(g) \quad \Delta H°_f = +24.5 \text{ kJ/mol}$$

(d) How much heat in kilojoules is released or absorbed in
the reaction of 25.0 g of X with a stoichiometric amount
of F_2, assuming 87.5% yield for the reaction?

chapter 10

Gases: Their Properties and Behavior

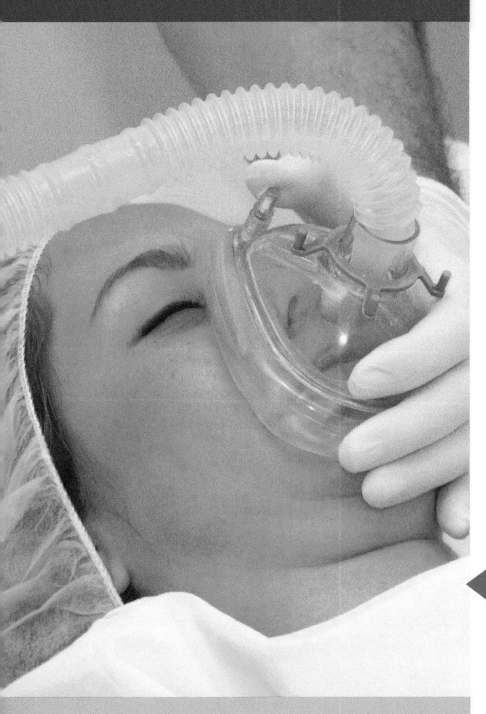

Contents

Volatile anesthetic compounds are liquids at room temperature but evaporate readily for administration by inhalation.

How do inhaled anesthetics work?

The answer to this question can be found on page 407 in the INQUIRY ?

A quick look around tells you that matter takes many forms. Most of the things around you are *solids*, substances whose constituent atoms, molecules, or ions are held rigidly together in a definite way, giving the solid a definite volume and shape. Other substances are *liquids*, whose constituent atoms or molecules are held together less strongly, giving the liquid a definite volume but a changeable and indefinite shape. Still other substances are *gases*, whose constituent atoms or molecules have little attraction for one another and are therefore free to move about in whatever volume is available.

Although gases are few in number—only about a hundred substances are gases at room temperature and atmospheric pressure—experimental studies of their properties were enormously important in the historical development of atomic theories. We'll look briefly at this historical development in the present chapter, and we'll see how the behavior of gases can be described.

10.1 GASES AND GAS PRESSURE

We live surrounded by a blanket of air—the mixture of gases that make up the Earth's atmosphere. As shown in **TABLE 10.1,** nitrogen and oxygen account for more than 99% by volume of dry air. The remaining 1% is largely argon, with trace amounts of several other substances also present. Carbon dioxide, about which there is so much current concern because of its relationship to climate change, is present in air only to the extent of about 0.040%, or 400 parts per million (ppm). Although small, this value has risen in the past 160 years from an estimated 290 ppm in 1850, as the burning of fossil fuels and the deforestation of tropical rain forests have increased.

TABLE 10.1   Composition of Dry Air at Sea Level

Constituent	% Volume	% Mass
N_2	78.08	75.52
O_2	20.95	23.14
Ar	0.93	1.29
CO_2	0.040	0.060
Ne	1.82×10^{-3}	1.27×10^{-3}
He	5.24×10^{-4}	7.24×10^{-5}
CH_4	1.7×10^{-4}	9.4×10^{-5}
Kr	1.14×10^{-4}	3.3×10^{-4}

Air is typical of gases in many respects, and its behavior illustrates several important points about gases. For instance, gas mixtures are always *homogeneous*, meaning that they are uniform in composition. Unlike liquids, which often fail to mix with one another and which may separate into distinct layers—oil and water, for example— gases always mix completely. Furthermore, gases are *compressible*. When pressure is applied, the volume of a gas contracts proportionately. Solids and liquids, however, are nearly incompressible, and even the application of great pressure changes their volume only slightly.

Homogeneous mixing and compressibility of gases both occur because the constituent particles—whether atoms or molecules—are far apart (**FIGURE 10.1**). Mixing occurs because individual gas particles have little interaction with their neighbors, so the chemical identity of those neighbors is irrelevant. In solids and liquids, by contrast, the constituent particles are packed closely together, where they are affected by various attractive and repulsive forces that can inhibit their mixing. Compressibility is possible in gases because less than 0.1% of the volume of a typical gas is taken up by the particles themselves under normal circumstances; the remaining 99.9% is empty space. By contrast, approximately 70% of a solid's or liquid's volume is taken up by the particles.

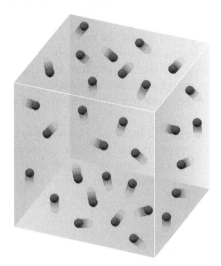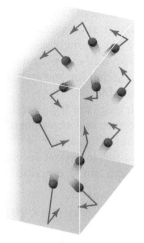

▶ **FIGURE 10.1**

Molecular view of a gas.

| A gas is a large collection of particles moving at random through a volume that is primarily empty space. | Collisions of randomly moving particles with the walls of the container exert a force per unit area that we perceive as gas pressure. |

One of the most obvious characteristics of gases is that they exert a measurable *pressure* on the walls of their container (Figure 10.1). We're all familiar with inflating a balloon or pumping up a bicycle tire and feeling the hardness that results from the pressure inside. In scientific terms, **pressure (P)** is defined as a force (*F*) exerted per unit area (*A*). Force, in turn, is defined as mass (*m*) times acceleration (*a*), which, on Earth, is usually the acceleration due to gravity, $a = 9.81$ m/s^2.

$$\text{Pressure } (P) = \frac{F}{A} = \frac{m \times a}{A}$$

The SI unit for force is the **newton (N)**, where $1 \text{ N} = 1 \text{ (kg} \cdot \text{m)/s}^2$, and the SI unit for pressure is the **pascal (Pa)**, where $1 \text{ Pa} = 1 \text{ N/m}^2 = 1 \text{ kg/(m} \cdot \text{s}^2)$. Expressed in more familiar units, a pascal is actually a very small amount—the pressure exerted by a mass of 10.2 mg resting on an area of 1.00 cm^2.

$$P = \frac{m \times a}{A} = \frac{(10.2 \text{ mg})\left(\dfrac{1 \text{ kg}}{10^6 \text{ mg}}\right)\left(9.81 \dfrac{\text{m}}{\text{s}^2}\right)}{(1.00 \text{ cm}^2)\left(\dfrac{1 \text{ m}}{10^2 \text{ cm}}\right)^2} = \frac{1.00 \times 10^{-4} \dfrac{\text{kg} \cdot \text{m}}{\text{s}^2}}{1.00 \times 10^{-4} \text{ m}^2}$$

$$= 1.00 \frac{\text{kg}}{\text{m} \cdot \text{s}^2} = 1.00 \text{ Pa}$$

In rough terms, a penny sitting on the tip of your finger exerts a pressure of about 250 Pa. Just as the air in a tire and a penny on your finger exert pressure, the mass of air in the atmosphere pressing down on the Earth's surface exerts what we call *atmospheric pressure*. In fact, a 1 m^2 column of air extending from the Earth's surface through the upper atmosphere has a mass of about 10,300 kg, producing an atmospheric pressure of approximately 101,000 Pa, or 101 kPa (**FIGURE 10.2**).

$$P = \frac{m \times a}{A} = \frac{10,300 \text{ kg} \times 9.81 \dfrac{\text{m}}{\text{s}^2}}{1.00 \text{ m}^2} = 101,000 \text{ Pa} = 101 \text{ kPa}$$

Pressure in everyday objects such as tires and basketballs is measured in units of pounds per square inch (lb/in.^{2} or psi). Atmospheric pressure can be expressed in lb/in.^{2}

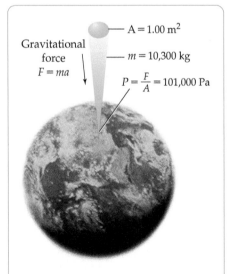

$A = 1.00$ m^2

Gravitational force
$F = ma$

$m = 10,300$ kg

$P = \dfrac{F}{A} = 101,000$ Pa

▲ **FIGURE 10.2**

Atmospheric pressure. A column of air 1 m^2 in cross-sectional area extending from the Earth's surface through the upper atmosphere has a mass of about 10,300 kg, producing an atmospheric pressure of approximately 101,000 Pa.

by converting the mass (10,300 kg) and the area (1 m²) of a column of the atmosphere from SI units to units of pounds and inches (Figure 10.2).

$$\frac{10{,}300 \text{ kg}}{1 \text{ m}^2} \times \frac{2.21 \text{ lb}}{1 \text{ kg}} \times \left(\frac{1 \text{ m}}{10^2 \text{ cm}}\right)^2 \times \left(\frac{2.54 \text{ cm}}{1 \text{ in.}}\right)^2 = 14.7 \frac{\text{lb}}{\text{in.}^2}$$

We don't feel the atmosphere pushing down on us because of an equivalent force within our bodies pushing outward. However, the force due to atmospheric pressure can be demonstrated by attaching a metal can to a vacuum pump. When the pump is turned on, the pressure inside the can decreases and atmospheric pressure is strong enough to crush the can (**FIGURE 10.3**)!

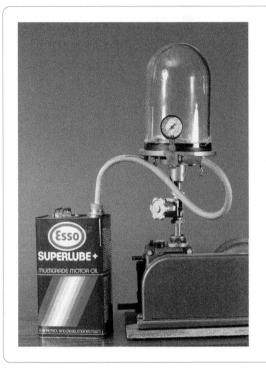 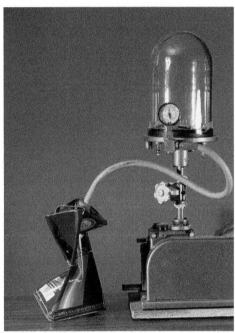

◀ **FIGURE 10.3**

Effect of atmospheric pressure on an evacuated can. (**a**) A vacuum pump is connected to a can. (**b**) Once the vacuum is turned on, pressure inside the can decreases and atmospheric pressure crushes the can.

As is frequently the case with SI units, which must serve many disciplines, the pascal is an inconvenient size for most chemical measurements. Thus, the alternative pressure units *millimeter of mercury (mm Hg)*, *atmosphere (atm)*, and *bar* are more often used.

The **millimeter of mercury**, also called a *torr* after the seventeenth-century Italian scientist Evangelista Torricelli (1608–1647), is based on atmospheric pressure measurements using a mercury *barometer*. As shown in **FIGURE 10.4**, a barometer consists of a long, thin tube that is sealed at one end, filled with mercury, and then inverted into a dish of mercury. Some mercury runs from the tube into the dish until the downward pressure of mercury due to the pull of gravity inside the column is exactly balanced by the outside atmospheric pressure, which presses on the mercury in the dish and pushes it up the column. The height of the mercury column varies slightly from day to day depending on the altitude and weather conditions, but atmospheric pressure at sea level is defined as exactly 760 mm Hg.

Knowing the density of mercury ($1.359\ 51 \times 10^4$ kg/m³ at 0 °C) and the acceleration due to gravity ($9.806\ 65$ m/s²), it's possible to calculate the pressure exerted by the column of mercury 760 mm (0.760 m) in height. Thus, 1 standard **atmosphere** (**atm**) of pressure (1 atm) is now defined as exactly 101,325 Pa:

$$P = (0.760 \text{ m})\left(1.359\ 51 \times 10^4 \frac{\text{kg}}{\text{m}^3}\right)\left(9.806\ 65 \frac{\text{m}}{\text{s}^2}\right) = 101{,}325 \text{ Pa}$$

$$1 \text{ atm} = 760 \text{ mm Hg} = 101{,}325 \text{ Pa}$$

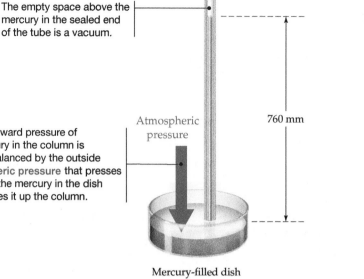

▶ **FIGURE 10.4**

A mercury barometer. The barometer measures atmospheric pressure by determining the height of a mercury column supported in a sealed glass tube.

▶ **Figure It out**

If the barometer were filled with liquid water (d = 1.00 g/mL) instead of liquid mercury (d = 13.6 g/mL), what would be the height of the water in the sealed tube?

Answer: Since mercury is 13.6 times more dense than water, the column of water would be 13.6 times higher (10,336 mm or 10.3 m). The large height makes water impractical for use in a barometer.

The empty space above the mercury in the sealed end of the tube is a vacuum.

The downward pressure of the mercury in the column is exactly balanced by the outside atmospheric pressure that presses down on the mercury in the dish and pushes it up the column.

Atmospheric pressure

760 mm

Mercury-filled dish

Although not strictly an SI unit, the **bar** is quickly gaining popularity as a unit of pressure because it is a convenient power of 10 of the SI unit pascal and because it differs from 1 atm by only about 1%:

$$1 \text{ bar} = 100{,}000 \text{ Pa} = 100 \text{ kPa} = 0.986\,923 \text{ atm}$$

TABLE 10.2 summarizes different units of pressure and the conversion factor between atmospheres and the specified unit.

Gas pressure inside a container is often measured using an open-end **manometer**, a simple instrument similar in principle to the mercury barometer. As shown in **FIGURE 10.5**, an open-end manometer consists of a U-tube filled with mercury, with one end connected

▶ **FIGURE 10.5**

Open-end manometers for measuring pressure in a gas-filled bulb.

▶ **Figure It out**

Describe the mercury level in the arm open to the bulb and the arm open to the atmosphere when the gas pressure in the bulb equals 1 atm.

Answer: The mercury levels in both arms are identical.

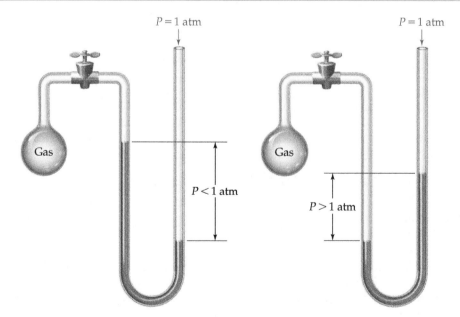

$P = 1$ atm

$P = 1$ atm

$P < 1$ atm

$P > 1$ atm

Gas

Gas

(a) The mercury level is higher in the arm open to the bulb because the pressure in the bulb is lower than atmospheric.

(b) The mercury level is higher in the arm open to the atmosphere because the pressure in the bulb is higher than atmospheric.

to a gas-filled container and the other end open to the atmosphere. The difference between the pressure of the gas in the container and the pressure of the atmosphere is equal to the difference between the heights of the mercury levels in the two arms of the U-tube. If the gas pressure inside the container is less than atmospheric, the mercury level is higher in the arm connected to the container (Figure 10.5a). If the gas pressure inside the container is greater than atmospheric, the mercury level is higher in the arm open to the atmosphere (Figure 10.5b).

TABLE 10.2 Conversions between Common Units of Pressure	
	1 atm
Pa (1 N/m^2)	*1.013 25 × 10^5
kPa	*101.325
bar	*1.013 25
mm Hg	*760
lb/in.^{2} or psi	14.7

*The following conversions are exact and do not limit the number of significant figures in a calculation.

— WORKED EXAMPLE 10.1

Converting Between Different Units of Pressure

Typical atmospheric pressure on top of Mt. Everest, whose official altitude is 8848 m, is 265 mm Hg. Convert this value to pascals, atmospheres, and bars.

STRATEGY

Use the conversion factors 101,325 Pa/760 mm Hg, 1 atm/760 mm Hg, and 1 bar/10^5 Pa to carry out the necessary calculations.

SOLUTION

$$(265 \text{ mm Hg})\left(\frac{101,325 \text{ Pa}}{760 \text{ mm Hg}}\right) = 3.53 \times 10^4 \text{ Pa}$$

$$(265 \text{ mm Hg})\left(\frac{1 \text{ atm}}{760 \text{ mm Hg}}\right) = 0.349 \text{ atm}$$

$$(3.53 \times 10^4 \text{ Pa})\left(\frac{1 \text{ bar}}{10^5 \text{ Pa}}\right) = 0.353 \text{ bar}$$

CHECK

One atmosphere equals 760 mm Hg pressure. Since 265 mm Hg is about one-third of 760 mm Hg, the air pressure on Mt. Everest is about one-third of standard atmospheric pressure—approximately 30,000 Pa, 0.3 atm, or 0.3 bar.

PRACTICE 10.1   Hurricane Irma (2017) had a central pressure of 914 mbar, the lowest recorded for an Atlantic hurricane. Calculate the pressure in units of mm Hg.

APPLY 10.2   At sea level, atmospheric pressure is 1 atm. The pressure increases by 1 atm for every 33 ft of water depth. Herbert Nitsch holds the world record for free-diving with a depth of 702 ft. What is the pressure in units of Pa at this depth?

All **PRACTICE** and **APPLY** problems are interactive in the eText.

▲ Atmospheric pressure decreases as altitude increases. On the top of Mt. Everest, typical atmospheric pressure is 265 mm Hg.

— CONCEPTUAL WORKED EXAMPLE 10.2

Using an Open-End Manometer to Measure Gas Pressure

What is the pressure of the gas inside the apparatus shown in mm Hg if the outside pressure is 750 mm Hg?

STRATEGY

The gas pressure in the bulb equals the difference between the outside pressure and the manometer reading. The pressure of the gas in the bulb is higher than atmospheric pressure because the liquid level is higher in the arm open to the atmosphere.

SOLUTION

$$P_{\text{gas}} = 750 \text{ mm Hg} + \left(25 \text{ cm Hg} \times \frac{10 \text{ mm}}{1 \text{ cm}}\right) = 1000 \text{ mm Hg} = 1.0 \times 10^3 \text{ mm Hg}$$

continued on next page

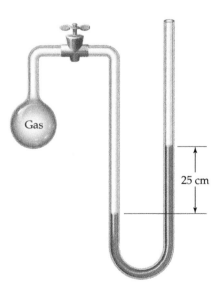

▲ Open-end manometer for Worked Example 10.2.

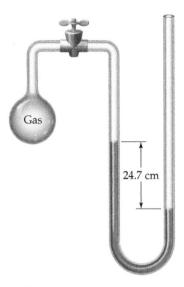

24.7 cm

▲ Open-end manometer for
Problem 10.3.

▶ **CONCEPTUAL PRACTICE 10.3**   What is the pressure of the gas inside the apparatus in mm Hg if the outside pressure is 0.975 atm?

▶ **CONCEPTUAL APPLY 10.4**   Assume that you are using an open-end manometer filled with mineral oil rather than mercury. The level of mineral oil in the arm connected to the bulb is 237 mm higher than the level in the arm connected to the atmosphere and atmospheric pressure is 746 mm Hg.

(a) Draw a picture of the manometer similar to Figure 10.5.
(b) What is the gas pressure in the bulb in mm of Hg? (The density of mercury is 13.6 g/mL, and the density of mineral oil is 0.822 g/mL.)

10.2 THE GAS LAWS

Unlike solids and liquids, different gases show remarkably similar physical behavior regardless of their chemical makeup. Helium and fluorine, for example, are vastly different in their chemical properties yet are almost identical in much of their physical behavior. Numerous observations made in the late 1600s showed that the properties of any gas can be defined by four variables: pressure (P), temperature (T), volume (V), and amount, or number of moles (n). The specific relationships among these four variables are called the **gas laws,** and a gas whose behavior follows the laws exactly is called an **ideal gas.**

Boyle's Law: The Relationship between Gas Volume and Pressure

Imagine that you have a sample of gas inside a cylinder with a movable piston at one end (**FIGURE 10.6**). What would happen if you were to increase the pressure on the gas by pushing down on the piston? Experience probably tells you that the volume of gas in the cylinder would decrease as you increase the pressure. According to **Boyle's law,** the volume of a fixed amount of gas at a constant temperature varies inversely with its pressure. If the gas pressure is doubled, the volume is halved; if the pressure is halved, the gas volume doubles.

> **Boyle's law $V \propto 1/P$ or $PV = k$ at constant n and T**   The volume of an ideal gas varies inversely with pressure. That is, P times V is constant when n and T are kept constant. (The symbol $\propto$ means "is proportional to," and k denotes a constant.)

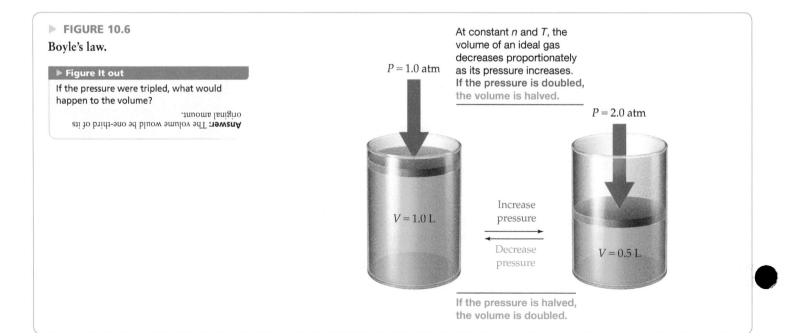

▶ **FIGURE 10.6**

Boyle's law.

▶ **Figure It out**

If the pressure were tripled, what would happen to the volume?

Answer: The volume would be one-third of its original amount.

At constant n and T, the volume of an ideal gas decreases proportionately as its pressure increases. **If the pressure is doubled, the volume is halved.**

$P = 1.0$ atm

$V = 1.0$ L

Increase pressure

Decrease pressure

$P = 2.0$ atm

$V = 0.5$ L

If the pressure is halved, the volume is doubled.

The validity of Boyle's law can be demonstrated by making a simple series of pressure–volume measurements on a gas sample (**TABLE 10.3**) and plotting them as in **FIGURE 10.7**. When V is plotted versus P, the result is a curve in the form of a hyperbola (Figure 10.7a). When V is plotted versus $1/P$, however, the result is a straight line (Figure 10.7b). Such graphical behavior is characteristic of mathematical equations of the form $y = mx + b$. In this case, $y = V$, $m = $ the slope of the line (the constant k in the present instance), $x = 1/P$, and $b = $ the y-intercept (a constant; 0 in the present instance). (See Appendix A.3 for a review of linear equations.)

$$V = k\left(\frac{1}{P}\right) + 0 \quad (\text{or } PV = k)$$
$$\uparrow \quad \uparrow\,\uparrow \qquad \uparrow$$
$$y = m\ x \quad + b$$

TABLE 10.3 Pressure–Volume Measurements on a Gas Sample at Constant n, T

Pressure (mm Hg)	Volume (L)
760	1
380	2
253	3
190	4
152	5
127	6
109	7
95	8
84	9
76	10

(a)

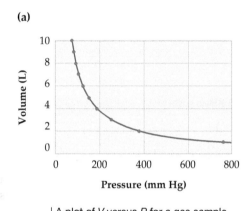

A plot of V versus P for a gas sample is a hyperbola.

(b)

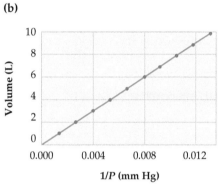

A plot of V versus $1/P$ is a straight line. Such a graph is characteristic of equations having the form $y = mx + b$.

◀ **FIGURE 10.7**
Boyle's law plot.

Charles's Law: The Relationship between Gas Volume and Temperature

Imagine again that you have a gas sample inside a cylinder with a movable piston at one end (**FIGURE 10.8**). What would happen if you were to raise the temperature of the sample while letting the piston move freely to keep the pressure constant? Experience tells you that the piston would move up because the volume of the gas in the cylinder would expand. You have most likely observed the relationship between temperature and volume of a gas when air in a hot-air balloon is heated, causing the air to expand and fill the balloon. According to **Charles's law,** the volume of a fixed amount of an ideal gas at a constant pressure varies directly with its absolute temperature. If the gas temperature in kelvins is doubled, the volume is doubled; if the gas temperature is halved, the volume is halved.

> **Charles's law $V \propto T$ or $V/T = k$ at constant n and P**   The volume of an ideal gas varies directly with absolute temperature. That is, V divided by T is constant when n and P are held constant.

The validity of Charles's law can be demonstrated by making a series of temperature–volume measurements on a gas sample, giving the results listed in **TABLE 10.4**. Like Boyle's law, Charles's law takes the mathematical form $y = mx + b$, where $y = V$, $m = $ the slope of the line (the constant k in the present instance), $x = T$, and $b = $ the y-intercept (0 in the present instance). A plot of V versus T is therefore a straight line whose slope is the constant k (**FIGURE 10.9**).

TABLE 10.4 Temperature–Volume Measurements on a Gas Sample at Constant n and P

Temperature (K)	Volume (L)
123	0.45
173	0.63
223	0.82
273	1.00
323	1.18
373	1.37

▶ **FIGURE 10.8**

Charles's law.

▶ **Figure It out**

If the absolute temperature were decreased by a factor of 3, what happens to the volume?

Answer: The volume is one-third of its initial amount.

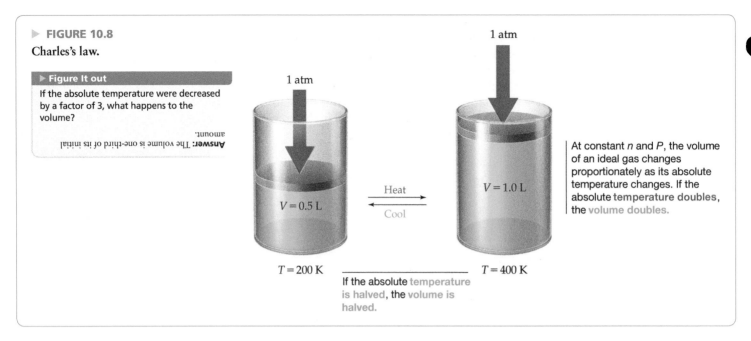

1 atm

1 atm

$V = 0.5$ L

Heat

Cool

$V = 1.0$ L

At constant n and P, the volume of an ideal gas changes proportionately as its absolute temperature changes. If the absolute temperature doubles, the volume doubles.

$T = 200$ K

$T = 400$ K

If the absolute temperature is halved, the volume is halved.

▶ **FIGURE 10.9**

Charles's law plot.

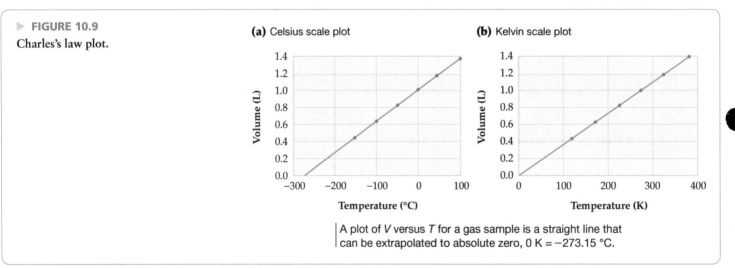

(a) Celsius scale plot

(b) Kelvin scale plot

A plot of V versus T for a gas sample is a straight line that can be extrapolated to absolute zero, 0 K = −273.15 °C.

$$V = kT + 0 \quad \left(\text{or } \frac{V}{T} = k\right)$$
$$\uparrow \quad \uparrow\uparrow \quad \uparrow$$
$$y = mx + b$$

The plots of volume versus temperature in Figure 10.9 demonstrate an interesting point. When temperature is plotted on the Celsius scale, the straight line can be extrapolated to $V = 0$ at $T = -273.15$ (Figure 10.9a). But because matter can't have a negative volume, this extrapolation suggests that −273.15 must be the lowest possible temperature, or *absolute zero* on the Kelvin scale (Figure 10.9b). In fact, the approximate value of absolute zero was first determined using this simple method.

Avogadro's Law: The Relationship between Volume and Amount

Imagine that you have two gas samples inside cylinders with movable pistons (**FIGURE 10.10**). One cylinder contains 1 mol of a gas and the other contains 2 mol of gas at the same temperature and pressure as the first. Common sense says that the gas in the second cylinder will have twice the volume of the gas in the first cylinder because there is twice as much of it. According to **Avogadro's law,** the volume of an ideal gas at a fixed pressure and temperature depends only on its molar amount. If the amount of the gas is doubled, the gas volume is doubled; if the amount is halved, the volume is halved.

Avogadro's law $V \propto n$ or $V/n = k$ **at constant T and P** The volume of an ideal gas varies directly with its molar amount. That is, V divided by n is constant when T and P are held constant.

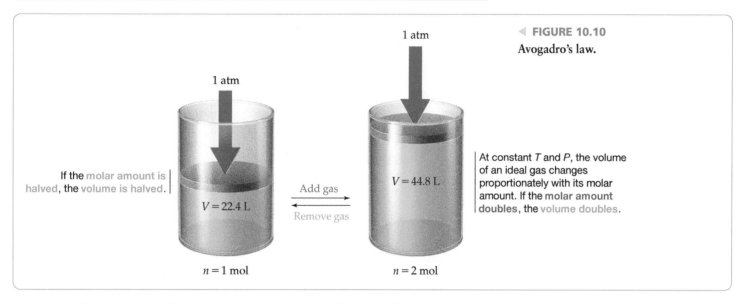

1 atm

1 atm

If the molar amount is halved, the volume is halved.

$V = 22.4$ L

Add gas

Remove gas

$V = 44.8$ L

At constant T and P, the volume of an ideal gas changes proportionately with its molar amount. If the molar amount doubles, the volume doubles.

$n = 1$ mol

$n = 2$ mol

◀ **FIGURE 10.10**

Avogadro's law.

Put another way, Avogadro's law also says that equal volumes of different gases at the same temperature and pressure contain the same molar amounts. A 1-L container of oxygen contains the same number of moles as a 1-L container of helium, fluorine, argon, or any other gas at the same T and P. Furthermore, 1 mol of an ideal gas occupies a volume, called the **standard molar volume**, of 22.414 L at 0 °C and exactly 1 atm pressure. For comparison, the standard molar volume is nearly identical to the volume of three basketballs.

Go to eText

CONCEPTUAL WORKED EXAMPLE 10.3

Visual Representations of Gas Laws

Show the approximate level of the movable piston in drawings **(a)** and **(b)** after the indicated changes have been made to the initial gas sample.

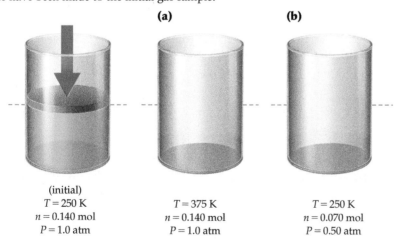

(a)

(b)

(initial)
$T = 250$ K
$n = 0.140$ mol
$P = 1.0$ atm

$T = 375$ K
$n = 0.140$ mol
$P = 1.0$ atm

$T = 250$ K
$n = 0.070$ mol
$P = 0.50$ atm

STRATEGY

Identify which of the variables P, n, and T have changed, and calculate the effect of each change on the volume according to the appropriate gas law.

SOLUTION

(a) The temperature T has increased by a factor of $375/250 = 1.5$, while the molar amount n and the pressure P are unchanged. Charles's law states that $V \propto T$; therefore, the volume will increase by a factor of 1.5.

continued on next page

(b) The temperature T is unchanged, while both the molar amount n and the pressure P are halved. Avogadro's laws states that $V \propto n$; therefore, halving the molar amount will halve the volume. Similarly, since $V \propto 1/P$ (Boyle's law), halving the pressure will double the volume. The two changes cancel, so the volume is unchanged.

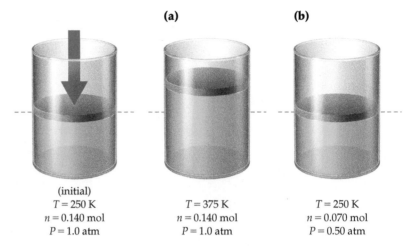

	(a)	**(b)**
(initial)		
$T = 250$ K	$T = 375$ K	$T = 250$ K
$n = 0.140$ mol	$n = 0.140$ mol	$n = 0.070$ mol
$P = 1.0$ atm	$P = 1.0$ atm	$P = 0.50$ atm

▶ **CONCEPTUAL PRACTICE 10.5**   Show the approximate level of the movable piston in the following drawing after the indicated changes have been made to the initial gas sample at a constant pressure of 1.0 atm.

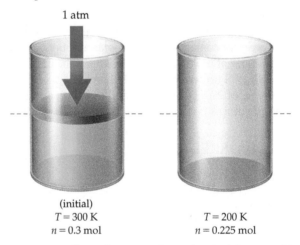

1 atm

(initial)
$T = 300$ K $T = 200$ K
$n = 0.3$ mol $n = 0.225$ mol

▶ **CONCEPTUAL APPLY 10.6**   Show the approximate level of the movable piston in drawings **(a)** and **(b)** after the indicated changes have been made to the initial gas sample.

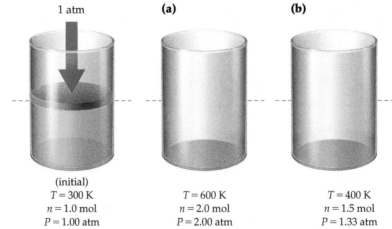

	(a)	**(b)**
1 atm		
(initial)		
$T = 300$ K	$T = 600$ K	$T = 400$ K
$n = 1.0$ mol	$n = 2.0$ mol	$n = 1.5$ mol
$P = 1.00$ atm	$P = 2.00$ atm	$P = 1.33$ atm

All **WORKED EXAMPLES** with this icon **Go to eText** have an interactive video in the eText.

10.3 THE IDEAL GAS LAW

All three gas laws discussed in the previous section can be combined into a single statement called the **ideal gas law,** which describes how the volume of a gas is affected by changes in pressure, temperature, and amount. When the values of any three of the variables P, V, T, and n are known, the value of the fourth can be calculated using the ideal gas law. The proportionality constant R in the equation is called the **gas constant (R)** and has the same value for all gases.

$$\text{Ideal gas law} \quad V = \frac{nRT}{P} \quad \text{or} \quad PV = nRT$$

Go to eText

BIG IDEA Question 1

Which of the following gases would have the largest volume at STP: 10 g of O_2, 10 g of Ar, 10 g of He? (Assume ideal behavior.)

The ideal gas law can be rearranged in different ways to take the form of Boyle's law, Charles's law, or Avogadro's law.

Boyle's law: $PV = nRT = k$ (When n and T are constant)

Charles's law: $\dfrac{V}{T} = \dfrac{nR}{P} = k$ (When n and P are constant)

Avogadro's law: $\dfrac{V}{n} = \dfrac{RT}{P} = k$ (When T and P are constant)

The value of the gas constant R can be calculated from knowledge of the standard molar volume of a gas. Since 1 mol of a gas occupies a volume of 22.414 L at 0 °C (273.15 K) and 1 atm pressure, the gas constant R is equal to 0.082 058 $(\text{L} \cdot \text{atm})/(\text{K} \cdot \text{mol})$, or 8.3145 J/(K·mol) in SI units:

$$R = \frac{P \cdot V}{n \cdot T} = \frac{(1 \text{ atm})(22.414 \text{ L})}{(1 \text{ mol})(273.15 \text{ K})} = 0.082 \ 058 \frac{\text{L} \cdot \text{atm}}{\text{K} \cdot \text{mol}}$$

$$= 8.3145 \text{ J/(K} \cdot \text{mol)} \quad \text{(When } P \text{ is in pascals and } V \text{ is in cubic meters)}$$

The specific conditions used in the calculation—0 °C (273.15 K) and 1 atm pressure—are said to represent **standard temperature and pressure,** abbreviated **STP.** These standard conditions are generally used when reporting measurements on gases. Note that the standard temperature for gas measurements (0 °C, or 273.15 K) is different from that usually assumed for thermodynamic measurements (25 °C, or 298.15 K; Section 9.5).

$$\text{Standard temperature and pressure (STP) for gases} \quad T = 0 \ ^\circ\text{C} \quad P = 1 \text{ atm}$$

We should also point out that the standard pressure for gas measurements, still listed here and in most other books as 1 atm (101,325 Pa), has actually been redefined to be 1 bar, or 100,000 Pa. This new standard pressure is now 0.986 923 atm, making the newly defined standard molar volume 22.711 L rather than 22.414 L. Like most other books, we'll continue for the present using 1 atm as the standard pressure.

The name *ideal* gas law implies that there must be some gases whose behavior is *nonideal*. In fact, there is no such thing as an ideal gas that obeys the equation perfectly under all circumstances. All real gases are nonideal to some extent and deviate slightly from the behavior predicted by the gas laws. As **TABLE 10.5** shows, for example, the actual molar volume of a real gas often differs slightly from the 22.414 L ideal value. Under most conditions, though, the deviations from ideal behavior are so slight as to make little difference. In Section 10.8, we'll discuss circumstances in which the deviations are greater.

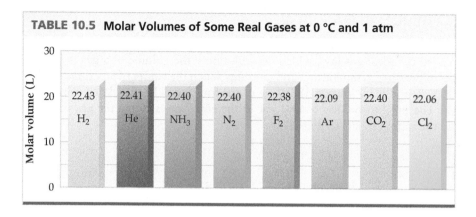

TABLE 10.5 Molar Volumes of Some Real Gases at 0 °C and 1 atm

22.43	22.41	22.40	22.40	22.38	22.09	22.40	22.06
H_2	He	NH_3	N_2	F_2	Ar	CO_2	Cl_2

Molar volume (L)

WORKED EXAMPLE 10.4

Using the Ideal Gas Law to Solve for an Unknown Variable

How many moles of gas (air) are in the lungs of an average adult with a lung capacity of 3.8 L? Assume that the lungs are at 1.00 atm pressure and a normal body temperature of 37 °C.

IDENTIFY

Known	Unknown
Volume ($V = 3.8$ L)	Moles of gas (n)
Pressure ($P = 1.00$ atm)	
Temperature ($T = 37$ °C)	

STRATEGY

This problem asks for a value of n when V, P, and T are given. Rearrange the ideal gas law to the form $n = PV/RT$, convert the temperature from degrees Celsius to kelvin, and substitute the given values of P, V, and T into the equation.

SOLUTION

$$n = \frac{PV}{RT} = \frac{(1.00 \text{ atm})(3.8 \text{ L})}{\left(0.082\ 06 \dfrac{\text{L} \cdot \text{atm}}{\text{K} \cdot \text{mol}}\right)(310 \text{ K})} = 0.15 \text{ mol}$$

The lungs of an average adult hold 0.15 mol of air.

CHECK

A lung volume of 4 L is about one-sixth of 22.4 L, the standard molar volume of an ideal gas. Thus, the lungs have a capacity of about one-sixth mol, or 0.17 mol.

▶ **PRACTICE 10.7** How many moles of methane gas, CH_4, are in a storage tank with a volume of 1.00×10^5 L at STP?

▶ **APPLY 10.8** An aerosol spray can with a volume of 350 mL contains 3.2 g of propane gas (C_3H_8) as propellant. What is the pressure in atmospheres of gas in the can at 20 °C?

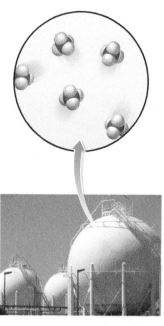

▲ How many moles of methane are in these tanks?

Go to
eText

— WORKED EXAMPLE 10.5

Using the Ideal Gas Law When Variables Change

In a typical automobile engine, the mixture of gasoline and air in a cylinder is compressed from 1.0 atm to 9.5 atm prior to ignition. If the uncompressed volume of the cylinder is 410 mL, what is the volume in milliliters when the mixture is fully compressed?

IDENTIFY

Known	Unknown
Initial pressure (P_i = 1.0 atm)	Final volume (V_f)
Final pressure (P_f = 9.5 atm)	
Initial volume (V_i = 410 mL)	

STRATEGY

Rearrange the ideal gas law so that variables that have changing values are on one side of the equation and constants are on the other side of the equation.

SOLUTION

In the ideal gas law, pressure and volume change and therefore remain on the left side of the equation, while n, R, and T remain constant.

$$PV = \underbrace{nRT}_{} = k$$

Variables   Constants
that change

Because pressure times volume is a constant (Boyle's law), the equation $P_iV_i = P_fV_f$ can be used to solve for the final volume:

$$V_f = \frac{P_iV_i}{P_f} = \frac{(1.0 \text{ atm})(410 \text{ mL})}{(9.5 \text{ atm})} = 43 \text{ mL}$$

CHECK

Because the pressure in the cylinder increases about 10-fold, the volume must decrease about 10-fold according to Boyle's law, from approximately 400 mL to 40 mL.

▶ **PRACTICE 10.9**   What final temperature (°C) is required for the pressure inside an automobile tire to increase from 2.15 atm at 0 °C to 2.37 atm, assuming the volume remains constant?

▶ **APPLY 10.10**   A weather balloon has a volume of 45.0 L when released under conditions of 745 mm Hg and 25.0°C What is the volume of the balloon at an altitude of 10,000 m where the pressure is 178 mm Hg and the temperature is 225 K?

10.4 STOICHIOMETRIC RELATIONSHIPS WITH GASES

Many chemical reactions, including some of the most important processes in the chemical industry, involve gases. Approximately 130 million metric tons of ammonia, for instance, is manufactured each year worldwide by the reaction of hydrogen with nitrogen according to the equation $3 H_2(g) + N_2(g) \longrightarrow 2 NH_3(g)$. Thus, it's necessary to be able to calculate amounts of gaseous reactants just as it's necessary to calculate amounts of solids, liquids, and solutions.

Most gas calculations are just applications of the ideal gas law in which three of the variables P, V, T, and n are known and the fourth variable must be calculated. The reaction used in the deployment of automobile air bags, for instance, is the high-temperature decomposition of sodium azide, NaN_3, to produce N_2 gas. (The sodium is then removed by a subsequent reaction.) Worked Example 10.6 shows how to calculate the volume of a gaseous product given the amount reactant.

FO4305OZ02

▲ Automobile air bags are inflated with N_2 gas produced by decomposition of sodium azide.

— WORKED EXAMPLE 10.6

Calculating the Volume of Gas Produced in a Chemical Reaction

How many liters of N_2 gas at 1.15 atm and 30 °C are produced by decomposition of 45.0 g of NaN_3?

$$2 NaN_3(s) \longrightarrow 2 Na(s) + 3 N_2(g)$$

continued on next page

IDENTIFY

Known	Unknown
Pressure ($P = 1.15$ atm)	Volume of N_2 (V)
Temperature ($T = 30\ °C$)	
Mass of NaN_3 (45.0 g)	

STRATEGY

Use stoichiometric relationships to find the number of moles of N_2 produced from 45.0 g of NaN_3 and then use the ideal gas law to find the volume of N_2.

SOLUTION

To find n, the number of moles of N_2 gas produced, we first need to find how many moles of NaN_3 are in 45.0 g:

$$\text{Molar mass of } NaN_3 = 65.0\ \text{g/mol}$$

$$\text{Moles of } NaN_3 = (45.0\ \text{g } NaN_3)\left(\frac{1\ \text{mol } NaN_3}{65.0\ \text{g } NaN_3}\right) = 0.692\ \text{mol } NaN_3$$

Next, find how many moles of N_2 are produced in the decomposition reaction. According to the balanced equation, 2 mol of NaN_3 yields 3 mol of N_2, so 0.692 mol of NaN_3 yields 1.04 mol of N_2:

$$\text{Moles of } N_2 = (0.692\ \text{mol } NaN_3)\left(\frac{3\ \text{mol } N_2}{2\ \text{mol } NaN_3}\right) = 1.04\ \text{mol } N_2$$

Finally, use the ideal gas law to calculate the volume of N_2. Remember to use the Kelvin temperature (303 K) rather than the Celsius temperature (30 °C) in the calculation.

▲ Carbonate-bearing rocks like limestone ($CaCO_3$) react with dilute acids such as HCl to produce bubbles of carbon dioxide.

$$V = \frac{nRT}{P} = \frac{(1.04\ \text{mol } N_2)\left(0.08206\ \dfrac{\text{L} \cdot \text{atm}}{\text{K} \cdot \text{mol}}\right)(303\ \text{K})}{1.15\ \text{atm}} = 22.5\ \text{L}$$

▶ **PRACTICE 10.11** Carbonate-bearing rocks like limestone ($CaCO_3$) react with dilute acids such as HCl to produce carbon dioxide, according to the equation

$$CaCO_3(s) + 2\ HCl(aq) \longrightarrow CaCl_2(aq) + CO_2(g) + H_2O(l)$$

What is the volume in liters of CO_2 at STP formed from complete reaction of 33.7 g of limestone ($CaCO_3$)?

▶ **APPLY 10.12** Approximately 83% of all ammonia produced is used as fertilizer for crops. Ammonia is synthesized from hydrogen and nitrogen gas according to the equation

$$3\ H_2(g) + N_2(g) \longrightarrow 2\ NH_3(g)$$

What volume of hydrogen and nitrogen gas is needed to synthesize 500.0 L of ammonia at STP?

Other applications of the ideal gas law make it possible to calculate such properties as density and molar mass. Densities are calculated by weighing a known volume of a gas at a known temperature and pressure, as shown in **FIGURE 10.11.** Gas density changes dramatically with temperature and pressure, so values for these variables must be specified. Gas density values are commonly reported at STP for consistency. The ideal gas law can be used to convert a density measured at any temperature and pressure to its value at STP. For example, if a sample of ammonia gas weighs 0.672 g and occupies a 1.000 L bulb at 25 °C and 733.4 mm Hg pressure, the density at STP can be calculated as follows. The density of any substance is mass divided by volume. For the ammonia sample, the mass is 0.672 g, but the volume of the gas is given under nonstandard

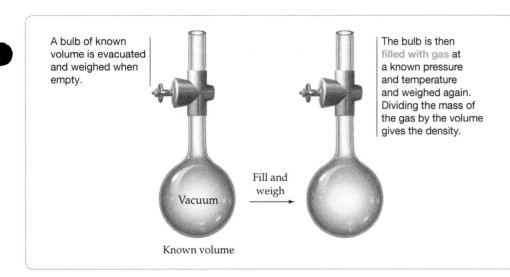

A bulb of known volume is evacuated and weighed when empty.

The bulb is then filled with gas at a known pressure and temperature and weighed again. Dividing the mass of the gas by the volume gives the density.

◀ **FIGURE 10.11**

Determining the density of an unknown gas.

Vacuum

Fill and weigh

Known volume

conditions and must first be converted to STP. Because the amount of sample n is constant, we can set the quantity PV/RT measured under nonstandard conditions equal to PV/RT at STP and then solve for V at STP.

$$n = \left(\frac{PV}{RT}\right)_{\text{measured}} = \left(\frac{PV}{RT}\right)_{\text{STP}} \quad \text{or} \quad V_{\text{STP}} = \left(\frac{PV}{RT}\right)_{\text{measured}} \left(\frac{RT}{P}\right)_{\text{STP}}$$

$$V_{\text{STP}} = \left(\frac{733.4 \text{ mm Hg} \times 1.000 \text{ L}}{298 \text{ K}}\right)\left(\frac{273 \text{ K}}{760 \text{ mm Hg}}\right) = 0.884 \text{ L}$$

The amount of gas in the 1.000 L bulb under the measured nonstandard conditions would have a volume of only 0.884 L at STP. Dividing the given mass by this volume gives the density of ammonia at STP:

$$\text{Density} = \frac{\text{Mass}}{\text{Volume}} = \frac{0.672 \text{ g}}{0.884 \text{ L}} = 0.760 \text{ g/L}$$

An equation relating gas density to molar mass can be found by rearranging the ideal gas law. Since density equals mass divided by volume, a term for mass must be added to the ideal gas law. Multiplying both sides of the equation by molar mass (M) incorporates mass into the equation because moles multiplied by molar mass equals mass as shown.

$$PV(M) = \underbrace{n(M)}_{\text{mass }(m)}RT$$

The equation can be rearranged to solve for density, which is defined as mass divided by volume:

$$PV(M) = mRT$$

$$\frac{P\cancel{V}(M)}{\cancel{V}} = \frac{mRT}{V} \quad \text{dividing both sides of the equation by } (V) \text{ yields:}$$

$$PM = \left(\frac{m}{V}\right)RT$$

$$\frac{PM}{RT} = \frac{\left(\frac{m}{V}\right)\cancel{RT}}{\cancel{RT}} \quad \text{dividing both sides of the equation by } (RT) \text{ yields:}$$

$$\boxed{\textbf{Gas density} \quad d = \frac{m}{V} = \frac{PM}{RT}}$$

 Go to eText

BIG IDEA Question 2

Which of the following gases has the highest density at a pressure of 1 atm and a temperature of 25 °C: H_2, He, CO_2, N_2?

If the density of an unknown gas is measured under conditions of known pressure and temperature the molar mass of the gas can be calculated. Worked Example 10.7 shows how to identify an unknown gas from a density measurement.

WORKED EXAMPLE 10.7

Identifying an Unknown by Using Gas Density to Find Molar Mass

An unknown gas found bubbling up in a swamp is collected, placed in a glass bulb, and found to have a density of 0.714 g/L at STP. What is the molar mass of the gas? What is a possible identity of the gas?

IDENTIFY

Known	Unknown
Density ($d = 0.714$ g/L)	Molar Mass (M)
Temperature ($T = 0\,°C$)	
Pressure ($P = 1$ atm)	

STRATEGY

Since d, T, and P are known, the equation for gas density can be rearranged to solve for molar mass. Once molar mass is known the identity of the gas can be suggested.

SOLUTION

Remember to use the Kelvin temperature (273 K) rather than the Celsius temperature (0 °C) in the gas density equation.

$$d = \frac{P(M)}{RT} \quad \text{or} \quad M = \frac{dRT}{P}$$

$$M = \frac{(0.714\text{g/L})\left(0.08206\ \dfrac{L \cdot atm}{mol \cdot K}\right)(273K)}{(1\ atm)} = 16.0\ \text{g/mol}$$

Thus, the molar mass of the unknown gas (actually methane, CH_4) is 16.0 g/mol.

▶ **PRACTICE 10.13**   A foul-smelling gas produced by the reaction of HCl with Na_2S was collected, and a 1.00 L sample was found to have a mass of 1.52 g at STP. What is the molar mass of the gas? What is its likely formula and name?

▶ **APPLY 10.14**   The image shows carbon dioxide gas generated by adding dry ice, CO_2 (s), to water. The carbon dioxide flows downward because it is denser than air.

(a) What is the density in g/L of carbon dioxide at 1 atm and 25 °C?

(b) Why is carbon dioxide denser than air?

10.5 MIXTURES OF GASES: PARTIAL PRESSURE AND DALTON'S LAW

Just as the gas laws apply to all pure gases, regardless of chemical identity, they also apply to *mixtures* of gases, such as air. The pressure, volume, temperature, and amount of a gas mixture are all related by the ideal gas law.

What is responsible for the pressure in a gas mixture? Because the pressure of a pure gas at constant temperature and volume is proportional to its amount ($P = nRT/V$), the pressure contribution from each individual gas in a mixture is also proportional to its amount in the mixture. In other words, the total pressure exerted by a mixture of gases in a container at constant V and T is equal to the sum of the pressures of each individual gas in the container, a statement known as **Dalton's law of partial pressures.**

BIG IDEA Question 3   Go to eText

The total pressure of a mixture of oxygen and nitrogen gas is 1.00 atm. If there are twice the number of moles of oxygen as nitrogen, what is the partial pressure of each gas?

Dalton's law of partial pressures $P_{total} = P_1 + P_2 + P_3 + \ldots$ at constant V and T, where $P_1, P_2, \ldots$ refer to the pressures each individual gas would have if it were alone.

The individual pressure contributions of the various gases in the mixture, P_1, P_2, and so forth, are called partial pressures and refer to the pressure each individual gas would exert if it were alone in the container. That is,

$$P_1 = n_1\left(\frac{RT}{V}\right) \quad P_2 = n_2\left(\frac{RT}{V}\right) \quad P_3 = n_3\left(\frac{RT}{V}\right) \; \dots \; \text{and so forth}$$

But because all the gases in the mixture have the same temperature and volume, we can rewrite Dalton's law to indicate that the total pressure depends only on the total molar amount of gas present and not on the chemical identities of the individual gases:

$$P_{total} = (n_1 + n_2 + n_3 + \; \dots \;)\left(\frac{RT}{V}\right)$$

The concentration of any individual component in a gas mixture is usually expressed as a **mole fraction** (X), which is defined simply as the number of moles of the component divided by the total number of moles in the mixture:

$$\text{Mole fraction } (X) = \frac{\text{Moles of component}}{\text{Total moles in mixture}}$$

The mole fraction of component 1, for example, is

$$X_1 = \frac{n_1}{n_1 + n_2 + n_3 + \; \dots \;} = \frac{n_1}{n_{total}}$$

But because $n = PV/RT$, we can also write

$$X_1 = \frac{P_1\left(\dfrac{V}{RT}\right)}{P_{total}\left(\dfrac{V}{RT}\right)} = \frac{P_1}{P_{total}}$$

which can be rearranged to solve for P_1.

The partial pressure of component 1 in a gas mixture is:

$$\text{Partial pressure} \quad P_1 = X_1 \cdot P_{total}$$

This equation says that the partial pressure exerted by each component in a gas mixture is equal to the mole fraction of that component times the total pressure. In air, for example, the mole fractions of N_2, O_2, Ar, and CO_2 are 0.7808, 0.2095, 0.0093, and 0.000 39, respectively (Table 10.1), and the total pressure of the air is the sum of the individual partial pressures:

$$P_{air} = P_{N_2} + P_{O_2} + P_{Ar} + P_{CO_2} + \; \dots$$

Thus, at a total air pressure of 1 atm (760 mm Hg), the partial pressures of the individual components are

$$
\begin{aligned}
P_{N_2} &= 0.7808 \times 1.00 \text{ atm} & &= 0.7808 \text{ atm} \\
P_{O_2} &= 0.2095 \times 1.00 \text{ atm} & &= 0.2095 \text{ atm} \\
P_{Ar} &= 0.0093 \times 1.00 \text{ atm} & &= 0.0093 \text{ atm} \\
P_{CO_2} &= 0.0004 \times 1.00 \text{ atm} & &= 0.0004 \text{ atm} \\
P_{air} &= P_{N_2} + P_{O_2} + P_{Ar} + P_{CO_2} & &= 1.0000 \text{ atm}
\end{aligned}
$$

There are numerous practical applications of Dalton's law, ranging from the use of anesthetic agents in hospital operating rooms, where partial pressures of both oxygen and anesthetic in the patient's lungs must be constantly monitored, to the composition of diving gases used for underwater exploration. Worked Example 10.8 shows how to calculate partial pressures in a mixture of gases.

Go to
eText

WORKED EXAMPLE 10.8

Calculating Partial Pressure

A 1.50-L steel container at 90 °C contains 5.50 g of H_2, 7.31 g of N_2, and 2.42 g of NH_3. What is the partial pressure of each gas and the total pressure in the container?

IDENTIFY

Known	Unknown
Mass of each gas	Partial pressure (P_{H_2}, P_{N_2}, P_{NH_3})
Volume ($V = 1.50$ L)	Total pressure (P_{tot})
Temperature ($T = 90$ °C)	

STRATEGY

Convert the mass of each gas into moles using molar mass. Find the partial pressure of each gas using the general formula $P_1 = n_1(RT/V)$. Find the total pressure of the gas mixture by summing the partial pressure of each gas.

SOLUTION

Moles of each gas:

$$5.50 \text{ g } H_2 \times \frac{1 \text{ mol } H_2}{2.02 \text{ g } H_2} = 2.72 \text{ mol } H_2 \quad 7.31 \text{ g } N_2 \times \frac{1 \text{ mol } N_2}{28.0 \text{ g } N_2} = 0.261 \text{ mol } N_2$$

$$2.42 \text{ g } NH_3 \times \frac{1 \text{ mol } NH_3}{17.0 \text{ g } NH_3} = 0.142 \text{ mol } NH_3$$

Partial pressure of each gas:

$$P_{H_2} = \frac{(2.72 \text{ mol } H_2)\left(0.08206 \frac{L \cdot atm}{K \cdot mol}\right)(363 \text{ K})}{1.50 \text{ L}} = 54.0 \text{ atm}$$

$$P_{N_2} = \frac{(0.261 \text{ mol } N_2)\left(0.08206 \frac{L \cdot atm}{K \cdot mol}\right)(363 \text{ K})}{1.50 \text{ L}} = 5.18 \text{ atm}$$

$$P_{NH_3} = \frac{(0.142 \text{ mol } NH_3)\left(0.08206 \frac{L \cdot atm}{K \cdot mol}\right)(363 \text{ K})}{1.50 \text{ L}} = 2.81 \text{ atm}$$

The total pressure is the sum of the partial pressure of each gas:

$$P_{tot} = P_{H_2} + P_{N_2} + P_{NH_3} = 54.0 \text{ atm} + 5.18 \text{ atm} + 2.81 \text{ atm} = 62.0 \text{ atm}$$

▶ **PRACTICE 10.15**   Nitrox is a gas mixture used by scuba divers to prevent nitrogen narcosis, a loss of mental and physical function, caused by increased levels of dissolved nitrogen in the blood. What is the partial pressure of nitrogen and oxygen in a 10.0 L tank that contains 235.5 g of oxygen and 366.8 g of nitrogen at a pressure of 50.0 atm and a temperature of 25 °C?

▶ **APPLY 10.16**   At an underwater depth of 250 ft, the pressure is 8.38 atm. What should the mole fraction of oxygen in the diving gas be for the partial pressure of oxygen in the gas to be 0.21 atm, the same as in air at 1.0 atm?

▲ The partial pressure of oxygen in the scuba tanks must be the same underwater as in air at atmospheric pressure.

10.6 THE KINETIC–MOLECULAR THEORY OF GASES

Thus far, we've concentrated on just describing the behavior of gases rather than on understanding the reasons for that behavior. Actually, the reasons are straightforward and were explained more than a century ago using a model called the **kinetic–molecular theory.** The kinetic–molecular theory is based on the following assumptions:

1. A gas consists of tiny particles, either atoms or molecules, moving about at random.
2. The volume of the particles themselves is negligible compared with the total volume of the gas. Most of the volume of a gas is empty space.
3. The gas particles act independently of one another; there are no attractive or repulsive forces between particles.
4. Collisions of the gas particles, either with other particles or with the walls of a container, are elastic. That is, they bounce off the walls at the same speed and therefore the same energy they hit with, so that the total kinetic energy of the gas particles is constant at constant T.
5. The average kinetic energy of the gas particles is proportional to the Kelvin temperature of the sample.

Beginning with these assumptions, it's possible not only to understand the behavior of gases but also to derive quantitatively the ideal gas law (though we'll not do so here). For example, look at how the individual gas laws follow from the five postulates of kinetic–molecular theory:

- **Boyle's law ($P \propto 1/V$):** Gas pressure is a measure of the number and forcefulness of collisions between gas particles and the walls of their container. The smaller the volume at constant n and T, the smaller the distance between the particles and the greater the frequency of collisions. Thus, pressure increases as volume decreases (**FIGURE 10.12a**).

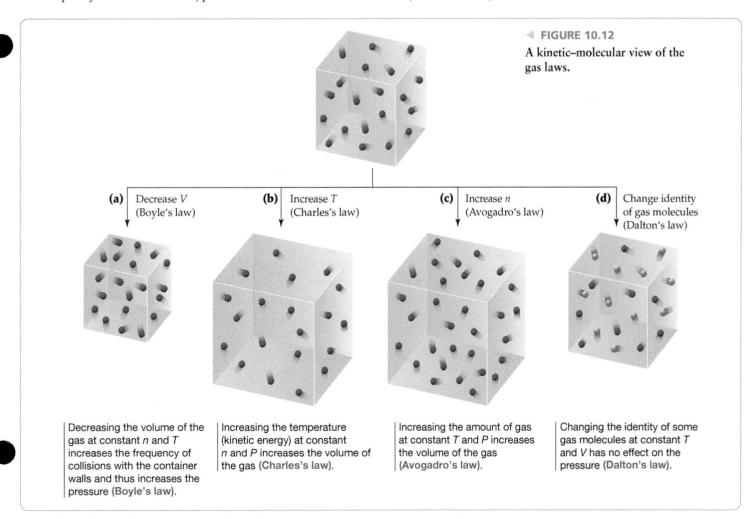

◀ **FIGURE 10.12**
A kinetic–molecular view of the gas laws.

(a) Decrease V (Boyle's law)

(b) Increase T (Charles's law)

(c) Increase n (Avogadro's law)

(d) Change identity of gas molecules (Dalton's law)

Decreasing the volume of the gas at constant n and T increases the frequency of collisions with the container walls and thus increases the pressure (**Boyle's law**).

Increasing the temperature (kinetic energy) at constant n and P increases the volume of the gas (**Charles's law**).

Increasing the amount of gas at constant T and P increases the volume of the gas (**Avogadro's law**).

Changing the identity of some gas molecules at constant T and V has no effect on the pressure (**Dalton's law**).

- **Charles's law ($V \propto T$):** Temperature is a measure of the average kinetic energy of the gas particles. The higher the temperature at constant n and P, the faster the gas particles move. A greater volume is required to avoid increasing the number of collisions with the walls of the container in order to maintain constant pressure. Thus, volume increases as temperature increases (**FIGURE 10.12b**).
- **Avogadro's law ($V \propto n$):** The more particles there are in a gas sample, the more volume the particles need at constant P and T to avoid increasing the number collisions with the walls of the container in order to maintain constant pressure. Thus, volume increases as molar amount increases (**FIGURE 10.12c**).
- **Dalton's law ($P_{total} = P_1 + P_2 + \dots$):** Because gas particles are far apart and act independently of one another, the chemical identity of the particles is irrelevant. Total pressure of a fixed volume of gas depends only on the temperature T and the total number of moles of gas n. The pressure exerted by a specific kind of particle thus depends on the mole fraction of that kind of particle in the mixture, not on the identity of the particle (**FIGURE 10.12d**).

One of the more important conclusions from kinetic–molecular theory comes from assumption 5—the relationship between temperature and E_K, the kinetic energy of molecular motion. Although we won't review the proof in this book, the total kinetic energy of a mole of gas particles equals $3RT/2$, and the average kinetic energy per particle is thus $3RT/2N_A$, where N_A is Avogadro's number. Knowing this relationship makes it possible to calculate the average speed u of a gas particle at a given temperature. To take a helium atom at room temperature (298 K), for instance, we can write

$$E_K = \frac{3\,RT}{2\,N_A} = \frac{1}{2}\,mu^2$$

which can be rearranged to give

Average speed of a gas particle (u).

$$u^2 = \frac{3\,RT}{mN_A}$$

$$\text{or} \quad u = \sqrt{\frac{3RT}{mN_A}} = \sqrt{\frac{3RT}{M}} \quad \text{where } M \text{ is the molar mass}$$

Substituting appropriate values for R $[8.314\ \text{J}/(\text{K}\cdot\text{mol})]$ and for M, the molar mass of helium (4.00×10^{-3} kg/mol), we have

$$u = \sqrt{\frac{(3)\left(8.314\ \dfrac{\text{J}}{\text{K}\cdot\text{mol}}\right)(298\ \text{K})}{4.00 \times 10^{-3}\ \dfrac{\text{kg}}{\text{mol}}}} = \sqrt{1.86 \times 10^6\ \frac{\text{J}}{\text{kg}}}$$

$$= \sqrt{1.86 \times 10^6\ \frac{\dfrac{\text{kg}\cdot\text{m}^2}{\text{s}^2}}{\text{kg}}} = 1.36 \times 10^3\ \text{m/s}$$

Thus, the average speed of a helium atom at room temperature is more than 1.3 km/s, or about 3000 mi/h! Average speeds of some other molecules at 25 °C are given in **TABLE 10.6**. The heavier the molecule, the slower the average speed.

Just because the average speed of helium atoms at 298 K is 1.36 km/s doesn't mean that all helium atoms are moving at that speed or that a given atom will travel from Maine to California in one hour. As shown in **FIGURE 10.13**, there is a broad distribution of speeds among particles in a gas. The distribution flattens out and the maximum speed moves higher as the temperature increases. This means that there is

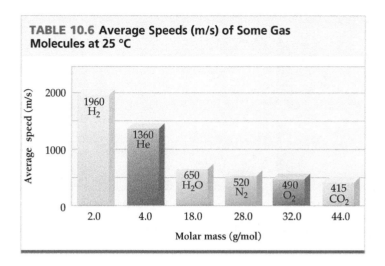

TABLE 10.6 Average Speeds (m/s) of Some Gas Molecules at 25 °C

Go to eText

BIG IDEA Question 4

What is the average speed of gaseous Ar atoms at 25 °C? Use Table 10.6 to make an estimate.

a greater range in molecular speed at higher temperatures. Furthermore, an individual gas particle is likely to travel only a very short distance before it collides with another particle and bounces off in a different direction. Thus, the actual path followed by a gas particle is a random zigzag.

For helium at room temperature and 1 atm pressure, the average distance between collisions, called the *mean free path*, is only about 2×10^{-7} m, or 1000 atomic diameters, and there are approximately 10^{10} collisions per second. For a larger O_2 molecule, the mean free path is about 6×10^{-8} m.

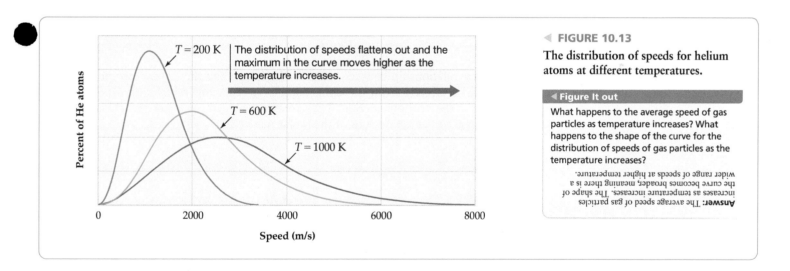

The distribution of speeds flattens out and the maximum in the curve moves higher as the temperature increases.

◀ **FIGURE 10.13**

The distribution of speeds for helium atoms at different temperatures.

◀ **Figure It out**

What happens to the average speed of gas particles as temperature increases? What happens to the shape of the curve for the distribution of speeds of gas particles as the temperature increases?

Answer: The average speed of gas particles increases as temperature increases. The shape of the curve becomes broader, meaning there is a wider range of speeds at higher temperature.

10.7 GAS DIFFUSION AND EFFUSION: GRAHAM'S LAW

The constant motion and high speeds of gas particles have some important practical consequences. One such consequence is that gases mix rapidly when they come in contact. Take the stopper off a bottle of perfume, for instance, and the odor will spread rapidly through a room as perfume molecules mix with the molecules in the air. This mixing of different molecules by random molecular motion with frequent collisions is called **diffusion.** A similar process in which gas molecules escape without collisions through a tiny hole into a vacuum is called **effusion** (**FIGURE 10.14**).

According to **Graham's law,** formulated in the mid-1800s by the Scottish chemist Thomas Graham (1805–1869), the rate of effusion of a gas is inversely proportional to the square root of its molar mass. In other words, the lighter the molecule, the more rapidly it effuses.

▶ **FIGURE 10.14**

Diffusion and effusion of gases.

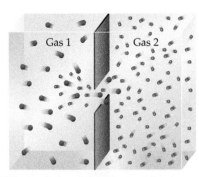
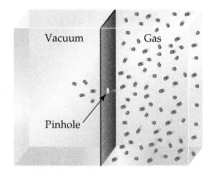

Diffusion is the mixing of gas molecules by random motion under conditions in which molecular collisions occur.

Effusion is the escape of a gas through a pinhole into a vacuum without molecular collisions.

Graham's law Rate of effusion $\propto \dfrac{1}{\sqrt{M}}$

The rate of effusion of a gas is inversely proportional to the square root of its molar mass, M.

In comparing two gases at the same temperature and pressure, we can set up an equation showing that the ratio of the effusion rates of the two gases is inversely proportional to the ratio of the square roots of their molar masses:

$$\frac{\text{Rate}_1}{\text{Rate}_2} = \frac{\sqrt{M_2}}{\sqrt{M_1}} = \sqrt{\frac{M_2}{M_1}}$$

The inverse relationship between the rate of effusion and the square root of the molar mass follows directly from the connection between temperature and kinetic energy described in the previous section. The average speed (u) for a gas particle is inversely proportional to the square root of the molar mass (M).

$$u = \sqrt{\frac{3RT}{M}}$$

$$u \propto \sqrt{\frac{1}{M}}$$

The rate of effusion is proportional to average speed of the molecules, resulting in Graham's law of effusion.

Diffusion is more complex than effusion because of the molecular collisions that occur, but Graham's law usually works as a good approximation. One of the most important practical consequences is that mixtures of gases can be separated into their pure components by taking advantage of the different rates of diffusion of the components. For example, naturally occurring uranium is a mixture of isotopes, primarily ^{235}U (0.72%) and ^{238}U (99.28%) . In uranium enrichment plants that purify the fissionable uranium-235 used for fuel in nuclear reactors, elemental uranium is converted into volatile uranium hexafluoride (bp 56 °C), and the UF_6 gas is allowed to diffuse from one chamber to another through a permeable membrane. The $^{235}UF_6$ and $^{238}UF_6$ molecules diffuse through the membrane at slightly different rates according to the square root of the ratio of their molar masses:

For $^{235}UF_6$, $M = 349.03$

For $^{238}UF_6$, $M = 352.04$

so $\dfrac{\text{Rate of } ^{235}UF_6 \text{ diffusion}}{\text{Rate of } ^{238}UF_6 \text{ diffusion}} = \sqrt{\dfrac{352.04}{349.03}} = 1.0043$

▲ Much of the uranium-235 used as a fuel in nuclear reactors is obtained by gas diffusion of UF_6 in cylinders like these.

The UF_6 gas that passes through the membrane is thus very slightly enriched in the lighter, faster-moving isotope. After repeating the process many thousands of times, a separation of isotopes can be achieved. Approximately 30% of the Western world's nuclear fuel supply—some 5000 tons per year—is produced by this gas diffusion method, although the percentage is dropping because better methods are now available.

WORKED EXAMPLE 10.9

Using Graham's Law to Calculate Diffusion Rates

Assume that you have a sample of hydrogen gas containing H_2, HD, and D_2 that you want to separate into pure components. What are the relative rates of diffusion of the three molecules according to Graham's law? $(H = {}^1H)$ and $(D = {}^2H)$

STRATEGY

First, find the molar masses of the three molecules: for H_2, $M = 2.016$; for HD, $M = 3.022$; for D_2, $M = 4.028$. Then apply Graham's law to different pairs of gas molecules.

SOLUTION

The relative rate of diffusion of each gas relative to the heaviest gas, D_2, can be calculated as shown.

Comparing HD with D_2, we have

$$\frac{\text{Rate of HD diffusion}}{\text{Rate of } D_2 \text{ diffusion}} = \sqrt{\frac{\text{molar mass of } D_2}{\text{molar mass of HD}}} = \sqrt{\frac{4.028}{3.022}} = 1.155$$

Comparing H_2 with D_2, we have

$$\frac{\text{Rate of } H_2 \text{ diffusion}}{\text{Rate of } D_2 \text{ diffusion}} = \sqrt{\frac{\text{molar mass of } D_2}{\text{molar mass of } H_2}} = \sqrt{\frac{4.028}{2.016}} = 1.414$$

Thus, the relative rates of diffusion are H_2 (1.414) > HD (1.155) > D_2 (1.000).

CHECK

The answer makes sense because the lower the molar mass of the gas, the higher the relative diffusion rate.

▸ **PRACTICE 10.17** Which gas diffuses faster, O_2 or Kr? How many times faster?

▸ **APPLY 10.18** An unknown gas is found to diffuse through a porous membrane 1.414 times faster than SO_2. What is the molecular weight of the gas? What is the likely identity of the gas?

10.8 THE BEHAVIOR OF REAL GASES

Here we expand on a point made earlier: The behavior of a real gas is often slightly different from that of an ideal gas. For instance, kinetic–molecular theory assumes that the volume of the gas particles themselves is negligible compared with the total gas volume. The assumption is valid at STP, where the volume taken up by molecules of a typical gas is less than 0.1% of the total volume, but the assumption is not valid at 500 atm and 0 °C, where the volume of the molecules is about 20% of the total volume (**FIGURE 10.15**). As a result, the volume of a real gas at high pressure is larger than predicted by the ideal gas law.

A second issue arising with real gases is the assumption that there are no attractive forces between particles. At lower pressures, this assumption is reasonable because the gas particles are so far apart. At higher pressures, however, the particles are much

▶ **FIGURE 10.15**

The volume of a real gas.

At lower pressure, the volume of the gas particles is negligible compared to the total volume.

At higher pressure, the volume of the gas particles is more significant compared to the total volume. As a result, the volume of a real gas at high pressure is somewhat larger than the ideal value.

▲ **FIGURE 10.16**

Molecules attract one another at distances up to about 10 molecular diameters. The result is a decrease in the actual volume of most real gases when compared with ideal gases at pressures up to 300 atm.

BIG IDEA Question 5   Go to eText

What temperature and pressure conditions are likely to result in a gas volume that is greater than the volume calculated by the ideal gas law?

closer together and the attractive forces between them become more important. In general, intermolecular attractions become significant at a distance of about ten molecular diameters and increase rapidly as the distance diminishes (**FIGURE 10.16**). The result is to draw the molecules of real gases together slightly, decreasing the volume at a given pressure (or decreasing the pressure for a given volume).

Note that the effect of molecular volume—to increase *V*—is opposite that of intermolecular attractions—to decrease *V*. The two factors therefore tend to cancel at intermediate pressures, but the effect of molecular volume dominates above about 300 atm.

Both deviations in the behavior of real gases can be dealt with mathematically by a modification of the ideal gas law called the **van der Waals equation,** which uses two correction factors, called *a* and *b*. The increase in *V*, caused by the volume of the individual gas particles, is corrected by subtracting the amount *nb* from the observed volume, where *n* is the number of moles of gas. The decrease in *V* at constant *P* (or, equivalently, the decrease in *P* at constant *V*), caused by the effect of attractive forces between gas particles, is corrected by adding an amount an^2/V^2 to the pressure.

van der Waals equation

Correction for intermolecular attractions

Correction for molecular volume

$$\left(P + \frac{an^2}{V^2}\right)(V - nb) = nRT$$

$$\text{or } P = \frac{nRT}{V - nb} - \frac{an^2}{V^2}$$

10.9 THE EARTH'S ATMOSPHERE AND THE GREENHOUSE EFFECT

The mantle of gases surrounding the Earth is far from the uniform mixture you might expect. Although atmospheric pressure decreases in a predictable way, the profile of temperature versus altitude is more complex (**FIGURE 10.17**). Four regions of the atmosphere have been defined based on this temperature curve. The temperature in the *troposphere,* the region nearest the Earth's surface, decreases regularly up to about

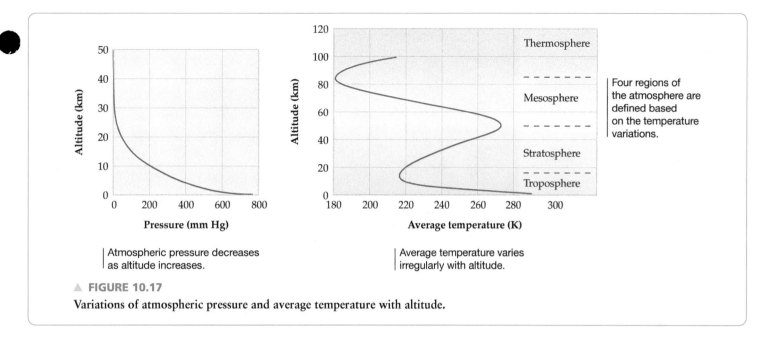

Atmospheric pressure decreases as altitude increases.

Average temperature varies irregularly with altitude.

▲ **FIGURE 10.17**

Variations of atmospheric pressure and average temperature with altitude.

12 km altitude, where it reaches a minimum value. In the *stratosphere,* the temperature then increases to about 50 km. Above the stratosphere, in the *mesosphere* (50–85 km), the temperature again decreases but then increases in the *thermosphere* (above 85 km). To give you a feeling for these altitudes, passenger jets normally fly near the top of the troposphere at altitudes of 10–12 km, and the world altitude record for jet aircraft is 37.65 km—roughly in the middle of the stratosphere.

One of the most well-known environmental problems related to the atmosphere is the increase in the concentration of greenhouse gases due to human activities. The **greenhouse effect** refers to the absorption of infrared (IR) radiation, also known as heat radiation, by gases in the atmosphere which causes an increase in planetary temperature. **Greenhouse gases** are gases that absorb infrared (IR) radiation. The greenhouse effect is a naturally occurring phenomenon that is critical in regulating climate. The average temperature of our planet would be only about 0 °F if the amount of solar radiation reaching the Earth was the only factor controlling climate. Most water would be frozen and the planet would not be suitable for maintaining life! The presence of greenhouse gases such as water, carbon dioxide, and methane increases the Earth's average global temperature to nearly 60 °F.

The principle of the greenhouse effect is illustrated in **FIGURE 10.18.** The Sun, Earth's primary energy source, emits radiation most strongly in the ultraviolet (UV) and visible regions of the electromagnetic spectrum (purple and yellow arrow). Most high-energy, biologically damaging UV radiation (purple arrow) is absorbed by ozone and oxygen in the stratosphere and therefore does not reach ground level. Consequently, most of the radiation reaching Earth is visible light (yellow arrow), which is absorbed by the Earth's surface (vegetation, rocks, water, concrete), causing it to heat up. The warm surface then re-emits infrared (IR) radiation toward space (red arrow). You can sense the IR radiation from Earth in the heat you feel well after dark from a black asphalt road that warmed up during the day. The IR radiation emitted from the Earth either escapes into space or is absorbed by greenhouse gases. The greenhouse gases, in turn, reradiate infrared energy and some of it returns to Earth, resulting in an increase in the temperature of the planet. The greenhouse effect is so named because the glass windows in a greenhouse act in a similar manner to the Earth's atmosphere. Visible light transmitted through the glass and absorbed by surfaces causes the inside to warm. Infrared radiation emitted by warm interior surfaces is absorbed by the glass, preventing some of the heat from escaping. Soon the greenhouse interior is much warmer than the temperature outside.

▶ **FIGURE 10.18**

The greenhouse effect.

▶ **Figure It out**

What is the primary region of electromagnetic radiation that strikes the surface of the Earth? What region of electromagnetic radiation is emitted from the surface of the Earth?

Answer: Visible radiation reaches Earth's surface; infrared radiation is emitted by the Earth.

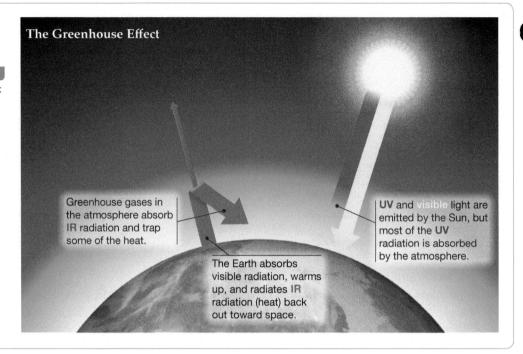

The Greenhouse Effect

Greenhouse gases in the atmosphere absorb IR radiation and trap some of the heat.

UV and visible light are emitted by the Sun, but most of the UV radiation is absorbed by the atmosphere.

The Earth absorbs visible radiation, warms up, and radiates IR radiation (heat) back out toward space.

FIGURE 10.19 illustrates differences in the nature of the electromagnetic radiation emitted by the Sun and Earth. Incoming solar radiation has a maximum intensity at 483 nm, which is in the visible region of the electromagnetic spectrum (blue light), while the Earth's emission has a maximum intensity near 10,000 nm in the IR. Why do the Earth and Sun emit different wavelengths of light? The answer is that the Sun

▶ **FIGURE 10.19**

Spectrum of solar radiation incident on the Earth's surface and spectrum of radiation emitted from Earth.

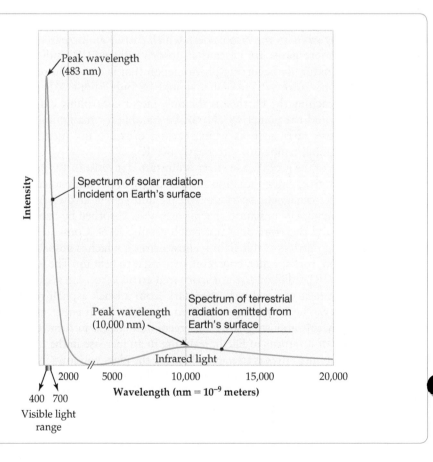

Peak wavelength (483 nm)

Spectrum of solar radiation incident on Earth's surface

Peak wavelength (10,000 nm)

Spectrum of terrestrial radiation emitted from Earth's surface

Infrared light

Intensity

2000   5000   10,000   15,000   20,000

400   700

Wavelength (nm = 10^{-9} meters)

Visible light range

is a lot hotter than the Earth. Both the Sun and Earth emit light as a *black-body*. A black-body emits a continuous spectrum, with wavelength dependent on temperature. A familiar example of black-body radiation is the reddish-orange glow of a hot electric burner. The peak wavelength of 483 nm in the Sun's emission spectrum corresponds to a very hot temperature, near 6000 K, while the peak wavelength in the Earth's emission spectrum, near 10,000 nm, corresponds to a temperature of 288 K (15 °C or 59 °F).

10.10 GREENHOUSE GASES

Why are some atmospheric gases classified as greenhouse gases and others are not? The two most concentrated gases in the atmosphere N_2 (~80% by volume) and O_2 (~20% by volume) do not absorb infrared radiation and therefore are not classified as greenhouse gases. In contrast, CO_2 with an atmospheric concentration of only 0.04% by volume is a greenhouse gas that plays an important role in the regulation of climate. In order to explain why CO_2 is a greenhouse gas and N_2 and O_2 are not, we must understand what occurs on a molecular level when a photon of IR radiation is absorbed.

Let's begin by revisiting the interaction of visible light with atoms. Recall from Section 5.3 that electrons falling from higher-energy to lower-energy orbitals emit a discrete series of colored lines called atomic line spectra. Conversely, when these same wavelengths of light are *absorbed* by an atom, electrons move from a lower-energy to higher-energy orbitals. The same absorption process occurs in molecules, but in this case electrons change positions not between atomic orbitals, but between molecular orbitals (Section 8.7).

Absorbed IR radiation, in the greenhouse effect, is not sufficiently energetic to cause electrons to jump to a higher energy orbital but does increase bond vibrations in molecules. To visualize a bond vibration, imagine two balls representing atoms on either end of a spring (**FIGURE 10.20**). The spring can stretch and retract. The absorption of IR radiation causes a molecule to reach an excited vibrational state in which stretching moves atoms further apart.

▲ A hot electric burner emits black-body radiation.

REMEMBER...
A line spectrum for an element consists of a few wavelengths separated by a dark background (Section 5.3).

REMEMBER...
A molecular orbital is a wave function whose square gives the probability of finding an electron within a given region of space in a molecule (Section 8.7).

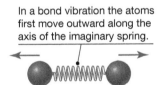

In a bond vibration the atoms first move outward along the axis of the imaginary spring.

Once the vibration has reached its maximum amplitude the atoms return to their original position.

◄ **FIGURE 10.20**
Visualization of a bond vibration in a diatomic molecule.

Let's examine some bond vibrations that occur in the greenhouse gas CO_2. CO_2 has two double bonds around the central carbon atom resulting in linear geometry. However, these bonds are not stationary and vibrate in three different ways, as shown below.

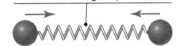
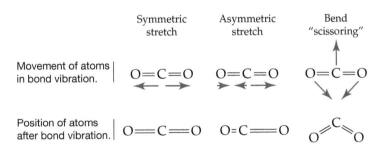

In the symmetric stretch, both C=O bonds lengthen and in the asymmetric stretch both oxygen atoms move in the same direction resulting in one longer and one shorter C=O bond. The bending vibration changes the linear geometry of CO_2 to a bent geometry. Once the atoms reach their new positions, the motion is reversed and the atoms oscillate around their original position. Every CO_2 molecule is constantly undergoing these vibrational motions.

A molecule will absorb a photon of IR radiation if two conditions are met:

REMEMBER...

In atoms and molecules, energy levels are **quantized**, meaning that certain specific energies are available for electrons (Section 5.4).

1. **The energy difference between the lower vibrational state and the excited vibrational state exactly matches the energy of the IR photon.** Recall that energy levels in atoms and molecules are **quantized** (Section 5.4). The photon can only be absorbed if its energy is exactly matched to the energy difference between states.
2. **The vibration results in a change in dipole moment** (Section 8.5).

Let's evaluate each vibration in CO_2 to determine if this requirement is met.

REMEMBER...

A **dipole moment** is a measure of the polarity of a molecule and is defined as the magnitude of charge at either end of the molecule times the distance between the charges (Section 8.5).

Symmetric stretch: Each C=O bond has a bond dipole because oxygen is more electronegative than carbon, but the dipoles exactly cancel each other out and CO_2 does not have a net dipole moment. When both C=O bonds elongate during the symmetric stretch, the bond dipoles still exactly cancel and there is no change in dipole moment. Thus the symmetric stretch vibration does not absorb IR radiation.

Asymmetric stretch: In the asymmetric stretch, both oxygen atoms carry a partial negative charge and they move in the same direction during the vibration. This results in shifting in negative charge to one side of the CO_2 molecule, creating a net dipole moment. Therefore, the asymmetric stretch absorbs IR radiation.

O=C≙O
⟶

O≙C=O
⟵

The asymmetric stretch produces an alternating dipole moment.

Bending (scissoring) vibration: The bending vibration changes CO_2 from a linear geometry with no dipole moment to a bent geometry with a dipole moment. Both bond dipoles point toward the partial negative charges on the oxygen atoms, giving the end of CO_2 with two oxygen atoms a partial negative charge and the end with the carbon atom a partial positive charge. Thus the bending vibration creates a change in dipole moment and absorbs IR radiation.

The bending vibration produces an alternating dipole moment.

FIGURE 10.21 shows the infrared absorption spectrum for CO_2. The vertical axis is the fraction of light transmitted by the sample, and the negative peaks indicate absorption of IR radiation. Notice that both the asymmetric stretch and the bending vibration have absorption peaks. The wavelength (frequency) for the symmetric stretch is indicated, but no absorption peak is present because this vibration does not cause a net change in dipole moment.

▶ **FIGURE 10.21**

Infrared absorption spectrum of carbon dioxide.

▶ **Figure It out**

Which vibration does not absorb IR radiation? Why?

Answer: The symmetric stretch because the vibration does not cause a change in dipole moment.

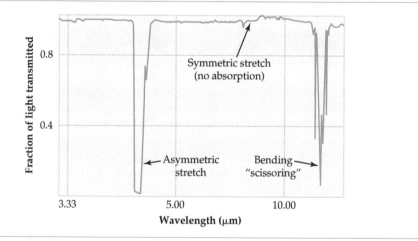

Molecules that absorb IR radiation are greenhouse gases, and water is one of the most important examples. The observation that clear nights are often cooler than cloudy ones is a familiar example of the greenhouse effect; water vapor in clouds strongly absorbs IR. Although water vapor is the largest contributor to the natural greenhouse effect, the amount in the atmosphere is mainly controlled by air temperature and not by emissions from human activities. The greenhouse gas emissions from human activities that are of greatest concern are carbon dioxide (CO_2), nitrous oxide (N_2O), methane (CH_4), and halogen-containing gases. Carbon dioxide is added to the atmosphere primarily from burning fossil fuels, but industrial processes and decaying organic matter also contribute significant amounts. Methane is emitted from fossil fuel mining and use, landfills, agricultural practices, livestock cultivation, and termites. Natural microbial activity in soil and oceans and the use of fertilizer in agricultural practices are sources of nitrous oxide. Halogenated gases, such as hydrofluorocarbons ($CHClF_2$) or sulfur hexafluoride (SF_6), are synthetic greenhouse gases emitted from a variety of industrial applications such as refrigeration and air conditioning.

Global-warming potential (GWP) is a relative measure of how much heat a greenhouse gas traps in the atmosphere on a per mass basis. Carbon dioxide is set to a reference value of 1. A GWP is calculated over a specific time interval, commonly 20, 100, or 500 years, and is influenced by how strongly a gas absorbs IR and how long it remains in the atmosphere. **TABLE 10.7** gives GWP values for the major greenhouse gases associated with human activities. Table 10.7 also includes radiative forcing values, which describe the relative contribution of each gas to the greenhouse effect. Higher positive values for radiative forcing indicate larger warming effects. Radiative forcing is defined as the net change in the energy balance of the Earth and is expressed in units of watts per square meter (W/m^2). Although CO_2 has the lowest GWP of the greenhouse gases listed in Table 10.7, it makes the greatest contribution to the greenhouse effect because it has a higher concentration than the other gases.

TABLE 10.7   Global-Warming Potentials and Radiative Forcing of Greenhouse Gases

Greenhouse Gas	Global Warming Potential (100-year value)	Atmospheric Concentration	Radiative Forcing (W/m^2)
CO_2	1	400 ppm	1.82
CH_4	21	1.8 ppm	0.48
N_2O	310	325 ppb	0.17
CFC-12	4600	0.52 ppb	0.17
SF_6	22,800	0.007 ppb	0.004
Halogenated gases (total)			0.360

Source: Intergovernmental Panel on Climate Change—Climate Change 2013: The Physical Science Basis.

10.11 CLIMATE CHANGE

Global warming describes an upset in the delicate thermal balance of incoming and outgoing radiation on Earth caused by increasing concentrations of greenhouse gases in the atmosphere. Rising levels of greenhouse gases will absorb more IR radiation and cause an enhanced warming effect. However, this phenomenon is more accurately called **climate change** because there will not be a uniform rise in temperature at all locations. Instead, different areas will warm by varying degrees and other areas may even experience cooling. Climate science is complex because it is global in scope and greenhouse gas concentration is only one of many factors that influence climate. Cloud cover, particulate matter, solar energy, and changing surface reflectivity due to melting polar ice caps or deforestation are just a few variables that influence global and regional temperature. In this section, we will describe how concentrations of greenhouse gases have changed over time and their impact on climate both today and in the future.

Since the Industrial Revolution began in the late 1700s, people have added significant quantities of greenhouse gases to the atmosphere by burning fossil fuels for energy, cutting down forests, and producing industrial goods such as cement or metals. Although natural mechanisms remove greenhouse gases from the atmosphere, concentrations are rising because the rate of addition exceeds the rate of removal. For example, careful measurements show that concentrations of atmospheric carbon dioxide have risen in the past 160 years from an estimated 290 parts per million (ppm) in 1850 to 400 ppm in 2014 (**FIGURE 10.22**).

▶ **FIGURE 10.22**

Annual concentration of atmospheric CO_2 since 1850.

Source: National Aeronautics and Space Administration Goddard Institute for Space Studies.

▶ **Figure It out**

Has the rate of CO_2 increase changed over the time period 1850–2017?

Answer: The rate of CO_2 increase has increased dramatically after 1970 and continues to rise each year.

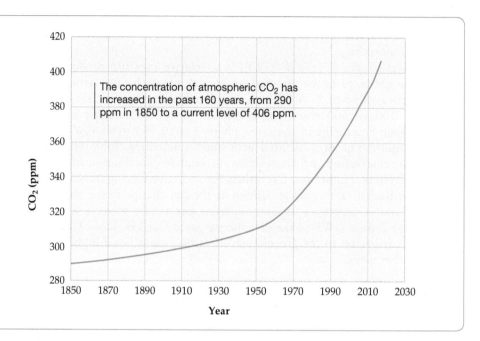

The concentration of atmospheric CO_2 has increased in the past 160 years, from 290 ppm in 1850 to a current level of 406 ppm.

It is also useful to examine greenhouse gas levels on a longer timescale to compare our current situation to other time periods. Scientists can create historical records from ice core samples taken from polar ice caps because the depth of the ice core can be correlated with time. Measurements of the greenhouse gases trapped in air bubbles in the ice show that current global atmospheric concentrations of CO_2 and CH_4 are unprecedented compared with their levels in the past 650,000 years (**FIGURE 10.23**). Concentrations of these greenhouse gases have fluctuated over time, but dramatic increases are evident in the last century. Note that CO_2 levels never exceeded 300 ppm in the long-term historical record, but burning of fossil fuels has caused the level to rise

▶ **FIGURE 10.23**

Long-term historical record of greenhouse gases obtained from ice core and atmospheric measurements.

Source: U.S. Environmental Protection Agency, Climate Change, Greenhouse Gases.

▶ **Figure It out**

How do current levels of CO_2 and CH_4 in the atmosphere compare to levels over the past 650,000 years?

Answer: The historical record shows that both CO_2 and CH_4 levels have risen dramatically in the past few centuries, and the current level is significantly higher than any time in the past 650,000 years.

(a) CO_2

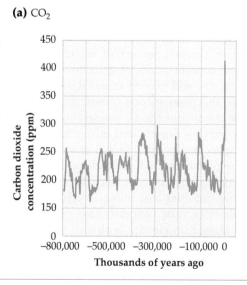

(b) CH_4

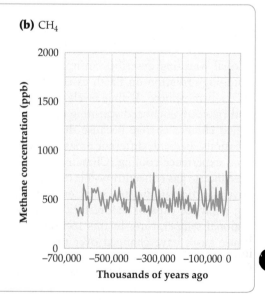

to more than 400 ppm. Methane concentrations did not exceed 0.8 ppm (800 ppb) in the historical record but have recently risen to 1.8 ppm (1,800 ppb). Similarly, N_2O levels have not exceeded 280 ppb over the past 100,000 years but have increased to a concentration of 325 ppb. Most of the halogenated gases are man-made, and their atmospheric concentration began rising as they were used in industrial processes over the past few decades.

Ever since the Earth formed, its climate has undergone dramatic shifts, from "ice ages" with high glacial coverage to relatively warm periods. Factors such as surface reflectivity, airborne dust, variations in the Earth's orbit, and solar intensity combined with greenhouse gas levels contributed to these periodic fluctuations in climate. Ice core samples can also be used to measure a long-term historical record of past climate. Analysis of the ratio of hydrogen isotopes (^2H:^1H) in H_2O in ice core samples helps scientists estimate long-term shifts in average global temperature. **FIGURE 10.24** shows the correlation between past carbon dioxide (CO_2) concentrations (top) and Antarctic temperature (bottom) over the past 800,000 years. Warmer periods coincide with higher CO_2 concentrations.

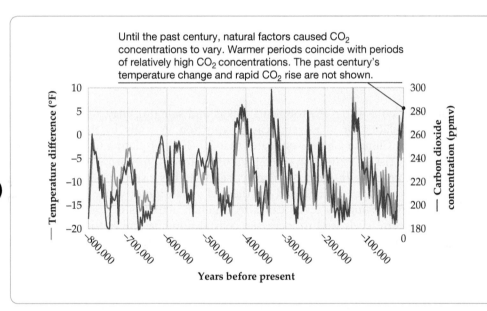

Until the past century, natural factors caused CO_2 concentrations to vary. Warmer periods coincide with periods of relatively high CO_2 concentrations. The past century's temperature change and rapid CO_2 rise are not shown.

◀ **FIGURE 10.24**

Correlation of CO_2 levels and past global temperatures measured in ice cores.

Source: National Research Council. Advancing the Science of Climate Change. The National Academies Press, 2010.

◀ **Figure It out**

Does this figure show that an increase in CO_2 concentration causes an increase in temperature?

Answer: No, it does not prove that the temperature changed in direct response to changes in CO_2 levels. It is possible that the opposite occurred: a temperature change could have caused changes in CO_2 levels. The factors are related in a complex way, but the data clearly shows that higher temperatures correspond with higher CO_2 levels.

Most atmospheric scientists believe the climate has changed rapidly in recent years due to human-induced changes to the atmosphere. **FIGURE 10.25** illustrates the change in global surface temperature (°C) from 1880 to present relative to the average temperature during the time period 1951–1980. Roughly two-thirds of the warming occurred since 1975. While the temperature increase may seem small (approximately 1 °C from 1990 to present), it is important to realize that warming is occurring at a much more rapid pace than past climate shifts, which occurred over thousands of years. Climate models predict that average global temperatures may increase by 3.5–4 °C (~6−7 °F) by 2100, depending on levels of future greenhouse gas emissions and that warming will continue at an accelerated pace.

Scientists track the effects of climate change using indicators that show trends in environmental conditions, such as the extent of sea ice, changes in sea level, changes in biodiversity, and ocean acidity. The Arctic is warming at a rate much faster than the rest of the Earth, and sea ice is melting at an astonishing rate. The area of Arctic sea ice in September 2012 was 1.3 million square miles (an area five times the size of Texas) less than the historical 1979–2000 average (**FIGURE 10.26**). Climate is an important influence on ecosystems, and the rate of temperature change may simply be too fast for many species to adapt in order to survive. Species that are especially climate sensitive, such as those dependent on sea ice (polar bears and seals) or those adapted

▶ **FIGURE 10.25**

Change in global surface temperature (°C) relative to 1951–1980 average temperatures. The dotted green line is the annual mean and the solid red line is the 5-year mean.

Source: National Aeronautics and Space Administration, GISS Surface Temperature Analysis.

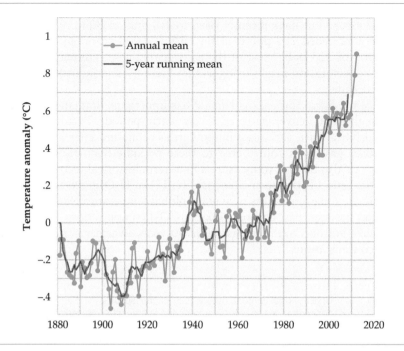

▲ Climate change may be the leading factor decreasing the populations of the American pika.

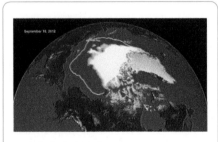

▲ **FIGURE 10.26**

Sea ice extent in the Arctic. September 2012 had the lowest sea ice extent on record, 49% below the 1979–2000 average for that month, shown by the yellow line.

to mountain or cold water environments (pikas and salmon), may suffer significant declines in population or in extreme cases extinction. A study by the Intergovernmental Panel on Climate Change estimates that 20–30% of plant and animal species evaluated are at risk of extinction if temperature reaches levels projected to occur by the end of the century. In the past century, sea level has risen by nearly ten inches due to the melting of ice on land and the thermal expansion of water as its temperature increases. Changing sea levels can affect human activities in coastal areas. For example, rising sea levels lead to increased coastal flooding and erosion and greater damage from storms. Sea level rise can alter ecosystems, transforming marshes and other freshwater systems into salt water. Coral reefs, home to much of the biodiversity in the oceans, are dying at a rapid rate due to higher temperatures and acidity caused by increased levels of dissolved carbon dioxide.

A great majority of climate scientists (more than 97% in a recent survey) agree that warming trends over the past century are very likely due to human activities, and most of the leading scientific organizations worldwide, such as the Intergovernmental Panel on Climate Change, The National Academy of Sciences, and the American Chemical Society, have issued public statements endorsing this position. Consequences of ongoing climate change have been scientifically documented, and humans must consider the increasing risk of extreme weather events (heat waves, droughts, and floods), loss of coastline, and changing weather patterns that impact agriculture and our food supply.

We must decide how to respond to this long-term environmental issue. Solutions can be complex, costly, and difficult to implement, but several options are being researched and tested. These include alternative energy sources (such as solar, wind, and nuclear), carbon sequestration—the trapping and storing of carbon dioxide—and conservation measures (increasing energy efficiency or decreasing use). Many scientists and policy makers believe that it is more economical to develop a sustainable energy system than to deal with the costs associated with the negative consequences of climate change.

INQUIRY ? How do inhaled anesthetics work?

nesthesia is a state of temporary induced loss of sensation or awareness in the practice of medicine. It may include analgesia (relief from or prevention of pain), paralysis (muscle relaxation), amnesia (loss of memory), or unconsciousness. William Morton's demonstration in 1846 of ether-induced anesthesia during dental surgery ranks as one of the most important medical breakthroughs of all time. Up to that point, all surgery had been carried out with the patient conscious. Use of chloroform as an anesthetic quickly followed Morton's work. Queen Victoria of England gave birth to a child in 1853 while anesthetized by chloroform.

◀ Morton's ether inhaler consisted of a large glass bulb. A sponge soaked with diethyl ether was placed inside the bulb, and the spout was placed in the patient's mouth. An opening on the opposite side from the patient allowed air to enter and be drawn over the ether-soaked sponge with each breath.

Throughout history, doctors and scientists have searched for the perfect inhaled anesthetic, one that rapidly induces anesthesia, smells pleasant, and is free of side effects. Hundreds of substances in addition to ether and chloroform have been shown to act as inhaled anesthetics. Halothane, isoflurane, sevoflurane, and desflurane are commonly used agents at present. All four are nontoxic, nonflammable, and potent at relatively low doses.

Despite their importance, surprisingly little is known about how inhaled anesthetics work in the body. Even the definition of anesthesia as a behavioral state is imprecise, and the nature of changes in brain function leading to anesthesia are unknown. Remarkably, the potency of different inhaled anesthetics correlates well with their solubility in olive oil: the more soluble in olive oil, the more potent as an anesthetic. This unusual observation has led many scientists to believe that anesthetics act by dissolving in the fatty membranes surrounding nerve cells. The resultant changes in the fluidity and shape of the membranes apparently decrease the ability of sodium ions to pass into the nerve cells, thereby blocking the firing of nerve impulses.

Depth of anesthesia is determined by the concentration of anesthetic agent that reaches the brain. Brain concentration, in turn, depends on the solubility and transport of the anesthetic agent in the bloodstream and on its partial pressure in inhaled air. Anesthetic potency is usually expressed as a minimum alveolar concentration (MAC), defined as the percent by volume of anesthetic in inhaled air that results in 50% of patients being unresponsive to surgical stimulus. The minimum alveolar concentration (MAC) can also be expressed as a partial pressure of the anesthetic gas. The ideal gas law shows that volume is

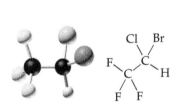

Halothane

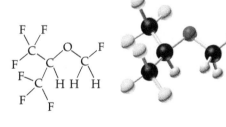

Sevoflurane

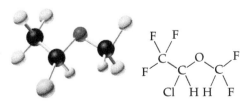

Isoflurane

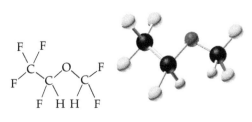

Desflurane

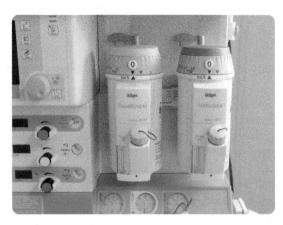

▲ A vaporizer holds a liquid anesthetic and converts it to gas for inhalation.

directly related to moles. Therefore, the volume fraction of a gas is the same as the mol fraction. **TABLE 10.11** indicates that desflurane has a MAC value of 6.2% by volume, which corresponds to a volume fraction of 0.062. The mole fraction of desflurane in the mixture is also 0.062. The total pressure inhaled by a patient is room pressure (~760 mm Hg), and the partial pressure of desflurane is calculated using the equation for partial pressure:

$$P_{desflurane} = X_{desflurane}P_{total} = (0.062)(760 \text{ mm Hg})$$
$$= 47 \text{ mm Hg}$$

As shown in Table 10.11, nitrous oxide (N_2O) is the least potent of common anesthetics. Fewer than 50% of patients are anesthetized by breathing an 80:20 mix of nitrous oxide and oxygen. Halothane is the most potent agent, and a partial pressure of only 5.7 mm Hg is sufficient to anesthetize 50% of patients.

TABLE 10.11   Relative Potency of Inhaled Anesthetics

Anesthetic	MAC (%)	MAC (partial pressure, mm Hg)
Nitrous oxide	—	>760
Desflurane	6.2	47
Sevoflurane	2.5	19
Isoflurane	1.4	11
Halothane	0.75	5.7

PROBLEM 10.19   A partial pressure of diethyl ether of 15 mm Hg results in anesthesia in 50% of patients. What is the MAC for diethyl ether? (Assume the total pressure is 760 mm Hg.)

PROBLEM 10.20   Chloroform has a MAC of 0.77%.
(a) What partial pressure of chloroform is required to anesthetize 50% of patients? (Assume the total pressure is 760 mm Hg.)
(b) What mass of chloroform in 10.0 L of air at STP will produce the appropriate MAC?

PROBLEM 10.21   A 70:30 by volume mix of $N_2O:O_2$ is used to reduce pain in children undergoing basic procedures such a venipuncture or lumbar puncture. A 10.0 L tank of the mix has a total a pressure of 50 atm at 25 °C.
(a) What is the partial pressure of N_2O and O_2 in the tank?
(b) How many grams of N_2O and O_2 are in the tank?

PROBLEM 10.22   A mixture of 8% sevoflurane ($C_4H_3F_7O$), 62% N_2O, and 30% O_2 was used to anesthetize cats for veterinary procedures. If a patient is breathing the mixture at a total pressure of 1 atm, what is the partial pressure of each gas in units of mm Hg?

STUDY GUIDE

Section	Concept Summary	Learning Objectives	Test Your Understanding
10.1 Gases and Gas Pressure	A gas is a collection of atoms or molecules moving independently through a volume that is largely empty space. Collisions of randomly moving particles with the walls of their container exert a force per unit area that we perceive as **pressure (P)**. The SI unit for pressure is the **pascal (Pa)**, but the **atmosphere (atm)**, the **millimeter of mercury (mm Hg)**, and the **bar** are more commonly used.	**10.1** Convert between different units of pressure.	Worked Example 10.1; Problems 10.38–10.39
		10.2 Determine the pressure from the height of a liquid mercury column in a barometer or manometer.	Worked Example 10.2; Problems 10.23–10.25, 10.40, 10.42
10.2 The Gas Laws	The physical condition of any gas is defined by four variables: pressure (P), temperature (T), volume (V), and molar amount (n). The specific relationships among these variables are called the **gas** laws: Boyle's law, Charles's law, and Avogadro's law.	**10.3** Use the individual gas laws to calculate pressure, volume, number of moles, or temperature of a gas sample when conditions change.	Worked Example 10.3; Problems 10.26, 10.48
10.3 The Ideal Gas Law	The three individual gas laws can be combined into a single **ideal gas law**, $PV = nRT$. If any three of the four variables P, V, T, and n are known, the fourth can be calculated. The constant R in the equation is called the **gas constant (R)** and has the same value for all gases. At **standard temperature and pressure** (**STP**; 1 atm and 0 °C), the **standard molar volume** of an ideal gas is 22.414 L.	**10.4** Use the ideal gas law to calculate pressure, volume, number of moles, or temperature of a gas sample.	Worked Examples 10.4–10.5; Problems 10.52–10.58
10.4 Stoichiometric Relationships with Gases	The ideal gas law can be used to calculate the number of moles in a gas sample. The number of moles in a gaseous reactant or product can be related to other substances in a chemical reaction using stoichiometry. Gas density can be used to identify unknown gases and depends on pressure, molar mass, and temperature.	**10.5** Calculate volumes of gases in chemical reactions.	Worked Example 10.6; Problems 10.70, 10.72, 10.74
		10.6 Calculate the density or molar mass of a gas using the formula for gas density.	Worked Example 10.7; Problems 10.66, 10.68
10.5 Mixtures of Gases: Partial Pressure and Dalton's Law	The gas laws apply to mixtures of gases as well as to pure gases. According to **Dalton's law of partial pressures**, the total pressure exerted by a mixture of gases in a container is equal to the sum of the pressures each individual gas would exert alone.	**10.7** Calculate the partial pressure, mole fraction, or amount of each gas in a mixture.	Worked Example 10.8; Problems 10.82, 10.86, 10.90
10.6 The Kinetic–Molecular Theory of Gases	The behavior of gases can be accounted for using a model called the **kinetic–molecular theory**, a group of five postulates: 1. A gas consists of tiny particles moving at random. 2. The volume of the gas particles is negligible compared with the total volume. 3. There are no forces between particles, either attractive or repulsive. 4. Collisions of gas particles are elastic. 5. The average kinetic energy of gas particles is proportional to their absolute temperature.	**10.8** Use the assumptions of kinetic–molecular theory to predict gas behavior.	Problems 10.27–10.28, 10.32
		10.9 Calculate the average molecular speed of a gas particle at a given temperature.	Problems 10.94, 10.98

Section	Concept Summary	Learning Objectives	Test Your Understanding
10.7 Gas Diffusion and Effusion: Graham's Law	The connection between temperature and kinetic energy obtained from the kinetic–molecular theory makes it possible to calculate the average speed of a gas particle at any temperature. An important practical consequence of this relationship is **Graham's law,** which states that the rate of a gas's **effusion,** or spontaneous passage through a pinhole in a membrane, depends inversely on the square root of the gas's molar mass.	**10.10** Interpret a molecular picture of effusion and diffusion. **10.11** Use Graham's law to estimate relative rates of diffusion for two gases.	Problem 10.33 Worked Example 10.9; Problems 10.100–10.103, 10.143
10.8 The Behavior of Real Gases	Real gases differ in their behavior from that predicted by the ideal gas law, particularly at high pressure, where gas particles are forced close together and intermolecular attractions become significant. The deviations from ideal behavior can be dealt with mathematically by the **van der Waals equation.**	**10.12** Identify temperature and pressure conditions that cause significant deviations from ideal behavior. **10.13** Use the van der Waals equation to calculate the properties of real gases.	Problems 10.110–10.111 Problems 10.112, 10.114
10.9–10.11 The Earth's Atmosphere, the Greenhouse Effect, and Climate Change	The **greenhouse effect** is the trapping of heat emitted from the Earth by gases that absorb infrared radiation (**greenhouse gases**). **Climate change** is occurring as a result of rising levels of greenhouse gases from human activities. Measurements and models show that certain regions such as the Arctic and land masses will warm more than other regions, such as the ocean.	**10.14** Explain the principle of the greenhouse effect. **10.15** Predict if a molecule is a greenhouse gas by examining molecular vibrations. **10.16** Describe the trends in greenhouse gas concentrations over time and measured and predicted effects of climate change.	Problems 10.120, 10.122 Problems 10.123–10.126 Problems 10.127–10.133

KEY TERMS

atmosphere (atm) *377*
Avogadro's law *382*
bar *378*
Boyle's law *380*
Charles's law *381*
climate change *403*
Dalton's law of partial
   pressures *390*

diffusion *395*
effusion *395*
gas constant (R) *385*
gas laws *380*
global warming *403*
Graham's law *395*
greenhouse effect *399*
greenhouse gas *399*

ideal gas *380*
ideal gas law *385*
kinetic–molecular theory *393*
manometer *378*
millimeter of mercury
   (mm Hg) *377*
mole fraction (X) *391*
newton (N) *376*

pascal (Pa) *376*
pressure (P) *376*
standard molar volume *383*
standard temperature and
   pressure (STP) *385*
van der Waals equation *398*

KEY EQUATIONS

- Boyle's Law (Section 10.2)

  $V \propto 1/P$ or $PV = k$ at constant n and T

- Charles's Law (Section 10.2)

  $V \propto T$ or $V/T = k$ at constant n and P

- Avogadro's Law (Section 10.2)

  $V \propto n$ or $V/n = k$ at constant T and P

- Ideal Gas Law (Section 10.3)

  $V = \dfrac{nRT}{P}$ or $PV = nRT$

- Standard Temperature and Pressure (STP) for Gases (Section 10.3)

  $T = 0\,°C \quad P = 1\,\text{atm}$

- Gas Density (Section 10.4)

  $$d = \frac{m}{V} = \frac{PM}{RT}$$

- Dalton's Law of Partial Pressures (Section 10.5)

  $P_{\text{total}} = P_1 + P_2 + P_3 + \ldots$
  where $P_1, P_2, \ldots$ are the pressures each individual gas would have if it were alone.

- Mole Fraction (Section 10.6)

  $$(X) = \frac{\text{Moles of component}}{\text{Total moles in mixture}}$$

- Partial Pressure (Section 10.6)

  $\mathbf{P_1} = X_1 \cdot P_{\text{total}}$

- Average Speed of gas Particle at Temperature T (Section 10.6)

  $$u = \sqrt{\frac{3RT}{mN_A}} = \sqrt{\frac{3RT}{M}} \quad \text{where } M \text{ is the molar mass.}$$

- Graham's Law (Section 10.7)

  $$\text{Rate of effusion} \propto \frac{1}{\sqrt{M}}$$

- Van der Waals Equation (Section 10.8)

  $$\left(P + \frac{an^2}{V^2}\right)(V - nb) = nRT \quad \text{or} \quad P = \frac{nRT}{V - nb} - \frac{an^2}{V^2}$$

  where a and b are correction factors to the ideal gas law.

PRACTICE TEST

After studying this chapter, you can assess your understanding with these practice test questions, which are correlated with chapter learning objectives. If you answer a question incorrectly, refer to the learning objectives in the end-of-chapter Study Guide for assistance. The Study Guide provides a conceptual summary, references a Worked Example to model how to solve the problem, and gives additional problems for more practice.

1. The recommended pressure for inflation of an automobile tire is 35 psi. What is the pressure in units of mm Hg? (1 atm = 14.7 psi; 1 atm = 760 mm Hg) **(LO 10.1)**

   (a) 3.9×10^5 mm Hg

   (b) 0.68 mm Hg

   (c) 3.2×10^2 mm Hg

   (d) 1.8×10^3 mm Hg

2. What is the pressure of the gas inside the bulb (atm) if the outside pressure is 0.92 atm? **(LO 10.2)**

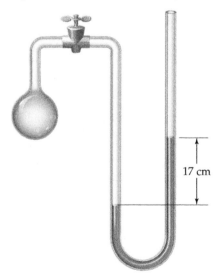

   (a) 1.1 atm   (b) 2.6 atm

   (c) 0.22 atm   (d) 0.70 atm

3. Assume that you have a gas cylinder with a movable piston filled with oxygen. The initial conditions are $T = 250$ K, $n = 0.140$ mol O_2, and $P = 1.00$ atm. If the initial volume is 1.0 L, what is the volume when the temperature is increased to 400 K and the pressure is decreased to 0.75 atm? **(LO 10.3)**

(initial)
$T = 250$ K
$n = 0.140$ mol
$P = 1.0$ atm

(a) 2.1 L   (b) 1.2 L
(c) 0.83 L   (d) 1.6 L

4. Many laboratory gases are sold in steel cylinders with a volume of 43.8 L. What is the mass in grams of argon inside a cylinder whose pressure is 17,180 kPa at 20 °C? **(LO 10.4)**
(a) 1.83×10^7 g   (b) 1.81×10^5 g
(c) 1.23×10^4 g   (d) 122 g

5. Propane gas (C_3H_8) is often used as fuel in rural areas. How many liters of CO_2 are formed at STP by the complete combustion of the propane in a container with a volume of 15.0 L and a pressure of 4.50 atm at 25.0 °C? The equation for the combustion of propane is: $C_3H_8(g) + 5\ O_2(g) \longrightarrow 3\ CO_2(g) + 4\ H_2O(l)$ **(LO 10.4, 10.5)**
(a) 61.8 L   (b) 186 L
(c) 20.6 L   (d) 2.21×10^3 L

6. A certain nonmetal reacts with hydrogen at 440 °C to form a poisonous, foul-smelling gas. A sample with a mass of 6.618 g was found to have a volume of 2.00 L at 25.0 °C and 1.00 atm. What is the molecular weight of the gas? **(LO 10.6)**
(a) 67.9 g/mol   (b) 162 g/mol
(c) 80.9 g/mol   (d) 193.6 g/mol

7. Trimix is a gas mixture consisting of oxygen, helium, and nitrogen used for deep scuba dives. The helium is included to reduce the effects of nitrogen narcosis and oxygen toxicity that occur when too much nitrogen and oxygen dissolve in the blood. A tank of Trimix has a total pressure of 200 atm, and the partial pressure of He is 34 atm. What is the percent by volume of He in the tank? **(LO 10.7)**
(a) 17%   (b) 38%
(c) 23%   (d) 83%

8. A sample of ammonia (NH_3) gas is completely decomposed to nitrogen and hydrogen over a heated iron catalyst. If the total pressure of the mixture of N_2 and H_2 is 1.80 atm, what is the partial pressure of N_2? **(LO 10.7)**
(a) 1.35 atm   (b) 0.45 atm
(c) 0.90 atm   (d) 0.63 atm

9. The apparatus shown consists of three bulbs connected by stopcocks. What is the pressure inside the system when the stopcocks are opened? Assume that the lines connecting the bulbs have zero volume and that the temperature remains constant. **(LO 10.3, 10.7)**
(a) 1.10 atm   (b) 1.73 atm
(c) 4.14 atm   (d) 1.41 atm

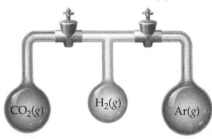

$P = 2.13$ atm $P = 0.861$ atm $P = 1.15$ atm
$V = 1.50$ L $V = 1.00$ L $V = 2.00$ L

10. A mixture of chlorine, hydrogen, and oxygen gas is in a container at STP. Which curve represents oxygen gas? **(LO 10.8)**
(a) Curve (a)   (b) Curve (b)   (c) Curve (c)

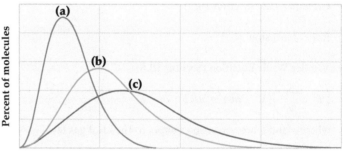

11. The coldest temperature recorded at ground level on Earth was −89.2 °C at the Vostok Station in Antarctica. What is the speed of a nitrogen molecule at this temperature? **(LO 10.9)**
(a) 933 m/s   (b) 40.2 m/s
(c) 404 m/s   (d) 12.8 m/s

12. An unknown gas is found to diffuse through a porous membrane 2.92 times more slowly than H_2. What is the molecular weight of the gas?
(a) 17.0 g/mol   (b) 5.84 g/mol
(c) 8.52 g/mol   (d) 30.0 g/mol

13. Identify the *true* statement about deviations from ideal gas behavior. **(LO 10.12)**
(a) The attractive forces between gas particles cause the true volume of the sample to be larger than predicted by the ideal gas law.
(b) The attractive forces between gas particles most influence the volume of a sample at low pressure.
(c) The volume of the gas particles themselves most influences the volume of the sample at low pressure.
(d) The volume of the gas particles themselves causes the true volume of the sample to be larger than predicted by the ideal gas law.

Answers:

1. d, 2. a, 3. a, 4. c, 5. b, 6. c, 7. a, 8. b, 9. d, 10. b, 11. c, 12. a, 13. d

> **Mastering Chemistry** provides end-of-chapter exercises, feedback-enriched tutorial problems, animations, and interactive activities to encourage problem-solving practice and deeper understanding of key concepts and topics.

RAN *Randomized in Mastering Chemistry*

CONCEPTUAL PROBLEMS

Problems 10.1–10.22 appear within the chapter.

10.23 A glass tube has one end in a dish of mercury and the other end closed by a stopcock. The distance from the surface of the mercury to the bottom of the stopcock is 850 mm. The apparatus is at 25 °C, and the mercury level in the tube is the same as that in the dish.

(1)   **(2)**   **(3)**

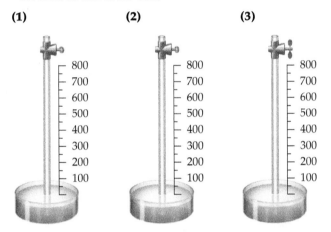

(a) Show on drawing (**1**) what the approximate level of mercury in the tube will be when the temperature of the entire apparatus is lowered from +25 °C to −25 °C.

(b) Show on drawing (**2**) what the approximate level of mercury in the tube will be when a vacuum pump is connected to the top of the tube, the stopcock is opened, the tube is evacuated, the stopcock is closed, and the pump is removed.

(c) Show on drawing (**3**) what the approximate level of mercury in the tube will be when the stopcock in drawing (**2**) is opened.

10.24 The apparatus shown is called a *closed-end* manometer because the arm not connected to the gas sample is closed to the atmosphere and is under vacuum. Explain how you can read the gas pressure in the bulb.

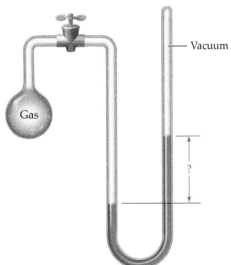

10.25 Redraw the following open-end manometer to show what it would look like when stopcock A is opened.

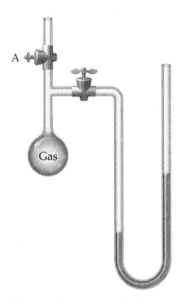

10.26 Assume that you have a sample of gas in a cylinder with a movable piston, as shown in the following drawing:

Redraw the apparatus to show what the sample will look like after (**a**) the temperature is increased from 300 K to 450 K at constant pressure, (**b**) the pressure is increased from 1 atm to 2 atm at constant temperature, and (**c**) the temperature is decreased from 300 K to 200 K and the pressure is decreased from 3 atm to 2 atm.

10.27 Assume that you have a sample of gas at 350 K in a sealed
RAN   container, as represented in (a). Which of the drawings
(b)–(d) represents the gas after the temperature is lowered
from 350 K to 150 K? The boiling point of the gas is 90 K.

(a)

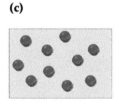

(b)   **(c)**   **(d)**

 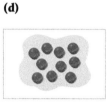

10.28 Assume that you have a mixture of He (atomic weight = 4)
and Xe (atomic weight = 131) at 300 K. Which of the drawings best represents the mixture? (Blue = He, red = Xe.)

(a)   **(b)**   **(c)**

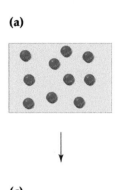

10.29 The following drawing represents a container holding a
RAN   mixture of four gases, red, blue, green, and black. If the total
pressure inside the container is 420 mm Hg, what is the partial pressure of each individual component?

10.30 Three bulbs, two of which contain different gases and one of
which is empty, are connected as shown in the following drawing.
Redraw the apparatus to represent the gases after the stopcocks
are opened and the system is allowed to come to equilibrium.

10.31 Show the approximate level of the movable piston in drawings (a), (b), and (c) after the indicated changes have been
made to the gas.

(a)

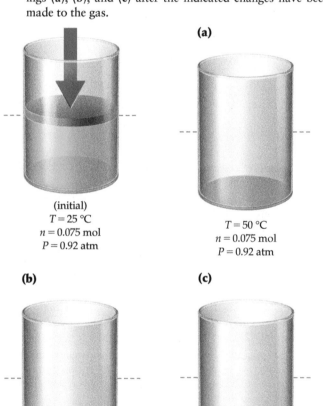

(initial)
T = 25 °C
n = 0.075 mol
P = 0.92 atm

T = 50 °C
n = 0.075 mol
P = 0.92 atm

(b)   **(c)**

T = 175 °C   T = 25 °C
n = 0.075 mol   n = 0.22 mol
P = 2.7 atm   P = 2.7 atm

10.32 A 1:1 mixture of helium (red) and argon (blue) at 300 K is portrayed below on the left. Draw the same mixture when the temperature is lowered to 150 K.

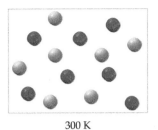

 300 K   150 K

10.33 Effusion of a 1:1 mixture of two gases through a small pinhole produces the results shown below.
RAN
(a) Which gas molecules—yellow or blue—have a higher average speed?
(b) If the yellow molecules have a molecular weight of 25, what is the molecular weight of the blue molecules?

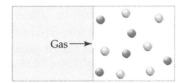

Gas⟶

SECTION PROBLEMS

Gases and Gas Pressure (Section 10.1)

10.34 Yet another common measure of pressure is the unit pounds
RAN per square inch (psi). How many pounds per square inch correspond to 1.00 atm? To 1.00 mm Hg?

10.35 If the density of water is 1.00 g/mL and the density of mercury is 13.6 g/mL, how high a column of water in meters can be supported by standard atmospheric pressure? By 1 bar?

10.36 Why do gases exert pressure?

10.37 Why are gases so much more compressible than solids or liquids?

10.38 Atmospheric pressure at the top of Pikes Peak in Colorado is approximately 480 mm Hg. Convert this value to atmospheres and to pascals.

10.39 Carry out the following conversions:
RAN
(a) 352 torr to kPa
(b) 0.255 atm to mm Hg
(c) 0.0382 mm Hg to Pa

10.40 What is the pressure in millimeters of mercury inside a con-
RAN tainer of gas connected to a mercury-filled open-end manometer of the sort shown in Figure 10.5 when the level in the arm connected to the container is 17.6 cm lower than the level in the arm open to the atmosphere and the atmospheric pressure reading outside the apparatus is 754.3 mm Hg?

10.41 What is the pressure in atmospheres inside a container of gas
RAN connected to a mercury-filled open-end manometer when the level in the arm connected to the container is 28.3 cm higher than the level in the arm open to the atmosphere and the atmospheric pressure reading outside the apparatus is 1.021 atm?

10.42 Assume that you have an open-end manometer filled with ethyl
RAN alcohol (density = 0.7893 g/mL at 20 °C) rather than mercury (density = 13.546 g/mL at 20 °C). What is the pressure in pascals if the level in the arm open to the atmosphere is 55.1 cm higher than the level in the arm connected to the gas sample and the atmospheric pressure reading is 752.3 mm Hg?

Ethyl alcohol

10.43 Assume that you have an open-end manometer filled with
RAN chloroform (density = 1.4832 g/mL at 20 °C) rather than mercury (density = 13.546 g/mL at 20 °C). What is the difference in height between the liquid in the two arms if the pressure in the arm connected to the gas sample is 0.788 atm and the atmospheric pressure reading is 0.849 atm? In which arm is the chloroform level higher?

Chloroform

10.44 What would the atmospheric pressure be in millimeters of mercury if our atmosphere were composed of pure CO_2 gas?

10.45 The surface temperature of Venus is about 1050 K, and the pressure is about 75 Earth atmospheres. Assuming that these conditions represent a Venusian "STP," what is the standard molar volume in liters of a gas on Venus?

10.46 Calculate the average molecular weight of air from the data given in Table 10.1.

10.47 What is the average molecular weight of a diving-gas mixture
RAN that contains 2.0% by volume O_2 and 98.0% by volume He?

The Gas Laws (Section 10.2–10.3)

10.48 Assume that you have a cylinder with a movable piston. What
RAN would happen to the gas pressure inside the cylinder if you were to do the following?
(a) Triple the Kelvin temperature while holding the volume constant
(b) Reduce the amount of gas by one-third while holding the temperature and volume constant
(c) Decrease the volume by 45% at constant T
(d) Halve the Kelvin temperature and triple the volume

10.49 Assume that you have a cylinder with a movable piston. What would happen to the gas volume of the cylinder if you were to do the following?

(a) Halve the Kelvin temperature while holding the pressure constant

(b) Increase the amount of gas by one-fourth while holding the temperature and pressure constant

(c) Decrease the pressure by 75% at constant T

(d) Double the Kelvin temperature and double the pressure

10.50 Which sample contains more molecules: 1.00 L of O_2 at STP, 1.00 L of air at STP, or 1.00 L of H_2 at STP?

10.51 Which sample contains more molecules: 2.50 L of air at 50 °C and 750 mm Hg pressure or 2.16 L of CO_2 at −10 °C and 765 mm Hg pressure?

10.52 Oxygen gas is commonly sold in 49.0-L steel containers at
RAN   a pressure of 150 atm. What volume in liters would the gas occupy at a pressure of 1.02 atm if its temperature remained unchanged? If its temperature was raised from 20 °C to 35 °C at constant P = 150 atm?

10.53 A compressed air tank carried by scuba divers has a volume
RAN   of 8.0 L and a pressure of 140 atm at 20 °C. What is the volume of air in the tank in liters at STP?

10.54 If 15.0 g of CO_2 gas has a volume of 0.30 L at 300 K, what
RAN   is its pressure in millimeters of mercury?

10.55 If 2.00 g of N_2 gas has a volume of 0.40 L and a pressure of
RAN   6.0 atm, what is its Kelvin temperature?

10.56 The matter in interstellar space consists almost entirely of hydrogen atoms at a temperature of 100 K and a density of approximately 1 atom/cm^3. What is the gas pressure in millimeters of mercury?

10.57 Methane gas, CH_4, is sold in a 43.8-L cylinder containing
RAN   5.54 kg. What is the pressure inside the cylinder in kilopascals at 20 °C?

10.58 A small cylinder of helium gas used for filling balloons has
RAN   a volume of 2.30 L and a pressure of 13,800 kPa at 25 °C. How many balloons can you fill if each one has a volume of 1.5 L and a pressure of 1.25 atm at 25 °C?

10.59 Dry ice (solid CO_2) has occasionally been used as an "explo-
RAN   sive" in mining. A hole is drilled, dry ice and a small amount of gunpowder are placed in the hole, a fuse is added, and the hole is plugged. When lit, the exploding gunpowder rapidly vaporizes the dry ice, building up an immense pressure. Assume that 500.0 g of dry ice is placed in a cavity with a volume of 0.800 L and the ignited gunpowder heats the CO_2 to 700 K. What is the final pressure inside the hole?

10.60 Which sample contains more molecules, 15.0 L of steam
RAN   (gaseous H_2O) at 123.0 °C and 0.93 atm pressure or a 10.5 g ice cube at −5.0 °C?

10.61 Which sample contains more molecules, 3.14 L of Ar at 85.0 °C and 1111 mm Hg pressure or 11.07 g of Cl_2?

10.62 Imagine that you have two identical flasks, one containing hydrogen at STP and the other containing oxygen at STP. How can you tell which is which without opening them?

10.63 Imagine that you have two identical flasks, one containing chlorine gas and the other containing argon at the same temperature and pressure. How can you tell which is which without opening them?

Gas Stoichiometry (Section 10.4)

10.64 What is the total mass in grams of oxygen in a room mea-
RAN   suring 4.0 m by 5.0 m by 2.5 m? Assume that the gas is at STP and that air contains 20.95% oxygen by volume.

10.65 The average oxygen content of arterial blood is approxi-
RAN   mately 0.25 g of O_2 per liter. Assuming a body temperature of 37 °C how many moles of oxygen are transported by each liter of arterial blood? How many milliliters?

10.66 One mole of an ideal gas has a volume of 22.414 L at STP. Assuming ideal behavior, what are the densities of the following gases in g/L at STP?

(a) CH_4 (b) CO_2 (c) O_2

10.67 What is the density in g/L of a gas mixture that contains 27.0%
RAN F_2 and 73.0% He by volume at 714 mm Hg and 27.5 °C?

10.68 An unknown gas is placed in a 1.500-L bulb at a pressure of
RAN   356 mm Hg and a temperature of 22.5 °C and is found to weigh 0.9847 g. What is the molecular weight of the gas?

10.69 What are the molecular weights of the gases with the follow-
RAN   ing densities:

(a) 1.342 g/L at STP

(b) 1.053 g/L at 25 °C and 752 mm Hg

10.70 Pure oxygen gas was first prepared by heating mercury(II)
RAN   oxide, HgO:

$$2 \, HgO(s) \longrightarrow 2 \, Hg(l) + O_2(g)$$

What volume in liters of oxygen at STP is released by heating 10.57 g of HgO?

10.71 How many grams of HgO would you need to heat if you
RAN   wanted to prepare 0.0155 mol of O_2 according to the equation in Problem 10.70?

10.72 Hydrogen gas can be prepared by reaction of zinc metal
RAN   with aqueous HCl:

$$Zn(s) + 2 \, HCl(aq) \longrightarrow ZnCl_2(aq) + H_2(g)$$

(a) How many liters of H_2 would be formed at 742 mm Hg and 15 °C if 25.5 g of zinc was allowed to react?

(b) How many grams of zinc would you start with if you wanted to prepare 5.00 L of H_2 at 350 mm Hg and 30.0 °C?

10.73 Ammonium nitrate can decompose explosively when heated
RAN   according to the equation

$$2 \, NH_4NO_3(s) \longrightarrow 2 \, N_2(g) + 4 \, H_2O(g) + O_2(g)$$

How many liters of gas would be formed at 450 °C and 1.00 atm pressure by explosion of 450 g of NH_4NO_3?

10.74 The reaction of sodium peroxide (Na_2O_2) with CO_2 is used
RAN   in space vehicles to remove CO_2 from the air and generate O_2 for breathing:

$$2 \, Na_2O_2(s) + 2 \, CO_2(g) \longrightarrow 2 \, Na_2CO_3(s) + O_2(g)$$

(a) Assuming that air is breathed at an average rate of 4.50 L/min (25 °C; 735 mm Hg) and that the concentration of CO_2 in expelled air is 3.4% by volume, how many grams of CO_2 are produced in 24 h?

(b) How many days would a 3.65 kg supply of Na_2O_2 last?

10.75 Titanium(III) chloride, a substance used in catalysts for preparing polyethylene, is made by high-temperature reaction of $TiCl_4$ vapor with H_2:

$$2\ TiCl_4(g) + H_2(g) \longrightarrow 2\ TiCl_3(s) + 2\ HCl(g)$$

(a) How many grams of $TiCl_4$ are needed for complete reaction with 155 L of H_2 at 435 °C and 795 mm Hg pressure?

(b) How many liters of HCl gas at STP will result from the reaction described in part (a)?

10.76 A typical high-pressure tire on a bicycle might have a volume of 365 mL and a pressure of 7.80 atm at 25 °C. Suppose the rider filled the tire with helium to minimize weight. What is the mass of the helium in the tire?

10.77 Assume that you have 1.00 g of nitroglycerin in a 500.0-mL steel container at 20.0 °C and 1.00 atm pressure. An explosion occurs, raising the temperature of the container and its contents to 425 °C. The balanced equation is

$$4\ C_3H_5N_3O_9(l) \longrightarrow$$
$$12\ CO_2(g) + 10\ H_2O(g) + 6\ N_2(g) + O_2(g)$$

(a) How many moles of nitroglycerin and how many moles of gas (air) were in the container originally?

(b) How many moles of gas are in the container after the explosion?

(c) What is the pressure in atmospheres inside the container after the explosion according to the ideal gas law?

10.78 One of the largest sources of SO_2 to the atmosphere is coal-fired power plants. Calculate the volume of SO_2 in L produced by burning 1 kg of coal that contains 2% sulfur. Assume all the sulfur is converted to SO_2.

10.79 Smelting of ores to produce pure metals is an atmospheric source of sulfur dioxide.

(a) Galena, the most common mineral of lead, is primarily lead(II) sulfide (PbS). The first step in the production of pure lead is to oxidize lead sulfide into lead(II) sulfite ($PbSO_3$). Lead(II) sulfite is then thermally decomposed into lead(II) oxide and sulfur dioxide gas. Balance the following equation.

$$PbSO_3(s) \xrightarrow{heat} PbO(s) + SO_2(g)$$

(b) How many liters of SO_2 are produced at 1 atm and 300 °C if 250 g of $PbSO_3$ is decomposed?

Dalton's Law and Mole Fraction (Section 10.5)

10.80 Use the information in Table 10.1 to calculate the partial pressure in atmospheres of each gas in dry air at STP.

10.81 Natural gas is a mixture of many substances, primarily CH_4, C_2H_6, C_3H_8, and C_4H_{10}. Assuming that the total pressure of the gases is 1.48 atm and that their mole ratio is 94:4.0:1.5:0.50, calculate the partial pressure in atmospheres of each gas.

10.82 A special gas mixture used in bacterial growth chambers contains 1.00% by weight CO_2 and 99.0% O_2. What is the partial pressure in atmospheres of each gas at a total pressure of 0.977 atm?

10.83 A gas mixture for use in some lasers contains 5.00% by weight HCl, 1.00% H_2, and 94% Ne. The mixture is sold in cylinders that have a volume of 49.0 L and a pressure of 13,800 kPa at 210 °C. What is the partial pressure in kilopascals of each gas in the mixture?

10.84 What is the mole fraction of each gas in the mixture described in Problem 10.83?

10.85 A mixture of Ar and N_2 gases has a density of 1.413 g/L at STP. What is the mole fraction of each gas?

10.86 A mixture of 14.2 g of H_2 and 36.7 g of Ar is placed in a 100.0-L container at 290 K.

(a) What is the partial pressure of H_2 in atmospheres?

(b) What is the partial pressure of Ar in atmospheres?

10.87 A 20.0-L flask contains 0.776 g of He and 3.61 g of CO_2 at 300 K.

(a) What is the partial pressure of He in mm Hg?

(b) What is the partial pressure of CO_2 in mm Hg?

10.88 A sample of magnesium metal reacts with aqueous HCl to yield H_2 gas:

$$Mg(s) + 2\ HCl(aq) \longrightarrow MgCl_2(aq) + H_2(g)$$

The gas that forms is found to have a volume of 3.557 L at 25 °C and a pressure of 747 mm Hg. Assuming that the gas is saturated with water vapor at a partial pressure of 23.8 mm Hg, what is the partial pressure in millimeters of mercury of the H_2? How many grams of magnesium metal were used in the reaction?

10.89 Chlorine gas was first prepared in 1774 by the oxidation of NaCl with MnO_2:

$$2\ NaCl(s) + 2\ H_2SO_4(l) + MnO_2(s) \longrightarrow$$
$$Na_2SO_4(s) + MnSO_4(s) + 2\ H_2O(g) + Cl_2(g)$$

Assume that the gas produced is saturated with water vapor at a partial pressure of 28.7 mm Hg and that it has a volume of 0.597 L at 27 °C and 755 mm Hg pressure.

(a) What is the mole fraction of Cl_2 in the gas?

(b) How many grams of NaCl were used in the experiment, assuming complete reaction?

10.90 Natural gas is a mixture of hydrocarbons, primarily methane (CH_4) and ethane (C_2H_6). A typical mixture might have $X_{methane} = 0.915$ and $X_{ethane} = 0.085$. Let's assume that we have a 15.50 g sample of natural gas in a volume of 15.00 L at a temperature of 20.00 °C.

(a) How many total moles of gas are in the sample?

(b) What is the pressure of the sample in atmospheres?

(c) What is the partial pressure of each component in the sample in atmospheres?

10.91 Gaseous compound **Q** contains only xenon and oxygen. When 0.100 g of **Q** is placed in a 50.0-mL steel vessel at 0 °C the pressure is 0.229 atm.

(a) What is the molar mass of **Q**, and what is a likely formula?

(b) When the vessel and its contents are warmed to 100 °C, **Q** decomposes into its constituent elements. What is the total pressure, and what are the partial pressures of xenon and oxygen in the container?

Kinetic–Molecular Theory (Section 10.6)

10.92 What are the basic assumptions of the kinetic–molecular theory?

10.93 What is the difference between heat and temperature?

10.94 The average temperature at an altitude of 20 km is 220 K. What is the average speed in m/s of an N_2 molecule at this altitude?

10.95 Calculate the average speed of a nitrogen molecule in m/s on a hot day in summer ($T = 37\,°C$) and on a cold day in winter ($T = -25\,°C$).

10.96 At what temperature (°C) will xenon atoms have the same
RAN average speed that Br_2 molecules have at 20° C?

10.97 At what temperature does the average speed of an oxygen
RAN molecule equal that of an airplane moving at 580 mph?

10.98 Which has a higher average speed, H_2 at 150 K or He at 375 °C?

10.99 Which has a higher average speed, a Ferrari at 145 mph or a gaseous UF_6 molecule at 145 °C?

10.100 A big-league fastball travels at about 45 m/s. At what tem-
RAN perature (°C) do helium atoms have this same average speed?

10.101 Traffic on the German autobahns reaches speeds of up to 230 km/h. At what temperature (°C) do oxygen molecules have this same average speed?

Graham's Law (Section 10.7)

10.102 What is the difference between effusion and diffusion?

10.103 Why does a helium-filled balloon lose pressure faster than an air-filled balloon?

10.104 What is the molecular weight of a gas that diffuses through
RAN a porous membrane 1.86 times faster than Xe? What might the gas be?

10.105 Chlorine occurs as a mixture of two isotopes, ^{35}Cl and ^{37}Cl. What is the ratio of the diffusion rates of the three species $(^{35}Cl)_2$, $^{35}Cl^{37}Cl$, and $(^{37}Cl)_2$?

10.106 Rank the following gases in order of their speed of diffu-
RAN sion through a membrane, and calculate the ratio of their diffusion rates: HCl, F_2, Ar.

10.107 Which will diffuse through a membrane more rapidly, CO or N_2? Assume that the samples contain only the most abundant isotopes of each element, ^{12}C, ^{16}O, and ^{14}N.

10.108 Two 112-L tanks are filled with gas at 330 K. One contains 5.00 mol of Kr, and the other contains 5.00 mol of O_2. Considering the assumptions of kinetic–molecular theory, rank the gases from low to high for each of the following properties.

   (a) Collision frequency   (b) Density (g/L)
   (c) Average speed   (d) Pressure

10.109 Two identical 732.0-L tanks each contain 212.0 g of gas at 293 K, with neon in one tank and nitrogen in the other. Based on the assumptions of kinetic–molecular theory, rank the gases from low to high for each of the following properties.

   (a) Average speed   (b) Pressure
   (c) Collision frequency   (d) Density (g/L)

Real Gases (Section 10.8)

10.110 (a) The volume of the gas particles themselves most affect the overall volume of the gas sample at _____ (high or low) pressure.

   (b) The volume of each particle causes the true volume of the gas sample to be _____ (larger or smaller) than the volume calculated by the ideal gas law.

10.111 (a) The attractive forces between particles most affect the overall volume of the gas sample at _____ (high or low) pressure.

   (b) The attractive forces between particles causes the true volume of the gas sample to be _____ (larger or smaller) than the volume calculated by the ideal gas law.

10.112 Assume that you have 0.500 mol of N_2 in a volume of 0.600 L
RAN at 300 K. Calculate the pressure in atmospheres using both the ideal gas law and the van der Waals equation. For N_2, $a = 1.35\,(L^2 \cdot atm)/mol^2$ and $b = 0.0387\,L/mol$.

10.113 Assume that you have 15.00 mol of N_2 in a volume of 0.600 L
RAN at 300 K. Calculate the pressure in atmospheres using both the ideal gas law and the van der Waals equation. For N_2, $a = 1.35\,(L^2 \cdot atm)/mol^2$ and $b = 0.0387\,L/mol$.

10.114 Uranium hexafluoride, a molecular solid used for purifica-
RAN tion of the uranium isotope needed to fuel nuclear power plants, sublimes at 56.5 °C. Assume that you have a 22.9 L vessel that contains 512.9 g of UF_6 at 70.0 °C.

   (a) What is the pressure in the vessel calculated using the ideal gas law?

   (b) What is the pressure in the vessel calculated using the van der Waals equation? (For UF_6, $a = 15.80\,(L^2 \cdot atm)/mol^2$; $b = 0.1128\,L/mol$.)

10.115 Use both the ideal gas law and the van der Waals equation
RAN to calculate the pressure in atmospheres of 45.0 g of NH_3 gas in a 1.000-L container at 0 °C, 50 °C, and 100 °C. For NH_3, $a = 4.17\,(L^2 \cdot atm)/mol^2$ and $b = 0.0371\,L/mol$.

The Earth's Atmosphere, Greenhouse Gases, and Climate Change (Sections 10.9–10.11)

10.116 Name the regions of the atmosphere. What property is used to distinguish between different regions of the atmosphere?

10.117 The Earth's atmosphere has a mass of approximately $5.15 \times$
RAN 10^{15} kg. If the average molar mass of air is 28.8 g/mol, how many moles of gas make up the atmosphere? What is the volume of the atmosphere in liters under condi-tions of STP? (Note: The average molar mass of air is the weighted average of the molar mass of nitrogen and oxygen. $0.20(32.0\,g/mol) + 0.80(28.0\,g/mol) = 28.8\,g/mol$.)

10.118 The troposphere contains about three quarters of the mass of the entire atmosphere. The troposphere is only 12 km thick while the whole atmosphere is about 120 km thick. Explain why the troposphere contains such a large fraction of the total mass.

10.119 The percent by volume of oxygen (20.95%) is constant thro-
RAN ughout the troposphere.

   (a) Express this percentage as a mole fraction.

   (b) Give the partial pressure of oxygen at sea level where the total atmospheric pressure = 1.0 atm.

   (c) Give the partial pressure oxygen at 11 km, the altitude where airplanes fly, if the total atmospheric pressure is 0.20 atm.

10.120 Fill in the blanks with the appropriate region of electro-magnetic radiation: UV, visible, infrared.

   (a) The Sun most strongly emits in the _____ and _____ regions of electromagnetic radiation.

   (b) The atmosphere filters out biologically damaging _____ radiation from incoming solar radiation and prevents it from reaching Earth.

   (c) The Earth most strongly emits _____ radiation.

   (d) Greenhouse gases absorb _____ radiation.

10.121 (a) The wavelength of maximum emission of solar radiation is 483 nm. Calculate the energy of one mole of photons with a wavelength of 483 nm.

(b) The wavelength of maximum emission of the Earth's radiation is 10,000 nm. Calculate the energy of one mole of photons with a wavelength of 10,000 nm.

(c) Which emits higher energy radiation, the Earth or the Sun?

10.122 Why do the Earth and Sun have different emission spectra?

10.123 Ozone (O_3) is a harmful pollutant in the troposphere. Draw the electron-dot structure for ozone (O_3), and evaluate the symmetric and asymmetric stretch to determine whether ozone is also a greenhouse gas.

10.124 Explain why nitrogen and oxygen, the two most abundant gases in the atmosphere, are not greenhouse gases.

10.125 The water molecule has similar bond vibrations to carbon dioxide. Decide whether the symmetric, asymmetric, and bending vibrations in water will result in the absorption of IR radiation.

10.126 Bond vibrations for the symmetric and asymmetric stretch in methane are illustrated below. Decide whether each vibration will result in the absorption of IR radiation. Arrows indicate the movement of atoms during the vibration.

Symmetric stretch   Asymmetric stretch

10.127 How many times larger is carbon dioxide's contribution to the greenhouse effect than methane? (Use radiative forcing values in Table 10.7 to compare the two gases.)

10.128 N_2O has a GWP value of 310 and CO_2 has a GWP value 1, but CO_2 makes a greater contribution to the greenhouse effect. Explain.

10.129 Although human activities do not have significant influence on the concentration of water vapor in the atmosphere, the surface warming caused by other greenhouse gases could increase rates of evaporation. If more water vapor entered the atmosphere, what would be the effect on climate based on the greenhouse effect?

10.130 What major greenhouse gases are associated with human activities?

10.131 What is the trend in atmospheric CO_2 and CH_4 concentrations over the past 150 years? Over several hundred thousand years?

10.132 Has the Earth's surface experienced warming as a result of increasing levels of greenhouse gases? If so, by how much?

10.133 Name several indicators that scientists use to evaluate the effects of climate change.

MULTICONCEPT PROBLEMS

10.134 A driver with a nearly empty fuel tank may say she is "running on fumes." If a 15.0-gallon automobile gas tank had only gasoline vapor remaining in it, what is the farthest the vehicle could travel if it gets 20.0 miles per gallon on liquid gasoline? Assume the average molar mass of molecules in gasoline is 105 g/mol, the density of liquid gasoline is 0.75 g/mL, the pressure is 743 mm Hg, and the temperature is 25 °C.

10.135 Pakistan's K2 is the world's second-tallest mountain, with an altitude of 28,251 ft. Its base camp, where climbers stop to acclimate, is located about 16,400 ft above sea level.

(a) Approximate atmospheric pressure P at different altitudes is given by the equation $P = e^{-h/7000}$, where P is in atmospheres and h is the altitude in meters. What is the approximate atmospheric pressure in mm Hg at K2 base camp?

(b) What is the atmospheric pressure in mm Hg at the summit of K2?

(c) Assuming the mole fraction of oxygen in air is 0.2095, what is the partial pressure of oxygen in mm Hg at the summit of K2?

10.136 Assume that you take a flask, evacuate it to remove all the air, and find its mass to be 478.1 g. You then fill the flask with argon to a pressure of 2.15 atm and reweigh it. What would the balance read in grams if the flask has a volume of 7.35 L and the temperature is 20.0 °C?

10.137 The apparatus shown consists of three temperature-jacketed 1.000-L bulbs connected by stopcocks. Bulb A contains a mixture of $H_2O(g)$, $CO_2(g)$, and $N_2(g)$ at 25 °C and a total pressure of 564 mm Hg. Bulb B is empty and is held at a temperature of −70 °C. Bulb C is also empty and is held at a temperature of −190 °C. The stopcocks are closed, and the volume of the lines connecting the bulbs is zero. CO_2 sublimes at −78 °C, and N_2 boils at −196 °C.

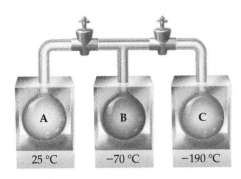

(a) The stopcock between A and B is opened, and the system is allowed to come to equilibrium. The pressure in A and B is now 219 mm Hg. What do bulbs A and B contain?

(b) How many moles of H_2O are in the system?

(c) Both stopcocks are opened, and the system is again allowed to come to equilibrium. The pressure throughout the system is 33.5 mm Hg. What do bulbs A, B, and C contain?

(d) How many moles of N_2 are in the system?

(e) How many moles of CO_2 are in the system?

10.138 When solid mercury(I) carbonate, Hg_2CO_3, is added to RAN nitric acid, HNO_3, a reaction occurs to give mercury(II) nitrate, $Hg(NO_3)_2$, water, and two gases A and B:

$$Hg_2CO_3(s) + HNO_3(aq) \longrightarrow$$
$$Hg(NO_3)_2(aq) + H_2O(l) + A(g) + B(g)$$

(a) When the gases are placed in a 500.0-mL bulb at 20 °C, the pressure is 258 mm Hg. How many moles of gas are present?

(b) When the gas mixture is passed over $CaO(s)$, gas A reacts, forming $CaCO_3(s)$:

$$CaO(s) + A(g) + B(g) \longrightarrow CaCO_3(s) + B(g)$$

The remaining gas B is collected in a 250.0-mL container at 20 °C and found to have a pressure of 344 mm Hg. How many moles of B are present?

(c) The mass of gas B collected in part (b) was found to be 0.218 g. What is the density of B in g/L?

(d) What is the molecular weight of B, and what is its formula?

(e) Write a balanced equation for the reaction of mercury(I) carbonate with nitric acid.

10.139 Consider the combustion reaction of 0.148 g of a hydrocarbon having formula C_nH_{2n+2} with an excess of O_2 in a 400.0-mL steel container. Before reaction, the gaseous mixture had a temperature of 25.0 °C and a pressure of 2.000 atm. After complete combustion and loss of considerable heat, the mixture of products and excess O_2 had a temperature of 125.0 °C and a pressure of 2.983 atm.

(a) What is the formula and molar mass of the hydrocarbon?

(b) What are the partial pressures in atmospheres of the reactants?

(c) What are the partial pressures in atmospheres of the products and the excess O_2?

10.140 A mixture of $CS_2(g)$ and excess $O_2(g)$ is placed in a 10.0-L RAN reaction vessel at 100.0 °C and a pressure of 3.00 atm. A spark causes the CS_2 to ignite, burning it completely, according to the equation

$$CS_2(g) + 3\ O_2(g) \longrightarrow CO_2(g) + 2\ SO_2(g)$$

After reaction, the temperature returns to 100.0 °C, and the mixture of product gases (CO_2, SO_2, and unreacted O_2) is found to have a pressure of 2.40 atm. What is the partial pressure of each gas in the product mixture?

10.141 When 10.0 g of a mixture of $Ca(ClO_3)_2$ and RAN $Ca(ClO)_2$ is heated to 700 °C in a 10.0-L vessel, both compounds decompose, forming $O_2(g)$ and $CaCl_2(s)$. The final pressure inside the vessel is 1.00 atm.

(a) Write balanced equations for the decomposition reactions.

(b) What is the mass of each compound in the original mixture?

10.142 A 5.00-L vessel contains 25.0 g of PCl_3 and 3.00 g of O_2 at RAN 15 °C. The vessel is heated to 200.0 °C, and the contents react to give $POCl_3$. What is the final pressure in the vessel, assuming that the reaction goes to completion and that all reactants and products are in the gas phase?

10.143 A steel container with a volume of 500.0 mL is evacuated, and 25.0 g of $CaCO_3$ is added. The container and contents are then heated to 1500 K, causing the $CaCO_3$ to decompose completely, according to the equation

$$CaCO_3(s) \longrightarrow CaO(s) + CO_2(g).$$

(a) Using the ideal gas law and ignoring the volume of any solids remaining in the container, calculate the pressure inside the container at 1500 K.

(b) Now make a more accurate calculation of the pressure inside the container. Take into account the volume of solid CaO (density = 3.34 g/mL) in the container, and use the van der Waals equation to calculate the pressure. The van der Waals constants for $CO_2(g)$ are $a = 3.59\ (L^2 \cdot atm)/mol^2$ and $b = 0.0427\ L/mol$.

10.144 Nitrogen dioxide dimerizes to give dinitrogen tetroxide: RAN $2\ NO_2(g) \longrightarrow N_2O_4(g)$. At 298 K, 9.66 g of an NO_2/N_2O_4 mixture exerts a pressure of 0.487 atm in a volume of 6.51 L. What are the mole fractions of the two gases in the mixture?

10.145 An empty 4.00-L steel vessel is filled with 1.00 atm of $CH_4(g)$ RAN and 4.00 atm of $O_2(g)$ at 300 °C. A spark causes the CH_4 to burn completely, according to the equation

$$CH_4(g) + 2\ O_2(g) \longrightarrow CO_2(g) + 2\ H_2O(g)$$
$$\Delta H° = -802\ kJ$$

(a) What mass of $CO_2(g)$ is produced in the reaction?

(b) What is the final temperature inside the vessel after combustion, assuming that the steel vessel has a mass of 14.500 kg, the mixture of gases has an average molar heat capacity of 21 J/(mol·°C), and the heat capacity of steel is 0.449 J/(g·°C)?

(c) What is the partial pressure of $CO_2(g)$ in the vessel after combustion?

10.146 When a gaseous compound X containing only C, H, and O is burned in O_2, 1 volume of the unknown gas reacts with 3 volumes of O_2 to give 2 volumes of CO_2 and 3 volumes of gaseous H_2O. Assume all volumes are measured at the same temperature and pressure.

(a) Calculate a formula for the unknown gas, and write a balanced equation for the combustion reaction.

(b) Is the formula you calculated an empirical formula or a molecular formula? Explain.

(c) Draw two different possible electron-dot structures for the compound X.

(d) Combustion of 5.000 g of X releases 144.2 kJ heat. Look up $\Delta H°_f$ values for $CO_2(g)$ and $H_2O(g)$ in Appendix B, and calculate $\Delta H°_f$ for compound X.

10.147 Isooctane, C_8H_{18}, is the component of gasoline from which RAN the term *octane rating* derives.

(a) Write a balanced equation for the combustion of isooctane to yield CO_2 and H_2O.

(b) Assuming that gasoline is 100% isooctane, that isooctane burns to produce only CO_2 and H_2O, and that the density of isooctane is 0.792 g/mL, what mass of

CO_2 in kilograms is produced each year by the annual U.S. gasoline consumption of 4.6×10^{10} L?

(c) What is the volume in liters of this CO_2 at STP?

(d) How many moles of air are necessary for the combustion of 1 mol of isooctane, assuming that air is 21.0% O_2 by volume? What is the volume in liters of this air at STP?

10.148 The *Rankine* temperature scale used in engineering is to the
RAN   Fahrenheit scale as the Kelvin scale is to the Celsius scale. That is, 1 *Rankine* degree is the same size as 1 Fahrenheit degree, and 0 °R = absolute zero.

(a) What temperature corresponds to the freezing point of water on the Rankine scale?

(b) What is the value of the gas constant R on the Rankine scale in $(L \cdot atm)/(°R \cdot mol)$?

(c) Use the van der Waals equation to determine the pressure inside a 400.0-mL vessel that contains 2.50 mol of CH_4 at a temperature of 525 °R. For CH_4, $a = 2.253$ $(L^2 \cdot atm)/mol^2$ and $b = 0.04278$ L/mol.

10.149 Chemical explosions are characterized by the instantaneous release of large quantities of hot gases, which set up a shock wave of enormous pressure (up to 700,000 atm) and velocity (up to 20,000 mi/h). For example, explosion of nitroglycerin $(C_3H_5N_3O_9)$ releases four gases, A, B, C, and D:

$$n \, C_3H_5N_3O_9(l) \longrightarrow a \, A(g) + b \, B(g) + c \, C(g) + d \, D(g)$$

Assume that the explosion of 1 mol (227 g) of nitroglycerin releases gases with a temperature of 1950 °C and a volume of 1323 L at 1.00 atm pressure.

(a) How many moles of hot gas are released by the explosion of 0.004 00 mol of nitroglycerin?

(b) When the products released by explosion of 0.004 00 mol of nitroglycerin were placed in a 500.0-mL flask and the flask was cooled to −10 °C, product A solidified and the pressure inside the flask was 623 mm Hg. How many moles of A were present, and what is its likely identity?

(c) When gases B, C, and D were passed through a tube of powdered Li_2O, gas B reacted to form Li_2CO_3.

The remaining gases, C and D, were collected in another 500.0-mL flask and found to have a pressure of 260 mm Hg at 25 °C. How many moles of B were present, and what is its likely identity?

(d) When gases C and D were passed through a hot tube of powdered copper, gas C reacted to form CuO. The remaining gas, D, was collected in a third 500.0-mL flask and found to have a mass of 0.168 g and a pressure of 223 mm Hg at 25 °C. How many moles each of C and D were present, and what are their likely identities?

(e) Write a balanced equation for the explosion of nitroglycerin.

10.150 Combustion analysis of 0.1500 g of methyl *tert*-butyl ether, an octane booster used in gasoline, gave 0.3744 g of CO_2 and 0.1838 g of H_2O. When a flask having a volume of 1.00 L was evacuated and then filled with methyl *tert*-butyl ether vapor at a pressure of 100.0 kPa and a temperature of 54.8 °C, the mass of the flask increased by 3.233 g.

Methyl *tert*-butyl ether

(a) What is the empirical formula of methyl *tert*-butyl ether?

(b) What is the molecular weight and molecular formula of methyl *tert*-butyl ether?

(c) Write a balanced equation for the combustion reaction.

(d) The enthalpy of combustion for methyl *tert*-butyl ether is $\Delta H°_{combustion} = -3368.7$ kJ/mol. What is its standard enthalpy of enthalpy of formation, $\Delta H°_f$?

chapter 11

Liquids and Phase Changes

Contents

Carbon dioxide (CO_2) exists in different phases at varying temperatures and pressures. At temperatures below -78.5 °C at a pressure of 1 atm, CO_2 is in the solid phase, which is commonly called dry ice. If the temperature is raised at a pressure of 1 atm, CO_2 becomes a gas. At higher temperatures and pressures, CO_2 forms the liquid and supercritical fluid phases that are used as alternatives to toxic solvents in applications such as decaffeinating coffee and dry cleaning.

How is caffeine removed from coffee?

The answer to this question can be found on page 439 in the INQUIRY ?

422

The kinetic-molecular theory developed in Chapter 10 accounts for the properties of gases by assuming that gas particles act independently of one another. **Intermolecular forces** between gas particles are so weak that particles in gases are free to move about at random and occupy the volume of space available. The same is not true in liquids and solids, which are distinguished from gases by the presence of substantial attractive forces between particles as well as small distances between particles. In this chapter, we'll examine how intermolecular forces influence the properties of liquids and the energy associated with phase changes. We'll look at what happens during the transitions between solid, liquid, gaseous, and supercritical fluid states and at the effects of temperature and pressure on these transitions. In addition, we'll learn about liquid crystals, an intermediate phase between the solid and liquid state, and applications of this unique crystalline phase.

REMEMBER...
Intermolecular forces are attractive forces that occur between all kinds of particles including molecules, ions, and atoms. Ion–dipole forces, dipole–dipole forces, London dispersion forces, and hydrogen bonding are all types of intermolecular forces (Section 8.6).

11.1 PROPERTIES OF LIQUIDS

Many familiar and observable properties of liquids can be explained by the presence of intermolecular forces. We all know, for instance, that some liquids, such as water or gasoline, flow easily when poured, whereas others, such as motor oil or maple syrup, flow sluggishly.

The measure of a liquid's resistance to flow is called its **viscosity**. Honey is a liquid that does not flow easily and has a high viscosity, and water is a liquid that flows readily and has a low viscosity. Viscosity is related to the ease with which individual molecules move around in the liquid and thus to the intermolecular forces present. Substances with weak intermolecular forces have relatively low viscosities, while substances with strong intermolecular forces have relatively high viscosities. Let's compare pentane (C_5H_{12}) with glycerol [$C_3H_5(OH)_3$]. Pentane, a nonpolar molecule, has London dispersion forces holding molecules together. The London dispersion forces are relatively weak, and therefore pentane has a low viscosity. In contrast, glycerol has three hydroxyl (OH) groups that can form hydrogen bonds with other glycerol molecules. The intermolecular forces between glycerol molecules are strong, resulting in a high viscosity. Glycerol is a clear liquid with a viscosity similar to honey.

BIG IDEA Question 1 Go to eText

Which of the following liquids has the highest viscosity: $CH_3CH_2CH_2CH_2OH$, $CH_3CH_2CH_2CH_2CH_3$, or $CH_3CH_2OCH_2CH_3$?

Pentane   Glycerol

Another familiar property of liquids is **surface tension,** the resistance of a liquid to spread out and increase its surface area. Surface tension is caused by the difference in intermolecular forces experienced by molecules at the surface of a liquid and those experienced by molecules in the interior. Molecules at the surface feel attractive forces on only one side and are thus pulled in toward the liquid, while molecules in the interior are surrounded and are pulled equally in all directions (**FIGURE 11.1**). The ability of a water strider to walk on water and the beading up of water on a newly waxed car are both due to surface tension.

Surface tension, like viscosity, is generally higher in liquids that have stronger intermolecular forces. Both properties are also temperature-dependent because molecules at higher temperatures have more kinetic energy to counteract the attractive forces holding them together. Data for some common substances are given in **TABLE 11.1**. Note that mercury has a particularly large surface tension, causing droplets to form beads (Figure 11.1) and giving the top of the mercury column in a barometer a rounded shape called a *meniscus*.

▲ Surface tension allows a water strider to walk on a pond without penetrating the surface.

▶ **FIGURE 11.1**

Surface tension. Surface tension is the resistance of a liquid to spread out and increase its surface area.

Surface tension causes these drops of liquid mercury to form beads.

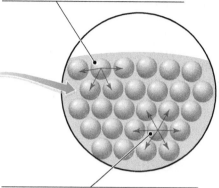

Molecules or atoms on the **surface** feel attractive forces on only one side and are thus drawn in toward the liquid.

Molecules or atoms in the **middle** of a liquid are attracted equally in all directions.

TABLE 11.1 Viscosities and Surface Tensions of Some Common Substances at 20 °C

Name	Formula	Viscosity ($N \cdot s/m^2$)	Surface Tension (J/m^2)
Pentane	C_5H_{12}	2.4×10^{-4}	1.61×10^{-2}
Benzene	C_6H_6	6.5×10^{-4}	2.89×10^{-2}
Water	H_2O	1.00×10^{-3}	7.29×10^{-2}
Ethanol	C_2H_5OH	1.20×10^{-3}	2.23×10^{-2}
Mercury	Hg	1.55×10^{-3}	4.6×10^{-1}
Glycerol	$C_3H_5(OH)_3$	1.49	6.34×10^{-2}

REMEMBER...

The gas pressure inside a container can be measured using an open-end **manometer**, which consists of a U-tube filled with mercury. The difference between the pressure of the gas and the pressure of the atmosphere is equal to the difference between the heights of the mercury levels in the two arms of the U-tube (Section 10.1).

▲ Because bromine is colored, it's possible to see its reddish vapor above the liquid.

11.2 VAPOR PRESSURE AND BOILING POINT

Vapor pressure and boiling point are two other properties of liquids that are influenced by the strength of intermolecular forces. We'll describe these properties and how they depend on molecular structure and intermolecular forces. The conversion of a liquid to a vapor is visible when the liquid boils, but it occurs under other conditions as well. Let's imagine the two experiments illustrated in **FIGURE 11.2**. In one experiment, we place a liquid in an open container; in the other experiment, we place the liquid in a closed container connected to a mercury **manometer**. After a certain amount of time has passed, the liquid in the first container has evaporated, while the liquid in the second container remains but the pressure has risen. At equilibrium and at a constant temperature, the pressure increase has a constant value called the **vapor pressure** of the liquid.

Molecules that enter the vapor phase in an open container can escape from the liquid and drift away until the liquid evaporates entirely, but molecules in a closed container are trapped. As more and more molecules pass from the liquid to the vapor, the chances increase that random motion will cause some of them to return occasionally to the liquid. Ultimately, the number of molecules returning to the liquid becomes equal to the number escaping, at which point a dynamic equilibrium exists. Although *individual* molecules are constantly passing back and forth from one phase to the other, the *total numbers* of molecules in both liquid and vapor phases remain constant.

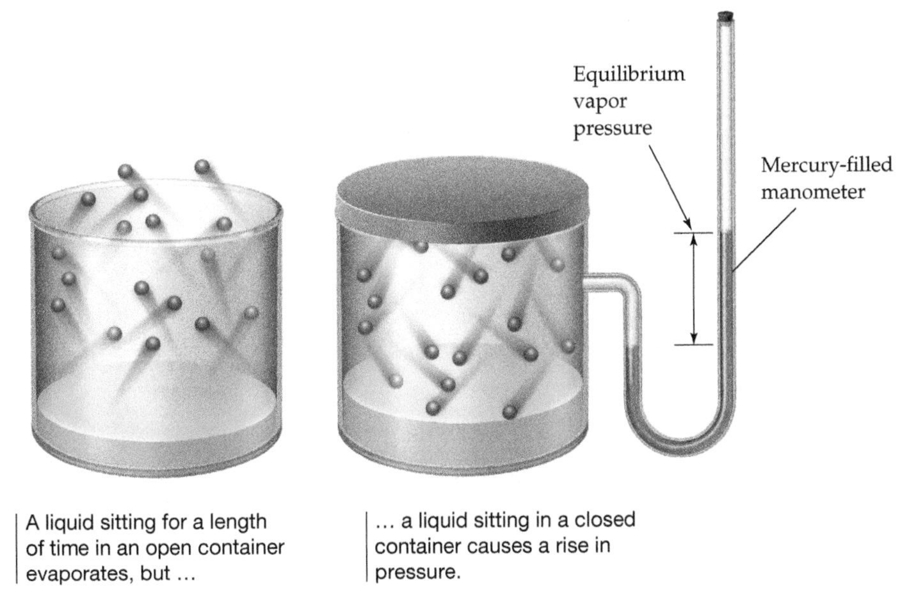

◀ **FIGURE 11.2**

The origin of vapor pressure.

◀ **Figure It Out**

Liquid water is placed in the manometer at 25 °C If the vapor pressure of water at 25 °C is 0.0313 atm, what is the height difference in the two ends of the mercury column in units of mm? (760 mm Hg = 1 atm.)

$$0.0313 \text{ atm} \times \frac{760 \text{ mm Hg}}{1 \text{ atm}} = 23.8 \text{ mm Hg}$$

Answer:

Equilibrium vapor pressure

Mercury-filled manometer

A liquid sitting for a length of time in an open container evaporates, but …

… a liquid sitting in a closed container causes a rise in pressure.

Evaporation and vapor pressure are both explained on a molecular level by the **kinetic–molecular theory,** developed in Section 10.6 to account for the behavior of gases. The molecules in a liquid are in constant motion but at a variety of speeds depending on the amount of kinetic energy they have. In considering a large sample, molecular kinetic energies follow a distribution curve like that shown in **FIGURE 11.3,** with the exact shape of the curve dependent on the temperature. The higher the temperature and the lower the boiling point of the substance, the greater the fraction of molecules in the sample that have sufficient kinetic energy to break free from the surface of the liquid and escape into the vapor.

REMEMBER…
The **kinetic–molecular theory** is a group of five postulates that can be used to account for the behavior of gases and to derive the ideal gas law. Temperature and kinetic energy are related according to the equation $E_K = (3/2)RT$ (Section 10.6).

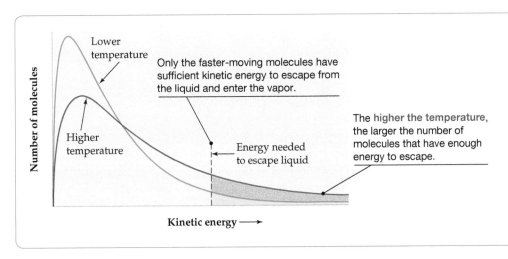

◀ **FIGURE 11.3**

The distribution of molecular kinetic energies in a liquid.

◀ **Figure It Out**

How does the area under the curve to the right of the dashed line change as temperature increases? How does an increase in temperature affect vapor pressure?

Answer: The area under the curve to the right of the dashed line increases with temperature. Compare the area under the red line (higher T) with the area under the blue line (lower T). At higher temperature, more molecules have sufficient kinetic energy to move from the liquid to the gas phase and vapor pressure increases.

The numerical value of a liquid's vapor pressure depends on the magnitude of the intermolecular forces present and on the temperature. The smaller the intermolecular forces, the higher the vapor pressure because loosely held molecules escape more easily. The higher the temperature, the higher the vapor pressure because a larger fraction of molecules have sufficient kinetic energy to escape.

The Clausius–Clapeyron Equation

If we measure vapor pressure at different temperatures, we can calculate the enthalpy of vaporization, or **heat of vaporization (ΔH_{vap}).** The heat of vaporization is the amount of energy needed to convert one mole of liquid into a gas. As indicated in **FIGURE 11.4,** the

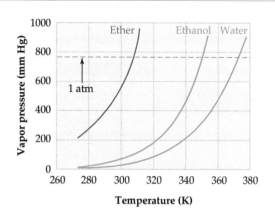

The vapor pressures of ether, ethanol, and water show a nonlinear rise when plotted as a function of temperature.

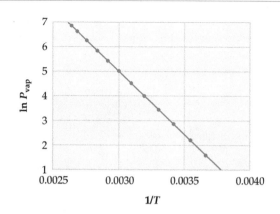

A plot of ln P_{vap} versus 1/T (kelvin) for water, prepared from the data in Table 11.2, shows a linear relationship.

▲ **FIGURE 11.4**

Vapor pressure of liquids at different temperatures.

vapor pressure of a liquid rises with temperature in a nonlinear way. We find a linear relationship, however, when we plot the natural logarithm of the vapor pressure, ln P_{vap}, against the inverse of the Kelvin temperature, 1/T. **TABLE 11.2** gives the appropriate data for water, and Figure 11.4 shows the plot. A linear graph is characteristic of mathematical equations of the form $y = mx + b$. In this case, $y = \ln P_{vap}$, $x = 1/T$, m is the slope of the line $(-\Delta H_{vap}/R)$, and b is the y-intercept (a constant, C). Thus, the data fit an expression known as the **Clausius–Clapeyron equation.**

Clausius–Clapeyron equation

$$\underset{y\ =}{\ln P_{vap}} = \underset{m}{\left(-\frac{\overset{\text{natural}}{\overset{\text{logarithm}}{\downarrow}}\Delta H_{vap}}{R}\right)}\underset{x}{\frac{1}{T}} + \underset{b}{C}$$

where ΔH_{vap} is the heat of vaporization of the liquid, R is the gas constant (Section 10.3), and C is a constant characteristic of each specific substance.

TABLE 11.2 Vapor Pressure of Water at Various Temperatures

Temp (K)	P_{vap} (mm Hg)	ln P_{vap}	1/T	Temp (K)	P_{vap} (mm Hg)	ln P_{vap}	1/T
273	4.58	1.522	0.003 66	333	149.4	5.007	0.003 00
283	9.21	2.220	0.003 53	343	233.7	5.454	0.002 92
293	17.5	2.862	0.003 41	353	355.1	5.872	0.002 83
303	31.8	3.459	0.003 30	363	525.9	6.265	0.002 75
313	55.3	4.013	0.003 19	373	760.0	6.633	0.002 68
323	92.5	4.527	0.003 10	378	906.0	6.809	0.002 65

The Clausius–Clapeyron equation makes it possible to calculate the heat of vaporization of a liquid by measuring its vapor pressure at several temperatures and then plotting the results to obtain the slope of the line. Alternatively, once we know the heat of vaporization and the vapor pressure at one temperature, we can calculate the vapor pressure of the liquid at any other temperature. We can derive a two-point form of the Clausius–Clapeyron equation that uses measurements of vapor pressure at two different temperatures to calculate ΔH_{vap}. Because C is a constant (its value is the same at any two pressures and temperatures), we can write the following expression:

$$C = \ln P_1 + \frac{\Delta H_{vap}}{RT_1} = \ln P_2 + \frac{\Delta H_{vap}}{RT_2}$$

We can also rearrange this equation so that pressure is on one side and temperature is on the other.

$$\ln P_1 - \ln P_2 = \frac{\Delta H_{vap}}{RT_2} - \frac{\Delta H_{vap}}{RT_1}$$

Further mathematical rearrangement yields a two-point form of the Clausius–Clapeyron equation:

$$\ln\left(\frac{P_1}{P_2}\right) = \frac{\Delta H_{vap}}{R}\left(\frac{1}{T_2} - \frac{1}{T_1}\right)$$

If we know ΔH_{vap} and the vapor pressure at one temperature, then we can calculate the vapor pressure of a liquid at a new temperature, as shown in Worked Example 11.1.

When the vapor pressure of a liquid rises to the point where it becomes equal to the external pressure, the liquid reaches its boiling point and changes into vapor. On a molecular level, you might picture boiling in the following way: Imagine that a few molecules in the interior of the liquid momentarily break free from their neighbors and form a microscopic bubble. If the external pressure from the atmosphere is greater than the vapor pressure inside the bubble, the bubble is immediately crushed. At the temperature at which the external pressure and the vapor pressure in the bubble are the same, however, the bubble is not crushed. Instead, it rises through the denser liquid, grows larger as more molecules join it, and appears as part of the vigorous action we associate with boiling.

The temperature at which a liquid boils when the external pressure is exactly 1 atm (760 mm Hg) is called the **normal boiling point**. On the plots in Figure 11.4, the normal boiling points of the three liquids are reached when the curves cross the dashed line representing 760 mm Hg: for ether, 34.6 °C (307.8 K); for ethanol, 78.3 °C (351.5 K); and for water, 100.0 °C (373.15 K)

If the external pressure is less than 1 atm, then the vapor pressure necessary for boiling is reached below 1 atm and the liquid boils at a lower than normal temperature. On top of Mt. Everest, for example, where the atmospheric pressure is only about 260 mm Hg, water boils at approximately 71 °C rather than 100 °C. Conversely, if the external pressure on a liquid is greater than 1 atm, the vapor pressure necessary for boiling is also higher than 1 atm and the liquid boils above its normal boiling point. Pressure cookers take advantage of this effect by raising the boiling point of water. When cooking a liquid such as a stew, the highest temperature is limited by the boiling point of water. High pressure inside the cooker can raise the boiling point of water to 250 °F (121 °C).

▲ **Figure It out**

What is the vapor pressure of the liquid at its boiling point?

Answer: The vapor pressure is equal to the external atmosphere at sea level, the external pressure is pressure at the boiling point. When open to the close to 1 atm.

Calculating a Vapor Pressure Using the Clausius–Clapeyron Equation

The vapor pressure of ethanol at 34.7 °C is 100.0 mm Hg, and the heat of vaporization of ethanol is 38.6 kJ/mol. What is the vapor pressure of ethanol in millimeters of mercury at 65.0 °C?

IDENTIFY

Known	Unknown
$\Delta H_{vap} = 38.6$ kJ/mol	Vapor pressure at 65.0 °C (P_2)

Vapor pressure at 34.7 °C ($P_1 = 100.0$ mm Hg)

$T_1 = 34.7$ °C

$T_2 = 65.0$ °C

STRATEGY

Use the rearranged form of the Clausius–Clapeyron equation that enables us to calculate the vapor pressure of the liquid at any other temperature if we know the heat of vaporization and the vapor pressure at one temperature. Convert temperature to Kelvin and ΔH_{vap} to units of joules.

$$\ln\left(\frac{P_1}{P_2}\right) = \frac{\Delta H_{vap}}{R}\left(\frac{1}{T_2} - \frac{1}{T_1}\right)$$

SOLUTION

Substitute all known values into the equation and solve for P_2.

$$\ln\left(\frac{100.0 \text{ mm Hg}}{P_2}\right) = \frac{38,600 \text{ J/mol}}{8.314 \text{ J/mol} \cdot \text{K}}\left(\frac{1}{338.2} - \frac{1}{307.9}\right)$$

$$\ln\left(\frac{100.0 \text{ mm Hg}}{P_2}\right) = -1.3509$$

$$\frac{100.0 \text{ mm Hg}}{P_2} = e^{-1.3509} = 0.2590$$

$$P_2 = 386 \text{ mm Hg}$$

CHECK

We would expect the vapor pressure to be larger at a higher temperature so the answer is physically reasonable. It is difficult to estimate the magnitude of the new vapor pressure due to the complexity of the calculation.

▶ **PRACTICE 11.1**   Bromine has $P_{vap} = 400$ mm Hg at 41.0 °C and a normal boiling point of 331.9 K. What is the heat of vaporization, ΔH_{vap}, of bromine in kJ/mol?

▶ **APPLY 11.2**   The normal boiling point of water is 100.0 °C, and the heat of vaporization is $\Delta H_{vap} = 40.7$ kJ/mol. What is the boiling point of water in °C on top of Pikes Peak in Colorado, where $P = 407$ mm Hg?

All **PRACTICE** and **APPLY** problems are interactive in the eText.

11.3 PHASE CHANGES BETWEEN SOLIDS, LIQUIDS, AND GASES

We're all familiar with seeing solid ice melt to liquid water, liquid water freeze to solid ice or evaporate to gaseous steam, and gaseous steam condense to liquid water. Such processes, in which the physical form but not the chemical identity of a substance changes, are called **phase changes,** or *changes of state.* Matter in any one state, or **phase,** can change into either of the other two. Solids can even change directly into gases, as occurs when dry ice (solid CO_2) undergoes *sublimation.* The names of the various phase changes are:

Fusion (melting)	solid → liquid
Freezing	liquid → solid
Vaporization	liquid → gas
Condensation	gas → liquid
Sublimation	solid → gas
Deposition	gas → solid

Like all naturally occurring processes, every phase change has associated with it a **free–energy change**, ΔG. As we saw in Section 9.13, ΔG is made up of two contributions, an enthalpy part (ΔH) and a temperature-dependent entropy part ($T\Delta S$) according to the equation $\Delta G = \Delta H - T\Delta S$. The enthalpy part is the heat flow associated with making or breaking the intermolecular attractions that hold liquids and solids together, while the entropy part is associated with the difference in molecular randomness between the various phases. Gases are more random and have more entropy than liquids, which in turn are more random and have more entropy than solids.

The melting of a solid to a liquid, the sublimation of a solid to a gas, and the vaporization of a liquid to a gas all involve an increase in randomness due to the increased mobility of particles. Heat must be absorbed to overcome intermolecular forces holding the particles together. Thus, both ΔS and ΔH are positive for these phase changes. By contrast, the freezing of a liquid to a solid, the deposition of a gas to a solid, and the condensation of a gas to a liquid all involve a decrease in randomness due to decreased mobility of particles. Attractive intermolecular forces form and heat is released in the process. Thus, both ΔS and ΔH have negative values for these phase changes. The situations are summarized in **FIGURE 11.5**.

REMEMBER...

The value of the **free–energy change (ΔG)** is a general criterion for the spontaneity of a chemical or physical process. If $\Delta G < 0$, the process is spontaneous; if $\Delta G = 0$, the process is at equilibrium; and if $\Delta G > 0$, the process is nonspontaneous (Section 9.13).

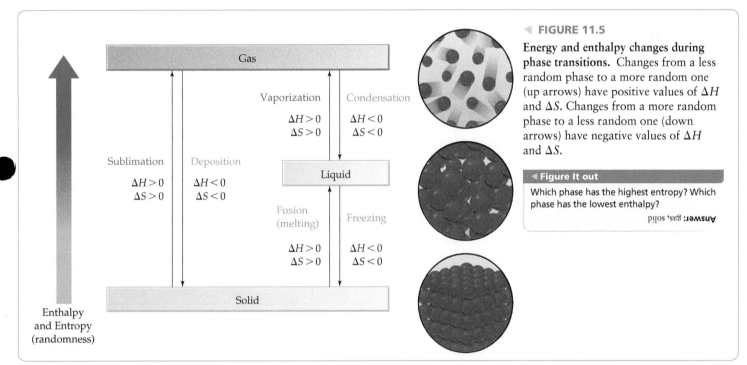

◀ **FIGURE 11.5**

Energy and enthalpy changes during phase transitions. Changes from a less random phase to a more random one (up arrows) have positive values of ΔH and ΔS. Changes from a more random phase to a less random one (down arrows) have negative values of ΔH and ΔS.

◀ **Figure It out**

Which phase has the highest entropy? Which phase has the lowest enthalpy?

Answer: gas, solid

Let's look at the transitions of solid ice to liquid water and liquid water to gaseous steam to see examples of energy relationships during phase changes.

Melting of ice to liquid water (fusion): $\Delta H = +6.01$ kJ/mol and $\Delta S = +22.0$ J/(K • mol)

Vaporization of liquid water to steam: $\Delta H = +40.67$ kJ/mol and $\Delta S = +109$ J/(K • mol)

Both ΔH and ΔS are larger for the liquid $\rightarrow$ vapor change than for the solid $\rightarrow$ liquid change because many more intermolecular attractions need to be overcome and much more randomness is gained in the change of liquid to vapor. This highly endothermic conversion of liquid water to gaseous water vapor is used by many organisms as a cooling mechanism. When our bodies perspire on a warm day, evaporation of the perspiration absorbs heat and leaves the skin feeling cooler.

For phase changes in the opposite direction, the numbers have the same absolute values but opposite signs.

▲ Evaporation of perspiration carries away heat and cools the body after exertion.

Freezing of liquid water to ice: $\Delta H = -6.01$ kJ/mol and $\Delta S = -22.0$ J/(K • mol)

Condensation of water vapor to liquid water: $\Delta H = -40.67$ kJ/mol and $\Delta S = -109$ J/(K • mol)

Citrus growers take advantage of the exothermic freezing of water when they spray their trees with water on cold nights to prevent frost damage. As water freezes on the leaves, it releases heat that protects the tree.

Knowing the values of ΔH and ΔS for a phase transition makes it possible to calculate the temperature at which the change occurs. Recall from Section 9.13 that ΔG is negative for a spontaneous process, positive for a nonspontaneous process, and zero for a process at equilibrium. Thus, by setting $\Delta G = 0$ and solving for T in the free-energy equation, we can calculate the temperature at which two phases are in equilibrium. For the solid $\rightarrow$ liquid phase change in water, for instance, we have

$$\Delta G = \Delta H - T\Delta S = 0 \quad \text{at equilibrium}$$

or $\quad T = \Delta H/\Delta S$

where $\Delta H = +6.01$ kJ/mol and $\Delta S = +22.0$ J/(K • mol)

so $\quad T = \dfrac{6.01 \dfrac{\text{kJ}}{\text{mol}}}{0.0220 \dfrac{\text{kJ}}{\text{K} \cdot \text{mol}}} = 273$ K

In other words, ice turns into liquid water, and liquid water turns into ice, at 273 K, or 0 °C, at 1 atm pressure—hardly a surprise. In practice, the calculation is more useful in the opposite direction. That is, the temperature at which a phase change occurs is measured and then used to calculate $\Delta S(= \Delta H/T)$.

▲ **Figure It out**

Why do citrus growers spray their trees with water on cold nights?

Answer: Heat is released when water freezes and prevents the fruit from freezing.

— **WORKED EXAMPLE 11.2**

Calculating an Entropy of Vaporization

The boiling point of water is 100 °C, and the enthalpy change for the conversion of water to steam is $\Delta H_{vap} = 40.67$ kJ/mol. What is the entropy change for vaporization, ΔS_{vap}, in J/(K • mol)?

IDENTIFY

Known	Unknown
Boiling point of water ($T = 100$ °C)	Entropy change (ΔS_{vap})
Enthalpy change ($\Delta H_{vap} = 40.67$ kJ/mol)	

STRATEGY

At the temperature where a phase change occurs, the two phases coexist in equilibrium and ΔG, the free-energy difference between the phases, is zero: $\Delta G = \Delta H - T\Delta S = 0$. Rearranging gives $\Delta S = \Delta H/T$, where both ΔH and T are known. Remember that T must be expressed in kelvin.

SOLUTION

$$\Delta S_{vap} = \frac{\Delta H_{vap}}{T} = \frac{40.67 \dfrac{\text{kJ}}{\text{mol}}}{373.15 \text{ K}} = 0.1090 \text{ kJ/(K} \cdot \text{mol)} = 109.0 \text{ J/(K} \cdot \text{mol)}$$

CHECK

Converting water from liquid to a gas should result in a large and positive entropy change. The calculation for ΔS_{vap} agrees with the expected sign and magnitude and is therefore a reasonable answer.

▶ **PRACTICE 11.3** The boiling point of ethanol is 78.4 °C, and the enthalpy change for the conversion of liquid to vapor is $\Delta H_{vap} = 38.56$ kJ/mol. What is the entropy change for vaporization, ΔS_{vap}, in J/(K • mol)?

▶ **APPLY 11.4** Chloroform ($CHCl_3$) has $\Delta H_{vap} = 29.2$ kJ/mol and $\Delta S_{vap} = 87.5$ J/ (K• mol). What is the boiling point of chloroform in kelvin?

11.4 ENERGY CHANGES DURING PHASE TRANSITIONS

The results of continuously adding heat to a substance can be displayed on a *heating curve* like that shown in **FIGURE 11.6** for H_2O. The curve has five different regions associated with five different processes involved when heating solid H_2O from −25.0 °C to gaseous H_2O at 125.0 °C.

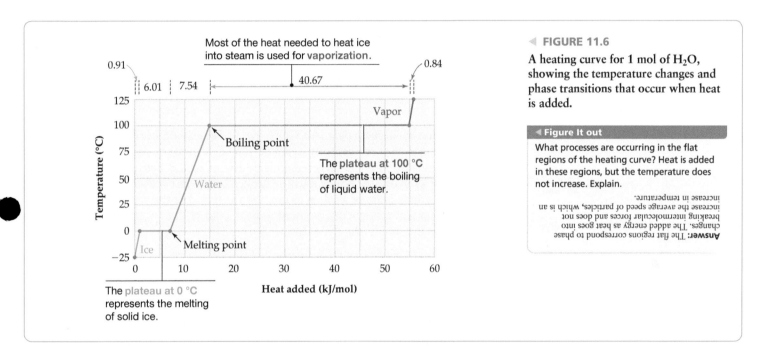

Most of the heat needed to heat ice into steam is used for **vaporization**.

◀ **FIGURE 11.6**

A heating curve for 1 mol of H_2O, showing the temperature changes and phase transitions that occur when heat is added.

◀ **Figure It out**

What processes are occurring in the flat regions of the heating curve? Heat is added in these regions, but the temperature does not increase. Explain.

Answer: The flat regions correspond to phase changes. The added energy as heat goes into breaking intermolecular forces and does not increase the average speed of particles, which is an increase in temperature.

1. **Heating solid H_2O:** Starting at an arbitrary temperature below the freezing point of water, say −25.0 °C, addition of heat raises the temperature of the ice until it reaches 0 °C. The molar heat capacity of ice (36.57 J/mol · C) and the temperature change (+25 °C) are used to calculate the amount of energy required to heat the ice:

$$\text{Energy to heat ice from } -25\,°C \text{ to } 0\,°C = \left(36.57\ \frac{J}{mol \cdot °C}\right)(25.0\,°C)$$
$$= 914\ J/mol = 0.914\ kJ/mol$$

REMEMBER...
The molar heat capacity (C_m) of a substance is the amount of heat necessary to raise the temperature of 1 mol of the substance by 1 °C (Section 9.7).

2. **Melting solid H_2O (fusion):** Once the temperature of the ice reaches 0 °C, addition of further heat goes into disrupting hydrogen bonds and other intermolecular forces rather than into increasing the temperature, as indicated by the plateau at 0 °C on the heating curve in Figure 11.6. At this temperature—the *melting point*—solid and liquid coexist in equilibrium as molecules break free from their positions in the ice crystals and enter the liquid phase. Not until the solid turns completely to liquid does the temperature again rise. The amount of energy required to overcome enough intermolecular forces to convert a solid into a liquid is the *enthalpy of fusion*, or **heat of fusion** (ΔH_{fusion}). For ice, $\Delta H_{fusion} = +6.01$ kJ/mol.

3. **Heating liquid H_2O:** Continued addition of heat to liquid water raises the temperature until it reaches its boiling point at 100 °C. The molar heat capacity of liquid

water (75.4 J/mol · C) and the temperature change (+100 °C) are used to calculate the amount of energy required to heat the liquid water:

$$\text{Energy to heat water from 0 °C to 100 °C} = \left(75.4\ \frac{J}{mol \cdot °C}\right)(100\ °C)$$

$$= 7.54 \times 10^3\ J/mol = 7.54\ kJ/mol$$

4. **Vaporizing liquid H_2O:** Once the temperature of the water reaches 100 °C addition of further heat again goes into overcoming intermolecular forces rather than into increasing the temperature, as indicated by the second plateau at 100 °C on the heating curve. At this temperature—the *boiling point*—liquid and vapor coexist in equilibrium as molecules break free from the surface of the liquid and enter the gas phase. The amount of energy necessary to convert a liquid into a gas is called the *enthalpy of vaporization,* or heat of vaporization (ΔH_{vap}). For water, $\Delta H_{vap} = +40.67\ kJ/mol$.

5. **Heating H_2O vapor:** Only after the liquid has been completely vaporized does the temperature again rise. The molar heat capacity of water vapor (33.6 J/mol · C) and the temperature change (+25 °C) are used to calculate the amount of energy required to heat the steam:

$$\text{Energy to heat steam from 100 °C to 125 °C} = \left(33.6\ \frac{J}{mol \cdot °C}\right)(25\ °C)$$

$$= 840\ J/mol = 0.840\ kJ/mol$$

Notice that the largest part (40.67 kJ/mol) of the 56.05 kJ/mol required to convert solid ice at −25 °C to gaseous steam at 125 °C is used for vaporization. The heat of vaporization for water is larger than the heat of fusion because only a relatively small number of hydrogen bonds must be broken to convert the solid to the liquid, but *all* hydrogen bonds must be broken to convert the liquid to the vapor.

The total amount of energy required to heat solid H_2O at −25 °C to gaseous H_2O at 125.0 °C is the sum of the five steps.

(1) Heating 1 mol of $H_2O(s)$ from −25 °C to 0 °C:		0.914 kJ
(2) Melting 1 mol of $H_2O(s)$ to form 1 mol $H_2O(l)$ at 0 °C:		6.01 kJ
(3) Heating 1 mol of $H_2O(l)$ from 0 °C to 100 °C:		7.54 kJ
(4) Vaporizing 1 mol of $H_2O(l)$ to form 1 mol $H_2O(g)$ at 100 °C:		40.67 kJ
(5) Heating 1 mol of $H_2O(g)$ from 100 °C to 125 °C:		0.840 kJ
Total energy required:		55.97 kJ

TABLE 11.3 gives further data on both heat of fusion and heat of vaporization for some common compounds. What is true for water is also true for other compounds: The heat of vaporization of a compound is always larger than its heat of fusion because all intermolecular forces must be overcome before vaporization can occur, but relatively fewer intermolecular forces must be overcome for a solid to change to a liquid.

BIG IDEA Question 2 Go to eText

Steam at 110 °C condenses on a person's skin and causes a mild burn. Which process releases the most heat to the skin: cooling steam from 110 °C to 100 °C; condensing steam to liquid water at 100 °C; or cooling liquid water from 100 °C to body temperature, 37 °C?

TABLE 11.3 Heats of Fusion and Heats of Vaporization for Some Common Compounds

Name	Formula	ΔH_{fusion}(kJ/mol)	ΔH_{vap}(kJ/mol)
Ammonia	NH_3	5.66	23.33
Benzene	C_6H_6	9.87	30.72
Ethanol	C_2H_5OH	4.93	38.56
Helium	He	0.02	0.08
Mercury	Hg	2.30	59.11
Water	H_2O	6.01	40.67

Go to
eText

WORKED EXAMPLE 11.3

Calculating the Amount of Heat of a Temperature Change

How much heat is required to convert 50.0 g of water at 25 °C to steam at 150 °C? The boiling point of water is 100 °C and $C_m[H_2O(l)] = 75.4$ J/(mol·°C), $\Delta H_{vap} = 40.67$ kJ/mol, $C_m[H_2O(g)] = 33.6$ J/(mol·°C).

IDENTIFY

Known	Unknown
Temperature change (25 °C to 150 °C)	
Amount of water (50.0 g)	Heat (q in kJ)
Boiling point 100 °C	
$\Delta H_{vap} = 40.67$ kJ/mol	
$C_m[H_2O(l)] = 75.4$ J/(mol·°C)	
$C_m[H_2O(g)] = 33.6$ J/(mol·°C)	

STRATEGY

There are three separate steps involved in heating the sample. Step 1 is heating the liquid water to its boiling point, step 2 is vaporizing the liquid water, and step 3 is heating the water vapor. First convert from grams to moles since the molar heat capacities are given and then calculate q for each step.

SOLUTION

First find the number of moles of water:

$$50.0 \text{ g } H_2O \times \frac{1 \text{ mol } H_2O}{18.0 \text{ g } H_2O} = 2.78 \text{ mol } H_2O$$

Step 1. Heating liquid H_2O from 25 °C to 100 °C:

$q = C_m \times$ Moles of a substance $\times \Delta T$

$$= \left(75.4 \frac{J}{mol \cdot °C}\right)(2.78 \text{ mol})(75 \text{ °C}) = 1.57 \times 10^4 \text{ J} = 15.7 \text{ kJ}$$

Step 2. Vaporizing liquid H_2O:

$q = \Delta H_{vap} \times$ Moles of a substance $= (40.67 \text{ kJ/mol})(2.78 \text{ mol})$

$$= 113.0 \text{ kJ}$$

Step 3. Heating gaseous H_2O from 100 °C to 150 °C:

$q = C_m \times$ Moles of a substance $\times \Delta T$

$$= \left(33.6 \frac{J}{mol \cdot °C}\right)(2.78 \text{ mol})(50 \text{ °C}) = 4.67 \times 10^3 \text{ J} = 4.67 \text{ kJ}$$

The total heat required is a sum of all three steps:
15.7 kJ + 113.0 kJ + 4.67 kJ = 133 kJ

CHECK

The amount of heat is large and positive, which is reasonable for a phase change. Also the magnitude of the heat for each step is reasonable. The phase change from liquid to vapor requires the most heat because strong hydrogen bonds between water molecules must be broken. Raising the temperature of liquid water requires more heat than raising the temperature of steam because the molar heat capacity and the temperature change are larger for the liquid.

▶ **PRACTICE 11.5** How much heat is required to convert 15.0 g of liquid benzene (C_6H_6) at 50 °C to gaseous benzene at 100 °C? The boiling point of benzene is 80.1 °C and $C_m[C_6H_6(l)] = 136.0$ J/(mol·°C), $\Delta H_{vap} = 30.72$ kJ/mol, $C_m[C_6H_6(g)] = 82.4$ J/(mol·°C)

▶ **APPLY 11.6** What is the sign and magnitude of q when 10.0 g of liquid water at 25 °C cools and freezes to form ice at −10 °C? The freezing point of water is 0 °C and $C_m[H_2O(l)] = 75.4$ J/(mol·°C), $\Delta H_{fus} = +6.01$ kJ/mol, $C_m[H_2O(s)] = 36.6$ J/(mol·°C).

All **WORKED EXAMPLES** with this icon [Go to eText] have an interactive video in the eText.

11.5 PHASE DIAGRAMS

Any one phase of matter can change spontaneously into either of the other two, depending on the temperature and pressure. A convenient way to picture the pressure–temperature dependency of a pure substance in a closed system without air present is to use what is called a **phase diagram.** As illustrated for water in **FIGURE 11.7,** a typical phase diagram shows which phase is stable at various combinations of pressure and temperature. When a boundary line between phases is crossed by changing either the temperature or the pressure, a phase change occurs.

Let's examine the phase diagram of water by starting in the lower left corner of Figure 11.7 and traveling up and right along the boundary line between solid on the left and gas on the right. Points on this line represent pressure/temperature combinations at which the two phases are in equilibrium in a closed system and a direct phase transition between solid and gas occurs. Further along the solid/gas line, an intersection is reached where two lines diverge to form the bounds of the liquid region. As shown in Figure 11.7, the solid/liquid boundary for H_2O goes up and slightly left, while the liquid/gas boundary continues curving up and to the right. Called the **triple point,** this three-way intersection represents a unique combination of pressure and

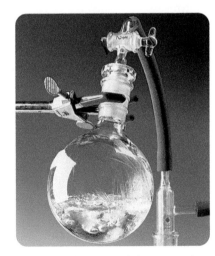

▲ At the triple point, solid exists in the boiling liquid. That is, solid, liquid, and gas coexist in equilibrium.

Go to
eText

▶ **FIGURE 11.7**

A phase diagram for H₂O. Various features of the diagram are discussed in the text. Note that the pressure and temperature axes are not drawn to scale.

▶ **Figure It out**

(a) What phase of water exists at a pressure of 1 atm and temperature of −1 °C?
(b) What phase change occurs if the pressure is decreased from 1 atm to 1 × 10⁻⁴ atm at a constant temperature of −1 °C?

Answer: (a) H₂O(s); (b) sublimation
H₂O(s) → H₂O(g)

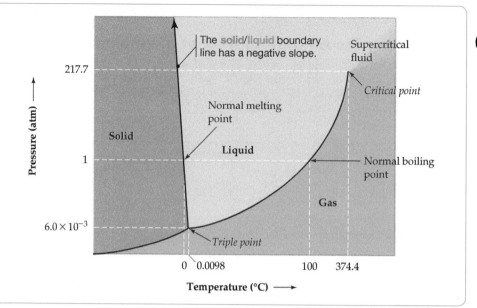

temperature at which all three phases coexist in equilibrium. For water, the triple-point temperature T_t is 0.0098 °C and the triple-point pressure P_t is 6.0×10^{-3} atm.

Continuing up and slightly left from the triple point, the solid/liquid boundary line represents the melting point of solid ice (or the freezing point of liquid water) at various pressures. When the pressure is 1 atm, the melting point—called the **normal melting point**—is exactly 0 °C. There is a slight negative slope to the line, indicating that the melting point of ice decreases as pressure increases. Water is unusual in this respect because most substances have a positive slope to their solid/liquid line, indicating that their melting points *increase* with pressure. For most substances, the solid phase is denser than the liquid because particles are packed closer together in the solid. Increasing the pressure pushes the molecules even closer together, thereby favoring the solid phase even more and giving the solid/liquid boundary line a positive slope. Water, however, becomes less dense when it freezes to a solid because large empty spaces are left between molecules due to the ordered three-dimensional network of hydrogen bonds in ice (**FIGURE 11.8**). As a result, increasing the pressure favors the liquid phase, giving the solid/liquid boundary a negative slope.

▶ **FIGURE 11.8**

Molecular picture of ice and liquid water.

▶ **Figure It out**

Which phase of H₂O has a greater density, solid or liquid?

Answer: The liquid phase has a greater density because there is less space between molecules than in the solid phase.

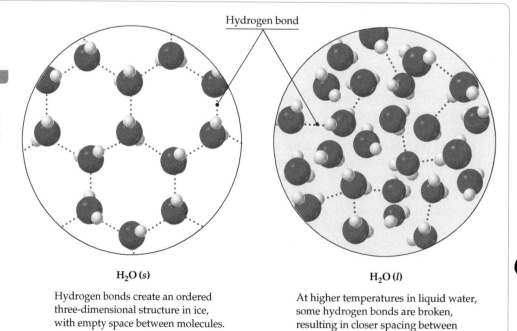

Hydrogen bond

H₂O (s)

Hydrogen bonds create an ordered three-dimensional structure in ice, with empty space between molecules.

H₂O (l)

At higher temperatures in liquid water, some hydrogen bonds are broken, resulting in closer spacing between molecules.

FIGURE 11.9 shows a simple demonstration of the effect of pressure on melting point. If a thin wire with heavy weights at each end is draped over a block of ice near 0 °C, the wire rapidly cuts through the block because the increased pressure lowers the melting point of the ice under the wire, causing the ice to liquefy.

Continuing up and right from the triple point, the liquid/gas boundary line represents the pressure/temperature combinations at which liquid and gas coexist and water vaporizes (or steam condenses). In fact, the part of the curve up to 1 atm pressure is simply the vapor pressure curve we saw previously in Figure 11.4. When the pressure is 1 atm, water is at its normal boiling point of 100 °C. Continuing along the liquid/gas boundary line, we reach the **critical point**, where the line abruptly ends. The critical temperature T_c is the temperature beyond which a gas cannot be liquefied, no matter how great the pressure, and the critical pressure P_c is the pressure beyond which a liquid cannot be vaporized, no matter how high the temperature. For water, $T_c = 374.4 \,°C$ and $P_c = 217.7$ atm.

We are accustomed to seeing solid/liquid and liquid/gas phase transitions, but behavior at the critical point lies so far outside our normal experiences that it's hard to imagine. Think of it this way: A *gas* at the critical point is under such high pressure, and its molecules are pressed so close together, that it becomes almost like a liquid. A *liquid* at the critical point is at such a high temperature, and its molecules are so relatively far apart, that it becomes almost like a gas. Thus, the two phases simply become the same and form a **supercritical fluid** that is neither liquid nor true gas. No distinct physical phase change occurs on going beyond the critical point. Rather, a whitish, pearly sheen momentarily appears, and the visible boundary between liquid and gas vanishes. Figure 11.11 in the Inquiry section shows an image of the supercritical phase transition of carbon dioxide.

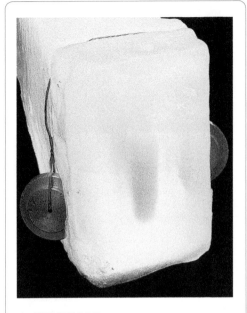

▲ **FIGURE 11.9**

Pressure and melting point. Why does the weighted wire cut through this block of ice?

CONCEPTUAL WORKED EXAMPLE 11.4

Interpreting a Phase Diagram

Freeze-dried foods are prepared by freezing the food and removing water by subliming the ice at low pressure. Look at the phase diagram of water in Figure 11.7, and give the maximum pressure in mm Hg at which ice and water vapor are in equilibrium.

STRATEGY

Solid and vapor are in equilibrium only below the triple-point pressure, $P_t = 6.0 \times 10^{-3}$ atm, which needs to be converted to millimeters of mercury.

SOLUTION

$$6.0 \times 10^{-3} \text{ atm} \times \frac{760 \text{ mm Hg}}{1 \text{ atm}} = 4.6 \text{ mm Hg}$$

▶ **CONCEPTUAL PRACTICE 11.7**   Look at the phase diagram of H_2O in Figure 11.7, and describe what happens to an H_2O sample when the pressure is first increased from 6×10^{-3} atm to 5 atm at 50 °C and the temperature is then increased from 50 °C to 375 °C at a presssure of 5 atm.

▶ **CONCEPTUAL APPLY 11.8**   The phase diagram for gallium metal is shown (the pressure axis is not to scale). In the region shown, gallium has two different solid phases.

(a) Where on the diagram are the solid, liquid, and vapor regions?

(b) How many triple points does gallium have? Circle each on the diagram.

(c) At 1 atm pressure, which phase is more dense, solid or liquid? Explain.

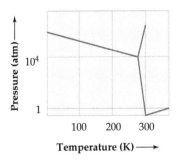

11.6 LIQUID CRYSTALS

You probably use items containing a liquid crystal display (LCD) every day. Laptop computer screens, smartphones, televisions, digital watches, microwave ovens, and other electronic devices display images using LCD technology. What is a liquid crystal? We normally think of a crystal as a solid material, like grain of salt (sodium chloride) or quartz rock. Solids have strong intermolecular forces, and their particles remain in fixed positions and orientations. The strong forces between molecules in solids lead to hard objects that retain their shape. In contrast, liquids flow and change shape because the molecules have weaker intermolecular forces that can be overcome to allow molecules to change positions and orientations. A **liquid crystal** is an intermediate phase that exhibits properties of both liquids and solids. Liquid crystals are ordered like solids but flow like liquids.

In 1888, the botanist and chemist Friedrich Reinitzer (1857–1927) was researching the chemical function of cholesterol in plants. He is credited for first describing the properties of the liquid crystal phase. Upon heating the compound cholesteryl benzoate, he was surprised to find that it had two distinct melting points. At 145.5 °C (293.9 °F), the solid melted to form a turbid (cloudy and opaque) liquid. When the turbid liquid was further heated, it turned clear at a temperature of 178.5 °C (353.3 °F). When the clear liquid was cooled, the phase transitions were reversed. The clear liquid became opaque at 178.5 °C, and the opaque liquid became solid at 145.5 °C. He concluded that he had discovered a new state of matter between the crystalline solid state and liquid state, called the liquid crystal state. **FIGURE 11.10** shows the solid, liquid crystal, and liquid states of cholesteryl benzoate created by heating the substance in a mineral oil bath.

▶ **FIGURE 11.10**

A macroscopic and molecular picture of the solid, liquid crystal, and liquid phases of cholesteryl benzoate.

Cholesteryl benzoate in the **solid phase** in a beaker of mineral oil at 25 °C.

Cholesteryl benzoate in the **liquid crystal phase** in a beaker of mineral oil at 150 °C.

Cholesteryl benzoate in the **liquid phase** in a beaker of mineral oil at 177 °C.

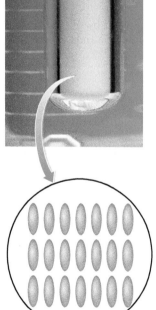

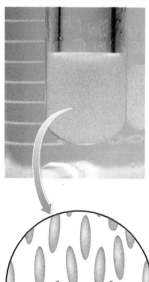

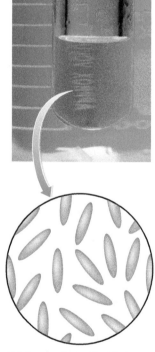

Molecules in the solid phase have fixed orientation and positions.

Molecules in the liquid crystal phase are oriented with long axes lined up approximately parallel to one another. The molecules can move positions, but the orientation remains the sames.

Molecules in the liquid phase have random orientations and move positions.

Liquid crystals are organic molecules that have elongated, rod-like shapes. A general structure for a liquid crystal molecule is:

Terminal group—⟨ ⟩—Linkage group—⟨ ⟩—Side chain

Liquid crystal molecules often contain six-membered carbon rings in which each carbon atom is sp^2 hybridized with trigonal planar geometry, creating a planar ring. The presence of rings provides a flat surface, which promotes stacking of the molecules. The rigidity of the molecule is enhanced when the linkage group contains a double bond such as (C=N), because there is no rotation around a double bond. The terminal group is often polar. This polarity creates strong intermolecular forces between molecules that lead to ordered arrangements. The side chain contains a chain of carbon atoms, which elongates the molecule. The structure for the molecule N-(4-Methoxybenzylidine)-4-butylaniline (MBBA), which exhibits a liquid crystal phase from 21–47 °C, is shown in **FIGURE 11.11**.

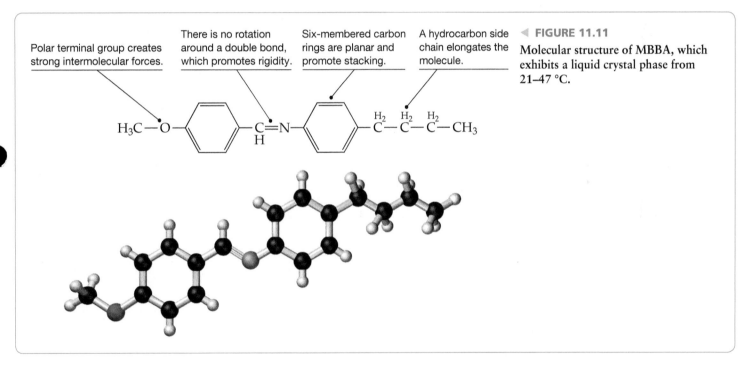

There is no rotation around a double bond, which promotes rigidity.

Polar terminal group creates strong intermolecular forces.

Six-membered carbon rings are planar and promote stacking.

A hydrocarbon side chain elongates the molecule.

◀ **FIGURE 11.11**
Molecular structure of MBBA, which exhibits a liquid crystal phase from 21–47 °C.

Go to
eText

BIG IDEA Question 3

Which of the following molecules is most likely to exhibit a liquid crystalline state?

(a) $CH_3CH_2CH_2CH_2CH_2CH_2CH_2CH_3$

(b)

(c) C_8H_{17}—O—⟨ ⟩—⟨ ⟩—C(=O)—O—⟨ ⟩—C_6H_{13}

Molecules that exhibit the liquid crystalline phase are classified based on their orientation patterns (**FIGURE 11.12**) In the **nematic liquid crystal phase**, the long axes of molecules are approximately parallel, but their ends do not line up. To picture the nematic phase, imagine a school of fish swimming in one direction. There is directional order, but the fish can change positions within the group. In the **smectic liquid crystal phase**, the long axes of molecules are parallel, but the ends are also aligned, creating rows of molecules. To picture the smectic phase, imagine soup cans stacked on shelves in the grocery store. Several variants of the smectic phase are known and depend on the angle formed between the long molecular axis and the plane of each layer. In smectic A, the angle between the rows of molecules and the long axis of the molecule is 90°. In smectic phases, molecules can rotate about their long axes but cannot readily slide past one another. In the **cholesteric liquid crystal phase**, the molecules are arranged such that one layer is placed on top of another, but the long axes of the molecules in each layer are rotated relative to the layers above and below. The stacking occurs in a helical pattern.

Liquid crystals have **anisotropic properties**, which depend on the directional orientation of the molecules. Electric fields can be used to change the orientation of polar molecules in a liquid crystal sample, which in turn changes their optical properties. Nematic liquid crystals are relatively translucent, but many of them become opaque when an electric field is applied. This behavior is ideal for producing dark images on a light background and is used in the liquid crystal displays of digital watches and handheld calculators. A cholesteric liquid crystal reflects different wavelengths of light depending on the pitch in the helical twist and on the viewing angle. Because the pitch of a cholesteric liquid crystal is temperature-dependent, observed color is a function of temperature. Liquid crystals can therefore serve as inexpensive, flexible thermometers. By mixing different compounds that form liquid crystals, thermometers can be customized to show temperature changes in a desired temperature range.

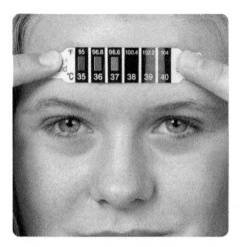

▲ An inexpensive liquid crystal thermometer for recording body temperature.

  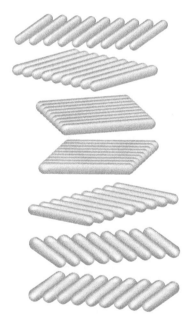

Nematic liquid crystal phase
The long axes of molecules are approximately parallel but their ends do not line up.

Smectic liquid crystal phase
The long axes of molecules are parallel and the ends are also aligned, creating rows of molecules. In the smectic A phase the angle between the long axis of the molecule and row is 90°.

Cholesteric liquid crystal phase
The molecules are arranged such that one layer is placed on top of another, but the long axes of the molecules are rotated at a slight angle to layer above and below it.

▲ **FIGURE 11.12**

Molecular orientation in nematic, smectic, and cholesteric liquid crystal phases. Liquid crystal phases exhibit various ordering patterns.

Organic compounds with carbon-hydrogen bonds are nonpolar. Caffeine has high solubility in the nonpolar solvent benzene because a significant portion of the molecule is nonpolar.

Caffeine   Benzene

Caffeine ($C_8H_{10}N_4O_2$) is a pesticide found naturally in the seeds and leaves of plants that kills or paralyzes certain insects that ingest it. In humans, caffeine acts a stimulant, and for this reason it is sometimes removed from coffee beans or tea leaves. *Extraction* is a process that refers to the separation of a substance from its surroundings, such as the removal of the caffeine molecule from a coffee bean. In 1905, Ludwig Roselius developed a method to extract caffeine from coffee using benzene (C_6H_6) as a solvent. Caffeine dissolves readily in the nonpolar solvent benzene because a significant portion of the caffeine molecule is nonpolar. If the polarity of solute and solvent are matched, then solubility will be high. In other words, nonpolar solvents dissolve nonpolar solutes and polar solvents dissolve polar solutes. However, in the food industry benzene is a poor choice for a solvent because it is highly toxic and carcinogenic (cancer causing). Residual benzene in the coffee can pose a severe health threat to those who consume it.

A much safer method uses supercritical CO_2 to extract caffeine from coffee beans. CO_2 is nontoxic, nonflammable, easily separated from a food sample, and recyclable. It is a nonpolar molecule and dissolves nonpolar solutes such as caffeine. However, at room temperature and pressure (25 °C and 1 atm), CO_2 is a gas and cannot be used as a solvent. Raising the temperature and pressure produces the supercritical phase of CO_2, which has unique properties between those of gases and liquids. Supercritical CO_2 has solvent properties like the liquid phase, but the extraction can be performed faster than with a conventional organic solvent because it diffuses rapidly and flows easily like a gas. Supercritical CO_2 also has low surface tension, allowing it to permeate into tiny pores in the coffee beans and dissolve caffeine on the inside.

The phase diagram of CO_2 shown in **FIGURE 11.13** shows that the supercritical phase of CO_2 can be reached at a relatively

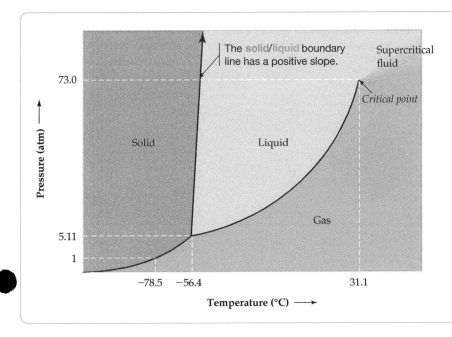

◀ **FIGURE 11.13**

A phase diagram for CO_2. The pressure and temperature axes are not to scale.

moderate temperature and pressure (31.1 °C and 73.0 atm). The easily attainable critical point for CO_2 makes it the most widely used supercritical fluid. Industrial and research applications use supercritical CO_2 as a solvent in environmentally friendly dry cleaning, analytical separations, and polymerization reactions. The phase diagram for CO_2 has many of the same features as that of water (Figure 11.7) but differs in several interesting respects. First, the triple point is at $P_t = 5.11$ atm, meaning that CO_2 can't be a liquid below this pressure, no matter what the temperature. At 1 atm pressure, CO_2 is a solid below -78.5 °C but a gas above this temperature. This means that carbon dioxide never exists in the liquid form at standard pressure. Second, the slope of the solid/liquid boundary is positive, meaning that the solid phase is favored as the pressure rises and that the melting point of solid CO_2 therefore increases with pressure.

We can observe the transition between a liquid and a supercritical fluid using a high pressure cell. As shown in **FIGURE 11.14,** initially, **CO_2** is present in the cell in the liquid phase and there is clear distinction between the gas and liquid phase. In the high pressure cell at 75 atm, increasing the temperature causes the liquid to become less dense, so that the separation between the liquid and gas phases becomes less distinct. Upon reaching the critical temperature, the density of the gas and liquid phase are identical and the boundary between them no longer exists.

PROBLEM 11.9

(a) Why was a new solvent needed for extracting caffeine from coffee beans?

(b) Why is the solubility of caffeine high in supercritical CO_2?

(c) Why is carbon dioxide the most widely used supercritical fluid?

PROBLEM 11.10 A fire extinguisher containing carbon dioxide has a pressure of 70 atm at 75 °F. What phase of CO_2 is present in the tank?

PROBLEM 11.11 Look at the phase diagram of CO_2 in Figure 11.13, and describe what happens to a CO_2 sample when the following changes are made:

(a) The temperature is increased from -100 °C to 0 °C at a constant pressure of 2 atm.

(b) The pressure is reduced from 72 atm to 5.0 atm at a constant temperature of 30 °C.

(c) The pressure is first increased from 3.5 atm to 76 atm at -10 °C, and the temperature is then increased from -10 °C to 45 °C.

PROBLEM 11.12 Liquid carbon dioxide is also used as nontoxic solvent in dry cleaning. Refer to the phase diagram for CO_2 (Figure 11.13) to answer the following questions.

(a) What is the minimum pressure at which liquid CO_2 can exist?

(b) What is the minimum temperature at which liquid CO_2 can exist?

(c) What is the maximum temperature at which liquid CO_2 can exist?

PROBLEM 11.13

(a) For the phase transition $CO_2(s) \longrightarrow CO_2(g)$, predict the sign of ΔS.

(b) At what temperature does CO_2 (s) spontaneously sublime at 1 atm? Use the phase diagram for CO_2 (Figure 11.13) to answer this question.

(c) If ΔH for the sublimation of 1 mol of CO_2 (s) is 26.1 kJ, calculate ΔS in (J/K·mol) for this phase transition. (*Hint:* Use the temperature found in part b to calculate the answer.)

PROBLEM 11.14 A sample of supercritical carbon dioxide was prepared by heating 100.0 g of $CO_2(s)$ at -78.5 °C to CO_2 (g) at 33 °C. Then the pressure was increased to 75.0 atm. How much heat was required to sublime the sample of $CO_2(s)$ and subsequently heat CO_2 (g)? ($\Delta H_{sub} = 26.1$ kJ/mol; C_m for CO_2 (g) $= 35.0$ J/mol · °C)

| At temperatures below the critical temperature, there is a clear boundary between liquid CO_2 and the gas phase. | Increasing temperature decreases the density of liquid CO_2, blurring the distinction between the liquid and gas phase. | At temperatures above the critical temperature (31.1°C), CO_2 is in the supercritical phase and the boundary disappears. | ◄ **FIGURE 11.14** **Visualization of phase transition between liquid and supercritical CO_2 in a high pressure cell.** |

STUDY GUIDE

Section	Concept Summary	Learning Objectives	Test Your Understanding
11.1 Properties of Liquids	Two important properties of liquids are **viscosity**, a measure of a liquid's resistance to flow, and **surface tension**, the resistance of a liquid to spread out and increase its surface area. These properties are influenced by the strength of the intermolecular forces in a liquid sample.	**11.1** Match an observation of a physical phenomenon with the correct liquid property.	Problems 11.20–11.21
		11.2 Predict which substance has higher viscosity or surface tension based on its molecular structure.	Problems 11.22–11.25
11.2 Vapor Pressure and Boiling Point	In a closed container, liquid molecules evaporate, and the resulting increase in pressure is called the **vapor pressure**. Vapor pressure increases as temperature increases because more molecules have sufficient energy to overcome intermolecular forces and enter the gas phase.	**11.3** Use the Clausius–Clapeyron equation to calculate ΔH_{vap} or vapor pressure at varying temperatures.	Worked Example 11.1; Problems 11.26, 11.28, 11.36
11.3 Phase Changes between Solids, Liquids, and Gases	Matter in any one **phase**—solid, liquid, or gas—can undergo a **phase change** to either of the other two phases. Like all naturally occurring processes, a phase change has an associated free-energy change, $\Delta G = \Delta H - T\Delta S$. The enthalpy component, ΔH, is a measure of the change in intermolecular forces; the entropy component, ΔS, is a measure of the change in molecular randomness accompanying the phase transition.	**11.4** Use the equation for free energy to calculate the temperature at which a phase change becomes spontaneous or ΔS for a phase change.	Worked Example 11.2; Problems 11.48–11.57
11.4 Energy Changes during Phase Transitions	When heat is added to a substance in the solid phase, the temperature increases until the melting point is reached. At the melting point, additional heat is added, but the temperature does not change because the energy is used to break some intermolecular forces between molecules. After all the solid has been converted to liquid, the temperature rises as more heat is added. At the melting point, additional heat is added, but the temperature does not change because the energy is used to overcome intermolecular forces and molecules enter the gas phase. The addition of further heat increases the temperature of the vapor.	**11.5** Draw or interpret a heating curve for a substance.	Problems 11.42, 11.44, 11.50
		11.6 Calculate the amount of heat associated with temperature and phase changes of substances.	Worked Example 11.3; Problems 11.46, 11.48
11.5 Phase Diagrams	The effects of temperature and pressure on phase changes can be displayed graphically on a **phase diagram**. A typical phase diagram has three regions—solid, liquid, and gas—separated by three boundary lines that represent pressure/temperature combinations at which two phases are in equilibrium and phase changes occur. At exactly 1 atm pressure, the temperature at the solid/liquid boundary corresponds to the **normal melting point** of the substance, and the temperature at the liquid/gas boundary corresponds to the **normal boiling point**. The three lines meet at the **triple point,** a unique combination of temperature and pressure at which all three phases coexist in equilibrium. The liquid/gas line runs from the triple point to the **critical point,** a pressure/temperature combination beyond which liquid and gas phases become a **supercritical fluid.**	**11.7** Use a phase diagram to identify the phase of a substance at a given temperature and pressure or the phase transition that occurs when the temperature and pressure are changed.	Worked Example 11.4; Problems 11.16–11.17, 11.55, 11.61
		11.8 Sketch a phase diagram given the appropriate data.	Problems 11.56–11.57, 11.62

Section	Concept Summary	Learning Objectives	Test Your Understanding
11.6 Liquid Crystals	A **liquid crystal** is an intermediate phase that exhibits properties of both liquids and solids. Molecules that form liquid crystals have a rigid, elongated shape and polar groups that align molecules through intermolecular forces. Liquid crystals have various ordered phases that depend on the molecular structure and the temperature. In the **nematic phase,** molecules are arranged with their long axes parallel to one another, but the ends of the molecules are not aligned. In the **smectic phase,** molecules are aligned in rows with the long axes parallel to one another and ends aligned. In the **cholesteric** phase, the molecules are arranged such that one layer is placed on top of another, but the long axes of the molecules are rotated at a slight angle to layer above and below it, creating a helix.	**11.9** Predict which molecules will exhibit a liquid crystal phase based on molecular structure.	Problems 11.68–11.69
		11.10 Classify the types of intermolecular forces between molecules in a liquid crystal.	Problems 11.70–11.73
		11.11 Match a molecular image with different states of matter, including solid, liquid, and common liquid crystal phases.	Problems 11.18–11.19

KEY TERMS

anisotropic properties *438*
Clausius–Clapeyron
 equation *426*
cholesteric liquid crystal
 phase *438*
critical point *435*

heat of fusion ($\mathbf{\Delta H_{fusion}}$) *431*
heat of vaporization
 ($\mathbf{\Delta H_{vap}}$) *425*
liquid crystal *436*
nematic liquid crystal
 phase *438*

normal boiling point *427*
normal melting point *434*
phase *428*
phase change *428*
phase diagram *433*
smectic liquid crystal phase *438*

supercritical fluid *435*
surface tension *423*
triple point *433*
vapor pressure *424*
viscosity *423*

KEY EQUATIONS

- **Clausius–Clapeyron Equation (Section 11.2)**

$$\ln P_{vap} = \left(-\frac{\Delta H_{vap}}{R} \right) \frac{1}{T} + C$$ where ΔH_{vap} is the heat of vaporization, R is the gas constant, and C is a constant characteristic of the specific substance.

- **Two-Point Form of the Clausius–Clapeyron Equation (Section 11.2)**

$$\ln \left(\frac{P_1}{P_2} \right) = \frac{\Delta H_{vap}}{R} \left(\frac{1}{T_2} - \frac{1}{T_1} \right)$$

PRACTICE TEST

After studying this chapter, you can assess your understanding with these practice test questions, which are correlated with chapter learning objectives. If you answer a question incorrectly, refer to the learning objectives in the end-of-chapter Study Guide for assistance. The Study Guide provides a conceptual summary, references a Worked Example to model how to solve the problem, and gives additional problems for more practice.

1. Three identical tubes are filled with different liquids; water, ethylene glycol, and olive oil. A small steel sphere was dropped into the tube, and the time it took to fall to the bottom was recorded.

Tube	Liquid	Time for Steel Sphere to Fall to Bottom
1	water	0.2 seconds
2	ethylene glycol	3.2 seconds
3	olive oil	16.5 seconds

Which property of liquids is responsible for the differences in time for the sphere to fall through the liquid? (**LO 11.1**)
(a) Surface tension
(b) Boiling point
(c) Viscosity
(d) Vapor pressure

2. Which organic compound has the lowest viscosity? (**LO 11.2**)

(a)

$$CH_3 - \underset{\underset{CH_3}{|}}{\overset{\overset{CH_3}{|}}{C}} - CH_3$$

(b)

H₃C—CH₂—CH₂—CH₂—CH₂—CH₃

(c)

H₃C—CH₂—CH₂—CH₂—CH₂—CH₂—NH₂

(d)

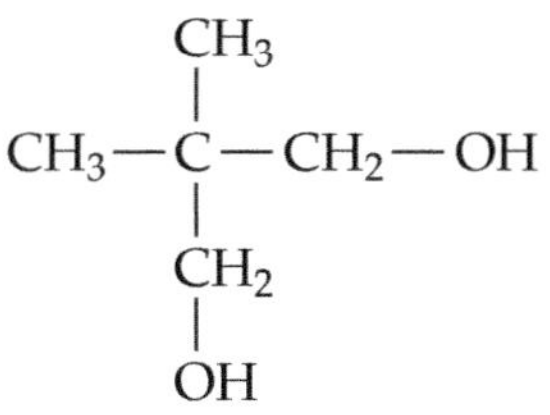

$$CH_3$$
$$|$$
$$CH_3-C-CH_2-OH$$
$$|$$
$$CH_2$$
$$|$$
$$OH$$

3. Nitrous oxide, occasionally used as an anesthetic by dentists and sometimes called "laughing gas," has P_{vap}=100.0 mm Hg at −111.3 °C and a normal boiling point of −88.5 °C. What is the heat of vaporization of nitrous oxide in kJ/mol? **(LO 11.3)**

(a) 22.1 kJ/mol   (b) −22.1 kJ/mol
(c) 218 kJ/mol   (d) 2.21 × 10⁴ kJ/mol

4. In Denver, the Mile-High City, water boils at 95 °C. What is atmospheric pressure in atmospheres in Denver? ΔH_{vap} for H₂O is 40.67 kJ/mol. **(LO 11.3)**

(a) 532 mm Hg   (b) 759 mm Hg
(c) 908 mm Hg   (d) 636 mm Hg

5. Magnesium metal has ΔH_{fusion} = 9.037 kJ/mol and ΔS_{fusion} = 9.79 J/(K·mol). What is the melting point of magnesium in °C? **(LO 11.4)**

(a) 0.923 °C   (b) 923 °C
(c) 650 °C   (d) 1.08 × 10³ °C

6. Consider a compound that has a melting point at 65 °C and a boiling point at 175 °C. Which of the following images represents a heating curve for the compound from 40 °C to 200 °C? **(LO 11.5)**

(a)

Time ⟶

(b)

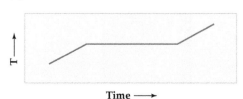

Time ⟶

(c)

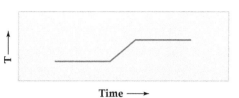

Time ⟶

(d)

Time ⟶

7. Mercury has a melting point of −38.8 °C, a molar heat capacity of 27.9 J/(K·mol) for the liquid and 28.2 J/(K·mol) for the solid, and ΔH_{fusion} = 2.33 kJ/mol. Assuming that the heat capacities don't change with temperature, how much energy in joules is needed to heat 7.50 g of Hg from a temperature of −50.0 °C to +50.0 °C? **(LO 11.6)**

(a) 236 J   (b) 192 J
(c) 105 J   (d) 111 J

Use the phase diagram for carbon dioxide to answer questions 8 and 9.

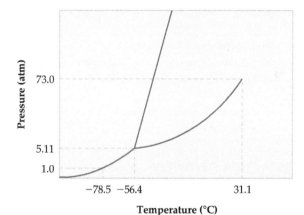

8. What phase transition occurs if a sample of carbon dioxide starts at 50 atm and 0 °C and the pressure is decreased to 1 atm? **(LO 11.7)**

(a) Solid to gas   (b) Liquid to solid
(c) Liquid to gas   (d) Gas to solid

9. Under which conditions will the supercritical phase exist?

(a) 5.11 atm and −56.4 °C   (b) 2 atm and 25 °C
(c) 50 atm and −50 °C   (d) 80 atm and 35 °C

10. Which of the following molecules is likely to have a liquid crystalline phase? **(LO 11.9)**

(a)

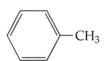

(b)

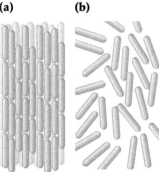

(c)

H_3CO—⬡—$N{=}N$—⬡—OCH_3

(d)

$$H{-}\underset{\underset{H}{|}}{\overset{\overset{H}{|}}{C}}{-}\underset{\underset{H}{|}}{\overset{\overset{H}{|}}{C}}{-}\underset{\underset{H}{|}}{\overset{\overset{H}{|}}{C}}{-}\underset{\underset{H}{|}}{\overset{\overset{H}{|}}{C}}{-}\underset{\underset{H}{|}}{\overset{\overset{H}{|}}{C}}{-}\underset{\underset{H}{|}}{\overset{\overset{H}{|}}{C}}{-}\underset{\underset{H}{|}}{\overset{\overset{H}{|}}{C}}{-}\underset{\underset{H}{|}}{\overset{\overset{H}{|}}{C}}{-}\underset{\underset{H}{|}}{\overset{\overset{H}{|}}{C}}{-}H$$

11. What types of intermolecular forces exist in a sample of a compound with the molecular structure shown?

⬡—⬡—$\overset{\overset{\textstyle H}{|}}{C}{=}N$—⬡—$\overset{\overset{\textstyle O}{\|}}{C}{-}O{-}C_2H_5$

(a) London dispersion forces
(b) London dispersion forces and dipole-dipole forces
(c) London dispersion forces, dipole-dipole forces, and hydrogen bonding
(d) London dispersion forces, hydrogen bonding, and ion-dipole forces

12. The following compound exists in the nematic liquid crystal phase from 14–28 °C. Which image represents the molecular arrangement of the molecules at 25 °C? **(LO 11.11)**

$N{\equiv}C$—⬡—$N{=}N$—⬡—C_6H_{13}

(a)   **(b)**

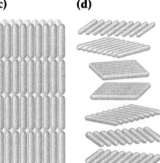

(c)   **(d)**

Answers:
1. c, 2. a, 3. a, 4. d, 5. c, 6. d, 7. b, 8. c, 9. d, 10. c, 11. b, 12. a

---

Mastering Chemistry   provides end-of-chapter exercises, feedback-enriched tutorial problems, animations, and interactive activities to encourage problem-solving practice and deeper understanding of key concepts and topics.

RAN   *Randomized in Mastering Chemistry*

CONCEPTUAL PROBLEMS

Problems 11.1–11.14 appear within the chapter.

11.15 Assume that you have a liquid in a cylinder equipped with a movable piston. There is no air in the cylinder, the volume of space above the liquid is 200 mL, and the equilibrium vapor pressure above the liquid is 28.0 mm Hg. What is the equilibrium pressure above the liquid when the volume of space is decreased from 200 mL to 100 mL at constant temperature?

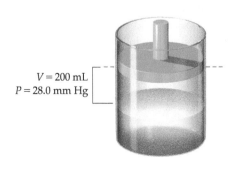

V = 200 mL
P = 28.0 mm Hg

Lower piston →

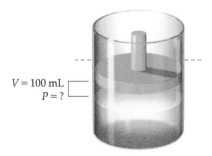

V = 100 mL
P = ?

11.16 The phase diagram of a substance is shown below.

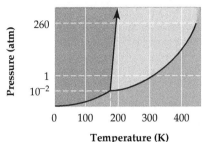

(a) Approximately what is the normal boiling point and what is the normal melting point of the substance?

(b) What is the physical state of the substance under the following conditions?
  (i) $T = 150$ K, $P = 0.5$ atm
  (ii) $T = 325$ K, $P = 0.9$ atm
  (iii) $T = 450$ K, $P = 265$ atm

11.17 The following phase diagram of elemental carbon has three different solid phases in the region shown.

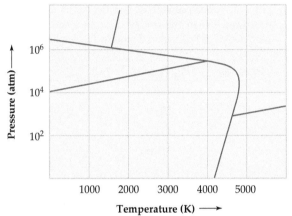

(a) Show where the solid, liquid, and vapor regions are on the diagram.

(b) How many triple points does carbon have? Circle each on the diagram.

(c) Graphite is the most stable solid phase under normal conditions. Identify the graphite phase on the diagram.

(d) On heating graphite to 2500 K at a pressure of 100,000 atm, it can be converted into diamond. Identify the diamond phase on the graph.

(e) Which phase is more dense, graphite or diamond? Explain.

11.18 A phase diagram for a substance that exhibits the liquid crystalline state is shown.

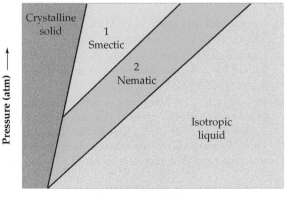

(a) Between which two phases are the smectic liquid crystal and nematic liquid crystal phases?

(b) If the substance is at a temperature and pressure represented by point 1 and the temperature is increased until the phase changes, what is the phase transition? Describe how the ordering of molecules changes, and propose a reason for the change.

(c) If the substance is at a temperature and pressure represented by point 2 and the pressure is increased until the phase changes, what is the phase transition? Describe how the ordering of molecules changes, and propose a reason for the change.

11.19 The following compound undergoes a phase transition from the solid to the smectic liquid crystal phase at 121 °C. Upon further heating, the compound undergoes another transition to liquid phase at 131 °C.

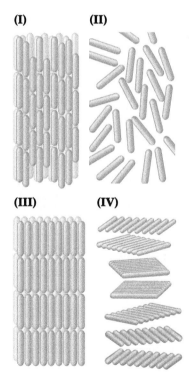

(a) Which image represents the arrangement of molecules at 125 °C?

(b) Which image represents the arrangement of molecules at 135 °C?

SECTION PROBLEMS

Properties of Liquids (Section 11.1)

11.20 A magnetized needle gently placed on the surface of a glass of water acts like a makeshift compass. Is it water's viscosity or its surface tension that keeps the needle on top?

11.21 Water flows quickly through the narrow neck of a bottle, but maple syrup flows sluggishly. Is this different behavior due to a difference in viscosity or in surface tension for the liquids?

11.22 Predict which substance in each pair has the highest surface tension.
RAN
  (a) CCl_4 or CH_2Br_2
  (b) Ethanol (CH_3CH_2OH) or ethylene glycol ($HOCH_2CH_2OH$)

11.23 Predict which substance in each pair has the highest viscosity.
RAN
  (a) Hexane ($CH_3CH_2CH_2CH_2CH_2CH_3$) or 1-hexanol ($CH_3CH_2CH_2CH_2CH_2CH_2OH$)
  (b) Pentane or neopentane

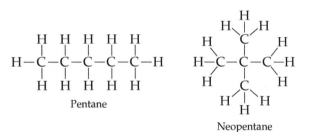

Pentane

Neopentane

11.24 The chemical structure for oleic acid, the primary component of olive oil, is shown. Explain why olive oil has a higher viscosity than water.

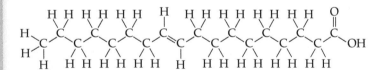

11.25 The viscosity of water at 20 °C is 1.00×10^{-3} $(N \cdot s/m^2)$ higher than dimethyl sulfide $[(CH_3)_2 S] \, 2.8 \times 10^{-4}$ $(N \cdot s/m^2)$. Explain the difference in viscosity based on chemical structure.

Vapor Pressure and Boiling Point (Section 11.2)

11.26 Carbon disulfide, CS_2, has $P_{vap} = 100$ mm Hg at -5.1 °C and a normal boiling point of 46.5 °C. What is ΔH_{vap} for carbon disulfide in kJ/mol?

Carbon disulfide

11.27 The vapor pressure of $SiCl_4$ is 100 mm Hg at 5.4 °C, and the normal boiling point is 57.7 °C. What is ΔH_{vap} for $SiCl_4$ in kJ/mol?

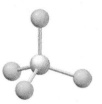

Silicon tetrachloride

11.28 What is the vapor pressure of CS_2 in mm Hg at 20.0 °C? (See Problem 11.26.)
RAN

11.29 What is the vapor pressure of $SiCl_4$ in mm Hg at 30.0 °C? (See Problem 11.27.)
RAN

11.30 Dichloromethane, CH_2Cl_2, is an organic solvent used for removing caffeine from coffee beans. The following table gives the vapor pressure of dichloromethane at various temperatures. Fill in the rest of the table, and use the data to plot curves of P_{vap} versus T and $\ln P_{vap}$ versus $1/T$.

Temp (K)	P_{vap} (mm Hg)	$\ln P_{vap}$	$1/T$
263	80.1	?	?
273	133.6	?	?
283	213.3	?	?
293	329.6	?	?
303	495.4	?	?
313	724.4	?	?

11.31 The following table gives the vapor pressure of mercury at various temperatures. Fill in the rest of the table, and use the data to plot curves of P_{vap} versus T and $\ln P_{vap}$ versus $1/T$.

Temp (K)	P_{vap} (mm Hg)	$\ln P_{vap}$	$1/T$
500	39.3	?	?
520	68.5	?	?
540	114.4	?	?
560	191.6	?	?
580	286.4	?	?
600	432.3	?	?

11.32 Use the plot you made in Problem 11.30 to find a value in kJ/mol for ΔH_{vap} for dichloromethane.

11.33 Use the plot you made in Problem 11.31 to find a value in kJ/mol for ΔH_{vap} for mercury. The normal boiling point of mercury is 630 K.

11.34 Choose any two temperatures and corresponding vapor pressures in the table given in Problem 11.30, and use those values to calculate ΔH_{vap} for dichloromethane in kJ/mol. How does the value you calculated compare to the value you read from your plot in Problem 11.32?

11.35 Choose any two temperatures and corresponding vapor pressures in the table given in Problem 11.31, and use those values to calculate ΔH_{vap} for mercury in kJ/mol. How does the value you calculated compare to the value you read from your plot in Problem 11.33?

11.36 Dichlorodifluoromethane, CCl_2F_2, one of the chlorofluorocarbon refrigerants responsible for destroying part of the Earth's ozone layer, has $P_{vap} = 40.0$ mm Hg at -81.6 °C and $P_{vap} = 400$ mm Hg at -43.9 °C. What is the normal boiling point of CCl_2F_2 in °C?

Dichlorodifluoromethane

11.37 Acetone, a common laboratory solvent, has $\Delta H_{vap} = 29.1$ kJ/
RAN   mol and a normal boiling point of 56.1 °C. At what temperature in °C does acetone have $P_{vap} = 105$ mm Hg?

Acetone

Phase Changes (Sections 11.3–11.4)

11.38 Why is ΔH_{vap} usually larger than ΔH_{fusion}?

11.39 Why is the heat of sublimation, ΔH_{subl}, equal to the sum of ΔH_{vap} and ΔH_{fusion} at the same temperature?

11.40 Naphthalene, better known as "mothballs," has bp = 218 °C and $\Delta H_{vap} = 43.3$ kJ/mol. What is the entropy of vaporization, ΔS_{vap} in J/(K·mol) for naphthalene?

11.41 Titanium tetrachloride, $TiCl_4$, has a melting point of -23.2 °C and has $\Delta H_{fusion} = 9.37$ kJ/mol. What is the entropy of fusion, ΔS_{fusion} in J/(K·mol), for $TiCl_4$?

11.42 Consider a compound that has melting point at 50 °C and a boiling point at 125 °C and a pressure of 1 atm. Draw a heating curve that represents heating the compound from 25 °C to 150 °C.

11.43 Consider a compound that sublimes at 50 °C and a pressure of 1 atm. Draw a heating curve that represents heating the compound from 25 °C to 75 °C.

11.44 Mercury has mp = -38.8 °C and bp = 356.6 °C. What, if any, phase changes take place under the following conditions at 1.0 atm pressure?

(a) The temperature of a sample is raised from -30 °C to 365 °C.

(b) The temperature of a sample is lowered from 291 K to 238 K.

(c) The temperature of a sample is lowered from 638 K to 231 K.

11.45 Iodine has mp = 113.7 °C and bp = 184.4 °C. What, if any, phase changes take place under the following conditions at 1.0 atm pressure?

(a) The temperature of a solid sample is held at 113.7 °C while heat is added.

(b) The temperature of a sample is lowered from 452 K to 389 K.

11.46 How much energy in kilojoules is needed to heat 5.00 g of
RAN   ice from -11.0 °C to °30.0 °C? The heat of fusion of water is 6.01 kJ/mol, and the molar heat capacity is 36.6 J/(K·mol) for ice and 75.4 J/(K·mol) for liquid water.

11.47 How much energy in kilojoules is released when 15.3 g of steam at 115.0 °C is condensed to give liquid water at 75.0 °C? The heat of vaporization of liquid water is 40.67 kJ/mol, and the molar heat capacity is 75.4 J/(K·mol) for the liquid and 33.6 J/(K·mol) for the vapor.

11.48 How much energy in kilojoules is released when 7.55 g of
RAN   water at 33.5 °C is cooled to -11.0 °C? (See Problem 11.46 for the necessary data.)

11.49 How much energy in kilojoules is released when 25.0 g of ethanol vapor at 93.0 °C is cooled to -11.0 °C? Ethanol has mp = -114.1 °C, bp = 78.3 °C, $\Delta H_{vap} = 38.56$ kJ/mol, and $\Delta H_{fusion} = 4.93$ kJ/mol. The molar heat capacity is 112.3 J/(K·mol) for the liquid and 65.6 J/(K·mol) for the vapor.

11.50 Draw a molar heating curve for ethanol, C_2H_5OH, similar to that shown for water in Figure 11.6. Begin with solid ethanol at its melting point, and raise the temperature to 100 °C. The necessary data are given in Problem 11.49.

11.51 Draw a molar heating curve for sodium similar to that shown for water in Figure 11.6. Begin with solid sodium at its melting point, and raise the temperature to 1000 °C. The necessary data are mp = 97.8 °C, bp = 883 °C, $\Delta H_{vap} = 89.6$ kJ/mol, and $\Delta H_{fusion} = 2.64$ kJ/mol. Assume that the molar heat capacity is 20.8 J/(K·mol) for both liquid and vapor phases and does not change with temperature.

Phase Diagrams (Section 11.5)

11.52 Water at room temperature is placed in a flask connected by rubber tubing to a vacuum pump, and the pump is turned on. After several minutes, the volume of the water has decreased, and what remains has turned to ice. Explain.

11.53 Ether at room temperature is placed in a flask connected by a rubber tube to a vacuum pump, the pump is turned on, and the ether begins boiling. Explain.

11.54 Look at the phase diagram of CO_2 in Figure 11.13, and tell
RAN   what phases are present under the following conditions.

(a) $T = -60$ °C, $P = 0.75$ atm

(b) $T = -35$ °C, $P = 18.6$ atm

(c) $T = -80$ °C, $P = 5.42$ atm

11.55 Look at the phase diagram of H_2O in Figure 11.7, and
RAN   tell what happens to an H_2O sample when the following changes are made.

(a) The temperature is reduced from 48 °C to -4.4 °C at a constant pressure of 6.5 atm.

(b) The pressure is increased from 85 atm to 226 atm at a constant temperature of 380 °C.

11.56 Bromine has $T_t = -7.3\,°C$, $P_t = 44$ mm Hg, $T_c = 315\,°C$, and $P_c = 102$ atm. The density of the liquid is 3.1 g/cm^3, and the density of the solid is 3.4 g/cm^3. Sketch a phase diagram for bromine, and label all points of interest.

11.57 Oxygen has $T_t = 54.3$ K, $P_t = 1.14$ mm Hg, $T_c = 154.6$ K, and $P_c = 49.77$ atm. The density of the liquid is 1.14 g/cm^3, and the density of the solid is 1.33 g/cm^3. Sketch a phase diagram for oxygen, and label all points of interest.

11.58 Refer to the bromine phase diagram you sketched in Problem 11.56, and tell what phases are present under the following conditions.
RAN
(a) $T = -10\,°C$, $P = 0.0075$ atm
(b) $T = 25\,°C$, $P = 16$ atm

11.59 Refer to the oxygen phase diagram you sketched in Problem 11.57, and tell what phases are present under the following conditions
RAN
(a) $T = -210\,°C$, $P = 1.5$ atm
(b) $T = -100\,°C$, $P = 66$ atm

11.60 Does solid oxygen (Problem 11.57) melt, as water does, when pressure is applied? Explain.

11.61 Assume that you have samples of the following three gases at 25 °C. Which of the three can be liquefied by applying pressure, and which cannot? Explain.
Ammonia: $T_c = 132.5\,°C$ and $P_c = 112.5$ atm
Methane: $T_c = -82.1\,°C$ and $P_c = 45.8$ atm
Sulfur dioxide: $T_c = 157.8\,°C$ and $P_c = 77.7$ atm

11.62 Benzene has a melting point of 5.53 °C and a boiling point of 80.09 °C at atmospheric pressure. Its density is 0.8787 g/cm^3 when liquid and 0.899 g/cm^3 when solid; it has $T_c = 289.01\,°C$, $P_c = 48.34$ atm, $T_t = 5.52\,°C$, and $P_t = 0.0473$ atm. Starting from a point at 200 K and 66.5 atm, trace the following path on a phase diagram.
(1) First, increase T to 585 K while keeping P constant.
(2) Next, decrease P to 38.5 atm while keeping T constant.
(3) Then, decrease T to 278.66 K while keeping P constant.
(4) Finally, decrease P to 0.0025 atm while keeping T constant.
What is your starting phase, and what is your final phase?

11.63 Refer to the oxygen phase diagram you drew in Problem 11.57, and trace the following path starting from a point at 0.0011 atm and −225 °C:
(1) First, increase P to 35 atm while keeping T constant.
(2) Next, increase T to −150 °C while keeping P constant.
(3) Then, decrease P to 1.0 atm while keeping T constant.
(4) Finally, decrease T to −215 °C while keeping P constant.
What is your starting phase, and what is your final phase?

11.64 How many phase transitions did you pass through in Problem 11.62, and what are they?

11.65 What phase transitions did you pass through in Problem 11.63?

11.66 Use the following data to sketch a phase diagram for krypton: $T_t = -169\,°C$, $P_t = 133$ mm Hg, $T_c = -63\,°C$, $P_c = 54$ atm, mp = −156.6 °C, bp = −152.3 °C. The density of solid krypton is 2.8 g/cm^3, and the density of the liquid is 2.4 g/cm^3. Can a sample of gaseous krypton at room temperature be liquefied by raising the pressure?

11.67 What is the physical phase of krypton (Problem 11.64) under the following conditions.
(a) $P = 5.3$ atm, $T = -153\,°C$
(b) $P = 65$ atm, $T = 250$ K

Liquid Crystals (Section 11.6)

11.68 State whether each of the following compounds is likely to have a liquid crystalline phase. Explain your reasoning.

(a)

$CH_3(CH_2)_7$ — ⬡ — $\overset{\overset{\textstyle O}{\|}}{C}$ — O — H

(b)

(benzene ring structure)

(c)

H_3C — O — ⬡ — N=N — ⬡ — C_4H_9

11.69 State whether each of the following compounds is likely to have a liquid crystalline phase. Explain your reasoning.

(a)

H — C—C—C—C—C—C—C—C—$\overset{\overset{\textstyle O}{\|}}{C}$ — O$^-$ Na$^+$
(chain with H atoms)

(b)

⬡—⬡ — $\overset{H}{\underset{}{C}}$=N—$\overset{\overset{\textstyle O}{\|}}{C}$—O—$C_2H_5$

(c)

H—C—C—C—C—C—C—C—C—C—H
(alkane chain with H atoms)

11.70 Intermolecular forces are important in creating ordered arrangements in liquid crystals. What types of intermolecular forces exist in a sample of the following compound?

C_8H_{17} — O — ⬡ — COOH

11.71 Intermolecular forces are important in creating ordered arrangements in liquid crystals. What types of intermolecular forces exist in a sample of the following compound?

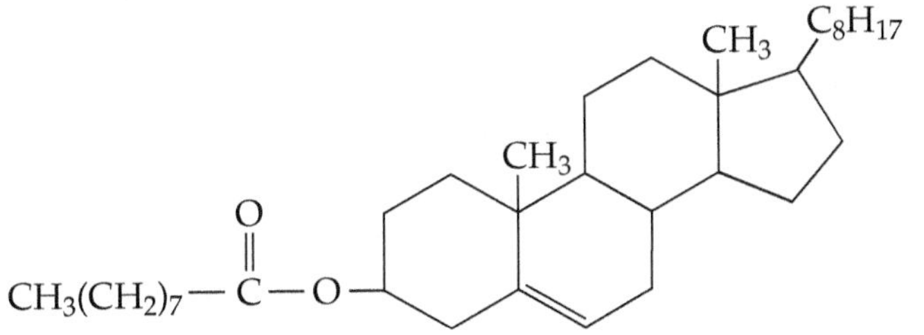

$CH_3(CH_2)_7$—$\overset{\overset{\displaystyle O}{\|}}{C}$—O— (steroid structure with CH₃, CH₃, CH₃, C₈H₁₇ groups)

11.72 Two compounds (labeled I and II) have a liquid crystal state. Predict which compound has a higher temperature associated with the solid to liquid crystal phase change. Explain your reasoning.

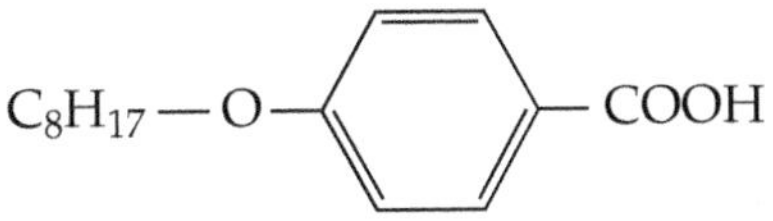

C_8H_{17}—O— —COOH

Compound I

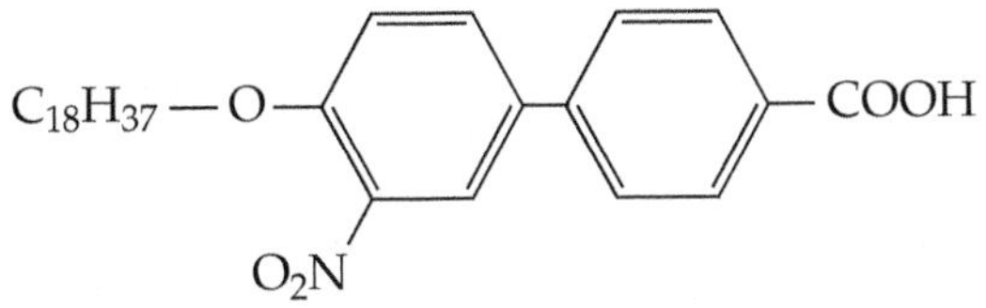

$C_{18}H_{37}$—O— —COOH, O_2N

Compound II

11.73 The compound cholesteryl benzoate is a rod-like molecule that undergoes a phase change from the solid to the liquid crystal phase at 145.5 °C. When cholesteryl benzoate is mixed with cholesteryl oleyl carbonate, a molecule with a curved shape, the temperature of the solid to liquid crystal transition changes. Predict if the transition temperature increases or decreases and explain your reasoning.

MULTICONCEPT PROBLEMS

11.74 For each of the following substances, identify the intermolecular force or forces that predominate. Using your knowledge of the relative strengths of the various forces, rank the substances in order of their normal boiling points.

$$Al_2O_3, F_2, H_2O, Br_2, ICl, NaCl$$

11.75 The chlorofluorocarbon refrigerant trichlorofluoromethane, CCl_3F, has $P_{vap} = 100.0$ mm Hg at -23 °C and $\Delta H_{vap} = 24.77$ kJ/mol.

(a) What is the normal boiling point of trichlorofluoromethane in °C?

(b) What is ΔS_{vap} for trichlorofluoromethane?

11.76 Look up thermodynamic data for ethanol (C_2H_5OH) in Appendix B, estimate the normal boiling point of ethanol, and calculate the vapor pressure of ethanol at 25 °C.
RAN

11.77 Substance **X** has a vapor pressure of 100 mm Hg at its triple point (48 °C). When 1 mol of **X** is heated at 1 atm pressure with a constant rate of heat input, the following heating curve is obtained:

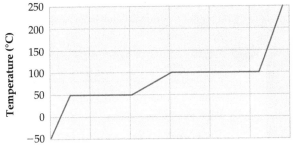

(a) Sketch the phase diagram for X, including labels for different phases, triple point, melting point, and boiling point.

(b) For each of the following, choose which phase of X (solid, liquid, or gas) fits the description.

  (i)   Is the most dense at 50 °C

  (ii) Is the least dense at 50 °C

  (iii) Has the greatest specific heat

  (iv) Predominates at 80 °C and 1 atm

  (v) Can have a vapor pressure of 20 mm Hg

chapter 12

Solids and Solid-State Materials

Contents

Solutions of quantum dots made from cadmiun selenide absorb ultraviolet light and emit visible light at specific wavelengths. Quantum dots are nanoparticles that could revolutionize applications from television displays to cancer treatment.

What are quantum dots, and what controls their color?

The answer to this question can be found on page 482 in the INQUIRY ?

The materials available for making tools, weapons, skyscrapers, computers, and lasers have had a profound effect on the development of human civilization. Materials play such a large role in human culture that archaeologists organize early human history in terms of materials—the Stone Age, in which only natural materials such as wood and stone were available; the Bronze Age, in which tools, weapons, art, and other objects were made of copper alloyed with tin; and the Iron Age, in which various artifacts were made of iron.

Copper and iron are still of enormous importance today. Copper is used to make electrical wiring, and iron is the main constituent of steel. Metals unknown in ancient times, such as aluminum and titanium, are widely used in the aircraft industry because of their low densities and high resistance to corrosion.

Modern technology is made possible by a host of *solid-state* materials, such as *semiconductors, superconductors, advanced ceramics, composites,* and *nanoparticles.* Semiconductors are used in miniature electronic devices found in computers. Superconductors are used to make the powerful electromagnets found in the magnetic resonance imaging (MRI) instruments employed in medical imaging. Advanced ceramics and composites have numerous engineering, electronic, and biomedical applications. Nanoparticles, whose unique properties depend on particle size, are used in applications ranging from sunscreen to treatments for cancer.

In this chapter, we'll look at the structure of solids and how different types of bonding interactions influence properties. We'll describe metallic bonding and how it is responsible for the properties of metals such as conductivity. We can then apply an understanding of the structure of solids to the bonding, properties, and exciting applications of modern solid-state materials.

12.1 TYPES OF SOLIDS

A brief look around tells you that most substances are solids rather than liquids or gases at room temperature. That brief look also shows that there are many different kinds of solids. Some solids, such as iron and aluminum, are hard and metallic. Others, such as sugar and table salt, are crystalline and easily broken. And still others, such as rubber and many plastics, are soft and amorphous.

The fundamental distinction between kinds of solids is that some are crystalline and others are amorphous. **Crystalline solids** are those whose constituent particles— atoms, ions, or molecules—have an ordered arrangement extending over a long range. This order on the atomic level is also seen on the visible level because crystalline solids usually have flat faces and distinct angles (**FIGURE 12.1a**). **Amorphous solids,** by contrast, are those whose constituent particles are randomly arranged and have no ordered long-range structure (**FIGURE 12.1b**). Rubber is an example.

(a) A crystalline solid, such as this amethyst, has flat faces and distinct angles. These regular macroscopic features reflect a similarly ordered arrangement of particles at the atomic level.

(b) An amorphous solid like rubber has a disordered arrangement of its constituent particles.

▲ FIGURE 12.1

Crystalline and amorphous solids.

451

Crystalline solids can be further categorized as *ionic, molecular, covalent network,* or *metallic*. **Ionic solids** are those like sodium chloride, whose constituent particles are ions. A crystal of sodium chloride is composed of alternating Na^+ and Cl^- ions ordered in a regular three-dimensional arrangement and held together by ionic bonds, as discussed in Section 2.12.

Molecular solids are those like sucrose or ice, whose constituent particles are molecules held together by the intermolecular forces discussed in Section 8.6. A crystal of ice, for instance, is composed of H_2O molecules held together in a regular way by hydrogen bonding (**FIGURE 12.2a**).

Covalent network solids are those like quartz (**FIGURE 12.2b**) or diamond, whose atoms are linked together by covalent bonds into a giant three-dimensional array. **FIGURE 12.2c** shows that individual SiO_4 units share oxygen atoms to make one *very large* molecule.

Metallic solids, such as silver or iron, also consist of large arrays of atoms, but their crystals have metallic properties such as electrical conductivity.

A summary of the different types of crystalline solids and their characteristics is given in **TABLE 12.1**.

(a) Ice consists of individual H_2O molecules held together in a regular manner by hydrogen bonds.

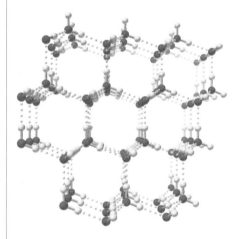

(b) Quartz (SiO_2) is essentially one very large molecule with Si–O covalent bonds. Each silicon atom has tetrahedral geometry and is bonded to four oxygens; each oxygen has approximately linear geometry and is bonded to two silicons.

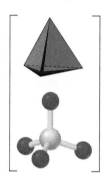

(c) This shorthand representation shows how SiO_4 tetrahedra join at their corners to share oxygen atoms.

▲ **FIGURE 12.2**

Crystal structures of ice, a molecular solid, and quartz, a covalent network solid.

▲ **Figure It Out**

Quartz has a much higher melting point than water. Explain the difference based on the types of forces that hold individual units together.

Answer: In ice, water molecules are held together by hydrogen bonds. In quartz, individual SiO_2 units are held together by covalent bonds. Since covalent bonds are stronger then hydrogen bonds, quartz has a higher melting point.

TABLE 12.1   Types of Crystalline Solids and Their Characteristics

Type of Solid	Intermolecular Forces	Properties	Examples
Ionic	Ion–ion forces	Brittle, hard, high-melting	$NaCl$, KBr, $MgCl_2$
Molecular	Dispersion forces, dipole–dipole forces, hydrogen bonds	Soft, low-melting, nonconducting	H_2O, Br_2, CO_2, CH_4
Covalent network	Covalent bonds	Hard, high-melting	C (diamond), SiO_2
Metallic	Metallic bonds	Variable hardness and melting point, conducting	Na, Zn, Cu, Fe

12.2 PROBING THE STRUCTURE OF SOLIDS: X-RAY CRYSTALLOGRAPHY

How can the structure of a solid be found experimentally? According to a principle of optics, the wavelength of light used to observe an object must be less than twice the length of the object itself. Since atoms have diameters of around 2×10^{-10} m and the visible light detected by our eyes has wavelengths of $4 - 7 \times 10^{-7}$ m, it's impossible to see atoms using even the finest optical microscope. To "see" atoms, we must use "light" with a wavelength of approximately 10^{-10} m, which is in the X-ray region of the **electromagnetic spectrum**.

The origins of X-ray crystallography go back to the work of Max von Laue in 1912. X rays were passed through a crystal of copper sulfate with a photographic plate on the other side. Laue observed a pattern of spots on the plate, indicating the X rays were being diffracted by the atoms in the crystal. **FIGURE 12.3** shows a diffraction pattern from a crystal of sodium chloride.

REMEMBER...
The **electromagnetic spectrum** is made up of all wavelengths of radiant energy, including visible light, infrared radiation, microwaves, radio waves, X rays, and so on (Section 5.1).

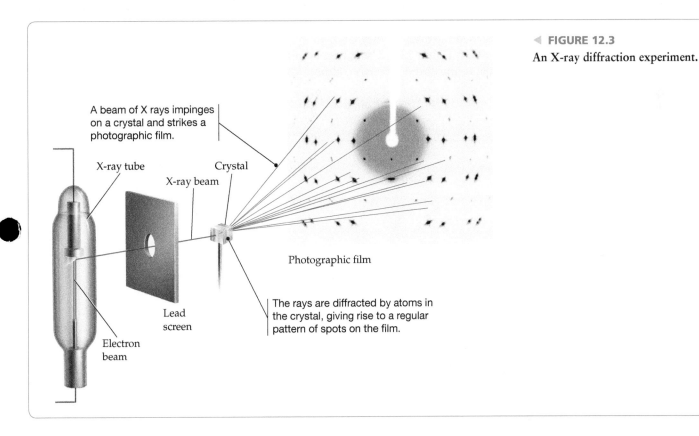

A beam of X rays impinges on a crystal and strikes a photographic film.

X-ray tube

X-ray beam

Crystal

Photographic film

Lead screen

Electron beam

The rays are diffracted by atoms in the crystal, giving rise to a regular pattern of spots on the film.

◀ **FIGURE 12.3**
An X-ray diffraction experiment.

Diffraction of electromagnetic radiation occurs when the paths of light rays are bent by an object containing regularly spaced lines (such as those in a diffraction grating) or points (such as the atoms in a crystal). This scattering happens only if the spacing between the lines or points is comparable to the wavelength of the radiation.

As shown schematically in **FIGURE 12.4**, the diffraction pattern is due to *interference* between two waves passing through the same region of space at the same time. If the

BIG IDEA Question 1

Why are X rays rather than visible light used to determine the position of atoms in a crystal?

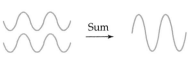

Constructive interference occurs if the waves are in-phase, producing a wave with increased intensity.

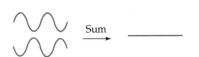

Destructive interference occurs if the waves are out-of-phase, resulting in cancellation.

◀ **FIGURE 12.4**
Interference of electromagnetic waves.

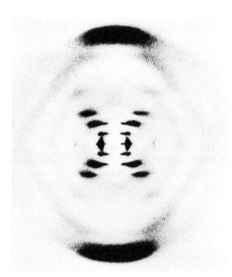

Rosalind Franklin's X-ray diffraction image of DNA taken in 1952 provided crucial data that led to the double helical model of DNA proposed by James Watson and Francis Crick.

waves are in-phase, peak to peak and trough to trough, the interference is constructive and the combined wave is increased in intensity. If the waves are out-of-phase, however, the interference is destructive and the wave is canceled. Constructive interference gives rise to the intense spots observed on Laue's photographic plate, while destructive interference causes the surrounding light areas.

How does the diffraction of X rays by atoms in a crystal give rise to the observed pattern of spots on a photographic plate? According to an explanation advanced in 1913 by the English physicist William H. Bragg (1862–1942) and his 22-year-old son, William L. Bragg (1890–1971), the X rays are diffracted by different layers of atoms in the crystal, leading to constructive interference in some instances but destructive interference in others.

To understand the Bragg analysis, imagine that incoming X rays with wavelength λ strike a crystal face at an angle θ and then bounce off at the same angle, just as light bounces off a mirror (**FIGURE 12.5**). Those rays that strike an atom in the top layer are all reflected at the same angle θ, and those rays that strike an atom in the second layer are also reflected at the angle θ. But because the second layer of atoms is farther from the X-ray source, the distance that the X rays have to travel to reach the second layer is farther than the distance they have to travel to reach the first layer by an amount indicated as BC in Figure 12.5. Using trigonometry, you can show that the extra distance BC is equal to the distance between atomic layers d (= AC) times the sine of the angle θ:

$$\sin \theta = \frac{BC}{d} \quad \text{so} \quad BC = d \sin \theta$$

▶ Go to eText

FIGURE 12.5

Diffraction of X rays of wavelength λ from atoms in the top two layers of a crystal.

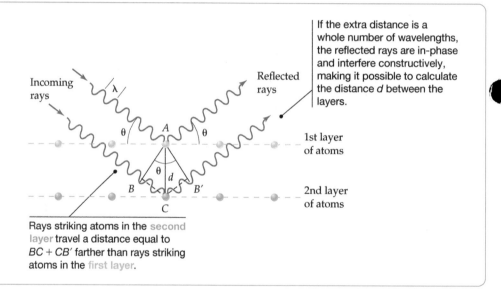

If the extra distance is a whole number of wavelengths, the reflected rays are in-phase and interfere constructively, making it possible to calculate the distance d between the layers.

Incoming rays

Reflected rays

1st layer of atoms

2nd layer of atoms

Rays striking atoms in the second layer travel a distance equal to BC + CB' farther than rays striking atoms in the first layer.

BIG IDEA Question 2

The wavelength of light used in X-ray crystallography is 1.54×10^{-10} m. What is the length of the distance $(BC + B'C)$ in Figure 12.5 that will result in constructive interference?

The extra distance $BC = CB'$ must also be traveled again by the *reflected* rays as they exit the crystal, making the total extra distance traveled equal to $2d \sin \theta$.

$$BC + CB' = 2d \sin \theta$$

The key to the Bragg analysis is the realization that the different rays striking the two layers of atoms are in-phase initially but can be in-phase after reflection only if the extra distance $BC + CB'$ is equal to a whole number of wavelengths $n\lambda$, where n is an integer (1, 2, 3, . . .). If the extra distance is not a whole number of wavelengths, then the reflected rays will be out-of-phase and will cancel. Setting the extra distance $2d \sin \theta = n\lambda$ and rearranging to solve for d gives the **Bragg equation**:

$$BC + CB' = 2d \sin \theta = n\lambda$$

$$\boxed{\text{Bragg equation} \quad d = \frac{n\lambda}{2 \sin \theta}}$$

Of the variables in the Bragg equation, the value of the wavelength λ is known, the value of sin θ can be measured, and the value of n is a small integer, usually 1. Thus, we can calculate the distance d between layers of atoms in a crystal. For their work, the Braggs shared the 1915 Nobel Prize in Physics. The younger Bragg was 25 years old at the time.

Computer-controlled X-ray diffractometers are now available that automatically rotate a crystal and measure the diffraction from all angles. Analysis of the X-ray diffraction pattern then makes it possible to measure the interatomic distance between any two nearby atoms in a crystal. For molecular substances, this knowledge of interatomic distances indicates which atoms are close enough to form a bond. X-ray analysis thus provides a means for determining the three dimensional structures of molecules (**FIGURE 12.6**).

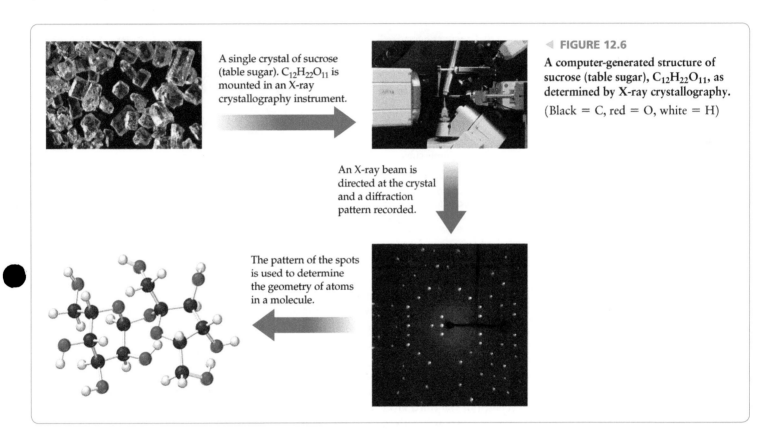

A single crystal of sucrose (table sugar). $C_{12}H_{22}O_{11}$ is mounted in an X-ray crystallography instrument.

An X-ray beam is directed at the crystal and a diffraction pattern recorded.

The pattern of the spots is used to determine the geometry of atoms in a molecule.

◄ **FIGURE 12.6**

A computer-generated structure of sucrose (table sugar), $C_{12}H_{22}O_{11}$, as determined by X-ray crystallography.

(Black = C, red = O, white = H)

12.3 THE PACKING OF SPHERES IN CRYSTALLINE SOLIDS: UNIT CELLS

How do particles—whether atoms, ions, or molecules—pack together in crystals? Let's look at metals, which are the simplest examples of crystal packing because the individual metal atoms are spheres. Not surprisingly, metal atoms (and other kinds of particles as well) generally pack together in crystals so that they can be as close as possible and maximize intermolecular attractions.

If you were to take a large number of uniformly sized marbles and arrange them in a box in some orderly way, you would have four options. You could arrange the marbles in orderly rows and stacks, with the spheres in one layer sitting directly on top of those in the previous layer so that all layers are identical (**FIGURE 12.7a**). In this arrangement, called **simple cubic packing**, each sphere is touched by six neighbors—four in its own layer, one above, and one below—and is thus said to have a **coordination number** of 6. Only 52% of the available volume is occupied by the spheres in simple cubic packing, making inefficient use of space and minimizing attractive forces. Of all the metals in the periodic table, only polonium crystallizes in this way.

Alternatively, you could slightly separate the spheres in a given layer and offset the alternating layers in an *a-b-a-b* arrangement so that the spheres in the *b* layers fit into the depressions between spheres in the *a* layers, and vice versa (**FIGURE 12.7b**). In this arrangement, called **body-centered cubic packing,** each sphere has a coordination number of 8—four neighbors above and four below—and space is used quite efficiently: 68% of the available volume is occupied. Iron, sodium, and 14 other metals crystallize in this way.

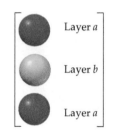

▶ **FIGURE 12.7**

Simple cubic packing and body-centered cubic packing.

▶ **Figure It out**

Which type of packing would result in a more dense substance, simple cubic packing or body-centered cubic packing?

Answer: Density is defined as mass per volume. In body-centered cubic packing the atoms have less space between them, resulting in a greater amount of mass per volume.

(a) Simple Cubic Packing: All layers are identical, and all atoms are lined up in stacks and rows.

(b) Body-Centered Cubic Packing: The spheres in layer *a* are separated slightly and the spheres in layer *b* are offset so that they fit into the depressions between atoms in layer *a*. The third layer is a repeat of the first.

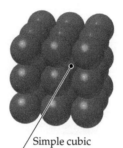

Simple cubic

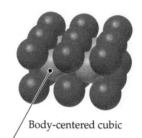

Body-centered cubic

Layer *a*

Layer *b*

Layer *a*

Coordination Number 6: Each sphere is touched by six neighbors, four in the same layer, one directly above, and one directly below.

Coordination Number 8: Each sphere is touched by eight neighbors, four in the layer below, and four in the layer above.

▲ What kind of packing arrangement do these oranges have?

▲ Just as these bricks are stacked together in a regular way, a crystal is made of many small repeating units called unit cells that stack together in a regular way.

The remaining two packing arrangements of spheres are both said to be *closest-packed*. The **hexagonal closest-packing** arrangement (**FIGURE 12.8a**) has two alternating layers, *a-b-a-b*. Each layer has a hexagonal arrangement of touching spheres, which are offset so that spheres in a *b* layer fit into the small triangular depressions between spheres in an *a* layer. Zinc, magnesium, and 19 other metals crystallize in this way.

The **cubic closest-packing** arrangement (**FIGURE 12.8b**) has *three* alternating layers, *a-b-c-a-b-c*. The *a-b* layers are identical to those in the hexagonal closest-packed arrangement, but the third layer is offset from both *a* and *b* layers. Silver, copper, and 16 other metals crystallize with this arrangement.

In both kinds of closest-packed arrangements, each sphere has a coordination number of 12—six neighbors in the same layer, three above, and three below—and 74% of the available volume is filled. The next time you're in a grocery store, look to see how the oranges or apples are stacked in their display box. They'll almost certainly have a closest-packed arrangement.

Having just taken a bulk view of how spheres can pack in a crystal, let's now take a close-up view. Just as a large brick wall is made up of many identical bricks stacked together in a repeating pattern, a crystal is made up of many small repeat units, called **unit cells,** stacked together in three dimensions.

Fourteen different unit-cell geometries occur in crystalline solids. All are parallelepipeds—six-sided geometric solids whose faces are parallelograms. We'll be concerned here only with the three unit cells that have cubic symmetry; that is, cells whose edges are equal in length and whose angles are 90°. The other 11 unit cells have noncubic geometry.

The three kinds of cubic unit cells are *primitive-cubic, body-centered cubic,* and *face-centered cubic.* As shown in **FIGURE 12.9a,** a **primitive-cubic unit cell** for a metal has an atom at each of its eight corners, where it is shared with seven neighboring cubes that come together at the same point. As a result, only 1/8 of each corner atom "belongs" to a given cubic unit. This primitive-cubic unit cell, with all atoms arranged in orderly rows and stacks, is the repeat unit found in simple cubic packing.

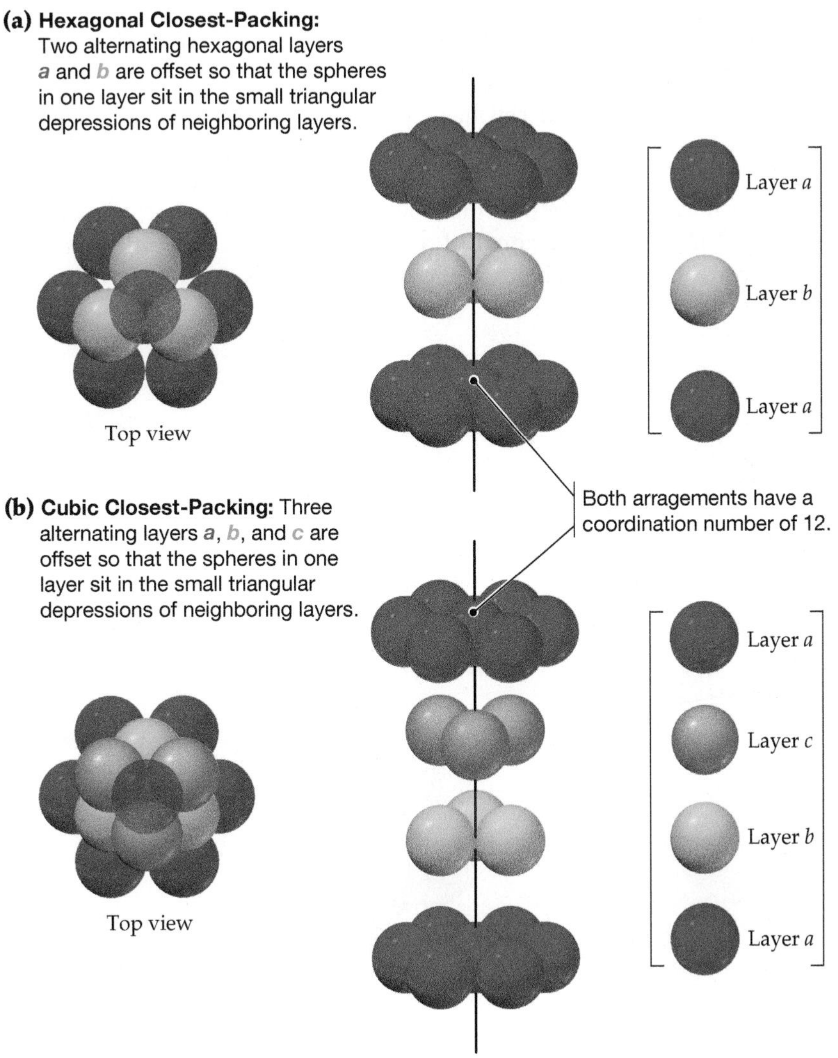
(a) Hexagonal Closest-Packing: Two alternating hexagonal layers *a* and *b* are offset so that the spheres in one layer sit in the small triangular depressions of neighboring layers.
Top view
Layer *a*
Layer *b*
Layer *a*
Both arragements have a coordination number of 12.
(b) Cubic Closest-Packing: Three alternating layers *a*, *b*, and *c* are offset so that the spheres in one layer sit in the small triangular depressions of neighboring layers.
Top view
Layer *a*
Layer *c*
Layer *b*
Layer *a*

FIGURE 12.8

Hexagonal closest-packing and cubic closest-packing. In both kinds of packing, each sphere is touched by 12 neighbors, 6 in the same layer, 3 in the layer above, and 3 in the layer below.

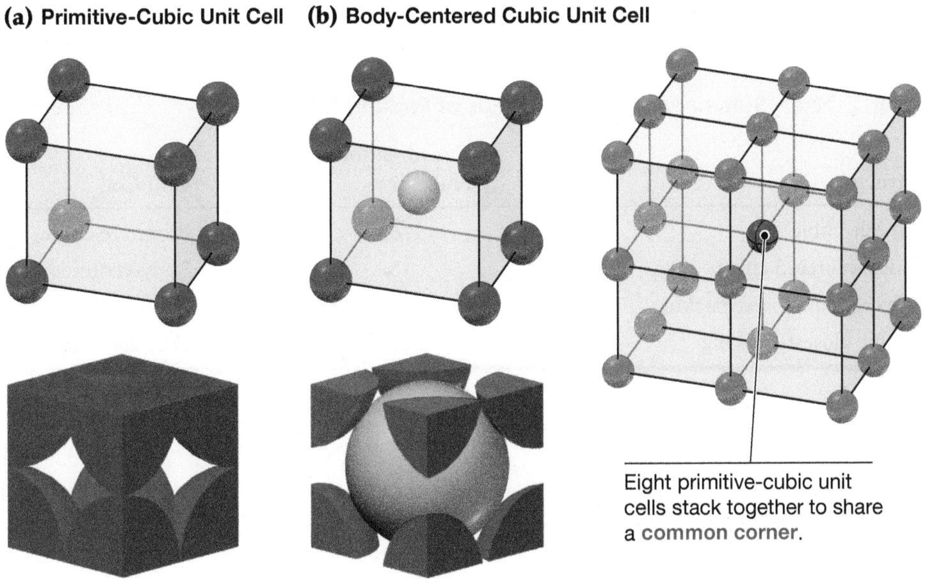
(a) Primitive-Cubic Unit Cell
(b) Body-Centered Cubic Unit Cell
Eight primitive-cubic unit cells stack together to share a common corner.

FIGURE 12.9

Geometries of (a) primitive-cubic and (b) body-centered cubic unit cells. Both skeletal (top) and space-filling views (bottom) are shown.

Figure It out

How many atoms are present in a primitive-cubic and body-centered cubic unit cell?

Answer: A primitive-cubic unit cell has an atom at each of its 8 corners, and each corner atom is shared by 8 cubes, so that only 1/8 of each atom "belongs" to a given unit cell. A primitive cubic unit cell has $1/8 \times 8$(corner atoms) = 1 atom. A body-centered unit cell has one atom inside and 8 corner atoms; [1(center atom)] + [$1/8 \times 8$ (corner atoms)] = 2 atoms.

Go to eText

BIG IDEA Question 3

Visualize a cubic unit cell. What fraction of an atom on the corner, an atom on an edge, and an atom on the face are in the unit cell?

A **body-centered cubic unit cell** has eight corner atoms plus an additional atom in the center of the cube (**FIGURE 12.9b**). This body-centered cubic unit cell, with two repeating offset layers and with the spheres in a given layer slightly separated, is the repeat unit found in body-centered cubic packing.

A **face-centered cubic unit cell** has eight corner atoms plus an additional atom in the center of each of its six faces that is shared with one other neighboring cube (**FIGURE 12.10a**). Thus, 1/2 of each face atom belongs to a given unit cell. This face-centered cubic unit cell is the repeat unit found in cubic closest-packing, as can be seen by looking down the body diagonal of a unit cell (**FIGURE 12.10b**). The faces of the unit-cell cube are at 54.790° angles to the layers of the atoms.

▶ **FIGURE 12.10**

Geometry of a face-centered cubic unit cell.

▶ **Figure It out**

How many atoms are present in a face-centered cubic unit cell?

Answer: A face-centered unit cell has 6 atoms on the faces that contribute 1/2 of an atom and 8 corner atoms that contribute 1/8 of an atom;
[1/2 × 6 (face atoms)] + [(1/8 × 8 (corner atoms)] = 4 atoms.

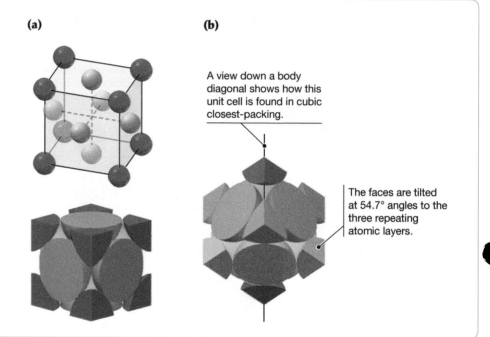

(a)   **(b)**

A view down a body diagonal shows how this unit cell is found in cubic closest-packing.

The faces are tilted at 54.7° angles to the three repeating atomic layers.

TABLE 12.2 summarizes the stacking patterns, coordination numbers, amount of space used, and unit cells for the four kinds of packing of spheres. Hexagonal closest-packing is the only one of the four that has a noncubic unit cell.

TABLE 12.2   Summary of the Four Kinds of Packing for Spheres

Structure	Stacking Pattern	Coordination Number	Space Used (%)	Unit Cell
Simple cubic	a-a-a-a-	6	52	Primitive-cubic
Body-centered cubic	a-b-a-b-	8	68	Body-centered cubic
Hexagonal closest-packing	a-b-a-b-	12	74	(Noncubic)
Cubic closest-packing	a-b-c-a-b-c-	12	74	Face-centered cubic

WORKED EXAMPLE 12.1

Using Unit-Cell Dimensions to Calculate the Radius of an Atom

Silver metal crystallizes in a cubic closest-packed arrangement with the edge of the unit cell having a length $d = 407$ pm. What is the radius in picometers of a silver atom?

STRATEGY AND SOLUTION

Cubic closest-packing uses a face-centered cubic unit cell. Looking at any one face of the cube head-on shows that the face atoms touch the corner atoms along the diagonal of the face but that corner atoms do not touch one another along the edges. Each diagonal is therefore equal to four atomic radii, $4r$:

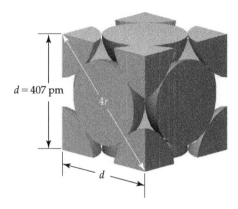

$d = 407$ pm

$4r$

d

Because the diagonal and two edges of the cube form a right triangle, we can use the Pythagorean theorem to set the sum of the squares of the two edges equal to the square of the diagonal, $d^2 + d^2 = (4r)^2$ and then solve for r, the radius of one atom:

$$d^2 + d^2 = (4r)^2$$

$$2d^2 = 16r^2 \quad \text{and} \quad r^2 = \frac{d^2}{8}$$

$$\text{thus} \quad r = \sqrt{\frac{d^2}{8}} = \sqrt{\frac{(407 \text{ pm})^2}{8}} = 144 \text{ pm}$$

The radius of a silver atom is 144 pm.

▶ **PRACTICE 12.1**   Calcium metal crystallizes in a cubic closest-packed arrangement. Calcium has an atomic radius of 197 pm. What is the edge length of a unit cell in pm?

▶ **APPLY 12.2**   Polonium metal crystallizes in a simple cubic arrangement, with the edge of a unit cell having a length $d = 334$ pm. What is the radius in picometers of a polonium atom?

All **PRACTICE** and **APPLY** problems are interactive in the eText.

 Go to eText

WORKED EXAMPLE 12.2

Using Unit-Cell Dimensions to Calculate the Density of a Metal

Nickel has a face-centered cubic unit cell with a length of 352.4 pm along an edge. What is the density of nickel in g/cm^3?

STRATEGY

Density is mass divided by volume. The mass of a single unit cell can be calculated by counting the number of atoms in the cell and multiplying by the mass of a single atom. The volume of a single cubic unit cell with edge d is $d^3 = (3.524 \times 10^{-8} \text{ cm})^3 = 4.376 \times 10^{-23} \text{ cm}^3$.

SOLUTION

Each of the eight corner atoms in a face-centered cubic unit cell is shared by eight unit cells, so that only $1/8 \times 8 = 1$ atom belongs to a single cell. In addition, each of the six face atoms is shared by two unit cells, so that $1/2 \times 6 = 3$ atoms belong to a single cell. Thus, a single cell has 1 corner atom and 3 face atoms, for a total of 4, and each atom has a mass equal to the molar mass of nickel (58.69 g/mol) divided by Avogadro's number (6.022×10^{23} atoms/mol). We can now calculate the density:

$$\text{Density} = \frac{\text{Mass}}{\text{Volume}}$$

$$= \frac{(4 \text{ atoms}) \left(\dfrac{58.69 \frac{\text{g}}{\text{mol}}}{6.022 \times 10^{23} \frac{\text{atoms}}{\text{mol}}} \right)}{4.376 \times 10^{-23} \text{ cm}^3}$$

$$= 8.909 \text{ g/cm}^3$$

The calculated density of nickel is 8.909 g/cm^3. (The measured value is 8.90 g/cm^3.)

▶ **PRACTICE 12.3**   Polonium metal crystallizes in a simple cubic arrangement, with the edge of a unit cell having a length $d = 334$ pm. What is the density of polonium?

▶ **APPLY 12.4**   The density of a sample of metal was measured to be 22.67 g/cm^3. An X-ray diffraction experiment measures the edge of a face-centered cubic cell as 383.3 pm. Calculate the atomic mass of the metal and identify it.

All **WORKED EXAMPLES** with this icon Go to eText have an interactive video in the eText.

12.4 STRUCTURES OF SOME IONIC SOLIDS

Simple ionic solids such as NaCl and KBr are like metals in that the individual ions are spheres that pack together in a regular way. They differ from metals, however, in that the spheres are not all the same size—anions generally have larger ionic radii than cations. As a result, ionic solids adopt a variety of different unit cells, depending

REMEMBER...

Atomic radius decreases when an atom is converted to a cation by loss of an electron and increases when the atom is converted to an anion by gain of an electron (Section 6.2).

on the size and charge of the ions. NaCl, KCl, and a number of other salts have a face-centered cubic unit cell in which the larger Cl^- anions occupy corners and faces while the smaller Na^+ cations fit into the holes between adjacent anions (**FIGURE 12.11**).

It's necessary, of course, that the unit cell of an ionic substance be electrically neutral, with equal numbers of positive and negative charges. In the NaCl unit cell, for instance, there are four Cl^- anions ($1/8 \times 8 = 1$ corner atom, plus $1/2 \times 6 = 3$ face atoms) and also four Na^+ cations ($1/4 \times 12 = 3$ edge atoms, plus 1 center atom). (Remember that each corner atom in a cubic unit cell is shared by eight cells, each face atom is shared by two cells, and each edge atom is shared by four cells.)

Two other common ionic unit cells are shown in **FIGURE 12.12**. Copper(I) chloride has a face-centered cubic arrangement of the larger Cl^- anions, with the smaller

▶ **FIGURE 12.11**

The unit cell of NaCl. Both a skeletal view (**a**) and a space-filling view (**b**) in which the unit cell is viewed edge-on are shown.

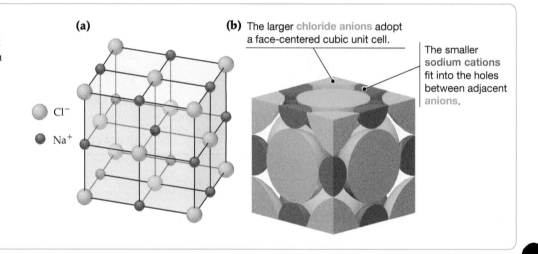

(a)

(b) The larger chloride anions adopt a face-centered cubic unit cell.

The smaller sodium cations fit into the holes between adjacent anions.

Cl^-

Na^+

▶ **FIGURE 12.12**

Unit cells of (a) CuCl and (b) $BaCl_2$.

▶ **Figure It out**

How many Cu and Cl ions are in the CuCl unit cell? What is the overall charge of the CuCl unit cell? How many Ba and Cl ions are in the $BaCl_2$ unit cell? What is the overall charge of the $BaCl_2$ unit cell?

Answer: CuCl unit cell: 4 Cu^+ ions and 4 Cl^- ions, overall charge is 0. $BaCl_2$ unit cell: 4 Ba^{2+} ions and 8 Cl^- ions, overall charge is 0.

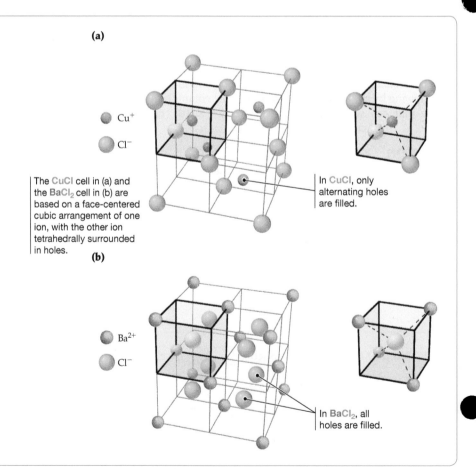

(a)

Cu^+

Cl^-

The CuCl cell in (a) and the $BaCl_2$ cell in (b) are based on a face-centered cubic arrangement of one ion, with the other ion tetrahedrally surrounded in holes.

In CuCl, only alternating holes are filled.

(b)

Ba^{2+}

Cl^-

In $BaCl_2$, all holes are filled.

Cu^+ cations in holes so that each is surrounded by a tetrahedron of four anions. Barium chloride, by contrast, has a face-centered cubic arrangement of the smaller Ba^{2+} *cations*, with the larger Cl^- anions surrounded tetrahedrally. As required for charge neutrality, there are twice as many Cl^- anions as Ba^{2+} cations.

CONCEPTUAL WORKED EXAMPLE 12.3

Using the Unit Cell to Determine Chemical Formula and Geometry

Rhenium oxide crystallizes in the following cubic unit cell:

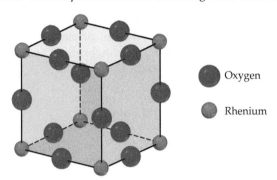

● Oxygen

● Rhenium

(a) How many rhenium ions and how many oxygen ions are in each unit cell?
(b) What is the formula of rhenium oxide?
(c) What is the oxidation state of rhenium?
(d) What is the geometry around each oxygen atom?
(e) What is the geometry around each rhenium atom?

STRATEGY AND SOLUTION

(a-b) Each corner atom in a cubic unit cell is shared by eight cells, each face atom is shared by two cells, and each edge atom is shared by four cells. In the unit cell there are rhenium atoms at each of the eight corners; $(8 \times 1/8) = 1$ rhenium atom. There are 12 oxygen atoms on the edges; $(12 \times 1/4) = 3$ oxygen atoms. Therefore, the formula is ReO_3.

(c) The oxidation state of oxygen is -2 and there are three O atoms for every one Re atom, therefore the oxidation state of Re must be $+6$ to make the unit cell of the ionic compound electrically neutral.

(d) Each oxygen forms two bonds. The diagram of the unit cell shows that the geometry is linear.

(e) If unit cells are linked together, Re forms 6 bonds with oxygen in an octahedral geometry.

▶ **CONCEPTUAL PRACTICE 12.5** Zinc sulfide crystallizes in the following cubic unit cell:

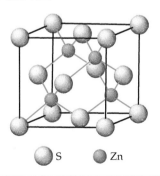

○ S   ● Zn

(a) How many sulfur ions and how many zinc ions are in each unit cell?
(b) What is the formula of zinc sulfide?
(c) What is the oxidation state of zinc?
(d) What is the geometry around each zinc atom?

▶ **CONCEPTUAL APPLY 12.6** Perovskite, a mineral containing calcium, oxygen, and titanium, crystallizes in the following cubic unit cell:

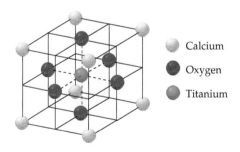

○ Calcium
● Oxygen
● Titanium

(a) What is the formula of perovskite?
(b) What is the oxidation number of the titanium ion in perovskite?
(c) What is the geometry around each titanium, oxygen, and calcium ion?

12.5 STRUCTURES OF SOME COVALENT NETWORK SOLIDS

The atoms in covalent network solids are held together by covalent bonds in a giant three-dimensional array. Covalent network solids are hard and have high melting points due to the strong covalent bonds between all atoms. Carbon and silicon form several different covalent network solids, including graphite, diamond, and glass.

Carbon

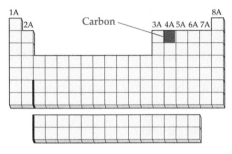

Carbon exists in more than 40 known structural forms, or **allotropes,** several of which are crystalline but most of which are amorphous. Graphite, the most common allotrope of carbon and the most stable under normal conditions, is a crystalline covalent network solid that consists of two-dimensional sheets of fused six-membered rings (**FIGURE 12.13a**). Each carbon atom is sp^2-hybridized and is bonded with trigonal planar geometry to three other carbons. The diamond form of elemental carbon is a covalent network solid in which each carbon atom is sp^3-hybridized and is bonded with tetrahedral geometry to four other carbons (**FIGURE 12.13b**).

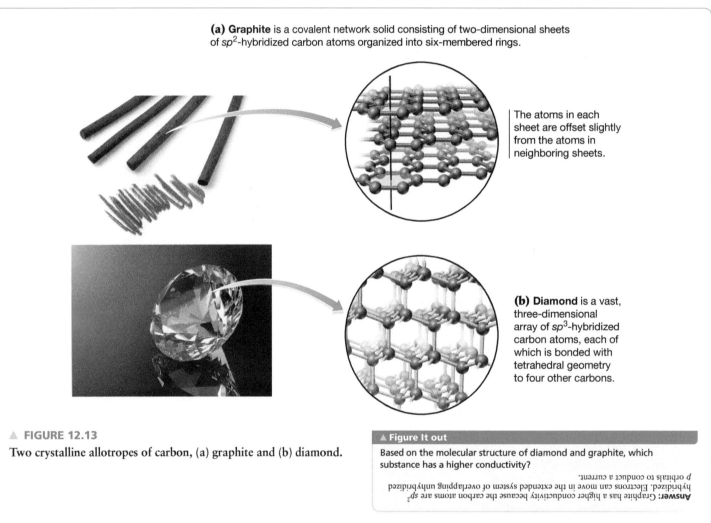

(a) Graphite is a covalent network solid consisting of two-dimensional sheets of sp^2-hybridized carbon atoms organized into six-membered rings.

The atoms in each sheet are offset slightly from the atoms in neighboring sheets.

(b) Diamond is a vast, three-dimensional array of sp^3-hybridized carbon atoms, each of which is bonded with tetrahedral geometry to four other carbons.

▲ **FIGURE 12.13**

Two crystalline allotropes of carbon, (a) graphite and (b) diamond.

▲ **Figure It out**

Based on the molecular structure of diamond and graphite, which substance has a higher conductivity?

Answer: Graphite has a higher conductivity because the carbon atoms are sp^2 hybridized. Electrons can move in the extended system of overlapping unhybridized p orbitals to conduct a current.

In addition to graphite and diamond, a third crystalline allotrope of carbon called *fullerene* was discovered in 1985 as a constituent of soot. Fullerene consists of spherical C_{60} molecules with the extraordinary shape of a soccer ball. The C_{60} ball has 12 pentagonal and 20 hexagonal faces, with each atom sp^2-hybridized and bonded to three other atoms (**FIGURE 12.14a**). Closely related to both graphite and fullerene are a group of carbon allotropes called *nanotubes*—tubular structures made of repeating six-membered carbon rings, as if a sheet of graphite were rolled up (**FIGURE 12.14b**). Typically, the tubes have a diameter of about 2–30 nm and a length of up to 1 mm.

12 pentagonal faces   20 hexagonal faces

(a) Fullerene is a molecular solid whose molecules have the shape of a soccer ball. The ball has 12 pentagonal and 20 hexagonal faces, and each carbon atom is sp^2-hybridized.

(b) Carbon nanotubes consist of sheets of graphite rolled into tubes of 2–30 nm diameter.

The different structures of the carbon allotropes lead to widely different properties. Diamond is the hardest known substance because of its three-dimensional network of strong single bonds that tie all atoms in a crystal together. In addition to its use in jewelry, diamond is widely used industrially for the tips of saw blades and drilling bits. It is an electrical insulator and has a melting point of about 8700 °C at a pressure of 6–10 million atm. Clear, colorless, and highly crystalline, diamonds are very rare and are found in only a few places in the world, particularly in central and southern Africa.

Graphite is the black, slippery substance used as the "lead" in pencils, as an electrode material in batteries, and as a lubricant in locks. All these properties result from its sheetlike structure. Air and water molecules can adsorb onto the flat faces of the sheets, allowing the sheets to slide over one another and giving graphite its greasy feeling and lubricating properties. Graphite is more stable than diamond at normal pressures but can be converted into diamond at very high pressure and temperature. In fact, approximately 120,000 kg of industrial diamonds are synthesized annually by applying 150,000 atm pressure to graphite at high temperature.

Fullerene, black and shiny like graphite, is the subject of much current research because of its interesting electronic properties. When fullerene is allowed to react with rubidium metal, a superconducting material called rubidium fulleride, Rb_3C_{60}, forms. Carbon nanotubes are being studied for use as fibers in the structural composites used to make golf clubs, bicycle frames, boats, and airplanes. Their tensile strength is approximately 50–60 times greater than that of steel. We'll look at the chemistry of carbon and some of its compounds in more detail in Section 22.6.

Silica

Just as living organisms are based on carbon compounds, most rocks and minerals are based on silicon compounds. Quartz and much sand, for instance, are nearly pure *silica,* SiO_2. Silicon and oxygen together, in fact, make up nearly 75% of the mass of the Earth's crust. Considering that silicon and carbon are both in group 4A of the periodic table, you might expect SiO_2 to be similar in its properties to CO_2. In fact, though, CO_2 is a molecular substance and a gas at room temperature, whereas SiO_2 (Figure 12.2b) is a covalent network solid with a melting point over 8700 °C.

The dramatic difference in properties between CO_2 and SiO_2 is due primarily to the difference in electronic structure between carbon and silicon. The π part of a *carbon–oxygen* **double bond** is formed by sideways overlap of a carbon 2p orbital with an oxygen 2p orbital (Section 8.4). If a similar *silicon*–oxygen double bond were to form, it would require overlap of an oxygen 2p orbital and a silicon 3p orbital. But because

LOOKING AHEAD...
Section 22.6 describes the structure and properties of elemental carbon and some of its compounds.

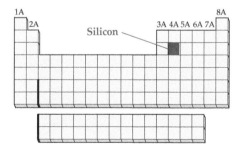

REMEMBER...
Two atoms form a **double bond** when they approach each other with their hybrid orbitals aligned head-on for σ bonding and with their unhybridized p orbitals aligned in a parallel, sideways manner to form a π bond (Section 8.4).

▲ Colored glasses contain transition metal ions.

▲ Because of its ductility, aluminum can be drawn into the wires used in electric power lines.

the Si—O bond distance is longer than the C—O distance and a 3*p* orbital is larger than a 2*p* orbital, overlap between the silicon and oxygen *p* orbitals is not as favorable. As a result, silicon forms four single bonds to four oxygens in a covalent network structure rather than two double bonds to two oxygens in a molecular structure.

Heating silica above about 8700 °C breaks many of its Si—O bonds and turns it from a crystalline solid into a viscous liquid. When this fluid is cooled, some of the Si—O bonds re-form in a random arrangement and a noncrystalline, amorphous solid called *quartz glass* is formed. If additives are mixed in before cooling, a wide variety of glasses can be prepared. Common window glass, for instance, is prepared by adding $CaCO_3$ and Na_2CO_3. Addition of various transition metal ions results in the preparation of colored glasses, and addition of B_2O_3 produces a high-melting *borosilicate glass* that is sold under the trade name Pyrex. Borosilicate glass is particularly useful for cooking utensils and laboratory glassware because it expands very little when heated and is thus unlikely to crack. We'll look further at the chemistry of silicon and silicon-containing minerals in Section 22.6.

12.6 BONDING IN METALS

Thus far, we've discussed different types of solids and examined their structures. Now we'll explain properties of metals using models of bonding. Some properties, such as hardness and melting point, vary considerably among metals, but other properties are characteristic of metals in general. For instance, all metals can be drawn into wires (ductility) or beaten into sheets (malleability) without breaking into pieces like glass or an ionic crystal. Furthermore, all metals have a high thermal and electrical conductivity. When you touch a metal, it feels cold because the metal efficiently conducts heat away from your hand, and when you connect a metal wire to the terminals of a battery, it conducts an electric current.

To understand those properties, we need to look at the bonding in metals. We'll consider two theoretical models that are commonly used: the *electron-sea model* and the *molecular orbital theory*.

Electron-Sea Model of Metals

If you try to draw an electron-dot structure for a metal, you'll quickly realize that there aren't enough valence electrons available to form an electron-pair bond between every pair of adjacent atoms. Sodium, for example, which has just one valence electron per atom ($3s^1$), crystallizes in a body-centered cubic structure in which each Na atom is surrounded by eight nearest neighbors. Consequently, the valence electrons can't be localized in a bond between any particular pair of atoms. Instead, they are delocalized and belong to the crystal as a whole.

In the **electron-sea model,** a metal crystal is viewed as a three-dimensional array of metal cations immersed in a sea of delocalized electrons that are free to move throughout the crystal (**FIGURE 12.15**). The continuum of delocalized, mobile valence electrons acts as an electrostatic glue that holds the metal cations together.

The electron-sea model affords a simple qualitative explanation for the electrical and thermal conductivity of metals. Because the electrons are mobile, they are free to move away from a negative electrode and toward a positive electrode when a metal is

▶ **FIGURE 12.15**

A two-dimensional representation of the electron-sea model of a metal. An ordered array of cations is immersed in a continuous distribution of delocalized, mobile valence electrons.

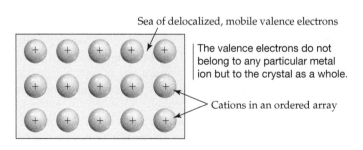

Sea of delocalized, mobile valence electrons

The valence electrons do not belong to any particular metal ion but to the crystal as a whole.

Cations in an ordered array

subjected to an electrical potential. The mobile electrons can conduct heat by carrying kinetic energy from one part of the crystal to another. Metals are malleable and ductile because the delocalized bonding extends in all directions; that is, it is not confined to oriented bond directions, as in covalent network solids like diamond or SiO_2. When a metallic crystal is deformed, no localized bonds are broken. Instead, the electron sea simply adjusts to the new distribution of cations, and the energy of the deformed structure is similar to that of the original. Thus, the energy required to deform a metal like sodium is relatively small. The energy required to deform a transition metal like iron is greater because iron has more valence electrons ($4s^2\,3d^6$), and the electrostatic "glue" is denser.

Molecular Orbital Theory for Metals

A more detailed understanding of the bonding in metals is provided by the molecular orbital theory, a model that is a logical extension of the **molecular orbital (MO)** description of small molecules discussed in Sections 8.7–8.9. Recall that in the H_2 molecule the 1s orbitals of the two H atoms overlap to give a σ bonding MO and a higher-energy σ^* antibonding MO. The bonding in the gaseous Na_2 molecule is similar: The 3s orbitals of the two Na atoms combine to give a σ and a σ^* MO. Because each Na atom has just one 3s valence electron, the lower-energy bonding orbital is filled and the higher-energy antibonding orbital is empty:

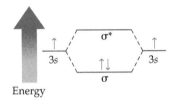

Now consider what happens if we bring together an increasingly larger number of Na atoms to build up a crystal of sodium metal. The key idea to remember from Section 8.7 is *that the number of molecular orbitals formed is the same as the number of atomic orbitals combined.* Thus, there will be three MOs for a triatomic Na_3 molecule, four MOs for Na_4, and so on as shown in **FIGURE 12.16**. A cubic crystal of sodium metal, 1.5 mm on an edge, contains about 10^{20} Na atoms and therefore has

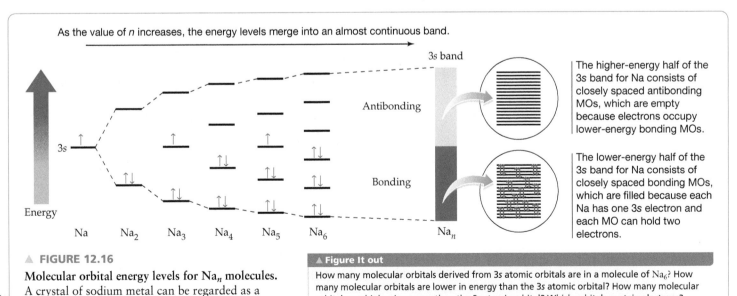

▲ **FIGURE 12.16**

Molecular orbital energy levels for Na$_n$ molecules. A crystal of sodium metal can be regarded as a giant Na$_n$. molecule, where n has a value of about 10^{20}. In this and subsequent figures, the red color denotes the filled portion of a band.

▲ **Figure It out**

How many molecular orbitals derived from 3s atomic orbitals are in a molecule of Na$_6$? How many molecular orbitals are lower in energy than the 3s atomic orbital? How many molecular orbitals are higher in energy than the 3s atomic orbital? Which orbitals contain electrons?

Answer: There are six total molecular orbitals with three lower in energy than the 3s atomic orbital and three higher in energy than the 3s atomic orbital. The lower energy orbitals contain electrons.

about 10^{20} MOs, each of which is delocalized over all the atoms in the crystal. Notice that the difference in energy between successive MOs in an Na_n molecule decreases as the number of Na atoms increases. In Figure 12.16, the energy gap between MOs is less in Na_3 than Na_2 and the trend continues as more Na atoms are added to the crystal. Eventually, the energy difference between MOs becomes so small that they merge into an almost continuous band of energy levels for large values of n. A **band** is a set of MOs that are very closely spaced in energy and consequently MO theory for metals is often called **band theory**. The bottom half of the 3s band for sodium metal in Figure 12.16 consists of bonding MOs that are lower in energy than the antibonding MOs in the top half of the band. The bonding MOs are filled because each Na atom has one 3s electron and each MO can hold two electrons.

How does band theory account for the electrical conductivity of metals? Because each of the MOs in a metal has a definite energy (Figure 12.16), each electron in a metal has a specific kinetic energy and a specific velocity. These values depend on the particular MO energy level and increase from the bottom of a band to the top. For a one-dimensional metal wire, electrons traveling in opposite directions at the same speed have the same kinetic energy. Thus, the energy levels within a band occur in degenerate pairs; one set of energy levels applies to electrons moving to the right, and the other set applies to electrons moving to the left.

In the absence of an electrical potential, the two sets of levels are equally populated. That is, for each electron moving to the right, another electron moves to the left with exactly the same speed (**FIGURE 12.17**). As a result, there is no net electric current in either direction. In the presence of an electrical potential, however, those electrons moving to the right (toward the positive terminal of a battery) are accelerated, those moving to the left (toward the negative terminal) are slowed down, and those moving to the left with very slow speeds undergo a change of direction. Thus, the number of electrons moving to the right is now greater than the number moving to the left and there is a net electric current.

> **REMEMBER...**
> Degenerate energy levels have the same energy (Section 5.11).

▶ **FIGURE 12.17**

Half-filled 3s band of MO energy levels for a one-dimensional sodium metal wire. The direction of electron motion for the two degenerate sets of energy levels is indicated by the horizontal arrows.

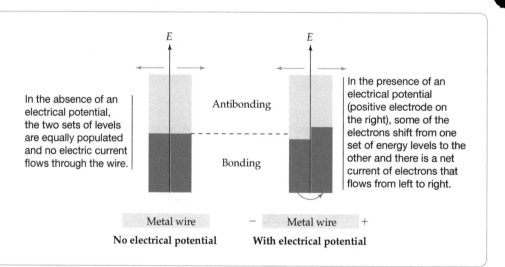

In the absence of an electrical potential, the two sets of levels are equally populated and no electric current flows through the wire.

In the presence of an electrical potential (positive electrode on the right), some of the electrons shift from one set of energy levels to the other and there is a net current of electrons that flows from left to right.

Metal wire

No electrical potential

Metal wire   − +

With electrical potential

Figure 12.17 shows that an electrical potential can shift electrons from one set of energy levels to the other only if the band is partially filled. If the band is completely filled, there are no available vacant energy levels to which electrons can be excited, and therefore the two sets of levels must remain equally populated, even in the presence of an electrical potential. This means that an electrical potential can't accelerate the electrons in a completely filled band, a result that applies to a three-dimensional crystal as well as to a one-dimensional wire. *Materials that have only completely filled bands are therefore electrical insulators. By contrast, materials that have partially filled bands are metallic conductors.*

Based on the preceding analysis, we would predict that magnesium should be an insulator because it has the electron configuration [Ar] $3s^2$ and should therefore have a completely filled 3s band. This prediction is wrong, however, because we have not yet considered the 3p valence orbitals. Just as the 3s orbitals combine to form a 3s band,

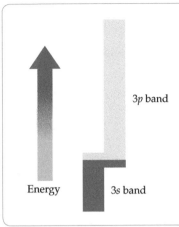

◀ **FIGURE 12.18**

The composite 3s–3p band in magnesium metal. The 3s and 3p bands have similar energies and overlap to give a composite band consisting of four MOs per Mg atom. The composite band can accommodate eight electrons per Mg atom but is only partially filled, since each Mg atom has just two valence electrons. (In this and subsequent figures, the separate sets of energy levels for the right- and left-moving electrons aren't shown.)

3p band

3s band

Energy

�── **Figure It out**

How would the level of filling in the 3s–3p band change if the metal were aluminum rather than magnesium?

Answer: The filled portion of the band (shown in red) would extend partially into the 3p band because each Al has two 3s electrons and one 3p electron.

so the 3p orbitals can combine to form a 3p band. If the 3s and 3p bands were widely separated in energy, the 3s band would be filled, the 3p band would be empty, and magnesium would be an insulator. In fact, though, the 3s and 3p bands overlap in energy, and the resulting composite band is only partially filled (**FIGURE 12.18**). Thus, magnesium and other alkaline earth elements are metallic conductors.

Transition metals have a d band that can overlap the s band to give a composite band consisting of six MOs per metal atom. Half of these MOs are bonding and half are antibonding. We might therefore expect maximum bonding for metals that have six valence electrons per metal atom because six electrons will just fill the bonding MOs and leave the antibonding MOs empty. Filled or nearly filled bonding orbitals for transition elements lead to high melting points within a series. In the first transition metal series, vanadium (5 valence electrons) has the highest melting point (1910 °C). In the second and third transition metal series, molybdenum and tungsten (both with 6 valence electrons) have the highest melting points. Molybdenum melts at 2623 °C, and tungsten melts at 3422 °C. In contrast, mercury has a low melting point (−39 °C) and is a liquid at room temperature. Mercury has 12 valence electrons and filled nonbonding orbitals, thus lowering the bonding interactions.

Go to
eText

— **WORKED EXAMPLE 12.4**

Using Band Theory to Account for Melting Points

The melting points of chromium and zinc are 1907 °C and 420 °C, respectively. Use band theory to account for the difference.

STRATEGY

The melting points will depend on the occupancy of the bonding and antibonding MOs. The greater the excess of bonding electrons relative to antibonding electrons, the stronger the bonding and the higher the melting point. Determine the electron configurations for each metal and the relative numbers of bonding and antibonding electrons.

SOLUTION

The electron configurations are [Ar] $3d^5 4s^1$ for Cr and [Ar] $3d^{10} 4s^2$ for Zn. Assume that the 3d and 4s bands overlap. The composite band, which can accommodate 12 valence electrons per metal atom, will be half-filled for Cr and completely filled for Zn. Strong bonding and a consequent high melting point are expected for Cr because all the bonding MOs are occupied and all the antibonding MOs are empty. Weak bonding and a low melting point are expected for Zn

because both bonding and antibonding MOs are occupied. (The fact that Zn has high electrical conductivity suggests that the 4p orbitals also contribute to the composite band.)

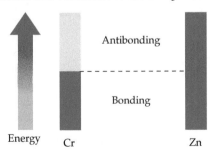

Antibonding

Bonding

Energy   Cr   Zn

▶ **PRACTICE 12.7** Mercury metal is a liquid at room temperature. Use band theory to suggest a reason for its low melting point (−39 °C).

continued on next page

▶ **CONCEPTUAL APPLY 12.8**   The following pictures represent the electron population of the composite *s–d* band for three metals—Hf, Pt, and Re:

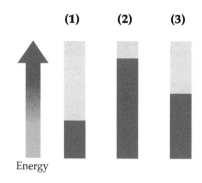

(a) Which picture corresponds to which metal? Explain
(b) Which metal has the highest melting point, and which is the hardest? Explain.
(c) Which metal has the lowest melting point, and which is the softest? Explain.

12.7 SEMICONDUCTORS

A **semiconductor,** such as silicon or germanium, is a material that has an electrical conductivity intermediate between that of a metal and that of an insulator. To understand the electrical properties of semiconductors, let's look first at the bonding in insulators. Take diamond, for example, a covalent network solid in which each C atom is bonded tetrahedrally to four other C atoms (Figure 12.13). In a localized description of the bonding, C—C electron-pair bonds result from the overlap of sp^3 hybrid orbitals. In a delocalized description, the 2s and 2p valence orbitals of all the C atoms combine to give bands of bonding and antibonding MOs—a total of four MOs per C atom. As is generally the case for insulators, the lower-energy bonding MOs, called the **valence band,** and the higher-energy antibonding MOs, called the **conduction band,** are separated in energy by a large **band gap.** The band gap in diamond is about 520 kJ/mol.

Each of the two bands in diamond can accommodate four electrons per C atom. Because carbon has just four valence electrons $(2s^2\,2p^2)$, the valence band is completely filled and the conduction band is completely empty. Diamond is therefore an electrical insulator because there are no vacant MOs in the valence band to which electrons can be excited by an electrical potential and because population of the vacant MOs of the conduction band is prevented by the large band gap. By contrast, metallic conductors have no energy gap between the highest occupied and lowest unoccupied MOs. The MOs of a semiconductor are similar to those of an insulator, but the band gap in a semiconductor is smaller. As a result, a few electrons have enough thermal energy to jump the gap and occupy the higher-energy conduction band. The conduction band is partially filled, and the valence band is partially empty because now it contains a few unoccupied MOs. When an electrical potential is applied to a semiconductor, it conducts a small amount of current because the potential can accelerate the electrons in the partially filled bands. **FIGURE 12.19** shows the difference in energy levels in conductors, insulators, and semiconductors. **TABLE 12.3** shows how the electrical properties of the group 4A elements vary with the size of the band gap.

The electrical conductivity of a semiconductor increases with increasing temperature because the number of electrons with sufficient thermal energy to occupy the conduction band increases as the temperature rises. At higher temperatures, there are more charge carriers (electrons) in the conduction band and more vacancies in the valence band. By contrast, the electrical conductivity of a metal decreases with increasing temperature. At higher temperatures, the metal cations undergo increased vibrational motion about their lattice sites and vibration of the cations disrupts the flow of electrons through the crystal. Thus, the temperature dependence of the electrical conductivity is the best criterion for distinguishing a metal from a semiconductor: As

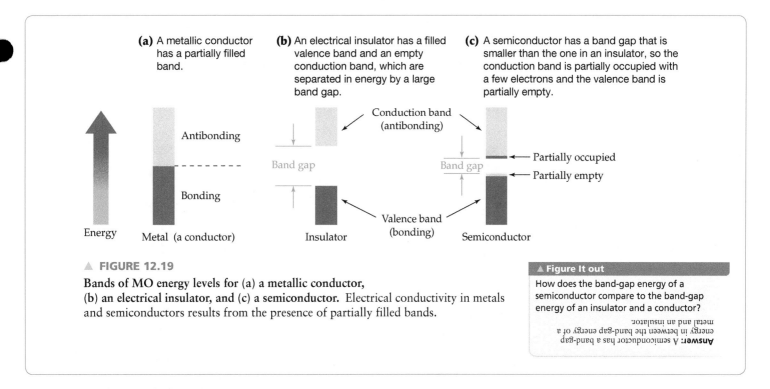

▲ **FIGURE 12.19**

**Bands of MO energy levels for (a) a metallic conductor,
(b) an electrical insulator, and (c) a semiconductor.** Electrical conductivity in metals
and semiconductors results from the presence of partially filled bands.

▲ **Figure It out**

How does the band-gap energy of a
semiconductor compare to the band-gap
energy of an insulator and a conductor?

Answer: A semiconductor has a band-gap
energy in between the band-gap energy of a
metal and an insulator.

TABLE 12.3   Band Gaps for the Group 4A Elements

Element*	Band Gap (kJ/mol)	Type of Material
C (diamond)	520	Insulator
Si	107	Semiconductor
Ge	65	Semiconductor
Sn (gray tin)	8	Semiconductor
Sn (white tin)	0	Metal
Pb	0	Metal

*Si, Ge, and gray Sn have the same structure as diamond.

---

the temperature increases, the conductivity of a metal decreases, whereas the conductivity of a semiconductor increases.

The conductivity of a semiconductor can be greatly increased by adding small amounts of certain impurities, a process called **doping.** Consider, for example, the addition of a group 5A element such as phosphorus to a group 4A semiconductor such as silicon. Like diamond, silicon has a structure in which each Si atom is surrounded tetrahedrally by four others and thus has a complete octet (**FIGURE 12.20a**). The added P atoms occupy normal Si positions in the structure, but each P atom has five valence electrons and therefore introduces an extra electron not needed for bonding (**FIGURE 12.20b**). In the MO picture, the extra electrons occupy the conduction band. The number of electrons in the conduction band of the silicon doped with P is much greater than the number in pure silicon, and the conductivity of the doped semiconductor is therefore correspondingly higher. When just one of every 1 million Si atoms is replaced by P, the number of electrons in the conduction band increases from $\sim 10^{10}/cm^3$ to $\sim 10^{17}/cm^3$ and the conductivity increases by a factor of $\sim 10^7$. Because the charge carriers are electrons, which are negatively charged, the silicon doped with a group 5A element is called an *n*-type semiconductor.

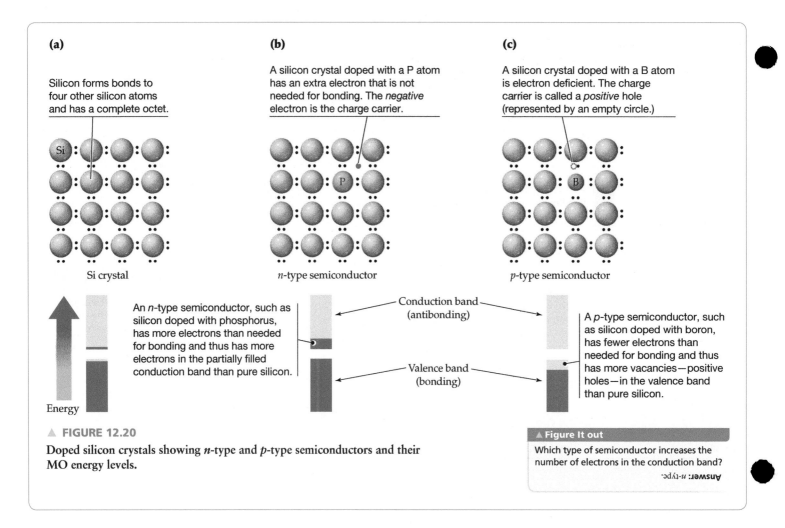

(a)

Silicon forms bonds to four other silicon atoms and has a complete octet.

Si crystal

(b)

A silicon crystal doped with a P atom has an extra electron that is not needed for bonding. The *negative* electron is the charge carrier.

n-type semiconductor

(c)

A silicon crystal doped with a B atom is electron deficient. The charge carrier is called a *positive* hole (represented by an empty circle.)

p-type semiconductor

An *n*-type semiconductor, such as silicon doped with phosphorus, has more electrons than needed for bonding and thus has more electrons in the partially filled conduction band than pure silicon.

Energy

Conduction band (antibonding)

Valence band (bonding)

A *p*-type semiconductor, such as silicon doped with boron, has fewer electrons than needed for bonding and thus has more vacancies—positive holes—in the valence band than pure silicon.

▲ **FIGURE 12.20**
Doped silicon crystals showing *n*-type and *p*-type semiconductors and their MO energy levels.

▲ **Figure It out**
Which type of semiconductor increases the number of electrons in the conduction band?

Answer: *n*-type.

Now let's consider a semiconductor in which silicon is doped with a group 3A element such as boron. Each B atom has just three valence electrons and therefore does not have enough electrons to form bonds to its four Si neighbors as shown in (**FIGURE 12.20c**). In the MO picture, the bonding MOs of the valence band are only partially filled.

The vacancies in the valence band can be thought of as positive holes in a filled band. When the electrons in the partially filled valence band move under the influence of an applied potential, the positive holes move in the opposite direction. Because the charge carriers can be regarded as the positive holes, silicon doped with a group 3A element is called a **p-type semiconductor**. In a localized picture, a positive hole is a missing electron in a B—Si electron-pair bond. When an electron from an adjacent Si—Si bond moves into the hole, the hole moves in the opposite direction.

WORKED EXAMPLE 12.5

Identifying the Type of a Doped Semiconductor

Consider a crystal of germanium that has been doped with a small amount of aluminum. Is the doped crystal an *n*-type or a *p*-type semiconductor? Compare the conductivity of the doped crystal with that of pure germanium.

STRATEGY

Consider the location of germanium and aluminum in the periodic table and the number of valence electrons in the doped crystal relative to the number in pure germanium. Doped semiconductors with more electrons than the pure semiconductor are *n*-type, and those with fewer electrons are *p*-type.

SOLUTION

Germanium, like silicon, is a group 4A semiconductor, and aluminum, like boron, is a group 3A element. The doped germanium is therefore a *p*-type semiconductor because each Al atom has one less valence electron than needed for bonding to the four neighboring Ge atoms. (Like silicon, germanium has the diamond structure.) The valence band is thus partially filled, which accounts for the electrical conductivity. The conductivity is greater than that of pure germanium because the doped germanium has many more positive holes in the valence band. That is, it has more vacant MOs available to which electrons can be excited by an electrical potential.

▶ **PRACTICE 12.9** Is germanium doped with arsenic an *n*-type or a *p*-type semiconductor? How is the conductivity affected by doping?

▶ **CONCEPTUAL APPLY 12.10** The diagrams show the electron population of the bands of MO energy levels for four materials–diamond, silicon, silicon doped with aluminum, and white tin:

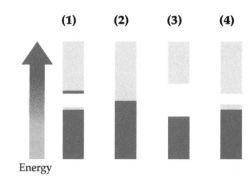

(a) Which picture corresponds to which material?
(b) Arrange the four materials in order of increasing electrical conductivity. Explain.

12.8 SEMICONDUCTOR APPLICATIONS

Doped semiconductors are essential components in the modern solid-state electronic devices found in numerous commercial products. These devices include

- diodes that convert alternating current to direct current.
- *light-emitting diodes (LEDs)* that serve as light sources in traffic signals, vehicle brake lights, digital clocks, and hospital pulse oximeters that monitor the oxygen saturation of hemoglobin in blood.
- *photovoltaic (solar) cells* that convert sunlight into electricity.
- *transistors* that control and amplify electrical signals in the integrated circuits of computers, cell phones, and many other consumer electronic products.

All these devices are made from *n*- and *p*-type semiconductors.

Diodes

A diode is a device that permits electric current to flow in one direction but is highly resistant to current flow in the opposite direction. It consists of a *p*-type semiconductor in contact with an *n*-type semiconductor to give a *p–n* junction (**FIGURE 12.21**).

If the *n*-type semiconductor is connected to the negative terminal of a battery and the *p*-type semiconductor is connected to the positive terminal (Figure 12.21a), electrons in the conduction band of the *n*-type semiconductor are repelled by the negative terminal and are attracted to the positive terminal. Consequently, they move into the region of the *p–n* junction from the *n*-side to the *p*-side where they fall into vacancies (positive holes) in the valence band of the *p*-type semiconductor. At the same time, the positive holes move in the opposite direction, from the *p*-side to the *n*-side where they combine with electrons in the conduction band of the *n*-type semiconductor. The motion of the electrons and holes constitutes an electric current that persists as long as the device is connected to the battery because the battery continues to pump electrons into the *n*-side and holes into the *p*-side. When the device is connected to the battery in this way, the *p–n* junction is said to be under a forward bias.

If the connections to the battery are reversed (*reverse bias*, Figure 12.21b), the charge carriers move in the reverse directions: Negative electrons move toward the positive battery terminal, and positive holes move toward the negative terminal. Because the charge carriers move away from the *p–n* junction, almost no electric current can flow through the junction; current flows only when the junction is under a forward bias. A *p–n* junction that is part of a circuit and subjected to an alternating potential acts as a *rectifier,* allowing current to flow in only one direction, thereby converting alternating current to direct current.

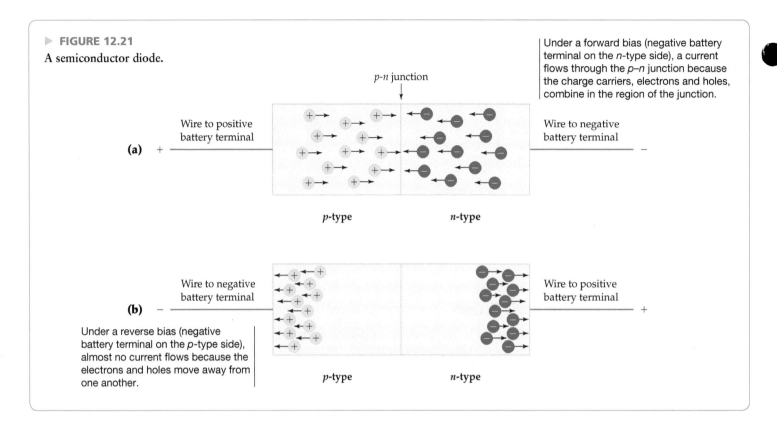

► **FIGURE 12.21**

A semiconductor diode.

p–n junction

Under a forward bias (negative battery terminal on the *n*-type side), a current flows through the *p–n* junction because the charge carriers, electrons and holes, combine in the region of the junction.

Wire to positive battery terminal

Wire to negative battery terminal

(a) +　　　　　　　　　　　　　　　　　　　　　　　　　　　　　－

p-type　　　　　　　　　*n*-type

Wire to negative battery terminal

Wire to positive battery terminal

(b) －　　　　　　　　　　　　　　　　　　　　　　　　　　　　　+

Under a reverse bias (negative battery terminal on the *p*-type side), almost no current flows because the electrons and holes move away from one another.

p-type　　　　　　　　　*n*-type

Light-Emitting Diodes (LEDs)

Energy is released as light when electrons and holes combine in the *p–n* junction of a diode under a forward bias because of an electronic transition from the conduction band to the valence band. A schematic of a light-emitting diode, or LED, is shown in **FIGURE 12.22**, along with the corresponding band theory energy-level diagram.

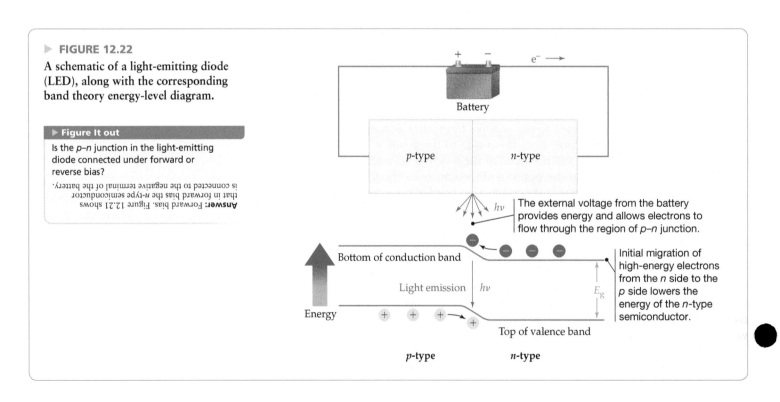

► **FIGURE 12.22**

A schematic of a light-emitting diode (LED), along with the corresponding band theory energy-level diagram.

► **Figure It out**

Is the *p–n* junction in the light-emitting diode connected under forward or reverse bias?

Answer: Forward bias. Figure 12.21 shows that in forward bias the *n*-type semiconductor is connected to the negative terminal of the battery.

e^- →

Battery

p-type　　　　　*n*-type

$h\nu$　The external voltage from the battery provides energy and allows electrons to flow through the region of *p–n* junction.

Bottom of conduction band

Initial migration of high-energy electrons from the *n* side to the *p* side lowers the energy of the *n*-type semiconductor.

Light emission $h\nu$

E_g

Energy

Top of valence band

p-type　　　　　*n*-type

In the energy-level diagram, the valence and conduction bands have different energies on the *n*- and *p*-type sides because a few electrons migrate across the junction from the *n* side to the *p* side. The decrease in energy on the n side is due to loss of high energy electrons. Soon there are no electrons left on the *n* side that have enough energy to get across the junction to the *p* side. In the presence of an external voltage, the energy required for electrons to climb the energy hill in the region of the *p–n* junction is provided by the forward bias from the battery. The energy of the light emitted by an LED is roughly equal to the band-gap energy, E_g, and is related to the wavelength of the light by the equation

$$E_g = h\nu = hc/\lambda$$

The semiconductors commonly used for making LEDs are 1:1 compounds of group 3A and 5A elements, such as GaAs, GaP, AlAs, and InP. These so-called 3–5 semiconductors have the same diamond structure and the same number of valence electrons as the elemental semiconductors Si and Ge. In GaAs, for example, each Ga atom is surrounded tetrahedrally by four As atoms and each As atom is surrounded tetrahedrally by four Ga atoms (**FIGURE 12.23**).

The average number of valence electrons per atom is four—three from Ga and five from As—so the valence band is nearly filled and the conduction band is nearly empty, as is the case for pure Si and Ge. *n*-type GaAs can be made by doping with an element having one more valence electron than As (for example, Se), and *p*-type GaAs can be made by doping with an element having one less valence electron than Ga (for example, Zn).

In one method for manufacturing LEDs, thin films of *n*-type and *p*-type semiconductors are layered onto a transparent substrate by a process called *chemical vapor deposition*. For example, a GaAs layer can be deposited by bringing highly purified vapors of trimethylgallium, $Ga(CH_3)_3$, and arsine, AsH_3, into contact with the hot substrate at about 1000 °C.

$$Ga(CH_3)_3(g) + AsH_3(g) \xrightarrow{\text{1000 °C}} GaAs(s) + 3\ CH_4(g)$$

The band gap and thus the color of the light emitted by an LED can be tuned by varying the semiconductor composition guided by period trends involving **atomic radii** and **electronegativity**. As atomic radii of atoms in a semiconductor crystal increase, the bonding electrons are held less strongly and the band-gap energy decreases. Consider the semiconductors of gallium (Group 3A) combined with the Group 5A elements P, As, and Sb. As the atomic radius increases from P to Sb, the band-gap energy becomes smaller; GaP (222 kJ/mol), GaAs (135 kJ/mol), and GaSb (67.5 kJ/mol).

When comparing the band-gap energy of the semiconductors with elements in the same period such as GaAs and ZnSe, the atomic radii are similar and there is only a small difference in bond length between atoms in the two different crystals. In this case, bond polarity has a greater influence on the band-gap energy. Electronegativity values given in Figure 7.4 can be used to calculate the difference in electronegativity between Ga and As ($2.0 - 1.6 = 0.40$) and Zn and Se ($2.4 - 1.6 = 0.80$). Since the difference in electronegativity between Zn and Se is larger than between Ga and As, the bonds between Zn and Se are stronger and the band-gap energy for ZnSe (261 kJ/mol) is higher than the band-gap energy for GaAs (135 kJ/mol).

We've seen that band-gap energy can be tuned by varying elemental composition of the semiconductor. *Solid solutions* in which one type of atom is substituted for a similar atom in the structure further expand the possible band-gap energies and colors for LED lights. For example, GaAs and GaP form solid solutions having composition GaP_xAs_{1-x} ($0 \le x \le 1$). The band gaps of these solutions vary from ~135 kJ/mol, corresponding to infrared light, to ~222 kJ/mol, corresponding to green light, as the value of x increases from 0 to 1. The familiar red light emitted by many commercial LEDs is produced by $GaP_{0.40}As_{0.60}$, which has a band gap of 181 kJ/mol. Light in the blue and green regions of the spectrum can be obtained from solid solutions of gallium and indium nitrides, $Ga_xIn_{1-x}N$.

BIG IDEA Question 4

Compare two different *p–n* semiconductor junctions with different band-gap energies. Will a larger or smaller band-gap energy result in emission of light with the longer wavelength in a light emitting diode?

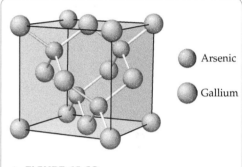

Arsenic

Gallium

▲ **FIGURE 12.23**
Unit cell of gallium arsenide showing the tetrahedral geometry around each As atom.

REMEMBER...
Atomic radius, half the distance between nuclei of two identical bonded atoms, decreases across a period and increases down a group (Section 5.14). **Electronegativity**, the ability of an atom to attract shared electrons in a bond, increases across a period and decreases down a group (Section 7.3).

▲ LED lights on a high-definition video board at the stadium of the Jacksonville Jaguars.

Red, green, and blue LEDs are combined to produce the numerous colors displayed on sports stadium video screens and large, outdoor message boards. For example, one of the world's largest high-definition video boards, 362 ft wide by 60 ft tall, at the Jacksonville Jaguars stadium contains more than 9 million LEDs that combine to produce a vast array of colors.

Compared with incandescent light bulbs, LEDs are smaller, brighter, longer lived, and more energy efficient, and they have faster switching times. For example, LED vehicle brake lights illuminate 0.2 seconds faster than conventional incandescent brake lights, an important safety feature. The faster illumination gives the driver of a trailing car an additional stopping distance of 19 feet at a speed of 65 mph.

Photovoltaic (Solar) Cells

Like an LED, a photovoltaic cell contains a *p–n* junction, but the two devices involve opposite processes. Whereas an LED converts electrical energy to light, a photovoltaic, or solar, cell converts light to electricity. When light with energy greater than or equal to the band gap shines on a *p–n* junction, electrons are excited from the valence band into the conduction band (**FIGURE 12.24**). If the *p–n* junction is made part of an electrical circuit, electrons will flow through the junction from the *p*-side to the *n*-side because of the negative (downhill) energy slope of the conduction band in the region of the junction. Positive holes left behind in the valence band move in the opposite direction. The resulting current can be used to charge a battery or power an electrical device, such as a calculator or a light bulb.

▶ **FIGURE 12.24**

A schematic of a photovoltaic (solar) cell, along with the corresponding band theory energy-level diagram.

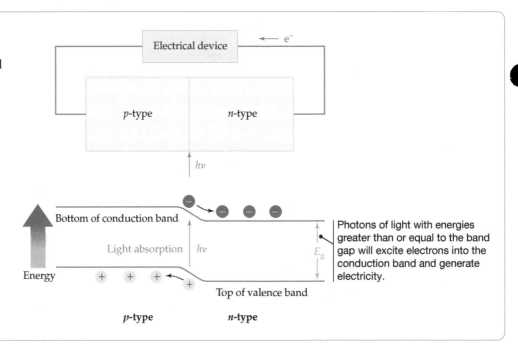

Silicon, which has a band gap of 107 kJ/mol, is commonly used to make solar cells; it absorbs sunlight with a wavelength less than about 1100 nm (ultraviolet, visible, and some of the infrared). Because the efficiency of light-to-electricity conversion in presently available solar cells is only about 20%, current research focuses on increasing the efficiency of these cells and reducing their cost by using cheaper materials and simpler manufacturing processes.

Transistors

The transistor, invented in 1947 at Bell Laboratories, is an essential component in radios, television sets, computers, cell phones, and a host of other electronic products. Transistors consist of *n–p–n* or *p–n–p* junctions that control and amplify electrical signals in modern integrated circuits. An amazing number of these extremely small devices can be packed into a small space, thus decreasing the size and increasing the speed of electrical equipment. For example, computer microprocessors now contain tens of billions of transistors on a silicon chip with a surface area of about 2 cm^2 and are able to execute more than 100 billion instructions per second.

▲ An Intel Core vPro Processor.

12.9 SUPERCONDUCTORS

The discovery of high-temperature superconductors is one of the more exciting scientific developments in the past few decades. It has stimulated an enormous amount of research in chemistry, physics, and materials science that could some day lead to a world of superfast computers, magnetically levitated trains, and power lines that carry electric current without loss of energy.

A **superconductor** is a material that loses all electrical resistance below a characteristic temperature called the **superconducting transition temperature**, T_c. This phenomenon was discovered in 1911 by the Dutch physicist Heike Kamerlingh Onnes, who found that mercury abruptly loses its electrical resistance when it is cooled with liquid helium to 4.2 K (**FIGURE 12.25**). Below its T_c, a superconductor becomes a perfect conductor and an electric current, once started, flows indefinitely without loss of energy.

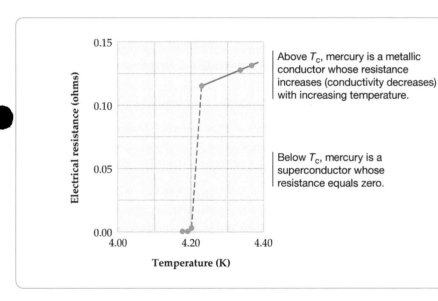

Above T_c, mercury is a metallic conductor whose resistance increases (conductivity decreases) with increasing temperature.

Below T_c, mercury is a superconductor whose resistance equals zero.

◀ **FIGURE 12.25**

The electrical resistance of mercury falls to zero at its superconducting transition temperature, $T_c = 4.2$ K.

Ever since the initial discovery in 1911, scientists have been searching for materials that superconduct at higher temperatures, and more than 6000 superconductors are now known. Until 1986, however, the record value of T_c was only 23.2 K, for the compound Nb_3Ge. The situation changed dramatically in 1986, when K. Alex Müller and J. Georg Bednorz of the IBM Zürich Research Laboratory reported a T_c of 35 K for the nonstoichiometric barium lanthanum copper oxide $Ba_xLa_{2-x}CuO_4$, where x has a value of about 0.1. Soon thereafter, scientists found even higher values of T_c for other copper-containing oxides: 90 K for $YBa_2Cu_3O_7$, 125 K for $Tl_2Ba_2Ca_2Cu_3O_{10}$, 133 K for $HgBa_2Ca_2Cu_3O_{8+x}$, and 138 K for $Hg_{0.8}Tl_{0.2}Ba_2Ca_2Cu_3O_{8.33}$. High values of T_c for these compounds were completely unexpected because most metal oxides— nonmetallic inorganic solids called *ceramics*—are electrical insulators. Within just one year of discovering the first ceramic superconductor, Müller and Bednorz were awarded the 1987 Nobel Prize in Physics.

One unit cell of $YBa_2Cu_3O_7$, the so-called 1-2-3 compound (1 yttrium atom, 2 barium atoms, and 3 copper atoms), is shown in **FIGURE 12.26**. Two-thirds of the Cu atoms are surrounded by a square pyramid of five O atoms, some of which are shared with neighboring CuO_5 groups to give two-dimensional layers of square pyramids.

▶ **FIGURE 12.26**

One unit cell of the crystal structure of $YBa_2Cu_3O_7$. The unit cell contains one Y atom, two Ba atoms, three Cu atoms, and seven O atoms.

▶ **Figure It out**

Prove that one unit cell of $YBa_2Cu_3O_7$ contains three Cu atoms and seven O atoms.

Answer: There are 8 Cu atoms at the corners $(8 \times 1/8) = 1$ Cu atom in unit cell and there are 8 Cu atoms at the edges $(8 \times 1/4) = 2$ Cu atoms in unit cell; giving a total of 3 Cu atoms in the unit cell. There are 12 O atoms on the edges $(12 \times 1/4) = 3$ O atoms in the unit cell and there are 8 O atoms on the faces $(8 \times 1/2) = 4$ O atoms in the unit cell; giving a total of 7 O atoms in the unit cell.

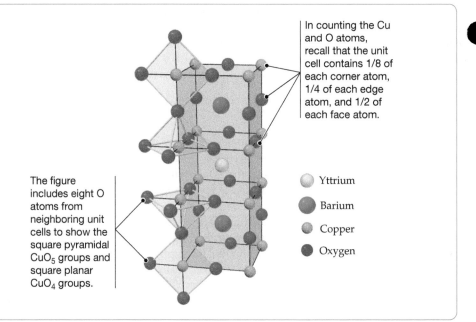

In counting the Cu and O atoms, recall that the unit cell contains 1/8 of each corner atom, 1/4 of each edge atom, and 1/2 of each face atom.

The figure includes eight O atoms from neighboring unit cells to show the square pyramidal CuO_5 groups and square planar CuO_4 groups.

○ Yttrium
● Barium
◐ Copper
● Oxygen

▲ **FIGURE 12.27**

Levitation of a magnet above a pellet of $YBa_2Cu_3O_7$ cooled to 77 K with liquid nitrogen. $YBa_2Cu_3O_7$ becomes a superconductor at approximately 90 K.

▲ **FIGURE 12.28**

A Maglev train is now in commercial operation in Shanghai, China, running at 430 km/h.

The remaining Cu atoms are surrounded by a square of four O atoms, two of which are shared with neighboring CuO_4 squares to give chains of CuO_4 groups. It's interesting to note that the Cu atoms have a fractional oxidation number of +2.33, based on the usual oxidation numbers of +3 for Y, +2 for Ba, and −2 for O. Both the infinitely extended layers of Cu and O atoms and the fractional oxidation number of Cu appear to play a role in the current flow, but a generally accepted theory of superconductivity in ceramic superconductors is not yet available. This is a field where experiment is far ahead of theory.

One of the most dramatic properties of a superconductor is its ability to levitate a magnet. When a superconductor is cooled below its T_c and a magnet is lowered toward it, the superconductor and the magnet repel each other, and the magnet hovers above the superconductor as though suspended in midair (**FIGURE 12.27**).

The force responsible for levitation arises in the following way: When the magnet moves toward the superconductor, it induces a supercurrent in the surface of the superconductor that continues to flow even after the magnet stops moving. The supercurrent, in turn, induces a magnetic field in the superconductor that exactly cancels the field from the magnet. Thus, the net magnetic field within the bulk of the superconductor is zero, a phenomenon called the *Meissner effect*. Outside the superconductor, however, the magnetic fields due to the magnet and the supercurrent repel each other, just as the north poles of two bar magnets do. The magnet therefore experiences an upward magnetic force as well as the usual downward gravitational force, and it remains suspended above the superconductor at the point where the two forces are equal. The Meissner effect is used in high-speed, magnetically levitated trains, such as the one now operating in Shanghai, China (**FIGURE 12.28**).

Other common applications of superconductors include the powerful superconducting magnets that are essential components in the magnetic resonance imaging (MRI) instruments widely used in medical diagnosis and the magnets that bend the path of charged particles in high-energy particle accelerators. These applications, however, use conventional superconductors ($T_c \leq 20$ K) that are cooled to 4.2 K with liquid helium, an expensive substance that requires sophisticated cryogenic (cooling) equipment. Much of the excitement surrounding the high-temperature superconductors arises because their T_c values are above the boiling point of liquid nitrogen (bp 77 K), an abundant refrigerant that is cheaper than milk. Of course, the search goes on for materials with still higher values of T_c. For applications such as long-distance electric power transmission, the goal is a material that superconducts at room temperature.

Presently known high-temperature superconductors are brittle powders with high melting points, so they are not easily fabricated into the wires and coils needed for electrical equipment. Nevertheless, commercial applications of high-temperature superconductors are beginning to emerge. Superconducting thin films are used as microwave filters in cell phone base stations, and superconducting wire 1 km in length is now commercially available.

In 1991, scientists at AT&T Bell Laboratories discovered a new class of high-temperature superconductors based on fullerene, the allotrope of carbon that contains C_{60} molecules (Section 12.5). These soccer ball-shaped C_{60} molecules react with potassium to give K_3C_{60}, a stable crystalline solid that contains a face-centered cubic array of C_{60}^{3-} ions, with K^+ ions in the cavities between them (**FIGURE 12.29**). At room temperature, K_3C_{60} is a metallic conductor, but it becomes a superconductor at 19.5 K. The related fullerides Rb_3C_{60} and Cs_3C_{60} have higher T_c values of 29.5 K and 40 K, respectively. In 2015, hydrogen sulfide under extremely high pressure (150 gigapascals) was found to undergo a superconducting transition near 203K.

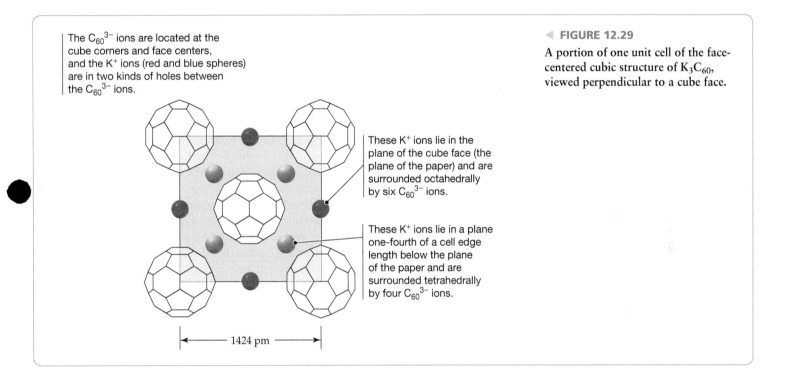

The C_{60}^{3-} ions are located at the cube corners and face centers, and the K^+ ions (red and blue spheres) are in two kinds of holes between the C_{60}^{3-} ions.

These K^+ ions lie in the plane of the cube face (the plane of the paper) and are surrounded octahedrally by six C_{60}^{3-} ions.

These K^+ ions lie in a plane one-fourth of a cell edge length below the plane of the paper and are surrounded tetrahedrally by four C_{60}^{3-} ions.

◀ **FIGURE 12.29**

A portion of one unit cell of the face-centered cubic structure of K_3C_{60}, viewed perpendicular to a cube face.

12.10 CERAMICS AND COMPOSITES

Ceramics are inorganic, nonmetallic, nonmolecular solids, including both crystalline materials such as quartz (SiO_2) and amorphous materials such as glasses. Known since ancient times, traditional silicate ceramics, such as pottery and porcelain, are made by heating aluminosilicate clays to high temperatures. Modern, so-called **advanced ceramics**—materials that have high-tech engineering, electronic, and biomedical applications—include *oxide ceramics*, such as alumina (Al_2O_3), and *nonoxide ceramics*, such as silicon carbide (SiC) and silicon nitride (Si_3N_4). Additional examples are listed in **TABLE 12.4,** which compares the properties of ceramics with those of aluminum and steel. Note that oxide ceramics are named by replacing the *-um* ending of the element name with an *-a;* thus, BeO is beryllia, and ZrO_2 is zirconia.

In many respects, the properties of ceramics are superior to those of metals: Ceramics have higher melting points and are stiffer, harder, and more resistant to wear and corrosion. Moreover, they maintain much of their strength at high temperatures, whereas metals either melt or corrode because of oxidation. Silicon nitride and silicon carbide, for example, are stable to oxidation in air up to 1400–1500 °C, and oxide ceramics

TABLE 12.4 Properties of Some Ceramic and Metallic Materials

Material	Melting Point (°C)	Density (g/cm³)	Elastic Modulus (GPa)[a]	Hardness (Mohs scale)[b]
Oxide ceramics				
Alumina, Al_2O_3	2054	3.99	380	9
Beryllia, BeO	2578	3.01	370	8
Zirconia, ZrO_2	2710	5.68	210	8
Nonoxide ceramics				
Boron carbide, B_4C	2350	2.50	280	9
Silicon carbide, SiC	2830	3.16	400	9
Silicon nitride, Si_3N_4	1900	3.17	310	9
Metals				
Aluminum	660	2.70	70	3
Plain carbon steel	1515	7.86	205	5

[a]The elastic modulus, measured in units of pressure (1 gigapascal = 1 GPa = 10^9 Pa) indicates the stiffness of a material when it is subjected to a load. The larger the value, the stiffer the material.

[b]Numbers on the Mohs hardness scale range from 1 for talc, a very soft material, to 10 for diamond, the hardest known natural substance

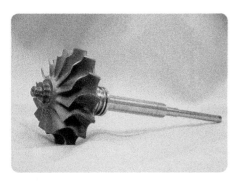

▲ A silicon nitride rotor for use in gas-turbine engines.

don't react with oxygen because they are already fully oxidized. Because ceramics are less dense than steel, they are attractive lightweight, high-temperature materials for replacing metal components in aircraft, space vehicles, and automotive engines.

Unfortunately, ceramics are brittle, as anyone who has dropped a coffee cup well knows. The brittleness, hardness, stiffness, and high melting points of ceramics are all due to strong chemical bonding. Take silicon carbide, for example, a covalent network solid that crystallizes in the diamond structure (**FIGURE 12.30**). Each Si atom is bonded tetrahedrally to four C atoms, and each C atom is bonded tetrahedrally to four Si atoms. The strong, highly directional covalent bonds ($D_{Si-C} = 435$ kJ/mol) prevent the planes of atoms from sliding over one another when the solid is subjected to the stress of a load or an impact. As a result, the solid can't deform to relieve the stress. It maintains its shape up to a point, but then the bonds give way suddenly, and the material fails catastrophically when the stress exceeds a certain threshold value. Oxide

▶ **FIGURE 12.30**

One unit cell of the cubic form of silicon carbide, SiC. The crystal can't deform under stress because the bonds are strong and highly directional.

▶ **Figure It out**

How many silicon and how many carbon atoms are in one unit cell of silicon carbide?

Answer: 4 C atoms and 4 Si atoms

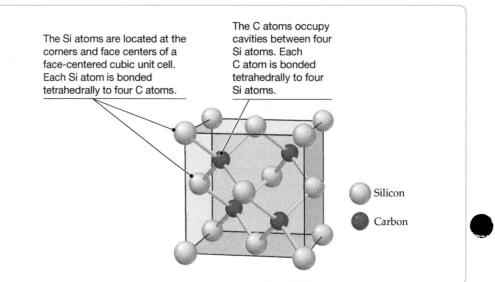

The Si atoms are located at the corners and face centers of a face-centered cubic unit cell. Each Si atom is bonded tetrahedrally to four C atoms.

The C atoms occupy cavities between four Si atoms. Each C atom is bonded tetrahedrally to four Si atoms.

○ Silicon
● Carbon

ceramics, in which the bonding is largely ionic, behave similarly. By contrast, metals are able to deform under stress because their planes of metal cations can slide easily in the electron sea (Section 12.6). As a result, metals dent but ceramics shatter.

Ceramic processing, the series of steps that leads from raw material to the finished ceramic object, determines properties such as strength and resistance to fracture of the final product. Processing often begins with a fine powder, which is combined with an organic binder, shaped, compacted, and finally *sintered* at temperatures of 1300–2000 °C. **Sintering,** which occurs below the melting point, is a process in which the particles of the powder are welded together without completely melting. During sintering, the crystal grains grow larger and the density of the material increases as the void spaces between particles disappear. Unfortunately, impurities and remaining voids can lead to microscopic cracks that cause the material to fail under stress. It is therefore important to minimize impurities and voids by beginning with high-purity, fine powders that can be tightly compacted prior to sintering.

One approach to making such powders is the **sol–gel method,** in which a metal oxide powder is synthesized from a metal alkoxide, a compound derived from a metal and an alcohol. In the synthesis of titania (TiO_2) from titanium ethoxide, $Ti(OCH_2CH_3)_4$, for example, the $Ti(OCH_2CH_3)_4$ starting material is made by the reaction of titanium(IV) chloride with ethanol and ammonia in a benzene solution:

$$TiCl_4 + 4\ \underset{\text{Ethanol}}{HOCH_2CH_3} + 4\ NH_3(g) \xrightarrow{\text{Benzene}} \underset{\text{Titanium ethoxide}}{Ti(OCH_2CH_3)_4} + 4\ NH_4Cl(s)$$

Pure $Ti(OCH_2CH_3)_4$ is then dissolved in an appropriate organic solvent, and water is added to bring about a *hydrolysis* reaction:

$$Ti(OCH_2CH_3)_4 + 4\ H_2O \xrightarrow{\text{Solution}} Ti(OH)_4(s) + 4\ HOCH_2CH_3$$

In this reaction, $Ti-OCH_2CH_3$ bonds are broken, $Ti-OH$ bonds are formed, and ethanol is regenerated. The $Ti(OH)_4$ forms in the solution as a colloidal dispersion called a sol, consisting of extremely fine particles having a diameter of only 0.001–0.1 μm. Subsequent reactions eliminate water molecules and form oxygen bridges between Ti atoms:

$$(HO)_3Ti-O-H + H-O-Ti(OH)_3 \longrightarrow (HO)_3Ti-O-Ti(OH)_3 + H_2O$$

Because all the OH groups can undergo this reaction, the particles of the sol link together through a three-dimensional network of oxygen bridges and the sol is converted to a more rigid, gelatinlike material called a *gel*. The remaining water and solvent are then removed by heating the gel, and TiO_2 is obtained as a fine powder consisting of high-purity particles with a diameter less than 1 μm (**FIGURE 12.31**).

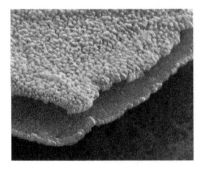 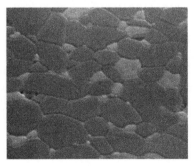

◀ **FIGURE 12.31**

Electron micrographs of a titania powder (left) and the dense ceramic produced by sintering the powder (right). The powder, consisting of tightly packed particles of TiO_2 with a diameter less than 1 μm, was made by the sol–gel method.

Oxide ceramics have many important uses. Alumina, for example, is the material of choice for making spark-plug insulators because of its high electrical resistance, high strength, high thermal stability, and chemical inertness. Because alumina is nontoxic

▲ Ceramic body armor plates are commonly used as inserts in bullet proof vests.

and essentially inert in biological systems, it is used in constructing dental crowns and the heads of artificial hips. Alumina is also used as a substrate material for electronic circuit boards. High purity silica (SiO_2) fibers were used to make the lightweight, heat-resistant ceramic tiles that protected the space shuttle on re-entry into the atmosphere. The shuttle was covered with more than 20,000 tiles with a density less than that of Styrofoam and able to withstand temperatures up to 1600 °C.

In the military, ceramics are widely used because they are lightweight and are strong enough to shatter and deflect bullets and other projectiles. For example, body armor, once made of steel (density, 7.86 g/cm^3), is now made of light ceramic materials such as boron carbide (density, 2.50 g/cm^3). Use of lightweight ceramic armor panels in place of steel in combat vehicles and military aircraft has increased maneuverability and decreased fuel consumption.

WORKED EXAMPLE 12.6

Synthesizing the 1-2-3 Superconductor by Hydrolysis of a Mixture of Metal Ethoxides

The 1-2-3 ceramic superconductor $YBa_2Cu_3O_7$ has been synthesized by the sol–gel method from a stoichiometric mixture of yttrium ethoxide, barium ethoxide, and copper(II) ethoxide in an appropriate organic solvent. The oxide product, before being heated in oxygen, has the formula $YBa_2Cu_3O_{6.5}$. Write a balanced equation for the hydrolysis of the stoichiometric mixture of metal ethoxides.

STRATEGY

First, determine the chemical formulas of the metal ethoxides, and then write a balanced equation for the reaction of the mixture of metal ethoxides with water.

SOLUTION

The ethoxide ligand is the anion of ethanol, $HOCH_2CH_3$, and has a charge of -1. Because yttrium is a group 3B element and has an oxidation number of $+3$, the formula of yttrium ethoxide must be $Y(OCH_2CH_3)_3$. Similarly, because both barium (in group 2A) and copper(II) have an oxidation number of $+2$, the formulas of barium ethoxide and copper ethoxide must be $Ba(OCH_2CH_3)_2$ and $Cu(OCH_2CH_3)_2$, respectively.

Hydrolysis of $Y(OCH_2CH_3)_3$, which breaks the $Y—OCH_2CH_3$ bonds and forms $Y—OH$ bonds, requires one H_2O molecule for each ethoxide ligand:

$$Y(OCH_2CH_3)_3 + 3\ H_2O \longrightarrow Y(OH)_3 + 3\ HOCH_2CH_3$$

Similarly, hydrolysis of $Ba(OCH_2CH_3)_2$ and $Cu(OCH_2CH_3)_2$ yields $Ba(OH)_2$ and $Cu(OH)_2$, respectively:

$$Ba(OCH_2CH_3)_2 + 2\ H_2O \longrightarrow Ba(OH)_2 + 2\ HOCH_2CH_3$$

$$Cu(OCH_2CH_3)_2 + 2\ H_2O \longrightarrow Cu(OH)_2 + 2\ HOCH_2CH_3$$

Because the three metal ethoxides are present together in a 1:2:3 ratio in the synthesis of the superconductor, we can write the product of the hydrolysis as $Y(OH)_3 \cdot 2\ Ba(OH)_2 \cdot 3\ Cu(OH)_2$,

or $YBa_2Cu_3(OH)_{13}$. Thus, the hydrolysis reaction requires 13 H_2O molecules for 13 OCH_2CH_3 ligands, and the balanced equation is

$$Y(OCH_2CH_3)_3 + 2\ Ba(OCH_2CH_3)_2 + 3\ Cu(OCH_2CH_3)_2$$
$$+ 13\ H_2O \longrightarrow YBa_2Cu_3(OH)_{13} + 13\ HOCH_2CH_3$$

Subsequent heating of $YBa_2Cu_3(OH)_{13}$ removes water, converting the mixed-metal hydroxide to the oxide $YBa_2Cu_3O_{6.5}$, which is then oxidized to $YBa_2Cu_3O_7$ by heating in O_2 gas.

▶ **PRACTICE 12.11** Silica glasses used in lenses, laser mirrors, and other optical components can be made by the sol–gel method. One step in the process is the hydrolysis of $Si(OCH_3)_4$. Write a balanced equation for the reaction.

▶ **APPLY 12.12** Crystals of the oxide ceramic barium titanate, $BaTiO_3$, have an unsymmetrical arrangement of ions, which gives the crystals an electric dipole moment. Such materials are called ferroelectrics and are used to make various electronic devices. Barium titanate can be made by the sol–gel method, which involves hydrolysis of a mixture of metal alkoxides. Write a balanced equation for the hydrolysis of a 1:1 mixture of barium isopropoxide and titanium isopropoxide, and explain how the resulting sol is converted to $BaTiO_3$. (The isopropoxide ligand is the anion of isopropyl alcohol, $HOCH(CH_3)_2$, also known as rubbing alcohol.)

Ceramics that are brittle and prone to fracture can be strengthened by mixing the ceramic powder prior to sintering with fibers of a second material, such as carbon, boron, or silicon carbide. The resulting hybrid material, called a **ceramic composite**, combines the advantageous properties of both components. An example is the composite consisting of fine grains of alumina reinforced with *whiskers* of silicon carbide. Whiskers are tiny, fiber-shaped particles, about 0.5 μm in diameter and 50 μm long, that are very strong because they are single crystals. Silicon carbide-reinforced alumina possesses high strength and high shock resistance, even at high temperatures, and has therefore been used to make high-speed cutting tools for machining very hard steels.

How do fibers and whiskers increase the strength and fracture toughness of a composite material? First, fibers have great strength along the fiber axis because most of the chemical bonds are aligned in that direction. Second, there are several ways in which fibers can prevent microscopic cracks from propagating to the point that they lead to the fracture of the material. The fibers can deflect cracks, thus preventing them from moving cleanly in one direction, and they can bridge cracks, thus holding the two sides of a crack together.

Silicon carbide–reinforced alumina is a composite in which both the fibers and the surrounding matrix are ceramics. There are other composites, however, in which the two phases are different types of materials. Examples are **ceramic–metal composites,** or **cermets,** such as aluminum metal reinforced with boron fiber, and **ceramic–polymer composites,** such as boron/epoxy and carbon/epoxy. (Epoxy is a resin consisting of long-chain organic molecules.) These materials are popular for aerospace and military applications because of their high strength-to-weight ratios. Boron-reinforced aluminum, for example, was used as a lightweight structural material in the space shuttle, and boron/epoxy and carbon/epoxy skins are used on military aircraft. Increased use of composite materials in commercial aircraft could result in weight savings of 20–30% and corresponding savings in fuel, which accounts for as much as 40% of the cost of operating an airline. Carbon/epoxy composites are also used to make strong, lightweight sports equipment, such as golf clubs, tennis racquets, fishing rods, and bicycles.

Ceramic fibers used in composites are usually made by high-temperature methods. Carbon (graphite) fiber, for example, can be made by the thermal decomposition of fibers of polyacrylonitrile, a long-chain organic molecule also used to make the textile Orlon:

▲ The skin of the B-2 advanced technology aircraft is a strong, lightweight composite material that contains carbon fibers.

$$-CH_2-CH\underbrace{-CH_2-CH}-CH_2-CH- \quad \text{Polyacrylonitrile}$$
$$\qquad\quad\; | \qquad\qquad\; | \qquad\qquad | $$
$$\qquad\quad CN \qquad\quad CN \qquad\quad CN$$

Repeating unit

In the final step of the multistep process, the carbon in the fiber is converted to graphite by heating at 400–2500 °C. Similarly, silicon carbide fiber can be made by heating fibers that contain long-chain molecules with alternating silicon and carbon atoms:

$$-SiH_2-CH_2\underbrace{-SiH_2-CH_2}-SiH_2-CH_2- \quad \xrightarrow[-H_2]{\text{Heat}} \quad SiC \text{ fiber}$$

Repeating unit

INQUIRY ? What are quantum dots, and what controls their color?

Nanoscience is the study of the unique properties of nanoscale structures. The nanoscale is defined as particles that have at least one dimension between 1 and 100 nanometers (nm). To appreciate the extremely small size of nanomaterials, it's helpful to note that 1 nm is the width of four H atoms laid side by side and about 50,000 times smaller than the diameter of a human hair. The Chapter 1 *Inquiry* describes some of the unique properties of nanoparticles that arise because of their small size. They tend to have lower melting points, different colors, and greater reactivity than bulk materials. For instance, gold nanoparticles are red and chemically reactive, whereas the bulk metal is yellow and inert. The lower melting points and greater reactivity of nanoparticles result because a substantial percentage of the atoms in a nanoparticle are on the surface or on edges and are therefore bound less tightly than those within the bulk solid; for a 5 nm particle, about 30% of the atoms are on the surface.

Quantum dots are examples of semiconductor nanoparticles whose color varies with the diameter of the particle. On the nanoscale, both particle size and chemical composition affect properties. Why do different sizes of quantum dots exhibit different colors? The answer has to do with the extremely small size of the particles. Because nanoparticles have dimensions intermediate between those of individual atoms and bulk solids, the spacing of their electronic energy levels is intermediate between the widely separated energy levels of individual atoms and the nearly continuous band of energy levels in a bulk metal. Moreover, the energy spacing depends on the size of the nanoparticle. **FIGURE 12.32** shows how orbital energy levels change as more atoms are bonded together in a nanoparticle. Overlap of two atomic orbitals (Figure 12.32a) creates two new molecular orbitals (MOs); a bonding MO which is lower in energy than the antibonding MO (Figure 12.32b). Inclusion of more atomic orbitals creates more lower-energy bonding MOs and more higher-energy antibonding MOs (Figure 12.32 c, d). As the number of bonding and antibonding MOs increases, the individual energy levels become closer together, forming a continuum called a band (Figure 12.32e). Notice that the collection of orbitals "spreads" as more atomic orbitals are added. When two bands spread, the band gap decreases.

Colors of quantum dots arise from electronic transitions. When a quantum dot that is about 1–10 nm in diameter is irradiated with ultraviolet light, it emits visible light with a wavelength that depends on the size of the nanoparticle. Irradiation with ultraviolet light causes an electron to be excited from the valence band to the conduction band. Subsequently, the electron returns to the valence band with emission of a visible photon having an energy roughly equal to the band-gap energy. The larger the nanoparticle, the smaller the band gap, and the greater the shift in the color of the emitted light from the higher-energy violet to the lower-energy red. A 3-nm cadmium selenide particle emits green light at 520 nm, whereas a 5.5-nm particle

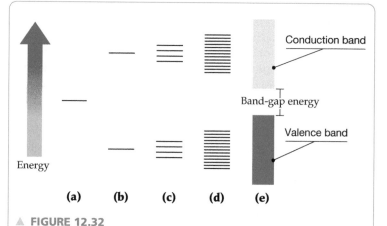

▲ **FIGURE 12.32**

Diagram showing the changes in energy levels as the number of atoms (*n*) in a nanoparticle increases. (a) atomic orbital, $n = 1$; (b) molecular orbitals, $n = 2$; (c) molecular orbitals, $n = 4$; (d) molecular orbitals, $n = 13$; (e) bands, $n = 6 \times 10^{23}$.

of the same substance emits red light at 620 nm. An entire rainbow of colors can be emitted by differently sized nanoparticles of a single substance, with the wavelength of the emitted light decreasing from the red to the blue part of the spectrum as the size of the particle decreases (**FIGURE 12.33**).

Because of their distinctive optical properties, quantum dots may have numerous applications. For example, quantum dots can be used to tag or trace biomolecules in the body. The surface of a quantum dot is chemically modified by attaching a

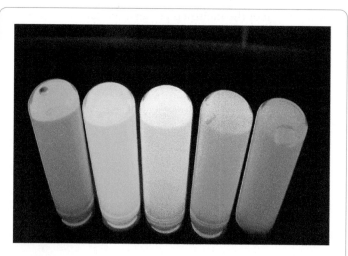

▲ **FIGURE 12.33**

The color of light emitted by these cadmium selenide nanoparticles, called quantum dots, varies from blue to red as the size of the nanoparticle increases. The average diameters of the particles are blue = 2.4 nm, green = 2.5 nm, yellow = 3.0 nm, orange = 2.9 nm, and red = 4.2 nm.

molecule that targets and binds to a specific biomolecule, such as DNA or a protein. When a quantum dot bonded to a biomolecule is irradiated with ultraviolet light, it serves as a fluorescent probe, or label, for that particular biomolecule, with the color of the emitted light dependent on the size of the quantum dot. Thus, irradiation with a single light source of a mixture of quantum dots having different sizes and different surface modifications allows simultaneous detection, tracking, and color imaging of a number of different biomolecules within a cell. Several types of nanoparticles, including quantum dots, are being investigated as potential weapons in the fight against cancer. For example, nanoparticles can be modified by attaching antibodies that target and bond to receptors found on the outside of cancer cell membranes but have little affinity for the membranes of healthy cells. Use of nanoparticles in imaging cancer cells may allow cancer to be detected at its earliest and most curable stages.

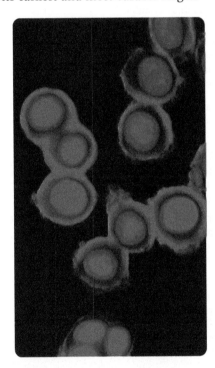

▲ Quantum dots are semiconductor nanoparticles that are replacing many conventional dyes used to stain cells. In this image, quantum dots emit red light tag proteins on the surface of breast cancer cells, while a conventional blue dye stains the cells' nuclei.

PROBLEM 12.13 Why are solutions of different-sized CdSe quantum dots different colors?

PROBLEM 12.14 Energy diagrams for three different-sized CdSe nanoparticles (2.2 nm, 3.5 nm, and 5.0 nm) are shown. The valence band is shown in red and the conduction band is shown in blue. Match each diagram to the correct particle size.

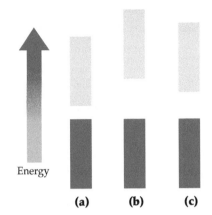

PROBLEM 12.15 A CdSe nanoparticle quantum dot is 2.0 nm in diameter. How many nanoparticles can fit across a human hair with a diameter of 50 μm?

PROBLEM 12.16 The absorption spectra of four different-sized CdSe quantum dot nanoparticles are shown. Particle size (a) is represented by the red trace, particle size (b) is orange, particle size (c) is yellow, and particle size (d) is green. Order particle sizes a–d from smallest to largest.

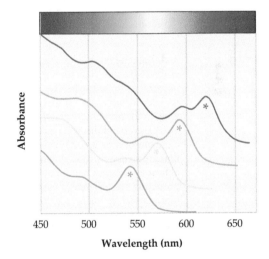

PROBLEM 12.17 You wish to track the motions of two proteins within a cell. To protein A you attach a quantum dot that emits yellow light, and to protein B, a quantum dot of the same substance that emits blue light. Which quantum dot is larger? Explain.

STUDY GUIDE

Section	Concept Summary	Learning Objectives	Test Your Understanding
12.1 Types of Solids	Solids can be characterized as **amorphous solids** if their particles are randomly arranged or **crystalline solids** if their particles are ordered. Crystalline solids can be further characterized as **ionic solids** if their particles are ions, **molecular solids** if their particles are molecules, **covalent network solids** if they consist of a covalently bonded array of atoms without discrete molecules, or **metallic solids** if their particles are metal atoms.	**12.1** Classify types of solids based on their chemical composition and properties.	Problems 12.28–12.31
12.2 Probing the Structure of Solids: X-Ray Crystallography	The structures of atoms in a solid can be measured by directing X rays at a crystal and analyzing the resulting diffraction pattern.	**12.2** Use the Bragg equation to calculate spacing between atomic layers in a crystal.	Problems 12.32–12.33
12.3–12.5 Packing of Spheres in Crystalline Solids and Unit Cells	The regular three-dimensional network of particles in a crystal is made up of small repeating units called **unit cells. Simple cubic packing** uses a **primitive-cubic unit cell**, with an atom at each corner of the cube. **Body-centered cubic packing** uses a **body-centered cubic unit cell**, with an atom at the center and at each corner of the cube. **Cubic closest-packing** uses a **face-centered cubic unit cell**, with an atom at the center of each face and at each corner of the cube. A fourth kind of packing, called **hexagonal closest-packing**, uses a noncubic unit cell.	**12.3** Identify the four kinds of spherical packing arrangements in crystalline solids and the three kinds of cubic unit cells.	Problem 12.18
		12.4 Calculate the density of a substance, atomic radii of its atoms, or molecular mass given its unit cell dimensions.	Worked Examples 12.1–12.2; Problems 12.36, 12.38, 12.42, 12.44
		12.5 Use an image of a unit cell to determine the formula and geometry of ionic compounds.	Worked Example 12.3; Problems 12.19–12.20
12.6 Bonding in Metals	Two bonding models are used for metals. The **electron-sea model** pictures a metal as an array of metal cations immersed in a sea of delocalized, mobile valence electrons that act as an electrostatic glue. In the molecular orbital theory for metals, also called **band theory**, the delocalized valence electrons occupy a vast number of MO energy levels that are so closely spaced that they merge into an almost continuous band. Both theories account for properties such as malleability, ductility, and high thermal and electrical conductivity, but band theory better explains how the number of valence electrons affects properties such as melting point and hardness.	**12.6** Describe the electron-sea model of metals, and explain how it accounts for their properties.	Problems 12.60–12.64
		12.7 Describe the molecular orbital theory for metals (band theory), draw MO energy-level diagrams for metals, and use band theory to account for their properties.	Worked Example 12.4; Problems 12.21, 12.68–12.72
12.7 Semiconductors	Band theory accounts for the electrical properties of metals, insulators, and semiconductors. Materials with partially filled bands are metallic conductors, and materials with only completely filled bands are electrical insulators. In insulators, the bonding MOs, called the **valence band**, and the antibonding MOs, called the **conduction band**, are separated in energy by a large **band gap,** In **semiconductors,** the band gap is smaller, and a few electrons have enough thermal energy to occupy the conduction band. The resulting partially filled valence and conduction bands give rise to a small conductivity. The conductivity can be increased by **doping**—adding a group 5A impurity to a group 4A element, which gives an *n*-type semiconductor, or adding a group 3A impurity to a group 4A element, which gives a *p*-type semiconductor.	**12.8** Draw MO diagrams for insulators, semiconductors, and conductors. Use band theory to explain why different substances have different conductivities.	Problems 12.22–12.24, 12.78
		12.9 Classify doped semiconductors as *n*-type or *p*-type and draw their MO diagrams.	Worked Example 12.5; Problems 12.82, 12.84, 12.86
		12.10 Predict relative band-gap energies based on periodic trends in atomic size and electronegativity.	Problems 12.98–12.101
		12.11 Relate the wavelength of light emitted or absorbed by a semiconductor to the band gap in its band theory energy-level diagram.	Problems 12.25, 12.96, 12.105

Section	Concept Summary	Learning Objectives	Test Your Understanding
12.8 Semiconductor Applications	Doped semiconductors are essential components in modern solid-state electronic devices, such as diodes, light-emitting diodes (LEDs), photovoltaic (solar) cells, and transistors.	**12.12** Describe some applications of semiconductors.	Problems 12.92–12.94
12.9 Superconductors	A **superconductor** is a material that loses all electrical resistance below a characteristic temperature called the **superconducting transition temperature,** T_c. An example is $YBa_2Cu_3O_7(T_c = 90$ K$)$. Below T_c, a superconductor can levitate a magnet, a consequence of the Meissner effect.	**12.13** Describe the appearance of a plot of electrical resistance versus temperature for a superconductor.	Problem 12.111
12.10 Ceramics and Composites	**Ceramics** are inorganic, nonmetallic, nonmolecular solids. Modern **advanced ceramics** include oxide ceramics such as Al_2O_3 and $YBa_2Cu_3O_7$ and nonoxide ceramics such as SiC and Si_3N_4. Ceramics are generally higher-melting, lighter, stiffer, harder, and more resistant to wear and corrosion than metals. One approach to ceramic processing is the **sol–gel method,** involving hydrolysis of a metal alkoxide. The strength and fracture toughness of ceramics can be increased by making **ceramic composites,** hybrid materials such as Al_2O_3/SiC in which fine grains of Al_2O_3 are reinforced by whiskers of SiC. Other types of composites include **ceramic–metal composites.**	**12.14** Describe the bonding in ceramics and account for their properties in terms of the bonding model.	Problems 12.116, 12.118, 12.120
		12.15 Describe the synthesis of ceramics, and write balanced equations for the reactions that occur in the sol–gel synthesis method.	Worked Example 12.6; Problems 12.126, 12.128

KEY TERMS

advanced ceramic 477
allotrope 462
amorphous solid 451
band 466
band gap 468
band theory 466
body-centered cubic packing 456
body-centered cubic unit cell 458
Bragg equation 454

ceramic 477
ceramic composite 480
ceramic–metal composite (cermet) 481
ceramic-polymer composite 481
conduction band 468
coordination number 455
covalent network solid 452
crystalline solid 451
cubic closest-packing 456
diffraction 453

doping 469
electron-sea model 464
face-centered cubic unit cell 458
hexagonal closest-packing 456
ionic solid 452
metallic solid 452
molecular solid 452
n-type semiconductor 469
p-type semiconductor 470
primitive-cubic unit cell 456
semiconductor 468

simple cubic packing 455
sintering 479
sol–gel method 479
superconducting transition temperature, T_c 475
superconductor 475
unit cell 456
valence band 468

KEY EQUATIONS

• Bragg Equation (Section 12.7)

$$d = \frac{n\lambda}{2\sin\theta}$$ where d is the distance between layers of atoms, λ is the wavelength of the X rays, and θ is the angle at which X rays strike

the atomic layers

PRACTICE TEST

After studying this chapter, you can assess your understanding with these practice test questions, which are correlated with chapter learning objectives. If you answer a question incorrectly, refer to the learning objectives in the end-of-chapter Study Guide for assistance. The Study Guide provides a conceptual summary, references a Worked Example to model how to solve the problem, and gives additional problems for more practice.

1. Silicon tetraiodide melts at 120.5 °C, diffracts X rays, and does not conduct electricity in either the solid or liquid phase. What type of solid is it? **(LO 12.1)**
   (a) Ionic solid
   (b) Covalent network solid
   (c) Molecular solid
   (d) Metallic solid

2. Diffraction of X rays with $\lambda = 131.5$ pm occurred at an angle of 25.5 degrees by a crystal of aluminum. Assuming first-order diffraction, what is the interplanar spacing in aluminum? (**LO 12.2**)

(a) 76.4 pm   (b) 183.1 pm

(c) 305.5 pm   (d) 152.7 pm

3. What is the atomic radius in picometers of an argon atom if solid argon has a density of 1.623 g/cm^3 and crystallizes at low temperature in a face-centered cubic unit cell? (**LO 12.3, 12.4**)

(a) 546.9 pm   (b) 273.4 pm

(c) 136.7 pm   (d) 193.4 pm

4. One form of silver telluride (Ag_2Te) crystallizes with a cubic unit cell and a density of 7.70 g/cm^3. X-ray crystallography shows that the edge of the cubic unit cell has a length of 529 pm. How many Ag atoms are in the unit cell? (**LO 12.3, 12.4**)

(a) 1   (b) 2   (c) 4   (d) 6

5. Niobium oxide crystallizes in the following cubic unit cell:

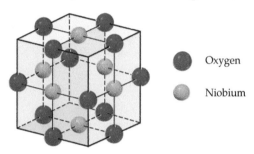

Oxygen

Niobium

What is the formula of niobium oxide, and what is the oxidation state of niobium? (**LO 12.5**)

(a) NbO, Nb = +2   (b) Nb_2O, Nb = +2

(c) NbO_2, Nb = +4   (d) Nb_2O_3, Nb = +3

6. The following diagrams represent the electron population of the composite s–d band for three metals—Ag, Mo, and Y:

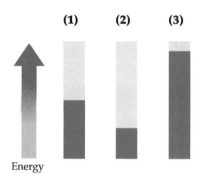

(1)   (2)   (3)

Energy

Which diagram corresponds to which metal? (**LO 12.7**)

(a) Ag = 3, Mo = 1, Y = 2

(b) Ag = 2, Mo = 1, Y = 3

(c) Ag = 2, Mo = 3, Y = 1

(d) Ag = 1, Mo = 2, Y = 3

7. Examine diagrams for the electron population of the composite s–d band for three metals in question 6. Which metal has the highest melting point? (**LO 12.7**)

(a) Metal 1

(b) Metal 2

(c) Metal 3

8. The following diagrams represent the electron population of molecular orbitals for different substances. What diagram corresponds to magnesium oxide, germanium, and tin? (**LO 12.8**)

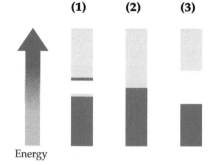

(1)   (2)   (3)

Energy

(a) Diagram 1 = tin, diagram 2 = magnesium oxide, diagram 3 = germanium

(b) Diagram 1 = germanium, diagram 2 = magnesium oxide, diagram 3 = tin

(c) Diagram 1 = germanium, diagram 2 = tin, diagram 3 = magnesium oxide

(d) Diagram 1 = magnesium oxide, diagram 2 = tin, diagram 3 = germanium

9. The molecular orbital diagram of a doped semiconductor is shown below. If the semiconductor is silicon, does the diagram represent *n*-type or *p*-type doping and which of the following elements could be dopant? (**LO 12.9**)

Energy

(a) *n*-type, As   (b) *n*-type, Ga

(c) *p*-type, As   (d) *p*-type, Ga

10. The small size of cellular telephones results from the use of sophisticated gallium arsenide (GaAs) semiconductors instead of ordinary silicon-based semiconductors. Which of the following semiconductors will have a *smaller* band-gap energy than GaAs? (**LO 12.10**)

(a) ZnSe   (b) GaP

(c) GaSb   (d) GaN

11. If the band-gap energy of a gallium phosphide (GaP) semiconductor is 222 kJ/mol, calculate the wavelength of light emitted in a GaP light-emitting diode (LED). (**LO 12.11**)

(a) 186 nm   (b) 245 nm

(c) 539 nm   (d) 854 nm

12. Identify the *false* statement about applications of semiconductors in the list below. (**LO 12.12**)

(a) A diode allows current to flow in one direction but not the other.

(b) A p–n junction under forward bias is used to create a solar cell.

(c) Transistors important in integrated circuits in computers consist of p–n–p junctions.

(d) A p–n junction under forward bias is used to create a light-emitting diode.

13. A superconductor is a material that loses all electrical resistance below a characteristic temperature called the superconducting transition temperature. Which graph represents the behavior of a superconductor? (LO 12.13)

(a)

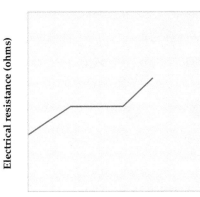

Temperature (K)

(b)

Temperature (K)

(c)

Temperature (K)

(d)

Temperature (K)

Answers:

1. c, 2. d, 3. d, 4. c, 5. a, 6. a, 7. a, 8. c, 9. d, 10. c, 11. c, 12. b, 13. c

---

Mastering Chemistry   provides end-of-chapter exercises, feedback-enriched tutorial problems, animations, and interactive activities to encourage problem-solving practice and deeper understanding of key concepts and topics.

RAN   *Randomized in Mastering Chemistry*

CONCEPTUAL PROBLEMS

Problems 12.1–12.17 appear within the chapter.

12.18 Identify each of the following kinds of packing:

(a)   (b)   (c)   (d)

12.19 Zinc sulfide, or sphalerite, crystallizes in the following cubic unit cell:

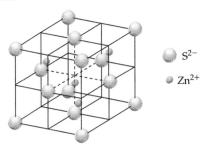

S^{2-}

Zn^{2+}

(a) What kind of packing do the sulfide ions adopt?

(b) How many S^{2-} ions and how many Zn^{2+} ions are in the unit cell?

12.20 Titanium oxide crystallizes in the following cubic unit cell:

RAN

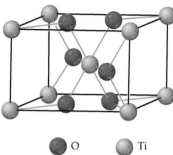

○ O   ○ Ti

(a) How many titanium ions and how many oxygen ions are in each unit cell?

(b) What is the formula of titanium oxide?

(c) What is the oxidation state of titanium?

12.21 The following diagrams show the electron populations of the composite $s - d$ bands for three different transition metals:

(1)   (2)   (3)

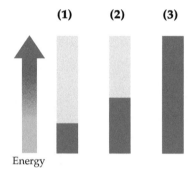

Energy

(a) Which metal has the highest melting point? Explain.

(b) Which metal has the lowest melting point? Explain.

(c) Arrange the metals in order of increasing hardness. Explain.

12.22 The following diagrams show the electron populations of the bands of MO energy levels for four different materials:

(1)   (2)   (3)   (4)

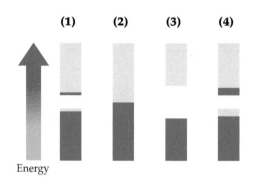

Energy

(a) Classify each material as an insulator, a semiconductor, or a conducting metal.

(b) Arrange the four materials in order of increasing electrical conductivity. Explain.

(c) Tell whether the conductivity of each material increases or decreases when the temperature increases.

12.23 The following diagram represents the electron population of the bands of MO energy levels for elemental silicon:

Energy

(a) Identify the valence band, conduction band, and band gap.

(b) In a drawing, show how the electron population changes when the silicon is doped with gallium.

(c) In a drawing, show how the electron population changes when the silicon is doped with arsenic.

(d) Compare the electrical conductivity of the doped silicon samples with that of pure silicon. Account for the differences.

12.24 Sketch the electron populations of the bands of MO energy levels for elemental carbon (diamond), silicon, germanium, gray tin, and white tin. (Band-gap data are given in Table 12.3.) Your sketches should show how the populations of the different bands vary with a change in the group 4A element.

12.25 Diagrams (1) and (2) are energy-level diagrams for two different LEDs. One LED emits red light, and the other emits blue light. Which one emits red, and which blue? Explain.

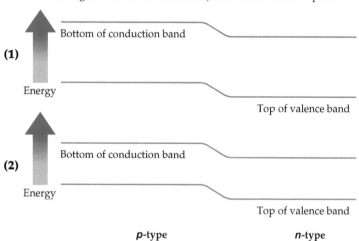

SECTION PROBLEMS

Types of Solids (Section 12.1)

12.26 List the four main classes of crystalline solids, and give a specific example of each.

12.27 What kinds of particles are present in each of the four main classes of crystalline solids?

12.28 Which of the substances Na_3PO_4, CBr_4, rubber, Au, and quartz best fits each of the following descriptions?

(a) Amorphous solid

(b) Ionic solid

(c) Molecular solid

(d) Covalent network solid

(e) Metallic solid

12.29 Which of the substances diamond, Hg, Cl_2, glass, and KCl best fits each of the following descriptions?

    (a) Amorphous solid

    (b) Ionic solid

    (c) Molecular solid

    (d) Covalent network solid

    (e) Metallic solid

12.30 Silicon carbide is very hard, has no known melting point, and diffracts X rays. What type of solid is it: amorphous, ionic, molecular, covalent network, or metallic?

12.31 Arsenic tribromide melts at 31.1 °C, diffracts X rays, and does not conduct electricity in either the solid or liquid phase. What type of solid is it: amorphous, ionic, molecular, covalent network, or metallic?

X-Ray Crystallography (Section 12.2)

12.32 Diffraction of X rays with $\lambda = 154.2$ pm occurred at an angle $\theta = 22.5\,°$ from a metal surface. What is the spacing (in pm) between the layers of atoms that diffracted the X rays?

12.33 Diffraction of X rays with $\lambda = 154.2$ pm occurred at an angle $\theta = 76.84\,°$ from a metal surface. What is the spacing (in pm) between layers of atoms that diffracted the X rays?

The Packing of Spheres in Crystalline Solids: Unit Cells (Section 12.3)

12.34 Which of the four kinds of packing used by metals makes the most efficient use of space, and which makes the least efficient use?

12.35 (a) What is a unit cell?

    (b) How many atoms are in one primitive cubic unit cell, one body-centered unit cell, and one face-centered unit cell?

12.36 Copper crystallizes in a face-centered cubic unit cell with an edge length of 362 pm. What is the radius of a copper atom in picometers? What is the density of copper in g/cm^3?

12.37 Lead crystallizes in a face-centered cubic unit cell with an edge length of 495 pm. What is the radius of a lead atom in picometers? What is the density of lead in g/cm^3?

12.38 Aluminum has a density of 2.699 g/cm^3 and crystallizes with a face-centered cubic unit cell. What is the edge length of a unit cell in picometers?

12.39 Tungsten crystallizes in a body-centered cubic unit cell with an edge length of 317 pm. What is the length in picometers of a unit-cell diagonal that passes through the center atom?

12.40 In light of your answer to Problem 12.39, what is the radius in picometers of a tungsten atom?
RAN

12.41 Sodium has a density of 0.971 g/cm^3 and crystallizes with a body-centered cubic unit cell. What is the radius of a sodium atom, and what is the edge length of the cell in picometers?

12.42 Titanium metal has a density of 4.506 g/cm^3 and an atomic radius of 144.8 pm. In what cubic unit cell does titanium crystallize?

12.43 Calcium metal has a density of 1.55 g/cm^3 and crystallizes in a cubic unit cell with an edge length of 558.2 pm.

    (a) How many Ca atoms are in one unit cell?

    (b) In which of the three cubic unit cells does calcium crystallize?

12.44 The atomic radius of Pb is 175 pm, and the density is 11.34 g/cm^3. Does lead have a primitive cubic structure or a face-centered cubic structure?
RAN

12.45 The density of a sample of metal was measured to be 6.84 g/cm^3. An X-ray diffraction experiment measures the edge of a face-centered cubic cell as 350.7 pm. What is the atomic weight, atomic radius, and identity of the metal?
RAN

12.46 If a protein can be induced to crystallize, its molecular structure can be determined by X-ray crystallography. Protein crystals, though solid, contain a large amount of water molecules along with the protein. The protein chicken egg-white lysozyme, for instance, crystallizes with a unit cell having angles of 90° and with edge lengths of 7.9×10^3 pm, 7.9×10^3 pm, and 3.8×10^3 pm. There are eight molecules in the unit cell. If the lysozyme molecule has a molecular weight of 1.44×10^4 and a density of 1.35 g/cm^3, what percent of the unit cell is occupied by the protein?
RAN

12.47 The molecular structure of a scorpion toxin, a small protein, was determined by X-ray crystallography. The unit cell has angles of 90°, contains 16 molecules, and has a volume of 1.019×10^2 nm^3. If the molecular weight of the toxin is 3336 and the density is about 1.35 g/cm^3, what percent of the unit cell is occupied by protein?
RAN

12.48 Iron crystallizes in a body-centered cubic unit cell with an edge length of 287 pm. Iron metal has a density of 7.86 g/cm^3 and a molar mass of 55.85 g. Calculate a value for Avogadro's number.
RAN

12.49 Silver metal crystallizes in a face-centered cubic unit cell with an edge length of 408 pm. The molar mass of silver is 107.9 g/mol. and its density is 11.50 g/cm^3. Use these data to calculate a value for Avogadro's number.

Structures of Ionic Solids (Section 12.4)

12.50 Sodium hydride, NaH, crystallizes in a face-centered cubic unit cell similar to that of NaCl (Figure 12.11). How many Na^+ ions touch each H^- ion, and how many H^- ions touch each Na^+ ion?
RAN

12.51 Cesium chloride crystallizes in a cubic unit cell with Cl^- ions at the corners and a Cs^+ ion in the center. Count the numbers of + and − charges, and show that the unit cell is electrically neutral.

12.52 If the edge length of an NaH unit cell is 488 pm, what is the length in picometers of an Na—H bond? (See Problem 12.50.)
RAN

12.53 The edge length of a CsCl unit cell (Problem 12.51) is 412.3 pm. What is the length in picometers of the Cs—Cl bond? If the ionic radius of a Cl^- ion is 181 pm, what is the ionic radius in picometers of a Cs^+ ion?

Structures of Covalent Network Solids (Section 12.5)

12.54 Silicon carbide, SiC, is a covalent network solid with a structure similar to that of diamond. Sketch a small portion of the SiC structure.

12.55 Based on the structure of SiC that you sketched in Problem 12.54, predict some of the properties of silicon carbide.

12.56 What is the hybridization of carbon atoms in carbon nanotubes? Explain why carbon nanotubes are good conductors of electricity.

12.57 Carbon and oxygen combine to form the molecular compound CO_2, while silicon and oxygen combine to form a covalent network solid with the formula unit SiO_2. Explain the difference in bonding between the two group 4A elements and oxygen.

Bonding in Metals (Section 12.6)

12.58 Potassium metal crystallizes in a body-centered cubic structure. Draw one unit cell, and try to draw an electron-dot structure for bonding of the central K atom to its nearest-neighbor K atoms. What is the problem?

12.59 Describe the electron-sea model of the bonding in cesium metal. Cesium has a body-centered cubic structure.

12.60 How does the electron-sea model account for the malleability and ductility of metals?

12.61 How does the electron-sea model account for the electrical and thermal conductivity of metals?

12.62 The melting point of sodium metal is 97.8 °C, and the melting point of sodium chloride is 801 °C. What can you infer about the relative strength of metallic and ionic bonding from these melting points?

12.63 Sodium metal is easily deformed by an applied force whereas sodium chloride is shattered. Explain.

12.64 Cesium metal is very soft, and tungsten metal is very hard. Explain the difference using the electron-sea model.

12.65 Sodium melts at 98 °C, and magnesium melts at 650 °C. Account for the higher melting point of magnesium using the electron-sea model.

12.66 Why is the molecular orbital theory for metals called band theory?

12.67 Draw an MO energy-level diagram that shows the population of the 4s band for potassium metal.

12.68 How does band theory account for the electrical conductivity of metals?

12.69 Materials with partially filled bands are metallic conductors, and materials with only completely filled bands are electrical insulators. Explain why the population of the bands affects the conductivity.

12.70 Draw an MO energy-level diagram for beryllium metal, and show the population of the MOs for the following two cases.
   (a) The 2s and 2p bands are well separated in energy.
   (b) The 2s and 2p bands overlap in energy.
   Which diagram agrees with the fact that beryllium has a high electrical conductivity? Explain.

12.71 Draw an MO energy-level diagram for calcium metal, and show the population of the MOs for the following two cases.
   (a) The 4s and 3d bands are well separated in energy.
   (b) The 4s and 3d bands overlap in energy.
   Which diagram agrees with the fact that calcium has a high electrical conductivity? Explain.

12.72 The melting points for the second-series transition elements increase from 1522 °C for yttrium to 2623 °C for molybdenum and then decrease to 321 °C for cadmium. Account for the trend using band theory.

12.73 Copper has a Mohs hardness value of 3, and iron has a Mohs hardness value of 5. Use band theory to explain why copper is softer than iron.

12.74 Tungsten is hard and has a very high melting point (3422 °C), and gold is soft and has a relatively low melting point (1064 °C). Are these facts in better agreement with the electron-sea model or the MO model (band theory)? Explain.

12.75 Explain why the enthalpy of vaporization of vanadium (460 kJ/mol) is much larger than that of zinc (114 kJ/mol).

Semiconductors (Section 12.7–12.8)

12.76 Define a semiconductor, and give three examples.

12.77 Tell what is meant by each of the following terms.
   (a) Valence band   (b) Conduction band
   (c) Band gap   (d) Doping

12.78 Draw the bands of MO energy levels and the electron population for:
   (a) A semiconductor   (b) An electrical insulator
   Explain why a semiconductor has the higher electrical conductivity.

12.79 Draw the bands of MO energy levels and the electron population for:
   (a) A semiconductor   (b) A metallic conductor
   Explain why a semiconductor has the lower electrical conductivity.

12.80 How does the electrical conductivity of a semiconductor change as the size of the band gap increases? Explain.

12.81 How does the electrical conductivity of a semiconductor change as the temperature increases? Explain.

12.82 Explain what an n-type semiconductor is, and give an example. Draw an MO energy-level diagram, and show the population of the valence band and the conduction band for an n-type semiconductor.

12.83 Explain what a p-type semiconductor is, and give an example. Draw an MO energy-level diagram, and show the population of the valence band and the conduction band for a p-type semiconductor.

12.84 Explain why germanium doped with phosphorus has a higher electrical conductivity than pure germanium.

12.85 Explain why silicon doped with gallium has a higher electrical conductivity than pure silicon.

12.86 Classify the following semiconductors as n-type or p-type.
   (a) Si doped with In
   (b) Ge doped with Sb
   (c) Gray Sn doped with As

12.87 Classify the following semiconductors as n-type or p-type.
   (a) Ge doped with As
   (b) Ge doped with B
   (c) Si doped with Sb

12.88 The so-called 2–6 semiconductors are 1:1 compounds of group 2B and 6A elements. Write a balanced equation for the synthesis of the 2–6 semiconductor CdSe by chemical vapor deposition from the molecular precursors $Cd(CH_3)_2(g)$ and $H_2Se(g)$.

12.89 Write a balanced equation for the synthesis of the 2–6 semiconductor ZnTe by chemical vapor deposition from dimethylzinc, $Zn(CH_3)_2$, and diethyltellurium, $Te(CH_2CH_3)_2$, assuming that the other product is gaseous propane ($CH_3CH_2CH_3$).

12.90 Arrange the following materials in order of increasing electrical conductivity.
   (a) Cu
   (b) Al_2O_3
   (c) Fe
   (d) Pure Ge
   (e) Ge doped with In

12.91 Arrange the following materials in order of increasing electrical conductivity.

(a) Pure gray Sn

(b) Gray Sn doped with Sb

(c) NaCl

(d) Ag

(e) Pure Si

12.92 Explain how a diode converts alternating current to direct current.

12.93 An LED produces light only when the *n*- and *p*-type semiconductors are connected to the proper terminals of a battery. Explain why no light is produced if you reverse the battery connections.

12.94 What are the main differences between an LED and a photovoltaic cell?

12.95 What determines the wavelength of light emitted by an LED?

12.96 What is the wavelength of the light emitted by a GaP_xAs_{1-x} LED
RAN with a band gap of 193 kJ/mol?

12.97 What band gap is needed for an LED to produce blue light
RAN with a wavelength of 470 nm?

12.98 Consider the two LED semiconductors GaN and InN.

(a) Which will have a smaller band-gap energy?

(b) One LED emits red light and the other emits ultraviolet light. Which is which?

12.99 Consider the two LED semiconductors GaN and GaP.

(a) Which will have a larger band-gap energy?

(b) One LED emits green light and the other emits ultraviolet light. Which is which?

12.100 Considering only atomic radii, rank the LED semiconductors made of solid solutions in order of increasing band-gap energy.

$$GaP_{1.00}As_{0.00}, \ GaP_{0.80}As_{0.20}, \ GaP_{0.50}As_{0.50}$$

12.101 Considering only electronegativity, rank the LED semiconductors made of solid solutions in order of increasing band-gap energy.

$$Al_{0.40}Ga_{0.60}As, \ Al_{0.25}Ga_{0.75}As, \ Al_{0.05}Ga_{0.95}As$$

12.102 A photovoltaic cell contains a *p–n* junction that converts
RAN solar light to electricity.

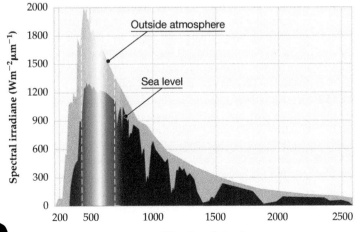

Wavelength (nm)

(a) Silicon semiconductors with a band-gap energy of 107 kJ/mol are commonly used to make photovoltaic cells. Calculate the wavelength that corresponds to the band-gap energy in silicon.

(b) Compare the wavelength absorbed by silicon to the wavelengths of light that reach the Earth's surface, as shown in the graph. Does the wavelength correspond to the highest intensity wavelength in the solar emission spectrum?

12.103 A photovoltaic cell contains a *p–n* junction that that con-
RAN verts solar light to electricity. An optimum semiconductor would have its band-gap energy matched to the wavelength of maximum solar intensity at the Earth's surface.

(a) What is the color and approximate wavelength of maximum solar intensity at the Earth's surface? Refer to the figure for Problem 12.102.

(b) Which of the following semiconductors absorb at a wavelength matched with maximum solar intensity? CdTe with a band-gap energy of 145 kJ/mol or ZnSe with a band-gap energy of 248 kJ/mol.

12.104 Gallium arsenide, a material used to manufacture laser printers and compact disc players, has a band gap of 130 kJ/mol. Is GaAs a metallic conductor, a semiconductor, or an electrical insulator? With what group 4A element is GaAs isoelectronic? (Isoelectronic substances have the same number of electrons.)

12.105 Wide band-gap semiconductors have a band gap between
RAN 2 and 7 electron volts (eV), where 1 eV = 96.485 kJ/mol. The wide band-gap semiconductor GaN, used to construct the laser in Blu-ray DVD players, has a band gap of 3.44 eV. The material in the laser, $Ga_xIn_{1-x}N$, has some indium substituted for gallium.

(a) What wavelength of light (in nm) would GaN emit, based on its band gap?

(b) If the light from the device is blue, does partial substitution of indium for gallium increase or decrease the band gap of $Ga_xIn_{1-x}N$ compared to GaN?

12.106 Does the conductivity increase, decrease, or remain the same when

(a) *n*-type InP is doped with Cd? Explain.

(b) *p*-type InP is doped with Se? Explain.

12.107 Zinc selenide (ZnSe) doped with Ga has some Ga atoms in place of Zn atoms and is an *n*-type semiconductor. Draw an MO energy-level diagram for doped ZnSe, show the population of the bands, and explain why Ga substitution gives an *n*-type semiconductor.

12.108 Consider a hypothetical material consisting of an infinite one-dimensional array of hydride ions (H^-).

(a) Show how a band of MOs could be formed from individual atomic orbitals, and explain why this material should be an electrical insulator.

(b) If some of the hydride ions in the above material were replaced with neutral hydrogen atoms, the resulting doped material should be a better electrical conductor. Draw an MO energy-level diagram for the doped material, show the population of the band, and explain why the doped material can conduct electricity.

(c) Would you describe the doped material as *n*-type or *p*-type? Explain.

12.109 Diamonds, a network covalent solid of carbon, have a band-
RAN gap energy of 530 kJ/mol, which corresponds to absorption of ultraviolet light of 225 nm. Since the band-gap energy of diamonds is in the ultraviolet region of the spectrum, a pure diamond does not absorb visible light and appears as a clear, colorless crystal. Colors in diamonds originate from

trace impurities, and nitrogen is the most common, resulting in a yellow-colored diamond.

(a) Classify a nitrogen-doped diamond as a *p*-type or *n*-type semiconductor.

(b) Draw a picture of the band structure of diamond to indicate the difference between pure diamond and nitrogen-doped diamond.

(c) Yellow diamonds absorb light at around 425 nm. Calculate the band-gap energy that corresponds to this wavelength.

Superconductors (Section 12.9)

12.110 What are the two most striking properties of a superconductor?

12.111 $YBa_2Cu_3O_7$ is a superconductor below its T_c of 90 K and a metallic conductor above 90 K. Make a rough plot of electrical resistance versus temperature for $YBa_2Cu_3O_7$.

12.112 What is the coordination environment of the K^+ ions in the fullerene-based superconductor K_3C_{60}?

12.113 Looking at Figure 12.27 identify the coordination numbers of the Cu, Y, and Ba atoms.

12.114 Superconductors with values of T_c above 77 K are of special interest. What's so special about 77 K?

12.115 The $YBa_2Cu_3O_7$ superconductor can be synthesized by the RAN sol–gel method from a stoichiometric mixture of metal ethoxides followed by heating in oxygen. How many grams of $Y(OCH_2CH_3)_3$ and how many grams of $Ba(OCH_2CH_3)_2$ are required to react with 75.4 g of $Cu(OCH_2CH_3)_2$ and an excess of water? Assuming a 100% yield, how many grams of $YBa_2Cu_3O_7$ are obtained?

Ceramics and Composites (Section 12.10)

12.116 What is a ceramic, and what properties distinguish a ceramic from a metal?

12.117 Contrast the bonding in ceramics with the bonding in metals.

12.118 Why are ceramics more wear-resistant than metals?

12.119 Why are oxide ceramics more corrosion-resistant than metals?

12.120 Silicon nitride (Si_3N_4), a high-temperature ceramic useful for making engine components, is a covalent network solid in which each Si atom is bonded to four N atoms and each N atom is bonded to three Si atoms. Explain why silicon nitride is more brittle than a metal like copper.

12.121 Magnesia (MgO), used as an insulator for electrical heating devices, has a face-centered cubic structure like that of NaCl. Draw one unit cell of the structure of MgO, and explain why MgO is more brittle than magnesium metal.

12.122 What is ceramic processing?

12.123 Describe what happens when a ceramic powder is sintered.

12.124 Zirconia (ZrO_2), an unusually tough oxide ceramic, has been used to make very sharp table knives. Write a balanced equation for the hydrolysis of zirconium isopropoxide in the sol–gel method for making zirconia powders. The isopropoxide ligand is the anion of isopropyl alcohol, $HOCH(CH_3)_2$.

12.125 Zinc oxide is a semiconducting ceramic used to make *varistors* (variable resistors). Write a balanced equation for the hydrolysis of zinc ethoxide in the sol–gel method for making ZnO powders.

12.126 Describe the reactions that occur when an $Si(OH)_4$ sol becomes a gel. What is the formula of the ceramic obtained when the gel is dried and sintered?

12.127 Describe the reactions that occur when a $Y(OH)_3$ sol becomes a gel. What is the chemical formula of the ceramic obtained when the gel is dried and sintered?

12.128 Some ceramics can be synthesized easily from molecular or ionic precursors. Titanium diboride (TiB_2) powder forms from the decomposition of titanium(III) borohydride dissolved in an organic solvent at 140 °C. Write a balanced equation for this reaction if the other products are diborane (B_2H_6) and molecular hydrogen. (The borohydride ion is BH_4^-.)

12.129 GeS_2 is a nonoxide ceramic accessible by the sol–gel method. Write a balanced equation for the preparation of GeS_2 from tetraethoxygermane, $Ge(OCH_2CH_3)_4$, and H_2S in an organic solvent.

12.130 Silicon nitride powder can be made by the reaction of silicon tetrachloride vapor with gaseous ammonia. The by-product is gaseous hydrogen chloride. Write a balanced equation for the reaction.

12.131 Boron, which is used to make composites, is deposited on a tungsten wire when the wire is heated electrically in the presence of boron trichloride vapor and gaseous hydrogen. Write a balanced equation for the reaction.

12.132 Explain why graphite/epoxy composites are good materials for making tennis rackets and golf clubs.

12.133 Explain why silicon carbide–reinforced alumina is stronger and tougher than pure alumina.

MULTICONCEPT PROBLEMS

12.134 The mineral magnetite is an iron oxide ore that has a density RAN of 5.20 g/cm³. At high temperature, magnetite reacts with carbon monoxide to yield iron metal and carbon dioxide. When 2.660 g of magnetite is allowed to react with sufficient carbon monoxide, the CO_2 product is found to have a volume of 1.136 L at 298 K and 751 mm Hg pressure.

(a) What mass of iron in grams is formed in the reaction?

(b) What is the formula of magnetite?

(c) Magnetite has a somewhat complicated cubic unit cell with an edge length of 839 pm. How many Fe and O atoms are present in each unit cell?

12.135 An 8.894 g block of aluminum was pressed into a thin RAN square of foil with 36.5 cm edge lengths.

(a) If the density of Al is 2.699 g/cm³, how thick is the foil in centimeters?

(b) How many unit cells thick is the foil? Aluminum crystallizes in a face-centered cubic structure and has an atomic radius of 143 pm.

12.136 A group 3A metal has a density of 2.70 g/cm³ and a cubic unit cell with an edge length of 404 pm. Reaction of a 1.07 cm³ chunk of the metal with an excess of hydrochloric acid gives a colorless gas that occupies 4.00 L at 23.0 °C and a pressure of 740 mm Hg.

(a) Identify the metal.

(b) Is the unit cell primitive, body-centered, or face-centered?

(c) What is the atomic radius of the metal atom in picometers?

12.137 A cube-shaped crystal of an alkali metal, 1.62 mm on an edge,
RAN was vaporized in a 500.0 mL evacuated flask. The resulting vapor pressure was 12.5 mm Hg at 802 °C. The structure of the solid metal is known to be body-centered cubic.

(a) What is the atomic radius of the metal atom in picometers?

(b) Use the data in Figure 5.19 to identify the alkali metal.

(c) What are the densities of the solid and the vapor in g/cm³?

12.138 Assume that 1.588 g of an alkali metal undergoes complete
RAN reaction with the amount of gaseous halogen contained in a 0.500 L flask at 298 K and 755 mm Hg pressure. In the reaction, 22.83 kJ is released ($\Delta H = -22.83$ kJ). The product, a binary ionic compound, crystallizes in a unit cell with anions in a face-centered cubic arrangement and with cations centered along each edge between anions. In addition, there is a cation in the center of the cube.

(a) What is the identity of the alkali metal?

(b) The edge length of the unit cell is 535 pm. Find the radius of the alkali metal cation from the data in Figure 6.1, and then calculate the radius of the halide anion. Identify the anion from the data in Figure 6.2.

(c) Sketch a space-filling, head-on view of the unit cell, labeling the ions. Are the anions in contact with one another?

(d) What is the density of the compound in g/cm³?

(e) What is the standard heat of formation for the compound?

12.139 Europium(II) oxide is a semiconductor with a band gap of 108 kJ/mol. Below 69 K, it is also ferromagnetic, meaning all the unpaired electrons on europium are aligned in the same direction. How many f electrons are present on each europium ion in EuO? (In lanthanide ions the $4f$ orbitals are lower in energy than the $6s$ orbitals.)

12.140 Red light with a wavelength of 660 nm from a 3.0 mW diode
RAN laser shines on a solar cell.

(a) How many photons per second are emitted by the laser? (1 W = 1 J/s)

(b) How much current (in amperes) flows in the circuit of the solar cell if all the photons are absorbed by the cell and each photon produces one electron?

12.141 The mineral wustite is a nonstoichiometric iron oxide with
RAN the empirical formula Fe_xO, where x is a number slightly less than 1. Wustite can be regarded as an FeO in which some of the Fe sites are vacant. It has a density of 5.75 g/cm³, a cubic unit cell with an edge length of 431 pm, and a face-centered cubic arrangement of oxygen atoms.

(a) What is the value of x in the formula Fe_xO?

(b) Based on the formula in part (a), what is the average oxidation state of Fe?

(c) Each Fe atom in wustite is in either the +2 or the +3 oxidation state. What percent of the Fe atoms are in the +3 oxidation state?

(d) Using X rays with a wavelength of 70.93 pm, at what angle would third-order diffraction be observed from the planes of atoms that coincide with the faces of the unit cells? Third-order diffraction means that the value of n in the Bragg equation is equal to 3.

(e) Wustite is a semiconducting iron(II) oxide in which some of the Fe^{2+} has been replaced by Fe^{3+}. Should it be described as an n-type or a p-type semiconductor? Explain.

12.142 The alkali metal fulleride superconductors M_3C_{60} have a cubic closest-packed (face-centered cubic) arrangement of nearly spherical C_{60}^{3-} anions with M^+ cations in the holes between the larger C_{60}^{3-} ions. The holes are of two types: octahedral holes, which are surrounded octahedrally by six C_{60}^{3-} ions; and tetrahedral holes, which are surrounded tetrahedrally by four C_{60}^{3-} ions.

(a) Sketch the three-dimensional structure of one unit cell.

(b) How many C_{60}^{3-} ions, octahedral holes, and tetrahedral holes are present per unit cell?

(c) Specify fractional coordinates for all the octahedral and tetrahedral holes. (Fractional coordinates are fractions of the unit cell edge lengths. For example, a hole at the center of the cell has fractional coordinates $\frac{1}{2}, \frac{1}{2}, \frac{1}{2}$.)

(d) The radius of a C_{60}^{3-} ion is about 500 pm. Assuming that the C_{60}^{3-} ions are in contact along the face diagonals of the unit cell, calculate the radii of the octahedral and tetrahedral holes.

(e) The ionic radii of Na^+, K^+, and Rb^+ are 102, 138, and 152 pm, respectively. Which of these ions will fit into the octahedral and tetrahedral holes? Which ions will fit only if the framework of C_{60}^{3-} ions expands?

12.143 A single-walled carbon nanotube can be regarded as a sin-
RAN gle sheet of graphite rolled into a cylinder (Figure 12.14). Calculate the number of C atoms in a single-walled nanotube having a length of 1.0 mm and a diameter of 1.08 nm. The C—C bond length in graphite is 141.5 pm.

12.144 Small molecules with C=C double bonds, called monomers, can join with one another to form long chain molecules called polymers. Thus, acrylonitrile, H_2C=CHCN, polymerizes under appropriate conditions to give polyacrylonitrile, a common starting material for producing the carbon fibers used in composites.

(a) Write electron-dot structures for acrylonitrile and polyacrylonitrile, and show how rearranging the electrons can lead to formation of the polymer.

(b) Use the bond dissociation energies in Table 7.1 to calculate ΔH per H_2C=CHCN unit for the conversion of acrylonitrile to polyacrylonitrile. Is the reaction endothermic or exothermic?

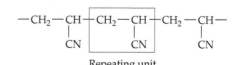

Repeating unit

Polyacrylonitrile

chapter 13

Solutions and Their Properties

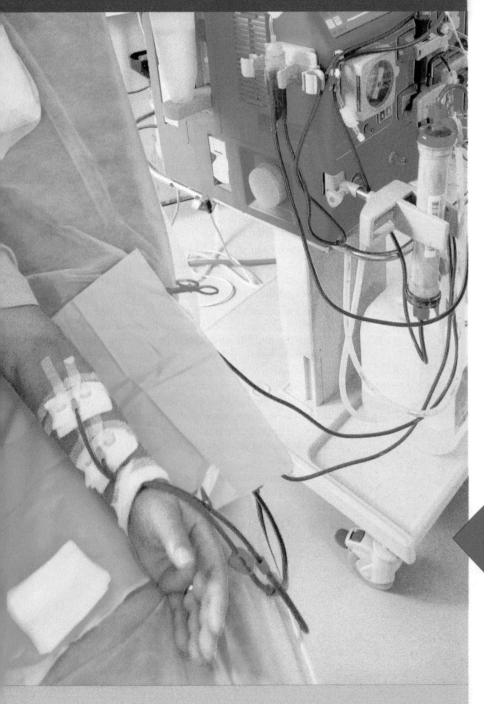

Contents

Hemodialysis on this artificial kidney machine cleanses the blood of individuals whose kidneys no longer function. The principles of osmosis discussed in this chapter are the basis for hemodialysis.

How does hemodialysis cleanse the blood of patients with kidney failure?

The answer to this question can be found on page 525 in the INQUIRY?

Thus far, we've been concerned mainly with pure substances, both elements and compounds. If you look around, though, most of the substances you see in day-to-day life are *mixtures*. Air is a gaseous mixture of (primarily) oxygen and nitrogen, gasoline is a liquid mixture of many different organic compounds, and rocks are solid mixtures of different minerals.

A **mixture** is any combination of two or more pure substances blended together in some arbitrary proportion without chemically changing the individual substances themselves. Mixtures can be classified as either *heterogeneous* or *homogeneous,* depending on their appearance. A heterogeneous mixture is one in which the mixing of components is visually nonuniform and that therefore has regions of different composition. Sugar with salt and oil with water are examples. A homogeneous mixture is one in which the mixing *is* uniform, at least to the naked eye, and that therefore has a constant composition throughout. Seawater (sodium chloride with water) and brass (copper with zinc) are examples. We'll explore the properties of some homogeneous mixtures in this chapter, with particular emphasis on the mixtures we call *solutions*.

13.1 SOLUTIONS

Homogeneous mixtures can be classified according to the size of their constituent particles as either *solutions* or *colloids*. **Solutions,** the most common class of homogeneous mixtures, contain particles with diameters in the range 0.1–2 nm, the size of a typical ion or small molecule. They are transparent, although they may be colored, and they don't separate on standing. **Colloids,** such as milk and fog, contain larger particles, with diameters in the range 2–500 nm. Although they are often murky or opaque to light, colloids don't separate on standing. Mixtures called *suspensions* also exist, having even larger particles than colloids. These are not truly homogeneous, however, because their particles separate out on standing and are visible with a low-power microscope. Blood, paint, and aerosol sprays are examples.

We usually think of a solution as a solid dissolved in a liquid or as a mixture of liquids, but there are many other kinds of solutions as well. In fact, any one state of matter can form a solution with any other state, and seven different kinds of solutions are possible (**TABLE 13.1**).

Even solutions of one solid with another and solutions of a gas in a solid are well known. Metal alloys, such as stainless steel (4–30% chromium in iron) and brass (10–40% zinc in copper), are solid/solid solutions, and hydrogen in palladium is a gas/solid solution. Metallic palladium, in fact, is able to absorb up to 935 times its own volume of H_2 gas!

For solutions in which a gas or solid is dissolved in a liquid, the dissolved substance is called the **solute** and the liquid is called the **solvent.** When one liquid is dissolved in another, the minor component is usually considered the solute and the major component is the solvent. Thus, ethyl alcohol is the solute and water the solvent in a mixture of 10% ethyl alcohol and 90% water, but water is the solute and ethyl alcohol the solvent in a mixture of 90% ethyl alcohol and 10% water.

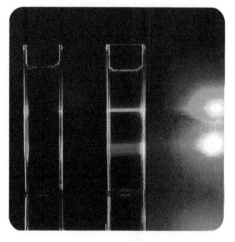

▲ The cuvette on the left contains a solution that does not scatter light from a laser. The cuvette on the right is a colloid, and the suspended particles scatter light from a laser.

TABLE 13.1	Some Different Kinds of Solutions
Kind of Solution	**Example**
Gas in gas	Air (O_2, N_2, Ar, and other gases)
Gas in liquid	Carbonated water (CO_2 in water)
Gas in solid	H_2 in palladium metal
Liquid in liquid	Gasoline (mixture of hydrocarbons)
Liquid in solid	Dental amalgam (mercury in silver)
Solid in liquid	Seawater (NaCl and other salts in water)
Solid in solid	Metal alloys, such as sterling silver (92.5% Ag, 7.5% Cu)

13.2 ENTHALPY CHANGES AND THE SOLUTION PROCESS

REMEMBER . . .
Intermolecular forces include ion–dipole, dipole–dipole, and London dispersion forces as well as hydrogen bonds (Section 8.6).

With the exception of gas/gas mixtures, such as air, the different kinds of solutions listed in Table 13.1 involve *condensed phases*, either liquid or solid. Thus, all the **intermolecular forces** described in Section 8.6 to explain the properties of pure liquids and solids are also important for explaining the properties of solutions. The situation is more complex for solutions than for pure substances, though, because three types of interactions among particles have to be taken into account: solute–solute interactions, solvent–solvent interactions, and solvent–solute interactions.

FIGURE 13.1 shows a molecular view of the solution-making process represented as three individual steps. The enthalpy change of each step is determined by the intermolecular forces involved in separating molecules or bringing molecules together.

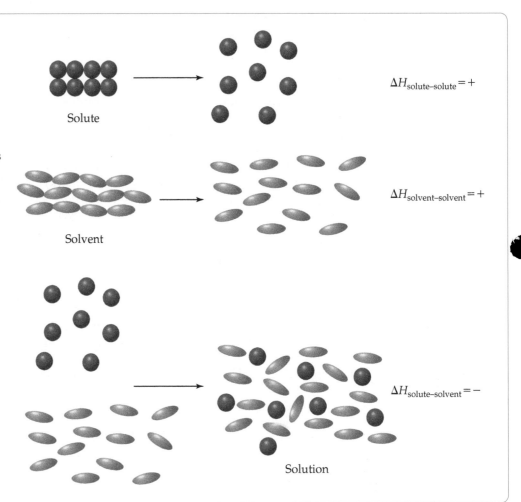

▶ **FIGURE 13.1**

A molecular view of the solution-making process portrayed as three individual steps. The enthalpy change for separating solute particles ($\Delta H_{\text{solute–solute}}$) is positive because energy is required to overcome intermolecular forces. The enthalpy change for separating solvent molecules ($\Delta H_{\text{solvent–solvent}}$) is positive because energy is required to overcome intermolecular forces. The enthalpy change for interacting solute and solvent ($\Delta H_{\text{solute–solvent}}$) is negative because energy is released when intermolecular forces form.

Solute $\Delta H_{\text{solute–solute}} = +$

Solvent $\Delta H_{\text{solvent–solvent}} = +$

$\Delta H_{\text{solute–solvent}} = -$

Solution

REMEMBER . . .
Lattice energy (U) is the sum of the electrostatic interaction energies between ions in a crystal. It is the amount of energy that must be supplied to break up an ionic solid into its individual gaseous ions (Section 6.8).

- **Step 1. Solute–solute interactions:** Energy is absorbed (positive ΔH) to overcome intermolecular forces holding solute particles together in a crystal. For an ionic solid, the amount of energy required to overcome solute–solute interactions is related to the **lattice energy**. As a result, substances with higher lattice energies tend to be less soluble than substances with lower lattice energies. Compounds with singly charged ions are thus more soluble than compounds with doubly or triply charged ions, as we saw when discussing solubility guidelines in Section 4.6.
- **Step 2. Solvent–solvent interactions:** Energy is absorbed (positive ΔH) to overcome intermolecular forces between solvent molecules because the molecules must be separated and pushed apart to make room for solute particles.
- **Step 3. Solvent–solute interactions:** Energy is released (negative ΔH) because intermolecular forces form when solvent molecules are attracted to solute particles.

The overall enthalpy change for solution formation is called the *heat of solution,* or **enthalpy of solution** (ΔH_{soln}), and is the sum of the enthalpy changes for each step.

$$\Delta H_{soln} = \Delta H_{solute-solute} + \Delta H_{solvent-solvent} + \Delta H_{solute-solvent}$$

The first two kinds of interactions are endothermic, requiring an input of energy to spread apart solvent molecules and to break apart crystals. Only the third interaction is exothermic, as attractive intermolecular forces develop between solvent and solute particles. The sum of the three interactions determines whether ΔH_{soln} is endothermic or exothermic. For some substances, the one exothermic interaction is sufficiently large to outweigh the two endothermic interactions, but for other substances, the opposite is true (**FIGURE 13.2**).

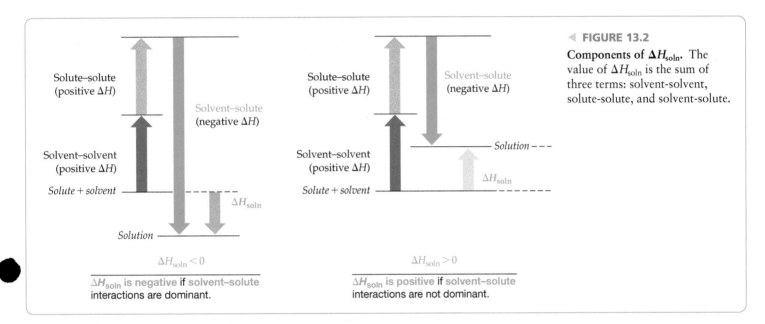

◀ **FIGURE 13.2**

Components of ΔH_{soln}. The value of ΔH_{soln} is the sum of three terms: solvent-solvent, solute-solute, and solvent-solute.

Athletes benefit from both negative and positive enthalpies of solution when they use instant hot packs or cold packs to treat injuries. Both kinds of instant packs consist of a pouch of water and a dry chemical, either $CaCl_2$ or $MgSO_4$ for hot packs and NH_4NO_3 for cold packs. When the pack is squeezed, the pouch breaks and the solid dissolves, either raising or lowering the temperature (**FIGURE 13.3**).

$$\text{Hot packs:} \quad CaCl_2(s) \quad \Delta H_{soln} = -81.3 \text{ kJ/mol}$$
$$MgSO_4(s) \quad \Delta H_{soln} = -91.2 \text{ kJ/mol}$$
$$\text{Cold pack:} \quad NH_4NO_3(s) \quad \Delta H_{soln} = +25.7 \text{ kJ/mol}$$

Let's examine the types of intermolecular forces that must break or form when sodium chloride (NaCl) dissolves in water. **FIGURE 13.4** shows a molecular representation of the process of dissolving sodium and chloride ions from a crystal of NaCl. When solid NaCl is placed in water, those ions that are less tightly held because of their position at a corner or an edge of the crystal are exposed to water molecules, which collide with them until an ion happens to break free. More water molecules then cluster around the ion, stabilizing it by means of ion–dipole attractions. A new edge or corner is thereby exposed on the crystal, and the process continues until the entire crystal has dissolved. The ions in solution are said to be *solvated*—more specifically, *hydrated,* when water is the solvent—meaning that they are surrounded and stabilized by an ordered shell of solvent molecules. For ionic substances in water, the amount of hydration energy released is generally greater for smaller cations than for larger ones because water molecules can approach the positive nuclei of smaller ions more closely and thus bind more tightly. In addition, hydration energy generally increases as the charge on the ion increases.

BIG IDEA Question 1

What is the sign of ΔH_{soln}, and how will the temperature of the solution change when a solute dissolves with the following enthalpy changes?

$$\Delta H_{solute-solute} = +90 \text{ kJ/mol}$$
$$\Delta H_{solvent-solvent} = +25 \text{ kJ/mol}$$
$$\Delta H_{solute-solvent} = -130 \text{ kJ/mol}$$

▶ **FIGURE 13.3**

Enthalpy of solution. Enthalpies of solution can be either negative (exothermic) or positive (endothermic).

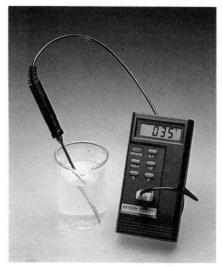 

Dissolution of $CaCl_2$ in water is **exothermic**, causing the temperature of the water to rise from its initial 25 °C value.

Dissolution of NH_4NO_3 is **endothermic**, causing the temperature of the water to fall from its initial 25 °C value.

▶ **FIGURE 13.4**

Dissolution of NaCl crystals in water.

▶ **Figure It out**

Name the strongest type of intermolecular force between (a) solvent–solvent, (b) solvent–solute, and (c) solute–solute when NaCl dissolves in water.

Answer: (a) hydrogen bonding, (b) ion–dipole, (c) ionic bond

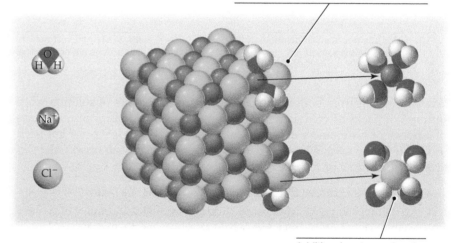

Water molecules surround an accessible edge or corner ion in a crystal and collide with it until the ion breaks free.

Additional water molecules then surround the ion and stabilize it by means of ion–dipole attractions.

13.3 PREDICTING SOLUBILITY

REMEMBER . . .
The sign of the Gibbs free-energy change (ΔG) is used to determine if a process occurs spontaneously. If ΔG is negative, the process is spontaneous, and if ΔG is positive, the process is nonspontaneous (Section 9.13).

Like all chemical and physical processes, the dissolution of a solute in a solvent has associated with it a **Gibbs free-energy change**, $\Delta G = \Delta H - T\Delta S$, whose value describes its spontaneity. If ΔG is negative, the process is spontaneous and the substance dissolves; if ΔG is positive, the process is nonspontaneous and the substance does not dissolve. The enthalpy term ΔH measures the heat flow into or out of the system during dissolution, and the temperature-dependent entropy term $T\Delta S$ measures the change in the amount of molecular randomness in the system. In the previous section, we defined the enthalpy of solution (ΔH_{soln}) as the heat released or absorbed upon solution formation, and similarly the **entropy of solution** (ΔS_{soln}) is the entropy change that occurs when the solute dissolves in the solvent.

What values might we expect for ΔH_{soln} and ΔS_{soln}? Let's take the entropy change first. Entropies of solution are usually positive because molecular randomness usually increases during dissolution: $+43.4$ J/(K·mol) for NaCl in water, for example. When a solid dissolves in a liquid, randomness increases upon going from a well-ordered crystal to a less-ordered state in which solvated ions or molecules are able to move freely in solution. When one liquid dissolves in another, randomness increases as the different molecules intermingle (**FIGURE 13.5**).

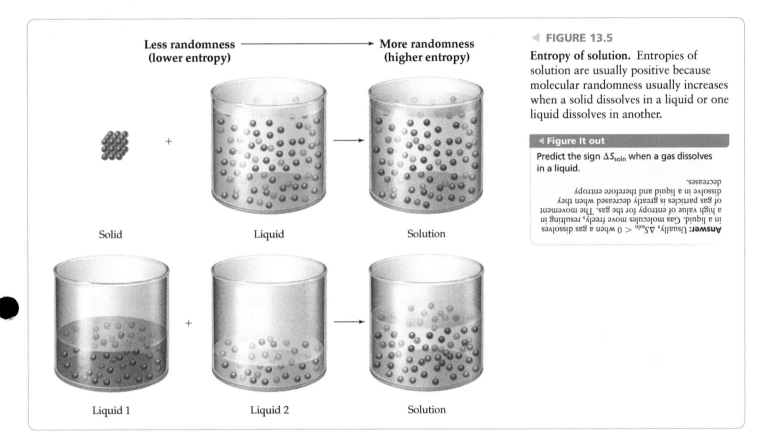

Less randomness   More randomness
(lower entropy)   (higher entropy)

Solid   Liquid   Solution

Liquid 1   Liquid 2   Solution

◀ **FIGURE 13.5**

Entropy of solution. Entropies of solution are usually positive because molecular randomness usually increases when a solid dissolves in a liquid or one liquid dissolves in another.

◀ **Figure It out**

Predict the sign ΔS_{soln} when a gas dissolves in a liquid.

Answer: Usually, $\Delta S_{soln} > 0$ when a gas dissolves in a liquid. Gas molecules move freely, resulting in a high value of entropy for the gas. The movement of gas particles is greatly decreased when they dissolve in a liquid and therefore entropy decreases.

As depicted in Figure 13.5, most solids and liquids dissolve with an increase in entropy, $\Delta S_{soln} = +$. However, values for enthalpies of solution, ΔH_{soln}, are more difficult to predict. Some solids dissolve exothermically and have a negative ΔH_{soln} (-37.0 kJ/mol for LiCl in water), but others dissolve endothermically and have a positive ΔH_{soln} ($+17.2$ kJ/mol for KCl in water). Figure 13.2 in the previous section shows how the interplay of intermolecular forces between solute and solvent particles influences ΔH_{soln}. **TABLE 13.2** lists values of ΔH_{soln} and ΔS_{soln} for some common ionic substances.

TABLE 13.2   Some Enthalpies and Entropies of Solution in Water at 25 °C

Substance	ΔH°_{soln}(kJ/mol)	ΔS°_{soln}[J/(K·mol)]
LiCl	-37.0	10.5
NaCl	3.9	43.4
KCl	17.2	75.0
LiBr	-48.8	21.5
NaBr	-0.6	54.6
KBr	19.9	89.0
KOH	-57.6	12.9

▲ Oil and water do not mix because water is polar and oil is nonpolar. The dissimilar types of intermolecular forces cause oil to be nearly insoluble in water.

How can we predict if a solute will be soluble in a given solvent? Thermodynamics tells us that if the free-energy change of solution formation is negative, then the process will be spontaneous. The free-energy change for dissolving sodium chloride at 25 °C is

$$\Delta G° = \Delta H° - T\Delta S°$$
$$\Delta G° = (+3,900 \text{ J/mol}) - (298 \text{ K})(43.4 \text{ J/K·mol})$$
$$= -9,033 \text{ J/mol} = -9.0 \text{ kJ/mol}$$

The negative sign of $\Delta G°$ indicates that process of dissolving sodium chloride in water is spontaneous. However, it is most convenient to predict solubility using the simple solubility rule summarized by the phrase "like dissolves like." Solutions will form when the types of intermolecular forces between solute particles, solvent molecules, and solvent and solute particles are similar in kind and in magnitude. Thus, ionic solids like NaCl dissolve in polar solvents like water because the strong ion–dipole attractions between Na^+ and Cl^- ions and polar H_2O molecules are similar in magnitude to the strong hydrogen bonding attractions between water molecules and to the strong ion–ion attractions between Na^+ and Cl^- ions. In the same way, nonpolar organic substances like cholesterol, $C_{27}H_{46}O$, dissolve in nonpolar organic solvents like benzene, C_6H_6, because of the similar London dispersion forces present among both kinds of molecules. Oil, however, does not dissolve appreciably in water because the two liquids have different kinds of intermolecular forces.

CONCEPTUAL WORKED EXAMPLE 13.1

Predicting Solubility from Chemical Structure

Pentane (C_5H_{12}) and 1-butanol (C_4H_9OH) are organic liquids with similar molecular weights but substantially different solubility behavior. Which of the two would you expect to be more soluble in water? Explain.

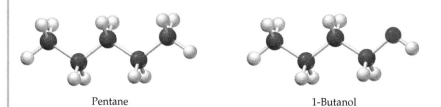

Pentane   1-Butanol

STRATEGY

Look at the two structures, and decide on the kinds of intermolecular forces present between molecules in each case. If the intermolecular forces between the three types of interactions—solvent–solvent interactions, solvent–solute interactions, and solute–solute interactions—are similar in magnitude, then a solution will likely form. The substance with intermolecular forces more like those in water will probably be more soluble in water.

SOLUTION

London dispersion forces are the intermolecular forces between two pentane molecules because pentane is nonpolar. In contrast, water is polar and forms hydrogen bonds. Pentane is unlikely to have strong intermolecular interactions with water. 1-Butanol, however, has an —OH group and can hydrogen bond to itself and to water. 1-Butanol forms 3 hydrogen bonds with water, resulting in strong solvent–solute interactions as shown. As a result, 1-butanol is more soluble in water.

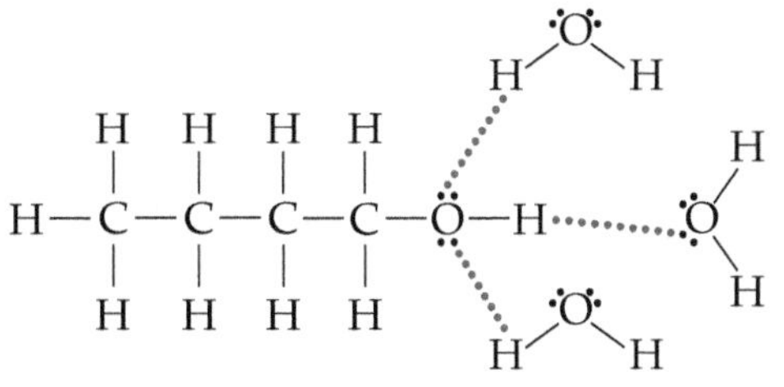

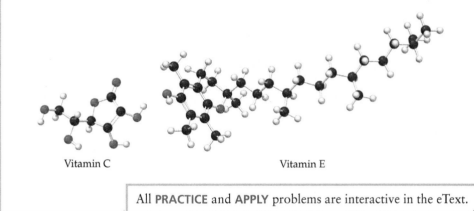

▶ **CONCEPTUAL PRACTICE 13.1**   Arrange the following three compounds in order of their expected increasing solubility in pentane (C_5H_{12}): 1,5 pentanediol ($C_5H_{12}O_2$), KBr, toluene (C_7H_8, a constituent of gasoline).

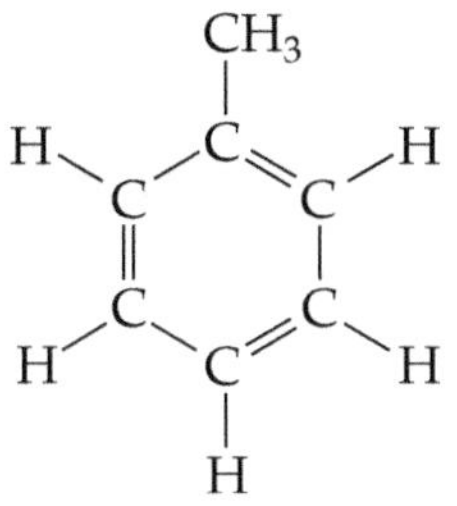

Toluene

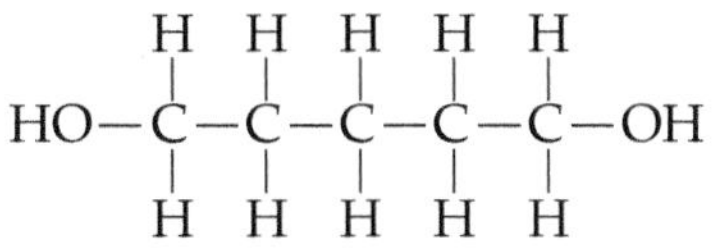

1,5 Pentanediol

▶ **CONCEPTUAL APPLY 13.2**   Many people take vitamin supplements to promote health, and the potential for overdose depends on whether a particular vitamin is fat soluble or water soluble. If a vitamin is soluble in nonpolar fatty tissues, it will accumulate in the body, whereas water-soluble vitamins are readily excreted. Use the ball-and-stick models of vitamin C and vitamin E to determine which molecule is more fat soluble.

Vitamin C

Vitamin E

All **PRACTICE** and **APPLY** problems are interactive in the eText.

13.4 CONCENTRATION UNITS FOR SOLUTIONS

In daily life, it's often sufficient to describe a solution as either *dilute* or *concentrated*. In scientific work, though, it's usually necessary to know the exact concentration of a solution—that is, to know the exact amount of solute dissolved in a certain amount of solvent. There are many ways of expressing concentration, each of which has its own advantages and disadvantages. The four most common units for concentration are **molarity**, *mole fraction*, *mass percent*, and *molality*.

Molarity (M)

The most common way of expressing concentration in a chemistry laboratory is to use molarity (M). As discussed in Section 4.1, a solution's molarity is given by the number of moles of solute per liter of solution (mol/L, abbreviated M). If, for example, you

REMEMBER . . .
A solution of a given **molarity** is prepared by dissolving a solute in a small amount of solvent and then diluting with more solvent to the desired volume. The solution is not made by dissolving the solute in the desired volume of solvent (Section 4.1).

dissolve 0.500 mol (20.0 g) of NaOH in enough water to give 1.000 L of solution, then the solution has a concentration of 0.500 M.

$$\text{Molarity (M)} = \frac{\text{Moles of solute}}{\text{Liters of solution}}$$

The advantages of using molarity are twofold: (1) Stoichiometry calculations are simplified because numbers of moles are used rather than mass, and (2) amounts of solution (and therefore of solute) are conveniently measured by volume (pouring) rather than by mass (weighing). As a result, titrations are particularly easy.

The disadvantages of using molarity are also twofold: (1) The exact solute concentration depends on the temperature because the volume of a solution expands or contracts as the temperature changes, and (2) the exact amount of solvent in a given volume can't be determined unless the density of the solution is known.

REMEMBER . . .
Titration is a procedure for determining the concentration of a solution by allowing a carefully measured amount of the solution to react with a standard solution of another substance whose concentration is known (Section 4.9).

Mole Fraction (*X*)

As discussed in Section 10.5, the mole fraction (*X*) of any component in a solution is given by the number of moles of the component divided by the total number of moles making up the solution (including solvent):

$$\text{Mole fraction (X)} = \frac{\text{Moles of component}}{\text{Total moles making up the solution}}$$

For example, a solution prepared by dissolving 1.00 mol (32.0 g) of methyl alcohol (CH_3OH) in 5.00 mol (90.0 g) of water has a methyl alcohol concentration $X = 1.00 \text{ mol}/(1.00 \text{ mol} + 5.00 \text{ mol}) = 0.167$. Note that mole fractions are dimensionless because the units cancel.

Mole fractions are independent of temperature and are particularly useful for calculations involving gas mixtures. Except in special situations, mole fractions are not often used for liquid solutions because other units are generally more convenient.

Mass Percent (Mass %)

As the name suggests, the **mass percent (mass %)** of any component in a solution is the mass of that component divided by the total mass of the solution times 100%:

$$\text{Mass percent} = \frac{\text{Mass of component}}{\text{Total mass of solution}} \times 100\%$$

For example, a solution prepared by dissolving 10.0 g of glucose in 100.0 g of water has a glucose concentration of 9.09 mass %:

$$\text{Mass \% glucose} = \frac{10.0 \text{ g}}{10.0 \text{ g} + 100.0 \text{ g}} \times 100\% = 9.09 \text{ mass \%}$$

Closely related to mass percent, and particularly useful for very dilute solutions, are the concentration units **parts per million (ppm)** and **parts per billion (ppb)**:

$$\text{Parts per million (ppm)} = \frac{\text{Mass of component}}{\text{Total mass of solution}} \times 10^6$$

$$\text{Parts per billion (ppb)} = \frac{\text{Mass of component}}{\text{Total mass of solution}} \times 10^9$$

A concentration of 1 ppm for a substance means that each kilogram (1 million mg) of solution contains 1 mg of solute. For dilute aqueous solutions near room temperature, where 1 kg has a volume of 1 L, 1 ppm also means that each liter of solution contains 1 mg of solute. In the same way, a concentration of 1 ppb means that each liter of an aqueous solution contains 0.001 mg of solute.

Values in ppm and ppb are frequently used for expressing the concentrations of trace amounts of impurities in air or water. Thus, you might express the maximum allowable concentration of lead in drinking water as 15 ppb, or about 1 g per 67,000 L.

The advantage of using mass percent (or ppm) for expressing concentration is that the values are independent of temperature because masses don't change when substances are heated or cooled. The disadvantage of using mass percent is that it is generally less convenient when working with liquid solutions to measure amounts by mass rather than by volume. Furthermore, the density of a solution must be known before a concentration in mass percent can be converted into molarity. Worked Example 13.5 shows how to make the conversion.

▲ Dissolving a quarter teaspoon of table sugar in this backyard swimming pool would give a sugar concentration of about 1 ppb.

WORKED EXAMPLE 13.2

Mass Percent Concentration

Assume that you have a 5.75 mass % solution of LiCl in water. What mass of solution in grams contains 1.60 g of LiCl?

IDENTIFY

Known	Unknown
Mass percent (5.75%)	Mass of solution (g)
Mass of solute (1.60 g LiCl)	

STRATEGY

Describing this concentration as 5.75 mass % means that 100.0 g of aqueous solution contains 5.75 g of LiCl (and 94.25 g of H_2O), a relationship that can be used as a conversion factor.

SOLUTION

$$\text{Mass of soln} = 1.60 \text{ g LiCl} \times \frac{100 \text{ g soln}}{5.75 \text{ g LiCl}} = 27.8 \text{ g soln}$$

▶ **PRACTICE 13.3** What is the mass percent concentration of a saline solution prepared by dissolving 1.00 mol of NaCl in 1.00 L of water? (Assume the density of water is 1.00 g/mL.)

▶ **APPLY 13.4** Calculate the amount of water (in grams) that must be added to 15.0 g of sucrose ($C_{12}H_{22}O_{11}$) to make a solution with a concentration of 7.50 mass %.

WORKED EXAMPLE 13.3

Using Parts per Million (ppm) as a Concentration Unit

Nitrate (NO_3^-) enters the water supply from fertilizer runoff or leaking septic tanks. The legal limit in drinking water is 10.0 ppm because infants who drink water exceeding this level can become seriously ill. What mass of nitrate (mg) is present in 1.5 L of water with a concentration of 10.0 ppm? (Assume the density of water is 1.0 g/mL.)

IDENTIFY

Known	Unknown
Concentration of nitrate (10.0 ppm)	Mass of nitrate (mg)
Volume of water (1.5 L)	

STRATEGY

Use density to convert from volume of water to mass of water (total mass of solution). Substitute known values into the equation for parts per million, and solve for the mass of NO_3^-.

SOLUTION

Find the mass of water: $1.5 \text{ L} \times \frac{1 \text{ mL}}{1 \times 10^{-3} \text{ L}} \times \frac{1.0 \text{ g}}{1 \text{ mL}} = 1500 \text{ g}$

Use the equation for ppm: $10.0 \text{ ppm} = \frac{\text{Mass of } NO_3^-}{1500 \text{ g}} \times 10^6$

Mass of $NO_3^- = 0.015 \text{ g} = 15 \text{ mg}$

CHECK

The mass of nitrate is very small, which is reasonable because ppm is used to express trace amounts of solutes. Since the

continued on next page

multiplication factor for ppm is 10^6, we would expect the amount of solute to be about 10^6 times less than the total mass of the solution.

▶ **PRACTICE 13.5** A 50.0 mL sample of drinking water was found to contain 1.25 μg of arsenic. Calculate the concentration in units of ppb, and determine whether the sample

exceeds the legal limit of 10.0 ppb. (Assume the density of water is 1.00 g/mL.)

▶ **APPLY 13.6** The legal limit for human exposure to carbon monoxide in the workplace is 35 ppm. Assuming that the density of air is 1.3 g/L, how many grams of carbon monoxide are in 1.0 L of air at the maximum allowable concentration?

Molality (m)

The **molality** (m) of a solution is defined as the number of moles of solute per kilogram of solvent (mol/kg):

$$\text{Molality } (m) = \frac{\text{Moles of solute}}{\text{Mass of solvent (kg)}}$$

To prepare a 1.000 m solution of KBr in water, for example, you might dissolve 1.000 mol of KBr (119.0 g) in 1.000 kg (1000 mL) of water. You can't say for sure what the final volume of the solution will be, although it will probably be a bit larger than 1000 mL. Although the names sound similar, note the differences between molarity and molality. Molarity is the number of moles of solute per *volume* (liter) of *solution*, whereas molality is the number of moles of solute per *mass* (kilogram) of *solvent*.

The main advantage of using molality is that it is temperature-independent because masses don't change when substances are heated or cooled. Thus, it is well suited for calculating certain properties of solutions that we'll discuss later in this chapter. The disadvantages of using molality are that amounts of solution must be measured by mass rather than by volume and that the density of the solution must be known to convert molality into molarity (see Worked Example 13.5).

A summary of the four most common units for expressing concentration, together with a comparison of their relative advantages and disadvantages, is given in **TABLE 13.3.**

TABLE 13.3   A Comparison of Various Concentration Units

Name	Units	Advantages	Disadvantages
Molarity (M)	$\dfrac{\text{mol solute}}{\text{L solution}}$	Useful in stoichiometry; by volume	Temperature-dependent; must know density to find solvent mass
Mole fraction (X)	none	Temperature-independent; useful in special applications	Measure by mass; must know density to convert to molarity
Mass %	%	Temperature-independent; useful for small amounts	Measure by mass; must know density to convert to molarity
Molality (m)	$\dfrac{\text{mol solute}}{\text{kg solvent}}$	Temperature-independent; useful in special applications	Measure by mass; must know density to convert to molarity

WORKED EXAMPLE 13.4

Calculating the Molality of a Solution

What is the molality of a solution made by dissolving 1.45 g of table sugar (sucrose, $C_{12}H_{22}O_{11}$) in 30.0 mL of water? The molar mass of sucrose is 342.3 g/mol, and the density of water is 1.00 g/mL.

IDENTIFY

Known	Unknown
Mass of sucrose (1.45 g)	Molality (m)
Volume of water (30.0 mL)	
Molar mass of sucrose (342.3 g/mol)	

STRATEGY

Molality is the number of moles of solute per kilogram of solvent. Thus, we need to find how many moles are in 1.45 g of sucrose and how many kilograms are in 30.0 mL of water.

SOLUTION

The number of moles of sucrose is

$$1.45 \text{ g sucrose} \times \frac{1 \text{ mol sucrose}}{342.3 \text{ g sucrose}} = 4.24 \times 10^{-3} \text{ mol sucrose}$$

Since the density of water is 1.00 g/mol, 30.0 mL of water has a mass of 30.0 g, or 0.0300 kg. Thus, the molality of the solution is

$$\text{Molality} = \frac{4.24 \times 10^{-3} \text{ mol}}{0.0300 \text{ kg}} = 0.141 \; m$$

▶ **PRACTICE 13.7**   What mass in grams of a 0.500 m solution of sodium acetate, CH_3CO_2Na, in water would you use to obtain 0.150 mol of sodium acetate?

▶ **APPLY 13.8**   What is the molality of a solution prepared by dissolving 0.385 g of cholesterol, $C_{27}H_{46}O$, in 40.0 g of chloroform, $CHCl_3$? What is the mole fraction of cholesterol in the solution?

WORKED EXAMPLE 13.5

Using Density to Convert Molarity to Other Measures of Concentration

A 0.750 M solution of H_2SO_4 in water has a density of 1.046 g/mL at 20 °C. What is the concentration of this solution in (a) mole fraction, (b) mass percent, and (c) molality? The molar mass of H_2SO_4 is 98.1 g/mol.

IDENTIFY

Known	Unknown
Concentration of H_2SO_4 solution (0.750M)	Mole fraction, mass percent, molality
Density of H_2SO_4 solution (1.046 g/mL)	
Molar mass of H_2SO_4 (98.1 g/mol)	

STRATEGY

Write the expressions for both the beginning and ending concentrations labeled with units. This will help you see how to use the known quantities of density and molar mass to change from the starting unit into the ending unit.

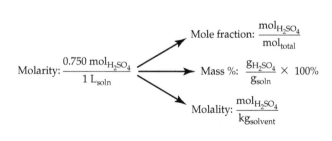

SOLUTION

(a) **Mole fraction:** To calculate mole fraction, the 1.00 L of solution needs to be converted to total number of moles making up the solution (mol H_2SO_4 + mol H_2O). The number of moles of H_2SO_4 in 1.00 L is 0.750, because the molarity is 0.750 M. To find the number of moles of H_2O, first find the total mass of the solution using the density.

$$\text{Mass of 1.00 L soln} = 1.00 \text{ L} \times \frac{1 \text{ mL}}{1 \times 10^{-3} \text{ L}} \times \frac{1.046 \text{ g}}{1 \text{ mL}} = 1046 \text{ g}$$

continued on next page

The total mass of solution is the mass of H_2SO_4 + the mass of H_2O. The mass of H_2SO_4 is found by converting from moles to grams:

$$\text{Mass of } H_2SO_4 \text{ in 1.00 L soln} = 0.750 \text{ mol} \times \frac{98.1 \text{ g}}{\text{mol}} = 73.6 \text{ g}$$

The mass of water is: 1046 g solution − 73.6 g H_2SO_4 = 972.4 g H_2O
The number of moles of water is found by converting mass into moles.

$$972.4 \text{ g } H_2O \times \frac{1 \text{ mol } H_2O}{18.0 \text{ g } H_2O} = 54.0 \text{ mol } H_2O$$

Thus, the mole fraction of H_2SO_4 is

$$X_{H_2SO_4} = \frac{0.750 \text{ mol } H_2SO_4}{0.750 \text{ mol } H_2SO_4 + 54.0 \text{ mol } H_2O} = 0.0137$$

(b) **Mass percent:** To calculate mass percent, convert 0.750 mol H_2SO_4 into grams using molar mass and convert 1.00 L of solution into grams using density. Both of these quantities were calculated in part (a). The mass % of H_2SO_4 is

$$\text{Mass \% of } H_2SO_4 = \frac{73.6 \text{ g } H_2SO_4}{1046 \text{ g solution}} \times 100\% = 7.04\%$$

(c) **Molality:** To calculate molality, convert 1.00 L of solution to kg of solvent. The mass of water was found in part (a) to be 972.4 g or 0.9724 kg.

$$\text{Molality of } H_2SO_4 = \frac{0.750 \text{ mol } H_2SO_4}{0.9724 \text{ kg } H_2O} = 0.771 \, m$$

▸ **PRACTICE 13.9**   The density at 20 °C of a 0.500 M solution of acetic acid in water is 1.0042 g/mL. What is the molality of the solution? The molar mass of acetic acid, CH_3CO_2H, is 60.05 g/mol.

▸ **APPLY 13.10**   The density at 20 °C of a 0.258 m solution of glucose in water is 1.0173 g/mL, and the molar mass of glucose is 180.2 g/mol. What is the molarity of the solution?

All **WORKED EXAMPLES** with this icon have an interactive video in the eText.

13.5 SOME FACTORS THAT AFFECT SOLUBILITY

If you take solid NaCl and add it to water, dissolution occurs rapidly at first but then slows down as more and more NaCl is added. Eventually the dissolution stops because a dynamic equilibrium is reached where the number of Na^+ and Cl^- ions leaving a crystal to go into solution is equal to the number of ions returning from the solution to the crystal. At this point, the maximum possible amount of NaCl has dissolved and the solution is said to be **saturated** in that solute.

$$\text{Solute + Solvent} \underset{\text{crystallize}}{\overset{\text{dissolve}}{\rightleftharpoons}} \text{Solution}$$

Note that this definition requires a saturated solution to be at *equilibrium* with undissolved solid at a given temperature. Some substances, however, can form what are called **supersaturated** solutions, which contain a greater-than-equilibrium amount of solute. For example, when a saturated solution of sodium acetate is prepared at high temperature and then cooled slowly, a supersaturated solution results, as shown in **FIGURE 13.6.** Such a solution is unstable, however, and precipitation occurs when a tiny seed crystal of sodium acetate is added to initiate crystallization.

Effect of Temperature on Solubility

The solubility of a substance—that is, the amount of the substance per unit volume of solvent needed to form a saturated solution at a given temperature—is a physical property characteristic of that substance. Different substances can have greatly different solubilities, as shown in **FIGURE 13.7.** Sodium chloride, for instance, has a solubility of

| A supersaturated solution of sodium acetate in water. | When a tiny seed crystal is added, larger crystals begin to grow and precipitate from the solution until equilibrium is reached. |

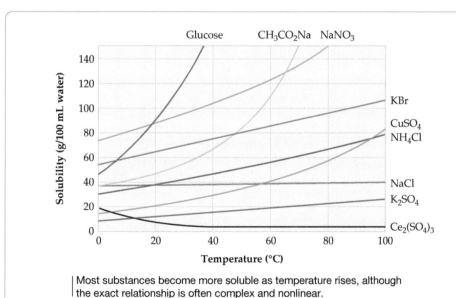

◀ **FIGURE 13.7**
Solubilities of some common solids in water as a function of temperature.

Most substances become more soluble as temperature rises, although the exact relationship is often complex and nonlinear.

35.9 g/100 mL of water at 20 °C, and sodium nitrate has a solubility of 87.3 g/100 mL of water at 20 °C. Sometimes, particularly when two liquids are involved, the solvent and solute are **miscible,** meaning that they are mutually soluble in all proportions. A solution of ethyl alcohol and water is an example.

Solubilities are temperature-dependent, so the temperature at which a specific measurement is made must be reported. As Figure 13.7 shows, there is no obvious correlation between structure and solubility or between solubility and temperature. The solubilities of most molecular and ionic solids increase with increasing temperature, but the solubilities of some (NaCl) are almost unchanged and the solubilities of others $[Ce_2(SO_4)_3]$ actually decrease.

The effect of temperature on the solubility of gases in water is more predictable than its effect on the solubility of solids: Gases become less soluble in water as the temperature increases (**FIGURE 13.8**). One consequence of this decreased solubility is that carbonated drinks bubble continuously as they warm up to room temperature after being refrigerated. Soon, they lose so much dissolved CO_2 that they go flat. A much more important consequence is the damage to aquatic life that can result from the decrease in concentration of dissolved oxygen in lakes and rivers when hot water is discharged from industrial plants, an effect known as thermal pollution.

▶ **FIGURE 13.8**

Solubilities of some gases in water as a function of temperature.

▶ **Figure It out**

Is the amount of dissolved oxygen higher in water at 10 °C or at 30 °C?

Answer: 10 °C

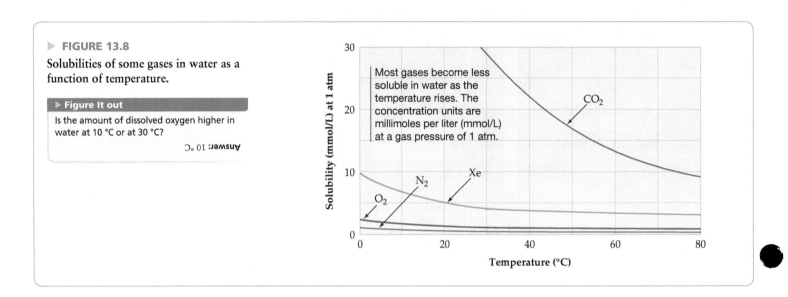

As shown by Figures 13.7 and 13.8, the solubility of most solids *increases* as temperature *increases,* while the solubility of most gases *decreases* as temperature *increases.* What is the reason for the different effect of temperature on the solubility of solids and gases? To answer this question, we must consider the change in entropy that occurs upon solution formation. Figure 13.5 shows that when a solid dissolves in a liquid, randomness increases and $\Delta S_{soln} > 0$. Particles in a solid crystal have a high degree of order and do not move positions relative to one another. When the crystal dissolves, solvated molecules or ions move freely in solution and have less order. Let's examine the effect of a positive value of ΔS_{soln} on solubility. The formula for Gibbs free energy is

$$\Delta G = \Delta H - T\Delta S$$

If ΔS is positive, then the term $(-T\Delta S)$ causes ΔG to become more negative as temperature increases. The more negative the value of ΔG, the greater the extent the reaction favors the soluble product.

In contrast, when gases dissolve in a liquid, the entropy decreases and $\Delta S_{soln} < 0$. Particles in the gas phase are moving rapidly and have a high degree of entropy. When gas particles dissolve in a liquid, their motion is greatly decreased and solvation increases order, thereby causing a decrease in entropy. If ΔS is negative, then the term $(-T\Delta S)$ causes ΔG to become more positive as temperature increases. A more positive value of ΔG means the reaction favors the undissolved reactant to a greater extent. In summary, substances that have a positive ΔS_{soln} become more soluble as temperature increases, while substances that have a negative ΔS_{soln} become less soluble as temperature increases.

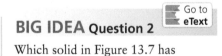

BIG IDEA Question 2

Go to eText

Which solid in Figure 13.7 has $\Delta S_{soln} < 0$?

Effect of Pressure on Solubility

Pressure has practically no effect on the solubility of liquids and solids but has a profound effect on the solubility of gases. According to **Henry's law,** the solubility of a gas in a liquid at a given temperature is directly proportional to the partial pressure of the gas over the solution.

> **Henry's law**   Solubility $= k \cdot P$

The constant k in this expression is characteristic of a specific gas, and P is the partial pressure of the gas over the solution. Doubling the partial pressure doubles the solubility, tripling the partial pressure triples the solubility, and so forth. Henry's law constants are usually given in units of $\text{mol}/(\text{L} \cdot \text{atm})$, and measurements are reported at 25 °C. Note that when the gas partial pressure P is 1 atm, the Henry's law constant k is numerically equal to the solubility of the gas in moles per liter.

Perhaps the most common example of Henry's law behavior occurs when you open a can of soda or other carbonated drink. Bubbles of gas immediately come fizzing out of solution because the pressure of CO_2 in the can drops and CO_2 suddenly becomes less soluble. A more serious example of Henry's law behavior occurs when a deep-sea diver surfaces too quickly and develops a painful and life-threatening condition called *decompression sickness (DCS)* or the *bends*. The bends occur because large amounts of nitrogen dissolve in the blood at high underwater pressures. When the diver ascends and pressure decreases too rapidly, bubbles of nitrogen form in the blood, blocking capillaries and inhibiting blood flow. The condition can be prevented by using an oxygen/helium mixture for breathing rather than air (oxygen/nitrogen) because helium has a much lower solubility in blood than nitrogen.

On a molecular level, the increase in gas solubility with increasing pressure occurs because of a change in the position of the equilibrium between dissolved and undissolved gas. At a given pressure, an equilibrium is established in which equal numbers of gas particles enter and leave the solution. When the pressure is increased, however, more particles are forced into solution than leave it, so gas solubility increases until a new equilibrium is established (**FIGURE 13.9**).

▲ Divers who ascend too quickly can develop the bends, a condition caused by the formation of nitrogen bubbles in the blood. Treatment involves placing the diver into a decompression chamber like this one and slowly lowering the pressure.

◀ **FIGURE 13.9**

A molecular view of Henry's law.

▶ **Figure It out**

After pressure is increased and equilibrium is restored, how does the concentration of gas in solution compare to its original level?

Answer: The concentration is higher since the pressure is higher.

Equilibrium	Pressure increase	Equilibrium restored
At a given pressure, an equilibrium exists in which equal numbers of gas particles enter and leave the solution.	When **pressure is increased** by pushing on the piston, more gas particles are temporarily forced into solution than are able to leave.	Solubility therefore increases until a new equilibrium is reached.

WORKED EXAMPLE 13.6

Using Henry's Law to Calculate Gas Solubility

The Henry's law constant of methyl bromide (CH_3Br), a gas used as a soil fumigating agent, is $k = 0.159$ mol/(L·atm) at 25 °C. What is the solubility in mol/L of methyl bromide in water at 25 °C and a partial pressure of 125 mm Hg?

IDENTIFY

Known	Unknown
$k = 0.159$ mol/(L·atm)	Solubility (mol/L)
T = 25 °C	
Partial pressure of $CH_3Br = 125$ mm Hg	

STRATEGY

According to Henry's law, solubility $= k \cdot P$.

SOLUTION

$$k = 0.159 \text{ mol/(L·atm)}$$

$$P = 125 \text{ mm Hg} \times \frac{1 \text{ atm}}{760 \text{ mm Hg}} = 0.164 \text{ atm}$$

Solubility $= k \cdot P = 0.159 \dfrac{\text{mol}}{\text{L·atm}} \times 0.164 \text{ atm} = 0.0261 \text{ M}$

The solubility of methyl bromide in water at a partial pressure of 125 mm Hg is 0.0261 M.

▶ **PRACTICE 13.11**   The solubility of CO_2 in water is 3.2×10^{-2} M at 25 °C and 1 atm pressure. What is the solubility of CO_2 in water at a partial pressure of CO_2 equal to 5.0 atm at 25 °C?

▶ **APPLY 13.12**   Use the Henry's law constant you calculated in Problem 13.11 to find the concentration of CO_2 in:

(a) A can of soda under a CO_2 pressure of 2.5 atm at 25 °C
(b) A can of soda open to the atmosphere at 25 °C (CO_2 is approximately 0.04% by volume in the atmosphere.)

13.6 PHYSICAL BEHAVIOR OF SOLUTIONS: COLLIGATIVE PROPERTIES

The behavior of solutions is qualitatively similar to that of pure solvents but is quantitatively different. Pure water boils at 100.0 °C and freezes at 0.0 °C, for instance, but a 1.00 m (molal) solution of NaCl in water boils at 101.0 °C and freezes at −3.7 °C.

The higher boiling point and lower freezing point observed for a solution compared to a pure solvent are examples of **colligative properties,** which depend only on the amount of dissolved solute but not on the solute's chemical identity. The word *colligative* means "bound together in a collection" and is used because a collection of solute particles is responsible for the observed effects. Other colligative properties are a lower vapor pressure for a solution compared with the pure solvent and *osmosis*, the migration of solvent and other small molecules through a semipermeable membrane.

Colligative properties   In comparing the properties of a pure solvent with those of a solution:

• The vapor pressure of the solution is lower.
• The boiling point of the solution is higher.
• The freezing (or melting) point of the solution is lower.
• The solution gives rise to *osmosis*, the migration of solvent molecules through a semipermeable membrane.

We'll look at each of the four colligative properties in more detail in Sections 13.7–13.9.

13.7 VAPOR-PRESSURE LOWERING OF SOLUTIONS: RAOULT'S LAW

Recall from Section 11.2 that a liquid in a closed container is in equilibrium with its vapor and that the amount of pressure exerted by the vapor is called the **vapor pressure**. When you compare the vapor pressure of a pure solvent with that of a solution at the same temperature, you find that the two values are different. If the solute is nonvolatile and has no appreciable vapor pressure of its own, as occurs when a solid is dissolved, then the vapor pressure of the solution is always lower than that of the pure solvent. If the solute *is* volatile and has a significant vapor pressure of its own, as occurs in a mixture of two liquids, then the vapor pressure of the mixture is intermediate between the vapor pressures of the two pure liquids.

Solutions with a Nonvolatile Solute

It's easy to demonstrate with **manometers** that a solution of a nonvolatile solute has a lower vapor pressure than the pure solvent (**FIGURE 13.10**). Alternatively, you can show the same effect by comparing the evaporation rate of a pure solvent with the evaporation rate of a solution. A solution always evaporates more slowly than a pure solvent does because its vapor pressure is lower and its molecules escape less readily.

> **REMEMBER . . .**
> At equilibrium, the rate of evaporation from the liquid to the vapor is equal to the rate of condensation from the vapor back to the liquid. The resulting **vapor pressure** is the partial pressure of the gas in the equilibrium (Section 11.2).

> **REMEMBER . . .**
> A **manometer** is a mercury-filled U-tube used for reading pressure. One end of the tube is connected to the sample container and the other end is open to the atmosphere (Section 10.1).

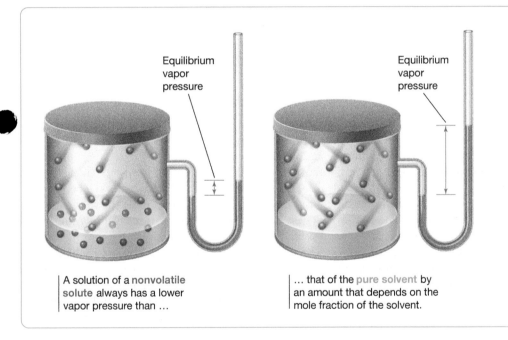

◀ **FIGURE 13.10**
The equilibrium vapor pressure of solutions.

A solution of a **nonvolatile solute** always has a lower vapor pressure than ...

... that of the **pure solvent** by an amount that depends on the mole fraction of the solvent.

According to **Raoult's law,** the vapor pressure of a solution containing a nonvolatile solute is equal to the vapor pressure of the pure solvent times the mole fraction of the solvent. That is,

Raoult's law	$P_{soln} = P_{solv} \times X_{solv}$

where P_{soln} is the vapor pressure of the solution, P_{solv} is the vapor pressure of pure solvent at the same temperature, and X_{solv} is the mole fraction of the solvent in the solution.

 Go to eText

BIG IDEA Question 3

100 mL of 1.0 M ethanol in water and 100 mL of 1.0 M NaCl in water are left out on the lab bench. Which solution will evaporate to dryness first?

What is the vapor pressure of a solution of 1.00 mol of glucose in 15.0 mol of water at 25 °C? The vapor pressure of pure water at 25 °C is 23.76 mm Hg so we expect the vapor pressure of the solution to be lower. The mole fraction of water in the solution is

$$X_{solv} = \frac{15.0 \text{ mol } H_2O}{15.0 \text{ mol } H_2O + 1.0 \text{ mol glucose}} = 0.938$$

Thus, Raoult's law predicts a vapor pressure for the solution of 22.3 mm Hg, which corresponds to a vapor-pressure lowering, ΔP_{soln}, of 1.5 mm Hg:

$$P_{soln} = P_{solv} \times X_{solv} = 23.76 \text{ mm Hg} \times 0.938 = 22.3 \text{ mm Hg}$$

$$\Delta P_{soln} = P_{solv} - P_{soln} = 23.76 \text{ mm Hg} - 22.3 \text{ mm Hg} = 1.5 \text{ mm Hg}$$

Alternatively, the amount of vapor-pressure lowering can be calculated directly by multiplying the mole fraction of the *solute* times the vapor pressure of the pure solvent. That is,

$$\Delta P_{soln} = P_{solv} \times X_{solute} = 23.76 \text{ mm Hg} \times \frac{1.00 \text{ mol glucose}}{1.00 \text{ mol glucose} + 15.0 \text{ mol } H_2O}$$

$$= 1.49 \text{ mm Hg}$$

If an ionic substance such as NaCl is the solute rather than a molecular substance, we have to calculate mole fractions based on the total concentration of solute *particles* (ions) rather than NaCl formula units. A solution of 1.00 mol of NaCl in 15.0 mol of water at 25 °C, for example, contains 2.00 mol of dissolved particles, assuming complete dissociation, resulting in a mole fraction for water of 0.882 and a solution vapor pressure of 21.0 mm Hg.

$$X_{water} = \frac{15.0 \text{ mol } H_2O}{1.00 \text{ mol Na}^+ + 1.00 \text{ mol Cl}^- + 15.0 \text{ mol } H_2O} = 0.882$$

$$P_{soln} = P_{solv} \times X_{solv} = 23.76 \text{ mm Hg} \times 0.882 = 21.0 \text{ mm Hg}$$

Because the mole fraction of water is smaller in the NaCl solution than in the glucose solution, the vapor pressure of the NaCl solution is lower: 21.0 mm Hg for NaCl versus 22.3 mm Hg for glucose at 25 °C.

Just as the ideal gas law discussed in Section 10.3 applies only to "ideal" gases, Raoult's law applies only to ideal solutions. Raoult's law approximates the behavior of most real solutions, but significant deviations from ideality occur as the solute concentration increases. The law works best when solute concentrations are low and when solute and solvent particles have similar intermolecular forces.

If the intermolecular forces between solute particles and solvent molecules are weaker than the forces between solvent molecules alone, then the solvent molecules are less tightly held in the solution and the vapor pressure is higher than Raoult's law predicts. Conversely, if the intermolecular forces between solute and solvent molecules are stronger than the forces between solvent molecules alone, then the solvent molecules are more tightly held in the solution and the vapor pressure is lower than predicted. Solutions of ionic substances, in particular, often have a vapor pressure significantly lower than predicted, because the ion-dipole forces between dissolved ions and polar water molecules are so strong.

A further complication is that ionic substances rarely dissociate completely, so a solution of an ionic compound usually contains fewer particles than the formula of the compound would suggest. The actual extent of dissociation can be expressed as a **van't Hoff factor (*i*)**.

Go to eText

BIG IDEA Question 4

The following diagram shows a close-up view of part of the vapor pressure curve for a pure solvent and a solution of a nonvolatile solute. Which curve represents the solution?

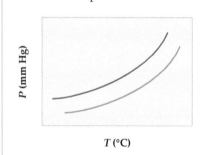

$$\text{van't Hoff factor } i = \frac{\text{Moles of particles in solution}}{\text{Moles of solute dissolved}}$$

Rearranging this equation shows that the number of moles of particles dissolved in a solution is equal to number of moles of solute dissolved times the van't Hoff factor.

$$\text{Moles of particles in solution} = i \times \text{Moles of solute dissolved}$$

For any nonelectrolyte such as sucrose ($C_{12}H_{22}O_{11}$), we can always assume $i = 1$. To take a solution of an electrolyte such as NaCl as an example, the experimentally determined van't Hoff factor for 0.05 m NaCl is 1.9, meaning that each mole of NaCl gives only 1.9 mol of particles rather than the 2.0 mol expected for complete dissociation. Of the 1.9 mol of particles, 0.9 mol is Cl^-, 0.9 mol is Na^+, and 0.1 mol is undissociated NaCl. Thus, NaCl is only $(0.9/1.0) \times 100\% = 90\%$ dissociated in a 0.05 m solution and the amount of vapor pressure lowering is less than expected.

Why do ionic compounds such as NaCl have incomplete dissociation in aqueous solution? **FIGURE 13.11** shows a molecular scale image of dissolved Na^+ and Cl^- ions. Each ion forms ion–dipole forces with surrounding water molecules in a hydration shell. In solution, the molecules and ions move around freely, and sometimes oppositely charged ions approach closely enough to attract one another and form an **ion pair**, which behaves as one dissolved particle instead of two. Due to ion pairing many ionic solids do not completely dissociate and have a measured van't Hoff factor lower than predicted.

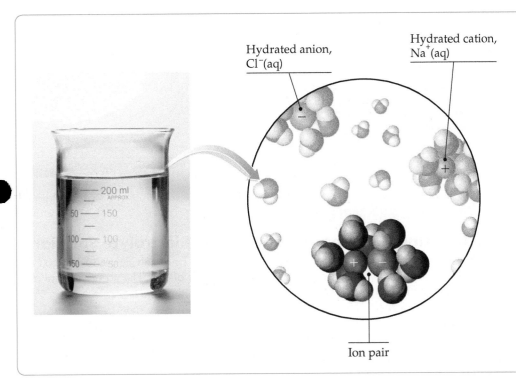

Hydrated anion, $Cl^-(aq)$

Hydrated cation, $Na^+(aq)$

Ion pair

◄ **FIGURE 13.11**

Ion pairing in an aqueous salt solution. In a solution of sodium chloride, hydrated cations and anions move freely and may temporarily attract one another to form an ion pair. An ion pair behaves as one particle and lowers the concentration of dissolved particles.

What accounts for the lowering of the vapor pressure when a nonvolatile solute is dissolved in a solvent? As we've noted on many prior occasions, a physical process such as the vaporization of a liquid to a gas is accompanied by a free-energy change, $\Delta G_{vap} = \Delta H_{vap} - T\Delta S_{vap}$. The more negative the value of ΔG_{vap}, the more favored the vaporization process. Thus, if we want to compare the ease of vaporization of a pure solvent with that of the solvent in a solution, we have to compare the signs and relative magnitudes of the ΔH_{vap} and ΔS_{vap} terms in the two cases.

The vaporization of a liquid to a gas is disfavored by enthalpy (positive ΔH_{vap}) because energy is required to overcome intermolecular attractions in the liquid. At the same time, however, vaporization is favored by entropy (positive ΔS_{vap}) because randomness increases when molecules go from a liquid state to a gaseous state.

Comparing a pure solvent with a solution, the *enthalpies* of vaporization for the pure solvent and solvent in a solution are similar because similar intermolecular forces must be overcome in both cases for solvent molecules to escape from the liquid. The *entropies* of vaporization for a pure solvent and a solvent in a solution are *not* similar, however.

Because a solution has more molecular randomness and higher entropy than a pure solvent does, the entropy *change* on going from liquid to vapor is smaller for the solvent in a solution than for the pure solvent. Subtracting a smaller $T\Delta S_{vap}$ from ΔH_{vap} thus results in a larger (less negative) ΔG_{vap} for the solution. As a result, vaporization is less favored for the solution and the vapor pressure of the solution at equilibrium is lower (**FIGURE 13.12**).

▶ **FIGURE 13.12**

Vapor-pressure lowering. The lower vapor pressure of a solution relative to that of a pure solvent is due to the difference in their entropies of vaporization, ΔS_{vap}.

▶ **Figure It out**

Which has a larger ΔS_{vap}, a pure solvent or a solution? How does a larger value of ΔS_{vap} affect the value of ΔG_{vap}?

Answer: ΔS_{vap} is larger for the pure solvent. A larger value of ΔS_{vap} makes ΔG_{vap} more negative, meaning the process is more product favored.

Because the entropy of the solvent in a solution is higher than that of pure solvent to begin with, ΔS_{vap} is smaller for the solution than for the pure solvent.

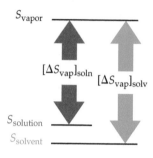

$[\Delta S_{vap}]_{soln}$ $[\Delta S_{vap}]_{solv}$

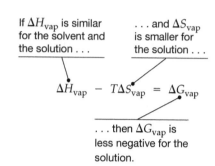

If ΔH_{vap} is similar for the solvent and the solution . . .   . . . and ΔS_{vap} is smaller for the solution . . .

$$\Delta H_{vap} - T\Delta S_{vap} = \Delta G_{vap}$$

. . . then ΔG_{vap} is less negative for the solution.

As a result, vaporization of the solvent from the solution is less favored (less negative ΔG_{vap}), and the vapor pressure of the solution is lower.

Go to eText

WORKED EXAMPLE 13.7

Calculating the Vapor Pressure of an Electrolyte Solution

What is the vapor pressure in mm Hg of a solution made by dissolving 18.30 g of NaCl in 500.0 g of H_2O at 70 °C, assuming a van't Hoff factor of 1.9? The vapor pressure of pure water at 70 °C is 233.7 mm Hg.

IDENTIFY

Known	Unknown
Mass of solute (18.3 g NaCl)	P_{soln}
Mass of solvent (500.0 g H_2O)	
$i = 1.9$	
$P_{H_2O} = 233.7$ mm Hg	

STRATEGY

According to Raoult's law, the vapor pressure of the solution (P_{soln}) equals the vapor pressure of pure solvent (P_{H_2O}) times the mole fraction of the solvent (X_{solv}) in the solution. Thus, we have to find the numbers of moles of solvent and solute particles and then calculate the mole fraction of solvent. Electrolytes dissociate in solution, and the van't Hoff factor is used to find the number of solute particles.

SOLUTION

First, use molar mass to calculate the number of moles of NaCl and H_2O.

$$\text{Moles of NaCl} = 18.3 \text{ g NaCl} \times \frac{1 \text{ mol NaCl}}{58.44 \text{ g NaCl}} = 0.313 \text{ mol NaCl}$$

$$\text{Moles of } H_2O = 500.0 \text{ g } H_2O \times \frac{1 \text{ mol } H_2O}{18.02 \text{ g } H_2O} = 27.75 \text{ mol } H_2O$$

Next, calculate the mole fraction of water in the solution. A van't Hoff factor of 1.9 means that the NaCl dissociates incompletely and gives only 1.9 particles per formula unit. Thus, the solution contains 1.9×0.313 mol $= 0.59$ mol of dissolved particles and the mole fraction of water is

$$\text{Mole fraction of } H_2O = \frac{27.75 \text{ mol}}{0.59 \text{ mol} + 27.75 \text{ mol}} = 0.9792$$

From Raoult's law, the vapor pressure of the solution is

$$P_{soln} = P_{solv} \times X_{solv} = 233.7 \text{ mm Hg} \times 0.9792 = 228.8 \text{ mm Hg}$$

▶ **PRACTICE 13.13** What is the vapor pressure in mm Hg of a solution made by dissolving 10.00 g of $CaCl_2$ in 100.0 g of H_2O at 70 °C, assuming a van't Hoff factor of 2.7? The vapor pressure of pure water at 70 °C is 233.7 mm Hg.

▶ **APPLY 13.14** A solution made by dissolving 8.110 g of $MgCl_2$ in 100.0 g of water has a vapor pressure 224.7 mm Hg at 70 °C. The vapor pressure of pure water at 70 °C is 233.7 mm Hg. What is the van't Hoff factor for $MgCl_2$?

Solutions with a Volatile Solute

According to **Dalton's law of partial pressures**, the overall vapor pressure P_{total} of a mixture of two volatile liquids, A and B, is the sum of the vapor-pressure contributions of the individual components, P_A and P_B:

$$P_{total} = P_A + P_B$$

The individual vapor pressures P_A and P_B are calculated by Raoult's law. That is, the vapor pressure of A is equal to the mole fraction of A (X_A) times the vapor pressure of pure A $(P°_A)$, and the vapor pressure of B is equal to the mole fraction of B (X_B) times the vapor pressure of pure B $(P°_B)$. Thus, the total vapor pressure of the solution is

$$P_{total} = P_A + P_B = (P°_A \cdot X_A) + (P°_B \cdot X_B)$$

Take a mixture of the two similar organic liquids benzene (C_6H_6, bp $= 80.1$ °C) and toluene (C_7H_8, bp $= 110.6$ °C), as an example. Pure benzene has a vapor pressure $P° = 96.0$ mm Hg at 25 °C, and pure toluene has $P° = 30.3$ mm Hg at the same temperature. In a 1:1 molar mixture of the two, where the mole fraction of each is $X = 0.500$, the vapor pressure of the solution is 63.2 mm Hg:

$$\begin{aligned} P_{total} &= (P°_{benzene})(X_{benzene}) + (P°_{toluene})(X_{toluene}) \\ &= (96.0 \text{ mm Hg} \times 0.500) + (30.3 \text{ mm Hg} \times 0.500) \\ &= 48.0 \text{ mm Hg} + 15.2 \text{ mm Hg} = 63.2 \text{ mm Hg} \end{aligned}$$

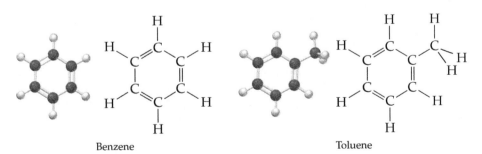

Benzene   Toluene

Note that the vapor pressure of the mixture is intermediate between the vapor pressures of the two pure liquids (**FIGURE 13.13**). Figure 13.13 shows that the vapor pressure of the solution is the sum of the vapor pressures of the individual components.

As with nonvolatile solutes, Raoult's law for a mixture of volatile liquids applies only to ideal solutions. Most real solutions show behaviors that deviate slightly from the ideal in either a positive or negative way, depending on the kinds and strengths of intermolecular forces present in the solution.

▶ **FIGURE 13.13**

Raoult's law for a mixture of volatile liquids. The vapor pressure of a solution of the two volatile liquids benzene and toluene at 25 °C is the sum of the two individual contributions, each calculated by Raoult's law.

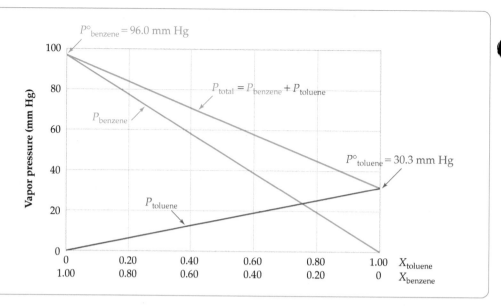

WORKED CONCEPTUAL EXAMPLE 13.8

Calculating the Vapor Pressure of a Solution with a Volatile Solute

What is the vapor pressure in mm Hg of a solution prepared by dissolving 25.0 g of ethyl alcohol (C_2H_5OH) in 100.0 g of water at 25 °C? The vapor pressure of pure water is 23.8 mm Hg, and the vapor pressure of ethyl alcohol is 61.2 mm Hg at 25 °C.

IDENTIFY

Known	Unknown
Mass of C_2H_5OH (25.0 g)	P_{soln}
Mass of H_2O (100.0 g)	
P_{H_2O} = 23.8 mm Hg	
$P_{C_2H_5OH}$ = 61.2 mm Hg	

STRATEGY

The total vapor pressure of a solution with a volatile solute is calculated using the general equation $P_{total} = P_A + P_B = (P°_A \cdot X_A) + (P°_B \cdot X_B)$. The mole fraction of each liquid is calculated by converting from grams to moles and then dividing by the total number of moles.

SOLUTION

Convert from grams to moles of each liquid:

$$\text{Mol } C_2H_5OH = 25.0 \text{ g } C_2H_5OH \times \frac{1 \text{ mol } C_2H_5OH}{46.0 \text{ g } C_2H_5OH}$$

$$= 0.543 \text{ mol } C_2H_5OH$$

$$\text{Mol } H_2O = 100.0 \text{ g } H_2O \times \frac{1 \text{ mol } H_2O}{18.0 \text{ g } H_2O} = 5.56 \text{ mol } H_2O$$

Calculate the mole fraction of each liquid:

$$X_{C_2H_5OH} = \frac{0.543 \text{ mol } C_2H_5OH}{0.543 \text{ mol } C_2H_5OH + 5.56 \text{ mol } H_2O} = 0.089$$

$$X_{H_2O} = \frac{5.56 \text{ mol } H_2O}{0.543 \text{ mol } C_2H_5OH + 5.56 \text{ mol } H_2O} = 0.911$$

The vapor pressure of the solution is:

$$P_{soln} = P_{H_2O} + P_{C_2H_5OH}$$

$$= (23.8 \text{ mm Hg} \times 0.911) + (61.2 \text{ mm Hg} \times 0.089)$$

$$= 27.1 \text{ mm Hg}$$

CHECK

The answer is reasonable because the vapor pressure of the solution, 27.1 mm Hg, is in between the vapor pressure of each pure liquid. The vapor pressure of the solution is closer to the vapor pressure of water than ethyl alcohol, which makes sense because water has a much higher mole fraction than ethyl alcohol in the mixture.

▶ **PRACTICE 13.15** What is the vapor pressure of the solution if 25.0 g of water is dissolved in 100.0 g of ethyl alcohol at 25 °C? The vapor pressure of pure water is 23.8 mm Hg, and the vapor pressure of ethyl alcohol is 61.2 mm Hg at 25 °C.

▶ **CONCEPTUAL APPLY 13.16** The following diagram shows a close-up view of part of the vapor-pressure curves for two pure liquids and a mixture of the two. Which curves represent pure liquids, and which represents the mixture?

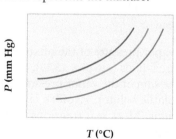

13.8 BOILING-POINT ELEVATION AND FREEZING-POINT DEPRESSION OF SOLUTIONS

We saw in Section 11.3 that the vapor pressure of a liquid rises with increasing temperature and that the liquid boils when its vapor pressure equals atmospheric pressure. Because a solution of a nonvolatile solute has a lower vapor pressure than a pure solvent has at a given temperature, the solution must be heated to a higher temperature to cause it to boil. Furthermore, the lower vapor pressure of the solution means that the liquid–vapor phase transition line on a **phase diagram** is always lower for the solution than for the pure solvent. As a result, the triple-point temperature T_t is lower for the solution, the solid–liquid phase transition line is shifted to a lower temperature for the solution, and the solution must be cooled to a lower temperature to freeze. **FIGURE 13.14** shows how the phase diagram for a solution is different than that of a pure solvent.

REMEMBER . . .
The **phase diagram** of a substance is a plot of pressure versus temperature, showing which phase is stable at any given combination of P and T (Section 11.9).

Because the liquid/vapor phase transition line is lower for the **solution** than for the **pure solvent**, the triple-point temperature T_t is lower and the solid/liquid phase transition line is shifted to a lower temperature. As a result, the freezing point of the solution is lower than that of the pure solvent by an amount ΔT_f.

Because the vapor pressure of the **solution** is lower than that of the **pure solvent** at a given temperature, the temperature at which the vapor pressure reaches atmospheric pressure is higher for the solution than for the solvent. Thus, the boiling point of the solution is higher by an amount ΔT_b.

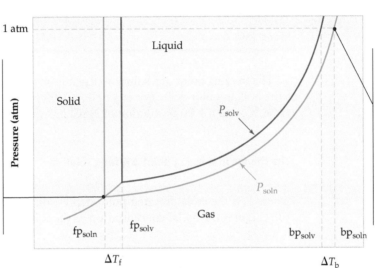

▲ **FIGURE 13.14**

Phase diagrams for a pure solvent (red) and a solution of a nonvolatile solute (green).

▲ **Figure It out**

How does the vapor pressure of a solution compare to the vapor pressure of a pure solvent? How is the freezing point and boiling point of the solution affected by the change in vapor pressure?

Answer: The vapor pressure of a solution is lower than the pure solvent. This causes the freezing point of the solution to be lower than that of the pure solvent and the boiling point of the solution to be higher than that of the pure solvent.

The boiling point of a solution depends on the concentration of dissolved particles, just as vapor pressure does. Thus, a $1.00\ m$ solution of glucose in water boils at approximately 100.51 °C at 1 atm pressure (0.51 °C above normal), but a $1.00\ m$ solution of NaCl in water boils at approximately 101.02 °C (1.02 °C above normal) because there are twice as many particles (ions) dissolved in the NaCl solution as there are in the glucose solution. As with vapor-pressure lowering described in the previous section, the actual amount of boiling-point elevation observed for a solution of an ionic substance depends on the extent of dissociation, as given by the van't Hoff factor.

The change in boiling point for a solution

$$\Delta T_b = K_b \cdot m \cdot i$$

where K_b is the **molal boiling-point-elevation constant (K_b)** characteristic of a given solvent, m is the molal concentration of the solute, and i is the van't Hoff factor.

Go to eText

BIG IDEA Question 5

Which solution has the highest boiling point: $0.5\ m$ $Ca(NO_3)_2$, $1.0\ m$ $C_{12}H_{22}O_{11}$, or $0.6\ m$ NaCl?

▲ Propylene glycol is one component of de-icer fluid that is sprayed on the wings of airplanes to prevent a build up of ice, which disrupts air flow and reduces lift. De-icing fluids melt ice and prevent it from reforming by lowering the freezing point of water.

The concentration must be expressed in molality—the number of moles of solute particles per kilogram of solvent—rather than molarity so that the solute concentration is independent of temperature. Molal boiling-point-elevation constants are given in TABLE 13.4 for some common substances.

TABLE 13.4 Molal Boiling-Point-Elevation Constants (K_b) and Molal Freezing-Point-Depression Constants (K_f) for Some Common Substances

Substance	$K_b[(°C \cdot kg)/mol]$	$K_f[(°C \cdot kg)/mol]$
Benzene (C_6H_6)	2.64	5.07
Camphor ($C_{10}H_{16}O$)	5.95	37.8
Chloroform ($CHCl_3$)	3.63	4.70
Diethyl ether ($C_4H_{10}O$)	2.02	1.79
Ethyl alcohol (C_2H_6O)	1.22	1.99
Water (H_2O)	0.51	1.86

The freezing point of a solution depends on the concentration of solute particles, just as the boiling point does. For example, a 1.00 m solution of glucose in water freezes at $-1.86\ °C$, and a 1.00 m solution of NaCl in water freezes at approximately $-3.72\ °C$.

The change in freezing point ΔT_f for a solution

$$\Delta T_f = -K_f \cdot m \cdot i$$

where K_f is the **molal freezing-point-depression constant (K_f)** characteristic of a given solvent, m is the molal concentration of the solute, and i is the van't Hoff factor.

Some molal freezing-point-depression constants are also given in Table 13.4.

Colligative properties may seem somewhat obscure, but in fact they have many practical uses, both in the chemical laboratory and in everyday life. Motorists in winter, for instance, take advantage of freezing-point lowering when they drive on streets where the snow has been melted by a sprinkling of salt. The antifreeze added to automobile radiators and the de-icer solution sprayed on airplane wings also work by lowering the freezing point of water. That same automobile antifreeze keeps radiator water from boiling over in summer by raising its boiling point.

The fundamental cause of boiling-point elevation and freezing-point depression in solutions is the same as the cause of vapor-pressure lowering: the entropy difference between the pure solvent and the solvent in solution. Let's take boiling-point elevations first. We know that liquid and vapor phases are in equilibrium at the boiling point (T_b) and that the free-energy difference between the two phases (ΔG_{vap}) is therefore zero.

Since $\Delta G_{vap} = \Delta H_{vap} - T_b \Delta S_{vap} = 0$

then $\Delta H_{vap} = T_b \Delta S_{vap}$ and $T_b = \dfrac{\Delta H_{vap}}{\Delta S_{vap}}$

When we compare the enthalpies of vaporization (ΔH_{vap}) for the pure solvent and for the solvent in a solution, we see that the values are similar because similar intermolecular forces holding the solvent molecules must be overcome in both cases. The values of the entropies of vaporization, however, are not similar. Because the solvent in solution has more molecular randomness than the pure solvent has, the entropy change between solution and vapor is smaller than the entropy change between pure solvent and vapor. But if ΔS_{vap} is smaller for the solution, then T_b must be correspondingly larger. In other words, the boiling point of the solution (T_b) is higher than that of the pure solvent (FIGURE 13.15).

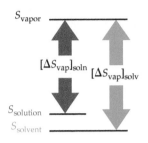

Because the solvent in a solution has a higher entropy to begin with, ΔS_{vap} is smaller for the solution than for the pure solvent.

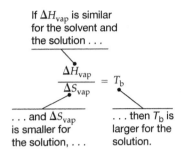

If ΔH_{vap} is similar for the solvent and the solution . . .

$$\frac{\Delta H_{vap}}{\Delta S_{vap}} = T_b$$

. . . and ΔS_{vap} is smaller for the solution, . . .

. . . then T_b is larger for the solution.

As a result, the boiling point of the solution T_b is higher than that of the pure solvent.

◀ **FIGURE 13.15**

Boiling-point elevation. The higher boiling point of a solution relative to that of a pure solvent is due to a difference in their entropies of vaporization, ΔS_{vap}.

A similar explanation accounts for freezing-point depression. Because liquid and solid phases are in equilibrium at the freezing point, the free-energy difference between the phases (ΔG_{fusion}) is zero:

$$\text{Since} \quad \Delta G_{fusion} = \Delta H_{fusion} - T_f \Delta S_{fusion} = 0$$

$$\text{then} \quad \Delta H_{fusion} = T_f \Delta S_{fusion} \quad \text{and} \quad T_f = \frac{\Delta H_{fusion}}{\Delta S_{fusion}}$$

When we compare the solvent in solution with pure solvent, we see that the values of the enthalpies of fusion (ΔH_{fusion}) are similar because similar intermolecular forces between solvent molecules are involved. The values of the entropies of fusion (ΔS_{fusion}), however, are not similar. Because the solvent in solution has more molecular randomness than the pure solvent has, the entropy change between the solvent in the solution and the solid is larger than the entropy change between pure solvent and the solid. With ΔS_{fusion} larger for the solution, T_f must be correspondingly smaller, meaning that the freezing point of the solution (T_f) is lower than that of the pure solvent (**FIGURE 13.16**).

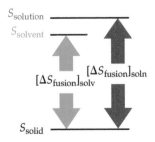

Because the solvent in a solution has a higher entropy level to begin with, ΔS_{fusion} is larger for the solution than for the pure solvent.

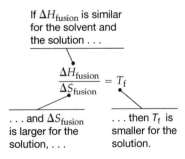

If ΔH_{fusion} is similar for the solvent and the solution . . .

$$\frac{\Delta H_{fusion}}{\Delta S_{fusion}} = T_f$$

. . . and ΔS_{fusion} is larger for the solution, . . .

. . . then T_f is smaller for the solution.

As a result, the freezing point of the solution T_f is lower than that of the pure solvent.

◀ **FIGURE 13.16**

Freezing-point lowering. The lower freezing point of a solution relative to that of a pure solvent is due to a difference in their entropies of fusion, ΔS_{fusion}.

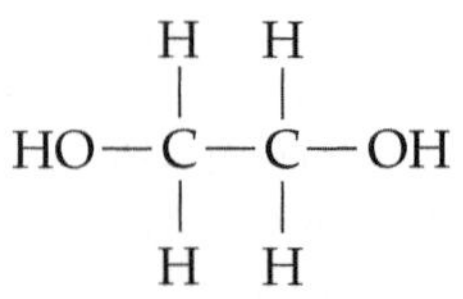

Ethylene glycol

Using Freezing-Point Depression to Calculate the Molality of a Solution

Ethylene glycol ($C_2H_6O_2$) is commonly dissolved in water to make antifreeze for your car's engine. What is the molality of an aqueous solution of ethylene glycol if the freezing point of the solution at 1 atm pressure is −37.0 °C? The molal freezing-point-depression constant for water is given in Table 13.4.

IDENTIFY

Known	Unknown
Freezing point (−37.0 °C)	Molality (m)
$K_f = 1.86$ (°C · kg)/mol	

STRATEGY

Ethylene glycol is a nonelectrolyte, therefore $i = 1$. Rearrange the equation for molal freezing-point depression to solve for m:

$$\Delta T_f = -K_f \cdot m \cdot i \quad \text{so} \quad m = \frac{-\Delta T_f}{K_f \cdot i}$$

where $K_f = 1.86$(°C · kg)/mol and $\Delta T_f = 0.0\ °C - 37.0\ °C = -37.0\ °C$.

SOLUTION

$$m = \frac{-(-37.0\ °C)}{1.86 \dfrac{(°C \cdot kg)}{mol}} = 19.9 \frac{mol}{kg} = 19.9\ m$$

The molality of the solution is 19.9 m.

▶ **PRACTICE 13.17**   What is the normal boiling point in °C of an antifreeze solution prepared by dissolving 616.9 g of ethylene glycol ($C_2H_6O_2$) in 500.0 g of water? The molal boiling-point-elevation constant for water is 0.51 (°C · kg)/mol.

▶ **APPLY 13.18**   The following phase diagram shows a close-up view of the liquid–vapor phase transition boundaries for pure chloroform and a solution of a nonvolatile solute in chloroform.

(a) What is the approximate boiling point of pure chloroform?
(b) What is the approximate molal concentration of the nonvolatile solute?
    See Table 13.4 to find K_b for chloroform.

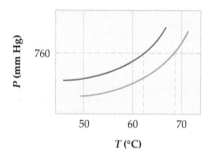

13.9 OSMOSIS AND OSMOTIC PRESSURE

Certain materials, including those that make up the membranes around living cells, are *semipermeable*. That is, they allow water or other smaller molecules to pass through, but they block the passage of larger solute molecules or solvated ions. When a solution and a pure solvent, or two solutions of different concentration, are separated by the right kind of semipermeable membrane, solvent molecules pass through the membrane in a process called **osmosis**. Although the passage of solvent through the membrane takes place in both directions, passage from the pure solvent side to the solution side is more favored and occurs at a greater rate. As a result, the amount of liquid on the pure solvent side of the membrane decreases, the amount of liquid on the solution side increases, and the concentration of the solution decreases.

Osmosis can be demonstrated with the experimental setup shown in **FIGURE 13.17**, in which a solution inside the bulb is separated from pure solvent in the beaker by a semi-permeable membrane. Solvent passes through the membrane from the beaker to the bulb, causing the liquid level in the attached tube to rise. The increased weight of liquid in the tube creates an increased pressure that pushes solvent back through the membrane until the rates of forward and reverse passage become equal and the liquid level stops rising.

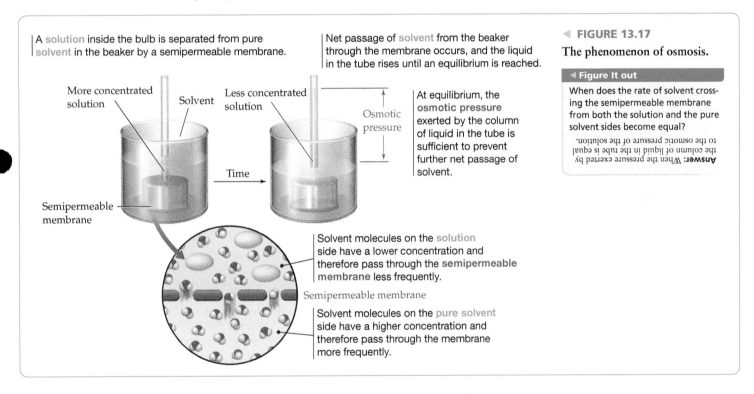

A solution inside the bulb is separated from pure solvent in the beaker by a semipermeable membrane.

Net passage of solvent from the beaker through the membrane occurs, and the liquid in the tube rises until an equilibrium is reached.

More concentrated solution

Solvent

Less concentrated solution

Osmotic pressure

Time

Semipermeable membrane

At equilibrium, the osmotic pressure exerted by the column of liquid in the tube is sufficient to prevent further net passage of solvent.

Solvent molecules on the solution side have a lower concentration and therefore pass through the semipermeable membrane less frequently.

Semipermeable membrane

Solvent molecules on the pure solvent side have a higher concentration and therefore pass through the membrane more frequently.

◄ **FIGURE 13.17**
The phenomenon of osmosis.

◄ **Figure It out**

When does the rate of solvent crossing the semipermeable membrane from both the solution and the pure solvent sides become equal?

Answer: When the pressure exerted by the column of liquid in the tube is equal to the osmotic pressure of the solution.

The amount of pressure necessary to achieve this equilibrium passage of solvent molecules through the membrane is called the solution's **osmotic pressure** (Π) (Greek capital pi). Osmotic pressures can be extremely high, even for relatively dilute solutions. The osmotic pressure of a 0.15 M NaCl solution at 25 °C, for example, is 7.3 atm, a value that will support a difference in water level of approximately 250 ft! The rise in water from the roots to leaves of trees, such as the giant coastal California redwoods, is due to osmotic pressure. The fluid inside the tree has a higher solute concentration than groundwater.

▲ The osmotic pressure that results when a cell membrane forms a boundary between a solution with low solute concentration and a solution with high solute concentration causes water to rise from roots to the leaves.

The amount of osmotic pressure at equilibrium depends on the concentration of solute particles in the solution according to the equation

> **Osmotic pressure**
>
> $$\Pi = MRTi$$
>
> where M is the molar concentration of solute, R is the gas constant $[0.082\ 06\ (L \cdot atm)/(K \cdot mol)]$, T is the temperature in kelvins, and i is the van't Hoff factor.

For example, a 1.00 M solution of glucose, a nonelectrolyte, in water at 300 K has an osmotic pressure of 24.6 atm:

$$\Pi = MRTi = \left(1.00\ \frac{mol}{L}\right)\left(0.082\ 06\ \frac{L \cdot atm}{K \cdot mol}\right)(300\ K)(1) = 24.6\ atm$$

Note that the solute concentration is given in *molarity* when calculating osmotic pressure rather than in molality as for other colligative properties. Because osmotic-pressure measurements are made at the specific temperature given in the equation $\Pi = MRTi$, it's not necessary to express concentration in a temperature-independent unit like molality.

Osmosis, like all colligative properties, results from a favorable increase in entropy of the pure solvent as it passes through the membrane and mixes with the solution. We can also explain osmosis on a molecular level by noting that molecules on the solvent side of the membrane, because of their greater concentration, approach the membrane a bit more frequently than molecules on the solution side, thereby passing through more often (Figure 13.17).

WORKED EXAMPLE 13.10

Calculating the Osmotic Pressure of a Solution

FIGURE 13.18 shows the transport of solvent across the membrane of red blood cells in solutions of varying concentration. (a) What is the osmotic pressure of a solution inside a red blood cell with a total concentration of dissolved particles of 0.30 M? (b) Red blood cells rupture if the pressure differential across their membrane is 5 atm or higher. Would the blood cells rupture if they were placed in pure water?

IDENTIFY

Known	Unknown
Molarity of particles in solution (0.30 M)	Osmotic pressure (Π)
Temperature (298 K)	

STRATEGY

If red blood cells were removed from plasma and placed in pure water, water would pass through the cell membrane, causing a pressure increase inside the cells. The maximum amount of this pressure would be

$$\Pi = MRTi$$

where $Mi = 0.30$ mol/L, $R = 0.082\ 06\ (L \cdot atm)/(K \cdot mol)$, and $T = 298$ K.

SOLUTION

(a) $\Pi = \left(0.30\ \dfrac{mol}{L}\right)\left(0.082\ 06\ \dfrac{L \cdot atm}{K \cdot mol}\right)(298\ K) = 7.3\ atm$

(b) Water molecules cross the cell membrane and cause the cell to swell because the solute concentration inside the cell is higher than on the outside. Since the osmotic pressure on the inside is 7.3 atm and on the outside is 0 atm, the differential exceeds 5 atm and the cells burst.

▶ **PRACTICE 13.19** What is the osmotic pressure of an intravenous solution prepared by dissolving 50.0 g of dextrose (sugar), $C_6H_{12}O_6$, in enough water to make 1.00 L of solution? Dextrose is a nonelectrolyte and body temperature is 37.0 °C.

▶ **APPLY 13.20** Cells in the human eye have an osmotic pressure of 8.0 atm at 25 °C. A saline solution, used to store contact lenses, is prepared by dissolving NaCl in water.

(a) What molar concentration of sodium chloride will give a solution with an equal osmotic pressure? The van't Hoff factor for NaCl is 1.9.

(b) A contact lens stored in saline solution with an osmotic pressure of 5.5 atm is placed in the eye. What is the net direction of water transfer across the cell membrane? What happens to the size of the cell?

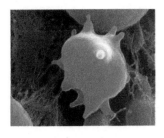
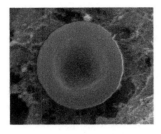
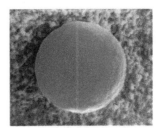

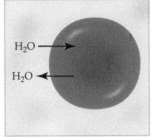
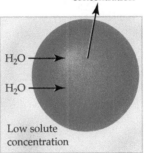

Low solute
concentration

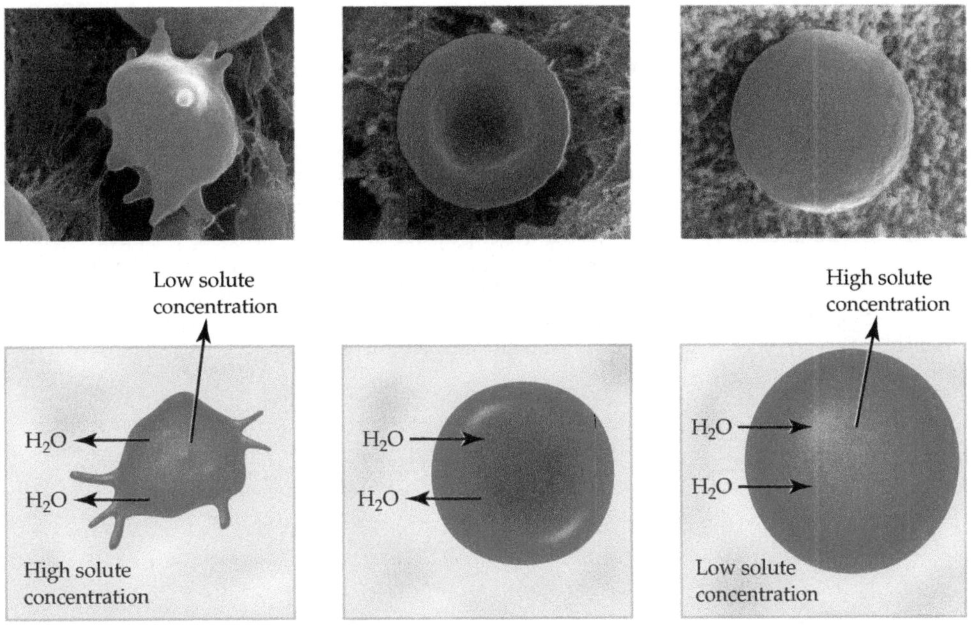

H_2O ←

H_2O ←

High solute
concentration

H_2O →

H_2O ←

High solute
concentration

H_2O →

H_2O →

Low solute
concentration

In a *hypertonic solution*, the solute concentration in the solution is greater than that inside the cell. There is a net transfer of water from the cell to the solution causing the cell to shrivel and shrink.

In a *isotonic solution*, the solute concentration in the solution is the same as that inside the cell. There is no net transfer of water across the cell membrane.

In a *hypotonic solution*, the solute concentration in the solution is less than that inside the cell. There is a net transfer of water across the cell membrane into the cell causing the cell to swell and burst if the internal pressure becomes too high.

In the laboratory, colligative properties are sometimes used for determining the molecular weight of an unknown substance. Any of the four colligative properties we've discussed can be used, but the most accurate values are obtained from osmotic-pressure measurements because the magnitude of the osmosis effect is so great. For example, a solution of 0.0200 M glucose in water at 300 K will give an osmotic-pressure reading of 374.2 mm Hg, a value that can easily be read to four significant figures. The same solution, however, will lower the freezing point by only 0.04 °C, a value that can be read to only one significant figure. Worked Example 13.11 shows how osmotic pressure can be used to find molecular weight.

WORKED EXAMPLE 13.11

Using Osmotic Pressure to Calculate the Molecular Weight of a Solute

A solution prepared by dissolving 20.0 mg of insulin, a nonelectrolyte, in water and diluting to a volume of 5.00 mL gives an osmotic pressure of 12.5 mm Hg at 300 K. What is the molecular weight of insulin?

IDENTIFY

Known	Unknown
Osmotic pressure ($\Pi = 12.5$ mm Hg)	Mol. wt. of insulin
Mass of insulin (20.0 mg)	
Volume of solution (5.00 mL)	

STRATEGY

To determine molecular weight, we need to know the number of moles of insulin represented by the 20.0 mg sample. We can do this by first rearranging the equation for osmotic pressure to find the molar concentration of the insulin solution and then multiplying by the volume of the solution to obtain the number of moles of insulin.

continued on next page

SOLUTION

Since $\Pi = MRTi$, then $M = \dfrac{\Pi}{RTi}$

$$M = \dfrac{12.5 \text{ mm Hg} \times \dfrac{1 \text{ atm}}{760 \text{ mm Hg}}}{0.082\,06 \dfrac{\text{L} \cdot \text{atm}}{\text{K} \cdot \text{mol}} \times 300 \text{ K} \times 1} = 6.68 \times 10^{-4}\ M$$

Since the volume of the solution is 5.00 mL, the number of moles of insulin is

$$\text{Moles insulin} = 6.68 \times 10^{-4}\ \dfrac{\text{mol}}{\text{L}} \times \dfrac{1 \times 10^{-3}\ \text{L}}{1\ \text{mL}} \times 5.00\ \text{mL}$$

$$= 3.34 \times 10^{-6}\ \text{mol}$$

Knowing both the mass and the number of moles of insulin, we can calculate the molar mass and thus the molecular weight:

$$\text{Molar mass} = \dfrac{\text{mass of insulin}}{\text{moles of insulin}} = \dfrac{0.0200\ \text{g insulin}}{3.34 \times 10^{-6}\ \text{mol insulin}}$$

$$= 5990\ \text{g/mol}$$

The molecular weight of insulin is 5990.

▶ **PRACTICE 13.21** A solution prepared by dissolving 0.8220 g of glucose, a nonelectrolyte, in enough water to produce 200.0 mL of solution has an osmotic pressure of 423.1 mm Hg at 298 K. What is the molecular weight of glucose?

▶ **APPLY 13.22** An unknown white powder is found on the table at a crime scene and the suspect claims that it is table sugar (sucrose, $C_{12}H_{22}O_{11}$). A forensic chemist dissolves 0.512 g of the unknown white powder in enough water to produce 100.0 mL of solution. The osmotic pressure is measured at 25 °C and found to be 278 mm Hg. Does the osmotic pressure measurement support the claim that the powder is sucrose?

▲ The Hadera seawater reverse-osmosis (SWRO) desalination plant in Israel is the world's largest, producing approximately 4.4 *billion* gallons per year of desalinated water.

One of the more interesting uses of colligative osmotic pressure is the desalination of seawater by *reverse osmosis*. When pure water and seawater are separated by a suitable membrane, the passage of water molecules from the pure side to the solution side is faster than passage in the reverse direction. As osmotic pressure builds up, though, the rates of forward and reverse water passage eventually become equal at an osmotic pressure of about 30 atm at 25 °C. If, however, a pressure even *greater* than 30 atm is applied to the solution side, then the reverse passage of water becomes favored. As a result, pure water can be obtained from seawater (**FIGURE 13.19**).

Desalination of seawater has many applications, from use on cruise ships, to use by troops on battlefields, to use for providing drinking water to entire countries. Israel, for instance, now obtains more than 35% of its domestic consumer water demand by reverse osmosis.

▶ **FIGURE 13.19**

Desalination of seawater by reverse osmosis at high pressure.

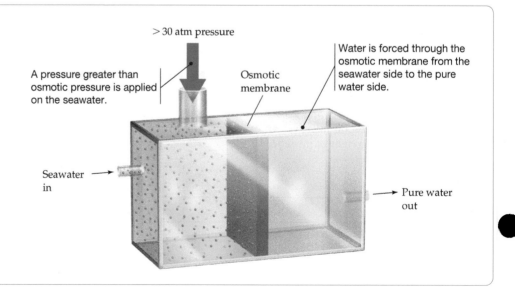

INQUIRY ? How does hemodialysis cleanse the blood of patients with kidney failure?

The primary function of the kidneys is to filter the blood and remove wastes by producing urine. Kidneys also have a number of other important functions on which life depends, including regulation of blood pressure, body pH, and electrolyte balance. In addition, they produce several important hormones and are responsible for the reabsorption from filtered blood of ions and small molecules such as glucose, amino acids, and water. Even though these functions are vital, more than 660,000 people in the United States currently suffer from end-stage renal disease, or kidney failure, for which there is no cure. For those afflicted, the only choices are regular dialysis treatments to replace some of the kidney's functions or organ transplantation.

The process of *dialysis* is similar to osmosis, except that *both* solvent molecules and small solute particles pass through the semipermeable dialysis membrane. Recall that in osmosis, only solvent molecules pass through a semipermeable membrane. In dialysis, only large colloidal particles such as cells and large molecules such as proteins can't pass (**FIGURE 13.20**). (The exact dividing line between a small molecule and a large one is imprecise, and dialysis membranes with a variety of pore sizes are available.) Because they don't dialyze, proteins can be separated from small ions and molecules, making dialysis a valuable procedure for purification of the proteins needed in laboratory studies.

The most important medical use of dialysis is in artificial kidney machines, where *hemodialysis* is used to cleanse the blood of patients, removing waste products like urea and controlling the potassium/sodium ion balance (**FIGURE 13.21**). Blood is diverted from the body and pumped through a dialysis tube suspended in a solution formulated to contain many of the same components as blood plasma. These substances—NaCl, NaHCO$_3$, KCl, and glucose—can have the same concentrations in the dialysis solution as they do in blood, so that they have no net passage through the membrane. However, it is sometimes desirable to adjust dialysis solutions to cause an increase in the concentration of a critical substance in the blood. For example, excess bicarbonate (HCO$_3{}^-$) in dialysis solution can be used to correct acid imbalances (that is, to adjust blood pH). **TABLE 13.5** shows the concentration and various components of a hemodialysis solution.

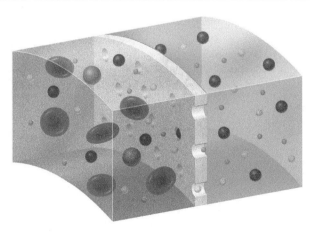

▲ **FIGURE 13.20**

Dialysis membrane. Solvent and small solute molecules (represented by pink, green, and yellow spheres) pass through tiny pores in the membrane. Cells and proteins (represented by red discs and blue spheres, respectively) are retained because they are too large to pass through the membrane.

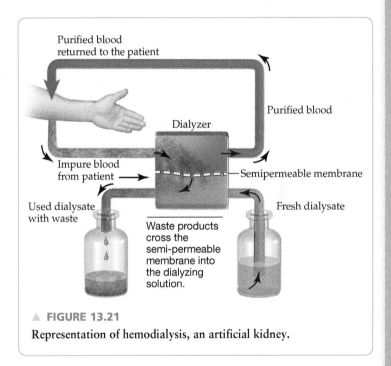

▲ **FIGURE 13.21**

Representation of hemodialysis, an artificial kidney.

TABLE 13.5	Composition of a Typical Hemodialysis Solution
Substance	Concentration
Na$^+$	137 mmol/L
Cl$^-$	105 mmol/L
Ca^{2+}	3.0 mmol/L
CH$_3$COO$^-$	4.0 mmol/L
K$^+$	2.0 mmol/L
HCO$_3{}^-$	33 mmol/L
Mg^{2+}	0.75 mmol/L
Dextrose (C$_6$H$_{12}$O$_6$)	1.11 mmol/dL

Small waste products pass through the dialysis membrane from the blood to the solution side, where they are washed away, but cells, proteins, and other important blood components are prevented by their size from passing through the membrane. A typical hemodialysis treatment lasts for approximately 3–4 hours, and treatments are usually repeated 3–4 times per week.

The water used to make dialysis solutions must be extremely pure because if a solute has a higher concentration in dialysis solution than in blood, it will pass through the membrane and enter the patient's body. Fluoride ion is of particular concern because over half of the drinking water in the United States is enriched with fluoride to promote dental health. The optimal level of fluoride in drinking water is 1.0 ppm, and the maximum legal limit is 4.0 ppm. Water from a municipal supply must be purified and the fluoride ion concentration reduced to less than 0.2 ppm before it can be used in dialysis. Fluoride poisoning in dialysis treatment has occurred when water purification systems malfunctioned. In a Chicago hospital in 1993, 12 patients became ill with symptoms of pain, vomiting, and itching, and 3 patients died from cardiac arrest. New standards are now in place in dialysis units to prevent cases of fluoride intoxication.

More than 95,000 Americans are waiting for a kidney transplant, according to the United Network for Organ Sharing, and there are not enough organs available to fill the need. Healthy people with two working kidneys can donate a kidney during their lifetime. However, most organ donations come from deceased donors. In most states, you can indicate your willingness to be an organ donor by signing the back of your driver's license.

PROBLEM 13.23 What is the difference between a dialysis membrane and the typical semipermeable membrane used for osmosis?

PROBLEM 13.24 Urea has a high solubility in blood serum and is one waste product filtered from blood in a dialysis treatment. Classify the strongest type of intermolecular force in the following interactions; solvent–solvent, solvent–solute, and solute–solute when urea is dissolved in aqueous blood serum (H_2O).

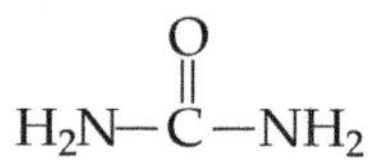

$$H_2N-\overset{\overset{\displaystyle O}{\|}}{C}-NH_2$$
Urea

PROBLEM 13.25

(a) Use Table 13.5 to calculate the osmotic pressure of the hemodialysis solution at 25 °C.

(b) If the osmotic pressure of blood at 25 °C is 7.70 atm, what is the direction of solvent movement across the semipermeable membrane in dialysis? (Blood to dialysis solution or dialysis solution to blood)

PROBLEM 13.26 The presence of fluoride ion in dialysis solution is dangerous because patients are exposed to much greater volumes of fluid than the average person. A typical 4-hour dialysis treatment exposes a patient to as much fluid as the average person drinks in one week (~12 L). What mass of fluoride ion is present in 12.0 L of water with a fluoride concentration of 2.0 ppm?

STUDY GUIDE

Section	Concept Summary	Learning Objectives	Test Your Understanding
13.1–13.2 Solutions and Enthalpy Changes	**Solutions** are homogeneous mixtures that contain particles the size of a typical ion or small molecule. For solutions in which a gas or solid is dissolved in a liquid, the dissolved substance is called the **solute** and the liquid is called the **solvent**. The **enthalpy of solution** (ΔH_{soln}) can be either positive or negative, depending on the relative strengths of solvent–solvent, solute–solute, and solvent–solute intermolecular forces.	**13.1** Classify the types of intermolecular forces involved in the process of forming a solution.	Problems 13.40–13.41, 13.48
		13.2 Determine the enthalpy of solution from the enthalpy changes of steps in the solution-making process.	Problems 13.42–13.43
13.3 Predicting Solubility	The dissolution of a solute in a solvent has an associated free-energy change, $\Delta G = \Delta H - T\Delta S$. The enthalpy change is the **enthalpy of solution** (ΔH_{soln}), and the entropy change is the **entropy of solution** (ΔS_{soln}). If $\Delta G < 0$, then the dissolving process is spontaneous. A simple rule for predicting solubility is "like dissolves like." A nonpolar solute will have a high solubility in a nonpolar solvent, and a polar solute will have a high solubility in a polar solvent.	**13.3** Predict the solubility of a substance based on the chemical structure of the solute and solvent.	Worked Example 13.1; Problems 13.27, 13.44–13.49, 13.52
13.4 Concentration Units for Solutions	The concentration of a solution can be expressed in many ways, including molarity (moles of solute per liter of solution), mole fraction (moles of solute per mole of solution), mass percent (mass of solute per mass of solution times 100%), and molality (moles of solute per kilogram of solvent).	**13.4** Calculate the concentration of a solution in units of mass percent.	Worked Example 13.2; Problem 13.66
		13.5 Calculate the concentration of a solution in units of parts per million (ppm) or parts per billion (ppb).	Worked Example 13.3; Problems 13.58, 13.69
		13.6 Calculate the concentration of a solution in units of molarity or molality.	Worked Example 13.4; Problems 13.60, 13.62, 13.70, 13.76
		13.7 Convert from one unit of concentration to another.	Worked Example 13.5; Problems 13.56, 13.64, 13.72, 13.74
13.5 Some Factors That Affect Solubility	When equilibrium is reached and no further solute dissolves in a given amount of solvent, a solution is said to be **saturated**. Solubilities are usually temperature-dependent. Gas solubilities usually decrease in water with increasing temperature, but the solubilities of solids can either increase or decrease. The solubilities of gases also depend on pressure. According to **Henry's law,** the solubility of a gas in a liquid at a given temperature is proportional to the partial pressure of the gas over the solution.	**13.8** Determine the solubility of a solute at a given temperature by reading a solubility graph.	Problems 13.80–13.81
		13.9 Use Henry's law to calculate the solubility of a gas.	Worked Example 13.6; Problems 13.82–13.86

Section	Concept Summary	Learning Objectives	Test Your Understanding
13.6–13.9 Physical Properties of Solutions: Colligative Properties (Vapor-Pressure Lowering, Boiling-Point Elevation, Freezing-Point Depression, Osmotic Pressure)	In comparison with a pure solvent, a solution has a lower vapor pressure at a given temperature, a lower freezing point, and a higher boiling point. In addition, a solution that is separated from a solvent by a semipermeable membrane gives rise to the phenomenon of **osmosis**. All four of these properties of solutions depend only on the concentration of dissolved solute particles rather than on the chemical identity of those particles and are therefore called **colligative properties**. The fundamental cause of all colligative properties is the same: the higher entropy of the solvent in a solution relative to that of the pure solvent.	**13.10** Qualitatively predict changes to the vapor pressure of a solution when a solute is added.	Problems 13.29–13.33, 13.88
		13.11 Calculate the vapor pressure of a solution containing a nonvolatile solute.	Worked Example 13.7; Problems 13.92, 13.102
		13.12 Calculate the vapor pressure of a solution containing a volatile solute.	Worked Example 13.8; Problems 13.108, 13.110
		13.13 Calculate the amount of boiling-point elevation and freezing-point depression for a solution.	Worked Example 13.9; Problems 13.90, 13.94, 13.114, 13.116, 13.132
		13.14 Calculate osmotic pressure for a solution.	Worked Example 13.10; Problems 13.34, 13.118, 13.120
		13.15 Use osmotic pressure or another colligative property to calculate the molecular weight of a solute.	Worked Example 13.11; Problems 13.124, 13.143

KEY TERMS

colligative property *510*
colloid *495*
enthalpy of solution
  (ΔH_{soln}) *497*
entropy of solution
  (ΔS_{soln}) *499*
Henry's law *509*

ion pair *513*
mass percent (mass %) *502*
miscible *507*
mixture *495*
molal boiling-point-elevation
  constant (K_b) *517*

molal freezing-point-depression
  constant (K_f) *518*
molality (m) *504*
osmosis *521*
osmotic pressure (Π) *521*
parts per billion (ppb) *502*
parts per million (ppm) *502*

Raoult's law *511*
saturated *506*
solute *495*
solution *495*
solvent *495*
supersaturated *506*
van't Hoff factor (i) *512*

KEY EQUATIONS

- Units of Solution Concentration (Section 13.4)

$$\text{Molarity (M)} = \frac{\text{Moles of solute}}{\text{Liters of solution}}$$

$$\text{Mole fraction (X)} = \frac{\text{Moles of component}}{\text{Total moles making up the solution}}$$

$$\text{Mass percent} = \frac{\text{Mass of component}}{\text{Total mass of solution}} \times 100\%$$

$$\text{Parts per million (ppm)} = \frac{\text{Mass of component}}{\text{Total mass of solution}} \times 10^6$$

$$\text{Parts per billion (ppb)} = \frac{\text{Mass of component}}{\text{Total mass of solution}} \times 10^9$$

$$\text{Molality } (m) = \frac{\text{Moles of solute}}{\text{Mass of solvent (kg)}}$$

- Henry's Law (Section 13.5)

$$\text{Solubility} = k \cdot P \qquad \text{where the constant } k \text{ is characteristic of a specific gas and } P \text{ is the partial pressure of the gas over the solution.}$$

- Raoult's Law and Vapor Pressure Lowering (Section 13.7)

$$P_{soln} = P_{solv} \times X_{solv}$$

$$\Delta P_{soln} = P_{solv} \times X_{solute}$$

- Van't Hoff Factor

$$i = \frac{\text{Moles of particles in solution}}{\text{Moles of solute dissolved}} \quad (i = 1 \text{ for nonelectrolytes})$$

- **Molal Boiling-Point Elevation (Section 13.8)**

  $\Delta T_b = K_b \cdot m \cdot i$ where K_b is the molal boiling-point-elevation constant and i is the van't Hoff factor

- **Molal Freezing-Point Depression (Section 13.8)**

  $\Delta T_f = -K_f \cdot m \cdot i$ where K_b is the molal freezing-point-depression constant and i is the van't Hoff factor

- **Osmotic Pressure (Section 13.9)**

  $\Pi = MRTi$

PRACTICE TEST

After studying this chapter, you can assess your understanding with these practice test questions, which are correlated with chapter learning objectives. If you answer a question incorrectly, refer to the learning objectives in the end-of-chapter Study Guide for assistance. The Study Guide provides a conceptual summary, references a Worked Example to model how to solve the problem, and gives additional problems for more practice.

1. Ethanol (CH_3CH_2OH) dissolves in hexane (C_6H_{14}). Give the strongest type of intermolecular force between two solute molecules, between two solvent molecules, and between a solute and solvent molecule. (**LO 13.1**)

	Solute–solute	Solvent–solvent	Solute–solvent
(a)	hydrogen bonding	dispersion forces	hydrogen bonding
(b)	dipole–dipole forces	dispersion forces	dipole–dipole forces
(c)	dispersion forces	dispersion forces	dispersion forces
(d)	hydrogen bonding	dispersion forces	dispersion forces

2. When lithium chloride dissolves in water, the temperature of the solution increases. Which diagram represents the enthalpy changes of the steps in the solution-making process? (**LO 13.2**)

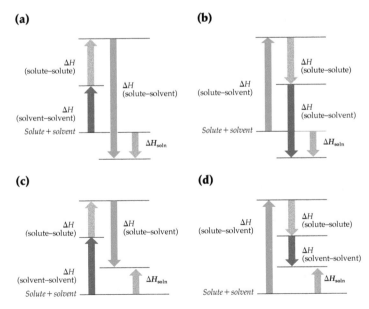

3. In which solvent does sodium acetate (CH_3COONa) have the highest solubility? (**LO 13.3**)
   - (a) Acetone CH_3OCH_3
   - (b) Methanol (CH_3OH)
   - (c) Ethanol (CH_3CH_2OH)
   - (d) Pentane ($CH_3CH_2CH_2CH_2CH_3$)

4. Rubbing alcohol is a 90% (by mass) solution of isopropyl alcohol, C_3H_8O, in water. How many moles of isopropyl alcohol are in 50.0 g of rubbing alcohol? (**LO 13.4**)
   - (a) 45 mol
   - (b) 0.75 mol
   - (c) 4.5 mol
   - (d) 1.3 mol

5. Fluoride ion is added to drinking water at low concentrations to prevent tooth decay. What mass of sodium fluoride (NaF) should be added to 750 L of water to make a solution that is 1.5 ppm in fluoride ion? (**LO 13.5**)
   - (a) 1.1 g
   - (b) 2.5 g
   - (c) 0.51 g
   - (d) 3.1 g

6. Concentrated hydrochloric acid solution contains 37.0% by mass HCl. If the density is 1.18 g/mL, what is the molar concentration? (**LO 13.6, 13.7**)
   - (a) 10.1 M
   - (b) 16.2 M
   - (c) 8.60 M
   - (d) 12.0 M

7. A 10.0 M aqueous solution of NaOH has a density of 1.33 g/cm³ at 25 °C. Calculate the mass percent of the NaOH in the solution. (**LO 13.4, 13.7**)
   - (a) 53.2%
   - (b) 30.1%
   - (c) 40.0%
   - (d) 13.3%

8. The molarity of a solution of sodium acetate (CH_3COONa) at 20 °C is 7.5 M. Use the graph showing the solubility of sodium acetate as a function of temperature to describe the solution. The solution is _____. (**LO 13.7, 13.8**)

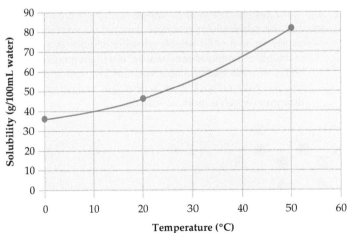

   - (a) Saturated
   - (b) Unsaturated
   - (c) Supersaturated
   - (d) Colloidal

9. A scuba diver is breathing compressed air that is 21% (by volume) oxygen at a depth of 66 ft. beneath the surface. The total pressure at this depth is 3.0 atm. The Henry's law constant (k) for O_2 in water at a normal body temperature $(37\,°C)$ is 1.93×10^{-3} mol/(L·atm). What is the solubility of O_2 in the blood of the diver at 66 ft. beneath the surface? **(LO 13.9)**

(a) 1.22×10^{-3} M   (b) 2.76×10^{-2} M

(c) 1.22×10^{-1} M   (d) 2.76×10^{-1} M

10. Which of the following solutions has the lowest vapor pressure? (Assume complete dissociation for ionic compounds.) **(LO 13.10)**

(a) An aqueous solution of 0.25 m $Ca(NO_3)_2$

(b) An aqueous solution of 0.30 m sodium chloride (NaCl)

(c) A mixture of 50 mL of water and 50 mL of ethanol (CH_3CH_2OH)

(d) A mixture of 25 mL of water and 50 mL of ethanol (CH_3CH_2OH)

11. A solution is prepared by dissolving 60.0 g KCl in 250.0 g of water at 25 °C. What is the vapor pressure of the solution if the vapor pressure of water at 25 °C is 23.76 mm Hg? **(LO 13.11)**

(a) 20.7 mm Hg   (b) 21.3 mm Hg

(c) 22.5 mm Hg   (d) 25.5 mm Hg

12. A solution contains 4.08 g of chloroform $(CHCl_3)$ and 9.29 g of acetone (CH_3COCH_3). The vapor pressures at 35 °C of pure chloroform and pure acetone are 295 torr and 332 torr, respectively. Assuming ideal behavior, calculate the vapor pressure above the solution. **(LO 13.12)**

(a) 256 torr   (b) 314 torr

(c) 325 torr   (d) 462 torr

13. How many kilograms of ethylene glycol (automobile antifreeze, $C_2H_6O_2$) dissolved in 3.55 kg of water are needed to lower the freezing point of water in an automobile radiator to $-22.0\,°C$? The molal freezing point depression constant for water is $K_f = 1.86$ $(°C·kg)/mol$. **(LO 13.13)**

(a) 0.865 kg   (b) 0.0420 kg

(c) 9.01 kg   (d) 2.61 kg

14. An intravenous glucose solution has the same osmotic pressure as blood serum (7.65 atm at 37 °C). What mass of glucose $(C_6H_{12}O_6)$ is dissolved in 0.50 L of the solution? **(LO 13.14)**

(a) 54 g   (b) 27 g

(c) 227 g   (d) 2.2 g

15. Hemoglobin is a large molecule that carries oxygen in the body. An aqueous solution that contains 2.61 g of hemoglobin in 100.0 mL has an osmotic pressure of 7.52 mmHg at 25 °C. What is the molar mass of the hemoglobin? Assume hemoglobin does not dissociate in water. **(LO 13.15)**

(a) 1.96×10^3 g/mol   (b) 84.8 g/mol

(c) 6.45×10^4 g/mol   (d) 3.65×10^3 g/mol

Answers:

1. d, 2. a, 3. b, 4. b, 5. b, 6. d, 7. b, 8. c, 9. a, 10. a, 11. b, 12. c, 13. d, 14. b, 15. c

Mastering Chemistry provides end-of-chapter exercises, feedback-enriched tutorial problems, animations, and interactive activities to encourage problem-solving practice and deeper understanding of key concepts and topics.

RAN   *Randomized in Mastering Chemistry*

CONCEPTUAL PROBLEMS

Problems 13.1–13.26 appear within the chapter.

13.27 Many people take vitamin supplements to promote health,
RAN   and the potential for overdose depends on whether a particular vitamin is fat soluble or water soluble. If a vitamin is soluble in nonpolar fatty tissues, it will accumulate in the body, whereas water-soluble vitamins are readily excreted. Use the ball-and-stick models of niacin and vitamin D to determine which is more fat soluble.

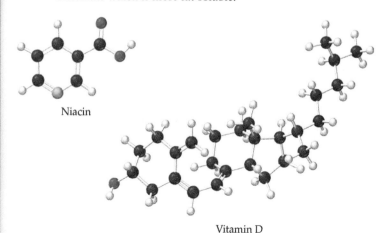

Niacin

Vitamin D

13.28 Rank the situations represented by the following drawings according to increasing entropy.

(a)   (b)

(c)

13.29 The following phase diagram shows part of the vapor-pressure curves for a pure liquid (green curve) and a solution of the first liquid with a second volatile liquid (red curve).

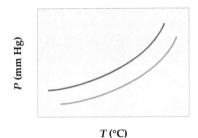

T (°C)

(a) Is the boiling point of the second liquid higher or lower than that of the first liquid?

(b) Draw on the diagram the approximate position of the vapor-pressure curve for the second liquid.

13.30 The following phase diagram shows part of the liquid–vapor phase-transition boundaries for pure ether and a solution of a nonvolatile solute in ether.

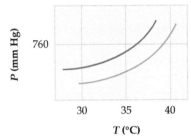

T (°C)

(a) What is the approximate normal boiling point of pure ether?

(b) What is the approximate molal concentration of the solute? [K_b for ether is 2.02(°C · kg)/mol.]

13.31 The following diagram shows a close-up view of part of the vapor-pressure curves for a solvent (red curve) and a solution of the solvent with a second liquid (green curve). Is the second liquid more volatile or less volatile than the solvent?

T (°C)

13.32 The following phase diagram shows part of the liquid–vapor phase-transition boundaries for two solutions of equal concentration, one containing a nonvolatile solute and the other containing a volatile solute whose vapor pressure at a given temperature is approximately half that of the pure solvent.

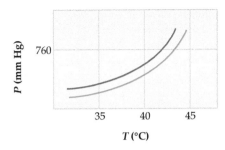

T (°C)

(a) Which curve, red or green, represents the solution of the nonvolatile solute and which represents the solution of the volatile solute?

(b) Draw on the diagram the approximate position of the vapor-pressure curve for the pure solvent.

(c) Based on your drawing, what is the approximate molal concentration of the nonvolatile solute, assuming the solvent has $K_b = 2.0 \, °C/m$?

(d) Based on your drawing, what is the approximate normal boiling point of the pure solvent?

13.33 Two beakers, one with pure water (blue) and the other with a solution of NaCl in water (green), are placed in a closed container as represented by drawing (a). Which of the drawings (b)–(d) represents what the beakers will look like after a substantial amount of time has passed?

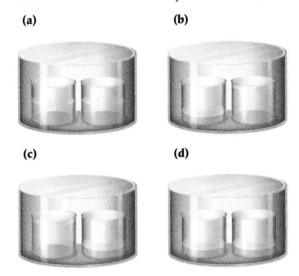

13.34 Assume that two liquids are separated by a semipermeable membrane. Make a drawing that shows the situation after equilibrium is reached.

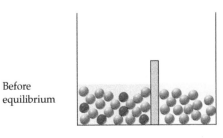

Before equilibrium

13.35 The following phase diagram shows a very small part of the solid–liquid phase-transition boundaries for two solutions of equal concentration. Substance A has $i = 1$, and substance B has $i = 3$.

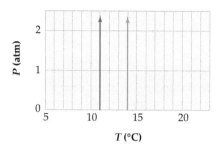

(a) Which line, red or blue, represents a solution of A, and which represents a solution of B?

(b) What is the approximate melting point of the pure liquid solvent?

(c) What is the approximate molal concentration of each solution, assuming the solvent has $K_f = 3.0\ °C/m$?

RAN

SECTION PROBLEMS

Solutions, Enthalpy Changes, and Solubility
(Sections 13.1–13.3)

13.36 If a single 5-g block of NaCl is placed in water, it dissolves slowly, but if 5 g of powdered NaCl is placed in water, it dissolves rapidly. Explain.

13.37 Why do ionic substances with higher lattice energies tend to be less soluble in water than substances with lower lattice energies?

13.38 Which would you expect to have the larger (more negative) hydration energy?

(a) Na^+ or Cs^+ (b) K^+ or Ba^{2+}

13.39 Which would you expect to have the larger hydration energy, SO_4^{2-} or ClO_4^-? Explain.

13.40 Classify the strongest type of intermolecular force in the following interactions: solvent–solvent, solvent–solute, and solute–solute when solid iodine (I_2) is placed in the water. Based on these interactions, predict whether I_2 is soluble in water.

13.41 Classify the strongest type of intermolecular force in the following interactions: solvent–solvent, solvent–solute, and solute–solute when solid glucose ($C_6H_{12}O_6$) is placed in the water. Based on these interactions predict whether glucose is soluble in water.

$$HO-\overset{\overset{\displaystyle H}{|}}{\underset{\underset{\displaystyle H}{|}}{C}}-\overset{\overset{\displaystyle OH}{|}}{\underset{\underset{\displaystyle H}{|}}{C}}-\overset{\overset{\displaystyle H}{|}}{\underset{\underset{\displaystyle OH}{|}}{C}}-\overset{\overset{\displaystyle OH}{|}}{\underset{\underset{\displaystyle H}{|}}{C}}-\overset{\overset{\displaystyle H}{|}}{\underset{\underset{\displaystyle OH}{|}}{C}}-\overset{\overset{\displaystyle O}{\|}}{C}-H$$

Glucose

13.42 What is the sign of ΔH_{soln}, and how will the temperature change when a solute dissolves with the following enthalpy changes?

$\Delta H_{solute-solute} = +642\ kJ/mol$

$\Delta H_{solvent-solvent} = +31\ kJ/mol$

$\Delta H_{solute-solvent} = -648\ kJ/mol$

13.43 What is the sign of ΔH_{soln}, and how will the temperature change when a solute dissolves with the following enthalpy changes?

$\Delta H_{solute-solute} = +56\ kJ/mol$

$\Delta H_{solvent-solvent} = +34\ kJ/mol$

$\Delta H_{solute-solvent} = -125\ kJ/mol$

13.44 Br_2 is much more soluble in tetrachloromethane, CCl_4, than in water. Explain.

13.45 Predict whether the solubility of formaldehyde, CH_2O, is greater in water or tetrachloromethane, CCl_4.

13.46 Predict whether the solubility of acetic acid, CH_3COOH, is greater in water or benzene, C_6H_6.

13.47 Predict whether the solubility of butane, C_4H_{10}, is greater in water or benzene, C_6H_6.

13.48 Arrange the following compounds in order of their expected increasing solubility in water: Br_2, KBr, toluene (C_7H_8, a constituent of gasoline).

13.49 Arrange the following compounds in order of their expected increasing solubility in hexane (C_6H_{14}): NaBr, CCl_4, CH_3COOH.

13.50 When ethyl alcohol, CH_3CH_2OH, dissolves in water, how many hydrogen bonds are formed between one ethyl alcohol molecule and surrounding water molecules? Sketch the hydrogen bonding interactions. (Hint: Add lone pairs of electrons to the structure before drawing hydrogen bonds.)

$$H-\overset{\overset{\displaystyle H}{|}}{\underset{\underset{\displaystyle H}{|}}{C}}-\overset{\overset{\displaystyle H}{|}}{\underset{\underset{\displaystyle H}{|}}{C}}-O-H$$

13.51 When the vitamin niacin ($C_6H_5O_2N$) dissolves in water, how many hydrogen bonds are formed between one niacin molecule and surrounding water molecules? Sketch the hydrogen bonding interactions. (Hint: Add lone pairs of electrons to the structure before drawing hydrogen bonds.)

Niacin

13.52 Ethyl alcohol, CH_3CH_2OH, is miscible with water at 20 °C, but pentyl alcohol, $CH_3CH_2CH_2CH_2CH_2OH$, is soluble in water only to the extent of 2.7 g/100 mL. Explain.

13.53 Pentyl alcohol ($CH_3CH_2CH_2CH_2CH_2OH$) is miscible with octane, C_8H_{18}, but methyl alcohol, CH_3OH, is insoluble in octane. Explain.

Units of Concentration (Section 13.4)

13.54 The dissolution of $CaCl_2$ (s) in water is exothermic, with $\Delta H_{soln} = -81.3$ kJ/mol. If you were to prepare a 1.00 m solution of $CaCl_2$ beginning with water at 25.0 °C, what would the final temperature of the solution be in °C? Assume that the specific heats of both pure H_2O and the solution are the same, 4.18 J/(K·g).

13.55 The dissolution of $NH_4ClO_4(s)$ in water is endothermic, with $\Delta H_{soln} = +33.5$ kJ/mol. If you prepare a 1.00 m solution of NH_4ClO_4 beginning with water at 25.0 °C, what is the final temperature of the solution in °C? Assume that the specific heats of both pure H_2O and the solution are the same, 4.18 J/(K·g).

13.56 Assuming that seawater is an aqueous solution of NaCl, what is its molarity? The density of seawater is 1.025 g/mL at 20 °C, and the NaCl concentration is 3.50 mass %.

13.57 Assuming that seawater is a 3.50 mass % aqueous solution of NaCl, what is the molality of seawater?

13.58 Propranolol ($C_{16}H_{21}NO_2$), a so-called beta-blocker that is used for treatment of high blood pressure, is effective at a blood plasma concentration of 50 ng/L. Express this concentration of propranolol in the following units:

(a) Parts per billion (assume a plasma density of 1.025 g/mL)

(b) Molarity

13.59 Residues of the herbicide atrazine ($C_8H_{14}ClN_5$) in water can be detected at concentrations as low as 0.050 μg/L. Express this concentration of atrazine in the following units:

(a) Parts per billion (assume a solution density of 1.00 g/mL)

(b) Molarity

13.60 How would you prepare each of the following solutions?

(a) A 0.150 M solution of glucose in water

(b) A 1.135 m solution of KBr in water

(c) A solution of methyl alcohol (methanol) and water in which $X_{methanol} = 0.15$ and $X_{water} = 0.85$

13.61 How would you prepare each of the following solutions?

(a) 100 mL of a 155 ppm solution of urea, CH_4N_2O, in water

(b) 100 mL of an aqueous solution whose K^+ concentration is 0.075 M

13.62 How would you prepare 165 mL of a 0.0268 M solution of benzoic acid ($C_7H_6O_2$) in chloroform ($CHCl_3$)?

13.63 How would you prepare 250 mL of a 0.325 M solution of benzoic acid ($C_7H_6O_2$) in chloroform ($CHCl_3$)?

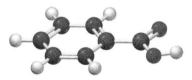

Benzoic acid

13.64 Which of the following solutions is more concentrated?
RAN
(a) 0.500 M KCl or 0.500 mass % KCl in water

(b) 1.75 M glucose or 1.75 m glucose in water

13.65 Which of the following solutions has the higher molarity?
RAN
(a) 10 ppm KI in water or 10,000 ppb KBr in water

(b) 0.25 mass % KCl in water or 0.25 mass % citric acid ($C_6H_8O_7$) in water

13.66 What is the mass percent concentration of the following
RAN solutions?

(a) Dissolve 0.655 mol of citric acid, $C_6H_8O_7$, in 1.00 kg of water.

(b) Dissolve 0.135 mg of KBr in 5.00 mL of water.

(c) Dissolve 5.50 g of aspirin, $C_9H_8O_4$, in 145 g of dichloromethane, CH_2Cl_2.

13.67 What is the molality of each solution prepared in Problem 13.66?

13.68 The ozone layer in the Earth's stratosphere has an average
RAN total pressure of 10 mm Hg (1.3×10^{-2} atm). The partial pressure of ozone in the layer is about 1.2×10^{-6} mm Hg (1.6×10^{-9} atm). What is the concentration of ozone in parts per million, assuming that the average molar mass of air is 29 g/mol?

13.69 A person is medically considered to have lead poisoning if
RAN he or she has a concentration of greater than 10 μg of lead per deciliter of blood. What is this concentration in parts per billion? Assume that the density of blood is the same as that of water.

13.70 What is the concentration of each of the following solutions?

(a) The molality of a solution prepared by dissolving 25.0 g of H_2SO_4 in 1.30 L of water

(b) The mole fraction of each component of a solution prepared by dissolving 2.25 g of nicotine, $C_{10}H_{14}N_2$, in 80.0 g of CH_2Cl_2

13.71 Household bleach is a 5.0 mass % aqueous solution of sodium hypochlorite, NaOCl. What is the molality of the bleach? What is the mole fraction of NaOCl in the bleach?

13.72 The density of a 16.0 mass % solution of sulfuric acid in
RAN water is 1.1094 g/mL at 25.0 °C. What is the molarity of the solution?

13.73 Ethylene glycol, $C_2H_6O_2$, is the principal constituent of auto-
RAN mobile antifreeze. If the density of a 40.0 mass % solution of ethylene glycol in water is 1.0514 g/mL at 20 °C, what is the molarity?

Ethylene glycol

13.74 What is the molality of the 40.0 mass % ethylene glycol
RAN ($C_2H_6O_2$) solution used for automobile antifreeze?

13.75 Ethylene glycol, $C_2H_6O_2$, is a colorless liquid used as auto-
RAN mobile antifreeze. If the density at 20 °C of a 4.028 m solution of ethylene glycol in water is 1.0241 g/mL, what is the molarity of the solution? The molar mass of ethylene glycol is 62.07 g/mol.

13.76 Nalorphine ($C_{19}H_{21}NO_3$), a relative of morphine, is used to combat withdrawal symptoms in narcotics users. How many grams of a 1.3×10^{-3} m aqueous solution of nalorphine are needed to obtain a dose of 1.5 mg?

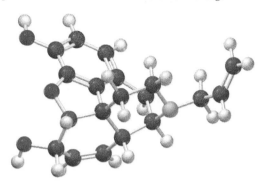

Nalorphine

13.77 How many grams of water should you add to 32.5 g of
RAN   sucrose, $C_{12}H_{22}O_{11}$, to get a 0.850 m solution?

13.78 A 0.944 M solution of glucose, $C_6H_{12}O_6$, in water has a
RAN   density of 1.0624 g/mL at 20 °C. What is the concentration of this solution in the following units?
   (a) Mole fraction
   (b) Mass percent
   (c) Molality

13.79 Lactose, $C_{12}H_{22}O_{11}$, is a naturally occurring sugar found in
RAN   mammalian milk. A 0.335 M solution of lactose in water has a density of 1.0432 g/L at 20 °C. What is the concentration of this solution in the following units?
   (a) Mole fraction
   (b) Mass percent
   (c) Molality

Solubility and Henry's Law (Section 13.5)

13.80 Look at the solubility graph in Figure 13.7, and estimate
RAN   which member of each of the following pairs has the higher molar solubility at the given temperature.
   (a) $CuSO_4$ or NH_4Cl at 60 °C
   (b) CH_3CO_2Na or glucose at 20 °C

13.81 Look at the solubility graph in Figure 13.7, and estimate
RAN   which member of each of the following pairs has the higher molar solubility at the given temperature.
   (a) NaCl or NH_4Cl at 40 °C
   (b) K_2SO_4 or $CuSO_4$ at 20 °C

13.82 Vinyl chloride ($H_2C = CHCl$), the starting material from which PVC polymer is made, has a Henry's law constant of at 0.091 mol/(L·atm) at 25 °C. What is the solubility of vinyl chloride in water in mol/L at 25 °C and a partial pressure of 0.75 atm?

13.83 Hydrogen sulfide, H_2S, is a toxic gas responsible for the odor of rotten eggs. The solubility of $H_2S(g)$ in water at STP is 0.195 M. What is the Henry's law constant of H_2S at 0 °C? What is the solubility of H_2S in water at 0 °C and a partial pressure of 25.5 mm Hg?

13.84 Fish generally need an O_2 concentration in water of at least
RAN   4 mg/L for survival. What partial pressure of oxygen above the water in atmospheres at 0 °C is needed to obtain this concentration? The solubility of O_2 in water at 0 °C and 1 atm partial pressure is 2.21×10^{-3} mol/L.

13.85 At an altitude of 10,000 ft, the partial pressure of oxygen in
RAN   the lungs is about 68 mm Hg. What is the concentration in mg/L of dissolved O_2 in blood (or water) at this partial pressure and a normal body temperature of 37 °C? The solubility of O_2 in water at 37 °C and 1 atm partial pressure is 1.93×10^{-3} mol/L.

13.86 Sulfur hexafluoride, which is used as a nonflammable insulator in high-voltage transformers, has a Henry's law constant of 2.4×10^{-4} mol/(L·atm) at 25 °C. What is the solubility in mol/L of sulfur hexafluoride in water at 25 °C and a partial pressure of 2.00 atm?

13.87 The nonstick polymer Teflon is made from tetrafluoroethylene, C_2F_4. If C_2F_4 is a gas that dissolves in water at 298 K to the extent of 1.01×10^{-3} M with a partial pressure of 0.63 atm, what is its Henry's law constant at 298 K?

Colligative Properties (Sections 13.6–13.9)

13.88 When 1 mL of toluene is added to 100 mL of benzene (bp 80.1 °C), the boiling point of the benzene solution rises, but when 1 mL of benzene is added to 100 mL of toluene (bp 110.6 °C), the boiling point of the toluene solution falls. Explain.

13.89 When solid $CaCl_2$ is added to liquid water, the temperature
RAN   rises. When solid $CaCl_2$ is added to ice at 0 °C, the temperature falls. Explain.

13.90 Rank the following aqueous solutions from lowest to highest freezing point: 0.10 m $FeCl_3$, 0.30 m glucose ($C_6H_{12}O_6$), 0.15 m $CaCl_2$. Assume complete dissociation.

13.91 Which of the following aqueous solutions has the **(a)** higher
RAN   freezing point, **(b)** higher boiling point, **(c)** lower vapor pressure: 0.50 m sucrose ($C_{12}H_{22}O_{11}$) or 0.35 m HNO_3?

13.92 What is the vapor pressure in mm Hg of a solution prepared
RAN   by dissolving 5.00 g of benzoic acid ($C_7H_6O_2$) in 100.00 g of ethyl alcohol (C_2H_6O) at 35 °C? The vapor pressure of pure ethyl alcohol at 35 °C is 100.5 mm Hg.

13.93 What is the normal boiling point in °C of a solution pre-
RAN   pared by dissolving 1.50 g of aspirin (acetylsalicylic acid, $C_9H_8O_4$) in 75.00 g of chloroform ($CHCl_3$)? The normal boiling point of chloroform is 61.7 °C, and K_b for chloroform is given in Table 13.4.

13.94 What is the freezing point in °C of a solution prepared by dissolving 7.40 g of $MgCl_2$ in 110 g of water? The value of K_f for water is given in Table 13.4, and the van't Hoff factor for $MgCl_2$ is $i = 2.7$.

13.95 Assuming complete dissociation, what is the molality of an
RAN   aqueous solution of KBr whose freezing point is −2.95 °C? The molal freezing-point-depression constant of water is given in Table 13.4.

13.96 When 9.12 g of HCl was dissolved in 190 g of water, the freezing point of the solution was −4.65 °C. What is the value of the van't Hoff factor for HCl?

13.97 The observed osmotic pressure for a 0.125 M solution of $MgCl_2$ at 310 K is 8.57 atm. Calculate the value of the van't Hoff factor for $MgCl_2$ under these conditions.

13.98 When 1 mol of NaCl is added to 1 L of water, the boiling point increases. When 1 mol of methyl alcohol is added to 1 L of water, the boiling point decreases. Explain.

13.99 When 100 mL of 9 M H_2SO_4 at 0 °C is added to 100 mL of liquid water at 0 °C, the temperature rises to 12 °C. When 100 mL of 9 M H_2SO_4 at 0 °C is added to 100 g of solid ice at 0 °C, the temperature falls to -12 °C. Explain the difference in behavior.

13.100 Draw a phase diagram showing how the phase boundaries
RAN differ for a pure solvent compared with a solution.

13.101 A solution concentration must be expressed in molality
RAN when considering boiling-point elevation or freezing-point depression but can be expressed in molarity when considering osmotic pressure. Why?

13.102 What is the vapor pressure in mm Hg of the following solu-
RAN tions, each of which contains a nonvolatile solute? The vapor pressure of water at 45.0 °C is 71.93 mm Hg.

(a) A solution of 10.0 g of urea, CH_4N_2O, in 150.0 g of water at 45.0 °C

(b) A solution of 10.0 g of LiCl in 150.0 g of water at 45.0 °C, assuming complete dissociation

13.103 What is the vapor pressure in mm Hg of a solution of 16.0 g of
RAN glucose ($C_6H_{12}O_6$) in 80.0 g of methanol (CH_3OH) at 27 °C? The vapor pressure of pure methanol at 27 °C is 140 mm Hg.

13.104 What is the boiling point in °C of each of the solutions in Problem 13.102? For water, $K_b = 0.51$ (°C·kg)/mol.

13.105 What is the freezing point in °C of each of the solutions in
RAN Problem 13.102? For water, $K_f = 1.86$ (°C·kg)/mol.

13.106 A 1.0 m solution of K_2SO_4 in water has a freezing point
RAN of -4.3 °C. What is the value of the van't Hoff factor i for K_2SO_4?

13.107 The van't Hoff factor for KCl is $i = 1.85$. What is the boil-
RAN ing point of a 0.75 m solution of KCl in water? For water, $K_b = 0.51$(°C·kg)/mol.

13.108 Heptane (C_7H_{16}) and octane (C_8H_{18}) are constituents of
RAN gasoline. At 80.0 °C, the vapor pressure of heptane is 428 mm Hg, and the vapor pressure of octane is 175 mm Hg. What is $X_{heptane}$ in a mixture of heptane and octane that has a vapor pressure of 305 mm Hg at 80.0 °C?

13.109 Cyclopentane (C_5H_{10}) and cyclohexane (C_6H_{12}) are vola-
RAN tile, nonpolar hydrocarbons. At 30.0 °C, the vapor pressure of cyclopentane is 385 mm Hg, and the vapor pressure of cyclohexane is 122 mm Hg. What is $X_{pentane}$ in a mixture of C_5H_{10} and C_6H_{12} that has a vapor pressure of 212 mm Hg at 30.0 °C?

13.110 Acetone, C_3H_6O, and ethyl acetate, $C_4H_8O_2$, are organic
RAN liquids often used as solvents. At 30 °C, the vapor pressure of acetone is 285 mm Hg, and the vapor pressure of ethyl acetate is 118 mm Hg. What is the vapor pressure in mm Hg at 30 °C of a solution prepared by dissolving 25.0 g of acetone in 25.0 g of ethyl acetate?

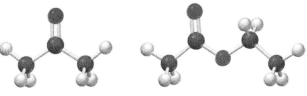

Acetone   Ethyl acetate

13.111 The industrial solvents chloroform, $CHCl_3$, and dichloro-
RAN methane, CH_2Cl_2, are prepared commercially by reaction of methane with chlorine, followed by fractional distillation of the product mixture. At 25 °C, the vapor pressure of $CHCl_3$ is 205 mm Hg, and the vapor pressure of CH_2Cl_2 is 415 mm Hg. What is the vapor pressure in mm Hg at 25 °C of a mixture of 15.0 g of $CHCl_3$ and 37.5 g of CH_2Cl_2?

13.112 What is the mole fraction of each component in the liquid mixture in Problem 13.110, and what is the mole fraction of each component in the vapor at 30 °C?

13.113 What is the mole fraction of each component in the liquid mixture in Problem 13.111, and what is the mole fraction of each component in the vapor at 25 °C?

13.114 A solution prepared by dissolving 5.00 g of aspirin,
RAN $C_9H_8O_4$, in 215 g of chloroform has a normal boiling point that is elevated by $\Delta T = 0.47$ °C over that of pure chloroform. What is the value of the molal boiling-point-elevation constant for chloroform?

13.115 A solution prepared by dissolving 3.00 g of ascorbic acid
RAN (vitamin C, $C_6H_8O_6$) in 50.0 g of acetic acid has a freezing point that is depressed by $\Delta T = 1.33$ °C below that of pure acetic acid. What is the value of the molal freezing-point-depression constant for acetic acid?

13.116 A solution of citric acid, $C_6H_8O_7$, in 50.0 g of acetic acid
RAN has a boiling point elevation of $\Delta T = 1.76$ °C. What is the molality of the solution if the molal boiling-point-elevation constant for acetic acid is $K_b = 3.07$ (°C·kg)/mol.

13.117 What is the normal boiling point in °C of ethyl alcohol if a
RAN solution prepared by dissolving 26.0 g of glucose ($C_6H_{12}O_6$) in 285 g of ethyl alcohol has a boiling point of 79.1 °C? See Table 13.4 to find K_b for ethyl alcohol.

13.118 What osmotic pressure in atmospheres would you expect
RAN for each of the following solutions?

(a) 5.00 g of NaCl in 350.0 mL of aqueous solution at 50 °C

(b) 6.33 g of sodium acetate, CH_3CO_2Na, in 55.0 mL of aqueous solution at 10 °C

13.119 What osmotic pressure in mm Hg would you expect for an
RAN aqueous solution of 11.5 mg of insulin (mol. weight = 5990) in 6.60 mL of solution at 298 K? What would be the height of the water column be in meters? The density of mercury is 13.534 g/mL at 298 K.

13.120 A solution of an unknown molecular substance in water at 300 K gives rise to an osmotic pressure of 4.85 atm. What is the molarity of the solution?

13.121 Human blood gives rise to an osmotic pressure of approximately 7.7 atm at body temperature, 37.0 °C. What must the molarity of an intravenous glucose solution be to give rise to the same osmotic pressure as blood?

13.122 When salt is spread on snow-covered roads at -2 °C, the
RAN snow melts. When salt is spread on snow-covered roads at -30 °C, nothing happens. Explain.

13.123 If cost per gram were not a concern, which of the follow-
RAN ing substances would be the most efficient per unit mass for melting snow from sidewalks and roads: glucose ($C_6H_{12}O_6$), LiCl, NaCl, or $CaCl_2$? Explain.

13.124 Cellobiose is a sugar obtained by degradation of cellulose. If 200.0 mL of an aqueous solution containing 1.500 g of cellobiose at 25.0 °C gives rise to an osmotic pressure of 407.2 mm Hg, what is the molecular weight of cellobiose?

13.125 Met-enkephalin is one of the so-called endorphins, a class of naturally occurring morphine-like chemicals in the brain.

What is the molecular weight of met-enkephalin if 20.0 mL of an aqueous solution containing 15.0 mg of met-enkephalin at 298 K supports a column of water 32.9 cm high? The density of mercury at 298 K is 13.534 g/mL.

13.126 The freezing point of a solution prepared by dissolving 1.00 mol of hydrogen fluoride, HF, in 500 g of water is -3.8 °C, but the freezing point of a solution prepared by dissolving 1.00 mol of hydrogen chloride, HCl, in 500 g of water is -7.4 °C. Explain.

13.127 The boiling point of a solution prepared by dissolving 71 g of Na_2SO_4 in 1.00 kg of water is 100.8 °C. Explain.

13.128 When a 2.850 g mixture of the sugars sucrose ($C_{12}H_{22}O_{11}$)
RAN and fructose ($C_6H_{12}O_6$) was dissolved in water to a volume of 1.50 L, the resultant solution gave an osmotic pressure of 0.1843 atm at 298.0 K. What is $X_{sucrose}$ of the mixture?

13.129 Glycerol ($C_3H_8O_3$) and diethylformamide ($C_5H_{11}NO$) are
RAN nonvolatile, miscible liquids. If the volume of a solution made by dissolving 10.208 g of a glycerol–diethylformamide mixture in water is 1.75 L and the solution has an osmotic pressure of 1.466 atm at 298.0 K, what is $X_{glycerol}$ of the mixture?

13.130 How many grams of naphthalene, $C_{10}H_8$ (commonly used as household mothballs), should be added to 150.0 g of benzene to depress its freezing point by 0.35 °C? See Table 13.4 to find K_f for benzene.

13.131 Bromine is sometimes used as a solution in tetrachloromethane, CCl_4. What is the vapor pressure in mm Hg of a solution of 1.50 g of Br_2 in 145.0 g of CCl_4 at 300 K? The vapor pressure of pure bromine at 300 K is 30.5 kPa, and the vapor pressure of CCl_4 is 16.5 kPa.

13.132 Assuming that seawater is a 3.5 mass % solution of NaCl
RAN and that its density is 1.00 g/mL, calculate both its boiling point and its freezing point in °C.

13.133 There's actually much more in seawater than just dissolved NaCl. Major ions present include 19,000 ppm Cl^-, 10,500 ppm Na^+, 2650 ppm SO_4^{2-}, 1350 ppm Mg^{2+}, 400 ppm Ca^{2+}, 380 ppm K^+, 140 ppm HCO_3^-, and 65 ppm Br^-.

(a) What is the total molality of all ions present in seawater?

(b) Assuming molality and molarity to be equal, what amount of osmotic pressure in atmospheres would seawater give rise to at 300 K?

13.134 The van't Hoff factor for $CaCl_2$ is 2.71. What is its mass %
RAN in an aqueous solution that has $T_f = -1.14$ °C?

13.135 What is the van't Hoff factor for K_2SO_4 in an aqueous solution that is 5.00% K_2SO_4 by mass and freezes at -1.21 °C?

13.136 If the van't Hoff factor for LiCl in a 0.62 m solution is 1.96, what is the vapor pressure depression in mm Hg of the solution at 298 K? (The vapor pressure of water at 298 K is 23.76 mm Hg.)

13.137 What is the value of the van't Hoff factor for KCl if a 1.00 m aqueous solution shows a vapor pressure depression of 0.734 mm Hg at 298 °C? (The vapor pressure of water at 298 K is 23.76 mm Hg.)

13.138 A solid mixture of KCl, KNO_3, and $Ba(NO_3)_2$ is 20.92 mass % chlorine, and a 1.000 g sample of the mixture in 500.0 mL of aqueous solution at 25 °C has an osmotic pressure of 744.7 mm Hg. What are the mass percents of KCl, KNO_3, and $Ba(NO_3)_2$ in the mixture?

13.139 A solution of LiCl in a mixture of water and methanol
RAN (CH_3OH) has a vapor pressure of 39.4 mm Hg at 17 °C and 68.2 mm Hg at 27 °C. The vapor pressure of pure water is 14.5 mm Hg at 17 °C and 26.8 mm Hg at 27 °C, and the vapor pressure of pure methanol is 82.5 mm Hg at 17 °C and 140.3 mm Hg at 27 °C. What is the composition of the solution in mass percent?

13.140 An aqueous solution of a certain organic compound has a
RAN density of 1.063 g/mL, an osmotic pressure of 12.16 atm at 25.0 °C, and a freezing point of -1.03 °C. The compound is known not to dissociate in water. What is the molar mass of the compound?

13.141 At 60 °C, compound X has a vapor pressure of 96 mm Hg,
RAN benzene (C_6H_6) has a vapor pressure of 395 mm Hg, and a 50:50 mixture by mass of benzene and X has a vapor pressure of 299 mm Hg. What is the molar mass of X?

13.142 Desert countries like Saudi Arabia have built reverse osmo-
RAN sis plants to produce freshwater from seawater. Assume that seawater has the composition 0.470 M NaCl and 0.068 M $MgCl_2$ and that both compounds are completely dissociated.

(a) What is the osmotic pressure of seawater at 25 °C?

(b) If the reverse osmosis equipment can exert a maximum pressure of 100.0 atm at 25.0 °C, what is the maximum volume of freshwater that can be obtained from 1.00 L of seawater?

13.143 A solution of 0.250 g of naphthalene (mothballs) in 35.00 g
RAN of camphor lowers the freezing point by 2.10 °C. What is the molar mass of naphthalene? The freezing-point-depression constant for camphor is 37.7 (°C · kg)/mol.

MULTICONCEPT PROBLEMS

13.144 Elemental analysis of β-carotene, a dietary source of vitamin
RAN A, shows that it contains 10.51% H and 89.49% C. Dissolving 0.0250 g of β-carotene in 1.50 g of camphor gives a freezing-point depression of 1.17 °C. What are the molecular weight and formula of β-carotene? [K_f for camphor is 37.7 (°C · kg)/mol.]

13.145 Lysine, one of the amino acid building blocks found in proteins, contains 49.29% C, 9.65% H, 19.16% N, and 21.89% O by elemental analysis. A solution prepared by dissolving 30.0 mg of lysine in 1.200 g of the organic solvent biphenyl gives a freezing-point depression of 1.37 °C. What are the molecular weight and formula of lysine? [K_f for biphenyl is 8.00 (°C · kg)/mol.]

13.146 The steroid hormone estradiol contains only C, H, and O; com-
RAN bustion analysis of a 3.47 mg sample yields 10.10 mg CO_2 and
2.76 mg H_2O. On dissolving 7.55 mg of estradiol in 0.500 g
of camphor, the melting point of camphor is depressed by
2.10 °C. What is the molecular weight of estradiol, and what is
a probable formula? [For camphor, $K_f = 37.7$ (°C·kg)/mol.]

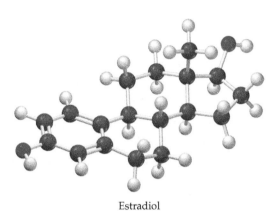

Estradiol

13.147 Many acids are partially dissociated into ions in aqueous
RAN solution. Trichloroacetic acid (CCl_3CO_2H), for instance, is
partially dissociated in water according to the equation

$$CCl_3CO_2H(aq) \rightleftharpoons H^+(aq) + CCl_3CO_2^-(aq)$$

What is the percentage of molecules dissociated if the freezing point of a 1.00 m solution of trichloroacetic acid in
water is −2.53 °C?

13.148 Addition of 50.00 mL of 2.238 m H_2SO_4 (solution density
= 1.1243 g/mL) to 50.00 mL of 2.238 M $BaCl_2$ gives a
white precipitate.
(a) What is the mass of the precipitate in grams?
(b) If you filter the mixture and add more H_2SO_4 solution to
the filtrate, would you obtain more precipitate? Explain.

13.149 Treatment of 1.385 g of an unknown metal M with an excess
of aqueous HCl evolved a gas that was found to have a volume
of 382.6 mL at 20.0 °C and 755 mm Hg pressure. Heating
the reaction mixture to evaporate the water and remaining
HCl then gave a white crystalline compound, MCl_x. After dissolving the compound in 25.0 g of water, the melting point of
the resulting solution was −3.53 °C.
(a) How many moles of H_2 gas are evolved?
(b) What mass of MCl_x is formed?
(c) What is the molality of particles (ions) in the solution
of MCl_x?
(d) How many moles of ions are in solution?
(e) What are the formula and molecular weight of MCl_x?
(f) What is the identity of the metal M?

13.150 A compound that contains only C and H was burned in
RAN excess O_2 to give CO_2 and H_2O. When 0.270 g of the compound was burned, the amount of CO_2 formed reacted
completely with 20.0 mL of 2.00 M NaOH solution
according to the equation

$$2 \text{ OH}^-(aq) + CO_2(g) \rightarrow CO_3^{2-}(aq) + H_2O(l)$$

When 0.270 g of the compound was dissolved in 50.0 g
of camphor, the resulting solution had a freezing point
of 177.9 °C. [Pure camphor freezes at 179.8 °C and has
$K_f = 37.7$ (°C·kg)/mol.]
(a) What is the empirical formula of the compound?
(b) What is the molecular weight of the compound?
(c) What is the molecular formula of the compound?

13.151 Combustion analysis of a 36.72-mg sample of the male
hormone testosterone gave 106.43 mg CO_2 and 32.10 mg
H_2O as the only combustion products. When 5.00 mg of
testosterone was dissolved in 15.0 mL of a suitable solvent
at 25 °C, an osmotic pressure of 21.5 mm Hg was measured. What is the molecular formula of testosterone?

13.152 When 8.900 g of a mixture of an alkali metal chloride
(**X**Cl) and an alkaline earth chloride (**Y**Cl$_2$) was dissolved
in 150.0 g of water, the freezing point of the resultant
solution was −4.42 °C. Addition of an excess of aqueous
AgNO$_3$ to the solution yielded a white precipitate with a
mass of 27.575 g. How much of each metal chloride was
present in the original mixture, and what are the identities
of the two metals **X** and **Y**?

13.153 Combustion analysis of a 3.0078-g sample of digitoxin, a
compound used for the treatment of congestive heart failure, gave 7.0950 g of CO_2 and 2.2668 g of H_2O. When
0.6617 g of digitoxin was dissolved in water to a total
volume of 0.800 L, the osmotic pressure of the solution at
298 K was 0.026 44 atm. What is the molecular formula of
digitoxin, which contains only C, H, and O?

13.154 A solution prepared by dissolving 100.0 g of a mixture
of sugar ($C_{12}H_{22}O_{11}$) and table salt (NaCl) in 500.0 g of
water has a freezing point of −2.25 °C. What is the mass
of each individual solute? Assume that NaCl is completely
dissociated.

chapter 14

Chemical Kinetics

Contents

Enzymes are large biological molecules that increase the rates of important reactions in the body. Amylase is the enzyme that speeds up the chemical reaction that breaks down starch into glucose, a first step in producing energy.

How do enzymes work?

The answer to this question can be found on page 583 in the INQUIRY

Chemists ask three fundamental questions when they study chemical reactions: What happens? To what extent does it happen? How fast and by what mechanism does it happen? The answer to the first question is given by the balanced chemical equation, which identifies the reactants, the products, and the stoichiometry of the reaction. The answer to the second question concerning the extent of reaction is addressed in Chapter 15, which deals with **chemical equilibrium.** In this chapter, we'll look at the answer to the third question—the speeds, or rates, and the mechanisms of chemical reactions. **Chemical kinetics** is the area of chemistry that describes reaction rates and the sequence of steps that occur in reactions.

Chemical kinetics is a subject of crucial environmental, biological, and economic importance. In the upper atmosphere, for example, maintenance or depletion of the ozone layer, which protects us from the sun's harmful ultraviolet radiation, depends on the relative rates of reactions that produce and destroy O_3 molecules. In our bodies, large protein molecules called *enzymes* increase the rates of numerous reactions essential to life processes. In the chemical industry, the profitability of many processes requires fast reaction rates. For example, the economical synthesis of ammonia, used as a fertilizer, depends on the rate at which gaseous N_2 and H_2 can be converted to NH_3. Like the biochemical reactions occurring in the body, industrial reactions can also be hastened by **catalysts.** Enzymes and other types of catalysts increase reaction rates without being consumed in the process.

In this chapter, we'll describe reaction rates and examine how they are affected by variables such as concentrations and temperature. We'll also see how chemists use rate data to propose a *mechanism,* or pathway, by which a reaction takes place. By understanding reaction mechanisms, we can control known reactions and predict new ones.

LOOKING AHEAD . . .
We'll see in Chapter 15 that **chemical equilibrium** is the state reached when the concentrations of reactants and products remain constant over time.

LOOKING AHEAD . . .
We'll see in Section 14.12 why **catalysts** speed up chemical reactions.

14.1 REACTION RATES

The rates of chemical reactions differ greatly. Some reactions, such as the combination of sodium and bromine, occur instantly. Other reactions, such as the breakdown of food for energy or the ripening of fruit, occur on the timescale of hours to days. Yet others, such as the rusting of iron or the formation of fossil fuels, occur over months or even thousands of years.

The reaction of sodium and bromine

Reactions that break down food to release energy

The rusting of iron

▲ Chemical reactions exhibit a wide range of reaction rates.

To describe a reaction rate quantitatively, we must specify how fast the concentration of a reactant or a product changes per unit time.

$$\text{Rate} = \frac{\text{Change in concentration}}{\text{Change in time}}$$

Look, for example, at the thermal decomposition of gaseous dinitrogen pentoxide, N_2O_5, to give the brown gas nitrogen dioxide, a common air pollutant, and molecular oxygen:

$$2\ N_2O_5(g) \longrightarrow 4\ NO_2(g) + O_2(g)$$

Colorless   Brown   Colorless

▲ Nitrogen dioxide (NO_2) is a brown gas that can be measured using absorbance spectrophotometry.

Changes in concentration as a function of time can be determined by measuring the increase in pressure as 2 gas molecules are converted to 5 gas molecules. Alternatively, concentration changes can be monitored by measuring the intensity of the brown color due to NO_2. Reactant and product concentrations as a function of time at 55 °C are listed in **TABLE 14.1**.

TABLE 14.1   Concentrations as a Function of Time at 55 °C for the Reaction $2 N_2O_5(g) \longrightarrow 4 NO_2(g) + O_2(g)$

| | Concentration (M) | | |
Time (s)	N_2O_5	NO_2	O_2
0	0.0200	0	0
100	0.0169	0.0063	0.0016
200	0.0142	0.0115	0.0029
300	0.0120	0.0160	0.0040
400	0.0101	0.0197	0.0049
500	0.0086	0.0229	0.0057
600	0.0072	0.0256	0.0064
700	0.0061	0.0278	0.0070

Note that the concentrations of NO_2 and O_2 increase as the concentration of N_2O_5 decreases.

The **reaction rate** is defined either as the *increase* in the concentration of a *product* per unit time or as the *decrease* in the concentration of a *reactant* per unit time. Let's look first at product formation. The rate of formation of O_2 is given by the equation

$$\text{Rate of formation of } O_2 = \frac{\Delta[O_2]}{\Delta t} = \frac{\text{Conc. of } O_2 \text{ at time } t_2 - \text{Conc. of } O_2 \text{ at time } t_1}{t_2 - t_1}$$

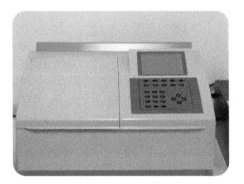

▲ The absorption of visible light, and thus the intensity of color, can be measured with a spectrophotometer.

where the square brackets surrounding O_2 denote its molar concentration, $\Delta[O_2]$ is the change in the molar concentration of O_2; Δt is the change in the time, and $\Delta[O_2]/\Delta t$ is the average rate of change in the molar concentration of O_2 during the interval from time t_1 to t_2. During the time period 300–400 s, for example, the average rate of formation of O_2 is 9.0×10^{-6} M/s:

$$\text{Rate of formation of } O_2 = \frac{\Delta[O_2]}{\Delta t} = \frac{0.0049 \text{ M} - 0.0040 \text{ M}}{400 \text{ s} - 300 \text{ s}}$$
$$= 9.0 \times 10^{-6} \text{ M/s}$$

The most common units of reaction rate are molar per second, M/s, or, equivalently, moles per liter second, mol/(L · s). We define reaction rate in terms of concentration (moles per liter) rather than amount (moles) because we want the rate to be independent of the scale of the reaction. When twice as much 0.0200 M N_2O_5 decomposes in a vessel of twice the volume, twice the number of moles of O_2 form per second, but the number of moles of O_2 *per liter* that form per second is unchanged.

Plotting the data of Table 14.1 to give the three curves in **FIGURE 14.1** affords additional insight into the concept of a reaction rate. Looking at the time period 300–400 s on the O_2 curve, $\Delta[O_2]$ and Δt are represented, respectively, by the vertical and horizontal sides of a right triangle. The slope of the third side, the hypotenuse of the triangle, is $\Delta[O_2]/\Delta t$, the average rate of O_2 formation during that time period.

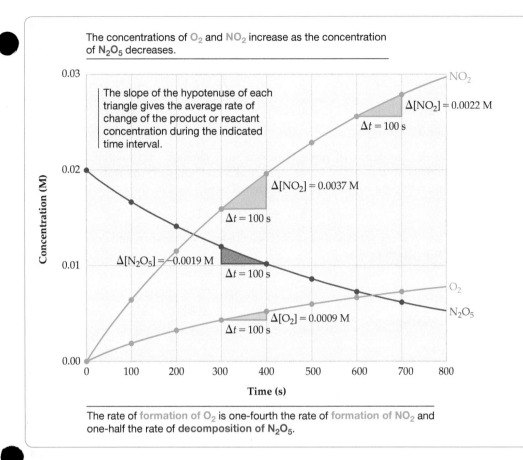

The concentrations of O_2 and NO_2 increase as the concentration of N_2O_5 decreases.

The slope of the hypotenuse of each triangle gives the average rate of change of the product or reactant concentration during the indicated time interval.

$\Delta[NO_2] = 0.0022$ M
$\Delta t = 100$ s

$\Delta[NO_2] = 0.0037$ M
$\Delta t = 100$ s

$\Delta[N_2O_5] = -0.0019$ M
$\Delta t = 100$ s

$\Delta[O_2] = 0.0009$ M
$\Delta t = 100$ s

The rate of formation of O_2 is one-fourth the rate of formation of NO_2 and one-half the rate of decomposition of N_2O_5.

▶ **FIGURE 14.1**

Concentrations as a function of time when gaseous N_2O_5 decomposes to gaseous NO_2 and O_2 at 55 °C.

◀ **Figure It out**

(a) Which product has the greatest rate of formation? (b) Describe how the average rate of formation of NO_2 changes over the course of reaction.

Answer: (a) Examine the triangles shown for a specific time period such as 300–400 s, NO_2 has the greatest rate of formation because it has the greatest change in concentration per unit of time. (b) The average rate of formation of NO_2 decreases as the reaction proceeds.

The steeper the slope of the hypotenuse, the faster the rate. Look, for example, at the triangle defined by $\Delta[NO_2]$ and Δt. The average rate of formation of NO_2 during the time period 300–400 s is 3.7×10^{-5} M/s, which is four times the rate of formation of O_2, in accord with the 4:1 ratio of the coefficients of NO_2 and O_2 in the chemical equation for the decomposition of N_2O_5.

$$\text{Rate of formation of } NO_2 = \frac{\Delta[NO_2]}{\Delta t} = \frac{0.0197 \text{ M} - 0.0160 \text{ M}}{400 \text{ s} - 300 \text{ s}}$$
$$= 3.7 \times 10^{-5} \text{ M/s}$$

As O_2 and NO_2 form, N_2O_5 disappears. Consequently, $\Delta[N_2O_5]/\Delta t$ is negative, in accord with the negative slope of the hypotenuse of the triangle defined by $\Delta[N_2O_5]$ and Δt in Figure 14.1. Because *reaction rate is defined as a positive quantity,* we must always introduce a minus sign when calculating the rate of disappearance of a reactant. During the time period 300–400 s, for example, the average rate of decomposition of N_2O_5 is 1.9×10^{-5} M/s:

$$\text{Rate of decomposition of } N_2O_5 = \frac{-\Delta[N_2O_5]}{\Delta t} = \frac{-(0.0101 \text{ M} - 0.0120 \text{ M})}{400 \text{ s} - 300 \text{ s}}$$
$$= 1.9 \times 10^{-5} \text{ M/s}$$

When quoting a reaction rate, it's important to specify the reactant or product on which the rate is based because the rates of product formation and reactant consumption may differ, depending on the coefficients in the balanced chemical equation. For the decomposition of N_2O_5, 4 mol of NO_2 form and 2 mol of N_2O_5 disappear for each

Go to
eText

BIG IDEA Question 1

Which statement about the change in concentration of NO for the following reaction is *true*?

$$2 \text{ NO } (g) + O_2 (g) \longrightarrow 2 \text{ NO}_2 (g)$$

(a) The concentration of NO increases with time.
(b) The rate of change of the NO concentration increases with time.
(c) The rate of loss of NO is twice the rate of the loss of O_2.
(d) The general reaction rate is the same as the rate of loss of NO.

mole of O_2 that forms. Therefore, the rate of formation of O_2 is one-fourth the rate of formation of NO_2 and one-half the rate of decomposition of N_2O_5:

$$\left(\begin{array}{c}\text{Rate of formation}\\\text{of } O_2\end{array}\right) = \frac{1}{4}\left(\begin{array}{c}\text{Rate of formation}\\\text{of } NO_2\end{array}\right) = \frac{1}{2}\left(\begin{array}{c}\text{Rate of decomposition}\\\text{of } N_2O_5\end{array}\right)$$

$$\text{or} \quad \frac{\Delta[O_2]}{\Delta t} = \frac{1}{4}\left(\frac{\Delta[NO_2]}{\Delta t}\right) = -\frac{1}{2}\left(\frac{\Delta[N_2O_5]}{\Delta t}\right)$$

To avoid the ambiguity of more than one rate, chemists have defined a *general* reaction rate equal to the rate of consumption of a reactant or formation of a product divided by its coefficient in the balanced chemical equation. For the general reaction,

$$a\,A + b\,B \longrightarrow c\,C + d\,D$$

where *a, b, c,* and *d* are stoichiometric coefficients for the reactants and products A, B, C, and D, the reaction rate is:

General reaction rate $\text{Rate} = -\dfrac{1}{a}\dfrac{\Delta[A]}{\Delta t} = -\dfrac{1}{b}\dfrac{\Delta[B]}{\Delta t} = \dfrac{1}{c}\dfrac{\Delta[C]}{\Delta t} = \dfrac{1}{d}\dfrac{\Delta[D]}{\Delta t}$

Thus, for the reaction

$$2\,N_2O_5(g) \longrightarrow 4\,NO_2(g) + O_2(g)$$

the reaction rate has a single value given by the equation

$$\text{Rate} = -\frac{1}{2}\left(\frac{\Delta[N_2O_5]}{\Delta t}\right) = \frac{1}{4}\left(\frac{\Delta[NO_2]}{\Delta t}\right) = \frac{\Delta[O_2]}{\Delta t}$$

In this chapter, when we describe the "rate" of a reaction, we are referring to the reaction rate as defined above.

We sometimes need to define a rate in terms of the *rate of formation* or the *rate of consumption* of a specific substance because the concentration of only one reactant or product is measured experimentally. For example, in the decomposition of N_2O_5 reaction, the rate of formation of NO_2 is likely to be measured because it is colored and absorbs visible light. The rate for a specific substance will be referred to as the consumption (or decomposition) of a reactant or formation of a product. For example,

$$\text{Rate of formation of } NO_2 = \frac{\Delta[NO_2]}{\Delta t} \text{ or Rate of decomposition of } N_2O_5 = -\left(\frac{\Delta[N_2O_5]}{\Delta t}\right)$$

It's important to specify the timeframe when quoting a rate because the rate changes as the reaction proceeds. For example, the average rate of formation of NO_2 is 3.7×10^{-5} M/s during the time period 300–400 s, but it is only 2.2×10^{-5} M/s during the period 600–700 s (Figure 14.1). Ordinarily, reaction rates depend on the concentrations of at least some of the reactants and therefore decrease as the reaction mixture runs out of reactants, as indicated by the decreasing slopes of the curves in Figure 14.1 as time passes.

Often, chemists want to know the rate of a reaction at a specific time *t* rather than the rate averaged over a time interval Δt. For example, what is the rate of formation of NO_2 at time $t = 300$ s?

The slope of the line tangent to a concentration-versus-time curve at time *t* is called the **instantaneous rate** at that particular time. **FIGURE 14.2** shows a line tangent to the curve at $t = 300$ s. The slope of line $= (0.017 \text{ M})/(400 \text{ s}) = 4.25 \times 10^{-5}$ M/s is the instantaneous rate of formation of NO_2 at 300 s. The slope of the line tangent to the curve at the beginning of the reaction ($t = 0$) is called the **initial rate.**

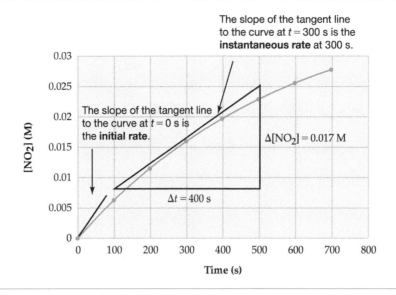

▶ **FIGURE 14.2**

Concentration of NO_2 versus time when N_2O_5 decomposes at 55 °C. The slope of the line tangent to the curve at $t = 300$ s is the instantaneous rate of formation of NO_2.

The slope of the tangent line to the curve at $t = 300$ s is the **instantaneous rate** at 300 s.

The slope of the tangent line to the curve at $t = 0$ s is the **initial rate**.

$\Delta[NO_2] = 0.017$ M

$\Delta t = 400$ s

WORKED EXAMPLE 14.1

Finding Relative Rates of Product Formation and Reactant Consumption

Ethanol (C_2H_5OH), the active ingredient in alcoholic beverages and an octane booster in gasoline, is produced by the fermentation of glucose. The balanced equation is

$$C_6H_{12}O_6(aq) \longrightarrow 2\ C_2H_5OH(aq) + 2\ CO_2(g)$$

(a) How is the rate of formation of ethanol related to the rate of consumption of glucose? Write this relationship in terms of $\Delta[C_2H_5OH]/\Delta t$ and $\Delta[C_6H_{12}O_6]/\Delta t$.

(b) If $-\Delta[C_6H_{12}O_6]/\Delta t = 1.7 \times 10^{-3}$ M/s, what is the value of $\Delta[C_2H_5OH]/\Delta t$ during the same time interval?

(c) If $\Delta[C_2H_5OH]/\Delta t = 3.6 \times 10^{-2}$ M/s, what is the reaction rate during the same time interval?

STRATEGY

To find the relative rates, look at the coefficients in the balanced chemical equation.

SOLUTION

(a) The expression for the reaction rate allows us to relate the rate of change in concentration of one reactant or product to another.

$$\text{Rate} = -\frac{1}{1}\frac{\Delta[C_6H_{12}O_6]}{\Delta t} = \frac{1}{2}\frac{\Delta[C_2H_5OH]}{\Delta t}$$

Solving for the rate of formation of C_2H_5OH yields:

$$\frac{\Delta[C_2H_5OH]}{\Delta t} = -2\frac{\Delta[C_6H_{12}O_6]}{\Delta t}$$

(b) $\dfrac{\Delta[C_2H_5OH]}{\Delta t} = 2\left(-\dfrac{\Delta[C_6H_{12}O_6]}{\Delta t}\right) = 2(1.7 \times 10^{-3}\ \text{M/s})$

$$= 3.4 \times 10^{-3}\text{M/s}$$

(c) $\text{Rate} = \dfrac{1}{2}\dfrac{\Delta[C_2H_5OH]}{\Delta t} = \dfrac{1}{2}(3.6 \times 10^{-2}\ \text{M/s})$

$$= 1.8 \times 10^{-2}\ \text{M/s}$$

CHECK

According to the balanced equation, 2 mol of ethanol are produced for each mole of glucose that reacts. Therefore, the rate of formation of ethanol is twice the rate of consumption of glucose.

▶ **PRACTICE 14.1**   The oxidation of iodide ion by arsenic acid, H_3AsO_4, is described by the balanced equation

$$3\ I^-(aq) + H_3AsO_4(aq) + 2\ H^+(aq)$$
$$\longrightarrow I_3^-(aq) + H_3AsO_3(aq) + H_2O(l)$$

(a) If $-\Delta[I^-]/\Delta t = 4.8 \times 10^{-4}$ M/s, what is the value of $\Delta[I_3^-]/\Delta t$ during the same time interval?

(b) What is the average rate of consumption of H^+ during that time interval?

▶ **CONCEPTUAL APPLY 14.2**

(a) Use the following graph of concentration as a function of time to calculate the average reaction rate for each reactant and product (A, B, and C) over the time period 0–500 s.

(b) Write the balanced reaction that corresponds to the data in graph.

continued on next page

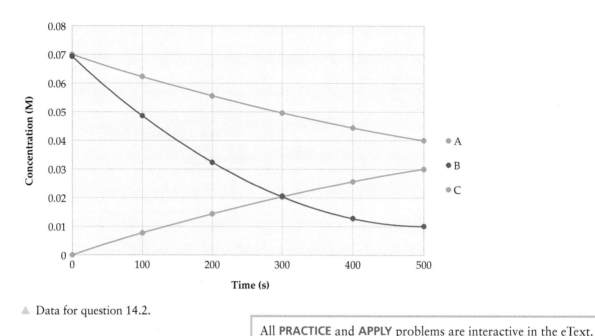

▲ Data for question 14.2.

All **PRACTICE** and **APPLY** problems are interactive in the eText.

14.2 RATE LAWS AND REACTION ORDER

We noted in Section 14.1 that the rate of decomposition of N_2O_5 depends on its concentration, slowing down as the N_2O_5 concentration decreases. To explore further how reaction rates depend on concentrations, let's consider the general reaction

$$a\,A + b\,B \longrightarrow \text{Products}$$

where A and B are the reactants and a and b are stoichiometric coefficients in the balanced chemical equation. The dependence of the reaction rate on the concentration of each reactant is given by an equation called the **rate law.**

> **Rate law** $\text{Rate} = k[\text{A}]^m[\text{B}]^n$

In this equation, k is a proportionality constant called the **rate constant.** The exponents m and n in the rate law indicate how sensitive the rate is to changes in [A] and [B], and they are *unrelated* to the coefficients a and b in the balanced equation. For the simple reactions discussed in this book, the exponents are usually small positive integers. For more complex reactions, however, the exponents can be negative, zero, or even fractions.

Let's consider reactant A to describe how exponents in the rate law affect reaction rate. We could also examine reactant B or both reactants, but for simplicity let's begin with only reactant A. **FIGURE 14.3** shows how the rate changes when the concentration of A is doubled for various values of the exponent m. In general, the rate is proportional to $[\text{A}]^m$.

- If $m = -1$, the rate is proportional to $[\text{A}]^{-1}$ or 1/[A]. If [A] is doubled, the rate decreases by a factor of 2.
- If $m = 0$, the rate is independent of [A] because any number raised to the zeroth power equals one ($[\text{A}]^0 = 1$).
- If $m = 1$, the rate is proportional to $[\text{A}]^1$ or [A]. If [A] is doubled, the rate doubles.
- If $m = 2$, the rate is proportional to $[\text{A}]^2$. If [A] is doubled, the rate increases by a factor of 4.

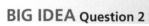

BIG IDEA Question 2

Go to eText

In the reaction $A + B \rightarrow$ Products, the rate law is rate $= k[\text{A}]^2$. By what factor does the rate change if the concentration of A is tripled?

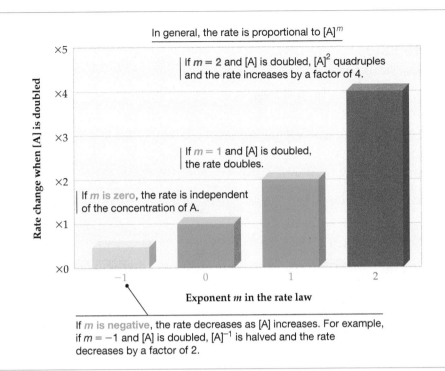

In general, the rate is proportional to $[A]^m$

If $m = 2$ and $[A]$ is doubled, $[A]^2$ quadruples and the rate increases by a factor of 4.

If $m = 1$ and $[A]$ is doubled, the rate doubles.

If m is zero, the rate is independent of the concentration of A.

Exponent m in the rate law

If m is negative, the rate decreases as $[A]$ increases. For example, if $m = -1$ and $[A]$ is doubled, $[A]^{-1}$ is halved and the rate decreases by a factor of 2.

◀ **FIGURE 14.3**

Change in reaction rate when the concentration of reactant A is doubled for different values of the exponent m in the rate law, rate $= k[A]^m[B]^n$.

The values of the exponents m and n determine the **reaction order** with respect to A and B, respectively. A reaction is *zeroth order* with respect to a reactant when its exponent in the rate law is zero. Similarly, a reaction is *first order* in a reactant when its exponent is one and *second order* in a reactant when its exponent is two. The sum of the exponents ($m + n$) defines the *overall* reaction order. Thus, if the rate law is

$$\text{Rate} = k[A]^2[B] \quad m = 2; n = 1; \text{ and } m + n = 3$$

we say that the reaction is *second order* in A, *first order* in B, and *third order* overall.

The values of the exponents in a rate law must be determined by experiment; they cannot be deduced from the stoichiometry of the reaction. As **TABLE 14.2** shows, there is no general relationship between the stoichiometric coefficients in the balanced chemical equation and the exponents in the rate law. In the first reaction in Table 14.2, for example, the coefficients of $(CH_3)_3CBr$ and H_2O in the balanced equation are both 1, but the exponents in the rate law are 1 for $(CH_3)_3CBr$, and 0 for H_2O:

$$\text{Rate} = k[(CH_3)_3CBr]^1[H_2O]^0 = k[(CH_3)_3CBr]$$

In Section 14.10, we'll see that the exponents in a rate law depend on the reaction mechanism.

TABLE 14.2 **Balanced Chemical Equations and Experimentally Determined Rate Laws for Some Reactions**

Reaction*	Rate Law
$(CH_3)_3CBr(soln) + H_2O(soln) \longrightarrow (CH_3)_3COH(soln) + H^+(soln) + Br^-(soln)$	$\text{Rate} = k[(CH_3)_3CBr]$
$HCO_2H(aq) + Br_2(aq) \longrightarrow 2\,H^+(aq) + 2\,Br^-(aq) + CO_2(g)$	$\text{Rate} = k[Br_2]$
$BrO_3^-(aq) + 5\,Br^-(aq) + 6\,H^+(aq) \longrightarrow 3\,Br_2(aq) + 3\,H_2O(l)$	$\text{Rate} = k[BrO_3^-][Br^-][H^+]^2$
$H_2(g) + I_2(g) \longrightarrow 2\,HI(g)$	$\text{Rate} = k[H_2][I_2]$

In general, the exponents in the rate law are not the same as the stoichiometric coefficients in the balanced chemical equation for the reaction.

*In the first reaction, "(*soln*)" denotes a nonaqueous solution.

WORKED EXAMPLE 14.2

Finding Reaction Order from a Rate Law

The second reaction in Table 14.2, shown in progress in **FIGURE 14.4**, is

$$HCO_2H(aq) + Br_2(aq) \longrightarrow 2\,H^+(aq) + 2\,Br^-(aq) + CO_2(g) \qquad \text{Rate} = k[Br]$$

Colorless   Red   Colorless

What is the order of the reaction with respect to each of the reactants? What is the overall reaction order?

▲ **FIGURE 14.4**

The reaction of formic acid (HCO_2H) and bromine (Br_2). As time passes (left to right), the red color of bromine disappears because Br_2 is reduced to the colorless Br^- ion. The concentration of Br_2 as a function of time, and thus the reaction rate, can be determined by measuring the intensity of the color.

STRATEGY

To find the reaction order with respect to each reactant, look at the exponents in the rate law, not the coefficients in the balanced chemical equation. Then sum the exponents to obtain the overall reaction order.

SOLUTION

The experimentally determined rate law for the reaction of formic acid with bromine is:

$$\text{Rate} = k[Br_2]$$

Because HCO_2H (formic acid) does not appear in the rate law, the rate is independent of the HCO_2H concentration, and so the reaction is zeroth order in HCO_2H. Because the exponent on $[Br_2]$ is 1 (it is understood to be 1 when no exponent is given), the reaction is first order in Br_2. The reaction is first order overall because the sum of the exponents is 1.

▶ **PRACTICE 14.3** The rate law for the reaction

$$BrO_3^-(aq) + 5\,Br^-(aq) + 6\,H^+(aq) \longrightarrow 3\,Br_2(aq) + 3\,H_2O(l)$$

is rate = $k[BrO_3^-][Br^-][H^+]^2$. What is the order of the reaction with respect to each reactant, and what is the overall order of the reaction?

▶ **APPLY 14.4** The order of each reactant in the following reaction was determined by varying concentration and measuring the change in rate.

$$2\,NO(g) + Cl_2(g) \longrightarrow 2\,NOCl(g)$$

When the concentration of NO was doubled while the concentration of Cl_2 was held constant, the rate increased by a factor of 2. In a separate experiment, when the concentration of Cl_2 was halved while the concentration of NO was held constant, the rate decreased by a factor of 4.

What is the order of the reaction with respect to each reactant? What is the overall reaction order?

14.3 METHOD OF INITIAL RATES: EXPERIMENTAL DETERMINATION OF A RATE LAW

One method of determining the values of the exponents in a rate law—the *method of initial rates*—is to carry out a series of experiments in which the initial rate of a reaction is measured as a function of different sets of initial concentrations. Take, for example, the oxidation of nitric oxide in air, one of the reactions that contributes to the formation of acid rain:

$$2\,NO(g) + O_2(g) \longrightarrow 2\,NO_2(g)$$

Some initial rate data are collected in **TABLE 14.3**.

TABLE 14.3 Initial Concentration and Rate Data for the Reaction
$2\,NO(g) + O_2(g) \longrightarrow 2\,NO_2(g)$

Experiment	Initial [NO]	Initial [O_2]	Initial Reaction Rate (M/s)
1	0.015	0.015	0.024
2	0.030	0.015	0.096
3	0.015	0.030	0.048
4	0.030	0.030	0.192

Note that pairs of experiments are designed to investigate the change in initial rate that occurs when the initial concentration of a single reactant is changed. In the first two experiments, for example, the concentration of NO is doubled from 0.015 M to 0.030 M while the concentration of O_2 is held constant. The initial rate increases by a factor of 4, from 0.024 M/s to 0.096 M/s, indicating that the rate depends on the concentration of NO squared, $[NO]^2$. When [NO] is held constant and $[O_2]$ is doubled (experiments 1 and 3), the initial rate doubles from 0.024 M/s to 0.048 M/s, indicating that the rate depends on the concentration of O_2 to the first power, $[O_2]^1$. Therefore, the rate law for the reaction is

$$Rate = k[NO]^2[O_2]$$

In accord with this rate law, which is second order in NO, first order in O_2, and third order overall, the initial rate increases by a factor of 8 when the concentrations of both NO and O_2 are doubled (experiments 1 and 4).

The preceding method uses initial rates rather than rates at a later stage of the reaction because chemical reactions are reversible and we want to avoid complications from the reverse reaction: reactants $\leftarrow$ products. As the product concentrations build up, the rate of the reverse reaction increases and the measured rate is affected by the concentrations of both reactants and products. At the beginning of the reaction, however, the product concentrations are zero, and therefore the products can't affect the measured rate. When we measure an initial rate, we are measuring the rate of only the forward reaction, so only reactants (and catalysts; see Section 14.12) can appear in the rate law.

One step in determining a rate law, as we've just seen, is to establish the reaction order. Another is to evaluate the numerical value of the rate constant k. Each reaction has its own characteristic value of the rate constant, which depends on temperature but not on concentrations. To evaluate k for the reaction

$$2\,NO(g) + O_2(g) \longrightarrow 2\,NO_2(g)$$

we can use the data from any one of the experiments in Table 14.3. Solving the rate law for k and substituting the initial rate and concentrations from the first experiment, we obtain

$$k = \frac{Rate}{[NO]^2[O_2]} = \frac{0.024\ M/s}{(0.015\ M)^2(0.015\ M)} = 7.1 \times 10^3/(M^2 \cdot s)$$

Try repeating the calculation for experiments 2–4, and show that you get the same value of k within the range of experimental error. Note that the units of k in this example are $1/(M^2 \cdot s)$ read as "one over molar squared second." The units of k depend on the overall order of the reaction and can be found by canceling units in the equation for k. For example, in the general second-order reaction, rate $= k[A][B]$, we can determine the units for k as follows:

$$k = \frac{Rate}{[A][B]} = \frac{M/s}{(M)(M)} = 1/(M \cdot s)\ \text{or}\ M^{-1}\,s^{-1}$$

Units for some common cases are:

Rate Law	Overall Reaction Order	Units for k
Rate $= k$	Zeroth order	M/s or M s^{-1}
Rate $= k[A]$	First order	1/s or s^{-1}
Rate $= k[A][B]$	Second order	1/(M·s) or M^{-1} s^{-1}
Rate $= k[A][B]^2$	Third order	1/(M^2·s) or M^{-2} s^{-1}

Be careful not to confuse the rate of a reaction and its rate constant. The *rate* depends on concentrations, whereas the rate constant does not—it has one value at a given temperature. The rate is usually expressed in units of M/s, whereas the units of the rate constant depend on the overall reaction order.

Worked Example 14.3 gives another instance of how a rate law can be determined from initial rates.

WORKED EXAMPLE 14.3

Go to eText

Determining a Rate Law from Initial Rates

Initial rate data are listed in the table for the reaction

$$2 \, NO(g) + 2 \, H_2(g) \longrightarrow N_2(g) + 2 \, H_2O(g)$$

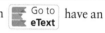
All **WORKED EXAMPLES** with this icon Go to eText have an interactive video in the eText.

Initial rate data at 950 °C are as follows:

Experiment	Initial [NO]	Initial [H$_2$]	Initial Reaction Rate (M/s)
1	0.200	0.122	0.0313
2	0.400	0.122	0.1250
3	0.200	0.244	0.0625

(a) What is the rate law?
(b) What is the value of the rate constant and its units?
(c) What is the initial rate of formation of N$_2$ when the initial concentrations are [NO] = 0.300 M and [H$_2$] = 0.150 M?

STRATEGY

(a) The rate law for the reaction can be written as

$$Rate = k[NO]^m[H_2]^n$$

where m is the order of the reaction in NO and n is the order of the reaction in H$_2$. To find the value of the exponent m, compare the change in initial rate with the change in the concentration of NO (when the concentration of H$_2$ is held constant). To find the value of the exponent n, compare the change in initial rate with the change in the concentration of H$_2$ (when the concentration of NO is held constant).
(b) To find the value of the rate constant k, solve the rate law for k and then substitute in the data any one of the three experiments.
(c) To calculate the initial rate of formation of N$_2$, substitute the rate constant found in part b and the given initial concentrations of NO (0.300 M) and H$_2$ (0.150 M) into the rate law.

SOLUTION

(a) **Determination of** m: Comparing experiments 1 and 2 shows that doubling the concentration of NO increases the initial rate of reaction by a factor of 4:

$$\frac{[NO]_2}{[NO]_1} = \frac{0.400 \text{ M}}{0.200 \text{ M}} = 2.00 \qquad \frac{(Rate)_2}{(Rate)_1} = \frac{0.1250 \text{ M/s}}{0.0313 \text{ M/s}} = 4.0$$

The initial rate is proportional to [NO]^2 and $m = 2$.
A more formal way to approach this problem is to write the rate law for each experiment:

$$(Rate)_1 = k[NO]^m[H_2]^n = k(0.200 \text{ M})^m(0.122 \text{ M})^n$$
$$(Rate)_2 = k[NO]^m[H_2]^n = k(0.400 \text{ M})^m(0.122 \text{ M})^n$$

If we then divide the second equation by the first, we obtain

$$\frac{(Rate)_2}{(Rate)_1} = \frac{k(0.400 \text{ M})^m(0.122 \text{ M})^n}{k(0.200 \text{ M})^m(0.122 \text{ M})^n} = \left(\frac{0.400 \text{ M}}{0.200 \text{ M}}\right)^m = (2.0)^m$$

The right-hand side of the equation equals the ratio of the experimental rates:

$$(2.0)^m = \frac{(0.1250 \text{ M/s})}{(0.0313 \text{ M/s})} = 4.0$$

It is relatively straightforward to solve the equation $(2.0)^m = 4.0$. Since $(2.0)^2 = 4.0$, $m = 2$. If the value of the exponent is not easily found, we can solve for m if we take the logarithm of both sides and apply the relationship $\log x^m = m \log x$ (see Appendix A.2 for a review of logarithms).

$$\log (2.0)^m = \log 4.0$$
$$m \log 2.0 = \log 4.0$$
$$m = \frac{\log 4.0}{\log 2.0} = 2$$

Since $m = 2$, the reaction is second order in NO.

Determination of n: Comparing experiments 1 and 3 shows that doubling the concentration of H_2 increases the initial rate by a factor of 2:

$$\frac{[H_2]_3}{[H_2]_1} = \frac{0.244 \text{ M}}{0.122 \text{ M}} = 2.00 \qquad \frac{(\text{Rate})_3}{(\text{Rate})_1} = \frac{0.0625 \text{ M/s}}{0.0313 \text{ M/s}} = 2.00$$

The initial rate is proportional to $[H_2]$ and $n = 1$. The formal approach can also be applied to experiments 1 and 3, yielding a value of $n = 1$. Therefore, the rate law for the formation of N_2 is

$$\text{Rate} = k[NO]^2[H_2]$$

The rate law is second order in NO, first order in H_2, and third order overall.

(b) Solving the rate law for k and substituting in the data from the first experiment gives

$$k = \frac{\text{Rate}}{[NO]^2[H_2]} = \frac{0.0313 \text{ M/s}}{(0.200 \text{ M})^2(0.122 \text{ M})} = 6.41/(M^2 \cdot s)$$

(c) Substituting the initial concentrations of NO (0.300 M) and H_2 (0.150 M) and the rate constant from part **(b)** $6.41/(M^2 \cdot s)$ into the rate law gives

$$\text{Rate} = \frac{\Delta[N_2]}{\Delta t} = k[NO]^2[H_2] = \left(\frac{6.41}{M^2 \cdot s}\right)(0.300 \text{ M})^2(0.150 \text{ M})$$

$$= 0.0865 \text{ M/s}$$

CHECK

(a) It's a good idea to check the units of the rate constant. The units of k, $1/(M^2 \cdot s)$, are the expected units for a third-order reaction.

(b) The initial concentration values are between the values used in experiment 1 and 2; therefore, the rate should be between the rates for experiments 1 and 2.

▶ **PRACTICE 14.5** The initial rates listed in the following table were measured for the reaction between nitrogen dioxide and carbon monoxide that occurs in engine exhaust.

$$NO_2(g) + CO(g) \longrightarrow NO(g) + CO_2(g)$$

Experiment	Initial $[NO_2]$	Initial $[CO]$	Initial Reaction Rate (M/s)
1	0.100	0.100	5.00×10^{-3}
2	0.150	0.100	1.13×10^{-2}
3	0.200	0.200	2.00×10^{-2}

(a) What is the rate law?
(b) What is the value of the rate constant and its units?

▶ **APPLY 14.6** The initial rates listed in the following table were measured in methanol solution for the reaction

$$C_2H_4Br_2 + 3\,I^- \longrightarrow C_2H_4 + 2\,Br^- + I_3^-$$

Experiment	Initial $[C_2H_4Br_2]$	Initial $[I^-]$	Initial Reaction Rate (M/s)
1	0.127	0.102	6.45×10^{-5}
2	0.343	0.102	174×10^{-4}
3	0.203	0.125	126×10^{-4}

(a) What is the rate law?
(b) What is the value of the rate constant and its units?
(c) What is the initial rate of formation of I_3^- when the concentrations of both reactants are 0.150 M?
(d) What is the initial rate of loss of I^- when the concentrations of both reactants are 0.150 M?

— **CONCEPTUAL WORKED EXAMPLE 14.4**

Using the Method of Initial Rates

The relative rates of the reaction $A + 2\,B \longrightarrow$ products in vessels (1)–(4) are $1:2:2:4$. Red spheres represent the initial number of A molecules, and blue spheres represent B molecules.

(1)   **(2)**   **(3)**   **(4)**

 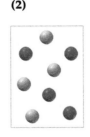  

(a) What is the order of the reaction in A and B? What is the overall reaction order?
(b) Write the rate law.

STRATEGY

(a) To find the reaction order, apply the method of initial rates. Count the number of A and B molecules in vessels (1)–(4) and compare the relative rates with the relative number of molecules of each type. Assume that all four

vessels have the same volume, so the concentrations are proportional to the number of molecules.

(b) The rate law can be written as rate $= k[A]^m[B]^n$, where the exponents m and n are the orders of the reaction in A and B, respectively.

continued on next page

SOLUTION

(a) Compare pairs of vessels in which the concentration of one reactant varies while the concentration of the other reactant remains constant. The concentration of A molecules in vessel (2) is twice that in vessel (1) while the concentration of B remains constant. Because the reaction rate in vessel (2) is twice that in vessel (1), the rate is proportional to [A], and therefore the reaction is first order in A. When the concentration of B is doubled while the concentration of A remains constant [compare vessels (1) and (3)], the rate doubles, so the reaction is first order in B. When the concentrations of both A and B are doubled, the rate increases by a factor of 4 [compare vessels (1) and (4)], in accord with a reaction that is first order in A and first order in B. The overall reaction order is the sum of the orders in A and B, or $1 + 1 = 2$.

(b) Since the reaction is first order in A and first order in B, the rate law is rate $= k[A][B]$. Note that the exponents in the rate law differ from the coefficients in the balanced chemical equation, $A + 2B \rightarrow$ products.

▶ **CONCEPTUAL PRACTICE 14.7** The relative rates of the reaction $A + B \rightarrow$ products in vessels (1)–(4) are $1:1:4:4$. Red spheres represent A molecules, and blue spheres represent B molecules.

(1)   **(2)**   **(3)**   **(4)**

 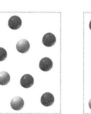  

(a) What is the order of the reaction in A and B? What is the overall reaction order?

(b) What is the rate law?

▶ **CONCEPTUAL APPLY 14.8** The rate law for the reaction $A + B \rightarrow C$ is Rate $= k[A]^2[B]$. The initial rate of the reaction in the vessel 1 is 0.01 M/s. Red spheres represent A molecules, and blue spheres represent B molecules. What is the initial rate of reaction in vessels (2) and (3)?

(1)   **(2)**   **(3)**

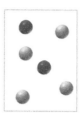  

14.4 INTEGRATED RATE LAW: ZEROTH-ORDER REACTIONS

Go to eText

BIG IDEA Question 3

How does the rate of a zeroth-order reaction change over time?

Thus far we've focused on the rate law, an equation that tells how a reaction rate depends on reactant concentrations. We're also interested in how reactant and product concentrations vary with time. For example, it's important to know the rate at which a drug reacts in the body, but we also want to know how much will remain after given amount of time (such as 12 hours) and how long it will take for the concentration to change by a given amount (say, 25%). Rate laws for zeroth-, first-, and second-order reactions can be integrated using calculus to convert the rate law to another form, called the **integrated rate law**. Integrated rate laws are *concentration–time equations* that let us calculate the concentration of reactant that remains at any time *t*. They can also be used to calculate the time required for the initial concentration of reactant to drop to a certain value or to a certain fraction of its initial concentration.

A **zeroth-order reaction** is one whose rate is independent of reactant concentration. A general reaction, $A \rightarrow$ Products, that is zeroth order in A will have the rate law:

$$\text{Rate of consumption of A} = -\frac{\Delta[A]}{\Delta t} = k[A]^0 = k(1) = k$$

Using calculus to integrate the rate law gives:

> **Integrated rate law for a zeroth-order reaction** $[A]_t = -kt + [A]_0$

In this equation, $[A]_0$ is the concentration of A at some initial time, arbitrarily considered to be $t = 0$, and $[A]_t$ is the concentration of A at any time t thereafter. Although you must be able to use the integrated rate law, you don't need to be able to derive it. Students who know some calculus, however, may be interested in the derivation shown in Appendix A.5.

The integrated rate law is an equation of the form $y = mx + b$, so a graph of [A] versus time is a straight line with a slope $= -k$ (**FIGURE 14.5**). Note that both the rate constant k and the rate of a zeroth-order reaction have a constant value equal to the negative slope of the [A] versus time plot.

Zeroth-order reactions are relatively uncommon, but they can occur under special circumstances. Take, for example, the decomposition of gaseous ammonia on a hot platinum surface:

$$NH_3(g) \xrightarrow[\text{Pt catalyst}]{1130\ K} 1/2\ N_2(g) + 3/2\ H_2(g)$$

The platinum surface is completely covered by a layer of NH_3 molecules (**FIGURE 14.6**), but the number of NH_3 molecules that can fit on the surface is limited by size constraints and is very small compared with the total number of NH_3 molecules. Most of the NH_3 is in the gas phase above the surface. Because only the NH_3 molecules on the surface can react, the reaction rate is constant and independent of the total concentration of NH_3:

$$Rate = k[NH_3]^0 = k$$

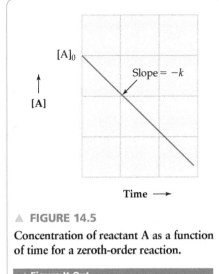

▲ FIGURE 14.5

Concentration of reactant A as a function of time for a zeroth-order reaction.

▲ Figure It Out

How is the rate constant (*k*) determined in a zeroth-order reaction?

Answer: In a plot of [A] versus time, the slope is equal to −*k*.

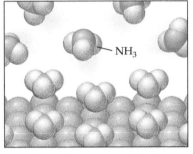

Most of the NH_3 molecules are in the gas phase above the surface and are unable to react.

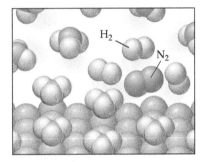

As NH_3 molecules on the surface decompose, they are replaced by molecules from the gas phase, so the number of NH_3 molecules on the surface remains constant.

H$_2$

N$_2$

NH$_3$

Because only the NH_3 molecules on the surface react under these conditions, the reaction rate is independent of the total concentration of NH_3.

◄ FIGURE 14.6

Decomposition of ammonia on a hot platinum surface. Only the NH_3 molecules attached to the surface can react.

WORKED EXAMPLE 14.5

Using the Integrated Rate Law for a Zeroth-Order Reaction

The rate constant for the zeroth-order decomposition of NH_3 on a platinum surface at 856 °C is 1.50×10^{-6} M/s. If the initial concentration of NH_3 is 1.50×10^{-2} M, calculate the concentration of NH_3 after 30 minutes.

IDENTIFY

Known	Unknown
$k = 1.50 \times 10^{-6}$ M/s, $[A]_0 = 1.50 \times 10^{-2}$ M, $t = 30.0$ min.	Final concentration $[A]_t$

STRATEGY

Convert units of time in minutes to seconds to match with units in the rate constant. Substitute known values into the integrated rate law for zeroth-order reactions and solve for $[A]_t$.

continued on next page

SOLUTION

$$[A]_t = -kt + [A]_0$$

$$[A]_t = -(1.50 \times 10^{-6} \text{ M/s})\left(30.0 \text{ min.} \times \frac{60 \text{ s}}{1 \text{ min.}}\right)$$

$$+ 1.50 \times 10^{-2} \text{ M} = 1.23 \times 10^{-2} \text{ M}$$

▶ **PRACTICE 14.9** The rate constant for the zeroth-order decomposition of NH_3 on a platinum surface at 856 °C is 1.50×10^{-6} M/s. How much time is required for the concentration of NH_3 to drop from 5.00×10^{-3} M to 1.00×10^{-3} M?

▶ **APPLY 14.10** Concentration-versus-time data were collected for the decomposition of ethanol (C_2H_5OH) on an alumina (Al_2O_3) surface at 600 K.

$$C_2H_5OH(g) \longrightarrow C_2H_4(g) + H_2O(g)$$

Time (s)	[C$_2$H$_5$OH]
0	5.0×10^{-2}
100	4.6×10^{-2}
200	4.2×10^{-2}
300	3.8×10^{-2}
400	3.4×10^{-2}
500	3.0×10^{-2}

Prepare a graph of $[C_2H_5OH]$ versus time.
(a) What is the order of the reaction? Explain.
(b) What is the rate constant for the decomposition of ethanol (include units)?
(c) What is the concentration of ethanol after 15.0 min?

14.5 INTEGRATED RATE LAW: FIRST-ORDER REACTIONS

A **first-order reaction** is one whose rate depends on the concentration of a single reactant raised to the first power. For the general reaction, A $\longrightarrow$ Products, the rate law is

$$\text{Rate of consumption of A} = -\frac{\Delta[A]}{\Delta t} = k[A]$$

Unlike in a zeroth-order reaction, in which the rate does not change when the quantity of [A] changes, in a first-order reaction, the rate changes and declines as [A] decreases. Using calculus to integrate the first-order rate law gives:

BIG IDEA Question 4

Go to eText

How does the rate of a first-order reaction change over time?

> **Integrated rate law for a first-order reaction** $\ln \dfrac{[A]_t}{[A]_0} = -kt.$

In this equation, ln denotes the natural logarithm; $[A]_0$ is the concentration of A at some initial time, arbitrarily considered to be $t = 0$; and $[A]_t$ is the concentration of A at any time t thereafter. (See Appendix A.2 for a review of logarithms.) The ratio $[A]_t/[A]_0$ is the fraction of A that remains at time t. Since $\ln([A]_t/[A]_0) = \ln[A]_t - \ln[A]_0$, we can rewrite the integrated rate law as

$$\ln[A]_t = -kt + \ln[A]_0$$

This equation is of the form $y = mx + b$, the equation for a straight line, so $\ln[A]_t$ is a linear function of time:

$$\underset{y}{\ln[A]_t} = \underset{m\ x}{(-k)t} + \underset{b}{\ln[A]_0}$$

A graph of $\ln[A]$ versus time is therefore a straight line having a slope $m = -k$ and an intercept $b = \ln[A]_0$ (**FIGURE 14.7b**). The value of the rate constant is simply equal to the negative slope of the straight line:

$$k = -(\text{Slope})$$

This graphical method of determining a rate constant, illustrated in Worked Example 14.7, is an alternative to the method of initial rates used in Worked Example 14.3. A plot of $\ln[A]$ versus time, however, will give a straight line only if the reaction is first order in A. Indeed, a good way of testing whether a reaction is first order is to examine the appearance of such a plot.

(a) Reactant concentration versus time

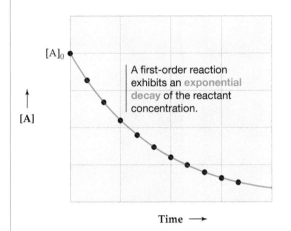

[A]$_0$

A first-order reaction exhibits an **exponential decay** of the reactant concentration.

[A]

Time →

(b) Natural logarithm of reactant concentration versus time

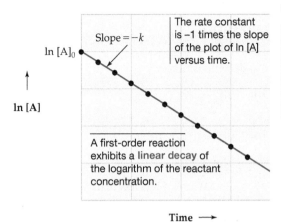

Slope $= -k$

ln [A]$_0$

ln [A]

A first-order reaction exhibits a **linear decay** of the logarithm of the reactant concentration.

Time →

◄ **FIGURE 14.7**

Plots for a first-order reaction.

The rate constant is –1 times the slope of the plot of ln [A] versus time.

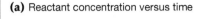

◄ **Figure It Out**

How is the rate constant (*k*) determined for a first-order reaction?

Answer: In a plot of ln[A] versus time, the slope is equal to −*k*.

Go to eText

WORKED EXAMPLE 14.6

Using the Integrated Rate Law for a First-Order Reaction

The decomposition of hydrogen peroxide in dilute sodium hydroxide solution is described by the equation

$$H_2O_2(aq) \longrightarrow H_2O(l) + 1/2\ O_2(g)$$

The reaction is first order in H_2O_2, the rate constant for the consumption of H_2O_2 at 20 °C is 1.8×10^{-5} s^{-1}, and the initial concentration of H_2O_2 is 0.30 M.

(a) What is the concentration of H_2O_2 after 4.00 h?
(b) How long will it take for the H_2O_2 concentration to drop to 0.12 M?
(c) How long will it take for 90% of the H_2O_2 to decompose?

IDENTIFY

Known	Unknown
(a) $k = 1.8 \times 10^{-5}$ s^{-1}, $[H_2O_2]_0 = 0.030$ M, $t = 4.00$ h	$[H_2O_2]_t$
(b) $k = 1.8 \times 10^{-5}$ s^{-1}, $[H_2O_2]_0 = 0.030$ M, $[H_2O_2]_t = 0.12$ M	time (t)
(c) $k = 1.8 \times 10^{-5}$ s^{-1}, 90% decomposed	time (t)

STRATEGY

Since this reaction has a first-order rate law, $-\Delta[H_2O_2]/\Delta t = k[H_2O_2]$, we can use the integrated rate law for a first-order reaction:

$$\ln \frac{[H_2O_2]_t}{[H_2O_2]_0} = -kt$$

In each part, we substitute the known quantities into the first-order integrated rate law and solve for the unknown.

SOLUTION

(a) Because k has units of s^{-1}, we must first convert the time from hours to seconds:

$$t = (4.00\ \text{h})\left(\frac{60\ \text{min}}{\text{h}}\right)\left(\frac{60\ \text{s}}{\text{min}}\right) = 1.44 \times 10^4\ \text{s}$$

Then, substitute the values of $[H_2O_2]_0$, k, and t into the first-order integrated rate law equation:

$$\ln \frac{[H_2O_2]_t}{0.30\ \text{M}} = -(1.8 \times 10^{-5}\ \text{s}^{-1})\,(1.44 \times 10^4\ \text{s}) = -0.259$$

Taking the natural antilogarithm (antiln) of both sides gives

$$\frac{[H_2O_2]_t}{0.30\ \text{M}} = e^{-0.259} = 0.772$$

where the number $e = 2.718\,28\ldots$ is the base of natural logarithms (see Appendix A.2). Therefore,

$$[H_2O_2]_t = (0.772)(0.30\ \text{M}) = 0.23\ \text{M}$$

(b) First, solve the integrated rate law for the time:

$$t = -\frac{1}{k}\ln \frac{[H_2O_2]_t}{[H_2O_2]_0}$$

Then calculate the time by substituting the concentrations and the value of k:

$$t = -\left(\frac{1}{1.8 \times 10^{-5}\ \text{s}^{-1}}\right)\left(\ln \frac{0.12\ \text{M}}{0.30\ \text{M}}\right)$$

$$= -\left(\frac{1}{1.8 \times 10^{-5}\ \text{s}^{-1}}\right)(-0.916) = 5.1 \times 10^4\ \text{s}$$

continued on next page

Thus, the H_2O_2 concentration reaches 0.12 M at a time of 5.1×10^4 s or 14 h.

(c) When 90% of the H_2O_2 has decomposed, 10% remains. Therefore,

$$\frac{[H_2O_2]_t}{[H_2O_2]_0} = \frac{(0.10)[H_2O_2]_0}{[H_2O_2]_0} = 0.10$$

The time required for 90% decomposition is

$$t = -\left(\frac{1}{1.8 \times 10^{-5}\ s^{-1}}\right)(\ln 0.10)$$

$$= -\left(\frac{1}{1.8 \times 10^{-5}\ s^{-1}}\right)(-2.30) = 1.3 \times 10^5\ s = 36\ h$$

CHECK

(a) The concentration of H_2O_2 (0.23 M) after 4 h is less than the initial concentration (0.30 M). (b) A longer period of time (14 h) is required for the concentration to drop to 0.12 M. (c) Still more time (36 h) is needed for the concentration to fall to 0.030 M (10% of the original concentration).

▶ **PRACTICE 14.11** In acidic aqueous solution, the purple complex ion $Co(NH_3)_5Br^{2+}$ undergoes a slow reaction in which the bromide ion is replaced by a water molecule, yielding the pinkish-orange complexion $Co(NH_3)_5(H_2O)^{3+}$:

$$Co(NH_3)_5Br^{2+}(aq) + H_2O(l) \longrightarrow$$
Purple

$$Co(NH_3)_5(H_2O)^{3+}(aq) + Br^-(aq)$$
Pinkish–orange

The reaction is first order in $Co(NH_3)_5Br^{2+}$, the rate constant at 25 °C is $6.3 \times 10^{-6}\ s^{-1}$, and the initial concentration of $Co(NH_3)_5Br^{2+}$ is 0.100 M.

▲ Aqueous solutions of $Co(NH_3)_5Br^{2+}$ (left) and $Co(NH_3)_5(H_2O)^{3+}$ (right).

(a) What is the molarity of $Co(NH_3)_5Br^{2+}$ after a reaction time of 10.0 h?

(b) How many hours are required for 75% of the $Co(NH_3)_5Br^{2+}$ to react?

▶ **APPLY 14.12** For a general first-order reaction, A ⟶ Products, 55.0% of A has reacted after 14.2 hours. (a) What is the value of the rate constant in units of h^{-1}? (b) How long will it take for 85.0% of A to react?

WORKED EXAMPLE 14.7

Plotting Data for a First-Order Reaction

Experimental concentration-versus-time data for the decomposition of gaseous N_2O_5 at 55 °C are listed in Table 14.1 and are plotted in Figure 14.1. Use those data to confirm that the decomposition of N_2O_5 is a first-order reaction. What is the value of the rate constant for the consumption of N_2O_5?

STRATEGY

To confirm that the reaction is first order, check to see whether a plot of $\ln [N_2O_5]$ versus time gives a straight line. The rate constant for a first-order reaction equals the negative slope of the straight line.

SOLUTION

Values of $\ln [N_2O_5]$ are listed in the following table and are plotted versus time in the graph:

Time (s)	$[N_2O_5]$	$\ln [N_2O_5]$
0	0.0200	−3.912
100	0.0169	−4.080
200	0.0142	−4.255
300	0.0120	−4.423
400	0.0101	−4.595
500	0.0086	−4.756
600	0.0072	−4.934
700	0.0061	−5.099

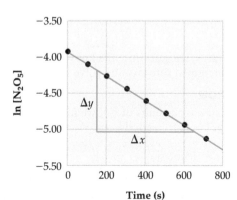

Because the data points lie on a straight line, the reaction is first order in N_2O_5. The slope of the line can be determined from the coordinates of any two widely separated points on the line, and the rate constant k can be calculated from the slope:

$$\text{Slope} = \frac{\Delta y}{\Delta x} = \frac{(-5.02) - (-4.17)}{650 \text{ s} - 150 \text{ s}} = \frac{-0.85}{500 \text{ s}} = -1.7 \times 10^{-3} \text{ s}^{-1}$$

$$k = -(\text{Slope}) = 1.7 \times 10^{-3} \text{ s}^{-1}$$

Note that the slope is negative and k is positive.

▶ **PRACTICE 14.13**   At high temperatures, cyclopropane is converted to propene, the material from which polypropylene plastics are made:

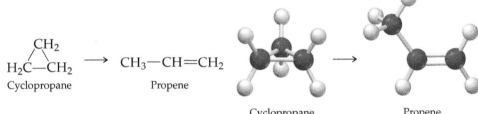

$$\underset{\text{Cyclopropane}}{\overset{\displaystyle \overset{\displaystyle CH_2}{\overset{\displaystyle \diagup \diagdown}{H_2C-CH_2}}}{}} \longrightarrow \underset{\text{Propene}}{CH_3-CH=CH_2}$$

Cyclopropane   Propene

Given the following concentration data, test whether the reaction is first order and calculate the value of the rate constant.

Time (min)	0	5.0	10.0	15.0	20.0
[Cyclopropane]	0.098	0.080	0.066	0.054	0.044

▶ **APPLY 14.14**   The gaseous decomposition of N_2O_5 was studied at 35 °C.

$$N_2O_5(g) \longrightarrow 2 \text{ NO}_2(g) + 1/2 \text{ O}_2(g)$$

A plot of $\ln[N_2O_5]$ versus time has a slope of $-9.8 \times 10^{-4} \text{ s}^{-1}$. If 0.100 mol of N_2O_5 is added to a 1.0 L flask at 35 °C, calculate the concentrations of N_2O_5, NO_2, and O_2 after 10.0 minutes.

The **half-life** of a reaction, symbolized by $t_{1/2}$, is the time required for the reactant concentration to drop to one-half of its initial value. Consider the first-order reaction

$$A \longrightarrow \text{Products}$$

To relate the reaction's half-life to the rate constant, let's begin with the integrated rate law:

$$\ln \frac{[A]_t}{[A]_0} = -kt$$

When $t = t_{1/2}$, the fraction of A that remains, $[A]_t/[A]_0$, is one-half. Therefore,

$$\ln \frac{1}{2} = -kt_{1/2}$$

$$\text{so} \quad t_{1/2} = \frac{-\ln \dfrac{1}{2}}{k} = \frac{\ln 2}{k}$$

$$\text{or} \quad t_{1/2} = \frac{0.693}{k}$$

Thus, we can calculate the half-life of a first-order reaction from the rate constant, and vice versa.

The half-life of a first-order reaction is a constant because it depends only on the rate constant and not on the reactant concentration. This point is worth noting because reactions that are not first order have half-lives that *do* depend on concentration; that is, the amount of time in one half-life changes as the reactant concentration changes for a non-first-order reaction.

Because the half-life of a first-order reaction is a constant, each successive half-life is an equal period of time in which the reactant concentration decreases by a factor of 2 (**FIGURE 14.8**).

▶ FIGURE 14.8

Concentration of a reactant A as a function of time for a first-order reaction. Each half-life represents an equal amount of time.

▶ **Figure It Out**

How many half-lives does it take for a reactant concentration to decrease to 1/8 of its initial value?

Answer: Three half-lives

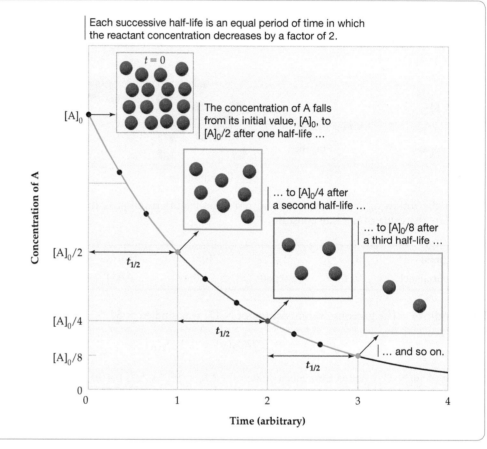

Each successive half-life is an equal period of time in which the reactant concentration decreases by a factor of 2.

The concentration of A falls from its initial value, $[A]_0$, to $[A]_0/2$ after one half-life ...

... to $[A]_0/4$ after a second half-life ...

... to $[A]_0/8$ after a third half-life ...

... and so on.

WORKED EXAMPLE 14.8

Determining the Half-Life for a First-Order Reaction

The rate constant for the first-order decomposition of cyclopropane to propene at a high temperature is 0.0400 min^{-1}.

$$\underset{\text{Cyclopropane}}{\overset{\displaystyle CH_2}{\underset{\displaystyle H_2C-CH_2}{\bigtriangleup}}} \longrightarrow \underset{\text{Propene}}{CH_3-CH=CH_2}$$

(a) Calculate the half-life of cyclopropane at the same temperature.
(b) How long will it take for cyclopropane to fall to 12.5% of its initial value?

STRATEGY AND SOLUTION

(a) For a first-order reaction, half-life can be calculated from the equation $t_{1/2} = 0.693/k$.

$$t_{1/2} = \frac{0.693}{0.0400(1/\text{min})} = 17.3 \text{ min}$$

(b) Since 12.5% of the initial concentration corresponds to 1/8 or $(1/2)^3$ of the initial concentration, the time required is three half-lives:

$$t = 3t_{1/2} = 3(17.3 \text{ min}) = 51.9 \text{ min}$$

▶ PRACTICE 14.15 Consider the first-order decomposition of H_2O_2 with a rate constant of 1.8×10^{-5} s^{-1} at 20 °C.

What is the half-life in hours and what is the molarity of H_2O_2 after four half-lives? The initial concentration of H_2O_2 is 0.30 M.

▶ CONCEPTUAL APPLY 14.16 Consider the first-order reaction $A \rightarrow B$ in which A molecules (red spheres) are converted to B molecules (blue spheres).

(a) Given the following pictures at $t = 0$ min and $t = 10$ min, what is the half-life of the reaction?

(b) Draw a picture that shows the number of A and B molecules present at $t = 15$ min.

$t = 0$ min $t = 10$ min

14.6 INTEGRATED RATE LAW: SECOND-ORDER REACTIONS

A **second-order reaction** is one whose rate depends either on the concentration of a single reactant raised to the second power or on the concentrations of two different reactants, each raised to the first power. For the simpler type, $A \rightarrow$ Products, the rate law is

$$\text{Rate of consumption of A} = -\frac{\Delta[A]}{\Delta t} = k[A]^2$$

An example is the thermal decomposition of nitrogen dioxide to yield NO and O_2:

$$NO_2(g) \longrightarrow NO(g) + 1/2\ O_2(g)$$

Using calculus, it's possible to convert the rate law to the integrated rate law:

$$\frac{1}{[A]_t} = kt + \frac{1}{[A]_0}$$

This integrated rate law allows us to calculate the concentration of A at any time t if the initial concentration $[A]_0$ is known.

Since the integrated rate law has the form $y = mx + b$, a graph of $1/[A]$ versus time is a straight line if the reaction is second order:

$$\underset{\substack{\uparrow \\ y}}{\frac{1}{[A]_t}} = \underset{\substack{\uparrow\uparrow \\ mx}}{kt} + \underset{\substack{\uparrow \\ b}}{\frac{1}{[A]_0}}$$

The slope of the straight line is the rate constant k, and the intercept is $1/[A]_0$. Thus, by plotting $1/[A]$ versus time, we can test whether the reaction is second order and can determine the value of the rate constant (see Worked Example 14.9). We can also obtain an expression for the half-life of a second-order reaction by substituting $[A]_t = [A]_0/2$ and $t = t_{1/2}$ into the integrated rate law:

$$\frac{1}{\left(\dfrac{[A]_0}{2}\right)} = kt_{1/2} + \frac{1}{[A]_0}$$

so $t_{1/2} = \dfrac{1}{k}\left(\dfrac{2}{[A]_0} - \dfrac{1}{[A]_0}\right)$

or $t_{1/2} = \dfrac{1}{k[A]_0}$

In contrast with a first-order reaction, the time required for the concentration of A to drop to one-half of its initial value in a second-order reaction depends on both the rate constant and the initial concentration. Thus, the value of $t_{1/2}$ increases as the reaction proceeds because the value of $[A]_0$ at the beginning of each successive half-life is smaller by a factor of 2. Consequently, each half-life for a second-order reaction is twice as long as the preceding one (**FIGURE 14.9**).

TABLE 14.4 summarizes some important differences between zeroth-order, first-order, and second-order reactions of the type A → Products.

▶ **FIGURE 14.9**

Concentration of a reactant A as a function of time for a second-order reaction.

▶ **Figure It Out**

In a second-order reaction, how does the half-life change as the reaction proceeds?

Answer: The time for each half-life increases as the reaction proceeds.

Each half-life is twice as long as the preceding one because $t_{1/2} = 1/k[A]_0$ and the concentration of A at the beginning of each successive half-life is smaller by a factor of 2.

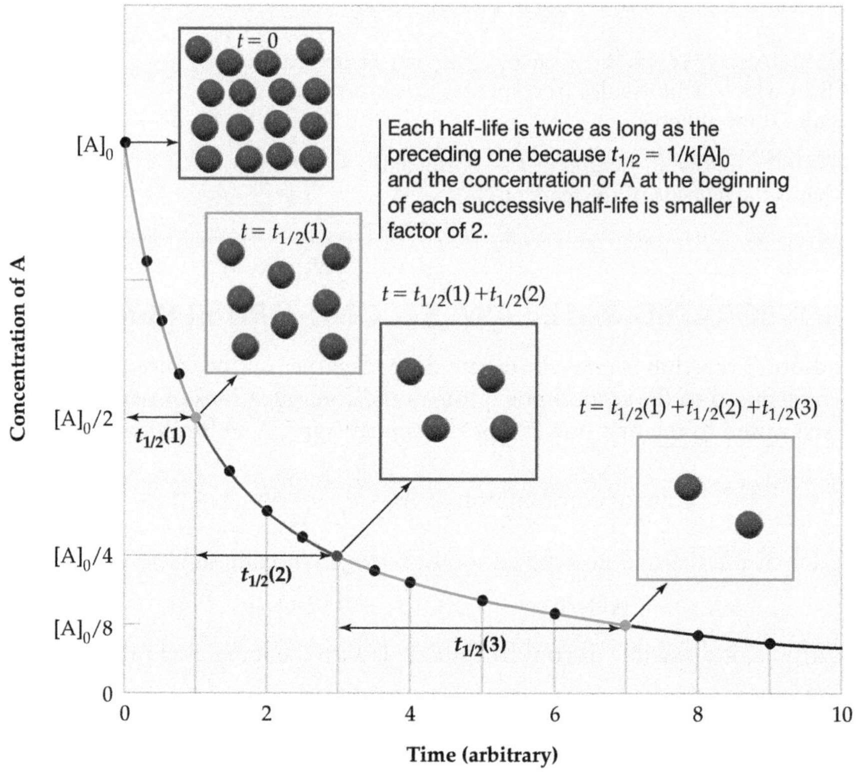

TABLE 14.4 Characteristics of zeroth-, first-, and second-order reactions of the Type A → Products			
	Zeroth-Order	**First-Order**	**Second-Order**
Rate law	$-\dfrac{\Delta[A]}{\Delta t} = k$	$-\dfrac{\Delta[A]}{\Delta t} = k[A]$	$-\dfrac{\Delta[A]}{\Delta t} = k[A]^2$
Integrated Rate Law	$[A]_t = -kt + [A]_0$	$\ln[A]_t = -kt + \ln[A]_0$	$\dfrac{1}{[A]_t} = kt + \dfrac{1}{[A]_0}$
Linear graph	$[A]$ versus t	$\ln[A]$ versus t	$\dfrac{1}{[A]}$ versus t
	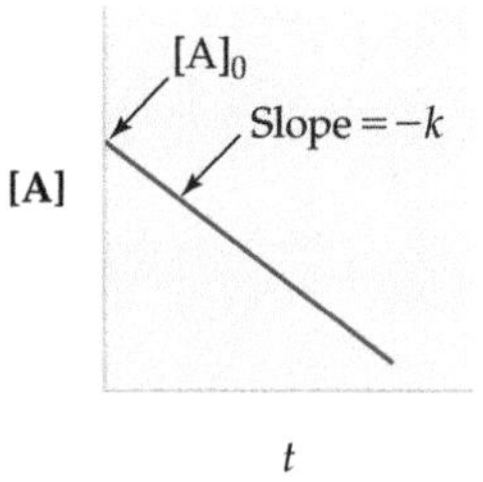	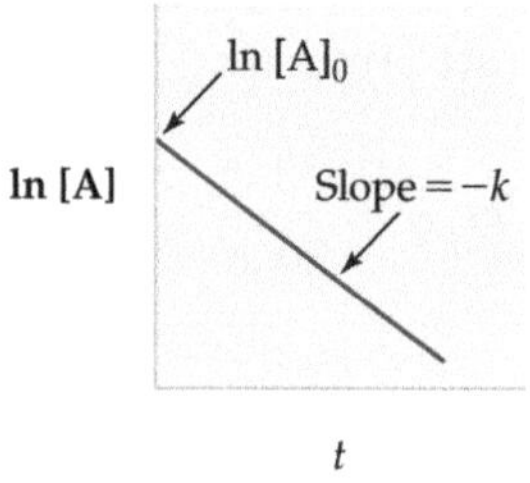	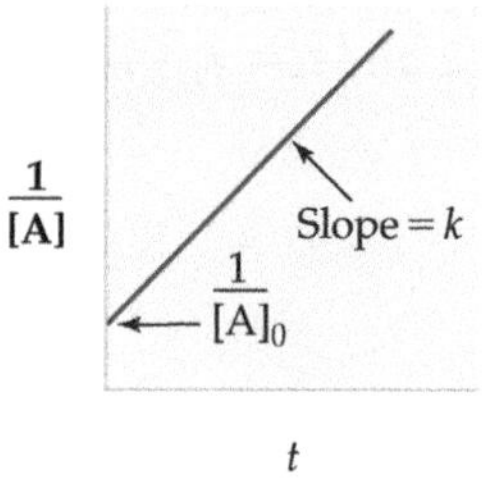
Graphical determination of k	$k = -(\text{Slope})$	$k = -(\text{Slope})$	$k = \text{Slope}$
Half-life	$t_{1/2} = \dfrac{[A]_0}{2k}$ (not constant)	$t_{1/2} = \dfrac{0.693}{k}$ (constant)	$t_{1/2} = \dfrac{1}{k[A]_0}$ (not constant)

WORKED EXAMPLE 14.9

Determining Reaction Order Graphically

At elevated temperatures, nitrogen dioxide decomposes to nitric oxide and molecular oxygen:

$$NO_2(g) \longrightarrow NO(g) + 1/2\ O_2(g)$$

Concentration–time data for the consumption of NO_2 at 300 °C are as follows:

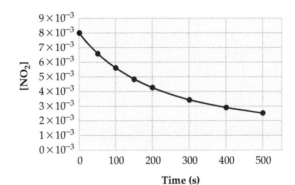

Time (s)	NO_2	Time (s)	$[NO_2]$
0	8.00×10^{-3}	200	4.29×10^{-3}
50	6.58×10^{-3}	300	3.48×10^{-3}
100	5.59×10^{-3}	400	2.93×10^{-3}
150	4.85×10^{-3}	500	2.53×10^{-3}

(a) What is the order of the reaction?
(b) What is the value of the rate constant for the decomposition of NO_2?
(c) What is the concentration of NO_2 at $t = 20.0$ min?
(d) What is the half-life of the reaction when the initial concentration of NO_2 is 6.00×10^{-3} M?
(e) What is $t_{1/2}$ when $[NO_2]_0$ is 3.00×10^{-3} M?

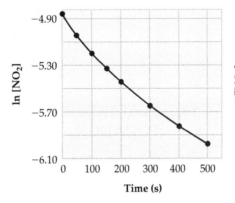

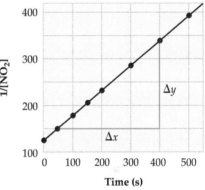

STRATEGY

To determine the order of the reaction, calculate values of ln $[NO_2]$ and $1/[NO_2]$ and then make graphs of $[NO_2]$, $\ln[NO_2]$, and $1/[NO_2]$ versus time. Obtain the rate constant from the slope of the straight-line plot, and calculate concentrations and half-lives using the appropriate equation in Table 14.4.

SOLUTION

Time (s)	$[NO_2]$	$\ln[NO_2]$	$1/[NO_2]$
0	8.00×10^{-3}	-4.828	125
50	6.58×10^{-3}	-5.024	152
100	5.59×10^{-3}	-5.187	179
150	4.85×10^{-3}	-5.329	206
200	4.29×10^{-3}	-5.451	233
300	3.48×10^{-3}	-5.661	287
400	2.93×10^{-3}	-5.833	341
500	2.53×10^{-3}	-5.980	395

(a) The plot of $[NO_2]$ versus time and ln $[NO_2]$ versus time is curved, but the plot of $1/[NO_2]$ versus time is a straight line. The reaction is, therefore, second order in NO_2.

(b) The rate constant equals the slope of the straight line in the plot of $1/[NO_2]$ versus time, which we can estimate from the coordinates of two widely separated points on the line:

$$k = \text{Slope} = \frac{\Delta y}{\Delta x} = \frac{340\ M^{-1} - 150\ M^{-1}}{400\ s - 50\ s}$$

$$= \frac{190\ M^{-1}}{350\ s} = 0.54/(M \cdot s)$$

(c) The concentration of NO_2 at $t = 20.0$ min $(1.20 \times 10^3\ s)$ can be calculated using the integrated rate law:

$$\frac{1}{[NO_2]_t} = kt + \frac{1}{[NO_2]_0}$$

Substituting the values of k, t, and $[NO_2]_0$ gives

$$\frac{1}{[NO_2]_t} = \left(\frac{0.54}{M \cdot s}\right)(1.20 \times 10^3\ s) + \frac{1}{8.00 \times 10^{-3}\ M} = \frac{773}{M}$$

$$[NO_2]_t = 1.3 \times 10^{-3}\ M$$

continued on next page

(d) The half-life of this second-order reaction when the initial concentration of NO_2 is 6.00×10^{-3} M can be calculated from the rate constant and the initial concentration:

$$t_{1/2} = \frac{1}{k[NO_2]_0} = \frac{1}{\left(\dfrac{0.54}{M \cdot s}\right)(6.00 \times 10^{-3} \text{ M})} = 3.1 \times 10^2 \text{ s}$$

(e) When $[NO_2]_0$ is 3.00×10^{-3} M, $t_{1/2} = 6.2 \times 10^2$ s (twice as long as when $[NO_2]_0$ is 6.00×10^{-3} M because $[NO_2]_0$ is now smaller by a factor of 2).

▶ **PRACTICE 14.17** Hydrogen iodide gas decomposes at 410 °C:

$$2 \text{ HI}(g) \longrightarrow H_2(g) + I_2(g)$$

The following data describe this decomposition:

Time (min)	0	20	40	60	80
[HI]	0.500	0.382	0.310	0.260	0.224

(a) What are the order of the reaction and the value of the rate constant for the decomposition of HI?
(b) At what time (in minutes) does the HI concentration reach 0.100 M?

▶ **APPLY 14.18** When exposed to ultraviolet light, a red dye molecule degrades to a colorless product represented by the equation

$$\text{Dye} \longrightarrow \text{Product}$$
$$\text{(red)} \qquad \text{(colorless)}$$

A solution of dye was prepared and its absorbance measured as a function of time. The absorbance of the dye is directly proportional to its concentration. Use the following data to determine the order and half-life for the reaction.

Time (s)	Absorbance	Time (s)	Absorbance
0	0.5000	40	0.1460
10	0.3680	60	0.0793
20	0.2710	80	0.0429
30	0.1990	100	0.0232

14.7 REACTION RATES AND TEMPERATURE: THE ARRHENIUS EQUATION

Thus far, our discussion has centered on experimental aspects of reaction rates. We've seen that the rate of a reaction depends on both reactant concentrations and the value of the rate constant.

Still another factor that affects reaction rates is temperature. Everyday experience tells us that the rates of chemical reactions increase with increasing temperature. Familiar fuels such as gas, oil, and coal are relatively inert at room temperature but burn rapidly at elevated temperatures. Many foods last almost indefinitely when stored in a freezer but spoil quickly at room temperature. Metallic magnesium is inert in cold water but reacts with hot water (**FIGURE 14.10**). As a rule of thumb, reaction rates tend to double when the temperature is increased by 10 °C.

To understand why reaction rates depend on temperature, we need a picture, or model, of how reactions take place. According to the **collision theory** model, a bimolecular reaction occurs when two properly oriented reactant molecules come together in a sufficiently energetic collision. To be specific, let's consider one of the simplest possible reactions, the reaction of an atom A with a diatomic molecule BC to give a diatomic molecule AB and an atom C:

$$A + BC \longrightarrow AB + C$$

An example from atmospheric chemistry is the reaction of an oxygen atom with an HCl molecule to give an OH molecule and a chlorine atom:

$$O(g) + HCl(g) \longrightarrow OH(g) + Cl(g)$$

If the reaction occurs in a single step, the electron distribution about the three nuclei must change in the course of the collision such that a new bond, A—B, develops at the same time the old bond, B—C, breaks. Between the reactant and product

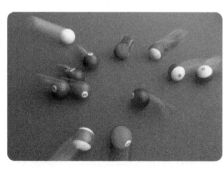

▲ Chemical reactions occur as a result of energetic collisions between atoms and molecules.

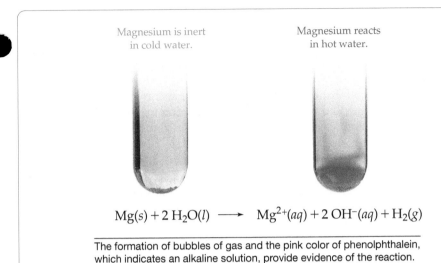

Magnesium is inert in cold water (left) but reacts with hot water (right).

Magnesium is inert in cold water.

Magnesium reacts in hot water.

$$Mg(s) + 2\,H_2O(l) \longrightarrow Mg^{2+}(aq) + 2\,OH^-(aq) + H_2(g)$$

The formation of bubbles of gas and the pink color of phenolphthalein, which indicates an alkaline solution, provide evidence of the reaction.

stages, the nuclei pass through a configuration in which all three atoms are weakly linked together. We can picture the progress of the reaction as

$$A + B{-}C \longrightarrow A{-}{-}{-}B{-}{-}{-}C \longrightarrow A{-}B + C$$

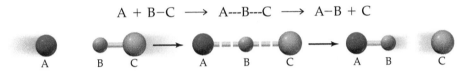

If A and BC have filled shells of electrons (no unpaired electrons or vacant, low-energy orbitals), they will repel each other. To achieve the configuration A—B—C, then, the atoms require energy to overcome this repulsion. The energy comes from the kinetic energy of the colliding particles and is converted to potential energy in A—B—C. In fact, A—B—C has more potential energy than either the reactants or the products. Thus, there is a potential energy barrier that must be surmounted before reactants can be converted to products, as depicted graphically on the potential energy profile in **FIGURE 14.11**.

The height of the barrier is called the **activation energy** (E_a). and the configuration of atoms at the maximum in the potential energy profile is called the **transition state,** or

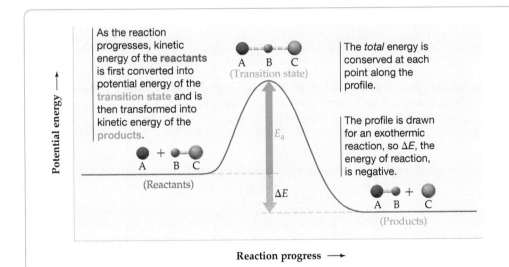

As the reaction progresses, kinetic energy of the **reactants** is first converted into potential energy of the transition state and is then transformed into kinetic energy of the products.

A   B   C
(Transition state)

The *total* energy is conserved at each point along the profile.

E_a

The profile is drawn for an exothermic reaction, so ΔE, the energy of reaction, is negative.

A + B C
(Reactants)

ΔE

A B + C
(Products)

Reaction progress ⟶

Potential energy ⟶

◀ **FIGURE 14.11**

Potential energy profile for the reaction $A + BC \rightarrow AB + C$.

◀ **Figure It Out**

What is present at the point of highest energy in the reaction? Where do molecules get the energy needed to surpass this point?

Answer: Transition state; if the energy of molecular collisions exceeds the energy of the transition state, the reactants will be converted to products.

the *activated complex*. Since energy is conserved in the collision, all the energy needed to climb the potential energy hill must come from the kinetic energy of the colliding molecules. If the collision energy is less than E_a, the reactant molecules can't surmount the barrier, and they simply bounce apart. If the collision energy is at least as great as E_a, however, the reactants can climb over the barrier and be converted to products.

Experimental evidence for the notion of an activation energy barrier comes from a comparison of collision rates and reaction rates. Collision rates in gases can be calculated from **kinetic–molecular theory** (Section 10.6). For a gas at room temperature (298 K) and 1 atm pressure, each molecule undergoes approximately 10^9 collisions per second, or 1 collision every 10^{-9} s. Thus, if every collision resulted in a reaction, every gas-phase reaction would be complete in about 10^{-9} s. By contrast, observed reactions often have half-lives of minutes or hours, so only a tiny fraction of the collisions lead to a reaction.

Very few collisions are productive because very few occur with a collision energy as large as the activation energy. The fraction of collisions with an energy equal to or greater than the activation energy E_a is represented in **FIGURE 14.12** at two different temperatures by the areas under the curves to the right of E_a.

▶ **FIGURE 14.12**

Plots of the fraction of collisions with a particular energy at two different temperatures.

▶ **Figure It Out**

Does the reaction rate increase or decrease when the temperature increases? Explain.

Answer: Increase; the larger area under the curve to the right of E_a shows that a larger fraction of molecules collide with energy greater than or equal to E_a.

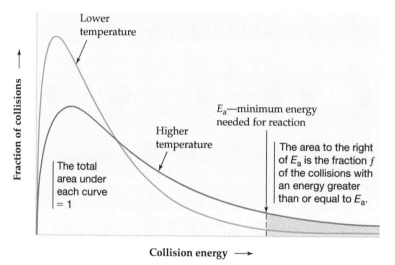

The fraction of collisions that are sufficiently energetic to result in reaction increases exponentially with increasing temperature.

When E_a is large compared to RT, this fraction f is approximated by the equation

$$f = e^{-E_a/RT}$$

where R is the gas constant $[8.314 \text{ J}/(\text{K} \cdot \text{mol})]$ and T is the absolute temperature in kelvin. Note that f is a very small number. For example, for a reaction having an activation energy of 75 kJ/mol, the value of f at 298 K is 7×10^{-14}:

$$f = \exp\left[\dfrac{-75,000\dfrac{\text{J}}{\text{mol}}}{\left(8.314\dfrac{\text{J}}{\text{K} \cdot \text{mol}}\right)(298 \text{ K})} \right] = e^{-30.3} = 7 \times 10^{-14}$$

Thus, only 7 collisions in 100 trillion are sufficiently energetic to convert reactants to products.

As the temperature increases, the distribution of collision energies broadens and shifts to higher energies (Figure 14.12), resulting in a rapid increase in the fraction of collisions that lead to products. At 308 K, for example, the calculated value of f for the reaction with $E_a = 75$ kJ/mol is 2×10^{-13}. Thus, a temperature increase of just 3%, from 298 K to 308 K, increases the value of f by a factor of 3. Collision theory therefore accounts nicely for the exponential dependence of reaction rates on reciprocal temperature. As T increases ($1/T$ decreases), $f = e^{-E_a/RT}$ increases exponentially. Collision theory also explains why reaction rates are so much lower than collision rates. Collision rates also increase with increasing temperature, but only by a small amount—less than 2% on going from 298 K to 308 K.

The fraction of collisions that lead to products is further reduced by an orientation requirement. Even if the reactants collide with sufficient energy, they won't react unless the orientation of the reaction partners is correct for formation of the transition state. For example, a collision of A with the C end of the molecule BC can't result in formation of AB:

$$A + C\text{–}B \longrightarrow A\text{---}C\text{---}B \not\longrightarrow A\text{–}B + C$$

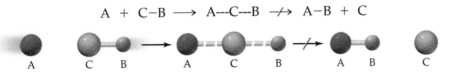

Go to eText

BIG IDEA Question 5

What is the main reason that higher temperatures increase reaction rate?

The reactant molecules would simply collide and then separate without reaction:

$$A + C\text{–}B \longrightarrow A\text{---}C\text{---}B \longrightarrow A + C\text{–}B$$

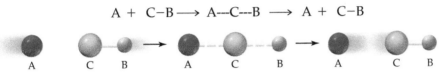

The fraction of collisions having proper orientation for the conversion of reactants to products is called the **steric factor** (p). For the reaction, $A + BC \rightarrow AB + C$, the value of p is expected to be about 0.5 because A has a nearly 1 : 1 probability of colliding with each of the B and C ends of BC. (This assumes that B and C have similar sizes and electronic properties.) For reactions of larger, more complex molecules, p is considerably less than 0.5.

Now let's see how the two parameters p and f enter into the rate law. Since bimolecular collisions between any two molecules—say, A and B—occur at a rate that is proportional to their concentrations, we can write

$$\text{Collision rate} = Z[A][B]$$

where Z is a constant related to the collision frequency and has units of a second-order rate constant, $1/(M \cdot s)$ or $M^{-1} s^{-1}$. The reaction rate is lower than the collision rate by a factor $p \times f$ because only a fraction of the colliding molecules have the correct orientation and the minimum energy needed for reaction:

$$\text{Reaction rate} = p \times f \times \text{Collision rate} = pfZ[A][B]$$

Since the rate law is

$$\text{Reaction rate} = k[A][B]$$

the rate constant predicted by collision theory is $k = pfZ$, or $k = pZe^{-E_a/RT}$:

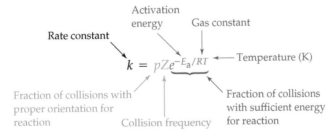

This expression is usually written in a form called the **Arrhenius equation,** named after Svante Arrhenius, the Swedish chemist who proposed it in 1889 on the basis of experimental studies of reaction rates.

Arrhenius equation $k = Ae^{-E_a/RT}$

The parameter $A(=pZ)$ is called the **frequency factor,** or pre-exponential factor. In accord with the minus sign in the exponent, the rate constant decreases as E_a increases and increases as T increases.

CONCEPTUAL WORKED EXAMPLE 14.10

Interpreting Potential Energy Profiles for Reactions

The potential energy profile for the one-step reaction $AB + CD \rightarrow AC + BD$ follows. The energies are in kJ/mol relative to an arbitrary zero of energy.

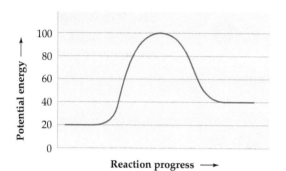

(a) What is the value of the activation energy for this reaction?
(b) Is the reaction exothermic or endothermic? What is ΔE for the reaction?
(c) What is the activation energy for the reverse reaction?
(d) Suggest a plausible structure for the transition state. Use dashed lines to indicate the atoms that are weakly linked together in the transition state.

STRATEGY AND SOLUTION

(a)–(c) The activation energy is the energy difference between the transition state and reactants, ΔE is the energy difference between products and reactants, and the activation energy for the reverse reaction is the energy difference between the transition state and products. ($E_a = 80$ kJ/mol, $\Delta E = 20$ kJ/mol (endothermic), and E_a (reverse) $= 60$ kJ/mol)

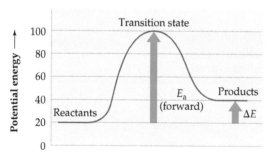

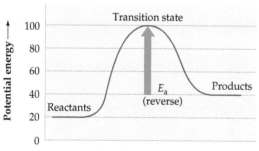

(d) In order for the reaction to occur, A must collide with C and B must collide with D. A plausible structure for the transition state is shown in the margin.

▶ **CONCEPTUAL PRACTICE 14.19**   The reaction

$$NO_2(g) + CO(g) \longrightarrow NO(g) + CO_2(g)$$

occurs in one step. The activation energy is 132 kJ/mol and ΔE is −226 kJ/mol.

(a) Is the reaction endothermic or exothermic?
(b) What is the activation energy for the reverse reaction?

▶ **CONCEPTUAL APPLY 14.20**   For the one-step reaction

$$NO_2(g) + CO(g) \longrightarrow NO(g) + CO_2(g),$$

draw four possible ways that the two reactant molecules can collide. Which is most likely to result in a successful reaction?

14.8 USING THE ARRHENIUS EQUATION

As we saw in Section 14.7, the activation energy E_a is one of the most important factors affecting the rate of a chemical reaction. We can determine its value using the Arrhenius equation if we know the values of the rate constant at different temperatures.

Taking the natural logarithm of both sides of the Arrhenius equation, we obtain the logarithmic form

> **Log form of the Arrhenius equation** $\ln k = \ln A - \dfrac{E_a}{RT}$

We can rearrange this equation into the form $y = mx + b$, so a graph of $\ln k$ versus $1/T$, called an **Arrhenius plot**, gives a straight line with slope $m = -E_a/R$ and intercept $b = \ln A$:

$$\ln k = \left(\frac{-E_a}{R}\right)\left(\frac{1}{T}\right) + \ln A$$

$$\underset{y}{\uparrow} \qquad \underset{m}{\uparrow}\ \underset{x}{\uparrow} \qquad \underset{b}{\uparrow}$$

We can determine the experimental value of the activation energy from the slope of the straight line, as shown in Worked Example 14.11.

$$E_a = -R(\text{Slope})$$

We can derive still another form of the Arrhenius equation that allows us to estimate the activation energy from rate constants at just two temperatures. At temperature T_1,

$$\ln k_1 = \left(\frac{-E_a}{R}\right)\left(\frac{1}{T_1}\right) + \ln A$$

and at temperature T_2,

$$\ln k_2 = \left(\frac{-E_a}{R}\right)\left(\frac{1}{T_2}\right) + \ln A$$

Subtracting the first equation from the second, and remembering that $(\ln k_2 - \ln k_1) = \ln (k_2/k_1)$, we obtain

> **Two point form of the Arrhenius equation** $\ln\left(\dfrac{k_2}{k_1}\right) = \left(\dfrac{-E_a}{R}\right)\left(\dfrac{1}{T_2} - \dfrac{1}{T_1}\right)$

This equation can be used to calculate E_a from rate constants k_1 and k_2 at temperatures T_1 and T_2. By the same token, if we know E_a and the rate constant k_1 at one temperature T_1, we can calculate the rate constant k_2 at another temperature T_2. Worked Example 14.11 shows how this is done.

WORKED EXAMPLE 14.11

Using the Arrhenius Equation

Rate constants for the gas-phase decomposition of hydrogen iodide, $HI(g) \rightarrow 1/2\ H_2(g) + 1/2\ I_2(g)$, are listed in the following table:

Temperature (°C)	$k(M^{-1}\,s^{-1})$	Temperature (°C)	$k(M^{-1}\,s^{-1})$
283	3.52×10^{-7}	427	1.16×10^{-3}
356	3.02×10^{-5}	508	3.95×10^{-2}
393	2.19×10^{-4}		

(a) Find the activation energy (in kJ/mol) using all five data points.
(b) Calculate E_a from the rate constants at 283 °C and 508 °C.
(c) Given the rate constant at 283 °C and the value of E_a obtained in part (b), what is the rate constant at 293 °C?

continued on next page

STRATEGY

(a) We can determine the activation energy E_a from the slope of a linear plot of ln k versus $1/T$.

(b) To calculate E_a from values of the rate constant at two temperatures, use the equation

$$\ln\left(\frac{k_2}{k_1}\right) = \left(\frac{-E_a}{R}\right)\left(\frac{1}{T_2} - \frac{1}{T_1}\right)$$

(c) Use the same equation and the known values of E_a and k_1 at T_1 to calculate k_2 at T_2.

SOLUTION

(a) Because the temperature in the Arrhenius equation is expressed in kelvin, we must first convert the Celsius temperatures to absolute temperatures. Then calculate values of $1/T$ and ln k, and plot ln k versus $1/T$. The results are shown in the following table and graph:

T (°C)	T (K)	$k(M^{-1}\,s^{-1})$	$1/T$ (1/K)	ln k
283	556	3.52×10^{-7}	0.001 80	−14.860
356	629	3.02×10^{-5}	0.001 59	−10.408
393	666	2.19×10^{-4}	0.001 50	−8.426
427	700	1.16×10^{-3}	0.001 43	−6.759
508	781	3.95×10^{-2}	0.001 28	−3.231

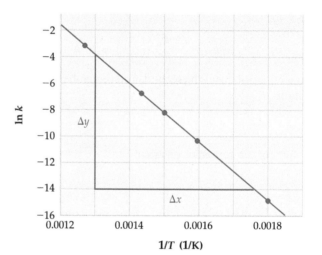

We can determine the slope of the straight-line plot from the coordinates of any two widely separated points on the line:

$$\text{Slope} = \frac{\Delta y}{\Delta x} = \frac{(-14.0) - (-3.9)}{(0.001\,75\,K^{-1}) - (0.001\,30\,K^{-1})} = \frac{-10.1}{0.000\,45\,K^{-1}} = -2.24 \times 10^4\,K$$

Finally, calculate the activation energy from the slope:

$$E_a = -R(\text{Slope}) = -\left(8.314\frac{J}{K \cdot mol}\right)(-2.24 \times 10^4\,K)$$
$$= 1.9 \times 10^5\,J/mol = 190\,kJ/mol$$

Note that the slope of the Arrhenius plot is negative and the activation energy is positive. The greater the activation energy for a particular reaction, the steeper the slope of the ln k versus $1/T$ plot and the greater the increase in the rate constant for a given increase in temperature.

(b) Substituting the values of $k_1 = 3.52 \times 10^{-7}\,M^{-1}\,s^{-1}$ at $T_1 = 556\,K$ (283 °C)) and $k_2 = 3.95 \times 10^{-2}\,M^{-1}\,s^{-1}$ at $T_2 = 781\,K$ (508 °C) into the equation

$$\ln\left(\frac{k_2}{k_1}\right) = \left(\frac{-E_a}{R}\right)\left(\frac{1}{T_2} - \frac{1}{T_1}\right)$$

gives

$$\ln \left(\frac{3.95 \times 10^{-2}\ M^{-1}\ s^{-1}}{3.52 \times 10^{-7}\ M^{-1}\ s^{-1}} \right) = \left(\frac{-E_a}{8.314 \dfrac{J}{K \cdot mol}} \right)\left(\frac{1}{781\ K} - \frac{1}{556\ K} \right)$$

Simplifying this equation gives

$$11.628 = \left(\frac{-E_a}{8.314 \dfrac{J}{K \cdot mol}} \right)\left(\frac{-5.18 \times 10^{-4}}{K} \right)$$

$$E_a = 1.87 \times 10^5\ J/mol = 187\ kJ/mol$$

(c) Use the same equation as in part (b), but now the known values are

$$k_1 = 3.52 \times 10^{-7}\ M^{-1}\ s^{-1}\ \text{at}\ T_1 = 556\ K\ (283\ ^\circ C)$$
$$E_a = 1.87 \times 10^5\ J/mol$$

and k_2 at $T_2 = 566\ K$ (293 °C) is the unknown.

$$\ln \left(\frac{k_2}{3.52 \times 10^{-7}\ M^{-1}\ s^{-1}} \right) = \left(\frac{-1.87 \times 10^5 \dfrac{J}{mol}}{8.314 \dfrac{J}{K \cdot mol}} \right)\left(\frac{1}{566\ K} - \frac{1}{556\ K} \right) = 0.715$$

Taking the antiln of both sides gives

$$\frac{k_2}{3.52 \times 10^{-7}\ M^{-1}\ s^{-1}} = e^{0.715} = 2.04$$

$$k_2 = 7.18 \times 10^{-7}\ M^{-1}\ s^{-1}$$

In this temperature range, a rise in temperature of 10 K doubles the rate constant.

▸ **PRACTICE 14.21**   Rate constants for the decomposition of gaseous dinitrogen pentoxide are $3.70 \times 10^{-5}\ s^{-1}$ at 25.0 °C and $1.70 \times 10^{-3}\ s^{-1}$ at 55.0 °C.

$$N_2O_5(g) \longrightarrow 2\ NO_2(g) + 1/2\ O_2(g)$$

(a) What is the activation energy for this reaction in kJ/mol?
(b) What is the rate constant at 35 °C?

▸ **APPLY 14.22**   The rate of the reaction $CH_3Br(aq) + I^-(aq) \rightarrow CH_3I(aq) + Br^-(aq)$ triples when the temperature is increased from 25 °C to 36 °C. What is the activation energy for this reaction?

14.9 REACTION MECHANISMS

In Section 14.7, we saw that many chemical reactions occur as a result of energetic collisions between properly oriented atoms and molecules. Often, reactions occur in two or more steps. The sequence of reaction steps that describes the pathway from reactants to products is called the **reaction mechanism**. Chemists want to know the sequence in which various reaction steps take place so that they can better control known reactions and predict new ones.

> **Reaction mechanism**   The sequence of reaction steps that describes the pathway from reactants to products.

A single step in a reaction mechanism is called an **elementary reaction**, or **elementary step**. To clarify the crucial distinction between an elementary reaction and an overall reaction, let's consider the gas-phase reaction of nitrogen dioxide and carbon monoxide to give nitric oxide and carbon dioxide:

$$NO_2(g) + CO(g) \longrightarrow NO(g) + CO_2(g) \quad \text{Overall reaction}$$

Experimental evidence suggests that this reaction takes place by a two-step mechanism:

Step 1. $NO_2(g) + NO_2(g) \longrightarrow NO(g) + NO_3(g)$ Elementary reaction
Step 2. $NO_3(g) + CO(g) \longrightarrow NO_2(g) + CO_2(g)$ Elementary reaction

In the first elementary step, two NO_2 molecules collide with enough energy to break one N—O bond and form another, resulting in the transfer of an oxygen atom from one NO_2 molecule to the other. In the second step, the NO_3 molecule formed in the first step collides with a CO molecule, and the transfer of an oxygen atom from NO_3 to CO yields an NO_2 molecule and a CO_2 molecule (**FIGURE 14.13**).

▶ **FIGURE 14.13**
Elementary steps in the reaction of NO_2 with CO.

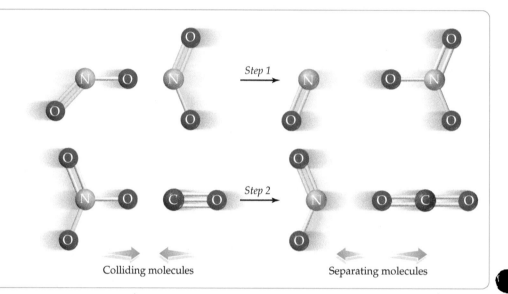

Colliding molecules   Separating molecules

The chemical equation for an elementary reaction is a description of an individual molecular event that involves breaking and/or making chemical bonds. By contrast, the balanced equation for an overall reaction describes only the stoichiometry of the overall process but provides no information about how the reaction occurs. The equation for the overall reaction of NO_2 with CO, for example, doesn't tell us that the reaction occurs by the direct transfer of an oxygen atom from an NO_2 molecule to a CO molecule.

Elementary reaction—describes an individual molecular event.
Overall reaction—describes the reaction stoichiometry.

The elementary steps in a proposed reaction mechanism must sum to give the overall reaction. When we sum the elementary steps in the reaction of NO_2 with CO and then cancel the molecules that appear on both sides of the resulting equation, for instance, we obtain the overall reaction:

Step 1. $NO_2(g) + NO_2(g) \longrightarrow NO(g) + NO_3(g)$ Elementary reaction
Step 2. $NO_3(g) + CO(g) \longrightarrow NO_2(g) + CO_2(g)$ Elementary reaction
_____
$NO_2(g) + \cancel{NO_2(g)} + \cancel{NO_3(g)} + CO(g) \longrightarrow NO(g) + \cancel{NO_3(g)} + \cancel{NO_2(g)} + CO_2(g)$
                $NO_2(g) + CO(g) \longrightarrow NO(g) + CO_2(g)$ Overall reaction

A species that is formed in one step of a reaction mechanism and consumed in a subsequent step, such as NO_3 in our example, is called a **reaction intermediate.** Reaction intermediates don't appear in the net equation for the overall reaction, and it's only by looking at the elementary steps that their presence is noticed.

Elementary reactions are classified on the basis of their **molecularity**, the number of molecules (or atoms) on the reactant side of the chemical equation. A **unimolecular reaction,**

▲ The beautiful northern lights, or aurora borealis, are often observed in the northern hemisphere at high latitudes. The light is partly produced by excited O atoms in the upper atmosphere.

for instance, is an elementary reaction that involves a single reactant molecule—for example, the unimolecular decomposition of ozone in the upper atmosphere:

$$O_3{}^*(g) \longrightarrow O_2(g) + O(g)$$

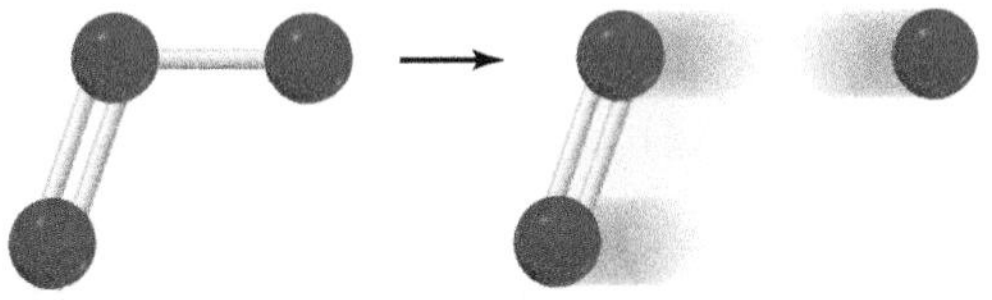

The asterisk on O_3 indicates that the ozone molecule is in an energetically excited state because it has absorbed ultraviolet light from the sun. The absorbed energy causes one of the two O—O bonds to break, with the loss of an oxygen atom.

A **bimolecular reaction** is an elementary reaction that results from an energetic collision between two reactant atoms or molecules. In the upper atmosphere, for example, an ozone molecule can react with an oxygen atom to yield two O_2 molecules:

$$O_3(g) + O(g) \longrightarrow 2\,O_2(g)$$

Both unimolecular and bimolecular reactions are common, but **termolecular reactions,** which involve three atoms or molecules, are rare. As any pool player knows, three-body collisions are much less probable than two-body collisions. There are some reactions, however, that require a three-body collision, notably the combination of two atoms to form a diatomic molecule. For example, oxygen atoms in the upper atmosphere combine as a result of collisions involving some third molecule M:

$$O(g) + O(g) + M(g) \longrightarrow O_2(g) + M(g)$$

In the atmosphere, M is most likely N_2, but in principle it could be any atom or molecule. The role of M is to carry away the energy that is released when the O—O bond is formed. If M were not involved in the collision, the two oxygen atoms would simply bounce off each other, and no reaction would occur.

WORKED EXAMPLE 14.12

Identifying Intermediates and Molecularity in a Reaction Mechanism

The following two-step mechanism has been proposed for the gas-phase decomposition of nitrous oxide (N_2O):

Step 1. $N_2O(g) \longrightarrow N_2(g) + O(g)$

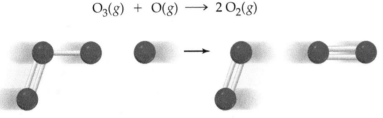

Step 2. $N_2O(g) + O(g) \longrightarrow N_2(g) + O_2(g)$

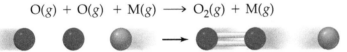

(a) Write the chemical equation for the overall reaction.
(b) Identify any reaction intermediates.
(c) What is the molecularity of each of the elementary reactions?
(d) What is the molecularity of the overall reaction?

continued on next page

STRATEGY

To find the overall reaction, sum the elementary steps. To identify intermediates and molecularity, look at the individual steps.

SOLUTION

(a) The overall reaction is the sum of the two elementary steps:

Step 1.	$N_2O(g) \longrightarrow N_2(g) + O(g)$	Elementary reaction
Step 2.	$N_2O(g) + O(g) \longrightarrow N_2(g) + O_2(g)$	Elementary reaction

$$2 N_2O(g) + \cancel{O(g)} \longrightarrow 2 N_2(g) + \cancel{O(g)} + O_2(g)$$

$$2 N_2O(g) \longrightarrow 2 N_2(g) + O_2(g) \qquad \text{Overall reaction}$$

(b) The oxygen atom is a reaction intermediate because it is formed in the first elementary step and consumed in the second step.

(c) The first elementary reaction is unimolecular because it involves a single reactant molecule. The second step is bimolecular because it involves two reactant atoms or molecules.

(d) It's inappropriate to use the word *molecularity* in connection with the overall reaction because the overall reaction does not describe an individual molecular event. The term *molecularity* only refers to elementary reactions.

▶ **PRACTICE 14.23** A suggested mechanism for the reaction of nitrogen dioxide and molecular fluorine is

Step 1. $NO_2(g) + F_2(g) \longrightarrow NO_2F(g) + F(g)$

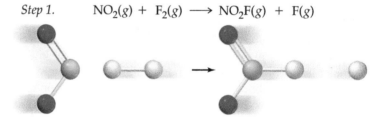

Step 2. $NO_2(g) + F(g) \longrightarrow NO_2F(g)$

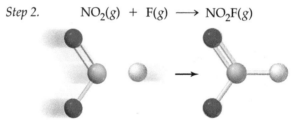

(a) Write the chemical equation for the overall reaction, and identify any reaction intermediates.

(b) What is the molecularity of each elementary reaction?

▶ **APPLY 14.24** A suggested mechanism for the decomposition of hydrogen peroxide is

Step 1. $H_2O_2(aq) \longrightarrow 2 OH(aq)$

Step 2. $H_2O_2(aq) + OH(aq) \longrightarrow H_2O(l) + HO_2(aq)$

Step 3. $HO_2(aq) + OH(aq) \longrightarrow H_2O(l) + O_2(g)$

(a) Write the chemical equation for the overall reaction, and identify reaction intermediates.

(b) What is the molecularity of each elementary reaction?

14.10 RATE LAWS FOR ELEMENTARY REACTIONS

Recall from Section 14.3 that the rate law for an overall chemical reaction must be determined experimentally. It can't be deduced from the stoichiometric coefficients in the balanced equation for the overall reaction. By contrast, the rate law for an elementary reaction follows directly from its molecularity because an elementary reaction is an individual molecular event. The concentration of each reactant in an elementary reaction appears in the rate law, with an exponent equal to its coefficient in the chemical equation for the elementary reaction.

Consider, for example, the unimolecular decomposition of ozone:

$$O_3(g) \longrightarrow O_2(g) + O(g)$$

The number of moles of O_3 per liter that decompose per unit time is directly proportional to the molar concentration of O_3:

$$\text{Rate of decomposition of } O_3 = -\frac{\Delta[O_3]}{\Delta t} = k[O_3]$$

The rate of a unimolecular reaction is always first order in the concentration of the reactant molecule.

For a bimolecular elementary reaction of the type $A + B \longrightarrow$ Products, the reaction rate depends on the frequency of collisions between A and B molecules. The frequency of AB collisions involving any *particular* A molecule is proportional to the molar concentration of B, and the total frequency of AB collisions involving *all* A molecules is proportional to the molar concentration of A times the molar concentration of B (**FIGURE 14.14**). Therefore, the reaction obeys the second-order rate law

$$\text{Rate} = k[A][B]$$

BIG IDEA Question 6

The rate law for which type of reaction can be deduced from the stoichiometry: overall reaction, elementary reaction, neither, or both?

(a) The frequency of AB collisions involving any one A molecule is proportional to the concentration of B molecules.

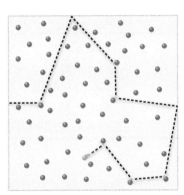

(b) Doubling the concentration of A molecules (from 1 to 2 per unit volume) doubles the total frequency of AB collisions.

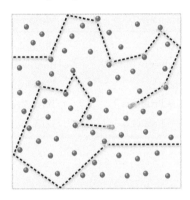

(c) Doubling the concentration of B molecules doubles the frequency of AB collisions involving any one A molecule.

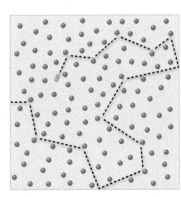

(d) Doubling the concentration of A molecules and doubling the concentration of B molecules quadruples the total frequency of AB collisions.

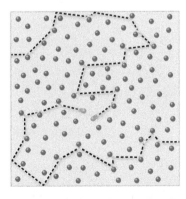

Thus, the total frequency of AB collisions is proportional to the concentration of A molecules times the concentration of B molecules.

◀ **FIGURE 14.14**

The effect of concentration on the frequency of collisions between A molecules (blue) and B molecules (red).

◀ **Figure It out**

What is the effect on collision frequency between A molecules and B molecules if the concentration of A is tripled and the concentration of B is halved?

Answer: The collision frequency increases by a factor of $3/2 = 1.5$.

An example is the conversion of bromomethane to methanol in a basic solution:

$$CH_3Br(aq) + OH^-(aq) \longrightarrow Br^-(aq) + CH_3OH(aq)$$

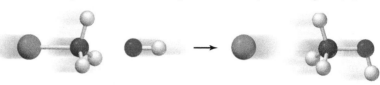

Bromomethane   Hydroxide ion   Bromide ion   Methanol

This reaction occurs in a single bimolecular step in which a new C—O bond forms at the same time as the C—Br bond breaks. The experimental rate law is

$$\text{Rate} = k[CH_3Br][OH^-]$$

By a similar line of reasoning, a bimolecular reaction of the type

$$A + A \longrightarrow \text{Products}$$

has the second-order rate law:

$$\text{Rate} = k[A][A] = k[A]^2$$

Rate laws for elementary reactions are summarized in **TABLE 14.5.** Note that the overall reaction order for an elementary reaction always equals its molecularity.

TABLE 14.5   Rate Laws for Elementary Reactions

Elementary Reaction	Molecularity	Rate Law
A → Products	Unimolecular	Rate $= k[A]$
A + A → Products	Bimolecular	Rate $= k[A]^2$
A + B → Products	Bimolecular	Rate $= k[A][B]$
A + A + B → Products	Termolecular	Rate $= k[A]^2[B]$
A + B + C → Products	Termolecular	Rate $= k[A][B][C]$

WORKED EXAMPLE 14.13

Writing Rate Laws for Elementary Reactions

The following two-step mechanism has been proposed for the gas-phase decomposition of nitrous oxide (N_2O):

Step 1. $N_2O(g) \longrightarrow N_2(g) + O(g)$

Step 2. $N_2O(g) + O(g) \longrightarrow N_2(g) + O_2(g)$

Write the rate law for the elementary reaction in *Step* 1 and *Step* 2.

STRATEGY

The molecularity of an elementary reaction determines the rate law. Unimolecular reactions are first order, bimolecular reactions are second order, and termolecular reactions are third order.

SOLUTION

Step 1. Rate $= k[N_2O]$ Step 2. Rate $= k[N_2O][O]$

▶ **PRACTICE 14.25** Write the rate law for each of the following elementary reactions:

(a) $Br(g) + Br(g) + Ar(g) \longrightarrow Br_2(g) + Ar(g)$

(b) $O_3(g) + O(g) \longrightarrow 2\ O_2(g)$

(c) $Co(CN)_5(H_2O)^{2-}(aq) \longrightarrow Co(CN)_5{}^{2-}(aq) + H_2O(l)$

▶ **APPLY 14.26** Select the statement about molecular collisions that best describes each of the elementary reactions in Problem 14.25.

(i) A reaction occurs without a collision between two reactant molecules.

(ii) A collision between two reactant molecules or atoms is required for a reaction to occur.

(iii) A simultaneous collision between three reactants is required for a reaction to occur.

14.11 RATE LAWS FOR OVERALL REACTIONS

We saw in the previous section that the rate law for a single-step, elementary reaction follows directly from its molecularity. By contrast, the rate law for a multistep, overall reaction depends on the reaction mechanism—that is, on the sequence of elementary steps and their relative rates. As we learned in Section 14.2, the rate law for the overall reaction must be determined by experiment; we'll refer to it hereafter as the observed, or experimental, rate law.

When an overall reaction occurs in two or more elementary steps, one of the steps is often much slower than the others. This slowest step in a reaction mechanism is called the **rate-determining step** because it acts as a bottleneck, limiting the rate at which reactants can be converted to products. In this respect, a chemical reaction is somewhat like using a funnel to pour a liquid into a container. The rate at which the liquid drains from the funnel depends entirely on the rate at which it flows through the narrow tube. As long as there is some liquid in the cone at the top, the rate at which it is poured in has no effect on the rate at which the liquid passes through the funnel. In this case, the rate of liquid draining from the narrow tube is the rate-determining step. The overall reaction can occur no faster than the speed of the rate-determining step.

Multistep Reactions with an Initial Slow Step

The rate-determining step might occur anywhere in the multistep sequence. In the reaction of nitrogen dioxide with carbon monoxide, for instance, the first step in the mechanism is slower and rate-determining, whereas the second step occurs more rapidly:

▲ When pouring liquid through a funnel, the rate-determining step is how fast the liquid can pass through the narrow tube.

$$NO_2(g) + NO_2(g) \xrightarrow{k_1} NO(g) + NO_3(g) \quad \text{Slower, rate-determining}$$
$$\underline{NO_3(g) + CO(g) \xrightarrow{k_2} NO_2(g) + CO_2(g) \quad \text{Faster}}$$
$$NO_2(g) + CO(g) \longrightarrow NO(g) + CO_2(g) \quad \text{Overall reaction}$$

The constants k_1 and k_2, written above the arrows in the preceding equations, are the rate constants for the elementary reactions. The rate of the overall reaction is determined by the rate of the first, slower step. In the second step, the unstable intermediate (NO_3) reacts as soon as it is formed.

Because the rate law for an overall reaction depends on the reaction mechanism, it provides important clues to the mechanism. A plausible mechanism must meet two criteria:

(1) The elementary steps must sum up to give the overall reaction.
(2) The mechanism must be consistent with the experimental rate law for the overall reaction.

For the overall reaction (reaction of NO_2 with CO), for example, the *experimental* rate law is

$$\text{Rate} = k[NO_2]^2$$

The rate law *predicted* by the proposed mechanism is that for the rate-determining step (first step) and follows directly from the molecularity of that step:

$$\text{Rate} = k_1[NO_2]^2$$

Because the experimental and predicted rate laws have the same form (second-order dependence on $[NO_2]$), the proposed mechanism is consistent with the experimental rate law. The observed rate constant k equals k_1, the rate constant for the first elementary step.

Multistep Reactions with an Initial Fast Step

In contrast to reactions in which the first step is slow and rate-determining, let's examine the proposed three-step mechanism for the reaction of nitric oxide with hydrogen:

$$2\,NO(g) \underset{k_{-1}}{\overset{k_1}{\rightleftharpoons}} N_2O_2(g) \qquad\qquad \text{Fast, reversible}$$
$$N_2O_2(g) + H_2(g) \xrightarrow{k_2} N_2O(g) + H_2O(g) \qquad \text{Slow, rate-determining}$$
$$\underline{N_2O(g) + H_2(g) \xrightarrow{k_3} N_2(g) + H_2O(g) \qquad \text{Fast}}$$
$$2\,NO(g) + 2\,H_2(g) \longrightarrow N_2(g) + 2\,H_2O(g) \quad \text{Overall reaction}$$

The first step, which is fast and reversible, produces a small concentration of the unstable intermediate, N_2O_2. This intermediate decomposes rapidly to NO in the reverse of the first step and reacts only slowly with H_2 in the second step, yielding a second intermediate, N_2O. The second step is the rate-determining step. In the third step, the N_2O reacts rapidly with H_2 to give N_2 and H_2O. On summing the three elementary steps, both intermediates cancel and we obtain the balanced equation for the overall reaction. The *predicted* rate law for the proposed mechanism is the rate law for the rate-determining step:

$$\text{Rate} = k_2[N_2O_2][H_2]$$

where k_2, is the rate constant for that step. Note that fast steps subsequent to the rate-determining step (the third step in our example) do not affect the rate of the reaction.

The *experimental* rate law for the overall reaction is

$$\text{Rate} = k[NO]^2[H_2]$$

where k is the observed rate constant for the overall reaction. To decide whether the proposed mechanism is plausible, we need to compare the experimental and predicted rate laws. The concentrations of reaction intermediates, such as N_2O_2, do not appear in the experimental rate law because their concentrations are usually very small and difficult to measure. Only reactants and products (and catalysts, if present) appear in the rate law for an overall reaction. Therefore, we must eliminate $[N_2O_2]$ from the predicted rate law, $\text{Rate} = k_2[N_2O_2][H_2]$.

To eliminate N_2O_2 from the right side of this equation, we assume that the fast, reversible, first step in the mechanism reaches a dynamic equilibrium. (We'll have more to say about chemical equilibrium in Chapter 15.) The rates of the forward and reverse reactions in the fast, reversible first step are given by

$$\text{Rate}_{\text{forward}} = k_1[NO]^2 \quad \text{Rate}_{\text{reverse}} = k_{-1}[N_2O_2]$$

Just as the rates of vaporization and condensation are equal for a liquid–vapor equilibrium (Section 11.3), the rates of the forward and reverse reactions are equal for a chemical equilibrium. Therefore,

$$k_1[NO]^2 = k_{-1}[N_2O_2] \quad \text{and} \quad [N_2O_2] = \frac{k_1}{k_{-1}}[NO]^2$$

Substituting this expression for $[N_2O_2]$ into the equation for the predicted rate law eliminates the intermediate and gives the final predicted rate law in terms of only reactants and products:

$$\text{Rate} = k_2[N_2O_2][H_2] = k_2\frac{k_1}{k_{-1}}[NO]^2[H_2]$$

The predicted and experimental rate laws now have the same form: Both are second order in NO and first order in H_2. Therefore, the proposed mechanism is consistent with the experiment and is a plausible mechanism for the reaction. Comparison of the two rate laws indicates that the observed rate constant k equals k_2k_1/k_{-1}.

Procedure Used in Studies of Reaction Mechanisms

Let's summarize the procedure that chemists use in establishing a reaction mechanism. First, determine the rate law by experiment. Then propose a series of elementary steps and work out the rate law predicted by the proposed mechanism. If the observed and predicted rate laws do not agree, discard the proposed mechanism, and devise another one. If the observed and predicted rate laws *do* agree, the proposed mechanism is a plausible (though not necessarily correct) pathway for the reaction. **FIGURE 14.15** summarizes the procedure.

LOOKING AHEAD . . .

In Chapter 15, we'll see that **chemical equilibrium** is reached when the rates of the forward and reverse reactions are equal.

REMEMBER . . .

In a **liquid–vapor equilibrium** the number of molecules escaping from the liquid per unit time (vaporization) equals the number returning to the liquid per unit time (condensation) (Section 11.3).

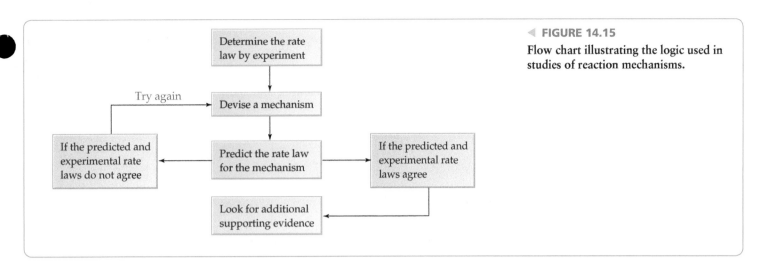

◀ **FIGURE 14.15**
Flow chart illustrating the logic used in studies of reaction mechanisms.

The case for a particular mechanism is strengthened considerably if we can isolate a reaction intermediate or detect an unstable intermediate. It's easy to disprove a mechanism, but it's seldom possible to finally prove a mechanism because there may be an alternative reaction pathway—not yet imagined—that also fits the experimental facts. The best we can do to establish a mechanism is to accumulate a convincing body of experimental evidence that supports it. Proving a reaction mechanism is more like proving a case in a court of law than like proving a theorem in mathematics.

WORKED EXAMPLE 14.14

Suggesting a Mechanism Given the Rate Law: Reactions with an Initial Slow Step

The following reaction has a second-order rate law:

$$H_2(g) + 2\,ICl(g) \longrightarrow I_2(g) + 2\,HCl(g) \quad Rate = k[H_2][ICl]$$

Devise a possible reaction mechanism.

STRATEGY

The reaction doesn't occur in a single elementary step because, if it did, the rate law would be third order: $Rate = k[H_2][ICl]^2$. The observed rate law will be obtained if the rate-determining step involves the bimolecular reaction of H_2 and ICl.

SOLUTION

A plausible sequence of elementary steps is

$$H_2(g) + ICl(g) \xrightarrow{k_1} HI(g) + HCl(g) \quad \text{Slower, rate-determining}$$
$$\underline{HI(g) + ICl(g) \xrightarrow{k_2} I_2(g) + HCl(g) \quad \text{Faster}}$$
$$H_2(g) + 2\,ICl(g) \longrightarrow I_2(g) + 2\,HCl(g) \quad \text{Overall reaction}$$

The rate law predicted by this mechanism, rate $= k_1[H_2][ICl]$, agrees with the observed rate law.

▶ **PRACTICE 14.27** The following reaction has a second-order rate law:

$$2\,NO(g) + Cl_2(g) \longrightarrow 2\,NOCl(g) \quad Rate = k[NO][Cl_2]$$

Devise a possible reaction mechanism.

continued on next page

▶ **APPLY 14.28** The following reaction has a first-order rate law:

$$Co(CN)_5(H_2O)^{2-}(aq) + I^-(aq) \longrightarrow Co(CN)_5I^{3-}(aq) + H_2O(l)$$

$$Rate = k[Co(CN)_5(H_2O)^{2-}]$$

Suggest a possible reaction mechanism, and show that your mechanism agrees with the observed rate law.

WORKED EXAMPLE 14.15

Supporting a Mechanism Given the Rate Law: Reactions with a Fast Initial Step

The experimental rate law for the decomposition of ozone is second order in ozone and inverse first order in molecular oxygen:

$$2 O_3(g) \longrightarrow 3 O_2(g) \quad Rate = k\frac{[O_3]^2}{[O_2]}$$

Show that the following mechanism is consistent with the experimental rate law, and relate the observed rate constant k to the rate constants for the elementary reactions:

$$O_3(g) \underset{k_{-1}}{\overset{k_1}{\rightleftharpoons}} O_2(g) + O(g) \quad \text{Fast, reversible}$$

$$\underline{O(g) + O_3(g) \overset{k_2}{\longrightarrow} 2 O_2(g) \qquad \text{Slow, rate-determining}}$$

$$2 O_3(g) \longrightarrow 3 O_2(g) \qquad \text{Overall reaction}$$

STRATEGY

To show that the mechanism is consistent with the experiment, we must derive the rate law predicted by the mechanism and compare it with the experimental rate law. If we assume that the faster, reversible step is at equilibrium, we can eliminate the concentration of the intermediate in the rate law for the rate-determining step.

SOLUTION

The first step in determining the rate law from the mechanism is to write the rate law for the rate-determining step:

$$Rate = k_2[O][O_3]$$

However, the O atom in the rate law is an intermediate. Therefore, the first fast equilibrium step must be used to find [O] in terms of the reactant O_3.

The rates of the forward and reverse reactions in the faster, reversible step are given by

$$Rate_{forward} = k_1[O_3] \quad Rate_{reverse} = k_{-1}[O_2][O]$$

Assuming that the first step is at equilibrium, we can equate the rates of the forward and reverse reactions and then solve for the concentration of the intermediate O atoms:

$$k_1[O_3] = k_{-1}[O_2][O] \quad \text{so, } [O] = \frac{k_1[O_3]}{k_{-1}[O_2]}$$

Substituting this expression for [O] into the predicted rate law for the overall reaction gives the predicted rate law in terms of only reactants and products:

$$Rate = k_2[O][O_3] = k_2\frac{k_1}{k_{-1}}\frac{[O_3]^2}{[O_2]}$$

Because the predicted and experimental rate laws have the same reaction orders in O_3 and O_2, the proposed mechanism is consistent with the experimental rate law and is a plausible mechanism

for the reaction. Comparison of the predicted and experimental rate laws indicates that the observed rate constant k equals k_2k_1/k_{-1}.

▶ **PRACTICE 14.29** The following mechanism has been proposed for the oxidation of nitric oxide to nitrogen dioxide:

$$NO(g) + O_2(g) \underset{k_{-1}}{\overset{k_1}{\rightleftharpoons}} NO_3(g) \qquad \text{Faster, reversible}$$

$$NO_3(g) + NO(g) \overset{k_2}{\longrightarrow} 2 NO_2(g) \quad \text{Slower, rate-determining}$$

The experimental rate law for the overall reaction is

$$Rate = k[NO]^2[O_2]$$

What is the rate law derived from the mechanism? (Relate the rate constant for the mechanism to the rate constants for the elementary reactions.)

▶ **APPLY 14.30** The following mechanism has been proposed for the reaction of hydrogen gas with iodine gas to produce hydrogen iodide.

$$I_2(g) \underset{k_{-1}}{\overset{k_1}{\rightleftharpoons}} 2I(g) \qquad\qquad \text{Faster, reversible}$$

$$H_2(g) + I(g) \underset{k_{-2}}{\overset{k_2}{\rightleftharpoons}} H_2I(g) \qquad \text{Faster, reversible}$$

$$H_2I(g) + I(g) \overset{k_3}{\longrightarrow} 2 HI(g) \qquad \text{Slower, rate-determining}$$

The experimental rate law for the overall reaction is

$$Rate = k[H_2][I_2]$$

(a) Write a balanced equation for the overall reaction.
(b) Show that the proposed mechanism is consistent with the experimental rate law.
(c) Relate the rate constant k to the rate constants for the elementary reactions.

14.12 CATALYSIS

Reaction rates are affected not only by reactant concentrations and temperature but also by the presence of *catalysts*. A **catalyst** is a substance that increases the rate of a transformation without itself being consumed in the process. An example is manganese dioxide, a black powder that speeds up the thermal decomposition of potassium chlorate:

$$2 \text{ KClO}_3(s) \xrightarrow[\text{Heat}]{\text{MnO}_2 \text{ catalyst}} 2 \text{ KCl}(s) + 3 \text{ O}_2(g)$$

In the absence of a catalyst, $KClO_3$ decomposes very slowly, even when heated, but when a small amount of MnO_2 is mixed with the $KClO_3$ before heating, rapid evolution of oxygen ensues. The MnO_2 can be recovered unchanged after the reaction is complete.

Catalysts are enormously important, both in the chemical industry and in living organisms. Nearly all industrial processes for the manufacture of essential chemicals use catalysts to favor formation of specific products and to lower reaction temperatures, thus reducing energy costs. In environmental chemistry, catalysts such as nitric oxide play a role in the formation of air pollutants, while other catalysts, such as platinum in automobile catalytic converters, are potent weapons in the battle to control air pollution. In living organisms, almost all the hundreds of thousands of chemical reactions that take place constantly are catalyzed by large molecules called *enzymes*. Nitrogenase, for example, an enzyme present in bacteria on the root nodules of leguminous plants such as peas and beans, catalyzes the conversion of atmospheric nitrogen to ammonia. The ammonia then serves as a fertilizer for plant growth.

How does a catalyst work? A catalyst accelerates the rate of a reaction by making available a different, lower-energy mechanism for the conversion of reactants to products. Take the decomposition of hydrogen peroxide in a basic, aqueous solution, for instance:

$$2 \text{ H}_2\text{O}_2(aq) \longrightarrow 2 \text{ H}_2\text{O}(l) + \text{O}_2(g)$$

Although unstable with respect to water and oxygen, hydrogen peroxide decomposes only very slowly at room temperature because the reaction has a high activation energy (76 kJ/mol). In the presence of iodide ion, however, the reaction is appreciably faster (**FIGURE 14.16**) because it can proceed by a different, lower-energy pathway:

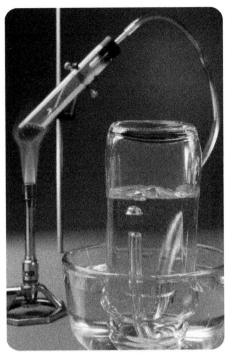

▲ MnO_2 catalyzes the thermal decomposition of $KClO_3$ to KCl and O_2

BIG IDEA Question 7

Why does the rate of a chemical reaction increase when a catalyst is added?

Step 1.	$H_2O_2(aq) + I^-(aq) \longrightarrow H_2O(l) + IO^-(aq)$	Slower, rate determining
Step 2.	$H_2O_2(aq) + IO^-(aq) \longrightarrow H_2O(l) + O_2(g) + I^-(aq)$	Faster
	$2 H_2O_2(aq) \longrightarrow 2 H_2O(l) + O_2(g)$	Overall reaction

The H_2O_2 first oxidizes the catalyst (I^-) to hypoiodite ion (IO^-) and then reduces the intermediate IO^- back to I^-. The catalyst does not appear in the overall reaction because it is consumed in step 1 and regenerated in step 2. The catalyst is, however, intimately involved in the reaction and appears in the observed rate law:

$$\text{Rate} = k[\text{H}_2\text{O}_2][\text{I}^-]$$

The rate law is consistent with the reaction of H_2O_2 and I^- as the rate-determining step. In general, a catalyst is consumed in one step of a reaction and is regenerated in a subsequent step, whereas an intermediate is formed in one step and is consumed in a subsequent step.

The catalyzed pathway for a reaction might have a faster rate than the uncatalyzed pathway either because of a larger frequency factor (A) or a smaller activation energy (E_a) in the Arrhenius equation. Usually, though, catalysts function by making available a reaction pathway with a lower activation energy (**FIGURE 14.17**). In the decomposition of hydrogen peroxide, for example, catalysis by I^- lowers E_a for the overall reaction by 19 kJ/mol.

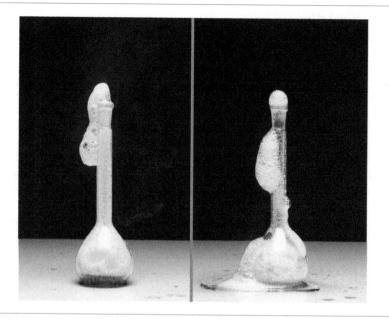

▶ **FIGURE 14.16**

▶ **FIGURE 14.16**

Catalytic decomposition of hydrogen peroxide. A 30% hydrogen peroxide solution (H_2O_2) undergoes a very slow decomposition into water (H_2O) and oxygen gas (O_2) under normal conditions. The rate of decomposition is greatly increased by addition of powdered solid potassium iodide (KI) which acts as a catalyst.

Because the reaction occurs in two steps, the energy profile for the catalyzed pathway shown in Figure 14.17 exhibits two maxima (two transition states) with a minimum between them that represents the energy of the intermediate species present after the first step. The first maximum is higher than the second because the first step is rate-determining, and the activation energy for the overall reaction is E_a for the first step. Maxima for both steps, though, are lower than E_a for the uncatalyzed pathway. Note that a catalyst does not affect the energies of the reactants and products, which are the same for both catalyzed and uncatalyzed pathways.

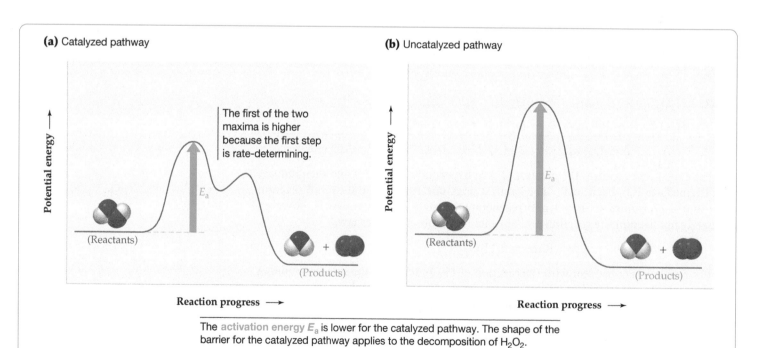

▲ **FIGURE 14.17**

Typical potential energy profiles for a reaction whose activation energy is lowered by the presence of a catalyst.

CONCEPTUAL WORKED EXAMPLE 14.16

Using the Method of Initial Rates for a Catalyzed Reaction

The relative rates of the reaction $A + B \rightarrow AB$ in vessels (1)—(4) are 1:2:1:2. Red spheres represent A molecules, blue spheres represent B molecules, and yellow spheres represent molecules of a third substance C.

(1)   **(2)**   **(3)**   **(4)**

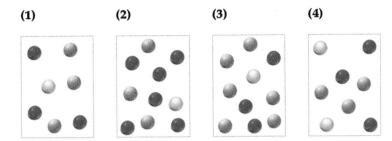

(a) What is the order of the reaction in A, B, and C?
(b) Write the rate law.
(c) Write a mechanism that agrees with the rate law.
(d) Why doesn't C appear in the equation for the overall reaction?

STRATEGY AND SOLUTION

(a) Count the number of molecules of each type in vessels (1)–(4), and compare the relative rates with the relative numbers of molecules. The concentration of A molecules in vessel (2) is twice that in vessel (1), while the concentrations of B and C remain constant. Because the reaction rate in vessel (2) is twice that in vessel (1), the rate is proportional to [A], and so the reaction is first order in A. When [B] is doubled [compare vessels (1) and (3)], the rate is unchanged, so the reaction is zeroth order in B. When [C] is doubled [compare vessels (1) and (4)], the rate doubles, so the reaction is first order in C.

(b) The rate law can be written as rate $= k[A]^m[B]^n[C]^p$, where the exponents m, n, and p specify the reaction orders in A, B, and C, respectively. Since the reaction is first order in A and C, and zeroth order in B, the rate law is rate $= k[A][C]$.

(c) The rate law tells us that A and C collide in the rate-determining step because the rate law for the overall reaction is the rate law for the rate-determining step. Subsequent steps in the mechanism are faster than the rate-determining step, and the various steps must sum up to give the overall reaction. Therefore, a plausible mechanism is

$$
\begin{array}{ll}
A + C \longrightarrow AC & \text{Slower, rate-determining} \\
\underline{AC + B \longrightarrow AB + C} & \text{Faster} \\
A + B \longrightarrow AB & \text{Overall reaction}
\end{array}
$$

(d) C doesn't appear in the overall reaction because it is consumed in the first step and regenerated in the second step. C is therefore a catalyst. AC is an intermediate because it is formed in the first step and consumed in the second step.

▶ **PRACTICE 14.31** The relative rates of the reaction $2 A + C_2 \rightarrow 2 AC$ in vessels (1)–(4) are 1:1:2:3. Red spheres represent A molecules, blue spheres represent B molecules, and connected yellow spheres represent C_2 molecules.

(1)   **(2)**   **(3)**   **(4)**

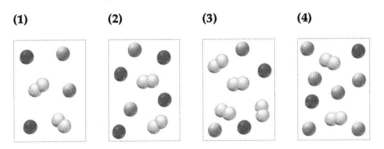

(a) What is the rate law?
(b) Does the following mechanism agree with the rate law?

$$
\begin{array}{lll}
\text{Step 1.} & C_2 + B \longrightarrow C + CB & \text{Slower, rate-determining} \\
\text{Step 2.} & C + A \longrightarrow AC & \text{Faster} \\
\text{Step 3.} & CB + A \longrightarrow AC + B & \text{Faster}
\end{array}
$$

(c) Identify all catalysts and intermediates in the mechanism.

▶ **APPLY 14.32**

(a) Draw a potential energy diagram for the mechanism proposed in Problem 14.31. Step 1 is endothermic, Step 2 is endothermic, and the overall reaction is exothermic. Label reactants, products, and activation energy.

(b) Propose a different mechanism that agrees with the rate law for Problem 14.31. Identify all catalysts and intermediates in the mechanism.

14.13 HOMOGENEOUS AND HETEROGENEOUS CATALYSTS

Catalysts are commonly classified as either *homogeneous* or *heterogeneous*. A **homogeneous catalyst** is one that exists in the same phase as the reactants. For example, iodide ion is a homogeneous catalyst for the decomposition of aqueous hydrogen peroxide because both I^- and H_2O_2 are present in the same aqueous solution phase.

In the atmosphere, nitric oxide is a gas-phase homogeneous catalyst for the conversion of molecular oxygen to ozone (O_3), a process described by the following series of reactions:

$$1/2\ O_2(g) + NO(g) \longrightarrow NO_2(g)$$
$$NO_2(g) \xrightarrow{\text{Sunlight}} NO(g) + O(g)$$
$$\underline{O(g) + O_2(g) \longrightarrow O_3(g)}$$
$$3/2\ O_2(g) \longrightarrow O_3(g) \qquad \text{Overall reaction}$$

Nitric oxide first reacts with atmospheric O_2 to give nitrogen dioxide (NO_2), a poisonous brown gas. Subsequently, NO_2 absorbs sunlight and dissociates to give an oxygen atom, which then reacts with O_2 to form ozone. As usual, the catalyst (NO) and the intermediates (NO_2 and O) do not appear in the chemical equation for the overall reaction.

A **heterogeneous catalyst** is one that exists in a different phase from that of the reactants. Ordinarily, the heterogeneous catalyst is a solid, and the reactants are either gases or liquids. In the Fischer–Tropsch process for manufacturing synthetic gasoline, for example, tiny particles of a metal such as iron or cobalt coated on alumina (Al_2O_3) catalyze the conversion of gaseous carbon monoxide and hydrogen to hydrocarbons such as octane (C_8H_{18}):

$$8\ CO(g) + 17\ H_2(g) \xrightarrow[\text{catalyst}]{\text{Co/Al}_2\text{O}_3} C_8H_{18}(l) + 8\ H_2O(l)$$

The mechanism of heterogeneous catalysis is often complex and not well understood. Important steps, however, frequently involve (1) attachment of reactants to the surface of the catalyst, a process called *adsorption*; (2) conversion of reactants to products on the surface; and (3) release of products from the surface, a process called desorption. The adsorption step is thought to involve chemical bonding of the reactants to the highly reactive metal atoms on the surface with accompanying breaking, or at least weakening, of bonds in the reactants.

To illustrate, take the catalytic hydrogenation of compounds with $C=C$ double bonds, a reaction used in the food industry to convert unsaturated vegetable oils to solid fats. The reaction is catalyzed by tiny particles of metals such as Ni, Pd, or Pt:

$$\underset{\text{Ethylene}}{H_2C{=}CH_2(g)} + H_2(g) \xrightarrow[\text{catalyst}]{\text{Metal}} \underset{\text{Ethane}}{H_3C{-}CH_3(g)}$$

As shown in **FIGURE 14.18,** the function of the metal surface is to adsorb the reactants and facilitate the rate-determining step by breaking the strong $H-H$ bond in the H_2 molecule. Because the $H-H$ bond breaking is accompanied by the simultaneous formation of bonds from the separating H atoms to the surface metal atoms, the activation energy for the process is lowered. The H atoms then move about on the surface until they encounter the C atoms of the adsorbed C_2H_4 molecule. Subsequent stepwise formation of two new $C-H$ bonds gives C_2H_6, which is finally desorbed from the surface.

Most of the catalysts used in industrial chemical processes are heterogeneous, in part because such catalysts can be easily separated from the reaction products. **TABLE 14.6** lists some commercial processes that employ heterogeneous catalysts.

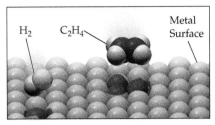

(a) H_2 and C_2H_4 are adsorbed on the metal surface.

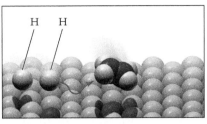

(b) The H–H bond breaks as H–metal bonds form, and the H atoms move about on the surface.

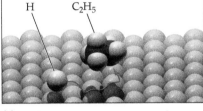

(c) One H atom forms a bond to a C atom of the adsorbed C_2H_4 to give a metal-bonded C_2H_5 group. A second H atom bonds to the C_2H_5 group.

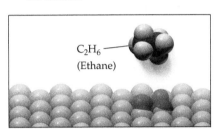

(d) The resulting C_2H_6 molecule is desorbed from the surface.

◀ **FIGURE 14.18**

Proposed mechanism for the catalytic hydrogenation of ethylene (C_2H_4) on a metal surface.

◀ **Figure It Out**

How does the catalytic surface lower the activation energy for the hydrogenation of ethylene?

Answer: In step (a), H_2 adsorbs on the surface which weakens the H—H bond.

TABLE 14.6   Some Heterogeneous Catalysts Used in Commercially Important Reactions

Reaction	Catalyst	Commercial Process	Product: Commercial Uses
$2\,SO_2 + O_2 \longrightarrow 2\,SO_3$	Pt or V_2O_5	Intermediate step in the contact process for synthesis of sulfuric acid	H_2SO_4: Manufacture of fertilizers, chemicals; oil refining
$4\,NH_3 + 5O_2 \longrightarrow 4\,NO + 6H_2O$	Pt and Rh	First step in the Ostwald process for synthesis of nitric acid	HNO_3: Manufacture of explosives, fertilizers, plastics, dyes, lacquers
$N_2 + 3H_2 \longrightarrow 2\,NH_3$	Fe, K_2O, and Al_2O_3	Haber process for synthesis of ammonia	NH_3: Manufacture of fertilizers, nitric acid
$H_2O + CH_4 \longrightarrow CO + 3\,H_2$	Ni	Steam–hydrocarbon re-forming process for synthesis of hydrogen	H_2: Manufacture of ammonia, methanol
$CO + H_2O \longrightarrow CO_2 + H_2$	ZnO and CuO	Water–gas shift reaction to improve yield in the synthesis of H_2	H_2: Manufacture of ammonia and methanol and generation of hydrogen in fuel cells
$CO + 2H_2 \longrightarrow CH_3OH$	Cu, ZnO, and Al_2O_3	Industrial synthesis of methanol	CH_3OH: Manufacture of plastics, adhesives, gasoline additives; industrial solvent
$\diagdown\!\!\diagup C{=}C\diagup\!\!\diagdown + H_2 \longrightarrow \diagdown\!\!\overset{H}{\underset{}{C}}{-}\overset{H}{\underset{}{C}}\diagup$	Ni, Pd, or Pt	Catalytic hydrogenation of compounds with C=C bonds as in conversion of unsaturated vegetable oils to solid fats	Food products: margarine, shortening

Another important application of heterogeneous catalysts is in automobile catalytic converters. Despite much work on engine design and fuel composition, automotive exhaust emissions contain air pollutants such as unburned hydrocarbons (C_xH_y), carbon monoxide, and nitric oxide. Carbon monoxide results from the incomplete combustion of hydrocarbon fuels, and nitric oxide is produced when atmospheric nitrogen and oxygen combine at the high temperatures present in an automobile engine. Catalytic converters help convert the offending pollutants to carbon dioxide, water, nitrogen, and oxygen (**FIGURE 14.19**):

$$C_xH_y(g) + (x + y/4)\, O_2(g) \longrightarrow x\, CO_2(g) + (y/2)\, H_2O(g)$$

$$2\, CO(g) + O_2(g) \longrightarrow 2\, CO_2(g)$$

$$2\, NO(g) \longrightarrow N_2(g) + O_2(g)$$

Typical catalysts for these reactions are the so-called noble metals Pt, Pd, and Rh and the transition metal oxides V_2O_5, Cr_2O_3, and CuO. The surface of the catalyst is rendered ineffective, or poisoned, by the adsorption of lead, which is why automobiles with catalytic converters (all those built since 1975) use unleaded gasoline.

▶ **FIGURE 14.19**

An automobile catalytic converter. The gases exhausted from an automobile engine pass through a catalytic converter, where air pollutants such as unburned hydrocarbons (C_xH_y), CO, and NO are converted to CO_2, H_2O, N_2, and O_2. The catalytic converter contains beads that are impregnated with the heterogeneous catalyst.

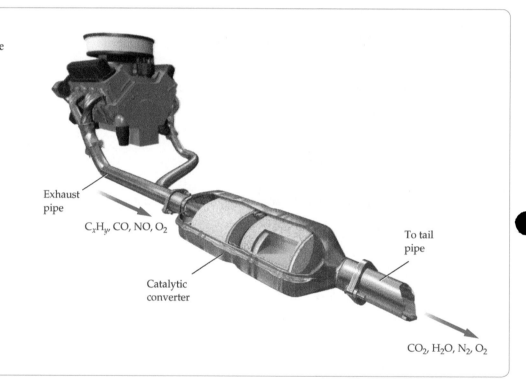

Exhaust pipe

C_xH_y, CO, NO, O_2

Catalytic converter

To tail pipe

CO_2, H_2O, N_2, O_2

INQUIRY ?   How do enzymes work?

Enzymes are large protein molecules that act as catalysts for biological reactions. They differ from simple inorganic catalysts in two important respects. First, enzymes have much larger and more complex structures than inorganic catalysts, with molar masses ranging from about 10^4 to greater than 10^6 g/mol. Second, enzymes are far more specific in their action than inorganic catalysts, often catalyzing only a single reaction of a single compound, called the enzyme's *substrate*. For example, the enzyme *amylase* found in human digestive systems is able to catalyze the breakdown of starch to yield glucose but has no effect on cellulose, even though starch and cellulose are structurally similar. Thus, humans can digest potatoes (starch) but not grass (cellulose).

$$\text{Starch} + H_2O \xrightarrow{\text{Amylase}} \text{Many glucose molecules}$$

$$\text{Cellulose} + H_2O \xrightarrow{\text{Amylase}} \text{No reaction}$$

The catalytic activity of an enzyme is measured by its *turnover number,* which is defined as the number of substrate molecules acted on by one molecule of enzyme per second. Most enzymes have turnover numbers in the range of 1–20,000, but some have much higher values. Carbonic anhydrase, which catalyzes the reaction of CO_2 with water to yield HCO_3^- ion, acts on *600,000* substrate molecules per second:

$$CO_2(aq) + H_2O(l) \rightleftharpoons H^+(aq) + HCO_3^-(aq)$$

The forward reaction occurs when the blood takes up CO_2 in the tissues, and the reverse reaction occurs when the blood releases CO_2 in the lungs. Remarkably, carbonic anhydrase increases the rate of these reactions by a factor of about 10^6!

How do enzymes work? According to the *lock-and-key model,* an enzyme is pictured as a large, irregularly shaped molecule with a cleft, or crevice, in its middle. Inside the crevice is an *active site,* a small region with the shape and chemical composition necessary to bind the substrate and catalyze the appropriate reaction. In other words, the active site acts like a lock into which only a specific key (substrate) can fit (**FIGURE 14.20**). An enzyme's active site is lined by various acidic, basic, and neutral amino acid side chains, all properly positioned for maximum interaction with the substrate.

Enzyme-catalyzed reactions begin when the substrate migrates into the active site to form an *enzyme–substrate complex.* Often, no covalent bonds are formed; the enzyme and substrate are held together only by hydrogen bonds and by weak intermolecular attractions. When enzyme and substrate are held together in a precisely defined arrangement, the appropriately positioned atoms in the active site facilitate a chemical reaction of the substrate molecule, and the enzyme plus the product then separate.

The reaction can be described by the following simplified mechanism:

$$E + S \underset{k_{-1}}{\overset{k_1}{\rightleftharpoons}} ES$$

$$ES \xrightarrow{k_2} E + P$$

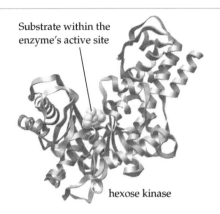

Substrate within the enzyme's active site

hexose kinase

▲ **FIGURE 14.20**

An enzyme's active site. An enzyme is a large, three-dimensional molecule containing a crevice with an active site. Only a substrate whose shape and structure are complementary to those of the active site can fit into the enzyme. This computer-generated structure depicts the active site of the enzyme hexose kinase. Hexose kinase catalyzes a reaction that adds a phosphoryl group $(-PO_3^{2-})$ to six-carbon sugar molecules, an important step in producing ATP to store energy in the body.

where E, S, and P represent the enzyme, substrate, and product, respectively, and ES represents the enzyme–substrate complex. The rate of product formation is given by

$$\frac{\Delta[P]}{\Delta t} = k_2[ES]$$

At low concentrations of the substrate, the concentration of the enzyme–substrate complex is proportional to the concentration of the substrate, and the rate of product formation is therefore first order in S. At high concentrations of the substrate, however, the enzyme becomes saturated with the substrate–that is, all the enzyme is in the form of the enzyme–substrate complex. At that point, the reaction rate reaches a maximum value and becomes independent of the concentration of the substrate (zeroth order in S) because only substrate bound to the enzyme can react. The dependence of the rate on substrate concentration is shown in **FIGURE 14.21**.

PROBLEM 14.33   Is an enzyme a homogeneous or heterogeneous catalyst?

PROBLEM 14.34   Given the mechanism for an enzyme-catalyzed reaction

$$E + S \underset{k_{-1}}{\overset{k_1}{\rightleftharpoons}} ES \qquad \text{Faster, reversible}$$

$$ES \xrightarrow{k_2} E + P \qquad \text{Slower, rate-determinning}$$

(a) Write a balanced equation for the overall reaction.
(b) Identify any catalysts and intermediates.

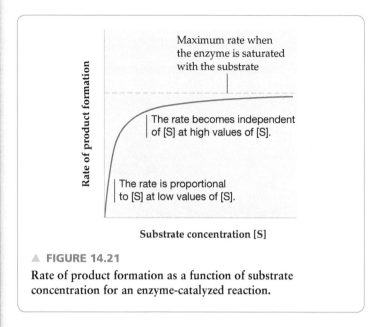

Maximum rate when the enzyme is saturated with the substrate

The rate becomes independent of [S] at high values of [S].

The rate is proportional to [S] at low values of [S].

Substrate concentration [S]

▲ **FIGURE 14.21**

Rate of product formation as a function of substrate concentration for an enzyme-catalyzed reaction.

(c) Determine the rate law predicted from the mechanism. Give the rate law in terms of reactant and catalyst. Do not include intermediates in the rate law.

(d) Relate the rate constant k to the rate constants for the elementary reactions.

PROBLEM 14.35 Sketch a potential energy profile for the enzyme-catalyzed reaction mechanism in Problem 14.34, and label the appropriate parts of your graph with E, S, P, ES, and the activation energy E_a. On the same graph, sketch a potential energy profile for the uncatalyzed reaction.

PROBLEM 14.36 At low substrate concentrations, the rate of product formation is first order in S. By what factor does the rate of an enzyme-catalyzed reaction change when the substrate concentration is changed from 1.4×10^{-5} M to 3.4×10^{-5} M?

PROBLEM 14.37 At high substrate concentrations, the rate of product formation is zeroth order in S.

(a) By what factor does the rate of an enzyme-catalyzed reaction change when the substrate concentration is changed from 2.8×10^{-3} M to 4.8×10^{-3} M?

(b) Why is the enzyme-catalyzed reaction zeroth order in S at high substrate concentrations?

PROBLEM 14.38 Chymotrypsin is a digestive enzyme component of pancreatic juice that acts to break down proteins. The rate constant for the reaction that occurs when chymotrypsin reacts with its substrate N-acetylvaline ethyl ester is 1.7×10^{-1} s^{-1}.

(a) What is the half-life for this reaction?

(b) How long does it take for an N-acetylvaline ethyl ester concentration of 5.5×10^{-3} M to reach a concentration of 1.5×10^{-4} M?

STUDY GUIDE

Section	Concept Summary	Learning Objectives	Test Your Understanding
14.1 Reaction Rates	**Chemical kinetics** is the area of chemistry concerned with reaction rates. A **reaction rate** is defined as the increase in the concentration of a product or the decrease in the concentration of a reactant per unit time. It can be expressed as the average rate during a given time interval, the **instantaneous rate** at a particular time, or the **initial rate** at the beginning of the reaction.	**14.1** Determine the average reaction rate over a specified period of time and estimate instantaneous reaction rate from a graph of concentration versus time.	Problems 14.39, 14.50, 14.52, 14.54
		14.2 Relate the rate of consumption of any reactant to the rate of formation of any product using reaction stoichiometry.	Worked Example 14.1; Problems 14.56, 14.58
14.2–14.3 Rate Laws, Reaction Order, and Method of Initial Rates	Reaction rates depend on reactant concentrations, temperature, and the presence of catalysts. The concentration dependence is given by the **rate law**, rate $= k[A]^m[B]^n$, where k is the **rate constant**, m and n specify the **reaction order** with respect to reactants A and B, and $m + n$ is the overall reaction order. The values of m and n must be determined by experiment; they cannot be deduced from the stoichiometry of the overall reaction. One experimental method for determining the rate law is the method of initial rates, which involves measuring the change in rate that occurs when the concentration of one reactant changes.	**14.3** Find the order of a reaction with respect to each reactant, the overall order of a reaction, and the units of the rate constant, given the rate law.	Worked Example 14.2; Problems 14.60–14.61
		14.4 Predict the change in reaction rate when the concentration of a reactant changes by a specified amount.	Problems 14.40, 14.62, 14.64
		14.5 Determine the rate law and rate constant using initial rate and concentration data.	Worked Examples 14.3–14.4; Problems 14.66–14.68
14.4–14.6 Integrated Rate Laws for Zeroth-, First-, and Second-Order Reactions	The **integrated rate law** is a concentration–time equation that lets us calculate concentrations at any time t or the time required for an initial concentration to reach any particular value. For a **zeroth-order reaction,** the integrated rate law is $[A]_t = -kt + [A]_0$, and a graph of [A] versus time is linear with a slope equal to $-k$. For a **first-order reaction,** the integrated rate law is ln $[A]_t = -kt + \ln [A]_0$. A graph of ln [A] versus time is a straight line with a slope equal to $-k$. For a **second-order reaction,** the integrated rate law is $1/[A]_t = kt + 1/[A]_0$. A graph of 1/[A] versus time is linear with a slope equal to k. The **half-life** $(t_{1/2})$ of a reaction is the time required for the reactant concentration to drop to one-half its initial value. The equation for half-life for an order reaction can be derived from the integrated rate law.	**14.6** Use the integrated rate law to determine the half-life and the concentrations remaining at various times for a zeroth-order reaction.	Worked Example 14.5; Problems 14.9–14.10
		14.7 Use the integrated rate law to determine the half-life and the concentrations remaining at various times for a first-order reaction.	Worked Examples 14.6, 14.8; Problems 14.41–14.42, 14.70, 14.72, 14.79, 14.85
		14.8 Use the integrated rate law to determine the half-life and the concentrations remaining at various times for a second-order reaction.	Worked Example 14.9; Problems 14.74–14.77
		14.9 Determine the reaction order and rate constant graphically.	Worked Examples 14.7, 14.9; Problems 14.81–14.83
14.7–14.8 Reaction Rates and Temperature: The Arrhenius Equation	The temperature dependence of rate constants is described by the **Arrhenius equation,** $k = Ae^{-E_a/RT}$, where A is the **frequency factor** and E_a is the **activation energy.** The value of E_a can be determined from the slope of a linear plot of ln k versus $1/T$, and it can be interpreted as the height of the potential energy barrier between the reactants and products. The configuration of atoms at the top of the barrier is called the **transition state.** According to **collision theory,** the rate constant is given by $k = pZe^{-E_a/RT}$, where p is a **steric factor** (the fraction of collisions in which the molecules have the proper orientation for reaction), Z is a constant related to the collision frequency, and $e^{-E_a/RT}$ is the fraction of collisions with energy equal to or greater than E_a.	**14.10** Interpret potential energy diagrams and suggest geometries for successful collisions and transition states using collision theory.	Worked Example 14.10; Problems 14.90–14.92
		14.11 Use the Arrhenius equation to calculate the activation energy of a reaction given rate constants at different temperatures.	Worked Example 14.11; Problems 14.94, 14.96, 14.98, 14.102

Section	Concept Summary	Learning Objectives	Test Your Understanding
14.9–14.10 Reaction Mechanisms and Rate Laws for Elementary Reactions	A **reaction mechanism** is the sequence of **elementary reactions,** or **elementary steps,** that defines the pathway from reactants to products. Elementary reactions are classified as **unimolecular reactions, bimolecular reactions,** or **termolecular reactions,** depending on whether one, two, or three reactant molecules (or atoms) are involved. The rate law for an elementary reaction follows directly from its **molecularity:** rate $= k[A]$ for a unimolecular reaction, and rate $= k[A]^2$ or rate $= k[A][B]$ for a bimolecular reaction.	**14.12** Given a reaction mechanism, write the overall reaction, identify intermediates, and determine the molecularity for each elementary step.	Worked Examples 14.12–14.13; Problems 14.106, 14.108, 14.110
14.11 Rate Laws for Overall Reactions	The observed rate law for a multistep, overall reaction depends on the sequence of elementary steps and their relative rates. The slowest step is called the **rate-determining step.** A chemical species that is formed in one elementary step and consumed in a subsequent step is called a **reaction intermediate.** An acceptable mechanism must meet two criteria: (1) The elementary steps must sum to give the overall reaction, and (2) the mechanism must be consistent with the observed rate law.	**14.13** For mechanisms with an initial slow step (a) predict the rate law for the overall reaction or (b) propose a mechanism that agrees with the experimental rate law.	Worked Example 14.14; Problems 14.46, 14.111–14.112
		14.14 For mechanisms with an initial fast equilibrium step, predict the rate law.	Worked Example 14.15; Problems 14.113–14.115
14.12–14.13 Catalysis	A **catalyst** is a substance that increases the rate of a reaction without being consumed in the reaction. It functions by making available an alternative reaction pathway that has a lower activation energy. A **homogeneous catalyst** is present in the same phase as the reactants, whereas a **heterogeneous catalyst** is present in a different phase. **Enzymes** are large protein molecules that act as catalysts for biological reactions.	**14.15** Use a molecular diagram for a catalyzed reaction to determine the rate law and propose a mechanism consistent with the rate law.	Worked Example 14.16; Problems 14.31–14.32, 14.47
		14.16 Identify catalysts and intermediates in the mechanism of a catalyzed reaction.	Problems 14.118–14.119, 14.121
		14.17 Interpret a potential energy diagram representing the mechanism in a catalyzed reaction.	Problem 14.48

KEY TERMS

activation energy (E_a) 561
Arrhenius equation 563
Arrhenius plot 565
bimolecular reaction 569
catalyst 577
chemical kinetics 539
collision theory 560
elementary reaction 567

elementary step 567
enzyme 583
first-order reaction 552
frequency factor 564
half-life 555
heterogeneous catalyst 580
homogeneous catalyst 580
initial rate 542

instantaneous rate 542
integrated rate law 550
molecularity 568
rate constant 544
rate-determining step 573
rate law 544
reaction intermediate 568
reaction mechanism 567

reaction order 545
reaction rate 540
second-order reaction 557
steric factor (p) 563
termolecular reaction 569
transition state 561
unimolecular reaction 568
zeroth-order reaction 550

KEY EQUATIONS

- Reaction Rate for the Reaction $a\,A + b\,B \longrightarrow c\,C + d\,D$ (Section 14.1)

- General Reaction Rate (Section 14.1)

$$\text{Rate} = -\frac{1}{a}\frac{\Delta[A]}{\Delta t} = -\frac{1}{b}\frac{\Delta[B]}{\Delta t} = \frac{1}{c}\frac{\Delta[C]}{\Delta t} = \frac{1}{d}\frac{\Delta[D]}{\Delta t}$$

where a, b, c, and d are stoichiometric coefficients for the reactants and products A, B, C, and D.
- Generalized Rate Law for the Reaction $a\,A + b\,B \longrightarrow$ Products (Section 14.2)

$$\text{Rate} = k[A]^m[B]^n$$

where k is the rate constant, m is the order of the reaction in A, and n is the order of the reaction in B.

- Integrated Rate Law and Half-Life ($t_{1/2}$) for a Zeroth-Order Reaction (Section 14.4)

$$[A]_t = -kt + [A]_0 \qquad t_{1/2} = \frac{[A]_0}{2k}$$

where $[A]_t$ is the concentration of A at time t and $[A]_0$ is the initial concentration of A.

- Integrated Rate Law and Half-Life ($t_{1/2}$) for a First-Order Reaction (Section 14.5)

$$\ln \frac{[A]_t}{[A]_0} = -kt \qquad t_{1/2} = \frac{0.693}{k}$$

- Integrated Rate Law and Half-Life ($t_{1/2}$) for a Second-Order Reaction (Section 14.6)

$$\frac{1}{[A]_t} = kt + \frac{1}{[A]_0} \qquad t_{1/2} = \frac{1}{k[A]_0}$$

- The Rate Constant Predicted by Collision Theory: The Arrhenius Equation (Section 14.7)

$$k = pZe^{-E_a/RT} = Ae^{-E_a/RT}$$

where p is the steric factor, Z is the collision frequency, E_a is the activation energy, and A $(=pZ)$ is the frequency factor

- Two Logarithmic Forms of the Arrhenius Equation (Section 14.8)

$$\ln k = \ln A - \frac{E_a}{RT} \qquad \ln \left(\frac{k_2}{k_1} \right) = \left(\frac{-E_a}{R} \right)\left(\frac{1}{T_2} - \frac{1}{T_1} \right)$$

where k_1 and k_2 are rate constants at temperature T_1 and T_2

PRACTICE TEST

After studying this chapter, you can assess your understanding with these practice test questions, which are correlated with chapter learning objectives. If you answer a question incorrectly, refer to the learning objectives in the end-of-chapter Study Guide for assistance. The Study Guide provides a conceptual summary, references a Worked Example to model how to solve the problem, and gives additional problems for more practice.

Use the following equation and graph to answer questions 1 and 2. Hydrogen iodide decomposes at 410 °C, according the reaction:

$$2\,HI(g) \longrightarrow H_2(g) + I_2(g)$$

The graph shows how the concentration of HI changes over time.

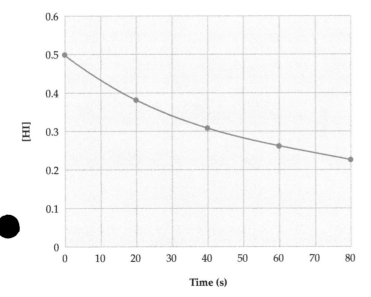

1. What is the average rate of loss of HI over the time period 0–40 s (LO 14.1)
   (a) 7.5×10^{-3} M/s
   (b) 4.8×10^{-3} M/s
   (c) 3.0×10^{-2} M/s
   (d) 3.5×10^{-3} M/s

2. What is the average rate of production of H_2 over the time period 0–40 s? (LO 14.2)
   (a) 3.8×10^{-3} M/s
   (b) 4.8×10^{-3} M/s
   (c) 2.4×10^{-3} M/s
   (d) 9.6×10^{-3} M/s

3. The rate of the reaction between CO and NO_2 was studied at 540 K and found to have an initial reaction rate of 6.8×10^{-8} M/min when $[CO] = 5.1 \times 10^{-4}$ M and $[NO_2] = 7.0 \times 10^{-5}$ M.

   $$CO(g) + NO_2(g) \longrightarrow CO_2(g) + NO(g)$$

   $$Rate = k[CO][NO_2]$$

   What is the rate constant for the reaction? (LO 14.3)
   (a) $1.9\ M^{-1}\ min^{-1}$
   (b) $0.53\ M^{-1}\ min^{-1}$
   (c) $0.53\ min^{-1}$
   (d) $1.9\ M^{-2}\ min^{-1}$

4. The following reaction is first order in A and B.

$$A + B \longrightarrow AB \qquad Rate = k[A][B]$$

(1)

(a)   **(b)**

(c)   **(d)**

In which image would the concentrations of A (red spheres) and B (blue spheres) produce a reaction rate that is 3 times higher than the rate of the reaction in image 1? (**LO 14.4**)

5. Reaction rates at various initial concentrations of reactants were measured at 900 °C for the following reaction.

$$2 NO(g) + 2 H_2(g) \longrightarrow N_2(g) + 2 H_2O(g)$$

Use the data in the table to determine the rate law for the reaction. (**LO 14.5**)

Initial [NO]	Initial $[H_2]$	Initial Rate (M/s)
0.300	0.100	5.69×10^{-2}
0.150	0.100	1.42×10^{-2}
0.250	0.200	7.90×10^{-2}

(a) Rate = $k[NO][H_2]$ (b) Rate = $k[NO][H_2]^2$
(c) Rate = $k[NO]^2[H_2]$ (d) Rate = $k[NO]^2[H_2]^2$

6. Sucrose, a sugar, decomposes in acidic solution to produce glucose and fructose. The reaction is first order in sucrose, and at 25 °C the rate constant is $k = 3.60 \times 10^{-3}$ h^{-1}. If the initial concentration of sucrose is 0.050 M, what is the concentration after 2 days? (**LO 14.7**)

(a) 0.013 M (b) 0.028 M
(c) 4.3×10^{-3} M (d) 0.042 M

7. Sucrose, a sugar, decomposes in acidic solution to produce glucose and fructose. The reaction is first order in sucrose, and at 25 °C the rate constant is $k = 3.60 \times 10^{-3}$ h^{-1}. How much time would it take for 80.0% of the sucrose in solution to react? (**LO 14.7**)

(a) 2.56 days (b) 9.78 days
(c) 18.6 days (d) 0.164 days

8. The gas phase decomposition of HI has the following rate law:

$$2 HI(g) \longrightarrow H_2(g) + I_2(g) \qquad Rate = k[HI]^2$$

At 443 °C, $k = 30.1$ M^{-1} min^{-1}. If the initial concentration of HI is 0.010 M, what is the concentration after 1.5 hours? (**LO 14.8**)

(a) 6.9×10^{-3} M (b) 1.8×10^{-3} M
(c) 3.6×10^{-4} M (d) 8.9×10^{-4} M

9. Chlorine monoxide (ClO) decomposes at room temperature according to the reaction

$$2 ClO(g) \longrightarrow Cl_2(g) + O_2(g)$$

The concentration of ClO was monitored over time, and three graphs were made:

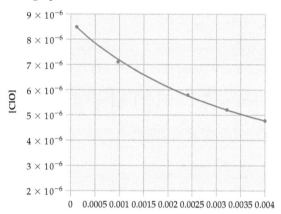

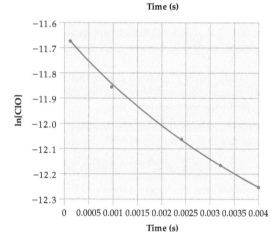

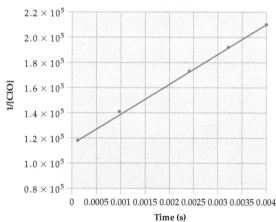

What is the rate law for the reaction? (**LO 14.9**)

(a) Rate = k (b) Rate = $k[ClO]$
(c) Rate = $k[ClO]^2$ (d) Rate = $k[ClO]^3$

10. The half-life for the first order decomposition of H_2O_2 is 10.7 hours. If the initial concentration of H_2O_2 is 0.56 M, what is the concentration after 3 half-lives? (**LO 14.7**)

   (a) 0.070 M   (b) 0.14 M
   (c) 0.047 M   (d) 0.018 M

11. A key reaction in the upper atmosphere is

$$O_3(g) \; + \; O(g) \; \longrightarrow \; 2\,O_2(g)$$

   For this process, the energy of activation for the forward reaction, $E_{a(fwd)}$, is 19 kJ/mol, and the enthalpy change for the reaction, ΔH_{rxn}, is -392 kJ/mol. What is the energy of activation for the reverse reaction, $E_{a(reverse)}$? (**LO 14.10**)

   (a) 411 kJ/mol   (b) 392 kJ/mol
   (c) 373 kJ/mol   (d) 196 kJ/mol

12. If a certain reaction has a rate constant of 9.51×10^{-9} L $mol^{-1}\,s^{-1}$ at 500 K and a rate constant of 1.10×10^{-5} L $mol^{-1}\,s^{-1}$ at 600 K, what is the value of E_a? (**LO 14.11**)

   (a) 1.00×10^2 kJ/mol   (b) 1.76×10^2 kJ/mol
   (c) 6.85×10^3 kJ/mol   (d) 1.72×10^5 kJ/mol

   To answer questions 13–15, refer to the mechanism:

$$H_2O_2(aq) + I^-(aq) \longrightarrow OH^-(aq) + HOI(aq)$$

   Slower, rate-determining

$$HOI(aq) + I^-(aq) \longrightarrow OH^-(aq) + I_2(aq) \qquad \text{Faster}$$

$$2\,OH^-(aq) + 2\,H_3O^+(aq) \longrightarrow 4\,H_2O(l) \qquad \text{Faster}$$

13. Identify the catalyst and intermediate(s) in the mechanism. (**LO 14.12, 14.16**)

   (a) Catalyst = I^-, intermediates = OH^-, HOI
   (b) Catalyst = H_3O^+, intermediate = HOI
   (c) No catalyst, intermediate = I_2
   (d) No catalyst, intermediates = OH^-, HOI

14. What is the rate law for the reaction? (**LO 14.13**)

   (a) Rate = $k[H_2O_2][I^-]^2$
   (b) Rate = $k[H_2O_2][I^-]$
   (c) Rate = $k[H_2O_2][I^-]^2[H_3O^+]^2$
   (d) Rate = $k[H_2O_2]^2[I^-]$

15. Select the correct potential energy diagram for the mechanism. (**LO 14.17**)

   (a)
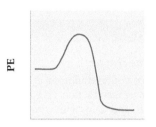
   Reaction progress

   (b)
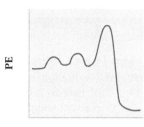
   Reaction progress

   (c)
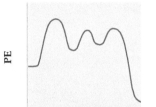
   Reaction progress

   (d)
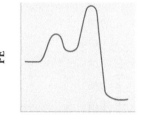
   Reaction progress

16. The reaction $H_2(g) + I_2(g) \longrightarrow 2\,HI(g)$ occurs by a two-step mechanism:

$$I_2(g) \underset{k_{-1}}{\overset{k_1}{\rightleftharpoons}} 2\,I(g) \qquad \text{Faster}$$

$$H_2(g) + 2\,I(g) \overset{k_2}{\longrightarrow} 2\,HI(g) \qquad \text{Slower, rate-determining}$$

   What is the rate law for the mechanism? Express the rate constant in terms of the rate constants for the elementary steps. (**LO 14.14**)

   (a) Rate = $k_1[I_2]$

   (b) Rate = $\dfrac{k_1 k_2}{k_{-1}}[H_2][I_2]^2$

   (c) Rate = $k_2[H_2][I]^2$

   (d) Rate = $\dfrac{k_1 k_2}{k_{-1}}[H_2][I_2]$

Answers:

1. b, 2. c, 3. a, 4. b, 5. c, 6. d, 7. c, 8. c, 9. c, 10. a, 11. a, 12. b, 13. d, 14. b, 15. c, 16. d

Mastering Chemistry   provides end-of-chapter exercises, feedback-enriched tutorial problems, animations, and interactive activities to encourage problem-solving practice and deeper understanding of key concepts and topics.

RAN   *Randomized in Mastering Chemistry*

CONCEPTUAL PROBLEMS

Problems 14.1–14.38 appear within the chapter.

14.39 (a) Use the following graph of concentration as a function
RAN   of time to calculate the average reaction rate for each
     reactant and product (A, B, and C) over the time period
     0–250 s.

 (b) Write the balanced reaction that corresponds to the data
     in the graph.

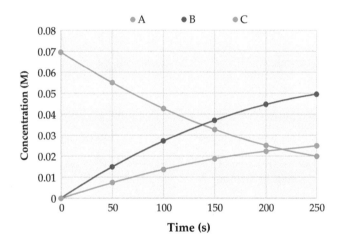

14.40 The following reaction is first order in A (red spheres) and
first order in B (blue spheres):

A + B ⟶ Products   Rate = $k[A][B]$

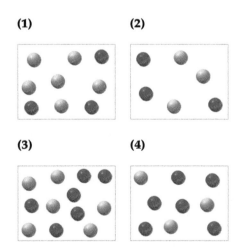

 (a) What are the relative rates of this reaction in vessels
     (1)–(4)? Each vessel has the same volume.
 (b) What are the relative values of the rate constant k for
     vessels (1)–(4)?

14.41 Consider the first-order decomposition of A molecules (red
spheres) in three vessels of equal volume.

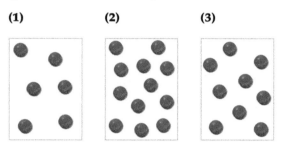

 (a) What are the relative rates of decomposition in vessels
     (1)–(3)?
 (b) What are the relative half-lives of the reactions in vessels
     (1)–(3)?
 (c) How will the rates and half-lives be affected if the
     volume of each vessel is decreased by a factor of 2?

14.42 Consider the first-order reaction A → B in which A molecules
(red spheres) are converted to B molecules (blue spheres).

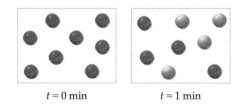

 (a) Given the pictures at $t = 0$ min and $t = 1$ min, draw
     pictures that show the number of A and B molecules
     present at $t = 2$ min and $t = 3$ min.
 (b) What is the half-life of the reaction?

14.43 The following pictures represent the progress of the reaction
RAN   A → B in which A molecules (red spheres) are converted to B
     molecules (blue spheres).

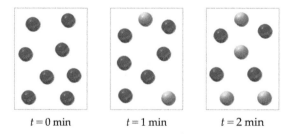

 (a) What is the order of the reaction?
 (b) Draw a picture that shows the number of A and B mol-
     ecules present at $t = 3$ min.

(c) Suppose that each sphere represents 6.0×10^{21} molecules and that the volume of the container is 1.0 L. What is the rate constant for the reaction in the usual units?

14.44 The following pictures represent the progress of a reaction in which two A molecules combine to give a more complex molecule A_2, $2\ A \rightarrow A_2$.

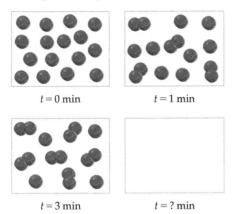

$t = 0$ min $t = 1$ min

$t = 3$ min $t = ?$ min

(a) Is the reaction first order or second order in A?

(b) What is the rate law?

(c) Draw an appropriate picture in the last box, and specify the time.

14.45 What is the molecularity of each of the following elementary reactions?

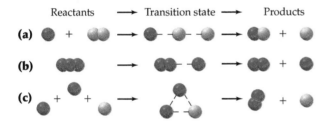

Reactants ⟶ Transition state ⟶ Products

(a)

(b)

(c)

14.46 The relative rates of the reaction $A + B \rightarrow AB$ in vessels (1)–(3) are 4:4:1. Red spheres represent A molecules, and blue spheres represent B molecules.

(1)   (2)   (3)

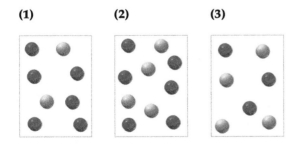

(a) What is the order of the reaction in A and B?

(b) Write the rate law.

(c) Write a mechanism that agrees with the rate law.

(d) Identify all intermediates in your mechanism.

14.47 The relative rates of the reaction $AB + B \rightarrow A + B_2$ in vessels (1)–(4) are 2:1:1:4. Red spheres represent A, blue spheres represent B, and green spheres represent C.

(1)   (2)

(3)   (4)

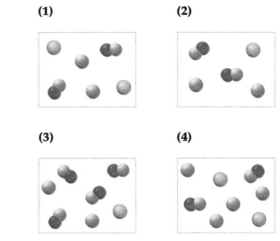

(a) Write the rate law.

(b) Write a mechanism that agrees with the rate law.

(c) Identify all intermediates and catalysts in your mechanism.

14.48 Consider a reaction that occurs by the following mechanism:

$$A + BC \rightarrow AC + B$$
$$AC + D \rightarrow A + CD$$

The potential energy profile for this reaction is as follows:

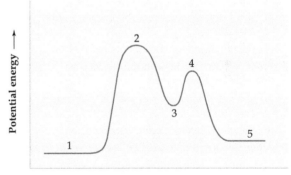

(a) What is the equation for the overall reaction?

(b) Write structural formulas for all species present at reaction stages 1–5. Identify each species as a reactant, product, catalyst, intermediate, or transition state.

(c) Which of the two steps in the mechanism is the rate-determining step? Write the rate law for the overall reaction.

(d) Is the reaction endothermic or exothermic? Add labels to the diagram that show the values of the energy of reaction ΔE and the activation energy E_a for the overall reaction.

14.49 Draw a plausible transition state for the bimolecular reaction of nitric oxide with ozone. Use dashed lines to indicate the atoms that are weakly linked together in the transition state.

$$NO(g) + O_3(g) \longrightarrow NO_2(g) + O_2(g)$$

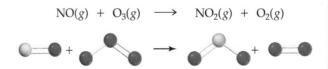

SECTION PROBLEMS

Reaction Rates (Section 14.1)

14.50 Use the data in Table 14.1 to calculate the average rate of
RAN decomposition of N_2O_5 and the average rate of formation of
O_2 during the time interval 200–300 s.

14.51 Use the data in Table 14.1 to calculate the average rate of
RAN decomposition of N_2O_5 and the average rate of formation of
NO_2 during the time interval 500–600 s.

14.52 Use the information in Table 14.1 and Figure 14.1 to estimate the instantaneous rate of appearance of NO_2 at
$t = 350$ s by calculating the average rate of appearance of
NO_2 over the following time intervals centered on $t = 350$ s.

(a) 0 to 700 s   (b) 100 to 600 s

(c) 200 to 500 s   (d) 300 to 400 s

Which is the best estimate, and why?

14.53 Concentration–time data for the decomposition of nitrogen
dioxide are given in Worked Example 14.9. Estimate the
instantaneous rate of disappearance of NO_2 at $t = 250$ s by
calculating the average rate of disappearance of NO_2 over
the following time intervals centered on $t = 250$ s.

(a) 0 to 500 s

(b) 100 to 400 s

(c) 200 to 300 s

Which is the best estimate, and why?

14.54 From the plot of concentration–time data in Figure 14.1,
RAN estimate:

(a) the instantaneous rate of decomposition of N_2O_5 at
$t = 200$ s.

(b) the initial rate of decomposition of N_2O_5.

14.55 From a plot of the concentration–time data in Worked
RAN Example 14.9, estimate:

(a) the instantaneous rate of decomposition of NO_2 at
$t = 100$ s.

(b) the initial rate of decomposition of NO_2.

14.56 Ammonia is manufactured in large amounts by the reaction

$$N_2(g) + 3 H_2(g) \rightarrow 2 NH_3(g)$$

(a) How is the rate of consumption of H_2 related to the rate
of consumption of N_2?

(b) How is the rate of formation of NH_3 related to the rate
of consumption of N_2?

14.57 In the first step of the Ostwald process for the synthesis of
nitric acid, ammonia is converted to nitric oxide by the high-
temperature reaction

$$4 NH_3(g) + 5 O_2(g) \rightarrow 4 NO(g) + 6 H_2O(g)$$

(a) How is the rate of consumption of O_2 related to the rate
of consumption of NH_3?

(b) How are the rates of formation of NO and H_2O related
to the rate of consumption of NH_3?

14.58 Chlorite is reduced by bromide in acidic solution according
RAN to the following balanced equation:

$$ClO_2^-(aq) + 4 Br^-(aq) + 4 H^+(aq) \rightarrow Cl^-(aq) +$$
$$2 Br_2(aq) + 2 H_2O(l)$$

(a) If $\Delta[Br_2]/\Delta t = 4.8 \times 10^{-6}$ M/s, what is the value of
$\Delta[ClO_2^-]/\Delta t$ during the same time interval?

(b) What is the average rate of consumption of Br^- during
the same time interval?

14.59 The oxidation of 2-butanone ($CH_3CO_2H_5$) by the cerium(IV)
RAN ion in aqueous solution to form acetic acid (CH_3CO_2H) occurs
according to the following balanced equation:

$$CH_3COC_2H_5(aq) + 6 Ce^{4+}(aq) + 3 H_2O(l) \rightarrow$$
$$2 CH_3CO_2H(aq) + 6 Ce^{3+}(aq) + 6 H^+(aq)$$

(a) If acetic acid appears at an average rate of 5.0×10^{-8} M/s,
what is $\Delta[H^+]/\Delta t$ during the same time interval?

(b) What is the average rate of consumption of Ce^{4+} during
the same time interval?

Rate Laws (Sections 14.2–14.3)

14.60 The gas-phase reaction of hydrogen and iodine monochloride is first order in H_2 and first order in ICl. What is the
rate law, and what are the units of the rate constant?

$$H_2(g) + 2 ICl(g) \rightarrow 2 HCl(g) + I_2(g)$$

14.61 The reaction $2NO(g) + 2 H_2(g) \rightarrow N_2(g) + 2 H_2O(g)$ is
first order in H_2 and second order in NO. Write the rate law,
and specify the units of the rate constant.

14.62 Bromomethane is converted to methanol in an alkaline solu-
RAN tion. The reaction is first order in each reactant.

$$CH_3Br(aq) + OH^-(aq) \rightarrow CH_3OH(aq) + Br^-(aq)$$

(a) Write the rate law.

(b) How does the reaction rate change if the OH^- concentration is decreased by a factor of 5?

(c) What is the change in rate if the concentrations of both
reactants are doubled?

14.63 The oxidation of Br^- by BrO_3^- in acidic solution is described
by the equation

$$5 Br^-(aq) + BrO_3^-(aq) + 6 H^+(aq) \rightarrow 3 Br_2(aq) + 3 H_2O(l)$$

The reaction is first order in Br^-, first order in BrO_3^-, and
second order in H^+.

(a) Write the rate law.

(b) What is the overall reaction order?

(c) How does the reaction rate change if the H^+concentra-
tion is tripled?

(d) What is the change in rate if the concentrations of both
Br^- and BrO_3^- are halved?

14.64 Oxidation of bis(bipyridine)copper(I) ion by molecular oxygen is described by the equation

$$Cu(C_{10}H_8N_2)_2^+(aq) + O_2(aq) \rightarrow products$$

The reaction is first order in oxygen and second order in
$Cu(C_{10}H_8N_2)_2^+$.

(a) Write the rate law.

(b) What is the overall reaction order?

(c) How does the reaction rate change if the concentration
of $Cu(C_{10}H_8N_2)_2^+$ is decreased by a factor of 4?

14.65 In aqueous solution the oxidation of nitric oxide occurs according to the equation

$$4 \, NO(aq) + O_2(aq) + 2 \, H_2O(l) \rightarrow 4 \, HNO_2(aq)$$

The reaction is second order in nitric oxide and first order in oxygen.

(a) Write the rate law.

(b) What is the overall reaction order?

(c) How does the reaction rate change when the concentration of oxygen is doubled?

(d) How does the reaction rate change when the concentration of nitric oxide is doubled and the concentration of oxygen is halved?

14.66 Initial rate data at 25 °C are listed in the table for the reaction
RAN

$$NH_4^+(aq) + NO_2^-(aq) \rightarrow N_2(g) + 2 \, H_2O(l)$$

Experiment	Initial [NH$_4^+$]	Initial [NO$_2^-$]	Initial Reaction Rate (M/s)
1	0.24	0.10	7.2×10^{-6}
2	0.12	0.10	3.6×10^{-6}
3	0.12	0.15	5.4×10^{-6}

(a) What is the rate law?

(b) What is the value of the rate constant?

(c) What is the initial rate when the initial concentrations are [NH$_4^+$] = 0.39 M and [NO$_2^-$] = 0.052 M?

14.67 The initial rates listed in the following table were determined
RAN for the reaction

$$2 \, NO(g) + Cl_2(g) \rightarrow 2 \, NOCl(g)$$

Experiment	Initial [NO]	Initial [Cl$_2$]	Initial Reaction Rate (M/s)
1	0.24	0.10	7.2×10^{-6}
2	0.12	0.10	3.6×10^{-6}
3	0.12	0.15	5.4×10^{-6}

(a) What is the rate law?

(b) What is the value of the rate constant?

(c) What is the initial rate when the initial concentrations of both reactants are 0.12 M?

14.68 The oxidation of iodide ion by hydrogen peroxide in an
RAN acidic solution is described by the balanced equation

$$H_2O_2(aq) + 3 \, I^-(aq) + 2 \, H^+(aq) \rightarrow I_3^-(aq) + 2 \, H_2O(l)$$

The rate of formation of the red triiodide ion, $\Delta[I_3^-]/\Delta t$, can be determined by measuring the rate of appearance of the color.

▲ A sequence of photographs showing the progress of the reaction of hydrogen peroxide (H$_2$O$_2$) and iodide ion (I$^-$). As time passes, the red color due to the triiodide ion (I$_3^-$) increases in intensity.

Initial rate data at 25 °C and a constant [H$^+$] are as follows:

Experiment	Initial [H$_2$O$_2$]	Initial [I$^-$]	Initial Reaction Rate (M/s)
1	0.100	0.100	1.15×10^{-4}
2	0.100	0.200	2.30×10^{-4}
3	0.200	0.100	2.30×10^{-4}
4	0.200	0.200	4.60×10^{-4}

(a) What is the rate law?

(b) What is the value of the rate constant?

(c) What is the initial rate when the initial concentrations are [H$_2$O$_2$] = 0.300 M and [I$^-$] = 0.400 M?

14.69 Trimethylamine and chlorine dioxide react in water in an electron transfer reaction to form the trimethylamine cation and chlorite ion:

$$(CH_3)_3 \, N(aq) + ClO_2(aq) + H_2O(l) \rightarrow$$
$$(CH_3)_3 \, NH^+(aq) + ClO_2^-(aq) + OH^-(aq)$$

Initial rate data obtained at 23 °C are listed in the following table.

Experiment	[(CH$_3$)$_3$ N]	[ClO$_2$]	Initial Reaction Rate (M/s)
1	3.25×10^{-3}	4.60×10^{-3}	0.90
2	6.50×10^{-3}	2.30×10^{-3}	0.90
3	1.30×10^{-2}	2.30×10^{-3}	1.79
4	2.60×10^{-2}	9.20×10^{-3}	14.4

(a) What is the rate law, including the value of the rate constant?

(b) What would be the initial rate in an experiment with initial concentrations [(CH$_3$)$_3$ N] = 4.2×10^{-2} M and [ClO$_2$] = 3.4×10^{-2} M?

Integrated Rate Law; Half-Life (Sections 14.4–14.6)

14.70 At 500 °C, cyclopropane (C_3H_6) rearranges to propene
RAN   (CH_3—CH=CH_2). The reaction is first order, and the rate
constant is $6.7 \times 10^{-4}\ s^{-1}$. If the initial concentration of
C_3H_6 is 0.0500 M:

(a) What is the molarity of C_3H_6 after 30 min?

(b) How many minutes does it take for the C_3H_6 concentration to drop to 0.0100 M?

(c) How many minutes does it take for 25% of the C_3H_6 to react?

14.71 The rearrangement of methyl isonitrile (CH_3NC) to aceto-
RAN   nitrile (CH_3CN) is a first-order reaction and has a rate constant of $5.11 \times 10^{-5}\ s^{-1}$ at 472 K.

$$CH_3-N\equiv C \longrightarrow CH_3-C\equiv N$$

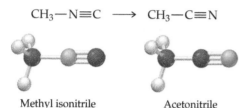

Methyl isonitrile   Acetonitrile

If the initial concentration of CH_3NC is 0.0340 M:

(a) What is the molarity of CH_3NC after 2.00 h?

(b) How many minutes does it take for the CH_3NC concentration to drop to 0.0300 M?

(c) How many minutes does it take for 20% of the CH_3NC to react?

14.72 What is the half-life (in minutes) of the reaction in Problem
RAN   14.70? How many minutes will it take for the concentration of cyclopropane to drop to 6.25% of its initial value?

14.73 What is the half-life (in hours) of the reaction in Problem
RAN   14.71? How many hours will it take for the concentration of methyl isonitrile to drop to 12.5% of its initial value?

14.74 Butadiene (C_4H_6) reacts with itself to form a dimer with
RAN   the formula C_8H_{12}. The reaction is second order in C_4H_6. Assume the rate constant at a particular temperature is $4.0 \times 10^{-2}\ M^{-1}\ s^{-1}$ and the initial concentration of C_4H_6 is 0.0200 M.

(a) What is its molarity after a reaction time of 1.00 h?

(b) What is the time (in hours) when the C_4H_6 concentration reaches a value of 0.0020 M?

14.75 Hydrogen iodide decomposes slowly to H_2 and I_2 at 600 K.
RAN   The reaction is second order in HI, and the rate constant is $9.7 \times 10^{-6}\ M^{-1}\ s^{-1}$. Assume the initial concentration of HI is 0.100 M.

(a) What is its molarity after a reaction time of 6.00 days?

(b) What is the time (in days) when the HI concentration reaches a value of 0.085 M?

14.76 What is the half-life (in minutes) of the reaction in Prob-
RAN   lem 14.74 when the initial C_4H_6 concentration is 0.0200 M? How many minutes does it take for the concentration of C_4H_6 to drop from 0.0100 M to 0.0050 M?

14.77 What is the half-life (in days) of the reaction in Problem
RAN   14.75 when the initial HI concentration is 0.100 M? How many days does it take for the concentration of HI to drop from 0.200 M to 0.100 M?

14.78 At 25 °C, the half-life of a certain first-order reaction is 248 s.
RAN   What is the value of the rate constant at this temperature?

14.79 The decomposition of N_2O_5 is a first-order reaction. At 25 °C,
RAN   it takes 5.2 h for the concentration to drop from 0.120 M to 0.060 M. How many hours does it take for the concentration to drop from 0.030 M to 0.015 M? From 0.480 M to 0.015 M?

14.80 You wish to determine the reaction order and rate constant for the following thermal decomposition reaction:

$$AB_2 \rightarrow 1/2\ A_2 + B_2$$

(a) What data would you collect?

(b) How would you use these data to determine whether the reaction is zeroth order, first order, or second order?

(c) Describe how you would determine the value of the rate constant.

14.81 At elevated temperatures, nitrous oxide decomposes according to the equation

$$N_2O(g) \rightarrow N_2(g) + 1/2\ O_2(g)$$

Given the following data, plot the appropriate graphs to determine whether the reaction is zeroth, first, or second order. What is the value of the rate constant for the consumption of N_2O?

Time (min)	0	60	90	300	600
[N_2O]	0.265	0.228	0.216	0.128	0.0630

14.82 Nitrosyl bromide decomposes at 10 °C.

$$NOBr(g) \rightarrow NO(g) + 1/2\ Br_2(g)$$

Use the following kinetic data to determine the order of the reaction and the value of the rate constant for consumption of NOBr.

Time (s)	0	10	40	120	320
[NOBr]	0.0390	0.0301	0.0175	0.00812	0.003 76

14.83 Consider the following concentration–time data for the decomposition reaction $AB \rightarrow A + B$.

Time (min)	0	20	40	120	220
[AB]	0.215	0.196	0.181	0.117	0.036

(a) Determine the order of the reaction and the value of the rate constant.

(b) What is the molarity of AB after a reaction time of 192 min?

(c) What is the time (in minutes) when the AB concentration reaches a value of 0.0250 M?

14.84 *Trans*-cycloheptene (C_7H_{12}), a strained cyclic hydrocar-
RAN   bon, converts to *cis*-cycloheptene at low temperatures. This molecular rearrangement is a second-order process with a rate constant of $0.030\ M^{-1}\ s^{-1}$ at 60 °C. If the initial concentration of *trans*-cycloheptene is 0.035 M:

(a) What is the concentration of *trans*-cycloheptene after a reaction time of 1600 s?

(b) At what time will the concentration drop to one-twentieth of its initial value?

(c) What is the half-life of *trans*-cycloheptene at an initial concentration of 0.075 M?

14.85 The light-stimulated conversion of 11-*cis*-retinal to 11-*trans*-
RAN retinal is central to the vision process in humans. This reaction also occurs (more slowly) in the absence of light. At
80.0 °C in heptane solution, the reaction is first order with a
rate constant of $1.02 \times 10^{-5}/s$.

(a) What is the molarity of 11-*cis*-retinal after 6.00 h if its
initial concentration is 3.50×10^{-3} M?

(b) How many minutes does it take for 25% of the 11-*cis*-retinal
to react?

(c) How many hours does it take for the concentration of
11-*trans*-retinal to reach 3.15×10^{-3} M?

The Arrhenius Equation (Sections 14.7–14.8)

14.86 Why don't all collisions between reactant molecules lead to
a chemical reaction?

14.87 Two reactions have the same activation energy, but their
rates at the same temperature differ by a factor of 10.
Explain.

14.88 When the temperature of a gas is raised by 10 °C, the collision frequency increases by only about 2%, but the reaction
rate increases by 100% (a factor of 2) or more. Explain.

14.89 What fraction of the molecules in a gas at 300 K collide with an
RAN energy equal to or greater than E_a when E_a equals 50 kJ/mol?
What is the value of this fraction when E_a is 100 kJ/mol?

14.90 The values of $E_a = 183$ kJ/mol and $\Delta E = 9$ kJ/mol have
been measured for the reaction

$$2 \, HI(g) \rightarrow H_2(g) + I_2(g)$$

Sketch a potential energy profile for this reaction that shows
the potential energy of reactants, products, and the transition state. Include labels that define E_a and ΔE.

14.91 The values of $E_a = 248$ kJ/mol and $\Delta E = 41$ kJ/mol have
been measured for the reaction

$$H_2(g) + CO_2(g) \rightarrow H_2O(g) + CO(g)$$

(a) Sketch a potential energy profile for this reaction that
shows the potential energy of reactants, products, and
the transition state. Include labels that define E_a and ΔE.

(b) Considering the geometry of the reactants and products,
suggest a plausible structure for the transition state.

14.92 Consider three reactions with different values of E_a and ΔE:

Reaction 1. $E_a = 20$ kJ/mol; $\Delta E = -60$ kJ/mol
Reaction 2. $E_a = 10$ kJ/mol; $\Delta E = -20$ kJ/mol
Reaction 3. $E_a = 40$ kJ/mol; $\Delta E = +15$ kJ/mol

(a) Sketch a potential energy profile for each reaction that
shows the potential energy of reactants, products, and
the transition state. Include labels that define E_a and ΔE.

(b) Assuming that all three reactions are carried out at the
same temperature and that all three have the same frequency factor A, which reaction is the fastest and which
is the slowest?

(c) Which reaction is the most endothermic, and which is
the most exothermic?

14.93 Consider the potential energy profile in Figure 14.17 for the
iodide ion-catalyzed decomposition of H_2O_2. What point on
the profile represents the potential energy of the transition
state for the first step in the reaction? What point represents
the potential energy of the transition state for the second

step? What point represents the potential energy of the intermediate products $H_2O(l) + IO^-(aq)$?

14.94 Rate constants for the reaction $NO_2(g) + CO(g) \rightarrow$
RAN $NO(g) + CO_2(g)$ are 1.3 $M^{-1} s^{-1}$ at 700 K and 23.0 $M^{-1} s^{-1}$
at 800 K.

(a) What is the value of the activation energy in kJ/mol?

(b) What is the rate constant at 750 K?

14.95 A certain first-order reaction has a rate constant of
$1.0 \times 10^{-3} \, s^{-1}$ at 25 °C.

(a) If the reaction rate doubles when the temperature is
increased to 35 °C, what is the activation energy for this
reaction in kJ/mol?

(b) What is the E_a (in kJ/mol) if the same temperature
change causes the rate to triple?

14.96 Reaction of the anti-cancer drug cisplatin, $Pt(NH_3)_2 \, Cl_2$,
with water is described by the equation

$$Pt(NH_3)_2Cl_2(aq) + H_2O(l) \longrightarrow$$
$$Pt(NH_3)_2(H_2O)Cl^+(aq) + Cl^-(aq)$$

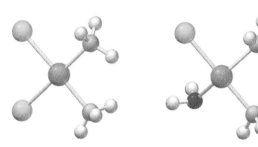

The rate of this reaction increases by a factor of 15 on raising the temperature from 25 °C to 50 °C. What is the value
of the activation energy in kJ/mol?

14.97 The widely used solvent ethyl acetate undergoes the following reaction in basic solution:

$$CH_3CO_2C_2H_5(aq) + OH^-(aq) \rightarrow$$
$$CH_3CO_2^-(aq) + C_2H_5OH(aq)$$

The rate of this reaction increases by a factor of 6.37 on
raising the temperature from 15 °C to 45 °C. Calculate the
value of the activation energy in kJ/mol.

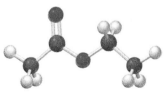

Ethyl acetate

14.98 If the rate of a reaction increases by a factor of 2.5 when the
RAN temperature is raised from 20 °C to 30 °C, what is the value
of the activation energy in kJ/mol? By what factor does the
rate of this reaction increase when the temperature is raised
from 120 °C to 130 °C?

14.99 *Mycobacterium avium,* a human pathogen responsible for respiratory infections, is sometimes found in hot tubs. *M. avium* can be inactivated by many disinfectants including chlorine, chlorine dioxide, and ozone. For inactivation by chlorine dioxide, the following rate constants were obtained. What is the activation energy (in kJ) for the inactivation of *M. avium* by chlorine dioxide?

Temperature (°C)	$k[\text{L}/(\text{mg} \cdot \text{min})]$
5	0.267
30	3.45

14.100 You wish to determine the activation energy for the following first-order reaction:

$$A \rightarrow B + C$$

(a) What data would you collect?

(b) How would you use these data to determine the activation energy?

14.101 Rate constants for the reaction $N_2O_5(g) \rightarrow 2\ NO_2(g) + 1/2\ O_2(g)$ exhibit the following temperature dependence:

Temperature (°C)	$k(\text{s}^{-1})$	Temperature (°C)	$k(\text{s}^{-1})$
25	3.7×10^{-5}	55	1.7×10^{-3}
45	5.1×10^{-4}	65	5.2×10^{-3}

Plot an appropriate graph of the data, and determine the activation energy for this reaction in kJ/mol.

14.102 The following rate constants describe the thermal decomposition of nitrogen dioxide:

$$NO_2(g) \longrightarrow NO(g) + 1/2\ O_2(g)$$

Temperature (°C)	$k(\text{M}^{-1}\,\text{s}^{-1})$	Temperature (°C)	$k(\text{M}^{-1}\,\text{s}^{-1})$
330	0.77	378	4.1
354	1.8	383	4.7

Plot an appropriate graph of the data, and calculate the value of E_a for this reaction in kJ/mol.

14.103 Poly(ethylene terephthalate) is a synthetic plastic used for fibers, bottles, and packaging. This material can be broken down for recycling by treatment with water at elevated temperatures (230-270 °C) and pressures according to the following equation:

Repeating unit

The reaction was carried out at different temperatures, and the following rate constants were obtained.

Temperature (°C)	$k(\text{M}^{-3/2}\,\text{s}^{-1})$
235	2.95×10^{-5}
250	5.77×10^{-5}
265	1.44×10^{-4}

Plot an appropriate graph of the data, and calculate the activation energy for this reaction in kJ/mol.

Reaction Mechanisms (Sections 14.9–14.11)

14.104 What is the relationship between the coefficients in a balanced chemical equation for an overall reaction and the exponents in the rate law?

14.105 What distinguishes the rate-determining step from the other steps in a reaction mechanism? How does the rate-determining step affect the observed rate law?

14.106 Consider the following mechanism for the reaction of hydrogen and iodine monochloride:

Step 1. $H_2(g) + ICl(g) \rightarrow HI(g) + HCl(g)$
Step 2. $HI(g) + ICl(g) \rightarrow I_2(g) + HCl(g)$

(a) Write the equation for the overall reaction.

(b) Identify any reaction intermediates.

(c) What is the molecularity of each elementary step?

14.107 The following mechanism has been proposed for the reaction of nitric oxide and chlorine:

Step 1. $NO(g) + Cl_2(g) \rightarrow NOCl_2(g)$
Step 2. $NOCl_2(g) + NO(g) \rightarrow 2\ NOCl(g)$

(a) What is the overall reaction?

(b) Identify any reaction intermediates.

(c) What is the molecularity of each elementary step?

14.108 Give the molecularity and the rate law for each of the following elementary reactions.

(a) $O_3(g) + Cl(g) \rightarrow O_2(g) + ClO(g)$

(b) $NO_2(g) \rightarrow NO(g) + O(g)$

(c) $ClO(g) + O(g) \rightarrow Cl(g) + O_2(g)$

(d) $Cl(g) + Cl(g) + N_2(g) \rightarrow Cl_2(g) + N_2(g)$

14.109 Identify the molecularity and write the rate law for each of the following elementary reactions.

(a) $I_2(g) \rightarrow 2\ I(g)$

(b) $2\ NO(g) + Br_2(g) \rightarrow 2\ NOBr(g)$

(c) $N_2O_5(g) \rightarrow NO_2(g) + NO_3(g)$

14.110 The thermal decomposition of nitryl chloride, NO_2Cl, is believed to occur by the following mechanism:

$$NO_2Cl(g) \xrightarrow{k_1} NO_2(g) + Cl(g)$$

$$Cl(g) + NO_2Cl(g) \xrightarrow{k_2} NO_2(g) + Cl_2(g)$$

(a) What is the overall reaction?

(b) What is the molecularity of each of the elementary steps?

(c) What rate law is predicted by this mechanism if the first step is rate-determining?

14.111 The substitution reactions of molybdenum hexacarbonyl, $Mo(CO)_6$, with a variety of other molecules L are believed to occur by the following mechanism:

$$Mo(CO)_6 \xrightarrow{k_1} Mo(CO)_5 + CO$$

$$Mo(CO)_5 + L \xrightarrow{k_2} Mo(CO)_5L$$

(a) What is the overall reaction?

(b) What is the molecularity of each of the elementary steps?

(c) Write the rate law, assuming that the first step is rate-determining.

14.112 The reaction $2\,NO_2(g) + F_2(g) \rightarrow 2\,NO_2F(g)$ has a second-order rate law, Rate $= k[NO_2][F_2]$. Suggest a mechanism that is consistent with this rate law.

14.113 The decomposition of ozone in the upper atmosphere is facilitated by NO. The overall reaction and the rate law are

$$O_3(g) + O(g) \rightarrow 2\,O_2(g) \qquad \text{Rate} = k[O_3][NO]$$

Write a mechanism that is consistent with the rate law.

14.114 A proposed mechanism for the oxidation of nitric oxide to nitrogen dioxide was described in Problem 14.29. Another possible mechanism for this reaction is

$$2\,NO(g) \underset{k_{-1}}{\overset{k_1}{\rightleftharpoons}} N_2O_2(g) \qquad \text{Faster, reversible}$$

$$N_2O_2(g) + O_2(g) \xrightarrow{k_2} 2NO_2(g) \quad \text{Slower, rate-determining}$$

(a) Write a balanced equation for the overall reaction.

(b) Show that this mechanism is consistent with the experimental rate law, Rate $= k[NO]^2[O_2]$.

(c) Relate the rate constant k to the rate constants for the elementary reactions.

14.115 The following mechanism has been proposed for the decomposition of dinitrogen pentoxide, which has the experimental rate law, Rate $= k[N_2O_5]$:

$$N_2O_5(g) \underset{k_{-1}}{\overset{k_1}{\rightleftharpoons}} NO_2(g) + NO_3(g) \qquad \text{Faster, reversible}$$

$$NO_2(g) + NO_3(g) \xrightarrow{k_2} NO(g) + NO_2(g) + O_2(g)$$
$$\text{Slow, rate-determining}$$

$$NO(g) + NO_3(g) \xrightarrow{k_3} 2\,NO_2(g) \qquad \text{Fast}$$

Note that the first step must be multiplied by 2 to obtain the overall reaction.

(a) Write a balanced equation for the overall reaction.

(b) Identify all reaction intermediates.

(c) Show that the proposed mechanism is consistent with the experimental rate law.

(d) Relate the rate constant k to the rate constants for the elementary reactions.

Catalysis (Sections 14.13)

14.116 Comment on the following statement: "A catalyst increases the rate of a reaction, but it is not consumed because it does not participate in the reaction."

14.117 Why doesn't a catalyst appear in the overall chemical equation for a reaction?

14.118 In the upper atmosphere, chlorofluorocarbons such as $CFCl_3$ absorb sunlight and subsequently fragment, yielding Cl radicals. The Inquiry in this chapter gave one mechanism for destruction of ozone by chlorine radicals. An alternative mechanism is shown:

$$Cl(g) + O_3(g) \rightarrow ClO(g) + O_2(g)$$
$$ClO(g) + O(g) \rightarrow Cl(g) + O_2(g)$$

(a) Write the chemical equation for the overall reaction.

(b) What is the role of the Cl atoms in this reaction?

(c) Is ClO a catalyst or a reaction intermediate?

(d) What distinguishes a catalyst from an intermediate?

14.119 Sulfur dioxide is oxidized to sulfur trioxide in the following sequence of reactions:

$$2\,SO_2(g) + 2\,NO_2(g) \rightarrow 2\,SO_3(g) + 2\,NO(g)$$
$$2\,NO(g) + O_2(g) \rightarrow 2\,NO_2(g)$$

(a) Write the chemical equation for the overall reaction.

(b) Identify any molecule that acts as a catalyst or intermediate in this reaction.

14.120 Consider the following mechanism for the decomposition of nitramide (NH_2NO_2) in aqueous solution:

$$NH_2NO_2(aq) + OH^-(aq) \rightarrow NHNO_2^-(aq) + H_2O(l)$$
$$NHNO_2^-(aq) \rightarrow N_2O(g) + OH^-(aq)$$

(a) Write the chemical equation for the overall reaction.

(b) Identify the catalyst and the reaction intermediate.

(c) How will the rate of the overall reaction be affected if HCl is added to the solution?

14.121 In Problem 14.113, you wrote a mechanism for the nitric oxide–facilitated decomposition of ozone. Does your mechanism involve a catalyst or a reaction intermediate? Explain.

14.122 The rate of the reaction $A + B_2 \rightarrow AB + B$ is directly proportional to the concentration of B_2, independent of the concentration of A, and directly proportional to the concentration of a substance C.

(a) What is the rate law?

(b) Write a mechanism that agrees with the experimental facts.

(c) What is the role of C in this reaction? Why doesn't C appear in the chemical equation for the overall reaction?

14.123 The half-life of a typical peptide bond (the C—N bond in a protein backbone) in neutral aqueous solution is about 500 years. When a protease enzyme acts on a peptide bond, the bond's half-life is about 0.010 s. Assuming that these half-lives correspond to first-order reactions, by what factor does the enzyme increase the rate of the peptide bond breaking reaction?

MULTICONCEPT PROBLEMS

14.124 Consider the reaction $H_2(g) + I_2(g) \rightarrow 2\ HI(g)$. The reac-
RAN tion of a fixed amount of H_2 and I_2 is studied in a cylinder
fitted with a movable piston. Indicate the effect of each of
the following changes on the rate of the reaction.

(a) An increase in temperature at constant volume

(b) An increase in volume at constant temperature

(c) The addition of a catalyst

(d) The addition of argon (an inert gas) at constant volume

14.125 Concentration–time data for the conversion of A and B to
D are listed in the following table.

Experiment	Time (s)	[A]	[B]	[C]	[D]
1	0	5.00	2.00	1.00	0.00
	60	4.80	1.90	1.00	0.10
2	0	10.00	2.00	1.00	0.00
	60	9.60	1.80	1.00	0.20
3	0	5.00	4.00	1.00	0.00
	60	4.80	3.90	1.00	0.10
4	0	5.00	2.00	2.00	0.00
	60	4.60	1.80	2.00	0.20

(a) Write a balanced equation for the reaction.

(b) What is the reaction order with respect to A, B, and C?
What is the overall reaction order?

(c) What is the rate law?

(d) Is a catalyst involved in this reaction? Explain.

(e) Suggest a mechanism that is consistent with the data.

(f) Calculate the rate constant for the formation of D.

14.126 Consider the following concentration–time data for the
RAN reaction of iodide ion and hypochlorite ion (OCl^-). The
products are chloride ion and hypoiodite ion (OI^-).

Experiment	Time(s)	$[I^-]$	$[OCl^-]$	$[OH^-]$
1	0	2.40×10^{-4}	1.60×10^{-4}	1.00
	10	2.17×10^{-4}	1.37×10^{-4}	1.00
2	0	1.20×10^{-4}	1.60×10^{-4}	1.00
	10	1.08×10^{-4}	1.48×10^{-4}	1.00
3	0	2.40×10^{-4}	4.00×10^{-5}	1.00
	10	2.34×10^{-4}	3.40×10^{-5}	1.00
4	0	1.20×10^{-4}	1.60×10^{-4}	2.00
	10	1.14×10^{-4}	1.54×10^{-4}	2.00

(a) Write a balanced equation for the reaction.

(b) Determine the rate law, and calculate the value of the
rate constant.

(c) Does the reaction occur by a single-step mechanism?
Explain.

(d) Propose a mechanism that is consistent with the rate
law, and express the rate constant in terms of the rate
constants for the elementary steps in your mechanism.
(Hint: Transfer of an H^+ ion between H_2O and OCl^- is
a rapid reversible reaction.)

14.127 Consider the reversible, first-order interconversion of two
RAN molecules A and B:

$$A \underset{k_r}{\overset{k_f}{\rightleftharpoons}} B$$

where $k_f = 3.0 \times 10^{-3}\ s^{-1}$ is the rate constant for the forward reaction and $k_r = 1.0 \times 10^{-3}\ s^{-1}$ is the rate constant
for the reverse reaction. We'll see in Chapter 15 that a reaction does not go to completion but instead reaches a state of
equilibrium with comparable concentrations of reactants and
products if the rate constants k_f and k_r have comparable values.

(a) What are the rate laws for the forward and reverse
reactions?

(b) Draw a qualitative graph that shows how the rates of
the forward and reverse reactions vary with time.

(c) What are the relative concentrations of B and A when
the rates of the forward and reverse reactions become
equal?

14.128 Assume that you are studying the first-order conversion of
RAN a reactant X to products in a reaction vessel with a constant volume of 1.000 L. At 1 p.m., you start the reaction
at 25 °C with 1.000 mol of X. At 2 p.m., you find that
0.600 mol of X remains, and you immediately increase the
temperature of the reaction mixture to 35 °C. At 3 p.m.,
you discover that 0.200 mol of X is still present. You want
to finish the reaction by 4 p.m. but need to continue it until
only 0.010 mol of X remains, so you decide to increase the
temperature once again. What is the minimum temperature required to convert all but 0.010 mol of X to products
by 4 p.m.?

14.129 The half-life for the first-order decomposition of N_2O_4 is
RAN 1.3×10^{-5} s.

$$N_2O_4(g) \rightarrow 2\ NO_2(g)$$

If N_2O_4 is introduced into an evacuated flask at a pressure
of 17.0 mm Hg, how many seconds are required for the
pressure of NO_2 to reach 1.3 mm Hg?

13.130 Some reactions are so rapid that they are said to be
RAN diffusion-controlled; that is, the reactants react as quickly
as they can collide. An example is the neutralization of
H_3O^+ by OH^-, which has a second-order rate constant of
$1.3 \times 10^{11}\ M^{-1}\ s^{-1}$ at 25 °C.

(a) If equal volumes of 2.0 M HCl and 2.0 M NaOH are
mixed instantaneously, how much time is required for
99.999% of the acid to be neutralized?

(b) Under normal laboratory conditions, would you expect
the rate of the acid–base neutralization to be limited by
the rate of the reaction or by the speed of mixing?

14.131 The reaction $2\,NO(g) + O_2(g) \rightarrow 2\,NO_2(g)$ has the third-
order rate law rate $= k[NO]^2[O_2]$, where $k = 25\ M^{-2}\ s^{-1}$.
Under the condition that $[NO] = 2\,[O_2]$, the integrated
rate law is

$$\frac{1}{[O_2]^2} = 8\,kt + \frac{1}{([O_2]_0)^2}$$

What are the concentrations of NO, O_2, and NO_2 after
100.0 s if the initial concentrations are $[NO] = 0.0200\ M$
and $[O_2] = 0.0100\ M$?

14.132 Consider the following data for the gas-phase decomposition of NO_2:

$$2\,NO_2(g) \rightarrow 2\,NO(g) + O_2(g)$$

Temperature (K)	Initial [NO$_2$]	Initial Rate of Decomposition of NO$_2$(M/s)
600	0.0010	5.4×10^{-7}
600	0.0020	2.2×10^{-6}
700	0.0020	5.2×10^{-5}

If 0.0050 mol of NO_2 is introduced into a 1.0 L flask and
allowed to decompose at 650 K, how many seconds does it
take for the NO_2 concentration to drop to 0.0010 M?

14.133 Use the following initial rate data to determine the activation energy (in kJ/mol) for the reaction $A + B \rightarrow C$.

Experiment	Temperature (K)	Initial [A]	Initial [B]	Initial Reaction Rate (M/s)
1	700	0.20	0.10	1.8×10^{-5}
2	700	0.40	0.10	3.6×10^{-5}
3	700	0.10	0.20	3.6×10^{-5}
4	600	0.50	0.50	4.3×10^{-5}

14.134 The following experimental data were obtained in a study
of the reaction $2\,HI(g) \rightarrow H_2(g) + I_2(g)$. Predict the concentration of HI that would give a rate of $1.0 \times 10^{-5}\ M/s$
at 650 K.

Experiment	Temperature (K)	Initial [HI]	Initial Reaction Rate (M/s)
1	700	0.10	1.8×10^{-5}
2	700	0.30	1.6×10^{-4}
3	800	0.20	3.9×10^{-3}
4	650	?	1.0×10^{-5}

14.135 Polytetrafluoroethylene (Teflon) decomposes when heated
above 500 °C. Rate constants for the decomposition are
$2.60 \times 10^{-4}\ s^{-1}$ at 530 °C and $9.45 \times 10^{-3}\ s^{-1}$ at 620 °C.
(a) What is the activation energy in kJ/mol?
(b) What is the half-life of this substance at 580 °C?

14.136 The reaction $A \rightarrow C$ is first order in the reactant A and is
known to go to completion. The product C is colored and
absorbs light strongly at 550 nm, while the reactant and

intermediates are colorless. A solution of A was prepared,
and the absorbance of C at 550 nm was measured as a
function of time. (Note that the absorbance of C is directly
proportional to its concentration.) Use the following data
to determine the half-life of the reaction.

Time (s)	Absorbance
0	0.000
10	0.444
20	0.724
100	1.188
200	1.200
500	1.200

14.137 Values of $E_a = 6.3\ kJ/mol$ and $A = 6.0 \times 10^8/(M \cdot s)$
have been measured for the bimolecular reaction:

$$NO(g) + F_2(g) \rightarrow NOF(g) + F(g)$$

(a) Calculate the rate constant at 25 °C.
(b) The product of the reaction is nitrosyl fluoride. Its formula is usually written as NOF, but its structure is actually ONF. Is the ONF molecule linear or bent?
(c) Draw a plausible transition state for the reaction. Use
dashed lines to indicate the atoms that are weakly
linked together in the transition state.
(d) Why does the reaction have such a low activation
energy?

14.138 A 1.50 L sample of gaseous HI having a density of
$0.0101\ g/cm^3$ is heated at 410 °C. As time passes, the
HI decomposes to gaseous H_2 and I_2. The rate law is
$-\Delta[HI]/\Delta t = k[HI]^2$, where $k = 0.031/(M \cdot min)$ at
410 °C.
(a) What is the initial rate of production of I_2 in molecules/min?
(b) What is the partial pressure of H_2 after a reaction time
of 8.00 h?

14.139 The rate constant for the decomposition of gaseous NO_2 to
NO and O_2 is $4.7/(M \cdot s)$ at 383 °C. Consider the decomposition of a sample of pure NO_2 having an initial pressure
of 746 mm Hg in a 5.00 L reaction vessel at 383 °C.
(a) What is the order of the reaction?
(b) What is the initial rate of formation of O_2 in g/(L $\cdot$ s)?
(c) What is the mass of O_2 in the vessel after a reaction time
of 1.00 min?

14.140 The rate constant for the first-order decomposition of gaseous N_2O_5 to NO_2 and O_2 is $1.7 \times 10^{-3}\ s^{-1}$ at 55 °C.
(a) If 2.70 g of gaseous N_2O_5 is introduced into an evacuated 2.00 L container maintained at a constant temperature of 55 °C, what is the total pressure in the
container after a reaction time of 13.0 minutes?
(b) Use the data in Appendix B to calculate the initial rate
at which the reaction mixture absorbs heat (in J/s). You
may assume that the heat of the reaction is independent
of temperature.
(c) What is the total amount of heat absorbed (in kilojoules) after a reaction time of 10.0 min?

14.141 For the thermal decomposition of nitrous oxide, $2 N_2O(g) \rightarrow 2 N_2(g) + O_2(g)$, values of the parameters in the Arrhenius equation are $A = 4.2 \times 10^9 \, s^{-1}$ and $E_a = 222 \, kJ/mol$. If a stream of N_2O is passed through a tube 25 mm in diameter and 20 cm long at a flow rate of 0.75 L/min at what temperature should the tube be maintained to have a partial pressure of 1.0 mm of O_2 in the exit gas? Assume that the total pressure of the gas in the tube is 1.50 atm.

14.142 A 0.500 L reaction vessel equipped with a movable piston is filled completely with a 3.00% aqueous solution of hydrogen peroxide. The H_2O_2 decomposes to water and O_2 gas in a first-order reaction that has a half-life of 10.7 h. As the reaction proceeds, the gas formed pushes the piston against a constant external atmospheric pressure of 738 mm Hg. Calculate the PV work done (in joules) after a reaction time of 4.02 h. (You may assume that the density of the solution is 1.00 g/mL and that the temperature of the system is maintained at 20 °C.)

14.143 At 791 K and relatively low pressures, the gas-phase decomposition of acetaldehyde (CH_3CHO) is second order in acetaldehyde.

$$CH_3CHO(g) \rightarrow CH_4(g) + CO(g)$$

The total pressure of a particular reaction mixture was found to vary as follows:

Time (min)	0	75	148	308	605
Total pressure (atm)	0.500	0.583	0.641	0.724	0.808

(a) Use the pressure data to determine the value of the rate constant in units of $atm^{-1} \, s^{-1}$.

(b) What is the rate constant in the usual units of $M^{-1} \, s^{-1}$?

(c) If the volume of the reaction mixture is 1.00 L, what is the total amount of heat liberated (in joules) after a reaction time of 605 s?

chapter 15

Chemical Equilibrium

Contents

Skiing the mountains may cause *altitude sickness* due to lower partial pressures of oxygen at high elevations. The transport of O_2 through the blood to muscle tissue depends on a chemical equilibrium involving hemoglobin, an oxygen-carrying protein.

How does high altitude affect oxygen transport in the body?

The answer to this question can be found on page 637 in the INQUIRY ?

At the beginning of Chapter 14, we raised three key questions about chemical reactions: What happens? How fast and by what mechanism does it happen? To what extent does it happen? The answer to the first question is given by the stoichiometry of the balanced chemical equation, and the answer to the second question is given by the kinetics of the reaction. In this chapter, we'll look at the answer to the third question: How far does a reaction proceed toward completion before it reaches a state of **chemical equilibrium**—a state in which the concentrations of reactants and products no longer change?

> **Chemical equilibrium**   The state reached when the concentrations of reactants and products remain constant over time.

We've already touched on the concept of equilibrium in connection with our study of the evaporation of liquids (Section 11.2). When a liquid evaporates in a closed container, it soon gives rise to a constant vapor pressure because of a dynamic equilibrium in which the number of molecules leaving the liquid equals the number returning from the vapor. Chemical reactions behave similarly. They can occur in both forward and reverse directions, and when the rates of the forward and reverse reactions become equal, the concentrations of reactants and products remain constant. At that point, the chemical system is at equilibrium.

Chemical equilibria are important in numerous industrial, biological, and environmental processes. For example, equilibria involving O_2 molecules and the protein hemoglobin play a crucial role in the transport and delivery of oxygen from our lungs to cells throughout our body. Similar equilibria involving CO molecules and hemoglobin account for the toxicity of carbon monoxide.

A mixture of reactants and products in the equilibrium state is called an **equilibrium mixture**. In this chapter, we'll address a number of important questions about the composition of equilibrium mixtures: What is the relationship between the concentrations of reactants and products in an equilibrium mixture? How can we determine equilibrium concentrations from initial concentrations? What factors can we exploit to alter the composition of an equilibrium mixture? This last question is particularly important when choosing conditions for the synthesis of industrial chemicals such as hydrogen, ammonia, and methanol (CH_3OH).

▲ In a closed container, liquid bromine and its vapor are in a dynamic equilibrium.

15.1 THE EQUILIBRIUM STATE

In previous chapters, we've generally assumed that chemical reactions result in complete conversion of reactants to products. Many reactions, however, do not go to completion. Take, for example, the decomposition of the colorless gas dinitrogen tetroxide (N_2O_4) to the dark brown gas nitrogen dioxide (NO_2).

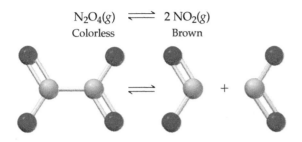

$$N_2O_4(g) \rightleftharpoons 2\,NO_2(g)$$
Colorless   Brown

FIGURE 15.1 shows the decomposition reaction of colorless dinitrogen tetroxide (N_2O_4) gas to form brown nitrogen dioxide (NO_2) gas. Initially, only N_2O_4 is present, and only the forward reaction of N_2O_4 decomposing into NO_2 occurs. As the concentration of the product (NO_2) increases, the reverse reaction of NO_2 molecules combining to make N_2O_4 begins to occur at an appreciable rate. The brown color caused by the formation of NO_2 continues to intensify until the rate of the forward reaction equals the rate of the reverse reaction and equilibrium has been reached.

▲ Hong Kong on a smoggy day. The brown color of the smog is due primarily to NO_2 formed from atmospheric reactions of combustion products.

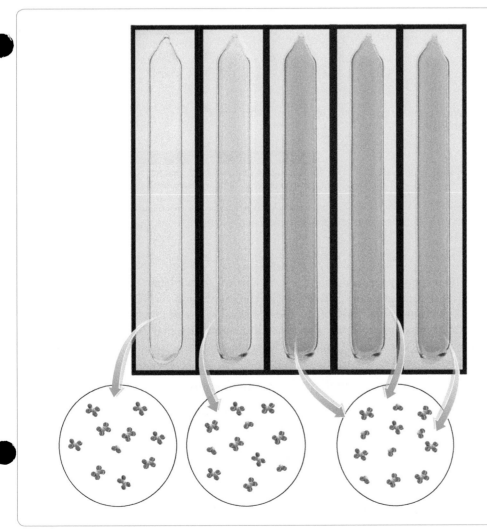

◀ **FIGURE 15.1**

The tubes show gases in the equilibrium reaction $N_2O_4(g) \rightleftharpoons NO_2(g)$. Initially, when the concentration of NO_2 is low, the forward reaction converts colorless N_2O_4 into brown NO_2. As NO_2 forms, the reverse reaction begins to occur. When the rates of the forward and reverse reactions become equal, equilibrium has been established, and a consistent brown color is observed.

FIGURE 15.2 shows a plot of the concentrations of N_2O_4 and NO_2 as a function of time in two different experiments carried out at 25 °C. In the first experiment (Figure 15.2a), we begin with N_2O_4 at an initial concentration of 0.0400 M. The formation of NO_2 is indicated by the appearance of a brown color, and its concentration can be monitored by measuring the intensity of the color with a spectrophotometer. According to the balanced equation, 2.0 mol of NO_2 forms for each mole of N_2O_4 that reacts, so the rate of formation of NO_2 is twice the rate of decomposition of N_2O_4. As time passes, the concentration of N_2O_4 decreases and the concentration of NO_2 increases until both concentrations level off at constant, equilibrium values: $[N_2O_4] = 0.0337$ M; $NO_2 = 0.0125$ M.

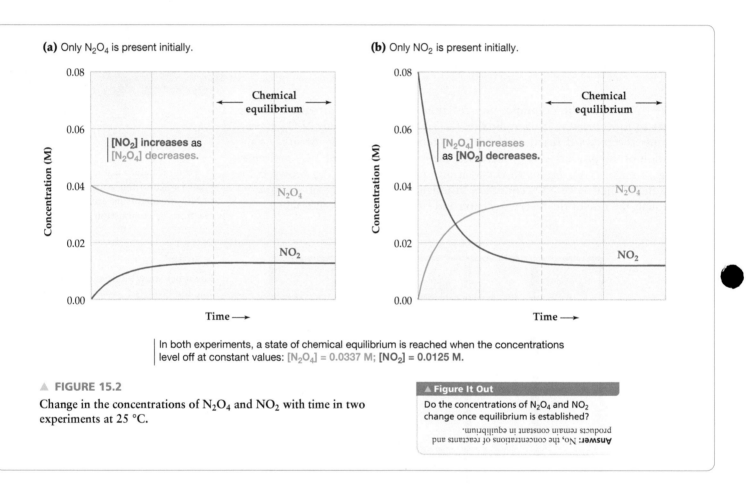

(a) Only N_2O_4 is present initially.

(b) Only NO_2 is present initially.

In both experiments, a state of chemical equilibrium is reached when the concentrations level off at constant values: $[N_2O_4] = 0.0337$ M; $[NO_2] = 0.0125$ M.

▲ **FIGURE 15.2**

Change in the concentrations of N_2O_4 and NO_2 with time in two experiments at 25 °C.

▲ **Figure It Out**

Do the concentrations of N_2O_4 and NO_2 change once equilibrium is established?

Answer: No, the concentrations of reactants and products remain constant in equilibrium.

In the second experiment, shown in Figure 15.2b, we begin with NO_2 as the sole reactant at a concentration of 0.0800 M. The conversion of NO_2 to N_2O_4 proceeds until the concentrations level off at the same values as obtained in the first experiment. Taken together, the two experiments demonstrate that the interconversion of N_2O_4 and NO_2 is reversible and that the same equilibrium state is reached starting from either substance.

$$N_2O_4(g) \rightleftharpoons 2\ NO_2(g) \quad \text{Reaction occurs in both directions}$$

To indicate that the reaction can proceed in both forward and reverse directions, we write the balanced equation with two arrows, one pointing from reactants to products and the other pointing from products to reactants. (The terms "reactants" and "products" could be confusing in this context because the products of the forward reaction are the reactants in the reverse reaction. To avoid confusion, we'll restrict the term *reactants* to the substances on the left side of the chemical equation and the term *products* to the substances on the right side of the equation.)

Strictly speaking, *all* chemical reactions are reversible. What we sometimes call irreversible reactions are simply those that proceed *nearly* to completion, so that the equilibrium mixture contains almost all products and almost no reactants. For such reactions, the reverse reaction is often too slow to be detected.

The reason why chemical reactions reach an equilibrium state follows from chemical kinetics (Chapter 14). Consider again the interconversion of N_2O_4 and NO_2. The rate of the forward reaction ($N_2O_4 \longrightarrow 2\,NO_2$) and the reverse reaction ($N_2O_4 \longleftarrow 2\,NO_2$) are given by the following **rate laws**:

$$\text{Rate forward} = k_f[N_2O_4] \qquad \text{Rate reverse} = k_r[NO_2]^2$$

The rate of the forward reaction decreases as the concentration of the reactant N_2O_4 decreases, while the rate of the reverse reaction increases as the concentration of the product NO_2 increases. Eventually, the decreasing rate of the forward reaction and the increasing rate of the reverse reaction become equal. At that point, there are no further changes in concentrations not because the reactions stop but because N_2O_4 and NO_2 both disappear as fast as they're formed. Thus, chemical equilibrium is a dynamic state in which forward and reverse reactions continue at equal rates so that there is no *net* conversion of reactants to products (**FIGURE 15.3**).

REMEMBER. . .

The reaction rate is given by the **rate law** and ordinarily depends on the concentrations of at least some of the reacting species, usually increasing with increasing concentration and decreasing with decreasing concentration (Section 14.2).

BIG IDEA Question 1

What two quantities are equal when a chemical reaction has reached equilibrium?

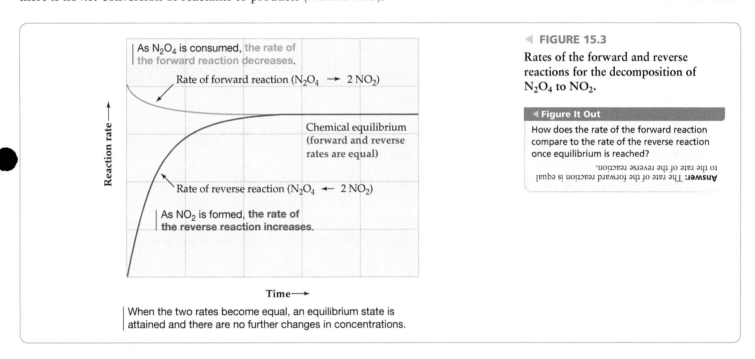

◀ **FIGURE 15.3**

Rates of the forward and reverse reactions for the decomposition of N_2O_4 to NO_2.

◀ **Figure It Out**

How does the rate of the forward reaction compare to the rate of the reverse reaction once equilibrium is reached?

Answer: The rate of the forward reaction is equal to the rate of the reverse reaction.

15.2 THE EQUILIBRIUM CONSTANT K_C

TABLE 15.1 lists concentration data for the experiments in Figure 15.2 along with data for three additional experiments. In experiments 1 and 2, the equilibrium mixtures have identical compositions because the initial concentration of N_2O_4 in experiment 1 is half the initial concentration of NO_2 in experiment 2; that is, the total number of N and O atoms is the same in both experiments. In experiments 3–5, different initial concentrations of N_2O_4 and/or NO_2 give different equilibrium concentrations. In all the experiments, however, the equilibrium concentrations are related. The last column of Table 15.1 shows that, at equilibrium, the expression $[NO_2]^2/[N_2O_4]$ has a constant value of, 4.64×10^{-3} M, within experimental error.

The expression $[NO_2]^2/[N_2O_4]$ appears to be related to the balanced equation for the reaction $N_2O_4(g) \rightleftharpoons 2\,NO_2(g)$ in that the concentration of the product is in the numerator, raised to the power of its coefficient in the balanced equation, and the concentration of the reactant is in the denominator. Is there an analogous expression with a constant value for every chemical reaction? If so, how is the form of that expression related to the balanced equation for the reaction?

Go to
eText

BIG IDEA Question 2

Does the value of the equilibrium constant depend on the initial concentrations of reactants and products?

▲ Another example of dynamic equilibrium: If the rate at which people move from the first floor to the second equals the rate at which people move from the second floor to the first, the number of people on each floor remains constant and the two populations are in dynamic equilibrium.

TABLE 15.1   Concentration Data at 25 °C for the Reaction $N_2O_4(g) \rightleftharpoons 2\,NO_2(g)$

	Initial Concentrations (M)		Equilibrium Concentrations (M)		Equilibrium Constant Expression
Experiment	$[N_2O_4]$	$[NO_2]$	$[N_2O_4]$	$[NO_2]$	$[NO_2]^2/[N_2O_4]$
1	0.0400	0.0000	0.0337	0.0125	4.64×10^{-3}
2	0.0000	0.0800	0.0337	0.0125	4.64×10^{-3}
3	0.0600	0.0000	0.0522	0.0156	4.66×10^{-3}
4	0.0000	0.0600	0.0246	0.0107	4.65×10^{-3}
5	0.0200	0.0600	0.0429	0.0141	4.63×10^{-3}

To answer those questions, let's consider a general reversible reaction:

$$a\,A + b\,B \rightleftharpoons c\,C + d\,D$$

where A and B are the reactants, C and D are the products, and a, b, c, and d are their respective stoichiometric coefficients in the balanced chemical equation. In 1864 on the basis of experimental studies of many reversible reactions, the Norwegian chemists Cato Maximilian Guldberg and Peter Waage proposed that the concentrations in an equilibrium mixture are related by the following **equilibrium equation**, where K_c is the *equilibrium constant* and the expression on the right side is called the *equilibrium constant expression*.

Equilibrium constant expression

$$K_c = \frac{[C]^c[D]^d}{[A]^a[B]^b}$$

$\longleftarrow$ Products
$\longleftarrow$ Reactants

Equilibrium constant $\nearrow$ $\nwarrow$ Equilibrium-constant expression

LOOKING AHEAD...
In Section 15.4, we'll see why **pure solids** and **pure liquids** are omitted from the equilibrium-constant expression.

As usual, square brackets indicate the molar concentration of the substance within the brackets, hence the subscript c for "concentration" in K_c. The substances in the equilibrium constant expression may be gases or molecules and ions in a solution but may not be **pure solids** or **pure liquids** for reasons that we'll explain in Section 15.4. The equilibrium equation is also known as the *law of mass action* because in the early days of chemistry, concentration was called "active mass."

The **equilibrium constant K_c** is the number obtained by multiplying the equilibrium concentrations of all the products and dividing by the product of the equilibrium concentrations of all the reactants, with the concentration of each substance raised to the power of its coefficient in the balanced chemical equation. No matter what the individual equilibrium concentrations may be in a particular experiment, *the equilibrium constant for a reaction at a particular temperature always has the same value*. Thus, the equilibrium equation for the decomposition reaction $N_2O_4(g) \rightleftharpoons 2\,NO_2(g)$ is

$$K_c = \frac{[NO_2]^2}{[N_2O_4]} = 4.64 \times 10^{-3} \qquad \text{at 25 °C}$$

where the equilibrium constant expression is $[NO_2]^2/[N_2O_4]$ and the equilibrium constant K_c has a value of 4.64×10^{-3} at 25 °C (Table 15.1).

Values of K_c are generally reported without units because the concentrations in the equilibrium constant expression are considered to be concentration ratios in which the molarity of each substance is divided by its molarity (1 M) in the **thermodynamic standard state** (Section 9.5). Because the units cancel, the concentration ratios and the values of K_c are dimensionless. For experiment 1 in Table 15.1, for example,

REMEMBER. . .
The **thermodynamic standard state** is the set of conditions under which thermodynamic measurements are reported: 1 M concentration for each solute in solution, 1 atm pressure for each gas, and a specified temperature, usually 25 °C (Section 9.5).

$$K_c = \frac{[NO_2]^2}{[N_2O_4]} = \frac{\left(\dfrac{0.0125 \ M}{1 \ M}\right)^2}{\left(\dfrac{0.0337 \ M}{1 \ M}\right)} = 4.64 \times 10^{-3} \qquad \text{at } 25 \ °C$$

Equilibrium constants are temperature-dependent, so the temperature must be given when citing a value of K_c. For example, K_c for the decomposition of N_2O_4 increases from 4.64×10^{-3} at 25 °C to 1.53 at 127 °C.

The form of the equilibrium constant expression and the numerical value of the equilibrium constant depend on the form of the balanced chemical equation. Look again at the chemical equation and the equilibrium equation for a general reaction:

$$a\,A + b\,B \rightleftharpoons c\,C + d\,D \qquad K_c = \frac{[C]^c[D]^d}{[A]^a[B]^b}$$

If we write the chemical equation in the *reverse direction,* the new equilibrium constant expression is the reciprocal of the original expression and the new equilibrium constant K_c (reverse) is the reciprocal of the original equilibrium constant K_c.

$$c\,C + d\,D \rightleftharpoons a\,A + b\,B \qquad K_c \text{ (reverse)} = \frac{[A]^a[B]^b}{[C]^c[D]^d} = \frac{1}{K_c}$$

If a chemical equation is *multiplied* by a common factor n, the new equilibrium constant expression is the original expression raised to the power of n and the new equilibrium constant K_c (new) is equal to $(K_c)^n$.

$$n(a\,A + b\,B \rightleftharpoons c\,C + d\,D) \qquad K_c \text{ (new)} = \frac{[C]^{nc}[D]^{nd}}{[A]^{na}[B]^{nb}} = \left(\frac{[C]^c[D]^d}{[A]^a[B]^b}\right)^n = (K_c)^n$$

It's important to specify the form of the balanced equation for a given K_c because the value of K_c will change when the reaction is reversed or when it is multiplied by an integer.

Whenever chemical equations for two (or more) reactions are *added* to get the equation for an overall reaction, the equilibrium constant for the overall reaction equals the product of the equilibrium constants for the individual reactions.

$$a\,A \rightleftharpoons b\,B \qquad K_{c1} = \frac{[B]^b}{[A]^a}$$

$$b\,B \rightleftharpoons c\,C \qquad K_{c2} = \frac{[C]^c}{[B]^b}$$

$$\overline{\phantom{a\,A \rightleftharpoons c\,C \qquad K_c\text{(overall)} = \frac{[C]^c}{[A]^a}}}$$

$$a\,A \rightleftharpoons c\,C \qquad K_c\text{(overall)} = \frac{[C]^c}{[A]^a}$$

When the K_c values for the two added reactions are multiplied, the resulting quantity is the equilibrium constant for the overall reaction.

$$K_{c1} \times K_{c2} = \frac{[B]^b}{[A]^a} \times \frac{[C]^c}{[B]^b} = \frac{[C]^c}{[A]^a} = K_c\text{(overall)}$$

WORKED EXAMPLE 15.1

Writing Equilibrium Constant Expressions for Gas-Phase and Solution Reactions

Write the equilibrium constant expression (K_c) for each of the following reactions. For part **(b)**, derive the mathematical relationship to the equilibrium constant in part **(a)**.

(a) Synthesis of ammonia: $N_2(g) + 3\,H_2(g) \rightleftharpoons 2\,NH_3(g)$
(b) Decomposition of ammonia: $NH_3(g) \rightleftharpoons 1/2\,N_2(g) + 3/2\,H_2(g)$
(c) Synthesis of the gasoline additive methyl *tert*-butyl ether (MTBE):

$$CH_3OH(soln) + C_4H_9OH(soln) \underset{\text{catalyst}}{\overset{H^+}{\rightleftharpoons}} C_4H_9OCH_3(soln) + H_2O(soln)$$

Methanol   *tert*-Butyl alcohol   Methyl *tert*-butyl ether

In this equation, *soln* denotes a largely organic solution that also contains water.

▲ MTBE is a component of some gasolines but is being phased out because it has been found to be an unsafe contaminant of groundwater.

STRATEGY

The rules for writing the equilibrium constant expression apply to reactions in liquid solutions as well as to gas-phase reactions (reactions in gaseous solutions). Put the concentrations of the products in the numerator and the concentrations of the reactants in the denominator. Then raise the concentration of each substance to the power of its coefficient in the balanced chemical reaction.

SOLUTION

(a) $K_c = \dfrac{[NH_3]^2}{[N_2][H_2]^3}$

     Coefficient of NH_3 — [$NH_3]^2$
     Coefficient of H_2 — $[H_2]^3$

(b) $K_c\,(\text{new}) = \dfrac{[N_2]^{1/2}[H_2]^{3/2}}{[NH_3]} = \dfrac{1}{\sqrt{K_c}}$

     Coefficient of N_2 Coefficient of H_2

The balanced equation in part **(b)** is the equation in part **(a)** multiplied by 1/2 and reversed. Multiplying an equation by 1/2 changes the original equilibrium constant by raising it to the power of 1/2 (square root). Reversing an equation changes the equilibrium constant to the reciprocal of the original equilibrium constant.

(c) $K_c = \dfrac{[C_4H_9OCH_3]\,[H_2O]}{[CH_3OH]\,[C_4H_9OH]}$ ← Products / ← Reactants

---

▶ **PRACTICE 15.1** The oxidation of sulfur dioxide to give sulfur trioxide is an important step in the industrial process for the synthesis of sulfuric acid.

What is the equilibrium constant expression, K_c, for the reaction below?

$$2\,SO_2(g) + O_2(g) \rightleftharpoons 2\,SO_3(g)$$

What is the equilibrium constant expression, K_c (reverse), for the reaction below?

$$2\,SO_3(g) \rightleftharpoons 2\,SO_2(g) + O_2(g)$$

How is the equilibrium constant K_c (reverse) related to K_c?

▶ **APPLY 15.2** Nitrogen dioxide, a pollutant that contributes to photochemical smog, is formed by a series of two reactions.

$$\begin{aligned} N_2(g) + O_2(g) &\rightleftharpoons 2\,NO(g) & K_{c1} &= 4.3 \times 10^{-25} \\ 2\,NO(g) + O_2(g) &\rightleftharpoons 2\,NO_2(g) & K_{c2} &= 6.4 \times 10^{9} \end{aligned}$$

$$N_2(g) + 2\,O_2(g) \rightleftharpoons 2\,NO_2(g) \qquad K_c(\text{overall}) = ?$$

(a) Write the equilibrium constant expression for the overall reaction.
(b) What is the value of K_c(overall)?

All **PRACTICE** and **APPLY** problems are interactive in the eText.

WORKED EXAMPLE 15.2

Calculating the Equilibrium Constant K_c

The following concentrations were measured for an equilibrium mixture at 500 K: $[N_2] = 3.0 \times 10^{-2}$ M; $[H_2] = 3.7 \times 10^{-2}$ M; $[NH_3] = 1.6 \times 10^{-2}$ M. Calculate the equilibrium constant at 500 K for each of the reactions:

(a) $N_2(g) + 3\,H_2(g) \rightleftharpoons 2\,NH_3(g)$

(b) $2\,NH_3(g) \rightleftharpoons N_2(g) + 3\,H_2(g)$

STRATEGY

To calculate the value of the equilibrium constant, substitute the equilibrium concentrations into the equilibrium equation.

SOLUTION

(a) $K_c = \dfrac{[NH_3]^2}{[N_2][H_2]^3} = \dfrac{(1.6 \times 10^{-2})^2}{(3.0 \times 10^{-2})(3.7 \times 10^{-2})^3} = 1.7 \times 10^2$

(b) $K_c \text{ (reverse)} = \dfrac{[N_2][H_2]^3}{[NH_3]^2} = \dfrac{(3.0 \times 10^{-2})(3.7 \times 10^{-2})^3}{(1.6 \times 10^{-2})^2} = 5.9 \times 10^{-3}$

Note that K_c (reverse) is the reciprocal of K_c. That is,

$$5.9 \times 10^{-3} = \dfrac{1}{1.7 \times 10^2}$$

▶ **PRACTICE 15.3** The following equilibrium concentrations were measured at 800 K: $[SO_2] = 3.0 \times 10^{-3}$ M; $[O_2] = 3.5 \times 10^{-3}$ M; $[SO_3] = 5.0 \times 10^{-2}$ M. What is the value of the equilibrium constants K_c and K_c (reverse) at 800 K for each of the reactions?

$$2\,SO_2(g) + O_2(g) \rightleftharpoons 2\,SO_3(g) \quad K_c = \text{?}$$
$$2\,SO_3(g) \rightleftharpoons 2\,SO_2(g) + O_2(g) \quad K_c \text{ (reverse)} = \text{?}$$

▶ **APPLY 15.4** Lactic acid, which builds up in muscle tissue upon strenuous exercise, is partially dissociated in aqueous solution:

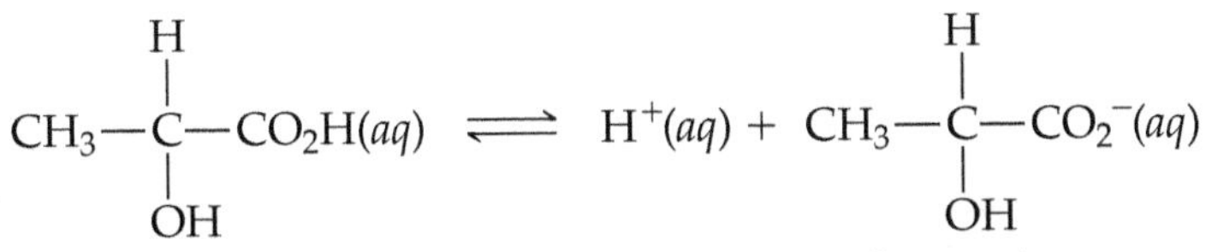

$$CH_3-\underset{\underset{OH}{|}}{\overset{\overset{H}{|}}{C}}-CO_2H(aq) \;\rightleftharpoons\; H^+(aq) + CH_3-\underset{\underset{OH}{|}}{\overset{\overset{H}{|}}{C}}-CO_2^{\,-}(aq)$$

Lactic acid   Lactate ion

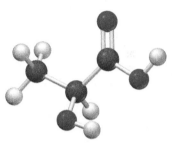

Lactic acid

The following equilibrium concentrations were measured at 25 °C:

Lactic acid $[C_3H_6O_3] = 9.64 \times 10^{-2}$ M; $[H^+] = 3.65 \times 10^{-3}$ M;
Lactate ion $[C_3H_5O_3^-] = 3.65 \times 10^{-3}$ M.

(a) Calculate the value of the equilibrium constant.
(b) What is the concentration of lactic acid ($C_3H_6O_3$) if at equilibrium

$$[C_3H_5O_3^-] = [H^+] = 1.17 \times 10^{-2} \text{ M?}$$

— **WORKED EXAMPLE 15.3**

Judging Whether a Mixture Is at Equilibrium

The following pictures represent mixtures of A molecules (red spheres) and B molecules (blue spheres), which interconvert according to the equation A $\rightleftharpoons$ B. If mixture (1) is at equilibrium, which of the other mixtures are also at equilibrium? Explain.

(1)   **(2)**   **(3)**   **(4)**

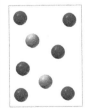   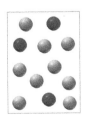

continued on next page

STRATEGY

The equilibrium constant for the reaction is given by $K_c = [B]/[A]$, where the concentrations are equilibrium concentrations in units of mol/L. Since the equilibrium constant expression has the same number of concentration terms in the numerator and denominator, the volume cancels and $K_c = $ (moles of B)/(moles of A). Because the number of moles is directly proportional to the number of molecules, $K_c = $ (molecules of B)/(molecules of A) in the mixture (1) at equilibrium. To determine whether the other mixtures are at equilibrium, count the number of molecules and compare the B/A ratio in mixtures (2)–(4) with the B/A ratio in the equilibrium mixture.

SOLUTION

For mixture (1), $K_c = [B]/[A] = 2/6 = 1/3$.
For mixture (2), $[B]/[A] = 4/4 = 1 \neq K_c$.
For mixture (3), $[B]/[A] = 3/9 = 1/3 = K_c$.
For mixture (4), $[B]/[A] = 9/3 = 3 \neq K_c$.
Mixture (3) is at equilibrium, but mixtures (2) and (4) are not at equilibrium because their equilibrium constant expression [B]/[A] does not equal K_c.

▶ **CONCEPTUAL PRACTICE 15.5** The following pictures represent mixtures that contain A atoms (red), B atoms (blue), and AB and B_2 molecules, which interconvert according to the equation $A + B_2 \rightleftharpoons AB + B$. If mixture (1) is at equilibrium, which of the other mixtures are also at equilibrium? Explain.

(1)    **(2)**    **(3)**    **(4)**

▶ **CONCEPTUAL APPLY 15.6** The equilibrium constant (K_c) for the reaction $B_2 \rightleftharpoons 2\,B$ is 4.5 at 25 °C. Draw a molecular picture similar to the ones shown in Worked Example 15.3 that represents a mixture of B_2 and B at equilibrium.

15.3 THE EQUILIBRIUM CONSTANT K_P

Because gas pressures are easily measured, equilibrium equations for gas-phase reactions are often written using **partial pressures** (Section 10.5) rather than molar concentrations. For example, the equilibrium equation for the decomposition of N_2O_4 can be written as

$$K_p = \frac{(P_{NO_2})^2}{P_{N_2O_4}} \qquad \text{for the reaction } N_2O_4(g) \rightleftharpoons 2\,NO_2(g)$$

where $P_{N_2O_4}$ and P_{NO_2} are the partial pressures (in atmospheres) of reactants and products at equilibrium, and the subscript p on K reminds us that the **equilibrium constant** K_p is defined using partial pressures. As is the case for K_c, values of K_p are dimensionless because the partial pressures in the equilibrium equation are actually ratios of partial pressures in atmospheres to the standard-state partial pressure of 1 atm. Thus, the units cancel. Note that the equilibrium equations for K_p and K_c have the same form except that the expression for K_p contains partial pressures instead of molar concentrations.

The constants K_p and K_c for the general gas-phase reaction $a\,A + b\,B \rightleftharpoons c\,C + d\,D$ are related because the pressure of each component in a mixture of ideal gases is directly proportional to its molar concentration. For component A, for example,

$$P_A V = n_A RT$$

so

$$P_A = \frac{n_A}{V} RT = [A]RT$$

Similarly, $P_B = [B]RT$, $P_C = [C]RT$, and $P_D = [D]RT$. The equilibrium equation for K_p is therefore given by

$$K_p = \frac{(P_C)^c(P_D)^d}{(P_A)^a(P_B)^b} = \frac{([C]RT)^c([D]RT)^d}{([A]RT)^a([B]RT)^b} = \frac{[C]^c[D]^d}{[A]^a[B]^b} \times (RT)^{(c+d)-(a+b)}$$

Because the first term on the right side equals K_c, the values of K_p and K_c are related by the equation

$$K_p = K_c(RT)^{\Delta n} \qquad \text{for the reaction } a\,A + b\,B \rightleftharpoons c\,C + d\,D$$

Here, R is the gas constant, 0.082 06 (L • atm)/(K • mol), T is the absolute temperature, and $\Delta n = (c + d) - (a + b)$ is the number of moles of gaseous products minus the number of moles of gaseous reactants.

For the decomposition of 1 mol of N_2O_4 to 2 mol of NO_2, $\Delta n = 2 - 1 = 1$, and $K_p = K_c(RT)$:

$$N_2O_4(g) \rightleftharpoons 2\ NO_2(g) \quad K_p = K_c(RT)$$

For the reaction of 1 mol of hydrogen with 1 mol of iodine to give 2 mol of hydrogen iodide, $\Delta n = 2 - (1 + 1) = 0$, and $K_p = K_c(RT)^0 = K_c$:

$$H_2(g) + I_2(g) \rightleftharpoons 2\ HI(g) \quad K_p = K_c$$

In general, K_p equals K_c only if the same number of moles of gases appear on both sides of the balanced chemical equation so that $\Delta n = 0$.

WORKED EXAMPLE 15.4

Determining the Equilibrium Constant K_p

Methane (CH_4) reacts with hydrogen sulfide to yield H_2 and carbon disulfide, a solvent used in manufacturing rayon and cellophane:

$$CH_4(g) + 2\ H_2S(g) \rightleftharpoons CS_2(g) + 4\ H_2(g)$$

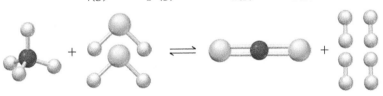

What is the value of K_p at 1000 K if the partial pressures in an equilibrium mixture at 1000 K are 0.20 atm of CH_4, 0.25 atm of H_2S, 0.52 atm of CS_2, and 0.10 atm of H_2?

STRATEGY

Write the equilibrium equation by setting K_p equal to the equilibrium constant expression using partial pressures. Put the partial pressures of products in the numerator and the partial pressures of reactants in the denominator, with the pressure of each substance raised to the power of its coefficient in the balanced chemical equation. Then substitute the partial pressures into the equilibrium equation and solve for K_p.

SOLUTION

$$K_p = \frac{(P_{CS_2})(P_{H_2})^4}{(P_{CH_4})(P_{H_2S})^2}$$

Coefficient of H_2

Coefficient of H_2S

$$K_p = \frac{(P_{CS_2})(P_{H_2})^4}{(P_{CH_4})(P_{H_2S})^2} = \frac{(0.52)(0.10)^4}{(0.20)(0.25)^2} = 4.2 \times 10^{-3}$$

Note that the partial pressures must be in units of atmospheres (not mm Hg) because the standard-state partial pressure for gases is 1 atm.

▶ **PRACTICE 15.7** In the industrial synthesis of hydrogen, mixtures of CO and H_2 are enriched in H_2 by allowing the CO to react with steam. The chemical equation for this water–gas shift reaction is

$$CO(g) + H_2O(g) \rightleftharpoons CO_2(g) + H_2(g)$$

What is the value of K_p at 700 K if the partial pressures in an equilibrium mixture at 700 K are 1.31 atm of CO, 10.0 atm of H_2O, 6.12 atm of CO_2, and 20.3 atm of H_2?

▶ **APPLY 15.8** At 25 °C, $K_p = 25$ for the reaction $H_2(g) + I_2(g) \rightleftharpoons 2\ HI(g)$. At equilibrium, the partial pressure of both H_2 and I_2 is 0.286 atm. What is the partial pressure of HI at equilibrium?

WORKED EXAMPLE 15.5

Relating the Equilibrium Constants K_p and K_c

Hydrogen is produced industrially by the steam–hydrocarbon re-forming process. The reaction that takes place in the first step of this process is

$$H_2O(g) + CH_4(g) \rightleftharpoons CO(g) + 3\,H_2(g)$$

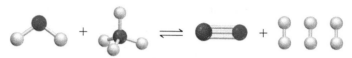

(a) If $K_c = 3.8 \times 10^{-3}$ at 1000 K, what is the value of K_p at the same temperature?
(b) If $K_p = 6.1 \times 10^4$ at 1125 °C, what is the value of K_c at 1125 °C?

IDENTIFY

Known	Unknown
(a) K_c, temperature (K), balanced reaction	K_p
(b) K_p, temperature (K), balanced reaction	K_c

STRATEGY

To calculate K_p from K_c, or vice versa, use the equation $K_p = K_c(RT)^{\Delta n}$, where R must be in units of $(L \cdot atm)/(K \cdot mol)$, T is the temperature in kelvin, and Δn is the number of moles of gaseous products minus the number of moles of gaseous reactants.

SOLUTION

(a) For this reaction, $\Delta n = (1 + 3) - (1 + 1) = 2$. Therefore,

$$K_p = K_c(RT)^{\Delta n} = K_c(RT)^2 = (3.8 \times 10^{-3})[(0.082\ 06)(1000)]^2 = 26$$

(b) Solving the equation $K_p = K_c(RT)^2$ for K_c gives

$$K_c = \frac{K_p}{(RT)^2} = \frac{6.1 \times 10^4}{[(0.082\ 06)(1398)]^2} = 4.6$$

Note that the temperature in these equations is the absolute temperature; 1125 °C corresponds to $1125 + 273 = 1398$ K.

▶ **PRACTICE 15.9** Nitric oxide reacts with oxygen to give nitrogen dioxide, an important reaction in the Ostwald process for the industrial synthesis of nitric acid:

$$2\,NO(g) + O_2(g) \rightleftharpoons 2\,NO_2(g)$$

If $K_c = 6.9 \times 10^5$ at 227 °C, what is the value of K_p at this temperature?

▶ **APPLY 15.10**   For the reaction

$$2\,SO_2(g) + O_2(g) \rightleftharpoons 2\,SO_3(g) \qquad K_c = 245\ (1000\ K)$$

What is K_p at 1000 K for the reaction $2\,SO_3(g) \rightleftharpoons 2\,SO_2(g) + O_2(g)$?

15.4 HETEROGENEOUS EQUILIBRIA

Thus far we've been discussing **homogeneous equilibria,** in which all reactants and products are in a single phase, usually either gaseous or solution. **Heterogeneous equilibria,** by contrast, are those in which reactants and products are present in more than one phase. Take, for example, the thermal decomposition of solid calcium carbonate, a reaction used in manufacturing cement:

$$\underset{\text{Limestone}}{CaCO_3(s)} \rightleftharpoons \underset{\text{Lime}}{CaO(s)} + CO_2(g)$$

When the reaction is carried out in a closed container, three phases are present at equilibrium: solid calcium carbonate, solid calcium oxide, and gaseous carbon dioxide. If pure solids and liquids are present in a chemical equation, then their concentrations are not included in the equilibrium constant expression because concentrations in the equilibrium equation are ratios of experimental concentrations to standard state concentrations. The ratio used in the equilibrium expression for a pure solid is equal to 1 because the standard state of a pure solid is the pure solid itself. Similarly, the standard state for a pure liquid is also the pure liquid, giving a ratio of 1. Therefore, the equilibrium equation simplifies, omitting concentrations of pure solids and liquids:

$$K_c = \frac{[CaO][CO_2]}{[CaCO_3]} = \frac{(1)[CO_2]}{(1)} = [CO_2]$$

The analogous equilibrium equation in terms of pressure is $K_p = P_{CO_2}$, where P_{CO_2} is the equilibrium pressure of CO_2 in atmospheres:

$$K_c = [CO_2] \qquad K_p = P_{CO_2}$$

As a general rule, the concentrations of pure solids and pure liquids are not included when writing an equilibrium equation. We include only the concentrations of gases and the concentrations of solutes in solutions because only those concentrations can be varied. The concentration of a pure solid or pure liquid is defined by its density, which can be expressed in units of molarity. The concentration of a pure solid and liquid will not change because its density is constant at a given temperature.

To establish equilibrium between solid $CaCO_3$, solid CaO, and gaseous CO_2, all three components must be present. It follows from the equations $K_c = [CO_2]$ and $K_p = P_{CO_2}$, however, that the concentration and pressure of CO_2 at equilibrium are constant, independent of how much solid CaO and $CaCO_3$ is present (**FIGURE 15.4**). If the temperature is changed, however, the concentration and pressure of CO_2 will also change because the values of K_c and K_p depend on temperature.

▲ The manufacture of cement begins with the thermal decomposition of limestone, $CaCO_3$ in large kilns.

Go to
eText

BIG IDEA Question 3

Which phases of substances are *not* included in the equilibrium expression?

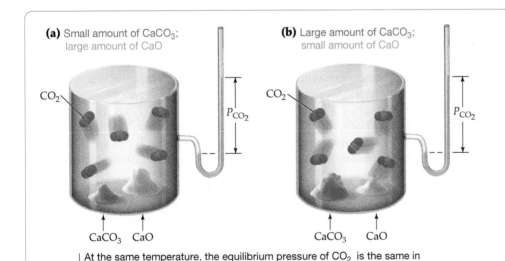

(a) Small amount of $CaCO_3$; large amount of CaO

(b) Large amount of $CaCO_3$; small amount of CaO

At the same temperature, the equilibrium pressure of CO_2 is the same in **(a)** and **(b)**, independent of how much solid $CaCO_3$ and CaO is present.

◀ **FIGURE 15.4**

Thermal decomposition of calcium carbonate:
$CaCO_3(s) \rightleftharpoons CaO(s) + CO_2(g)$.

◀ **Figure It Out**

What would happen to the equilibrium partial pressure of CO_2 in container (b) if the amount of $CaCO_3$ is doubled and the amount of CaO is halved? (Assume constant temperature.)

Answer: The partial pressure of CO_2 will not change as long as there is solid $CaCO_3$ and CaO present.

WORKED EXAMPLE 15.6

Writing Equilibrium Equations for Heterogeneous Equilibria

Write the equilibrium constant expression for each of the following reactions:

(a) $CO_2(g) + C(s) \rightleftharpoons 2\,CO(g)$
(b) $Hg(l) + Hg^{2+}(aq) \rightleftharpoons Hg_2^{2+}(aq)$

continued on next page

STRATEGY

Write the usual equilibrium constant expressions but omit the pure solid carbon in part (a) and the pure liquid mercury in part (b) because the ratio of their concentrations to their concentrations in the standard state is equal to 1.

SOLUTION

(a) $K_c = \dfrac{[CO]^2}{[CO_2]}$

Alternatively, because CO and CO_2 are gases, the equilibrium equation can be written using partial pressures:

$$K_p = \frac{(P_{CO})^2}{P_{CO_2}}$$

(b) $K_c = \dfrac{[Hg_2^{2+}]}{[Hg^{2+}]}$

In this case, it's not appropriate to write an expression for K_p because the reactants and products are not in the gas phase.

▶ **PRACTICE 15.11** Write the equilibrium constant expression (K_p) for the reaction:

$$SiCl_4(g) + 2\,H_2(g) \rightleftharpoons Si(s) + 4\,HCl(g)$$

▶ **APPLY 15.12** Magnesium hydroxide is the active ingredient in the antacid milk of magnesia. What is the value of K_c if solid magnesium hydroxide is dissolved in water and equilibrium concentrations are $[Mg^{2+}] = 1.65 \times 10^{-4}$ M and $[OH^-] = 3.30 \times 10^{-4}$ M?

15.5 USING THE EQUILIBRIUM CONSTANT

Knowing the value of the equilibrium constant for a chemical reaction lets us judge the extent of the reaction, predict the direction of the reaction, and calculate equilibrium concentrations from initial concentrations. Let's look at each possibility.

Judging the Extent of Reaction

The numerical value of the equilibrium constant for a reaction indicates the extent to which reactants are converted to products; that is, it measures how far the reaction proceeds before the equilibrium state is reached. Consider, for example, the reaction of H_2 with O_2, which has a very large equilibrium constant ($K_c = 2.4 \times 10^{47}$ at 500 K):

$$2\,H_2(g) + O_2(g) \rightleftharpoons 2\,H_2O(g)$$

$$K_c = \frac{[H_2O]^2}{[H_2]^2[O_2]} = 2.4 \times 10^{47} \qquad \text{at 500 K}$$

Because products appear in the numerator of the equilibrium constant expression and reactants are in the denominator, a very large value of K_c means that the equilibrium ratio of products to reactants is very large. In other words, the reaction proceeds nearly to completion. For example, if stoichiometric amounts of H_2 and O_2 are allowed to react and $[H_2O] = 0.10$ M at equilibrium, then the concentrations of H_2 and O_2 that remain at equilibrium are negligibly small: $[H_2] = 4.4 \times 10^{-17}$ M and $[O_2] = 2.2 \times 10^{-17}$ M. (Try substituting these concentrations into the equilibrium equation to show that they satisfy the equation.)

By contrast, if a reaction has a very small value of K_c, the equilibrium ratio of products to reactants is very small and the reaction proceeds hardly at all before equilibrium is reached. For example, the reverse of the reaction of H_2 with O_2 gives the same equilibrium mixture as obtained from the forward reaction ($[H_2] = 4.4 \times 10^{-17}$ M, $[O_2] = 2.2 \times 10^{-17}$ M, $[H_2O] = 0.10$ M). The reverse reaction does not occur to any appreciable extent, however, because its equilibrium constant is so small: $K_c(\text{reverse}) = 1/K_c = 1/(2.4 \times 10^{47}) = 4.2 \times 10^{-48}$.

$$2\,H_2O(g) \rightleftharpoons 2\,H_2(g) + O_2(g)$$

$$K_c = \frac{[H_2]^2[O_2]}{[H_2O]^2} = 4.2 \times 10^{-48} \qquad \text{at 500 K}$$

If a reaction has an intermediate value of K_c—say, a value in the range of 10^3 to 10^{-3}— then appreciable concentrations of both reactants and products are present in the equilibrium mixture. The reaction of hydrogen with iodine, for example, has $K_c = 57.0$ at 700 K:

$$H_2(g) + I_2(g) \rightleftharpoons 2\,HI(g) \qquad K_c = \frac{[HI]^2}{[H_2][I_2]} = 57.0 \quad \text{at 700 K}$$

If the equilibrium concentrations of H_2 and I_2 are both 0.010 M, then the concentration of HI at equilibrium is 0.075 M:

$$[HI]^2 = K_c[H_2][I_2]$$
$$[HI] = \sqrt{K_c[H_2][I_2]} = \sqrt{(57.0)(0.010)(0.010)} = 0.075 \text{ M}$$

Thus, the concentrations of both reactants and products—0.010 M and 0.075 M—are appreciable.

(Note that pressing the $\sqrt{x}$ key on a calculator gives a positive number. Remember, though, that the square root of a positive number can be positive or negative. Of the two roots for the concentration of HI, ±0.075 M, we choose the positive one because the concentration of a chemical substance is always a positive quantity.)

The gas-phase decomposition of N_2O_4 to NO_2 is another reaction with a value of K_c that is neither large nor small: $K_c = 4.64 \times 10^{-3}$ at 25 °C. Accordingly, equilibrium mixtures contain appreciable concentrations of both N_2O_4 and NO_2, as shown previously in Table 15.1.

We can make the following generalizations concerning the composition of equilibrium mixtures:

- If $K_c > 10^3$, products predominate over reactants. If K_c is very large, the reaction proceeds nearly to completion.
- If $K_c < 10^{-3}$, reactants predominate over products. If K_c is very small, the reaction proceeds hardly at all.
- If K_c is in the range 10^{-3} to 10^3, appreciable concentrations of both reactants and products are present.

These points are illustrated in **FIGURE 15.5**.

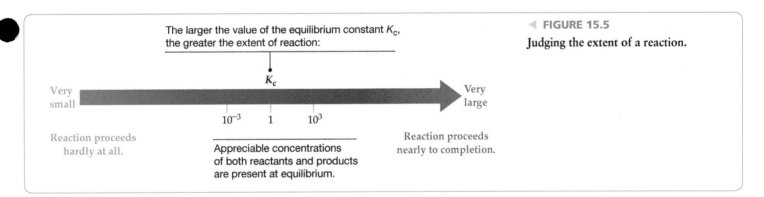

◀ **FIGURE 15.5**

Judging the extent of a reaction.

WORKED EXAMPLE 15.7

Judging the Extent of a Reaction

When the following reactions come to equilibrium, does the equilibrium mixture contain mostly reactants or mostly products?

(a) $NO(g) + O(g) \rightleftharpoons NO_2(g)$ $K_c = 1.5 \times 10^{48}$ at 25 °C
(b) $N_2(g) + O_2(g) \rightleftharpoons 2 NO(g)$ $K_c = 1.7 \times 10^{-3}$ at 2025 °C
(c) $N_2(g) + O_2(g) \rightleftharpoons 2 NO(g)$ $K_c = 4.5 \times 10^{-31}$ at 25 °C

STRATEGY

Examine the value of K_c. If K_c is large ($>10^3$), nearly all the reactants are converted to products. If K_c is small ($<10^{-3}$), only a small amount of reactants are converted to product. If K_c has an intermediate value ($10^{-3} < K_c < 10^3$), then appreciable amounts of both reactants and products are present at equilibrium.

SOLUTION

(a) K_c is large, mostly products present at equilibrium.

continued on next page

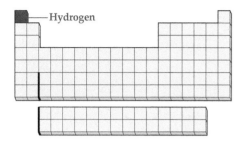
—Hydrogen

(b) K_c has an intermediate value therefore appreciable amounts of both reactants and products are present at equilibrium.

(c) K_c is small, mostly reactants present at equilibrium.

▶ **PRACTICE 15.13** When wine spoils, ethanol is oxidized to acetic acid as O_2 from the air reacts with the wine:

$$CH_3CH_2OH(aq) + O_2(aq) \rightleftharpoons CH_3CO_2H(aq) + H_2O(l)$$

Ethanol   Acetic acid

The value of K_c for this reaction at 25 °C is 1.2×10^{82}. What are the relative amounts of ethanol and acetic acid once the reaction has reached equilibrium?

▶ **APPLY 15.14** The value of K_c for the dissociation reaction $H_2(g) \rightleftharpoons 2 H(g)$ is 1.2×10^{-42} at 500 K. Does the equilibrium mixture contain mainly H_2 molecules or H atoms? Use the electron configuration of hydrogen to explain the equilibrium position.

Predicting the Direction of Reaction

Let's look again at the gaseous reaction of hydrogen with iodine:

$$H_2(g) + I_2(g) \rightleftharpoons 2HI(g) \quad K_c = 57.0 \text{ at } 700 \text{ K}$$

Suppose that we have a mixture of $H_2(g)$, $I_2(g)$, and $HI(g)$ at 700 K with concentrations $[H_2]_t = 0.10$ M, $[I_2]_t = 0.20$ M, and $[HI]_t = 0.40$ M. (The subscript t on the concentration symbols means that the concentrations were measured at some arbitrary time t, not necessarily at equilibrium.) If we substitute these concentrations into the equilibrium constant expression, we obtain a value called the **reaction quotient Q_c**.

Reaction quotient $Q_c = \dfrac{[HI]_t^2}{[H_2]_t[I_2]_t} = \dfrac{(0.40)^2}{(0.10)(0.20)} = 8.0$

The reaction quotient Q_c is defined in the same way as the equilibrium constant K_c except that the concentrations in Q_c are not necessarily equilibrium values.

For the case at hand, the numerical value of Q_c (8.0) does not equal K_c (57.0), so the mixture of $H_2(g)$, $I_2(g)$, and $HI(g)$ is not at equilibrium. As time passes, though, reaction will occur, changing the concentrations and thus changing the value of Q_c in the direction of K_c. After a sufficiently long time, an equilibrium state will be reached and $Q_c = K_c$.

The reaction quotient Q_c is useful because it lets us predict the direction of reaction by comparing the values of Q_c and K_c. If Q_c is less than K_c, movement toward equilibrium increases Q_c by converting reactants to products (that is, net reaction proceeds from left to right). If Q_c is greater than K_c, movement toward equilibrium decreases Q_c by converting products to reactants (that is, net reaction proceeds from right to left). If Q_c equals K_c, the reaction mixture is already at equilibrium, and no net reaction occurs.

Thus, we can make the following generalizations concerning the direction of the reaction:

- If $Q_c < K_c$, net reaction goes from left to right (reactants to products).
- If $Q_c > K_c$, net reaction goes from right to left (products to reactants).
- If $Q_c = K_c$, no net reaction occurs.

These points are illustrated in **FIGURE 15.6**.

Go to eText

BIG IDEA Question 4

For a reaction, $Q_c = 25$ and $K_c = 2.5$. Which direction will the reaction shift to reach equilibrium?

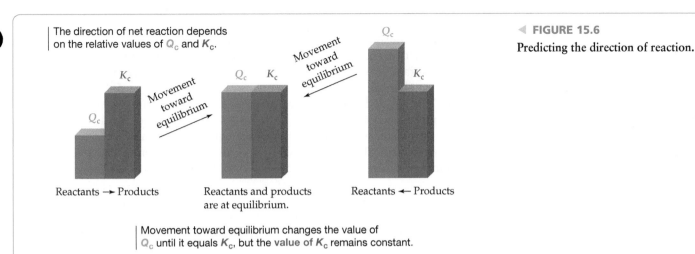

The direction of net reaction depends on the relative values of Q_c and K_c.

K_c

Movement toward equilibrium

Q_c

Q_c K_c

Movement toward equilibrium

Q_c

K_c

Reactants → Products

Reactants and products are at equilibrium.

Reactants ← Products

◀ **FIGURE 15.6**

Predicting the direction of reaction.

Movement toward equilibrium changes the value of Q_c until it equals K_c, but the value of K_c remains constant.

WORKED EXAMPLE 15.8

Predicting the Direction of Reaction

A mixture of 1.57 mol of N_2, 1.92 mol of H_2, and 8.13 mol of NH_3 is introduced into a 20.0 L reaction vessel at 500 K. At this temperature, the equilibrium constant K_c for the reaction $N_2(g) + 3 H_2(g) \rightleftharpoons 2 NH_3(g)$ is 1.7×10^2. Is the reaction mixture at equilibrium? If not, what is the direction of the net reaction?

STRATEGY

To determine whether the reaction mixture is at equilibrium, we need to calculate the value of the reaction quotient Q_c and then compare it with the equilibrium constant K_c. If the mixture is not at equilibrium, the relative values of Q_c and K_c tell us the direction of the net reaction. Because we are given amounts in moles, we must first convert moles to molar concentrations before substituting into the expression for Q_c.

SOLUTION

The initial concentration of N_2 is $(1.57 \text{ mol})/(20.0 \text{ L}) = 0.0785$ M. Similarly, $[H_2] = 0.0960$ M and $[NH_3] = 0.406$ M. Substituting these concentrations into the equilibrium constant expression gives

$$Q_c = \frac{[NH_3]_t^2}{[N_2]_t[H_2]_t^3} = \frac{(0.406)^2}{(0.0785)(0.0960)^3} = 2.37 \times 10^3$$

Because Q_c does not equal K_c (1.7×10^2), the reaction mixture is not at equilibrium. Because Q_c is greater than K_c, net reaction will proceed from products to reactants, decreasing the concentration of NH_3 and increasing the concentrations of N_2 and H_2 until $Q_c = K_c = 1.7 \times 10^2$.

CHECK

Approximate initial concentrations can be calculated by dividing rounded values of the number of moles of each substance by the volume; $[N_2] \approx (1.6 \text{ mol})/(20 \text{ L}) \approx 0.08$ M, $[H_2] \approx (2 \text{ mol})/(20 \text{ L}) \approx 0.1$ M, and $[NH_3] \approx (8 \text{ mol})/(20 \text{ L}) \approx 0.4$ M. Substituting these concentrations into the expression for Q_c gives a ballpark estimate of Q_c.

$$Q_c = \frac{[NH_3]_t^2}{[N_2]_t[H_2]_t^3} \approx \frac{(0.4)^2}{(0.08)(0.1)^3} \approx 2 \times 10^3$$

You can calculate this value without a calculator because it equals $(16 \times 10^{-2})/(8 \times 10^{-2})$. The ballpark estimate of Q_c, like the more exact value (2.37×10^3), exceeds K_c, so the reaction mixture is not at equilibrium.

▶ **PRACTICE 15.15** The equilibrium constant K_c for the reaction $2 NO(g) + O_2(g) \rightleftharpoons 2 NO_2(g)$ is 6.9×10^5 at 500 K. A 5.0 L reaction vessel at 500 K was filled with 0.060 mol of NO, 1.0 mol of O_2, and 0.80 mol of NO_2. What is the value of Q_c, and in which direction does the net reaction proceed?

▶ **CONCEPTUAL APPLY 15.16** The reaction $A_2 + B_2 \rightleftharpoons 2 AB$ has an equilibrium constant $K_c = 4$. The following pictures represent reaction mixtures that contain A_2 molecules (red), B_2 molecules (blue), and AB molecules:

(1)   **(2)**   **(3)**

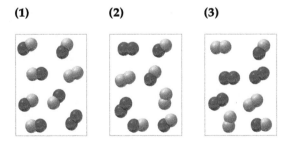

(a) Which reaction mixture is at equilibrium?
(b) For those reaction mixtures that are not at equilibrium, will the net reaction go in the forward or reverse direction to reach equilibrium?

Calculating Equilibrium Concentrations

If the equilibrium constant and all the equilibrium concentrations but one are known, the unknown concentration can be calculated directly from the equilibrium equation. To illustrate, let's consider the following problem: What is the concentration of NO in an equilibrium mixture of gaseous NO, O_2, and NO_2 at 500 K that contains 1.0×10^{-3} M O_2 and 5.0×10^{-2} M NO_2? At this temperature, the equilibrium constant K_c for the reaction $2\,NO(g) + O_2(g) \rightleftharpoons 2\,NO_2(g)$ is 6.9×10^5.

In this problem, we know K_c and all the equilibrium concentrations except one, and we're asked to calculate the unknown equilibrium concentration. First, we write the equilibrium equation for the reaction and solve for the unknown concentration:

$$K_c = \frac{[NO_2]^2}{[NO]^2[O_2]} \qquad [NO] = \sqrt{\frac{[NO_2]^2}{[O_2]K_c}}$$

Then we substitute the known values of K_c, $[O_2]$, and $[NO_2]$ into the expression for [NO]. Taking the positive square root because the concentration of NO is a positive quantity, we obtain:

$$[NO] = \sqrt{\frac{(5.0 \times 10^{-2})^2}{(1.0 \times 10^{-3})(6.9 \times 10^5)}} = \sqrt{3.6 \times 10^{-6}} = 1.9 \times 10^{-3}\ \text{M}$$

To be sure that we haven't made any errors, it's a good idea to check the result by substituting it into the equilibrium equation:

$$K_c = 6.9 \times 10^5 = \frac{[NO_2]^2}{[NO]^2[O_2]} = \frac{(5.0 \times 10^{-2})^2}{(1.9 \times 10^{-3})^2(1.0 \times 10^{-3})} = 6.9 \times 10^5$$

Another type of problem is one in which we know the initial concentrations but do not know any of the equilibrium concentrations. To solve this kind of problem, follow the series of steps summarized in **FIGURE 15.7** and illustrated in Worked Examples 15.9 and 15.10. The same approach can be used to calculate equilibrium partial pressures from initial partial pressures and K_p, as shown in Worked Examples 15.11.

▶ **FIGURE 15.7**

Steps to follow in calculating equilibrium concentrations from initial concentrations.

Step 1. Compute Q_c and compare to K_c to determine the direction of reaction.

Step 2. Write the balanced equation for the reaction. Under the balanced equation, make a table that lists for each substance involved in the reaction:
   (a) The initial concentration
   (b) The change in concentration on going to equilibrium
   (c) The equilibrium concentration
In constructing the table, define x as the concentration (mol/L) of one of the substances that reacts on going to equilibrium and then use the stoichiometry of the reaction to determine the concentrations of the other substances in terms of x.

Step 3. Substitute the equilibrium concentrations into the equation for the equilibrium constant (K_c) and solve for x. If you must solve a quadratic equation, choose the mathematical solution that makes chemical sense.

Step 4. Calculate the equilibrium concentrations from the calculated value of x.

Step 5. Check your results by substituting them into the equation for the equilibrium constant (K_c).

Go to eText

WORKED EXAMPLE 15.9

Calculating Equilibrium Concentrations (Solving the Equilibrium Equation by Taking the Square Root of Both Sides of the Equation)

The equilibrium constant K_c for the reaction of H_2 with I_2 is 57.0 at 700 K:

$$H_2(g) + I_2(g) \rightleftharpoons 2\,HI(g) \quad K_c = 57.0 \text{ at } 700 \text{ K}$$

If 1.00 mol of H_2 is allowed to react with 1.00 mol of I_2 in a 10.0 L reaction vessel at 700 K, what are the concentrations of H_2, I_2, and HI at equilibrium? What is the concentration of the reactants and products in equilibrium?

STRATEGY

We need to calculate equilibrium concentrations from initial concentrations, so we use the method outlined in Figure 15.7.

SOLUTION

Step 1. $\quad Q_c = \dfrac{[HI]_t^2}{[H_2]_t[I_2]_t} = \dfrac{[0]^2}{[0.100][0.100]} = 0$

$Q_c(0) < K_c\,(57.0)$; therefore the reaction proceeds in the direction of the products.

Step 2. The balanced equation is given: $H_2(g) + I_2(g) \rightleftharpoons 2\,HI(g)$.

The initial concentrations are $[H_2] = [I_2] = (1.00 \text{ mol})/(10.0 \text{ L}) = 0.100$ M. For convenience, define an unknown, x, as the concentration (mol/L) of H_2 that reacts. According to the balanced equation for the reaction and the direction of the reaction, x mol/L of H_2 reacts with x mol/L of I_2 to give $2x$ mol/L of HI. This reduces the initial concentrations of H_2 and I_2 from 0.100 mol/L to $(0.100 - x)$ mol/L at equilibrium. Let's summarize these results in a table under the balanced equation:

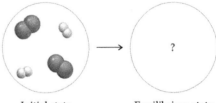

Initial state   Equilibrium state

	$H_2(g)$	+	$I_2(g)$	$\rightleftharpoons$	$2\,HI(g)$
Initial concentration (M)	0.100		0.100		0
Change (M)	$-x$		$-x$		$+2x$
Equilibrium concentration (M)	$(0.100 - x)$		$(0.100 - x)$		$2x$

Step 3. Substitute the equilibrium concentrations into the equation for the equilibrium constant:

$$K_c = 57.0 = \frac{[HI]^2}{[H_2][I_2]} = \frac{(2x)^2}{(0.100 - x)(0.100 - x)} = \left(\frac{2x}{0.100 - x}\right)^2$$

Because the right side of this equation is a perfect square, we can take the square root of both sides:

$$\sqrt{57.0} = \pm\,7.55 = \frac{2x}{0.100 - x}$$

Solving for x, we obtain two solutions. The equation with the positive square root of 57.0 gives

$$+7.55(0.100 - x) = 2x$$
$$0.755 = 2x + 7.55x$$
$$x = \frac{0.755}{9.55} = 0.0791 \text{ M}$$

The equation with the negative square root of 57.0 gives

$$-7.55(0.100 - x) = 2x$$
$$-0.755 = 2x - 7.55x$$
$$x = \frac{-0.755}{-5.55} = 0.136 \text{ M}$$

continued on next page

Because the initial concentrations of H_2 and I_2 are 0.100 M, x can't exceed 0.100 M. Therefore, discard $x = 0.136$ M as chemically unreasonable and choose the first solution, $x = 0.0791$ M.

Step 4. Calculate the equilibrium concentrations from the calculated value of x:

$$[H_2] = [I_2] = 0.100 - x = 0.100 - 0.0791 = 0.021 \text{ M}$$
$$[HI] = 2x = (2)(0.0791) = 0.158 \text{ M}$$

Step 5. Check the results by substituting them into the equilibrium equation:

$$K_c = 57.0 = \frac{[HI]^2}{[H_2][I_2]} = \frac{(0.158)^2}{(0.021)(0.021)} = 57$$

▶ **PRACTICE 15.17**   The H_2/CO ratio in mixtures of carbon monoxide and hydrogen (called *synthesis gas*) is increased by the water–gas shift reaction $CO(g) + H_2O(g) \rightleftharpoons CO_2(g) + H_2(g)$, which has an equilibrium constant $K_c = 4.24$ at 800 K. Calculate the equilibrium concentrations of CO_2, H_2, CO, and H_2O at 800 K if only CO and H_2O are present initially at concentrations of 0.150 M.

▶ **APPLY 15.18**   Calculate the equilibrium concentrations at 800 K for the reaction in Problem 15.17 if CO, H_2O, CO_2, and H_2 are added to a reaction vessel with initial concentrations of 0.100 M.

All **WORKED EXAMPLES** with this icon have an interactive video in the eText.

WORKED EXAMPLE 15.10

Calculating Equilibrium Concentrations (Solving the Equilibrium Equation Using the Quadratic Equation)

Calculate the equilibrium concentrations of H_2, I_2, and HI at 700 K if the initial concentrations are $[H_2] = 0.100$ M and $[I_2] = 0.200$ M. The equilibrium constant K_c for the reaction $H_2(g) + I_2(g) \rightleftharpoons 2\,HI(g)$ is 57.0 at 700 K.

STRATEGY

This problem is similar to a Worked Examples 15.9 except that the initial concentrations of H_2 and I_2 are unequal. Again, we follow the steps in Figure 15.7.

SOLUTION

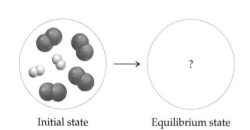

Initial state   Equilibrium state

Step 1. $Q_c = \dfrac{[HI]_t^2}{[H_2]_t[I_2]_t} = \dfrac{[0]^2}{[0.100][0.200]} = 0$

$Q_c\,(0) < K_c\,(57.0)$; therefore the reaction proceeds in the direction of the products.

Step 2.   The balanced equation is $H_2(g) + I_2(g) \rightleftharpoons 2\,HI(g)$. Again, define x as the concentration of H_2 that reacts. Set up a table of concentrations under the balanced equation:

	$H_2(g)$	$+$	$I_2(g)$	$\rightleftharpoons$	$2\,HI(g)$
Initial concentration (M)	0.100		0.200		0
Change (M)	$-x$		$-x$		$+2x$
Equilibrium concentration (M)	$(0.100 - x)$		$(0.200 - x)$		$2x$

Step 3.   Substitute the equilibrium concentrations into the equilibrium equation:

$$K_c = 57.0 = \frac{[HI]^2}{[H_2][I_2]} = \frac{(2x)^2}{(0.100 - x)(0.200 - x)}$$

Because the right side of this equation is not a perfect square, we must put the equation into the standard quadratic form, $ax^2 + bx + c = 0$, and then solve for x using the quadratic formula (Appendix A.4):

$$x = \frac{-b \pm \sqrt{b^2 - 4ac}}{2a}$$

Rearranging the equilibrium equation gives

$$(57.0)(0.0200 - 0.300x + x^2) = 4x^2$$

or

$$53.0x^2 - 17.1x + 1.14 = 0$$

Substituting the values of a, b, and c into the quadratic formula gives two solutions:

$$x = \frac{17.1 \pm \sqrt{(-17.1)^2 - 4(53.0)(1.14)}}{2(53.0)} = \frac{17.1 \pm 7.1}{106} = 0.228 \quad \text{and} \quad 0.0943$$

Discard the solution ($x = 0.228$) because the H_2 concentration can't change by more than its initial value (0.100 M). Therefore, choose the solution ($x = 0.0943$).

Step 4. Calculate the equilibrium concentrations from the calculated value of x:

$$[H_2] = 0.100 - x = 0.100 - 0.0943 = 0.006 \text{ M}$$
$$[I_2] = 0.200 - x = 0.200 - 0.0943 = 0.106 \text{ M}$$
$$[HI] = 2x = (2)(0.0943) = 0.189 \text{ M}$$

Step 5. Check the results by substituting them into the equilibrium equation:

$$K_c = 57.0 = \frac{[HI]^2}{[H_2][I_2]} = \frac{(0.189)^2}{(0.006)(0.106)} = 56.2$$

The calculated value of $K_c(56.2)$, which should be rounded to one significant figure (6×10^1), agrees with the value given in the problem (57.0).

▶ **PRACTICE 15.19** Calculate the equilibrium concentrations of H_2, I_2, and HI at 700 K if the initial concentrations are $[H_2] = 0.100$ M and $[I_2] = 0.300$ M. The equilibrium constant K_c for the reaction $H_2(g) + I_2(g) \rightleftharpoons 2\,HI(g)$ is 57.0 at 700 K.

▶ **APPLY 15.20** Calculate the equilibrium concentrations of N_2O_4 and NO_2 at 25 °C in a vessel that contains an initial N_2O_4 concentration of 0.500 M. The equilibrium constant K_c for the reaction $N_2O_4(g) \rightleftharpoons 2\,NO_2(g)$ is 4.64×10^{-3} at 25 °C.

— **WORKED EXAMPLE 15.11**

Calculating Equilibrium Concentrations (Solving the Equilibrium Equation Using the Small x Approximation)

Calculate the equilibrium concentrations of CO_2, H_2, CO, and H_2O at 298 K if the initial concentrations are $[CO_2] = 1.50$ M and $[H_2] = 1.10$ M. The equilibrium constant K_c for the reaction $CO_2(g) + H_2(g) \rightleftharpoons CO(g) + H_2O(g)$ is 9.71×10^{-6} at 298 K.

STRATEGY

Use the method outlined in Figure 15.7.

SOLUTION

Step 1. $Q_c = \dfrac{[CO][H_2O]}{[CO_2][H_2]} = \dfrac{(0)(0)}{(1.50)(1.10)} = 0$

$Q_c(0) < K_c(9.71 \times 10^{-6})$; therefore, the reaction proceeds in the direction of the products.

continued on next page

Step 2. Use the balanced equation to set up a table of concentrations representing initial concentrations, changes in concentrations, and equilibrium concentrations. Define x as the concentration of CO_2 that reacts.

	$CO_2(g)$	+	$H_2(g)$	$\rightleftharpoons$ $CO(g)$	+	$H_2O(g)$
Initial concentration (M)	1.50		1.10	0		0
Change (M)	$-x$		$-x$	$+x$		$+x$
Equilibrium concentration (M)	$(1.50 - x)$		$(1.10 - x)$	x		x

Step 3. Substitute equilibrium concentrations into the equilibrium equation.

$$K_c = 9.71 \times 10^{-6} = \frac{[CO][H_2O]}{[CO_2][H_2]} = \frac{(x)(x)}{(1.50 - x)(1.10 - x)}$$

For reactions in which the value of the equilibrium constant K_c is very small, the reaction proceeds to the products to a very small degree, and x will be negligible compared to the initial concentrations (1.50 M and 1.10 M). Therefore, we can make two assumptions that greatly simplify the solution. The assumptions are:

$$(1.50 - x) \approx 1.50 \qquad (1.10 - x) \approx 1.10$$

Using the assumptions, we can now write the equilibrium equation as:

$$K_c = 9.71 \times 10^{-6} = \frac{(x)(x)}{(1.50)(1.10)} = \frac{x^2}{1.65}$$

$$x^2 = 1.60 \times 10^{-5}$$
$$x = 4.00 \times 10^{-3}$$

We must check the validity of the assumption that x is small compared to the numbers it is subtracted from. Calculate the percentage that the calculated value of x is of the initial concentration.

$$\frac{4.00 \times 10^{-3}}{1.50} \times 100\% = 0.267\% \qquad \frac{4.00 \times 10^{-3}}{1.10} \times 100\% = 0.364\%$$

If x is less than 5% of the initial concentration, the assumption is valid and does not cause significant error in the equilibrium concentrations. In this case, the assumption is valid because 0.267% and 0.356% are less than 5%.

Step 4. Calculate the equilibrium concentrations using the expressions in the last row of the table and the value of x.

$$[CO_2] = 1.50 - x = 1.50 - 4.00 \times 10^{-3} = 1.50 \text{ M}$$
$$[H_2] = 1.10 - x = 1.10 - 4.00 \times 10^{-3} = 1.10 \text{ M}$$
$$[CO] = [H_2O] = x = 4.00 \times 10^{-3} \text{ M}$$

Step 5. We can check the equilibrium concentrations by substituting the values in the equilibrium equation and calculating the equilibrium constant.

$$K_c = \frac{[CO][H_2O]}{[CO_2][H_2]} = \frac{(4.00 \times 10^{-3})(4.00 \times 10^{-3})}{(1.50)(1.10)} = 9.70 \times 10^{-6}$$

The value of K_c calculated using the equilibrium concentrations (9.70×10^{-6}) matches the value of K_c given in the problem (9.71×10^{-6}) within rounding error.

▶ **PRACTICE 15.21** Calculate the equilibrium concentrations of CO_2, H_2, CO, and H_2O at 298 K if the initial concentrations are $[CO_2] = 0.750$ M and $(H_2) = 0.550$ M. The equilibrium constant K_c for the reaction $CO_2(g) + H_2(g) \rightleftharpoons CO(g) + H_2O(g)$ is 9.71×10^{-6} at 298 K.

▶ **APPLY 15.22** Calculate the equilibrium concentrations of $H_2O(g)$, $Cl_2(g)$, $HCl (g)$, and $O_2(g)$ at 298 K if the initial concentrations are $[H_2O] = 0.130$ M and $[Cl_2] = 0.250$ M. The equilibrium constant K_c for the reaction $H_2O(g) + Cl_2(g) \rightleftharpoons 2 HCl(g) + O_2(g)$ is 8.96×10^{-9} at 298 K.

WORKED EXAMPLE 15.12

Calculating Equilibrium Partial Pressures from Initial Partial Pressures

One reaction that occurs in producing steel from iron ore is the reduction of iron(II) oxide by carbon monoxide to give iron metal and carbon dioxide. The equilibrium constant K_p for the reaction at 1000 K is 0.259.

$$\text{FeO}(s) + \text{CO}(g) \rightleftharpoons \text{Fe}(s) + \text{CO}_2(g) \quad K_p = 0.259 \text{ at } 1000 \text{ K}$$

What are the equilibrium partial pressures of CO and CO_2 at 1000 K if the initial partial pressures are $P_{CO} = 0.500$ atm and $P_{CO_2} = 1.000$ atm?

▲ Iron ore reacts with carbon monoxide in a blast furnace to extract iron used for producing steel.

STRATEGY

We can calculate equilibrium partial pressures from initial partial pressures and K_p in the same way that we calculate equilibrium concentrations from initial concentrations and K_c. Follow the steps in Figure 15.7, but substitute partial pressures for concentrations.

SOLUTION

Step 1. $Q_p = \dfrac{(P_{CO_2})_t}{(P_{CO})_t} = \dfrac{1.000 \text{ atm}}{0.500 \text{ atm}} = 2.00$

Q_p (2.00) > K_p (0.259); therefore the reaction proceeds in the direction of the reactants.

Step 2. The balanced equation is $\text{FeO}(s) + \text{CO}(g) \rightleftharpoons \text{Fe}(s) + \text{CO}_2(g)$. Define x as the partial pressure of CO_2 that reacts. Since the reaction is shifting *toward reactants,* the reactants will have values of $+x$ and the products will have values of $-x$. Set up a table of partial pressures of the gases under the balanced equation:

	FeO(s)	+	CO(g)	⇌ Fe(s) +	CO₂(g)
Initial pressure (atm)			0.500		1.000
Change (atm)			+x		−x
Equilibrium pressure (atm)			(0.500 + x)		(1.000 − x)

Step 3. Substitute the equilibrium partial pressures into the equilibrium equation for K_p:

$$K_p = 0.259 = \frac{P_{CO_2}}{P_{CO}} = \frac{1.000 - x}{0.500 + x}$$

As usual for a heterogeneous equilibrium, we omit the pure solids from the equilibrium equation. Rearranging the equilibrium equation and solving for x gives

$$0.1295 + 0.259x = 1.000 - x$$

$$x = \frac{0.8705}{1.259} = 0.691$$

Step 4. Calculate the equilibrium partial pressures from the calculated value of x:

$$P_{CO} = 0.500 + x = 0.500 + (0.691) = 1.191 \text{ atm}$$
$$P_{CO_2} = 1.000 - x = 1.000 - (0.691) = 0.309 \text{ atm}$$

Step 5. Check the results by substituting them into the equilibrium equation:

$$K_p = 0.259 = \frac{P_{CO_2}}{P_{CO}} = \frac{0.309}{1.191} = 0.259$$

▶ **PRACTICE 15.23** The equilibrium constant K_p for the reaction $\text{C}(s) + \text{H}_2\text{O}(g) \rightleftharpoons \text{CO}(g) + \text{H}_2(g)$ is 2.44 at 1000 K. What are the equilibrium partial pressures of H_2O, CO, and H_2 if the initial partial pressures are $P_{CO} = 1.00$ atm and $P_{H_2} = 1.00$ atm?

▶ **APPLY 15.24** The equilibrium constant K_p for the reaction $\text{C}(s) + \text{H}_2\text{O}(g) \rightleftharpoons \text{CO}(g) + \text{H}_2(g)$ is 2.44 at 1000 K. What are the equilibrium partial pressures of H_2O, CO, and H_2 if the initial partial pressures are $P_{H_2O} = 1.20$ atm, $P_{CO} = 1.00$ atm, and $P_{H_2} = 1.40$ atm?

15.6 FACTORS THAT ALTER THE COMPOSITION OF AN EQUILIBRIUM MIXTURE: LE CHÂTELIER'S PRINCIPLE

One of the main goals of chemical synthesis is to maximize the conversion of reactants to products while minimizing the expenditure of energy. This objective is achieved easily if the reaction goes nearly to completion at mild temperature and pressure. If the reaction gives an equilibrium mixture that is rich in reactants and poor in products, however, then the experimental conditions must be adjusted. Consider the process for the synthesis of ammonia from its elements developed by German chemist Fritz Haber (1868–1934) in 1913 (**FIGURE 15.8**):

$$N_2(g) + 3 H_2(g) \rightleftharpoons 2 NH_3(g)$$

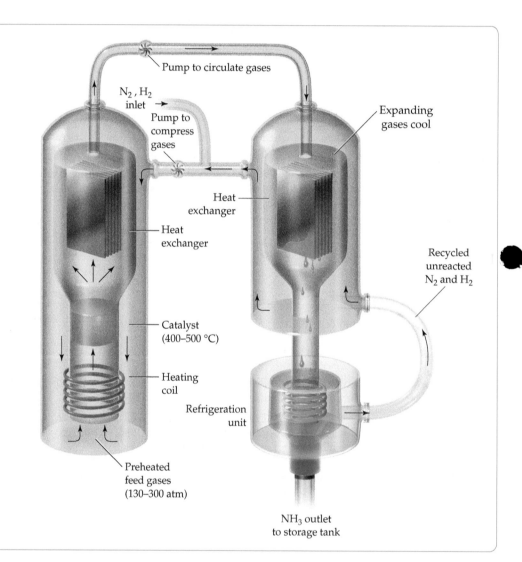

▶ **FIGURE 15.8**

Representation of the Haber–Bosch process for the industrial production of ammonia. A mixture of gaseous N_2 and H_2 at 130–300 atm pressure is passed over a catalyst at 400–500 °C. The reaction $N_2(g) + 3 H_2(g) \rightleftharpoons 2 NH_3(g)$ produces ammonia. The NH_3 in the gaseous mixture of reactants and products is liquefied, and the unreacted N_2 and H_2 are recycled.

The choice of experimental conditions in the Haber process is of real economic importance because annual worldwide production of ammonia is about 130 million metric tons, primarily for use as fertilizer. Industrial production of ammonia is also known as the Haber–Bosch process to recognize the work of Karl Bosch, who engineered the technology for scaling up production. The Haber–Bosch process is one of the most important inventions of the twentieth century because without ammonia-based fertilizer the world's population could not have grown from 1.6 billion in 1900 to today's 7 billion. It has been estimated that almost half of the world's current population subsists on crops grown with the output of the Haber–Bosch process.

We can exploit several factors to alter the composition of an equilibrium mixture:

- The concentration of reactants or products can be changed.
- The pressure and volume can be changed.
- The temperature can be changed.

A possible fourth factor, addition of a catalyst, increases only the rate at which equilibrium is reached. As we'll see in Section 15.10, a **catalyst** does not affect the equilibrium concentrations.

We can predict the qualitative effect of the listed changes on the composition of an equilibrium mixture using a principle first described by the French chemist Henri-Louis Le Châtelier (1850–1936; pronounced Li Sha–tell–**yea**).

> **Le Châtelier's principle** If a stress is applied to a reaction mixture at equilibrium, net reaction occurs in the direction that relieves the stress.

The word "stress" in this context means a change in concentration, pressure, volume, or temperature that disturbs the original equilibrium. Reaction then occurs to change the composition of the mixture until a new state of equilibrium is reached. The direction that the reaction takes (reactants to products or vice versa) is the one that reduces the stress. In the next three sections, we'll look at the different kinds of stress that can change the composition of an equilibrium mixture.

15.7 ALTERING AN EQUILIBRIUM MIXTURE: CHANGES IN CONCENTRATION

Let's consider the equilibrium that occurs in the Haber process for the synthesis of ammonia:

$$N_2(g) + 3 H_2(g) \rightleftharpoons 2 NH_3(g) \qquad K_c = 0.291 \text{ at } 700 \text{ K}$$

Suppose that we have an equilibrium mixture of 0.50 M N_2, 3.00 M H_2, and 1.98 M NH_3 at 700 K and that we disturb the equilibrium by increasing the N_2 concentration to 1.50 M. Le Châtelier's principle tells us that reaction will occur to relieve the stress of the increased concentration of N_2 by converting some of the N_2 to NH_3. As the N_2 concentration decreases, the H_2 concentration must also decrease and the NH_3 concentration must increase in accord with the stoichiometry of the balanced equation. These changes are illustrated in **FIGURE 15.9.**

In general, when an equilibrium is disturbed by the addition or removal of any reactant or product, Le Châtelier's principle predicts that

- The concentration stress of an *added* reactant or product is relieved by net reaction in the direction that *consumes* the added substance.
- The concentration stress of a *removed* reactant or product is relieved by net reaction in the direction that *replenishes* the removed substance.

If these rules are applied to the equilibrium $N_2(g) + 3 H_2(g) \rightleftharpoons 2 NH_3(g)$, then the yield of ammonia is increased by an increase in the N_2 or H_2 concentration or by a decrease in the NH_3 concentration (**FIGURE 15.10**). In the industrial production of ammonia, the concentration of gaseous NH_3 is decreased by liquefying the ammonia (bp -33 °C) as it's formed, and so more ammonia is produced.

LOOKING AHEAD . . .
We'll see in Section 15.10 that a **catalyst** increases the rates of both the forward and reverse reactions but does not change the composition of the equilibrium mixture.

BIG IDEA Question 5

How can the net reaction direction be shifted to make more product?

▶ **FIGURE 15.9**

Changes in concentrations when N_2 is added to an equilibrium mixture of N_2, H_2, and NH_3.

▶ **Figure It Out**

How do the equilibrium concentrations of H_2, NH_3, and N_2 change after the addition of N_2? What happens to the value of K_c?

Answer: [H₂] decreases, [NH₃] increases, [N₂] increases, and the value of the equilibrium constant K_c remains unchanged.

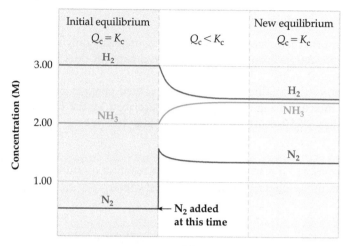

Net conversion of N_2 and H_2 to NH_3 occurs until a new equilibrium is established. That is, the N_2 and H_2 concentrations decrease, while the NH_3 concentration increases.

▶ **FIGURE 15.10**

Effect of concentration changes on the equilibrium $N_2(g) + 3 H_2(g) \rightleftharpoons 2 NH_3(g)$. An increase in the N_2 or H_2 concentration or a decrease in the NH_3 concentration shifts the equilibrium from left to right. A decrease in the N_2 or H_2 concentration or an increase in the NH_3 concentration shifts the equilibrium from right to left.

Any of the changes marked in blue shifts the equilibrium to the left.

Any of the changes marked in pink shifts the equilibrium to the right.

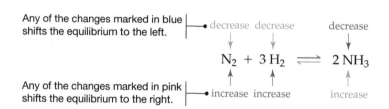

Le Châtelier's principle is a handy rule for predicting changes in the composition of an equilibrium mixture, but it doesn't explain *why* those changes occur. To see why Le Châtelier's principle works, let's look again at the reaction quotient Q_c. For the initial equilibrium mixture of 0.50 M N_2, 3.00 M H_2, and 1.98 M NH_3, at 700 K, Q_c equals the equilibrium constant K_c (0.291) because the system is at equilibrium:

$$Q_c = \frac{[NH_3]_t^2}{[N_2]_t[H_2]_t^3} = \frac{(1.98)^2}{(0.50)(3.00)^3} = 0.29 = K_c$$

When we disturb the equilibrium by increasing the N_2 concentration to 1.50 M, the denominator of the equilibrium constant expression increases and Q_c decreases to a value less than K_c:

$$Q_c = \frac{[NH_3]_t^2}{[N_2]_t[H_2]_t^3} = \frac{(1.98)^2}{(1.50)(3.00)^3} = 0.0968 < K_c$$

For the system to move to a new state of equilibrium, Q_c must increase; that is, the numerator of the equilibrium constant expression must increase and the denominator must decrease. This implies the net conversion of N_2 and H_2 to NH_3, just as predicted by Le Châtelier's principle. When the new equilibrium is established (Figure 15.9), the concentrations are 1.31 M N_2, 2.43 M H_2, and 2.36 M NH_3, and Q_c again equals K_c:

$$Q_c = \frac{[NH_3]_t^2}{[N_2]_t[H_2]_t^3} = \frac{(2.36)^2}{(1.31)(2.43)^3} = 0.296 = K_c$$

As another example of how a change in concentration affects an equilibrium, let's consider the reaction in aqueous solution of iron(III) ions and thiocyanate (SCN^-) ions to give an equilibrium mixture that contains the red complex ion $FeNCS^{2+}$:

$$Fe^{3+}(aq) + SCN^-(aq) \rightleftharpoons FeNCS^{2+}(aq)$$
$$\text{Pale yellow} \qquad \text{Colorless} \qquad\qquad \text{Red}$$

We can detect shifts in the position of this equilibrium by observing how the color of the solution changes when we add various reagents (**FIGURE 15.11**). If we add aqueous $FeCl_3$, the red color gets darker, as predicted by Le Châtelier's principle. The concentration stress of added Fe^{3+} is relieved by net reaction from left to right, which consumes some of the Fe^{3+} and increases the concentration of $FeNCS^{2+}$. (Note that the Cl^- ions are not involved in the reaction.) Similarly, if we add aqueous KSCN, the stress of added SCN^- shifts the equilibrium from left to right and again the red color gets darker.

The equilibrium can be shifted in the opposite direction by adding reagents that remove Fe^{3+} or SCN^- ions. For example, oxalic acid ($H_2C_2O_4$), a poisonous substance present in the leaves of plants such as rhubarb, reacts with Fe^{3+} to form the stable, yellow complex ion $Fe(C_2O_4)_3{}^{3-}$, thus decreasing the concentration of free $Fe^{3+}(aq)$. In accord with Le Châtelier's principle, the concentration stress of removed Fe^{3+} is relieved by the dissociation of $FeNCS^{2+}$ to replenish the Fe^{3+} ions. Because the concentration of $FeNCS^{2+}$ decreases, the red color disappears.

$$3 H_2C_2O_4(aq) + Fe^{3+}(aq) \longrightarrow Fe(C_2O_4)_3{}^{3-}(aq) + 6 H^+(aq)$$
$$FeNCS^{2+}(aq) \longrightarrow Fe^{3+}(aq) + SCN^-(aq)$$

▲ The extremely sour leaves of rhubarb contain toxins such as oxalic acid, but the stalks and roots are nutritious.

Addition of aqueous $HgCl_2$ also eliminates the red color because $HgCl_2$ reacts with SCN^- ions to form the stable, colorless, complex ion $Hg(SCN)_4{}^{2-}$. Removal of free $SCN^-(aq)$ results in dissociation of the red $FeNCS^{2+}$ ions so as to replenish the SCN^- ions.

$$HgCl_2(aq) + 4 SCN^-(aq) \longrightarrow Hg(SCN)_4{}^{2-}(aq) + 2 Cl^-(aq)$$
$$FeNCS^{2+}(aq) \longrightarrow Fe^{3+}(aq) + SCN^-(aq)$$

(a) Original solution: Fe^{3+}(pale yellow), SCN^-(colorless), and $FeNCS^{2+}$(red).

(b) After adding $FeCl_3$ to **(a)**: [$FeNCS^{2+}$] increases.

(c) After adding KSCN to **(a)**: [$FeNCS^{2+}$] increases.

(d) After adding $H_2C_2O_4$ to **(a)**: [$FeNCS^{2+}$] decreases as [$Fe(C_2O_4)_3{}^{3-}$] increases.

(e) After adding $HgCl_2$ to **(a)**: [$FeNCS^{2+}$] decreases as [$Hg(SCN)_4{}^{2-}$] increases.

▲ **FIGURE 15.11**
Color changes produced by adding various reagents to an equilibrium mixture of Fe^{3+} (pale yellow), SCN^- (colorless), and $FeNCS^{2+}$ (red).

WORKED EXAMPLE 15.13

Applying Le Châtelier's Principle to Concentration Changes

The reaction of iron(III) oxide with carbon monoxide occurs in a blast furnace when iron ore is reduced to iron metal:

$$Fe_2O_3(s) + 3\ CO(g) \rightleftharpoons 2\ Fe(l) + 3\ CO_2(g)$$

Use Le Châtelier's principle to predict the direction of the net reaction when an equilibrium mixture is disturbed by:

(a) Adding Fe_2O_3
(b) Removing CO_2
(c) Removing CO; also account for the change using the reaction quotient Q_c.

STRATEGY

To predict the direction of net reaction, recall that a concentration stress is relieved by reaction in the direction that consumes an added substance or replenishes a removed substance.

SOLUTION

(a) Because Fe_2O_3 is a pure solid, it's concentration does not appear in the equilibrium expression. Therefore, there is no concentration stress and the original equilibrium is undisturbed.
(b) Le Châtelier's principle predicts that the concentration stress of removed CO_2 will be relieved by net reaction from reactants to products to replenish the CO_2.
(c) Le Châtelier's principle predicts that the concentration stress of removed CO will be relieved by net reaction from products to reactants to replenish the CO. The reaction quotient is

$$Q_c = \frac{[CO_2]_t^3}{[CO]_t^3}$$

When the equilibrium is disturbed by reducing [CO], Q_c increases, so that $Q_c > K_c$. For the system to move to a new state of equilibrium, Q_c must decrease—that is, $[CO_2]$ must decrease and [CO] must increase. Therefore, the net reaction goes from products to reactants, as predicted by Le Châtelier's principle.

▶ **PRACTICE 15.25**   Consider the equilibrium for the water–gas shift reaction:

$$CO(g) + H_2O(g) \rightleftharpoons CO_2(g) + H_2(g)$$

Which of the following changes will increase the concentration of H_2?

(a) Adding CO
(b) Adding CO_2
(c) Removing H_2O
(d) Removing CO_2
(e) Both (a) and (d)

▶ **APPLY 15.26**   Solid particles that form in the kidney are called kidney stones and frequently cause acute pain. One common type of kidney stone is formed from a precipitation reaction of calcium and oxalate:

$$Ca^{2+}(aq) + C_2O_4^{2-}(aq) \rightleftharpoons CaC_2O_4(s)$$

Use Le Châtelier's principle to explain the following statements.

(a) A person taking diuretics, medicines that help kidneys remove fluids, may be at increased risk for developing kidney stones.
(b) A person diagnosed with hypercalciuria, a genetic condition causing elevated levels of calcium in the urine, has an increased risk for developing kidney stones.
(c) One simple treatment for kidney stones is to avoid foods high in oxalate such as spinach, rhubarb, and nuts.
(d) Another simple treatment for kidney stones is to increase consumption of water.

15.8 ALTERING AN EQUILIBRIUM MIXTURE: CHANGES IN PRESSURE AND VOLUME

To illustrate how an equilibrium mixture is affected by a change in pressure as a result of a change in the volume, let's return to the Haber process for the synthesis of ammonia. The balanced equation for the reaction has 4 mol of gas on the reactant side of the equation and 2 mol on the product side:

$$N_2(g) + 3 H_2(g) \rightleftharpoons 2 NH_3(g) \quad K_c = 0.291 \text{ at } 700 \text{ K}$$

What happens to the composition of the equilibrium mixture if we increase the pressure by decreasing the volume? The **ideal gas law** states that the pressure of a gas sample is inversely proportional to the volume at a constant temperature and constant number of moles. Therefore, decreasing volume will increase pressure. According to Le Châtelier's principle, net reaction will occur in the direction that relieves the stress of the increased pressure, which means that the number of moles of gas must decrease. Therefore, we predict that the net reaction will proceed from left to right because the forward reaction converts 4 mol of gaseous reactants to 2 mol of gaseous products.

In general, Le Châtelier's principle predicts that

- An *increase* in pressure by reducing the volume will bring about net reaction in the direction that *decreases* the number of moles of gas.
- A *decrease* in pressure by expanding the volume will bring about net reaction in the direction that *increases* the number of moles of gas.

To see why Le Châtelier's principle works for pressure (volume) changes, let's look again at the reaction quotient for the equilibrium mixture of 0.50 M N_2, 3.00 M H_2, and 1.98 M NH_3 at 700 K:

$$Q_c = \frac{[NH_3]_t^2}{[N_2]_t[H_2]_t^3} = \frac{(1.98)^2}{(0.50)(3.00)^3} = 0.29 = K_c$$

If we disturb the equilibrium by reducing the volume by a factor of 2, we not only double the total pressure, we also double the partial pressure and thus the molar concentration of each reactant and product (because molarity = n/V). Because the balanced equation has more moles of gaseous reactants than gaseous products, the increase in the denominator of the equilibrium constant expression is greater than the increase in the numerator, and the new value of Q_c is less than the equilibrium constant K_c:

$$Q_c = \frac{[NH_3]_t^2}{[N_2]_t[H_2]_t^3} = \frac{(3.96)^2}{(1.00)(6.00)^3} = 0.0726 < K_c$$

For the system to move to a new state of equilibrium, Q_c must increase, which means that the net reaction must go from reactants to products, as predicted by Le Châtelier's principle (**FIGURE 15.12**). In practice, the yield of ammonia in the Haber process is increased by running the reaction at high pressure, typically 130–300 atm.

The composition of an equilibrium mixture is unaffected by a change in pressure if the reaction involves no change in the number of moles of gas. For example, the reaction of hydrogen with gaseous iodine has 2 mol of gas on both sides of the balanced equation:

$$H_2(g) + I_2(g) \rightleftharpoons 2 HI(g)$$

If we double the pressure by halving the volume, the numerator and denominator of the reaction quotient change by the same factor and Q_c remains unchanged:

$$Q_c = \frac{[HI]_t^2}{[H_2]_t[I_2]_t}$$

In applying Le Châtelier's principle to a heterogeneous equilibrium, the effect of pressure changes on solids and liquids can be ignored because the volume (and concentration) of a solid or a liquid is nearly independent of pressure. Consider, for example,

> **REMEMBER. . .**
> The **ideal gas law** describes how the pressure of a gas is affected by changes in volume, temperature, and amount; $P = nRT/V$ (Section 10.3).

(a) A mixture of gaseous N_2, H_2, and NH_3 at equilibrium ($Q_c = K_c$).

(b) When the pressure is increased by decreasing the volume, the mixture is no longer at equilibrium ($Q_c < K_c$).

(c) Net reaction occurs from reactants to products, decreasing the total number of gaseous molecules until equilibrium is re-established ($Q_c = K_c$).

$$\xrightarrow[\text{V decreases}]{\text{P increases as}}$$

$$\xrightarrow[\text{to form products}]{\text{Net reaction}}$$

= N_2

= H_2

= NH_3

▲ **FIGURE 15.12**

Qualitative effect of pressure and volume on the equilibrium $N_2(g) + 3\,H_2(g) \rightleftharpoons 2\,NH_3(g)$.

▲ **Figure It Out**

Why does the reaction shift toward the products when the volume is decreased?

Answer: The change caused by the decrease in volume is an increase in pressure. The reaction must shift to lower the number of moles of gas to decrease pressure. In the synthesis of ammonia, there are two moles of gas in the products and four moles of gas in the reactants.

the high-temperature reaction of carbon with steam, the first step in converting coal to gaseous fuels:

$$C)(s) + H_2O(g) \rightleftharpoons CO(g) + H_2(g)$$

Ignoring the carbon because it's a solid, we predict that a decrease in volume (increase in pressure) will shift the equilibrium from products to reactants because the reverse reaction decreases the amount of gas from 2 mol to 1 mol.

Throughout this section, we've been careful to limit the application of Le Châtelier's principle to pressure changes that result from a change in *volume*. What happens, though, if we keep the volume constant but increase the total pressure by adding a gas that is not involved in the reaction—say, an inert gas such as argon? In that case, the equilibrium remains undisturbed because adding an inert gas at constant volume does not change the partial pressures or the molar concentrations of the substances involved in the reaction. Only if the added gas is a reactant or product does the reaction quotient change.

WORKED EXAMPLE 15.14

Applying Le Châtelier's Principle to Pressure and Volume Changes

Does the number of moles of products increase, decrease, or remain the same when each of the following equilibria is subjected to a decrease in pressure by increasing the volume?

(a) $PCl_5(g) \rightleftharpoons PCl_3(g) + Cl_2(g)$
(b) $CaO(s) + CO_2(g) \rightleftharpoons CaCO_3(s)$
(c) $3\,Fe(s) + 4\,H_2O(g) \rightleftharpoons Fe_3O_4(s) + 4\,H_2(g)$

STRATEGY

According to Le Châtelier's principle, the stress of a decrease in pressure is relieved by net reaction in the direction that increases pressure by increasing the number of moles of gas.

SOLUTION

(a) Because the forward reaction converts 1 mol of gas to 2 mol of gas, net reaction will go from reactants to products, thus increasing the number of moles of PCl_3 and Cl_2.

(b) Because there is 1 mol of gas on the reactant side of the balanced equation and none on the product side, the stress of a decrease in pressure is relieved by net reaction from products to reactants. The number of moles of $CaCO_3$ therefore decreases.

(c) Because there are 4 mol of gas on both sides of the balanced equation, the composition of the equilibrium mixture is unaffected by a change in pressure. The number of moles of Fe_3O_4 and H_2 remains the same.

▶ **PRACTICE 15.27** Does the number of moles of products increase, decrease, or remain the same when the following equilibrium is subjected to an increase in pressure by decreasing the volume?

$$2\ CO(g) \rightleftharpoons C(s) + CO_2(g)$$

▶ **CONCEPTUAL APPLY 15.28** The following picture represents the equilibrium mixture for the gas-phase reaction $A_2 \rightleftharpoons 2\ A$:

(a) Draw a picture that shows how the concentrations change when the pressure is increased by reducing the volume.
(b) Does the value of K_c stay the same, increase, or decrease when the pressure is increased by reducing the volume?

15.9 ALTERING AN EQUILIBRIUM MIXTURE: CHANGES IN TEMPERATURE

When an equilibrium is disturbed by a change in concentration, pressure, or volume, the composition of the equilibrium mixture changes because the reaction quotient Q_c no longer equals the equilibrium constant K_c. As long as the temperature remains constant, however, concentration, pressure, or volume changes don't change the value of the equilibrium constant. By contrast, a change in temperature nearly always changes the value of the equilibrium constant. For the synthesis of ammonia in the Haber process, which is an exothermic reaction, the equilibrium constant K_c decreases by a factor of 10^{11} over the temperature range 300–1000 K (**FIGURE 15.13**).

$$N_2(g) + 3\ H_2(g) \rightleftharpoons 2\ NH_3(g) + 92.2\ kJ \quad \Delta H° = -92.2\ kJ$$

Go to eText

BIG IDEA Question 6

What causes the net reaction direction to shift when the temperature of an equilibrium mixture is changed?

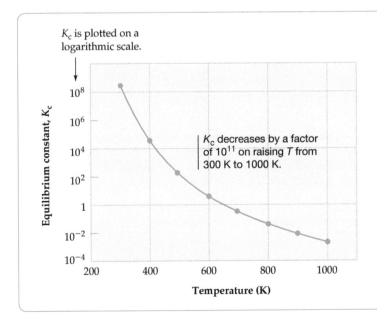

K_c is plotted on a logarithmic scale.

Temp (K)	K_c
300	2.6×10^8
400	3.9×10^4
500	1.7×10^2
600	4.2
700	2.9×10^{-1}
800	3.9×10^{-2}
900	8.1×10^{-3}
1000	2.3×10^{-3}

◀ **FIGURE 15.13**

Temperature dependence of the equilibrium constant for the reaction $N_2(g) + 3\ H_2(g) \rightleftharpoons 2\ NH_3(g)$.

K_c decreases by a factor of 10^{11} on raising T from 300 K to 1000 K.

At low temperatures, the equilibrium mixture is rich in NH_3 because K_c is large. At high temperatures, the equilibrium shifts in the direction of N_2 and H_2.

In general, the temperature dependence of an equilibrium constant depends on the sign of $\Delta H°$ for the reaction.

$N_2O_4(g) \rightleftharpoons 2\,NO_2(g); \Delta H° > 0$

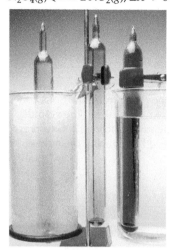

K_c increases as

T increases

▲ **FIGURE 15.14**

Sample tubes containing an equilibrium mixture of N_2O_4 and NO_2 immersed in ice water (left), at room temperature (center), and immersed in hot water (right). The darker brown color of the sample at the highest temperature indicates that the equilibrium $N_2O_4(g) \rightleftharpoons 2\,NO_2(g)$ shifts from reactants to products with increasing temperature, as expected for an endothermic reaction.

▲ Fingernail polish can be removed by dissolving it in ethyl acetate.

- The equilibrium constant for an exothermic reaction (negative $\Delta H°$) decreases as the temperature increases.
- The equilibrium constant for an endothermic reaction (positive $\Delta H°$) increases as the temperature increases.

You can predict the way in which K_c depends on temperature by using Le Châtelier's principle. If a reaction is endothermic, add heat to the reactant side of the equation. If a reaction is exothermic, add heat to the product side of the reaction. Take the endothermic decomposition of N_2O_4, for example:

add heat

$$N_2O_4(g) + 55.3\ \text{kJ} \rightleftharpoons 2\,NO_2(g) \qquad \Delta H° = +55.3\ \text{kJ}$$
Colorless · · · · · · · · · · · · · · · · · · · Brown

remove heat

Le Châtelier's principle says that if heat is added to an equilibrium mixture, thus increasing its temperature, net reaction occurs in the direction that relieves the stress of the added heat. For an endothermic reaction, such as the decomposition of N_2O_4, heat is absorbed by reaction in the forward direction. The equilibrium therefore shifts to the product side at the higher temperature, which means that K_c increases with increasing temperature. Because N_2O_4 is colorless and NO_2 has a brown color, the effect of temperature on the $N_2O_4 - NO_2$ equilibrium is readily apparent from the color of the mixture (**FIGURE 15.14**). For an exothermic reaction, such as the synthesis of NH_3, heat is absorbed by net reaction in the reverse direction, so K_c decreases with increasing temperature.

remove heat

$$N_2(g) + 3\,H_2(g) \rightleftharpoons 2\,NH_3(g) + 92.2\ \text{kJ} \qquad \Delta H° = -92.2\ \text{kJ}$$

add heat

WORKED EXAMPLE 15.15

Applying Le Châtelier's Principle to Temperature Changes

In the first step of the Ostwald process for the synthesis of nitric acid, ammonia is oxidized to nitric oxide by the reaction

$$4\,NH_3(g) + 5\,O_2(g) \rightleftharpoons 4\,NO(g) + 6\,H_2O(g) \quad \Delta H° = -901.2\ \text{kJ}$$

How does the equilibrium amount of NO vary with an increase in temperature?

STRATEGY

Le Châtelier's principle predicts that the stress of added heat when the temperature is increased will be relieved by net reaction in the direction that absorbs the heat. It's helpful to include the heat in the balanced equation—on the reactant side if the reaction is endothermic or on the product side if the reaction is exothermic.

SOLUTION

Because the oxidation of ammonia is exothermic, we include the heat (901.2 kJ) on the product side:

$$4\,NH_3(g) + 5\,O_2(g) \rightleftharpoons 4\,NO(g) + 6\,H_2O(g) + 901.2\ \text{kJ}$$

The stress of added heat when the temperature is increased will be relieved by net reaction from products to reactants, which absorbs the added heat. The equilibrium will therefore shift to the reactant side (K_c will decrease) with an increase in temperature. Consequently, the equilibrium mixture will contain less NO at higher temperatures.

▶ **PRACTICE 15.29** When air is heated at very high temperatures in an automobile engine, the air pollutant nitric oxide is produced by the reaction

$$N_2(g) + O_2(g) \rightleftharpoons 2\,NO(g) \quad \Delta H° = +182.6\ \text{kJ}$$

How does the equilibrium amount of NO vary with an increase in temperature?

▶ **APPLY 15.30**   Ethyl acetate, a solvent used in many fingernail-polish removers, is made by the reaction of acetic acid with ethanol:

$$CH_3CO_2H(soln) + CH_3CH_2OH(soln) \rightleftharpoons CH_3CO_2CH_2CH_3(soln) + H_2O(soln)$$

Acetic acid   Ethanol   Ethyl acetate

$$\Delta H° = -2.9 \text{ kJ}$$

Does the amount of ethyl acetate in an equilibrium mixture increase or decrease when the temperature is increased? How does K_c change when the temperature is decreased? Justify your answers using Le Châtelier's principle.

CONCEPTUAL WORKED EXAMPLE 15.16

Applying Le Châtelier's Principle to Temperature Changes

The pictures represent the composition of the equilibrium mixture at 600 K and 650 K for the combination of two A molecules, $2 A(g) \rightleftharpoons A_2(g)$.

    Is the reaction endothermic or exothermic? Explain using Le Châtelier's principle.

$T = 600$ K $T = 650$ K

STRATEGY

We can determine the direction of net reaction on raising the temperature by counting the number of A and A_2 molecules at each temperature. According to Le Châtelier's principle, if the net reaction converts reactants to products on raising the temperature, heat is on the reactant side of the chemical equation and the reaction is endothermic. Conversely, if the net reaction converts products to reactants on raising the temperature, heat is on the product side and the reaction is exothermic.

SOLUTION

Two A and five A_2 molecules are present at 600 K, and six A and three A_2 molecules are present at 650 K. On raising the temperature, the net reaction converts products to reactants, and so heat is on the product side of the chemical equation:

$$2 A(g) \rightleftharpoons A_2(g) + \text{heat}$$

The reaction is therefore exothermic, as expected for a reaction in which a chemical bond is formed. Note that Le Châtelier's principle predicts that net reaction will occur in the direction that uses up the added heat.

▶ **CONCEPTUAL PRACTICE 15.31**   The following pictures represent the composition of the equilibrium mixture for the reaction $A(g) + B(s) \rightleftharpoons AB(g)$ at 400 K and 500 K:

$T = 400$ K $T = 500$ K

Is the reaction endothermic or exothermic? Explain using Le Châtelier's principle.

▶ **CONCEPTUAL APPLY 15.32**   The following picture represents an equilibrium mixture of solid $BaCO_3$, solid BaO, and gaseous CO_2 obtained as a result of the endothermic decomposition of $BaCO_3$:

(a) Draw a picture that represents the equilibrium mixture after addition of four more CO_2 molecules.
(b) Draw a picture that represents the equilibrium mixture at a higher temperature.

15.10 THE LINK BETWEEN CHEMICAL EQUILIBRIUM AND CHEMICAL KINETICS

We emphasized in Section 15.1 that the equilibrium state is a dynamic one in which reactant and product concentrations remain constant, not because the reaction stops but because the rates of the forward and reverse reactions are equal. To explore this idea further, let's consider the general, reversible reaction

$$A + B \rightleftharpoons C + D$$

Let's assume that the forward and reverse reactions occur in a single bimolecular step; that is, they are **elementary reactions** (Section 14.10). We can then write the following rate laws:

$$\text{Rate forward} = k_f[A][B]$$

$$\text{Rate reverse} = k_r[C][D]$$

REMEMBER. . .
Because an **elementary reaction** describes an individual molecular event, its rate law follows directly from its stoichiometry (Section 14.10).

If we begin with a mixture that contains all reactants and no products, the initial rate of the reverse reaction is zero because $[C] = [D] = 0$. As A and B are converted to C and D by the forward reaction, the rate of the forward reaction decreases because [A] and [B] are getting smaller. At the same time, the rate of the reverse reaction increases because [C] and [D] are getting larger. Eventually, the decreasing rate of the forward reaction and the increasing rate of the reverse reaction become equal, and thereafter the concentrations remain constant; that is, the system is at chemical equilibrium (Figure 15.3).

Because the forward and reverse rates are equal at equilibrium, we can write

$$k_f[A][B] = k_r[C][D] \qquad \text{at equilibrium}$$

which can be rearranged to give

$$\frac{k_f}{k_r} = \frac{[C][D]}{[A][B]}$$

The right side of this equation is the equilibrium constant expression for the forward reaction, which equals the equilibrium constant K_c since the reaction mixture is at equilibrium.

$$K_c = \frac{[C][D]}{[A][B]}$$

Therefore, the equilibrium constant is simply the ratio of the rate constants for the forward and reverse reactions:

$$K_c = \frac{k_f}{k_r}$$

In deriving this equation for K_c, we have assumed a single-step mechanism. For a multistep mechanism, each step has a characteristic rate constant ratio, k_f/k_r. When equilibrium is reached, each step in the mechanism must be at equilibrium, and K_c for the overall reaction is equal to the product of the rate constant ratios for the individual steps.

REMEMBER. . .
Because the fraction of collisions with sufficient energy for reaction is given by $e^{-E_a/RT}$, the **Arrhenius equation** indicates that the rate constant decreases as E_a increases and increases as T increases (Section 14.7).

The equation relating K_c to k_f and k_r provides a fundamental link between chemical equilibrium and chemical kinetics: The relative values of the rate constants for the forward and reverse reactions determine the composition of the equilibrium mixture. When k_f is much larger than k_r, K_c is very large and the reaction goes almost to completion. Such a reaction is sometimes said to be irreversible because the reverse reaction is often too slow to be detected. When k_f and k_r have comparable values, K_c has a value near 1, and comparable concentrations of both reactants and products are present at equilibrium. This is the usual situation for a reversible reaction.

REMEMBER. . .
The greater the **activation energy**, the steeper the slope of an Arrhenius plot (a graph of ln k versus $1/T$) and the greater the increase in k for a given increase in T (Section 14.8).

The equation $K_c = k_f/k_r$ also helps explain why equilibrium constants depend on temperature. Recall from Section 14.7 that the rate constant increases as the temperature increases, in accord with the **Arrhenius equation** $k = Ae^{-E_a/RT}$. In general, the forward and reverse reactions have different values of the **activation energy**, so k_f and k_r increase by different amounts as the temperature increases. The ratio $k_f/k_r = K_c$ is therefore temperature-dependent. For an exothermic reaction, E_a for the reverse reaction is greater than E_a for the forward reaction. Consequently, as the temperature increases, k_r increases by more than k_f increases, and so $K_c = k_f/k_r$ for an exothermic reaction decreases as the temperature increases. Conversely, K_c for an endothermic reaction increases as the temperature increases. These results are in accord with Le Châtelier's principle (Section 15.9).

Chemical kinetics also explains the effect of a catalyst on a reaction in equilibrium. Recall from Section 14.12 that a **catalyst** increases the rate of a chemical reaction by making available a new, lower-energy pathway for the conversion of reactants to products. Because the forward and reverse reactions pass through the same transition state, a catalyst lowers the activation energy for the forward and reverse reactions by exactly the same amount. As a result, the rates of the forward and reverse reactions increase by the same factor (**FIGURE 15.15**). Because k_f and k_r increase by the same factor the ratio k_f/k_r is unaffected, and the value of the equilibrium constant $K_c = k_f/k_r$ remains unchanged. Thus, addition of a catalyst does not alter the composition of an equilibrium mixture.

If a reaction mixture is at equilibrium in the absence of a catalyst (that is, the forward and reverse rates are equal), it will still be at equilibrium after a catalyst is added because the forward and reverse rates, though faster, remain equal. If a reaction mixture is not at equilibrium, a catalyst accelerates the rate at which equilibrium is reached. Because a catalyst has no effect on the equilibrium concentrations, it does not appear in the balanced chemical equation or in the equilibrium constant expression.

Even though a catalyst doesn't change the position of an equilibrium, it can nevertheless significantly influence the choice of optimum conditions for a reaction. Look again at the Haber synthesis of ammonia. Because the reaction $N_2(g) + 3\,H_2(g) \rightleftharpoons 2\,NH_3(g)$ is exothermic, its equilibrium constant decreases with increasing temperature, and optimum yields of NH_3 are obtained at low temperatures. At those low temperatures, however, the rate at which equilibrium is reached is too slow for the reaction to be practical. We thus have what appears to be a no-win situation: Low temperatures give good yields but slow rates, whereas high temperatures give satisfactory rates but poor yields. The answer to the dilemma is to find a catalyst.

REMEMBER. . .
A **catalyst** is a substance that increases the rate of a transformation without being consumed in the process (Section 14.12).

◀ **FIGURE 15.15**

Potential energy profiles for a reaction whose activation energy is lowered by the presence of a catalyst.

◀ Figure It Out

What is the effect of adding a catalyst on the equilibrium constant of a reaction?

Answer: The equilibrium constant is unchanged when a catalyst is added because the activation energy barrier for the forward and reverse reactions by the same amount.

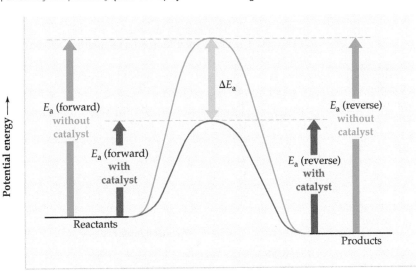

The activation energy for the **catalyzed pathway** (red curve) is lower than that for the **uncatalyzed pathway** (blue curve) by an amount ΔE_a.

A catalyst lowers the activation energy barrier for the forward and reverse reactions by the same amount. The catalyst therefore accelerates the forward and reverse reactions by the same factor, and the composition of the equilibrium mixture is unchanged.

The introduction of a catalyst is what makes the Haber-Bosch process so efficient. Haber discovered that a catalyst consisting of iron mixed with certain metal oxides causes the reaction to occur at a satisfactory rate at temperatures where the equilibrium concentration of NH_3 is reasonably favorable. The yield of NH_3 can be improved further by running the reaction at high pressures. Typical reaction conditions for the industrial synthesis of ammonia are 400–500 °C and 130–300 atm.

WORKED EXAMPLE 15.17

Exploring the Link between Equilibrium and Kinetics

The equilibrium constant K_c for the reaction of hydrogen with iodine is 57.0 at 700 K, and the reaction is endothermic ($\Delta H = 9$ kJ).

$$H_2(g) + I_2(g) \underset{k_r}{\overset{k_f}{\rightleftharpoons}} 2\,HI(g) \quad K_c = 57.0 \text{ at } 700 \text{ K}$$

(a) Is the rate constant k_f for the formation of HI larger or smaller than the rate constant k_r for the decomposition of HI?

(b) The value of k_r at 700 K is $1.16 \times 10^{-3}\,M^{-1}\,s^{-1}$. What is the value of k_f at the same temperature?

(c) How are the values of k_f, k_r, and K_c affected by the addition of a catalyst?

(d) How are the values of k_f, k_r, and K_c affected by an increase in temperature?

STRATEGY

To answer these questions, make use of the relationship $K_c = k_f/k_r$. Remember that a catalyst increases k_f and k_r by the same factor, and recall that the temperature dependence of a rate constant increases with increasing value of the activation energy (Section 14.8).

SOLUTION

(a) Because $K_c = k_f/k_r = 57.0$, the rate constant for the formation of HI (forward reaction) is larger than the rate constant for the decomposition of HI (reverse reaction) by a factor of 57.0.

(b) Because $K_c = k_f/k_r$,

$$k_f = (K_c)(k_r) = (57.0)(1.16 \times 10^{-3}\,M^{-1}\,s^{-1}) =$$
$$6.61 \times 10^{-2}\,M^{-1}\,s^{-1}$$

(c) A catalyst lowers the activation energy barrier for the forward and reverse reactions by the same amount, thus increasing the rate constants k_f and k_r by the same factor. Because the equilibrium constant K_c equals the ratio of k_f to k_r, the value of K_c is unaffected by the addition of a catalyst.

(d) Because the reaction is endothermic, E_a for the forward reaction is greater than E_a for the reverse reaction. Consequently, as the temperature increases, k_f increases by more than k_r increases, and therefore $K_c = k_f/k_r$ increases, consistent with Le Châtelier's principle.

▶ **PRACTICE 15.33** Nitric oxide emitted from the engines of supersonic aircraft can contribute to the destruction of stratospheric ozone:

$$NO(g) + O_3(g) \underset{k_r}{\overset{k_f}{\rightleftharpoons}} NO_2(g) + O_2(g)$$

▲ Nitric oxide emissions from supersonic aircraft can contribute to destruction of the ozone layer.

This reaction is highly exothermic ($\Delta H = -201$ kJ), and its equilibrium constant K_c is 3.4×10^{34} at 300 K.

(a) Which rate constant is larger, k_f or k_r?

(b) The value of k_f at 300 K is $8.5 \times 10^6\,M^{-1}\,s^{-1}$. What is the value of k_r at the same temperature?

▶ **APPLY 15.34** The energy profile of the reaction $A \underset{k_r}{\overset{k_f}{\rightleftharpoons}} B$ is shown. The frequency factor (A) is similar for the forward and reverse reaction.

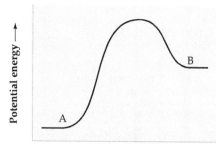

(a) Which rate constant is larger, k_f or k_r?

(b) Predict whether the equilibrium constant is greater or less than 1.

(c) What would be the effect of decreasing the temperature on the rate of the forward and reverse reaction? What would be the effect of decreasing temperature on the value of the equilibrium constant?

Humans, like all animals, need oxygen. The oxygen comes from breathing: about 500 mL of air is drawn into the lungs of an average person with each breath. When the freshly inspired air travels through the bronchial passages and enters the approximately 150 million alveolar sacs of the lungs, it picks up moisture and mixes with air remaining from the previous breath. As it mixes, the concentrations of both water vapor and carbon dioxide increase. These gas concentrations are measured by their partial pressures, with the partial pressure of oxygen in the lungs usually around 100 mm Hg (**TABLE 15.2**). Oxygen then diffuses through the delicate walls of the lung alveoli and into arterial blood, which transports it to all body tissues.

Only about 3% of the oxygen in blood is dissolved as $O_2(aq)$; the rest is chemically bound to *hemoglobin* molecules (Hb), large proteins that contain *heme* groups embedded in them. Each hemoglobin molecule contains four heme groups, and each heme group contains an iron atom that can bind to one O_2 molecule. Thus, a single hemoglobin molecule can bind four molecules of oxygen.

The entire system of oxygen transport and delivery in the body depends on the pickup and release of O_2 by hemoglobin according to the following series of equilibria:

$$Hb + O_2 \rightleftharpoons Hb(O_2)$$
$$Hb(O_2) + O_2 \rightleftharpoons Hb(O_2)_2$$
$$Hb(O_2)_2 + O_2 \rightleftharpoons Hb(O_2)_3$$
$$Hb(O_2)_3 + O_2 \rightleftharpoons Hb(O_2)_4$$

The positions of the different equilibria depend on the partial pressures of oxygen (P_{O_2}) in the various tissues. In hardworking, oxygen-starved muscles, where P_{O_2} is low, oxygen is released from hemoglobin as the equilibria shift toward the left, according to Le Châtelier's principle. In the lungs, where P_{O_2} is high, oxygen is absorbed by hemoglobin as the equilibria shift toward the right.

The amount of oxygen carried by hemoglobin at any given value of P_{O_2} is usually expressed as the percentage of possible binding sites that contain bound oxygen, called percent saturation. The curve in **FIGURE 15.16** shows the saturation is 97.5% in the lungs, where $P_{O_2} = 100$ mm Hg, meaning that each hemoglobin molecule is carrying close to its maximum possible amount of 4 O_2 molecules. When $P_{O_2} = 26$ mm Hg, however, the saturation drops to 50%.

What about people who live or vacation at high altitude? At the top of a mountain ski resort, where the altitude is 10,156 ft, the partial pressure of O_2 in the lungs is only about 68 mm Hg. Hemoglobin is only 90% saturated with O_2 at this pressure, so less oxygen is available for delivery to the tissues. People who ascend suddenly from sea level to high altitude thus experience a feeling of oxygen deprivation, or *hypoxia*, as their bodies are unable to supply enough oxygen to their tissues. The

TABLE 15.2   Partial Pressure of Oxygen in Human Lungs and Blood at Sea Level

Source	P_{O_2}(mm Hg)
Dry air	159
Alveolar air	100
Arterial blood	95
Venous blood	40

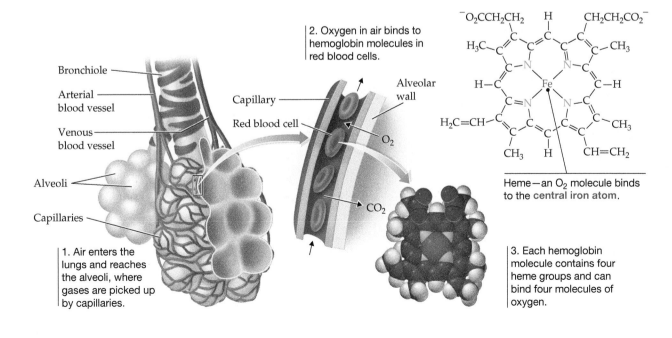

2. Oxygen in air binds to hemoglobin molecules in red blood cells.

Bronchiole

Arterial blood vessel

Capillary

Venous blood vessel

Red blood cell

Alveolar wall

O_2

Alveoli

Capillaries

CO_2

1. Air enters the lungs and reaches the alveoli, where gases are picked up by capillaries.

Heme—an O_2 molecule binds to the **central iron atom**.

3. Each hemoglobin molecule contains four heme groups and can bind four molecules of oxygen.

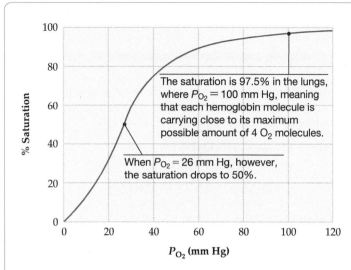

▲ FIGURE 15.16

An oxygen-carrying curve for hemoglobin. The percent saturation of the oxygen-binding sites on hemoglobin depends on the partial pressure of oxygen (P_{O_2}).

body responds to lower oxygen levels by producing more hemoglobin molecules, which both provide more capacity for O_2 transport and also drive the Hb + O_2 equilibrium to the right. However, the time required to adapt the lower O_2 pressures is typically days to weeks, so many vacationers in the mountains experience mild to moderate *altitude sickness*. Symptoms include headache, fatigue, upset stomach, and dizziness. Although discomfort usually subsides in a few days, some cases progress to more serious complications such as fluid in the lungs and swelling in the brain. The best treatment for altitude sickness is to return to lower elevations where P_{O_2} increases.

▲ The bodies of athletes who compete in high-altitude sports produce increased amounts of hemoglobin to cope with the low O_2 pressures at high altitudes.

PROBLEM 15.35   Refer to Figure 15.16 to answer the following questions:

(a) Estimate the percent saturation of hemoglobin in resting muscle tissue where the partial pressure of oxygen is 40 mm Hg.

(b) Estimate the percent saturation of hemoglobin in working muscle tissues where the partial pressure of oxygen is 15 mm Hg.

(c) The steepest part of the percent saturation curve is in the range 10–40 mm Hg. Why is it important to have a steep curve over this partial pressure range of oxygen?

PROBLEM 15.36   Which direction will the equilibrium reaction of hemoglobin and oxygen shift as temperature increases in the muscles during intense exercise? Does the effect of temperature on equilibrium position help muscles acquire the oxygen they need?

$$Hb + 4\,O_2 \rightleftharpoons Hb(O_2)_4 \qquad \Delta H = -200 \text{ kJ/mol}$$

PROBLEM 15.37   Equilibrium constants at 25 °C are given for the sequential equilibrium reactions in the binding of oxygen to hemoglobin.

$$Hb + O_2 \rightleftharpoons Hb(O_2) \qquad K_{c1} = 1.5 \times 10^4$$
$$Hb(O_2) + O_2 \rightleftharpoons Hb(O_2)_2 \qquad K_{c2} = 3.5 \times 10^4$$
$$Hb(O_2)_2 + O_2 \rightleftharpoons Hb(O_2)_3 \qquad K_{c3} = 5.9 \times 10^4$$
$$Hb(O_2)_3 + O_2 \rightleftharpoons Hb(O_2)_4 \qquad K_{c4} = 1.7 \times 10^6$$

(a) How does the value of the equilibrium constant change for each sequential binding step (K_{c1} through K_{c4})?

(b) Do the equilibrium reactions become more product- or reactant-favored with the binding of each oxygen?

(c) How do the sequential equilibrium steps affect the oxygen-carrying capacity of hemoglobin?

PROBLEM 15.38   Use the equilibrium constants provided in Problem 15.37 to answer the following questions:

(a) Calculate the equilibrium constant at 25 °C for the reaction

$$Hb + 4\,O_2 \rightleftharpoons Hb(O_2)_4 \quad K_c = ?$$

(b) Calculate the ratio of $[Hb(O_2)_4/[Hb]]$ in a solution similiar to arterial blood with a partial pressure of oxygen of 95 mmHg. (The solubility of oxygen at 25 °C is 1.61 μM/mm Hg.)

PROBLEM 15.39   The equilibrium constant for the reaction of hemoglobin with CO (Hb + CO) is greater than the equilibrium constant for the reaction of Hb + O_2. Use Le Châtelier's principle to predict how CO affects the equilibrium Hb + $O_2 \rightleftharpoons Hb(O_2)$. Suggest a possible reason for the toxicity of CO.

PROBLEM 15.40   The following reaction shows the equilibrium reaction that occurs when carbon monoxide enters the blood. (Note: The K value uses partial pressures of O_2 and CO in the equilibrium expression.)

$$Hb(O_2)(aq) + CO(g) \rightleftharpoons Hb(CO)(aq) + O_2(g)$$
$$K = 207 \text{ at } 37 \text{ °C}$$

(a) What is the ratio of $[Hb(CO)]$ to $[Hb(O_2)]$ in air that contains 20% O_2 and 0.15% CO by volume?

(b) The treatment for mild carbon monoxide poisoning is breathing pure oxygen. Use Le Châtelier's principle to explain why this treatment is effective.

STUDY GUIDE

Section	Concept Summary	Learning Objectives	Test Your Understanding
15.1 The Equilibrium State	**Chemical equilibrium** is a dynamic state in which the concentrations of reactants and products remain constant because the rates of the forward and reverse reactions are equal.	**15.1** Describe the characteristics of a reaction in chemical equilibrium.	Problems 15.41, 15.52–15.53
15.2 The Equilibrium Constant K_c	For the general reaction $a\text{ A} + b\text{ B} \rightleftharpoons c\text{ C} + d\text{ D}$, concentrations in the **equilibrium mixture** are related by the **equilibrium equation:** $$K_c = \frac{[\text{C}]^c[\text{D}]^d}{[\text{A}]^a[\text{B}]^b}$$ The quotient on the right side of the equation is called the *equilibrium constant expression*. The **equilibrium constant K_c** is the number obtained when equilibrium concentrations (in mol/L) are substituted into the equilibrium constant expression. The value of K_c varies with temperature and depends on the form of the balanced chemical equation.	**15.2** Write an equilibrium constant expression K_c for gas and solution phase reactions.	Worked Example 15.1; Problems 15.54, 15.56
		15.3 Determine a new equilibrium constant expression when reactions are manipulated or added.	Problems 15.60, 15.62, 15.64
		15.4 Calculate the value of the equilibrium constant K_c given concentrations of reactant and products.	Worked Example 15.2; Problems 15.66, 15.72
		15.5 Calculate the equilibrium constant from a molecular representation of a reaction mixture.	Worked Example 15.3; Problem 15.43
15.3 The Equilibrium Constant K_p	The **equilibrium constant K_p** can be used for gas-phase reactions. It is defined in the same way as K_c except that the equilibrium constant expression contains partial pressures (in atmospheres) instead of molar concentrations. The constants K_p and K_c are related by the equation $K_p = K_c(RT)^{\Delta n}$, where $\Delta n = (c + d) - (a + b)$.	**15.6** Write the equilibrium expression K_p in terms of partial pressures of reactants and products. Relate the equilibrium constants K_p and K_c.	Worked Example 15.4–15.5 Problems 15.68, 15.70
15.4 Heterogeneous Equilibria	**Homogeneous equilibria** are those in which all reactants and products are in a single phase; **heterogeneous equilibria** are those in which reactants and products are present in more than one phase. The equilibrium equation for a heterogeneous equilibrium does not include concentrations of pure solids or pure liquids.	**15.7** Write equilibrium constant expressions for reactions involving heterogeneous equilibria.	Worked Example 15.6; Problems 15.74–75
15.5 Using the Equilibrium Constant	The value of the equilibrium constant for a reaction makes it possible to judge the extent of reaction, predict the direction of reaction, and calculate equilibrium concentrations (or partial pressures) from initial concentrations (or partial pressures). The farther the reaction proceeds toward completion, the larger the value of K_c. The direction of a reaction not at equilibrium depends on the relative values of K_c and the **reaction quotient Q_c**, which is defined in the same way as K_c except that the concentrations in the equilibrium constant expression are not necessarily equilibrium concentrations. If $Q_c < K_c$, the net reaction goes from reactants to products to attain equilibrium; if $Q_c < K_c$, the net reaction goes from products to reactants; if $Q_c = K_c$, the system is at equilibrium.	**15.8** Determine the extent of a reaction given the equilibrium constant.	Worked Example 15.7; Problems 15.76, 15.78
		15.9 Predict the direction a reaction will shift to reach equilibrium given non-equilibrium concentrations.	Worked Example 15.8; Problems 15.80, 15.82
		15.10 Calculate concentrations or partial pressures of products and reactants in equilibrium.	Worked Examples 15.9, 15.11; Problems 15.84, 15.86, 15.90, 15.116
		15.11 Calculate concentrations of reactants and products in equilibrium using the quadratic equation to solve the mathematical operations.	Worked Examples 15.10, 15.12; Problems 15.88, 15.92

Section	Concept Summary	Learning Objectives	Test Your Understanding
15.6–15.9 Factors That Alter the Composition of an Equilibrium Mixture: Le Châtelier's Principle (Concentration, Pressure, Volume, Temperature)	The composition of an equilibrium mixture can be altered by changes in concentration, pressure (volume), or temperature. The qualitative effect of these changes is predicted by **Le Châtelier's principle**, which says that if a stress is applied to a reaction mixture at equilibrium, net reaction occurs in the direction that relieves the stress. Temperature changes affect equilibrium concentrations because K_c is temperature-dependent. As the temperature increases, K_c for an exothermic reaction decreases and K_c for an endothermic reaction increases.	**15.12** Predict the direction a reaction in equilibrium will shift when concentrations of reactants or products are changed.	Worked Example 15.13; Problems 15.45, 15.124, 15.128
		15.13 Predict the direction a reaction in equilibrium will shift when volume or pressure is changed.	Worked Example 15.14; Problems 15.44, 15.120, 15.131
		15.14 Predict the direction a reaction in equilibrium will shift when temperature is changed.	Worked Examples 15.15–15.16; Problems 15.46, 15.51, 15.122, 15.156
15.10 The Link between Chemical Equilibrium and Chemical Kinetics	The equilibrium constant for a single-step reaction equals the ratio of the rate constants for the forward and reverse reactions: $K_c = k_f/k_r$. A catalyst increases the rate at which chemical equilibrium is reached, but it does not affect the equilibrium constant or the equilibrium concentrations.	**15.15** Describe the effect of a catalyst on the equilibrium constant for a reaction.	Problems 15.50, 15.132
		15.16 Relate the equilibrium constant to the rate of the forward and reverse reactions.	Worked Example 15.17; Problems 15.134, 15.137–15.138

KEY TERMS

chemical equilibrium *602*
equilibrium constant K_c *606*
equilibrium constant K_p *610*

equilibrium equation *606*
equilibrium mixture *602*
heterogeneous equilibria *612*

homogeneous equilibria *612*
Le Châtelier's principle *625*
reaction quotient Q_c *616*

KEY EQUATIONS

- Equilibrium Equation for the Reaction $a\,A + b\,B \rightleftharpoons c\,C + d\,D$ in Terms of Molar Concentrations (Section 15.2)

$$K_c = \frac{[C]^c[D]^d}{[A]^a[B]^b}$$

- Equilibrium Equation for the Gas Phase Reaction $a\,A + b\,B \rightleftharpoons c\,C + d\,D$ in Terms of Partial Pressure (Section 15.3)

$$K_p = \frac{(P_C)^c(P_D)^d}{(P_A)^a(P_B)^b}$$

- Relationship between the Equilibrium Constants K_p and K_c for the Reaction $a\,A + b\,B \rightleftharpoons c\,C + d\,D$ (Section 15.3)

$$K_p = K_c(RT)^{\Delta n}$$

where R is the gas constant, 0.082 06 (L · atm)/(K · mol), T is the absolute temperature, and $\Delta n = (c + d) - (a + b)$ is the number of moles of gaseous products minus the number of moles of gaseous reactants.

- Reaction Quotient Q_c for the Reaction $a\,A + b\,B \rightleftharpoons c\,C + d\,D$ (Section 15.5)

$$Q_c = \frac{[C]_t^c[D]_t^d}{[A]_t^a[B]_t^b}$$

where the subscript t denotes concentration at an arbitrary time t, not necessarily equilibrium.

- Relationship between the Equilibrium Constant K_c and the Rate Constant for the Forward and Reverse Reaction (Section 15.10)

$$K_c = k_f/k_r.$$

PRACTICE TEST

After studying this chapter, you can assess your understanding with these practice test questions, which are correlated with chapter learning objectives. If you answer a question incorrectly, refer to the learning objectives in the end-of-chapter Study Guide for assistance. The Study Guide provides a conceptual summary, references a Worked Example to model how to solve the problem, and gives additional problems for more practice.

1. Hydrogen is synthesized by the following reaction:

$$CO(g) + H_2O(g) \rightleftharpoons CO_2(g) + H_2(g)$$

The graph shows how the concentrations of $H_2O(g)$ and $H_2(g)$ change with time. At which point (a–e) is equilibrium *first* reached? (**LO 15.1**)

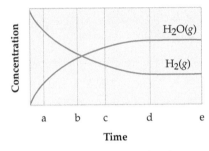

2. What is the equilibrium expression for the reaction of iron metal with water vapor? (**LO 15.2 and 15.6**)

$$2 Fe(s) + 3 H_2O(g) \rightleftharpoons Fe_2O_3(s) + 3 H_2(g)$$

(a) $K_c = \dfrac{[H_2]^3[Fe_2O_3]}{[Fe][H_2O]^3}$

(b) $K_c = \dfrac{[Fe][H_2O]^3}{[H_2]^3[Fe_2O_3]}$

(c) $K_c = \dfrac{[H_2O]^3}{[H_2]^3}$

(d) $K_c = \dfrac{[H_2]^3}{[H_2O]^3}$

3. Given the data for the following reactions at 298 K

$$N_2(g) + O_2(g) \rightleftharpoons 2 NO(g) \quad K_p = 4.4 \times 10^{-31}$$

$$NO(g) + \frac{1}{2} O_2(g) \rightleftharpoons NO_2(g) \quad K_p = 1.5 \times 10^6$$

calculate the value of the equilibrium constant K_p at 298 K for the reaction

$$N_2(g) + 2 O_2(g) \rightleftharpoons 2 NO_2(g) \quad K_p = ? \quad (\textbf{LO 15.3})$$

(a) $K_p = 6.6 \times 10^{-25}$ (b) $K_p = 1.3 \times 10^{-24}$
(c) $K_p = 9.9 \times 10^{-19}$ (d) $K_p = 5.4 \times 10^{-28}$

4. Sulfur dioxide reacts with oxygen in a step in the production of sulfuric acid.

$$2 SO_2(g) + O_2(g) \rightleftharpoons 2 SO_3(g) \quad K_c = 7.9 \times 10^4 \quad (800\ K)$$

For an equilibrium mixture in which $[SO_2] = 4.5 \times 10^{-3}$ M and $[O_2] = 1.5 \times 10^{-3}$ M, what is $[SO_3]$? (**LO 15.4**)

(a) $[SO_3] = 6.2 \times 10^{-7}$ M
(b) $[SO_3] = 4.9 \times 10^{-2}$ M
(c) $[SO_3] = 0.73$ M
(d) $[SO_3] = 2.4 \times 10^{-3}$ M

5. The reaction $A_2(g) + B_2(g) \rightleftharpoons 2 AB(g)$ has an equilibrium constant $K_c = 9$. The following figure represents a reaction mixture that contains A_2 molecules (red), B_2 molecules (blue), and AB molecules. What statement about the mixture is true? (**LO 15.5**)

(a) The mixture is at equilibrium, and there will be no net shift in reaction direction.
(b) The reaction will shift toward the reactants to reach equilibrium.
(c) The reaction will shift toward the products to reach equilibrium.
(d) More information is needed to answer this question.

6. For the reaction, $2 NH_3(g) \rightleftharpoons N_2(g) + 3 H_2(g)$, $K_p = 1.5 \times 10^3$ at 400 °C, what is the value of K_c for the reaction at the same temperature? (**LO 15.6**)

(a) 4.6×10^6 (b) 1.4
(c) 1.8×10^{-1} (d) 4.9×10^{-1}

7. The gas-phase reaction $2 SO_2(g) + O_2(g) \rightleftharpoons 2 SO_3(g)$, has an equilibrium constant $K_c = 5.8 \times 10^3$ at 600 °C.

A mixture contains $[SO_2] = 0.10$ M, $[O_2] = 0.100$ M, and $[SO_3] = 0.200$ M. Which statement is true about the reaction direction and equilibrium mixture? (**LO 15.8, 15.9**)

(a) The mixture is at equilibrium and contains appreciable amount of reactants and products.
(b) The reaction will shift to make more product, and the equilibrium mixture contains appreciable amount of reactants and products.
(c) The reaction will shift to make more reactants, and the equilibrium mixture contains mostly product.
(d) The reaction will shift to make more product, and the equilibrium mixture contains mostly reactant.

8. Consider the reaction $2 NOCl(g) \rightleftharpoons 2 NO(g) + Cl_2(g)$. To start the reaction, NOCl (g) was added to a flask at a concentration of 2.5 M. After equilibrium was reached, the concentration of $Cl_2(g)$ was 0.60 M. Calculate the equilibrium constant (K_c) for this reaction at the temperature of this experiment. (**LO 15.10**)

(a) 0.19   (b) 0.51
(c) 0.66   (d) 0.17

9. At a temperature of 430 °C, the reaction $H_2(g) + I_2(g) \rightleftharpoons 2 HI(g)$ has an equilibrium constant $K_c = 54.3$. Suppose that a mixture of 0.500 mol of $H_2(g)$ and 0.500 mol of $I_2(g)$ is placed into a 1.00-L stainless-steel flask at 430 °C. Calculate the concentration of $HI(g)$ when equilibrium is reached. (**LO 15.10**)

(a) 0.393 M   (b) 0.107 M
(c) 0.500 M   (d) 0.786 M

10. Nitrogen dioxide decomposes to nitric oxide and oxygen according the equation:

$$2\,NO_2(g) \rightleftharpoons 2\,NO(g) + O_2(g)$$

Initially the only gas in the system is NO_2 at a pressure of 7.82 atm. After the system reaches equilibrium, the total pressure is 9.86 atm. What is the value of K_p for the reaction at this temperature? **(LO 15.10)**

(a) 2.43   (b) 0.124

(c) 2.22   (d) 1.36

11. Phosphorus pentachloride decomposes to phosphorus trichloride and chlorine at high temperatures according to the equation:

$$PCl_5(g) \rightleftharpoons PCl_3(g) + Cl_2(g)$$

At 250 °C, 0.250 M PCl_5 is added to the flask. If $K_c = 1.80$, what are the equilibrium concentrations of each gas? **(LO 15.11)**

(a) $[PCl_5] = 0.028$ M, $[PCl_3] = 0.222$ M, and $[Cl_2] = 0.222$ M

(b) $[PCl_5] = 0.125$ M, $[PCl_3] = 0.474$ M, and $[Cl_2] = 0.474$ M

(c) $[PCl_5] = 1.80$ M, $[PCl_3] = 1.80$ M, and $[Cl_2] = 1.80$ M

(d) $[PCl_5] = 2.27$ M, $[PCl_3] = 2.02$ M, and $[Cl_2] = 2.02$ M

12. Suppose that the following *endothermic* reaction is at equilibrium:

$$2\,CO(g) + O_2(g) \rightleftharpoons 2\,CO_2(g)$$

Which of the following changes would cause the reaction to shift toward products? **(LO 15.12 – 15.15)**

(a) Remove $CO(g)$.

(b) Increase the temperature.

(c) Add a catalyst.

(d) Increase the volume of the container.

13. For the general, single-step reaction $A(g) + B(g) \rightleftharpoons AB(g)$, $K_c = 4.5 \times 10^{-6}$, which of the following statements is true? **(LO 15.16)**

(a) E_a (forward) $< E_a$ (reverse)

(b) The equilibrium mixture contains mostly products.

(c) $k_r > k_f$

(d) The reaction is exothermic.

Answers:

1. d, 2. d, 3. c, 4. b, 5. c, 6. d, 7. b, 8. b, 9. d, 10. a, 11. a, 12. b, 13. c

---

Mastering Chemistry   provides end-of-chapter exercises, feedback-enriched tutorial problems, animations, and interactive activities to encourage problem-solving practice and deeper understanding of key concepts and topics.

RAN   *Randomized in Mastering Chemistry*

CONCEPTUAL PROBLEMS

Problems 15.1–15.40 appear within the chapter.

15.41 Consider the interconversion of A molecules (red spheres) and B molecules (blue spheres) according to the reaction A $\rightleftharpoons$ B. Each of the series of pictures at the right represents a separate experiment in which time increases from left to right:

(a) Which of the experiments has resulted in an equilibrium state?

(b) What is the value of the equilibrium constant K_c for the reaction A $\rightleftharpoons$ B?

(c) Explain why you can calculate K_c without knowing the volume of the reaction vessel.

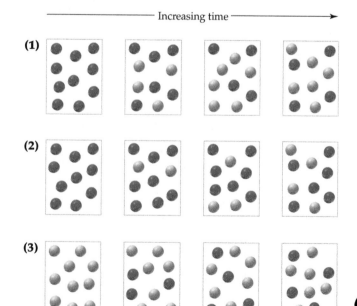

Increasing time ⟶

(1)

(2)

(3)

15.42 The following pictures represent the equilibrium state for three different reactions of the type $A_2 + X_2 \rightleftharpoons 2AX$ (X = B, C, or D).

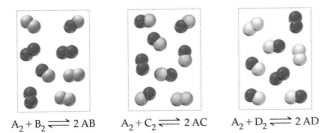

$$A_2 + B_2 \rightleftharpoons 2\,AB \qquad A_2 + C_2 \rightleftharpoons 2\,AC \qquad A_2 + D_2 \rightleftharpoons 2\,AD$$

(a) Which reaction has the largest equilibrium constant?

(b) Which reaction has the smallest equilibrium constant?

15.43 The reaction $A_2 + B \rightleftharpoons A + AB$ has an equilibrium con-
RAN stant $K_c = 2$. The following pictures represent reaction mixtures that contain A atoms (red), B atoms (blue), and A_2 and AB molecules.

(1)   **(2)**   **(3)**

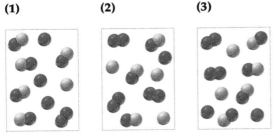

(a) Which reaction mixture is at equilibrium?

(b) For those mixtures that are not at equilibrium, will the reaction go in the forward or reverse direction to reach equilibrium?

15.44 The following pictures represent the initial state and the equilibrium state for the reaction of A_2 molecules (red) with B atoms (blue) to give AB molecules.

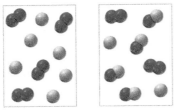

Initial state   Equilibrium state

(a) Write a balanced chemical equation for the reaction.

(b) If the volume of the equilibrium mixture is decreased, will the number of AB molecules increase, decrease, or remain the same? Explain.

15.45 Consider the reaction $A + B \rightleftharpoons AB$. The vessel on the right contains an equilibrium mixture of A molecules (red spheres), B molecules (blue spheres), and AB molecules. If the stopcock is opened and the contents of the two vessels are allowed to mix, will the reaction go in the forward or reverse direction? Explain.

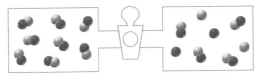

15.46 The following pictures represent the composition of the equilibrium mixture for the reaction $A + B \rightleftharpoons AB$ at 300 K and at 400 K.

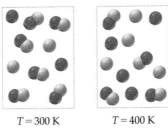

$T = 300\ \text{K}$ $T = 400\ \text{K}$

Is the reaction exothermic or endothermic? Explain using Le Châtelier's principle.

15.47 The following pictures represent equilibrium mixtures at 325 K and 350 K for a reaction involving A atoms (red), B atoms (blue), and AB molecules.

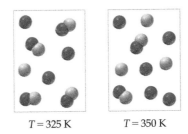

$T = 325\ \text{K}$ $T = 350\ \text{K}$

(a) Write a balanced equation for the reaction that occurs on raising the temperature.

(b) Is the reaction exothermic or endothermic? Explain using Le Châtelier's principle.

(c) If the volume of the container is increased, will the number of A atoms increase, decrease, or remain the same? Explain.

15.48 The following picture represents the composition of the equilibrium mixture for the endothermic reaction $A_2 \rightleftharpoons 2\,A$ at 500 K.

Draw a picture that represents the equilibrium mixture after each of the following changes.

(a) Adding a catalyst

(b) Increasing the volume

(c) Decreasing the temperature

15.49 The following picture represents the equilibrium state for the reaction $2\,AB \rightleftharpoons A_2 + B_2$.

Which rate constant is larger, k_f or k_r? Explain.

15.50 The following pictures represent the initial and equilibrium states for the exothermic decomposition of gaseous A molecules (red) to give gaseous B molecules (blue).

Initial state   Equilibrium state

(a) Write a balanced chemical equation for the reaction.

(b) Will the number of A molecules in the equilibrium mixture increase, decrease, or remain the same after each of the following changes? Explain.
   (1) Increasing the temperature
   (2) Decreasing the volume
   (3) Increasing the pressure by adding an inert gas
   (4) Adding a catalyst

15.51 The following pictures represent the initial and equilibrium states for the exothermic reaction of solid A (red) with gaseous B_2 (blue) to give gaseous AB.

Initial state   Equilibrium state

(a) Write a balanced chemical equation for the reaction.

(b) Will the number of AB molecules in the equilibrium mixture increase, decrease, or remain the same after each of the following changes? Explain.
   (1) Increasing the partial pressure of B_2
   (2) Adding more solid A
   (3) Increasing the volume
   (4) Increasing the temperature

SECTION PROBLEMS

The Equilibrium State (Section 15.1)

15.52 Identify the *true* statement about the rate of the forward and reverse reaction once a reaction has reached equilibrium.

(a) The rate of the forward reaction and the reverse reaction is zero.

(b) The rate of the forward reaction is greater than the rate of the reverse reaction.

(c) The rate of the reverse reaction is greater than the rate of the forward reaction.

(d) The rate of the forward reaction is equal to the rate of the reverse reaction.

15.53 Identify the *true* statement about the concentrations of A and B once the reaction A $\rightleftharpoons$ B has reached equilibrium.

(a) The concentration A equals the concentration of B.

(b) The concentrations of A and B are constant.

(c) The concentration of A decreases and the concentration of B increases.

(d) The concentration of B decreases and the concentration of A increases.

Equilibrium Constant Expressions and Equilibrium Constants (Section 15.2–15.4)

15.54 Diethyl ether, used as an anesthetic, is synthesized by heating ethanol with concentrated sulfuric acid. Write the equilibrium constant expression for K_c.

$$2\ C_2H_5OH(soln) \rightleftharpoons C_2H_5OC_2H_5(soln) + H_2O(soln)$$

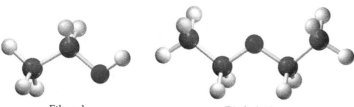

Ethanol   Diethyl ether

15.55 Ethylene glycol, used as antifreeze in automobile radiators, is manufactured by the hydration of ethylene oxide. Write the equilibrium constant expression for K_c.

$$H_2C\!-\!CH_2(soln) + H_2O(soln) \rightleftharpoons HOCH_2CH_2OH(soln)$$

Ethylene oxide   Ethylene glycol

15.56 For each of the following equilibria, write the equilibrium RAN constant expression for K_c.

(a) $CH_4(g) + H_2O(g) \rightleftharpoons CO(g) + 3\ H_2(g)$

(b) $3\ F_2(g) + Cl_2(g) \rightleftharpoons 2\ ClF_3(g)$

(c) $H_2(g) + F_2(g) \rightleftharpoons 2\ HF(g)$

15.57 For each of the following equilibria, write the equilibrium constant expression for K_c.

(a) $2\ C_2H_4(g) + O_2(g) \rightleftharpoons 2CH_3CHO(g)$

(b) $CO(g) + H_2O(g) \rightleftharpoons CO_2(g) + H_2(g)$

(c) $4\ NH_3(g) + 5\ O_2(g) \rightleftharpoons 4\ NO(g) + 6\ H_2O(g)$

15.58 For each of the equilibria in Problem 15.56, write the equi- RAN librium constant expression for K_p and give the equation that relates K_p and K_c.

15.59 For each of the equilibria in Problem 15.57, write the equi- RAN librium constant expression for K_p and give the equation that relates K_p and K_c.

15.60 If $K_c = 7.5 \times 10^{-9}$ at 1000 K for the reaction $N_2(g) +$
RAN $O_2(g) \rightleftharpoons 2\,NO(g)$, give the value of K_c at 1000 K for the reaction

(a) $2\,NO(g) \rightleftharpoons N_2(g) + O_2(g)$

(b) $NO(g) \rightleftharpoons 1/2\,N_2(g) + 1/2\,O_2(g)$

(c) $2\,N_2(g) + 2\,O_2(g) \rightleftharpoons 4\,NO(g)$

15.61 At 400 K, $K_p = 50.2$ for the reaction $N_2O_4(g) \rightleftharpoons$
RAN $2\,NO_2(g)$, what is K_p at 400 K for the reaction?

(a) $2\,NO_2(g) \rightleftharpoons N_2O_4(g)$

(b) $2\,N_2O_4(g) \rightleftharpoons 4\,NO_2(g)$

(c) $NO_2(g) \rightleftharpoons 1/2\,N_2O_4(g)$

15.62 The reaction $2\,AsH_3(g) \rightleftharpoons As_2(g) + 3\,H_2(g)$ has $K_p = 7.2 \times 10^7$ at 1073 K. At the same temperature, what is K_p for each of the following reactions?

(a) $As_2(g) + 3\,H_2(g) \rightleftharpoons 2\,AsH_3(g)$

(b) $4\,AsH_3(g) \rightleftharpoons 2\,As_2(g) + 6\,H_2(g)$

(c) $9\,H_2(g) + 3\,As_2(g) \rightleftharpoons 6\,AsH_3(g)$

15.63 The reaction

$$2\,PH_3(g) + As_2(g) \rightleftharpoons 2\,AsH_3(g) + P_2(g)$$

has $K_p = 2.9 \times 10^{-5}$ at 873 K. At the same temperature, what is K_p for each of the following reactions?

(a) $2\,AsH_3(g) + P_2(g) \rightleftharpoons 2\,PH_3(g) + As_2(g)$

(b) $6\,PH_3(g) + 3\,As_2(g) \rightleftharpoons 3\,P_2(g) + 6\,AsH_3(g)$

(c) $2\,P_2(g) + 4\,AsH_3(g) \rightleftharpoons 2\,As_2(g) + 4\,PH_3(g)$

15.64 Calculate the value of the equilibrium constant at 427 °C for
RAN the reaction

$$Na_2O(s) + 1/2\,O_2(g) \rightleftharpoons Na_2O_2(s)$$

given the following equilibrium constants at 427 °C.

$Na_2O(s) \rightleftharpoons 2\,Na(l) + 1/2\,O_2(g)$ $K_c = 2 \times 10^{-25}$
$Na_2O_2(s) \rightleftharpoons 2\,Na(l) + O_2(g)$ $K_c = 5 \times 10^{-29}$

15.65 Calculate the value of the equilibrium constant for the reaction
RAN
$$4\,NH_3(g) + 3\,O_2(g) \rightleftharpoons 2\,N_2(g) + 6\,H_2O(g)$$

given the following equilibrium constants at a certain temperature.

$2\,H_2(g) + O_2(g) \rightleftharpoons 2\,H_2O(g)$ $K_c = 3.2 \times 10^{81}$
$N_2(g) + 3\,H_2(g) \rightleftharpoons 2\,NH_3(g)$ $K_c = 3.5 \times 10^{8}$

15.66 An equilibrium mixture of PCl_5, PCl_3, and Cl_2 at a certain tem-
RAN perature contains 8.3×10^{-3} M PCl_5, 1.5×10^{-2} M PCl_3, and 3.2×10^{-2} M Cl_2. Calculate the equilibrium constant K_c for the reaction $PCl_5(g) \rightleftharpoons PCl_3(g) + Cl_2(g)$.

15.67 The partial pressures in an equilibrium mixture of NO, Cl_2, and NOCl at 500 K are as follows: $P_{NO} = 0.240$ atm, $P_{Cl_2} = 0.608$ atm, $P_{NOCl} = 1.35$ atm. What is K_p at 500 K for the reaction $2\,NO(g) + Cl_2(g) \rightleftharpoons 2\,NOCl(g)$?

15.68 At 298 K, K_c is 2.2×10^5 for the reaction $F(g) + O_2(g) \rightleftharpoons O_2F(g)$. What is the value of K_p at this temperature?

15.69 At 298 K, K_p is 1.6×10^{-6} for the reaction $2\,NOCl(g) \rightleftharpoons 2\,NO(g) + Cl_2(g)$. What is the value of K_c at this temperature?

15.70 The vapor pressure of water at 25 °C is 0.0313 atm. Calculate the values of K_p and K_c at 25 °C for the equilibrium $H_2O(l) \rightleftharpoons H_2O(g)$.

15.71 Naphthalene, a white solid used to make mothballs, has a vapor pressure of 0.10 mm Hg at 27 °C. Calculate the values of K_p and K_c at 27 °C for the equilibrium $C_{10}H_8(s) \rightleftharpoons C_{10}H_8(g)$.

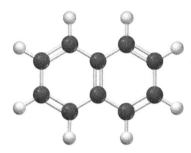

Naphthalene

15.72 Chlorine monoxide and dichlorine dioxide are involved in the catalytic destruction of stratospheric ozone. They are related by the equation

$$2\,ClO(g) \rightleftharpoons Cl_2O_2(g)$$

for which K_c is 4.96×10^{11} at 253 K. For an equilibrium mixture in which $[Cl_2O_2]$ is 6.00×10^{-6} M, what is $[ClO]$?

15.73 Sulfur dioxide reacts with oxygen in a step in the production
RAN of sulfuric acid.

$2\,SO_2(g) + O_2(g) \rightleftharpoons 2\,SO_3(g)$ $K_c = 7.9 \times 10^4$ (800 K)

For an equilibrium mixture in which $[SO_2] = 1.5 \times 10^{-3}$ M and $[O_2] = 3.0 \times 10^{-3}$ M, what is $[SO_3]$?

15.74 For each of the following equilibria, write the equilibrium
RAN constant expression for K_c. Where appropriate, also write the equilibrium constant expression for K_p.

(a) $Fe_2O_3(s) + 3\,CO(g) \rightleftharpoons 2\,Fe(l) + 3\,CO_2(g)$

(b) $4\,Fe(s) + 3\,O_2(g) \rightleftharpoons 2\,Fe_2O_3(s)$

(c) $BaSO_4(s) \rightleftharpoons BaO(s) + SO_3(g)$

(d) $BaSO_4(s) \rightleftharpoons Ba^{2+}(aq) + SO_4^{2-}(aq)$

15.75 For each of the following equilibria, write the equilibrium constant expression for K_c. Where appropriate, also write the equilibrium constant expression for K_p.

(a) $WO_3(s) + 3\,H_2(g) \rightleftharpoons W(s) + 3\,H_2O(g)$

(b) $Ag^+(aq) + Cl^-(aq) \rightleftharpoons AgCl(s)$

(c) $2\,FeCl_3(s) + 3\,H_2O(g) \rightleftharpoons Fe_2O_3(g) + 6\,HCl(g)$

(d) $MgCO_3(s) \rightleftharpoons MgO(s) + CO_2(g)$

Using the Equilibrium Constant (Section 15.5)

15.76 When the following reactions come to equilibrium, does the equilibrium mixture contain mostly reactants or mostly products?

(a) $H_2(g) + S(s) \rightleftharpoons H_2S(g)$; $K_c = 7.8 \times 10^5$

(b) $N_2(g) + 2\,H_2(g) \rightleftharpoons N_2H_4(g)$; $K_c = 7.4 \times 10^{-26}$

15.77 Which of the following reactions yield appreciable equilibrium concentrations of both reactants and products?

(a) $2\,Cu(s) + O_2(g) \rightleftharpoons 2\,CuO(s)$; $K_c = 4 \times 10^{45}$

(b) $H_3PO_4(aq) \rightleftharpoons H^+(aq) + H_2PO_4^-(aq)$; $K_c = 7.5 \times 10^{-3}$

(c) $2\,HBr(g) \rightleftharpoons H_2(g) + Br_2(g)$; $K_c = 2 \times 10^{-19}$

15.78 A chemical engineer is studying reactions to produce SO_3 as a step in the manufacture of sulfuric acid. The value of K_p for the reaction $2\,SO_2(g) + O_2(g) \rightleftharpoons 2\,SO_3(g)$ is 2.5×10^{10} at 500 K. Will a mixture of SO_2 and O_2 produce much SO_3 when equilibrium is reached?

15.79 The value of K_c for the reaction $3 O_2(g) \rightleftharpoons 2 O_3(g)$ is 1.7×10^{-56} at 25°C. Do you expect pure air at 25 °C to contain much O_3 (ozone) when O_2 and O_3 are in equilibrium? If the equilibrium concentration of O_2 in air at 25 °C is 8×10^{-3} M, what is the equilibrium concentration of O_3?

15.80 At 1400 K, $K_c = 2.5 \times 10^{-3}$ for the reaction $CH_4(g) + 2 H_2S(g) \rightleftharpoons CS_2(g) + 4 H_2(g)$. A 10.0-L reaction vessel at 1400 K contains 2.0 mol of CH_4, 3.0 mol of CS_2, 3.0 mol of H_2, and 4.0 mol of H_2S. Is the reaction mixture at equilibrium? If not, in which direction does the reaction proceed to reach equilibrium?

15.81 The first step in the industrial synthesis of hydrogen is the reaction of steam and methane to give synthesis gas, a mixture of carbon monoxide and hydrogen:

$$H_2O(g) + CH_4(g) \rightleftharpoons CO(g) + 3 H_2(g) \quad K_c = 4.7 \text{ at } 1400 \text{ K}$$

A mixture of reactants and products at 1400 K contains 0.035 M H_2O, 0.050 M CH_4, 0.15 M CO, and 0.20 M H_2. In which direction does the reaction proceed to reach equilibrium?

15.82 Phosphine (PH_3) decomposes at elevated temperatures, RAN yielding gaseous P_2 and H_2:

$$2 PH_3(g) \rightleftharpoons P_2(g) + 3 H_2(g) \quad K_p = 398 \text{ at } 873 \text{ K}$$

(a) If the initial partial pressures are $P_{PH_3} = 0.0260$ atm, $P_{P_2} = 0.871$ atm, $P_{H_2} = 0.517$ atm, calculate Q_p and determine the direction of reaction to attain equilibrium.

(b) When a mixture of PH_3, P_2, and H_2 comes to equilibrium at 873 K, $P_{P_2} = 0.412$ atm and $P_{H_2} = 0.822$ atm. What is P_{PH_3}?

15.83 At 500 K, the equilibrium constant for the dissociation reac- RAN tion $H_2(g) \rightleftharpoons 2 H(g)$ is very small ($K_c = 1.2 \times 10^{-42}$).

(a) What is the molar concentration of H atoms at equilibrium if the equilibrium concentration of H_2 is 0.10 M?

(b) How many H atoms and H_2 molecules are present in 1.0 L of 0.10 M H_2 at 500 K?

15.84 Calculate the equilibrium concentrations of N_2O_4 and NO_2 RAN at 25 °C in a vessel that contains an initial N_2O_4 concentration of 0.0500 M. The equilibrium constant K_c for the reaction $N_2O_4(g) \rightleftharpoons 2 NO_2(g)$ is 4.64×10^{-3} at 25 °C.

15.85 Calculate the equilibrium concentrations at 25 °C for the RAN reaction in Problem 15.84 if the initial concentrations are $[N_2O_4] = 0.0200$ M and $[NO_2] = 0.0300$ M.

15.86 A sample of HI (9.30×10^{-3} mol) was placed in an empty 2.00-L container at 1000 K. After equilibrium was reached, the concentration of I_2 was 6.29×10^{-4} M. Calculate the value of K_c at 1000 K for the reaction $H_2(g) + I_2(g) \rightleftharpoons 2 HI(g)$.

15.87 The industrial solvent ethyl acetate is produced by the reac- RAN tion of acetic acid with ethanol:

$$\underset{\text{Acetic acid}}{CH_3CO_2H(soln)} + \underset{\text{Ethanol}}{CH_3CH_2OH(soln)} \rightleftharpoons$$

$$\underset{\text{Ethyl acetate}}{CH_3CO_2CH_2CH_3(soln)} + H_2O(soln)$$

(a) Write the equilibrium constant expression for K_c.

(b) A solution prepared by mixing 1.00 mol of acetic acid and 1.00 mol of ethanol contains 0.65 mol of ethyl acetate at equilibrium. Calculate the value of K_c. Explain why you can calculate K_c without knowing the volume of the solution.

15.88 A characteristic reaction of ethyl acetate is hydrolysis, the RAN reverse of the reaction in Problem 15.87. Write the equilibrium equation for the hydrolysis of ethyl acetate, and use the data in Problem 15.87 to calculate K_c for the hydrolysis reaction.

15.89 Gaseous indium dihydride is formed from the elements at RAN elevated temperature:

$$In(g) + H_2(g) \rightleftharpoons InH_2(g) \quad K_p = 1.48 \text{ at } 973 \text{ K}$$

Partial pressures measured in a reaction vessel are: $P_{In} = 0.0600$ atm, $P_{H_2} = 0.0350$ atm, $P_{InH_2} = 0.0760$ atm.

(a) Calculate Q_p, and determine the direction of reaction to attain equilibrium.

(b) Determine the equilibrium partial pressures of all the gases.

15.90 The following reaction, which has $K_c = 0.145$ at 298 K, takes place in carbon tetrachloride solution:

$$2 BrCl(soln) \rightleftharpoons Br_2(soln) + Cl_2(soln)$$

A measurement of the concentrations shows $[BrCl] = 0.050$ M, $[Br_2] = 0.035$ M, and $[Cl_2] = 0.030$ M.

(a) Calculate Q_c, and determine the direction of reaction to attain equilibrium.

(b) Determine the equilibrium concentrations of BrCl, Br_1, and Cl_2.

15.91 An equilibrium mixture of N_2, H_2, and NH_3 at 700 K contains 0.036 M N_2 and 0.15 M H_2. At this temperature, K_c for the reaction $N_2(g) + 3 H_2(g) \rightleftharpoons 2 NH_3(g)$ is 0.29. What is the concentration of NH_3?

15.92 An equilibrium mixture of O_2, SO_2, and SO_3 contains equal RAN concentrations of SO_2 and SO_3. Calculate the concentration of O_2 if $K_c = 2.7 \times 10^2$ for the reaction $2 SO_2(g) + O_2(g) \rightleftharpoons 2 SO_3(g)$.

15.93 The air pollutant NO is produced in automobile engines from RAN the high-temperature reaction $N_2(g) + O_2(g) \rightleftharpoons 2 NO(g)$; $K_c = 1.7 \times 10^{-3}$ at 2300 K. If the initial concentrations of N_2 and O_2 at 2300 K are both 1.40 M, what are the concentrations of NO, N_2, and O_2 when the reaction mixture reaches equilibrium?

15.94 Recalculate the equilibrium concentrations in Problem 15.93 RAN if the initial concentrations are 2.24 M N_2 and 0.56 M O_2. (This N_2/O_2 concentration ratio is the ratio found in air.)

15.95 The interconversion of L-α-lysine and L-β-lysine, for which $K_c = 7.20$ at 333 K, is catalyzed by the enzyme lysine 2,3-aminomutase.

L-α-Lysine occurs in proteins while L-β-lysine is a precursor to certain antibiotics. At 333 K, a solution of L-α-lysine at a concentration of 3.00×10^{-3} M is placed in contact with lysine 2,3-aminomutase. What are the equilibrium concentrations of L-α-lysine and L-β-lysine?

15.96 The value of K_c for the reaction of acetic acid with ethanol is
RAN   3.4 at 25°C:

$$CH_3CO_2H(soln) + CH_3CH_2OH(soln) \rightleftharpoons$$

<small>Acetic acid   Ethanol</small>

$$CH_3CO_2CH_2CH_3(soln) + H_2O(soln) \quad K_c = 3.4$$

<small>Ethyl acetate</small>

(a) How many moles of ethyl acetate are present in an equilibrium mixture that contains 4.0 mol of acetic acid, 6.0 mol of ethanol, and 12.0 mol of water at 25 °C?

(b) Calculate the number of moles of all reactants and products in an equilibrium mixture prepared by mixing 1.00 mol of acetic acid and 10.00 mol of ethanol.

15.97 In a basic aqueous solution, chloromethane undergoes a
RAN   substitution reaction in which Cl^- is replaced by OH^-:

$$CH_3Cl(aq) + OH^-(aq) \rightleftharpoons CH_3OH(aq) + Cl^-(aq)$$

<table>
<tr><td>Chloromethane</td><td>Methanol</td></tr>
</table>

The equilibrium constant K_c is 1×10^{16}. Calculate the equilibrium concentrations of CH_3Cl, CH_3OH, OH^-, and Cl^- in a solution prepared by mixing equal volumes of 0.1 M CH_3Cl and 0.2 M NaOH. (Hint: In defining x, assume that the reaction goes 100% to completion, and then take account of a small amount of the reverse reaction.)

15.98 At 700 K, $K_p = 0.140$ for the reaction $ClF_3(g) \rightleftharpoons$
RAN $ClF(g) + F_2(g)$. Calculate the equilibrium partial pressures of ClF_3, ClF, and F_2 if only ClF_3 is present initially, at a partial pressure of 1.47 atm.

15.99 The reaction of iron(III) oxide with carbon monoxide is
RAN   important in making steel. At 1200 K, $K_p = 19.9$ for the reaction

$$Fe_2O_3(s) + 3\,CO(g) \rightleftharpoons 2\,Fe(l) + 3\,CO_2(g)$$

What are the equilibrium partial pressures of CO and CO_2 if CO is the only gas present initially, at a partial pressure of 0.978 atm?

15.100 The equilibrium concentrations in a gas mixture at a par-
RAN   ticular temperature are 0.13 M H_2, 0.70 M I_2, and 2.1 M HI. What equilibrium concentrations are obtained at the same temperature when 0.20 mol of HI is injected into an empty 500.0-mL container?

15.101 A 5.00-L reaction vessel is filled with 1.00 mol of
RAN H_2, 1.00 mol of I_2, and 2.50 mol of HI. Calculate the equilibrium concentrations of H_2, I_2, and HI at 500 K. The equilibrium constant K_c at 500 K for the reaction $H_2(g) + I_2(g) \rightleftharpoons 2\,HI(g)$ is 129.

15.102 At 1000 K, the value of K_c for the reaction $C(s) +$
RAN $H_2O(g) \rightleftharpoons CO(g) + H_2(g)$ is 3.0×10^{-2}. Calculate the equilibrium concentrations of H_2O, CO_2, and H_2 in a reaction mixture obtained by heating 6.00 mol of steam and an excess of solid carbon in a 5.00-L container. What is the molar composition of the equilibrium mixture?

15.103 When 1.000 mol of PCl_5 is introduced into a 5.000-L con-
RAN   tainer at 500 K, 78.50% of the PCl_5 dissociates to give an equilibrium mixture of PCl_5, PCl_3, and Cl_2:

$$PCl_5(g) \rightleftharpoons PCl_3(g) + Cl_2(g)$$

(a) Calculate the values of K_c and K_p.

(b) If the initial concentrations in a particular mixture of reactants and products are $[PCl_5] = 0.500$ M, $[PCl_3] = 0.150$ M, and $[Cl_2] = 0.600$ M, in which direction does the reaction proceed to reach equilibrium? What are the concentrations when the mixture reaches equilibrium?

15.104 Consider the reaction $C(s) + CO_2(g) \rightleftharpoons 2\,CO(g)$.
RAN   When 1.50 mol of CO_2 and an excess of solid carbon are heated in a 20.0-L container at 1100 K, the equilibrium concentration of CO is 7.00×10^{-2} M.

(a) What is the equilibrium concentration of CO_2?

(b) What is the value of the equilibrium constant K_c at 1100 K?

15.105 The equilibrium constant K_p for the gas-phase thermal decom-
RAN   position of *tert*-butyl chloride is 3.45 at 500 K:

$$(CH_3)_3CCl(g) \rightleftharpoons (CH_3)_2C = CH_2(g) + HCl(g)$$

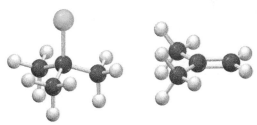

<table>
<tr><td>*tert*-Butyl chloride</td><td>Isobutylene</td></tr>
</table>

(a) Calculate the value of K_c at 500 K.

(b) Calculate the molar concentrations of reactants and products in an equilibrium mixture obtained by heating 1.00 mol of *tert*-butyl chloride in a 5.00-L vessel at 500 K.

(c) A mixture of isobutylene (0.400 atm partial pressure at 500 K) and HCl (0.600 atm partial pressure at 500 K) is allowed to reach equilibrium at 500 K. What are the equilibrium partial pressures of *tert*-butyl chloride, isobutylene, and HCl?

15.106 At 100 °C, $K_c = 4.72$ for the reaction $2\,NO_2(g) \rightleftharpoons$
RAN $N_2O_4(g)$. An empty 10.0-L flask is filled with 4.60 g of NO_2 at 100 °C. What is the total pressure in the flask at equilibrium?

15.107 At 25 °C, $K_c = 216$ for the reaction $2\,NO_2(g) \rightleftharpoons$
RAN $N_2O_4(g)$. A 1.00-L flask containing a mixture of NO_2 and N_2O_4 at 25 °C has a total pressure of 1.50 atm. What is the partial pressure of each gas?

15.108 At 500 °C, F_2 gas is stable and does not dissociate, but at
RAN   840 °C, some dissociation occurs: $F_2(g) \rightleftharpoons 2\,F(g)$. A flask filled with 0.600 atm of F_2 at 500 °C was heated to 840 °C, and the pressure at equilibrium was measured to be 0.984 atm. What is the equilibrium constant K_p for the dissociation of F_2 gas at 840 °C?

15.109 The reaction $NO(g) + NO_2(g) \rightleftharpoons N_2O_3(g)$ takes place
RAN   in the atmosphere with $K_c = 13$ at 298 K. A gas mixture is
prepared with 2.0 mol NO and 3.0 mol NO_2 and an initial
total pressure of 1.65 atm.

(a) What are the equilibrium partial pressures of NO,
NO_2, and N_2O_3 at 298 K?

(b) What is the volume of the container?

15.110 Phosgene ($COCl_2$) is a toxic gas that damages the lungs. At
RAN   360 °C, $K_c = 8.4 \times 10^{-4}$ for the decomposition of phosgene:

$$COCl_2(g) \rightleftharpoons CO(g) + Cl_2(g)$$

If an empty 50.0-L container is charged with 1.00 mol of
phosgene at 360 °C, what is the total pressure in the container after the system comes to equilibrium?

15.111 The equilibrium constant K_c for the reaction $N_2(g) +$
RAN $3 H_2(g) \rightleftharpoons 2 NH_3(g)$ is 4.20 at 600 K. When a quantity
of gaseous NH_3 was placed in a 1.00-L reaction vessel at
600 K and the reaction was allowed to reach equilibrium,
the vessel was found to contain 0.200 mol of N_2. How
many moles of NH_3 were placed in the vessel?

15.112 At 45 °C, $K_c = 0.619$ for the reaction $N_2O_4(g) \rightleftharpoons$
RAN $2 NO_2(g)$. If 46.0 g of N_2O_4 is introduced into an empty
2.00-L container, what are the partial pressures of NO_2 and
N_2O_4 after equilibrium has been achieved at 45 °C?

15.113 When 9.25 g of ClF_3 was introduced into an empty 2.00-L
RAN   container at 700.0 K, 19.8% of the ClF_3 decomposed to
give an equilibrium mixture of ClF_3, ClF, and F_2.

$$ClF_3(g) \rightleftharpoons ClF(g) + F_2(g)$$

(a) What is the value of the equilibrium constant K_c at 700.0 K?

(b) What is the value of the equilibrium constant K_p at
700.0 K?

(c) In a separate experiment, 39.4 g of ClF_3 was introduced
into an empty 2.00-L container at 700.0 K. What are
the concentrations of ClF_3, ClF, and F_2 when the mixture reaches equilibrium?

15.114 The following reaction in aqueous solution is catalyzed by
RAN   the enzyme aspartase and has $K_c = 6.95 \times 10^{-3}$ at 37 °C:

L-Aspartate   Fumarate

If the initial concentration of L-aspartate is 8.32×10^{-3} M,
what are the equilibrium concentrations of L-aspartate,
fumarate, and ammonium ion at 37°C?

15.115 The reaction of fumarate with water to form L-malate is
RAN   catalyzed by the enzyme fumarase; $K_c = 3.3$ at 37°C.

Fumarate   L-Malate

When a reaction mixture with [fumarate] $= 1.56 \times 10^{-3}$ M
and [L-malate] $= 2.27 \times 10^{-3}$ M comes to equilibrium in
the presence of fumarase at 37 °C, what are the equilibrium
concentrations of fumarate and L-malate? (Water can be omitted from the equilibrium equation because its concentration in
dilute solutions is essentially the same as that in pure water.)

15.116 Calculate the equilibrium concentrations of SO_2, Cl_2,
and SO_2Cl_2 at 298 K if the initial concentrations are
$[SO_2] = 1.50$ M and $[Cl_2] = 0.85$ M. The equilibrium constant K_c for the reaction $SO_2(g) + Cl_2(g) \rightleftharpoons SO_2Cl_2(g)$
is 8.40×10^{-3} at 298 K.

15.117 Calculate the equilibrium concentrations of $H_2O(g)$, $Cl_2(g)$,
HCl (g), and $O_2(g)$ at 298 K if the initial concentrations
are $[H_2O] = 0.050$ and $[Cl_2] = 0.100$ M. The equilibrium constant K_c for the reaction $H_2O(g) + Cl_2(g) \rightleftharpoons$
$2 HCl(g) + O_2(g)$ is 8.96×10^{-9} at 298 K.

Le Châtelier's Principle (Sections 15.6–15.9)

15.118 Consider the following equilibrium:

$$Ag^+(aq) + Cl^-(aq) \rightleftharpoons AgCl(s)$$

Use Le Châtelier's principle to predict how the amount of
solid silver chloride will change when the equilibrium is
disturbed by:

(a) Adding NaCl

(b) Adding $AgNO_3$

(c) Adding NO_3, which reacts with Ag^+ to form the complex ion $Ag(NH_3)_2^+$

(d) Removing Cl^-; also account for the change using the
reaction quotient Q_c

15.119 Will the concentration of NO_2 increase, decrease, or remain
the same when the equilibrium

$$NO_2Cl(g) + NO(g) \rightleftharpoons NOCl(g) + NO_2(g)$$

is disturbed by the following changes?

(a) Adding NOCl

(b) Adding NO

(c) Removing NO

(d) Adding NO_2Cl; also account for the change using the
reaction quotient Q_c

15.120 When each of the following equilibria is disturbed by
increasing the pressure as a result of decreasing the volume,
does the number of moles of reaction products increase,
decrease, or remain the same?

(a) $2 CO_2(g) \rightleftharpoons 2 CO(g) + O_2(g)$

(b) $N_2(g) + O_2(g) \rightleftharpoons 2 NO(g)$

(c) $Si(s) + 2 Cl_2(g) \rightleftharpoons SiCl_4(g)$

15.121 For each of the following equilibria, use Le Châtelier's principle to predict the direction of reaction when the volume is
increased.

(a) $C(s) + H_2O(g) \rightleftharpoons CO(g) + H_2(g)$

(b) $2 H_2(g) + O_2(g) \rightleftharpoons 2 H_2O(g)$

(c) $2 Fe(s) + 3 H_2O(g) \rightleftharpoons Fe_2O_3(s) + 3 H_2(g)$

15.122 For the water–gas shift reaction

$$CO(g) + H_2O(g) \rightleftharpoons CO_2(g) + H_2(g), \Delta H° = -41.2 \text{ kJ}$$

does the amount of H_2 in an equilibrium mixture increase or decrease when the temperature is increased? How does K_c change when the temperature is decreased? Justify your answers using Le Châtelier's principle.

15.123 The value of $\Delta H°$ for the reaction $3 O_2(g) \rightleftharpoons 2 O_3(g)$ is +285 kJ. Does the equilibrium constant for this reaction increase or decrease when the temperature increases? Justify your answer using Le Châtelier's principle.

15.124 Consider the exothermic reaction

$$CoCl_4^{2-}(aq) + 6 H_2O(l) \rightleftharpoons Co(H_2O)_6^{2+}(aq) + 4 Cl^-(aq)$$

which interconverts the blue $CoCl_4^{2-}$ ion and the pink $Co(H_2O)_6^{2+}$ ion. Will the equilibrium concentration of $CoCl_4^{2-}$ increase or decrease when the following changes occur?

(a) HCl is added.

(b) $Co(NO_3)_2$ is added.

(c) The solution is diluted with water.

(d) The temperature is increased.

15.125 Consider the endothermic reaction

$$Fe^{3+}(aq) + Cl^-(aq) \rightleftharpoons FeCl^{2+}(aq)$$

Use Le Châtelier's principle to predict how the equilibrium concentration of the complex ion $FeCl^{2+}$ will change when:

(a) $Fe(NO_3)_3$ is added.

(b) Cl^- is precipitated as AgCl by addition of $AgNO_3$.

(c) The temperature is increased.

(d) A catalyst is added.

15.126 Methanol (CH_3OH) is manufactured by the reaction of carbon monoxide with hydrogen in the presence of a $Cu/ZnO/Al_2O_3$ catalyst:

$$CO(g) + 2 H_2(g) \xrightarrow{\underset{\text{catalyst}}{Cu/ZnO/Al_2O_3}} CH_3OH(g) \quad \Delta H° = -91 \text{ kJ}$$

Does the amount of methanol increase, decrease, or remain the same when an equilibrium mixture of reactants and products is subjected to the following changes?

(a) The temperature is increased.

(b) The volume is decreased.

(c) Helium is added.

(d) CO is added.

(e) The catalyst is removed.

15.127 In the gas phase at 400 °C, isopropyl alcohol (rubbing alcohol) decomposes to acetone, an important industrial solvent:

$$\underset{\text{Isopropyl alcohol}}{(CH_3)_2CHOH(g)} \rightleftharpoons \underset{\text{Acetone}}{(CH_3)_2CO(g)} + H_2(g) \quad \Delta H° = +57.3 \text{ kJ}$$

Does the amount of acetone increase, decrease, or remain the same when an equilibrium mixture of reactants and products is subjected to the following changes?

(a) The temperature is increased.

(b) The volume is increased.

(c) Argon is added.

(d) H_2 is added.

(e) A catalyst is added.

15.128 The following reaction is important in gold mining:

$$4 Au(s) + 8 CN^-(aq) + O_2(g) + 2 H_2O(l) \rightleftharpoons$$
$$4 Au(CN)_2^-(aq) + 4 OH^-(aq)$$

For a reaction mixture at equilibrium, in which direction would the reaction go to reestablish equilibrium after each of the following changes?

(a) Adding gold

(b) Increasing the hydroxide concentration

(c) Increasing the partial pressure of oxygen

(d) Adding $Fe^{3+}(aq)$, which reacts with cyanide to form $Fe(CN)_6^{3-}(aq)$

15.129 The following reaction, catalyzed by iridium, is endothermic at 700 K:

$$CaO(s) + CH_4(g) + 2 H_2O(g) \rightleftharpoons CaCO_3(s) + 4 H_2(g)$$

For a reaction mixture at equilibrium at 700 K, how would the following changes affect the total quantity of $CaCO_3$ in the reaction mixture once equilibrium is reestablished?

(a) Increasing the temperature

(b) Adding calcium oxide

(c) Removing methane (CH_4)

(d) Increasing the total volume

(e) Adding iridium

15.130 The equilibrium constant K_p for the reaction $PCl_5(g) \rightleftharpoons PCl_3(g) + Cl_2(g)$ is 3.81×10^2 at 600 K and 2.69×10^3 at 700 K.

(a) Is the reaction endothermic or exothermic?

(b) How are the equilibrium amounts of reactants and products affected by (i) an increase in volume, (ii) addition of an inert gas, and (iii) addition of a catalyst?

15.131 Baking soda (sodium bicarbonate) decomposes when it is heated:

$$2 NaHCO_3(s) \rightleftharpoons Na_2CO_3(s) + CO_2(g) + H_2O(g)$$
$$\Delta H° = +136 \text{ kJ}$$

Consider an equilibrium mixture of reactants and products in a closed container. How does the number of moles of CO_2 change when the mixture is disturbed by the following:

(a) Adding solid $NaHCO_3$

(b) Adding water vapor

(c) Decreasing the volume of the container

(d) Increasing the temperature

Chemical Equilibrium and Chemical Kinetics (Section 15.10)

15.132 A platinum catalyst is used in automobile catalytic converters to hasten the oxidation of carbon monoxide:

$$2\,CO(g) + O_2(g) \xrightleftharpoons{Pt} 2\,CO_2(g) \quad \Delta H° = -566\,kJ$$

Suppose that you have a reaction vessel containing an equilibrium mixture of $CO(g)$, $O_2(g)$, and $CO_2(g)$. Under the following conditions, will the amount of CO increase, decrease, or remain the same?

(a) A platinum catalyst is added.
(b) The temperature is increased.
(c) The pressure is increased by decreasing the volume.
(d) The pressure is increased by adding argon gas.
(e) The pressure is increased by adding O_2 gas.

15.133 Consider the following gas-phase reaction: $2\,A(g) + B(g) \rightleftharpoons C(g) + D(g)$. An equilibrium mixture of reactants and products is subjected to the following changes:

(a) A decrease in volume
(b) An increase in temperature
(c) Addition of reactants
(d) Addition of a catalyst
(e) Addition of an inert gas

Which of these changes affect the composition of the equilibrium mixture but leave the value of the equilibrium constant K_c unchanged? Which of the changes affect the value of K_c? Which affect neither the composition of the equilibrium mixture nor K_c?

15.134 For the reaction $A_2 + 2B \rightleftharpoons 2\,AB$, the rate of the forward reaction is 18 M/s and the rate of the reverse reaction is 12 M/s. The reaction is not at equilibrium. Will the reaction proceed in the forward or reverse direction to attain equilibrium?

15.135 For the reaction $2\,A_3 + B_2 \rightleftharpoons 2\,A_3B$, the rate of the forward reaction is 0.35 M/s and the rate of the reverse reaction is 0.65 M/s. The reaction is not at equilibrium. Will the reaction proceed in the forward or reverse direction to attain equilibrium?

15.136 Consider a general, single-step reaction of the type $A + B \rightleftharpoons C$. Show that the equilibrium constant is equal to the ratio of the rate constants for the forward and reverse reactions, $K_c = k_f/k_r$.

15.137 Which of the following relative values of k_f and k_r results in an equilibrium mixture that contains large amounts of reactants and small amounts of products?

(a) $k_f > k_r$ (b) $k_f = k_r$ (c) $k_f < k_r$

15.138 Consider the gas-phase hydration of hexafluoroacetone, $(CF_3)_2CO$:

$$(CF_3)_2CO(g) + H_2O(g) \xrightleftharpoons[k_r]{k_f} (CF_3)_2C(OH)_2(g)$$

At 76 °C, the forward and reverse rate constants are $k_f = 0.13\,M^{-1}\,s^{-1}$ and $k_r = 6.2 \times 10^{-4}\,s^{-1}$. What is the value of the equilibrium constant K_c?

15.139 Consider the reaction of chloromethane with OH^- in aqueous solution:

$$CH_3Cl(aq) + OH^-(aq) \xrightleftharpoons[k_r]{k_f} CH_3OH(aq) + Cl^-(aq)$$

At 25 °C, the rate constant for the forward reaction is $6 \times 10^{-6}\,M^{-1}\,s^{-1}$, and the equilibrium constant K_c is 1×10^{16}. Calculate the rate constant for the reverse reaction at 25 °C.

15.140 In automobile catalytic converters, the air pollutant nitric oxide is converted to nitrogen and oxygen. Listed in the table are forward and reverse rate constants for the reaction $2\,NO(g) \rightleftharpoons N_2(g) + O_2(g)$.

Temperature (K)	$k_f(M^{-1}\,s^{-1})$	$k_r(M^{-1}\,s^{-1})$
1400	0.29	1.1×10^{-6}
1500	1.3	1.4×10^{-5}

Is the reaction endothermic or exothermic? Explain in terms of kinetics.

15.141 Forward and reverse rate constants for the reaction $CO_2(g) + N_2(g) \rightleftharpoons CO(g) + N_2O(g)$ exhibit the following temperature dependence:

Temperature (K)	$k_f(M^{-1}\,s^{-1})$	$k_r(M^{-1}\,s^{-1})$
1200	9.1×10^{-11}	1.5×10^5
1300	2.7×10^{-9}	2.6×10^5

Is the reaction endothermic or exothermic? Explain in terms of kinetics.

15.142 As shown in Figure 15.15 a catalyst lowers the activation energy for the forward and reverse reactions by the same amount, ΔE_a.

(a) Apply the Arrhenius equation, $k = Ae^{-E_a/RT}$, to the forward and reverse reactions, and show that a catalyst increases the rates of both reactions by the same factor.
(b) Use the relation between the equilibrium constant and the forward and reverse rate constants, $K_c = k_f/k_r$, to show that a catalyst does not affect the value of the equilibrium constant.

15.143 Given the Arrhenius equation, $k = Ae^{-E_a/RT}$, and the relation between the equilibrium constant and the forward and reverse rate constants, $K_c = k_f/k_r$, explain why K_c for an exothermic reaction decreases with increasing temperature.

MULTICONCEPT PROBLEMS

15.144 Vinegar contains acetic acid, a weak acid that is partially dissociated in aqueous solution:

$$CH_3CO_2H(aq) \rightleftharpoons H^+(aq) + CH_3CO_2^-(aq)$$

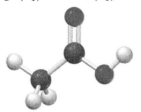

Acetic acid

(a) Write the equilibrium constant expression for K_c.

(b) What is the value of K_c if the extent of dissociation in 1.0 M CH_3CO_2H is 0.42%?

15.145 Heavy water, symbolized D_2O (D = ^2H) finds use as a
RAN neutron moderator in nuclear reactors. In a mixture with ordinary water, exchange of isotopes occurs according to the following equation:

$$H_2O + D_2O \rightleftharpoons 2\ HDO \quad K_c = 3.86\ at\ 298\ K$$

When 1.00 mol of H_2O is combined with 1.00 mol of D_2O, what are the equilibrium amounts of H_2O, D_2O, and HDO (in moles) at 298 K? Assume the density of the mixture is constant at 1.05 g/cm^3.

15.146 Refining petroleum involves cracking large hydrocarbon molecules into smaller, more volatile pieces. A simple example of hydrocarbon cracking is the gas-phase thermal decomposition of butane to give ethane and ethylene:

$$
\begin{array}{c}
\ce{H} \\
| \\
\ce{H-C-H} \\
| \\
\ce{H-C-H} \\
| \\
\ce{H-C-H} \\
| \\
\ce{H-C-H} \\
| \\
\ce{H}
\end{array}
\rightleftharpoons
\begin{array}{c}
\ce{H} \\
| \\
\ce{H-C-H} \\
| \\
\ce{H-C-H} \\
| \\
\ce{H}
\end{array}
+
\begin{array}{c}
\ce{H} \quad\ \ \ce{H} \\
\ce{\backslash}\ \ \ce{/} \\
\ce{C} \\
\| \\
\ce{C} \\
\ce{/}\ \ \ce{\backslash} \\
\ce{H} \quad\ \ \ce{H}
\end{array}
$$

Butane, C_4H_{10} Ethane, C_2H_6 Ethylene, C_2H_4

(a) Write the equilibrium constant expressions for K_p and K_c.

(b) The value of K_p at 500 °C is 12. What is the value of K_c?

(c) A sample of butane having a pressure of 50 atm is heated at 500 °C in a closed container at constant volume. When equilibrium is reached, what percentage of the butane has been converted to ethane and ethylene? What is the total pressure at equilibrium?

(d) How would the percent conversion in part (c) be affected by a decrease in volume?

15.147 At 1000 K, $K_p = 2.1 \times 10^6$ and $\Delta H° = -107.7$ kJ for the
RAN reaction $H_2(g) + Br_2(g) \rightleftharpoons 2\ HBr(g)$.

(a) A 0.974 mol quantity of Br_2 is added to a 1.00-L reaction vessel that contains 1.22 mol of H_2 gas at 1000 K. What are the partial pressures of H_2, Br_2, and HBr at equilibrium?

(b) For the equilibrium in part (a), each of the following changes will increase the equilibrium partial pressure of HBr. Choose the change that will cause the greatest increase in the pressure of HBr, and explain your choice.

  (i) Adding 0.10 mol of H_2

  (ii) Adding 0.10 mol of Br_2

  (iii) Decreasing the temperature to 700 K

15.148 Consider the gas-phase decomposition of NOBr:

$$2\ NOBr(g) \rightleftharpoons 2\ NO(g) + Br_2(g)$$

(a) When 0.0200 mol of NOBr is added to an empty 1.00-L flask and the decomposition reaction is allowed to reach equilibrium at 300 K, the total pressure in the flask is 0.588 atm. What is the equilibrium constant K_c for this reaction at 300 K?

(b) What is the value of K_p for this reaction at 300 K?

15.149 Halogen lamps are ordinary tungsten filament lamps in which the lamp bulb contains a small amount of a halogen (often bromine). At the high temperatures of the lamp, the halogens dissociate and exist as single atoms.

(a) In an ordinary tungsten lamp, the hot tungsten filament is constantly evaporating and the tungsten condenses on the relatively cool walls of the bulb. In a Br-containing halogen lamp, the tungsten reacts with the Br atoms to give gaseous WBr_4:

$$W(s) + 4\ Br(g) \rightleftharpoons WBr_4(g)$$

At the walls of the lamp, where the temperature is about 900 K, this reaction has an equilibrium constant K_p of about 100. If the equilibrium pressure of $Br(g)$ is 0.010 atm, what is the equilibrium pressure of $WBr_4(g)$ near the walls of the bulb?

(b) Near the tungsten filament, where the temperature is about 2800 K, the reaction in part (a) has a K_p value of about 5.0. Is the reaction exothermic or endothermic?

(c) When the $WBr_4(g)$ diffuses back toward the filament, it decomposes, depositing tungsten back onto the filament. Show quantitatively that the pressure of WBr_4 from part (a) will cause the reaction in part (a) to go in reverse direction at 2800 K. [The pressure of $Br(g)$ is still 0.010 atm.] Thus, tungsten is continually recycled from the walls of the bulb back to the filament, allowing the bulb to last longer and burn brighter.

15.150 The decomposition of solid ammonium carbamate, (NH_4)
RAN (NH_2CO_2), to gaseous ammonia and carbon dioxide is an endothermic reaction.

$$(NH_4)(NH_2CO_2)(s) \rightleftharpoons 2\ NH_3(g) + CO_2(g)$$

(a) When solid $(NH_4)(NH_2CO_2)$ is introduced into an evacuated flask at 25 °C, the total pressure of gas at equilibrium is 0.116 atm. What is the value of K_p at 25 °C?

(b) Given that the decomposition reaction is at equilibrium, how would the following changes affect the total quantity of NH_3 in the flask once equilibrium is reestablished?

  (i) Adding CO_2

  (ii) Adding $(NH_4)(NH_2CO_2)$

(iii) Removing CO_2

(iv) Increasing the total volume

(v) Adding neon

(vi) Increasing the temperature

15.151 The F—F bond in F_2 is relatively weak because the lone pairs of electrons on one F atom repel the lone pairs on the other F atom; $K_p = 7.83$ at 1500 K for the reaction $F_2(g) \rightleftharpoons 2\ F(g)$.

(a) If the equilibrium partial pressure of F_2 molecules at 1500 K is 0.200 atm, what is the equilibrium partial pressure of F atoms in atm?

(b) What fraction of the F_2 molecules dissociate at 1500 K?

(c) Why is the F—F bond in F_2 weaker than the Cl—Cl bond in Cl_2?

15.152 When 0.500 mol of N_2O_4 is placed in a 4.00-L reaction vessel and heated at 400 K, 79.3% of the N_2O_4 decomposes to NO_2.

(a) Calculate K_c and K_p at 400 K for the reaction $N_2O_4(g) \rightleftharpoons 2\ NO_2(g)$.

(b) Draw an electron-dot structure for NO_2, and rationalize the structure of N_2O_4.

15.153 The equilibrium constant K_c for the gas-phase thermal decomposition of cyclopropane to propene is 1.0×10^5 at 500 K:

RAN

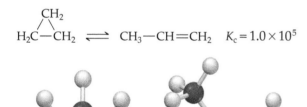

$$H_2\overset{\overset{\displaystyle CH_2}{\diagup\ \diagdown}}{C—CH_2} \rightleftharpoons CH_3—CH=CH_2 \quad K_c = 1.0 \times 10^5$$

Cyclopropane   Propene

(a) What is the value of K_p at 500 K?

(b) What is the equilibrium partial pressure of cyclopropane at 500 K when the partial pressure of propene is 5.0 atm?

(c) Can you alter the ratio of the two concentrations at equilibrium by adding cyclopropane or by decreasing the volume of the container? Explain.

(d) Which has the larger rate constant, the forward reaction or the reverse reaction?

(e) Why is cyclopropane so reactive? (Hint: Consider the hybrid orbitals used by the C atoms.)

15.154 Acetic acid tends to form dimers, $(CH_3CO_2H)_2$, because of hydrogen bonding:

$$2\ CH_3—C\overset{\displaystyle O}{\underset{\displaystyle O—H}{\diagup\!\!\parallel}} \rightleftharpoons CH_3—C\overset{\displaystyle O\cdots\!\cdot H—O}{\underset{\displaystyle O—H\cdots\!\cdot O}{\diagup\!\!\parallel \qquad \diagdown\!\!\parallel}}C—CH_3$$

Monomer   Dimer

The equilibrium constant K_c for this reaction is 1.51×10^2 in benzene solution but only 3.7×10^{-2} in water solution.

(a) Calculate the ratio of dimers to monomers for 0.100 M acetic acid in benzene.

(b) Calculate the ratio of dimers to monomers for 0.100 M acetic acid in water.

(c) Why is K_c for the water solution so much smaller than K_c for the benzene solution?

15.155 A 125.4 g quantity of water and an equal molar amount of carbon monoxide were placed in an empty 10.0-L vessel, and the mixture was heated to 700 K. At equilibrium, the partial pressure of CO was 9.80 atm. The reaction is

$$CO(g) + H_2O(g) \rightleftharpoons CO_2(g) + H_2(g)$$

(a) What is the value of K_p at 700 K?

(b) An additional 31.4 g of water was added to the reaction vessel, and a new state of equilibrium was achieved. What are the equilibrium partial pressures of each gas in the mixture? What is the concentration of H_2 in molecules/cm^3?

15.156 A 79.2 g chunk of dry ice (solid CO_2) and 30.0 g of graphite (carbon) were placed in an empty 5.00-L container, and the mixture was heated to achieve equilibrium. The reaction is

$$CO_2(g) + C(s) \rightleftharpoons 2\ CO(g)$$

(a) What is the value of K_p at 1000 K if the gas density at 1000 K is 16.3 g/L?

(b) What is the value of K_p at 1100 K if the gas density at 1100 K is 16.9 g/L?

(c) Is the reaction exothermic or endothermic? Explain.

15.157 The amount of carbon dioxide in a gaseous mixture of CO_2 and CO can be determined by passing the gas into an aqueous solution that contains an excess of $Ba(OH)_2$. The CO_2 reacts, yielding a precipitate of $BaCO_3$, but the CO does not react. This method was used to analyze the equilibrium composition of the gas obtained when 1.77 g of CO_2 reacted with 2.0 g of graphite in a 1.000-L container at 1100 K. The analysis yielded 3.41 g of $BaCO_3$. Use these data to calculate K_p at 1100 K for the reaction

$$CO_2(g) + C(s) \rightleftharpoons 2\ CO(g)$$

15.158 A 14.58 g quantity of N_2O_4 was placed in a 1.000-L reaction vessel at 400 K. The N_2O_4 decomposed to an equilibrium mixture of N_2O_4 and NO_2 that had a total pressure of 9.15 atm.

(a) What is the value of K_c for the reaction $N_2O_4(g) \rightleftharpoons 2\ NO_2(g)$ at 400 K?

(b) How much heat (in kilojoules) was absorbed when the N_2O_4 decomposed to give the equilibrium mixture? (Standard heats of formation may be found in Appendix B.)

15.159 Consider the sublimation of mothballs at 27 °C in a room having dimensions 8.0 ft × 10.0 ft × 8.0 ft. Assume that the mothballs are pure solid naphthalene (density 1.16 g/cm^3) and that they are spheres with a diameter of 12.0 mm. The equilibrium constant K_c for the sublimation of naphthalene is 5.40×10^{-6} at 27 °C.

RAN

$$C_{10}H_8(s) \rightleftharpoons C_{10}H_8(g)$$

(a) When excess mothballs are present, how many gaseous naphthalene molecules are in the room at equilibrium?

(b) How many mothballs are required to saturate the room with gaseous naphthalene?

15.160 Ozone is unstable with respect to decomposition to ordinary oxygen:

$$2\,O_3(g) \rightleftharpoons 3\,O_2(g) \quad K_p = 1.3 \times 10^{57}$$

How many O_3 molecules are present at equilibrium in 10 million cubic meters of air at 25 °C and 720 mm Hg pressure?

15.161 The equilibrium constant for the dimerization of acetic acid in benzene solution is 1.51×10^2 at 25 °C.

$$2\,CH_3CO_2H \rightleftharpoons (CH_3CO_2H)_2 \quad K_c = 1.51 \times 10^2 \text{ at } 25 \text{ °C}$$

(a) What are the equilibrium concentrations of monomer and dimer at 25 °C in a solution prepared by dissolving 0.0300 mol of pure acetic acid in enough benzene to make 250.0 mL of solution?

(b) What is the osmotic pressure of the solution at 25 °C?

15.162 For the decomposition reaction $PCl_5(g) \rightleftharpoons PCl_3(g) + Cl_2(g)$, $K_p = 381$ at 600 K and $K_c = 46.9$ at 700 K.

(a) Is the reaction endothermic or exothermic? Explain. Does your answer agree with what you would predict based on bond energies?

(b) If 1.25 g of PCl_5 is introduced into an evacuated 0.500-L flask at 700 K and the decomposition reaction is allowed to reach equilibrium, what percent of the PCl_5 will decompose and what will be the total pressure in the flask?

(c) Write electron-dot structures for PCl_5 and PCl_3, and indicate whether these molecules have a dipole moment. Explain.

15.163 Propanol (PrOH) and methyl methacrylate (MMA) associate in solution by an intermolecular force, forming an adduct represented as PrOH·MMA. The equilibrium constant for the association reaction is $K_c = 0.701$ at 298 K.

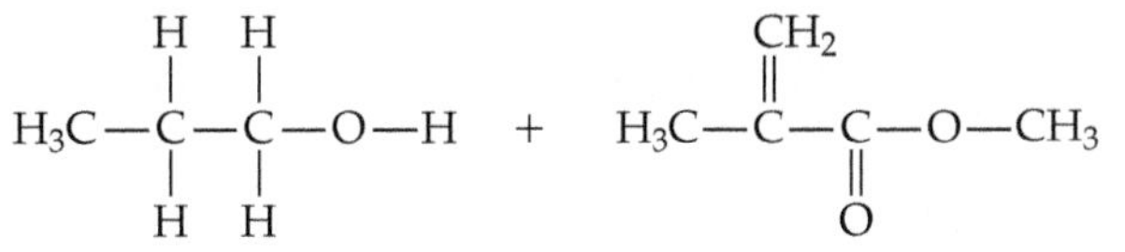

$$\rightleftharpoons \text{PrOH} \cdot \text{MMA}$$

(a) What is the predominant intermolecular force accounting for the interaction between PrOH and MMA?

(b) Draw a plausible structure for the PrOH·MMA adduct. Use $\cdots$ to signify an intermolecular interaction.

(c) If the initial concentrations are $[\text{PrOH}] = 0.100$ M and $[\text{MMA}] = 0.0500$ M, what are the equilibrium concentrations of PrOH, MMA, and PrOH·MA?

chapter 16

Aqueous Equilibria: Acids and Bases

Sulfuric acid and nitric acid formed in the atmosphere from industrial emissions dramatically lowered the pH of rain in the latter half of the 20th century. Acid rain killed pine trees and aquatic life in mountain lakes and disrupted entire ecosystems.

Has the problem of acid rain been solved?

The answer to this question can be found on page 694 in the INQUIRY ?

Contents

In the previous chapter, we discussed the principles of chemical equilibria. We'll now apply those principles to solutions of acids and bases. Acids and bases are common in daily life, including the acetic acid in vinegar, citric acid in lemons and other citrus fruits, magnesium hydroxide in commercial antacids, and ammonia in household cleaning products. Hydrochloric acid is the acid in gastric juice, also known as stomach acid. Without the more than 1.0 liter of HCl secreted by the lining of the stomach every day, we would not be able to digest our food.

Humans have known the characteristic properties of acids and bases for centuries. Acids react with metals such as iron and zinc to yield H_2 gas, and they change the color of the plant dye *litmus* from blue to red. By contrast, bases feel slippery, and they change the color of litmus from red to blue. When acids and bases are mixed in the right proportion, the characteristic acidic and basic properties disappear and are replaced with new substances known as *salts*.

What is it that makes an acid an acid and a base a base? We first raised those questions in Section 4.7, and we'll now take a closer look at some of the concepts that chemists have developed to describe the chemical behavior of acids and bases. We'll also apply the principles of chemical equilibrium discussed in Chapter 15 to determine the concentrations of the substances present in aqueous solutions of acids and bases. We can understand an enormous amount of chemistry in terms of acid–base reactions, perhaps the most important reaction type in all of chemistry.

16.1 ACID–BASE CONCEPTS: THE BRØNSTED–LOWRY THEORY

Thus far, we've been using the Arrhenius theory of acids and bases (Section 4.7). According to Arrhenius, acids are substances that dissociate in water to produce hydrogen ions (H^+) and bases are substances that dissociate in water to yield hydroxide ions (OH^-). Thus, HCl and H_2SO_4 are acids, and NaOH and $Ba(OH)_2$ are bases.

> **A generalized Arrhenius acid** $HA(aq) \rightleftharpoons H^+(aq) + A^-(aq)$
>
> **A generalized Arrhenius base** $MOH(aq) \rightleftharpoons M^+(aq) + OH^-(aq)$

The Arrhenius theory accounts for the properties of many common acids and bases, but it has important limitations. For one thing, the Arrhenius theory is restricted to aqueous solutions; for another, it doesn't account for the basicity of substances like ammonia (NH_3) that don't contain OH groups. In 1923, the Danish chemist Johannes Brønsted (1879–1947) and the English chemist Thomas Lowry (1843–1909) independently proposed a more general theory of acids and bases. According to the **Brønsted–Lowry theory**, an acid is any substance (molecule or ion) that can transfer a proton (H^+ ion) to another substance, and a base is any substance that can accept a proton. In short, acids are proton donors, bases are proton acceptors, and acid–base reactions are proton-transfer reactions:

> **Brønsted–Lowry acid** A substance that can transfer H^+
>
> **Brønsted–Lowry base** A substance that can accept H^+

It follows from this equation that the products of a Brønsted–Lowry acid–base reaction, BH^+ and A^-, are themselves, respectively, acids and bases. The species BH^+ produced when the base B accepts a proton from HA can itself donate a proton back to A^-, meaning that it is a Brønsted–Lowry acid. Similarly, the species A^- produced when HA loses a proton can itself accept a proton back from BH^+, meaning that it is a Brønsted–Lowry base. Chemical species whose formulas differ only by one proton are said to be **conjugate acid–base pairs**. Thus, A^- is the **conjugate base** of the acid HA, and HA is the **conjugate acid** of the base A⁻. Similarly, B is the conjugate base of the acid BH⁺, and BH⁺ is the conjugate acid of the base B.

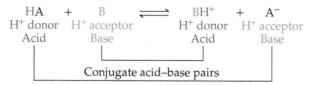

$$\underset{\substack{\text{H⁺ donor} \\ \text{Acid}}}{\text{HA}} \quad + \quad \underset{\substack{\text{H⁺ acceptor} \\ \text{Base}}}{\text{B}} \quad \rightleftharpoons \quad \underset{\substack{\text{H⁺ donor} \\ \text{Acid}}}{\text{BH⁺}} \quad + \quad \underset{\substack{\text{H⁺ acceptor} \\ \text{Base}}}{\text{A⁻}}$$

Conjugate acid–base pairs

To see what's going on in an acid–base reaction, keep your eye on the proton. For example, when a Brønsted–Lowry acid HA dissolves in water, it reacts reversibly with water in an *acid-dissociation equilibrium*. The acid transfers a proton to the solvent, which acts as a base (a proton acceptor). The products are the **hydronium ion, H_3O^+** (the conjugate acid of H_2O), and A⁻ (the conjugate base of HA):

Go to eText

BIG IDEA Question 1

What is the conjugate acid of the Brønsted–Lowry base OH⁻?

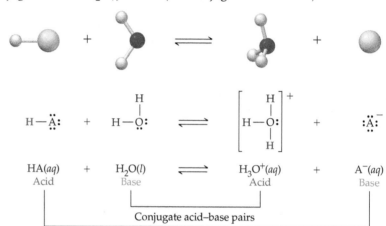

$$\underset{\substack{\text{HA}(aq) \\ \text{Acid}}}{} \quad + \quad \underset{\substack{\text{H}_2\text{O}(l) \\ \text{Base}}}{} \quad \rightleftharpoons \quad \underset{\substack{\text{H}_3\text{O}^+(aq) \\ \text{Acid}}}{} \quad + \quad \underset{\substack{\text{A}^-(aq) \\ \text{Base}}}{}$$

Conjugate acid–base pairs

In the reverse reaction, H_3O^+ acts as the proton donor (acid) and A⁻ acts as the proton acceptor (base). Typical examples of Brønsted–Lowry acids include not only electrically neutral molecules, such as HCl, HNO_3, and HF, but also cations and anions of salts that contain transferable protons, such as NH_4^+, HSO_4^-, and HCO_3^-.

When a Brønsted–Lowry base such as NH_3 dissolves in water, it accepts a proton from the solvent, which acts as an acid. The products are the hydroxide ion, OH⁻ (the conjugate base of water), and the ammonium ion, NH_4^+ (the conjugate acid of NH_3). In the reverse reaction, NH_4^+ acts as the proton donor and OH⁻ acts as the proton acceptor:

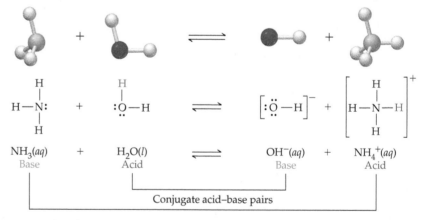

$$\underset{\substack{\text{NH}_3(aq) \\ \text{Base}}}{} \quad + \quad \underset{\substack{\text{H}_2\text{O}(l) \\ \text{Acid}}}{} \quad \rightleftharpoons \quad \underset{\substack{\text{OH}^-(aq) \\ \text{Base}}}{} \quad + \quad \underset{\substack{\text{NH}_4^+(aq) \\ \text{Acid}}}{}$$

Conjugate acid–base pairs

For a molecule or ion to accept a proton, it must have at least one unshared pair of electrons that it can use for bonding to the proton. As shown by the following electron-dot structures, all Brønsted–Lowry bases have one or more lone pairs of electrons:

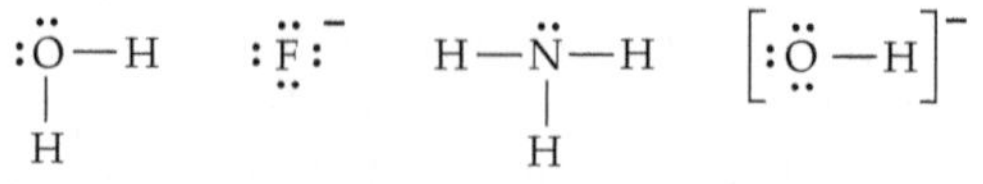

Some Brønsted–Lowry bases

As we've seen, the proton is fundamental to both the Arrhenius and the Brønsted–Lowry definitions of an acid. In the Arrhenius definition, an acid (HA) dissociates to give an aqueous hydrogen ion, or hydrated proton, written as $H^+(aq)$:

$$HA(aq) \rightleftharpoons H^+(aq) + A^-(aq)$$

As a bare proton, the positively charged H^+ ion is too reactive to exist in aqueous solution, and so it bonds to the oxygen atom of a solvent water molecule to give the trigonal pyramidal hydronium ion, H_3O^+. The H_3O^+ ion, which can be regarded as the simplest hydrate of the proton, $[H(H_2O)]^+$, can associate through hydrogen bonding with additional water molecules to give higher hydrates with the general formula $[H(H_2O)_n]^+$ (n = 2, 3, or 4), such as $H_5O_2^+$, $H_7O_3^+$, and $H_9O_4^+$. It's likely that acidic aqueous solutions contain a distribution of $[H(H_2O)_n]^+$ ions having different values of n. In this book, though, we'll use the symbols $H^+(aq)$ and $H_3O^+(aq)$ to mean the same thing—namely, a proton hydrated by an unspecified number of water molecules. Ordinarily, we use H_3O^+ in acid–base reactions to emphasize the proton-transfer character of those reactions.

The hydronium ion, H_3O^+

WORKED EXAMPLE 16.1

Explaining Acidity with the Arrhenius and Brønsted–Lowry Theories

Account for the acidic properties of nitrous acid (HNO_2) using the Arrhenius theory and the Brønsted–Lowry theory, and identify the conjugate base of HNO_2.

STRATEGY

To account for the acidity of a substance, consider how it can produce H^+ ions in water (Arrhenius theory) and how it can act as a proton donor (Brønsted–Lowry theory).

Nitrous acid

SOLUTION

HNO_2 is an Arrhenius acid because it dissociates in water to produce H^+ ions:

$$HNO_2(aq) \rightleftharpoons H^+(aq) + NO_2^-(aq)$$

Nitrous acid is a Brønsted–Lowry acid because it acts as a proton donor when it dissociates, transferring a proton to water to give the hydronium ion, H_3O^+:

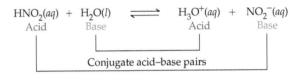

The conjugate base of HNO_2 is NO_2^-, the species that remains after HNO_2 has lost a proton.

▶ **PRACTICE 16.1** Write a balanced equation for the dissociation of the Brønsted–Lowry acid, HSO_4^-, in water.

▶ **APPLY 16.2** Write the reaction between the carbonate ion (CO_3^{2-}) and water.

All **PRACTICE** and **APPLY** problems are interactive in the eText.

CONCEPTUAL WORKED EXAMPLE 16.2

Identifying Brønsted–Lowry Acids, Bases, and Conjugate Acid–Base Pairs

For the following reaction in aqueous solution, identify the Brønsted–Lowry acids, bases, and conjugate acid–base pairs:

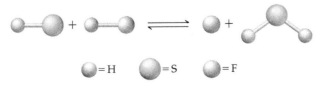

continued on next page

STRATEGY

The simplest approach is to identify the conjugate acid–base pairs, the species whose formulas differ by just one proton.

SOLUTION

The second reactant is HF, and the first product is its conjugate base F⁻. The second product is H₂S, and the first reactant is its conjugate base HS⁻. Therefore, the Brønsted–Lowry acids, bases, and conjugate acid–base pairs are as follows:

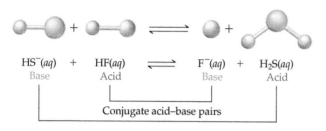

$$HS^-(aq) + HF(aq) \rightleftharpoons F^-(aq) + H_2S(aq)$$

Base · · · · · · · · · Acid · · · · · · · · · Base · · · · · · · · · Acid

Conjugate acid–base pairs

▶ **CONCEPTUAL PRACTICE 16.3** For the following reaction in aqueous solution, identify the Brønsted–Lowry acids, bases, and conjugate acid–base pairs:

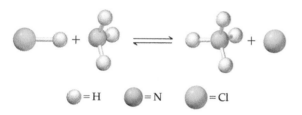

◯ = H   ⬤ = N   ⬤ = Cl

▶ **CONCEPTUAL APPLY 16.4** For the following reactions in aqueous solution, identify the Brønsted–Lowry acids, bases, and conjugate acid–base pairs. In which reaction is water an acid and in which reaction is water a base?

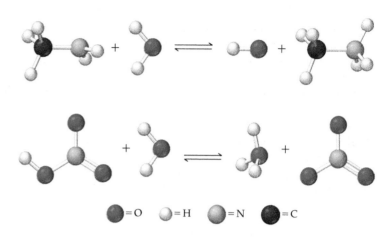

⬤ = O   ◯ = H   ◯ = N   ⬤ = C

16.2 ACID STRENGTH AND BASE STRENGTH

A helpful way of viewing an acid-dissociation equilibrium is to realize that the two bases, H₂O and A⁻, are competing for protons:

$$HA(aq) + H_2O(l) \rightleftharpoons H_3O^+(aq) + A^-(aq)$$

Acid · · · · · Base · · · · · Acid · · · · · Base

In the equation, the two bases are H₂O and A⁻. In general, a base is often represented in a chemical equation as A⁻ because an anion can accept a proton (H⁺). If H₂O is a stronger base (a stronger proton acceptor) than A⁻, the majority of protons will be transferred from HA to H₂O, and the solution will contain mainly H₃O⁺ and A⁻ ions.

$$HA(aq) + H_2O(l) \rightleftharpoons H_3O^+(aq) + A^-(aq)$$

Solution contains mainly products when H₂O is a stronger base than A⁻.

If A⁻ is a stronger base than H₂O, the A⁻ ions will get the protons, and the solution will contain mainly HA and H₂O.

$$HA(aq) + H_2O(l) \rightleftharpoons H_3O^+(aq) + A^-(aq)$$

Solution contains mainly reactants when A⁻ is a stronger base than H₂O.

When beginning with equal concentrations of reactants and products, *the proton is always transferred to the stronger base.* This means that the direction of reaction to reach equilibrium is proton transfer from the stronger acid to the stronger base to give the weaker acid and the weaker base:

$$\text{Stronger acid} + \text{Stronger base} \longrightarrow \text{Weaker acid} + \text{Weaker base}$$

Different acids differ in their ability to donate protons. A **strong acid** is one that is almost completely dissociated in water and is therefore a strong electrolyte (Section 4.3). Thus, the acid-dissociation equilibrium of a strong acid lies nearly 100% to the right, and the solution contains almost entirely H_3O^+ and A^- ions with only a negligible amount of undissociated HA molecules. Typical strong acids are perchloric acid ($HClO_4$), hydrochloric acid (HCl), hydrobromic acid (HBr), hydroiodic acid (HI), nitric acid (HNO_3), and sulfuric acid (H_2SO_4). It follows from this definition that strong acids have very weak conjugate bases. The ions ClO_4^-, Cl^-, Br^-, I^-, NO_3^-, and HSO_4^- have a negligible tendency to combine with a proton in aqueous solution, and they are therefore much weaker bases than H_2O.

A **weak acid** is one that is only partially dissociated in water and is thus a weak electrolyte. Only a small fraction of the weak acid molecules transfer a proton to water, and the solution therefore contains mainly undissociated HA molecules along with small amounts of H_3O^+ and the conjugate base A^-. Typical weak acids are nitrous acid (HNO_2), hydrofluoric acid (HF), and acetic acid (CH_3CO_2H). In the case of very weak acids, such as NH_3, OH^-, and H_2, the acid has practically no tendency to transfer a proton to water and the acid-dissociation equilibrium lies essentially 100% to the left. It follows from this definition that very weak acids have strong conjugate bases. For example, the NH_2^-, O^{2-}, and H^- ions are essentially 100% protonated in aqueous solution and are much stronger bases than H_2O.

The equilibrium concentrations of HA, H_3O^+, and A^- for strong acids, weak acids, and very weak acids are represented in **FIGURE 16.1**. The inverse relationship between the strength of an acid and the strength of its conjugate base is illustrated in **TABLE 16.1**.

BIG IDEA Question 2

HOCl is a stronger acid than HOI. Which is a stronger base: OCl^-, or OI^-?

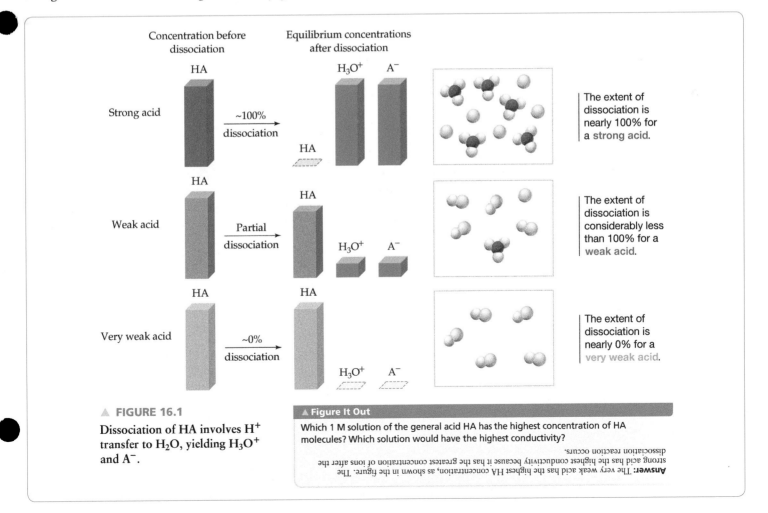

▲ **FIGURE 16.1**

Dissociation of HA involves H^+ transfer to H_2O, yielding H_3O^+ and A^-.

▲ Figure It Out

Which 1 M solution of the general acid HA has the highest concentration of HA molecules? Which solution would have the highest conductivity?

Answer: The very weak acid has the highest HA concentration, as shown in the figure. The strong acid has the highest conductivity because it has the greatest concentration of ions after the dissociation reaction occurs.

TABLE 16.1   Relative Strengths of Conjugate Acid–Base Pairs

	Acid, HA			Base, A⁻		
Stronger acid	$HClO_4$ HCl H_2SO_4 HNO_3	Strong acids: 100% dissociated in aqueous solution.		ClO_4^- Cl^- HSO_4^- NO_3^-	Very weak bases: Negligible tendency to be protonated in aqueous solution.	Weaker base
	H_3O^+			H_2O		
	HSO_4^- H_3PO_4 HNO_2 HF CH_3CO_2H H_2CO_3 H_2S NH_4^+ HCN HCO_3^-	Weak acids: Exist in solution as a mixture of HA, A⁻, and H_3O^+.		SO_4^{2-} $H_2PO_4^-$ NO_2^- F^- $CH_3CO_2^-$ HCO_3^- HS^- NH_3 CN^- CO_3^{2-}	Weak bases: Moderate tendency to be protonated in aqueous solution.	
	H_2O			OH^-		
Weaker acid	NH_3 OH^- H_2	Very weak acids: Negligible tendency to dissociate.		NH_2^- O^{2-} H^-	Strong bases: 100% protonated in aqueous solution.	Stronger base

WORKED EXAMPLE 16.3

Predicting the Direction of Acid–Base Reactions

If you mix equal concentrations of reactants and products, which of the following reactions proceed to the right and which proceed to the left?

(a) $H_2SO_4(aq) + NH_3(aq) \rightleftharpoons NH_4^+(aq) + HSO_4^-(aq)$
(b) $HCO_3^-(aq) + SO_4^{2-}(aq) \rightleftharpoons HSO_4^-(aq) + CO_3^{2-}(aq)$

STRATEGY

To predict the direction of reaction, use the balanced equation to identify the acids and bases, and then use Table 16.1 to identify the stronger acid and the stronger base. When equal concentrations of reactants and products are present, proton transfer always occurs from the stronger acid to the stronger base.

SOLUTION

(a) In this reaction, H_2SO_4 and NH_4^+ are the acids, and NH_3 and HSO_4^- are the bases. According to Table 16.1, H_2SO_4 is a stronger acid than NH_4^+ and NH_3 is a stronger base than HSO_4^-. Therefore, NH_3 gets the proton and the reaction proceeds from left to right.

$H_2SO_4(aq) + NH_3(aq) \longrightarrow NH_4^+(aq) + HSO_4^-(aq)$
Stronger acid   Stronger base   Weaker acid   Weaker base

(b) HCO_3^- and HSO_4^- are the acids, and SO_4^{2-} and CO_3^{2-} are the bases. Table 16.1 indicates that HSO_4^- is the stronger acid and CO_3^{2-} is the stronger base. Therefore, CO_3^{2-} gets the proton and the reaction proceeds from right to left.

$HCO_3^-(aq) + SO_4^{2-}(aq) \longleftarrow HSO_4^-(aq) + CO_3^{2-}(aq)$
Weaker acid   Weaker base   Stronger acid   Stronger base

▶ **PRACTICE 16.5**   If you mix equal concentrations of reactants and products, will the reaction proceed to the right or the left?

$HF(aq) + NO_3^-(aq) \rightleftharpoons HNO_3(aq) + F^-(aq)$

▶ **CONCEPTUAL APPLY 16.6**   The following pictures represent aqueous solutions of two acids HA (A = X or Y); water molecules have been omitted for clarity.

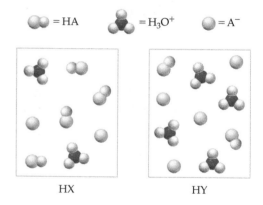

HX   HY

(a) Which is the stronger acid, HX or HY?
(b) Which is the stronger base, X⁻ or Y⁻?
(c) If you mix equal concentrations of reactants and products, will the following reaction proceed to the right or to the left?

$HX + Y^- \rightleftharpoons HY + X^-$

16.3 FACTORS THAT AFFECT ACID STRENGTH

Why is one acid stronger than another? Although a complete analysis of the factors that determine the strength of an acid is complex, the extent of dissociation of an acid HA is often determined by the strength and polarity of the H—A bond. The strength of the H—A bond, as we saw in Section 7.2, is given by the **bond dissociation energy (D)**, which is the amount of energy required to dissociate HA into an H atom and an A atom. The polarity of the H—A bond increases with an increase in the electronegativity of A and is related to the ease of electron transfer from an H atom to an A atom to give an H^+ cation and an A^- anion. In general, the weaker and more polar the H—A bond, the stronger the acid.

Let's look first at the hydrohalic acids HF, HCl, HBr, and HI. **Electrostatic potential maps** (Section 7.3) show that all these molecules are polar, with the halogen atom being electron rich (red) and the H atom being electron poor (blue).

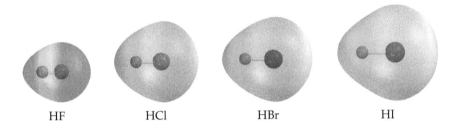

HF   HCl   HBr   HI

The variation in polarity in this series, however, is much less important than the variation in bond strength, which decreases markedly from 570 kJ/mol for HF to 298 kJ/mol for HI.

In general, for binary acids of elements in the *same group* of the periodic table, the H—A bond strength is the most important determinant of acidity. The H—A bond strength generally decreases with increasing size of element A down a group, so acidity increases. For HA (A = F, Cl, Br, or I), for example, the size of A increases from F to I, so bond strength decreases and acidity increases from HF to HI. Hydrofluoric acid is a weak acid, whereas HCl, HBr, and HI are strong acids.

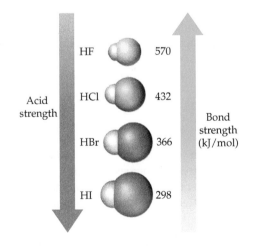

Acid strength

HF	570
HCl	432
HBr	366
HI	298

Bond strength (kJ/mol)

As a further example of this effect, H_2S is a stronger acid than H_2O.

For binary acids of elements in the *same row* of the periodic table, changes in the H—A bond strength are smaller and the polarity of the H—A bond is the most important determinant of acid strength. The strengths of binary acids of the second-row elements, for example, increase as the electronegativity of A increases.

REMEMBER . . .
The strength of a bond is measured by the **bond dissociation energy (D)**, the amount of energy that must be supplied to break a chemical bond in a molecule in the gaseous state represented by the reaction $HA(g) \rightarrow H(g) + A(g)$ (Section 7.2).

REMEMBER . . .
Electrostatic potential maps use color to portray the calculated electron distribution in a molecule. Electron-rich regions are red, and electron-poor regions are blue. Intermediate regions may be yellow, orange, or green (Section 7.3).

Go to eText

BIG IDEA Question 3

Which properties of the H—A bond result in the strongest acid?

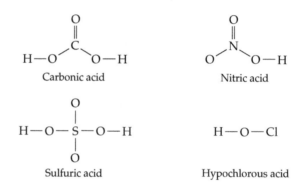

As the electrostatic potential maps show, the C—H bond is relatively nonpolar, and methane has no tendency to dissociate in water into H_3O^+ and CH_3^- ions. The N—H bond is more polar, but dissociation of NH_3 into H_3O^+ and NH_2^- ions is still negligibly small. Water and hydrofluoric acid, however, are increasingly stronger acids. Periodic trends in the strength binary acids are summarized in **FIGURE 16.2**.

Oxoacids, such as H_2CO_3, HNO_3, H_2SO_4, and $HClO$, have the general formula H_nYO_m, where Y is a nonmetallic atom, such as C, N, S, or Cl, and n and m are integers. The atom Y is always bonded to one or more hydroxyl (OH) groups and can be bonded, in addition, to one or more oxygen atoms:

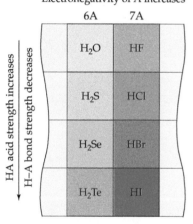

▲ **FIGURE 16.2**

Periodic trends in the strength of binary acids. The acid strength of a binary acid HA increases from left to right in the periodic table, with increasing electronegativity of A, and from top to bottom, with decreasing H—A bond strength.

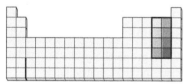

▲ Figure It Out
Which binary acid of 6A and 7A elements is the strongest and why?

Answer: HI because binary acid strength increases down a group and across a period.

Because the dissociation of an oxoacid requires breaking an O—H bond, any factor that weakens the O—H bond or increases its polarity increases the strength of the acid. Two such factors are the electronegativity of Y and the oxidation number of Y in the general reaction

$$-\overset{|}{\underset{|}{Y}}-O-H + H_2O \rightleftharpoons H_3O^+ + -\overset{|}{\underset{|}{Y}}-O^-$$

- **For oxoacids that contain the same number of OH groups and the same number of O atoms, acid strength increases as the electronegativity of Y increases.** For example, the acid strength of the hypohalous acids HOY (Y = Cl, Br, or I) increases as the electronegativity of the halogen increases:

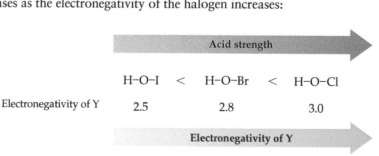

As the halogen becomes more electronegative, an increasing amount of electron density shifts from the O—H bond toward the halogen, thus weakening the O—H bond and increasing its polarity. As a result, the proton is more easily transferred to a solvent water molecule, and so the acid strength increases.

• **For oxoacids that contain the same atom Y but different numbers of oxygen atoms, acid strength increases as the oxidation number of Y increases.** The oxidation number of Y increases, in turn, as the number of oxygen atoms increases. This effect is illustrated by the oxoacids of chlorine:

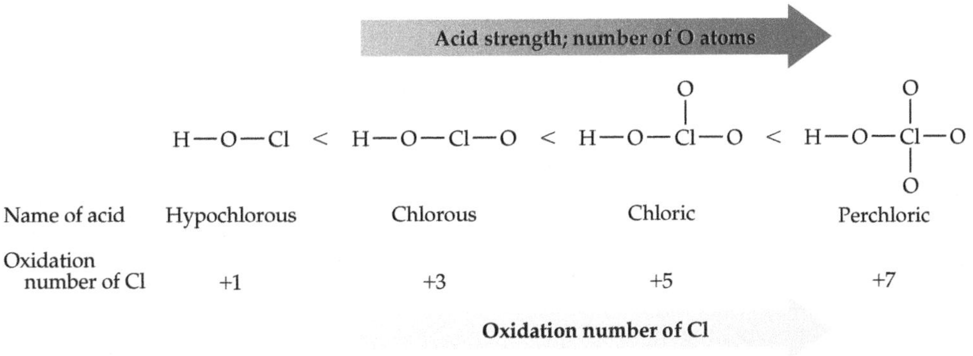

Acid strength; number of O atoms →

$$H-O-Cl \quad < \quad H-O-Cl-O \quad < \quad H-O-\overset{\displaystyle O}{\underset{}{Cl}}-O \quad < \quad H-O-\overset{\displaystyle O}{\underset{\displaystyle O}{Cl}}-O$$

Name of acid	Hypochlorous	Chlorous	Chloric	Perchloric
Oxidation number of Cl	+1	+3	+5	+7

Oxidation number of Cl

As the number of O atoms in $HClO_m$ increases, an increasing amount of electron density shifts from the Cl atom toward the more electronegative O atoms. The amount of positive charge on the Cl atom therefore increases as its oxidation number increases. The increased positive charge on the Cl atom in turn attracts an increasing amount of electron density from the O—H bond, thus weakening the O—H bond and increasing its polarity. As a result, the proton is more easily transferred to a solvent water molecule.

Another factor that affects the acid strength of oxoacids is the relative stability of the corresponding oxoanions. The ClO_m^- anion becomes more stable as the number of O atoms increases in the series $ClO^- < ClO_2^- < ClO_3^- < ClO_4^-$ because a larger number of electronegative O atoms can better accommodate the anion's negative charge. As the stability of the anion increases, the corresponding acid has a greater tendency to dissociate. The increase in acid strength with increasing number of O atoms is further illustrated by the oxoacids of sulfur: H_2SO_4 is a stronger acid than H_2SO_3.

WORKED EXAMPLE 16.4

Evaluating Acid Strength Based upon Molecular Structure

Identify the stronger acid in each of the following pairs:

(a) H_2S or HCl   (b) $HClO_3$ or $HBrO_3$ (c) H_2SO_4 or H_2SO_3

STRATEGY

For a given acid, where an element A is bonded to an acidic hydrogen atom, evaluate the strength and polarity of the H—A bond. Acid strength increases as the strength of the H—A bond decreases and the polarity of the H—A bond increases.

SOLUTION

(a) In comparing H_2S and HCl, note that S and Cl are in the same row of the periodic table. Therefore, the H—S bond should have similar strength to the H—Cl bond and bond polarity is the most important determinant of acid strength. Cl is more electronegative than S, making the H—Cl bond more polar than the H—S bond. HCl is the stronger acid.

(b) $HClO_3$ and $HBrO_3$ are oxoacids that contain the same number of O atoms and acid strength is determined by the electronegativity of the central atom. Because Cl is more electronegative than Br it can pull more electron density from the O—H bond resulting in a weaker and more polar bond in $HClO_3$. Therefore, $HClO_3$ is the stronger acid.

continued on next page

(c) H_2SO_4 and H_2SO_3 are oxoacids with different numbers of oxygen atoms and different oxidation states of sulfur. The oxidation state of sulfur in H_2SO_4 is +6 which is higher than +4 in H_2SO_3. The higher positive charge on the S atom is more effective in pulling electron density from the O—H bond, resulting in a weaker and more polar bond in H_2SO_4. Therefore, H_2SO_4 is the stronger acid.

PRACTICE 16.7 Which pair has the stronger acid listed first?

(a) H_2S and H_2Se
(b) HNO_2 and HNO_3
(c) H_2Te and HI
(d) H_2SO_3 and H_2SeO_3

APPLY 16.8 Which acid is stronger, H_3PO_4 or H_3AsO_4?

16.4 DISSOCIATION OF WATER

One of the most important properties of water is its ability to act both as an acid and as a base. In the presence of an acid, water acts as a base, whereas in the presence of a base, water acts as an acid. It's not surprising, therefore, that in pure water one molecule can donate a proton to another in a reaction in which water acts as both an acid and a base in the same reaction:

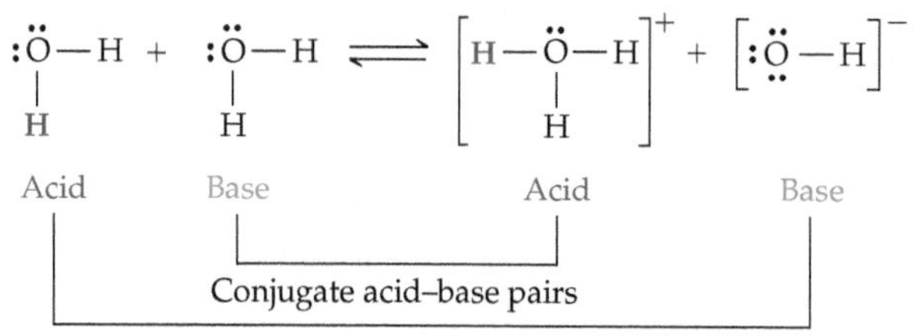

Called the ***dissociation of water,*** this reaction is characterized by the equilibrium equation $K_w = [H_3O^+][OH^-]$, where the equilibrium constant K_w is called the *ion-product constant for water.*

Dissociation of water $2\ H_2O(l) \rightleftharpoons H_3O^+(aq) + OH^-(aq)$

Ion-product constant for water $K_w = [H_3O^+][OH^-]$

REMEMBER . . .
The **concentration** of a pure liquid or a pure solid is omitted from the equilibrium constant expression because the ratio of its actual concentration to its concentration in the thermodynamic standard state is equal to 1 (Section 15.4).

As discussed in Section 15.4, the **concentration** of water is omitted from the equilibrium constant expression because water is a pure liquid.

Note two important aspects of the dynamic equilibrium in the dissociation of water. First, the forward and reverse reactions are rapid: H_2O molecules, H_3O^+ ions, and OH^- ions continually interconvert as protons transfer quickly from one species to another. Second, the position of the equilibrium lies far to the left: At any given instant, only a tiny fraction of the water molecules are dissociated into H_3O^+ and OH^- ions. The vast majority of the H_2O molecules are undissociated.

We can calculate the extent of the dissociation of the water molecules starting from experimental measurements that show the H_3O^+ concentration in pure water to be 1.0×10^{-7} M at 25 °C:

$$[H_3O^+] = 1.0 \times 10^{-7}\ \text{M} \quad \text{at } 25\ °\text{C}$$

Since the dissociation reaction of water produces equal concentrations of H_3O^+ and OH^- ions, the OH^- concentration in pure water is also 1.0×10^{-7} M at 25 °C:

$$[H_3O^+] = [OH^-] = 1.0 \times 10^{-7}\ \text{M} \quad \text{at } 25\ °\text{C}$$

Furthermore, we know that the molar concentration of pure water, calculated from its density and molar mass, is 55.4 M at 25 °C:

$$[H_2O] = \left(\frac{997\text{ g}}{L}\right)\left(\frac{1\text{ mol}}{18.0\text{ g}}\right) = 55.4\text{ mol/L}\quad\text{at 25 °C}$$

From these facts, we conclude that the ratio of dissociated to undissociated water molecules is about 2 in 10^9, a very small number indeed:

$$\frac{[H_2O]_{\text{dissociated}}}{[H_2O]_{\text{undissociated}}} = \frac{1.0\times10^{-7}\text{M}}{55.4\text{ M}} = 1.8\times10^{-9}\quad\text{about 2 in }10^9$$

In addition, we can calculate that the numerical value of K_w at 25 °C is 1.0×10^{-14}:

$$K_w = [H_3O^+][OH^-] = (1.0\times10^{-7})(1.0\times10^{-7})$$
$$= 1.0\times10^{-14}\quad\text{at 25 °C}$$

In very dilute solutions, the water is almost a pure liquid and the product of the H_3O^+ and OH^- concentrations is unaffected by the presence of solutes. This is not true in more concentrated solutions, but we'll neglect that complication and assume that the product of the H_3O^+ and OH^- concentrations is always 1.0×10^{-14} at 25 °C in any aqueous solution.

We can distinguish acidic, neutral, and basic aqueous solutions by the relative values of the H_3O^+ and OH^- concentrations:

Acidic: $[H_3O^+] > [OH^-]$
Neutral: $[H_3O^+] = [OH^-]$
Basic: $[H_3O^+] < [OH^-]$

At 25 °C, $[H_3O^+] > 1.0\times10^{-7}$M in an acidic solution, $[H_3O^+] = [OH^-] = 1.0\times10^{-7}$M in a neutral solution, and $[H_3O^+] < 1.0\times10^{-7}$M in a basic solution (**FIGURE 16.3**). If one of the concentrations, $[H_3O^+]$ or $[OH^-]$, is known, the other is readily calculated:

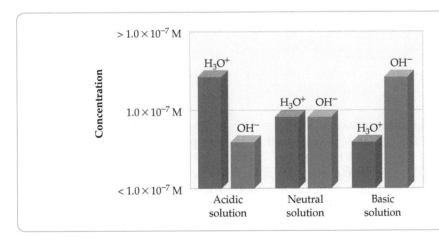

◀ **FIGURE 16.3**

Values of the H_3O^+ and OH^- concentrations at 25 °C in acidic, neutral, and basic solutions.

Since

$$[H_3O^+][OH^-] = K_w = 1.0\times10^{-14}$$

then

$$[H_3O^+] = \frac{1.0\times10^{-14}}{[OH^-]}\quad\text{and}\quad[OH^-] = \frac{1.0\times10^{-14}}{[H_3O^+]}$$

In the previous discussion, we were careful to emphasize that the value of $K_w = 1.0\times10^{-14}$ applies only at 25 °C. This is because K_w, like all equilibrium constants, is affected by temperature and the H_3O^+ and OH^- concentrations in neutral aqueous solutions at temperatures other than 25 °C deviate from 1.0×10^{-7} M (see Problem 16.10). Unless otherwise indicated, we'll always assume a temperature of 25 °C.

Calculating $[OH^-]$ from K_w and $[H_3O^+]$

The concentration of H_3O^+ ions in a sample of lemon juice is 2.5×10^{-3} M. Calculate the concentration of OH^- ions, and classify the solution as acidic, neutral, or basic.

IDENTIFY

Known	Unknown
$K_w = 1.0 \times 10^{-14}$	$[OH^-]$
$[H_3O^+] = 2.5 \times 10^{-3}$ M	

STRATEGY

When $[H_3O^+]$ is known, the OH^- concentration can be found from the expression $[OH^-] = K_w/[H_3O^+]$.

SOLUTION

$$[OH^-] = \frac{K_w}{[H_3O^+]} = \frac{1.0 \times 10^{-14}}{2.5 \times 10^{-3}} = 4.0 \times 10^{-12} \text{ M}$$

Because $[H_3O^+] > [OH^-]$, the solution is acidic.

CHECK

Because the product of the H_3O^+ and OH^- concentrations must equal 10^{-14} and because the H_3O^+ concentration is in the range 10^{-3} M to 10^{-2} M, the OH^- concentration must be in the range 10^{-11} M to 10^{-12} M. The ballpark check and the solution agree.

▶ **PRACTICE 16.9** The concentration of H_3O^+ ions in the runoff from a coal mine is 1.4×10^{-4} M. Calculate the concentration of OH^- ions, and classify the solution as acidic, neutral, or basic.

▶ **APPLY 16.10** At 50 °C the value of K_w is 5.5×10^{-14}. What are the concentrations of H_3O^+ and OH^- in a neutral solution at 50 °C?

16.5 THE pH SCALE

Rather than write hydronium ion concentrations in units of molarity, it's more convenient to express them on a logarithmic scale known as the *pH scale*. The term **pH** is derived from the French *puissance d'hydrogène* ("power of hydrogen") and refers to the power of 10 (the exponent) used to express the molar H_3O^+ concentration. The pH of a solution is defined as the negative base-10 logarithm (log) of the molar hydronium ion concentration:

$$\textbf{pH} = -\log[H_3O^+] \quad \text{or} \quad [H_3O^+] = \text{antilog}(-pH) = 10^{-pH}$$

Thus, an acidic solution having $[H_3O^+] = 10^{-2}$ M has a pH of 2, a basic solution having $[OH^-] = 10^{-2}$ M and $[H_3O^+] = 10^{-12}$ M has a pH of 12, and a neutral solution having $[H_3O^+] = 10^{-7}$ M has a pH of 7. Note that we can take the log of $[H_3O^+]$ because $[H_3O^+]$ is a dimensionless ratio of the actual concentration to the concentration (1 M) in the standard state. Although much less frequently used than pH, a pOH can be defined in the same way as pH and used to express the molar OH^- concentration.

$$\textbf{pOH} = -\log[OH^-]$$

It follows from the equation $[H_3O^+][OH^-] = 1.0 \times 10^{-14}$ that pH+ pOH = 14.00.

If you use a calculator to find the pH from the H_3O^+ concentration, your answer will have more decimal places than the proper number of significant figures. For example, the pH of the lemon juice in Worked Example 16.5 ($[H_3O^+] = 2.5 \times 10^{-3}$M) is found on a calculator to be

$$pH = -\log(2.5 \times 10^{-3}) = 2.60206$$

This result should be rounded to pH 2.60 (two significant figures) because $[H_3O^+]$ has only two significant figures. Note that the only significant figures in a logarithm are the digits to the right of the decimal point; the number to the left of the decimal point is an exact number related to the integral power of 10 in the exponential expression for $[H_3O^+]$:

$$pH = -\log(\underbrace{2.5} \times 10^{-3}) = -\log 10^{-3} - \log 2.5 = 3 - \underbrace{0.40} = 2.60$$

2 significant figures (2 SFs)   Exact number   Exact number   2 SFs   Exact number   2 SFs

Because the pH scale is logarithmic, the pH changes by 1 unit when $[H_3O^+]$ changes by a factor of 10, by 2 units when $[H_3O^+]$ changes by a factor of 100, and by 6 units when $[H_3O^+]$ changes by a factor of 1,000,000. To appreciate the extent to which the pH scale is a compression of the $[H_3O^+]$ scale, compare the amounts of 12 M HCl required to change the pH of the water in a backyard swimming pool: Only about 100 mL of 12 M HCl is needed to change the pH from 7 to 6, but a 10,000 L truckload of 12 M HCl is needed to change the pH from 7 to 1.

FIGURE 16.4 shows the pH scale and pH values for some common substances. Because the pH is the *negative* log of $[H_3O^+]$, the pH decreases as $[H_3O^+]$ increases. Thus, when $[H_3O^+]$ increases from 10^{-7}M to 10^{-6}M, the pH decreases from 7 to 6. As a result, acidic solutions have pH less than 7, and basic solutions have pH greater than 7.

Acidic solution: pH < 7
Neutral solution: pH = 7
Basic solution: pH > 7

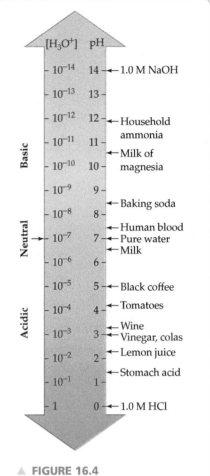

▲ **FIGURE 16.4**
The pH scale and pH values for some common substances.

— **WORKED EXAMPLE 16.6**

Calculating the pH from the H_3O^+ Concentration

Calculate the pH of an aqueous ammonia solution that has an OH^- concentration of 1.9×10^{-3} M.

IDENTIFY

Known	Unknown
$[OH^-] = 1.9 \times 10^{-3}$	pH
$K_w = 1.0 \times 10^{-14}$	

STRATEGY

First, calculate the H_3O^+ concentration from the OH^- concentration, and then take the negative logarithm of $[H_3O^+]$ to convert to pH.

SOLUTION

$$[H_3O^+] = \frac{K_w}{[OH^-]} = \frac{1.0 \times 10^{-14}}{1.9 \times 10^{-3}} = 5.3 \times 10^{-12} M$$

$$pH = -\log [H_3O^+] = -\log (5.3 \times 10^{-12}) = 11.28$$

The pH is quoted to two significant figures (.28) because $[H_3O^+]$ is known to two significant figures (5.3).

continued on next page

▲ The copper mines at Iron Mountain produce highly acidic runoff. The orange-brown precipitate is $Fe(OH)_3(s)$.

CHECK

Because $[OH^-]$ is between 10^{-3} M and 10^{-2} M, $[H_3O^+]$ is between 10^{-11} M and 10^{-12} M.

Therefore, the pH is between 11 and 12, in agreement with the solution.

▶ **PRACTICE 16.11**   Calculate the pH of a sample of seawater that has an OH^- concentration of 1.58×10^{-6} M.

▶ **APPLY 16.12**   During mining operations, the mineral pyrite (FeS_2) is exposed to air and oxygen and reacts to produce sulfuric acid (H_2SO_4). The drainage from the Iron Mountain Mine in California was found to have $[H_3O^+] = 6.3$ M. Calculate the pH of the water seeping from the mine.

WORKED EXAMPLE 16.7

Calculating the H₃O⁺ Concentration from the pH

Acid rain is a matter of serious concern because most species of fish die in waters having a pH lower than 4.5–5.0. Calculate the H_3O^+ concentration in a lake that has a pH of 4.50.

IDENTIFY

Known	Unknown
pH = 4.50	$[H_3O^+]$

STRATEGY

Calculate the H_3O^+ concentration by taking the antilogarithm of the negative of the pH.

SOLUTION

$$[H_3O^+] = \text{antilog}\,(-pH) = 10^{-pH} = 10^{-4.50} = 3.2 \times 10^{-5}\ M$$

$[H_3O^+]$ is reported to two significant figures because the pH has two digits beyond the decimal point. (If you need help in finding the antilog of a number, see Appendix A.2.)

CHECK

Because a pH of 4.50 is between 4 and 5, $[H_3O^+]$ is between 10^{-4} M and 10^{-5} M, in agreement with the solution.

▶ **PRACTICE 16.13**   Calculate the concentrations of H_3O^+ and OH^- in a cola beverage (pH 2.80).

▶ **APPLY 16.14**   The pH of milk is 6.6 and the pH of black coffee is 5.0. How many times greater is the $[H_3O^+]$ concentration in coffee?

16.6 MEASURING pH

The approximate pH of a solution can be determined by using an **acid–base indicator**, a substance that changes color in a specific pH range (**FIGURE 16.5**). Indicators (abbreviated HIn) exhibit pH-dependent color changes because they are weak acids and have different colors in their acid (HIn) and conjugate base (In^-) forms:

$$\underset{\text{Color A}}{HIn(aq)} + H_2O(l) \rightleftharpoons H_3O^+(aq) + \underset{\text{Color B}}{In^-(aq)}$$

Bromthymol blue, for example, changes color in the pH range 6.0–7.6, from yellow in its acid form to blue in its base form.

Especially convenient for making approximate pH measurements is a commercially available mixture of indicators known as *universal indicator* that exhibits various

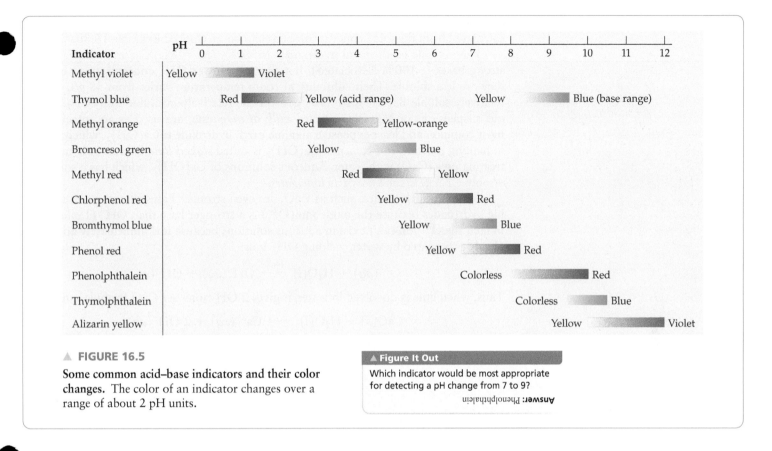

Indicator	pH	0	1	2	3	4	5	6	7	8	9	10	11	12

Methyl violet — Yellow → Violet

Thymol blue — Red → Yellow (acid range)   Yellow → Blue (base range)

Methyl orange — Red → Yellow-orange

Bromcresol green — Yellow → Blue

Methyl red — Red → Yellow

Chlorphenol red — Yellow → Red

Bromthymol blue — Yellow → Blue

Phenol red — Yellow → Red

Phenolphthalein — Colorless → Red

Thymolphthalein — Colorless → Blue

Alizarin yellow — Yellow → Violet

▲ **FIGURE 16.5**

Some common acid–base indicators and their color changes. The color of an indicator changes over a range of about 2 pH units.

▲ **Figure It Out**

Which indicator would be most appropriate for detecting a pH change from 7 to 9?

Answer: Phenolphthalein

colors depending on the pH (see Figure 16.9). More accurate pH values can be determined with an electronic instrument called a *pH meter* (**FIGURE 16.6**), a device that measures the pH-dependent electrical potential of the test solution.

16.7 THE pH IN SOLUTIONS OF STRONG ACIDS AND STRONG BASES

The commonly encountered strong acids listed in Table 16.1 include three *monoprotic acids* ($HClO_4$, HCl, and HNO_3), which contain a single dissociable proton, and one *diprotic acid* (H_2SO_4), which has two dissociable protons. Because strong monoprotic acids are nearly 100% dissociated in aqueous solution, the H_3O^+ and A^- concentrations are equal to the initial concentration of the acid and the concentration of undissociated HA molecules is essentially zero.

$$HA(aq) + H_2O(l) \xrightarrow{100\%} H_3O^+(aq) + A^-(aq)$$

The pH of a solution of a strong monoprotic acid is easily calculated from the H_3O^+ concentration, as shown in Worked Example 16.8a. Calculation of the pH of a H_2SO_4 solution is more complicated because essentially 100% of the H_2SO_4 molecules dissociate to give H_3O^+ and HSO_4^- ions but much less than 100% of the resulting HSO_4^- ions dissociate to give H_3O^+ and SO_4^{2-} ions. In Section 16.11, we'll calculate the pH of diprotic acids like H_2SO_4.

The most familiar examples of **strong bases** are alkali metal hydroxides, MOH, such as NaOH (*caustic soda*) and KOH (*caustic potash*). These compounds are water-soluble ionic solids that exist in aqueous solution as alkali metal cations (M^+) and OH^- anions:

$$MOH(s) \xrightarrow{H_2O} MOH(aq) \xrightarrow{100\%} M^+(aq) + OH^-(aq)$$

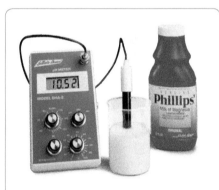

▲ **FIGURE 16.6**

A pH meter with its electrical probe dipping into milk of magnesia. An accurate value of the pH (10.52) is shown on the meter.

Thus, 0.10 M NaOH contains 0.10 M Na^+ and 0.10 M OH^-, and the pH is readily calculated from the OH^- concentration, as shown in Worked Example 16.8b.

The alkaline earth metal hydroxides $M(OH)_2$ (M = Mg, Ca, Sr, or Ba) are also strong bases (~100% dissociated), but they give lower OH^- concentrations because they are less soluble. Their solubility at room temperature varies from 38 g/L for the relatively soluble $Ba(OH)_2$ to $\sim 10^{-2}$ g/L for the relatively insoluble $Mg(OH)_2$. Aqueous suspensions of $Mg(OH)_2$, called *milk of magnesia,* are used as an antacid. The most common and least expensive alkaline earth hydroxide is $Ca(OH)_2$, which is used in making mortars and cements. $Ca(OH)_2$ is called *slaked lime* because it is made by treating *lime* (CaO) with water. Aqueous solutions of $Ca(OH)_2$, which has a solubility of only ~1.3 g/L, are known as *limewater.*

Alkaline earth oxides, such as CaO, are even stronger bases than the corresponding hydroxides because the oxide ion (O^{2-}) is a stronger base than OH^- (Table 16.1). In fact, the O^{2-} ion can't exist in aqueous solutions because it is immediately and completely protonated by water, yielding OH^- ions:

$$O^{2-}(aq) + H_2O(l) \xrightarrow{100\%} OH^-(aq) + OH^-(aq)$$

Thus, when lime is dissolved in water, it gives 2 OH^- ions per CaO formula unit:

$$CaO(s) + H_2O(l) \longrightarrow Ca^{2+}(aq) + 2\,OH^-(aq)$$

Lime is the world's most important strong base. Annual worldwide production is around 283 million metric tons for use in steelmaking, water purification, and chemical manufacture. Lime is made by the decomposition of limestone, $CaCO_3$, at temperatures of 800–1000 °C:

$$CaCO_3(s) \xrightarrow{\text{Heat}} CaO(s) + CO_2(g)$$

WORKED EXAMPLE 16.8

Calculating the pH of Strong Acid and Strong Base Solutions

Calculate the pH of each of the following solutions:

(a) A 0.025 M HNO_3 solution
(b) A 0.10 M solution of NaOH
(c) A solution prepared by dissolving 0.185 g of slaked lime $[Ca(OH)_2]$ in enough water to produce 0.50 L.

IDENTIFY

Known	Unknown
Concentration of strong acid or base	pH

STRATEGY

Write the dissociation reaction for each acid or base and use the stoichiometry of the balanced equation to find the $[H_3O^+]$ or $[OH^-]$ in solution. Strong acids and strong bases are essentially 100% dissociated in aqueous solution. The pH equals the negative log of the $[H_3O^+]$. For solutions of strong bases, $[H_3O^+]$ can be found from $[OH^-]$ using the expression for $K_w, [H_3O^+] = K_w/[OH^-]$.

SOLUTION

(a) Because strong acids are completely dissociated, $[H_3O^+]$ equals the initial concentration of HNO_3.

$$HNO_3(aq) + H_2O(l) \xrightarrow{100\%} H_3O^+(aq) + NO_3^-(aq)$$

$$pH = -\log[H_3O^+] = -\log(2.5 \times 10^{-2}) = 1.60$$

(b) Because strong bases are completely dissociated, $[OH^-]$ equals the initial concentration of NaOH (0.10 M).

$$NaOH(s) \xrightarrow{H_2O} NaOH(aq) \xrightarrow{100\%} Na^+(aq) + OH^-(aq)$$

$$[H_3O^+] = \frac{K_w}{[OH^-]} = \frac{1.0 \times 10^{-14}}{0.10} = 1.0 \times 10^{-13} \, M$$

$$pH = -\log(1.0 \times 10^{-13}) = 13.00$$

(c) First calculate the number of moles of $Ca(OH)_2$ dissolved from the given mass of $Ca(OH)_2$ and its molar mass (74.1 g/mol).

$$0.185 \text{ g } Ca(OH)_2 \times \frac{1 \text{ mol } Ca(OH)_2}{74.1 \text{ g } Ca(OH)_2} = 2.50 \times 10^{-3} \text{mol}$$

Divide the number of moles by volume to find the molar concentration of $Ca(OH)_2$.

$$[Ca(OH)_2] = \frac{2.50 \times 10^{-3} \text{ mol } Ca(OH)_2}{0.50 \text{ L}} = 5.0 \times 10^{-3} M$$

Because slaked lime is a strong base and completely dissociated, it provides 2 OH^- ions per $Ca(OH)_2$ formula unit. Therefore, $[OH^-] = 2(0.0050 \text{ M}) = 0.010 \text{ M}$.

$$Ca(OH)_2(s) \xrightarrow{H_2O} Ca(OH)_2(aq) \xrightarrow{100\%} Ca^{2+}(aq) + 2OH^-(aq)$$

$$[H_3O^+] = \frac{K_w}{[OH^-]} = \frac{1.0 \times 10^{-14}}{0.010} = 1.0 \times 10^{-12} \, M$$

$$pH = -\log(1.0 \times 10^{-12}) = 12.00$$

CHECK

A solution of an acid will result in pH < 7 and a solution of a base will result in pH > 7, so calculated pH values are reasonable. In addition, the pH should be close to the negative exponent of the $[H_3O^+]$.

▶ **PRACTICE 16.15**   Calculate the pH of the following solutions:

(a) 0.050 M $HClO_4$ (b) 0.010 M $Ba(OH)_2$

▶ **APPLY 16.16**   Calculate the pH of a solution prepared by dissolving 0.25 g of CaO in enough water to make 1.50 L of solution.

▲ Lime is spread on lawns and gardens to raise the pH of acidic soils.

16.8 EQUILIBRIA IN SOLUTIONS OF WEAK ACIDS

It's important to realize that a weak acid is not the same thing as a dilute solution of a strong acid. Whereas a strong acid is 100% dissociated in aqueous solution, a weak acid is only partially dissociated. It is possible that the H_3O^+ concentration from complete dissociation of a dilute strong acid is the same as that from partial dissociation of a more concentrated weak acid.

Like the equilibrium reactions discussed in Chapter 15, the dissociation of a weak acid in water is characterized by an equilibrium equation. The equilibrium constant for the dissociation reaction, denoted K_a, is called the **acid-dissociation constant**:

The acid-dissociation constant K_a

$$HA(aq) + H_2O(l) \rightleftharpoons H_3O^+(aq) + A^-(aq) \quad K_a = \frac{[H_3O^+][A^-]}{[HA]}$$

Note that water has been omitted from the equilibrium equation because its concentration in dilute solutions is essentially the same as that in pure water (55.4 M) and pure liquids are always omitted from equilibrium equations (Section 15.4).

Values of K_a and pK_a for some typical weak acids are listed in **TABLE 16.2**. Just as the pH is defined as $-\log[H_3O^+]$, the pK_a of an acid is defined as $-\log K_a$. Note that the pK_a

Go to
eText

BIG IDEA Question 4

Which solution would have a lower pH: 1 M HOI ($pK_a = 10.6$) or 1 M HOCl ($pK_a = 7.46$)?

decreases as K_a increases. As indicated by the equilibrium equation, the larger the value of K_a, the stronger the acid. Thus, methanol ($K_a = 2.9 \times 10^{-16}$; $pK_a = 15.54$) is the weakest of the acids listed in Table 16.2, and nitrous acid ($K_a = 4.5 \times 10^{-4}$; $pK_a = 3.35$) is the strongest of the weak acids. Strong acids, such as HCl, have K_a values that are much greater than 1 and pK_a values that are negative. A more complete list of K_a values for weak acids is given in Appendix C. Numerical values of acid-dissociation constants are determined from pH measurements, as shown in Worked Example 16.9.

TABLE 16.2   Acid-Dissociation Constants at 25 °C

	Acid	Molecular Formula	Structural Formula*	K_a	$pK_a^\dagger$
Stronger acid	Hydrochloric	HCl	H—Cl	2×10^6	−6.3
	Nitrous	HNO_2	H—O—N==O	4.5×10^{-4}	3.35
	Hydrofluoric	HF	H—F	3.5×10^{-4}	3.46
	Formic	HCO_2H	H—C(=O)—O—H	1.8×10^{-4}	3.74
	Ascorbic (vitamin C)	$C_6H_8O_6$	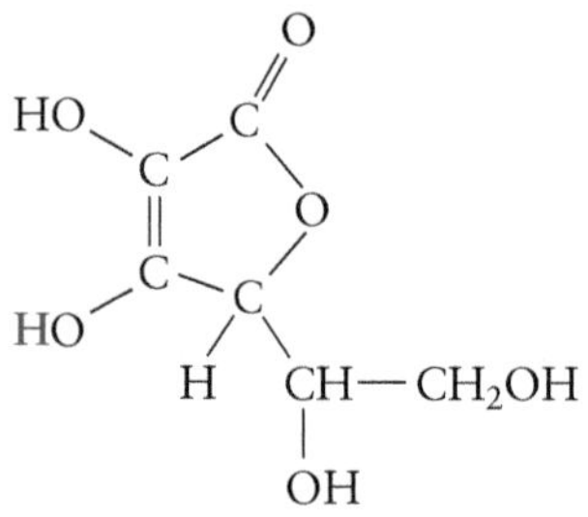	8.0×10^{-5}	4.10
	Acetic	CH_3CO_2H	CH_3—C(=O)—O—H	1.8×10^{-5}	4.74
	Hypochlorous	HOCl	H—O—Cl	3.5×10^{-8}	7.46
Weaker acid	Hydrocyanic	HCN	H—C≡N	4.9×10^{-10}	9.31
	Methanol	CH_3OH	CH_3—O—H	2.9×10^{-16}	15.54

* The proton that is transferred to water when the acid dissociates is shown in red.
† $pK_a = -\log K_a$.

WORKED EXAMPLE 16.9

Calculating K_a and pK_a for a Weak Acid from the pH of the Solution

The pH of 0.250 M HF is 2.036. What are the values of K_a and pK_a for hydrofluoric acid?

IDENTIFY

Known	Unknown
Concentration of HF (0.250 M) pH = 2.036	K_a, pK_a

STRATEGY

Use the general procedure for solving equilibrium problems described in Section 15.5. First, write the balanced equation for the dissociation equilibrium. Then, add known quantities to the "initial," "change," and "equilibrium" rows in the equilibrium table. Define x as the concentration of HF that dissociates. Because x equals the H_3O^+ concentration, its value can be calculated from the pH. Finally, substitute the equilibrium concentrations into the equilibrium equation to obtain the value of K_a and take the negative log of K_a to obtain the pK_a.

SOLUTION

	$HF(aq) + H_2O(l) \rightleftharpoons H_3O^+(aq) + F^-(aq)$		
Initial concentration (M)	0.250	~0*	0
Change (M)	$-x$	$+x$	$+x$
Equilibrium concentration (M)	$(0.250 - x)$	x	x

*A very small concentration of H_3O^+ is present initially because of the dissociation of water.

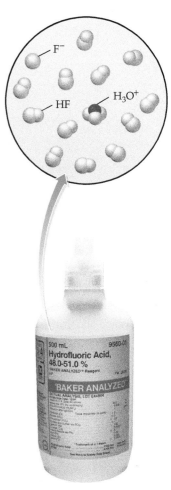

We can calculate the value of x from the pH:

$$x = [H_3O^+] = \text{antilog}\,(-pH) = 10^{-pH} = 10^{-2.036} = 9.20 \times 10^{-3}\,M$$

The other equilibrium concentrations are

$$[F^-] = x = 9.20 \times 10^{-3}\,M$$

$$[HF] = 0.250 - x = 0.250 - 0.00920 = 0.241\,M$$

Substituting these concentrations into the equilibrium equation gives the value of K_a:

$$K_a = \frac{[H_3O^+][F^-]}{[HF]} = \frac{(x)(x)}{(0.250 - x)} = \frac{(9.20 \times 10^{-3})(9.20 \times 10^{-3})}{0.241} = 3.51 \times 10^{-4}$$

$$pK_a = -\log K_a = -\log\,(3.51 \times 10^{-4}) = 3.455$$

CHECK

Because the pH is about 2, $[H_3O^+]$ and $[F^-]$ are about 10^{-2} M, and [HF] is about 0.25 M $(0.250\,M - 10^{-2}\,M)$. The value of K_a is therefore about $(10^{-2})(10^{-2})/0.25$, or 4×10^{-4}, and the pK_a is between 3 and 4. The estimate and the solution agree.

▶ **PRACTICE 16.17** The pH of a 0.10 M solution of lactic acid $(HC_3H_5O_3)$ is 2.42. Calculate the K_a and pK_a for lactic acid.

▶ **CONCEPTUAL APPLY 16.18** The following pictures represent aqueous solutions of three acids HA (A = X, Y, or Z); water molecules have been omitted for clarity:

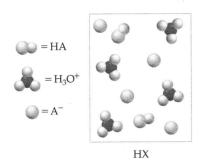

HX   HY   HZ

(a) Arrange the three acids in order of increasing value of K_a.
(b) Which acid, if any, is a strong acid?
(c) Which solution has the highest pH, and which has the lowest pH?

▲ Of the hydrohalic acids HF, HCl, HBr, and HI, hydrofluoric acid is the only weak acid. HF has a positive pK_a value, while the strong acids HCl, HBr, and HI have negative pK_a values.

16.9 CALCULATING EQUILIBRIUM CONCENTRATIONS IN SOLUTIONS OF WEAK ACIDS

Once we have measured the K_a value for a weak acid, we can use it to calculate equilibrium concentrations and the pH in the acid solution. We'll illustrate the approach to such a problem by calculating the concentrations of all species present (H_3O^+, CN^-, HCN, OH^-) and the pH in a 0.10 M HCN solution. The approach we'll take is quite general and will be useful on numerous later occasions.

The key to solving acid–base equilibrium problems is to think about the chemistry—that is, to consider the possible proton-transfer reactions that can take place between Brønsted–Lowry acids and bases.

Step 1. Write the balanced equation for the principal acid–base reaction. The **principal acid–base reaction** is the one that has the largest equilibrium constant. Any other proton transfer reactions are called **subsidiary reactions.** We begin by listing the species present initially before any dissociation and by classifying them as acids or bases. Since water can behave either as an acid or a base, our list of species present initially is

$$\text{HCN} \qquad \text{H}_2\text{O}$$
$$\text{Acid} \qquad \text{Acid or base}$$

Because we have two acids (HCN and H_2O) and just one base (H_2O), two proton-transfer reactions are possible:

$$\text{HCN}(aq) + \text{H}_2\text{O}(l) \rightleftharpoons \text{H}_3\text{O}^+(aq) + \text{CN}^-(aq) \quad K_a = 4.9 \times 10^{-10}$$
$$\text{H}_2\text{O}(l) + \text{H}_2\text{O}(l) \rightleftharpoons \text{H}_3\text{O}^+(aq) + \text{OH}^-(aq) \quad K_w = 1.0 \times 10^{-14}$$

Since K_a for HCN is more than 10,000 times greater than K_w, we can make the assumption that the equilibrium concentration of H_3O^+ ions is determined by the dissociation of the stronger acid HCN.

$$[\text{H}_3\text{O}^{+1}]\,(\text{total}) \approx [\text{H}_3\text{O}^+]\,(\text{from HCN})$$

Therefore, the balanced equation for the principal reaction is the dissociation of the weak acid HCN:

$$\text{HCN}(aq) + \text{H}_2\text{O}(l) \rightleftharpoons \text{H}_3\text{O}^+(aq) + \text{CN}^-(aq) \quad K_a = 4.9 \times 10^{-10}$$

Step 2. Make a table that lists initial concentration, change in concentration, and equilibrium concentration for each of the reactants and products in the balanced chemical equation. We can express the concentrations of the species involved in the reaction in terms of the concentration of HCN that dissociates—say, x mol/L. According to the balanced equation for the dissociation of HCN, if x mol/L of HCN dissociates, then x mol/L of H_3O^+ and x mol/L of CN^- are formed and the initial concentration of HCN before dissociation (0.10 mol/L in our example) is reduced to $(0.10-x)$ mol/L at equilibrium. Let's summarize these considerations in the table under the reaction:

Principal reaction	$\text{HCN}(aq) + \text{H}_2\text{O}(l) \rightleftharpoons \text{H}_3\text{O}^+(aq) + \text{CN}^-(aq)$		
Initial concentration (M)	0.10	~0	0
Change (M)	$-x$	$+x$	$+x$
Equilibrium concentration (M)	$0.10 - x$	x	x

Step 3. Substitute the equilibrium concentrations into the equilibrium equation for the principal reaction, and solve for x.

$$K_a = 4.9 \times 10^{-10} = \frac{[\text{H}_3\text{O}^+][\text{CN}^-]}{[\text{HCN}]} = \frac{(x)(x)}{(0.10 - x)}$$

Because K_a is very small, the reaction will not proceed very far to the right and x will be negligibly small compared to 0.10. Therefore, we can make the approximation that $(0.10 - x) \approx 0.10$, which greatly simplifies the solution:

$$4.9 \times 10^{-10} = \frac{(x)(x)}{(0.10 - x)} = \frac{x^2}{0.10}$$

$$x^2 = 4.9 \times 10^{-11}$$

$$x = 7.0 \times 10^{-6}$$

We must check the validity of the assumption that x is small compared to the number it is subtracted from. To check the assumption, calculate the percentage that the calculated value of x is of the initial concentration.

$$\frac{7.0 \times 10^{-6}}{0.10} \times 100\% = 7.0 \times 10^{-3}\%$$

It's important to check the validity of the simplifying approximation in every problem because x is not always negligible compared to the initial concentration of the acid. Worked Example 16.10 illustrates such a case.

Step 4. Calculate the concentrations of all reactants and products involved in the principal reaction. We use the calculated value of x to obtain the equilibrium concentration of all substances involved in the reaction:

$$[H_3O^+] = [CN^-] = x = 7.0 \times 10^{-6} \text{ M}$$
$$[HCN] = 0.10 - x = 0.10 - (7.0 \times 10^{-6}) = 0.10 \text{ M}$$

The small concentration, $[OH^-]$, is obtained from the subsidiary equilibrium, the dissociation of water:

$$[OH^-] = \frac{K_w}{[H_3O^+]} = \frac{1.0 \times 10^{214}}{7.0 \times 10^{-6}} = 1.4 \times 10^{-9} \text{ M}$$

Step 5. Calculate the pH. Use the equation

$$pH = -\log [H_3O^+] = -\log (7.0 \times 10^{-6}) = 5.15$$

FIGURE 16.7 summarizes the steps followed in solving this problem. This same systematic approach is applied to all acid–base equilibrium problems in this chapter and in Chapter 17.

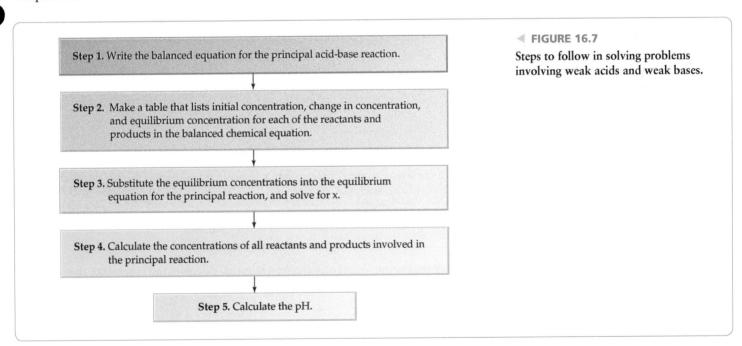

Step 1. Write the balanced equation for the principal acid-base reaction.

Step 2. Make a table that lists initial concentration, change in concentration, and equilibrium concentration for each of the reactants and products in the balanced chemical equation.

Step 3. Substitute the equilibrium concentrations into the equilibrium equation for the principal reaction, and solve for x.

Step 4. Calculate the concentrations of all reactants and products involved in the principal reaction.

Step 5. Calculate the pH.

◄ **FIGURE 16.7**

Steps to follow in solving problems involving weak acids and weak bases.

 Go to eText

WORKED EXAMPLE 16.10

Calculating the pH and the Equilibrium Concentrations in a Solution of a Weak Acid

Calculate the pH and the concentrations of all species present (H_3O^+, F^-, HF, and OH^-) in 0.050 M HF.

continued on next page

IDENTIFY

Known	Unknown
Concentration of HF (0.050 M)	pH
$K_a = 3.5 \times 10^{-4}$	equilibrium concentrations of H_3O^+, F^-, HF, and OH^-

STRATEGY

Follow the five-step sequence outlined in Figure 16.7.

SOLUTION

Step 1. Write the balanced equation for the principal acid–base reaction. The species present initially are:

$$\underset{\text{Acid}}{\text{HF}} \qquad \underset{\text{Acid or base}}{\text{H}_2\text{O}}$$

The possible proton-transfer reactions are

$$HF(aq) + H_2O(l) \rightleftharpoons H_3O^+(aq) + F^-(aq) \qquad K_a = 3.5 \times 10^{-4}$$
$$H_2O(l) + H_2O(l) \rightleftharpoons H_3O^+(aq) + OH^-(aq) \qquad K_w = 1.0 \times 10^{-14}$$

Since $K_a \gg K_w$, the principal reaction is the dissociation of HF.

$$HF(aq) + H_2O(l) \rightleftharpoons H_3O^+(aq) + F^-(aq) \qquad K_a = 3.5 \times 10^{-4}$$

Step 2. Make a table that lists initial concentration, change in concentration, and equilibrium concentration for each of the reactants and products in the balanced chemical equation.

Principal reaction	$HF(aq)\ +\ H_2O(l) \rightleftharpoons H_3O^+(aq)\ +\ F^-(aq)$		
Initial concentration (M)	0.050	~0	0
Change (M)	$-x$	$+x$	$+x$
Equilibrium concentration (M)	$0.050 - x$	x	x

Step 3. Substitute the equilibrium concentrations into the equilibrium equation for the principal reaction and solve for x.

$$K_a = 3.5 \times 10^{-4} = \frac{[H_3O^+][F^-]}{[HF]} = \frac{(x)(x)}{(0.050 - x)}$$

Making the usual approximation that x is negligible compared with the initial concentration of the acid, we assume that $(0.050-x) \approx 0.050$ and then solve for an approximate value of x:

$$x^2 \approx (3.5 \times 10^{-4})(0.050)$$
$$x \approx 4.2 \times 10^{-3}$$

To check the assumption that x can be neglected, calculate the percentage that the calculated value of x is of the initial concentration.

$$\frac{4.2 \times 10^{-3}}{0.050} \times 100\% = 8.4\%$$

If x is less than 5% of the initial concentration, the assumption is valid and does not cause significant error in the equilibrium concentrations. In this case, the assumption is invalid because 8.4% is greater than 5%. We must solve for x without making approximations using the quadratic equation:

$$3.5 \times 10^{-4} = \frac{x^2}{(0.050 - x)}$$
$$x^2 + (3.5 \times 10^{-4})x - (1.75 \times 10^{-5}) = 0$$

We use the standard quadratic formula (Appendix A.4):

$$x = \frac{-b \pm \sqrt{b^2 - 4ac}}{2a}$$

$$= \frac{-(3.5 \times 10^{-4}) \pm \sqrt{(3.5 \times 10^{-4})^2 - 4(1)(-1.75 \times 10^{-5})}}{2(1)}$$

$$= \frac{(-3.5 \times 10^{-4}) \pm (8.37 \times 10^{-3})}{2}$$

$$= +4.0 \times 10^{-3} \quad \text{or} \quad -4.4 \times 10^{-3}$$

Of the two solutions for x, only the positive value has physical meaning, since x is the H_3O^+ concentration. Therefore,

$$x = 4.0 \times 10^{-3}$$

Step 4. Calculate the concentrations of all reactants and products involved in the principal reaction.

$$[H_3O^+] = [F^-] = x = 4.0 \times 10^{-3}\,M$$

$$[HF] = (0.050 - x) = (0.050 - 0.0040) = 0.046\,M$$

The small concentration, $[OH^-]$, is obtained from the subsidiary equilibrium, the dissociation of water:

$$[OH^-] = \frac{K_w}{[H_3O^+]} = \frac{1.0 \times 10^{-14}}{4.0 \times 10^{-3}} = 2.5 \times 10^{-12}\,M$$

Step 5. Calculate the pH. Use the equation

$$pH = -\log[H_3O^+] = -\log(4.0 \times 10^{-3}) = 2.40$$

CHECK

Check your arithmetic by substituting the equilibrium concentrations into the expression for K_a to make sure it is equal to the value of $K_a(3.5 \times 10^{-4})$ given in the problem.

$$K_a = \frac{[H_3O^+][F^-]}{[HF]} = \frac{(4.0 \times 10^{-3})(4.0 \times 10^{-3})}{(0.046)} = 3.5 \times 10^{-4}$$

▶ **PRACTICE 16.19** Acetic acid, CH_3CO_2H, is the solute that gives vinegar its characteristic odor and sour taste. Calculate the pH and the concentrations of all species present (H_3O^+, $CH_3CO_2^-$, CH_3CO_2H, and OH^-), in 0.100 M CH_3CO_2H ($K_a = 1.8 \times 10^{-5}$).

▶ **APPLY 16.20** What concentration of formic acid will result in a solution with pH = 2.00? Refer to Table 16.2 to find the K_a value for formic acid.

All **WORKED EXAMPLES** with this icon [Go to eText] have an interactive video in the eText.

16.10 PERCENT DISSOCIATION IN SOLUTIONS OF WEAK ACIDS

In addition to K_a, another useful measure of the strength of a weak acid is the **percent dissociation**, defined as the concentration of the acid that dissociates divided by the initial concentration of the acid times 100%.

$$\text{Percent dissociation} = \frac{[HA]_{\text{dissociated}}}{[HA]_{\text{initial}}} \times 100\%$$

Using the procedure for solving a weak acid problem shown in Worked Example 16.10, we can find $[H_3O^+]$ concentrations for various initial concentrations of acetic acid. A 1.00 M solution of CH_3CO_2H has $[H_3O^+] = 4.2 \times 10^{-3}$ M, and because $[H_3O^+]$ equals the concentration of CH_3CO_2H that dissociates, the percent dissociation is 0.42%.

$$\text{Percent dissociation} = \frac{[CH_3CO_2H]_{\text{dissociated}}}{[CH_3CO_2H]_{\text{initial}}} \times 100\%$$

$$= \frac{4.2 \times 10^{-3}\text{M}}{1.00\text{ M}} \times 100\% = 0.42\%$$

In general, the percent dissociation depends on the acid and increases with increasing values of K_a. For a given weak acid, the percent dissociation increases with decreasing concentration, as shown in **FIGURE 16.8**. A 0.0100 M CH_3CO_2H solution has $[H_3O^+] = 4.2 \times 10^{-4}$ M, and the percent dissociation is 4.2%:

$$\text{Percent dissociation} = \frac{[CH_3CO_2H]_{\text{dissociated}}}{[CH_3CO_2H]_{\text{initial}}} \times 100\%$$

$$= \frac{4.2 \times 10^{-4}\text{M}}{0.0100\text{ M}} \times 100\% = 4.2\%$$

▷ **FIGURE 16.8**

The percent dissociation of acetic acid increases as the concentration of the acid decreases.

▶ **Figure It Out**

Predict which solution has a higher percent dissociation, 0.500 M HF or 0.050 M HF.

Answer: The HF solution with a concentration of 0.05 M will have a higher percent dissociation because its concentration is lower than the 0.50 M HF solution.

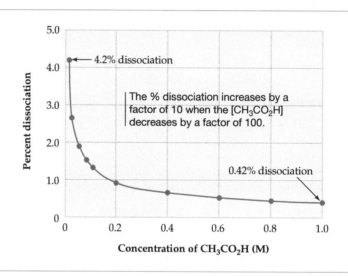

The % dissociation increases by a factor of 10 when the $[CH_3CO_2H]$ decreases by a factor of 100.

4.2% dissociation

0.42% dissociation

Concentration of CH_3CO_2H (M)

Percent dissociation

Go to eText

BIG IDEA Question 5

For a solution of the weak acid sulfurous acid (H_2SO_3), is the pH determined primarily from the first or second dissociation? Table 16.3 lists the K_a values for H_2SO_3.

16.11 POLYPROTIC ACIDS

Acids that contain more than one dissociable proton are called **polyprotic acids**. Polyprotic acids dissociate in a stepwise manner, and each dissociation step is characterized by its own acid-dissociation constant, K_{a1}, K_{a2}, and so forth. For example, carbonic acid (H_2CO_3), the diprotic acid that forms when gaseous carbon dioxide dissolves in water, is important in maintaining a constant pH in human blood. It undergoes the following dissociation reactions:

$$H_2CO_3(aq) + H_2O(l) \rightleftharpoons H_3O^+(aq) + HCO_3^-(aq) \quad K_{a1} = \frac{[H_3O^+][HCO_3^-]}{[H_2CO_3]} = 4.3 \times 10^{-7}$$

$$HCO_3^-(aq) + H_2O(l) \rightleftharpoons H_3O^+(aq) + CO_3^{2-}(aq) \quad K_{a2} = \frac{[H_3O^+][CO_3^{2-}]}{[HCO_3^-]} = 5.6 \times 10^{-11}$$

As shown in **TABLE 16.3,** the values of stepwise dissociation constants of polyprotic acids decrease, typically by a factor of 10^4 to 10^6, in the order $K_{a1} > K_{a2} > K_{a3}$. Because of electrostatic forces, it's more difficult to remove a positively charged proton from a negative ion, such as HCO_3^-, than from an uncharged molecule, such as H_2CO_3, so $K_{a2} < K_{a1}$. In the case of triprotic acids (such as H_3PO_4), it's more difficult to remove H^+ from an anion with a double negative charge (such as HPO_4^{2-}) than from an anion with a single negative charge (such as $H_2PO_4^-$), so $K_{a3} < K_{a2}$.

TABLE 16.3   **Stepwise Dissociation Constants for Polyprotic Acids at 25 °C**

Name	Formula	K_{a1}	K_{a2}	K_{a3}
Carbonic acid	H_2CO_3	4.3×10^{-7}	5.6×10^{-11}	
Hydrogen sulfide[a]	H_2S	1.0×10^{-7}	$\sim 10^{-19}$	
Oxalic acid	$H_2C_2O_4$	5.9×10^{-2}	6.4×10^{-5}	
Phosphoric acid	H_3PO_4	7.5×10^{-3}	6.2×10^{-8}	4.8×10^{-13}
Sulfuric acid	H_2SO_4	Very large	1.2×10^{-2}	
Sulfurous acid	H_2SO_3	1.5×10^{-2}	6.3×10^{-8}	

[a]Because of its very small size, K_{a2} for H_2S is difficult to measure and its value is uncertain.

Polyprotic acid solutions contain a mixture of acids—H_2A, HA^-, and H_2O in the case of a diprotic acid. Because H_2A is by far the strongest acid, the principal reaction is the dissociation of H_2A and essentially all the H_3O^+ in the solution comes from the first dissociation step. Worked Example 16.11 shows how calculations are done.

▲ Carbonated beverages contain polyprotic acids such as carbonic acid and phosphoric acid.

WORKED EXAMPLE 16.11

Calculating the pH and the Equilibrium Concentrations in a Solution of a Diprotic Acid

Calculate the pH and the concentrations of all species present (H_2CO_3, HCO_3^-, CO_3^{2-}, H_3O^+, and OH^-) in a 0.020 M carbonic acid solution.

IDENTIFY

Known	Unknown
Concentration of H_2CO_3 (0.020 M)	pH
$K_{a1} = 4.3 \times 10^{-7}$, $K_{a2} = 5.6 \times 10^{-11}$ (Table 16.3)	equilibrium concentrations

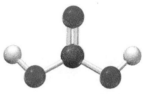

Carbonic acid

STRATEGY

Use the five-step procedure summarized in Figure 16.7.

SOLUTION

Step 1. Write the balanced equation for the principal acid–base reaction. The species present initially are H_2CO_3 (acid) and H_2O (acid or base). Because $K_{a1}(4.3 \times 10^{-7}) \gg K_w(1.0 \times 10^{-14})$, the principal reaction is the dissociation of H_2CO_3.

Step 2. Make a table that lists initial concentration, change in concentration, and equilibrium concentration for each of the reactants and products in the balanced chemical equation.

Principal reaction	$H_2CO_3(aq) + H_2O(l) \rightleftharpoons H_3O^+(aq) + HCO_3^-(aq)$		
Initial concentration (M)	0.020	~ 0	0
Change (M)	$-x$	$+x$	$+x$
Equilibrium concentration (M)	$0.020 - x$	x	x

continued on next page

▲ Dry ice (solid CO_2) acidifies water. When added to water, CO_2 sublimes and reacts with water to form carbonic acid, H_2CO_3. This sequence of photos shows how the pH of the solution changes as CO_2 dissolves. Initially the water has a neutral pH, as shown by the green color of universal indicator, but as carbon dioxide dissolves, the color changes to orange-red, signifying acid formation. The white fog is a mist of water droplets condensing from the air.

Step 3. Substitute the equilibrium concentrations into the equilibrium equation for the principal reaction, and solve for x.

$$K_{a1} = 4.3 \times 10^{-7} = \frac{[H_3O^+][HCO_3^-]}{[H_2CO_3]} = \frac{(x)(x)}{(0.020 - x)}$$

Assuming that $(0.020 - x) \approx 0.020$,

$$x^2 = (4.3 \times 10^{-7})(0.020)$$

$$x = 9.3 \times 10^{-5} \quad \text{Approximation } (0.020 - x) \approx 0.020 \text{ is justified.}$$

Step 4. Calculate the concentrations of all reactants and products involved in the reactions.

$$[H_3O^+] = [HCO_3^-] = x = 9.3 \times 10^{-5} \text{ M}$$

$$[H_2CO_3] = 0.020 - x = 0.020 - 0.000\,093 = 0.020 \text{ M}$$

The concentrations of CO_3^{2-} and OH^- are obtained from the subsidiary equilibria—(1) dissociation of HCO_3^- and (2) dissociation of water—and from the concentrations already determined:

(1) $HCO_3^-(aq) + H_2O(l) \rightleftharpoons H_3O^+(aq) + CO_3^{2-}(aq)$

$$K_{a2} = 5.6 \times 10^{-11} = \frac{[H_3O^+][CO_3^{2-}]}{[HCO_3^-]} = \frac{(9.3 \times 10^{-5})[CO_3^{2-}]}{(9.3 \times 10^{-5})}$$

$$[CO_3^{2-}] = K_{a2} = 5.6 \times 10^{-11} \text{ M}$$

(In general, for a solution of a weak diprotic acid that has a very small value of K_{a2}, $[A^{2-}] = K_{a2}$.)

(2) $$[OH^-] = \frac{K_w}{[H_3O^+]} = \frac{1.0 \times 10^{-14}}{9.3 \times 10^{-5}} = 1.1 \times 10^{-10} \text{ M}$$

The second dissociation of H_2CO_3 produces a negligible amount of H_3O^+ compared with the H_3O^+ obtained from the first dissociation. Of the 9.3×10^{-5} mol/L of HCO_3^- produced by the first dissociation, only 5.6×10^{-11} mol/L dissociates to form H_3O^+ and CO_3^{2-}.

Step 5. Calculate the pH.

$$\text{pH} = -\log[H_3O^+] = -\log(9.3 \times 10^{-5}) = 4.03$$

CHECK

Check your arithmetic by substituting the equilibrium concentrations into the expression for K_{a1} to make sure the value is equal to the one given in Table 16.3.

$$K_{a1} = \frac{[H_3O^+][HCO_3^-]}{[H_2CO_3]} = \frac{(9.3 \times 10^{-5})(9.3 \times 10^{-5})}{(0.020)} = 4.3 \times 10^{-7}$$

▶ **PRACTICE 16.21**   Calculate the pH and the concentrations of all species present in 0.10 M ascorbic acid ($H_2C_6H_6O_6$) solution. ($K_{a1} = 8.0 \times 10^{-5}$, $K_{a2} = 1.6 \times 10^{-12}$)

▶ **APPLY 16.22**   Carbonated drinks are prepared by dissolving CO_2 under high pressure. CO_2 reacts with water to produce carbonic acid according to the equation

$$CO_2(g) + H_2O(l) \rightleftharpoons CO_2(aq) + H_2O(l) \rightleftharpoons H_2CO_3(aq)$$

Calculate the pH and the concentrations of all species present (H_2CO_3, HCO_3^-, CO_3^{2-}, H_3O^+, and OH^-) in a can of carbonated water with $[H_2CO_3] = 0.16$ M. K_a values for H_2CO_3 are in Table 16.3.

Sulfuric acid (H_2SO_4), a strong acid, differs from most other polyprotic acids because it has a very large value of K_{a1}. Essentially all the H_2SO_4 molecules dissociate to give H_3O^+ and HSO_4^- ions, but only a fraction of the resulting HSO_4^- ions dissociate to give additional H_3O^+ ions and SO_4^{2-} ions. The pH and concentrations of all species present (HSO_4^-, SO_4^{2-}, H_3O^+, and OH^-) in 0.10 M H_2SO_4 can be calculated using the five-step procedure summarized in Figure 16.7.

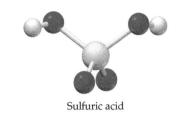

Sulfuric acid

Step 1. Write the balanced equation for the principal acid–base reaction. Due to the large value of K_{a1}, H_2SO_4 is essentially 100% dissociated to give H_3O^+ and HSO_4^-. After the first dissociation the species present are $[H_3O^+] = 0.10$ M, $[HSO_4^-] = 0.10$ M, and H_2O. Because K_{a2} (1.2×10^{-2}) $\gg$ K_w (1.0×10^{-14}), we consider the principal reaction to be dissociation of HSO_4^-.

$$HSO_4^-(aq) + H_2O(aq) \rightleftharpoons H_3O^+(aq) + SO_4^{2-}(aq) \quad K_{a2} = 1.2 \times 10^{-2}$$

Step 2. Make a table that lists initial concentration, change in concentration, and equilibrium concentration for each of the reactants and products in the balanced chemical equation.

Principal reaction	$HSO_4^-(aq)$ + $H_2O(l)$ $\rightleftharpoons$ $H_3O^+(aq)$ + $SO_4^{2-}(aq)$		
Initial concentration (M)	0.10	0.10	0
Change (M)	$-x$	$+x$	$+x$
Equilibrium concentration (M)	$0.10 - x$	$0.10 + x$	x

Note that the second dissociation step takes place in the presence of 0.10 M H_3O^+ from the first dissociation step.

Step 3. Substitute the equilibrium concentrations into the equilibrium equation for the principal reaction, and solve for x.

$$K_{a2} = 1.2 \times 10^{-2} = \frac{[H_3O^+][SO_4^{2-}]}{[HSO_4^-]} = \frac{(0.10 + x)(x)}{(0.10 - x)}$$

Neglecting x compared with 0.10 and solving this equation would give $x = K_{a2} = 0.012$, which is not negligible compared with 0.10. Therefore, we use the quadratic equation to obtain the value of x:

$$0.0012 - 0.012x = 0.10x + x^2$$

$$x^2 + 0.112x - 0.0012 = 0$$

$$x = \frac{-b \pm \sqrt{b^2 - 4ac}}{2a} = \frac{-0.112 \pm \sqrt{(0.112)^2 - 4(1)(-0.0012)}}{2(1)}$$

$$= \frac{-0.112 \pm 0.132}{2}$$

$$= +0.010 \quad \text{or} \quad -0.122$$

Because x is the SO_4^{2-} concentration, it must be positive. Therefore,

$$x = 0.010$$

Step 4. Calculate the concentrations of all reactants and products involved in the reactions.

$$[SO_4^{2-}] = x = 0.010 \text{ M}$$

$$[HSO_4^-] = 0.10 - x = 0.10 - 0.010 = 0.09 \text{ M}$$

$$[H_3O^+] = 0.10 + x = 0.10 + 0.010 = 0.11 \text{ M}$$

$$[OH^-] = \frac{K_w}{[H_3O^+]} = \frac{1.0 \times 10^{-14}}{0.11} = 9.1 \times 10^{-14} \text{ M}$$

▲ The brightness of the planet Venus is due in part to thick, highly reflective clouds in its upper atmosphere. These clouds consist of sulfur dioxide and droplets of sulfuric acid.

Step 5. Calculate the pH.

$$pH = -\log[H_3O^+] = -\log 0.11 = 0.96$$

The easiest check in this case is to substitute the concentrations obtained in Step 4 into the equilibrium equation for the second dissociation step and show that the equilibrium constant expression equals K_{a2}:

$$\frac{[H_3O^+][SO_4^{2-}]}{[HSO_4^-]} = \frac{(0.11)(0.010)}{0.09} = 0.012 = K_{a2}$$

Since $[H_3O^+]$ is approximately 10^{-1} M, the pH should be about 1, in agreement with the detailed solution.

16.12 EQUILIBRIA IN SOLUTIONS OF WEAK BASES

Weak bases, such as ammonia, accept a proton from water to give the conjugate acid of the base and OH^- ions:

$$NH_3(aq) + H_2O(l) \rightleftharpoons NH_4^+(aq) + OH^-(aq)$$

The equilibrium reaction of any general base B with water is characterized by the equilibrium equation and the equilibrium constant is called the **base-dissociation constant, K_b.**

Base-dissociation constant, K_b

$$B(aq) + H_2O(l) \rightleftharpoons BH^+(aq) + OH^-(aq) \quad K_b = \frac{[BH^+][OH^-]}{[B]}$$

As usual, $[H_2O]$ is omitted from the equilibrium constant expression. **TABLE 16.4** lists some typical weak bases and gives their K_b values at 25 °C (The term *base-protonation constant* might be a more descriptive name for K_b, but the term *base-dissociation constant* is still widely used.)

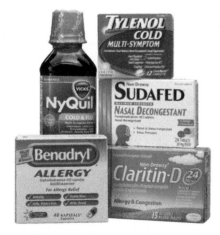

▲ Many over-the-counter drugs contain salts formed from amines and hydrochloric acid.

TABLE 16.4 K_b Values for Some Weak Bases and K_a Values for Their Conjugate Acids at 25 °C

Base	Formula, B	K_b	Conjugate Acid, BH$^+$	K_a
Ammonia	NH_3	1.8×10^{-5}	NH_4^+	5.6×10^{-10}
Aniline	$C_6H_5NH_2$	4.3×10^{-10}	$C_6H_5NH_3^+$	2.3×10^{-5}
Dimethylamine	$(CH_3)_2NH$	5.4×10^{-4}	$(CH_3)_2NH_2^+$	1.9×10^{-11}
Hydrazine	N_2H_4	8.9×10^{-7}	$N_2H_5^+$	1.1×10^{-8}
Hydroxylamine	NH_2OH	9.1×10^{-9}	NH_3OH^+	1.1×10^{-6}
Methylamine	CH_3NH_2	3.7×10^{-4}	$CH_3NH_3^+$	2.7×10^{-11}

Many weak bases are organic compounds called *amines,* derivatives of ammonia in which one or more hydrogen atoms are replaced by an organic, carbon-based group, such as a methyl group (CH_3). Methylamine (CH_3NH_2), for example, is an organic amine responsible for the odor of rotting fish.

Amines are organic compounds in which one, two, or three substituents, such as a methyl group, are attached to a basic nitrogen atom.

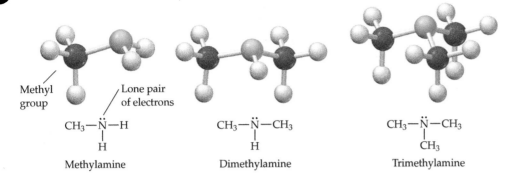

Methyl group

Lone pair of electrons

$$CH_3—\overset{..}{N}—H$$
$$|$$
$$H$$

Methylamine

$$CH_3—\overset{..}{N}—CH_3$$
$$|$$
$$H$$

Dimethylamine

$$CH_3—\overset{..}{N}—CH_3$$
$$|$$
$$CH_3$$

Trimethylamine

The basicity of an amine is due to the lone pair of electrons on the nitrogen atom, which can form a coordinate covalent bond with a proton.

Equilibria in solutions of weak bases are treated by the same procedure used for solving problems involving weak acids. Worked Example 16.12 illustrates the procedure.

Go to eText

WORKED EXAMPLE 16.12

Calculating the pH and the Equilibrium Concentrations in a Solution of a Weak Base

Codeine ($C_{18}H_{21}NO_3$), a drug used in painkillers and cough medicines, is a naturally occurring amine that has $K_b = 1.6 \times 10^{-6}$. Calculate the pH and the concentrations of all species present in a 0.0012 M solution of codeine.

IDENTIFY

Known	Unknown
Concentration of codeine (0.0012 M)	pH
$K_b = 1.6 \times 10^{-6}$	Equilibrium concentrations

STRATEGY

Use the procedure outlined in Figure 16.7.

SOLUTION

Step 1. Write the balanced equation for the principal acid–base reaction. Let's use Cod as an abbreviation for codeine and CodH$^+$ for its conjugate acid. The species present initially are Cod (base) and H_2O (acid or base).

There are two possible proton-transfer reactions:

$$Cod(aq) + H_2O(l) \rightleftharpoons CodH^+(aq) + OH^-(aq) \qquad K_b = 1.6 \times 10^{-6}$$

$$H_2O(l) + H_2O(l) \rightleftharpoons H_3O^+(aq) + OH^-(aq) \qquad K_w = 1.0 \times 10^{-14}$$

Since Cod is a much stronger base than $H_2O(K_b \gg K_w)$, the principal reaction involves the protonation of codeine.

Step 2. Make a table that lists initial concentration, change in concentration, and equilibrium concentration for each of the reactants and products in the balanced chemical equation.

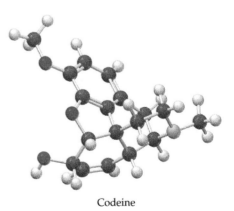

Codeine

Principal reaction	$Cod(aq)$	$+ H_2O(l)$	$\rightleftharpoons CodH^+(aq)$	$+ OH^-(aq)$
Initial concentration (M)	0.0012		0	~0
Change (M)	$-x$		$+x$	$+x$
Equilibrium concentration (M)	$0.0012 - x$		x	x

continued on next page

Step 3. Substitute the equilibrium concentrations into the equilibrium equation for the principal reaction, and solve for x. The value of x is obtained from the equilibrium equation:

$$K_b = 1.6 \times 10^{-6} = \frac{[CodH^+][OH^-]}{[Cod]} = \frac{(x)(x)}{(0.0012 - x)}$$

Assuming that $(0.0012 - x) \approx 0.0012$,

$$x^2 = (1.6 \times 10^{-6})(0.0012)$$
$$x = 4.4 \times 10^{-5}$$

To check the assumption that x can be neglected, calculate the percentage that the calculated value of x is of the initial concentration.

$$\frac{4.4 \times 10^{-5}}{0.0012} \times 100\% = 3.7\%$$

In this case, the assumption is valid because 3.7% is less than 5%.

Step 4. Calculate the concentrations of all reactants and products involved in the reactions.

$$[CodH^+] = [OH^-] = x = 4.4 \times 10^{-5} \text{ M}$$
$$[Cod] = 0.0012 - x = 0.0012 - 0.000044 = 0.0012 \text{ M}$$

The dissociation of water equilibrium is used to calculate $[H_3O^+]$.

$$[H_3O^+] = \frac{K_w}{[OH^-]} = \frac{1.0 \times 10^{-14}}{4.4 \times 10^{-5}} = 2.3 \times 10^{-10} \text{ M}$$

Step 5. Calculate the pH. $pH = -\log[H_3O^+] = -\log(2.3 \times 10^{-10}) = 9.64$

CHECK

The pH is greater than 7, as expected for a solution of a weak base. Check your arithmetic by substituting the equilibrium concentrations into the expression for K_b to make sure the value is the same as the given value (1.6×10^{-6}).

$$K_b = \frac{[CodH^+][OH^-]}{[Cod]} = \frac{(4.4 \times 10^{-5})(4.4 \times 10^{-5})}{(0.0012)} = 1.6 \times 10^{-6}$$

▶ **PRACTICE 16.23**   Calculate the pH and the concentrations of all species present in 0.40 M NH_3 ($K_b = 1.8 \times 10^{-5}$).

▶ **APPLY 16.24**   Lactated Ringer's solution is given intravenously to replenish fluids in patients who have experienced significant blood loss. The solution contains several different ions including sodium, potassium, chloride, calcium, and lactate ($C_3H_5O_3^-$). Lactate is the only species that affects the pH. The solution has a lactate concentration of 0.028 M and pH of 8.16. What is the value of K_b for lactate?

16.13 RELATION BETWEEN K_a AND K_b

We've seen in previous sections that the strength of an acid can be expressed by its value of K_a and the strength of a base can be expressed by its value of K_b. For a conjugate acid–base pair, the two equilibrium constants are related in a simple way that makes it possible to calculate either one from the other. Let's consider the conjugate acid–base pair NH_4^+ and NH_3, for example, where K_a refers to proton transfer from the acid NH_4^+ to water and K_b refers to proton transfer from water to the base NH_3. The sum of the two reactions is simply the dissociation of water:

$$NH_4^+(aq) + H_2O(l) \rightleftharpoons H_3O^+(aq) + NH_3(aq) \quad K_a = \frac{[H_3O^+][NH_3]}{[NH_4^+]} = 5.6 \times 10^{-10}$$

$$NH_3(aq) + H_2O(l) \rightleftharpoons NH_4^+(aq) + OH^-(aq) \quad K_b = \frac{[NH_4^+][OH^-]}{[NH_3]} = 1.8 \times 10^{-5}$$

Net: $\quad 2\,H_2O(l) \rightleftharpoons H_3O^+(aq) + OH^-(aq) \quad K_w = [H_3O^+][OH^-] = 1.0 \times 10^{-14}$

The equilibrium constant for the net reaction equals the product of the equilibrium constants for the reactions added:

$$K_a \times K_b = \frac{[H_3O^+][NH_3]}{[NH_4^+]} \times \frac{[NH_4^+][OH^-]}{[NH_3]} = [H_3O^+][OH^-] = K_w$$

$$= (5.6 \times 10^{-10})(1.8 \times 10^{-5}) = 1.0 \times 10^{-14}$$

For any conjugate acid–base pair, the product of the acid-dissociation constant for the acid and the base-dissociation constant for the base always equals the ion-product constant for water:

$$\boxed{K_a \times K_b = K_w}$$

As the strength of an acid increases (larger K_a), the strength of its conjugate base decreases (smaller K_b) because the product $K_a \times K_b$ remains constant at 1.0×10^{-14}. This inverse relationship between the strength of an acid and the strength of its conjugate base was illustrated qualitatively in Table 16.1. Taking the negative base-10 logarithm of both sides of the equation $K_a \times K_b = K_w$ gives another useful relationship:

$$pK_a + pK_b = pK_w = 14.00$$

where $pK_a = -\log K_a$, $pK_b = -\log K_b$, and $pK_w = -\log K_w$.

Compilations of equilibrium constants, such as Appendix C, generally list either K_a or K_b, but not both, because K_a is easily calculated from K_b and vice versa:

$$K_a = \frac{K_w}{K_b} \quad \text{and} \quad K_b = \frac{K_w}{K_a}$$

Similarly, pK_a can be calculated from pK_b and vice versa:

$$pK_a = 14.00 - pK_b \quad \text{and} \quad pK_b = 14.00 - pK_a$$

WORKED EXAMPLE 16.13

Relating K_a, K_b, pK_a, and pK_b

(a) K_b for trimethylamine is 6.5×10^{-5}. Calculate K_a for the trimethylammonium ion, $(CH_3)_3NH^+$.

(b) K_a for HCN is 4.9×10^{-10}. Calculate K_b for CN^-.

(c) Pyridine (C_5H_5N), an organic solvent, has $pK_b = 8.74$. What is the value of pK_a for the pyridinium ion, $C_5H_5NH^+$?

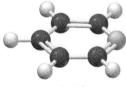

Pyridine

STRATEGY

To calculate K_a from K_b (or vice versa), use the equation $K_a = K_w/K_b$ or $K_b = K_w/K_a$. To calculate pK_a from pK_b, use the equation $pK_a = 14.00 - pK_b$.

SOLUTION

(a) K_a for $(CH_3)_3NH^+$ is the equilibrium constant for the acid-dissociation reaction

$$(CH_3)_3NH^+(aq) + H_2O(l) \rightleftharpoons H_3O^+(aq) + (CH_3)_3N(aq)$$

continued on next page

Because $K_a = K_w/K_b$, we can find K_a for $(CH_3)_3NH^+$ from K_b for its conjugate base $(CH_3)_3N$:

$$K_a = \frac{K_w}{K_b} = \frac{1.0 \times 10^{-14}}{6.5 \times 10^{-5}} = 1.5 \times 10^{-10}$$

(b) K_b for CN^- is the equilibrium constant for the base-protonation reaction

$$CN^-(aq) + H_2O(l) \rightleftharpoons HCN(aq) + OH^-(aq)$$

Because $K_b = K_w/K_a$, we can find K_b for CN^- from K_a for its conjugate acid HCN:

$$K_b = \frac{K_w}{K_a} = \frac{1.0 \times 10^{-14}}{4.9 \times 10^{-10}} = 2.0 \times 10^{-5}$$

(c) We can find pK_a for $C_5H_5NH^+$ from pK_b for C_5H_5N:

$$pK_a = 14.00 - pK_b = 14.00 - 8.74 = 5.26$$

▶ **PRACTICE 16.25** Piperidine $(C_5H_{11}N)$ is an amine found in black pepper. Find K_b for piperidine in Appendix C, and then calculate K_a for the $C_5H_{11}NH^+$ cation.

Piperidine

▶ **CONCEPTUAL APPLY 16.26** The following pictures represent aqueous solutions of three acids HA(A = X, Y, or Z); water molecules have been omitted for clarity:

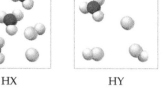

HX   HY   HZ

(a) Which acid has the largest value of K_a?
(b) Which conjugate base $(A^- = X^-, Y^-, \text{ or } Z^-)$ has the largest value of K_b?
(c) Which A^- ion is the weakest base?

16.14 ACID–BASE PROPERTIES OF SALTS

REMEMBER . . .
A neutralization reaction between an acid and a base forms H_2O and a **salt**, which contains the cation from the base and the anion from the acid (Section 4.7).

Recall from Section 4.7 that when an acid neutralizes a base an ionic compound called a **salt** is formed. Salt solutions can be neutral, acidic, or basic, depending on the acid–base properties of the constituent cations and anions (**FIGURE 16.9**). As a general rule, salts formed by the reaction of a strong acid with a strong base give neutral solutions,

▶ **FIGURE 16.9**

Aqueous salt solutions (0.10 M) of NaCl, NH₄Cl, AlCl₃, NaCN, and (NH₄)₂CO₃. A few drops of universal indicator have been added to each solution. Universal indicator is red at pH 3, yellow at pH 5, green at pH 7, and blue at pH ≥ 10.

$NH_4Cl(aq)$: pH = 5
$AlCl_3(aq)$: pH = 3
$NaCN(aq)$: pH > 10
$NaCl(aq)$: pH = 7
$(NH_4)_2CO_3(aq)$: pH > 10

salts formed by the reaction of a strong acid with a weak base give acidic solutions, and salts formed by the reaction of a weak acid with a strong base give basic solutions. It's as if the influence of the stronger partner dominates:

$$\text{Strong acid} + \text{Strong base} \longrightarrow \textbf{Neutral solution}$$
$$\text{Strong acid} + \text{Weak base} \longrightarrow \textbf{Acidic solution}$$
$$\text{Weak acid} + \text{Strong base} \longrightarrow \text{Basic solution}$$

Salts That Yield Neutral Solutions

Salts such as NaCl that are derived from a strong base (NaOH) and a strong acid (HCl) yield neutral solutions because neither the cation nor the anion reacts appreciably with water to produce H_3O^+ or OH^- ions. As the conjugate base of a strong acid, Cl^- has no tendency to make the solution basic by picking up a proton from water. As the cation of a strong base, the hydrated Na^+ ion has only a negligible tendency to make the solution acidic by transferring a proton to a solvent water molecule.

The following ions do not react appreciably with water to produce either H_3O^+ or OH^- ions:

- **Cations from strong bases:**

  Alkali metal cations of group 1A (Li^+, Na^+, K^+)

  Alkaline earth cations of group 2A (Mg^{2+}, Ca^{2+}, Sr^{2+}, Ba^{2+}), except for Be^{2+}

- **Anions from strong monoprotic acids:**

  Cl^-, Br^-, I^-, NO_3^-, and ClO_4^-

Salts that contain only these ions give neutral solutions in pure water (pH = 7).

Salts That Yield Acidic Solutions

Salts such as NH_4Cl that are derived from a weak base (NH_3) and a strong acid (HCl) produce acidic solutions. In such a case, the anion is neither an acid nor a base, but the cation is a weak acid:

$$NH_4^+(aq) + H_2O(l) \rightleftharpoons H_3O^+(aq) + NH_3(aq)$$

Related ammonium salts derived from amines, such as $[CH_3NH_3]Cl$, $[(CH_3)_2NH_2]Cl$, and $[(CH_3)_3NH]Cl$, also give acidic solutions because they too have cations with at least one dissociable proton. The pH of a solution that contains an acidic cation can be calculated by the standard procedure outlined in Figure 16.7. For a 0.10 M NH_4Cl solution, the pH is 5.12. Although the reaction of a cation or anion of a salt with water to produce H_3O^+ or OH^- ions is sometimes called a *salt hydrolysis reaction*, there is no fundamental difference between a salt hydrolysis reaction and any other Brønsted–Lowry acid–base reaction.

Another type of acidic cation is the hydrated cation of a small, highly charged metal ion, such as Al^{3+}. In aqueous solution, the Al^{3+} ion bonds to six water molecules to give the hydrated cation $Al(H_2O)_6^{3+}$. As shown in **FIGURE 16.10**, the negative (oxygen) end of each dipolar water molecule bonds to the positive metal cation, and the six water molecules are located at the vertices of a regular octahedron.

All metal ions exist in aqueous solution as hydrated cations, but the acidity of these cations varies greatly depending on the charge and size of the unhydrated metal ion. Because of the high (3+) charge on the Al^{3+} ion, electrons in the O—H bonds of the bound water molecules are attracted toward the Al^{3+} ion. The attraction is strong because the Al^{3+} ion is small and the electrons in the O—H bonds are relatively close to the center of positive charge. As a result, electron density shifts from the O—H bonds toward the Al^{3+} ion, thus weakening the O—H bonds and

▶ **FIGURE 16.10**

Octahedral structure of the $Al(H_2O)_6{}^{3+}$ cation.

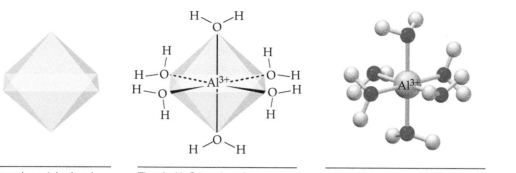

A regular octahedron has eight equilateral triangular faces and six vertices.	The six Al–O bonds point toward the six vertices of the octahedron.	A model of the $Al(H_2O)_6{}^{3+}$ cation, showing the octahedral arrangement of bonds to the six H_2O molecules.

increasing their polarity, which in turn eases the transfer of a proton to a solvent water molecule:

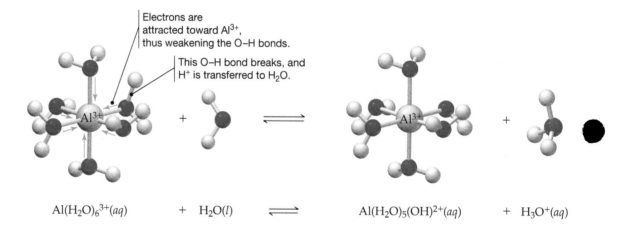

Electrons are attracted toward Al^{3+}, thus weakening the O–H bonds.

This O–H bond breaks, and H^+ is transferred to H_2O.

$$Al(H_2O)_6{}^{3+}(aq) \quad + \quad H_2O(l) \quad \rightleftharpoons \quad Al(H_2O)_5(OH)^{2+}(aq) \quad + \quad H_3O^+(aq)$$

The acid-dissociation constant for $Al(H_2O)_6{}^{3+}$, $K_a = 1.4 \times 10^{-5}$, is much larger than $K_w = 1.0 \times 10^{-14}$, which means that the water molecules in the hydrated cation are much stronger proton donors than are free solvent water molecules. In fact, the acid strength of $Al(H_2O)_6{}^{3+}$ is comparable to that of acetic acid, which has $K_a = 1.8 \times 10^{-5}$. In general, the acidity of hydrated main-group cations increases from left to right in the periodic table as the metal ion charge increases and the metal ion size decreases ($Li^+ < Be^{2+}$; $Na^+ < Mg^{2+} < Al^{3+}$). Transition metal cations, such as Zn^{2+}, Cr^{3+}, and Fe^{3+}, also give acidic solutions; their K_a values are listed in Table C.2 of Appendix C.

Salts That Yield Basic Solutions

Salts such as NaCN that are derived from a strong base (NaOH) and a weak acid (HCN) yield basic solutions. In this case, the cation is neither an acid nor a base but the anion is a weak base:

$$CN^-(aq) + H_2O(l) \rightleftharpoons HCN(aq) + OH^-(aq)$$

Other anions that exhibit basic properties are listed in Table 16.1 and include $NO_2{}^-$, F^-, $CH_3CO_2{}^-$, and $CO_3{}^{2-}$. The pH of an acidic or basic salt solution can be calculated by the standard procedure, as shown in Worked Example 16.14.

── WORKED EXAMPLE 16.14

Calculating the pH of a Salt Solution

Calculate the pH of

(a) a 0.10 M solution of $AlCl_3$; K_a for $Al(H_2O)_6^{3+}$ is 1.4×10^{-5}.
(b) a 0.10 M solution of NaCN; K_a for HCN is 4.9×10^{-10}.

STRATEGY

Determine if the salt will be acidic or basic by examining the cations and anions. The pH can be calculated by using the strategy for solving weak acid and base problems outlined in Figure 16.7.

SOLUTION (a)

Step 1 and 2. Write the balanced equation for the principal acid–base reaction, and set up the equilibrium table. The species present initially are $Al(H_2O)_6^{3+}$ (acid), Cl^- (inert), and H_2O (acid or base). Because $Al(H_2O)_6^{3+}$ is a much stronger acid than water ($K_a \gg K_w$), the principal reaction is dissociation of $Al(H_2O)_6^{3+}$:

Principal reaction	$Al(H_2O)_6^{3+}(aq) + H_2O(l) \rightleftharpoons H_3O^+(aq) + Al(H_2O)_5(OH)^{2+}(aq)$		
Initial concentration (M)	0.10	~0	0
Change (M)	$-x$	$+x$	$+x$
Equilibrium concentration (M)	$0.10 - x$	x	x

Step 3. Substitute the equilibrium concentrations into the equilibrium equation for the principal reaction, and solve for x.

$$K_a = 1.4 \times 10^{-5} = \frac{[H_3O^+][Al(H_2O)_5(OH)^{2+}]}{[Al(H_2O)_6^{3+}]} = \frac{(x)(x)}{(0.10-x)} \approx \frac{x^2}{0.10}$$

$$x = 1.2 \times 10^{-3} \text{ M}$$

Step 4 and 5. Find the equilibrium concentrations and calculate pH. The equilibrium table shows that in equilibrium $[H_3O^+] = x$.

$$[H_3O^+] = x = 1.2 \times 10^{-3} \text{ M} \quad \text{pH} = -\log (1.2 \times 10^{-3}) = 2.92$$

Thus, $Al(H_2O)_6^{3+}$ is a much stronger acid than NH_4^+, which agrees with the colors of the indicator in Figure 16.9.

SOLUTION (b)

Step 1 and 2. Write the balanced equation for the principal acid–base reaction and the initial, change, and equilibrium table. The species present initially are Na^+ (inert), CN^- (base), and H_2O (acid or base).

There are two possible proton-transfer reactions:

$$CN^-(aq) + H_2O(l) \rightleftharpoons HCN(aq) + OH^-(aq) \quad K_b$$
$$H_2O(l) + H_2O(l) \rightleftharpoons H_3O^+(aq) + OH^-(aq) \quad K_w$$

$K_b = K_w/(K_a \text{ for HCN}) = 2.0 \times 10^{-5}$. Because $K_b \gg K_w$, CN^- is a stronger base than H_2O and the principal reaction is proton transfer from H_2O to CN^-.

Principal reaction	$CN^-(aq) + H_2O(l) \rightleftharpoons HCN(aq) + OH^-(aq)$		
Initial concentration (M)	0.10	0	~0
Change (M)	$-x$	$+x$	$+x$
Equilibrium concentration (M)	$0.10 - x$	x	x

continued on next page

Step 3. Substitute the equilibrium concentrations into the equilibrium equation for the principal reaction, and solve for x.

$$K_b = 2.0 \times 10^{-5} = \frac{[\text{HCN}][\text{OH}^-]}{[\text{CN}^-]} = \frac{(x)(x)}{(0.10-x)} \approx \frac{x^2}{0.10}$$

$$x = 1.4 \times 10^{-3} \text{ M}$$

Step 4 and 5. Find equilibrium concentrations, and calculate pH. The equilibrium table shows that in equilibrium $x = [\text{OH}^-]$.

$$x = [\text{OH}^-] = 1.4 \times 10^{-3} \text{ M}$$

$$[\text{H}_3\text{O}^+] = \frac{K_w}{[\text{OH}^-]} = \frac{1.0 \times 10^{-14}}{1.4 \times 10^{-3}} = 7.1 \times 10^{-12}$$

$$\text{pH} = -\log(7.1 \times 10^{-12}) = 11.15$$

The solution is basic, which agrees with the color of the indicator in Figure 16.9.

▶ **PRACTICE 16.27** Predict whether a solution of 0.20 M NaNO_2 is neutral, acidic, or basic, and calculate the pH. (K_a for HNO_2 is 4.6×10^{-4})

▶ **APPLY 16.28**

(a) Calculate the pH and percent dissociation of $\text{Zn}(\text{H}_2\text{O})_6^{2+}$ prepared from a 0.40 M ZnCl_2 solution. K_a for $\text{Zn}(\text{H}_2\text{O})_6^{2+}$ is 2.5×10^{-10}.
(b) Which solution will have a higher percent dissociation 0.40 M $\text{Zn}(\text{H}_2\text{O})_6^{2+}$ or 0.40 M $\text{Fe}(\text{H}_2\text{O})_6^{3+}$?

Salts That Contain Acidic Cations and Basic Anions

Finally, let's look at a salt such as $(\text{NH}_4)_2\text{CO}_3$ in which both the cation and the anion can undergo proton-transfer reactions. Because NH_4^+ is a weak acid and CO_3^{2-} is a weak base, the pH of an $(\text{NH}_4)_2\text{CO}_3$ solution depends on the relative acid strength of the cation and base strength of the anion:

$$\text{NH}_4^+ (aq) + \text{H}_2\text{O}(l) \rightleftharpoons \text{H}_3\text{O}^+(aq) + \text{NH}_3(aq) \quad \text{Acid strength } (K_a)$$
$$\text{CO}_3^{2-} (aq) + \text{H}_2\text{O}(l) \rightleftharpoons \text{HCO}_3^-(aq) + \text{OH}^-(aq) \quad \text{Base strength } (K_b)$$

We can distinguish three possible cases:

- $K_a > K_b$: If K_a for the cation is greater than K_b for the anion, the solution will contain an excess of H_3O^+ ions (pH < 7).
- $K_a < K_b$: If K_a for the cation is less than K_b for the anion, the solution will contain an excess of OH^- ions (pH > 7).
- $K_a \approx K_b$: If K_a for the cation and K_b for the anion are comparable, the solution will contain approximately equal concentrations of H_3O^+ and OH^- ions (pH $\approx$ 7).

To determine whether an $(\text{NH}_4)_2\text{CO}_3$ solution is acidic, basic, or neutral, let's work out the values of K_a for NH_4^+ and K_b for CO_3^{2-}:

$$K_a \text{ for } \text{NH}_4^+ = \frac{K_w}{K_b \text{ for } \text{NH}_3} = \frac{1.0 \times 10^{-14}}{1.8 \times 10^{-5}} = 5.6 \times 10^{-10}$$

$$K_b \text{ for } \text{CO}_3^{2-} = \frac{K_w}{K_a \text{ for } \text{HCO}_3^-} = \frac{K_w}{K_{a2} \text{ for } \text{H}_2\text{CO}_3} = \frac{1.0 \times 10^{-14}}{5.6 \times 10^{-11}} = 1.8 \times 10^{-4}$$

Because $K_a < K_b$, the solution is basic (pH > 7), in accord with the color of the indicator in Figure 16.9.

TABLE 16.5 gives a summary of the acid–base properties of salts.

TABLE 16.5 ACID–BASE PROPERTIES OF SALTS

Type of Salt	Examples	Ions That React with Water	pH of Solution
Cation from strong base; anion from strong acid	$NaCl$, KNO_3, BaI_2	None	~7
Cation from weak base; anion from strong acid	NH_4Cl, NH_4NO_3, $[(CH_3)_3NH]Cl$	Cation	<7
Small, highly charged, cation; anion from strong acid	$AlCl_3$, $Cr(NO_3)_3$, $Fe(ClO_4)_3$	Hydrated cation	<7
Cation from strong base; anion from weak acid	$NaCN$, KF, Na_2CO_3	Anion	>7
Cation from weak base; anion from weak acid	NH_4CN, NH_4F, $(NH_4)_2CO_3$	Cation and anion	<7 if $K_a > K_b$ >7 if $K_a < K_b$ ~7 if $K_a \approx K_b$

16.15 LEWIS ACIDS AND BASES

In 1923, the same year in which Brønsted and Lowry defined acids and bases in terms of their proton donor/acceptor properties, the American chemist G. N. Lewis (1875–1946) proposed an even more general concept of acids and bases. Lewis noticed that when a base accepts a proton, it does so by sharing a lone pair of electrons with the proton to form a new covalent bond. Using ammonia as an example, the reaction can be written in the following format, in which the curved arrow represents the donation of the nitrogen lone pair to form a bond with H^+:

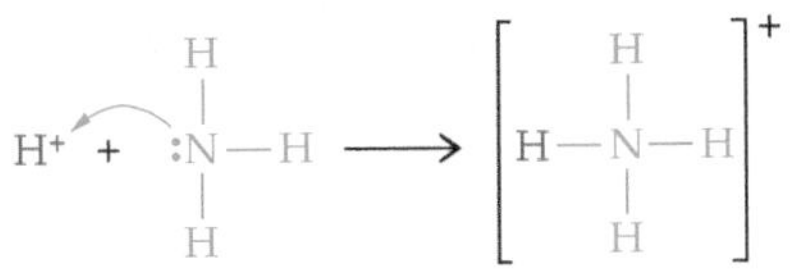

In this reaction, the proton behaves as an electron-pair acceptor and the ammonia molecule behaves as an electron-pair donor. Consequently, the Lewis definition of acids and bases states that a *Lewis acid* is an electron-pair acceptor and a *Lewis base* is an electron-pair donor.

> **Lewis acid**   An electron-pair acceptor
>
> **Lewis base**   An electron-pair donor

Because all proton acceptors have an unshared pair of electrons and all electron-pair donors can accept a proton, the Lewis and the Brønsted–Lowry definitions of a base are simply different ways of looking at the same property. All Lewis bases are Brønsted–Lowry bases, and all Brønsted–Lowry bases are Lewis bases. The Lewis definition of an acid, however, is considerably more general than the Brønsted–Lowry definition. Lewis acids include not only H^+ but also other cations and neutral molecules having vacant valence orbitals that can accept a share in a pair of electrons donated by a Lewis base.

Common examples of cationic Lewis acids are metal ions, such as Al^{3+} and Cu^{2+}. Hydration of the Al^{3+} ion, for example, is a Lewis acid–base reaction in which each of six H_2O molecules donates a pair of electrons to Al^{3+} to form the hydrated cation $Al(H_2O)_6{}^{3+}$:

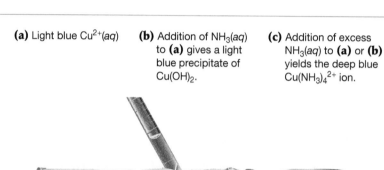

Lewis acid   Lewis base

Similarly, the reaction of Cu^{2+} ion with ammonia is a Lewis acid–base reaction in which each of four NH_3 molecules donates a pair of electrons to Cu^{2+} to form the deep blue $Cu(NH_3)_4{}^{2+}$ ion (**FIGURE 16.11**).

$$Cu^{2+} + 4:NH_3 \longrightarrow Cu(NH_3)_4{}^{2+}$$

Lewis acid   Lewis base

▶ **FIGURE 16.11**

The addition of aqueous ammonia to a solution of the $Cu^{2+}(aq)$ ion is a Lewis acid–base reaction.

(a) Light blue $Cu^{2+}(aq)$

(b) Addition of $NH_3(aq)$ to **(a)** gives a light blue precipitate of $Cu(OH)_2$.

(c) Addition of excess $NH_3(aq)$ to **(a)** or **(b)** yields the deep blue $Cu(NH_3)_4{}^{2+}$ ion.

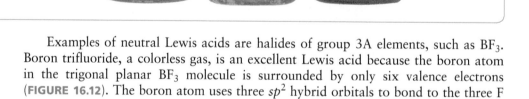

Examples of neutral Lewis acids are halides of group 3A elements, such as BF_3. Boron trifluoride, a colorless gas, is an excellent Lewis acid because the boron atom in the trigonal planar BF_3 molecule is surrounded by only six valence electrons (**FIGURE 16.12**). The boron atom uses three sp^2 hybrid orbitals to bond to the three F

▶ **FIGURE 16.12**

The reaction of the Lewis acid BF_3 with the Lewis base NH_3.

The electrostatic potential maps show that the B atom is electron poor (blue) and the N atom is electron rich (red).

The geometry about boron changes from trigonal planar in BF_3 to tetrahedral in the adduct. Both boron and nitrogen use sp^3 hybrid orbitals in the adduct.

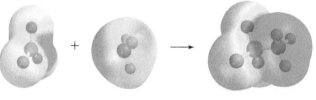

Lewis acid   Lewis base   Acid–base adduct

atoms and has a vacant $2p$ valence orbital that can accept a pair of electrons from a Lewis base, such as NH_3.

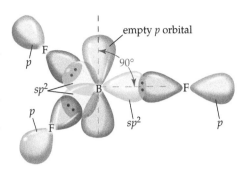

The Lewis acid and base sites are evident in electrostatic potential maps, which show the electron-poor B atom (blue) and the electron-rich N atom (red). In the product, called an *acid–base adduct,* the boron atom has acquired a stable octet of electrons.

Additional examples of neutral Lewis acids are oxides of nonmetals, such as CO_2, SO_2, and SO_3. The reaction of SO_3 with water, for example, can be viewed as a Lewis acid–base reaction in which SO_3 accepts a lone pair of electrons from a water molecule:

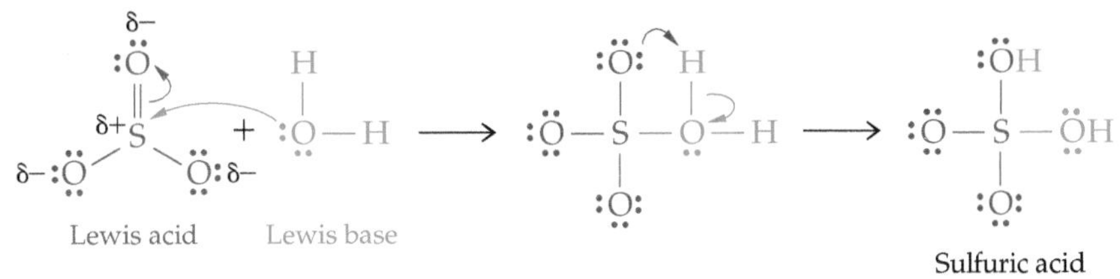

Because the sulfur–oxygen bonds are polar, with a partial positive charge ($\delta+$) on the less electronegative S atom, the S atom attracts an electron pair from H_2O. Formation of a bond from the water O atom to the S atom in the first step is helped along by the shift of a shared pair of electrons to oxygen. In the second step, a proton shifts from one oxygen atom to another, yielding sulfuric acid (H_2SO_4).

WORKED EXAMPLE 16.15

Identifying Lewis Acids and Bases

For each of the following reactions, identify the Lewis acid and the Lewis base.

(a) $CO_2 + OH^- \longrightarrow HCO_3^-$
(b) $B(OH)_3 + OH^- \longrightarrow B(OH)_4^-$
(c) $6\ CN^- + Fe^{3+} \longrightarrow Fe(CN)_6^{3-}$

STRATEGY

To identify the Lewis acid and the Lewis base, determine which molecule or ion can accept an electron pair (the Lewis acid) and which can donate an electron pair (the Lewis base).

SOLUTION

(a) The carbon atom of $O=C=O$ bears a partial positive charge ($\delta+$) because carbon is less electronegative than oxygen. Therefore, the carbon atom attracts an electron pair from OH^-. Formation of a covalent bond from OH^- to CO_2 is helped along by a shift of a shared electron pair to oxygen:

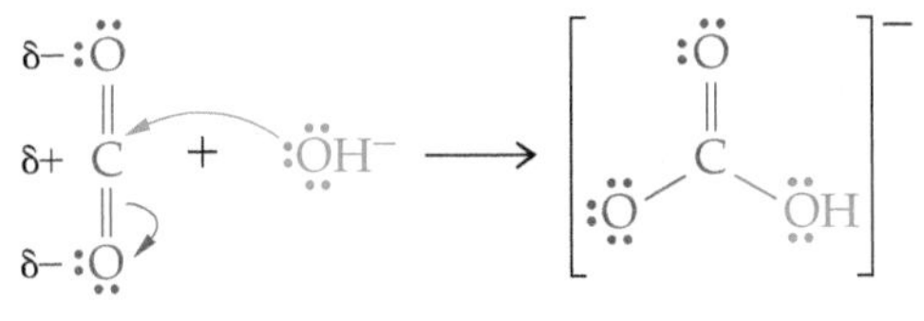

The Lewis acid (electron-pair acceptor) is CO_2; the Lewis base (electron-pair donor) is OH^-.

(b) The Lewis acid is boric acid, $B(OH)_3$, a weak acid and mild antiseptic used in eyewash. The boron atom in $B(OH)_3$ has a vacant valence orbital and completes its octet by accepting a pair of electrons from the Lewis base, OH^-.
(c) The Lewis acid is Fe^{3+}, and the Lewis base is CN^-. Each of the six $:C{\equiv}N:^-$ ions bond to the Fe^{3+} ion by donating a lone pair of electrons on the C atom.

▶ **PRACTICE 16.29** Identify the Lewis acid and Lewis base in the reaction $AlCl_3 + Cl^- \longrightarrow AlCl_4^-$.

▶ **CONCEPTUAL APPLY 16.30** For the following Lewis acid–base reaction, draw electron-dot structures for the reactants and products, and use the curved arrow notation to represent the donation of a lone pair of electrons from the Lewis base to the Lewis acid.

$$BeCl_2 + 2\ Cl^- \rightarrow BeCl_4^{2-}$$

INQUIRY  Has the problem of acid rain been solved?

Acid rain is one of the more important environmental issues of recent times. An aqueous solution is classified as acidic if it has a pH less than 7, but precipitation is classified as **acid rain** if the pH is less than 5.6. Under normal conditions, rain is slightly acidic (pH = 5.6) because carbon dioxide (CO_2) dissolves in raindrops and reacts with water to produce carbonic acid (H_2CO_3). However, because of power production, manufacturing, and automotive pollution, the pH of rainfall in many industrialized regions has decreased to between 3 and 4.5.

One major contributor to acid rain is sulfuric acid (H_2SO_4). Large power plants and smelters that burn sulfur-containing fossil fuels emit millions of tons of sulfur dioxide (SO_2) gas into the atmosphere, where some is oxidized by air to produce sulfur trioxide (SO_3). Sulfur oxides then dissolve in rain to form dilute sulfurous acid and sulfuric acid.

$$SO_2(g) + H_2O(l) \rightarrow H_2SO_3(aq) \quad \text{Sulfurous acid}$$

$$SO_3(g) + H_2O(l) \rightarrow H_2SO_4(aq) \quad \text{Sulfuric acid}$$

The other main contributor to acid rain is nitric acid (HNO_3). Nitrogen oxides are produced by the high-temperature reaction of N_2 and O_2 in coal-burning plants and in automobile engines. Nitrogen dioxide (NO_2) reacts in moist air to form nitric acid represented by a simplified reaction.

$$4\,NO_2(g) + 2\,H_2O(l) + O_2(g) \rightarrow 4\,HNO_3(aq)$$

Oxides of both sulfur and nitrogen have always been present in the atmosphere, produced by such natural sources as volcanoes and lightning, but the quantity of sulfur and nitrogen has increased dramatically over the past century as a result of industrialization.

The environmental consequences of acid rain can be devastating because many processes in nature do not operate outside a narrow pH range. Thousands of lakes in the Adirondack region of New York State and in southeastern Canada became so acidic that all fish life disappeared. Massive tree die-offs occurred throughout central and eastern Europe as acid rain lowered the pH and dissolved toxic metals in the soil. Countless marble statues have been slowly destroyed as acid rain dissolves calcium carbonate.

$$CaCO_3(s) + 2\,H^+(aq) \rightarrow Ca^{2+}(aq) + H_2O(l) + CO_2(g)$$

Fortunately, acidic emissions from automobiles and power plants have been greatly reduced in recent years as a result of the Clean Air Act and Acid Rain Program, which started in 1990. Nitrogen oxide emissions have been lowered by equipping automobiles with catalytic converters (Section 14.13), which catalyze the decomposition of nitrogen oxides to N_2 and O_2. Sulfur dioxide emissions from power plants have been reduced by scrubbing combustion products before they

▲ Before and after images showing the effects of acid rain on decorative building details from a city block of identical nineteenth-century brownstones. The building on the left has been restored to its original state, while the building on right shows acid rain damage.

are emitted from smokestacks. In this process an aqueous suspension of lime (CaO) is added to the combustion chamber and the stack. The lime reacts with SO_2 to give calcium sulfite ($CaSO_3$).

$$CaO(s) + SO_2(g) \rightarrow CaSO_3(s)$$

In some scrubbers, $CaSO_3$ is oxidized to $CaSO_4 \cdot 2\,H_2O$ (gypsum), which is used to make drywall in the building industry. The reduction in SO_2 emissions raised the average pH of rain in the eastern United States significantly from 1985 to 2016, as shown in **FIGURE 16.13**.

Unfortunately, it takes many years for lakes and streams to recover, even after the pH of rainwater returns to normal levels. One surprising and unanticipated long-term consequence of acid rain is the "jellification" of lakes. Jellification refers to the rising dominance of small jelly-clad invertebrates called *Holopedium glacialis*. Acid rain decreased the calcium levels in many lakes across eastern North America, which led to a decline in traditional plankton such as *Daphnia* (water fleas). Lower calcium levels hinder the ability of *Daphnia* to build their calcium-rich exoskeletons for protection. In contrast, *Holopedium glacialis* are able to use their jelly-like outer coating to shield themselves from predators and thrive in water with low calcium levels.

Although we have greatly reduced acid rain through environmental regulations, we cannot claim complete recovery from acidification. Lakes may have been pushed into an entirely new ecological state, and scientists predict that the rise in jelly-coated

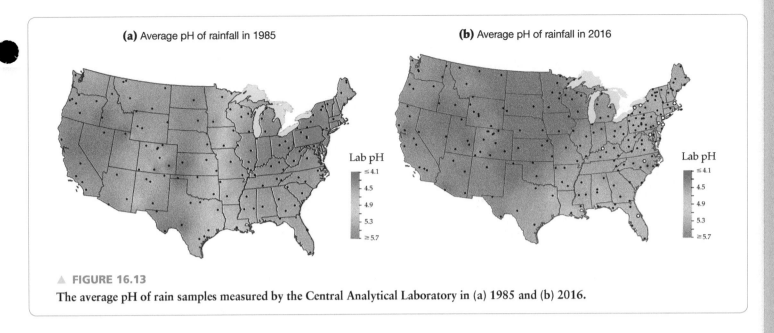

(a) Average pH of rainfall in 1985

(b) Average pH of rainfall in 2016

Lab pH

≤ 4.1
4.5
4.9
5.3
≥ 5.7

▲ FIGURE 16.13

The average pH of rain samples measured by the Central Analytical Laboratory in (a) 1985 and (b) 2016.

organisms will prevent vital nutrients from being passed up the food chain to fish stocks and clog filtration systems for drinking water. It will likely take thousands of years to return to historic calcium concentrations from natural weathering of rocks and minerals. Furthermore, acid rain continues to pose a large threat in countries like India and China where pollution-reducing technologies are not used.

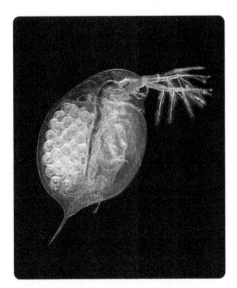

▲ A jelly-coated invertebrate (*Holopedium glacialis*) collected in a lake in Ontario, Canada.

PROBLEM 16.31   What are the chemical formulas and names of the two acids that are the largest contributors to acid rain?

PROBLEM 16.32   What were the average pH ranges for rainfall in the eastern United States in 1985 and 2016? Refer to Figure 16.13 to answer the question.

PROBLEM 16.33   One approach to mitigating the effects of acid rain is the addition of lime (CaO) to lakes and streams. This approach, called liming, has been widely used in Norway and Sweden but only infrequently in the United States. Write the reaction that occurs when CaO is added to acid.

PROBLEM 16.34

(a) Natural or "unpolluted" rain has a pH of 5.6. By what factor has the concentration of H_3O^+ increased in acid rain with a pH of 3.6?

(b) In southeastern Canada, the average pH of lakes is 5.1, well below the pH of 6.5 required for a healthy aquatic ecosystem. Calculate $[H_3O^+]$ and $[OH^-]$ in a lake with pH = 5.10.

PROBLEM 16.35   What is the pH of 1.00 L of rainwater that has dissolved 5.50 mg of NO_2? Assume that all of the NO_2 has reacted with water to give nitric acid, according to the reaction given in the text of the Inquiry section.

PROBLEM 16.36   The reaction of lime (CaO) with SO_2 in the scrubber of a power plant is a Lewis acid–base reaction. Explain.

PROBLEM 16.37   Because sulfur and nitrogen oxides are the main causes of acid rain, elevated levels of NO_3^- and SO_4^{2-} exist in droplets in the atmosphere. During dry weather, solid aerosol particles containing the salts NH_4NO_3 and $(NH_4)_2SO_4$ "dry" deposit on land and water. Are these salts acidic, basic, or neutral? (K_a and K_b values are in Appendix C.)

STUDY GUIDE

Section	Concept Summary	Learning Objectives	Test Your Understanding
16.1 Acid–Base Concepts: The Brønsted–Lowry Theory	According to the **Arrhenius** theory, acids (HA) are substances that dissociate in water to produce $H^+(aq)$ and bases (MOH) are substances that dissociate to yield $OH^-(aq)$. The more general **Brønsted–Lowry theory** defines an acid as a proton donor, a base as a proton acceptor, and an acid–base reaction as a proton-transfer reaction. Examples of Brønsted–Lowry acids are HCl, NH_4^+, and HSO_4^-; examples of Brønsted–Lowry bases are OH^-, F^-, and NH_3.	**16.1** Identify Arrhenius acids and bases and Brønsted–Lowry acids and bases and conjugate acid–base pairs.	Worked Examples 16.1–16.2; Problems 16.38, 16.48, 16.52
16.2 Acid Strength and Base Strength	A **strong acid** (HA) is nearly 100% dissociated, whereas a **weak acid** (HA) is only partially dissociated, existing as an equilibrium mixture of HA, H_3O^+, and A^-: $$HA(aq) + H_2O(l) \rightleftharpoons H_3O^+(aq) + A^-(aq)$$ The strength of an acid (HA) and the strength of its **conjugate base** (A^-) are inversely related. The H_3O^+ ion, a hydrated proton, is called the **hydronium ion**.	**16.2** Rank acids in order of increasing strength given a molecular image representing dissociation in solution.	Problem 16.39
		16.3 Predict the direction of a reaction based upon relative strengths of the conjugate acid–base pairs.	Worked Example 16.3; Problems 16.56–16.57
16.3 Factors That Affect Acid Strength	The acid strength of a binary acid (HA) increases down a group of the periodic table with decreasing strength of the H—A bond and increases from left to right across a row with increasing polarity of the H—A bond. The acid strength of an oxoacid, $H_nYO_m(Y = C, N, S, Cl)$, increases with increasing electronegativity and increasing oxidation number of the central atom Y.	**16.4** Predict the relative strengths of binary acids (HA) and oxoacids (H_nYO_m) based on their chemical structure.	Worked Example 16.4; Problems 16.40, 16.60, 16.62, 16.66
16.4 Dissociation of Water	Water, which can act both as an acid and as a base, undergoes the dissociation reaction $H_2O + H_2O \rightleftharpoons H_3O^+ + OH^-$. In pure water at 25 °C, $[H_3O^+] = [OH^-] = 1.0 \times 10^{-7}M$. The **ion-product constant for water**, K_w, is given by $K_w = [H_3O^+][OH^-] = 1.0 \times 10^{-14}$.	**16.5** Calculate the concentration of $[H_3O^+]$ or $[OH^-]$ using the value of K_w.	Worked Example 16.5; Problems 16.68, 16.70
16.5–16.6 The pH Scale and Measuring pH	The acidity of an aqueous solution is expressed on the **pH scale**, where $pH = -\log[H_3O^+]$. At 25 °C acidic solutions have $pH < 7$, basic solutions have $pH > 7$, and neutral solutions have $pH = 7$. The pH of a solution can be determined using an **acid–base indicator** or a pH meter.	**16.6** Calculate the pH of a solution given $[H_3O^+]$ or $[OH^-]$, or calculate $[H_3O^+]$ or $[OH^-]$ given the pH of a solution.	Worked Examples 16.6–16.7; Problems 16.74, 16.76
16.7 The pH in Solutions of Strong Acids and Strong Bases	**Strong acids** and **strong bases** are nearly 100% dissociated in aqueous solution. Therefore, $[H_3O^+]$ or $[OH^-]$ can be determined from the stoichiometry of the balanced dissociation reaction for the acid or base.	**16.7** Calculate the pH of a strong acid or base solution.	Worked Example 16.8; Problems 16.82, 16.84
16.8–16.10 Equilibria in Solutions of Weak Acids	The extent of dissociation of a weak acid HA is measured by its **acid-dissociation constant**, (K_a). The five-step process outlined in Figure 16.7 is used to solve equilibrium problems involving weak acids. The quadratic equation is used to solve for $[H_3O^+]$ when HA dissociates to an extent that significantly alters its initial concentration. This is the case when the initial concentration of HA is small or K_a is large.	**16.8** Calculate the value for K_a of a weak acid given the initial concentration of the acid and its equilibrium pH.	Worked Example 16.9; Problems 16.90, 16.92
		16.9 Given the K_a value of a weak acid and its initial concentration, calculate the pH, percent dissociation, and the concentration of all species present in solution.	Worked Example 16.10; Problems 16.94, 16.96

Section	Concept Summary	Learning Objectives	Test Your Understanding
16.11 Polyprotic Acids	Polyprotic acids contain more than one dissociable proton and dissociate in a stepwise manner. Because the stepwise dissociation constants decrease in the order $K_{a1} \gg K_{a2} \gg K_{a3}$, nearly all the H_3O^+ in a polyprotic acid solution comes from the first dissociation step.	**16.10** Calculate the pH and the concentration of all species present in a solution of a diprotic acid.	Worked Example 16.11; Problems 16.106, 16.108
16.12–16.13 Equilibria in Solutions of Weak Bases and the Relation between K_a and K_b	The extent of dissociation of a weak base B is measured by its **base-dissociation constant, K_b**. Examples of weak bases are NH_3 and derivatives of NH_3 called *amines*. The five-step process outlined in Figure 16.7 is used to solve equilibrium problems involving weak bases. For any conjugate acid–base pair, (K_a for the acid) × (K_b for the base) = K_w.	**16.11** Calculate the pH and equilibrium concentrations in a solution of a weak base.	Worked Example 16.12; Problems 16.114, 16.116, 16.118
		16.12 Relate K_a, K_b, pK_a, and pK_b for a conjugate acid–base pair.	Worked Example 16.13; Problems 16.120, 16.122
16.14 Acid–Base Properties of Salts	Aqueous solutions of salts can be neutral, acidic, or basic, depending on the acid–base properties of the constituent ions. Group 1A and 2A cations (except Be^{2+}) and anions that are conjugate bases of strong acids, such as Cl^-, do not react appreciably with water to produce H_3O^+ or OH^- ions. Cations that are conjugate acids of weak bases, such as NH_4^+, and hydrated cations of small, highly charged metal ions, such as Al^{3+}, yield acidic solutions. Anions that are conjugate bases of weak acids, such as CN^-, yield basic solutions.	**16.13** Predict whether a salt is acidic, basic, or neutral and calculate the pH of the salt solution.	Worked Example 16.14; Problems 16.44, 16.124, 16.128, 16.133–16.134
16.15 Lewis Acids and Bases	A **Lewis acid** is an electron-pair acceptor, and a **Lewis base** is an electron-pair donor. Lewis acids include not only H^+ but also other cations and neutral molecules that can accept a share in a pair of electrons from a Lewis base. Examples of Lewis acids are Al^{3+}, Cu^{2+}, BF_3, SO_3, and CO_2.	**16.14** Identify the Lewis acid and Lewis base and use curved arrow notation to indicate donation of a lone pair of elections in Lewis acid–base reaction.	Worked Example 16.15; Problems 16.45–16.47, 16.142

KEY TERMS

acid–base indicator *668*
acid-dissociation
 constant (K_a) *671*
acid rain *694*
Arrhenius acid *655*
Arrhenius base *655*
base-dissociation
 constant (K_b) *682*

Brønsted–Lowry acid *655*
Brønsted–Lowry base *655*
Brønsted–Lowry theory *655*
conjugate acid *656*
conjugate acid–base pair *655*
conjugate base *655*
dissociation of water *664*
hydronium ion, H_3O^+ *656*

ion-product constant for
 water (K_w) *664*
Lewis acid *691*
Lewis base *691*
percent dissociation *677*
pH *666*
polyprotic acid *678*
principal acid-base reaction *674*

strong acid *659*
strong bases *669*
subsidiary reaction *674*
weak acid *659*

KEY EQUATIONS

- **Ion–Product Constant for Water (Section 16.4)**

  $K_w = [H_3O^+][OH^-] = 1.0 \times 10^{-14}$ at 25 °C

- **Definition of pH (Section 16.5)**

  $pH = -\log [H_3O^+]$ or $[H_3O^+] = 10^{-pH}$

- **Acid–Dissociation Constant for a Weak Acid, HA (Section 16.8)**

  $K_a = \dfrac{[H_3O^+][A^-]}{[HA]}$ $pK_a = -\log K_a$

- **Percent Dissociation of a Weak Acid (Section 16.10)**

  $\text{Percent dissociation} = \dfrac{[HA]_{\text{dissociated}}}{[HA]_{\text{initial}}} \times 100\%$

- Base–Dissociation or Base–Protonation Constant for a Weak Base, B (Section 16.12)

$$K_b = \frac{[BH^+][OH^-]}{[B]} \quad pK_b = -\log K_b$$

- Relation between K_a and K_b for a Conjugate Acid–Base Pair (Section 16.13)

$$K_a \times K_b = K_w \quad pK_a + pK_b = pK_w = 14.00$$

PRACTICE TEST

After studying this chapter, you can assess your understanding with these practice test questions, which are correlated with chapter learning objectives. If you answer a question incorrectly, refer to the learning objectives in the end-of-chapter Study Guide for assistance. The Study Guide provides a conceptual summary, references a Worked Example to model how to solve the problem, and gives additional problems for more practice.

1. Which of the following is a Brønsted-Lowry base, but not an Arrhenius base? (**LO 16.1**)
   (a) HNO_3 (b) CsOH
   (c) CH_3NH_2 (d) CH_3OH

2. The following pictures represent equal volumes of aqueous solutions of three acids HA (A = X, Y, or Z); water molecules have been omitted for clarity. Which is the strongest acid? (**LO 16.2**)

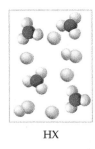 

     HX   HY   HZ

   (a) HX
   (b) HY
   (c) HZ
   (d) All three acids are strong acids and have equal strength.

3. Consider the conjugate bases, (X^-, Y^-, Z^-) in Problem 2. If you mix equal concentrations of reactants and products, which of the following reactions will proceed to the left? (**LO 16.3**)
   (a) $HX + Y^- \rightleftharpoons HY + X^-$
   (b) $HX + Z^- \rightleftharpoons HZ + X^-$
   (c) $HY + X^- \rightleftharpoons HX + Y^-$
   (d) $HZ + Y^- \rightleftharpoons HY + Z^-$

4. Which is the strongest acid? (**LO 16.4**)
   (a) $HClO_3$ (b) $HBrO_3$
   (c) H_2SO_3 (d) H_2TeO_3

5. What is the concentration of hydroxide ions $[OH^-]$ in a glass of wine with pH = 3.64? (**LO 16.5, 16.6**)
   (a) 2.3×10^{-4} M   (b) 6.4×10^{-3} M
   (c) 6.8×10^{-9} M   (d) 4.4×10^{-11} M

6. What is the pH of an aqueous solution of $Ca(OH)_2$ at 25.0 °C with a concentration of 6.3×10^{-5} M? (**LO 16.7**)
   (a) 4.20   (b) 10.10
   (c) 11.36   (d) 9.80

7. An acid solution with a concentration of 0.500 M has a pH = 3.21. What is the K_a of the acid? (**LO 16.8**)
   (a) 1.2×10^{-5} (b) 1.7×10^{-6}
   (c) 7.6×10^{-7} (d) 5.4×10^{-3}

8. What is the pH of a 1.5 M $C_6H_5CO_2H$ solution if K_a of $C_6H_5CO_2H$ is 6.5×10^{-5}? (**LO 16.9**)
   (a) 4.01   (b) 5.32
   (c) 3.14   (d) 2.01

9. Determine the following concentrations for a 0.40 M H_2Se solution that has the stepwise dissociation constants of $K_{a1} = 1.3 \times 10^{-4}$ and $K_{a2} = 1.0 \times 10^{-11}$. (**LO 16.10**)
   (a) $[H_2Se] = 0.35, [HSe^-] = 5.0 \times 10^{-2}, [H_3O^+] = 3.0 \times 10^{-3}, [Se^{2-}] = 1.3 \times 10^{-4}$
   (b) $[H_2Se] = 0.39, [HSe^-] = 7.2 \times 10^{-3}, [H_3O^+] = 7.2 \times 10^{-3}, [Se^{2-}] = 1.0 \times 10^{-11}$
   (c) $[H_2Se] = 0.31, [HSe^-] = 9.0 \times 10^{-2}, [H_3O^+] = 9.0 \times 10^{-2}, [Se^{2-}] = 1.0 \times 10^{-11}$
   (d) $[H_2Se] = 0.40, [HSe^-] = 1.3 \times 10^{-4}, [H_3O^+] = 1.3 \times 10^{-4}, [Se^{2-}] = 1.0 \times 10^{-11}$

10. Ammonia (NH_3) has base dissociation constant (K_b) of 1.8×10^{-5}. What is the concentration of an aqueous ammonia solution that has a pH of 11.68? (**LO 16.11**)
    (a) 0.28 M   (b) 3.6 M
    (c) 9.0×10^{-3} M   (d) 1.3 M

11. Which one of the following salts will form an acidic solution when dissolved in water? (**LO 16.13**)
    (a) NaF   (b) $NaNO_3$
    (c) $FeCl_3$ (d) NaCl

12. Calculate the pH of a 0.50 M solution of $NaNO_2$. (K_a for $HNO_2 = 4.5 \times 10^{-4}$) (**LO 16.12, 16.13**)
    (a) 1.82   (b) 3.64
    (c) 10.35   (d) 8.52

13. Consider the reaction: $SO_2 + OH^- \rightarrow HSO_3^-$. Which reaction scheme shows the correct use of the curved arrow notation representing the donation of an electron pair and the correct labeling of the Lewis acid and Lewis base? **(LO 16.14)**

(a)

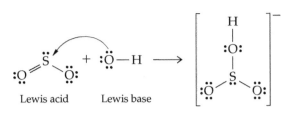

(b)

(c)

(d)

Answers:

1. c, 2. a, 3. c, 4. a, 5. d, 6. b, 7. c, 8. d, 9. b, 10. d, 11. c, 12. d, 13. a

Mastering Chemistry   provides end-of-chapter exercises, feedback-enriched tutorial problems, animations, and interactive activities to encourage problem-solving practice and deeper understanding of key concepts and topics.

RAN   *Randomized in Mastering Chemistry*

CONCEPTUAL PROBLEMS

Problems 16.1–16.37 appear within the chapter.

16.38 For each of the following reactions, identify the Brønsted–Lowry acids and bases.

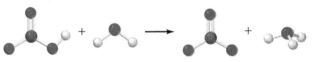

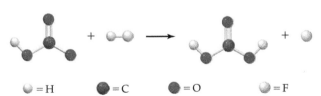

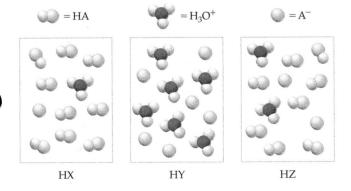

○ = H   ● = C   ● = O   ○ = F

16.39 The following pictures represent aqueous solutions of three acids HA(A = X, Y, or Z); water molecules have been omitted for clarity.

HX   HY   HZ

(a) What is the conjugate base of each acid?

(b) Arrange the three acids in order of increasing acid strength.

(c) Which acid, if any, is a strong acid?

(d) Which acid has the smallest value of K_a?

(e) What is the percent dissociation in the solution of HZ?

16.40 Locate sulfur, selenium, chlorine, and bromine in the periodic table:

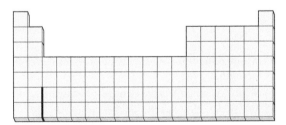

(a) Which binary acid (H_2S, H_2Se, HCl, or HBr) is the strongest? Which is the weakest? Explain.

(b) Which oxoacid (H_2SO_3, H_2SeO_3, $HClO_3$, or $HBrO_3$) is the strongest? Which is the weakest? Explain.

16.41 Which of the following pictures represents a solution of a weak diprotic acid, H_2A? (Water molecules have been omitted for clarity.) Which pictures represent an impossible situation? Explain.

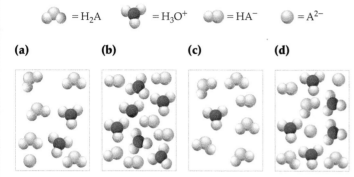

= H_2A = H_3O^+ = HA^- = A^{2-}

(a)   (b)   (c)   (d)

16.42 Which of the following pictures best represents an aqueous solution of sulfuric acid? Explain. (Water molecules have been omitted for clarity.)

= H_2SO_4 = H_3O^+ = HSO_4^- = SO_4^{2-}

(a)   (b)   (c)

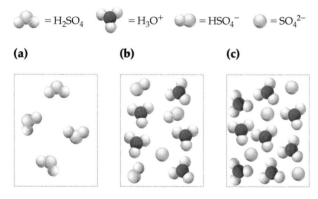

16.43 The following pictures represent solutions of three salts NaA ($A^- = X^-$, Y^-, or Z^-); water molecules and Na^+ ions have been omitted for clarity.

= A^- = HA   = OH^-

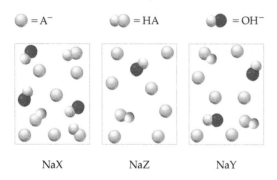

NaX   NaZ   NaY

(a) Arrange the three A^- anions in order of increasing base strength.
(b) Which A^- anion has the strongest conjugate acid?
(c) Which A^- anion has the smallest value of pK_b?
(d) Why, within each box, is the number of HA molecules and OH^- anions the same?

16.44 The following picture represents the hydrated metal cation $M(H_2O)_6^{n+}$, where $n = 1, 2,$ or 3.

(a) Write a balanced equation for the reaction of $M(H_2O)_6^{n+}$ with water and write the equilibrium equation for the reaction.
(b) Does the equilibrium constant increase, decrease, or remain the same as the value of n increases? Explain.
(c) Which $M(H_2O)_6^{n+}$ ion ($n = 1, 2,$ or 3) is the strongest acid, and which has the strongest conjugate base?

16.45 Look at the electron-dot structures of the following molecules and ions:

(a) Which of these molecules and ions can behave as a Brønsted–Lowry acid? Which can behave as a Brønsted–Lowry base?
(b) Which can behave as a Lewis acid? Which can behave as a Lewis base?

16.46 Boric acid (H_3BO_3) is a weak monoprotic acid that yields H_3O^+ ions in water. H_3BO_3 might behave either as a Brønsted–Lowry acid or as a Lewis acid, though it is, in fact, a Lewis acid.

Boric acid

(a) Write a balanced equation for the reaction with water in which H_3BO_3 behaves as a Brønsted–Lowry acid.
(b) Write a balanced equation for the reaction with water in which H_3BO_3 behaves as a Lewis acid. Hint: One of the reaction products contains a tetrahedral boron atom.

16.47 The reaction of PCl_4^+ with Cl^- is a Lewis acid–base reaction. Draw electron-dot structures for the reactants and products, and use the curved arrow notation (Section 16.15) to represent the donation of a lone pair of electrons from the Lewis base to the Lewis acid.

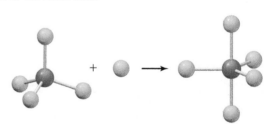

SECTION PROBLEMS

Acid–Base Concepts (Section 16.1)

16.48 Which of the following are Brønsted–Lowry bases but not
RAN   Arrhenius bases?
   (a) $Ca(OH)_2$ (b) NH_3
   (c) HCO_3^- (d) $NaOH$

16.49 Which of the following can behave both as a Brønsted–
RAN   Lowry acid and as a Brønsted–Lowry base?
   (a) HCO_3^- (b) CN^-
   (c) H_2O (d) H_2CO_3

16.50 Give the formula for the conjugate base of each of the following Brønsted–Lowry acids.
   (a) HSO_4^- (b) H_2SO_3 (c) $H_2PO_4^-$
   (d) NH_4^+ (e) H_2O (f) NH_3

16.51 Give the formula for the conjugate acid of each of the following Brønsted–Lowry bases.
   (a) SO_3^{2-} (b) H_2O (c) CH_3NH_2
   (d) OH^- (e) HCO_3^- (f) H^-

16.52 For each of the following reactions, identify the Brønsted–Lowry acids and bases and the conjugate acid–base pairs.
   (a) $CH_3CO_2H(aq) + NH_3(aq) \rightleftharpoons$
                      $NH_4^+(aq) + CH_3CO_2^-(aq)$
   (b) $CO_3^{2-}(aq) + H_3O^+(aq) \rightleftharpoons H_2O(l) + HCO_3^-(aq)$
   (c) $HSO_3^-(aq) + H_2O(l) \rightleftharpoons H_3O^+(aq) + SO_3^{2-}(aq)$
   (d) $HSO_3^-(aq) + H_2O(l) \rightleftharpoons H_2SO_3(aq) + OH^-(aq)$

16.53 For each of the following reactions, identify the Brønsted–Lowry acids and bases and the conjugate acid–base pairs.
   (a) $CN^-(aq) + H_2O(l) \rightleftharpoons OH^-(aq) + HCN(aq)$
   (b) $H_2PO_4^-(aq) + H_2O(l) \rightleftharpoons H_3O^+(aq) + HPO_4^{2-}(aq)$
   (c) $HPO_4^{2-}(aq) + H_2O(l) \rightleftharpoons OH^-(aq) + H_2PO_4^-(aq)$
   (d) $NH_4^+(aq) + NO_2^-(aq) \rightleftharpoons HNO_2(aq) + NH_3(aq)$

16.54 Aqueous solutions of hydrogen sulfide contain H_2S, HS^-, S^{2-}, H_3O^+, OH^-, and H_2O in varying concentrations. Which of these species can act only as an acid? Which can act only as a base? Which can act both as an acid and as a base?

16.55 The hydronium ion H_3O^+ is the strongest acid that can exist in aqueous solution because stronger acids dissociate by transferring a proton to water. What is the strongest base that can exist in aqueous solution?

Acid and Base Strength (Section 16.2)

16.56 Choose from the conjugate acid–base pairs HSO_4^-/SO_4^{2-},
RAN HF/F^-, and NH_4^+/NH_3 to complete the following equation with the pair that gives an equilibrium constant $K_c > 1$.

$$\underline{\hspace{1cm}} + NO_2^- \rightarrow \underline{\hspace{1cm}} + HNO_2$$

16.57 Choose from the conjugate acid–base pairs F^-/HF, $NH_3/$
RAN NH_4^+, and NO_3^-/HNO_3, to complete the following equation with the pair that gives an equilibrium constant $K_c > 1$.

$$\underline{\hspace{1cm}} + H_2CO_3 \rightarrow \underline{\hspace{1cm}} + HCO_3^-$$

16.58 Which acid in each of the following pairs has the stron-
RAN   ger conjugate base? See Table 16.1 to compare the relative strengths of conjugate acid-base pairs.
   (a) HCl or HF
   (b) HF or NH_4^+
   (c) HCN or HSO_4^-

16.59 Which base in each of the following pairs has the stronger conjugate acid? See Table 16.1 to compare the relative strengths of conjugate acid-base pairs.
   (a) Cl^- or CO_3^{2-}
   (b) CN^- or NH_3
   (c) HS^- or O^{2-}

Factors That Affect Acid Strength (Section 16.3)

16.60 Arrange each group of compounds in order of increasing acid strength. Explain your reasoning.
   (a) HCl, H_2S, PH_3
   (b) NH_3, PH_3, AsH_3
   (c) $HBrO$, $HBrO_3$, $HBrO_4$

16.61 Arrange each group of compounds in order of decreasing acid strength. Explain your reasoning.
   (a) H_2O, H_2S, H_2Se
   (b) $HClO_3$, $HBrO_3$, HIO_3
   (c) PH_3, H_2S, HCl

16.62 Identify the strongest acid in each of the following sets. Explain your reasoning.
   (a) H_2O, HF, or HCl
   (b) $HClO_2$, $HClO_3$, or $HBrO_3$
   (c) HBr, H_2S, or H_2Se

16.63 Identify the weakest acid in each of the following sets. Explain your reasoning.
   (a) H_2SO_3, $HClO_3$, $HClO_4$
   (b) NH_3, H_2O, H_2S
   (c) $B(OH)_3$, $Al(OH)_3$, $Ga(OH)_3$

16.64 Identify the stronger acid in each of the following pairs. Explain your reasoning.
   (a) H_2Se or H_2Te (b) H_3PO_4 or H_3AsO_4
   (c) $H_2PO_4^-$ or HPO_4^{2-} (d) CH_4 or NH_4^+

16.65 Identify the stronger base in each of the following pairs. Explain your reasoning.
   (a) ClO_2^- or ClO_3^- (b) HSO_4^- or $HSeO_4^-$
   (c) HS^- or OH^- (d) HS^- or Br^-

16.66 The following organic compounds both contain an O—H
RAN   bond. Which compound is more acidic?

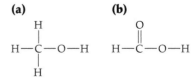

(a)   **(b)**

16.67 The following organic compounds contain the acetic acid group (COOH). Rank them from weakest to strongest acid.

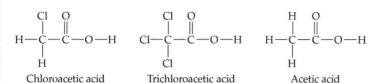

Chloroacetic acid   Trichloroacetic acid   Acetic acid

Dissociation of Water; pH (Sections 16.4–16.6)

16.68 The concentration of OH^- in a sample of seawater is
RAN 2.0×10^{-6} M. Calculate the concentration of H_3O^+ ions, and classify the solution as acidic, neutral, or basic.

16.69 The concentration of OH^- in human blood is 2.24×10^{-7} M.
RAN Calculate the concentration of H_3O^+ ions, and classify the solution as acidic, neutral, or basic.

16.70 For each of the following solutions, calculate $[OH^-]$ from
RAN $[H_3O^+]$, or $[H_3O^+]$ from $[OH^-]$. Classify each solution as acidic, basic, or neutral.
  (a) $[H_3O^+] = 3.4 \times 10^{-9}$ M
  (b) $[OH^-] = 0.010$ M
  (c) $[OH^-] = 1.0 \times 10^{-10}$ M
  (d) $[H_3O^+] = 1.0 \times 10^{-7}$ M
  (e) $[H_3O^+] = 8.6 \times 10^{-5}$ M

16.71 For each of the following solutions, calculate $[OH^-]$ from
RAN $[H_3O^+]$, or $[H_3O^+]$ from $[OH^-]$. Classify each solution as acidic, basic, or neutral.
  (a) $[H_3O^+] = 2.5 \times 10^{-4}$ M
  (b) $[H_3O^+] = 2.0$ M
  (c) $[OH^-] = 5.6 \times 10^{-9}$ M
  (d) $[OH^-] = 1.5 \times 10^{-3}$ M
  (e) $[OH^-] = 1.0 \times 10^{-7}$ M

16.72 Water superheated under pressure to 200 °C and 750 atm has $K_w = 1.5 \times 10^{-11}$. What is $[H_3O^+]$ and $[OH^-]$ at 200 °C? Is the water acidic, basic, or neutral?

16.73 Water at 500 °C and 250 atm is a supercritical fluid. Under these conditions, K_w is approximately 1.7×10^{-19}. Estimate $[H_3O^+]$ and $[OH^-]$ at 500 °C Is the water acidic, basic, or neutral?

16.74 Calculate the pH to the correct number of significant figures for
RAN solutions with the following concentrations of H_3O^+ or OH^-.
  (a) $[H_3O^+] = 2.0 \times 10^{-5}$ M
  (b) $[OH^-] = 4 \times 10^{-3}$ M
  (c) $[H_3O^+] = 3.56 \times 10^{-9}$ M
  (d) $[H_3O^+] = 10^{-3}$ M
  (e) $[OH^-] = 12$ M

16.75 What is the pH to the correct number of significant figures
RAN for solutions with the following concentrations of H_3O^+ or OH^-?
  (a) $[OH^-] = 7.6 \times 10^{-3}$ M
  (b) $[H_3O^+] = 10^{-8}$ M
  (c) $[H_3O^+] = 5.0$ M
  (d) $[OH^-] = 1.0 \times 10^{-7}$ M
  (e) $[H_3O^+] = 2.18 \times 10^{-10}$ M

16.76 Calculate the H_3O^+ concentration to the correct number of
RAN significant figures for solutions with the following pH values.
  (a) 4.1   (b) 10.82   (c) 0.00
  (d) 14.25   (e) −1.0   (f) 5.238

16.77 What is the H_3O^+ concentration to the correct number
RAN of significant figures for solutions with the following pH values?
  (a) 9.0   (b) 7.00   (c) −0.3
  (d) 15.18   (e) 2.63   (f) 10.756

16.78 Which of the indicators given in Figure 16.5, thymol blue,
RAN alizarin yellow, chlorphenol red, or methyl orange, would be most appropriate to detect a pH change from:
  (a) 7 to 5?   (b) 8 to 10?   (c) 3 to 5?

16.79 Which of the indicators given in Figure 16.5, methyl violet, bromcresol green, phenol red, or thymolphthalein, would be most appropriate to detect a pH change from:
  (a) 4 to 6?   (b) 8 to 10?   (c) 2 to 0?

Strong Acids and Strong Bases (Section 16.7)

16.80 Which of the following species behave as strong acids or as
RAN strong bases in aqueous solution?
  (a) HNO_2 (b) HNO_3
  (c) NH_4^+ (d) Cl^-

16.81 Which of the following species behave as strong acids or as strong bases in aqueous solution?
  (a) H^- (b) O^{2-}
  (c) H_2SO_4 (d) CsOH

16.82 Calculate the pH of the following solutions:
RAN (a) 1.0×10^{-3}M $Sr(OH)_2$
  (b) 0.015 MHNO_3
  (c) 0.035 M NaOH

16.83 Calculate the pH of the following solutions:
RAN (a) 0.48 M HCl
  (b) 2.5×10^{-3}M $Ba(OH)_2$
  (c) 0.075 M NaOH

16.84 Calculate the pH of solutions prepared by:
RAN (a) Dissolving 4.8 g of lithium hydroxide in water to give 250 mL of solution.
  (b) Dissolving 0.93 g of hydrogen chloride in water to give 0.40 L of solution.
  (c) Diluting 50.0 mL of 0.10 M HCl to a volume of 1.00 L.
  (d) Mixing 100.0 mL of 2.0×10^{-3} M HCl and 400.0 mL of 1.0×10^{-3} M $HClO_4$. (Assume that volumes are additive.)

16.85 Calculate the pH of solutions prepared by:
RAN (a) Dissolving 0.20 g of sodium oxide in water to give 100.0 mL of solution.
  (b) Dissolving 1.26 g of pure nitric acid in water to give 0.500 L of solution.
  (c) Diluting 40.0 mL of 0.075 M $Ba(OH)_2$ to a volume of 300.0 mL.
  (d) Mixing equal volumes of 0.20 M HCl and 0.50 M HNO_3. (Assume that volumes are additive.)

16.86 How many grams of CaO should be dissolved in sufficient
RAN water to make 1.00 L of a solution with a pH of 10.50?

16.87 How many grams of SrO should be dissolved in sufficient
RAN water to make 2.00 L of a solution with a pH = 10.0?

Weak Acids (Sections 16.8–16.10)

16.88 Look up the values of K_a in Appendix C for C_6H_5OH, HNO_3, CH_3CO_2H, and HOCl, and arrange these acids in order of:

(a) Increasing acid strength.

(b) Decreasing percent dissociation.

(c) Also estimate $[H_3O^+]$ in a 1.0 M solution of each acid.

16.89 Look up the values of K_a in Appendix C for HCO_2H, HCN,
RAN $HClO_4$, and HOBr, and arrange these acids in order of:

(a) Increasing acid strength.

(b) Decreasing percent dissociation.

(c) Also estimate $[H_3O^+]$ in a 1.0 M solution of each acid.

16.90 The pH of 0.040 M hypobromous acid (HOBr) is 5.05. Set
RAN up the equilibrium equation for the dissociation of HOBr, and calculate the value of the acid-dissociation constant (K_a).

16.91 Lactic acid ($C_3H_6O_3$), which occurs in sour milk and foods
RAN such as sauerkraut, is a weak monoprotic acid. The pH of a 0.10 M solution of lactic acid is 2.43. What are the values of K_a and pK_a for lactic acid?

Lactic acid

16.92 The pH of 0.050 M gallic acid, an acid found in tea leaves, is 2.86. Calculate K_a and pK_a for gallic acid.

HO   H
  \\   /
   C = C
  /   \\   O
HO — C   C — C Acidic hydrogen
      \\ /   \\
       C — C   OH
      /   \\
    HO   H

Gallic acid

16.93 The pH of 0.040 M pyruvic acid, a compound involved in metabolic pathways, is 1.96. Calculate K_a and pK_a for pyruvic acid.

H₃C   O
   \\ ||
    C — C   Acidic hydrogen
   || \\
    O   OH

Pyruvic acid

16.94 A vitamin C tablet containing 250 mg of ascorbic acid ($C_6H_8O_6$; $K_a = 8.0 \times 10^{-5}$) is dissolved in a 250 mL glass of water. What is the pH of the solution?

16.95 Acetic acid (CH_3COOH; $K_a = 1.8 \times 10^{-5}$) has a concen-
RAN tration in vinegar of 3.50% by mass. What is the pH of vinegar? (The density of vinegar is 1.02 g/mL.)

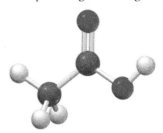

Acetic acid

16.96 Acrylic acid ($HC_3H_3O_2$) is used in the manufacture of paints
RAN and plastics. The pK_a of acrylic acid is 4.25.

(a) Calculate the pH and the concentrations of all species (H_3O^+, $C_3H_3O_2^-$, $HC_3H_3O_2$, and OH^-) in 0.150 M acrylic acid.

(b) Calculate the percent dissociation in 0.0500 M acrylic acid.

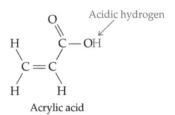

Acrylic acid

16.97 Hippuric acid ($HC_9H_8NO_3$), found in horse urine, has
RAN $pK_a = 3.62$.

(a) Calculate the pH and the concentrations of all species (H_3O^+, $C_9H_8NO_3^-$, $HC_9H_8NO_3$, and OH^-) in 0.100 M hippuric acid.

(b) Calculate the percent dissociation in 0.0750 M hippuric acid.

Hippuric acid

16.98 Calculate the pH and the percent dissociation in 1.5 M
RAN HNO_2 ($K_a = 4.5 \times 10^{-4}$).

16.99 A typical aspirin tablet contains 324 mg of aspirin (acetylsalicy-
RAN lic acid, $C_9H_8O_4$), a monoprotic acid having $K_a = 3.0 \times 10^{-4}$. If you dissolve two aspirin tablets in a 300 mL glass of water, what is the pH of the solution and the percent dissociation?

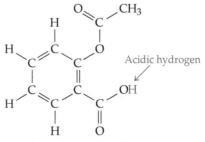

Acetylsalicylic acid (Aspirin)

16.100 Calculate the percent dissociation of HF ($K_a = 3.5 \times 10^{-4}$) in
(a) 0.050 M HF   (b) 0.50 M HF

16.101 Calculate the percent dissociation of HNO_2 ($K_a = 4.5 \times 10^{-4}$) in
RAN
(a) 0.010 M HNO_2 (b) 1.00 M HNO_2

Polyprotic Acids (Section 16.11)

16.102 Write balanced net ionic equations and the corresponding equilibrium equations for the stepwise dissociation of the diprotic acid H_2SeO_4.

16.103 Write balanced net ionic equations and the corresponding equilibrium equations for the stepwise dissociation of the triprotic acid H_3PO_4.

16.104 Calculate the pH and the concentrations of all species present (H_2CO_3, HCO_3^-, CO_3^{2-}, H_3O^+, and OH^-) in 0.010 M
RAN
H_2CO_3 ($K_{a1} = 4.3 \times 10^{-7}$; $K_{a2} = 5.6 \times 10^{-11}$).

16.105 Calculate the pH and the concentrations of H_2SO_3, HSO_3^-,
RAN
SO_3^{2-}, H_3O^+, and OH^- in 0.025 M H_2SO_3($K_{a1} = 1.5 \times 10^{-2}$; $K_{a2} = 6.3 \times 10^{-8}$).

16.106 Oxalic acid ($H_2C_2O_4$) is a diprotic acid that occurs in plants
RAN
such as rhubarb and spinach. Calculate the pH and the concentration of $C_2O_4^{2-}$ ions in 0.20 M $H_2C_2O_4$ ($K_{a1} = 5.9 \times 10^{-2}$; $K_{a2} = 6.4 \times 10^{-5}$).

Oxalic acid

16.107 Tartaric acid ($C_4H_6O_6$) is a diprotic acid that plays an
RAN
important role in lowering the pH of wine to a level that bacteria cannot survive. Calculate the pH of a 0.50 M tartaric acid solution ($pK_{a1} = 2.89$; $pK_{a2} = 4.40$).

Tartaric acid

16.108 Like sulfuric acid, selenic acid (H_2SeO_4) is a diprotic acid
RAN
that has a very large value of K_{a1}. Calculate the pH and the concentrations of all species present in 0.50 M H_2SeO_4 ($K_{a2} = 1.2 \times 10^{-2}$).

16.109 Calculate the concentrations of H_3O^+ and SO_4^{2-} in a solu-
RAN
tion prepared by mixing equal volumes of 0.2 M HCl and 0.6 M H_2SO_4(K_{a2} for H_2SO_4 is 1.2×10^{-2}).

16.110 Quinolinic acid, $H_2C_7H_3NO_4$($pK_{a1} = 2.43$; $pK_{a2} = 4.78$),
RAN
has been implicated in the progression of Alzheimer's disease. Calculate the pH and the concentrations of all species

present ($H_2C_7H_3NO_4$, $HC_7H_3NO_4^-$, H_3O^+, and OH^-) in a 0.050 M solution of quinolinic acid.

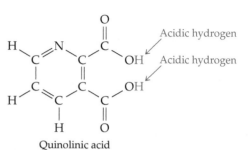

Quinolinic acid

16.111 Calculate the pH and the concentrations of all species present ($H_2C_8H_4O_4$, $HC_8H_4O_4^-$, $C_8H_4O_4^{2-}$, H_3O^+, and OH^-) in a 0.0250 M solution of phthalic acid, $H_2C_8H_4O_4$ ($pK_{a1} = 2.89$; $pK_{a2} = 5.51$).

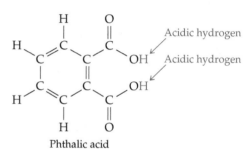

Phthalic acid

Weak Bases (Section 16.12)

16.112 Write a balanced net ionic equation and the corresponding equilibrium equation for the reaction of the following weak bases with water.
(a) Dimethylamine, $(CH_3)_2NH$
(b) Aniline, $C_6H_5NH_2$
(c) Cyanide ion, CN^-

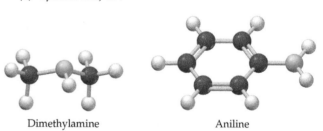

Dimethylamine   Aniline

16.113 Write a balanced net ionic equation and the corresponding equilibrium equation for the reaction of the following weak bases with water.
(a) Pyridine, C_5H_5N
(b) Ethylamine, $C_2H_5NH_2$
(c) Acetate ion, $CH_3CO_2^-$

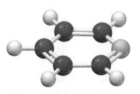

Pyridine   Ethylamine

16.114 Strychnine ($C_{21}H_{22}N_2O_2$), a deadly poison used for killing rodents, is a weak base having $K_b = 1.8 \times 10^{-6}$. Calculate the pH of a saturated solution of strychnine (16 mg/100 mL).

16.115 What is the pH of 0.5 M ammonia (NH_3)?
RAN ($K_b = 1.8 \times 10^{-5}$)

16.116 Morphine ($C_{17}H_{19}NO_3$), a narcotic used in painkillers, is a weak organic base. If the pH of a 7.0×10^{-4} M solution of morphine is 9.50, what are the values of K_b and pK_b?

16.117 A 1.00×10^{-3} M solution of quinine, a drug used in treat-
RAN ing malaria, has a pH of 9.75. What are the values of K_b and pK_b?

16.118 Oxycodone ($C_{18}H_{21}NO_4$), a narcotic analgesic, is a weak base
RAN with $pK_b = 5.47$. Calculate the pH and the concentrations of all species present ($C_{18}H_{21}NO_4$, $HC_{18}H_{21}NO_4^+$, H_3O^+, and OH^-) in a 0.002 50 M oxycodone solution.

16.119 Morpholine (C_4H_9NO) is a weak organic base with
RAN $pK_b = 5.68$. Calculate the pH and the concentrations of all species present (C_4H_9NO, $HC_4H_9NO^+$, and OH^-) in a 0.0100 M morpholine solution.

Relation between K_a and K_b (Section 16.13)

16.120 Using values of K_b in Appendix C, calculate values of K_a for each of the following ions.
   (a) Propylammonium ion, $C_3H_7NH_3^+$
   (b) Hydroxylammonium ion, NH_3OH^+
   (c) Anilinium ion, $C_6H_5NH_3^+$
   (d) Pyridinium ion, $C_5H_5NH^+$

16.121 Using values of K_a in Appendix C, calculate values of K_b for each of the following ions.
   (a) Fluoride ion, F^-
   (b) Hypobromite ion, OBr^-
   (c) Hydrogen sulfide ion, HS^-
   (d) Sulfide ion, S^{2-}

16.122 Nicotine ($C_{10}H_{14}N_2$) can accept two protons because it has two basic N atoms ($K_{b1} = 1.0 \times 10^{-6}$; $K_{b2} = 1.3 \times 10^{-11}$). Calculate the values of K_a for the conjugate acids $C_{10}H_{14}N_2H^+$ and $C_{10}H_{14}N_2H_2^{2+}$.

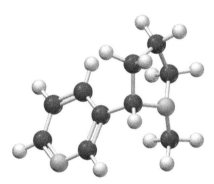

Nicotine

16.123 Sodium benzoate ($C_6H_5CO_2Na$) is used as a food preserva-
RAN tive. Calculate the pH and the concentrations of all species present (Na^+, $C_6H_5CO_2^-$, $C_6H_5CO_2H$, H_3O^+, and OH^-) in 0.050 M sodium benzoate; K_a for benzoic acid ($C_6H_5CO_2H$) is 6.5×10^{-5}.

Acid–Base Properties of Salts (Section 16.14)

16.124 Write a balanced net ionic equation for the reaction of each of the following ions with water. In each case, identify the Brønsted–Lowry acids and bases and the conjugate acid–base pairs.
   (a) $CH_3NH_3^+$ (b) $Cr(H_2O)_6^{3+}$
   (c) $CH_3CO_2^-$ (d) PO_4^{3-}

16.125 Write a balanced net ionic equation for the principal reaction in solutions of each of the following salts. In each case, identify the Brønsted–Lowry acids and bases and the conjugate acid–base pairs.
   (a) Na_2CO_3 (b) NH_4NO_3
   (c) NaCl   (d) $ZnCl_2$

16.126 Classify each of the following ions according to whether they react with water to give a neutral, acidic, or basic solution.
   (a) F^- (b) Br^-
   (c) NH_4^+ (d) $K(H_2O)_6^+$
   (e) SO_3^{2-} (f) $Cr(H_2O)_6^{3+}$

16.127 Classify each of the following salt solutions as neutral, acidic, or basic. See Appendix C for values of equilibrium constants.
   (a) $Fe(NO_3)_3$ (b) $Ba(NO_3)_2$
   (c) NaOCl   (d) NH_4I
   (e) NH_4NO_2 (f) $(CH_3NH_3)Cl$

16.128 Calculate the concentrations of all species present and the
RAN pH in 0.10 M solutions of the following substances. See Appendix C for values of equilibrium constants.
   (a) Ethylammonium nitrate, $(C_2H_5NH_3)NO_3$
   (b) Sodium acetate, $Na(CH_3CO_2)$
   (c) Sodium nitrate, $NaNO_3$

16.129 Calculate the pH and the percent dissociation of the hyd-
RAN rated cation in 0.020 M solutions of the following sub-stances. See Appendix C for values of equilibrium constants.
   (a) $Fe(NO_3)_2$ (b) $Fe(NO_3)_3$

16.130 Calculate K_a for the cation and K_b for the anion in an aqueous NH_4CN solution. Is the solution acidic, basic, or neutral?

16.131 Classify each of the following salt solutions as acidic, basic, or neutral.
   (a) KBr   (b) $NaNO_2$
   (c) NH_4Br (d) $ZnCl_2$
   (e) NH_4F

16.132 The hydrated cation $M(H_2O)_6^{3+}$ has $K_a = 10^{-4}$, and the acid HA has $K_a = 10^{-5}$. Identify the principal reaction in an aqueous solution of each of the following salts, and clas-sify each solution as acidic, basic, or neutral.
   (a) NaA   (b) $M(NO_3)_3$
   (c) $NaNO_3$ (d) MA_3

16.133 Classify each of the following salt solutions as neutral, acidic, or basic. See Appendix C for values of equilibrium constants.
   (a) NH_4F (b) $(NH_4)_2SO_3$

16.134 Calculate the pH and the concentrations of all species present
RAN   in 0.25 M solutions of each of the salts in Problem 16.133.
(Hint: The principal reaction is proton transfer from the
cation to the anion.)

16.135 Calculate the pH and the percent dissociation of the hydrated cation in the following solutions. See Appendix C for
the value of the equilibrium constant.
(a) 0.010 M $Cr(NO_3)_3$
(b) 0.0050 M $Cr(NO_3)_3$

16.136 Baking powder contains baking soda ($NaHCO_3$) and an
acidic substance such as sodium alum, $NaAl(SO_4)_2 \cdot 12\ H_2O$.
These components react in an aqueous medium to produce
CO_2 gas, which "raises" the dough. Write a balanced net
ionic equation for the reaction.

16.137 Arrange the following substances in order of increasing
$[H_3O^+]$ for a 0.10 M solution of each.
(a) $Zn(NO_3)_2$ (b) Na_2O
(c) $NaOCl$ (d) $NaClO_4$
(e) $HClO_4$

Lewis Acids and Bases (Section 16.15)

16.138 For each of the following reactions, identify the Lewis acid
and the Lewis base.
(a) $SiF_4 + 2\ F^- \rightarrow SiF_6^{2-}$
(b) $4\ NH_3 + Zn^{2+} \rightarrow Zn(NH_3)_4^{2+}$
(c) $2\ Cl^- + HgCl_2 \rightarrow HgCl_4^{2-}$
(d) $CO_2 + H_2O \rightarrow H_2CO_3$

16.139 For each of the following reactions, identify the Lewis acid
and the Lewis base.
(a) $2\ Cl^- + BeCl_2 \rightarrow BeCl_4^{2-}$
(b) $Mg^{2+} + 6\ H_2O \rightarrow Mg(H_2O)_6^{2+}$
(c) $SO_3 + OH^- \rightarrow HSO_4^-$
(d) $F^- + BF_3 \rightarrow BF_4^-$

16.140 For each of the Lewis acid–base reactions in Problem
16.138, draw electron-dot structures for the reactants and
products, and use the curved arrow notation to represent
the donation of a lone pair of electrons from the Lewis base
to the Lewis acid.

16.141 For each of the Lewis acid–base reactions in Problem
16.139, draw electron-dot structures for the reactants and
products, and use the curved arrow notation to represent
the donation of a lone pair of electrons from the Lewis base
to the Lewis acid.

16.142 Classify each of the following as a Lewis acid or a Lewis
base.
(a) CN^- (b) H^+
(c) H_2O (d) Fe^{3+}
(e) OH^- (f) CO_2
(g) $P(CH_3)_3$ (h) $B(CH_3)_3$

16.143 Which would you expect to be the stronger Lewis acid in
each of the following pairs? Explain.
(a) BF_3 or BH_3 (b) SO_2 or SO_3
(c) Sn^{2+} or Sn^{4+} (d) CH_3^+ or CH_4

MULTICONCEPT PROBLEMS

16.144 At 0 °C, the density of liquid water is 0.9998 g/mL and the
value of K_w is 1.14×10^{-15}. What fraction of the molecules
in liquid water are dissociated at 0 °C? What is the percent dissociation at 0 °C? What is the pH of a neutral solution at 0 °C?

16.145 What is the pH and the principal source of H_3O^+ ions in
1.0×10^{-10} M HCl? (Hint: The pH of an acid solution
can't exceed 7.) What is the pH of 1.0×10^{-7} M HCl?

16.146 Calculate the pH and the concentrations of all species pres-
RAN   ent (H_3O^+, F^-, HF, Cl^-, and OH^-) in a solution that contains 0.10 M HF ($K_a = 3.5 \times 10^{-4}$) and 0.10 M HCl.

16.147 When NO_2 is bubbled into water, it is completely converted
RAN   to HNO_3 and HNO_2:

$$2\ NO_2(g) + H_2O(l) \rightarrow HNO_3(aq) + HNO_2(aq)$$

Calculate the pH and the concentrations of all species present (H_3O^+, OH^-, HNO_2, NO_2^-, and NO_3^-) in a solution
prepared by dissolving 0.0500 mol of NO_2 in 1.00 L of
water. K_a for HNO_2 is 4.5×10^{-4}.

16.148 Normal rain has a pH of 5.6 due to dissolved atmospheric
carbon dioxide at a current level of 400 ppm. Various models predict that burning fossil fuels will increase the atmospheric CO_2 concentration to between 500 and 1000 ppm
(by volume) by the year 2100. Calculate the pH of rain in a
scenario where the CO_2 concentration is 750 ppm.
(a) First use Henry's Law to calculate the concentration
of dissolved CO_2. Solubility $= k \times P$ (Section 12.4)
and the Henry's Law constant (k) for CO_2 at 25 °C is
3.2×10^{-2} mol/(L·atm).

(b) Next calculate the pH of the rain. CO_2 reacts with water
to produce carbonic acid according to the equation:

$$CO_2(aq) + H_2O(l) \rightleftharpoons H_2CO_3(aq)$$

Assume all the dissolved CO_2 is converted to H_2CO_3.
Acid dissociation constants for H_2CO_3 are $K_{a1} = 4.3 \times 10^{-7}$; $K_{a2} = 5.6 \times 10^{-11}$. (Worked Example Worked
Example 16.11 is a model for this calculation.)
(c) Will rising CO_2 levels affect the acidity of rainfall?

16.149 Sulfur dioxide is quite soluble in water:
RAN

$$SO_2(g) + H_2O(l) \rightleftharpoons H_2SO_3(aq)\quad K = 1.33$$

The H_2SO_3 produced is a weak diprotic acid ($K_{a1} = 1.5 \times 10^{-2}$; $K_{a2} = 6.3 \times 10^{-8}$). Calculate the pH and the concentrations of H_2SO_3, HSO_3^-, and SO_3^{2-} in a solution
prepared by continuously bubbling SO_2 at a pressure of
1.00 atm into pure water.

16.150 For a solution of two weak acids with comparable values
RAN   of K_a, there is no single principal reaction. The two acid-
dissociation equilibrium equations must therefore be solved
simultaneously. Calculate the pH in a solution that is 0.10 M
in acetic acid (CH_3CO_2H, $K_a = 1.8 \times 10^{-5}$) and 0.10 M
in benzoic acid ($C_6H_5CO_2H$, $K_a = 6.5 \times 10^{-5}$). (Hint: Let
$x = [CH_3CO_2H]$ that dissociates and $y = [C_6H_5CO_2H]$
that dissociates; then $[H_3O^+] = x + y$.)

16.151 Acid and base behavior can be observed in solvents other
than water. One commonly used solvent is dimethyl

sulfoxide (DMSO), which can be treated as a monoprotic acid "HSol." Just as water can behave either as an acid or a base, so HSol can behave either as a Brønsted–Lowry acid or base.

(a) The equilibrium constant for self-dissociation of HSol (call it K_{HSol}) is 1×10^{-35}. Write the chemical equation for the self-dissociation reaction and the corresponding equilibrium equation. (Hint: The equilibrium equation is analogous to the equilibrium equation for K_w in the case of water.)

(b) The weak acid HCN has an acid dissociation constant $K_a = 1.3 \times 10^{-13}$ in the solvent HSol. If 0.010 mol of NaCN is dissolved in 1.00 L of HSol, what is the equilibrium concentration of H_2Sol^+?

16.152 A 7.0 mass % solution of H_3PO_4 in water has a density of 1.0353 g/mL. Calculate the pH and the molar concentrations of all species present (H_3PO_4, $H_2PO_4^-$, PO_4^{3-}, H_3O^+, and OH^-) in the solution. Values of equilibrium constants are listed in Appendix C.

16.153 In the case of very weak acids, $[H_3O^+]$ from the dissociation of water is significant compared with $[H_3O^+]$ from the dissociation of the weak acid. The sugar substitute saccharin ($C_7H_5NO_3S$), for example, is a very weak acid having $K_a = 2.1 \times 10^{-12}$ and a solubility in water of 348 mg/100 mL. Calculate $[H_3O^+]$ in a saturated solution of saccharin. (Hint: Equilibrium equations for the dissociation of saccharin and water must be solved simultaneously.)

16.154 In aqueous solution, sodium acetate behaves as a strong electrolyte, yielding Na^+ cations and $CH_3CO_2^-$ anions. A particular solution of sodium acetate has a pH of 9.07 and a density of 1.0085 g/mL. What is the molality of this solution, and what is its freezing point?

16.155 During a certain time period, 4.0 million tons of SO_2 were
RAN released into the atmosphere and subsequently oxidized to SO_3. As explained in the Inquiry, the acid rain produced when the SO_3 dissolves in water can damage marble statues:

$$CaCO_3(s) + H_2SO_4(aq) \rightarrow CaSO_4(aq) + CO_2(g) + H_2O(l)$$

(a) How many 500 pound marble statues could be damaged by the acid rain? (Assume that the statues are pure $CaCO_3$ and that a statue is damaged when 3.0% of its mass is dissolved.)

(b) How many liters of CO_2 gas at 20 °C and 735 mm Hg is produced as a byproduct?

(c) The cation in aqueous H_2SO_4 is trigonal pyramidal rather than trigonal planar. Explain.

16.156 Neutralization reactions involving either a strong acid or a
RAN strong base go essentially to completion, and therefore we must take such neutralizations into account before calculating concentrations in mixtures of acids and bases. Consider a mixture of 3.28 g of Na_3PO_4 and 300.0 mL of 0.180 M HCl. Write balanced net ionic equations for the neutralization reactions and calculate the pH of the solution.

16.157 We've said that alkali metal cations do not react appreciably with water to produce H_3O^+ ions, but in fact, all cations are acidic to some extent. The most acidic alkali metal cation is the smallest one, Li^+, which has $K_a = 2.5 \times 10^{-14}$ for the reaction

$$Li(H_2O)_4^+(aq) + H_2O(l) \rightleftharpoons H_3O^+(aq) + Li(H_2O)_3(OH)(aq)$$

This reaction and the dissociation of water must be considered simultaneously in calculating the pH of Li^+ solutions, which nevertheless have pH $\approx$ 7. Check this by calculating the pH of 0.10 M LiCl.

16.158 A 1.000 L sample of HF gas at 20.0 °C and 0.601 atm pres-
RAN sure was dissolved in enough water to make 50.0 mL of hydrofluoric acid.

(a) What is the pH of the solution?

(b) To what volume must you dilute the solution to triple the percent dissociation?

16.159 A 200.0 mL sample of 0.350 M acetic acid (CH_3CO_2H)
RAN was allowed to react with 2.000 L of gaseous ammonia at 25 °C and a pressure of 650.8 mm Hg. Assuming no change in the volume of the solution, calculate the pH and the equilibrium concentrations of all species present (CH_3CO_2H, $CH_3CO_2^-$, NH_3, NH_4^+, H_3O^+, and OH^-). Values of equilibrium constants are listed in Appendix C.

16.160 You may have been told not to mix bleach and ammonia. The reason is that bleach (sodium hypochlorite) reacts with ammonia to produce toxic chloramines, such as NH_2Cl. For example, in basic solution:

$$OCl^-(aq) + NH_3(aq) \rightarrow OH^-(aq) + NH_2Cl(aq)$$

(a) The following initial rate data for this reaction were obtained in basic solution at 25 °C

pH	Initial $[OCl^-]$	Initial $[NH_3]$	Initial Rate (M/s)
12	0.001	0.01	0.017
12	0.002	0.01	0.033
12	0.002	0.03	0.100
13	0.002	0.03	0.010

What is the rate law for the reaction? What is the numerical value of the rate constant k, including the correct units?

(b) The following mechanism has been proposed for this reaction in basic solution:

$$H_2O + OCl^- \rightleftharpoons HOCl + OH^- \quad \text{Fast, equilibrium constant } K_1$$
$$HOCl + NH_3 \rightarrow H_2O + NH_2Cl \quad \text{Slow, rate constant } k_2$$

Assuming that the first step is in equilibrium and the second step is rate-determining, calculate the value of the rate constant k_2 for the second step. K_a for HOCl is 3.5×10^{-8}.

chapter 17

Applications of Aqueous Equilibria

Contents

The pteropod, or "sea butterfly," is a tiny sea creature that makes its calcium carbonate ($CaCO_3$) shell from Ca^{2+} and CO_3^{2-} ions in seawater in a process called calcification. The ability of calcifying organisms to build their protective shells is greatly affected by the pH and buffering capacity of the water. These organisms are vital in the marine food chain and could be devastated by ocean acidification.

What is causing ocean acidification?

The answer to this question can be found on page 754 in the INQUIRY ?

Aqueous equilibria play a crucial role in many environmental and biological processes. For example, the pH of many lakes and streams must be in the range 6.0–8.0 for plant and aquatic life to flourish. The pH of human blood is carefully controlled at a value of 7.4 by equilibria involving primarily the conjugate acid–base pair H_2CO_3 and HCO_3^-. At lower pH values, the affinity of hemoglobin in red blood cells for O_2 molecules decreases and the blood carries less oxygen to the tissues. At higher pH values, the affinity of hemoglobin for O_2 molecules increases, making it more difficult to release the oxygen needed in the tissues.

We began our study of the principles of chemical equilibria in Chapter 15 and learned how to write equilibrium equations and calculate the concentrations of reactants and products in an equilibrium mixture. In Chapter 16, we applied those principles to equilibria in aqueous solutions of a weak acid or a weak base. In this chapter, we'll continue our study by examining some applications of aqueous equilibria. First we'll see how to calculate the pH of *mixtures* of acids and bases. Then we'll look at the dissolution and precipitation of slightly soluble ionic compounds and the factors that affect solubility. Aqueous equilibria involving the dissolution and precipitation of ionic compounds are important in a great many natural processes, from tooth decay to the formation of coral reefs and limestone caves.

17.1 NEUTRALIZATION REACTIONS

We've seen on numerous occasions that the neutralization reaction of an acid with a base produces water and a salt. But to what extent does a neutralization reaction go to completion? We must answer that question before we can make pH calculations on mixtures of acids and bases. We'll look at four types of neutralization reactions: (1) strong acid–strong base, (2) weak acid–strong base, (3) strong acid–weak base, and (4) weak acid–weak base.

Strong Acid–Strong Base

As an example of a strong acid–strong base reaction, let's consider the reaction of hydrochloric acid with aqueous sodium hydroxide to give water and an aqueous solution of sodium chloride:

$$HCl(aq) + NaOH(aq) \longrightarrow H_2O(l) + NaCl(aq)$$

Because $HCl(aq)$, $NaOH(aq)$, and $NaCl(aq)$ are all completely dissociated, the net ionic equation for the neutralization reaction is

$$H_3O^+(aq) + OH^-(aq) \rightleftharpoons 2 H_2O(l)$$

If we mix equal numbers of moles of $HCl(aq)$ and $NaOH(aq)$, the concentrations of H_3O^+ and OH^- remaining in the NaCl solution after neutralization will be the same as those in pure water, $[H_3O^+] = [OH^-] = 1.0 \times 10^{-7}$ M. In other words, the reaction of HCl with NaOH proceeds far to the right.

We come to the same conclusion by looking at the equilibrium constant for the reaction. Because the neutralization reaction of any strong acid with a strong base is the reverse of the dissociation of water, its equilibrium constant, K_n ("n" for neutralization), is the reciprocal of the ion-product constant for water, $K_n = 1/K_w$:

$$H_3O^+(aq) + OH^-(aq) \rightleftharpoons 2 H_2O(l)$$

$$K_n = \frac{1}{[H_3O^+][OH^-]} = \frac{1}{K_w} = \frac{1}{1.0 \times 10^{-14}} = 1.0 \times 10^{14}$$

The value of $K_n (1.0 \times 10^{14})$ for a strong acid–strong base reaction is a very large number, which means that the neutralization reaction proceeds essentially 100% to completion. After neutralization of equal molar amounts of acid and base, the solution contains a salt derived from a strong base and a strong acid. Because neither the cation nor the anion of the salt has acidic or basic properties, the pH is 7 (Section 16.14).

Weak acid–strong base neutralization
$$CH_3CO_2H(aq) + OH^-(aq) \rightleftharpoons$$
$$H_2O(l) + CH_3CO_2^-(aq)$$

▲ When $NaOH(aq)$ is added to $CH_3CO_2H(aq)$ containing the acid–base indicator phenolphthalein, the color of the indicator changes from colorless to pink in the pH range 8.2–9.8 because neutralization of the acetic acid gives a weakly basic solution.

Weak Acid–Strong Base

Because a weak acid HA is largely undissociated, the net ionic equation for the neutralization reaction of a weak acid with a strong base involves proton transfer from HA to the strong base, OH⁻:

$$HA(aq) + OH^-(aq) \rightleftharpoons H_2O(l) + A^-(aq)$$

Acetic acid (CH_3CO_2H), for example, reacts with aqueous NaOH to give water and aqueous sodium acetate (CH_3CO_2Na):

$$CH_3CO_2H(aq) + OH^-(aq) \rightleftharpoons H_2O(l) + CH_3CO_2^-(aq)$$

Na^+ ions do not appear in the net ionic equation because both NaOH and CH_3CO_2Na are completely dissociated.

To obtain the equilibrium constant K_n for the neutralization of acetic acid, we multiply equilibrium constants for reactions that add to give the net ionic equation for the neutralization reaction. Because CH_3CO_2H is on the left side of the equation and $CH_3CO_2^-$ is on the right side, we surmise that one of the reactions must be the dissociation of CH_3CO_2H. Because H_2O is on the right side of the equation and OH⁻ is on the left side, the other reaction must be the reverse of the dissociation of H_2O. Note that H_3O^+ and one H_2O molecule cancel when the two equations are added:

$$CH_3CO_2H(aq) + H_2O(l) \rightleftharpoons H_3O^+(aq) + CH_3CO_2^-(aq) \qquad K_a = 1.8 \times 10^{-5}$$
$$H_3O^+(aq) + OH^-(aq) \rightleftharpoons H_2O(l) + H_2O(l) \qquad 1/K_w = 1.0 \times 10^{14}$$

Net: $\;\;CH_3CO_2H(aq) + OH^-(aq) \rightleftharpoons H_2O(l) + CH_3CO_2^-(aq) \quad K_n = (K_a)(1/K_w) = (1.8 \times 10^{-5})(1.0 \times 10^{14})$
$$= 1.8 \times 10^9$$

As we saw in Section 15.2, the equilibrium constant for the net reaction equals the product of the equilibrium constants for the reactions added. Therefore, we multiply K_a for CH_3CO_2H by the reciprocal of K_w to get K_n for the neutralization reaction. (We use $1/K_w$ because the H_2O dissociation reaction is written in the reverse direction.) The resulting large value of $K_n (1.8 \times 10^9)$ means that the neutralization reaction proceeds nearly 100% to completion.

As a general rule, the neutralization of any weak acid with a strong base will go 100% to completion because OH⁻ has a great affinity for protons. After neutralization of equal molar amounts of CH_3CO_2H and NaOH, the solution contains Na^+, which has no acidic or basic properties, and $CH_3CO_2^-$, which is a weak base. Therefore, the pH is greater than 7.

Strong Acid–Weak Base

A strong acid HA is completely dissociated into H_3O^+ and A^- ions, and its neutralization reaction with a weak base therefore involves proton transfer from H_3O^+ to the weak base B:

$$H_3O^+(aq) + B(aq) \rightleftharpoons H_2O(l) + BH^+(aq)$$

For example, the net ionic equation for the neutralization of hydrochloric acid with aqueous ammonia is

$$H_3O^+(aq) + NH_3(aq) \rightleftharpoons H_2O(l) + NH_4^+(aq)$$

As in the weak acid–strong base case, we can obtain the equilibrium constant for the neutralization reaction by multiplying equilibrium constants for reactions that add to give the net ionic equation:

$$NH_3(aq) + H_2O(l) \rightleftharpoons NH_4^+(aq) + OH^-(aq) \qquad K_b = 1.8 \times 10^{-5}$$
$$H_3O^+(aq) + OH^-(aq) \rightleftharpoons H_2O(l) + H_2O(l) \qquad 1/K_w = 1.0 \times 10^{14}$$

Net: $\;\;H_3O^+(aq) + NH_3(aq) \rightleftharpoons H_2O(l) + NH_4^+(aq) \qquad K_n = (K_b)(1/K_w)$
$$= (1.8 \times 10^{-5})(1.0 \times 10^{14})$$
$$= 1.8 \times 10^9$$

Strong acid–weak base neutralization
$$H_3O^+(aq) + NH_3(aq) \rightleftharpoons H_2O(l) + NH_4^+(aq)$$

▲ When hydrochloric acid is added to aqueous ammonia containing the acid–base indicator methyl red, the color of the indicator changes from yellow to red in the pH range 4.2–6.0 because the neutralization of the NH_3 gives a weakly acidic solution.

Again, because the equilibrium constant K_n is a very large number (1.8×10^9), we know the neutralization reaction proceeds nearly 100% to the right. (It's purely coincidental that the neutralization reactions of CH_3CO_2H with NaOH and of HCl with NH_3 have the same value of K_n. The two K_n values are the same because K_a for CH_3CO_2H happens to have the same value as K_b for NH_3.)

The neutralization of any weak base with a strong acid generally goes 100% to completion because H_3O^+ is a powerful proton donor. After the neutralization of equal molar amounts of NH_3 and HCl, the solution contains NH_4^+, which is a weak acid, and Cl^-, which has no acidic or basic properties. Therefore, the pH is less than 7.

Weak Acid–Weak Base

Both a weak acid HA and a weak base B are largely undissociated, and the neutralization reaction between them therefore involves proton transfer from the weak acid to the weak base. For example, the net ionic equation for the neutralization of acetic acid with aqueous ammonia is

$$CH_3CO_2H(aq) + NH_3(aq) \rightleftharpoons NH_4^+(aq) + CH_3CO_2^-(aq)$$

The equilibrium constant K_n can be obtained by multiplying equilibrium constants for (1) the acid dissociation of acetic acid, (2) the base protonation of ammonia, and (3) the reverse of the dissociation of water:

$$CH_3CO_2H(aq) + H_2O(l) \rightleftharpoons H_3O^+(aq) + CH_3CO_2^-(aq) \qquad K_a = 1.8 \times 10^{-5}$$
$$NH_3(aq) + H_2O(l) \rightleftharpoons NH_4^+(aq) + OH^-(aq) \qquad K_b = 1.8 \times 10^{-5}$$
$$H_3O^+(aq) + OH^-(aq) \rightleftharpoons 2 H_2O(l) \qquad 1/K_w = 1.0 \times 10^{14}$$

Net: $CH_3CO_2H(aq) + NH_3(aq) \rightleftharpoons NH_4^+(aq) + CH_3CO_2^-(aq) \qquad K_n = (K_a)(K_b)(1/K_w)$

$$K_n = (K_a)(K_b)\left(\frac{1}{K_w}\right) = (1.8 \times 10^{-5})(1.8 \times 10^{-5})(1.0 \times 10^{14}) = 3.2 \times 10^4$$

The value of K_n in this case is smaller than it is for the preceding three cases, indicating that the neutralization does not proceed as far toward completion.

In general, weak acid–weak base neutralizations have less tendency to proceed to completion than neutralizations involving strong acids or strong bases. The neutralization of HCN with aqueous ammonia, for example, has a value of K_n less than one, which means that the reaction proceeds less than halfway to completion:

$$HCN(aq) + NH_3(aq) \rightleftharpoons NH_4^+(aq) + CN^-(aq) \quad K_n = 0.88$$

WORKED EXAMPLE 17.1

Writing an Equation for a Neutralization Reaction and Estimating the pH of the Resulting Solution

Write a balanced net ionic equation for the neutralization of equal molar amounts of nitric acid and the weak base methylamine (CH_3NH_2). Indicate whether the pH after neutralization is greater than, equal to, or less than 7.

STRATEGY

The formulas that should appear in the net ionic equation depend on whether the acid and base are strong (completely dissociated) or weak (largely undissociated). The pH after neutralization depends on the acid–base properties of the cation and anion in the resulting salt solution (Section 16.14).

SOLUTION

Because HNO_3 is a strong acid and CH_3NH_2 is a weak base, the neutralization proceeds to completion and the net ionic equation is

$$H_3O^+(aq) + CH_3NH_2(aq) \longrightarrow H_2O(l) + CH_3NH_3^+(aq)$$

continued on next page

In this reaction methylamine uses the lone pair of electrons on its N atom to accept a proton from H_3O^+:

$$\begin{bmatrix} H-\overset{\displaystyle H}{\underset{\displaystyle H}{\overset{|}{\underset{|}{\ddot{O}}}}}-H \end{bmatrix}^+ + \;:\overset{\displaystyle H}{\underset{\displaystyle H}{\overset{|}{\underset{|}{N}}}}-CH_3 \longrightarrow H-\ddot{O}: + \begin{bmatrix} H-\overset{\displaystyle H}{\underset{\displaystyle H}{\overset{|}{\underset{|}{N}}}}-CH_3 \end{bmatrix}^+$$

After neutralization, the solution contains $CH_3NH_3^+$ (a weak acid) and NO_3^-, which has no acidic or basic properties. Therefore, the pH is less than 7.

▶ **PRACTICE 17.1**   Write a balanced net ionic equation for the neutralization of equimolar amounts of HNO_2 and KOH. Indicate whether the pH after neutralization is greater than, equal to, or less than 7. Values of K_a and K_b are listed in Appendix C.

▶ **APPLY 17.2**   Write balanced net ionic equations for the neutralization of the following acids and bases, calculate the value of K_n for each neutralization reaction, and arrange the reactions in order of increasing tendency to proceed to completion. Values of K_a and K_b are listed in Appendix C.

(a) HF and NaOH   (b) HCl and KOH   (c) HF and NH_3

All **PRACTICE** and **APPLY** problems are interactive in the eText.

17.2 THE COMMON-ION EFFECT

A solution of a weak acid and its conjugate base is an important acid–base mixture because such mixtures regulate the pH in biological systems. If a conjugate base (A^-) is added to a solution of a weak acid (HA), then A^- is considered to be a **common ion** because it is already present in the mixture as a product of the acid-dissociation reaction.

$$HA(aq) + H_2O(l) \rightleftharpoons H_3O^+(aq) + \underset{\underset{\text{common ion}}{\uparrow}}{A^-(aq)}$$

Let's examine **FIGURE 17.1** to discover the meaning of the common ion effect. The pink solution on the right-hand side of the figure is a 0.10 M solution of acetic acid (CH_3CO_2H) with $[H_3O^+] = 1.3 \times 10^{-3}$ M and a pH of 2.89. The yellow solution on the left-hand side of the figure contains a solution that is 0.10 M in both the weak acid CH_3CO_2H and its conjugate base $CH_3CO_2^-$ with $[H_3O^+] = 1.8 \times 10^{-5}$ M and a pH of 4.74.

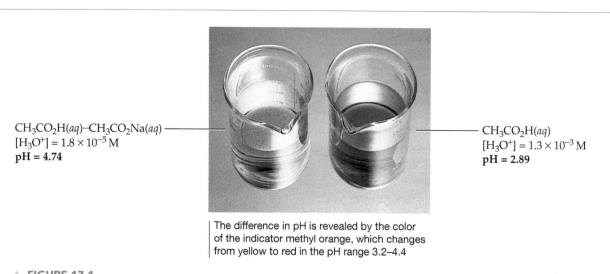

$CH_3CO_2H(aq)$–$CH_3CO_2Na(aq)$
$[H_3O^+] = 1.8 \times 10^{-5}$ M
pH = 4.74

$CH_3CO_2H(aq)$
$[H_3O^+] = 1.3 \times 10^{-3}$ M
pH = 2.89

The difference in pH is revealed by the color of the indicator methyl orange, which changes from yellow to red in the pH range 3.2–4.4

▲ **FIGURE 17.1**

A demonstration of the common-ion effect. The 0.10 M acetic acid–0.10 M sodium acetate solution on the left has a lower H_3O^+ concentration and a higher pH than the 0.10 M acetic acid solution on the right.

The decrease in $[H_3O^+]$ that occurs when acetate ions have been added to the acetic acid solution is an example of the **common-ion effect**, the shift in equilibrium that occurs when adding a substance that increases the concentration of an ion already involved in the equilibrium. Thus, added acetate ions shift the acetic acid-dissociation equilibrium to the left as shown in **FIGURE 17.2**.

$$CH_3CO_2H(aq) + H_2O(l) \rightleftharpoons H_3O^+(aq) + CH_3CO_2^-(aq)$$

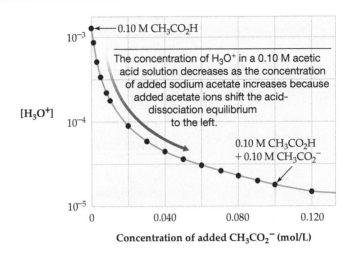

▶ FIGURE 17.2

The common-ion effect. Note that $[H_3O^+]$ is plotted on a logarithmic scale.

The common-ion effect is another example of **Le Chatelier's principle** (Section 15.6), in which the stress of the equilibrium of raising one of the product concentrations is relieved by shifting the equilibrium to the reactant side. Worked Example 17.2 shows how to calculate the pH of a solution containing a weak acid and its conjugate base (a common ion).

> **REMEMBER . . .**
> **Le Châtelier's principle** says that if a concentration, pressure, or temperature stress is applied to a reaction mixture at equilibrium, net reaction occurs in the direction that relieves the stress. Net reaction continues until a new state of equilibrium is achieved (Section 15.6).

WORKED EXAMPLE 17.2

Calculating the Effect of a Common Ion on Concentrations, pH, and Percent Dissociation

Calculate the concentrations of all species present, the pH, and the percent dissociation of acetic acid in a solution that is 0.10 M in CH_3CO_2H ($K_a = 1.8 \times 10^{-5}$) and 0.10 M in $NaCH_3CO_2$. (For comparison, the pH of a 0.10 M CH_3CO_2H solution that contains no $NaCH_3CO_2$ is 2.89 and the percent dissociation is 1.3%.)

BIG IDEA Question 1

How does the pH change when 1.0 mol of solid NH_4Cl is added to 1.0 L of a 1.0 M solution of NH_3?

IDENTIFY

Known	Unknown
0.10 M CH_3CO_2H and 0.10 M $NaCH_3CO_2$	pH
For CH_3CO_2H, $K_a = 1.8 \times 10^{-5}$	Concentration of all species Percent dissociation

STRATEGY

This problem can be solved by thinking about the chemistry involved and identifying the acid–base properties of the various species in solution. The procedure for solving acid–base problems outlined in Figure 16.7 can be used.

SOLUTION

Step 1. Write the balanced equation for the acid–base reaction. Because acetic acid is largely undissociated in aqueous solution and the salt sodium acetate is essentially 100% dissociated, the species present initially are

$$CH_3CO_2H \quad Na^+ \quad CH_3CO_2^- \quad H_2O$$
$$\text{Acid} \qquad \text{Inert} \quad \text{Base} \qquad \text{Acid or base}$$

Since K_a for the weak acid CH_2CO_2H is much larger than K_w, we consider the dissociation of acetic acid to be the principal reaction.

$$CH_3CO_2H(aq) + H_2O(l) \rightleftharpoons H_3O^+(aq) + CH_3CO_2^-(aq) \qquad K_a = 1.8 \times 10^{-5}$$

Step 2. Make a table that lists initial concentration, change in concentration, and equilibrium concentration for each of the reactants and products in the balanced chemical equation.

continued on next page

We define x as the concentration of acid that dissociates—here acetic acid—but we need to remember that the acetate ions come from two sources: 0.10 mol/L of acetate comes from the sodium acetate present *initially,* and x mol/L comes from the dissociation of acetic acid.

Principal Reaction	$CH_3CO_2H(aq) + H_2O(l) \rightleftharpoons H_3O^+(aq) + CH_3CO_2^-(aq)$		
Initial concentration (M)	0.10	~0	0.10
Change (M)	$-x$	$+x$	$+x$
Equilibrium concentration (M)	$0.10 - x$	x	$0.10 + x$

Step 3. Substitute the equilibrium concentrations into the equilibrium equation for the reaction, and solve for x.

$$K_a = 1.8 \times 10^{-5} = \frac{[H_3O^+][CH_3CO_2^-]}{[CH_3CO_2H]} = \frac{(x)(0.10 + x)}{(0.10 - x)}$$

Because K_a is small, x is small compared to 0.10 we can make the approximation that $(0.10 + x) \approx (0.10 - x) \approx 0.10$, which simplifies the solution of the equation:

$$1.8 \times 10^{-5} = \frac{(x)(0.10 + x)}{(0.10 - x)} \approx \frac{(x)(0.10)}{0.10}$$

$$x = 1.8 \times 10^{-5} \text{ M}$$

To check the assumption, calculate the percentage that the calculated value of x is of the initial concentration.

$$\frac{1.8 \times 10^{-5}}{0.10} \times 100\% = 0.018\%$$

If x is less than 5% of the initial concentration, the assumption is valid and does not cause significant error in the calculated equilibrium concentrations. In this case, the assumption is valid because 0.018% is less than 5%. *It's important to check the validity of the simplifying approximation in every problem* because x is not always negligible compared to the initial concentration of the acid.

Step 4. Calculate the concentrations of all reactants and products involved in the reaction.

We use the calculated value of x to obtain the equilibrium concentration of all substances involved in the reaction:

$$[CH_3CO_2H] = 0.10 \text{ M} - x = 0.10 \text{ M} - 1.8 \times 10^{-5} = 0.10 \text{ M}$$
$$[H_3O^+] = x = 1.8 \times 10^{-5}$$
$$[CH_3CO_2^-] = 0.10 \text{ M} + x = 0.10 \text{ M} + 1.8 \times 10^{-5} = 0.10 \text{ M}$$

Step 5. Calculate the pH.

$$pH = -\log(1.8 \times 10^{-5}) = 4.74$$

The percent dissociation of acetic acid is

$$\text{Percent dissociation} = \frac{[CH_3CO_2H]_{\text{dissociated}}}{[CH_3CO_2H]_{\text{initial}}} \times 100\% = \frac{1.8 \times 10^{-5}}{0.10} \times 100\% = 0.018\%$$

CHECK

In the $CH_3CO_2H/CH_3CO_2^-$ solution, $CH_3CO_2^-$ is the common ion, and raising its concentration shifts the equilibrium for the principal reaction to the left.

$$CH_3CO_2H(aq) + H_2O(l) \rightleftharpoons H_3O^+(aq) + CH_3CO_2^-(aq)$$

Consequently, the H_3O^+ concentration and the percent dissociation should decrease. As expected, the calculated value of the pH (4.74) is higher than the pH (2.89) of the 0.10 M CH_3CO_2H solution that contains no $CH_3CO_2^-$ ions.

▶ **PRACTICE 17.3** Calculate the pH and the percent dissociation of HCN ($K_a = 4.9 \times 10^{-10}$) in a solution that is 0.025 M in HCN and 0.010 M in NaCN.

▶ **APPLY 17.4** Calculate the pH of a solution prepared by mixing equal volumes of 0.20 M methylamine (CH_3NH_2, $K_b = 3.7 \times 10^{-4}$) and 0.60 M CH_3NH_3Cl.

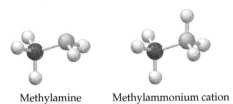

Methylamine   Methylammonium cation

CONCEPTUAL WORKED EXAMPLE 17.3

Determining the Effect of a Common Ion on pH and Percent Dissociation

The following pictures represent initial concentrations in solutions of a weak acid HA that may also contain the sodium salt NaA. Which solution has the highest pH? Which has the largest percent dissociation of HA? (Na^+ and H_3O^+ ions and solvent water molecules have been omitted for clarity.)

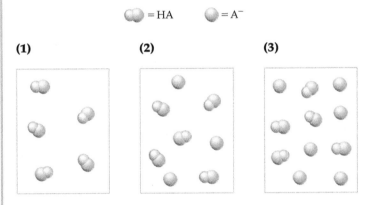

= HA   = A⁻

(1)   (2)   (3)

STRATEGY

The pH and the percent dissociation of HA are determined by the extent of the reaction $HA(aq) + H_2O(l) \rightleftharpoons H_3O^+(aq) + A^-(aq)$. The dissociation equilibrium shifts to the left upon adding more of the common ion A^-, thus decreasing the H_3O^+ concentration (increasing the pH) and decreasing the percent dissociation. To answer the questions posed, simply count the number of A^- ions.

SOLUTION

All three solutions contain the same number of HA molecules, but different numbers of A^- ions—none for solution 1, three for solution 2, and six for solution 3. The dissociation equilibrium lies farthest to the left for solution 3, and therefore solution 3 has the lowest H_3O^+ concentration and the highest pH. For solution 1, no common ion is present to suppress the dissociation of HA, and therefore solution 1 has the largest percent dissociation.

▶ **CONCEPTUAL PRACTICE 17.5** The following pictures represent initial concentrations in solutions of the weak acid HF that may also contain the sodium salt NaF. Which solution has the highest pH? Which has the largest percent dissociation of HF?

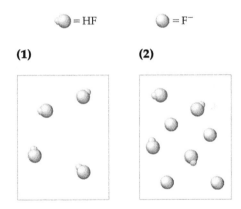

= HF   = F⁻

(1)   (2)

▶ **CONCEPTUAL APPLY 17.6** The following pictures represent initial concentrations in solutions of a weak base B that may also contain the chloride salt BH^+Cl^-. Which solution has the lowest pH? Which has the largest percent dissociation of B? (Cl^- and OH^- ions and solvent water molecules have been omitted for clarity.)

= B   = BH⁺

(1)   (2)   (3)

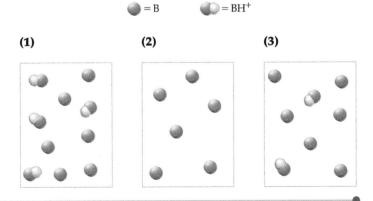

17.3 BUFFER SOLUTIONS

Solutions like those discussed in Section 17.2 that contain a weak acid and its conjugate base are called **buffer solutions** because they resist large changes in pH.

Buffer solution

Weak acid (HA)

+

Conjugate base(A⁻)

For example: $\begin{cases} CH_3CO_2H + CH_3CO_2^- \\ HF + F^- \\ NH_4^+ + NH_3 \\ H_2PO_4^- + HPO_4^{2-} \end{cases}$

BIG IDEA Question 2   [Go to eText]

The following pictures represent *initial* concentrations in solutions that contain a weak acid HA and/or its sodium salt NaA. Which of the solutions are buffer solutions? (Na⁺ ions and solvent water molecules have been omitted for clarity.)

⬤ = HA   ● = A⁻

(1)

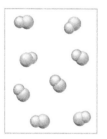

(2)

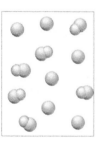

(3)

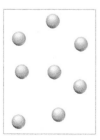

How do buffer solutions resist changes in pH? If a small amount of OH⁻ is added to a buffer solution, the pH increases, but not by much because the acid component of the buffer solution neutralizes the added OH⁻. The hydroxide ion (OH⁻), which has a large effect on pH, reacts with HA to produce the A⁻ ion, which has a small effect on pH because it is a weak base.

$$HA(aq) + OH^-(aq) \longrightarrow H_2O(l) + A^-(aq)$$

If a small amount of H₃O⁺ is added to a buffer solution, the pH decreases, but again not by much because the base component of the buffer solution neutralizes the added H₃O⁺.

$$A^-(aq) + H_3O^+(aq) \longrightarrow H_2O(l) + HA(aq)$$

The hydronium ion (H₃O⁺), which has a large effect on pH, reacts with A⁻ to produce the HA molecule, which has a small effect on pH because it is a weak acid.

Buffer solutions are very important in biological systems. Blood, for example, is a buffer solution that can reduce pH changes when acids and bases are produced in biological reactions. Cardiac arrest is one condition that can add acid to blood due to the buildup of carbon dioxide that occurs when the heart stops circulating blood. In contrast, hyperventilation increases the amount of CO_2 removed from the body and can raise blood pH. The pH of human blood is carefully controlled in the pH range 7.35 to 7.45 by conjugate acid–base pairs, primarily H_2CO_3 and its conjugate base HCO_3^-. The oxygen-carrying ability of hemoglobin in blood depends on control of the pH to within 0.1 pH unit.

To appreciate the ability of a buffer solution to maintain a nearly constant pH, let's contrast the behavior of the 0.10 M acetic acid–0.10 M sodium acetate buffer with that of a 1.8×10^{-5} M HCl solution. As shown in **FIGURE 17.3**, the dilute solution of a strong acid has the same pH (4.74) as the buffer solution, but it doesn't have the capacity to soak up added acid or base. For example, if we add 0.01 mol of solid NaOH to 1.00 L of 1.8×10^{-5} M HCl, a negligible amount of the OH⁻(1.8×10^{-5} mol) is neutralized and the concentration of OH⁻ after neutralization is 0.01 mol/1.00 L = 0.01 M. As a result, the pH rises from 4.74 to 12.0.

$$[H_3O^+] = \frac{K_w}{[OH^-]} = \frac{1.0 \times 10^{-14}}{0.01} = 1 \times 10^{-12} \, M$$

$$pH = 12.0$$

This pH change is indicated by the color change from red in Figure 17.3a to yellow in Figure 17.3b. In contrast, when 0.01 mole of solid NaOH is added to the 0.10 M acetic acid–0.10 M sodium acetate buffer, the pH barely changes, as indicated by the constant red color in 17.3c and d.

(a) 1.00 L of 1.8 × 10⁻⁵ M HCl (pH = 4.74)

(b) The solution from **(a)** turns yellow (pH > 5.4) after addition of only a few drops of 0.10 M NaOH.

(c) 1.00 L of a 0.10 M acetic acid–0.10 M sodium acetate buffer solution (pH = 4.74)

(d) The solution from **(c)** is still red (pH < 5.4) after addition of 100 mL of 0.10 M NaOH.

▲ **FIGURE 17.3**

The abilities of a strong acid solution and a buffer solution to absorb added base. The color of each solution is due to the presence of a few drops of methyl red, an acid–base indicator that is red at pH less than about 5.4 and yellow at pH greater than about 5.4.

FIGURE 17.4 illustrates how a buffer functions. The pH or $[H_3O^+]$ is influenced by the ratio of weak acid to conjugate base since the K_a value for the weak acid remains constant.

$$[H_3O^+] = K_a \frac{[HA]}{[A^-]}$$

When OH^- or H_3O^+ is added to the buffer solution, the ratio of weak acid to conjugate base changes. If the amount of OH^- or H_3O^+ added is small compared to the amount of weak acid and base in the buffer, then the ratio and pH will only change slightly.

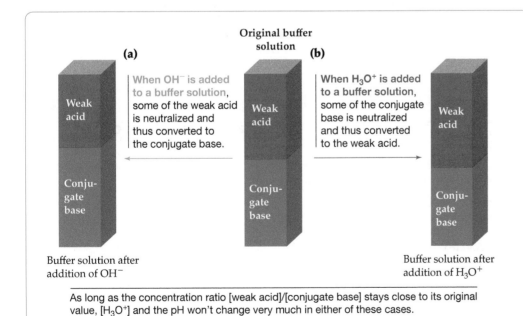

(a)

Weak acid

When OH⁻ is added to a buffer solution, some of the weak acid is neutralized and thus converted to the conjugate base.

Original buffer solution

Weak acid

(b)

When H₃O⁺ is added to a buffer solution, some of the conjugate base is neutralized and thus converted to the weak acid.

Weak acid

Conjugate base

Conjugate base

Conjugate base

Buffer solution after addition of OH⁻

Buffer solution after addition of H₃O⁺

As long as the concentration ratio [weak acid]/[conjugate base] stays close to its original value, [H₃O⁺] and the pH won't change very much in either of these cases.

◀ **FIGURE 17.4**

The addition of (a) OH^- and (b) H_3O^+ to a buffer solution.

◀ **Figure It Out**

Under what conditions will the addition of OH^- and H_3O^+ change the pH of the buffer solution by a large amount?

Answer: The pH of the solution will change significantly if the amount of OH⁻ and H₃O⁺ added changes the ratio of weak acid to conjugate base by a large amount.

We sometimes talk about the buffering ability of a solution using the term **buffer capacity** as a measure of the amount of acid or base that the solution can absorb without a significant change in pH. Buffer capacity is also a measure of how little the pH changes with the addition of a given amount of acid or base. Buffer capacity depends on how many moles of weak acid and conjugate base are present. For equal volumes of solution, the more concentrated the solution, the greater the buffer capacity. For solutions having the same concentration, the greater the volume, the greater the buffer capacity.

To see how a buffer solution works, let's return to the 0.10 M acetic acid–0.10 M sodium acetate solution discussed in Section 17.2. The principal reaction and the equilibrium concentrations for the solution are

Principal Reaction	$CH_3CO_2H(aq) + H_2O(l) \rightleftharpoons H_3O^+(aq) + CH_3CO_2^-(aq)$		
Equilibrium concentration (M)	$0.10 - x$	x	$0.10 + x$

If we solve the equilibrium equation for $[H_3O^+]$, we obtain

$$K_a = \frac{[H_3O^+][CH_3CO_2^-]}{[CH_3CO_2H]}$$

$$[H_3O^+] = K_a \frac{[CH_3CO_2H]}{[CH_3CO_2^-]}$$

Thus, the H_3O^+ concentration in a buffer solution has a value close to the value of K_a for the weak acid but differs by a factor equal to the concentration ratio [weak acid]/[conjugate base]. In the 0.10 M acetic acid–0.10 M sodium acetate solution, in which the concentration ratio is 1.0, $[H_3O^+]$ equals K_a:

$$[H_3O^+] = K_a \frac{(0.10 - x)}{(0.10 + x)} = K_a\left(\frac{0.10}{0.10}\right) = K_a = 1.8 \times 10^{-5}\ M$$

$$pH = pK_a = -\log(1.8 \times 10^{-5}) = 4.74$$

Note that in calculating this result we have set the *equilibrium* concentrations, $(0.10 - x)$ and $(0.10 + x)$, equal to the *initial* concentrations, 0.10, because x is negligible compared with the initial concentrations. For commonly used buffer solutions, K_a is small and the initial concentrations are relatively large. As a result, x is generally negligible compared with the initial concentrations, and we can use initial concentrations in the calculations. Now let's consider what happens when we add H_3O^+ or OH^- to a buffer solution as shown in Worked Example 17.4.

WORKED EXAMPLE 17.4

Go to
eText

Calculating the pH of a Buffer after Addition of OH⁻ or H₃O⁺

Calculate the pH of 1.00 L of a 0.10 M acetic acid–0.10 M sodium acetate solution with an initial pH of 4.74 after the addition of

(a) 0.01 mol of OH^-
(b) 0.01 mol of H_3O^+

IDENTIFY

Known	Unknown
Volume of buffer solution (1.00 L)	Final pH
Initial concentration of acetic acid and acetate ion (0.10 M)	
Amount of acid or base added (0.01 mol)	

STRATEGY

Because neutralization reactions involving strong acids or strong bases go essentially 100% to completion (Section 17.1), we must write the reaction and set up an equilibrium table to find $[H_3O^+]$.

SOLUTION (a)

When 0.01 mol of OH^- is added to the buffer solution, it will react with the weak acid (CH_3CO_2H). The neutralization reaction will alter the numbers of moles. We set up an equilibrium table to keep track of the change, putting the initial number of moles of acetic acid and acetate ion in the first row. Initially, we have (1.00 L) (0.10 mol/L) = 0.10 mol of acetic acid and an equal amount of acetate ion.

Neutralization Reaction	$CH_3CO_2H(aq)$ +	$OH^-(aq)$	$\xrightarrow{100\%}$	$H_2O(l)$ +	$CH_3CO_2^-(aq)$
Before reaction (mol)	0.10	0.01			0.10
Change (mol)	−0.01	−0.01			+0.01
After reaction (mol)	0.09	~0			0.11

If we assume that the solution volume remains constant at 1.00 L, the concentrations of the buffer components after neutralization are

$$[CH_3CO_2H] = \frac{0.09 \text{ mol}}{1.00 \text{ L}} = 0.09 \text{ M}$$

$$[CH_3CO_2^-] = \frac{0.11 \text{ mol}}{1.00 \text{ L}} = 0.11 \text{ M}$$

Substituting these concentrations into the expression for $[H_3O^+]$, we can then calculate the pH:

$$[H_3O^+] = K_a\frac{[CH_3CO_2H]}{[CH_3CO_2^-]}$$

$$= (1.8 \times 10^{-5})\left(\frac{0.09}{0.11}\right) = 1.5 \times 10^{-5} \text{ M}$$

$$pH = 4.82$$

CHECK (a)

Adding 0.01 mol of NaOH changes $[H_3O^+]$ by only a small amount because the concentration ratio [weak acid]/[conjugate base] changes by only a small amount, from 1.0 to 0.818 (0.09/0.11). The corresponding change in pH, from 4.74 to 4.82, is only 0.08 pH unit.

SOLUTION (b)

When 0.01 mol of H_3O^+ is added to the buffer solution, it will react with the weak base ($CH_3CO_2^-$). The added strong acid will convert 0.01 mol of acetate ions to 0.01 mol of acetic acid because of the neutralization reaction.

Neutralization Reaction	$CH_3CO_2^-(aq)$ +	$H_3O^+(aq)$	$\xrightarrow{100\%}$	$H_2O(l)$ +	$CH_3CO_2H(aq)$
Before reaction (mol)	0.10	0.01			0.10
Change (mol)	−0.01	−0.01			+0.01
After reaction (mol)	0.09	~0			0.11

The concentrations after neutralization will be $[CH_3CO_2H] = 0.11$ M and $[CH_3CO_2^-] = 0.09$ M, and the pH of the solution will be 4.66:

$$[H_3O^+] = K_a\frac{[CH_3CO_2H]}{[CH_3CO_2^-]}$$

$$= (1.8 \times 10^{-5})\left(\frac{0.11}{0.09}\right) = 2.2 \times 10^{-5} \text{ M}$$

$$pH = 4.66$$

CHECK (b)

Adding 0.01 mol of H_3O^+ changes the pH by only a small amount because the concentration ratio [weak acid]/[conjugate base] changes by only a small amount, from 1.0 to 1.2 (0.11/0.09). The corresponding change in pH, from 4.74 to 4.66, is only −0.08 pH unit.

continued on next page

▶ **PRACTICE 17.7**   Calculate the pH of 0.100 L of a buffer solution that is 0.25 M in HF and 0.50 M in NaF with an initial pH of 3.76 after the addition of 0.004 mol of KOH. ($K_a = 3.5 \times 10^{-4}$)

▶ **APPLY 17.8** (a) Calculate the change in pH when 0.002 mol of HNO_3 is added to 0.100 L of a buffer solution that is 0.050 M in HF and 0.100 M in NaF.

(b) Will the pH change if the solution is diluted by a factor of 2?

17.4 THE HENDERSON–HASSELBALCH EQUATION

We saw in Section 17.3 that the H_3O^+ concentration in a buffer solution depends on the dissociation constant of the weak acid and on the concentration ratio [weak acid]/[conjugate base]:

$$[H_3O^+] = K_a \frac{[\text{Acid}]}{[\text{Base}]}$$

This equation can be rewritten in logarithmic form by taking the negative base-10 logarithm of both sides:

$$pH = -\log[H_3O^+] = -\log\left(K_a \frac{[\text{Acid}]}{[\text{Base}]}\right) = -\log K_a - \log \frac{[\text{Acid}]}{[\text{Base}]}$$

Because $pK_a = -\log K_a$ and

$$-\log \frac{[\text{Acid}]}{[\text{Base}]} = \log \frac{[\text{Base}]}{[\text{Acid}]}$$

we obtain an expression called the **Henderson–Hasselbalch equation.**

> **Henderson–Hasselbalch equation** $pH = pK_a + \log \dfrac{[\text{Base}]}{[\text{Acid}]}$

The Henderson–Hasselbalch equation says that the pH of a buffer solution has a value close to the pK_a of the weak acid, differing only by the amount log [base]/[acid]. When [base]/[acid] = 1, then log [base]/[acid] = 0 and the pH equals the pK_a.

The real importance of the Henderson–Hasselbalch equation, particularly in biochemistry, is that it tells us how the pH affects the percent dissociation of a weak acid. Suppose, for example, that you have a solution containing the amino acid glycine, one of the molecules from which proteins are made, and that the pH of the solution is 2.00 pH units greater than the pK_a of the amine group of glycine:

Glycine

$$\overset{+}{H_3}NCH_2\overset{O}{\overset{\|}{C}}O^-(aq) + H_2O(l) \rightleftharpoons H_2NCH_2\overset{O}{\overset{\|}{C}}O^-(aq) + H_3O^+(aq) \qquad K_a = 2.5 \times 10^{-10}$$
$$pK_a = 9.60$$

Glycine

$$pH = pK_a + 2.00 = 11.60$$

Since $pH = pK_a + 2.00$, then log [base]/[acid] = 2.00 and [base]/[acid] = $1.0 \times 10^2 = 100/1$. According to the Henderson–Hasselbalch equation, therefore, 100 of every 101 glycine molecules are dissociated, which corresponds to 99% dissociation:

$$\log \frac{[\text{Base}]}{[\text{Acid}]} = pH - pK_a = 2.00$$

$$\frac{[\text{Base}]}{[\text{Acid}]} = 1.0 \times 10^2 = \frac{100}{1} \qquad 99\% \text{ dissociation}$$

The Henderson–Hasselbalch equation thus gives the following relationships:

At pH = pK_a + 2.00 $\dfrac{[\text{Base}]}{[\text{Acid}]} = 1.0 \times 10^2 = \dfrac{100}{1}$ 99% dissociation

At pH = pK_a + 1.00 $\dfrac{[\text{Base}]}{[\text{Acid}]} = 1.0 \times 10^1 = \dfrac{10}{1}$ 91% dissociation

At pH = pK_a + 0.00 $\dfrac{[\text{Base}]}{[\text{Acid}]} = 1.0 \times 10^0 = \dfrac{1}{1}$ 50% dissociation

At pH = pK_a − 1.00 $\dfrac{[\text{Base}]}{[\text{Acid}]} = 1.0 \times 10^{-1} = \dfrac{1}{10}$ 9% dissociation

At pH = pK_a − 2.00 $\dfrac{[\text{Base}]}{[\text{Acid}]} = 1.0 \times 10^{-2} = \dfrac{1}{100}$ 1% dissociation

The Henderson–Hasselbalch equation also tells us how to prepare a buffer solution with a given pH. The general idea is to select a weak acid whose pK_a is close to the desired pH and then adjust the [base]/[acid] ratio to the value specified by the Henderson–Hasselbalch equation. For example, to prepare a buffer having a pH near 7, we might use the $H_2PO_4^-$ − HPO_4^{2-} conjugate acid–base pair because the pK_a for $H_2PO_4^-$ is $-\log (6.2 \times 10^{-8}) = 7.21$. Similarly, a mixture of NH_4Cl and NH_3 would be a good choice for a buffer having a pH near 9 because the pK_a for NH_4^+ is $-\log (5.6 \times 10^{-10}) = 9.25$. As a rule of thumb, the pK_a of the weak acid component of a buffer should be within ± 1 pH unit of the desired pH.

Because buffer solutions are widely used in the laboratory and in medicine, prepackaged buffers having a variety of precisely known pH values are commercially available (**FIGURE 17.5**). The manufacturer prepares these buffers by choosing a buffer system having an appropriate pK_a value and then adjusting the amounts of the ingredients so that the [base]/[acid] ratio has the proper value.

The pH of a buffer solution does not depend on the volume of the solution because a change in solution volume changes the concentrations of the acid and base by the same amount. Thus, the [base]/[acid] ratio and the pH remain unchanged. As a result, the volume of water used to prepare a buffer solution is not critical and you can dilute a buffer without changing its pH. The pH depends only on pK_a and on the relative molar amounts of the weak acid and the conjugate base.

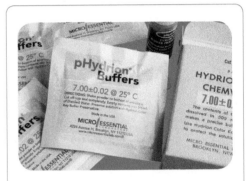

▲ **FIGURE 17.5**

Prepackaged buffer solutions of known pH and solid ingredients for preparing buffer solutions of known pH.

WORKED EXAMPLE 17.5

Using the Henderson–Hasselbalch Equation to Calculate the pH of Buffer Solution

Use the Henderson–Hasselbalch equation to calculate the pH of a buffer solution that is 0.45 M in NH_4Cl and 0.15 M in NH_3.

IDENTIFY

Known	Unknown
Concentration of weak acid and conjugate base (0.45 M NH_4Cl and 0.15 M NH_3)	pH

STRATEGY

Since NH_4^+ is the weak acid in the NH_4^+−NH_3 buffer solution, we need to find the pK_a for NH_4^+ from the tabulated K_b value for NH_3 (Appendix C). Then, substitute $[NH_3]$, $[NH_4^+]$, and the pK_a value into the Henderson–Hasselbalch equation to find the pH.

SOLUTION

Since K_b for NH_3 is 1.8×10^{-5}, K_a for NH_4^+ is 5.6×10^{-10} and $pK_a = 9.25$.

$$K_a = \frac{K_w}{K_b} = \frac{1.0 \times 10^{-14}}{1.8 \times 10^{-5}} = 5.6 \times 10^{-10}$$

$$pK_a = -\log K_a = -\log (5.6 \times 10^{-10}) = 9.25$$

continued on next page

Since $[base] = [NH_3] = 0.15$ M and $[acid] = [NH_4^+] = 0.45$ M,

$$pH = pK_a + \log \frac{[Base]}{[Acid]} = 9.25 + \log \left(\frac{0.15}{0.45}\right) = 9.25 - 0.48 = 8.77$$

The pH of the buffer solution is 8.77.

CHECK

A common error in using the Henderson–Hasselbalch equation is to invert the [base]/[acid] ratio, so it's wise to check that your answer makes chemical sense. If the concentrations of the acid and its conjugate base are equal, the pH will equal the pK_a. If the acid predominates, the pH will be less than the pK_a, and if the conjugate base predominates, the pH will be greater than the pK_a. The $[acid] = [NH_4^+]$ is greater than $[base] = [NH_3]$ and so the calculated pH (8.77) should be less than the pK_a (9.25).

▶ **PRACTICE 17.9** Use the Henderson–Hasselbalch equation to calculate the pH of a buffer solution prepared by mixing equal volumes of 0.20 M $NaHCO_3$ and 0.10 M Na_2CO_3. (For HCO_3^-, $K_a = 5.6 \times 10^{-11}$.)

▶ **APPLY 17.10** The pK_a of the amine group of the amino acid serine is 9.15.

(a) What is the pH of a solution in which the ratio of the acidic form to the basic form of the amine group in serine is 50.0?
(b) What is the percent dissociation of serine at the pH in part (a)?

Serine

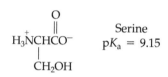

Serine
$pK_a = 9.15$

WORKED EXAMPLE 17.6

Using the Henderson–Hasselbalch Equation to Prepare a Buffer Solution

What $[NH_3]/[NH_4^+]$ ratio is required for a buffer solution that has pH = 7.00? Is a mixture of NH_3 and NH_4Cl a good choice for a buffer having pH = 7.00?

IDENTIFY

Known	Unknown
pH = 7.00	Ratio of NH_3/NH_4^+
Identity of weak acid and conjugate base (NH_4Cl–NH_3)	

STRATEGY

Find the pK_a value for NH_4^+ from the tabulated K_b value for NH_3 (Appendix C). Use the Henderson–Hasselbalch equation to calculate the $[NH_3]/[NH_4^+]$ ratio for the desired pH.

SOLUTION

Since K_b for NH_3 is 1.8×10^{-5}, K_a for NH_4^+ is 5.6×10^{-10} and $pK_a = 9.25$.

$$K_a = \frac{K_w}{K_b} = \frac{1.0 \times 10^{-14}}{1.8 \times 10^{-5}} = 5.6 \times 10^{-10}$$

$$pK_a = -\log K_a = -\log (5.6 \times 10^{-10}) = 9.25$$

Rearrange the Henderson–Hasselbalch equation to obtain an expression for the relative amounts of NH_3 and NH_4^+ in a solution having pH = 7.00:

$$\log \frac{[Base]}{[Acid]} = pH - pK_a = 7.00 - 9.25 = -2.25$$

Therefore,

$$\frac{[NH_3]}{[NH_4^+]} = \text{antilog} (-2.25) = 10^{-2.25} = 5.6 \times 10^{-3}$$

For a typical value of $[NH_4^+]$—say, 1.0 M—the NH_3 concentration would have to be very small (0.0056 M). Such a solution is a poor buffer because it has little capacity to absorb added acid. Also, because the $[NH_3]/[NH_4^+]$ ratio is far from 1.0, addition of a small amount of H_3O^+ or OH^- will result in a large change in the pH.

CHECK

Because the desired pH (7.00) is much less than the pK_a value (9.25), the buffer must contain much more weak acid than conjugate base, in agreement with the solution.

▶ **PRACTICE 17.11** How would you prepare an $NaHCO_3$–Na_2CO_3 buffer solution that has pH = 10.40?

▶ **APPLY 17.12** Suppose you are performing an experiment that requires a constant pH of 7.00.

(a) Which buffer system would be most appropriate? $HOCl$–$NaOCl$ or $HOBr$–$NaOBr$? Refer to K_a values in Appendix C to explain your answer.
(b) Calculate the ratio of conjugate base to weak acid required to prepare a buffer with pH = 7.00.

17.5 pH TITRATION CURVES

Let's continue our study of equilibria in mixtures of acids and bases by looking at the solutions obtained in the course of an acid–base titration. In a typical acid–base **titration** (Section 4.9), a solution containing a known concentration of base (or acid) is added slowly from a buret to a second solution containing an unknown concentration of acid (or base). The progress of the titration is monitored, either by using a pH meter (**FIGURE 17.6a**) or by observing the color of a suitable acid–base indicator. With a pH meter, you can record data to produce a **pH titration curve,** a plot of the pH of the solution as a function of the volume of added titrant (**FIGURE 17.6b**).

> **REMEMBER . . .**
> Titration is a procedure for determining the concentration of a solution by allowing a carefully measured volume to react with a standard solution of another substance, whose concentration is known (Section 4.9).

(a) A pH titration in which 0.100 M NaOH is added slowly from a buret to an HCl solution of unknown concentration. The pH of the solution is measured with a pH meter and is recorded as a function of the volume of NaOH added.

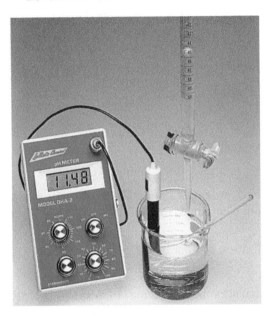

(b) The pH titration curve for titration of 40.0 mL of 0.100 M HCl with 0.100 M NaOH.

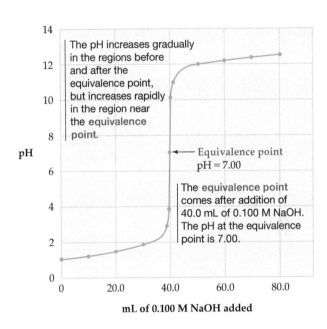

The pH increases gradually in the regions before and after the equivalence point, but increases rapidly in the region near the equivalence point.

← Equivalence point pH = 7.00

The equivalence point comes after addition of 40.0 mL of 0.100 M NaOH. The pH at the equivalence point is 7.00.

mL of 0.100 M NaOH added

▲ **FIGURE 17.6**
A strong acid–strong base titration.

Why study titration curves? The shape of a pH titration curve makes it possible to identify the **equivalence point** in a titration, the point at which stoichiometrically equivalent quantities of acid and base have been mixed together. Knowing the shape of the titration curve is also useful in selecting a suitable indicator to signal the equivalence point. We'll explore both of these points later.

We can calculate pH titration curves using the principles of aqueous solution equilibria discussed in Chapter 16. To understand why titration curves have certain characteristic shapes, let's calculate these curves for four common types of titration: (1) strong acid–strong base, (2) weak acid–strong base, (3) weak base–strong acid, and (4) polyprotic acid–strong base. Because titrations typically use milliliter volumes, we'll express solution volumes in milliliters (mL) and amounts of solute in millimoles (mmol). Molar concentration can thus be expressed in mmol/mL, a unit that is equivalent to mol/L:

$$\text{Molarity} = \frac{\text{mmol of solute}}{\text{mL of solution}} = \frac{10^{-3}\ \text{mol of solute}}{10^{-3}\ \text{L of solution}} = \frac{\text{mol of solute}}{\text{L of solution}}$$

17.6 STRONG ACID–STRONG BASE TITRATIONS

As an example of a strong acid–strong base titration, let's consider the titration of 40.0 mL of 0.100 M HCl with 0.100 M NaOH shown in Figure 17.6. We'll calculate the pH at selected points in the course of the titration to illustrate the procedures we use to calculate the entire curve.

1. **Before addition of any NaOH.** Since HCl is a strong acid, the initial concentration of H_3O^+ is 0.100 M and the pH is 1.00. (We've rounded the value of the pH to two significant figures.)

2. **Before the equivalence point.** Let's calculate the pH after addition of 10.0 mL of 0.100 M NaOH. The added OH^- ions will decrease $[H_3O^+]$ because of the neutralization reaction.

$$H_3O^+(aq) + OH^-(aq) \xrightarrow{100\%} 2\,H_2O(l)$$

Prepare a reaction table with the number of moles of each reactant.

Neutralization Reaction	$H_3O^+(aq)$ +	$OH^-(aq)$ $\xrightarrow{100\%}$	$2\,H_2O(l)$
Before reaction (mmol)	4.00	1.00	
Change (mmol)	−1.00	−1.00	
After reaction (mmol)	3.00	~0	

The number of millimoles of H_3O^+ present initially is the product of the initial volume of HCl and its molarity:

$$\text{mmol } H_3O^+ \text{ initial} = (40.0 \text{ mL})(0.100 \text{ mmol/mL}) = 4.00 \text{ mmol}$$

Similarly, the number of millimoles of OH^- added is the product of the volume of NaOH added and its molarity:

$$\text{mmol } OH^- \text{ added} = (10.0 \text{ mL})(0.100 \text{ mmol/mL}) = 1.00 \text{ mmol}$$

For each mmol of OH^- added, an equal amount of H_3O^+ will disappear because of the neutralization reaction. The number of millimoles of H_3O^+ remaining after neutralization is therefore

$$\text{mmol } H_3O^+ \text{ after neutralization} = \text{mmol } H_3O^+_{\text{initial}} - \text{mmol } OH^-_{\text{added}}$$

$$= 4.00 \text{ mmol} - 1.00 \text{ mmol} = 3.00 \text{ mmol}$$

We've carried out this calculation using *amounts* of acid and base (mmol) rather than concentrations (molarity) because the volume changes as the titration proceeds. If we divide the number of millimoles of H_3O^+ after neutralization by the total volume (now 40.0 + 10.0 = 50.0 mL), we obtain $[H_3O^+]$ after neutralization:

$$[H_3O^+] \text{ after neutralization} = \frac{3.00 \text{ mmol}}{50.0 \text{ mL}} = 6.00 \times 10^{-2} \text{ M}$$

$$pH = -\log(6.00 \times 10^{-2}) = 1.22$$

This same procedure can be used to calculate the pH at other points prior to the equivalence point, giving the results summarized in the top part of **TABLE 17.1**.

3. **At the equivalence point.** After addition of 40.0 mL of 0.100 M NaOH, we have added (40.0 mL) (0.100 mmol/mL) = 4.00 mmol of NaOH, which is just enough OH^- to neutralize all the 4.00 mmol of HCl present initially as shown in the reaction table.

Neutralization Reaction	$H_3O^+(aq)$ +	$OH^-(aq)$ $\xrightarrow{100\%}$	$2\,H_2O(l)$
Before reaction (mmol)	4.00	4.00	
Change (mmol)	−4.00	−4.00	
After reaction (mmol)	~0	~0	

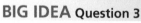

BIG IDEA Question 3 Go to eText

Why are moles used in reaction tables for titration problems instead of concentrations?

This is the equivalence point of the titration, and the pH is 7.00 because the solution contains only water and NaCl, a neutral salt derived from a strong base and a strong acid.

4. **Beyond the equivalence point.** After addition of 60.0 mL of 0.100 M NaOH, we have added $(60.0 \text{ mL})(0.100 \text{ mmol/mL}) = 6.00$ mmol of NaOH, which is more than enough to neutralize the 4.00 mmol of HCl present initially. Consequently, an excess of $OH^-(6.00 - 4.00 = 2.00$ mmol) is present.

Neutralization Reaction	$H_3O^+(aq) + OH^-(aq) \xrightarrow{100\%} 2 H_2O(l)$	
Before reaction (mmol)	4.00	6.00
Change (mmol)	−4.00	−4.00
After reaction (mmol)	~0	2.00

Since the total volume is now $40.0 + 60.0 = 100.0$ mL, the concentration of OH^- is

$$[OH^-] \text{ after neutralization} = \frac{2.00 \text{ mmol}}{100.0 \text{ mL}} = 2.00 \times 10^{-2} \text{ M}$$

The H_3O^+ concentration and the pH are

$$[H_3O^+] = \frac{K_w}{[OH^-]} = \frac{1.0 \times 10^{-14}}{2.00 \times 10^{-2}} = 5.0 \times 10^{-13} \text{ M}$$

$$pH = -\log (5.0 \times 10^{-13}) = 12.30$$

Sample results for pH calculations at other places beyond the equivalence point are also included in the bottom part of Table 17.1.

TABLE 17.1   Sample Results for pH Calculations at Various Points in the Titration of 40.0 mL of 0.100 M HCl with 0.100 M NaOH

mL NaOH Added	mmol OH⁻ Added	mmol H_3O^+ after Neutralization	Total Volume (mL)		$[H_3O^+]$ after Neutralization	pH
Before the equivalence point:						
0.0	0.0	4.00	40.0		1.00×10^{-1}	1.00
10.0	1.00	3.00	50.0		6.00×10^{-2}	1.22
20.0	2.00	2.00	60.0		3.33×10^{-2}	1.48
30.0	3.00	1.00	70.0		1.43×10^{-2}	1.84
39.0	3.90	0.10	79.0		1.27×10^{-3}	2.90
39.9	3.99	0.01	79.9		1.25×10^{-4}	3.90
At the equivalence point:						
40.0	4.00	0.00	80.0		1.00×10^{-7}	7.00
Beyond the equivalence point:						
		mmol OH⁻ after Neutralization		$[OH^-]$ after Neutralization		
40.1	4.01	0.01	80.1	1.25×10^{-4}	8.01×10^{-11}	10.10
41.0	4.10	0.10	81.0	1.23×10^{-3}	8.10×10^{-12}	11.09
50.0	5.00	1.00	90.0	1.11×10^{-2}	9.00×10^{-13}	12.05
60.0	6.00	2.00	100.0	2.00×10^{-2}	5.00×10^{-13}	12.30
70.0	7.00	3.00	110.0	2.73×10^{-2}	3.67×10^{-13}	12.43
80.0	8.00	4.00	120.0	3.33×10^{-2}	3.00×10^{-13}	12.52

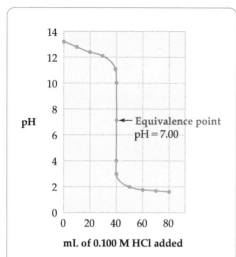

▲ FIGURE 17.7

A strong base–strong acid titration curve. The curve shown is for the titration of 40.0 mL of 0.100 M NaOH with 0.100 M HCl.

Plotting the pH data in Table 17.1 as a function of milliliters of NaOH added gives the pH titration curve shown previously in Figure 17.6b. This curve exhibits a gradual increase in pH in the regions before and after the equivalence point but a very sharp increase in pH in the region near the equivalence point. Thus, when the volume of added NaOH increases from 39.9 to 40.1 mL (0.2 mL is only about 4 drops), the pH increases from 3.9 to 10.1 (Table 17.1). This very sharp increase in pH in the region of the equivalence point is characteristic of the titration curve for any strong acid–strong base reaction, a feature that allows us to identify the equivalence point at which the concentration of the acid is unknown.

The pH curve for the titration of a strong base with a strong acid is similar except that the initial pH is high and then decreases as acid is added (**FIGURE 17.7**).

WORKED EXAMPLE 17.7

Calculating the pH in the Titration of a Strong Acid with a Strong Base

A 10.0 mL volume of 0.250 M HNO_3 is titrated with 0.150 M NaOH. Calculate the pH of the solution after addition of 15.0 mL of the NaOH solution.

IDENTIFY

Known	Unknown
Volume and molarity of HNO_3 (10.0 mL, 0.250 M)	pH
Volume and molarity of NaOH (15.0 mL, 0.150 M)	

STRATEGY

Write the neutralization reaction and set up a reaction table with amounts in moles (or millimoles). Determine if there is excess acid *(before equivalence point)*, water *(at equivalence point)*, or excess base *(after equivalence point)* once the neutralization reaction has occurred. Find the concentrations of the species remaining after the neutralization reaction by dividing the number of moles by the final volume.

SOLUTION

Prepare a reaction table with the number of moles of each reactant.

Neutralization Reaction	$H_3O^+(aq)$ +	$OH^-(aq)$ $\xrightarrow{100\%}$	$2\,H_2O(l)$
Before reaction (mmol)	2.50	2.25	
Change (mmol)	−2.25	−2.25	
After reaction (mmol)	0.25	~0	

The number of millimoles of H_3O^+ present initially is the product of the initial volume of HNO_3 and its molarity: mmol H_3O^+ initial = (10.0 mL)(0.250 mmol/mL) = 2.50 mmol.

Similarly, the number of millimoles of OH^- added is the product of the volume of NaOH added and its molarity: mmol OH^- added = (15.0 mL)(0.150 mmol/mL) = 2.25 mmol. For each mmol of OH^- added, an equal amount of H_3O^+ will disappear because of the neutralization reaction. The number of millimoles of H_3O^+ remaining after neutralization is therefore

$$\text{mmol } H_3O^+ \text{ after neutralization} = \text{mmol } H_3O^+_{\text{initial}} - \text{mmol } OH^-_{\text{added}}$$

$$= 2.50 \text{ mmol} - 2.25 \text{ mmol} = 0.25 \text{ mmol}$$

To find the concentration H_3O^+ divide the number of millimoles of H_3O^+ after neutralization by the total volume (now 40.0 + 10.0 = 50.0 mL), we obtain $[H_3O^+]$ after neutralization:

$$[H_3O^+] \text{ after neutralization} = \frac{0.25 \text{ mmol}}{50.0 \text{ mL}} = 5.00 \times 10^{-3} \text{ M}$$

$$pH = -\log(5.00 \times 10^{-3}) = 2.30$$

▶ **PRACTICE 17.13**   A 40.0 mL volume of 0.100 M HCl is titrated with 0.100 M NaOH. Calculate the pH of the solution after addition of 45.0 mL of base.

▶ **APPLY 17.14**   A 40.0 mL volume of 0.100 M NaOH is titrated with 0.0500 M HCl. Calculate the pH after addition of the following volumes of acid:

(a) 60.0 mL   (b) 80.2 mL   (c) 100.0 mL

17.7 WEAK ACID–STRONG BASE TITRATIONS

As an example of a weak acid–strong base titration, let's consider the titration of 40.0 mL of 0.100 M acetic acid ($K_a = 1.8 \times 10^{-5}$; $pK_a = 4.74$) with 0.100 M NaOH. Calculating the pH at selected points along the titration curve is straightforward because we've already met all the equilibrium problems that arise.

1. **Before addition of any NaOH.** The equilibrium problem at this point is the familiar one of calculating the pH of a solution of a **weak acid.** The calculated pH of 0.100 M acetic acid is 2.89.

2. **Before the equivalence point.** Since acetic acid is largely undissociated and NaOH is completely dissociated, the neutralization reaction is

$$CH_3CO_2H(aq) + OH^-(aq) \xrightarrow{100\%} H_2O(l) + CH_3CO_2^-(aq)$$

Set up a reaction table with initial amounts of reactants in moles to determine which species remain after the neutralization reaction.

Neutralization Reaction	$CH_3CO_2H(aq)$	+ $OH^-(aq)$	$\xrightarrow{100\%}$ $H_2O(l)$ +	$CH_3CO_2^-(aq)$
Before reaction (mmol)	4.00	1.00		0
Change (mmol)	−1.00	−1.00		+1.00
After reaction (mmol)	3.00	~0		1.00

After addition of 10.0 mL of 0.100 M NaOH, we have added (10.0 mL) (0.100 mmol/mL)=1.00 mmol of NaOH, which is enough OH⁻ to neutralize 1.00 mmol of the 4.00 mmol of CH_3CO_2H present initially. Neutralization gives a buffer solution that contains 1.00 mmol of $CH_3CO_2^-$ and 4.00 − 1.00 = 3.00 mmol of CH_3CO_2H. Consequently, the [base]/[acid] ratio is 1.00/3.00 and the Henderson–Hasselbalch equation gives a pH of 4.26.

$$pH = pK_a + \log \frac{[\text{Base}]}{[\text{Acid}]} = 4.74 + \log \frac{1.00}{3.00} = 4.26$$

This same procedure can be used to calculate the pH at other points prior to the equivalence point. Some results are summarized in the top part of **TABLE 17.2**.

3. **Halfway to the equivalence point.** After addition of 20.0 mL of 0.100 M NaOH, we have added 2.00 mmol of NaOH, which is enough OH⁻ to neutralize exactly half the 4.00 mmol of CH_3CO_2H present initially.

Neutralization Reaction	$CH_3CO_2H(aq)$	+ $OH^-(aq)$	$\xrightarrow{100\%}$ $H_2O(l)$ +	$CH_3CO_2^-(aq)$
Before reaction (mmol)	4.00	2.00		0
Change (mmol)	−2.00	−2.00		+2.00
After reaction (mmol)	2.00	~0		2.00

Neutralization gives a buffer solution that contains 2.00 mmol of $CH_3CO_2^-$ and 4.00 − 2.00 = 2.00 mmol of CH_3CO_2H. Consequently, the [base]/[acid] ratio is 1.00 and pH = pK_a:

$$pH = pK_a + \log \frac{[\text{Base}]}{[\text{Acid}]} = pK_a + \log \frac{[2.00]}{[2.00]} = pK_a = 4.74$$

REMEMBER . . .
The step-by-step procedure for solving equilibrium problems involving **weak acids** is outlined in Figure 16.7 (Section 16.9).

TABLE 17.2   Sample Results for pH Calculations at Various Points in the Titration of 40.0 mL of 0.100 M Acetic Acid with 0.100 M NaOH

mL NaOH Added	mmol OH⁻ Added	mmol $CH_3CO_2^-$ after Neutralization	mmol CH_3CO_2H after Neutralization	pH
Before the equivalence point:				
0.0	0.0	0.0	4.00	2.89
10.0	1.00	1.00	3.00	4.26
20.0	2.00	2.00	2.00	4.74
30.0	3.00	3.00	1.00	5.22
At the equivalence point:				
40.0	4.00	4.00	0.00	8.72

Beyond the equivalence point:

mL NaOH Added	mmol OH⁻ Added	mmol OH⁻ after Neutralization	Total Volume (mL)	$[OH^-]$ after Neutralization	$[H_3O^+]$ after Neutralization	pH
50.0	5.00	1.00	90.0	1.11×10^{-2}	9.0×10^{-13}	12.05
60.0	6.00	2.00	100.0	2.00×10^{-2}	5.0×10^{-13}	12.30

 Go to eText

BIG IDEA Question 4

At which point on the titration curve of a weak acid (HA) with a strong base (NaOH) does a buffer exist?

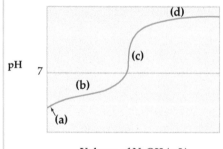

Volume of NaOH (mL)

REMEMBER . . .
Acetate ion ($CH_3CO_2^-$) gives a **basic salt solution** because it can accept a proton from water, yielding OH^-: $CH_3CO_2^-(aq) + H_2O(l) \rightleftharpoons CH_3CO_2H(aq) + OH^-(aq)$ (Section 16.14).

Halfway to the equivalence point is a special point in the titration of any weak acid with a strong base because at that point the pH always equals the pK_a of the weak acid.

4. **At the equivalence point.** The equivalence point is reached after adding 40.0 mL of 0.100 M NaOH (4.00 mmol), which is just enough OH⁻ to neutralize all the 4.00 mmol of CH_3CO_2H present initially.

Neutralization Reaction	$CH_3CO_2H(aq)$ + $OH^-(aq)$ $\xrightarrow{100\%}$ $H_2O(l)$ + $CH_3CO_2^-(aq)$		
Before reaction (mmol)	4.00	4.00	0
Change (mmol)	−4.00	−4.00	+4.00
After reaction (mmol)	~0	~0	4.00

After neutralization, the solution contains 0.0500 M $CH_3CO_2^-$:

$$[Na^+] = [CH_3CO_2^-] = \frac{4.00 \text{ mmol}}{40.0 \text{ mL} + 40.0 \text{ mL}} = 0.0500 \text{ M}$$

Because Na^+ is neither an acid nor a base and $CH_3CO_2^-$ is a weak base, we have a **basic salt solution** (Section 16.14), whose pH can be calculated as 8.72 by the method outlined in Worked Example 16.14. For a weak monoprotic acid–strong base titration, the pH at the equivalence point is *always greater than 7* because the anion of the weak acid is a base.

5. **After the equivalence point.** After addition of 50.0 mL of 0.100 M NaOH, we have added 5.00 mmol of NaOH, which is more than enough OH⁻ to neutralize the 4.00 mmol of CH_3CO_2H present initially.

Neutralization Reaction	$CH_3CO_2H(aq)$ + $OH^-(aq)$ $\xrightarrow{100\%}$ $H_2O(l)$ + $CH_3CO_2^-(aq)$		
Before reaction (mmol)	4.00	5.00	0
Change (mmol)	−4.00	−4.00	+4.00
After reaction (mmol)	~0	1.00	4.00

The total volume is $40.0 + 50.0 = 90.0$ mL, and the concentrations after neutralization are

$$[CH_3CO_2^-] = \frac{4.00 \text{ mmol}}{90.0 \text{ mL}} = 0.0444 \text{ M}$$

$$[OH^-] = \frac{1.00 \text{ mmol}}{90.0 \text{ mL}} = 0.0111 \text{ M}$$

The principal reaction is the same as that at the equivalence point:

$$CH_3CO_2^-(aq) + H_2O(l) \rightleftharpoons CH_3CO_2H(aq) + OH^-(aq)$$

In this case, however, $[OH^-]$ from the principal reaction is negligible compared with $[OH^-]$ from the excess NaOH. The hydronium ion concentration and the pH can be calculated from the excess $[OH^-]$:

$$[H_3O^+] = \frac{K_w}{[OH^-]} = \frac{1.0 \times 10^{-14}}{0.0111} = 9.0 \times 10^{-13} \text{ M}$$

$$pH = 12.05$$

In general, $[OH^-]$ from the reaction of the anion of a weak acid with water is negligible beyond the equivalence point, and the pH is determined by the concentration of OH^- from the excess NaOH.

The results of pH calculations for the titration of 0.100 M CH_3CO_2H with 0.100 M NaOH are plotted in **FIGURE 17.8**. Comparison of the titration curves for the weak acid–strong base titration and the strong acid–strong base case shows several significant differences:

- The initial rise in pH is greater for the titration of the weak acid than for the strong acid, but the curve then becomes more level in the region midway to the equivalence

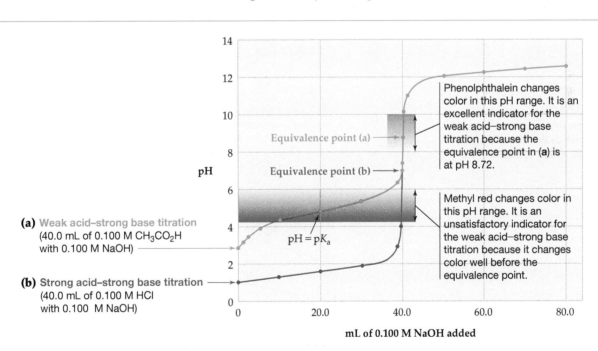

(a) Weak acid–strong base titration
(40.0 mL of 0.100 M CH_3CO_2H
with 0.100 M NaOH)

(b) Strong acid–strong base titration
(40.0 mL of 0.100 M HCl
with 0.100 M NaOH)

Equivalence point (a)

Equivalence point (b)

$pH = pK_a$

pH

mL of 0.100 M NaOH added

Phenolphthalein changes color in this pH range. It is an excellent indicator for the weak acid–strong base titration because the equivalence point in (a) is at pH 8.72.

Methyl red changes color in this pH range. It is an unsatisfactory indicator for the weak acid–strong base titration because it changes color well before the equivalence point.

Either phenolphthalein or methyl red can be used for the strong acid–strong base titration because the curve rises very steeply in the region of the equivalence point in (b) at pH 7.00.

▲ **FIGURE 17.8**

(a) A weak acid–strong base titration curve (blue) compared with (b) a strong acid–strong base curve (red). The pH ranges in which the acid–base indicators phenolphthalein and methyl red change color are indicated.

▲ **Figure It Out**

Why is the pH at the equivalence point higher in the titration of a weak acid with a strong base than in the titration of a strong acid with a strong base?

Answer: In the strong acid–strong base titration, H_2O is present at the equivalence point and the pH = 7. In the weak acid (HA)–strong base titration, the conjugate base A^- is present at the equivalence point and pH > 7.

point. Both effects are due to the buffering action of the weak acid–conjugate base mixture. The curve has its minimum slope exactly halfway to the equivalence point, where the buffering action is maximized and the $pH = pK_a$ for the weak acid.

- The increase in pH in the region near the equivalence point is smaller in the weak acid case than in the strong acid case.

- The pH at the equivalence point is greater than 7 in the weak acid titration because the anion of a weak acid is a base.

Beyond the equivalence point, the curves for the weak acid–strong base and strong acid–strong base titrations are identical because the pH in both cases is determined by the concentration of OH^- from the excess NaOH.

Figure 17.8 shows how knowing the shape of a pH titration curve makes it possible to select a suitable acid–base indicator to signal the equivalence point of a titration. Phenolphthalein is an excellent indicator for the CH_3CO_2H–NaOH titration because the pH at the equivalence point (8.72) falls within the pH range (8.2–9.8) in which phenolphthalein changes color. Methyl red, however, is an unacceptable indicator for this titration because it changes color in a pH range (4.2–6.0) well before the equivalence point. Anyone who tried to determine the acetic acid content in a solution of unknown concentration would badly underestimate the amount of acid present if methyl red were used as the indicator.

Either phenolphthalein or methyl red is a suitable indicator for a strong acid–strong base titration. The increase in pH in the region of the equivalence point is so steep that any indicator changing color in the pH range 4–10 can be used without making a significant error in locating the equivalence point.

FIGURE 17.9 shows the titration curve of 40.0 mL of 0.100 M solutions of various weak acids with 0.100 M NaOH. The equivalence point occurs after the addition of 40.0 mL of the NaOH solution for each acid. Notice that as the acid strength decreases, the pH at the equivalence point increases. This trend occurs because at the equivalence point, the conjugate base dissociation is the principal reaction.

$$A^-(aq) + H_2O(l) \rightleftharpoons HA(aq) + OH^-(aq)$$

▶ **FIGURE 17.9**

Changes in weak acid–strong base pH titration curves as the strength of the weak acid (K_a value) changes.

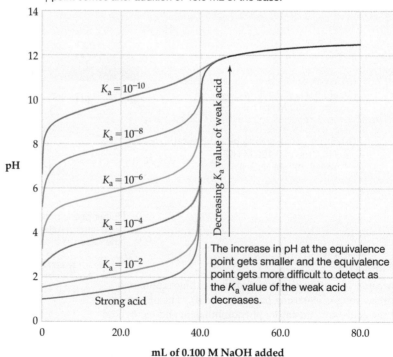

The curves shown are for titration of 40.0 mL of 0.100 M solutions of various weak acids with 0.100 M NaOH. In each case, the equivalence point comes after addition of 40.0 mL of the base.

$K_a = 10^{-10}$

$K_a = 10^{-8}$

$K_a = 10^{-6}$

$K_a = 10^{-4}$

$K_a = 10^{-2}$

Decreasing K_a value of weak acid

The increase in pH at the equivalence point gets smaller and the equivalence point gets more difficult to detect as the K_a value of the weak acid decreases.

Strong acid

mL of 0.100 M NaOH added

The weaker the acid, the stronger the conjugate base, and the higher the pH. Also note the weaker the acid, the smaller the increase in pH near the equivalence point in the titration curve.

WORKED EXAMPLE 17.8

Calculating the pH at the Equivalence Point in the Titration of a Weak Acid with a Strong Base

A 30.0 mL volume of 0.150 M HCO_2H (formic acid) is titrated with 0.300 M NaOH. Calculate the pH of the solution at the equivalence point and choose the most suitable indicator. The following acid–base indicators change color in the indicated pH ranges: bromthymol blue (6.0–7.6), phenolphthalein (8.1–9.8), and alizarin yellow (10.1–12.0).

IDENTIFY

Known	Unknown
Molarity of NaOH (0.300 M NaOH)	pH at equivalence point
Volume and molarity of HCO_2H (30.0 mL, 0.150 M)	Indicator

STRATEGY

Write the neutralization reaction and set up a reaction table with amounts in moles. The number of moles of NaOH *equals* the number of moles of HCO_2H at the equivalence point. Determine the concentration of formate ion (HCO_2^-) after the neutralization reaction has occurred and calculate the pH by solving a weak base equilibria problem.

SOLUTION

Prepare a reaction table with the number of moles of each reactant.

Neutralization Reaction	$HCO_2H(aq)$	$+ OH^-(aq)$	$\xrightarrow{100\%}$ $H_2O(l)$	$+ HCO_2^-(aq)$
Before reaction (mmol)	4.50	4.50		0
Change (mmol)	−4.50	−4.50		+4.50
After reaction (mmol)	~0	~0		4.50

After neutralization, the solution contains 4.50 mmol of HCO_2^- (formate ion). The total volume of solution at the equivalence point is the initial volume of formic acid (30.0 mL) plus the volume of NaOH added (15.0 mL). The volume of NaOH is found from the number of moles at the equivalence point (4.50 mmol) and the molarity.

$$\text{Volume of NaOH} = 4.50 \text{ mmol} \times \frac{\text{mL}}{0.300 \text{ mmol}} = 15.0 \text{ mL}$$

After neutralization, the solution contains 0.100 M HCO_2^-:

$$[HCO_2^-] = \frac{4.50 \text{ mmol}}{30.0 \text{ mL} + 15.0 \text{ mL}} = 0.100 \text{ M}$$

The pH of a basic anion can be found following the procedure in Worked Example 16.14 and an abbreviated solution is given here.

The primary reaction is

	$HCO_2^-(aq) + H_2O(l) \rightleftharpoons HCO_2H(aq) + OH^-(aq)$		
Equilibrium concentration (M)	$0.100 - x$	x	x

The value of K_b is found from the K_a value of formic acid listed in Appendix C.

$$K_b = K_w/(K_a \text{ for } HCO_2H) = (1.0 \times 10^{-14})/(1.8 \times 10^{-4})$$
$$= 5.6 \times 10^{-11}$$

The value of x is obtained from the equilibrium equation:

$$K_b = 5.6 \times 10^{-11} = \frac{[HCO_2H][OH^-]}{[HCO_2^-]} = \frac{(x)(x)}{(0.100 - x)} \approx \frac{x^2}{0.10}$$

$$x = [OH^-] = 2.4 \times 10^{-6} \text{ M}$$

The pH is calculated:

$$[H_3O^+] = \frac{K_w}{[OH^-]} = \frac{1.0 \times 10^{-14}}{2.4 \times 10^{-6}} = 4.2 \times 10^{-9}$$

$$pH = -\log(4.2 \times 10^{-9}) = 8.38$$

Based on the pH at the equivalence point, phenolphthalein, which changes color in the pH range 8.1–9.8 is the best indicator for this titration.

CHECK

The pH at the equivalence point in the titration is greater than 7, which is chemically reasonable for the titration of a weak acid with a strong base that produces the weak conjugate base at the equivalence point.

▶ **PRACTICE 17.15** What is the pH at the equivalence point in the titration of 100.0 mL of 0.0500 M HOCl ($K_a = 3.5 \times 10^{-8}$) with 0.100 M NaOH?

continued on next page

CONCEPTUAL APPLY 17.16 The following pictures represent solutions at various points in the titration of a weak acid HA with aqueous NaOH. (Na^+ ions and solvent water molecules have been omitted for clarity.)

Which picture corresponds to each of the following points in the titration?

(a) Before the addition of any NaOH
(b) Before the equivalence point
(c) At the equivalence point
(d) After the equivalence point

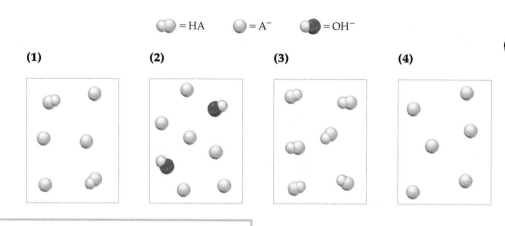

All **WORKED EXAMPLES** with this icon have an interactive video in the eText.

17.8 WEAK BASE–STRONG ACID TITRATIONS

FIGURE 17.10 shows the pH titration curve for a typical weak base–strong acid titration, the titration of 40.0 mL of 0.100 M NH_3 with 0.100 M HCl. The pH calculations for each region of the graph are outlined generally because all the types of acid-base problems have been shown in detail in previous sections.

▶ **FIGURE 17.10**

A weak base–strong acid titration curve.

▶ **Figure It Out**

Why is the pH at the equivalence point less than 7?

Answer: At the equivalence point, the weak base (NH_3) has been converted to its conjugate acid (NH_4^+).

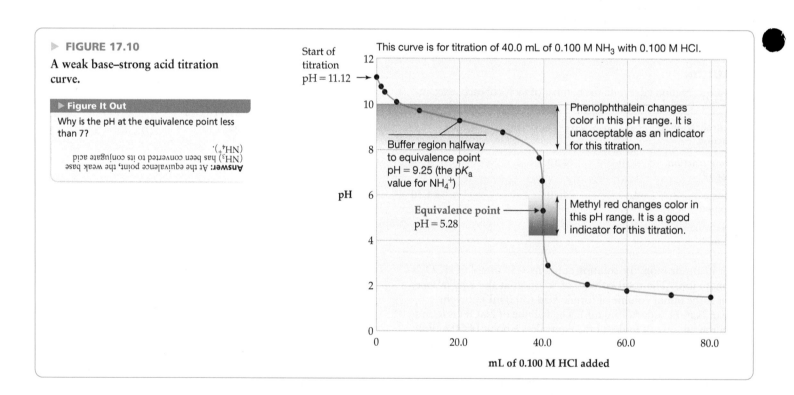

REMEMBER . . .

Equilibria in **solutions of weak bases** are treated by the same procedure used for solving problems involving weak acids (Section 16.12).

1. **Before addition of any HCl.** The equilibrium problem at the start of the titration is the familiar one of calculating the pH of a **solution of a weak base** (Section 16.12). The principal reaction at this point is the reaction of ammonia with water:

$$NH_3(aq) + H_2O(l) \rightleftharpoons NH_4^+(aq) + OH^-(aq) \quad K_b = 1.8 \times 10^{-5}$$

The initial pH is 11.12.

2. **Before the equivalence point.** As HCl is added to the NH_3 solution, NH_3 is converted to NH_4^+ because of the neutralization reaction

$$NH_3(aq) + H_3O^+(aq) \xrightarrow{100\%} NH_4^+(aq) + H_2O(l)$$

The neutralization reaction goes to completion, but the amount of H_3O^+ added before the equivalence point is insufficient to convert all the NH_3 to NH_4^+. We therefore have an NH_4^+–NH_3 buffer solution, which accounts for the leveling of the titration curve in the buffer region between the start of the titration and the equivalence point. The pH at any specific point can be calculated from the Henderson–Hasselbalch equation. After addition of 20.0 mL of 0.100 M HCl (halfway to the equivalence point), $[NH_3] = [NH_4^+]$ and the pH equals pK_a for NH_4^+ (9.25).

3. **At the equivalence point.** The equivalence point is reached after adding 40.0 mL of 0.100 M HCl (4.00 mmol). At this point, the 4.00 mmol of NH_3 present initially has been converted to 4.00 mmol of NH_4^+; $[NH_4^+] = (4.00 \text{ mmol})/(80.0 \text{ mL}) = 0.0500$ M. Since NH_4^+ is a weak acid and Cl^- is neither an acid nor a base, we have an acidic salt solution (Section 16.14). The principal reaction is

$$NH_4^+(aq) + H_2O(l) \rightleftharpoons H_3O^+(aq) + NH_3(aq) \quad K_a = K_w/K_b = 5.6 \times 10^{-10}$$

The pH at the equivalence point is 5.28.

4. **Beyond the equivalence point.** In this region, all the NH_3 has been converted to NH_4^+, and excess H_3O^+ is present from the excess HCl. Because the acid dissociation of NH_4^+ produces a negligible $[H_3O^+]$ compared with $[H_3O^+]$ from the excess HCl, the pH can be calculated directly from the concentration of the excess HCl. For example, after addition of 60.0 mL of 0.100 M HCl:

$$[H_3O^+] = \frac{6.00 \text{ mmol} - 4.00 \text{ mmol}}{100.0 \text{ mL}} = 0.0200 \text{ M}$$

$$pH = 1.70$$

17.9 POLYPROTIC ACID–STRONG BASE TITRATIONS

As a final example of an acid–base titration, let's consider the gradual addition of NaOH to the protonated form of the amino acid alanine (H_2A^+), a substance that behaves as a diprotic acid. Amino acids are both acidic and basic and can be protonated by strong acids such as HCl, yielding salts such as $H_2A^+Cl^-$. The protonated form of the amino acid has two dissociable protons and can react with two molar amounts of OH^- to give first the neutral form and then the anionic form:

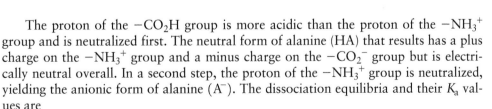

| Protonated form (H_2A^+) of alanine | Neutral form (HA) of alanine | Anionic form (A^-) of alanine |

Alanine cation (H_2A^+)

The proton of the $-CO_2H$ group is more acidic than the proton of the $-NH_3^+$ group and is neutralized first. The neutral form of alanine (HA) that results has a plus charge on the $-NH_3^+$ group and a minus charge on the $-CO_2^-$ group but is electrically neutral overall. In a second step, the proton of the $-NH_3^+$ group is neutralized, yielding the anionic form of alanine (A^-). The dissociation equilibria and their K_a values are

$$H_2A^+(aq) + H_2O(l) \rightleftharpoons H_3O^+(aq) + HA(aq) \quad K_{a1} = 4.6 \times 10^{-3}; \, pK_{a1} = 2.34$$

$$HA(aq) + H_2O(l) \rightleftharpoons H_3O^+(aq) + A^-(aq) \quad K_{a2} = 2.0 \times 10^{-10}; \, pK_{a2} = 9.69$$

FIGURE 17.11 shows the pH titration curve for the addition of solid NaOH to 1.00 L of a 1.00 M solution of H_2A^+(1.00 mol). Because K_{a1} and K_{a2} are separated by several powers of 10, the titration curve exhibits two well-defined equivalence points and two buffer regions.

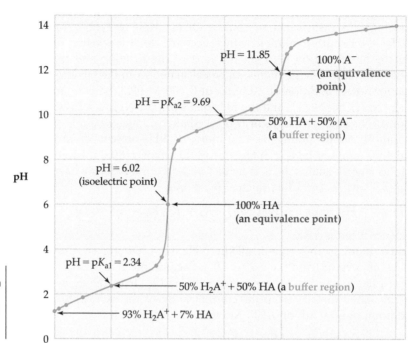

pH = 11.85   100% A⁻ (an equivalence point)

pH = pK_{a2} = 9.69   50% HA + 50% A⁻ (a buffer region)

pH = 6.02 (isoelectric point)   100% HA (an equivalence point)

pH = pK_{a1} = 2.34   50% H₂A⁺ + 50% HA (a buffer region)

93% H₂A⁺ + 7% HA

Because K_{a1} and K_{a2} are separated by several powers of 10, the titration curve exhibits two well–defined equivalence points, at pH 6.02 and pH 11.85, and two buffer regions, near pH 2.34 and pH 9.69.

▲ **FIGURE 17.11**

A diprotic acid–strong base titration curve. The change in the pH of 1.00 L of a 1.00 M solution of the protonated form of alanine (H_2A^+) on the addition of solid NaOH.

▲ **Figure It Out**

At what points in the titration curve does the pH equal pK_{a1} and pK_{a2}?

Answer: The pH equals pK_{a1} halfway to the first equivalence point, and the pH equals pK_{a2} halfway to the second equivalence point.

Calculations for Diprotic Acid–Strong Base Titrations

As a review of many of the acid–base problems we've encountered in this chapter and in Chapter 16, let's calculate the pH at several points on the curve for titration of the protonated form of alanine, H_2A^+, with NaOH. We've met all these equilibrium problems previously, except for the situation at the first equivalence point. To simplify the calculations, we'll assume that the added base is solid NaOH so that we can neglect volume changes in the course of the titration.

1. **Before addition of any NaOH.** The equilibrium problem at the start of the titration is the familiar one of calculating the pH of a diprotic acid (Section 16.11). The principal reaction is the dissociation of H_2A^+, and $[H_3O^+]$ can be calculated from the equilibrium equation

$$K_{a1} = 4.6 \times 10^{-3} = \frac{[H_3O^+][HA]}{[H_2A^+]} = \frac{(x)(x)}{1.00 - x}$$

Solving the quadratic equation gives $[H_3O^+] = 0.068$ M and pH = 1.17.

2. **Halfway to the first equivalence point.** As NaOH is added, H_2A^+ is converted to HA because of the neutralization reaction

$$H_2A^+(aq) + OH^-(aq) \xrightarrow{100\%} H_2O(l) + HA(aq)$$

The addition of 0.50 mol NaOH is enough OH⁻ to neutralize exactly half of the H_2A^+ present initially.

Neutralization Reaction	$H_2A^+(aq)$ + $OH^-(aq)$ $\xrightarrow{100\%}$ $H_2O(l)$ + $HA(aq)$			
Before reaction (mol)	1.00	0.50		0
Change (mol)	−0.50	−0.50		+0.50
After reaction (mol)	0.50	~0		0.50

Halfway to the first equivalence point, we have an H_2A^+–HA buffer solution with $[H_2A^+] = [HA]$. The Henderson–Hasselbalch equation gives pH = pK_{a1} = 2.34.

3. **At the first equivalence point.** At this point, we have added just enough NaOH to convert all the H_2A^+ to HA.

Neutralization Reaction	$H_2A^+(aq)$ + $OH^-(aq)$ $\xrightarrow{100\%}$ $H_2O(l)$ + $HA(aq)$			
Before reaction (mol)	1.00	1.00		0
Change (mol)	−1.00	−1.00		+1.00
After reaction (mol)	~0	~0		1.00

The principal reaction at the first equivalence point is proton transfer between HA molecules:

$$2\,HA(aq) \rightleftharpoons H_2A^+(aq) + A^-(aq) \qquad K = K_{a2}/K_{a1} = 4.3 \times 10^{-8}$$

It can be shown that the pH at this point equals the average of the two pK_a values:

$$\text{pH at first equivalence point} \quad \frac{pK_{a1} + pK_{a2}}{2}$$

Since H_2A^+ has pK_{a1} = 2.34 and pK_{a2} = 9.69, the pH at the first equivalence point is $(2.34 + 9.69)/2$ = 6.02. The same situation holds for most polyprotic acids: The pH at the first equivalence point equals the average of pK_{a1} and pK_{a2}.

For an amino acid, the pH value, $(pK_{a1} + pK_{a2})/2$, is called the *isoelectric point* (Figure 17.11). At that point, the concentration of the neutral HA is at a maximum, and the very small concentrations of H_2A^+ and A^- are equal. Biochemists use isoelectric points to separate mixtures of amino acids and proteins.

4. **Halfway between the first and second equivalence points.** At this point, half the HA has been converted to A^- because of the neutralization reaction

$$HA(aq) + OH^-(aq) \xrightarrow{100\%} H_2O(l) + A^-(aq)$$

Neutralization Reaction	HA(aq) + $OH^-(aq)$ $\xrightarrow{100\%}$ $H_2O(l)$ + $A^-(aq)$			
Before reaction (mol)	1.00	0.50		0
Change (mol)	−0.50	−0.50		+0.50
After reaction (mol)	0.50	~0		0.50

We thus have an HA–A^- buffer solution with [HA] = $[A^-]$. Therefore, pH = pK_{a2} = 9.69.

5. **At the second equivalence point.** At this point, we have added enough NaOH to convert all the HA to A^- and we have a 1.00 M solution of a **basic salt** (Section 16.14).

REMEMBER . . .
A **basic salt** contains the anion of a weak acid, so the anion is a proton acceptor (Section 16.14).

Neutralization Reaction	HA(aq) + $OH^-(aq)$ $\xrightarrow{100\%}$ $H_2O(l)$ + $A^-(aq)$			
Before reaction (mol)	1.00	1.00		0
Change (mol)	−1.00	−1.00		+1.00
After reaction (mol)	~0	~0		1.00

The principal reaction is

$$A^-(aq) + H_2O(l) \rightleftharpoons HA(aq) + OH^-(aq)$$

and its equilibrium constant is

$$K_b = \frac{K_w}{K_a \text{ for HA}} = \frac{K_w}{K_{a2}} = \frac{1.0 \times 10^{-14}}{2.0 \times 10^{-10}} = 5.0 \times 10^{-5}$$

We can obtain $[OH^-]$ from the equilibrium equation for the principal reaction and then calculate $[H_3O^+]$ and the pH in the usual way:

$$K_b = 5.0 \times 10^{-5} = \frac{[HA][OH^-]}{[A^-]} = \frac{(x)(x)}{1.00 - x} \quad x = OH^- = 7.1 \times 10^{-3} \text{ M}$$

$$[H_3O^+] = \frac{K_w}{[OH^-]} = \frac{1.0 \times 10^{-14}}{7.1 \times 10^{-3}} = 1.4 \times 10^{-12} \text{ M}$$

$$pH = 11.85$$

Since the initial solution of H_2A^+ contained 1.00 mol of H_2A^+, the amount of NaOH required to reach the second equivalence point is 2.00 mol. Beyond the second equivalence point, the pH is determined by $[OH^-]$ from the excess NaOH.

WORKED EXAMPLE 17.9

Calculating the pH for a Diprotic Acid–Strong Base Titration

Assume that you are titrating 30.0 mL of a 0.0600 M solution of the protonated form of the amino acid methionine (H_2A^+) with 0.0900 M NaOH. Calculate the pH after addition of the following volumes of 0.0900 M NaOH:

(a) 20.0 mL   (b) 30.0 mL   (c) 35.0 mL

Methionine cation

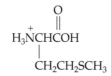

Methionine cation (H_2A^+)
$K_{a1} = 5.2 \times 10^{-3}$
$K_{a2} = 6.2 \times 10^{-10}$

IDENTIFY

Known	Unknown
Volume and molarity of H_2A^+ (30.0 mL, 0.0600 M)	pH at various points in the titration
Molarity of NaOH (0.0900 M)	
Volume of NaOH (20.0 mL, 30.0 mL, and 35.0 mL)	

STRATEGY

First calculate the number of millimoles of H_2A^+ present initially and the number of millimoles of NaOH added. Then you can tell which species remain after neutralization and therefore what type of equilibrium problem you must solve.

SOLUTION

(a) The number of millimoles of H_2A^+ present initially and the number of millimoles NaOH added are

$$\text{mmol } H_2A^+ \text{ initial} = (30.0 \text{ mL})(0.0600 \text{ mmol/mL}) = 1.80 \text{ mmol}$$

$$\text{mmol NaOH added} = (20.0 \text{ mL})(0.0900 \text{ mmol/mL}) = 1.80 \text{ mmol}$$

The added base is just enough to reach the first equivalence point, converting all the H_2A^+ to HA because of the neutralization reaction.

Neutralization Reaction	$H_2A^+(aq)$ + $OH^-(aq)$ $\xrightarrow{100\%}$ $H_2O(l)$ + $HA(aq)$		
Before reaction (mol)	1.80	1.80	0
Change (mol)	-1.80	-1.80	$+1.80$
After reaction (mol)	~ 0	~ 0	1.80

Therefore, the pH equals the average of pK_{a1} and pK_{a2}:

$$pH = \frac{pK_{a1} + pK_{a2}}{2} = \frac{[-\log(5.2 \times 10^{-3})] + [-\log(6.2 \times 10^{-10})]}{2}$$

$$= \frac{2.28 + 9.21}{2} = 5.74$$

(b) The amount of NaOH added to the 1.80 mmol of H_2A^+ present initially is $(30.0 \text{ mL}) \times (0.0900 \text{ mmol/mL}) = 2.70$ mmol. The added base is enough to convert all of the H_2A^+ to 1.80 mmol of HA and then convert $2.70 - 1.80 = 0.90$ mmol of the resulting HA to 0.90 mmol of A^- because of the neutralization reaction.

Neutralization Reaction	$HA(aq)$ + $OH^-(aq)$ $\xrightarrow{100\%}$ $H_2O(l)$ + $A^-(aq)$		
Before reaction (mol)	1.80	0.90	0
Change (mol)	-0.90	-0.90	$+0.90$
After reaction (mol)	0.90	~ 0	0.90

After neutralization, we have an HA–A^- buffer solution that contains equal amounts of HA and A^- (0.90 mmol), and so the pH equals pK_{a2} of the methionine cation H_2A^+.

$$pH = pK_{a2} = -\log(6.2 \times 10^{-10}) = 9.21$$

(c) The amount of NaOH added to the 1.80 mmol of H_2A^+ present initially is $(35.0 \text{ mL}) \times (0.0900 \text{ mmol/mL}) = 3.15$ mmol. The added base is enough to convert all of the H_2A^+ to 1.80 mmol of HA and then convert $3.15 - 1.80 = 1.35$ mmol of the resulting HA to 1.35 mmol of A^-.

Neutralization Reaction	$HA(aq)$ + $OH^-(aq)$ $\xrightarrow{100\%}$ $H_2O(l)$ + $A^-(aq)$		
Before reaction (mol)	1.80	1.35	0
Change (mol)	-1.35	-1.35	$+1.35$
After reaction (mol)	0.45	~ 0	1.35

After neutralization, we have an HA–A^- buffer solution that contains 0.45 mmol of HA and 1.35 mmol of A^-. We can use the Henderson–Hasselbalch equation to calculate the pH:

$$pH = pK_{a2} + \log \frac{[\text{Base}]}{[\text{Acid}]} = 9.21 + \log \frac{1.35}{0.45} = 9.69$$

CHECK

Because the conjugate base predominates in part (c), the pH should be greater than pK_{a2}, in agreement with the solution. Also note that the pH increases as NaOH is added [from 5.74 in part (a) to 9.21 in part (b) to 9.69 in part (c)].

▶ **PRACTICE 17.17** Assume that 40.0 mL of 0.0800 M H_2SO_3 ($K_{a1} = 1.5 \times 10^{-2}$, $K_{a2} = 6.3 \times 10^{-8}$) is titrated with 0.160 M NaOH. Calculate the pH after addition of the following volumes of 0.160 M NaOH:

(a)   20.0 mL

(b) 30.0 mL

continued on next page

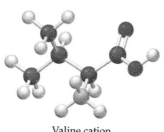

Valine cation

▶ **APPLY 17.18** Assume that 40.0 mL of a 0.0250 M solution of the protonated form of the amino acid valine (H_2A^+) is titrated with 0.100 M NaOH. Calculate the pH after addition of the following volumes of 0.100 M NaOH:

(a) 10.0 mL   (b) 15.0 mL   (c) 20.0 mL

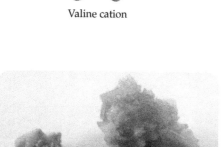

Valine cation (H_2A^+)
$K_{a1} = 4.8 \times 10^{-3}$
$K_{a2} = 2.4 \times 10^{-10}$

▲ Most kidney stones consist of insoluble calcium salts, such as calcium oxalate.

17.10 SOLUBILITY EQUILIBRIA FOR IONIC COMPOUNDS

Many biological and environmental processes involve the dissolution or precipitation of a sparingly soluble ionic compound. Tooth decay, for example, begins when tooth enamel, composed of the mineral hydroxyapatite, $Ca_5(PO_4)_3OH$, dissolves on reaction with organic acids produced by the bacterial decomposition of foods rich in sugar. Kidney stones form when moderately insoluble calcium salts, such as calcium oxalate, CaC_2O_4, precipitate slowly over a long period of time. To understand the quantitative aspects of such solubility and precipitation phenomena, we must examine the principles of solubility equilibria.

Let's consider the solubility equilibrium in a saturated solution of calcium fluoride in contact with an excess of solid calcium fluoride. Like most sparingly soluble ionic solutes, calcium fluoride is a strong electrolyte in water and exists in the aqueous phase as dissociated hydrated ions, $Ca^{2+}(aq)$ and $F^-(aq)$. At equilibrium, the ion concentrations remain constant because the rate at which solid CaF_2 dissolves to give $Ca^{2+}(aq)$ and $F^-(aq)$ exactly equals the rate at which the ions crystallize to form solid CaF_2:

$$CaF_2(s) \rightleftharpoons Ca^{2+}(aq) + 2\,F^-(aq)$$

The equilibrium equation for the dissolution reaction is

$$K_{sp} = [Ca^{2+}][F^-]^2$$

where the equilibrium constant K_{sp} is called the *solubility product constant,* or simply the **solubility product K_{sp}.** As usual for a heterogeneous equilibrium, the concentration of the solid, CaF_2, is omitted from the equilibrium equation (Section 15.4).

For the general solubility equilibrium

$$M_mX_x(s) \rightleftharpoons m\,M^{n+}(aq) + x\,X^{y-}(aq)$$

the equilibrium-constant expression for K_{sp} is

> **Solubility product** $K_{sp} = [M^{n+}]^m[X^{y-}]^x$

Thus, K_{sp} always equals the product of the equilibrium concentrations of all the ions on the right side of the chemical equation, with the concentration of each ion raised to the power of its coefficient in the balanced equation.

REMEMBER . . .
The concentration of a pure solid or a pure liquid is omitted from the equilibrium-constant expression because the ratio of its actual concentration to its concentration in the thermodynamic standard state is equal to 1 (Section 15.4).

WORKED EXAMPLE 17.10

Writing Equilibrium-Constant Expressions for K_{sp}

Write the equilibrium-constant expression for the solubility product of silver chromate, Ag_2CrO_4.

STRATEGY AND SOLUTION

First write the balanced equation for the solubility equilibrium:

$$Ag_2CrO_4(s) \rightleftharpoons 2\,Ag^+(aq) + CrO_4{}^{2-}(aq)$$

The exponents in the equilibrium-constant expression for K_{sp} are the coefficients in the balanced equation. Therefore,

$$K_{sp} = [Ag^+]^2[CrO_4{}^{2-}]$$

▶ **PRACTICE 17.19** Write the equilibrium-constant expression for K_{sp} of $Ca_3(PO_4)_2$.

▶ **CONCEPTUAL APPLY 17.20** The following pictures represent solutions of three silver salts: AgX, AgY, and AgZ. The pictures represent saturated solutions. (Other ions and solvent water molecules have been omitted for clarity.)

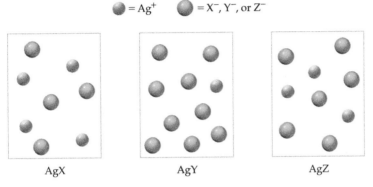

(a) Which salt has the largest value of K_{sp}? (b) Which salt has the smallest value of K_{sp}?

17.11 MEASURING K_{sp} AND CALCULATING SOLUBILITY FROM K_{sp}

The numerical value of a solubility product K_{sp} is measured by experiment. For example, we could determine K_{sp} for CaF_2 by adding an excess of solid CaF_2 to water, stirring the mixture to give a saturated solution of CaF_2, and then measuring the concentrations of Ca^{2+} and F^- in the saturated solution. To make sure that the concentrations had reached constant equilibrium values, we would want to stir the mixture for an additional period of time and then repeat the measurements. Suppose that we found $[Ca^{2+}] = 2.1 \times 10^{-4}$ M and $[F^-] = 4.1 \times 10^{-4}$ M. (The value of $[F^-]$ is twice the value of $[Ca^{2+}]$ because each mole of CaF_2 that dissolves yields 1 mol of Ca^{2+} ions and 2 mol of F^- ions.) We could then calculate K_{sp} for CaF_2:

$$K_{sp} = [Ca^{2+}][F^-]^2 = (2.1 \times 10^{-4})(4.1 \times 10^{-4})^2 = 3.5 \times 10^{-11}$$

Another way to measure K_{sp} for CaF_2 is to approach the equilibrium from the opposite direction—that is, by mixing sources of Ca^{2+} and F^- ions to give a precipitate of solid CaF_2 and a saturated solution of CaF_2. Suppose, for example, that we mix solutions of $CaCl_2$ and NaF, allow time for equilibrium to be reached, and then measure $[Ca^{2+}] = 3.5 \times 10^{-5}$ M and $[F^-] = 1.0 \times 10^{-3}$ M. These ion concentrations yield the same value of K_{sp}:

$$K_{sp} = [Ca^{2+}][F^-]^2 = (3.5 \times 10^{-5})(1.0 \times 10^{-3})^2 = 3.5 \times 10^{-11}$$

The value of K_{sp} is unaffected by the presence of other ions in solution, such as Na^+ from NaF and Cl^- from $CaCl_2$, as long as the solution is very dilute. As ion concentrations increase, K_{sp} values are somewhat modified because of electrostatic interactions between ions, but we'll ignore that complication here.

If the saturated solution is prepared by a method other than the dissolution of CaF_2 in pure water, there are no separate restrictions on $[Ca^{2+}]$ and $[F^-]$; the only

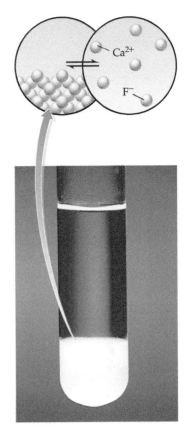

▲ A saturated solution of calcium fluoride in contact with solid CaF_2 contains constant equilibrium concentrations of $Ca^{2+}(aq)$ and $F^-(aq)$ because at equilibrium the ions crystallize at the same rate as the solid dissolves.

restriction on the ion concentrations is that the value of the equilibrium-constant expression $[Ca^{2+}][F^-]^2$ must equal the K_{sp}. That condition is satisfied by an infinite number of combinations of $[Ca^{2+}]$ and $[F^-]$, and therefore we can prepare many different solutions that are saturated with respect to CaF_2. For example, if $[F^-]$ is 1.0×10^{-2} M, then $[Ca^{2+}]$ must be 3.5×10^{-7} M:

$$[Ca^{2+}] = \frac{K_{sp}}{[F^-]^2} = \frac{3.5 \times 10^{-11}}{(1.0 \times 10^{-2})^2} = 3.5 \times 10^{-7}$$

Selected values of K_{sp} for various ionic compounds at 25°C are listed in **TABLE 17.3**, and additional values can be found in Appendix C. Like all equilibrium constants, values of K_{sp} depend on temperature (Section 15.8).

Go to
eText

BIG IDEA Question 5

Compare the K_{sp} values for CaF_2 and $PbCl_2$ listed in Table 17.3. Which salt has a higher molar solubility?

TABLE 17.3 K_{sp} Values for Some Ionic Compounds at 25 °C

Name	Formula	K_{sp}
Aluminum hydroxide	$Al(OH)_3$	1.9×10^{-33}
Barium carbonate	$BaCO_3$	2.6×10^{-9}
Calcium carbonate	$CaCO_3$	5.0×10^{-9}
Calcium fluoride	CaF_2	3.5×10^{-11}
Lead(II) chloride	$PbCl_2$	1.2×10^{-5}
Lead(II) chromate	$PbCrO_4$	2.8×10^{-13}
Silver chloride	$AgCl$	1.8×10^{-10}
Silver sulfate	Ag_2SO_4	1.2×10^{-5}

Once the K_{sp} value for a compound has been measured, you can use it to calculate the solubility of the compound—the amount of compound that dissolves per unit volume of saturated solution. Because of two complications, however, calculated solubilities are often approximate. First, K_{sp} values can be difficult to measure, and values listed in different sources might differ by as much as a factor of 10 or more. Second, calculated solubilities can be less than observed solubilities because of side reactions. For example, dissolution of $PbCl_2$ gives both Pb^{2+} and $PbCl^+$ because of some ion association between Pb^{2+} and Cl^- ions:

(1) $PbCl_2(s) \rightleftharpoons Pb^{2+}(aq) + 2\,Cl^-(aq)$

(2) $Pb^{2+}(aq) + Cl^-(aq) \rightleftharpoons PbCl^+(aq)$

In this book, we will calculate approximate solubilities assuming that ionic solutes are completely dissociated (reaction 1). In the case of $PbCl_2$, ignoring the second equilibrium gives a calculated solubility that is too low by a factor of about 2.

WORKED EXAMPLE 17.11

Calculating K_{sp} from Ion Concentrations

A particular saturated solution of silver chromate, Ag_2CrO_4, has $[Ag^+] = 5.0 \times 10^{-5}$ M and $[CrO_4^{2-}] = 4.4 \times 10^{-4}$ M. What is the value of K_{sp} for Ag_2CrO_4?

IDENTIFY

Known	Unknown
$[CrO_4^{2-}] = 4.4 \times 10^{-4}$ M	K_{sp}
$[Ag^+] = 5.0 \times 10^{-5}$ M	

STRATEGY

Substituting the equilibrium concentrations into the expression for K_{sp} of Ag_2CrO_4 gives the value of K_{sp}.

SOLUTION

$$K_{sp} = [Ag^+]^2[CrO_4^{2-}] = (5.0 \times 10^{-5})^2(4.4 \times 10^{-4}) = 1.1 \times 10^{-12}$$

CHECK

K_{sp} is approximately $(5)^2(4) = 100$ times $(10^{-5})^2(10^{-4})$. So K_{sp} is about 10^{-12}, in agreement with the solution.

▶ **PRACTICE 17.21** A saturated solution of $Ca_3(PO_4)_2$ has $[Ca^{2+}] = 2.01 \times 10^{-8}$ M and $[PO_4^{3-}] = 1.6 \times 10^{-5}$ M. Calculate K_{sp} for $Ca_3(PO_4)_2$.

▶ **APPLY 17.22** Ca^{2+}, which causes clotting, is removed from donated blood by precipitation with sodium oxalate ($Na_2C_2O_4$). CaC_2O_4 is a sparingly soluble salt ($K_{sp} = 2.3 \times 10^{-9}$). If the desired $[Ca^{2+}]$ is less than 3.0×10^{-8} M, what must be the minimum concentration of $Na_2C_2O_4$ in the blood sample?

▲ Addition of aqueous K_2CrO_4 to aqueous $AgNO_3$ gives a red precipitate of Ag_2CrO_4 and a saturated solution of Ag_2CrO_4.

─── **WORKED EXAMPLE 17.12**

Calculating Solubility from K_{sp}

Calculate the solubility of MgF_2 in water at 25 °C in units of:

(a) Moles per liter   (b) Grams per liter

IDENTIFY

Known	Unknown
K_{sp} at 25 °C (7.4×10^{-11}) (Appendix C)	Solubility (M, g/L)

STRATEGY

(a) Write the balanced equation for the solubility equilibrium assuming the complete dissociation of MgF_2 and set up an equilibrium table. If we define x as the number of moles per liter of MgF_2 that dissolves, then the saturated solution contains x mol/L of Mg^{2+} and $2x$ mol/L of F^-. Substituting these equilibrium concentrations into the expression for K_{sp} and solving for x gives the molar solubility.
(b) To convert the solubility from units of moles per liter to units of grams per liter, multiply the molar solubility of MgF_2 by its molar mass (62.3 g/mol).

SOLUTION

(a) The equilibrium table for the dissolution of MgF_2 is

Solubility Equilibrium	$MgF_2(s) \rightleftharpoons Mg^{2+}(aq) + 2 F^-(aq)$	
Initial concentration (M)	0	0
Change (M)	$+x$	$+2x$
Equilibrium concentration (M)	x	$2x$

Substituting the equilibrium concentrations into the expression for K_{sp} gives

$$K_{sp} = 7.4 \times 10^{-11} = [Mg^{2+}][F^-]^2 = (x)(2x)^2$$

$$4x^3 = 7.4 \times 10^{-11}$$

$$x^3 = 1.8 \times 10^{-11}$$

$$x = [Mg^{2+}] = \text{Molar solubility} = 2.6 \times 10^{-4} \text{ mol/L}$$

continued on next page

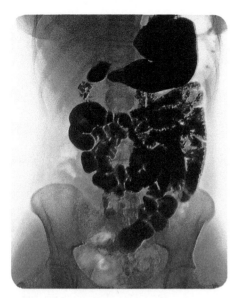

▲ The contrast in this X-ray photograph of the stomach and intestines is due to barium sulfate.

Thus, the molar solubility of MgF_2 in water at 25 °C is 2.6×10^{-4} M.

Note that the number 2 appears twice in the expression $(x)(2x)^2$. The exponent 2 is required because of the equilibrium equation, $K_{sp} = [Mg^{2+}][F^-]^2$. The coefficient 2 in $2x$ is required because each mole of MgF_2 that dissolves gives 2 mol of $F^-(aq)$.

(b) Solubility (in g/L) = $\dfrac{2.6 \times 10^{-4}\,\text{mol}}{\text{L}} \times \dfrac{62.3\ \text{g}}{\text{mol}} = 1.6 \times 10^{-2}$ g/L

CHECK

(a) The molar solubility of MgF_2 is the cube root of $K_{sp}/4$, which equals the cube root of approximately 20×10^{-12}. Because $(2)^3 = 8$, $(3)^3 = 27$, and $(10^{-4})^3 = 10^{-12}$, the molar solubility is between 2×10^{-4} M and 3×10^{-4} M. The estimate and the solution agree.

▶ **PRACTICE 17.23** What is the molar solubility of Ag_2CrO_4 in water at 25 °C? ($K_{sp} = 1.1 \times 10^{-12}$)

▶ **APPLY 17.24** Prior to having an X-ray exam of the upper gastrointestinal tract, a patient drinks an aqueous suspension of solid $BaSO_4$. (Scattering of X rays by barium greatly enhances the quality of the photograph.) Although Ba^{2+} is toxic, ingestion of $BaSO_4$ is safe because it is quite insoluble. If a saturated solution prepared by dissolving solid $BaSO_4$ in water has $[Ba^{2+}] = 1.05 \times 10^{-5}$ M, what is the value of K_{sp} for $BaSO_4$?

17.12 FACTORS THAT AFFECT SOLUBILITY

Solubility and the Common-Ion Effect

We've already discussed the common-ion effect in connection with the dissociation of weak acids and bases (Section 17.2). To see how a common ion affects the position of a solubility equilibrium, let's look again at the solubility of MgF_2:

$$MgF_2(s) \rightleftharpoons Mg^{2+}(aq) + 2\,F^-(aq)$$

In Worked Example 17.12, we found that the molar solubility of MgF_2 in pure water at 25 °C is 2.6×10^{-4} M. Thus,

$$[Mg^{2+}] = 2.6 \times 10^{-4}\ \text{M} \qquad [F^-] = 2[Mg^{2+}] = 5.2 \times 10^{-4}\ \text{M}$$

When MgF_2 dissolves in a solution that contains a common ion from another source—say, F^- from NaF—the position of the solubility equilibrium is shifted to the left by the common-ion effect. If $[F^-]$ is larger than 5.2×10^{-4} M, then $[Mg^{2+}]$ must be correspondingly smaller than 2.6×10^{-4} M to maintain the equilibrium-constant expression $[Mg^{2+}][F^-]^2$ at a constant value of $K_{sp} = 7.4 \times 10^{-11}$. A smaller value of $[Mg^{2+}]$ means that MgF_2 is less soluble in a sodium fluoride solution than it is in pure water. Similarly, the presence of Mg^{2+} from another source—say, $MgCl_2$—shifts the solubility equilibrium to the left and decreases the solubility of MgF_2.

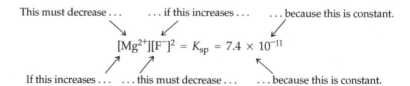

In general, the solubility of a slightly soluble ionic compound is decreased by the presence of a common ion in the solution, as predicted by Le Châtelier's principle (Section 15.6). This effect is illustrated in **FIGURE 17.12.** Quantitative aspects of the common-ion effect are explored in Worked Example 17.13.

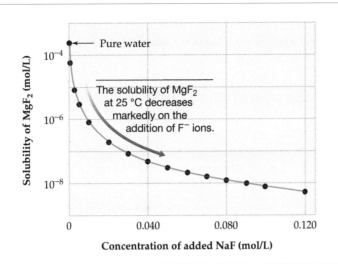

The common-ion effect. The calculated solubility of MgF_2 is plotted on a logarithmic scale.

◀ **Figure It Out**

By what factor does the solubility of MgF_2 change between pure water and a solution that contains 0.080 M NaF?

Answer: The solubility decreases by a factor of approximately 10^4. (It is approximately 10,000 times less soluble.)

WORKED EXAMPLE 17.13

Calculating Solubility in a Solution That Contains a Common Ion

Calculate the molar solubility of MgF_2 in 0.10 M NaF at 25 °C.

IDENTIFY

Known	Unknown
K_{sp} at 25 °C (7.4 × 10^{-11}) (Appendix C)	Solubility (M)
Concentration of common ion (0.10 M NaF)	

STRATEGY

Write the balanced equation for the solubility equilibrium assuming the complete dissociation of MgF_2 and set up an equilibrium table. Once again, we define x as the molar solubility of MgF_2. Substituting the equilibrium concentrations into the expression for K_{sp} and solving for x gives the molar solubility.

SOLUTION

The equilibrium table for the dissolution is

Solubility Equilibrium	$MgF_2(s) \rightleftharpoons Mg^{2+}(aq) + 2 F^-(aq)$	
Initial concentration (M)	0	0.10
Change (M)	+x	+2x
Equilibrium concentration (M)	x	0.10 + 2x

The dissolution of x mol/L of MgF_2 provides x mol/L of Mg^{2+} and $2x$ mol/L of F^-, but the total concentration of F^- is $(0.10 + 2x)$ mol/L because the solution already contains 0.10 mol/L of F^- from the completely dissociated NaF.

Substituting the equilibrium concentrations into the expression for K_{sp} gives

$$K_{sp} = 7.4 \times 10^{-11} = [Mg^{2+}][F^-]^2 = (x)(0.10 + 2x)^2$$

Because K_{sp} is small, $2x$ will be small compared to 0.10 and we can make the approximation that $(0.10 + 2x) \approx 0.10$. Therefore,

$$7.4 \times 10^{-11} = (x)(0.10 + 2x)^2 \approx (x)(0.10)^2$$

$$x = [Mg^{2+}] = \text{Molar solubility} = \frac{7.4 \times 10^{-11}}{(0.10)^2} = 7.4 \times 10^{-9} \text{ M}$$

In accord with Le Châtelier's principle, the calculated solubility of MgF_2 in 0.10 M NaF is less than that in pure water (by a factor of about 35,000!). (See Figure 17.12.)

CHECK

We can check our results by substituting the calculated equilibrium concentrations into the expression for K_{sp}:

$$K_{sp} = 7.4 \times 10^{-11} = [Mg^{2+}][F^-]^2 = (7.4 \times 10^{-9})(0.10)^2$$
$$= 7.4 \times 10^{-11}$$

▶ **PRACTICE 17.25** Calculate the molar solubility of MgF_2 in 0.10 M $MgCl_2$ at 25 °C.

▶ **APPLY 17.26** Calculate the molar solubility of $Zn(OH)_2$ in a solution buffered at pH = 11 at 25 °C. ($K_{sp} = 4.1 \times 10^{-17}$)

Solubility and the pH of the Solution

An ionic compound that contains a basic anion becomes more soluble as the acidity of the solution increases. The solubility of calcium carbonate ($CaCO_3$), for example, increases with decreasing pH (**FIGURE 17.13**) because the carbonate ions (CO_3^{2-})

combine with protons to give bicarbonate ions (HCO_3^-). As CO_3^{2-} ions are removed from the solution, the solubility equilibrium shifts to the right, as predicted by Le Châtelier's principle. The net reaction is the dissolution of $CaCO_3$ in acidic solution to give Ca^{2+} ions and HCO_3^- ions:

$$CaCO_3(s) \rightleftharpoons Ca^{2+}(aq) + \cancel{CO_3^{2-}(aq)}$$
$$H_3O^+(aq) + \cancel{CO_3^{2-}(aq)} \rightleftharpoons HCO_3^-(aq) + H_2O(l)$$

$$\text{Net} \quad CaCO_3(s) + H_3O^+(aq) \rightleftharpoons Ca^{2+}(aq) + HCO_3^-(aq) + H_2O(l)$$

Other salts that contain basic anions, such as CN^-, PO_4^{3-}, S^{2-}, or F^-, behave similarly. By contrast, pH has no effect on the solubility of salts that contain anions of strong acids (Cl^-, Br^-, I^-, NO_3^-, and ClO_4^-) because these anions are not protonated by H_3O^+.

▶ **FIGURE 17.13**

Plot of the solubility of $CaCO_3$ at 25 °C versus the pH of the solution. The solubility is plotted on a logarithmic scale.

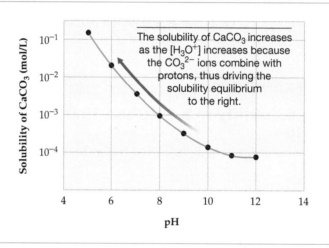

The solubility of $CaCO_3$ increases as the $[H_3O^+]$ increases because the CO_3^{2-} ions combine with protons, thus driving the solubility equilibrium to the right.

The effect of pH on the solubility of $CaCO_3$ has important environmental consequences. For instance, the formation of limestone caves, such as Mammoth Cave in Kentucky, is due to the slow dissolution of limestone ($CaCO_3$) in the slightly acidic natural water of underground streams, and the formation of stalactites and stalagmites in these caves is due to the slow precipitation of $CaCO_3$ from dripping water. Marble, another form of $CaCO_3$, also dissolves in acid, which accounts for the deterioration of marble monuments on exposure to acid rain (Inquiry Chapter 16).

The effect of pH on solubility is also important in understanding how acidity affects tooth decay. Tooth enamel is hydroxyapatite, $Ca_5(PO_4)_3(OH)$, a very insoluble compound, with $K_{sp} = 2.3 \times 10^{-59}$. Despite its small K_{sp}, hydroxyapatite can be demineralized, or dissolved, by reaction with acid:

$$Ca_5(PO_4)_3(OH)(s) + 4\,H^+(aq) \underset{\text{Remineralization}}{\overset{\text{Demineralization}}{\rightleftharpoons}} 5\,Ca^{2+}(aq) + 3\,HPO_4^{2-}(aq) + H_2O(l)$$

Both acidic beverages like sodas and bacterial metabolism of the sugars and starches left on teeth after eating are common sources of acid. The simplest and most common way to protect teeth from cavities is to brush them regularly, which physically removes the food that can feed bacteria. In addition, teeth can be protected by chemical treatment with a dilute solution of fluoride ion, which makes the enamel more resistant to attack by acid. When hydroxyapatite, $Ca_5(PO_4)_3OH$, comes in contact with F^- ions in drinking water or fluoride-containing toothpaste, OH^- ions in $Ca_5(PO_4)_3OH$ are replaced by F^- ions, giving the mineral fluorapatite, $Ca_5(PO_4)_3F$.

$$Ca_5(PO_4)_3(OH)(s) + F^-(aq) \rightleftharpoons Ca_5(PO_4)_3(F)(s) + OH^-(aq)$$

Because F^- is a much weaker base than OH^-, $Ca_5(PO_4)_3F$ is much more resistant than $Ca_5(PO_4)_3OH$ to dissolving in acids.

▲ These downward-growing, icicle-shaped structures, called stalactites, and the upward growing columns, called stalagmites, are formed in limestone caves by the slow precipitation of calcium carbonate from dripping water.

Solubility and the Formation of Complex Ions

The solubility of an ionic compound increases dramatically if the solution contains a **Lewis base** that can form a **coordinate covalent bond** to the metal cation (Section 7.5). Silver chloride, for example, is quite insoluble in water and in acid, but it dissolves in an excess of aqueous ammonia, forming the *complex ion* $Ag(NH_3)_2^+$ **(FIGURE 17.14)**. A **complex ion** is an ion that contains a metal cation bonded to one or more small molecules or ions, such as NH_3, CN^-, or OH^-. In accord with Le Châtelier's principle, ammonia shifts the solubility equilibrium to the right by tying up the Ag^+ ion in the form of the complex ion:

$$AgCl(s) \rightleftharpoons Ag^+(aq) + Cl^-(aq)$$
$$Ag^+(aq) + 2\,NH_3(aq) \rightleftharpoons Ag(NH_3)_2^+(aq)$$

The formation of a complex ion is a stepwise process, and each step has its own characteristic equilibrium constant. For the formation of $Ag(NH_3)_2^+$, the reactions are

$$Ag^+(aq) + NH_3(aq) \rightleftharpoons Ag(NH_3)^+(aq) \quad K_1 = 2.1 \times 10^3$$
$$\underline{Ag(NH_3)^+(aq) + NH_3(aq) \rightleftharpoons Ag(NH_3)_2^+(aq) \quad K_2 = 8.1 \times 10^3}$$
$$\text{Net:} \quad Ag^+(aq) + 2\,NH_3(aq) \rightleftharpoons Ag(NH_3)_2^+(aq) \quad K_f = 1.7 \times 10^7$$

$$K_1 = \frac{[Ag(NH_3)^+]}{[Ag^+][NH_3]} = 2.1 \times 10^3 \qquad K_2 = \frac{[Ag(NH_3)_2^+]}{[Ag(NH_3)^+][NH_3]} = 8.1 \times 10^3$$

$$K_f = K_1 K_2 = \frac{[Ag(NH_3)_2^+]}{[Ag^+][NH_3]^2} = 1.7 \times 10^7 \quad \text{at } 25\,°C$$

The stability of a complex ion is measured by its **formation constant K_f** (or *stability constant*), the equilibrium constant for the formation of the complex ion from the hydrated metal cation. The large value of K_f for $Ag(NH_3)_2^+$ means that this complex

---

REMEMBER . . .
A **coordinate covalent bond** forms when one atom donates two electrons (a lone pair) to another atom that has a vacant orbital (Section 7.5).
The electron-pair donor is a **Lewis base** (Section 16.15).

LOOKING AHEAD. . .
We'll see many more examples of **complex ions** in Chapter 21.

BIG IDEA Question 6

Which of the following compounds are more soluble in acidic solution than in pure water: AgCN, PbI$_2$, Al(OH)$_3$, ZnS?

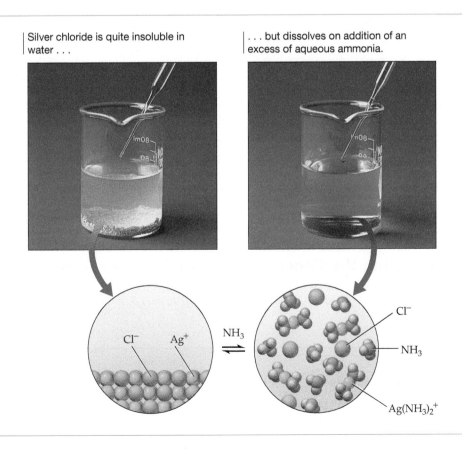

Silver chloride is quite insoluble in water . . .

. . . but dissolves on addition of an excess of aqueous ammonia.

◀ **FIGURE 17.14**

The reaction of silver chloride with aqueous ammonia.

◀ **Figure It Out**

Why does the addition of ammonia cause silver chloride to dissolve?

Answer: Ammonia forms a soluble complex ion with silver, thus removing Ag$^+$ from solution and shifting the solubility equilibrium of AgCl(s) to the right.

ion is quite stable, and nearly all the Ag^+ ion in an aqueous ammonia solution is therefore present in the form of $Ag(NH_3)_2^+$ (see Worked Example 17.14).

The net reaction for dissolution of AgCl in aqueous ammonia is the sum of the equations for the dissolution of AgCl in water and the reaction of $Ag^+(aq)$ with $NH_3(aq)$ to give $Ag(NH_3)_2^+$:

$$AgCl(s) \rightleftharpoons Ag^+(aq) + Cl^-(aq) \qquad\qquad K_{sp} = 1.8 \times 10^{-10}$$

$$Ag^+(aq) + 2\,NH_3(aq) \rightleftharpoons Ag(NH_3)_2^+(aq) \qquad K_f = 1.7 \times 10^7$$

Net: $\quad AgCl(s) + 2\,NH_3(aq) \rightleftharpoons Ag(NH_3)_2^+(aq) + Cl^-(aq) \quad K = 3.1 \times 10^{-3}$

The equilibrium constant K for the net reaction is the product of the equilibrium constants for the reactions added:

$$K = \frac{[Ag(NH_3)_2^+][Cl^-]}{[NH_3]^2} = (K_{sp})(K_f) = (1.8 \times 10^{-10})(1.7 \times 10^7) = 3.1 \times 10^{-3}$$

Because K is much larger than K_{sp}, the solubility equilibrium for AgCl lies much farther to the right in the presence of ammonia than it does in the absence of ammonia. The increase in the solubility of AgCl on the addition of ammonia is shown graphically in **FIGURE 17.15**. In general, the solubility of an ionic compound increases when the metal cation is tied up in the form of a complex ion. The quantitative effect of complex formation on the solubility of AgCl is explored in Worked Example 17.14.

▶ **FIGURE 17.15**

Plot of the solubility of AgCl in aqueous ammonia at 25 °C versus the concentration of ammonia. The solubility is plotted on a logarithmic scale.

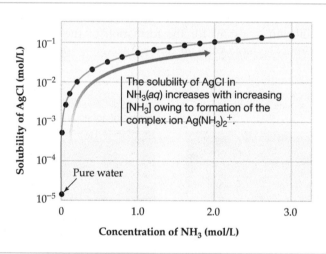

WORKED EXAMPLE 17.14

Calculating the Concentrations of Complex Ions

What are the concentrations of Ag^+, $Ag(NH_3)^+$, and $Ag(NH_3)_2^+$ in a solution prepared by adding 0.10 mol of $AgNO_3$ to 1.0 L of 3.0 M NH_3? $K_f = 1.7 \times 10^7$, $K_1 = 2.1 \times 10^3$, and $K_2 = 8.1 \times 10^3$.

IDENTIFY

Known	Unknown
Molarity and volume of NH_3 (3.0 M, 1.0 L)	$[Ag^+]$, $[Ag(NH_3)^+]$, $[Ag(NH_3)_2^+]$
Amount of $AgNO_3$ (0.10 mol)	
Equilibrium constants and reactions	

STRATEGY

Because K_1, K_2, and K_f for $Ag(NH_3)_2^+$ are all large numbers, we surmise that nearly all the Ag^+ from $AgNO_3$ will be converted to $Ag(NH_3)_2^+$:

$$Ag^+(aq) + 2\,NH_3(aq) \rightleftharpoons Ag(NH_3)_2^+(aq) \quad K_f = 1.7 \times 10^7$$

To calculate the concentrations, it's convenient to imagine that 100% of the Ag^+ is converted to $Ag(NH_3)_2^+$, followed by a tiny amount of back-reaction [dissociation of $Ag(NH_3)_2^+$] to give a small equilibrium concentration of Ag^+.

SOLUTION

First, determine concentrations assuming 100% conversion to $Ag(NH_3)_2^+$. The conversion of 0.10 mol/L of Ag^+ to $Ag(NH_3)_2^+$ consumes 0.20 mol/L of NH_3 and the following concentrations are obtained:

$$[Ag^+] = 0 \text{ M}$$
$$[Ag(NH_3)_2^+] = 0.10 \text{ M}$$
$$[NH_3] = 3.0 - 0.20 = 2.8 \text{ M}$$

Next, set up an equilibrium table for the dissociation of $Ag(NH_3)_2^+$:

	$Ag^+(aq)$ +	$2\,NH_3(aq)$ $\rightleftharpoons$	$Ag(NH_3)_2^+(aq)$
Initial concentration (M)	0	2.8	0.10
Change (M)	$+x$	$+2x$	$-x$
Equilibrium concentration (M)	x	$2.8 + 2x$	$0.10 - x$

The dissociation of x mol/L of $Ag(NH_3)_2^+$ in the back-reaction produces x mol/L of Ag^+ and $2x$ mol/L of NH_3. Therefore, the equilibrium concentrations (in mol/L) are

$$[Ag(NH_3)_2^+] = 0.10 - x$$
$$[Ag^+] = x$$
$$[NH_3] = 2.8 + 2x$$

Substituting the equilibrium concentrations into the expression for K_f and making the approximation that x is negligible compared to 0.10 (and to 2.8) gives

$$K_1 = 1.7 \times 10^7 = \frac{[Ag(NH_3)_2^+]}{[Ag^+][NH_3]^2} = \frac{0.10 - x}{(x)(2.8 + 2x)^2} = \frac{0.10}{(x)(2.8)^2}$$

$$[Ag^+] = x = \frac{0.10}{(1.7 \times 10^7)(2.8)^2} = 7.5 \times 10^{-10} \text{ M}$$

$$[Ag(NH_3)_2^+] = 0.10 - x = 0.10 - (7.5 \times 10^{-10}) = 0.10 \text{ M}$$

The concentration of $Ag(NH_3)^+$ can be calculated from either of the stepwise equilibria. Let's use the equilibrium equation for the formation of $Ag(NH_3)^+$ from Ag^+:

$$K_1 = \frac{[Ag(NH_3)^+]}{[Ag^+][NH_3]} = 2.1 \times 10^3$$

$$[Ag(NH_3)^+] = K_1[Ag^+][NH_3] = (2.1 \times 10^3)(7.5 \times 10^{-10})(2.8) = 4.4 \times 10^{-6} \text{ M}$$

Thus, nearly all the Ag^+ is in the form of $Ag(NH_3)_2^+$.

CHECK

The approximate equilibrium concentrations are $[Ag^+] = 7 \times 10^{-10}$ M, $[Ag(NH_3)^+] = 4 \times 10^{-6}$ M, $[Ag(NH_3)_2^+] = 0.1$ M, and $[NH_3] = 3$ M. We can check these results by substituting them into the equilibrium-constant expressions for K_f and K_1:

$$K_f = 1.7 \times 10^7 = \frac{[Ag(NH_3)_2^+]}{[Ag^+][NH_3]^2} \approx \frac{0.1}{(7 \times 10^{-10})(3)^2} = 2 \times 10^7$$

$$K_1 = 2.1 \times 10^3 = \frac{[Ag(NH_3)_2^+]}{[Ag^+][NH_3]} \approx \frac{4 \times 10^{-6}}{(7 \times 10^{-10})(3)} = 2 \times 10^3$$

The estimates of K_f and K_1 agree with the experimental values.

continued on next page

▶ **PRACTICE 17.27**   In an excess of $NH_3(aq)$, Cu^{2+} ion forms a deep blue complex ion, $Cu(NH_3)_4^{2+}$, which has a formation constant $K_f = 5.6 \times 10^{11}$. Calculate the concentration of Cu^{2+} in a solution prepared by adding 5.0×10^{-3} mol of $CuSO_4$ to 0.500 L of 0.400 M NH_3.

▶ **APPLY 17.28**   Cyanide ion is used in gold mining because it forms a soluble complex ion with Au^+. To study the feasibility of using cyanide to extract gold from ore, a chemist mixes 25.0 mL of 3.0×10^{-2} M $AuNO_3$ with 35.0 mL of 1.0 M NaCN. What is the final concentration of Au^+? (K_f of $Au(CN)_2^-$ is 2×10^{38})

WORKED EXAMPLE 17.15

Exploring the Effect of Complex Formation on Solubility

Calculate the molar solubility of AgCl at 25 °C in:

(a) Pure water
(b) 3.0 M NH_3

IDENTIFY

Known	Unknown
K_{sp} of AgCl at 25 °C (1.8×10^{-10}) (Appendix C)	Solubility (M)
Concentration of NH_3 (3.0 M)	
K_f of $Ag(NH_3)_2^+$ is 1.7×10^7 (Appendix C)	

STRATEGY

Write the balanced equation for the dissolution reaction, and define x as the number of moles per liter of AgCl that dissolves. Then express the equilibrium concentrations in terms of x, and substitute them into the appropriate equilibrium equation. Solving for x gives the molar solubility.

SOLUTION

(a) In pure water, the solubility equilibrium is

$$AgCl(s) \rightleftharpoons Ag^+(aq) + Cl^-(aq)$$

Substituting the equilibrium concentrations (x mol/L) into the expression for K_{sp} gives

$$K_{sp} = 1.8 \times 10^{-10} = [Ag^+][Cl^-] = (x)(x)$$
$$x = \text{Molar solubility} = \sqrt{1.8 \times 10^{-10}} = 1.3 \times 10^{-5} \text{ M}$$

(b) The balanced equation for the dissolution of AgCl in aqueous NH_3 is

$$
\begin{array}{lr}
AgCl(s) \rightleftharpoons Ag^+(aq) + Cl^-(aq) & K_{sp} = 1.8 \times 10^{-10} \\
\underline{Ag^+(aq) + 2\,NH_3(aq) \rightleftharpoons Ag(NH_3)_2^+(aq)} & \underline{K_f = 1.7 \times 10^7} \\
\text{Net:}\quad AgCl(s) + 2\,NH_3(aq) \rightleftharpoons Ag(NH_3)_2^+(aq) + Cl^-(aq) & K = 3.1 \times 10^{-3}
\end{array}
$$

Setting up an equilibrium table for the reaction:

Solubility Equilibrium	$AgCl(s) + 2\,NH_3(aq) \rightleftharpoons Ag(NH_3)_2^+(aq) + Cl^-(aq)$		
Initial concentration (M)	3.0	0	0
Change (M)	$-2x$	$+x$	$+x$
Equilibrium concentration (M)	$3.0 - 2x$	x	x

If we define x as the number of moles per liter of AgCl that dissolves, then the saturated solution contains x mol/L of $Ag(NH_3)_2^+$, x mol/L of Cl^-, and $(3.0 - 2x)$ mol/L of NH_3. (We're assuming that essentially all the Ag^+ is in the form of $Ag(NH_3)_2^+$, as proved in Worked Example 17.14.) Substituting the equilibrium concentrations into the equilibrium equation gives

$$K = 3.1 \times 10^{-3} = \frac{[Ag(NH_3)_2^+][Cl^-]}{[NH_3]^2} = \frac{(x)(x)}{(3.0 - 2x)^2}$$

Taking the square root of both sides, we obtain

$$5.6 \times 10^{-2} = \frac{x}{3.0 - 2x}$$

$$x = (5.6 \times 10^{-2})(3.0 - 2x) = 0.17 - 0.11x$$

$$x = \frac{0.17}{1.11} = 0.15 \text{ M}$$

The molar solubility of AgCl in 3.0 M NH_3 is 0.15 M. Thus, AgCl is much more soluble in aqueous NH_3 than in pure water, as shown in Figure 17.15.

CHECK

Check the calculated equilibrium concentrations by substituting them into the appropriate equilibrium equation. For part **(b)**, for example, $[Ag(NH_3)_2^+] = [Cl^-] = x = 0.15$ M and $[NH_3] = 3.0 - 2x = 2.7$ M. Since $[NH_3]$ is 10% less than 3 M, $K = [Ag(NH_3)_2^+]$ $[Cl^-]/[NH_3]^2$ is about 20% greater than $(0.15/3)^2 = (0.05)^2 = 2.5 \times 10^{-3}$. The estimate and the experimental value of $K = 3.1 \times 10^{-3}$ agree.

▶ **PRACTICE 17.29** Silver bromide dissolves in aqueous sodium thiosulfate, $Na_2S_2O_3$, yielding the complex ion $Ag(S_2O_3)_2^{3-}$:

$$AgBr(s) + 2 S_2O_3^{2-}(aq) \rightleftharpoons Ag(S_2O_3)_2^{3-}(aq) + Br^-(aq) \quad K = ?$$

Calculate the equilibrium constant K for the dissolution reaction, and calculate the molar solubility of AgBr in 0.10 M $Na_2S_2O_3$. $K_{sp} = 5.4 \times 10^{-13}$ for AgBr and $K_f = 4.7 \times 10^{13}$ for $Ag(S_2O_3)_2^{3-}$.

▶ **APPLY 17.30** In laboratory, a student adds 3 drops of 3 M NH_3 to 3 mL of a pale blue solution of 0.1 M $Cu(NO_3)_2$ and a precipitate forms. (The presence of Cu^{2+} ion gives a pale blue solution.) The student then adds 15 more drops of 3 M NH_3 and the precipitate dissolves and a dark blue solution forms. Explain the observations by writing equilibrium reactions for each step. Refer to Appendix C for solubility and complex ion equilibrium reactions and constants.

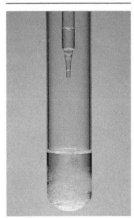

Aluminum hydroxide, a gelatinous white precipitate, forms on addition of aqueous NaOH to $Al^{3+}(aq)$.

The precipitate dissolves on addition of excess aqueous NaOH, yielding the colorless $Al(OH)_4^-$ ion. The precipitate also dissolves in aqueous HCl, yielding the colorless Al^{3+} ion.

Solubility and Amphoterism

Certain metal hydroxides, such as aluminum hydroxide (**FIGURE 17.16**), are soluble both in strongly acidic and in strongly basic solutions:

In acid: $Al(OH)_3(s) + 3 H_3O^+(aq) \rightleftharpoons Al^{3+}(aq) + 6 H_2O(l)$

In base: $Al(OH)_3(s) + OH^-(aq) \rightleftharpoons Al(OH)_4^-(aq)$

Such hydroxides are said to be **amphoteric** (am-fo-**tare**-ic), a term that comes from the Greek word *amphoteros*, meaning "in both ways."

The dissolution of $Al(OH)_3$ in excess base is just a special case of the effect of complex-ion formation on solubility: $Al(OH)_3$ dissolves because excess OH^- ions convert it to the soluble complex ion $Al(OH)_4^-$ (aluminate ion). The effect of pH on the solubility of $Al(OH)_3$ is shown in **FIGURE 17.17**.

Other examples of amphoteric hydroxides include $Zn(OH)_2$, $Cr(OH)_3$, $Sn(OH)_2$, and $Pb(OH)_2$, which react with excess OH^- ions to form the soluble complex ions $Zn(OH)_4^{2-}$ (zincate ion), $Cr(OH)_4^-$ (chromite ion), $Sn(OH)_3^-$ (stannite ion), and $Pb(OH)_3^-$ (plumbite ion), respectively. By contrast, basic hydroxides, such as $Mn(OH)_2$, $Fe(OH)_2$, and $Fe(OH)_3$, dissolve in strong acid but not in strong base.

▲ **FIGURE 17.16**

The amphoteric behavior of $Al(OH)_3$.

▶ **FIGURE 17.17**

A plot of solubility versus pH shows that Al(OH)₃ is an amphoteric hydroxide.

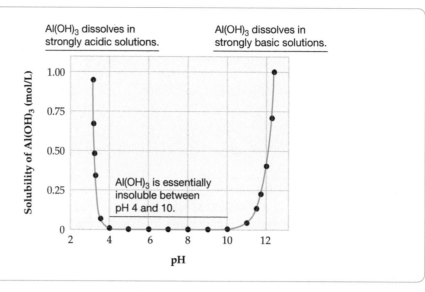

Al(OH)₃ dissolves in strongly acidic solutions.

Al(OH)₃ dissolves in strongly basic solutions.

Al(OH)₃ is essentially insoluble between pH 4 and 10.

REMEMBER . . .
The **reaction quotient** Q_c is defined in the same way as the equilibrium constant K_c except that the concentrations in the equilibrium-constant expression are not necessarily equilibrium values (Section 15.5).

17.13 PRECIPITATION OF IONIC COMPOUNDS

A common problem in chemistry is to decide whether a precipitate of an ionic compound will form when solutions that contain the constituent ions are mixed. For example, will CaF_2 precipitate on mixing solutions of $CaCl_2$ and NaF? In other words, will the dissolution reaction proceed in the reverse direction, from right to left?

$$CaF_2(s) \rightleftharpoons Ca^{2+}(aq) + 2\,F^-(aq)$$

We touched on this question briefly in Section 4.6 when we looked at solubility guidelines, but we can now get a more quantitative view. The answer depends on the value of the **ion product (IP)**, a number defined by the expression

$$IP = [Ca^{2+}]_t[F^-]_t^{\,2}$$

The IP is defined in the same way as K_{sp}, except that the concentrations in the expression for IP are initial concentrations—that is, arbitrary concentrations at time t, not necessarily equilibrium concentrations. Thus, the IP is actually a **reaction quotient** Q_c (Section 15.5), but the term *ion product* is more descriptive because, as usual, solid CaF_2 is omitted from the equilibrium-constant expression. For the general solubility equilibrium

$$M_mX_x(s) \rightleftharpoons m\,M^{n+}(aq) + x\,X^{y-}(aq)$$

the ion product is given by

Ion product $IP = [M^{n+}]_t^{\,m}[X^{y-}]_t^{\,x}$

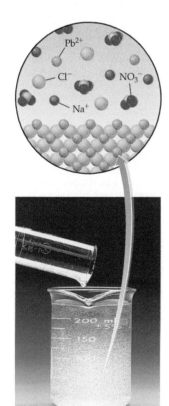

▲ When 0.150 L of 0.10 M Pb(NO₃)₂ and 0.100 L of 0.20 M NaCl are mixed, a white precipitate of PbCl₂ forms because the ion product is greater than K_{sp}.

If the value of IP is greater than K_{sp}, the solution is supersaturated with respect to CaF_2—a nonequilibrium situation. In that case, CaF_2 will precipitate, thus reducing the ion concentrations until IP equals K_{sp}. At that point, solubility equilibrium is reached and the solution is saturated.

In general, we only need to calculate the value of IP and then compare it with K_{sp} to decide whether an ionic compound will precipitate. Three cases arise:

1. If $IP > K_{sp}$, the solution is supersaturated and precipitation will occur.
2. If $IP = K_{sp}$, the solution is saturated and equilibrium exists already.
3. If $IP < K_{sp}$, the solution is unsaturated and precipitation will not occur.

WORKED EXAMPLE 17.16

Deciding Whether a Precipitate Will Form When Solutions Are Mixed

Will a precipitate form when 0.150 L of 0.10 M $Pb(NO_3)_2$ and 0.100 L of 0.20 M NaCl are mixed?

IDENTIFY

Known	Unknown
Molarity and volume of reactants (0.150 L of 0.10 M $Pb(NO_3)_2$)	IP
(0.100 L of 0.20 M NaCl)	
$K_{sp} = 1.2 \times 10^{-5}$ (Appendix C)	

STRATEGY

Since ionic compounds are strong electrolytes, $Pb(NO_3)_2$ and NaCl exist in solution as separate cations and anions. Use the solubility guidelines discussed in Section 4.6 to decide which ions might form a precipitate, and calculate their concentrations after mixing. Then calculate the IP for the possible precipitate and compare it with the value of K_{sp}.

SOLUTION

After the two solutions are mixed, the combined solution contains Pb^{2+}, NO_3^-, Na^+, and Cl^- ions and has a volume of 0.150 L + 0.100 L = 0.250 L. Because sodium salts and nitrate salts are soluble in water, the only compound that might precipitate is $PbCl_2$, which has $K_{sp} = 1.2 \times 10^{-5}$ (Appendix C). To calculate the value of IP for $PbCl_2$, first calculate the number of moles of Pb^{2+} and Cl^- in the combined solution:

Moles Pb^{2+} = (0.150 L)(0.10 mol/L) = 1.5×10^{-2} mol

Moles Cl^- = (0.100 L)(0.20 mol/L) = 2.0×10^{-2} mol

Then convert moles to molar concentrations:

$$[Pb^{2+}] = \frac{1.5 \times 10^{-2} \text{ mol}}{0.250 \text{ L}} = 6.0 \times 10^{-2} \text{ M}$$

$$[Cl^-] = \frac{2.0 \times 10^{-2} \text{ mol}}{0.250 \text{ L}} = 8.0 \times 10^{-2} \text{ M}$$

The ion product is

$$IP = [Pb^{2+}]_t[Cl^-]_t{}^2 = (6.0 \times 10^{-2})(8.0 \times 10^{-2})^2 = 3.8 \times 10^{-4}$$

Since $K_{sp} = 1.2 \times 10^{-5}$, IP is greater than K_{sp} and $PbCl_2$ will precipitate.

▶ **PRACTICE 17.31** Will a precipitate form on mixing equal volumes of 1.0×10^{-5} M $Ba(NO_3)_2$ and 4.0×10^{-5} M Na_2CO_3?

▶ **APPLY 17.32** Will a precipitate form on mixing 25 mL of 1.0×10^{-3} M $MnSO_4$, 25 mL of 1.0×10^{-3} M $FeSO_4$, and 200 mL of a buffer solution that is 0.20 M in NH_4Cl and 0.20 M in NH_3? Values of K_{sp} can be found in Appendix C.

17.14 SEPARATION OF IONS BY SELECTIVE PRECIPITATION

A convenient method for separating a mixture of ions is to add a solution that will precipitate some of the ions but not others. The anions SO_4^{2-} and Cl^-, for example, can be separated by addition of a solution of barium nitrate, $Ba(NO_3)_2$. Insoluble $BaSO_4$ precipitates, but Cl^- remains in solution because $BaCl_2$ is soluble. Similarly, the cations Ag^+ and Zn^{2+} can be separated by addition of dilute HCl. Silver chloride, AgCl, precipitates, but Zn^{2+} stays in solution because $ZnCl_2$ is soluble.

In Section 17.15, we'll see that mixtures of metal cations, M^{2+}, can be separated into two groups by the selective precipitation of metal sulfides, MS. For example, Pb^{2+}, Cu^{2+}, and Hg^{2+}, which form very insoluble sulfides, can be separated from Mn^{2+}, Fe^{2+}, Co^{2+}, Ni^{2+}, and Zn^{2+}, which form more soluble sulfides. The separation is carried out in an acidic solution and makes use of the following solubility equilibrium:

$$MS(s) + 2 H_3O^+(aq) \rightleftharpoons M^{2+}(aq) + H_2S(aq) + 2 H_2O(l)$$

The equilibrium constant for this reaction, called the *solubility product in acid*, is given the symbol K_{spa}:

$$\boxed{\text{Solubility product in acid} \quad K_{spa} = \frac{[M^{2+}][H_2S]}{[H_3O^+]^2}}$$

▲ Adding H_2S to an acidic solution of Hg^{2+} and Ni^{2+} precipitates Hg^{2+} as black HgS but leaves green Ni^{2+} in solution.

TABLE 17.4 Solubility Products in Acid (K_{spa}) at 25 °C for Metal Sulfides	
Metal Sulfide, MS	K_{spa}
MnS	3×10^7
FeS	6×10^2
CoS	3
NiS	8×10^{-1}
ZnS	3×10^{-2}
PbS	3×10^{-7}
CuS	6×10^{-16}
HgS	2×10^{-32}

The separation depends on adjusting the H_3O^+ concentration so that the reaction quotient Q_c exceeds K_{spa} for the very insoluble sulfides but not for the more soluble ones (**TABLE 17.4**).

We use K_{spa} for metal sulfides rather than K_{sp} for two reasons. First, the ion separations are carried out in acidic solution, so use of K_{spa} is more convenient. Second, because the value of K_{a2} for H_2S ($\sim 10^{-19}$) is very small, S^{2-}, like O^{2-}, is highly basic and is not an important species in aqueous solutions. The principal sulfide-containing species in aqueous solutions are H_2S in acidic solutions and HS^- in basic solutions.

In a typical experiment, the M^{2+} concentrations are about 0.01 M, and the H_3O^+ concentration is adjusted to about 0.3 M by adding HCl. The solution is then saturated with H_2S gas, which gives an H_2S concentration of about 0.10 M. Substituting these concentrations into the equilibrium-constant expression, we find that the reaction quotient Q_c is 1×10^{-2}:

$$Q_c = \frac{[M^{2+}]_t[H_2S]_t}{[H_3O^+]_t^{\,2}} = \frac{(0.01)(0.10)}{(0.3)^2} = 1 \times 10^{-2}$$

This value of Q_c exceeds K_{spa} for PbS, CuS, and HgS (Table 17.4) but does not exceed K_{spa} for MnS, FeS, CoS, NiS, or ZnS. As a result, PbS, CuS, and HgS precipitate under these acidic conditions, but Mn^{2+}, Fe^{2+}, Co^{2+}, Ni^{2+}, and Zn^{2+} remain in solution.

17.15 QUALITATIVE ANALYSIS

Qualitative analysis is a procedure for identifying the ions present in an unknown solution. The ions are identified by specific chemical tests, but because one ion can interfere with the test for another, the ions must first be separated. In the traditional scheme of analysis for metal cations, some 20 cations are separated initially into five groups by selective precipitation (**FIGURE 17.18**).

▶ **FIGURE 17.18**

Flowchart for the separation of metal cations in qualitative analysis.

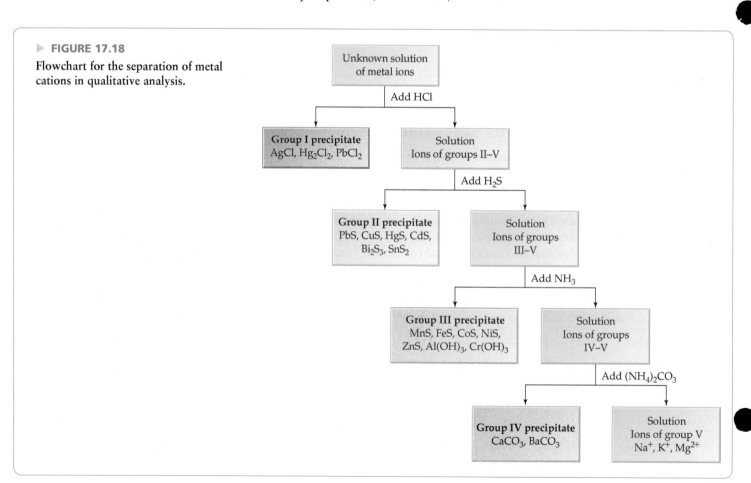

- **Group I: Ag⁺, Hg₂²⁺, and Pb²⁺.** When aqueous HCl is added to the unknown solution, the cations of group I precipitate as insoluble chlorides: $AgCl$, Hg_2Cl_2, and $PbCl_2$. The cations of groups II–V, which form soluble chlorides, remain in solution. A small amount of the Pb^{2+} also remains in solution because $PbCl_2$ is slightly soluble.

- **Group II: Pb²⁺, Cu²⁺, Hg²⁺, Cd²⁺, Bi³⁺, and Sn⁴⁺.** After the insoluble chlorides have been removed, the solution is treated with H_2S to precipitate the cations of group II as insoluble sulfides: PbS, CuS, HgS, CdS, Bi_2S_3, and SnS_2. Because the solution is strongly acidic at this point ($[H_3O^+] \approx 0.3$ M), only the most insoluble sulfides precipitate. The acid-insoluble sulfides are then removed from the solution.

- **Group III: Mn²⁺, Fe²⁺, Co²⁺, Ni²⁺, Zn²⁺, Al³⁺, and Cr³⁺.** At this point, aqueous NH_3 is added, neutralizing the acidic solution and giving an $NH_4^+ - NH_3$ buffer that is slightly basic (pH ≈ 8). The decrease in $[H_3O^+]$ shifts the metal sulfide solubility equilibrium to the left, thus precipitating the 2+ cations of group III as insoluble sulfides: MnS, FeS, CoS, NiS, and ZnS. The 3+ cations precipitate from the basic solution, not as sulfides but as insoluble hydroxides: $Al(OH)_3$ and $Cr(OH)_3$.

- **Group IV: Ca²⁺ and Ba²⁺.** After the base-insoluble sulfides and the insoluble hydroxides have been removed, the solution is treated with $(NH_4)_2CO_3$ to precipitate the cations of group IV as insoluble carbonates: $CaCO_3$ and $BaCO_3$. Magnesium carbonate does not precipitate at this point because $[CO_3^{2-}]$ in the NH_4^+–NH_3 buffer is maintained at a low value.

- **Group V: Na⁺, K⁺, and Mg²⁺.** The only ions remaining in solution at this point are those whose chlorides, sulfides, and carbonates are soluble under the conditions of the previous reactions. Magnesium ion is separated and identified by the addition of a solution of $(NH_4)_2HPO_4$; if Mg^{2+} is present, a white precipitate of $Mg(NH_4)PO_4$ forms. The alkali metal ions are usually identified by the characteristic colors that they impart to a Bunsen flame (**FIGURE 17.19**).

Once the cations have been separated into groups, further separations and specific tests are carried out to determine the presence or absence of the ions in each group. In group I, for example, lead can be separated from silver and mercury by treating the precipitate with hot water. The more soluble $PbCl_2$ dissolves, but the less soluble $AgCl$ and Hg_2Cl_2 do not. To test for Pb^{2+}, the solid chlorides are removed, and the solution is treated with a solution of K_2CrO_4. If Pb^{2+} is present, a yellow precipitate of $PbCrO_4$ forms.

Detailed procedures for separating and identifying all the ions can be found in general chemistry laboratory manuals. Although modern methods of metal-ion analysis employ sophisticated analytical instruments, qualitative analysis is still included in many general chemistry laboratory courses because it is an excellent vehicle for developing laboratory skills and for learning about acid–base, solubility, and complex-ion equilibria.

▲ When aqueous potassium chromate is added to a solution that contains Pb^{2+}, a yellow precipitate of $PbCrO_4$ forms.

▶ **FIGURE 17.19**
Flame tests for sodium and potassium.

Sodium imparts a persistent yellow color to the flame.

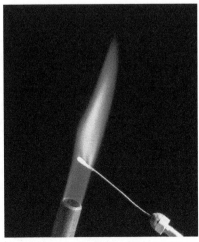

Potassium turns the flame a fleeting violet color.

The rapidly increasing concentration of CO_2 in the atmosphere is of concern due to its ability to absorb infrared radiation and cause global climate change (Sections 10.10–10.11). In the 400,000 years leading up to the Industrial Revolution, the concentration of carbon dioxide did not exceed 280 parts per million by volume (ppmv), but burning fossil fuels for energy caused levels to rise to 407 ppmv by 2017. Oceans currently absorb approximately one-fourth of the CO_2 emitted each year, which at first glance brings hope for mitigating the atmospheric CO_2 problem. However, an undesirable consequence of ocean uptake occurs when dissolved carbon dioxide undergoes several equilibrium reactions, increasing acidity, and shifting the fraction of bicarbonate (HCO_3^-) and carbonate (CO_3^{2-}) ions present.

$$CO_2(g) + H_2O(l) \rightleftharpoons H_2CO_3(aq)$$

$$H_2CO_3(aq) + H_2O(l) \rightleftharpoons H_3O^+(aq) + HCO_3^-(aq)$$
$$K_{a1} = 4.3 \times 10^{-7}$$

$$HCO_3^-(aq) + H_2O(l) \rightleftharpoons H_3O^+(aq) + CO_3^{2-}(aq)$$
$$K_{a2} = 5.6 \times 10^{-11}$$

Henry's Law (Section 13.5) states that as the partial pressure of carbon dioxide increases in the atmosphere, the amount of dissolved CO_2 will increase. Furthermore, dissolved CO_2 levels exceed equilibrium predictions due to natural processes such as the removal of carbonate from the equilibrium system by organisms that produce calcium carbonate shells and transport of surface layers to great depths. Decades of measurements by the National Oceanic and Atmospheric Association (NOAA) have shown that surface ocean pH has dropped from a pre-industrial value of 8.2 to a current value of 8.1. Although the pH change seems small, it represents a 26% increase in the concentration of H_3O^+ ions. Acidification can be even more severe in sensitive polar regions, where colder water temperatures dissolve more CO_2. FIGURE 17.20 shows the correlation between atmospheric CO_2 levels and ocean pH taken at Mauna Loa research station in Hawaii.

Ocean acidification is expected to affect species to varying degrees. Sea grasses and algae may thrive due to the accelerated rate of photosynthesis in a high CO_2 environment. In contrast, detrimental effects to marine life include alteration of metabolic reaction rates that have been optimized for a narrow pH range and hindered uptake of essential minerals. One of the most serious consequences is the decrease in CO_3^{2-} in seawater necessary for organisms to build calcium carbonate ($CaCO_3$) shells and skeletons in a process called calcification. FIGURE 17.21a shows the distribution of species in the carbonate equilibrium as a function of pH. Bicarbonate ion (HCO_3^-) is the major species at the current ocean pH (8.1), with minor contributions from CO_2 and CO_3^{2-}. FIGURE 17.21b is a magnified view of the minor species in the equilibrium and shows the dramatic effect of pH on the fractional amount of CO_3^{2-}. As pH decreases excess hydronium ions react with carbonate to lower its concentration.

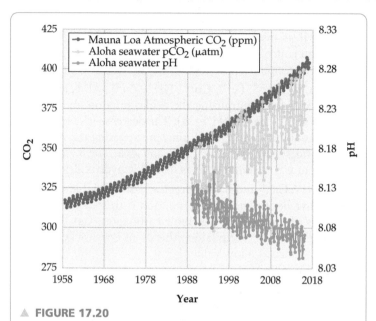

▲ **FIGURE 17.20**

Correlation between rising levels of CO_2 in the atmosphere at Mauna Loa station in Hawaii and rising levels of CO_2 in the nearby ocean. The higher the level of dissolved CO_2, the lower the pH of the ocean.

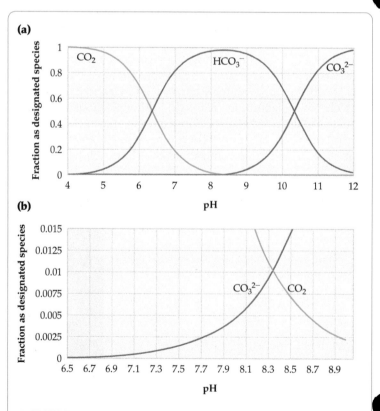

▲ **FIGURE 17.21**

(a) Distribution of species in the $CO_2/HCO_3^-/CO_3^{2-}$ equilibrium as a function of pH. (b) Magnified view of the minor species present in the pH range of seawater.

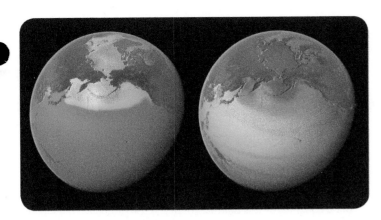

▲ Models showing past and future states of ocean acidification in the Arctic Ocean and the northern Pacific Ocean. This model tracked the change over time in calcium carbonate saturation, from 1850 to 1860 (left) to 2090 to 2100 (right). Coral growth is maximal in dark blue areas, while red areas show conditions that are corrosive for coral shells and skeletons. The data is from the Max Planck Institute Earth System Model (MPI-ERM), launched in 2013.

▲ These photos show the dissolution of a pteropod's shell when placed in sea water with pH and carbonate levels projected for the year 2100.

As water becomes more acidic, the solubility of calcium carbonate increases and structures such as coral and the shells of marine organisms dissolve to a greater degree. Studies indicate that as the ocean becomes more acidic, populations of certain species of shellfish and coral could decline and even go extinct. Small changes to the intricate network of the marine ecosystem could have a cascading effect and potentially wreak havoc on sea life, the fishing industry, food supply, economy, and coastal recreation.

PROBLEM 17.33

(a) What is the chemical formula of the acid formed when carbon dioxide dissolves in water?

(b) Write balanced equations for the first and second acid dissociation reactions.

PROBLEM 17.34

(a) Examine Figure 17.20. How did the ocean pH measured in Hawaii change from the year 1988 to the year 2017?

(b) Estimates of future carbon dioxide levels, based on business-as-usual emission scenarios, indicate that by the end of this century, the concentration of hydronium ions in the surface waters of the ocean could increase by 150% from the initial level in 1880. If the initial pH was 8.2, calculate what the pH will be in 2100.

$$\text{Percent change} = \frac{[H_3O^+]_{final} - [H_3O^+]_{initial}}{[H_3O^+]_{initial}} \times 100\%$$

PROBLEM 17.35 HCO_3^- and CO_3^{2-} are the primary ions in the ocean that act as a buffer against pH change due to acid rain and dissolved carbon dioxide. Use K_a values provided in the Inquiry section to calculate the ratio of $[HCO_3^-]/[CO_3^{2-}]$ ions at pH = 8.2 and pH = 7.9.

PROBLEM 17.36 Coral and the shells of marine organisms are made of calcium carbonate. Calculate the molar solubility of $CaCO_3$ in pure water. (K_{sp} for $CaCO_3 = 5.0 \times 10^{-9}$.)

PROBLEM 17.37 The following reactions represent the dissolution of $CaCO_3$ due to increased acidity from CO_2 dissolving in water to form carbonic acid (H_2CO_3).

$$CaCO_3(s) \rightleftharpoons Ca^{2+}(aq) + CO_3^{2-}(aq)$$

$$H_2CO_3(aq) + H_2O(l) \rightleftharpoons H_3O^+(aq) + HCO_3^-(aq)$$

$$\underline{CO_3^{2-}(aq) + H_3O^+(aq) \rightleftharpoons HCO_3^-(aq) + H_2O(l)}$$

$$CaCO_3(s) + H_2CO_3(aq) \rightleftharpoons Ca^{2+}(aq) + 2\,HCO_3^-(aq) \quad K = ?$$

(a) Use K_a values provided in the Inquiry section and the K_{sp} value for $CaCO_3$ (5.0×10^{-9}) to determine K for the overall reaction.

$$CaCO_3(s) + H_2CO_3(aq) \rightleftharpoons Ca^{2+}(aq) + 2\,HCO_3^-(aq) \quad K = ?$$

(b) Predict the effect of increasing levels of atmospheric CO_2 on molar solubility of $CaCO_3$.

STUDY GUIDE

Section	Concept Summary	Learning Objectives	Test Your Understanding
17.1 Neutralization Reactions	Neutralization reactions involving a strong acid and/or a strong base have very large equilibrium constants (K_n) and proceed nearly 100% to completion. Weak acid–weak base neutralizations do not go to completion.	**17.1** Write a balanced equation for a neutralization reaction and predict whether the pH after neutralization is greater than, equal to, or less than 7.	Worked Example 17.1; Problems 17.38, 17.47, 17.49, 17.51
17.2 The Common-Ion Effect	The **common-ion effect** is the shift in the position of an equilibrium that occurs when a substance is added that provides more of an ion already involved in the equilibrium. An example is the decrease in the percent dissociation of a weak acid on the addition of its conjugate base.	**17.2** Calculate the equilibrium concentrations of all species, pH, and percent dissociation in a solution of a weak acid when a common ion is present.	Worked Examples 17.2–17.3; Problems 17.57, 17.59, 17.61
17.3–17.4 Buffer Solutions and the Henderson–Hasselbalch Equation	A solution of a weak acid and its conjugate base is called a **buffer solution** because it resists drastic changes in pH. **Buffer capacity**, the ability of a buffer solution to absorb small amounts of added H_3O^+ or OH^- without a significant change in pH, increases with increasing amounts of weak acid and conjugate base. The pH of a buffer solution has a value close to the pK_a ($-\log K_a$) of the weak acid and can be calculated from the **Henderson–Hasselbalch equation:** $$pH = pK_a + \log \frac{[\text{Conjugate base}]}{[\text{Weak acid}]}$$	**17.3** Calculate the initial pH of a buffer solution and the pH after the addition of a strong acid or a strong base.	Worked Example 17.4; Problems 17.67, 17.69
		17.4 Use the Henderson–Hasselbalch equation to calculate the pH of a buffer solution.	Worked Examples 17.5–17.6; Problems 17.71, 17.73, 17.75, 17.77
17.5–17.9 pH Titration Curves: Strong Acid–Strong Base; Weak Acid–Strong Base; Weak Base–Strong Acid, Diprotic Acid–Strong Base	A **pH titration curve** is a plot of the pH of a solution as a function of the volume of base (or acid) added in the course of an acid–base titration. For a strong acid–strong base titration, the titration curve exhibits a sharp change in pH in the region of the **equivalence point,** the point at which stoichiometrically equivalent amounts of acid and base have been mixed together. For weak acid–strong base and weak base–strong acid titrations, the titration curves display a relatively flat buffer region midway to the equivalence point, a steep change in pH in the region of the equivalence point, and a pH at the equivalence point that is not equal to 7.00. The titration curve for a diprotic acid–strong base titration exhibits two equivalence points and two buffer regions.	**17.5** Calculate the pH at various points in a strong acid–strong base titration.	Worked Example 17.7; Problems 17.81, 17.83
		17.6 Calculate the pH at various points in a weak acid–strong base titration or weak base–strong acid titration.	Worked Example 17.8; Problems 17.41, 17.85, 17.89
		17.7 Calculate the pH at various points in a diprotic acid–strong base titration.	Worked Example 17.9; Problem 17.95
		17.8 Visualize the molecular species present during a titration and interpret titration curves.	Problems 17.41–17.44
17.10–17.11 Solubility Equilibria for Ionic Compounds	The **solubility product** (K_{sp}) for an ionic compound is the equilibrium constant for dissolution of the compound in water.	**17.9** Write the equilibrium-constant expression for dissolution of an ionic compound, and calculate the value of its K_{sp}.	Worked Examples 17.10–17.11; Problems 17.99, 17.101, 17.103
		17.10 Calculate equilibrium concentrations of ions and the solubility of an ionic compound from its K_{sp}.	Worked Example 17.12; Problem 17.107

Section	Concept Summary	Learning Objectives	Test Your Understanding
17.12 Factors That Affect Solubility	The solubility of the compound and K_{sp} are related by the equilibrium equation for the dissolution reaction. The solubility of an ionic compound is (1) suppressed by the presence of a common ion in the solution, (2) increased by decreasing the pH if the compound contains a basic anion, such as OH^-, S^{2-}, or CO_3^{2-}, and (3) increased by the presence of a Lewis base, such as NH_3, CN^-, or OH^-, that can bond to the metal cation to form a **complex ion**. The stability of a complex ion is measured by its **formation constant (K_f)**.	**17.11** Predict how the presence of molecular and ionic species will affect the solubility of an ionic compound.	Problems 17.109, 17.113, 17.115, 17.117
		17.12 Calculate solubility in a solution that contains a common ion.	Worked Example 17.13; Problem 17.111
		17.13 Use the formation constant K_f to calculate ion concentrations in a solution that contains a complex ion.	Worked Example 17.14; Problem 17.119
		17.14 Use the formation constant K_f to calculate the solubility of an ionic compound when the cation forms a complex ion.	Worked Example 17.15; Problems 17.123, 17.125
17.13–17.15 Precipitation of Ionic Compounds and Separation of Ions in Qualitative Analysis	When solutions of soluble ionic compounds are mixed, an insoluble compound will precipitate if the **ion product (IP)** for the insoluble compound exceeds its K_{sp}. The IP is defined in the same way as K_{sp}, except that the concentrations in the expression for IP are not necessarily equilibrium concentrations. Certain metal cations can be separated by the selective precipitation of metal sulfides. Selective precipitation is important in **qualitative analysis**, a procedure for identifying the ions present in an unknown solution.	**17.15** Calculate the ion product IP for an ionic compound, and determine whether a precipitate will form when various solutions are mixed.	Worked Example 17.16; Problems 17.127, 17.130
		17.16 Design a scheme for separating ions is a mixture based on selective precipitation.	Problem 17.136–17.137

KEY TERMS

amphoteric *749*
buffer capacity *717*
buffer solutions *716*
common ion *712*

common-ion effect *712*
complex ion *745*
equivalence point *723*
formation constant (K_f) *745*

Henderson–Hasselbalch equation *720*
ion product (IP) *750*

pH titration curve *723*
qualitative analysis *752*
solubility product (K_{sp}) *738*

KEY EQUATIONS

- Henderson–Hasselbalch Equation for Calculating the pH of a Buffer Solution (Section 17.4)

$$pH = pK_a + \log \frac{[\text{Conjugate base}]}{[\text{Weak acid}]}$$

- pH at the First Equivalence Point in a Diprotic Acid–Strong base Titration (Section 17.9)

$$pH \text{ at first equivalence point} = \frac{pK_{a1} + pK_{a2}}{2}$$

- Solubility Product K_{sp} for the Dissolution Reaction $M_mX_x(s) \rightleftharpoons m\ M^{n+}(aq) + x\ X^{y-}(aq)$ (Section 17.10)

$$K_{sp} = [M^{n+}]^m[X^{y-}]^x$$

- Ion Product for the Reaction $M_mX_x(s) \rightleftharpoons m\ M^{n+}(aq) + x\ X^{y-}(aq)$ (Section 17.13)

$$IP = [M^{n+}]_t^m[X^{y-}]_t^x$$

where the subscript t denotes concentrations at an arbitrary time t, not necessarily at equilibrium.

- Solubility Product in Acid for the Reaction $MS(s) + 2H_3O^+(aq) \rightleftharpoons M^{2+}(aq) + H_2S(aq) + 2\ H_2O(l)$ (Section 17.14)

$$K_{spa} = \frac{[M^{2+}][H_2S]}{[H_3O^+]^2}$$

PRACTICE TEST

After studying this chapter, you can assess your understanding with these practice test questions, which are correlated with chapter learning objectives. If you answer a question incorrectly, refer to the learning objectives in the end-of-chapter Study Guide for assistance. The Study Guide provides a conceptual summary, references a Worked Example to model how to solve the problem, and gives additional problems for more practice.

1. Which of the following mixtures has the highest pH? **(LO 17.1)**
   (a) Equal volumes of 1.0 M HCl and 1.0 M NaOH
   (b) Equal volumes of 0.1 M HNO_3 and 0.1 M KOH
   (c) Equal volumes of 0.1 M HCN and 0.1 M NaOH
   (d) Equal volumes of 0.1 M NaF and 0.1 M HCl

2. What is the percent dissociation in a solution that is 0.50 M in acetic acid (CH_3CO_2H) and 0.10 M in sodium acetate ($NaCH_3CO_2$)? ($K_a = 1.8 \times 10^{-5}$) **(LO 17.2)**
   (a) 0.018%   (b) 0.090%
   (c) 7.2×10^{-3}%   (d) 2.3%

3. A 1.00 L buffer solution is 0.250 M in HF and 0.250 M in NaF. Calculate the pH of the solution after the addition of 100.0 mL of 1.00 M HCl. ($K_a = 3.5 \times 10^{-4}$) **(LO 17.3)**
   (a) 4.11   (b) 3.82
   (c) 3.46   (d) 3.09

4. What is the pH of a buffer solution prepared by dissolving 0.250 mol of NaH_2PO_4 and 0.075 mol of NaOH in enough water to make 1.00 L of solution? (K_a ($H_2PO_4^-$) = 6.2×10^{-8}) **(LO 17.4)**
   (a) 6.32   (b) 6.83
   (c) 7.21   (d) 7.71

5. Which is the best acid-base pair to use in the preparation of a buffer with pH = 10.5? **(LO 17.4)**
   (a) HOI and OI^- $K_a = 2.0 \times 10^{-11}$
   (b) HNO_2 and NO_2^- $K_a = 4.5 \times 10^{-4}$
   (c) HIO_3 and IO_3^- $K_a = 1.7 \times 10^{-1}$
   (d) $H_2PO_4^-$ and HPO_4^{2-} $K_a = 6.2 \times 10^{-8}$

6. A 25.0 mL sample of 0.250 M HNO_3 is titrated with 0.100 M KOH. What is the pH after addition of 62.5 mL of the KOH solution? **(LO 17.5)**
   (a) 4.52   (b) 5.67
   (c) 7.00   (d) 8.95

7. A 50.00 mL sample of 1.00 M CH_3COOH is titrated with 1.00 M NaOH. Calculate the pH after 15.00 mL of NaOH has been added.($K_a = 1.8 \times 10^{-5}$) **(LO 17.6)**
   (a) 4.37   (b) 4.74
   (c) 5.10   (d) 6.78

8. Calculate the pH at the equivalence point in the titration of 50.0 mL of 0.50 M HCO_2H with 1.0 M KOH. ($K_a = 1.8 \times 10^{-4}$) **(LO 17.6)**
   (a) 3.53   (b) 7.00
   (c) 8.63   (d) 9.78

9. A 25.00 mL sample of 1.00 M CH_3CO_2H is titrated with 1.00 M NaOH. Calculate the pH after 30.00 mL of NaOH has been added.($K_a = 1.8 \times 10^{-5}$) **(LO 17.6)**
   (a) 6.90   (b) 9.36
   (c) 11.54   (d) 12.95

10. You are titrating 30.0 mL of a 0.100 M solution of the protonated form of the amino acid glycine (H_2A^+) with 0.15 M NaOH. Calculate the pH after the addition of 28.0 mL of the NaOH solution. ($K_{a1} = 4.57 \times 10^{-3}$, $K_{a2} = 2.51 \times 10^{-10}$) **(LO 17.7)**
    (a) 10.60   (b) 9.42
    (c) 4.69   (d) 2.34

11. The pH titration curve applies to the titration of 40.0 mL of a 0.100 M solution of an acid with 0.100 M NaOH. What are the approximate pK_a values for this acid? **(LO 17.8)**

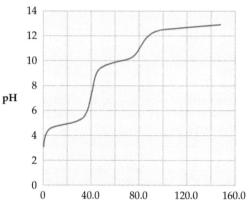

mL of 0.100 M NaOH added

    (a) pK_{a1} = 5, pK_{a2} = 10
    (b) pK_{a1} = 7, pK_{a2} = 11
    (c) pK_{a1} = 5, pK_{a2} = 10, pK_{a3} = 13
    (d) pK_{a1} = 5, pK_{a2} = 7, pK_{a3} = 10

12. What is the solubility-constant expression for $Ca_3(PO_4)_2$ (s)? **(LO 17.9)**
    (a) $[Ca^{2+}]^3[PO_4^{3-}]^2$ (b) $\dfrac{[Ca^{2+}]^3[PO_4^{3-}]^2}{[Ca_3(PO_4)_2]}$
    (c) $\dfrac{[Ca_3(PO_4)_2]}{[Ca^{2+}]^3[PO_4^{3-}]^2}$ (d) $[Ca^{2+}]^2[PO_4^{3-}]^3$

13. What is the molar solubility of Ag_2SO_3 in water? The solubility-product constant for silver sulfite is 1.5×10^{-14} at 25 °C. **(LO 17.10)**
    (a) 1.2×10^{-7} M   (b) 2.0×10^{-5} M
    (c) 8.7×10^{-8} M   (d) 1.6×10^{-5} M

14. Consider a saturated solution of the slightly soluble salt $BaCO_3$. Adding which of the following substances will increase the solubility? **(LO 17.11)**
    (a) HNO_3 (b) Na_2CO_3
    (c) $Ba(NO_3)_2$ (d) KOH

15. What is the molar solubility of BaF_2 in a solution containing 0.0750 M LiF? ($K_{sp} = 1.7 \times 10^{-6}$) **(LO 17.12)**
    (a) 2.3×10^{-5} M   (b) 3.0×10^{-4} M
    (c) 1.2×10^{-2} M   (d) 1.3×10^{-3} M

16. In excess of $NH_3(aq)$, Zn^{2+} forms a complex ion, $[Zn(NH_3)_4]^{2+}$ which has a formation constant $K_f = 7.8 \times 10^8$. Calculate the concentration of Zn^{2+} in a solution prepared by adding 1.00×10^{-2} mol $Zn(NO_3)_2$ to 1.00 L of 0.250 M NH_3. **(LO 17.13)**

(a) 7.9×10^{-4} M   (b) 2.8×10^{-6} M

(c) 3.9×10^{-9} M   (d) 6.4×10^{-11} M

17. What is the molar solubility of AgI in 0.20 M NaCN? **(LO 17.14)**

$$K_{sp}(AgI) = 8.5 \times 10^{-17} \quad K_f[Ag(CN)_2]^- = 3.0 \times 10^{20}$$

(a) 6.2×10^{-4} M   (b) 1.0×10^{-1} M

(c) 7.6×10^{-2} M   (d) 2.1×10^{-3} M

18. A solution containing sulfide ions is added to a solution of 0.036 M Cu^{2+} and 0.044 M Fe^{2+}. At what concentration of sulfide ion will a precipitate begin to form? What is the identity of the precipitate? **(LO 17.15, 17.16)**

$$K_{sp}(CuS) = 1.3 \times 10^{-36}, \quad K_{sp}(FeS) = 6.3 \times 10^{-18}$$

(a) 1.4×10^{-16} M, FeS   (b) 3.6×10^{-35} M, CuS

(c) 3.6×10^{-35} M, FeS   (d) 1.4×10^{-16} M, C

Answers:

1. c, 2. a, 3. d, 4. b, 5. a, 6. c, 7. a, 8. c, 9. d, 10. b, 11. a, 12. a, 13. d, 14. a, 15. b, 16. c, 17. b, 18. b

---

MASTERING CHEMISTRY   provides end-of-chapter exercises, feedback-enriched tutorial problems, animations, and interactive activities to encourage problem-solving practice and deeper understanding of key concepts and topics.

RAN   *Randomized in Mastering Chemistry*

CONCEPTUAL PROBLEMS

Problem 17.1–17.37 appear within the chapter.

17.38 The strong acid HA is mixed with an equal molar amount of aqueous NaOH. Which of the following pictures represents the equilibrium state of the solution? (Na^+ ions and solvent water molecules have been omitted for clarity.)

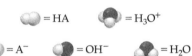

(1)   **(2)**   **(3)**   **(4)**

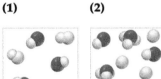

17.39 The following pictures represent initial concentrations in solutions that contain a weak acid HA ($pK_a = 6.0$) and its sodium salt NaA. (Na^+ ions and solvent water molecules have been omitted for clarity.)

(1)   **(2)**   **(3)**   **(4)**

(a) Which solution has the highest pH? Which has the lowest pH?

(b) Draw a picture that represents the equilibrium state of solution (1) after the addition of two H_3O^+ ions.

(c) Draw a picture that represents the equilibrium state of solution (1) after the addition of two OH^- ions.

17.40 The following pictures represent solutions that contain one or more of the compounds H_2A, $NaHA$, and Na_2A, where H_2A is a weak diprotic acid. (Na^+ ions and solvent water molecules have been omitted for clarity.)

(a) Which of the solutions are buffer solutions?

(b) Which solution has the greatest buffer capacity?

(1)   **(2)**   **(3)**   **(4)**

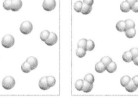  

17.41 The following plot shows two pH titration curves, each representing the titration of 50.0 mL of 0.100 M acid with 0.100 M NaOH:

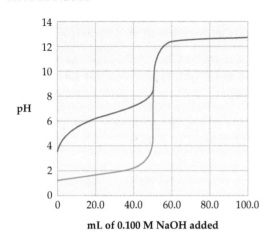

mL of 0.100 M NaOH added

(a) Which of the two curves represents the titration of a strong acid? Which represents a weak acid?

(b) What is the approximate pH at the equivalence point for each of the acids?

(c) What is the approximate pK_a of the weak acid?

17.42 The following pictures represent solutions at various stages in the titration of a weak base B with aqueous HCl. (Cl⁻ ions and solvent water molecules have been omitted for clarity.)

● = B   ◐ = BH⁺   ◓ = H₃O⁺

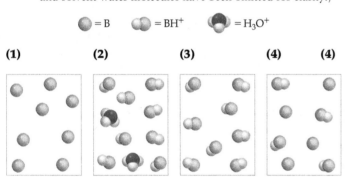

(1)   (2)   (3)   (4)   (4)

(a) To which of the following stages do solutions 1–4 correspond?

(i) The initial solution before addition of any HCl

(ii) Halfway to the equivalence point

(iii) At the equivalence point

(iv) Beyond the equivalence point

(b) Is the pH at the equivalence point more or less than 7?

17.43 The following pictures represent solutions at various stages in the titration of a weak diprotic acid H_2A with aqueous NaOH. (Na⁺ ions and water molecules have been omitted for clarity.)

◓ = OH⁻   ◑ = HA⁻   ● = A²⁻   ◔ = H₂A

(1)   (2)   (3)   (4)

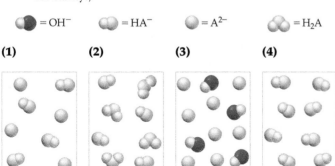

(a) To which of the following stages do solutions 1–4 correspond?

(i) Halfway to the first equivalence point

(ii) At the first equivalence point

(iii) Halfway between the first and second equivalence points

(iv) Beyond the second equivalence point

(b) Which solution has the highest pH? Which has the lowest pH?

17.44 The following pictures represent solutions at various stages in the titration of sulfuric acid H_2A ($A^{2-} = SO_4^{2-}$) with aqueous NaOH. (Na⁺ ions and water molecules have been omitted for clarity.)

◔ = H₂A   ◑ = HA⁻   ● = A²⁻   ◐ = OH⁻   ◓ =H₃O⁺

(1)   (2)   (3)   (4)

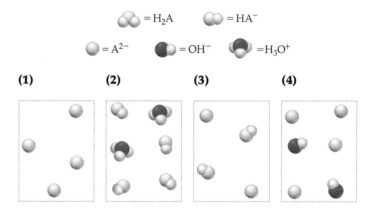

(a) To which of the following stages do solutions 1–4 correspond?

(i) Halfway to the first equivalence point

(ii) Halfway between the first and second equivalence points

(iii) At the second equivalence point

(iv) Beyond the second equivalence point

(b) Draw a picture that represents the solution prior to addition of any NaOH.

17.45 The following pictures represent solutions of AgCl, which also may contain ions other than Ag⁺ and Cl⁻ that are not shown. If solution 1 is a saturated solution of AgCl, classify solutions 2–4 as unsaturated, saturated, or supersaturated.

● = Ag⁺   ● = Cl⁻

(1)   (2)   (3)   (4)

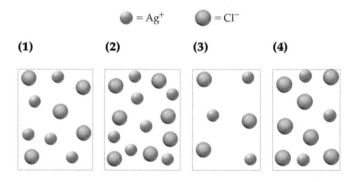

17.46 The following pictures represent solutions of Ag_2CrO_4, which also may contain ions other than Ag^+ and CrO_4^{2-} that are not shown. Solution 1 is in equilibrium with solid Ag_2CrO_4. Will a precipitate of solid Ag_2CrO_4 form in solutions 2–4? Explain.

(1)   **(2)**   **(3)**   **(4)**

SECTION PROBLEMS

Neutralization Reactions (Section 17.1)

17.47 Is the pH greater than, equal to, or less than 7 after the neu-
RAN tralization of each of the following pairs of acids and bases?
  (a) HNO_3 and KOH
  (b) HOI and $Ba(OH)_2$
  (c) HBr and aniline $(C_6H_5NH_2)$
  (d) HNO_2 and KOH

17.48 Is the pH greater than, equal to, or less than 7 after the neu-
RAN tralization of each of the following pairs of acids and bases?
  (a) NaOH and benzoic acid $(C_6H_5CO_2H)$
  (b) NH_3 and $HClO_4$
  (c) KOH and HI
  (d) $(CH_3)_3N$ and HOBr

17.49 Which of the following mixtures has the higher pH?
RAN (a) Equal volumes of 0.10 M HCN and 0.10 M NaOH
  (b) Equal volumes of 0.10 M $HClO_4$ and 0.10 M NaOH

17.50 Which of the following mixtures has the lower pH?
RAN (a) Equal volumes of 0.10 M HI and 0.10 M KOH
  (b) Equal volumes of 0.10 M HI and 0.10 M NH_3

17.51 Phenol $(C_6H_5OH, K_a = 1.3 \times 10^{-10})$ is a weak acid used in mouthwashes, and pyridine $(C_5H_5N, K_b = 1.8 \times 10^{-9})$ is a weak base used as a solvent. Calculate the value of K_n for the neutralization of phenol by pyridine. Does the neutralization reaction proceed very far toward completion?

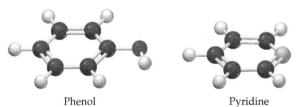

Phenol   Pyridine

17.52 Aniline $(C_6H_5NH_2, K_b = 4.3 \times 10^{-10})$ is a weak base used in the manufacture of dyes. Calculate the value of K_n for the neutralization of aniline by vitamin C (ascorbic acid, $C_6H_8O_6$, $K_a = 8.0 \times 10^{-5}$). Does much aniline remain at equilibrium?

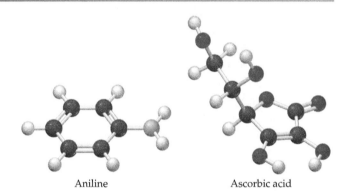

Aniline   Ascorbic acid

17.53 The equilibrium constant K_n for the neutralization of lactic acid $(C_3H_6O_3)$ and urea (CH_4N_2O) is 2.1×10^{-4}. What is K_b for urea? See Appendix C for the K_a of lactic acid.

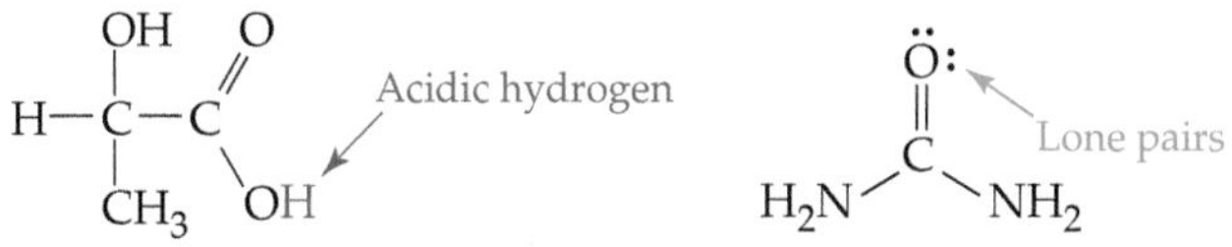

Lactic acid   Urea

17.54 The equilibrium constant K_n for the neutralization of boric acid (H_3BO_3) and caffeine $(C_8H_{10}N_4O_2)$ is 24. What is K_b for caffeine? See Appendix C for the K_a of boric acid.

The Common-Ion Effect (Section 17.2)

17.55 Does the pH increase, decrease, or remain the same when the substances are added to the solutions?
  (a) LiF to an HF solution
  (b) KI to an HI solution
  (c) NH_4Cl to an NH_3 solution

17.56 Does the pH increase, decrease, or remain the same on the addition of each of the following?
  (a) NH_4NO_3 to an NH_3 solution
  (b) Na_2CO_3 to an $NaHCO_3$ solution
  (c) $NaClO_4$ to an NaOH solution

17.57 Calculate the pH of a solution that is 0.25 M in HF and
RAN 0.10 M in NaF.

17.58 Calculate the pH in a solution prepared by dissolving
RAN   0.10 mol of solid NH_4Cl in 0.500 L of 0.40 M NH_3. Assume that there is no volume change.

17.59 The pH of a solution of HN_3 ($K_a = 1.9 \times 10^{-5}$) and
RAN NaN_3 is 4.86. What is the molarity of NaN_3 if the molarity of HN_3 is 0.016 M?

17.60 The pH of a solution of NH_3 and NH_4Br is 8.90. What is
RAN   the molarity of NH_4Br if the molarity of NH_3 is 0.016 M?

17.61 Calculate the percent dissociation of 0.10 M hydrazoic acid
RAN   (HN_3, $K_a = 1.9 \times 10^{-5}$). Recalculate the percent dissociation of 0.10 M HN_3 in the presence of 0.10 M HCl, and explain the change.

17.62 Calculate the pH of 100.0 mL of 0.30 M NH_3 before and
RAN   after the addition of 4.0 g of NH_4NO_3, and account for the change. Assume that the volume remains constant.

Buffer Solutions (Sections 17.3–17.4)

17.63 Which of the following gives a buffer solution when equal volumes of the two solutions are mixed?
   (a) 0.10 M HF and 0.10 M NaF
   (b) 0.10 M HF and 0.10 M NaOH
   (c) 0.20 M HF and 0.10 M NaOH
   (d) 0.10 M HCl and 0.20 M NaF

17.64 Which of the following gives a buffer solution when equal volumes of the two solutions are mixed?
   (a) 0.10 M NH_3 and 0.10 M HCl
   (b) 0.20 M NH_3 and 0.10 M HCl
   (c) 0.10 M NH_4Cl and 0.10 M NH_3
   (d) 0.20 M NH_4Cl and 0.10 M NaOH

17.65 Which of the following solutions has the greater buffer capacity:
   100 mL of 0.30 M HNO_2–0.30 M $NaNO_2$ or
   100 mL of 0.10 M HNO_2–0.10 M $NaNO_2$? Explain.

17.66 Which of the following solutions has the greater buffer capacity:
   50 mL of 0.20 M NH_4Br–0.30 M NH_3 or
   50 mL of 0.40 M NH_4Br–0.60 M NH_3? Explain.

17.67 Calculate the pH of a buffer solution that is 0.20 M in HCN
RAN   and 0.12 M in NaCN. Will the pH change if the solution is diluted by a factor of 2? Explain.

17.68 Calculate the pH of a buffer solution prepared by dissolving
RAN   4.2 g of $NaHCO_3$ and 5.3 g of Na_2CO_3 in 0.20 L of water. Will the pH change if the solution volume is increased by a factor of 10? Explain.

17.69 Calculate the pH of 0.250 L of a 0.36 M formic acid–0.30 M sodium formate buffer before and after the addition of (a) 0.0050 mol of NaOH and (b) 0.0050 mol of HCl. Assume that the volume remains constant.

17.70 Calculate the pH of 0.375 L of a 0.18 M acetic acid–0.29 M
RAN   sodium acetate buffer before and after the addition of (a) 0.0060 mol of KOH and (b) 0.0060 mol of HBr. Assume that the volume remains constant.

17.71 Use the Henderson–Hasselbalch equation to calculate the pH
RAN   of a buffer solution that is 0.25 M in formic acid (HCO_2H) and 0.50 M in sodium formate (HCO_2Na).

17.72 A food chemist studying the formation of lactic acid in sour
RAN   milk prepares buffer that is 0.58 M in lactic acid ($HC_3H_5O_3$) and 0.36 M in sodium lactate ($NaC_3H_5O_3$). Use the Henderson–Hasselbalch equation to calculate the pH of the buffer solution.

17.73 Use the Henderson–Hasselbalch equation to calculate the
RAN   ratio of HCO_3^- to H_2CO_3 in blood having a pH of 7.40. The value of K_a for carbonic acid at body temperature (37 °C) is 7.9×10^{-7}.

17.74 The ratio of HCO_3^- to H_2CO_3 in blood is called the "bicarb
RAN   number" and is used as a measure of blood pH in hospital emergency rooms. A newly diagnosed diabetic patient is admitted to the emergency room with ketoacidosis and a bicarb number of 10. Calculate the blood pH. K_a for carbonic acid at body temperature (37 °C) is 7.9×10^{-7}.

17.75 In what volume ratio should you mix 1.0 M solutions of
RAN NH_4Cl and NH_3 to produce a buffer solution having pH = 9.80?

17.76 Give a recipe for preparing a CH_3CO_2H–CH_3CO_2Na buf-
RAN   fer solution that has pH = 4.44.

17.77 You need a buffer solution that has pH = 7.00. Which of the following buffer systems should you choose? Explain.
   (a) H_3PO_4 and $H_2PO_4^-$
   (b) $H_2PO_4^-$ and HPO_4^{2-}
   (c) HPO_4^{2-} and PO_4^{3-}

17.78 Which of the following conjugate acid–base pairs should you choose to prepare a buffer solution that has pH = 4.50? Explain.
   (a) HSO_4^- and SO_4^{2-}
   (b) HOCl and OCl^-
   (c) $C_6H_5CO_2H$ and $C_6H_5CO_2^-$

17.79 Consider a buffer solution that contains equal concentrations of $H_2PO_4^-$ and HPO_4^{2-}. Will the pH increase, decrease, or remain the same when each of the following substances is added?
   (a) Na_2HPO_4 (b) HBr   (c) KOH
   (d) KI   (e) H_3PO_4 (f) Na_3PO_4

17.80 Calculate the concentrations of NH_4^+ and NH_3 and the
RAN   pH in a solution prepared by mixing 20.0 g of NaOH and 0.500 L of 1.5 M NH_4Cl. Assume that the volume remains constant.

Strong Acid–Strong Base Titrations (Sections 17.5–17.6)

17.81 Consider the titration of 60.0 mL of 0.150 M HNO_3 with
RAN   0.450 M NaOH.
   (a) How many millimoles of HNO_3 are present at the start of the titration?
   (b) How many milliliters of NaOH are required to reach the equivalence point?
   (c) What is the pH at the equivalence point?
   (d) Sketch the general shape of the pH titration curve.

17.82 Make a rough plot of pH versus milliliters of acid added for
RAN   the titration of 50.0 mL of 1.0 M NaOH with 1.0 M HCl. Indicate the pH at the following points, and tell how many milliliters of acid are required to reach the equivalence point.
   (a) At the start of the titration
   (b) At the equivalence point
   (c) After the addition of a large excess of acid

17.83 Consider the titration of 25.0 mL of 0.125 M HCl with 0.100 M KOH. Calculate the pH after the addition of each of the following volumes of base:

(a) 3.0 mL   (b) 20.0 mL   (c) 65.0 mL

17.84 Consider the titration of 50.0 mL of 0.116 M NaOH with
RAN 0.0750 M HCl. Calculate the pH after the addition of each of the following volumes of acid:

(a) 5.0 mL   (b) 50.0 mL   (c) 0.10 L

Weak Acid–Strong Base Titrations (Section 17.7)

17.85 Consider the titration of 40.0 mL of 0.250 M HF with
RAN 0.200 M NaOH. How many milliliters of base are required to reach the equivalence point? Calculate the pH at each of the following points.

(a) After the addition of 10.0 mL of base

(b) Halfway to the equivalence point

(c) At the equivalence point

(d) After the addition of 80.0 mL of base

17.86 Consider the titration of 25.0 mL of 0.200 M HCO_2H with
RAN 0.250 M NaOH. How many milliliters of base are required to reach the equivalence point? Calculate the pH at each of the following points.

(a) After the addition of 7.0 mL of base

(b) Halfway to the equivalence point

(c) At the equivalence point

(d) After the addition of 25.0 mL of base

17.87 On the same graph, sketch pH titration curves for the titration of (1) a strong acid with a strong base and (2) a weak acid with a strong base. How do the two curves differ with respect to the following?

(a) The initial pH

(b) The pH in the region between the start of the titration and the equivalence point

(c) The pH at the equivalence point

(d) The pH beyond the equivalence point

(e) The volume of base required to reach the equivalence point

17.88 Consider the titration of 50.0 mL of 0.010 M HA($K_a = 1.0 \times 10^{-4}$) with 0.010 M NaOH.

(a) Sketch the pH titration curve, and label the equivalence point.

(b) How many milliliters of 0.010 M NaOH are required to reach the equivalence point?

(c) Is the pH at the equivalence point greater than, equal to, or less than 7?

(d) What is the pH exactly halfway to the equivalence point?

Weak Base–Strong Acid Titrations (Section 17.8)

17.89 A 100.0 mL sample of 0.100 M methylamine (CH_3NH_2,
RAN $K_b = 3.7 \times 10^{-4}$) is titrated with 0.250 M HNO_3. Calculate the pH after the addition of each of the following volumes of acid.

(a) 0.0 mL   (b) 20.0 mL

(c) 40.0 mL   (d) 60.0 mL

17.90 A 50.0 mL sample of 0.250 M ammonia (NH_3, $K_b =
RAN 1.8 \times 10^{-5}$) is titrated with 0.250 M HNO_3. Calculate the pH after the addition of each of the following volumes of acid.

(a) 0.0 mL   (b) 25.0 mL

(c) 50.0 mL   (d) 60.0 mL

17.91 The equivalence point was reached in titrations of three unknown acids at pH 9.16 (acid A), 8.88 (acid B), and 8.19 (acid C).

(a) Which is the strongest acid?

(b) Which is the weakest acid?

17.92 The equivalence point was reached in titrations of three unknown bases at pH 5.53 (base A), 4.11 (base B), and 6.00 (base C).

(a) Which is the strongest base?

(b) Which is the weakest base?

17.93 What is the pH at the equivalence point for the titration
RAN of 0.10 M solutions of the following acids and bases, and which of the indicators in Figure 17.5 would be suitable for each titration?

(a) HNO_2 and NaOH

(b) HI and NaOH

(c) CH_3NH_2 (methylamine) and HCl

17.94 What is the pH at the equivalence point for the titration of
RAN 0.20 M solutions of the following acids and bases? Which of the indicators in Figure 17.5 would be suitable for each titration?

(a) $C_5H_{11}N$ (piperidine) and HNO_3

(b) $NaHSO_3$ and NaOH

(c) $Ba(OH)_2$ and HBr

Polyprotic Acid–Strong Base Titrations (Section 17.9)

17.95 Consider the titration of 50.0 mL of a 0.100 M solution of
RAN the protonated form of the amino acid alanine (H_2A^+; $K_{a1} = 4.6 \times 10^{-3}$, $K_{a2} = 2.0 \times 10^{-10}$) with 0.100 M NaOH. Calculate the pH after the addition of each of the following volumes of base.

(a) 10.0 mL   (b) 25.0 mL   (c) 50.0 mL

(d) 75.0 mL   (e) 100.0 mL

17.96 Consider the titration of 25.0 mL of 0.0200 M H_2CO_3 with 0.0250 M KOH. Calculate the pH after the addition of each of the following volumes of base.

(a) 10.0 mL   (b) 20.0 mL   (c) 30.0 mL

(d) 40.0 mL   (e) 50.0 mL

17.97 Consider the titration of 50.0 mL of 1.00 M H_3PO_4 with 1.00 M KOH. Calculate the pH after the addition of each of the following volumes of base.

(a) 25.0 mL   (b) 50.0 mL

(c) 75.0 mL   (d) 100.0 mL

17.98 The titration of 0.02500 L of a diprotic acid solution with
RAN 0.1000 M NaOH requires 34.72 mL of titrant to reach the second equivalence point. The pH is 3.95 at the first equivalence point and 9.27 at the second equivalence point. If the acid solution contained 0.2015 g of the acid, what is the molar mass, pK_{a1}, and pK_{a2} of the acid?

Solubility Equilibria (Sections 17.10–17.11)

17.99 For each of the following compounds, write a balanced net ionic equation for the dissolution of the compound in water, and write the equilibrium-constant expression for K_{sp}.

(a) Ag_2CO_3 (b) $PbCrO_4$

(c) $Al(OH)_3$ (d) Hg_2Cl_2

17.100 For each of the following, write the equilibrium-constant
RAN expression for K_{sp}.

(a) $Ca(OH)_2$ (b) Ag_3PO_4

(c) $BaCO_3$ (d) $Ca_5(PO_4)_3OH$

17.101 A particular saturated solution of PbI_2 has $[Pb^{2+}] = 5.0 \times$
RAN 10^{-3} M and $[I^-] = 1.3 \times 10^{-3}$ M.

(a) What is the value of K_{sp} for PbI_2?

(b) What is $[I^-]$ in a saturated solution of PbI_2 that has $[Pb^{2+}] = 2.5 \times 10^{-4}$ M?

(c) What is $[Pb^{2+}]$ in a saturated solution that has $[I^+] = 2.5 \times 10^{-4}$ M?

17.102 A particular saturated solution of $Ca_3(PO_4)_2$ has $[Ca^{2+}] =$
RAN $[PO_4^{3-}] = 2.9 \times 10^{-7}$ M.

(a) What is the value of K_{sp} for $Ca_3(PO_4)_2$?

(b) What is $[Ca^{2+}]$ in a saturated solution of $Ca_3(PO_4)_2$ that has $[PO_4^{3-}] = 0.010$ M?

(c) What is $[PO_4^{3-}]$ in a saturated solution that has $[Ca^{2+}] = 0.010$ M?

17.103 If a saturated solution prepared by dissolving Ag_2CO_3 in
RAN water has $[Ag^+] = 2.56 \times 10^{-4}$ M, what is the value of K_{sp} for Ag_2CO_3?

17.104 If a saturated aqueous solution of the shock-sensitive compound lead(II) azide, $Pb(N_3)_2$, has $[Pb^{2+}] = 8.5 \times 10^{-4}$ M, what is the value of K_{sp} for $Pb(N_3)_2$?

17.105 Use the following solubility data to calculate a value of K_{sp} for each compound.

(a) SrF_2; 1.03×10^{-3} M

(b) CuI; 1.05×10^{-6} M

(c) MgC_2O_4; 0.094 g/L

(d) $Zn(CN)_2$; 4.95×10^{-4} g/L

17.106 Use the following solubility data to calculate a value of K_{sp} for each compound.

(a) $CdCO_3$; 1.0×10^{-6} M

(b) $Ca(OH)_2$; 1.06×10^{-2} M

(c) $PbBr_2$; 4.34 g/L

(d) $BaCrO_4$; 2.8×10^{-3} g/L

17.107 Use the values of K_{sp} in Appendix C to calculate the molar solubility of the following compounds:

(a) $BaCrO_4$ (b) $Mg(OH)_2$ (c) Ag_2SO_3

17.108 Use the values of K_{sp} in Appendix C to calculate the solubil-
RAN ity of the following compounds (in g/L):

(a) Ag_2CO_3 (b) $CuBr$ (c) $Cu_3(PO_4)_2$

Factors That Affect Solubility (Section 17.12)

17.109 Use Le Châtelier's principle to explain the following changes in the solubility of Ag_2CO_3 in water.

(a) Decrease on addition of $AgNO_3$

(b) Increase on addition of HNO_3

(c) Decrease on addition of Na_2CO_3

(d) Increase on addition of NH_3

17.110 Use Le Châtelier's principle to predict whether the solubility of BaF_2 will increase, decrease, or remain the same on addition of each of the following substances.

(a) HCl   (b) KF

(c) $NaNO_3$ (d) $Ba(NO_3)_2$

17.111 Calculate the molar solubility of $PbCrO_4$ in:
RAN
(a) Pure water

(b) 1.0×10^{-3} M K_2CrO_4

17.112 Calculate the molar solubility of SrF_2 in:
RAN
(a) 0.010 M $Sr(NO_3)_2$

(b) 0.010 M NaF

17.113 Which of the following compounds are more soluble in acidic
RAN solution than in pure water? Write a balanced net ionic equation for each dissolution reaction.

(a) AgBr   (b) $CaCO_3$

(c) $Ni(OH)_2$ (d) $Ca_3(PO_4)_2$

17.114 Which of the following compounds are more soluble in acidic
RAN solution than in pure water? Write a balanced net ionic equation for each dissolution reaction.

(a) MnS   (b) $Fe(OH)_3$

(c) AgCl   (d) $BaCO_3$

17.115 Consider saturated solutions of the slightly soluble salt AgBr. Is the solubility of AgBr increased, decreased, or unaffected by the addition of each of the following substances?

(a) HBr   (b) HNO_3

(c) $AgNO_3$ (d) NH_3

17.116 Consider saturated solutions of the slightly soluble salt $BaCO_3$. Is the solubility of $BaCO_3$ increased, decreased, or unaffected by the addition of each of the following substances?

(a) HNO_3 (b) $Ba(NO_3)_2$

(c) Na_2CO_3 (d) CH_3CO_2H

17.117 Is the solubility of $Zn(OH)_2$ increased, decreased, or unchanged on addition of each of the following substances? Write a balanced net ionic equation for each dissolution reaction. (See Appendix C.6 for formulas of complex ions.)

(a) HCl   (b) KOH   (c) NaCN

17.118 Is the solubility of $Fe(OH)_3$ increased, decreased, or unchanged on addition of each of the following substances? Write a balanced net ionic equation for each dissolution reaction. (See Appendix C.6 for formulas of complex ions.)

(a) $HBr(aq)$ (b) $NaOH(aq)$

(c) $KCN(aq)$

17.119 Silver ion reacts with excess CN^- to form a colorless
RAN complex ion, $[Ag(CN)_2]^-$, which has a formation constant $K_f = 3.0 \times 10^{20}$. Calculate the concentration of Ag^+ in a solution prepared by mixing equal volumes of 2.0×10^{-3} M $AgNO_3$ and 0.20 M NaCN.

17.120 Dissolution of 5.0×10^{-3} mol of $Cr(OH)_3$ in 1.0 L of 1.0 M NaOH gives a solution of the complex ion $[Cr(OH)_4]^-$ ($K_f = 8 \times 10^{29}$). What fraction of the chromium in such a solution is present as uncomplexed Cr^{3+}?

17.121 Write a balanced net ionic equation for each of the following dissolution reactions, and use the appropriate K_{sp} and K_f values in Appendix C to calculate the equilibrium constant for each.

(a) AgI in aqueous NaCN to form $[Ag(CN)_2]^-$

(b) $Al(OH)_3$ in aqueous NaOH to form $[Al(OH)_4]^-$

(c) $Zn(OH)_2$ in aqueous NH_3 to form $[Zn(NH_3)_4]^{2+}$

17.122 Write a balanced net ionic equation for each of the following dissolution reactions, and use the appropriate K_{sp} and

K_f values in Appendix C to calculate the equilibrium constant for each.

(a) $Zn(OH)_2$ in aqueous NaOH to form $[Zn(OH)_4]^{2-}$

(b) $Cu(OH)_2$ in aqueous NH_3 to form $[Cu(NH_3)_4]^{2+}$

(c) AgBr in aqueous NH_3 to form $[Ag(NH_3)_2]^+$

17.123 Calculate the molar solubility of AgI in:

RAN

(a) Pure water

(b) 0.10 M NaCN; K_f for $[Ag(CN)_2]^-$ is 3.0×10^{20}

17.124 Calculate the molar solubility of $Cr(OH)_3$ in 0.50 M NaOH;

RAN K_f for $Cr(OH)_4^-$ is 8×10^{29}.

17.125 Zinc hydroxide, $Zn(OH)_2$ ($K_{sp} = 4.1 \times 10^{-17}$), is nearly

RAN insoluble in water but is more soluble in strong base because Zn^{2+} forms the soluble complex ion $[Zn(OH)_4]^{2-}$ ($K_f = 3 \times 10^{15}$).

(a) What is the molar solubility of $Zn(OH)_2$ in pure water? (You may ignore OH^- from the self-dissociation of water.)

(b) What is the pH of the solution in part (a)?

(c) What is the molar solubility of $Zn(OH)_2$ in 0.10 M NaOH?

17.126 Citric acid (H_3Cit) can be used as a household cleaning agent

RAN to dissolve rust stains. The rust, represented as $Fe(OH)_3$, dissolves because the citrate ion forms a soluble complex with Fe^{3+}:

$$Fe(OH)_3(s) + H_3Cit(aq) \rightleftharpoons Fe(Cit)(aq) + 3 H_2O(l)$$

(a) Using the equilibrium constants in Appendix C and $K_f = 6.3 \times 10^{11}$ for Fe(Cit), calculate the equilibrium constant K for the reaction.

(b) Calculate the molar solubility of $Fe(OH)_3$ in a 0.500 M solution of H_3Cit.

Precipitation; Qualitative Analysis (Sections 17.13–17.15)

17.127 What compound, if any, will precipitate when 80.0 mL of 1.0×10^{-5} M $Ba(OH)_2$ is added to 20.0 mL of 1.0×10^{-5} M $Fe_2(SO_4)_3$?

17.128 "Hard" water contains alkaline earth cations such as Ca^{2+}, which reacts with CO_3^{2-} to form insoluble deposits of $CaCO_3$. Will a precipitate of $CaCO_3$ form if a 250 mL sample of hard water having $[Ca^{2+}] = 8.0 \times 10^{-4}$ M is treated with the following?

(a) 0.10 mL of 2.0×10^{-3} M Na_2CO_3

(b) 10 mg of solid Na_2CO_3

17.129 The pH of a sample of hard water (Problem 17.128) hav-

RAN ing $[Mg^{2+}] = 2.5 \times 10^{-4}$ M is adjusted to pH 10.80. Will $Mg(OH)_2$ precipitate?

17.130 Fluoride ion is added to municipal water supplies to pre-

RAN vent tooth decay. The concentration of Ca^{2+} in water equilibrated with the atmosphere and limestone ($CaCO_3$) is 5.0×10^{-4} M. If fluoride ion were added to the water until a concentration of 1.0 ppm was reached, would a precipitate of CaF_2 ($K_{sp} = 3.5 \times 10^{-11}$) form?

17.131 In qualitative analysis, Al^{3+} and Mg^{2+} are separated in an $NH_4^+-NH_3$ buffer having pH ≈ 8. Assuming cation concentrations of 0.010 M, show why $Al(OH)_3$ precipitates but $Mg(OH)_2$ does not.

17.132 Can Fe^{2+} be separated from Sn^{2+} by bubbling H_2S through a 0.3 M HCl solution that contains 0.01 M Fe^{2+} and 0.01 M Sn^{2+}? A saturated solution of H_2S has $[H_2S] \approx$ 0.10 M. Values of K_{spa} are 6×10^2 for FeS and 1×10^{-5} for SnS.

17.133 Can Co^{2+} be separated from Zn^{2+} by bubbling H_2S through a 0.3 M HCl solution that contains 0.01 M Co^{2+} and 0.01 M Zn^{2+}? A saturated solution of H_2S has $[H_2S] \approx$ 0.10 M. Values of K_{spa} are 3 for CoS and 3×10^{-2} for ZnS.

17.134 Will FeS precipitate in a solution that is 0.10 M in $Fe(NO_3)_2$, 0.4 M in HCl, and 0.10 M in H_2S? Will FeS precipitate if the pH of the solution is adjusted to pH 8 with an $NH_4^+-NH_3$ buffer? $K_{spa} = 6 \times 10^2$ for FeS.

17.135 Will CoS precipitate in a solution that is 0.10 M in $Co(NO_3)_2$, 0.5 M in HCl, and 0.10 M in H_2S? Will CoS precipitate if the pH of the solution is adjusted to pH 8 with an $NH_4^+-NH_3$ buffer? $K_{spa} = 3$ for CoS.

17.136 Using the qualitative analysis flowchart in Figure 17.18,

RAN tell how you could separate the following pairs of ions.

(a) Ag^+ and Cu^{2+} (b) Na^+ and Ca^{2+}

(c) Mg^{2+} and Mn^{2+} (d) K^+ and Cr^{3+}

17.137 Give a method for separating the following pairs of ions by the addition of no more than two substances.

(a) Hg_2^{2+} and Co^{2+} (b) Na^+ and Mg^{2+}

(c) Fe^{2+} and Hg^{2+} (d) Ba^{2+} and Pb^{2+}

17.138 Assume that you have three white solids: NaCl, KCl, and $MgCl_2$. What tests could you do to tell which is which?

17.139 In qualitative analysis, Ag^+, Hg_2^{2+}, and Pb^{2+} are separated

RAN from other cations by the addition of HCl. Calculate the concentration of Cl^- ions required to just begin the precipitation of (a) AgCl, (b) Hg_2Cl_2, (c) $PbCl_2$ in a solution having metal-ion concentrations of 0.030 M. What fraction of the Pb^{2+} remains in solution when the Ag^+ just begins to precipitate?

MULTICONCEPT PROBLEMS

17.140 Teeth can be protected from decay by chemical treatment with a dilute solution of fluoride ion, which makes the enamel more resistant to attack by acid. Fluoride functions both by increasing the rate at which enamel remineralizes and by causing the partial conversion of hydroxyapatite to fluorapatite through exchange of F^- for OH^- in healthy enamel.

$$Ca_5(PO_4)_3(OH)(s) + F^-(aq) \rightleftharpoons Ca_5(PO_4)_3(F)(s) + OH^-(aq) \quad K = ?$$

Use the K_{sp} values provided to calculate the equilibrium constant of the above reaction.

$$Ca_5(PO_4)_3(OH)(s) \rightleftharpoons 5 Ca^{2+}(aq) + 3 PO_4^{3-}(aq) + OH^-(aq)$$
$$K_{sp} = 2.3 \times 10^{-59}$$

$$Ca_5(PO_4)_3(F)(s) \rightleftharpoons 5 Ca^{2+}(aq) + 3 PO_4^{3-}(aq) + F^-(aq)$$
$$K_{sp} = 3.2 \times 10^{-60}$$

17.141 Calculate the molar solubility of MnS in a 0.30 M NH_4Cl
RAN −0.50 M NH_3 buffer solution that is saturated with H_2S
($[H_2S] \approx 0.10$ M). What is the solubility of MnS (in g/L)?
(K_{spa} for MnS is 3×10^7.)

17.142 The acidity of lemon juice is derived primarily from citric
RAN acid (H_3Cit), a triprotic acid. What are the concentrations
of H_3Cit, H_2Cit^-, $HCit^{2-}$, and Cit^{3-} in a sample of lemon
juice that has a pH of 2.37 and a total concentration of the
four citrate-containing species of 0.350 M?

Citric acid

17.143 A 100.0 mL sample of a solution that is 0.100 M in HCl
RAN and 0.100 M in HCN is titrated with 0.100 M NaOH. Cal-
culate the pH after the addition of the following volumes of
NaOH:

(a) 0.0 mL   (b) 75.0 mL

(c) 100.0 mL   (d) 125.0 mL

17.144 A 0.0100 mol sample of solid $Cd(OH)_2$ ($K_{sp} = 5.3 \times 10^{-15}$)
RAN in 100.0 mL of water is titrated with 0.100 M HNO_3.

(a) What is the molar solubility of $Cd(OH)_2$ in pure
water? What is the pH of the solution before the addition of any HNO_3?

(b) What is the pH of the solution after the addition of
90.0 mL of 0.100 M HNO_3?

(c) How many milliliters of 0.100 M HNO_3 must be added
to completely neutralize the $Cd(OH)_2$?

17.145 One type of kidney stone is a precipitate of calcium oxalate
RAN (CaC_2O_4, $K_{sp} = 2.3 \times 10^{-9}$). A urine sample has a Ca^{2+}
concentration of 2.5×10^{-3} M and an oxalic acid ($H_2C_2O_4$,
$K_{a1} = 5.9 \times 10^{-2}$, $K_{a2} = 6.4 \times 10^{-5}$) concentration of
1.1×10^{-4} M.

(a) A typical pH for urine is 5.5. Will a precipitate of calcium oxalate form under these conditions?

(b) A vegetarian diet results in a higher pH for urine, typically greater than 7. Would kidney stones be more or
less likely to form in urine with a higher pH?

17.146 When a typical diprotic acid H_2A ($K_{a1} = 10^{-4}$; $K_{a2} = 10^{-10}$)
RAN is titrated with NaOH, the principal A-containing species
at the first equivalence point is HA^-.

(a) By considering all four proton-transfer reactions that
can occur in an aqueous solution of HA^-, show that the
principal reaction is $2\ HA^- \rightleftharpoons H_2A + A^{2-}$.

(b) Assuming that this is the principal reaction, show that
the pH at the first equivalence point equals the average
of pK_{a1} and pK_{a2}.

(c) How many A^{2-} ions are present in 50.0 mL of 1.0 M
NaHA?

17.147 Ethylenediamine ($NH_2CH_2CH_2NH_2$, abbreviated en) is an
organic base that can accept two protons:

$$en(aq) + H_2O(l) \rightleftharpoons enH^+(aq) + OH^-(aq)$$
$$K_{b1} = 5.2 \times 10^{-4}$$
$$enH^+(aq) + H_2O(l) \rightleftharpoons enH_2^{2+}(aq) + OH^-(aq)$$
$$K_{b2} = 3.7 \times 10^{-7}$$

(a) Consider the titration of 30.0 mL of 0.100 M ethylenediamine with 0.100 M HCl. Calculate the pH after
the addition of the following volumes of acid, and construct a qualitative plot of pH versus milliliters of HCl
added:

(i) 0.0 mL   (ii) 15.0 mL

(iii) 30.0 mL   (iv) 45.0 mL

(v) 60.0 mL   (vi) 75.0 mL

(b) Draw the structure of ethylenediamine, and explain
why it can accept two protons.

(c) What hybrid orbitals do the N atoms use for bonding?

17.148 A 40.0 mL sample of a mixture of HCl and H_3PO_4 was
titrated with 0.100 M NaOH. The first equivalence point
was reached after 88.0 mL of base, and the second equivalence point was reached after 126.4 mL of base.

(a) What is the concentration of H_3O^+ at the first equivalence point?

(b) What are the initial concentrations of HCl and
H_3PO_4 in the mixture?

(c) What percent of the HCl is neutralized at the first
equivalence point?

(d) What is the pH of the mixture before the addition of
any base?

(e) Sketch the pH titration curve, and label the buffer
regions and equivalence points.

(f) What indicators would you select to signal the equivalence points?

17.149 A 1.000 L sample of HCl gas at 25 °C and 732.0 mm Hg
RAN was absorbed completely in an aqueous solution that contained 6.954 g of Na_2CO_3 and 250.0 g of water.

(a) What is the pH of the solution?

(b) What is the freezing point of the solution?

(c) What is the vapor pressure of the solution? (The vapor
pressure of pure water at 25 °C is 23.76 mm Hg.)

17.150 A saturated solution of an ionic salt MX exhibits an osmotic
RAN pressure of 74.4 mm Hg at 25 °C. Assuming that MX is
completely dissociated in solution, what is the value of its
K_{sp}?

17.151 Consider the reaction that occurs on mixing 50.0 mL of
RAN 0.560 M $NaHCO_3$ and 50.0 mL of 0.400 M NaOH at 25 °C.

(a) Write a balanced net ionic equation for the reaction.

(b) What is the pH of the resulting solution?

(c) How much heat (in joules) is liberated by the reaction?
(Standard heats of formation are given in Appendix B.)

(d) What is the final temperature of the solution to the
nearest 0.1 °C. You may assume that all the heat liberated is absorbed by the solution, the mass of the solution is 100.0 g, and its specific heat is 4.18 J/(g·°C).

17.152 In qualitative analysis, Ca^{2+} and Ba^{2+} are separated from Na^+, K^+, and Mg^{2+} by adding aqueous $(NH_4)_2CO_3$ to a solution that also contains aqueous NH_3 (Figure 17.18). Assume that the concentrations after mixing are 0.080 M $(NH_4)_2CO_3$ and 0.16 M NH_3.

(a) List all the Brønsted–Lowry acids and bases present initially, and identify the principal reaction.

(b) Calculate the pH and the concentrations of all species present in the solution.

(c) In order for the human eye to detect the appearance of a precipitate, a very large number of ions must come together to form solid particles. For this and other reasons, the ion product must often exceed K_{sp} by a factor of about 10^3 before a precipitate can be detected in a typical qualitative analysis experiment. Taking this fact into account, show quantitatively that the CO_3^{2-} concentration is large enough to give observable precipitation of $CaCO_3$ and $BaCO_3$, but not $MgCO_3$. Assume that the metal-ion concentrations are 0.010 M.

(d) Show quantitatively which of the Mg^{2+}, Ca^{2+}, and Ba^{2+} ions, if any, should give an observable precipitate of the metal hydroxide.

(e) Could the separation of Ca^{2+} and Ba^{2+} from Mg^{2+} be accomplished using 0.80 M Na_2CO_3 in place of 0.080 M $(NH_4)_2CO_3$? Show quantitatively why or why not.

17.153 A railroad tank car derails and spills 36 tons of concen-
RAN trated sulfuric acid. The acid is 98.0 mass% H_2SO_4 and has a density of 1.836 g/mL.

(a) What is the molarity of the acid?

(b) How many kilograms of sodium carbonate are needed to completely neutralize the acid?

(c) How many liters of carbon dioxide at 18 °C and 745 mm Hg are produced as a by-product?

17.154 Some progressive hair coloring products marketed to men, such
RAN as Grecian Formula 16, contain lead acetate, $Pb(CH_3CO_2)_2$. As the coloring solution is rubbed on the hair, the Pb^{2+} ions react with the sulfur atoms in hair proteins to give lead(II) sulfide (PbS), which is black. A typical coloring solution contains 0.3 mass% $Pb(CH_3CO_2)_2$, and about 2 mL of the solution is used per application.

(a) Assuming that 30% of the $Pb(CH_3CO_2)_2$ is converted to PbS, how many milligrams of PbS are formed per application of the coloring solution?

(b) Suppose the hair is washed with shampoo and water that has pH = 5.50. How many washings would be required to remove 50% of the black color? Assume that 3 gal of water is used per washing and that the water becomes saturated with PbS.

(c) Does the calculated number of washings look reasonable, given that frequent application of the coloring solution is recommended? What process(es) in addition to dissolution might contribute to the loss of color?

chapter 18

Thermodynamics: Entropy, Free Energy, and Spontaneity

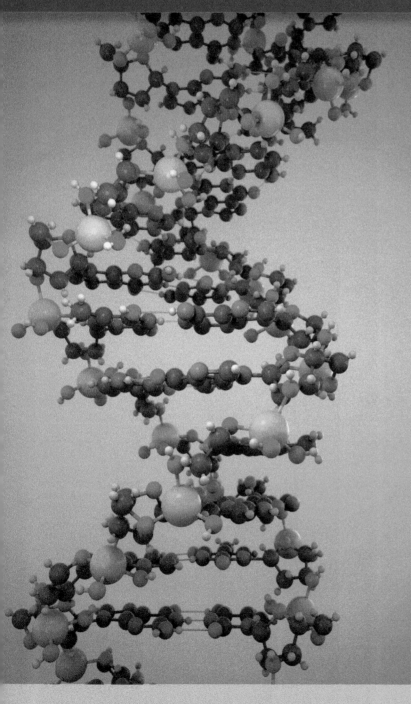

Contents

The second law of thermodynamics states that all spontaneous processes lead to increased randomness. Yet living organisms create highly ordered structures, such as DNA. A DNA molecule has two strands, each with a specific order of nucleotide bases.

Does the formation of highly ordered molecules violate the second law of thermodynamics?

The answer to this question can be found on page 798 in the **INQUIRY** ?

Each day we observe familiar processes such as ice cubes melting in a glass of water, a punctured tire losing air, and metal that has rusted after exposure to air and water for a long period of time. Throughout this book we have studied various aspects of these chemical and physical changes, such as the amount of heat transfer (thermochemistry), the rate of the reaction (kinetics), and the extent of the reaction (equilibrium). However, we likely have never considered why these changes, and not the opposite, occur in the first place. Why doesn't water spontaneously turn to ice at room temperature? Why doesn't a flat tire refill itself with air and rusted metal return to its original shiny state? In other words, what factors determine the direction and the extent of a process?

The *direction* in a chemical or physical process describes whether the reactants turn into products or vice versa. We know that ice in a glass of water at room temperature will always melt and that the reverse process, liquid water freezing back into solid cubes, will not occur. The *extent* of the reaction refers to the position of the equilibrium, which is described by the numerical value of the equilibrium constant K. Reactions that lead to an equilibrium state consisting mainly of products have a large value of K, and reactions that produce mainly reactants in equilibrium have a small value of K. We observe that the ice cubes melt completely in the water and the physical change of $H_2O(s)$ converting to $H_2O(l)$ at 25 °C will have a very large value of K.

In order to understand the fundamental reason that chemical and physical changes occur, we turn to **thermodynamics**, a field that deals with relationships between forms of energy. We will find that entropy, which is related to molecular randomness, is the driving force for these changes.

18.1 SPONTANEOUS PROCESSES

In Section 9.12 we defined a **spontaneous process** as one that proceeds on its own without any external influence. The reverse of a spontaneous process is always nonspontaneous and takes place only in the presence of some continuous external influence. Take, for example, the expansion of a gas into a vacuum. When the stopcock in the apparatus shown in **FIGURE 18.1** is opened, the gas in bulb A expands spontaneously into the evacuated bulb B until the gas pressure in the two bulbs is the same. The reverse process, migration of all the gas molecules into one bulb, does not occur spontaneously. To compress a gas from a larger to a smaller volume, we would have to push on the gas with a piston.

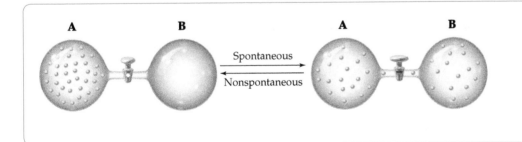

◀ FIGURE 18.1

The expansion of a gas into a vacuum. When the stopcock is opened, the gas in bulb A expands spontaneously into evacuated bulb B to fill all the available volume. The reverse process, compression of the gas into a single bulb, is nonspontaneous.

As a second example, consider the combination of hydrogen and oxygen in the presence of a platinum catalyst:

$$2\,H_2(g) + O_2(g) \xrightarrow{\text{Catalyst}} 2\,H_2O(l)$$

The forward reaction occurs spontaneously, but the reverse reaction, decomposition of water into its elements, does not occur no matter how long we wait. We'll see in Section 19.12 that we can force the reverse reaction to occur by **electrolysis**, but that reverse process is nonspontaneous and requires a continuous input of electrical energy.

In general, whether the forward or reverse reaction is spontaneous depends on temperature, pressure, and composition of the reaction mixture. Consider, for example, the Haber synthesis of ammonia:

$$N_2(g) + 3\,H_2(g) \xrightleftharpoons{\text{Catalyst}} 2\,NH_3(g)$$

LOOKING AHEAD . . .

As we'll see in Section 19.12, **electrolysis** is the process of using an electric current to bring about a chemical change.

A mixture of gaseous N_2, H_2, and NH_3, each at a partial pressure of 1 atm, reacts spontaneously at 300 K to convert some of the N_2 and H_2 to NH_3. We can predict the direction of spontaneous reaction from the relative values of the equilibrium constant K and the **reaction quotient** Q (Section 15.5). Since $K_p = 4.4 \times 10^5$ at 300 K and $Q_p = 1$ for partial pressures of 1 atm, the reaction will proceed in the forward direction because Q_p is less than K_p. Under these conditions, the reverse reaction is nonspontaneous. At 700 K, however, $K_p = 8.8 \times 10^{-5}$ and the reverse reaction is spontaneous because Q_p is greater than K_p.

A spontaneous reaction always moves a reaction mixture toward equilibrium. By contrast, a nonspontaneous reaction moves the composition of a mixture away from the equilibrium composition. Remember, though, that the word *spontaneous* doesn't necessarily mean "fast." A spontaneous reaction can be either fast or slow—for example, the gradual rusting of iron metal is a slow spontaneous reaction. Thermodynamics tells us where a reaction is headed, but it says nothing about how long it takes to get there. As discussed in Section 14.7, the rate of a reaction depends on factors such as temperature, concentration, and the height of the **activation energy** barrier between the reactants and products (**FIGURE 18.2**). Large values of activiation energy lead to slow reaction rates.

The rusting of these wheel rims is a slow but spontaneous reaction.

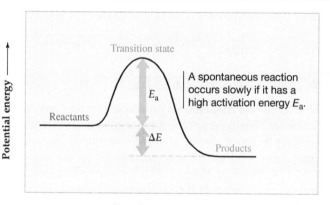

Reaction progress ⟶

A spontaneous reaction occurs slowly if it has a high activation energy E_a.

▲ **FIGURE 18.2**

A familiar example of a slow spontaneous reaction and the potential energy profile for the reaction. The rate of the reaction depends on kinetics.

BIG IDEA Question 1 Go to eText

Which of the following is a non-spontaneous process?
(a) Heat flows from a hot object to a cold object.
(b) For the reaction A ⇌ B, K = 0.20. Substance A is converted to substance B in a mixture in which [A] = 2.0 M and [B] = 1.0 M.
(c) Steam condenses to liquid water on a mirror at 28 °C.

18.2 ENTHALPY, ENTROPY, AND SPONTANEOUS PROCESSES

Let's look more closely at spontaneous processes and at the thermodynamic factors that cause them to occur. We saw in Chapter 9 that most spontaneous chemical reactions are accompanied by the conversion of potential energy to heat. For example, when methane burns in air, the potential energy stored in the chemical bonds of CH_4 and O_2 is partly converted to heat, which flows from the system (reactants plus products) to the surroundings:

$$CH_4(g) + 2\,O_2(g) \longrightarrow CO_2(g) + 2\,H_2O(l) \quad \Delta H° = -890.3 \text{ kJ}$$

Because heat is lost by the system, the reaction is exothermic and the standard enthalpy of reaction is negative ($\Delta H° = -890.3$ kJ). The total energy is conserved, so all the energy lost by the system shows up as heat gained by the surroundings.

Because spontaneous reactions so often give off heat, the nineteenth-century French chemist Marcellin Berthelot proposed that spontaneous chemical or physical changes are *always* exothermic. But Berthelot's proposal can't be correct. Ice, for example, spontaneously absorbs heat from the surroundings and melts at temperatures above 0 °C. Similarly, liquid water absorbs heat and spontaneously boils at temperatures above

100 °C. As further examples, gaseous N_2O_4 absorbs heat when it decomposes to NO_2 at 400 K, and table salt absorbs heat when it dissolves in water at room temperature:

$$H_2O(s) \longrightarrow H_2O(l) \qquad \Delta H_{fusion} = +6.01 \text{ kJ}$$
$$H_2O(l) \longrightarrow H_2O(g) \qquad \Delta H_{vap} = +40.7 \text{ kJ}$$
$$N_2O_4(g) \longrightarrow 2\,NO_2(g) \qquad \Delta H° = +55.3 \text{ kJ}$$
$$NaCl(s) \longrightarrow Na^+(aq) + Cl^-(aq) \qquad \Delta H° = +3.88 \text{ kJ}$$

All these processes are endothermic, yet all are spontaneous. In all cases, the system moves spontaneously to a state of *higher* enthalpy by absorbing heat from the surroundings.

Because some spontaneous reactions are exothermic and others are endothermic, enthalpy alone can't account for the direction of spontaneous change; a second factor must be involved. This second determinant of spontaneous change is nature's tendency to move to a condition of maximum randomness (Section 9.12).

Molecular randomness is called **entropy** and is denoted by the symbol S. Entropy is a **state function** (Section 9.2), and the entropy change ΔS for a process thus depends only on the initial and final states of the system:

$$\Delta S = S_{final} - S_{initial}$$

When the randomness of a system increases, ΔS has a positive value; when randomness decreases, ΔS is negative.

The randomness (or disorder) of a system comes about because the particles in the system (atoms, ions, and molecules) are in constant motion, moving about in the accessible volume, colliding with each other and continually exchanging energy. Randomness—and thus entropy—is a probability concept, related to the number of ways that a particular state of a system can be achieved. A particular state of a macroscopic system, characterized by its temperature, pressure, volume, and number of particles, can be achieved in a vast number of ways in which the fluctuating positions and energies of the individual particles differ but the volume and total energy are constant.

We'll examine the relationship between entropy and probability in the next section, but first let's take a qualitative look at the four spontaneous endothermic processes mentioned previously (melting of ice, boiling of liquid water, decomposition of N_2O_4, and dissolving of NaCl in water). Each of these processes involves an increase in the randomness of the system. When ice melts, for example, randomness increases because the highly ordered crystalline arrangement of tightly held water molecules collapses and the molecules become free to move about in the liquid. When liquid water vaporizes, randomness further increases because the molecules can now move independently in the much larger volume of the gas. In general, processes that convert a solid to a liquid or a liquid to a gas involve an increase in randomness and thus an increase in entropy (**FIGURE 18.3**).

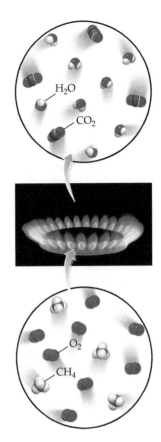

▲ The combustion of natural gas (mainly CH_4) in air is a spontaneous, exothermic reaction.

REMEMBER ...
A **state function** is a function or property whose value depends only on the present state (condition) of the system, not on the path used to arrive at that condition. Pressure, volume, temperature, enthalpy, and entropy are state functions (Section 9.2).

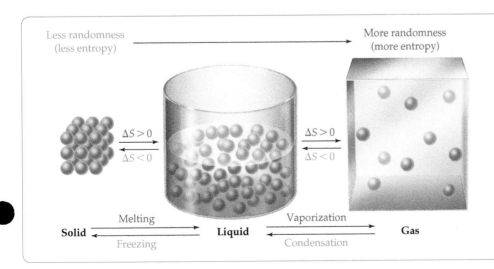

◀ FIGURE 18.3

How molecular randomness—and thus entropy—changes when solids, liquids, and gases interconvert.

◀ Figure It Out
Which phase of matter results in the greatest number of ways that particles can be arranged?

Answer: Gas

The decomposition of N_2O_4 ($O_2N—NO_2$) is accompanied by an increase in randomness because breaking the N—N bond allows the two gaseous NO_2 fragments to move independently. Whenever a molecule breaks into two or more pieces, the amount of molecular randomness increases. More generally, randomness—and thus entropy—increases whenever a reaction results in an increase in the number of gaseous particles (**FIGURE 18.4**).

▶ **FIGURE 18.4**

How molecular randomness—and thus entropy—changes when the number of gaseous particles changes.

▶ **Figure It Out**

Why does increasing the number of gaseous particles increase the entropy?

Answer: There are 5 N_2O_4 molecules in the reactants and 10 NO_2 molecules in products. There are more ways to arrange 10 molecules relative to one another than there are ways to arrange 5 molecules relative to one another. As the number of possible arrangements increases, the randomness increases.

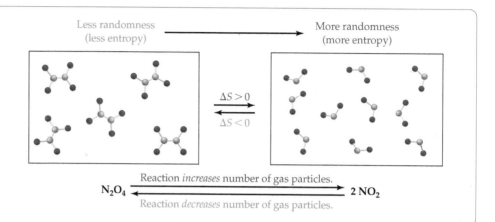

Less randomness (less entropy) → More randomness (more entropy)

$\Delta S > 0$
$\Delta S < 0$

Reaction *increases* number of gas particles.
$$N_2O_4 \rightleftharpoons 2\,NO_2$$
Reaction *decreases* number of gas particles.

REMEMBER . . .

Ion–dipole forces are the result of electrostatic attractions between an ion and partial charges on a polar molecule (Section 8.6)

The entropy change that occurs on dissolving sodium chloride in water occurs because the crystal structure of solid NaCl is disrupted and the Na^+ and Cl^- ions become *hydrated*, or surrounded by an ordered shell of solvent water molecules. Recall that when an ionic salt dissolves in water **ion–dipole forces** form between a polar water molecule and a charged ion (Section 8.6). Disruption of the crystal increases randomness because the Na^+ and Cl^- ions are tightly held in the solid but are free to move about in the liquid. The hydration process, however, *decreases* randomness because the polar, hydrating water molecules adopt an orderly arrangement about the Na^+ and Cl^- ions. It turns out that the overall dissolution process for NaCl results in a net increase in randomness, and ΔS is thus positive (**FIGURE 18.5**). This is usually the case for the dissolution of molecular solids, such as $HgCl_2$, and salts that contain +1 cations and −1 anions. For salts

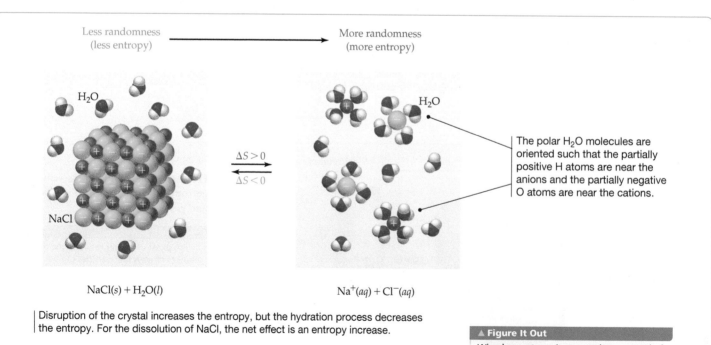

Less randomness (less entropy) → More randomness (more entropy)

H_2O

$\Delta S > 0$
$\Delta S < 0$

H_2O

The polar H_2O molecules are oriented such that the partially positive H atoms are near the anions and the partially negative O atoms are near the cations.

NaCl

$NaCl(s) + H_2O(l)$ $Na^+(aq) + Cl^-(aq)$

Disruption of the crystal increases the entropy, but the hydration process decreases the entropy. For the dissolution of NaCl, the net effect is an entropy increase.

▲ **FIGURE 18.5**

How molecular randomness—and thus entropy—changes when NaCl dissolves in water. When NaCl dissolves, the crystal breaks up and the Na^+ and Cl^- ions become surrounded by hydrating water molecules.

▲ **Figure It Out**

Why does entropy increase when a crystal of NaCl dissolves in water?

Answer: The ordered arrangement of alternating Na^+ and Cl^- ions in the crystal is destroyed, and the ions move randomly in the solution.

such as $CaSO_4$, which contain more highly charged ions, the hydrating water molecules are more strongly attached to the ions and the dissolution process often results in a net decrease in entropy. The following dissolution reactions illustrate the point:

$$HgCl_2(s) \longrightarrow HgCl_2(aq) \qquad\qquad \Delta S = +9 \text{ J}/(\text{K} \cdot \text{mol})$$

$$NaCl(s) \longrightarrow Na^+(aq) + Cl^-(aq) \qquad \Delta S = +43 \text{ J}/(\text{K} \cdot \text{mol})$$

$$CaSO_4(s) \longrightarrow Ca^{2+}(aq) + SO_4^{2-}(aq) \qquad \Delta S = -140 \text{ J}/(\text{K} \cdot \text{mol})$$

WORKED EXAMPLE 18.1

Predicting the Sign of ΔS

Predict the sign of ΔS in the system for each of the following processes:

(a) $CO_2(s) \longrightarrow CO_2(g)$ (sublimation of dry ice)
(b) $CaSO_4(s) \longrightarrow CaO(s) + SO_3(g)$
(c) $N_2(g) + 3 H_2(g) \longrightarrow 2 NH_3(g)$
(d) $I_2(s) \longrightarrow I_2(aq)$ (dissolution of iodine in water)

STRATEGY

To predict the sign of ΔS, look to see whether the process involves a phase change, a change in the number of gaseous molecules, or the dissolution (or precipitation) of a solid. Entropy generally increases for phase transitions that convert a solid to a liquid or a liquid to a gas, for reactions that increase the number of gaseous molecules, and for the dissolution of molecular solids or salts with +1 cations and −1 anions.

SOLUTION

(a) The molecules in a gas are free to move about randomly, whereas the molecules in a solid are tightly held in a highly ordered arrangement. Therefore, randomness increases when a solid sublimes and ΔS is positive.
(b) One mole of gaseous molecules appears on the product side of the equation and none appears on the reactant side. Because the reaction increases the number of gaseous molecules, the entropy change is positive.
(c) The entropy change is negative because the reaction decreases the number of gaseous molecules from 4 mol to 2 mol. Fewer particles move independently after reaction than before.
(d) Iodine molecules are electrically neutral and form a molecular solid. The dissolution process destroys the order of

the crystal and enables the iodine molecules to move about randomly in the liquid. Therefore, ΔS is positive.

▶ **PRACTICE 18.1** Which of the following reactions has a decrease in entropy ($\Delta S < 0$)?

(a) $CO_2(s) \longrightarrow CO_2(g)$
(b) $I_2(g) \longrightarrow 2 I(g)$
(c) $CaCO_3(s) \longrightarrow CaO(s) + CO_2(g)$
(d) $Ag^+(aq) + Br^-(aq) \longrightarrow AgBr(s)$

▶ **CONCEPTUAL APPLY 18.2** Consider the gas-phase reaction of A_2 molecules (red) with B atoms (blue):

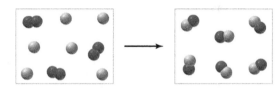

(a) Write a balanced equation for the reaction.
(b) Predict the sign of ΔS for the reaction.

All **WORKED EXAMPLES** with this icon have an interactive video in the eText.

18.3 ENTROPY AND PROBABILITY

We saw in Section 18.2 that the driving force for spontaneous processes that absorb heat (endothermic processes) is the accompanying increase in entropy or randomness. Why do systems tend to move spontaneously to a state of maximum randomness? The answer is that a random arrangement of particles is more probable than an ordered arrangement because a random arrangement can be achieved in more ways. The arrangement of particles refers to the positions and/or energy levels that are available to a system in a given state. Let's begin with a simple example. Suppose that you shake a box containing 20 identical coins and then count the number of heads (H) and tails (T). It's very unlikely that all 20 coins will come up heads; that is, a perfectly ordered arrangement of 20 heads (or 20 tails) is much less probable than a random mixture of heads and tails.

The probabilities of the ordered and random arrangements are proportional to the number of ways that the arrangements can be achieved. The perfectly ordered arrangement

▲ Shaking a box that contains 20 quarters gives a random arrangement of heads and tails.

(a) The perfectly ordered "heads-up" structure.

20 "heads"
0 "tails"

(b) The molecules arranged randomly in one of the 2^{20} ways in which a disordered structure can be obtained.

9 "heads"
11 "tails"

▲ **FIGURE 18.6**

A hypothetical crystal containing 20 CO molecules.

of 20 heads can be achieved in only one way because it consists of a single configuration. In how many ways, though, can a random arrangement be achieved? If there were just two coins in the box, each of them could come up in one of two ways (H or T), and the two together could come up in $2 \times 2 = 2^2 = 4$ ways (HH, HT, TH, or TT). Three coins could come up in $2 \times 2 \times 2 = 2^3 = 8$ ways (HHH, THH, HTH, HHT, HTT, THT, TTH, or TTT), and so on. For the case of 20 coins, the number of possible arrangements is $2^{20} = 1,048,576$.

Because the ordered arrangement of 20 heads (or 20 tails) can be achieved in only one way and a random mixture of heads and tails can be achieved in $2^{20} - 2 \approx 2^{20}$ ways, a random arrangement is 2^{20} times more probable than a perfectly ordered arrangement. If you begin with an ordered arrangement of 20 heads and shake the box, the system will move to a state with a random mixture of heads and tails because that state is more probable. (Note that the state with a random arrangement of coins includes all possible arrangements except the two perfectly ordered arrangements.)

An analogous chemical example is a crystal containing diatomic molecules such as carbon monoxide in which the two distinct ends of the CO molecule correspond to the heads and tails of a coin. Let's suppose that the long dimensions of the molecules are oriented vertically (**FIGURE 18.6**) and that the temperature is 0 K, so that the molecules are locked into a fixed arrangement. The state in which the molecules pack together in a perfectly ordered "heads-up" arrangement (Figure 18.6a) can be achieved in only one way, whereas the state in which the molecules are arranged randomly with respect to the vertical direction can be achieved in many ways—2^{20} ways for a hypothetical crystal containing 20 CO molecules (Figure 18.6b). Therefore, a structure in which the molecules are arranged randomly is 2^{20} times more probable than the perfectly ordered heads-up structure.

Boltzmann's Equation

The Austrian physicist Ludwig Boltzmann proposed that the entropy of a particular state is related to the number of ways that the state can be achieved, according to the equation

> **Boltzmann's equation** $S = k \ln W$

where S is the entropy of the state, $\ln W$ is the natural logarithm of the number of ways that the state can be achieved, and k, now known as *Boltzmann's constant*, is a universal constant equal to the gas constant R divided by Avogadro's number $(k = R/N_A = 1.38 \times 10^{-23} \, \text{J/K})$. Because a logarithm is dimensionless, the Boltzmann equation implies that entropy has the same units as the constant k, joules per kelvin.

Now let's apply Boltzmann's equation to our hypothetical crystal containing 20 CO molecules. Because a perfectly ordered state can be achieved in only one way ($W = 1$ in the Boltzmann equation) and because $\ln 1 = 0$, the entropy of the perfectly ordered state is zero:

$$S = k \ln W = k \ln 1 = 0$$

The more probable state in which the molecules are arranged randomly can be achieved in 2^{20} ways and thus has a higher entropy:

$$S = k \ln W = k \ln 2^{20}$$
$$= (1.38 \times 10^{-23} \, \text{J/K})(20)(\ln 2)$$
$$= 1.91 \times 10^{-22} \, \text{J/K}$$

where we have made use of the relation $\ln x^a = a \ln x$ (Appendix A.2).

If our crystal contained 1 mol of CO molecules, the entropy of the perfectly ordered state (6.02×10^{23} C atoms up) would still be zero, but the entropy of the state with a random arrangement of CO molecules would be much higher because Avogadro's number of molecules can be arranged randomly in a huge number of ways ($W = 2^{N_A} = 2^{6.02 \times 10^{23}}$).

According to Boltzmann's equation, the entropy of the state with a random arrangement of CO molecules is

$$S = k \ln W = k \ln 2^{N_A} = kN_A \ln 2$$

Because $k = R/N_A$,

$$S = R \ln 2 = (8.314 \text{ J/K})(0.693)$$
$$= 5.76 \text{ J/K}$$

Based on experimental measurements, the entropy of 1 mol of solid carbon monoxide near 0 K is about 5 J/K, indicating that the CO molecules adopt a nearly random arrangement. Entropy associated with a random arrangement of molecules in space is sometimes called positional, or configurational, entropy.

The nearly random arrangement of CO molecules in crystalline carbon monoxide is unusual but can be understood in terms of molecular structure. Because CO molecules have a dipole moment of only 0.11 D, intermolecular dipole–dipole forces are unusually weak (Section 8.6), and the molecules therefore have little preference for a slightly lower energy, completely ordered arrangement. By contrast, HCl, with a larger dipole moment of 1.11 D, forms an ordered crystalline solid, and so the entropy of 1 mol of solid HCl at 0 K is 0 J/K.

Expansion of an Ideal Gas

Boltzmann's equation also explains why a gas expands into a vacuum. If the two bulbs in Figure 18.1 have equal volumes, each molecule has one chance in two of being in bulb A (heads, in our coin example) and one chance in two of being in bulb B (tails) when the stopcock is opened. It's exceedingly unlikely that all the molecules in 1 mol of gas will be in bulb A because that state can be achieved in only one way. The state in which Avogadro's number of molecules are randomly distributed between bulbs A and B can be achieved in $2^{6.02 \times 10^{23}}$ ways, and the entropy of that state is therefore higher than the entropy of the ordered state by the now familiar amount, $R \ln 2 = 5.76$ J/K. Thus, a gas expands spontaneously because the state of greater volume is more probable.

We can derive a general equation for the entropy change that occurs on the expansion of an ideal gas at constant temperature by considering the distribution of N molecules among B hypothetical boxes, or cells, each having an equal volume v.

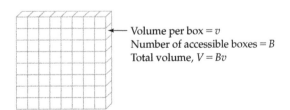

Volume per box $= v$
Number of accessible boxes $= B$
Total volume, $V = Bv$

Since the energy of an ideal gas depends only on the temperature (per the kinetic–molecular theory of gases, Section 10.6), ΔE for the expansion of an ideal gas at constant temperature is zero. To calculate the entropy change $\Delta S = S_{\text{final}} - S_{\text{initial}}$ using the Boltzmann equation, we have only to find the number of ways N molecules can be distributed among the B boxes.

A single molecule can go into any one of the boxes and can thus be assigned to B boxes in B ways. Two molecules can occupy the boxes in $B \times B = B^2$ ways, three molecules can fill the boxes in $B \times B \times B = B^3$ ways, and so on. The number of ways that N molecules can occupy B boxes is $W = B^N$.

REMEMBER . . .
The dipole moment (μ) is a measure of the net polarity of a molecule and is defined as $\mu = Q \times r$, where Q is the magnitude of the charge at either end of the molecular dipole and r is the distance between the charges (Section 8.5).
Dipole–dipole forces result from electrical interactions among neighboring polar molecules (Section 8.6).

BIG IDEA Question 2

Consider a disordered crystal of phosgene ($COCl_2$), in which each trigonal planar molecule is oriented randomly in one of three possible ways. What is the entropy of the disordered state of the crystal if the crystal contains 1 mol of $COCl_2$ molecules?

REMEMBER . . .
According to the kinetic–molecular theory, the kinetic energy of 1 mol of an ideal gas equals 3 $RT/2$ and is independent of pressure and volume (Section 10.6).

Now suppose that the initial volume comprises $B_{initial}$ boxes and the final volume consists of B_{final} boxes:

$$V_{initial} = B_{initial}v \quad \text{and} \quad V_{final} = B_{final}v$$

Then the probabilities of the initial and final states—that is, the number of ways they can be achieved—are

$$W_{initial} = (B_{initial})^N \quad \text{and} \quad W_{final} = (B_{final})^N$$

According to the Boltzmann equation, the entropy change due to a change in volume is

$$\Delta S = S_{final} - S_{initial} = k \ln W_{final} - k \ln W_{initial} = k \ln \frac{W_{final}}{W_{initial}}$$

$$= k \ln \left(\frac{B_{final}}{B_{initial}}\right)^N = kN \ln \frac{B_{final}}{B_{initial}}$$

Because $V_{initial} = B_{initial}v$ and $V_{final} = B_{final}v$

$$\Delta S = kN \ln \left(\frac{V_{final}/v}{V_{initial}/v}\right) = kN \ln \frac{V_{final}}{V_{initial}}$$

Finally, because $k = R/N_A$ and the number of particles (N) equals the number of moles of gas times Avogadro's number ($N = nN_A$), then

$$kN = \left(\frac{R}{N_A}\right)(nN_A) = nR$$

and so the entropy change for expansion (or compression) of n moles of an ideal gas at constant temperature is

> **Entropy change upon volume change of a gas** $\Delta S = nR \ln \dfrac{V_{final}}{V_{initial}}$

For a twofold expansion of 1 mol of an ideal gas at constant temperature, $\Delta S = R \ln 2$, the same result as we obtained previously.

Because the pressure and volume of an ideal gas are related inversely ($P = nRT/V$), we can also write

> $$\Delta S = nR \ln \frac{P_{initial}}{P_{final}}$$

Thus, the entropy of a gas *increases* when its pressure *decreases* at constant temperature, and the entropy *decreases* when its pressure *increases*. Common sense tells us that the more we squeeze the gas, the less space the gas molecules have and so randomness decreases.

BIG IDEA Question 3   Go to eText

Which container represents a system with higher entropy?

(a)   (b)

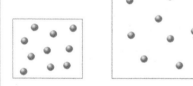

WORKED EXAMPLE 18.2

Calculating the Entropy (S) of a State

Consider the distribution of 1000 ideal gas molecules among three bulbs (A, B, and C) of equal volume. For each of the following states, determine the number of ways (W) that the state can be achieved, and use Boltzmann's equation to calculate the entropy of the state.

(a) 1000 molecules in bulb A

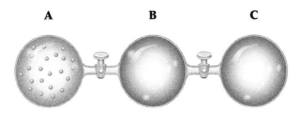

(b) 1000 molecules randomly distributed among bulbs A, B, and C

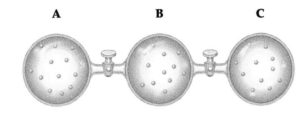

IDENTIFY

Known	Unknown
Number of molecules ($N = 1000$)	Number of ways (W)
Number of bulbs or "boxes" ($B = 3$)	Entropy (S)
Boltzmann's equation: $S = k \ln W$	

STRATEGY

Use the equation $W = B^N$ to calculate the number of ways that N molecules can occupy B boxes. Then use Boltzmann's equation to calculate entropy.

SOLUTION

(a) $W = B^N = 1^{1000} = 1$

$S = k \ln W = (1.38 \times 10^{-23} \text{ J/K})(\ln 1) = 0$

(b) $W = B^N = 3^{1000}$

$S = k \ln W = (1.38 \times 10^{-23} \text{ J/K})(\ln 3^{1000})$

$= (1.38 \times 10^{-23} \text{ J/K})(1000)(\ln 3)$

$= 1.52 \times 10^{-20} \text{ J/K}$

▶ **PRACTICE 18.3** Consider the distribution of ideal gas molecules among three bulbs (A, B, and C) of equal volume. Which of the following states has an entropy of 1.52×10^{-22} J/K?

(a) 10 molecules in bulb A
(b) 10 molecules randomly distributed in bulbs A and B
(c) 10 molecules randomly distributed in bulbs A, B, and C

▶ **APPLY 18.4**

(a) Which state has the higher entropy? Explain in terms of probability.
State A: 1 mol of N_2 gas at STP
State B: 1 mol of N_2 gas at 273 K in a volume of 11.2 L
(b) Calculate the change in entropy (ΔS) from State A to State B.

All **PRACTICE** and **APPLY** problems are interactive in the eText.

18.4 ENTROPY AND TEMPERATURE

Thus far we've seen that entropy is associated with the orientation and distribution of molecules in space. Disordered crystals have higher entropy than ordered crystals, and expanded gases have higher entropy than compressed gases.

Entropy is also associated with molecular motion. As the temperature of a substance increases, random molecular motion increases and there is a corresponding increase in the average kinetic energy of the molecules. But not all the molecules have the same energy. As we saw in Section 10.6, there is a distribution of molecular speeds in a gas, a distribution that broadens and shifts to higher speeds with increasing temperature (Figure 10.13). In solids, liquids, and gases, the total energy of a substance can be distributed among the individual molecules in a number of ways that increases as the total energy increases. According to Boltzmann's equation, the more ways that the energy can be distributed, the greater the randomness of the state and the higher its entropy. Therefore, the entropy of a substance increases with increasing temperature (**FIGURE 18.7**).

▶ **FIGURE 18.7**

A substance at a higher temperature has greater entropy than the same substance at a lower temperature.

▶ **Figure It Out**

Why does a substance at higher temperature have a greater entropy than a substance at lower temperature?

Answer: At higher temperatures, the molecules have a wider range of kinetic energies, which means there are a greater number of ways to distribute the energy.

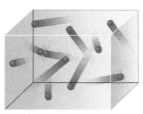

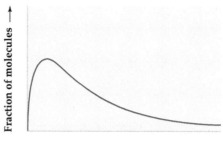

Higher temperature:
• Higher average molecular speed
• Broader distribution of individual kinetic energies
• More randomness
• Higher entropy

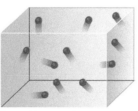

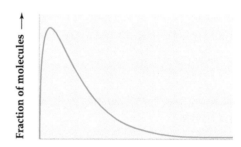

Lower temperature:
• Lower average molecular speed
• Narrower distribution of individual kinetic energies
• Less randomness
• Lower entropy

A typical plot of entropy versus temperature is shown in **FIGURE 18.8.** At absolute zero, every substance is a solid whose particles are tightly held in a crystalline structure. If there is no residual orientational disorder, like that in carbon monoxide (Figure 18.6b), the entropy of the substance at 0 K will be zero, a general result summarized in the **third law of thermodynamics.**

The third law of thermodynamics   The entropy of a perfectly ordered crystalline substance at 0 K is zero.

(The first law of thermodynamics was discussed in Section 9.1. We'll review the first law and discuss the second law in Section 18.6.)

As the temperature of a solid is raised, the added energy increases the vibrational motion of the molecules about their equilibrium positions in the crystal. The number of ways in which the vibrational energy can be distributed increases with rising temperature, and the entropy of the solid thus increases steadily as the temperature increases.

At the melting point, there is a discontinuous jump in entropy because there are many more ways of arranging the molecules in the liquid than in the solid in which positions are fixed. Furthermore, the molecules in the liquid can undergo translational and rotational motion as well as vibrational motion (**FIGURE 18.9**), and so there are many more ways of distributing the total energy in the liquid. An even greater jump in entropy is observed at the boiling point because molecules in the gas have much higher speeds and occupy a much larger volume. Between the melting point and the boiling point, the entropy of a liquid increases steadily as molecular motion increases and the number of ways of distributing the total energy among the individual molecules increases. For the same reason, the entropy of a gas rises steadily as its temperature increases.

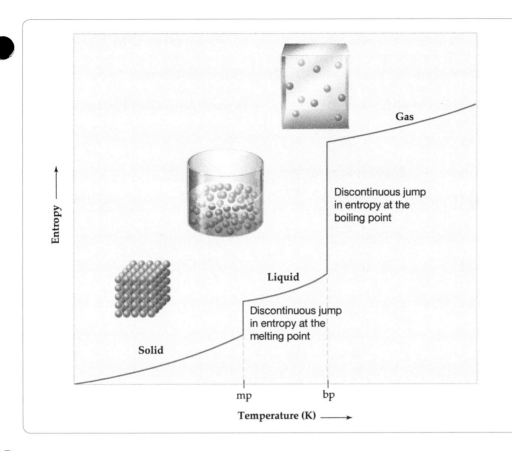

◀ **FIGURE 18.8**

Entropy versus temperature. The entropy of a pure substance, equal to zero at 0 K, shows a steady increase with rising temperature, punctuated by discontinuous jumps in entropy at the temperatures of the phase transitions.

◀ Figure It Out

Which phase transition, solid to liquid or liquid to gas, results in the largest increase in entropy? Explain.

Answer: The liquid-to-gas phase transition has a larger increase in entropy than the solid-to-liquid phase transition because particles are completely separated during vaporization, leading to a much larger number of ways to arrange the particles.

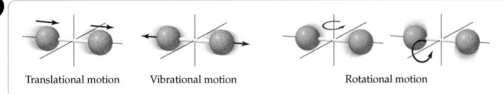

Translational motion   Vibrational motion   Rotational motion

◀ **FIGURE 18.9**

Types of molecular motion, illustrated for a diatomic molecule. Translational motion is motion of the center of mass. Vibrational motion is the stretching and compression of chemical bonds. Rotational motion is rotation of the molecule about fixed coordinate axes.

18.5 STANDARD MOLAR ENTROPIES AND STANDARD ENTROPIES OF REACTION

We won't describe how the entropy of a substance is determined, except to note that two approaches are available: (1) calculations based on Boltzmann's equation and (2) experimental measurements of **heat capacities** (Section 9.7) down to very low temperatures. Suffice it to say that **standard molar entropies**, denoted by $S°$, are known for many substances.

REMEMBER . . .
The molar **heat capacity** is the amount of heat needed to raise the temperature of 1 mol of a substance by 1°C (Section 9.7).

Standard molar entropy, $S°$ The entropy of 1 mol of the pure substance at 1 atm pressure and a specified temperature, usually 25 °C.

Values of $S°$ for some common substances at 25 °C are listed in **TABLE 18.1,** and additional values are given in Appendix B. Note that the units of $S°$ are *joules* (not kilojoules) per kelvin mole [J/(K·mol)]. Standard molar entropies are often called *absolute entropies* because they are measured with respect to an absolute reference point— the entropy of the perfectly ordered crystalline substance at 0 K [$S° = 0$ J/(K·mol) at $T = 0$ K].

Standard molar entropies make it possible to compare the entropies of different substances under the same conditions of temperature and pressure. It's apparent from Table 18.1, for example, that the entropies of gaseous substances tend to be larger than those of liquids, which, in turn, tend to be larger than those of solids. Table 18.1 also shows that $S°$ values increase with increasing molecular complexity. Compare, for example, CH_3OH, which has $S° = 127$ J/(K·mol), and CH_3CH_2OH, which has $S° = 161$ J/(K·mol).

TABLE 18.1　Standard Molar Entropies for Some Common Substances at 25 °C

Substance	Formula	$S°$ [J/(K·mol)]	Substance	Formula	$S°$ [J/(K·mol)]
Gases			**Liquids**		
Acetylene	C_2H_2	200.8	Acetic acid	CH_3CO_2H	160
Ammonia	NH_3	192.3	Ethanol	CH_3CH_2OH	161
Carbon dioxide	CO_2	213.6	Methanol	CH_3OH	127
Carbon monoxide	CO	197.6	Water	H_2O	69.9
Ethylene	C_2H_4	219.5	**Solids**		
Hydrogen	H_2	130.6	Calcium carbonate	$CaCO_3$	91.7
Methane	CH_4	186.2	Calcium oxide	CaO	38.1
Nitrogen	N_2	191.5	Diamond	C	2.4
Nitrogen dioxide	NO_2	240.0	Graphite	C	5.7
Dinitrogen tetroxide	N_2O_4	304.3	Iron	Fe	27.3
Oxygen	O_2	205.0	Iron(III) oxide	Fe_2O_3	87.4

BIG IDEA Question 4

Go to eText

Order the following gases from lowest to highest entropy.

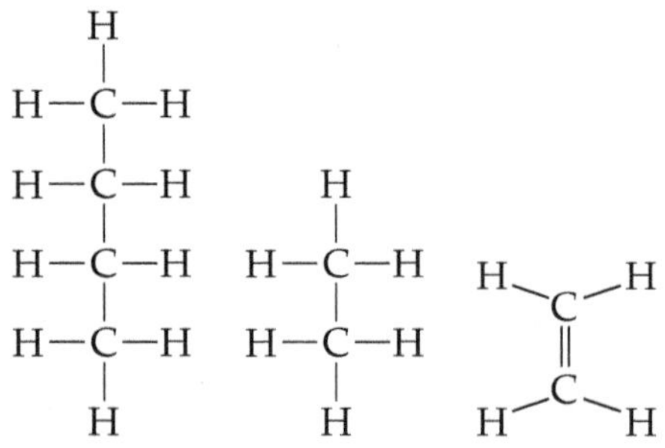

Butane, C_4H_{10}　Ethane, C_2H_6　Ethylene, C_2H_4

Once we have values for standard molar entropies, it's easy to calculate the entropy change for a chemical reaction. The **standard entropy of reaction** ($\Delta S°$) can be obtained simply by subtracting the standard molar entropies of all the reactants from the standard molar entropies of all the products:

> **Standard entropy change of reaction**　$\Delta S° = S°(\text{products}) - S°(\text{reactants})$

Because $S°$ values are quoted on a per-mole basis, the $S°$ value for each substance must be multiplied by the stoichiometric coefficient of that substance in the balanced chemical equation. Thus, for the general reaction

$$a\,A + b\,B \longrightarrow c\,C + d\,D$$

the standard entropy of reaction is

> **Standard entropy change of reaction**
> $$\Delta S° = [c\,S°(C) + d\,S°(D)] - [a\,S°(A) + b\,S°(B)]$$

In this equation, the units of the coefficients are moles, the units of $S°$ are J/(K·mol), and the units of $\Delta S°$ are J/K.

As an example, let's calculate the standard entropy change for the reaction

$$N_2O_4(g) \longrightarrow 2\,NO_2(g)$$

Using the appropriate $S°$ values from Table 18.1, we find that $\Delta S° = 175.7$ J/K:

$$\Delta S° = 2\,S°(NO_2) - S°(N_2O_4)$$

$$= (2\text{ mol})\left(240.0\,\frac{J}{K\cdot mol}\right) - (1\text{ mol})\left(304.3\,\frac{J}{K\cdot mol}\right)$$

$$= 175.7\text{ J/K}$$

Although the standard molar entropy of N_2O_4 is larger than that of NO_2, as expected for a more complex molecule, $\Delta S°$ for the reaction is positive because 1 mol of N_2O_4 is converted to 2 mol of NO_2. As noted in Figure 18.4, we expect an increase in entropy whenever a molecule breaks into two or more pieces.

WORKED EXAMPLE 18.3

Calculating the Standard Entropy of Reaction

Calculate the standard entropy of reaction at 25 °C for the Haber synthesis of ammonia:

$$N_2(g) + 3\,H_2(g) \longrightarrow 2\,NH_3(g)$$

IDENTIFY

Known	Unknown
$S°$ values for reactants and products (Appendix B)	Standard entropy of reaction ($\Delta S°$)

STRATEGY

To calculate $\Delta S°$ for the reaction, subtract the standard molar entropies of all the reactants from the standard molar entropies of all the products. Remember to multiply the $S°$ value for each substance by its coefficient in the balanced chemical equation.

SOLUTION

$$\Delta S° = 2\,S°(NH_3) - [S°(N_2) + 3\,S°(H_2)]$$

$$= (2\text{ mol})\left(192.3\,\frac{J}{K\cdot mol}\right) - \left[(1\text{ mol})\left(191.5\,\frac{J}{K\cdot mol}\right) + (3\text{ mol})\left(130.6\,\frac{J}{K\cdot mol}\right)\right]$$

$$= -198.7\text{ J/K}$$

CHECK

$\Delta S°$ should be negative because the reaction decreases the number of gaseous molecules from 4 mol to 2 mol.

▶ **PRACTICE 18.5**   Calculate the standard entropy of reaction for hydrogen gas reacting with oxygen gas to form liquid water at 25 °C.

$$2\,H_2(g) + O_2(g) \longrightarrow 2\,H_2O(l)$$

▶ **APPLY 18.6**   The *unbalanced* reaction for the combustion of gaseous propane (C_3H_8) at 25 °C is:

$$C_3H_8(g) + O_2(g) \longrightarrow CO_2(g) + H_2O(l) \quad \Delta S° = -376.6\text{ J/K}$$

(a) Balance the reaction and explain why $\Delta S°$ is negative.
(b) Use values in Appendix B to calculate $S°[J/(K\cdot mol)]$ for $C_3H_8\,(g)$.

18.6 ENTROPY AND THE SECOND LAW OF THERMODYNAMICS

We've seen thus far that molecular systems tend to move spontaneously toward a state of minimum enthalpy and maximum entropy. In any particular reaction, though, the enthalpy of the system can either increase or decrease. Similarly, the entropy of the system can either increase or decrease. How, then, can we decide whether a reaction will occur spontaneously? In Section 9.13, we said that it is the value of the *free-energy change*, ΔG, that is the criterion for spontaneity, where $\Delta G = \Delta H - T\Delta S$. If $\Delta G < 0$, the reaction is spontaneous; if $\Delta G > 0$, the reaction is nonspontaneous; and if $\Delta G = 0$, the reaction is at equilibrium. In this section and the next, we'll see how that conclusion was reached. Let's begin by looking at the first and second laws of thermodynamics.

First law of thermodynamics   In any process, spontaneous or nonspontaneous, the total energy of a system and its surroundings is constant.

Second law of thermodynamics   In any *spontaneous* process, the total entropy of a system and its surroundings always increases.

The first law is simply a statement of the conservation of energy (Section 9.1). It says that energy (or enthalpy) can flow between a system and its surroundings but the total energy of the system plus the surroundings always remains constant. In an exothermic reaction, the system loses heat to the surroundings; in an endothermic reaction, the system gains heat from the surroundings. Because energy is conserved in all chemical processes, spontaneous and nonspontaneous, the first law helps us keep track of energy flow between the system and the surroundings but it doesn't tell us whether a particular reaction will be spontaneous or nonspontaneous. For example, the first law says that when a hot cup of cocoa cools down by transferring energy to the surrounding air molecules, the total amount of energy is conserved. It does not explain why a hot cup of cocoa cools by spontaneously transferring heat to the surroundings and the reverse process of room temperature air transferring heat and warming a cup of cool liquid never happens.

The second law, however, provides a clear-cut criterion of spontaneity. It says that the direction of spontaneous change is always determined by the sign of the total entropy change:

Total entropy change $\Delta S_{\text{total}} = \Delta S_{\text{system}} + \Delta S_{\text{surroundings}}$

Specifically,

If $\Delta S_{\text{total}} > 0$, the reaction is spontaneous.

If $\Delta S_{\text{total}} < 0$, the reaction is nonspontaneous.

If $\Delta S_{\text{total}} = 0$, the reaction mixture is at equilibrium.

All reactions proceed spontaneously in the direction that increases the entropy of the system plus surroundings. A reaction that is nonspontaneous in the forward direction is spontaneous in the reverse direction because ΔS_{total} for the reverse reaction equals $-\Delta S_{\text{total}}$ for the forward reaction. If ΔS_{total} is zero, the reaction doesn't go spontaneously in either direction, and so the reaction mixture is at equilibrium.

To determine the value of ΔS_{total}, we need values for the entropy changes in the system and the surroundings. The entropy change in the system, ΔS_{sys}, is just the entropy of reaction, which can be calculated from standard molar entropies (Table 18.1), as described in Worked Worked Example 18.3. For a reaction that occurs at constant pressure, the entropy change in the surroundings is directly proportional to the enthalpy change for the reaction (ΔH) and inversely proportional to the kelvin temperature (T) of the surroundings, according to the equation

---

Entropy change in the surroundings $\Delta S_{\text{surr}} = \dfrac{-\Delta H}{T}$

Although we won't derive this equation to calculate ΔS_{surr}, we can nevertheless justify its form. To see why ΔS_{surr} is proportional to $-\Delta H$, recall that for an exothermic reaction ($\Delta H < 0$), the system loses heat to the surroundings (**FIGURE 18.10a**). As a result, the random motion of the molecules in the surroundings increases and the entropy of the surroundings also increases ($\Delta S_{surr} > 0$). Conversely, for an endothermic reaction ($\Delta H > 0$), the system gains heat from the surroundings (**FIGURE 18.10b**) and the entropy of the surroundings therefore decreases ($\Delta S_{surr} < 0$). Because ΔS_{surr} is positive when ΔH is negative and vice versa, ΔS_{surr} is proportional to $-\Delta H$:

$$\Delta S_{surr} \propto -\Delta H$$

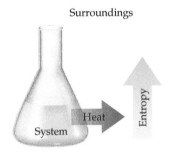 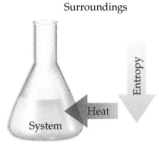

(a) When an exothermic reaction occurs in the system ($\Delta H < 0$), the surroundings gains heat and its entropy increases ($\Delta S_{surr} > 0$).

(b) When an endothermic reaction occurs in the system ($\Delta H > 0$), the surroundings loses heat and its entropy decreases ($\Delta S_{surr} < 0$).

◀ **FIGURE 18.10**
The relationship between the enthalpy change in the system (ΔH) and the entropy change in the surroundings (ΔS_{surr}).

◀ **Figure It Out**

What is the sign of ΔS_{surr} when an ice cube melts?

$$\Delta S_{surr} = \frac{-\Delta H}{T}, \text{(endothermic process, } +\text{ } = \Delta H)$$
To melt an ice cube, heat must be absorbed
Answer: $\Delta S_{surr} < 0.$

The reason why ΔS_{surr} is inversely proportional to the absolute temperature T is more subtle. We can think of the surroundings as an infinitely large constant-temperature bath to which heat can be added without changing its temperature. If the surroundings has a low temperature, it has only a small amount of randomness, in which case the addition of a given quantity of heat results in a substantial increase in the amount of randomness (a relatively large value of ΔS_{surr}). If the surroundings has a high temperature, it already has a large amount of randomness, and the addition of the same quantity of heat produces only a marginal increase in the amount of randomness (a relatively small value of ΔS_{surr}). Thus, ΔS_{surr} varies inversely with temperature:

$$\Delta S_{surr} \propto \frac{1}{T}$$

Adding heat to the surroundings is somewhat analogous to tossing a rock into a lake. If the lake exhibits little motion (calm, smooth surface) before the rock is thrown, the rock's impact produces considerable motion, evident in a circular pattern of waves. If the lake already exhibits appreciable motion (rough, choppy surface), the additional motion produced when the rock hits the water is hardly noticeable.

▲ Adding heat to cold surroundings is analogous to tossing a rock into calm waters. Both processes produce a considerable increase in the amount of motion and thus a relatively large increase in entropy.

WORKED EXAMPLE 18.4

Determining Whether a Reaction Is Spontaneous

Consider the oxidation of iron metal:

$$4\,Fe(s) + 3\,O_2(g) \longrightarrow 2\,Fe_2O_3(s)$$

Calculate the value of ΔS_{total}, and decide whether the reaction is spontaneous at 25 °C.

IDENTIFY

Known	Unknown
Values of $\Delta H°_f$ and $\Delta S°$ for reactants and products (Appendix B)	ΔS_{total}

continued on next page

▲ Adding heat to hot surroundings is analogous to tossing a rock into rough waters. Both processes produce a relatively small increase in the amount of motion and thus a relatively small increase in entropy.

STRATEGY

To determine the sign of $\Delta S_{total} = \Delta S_{sys} + \Delta S_{surr}$, we need to calculate the values of ΔS_{sys} and ΔS_{surr}. The entropy change in the system equals the standard entropy of reaction and can be calculated using the standard molar entropies. To obtain $\Delta S_{surr} = -\Delta H°/T$, calculate $\Delta H°$ for the reaction from standard heats of formation (Section 9.9).

SOLUTION

$\Delta S_{sys} = \Delta S° = 2\, S°(Fe_2O_3) - [4\, S°(Fe) + 3\, S°(O_2)]$

$= (2\text{ mol})\left(87.4\,\frac{J}{K \cdot mol}\right) - \left[(4\text{ mol})\left(27.3\,\frac{J}{K \cdot mol}\right) + (3\text{ mol})\left(205.0\,\frac{J}{K \cdot mol}\right)\right]$

$= -549.5\text{ J/K}$

$\Delta H° = 2\, \Delta H°_f(Fe_2O_3) - [4\, \Delta H°_f(Fe) + 3\, \Delta H°_f(O_2)]$

Because $\Delta H°_f = 0$ for elements and $\Delta H°_f = -824.2$ kJ/mol for Fe_2O_3, $\Delta H°$ for the reaction is

$\Delta H° = 2\, \Delta H°_f(Fe_2O_3) = (2\text{ mol})(-824.2\text{ kJ/mol}) = -1648.4\text{ kJ}$

Therefore, at 25 °C = 298.15 K,

$\Delta S_{surr} = \frac{-\Delta H°}{T} = \frac{-(-1,648,400\text{ J})}{298.15\text{ K}} = 5529\text{ J/K}$

$\Delta S_{total} = \Delta S_{sys} + \Delta S_{surr} = -549.5\text{ J/K} + 5529\text{ J/K} = 4980\text{ J/K}$

Because the total entropy change is positive, the reaction is spontaneous under standard-state conditions at 25 °C.

CHECK

Since the reaction consumes 3 mol of gas, ΔS_{sys} is negative. Because the oxidation of iron metal is highly exothermic, $\Delta S_{surr} = -\Delta H°/T$ is positive and very large. The value of ΔS_{surr} is greater than the absolute value of ΔS_{sys}, and so ΔS_{total} is positive, in agreement with the solution.

▶ **PRACTICE 18.7** Calculate the value of ΔS_{total}, and decide whether the decomposition of calcium carbonate is spontaneous under standard-state conditions at 25 °C (Hint: Look up values of $\Delta H°_f$ and $\Delta S°$ for each reactant and product in Appendix B.).

$$CaCO_3(s) \longrightarrow CaO(s) + CO_2(g)$$

▶ **APPLY 18.8** Use the values of $\Delta H°_f$ and $\Delta S°$ in Appendix B to estimate the boiling point of Br_2 in degrees Celsius.

18.7 FREE ENERGY AND THE SPONTANEITY OF CHEMICAL REACTIONS

Chemists are generally more interested in the system (the reaction mixture) than the surroundings, and it's therefore convenient to restate the second law in terms of the thermodynamic properties of the system, without regard to the surroundings. For this purpose, we use the thermodynamic property called **free energy**, denoted by G in honor of J. Willard Gibbs (1839–1903), the American mathematical physicist who laid the foundations of chemical thermodynamics. The free energy G of a system is defined as

Free energy $\quad G = H - TS$

As you might expect from its name, free energy has units of energy (J or kJ).

Because enthalpy, entropy, and temperature are state functions, free energy is also a state function and the change in free energy (ΔG) for a process is independent of path. ΔG for a reaction at constant temperature equals the change in enthalpy minus the product of temperature times the change in entropy.

> **Free-energy change** $\Delta G = \Delta H - T\Delta S$

To see what this equation for free-energy change has to do with spontaneity, let's return to the relationship

$$\Delta S_{total} = \Delta S_{sys} + \Delta S_{surr} = \Delta S + \Delta S_{surr}$$

where we have now dropped the subscript *sys*. (*It's generally understood that symbols without a subscript refer to the system, not the surroundings.*) Since $\Delta S_{surr} = -\Delta H/T$, where ΔH is the heat gained by the system at constant pressure, we can also write

$$\Delta S_{total} = \Delta S - \frac{\Delta H}{T}$$

Multiplying both sides by $-T$ gives

$$-T\Delta S_{total} = \Delta H - T\Delta S$$

The right side of this equation is just ΔG, the change in the free energy of the system at constant temperature and pressure. Therefore,

$$-T\Delta S_{total} = \Delta G$$

Note that ΔG and ΔS_{total} have opposite signs because the absolute temperature T is always positive.

According to the second law of thermodynamics, a reaction is spontaneous if ΔS_{total} is positive, nonspontaneous if ΔS_{total} is negative, and at equilibrium if ΔS_{total} is zero. Because $-T\Delta S_{total} = \Delta G$, and because ΔG and ΔS_{total} have opposite signs, we can restate the thermodynamic criterion for the spontaneity of a reaction carried out at constant temperature and pressure in the following way:

> If $\Delta G < 0$, the reaction is spontaneous.
>
> If $\Delta G > 0$, the reaction is nonspontaneous.
>
> If $\Delta G = 0$, the reaction mixture is at equilibrium.

In other words, *in any spontaneous process at constant temperature and pressure, the free energy of the system always decreases.*

The relationship between the signs of the enthalpy and entropy contributions to ΔG in the free-energy equation $\Delta G = \Delta H - T\Delta S$ is summarized in four different cases in **TABLE 18.2**.

TABLE 18.2   Signs of Enthalpy, Entropy, and Free-Energy Changes and Reaction Spontaneity for a Reaction at Constant Temperature and Pressure

ΔH	ΔS	$\Delta G + \Delta H - T\Delta S$	Reaction Spontaneity	Example
−	+	−	Spontaneous at all temperatures	$2\,NO_2(g) \longrightarrow N_2(g) + 2\,O_2(g)$
+	−	+	Nonspontaneous at all temperatures	$3\,O_2(g) \longrightarrow 2\,O_3(g)$
−	−	− or +	Spontaneous at low temperatures where ΔH outweighs $T\Delta S$	$N_2(g) + 3\,H_2(g) \longrightarrow 2\,NH_3(g)$
			Nonspontaneous at high temperatures where $T\Delta S$ outweighs ΔH	
+	+	− or +	Spontaneous at high temperatures where $T\Delta S$ outweighs ΔH	$2\,HgO(s) \longrightarrow 2\,Hg(l) + O_2(g)$
			Nonspontaneous at low temperatures where ΔH outweighs $T\Delta S$	

BIG IDEA Question 5   Go to eText

BIG IDEA Question 5

What are the signs of ΔS and ΔH in a reaction that is nonspontaneous at low temperatures and spontaneous at high temperatures?

Case 1. When ΔH is negative and ΔS is positive, ΔG is negative at all temperatures. Therefore, exothermic reactions with a positive entropy change are always spontaneous.

Case 2. When ΔH is positive and ΔS is negative, ΔG is positive at all temperatures. Therefore, endothermic reactions with a negative entropy change are always nonspontaneous, no matter what the temperature.

If, however, ΔH and ΔS are either both negative or both positive, the sign of ΔG (and therefore the spontaneity of the reaction) depends on the temperature. Temperature T in the equation $\Delta G = \Delta H - T\Delta S$ acts as a weighting factor that determines the relative importance of the enthalpy and entropy contributions to ΔG.

Case 3. If ΔH and ΔS are both negative, the reaction will be spontaneous only if the absolute value of ΔH is larger than the absolute value of $T\Delta S$. This is most likely at low temperatures, where the weighting factor T in $T\Delta S$ is small.

Case 4. If ΔH and ΔS are both positive, the reaction will be spontaneous only if $T\Delta S$ is larger than ΔH, which is most likely at high temperatures. We've already seen several examples of this situation in **phase changes** such as melting or vaporization (Section 11.2).

> REMEMBER . . .
> The phase change associated with the melting of ice is spontaneous above 0 °C where $T\Delta S$ outweighs ΔH and ΔG is negative but is nonspontaneous below 0 °C where ΔH outweighs $T\Delta S$ and ΔG is positive (Section 11.2).

In cases 3 and 4, when ΔH and ΔS are either both negative or both positive, we can estimate the temperature at which a nonspontaneous reaction becomes spontaneous, or a spontaneous reaction becomes nonspontaneous, because that's the temperature at which the reaction is at equilibrium. At equilibrium, $\Delta G = \Delta H - T\Delta S = 0$ and so the crossover temperature at which a reaction cvhanges between spontaneous and nonspontaneous is given by $T = \Delta H/\Delta S$.

$$\Delta G = \Delta H - T\Delta S = 0 \qquad \text{At equilibrium}$$

$$T = \frac{\Delta H}{\Delta S} \qquad \text{Crossover temperature}$$

This equation provides only an estimate, not an exact value, of the crossover temperature because the values of ΔH and ΔS are somewhat dependent on temperature. In general, the enthalpies and entropies of both reactants and products increase with increasing temperature, but the increases for the products tend to cancel the increases for the reactants. As a result, values of ΔH and ΔS for a reaction are relatively independent of temperature, at least over a small temperature range.

WORKED EXAMPLE 18.5

Determining the Temperature at Which a Reaction Becomes Spontaneous

Iron metal can be produced by reducing iron(III) oxide with hydrogen:

$$\text{Fe}_2\text{O}_3(s) + 3\,\text{H}_2(g) \longrightarrow 2\,\text{Fe}(s) + 3\,\text{H}_2\text{O}(g) \quad \Delta H° = +98.8 \text{ kJ}; \Delta S° = +141.5 \text{ J/K}$$

(a) Is this reaction spontaneous under standard-state conditions at 25 °C?
(b) At what temperature will the reaction become spontaneous?

IDENTIFY

Known	Unknown
$\Delta H° = +98.8 \text{ kJ}$	ΔG at 25 °C
$\Delta S° = +141.5 \text{ J/K}$	Temperature (T) for spontaneity

STRATEGY

To determine whether the reaction is spontaneous at 25 °C, we need to determine the sign of $\Delta G = \Delta H - T\Delta S$. To find the crossover temperature at which the reaction becomes spontaneous, we use the equation $T = \Delta H/\Delta S$.

SOLUTION

(a) At 25 °C (298 K), ΔG for the reaction is

$$\Delta G = \Delta H - T\Delta S = (98.8 \text{ kJ}) - (298 \text{ K})(0.1415 \text{ kJ/K})$$

$$= (98.8 \text{ kJ}) - (42.2 \text{ kJ})$$

$$= 56.6 \text{ kJ}$$

Because the positive ΔH term is larger than the positive $T\Delta S$ term, ΔG is positive and the reaction is nonspontaneous at 298 K.

(b) We can estimate the crossover temperature at which ΔG changes from positive to negative by substituting the values of ΔH and ΔS into the equation $T = \Delta H / \Delta S$:

$$T = \frac{\Delta H}{\Delta S} = \frac{98.8 \text{ kJ}}{0.1415 \text{ kJ/K}} = 698 \text{ K}$$

Because this calculation assumes the values of ΔH and ΔS are unchanged on going from the standard-state temperature of 298 K to 698 K, the calculated value of T is only an estimate.

CHECK

In the case in which $\Delta H°$ is positive and $\Delta S°$ is positive, the reaction becomes spontaneous when $T\Delta S$ becomes larger than ΔH. Higher temperatures increase the magnitude of the $T\Delta S$ term, ΔG becomes negative, and the reaction becomes spontaneous. We would predict that the temperature at which the reaction becomes spontaneous will be higher than the temperature at which the reaction is nonspontaneous (298 K). The crossover temperature (698 K) is higher than 298 K.

▶ **PRACTICE 18.9**   Consider the decomposition of gaseous N_2O_4:

$$N_2O_4(g) \longrightarrow 2 \text{ NO}_2(g) \quad \Delta H° = +55.3 \text{ kJ}; \Delta S° = +175.7 \text{ J/K}$$

Estimate the temperature conditions for a spontaneous reaction.

▶ **CONCEPTUAL APPLY 18.10**   What are the signs $(+, -, \text{ or } 0)$ of ΔH, ΔS, and ΔG for the following spontaneous reaction of A atoms (red) and B atoms (blue)?

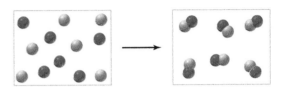

18.8 STANDARD FREE-ENERGY CHANGES FOR REACTIONS

The free energy of a substance, like its enthalpy and entropy, depends on its temperature, pressure, physical state (solid, liquid, or gas), and concentration (in the case of solutions). As a result, free-energy changes for chemical reactions must be compared under a well-defined set of standard-state conditions.

> **Standard-state conditions**
> - Solids, liquids, and gases in pure form at 1 atm pressure
> - Solutes at 1 M concentration
> - A specified temperature, usually 25 °C

The **standard free-energy change ($\Delta G°$)** for a reaction is the change in free energy that occurs when reactants in their standard states are converted to products in their standard states. As with $\Delta H°$, the value of $\Delta G°$ is an **extensive property** (Section 2.3) that refers to the number of moles indicated in the balanced chemical equation. For example, $\Delta G°$ at 25 °C for the reaction

$$2 \text{ Na}(s) + 2 \text{ H}_2\text{O}(l) \longrightarrow \text{H}_2(g) + 2 \text{ Na}^+(aq) + 2 \text{ OH}^-(aq)$$

is the change in free energy that occurs when 2 mol of solid sodium reacts completely with 2 mol of liquid water to give 1 mol of hydrogen gas at 1 atm pressure, along with an aqueous solution that contains 2 mol of Na^+ ions and 2 mol of OH^- ions at 1 M concentrations, with all reactants and products at a temperature of 25 °C. For this reaction, $\Delta G° = -364$ kJ.

Because the free-energy change for any process at constant temperature and pressure is $\Delta G = \Delta H - T\Delta S$, we can calculate the standard free-energy change $\Delta G°$ for a reaction from the standard enthalpy change $\Delta H°$ and the standard entropy change $\Delta S°$. Consider again the Haber synthesis of ammonia:

$$N_2(g) + 3 \text{ H}_2(g) \longrightarrow 2 \text{ NH}_3(g) \quad \Delta H° = -92.2 \text{ kJ}; \Delta S° = -198.7 \text{ J/K}$$

The standard free-energy change at 25 °C (298 K) is

$$\Delta G° = \Delta H° - T\Delta S° = (-92.2 \times 10^3 \text{ J}) - (298 \text{ K})(-198.7 \text{ J/K})$$
$$= (-92.2 \times 10^3 \text{ J}) - (-59.2 \times 10^3 \text{ J})$$
$$= -33.0 \times 10^3 \text{ J} \quad \text{or} \quad -33.0 \text{ kJ}$$

Because the negative $\Delta H°$ term is larger than the negative $T\Delta S°$ term at 25 °C, $\Delta G°$ is negative and the reaction is spontaneous under standard-state conditions.

REMEMBER . . .
Extensive properties, like mass and volume, have values that depend on the sample size. *Intensive properties*, like temperature and melting point, have values that do not depend on the sample size (Section 2.3).

It's important to note that a standard free-energy change applies to a hypothetical process rather than an actual process. In the hypothetical process, separate reactants in their standard states are completely converted to separate products in their standard states. In an actual process, however, reactants and products are mixed together and the reaction may not go to completion.

Take the synthesis of ammonia, for example. The hypothetical process is

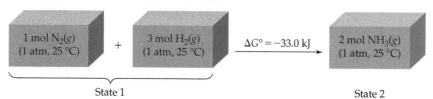

where $\Delta G° = -33.0$ kJ is the change in the free energy of the system on going from state 1 to state 2. In an actual synthesis of ammonia, however, the reactants N_2 and H_2 are not separate but are mixed together. Moreover, the reaction doesn't go to completion; it reaches an equilibrium state in which both reactants and products are present together.

How, then, should we think about the meaning of $\Delta G°$ in the context of an actual reaction? One way would be to suppose that we have a mixture of N_2, H_2, and NH_3 in a reaction vessel, each substance present at a partial pressure of 1 atm.

The reaction vessel has an inlet that allows the addition of $N_2(g)$ and $H_2(g)$ at the rate at which they react, to maintain a constant pressure of 1 atm for both reactant gases. The reaction vessel also has an outlet that removes NH_3 at the rate at which it is formed, thus maintaining a constant pressure of 1 atm for $NH_3(g)$. (One method of removing NH_3 is by cooling the reaction chamber and selectively condensing NH_3, which has a higher boiling point than N_2 and H_2.)

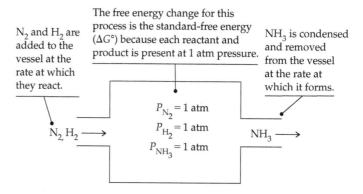

Under these conditions, the free-energy change for the reaction of 1 mol of N_2 with 3 mol of H_2 to form 2 mol of NH_3 is the standard free-energy change ($\Delta G° = -33.0$ kJ) because each reactant and product is present at 1 atm pressure.

As always, the value of $\Delta G°$ indicates only whether the reaction is spontaneous and provides no information about its rate.

WORKED EXAMPLE 18.6

Calculating $\Delta G°$ for a Reaction from $\Delta H°$ and $\Delta S°$

Iron metal is produced commercially by reducing iron(III) oxide in iron ore with carbon monoxide:

$$Fe_2O_3(s) + 3\,CO(g) \longrightarrow 2\,Fe(s) + 3\,CO_2(g)$$

(a) Calculate the standard free-energy change for this reaction at 25 °C.
(b) Is the reaction spontaneous under standard-state conditions at 25 °C?
(c) Does the reverse reaction become spontaneous at higher temperatures? Explain.

IDENTIFY

Known	Unknown
Temperature (25 °C)	$\Delta G°$
Values of $\Delta H°_f$ and $S°$ (Appendix B)	Spontaneous or nonspontaneous at 25 °C Conditions for spontaneity

STRATEGY

(a) We can calculate the standard free-energy change from the relation $\Delta G° = \Delta H° - T\Delta S°$, but first we must find $\Delta H°$ and $\Delta S°$ from standard heats of formation $(\Delta H°_f)$ and standard molar entropies $(S°)$.

(b) The reaction is spontaneous under standard-state conditions if $\Delta G°$ is negative.

(c) The spontaneity of the reaction at higher temperatures depends on the signs and magnitudes of $\Delta H°$ and $\Delta S°$.

SOLUTION

(a) The following values of $\Delta H°_f$ and $S°$ are found in Appendix B:

	$Fe_2O_3(s)$	$CO(g)$	$Fe(s)$	$CO_2(g)$
$\Delta H°_f$(kJ/mol)	−824.2	−110.5	0	−393.5
$S°[J/(K \cdot mol)]$	87.4	197.6	27.3	213.6

So we have

$$\Delta H° = [2\ \Delta H°_f(Fe) + 3\ \Delta H°_f(CO_2)] - [\Delta H°_f(Fe_2O_3)$$
$$+\ 3\ \Delta H°_f(CO)]$$

$$= [(2\ mol)(0\ kJ/mol) + (3\ mol)(-393.5\ kJ/mol)]$$

$$-[(1\ mol)(-824.2\ kJ/mol) + (3\ mol)(-110.5\ kJ/mol)]$$

$$\Delta H° = -24.8\ kJ$$

and

$$\Delta S° = [2\ S°(Fe) + 3\ S°(CO_2)] - [S°(Fe_2O_3) + 3\ S°(CO)]$$

$$= \left[(2\ mol)\left(27.3\ \frac{J}{K \cdot mol}\right) + (3\ mol)\left(213.6\ \frac{J}{K \cdot mol}\right)\right]$$

$$-\left[(1\ mol)\left(87.4\ \frac{J}{K \cdot mol}\right) + (3\ mol)\left(197.6\ \frac{J}{K \cdot mol}\right)\right]$$

$$\Delta S° = +15.2\ J/K \quad or \quad 0.0152\ kJ/K$$

Therefore,

$$\Delta G° = \Delta H° - T\Delta S°$$
$$= (-24.8\ kJ) - (298\ K)(0.0152\ kJ/K)$$
$$\Delta G° = -29.3\ kJ$$

(b) Because $\Delta G°$ is negative, the reaction is spontaneous at 25 °C. This means that a mixture of $Fe_2O_3(s)$, $CO(g)$, $Fe(s)$, and $CO_2(g)$, with each gas at a partial pressure of 1 atm, will react at 25 °C to produce more iron metal.

(c) Because $\Delta H°$ is negative and $\Delta S°$ is positive, $\Delta G°$ will be negative at all temperatures. The forward reaction is therefore spontaneous at all temperatures, and the reverse reaction does not become spontaneous at higher temperatures.

▶ **PRACTICE 18.11** Consider the thermal decomposition of calcium carbonate:

$$CaCO_3(s) \longrightarrow CaO(s) + CO_2(g)$$

(a) Use the data in Appendix B to calculate the standard free-energy change for this reaction at 25 °C.

(b) Will a mixture of solid $CaCO_3$, solid CaO, and gaseous CO_2 at 1 atm pressure react spontaneously at 25 °C to produce more CaO and CO_2?

▶ **CONCEPTUAL APPLY 18.12** Consider the following endothermic decomposition of AB_2 molecules:

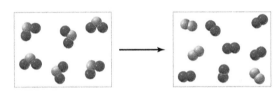

(a) What is the sign (+, −, or 0) of $\Delta S°$ for the reaction?

(b) Is the reaction more likely to be spontaneous at high temperatures or at low temperatures? Explain.

18.9 STANDARD FREE ENERGIES OF FORMATION

The **standard free energy of formation** $(\Delta G°_f)$ of a substance is the free-energy change for the formation of 1 mol of the substance in its standard state from the most stable form of its constituent elements in their standard states. For example, we found in Section 18.8 that the standard free-energy change $\Delta G°$ for the synthesis of 2 mol of NH_3 from its constituent elements is −33.0 kJ:

$$N_2(g) + 3\ H_2(g) \longrightarrow 2\ NH_3(g) \quad \Delta G° = -33.0\ kJ$$

Therefore, $\Delta G°_f$ for ammonia is −33.0 kJ/2 mol, or −16.5 kJ/mol.

Values of $\Delta G°_f$ at 25 °C for some common substances are listed in **TABLE 18.3**, and additional values are given in Appendix B. Note that $\Delta G°_f$ for an element in its most stable form at 25 °C is defined to be zero. Thus, solid graphite has $\Delta G°_f = 0$ kJ/mol, but diamond, a less stable form of solid carbon at 25 °C, has $\Delta G°_f = 2.9$ kJ/mol.

TABLE 18.3   Standard Free Energies of Formation for Some Common Substances at 25 °C

Substance	Formula	$\Delta G_f°$ [kJ/mol]	Substance	Formula	$\Delta G_f°$ [kJ/mol]
Gases			**Liquids**		
Acetylene	C_2H_2	209.9	Acetic acid	CH_3CO_2H	−390
Ammonia	NH_3	−16.5	Ethanol	CH_3CH_2OH	−174.9
Carbon dioxide	CO_2	−394.4	Methanol	CH_3OH	−166.6
Carbon monoxide	CO	−137.2	Water	H_2O	−237.2
Ethylene	C_2H_4	68.1	**Solids**		
Hydrogen	H_2	0	Calcium carbonate	$CaCO_3$	−1129.1
Methane	CH_4	−50.8	Calcium oxide	CaO	−603.3
Nitrogen	N_2	0	Diamond	C	2.9
Nitrogen dioxide	NO_2	51.3	Graphite	C	0
Dinitrogen tetroxide	N_2O_4	99.8	Iron(III) oxide	Fe_2O_3	−742.2

As with standard enthalpies of formation, $\Delta H°_f$, a zero value of $\Delta G°_f$ for elements in their most stable form establishes a thermochemical "sea level," or reference point, with respect to which the standard free energies of other substances are measured. We can't measure the absolute value of a substance's free energy (as we can the entropy), but that's not a problem because we are interested only in free-energy *differences* between reactants and products.

The standard free energy of formation of a substance measures its thermodynamic stability with respect to its constituent elements. Substances that have a negative value of $\Delta G°_f$, such as carbon dioxide and water, are stable and do not decompose to their constituent elements under standard-state conditions. Substances that have a positive value of $\Delta G°_f$, such as ethylene and nitrogen dioxide, are thermodynamically unstable with respect to their constituent elements and can, in principle, decompose. In fact, however, such substances can exist for long periods of time if the rate of their decomposition is slow.

There's no point in trying to synthesize a substance from its elements under standard-state conditions if the substance has a positive value of $\Delta G°_f$. Such a substance would have to be prepared at other temperatures and/or pressures, or it would have to be made from alternative starting materials using a reaction that has a negative free-energy change. Thus, a knowledge of thermodynamics can save considerable time in chemical synthesis.

In Section 18.8, we calculated standard free-energy changes for reactions from the equation $\Delta G° = \Delta H° - T\Delta S°$, using tabulated values of $\Delta H°_f$ and $S°$ to find $\Delta H°$ and $\Delta S°$. Alternatively, we can calculate $\Delta G°$ more directly by subtracting the standard free energies of formation of all the reactants from the standard free energies of formation of all the products.

> **Standard free-energy change of a reaction**
> $$\Delta G° = \Delta G°_f(\text{products}) - \Delta G°_f(\text{reactants})$$

For the general reaction

$$a\,A + b\,B \longrightarrow c\,C + d\,D$$

the standard free-energy change is

> **Standard free-energy change of a reaction**
> $$\Delta G° = [c\,\Delta G°_f(C) + d\,\Delta G°_f(D)] - [a\,\Delta G°_f(A) + b\,\Delta G°_f(B)]$$

To illustrate, let's calculate the standard free-energy change for the reaction in Worked Example 18.6—the reduction of iron(III) oxide by carbon monoxide:

$$Fe_2O_3(s) + 3\,CO(g) \longrightarrow 2\,Fe(s) + 3\,CO_2(g)$$

Using the $\Delta G°_f$ values in Table 18.3, we obtain

$$\Delta G° = [2\,\Delta G°_f(Fe) + 3\,\Delta G°_f(CO_2)] - [\Delta G°_f(Fe_2O_3) + 3\,\Delta G°_f(CO)]$$
$$= [(2\,mol)(0\,kJ/mol) + (3\,mol)(-394.4\,kJ/mol)]$$
$$-[(1\,mol)(-742.2\,kJ/mol) + (3\,mol)(-137.2\,kJ/mol)]$$
$$\Delta G° = -29.4\,kJ$$

This result agrees well with the value of -29.3 kJ calculated from $\Delta G° = \Delta H° - T\Delta S°$ in Worked Example 18.6.

WORKED EXAMPLE 18.7

Calculating $\Delta G°$ for a Reaction from $\Delta G°_f$ Values

(a) Calculate the standard free-energy change for the oxidation of ammonia to give nitric oxide (NO) and water. Is it worth trying to find a catalyst for this reaction under standard-state conditions at 25 °C?

$$4\,NH_3(g) + 5\,O_2(g) \longrightarrow 4\,NO(g) + 6\,H_2O(l)$$

(b) Is it worth trying to find a catalyst for the synthesis of NO from gaseous N_2 and O_2 under standard-state conditions at 25 °C?

IDENTIFY

Known	Unknown
Standard free energies of formation $\Delta G°_f$ (Appendix B)	$\Delta G°$

STRATEGY

We can calculate $\Delta G°$ most easily from standard free energies of formation using the formula:

$$\Delta G° = \Delta G°_f(products) - \Delta G°_f(reactants)$$

Remember to multiply the $\Delta G°$ value for each substance by its coefficient in the balanced chemical reaction. It's worth trying to find a catalyst for a reaction only if the reaction has a negative free-energy change.

SOLUTION

(a) $\Delta G° = [4\,\Delta G°_f(NO) + 6\,\Delta G°_f(H_2O)]$
$$- [4\,\Delta G°_f(NH_3) + 5\,\Delta G°_f(O_2)]$$
$$= [(4\,mol)(87.6\,kJ/mol) + (6\,mol)(-237.2\,kJ/mol)]$$
$$-[(4\,mol)(-16.5\,kJ/mol) + (5\,mol)(0\,kJ/mol)]$$
$$\Delta G° = -1006.8\,kJ$$

It is worth looking for a catalyst because the negative value of $\Delta G°$ indicates that the reaction is spontaneous under standard-state conditions. (This reaction is the first step in the Ostwald process for the production of nitric acid. In industry, the reaction is carried out using a platinum–rhodium catalyst.)

(b) It's not worth looking for a catalyst for the reaction $N_2(g) + O_2(g) \longrightarrow 2\,NO(g)$ because the standard free energy of formation of NO is positive

($\Delta G°_f = 87.6$ kJ/mol). This means that NO is unstable and will decompose to N_2 and O_2 under standard-state conditions at 25 °C. A catalyst would only increase the rate of its decomposition. A catalyst can't affect the composition of the equilibrium mixture (Section 15.11), and so it can't reverse the direction of the reaction.

▶ **PRACTICE 18.13**   Using values of $\Delta G°_f$ in Appendix B, calculate the standard free-energy change for the reaction of calcium carbide (CaC_2) with water. Might this reaction be used for synthesis of acetylene (HC≡CH, or C_2H_2) ?

$$CaC_2(s) + 2\,H_2O(l) \longrightarrow C_2H_2(g) + Ca(OH)_2(s)$$

▶ **APPLY 18.14**

(a) Using values of $\Delta G°_f$ in Table 18.3, calculate the standard free-energy change $\Delta G°$ for the reaction:
$$C_{diamond}(s) \longrightarrow C_{graphite}(s)$$

(b) Will a diamond spontaneously convert to graphite under standard-state conditions at 25 °C?

(c) We do not observe the diamond converting to graphite. Explain.

18.10 FREE-ENERGY CHANGES FOR REACTIONS UNDER NONSTANDARD-STATE CONDITIONS

The sign of the standard free-energy change $\Delta G°$ tells us the direction of spontaneous reaction when both reactants and products are present at standard-state conditions. In actual reactions, however, the composition of the reaction mixture seldom corresponds to standard-state pressures and concentrations. Moreover, the partial pressures and concentrations change as a reaction proceeds. How, then, do we calculate the free-energy change ΔG for a reaction when the reactants and products are present at non-standard-state pressures and concentrations?

The answer is given by the relation

> **Free-energy change** under nonstandard-state conditions $\Delta G = \Delta G° + RT \ln Q$

where ΔG is the free-energy change under nonstandard-state conditions, $\Delta G°$ is the free-energy change under standard-state conditions, R is the gas constant, T is the temperature in kelvin, and Q is the reaction quotient (Q_p for reactions involving gases because the standard state for gases is 1 atm pressure, or Q_c for reactions involving solutes in solution because the standard state for solutes is 1 M concentration). Recall from Section 15.5 that the reaction quotient Q_c is an expression having the same form as the equilibrium-constant expression K_c except that the concentrations do not necessarily have equilibrium values. Similarly, Q_p has the same form as K_p except that the partial pressures have arbitrary values. For example, for the synthesis of ammonia:

$$N_2(g) + 3\,H_2(g) \rightleftharpoons 2\,NH_3(g) \quad Q_p = \frac{(P_{NH_3})^2}{(P_{N_2})(P_{H_2})^3}$$

For reactions that involve both gases and solutes in solution, the reaction quotient Q contains partial pressures of gases in atmospheres and molar concentrations of solutes.

We won't derive the equation for ΔG under nonstandard-state conditions. It's hardly surprising, however, that ΔG and Q should turn out to be related because both predict the direction of a reaction. Worked Example 18.8 shows how to use this equation.

 Go to eText

WORKED EXAMPLE 18.8

Calculating ΔG for a Reaction under Nonstandard-State Conditions

Calculate the free-energy change for ammonia synthesis at 25 °C (298 K) given the following sets of partial pressures:

(a) 1.0 atm N_2, 3.0 atm H_2, 0.020 atm NH_3
(b) 0.010 atm N_2, 0.030 atm H_2, 2.0 atm NH_3

$$N_2(g) + 3\,H_2(g) \rightleftharpoons 2\,NH_3(g) \quad \Delta G° = -33.0 \text{ kJ}$$

IDENTIFY

Known	Unknown
Partial pressures of reactants and products	ΔG
$T = 298$ K	
$\Delta G° = -33.0$ kJ	

STRATEGY

We can calculate ΔG from the relation $\Delta G = \Delta G° + RT \ln Q$, where Q is Q_p for the reaction $N_2(g) + 3\,H_2(g) \rightleftharpoons 2\,NH_3(g)$.

SOLUTION

(a) The value of Q_p is

$$Q_p = \frac{(P_{NH_3})^2}{(P_{N_2})(P_{H_2})^3} = \frac{(0.020)^2}{(1.0)(3.0)^3} = 1.5 \times 10^{-5}$$

Substituting this value of Q_p into the equation for ΔG gives

$$\begin{aligned} \Delta G &= \Delta G° + RT \ln Q_p \\ &= (-33.0 \times 10^3 \text{ J/mol}) + [8.314 \text{ J/(K·mol)}](298 \text{ K})(\ln 1.5 \times 10^{-5}) \\ &= (-33.0 \times 10^3 \text{ J/mol}) + (-27.5 \times 10^3 \text{ J/mol}) \\ \Delta G &= -60.5 \text{ kJ/mol} \end{aligned}$$

To maintain consistent units in this calculation, we have expressed ΔG and $\Delta G°$ in units of J/mol because R has units of J/(K·mol). The phrase *per mole* in this context means per molar amounts of reactants and products indicated by the coefficients in the balanced equation. Thus, the free-energy change is -60.5 kJ when 1 mol of N_2 and 3 mol of H_2 are converted to 2 mol of NH_3 under the specified conditions.

ΔG is more negative than $\Delta G°$ because Q_p is less than 1 and $\ln Q_p$ is therefore a negative number. Thus, the reaction has a greater thermodynamic tendency to occur under the cited conditions than it does under standard-state conditions. When each reactant and product is present at a partial pressure of 1 atm, $Q_p = 1$, $\ln Q_p = 0$, and $\Delta G = \Delta G°$.

(b) The value of Q_p is

$$Q_p = \frac{(P_{NH_3})^2}{(P_{N_2})(P_{H_2})^3} = \frac{(2.0)^2}{(0.010)(0.030)^3} = 1.5 \times 10^7$$

The corresponding value of ΔG is

$$\begin{aligned} \Delta G &= \Delta G° + RT \ln Q_P \\ &= (-33.0 \times 10^3 \text{ J/mol}) + [8.314 \text{ J/(K·mol)}](298 \text{ K})(\ln 1.5 \times 10^7) \\ &= (-33.0 \times 10^3 \text{ J/mol}) + (40.9 \times 10^3 \text{ J/mol}) \\ \Delta G &= 7.9 \text{ kJ/mol} \end{aligned}$$

Because Q_p is large enough to give a positive value for ΔG, the reaction is nonspontaneous in the forward direction but spontaneous in the reverse direction. Thus, the direction in which a reaction proceeds spontaneously depends on the composition of the reaction mixture.

CHECK

(a) Based on Le Châtelier's principle (Section 15.6), we expect that the reaction will have a greater tendency to occur under the cited conditions than under standard-state conditions because one of the reactant partial pressures is greater than 1 atm and the product partial pressure is less than 1 atm. We therefore predict that Q_p will be less than 1 and ΔG will be more negative than $\Delta G°$, in agreement with the solution.

(b) In this case, the reaction mixture is rich in the product and poor in the reactants. Therefore, Q_p is expected to be greater than 1 and ΔG should be more positive than $\Delta G°$, in agreement with the solution.

▸ **PRACTICE 18.15** Calculate ΔG for the formation of ethylene (C_2H_4) from carbon and hydrogen at 25 °C when the partial pressures are 100 atm H_2 and 0.100 atm C_2H_4.

$$2 \text{ C}(s) + 2 \text{ H}_2(g) \longrightarrow C_2H_4(g) \qquad \Delta G° = 68.1 \text{ kJ}$$

Is the reaction spontaneous in the forward or the reverse direction?

▸ **CONCEPTUAL APPLY 18.16** Consider the following gas-phase reaction of A_2 (red) and B_2 (blue) molecules:

$$A_2 + B_2 \rightleftharpoons 2 \text{ AB} \qquad \Delta G° = 15 \text{ kJ}$$

continued on next page

(a) Which of the following reaction mixtures has the largest ΔG of reaction? Which has the smallest?

(1)   **(2)**   **(3)**

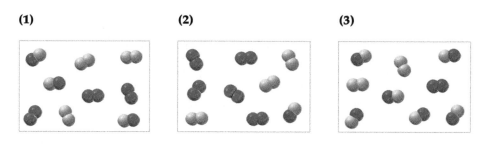

(b) If the partial pressure of each reactant and product in reaction mixture 1 is equal to 1 atm, what is the value of ΔG for the reaction in mixture 1?

18.11 FREE ENERGY AND CHEMICAL EQUILIBRIUM

Now that we've seen how ΔG for a reaction depends on the composition of a reaction mixture, we can understand how the total free energy of a reaction mixture changes as the reaction progresses toward equilibrium. Look again at the expression for calculating ΔG:

$$\Delta G = \Delta G° + RT \ln Q$$

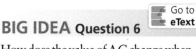

BIG IDEA Question 6

How does the value of ΔG change when the product of a reaction is removed from an equilibrium mixture?

If the reaction mixture contains mainly reactants and almost no products, Q will be much less than 1 and $RT \ln Q$ will be a very large negative number (minus infinity when $Q = 0$). Consequently, no matter what the value of $\Delta G°$ (positive or negative), the negative $RT \ln Q$ term will dominate the $\Delta G°$ term and ΔG will be negative. This means that the forward reaction is always spontaneous when the concentration of products is very small. Conversely, if the reaction mixture contains mainly products and almost no reactants, Q will be much greater than 1 and $RT \ln Q$ will be a very large positive number (plus infinity when no reactants are present). Consequently, the positive $RT \ln Q$ term will dominate the $\Delta G°$ term and ΔG will be positive. Thus, the reverse reaction is always spontaneous when the concentration of reactants is very small. These conditions are summarized as follows:

- When the reaction mixture is mostly reactants,

$$Q \ll 1 \qquad RT \ln Q \ll 0 \qquad \Delta G < 0$$

the total free energy decreases as the reaction proceeds spontaneously in the forward direction.

- When the reaction mixture is mostly products,

$$Q \gg 1 \qquad RT \ln Q \gg 0 \qquad \Delta G > 0$$

the total free energy decreases as the reaction proceeds spontaneously in the reverse direction.

FIGURE 18.11 shows how the total free energy of a reaction mixture changes as the reaction progresses. Because the free energy decreases as pure reactants form products and also decreases as pure products form reactants, the free-energy curve must go through a minimum somewhere between pure reactants and pure products. At that minimum free-energy composition, the system is at equilibrium because the conversion of either reactants to products or products to reactants would increase the free energy. The equilibrium composition persists indefinitely unless the system is disturbed by an external influence.

The sign of ΔG for the reaction is the same as the sign of the slope of the free-energy curve (Figure 18.11). To the left of the equilibrium composition, ΔG and the slope of the curve are negative and the free energy decreases as reactants are converted to products. To the right of the equilibrium composition, ΔG and the slope of the curve are positive. Exactly at the equilibrium composition, ΔG and the slope of the curve are zero and no net reaction occurs.

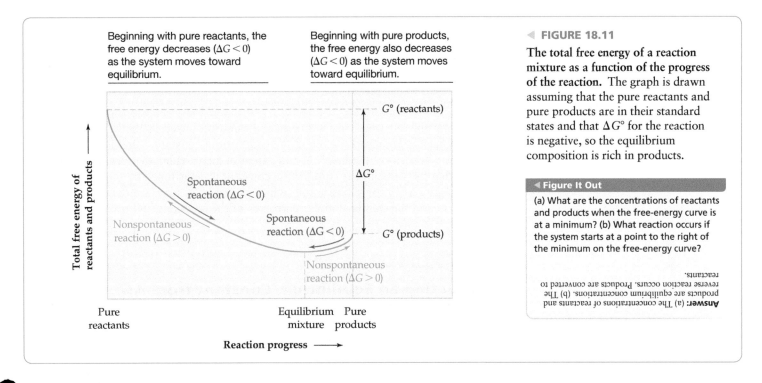

Beginning with pure reactants, the free energy decreases ($\Delta G < 0$) as the system moves toward equilibrium.

Beginning with pure products, the free energy also decreases ($\Delta G < 0$) as the system moves toward equilibrium.

◀ **FIGURE 18.11**

The total free energy of a reaction mixture as a function of the progress of the reaction. The graph is drawn assuming that the pure reactants and pure products are in their standard states and that $\Delta G°$ for the reaction is negative, so the equilibrium composition is rich in products.

◀ **Figure It Out**

(a) What are the concentrations of reactants and products when the free-energy curve is at a minimum? (b) What reaction occurs if the system starts at a point to the right of the minimum on the free-energy curve?

Answer: (a) The concentrations of reactants and products are equilibrium concentrations. (b) The reverse reaction occurs. Products are converted to reactants.

We can now derive a relationship between free energy and the equilibrium constant. At equilibrium, ΔG for a reaction is zero and the reaction quotient Q equals the equilibrium constant K. Substituting $\Delta G = 0$ and $Q = K$ into the equation

$$\Delta G = \Delta G° + RT \ln Q$$

gives $0 = \Delta G° + RT \ln K$ or

Relationship between standard free-energy change and equilibrium constant

$$\Delta G° = -RT \ln K$$

In this equation, K is K_p for reactions involving gases and K_c for reactions involving solutes in solution. For reactions involving both gases and solutes in solution, the equilibrium-constant expression contains partial pressures of gases and molar concentrations of solutes.

The equation $\Delta G° = -RT \ln K$ is one of the most important relationships in chemical thermodynamics because it allows us to calculate the equilibrium constant for a reaction from the standard free-energy change, or vice versa. This relationship is especially useful when K is difficult to measure. Consider a reaction so slow that it takes more than an experimenter's lifetime to reach equilibrium or a reaction that goes essentially to completion, so that the equilibrium concentrations of the reactants are extremely small and hard to measure. We can't measure K directly in such cases, but we can calculate its value from $\Delta G°$.

The relationship between $\Delta G°$ and the equilibrium constant K is summarized in **TABLE 18.4**. A reaction with a negative value of $\Delta G°$ has an equilibrium constant greater than 1, which corresponds to a minimum in the free-energy curve of Figure 18.11 at a composition rich in products. Conversely, a reaction that has a positive value of $\Delta G°$ has an equilibrium constant less than 1 and a minimum in the free-energy curve at a composition rich in reactants. Try redrawing Figure 18.11 for the case in which $\Delta G° > 0$.

TABLE 18.4   Relationship between the Standard Free-Energy Change and the Equilibrium Constant for a Reaction: $\Delta G° = -RT \ln K$

$\Delta G°$	$\ln K$	K	Comment
$\Delta G° < 0$	$\ln K > 0$	$K > 1$	The equilibrium mixture is mainly products.
$\Delta G° > 0$	$\ln K < 0$	$K < 1$	The equilibrium mixture is mainly reactants.
$\Delta G° = 0$	$\ln K = 0$	$K = 1$	The equilibrium mixture contains comparable amounts of reactants and products.

We have now answered the fundamental question posed at the beginning of this chapter: What determines the value of the equilibrium constant—that is, what properties of nature determine the direction and extent of a particular chemical reaction? The answer is that the value of the equilibrium constant is determined by the *standard* free-energy change ($\Delta G°$) for the reaction, which depends, in turn, on the standard heats of formation and the standard molar entropies of the reactants and products.

WORKED EXAMPLE 18.9

Calculating an Equilibrium Constant from $\Delta G°$ for a Reaction

Methanol (CH_3OH), an important alcohol used in the manufacture of adhesives, fibers, and plastics, is synthesized industrially by the reaction

$$CO(g) + 2\,H_2(g) \rightleftharpoons CH_3OH(g)$$

Use the thermodynamic data in Appendix B to calculate the equilibrium constant for this reaction at 25 °C.

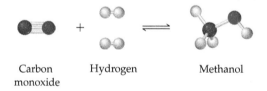

Carbon monoxide   Hydrogen   Methanol

IDENTIFY

Known	Unknown
Standard free energies of formation ($\Delta G°_f$)	Equilibrium constant (K_p)
$T = 25\,°C$	

STRATEGY

First calculate $\Delta G°$ for the reaction from the tabulated values of $\Delta G°_f$ for the reactants and products. Then use the equation $\Delta G° = -RT \ln K$ to find the value of the equilibrium constant.

SOLUTION

$$\Delta G° = \Delta G°_f(CH_3OH) - [\Delta G°_f(CO) + 2\,\Delta G°_f(H_2)]$$
$$= (1\ mol)(-162.3\ kJ/mol) - [(1\ mol)(-137.2\ kJ/mol) + (2\ mol)(0\ kJ/mol)]$$
$$\Delta G° = -25.1\ kJ$$

Solving the equation $\Delta G° = -RT \ln K$

for $\ln K$ gives $\ln K = \dfrac{-\Delta G°}{RT} = \dfrac{-(-25.1 \times 10^3\ J/mol)}{[8.314\ J/(K \cdot mol)](298\ K)} = 10.1$

Therefore, $K = K_p = \text{antiln } 10.1 = e^{10.1} = 2 \times 10^4$

The equilibrium constant obtained by this procedure is K_p because the reactants and products are gases and their standard states are defined in terms of pressure. If we want the value of K_c, we must calculate it from the relation $K_p = K_c(RT)^{\Delta n}$ (Section 15.3), where R must be expressed in the proper units $[R = 0.082\ 06\ (\text{L} \cdot \text{atm})/(\text{K} \cdot \text{mol})]$.

CHECK

$\Delta G°$ is negative, so the equilibrium mixture should be mainly products ($K > 1$), in agreement with the solution.

▶ **PRACTICE 18.17**   Given the data in Appendix B, calculate K_p at 25 °C for the reaction

$$CaCO_3(s) \rightleftharpoons CaO(s) + CO_2(g).$$

▶ **APPLY 18.18**   At 25°C, K_w for the dissociation of water is 1.0×10^{-14}. Calculate $\Delta G°$ for the reaction $2\,H_2O(l) \rightleftharpoons H_3O^+(aq) + OH^-(aq)$.

WORKED EXAMPLE 18.10

Calculating a Vapor Pressure from $\Delta G°$

The value of $\Delta G°_f$ at 25 °C for gaseous mercury is 31.85 kJ/mol. What is the vapor pressure of mercury at 25 °C?

IDENTIFY

Known	Unknown
Hg (g), $\Delta G°_f$ = 31.85 kJ/mol	P_{Hg}
$T = 25\,°C$	

STRATEGY

The vapor pressure (in atm) equals K_p for the reaction

$$Hg(l) \rightleftharpoons Hg(g) \quad K_p = P_{Hg}$$

Hg(l) is omitted from the equilibrium-constant expression because it is a pure liquid. Because the standard state for elemental mercury is the pure liquid, $\Delta G°_f = 0$ for Hg(l) and $\Delta G°$ for the vaporization reaction simply equals $\Delta G°_f$ for Hg(g) 31.85 kJ/mol). We can calculate K_p from the equation $\Delta G° = -RT \ln K_p$, as in Worked Example 18.9.

SOLUTION

$$\ln K_p = \frac{-\Delta G°}{RT} = \frac{-(31.85 \times 10^3\,\text{J/mol})}{[8.314\,\text{J}/(\text{K} \cdot \text{mol})](298\,\text{K})} = -12.86$$

$$K_p = \text{antiln}\,(-12.86) = e^{-12.86} = 2.6 \times 10^{-6}$$

Since K_p is defined in units of atmospheres, the vapor pressure of mercury at 25 °C is 2.6×10^{-6} atm (0.0020 mm Hg). Because the vapor pressure is appreciable and mercury is toxic in the lungs, mercury should not be handled without adequate ventilation.

CHECK

$\Delta G°$ is positive, so the vaporization reaction should not proceed very far before reaching equilibrium. Thus, K_p should be less than 1, in agreement with the solution.

▲ Mercury has an appreciable vapor pressure at room temperature, and its handling requires adequate ventilation.

▶ **PRACTICE 18.19**   Use the data in Appendix B to calculate the vapor pressure of water at 25 °C.

▶ **APPLY 18.20**   If the vapor pressure of ethanol (CH_3CH_2OH) at 25 °C is 60.6 mm Hg, calculate the values of K_p and $\Delta G°$.

INQUIRY ? Does the formation of highly ordered molecules violate the second law of thermodynamics?

At first glance, living organisms seem to violate the second law of thermodynamics which states that all spontaneous processes increase the entropy of the universe. The more complex the organism, the lower the amount of randomness, and the *lower* the entropy. How is it possible for enormously complex living organisms to grow and evolve if the randomness of the universe is always increasing? The dilemma of biological order can be explained by examining *both* the system and surroundings. It's perfectly possible for the randomness of any *system* to decrease spontaneously as long as the randomness of the *surroundings* increases by an even greater amount.

Let's consider a molecular example of biological order; the formation of a double helix of **deoxyribonucleic acid (DNA)** from two single strands of DNA. DNA stores our genetic code and is composed of two complementary molecular strands that spontaneously link through hydrogen bonds (Section 8.6). The DNA bases adenine (A) and thymine (T) form a pair and guanine (G) and cytosine (C) form a pair as shown:

$$\text{ATTGC} + \text{TAACG} \longrightarrow \begin{matrix} \text{TAACG} \\ \vdots \\ \text{ATTGC} \end{matrix}$$

The formation of the double strand results in a decrease in entropy of the system because the two individual strands that were free to diffuse and rotate in solution are now bound into one molecule with fixed positions. As the two DNA strands come together, the formation of intermolecular forces (specifically hydrogen bonds) between DNA bases releases a substantial amount of heat (250 kJ/mol of base pairs). The heat released increases the entropy of the surroundings. If we examine the total entropy, we find that the increase in entropy of the surroundings is greater than the decrease in entropy of the system leading to a spontaneous process, ($\Delta S_{total} = \Delta S_{system} + \Delta S_{surroundings} > 0$).

In contrast to the spontaneous formation of double-stranded DNA from two complementary single strands, many biological reactions are nonspontaneous because they do not cause an increase in total entropy ($\Delta S_{total} = \Delta S_{system} + \Delta S_{surroundings} < 0$). For example, **proteins** are highly ordered biological molecules that consist of amino

acids connected in a specific order and folded into the exact shape required to carry out a biological function, such as catalyzing a reaction. Synthesis of a protein involves linking amino acids molecules together. A representative reaction between two amino acid molecules of glycine (Gly), for example, is

$$\text{Gly}_1(aq) + \text{Gly}_2(aq) \rightleftharpoons$$
$$\text{Gly}_1\text{—Gly}_2(aq) + H_2O(l) \quad \Delta G^{\circ\prime} = +15 \text{ kJ}$$

In biochemical systems, the standard-state concentration of H_3O^+ is 1×10^{-7}M, not 1 M, and the standard free-energy change has the symbol $\Delta G^{\circ\prime}$. The positive value of $\Delta G^{\circ\prime}$ means that the reaction is reactant favored and only a small amount of product is formed at equilibrium. Therefore, the synthesis of a protein from its component amino acids requires an input of free energy.

Where does the energy needed to build complex biological molecules and sustain living organisms come from? Photosynthetic cells in plants use the Sun's energy to make glucose, which is then used by animals as their primary source of energy. Glucose ($C_6H_{12}O_6$) is oxidized for energy in the body according to the equation:

$$C_6H_{12}O_6(s) + 6\,O_2(g) \rightleftharpoons$$
$$6\,CO_2(g) + 6\,H_2O(l) \quad \Delta G^{\circ\prime} = -2870 \text{ kJ}$$

This reaction increases the entropy of the surroundings as the animal releases heat and the simple waste products CO_2 and H_2O to the environment. The free energy obtained from glucose oxidation can be used to build and organize molecules such as proteins. Rather than directly using all the free energy from glucose, the molecule adenosine triphosphate (ATP^{4-}, **FIGURE 18.12**) is synthesized and used to store energy in convenient amounts.

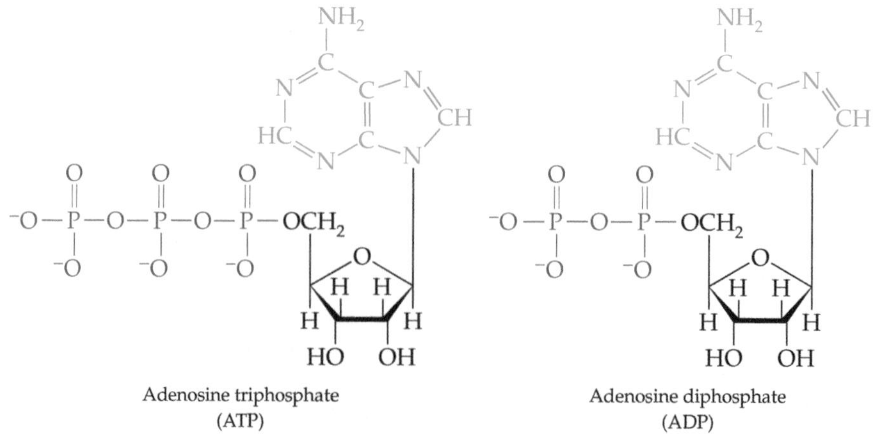

LOOKING AHEAD . . .
We will discuss the chemical structure and function of the biological molecules **deoxyribonucleic acid (DNA)** and **proteins** in Chapter 23.

▲ **FIGURE 18.12**
The structures of ATP^{4-} and ADP^{3-}. The breakdown of ATP into ADP and phosphate releases energy.

Free energy is liberated when ATP^{4-} reacts with water (hydrolysis) to form adenosine diphosphate (ADP^{3-}, Figure 18.12) as shown:

$$ATP^{4-}(aq) + H_2O(l) \rightleftharpoons$$
$$ADP^{3-}(aq) + H_2PO_4^-(aq) \quad \Delta G^{\circ\prime} = -30.5 \text{ kJ}$$

The reaction for linking the two glycine molecules can be coupled to ATP hydrolysis to make the overall reaction spontaneous and shift the equilibrium toward the formation of products.

$$Gly_1(aq) + Gly_2(aq) \rightleftharpoons$$
$$Gly_1—Gly_2(aq) + H_2O(l) \, \Delta G^{\circ\prime} = +15.0 \text{ kJ}$$
$$ATP^{4-}(aq) + H_2O(l) \rightleftharpoons$$
$$\underline{ADP^{3-}(aq) + H_2PO_4^-(aq) \, \Delta G^{\circ\prime} = -30.5 \text{ kJ}}$$
$$Gly_1(aq) + Gly_2(aq) + ATP^{4-}(aq) \rightleftharpoons$$
$$Gly_1—Gly_2(aq) + ADP^{3-}(aq) + H_2PO_4^-(aq) \, \Delta G^{\circ\prime} = -15.5 \text{ kJ}$$

Coupling reactions requires that the reactants be in close proximity and many of these reactions are facilitated by enzymes that bind reactants in its active site (Section 14.14). FIGURE 18.13 summarizes the free-energy relationships between the oxidation of carbon-based foods, the cycling of ADP and ATP, and the synthesis of complex biological molecules.

▲ **Many scientists call ATP the energy currency of cells.** Just as cash drives many commercial transactions, ATP is the principal medium of energy exchange in biological systems.

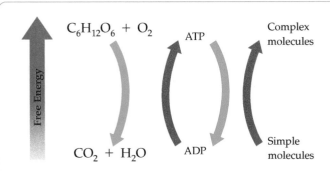

▲ FIGURE 18.13

Interconversion of free energy in living organisms. Free energy released in the oxidation of glucose is used to synthesize ATP from ADP. As ATP is hydrolyzed and converted back into ADP, free energy is released and used to build complex biological molecules such as proteins.

PROBLEM 18.21 Two complementary strands of DNA are placed in solution and spontaneously form a double strand.
(a) What is the sign of ΔS_{sys}?
(b) What is the sign of ΔG?
(c) Does forming a highly ordered molecule violate the second law of thermodynamics? Explain.

PROBLEM 18.22 In a human cell, 32 mol of ATP^{4-} are formed from ADP^{3-} every time one mole of glucose is oxidized to CO_2 and H_2O.
(a) Write the overall reaction and calculate $\Delta G^{\circ\prime}$.
(b) Is the reaction spontaneous or nonspontaneous?

PROBLEM 18.23 The first step in the oxidation of glucose for energy in the body is the conversion of glucose to glucose 6-phosphate:

$$Glucose(aq) + H_2PO_4^-(aq) \xrightarrow{H_2O} Glucose\ 6-phosphate(aq) +$$
$$H_2O(l) + H_3O^+(aq) \quad \Delta G^{\circ\prime} = 13.8 \text{ kJ}$$

(a) Is the reaction spontaneous or nonspontaneous?
(b) Couple the reaction to the hydrolysis of 1 mol of ATP^{4-} and calculate $\Delta G^{\circ\prime}$. Is the overall reaction spontaneous or nonspontaneous?
(c) Calculate the equilibrium constant for the overall reaction. (The temperature of the human body is 37 °C.)

PROBLEM 18.24 ATP^{4-} is the source of energy for muscle contraction.
(a) Calculate $\Delta G^{\prime}$ for the hydrolysis of ATP^{4-} under conditions in active muscle with the following concentrations: ATP^{4-} = 8.0 mM, ADP^{3-} = 0.9 mM, and $H_2PO_4^-$ = 8.0 mM. Use the value of $\Delta G^{\circ\prime}$ provided in the Inquiry, and the temperature of the human body is 37 °C.
(b) How does the amount of free energy released by hydrolysis of ATP^{4-} under physiological conditions in muscles compare to the standard free energy?

PROBLEM 18.25 The amount of ATP^{4-} in the body sustains muscle contraction for less than 1 second. After stored ATP^{4-} is used up in a burst of intense exercise, creatine phosphate serves to convert ADP^{3-} to ATP^{4-}. The hydrolysis reaction for creatine phosphate is:

$$creatine\ phosphate(aq) + H_2O(l) \rightleftharpoons$$
$$creatine(aq) + H_2PO_4^-(aq) \quad \Delta G^{\circ\prime} = -43.1 \text{ kJ}$$

Calculate $\Delta G^{\circ\prime}$ for the reaction:

$$creatine\ phosphate(aq) + ADP^{3-}(aq) \rightleftharpoons$$
$$creatine(aq) + ATP^{4-}(aq)$$

STUDY GUIDE

Section	Concept Summary	Learning Objectives	Test Your Understanding
18.1 Spontaneous Processes	**Thermodynamics** deals with the interconversion of heat and other forms of energy and allows us to predict the direction and extent of chemical reactions and other spontaneous processes. A **spontaneous process** proceeds on its own without any external influence. All spontaneous reactions move toward equilibrium.	**18.1** Define a spontaneous process and classify various physical processes and chemical reactions as spontaneous or nonspontaneous.	Problems 18.38, 18.40
18.2–18.4 Entropy and Spontaneous Processes	**Entropy**, denoted by S, is a state function that measures molecular randomness. A state of a system with a random distribution of molecules and molecular energies (a state of high entropy) can be achieved in more ways (W) than an ordered state and is therefore more probable. The entropy of a state can be calculated from **Boltzmann's equation**, $S = k \ln W$. The entropy of a system (reactants plus products) increases (ΔS is positive) for the following processes: phase transitions that convert a solid to a liquid or a liquid to a gas, reactions that increase the number of gaseous particles, dissolution of molecular solids and certain salts in water, raising the temperature of a substance, and the expansion of a gas at constant temperature. According to the **third law of thermodynamics**, the entropy of a pure, perfectly ordered crystalline substance at 0 K is zero.	**18.2** Predict the sign of ΔS for physical processes and chemical reactions.	Worked Example 18.1; Problems 18.26, 18.44–18.45
		18.3 Determine the number of arrangements in a system and use Boltzmann's equation to calculate the entropy.	Worked Example 18.2; Problems 18.49–18.50
		18.4 Calculate ΔS for the expansion or compression of an ideal gas at constant temperature.	Problem 18.58
		18.5 Predict how the phase of matter, or conditions such as temperature, volume, and pressure affect the entropy of a substance.	Problems 18.54, 18.56
18.5 Standard Molar Entropies and Standard Entropies of Reaction	The **standard molar entropy** ($S°$) of a substance is the absolute entropy of 1 mol of the pure substance at 1 atm pressure and a specified temperature, usually 25 °C. The **standard entropy of reaction** ($\Delta S°$) can be calculated from the relation: $$\Delta S° = S°(\text{products}) - S°(\text{reactants}).$$ The standard entropy change for the reaction $a\,A + b\,B \rightarrow c\,C + d\,D$ can be calculated from the formula: $$\Delta S° = [c\,S°(C) + d\,S°(D)] - [a\,S°(A) + b\,S°(B)]$$	**18.6** Predict how atomic structure affects the entropy of a substance.	Problems 18.62–18.63
		18.7 Use standard molar entropies ($S°$) to calculate the standard entropy of reaction ($\Delta S°$).	Worked Example 18.3; Problems 18.64, 18.66
18.6 Entropy and the Second Law of Thermodynamics	The **first law of thermodynamics** states that in any process, the total energy of a system and its surroundings remains constant. The **second law of thermodynamics** says that in any *spontaneous* process, the total entropy of a system and its surroundings ($\Delta S_{total} = \Delta S_{sys} + \Delta S_{surr}$) always increases. A chemical reaction is spontaneous if $\Delta S_{total} > 0$, nonspontaneous if $\Delta S_{total} < 0$, and at equilibrium if $\Delta S_{total} = 0$. Reactions that are nonspontaneous in the forward direction are spontaneous in the reverse direction. For a reaction at constant pressure, $\Delta S_{surr} = -\Delta H/T$, and ΔS_{sys} is ΔS for the reaction.	**18.8** Describe the factors that affect the spontaneity of a reaction.	Problems 18.68, 18.70
		18.9 Calculate values of ΔS_{sys}, ΔS_{surr}, and ΔS_{total} for a reaction, and use these values to determine whether the reaction is spontaneous.	Worked Example 18.4; Problems 18.74, 18.76

Section	Concept Summary	Learning Objectives	Test Your Understanding
18.7 Free Energy and Spontaneity of Chemical Reactions	Free energy, $G = H - TS$, is a state function that indicates whether a reaction is spontaneous or nonspontaneous. A reaction at constant temperature and pressure is spontaneous if $\Delta G < 0$, nonspontaneous if $\Delta G > 0$, and at equilibrium if $\Delta G = 0$. In the equation $\Delta G = \Delta H - T\Delta S$, temperature is a weighting factor that determines the relative importance of the enthalpy and entropy contributions to ΔG.	**18.10** Describe how the sign and magnitude of ΔS, ΔH, and temperature affect the value of ΔG and the spontaneity of the reaction.	Problems 18.27–18.31, 18.82, 18.84, 18.86
		18.11 Estimate the temperature at which a reaction changes between spontaneous and nonspontaneous.	Worked Example 18.5; Problems 18.88, 18.90
18.8–18.9 Standard Free-Energy Changes and Standard Free Energies of Formation	The **standard free-energy change ($\Delta G°$)** for a reaction is the change in free energy that occurs when reactants in their standard states are completely converted to products in their standard states. The **standard free energy of formation ($\Delta G°_f$)** of a substance is the free-energy change for formation of 1 mol of the substance in its standard state from the most stable form of the constituent elements in their standard states. Substances with a negative value of $\Delta G°_f$ are thermodynamically stable with respect to the constituent elements. We can calculate $\Delta G°$ for a reaction in either of two ways: (1) $\Delta G° = \Delta G°_f(\text{products}) - \Delta G°_f(\text{reactants})$ or (2) $\Delta G° = \Delta H° - T\Delta S°$.	**18.12** Calculate the standard free-energy change ($\Delta G°$) from $\Delta H°$ and $\Delta S°$ and interpret the meaning of the sign of $\Delta G°$.	Worked Example 18.6; Problems 18.94–18.95
		18.13 Calculate $\Delta G°$ for a reaction from values of $\Delta G°_f$ and determine if the reaction is spontaneous under standard-state conditions and how temperature will affect spontaneity.	Worked Example 18.7; Problems 18.96, 18.98, 18.100, 18.104
18.10–18.11 Free Energy, Nonstandard Conditions, and Chemical Equilibrium	The free-energy change, ΔG, for a reaction under nonstandard-state conditions is given by $\Delta G = \Delta G° + RT \ln Q$, where Q is the reaction quotient. At equilibrium, $\Delta G = 0$ and $Q = K$. As a result, $\Delta G° = -RT \ln K$, which allows us to calculate the equilibrium constant from $\Delta G°$ and vice versa.	**18.14** Calculate ΔG for a reaction under nonstandard-state conditions.	Worked Example 18.8; Problems 18.108, 18.110
		18.15 Use the relationship between $\Delta G°$ and the equilibrium constant for a reaction, and be able to calculate one from the other.	Worked Examples 18.9–18.10; Problems 18.37, 18.112, 18.118
		18.16 Interpret the shape of the curve in a plot of free energy versus reaction progress.	Problem 18.34

KEY TERMS

entropy (S) 771
first law of thermodynamics 782
free energy (G) 784
second law of thermodynamics 782

spontaneous process 769
standard entropy of reaction ($\Delta S°$) 780
standard free-energy change ($\Delta G°$) 787

standard free energy of formation ($\Delta G°_f$) 789
standard molar entropy ($S°$) 779
thermodynamics 769

third law of thermodynamics 778

KEY EQUATIONS

- Boltzmann equation for calculating the entropy S of a state of a system from the number of ways W that the state can be achieved (Section 18.3)

$S = k \ln W$ where the Boltzmann constant $k = 1.38 \times 10^{-23}$ J/K

- Entropy change for the expansion (or compression) of n moles of an ideal gas at constant temperature (Section 18.3)

$$\Delta S = nR \ln\frac{V_{\text{final}}}{V_{\text{initial}}} \qquad \Delta S = nR \ln\frac{P_{\text{initial}}}{P_{\text{final}}}$$

- Standard entropy change for a reaction (Section 18.5)

$\Delta S° = S°(\text{products}) - S°(\text{reactants})$ where $S°$ is the standard molar entropy

- Standard entropy change for the reaction $a\,A + b\,B \longrightarrow c\,C + d\,D$ (Section 18.5)

$\Delta S° = [c\,S°(C) + d\,S°(D)] - [a\,S°(A) + b\,S°(B)]$

- Total entropy change for a reaction (Section 18.6)

$$\Delta S_{total} = \Delta S_{system} + \Delta S_{surroundings}$$

where ΔS_{system} is the entropy of reaction and $\Delta S_{surroundings}$ is the entropy change in the surroundings

- Entropy change in the surroundings for a reaction at constant pressure (Section 18.6)

$$\Delta S_{surroundings} = -\Delta H/T$$

where ΔH is the enthalpy of reaction and T is the kelvin temperature of the surroundings

- Definition of free energy (Section 18.7)

$$G = H - TS$$

- Free-energy change for a reaction at constant temperature and pressure (Section 18.7)

$$\Delta G = \Delta H - T\Delta S$$

- Standard free-energy change for a reaction (Section 18.9)

$$\Delta G° = \Delta G°_f(products) - \Delta G°_f(reactants) \text{ where } \Delta G°_f \text{ is the standard free energy of formation}$$

- Standard free-energy change for the reaction $a\,A + b\,B \longrightarrow c\,C + d\,D$ (Section 18.9)

$$\Delta G° = [c\,\Delta G°_f(C) + d\,\Delta G°_f(D)] - [a\,\Delta G°_f(A) + b\,\Delta G°_f(B)]$$

- Free-energy change for a reaction under nonstandard-state conditions (Section 18.10)

$$\Delta G = \Delta G° + RT\,\ln Q \text{ where } Q \text{ is the reaction quotient}$$

- Relationship between $\Delta G°$ and the equilibrium constant for a reaction (Section 18.11)

$$\Delta G° = -RT\,\ln K$$

PRACTICE TEST

After studying this chapter, you can assess your understanding with these practice test questions, which are correlated with chapter learning objectives. If you answer a question incorrectly, refer to the learning objectives in the end-of-chapter Study Guide for assistance. The Study Guide provides a conceptual summary, references a Worked Example to model how to solve the problem, and gives additional problems for more practice.

1. Which of the following processes is nonspontaneous? **(LO 18.1)**
   (a) Transferring heat from a block of ice to air in a room maintained at 25 °C.
   (b) Water evaporating from an open beaker.
   (c) Dissolving a teaspoon of salt in a warm cup of water.
   (d) Liquid nitrogen boiling when poured onto the floor in a room at 25 °C and 1 atm.

2. Which of the following reactions has $\Delta S_{sys} > 0$? **(LO 18.2)**
   (a) $N_2(g) + 3\,H_2(g) \longrightarrow 2\,NH_3(g)$
   (b) $Ag^+(aq) + Cl^-(aq) \longrightarrow AgCl(s)$
   (c) $2\,H_2O_2(aq) \longrightarrow 2\,H_2O(l) + O_2(g)$
   (d) $2\,I(g) \longrightarrow I_2(g)$

3. What is the change in entropy (ΔS) when 1.0×10^{21} gas molecules initially in one bulb are distributed evenly between two bulbs as shown? **(LO 18.3)**

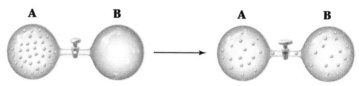

   (a) $\Delta S = +9.6 \times 10^{-3}$ J/K
   (b) $\Delta S = -9.6 \times 10^{-3}$ J/K
   (c) $\Delta S = +1.3 \times 10^{-21}$ J/K
   (d) $\Delta S = -1.3 \times 10^{-21}$ J/K

4. What is the entropy change (ΔS) when 1.32 g of propane (C_3H_8) at 0.100 atm pressure is compressed by a factor of five at a constant temperature of 20 °C? Assume that propane behaves as an ideal gas. **(LO 18.4)**
   (a) $\Delta S = +13$ J/K
   (b) $\Delta S = -13$ J/K
   (c) $\Delta S = -0.40$ J/K
   (d) $\Delta S = +0.40$ J/K

5. Which pair has the higher entropy state listed first? **(LO 18.5, 18.6)**
   (a) 1 mol of NaCl(s) at 25 °C or 1 mol of NaCl(s) at 100 °C
   (b) 1 mol of $C_2H_4(g)$ at 50 °C or 1 mol of Ar(g) at 50 °C
   (c) 1 mol of $N_2(l)$ or 1 mol of $N_2(g)$
   (d) 1 mol of $O_2(g)$ at 100 °C and 1 atm or 1 mol of $O_2(g)$ at 100 °C and 5 atm

6. Calculate $\Delta S°$ for the following reaction. The standard molar entropy ($S°$) for each species is shown below the reaction. (**LO 18.7**)

$$4\,NH_3(g) + 5\,O_2(g) \longrightarrow 4\,NO(g) + 6\,H_2O(g)$$

$S°$(J/K·mol)   192.8   205.2   210.8   188.8

(a) +287.4 J/K   (b) −401.2 J/K
(c) +160.0 J/K   (d) +178.8 J/K

7. Identify the true statement about a spontaneous process. (**LO 18.8**)

(a) A reaction that is nonspontaneous in the forward direction is spontaneous in the reverse direction.

(b) Adding a catalyst will cause a nonspontaneous reaction to become spontaneous.

(c) In a spontaneous process, the entropy of the system always decreases.

(d) An endothermic reaction is always spontaneous.

8. Calculate ΔS_{total}, and determine whether the reaction is spontaneous or nonspontaneous under standard-state conditions. (**LO 18.9**)

$$2\,Br^-(aq) + Cl_2(g) \longrightarrow Br_2(l) + 2\,Cl^-(aq)$$

$\Delta H°_f$(kJ/mol)   −121.5   0   0   −167.2
$S°$ (J/K·mol)   82.4   223.0   152.2   56.5

(a) −429 J/K; nonspontaneous
(b) −123 J/K; spontaneous
(c) + 3,530 J/K; nonspontaneous
(d) + 184 J/K; spontaneous

9. Consider the following endothermic reaction of gaseous AB_3 molecules with A_2 molecules.

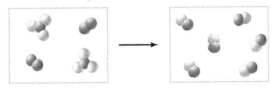

Identify the true statement about the spontaneity of the reaction. (**LO 18.10**)

(a) The reaction is likely to be spontaneous at high temperatures.

(b) The reaction is likely to be spontaneous at low temperatures.

(c) The reaction is always spontaneous.

(d) The reaction is never spontaneous.

10. The Haber process is used for the industrial synthesis of ammonia:

$$N_2(g) + 3\,H_2(g) \longrightarrow 2\,NH_3(g) \quad \Delta H° = -92.2\ kJ,$$
$$\Delta S° = -199\ J/K$$

Under what temperature conditions is the reaction spontaneous? (**LO 18.11**)

(a) $T < 463\ K$ (b) $T > 463\ K$
(c) $T > 291\ K$ (d) $T < 291\ K$

11. When gaseous carbon monoxide contacts solid nickel, liquid nickel tetracarbonyl is formed.

$$Ni(s) + 4\,CO(g) \longrightarrow Ni(CO)_4(l)$$

$\Delta H°_f$(kJ/mol)   0   − 110.5   − 633.0
$S°$ (J/K·mol)   29.9   197.6   313.4

Use standard enthalpy of formation $\Delta H°_f$ values and standard molar entropy ($S°$) values to calculate $\Delta G°$. Is the reaction spontaneous or nonspontaneous under standard-state conditions at 25 °C? (**LO 18.12**)

(a) + 151 kJ; nonspontaneous
(b) −203 kJ; nonspontaneous
(c) + 12.5 kJ; nonspontaneous
(d) −40.0 kJ; spontaneous

12. Nitrogen reacts with fluorine to form nitrogen trifluoride:

$$N_2(g) + 3\,F_2(g) \longrightarrow 2\,NF_3(g) \quad \Delta H° = -249\ kJ\ and$$
$$\Delta S° = -278\ J/K$$

Calculate $\Delta G°$, and determine whether the equilibrium composition should favor reactants or products at 25 °C. (**LO 18.12**)

(a) $\Delta G° = -6.7\ kJ$; the equilibrium composition should favor products.

(b) $\Delta G° = -332\ kJ$; the equilibrium composition should favor reactants.

(c) $\Delta G° = -166\ kJ$; the equilibrium composition should favor products.

(d) $\Delta G° = + 82.6\ kJ$; the equilibrium composition should favor reactants.

13. What is the equilibrium constant (K) for the following reaction at 298 K? The standard free energy of formation ($\Delta G°_f$) for each species is provided. (**LO 18.13, 18.15**)

$$2\,CO(g) + 2\,H_2(g) \longrightarrow CH_3OH(l)$$

$\Delta G°_f$(kJ/mol)   −137.2   0   −166.6

(a) 1.4×10^5 (b) 1.3×10^{-19}
(c) 0.96   (d) 7.9×10^{18}

14. Sulfuric acid is produced in larger amounts by weight than any other chemical. It is used in manufacturing fertilizers, for oil refining, and in hundreds of other processes. An intermediate step in the industrial process for the synthesis of H_2SO_4 is the catalytic oxidation of sulfur dioxide:

$$2\,SO_2(g) + O_2(g) \longrightarrow 2\,SO_3(g) \quad \Delta G° = -141.8\ kJ$$

Calculate ΔG at 25 °C, given the following sets of partial pressures: 2.00 atm SO_2, 1.00 atm O_2, and 10.0 atm SO_3. (**LO 18.14**)

(a) +527 kJ   (b) +7.83 kJ
(c) −150 kJ   (d) −134 kJ

15. Ammonium hydrogen sulfide, a stink bomb ingredient, decomposes to ammonia and hydrogen sulfide.

$$NH_4HS(s) \longrightarrow NH_3(g) + H_2S(g)$$

Calculate the standard free-energy change for the reaction at 25 °C if the total pressure resulting from solid NH_4HS placed in an evacuated container is 0.658 atm at 25 °C. (**LO 18.15**)

(a) −43.8 kJ   (b) + 1.04 kJ
(c) −462 kJ   (d) + 5.51 kJ

16. Consider the following graph of total free energy of reactants and products versus reaction progress for the general reaction, Reactants ⟶ Products.

At which of the four points (labeled *a*, *b*, *c*, and *d*) is $Q < K$?
(LO 18.16)

(a) Point a   (b) Points c and d

(c) Points a, c, and d   (d) Point b

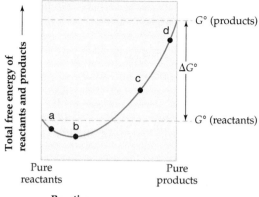

Answers:
1. a, 2. c, 3. a, 4. c, 5. b, 6. d, 7. a, 8. d, 9. a, 10. a, 11. d, 12. c, 13. b, 14. d, 15. d, 16. a

---

Mastering Chemistry   provides end-of-chapter exercises, feedback-enriched tutorial problems, animations, and interactive activities to encourage problem-solving practice and deeper understanding of key concepts and topics.

RAN   *Randomized in Mastering Chemistry*

CONCEPTUAL PROBLEMS

Problems 18.1–18.25 appear within the chapter.

18.26 Consider the gas-phase reaction of AB_3 and A_2 molecules:

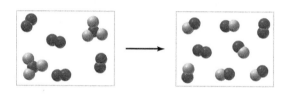

(a) Write a balanced equation for the reaction.

(b) What is the sign of the entropy change for the reaction?

18.27 Ideal gases A (red spheres) and B (blue spheres) occupy two separate bulbs. The contents of both bulbs constitute the initial state of an isolated system. Consider the process that occurs when the stopcock is opened.

(a) Sketch the final (equilibrium) state of the system.

(b) What are the signs (+ , −, or 0) of ΔH, ΔS, and ΔG for this process? Explain.

(c) How does this process illustrate the second law of thermodynamics?

(d) Is the reverse process spontaneous or nonspontaneous? Explain.

18.28 What are the signs (+ , −, or 0) of ΔH, ΔS, and ΔG for the spontaneous sublimation of a crystalline solid? Explain.

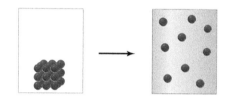

18.29 What are the signs (+ , −, or 0) of ΔH, ΔS, and ΔG for the spontaneous condensation of a vapor to a liquid? Explain.

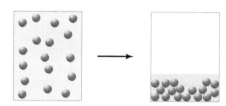

18.30 An ideal gas is compressed at constant temperature. What are the signs (+ , −, or 0) of ΔH, ΔS, and ΔG for the process? Explain.

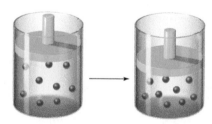

18.31 Consider the following spontaneous reaction of A_2 molecules (red) and B_2 molecules (blue):

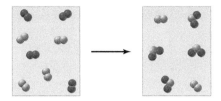

(a) Write a balanced equation for the reaction.

(b) What are the signs (+ , −, or 0) of ΔH, ΔS, and ΔG for the reaction? Explain.

18.32 Consider the dissociation reaction $A_2(g) \rightleftharpoons 2\,A(g)$. The following pictures represent two possible initial states and the equilibrium state of the system:

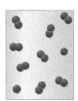

Initial state 1   Initial state 2   Equilibrium state

(a) Is the reaction quotient Q_p for initial state 1 greater than, less than, or equal to the equilibrium constant K_p? Is Q_p for initial state 2 greater than, less than, or equal to K_p?

(b) What are the signs (+ , −, or 0) of ΔH, ΔS, and ΔG when the system goes from initial state 1 to the equilibrium state? Explain. Is this a spontaneous process?

(c) What are the signs (+ , −, or 0) of ΔH, ΔS, and ΔG when the system goes from initial state 2 to the equilibrium state? Explain. Is this a spontaneous process?

(d) Relate each of the pictures to the graph in Figure 18.11.

18.33 Consider again the dissociation reaction $A_2(g) \rightleftharpoons 2\,A(g)$ (Problem 18.32).

(a) What are the signs (+ , −, or 0) of the standard enthalpy change, $\Delta H°$, and the standard entropy change, $\Delta S°$, for the forward reaction?

(b) Distinguish between the meaning of $\Delta S°$ for the dissociation reaction and ΔS for the process in which the system goes from initial state 1 to the equilibrium state (pictured in Problem 18.32).

(c) What can you say about the sign of $\Delta G°$ for the dissociation reaction? How does $\Delta G°$ depend on temperature? Will $\Delta G°$ increase, decrease, or remain the same if the temperature increases?

(d) Will the equilibrium constant K_p increase, decrease, or remain the same if the temperature increases? How will the picture for the equilibrium state (Problem 18.32) change if the temperature increases?

(e) What is the value of ΔG for the dissociation reaction when the system is at equilibrium?

18.34 Consider the following graph of the total free energy of reactants and products versus reaction progress for a general reaction, Reactants → Products:

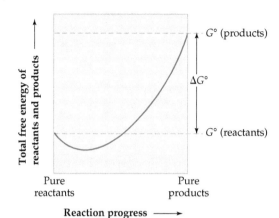

(a) Account for the shape of the curve, and identify the point at which $\Delta G = 0$. What is the significance of that point?

(b) Why is the minimum in the plot on the left side of the graph?

18.35 The following pictures represent equilibrium mixtures for the interconversion of A molecules (red) and X, Y, or Z molecules (blue):

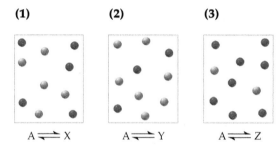

What is the sign of $\Delta G°$ for each of the three reactions?

18.36 The following pictures represent the composition of the equilibrium mixture at 25 °C and 45 °C for the reaction A $\rightleftharpoons$ B, where A molecules are represented by red spheres and B molecules by blue spheres:

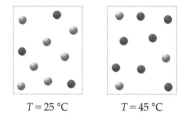

$T = 25\ °C$ $T = 45\ °C$

What are the signs of $\Delta H°$ and $\Delta S°$ for the forward reaction A → B? Explain. (Assume that $\Delta H°$ and $\Delta S°$ are independent of temperature.)

18.37 The following pictures represent mixtures of A_2 (red), B_2
RAN (blue), and AB molecules at 25 °C. The reaction is: $A_2(g) +$
$B_2(g) \rightleftharpoons 2\ AB(g)$

(1)   **(2)**

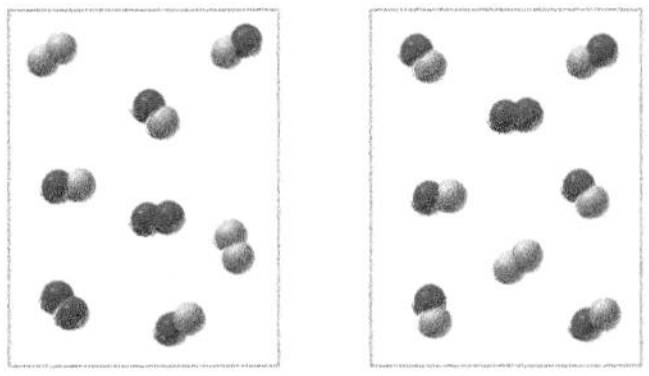

(a) If mixture 1 is at equilibrium, calculate K_p and $\Delta G°$.
Assume that each molecule represents 0.5 atm.

(b) Calculate the value of ΔG for mixture 2. Does the reaction proceed toward products or reactants?

SECTION PROBLEMS

Spontaneous Processes (Section 18.1)

18.38 Which of the following processes are spontaneous, and which are nonspontaneous?

(a) Freezing of water at 2 °C

(b) Corrosion of iron metal

(c) Expansion of a gas to fill the available volume

(d) Separation of an unsaturated aqueous solution of potassium chloride into solid KCl and liquid water

18.39 Tell whether the following processes are spontaneous or nonspontaneous.

(a) Dissolving sugar in hot coffee

(b) Decomposition of NaCl to solid sodium and gaseous chlorine at 25 °C and 1 atm pressure

(c) Uniform mixing of bromine vapor and nitrogen gas

(d) Boiling of gasoline at 25 °C and 1 atm pressure

18.40 Assuming that gaseous reactants and products are present at 1 atm partial pressure, which of the following reactions are spontaneous in the forward direction?

(a) $N_2(g) + 2\ H_2(g) \rightarrow N_2H_4(l)$; $K_p = 7 \times 10^{-27}$

(b) $2\ Mg(s) + O_2(g) \rightarrow 2\ MgO(s)$; $K_p = 2 \times 10^{198}$

(c) $MgCO_3(s) \rightarrow MgO(s) + CO_2(g)$; $K_p = 9 \times 10^{-10}$

(d) $2\ CO(g) + O_2(g) \rightarrow 2\ CO_2(g)$; $K_p = 1 \times 10^{90}$

18.41 Assuming that dissolved reactants and products are present at 1 M concentrations, which of the following reactions are nonspontaneous in the forward direction?

(a) $HCN(aq) + H_2O(l) \rightarrow H_3O^+(aq) + CN^-(aq)$;
$K = 4.9 \times 10^{-10}$

(b) $H_3O^+(aq) + OH^-(aq) \rightarrow 2\ H_2O(l)$; $K = 1.0 \times 10^{14}$

(c) $Ba^{2+}(aq) + CO_3{}^{2-}(aq) \rightarrow BaCO_3(s)$; $K = 3.8 \times 10^8$

(d) $AgCl(s) \rightarrow Ag^+(aq) + Cl^-(aq)$; $K = 1.8 \times 10^{-10}$

Enthalpy, Entropy, and Spontaneous Processes (Section 18.2)

18.42 Define entropy, and give an example of a process in which the entropy of a system increases.

18.43 Comment on the following statement: Exothermic reactions are spontaneous, but endothermic reactions are nonspontaneous.

18.44 Predict the sign of the entropy change in the system for each of the following processes.

(a) A solid sublimes.

(b) A liquid freezes.

(c) AgI precipitates from a solution containing Ag^+ and I^- ions.

(d) Gaseous CO_2 bubbles out of a carbonated beverage.

18.45 Predict the sign of ΔS in the system for each of the following reactions.

(a) $PCl_5(s) \rightarrow PCl_3(l) + Cl_2(g)$

(b) $CH_4(g) + 2\ O_2(g) \rightarrow CO_2(g) + 2\ H_2O(l)$

(c) $2\ H_3O^+(aq) + CO_3{}^{2-}(aq) \rightarrow CO_2(g) + 3\ H_2O(l)$

(d) $Mg(s) + Cl_2(g) \rightarrow MgCl_2(s)$

18.46 Predict the sign of ΔS for each process in Problem 18.38.
RAN

18.47 Predict the sign of ΔS for each process in Problem 18.39.
RAN

Entropy (Sections 18.3–18.4)

18.48 When rolling a pair of dice, there are two ways to get a point total of 3 $(1 + 2; 2 + 1)$ but only one way to get a point total of 2 $(1 + 1)$. How many ways are there of getting point totals of 4 to 12? What is the most probable point total?

18.49 Consider a disordered crystal of monodeuteriomethane in
RAN which each tetrahedral CH_3D molecule is oriented randomly in one of four possible ways. Use Boltzmann's equation to calculate the entropy of the disordered state of the crystal if the crystal contains:

(a) 12 molecules

(b) 120 molecules

(c) 1 mol of molecules

What is the entropy of the crystal if the C–D bond of each of the CH_3D molecules points in the same direction?

18.50 Consider the distribution of ideal gas molecules among three
RAN bulbs (A, B, and C) of equal volume. For each of the following states, determine the number of ways (W) that the state can be achieved, and use Boltzmann's equation to calculate the entropy of the state.

(a) 2 molecules in bulb A

(b) 2 molecules randomly distributed among bulbs A, B, and C

(c) 3 molecules in bulb A

(d) 3 molecules randomly distributed among bulbs A, B, and C

(e) 1 mol of molecules in bulb A

(f) 1 mol of molecules randomly distributed among bulbs A, B, and C

What is ΔS on going from state **(e)** to state **(f)**? Compare your result with ΔS calculated from the equation $\Delta S = nR \ln (V_{final}/V_{initial})$.

18.51 What is the entropy of 100 molecules in a system of
RAN 1000 boxes? What is the entropy of 100 molecules in a system of 10,000 boxes?

18.52 By what factor does the entropy increase for a collection of
RAN 1000 molecules moved from 1.00×10^6 boxes to 1.00×10^7 boxes? For a move from 1.00×10^{16} to 1.00×10^{17} boxes? (Express your answers to three significant figures.)

18.53 If the entropy of a collection of molecules in 50,000 boxes is 3.73×10^{-20} J/K, how many molecules are there?

18.54 Which state has higher entropy? Explain in terms of probability.
RAN (a) A perfectly ordered crystal of solid nitrous oxide ($N \equiv N-O$) or a disordered crystal in which the molecules are oriented randomly

  (b) Quartz glass (an amorphous solid) or a quartz crystal

18.55 Which state has higher entropy? Explain in terms of probability.
RAN (a) 1 mol of N_2 gas at STP or 1 mol of N_2 gas at 273 K and 0.10 atm

  (b) A crystal of NaCl at 0 °C or the same crystal at 50 °C

18.56 Which state in each of the following pairs has the higher
RAN entropy per mole of substance?

  (a) H_2 at 25 °C in a volume of 10 L or H_2 at 25 °C in a volume of 50 L

  (b) O_2 at 25 °C and 1 atm or O_2 at 25 °C and 10 atm

  (c) H_2 at 25 °C and 1 atm or H_2 at 100 °C and 1 atm

  (d) CO_2 at STP or CO_2 at 100 °C and 0.1 atm

18.57 Which state in each of the following pairs has the higher entropy per mole of substance?

  (a) Ice at -40 °C or ice at 0 °C

  (b) N_2 at STP or N_2 at 0 °C and 10 atm

  (c) N_2 at STP or N_2 at 0 °C in a volume of 50 L

  (d) Water vapor at 150 °C and 1 atm or water vapor at 100 °C and 2 atm

18.58 What is the entropy change when the volume of 1.6 g of
RAN O_2 increases from 2.5 L to 3.5 L at a constant temperature of 75 °C? Assume that O_2 behaves as an ideal gas.

18.59 What is the value of ΔS when 2.4 g of CH_4 is compressed
RAN from 30.0 L to 20.0 L at a constant temperature of 100 °C? Assume that CH_4 behaves as an ideal gas.

18.60 Make a rough, qualitative plot of the standard molar entropy versus temperature for methane from 0 K to 298 K. Incorporate the following data into your plot: mp $= -182$ °C; bp $= -164$ °C; $S° = 186.2$ J/(K · mol) at 25 °C.

18.61 Make a rough, qualitative plot of the fraction of molecules versus kinetic energy for methane gas at 10 °C and 100 °C. Use the plot to explain why a sample of methane gas has higher entropy at 100 °C.

Standard Molar Entropies and Standard Entropies of Reaction (Section 18.5)

18.62 Which substance in each of the following pairs would you expect to have the higher standard molar entropy? Explain.

  (a) $C_2H_2(g)$ or $C_2H_6(g)$

  (b) $CO_2(g)$ or $CO(g)$

  (c) $I_2(s)$ or $I_2(g)$

  (d) $CH_3OH(g)$ or $CH_3OH(l)$

18.63 Which substance in each of the following pairs would you expect to have the higher standard molar entropy? Explain.

  (a) $NO(g)$ or $NO_2(g)$

  (b) $CH_3CO_2H(l)$ or $HCO_2H(l)$

  (c) $Br_2(l)$ or $Br_2(s)$

  (d) $S(s)$ or $SO_3(g)$

18.64 Use the standard molar entropies in Appendix B to calculate
RAN the standard entropy of reaction for the oxidation of carbon monoxide to carbon dioxide:

$$2\,CO(g) + O_2(g) \rightarrow 2\,CO_2(g)$$

18.65 Use the standard molar entropies in Appendix B to calculate the standard entropy of reaction for the oxidation of graphite to carbon dioxide:

$$C(s) + O_2(g) \rightarrow 2\,CO_2(g)$$

18.66 Use the standard molar entropies in Appendix B to calculate
RAN $\Delta S°$ at 25 °C for each of the following reactions. Account for the sign of the entropy change in each case.

  (a) $2\,H_2O_2(l) \rightarrow 2\,H_2O(l) + O_2(g)$

  (b) $2\,Na(s) + Cl_2(g) \rightarrow 2\,NaCl(s)$

  (c) $2\,O_3(g) \rightarrow 3\,O_2(g)$

  (d) $4\,Al(s) + 3\,O_2(g) \rightarrow 2\,Al_2O_3(s)$

18.67 Use the $S°$ values in Appendix B to calculate $\Delta S°$ at 25 °C for each of the following reactions. Suggest a reason for the sign of $\Delta S°$ in each case.

  (a) $2\,S(s) + 3\,O_2(g) \rightarrow 2\,SO_3(g)$

  (b) $SO_3(g) + H_2O(l) \rightarrow H_2SO_4(aq)$

  (c) $AgCl(s) \rightarrow Ag^+(aq) + Cl^-(aq)$

  (d) $NH_4NO_3(s) \rightarrow N_2O(g) + 2\,H_2O(g)$

Entropy and the Second Law of Thermodynamics (Section 18.6)

18.68 State the second law of thermodynamics.

18.69 An isolated system is one that exchanges neither matter nor energy with the surroundings. What is the entropy criterion for spontaneous change in an isolated system? Give an example of a spontaneous process in an isolated system.

18.70 Give an equation that relates the entropy change in the surroundings to the enthalpy change in the system. What is the sign of ΔS_{surr} for the following?

  (a) An exothermic reaction

  (b) An endothermic reaction

18.71 When heat is added to the surroundings, the entropy of the surroundings increases. How does ΔS_{surr} depend on the temperature of the surroundings? Explain.

18.72 Reduction of mercury(II) oxide with zinc gives metallic mercury:

$$HgO(s) + Zn(s) \rightarrow ZnO(s) + Hg(l)$$

  (a) If $\Delta H° = -259.7$ kJ/mol and $\Delta S° = +7.8$ J/K, what is ΔS_{total} for this reaction? Is the reaction spontaneous under standard-state conditions at 25 °C?

  (b) Estimate at what temperature, if any, the reaction will become nonspontaneous.

18.73 Elemental sulfur is formed by the reaction of zinc sulfide
RAN with oxygen:

$$2\,ZnS(s) + O_2(g) \rightarrow 2\,ZnO(s) + 2\,S(s)$$

(a) If $\Delta H° = -289.0\,kJ/mol$ and $\Delta S° = -169.4\,J/K$, what is ΔS_{total} for this reaction? Is the reaction spontaneous under standard-state conditions at 25 °C?

(b) At what temperature, if any, will the reaction become nonspontaneous?

18.74 In lightning storms, oxygen is converted to ozone:

$$3\,O_2(g) \rightarrow 2\,O_3(g)$$

By determining the sign of ΔS_{total}, show whether the reaction is spontaneous at 25 °C.

18.75 Sulfur dioxide emitted from coal-fired power plants is oxidized to sulfur trioxide in the atmosphere:

$$2\,SO_2(g) + O_2(g) \rightarrow 2\,SO_3(g)$$

By determining the sign of ΔS_{total}, show whether the reaction is spontaneous at 25 °C.

18.76 Elemental mercury can be produced from its oxide:

$$2\,HgO(s) \rightarrow 2\,Hg(l) + O_2(g)$$

(a) Use data in Appendix B to calculate ΔS_{sys}, ΔS_{surr}, and ΔS_{total} for this reaction. Is the reaction spontaneous under standard-state conditions at 25 °C?

(b) Estimate the temperature at which the reaction will become spontaneous.

18.77 Phosphorus pentachloride forms from phosphorus trichlo-
RAN ride and chlorine:

$$PCl_3(g) + Cl_2(g) \rightarrow PCl_5(g)$$

(a) Use data in Appendix B to calculate ΔS_{sys}, ΔS_{surr}, and ΔS_{total} for this reaction. Is the reaction spontaneous under standard-state conditions at 25 °C?

(b) Estimate the temperature at which the reaction will become nonspontaneous.

18.78 For the vaporization of benzene, $\Delta H_{vap} = 30.7\,kJ/mol$ and
RAN $\Delta S_{vap} = 87.0\,J/(K \cdot mol)$. Calculate ΔS_{surr} and ΔS_{total} at:
(a) 70 °C (b) 80 °C (c) 90 °C
Does benzene boil at 70°C and 1 atm pressure? Calculate the normal boiling point of benzene.

18.79 For the melting of sodium chloride, $\Delta H_{fusion} = 28.16\,kJ/mol$
RAN and $\Delta S_{fusion} = 26.22\,J/(K \cdot mol)$. Calculate ΔS_{surr} and ΔS_{total} at:
(a) 1050 K
(b) 1074 K
(c) 1100 K
(d) Does NaCl melt at 1100 K? Calculate the melting point of NaCl.

Free Energy (Section 18.7)

18.80 Describe how the signs of ΔH and ΔS determine whether a reaction is spontaneous or nonspontaneous at constant temperature and pressure.

18.81 The entropy change for a certain nonspontaneous reaction
RAN at 50 °C is 104 J/K.

(a) Is the reaction endothermic or exothermic?

(b) What is the minimum value of ΔH (in kJ) for the reaction?

18.82 Identify the true statement. A spontaneous reaction must have
RAN (a) a negative enthalpy change.

(b) a positive enthalpy change.

(c) a positive entropy change.

(d) a positive free-energy change.

(e) a negative free-energy change.

18.83 Which of the following reactions will be spontaneous at any
RAN temperature?

(a) $\Delta H = +$, $\Delta S = -$ (b) $\Delta H = -$, $\Delta S = -$
(c) $\Delta H = +$, $\Delta S = +$ (d) $\Delta H = -$, $\Delta S = +$

18.84 The melting point of benzene is 5.5 °C. Predict the signs of ΔH, ΔS, and ΔG for melting of benzene at:
(a) 0 °C. (b) 15 °C.

18.85 Consider a twofold expansion of 1 mol of an ideal gas at
RAN 25 °C in the isolated system shown in Figure 18.1.

(a) What are the values of ΔH, ΔS, and ΔG for the process?

(b) How does this process illustrate the second law of thermodynamics?

18.86 Given the data in Problem 18.78, calculate ΔG for the vaporiza-
RAN tion of benzene at:
(a) 70 °C. (b) 80 °C. (c) 90 °C.
Predict whether benzene will boil at each of these temperatures and 1 atm pressure.

18.87 Given the data in Problem 18.79, calculate ΔG for the melt-
RAN ing of sodium chloride at:
(a) 1050 K. (b) 1074 K. (c) 100 K.
Predict whether NaCl will melt at each of these temperatures and 1 atm pressure.

18.88 Calculate the melting point of benzoic acid ($C_6H_5CO_2H$), given the following data: $\Delta H_{fusion} = 18.02\,kJ/mol$ and $\Delta S_{fusion} = 45.56\,J/(K \cdot mol)$.

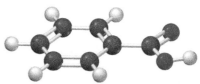

Benzoic acid

18.89 Calculate the enthalpy of fusion of naphthalene ($C_{10}H_8$), given that its melting point is 128 °C and its entropy of fusion is 47.7 J/(K · mol).

Naphthalene

18.90 Calculate the normal boiling point of ethanol (CH_3CH_2OH), given that its enthalpy of vaporization is 38.6 kJ/mol and its entropy of vaporization is 110 J/(K · mol).

18.91 Chloroform ($CHCl_3$) has a normal boiling point of 61 °C and an enthalpy of vaporization of 29.24 kJ/mol. What are its values of ΔG_{vap} and ΔS_{vap} at 61 °C?

Standard Free-Energy Changes and Standard Free Energies of Formation (Sections 18.8–18.9)

18.92 Define (a) the standard free-energy change, $\Delta G°$, for a reaction and (b) the standard free energy of formation, $\Delta G°_f$, of a substance.

18.93 What is meant by the standard state of a substance?

18.94 Use the data in Appendix B to calculate $\Delta H°$ and $\Delta S°$ for each of the following reactions. From the values of $\Delta H°$ and $\Delta S°$, calculate $\Delta G°$ at 25 °C, and predict whether each reaction is spontaneous under standard-state conditions.
(a) $N_2(g) + 2\,O_2(g) \rightarrow 2\,NO_2(g)$
(b) $CH_3CH_2OH(l) + O_2(g) \rightarrow CH_3CO_2H(l) + H_2O(l)$

18.95 Use the data in Appendix B to calculate $\Delta H°$ and $\Delta S°$ for each of the following reactions. From the values of $\Delta H°$ and $\Delta S°$, calculate $\Delta G°$ at 25 °C and predict whether each reaction is spontaneous under standard-state conditions.
(a) $2\,SO_2(g) + O_2(g) \rightarrow 2\,SO_3(g)$
(b) $CH_3OH(l) + O_2(g) \rightarrow HCO_2H(l) + H_2O(l)$

18.96 Use the standard free energies of formation in Appendix B
RAN to calculate $\Delta G°$ at 25 °C for the reaction

$$2\,KClO_3(s) \rightarrow 2\,KCl(s) + 3\,O_2(g)$$

18.97 Use the standard free energies of formation in Appendix B
RAN to calculate $\Delta G°$ at 25 °C for each reaction in

$$N_2(g) + 2\,H_2(g) \rightarrow N_2H_4(l)$$

18.98 Use the data in Appendix B to tell which of the following
RAN compounds are thermodynamically stable with respect to their constituent elements at 25 °C.
(a) $BaCO_3(s)$ (b) $HBr(g)$
(c) $N_2O(g)$ (d) $C_2H_4(g)$

18.99 Use the data in Appendix B to decide whether synthesis of the following compounds from their constituent elements is thermodynamically feasible at 25 °C.
(a) $C_6H_6(l)$ (b) $NO(g)$
(c) $PH_3(g)$ (d) $FeO(s)$

18.100 Use the values of $\Delta G°_f$ in Appendix B to calculate the stan-
RAN dard free-energy change for the synthesis of dichloroethane from ethylene and chlorine:

$$C_2H_4(g) + Cl_2(g) \rightarrow CH_2ClCH_2Cl(l)$$

Is it possible to synthesize dichloroethane from gaseous C_2H_4 and Cl_2, each at 25 °C and 1 atm pressure?

18.101 Use the values of $\Delta G°_f$ in Appendix B to calculate the standard free-energy change for the conversion of ammonia to hydrazine:

$$2\,NH_3(g) \rightarrow H_2(g) + N_2H_4(l)$$

Is it worth trying to find a catalyst for this reaction under standard-state conditions at 25 °C?

18.102 Ethanol is manufactured in industry by the hydration of ethylene:

$$CH_2 = CH_2(g) + H_2O(l) \rightarrow CH_3CH_2OH(l)$$

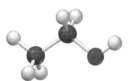

Ethylene   Ethanol

Using the data in Appendix B, calculate $\Delta G°$ and show that this reaction is spontaneous at 25 °C. Why does this reaction become nonspontaneous at higher temperatures? Estimate the temperature at which the reaction becomes nonspontaneous.

18.103 Sulfur dioxide in the effluent gases from coal-burning electric power plants is one of the principal causes of acid rain. One method for reducing SO_2 emissions involves partial conversion of SO_2 to H_2S, followed by catalytic conversion of the H_2S and the remaining SO_2 to elemental sulfur:

$$2\,H_2S(g) + SO_2(g) \rightarrow 3\,S(s) + 2\,H_2O(g)$$

Using the data in Appendix B, calculate $\Delta G°$, and show that this reaction is spontaneous at 25 °C. Why does this reaction become nonspontaneous at high temperatures? Estimate the temperature at which the reaction becomes nonspontaneous.

18.104 Consider the conversion of acetylene to benzene:

$$3\,C_2H_2(g) \rightarrow C_6H_6(l)$$

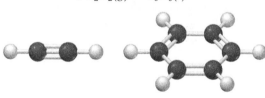

Acetylene   Benzene

Is a catalyst for this reaction possible? Is it possible to synthesize benzene from graphite and gaseous H_2 at 25 °C and 1 atm pressure?

18.105 Consider the conversion of 1,2-dichloroethane to vinyl chloride, the starting material for manufacturing poly(vinyl chloride) (PVC) plastics:

$$CH_2ClCH_2Cl(l) \rightarrow CH_2 = CHCl(g) + HCl(g)$$

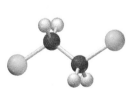

1,2-Dichloroethane   Vinyl chloride

Is this reaction spontaneous under standard-state conditions? Would it help to carry out the reaction in the presence of base to remove HCl? Explain. Is it possible to synthesize vinyl chloride from graphite, gaseous H_2, and gaseous Cl_2 at 25 °C and 1 atm pressure?

Free Energy for Reactions under Nonstandard-State Conditions (Section 18.10)

18.106   What is the relationship between the free-energy change under nonstandard-state conditions, ΔG, the free-energy change under standard-state conditions, $\Delta G°$, and the reaction quotient, Q?

18.107   Compare the values of ΔG and $\Delta G°$ when:
RAN   **(a)** $Q < 1$.   **(b)** $Q = 1$.   **(c)** $Q > 1$.

Does the thermodynamic tendency for the reaction to occur increase or decrease as Q increases?

18.108   Use the data in Appendix B to calculate ΔG for the decomRAN   position of nitrosyl chloride at 25 °C when the partial pressures are 2.00 atm of NOCl, 1.00×10^{-3} atm of NO, and 1.00×10^{-3} atm of Cl_2:

$$2 \, NO(g) + Cl_2(g) \rightarrow 2 \, NOCl(g)$$

Is the reaction spontaneous in the forward or the reverse direction under these conditions?

18.109   What is ΔG for the formation of solid uranium hexafluoride from uranium and fluorine at 25 °C when the partial pressure of F_2 is 0.045 atm? The standard free energy of formation of $UF_6(s)$ is -2068 kJ/mol.

$$U(s) + 3 \, F_2(g) \rightarrow UF_6(s)$$

Is the reaction spontaneous in the forward or the reverse direction under these conditions?

18.110   Sulfuric acid is produced in larger amounts by weight than
RAN   any other chemical. It is used in manufacturing fertilizers, oil refining, and hundreds of other processes. An intermediate step in the industrial process for the synthesis of H_2SO_4 is the catalytic oxidation of sulfur dioxide:

$$2 \, SO_2(g) + O_2(g) \rightarrow 2 \, SO_3(g) \quad \Delta G° = -141.8 \, kJ$$

Calculate ΔG at 25 °C, given the following sets of partial pressures.

(a) 100 atm SO_2, 100 atm O_2, 1.0 atm SO_3
(b) Each reactant and product at a partial pressure of 1.0 atm

18.111   Urea (NH_2CONH_2), an important nitrogen fertilizer, is
RAN   produced industrially by the reaction

$$2 \, NH_3(g) + CO_2(g) \rightarrow NH_2CONH_2(aq) + H_2O(l)$$

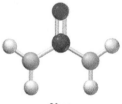

Urea

Given that $\Delta G° = -13.6$ kJ, calculate ΔG at 25 °C for the following sets of conditions.

(a) 10 atm NH_3, 10 atm CO_2, 1.0 M NH_2CONH_2
(b) 0.10 atm NH_3, 0.10 atm CO_2, 1.0 M NH_2CONH_2

Is the reaction spontaneous for the conditions in part (a) and/or part (b)?

Free Energy and Chemical Equilibrium (Section 18.11)

18.112   What is the relationship between the standard free-energy change, $\Delta G°$, for a reaction and the equilibrium constant, K? What is the sign of $\Delta G°$ when:
(a) $K > 1$?
(b) $K = 1$?
(c) $K < 1$?

18.113   Do you expect a large or small value of the equilibrium constant for a reaction with the following values of $\Delta G°$?
(a) $\Delta G°$ is positive.
(b) $\Delta G°$ is negative.

18.114   Given values of $\Delta G°_f$ at 25 °C for liquid ethanol (-174.9 kJ/mol) and gaseous ethanol (-167.9 kJ/mol), calculate the vapor pressure of ethanol at 25 °C.

18.115   At 25 °C, K_a for acid dissociation of aspirin ($C_9H_8O_4$) is 3.0×10^{-4}. Calculate $\Delta G°$ for the reaction $C_9H_8O_4(aq) + H_2O(l) \rightleftharpoons H_3O^+(aq) + C_9H_7O_4^-(aq)$.

18.116   If $\Delta G°_f$ for gaseous bromine is 3.14 kJ/mol at 25 °C, what is the vapor pressure of bromine at 25 °C?

18.117   Calculate the equilibrium partial pressure of iodine vapor
RAN   above solid iodine at 25 °C if $\Delta G°_f$ for gaseous iodine is 19.4 kJ/mol at 25 °C.

18.118   Ethylene oxide, C_2H_4O, is used to make antifreeze (ethylene glycol, $HOCH_2CH_2OH$). It is produced industrially by the catalyzed air oxidation of ethylene:

$$2 \, CH_2{=}CH_2(g) + O_2(g) \longrightarrow 2 \, CH_2{-}CH_2(g)$$

Ethylene oxide

Use the data in Appendix B to calculate $\Delta G°$ and K_p for this reaction at 25 °C.

18.119   The first step in the commercial production of titanium metal is the reaction of rutile (TiO_2) with chlorine and graphite:

$$TiO_2(s) + 2 \, Cl_2(g) + 2 \, C(s) \rightarrow TiCl_4(l) + 2 \, CO(g)$$

Use the data in Appendix B to calculate $\Delta G°$ and the equilibrium constant for this reaction at 25 °C.

18.120   Ammonium nitrate is dangerous because it decomposes
RAN   (sometimes explosively) when heated:

$$NH_4NO_3(s) \rightarrow N_2O(g) + 2 \, H_2O(g)$$

(a) Using the data in Appendix B, show that this reaction is spontaneous at 25 °C.
(b) How does $\Delta G°$ for the reaction change when the temperature is raised?
(c) Calculate the equilibrium constant K_p at 25 °C.
(d) Calculate ΔG for the reaction when the partial pressure of each gas is 30 atm.

18.121   At 25 °C, K_{sp} for $PbCrO_4$ is 2.8×10^{-13}. Calculate the
RAN   standard free-energy change at 25 °C for the reaction $PbCrO_4(s) \rightleftharpoons Pb^{2+}(aq) + CrO_4^{2-}(aq)$.

18.122 Use the data in Appendix B to calculate the equilibrium
RAN   pressure of CO_2 in a closed 1 L vessel that contains each of
the following samples:

(a) 15 g of $MgCO_3$ and 1.0 g of MgO at 25 °C

(b) 15 g of $MgCO_3$ and 1.0 g of MgO at 280 °C

(c) 30 g of $MgCO_3$ and 1.0 g of MgO at 280 °C

Assume that $\Delta H°$ and $\Delta S°$ are independent of temperature.

18.123 The equilibrium constant K_b for dissociation of aqueous
ammonia is 1.710×10^{-5} at 20 °C and 1.892×10^{-5} at
50 °C. What are the values of $\Delta H°$ and $\Delta S°$ for the reaction

$$NH_3(aq) + H_2O(l) \rightleftharpoons NH_4^+(aq) + OH^-(aq)?$$

18.124 Consider the Haber synthesis of gaseous
RAN NH_3 ($\Delta H°_f = -46.1$ kJ/mol; $\Delta G°_f = -16.5$ kJ/mol):

$$N_2(g) + 3 H_2(g) \rightarrow 2 NH_3(g)$$

(a) Use only these data to calculate $\Delta H°$ and $\Delta S°$ for the
reaction at 25 °C.

(b) Account for the sign of $\Delta S°$.

(c) Is the reaction spontaneous under standard-state conditions at 25 °C? Explain.

(d) What are the equilibrium constants K_p and K_c for the
reaction at 350 K? Assume that $\Delta H°$ and $\Delta S°$ are independent of temperature.

18.125 Consider the dissolution of AgBr in water at 25 °C:

$$AgBr(s) \rightleftharpoons Ag^+(aq) + Br^-(aq)$$

(a) Use the standard heats of formation and standard
molar entropies in Appendix B to calculate $\Delta G°$ for the
reaction at 25 °C.

(b) Calculate K_{sp} for AgBr at 25 °C.

(c) Calculate ΔG for the dissolution of AgBr at 25 °C when
$[Ag^+] = [Br^-] = 1.00 \times 10^{-5}$ M.
Is your result consistent with the relative values of Q and K_{sp}?

MULTICONCEPT PROBLEMS

18.126 *Trouton's rule* says that the ratio of the molar heat of vaporization of a liquid to its normal boiling point (in kelvin)
is approximately the same for all liquids: $\Delta H_{vap}/T_{bp} \approx$
88 J/(K • mol)

(a) Check the reliability of Trouton's rule for the liquids
listed in the following table.

(b) Explain why liquids tend to have the same value of
$\Delta H_{vap}/T_{bp}$.

(c) Which of the liquids in the table deviate(s) from Trouton's
rule? Explain.

Liquid	bp (°C)	ΔH_{vap}(kJ/mol)
Ammonia	−33.3	23.4
Benzene	80.1	30.8
Carbon tetrachloride	76.8	29.8
Chloroform	61.1	29.2
Mercury	356.6	59.11

18.127 The temperature dependence of the equilibrium constant is
RAN   given by the equation

$$\ln K = \frac{-\Delta H°}{R}\left(\frac{1}{T}\right) + \frac{\Delta S°}{R}$$

where $\Delta H°$ and $\Delta S°$ are assumed to be independent of
temperature.

(a) Derive this equation from equations given in this chapter.

(b) Explain how this equation can be used to determine
experimental values of $\Delta H°$ and $\Delta S°$ from values of K
at several different temperatures.

(c) Use this equation to predict the sign of $\Delta H°$ for a reaction
whose equilibrium constant increases with increasing
temperature. Is the reaction endothermic or exothermic?
Is your prediction in accord with Le Châtelier's principle?

18.128 The normal boiling point of bromine is 58.8 °C, and the
standard entropies of the liquid and vapor are $S°[Br_2(l)] =$
152.2 J/(K • mol); $S°[Br_2(g)] = 245.4$ J/(K • mol). At what
temperature does bromine have a vapor pressure of
227 mm Hg?

18.129 The molar solubility of lead iodide is 1.45×10^{-3} M at
20 °C and 6.85×10^{-3} M at 80 °C. What are the values of
$\Delta H°$ and $\Delta S°$ for dissolution of PbI_2?

$$PbI_2(s) \rightarrow Pb^{2+}(aq) + 2 I^-(aq)$$

Assume that $\Delta H°$ and $\Delta S°$ are independent of temperature.

18.130 Use the data in Appendix B to calculate the equilibrium
constant K for the following reaction at 80 °C:

$$2 Br^-(aq) + Cl_2(g) \rightarrow Br_2(l) + 2 Cl^-(aq)$$

Assume that $\Delta H°$ and $\Delta S°$ are independent of temperature.

18.131 Use the data from Appendix B to determine the normal
RAN   boiling point of carbon disulfide (CS_2).

18.132 A humidity sensor consists of a cardboard square that is
colored blue in dry weather and red in humid weather. The
color change is due to the reaction:

$$CoCl_2(s) + 6 H_2O(g) \rightleftharpoons [Co(H_2O)_6]Cl_2(s)$$
$$\text{Blue} \qquad\qquad\qquad\qquad \text{Red}$$

For this reaction at 25 °C, $\Delta H° = -352$ kJ/mol and $\Delta S° =$
−899 J/(K • mol). Assuming that $\Delta H°$ and $\Delta S°$ are independent of temperature, what is the vapor pressure of
water (in mm Hg) at equilibrium for the above reaction at
35 °C on a hot summer day?

18.133 The following reaction, sometimes used in the laboratory to generate small quantities of oxygen gas, has
$\Delta G° = -224.4$ kJ/mol at 25 °C:

$$2 KClO_3(s) \rightarrow 2 KCl(s) + 3 O_2(g)$$

Use the following additional data at 25 °C to calculate the standard molar entropy $S°$ of O_2 at 25 °C: $\Delta H°_f(KClO_3)=$ -397.7 kJ/mol, $\Delta H°_f(KCl) = -436.5$ kJ/mol, $S°(KClO_3) =$ 143.1 J/(K • mol), and $S°(KCl) = 82.6$ J/(K • mol).

18.134 Consider the equilibrium $N_2O_4(g) \rightleftharpoons 2\,NO_2(g)$.

(a) Use the thermodynamic data in Appendix B to determine the temperature at which an equilibrium mixture with a total pressure of 1.00 atm will contain twice as much NO_2 as N_2O_4. Assume that $\Delta H°$ and $\Delta S°$ are independent of temperature.

(b) At what temperature will an equilibrium mixture with a total pressure of 1.00 atm contain equal amounts of NO_2 and N_2O_4?

18.135 Sorbitol ($C_6H_{14}O_6$), a substance used as a sweetener in foods, is prepared by the reaction of glucose with hydrogen in the presence of a catalyst:

$$C_6H_{12}O(aq) + H_2(g) \rightarrow C_6H_{14}O_6(aq)$$

Which of the following quantities are affected by the catalyst?

(a) Rate of the forward reaction

(b) Rate of the reverse reaction

(c) Spontaneity of the reaction

(d) $\Delta H°$

(e) $\Delta S°$

(f) $\Delta G°$

(g) The equilibrium constant

(h) Time required to reach equilibrium

18.136 A mixture of 14.0 g of N_2 and 3.024 g of H_2 in a 5.00 L container is heated to 400 °C. Use the data in Appendix B to calculate the molar concentrations of N_2, H_2, and NH_3 at equilibrium. Assume that $\Delta H°$ and $\Delta S°$ are independent of temperature, and remember that the standard state of a gas is defined in terms of pressure.

$$N_2(g) + 3\,H_2(g) \rightleftharpoons 2\,NH_3(g)$$

18.137 One step in the commercial synthesis of sulfuric acid is the catalytic oxidation of sulfur dioxide:

$$2\,SO_2(g) + O_2(g) \rightleftharpoons 2\,SO_3(g)$$

(a) A mixture of 192 g of SO_2, 48.0 g of O_2, and a V_2O_5 catalyst is heated to 800 K in a 15.0 L vessel. Use the data in Appendix B to calculate the partial pressures of SO_3, SO_2, and O_2 at equilibrium. Assume that $\Delta H°$ and $\Delta S°$ are independent of temperature.

(b) Does the percent yield of SO_3 increase or decrease on raising the temperature from 800 K to 1000 K? Explain.

(c) Does the total pressure increase or decrease on raising the temperature from 800 K to 1000 K? Calculate the total pressure (in atm) at 1000 K.

18.138 The lead storage battery uses the reaction

$$Pb(s) + PbO_2(s) + 2\,H^+(aq) + 2\,HSO_4^-(aq) \rightarrow$$
$$2\,PbSO_4(s) + 2\,H_2O(l)$$

(a) Use the data in Appendix B to calculate $\Delta G°$ for this reaction.

(b) Calculate ΔG for this reaction on a cold winter's day (10 °F) in a battery that has run down to the point where the sulfuric acid concentration is only 0.100 M.

18.139 What is the molar solubility of $CaCO_3$ at 50 °C in a solution prepared by dissolving 1.000 L of CO_2 gas (at 20 °C and 731 mm Hg) and 3.335 g of solid $Ca(OH)_2$ in enough water to make 500.0 mL of solution at 50 °C? Is the solubility of $CaCO_3$ at 50 °C larger or smaller than at 25 °C? Explain. You may assume that $\Delta H°$ and $\Delta S°$ are independent of temperature.
RAN

18.140 A 1.00 L volume of gaseous ammonia at 25.0 °C and 744 mm Hg was dissolved in enough water to make 500.0 mL of aqueous ammonia at 2.0 °C. What is K_b for NH_3 at 2.0 °C, and what is the pH of the solution? Assume that $\Delta H°$ and $\Delta S°$ are independent of temperature.

18.141 Consider the unbalanced equation: $I_2(s) \rightarrow I^-(aq) + IO_3^-(aq)$.
RAN

(a) Balance the equation for this reaction in basic solution.

(b) Use the data in Appendix B and $\Delta G°_f$ for $IO_3^-(aq) =$ -128.0 kJ/mol to calculate $\Delta G°$ for the reaction at 25 °C.

(c) Is the reaction spontaneous or nonspontaneous under standard-state conditions?

(d) What pH is required for the reaction to be at equilibrium at 25 °C when $[I^-] = 0.10$ M and $[IO_3^-] = 0.50$ M?

18.142 A mixture of NO_2 and N_2O_4, each at an initial partial pressure of 1.00 atm and a temperature of 100 °C, is allowed to react.

(a) Use the data in Appendix B to calculate the partial pressure of each gas at equilibrium. Assume that $\Delta H°$ and $\Delta S°$ are independent of temperature.

$$N_2O_4(g) \rightleftharpoons 2\,NO_2(g)$$

(b) What is the shape of the N_2O_4 molecule, and what hybrid orbitals do the N atoms use in bonding?

chapter 19

Electrochemistry

Contents

The Toyota Mirai is a hydrogen-powered vehicle that generates electricity using a zero-emission fuel cell. A fuel cell is similar to a battery and converts chemical energy into electrical energy. Toyota plans to increase production of fuel cell vehicles to 30,000 per year by 2020.

How do hydrogen fuel cells work?

The answer to this question can be found on page 854 in the INQUIRY ?

B atteries are everywhere. Pause for a moment and count the number of batteries you have used today. Batteries provide the electric current to start our automobiles and power products such as cell phones, watches, laptop computers, calculators, and medical devices such as heart pacemakers or insulin pumps. In hybrid and all-electric vehicles, batteries power an electric motor that supplements or replaces a gasoline engine. A battery is an *electrochemical cell,* a device for converting chemical energy to electrical energy. A battery takes the energy released by a spontaneous chemical reaction and uses it to produce electricity.

Electrochemistry, the area of chemistry concerned with the interconversion of chemical and electrical energy, is enormously important in modern science and technology not only because of batteries but also because it makes possible the manufacture of essential industrial chemicals and materials. Sodium hydroxide, for example, which is used in the manufacture of paper, textiles, soaps, and detergents, is produced by passing an electric current through an aqueous solution of sodium chloride. Chlorine, essential to the manufacture of plastics such as poly(vinyl chloride) (PVC), is obtained in the same process. Aluminum metal is also produced in an electrochemical process, as is pure copper for use in electrical wiring.

In this chapter, we'll look at the principles involved in the design and operation of electrochemical cells. In addition, we'll explore some important connections between electrochemistry and thermodynamics.

19.1 BALANCING REDOX REACTIONS BY THE HALF-REACTION METHOD

Oxidation–reduction (redox) reactions are processes that involve a transfer of electrons from one substance to another and are the basis for electrochemical cells. Recall from Section 4.10 that an oxidation is defined as a loss of one or more electrons by a substance and a reduction is the gain of one or more electrons by a substance. We'll describe electrochemical cells in the next section, but first let's review redox reactions and learn a method for balancing them.

As an example, let's look at the reaction of aqueous potassium dichromate ($K_2Cr_2O_7$) with aqueous NaCl. The reaction occurs in acidic solution according to the unbalanced net ionic equation

$$Cr_2O_7{}^{2-}(aq) + Cl^-(aq) \longrightarrow Cr^{3+}(aq) + Cl_2(aq) \quad \textbf{Unbalanced}$$

REMEMBER . . .
An **oxidation number** indicates whether an atom is neutral, electron-rich, or electron-poor and is a convenient number for keeping track of electrons during redox reactions. We covered the rules for assigning oxidation numbers to atoms in Section 4.10.

Recall from Section 4.10 that **oxidation numbers** are used to determine which atoms get oxidized and reduced in a redox reaction. Before continuing, it's a good idea to review the guidelines for assigning oxidation numbers as discussed in Section 4.10. An increase in the oxidation number of an element indicates it has been oxidized, while a decrease in the oxidation number of an element indicates reduction. In the reaction, the chloride ion is oxidized from -1 to 0, and the chromium atom is reduced from $+6$ to $+3$.

Undergoes reduction
from +6 to +3

Undergoes oxidation
from −1 to 0

$$Cr_2O_7{}^{2-}(aq) + Cl^-(aq) \longrightarrow Cr^{3+}(aq) + Cl_2(aq)$$

+6   −1   +3   0

BIG IDEA Question 1

Go to
eText

Which element gets oxidized in the following reaction?

$$TiO_2(s) + C(s) + 2\,Cl_2(g) \longrightarrow$$
$$TiCl_4(l) + CO_2(g)$$

Simple redox reactions can often be balanced by the trial-and-error method, but many reactions are so complex that we need to use a more systematic approach. Although a number of different methods are available for balancing equations, we'll look only

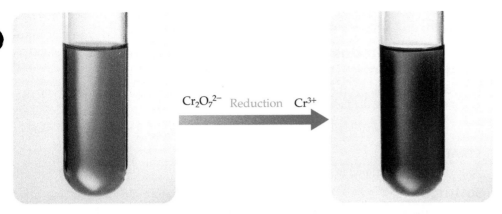

$Cr_2O_7^{2-}$ Reduction Cr^{3+}

▲ The orange **dichromate ion** is reduced by addition of Cl^- to give the green Cr^{3+} **ion.**

at the **half-reaction method,** which focuses on the transfer of electrons. The key to the half-reaction method is to realize that a redox reaction can be broken into two parts, or **half-reactions.** One half-reaction describes the oxidation part of the process, and the other describes the reduction part. Each half is balanced separately, and the two halves are then added to obtain the final equation. **FIGURE 19.1** summarizes the six steps for balancing a redox reaction in acidic solution by the half-reaction method. Worked Example 19.1 demonstrates how to balance a redox reaction in acidic solution. Worked Example 19.2 shows how to use the method for balancing an equation for a reaction that takes place in basic solution. The procedure is the same as that used for balancing a reaction in acidic solution, but OH^- ions are added as a final step to neutralize any H^+ ions that appear in the equation. This simply reflects the fact that basic solutions contain negligibly small amounts of H^+ but relatively large amounts of OH^-.

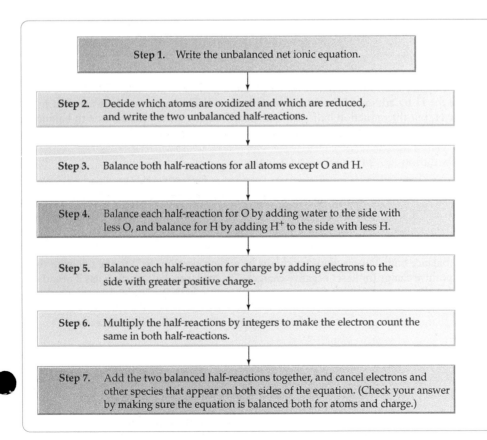

Step 1. Write the unbalanced net ionic equation.

Step 2. Decide which atoms are oxidized and which are reduced, and write the two unbalanced half-reactions.

Step 3. Balance both half-reactions for all atoms except O and H.

Step 4. Balance each half-reaction for O by adding water to the side with less O, and balance for H by adding H^+ to the side with less H.

Step 5. Balance each half-reaction for charge by adding electrons to the side with greater positive charge.

Step 6. Multiply the half-reactions by integers to make the electron count the same in both half-reactions.

Step 7. Add the two balanced half-reactions together, and cancel electrons and other species that appear on both sides of the equation. (Check your answer by making sure the equation is balanced both for atoms and charge.)

◀ **FIGURE 19.1**

Steps for balancing redox equations for reactions in acidic solution using the half-reaction method.

WORKED EXAMPLE 19.1

Balancing a Redox Equation for a Reaction in Acidic Solution

Balance the equation for the reaction between the dichromate ion and chloride ion in acidic solution using the half-reaction method.

$$Cr_2O_7^{2-}(aq) + Cl^-(aq) \longrightarrow Cr^{3+}(aq) + Cl_2(aq) \quad \text{Unbalanced}$$

STRATEGY

Follow the steps outlined in Figure 19.1.

SOLUTION

Step 1. Write the unbalanced net ionic equation.

$$Cr_2O_7^{2-}(aq) + Cl^-(aq) \longrightarrow Cr^{3+}(aq) + Cl_2(aq)$$

Step 2. Decide which atoms are oxidized and which are reduced, and write the two unbalanced half-reactions. As shown earlier in this section, the chloride ion is oxidized from -1 to 0, and the chromium atom is reduced from $+6$ to $+3$. Thus, we can write two unbalanced half-reactions that show the separate parts:

$$\text{Oxidation half-reaction} \quad Cl^-(aq) \longrightarrow Cl_2(aq)$$
$$\text{Reduction half-reaction} \quad Cr_2O_7^{2-}(aq) \longrightarrow Cr^{3+}(aq)$$

Step 3. Balance both half-reactions for all atoms except O and H. The oxidation half-reaction needs a coefficient of 2 before the Cl^-, and the reduction half-reaction needs a coefficient of 2 before the Cr^{3+}.

Add this coefficient to balance for Cl.

$$\text{Oxidation} \quad 2\,Cl^-(aq) \longrightarrow Cl_2(aq)$$

Add this coefficient to balance for Cr.

$$\text{Reduction} \quad Cr_2O_7^{2-}(aq) \longrightarrow 2\,Cr^{3+}(aq)$$

Step 4. Balance each half-reaction for O by adding H_2O to the side with less O, and balance for H by adding H^+ to the side with less H. The oxidation half-reaction has no O or H, but the reduction half-reaction needs 7 H_2O on the product side to balance for O and then 14 H^+ on the reactant side to balance for H:

$$\text{Oxidation} \quad 2\,Cl^-(aq) \longrightarrow Cl_2(aq)$$

Then add 14 H^+ to balance for H.

First, add 7 H_2O to balance for O.

$$\text{Reduction} \quad Cr_2O_7^{2-}(aq) + 14\,H^+(aq) \longrightarrow 2\,Cr^{3+}(aq) + 7\,H_2O(l)$$

Step 5. Balance both half-reactions for charge by adding electrons (e^-) to the side with the greater positive charge (or lesser negative charge). The oxidation half-reaction has 2 negative charges on the reactant side (2 Cl^-) and must therefore have 2 e^- added to the product side. The reduction half-reaction has a net of 12 positive charges on the reactant side and 6 positive charges on the product side and so must have 6 e^- added to the reactant side:

Add these electrons to balance for charge.

$$\text{Oxidation} \quad 2\,Cl^-(aq) \longrightarrow Cl_2(aq) + 2\,e^-$$

Add these electrons to balance for charge.

$$\text{Reduction} \quad Cr_2O_7^{2-}(aq) + 14\,H^+(aq) + 6\,e^- \longrightarrow 2\,Cr^{3+}(aq) + 7\,H_2O(l)$$

Step 6. Multiply the half-reactions by integers to make the electron count the same in both half-reactions. The number of electrons released in the oxidation half-reaction must be the same as the number consumed in the reduction half-reaction. Because the reduction half-reaction has 6 e^- but the oxidation half-reaction has only 2 e^-, the oxidation half-reaction must be multiplied by 3:

Multiply by this coefficient to equalize the numbers of electrons in the two half-reactions.

Oxidation $3 \times [2\, Cl^-(aq) \longrightarrow Cl_2(aq) + 2\, e^-]$

or $6\, Cl^-(aq) \longrightarrow 3\, Cl_2(aq) + 6\, e^-$

Reduction $Cr_2O_7{}^{2-}(aq) + 14\, H^+(aq) + 6\, e^- \longrightarrow 2\, Cr^{3+}(aq) + 7\, H_2O(l)$

Step 7. Add the two half-reactions together, and cancel electrons and other species that occur on both sides of the equation.

$$6\, Cl^-(aq) \longrightarrow 3\, Cl_2(aq) + \cancel{6\, e^-}$$
$$\underline{Cr_2O_7{}^{2-}(aq) + 14\, H^+(aq) + \cancel{6\, e^-} \longrightarrow 2\, Cr^{3+}(aq) + 7\, H_2O(l)}$$
$$Cr_2O_7{}^{2-}(aq) + 14\, H^+(aq) + 6\, Cl^-(aq) \longrightarrow 3\, Cl_2(aq) + 2\, Cr^{3+}(aq) + 7\, H_2O(l)$$

Check the answer to make sure it is balanced for both atoms and charge.

$$Cr_2O_7{}^{2-}(aq) + 14\, H^+(aq) + 6\, Cl^-(aq) \longrightarrow 3\, Cl_2(aq) + 2\, Cr^{3+}(aq) + 7\, H_2O(l)$$

Charge: $(-2) + (+14) + (6 \times -1) = +6$ Charge: $(2 \times +3) = +6$

Cr = 2	Cr = 2
O = 7	O = 7
H = 14	H = 14
Cl = 6	Cl = 6

▶ **PRACTICE 19.1** Balance the following net ionic equation by the half-reaction method. The reaction takes place in acidic solution.

$$NO_3{}^-(aq) + Cu(s) \longrightarrow NO(g) + Cu^{2+}(aq) \qquad \text{Unbalanced}$$

▶ **APPLY 19.2** Balance the following net ionic equation by the half-reaction method. The reaction takes place in acidic solution.

$$I^-(aq) + IO_3{}^-(aq) \longrightarrow I_3{}^-(aq) \qquad \text{Unbalanced}$$

All **WORKED EXAMPLES** with this icon [Go to eText] have an interactive video in the eText.

WORKED EXAMPLE 19.2

Balancing a Redox Equation for a Reaction in Basic Solution

Aqueous sodium hypochlorite (NaOCl; household bleach) is a strong oxidizing agent that reacts with $[Cr(OH)_4{}^-]$ in basic solution to yield $(CrO_4{}^{2-})$ and chloride ion. The net ionic equation is

$$ClO^-(aq) + Cr(OH)_4{}^-(aq) \longrightarrow CrO_4{}^{2-}(aq) + Cl^-(aq) \qquad \text{Unbalanced}$$

Balance the equation using the half-reaction method.

STRATEGY

Follow the steps outlined in Figure 19.1, and add OH^- ions as a final step to neutralize any H^+ ions that appear in the equation.

SOLUTION

Step 1. Write the unbalanced net ionic equation.

$$ClO^-(aq) + Cr(OH)_4{}^-(aq) \longrightarrow CrO_4{}^{2-}(aq) + Cl^-(aq) \qquad \text{Unbalanced}$$

continued on next page

Step 2. Decide which atoms are oxidized and which are reduced, and write the two unbalanced half-reactions. The unbalanced net ionic equation shows that chromium is oxidized (from +3 to +6) and chlorine is reduced (from +1 to −1) Thus, we can write the following half-reactions:

$$\text{Oxidation half-reaction} \quad Cr(OH)_4^-(aq) \longrightarrow CrO_4^{2-}(aq)$$
$$\text{Reduction half-reaction} \quad ClO^-(aq) \longrightarrow Cl^-(aq)$$

Step 3. Balance both half-reactions for all atoms except O and H. The half-reactions are already balanced for atoms other than O and H.

Step 4. Balance each half-reaction for O by adding H_2O to the side with less O, and balance for H by adding H^+ to the side with less H.

$$\text{Oxidation} \quad Cr(OH)_4^-(aq) \longrightarrow CrO_4^{2-}(aq) + 4\,H^+(aq)$$
$$\text{Reduction} \quad ClO^-(aq) + 2\,H^+(aq) \longrightarrow Cl^-(aq) + H_2O(l)$$

Step 5. Balance both half-reactions for charge by adding electrons (e^-) to the side with the greater positive charge.

$$\text{Oxidation} \quad Cr(OH)_4^-(aq) \longrightarrow CrO_4^{2-}(aq) + 4\,H^+(aq) + 3\,e^-$$
$$\text{Reduction} \quad ClO^-(aq) + 2\,H^+(aq) + 2\,e^- \longrightarrow Cl^-(aq) + H_2O(l)$$

Step 6. Multiply the half-reactions by integers to make the electron count the same in both half-reactions. The oxidation half-reaction must be multiplied by 2, and the reduction half-reaction must be multiplied by 3 to give 6 e^- in both:

$$\text{Oxidation} \quad 2 \times [Cr(OH)]_4^-(aq) \longrightarrow CrO_4^{2-}(aq) + 4\,H^+(aq) + 3\,e^-]$$
$$\text{or} \quad 2\,Cr(OH)_4^-(aq) \longrightarrow 2\,CrO_4^{2-}(aq) + 8\,H^+(aq) + 6\,e^-$$
$$\text{Reduction} \quad 3 \times [ClO^-(aq) + 2\,H^+(aq) + 2\,e^- \longrightarrow Cl^-(aq) + H_2O(l)]$$
$$\text{or} \quad 3\,ClO^-(aq) + 6\,H^+(aq) + 6\,e^- \longrightarrow 3\,Cl^-(aq) + 3\,H_2O(l)$$

Step 7. Add the two half-reactions together, and cancel electrons and other species that occur on both sides of the equation. The electrons must always cancel.

$$2\,Cr(OH)_4^-(aq) \longrightarrow 2\,CrO_4^{2-}(aq) + 8\,H^+(aq) + 6\,e^-$$
$$\underline{3\,ClO^-(aq) + 6\,H^+(aq) + 6\,e^- \longrightarrow 3\,Cl^-(aq) + 3\,H_2O(l)}$$
$$2\,Cr(OH)_4^-(aq) + 3\,ClO^-(aq) + 6\,H^+(aq) \longrightarrow$$
$$2\,CrO_4^{2-}(aq) + 3\,Cl^-(aq) + 3\,H_2O(l) + 8\,H^+(aq)$$

Now, cancel the other species that appear on both sides of the equation. Since there are 8 H^+ in the products and 6 H^+ in the reactants, the result after cancellation is 2 H^+ in the products.

$$2\,Cr(OH)_4^-(aq) + 3\,ClO^-(aq) \longrightarrow 2\,CrO_4^{2-}(aq) + 3\,Cl^-(aq) + 3\,H_2O(l) + 2\,H^+(aq)$$

Finally, since we know that the reaction takes place in basic solution, we must add 2 OH^- ions to *both* sides of the equation to neutralize the 2 H^+ ions on the right, giving 2 additional H_2O.

$$2\,OH^-(aq) + 2\,Cr(OH)_4^-(aq) + 3\,ClO^-(aq) \longrightarrow 2\,CrO_4^{2-}(aq) + 3\,Cl^-(aq)$$
$$+ 3\,H_2O(l) + 2\,H^+(aq) + 2OH^-(aq)$$

$$2\,OH^-(aq) + 2\,Cr(OH)_4^-(aq) + 3\,ClO^-(aq) \longrightarrow 2\,CrO_4^{2-}(aq) + 3\,Cl^-(aq)$$
$$+ 3\,H_2O(l) + 2\,H_2O(l)$$

The final net ionic equation, balanced for both atoms and charge, is

$$2\,Cr(OH)_4^-(aq) + 3\,ClO^-(aq) + 2\,OH^-(aq) \longrightarrow 2\,CrO_4^{2-}(aq) + 3\,Cl^-(aq) + 5\,H_2O(l)$$

Charge: $(2 \times -1) + (3 \times -1) + (2 \times -1) = -7$ Charge: $(2 \times -2) + (3 \times -1) = -7$

Cr = 2	Cr = 2
Cl = 3	Cl = 3
O = 13	O = 13
H = 10	H = 10

▶ **PRACTICE 19.3** Balance the following equation by the half-reaction method. The reaction takes place in basic solution.

$$I^-(aq) + MnO_4^-(aq) \longrightarrow I_2(aq) + MnO_2(s) \qquad \text{Unbalanced}$$

▶ **APPLY 19.4** Balance the following equation by the half-reaction method. The reaction takes place in basic solution.

$$Fe(OH)_2(s) + H_2O(l) + O_2(g) \longrightarrow Fe(OH)_3(s) \qquad \text{Unbalanced}$$

All **PRACTICE** and **APPLY** problems are interactive in the eText.

19.2 GALVANIC CELLS

Electrochemical cells are of two types: **galvanic cells** (also called *voltaic cells* and commonly known as batteries) and **electrolytic cells.** The names "galvanic" and "voltaic" honor the Italian scientists Luigi Galvani (1737–1798) and Alessandro Volta (1745–1827), who conducted pioneering work in the field of electrochemistry. In a galvanic cell, a spontaneous chemical reaction generates an electric current. In an electrolytic cell, an electric current drives a nonspontaneous reaction. The two types are therefore the reverse of each other. We'll describe galvanic cells in this section and examine electrolytic cells later.

Let's look at a spontaneous redox reaction, the type of reaction required for building a galvanic cell. If you immerse a strip of zinc metal in an aqueous solution of copper sulfate, you will observe dark-colored solid deposits on the surface of the zinc and that the blue color characteristic of the Cu^{2+} ion slowly disappears from the solution (**FIGURE 19.2**).

Chemical analysis shows that the dark-colored deposit consists of tiny particles of copper metal and that the solution now contains zinc ions. Therefore, the reaction is

$$Zn(s) + Cu^{2+}(aq) \longrightarrow Zn^{2+}(aq) + Cu(s)$$

This is a redox reaction in which Zn is oxidized to Zn^{2+} and Cu^{2+} is reduced to Cu. Remember that an *oxidation* is a loss of electrons (an increase in oxidation number) and a *reduction* is a gain of electrons (a decrease in oxidation number).

We can represent the oxidation and reduction aspects of the reaction by separating the overall process into *half-reactions,* one representing the oxidation reaction and the other representing the reduction:

$$\text{Oxidation half-reaction:} \quad Zn(s) \longrightarrow Zn^{2+}(aq) + 2\ e^-$$
$$\text{Reduction half-reaction:} \quad Cu^{2+}(aq) + 2\ e^- \longrightarrow Cu(s)$$

We say that Cu^{2+} is the *oxidizing agent* because, in gaining electrons from Zn, it causes the oxidation of Zn to Zn^{2+}. Similarly, we say that Zn is the *reducing agent* because, in losing electrons to Cu^{2+}, it causes the reduction of Cu^{2+} to Cu.

If the reaction is carried out as shown in Figure 19.2, electrons are transferred directly from Zn to Cu^{2+} and the enthalpy of reaction is lost to the surroundings as heat. If the reaction is carried out using the electrochemical cell depicted in **FIGURE 19.3,** however, some of the chemical energy released by the reaction is converted to electrical energy, which can be used to light a light bulb or run an electric motor.

The apparatus shown in Figure 19.3 is a type of galvanic cell called a *Daniell cell,* after John Frederick Daniell (1790–1845), the English chemist who first constructed it in 1836. The key to generating electricity is the presence of two separate *half-cells,* a beaker containing a strip of zinc that dips into an aqueous solution of zinc sulfate and a second beaker containing a strip of copper that dips into aqueous copper sulfate. The strips of zinc and copper are called **electrodes** and are connected by an electrically conducting wire. In addition, the two solutions are connected by a **salt bridge,** a U-shaped tube that contains a gel permeated with a solution of an inert electrolyte, such as sodium sulfate, Na_2SO_4. The ions of the inert electrolyte do not react with the other ions in the solutions, and they are not oxidized or reduced at the electrodes.

▶ **FIGURE 19.2**

The redox reaction of zinc metal with aqueous Cu^{2+} ions.

▶ **Figure It Out**

Which species is reduced: Zn(*s*) or Cu^{2+} (*aq*)?

Answer: Cu^{2+}(*aq*) is reduced to Cu(*s*). Cu(*s*) deposits on the surface of the Zn metal.

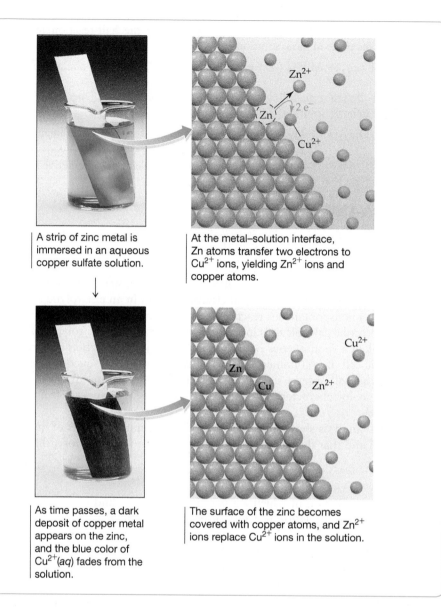

A strip of zinc metal is immersed in an aqueous copper sulfate solution.

At the metal–solution interface, Zn atoms transfer two electrons to Cu^{2+} ions, yielding Zn^{2+} ions and copper atoms.

As time passes, a dark deposit of copper metal appears on the zinc, and the blue color of Cu^{2+}(*aq*) fades from the solution.

The surface of the zinc becomes covered with copper atoms, and Zn^{2+} ions replace Cu^{2+} ions in the solution.

The reaction that occurs in the Daniell cell is the same one that occurs when Zn reacts directly with Cu^{2+}, but now, because the Zn metal and Cu^{2+} ions are in separate compartments, the electrons are transferred from Zn to Cu^{2+} *through the wire*. Consequently, the oxidation and reduction half-reactions occur at separate electrodes and an electric current flows through the wire. Electrons are not transferred through the solution because the metal wire is a much better conductor of electrons than is water. In fact, free electrons react rapidly with water and are therefore unstable in aqueous solutions.

The electrode at which oxidation takes place is called the **anode** (the zinc strip in this example), and the electrode at which reduction takes place is called the **cathode** (the copper strip). The anode and cathode half-reactions must add to give the overall cell reaction:

Anode (oxidation) half-reaction:	$Zn(s) \longrightarrow Zn^{2+}(aq) + 2\ e^-$
Cathode (reduction) half-reaction:	$Cu^{2+}(aq) + 2\ e^- \longrightarrow Cu(s)$
Overall cell reaction:	$Zn(s) + Cu^{2+}(aq) \longrightarrow Zn^{2+}(aq) + Cu(s)$

The salt bridge is necessary to complete the electrical circuit. Without it, the solution in the anode compartment would become positively charged as Zn^{2+} ions appeared

(a) A galvanic cell

The negative particles (electrons in the wire and anions in solution) travel around the circuit in the same clockwise direction.

The resulting electric current can be used to power a light bulb.

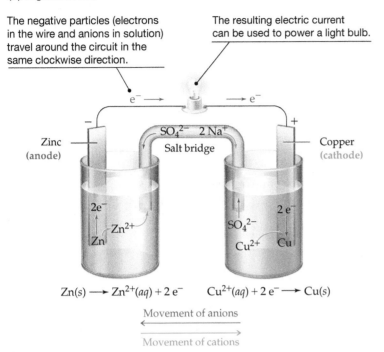

Zinc (anode)

SO_4^{2-} 2 Na$^+$
Salt bridge

Copper (cathode)

$2e^-$

Zn^{2+}
Zn

SO_4^{2-}

$2\,e^-$

Cu^{2+} Cu

$$Zn(s) \longrightarrow Zn^{2+}(aq) + 2\,e^- \qquad Cu^{2+}(aq) + 2\,e^- \longrightarrow Cu(s)$$

Movement of anions

Movement of cations

(b) An operating Daniell cell

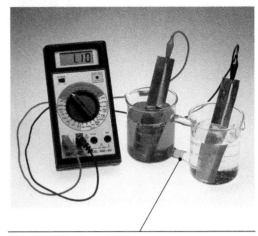

The salt bridge in part (**a**) is replaced by a porous glass disk that allows ions to flow between the anode and cathode compartments but prevents bulk mixing. If Cu^{2+} crossed the membrane and came into direct contact with zinc, then electrons would be transferred directly and would not flow through the wire.

▲ **FIGURE 19.3**

A Daniell cell, an example of a galvanic cell. (a) A Daniell cell uses the oxidation of zinc metal to **Zn^{2+}** ions and the reduction of **Cu^{2+}** ions to copper metal. (b) In this operating Daniell cell, the light bulb in part (a) is replaced with a digital voltmeter (more about that in Section 19.4).

▲ **Figure It Out**

(a) Do electrons flow toward the anode or the cathode? (b) Which electrode will lose mass and which will gain mass in the Daniell cell?

Answer: (a) Electrons flow toward the cathode where they are needed for reduction of Cu^{2+}. (b) The mass of the copper electrode will increase as Cu^{2+}(aq) is reduced to Cu(s) and the mass of the zinc electrode will decrease as Zn(s) is oxidized to Zn^{2+}(aq).

in it, and the solution in the cathode compartment would become negatively charged as Cu^{2+} ions were removed from it. Because of the charge imbalance, the electrode reactions would quickly come to a halt and electron flow through the wire would cease.

With the salt bridge in place, electrical neutrality is maintained in both compartments by a flow of ions. Anions (in this case SO$_4^{2-}$) flow through the salt bridge from the cathode compartment to the anode compartment, and cations migrate through the salt bridge from the anode compartment to the cathode compartment. For the cell shown in Figure 19.3, Na$^+$ ions move out of the salt bridge into the cathode compartment and Zn^{2+} ions move into the salt bridge from the anode compartment. The anode and cathode get their names from the direction of ion flow between the two compartments: *An*ions move toward the *an*ode, and *cat*ions move toward the *cat*hode in a galvanic cell.

The electrodes of commercial galvanic cells are generally labeled with plus (+) and minus (−) signs, although the magnitude of the actual charge on the electrodes is infinitesimally small and the sign of the charge associated with each electrode depends on the point of view. From the perspective of the wire, the anode looks negative because a stream of negatively charged electrons comes from it. From the perspective of the solution, however, the anode looks positive because a stream of positively charged Zn^{2+} ions move from it. Because galvanic cells are used to supply electric current to an external circuit, it makes sense to adopt the perspective of the wire. Consequently, we regard the anode as the negative (−) electrode and the cathode as the positive (+) electrode. Thus, electrons move through the external circuit from the negative electrode, where they are produced by the anode half-reaction, to the positive electrode, where they are consumed by the cathode half-reaction.

Go to
eText

BIG IDEA Question 2

In a galvanic cell, electrons flow toward the electrode with a _____ sign, which is called the _____. At this electrode, the _____ half-reaction occurs.

Anode: $\left\{\begin{array}{l}\text{Is where oxidation occurs}\\ \text{Is where electrons are produced}\\ \text{Is what anions migrate toward}\\ \text{Has a negative sign}\end{array}\right.$ Cathode: $\left\{\begin{array}{l}\text{Is where reduction occurs}\\ \text{Is where electrons are consumed}\\ \text{Is what cations migrate toward}\\ \text{Has a positive sign}\end{array}\right.$

In some galvanic cells, one (or both) of the half-cell reactions occurs at an inert electrode. Take, for instance, a cell that uses the reaction

$$Fe(s) + 2\ Fe^{3+}(aq) \longrightarrow 3\ Fe^{2+}(aq)$$

In this reaction, iron metal is oxidized to iron(II) ions and iron(III) ions are reduced to iron(II) ions. Therefore, the cell half-reactions are

Anode (oxidation):	$Fe(s) \longrightarrow Fe^{2+}(aq) + 2\ e^-$
Cathode (reduction):	$2 \times [Fe^{3+}(aq) + e^- \longrightarrow Fe^{2+}(aq)]$
Overall cell reaction:	$Fe(s) + 2\ Fe^{3+}(aq) \longrightarrow 3\ Fe^{2+}(aq)$

Notice that the cathode half-reaction is multiplied by a factor of 2 so that the two half-reactions will add to give the overall cell reaction. Whenever half-reactions are added, the electrons must cancel. No electrons can appear in the overall reaction because all the electrons lost by the reducing agent are gained by the oxidizing agent.

A possible experimental setup for this cell is shown in **FIGURE 19.4.** The anode compartment consists of an iron metal electrode dipping into an aqueous solution of $Fe(NO_3)_2$. Note, though, that any inert electrolyte can be used to carry the current in the anode compartment; Fe^{2+} does not need to be present initially because it's not a reactant in the anode half-reaction.

▶ **FIGURE 19.4**

A galvanic cell consisting of an Fe/Fe^{2+} half-cell and an Fe^{3+}/Fe^{2+} half-cell. The cathode reaction takes place at an inert platinum electrode.

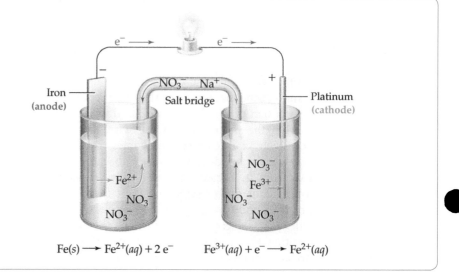

Since Fe^{3+} is a reactant in the cathode half-reaction, $Fe(NO_3)_3$ is a good electrolyte for the cathode compartment. The cathode can be any electrical conductor that doesn't react with the ions in the solution. A platinum wire is a common inert electrode, but iron metal can't be used because it would react directly with the Fe^{3+} ions, thus short-circuiting the cell. The salt bridge contains $NaNO_3$, but any inert electrolyte would do. Electrons flow through the wire from the iron anode $(-)$ to the platinum cathode $(+)$. Anions move from the cathode compartment toward the anode while cations migrate from the anode compartment toward the cathode.

CONCEPTUAL WORKED EXAMPLE 19.3

Designing a Galvanic Cell

Design a galvanic cell that uses the redox reaction

$$2\ Cr(s) + 3\ Sn^{2+}(aq) \longrightarrow 2\ Cr^{3+}(aq) + 3\ Sn(s)$$

Identify the anode and cathode half-reactions, and sketch the experimental setup. Label the anode and cathode, indicate the direction of electron and ion flow, and identify the sign of each electrode.

STRATEGY

First, separate the overall cell reaction into balanced anode (oxidation) and cathode (reduction) half-reactions. Then, set up two half-cells that use these half-reactions and connect the half-cells with a conducting wire and a salt bridge.

SOLUTION

The cell half-reactions involve oxidation of chromium metal to chromium(III) ions at the anode and reduction of tin(II) ions to tin metal at the cathode:

Anode (oxidation): $2 \times [Cr(s) \longrightarrow Cr^{3+}(aq) + 3\ e^-]$
Cathode (reduction): $3 \times [Sn^{2+}(aq) + 2\ e^- \longrightarrow Sn(s)]$

Overall cell reaction: $2\ Cr(s) + 3\ Sn^{2+}(aq) \longrightarrow 2\ Cr^{3+}(aq) + 3\ Sn(s)$

The anode half-reaction is multiplied by a factor of 2 and the cathode half-reaction is multiplied by a factor of 3 so that the electrons will cancel when the two half-reactions are added to give the overall cell reaction.

A possible experimental setup is shown below. The anode compartment consists of a chromium metal electrode dipping into an aqueous solution of $Cr(NO_3)_3$, and the cathode compartment consists of a tin metal electrode dipping into an aqueous solution of $Sn(NO_3)_2$. The salt bridge contains $NaNO_3$, but any inert electrolyte can be used. Electrons flow through the wire from the chromium anode $(-)$ to the tin cathode $(+)$. Anions move from the cathode compartment toward the anode, while cations migrate from the anode compartment toward the cathode.

▶ **CONCEPTUAL PRACTICE 19.5**   Describe a galvanic cell that uses the reaction

$$2\ Ag^+(aq) + Ni(s) \longrightarrow 2\ Ag(s) + Ni^{2+}(aq)$$

Identify the anode and cathode half-reactions, and sketch the experimental setup. Label the anode and cathode, indicate the direction of electron and ion flow, and identify the sign of each electrode.

▶ **CONCEPTUAL APPLY 19.6**   Use the diagram of the galvanic cell to identify the anode and cathode half-reactions and write an overall balanced reaction. Label the anode and the cathode, and identify the sign of each electrode.

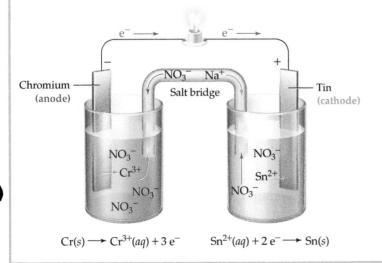

$Cr(s) \longrightarrow Cr^{3+}(aq) + 3\ e^-$ $Sn^{2+}(aq) + 2\ e^- \longrightarrow Sn(s)$

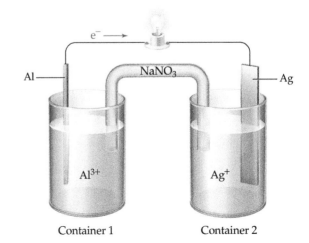

Container 1   Container 2

19.3 SHORTHAND NOTATION FOR GALVANIC CELLS

Rather than describing a galvanic cell in words, it's convenient to use a shorthand notation for representing the cell. For the Daniell cell in Figure 19.3, which uses the reaction

$$Zn(s) + Cu^{2+}(aq) \longrightarrow Zn^{2+}(aq) + Cu(s)$$

we can write the following expression:

$$Zn(s)\,|\,Zn^{2+}(aq)\,\|\,Cu^{2+}(aq)\,|\,Cu(s)$$

In this notation, a single vertical line ($|$) represents a phase boundary, such as that between a solid electrode and an aqueous solution, and the double vertical line ($\|$) denotes a salt bridge. The shorthand for the anode half-cell is always written on the left of the salt-bridge symbol, followed on the right by the shorthand for the cathode half-cell. The electrodes are written on the extreme left (anode) and the extreme right (cathode), and the reactants in each half-cell are always written first, followed by the products. With these arbitrary conventions, electrons move through the external circuit from left to right (from anode to cathode). Reading the shorthand thus suggests the overall cell reaction: Zn is oxidized to Zn^{2+}, and Cu^{2+} is reduced to Cu.

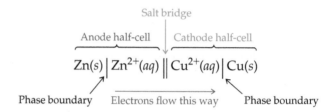

For the galvanic cell in Figure 19.4, based on the reaction

$$Fe(s) + 2\ Fe^{+3}(aq) \longrightarrow 3\ Fe^{2+}(aq)$$

the shorthand notation is

$$Fe(s)\,|\,Fe^{2+}(aq)\,\|\,Fe^{3+}(aq),\,Fe^{2+}(aq)\,|\,Pt(s)$$

The shorthand for the cathode half-cell includes both reactant (Fe^{3+}) and product (Fe^{2+}) as well as the electrode (Pt). The two ions $Fe^{3+}(aq)$ and $Fe^{2+}(aq)$ are separated by a comma rather than a vertical line because they are in the same phase.

The notation for a cell involving a gas has an additional vertical line because an additional phase is present. Thus, the notation

$$Cu(s)\,|\,Cu^{2+}(aq)\,\|\,Cl_2(g)\,|\,Cl^-(aq)\,|\,C(s)$$

specifies a cell in which copper is oxidized to Cu^{2+} at a copper anode and Cl_2 gas is reduced to Cl^- at an inert graphite (carbon) cathode. The cell reaction is

$$Cu(s) + Cl_2(g) \longrightarrow Cu^{2+}(aq) + 2\ Cl^-(aq)$$

A more detailed notation would include ion concentrations and gas pressures, as in the following:

$$Cu(s)\,|\,Cu^{2+}(1.0\ M)\,\|\,Cl_2(1\ atm)\,|\,Cl^-(1.0\ M)\,|\,C(s)$$

Although the anode half-cell always appears on the left in the shorthand notation, its location in a cell drawing is arbitrary. This means that you can't infer which electrode is the anode and which is the cathode from the location of the electrodes in a cell drawing. You must identify the electrodes based on whether each electrode half-reaction is an oxidation or a reduction.

WORKED EXAMPLE 19.4

Interpreting the Shorthand Notation for a Galvanic Cell

Given the following shorthand notation

$$Pt(s)\,|\,Sn^{2+}(aq),\,Sn^{4+}(aq)\,\|\,Ag^{+}(aq)\,|\,Ag(s)$$

write a balanced equation for the cell reaction, and give a brief description of the cell.

STRATEGY

We can obtain the cell half-reactions simply by reading the shorthand notation. To find the balanced equation for the cell reaction, add the two half-reactions after multiplying each by an appropriate factor so that the electrons will cancel. The shorthand notation specifies the anode on the left, the cathode on the right, and the reactants in the half-cell compartments.

SOLUTION

Because the anode always appears at the left in the shorthand notation, the anode (oxidation) half-reaction is

$$Sn^{2+}(aq) \longrightarrow Sn^{4+}(aq) + 2\,e^{-}$$

The platinum electrode is inert and serves only to conduct electrons. The cathode (reduction) half-reaction is

$$2 \times [Ag^{+}(aq) + e^{-} \longrightarrow Ag(s)]$$

The cathode half-reaction is multiplied by a factor of 2 so that the electrons will cancel when the two half-reactions are summed to give the cell reaction:

$$Sn^{2+}(aq) + 2\,Ag^{+}(aq) \longrightarrow Sn^{4+}(aq) + 2\,Ag(s)$$

The cell consists of a platinum wire anode dipping into an Sn^{2+} solution—say, $Sn(NO_3)_2(aq)$—and a silver cathode dipping into an Ag^{+} solution—say, $AgNO_3(aq)$. As usual, the anode and cathode half-cells must be connected by a wire and a salt bridge containing inert ions.

▶ **PRACTICE 19.7** Write a balanced equation for the overall cell reaction, and sketch a diagram of a galvanic cell represented by the following shorthand notation:

$$Pb(s)\,|\,Pb^{2+}(aq)\,\|\,Br_2(aq),\,Br^{-}(aq)\,|\,Pt(s)$$

▶ **CONCEPTUAL APPLY 19.8** Consider the following galvanic cell with containers arranged on the lab bench as shown. Use the direction of electron flow to complete parts (a)–(d).

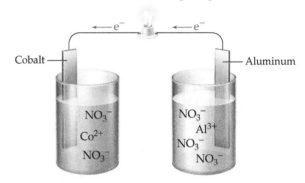

(a) Complete the drawing by adding any components essential for a functioning cell.
(b) Label the anode and cathode, and indicate the direction of ion flow.
(c) Write a balanced equation for the cell reaction.
(d) Write the shorthand notation for the cell.

19.4 CELL POTENTIALS AND FREE-ENERGY CHANGES FOR CELL REACTIONS

Let's return to the Daniell cell shown in Figure 19.3 to find an electrical measure of the tendency for a cell reaction to occur. Electrons move through the external circuit from the zinc anode to the copper cathode because they have lower energy when on copper than on zinc. The driving force for the movement of the negatively charged electrons away from the anode ($-$ electrode) and toward the cathode ($+$electrode) is an electrical potential called the **electromotive force** (**emf**), also known as the **cell potential** (E) or the **cell voltage**. The SI unit of electrical potential is the volt (V), and the potential of a galvanic cell is defined as a positive quantity.

The relationship between the volt and the SI units of energy (joule, J) and electric charge (coulomb, C) is given by the equation

$$1\,J = 1\,C \times 1\,V$$

where 1 C is the amount of charge transferred when a current of 1 **ampere** (**A**) flows for 1 second (s). When 1 C of charge moves between two electrodes that differ in electrical potential by 1 V, 1 J of energy is released by the cell and can be used to do electrical work.

The relationship between the watt (W), the ampere, and the volt is given by

$$1\,W = \frac{1\,J}{s} = \frac{1\,C \times 1\,V}{s} = 1\,A \times 1\,V$$

> **REMEMBER . . .**
> The **ampere** (1 A = 1 C/s) is the SI unit of electric current (Table 1.2).

where 1 W is 1 J/s and 1 A is 1 C/s. Thus, the current passing through a 75 W household light bulb at a household voltage of 110 V is about 0.7 A, which means that the electric charge of the electrons passing through the bulb in 1 s is 0.7 C.

A cell potential is measured with an electronic instrument called a *voltmeter* (Figure 19.3b), which is designed to give a positive reading when the + and − terminals of the voltmeter are connected to the + (cathode) and − (anode) electrodes of the cell, respectively. Thus, the voltmeter–cell connections required to get a positive reading on the voltmeter indicate which electrode is the anode and which is the cathode.

We've now seen two quantitative measures of the tendency for a chemical reaction to occur: the cell potential E, an electrochemical quantity, and the free-energy change ΔG, a thermochemical quantity (Section 18.7). The values of ΔG and E are directly proportional and are related by the equation

> **REMEMBER . . .**
> If the free-energy change ΔG for a reaction is negative, the reaction is spontaneous. The greater the negative value of ΔG, the greater the tendency for the reaction to occur (Section 18.7).

> **Relationship between free-energy change and cell potential** $\Delta G = -nFE$

where n is the number of moles of electrons transferred in the reaction and F is the **faraday** (or *Faraday constant*), the electric charge on 1 mol of electrons (96,485 C/mol e⁻). In our calculations, we'll round the value of F to three significant figures:

$$F = 96{,}500 \text{ C/mol e}^-$$

The faraday is named in honor of Michael Faraday (1791–1867), the nineteenth-century English scientist who laid the foundations for our current understanding of electricity.

Two features of the equation $\Delta G = -nFE$ are worth noting: the units and the minus sign. When we multiply the charge transferred (nF) in coulombs by the cell potential (E) in volts, we obtain an energy (ΔG) in joules, in accord with the relationship $1 \text{ J} = 1 \text{ C} \times 1 \text{ V}$. The minus sign is required because E and ΔG have opposite signs: The spontaneous reaction in a galvanic cell has a positive cell potential but a negative free-energy change (Section 18.7).

Later in this chapter, we'll see that cell potentials, like free-energy changes, depend on the composition of the reaction mixture. The **standard cell potential** ($E°$) is the cell potential when both reactants and products are in their standard states—solutes at 1 M concentrations, gases at a partial pressure of 1 atm, solids and liquids in pure form, with all at a specified temperature, usually 25 °C. For example, $E°$ for the reaction

$$\text{Zn}(s) + \text{Cu}^{2+}(aq) \longrightarrow \text{Zn}^{2+}(aq) + \text{Cu}(s) \quad E° = 1.10 \text{ V}$$

is the cell potential measured at 25 °C for a cell that has pure Zn and Cu metal electrodes and 1 M concentrations of Zn^{2+} and Cu^{2+}.

The standard free-energy change and the standard cell potential are related by the equation

$$\Delta G° = -nFE°$$

Because $\Delta G°$ and $E°$ are directly proportional, a voltmeter can be regarded as a "free-energy meter." When a voltmeter measures $E°$, it also indirectly measures $\Delta G°$.

> **Go to eText**
>
> **BIG IDEA Question 3**
>
> In a galvanic cell (a battery), what is the sign of the free-energy change (ΔG) of the reaction and the sign of the cell potential (E)?

— WORKED EXAMPLE 19.5

Calculating a Standard Free-Energy Change from a Standard Cell Potential

The standard cell potential at 25 °C is 0.90 V for the reaction

$$2 \text{ Al}(s) + 3 \text{ Zn}^{2+}(aq) \longrightarrow 2 \text{ Al}^{3+}(aq) + 3 \text{ Zn}(s)$$

Calculate the standard free-energy change for this reaction at 25 °C.

IDENTIFY

Known	Unknown
Standard cell potential ($E° = 0.90$ V)	Standard free-energy change($\Delta G°$)

STRATEGY

To calculate $\Delta G°$, we use the equation $\Delta G° = -nFE°$, where n can be determined from the balanced chemical equation, F is 96,500 C/mol e⁻, and $E°$ is given.

SOLUTION

When two moles of Al are oxidized to 2 moles of Al^{3+} ions, six moles of electrons are transferred to three moles of Zn^{2+} ions, which are reduced to three moles of Zn. Therefore, $n = 6$ mol e⁻ for this reaction, and the standard free-energy change is

$$\Delta G° = -nFE° = -(6 \text{ mol e}^-)\left(\frac{96,500 \text{ C}}{\text{mol e}^-}\right)(0.90 \text{ V})\left(\frac{1 \text{ J}}{1 \text{ C} \cdot \text{V}}\right)$$

$$= -5.2 \times 10^5 \text{ J} = -520 \text{ kJ}$$

CHECK

The sign of $\Delta G°$ is negative, as expected for a spontaneous reaction. F is approximately 10^5 C/mol e⁻ and $E°$ is approximately 1 V, so $\Delta G° = -nFE°$ is approximately $-(6 \text{ mol e}^-)$ (10^5 C/mol e^-) (1 V) $= -6 \times 10^5$ J, or about -600 kJ. The estimate of -600 kJ has the same magnitude and sign as the solution, -520 kJ.

▶ **PRACTICE 19.9**   The standard cell potential at 25 °C is 1.20 V for the reaction:

$$2 \text{ Ag}^+(aq) + \text{Cd}(s) \longrightarrow 2 \text{ Ag}(s) + \text{Cd}^{2+}(aq)$$

What is the standard free-energy change for this reaction at 25 °C?

▶ **APPLY 19.10**   The standard free-energy change is 59.8 kJ for the reaction:

$$\text{Hg}(l) + \text{I}_2(s) \longrightarrow \text{Hg}^{2+}(aq) + 2 \text{ I}^-(aq)$$

Is the reaction spontaneous or nonspontaneous? What is the standard cell potential at 25 °C?

19.5 STANDARD REDUCTION POTENTIALS

The standard potential of any galvanic cell is the sum of the standard half-cell potentials for the oxidation half-reaction at the anode and the reduction half-reaction at the cathode:

$$E°_{cell} = E°_{ox} + E°_{red}$$

Consider, for example, a cell in which H_2 gas is oxidized to H^+ ions at the anode and Cu^{2+} ions are reduced to copper metal at the cathode (**FIGURE 19.5**):

Anode (oxidation):	$H_2(g) \longrightarrow 2 \text{ H}^+(aq) + 2 \text{ e}^-$
Cathode (reduction):	$Cu^{2+}(aq) + 2 \text{ e}^- \longrightarrow Cu(s)$
Overall cell reaction:	$H_2(g) + Cu^{2+}(aq) \longrightarrow 2 \text{ H}^+(aq) + Cu(s)$

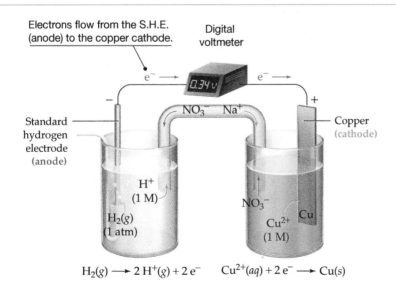

Electrons flow from the S.H.E. (anode) to the copper cathode.

Digital voltmeter

Standard hydrogen electrode (anode)

Copper (cathode)

H^+ (1 M)

$H_2(g)$ (1 atm)

NO_3^-

Cu^{2+} (1 M)   Cu

$H_2(g) \longrightarrow 2 \text{ H}^+(g) + 2 \text{ e}^-$ $Cu^{2+}(aq) + 2 \text{ e}^- \longrightarrow Cu(s)$

The standard hydrogen electron is a piece of platinum foil that is in contact with bubbles of $H_2(g)$ at 1 atm pressure and with $H^+(aq)$ at 1 M concentration.

◀ **FIGURE 19.5**

A galvanic cell consisting of a Cu^{2+} (1 M)/Cu half-cell and a standard hydrogen electrode (S.H.E.). The measured standard cell potential at 25 °C is 0.34 V.

[In this chapter we represent the hydrated proton as $H^+(aq)$ rather than $H_3O^+(aq)$ because we're interested here in electron transfer, not proton transfer as in Chapter 16.] The standard potential for this cell, 0.34 V at 25 °C, measures the sum of the reaction tendencies of the oxidation and reduction half-reactions:

$$E°_{cell} = E°_{ox} + E°_{red} = E°_{H_2 \longrightarrow H^+} + E°_{Cu^{2+} \longrightarrow Cu} = 0.34 \text{ V}$$

If we could determine $E°$ values for individual half-reactions, we could combine those values to obtain $E°$ values for a host of cell reactions. Unfortunately, it's not possible to measure the potential of a single electrode; we can measure only a potential *difference* by placing a voltmeter between *two* electrodes. Nevertheless, we can develop a set of standard half-cell potentials by choosing an arbitrary standard half-cell as a reference point, assigning it an arbitrary potential, and then expressing the potential of all other half-cells relative to the reference half-cell. Recall that this same approach was used in Section 9.9 for determining **standard enthalpies of formation,** $\Delta H°_f$.

To define an electrochemical "sea level," chemists have chosen a reference half-cell called the **standard hydrogen electrode (S.H.E.)**, shown as the anode in Figure 19.5 and in more detail in **FIGURE 19.6.** It consists of a platinum electrode in contact with H_2 gas and aqueous H^+ ions at standard-state conditions [1 atm $H_2(g)$, 1 M $H^+(aq)$, 25 °C]. The corresponding half-reaction, written in either direction, is assigned an arbitrary potential of exactly 0 V:

$$2 H^+(aq, 1 \text{ M}) + 2 e^- \longrightarrow H_2(g, 1 \text{ atm}) \quad E° = 0 \text{ V}$$
$$H_2(g, 1 \text{ atm}) \longrightarrow 2 H^+(aq, 1 \text{ M}) + 2 e^- \quad E° = 0 \text{ V}$$

> **REMEMBER . . .**
>
> The **standard enthalpy of formation,** $\Delta H°_f$, of a substance is expressed relative to an arbitrary reference of $\Delta H°_f = 0$ kJ/mol for its constituent elements in their standard states (Section 9.9).

▶ **FIGURE 19.6**

The standard hydrogen electrode (S.H.E.) is used as reference electrode and is assigned an arbitrary potential of 0 V. It can serve as either the cathode or anode in a galvanic cell.

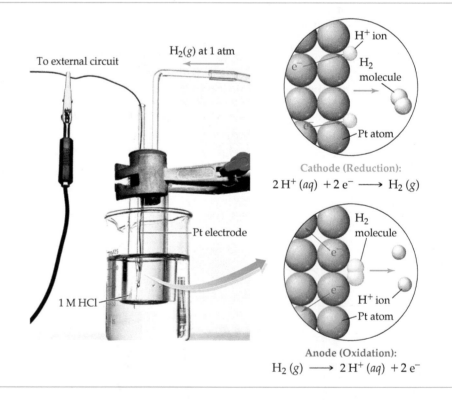

Cathode (Reduction):
$$2 H^+(aq) + 2 e^- \longrightarrow H_2(g)$$

Anode (Oxidation):
$$H_2(g) \longrightarrow 2 H^+(aq) + 2 e^-$$

With the standard hydrogen electrode as the standard reference electrode, the entire potential of the cell

$$Pt(s) | H_2(1 \text{ atm}) | H^+(1 \text{ M}) \| Cu^{2+}(1 \text{ M}) | Cu(s)$$

can be attributed to the Cu^{2+}/Cu half-cell:

$$E°_{cell} = E°_{H_2 \rightarrow H^+} + E°_{Cu^{2+} \rightarrow Cu} = 0.34 \text{ V}$$
$$\phantom{E°_{cell} =} \underset{0.34 \text{ V}}{\uparrow} \quad \underset{0 \text{ V}}{\uparrow} \quad \underset{0.34 \text{ V}}{\uparrow}$$

Because the Cu^{2+}/Cu half-reaction is a reduction, the corresponding half-cell potential, $E° = 0.34$ V, is called a **standard reduction potential:**

$$Cu^{2+}(aq) + 2\,e^- \longrightarrow Cu(s) \quad \text{Standard reduction potential: } E° = 0.34 \text{ V}$$

In a cell in which this half-reaction occurs in the opposite direction, the corresponding half-cell potential has the same magnitude but opposite sign:

$$Cu(s) \longrightarrow Cu^{2+}(aq) + 2\,e^- \quad E° = -0.34 \text{ V}$$

Whenever the direction of a half-reaction is *reversed,* the sign of $E°$ must also be *reversed.* Thus, the standard potential for an oxidation half-reaction is the negative of the standard reduction potential.

Hundreds of potentials for half-reactions have been measured using the standard hydrogen reference electrode or other reference electrodes. A short list is presented in **TABLE 19.1,** and a more complete tabulation is given in Appendix D. The following conventions are observed when constructing a table of half-cell potentials:

- **The half-reactions are written as reductions** rather than as oxidations. This means that oxidizing agents and electrons are on the left side of each half-reaction and reducing agents are on the right side.

- **The listed half-cell potentials are standard reduction potentials,** also known as standard electrode potentials.

- **The half-reactions are listed in order of decreasing standard reduction potential,** meaning a decreasing tendency to occur in the forward direction and an increasing

TABLE 19.1   Standard Reduction Potentials at 25 °C

	Reduction Half-Reaction		$E°$ (V)	
Stronger oxidizing agent	$F_2(g) + 2\,e^-$	$\longrightarrow 2\,F^-(aq)$	2.87	Weaker reducing agent
	$H_2O_2(aq) + 2\,H^+(aq) + 2\,e^-$	$\longrightarrow 2\,H_2O(l)$	1.78	
	$MnO_4^-(aq) + 8\,H^+(aq) + 5\,e^-$	$\longrightarrow Mn^{2+}(aq) + 4\,H_2O(l)$	1.51	
	$Cl_2(g) + 2\,e^-$	$\longrightarrow 2\,Cl^-(aq)$	1.36	
	$Cr_2O_7^{2-}(aq) + 14\,H^+(aq) + 6\,e^-$	$\longrightarrow 2\,Cr^{3+}(aq) + 7\,H_2O(l)$	1.36	
	$O_2(g) + 4\,H^+(aq) + 4\,e^-$	$\longrightarrow 2\,H_2O(l)$	1.23	
	$Br_2(aq) + 2\,e^-$	$\longrightarrow 2\,Br^-(aq)$	1.09	
	$Ag^+(aq) + e^-$	$\longrightarrow Ag(s)$	0.80	
	$Fe^{3+}(aq) + e^-$	$\longrightarrow Fe^{2+}(aq)$	0.77	
	$O_2(g) + 2\,H^+(aq) + 2\,e^-$	$\longrightarrow H_2O_2(aq)$	0.70	
	$I_2(s) + 2\,e^-$	$\longrightarrow 2\,I^-(aq)$	0.54	
	$O_2(g) + 2\,H_2O(l) + 4\,e^-$	$\longrightarrow 4\,OH^-(aq)$	0.40	
	$Cu^{2+}(aq) + 2\,e^-$	$\longrightarrow Cu(s)$	0.34	
	$Sn^{4+}(aq) + 2\,e^-$	$\longrightarrow Sn^{2+}(aq)$	0.15	
	$2\,H^+(aq) + 2\,e^-$	$\longrightarrow H_2(g)$	0	
	$Pb^{2+}(aq) + 2\,e^-$	$\longrightarrow Pb(s)$	-0.13	
	$Ni^{2+}(aq) + 2\,e^-$	$\longrightarrow Ni(s)$	-0.26	
	$Cd^{2+}(aq) + 2\,e^-$	$\longrightarrow Cd(s)$	-0.40	
	$Fe^{2+}(aq) + 2\,e^-$	$\longrightarrow Fe(s)$	-0.45	
	$Zn^{2+}(aq) + 2\,e^-$	$\longrightarrow Zn(s)$	-0.76	
	$2\,H_2O(l) + 2\,e^-$	$\longrightarrow H_2(g) + 2\,OH^-(aq)$	-0.83	
	$Al^{3+}(aq) + 3\,e^-$	$\longrightarrow Al(s)$	-1.66	
	$Mg^{2+}(aq) + 2\,e^-$	$\longrightarrow Mg(s)$	-2.37	Stronger reducing agent
Weaker oxidizing agent	$Na^+(aq) + e^-$	$\longrightarrow Na(s)$	-2.71	
	$Li^+(aq) + e^-$	$\longrightarrow Li(s)$	-3.04	

tendency to occur in the reverse direction. Consequently, the strongest oxidizing agents are located in the upper left of the table (F_2, H_2O_2, MnO_4^-, and so forth), and the strongest reducing agents are found in the lower right of the table (Li, Na, Mg, and so forth).

By choosing $E° = 0$ V for the standard hydrogen electrode, we obtain standard reduction potentials that range from about $+3$ V to -3 V.

Note that the strongest oxidizing agents in Table 19.1 tend to be either elements on the right side of the periodic table that have large electron affinities, such as F_2 and Cl_2, or oxoanions that have a metal atom in a high oxidation state, such as MnO_4^- and $Cr_2O_7^{2-}$. Metal atoms in high oxidation states have a strong affinity for electrons and are relatively easily reduced to lower oxidation states. Note also that the strongest reducing agents in Table 19.1 are the main-group metals on the left side of the periodic table that have relatively low ionization energies and are therefore relatively easily oxidized.

Finally, note how the ordering of the half-reactions in Table 19.1 corresponds to their ordering in the activity series in Table 4.5. The more active metals at the top of the activity series have the more positive oxidation potentials and therefore the more negative standard reduction potentials.

19.6 USING STANDARD REDUCTION POTENTIALS

A table of standard reduction potentials summarizes an enormous amount of chemical information in a very small space. It enables us to arrange oxidizing or reducing agents in order of increasing strength, and it permits us to predict the spontaneity or non-spontaneity of thousands of redox reactions. Suppose, for example, that the table contains just 100 half-reactions. Each reduction half-reaction can be paired with any one of the remaining 99 oxidation half-reactions to give a total of $100 \times 99 = 9900$ cell reactions. By calculating the $E°$ values for these cell reactions, we would find that half of them are spontaneous and the other half are nonspontaneous. (Can you see why?)

To illustrate how to use the tabulated $E°$ values, let's calculate $E°$ for the oxidation of $Zn(s)$ by $Ag^+(aq)$:

$$2\,Ag^+(aq) + Zn(s) \longrightarrow 2\,Ag(s) + Zn^{2+}(aq)$$

First, find the relevant half-reactions in Table 19.1 and write them in the appropriate direction for reduction of Ag^+ and oxidation of Zn. Next, add the half-reactions to get the overall reaction. Before adding, though, multiply the Ag^+/Ag half-reaction by a factor of 2 so that the electrons will cancel:

Reduction:	$2 \times [Ag^+(aq) + e^- \longrightarrow Ag(s)]$	$E° = 0.80$ V
Oxidation:	$Zn(s) \longrightarrow Zn^{2+}(aq) + 2\,e^-$	$E° = 0.76$ V
Overall reaction:	$2\,Ag^+(aq) + Zn(s) \longrightarrow 2\,Ag(s) + Zn^{2+}(aq)$	$E° = 1.56$ V

Then, tabulate the $E°$ values for the half-reactions, remembering that $E°$ for oxidation of zinc is the negative of the standard reduction potential (-0.76 V) given in Table 19.1. Do not multiply the $E°$ value for reduction of $Ag^+(0.80$ V) by a factor of 2, however, because an electrical potential does not depend on amounts of reactants and products. The potential for the half-reaction is determined by the identity of the reactants and products and is an indication of the tendency of the electrons to be transferred. A familiar example is large and small batteries that have the same redox reaction and the same potential.

The reason why $E°$ values are independent of the amount of reactant can be understood by looking at the equation $\Delta G° = -nFE°$. Free energy is an **extensive property** (Section 2.3) because it depends on the amount of substance. If twice as much Ag^+ is reduced, the free-energy change, $\Delta G°$, doubles. The number of electrons transferred, n, also doubles, however, so the ratio $E° = -\Delta G°/nF$ is constant. Electrical potential is therefore an **intensive property**, which does not depend on the amount of substance.

▲ The potential ($E°$) of a redox reaction depends on the identity of the reactants and products but does not depend on the amount of reactants and products. For example, a small battery and a large battery that have the same redox reaction will have the same potential (9 V). The large battery will last longer because it has a larger amount of reactants.

REMEMBER...

Extensive properties, like mass and volume, have values that depend on the sample size. Intensive properties, like temperature and melting point, have values that do not depend on the sample size (Section 2.3).

The $E°$ value for the overall reaction of Zn(s) with Ag^+ (aq) is the sum of the $E°$ values for the two half-reactions: 0.80 V + 0.76 V = 1.56 V. Because $E°$ is positive (and $\Delta G°$ is negative), oxidation of zinc by Ag^+ is a spontaneous reaction under standard-state conditions. Just as Ag^+ can oxidize Zn, it's evident from Table 19.1 that Ag^+ can oxidize any reducing agent that lies below it in the table (Fe^{2+}, H_2O_2, Cu, and so forth). Summing $E°$ for the Ag^+/Ag reduction (0.80 V) and $E°$ for any oxidation half-reaction that lies below the Ag^+/Ag half-reaction always gives a positive $E°$ for the overall reaction.

In general, an oxidizing agent can oxidize any reducing agent that lies below it in the table but can't oxidize a reducing agent that appears above it in the table. Thus, Ag^+ can't oxidize Br^-, H_2O, Cr^{3+}, and so forth because $E°$ for the overall reaction is negative. Simply by glancing at the locations of the oxidizing and reducing agents in the table, we can predict whether a reaction is spontaneous or nonspontaneous.

WORKED EXAMPLE 19.6

Arranging Oxidizing and Reducing Agents in Order of Increasing Strength

(a) Arrange the following oxidizing agents in order of increasing strength under standard-state conditions: $Br_2(aq)$, $Fe^{3+}(aq)$, $Cr_2O_7^{2-}(aq)$.
(b) Arrange the following reducing agents in order of increasing strength under standard-state conditions: Al(s), Na(s), Zn(s).

STRATEGY

Pick out the half-reactions in Table 19.1 that involve the given oxidizing or reducing agents, and list them, along with their $E°$ values, in the order in which they occur in the table. The strength of an oxidizing agent increases as the $E°$ value increases, and the strength of a reducing agent increases as the $E°$ value decreases.

SOLUTION

(a) List the half-reactions that involve Br_2, Fe^{3+}, and $Cr_2O_7^{2-}$ in the order in which they occur in Table 19.1:

$$Cr_2O_7^{2-}(aq) + 14\,H^+(aq) + 6\,e^- \longrightarrow$$
$$2\,Cr^{3+}(aq) + 7\,H_2O(l) \qquad E° = 1.36\text{ V}$$
$$Br_2(aq) + 2\,e^- \longrightarrow 2\,Br^-(aq) \qquad E° = 1.09\text{ V}$$
$$Fe^{3+}(aq) + e^- \longrightarrow Fe^{2+}(aq) \qquad E° = 0.77\text{ V}$$

You can see that $Cr_2O_7^{2-}$ has the greatest tendency to be reduced (largest $E°$) and Fe^{3+} has the least tendency to be reduced (smallest $E°$). The species that has the greatest tendency to be reduced is the strongest oxidizing agent, so oxidizing strength increases in the order $Fe^{3+} < Br_2 < Cr_2O_7^{2-}$. As a shortcut, simply note that the strength of the oxidizing agents, listed on the left side of Table 19.1, increases on moving up in the table.

(b) List the reduction half-reactions that involve Al(s), Na(s), and Zn(s) in the order in which they occur in Table 19.1:

$$Zn^{2+}(aq) + 2\,e^- \longrightarrow Zn(s) \qquad E° = -0.76\text{ V}$$
$$Al^{3+}(aq) + 3\,e^- \longrightarrow Al(s) \qquad E° = -1.66\text{ V}$$
$$Na^+(aq) + e^- \longrightarrow Na(s) \qquad E° = -2.71\text{ V}$$

The last half-reaction has the least tendency to occur in the forward direction (most negative $E°$) and the greatest tendency to occur in the reverse direction.

$$Zn(s) \longrightarrow Zn^{2+}(aq) + 2\,e^- \qquad E° = +0.76\text{ V}$$
$$Al(s) \longrightarrow Al^{3+}(aq) + 3\,e^- \qquad E° = +1.66\text{ V}$$
$$Na(s) \longrightarrow Na^+(aq) + e^- \qquad E° = +2.71\text{ V}$$

Therefore, Na is the strongest reducing agent, and reducing strength increases in the order Zn < Al < Na. As a shortcut, note that the strength of the reducing agents, listed on the right side of Table 19.1, increases on moving down in the table.

▶ **PRACTICE 19.11** Which substance is the strongest reducing agent: $Cl_2(g)$, Al(s), Cu(s), or $Zn^{2+}(aq)$?

▶ **APPLY 19.12** Consider the following table of standard reduction potentials:

Reduction Half-Reaction	$E°$ (V)
$A^{3+} + 2\,e^- \longrightarrow A^+$	1.47
$B^{2+} + 2\,e^- \longrightarrow B$	0.60
$C^{2+} + 2\,e^- \longrightarrow C$	-0.21
$D^+ + e^- \longrightarrow D$	-1.38

(a) Which substance is the strongest reducing agent? Which is the strongest oxidizing agent?
(b) Which substances can be oxidized by B^{2+}? Which can be reduced by C?
(c) Write a balanced equation for the overall cell reaction that delivers the highest voltage, and calculate $E°$ for the reaction.

WORKED EXAMPLE 19.7

Predicting Whether a Redox Reaction Is Spontaneous

Predict from Table 19.1 whether $Pb^{2+}(aq)$ can oxidize $Al(s)$ or $Cu(s)$ under standard-state conditions. Calculate $E°$ for each reaction at 25 °C.

STRATEGY

To predict whether a redox reaction is spontaneous, remember that an oxidizing agent can oxidize any reducing agent that lies below it in the table but can't oxidize one that lies above it. To calculate $E°$ for a redox reaction, sum the $E°$ values for the reduction and oxidation half-reactions. If the value of $E°$ is positive then the reaction is spontaneous.

SOLUTION

$Pb^{2+}(aq)$ is above $Al(s)$ in the table but below $Cu(s)$. Therefore, $Pb^{2+}(aq)$ can oxidize $Al(s)$ but can't oxidize $Cu(s)$. To confirm these predictions, calculate $E°$ values for the overall reactions.

For the oxidation of Al by Pb^{2+}, $E°$ is positive (1.53 V), and the reaction is therefore spontaneous:

$$
\begin{array}{ll}
3 \times [Pb^{2+}(aq) + 2\,e^- \longrightarrow Pb(s)] & E° = -0.13 \text{ V} \\
\underline{2 \times [Al(s) \longrightarrow Al^{3+}(aq) + 3\,e^-} & \underline{E° = 1.66 \text{ V}} \\
3\,Pb^{2+}(aq) + 2\,Al(s) \longrightarrow 3\,Pb(s) + 2\,Al^{3+}(aq) & E° = 1.53 \text{ V}
\end{array}
$$

Note that we have multiplied the Pb^{2+}/Pb half-reaction by a factor of 3 and the Al/Al^{3+} half-reaction by a factor of 2 so that the electrons will cancel, but we do not multiply the $E°$ values by these factors because electrical potential is an intensive property.

For the oxidation of Cu by Pb^{2+}, $E°$ is negative (-0.47 V) and the reaction is therefore nonspontaneous:

$$
\begin{array}{ll}
Pb^{2+}(aq) + 2\,e^- \longrightarrow Pb(s) & E° = -0.13 \text{ V} \\
\underline{Cu(s) \longrightarrow Cu^{2+}(aq) + 2\,e^-} & \underline{E° = -0.34 \text{ V}} \\
Pb^{2+}(aq) + Cu(s) \longrightarrow Pb(s) + Cu^{2+}(aq) & E° = -0.47 \text{ V}
\end{array}
$$

▶ **PRACTICE 19.13**   Use Table 19.1 to calculate the value of $E°$ for the reaction and determine if the reaction occurs spontaneously.

$$2\,Fe^{3+}(aq) + 2\,I^-(aq) \longrightarrow 2\,Fe^{2+}(aq) + I_2(s)$$

▶ **APPLY 19.14**   Use Table 19.1 to predict if a reaction occurs when a piece of $Ag(s)$ is placed in a 1.0 M solution of $Mg(NO_3)_2(aq)$.

Go to eText

WORKED EXAMPLE 19.8

Calculating the Voltage of a Galvanic Cell under Standard-State Conditions

What is the voltage of the galvanic cell shown in the illustration? What is the anode half-reaction, and what is the cathode half-reaction? Which direction do the electrons flow?

Digital voltmeter

Cd(s)

NaNO₃

Ag(s)

Cd(NO₃)₂
(1 M)

AgNO₃
(1 M)

STRATEGY

In a galvanic cell, there must be a reduction half-reaction and an oxidation half-reaction whose values of $E°$ sum to give a positive value of $E°$ for the cell. Any spontaneous redox reaction must have a positive value of $E°$. Use Table 19.1 to look up the values of $E°$ for half-reactions and then determine which half-reaction is the reduction and which is the oxidation.

SOLUTION

Table 19.1 gives the standard reduction potentials for the two half-reactions we must consider.

$$Cd^{2+}(aq) + 2\,e^- \longrightarrow Cd(s) \qquad E° = -0.40\ V$$
$$Ag^+(aq) + e^- \longrightarrow Ag(s) \qquad E° = +0.80\ V$$

One reaction must be reversed to an oxidation because any redox reaction must have one reduction half-reaction and one oxidation half-reaction. The value of $E°$ for the galvanic cell must be positive; therefore, the oxidation half-reaction is the oxidation of Cd(s) to $Cd^{2+}(aq)$.

Anode (oxidation): $Cd(s) \longrightarrow Cd^{2+}(aq) + 2\,e^-$ $E° = +0.40\ V$
Cathode (reduction): $2 \times [Ag^+(aq) + e^- \longrightarrow Ag(s)]$ $E° = +0.80\ V$
Overall cell reaction: $2\,Ag^+(aq) + Cd(s) \longrightarrow 2\,Ag(s) + Cd^{2+}(aq)$ $E° = +1.20\ V$

Note that the sign of $E°$ was changed when the cadmium half-reaction was changed from a reduction ($-0.40\ V$) to an oxidation ($+0.40\ V$). Also, when the reduction half-reaction for silver was multiplied by the 2, the value of $E°$ did not change. The value of $E°$ for the overall cell reaction is the sum of $E°$ values for the oxidation and reduction half-reactions, $0.80\ V + 0.40\ V = +1.20\ V$. Electrons flow from the cadmium half-cell to the silver half-cell where $Ag^+(aq)$ is reduced to Ag(s).

▶ **PRACTICE 19.15** You have the following materials that can be used to construct a galvanic cell: 1.0 M $NiCl_2$, 1.0 M $Zn(NO_3)_2$, Ni(s), Zn(s), and a salt bridge.

(a) What is the voltage of the galvanic cell?
(b) What is the shorthand notation for the galvanic cell?

▶ **APPLY 19.16** The standard potential for the following galvanic cell is 0.92 V:

$$Al(s)\,|\,Al^{3+}(1\ M)\,\|\,Cr^{3+}(1\ M)\,|\,Cr(s)$$

Look up the standard reduction potential for the Al^{3+}/Al half-cell in Table 19.1, and calculate the standard reduction potential for the Cr^{3+}/Cr half-cell.

19.7 CELL POTENTIALS UNDER NONSTANDARD-STATE CONDITIONS: THE NERNST EQUATION

Cell potentials, like free-energy changes, depend on temperature and on the composition of the reaction mixture—that is, on the concentrations of solutes and the partial pressures of gases. This dependence can be derived from the equation

$$\Delta G = \Delta G° + RT \ln Q$$

Recall from Section 18.10 that ΔG is the free-energy change for a reaction under nonstandard-state conditions, $\Delta G°$ is the free-energy change under standard-state conditions, and Q is the reaction quotient. Since $\Delta G = -nFE$ and $\Delta G° = -nFE°$ we can rewrite the equation for ΔG in the form

$$-nFE = -nFE° + RT \ln Q$$

Dividing by $-nF$, we obtain the **Nernst equation,** named after Walther Nernst (1864–1941), the German chemist who first derived it:

REMEMBER . . .
The reaction quotient Q is an expression having the same form as the equilibrium constant expression K except that the concentrations and partial pressures do not necessarily have equilibrium values (Section 18.10).

Nernst equation $E = E° - \dfrac{RT}{nF} \ln Q$ or $E = E° - \dfrac{2.303\,RT}{nF} \log Q$

Go to
eText

BIG IDEA Question 4

How would the potential of a galvanic cell change if the concentration of reactants was increased while holding the concentration of the products constant?

Because of an intimate connection between the cell voltage and pH (Section 19.8), we will write the Nernst equation in terms of base-10 logarithms: $\log Q$ rather than $\ln Q$. At 25 °C, $2.303RT/F$ has a value of 0.0592 V, and therefore

$$E = E° - \frac{0.0592 \text{ V}}{n}\log Q \quad \text{in volts, at 25 °C}$$

In actual galvanic cells, the concentrations and partial pressures of reactants and products seldom have standard-state values, and the values change as the cell reaction proceeds. The Nernst equation is useful because it enables us to calculate cell potentials under nonstandard-state conditions, as shown in Worked Example 19.9.

WORKED EXAMPLE 19.9

Using the Nernst Equation to Calculate the Cell Potential under Nonstandard-State Conditions

Consider a galvanic cell that uses the reaction

$$Zn(s) + 2 H^+(aq) \longrightarrow Zn^{2+}(aq) + H_2(g)$$

Calculate the cell potential at 25 °C when $[H^+] = 1.0$ M, $[Zn^{2+}] = 0.0010$ M, and $P_{H_2} = 0.10$ atm.

IDENTIFY

Known	Unknown
Concentrations ($[H^+] = 1.0$ M, $[Zn^{2+}] = 0.0010$ M, $P_{H_2} = 0.10$ atm)	E
Standard reduction potentials (Table 19.1)	

STRATEGY

We can calculate the standard cell potential $E°$ from the standard reduction potentials in Table 19.1 and then use the Nernst equation to find the cell potential E under the cited conditions.

SOLUTION

The standard cell potential is

$$E° = E°_{Zn \longrightarrow Zn^{2+}} + E°_{H^+ \longrightarrow H_2} = (0.76 \text{ V}) + 0 \text{ V} = 0.76 \text{ V}$$

The cell potential at 25 °C under nonstandard-state conditions is given by the Nernst equation:

$$E = E° - \frac{0.0592 \text{ V}}{n}\log Q = E° - \left(\frac{0.0592 \text{ V}}{n}\right)\left(\log\frac{[Zn^{2+}](P_{H_2})}{[H^+]^2}\right)$$

where the reaction quotient contains both molar concentrations of solutes and the partial pressure of a gas (in atm). As usual, zinc has been omitted from the reaction quotient because it is a pure solid. For this reaction, 2 mol of electrons are transferred, so $n = 2$. Substituting into the Nernst equation the appropriate values of $E°$, n, $[H^+]$, $[Zn^{2+}]$, and P_{H_2} gives

$$E = (0.76 \text{ V}) - \left(\frac{0.0592 \text{ V}}{2}\right)\left(\log\frac{(0.0010)(0.10)}{(1.0)^2}\right)$$

$$= (0.76 \text{ V}) - \left(\frac{0.0592 \text{ V}}{2}\right)(-4.0)$$

$$= 0.76 \text{ V} + 0.12 \text{ V} = 0.88 \text{ V} \quad \text{at 25 °C}$$

CHECK

We expect that the reaction will have a greater tendency to occur under the cited conditions than under standard-state conditions because the product concentrations are lower than standard-state values. We therefore predict that the cell potential E will be greater than the standard cell potential $E°$, in agreement with the solution.

▶ **PRACTICE 19.17** Consider a galvanic cell that uses the reaction

$$Cu(s) + 2 Fe^{3+}(aq) \longrightarrow Cu^{2+}(aq) + 2 Fe^{2+}(aq)$$

What is the potential of a cell at 25 °C that has the following ion concentrations?

$$[Fe^{3+}] = 1.0 \times 10^{-4} \text{ M} \quad [Cu^{2+}] = 0.25 \text{ M} \quad [Fe^{2+}] = 0.20 \text{ M}$$

▶ **APPLY 19.18** Accidentally chewing on a stray fragment of aluminum foil can cause a sharp tooth pain if the aluminum comes in contact with an amalgam filling. The filling, an alloy of silver, tin, and mercury, acts as the cathode of a tiny galvanic cell, the aluminum behaves as the anode, and saliva serves as the electrolyte. When the aluminum and the filling come in contact, an electric current passes from the aluminum to the filling, which is sensed by a nerve in the tooth. Aluminum is oxidized at the anode, and O_2 gas is reduced to water at the cathode.

(a) Write balanced equations for the anode, cathode, and overall cell reactions.
(b) Write the Nernst equation in a form that applies at body temperature 37 °C.
(c) Calculate the cell voltage at 37 °C. You may assume that $[Al^{3+}] = 1.0 \times 10^{-9}$ M, $P_{O_2} = 0.20$ atm, and that saliva has a pH of 7.0. Also assume that the $E°$ values in Table 19.1 apply at 37 °C.

CONCEPTUAL WORKED EXAMPLE 19.10

Exploring How Changes in Concentrations Affect Cell Voltage

Consider the following galvanic cell:

(a) What is the change in the cell voltage on increasing the ion concentrations in the anode compartment by a factor of 10?
(b) What is the change in the cell voltage on increasing the ion concentrations in the cathode compartment by a factor of 10?

IDENTIFY

Known	Unknown
Concentration changes	Change in E
Cell diagram (used to determine overall reaction)	

STRATEGY

The direction of electron flow in the picture indicates that lead is the anode and silver is the cathode. Therefore, the cell reaction is

$$Pb(s) + 2\,Ag^+(aq) \longrightarrow Pb^{2+}(aq) + 2\,Ag(s)$$

The cell potential at 25 °C is given by the Nernst equation, where $n = 2$ and $Q = [Pb^{2+}]/[Ag^+]^2$:

$$E = E° - \frac{0.0592\ \text{V}}{n}\log Q = E° - \left(\frac{0.0592\ \text{V}}{2}\right)\left(\log\frac{[Pb^{2+}]}{[Ag^+]^2}\right)$$

The change in E on changing the ion concentrations will be determined by the change in the log term in the Nernst equation.

SOLUTION

(a) Pb^{2+} is in the anode compartment, and Ag^+ is in the cathode compartment. Suppose that the original concentrations of Pb^{2+} and Ag^+ are 1 M, so that $E = E°$. Increasing $[Pb^{2+}]$ to 10 M gives

$$E = E° - \left(\frac{0.0592\ \text{V}}{2}\right)\left(\log\frac{(10)}{(1)^2}\right)$$

Because $\log 10 = 1.0$, $E = E° - 0.03$ V. Thus, increasing the Pb^{2+} concentration by a factor of 10 decreases the cell voltage by 0.03 V.

(b) Increasing $[Ag^+]$ to 10 M gives

$$E = E° - \left(\frac{0.0592\ \text{V}}{2}\right)\left(\log\frac{(1)}{(10)^2}\right)$$

Because $\log (10)^{-2} = -2.0$, $E = E° + 0.06$ V. Thus, increasing the Ag^+ concentration by a factor of 10 increases the cell voltage by 0.06 V.

CHECK

We expect that the reaction will have a lesser tendency to occur when the product ion concentration, $[Pb^{2+}]$, is increased and a greater tendency to occur when the reactant ion concentration, $[Ag^+]$, is increased. Therefore, the cell voltage E will decrease when $[Pb^{2+}]$ is increased and will increase when $[Ag^+]$ is increased. The prediction and the solution agree.

continued on next page

▸ **CONCEPTUAL PRACTICE 19.19**   Consider the following galvanic cell:

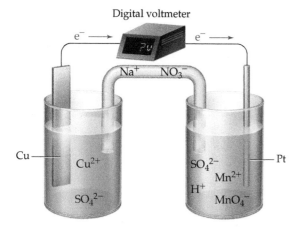

Digital voltmeter

What is the change in the cell voltage when the ion concentration in the anode compartment is increased by a factor of 100?

▸ **CONCEPTUAL APPLY 19.20**   Consider the following galvanic cell:

Digital voltmeter

What is the Zn^{2+}:Cu^{2+} concentration ratio at 25 °C if the measured cell potential is 1.16 V?

19.8 ELECTROCHEMICAL DETERMINATION OF pH

The electrochemical determination of pH using a pH meter is a particularly important application of the Nernst equation. Consider, for example, a cell with a hydrogen electrode as the anode and a second reference electrode as the cathode:

$$Pt\,|\,H_2(1\text{ atm})\,|\,H^+(?\text{ M})\,\|\,\text{Reference cathode}$$

The hydrogen electrode consists of a platinum wire that is in contact with H_2 at 1 atm and dips into a solution of unknown pH. The potential of this cell is

$$E_{cell} = E_{H_2 \longrightarrow H^+} + E_{ref}$$

We can calculate the potential for the hydrogen electrode half-reaction

$$H_2(g) \longrightarrow 2\,H^+(aq) + 2\,e^-$$

by applying the Nernst equation to this half-reaction:

$$E_{H_2 \longrightarrow H^+} = (E°_{H_2 \longrightarrow H^+}) - \left(\frac{0.0592\text{ V}}{n}\right)\left(\log\frac{[H^+]^2}{P_{H_2}}\right)$$

Since $E° = 0$ V for the standard hydrogen electrode, $n = 2$, and $P_{H_2} = 1$ atm, we can rewrite this equation as

$$E_{H_2 \longrightarrow H^+} = -\left(\frac{0.0592\text{ V}}{2}\right)(\log[H^+]^2)$$

Further, because $\log[H^+]^2 = 2\log[H^+]$ and $-\log[H^+] = pH$, the half-cell potential for the hydrogen electrode is directly proportional to the pH:

$$E_{H_2 \longrightarrow H^+} = -\left(\frac{0.0592\text{ V}}{2}\right)(2)(\log[H^+]) = (0.0592\text{ V})(-\log[H^+])$$

$$= (0.0592\text{ V})(pH)$$

The overall cell potential is

$$E_{cell} = (0.0592\text{ V})(pH) + E_{ref}$$

and the pH is therefore a linear function of the cell potential:

$$pH = \frac{E_{cell} - E_{ref}}{0.0592 \text{ V}}$$

Thus, a higher cell potential indicates a higher pH, meaning that we can measure the pH of a solution simply by measuring E_{cell}.

In actual pH measurements, a *glass electrode* replaces the cumbersome hydrogen electrode and a *calomel electrode* is used as the reference. A glass electrode consists of a silver wire coated with silver chloride that dips into a reference solution of dilute hydrochloric acid (**FIGURE 19.7**). The hydrochloric acid is separated from the test solution of unknown pH by a thin glass membrane. For rugged applications, the exterior of the electrode is made of an epoxy resin, which protects the glass tip. A calomel electrode consists of mercury(I) chloride (Hg_2Cl_2, commonly called calomel) in contact with liquid mercury and aqueous KCl. The cell half-reactions are

$$2 \times [Ag(s) + Cl^-(aq) \longrightarrow AgCl(s) + e^-] \quad E° = -0.22 \text{ V}$$
$$Hg_2Cl_2(s) + 2 \text{ e}^- \longrightarrow 2 \text{ Hg}(l) + 2 \text{ Cl}^-(aq) \quad E° = 0.28 \text{ V}$$

The overall cell potential, E_{cell}, depends not only on the potentials of these two half-reactions but also on the boundary potential that develops across the thin glass membrane separating the reference HCl solution from the test solution. Because the boundary potential depends linearly on the difference in the pH of the solutions on the two sides of the membrane, the pH of the test solution can be determined by measuring E_{cell}. The cell potential is measured with a pH meter, a voltage-measuring device that electronically converts E_{cell} to pH and displays the result in pH units.

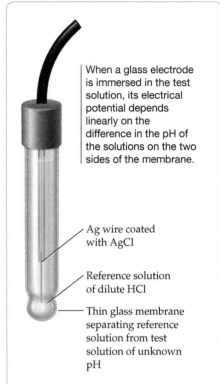

When a glass electrode is immersed in the test solution, its electrical potential depends linearly on the difference in the pH of the solutions on the two sides of the membrane.

Ag wire coated with AgCl

Reference solution of dilute HCl

Thin glass membrane separating reference solution from test solution of unknown pH

▲ **FIGURE 19.7**
A glass electrode consists of a silver wire coated with silver chloride that dips into a reference solution of dilute hydrochloric acid.

WORKED EXAMPLE 19.11

Using the Nernst Equation in the Electrochemical Determination of pH

The following cell has a potential of 0.55 V at 25 °C:

$$Pt(s)\,|\,H_2(1 \text{ atm})\,|\,H^+(? \text{ M})\,\|\,Cl^-(1 \text{ M})\,|\,Hg_2Cl_2(s)\,|\,Hg(l)$$

What is the pH of the solution in the anode compartment?

IDENTIFY

Known	Unknown
$E_{cell} = 0.55$ V	pH
Redox reaction (from shorthand cell notation)	
Standard reduction potentials for half cells (Appendix D)	

STRATEGY

First, read the shorthand notation to obtain the cell reaction. Then, calculate the standard cell potential by looking up the standard reduction potentials in Appendix B. Finally, apply the Nernst equation to find the pH.

SOLUTION

The cell reaction is

$$H_2(g) + Hg_2Cl_2(s) \longrightarrow 2 \text{ H}^+(aq) + 2 \text{ Hg}(l) + 2 \text{ Cl}^-(aq)$$

and the cell potential is

$$E_{cell} = E_{H_2 \longrightarrow H^+} + E_{Hg_2Cl_2 \longrightarrow Hg} = 0.55 \text{ V}$$

Because the reference electrode is the standard calomel electrode, which has $E = E° = 0.28$ V (Appendix D), the half-cell potential for the hydrogen electrode is 0.27 V:

$$E_{H_2 \longrightarrow H^+} = E_{cell} - E_{Hg_2Cl_2 \longrightarrow Hg} = 0.55 \text{ V} - 0.28 \text{ V} = 0.27 \text{ V}$$

We can then apply the Nernst equation to the half-reaction $H_2(g) \longrightarrow 2 \text{ H}^+(aq) + 2 \text{ e}^-$:

$$E_{H_2 \longrightarrow H^+} = (E°_{H_2 \longrightarrow H^+}) - \left(\frac{0.0592 \text{ V}}{n}\right)\left(\log \frac{[H^+]^2}{P_{H_2}}\right)$$

Substituting in the values of E, $E°$, n, and P_{H_2} gives

$$0.27 \text{ V} = (0 \text{ V}) - \left(\frac{0.0592 \text{ V}}{2}\right)\left(\log \frac{[H^+]^2}{1}\right) = (0.0592 \text{ V})(\text{pH})$$

Therefore, the pH is

$$pH = \frac{0.27 \text{ V}}{0.0592 \text{ V}} = 4.6$$

▶ **PRACTICE 19.21** What is the pH of the solution in the anode compartment of the following cell if the measured cell potential at 25 °C is 0.28 V?

$$Pt(s)\,|\,H_2(1 \text{ atm})\,|\,H^+(? \text{ M})\,\|\,Pb^{2+}(1 \text{ M})\,|\,Pb(s)$$

▶ **APPLY 19.22** If the pH in the anode compartment is 7.0, what is the cell potential at 25 °C?

$$Pt(s)\,|\,H_2(1 \text{ atm})\,|\,H^+(? \text{ M})\,\|\,Cl^-(1 \text{ M})\,|\,Hg_2Cl_2(s)\,|\,Hg(l)$$

19.9 STANDARD CELL POTENTIALS AND EQUILIBRIUM CONSTANTS

We saw in Section 19.4 that the standard free-energy change for a reaction is related to the standard cell potential by the equation

$$\Delta G° = -nFE°$$

In addition, we showed in Section 18.11 that the standard free-energy change is also related to the equilibrium constant for the reaction:

$$\Delta G° = -RT \ln K$$

Combining these two equations, we obtain

$$-nFE° = -RT \ln K$$

Or

> **Relationship between standard cell potential and equilibrium constant**
>
> $$E° = \frac{RT}{nF} \ln K = \frac{2.303RT}{nF} \log K$$

Since $2.303\,RT/F$ has a value of 0.0592 V at 25 °C, we can rewrite this equation in the following simplified form, which relates the standard cell potential and the equilibrium constant for any redox reaction:

> **Simplified equation relating standard cell potential and equilibrium constant**
>
> $$E° = \frac{0.0592\ \text{V}}{n} \log K \quad \text{in volts, at 25°C}$$

Very small concentrations are difficult to measure, so the determination of an equilibrium constant from concentration measurements is not feasible when K is either very large or very small. Standard cell potentials, however, are relatively easy to measure. Consequently, the most common use of this equation is in calculating equilibrium constants from standard cell potentials.

As an example, let's calculate the value of K for the reaction in the Daniell cell:

$$\text{Zn}(s) + \text{Cu}^{2+}(aq) \longrightarrow \text{Zn}^{2+}(aq) + \text{Cu}(s) \quad E° = 1.10 \text{ V}$$

Solving for $\log K$ in the simplified equation that relates $E°$ and $\log K$ and substituting in the appropriate values of $E°$ and n, we obtain

$$E° = \frac{0.0592\ \text{V}}{n} \log K$$

$$\log K = \frac{nE°}{0.0592\ \text{V}} = \frac{(2)(1.10\ \text{V})}{0.0592\ \text{V}} = 37.2$$

$$K = \text{antilog } 37.2 = 10^{37.2} = 2 \times 10^{37} \quad \text{at 25 °C}$$

Because $K = [\text{Zn}^{2+}]/[\text{Cu}^{2+}]$ is a very large number, the reaction goes essentially to completion. For example, in equilibrium when $[\text{Zn}^{2+}] = 1$ M, $[\text{Cu}^{2+}]$ is less than 10^{-37} M.

The preceding calculation shows that even a relatively small value of $E°$ (+1.10 V) corresponds to a huge value of K (2×10^{37}). A positive value of $E°$ corresponds to a positive value of $\log K$ and therefore $K > 1$, and a negative value of $E°$ corresponds to a negative value of $\log K$ and therefore $K < 1$. Because the standard reduction potentials in Table 19.1 span a range of about 6 V, $E°$ for a redox reaction can range from +6 V

REMEMBER . . .
The equation $\Delta G° = -RT \ln K$ follows directly from the equation $\Delta G = \Delta G° + RT \ln Q$ because, at equilibrium, $\Delta G = 0$ and $Q = K$ (Section 18.11).

Go to eText

BIG IDEA Question 5

A redox reaction has a standard cell potential ($E°$) of −0.78 V. How do the concentrations of reactants and products compare in equilibrium?

for reaction of the strongest oxidizing agent with the strongest reducing agent to -6 V for reaction of the weakest oxidizing agent with the weakest reducing agent. However, $E°$ values outside the range $+3$ V to -3 V are uncommon. For the case of $n = 2$, the correspondence between the values of $E°$ and K is indicated in **FIGURE 19.8.** Equilibrium constants for redox reactions tend to be either very large or very small compared to equilibrium constants for acid–base reactions, which are in the range of 10^{14} to 10^{-14}. Thus, redox reactions typically go either essentially to completion or almost not at all.

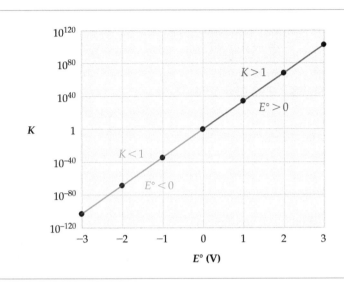

◀ **FIGURE 19.8**
The relationship between the equilibrium constant K for a redox reaction with $n = 2$ and the standard cell potential $E°$. Note that K is plotted on a logarithmic scale.

In previous chapters, we discussed two different ways to determine the value of an equilibrium constant K: from concentration data (Section 15.2) and from thermochemical data (Section 18.11). In this section, we've added a third way: from electrochemical data. The following are the key relationships needed for each approach:

1. K from concentration data for solutes: $K = \dfrac{[C]^c[D]^d}{[A]^a[B]^b}$

2. K from thermochemical data: $\Delta G° = -RT \ln K$; $\ln K = \dfrac{-\Delta G°}{RT}$

3. K from electrochemical data: $E° = \dfrac{RT}{nF} \ln K$; $\ln K = \dfrac{nFE°}{RT}$

WORKED EXAMPLE 19.12

Using Standard Reduction Potentials to Calculate an Equilibrium Constant

Use the standard reduction potentials in Table 19.1 to calculate the equilibrium constant at 25 °C for the reaction

$$6\,Br^-(aq) + Cr_2O_7^{2-}(aq) + 14\,H^+(aq) \longrightarrow 3\,Br_2(aq) + 2\,Cr^{3+}(aq) + 7\,H_2O(l)$$

IDENTIFY

Known	Unknown
Standard reduction potentials ($E°$) for half-reactions	Equilibrium constant (K)

STRATEGY

Calculate $E°$ for the reaction from standard reduction potentials, as in Worked Example 19.7. Then use the equation $\log K = nE°/0.0592$ V to determine the equilibrium constant.

continued on next page

SOLUTION

Find the relevant half-reactions in Table 19.1, and write them in the proper direction for the oxidation of Br^- and reduction of $Cr_2O_7^{2-}$. Before adding the half-reactions to get the overall reaction, multiply the Br^-/Br_2 half-reaction by a factor of 3 so that the electrons will cancel:

$$3 \times [2\,Br^-(aq) \longrightarrow Br_2(aq) + 2\,e^-] \qquad\qquad\qquad E° = -1.09\,V$$

$$\underline{Cr_2O_7^{2-}(aq) + 14\,H^+(aq) + 6\,e^- \longrightarrow 2\,Cr^{3+}(aq) + 7\,H_2O(l) \qquad E° = 1.36\,V}$$

$$6\,Br^-(aq) + Cr_2O_7^{2-}(aq) + 14\,H^+(aq) \longrightarrow 3\,Br_2(aq) + 2\,Cr^{3+}(aq) + 7\,H_2O(l) \quad E° = 0.27\,V$$

Note that $E°$ for the Br^-/Br_2 oxidation is the negative of the tabulated standard reduction potential (1.09 V), and remember that we don't multiply this $E°$ value by a factor of 3 because electrical potential is an intensive property. The $E°$ value for the overall reaction is the sum of the $E°$ values for the half-reactions: $-1.09\,V + 1.36\,V = 0.27\,V$. To calculate the equilibrium constant, use the relation between $\log K$ and $nE°$, with $n = 6$:

$$\log K = \frac{nE°}{0.0592\,V} = \frac{(6)(0.27\,V)}{0.0592\,V} = 27 \quad K = 1 \times 10^{27} \quad \text{at } 25\,°C$$

CHECK

$E°$ is positive, so K should be greater than 1, in agreement with the solution.

▶ **PRACTICE 19.23** Use the data in Table 19.1 to calculate the equilibrium constant at 25 °C for the reaction

$$4\,Fe^{2+}(aq) + O_2(g) + 4\,H^+(aq) \longrightarrow 4\,Fe^{3+}(aq) + 2\,H_2O(l)$$

▶ **APPLY 19.24** What is the value of $E°$ for a redox reaction involving the transfer of 2 mol of electrons if its equilibrium constant is 1.8×10^{-5}?

19.10 BATTERIES

By far the most important practical application of galvanic cells is their use as *batteries*. In multicell batteries, such as those in automobiles, the individual galvanic cells are linked in series, with the anode of each cell connected to the cathode of the adjacent cell. The voltage provided by the battery is the sum of the individual cell voltages.

The features required in a battery depend on the application. In general, though, a commercially successful battery should be compact, lightweight, physically rugged, and inexpensive, and it must provide a stable source of power for relatively long periods of time. Battery design is an active area of research that requires considerable ingenuity as well as a solid understanding of electrochemistry. Let's look at several of the most common types of commercial batteries.

Lead Storage Battery

The *lead storage battery* is perhaps the most familiar of all galvanic cells because it has been used as a reliable source of power for starting automobiles for nearly a century. A typical 12 V battery consists of six cells connected in series, each cell providing a potential of about 2 V. The cell design is illustrated in **FIGURE 19.9**. The anode, a series of lead grids packed with spongy lead, and the cathode, a second series of grids packed with lead dioxide, dip into the electrolyte, an aqueous solution of sulfuric acid (38% by weight). When the cell is discharging (providing current) the electrode half-reactions and the overall cell reaction are:

Anode: $\quad Pb(s) + HSO_4^-(aq) \longrightarrow PbSO_4(s) + H^+(aq) + 2\,e^-$	$E° = 0.296\,V$
Cathode: $\quad PbO_2(s) + 3\,H^+(aq) + HSO_4^-(aq) + 2\,e^- \longrightarrow PbSO_4(s) + 2\,H_2O(l)$	$E° = 1.628\,V$
Overall: $\quad Pb(s) + PbO_2(s) + 2\,H^+(aq) + 2\,HSO_4^-(aq) \longrightarrow 2\,PbSO_4(s) + 2\,H_2O(l)$	$E° = 1.924\,V$

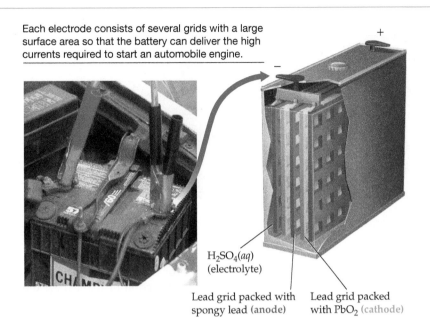

Each electrode consists of several grids with a large surface area so that the battery can deliver the high currents required to start an automobile engine.

+

−

$H_2SO_4(aq)$
(electrolyte)

Lead grid packed with spongy lead (anode)   Lead grid packed with PbO_2 (cathode)

◀ FIGURE 19.9

A lead storage battery and a cutaway view of one cell.

Note that these equations contain HSO_4^- ions rather than SO_4^{2-} because SO_4^{2-} is protonated in strongly acidic solutions.

Lead is oxidized to lead sulfate at the anode, and lead dioxide is reduced to lead sulfate at the cathode. The cell doesn't need to have separate anode and cathode compartments because the oxidizing and reducing agents are both solids (PbO_2 and Pb) that are kept from coming in contact by the presence of insulating spacers between the grids.

Because the reaction product, solid $PbSO_4$, adheres to the surface of the electrodes, a run-down lead storage battery can be recharged by using an external source of direct current to drive the cell reaction in the reverse, nonspontaneous direction. In an automobile, the battery is continuously recharged by a device called an *alternator,* which is driven by the engine.

A lead storage battery typically provides good service for several years, but eventually the spongy $PbSO_4$ deposits on the electrodes turn into a hard, crystalline form that can't be converted back to Pb and PbO_2. Then it's no longer possible to recharge the battery and it must be replaced. Because lead is extremely toxic, replaced batteries should be recycled, as is mandated by most states.

Dry-Cell Batteries

The *dry cell,* or *Leclanché cell,* was patented in 1866 by the Frenchman Georges Leclanché (1839–1882). Used for many years as a power source for flashlights and portable radios, it consists of a zinc metal can, which serves as the anode, and an inert graphite rod surrounded by a paste of solid manganese dioxide and carbon black, which functions as the cathode (FIGURE 19.10). Surrounding the MnO_2–containing paste is the electrolyte, a moist paste of ammonium chloride and zinc chloride in starch. A dry cell is not completely dry but gets its name because the electrolyte is a viscous, aqueous paste rather than a liquid solution. The electrode reactions, which are rather complicated, can be represented in simplified form by the following equations:

Anode: $Zn(s) \longrightarrow Zn^{2+}(aq) + 2\,e^-$

Cathode: $2\,MnO_2(s) + 2\,NH_4^+(aq) + 2\,e^- \longrightarrow Mn_2O_3(s) + 2\,NH_3(aq) + H_2O(l)$

The Leclanché cell is largely of historical interest and has been displaced by the *alkaline dry cell,* a modified version in which the acidic NH_4Cl electrolyte of the Leclanché

> **FIGURE 19.10**
> Leclanché dry cell and a cutaway view.

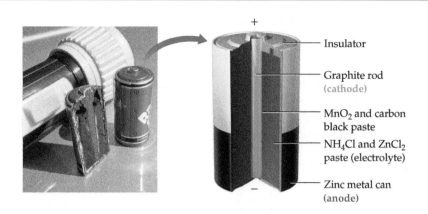

Insulator

Graphite rod
(cathode)

MnO_2 and carbon
black paste

NH_4Cl and $ZnCl_2$
paste (electrolyte)

Zinc metal can
(anode)

cell is replaced by a basic electrolyte, either NaOH or KOH. As in the Leclanché cell, the electrode reactions involve oxidation of zinc and reduction of manganese dioxide, but the oxidation product is zinc oxide, as is appropriate to the basic conditions:

Anode: $Zn(s) + 2\,OH^-(aq) \longrightarrow ZnO(s) + H_2O(l) + 2\,e^-$

Cathode: $2\,MnO_2(s) + H_2O(l) + 2\,e^- \longrightarrow Mn_2O_3(s) + 2\,OH^-(aq)$

Corrosion of the zinc anode is a significant side reaction under acidic conditions because zinc reacts with H^+ (aq) to give Zn^{2+} (aq) and H_2 (g). Under basic conditions, however, the cell has a longer life because zinc corrodes more slowly. The alkaline cell also produces higher power and a more stable current and voltage because of more efficient ion transport in the alkaline electrolyte. The voltage of an alkaline cell is about 1.5 V.

Nickel–Cadmium Batteries

Nickel–cadmium, or *"ni–cad," batteries* are popular for use in calculators and portable power tools because, unlike most other dry-cell batteries, they are rechargeable. The anode of a ni–cad battery is cadmium metal, and the cathode is the nickel(III) compound NiO(OH) supported on nickel metal. The electrode reactions are

Anode: $Cd(s) + 2\,OH^-(aq) \longrightarrow Cd(OH)_2(s) + 2\,e^-$

Cathode: $NiO(OH)(s) + H_2O(l) + e^- \longrightarrow Ni(OH)_2(s) + OH^-(aq)$

Ni–cad batteries can be recharged hundreds of times because the solid products of the electrode reactions adhere to the surface of the electrodes.

Nickel–Metal Hydride Batteries

Because cadmium is an expensive, toxic, heavy metal, ni–cad batteries have been replaced in many applications by lighter, more environmentally friendly *nickel–metal hydride,* or *NiMH, batteries.* An NiMH battery delivers about the same voltage (1.2 V) as a ni–cad battery but has about twice the energy density, the amount of energy stored per unit mass. Though both batteries have the same cathode, they have different anodes. In an NiMH battery, the anode is a special metal alloy, such as $LaNi_5$, that is capable of absorbing and releasing large amounts of hydrogen at ordinary temperatures and pressures. When the cell is discharging, the electrode and overall cell reactions can be written as

▲ Rechargeable nickel–metal hydride batteries

Anode: $MH_{ab}(s) + OH^-(aq) \longrightarrow M(s) + H_2O(l) + e^-$

Cathode: $NiO(OH)(s) + H_2O(l) + e^- \longrightarrow Ni(OH)_2(s) + OH^-(aq)$

Overall: $MH_{ab}(s) + NiO(OH)(s) \longrightarrow M(s) + Ni(OH)_2(s)$

where M represents the hydrogen-absorbing metal alloy and H_{ab} denotes the absorbed hydrogen. Thus, the metal hydride is oxidized at the anode, $NiO(OH)$ is reduced at the cathode, and the overall cell reaction transfers hydrogen from the anode to the cathode with no change in the KOH electrolyte.

NiMH batteries are used in a host of consumer electronic products and also in hybrid gas–electric automobiles, which are powered by both a gasoline engine and an electric motor. Battery packs consisting of many NiMH batteries supply energy to the electric motor and also store energy that is captured on braking. When the brakes are applied, the electric motor acts as a generator and recharges the batteries. Some of the car's kinetic energy, which would otherwise be dissipated as heat, is converted to electrical energy, thus reducing use of the gasoline engine and increasing mileage.

Lithium and Lithium-Ion Batteries

Lithium and lithium-ion batteries are even more popular for portable electronic products than nickel-based batteries because of their light weight and high voltage (3–4 V). Lithium has a higher $E°$ value for oxidation and a lower atomic weight than any other metal; only 6.94 g of lithium is needed to provide 1 mol of electrons.

Lithium batteries use a lithium metal anode and an electrolyte consisting of a lithium salt, such as $LiClO_4$, in an organic solvent. Most lithium batteries have a manganese dioxide cathode, which can absorb a variable number of Li^+ ions into its solid-state structure. The electrode reactions involve oxidation of lithium at the anode and reduction of MnO_2 at the cathode:

$$\text{Anode:} \quad x\,Li(s) \longrightarrow x\,Li^+(soln) + x\,e^-$$
$$\text{Cathode:} \quad MnO_2(s) + x\,Li^+(soln) + x\,e^- \longrightarrow Li_xMnO_2(s)$$

Li^+ ions migrate through the electrolyte from the anode to the cathode, while electrons move from the anode to the cathode through the external circuit. Lithium batteries are used in watches, calculators, and other small consumer devices.

The distinction between a lithium battery and a lithium-ion battery lies in the nature of the anode. Whereas a lithium battery uses a lithium metal anode, a lithium-ion battery uses a graphite anode that has lithium atoms inserted between its layers of carbon atoms. This so-called lithiated graphite, written as Li_xC_6, contains no lithium metal as such, which is why a lithium-ion battery isn't called a lithium battery. The usual cathode in a lithium-ion battery is CoO_2, another metal oxide that can incorporate Li^+ ions into its structure. The electrolyte is either a solution of a lithium salt in an organic solvent or a solid-state polymer electrolyte that can transport Li^+ ions.

Because an uncharged lithium-ion battery has a graphite anode and an $LiCoO_2$ cathode, the electrode reactions on discharging the battery can be written as

$$\text{Anode:} \quad Li_xC_6(s) \longrightarrow x\,Li^+(soln) + 6\,C(s) + x\,e^-$$
$$\text{Cathode:} \quad Li_{1-x}CoO_2(s) + x\,Li^+(soln) + x\,e^- \longrightarrow LiCoO_2(s)$$

When the battery is charged initially or recharged subsequently, the electrode reactions occur in the reverse direction.

Lithium-ion batteries are used in cell phones, laptop computers, digital cameras, power tools, and electric cars. Recent concerns about the possible overheating and explosion of lithium ion batteries are being addressed by the development of alternative electrode materials, such as lithium titanate (Li_2TiO_3) and lithium iron phosphate $(LiFePO_4)$, which have greater thermal stability than $LiCoO_2$.

19.11 CORROSION

Corrosion is the oxidative deterioration of a metal, such as the conversion of iron to rust, a hydrated iron(III) oxide of approximate composition $Fe_2O_3 \cdot H_2O$. The rusting of iron has enormous economic consequences. It has been estimated that as much as

▲ The Tesla Motors Model 3 is marketed as an affordable zero-emission electric vehicle, and the price starts at \$36,000. It has a top speed of 141 mph, a 0-to-60 mph acceleration time of 5.1 s, and a 220-mile range. The vehicle is powered by 6831 lithium ion batteries.

▲ This steel railroad trestle collapsed during a tornado due to corrosion of the bolts that connected the structure to its concrete footings.

one-fourth of the steel produced in the United States goes to replace steel structures and products that have been destroyed by corrosion.

To prevent corrosion, we first have to understand how it occurs. One important fact is that the rusting of iron requires both oxygen and water; it doesn't occur in oxygen-free water or in dry air. Another clue is the observation that rusting involves pitting of the metal surface, but the rust is deposited at a location physically separated from the pits. This suggests that rust does not form by direct reaction of iron and oxygen but rather by an electrochemical process in which iron is oxidized in one region of the surface and oxygen is reduced in another region.

A possible mechanism for rusting, consistent with the known facts, is illustrated in **FIGURE 19.11.** The surface of the iron and a droplet of surface water constitute a tiny galvanic cell in which different regions of the surface act as anode and cathode while the aqueous phase serves as the electrolyte. Iron is oxidized more readily in some regions (anode regions) than in others (cathode regions) because the composition of the metal is somewhat inhomogeneous and the surface is irregular. Factors such as impurities, phase boundaries, and mechanical stress may influence the ease of oxidation in a particular region of the surface.

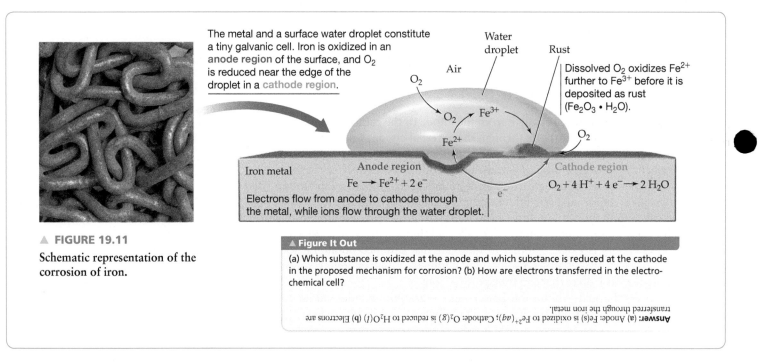

▲ **FIGURE 19.11**

Schematic representation of the corrosion of iron.

▲ **Figure It Out**

(a) Which substance is oxidized at the anode and which substance is reduced at the cathode in the proposed mechanism for corrosion? (b) How are electrons transferred in the electrochemical cell?

Answer: (a) Anode: Fe(s) is oxidized to $Fe^{2+}(aq)$; Cathode: $O_2(g)$ is reduced to $H_2O(l)$ (b) Electrons are transferred through the iron metal.

At an anode region, iron is oxidized to Fe^{2+} ions,

$$Fe(s) \longrightarrow Fe^{2+}(aq) + 2\ e^- \quad E° = 0.45\ V$$

while at a cathode region, oxygen is reduced to water:

$$O_2(g) + 4\ H^+(aq) + 4\ e^- \longrightarrow 2\ H_2O(l) \quad E° = 1.23\ V$$

The actual potential for the reduction half-reaction is less than the standard potential (1.23 V) because the water droplet is not 1 M in H^+ ions. (In fact, the water is only slightly acidic because the main source of H^+ ions is the reaction of water with dissolved atmospheric carbon dioxide.) Even at pH 7, however, the potential for the reduction half-reaction is 0.81 V, which means that the cell potential is highly positive, indicative of a spontaneous reaction.

The electrons required for reduction of O_2 at the cathode region are supplied by a current that flows through the metal from the more easily oxidized anode region (Figure 19.11). The electrical circuit is completed by migration of ions in the water droplet. When Fe^{2+} ions migrate away from the pitted anode region, they come in contact with O_2 dissolved in the surface portion of the water droplet and are further oxidized to Fe^{3+} ions:

$$4\,Fe^{2+}(aq) + O_2(g) + 4\,H^+(aq) \longrightarrow 4\,Fe^{3+}(aq) + 2\,H_2O(l)$$

Iron(III) forms a very insoluble hydrated oxide even in moderately acidic solutions, and so the iron(III) is deposited as the familiar red-brown material that we call rust:

$$\underset{\text{Rust}}{2\,Fe^{3+}(aq) + 4\,H_2O(l) \longrightarrow Fe_2O_3 \cdot H_2O(s) + 6\,H^+(aq)}$$

An electrochemical mechanism for corrosion also explains nicely why automobiles rust more rapidly in places where road salt is used to melt snow and ice. Dissolved salts in the water droplet greatly increase the conductivity of the electrolyte, thus accelerating the pace of corrosion.

A glance at a table of standard reduction potentials indicates that the O_2/H_2O half-reaction lies above the M^{n+}/M half-reaction for nearly all metals, so O_2 can oxidize all metals except a few, such as gold and platinum. Aluminum, for example, has $E° = -1.66$ V for the Al^{3+}/Al half-reaction and is oxidized more readily than iron. In other words, the corrosion of aluminum products such as aircraft and automobile parts, window frames, cooking utensils, and soda cans should be a serious problem. Fortunately, it isn't, because oxidation of aluminum gives a very hard, almost impenetrable film of Al_2O_3 that adheres to the surface of the metal and protects it from further contact and reaction with oxygen. Other metals such as magnesium, chromium, titanium, and zinc form similar protective oxide coatings. In the case of iron, however, rust is too porous to shield the underlying metal from further oxidation.

Prevention of Corrosion

Corrosion of iron can be prevented, or at least minimized, by shielding the metal surface from oxygen and moisture. A coat of paint is effective for a while, but rust begins to form as soon as the paint is scratched or chipped. Metals such as tin or zinc afford a more durable surface coating for iron. The steel used in making automobiles, for example, is coated by dipping it into a bath of molten zinc, a process known as **galvanizing**. As the potentials indicate, zinc is oxidized more easily than iron, and therefore, when the metal corrodes, zinc is oxidized instead of iron.

$$Fe^{2+}(aq) + 2\,e^- \longrightarrow Fe(s) \qquad E° = -0.45\text{ V}$$
$$Zn^{2+}(aq) + 2\,e^- \longrightarrow Zn(s) \qquad E° = -0.76\text{ V}$$

Any incipient oxidation of iron would be reversed immediately because Zn can reduce Fe^{2+} to Fe.

Oxidation:	$Zn(s) \longrightarrow Zn^{2+}(aq) + 2\,e^-$	$E° = 0.76$ V
Reduction:	$Fe^{2+}(aq) + 2\,e^- \longrightarrow Fe(s)$	$E° = -0.45$ V
Overall:	$Fe^{2+}(aq) + Zn(s) \longrightarrow Fe(s) + Zn^{2+}(aq)$	$E° = 0.31$ V

As long as the zinc and iron are in contact, the zinc protects the iron from oxidation even if the zinc layer becomes scratched (**FIGURE 19.12**).

The technique of protecting a metal from corrosion by connecting it to a second metal that is more easily oxidized is called **cathodic protection**. It's unnecessary to cover the entire surface of the metal with a second metal, as in galvanizing iron. All that's required is electrical contact with the second metal. An underground steel pipeline, for example, can be protected by connecting it through an insulated wire to a

▲ This titanium bicycle doesn't corrode because a hard, impenetrable layer of TiO_2 adheres to the surface and protects the metal from further oxidation.

BIG IDEA Question 6

If zinc could not be used to produce galvanized steel, what other metal could serve as a substitute: Ni or Al? (Refer to Table 19.1 for standard reduction potentials.)

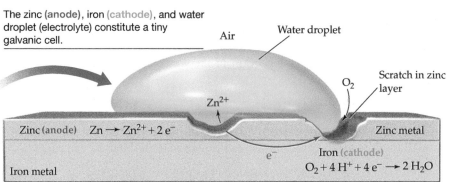

The zinc (anode), iron (cathode), and water droplet (electrolyte) constitute a tiny galvanic cell.

Air

Water droplet

Zn^{2+}

O_2

Scratch in zinc layer

Zinc (anode) $Zn \longrightarrow Zn^{2+} + 2\,e^-$

Zinc metal

Iron metal

e^-

Iron (cathode)
$O_2 + 4\,H^+ + 4\,e^- \longrightarrow 2\,H_2O$

▲ **FIGURE 19.12**

Galvanized iron. A layer of zinc protects iron from oxidation, even when the zinc layer becomes scratched.

Oxygen is reduced at the cathode, and zinc is oxidized at the anode, thus protecting the iron from oxidation.

▲ **Figure It Out**

How does zinc protect iron from rusting?

Answer: Zinc is more easily oxidized than iron and therefore zinc is oxidized by oxygen instead of iron.

▲ A ship's steel propeller and housing with an attached strip of zinc that serves as a sacrificial anode. The zinc corrodes instead of the iron.

stake of magnesium, which acts as a *sacrificial anode* and corrodes instead of the iron. In effect, the arrangement is a galvanic cell in which the easily oxidized magnesium acts as the anode, the pipeline behaves as the cathode, and moist soil is the electrolyte. The cell half-reactions are

Anode: $Mg(s) \longrightarrow Mg^{2+}(aq) + 2\,e^-$ $E° = 2.37\ V$

Cathode: $O_2(g) + 4\,H^+(aq) + 4\,e^- \longrightarrow 2\,H_2O(l)$ $E° = 1.23\ V$

For large steel structures such as pipelines, storage tanks, bridges, and ships, cathodic protection is the best defense against premature rusting.

19.12 ELECTROLYSIS AND ELECTROLYTIC CELLS

Thus far, we've been concerned only with galvanic cells—electrochemical cells in which a spontaneous redox reaction produces an electric current. A second important kind of electrochemical cell is the electrolytic cell, in which an electric current is used to drive a nonspontaneous reaction. Thus, the processes occurring in galvanic and electrolytic cells are the reverse of each other: A galvanic cell converts chemical energy to electrical energy when a reaction with a positive value of E (and a negative value of ΔG) proceeds toward equilibrium; an electrolytic cell converts electrical energy to chemical energy when an electric current drives a reaction with a negative value of E (and a positive value of ΔG) in a direction away from equilibrium.

The process of using an electric current to bring about chemical change is called **electrolysis.** The opposite signs of E and $\Delta G = -nFE$ for the two kinds of cells are summarized in **TABLE 19.2,** along with the situation for a reaction that has reached equilibrium—a dead battery!

TABLE 19.2   Relationship between Cell Potentials E and Free-Energy Changes ΔG

Reaction Type	E	ΔG	Cell Type
Spontaneous	+	−	Galvanic (battery)
Nonspontaneous	−	+	Electrolytic
Equilibrium	0	0	Dead battery

Electrolysis of Molten Sodium Chloride

An electrolytic cell has two electrodes that dip into an electrolyte and are connected to a battery or some other source of direct electric current. A cell for electrolysis of molten sodium chloride, for example, is illustrated in **FIGURE 19.13**. The battery serves as an electron pump, pushing electrons into one electrode and pulling them out of the other. The negative electrode attracts Na^+ cations, which combine with the electrons supplied by the battery and are thereby reduced to liquid sodium metal. Similarly, the positive electrode attracts Cl^- anions, which replenish the electrons removed by the battery and are thereby oxidized to chlorine gas. The electrode reactions and overall cell reaction are

Anode (oxidation):	$2\ Cl^-(l) \longrightarrow Cl_2(g) + 2\ e^-$
Cathode(reduction):	$2\ Na^+(l) + 2\ e^- \longrightarrow 2\ Na(l)$
Overall cell reaction:	$2\ Na^+(l) + 2\ Cl^-(l) \longrightarrow 2\ Na(l) + Cl_2(g)$

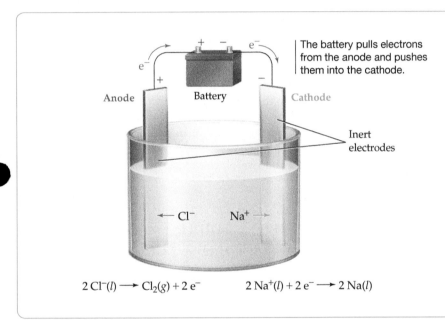

$$2\ Cl^-(l) \longrightarrow Cl_2(g) + 2\ e^- \qquad 2\ Na^+(l) + 2\ e^- \longrightarrow 2\ Na(l)$$

◀ **FIGURE 19.13**

Electrolysis of molten sodium chloride. Chloride ions are oxidized to Cl_2 gas at the anode, and Na^+ ions are reduced to sodium metal at the cathode.

◀ **Figure It Out**

Compare the direction of electron flow and the charges on the electrodes in this electrolytic cell with a galvanic cell (Figure 19.3). How do the two cells differ?

Answer: In both a galvanic and an electrolytic cell, electrons flow from the anode to cathode. In a galvanic cell the anode is negative and the cathode is positive, but in an electrolytic cell the cathode is positive and the anode is negative.

As in a galvanic cell, the anode is the electrode where oxidation takes place and the cathode is the electrode where reduction takes place. The signs of the electrodes, however, are opposite for galvanic and electrolytic cells. In a galvanic cell, the anode is considered to be negative because it supplies electrons to the external circuit, but in an electrolytic cell, the anode is considered to be positive because electrons are pulled out of it by the battery. The sign of each electrode in the electrolytic cell is the same as the sign of the battery electrode to which it is attached.

Electrolysis of Aqueous Sodium Chloride

When an aqueous salt solution is electrolyzed, the electrode reactions may differ from those for electrolysis of the molten salt because water may be involved. In the electrolysis of aqueous sodium chloride, for example, the cathode half-reaction might be either the reduction of Na^+ to sodium metal, as in the case of molten sodium chloride, or the reduction of water to hydrogen gas:

$$Na^+(aq) + e^- \longrightarrow Na(s) \qquad\qquad E° = -2.71\ V$$

$$2\ H_2O(l) + 2\ e^- \longrightarrow H_2(g) + 2\ OH^-(aq) \qquad E° = -0.83\ V$$

Because the standard potential is much less negative for the reduction of water than for the reduction of Na^+, water is reduced preferentially and bubbles of hydrogen gas are produced at the cathode.

The anode half-reaction might be either oxidation of Cl^- to Cl_2 gas, as in the case of molten sodium chloride, or oxidation of water to oxygen gas:

$$2\ Cl^-(aq) \longrightarrow Cl_2(g) + 2\ e^- \qquad\qquad E° = -1.36\ V$$
$$2\ H_2O(l) \longrightarrow O_2(g) + 4\ H^+(aq) + 4\ e^- \qquad E° = -1.23\ V$$

Based on the $E°$ values, we might expect a slight preference for oxidation of water in a solution having 1 M ion concentrations. For a neutral solution ($[H^+] = 10^{-7}\ M$), the preference for water oxidation will be even greater because its oxidation potential at pH 7 is $-0.81\ V$. The observed product at the anode, however, is Cl_2, not O_2, because of a phenomenon called *overpotential*.

Experiments indicate that the applied voltage required for an electrolysis is always greater than the voltage calculated from standard reduction potentials. The additional voltage required is the **overpotential**, also called the *overvoltage*. The overpotential is needed because the rate of electron transfer at the electrode–solution interface for one or both of the cell half-reactions is often slow, thus limiting the amount of current that passes through an electrolytic cell. For electrode half-reactions involving solution or deposition of metals, the overpotential is quite small, but for half-reactions that produce O_2 (or H_2) gas, the overpotential can be as large as 1 V. Present theory is unable to predict the magnitude of the overpotential, but it's known that the overpotential for the formation of O_2 is much larger than that for the formation of Cl_2. Because of overpotential, it's sometimes difficult to predict which half-reaction will occur when $E°$ values for the competing half-reactions are similar. In such cases, only experiment can tell what actually happens.

The observed electrode reactions and overall cell reaction for electrolysis of aqueous sodium chloride are

Anode (oxidation):	$2\ Cl^-(aq) \longrightarrow Cl_2(g) + 2\ e^-$	$E° = -1.36\ V$
Cathode (reduction):	$2\ H_2O(l) + 2\ e^- \longrightarrow H_2(g) + 2\ OH^-(aq)$	$E° = -0.83\ V$

Overall cell reaction: $2\ Cl^-(aq) + 2\ H_2O(l) \longrightarrow Cl_2(g) + H_2(g) + 2\ OH^-(aq)$ $E° = -2.19\ V$

Sodium ion acts as a spectator ion and is not involved in the electrode reactions. Thus, the sodium chloride solution is converted to a sodium hydroxide solution as the electrolysis proceeds. The minimum potential required to force this nonspontaneous reaction to occur under standard-state conditions is 2.19 V plus the overpotential.

Electrolysis of Water

The electrolysis of any aqueous solution requires the presence of an electrolyte to carry the current in the solution. If the ions of the electrolyte are less easily oxidized and reduced than water, however, then water will react at both electrodes. Take, for instance, the electrolysis of an aqueous solution of the inert electrolyte Na_2SO_4. Water is oxidized at the anode in preference to SO_4^{2-} ions and is reduced at the cathode in preference to Na^+ ions. The electrode and overall cell reactions are

Anode (oxidation):	$2\ H_2O(l) \longrightarrow O_2(g) + 4\ H^+(aq) + 4\ e^-$
Cathode (reduction):	$4\ H_2O(l) + 4\ e^- \longrightarrow 2\ H_2(g) + 4\ OH^-(aq)$

Overall cell reaction: $6\ H_2O(l) \longrightarrow 2\ H_2(g) + O_2(g) + 4\ H^+(aq) + 4\ OH^-(aq)$

If the anode and cathode solutions are mixed, the H^+ and OH^- ions react to form water:

$$4\ H^+(aq) + 4\ OH^-(aq) \longrightarrow 4\ H_2O(l)$$

The net electrolysis reaction is therefore the decomposition of water, a process sometimes used in the laboratory to produce small amounts of pure H_2 and O_2:

$$2\ H_2O(l) \longrightarrow 2\ H_2(g) + O_2(g)$$

19.13 COMMERCIAL APPLICATIONS OF ELECTROLYSIS

Electrolysis is used in the manufacture of many important chemicals and in numerous processes for purification and electroplating of metals. Let's look at some examples.

Manufacture of Sodium

Sodium metal is produced commercially in a *Downs cell* by electrolysis of a molten mixture of sodium chloride and calcium chloride (**FIGURE 19.14**). The presence of $CaCl_2$ allows the cell to operate at a lower temperature because the melting point of the $NaCl$–$CaCl_2$ mixture (about 580 °C) is depressed well below that of pure $NaCl$ (801 °C). The liquid sodium produced at the cylindrical steel cathode is less dense than the molten salt and thus floats to the top part of the cell, where it is drawn off into a suitable container. Chlorine gas forms at the graphite anode, which is separated from the cathode by an iron screen to keep the highly reactive sodium and chlorine away from each other. Because the Downs process requires high currents (typically 25,000–40,000 A), plants for producing sodium are located near sources of inexpensive hydroelectric power, such as Niagara Falls, New York.

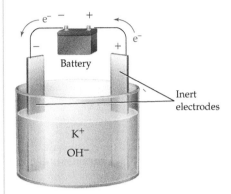

BIG IDEA Question 7

Go to eText

Metallic potassium was first prepared by Humphrey Davy in 1807 by electrolysis of molten potassium hydroxide:

What reaction occurs at the anode, and what is the sign of the anode? (Refer to Appendix D for standard reduction potentials.)

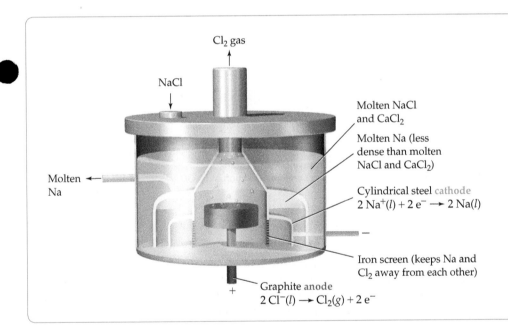

◄ **FIGURE 19.14**

Cross-sectional view of a Downs cell for commercial production of sodium metal by electrolysis of molten sodium chloride. The cell design keeps the sodium and chlorine apart from each other and out of contact with air.

Cl₂ gas

NaCl

Molten Na

Molten NaCl and $CaCl_2$

Molten Na (less dense than molten NaCl and $CaCl_2$)

Cylindrical steel cathode
$2\ Na^+(l) + 2\ e^- \longrightarrow 2\ Na(l)$

Iron screen (keeps Na and Cl₂ away from each other)

Graphite anode
$2\ Cl^-(l) \longrightarrow Cl_2(g) + 2\ e^-$

Manufacture of Chlorine and Sodium Hydroxide

Production of chlorine and sodium hydroxide by electrolysis of aqueous sodium chloride is the basis of the *chlor-alkali industry,* a business that generates annual sales of approximately $4 billion in the United States alone. Annual U.S. production is about 10 million metric tons for chlorine and 8 million metric tons for sodium hydroxide. Chlorine is used in water and sewage treatment, as a bleaching agent in manufacturing paper, and in the manufacture of plastics such as poly(vinyl chloride) (PVC). Sodium hydroxide is employed in making paper, textiles, soaps, and detergents.

FIGURE 19.15 shows the essential features of a membrane cell for commercial production of chlorine and sodium hydroxide. A saturated aqueous solution of sodium chloride (*brine*) flows into the anode compartment, where Cl^- is oxidized to Cl_2 gas, and water enters the cathode compartment, where it is converted to H_2 gas and OH^- ions. Between the anode and cathode compartments is a special polymer membrane that is permeable to cations but not to anions or water. The membrane keeps the Cl_2 and OH^- ions apart but allows a current of Na^+ ions to flow into the cathode compartment, thus carrying the current in the solution and maintaining electrical neutrality in both compartments. The Na^+ and OH^- ions flow out of the cathode compartment as an aqueous solution of NaOH.

▶ **FIGURE 19.15**

A membrane cell for electrolytic production of Cl_2 and NaOH. Chloride ion is oxidized to Cl_2 gas at the anode, and water is converted to H_2 gas and OH^- ions at the cathode.

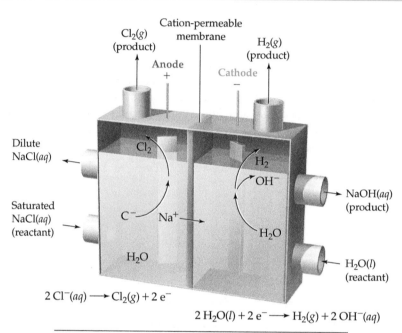

$$2\,Cl^-(aq) \longrightarrow Cl_2(g) + 2\,e^-$$

$$2\,H_2O(l) + 2\,e^- \longrightarrow H_2(g) + 2\,OH^-(aq)$$

Sodium ions move from the anode compartment to the cathode compartment through a cation-permeable membrane and then flow out of the cell as an aqueous solution of NaOH.

Manufacture of Aluminum

Although aluminum is the third most abundant element in the Earth's crust at 8.23% by mass, it remained a rare and expensive metal until 1886, when a 22-year-old American, Charles Martin Hall (1863–1914), and a 23-year-old Frenchman, Paul Héroult (1863–1914), independently devised a practical process for the electrolytic production of aluminum. Still used today, the **Hall–Héroult process** involves electrolysis of a molten mixture of aluminum oxide (Al_2O_3) and cryolite (Na_3AlF_6) at about 1000 °C in a cell with graphite electrodes (**FIGURE 19.16**). Electrolysis of pure Al_2O_3 is impractical because it melts at a very high temperature (2045 °C), and electrolysis of aqueous Al^{3+} solutions is not feasible because water is reduced in preference to Al^{3+} ions. Thus, the use of cryolite as a solvent for Al_2O_3 is the key to the success of the Hall–Héroult process.

The electrode reactions are still not fully understood, but they probably involve complex anions of the type $AlF_xO_y^{+3-x-2y}$, formed by the reaction of Al_2O_3 and Na_3AlF_6. The complex anions are reduced at the cathode to molten aluminum metal and are oxidized at the anode to O_2 gas, which reacts with the graphite anodes to give CO_2 gas.

▲ Alcoa aluminum plant along the Columbia River at Wenatchee, Washington.

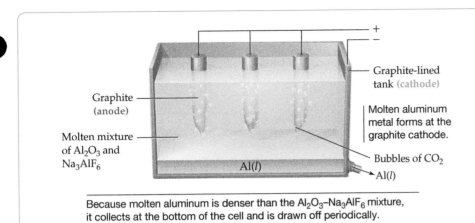

An electrolytic cell for production of aluminum by the Hall–Héroult process.

Graphite (anode)

Molten mixture of Al_2O_3 and Na_3AlF_6

Graphite-lined tank (cathode)

Molten aluminum metal forms at the graphite cathode.

Bubbles of CO_2

Al(l)

Al(l)

Because molten aluminum is denser than the Al_2O_3–Na_3AlF_6 mixture, it collects at the bottom of the cell and is drawn off periodically.

As a result, the anodes are chewed up rapidly and must be replaced frequently. The cell operates at a low voltage (5–6 V) but with very high currents (up to 250,000 A) because 1 mol of electrons produces only 9.0 g of aluminum. Electrolytic production of aluminum is the largest single consumer of electricity in the United States, making recycling of aluminum products cost-effective. As with the manufacture of sodium, aluminum production generally takes place near hydroelectric plants, such as those along the Columbia River in Washington.

Electrorefining and Electroplating

The purification of a metal by means of electrolysis is called **electrorefining**. For example, impure copper obtained from ores is converted to pure copper in an electrolytic cell that has impure copper as the anode and pure copper as the cathode (**FIGURE 19.17**). The electrolyte is an aqueous solution of copper sulfate.

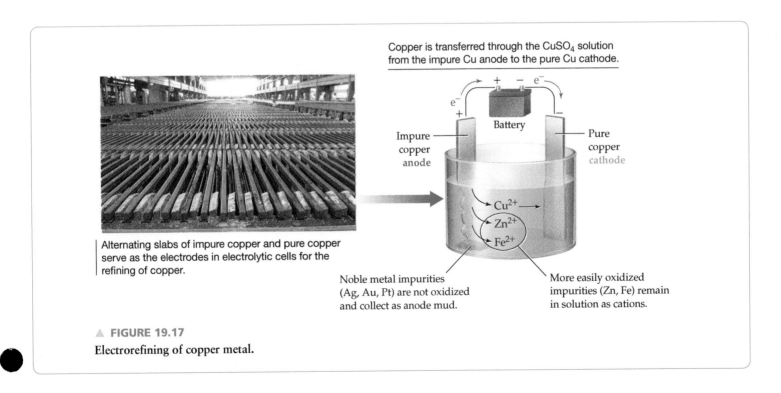

Copper is transferred through the $CuSO_4$ solution from the impure Cu anode to the pure Cu cathode.

Battery

Impure copper anode

Pure copper cathode

Cu^{2+}

Zn^{2+}

Fe^{2+}

Alternating slabs of impure copper and pure copper serve as the electrodes in electrolytic cells for the refining of copper.

Noble metal impurities (Ag, Au, Pt) are not oxidized and collect as anode mud.

More easily oxidized impurities (Zn, Fe) remain in solution as cations.

▲ **FIGURE 19.17**
Electrorefining of copper metal.

▲ Gold-plated objects are made by electroplating pure gold metal onto another metal.

BIG IDEA Question 8   📋 Go to eText

The Oscar statuette awarded by the Academy of Motion Pictures Arts and Sciences for excellence in acting and film production is made of bronze and coated in gold. In the electroplating process, to what electrode should the statue be connected, and what is the half-reaction that occurs at this electrode?

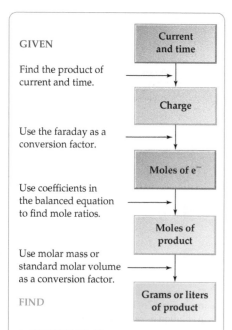

▲ **FIGURE 19.18**

Sequence of conversions used to calculate the amount of product produced by passing a current through an electrolytic cell for a fixed period of time.

At the impure Cu anode, copper is oxidized along with more easily oxidized metallic impurities such as zinc and iron. Less easily oxidized impurities such as silver, gold, and platinum fall to the bottom of the cell as *anode mud,* which is reprocessed to recover the precious metals. At the pure Cu cathode, Cu^{2+} ions are reduced to pure copper metal, but the less easily reduced metal ions (Zn^{2+}, Fe^{2+}, and so forth) remain in the solution.

$$\text{Anode (oxidation):} \quad M(s) \longrightarrow M^{2+}(aq) + 2\ e^- \quad (M = Cu, Zn, Fe)$$
$$\text{Cathode (reduction):} \quad Cu^{2+}(aq) + 2\ e^- \longrightarrow Cu(s)$$

Thus, the net cell reaction simply transfers copper metal from the impure anode to the pure cathode. The copper obtained by this process is 99.95% pure.

Closely related to electrorefining is **electroplating,** the coating of one metal on the surface of another using electrolysis. For example, steel motorcycle parts are often plated with chromium to protect them from corrosion, and silver-plating is commonly used to make items of fine table service. The object to be plated is carefully cleaned and then set up as the cathode of an electrolytic cell that contains a solution of ions of the metal to be deposited.

19.14 QUANTITATIVE ASPECTS OF ELECTROLYSIS

In the 1830s, Michael Faraday showed that the amount of substance produced at an electrode by electrolysis depends on the quantity of charge passed through the cell. For example, passing 1 mol of electrons through a Downs cell yields 1 mol (23.0 g) of sodium at the cathode:

$$\underset{\text{1 mol}}{Na^+(l)} + \underset{\text{1 mol}}{e^-} \longrightarrow \underset{\text{1 mol (23.0 g)}}{Na(l)}$$

Similarly, passing 1 mol of electrons in the Hall–Héroult process produces 1/3 mol (9.0 g) of aluminum, because 3 mol of electrons are required to reduce 1 mol of Al^{3+} to aluminum metal:

$$\underset{\text{1/3 mol}}{Al^{3+}(l)} + \underset{\text{1 mol}}{3\ e^-} \longrightarrow \underset{\text{1/3 mol (9.0 g)}}{Al(l)}$$

In general, the amount of product formed in an electrode reaction follows directly from the stoichiometry of the reaction and the molar mass of the product.

To find out how many moles of electrons pass through a cell in a particular experiment, we need to measure the electric current and the time that the current flows. The number of coulombs of charge passed through the cell equals the product of the current in amperes (coulombs per second) and the time in seconds:

$$\text{Charge (C)} = \text{Current (A)} \times \text{Time (s)}$$

Because the charge on 1 mol of electrons is 96,500 C (Section 19.4), the number of moles of electrons passed through the cell is

$$\text{Moles of } e^- = \text{Charge (C)} \times \frac{1 \text{ mol } e^-}{96,500 \text{ C}}$$

The sequence of conversions in **FIGURE 19.18** is used to calculate the mass or volume of product produced by passing a known current through a cell for a fixed period of time. The key is to think of the electrons as a reactant in a balanced chemical equation and then to proceed as with any other stoichiometry problem. Worked Example 19.13 illustrates the calculations. Alternatively, we can calculate the current (or time) required to produce a given amount of product by working through the sequence in Figure 19.18 in the reverse direction.

Go to
eText

WORKED EXAMPLE 19.13

Calculating the Amount of Product Produced by Electrolysis

A constant current of 30.0 A is passed through an aqueous solution of NaCl for 1.00 h. How many grams of NaOH and how many liters of Cl_2 gas at STP are produced?

IDENTIFY

Known	Unknown
NaCl (aq)	Amount of products (g NaOH and L of Cl_2)
Time (1.00 h)	
Current (30.0 A = 30.0 C/s)	
STP (at STP 1 mol of gas has a volume of 22.4 L)	

STRATEGY

To convert the current and time to grams or liters of product, carry out the sequence of conversions in Figure 19.18. Another strategy is to start with the known unit and use conversions to reach the desired unit.

SOLUTION

Because electrons can be thought of as a reactant in the electrolysis process, the first step is to calculate the charge and the number of moles of electrons passed through the cell:

$$\text{Charge} = \left(30.0\frac{C}{s}\right)(1.00\text{ h})\left(\frac{60\text{ min}}{1\text{ h}}\right)\left(\frac{60\text{ s}}{1\text{ min}}\right) = 1.08 \times 10^5\,C$$

$$\text{Moles of e}^- = (1.08 \times 10^5\text{ C})\left(\frac{1\text{ mol e}^-}{96,500\text{ C}}\right) = 1.12\text{ mol e}^-$$

The cathode reaction yields 2 mol of OH^- per 2 mol of electrons (Section 19.12: Electrolysis of Aqueous Sodium Chloride), so 1.12 mol of NaOH will be obtained:

$$2\text{ H}_2\text{O}(l) + 2\text{ e}^- \longrightarrow \text{H}_2(g) + 2\text{ OH}^-(aq)$$

$$\text{Moles of NaOH} = (1.12\text{ mol e}^-)\left(\frac{2\text{ mol NaOH}}{2\text{ mol e}^-}\right) = 1.12\text{ mol NaOH}$$

Converting the number of moles of NaOH to grams gives 44.8 g of NaOH:

$$\text{Grams of NaOH} = (1.12\text{ mol NaOH})\left(\frac{40.0\text{ g NaOH}}{1\text{ mol NaOH}}\right) = 44.8\text{ g NaOH}$$

As a shortcut, the entire sequence of conversions can be carried out in one step. For example, the mass of NaOH produced at the cathode is

$$\left(30.0\frac{C}{s}\right)(1.00\text{ h})\left(\frac{3600\text{ s}}{h}\right)\left(\frac{1\text{ mol e}^-}{96,500\text{ C}}\right)\left(\frac{2\text{ mol NaOH}}{2\text{ mol e}^-}\right)\left(\frac{40.0\text{ g NaOH}}{1\text{ mol NaOH}}\right)$$

$$= 44.8\text{ g NaOH}$$

The anode reaction gives 1 mol of Cl_2 per 2 mol of electrons, so 0.560 mol of Cl_2 will be obtained:

$$2\text{ Cl}^-(aq) \longrightarrow \text{Cl}_2(g) + 2\text{ e}^-$$

$$\text{Moles of Cl}_2 = (1.12\text{ mol e}^-)\left(\frac{1\text{ mol Cl}_2}{2\text{ mol e}^-}\right) = 0.560\text{ mol Cl}_2$$

Since 1 mol of an ideal gas occupies 22.4 L at STP, the volume of Cl_2 obtained is 12.5 L:

$$\text{Liters of Cl}_2 = (0.560\text{ mol Cl}_2)\left(\frac{22.4\text{ L Cl}_2}{\text{mol Cl}_2}\right) = 12.5\text{ L Cl}_2$$

As a shortcut, the entire sequence of conversions can be carried out in one step. For example, the volume of Cl_2 produced at the anode is

$$\left(30.0\frac{C}{s}\right)(1.00\text{ h})\left(\frac{3600\text{ s}}{h}\right)\left(\frac{1\text{ mol e}^-}{96,500\text{ C}}\right)\left(\frac{1\text{ mol Cl}_2}{2\text{ mol e}^-}\right)\left(\frac{22.4\text{ L Cl}_2}{\text{mol Cl}_2}\right)$$

$$= 12.5\text{ L Cl}_2$$

CHECK

Since approximately 1 mol of electrons is passed through the cell and the electrode reactions yield 1 mol of NaOH and 0.5 mol of Cl_2 per mole of electrons, 1 mol of NaOH (~40 g) and 0.5 mol of Cl_2 (~11 L at STP) will be formed. The estimate and the solution agree.

▶ **PRACTICE 19.25** How many kilograms of aluminum can be produced in 8.00 h by passing a constant current of 1.00×10^5 A through a molten mixture of aluminum oxide and cryolite?

▶ **APPLY 19.26** A layer of silver is electroplated on a coffee server using a constant current of 0.100 A. How much time is required to deposit 3.00 g of silver? (Hint: Work through the steps in Figure 19.18 in reverse order.)

INQUIRY ? How do hydrogen fuel cells work?

The need for nonpolluting energy sources has become more urgent as scientific evidence accumulates that climate change is caused by greenhouse gas emissions (Section 10.11) and that pollution has a wide range of negative health effects. Hydrogen produced from renewable sources such as wind and solar can be a clean energy solution. The direct combustion of hydrogen has been used by NASA in its space program to launch the space shuttle and other rockets into orbit, while an electrochemical reaction of hydrogen in a fuel cell is used to generate electricity onboard spacecraft. The only waste product of the fuel cell is water—pure enough for the astronauts to drink!

What is a fuel cell? A **fuel cell** is a galvanic cell in which one of the reactants is a fuel such as hydrogen or methanol. A fuel cell differs from an ordinary battery in that the reactants are not contained within the cell but instead are continuously supplied from an external reservoir. The hydrogen–oxygen fuel cell that is used as a source of electric power in space vehicles contains porous carbon electrodes impregnated with metallic catalysts and an electrolyte consisting of hot, aqueous KOH (**FIGURE 19.19**). The fuel, gaseous H_2, and the oxidizing agent, gaseous O_2, don't react directly but instead flow into separate cell compartments, where H_2 is oxidized at the anode and O_2 is reduced at the cathode. The overall cell reaction is simply the conversion of hydrogen and oxygen to water:

Anode: $\quad 2\,H_2(g) + 4\,OH^-(aq) \longrightarrow 4\,H_2O(l) + 4\,e^-$
Cathode: $O_2(g) + 2\,H_2O(l) + 4\,e^- \longrightarrow 4\,OH^-(aq)$
Overall: $\qquad 2\,H_2(g) + O_2(g) \longrightarrow 2\,H_2O(l)$

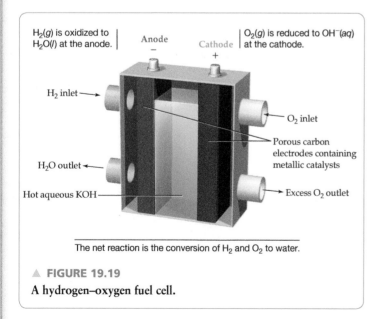

$H_2(g)$ is oxidized to $H_2O(l)$ at the anode. | $O_2(g)$ is reduced to $OH^-(aq)$ at the cathode.

The net reaction is the conversion of H_2 and O_2 to water.

▲ **FIGURE 19.19**

A hydrogen–oxygen fuel cell.

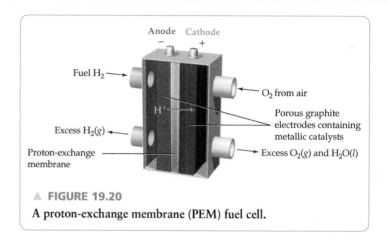

▲ **FIGURE 19.20**

A proton-exchange membrane (PEM) fuel cell.

In fuel cells designed to generate the electricity needed to power cars, buses, and power plants, the aqueous KOH electrolyte is replaced by a special polymer membrane that conducts protons but not electrons (**FIGURE 19.20**). The electrode and overall cell reactions in the *proton-exchange membrane (PEM) fuel cells* are

Anode: $\qquad\qquad\qquad 2\,H_2(g) \longrightarrow 4\,H^+(aq) + 4\,e^-$
Cathode: $\quad O_2(g) + 4\,H^+(aq) + 4\,e^- \longrightarrow 2\,H_2O(l)$
Overall: $\qquad\qquad 2\,H_2(g) + O_2(g) \longrightarrow 2\,H_2O(l)$

Protons pass through the membrane from the anode to the cathode, while electrons move through the external circuit from the anode to the cathode, generating electricity. Most fuel cells produce less than 1.2 V, and therefore a stack of cells is needed to drive an electric motor that powers the vehicle. Vehicles with fuel cells are quiet and emission free because the only reaction product is water. The vehicle itself doesn't produce environmental pollutants or greenhouse gases such as CO_2, although CO_2 may be generated in the production of hydrogen.

Fuel-cell-powered demonstration vehicles (FCVs) have been in existence for several years: Buses have been tested in several North American and European cities, and the first fuel-cell-powered truck has been delivered to the U.S. Army. Three types of FCVs—the Honda Clarity, the Hyundai Tucson, and the Toyota Mirai—are available for sale and lease in select markets such as California, which has 34 hydrogen fueling stations statewide. It takes only five minutes to fuel a vehicle with hydrogen, and these models can drive more than 300 miles on one tank. The main obstacles to large-scale commercialization are the relatively high cost of the FCVs, the lack of low cost, nonpolluting sources of hydrogen, and development of a hydrogen-fuel infrastructure. Fossil fuels are still the main source of hydrogen, but electrolysis of water using electricity

▲ Refueling a hydrogen fuel cell car. Hydrogen can be produced from electrolysis of H_2O using electricity from renewable sources such as wind and solar. This hydrogen has been produced using electricity from wind turbines. Photographed at the National Renewable Energy Laboratory's Wind Technology Center in Boulder, Colorado, USA.

produced from solar or wind is a cleaner production method. It costs close to $2 million to build a hydrogen fueling station, and many investors are not willing to take the risk with so few vehicles on the road.

Consumer applications of fuel cells might also become available in small electronic products, such as cell phones and laptop computers. The cell proposed for use in these products is the *direct methanol fuel cell* (DMFC). It is similar to the PEM fuel cell but uses aqueous methanol (CH_3OH) as the fuel rather than gaseous hydrogen:

Anode: $2\ CH_3OH(aq) + 2\ H_2O(l) \longrightarrow$
$2\ CO_2(g) + 12\ H^+(aq) + 12\ e^-$

Cathode: $3\ O_2(g) + 12\ H^+(aq) + 12\ e^- \longrightarrow 6\ H_2O(l)$

Overall: $2\ CH_3OH(aq) + 3\ O_2(g) \longrightarrow$
$2\ CO_2(g) + 4\ H_2O(l)$

The DMFC is lighter and has a higher energy density than conventional batteries. Moreover, methanol is more readily available, safer, and easier to store than hydrogen.

Fuel cells are also likely to find use as power generators for hospitals, hotels, and apartment buildings. Indeed, for some time now, the Tokyo Electric Power Company in Japan has been operating an 11-megawatt fuel-cell power plant capable of supplying electricity to about 4000 households.

PROBLEM 19.27 In what ways are fuel cells and batteries similar, and in what ways are they different?

PROBLEM 19.28 What are the main obstacles to large-scale commercialization of fuel cell vehicles?

PROBLEM 19.29 The cell reaction in a hydrogen–oxygen fuel cell in space vehicles (Figure 19.19) is $2\ H_2(g) + O_2(g) \longrightarrow 2\ H_2O(l)$.

(a) Use the standard reduction potential data in Appendix D to calculate the standard potential for this fuel cell.

(b) Calculate the value of $\Delta G°$ (in kilojoules) and K at 25 °C.

(c) What is the cell voltage at 25 °C if the partial pressure of each gas is 25 atm?

PROBLEM 19.30

(a) Use the standard reduction potential data in Appendix D to calculate the standard potential for the cell reaction in the PEM fuel cell used in electric automobiles (Figure 19.20).

(b) What is the cell voltage at 25 °C if the partial pressure of H_2 is 6 atm and the partial pressure of O_2 is 0.20 atm?

PROBLEM 19.31 Use the thermodynamic data in Appendix B to calculate the ΔG, standard cell potential ($E°$) and the equilibrium constant (K) at 25 °C for the cell reaction in a direct methanol fuel cell:

$$2\ CH_3OH(l) + 3\ O_2(g) \longrightarrow 2\ CO_2(g) + 4\ H_2O(l)$$

PROBLEM 19.32 A steam–hydrocarbon reforming process is one method for producing hydrogen from fossil fuels for use in a fuel cell. In the first step, steam reacts with hydrocarbons, such as CH_4 at high temperatures in the presence of a catalyst, yielding H_2 and CO. In the second step, the reaction of CO and H_2O called the water-gas shift reaction removes toxic carbon monoxide and produces more hydrogen.

Step 1: $H_2O(g) + CH_4(g) \xrightarrow{1100°C,\ Ni\ catalyst} CO(g) + 3\ H_2(g)$

Step 2: $CO(g) + H_2O(g) \xrightarrow{400°C,\ catalyst} CO_2(g) + H_2(g)$

(a) In step 1, which element is oxidized and which is reduced?

(b) In step 2, which element is oxidized and which is reduced? What is the oxidizing agent and reducing agent?

(c) What are drawbacks of steam reforming in the production of hydrogen?

PROBLEM 19.33 Another method of hydrogen production is the electrolysis of water (Section 19.12). If the electricity is generated from a clean energy source such as solar energy instead of fossil fuels, then hydrogen production could be less polluting.

Anode (oxidation): $2\ H_2O(l) \longrightarrow O_2(g) + 4\ H^+(aq) + 4\ e^-$

Cathode (reduction): $4\ H_2O(l) + 4\ e^- \longrightarrow$
$2\ H_2(g) + 4\ OH^-(aq)$

Overall cell reaction: $6\ H_2O(l) \longrightarrow$
$2\ H_2(g) + O_2(g) + 4\ H^+(aq) + 4\ OH^-(aq)$

(a) How many grams of $H_2(g)$ will be produced if 250.0 A is applied for 30.0 minutes?

(b) How long will it take to produce 25.0 mol of O_2 using a current of 500.0 A?

STUDY GUIDE

Section	Concept Summary	Learning Objectives	Test Your Understanding
19.1 Balancing Redox Reactions by the Half-Reaction Method	Oxidation–reduction reactions, **or redox reactions,** are processes in which one or more electrons are transferred between reaction partners. An oxidation is the loss of one or more electrons; a reduction is the gain of one or more electrons. Redox reactions can be balanced using the **half-reaction method** summarized in Figure 19.1.	**19.1** Assign oxidation numbers and classify half-reactions as oxidation or reduction reactions.	Problems 19.42, 19.46
		19.2 Use the half-reaction method to balance a redox reaction in acidic solution.	Worked Example 19.1; Problems 19.50–19.51
		19.3 Use the half-reaction method to balance a redox reaction in basic solution.	Worked Example 19.2; Problems 19.52–19.53
19.2–19.3 Galvanic Cells	**Electrochemistry** is the area of chemistry concerned with the interconversion of chemical and electrical energy. Chemical energy is converted to electrical energy in a **galvanic cell,** a device in which a spontaneous redox reaction is used to produce an electric current. It's convenient to separate cell reactions into half-reactions because oxidation and reduction occur at separate **electrodes.** The electrode at which oxidation occurs is called the **anode,** and the electrode at which reduction occurs is called the **cathode.**	**19.4** Draw a galvanic cell. Label the anode and cathode, indicate the direction of electron and ion flow, and write balanced equations for the half-reactions and overall cell reaction.	Worked Example 19.3; Problems 19.34, 19.56
		19.5 Use shorthand notation to represent a galvanic cell.	Worked Example 19.4; Problems 19.58, 19.60, 19.62
19.4 Cell Potentials and Standard Free-Energy Changes for Cell Reactions	The **cell potential E** (also called the **cell voltage** or **electromotive force**) is an electrical measure of the cell reaction's tendency to occur. Cell potentials depend on temperature, ion concentrations, and gas pressures. The **standard cell potential $E°$** is the cell potential when reactants and products are in their standard states. Cell potentials are related to free-energy changes by the equations $\Delta G = -nFE$ and $\Delta G° = -nFE°$, where n is the number of moles of electrons transferred in the cell reaction and $F = 96,500$ C/mol e$^-$ is the **faraday,** the charge on 1 mol of electrons.	**19.6** Interconvert between cell potentials and standard free-energy changes for electrochemical reactions.	Worked Example 19.5; Problems 19.68, 19.70, 19.72
19.5–19.6 Standard Reduction Potentials	The **standard reduction potential** ($E°$) for a half-reaction is defined relative to an arbitrary value of 0 V for the **standard hydrogen electrode (S.H.E.):** $$2\,H^+(aq, 1\,M) + 2\,e^- \longrightarrow H_2(g, 1\,atm) \quad E° = 0\,V$$ Tables of standard reduction potentials—also called **standard electrode potentials**—are arranged by strength of the oxidizing agent. The sign of $E°$ changes when a half-reaction is reversed. $E°$ values for cell reactions can be calculated by adding the $E°$ values for the oxidation and reduction half-reactions. Positive values of $E°$ for the cell indicate a spontaneous reaction and negative values of $E°$ indicate a nonspontaneous reaction.	**19.7** Use tabulated values of standard reduction potentials for half-reactions (Appendix D) to order substances by increasing strength as an oxidizing or reducing agent.	Worked Example 19.6; Problems 19.74, 19.76, 19.78
		19.8 Use standard reduction potentials for half-reactions to calculate the potential of a redox reaction and determine spontaneity.	Worked Example 19.7; Problems 19.80, 19.82
		19.9 Use standard reduction potentials for half-reactions to calculate the voltage of a galvanic cell.	Worked Example 19.8; Problems 19.84, 19.86
19.7–19.8 Cell Potentials under Nonstandard-State Conditions: The Nernst Equation	Cell potentials under nonstandard-state conditions can be calculated using the **Nernst equation:** $$E = E° - \frac{0.0592\,V}{n}\log Q \quad \text{in volts, at 25 °C}$$ where Q is the reaction quotient.	**19.10** Calculate the cell potential under nonstandard-state conditions using the Nernst equation.	Worked Examples 19.9–19.10; Problems 19.41 19.94, 19.98
		19.11 Use the cell potential to calculate solution concentration.	Worked Example 19.11; Problems 19.104, 19.106

Section	Concept Summary	Learning Objectives	Test Your Understanding
19.9 Standard Cell Potential and Equilibrium Constants	The equilibrium constant K and the standard cell potential $E°$ are related by the equation $$E° = \frac{0.0592\ \text{V}}{n}\log K \quad \text{in volts, at 25 °C}$$	**19.12** Use standard reduction potentials to calculate the equilibrium constant.	Worked Example 19.12; Problems 19.110, 19.112, 19.116
19.10–19.11 Batteries and Corrosion	A **battery** consists of one or more galvanic cells. **Corrosion** of iron (rusting) is an electrochemical process in which iron is oxidized in an anode region of the metal surface and oxygen is reduced in a cathode region. Corrosion can be prevented by covering iron with another metal, such as zinc in the process called **galvanizing**, or simply by putting the iron in electrical contact with a second metal that is more easily oxidized, a process called **cathodic protection**.	**19.13** Describe batteries and fuel cells, and calculate values of $E°$, $\Delta G°$, and K for the cell reaction.\n\n**19.14** Describe how corrosion occurs and methods used to prevent it.	Problems 19.118, 19.120, 19.122\n\nProblems 19.127–19.128, 19.130
19.12–19.14 Electrolysis	Electrical energy is converted to chemical energy in an **electrolytic cell**, a cell in which an electric current drives a nonspontaneous reaction. **Electrolysis**, the process of using an electric current to bring about chemical change, is employed to produce sodium, chlorine, sodium hydroxide, and aluminum (**Hall–Héroult process**) and is used in **electrorefining** and **electroplating**. The product obtained at an electrode depends on the reduction potentials and **overpotential**. The amount of product obtained is related to the number of moles of electrons passed through the cell, which depends on the current and the time that the current flows.	**19.15** Draw an electrolytic cell. Label the anode and cathode, indicate the direction of electron and ion flow, and write balanced equations for the electrode and overall cell reactions.\n\n**19.16** Predict the products formed and the electrode and overall cell reactions when an aqueous solution of an ionic compound is electrolyzed.\n\n**19.17** Relate the current, time, and amount of product produced in an electrolytic cell.	Problems 19.38–19.39, 19.134\n\nProblems 19.136–19.137\n\nWorked Example 19.13; Problems 19.138, 19.142, 19.144

KEY TERMS

anode *820*
cathode *820*
cathodic protection *845*
cell potential (*E*) *825*
cell voltage *825*
corrosion *843*
electrochemical cell *819*
electrochemistry *814*

electrode *819*
electrolysis *846*
electrolytic cell *819*
electromotive force (emf) *825*
electroplating *852*
electrorefining *851*
faraday *826*
fuel cell *854*

galvanic cell *819*
galvanizing *845*
Hall–Héroult process *850*
half-reaction method *815*
half-reactions *815*
Nernst equation *833*
overpotential *848*
salt bridge *819*

standard cell
    potential (*E°*) *826*
standard electrode
    potential *829*
standard hydrogen electrode
    (S.H.E.) *828*
standard reduction
    potential *829*

KEY EQUATIONS

- **Relationships between the free-energy change and the cell potential for a redox reaction (Section 19.4)**
  $\Delta G = -nFE$ and $\Delta G° = -nFE°$, where n is the number of moles of electrons transferred in the cell reaction and $F = 96{,}500$ C/mol e$^-$ is the faraday

- **The Nernst equation, which relates the cell potential under nonstandard-state conditions E to the standard cell potential $E°$ and the reaction quotient Q (Section 19.7)**

$$E = E° - \frac{RT}{nF}\ln Q \quad \text{or} \quad E = E° - \frac{2.303\ RT}{nF}\log Q$$

which can be simplified to:

$$E = E° - \frac{0.0592\ \text{V}}{n}\log Q \quad \text{in volts, at 25 °C}$$

- **Relationship between the standard cell potential and the equilibrium constant (Section 19.9)**

$$E° = \frac{RT}{nF}\ln K = \frac{2.303\ RT}{nF}\log K$$

which can be simplified to:

$$E° = \frac{0.0592\ \text{V}}{n}\log K \quad \text{in volts, at 25 °C}$$

PRACTICE TEST

After studying this chapter, you can assess your understanding with these practice test questions, which are correlated with chapter learning objectives. If you answer a question incorrectly, refer to the learning objectives in the end-of-chapter Study Guide for assistance. The Study Guide provides a conceptual summary, references a Worked Example to model how to solve the problem, and gives additional problems for more practice.

1. Which of the following unbalanced half-reactions is correctly labeled as an oxidation or reduction? (**LO 19.1**)
   (a) $NO_3^-(aq) \longrightarrow NO(g)$; reduction
   (b) $Zn(s) \longrightarrow Zn^{2+}(aq)$; reduction
   (c) $ClO_3^-(aq) \longrightarrow Cl_2(g)$; oxidation
   (d) $Br^-(aq) \longrightarrow Br_2(l)$; reduction

2. What is the coefficient on Sn^{2+} when the following reaction is balanced in acidic solution? (**LO 19.2**)

   $$Sn^{2+}(aq) + IO_4^-(aq) \longrightarrow Sn^{4+}(aq) + I^-(aq)$$

   (a) 2   (b) 4
   (c) 5   (d) 7

3. Balance the redox reaction in basic solution. What is the coefficient on the hydroxide ion, and on which side of the equation does it appear? (**LO 19.3**)

   $$Mn^{2+}(aq) + H_2O_2(aq) \longrightarrow MnO_2(s) + H_2O(l)$$

   (a) 2 OH^- in reactants   (b) 4 OH^- in products
   (c) 4 OH^- in reactants   (d) 3 OH^- in reactants

   Use the diagram of a galvanic cell to answer questions 4 and 5.

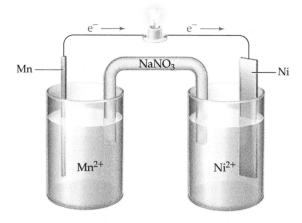

4. What is the reaction occurring at the anode? (**LO 19.4**)
   (a) $Ni^{2+}(aq) + 2 e^- \longrightarrow Ni(s)$
   (b) $Ni(s) \longrightarrow Ni^{2+}(aq) + 2 e^-$
   (c) $Mn^{2+}(aq) + 2 e^- \longrightarrow Mn(s)$
   (d) $Mn(s) \longrightarrow Mn^{2+}(aq) + 2 e^-$

5. What is the line notation for the galvanic cell? (**LO 19.5**)
   (a) $Mn(s)|Mn^{2+}(aq)\|Ni^{2+}(aq)|Ni(s)$
   (b) $Mn^{2+}(aq)|Mn(s)\|Ni^{2+}(aq)|Ni(s)$
   (c) $Ni(s)|Ni^{2+}(aq)\|Mn^{2+}(aq)|Mn(s)$
   (d) $Ni^{2+}(aq)|Ni(s)\|Mn(s)|Mn^{2+}(aq)$

6. The nickel–cadmium battery used in power tools delivers a voltage of 1.20 V. Calculate the standard free-energy change (in kilojoules) for the cell reaction. (**LO 19.6**)

   $$Cd(s) + 2\,NiO(OH)(s) + 2\,H_2O(l) \longrightarrow Cd(OH)_2(s) + 2\,Ni(OH)_2(s)$$

   (a) +347 kJ   (b) −232 kJ
   (c) −463 kJ   (d) +115 kJ

7. Consider the following table of standard reduction potentials:

Reduction Half-Reaction	$E°$ (V)
$D^{3+} + 3\,e^- \longrightarrow D$	1.50
$B^{2+} + 2\,e^- \longrightarrow B$	0.48
$C_2 + 2\,e^- \longrightarrow 2\,C^-$	0.17
$A + e^- \longrightarrow A^-$	−0.89

   Which substance(s) can be reduced by C^-? (**LO 19.7**)
   (a) D and B   (b) A^-
   (c) D^{3+} and B^{2+} (d) A

8. Consider the following galvanic cell with half-reactions in two containers.

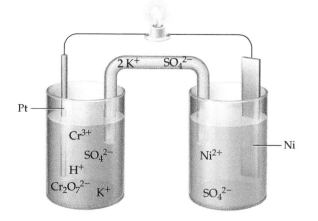

   What is the reaction occurring at the anode, and what voltage is produced by the cell? (Refer to Table 19.1 for standard reduction potentials.) (**LO 19.8, 19.9, 19.13**)
   (a) $Cr_2O_7^{2-}(aq) + 14\,H^+(aq) + 6\,e^- \longrightarrow 2\,Cr^{3+}(aq) + 7\,H_2O(l)$, 1.62 V
   (b) $2\,Cr^{3+}(aq) + 7\,H_2O(l) \longrightarrow Cr_2O_7^{2-}(aq) + 14\,H^+(aq) + 6\,e^-$, 1.10 V
   (c) $Ni^{2+}(aq) + 2\,e^- \longrightarrow Ni(s)$, 1.10 V
   (d) $Ni(s) \longrightarrow Ni^{2+}(aq) + 2\,e^-$, 1.62 V

9. Consider a galvanic cell that uses the reaction

   $$2\,Ag^+(aq) + Sn(s) \longrightarrow 2\,Ag(s) + Sn^{2+}(aq)$$

   Calculate the potential at 25 °C for a cell that has the following ion concentrations: $[Ag^+] = 0.010$ M, $[Sn^{2+}] = 0.020$ M. (Refer to Appendix B for standard reduction potentials.) (**LO 19.10**)
   (a) 0.94 V   (b) 0.93 V
   (c) 1.01 V   (d) 0.87 V

10. What is the pH of the solution in the cathode compartment of the following cell if the measured cell potential at 25 °C is 0.58 V? **(LO 19.11)**

$$Zn(s)\,|\,Zn^{2+}(1\ M)\,\|\,H^+(?\ M)\,|\,H_2(1\ atm)\,|\,Pt(s)$$

(Refer to Table 19.1 for standard reduction potentials.)

(a) 8.0   (b) 4.5   (c) 2.2   (d) 3.0

11. The following galvanic cell has a potential of 0.578 V at 25 °C

$$Ag(s)\,|\,AgCl(s)\,|\,Cl^-(1.0\ M)\,\|\,Ag^+(1.0\ M)\,|\,Ag(s)$$

Use this information to calculate K_{sp} for AgCl at 25 °C. **(LO 19.12)**

(a) 5.91×10^9 (b) 7.83×10^{-6}

(c) 1.69×10^{-10} (d) 3.85×10^{-3}

12. The technique of protecting a metal from corrosion by connecting it to a second metal that is more easily oxidized is called cathodic protection. Which of the following metals would be suitable for the cathodic protection of iron? (Refer to Table 19.1 for standard reduction potentials.) **(LO 19.14)**

(a) Al(s)   (b) Cu(s)   (c) Ni(s)   (d) Ag(s)

13. Is the following cell a galvanic or electrolytic cell? What is the half-reaction that occurs at the cathode? **(LO 19.15)**

(a) galvanic; $K^+(l) + e^- \longrightarrow K(l)$

(b) electrolytic; $K^+(l) + e^- \longrightarrow K(l)$

(c) galvanic; $2\,I^-(l) \longrightarrow I_2(l) + 2\,e^-$

(d) electrolytic; $I_2(l) + 2\,e^- \longrightarrow 2\,I^-(l)$

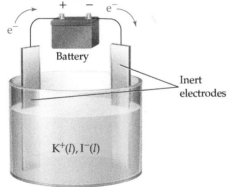

14. What are the products of the overall reaction in the electrolysis of an aqueous solution of sodium hydroxide? (Refer to Table 19.1 for standard reduction potentials.) **(LO 19.16)**

(a) Na(s) and $O_2(g)$ (b) $H_2(g)$ and $O_2(g)$

(c) Na(s) and $H_2(g)$ (d) Na(s) and $H_2O_2(aq)$

15. Adiponitrile, a key intermediate in the manufacture of nylon, is made industrially by an electrolytic process that reduces acrylonitrile:

Anode (oxidation): $2\,H_2O \longrightarrow O_2 + 4\,H^+ + 4\,e^-$

Cathode (reduction):

$$2\,CH_2CH_2CN + 2\,H^+ + 2\,e^- \longrightarrow NC(CH_2)_4CN$$

$\quad$ Acrylonitrile $\qquad\qquad\qquad\qquad\qquad$ Adiponitrile

How many kilograms of adiponitrile ($M = 108.14$ g/mol) are produced in 10.0 h in a cell that has a constant current of 3.00×10^3 A? **(LO 19.17)**

(a) 289 kg   (b) 579 kg

(c) 30.3 kg   (d) 60.5 kg

Answers:

1. a, 2. b, 3. a, 4. d, 5. a, 6. b, 7. c, 8. d, 9. d, 10. d, 11. c, 12. a, 13. b, 14. b, 15. d

---

Mastering Chemistry   provides end-of-chapter exercises, feedback-enriched tutorial problems, animations, and interactive activities to encourage problem-solving practice and deeper understanding of key concepts and topics.

RAN   *Randomized in Mastering Chemistry*

CONCEPTUAL PROBLEMS

Problems 19.1–19.33 appear within the chapter.

19.34 The following picture of a galvanic cell has lead and zinc electrodes:

(a) Label the electrodes, and identify the ions present in the solutions.

(b) Label the anode and cathode.

(c) Indicate the direction of electron flow in the wire and ion flow in the solutions.

(d) Which electrolyte could be used in the salt bridge, and indicate the direction of ion flow.

(e) Write balanced equations for the electrode and overall cell reactions.

19.35 Consider the following table of standard reduction potentials:

Reduction Half-Reaction	$E°$ (V)
$A^+ + e^- \longrightarrow A$	0.80
$B^{2+} + 2\,e^- \longrightarrow B$	0.38
$C_2 + 2\,e^- \longrightarrow 2\,C^-$	0.17
$D^{3+} + 3\,e^- \longrightarrow D$	-1.36

(a) Which substance is the strongest oxidizing agent? Which is the strongest reducing agent?

(b) Which substances can be oxidized by B^{2+}? Which can be reduced by D?

(c) Write a balanced equation for the overall cell reaction that delivers a voltage of 1.53 V under standard-state conditions.

19.36 Consider a Daniell cell with 1.0 M ion concentrations:

Does the cell voltage increase, decrease, or remain the same when each of the following changes is made? Explain.

(a) 5.0 M $CuSO_4$ is added to the cathode compartment.

(b) 5.0 M H_2SO_4 is added to the cathode compartment.

(c) 5.0 M $Zn(NO_3)_2$ is added to the anode compartment.

(d) 1.0 M $Zn(NO_3)_2$ is added to the anode compartment.

19.37 Consider the following galvanic cells:

1. $Cu(s)\,|\,Cu^{2+}(1\text{ M})\,\|\,Fe^{3+}(1\text{ M}),\,Fe^{2+}(1\text{ M})\,|\,Pt(s)$
2. $Cu(s)\,|\,Cu^{2+}(1\text{ M})\,\|\,Fe^{3+}(1\text{ M}),\,Fe^{2+}(5\text{ M})\,|\,Pt(s)$
3. $Cu(s)\,|\,Cu^{2+}(0.1\text{ M})\,\|\,Fe^{3+}(0.1\text{ M}),\,Fe^{2+}(0.1\text{ M})\,|\,Pt(s)$

(a) Write a balanced equation for each cell reaction.

(b) Sketch each cell. Label the anode and cathode, and indicate the direction of electron and ion flow.

(c) Which of the three cells has the largest cell potential? Which has the smallest cell potential? Explain.

19.38 Sketch a cell with inert electrodes suitable for electrolysis of aqueous $CuBr_2$.

(a) Label the anode and cathode.

(b) Indicate the direction of electron and ion flow.

(c) Write balanced equations for the anode, cathode, and overall cell reactions.

19.39 Consider the following electrochemical cell:

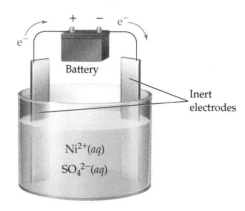

(a) Is the cell a galvanic or an electrolytic cell? Explain.

(b) Label the anode and cathode, and show the direction of ion flow.

(c) Write balanced equations for the anode, cathode, and overall cell reactions.

19.40 Porous pellets of TiO_2 can be reduced to titanium metal at the cathode of an electrochemical cell containing molten $CaCl_2$ as the electrolyte. When the TiO_2 is reduced, the O^{2-} ions dissolve in the $CaCl_2$ and are subsequently oxidized to O_2 gas at the anode. This approach may be the basis for a less expensive process than the one currently used for producing titanium.

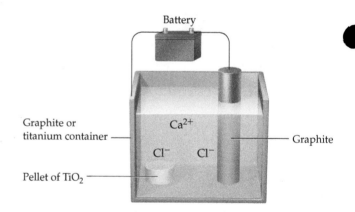

(a) Label the anode and cathode, and indicate the signs of the electrodes.

(b) Indicate the direction of electron and ion flow.

(c) Write balanced equations for the anode, cathode, and overall cell reactions.

19.41 Consider the following galvanic cell with 0.10 M concentrations:

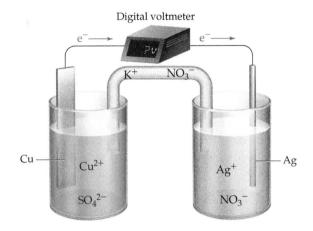

Digital voltmeter

K^+ NO_3^-

Cu — Cu^{2+} SO_4^{2-}

Ag — Ag^+ NO_3^-

Does the cell voltage increase, decrease, or remain the same when each of the following changes is made? Explain.

(a) 0.10 M NaCl is added to the cathode compartment.
(b) 0.10 M NaCl is added to the anode compartment.
(c) 1.0 M NH_3 is added to the cathode compartment.
(d) 1.0 M NH_3 is added to the anode compartment.

SECTION PROBLEMS

Balancing Redox Reactions (Section 19.1)

19.42 Classify each of the following unbalanced half-reactions as either an oxidation or a reduction.
(a) $HClO(aq) \longrightarrow Cl_2(g)$
(b) $Pt^{2+}(aq) \longrightarrow Pt(s)$
(c) $Cr(s) \longrightarrow Cr^{3+}(aq)$
(d) $SbCl_4^-(aq) \longrightarrow SbCl_6^-(aq)$

19.43 Classify each of the following unbalanced half-reactions as either an oxidation or a reduction.
(a) $O_2(g) \longrightarrow OH^-(aq)$
(b) $H_2O_2(aq) \longrightarrow O_2(g)$
(c) $MnO_4^-(aq) \longrightarrow MnO_4^{2-}(aq)$
(d) $CH_3OH(aq) \longrightarrow CH_2O(aq)$

19.44 Balance the half-reactions in Problem 19.42, assuming that they occur in acidic solution.

19.45 Balance the half-reactions in Problem 19.43, assuming that they occur in basic solution.

19.46 Write unbalanced oxidation and reduction half-reactions for the following processes.
(a) $Te(s) + NO_3^-(aq) \longrightarrow TeO_2(s) + NO(g)$
(b) $H_2O_2(aq) + Fe^{2+}(aq) \longrightarrow Fe^{3+}(aq) + H_2O(l)$

19.47 Write unbalanced oxidation and reduction half-reactions for the following processes.
(a) $Mn(s) + NO_3^-(aq) \longrightarrow Mn^{2+}(aq) + NO_2(g)$
(b) $Mn^{3+}(aq) \longrightarrow MnO_2(s) + Mn^{2+}(aq)$

19.48 Balance the following half-reactions.
(a) (acidic) $Cr_2O_7^{2-}(aq) \longrightarrow Cr^{3+}(aq)$
(b) (basic) $CrO_4^{2-}(aq) \longrightarrow Cr(OH)_4^-(aq)$
(c) (basic) $Bi^{3+}(aq) \longrightarrow BiO_3^-(aq)$
(d) (basic) $ClO^-(aq) \longrightarrow Cl^-(aq)$

19.49 Balance the following half-reactions.
(a) (acidic) $VO^{2+}(aq) \longrightarrow V^{3+}(aq)$
(b) (basic) $Ni(OH)_2(s) \longrightarrow Ni_2O_3(s)$
(c) (acidic) $NO_3^-(aq) \longrightarrow NO_2(aq)$
(d) (basic) $Br_2(aq) \longrightarrow BrO_3^-(aq)$

19.50 Write balanced net ionic equations for the following reactions in acidic solution.
(a) $Zn(s) + VO^{2+}(aq) \longrightarrow Zn^{2+}(aq) + V^{3+}(aq)$
(b) $TeO_2(s) + Cr^{2+}(aq) \longrightarrow Te(s) + Cr^{3+}(aq)$
(c) $I^-(aq) + IO_3^-(aq) \longrightarrow I_3^-(aq)$

19.51 Write balanced net ionic equations for the following reactions in acidic solution.
(a) $MnO_4^-(aq) + C_2H_5OH(aq) \longrightarrow$
$Mn^{2+}(aq) + CH_3CO_2H(aq)$
(b) $ClO_4^-(aq) + Co(s) \longrightarrow Co^{2+}(aq) + ClO_3^-(aq)$
(c) $PbO_2(s) + Cl^-(aq) \longrightarrow PbCl_2(s) + O_2(g)$

19.52 Write balanced net ionic equations for the following reactions in basic solution.
(a) $MnO_4^-(aq) + IO_3^-(aq) \longrightarrow MnO_2(s) + IO_4^-(aq)$
(b) $N_2(g) + NO_2^-(aq) \longrightarrow N_2H_4(aq) + NO_3^-(aq)$
(c) $H_2O_2(aq) + ClO_4^-(aq) \longrightarrow ClO_2^-(aq) + O_2(g)$

19.53 Write balanced net ionic equations for the following reactions in basic solution.
(a) $S_2O_3^{2-}(aq) + I_2(aq) \longrightarrow S_4O_6^{2-}(aq) + I^-(aq)$
(b) $ClO^-(aq) + MnO_2(s) \longrightarrow Cl^-(aq) + MnO_4^-(aq)$
(c) $Zn(s) + NO_3^-(aq) \longrightarrow NH_3(aq) + Zn(OH)_4^{2-}(aq)$

Galvanic Cells (Sections 19.2–19.3)

19.54 Why is the cathode of a galvanic cell considered to be the positive electrode?

19.55 What is the function of a salt bridge in a galvanic cell?

19.56 Describe galvanic cells that use the following reactions. In each case, write the anode and cathode half-reactions and sketch the experimental setup. Label the anode and cathode, identify the sign of each electrode, and indicate the direction of electron and ion flow.

(a) $Cd(s) + Sn^{2+}(aq) \longrightarrow Cd^{2+}(aq) + Sn(s)$

(b) $2 Al(s) + 3 Cd^{2+}(aq) \longrightarrow 2 Al^{3+}(aq) + 3 Cd(s)$

(c) $Cr_2O_7^{2-}(aq) + 6 Fe^{2+}(aq) + 14 H^+(aq) \longrightarrow$
$2 Cr^{3+}(aq) + 6 Fe^{3+}(aq) + 7 H_2O(l)$

19.57 Describe galvanic cells that use the following reactions. In each case, write the anode and cathode half-reactions and sketch the experimental setup. Label the anode and cathode, identify the sign of each electrode, and indicate the direction of electron and ion flow.

(a) $3 Cu^{2+}(aq) + 2 Cr(s) \longrightarrow 3 Cu(s) + 2 Cr^{3+}(aq)$

(b) $Pb(s) + 2 H^+(aq) \longrightarrow Pb^{2+}(aq) + H_2(g)$

(c) $Cl_2(g) + Sn^{2+}(aq) \longrightarrow Sn^{4+}(aq) + 2 Cl^-(aq)$

19.58 Write a balanced equation for the overall cell reaction in the following galvanic cell, and tell why inert electrodes are required at the anode and cathode.

$$Pt(s) | Br^-(aq) | Br_2(l) \| Cl_2(g) | Cl^-(aq) | Pt(s)$$

19.59 Write the shorthand notation for a galvanic cell that uses the
RAN　following cell reaction. Include inert electrodes if necessary.

$$2 Fe(s) + Cr_2O_7^{2-}(aq) + 14 H^+(aq) \longrightarrow$$
$$2 Fe^{3+}(aq) + 2 Cr^{3+}(aq) + 7 H_2O(l)$$

19.60 Write the standard shorthand notation for a galvanic cell
RAN　that uses the following cell reaction. Include inert electrodes if necessary.

$$2 Al(s) + 3 Cd^{2+}(aq) \longrightarrow 2 Al^{3+}(aq) + 3 Cd(s)$$

19.61 Write the standard shorthand notation for a galvanic cell that uses the following cell reaction. Include inert electrodes if necessary.

$$Fe(s) + I_2(s) \longrightarrow Fe^{2+}(aq) + 2 I^-(aq)$$

19.62 An H_2/H^+ half-cell (anode) and an Ag^+/Ag half-cell (cathode) are connected by a wire and a salt bridge.

(a) Sketch the cell, indicating the direction of electron and ion flow.

(b) Write balanced equations for the electrode and overall cell reactions.

(c) Give the shorthand notation for the cell.

19.63 A galvanic cell is constructed from a Zn/Zn^{2+} half-cell (anode) and a Cl_2/Cl^- half-cell (cathode).

(a) Sketch the cell, indicating the direction of electron and ion flow.

(b) Write balanced equations for the electrode and overall cell reactions.

(c) Give the shorthand notation for the cell.

19.64 Write balanced equations for the electrode and overall cell reactions in the following galvanic cells. Sketch each cell, labeling the anode and cathode and showing the direction of electron and ion flow.

(a) $Co(s) | Co^{2+}(aq) \| Cu^{2+}(aq) | Cu(s)$

(b) $Fe(s) | Fe^{2+}(aq) \| O_2(g) | H^+(aq), H_2O(l) | Pt(s)$

19.65 Write balanced equations for the electrode and overall cell reactions in the following galvanic cells. Sketch each cell, labeling the anode and cathode and showing the direction of electron and ion flow.

(a) $Mn(s) | Mn^{2+}(aq) \| Pb^{2+}(aq) | Pb(s)$

(b) $Pt(s) | H_2(g) | H^+(aq) \| Cl^-(aq) | AgCl(s) | Ag(s)$

Cell Potentials and Free-Energy Changes (Section 19.4)

19.66 What conditions must be met for a cell potential E to qualify as a standard cell potential $E°$?

19.67 How are standard reduction potentials defined?

19.68 The silver oxide–zinc battery used in watches delivers a voltage of 1.60 V. Calculate the free-energy change (in kilojoules) for the cell reaction

$$Zn(s) + Ag_2O(s) \longrightarrow ZnO(s) + 2 Ag(s)$$

19.69 The standard cell potential for a lead storage battery is 1.924 V. Calculate $\Delta G°$ (in kilojoules) for the cell reaction

$$Pb(s) + PbO_2(s) + 2 H^+(aq) + 2 HSO_4^-(aq) \longrightarrow$$
$$2 PbSO_4(s) + 2 H_2O(l)$$

19.70 What is the value of x for the following reaction if $E° = 1.43$ V
RAN　and $\Delta G° = -414$ kJ?

$$A + B^{x+} \longrightarrow A^{x+} + B$$

19.71 What are the values of x and y for the following reaction if $E° = 0.91$ V and $\Delta G° = -527$ kJ?

$$2 A^{x+} + 3 B \longrightarrow 2 A + 3 B^{y+}$$

19.72 Use the standard free energies of formation in Appendix B to calculate the standard cell potential for the reaction in the hydrogen–oxygen fuel cell:

$$2 H_2(g) + O_2(g) \longrightarrow 2 H_2O(l)$$

19.73 Consider a fuel cell that uses the reaction

$$CH_4(g) + 2 O_2(g) \longrightarrow CO_2(g) + 2 H_2O(l)$$

Given the standard free energies of formation in Appendix B, what is the value of $E°$ for the cell reaction?

Standard Reduction Potentials (Sections 19.5–19.6)

19.74 Arrange the following oxidizing agents in order of increas-
RAN　ing strength under standard-state conditions: $Br_2(aq)$, $MnO_4^+(aq)$, $Sn^{4+}(aq)$.

19.75 List the following reducing agents in order of increasing strength under standard-state conditions: $Al(s)$, $Pb(s)$, $Fe(s)$.

19.76 Consider the following substances: $I_2(s)$, $Fe^{2+}(aq)$, $Cr_2O_7^{2-}$
RAN (aq). Which is the strongest oxidizing agent? Which is the weakest oxidizing agent?

19.77 Consider the following substances: $Fe^{2+}(aq)$, $Sn^{2+}(aq)$, $I^-(aq)$. Identify the strongest reducing agent and the weakest reducing agent.

19.78 Consider the following substances: $Fe(s)$, $PbO_2(s)$, $H^+(aq)$, $Al(s)$, $Ag(s)$, $Cr_2O_7^{2-}(aq)$.

(a) Look at the $E°$ values in Appendix D, and classify each substance as an oxidizing agent or a reducing agent.

(b) Which is the strongest oxidizing agent? Which is the weakest oxidizing agent?

(c) Which is the strongest reducing agent? Which is the weakest reducing agent?

(d) Which substances can be oxidized by $Cu^{2+}(aq)$? Which can be reduced by $H_2O_2(aq)$?

19.79 The following cell reactions occur spontaneously:

$$B + A^+ \rightarrow B^+ + A$$
$$C + A^+ \rightarrow C^+ + A$$
$$B + C^+ \rightarrow B^+ + C$$

(a) Arrange the following reduction half-reactions in order of decreasing tendency to occur: $A^+ + e^- \longrightarrow A$, $B^+ + e^- \longrightarrow B$, and $C^+ + e^- \longrightarrow C$.

(b) Which of these substances (A, A^+, B, B^+, C, C^+) is the strongest oxidizing agent? Which is the strongest reducing agent?

(c) Which of the three cell reactions delivers the highest voltage?

19.80 Use the data in Appendix D to predict whether the following reactions can occur under standard-state conditions.

(a) Oxidation of $Sn^{2+}(aq)$ by $Br_2(aq)$

(b) Reduction of $Ni^{2+}(aq)$ by $Sn^{2+}(aq)$

(c) Oxidation of $Ag(s)$ by $Pb^{2+}(aq)$

(d) Reduction of $I_2(s)$ by $H_2SO_3(aq)$

19.81 Use the data in Appendix D to predict whether the following reactions can occur under standard-state conditions.

(a) Reduction of $Pb^{2+}(aq)$ by $Ni(s)$

(b) Oxidation of $Au^+(aq)$ by $Mn^{2+}(aq)$

(c) Reduction of $I_2(s)$ by $Mn(s)$

(d) Oxidation of $Fe^{2+}(aq)$ by $Br_2(aq)$

19.82 What reaction can occur, if any, when the following experiments are carried out under standard-state conditions?

(a) Oxygen gas is bubbled through an acidic solution of $Cr(NO_3)_3$.

(b) A strip of lead is dipped into an aqueous solution of $AgNO_3$.

(c) Chlorine gas is bubbled through aqueous $H_2C_2O_4$.

(d) A nickel wire is dipped into an aqueous solution of $HClO$.

19.83 What reaction can occur, if any, when the following experiments are carried out under standard-state conditions?

(a) A strip of zinc is dipped into an aqueous solution of $Pb(NO_3)_2$.

(b) An acidic solution of $FeSO_4$ is exposed to oxygen.

(c) A silver wire is immersed in an aqueous solution of $NiCl_2$.

(d) Hydrogen gas is bubbled through aqueous $Cd(NO_3)_2$.

19.84 The standard potential for the following galvanic cell is
RAN 0.40 V:

$$Zn(s)\,|\,Zn^{2+}(aq)\,\|\,Eu^{3+}(aq), Eu^{2+}(aq)\,|\,Pt(s)$$

(Europium, Eu, is one of the lanthanide elements.) Use the data in Table 19.1 to calculate the standard reduction potential for the Eu^{3+}/Eu^{2+} half-cell.

19.85 The following reaction has an $E°$ value of 0.27 V:
RAN
$$Cu^{2+}(aq) + 2\,Ag(s) + 2\,Br^-(aq) \longrightarrow Cu(s) + 2\,AgBr(s)$$

Use the data in Table 19.1 to calculate the standard reduction potential for the half-reaction

$$AgBr(s) + e^- \longrightarrow Ag(s) + Br^-(aq)$$

19.86 Given the following half-reactions, combine the two that give
RAN the cell reaction with the most positive $E°$. Write a balanced equation for the cell reaction, and calculate $E°$ and $\Delta G°$.

$$
\begin{aligned}
&Co^{2+}(aq) + 2\,e^- \longrightarrow Co(s) && E° = -0.28\text{ V}\\
&I_2(s) + 2\,e^- \longrightarrow 2\,I^-(aq) && E° = 0.54\text{ V}\\
&Cu^{2+}(aq) + 2\,e^- \longrightarrow Cu(s) && E° = 0.34\text{ V}
\end{aligned}
$$

19.87 Combine the two half-reactions in Problem 19.86 that give
RAN the spontaneous cell reaction with the smallest $E°$. Write a balanced equation for the cell reaction, and calculate $E°$ and $\Delta G°$.

19.88 Calculate the standard cell potential and the standard free-
RAN energy change (in kilojoules) for the reaction below. (See Appendix D for standard reduction potentials.)

$$2\,Al(s) + 3\,Cd^{2+}(aq) \longrightarrow 2\,Al^{3+}(aq) + 3\,Cd(s)$$

19.89 Calculate $E°$ and $\Delta G°$ (in kilojoules) for the reaction below.
RAN (See Appendix D for standard reduction potentials.)

$$Cr_2O_7^{2-}(aq) + 6\,Fe^{2+}(aq) + 14\,H^+(aq) \longrightarrow$$
$$2\,Cr^{3+}(aq) + 6\,Fe^{3+}(aq) + 7\,H_2O(l)$$

19.90 Calculate $E°$ for each of the following reactions, and tell which are spontaneous under standard-state conditions.

(a) $2\,Fe^{2+}(aq) + Pb^{2+}(aq) \longrightarrow 2\,Fe^{3+}(aq) + Pb(s)$

(b) $Mg(s) + Ni^{2+}(aq) \longrightarrow Mg^{2+}(aq) + Ni(s)$

19.91 Calculate $E°$ for each of the following reactions, and tell which are spontaneous under standard-state conditions.

(a) $5\,Ag^+(aq) + Mn^{2+}(aq) + 4\,H_2O(l) \longrightarrow$
    $5\,Ag(s) + MnO_4^-(aq) + 8\,H^+(aq)$

(b) $2\,H_2O_2(aq) \longrightarrow O_2(g) + 2\,H_2O(l)$

19.92 Consider a galvanic cell that uses the following half-reactions:

$$MnO_4^-(aq) + 8\,H^+(aq) + 5\,e^- \longrightarrow Mn^{2+}(aq)$$
$$+ 4\,H_2O(l) + Sn^{4+}(aq) + 2\,e^- \longrightarrow Sn^{2+}(aq)$$

(a) Write a balanced equation for the overall cell reaction.

(b) What is the oxidizing agent, and what is the reducing agent?

(c) Calculate the standard cell potential.

19.93 Given the following half-reactions and $E°$ values,

$$
\begin{aligned}
&Mn^{3+}(aq) + e^- \longrightarrow Mn^{2+}(aq) && E° = 1.54\text{ V}\\
&MnO_2(s) + 4\,H^+(aq) + e^- \longrightarrow Mn^{3+}(aq) + 2\,H_2O(l) && E° = 0.95\text{ V}
\end{aligned}
$$

write a balanced equation for the formation of Mn^{2+} and MnO_2 from Mn^{3+}, and calculate the value of $E°$ for this reaction. Is the reaction spontaneous under standard-state conditions?

The Nernst Equation (Sections 19.7–19.8)

19.94 Consider a galvanic cell that uses the reaction

$$2 \, Ag^+(aq) + Ni(s) \longrightarrow 2 \, Ag(s) + Ni^{2+}(aq)$$

Calculate the potential at 25 °C for a cell that has the following ion concentrations: $[Ag^+] = 0.010$ M, $[Ni^{2+}] = 0.100$ M.

19.95 Consider a galvanic cell based on the reaction
RAN

$$2 \, Fe^{2+}(aq) + Cl_2(g) \longrightarrow 2 \, Fe^{3+}(aq) + 2 \, Cl^-(aq)$$

Calculate the cell potential at 25 °C when $[Fe^{2+}] = 1.0$ M, $[Fe^{3+}] = 1.0 \times 10^{-3}$ M, $[Cl^-] = 3.0 \times 10^{-3}$ M, and $P_{Cl_2} = 0.50$ atm.

19.96 What is the cell potential at 25 °C for the following galvanic
RAN cell?

$$Pb(s) \, | \, Pb^{2+}(1.0 \text{ M}) \, \| \, Cu^{2+}(1.0 \times 10^{-4} \text{ M}) \, | \, Cu(s)$$

If the Pb^{2+} concentration is maintained at 1.0 M, what is the Cu^{2+} concentration when the cell potential drops to zero?

19.97 A galvanic cell has an iron electrode in contact with 0.10 M
RAN $FeSO_4$ and a copper electrode in contact with a $CuSO_4$ solution. If the measured cell potential at 25 °C is 0.67 V, what is the concentration of Cu^{2+} in the $CuSO_4$ solution?

19.98 What is the $Zn^{2+}:Cu^{2+}$ concentration ratio in the following
RAN cell at 25 °C if the measured cell potential is 1.07 V?

$$Zn(s) \, | \, Zn^{2+}(aq) \, \| \, Cu^{2+}(aq) \, | \, Cu(s)$$

19.99 What is the $Fe^{2+}:Sn^{2+}$ concentration ratio in the following
RAN cell at 25 °C if the measured cell potential is 0.35 V?

$$Fe(s) \, | \, Fe^{2+}(aq) \, \| \, Sn^{2+}(aq) \, | \, Sn(s)$$

19.100 The Nernst equation applies to both cell reactions and half-
RAN reactions. For the conditions specified, calculate the potential for the following half-reactions at 25 °C.

(a) $I_2(s) + 2 \, e^- \longrightarrow 2 \, I^-(aq)$; $[I^-] = 0.020$ M

(b) $Fe^{3+}(aq) + e^- \longrightarrow Fe^{2+}(aq)$; $[Fe^{3+}] = [Fe^{2+}] = 0.10$ M

(c) $Sn^{2+}(aq) \longrightarrow Sn^{4+}(aq) + 2 \, e^-$; $[Sn^{2+}] = 1.0 \times 10^{-3}$ M, $[Sn^{4+}] = 0.40$ M

(d) $2 \, Cr^{3+}(aq) + 7 \, H_2O(l) \longrightarrow$ $Cr_2O_7^{2-}(aq) + 14 \, H^+(aq) + 6 \, e^-$; $[Cr^{3+}] = [Cr_2O_7^{2-}] = 1.0$ M, $[H^+] = 0.010$ M

19.101 When suspected drunk drivers are tested with a Breathalyzer,
RAN the alcohol (ethanol) in the exhaled breath is oxidized to acetic acid with an acidic solution of potassium dichromate:

$$3 \, CH_3 CH_2 OH(aq) + 2 \, Cr_2O_7^{2-}(aq) + 16 \, H^+(aq) \longrightarrow$$
$$\text{Ethanol}$$
$$3 \, CH_3CO_2H(aq) + 4 \, Cr^{3+}(aq) + 11 \, H_2O(l)$$
$$\text{Acetic acid}$$

The color of the solution changes because some of the orange $Cr_2O_7^{2-}$ is converted to the green Cr^{3+} The Breathalyzer measures the color change and produces a meter reading calibrated in blood alcohol content.

(a) What is $E°$ for the reaction if the standard half-cell potential for the reduction of acetic acid to ethanol is 0.058 V?

(b) What is the value of E for the reaction when the concentrations of ethanol, acetic acid, $Cr_2O_7^{2-}$, and Cr^{3+} are 1.0 M and the pH is 4.00?

19.102 What is the reduction potential at 25 °C for the hydrogen elec-
RAN trode in each of the following solutions? The half-reaction is

$$2 \, H^+(aq) + 2 \, e^- \longrightarrow H_2(g, 1 \text{ atm})$$

(a) 1.0 M HCl

(b) A solution having pH 4.00

(c) Pure water

(d) 1.0 M NaOH

19.103 At one time on Earth, iron was present mostly as iron(II). Later, once plants had produced a significant quantity of oxygen in the atmosphere, the iron became oxidized to iron(III). Show that $Fe^{2+}(aq)$ can be spontaneously oxidized to $Fe^{3+}(aq)$ by $O_2(g)$ at 25 °C assuming the following reasonable environmental conditions:

$$[Fe^{2+}] = [Fe^{3+}] = 1 \times 10^{-7} \text{M}; \text{ pH} = 7.0; P_{O_2} = 160 \text{ mm Hg}.$$

19.104 Standard reduction potentials for the Pb^{2+}/Pb and Cd^{2+}/Cd half-reactions are -0.13 V and -0.40 V, respectively. At what relative concentrations of Pb^{2+} and Cd^{2+} will these half-reactions have the same reduction potential?

19.105 Copper reduces dilute nitric acid to nitric oxide (NO) but
RAN reduces concentrated nitric acid to nitrogen dioxide (NO_2):

(1) $3 \, Cu(s) + 2 \, NO_3^-(aq) + 8 \, H^+(aq) \longrightarrow$ $3 \, Cu^{2+}(aq) + 2 \, NO(g) + 4 \, H_2O(l)$ $E° = 0.62$ V

(2) $Cu(s) + 2 \, NO_3^-(aq) + 4 \, H^+(aq) \longrightarrow$ $Cu^{2+}(aq) + 2 \, NO_2(g) + 2 \, H_2O(l)$ $E° = 0.45$ V

Assuming that $[Cu^{2+}] = 0.10$ M and that the partial pressures of NO and NO_2 are 1.0×10^{-3} atm, calculate the potential (E) for reactions (1) and (2) at 25 °C and show which reaction has the greater thermodynamic tendency to occur when the concentration of HNO_3 is

(a) 1.0 M

(b) 10.0 M

(c) At what HNO_3 concentration do reactions (1) and (2) have the same value of E?

19.106 The following cell has a potential of 0.15 V at 25 °C:
RAN

$$Pt(s) \, | \, H_2(1 \text{ atm}) \, | \, H^+(? \text{ M}) \, \| \, Ni^{2+}(1 \text{ M}) \, | \, Ni(s)$$

What is the pH of the solution in the anode compartment?

19.107 What is the pH of the solution in the cathode compartment
RAN of the following cell if the measured cell potential at 25 °C is 0.17 V?

$$Zn(s) \, | \, Zn^{2+}(1 \text{ M}) \, \| \, H^+(? \text{ M}) \, | \, H_2(1 \text{ atm}) \, | \, Pt(s)$$

Standard Cell Potentials and Equilibrium Constants (Section 19.9)

19.108 Beginning with the equations that relate $E°$, $\Delta G°$, and K, show that $\Delta G°$ is negative and $K > 1$ for a reaction that has a positive value of $E°$.

19.109 If a reaction has an equilibrium constant $K < 1$, is $E°$ positive or negative? What is the value of K when $E° = 0$ V?

19.110 Use the data in Table 19.1 to calculate the equilibrium con-
RAN stant at 25 °C for the reaction

$$Ni(s) + 2\,Ag^+(aq) \longrightarrow Ni^{2+}(aq) + 2\,Ag(s)$$

19.111 From standard reduction potentials, calculate the equilibrium constant at 25 °C for the reaction

$$2\,MnO_4^-(aq) + 10\,Cl^-(aq) + 16\,H^+(aq) \longrightarrow$$
$$2\,Mn^{2+}(aq) + 5\,Cl_2(g) + 8\,H_2O(l)$$

19.112 Calculate the equilibrium constant at 25 °C for the reaction
RAN $Cd(s) + Sn^{2+}(aq) \longrightarrow Cd^{2+}(aq) + Sn(s)$.

19.113 Calculate the equilibrium constant at 25 °C for the reaction
RAN $Cl_2(g) + Sn^{2+}(aq) \longrightarrow Sn^{4+}(aq) + 2\,Cl^-(aq)$.

19.114 Calculate the equilibrium constant at 25 °C for the reaction $Hg_2^{2+}(aq) \longrightarrow Hg(l) + Hg^{2+}(aq)$. See Appendix D for standard reduction potentials.

19.115 Use standard reduction potentials to calculate the equi-
RAN librium constant at 25 °C for decomposition of hydrogen peroxide:

$$2\,H_2O_2(l) \longrightarrow 2\,H_2O(l) + O_2(g)$$

19.116 The following galvanic cell has a potential of 1.214 V at 25 °C:

$$Hg(l)\,|\,Hg_2Br_2(s)\,|\,Br^-(0.10\ M)\,\|\,MnO_4^-(0.10\ M),$$
$$Mn^{2+}(0.10\ M),\,H^+(0.10\ M)\,|\,Pt(s)$$

Calculate the value of K_{sp} for Hg_2Br_2 at 25 °C.

19.117 For the following half-reaction, $E° = 1.103$ V:

$$Cu^{2+}(aq) + 2\,CN^-(aq) + e^- \longrightarrow Cu(CN)_2^-(aq)$$

Calculate the formation constant K_f for $Cu(CN)_2^-$.

Batteries (Section 19.10)

19.118 Write a balanced equation for the overall cell reaction when a lead storage battery is being charged. Refer to Section 19.10.

19.119 Write a balanced equation for the overall cell reaction when a nickel-cadmium battery is producing current. Refer to Section 19.10.

19.120 You are on your dream vacation at the beach when a major storm knocks out the power for days. Your cell phone is dead, and you want to make a battery to charge it. You find the following materials in the beach house.

blue stone algaecide for pools, which can be used to make a 1.0 M Cu^{2+} solution

alum in the kitchen, which can be used to make a 1.0 M Al^{3+} solution

aluminum foil, copper wire, and bologna, which can be used as a salt bridge

(a) What are the half reactions and overall reaction in the battery?

(b) What voltage can be generated?

(c) Draw a diagram using beakers, a voltmeter, and salt bridge to show how a battery can be constructed. Label the anode, cathode, and direction of electron flow.

(d) An iPhone requires 5.0 V for charging. Can this battery charge the phone? Explain.

19.121 A storm has knocked out power to your beach house, and you would like to build a battery from household items to charge your iPhone. You have the following materials.

alum in the kitchen, which can be used to make a 1.0 M Al^{3+} solution

bleach, which is a solution that is approximately a 1.0 M in ClO^-

aluminum foil, a platinum necklace and bologna, which can be used as a salt bridge

(a) What are the half-reactions and overall reaction in the battery?

(b) What voltage can be generated?

(c) Draw a diagram using beakers, voltmeter, and salt bridge to show how a battery can be constructed. Label the anode, cathode, and direction of electron flow.

(d) An iPhone requires 5.0 V for charging. Can this battery charge the phone? Explain.

19.122 For a lead storage battery:

(a) Sketch one cell that shows the anode, cathode, electrolyte, direction of electron and ion flow, and sign of the electrodes.

(b) Write the anode, cathode, and overall cell reactions.

(c) Calculate the equilibrium constant for the cell reaction ($E° = 1.924$ V).

(d) What is the cell voltage when the cell reaction reaches equilibrium?

19.123 A mercury battery uses the following electrode half-reactions:
RAN

$$HgO(s) + H_2O(l) + 2\,e^- \longrightarrow$$
$$Hg(l) + 2\,OH^-(aq) \quad E° = 0.098\ V$$

$$ZnO(s) + H_2O(l) + 2\,e^- \longrightarrow$$
$$Zn(s) + 2\,OH^-(aq) \quad E° = -1.260\ V$$

(a) Write a balanced equation for the overall cell reaction.

(b) Calculate $\Delta G°$ (in kilojoules) and K at 25 °C for the cell reaction.

(c) What is the effect on the cell voltage of a tenfold change in the concentration of KOH in the electrolyte? Explain.

Corrosion (Section 19.11)

19.124 What is rust? What causes it to form? What can be done to prevent its formation?

19.125 How does the pH of the solution affect the formation of rust?

19.126 The standard oxidation potential for the reaction $Cr(s) \longrightarrow Cr^{3+}(aq) + 3\,e^-$ is 0.74 V. Despite the large, positive oxidation potential, chromium is sometimes used as a protective coating on steel. Why doesn't the chromium corrode?

19.127 Which of the following describes the process of galvanization that protects steel from rusting?
RAN
(a) Steel is coated with a layer of paint.
(b) Iron in steel is oxidized to form a protective oxide coating.
(c) Steel is coated with zinc because zinc is more easily oxidized than iron.
(d) A strip of magnesium is attached to steel because the magnesium is more easily oxidized than iron.

19.128 What is meant by cathodic protection?
RAN
(a) Steel is coated with a layer of paint.
(b) Iron in steel is oxidized to form a protective oxide coating.
(c) Steel is coated with zinc because zinc is more easily oxidized than iron.
(d) A strip of magnesium is attached to steel because the magnesium is more easily oxidized than iron.

19.129 Zinc is attached to a ship's steel propeller to prevent the steel from rusting. Write balanced equations for the corrosion reactions that occur (a) in the presence of Zn and (b) in the absence of Zn.
RAN

19.130 Which of the following metals can offer cathodic protection to iron? Select all the correct choices.
RAN

Mn, Ni, Pb, Sn, Al

19.131 If the metal zinc were not available for the galvanization process, which metal would be a reasonable alternative based on standard reduction potentials? Select all the correct choices.
RAN

Mn, Cd, Mg, Co, Cr

Electrolysis (Sections 19.12–19.14)

19.132 Magnesium metal is produced by the electrolysis of molten magnesium chloride using inert electrodes.
(a) Sketch the cell, label the anode and cathode, indicate the sign of the electrodes, and show the direction of electron and ion flow.
(b) Write balanced equations for the anode, cathode, and overall cell reactions.

19.133 (a) Sketch a cell with inert electrodes suitable for the electrolysis of an aqueous solution of sulfuric acid. Label the anode and cathode, and indicate the direction of electron and ion flow. Identify the positive and negative electrodes.
(b) Write balanced equations for the anode, cathode, and overall cell reactions.

19.134 List the anode and cathode half-reactions that might occur when an aqueous solution of $MgCl_2$ is electrolyzed in a cell having inert electrodes. Predict which half-reactions will occur, and justify your answers.

19.135 What products should be formed when the following reactants are electrolyzed in a cell having inert electrodes? Account for any differences.
(a) Molten KCl   (b) Aqueous KCl

19.136 Predict the anode, cathode, and overall cell reactions when an aqueous solution of each of the following salts is electrolyzed in a cell having inert electrodes.
(a) NaBr   (b) $CuCl_2$ (c) LiOH

19.137 Predict the anode, cathode, and overall cell reactions when an aqueous solution of each of the following salts is electrolyzed in a cell having inert electrodes.
(a) Ag_2SO_4 (b) $Ca(OH)_2$ (c) KI

19.138 How many grams of silver will be obtained when an aqueous silver nitrate solution is electrolyzed for 20.0 min with a constant current of 2.40 A?
RAN

19.139 A constant current of 100.0 A is passed through an electrolytic cell having an impure copper anode, a pure copper cathode, and an aqueous $CuSO_4$ electrolyte. How many kilograms of copper are refined by transfer from the anode to the cathode in a 24.0-h period?
RAN

19.140 How many hours are required to produce 1.00×10^3 kg of sodium by the electrolysis of molten NaCl with a constant current of 3.00×10^4 A? How many liters of Cl_2 at STP will be obtained as a by-product?
RAN

19.141 What constant current (in amperes) is required to produce aluminum by the Hall–Héroult process at a rate of 40.0 kg/h?
RAN

19.142 Electrolysis of a metal nitrate solution $M(NO_3)_2(aq)$ for 325 min with a constant current of 20.0 A gives 111 g of the metal. Identify the metal ion M^{2+}.
RAN

19.143 What is the metal ion in a metal nitrate solution $M(NO_3)_3(aq)$ if 90.52 g of metal was recovered from a 4.00 h electrolysis at a constant current of 35.0 A?
RAN

19.144 Aluminum, titanium, and several other metals can be colored by an electrochemical process called *anodizing*. Anodizing *oxidizes* a metal anode to yield a porous metal oxide coating that can incorporate dye molecules to give brilliant colors. In the oxidation of aluminum, for instance, the electrode reactions are
RAN

Cathode (reduction): $6\,H^+(aq) + 6\,e^- \longrightarrow 3\,H_2(g)$

Anode (oxidation): $2\,Al(s) + 3\,H_2O(l) \longrightarrow$
$$Al_2O_3(s) + 6\,H^+(aq) + 6\,e^-$$

Overall reaction: $2\,Al(s) + 3\,H_2O(l) \longrightarrow Al_2O_3(s) + 3\,H_2(g)$

The thickness of the aluminum oxide coating that forms on the anode can be controlled by varying the current flow during the electrolysis. How many minutes are required to produce a 0.0100-mm-thick coating of Al_2O_3 (density 3.97 g/cm^3) on a square piece of aluminum metal 10.0 cm on an edge if the current passed through the piece is 0.600 A?

▲ Anodized aluminum sports bottles.

19.145 Titanium anodizing proceeds much like that of aluminum, but
RAN the resultant coat of TiO_2 is much thinner (10^{-4}mm) than the corresponding coat of Al_2O_3 (10^{-2}mm). Furthermore, the iridescent colors of anodized titanium result not from the absorption of organic dyes but from the interference of light as it is reflected by the anodized surface. The cathode half-reaction is the same as the one for anodizing aluminum in Problem 19.144. What is the overall cell reaction and cell potential for anodizing titanium if the anode half-reaction is

$$Ti(s) + 2 H_2O(l) \longrightarrow$$
$$TiO_2(s) + 4 H^+(aq) + 4 e^- \quad E° = +1.066 \text{ V}$$

▲ Anodized titanium medical implant components such as bone screws are color coded to assist the surgeon during surgery.

19.146 In order to charge a lead storage battery (Section 19.10)
RAN 500.0 g of PbSO4 (s) must be converted into $PbO_2(s)$ and Pb(s).
(a) Does the reaction represent an electrolytic or galvanic cell?
(b) How many coulombs of electrical charge are needed?
(c) If a current of 500 A is used, how long will it take?

19.147 When the nickel–zinc battery, used in digital cameras, is
RAN recharged, the following cell reaction occurs:

$$2 Ni(OH)_2(s) + Zn(OH)_2(s) \longrightarrow 2 Ni(OH)_3(s) + Zn(s)$$

(a) How many grams of zinc are formed when $3.35 × 10^{-2}$ g of $Ni(OH)_2$ are consumed?
(b) How many minutes are required to fully recharge a dead battery that contains $6.17 × 10^{-2}$g of Zn with a constant current of 0.100 A?

MULTICONCEPT PROBLEMS

19.148 Consider the following half-reactions and $E°$ values:
RAN

$$Ag^+(aq) + e^- \longrightarrow Ag(s) \qquad E° = 0.80 \text{ V}$$
$$Cu^{2+}(aq) + 2 e^- \longrightarrow Cu(s) \qquad E° = 0.34 \text{ V}$$
$$Pb^{2+}(aq) + 2 e^- \longrightarrow Pb(s) \qquad E° = -0.13 \text{ V}$$

(a) Which of these metals or ions is the strongest oxidizing agent? Which is the strongest reducing agent?
(b) The half-reactions can be used to construct three different galvanic cells. Tell which cell delivers the highest voltage, identify the anode and cathode, and tell the direction of electron and ion flow.
(c) Write the cell reaction for part (b), and calculate the values of $E°$, $\Delta G°$ (in kilojoules), and K for this reaction at 25 °C.
(d) Calculate the voltage for the cell in part (b) if both ion concentrations are 0.010 M.

19.149 Consider a galvanic cell that uses the following half-reactions:
RAN

$$2 H^+(aq) + 2 e^- \longrightarrow H_2(g)$$
$$Al^{3+}(aq) + 3 e^- \longrightarrow Al(s)$$

(a) What materials are used for the electrodes? Identify the anode and cathode, and indicate the direction of electron and ion flow.
(b) Write a balanced equation for the cell reaction, and calculate the standard cell potential.
(c) Calculate the cell potential at 25 °C if the ion concentrations are 0.10 M and the partial pressure of H_2 is 10.0 atm.

(d) Calculate $\Delta G°$ (in kilojoules) and K for the cell reaction at 25 °C.
(e) Calculate the mass change (in grams) of the aluminum electrode after the cell has supplied a constant current of 10.0 A for 25.0 min.

19.150 Chlorine can be prepared in the laboratory by the reaction
RAN of hydrochloric acid and potassium permanganate.
(a) Use data in Appendix D to write a balanced equation for the reaction. The reduction product is Mn^{2+}.
(b) Calculate $E°$ and $\Delta G°$ for the reaction.
(c) How many liters of Cl_2 at 1.0 atm and 25 °C will result from the reaction of 179 g $KMnO_4$ with an excess of HCl?

19.151 The sodium–sulfur battery has molybdenum electrodes with anode and cathode compartments separated by β-alumina, a ceramic through which sodium ions can pass. Because the battery operates at temperatures above 300 °C, all the reactants and products are present in a molten solution. The cell voltage is about 2.0 V.
(a) What is the cell reaction if the shorthand notation is

$$Mo(s)\,|\,Na(soln),\, Na^+(soln)\,\|\,S(soln),\, S^{2-}(soln)\,|\,Mo(s)?$$

(b) How many kilograms of sodium are consumed when a 25 kW sodium–sulfur battery produces current for 32 min?

19.152 Consider the addition of the following half-reactions:

$$(1) \ Fe^{3+}(aq) + 3 \ e^- \longrightarrow Fe(s) \qquad E°_1 = -0.04 \ V$$
$$\underline{(2) \ Fe(s) \longrightarrow Fe^{2+}(aq) + 2 \ e^- \qquad E°_2 = 0.45 \ V}$$
$$(3) \ Fe^{3+}(aq) + e^- \longrightarrow Fe^{2+}(aq) \qquad E°_3 = ?$$

Because half-reactions (1) and (2) contain a different number of electrons, the net reaction (3) is another half-reaction, and $E°_3$ can't be obtained simply by adding $E°_1$ and $E°_2$. The free-energy changes, however, are additive because G is a state function:

$$\Delta G°_3 = \Delta G°_1 + \Delta G°_2$$

(a) Starting with the relationship between $\Delta G°$ and $E°$, derive a general equation that relates the $E°$ values for half-reactions (1), (2), and (3).

(b) Calculate the value of $E°_3$ for the Fe^{3+}/Fe^{2+} half-reaction.

(c) Explain why the $E°$ values would be additive $(E°_3 = E°_1 + E°_2)$ if reaction (3) were an overall cell reaction rather than a half-reaction.

19.153 A galvanic cell has a silver electrode in contact with
RAN   0.050 M $AgNO_3$ and a copper electrode in contact with 1.0 M $Cu(NO_3)_2$.

(a) Write a balanced equation for the cell reaction, and calculate the cell potential at 25 °C.

(b) Excess $NaBr(aq)$ is added to the $AgNO_3$ solution to precipitate AgBr. What is the cell potential at 25 °C after the precipitation of AgBr if the concentration of excess Br^- is 1.0 M? Write a balanced equation for the cell reaction under these conditions. (K_{sp} for AgBr at 25 °C is 5.4×10^{-13}.)

(c) Use the result in part (b) to calculate the standard reduction potential $E°$ for the half-reaction

$$AgBr(s) + e^- \longrightarrow Ag(s) + Br^-(aq).$$

19.154 Given the following standard reduction potentials at 25 °C, **(a)** balance the equation for the reaction of H_2MoO_4 with elemental arsenic in acidic solution to give Mo^{3+} and H_3AsO_4 and **(b)** calculate $E°$ for this reaction.

Half-Reaction	$E°$ (V)
$H_3AsO_4(aq) + 2 \ H^+(aq) + 2 \ e^- \longrightarrow H_3AsO_3(aq) + H_2O(l)$	+0.560
$H_3AsO_3(aq) + 3 \ H^+(aq) + 3 \ e^- \longrightarrow As(s) + 3 \ H_2O(l)$	+0.240
$H_2MoO_4(aq) + 2 \ H^+(aq) + 2 \ e^- \longrightarrow MoO_2(s) + 2 \ H_2O(l)$	+0.646
$MoO_2(s) + 4 \ H^+(aq) + e^- \longrightarrow Mo^{3+}(aq) + 2 \ H_2O(l)$	-0.008

19.155 The reaction of MnO_4^- with oxalic acid $(H_2C_2O_4)$ in
RAN   acidic solution, yielding Mn^{2+} and CO_2 gas, is widely used to determine the concentration of permanganate solutions.

(a) Write a balanced net ionic equation for the reaction.

(b) Use the data in Appendix D to calculate $E°$ for the reaction.

(c) Show that the reaction goes to completion by calculating the values of $\Delta G°$ and K at 25 °C.

(d) A 1.200 g sample of sodium oxalate $(Na_2C_2O_4)$ is dissolved in dilute H_2SO_4 and then titrated with a $KMnO_4$ solution. If 32.50 mL of the $KMnO_4$ solution is required to reach the equivalence point, what is the molarity of the $KMnO_4$ solution?

19.156 Calculate the standard reduction potential for $Ba^{2+}(aq) +$ $2 \ e^- \longrightarrow Ba(s)$ given that $\Delta G° = 16.7$ kJ for the reaction $Ba^{2+}(aq) + 2 \ Cl^-(aq) \longrightarrow BaCl_2(s)$. Use any data needed from Appendixes B and D.

19.157 A concentration cell has the same half-reactions at the
RAN   anode and cathode, but a voltage results from different concentrations in the two electrode compartments.

(a) What is x in the concentration cell $Cu(s) | \ Cu^{2+}$ $(x \ M) \| Cu^{2+}(0.10 \ M) | Cu(s)$ if the measured cell potential is 0.0965 V?

(b) A similar cell has 0.10 M Cu^{2+} in both compartments. When a stoichiometric amount of ethylenediamine $(NH_2CH_2CH_2NH_2)$ is added to one compartment, the measured cell potential is 0.179 V. Calculate the formation constant K_f for the complex ion $Cu(NH_2CH_2CH_2NH_2)_2^{2+}$. Assume there is no volume change.

19.158 Consider the redox titration (Section 4.13) of 120.0 mL of
RAN   0.100 M $FeSO_4$ with 0.120 M $K_2Cr_2O_7$ at 25 °C, assuming that the pH of the solution is maintained at 2.00 with a suitable buffer. The solution is in contact with a platinum electrode and constitutes one half-cell of an electrochemical cell. The other half-cell is a standard hydrogen electrode. The two half-cells are connected with a wire and a salt bridge, and the progress of the titration is monitored by measuring the cell potential with a voltmeter.

(a) Write a balanced net ionic equation for the titration reaction, assuming that the products are Fe^{3+} and Cr^{3+}.

(b) What is the cell potential at the equivalence point?

19.159 Consider a galvanic cell that utilizes the following half-
RAN   reactions:

Anode:	$Zn(s) + H_2O(l) \longrightarrow ZnO(s) + 2 \ H^+(aq) + 2 \ e^-$
Cathode:	$Ag^+(aq) + e^- \longrightarrow Ag(s)$

(a) Write a balanced equation for the cell reaction, and use the thermodynamic data in Appendix B to calculate the values of $\Delta H°$, $\Delta S°$, and $\Delta G°$ for the reaction.

(b) What are the values of $E°$ and the equilibrium constant K for the cell reaction at 25 °C?

(c) What happens to the cell voltage if aqueous ammonia is added to the cathode compartment? Calculate the cell voltage assuming that the solution in the cathode compartment was prepared by mixing 50.0 mL of 0.100 M $AgNO_3$ and 50.0 mL of 4.00 M NH_3.

(d) Will AgCl precipitate if 10.0 mL of 0.200 M NaCl is added to the solution in part (c)? Will AgBr precipitate if 10.0 mL of 0.200 M KBr is added to the resulting solution?

19.160 The nickel–iron battery has an iron anode, an NiO(OH)
RAN   cathode, and a KOH electrolyte. This battery uses the following half-reactions and has an $E°$ value of 1.37 V at 25 °C.

$$Fe(s) + 2 \ OH^-(aq) \longrightarrow Fe(OH)_2(s) + 2 \ e^-$$
$$NiO(s) + H_2O(l) + e^- \longrightarrow Ni(OH)_2(s) + OH^-(aq)$$

(a) Write a balanced equation for the cell reaction.

(b) Calculate $\Delta G°$ (in kilojoules) and the equilibrium constant K for the cell reaction at 25 °C.

(c) What is the cell voltage at 25 °C when the concentration of KOH in the electrolyte is 5.0 M?

(d) How many grams of $Fe(OH)_2$ are formed at the anode when the battery produces a constant current of 0.250 A for 40.0 min? How many water molecules are consumed in the process?

19.161 Experimental solid-oxide fuel cells that use butane (C_4H_{10})
RAN as the fuel have been reported recently. These cells contain
composite metal/metal oxide electrodes and a solid metal
oxide electrolyte. The cell half-reactions are

Anode: $C_4H_{10}(g) + 13\ O^{2-}(s) \longrightarrow 4\ CO_2(g) + 5\ H_2O(l) + 26\ e^-$

Cathode: $O_2(g) + 4\ e^- \longrightarrow 2\ O^{2-}(s)$

(a) Write a balanced equation for the cell reaction.

(b) Use the thermodynamic data in Appendix B to calculate
the values of $E°$ and the equilibrium constant K for the
cell reaction at 25 °C. Will $E°$ and K increase, decrease,
or remain the same on raising the temperature?

(c) How many grams of butane are required to produce a
constant current of 10.5 A for 8.00 h? How many liters
of gaseous butane at 20 °C and 815 mm Hg pressure are
required?

19.162 The half-reactions that occur in ordinary alkaline batteries
RAN can be written as

Cathode: $MnO_2(s) + H_2O(l) + e^- \longrightarrow MnO(OH)(s) + OH^-(aq)$

Anode: $Zn(s) + 2\ OH^-(aq) \longrightarrow Zn(OH)_2(s) + 2\ e^-$

In 1999, researchers in Israel reported a new type of alkaline
battery, called a "super-iron" battery. This battery uses the same
anode reaction as an ordinary alkaline battery but involves the
reduction of FeO_4^{2-} ion (from K_2FeO_4) to solid $Fe(OH)_3$ at the
cathode.

(a) Use the following standard reduction potential and any
data from Appendixes C and D to calculate the standard cell potential expected for an ordinary alkaline
battery:

$$MnO(OH)(s) + H_2O(l) + e^- \longrightarrow Mn(OH)_2(s) + OH^-(aq)$$
$$E° = -0.380\ V$$

(b) Write a balanced equation for the cathode half-reaction
in a super-iron battery. The half-reaction occurs in a
basic environment.

(c) A super-iron battery should last longer than an ordinary
alkaline battery of the same size and weight because its
cathode can provide more charge per unit mass. Quantitatively compare the number of coulombs of charge
released by the reduction of 10.0 g of K_2FeO_4 to $Fe(OH)_3$
with the number of coulombs of charge released by the
reduction of 10.0 g of MnO_2 to $MnO(OH)$.

19.163 Gold metal is extracted from its ore by treating the crushed
rock with an aerated cyanide solution. The unbalanced
equation for the reaction is

$$Au(s) + CN^-(aq) + O_2(g) \longrightarrow Au(CN)_2^-(aq)$$

(a) Balance the equation for this reaction in basic solution.

(b) Use any of the following data at 25 °C to calculate $\Delta G°$ for
this reaction at 25 °C: K_f for $Au(CN)_2^- = 6.2 \times 10^{38}$, K_a
for $HCN = 4.9 \times 10^{-10}$, and standard reduction potentials are

$$O_2(g) + 4\ H^+(aq) + 4\ e^- \longrightarrow 2\ H_2O(l) \quad E° = 1.229\ V$$
$$Au^{3+}(aq) + 3\ e^- \longrightarrow Au(s) \quad E° = 1.498\ V$$
$$Au^{3+}(aq) + 2\ e^- \longrightarrow Au^+(aq) \quad E° = 1.401\ V$$

19.164 Consider the redox titration of 100.0 mL of a solution of
RAN 0.010 M Fe^{2+} in 1.50 M H_2SO_4 with a 0.010 M solution
of $KMnO_4$, yielding Fe^{3+} and Mn^{2+}. The titration is carried out in an electrochemical cell equipped with a platinum electrode and a calomel reference electrode consisting
of an Hg_2Cl_2/Hg electrode in contact with a saturated KCl
solution having $[Cl^-] = 2.9$ M. Using any data in Appendixes C and D, calculate the cell potential after addition of
(a) 5.0 mL, (b) 10.0 mL, (c) 19.0 mL, and (d) 21.0 mL of
the $KMnO_4$ solution.

chapter 20

Nuclear Chemistry

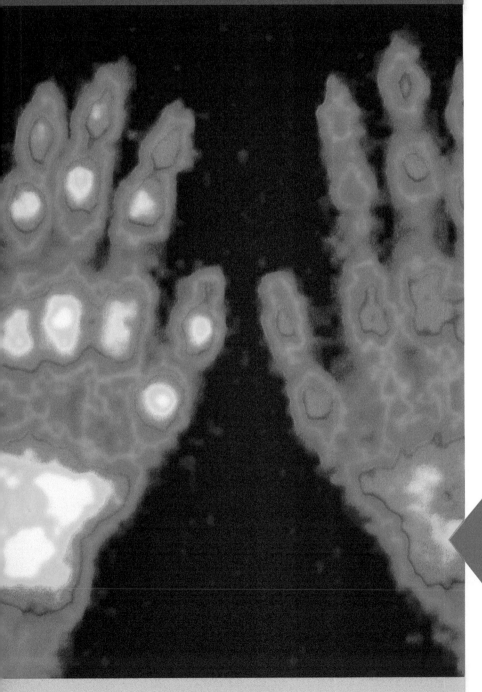

Contents

Radioisotopes are used to create images to gain information about the body and diagnose disease. The injected radioisotope accumulates in inflamed tissue in arthritic joints. Emitted gamma rays are used to create an image showing extensive arthritis in the fingers and wrists.

How are radioisotopes used in medicine?

The answer to this question can be found on page 894 in the INQUIRY ?

Nuclear chemistry, the study of properties and reactions of atomic nuclei, is a topic of high societal importance. Due to the incredible strength and long-term radiation effects of atomic weapons, 190 countries worldwide have joined the International Non-Proliferation Treaty, which aims to prevent the spread of weapons technology, disarm weapons stockpiles, and promote peaceful uses of nuclear technology such as nuclear power and medicine. Nuclear power plants operate in 30 countries worldwide, and approximately 20% of electricity generated in the United States comes from **nuclear reactions.** Some people consider nuclear energy a "clean" source of electricity because it does not emit air pollutants or the greenhouse gas carbon dioxide. However, drawbacks of nuclear power include the need for a long-term waste disposal plan and the potential for serious accidents such as the Chernobyl plant in Ukraine in 1986 and the Fukushima Daiichi plant in Japan in 2011. Although radiation from nuclear accidents is a public health concern, nuclear medicine uses radioactive isotopes to diagnose and treat conditions ranging from appendicitis to cancer. In this chapter we'll study various aspects of nuclear reactions including changes in atomic nuclei, the energy of nuclear reactions, and rates of radioactive decay.

20.1 NUCLEAR REACTIONS AND THEIR CHARACTERISTICS

One way that nuclear reactions differ from chemical reactions is that one element is converted into another. For example, our Sun's energy comes from a fusion reaction in which hydrogen is converted to helium.

$$_1^1H + {}_1^2H \longrightarrow {}_2^3He$$

In contrast, the identities of the atoms remain the same in a chemical reaction; only the bonds between atoms change. When natural gas (methane; CH_4) burns in oxygen, for instance, the C, H, and O atoms combine in a different way to yield carbon dioxide (CO_2) and water (H_2O), but they still remain C, H, and O atoms.

Recall from Section 2.8 that an atom is characterized by its atomic number Z, and its mass number, A. The atomic number, written as a subscript to the left of the element symbol, gives the number of protons in the nucleus. The mass number, written as a superscript to the left of the elemental symbol, gives the total number of particles in the nucleus, or **nucleons,** a general term for both protons (p) and neutrons (n). The most abundant isotope of carbon, for example, has 6 protons and 6 neutrons for a total of 12 nucleons.

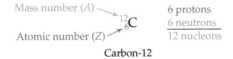

Carbon-12

Atoms with identical atomic numbers but different mass numbers are called isotopes (Section 2.8), and the nucleus of a given isotope is called a **nuclide.** Take the element carbon, for example. There are 15 known isotopes of carbon, two of which occur commonly (^{12}C and ^{13}C) and one of which (^{14}C) is produced in small amounts in the upper atmosphere by the action of cosmic rays on ^{14}N atoms. The remaining 12 carbon isotopes have been produced artificially using reactions similar to those we will describe in Section 20.7. Only the two commonly occurring isotopes are stable; the other 13 undergo spontaneous changes to their nuclei. Carbon-14, for instance, slowly decays to give nitrogen-14 plus an electron, a process we can write as the following *nuclear equation:*

$$_6^{14}C \longrightarrow {}_7^{14}N + {}_{-1}^0e$$

In a **nuclear equation,** the element symbols represent only the *nuclei* of atoms rather than the entire neutral atoms, so the subscript represents only the number of nuclear charges (protons). An emitted electron is written as $_{-1}^0e$, where the superscript 0 indicates that the mass of an electron is essentially zero when compared to that of a proton or neutron, and the subscript indicates that the charge is −1.

The equation is *balanced* because the total number of neutrons and protons, collectively called *nucleons,* or nuclear particles, is the same on both sides of the equation and the number of charges on the nuclei and on any elementary particles (protons and electrons) is the same on both sides. A nuclear equation is balanced when the *sum of the mass numbers of reactants equals the sum of the mass numbers of the products.* Likewise, the reaction is balanced when the *sum of the atomic numbers of the reactants equals the sum of the atomic numbers of the products.*

$$^{14}_{6}\text{C} \longrightarrow {}^{14}_{7}\text{N} + {}^{0}_{-1}\text{e}$$

Mass number: 14   =   14 + 0
Atomic number: 6   =   7 + (−1)

Nuclear reactions, such as the spontaneous change of ^{14}C to ^{14}N, are distinguished from chemical reactions in several ways:

- A nuclear reaction involves a change in an atom's nucleus, usually producing a different element. A chemical reaction, by contrast, involves only a change in the way that different atoms are combined. A chemical reaction never changes the nuclei themselves or produces a different element.

- Different isotopes of an element have essentially the same behavior in chemical reactions but often have completely different behavior in nuclear reactions.

- The energy change accompanying a nuclear reaction is far greater than that accompanying a chemical reaction. The nuclear transformation of 1.0 g of uranium-235 ($^{235}_{92}\text{U}$) releases more than one million times as much energy as the chemical combustion of 1.0 g of methane.

BIG IDEA Question 1

Go to eText

Which reaction releases more energy: the combustion of 1 g of hydrogen gas or the fusion of 1 g of hydrogen gas to form helium gas?

20.2 RADIOACTIVITY

Scientists have known since 1896 that many nuclei are **radioactive**—they undergo a spontaneous decay and emit some form of *radiation.* Early studies of radioactive isotopes, or **radioisotopes,** by Ernest Rutherford in 1897 showed that there are three common types of radiation with markedly different properties: *alpha* (α), *beta* (β), and *gamma* (γ) *radiation,* named after the first three letters of the Greek alphabet.

Alpha (α) Radiation

Using the simple experiment shown in **FIGURE 20.1,** Rutherford found that **α radiation** consists of a stream of particles that are repelled by a positively charged electrode, attracted by a negatively charged electrode, and have a mass-to-charge ratio which identifies them as helium nuclei, $^{4}_{2}\text{He}^{2+}$. Alpha particles thus consist of two protons and two neutrons.

Experiment

The radioactive source in the shielded box emits radiation, which passes between two electrodes.

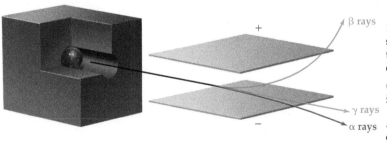

Results

Beta radiation is strongly deflected toward the positive electrode.

Gamma radiation is undeflected.

Alpha radiation is deflected toward the negative electrode.

Conclusion

Alpha particles and beta particles are both electrically charged, but they carry charges with opposite signs. Alpha particles have a positive charge, while beta particles have a negative charge. Gamma rays are not affected by an electric field and therefore have no electric charge.

▲ **FIGURE 20.1**
The effect of an electric field on α, β, and γ radiation.

Because the emission of an α particle from a nucleus results in a loss of two protons and two neutrons, it reduces the mass number of the nucleus by 4 and reduces the atomic number by 2. Alpha emission is particularly common for heavy radioactive isotopes. Uranium-238, for example, spontaneously emits an α particle and forms thorium-234.

$$^{238}_{92}\text{U} \longrightarrow {}^{4}_{2}\text{He} + {}^{234}_{90}\text{Th}$$

Mass number: $238 \;=\; 4 \;+ 234$
Atomic number: $92 \;=\; 2 \;+\; 90$

The nuclear equation for the radioactive decay of uranium-238 is balanced because the sum of the mass numbers and atomic numbers on both sides of the equation are equal. In the decay of $^{238}_{92}\text{U}$ to give $^{4}_{2}\text{He}$ and $^{234}_{90}\text{Th}$, there are 238 nucleons and 92 nuclear charges on both sides of the equation.

Beta (β) Radiation

Further work by Rutherford in the late 1800s showed that **β radiation** consists of a stream of particles that are attracted to a positive electrode (Figure 20.1), repelled by a negative electrode, and have a mass-to-charge ratio identifying them as electrons, $^{0}_{-1}\text{e}$ or β^-. Beta emission occurs when a neutron in the nucleus spontaneously decays into a proton plus an electron, which is then ejected. The product nucleus has the same mass number as the starting nucleus because a neutron has turned into a proton, but it has a higher atomic number because it has the newly created proton. The reaction of ^{131}I to give ^{131}Xe is an example:

$$^{131}_{53}\text{I} \longrightarrow {}^{131}_{54}\text{Xe} + {}^{0}_{-1}\text{e}$$

Mass number: $131 \;=\; 131 \;+\; 0$
Atomic number: $53 \;=\; 54 \;+ (-1)$

Writing the emitted β particle as $^{0}_{-1}\text{e}$ in the nuclear equation makes clear the charge balance of the nuclear reaction. The subscript in the $^{131}_{53}\text{I}$ nucleus on the left (53) is balanced by the sum of the two subscripts on the right ($54 - 1 = 53$).

Gamma (γ) Radiation

Gamma (γ) radiation is unaffected by either electric or magnetic fields (Figure 20.1) and has no mass. Like visible light, ultraviolet rays, and X rays, γ radiation is simply electromagnetic radiation of very high energy. Gamma radiation almost always accompanies α and β emission as a mechanism for the release of energy, but it is often not shown when writing nuclear equations because it changes neither the mass number nor the atomic number of the product nucleus.

Positron Emission and Electron Capture

In addition to α, β, and γ radiation, two other types of radioactive decay processes also occur commonly: *positron emission* and *electron capture*. **Positron emission** occurs when a proton in the nucleus changes into a neutron plus an ejected *positron* ($^{0}_{+1}\text{e}$ or β^+), a particle that can be thought of as a positive electron. A positron has the same mass as an electron but an opposite charge.

The result of positron emission is a decrease in the atomic number of the product nucleus but no change in the mass number. Potassium-40, for example, undergoes positron emission to yield argon-40, a nuclear reaction important in geology for dating rocks. Note once again that the sum of the two subscripts on the right of the nuclear equation ($18 + 1 = 19$) is equal to the subscript in the $^{40}_{19}\text{K}$ nucleus on the left.

$$^{40}_{19}\text{K} \longrightarrow {}^{40}_{18}\text{Ar} + {}^{0}_{1}\text{e}$$

Mass number: $40 \;=\; 40 \;+ 0$
Atomic number: $19 \;=\; 18 \;+ 1$

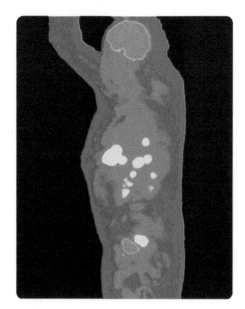

You might already know that the acronym PET used in medical imaging stands for *positron emission tomography*. PET can be used to detect tumors by injecting a chemical compound containing a positron-emitting isotope such as ^{18}F into the body. The compound binds specifically to a protein on the surface of a cancer cell and accumulates in a tumor. When decay occurs, the emitted positron reacts with a nearby electron and is instantly annihilated, releasing gamma rays whose position in the body can be detected.

Electron capture is a process in which the nucleus captures one of the surrounding electrons in an atom, thereby converting a proton into a neutron. The mass number of the product nucleus is unchanged, but the atomic number decreases by 1, just as in positron emission. The conversion of mercury-197 into gold-197 is an example:

$$^{197}_{80}Hg + {}^{0}_{-1}e \longrightarrow {}^{197}_{79}Au$$

Mass number: $197 + 0 = 197$
Atomic number: $80 + (-1) = 79$

Characteristics of the different kinds of radioactive decay processes are summarized in **TABLE 20.1**.

▲ A PET scan of this patient shows the presence of recurrent lung carcinoma.

TABLE 20.1   A Summary of Radioactive Decay Processes

Process	Symbol	Change in Atomic Number	Change in Mass Number	Change in Neutron Number
Alpha emission	$^{4}_{2}He$ or α	-2	-4	-2
Beta emission	$^{0}_{-1}e$ or β^{-}	$+1$	0	-1
Gamma emission	$^{0}_{0}\gamma$ or γ	0	0	0
Positron emission	$^{0}_{1}e$ or β^{+}	-1	0	$+1$
Electron capture	E. C.	-1	0	$+1$

WORKED EXAMPLE 20.1

Balancing Nuclear Equations

Write a balanced nuclear equation for each of the following processes:

(a) Alpha emission from curium-242: $^{242}_{96}Cm \longrightarrow {}^{4}_{2}He + ?$
(b) Beta emission from magnesium-28: $^{28}_{12}Mg \longrightarrow {}^{0}_{-1}e + ?$
(c) Positron emission from xenon-118: $^{118}_{54}Xe \longrightarrow {}^{0}_{1}e + ?$

STRATEGY

The key to writing nuclear equations is to make sure that sum of the mass numbers of reactants equals the sum of the mass numbers of the products and that the sum of the atomic numbers of the reactants equals the sum of the atomic numbers of the products.

SOLUTION

(a) In α emission, the mass number decreases by 4 and the atomic number decreases by 2, giving plutonium-238:

$$^{242}_{96}Cm \longrightarrow {}^{4}_{2}He + {}^{238}_{94}Pu$$

Mass number: $242 = 4 + 238$
Atomic number: $96 = 2 + 94$

(b) In β emission, the mass number is unchanged and the atomic number increases by 1, giving aluminum-28:

$$^{28}_{12}Mg \longrightarrow {}^{0}_{-1}e + {}^{28}_{13}Al$$

Mass number: $28 = 0 + 28$
Atomic number: $12 = (-1) + 13$

(c) In positron emission, the mass number is unchanged and the atomic number decreases by 1, giving iodine-118:

$$^{118}_{54}Xe \longrightarrow {}^{0}_{1}e + {}^{118}_{53}I$$

Mass number: $118 = 0 + 118$
Atomic number: $54 = 1 + 53$

▶ **PRACTICE 20.1**   Write a balanced nuclear equation for each of the following processes:

(a) Beta emission from ruthenium-106: $^{106}_{44}Ru \longrightarrow {}^{0}_{-1}e + ?$
(b) Alpha emission from bismuth-189: $^{189}_{83}Bi \longrightarrow {}^{4}_{2}He + ?$
(c) Electron capture by polonium-204: $^{204}_{84}Po + {}^{0}_{-1}e \longrightarrow ?$

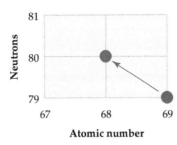

▶ **CONCEPTUAL APPLY 20.2** Identify the isotopes involved, and tell what type of decay process is occurring in the following nuclear reaction:

All **PRACTICE** and **APPLY** problems are interactive in the eText.

20.3 NUCLEAR STABILITY

Why do some nuclei undergo spontaneous radioactive decay while others do not? Why, for instance, does a carbon-14 nucleus, with 6 protons and 8 neutrons, spontaneously emit a β particle, whereas a carbon-13 nucleus, with 6 protons and 7 neutrons, is nonradioactive?

The answer has to do with the neutron/proton ratio in the nucleus and with the forces holding the nucleus together. To see the effect of the neutron/proton ratio on nuclear stability, look at the graph in **FIGURE 20.2.** The y-axis gives the number of neutrons, and the x-axis gives the number of protons. The first 92 elements are naturally occurring, while the remainder are the artificially produced **transuranium elements.** (Technetium and promethium cannot be found on Earth because all their isotopes are radioactive and have very short lifetimes. Francium and astatine occur on Earth only in very tiny amounts.)

When the more than 3600 known isotopes are plotted on the neutron/proton graph in Figure 20.2, they fall in a curved band sometimes called the *band of nuclear stability.* Even within the band, only 264 of the isotopes are nonradioactive. The others decay spontaneously, although their rates of decay vary enormously. On either side of the band is a so-called sea of instability representing the large number of unstable neutron–proton combinations that have never been detected. Particularly interesting is the island of stability predicted to exist for a few superheavy isotopes near 114 protons and 184 neutrons. The first members of this group—$^{287}_{114}\text{Fl}$, $^{288}_{114}\text{Fl}$, and $^{289}_{114}\text{Fl}$—were prepared by bombarding

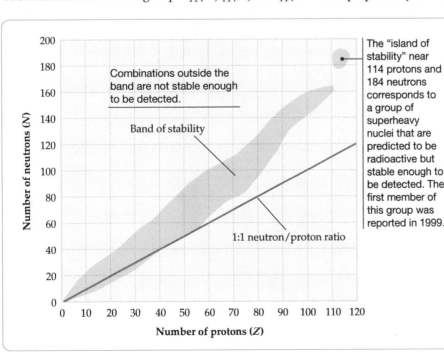

The "island of stability" near 114 protons and 184 neutrons corresponds to a group of superheavy nuclei that are predicted to be radioactive but stable enough to be detected. The first member of this group was reported in 1999.

◀ **FIGURE 20.2**

The band of nuclear stability. The band indicates various neutron/proton combinations that give rise to nuclei that are either nonradioactive or that are radioactive but decay slowly enough to exist for a measurable time.

◀ **Figure It Out**

Would a nuclide with 100 neutrons and 90 protons exist long enough to be detected?

Answer: No, the point on the graph at 100 neutrons and 90 protons is not in the band of stability.

a target of plutonium-244 with a nuclei of calcium-48 in a particle reactor in 1999. The element flerovium (symbol, Fl) was officially named in 2012 after the Flerov Laboratory of Nuclear Reactions where it was discovered. Other isotopes of superheavy elements in the island of stability (Z = 113, 115, 117 and 118) were discovered during the time period 2003 to 2005. The names and symbols were approved in 2016, and now these elements appear in the periodic table. While practical uses of these superheavy elements are not yet known, their existence upholds nuclear theory about the "island of stability" and helps scientists better understand how nuclei are held together.

We can make several generalizations about nuclear stability:

- Every element in the periodic table has at least one radioactive isotope.
- Hydrogen is the only element whose most abundant isotope ($_1^1H$) contains more protons (1) than neutrons (0).
- The ratio of neutrons to protons gradually increases in the band of stability.
- All isotopes heavier than bismuth-209 are radioactive, even though they may decay slowly and be stable enough to occur naturally.

A close-up look at a segment of the band of nuclear stability (**FIGURE 20.3**) shows the interesting trend that radioactive nuclei with higher neutron/proton ratios (top side of the band) tend to emit β particles while nuclei with lower neutron/proton ratios (bottom side of the band) tend to undergo nuclear decay by positron emission, electron capture, or α emission.

The trend shown in Figure 20.3 makes sense if you think about it: The nuclei on the top side of the band are neutron-rich and therefore undergo a process—β emission—that *decreases* the neutron/proton ratio by converting a neutron into a proton. The resulting new isotope is closer to the center of the band of stability instead of along the top edge. The nuclei on the bottom side of the band, by contrast, are neutron-poor

FIGURE 20.3

A close-up look at the band of nuclear stability. This look at the region from Z = 66 (dysprosium) through Z = 79 (gold) shows the types of radioactive processes that various radioisotopes undergo.

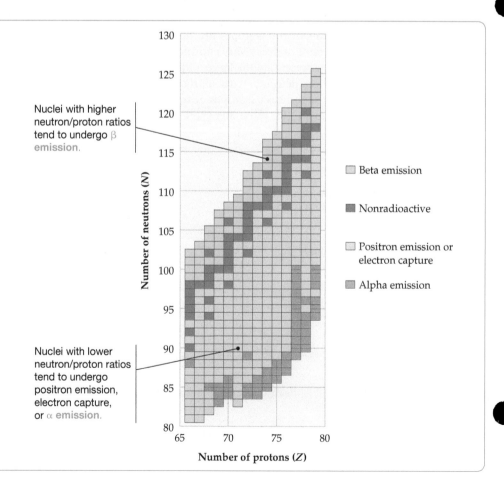

and therefore undergo processes that *increase* the neutron/proton ratio. Take a minute to convince yourself that α emission does, in fact, increase the neutron/proton ratio for heavy nuclei in which n > p.

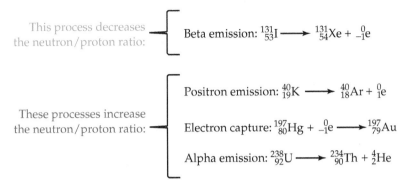

This process decreases the neutron/proton ratio:

Beta emission: $^{131}_{53}\text{I} \longrightarrow \,^{131}_{54}\text{Xe} + \,^{0}_{-1}\text{e}$

These processes increase the neutron/proton ratio:

Positron emission: $^{40}_{19}\text{K} \longrightarrow \,^{40}_{18}\text{Ar} + \,^{0}_{1}\text{e}$

Electron capture: $^{197}_{80}\text{Hg} + \,^{0}_{-1}\text{e} \longrightarrow \,^{197}_{79}\text{Au}$

Alpha emission: $^{238}_{92}\text{U} \longrightarrow \,^{234}_{90}\text{Th} + \,^{4}_{2}\text{He}$

Go to eText

BIG IDEA Question 2

Of the two isotopes ^{173}Au and ^{199}Au, one decays by β emission and one decays by α emission. Which is which?

Many radioactive nuclei cannot reach a stable state through a single decay process. Consequently, a series of nuclear decays, called a **radioactive decay chain**, will occur until a stable nucleus is formed. **FIGURE 20.4** shows the decay chain for the radioactive isotope Pu-239 produced in nuclear reactors.

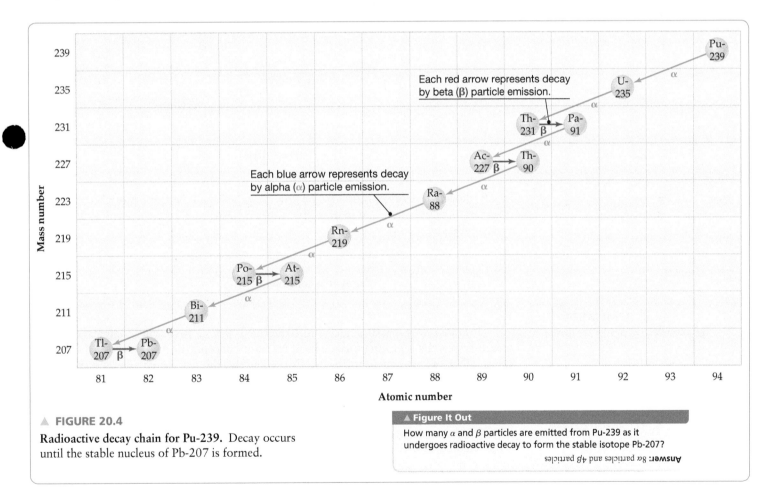

▲ **FIGURE 20.4**

Radioactive decay chain for Pu-239. Decay occurs until the stable nucleus of Pb-207 is formed.

▲ **Figure It Out**

How many α and β particles are emitted from Pu-239 as it undergoes radioactive decay to form the stable isotope Pb-207?

Answer: 8α particles and 4β particles

20.4 RADIOACTIVE DECAY RATES

In Section 20.3 we saw that many nuclei are radioactive. The rate at which a radioactive nucleus decays is characterized by the time it takes for half the sample to disappear, its half-life, $t_{1/2}$. For example, phosphorus-32, a radioisotope used in leukemia therapy, has a half-life of 14.26 days. Since radioactive decay is a **first-order process**, the

REMEMBER . . .

A **first-order process** is one whose rate depends on the concentration of only one substance raised to the first power (Section 14.5).

half-life is constant and after each time period of 14.26 days, the number of radioac-tive phosphorus-32 nuclei decreases by a factor of 2 (**FIGURE 20.5**).

The half-lives of different radioisotopes vary enormously. Some, such as uranium-238, decay at a barely perceptible rate over billions of years, while others, such as carbon-21, decay within microseconds. For example, carbon-14 decays to nitrogen-14 by emission of a β^- particle with a half-life of 5715 years:

$$^{14}_{6}C \longrightarrow {}^{14}_{7}N + {}^{0}_{-1}e \qquad t_{1/2} = 5715 \text{ y}$$

▶ **FIGURE 20.5**

Number of radioactive phosphorus-32 nuclei as a function of time. The half-life of phosphorus-32 is 14.26 days.

▶ **Figure It Out**

A sample of phosphorus-32 contains 1000 nuclei. How much time has elapsed when 250 phosphorus-32 nuclei remain?

Answer: When 250 phosphorus-32 nuclei remain, two half-lives have passed. Since each half-life is 14.26 days, the total time would be 28.52 days.

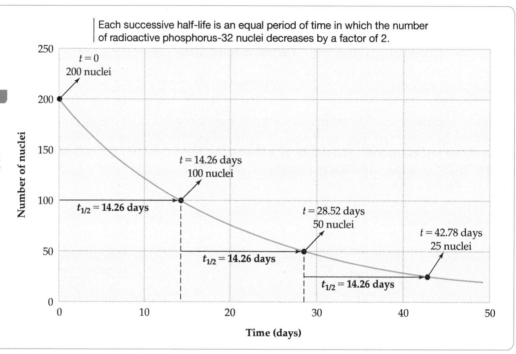

Each successive half-life is an equal period of time in which the number of radioactive phosphorus-32 nuclei decreases by a factor of 2.

$t = 0$
200 nuclei

$t = 14.26$ days
100 nuclei

$t_{1/2} = 14.26$ **days**

$t = 28.52$ days
50 nuclei

$t_{1/2} = 14.26$ **days**

$t = 42.78$ days
25 nuclei

$t_{1/2} = 14.26$ **days**

Number of nuclei

Time (days)

Radioactive decay is a first-order process, whose rate is proportional to the num-ber of radioactive nuclei N in a sample times a first-order rate constant k, called the decay constant.

$$\text{Decay rate constant} = -\frac{\Delta N}{\Delta t} = kN$$

REMEMBER . . .

The integrated rate law for the general reaction, A $\longrightarrow$ B, is

$$\ln\left(\frac{[A]_t}{[A]_0}\right) = -kt$$

The equation can be used to calculate the amount of reactant remaining after given amount of time (Section 14.5).

This rate law is strictly analogous to the rate law for a first-order chemical reaction, as can be seen by replacing the chemical reactant concentration [A] (Section 14.5) with the number of radioactive nuclei N. Consequently, the integrated rate law for radio-active decay and the relationship between the half-life and the decay constant are—analogous to the equations given in Section 14.5 for a first-order chemical reaction:

The integrated rate law for radioactive decay $\ln\left(\dfrac{N_t}{N_0}\right) = -kt$

where N_0 is the number of radioactive nuclei originally present in a sample and N_t is the number remaining at time t.

The relationship between the half-life and the decay constant

$$t_{1/2} = \frac{\ln 2}{k} \quad \text{and} \quad k = \frac{\ln 2}{t_{1/2}}$$

Thus, if we know the value of either the decay constant k or the half-life $t_{1/2}$, we can calculate the value of the other. Furthermore, if we know the value of $t_{1/2}$, we can calculate the ratio of remaining and initial amounts of a radioactive sample N_t/N_0 at any time t by substituting the expression for k into the integrated rate law. Since

$$\ln\left(\frac{N_t}{N_0}\right) = -kt \quad \text{and} \quad k = \frac{\ln 2}{t_{1/2}}$$

then

$$\ln\left(\frac{N_t}{N_0}\right) = (-\ln 2)\left(\frac{t}{t_{1/2}}\right)$$

Worked Example 20.2 shows how to calculate a half-life from a decay constant, Worked Example 20.3 shows how to determine the percentage of a radioactive sample remaining at time t, and Worked Example 20.4 shows how to use decay rates to calculate half-life.

WORKED EXAMPLE 20.2

Calculating a Half-Life from a Decay Constant

The decay constant for sodium-24, a radioisotope used medically in blood studies, is $4.63 \times 10^{-2} \, \text{h}^{-1}$. What is the half-life of ^{24}Na?

IDENTIFY

Known	Unknown
$k = 4.63 \times 10^{-2} \, \text{h}^{-1}$	$t_{1/2}$

STRATEGY

Half-life can be calculated from the decay constant by using the equation

$$t_{1/2} = \frac{\ln 2}{k}$$

SOLUTION

Substituting the values $k = 4.63 \times 10^{-2} \, \text{h}^{-1}$ and $\ln 2 = 0.693$ into the equation gives

$$t_{1/2} = \frac{0.693}{4.63 \times 10^{-2} \, \text{h}^{-1}} = 15.0 \, \text{h}$$

▶ **PRACTICE 20.3** The decay constant for mercury-197, a radioisotope used medically in kidney scans, is $1.08 \times 10^{-2} \, \text{h}^{-1}$. What is the half-life of mercury-197?

▶ **APPLY 20.4** The half-life of radon-222, a radioactive gas of concern as a health hazard in some homes, is 3.82 days. What is the decay constant of ^{222}Rn?

Go to
eText

WORKED EXAMPLE 20.3

Using Half-Life to Calculate an Amount Remaining

Phosphorus-32, a radioisotope used in leukemia therapy, has a half-life of 14.26 days. What percent of a sample remains after 35.0 days?

IDENTIFY

Known	Unknown
$t_{1/2} = 14.26$ days	Percent of sample remaining $(N_t/N_0) \times 100$
$t = 35.0$ days	

STRATEGY

The ratio of remaining (N_t) and initial (N_0) amounts of a radioactive sample at time t is given by the equation

$$\ln\left(\frac{N_t}{N_0}\right) = -kt$$

Taking N_0 as 100%, we can obtain N_t. The value of the rate constant can be found from the equation $k = 0.693/t_{1/2}$.

SOLUTION

We calculate the value of the rate constant using the half-life.

$$k = \frac{0.693}{14.26 \, \text{days}} = 4.860 \times 10^{-2} \, \text{days}^{-1}$$

continued on next page

Substituting values for t and for k into the equation gives

$$\ln\left(\frac{N_t}{N_0}\right) = (-4.860 \times 10^{-2} \text{ days}^{-1})(35.0 \text{ days}) = -1.70$$

Taking the natural antilogarithm of -1.70 then gives the ratio N_t/N_0:

$$\frac{N_t}{N_0} = \text{antiln } (-1.70) = e^{-1.70} = 0.183$$

Since the initial amount of ^{32}P was 100%, we can set $N_0 = 100\%$ and solve for N_t:

$$\frac{N_t}{100\%} = 0.183 \quad \text{so} \quad N_t = (0.183)(100\%) = 18.3\%$$

After 35.0 days, 18.3% of a ^{32}P sample remains and $100\% - 18.3\% = 81.7\%$ has decayed.

CHECK

We can estimate the answer by considering half-life. Since phosphorus-32 has a half-life of 14.26 days and the time is 35 days, we know the time of the reaction is more than two half-lives. After two half-lives, 75% has reacted and 25% remains. Since the time is over two half-lives, we would estimate that less than 25% remains, which agrees with the answer of 18.3%.

▶ **PRACTICE 20.5** What percentage of $^{14}_{6}C$ ($t_{1/2} = 5715$ years) remains in a sample estimated to be 16,230 years old?

▶ **APPLY 20.6** Cesium-137 is a radioactive isotope released as a result of the Fukushima Daiichi nuclear disaster in Japan in 2011. If 89.2% remains after 5.00 years, what is the half-life?

All **WORKED EXAMPLES** with this icon  have an interactive video in the eText.

WORKED EXAMPLE 20.4

Using Decay Rates to Calculate a Half-Life

A sample of ^{41}Ar, a radioisotope used to measure the flow of gases from smokestacks, decays initially at a rate of 34,500 disintegrations/min, but the decay rate falls to 21,500 disintegrations/min after 75.0 min. What is the half-life of ^{41}Ar?

IDENTIFY

Known	Unknown
Rate at $t = 0$ (34,500 disintegrations/min)	$t_{1/2}$
Rate at $t = 75.0$ min (21,500 disintegrations/min)	

STRATEGY

Use the integrated rate law to find the decay constant (k):

$$\ln\left(\frac{N_t}{N_0}\right) = -kt$$

In this problem, we are given decay rates rather than values of N_t and N_0. For a first-order process, in which rate $= kN$, the decay rate is directly proportional to N. Therefore we can use the decay rate at time (t) in the equation for N_t and the decay rate at time $= 0$ for N_0 in the integrated rate law. Once k is determined, the relationship between half-life and decay constant is used to calculate half-life.

$$t_{1/2} = \frac{0.693}{k}$$

SOLUTION

Since the decay rate is directly proportional to the number of nuclei (N), the initial decay rate can be substituted for N_0 and the decay rate at time t can be substituted for N_t.

$$\ln\left(\frac{21,500}{34,500}\right) = -k(75.0 \text{ min})$$

$$k = 6.31 \times 10^{-3} \text{ min}^{-1}$$

The relationship between half-life and decay constant is used to find half-life.

$$t_{1/2} = \frac{0.693}{k} = \frac{0.693}{6.31 \times 10^{-3}\text{min}^{-1}} = 110 \text{ min}$$

▲ The flow of gases from a smokestack can be measured by releasing ^{41}Ar and monitoring its passage.

▶ **PRACTICE 20.7** What is the half-life of iron-59, a radioisotope used medically in the diagnosis of anemia, if a sample with an initial decay rate of 16,800 disintegrations/min decays at a rate of 10,860 disintegrations/min after 28.0 days?

▶ **APPLY 20.8** What is the decay rate (disintegrations/min) of iron-59 after exactly 40 days? Refer to Practice Problem 20.7 for relevant information.

20.5 DATING WITH RADIOISOTOPES

Biblical scrolls are found in a cave near the Dead Sea. How old are they? A mummified body is discovered in a peat bog near Ramten, Denmark. How old is it? The bones of a 17- to 19-year-old Paleo-American woman are found in an archeological excavation near Arch, New Mexico. How long have humans lived in the area? These and many other questions can be answered by archaeologists using a technique called *radiocarbon dating*. The results of radiocarbon measurements date the Dead Sea Scrolls to 250–130 BC, the bog mummy to 375–250 BC, and the human remains found in New Mexico to ~8000 BC.

Radiocarbon dating of archaeological artifacts depends on the slow and constant production of radioactive carbon-14 in the upper atmosphere by neutron bombardment of nitrogen atoms. (The neutrons come from the bombardment of other atoms by cosmic rays.)

▲ Radiocarbon measurements date this bog mummy to 375–250 BC.

$$^{14}_{7}\text{N} + ^{1}_{0}\text{n} \longrightarrow ^{14}_{6}\text{C} + ^{1}_{1}\text{H}$$

Carbon-14 atoms produced in the upper atmosphere combine with oxygen to yield $^{14}\text{CO}_2$, which slowly diffuses into the lower atmosphere, where it mixes with ordinary $^{12}\text{CO}_2$ and is taken up by plants during photosynthesis. When these plants are eaten, carbon-14 enters the food chain and is ultimately distributed throughout all living organisms.

As long as a plant or animal is living, a dynamic equilibrium exists in which an organism excretes or exhales the same ratio of ^{14}C to ^{12}C that it takes in. As a result, the ratio of ^{14}C to ^{12}C in the living organism remains constant and is the same as that in the atmosphere—about 1 part in 10^{12}. When the plant or animal dies, however, it no longer takes in more ^{14}C and the $^{14}\text{C}/^{12}\text{C}$ ratio in the organism slowly decreases as ^{14}C undergoes radioactive decay by β emission, with $t_{1/2} = 5715$ years.

▲ The rate of radioactive decay of carbon-14 was used to date the Dead Sea Scrolls to 250–130 BC.

$$^{14}_{6}\text{C} \longrightarrow ^{14}_{7}\text{N} + ^{0}_{-1}\text{e}$$

At 5715 years (one ^{14}C half-life) after the death of the organism, the $^{14}\text{C}/^{12}\text{C}$ ratio has decreased by a factor of 2; at 11,430 years after death, the $^{14}\text{C}/^{12}\text{C}$ ratio has decreased by a factor of 4; and so on. By measuring the present $^{14}\text{C}/^{12}\text{C}$ ratio in the traces of any once-living organism, archaeologists can determine how long ago the organism died. Human or animal hair from well-preserved remains, charcoal or wood fragments from once-living trees, and cotton or linen from once-living plants are all useful sources for radiocarbon dating. The technique becomes less accurate as samples get older and the amount of ^{14}C they contain diminishes, but artifacts with an age of 1000–20,000 years can be dated with reasonable accuracy. The outer limit of the technique is about 50,000 years.

Just as radiocarbon measurements allow the dating of once-living organisms, similar measurements on other radioisotopes make possible the dating of rocks. Uranium-238, for example, has a half-life of 4.47×10^9 years and decays through a series of events to yield lead-206. The age of a uranium-containing rock can therefore be determined by measuring the $^{238}\text{U}/^{206}\text{Pb}$ ratio. Similarly, potassium-40 has a half-life of 1.25×10^9 years and decays through electron capture and positron emission to yield argon-40. (Both processes yield the same product.)

$$^{40}_{19}\text{K} + ^{0}_{-1}\text{e} \longrightarrow ^{40}_{18}\text{Ar}$$
$$^{40}_{19}\text{K} \longrightarrow ^{40}_{18}\text{Ar} + ^{0}_{1}\text{e}$$

The age of a rock can be found by crushing a sample, measuring the amount of ^{40}Ar gas that escapes, and comparing the amount of ^{40}Ar with the amount of ^{40}K remaining in the sample. It is through techniques such as these that the age of the Earth has been estimated at 4.54 billion years.

WORKED EXAMPLE 20.5

Using Radiocarbon Dating

Radiocarbon measurements made in 1988 on the Shroud of Turin, a religious artifact thought by some to be the burial shroud of Christ, showed a ^{14}C decay rate of 14.2 disintegrations/min per gram of carbon. How old is the artifact implied by this result, if currently living organisms decay at the rate of 15.3 disintegrations/min per gram of carbon? The half-life of ^{14}C is 5715 years.

IDENTIFY

Known	Unknown
$t_{1/2} = 5715$ years	Age of the artifact (t)
Rate at $t = 0$ (15.3 disintegrations/min)	
Rate at $t = ?$ (14.2 disintegrations/min)	

STRATEGY

Use the half-life to find the decay constant. Then use the integrated rate law to find (t), which represents the age of the artifact. Since the decay rate is directly proportional to the number of nuclei (N), the initial decay rate can be substituted for N_0 and the decay rate at time t can be substituted for N_t.

SOLUTION

Use the half-life to find the decay constant.

$$k = \frac{0.693}{t_{1/2}} = \frac{0.693}{5715 \text{ yr}} = 1.213 \times 10^{-4} \text{ yr}^{-1}$$

Use the integrated rate law to find (t), which represents the age of the artifact.

$$\ln\left(\frac{N_t}{N_0}\right) = -kt$$

$$\ln\left(\frac{14.2}{15.3}\right) = -(1.213 \times 10^{-4} \text{ y}^{-1})t$$

$$t = 615 \text{ yr}$$

To date the object, subtract the age from the year the measurement was made. An age of 615 years in 1988 dates the object to the year 1373.

$$1988 - 615 = 1373$$

This result implies the Shroud of Turin comes from medieval times.

▲ What is the age of this painting from the Lascaux cave in France?

▶ **PRACTICE 20.9**   Charcoal found in the Lascaux cave in France, site of many prehistoric cave paintings, was observed in 1950 to decay at a rate of 2.4 disintegrations/min per gram of carbon. Date the charcoal-based radiocarbon measurements. Living organisms decay at the rate of 15.3 disintegrations/min per gram of carbon and the half-life of ^{14}C is 5715 years.

▶ **APPLY 20.10**   The age of an igneous rock that has solidified from magma can be found by analyzing the amount of ^{40}K and ^{40}Ar. If the rock contains 1.20 mmol of ^{40}K and 0.95 mmol of ^{40}Ar, how long ago did the rock cool? The half-life of potassium-40 is 1.25×10^9 years.

$$^{40}_{19}K \longrightarrow {}^{40}_{18}Ar + {}^{0}_{1}e$$

20.6 ENERGY CHANGES DURING NUCLEAR REACTIONS

We saw in Figure 20.2 that the band of nuclear stability obtained by plotting the number of protons versus the number of neutrons for observable nuclei is slightly curved. Lighter elements have an approximately 1:1 neutron/proton ratio, but heavier elements need a substantially higher ratio of neutrons to protons for stability. Such observations suggest that neutrons function as a kind of nuclear glue to overcome the proton–proton repulsions that would otherwise cause a nucleus to fly apart. The more protons there are in the nucleus, the more glue is needed.

In principle, it should be possible to measure the strength of the forces holding a nucleus together by measuring the amount of energy necessary to tear the nucleus apart into its component protons and neutrons. That is, the amount of energy that must be *added* to tear the nucleus apart is numerically equal to the amount of energy that is *released* when a stable nucleus forms from its constituent particles. The two processes are simply the reverse of one another. With a helium-4 nucleus, for instance, the energy change associated with combining two neutrons and two protons to form $_2^4$He is a direct measure of helium-4 nuclear stability.

$$2\,_1^1\text{H} + 2\,_0^1\text{n} \longrightarrow\ _2^4\text{He} \qquad \Delta E = \ ?$$

Unfortunately, there is a problem with actually carrying out the measurement, because temperatures rivaling those in the interior of the Sun (10^7 K) are necessary for nuclei to form. That is, the activation energy required to force the elementary particles close enough for reaction is enormous. Nevertheless, the energy change for the process can be calculated using the now-famous equation $\Delta E = \Delta mc^2$, which was proposed by Albert Einstein in 1905 as part of his special theory of relativity and which relates the energy change ΔE of a process to a corresponding mass change Δm. (Recall that the symbol c represents the speed of light; the value is $c = 299\ 792\ 458$ m/s.)

> REMEMBER . . .
> Activation energy is the potential-energy barrier that must be surmounted before reactants can be converted to products. The height of the barrier controls the rate of the reaction (Section 14.7).

> Einstein's famous equation relating mass and energy $\Delta E = \Delta mc^2$

Using the helium-4 nucleus for an example calculation, the mass of the two neutrons and two protons that combine to form the nucleus is 4.031 90 unified atomic mass units (u):

> REMEMBER . . .
> Atomic masses are given using the unified atomic mass unit (u), defined as 1/12 the mass of a ^{12}C atom. In practice, the unit is not often specified and atomic masses are written as dimensionless (Section 2.9).

$$
\begin{array}{lll}
\text{Mass of 2 neutrons} = (2)(1.008\ 67) & = 2.017\ 34 \\
\text{Mass of 2 protons}\ \ = (2)(1.007\ 28) & = 2.014\ 56 \\
\hline
\text{Total mass of 2 n} + \text{2 p} & = 4.031\ 90
\end{array}
$$

The mass of the helium-4 *nucleus* can be found by subtracting the mass of two electrons from the experimentally measured mass of a helium-4 *atom* (4.002 60).

$$
\begin{array}{lll}
\text{Mass of helium-4 atom} & = & 4.002\ 60 \\
-\text{Mass of 2 electrons} = -(2)(5.486 \times 10^{-4}) & = & -0.001\ 10 \\
\hline
\text{Mass of helium-4 nucleus} & = & 4.001\ 50
\end{array}
$$

Subtracting the mass of the helium nucleus (4.001 50) from the combined mass of its constituent particles (4.031 90) shows a difference of 0.030 40. That is, 0.030 40 u (or 0.030 40 g/mol) is lost when two protons and two neutrons combine to form a helium-4 nucleus:

BIG IDEA Question 3

How does the mass of a nucleus compare to the combined masses of its component protons and neutrons?

$$
\begin{array}{lll}
\text{Mass of 2 n} + \text{2 p} & = & 4.031\ 90 \\
-\text{Mass of } ^4\text{He nucleus} & = & -4.001\ 50 \\
\hline
\text{Mass difference} & = & 0.030\ 40
\end{array}
$$

$$\text{Mass difference in grams} = \left(0.030\ 40\frac{\text{u}}{\text{atom}}\right)\left(1.6605 \times 10^{-24}\frac{\text{g}}{\text{u}}\right) = 5.048 \times 10^{-26}\text{g/atom}$$

$$\text{Mass difference in g/mol} = \left(5.048 \times 10^{-26}\frac{\text{g}}{\text{atom}}\right)\left(6.022 \times 10^{23}\frac{\text{atom}}{\text{mol}}\right) = 3.040 \times 10^{-2}\frac{\text{g}}{\text{mol}}$$

The loss in mass that occurs when protons and neutrons combine to form a nucleus is called the **mass defect** of the nucleus. This lost mass is converted into energy that is released during the nuclear reaction and is thus a direct measure of the **binding energy** holding the nucleus together. The larger the binding energy, the more stable the nucleus. For a helium-4 nucleus, the binding energy is

$$\Delta E = \Delta mc^2 = (5.048 \times 10^{-26}\ \text{g})\left(10^{-3}\frac{\text{kg}}{\text{g}}\right)\left(3.00 \times 10^{8}\frac{\text{m}}{\text{s}}\right)^2$$

$$= 4.54 \times 10^{-12}\left(\frac{\text{kg} \cdot \text{m}^2}{\text{s}^2}\right) = 4.54 \times 10^{-12}\ \text{J}$$

The binding energy can be expressed on a kJ/mol basis by converting joules to kilojoules and multiplying by Avogadro's number:

$$\text{Binding energy} = \left(4.54 \times 10^{-12}\frac{\text{J}}{\text{atom}}\right)\left(\frac{1\ \text{kJ}}{1000\ \text{J}}\right)\left(6.022 \times 10^{23}\frac{\text{atom}}{\text{mol}}\right)$$

$$= 2.73 \times 10^{9}\frac{\text{kJ}}{\text{mol}}$$

The binding energy of a helium-4 nucleus is 4.54×10^{-12} J, or 2.73×10^{9} kJ/mol. In other words, 4.54×10^{-12} J is released when a helium-4 nucleus is formed from 2 protons and 2 neutrons, and 4.54×10^{-12} J must be supplied to disintegrate a helium-4 nucleus into isolated protons and neutrons. This amount of energy is more than *10 million times* the energy change associated with a typical chemical process.

To make comparisons among different nuclei easier, binding energies are usually expressed on a per-**nucleon** basis using *electron volts* (*eV*) as the energy unit, where 1 eV = 1.60×10^{-19} J and 1 million electron volts (1 MeV) = 1.60×10^{-13} J. Thus, the helium-4 binding energy is 7.09 MeV/nucleon:

$$\text{Helium-4 binding energy} = \left(4.54 \times 10^{-12}\frac{\text{J}}{\text{nucleus}}\right)\left(\frac{1\ \text{MeV}}{1.60 \times 10^{-13}\ \text{J}}\right)\left(\frac{1\ \text{nucleus}}{4\ \text{nucleons}}\right)$$

$$= 7.09\ \text{MeV/nucleon}$$

A plot of binding energy per nucleon for the most stable isotope of each element is shown in **FIGURE 20.6.** Since a larger binding energy per nucleon corresponds to greater stability, the most stable nuclei are at the top of the curve and the least stable nuclei are at the bottom. Iron-56, with a binding energy of 8.79 MeV/nucleon, is the most stable isotope known.

The idea that mass and energy are interconvertible is potentially disturbing because it seems to overthrow two of the fundamental principles on which chemistry is based—the **law of mass conservation** and the **law of energy conservation.** In

> **REMEMBER . . .**
> A **nucleon** is the general term for a nuclear particle, either proton or neutron (Section 20.1).

> **REMEMBER . . .**
> The **law of mass conservation** says that mass is neither created nor destroyed in chemical reactions (Section 2.4).

> **REMEMBER . . .**
> The **law of energy conservation** says that energy can neither be created nor destroyed but only converted from one form to another (Section 9.1).

▶ **FIGURE 20.6**

The binding energy per nucleon for the most stable isotope of each naturally occurring element.

▶ **Figure It Out**

Which nucleus has the highest binding energy per nucleon? Which nucleus is most stable?

Answer: ^{56}Fe has the highest binding energy per nucleon and is therefore the most stable nucleus.

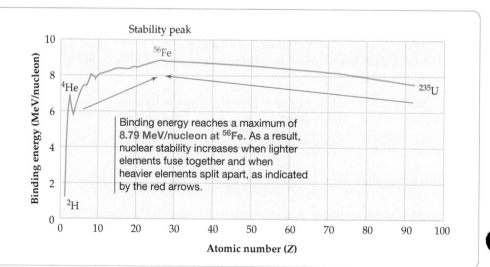

fact, what the mass–energy interconversion means is that neither mass nor energy is conserved independently; rather, it is the *combination* of mass and energy that is conserved. Every time we do a reaction, whether nuclear *or* chemical, mass and energy are interconverted, but the combination of the two is conserved. For the energy change involved in a typical chemical reaction, however, the effect is so small that the mass change can't be detected by even the best analytical balance. Worked Example 20.7 illustrates such a mass–energy calculation for a chemical reaction.

Go to
eText

WORKED EXAMPLE 20.6

Calculating a Mass Defect and a Binding Energy

Helium-6 is a radioactive isotope with $t_{1/2} = 0.807$ s. Calculate the mass defect for the formation of a ^{6}He nucleus in both grams and g/mol, and calculate the binding energy in both MeV/nucleon and kJ/mol. Is a ^{6}He nucleus more stable or less stable than a ^{4}He nucleus? The mass of a ^{6}He atom is 6.018 89.

IDENTIFY

Known	Unknown
Mass of ^{6}He atom (6.018 89 u)	Binding energy (MeV/nucleon and kJ/mol)
$t_{1/2} = 0.807$ s (not relevant for this problem)	

STRATEGY

Find the mass defect by subtracting the mass of the ^{6}He nucleus from the mass of its constituent nucleons, and use Einstein's famous equation $\Delta E = \Delta mc^2$ to find the ^{6}He binding energy. Then use Avogadro's number $(6.022 \times 10^{23}\ \text{mol}^{-1})$ to find the per-mole value.

SOLUTION

First, calculate the total mass of the nucleons (4 n + 2 p):

Mass of 4 neutrons = (4)(1.008 67)	= 4.034 68
Mass of 2 protons = (2)(1.007 28)	= 2.014 56
Mass of 4 n + 2 p	= 6.049 24

Next, calculate the mass of a ^{6}He nucleus by subtracting the mass of two electrons from the mass of a ^{6}He atom:

Mass of helium-6 atom	=	6.018 89
−Mass of 2 electrons = −(2)(5.486 × 10⁻⁴)	=	−0.001 10
Mass of helium-6 nucleus	=	6.017 79

Then subtract the mass of the ^{6}He nucleus from the mass of the constituent nucleons and convert unified atomic mass units to grams to find the mass defect:

Mass defect in u = Mass of nucleons − Mass of nucleus

$$= (6.049\ 24\ \text{u}) - (6.017\ 79\ \text{u}) = 0.031\ 45\ \text{u}$$

$$\text{Mass defect in grams} = \left(0.031\ 45\ \frac{\text{u}}{\text{atom}}\right)\left(1.6605 \times 10^{-24}\ \frac{\text{g}}{\text{u}}\right)$$

$$= 5.222 \times 10^{-26}\ \text{g/atom}$$

$$\text{Mass defect in g/mol} = \left(5.222 \times 10^{-26}\ \frac{\text{g}}{\text{atom}}\right)\left(6.022 \times 10^{23}\ \frac{\text{atom}}{\text{mol}}\right)$$

$$= 3.145 \times 10^{-2}\ \frac{\text{g}}{\text{mol}}$$

Now, use Einstein's equation to convert the mass defect into binding energy:

$$\Delta E = \Delta mc^2 = (5.222 \times 10^{-26}\ \text{g})\left(\frac{1\ \text{kg}}{1000\ \text{g}}\right)\left(3.00 \times 10^8\ \frac{\text{m}}{\text{s}}\right)^2$$

$$= 4.70 \times 10^{-12}\ \text{J}$$

Binding energy in Mev/nucleon

$$= \left(4.70 \times 10^{-12}\ \frac{\text{J}}{\text{nucleus}}\right)\left(\frac{1\ \text{MeV}}{1.60 \times 10^{-13}\ \text{J}}\right)\left(\frac{1\ \text{nucleus}}{6\ \text{nucleons}}\right)$$

$$= 4.90\ \text{MeV/nucleon}$$

Binding energy in kJ/mol

$$= \left(4.70 \times 10^{-12}\ \frac{\text{J}}{\text{atom}}\right)\left(\frac{1\ \text{kJ}}{1000\ \text{J}}\right)\left(6.022 \times 10^{23}\ \frac{\text{atom}}{\text{mol}}\right)$$

$$= 2.83 \times 10^9\ \frac{\text{kJ}}{\text{mol}}$$

The binding energy of a radioactive ^{6}He nucleus is 4.90 MeV/nucleon, making ^{6}He less stable than ^{4}He, whose binding energy is 7.09 MeV/nucleon.

▶ **PRACTICE 20.11** Calculate the mass defect for the formation of an oxygen-16 nucleus in g/mol, and calculate the binding energy in kJ/mol. The mass of an ^{16}O atom is 15.994 91.

▶ **APPLY 20.12** The mass defect for a Li-6 nucleus is 0.034 37 g/mol. What is the atomic mass of Li-6? What is the binding energy for a Li-6 nucleus in units of MeV/nucleon?

WORKED EXAMPLE 20.7

Calculating the Mass Defect in a Chemical Reaction

What is the mass change in g/mol of hydrogen when hydrogen atoms (H) combine to form hydrogen molecules (H_2)?

$$2\,H \longrightarrow H_2 \quad \Delta E = -436 \text{ kJ/mol}$$

IDENTIFY

Known	Unknown
$\Delta E = -436$ kJ/mol	Δm

STRATEGY

The problem asks us to calculate a mass defect Δm when the energy change ΔE is known. To do this, we have to rearrange the Einstein equation to solve for mass, remembering that $1\,J = 1 \text{ kg} \cdot m^2/s^2$.

SOLUTION

$$\Delta m = \frac{\Delta E}{c^2} = \frac{\left(-436\dfrac{kJ}{mol}\right)\left(\dfrac{1000\,J}{1\,kJ}\right)\left(\dfrac{1 \text{ kg} \cdot m^2/s^2}{J}\right)\left(\dfrac{1000\,g}{1\,kg}\right)}{\left(3.00 \times 10^8\,\dfrac{m}{s}\right)^2}$$

$$= -4.84 \times 10^{-9}\,\frac{g}{mol}$$

The loss in mass accompanying formation of H_2 molecules from their constituent H atoms is 4.84×10^{-9} g/mol, much too small an amount to be detectable by any balance and far smaller than the mass defect of a stable nucleus, which we calculated to be approximately 10^{-2} g/mol.

▶ **PRACTICE 20.13** What is the mass change in g/mol of NaCl for the reaction of sodium metal with chlorine gas (Cl_2) to give sodium chloride?

$$2\,Na(s) + Cl_2(g) \longrightarrow 2\,NaCl(s) \quad \Delta E = -820 \text{ kJ}$$

▶ **APPLY 20.14** An average household uses 3.9×10^7 kJ of energy per year produced primarily from the combustion of fossil fuels such as coal and methane (CH_4). Calculate the mass lost in grams when these fuels are burned to provide energy for one home.

20.7 NUCLEAR FISSION AND FUSION

A careful look at the plot of atomic number versus binding energy per nucleon in Figure 20.5 leads to some interesting and important conclusions. The fact that binding energy per nucleon begins at a relatively low value for ^2_1H, reaches a maximum for $^{56}_{26}Fe$, and then gradually tails off implies that both lighter and heavier nuclei are less stable than midweight nuclei near iron-56. Very heavy nuclei can therefore gain stability and release energy if they fragment to yield midweight nuclei, while very light nuclei can gain stability and release energy if they fuse together. The two resultant processes—**fission** for the fragmenting of heavy nuclei and **fusion** for the joining together of light nuclei—have changed the world since their discovery in the late 1930s and early 1940s.

Nuclear Fission

Certain nuclei—uranium-233, uranium-235, and plutonium-239, for example—do more than undergo simple radioactive decay; they break into fragments when struck by neutrons. As illustrated in **FIGURE 20.7**, an incoming neutron causes the nucleus to split into two smaller pieces of roughly similar size.

The fission of a given nucleus does not occur in exactly the same way each time, and nearly 800 different fission products have been identified from uranium-235. One of the more frequently occurring pathways generates barium-142 and krypton-91, along with two additional neutrons plus the one neutron that initiated the fission:

$$^1_0n + {}^{235}_{92}U \longrightarrow {}^{142}_{56}Ba + {}^{91}_{36}Kr + 3\,{}^1_0n$$

The three neutrons released by fission of a ^{235}U nucleus can induce up to three more fissions yielding as many as nine neutrons, which can induce up to nine more fissions yielding 27 neutrons, and so on indefinitely. The result is a **chain reaction** that continues to occur even if the external supply of bombarding neutrons is cut off. If the sample size

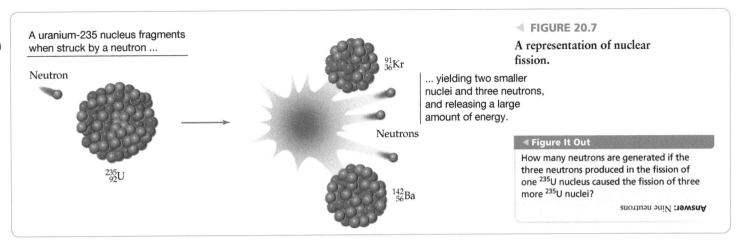

A uranium-235 nucleus fragments when struck by a neutron ...

Neutron

$^{235}_{92}U$

$^{91}_{36}Kr$

... yielding two smaller nuclei and three neutrons, and releasing a large amount of energy.

Neutrons

$^{142}_{56}Ba$

◀ **FIGURE 20.7**

A representation of nuclear fission.

◀ **Figure It Out**

How many neutrons are generated if the three neutrons produced in the fission of one ^{235}U nucleus caused the fission of three more ^{235}U nuclei?

Answer: Nine neutrons

is small, many of the neutrons escape before initiating additional fission events, and the chain reaction soon stops. If there is a sufficient amount of ^{235}U, though—an amount called the **critical mass**—enough neutrons remain for the chain reaction to become self-sustaining. Under high-pressure conditions that confine the ^{235}U to a small volume, the chain reaction may even occur so rapidly that a nuclear explosion results. For ^{235}U, the critical mass is about 56 kg, although the amount can be reduced to 15 kg by placing a coating of ^{238}U around the ^{235}U to reflect back some of the escaping neutrons.

Uranium used in atomic weapons based on fission of ^{235}U must be processed because naturally occurring uranium is a mixture of two major isotopes. The nonfissionable ^{238}U isotope has a natural abundance of 99.3%, while the fissionable ^{235}U isotope is present only to the extent of 0.7%. Weapons-grade uranium must contain greater than 85% of the fissionable isotope ^{235}U. The process of increasing the percentage of ^{235}U in a sample is called enrichment. One of the early techniques for enriching uranium was based on the difference in diffusion rates $^{235}UF_6(g)$ and $^{238}UF_6(g)$ as described in Section 10.7. The modern method of uranium enrichment is gas centrifugation, and it requires less energy than gaseous diffusion. Heavier gas molecules containing ^{238}U preferentially move to the outer edges of the rotating cylinder, while lighter gas molecules containing ^{235}U concentrate closer to the center.

We can calculate the amount of energy released during nuclear fission as in Worked Example 20.8 by finding the accompanying mass change and then using the Einstein mass–energy relationship. (When calculating the mass change for a nuclear reaction, it's simplest to use the masses of the *atoms* corresponding to the relevant nuclei, rather than the masses of the nuclei themselves, because the number of electrons is the same in both reactants and products and thus cancels from the calculation.)

▲ An enormous amount of energy is released in the explosion that accompanies an uncontrolled nuclear chain reaction.

WORKED EXAMPLE 20.8

Calculating the Energy Released in a Nuclear Reaction

How much energy in both joules and kJ/mol is released by the fission of uranium-235 to form barium-142 and krypton-91? The atomic masses are ^{235}U (235.0439), ^{142}Ba (141.9164), ^{91}Kr (90.9234), and ^1_0n (1.008 67).

$$^1_0n + ^{235}_{92}U \longrightarrow ^{142}_{56}Ba + ^{91}_{36}Kr + 3\,^1_0n$$

IDENTIFY

Known	Unknown
Masses of all reactants and products	ΔE in J and kJ/mol

STRATEGY

Calculate the change in mass (Δm) by subtracting the masses of the products from the mass of the ^{235}U reactant, and then use the Einstein equation to convert mass to energy.

continued on next page

SOLUTION

Mass of ^{235}U		=	235.0439
−Mass of ^{142}Ba		=	−141.9164
−Mass of ^{91}Ba		=	−90.9234
−Mass of $2{}_0^1n = -(2)(1.008\ 67)$		=	−2.0173
Mass change (Δm):		=	0.1868

Mass difference in grams

$$= \left(0.1868\ \frac{u}{\text{atom}}\right)\left(1.6605 \times 10^{-24}\ \frac{g}{u}\right) = 3.102 \times 10^{-25}\ \text{g/atom}$$

$$\Delta E = \Delta mc^2 = (3.102 \times 10^{-25}\ \text{g})\left(\frac{1\ \text{kg}}{1000\ \text{g}}\right)\left(3.00 \times 10^8\ \frac{\text{m}}{\text{s}}\right)^2$$

$$= 2.79 \times 10^{-11}\ \text{J per atom}$$

$$\Delta E \text{ in kJ/mol} = \left(2.79 \times 10^{-11}\frac{\text{J}}{\text{atom}}\right)\left(\frac{1\ \text{kJ}}{1000\ \text{J}}\right)\left(6.022 \times 10^{23}\ \frac{\text{atom}}{\text{mol}}\right)$$

$$= 1.68 \times 10^{10}\ \frac{\text{kJ}}{\text{mol}}$$

Nuclear fission of a ^{235}U atom releases 2.79×10^{-11} J, or 1.68×10^{10} kJ/mol.

CHECK

A highly exothermic chemical reaction releases approximately 10^3 kJ/mol. The fission reaction releases 10^7 (10 million) times more energy, which is reasonable.

▶ **PRACTICE 20.15**   An alternative pathway for the nuclear fission of ^{235}U produces tellurium-137 and zirconium-97. How much energy in kJ/mol is released in this fission pathway?

$${}_0^1n + {}_{92}^{235}U \longrightarrow {}_{52}^{137}Te + {}_{40}^{97}Zr + 2{}_0^1n$$

The masses are ^{235}U (235.0439), ^{137}Te (136.9254), ^{97}Zr (96.9110), and ${}_0^1n$ (1.008 67).

▶ **APPLY 20.16**   Uranium-238 undergoes alpha decay to thorium-234.

(a) Write a balanced nuclear equation.
(b) Calculate the mass change in (g/atom) and energy change in (kJ/mol).
(c) Is energy absorbed or released when uranium undergoes radioactive decay?

The masses are ^{238}U (238.0508), ^{234}Th (234.0436), ^4He (4.0026).

Nuclear Reactors

The same fission process that leads to a nuclear explosion under uncontrolled conditions can be used to generate electric power when carried out in a controlled manner in a nuclear reactor (**FIGURE 20.8**). The principle behind a nuclear reactor is straightforward: Uranium fuel is placed in a pressurized containment vessel surrounded by water, and *control rods* made of boron and cadmium are added. The water slows the neutrons released during fission so their escape is more difficult, and the control rods absorb and thus regulate the flow of neutrons when raised and lowered. As a result, the fission is maintained at a barely self-sustainable rate to prevent overheating. Energy from the controlled fission then heats a circulating coolant, which in turn produces steam to drive a turbine and produce electricity.

In a typical nuclear reactor, the fuel is made of compressed pellets of UO_2 that have been isotopically enriched to a 3% concentration of ^{235}U and then encased in zirconium rods. The rods are placed in a pressure vessel filled with water, which acts as a moderator to slow the neutrons so they can be captured more readily. No nuclear explosion can occur in a reactor because the amount and concentration of fissionable fuel is too low and because the fuel is not confined by pressure into a small volume. The level of enrichment in fuel rods (3%) is significantly lower than the level of enrichment required for a nuclear weapon (>85%). In a worst-case accident, however, uncontrolled fission can lead to enormous overheating that can melt the reactor and surrounding containment vessel, thereby releasing large amounts of radioactivity to the environment. This is exactly what happened in the Fukushima Daiichi nuclear disaster

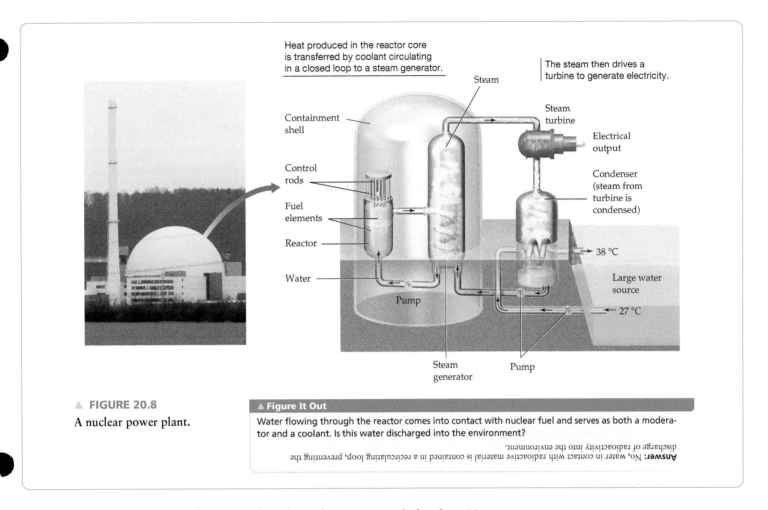

Heat produced in the reactor core is transferred by coolant circulating in a closed loop to a steam generator.

The steam then drives a turbine to generate electricity.

Steam

Steam turbine

Containment shell

Electrical output

Control rods

Condenser (steam from turbine is condensed)

Fuel elements

Reactor

38 °C

Water

Large water source

Pump

27 °C

Steam generator

Pump

▲ **FIGURE 20.8**

A nuclear power plant.

▲ **Figure It Out**

Water flowing through the reactor comes into contact with nuclear fuel and serves as both a moderator and a coolant. Is this water discharged into the environment?

Answer: No, water in contact with radioactive material is contained in a recirculating loop, preventing the discharge of radioactivity into the environment.

in Japan on March 11, 2011, when an earthquake and tsunami struck the plant. Newer reactors get around this problem by adopting what are called *passive safety* designs that automatically slow runaway reactors without any electronic feedback or operator input, but the Fukushima reactors were too old to have this technology.

Thirty countries around the world now obtain some of their electricity from nuclear energy (**FIGURE 20.9**). The United States produces more nuclear power than any other country—approximately 100,000 megawatts per year from 99 commercial generators. France produces the highest percentage of its electricity from nuclear reactors—75% from 58 plants. Worldwide, 451 nuclear plants were in operation in 2017, with an additional 58 under construction and more being planned. The total output was 395,000 megawatts, or approximately 15% of the world's electrical power.

One major problem holding back future development is the yet unsolved matter of how to dispose of the radioactive wastes generated by the plants. It will take at least 600 years for waste strontium-90 to decay to safe levels, and at least 200,000 years for plutonium-239 to decay.

Nuclear Fusion

Just as heavy nuclei such as ^{235}U release energy when they undergo *fission,* very light nuclei such as the isotopes of hydrogen release enormous amounts of energy when they undergo *fusion.* In fact, the fusion of hydrogen nuclei into helium is what powers our Sun and other stars. Among the processes thought to occur in the Sun are those in the following sequence leading to helium-4:

$$^{1}_{1}H + {}^{1}_{1}H \longrightarrow {}^{2}_{1}H + {}^{0}_{1}e$$

$$^{1}_{1}H + {}^{2}_{1}H \longrightarrow {}^{3}_{2}He$$

$$^{3}_{2}He + {}^{3}_{2}He \longrightarrow {}^{4}_{2}He + 2\,{}^{1}_{1}H$$

$$^{3}_{2}He + {}^{1}_{1}H \longrightarrow {}^{4}_{2}He + {}^{0}_{1}e$$

The main appeal of nuclear fusion as a power source is that the hydrogen isotopes used as fuel are cheap and plentiful and that the fusion products are nonradioactive and nonpolluting. The technical problems that must be solved before achieving a practical and controllable fusion method are staggering, however. Not the least of the problems is that a temperature of approximately 40 *million* kelvin is needed to initiate the fusion process.

▶ **FIGURE 20.9**

Megawatts of electricity generated by nuclear power in 2017.

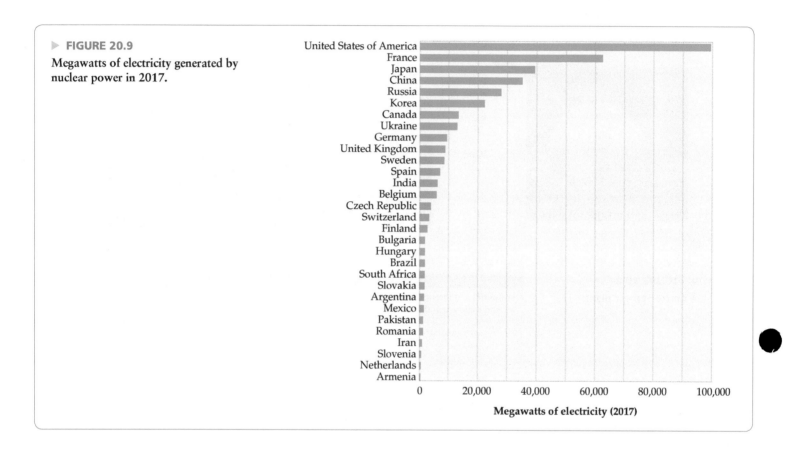

Megawatts of electricity (2017)

20.8 NUCLEAR TRANSMUTATION

Fewer than 300 of the more than 3600 known isotopes occur naturally. The remainder have been made by **nuclear transmutation,** the change of one element into another. Such transmutation is often brought about by bombardment of an atom with a high-energy particle such as a proton, neutron, or α particle. In the ensuing collision between particle and atom, an unstable nucleus is momentarily created, a nuclear change occurs, and a different element is produced. The first nuclear transmutation was accomplished in 1917 by Ernest Rutherford, who bombarded ^{14}N nuclei with α particles and found that ^{17}O was produced:

$$^{14}_{7}\text{N} + ^{4}_{2}\text{He} \longrightarrow ^{17}_{8}\text{O} + ^{1}_{1}\text{H}$$

Other nuclear transmutations can lead to the synthesis of entirely new elements never before seen on Earth. In fact, all the transuranium elements—those elements with atomic numbers greater than 92—have been produced by transmutation. Plutonium, for example, can be made by bombarding uranium-238 with α particles:

$$^{238}_{92}\text{U} + ^{4}_{2}\text{He} \longrightarrow ^{241}_{94}\text{Pu} + ^{1}_{0}\text{n}$$

The plutonium-241 that results from U-238 bombardment is itself radioactive with a half-life of 14.3 years, decaying by β emission to yield americium-241. (If that name sounds familiar, it's because americium is used commercially in making smoke detectors.) Americium-241 is also radioactive, decaying by α emission with a half-life of 433 years.

$$^{241}_{94}\text{Pu} \longrightarrow {}^{241}_{95}\text{Am} + {}^{0}_{-1}\text{e}$$
$$^{241}_{95}\text{Am} \longrightarrow {}^{237}_{93}\text{Np} + {}^{4}_{2}\text{He}$$

Still other nuclear transmutations are carried out using neutrons, protons, or even whole atoms for bombardment. The cobalt-60 used in radiation therapy for cancer patients can be prepared by neutron bombardment of iron-58. Iron-58 first absorbs a neutron to yield iron-59, the iron-59 undergoes β decay to yield cobalt-59, and the cobalt-59 then absorbs a second neutron to yield cobalt-60:

$$^{58}_{26}\text{Fe} + {}^{1}_{0}\text{n} \longrightarrow {}^{59}_{26}\text{Fe}$$
$$^{59}_{26}\text{Fe} \longrightarrow {}^{59}_{27}\text{Co} + {}^{0}_{-1}\text{e}$$
$$^{59}_{27}\text{Co} + {}^{1}_{0}\text{n} \longrightarrow {}^{60}_{27}\text{Co}$$

The overall change can be written as

$$^{58}_{26}\text{Fe} + 2\,{}^{1}_{0}\text{n} \longrightarrow {}^{60}_{27}\text{Co} + {}^{0}_{-1}\text{e}$$

In 2016, the International Union of Pure and Applied Chemistry (IUPAC) added four new elements to the periodic table, nihonium (Nh) ($Z = 113$), moscovium (Mc) ($Z = 115$), tennessine (Ts) ($Z = 117$), and oganesson (Og) ($Z = 118$). The nuclear transmutation experiments to synthesize and characterize these elements were carried out at the Flerov Laboratory of Nuclear Reactions in Russia in collaboration with the Analytical and Nuclear Chemistry Division of the Lawrence Livermore National Laboratory in the United States. To synthesize moscovium (Mc) ($Z = 115$), a cyclotron accelerated a beam of neutron-rich calcium ions (^{48}Ca) to nearly 10% the speed of light and aimed the beam at a target consisting of ^{243}Am. The result of the nuclear transmutation reaction was ^{287}Mc and four neutrons. Moscovium-287 then undergoes five alpha decays with the first decay producing nihonium (Nh) ($Z = 113$).

$$^{243}_{95}\text{Am} + {}^{48}_{20}\text{Ca} \rightarrow {}^{287}_{115}\text{Mc} + 4\,{}^{1}_{0}\text{n}$$
$$^{287}_{115}\text{Mc} \rightarrow {}^{283}_{113}\text{Nh} + {}^{4}_{2}\text{He}$$

▲ The four newest elements added to the periodic table ($Z = 113$, 115, 117, and 118) were synthesized in the particle accelerator at the Flerov Laboratory for Nuclear Reactions in Dubna, Russia. The ^{48}Ca ions were accelerated to 246 MeV.

WORKED EXAMPLE 20.9

Balancing a Nuclear Transmutation Equation

The element berkelium was first prepared at the University of California at Berkeley in 1949 by α bombardment of $^{241}_{95}\text{Am}$. Two neutrons are also produced during the reaction. What isotope of berkelium results from this transmutation? Write a balanced nuclear equation.

STRATEGY AND SOLUTION

According to the periodic table, berkelium has $Z = 97$. Since the sum of the reactant mass numbers is $241 + 4 = 245$ and 2 neutrons are produced, the berkelium isotope must have a mass number of $245 - 2 = 243$.

$$^{241}_{95}\text{Am} + {}^{4}_{2}\text{He} \longrightarrow {}^{243}_{97}\text{Bk} + 2\,{}^{1}_{0}\text{n}$$

▶ **PRACTICE 20.17** A new element was synthesized in 1996 when ^{70}Zn nuclei were fired at a target made of ^{208}Pb in a heavy ion accelerator. The new element and one neutron were formed in the reaction. What new isotope was formed?

▶ **APPLY 20.18** Californium-246 is formed by bombardment of uranium-238 atoms. If four neutrons are formed as by-products, what particle is used for the bombardment?

20.9 DETECTING AND MEASURING RADIOACTIVITY

Radioactive emissions are invisible. We can't see, hear, smell, touch, or taste them, no matter how high the dose. We can, however, measure and detect radioactivity using a suitable instrument, such as a Geiger counter (**FIGURE 20.10**), and the effects of that radiation on living organisms can be readily observed.

Radiation intensity is expressed in different ways, depending on what is being measured. Some units measure the number of nuclear decay events, while others measure the amount of exposure to radiation or the biological consequences of radiation (**TABLE 20.2**).

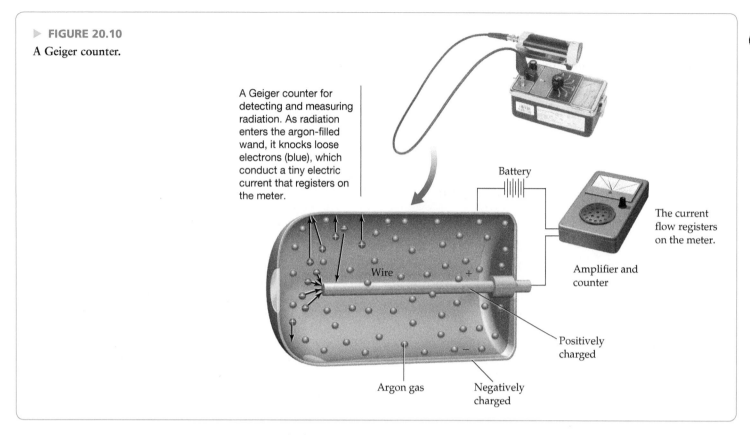

A Geiger counter for detecting and measuring radiation. As radiation enters the argon-filled wand, it knocks loose electrons (blue), which conduct a tiny electric current that registers on the meter.

Battery

The current flow registers on the meter.

Amplifier and counter

Wire

Positively charged

Argon gas

Negatively charged

- The *becquerel* (Bq) is the SI unit for measuring the number of nuclear disintegrations occurring per second in a sample: 1 Bq = 1 disintegration/s. The *curie* (Ci) and *millicurie* (mCi) also measure disintegrations per unit time, but they are far larger units than the becquerel and are more often used, particularly in medicine and biochemistry. One curie is the decay rate of 1 g of radium, equal to 3.7×10^{10} Bq:

$$1 \text{ Bq} = 1 \text{ disintegration/s}$$

$$1 \text{ Ci} = 3.7 \times 10^{10} \text{ Bq} = 3.7 \times 10^{10} \text{ disintegrations/s}$$

As an example, a 1.5 mCi sample of tritium is equal to 5.6×10^7 Bq, meaning that it undergoes 5.6×10^7 disintegrations/s.

$$(1.5 \text{ mCi})\left(10^{-3}\frac{\text{Ci}}{\text{mCi}}\right)\left(3.7 \times 10^{10}\frac{\text{Bq}}{\text{Ci}}\right) = 5.6 \times 10^7 \text{ Bq}$$

- The *gray* (Gy) is the SI unit for measuring the amount of energy absorbed per kilogram of tissue exposed to a radiation source: 1 Gy = 1 J/kg. The *rad* (*radiation absorbed dose*) also measures tissue exposure and is more often used in medicine.

$$1 \text{ Gy} = 1 \text{ J/kg} \qquad 1 \text{ rad} = 0.01 \text{ Gy}$$

TABLE 20.2   Units for Measuring Radiation

Unit	Quantity Measured	Description
Becquerel (Bq)	Decay events	1 Bq = 1 disintegration/s
Curie (Ci)	Decay events	1 Ci = 3.7×10^{10} disintegrations/s
Gray (Gy)	Energy absorbed per kilogram of tissue	1 Gy = 1 J/kg tissue
Rad	Energy absorbed per kilogram of tissue	1 rad = 0.01 Gy
Sievert (Sv)	Tissue damage	1 Sv = 1 J/kg
Rem	Tissue damage	1 rem = 0.01 Sv

- The *sievert* (Sv) is the SI unit that measures the amount of tissue damage caused by radiation. It takes into account not just the energy absorbed per kilogram of tissue but also the different biological effects of different kinds of radiation. For example, 1 Gy of α radiation causes 20 times more tissue damage than 1 Gy of γ rays, but 1 Sv of α radiation and 1 Sv of γ rays cause the same amount of damage. The *rem (roentgen equivalent for man)* is an analogous non-SI unit that is more frequently used in medicine.

$$1 \text{ rem} = 0.01 \text{ Sv}$$

The effects of radiation on the human body vary with the energy and kind of radiation, as well as with the length of exposure and whether the radiation is from an external or internal source. When coming from an external source, X rays and γ radiation are more harmful than α and β particles because they penetrate clothing and skin. When coming from an internal source, however, α and β particles are particularly dangerous because all their energy is given up to surrounding tissue. Alpha emitters, although relatively harmless externally, are especially hazardous internally and are almost never used in medical applications.

Because of their relatively large mass, α particles move slowly—up to only one-tenth the speed of light—and can be stopped by a few sheets of paper or by the top layer of skin. Beta particles, because they are much lighter, move at up to nine-tenths the speed of light and have about 100 times the penetrating power of α particles. A block of wood or heavy protective clothing is necessary to stop β radiation, which would otherwise penetrate and burn the skin.

Gamma rays and X rays move at the speed of light and have about 1000 times the penetrating power of α particles. A lead block several inches thick is needed to stop γ and X radiation, which could otherwise penetrate and damage the body's internal organs. Some properties of different kinds of radiation are summarized in **TABLE 20.3**.

BIG IDEA Question 4

Americinium-241 is an alpha emitter used in smoke detectors. Why is it safe to place this radioactive isotope in homes?

TABLE 20.3   Some Properties of Radiation

Type of Radiation	Energy Range	Penetrating Distance in Water[a]
α	3–9 MeV	0.02–0.04 mm
β	0–3 MeV	0–4 mm
X	100 eV–10 keV	0.01–1 cm
γ	10 keV–10 MeV	1–20 cm

[a]Distances at which one-half of the radiation has been stopped.

The biological effects of different radiation doses are given in **TABLE 20.4**. Although the effects sound fearful, the average radiation dose received annually by most people is only about 120 mrem. About 70% of this radiation comes from natural sources such as rocks and cosmic rays—energetic particles coming from interstellar space. The remaining 30% comes from medical procedures such as X rays. The amount due to emissions from nuclear power plants and to fallout from atmospheric testing of nuclear weapons in the 1950s is barely detectable.

TABLE 20.4   Biological Effects of Short-Term Radiation on Humans

Dose (rem)	Biological Effects
0–25	No detectable effects
25–100	Temporary decrease in white blood cell count
100–200	Nausea, vomiting, longer-term decrease in white blood cells
200–300	Vomiting, diarrhea, loss of appetite, listlessness
300–600	Vomiting, diarrhea, hemorrhaging, eventual death in some cases
Above 600	Eventual death in nearly all cases

INQUIRY ? How are radioisotopes used in medicine?

The origins of nuclear medicine date to 1901, when the French physician Henri Danlos first used radium in the treatment of a tuberculous skin lesion. Since that time, uses of radioactivity have become a crucial part of modern medical care, both diagnostic and therapeutic. Current nuclear techniques can be grouped into three classes: (1) *in vivo* procedures, (2) therapeutic procedures, and (3) imaging procedures.

In Vivo Procedures

In vivo studies—those that take place *inside* the body—are carried out to assess the functioning of a particular organ or body system. A radiopharmaceutical agent is administered, and its path in the body—whether it is absorbed, excreted, diluted, or concentrated—is determined by analysis of blood or urine samples.

An example of the many *in vivo* procedures using radioactive agents is the determination of whole-blood volume by injecting a known quantity of red blood cells labeled with radioactive chromium-51. After a suitable interval to allow the labeled cells to be distributed evenly throughout the body, a blood sample is taken, the amount of dilution of the ^{51}Cr is measured, and the blood volume is calculated.

Therapeutic Procedures

Therapeutic procedures—those in which radiation is used to kill diseased tissue—can involve either external or internal sources of radiation. External radiation therapy for the treatment of cancer is often carried out with γ rays from a cobalt-60 source. The highly radioactive source is shielded by a thick lead container and has a small opening directed toward the site of the tumor. By focusing the radiation beam on the tumor and rotating the patient's body, the tumor receives the full exposure while the exposure of surrounding parts of the body is minimized. Nevertheless, sufficient exposure occurs so that most patients suffer some effects of radiation sickness.

Internal radiation therapy is a much more selective technique than external therapy. Boron neutron capture therapy (BCNT) is a method used to treat brain tumors or other tumors that are difficult to treat with external sources of radiation because too much damage to the surrounding tissue will occur. A boron-10 compound that is selectively taken up by tumor cells is injected into the body, and then a beam of low-energy neutrons is directed at the tumor. Boron-10 captures a neutron to form an excited state of boron-11 (depicted by the superscript *m*), which emits an alpha particle to form lithium-7.

$$^{10}_{5}\text{B} + ^{1}_{0}\text{n} \longrightarrow ^{11m}\text{B} \longrightarrow ^{7}_{3}\text{Li} + ^{4}_{2}\text{He}$$

The alpha particles only penetrate a few micrometers and therefore preferentially kill boron-containing tumor cells. The success of this treatment depends on the selective delivery of sufficient amounts of boron-10 to the tumor with only small amounts taken up in the surrounding normal tissue.

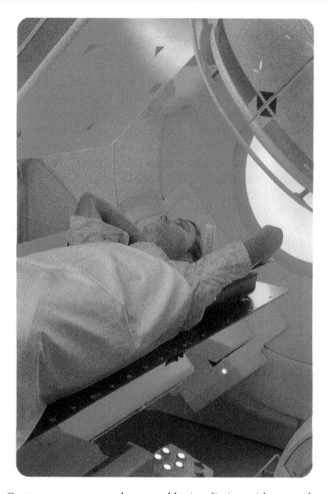

▲ Cancerous tumors can be treated by irradiation with γ rays from this cobalt-60 source.

In contrast to the chemical localization of BCNT radiation treatment, brachytherapy is a technique that treats cancer by placing a radiation source such as ^{125}I in a sealed capsule or wire inside the body near the area requiring treatment. ^{125}I captures an electron to produce an excited state of ^{125}Te, which emits gamma rays to supply a high dose of localized radiation to tumor.

The half-lives of some medically useful radioisotopes are listed in **TABLE 20.5.** As you might expect, the radioisotopes used internally in medical applications have fairly short half-lives so that they decay rapidly and don't cause long-term health hazards.

Imaging Procedures

Imaging procedures give diagnostic information about the health of body organs by analyzing the distribution pattern of radioisotopes introduced into the body. A radiopharmaceutical agent that is known to concentrate in a specific tissue or organ is injected into the body, and its distribution pattern is monitored by external radiation detectors. Depending on the disease and the organ, a diseased organ might concentrate more of the radiopharmaceutical than a normal organ and thus show

TABLE 20.5   Half-Lives of Some Useful Radioisotopes

Radioisotope	Symbol	Radiation	Half-life	Use
Fluorine-18	$^{18}_{9}\text{F}$	β^+	110 minutes	PET scans
Phosphorus-32	$^{32}_{15}\text{P}$	β^-	14.28 days	Leukemia therapy
Cobalt-60	$^{60}_{27}\text{Co}$	β^-, γ	5.27 years	Cancer therapy
Technetium-99m*	$^{99m}_{43}\text{Tc}$	γ	6.01 hours	Brain scans
Iodine-123	$^{123}_{53}\text{I}$	γ	13.2 hours	Thyroid therapy

*The *m* in technetium-99*m* stands for *metastable*, meaning that it undergoes γ emission but does not change its mass number or atomic number.

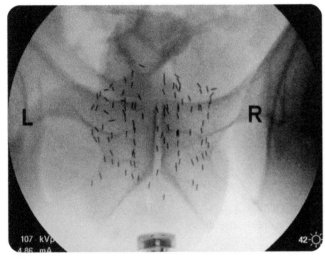

▲ X ray of brachytherapy seeds that have been inserted into a prostate gland to treat cancer. The seeds are made of titanium and radioactive ^{125}I. The γ-ray radiation kills the cancerous cells.

up as a radioactive "hot" spot against a "cold" background. Alternatively, the diseased organ might concentrate less of the radiopharmaceutical than a normal organ and thus show up as a cold spot on a hot background. The radioisotope most widely used today is technetium-99*m*, whose short half-life of 6.01 hours minimizes a patient's exposure to harmful effects. Gamma scanning, named for the emitted gamma rays, using the radioactive tracer ^{99m}Tc is an important tool in the diagnosis of stress fractures, cancer, and other conditions.

PROBLEM 20.19   Sodium-24 in saline solution can be injected into the blood stream to trace the flow of blood and detect obstructions in the circulatory system. Write the balanced nuclear equation for the β decay of sodium-24.

PROBLEM 20.20   Fluorine-18 used in positron emission tomography (PET) scans is produced in particle accelerators.

(a) One way to make fluorine-18 is to bombard oxygen-18 with a high energy proton. What other particle is produced in this nuclear reaction?

$$^{18}_{8}\text{O} + ^{1}_{1}\text{H} \longrightarrow ^{18}_{9}\text{F} + ?$$

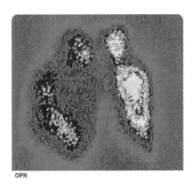

▲ A gamma scan of human lungs created by injecting ^{99m}Tc into the bloodstream. A pulmonary embolism, the formation of a blood clot, is shown in the left lung (right on image). The embolism is revealed by the notch on the side of the lung.

(b) Another way to make fluorine-18 is to bombard oxygen-16 with ^{3}He. What other particle is produced in this nuclear reaction?

$$^{16}_{8}\text{O} + ^{3}_{2}\text{He} \longrightarrow ^{18}_{9}\text{F} + ?$$

PROBLEM 20.21   Iodine-125 used in brachytherapy treatments has a half-life of 59.5 days. How long will it take before the dose of radiation reaches 10% of the initial level?

PROBLEM 20.22   A blood-volume determination was carried out on a patient by injection with 20.0 mL of blood that had been radioactively labeled with Cr-51 to an activity of 4.10 μCi/mL. After a brief period to allow for mixing in the body, blood was drawn from the patient for analysis. Unfortunately, a mix-up in the laboratory prevented an immediate analysis, and it was not until 17.0 days later that a measurement on the blood was made. The radiation level was then determined to be 0.00935 μCi/mL. If ^{51}Cr has a half-life to 27.7 days, what is the volume of blood in the patient?

PROBLEM 20.23   In a cancer treatment called boron neutron-capture therapy, a compound containing boron-10 is injected into a patient where it selectively binds to cancer cells. Irradiating the affected area with neutrons then induces the following reaction:

$$^{10}\text{B} + ^{1}\text{n} \rightarrow ^{4}\text{He} + ^{7}\text{Li} + \gamma$$

The α radiation kills the cancer cells, leaving the surrounding tissue unharmed. The reactants in this nuclear process have essentially no kinetic energy, but the products have a total kinetic energy of 2.31 MeV. What is the energy of the γ photon released? Relevant masses are: ^{4}He(4.002 603), ^{7}Li (7.016 004), ^{10}B (10.012 937), e^-(0.000 548 6), $^{1}_{0}\text{n}$ (1.008 665).

PROBLEM 20.24   To treat a brain tumor with gamma knife radiosurgery, the patient's head is positioned within a hemispherical dome covered by 201 individual ^{60}Co sources, whose narrow beams are directed inward toward the tumor target point. The tumor receives a very high dose of radiation because all the beams converge on it, while any irradiated healthy tissue receives only the radiation of a single beam. For a prescribed dose of 1800 rad, how long should the radiation treatment go if 2.2×10^{11} disintegrations of ^{60}Co are required to give a dose of 1.0 rad? Assume that all the sources are directed at the tumor with an activity of 30 Ci each.

STUDY GUIDE

Section	Concept Summary	Learning Objectives	Test Your Understanding
20.1–20.2 Nuclear Reactions and Radioactivity	**Nuclear reactions** differ from chemical reactions in that they involve a change in an atom's nucleus, producing a different element. Nuclear reactions are written using balanced **nuclear equations,** in which the element symbols represent only the nuclei rather than neutral atoms. **Radioactivity** is the spontaneous emission of radiation from an unstable nucleus. **Alpha (α) radiation** consists of helium nuclei, small particles containing two protons and two neutrons ($^4_2\text{He}^{2+}$). **Beta (β) radiation** consists of electrons ($^0_{-1}\text{e}$), and **gamma (γ) radiation** consists of high-energy electromagnetic radiation that has no mass. **Positron emission** is the ejection of a *positron* (^0_1e or β^+), a particle that has the same mass as an electron but a positive charge. **Electron capture** is the capture of a surrounding electron by a proton in the nucleus.	20.1 Write balanced nuclear reactions for different types of radioactive decay.	Worked Example 20.1; Problems 20.30, 20.34, 20.36
20.3 Nuclear Stability	The stability of a given nucleus is related to its neutron/proton ratio. As the number of protons in the nucleus increases, the ratio of neutrons to protons also increases for stable nuclei, giving a curved appearance to the band of stability. All isotopes heavier than bismuth-209 are radioactive, even though they may decay slowly and be stable enough to occur naturally. Many radioactive isotopes undergo a series of emissions called **radioactive decay series** to reach a stable nucleus.	20.2 Predict the type of radioactive decay for a given isotope. 20.3 Write a balanced nuclear reaction for a radioactive decay series.	Problems 20.40, 20.42 Problems 20.44, 20.46, 20.48
20.4 Radioactive Decay Rates	Radioactive decay is a first-order process, whose rate is proportional to the number of radioactive nuclei N in a sample times a first-order rate constant k, called the **decay constant.** The number N_t of radioactive nuclei remaining at any time t can be calculated from the integrated rate law: $\ln(N_t/N_0) = -kt$, where N_0 is the number of radioactive nuclei originally present.	20.4 Relate half-life and decay constant. 20.5 Calculate the amount of radioactive isotope remaining after a given amount of time. 20.6 Relate decay rates to decay constant, half-life, or amount remaining.	Worked Example 20.2; Problems 20.50, 20.56 Worked Example 20.3; Problems 20.52, 20.54, 20.62 Worked Example 20.4; Problems 20.58–20.59
20.5 Dating with Radioisotopes	Radioisotopic dating is a technique used to determine the age of rocks or carbon-based materials by measuring the amounts of the initial radioactive isotope and its decay product.	20.7 Calculate the age of an object using radioisotopic dating.	Worked Example 20.5; Problems 20.64, 20.66, 20.70
20.6 Energy Changes during Nuclear Reactions	The strength of the forces involved can be measured by calculating an atom's **mass defect**—the difference in mass between a given nucleus and the total mass of its constituent nucleons (protons and neutrons). Applying the Einstein equation $\Delta E = \Delta mc^2$ then allows calculation of the nuclear **binding energy.** The energy change accompanying a nuclear reaction is far greater than that accompanying any chemical reaction.	20.8 Calculate the mass defect and binding energy of a nucleus. 20.9 Calculate the mass and energy change of a chemical reaction.	Worked Example 20.6; Problems 20.76, 20.78, 20.86 Worked Example 20.7; Problems 20.84–20.85

Section	Concept Summary	Learning Objectives	Test Your Understanding
20.7 Nuclear Fission and Fusion	Certain heavy nuclei, such as uranium-235, undergo **nuclear fission** when struck by neutrons, breaking apart into fragment nuclei and releasing enormous amounts of energy. Light nuclei, such as the isotopes of hydrogen, undergo **nuclear fusion** when heated to sufficiently high temperatures, forming heavier nuclei and releasing energy.	**20.10** Identify key aspects of fission and fusion reactions and their role in nuclear power and weapons.	Problems 20.88, 20.93
		20.11 Calculate the energy change associated with a fission or fusion reaction.	Worked Example 20.8; Problems 20.94–20.96
20.8 Nuclear Transmutation	Of the more than 3600 known isotopes, most have been made by **nuclear transmutation,** the change of one element into another. Such transmutation is often brought about by bombardment of an atom with a high-energy particle such as a proton, neutron, α particle, or neutron-rich nucleus such as ^{48}Ca.	**20.12** Write balanced equations for nuclear transmutation reactions.	Worked Example 20.9; Problems 20.98, 20.100, 20.104
20.9 Detecting and Measuring Radioactivity	Radiation intensity is expressed in different ways, according to what property is being measured. The becquerel (Bq) and the curie (Ci) measure the number of radioactive disintegrations per second in a sample. The gray (Gy) and the rad measure the amount of radiation absorbed per kilogram of tissue. The sievert (Sv) and the rem measure the amount of tissue damage caused by radiation. Radiation effects become noticeable with a human exposure of 25 rem and become lethal at an exposure above 600 rem.	**20.13** Calculate the radiation intensity in a radioactive sample and convert between units of radiation intensity.	Problems 20.108, 20.110, 20.112

KEY TERMS

alpha (α) radiation 872	electron capture 874	nuclear equation 871	radioactive 872
binding energy 883	fission 886	nuclear reaction 871	radioactive decay chain 877
beta (β) radiation 873	fusion 886	nuclear transmutation 890	radioisotopes 872
chain reaction 886	gamma (γ) radiation 873	nucleon 871	transuranium element 875
critical mass 887	mass defect 883	nuclide 871	
decay constant 878	nuclear chemistry 871	positron emission 873	

KEY EQUATIONS

- First-order process for radioactive decay rate (Section 20.4)

$$\text{Decay rate} = -\frac{\Delta N}{\Delta t} = kN$$

- Integrated rate law for radioactive decay (Section 20.4)

$$\ln\left(\frac{N_t}{N_0}\right) = -kt$$

where N_0 is the number of radioactive nuclei originally present in a sample and N_t is the number remaining at time t.

- Relationship between the half-life and decay constant (Section 20.4)

$$t_{1/2} = \frac{\ln 2}{k} \quad \text{and} \quad k = \frac{\ln 2}{t_{1/2}}$$

- Relationship between the energy change ΔE of a process to a corresponding mass change Δm (Section 20.6)

$\Delta E = \Delta mc^2$, where c represents the speed of light; its value is $c = 299\ 792\ 458$ m/s.

PRACTICE TEST

After studying this chapter, you can assess your understanding with these practice test questions, which are correlated with chapter learning objectives. If you answer a question incorrectly, refer to the learning objectives in the end-of-chapter Study Guide for assistance. The Study Guide provides a conceptual summary, references a Worked Example to model how to solve the problem, and gives additional problems for more practice.

1. What is the balanced nuclear equation for the alpha decay of plutonium-238? **(LO 20.1)**

   (a) $^{238}_{94}Pu \longrightarrow {}^{234}_{92}U + {}^{4}_{2}He$

   (b) $^{238}_{94}Pu + {}^{4}_{2}He \longrightarrow {}^{242}_{96}Cm$

   (c) $^{238}_{94}Pu \rightarrow {}^{238}_{95}Am + {}^{0}_{-1}e$

   (d) $^{238}_{94}Pu \rightarrow {}^{238}_{93}Np + {}^{0}_{1}e$

2. Which of the following radioisotopes will most likely undergo beta emission? (Refer to Figure 20.2 to answer this question.) **(LO 20.2)**

   (a) ^{196}Rn (b) ^{27}P

   (c) ^{90}Ru (d) ^{109}Mo

3. Uranium-235 undergoes a radioactive decay series involving sequential loss of $\alpha, \beta, \alpha, \alpha, \alpha, \alpha, \beta, \alpha, \beta, \alpha$, and β particles. What is the final stable nucleus? **(LO 20.3)**

   (a) ^{211}W (b) ^{211}Pt

   (c) ^{207}Pb (d) ^{207}Pt

4. Fluorine-18 undergoes positron emission with a half-life of 1.10×10^2 minutes. If a patient is given a 250 mg dose for a PET scan, how long will it take for the amount of fluorine-18 to drop to 75 mg? **(LO 20.4 and 20.5)**

   (a) 56 minutes   (b) 96 minutes

   (c) 132 minutes   (d) 191 minutes

5. A sample of ^{201}Tl, a radioisotope used to determine the function of the heart, decays initially at a rate of 25,700 disintegrations/min, but the decay rate falls to 15,990 disintegrations/min after 50.0 hours. What is the half-life of ^{201}Tl, in hours? **(LO 20.6)**

   (a) 73.0 hours   (b) 105 hours

   (c) 1.56×10^{-2} hours   (d) 3.84×10^2 hours

6. In a cave in Oregon, archaeologists found bones, plant remains, and fossilized feces. DNA remaining in the feces indicates their human origin but not their age. To date the remains, the decay rate was measured and found to be 2.71 disintegrations/min per gram of carbon. Currently living organisms have a decay rate of 15.3 disintegrations/min per gram of carbon, and the half-life of ^{14}C is 5715 years. How old are the remains? **(LO 20.7)**

   (a) 1460 years   (b) 9900 years

   (c) 14,300 years   (d) 18,600 years

7. Calculate the binding energy a uranium-235 nucleus in units of MeV/nucleon. The mass of an ^{235}U atom is 235.043 929, the mass of a proton is 1.007 28, the mass of a neutron is 1.008 67, and the mass of an electron is 5.486×10^{-4}. $(1 \text{ MeV} = 1.60 \times 10^{-13} \text{ J})$ **(LO 208)**

   (a) 2.84 MeV/nucleon

   (b) 1.70×10^3 MeV/nucleon

   (c) 11.3 MeV/nucleon

   (d) 7.62 MeV/nucleon

8. Identify the *true* statement about nuclear power plants and nuclear weapons. **(LO 20.10)**

   (a) Nuclear power plants and nuclear weapons both use uranium enriched to about 90% U-235.

   (b) Nuclear power plants emit large amounts of CO_2 just like coal burning power plants.

   (c) The United States produces less than 1% of its electrical power from nuclear energy.

   (d) A nuclear weapon explodes when two pieces of fissionable uranium-235 are pushed together to reach a critical mass.

9. How much energy in kJ/mol is released by the fission of uranium-235 to form barium-140 and krypton-93? The atomic masses are ^{235}U (235.0439), ^{140}Ba (139.9106), ^{93}Kr (92.9313) and $^{1}_{0}n$ (1.00867). $(c = 3.00 \times 10^8 \text{ m/s})$ **(LO 20.10)**

   $$^{1}_{0}n + {}^{235}_{92}U \longrightarrow {}^{140}_{56}Ba + {}^{93}_{36}Kr + 3\,{}^{1}_{0}n$$

   (a) 6.59×10^9 kJ/mol   (b) 1.66×10^{10} kJ/mol

   (c) 1.98×10^{11} kJ/mol   (d) 1.66×10^{16} kJ/mol

10. What is the product in the balanced nuclear equation for the reaction of argon-40 with a proton? **(LO 20.12)**

    $$^{40}_{18}Ar + {}^{1}_{1}H \longrightarrow \; ? \; + {}^{1}_{0}n$$

    (a) ^{42}K (b) ^{40}K

    (c) ^{41}Cl (d) ^{40}Cl

11. A sample ^{99m}Tc used for a whole body bone scan has an activity of 600 MBq. If the half-life is 6.01 hours, what mass of ^{99m}Tc was injected? **(LO 20.13)**

    (a) 3.1 ng   (b) 8.4 μg

    (c) 67 μg   (d) 2.7 mg

Answers:

1. a, 2. d, 3. c, 4. d, 5. a, 6. c, 7. d, 8. d, 9. b, 10. b, 11. a

Mastering **Chemistry**   provides end-of-chapter exercises, feedback-enriched tutorial problems, animations, and interactive
activities to encourage problem-solving practice and deeper understanding of key concepts and topics.

RAN   *Randomized in Mastering Chemistry*

CONCEPTUAL PROBLEMS

Problems 20.1–20.24 appear within the chapter.

20.25 Isotope A decays to isotope E through the following series of steps, in which the products of the individual decay events are themselves radioactive and undergo further decay until a stable nucleus is ultimately reached. Two kinds of processes are represented, one by the shorter arrows pointing right and the other by the longer arrows pointing left.

(a) To what kind of nuclear decay process does each kind of arrow correspond?

(b) Identify and write the symbol $_Z^A X$ for each isotope in the series:

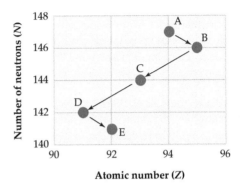

SECTION PROBLEMS

Nuclear Reactions and Radioactivity (Sections 20.1–20.2)

20.26 Positron emission and electron capture both give a product nucleus whose atomic number is 1 less than the starting nucleus. Explain.

20.27 What is the difference between an α particle and a helium atom?

20.28 Why does beta emission *raise* the atomic number of the product while positron emission *lowers*. the atomic number?

20.29 Why does gamma emission not change the atomic number of the product?

20.30 Complete and balance the following nuclear equations.

RAN   (a) $_{50}^{126}Sn \longrightarrow _{-1}^{0}e + ?$ (b) $_{88}^{210}Ra \rightarrow _2^4He + ?$

(c) $_{37}^{77}Rb \rightarrow _1^0e + ?$ (d) $_{36}^{76}Kr + _{-1}^{0}e \rightarrow ?$

20.31 Complete and balance the following nuclear equations.

(a) $_{38}^{90}Sr \longrightarrow _{-1}^{0}e + ?$ (b) $_{100}^{247}Fm \rightarrow _2^4He + ?$

(c) $_{25}^{49}Mn \rightarrow _1^0e + ?$ (d) $_{18}^{37}Ar + _{-1}^{0}e \longrightarrow ?$

20.32 What particle is produced by the decay of thorium-214 to
RAN   radium-210?

20.33 What particle is produced by the decay of uranium-239 to
RAN   neptunium-239?

20.34 What particle is produced in each of the following decay
RAN   reactions?

(a) $_{80}^{188}Hg \longrightarrow _{79}^{188}Au + ?$ (b) $_{85}^{288}At \longrightarrow _{83}^{214}Bi + ?$

(c) $_{90}^{234}Th \longrightarrow _{91}^{234}Pa + ?$

20.35 What particle is produced in each of the following decay reactions?

(a) $_{11}^{24}Na \longrightarrow _{12}^{24}Mg + ?$ (b) $_{60}^{135}Nd \longrightarrow _{59}^{135}Pr + ?$

(c) $_{78}^{170}Pt \rightarrow _{76}^{166}Os + ?$

20.36 Write balanced nuclear equations for the following processes.
RAN   (a) Alpha emission of ^{162}Re

(b) Electron capture of ^{138}Sm

(c) Beta emission of ^{188}W

(d) Positron emission of ^{165}Ta

20.37 Write balanced nuclear equations for the following processes:

(a) Beta emission of ^{157}Eu

(b) Electron capture of ^{126}Ba

(c) Alpha emission of ^{146}Sm

(d) Positron emission of ^{125}Ba

20.38 Radioactive ^{100}Tc decays to form ^{100}Mo. There are two possible pathways for this decay. Write balanced equations for both.

20.39 ^{226}Ac can decay by any of three different nuclear processes: α emission, β emission, or electron capture. Write a balanced nuclear equation for the decay of ^{226}Ac by each process.

Nuclear Stability (Section 20.3)

20.40 Of the two isotopes of tungsten, ^{160}W and ^{185}W, one decays by β emission and one decays by α emission. Which does which? Explain.

20.41 Of the two isotopes of iodine, ^{136}I and ^{122}I, one decays by β emission and one decays by positron emission. Which does which? Explain.

20.42 Of the two isotopes of iodine, ^{196}Pb and ^{206}Pb, one is nonradioactive and one decays by positron emission. Which does which? Explain.

20.43 Why do nuclei that are neutron-rich emit β particles? Why do nuclei that are neutron-poor emit α particles or positrons or undergo electron capture?

20.44 Americium-241, a radioisotope used in smoke detectors, decays by a series of 12 reactions involving sequential loss of α, α, β, α, α, β, α, α, α, β, α, and β particles. Identify each intermediate nucleus and the final stable product nucleus.

20.45 Radon-222 decays by a series of three α emissions and two β emissions. What is the final stable nucleus?

20.46 Thorium-232 decays by a 10-step series, ultimately yielding lead-208. How many α particles and how many β particles are emitted?

20.47 How many α particles and how many β particles are emitted in the 11-step decay of ^{235}U into ^{207}Pb?

20.48 Neptunium-237 decays by a series of steps to bismuth-209. How many α and β particles are produced by this decay process?

20.49 Naturally occurring uranium-238 undergoes a radioactive
RAN   decay series and emits 8 α particles and 6 β particles. What is the stable nucleus at the end of the series?

Radioactive Decay Rates (Section 20.4)

20.50 The half-life of indium-111, a radioisotope used in studying the distribution of white blood cells, is $t_{1/2} = 2.805$ days. What is the decay constant of ^{111}In?

20.51 The decay constant of plutonium-239, a waste product from nuclear reactors, is 2.88×10^{-5} year^{-1}. What is the half-life of ^{239}Pu?

20.52 The decay constant of ^{35}S is 7.95×10^{-3} day^{-1}. What per-
RAN   centage of an ^{35}S sample remains after 185 days?

20.53 Plutonium-239 has a decay constant of 2.88×10^{-5} year^{-1}.
RAN   What percentage of a ^{239}Pu sample remains after 1000 years? After 25,000 years? After 100,000 years?

20.54 Polonium-209, an α emitter, has a half-life of 102 years. How
RAN   many α particles are emitted in 1.0 s from a 1.0 ng sample of ^{209}Po?

20.55 Chlorine-36 is a β emitter, with a half-life of 3.0×10^5
RAN   years. How many β particles are emitted in 1.0 min from a 5.0 mg sample of ^{36}Cl?

20.56 A 1.0 mg sample of ^{79}Se decays initially at a rate of 1.5×10^5
RAN   disintegrations/s. What is the half-life of ^{79}Se in years?

20.57 What is the half-life (in years) of ^{44}Ti if a 1.0 ng sample decays initially at a rate of 4.8×10^3 disintegrations/s?

20.58 A sample of ^{37}Ar undergoes 8540 disintegrations/min
RAN   initially but undergoes 6990 disintegrations/min after 10.0 days. What is the half-life of ^{37}Ar in days?

20.59 A sample of ^{28}Mg decays initially at a rate of 53,500 disintegra-tions/min, but the decay rate falls to 10,980 disintegrations/min after 48.0 hours. What is the half-life of ^{28}Mg in hours?

20.60 Radioactive decay exhibits a first-order rate law, rate =
RAN $-\Delta N/\Delta t = kN$, where N denotes the number of radio-active nuclei present at time t. The half-life of strontium-90, a dangerous nuclear fission product, is 29 years.
  (a) What fraction of the strontium-90 remains after three half-lives?
  (b) What is the value of the decay constant for strontium-90?
  (c) How many years are required for 99% of the strontium-90 to disappear?

20.61 Potassium ion, K$^+$, is present in most foods and is an essen-
RAN   tial nutrient in the human body. Potassium-40, however, which has a natural abundance of 0.0117%, is radioactive with $t_{1/2} = 1.25 \times 10^9$ years. What is the decay constant of ^{40}K? How many ^{40}K$^+$ ions are present in 1.00 g of KCl? How many disintegrations/s does 1.00 g of KCl undergo?

20.62 The electronic systems on the New Horizons spacecraft, which launched on January 19, 2006, and reached its closest approach to Pluto on July 14, 2015, were powered by elec-tricity generated by heat. The heat came from the radioac-tive decay of ^{238}Pu in the 11 kg of ^{238}PuO$_2$ fuel onboard. The generator provided 240 W when the spacecraft was launched. If the power output is directly proportional to the amount of ^{238}Pu in the generator, what was the power output when the spacecraft reached Pluto? The half-life of ^{238}Pu is 87.7 y.

20.63 The radioisotope ^{226}Ac can decay by any of three different
RAN   nuclear processes: α emission, β emission, or electron capture.
  (a) Write a balanced nuclear equation for the decay of ^{226}Ac by each decay mode.
  (b) For the decay of ^{226}Ac by all processes combined, the first-order rate constant is $k = 0.556$ d^{-1}. How many days are required for 80.0% of a sample of ^{226}Ac to decay?

Radiocarbon Dating (Section 20.5)

20.64 The age of any remains from a once-living organism can be
RAN   determined by radiocarbon dating, a procedure that works by determining the concentration of radioactive ^{14}C in the remains. All living organisms contain an equilibrium con-centration of radioactive ^{14}C that gives rise to an average of 15.3 nuclear decay events per minute per gram of carbon. At death, however, no additional ^{14}C is taken in, so the con-centration slowly drops as radioactive decay occurs. What is the age of a bone fragment from an archaeological dig if the bone shows an average of 2.3 radioactive events per minute per gram of carbon? For ^{14}C, $t_{1/2} = 5715$ years.

20.65 What is the ^{14}C activity in decay events per minute of 1.0 g
RAN   of a 3000-year-old wooden object? All living organisms con-tain an equilibrium concentration of radioactive ^{14}C that gives rise to an average of 15.3 nuclear decay events per minute per gram of carbon. (^{14}C, $t_{1/2} = 5715$ years)

20.66 The Voynich manuscript is currently in the possession of the Beineke Rare Books Library at Yale University. The document was once in the possession of the Holy Roman Emperor Rudolph II and is written in an encrypted, indeci-pherable language. The document consists of strange draw-ings of astronomical features and unknown plants, and its purpose remains unclear. Radiocarbon measurements made in 1990 on this manuscript show a ^{14}C decay rate of 14.4 disintegrations/min per gram of carbon. What age is implied by this result if currently living organisms decay at the rate of 15.3 disintegrations/min per gram of carbon? The half-life of ^{14}C is 5715 years.

20.67 The remains of the Arlington Springs Man were discov-ered in 1959 from an eroding wall of Arlington Canyon on Santa Rosa island off the coast of Southern California. In 1987, new radiocarbon dating methods determined that the Arlington Springs Man was 13,000 years old. If currently living organisms decay at the rate of 15.3 disintegrations/min per gram of carbon, what is the expected decay rate for the artifact? The half-life of ^{14}C is 5715 years.

20.68 A mammoth bone found in Niederweningen, Switzer-land, was analyzed and determined to be approximately 46,000 years old. What percentage of the ^{14}C remains in the sample? The half-life of ^{14}C is 5715 years.

20.69 What is the age of a bone fragment that shows an average of 5.6 disintegrations/min per gram of carbon? The carbon in liv-ing organisms undergoes an average of 15.3 disintegrations/min per gram of carbon, and the half-life of ^{14}C is 5715 years.

20.70 Uranium-238 has a half-life of 4.47×10^9 years and decays
RAN   through a series of events to yield lead-206. Estimate the age of a rock that contains 105 μmol of ^{238}U and 33 μmol of ^{206}Pb. Assume all the ^{206}Pb is from the decay of ^{238}U.

20.71 Uranium-238 has a half-life of 4.47×10^9 years and decays through a series of events to yield lead-206. Estimate the age of a rock that contains 125 μmol of ^{238}U and 38 μmol of ^{206}Pb. Assume all the ^{206}Pb is from the decay of ^{238}U.

20.72 The age of an igneous rock that has solidified from magma
RAN can be found by analyzing the amount of ^{40}K and ^{40}Ar.
Potassium-40 emits a positron to produce argon-40 and the
half-life of ^{40}K is 1.25×10^9 years.

$$^{40}_{19}K \rightarrow \,^{40}_{18}Ar + \,^{0}_{1}e$$

If the rock contains 3.35 mmol of ^{40}K and 0.25 mmol of
^{40}Ar, how long ago did the rock cool?

20.73 The age of an igneous rock that has solidified from magma
can be found by analyzing the amount of ^{40}K and ^{40}Ar.
Potassium-40 emits a positron to produce argon-40 and the
half-life of ^{40}K is 1.25×10^9 years. What is the age of a rock
whose $^{40}Ar/^{40}K$ ratio is 1.42?

Energy Changes during Nuclear Reactions (Section 20.6)

20.74 Why does a given nucleus have less mass than the sum of its
constituent protons and neutrons?

20.75 In an endothermic chemical reaction, do the products have
RAN more mass, less mass, or the same mass as the reactants?
Explain.

20.76 Calculate the mass defect (in g/mol) for the following nuclei.
RAN (a) ^{52}Fe (atomic mass = 51.948 11)
(b) ^{92}Fe (atomic mass = 91.906 81)

20.77 Calculate the mass defect (in g/mol) for the following nuclei.
(a) ^{32}S (atomic mass = 11.972 07)
(b) ^{40}Ca (atomic mass = 39.962 59)

20.78 Calculate the mass defect (in g/mol) and the binding energy
RAN (in MeV/nucleon) for the following nuclei. Which of the two
is more stable?
(a) ^{50}Cr (atomic mass = 49.946 05)
(b) ^{64}Zn (atomic mass = 63.929 15)

20.79 Calculate the mass defect (in g/mol) and the binding energy
(in MeV/nucleon) for the following nuclei. Which of the two
is more stable?
(a) ^{7}Li (atomic mass = 7.016 004)
(b) ^{39}K (atomic mass = 38.963 706)

20.80 Calculate the binding energy (in MeV/nucleon) for the fol-
RAN lowing nuclei.
(a) ^{58}Ni (atomic mass = 57.935 35)
(b) ^{84}Kr (atomic mass = 83.911 51)

20.81 Calculate the binding energy (in MeV/nucleon) for the fol-
RAN lowing nuclei.
(a) ^{63}Cu (atomic mass = 62.939 60)
(b) ^{84}Sr (atomic mass = 83.913 43)

20.82 What is the energy change ΔE (in kJ/mol) when an α particle
RAN is emitted from ^{174}Ir? The atomic mass of ^{174}Ir is 173.966
66 the atomic mass of ^{170}Re is 169.958 04, and the atomic
mass of a ^{4}He atom is 4.002 60.

$$^{174}_{77}Ir \rightarrow \,^{170}_{75}Re + \,^{4}_{2}He \quad \Delta E = ?$$

20.83 Magnesium-28 is a β emitter that decays to aluminum-28.
How much energy is released in kJ/mol? The atomic mass of
^{28}Mg is 27.983 88, and the atomic mass of ^{28}Al is 27.981 91.

20.84 What is the mass change in grams per mole of NH_3 accom-
RAN panying the formation of NH_3 from H_2 and N_2?

$$N_2(g) + 3H_2(g) \rightarrow 2\,NH_3(g) \quad \Delta H° = -92.2 \text{ kJ}$$

20.85 What is the mass change in grams accompanying the formation of 1 mol of CO and 1 mol of H_2 in the water–gas
reaction?

$$C(s) + H_2O(g) \rightarrow CO(g) + H_2(g) \quad \Delta H° = +131 \text{ kJ}$$

20.86 Thorium-232 decays by a 10-step series of nuclear reactions,
RAN ultimately yielding lead-208, along with 6 α particles and 4 β
particles. How much energy (in kJ/mol) is released during the
overall process? The relevant masses are ^{232}Th = 232.038 054,
^{208}Pb = 207.976 627, electron= 0.000 548 6, and ^{4}He=
4.002 603.

20.87 The radioactive isotope ^{100}Tc decays to form the stable isotope ^{100}Mo.
(a) There are two possible pathways for this decay. Write
balanced equations for both.
(b) Only one of the pathways is observed. Calculate the energy
released by both pathways, and explain why only one is
observed. Relevant masses are: ^{100}Tc = 99.907 657,
^{100}Mo = 99.907 48, electron = 0.000 548 6.

Fission and Fusion (Section 20.7)

20.88 Identify the *false* statement about nuclear fission.
RAN (a) Nuclear fission is induced by bombarding a U-235 sample with beta particles.
(b) Nuclear fission is the splitting of a heavy element into
lighter elements.
(c) Mass is converted to energy in nuclear fission.
(d) Nuclear fission releases huge amounts of energy compared to chemical reactions.
(e) Nuclear fission can cause a chain reaction because additional neutrons are produced with each fission of a nucleus.

20.89 On September 6, 1954, President Dwight David Eisenhower
waved a ceremonial "neutron wand" over a neutron counter
in Denver, Colorado, to signal a bulldozer in Shippingport,
Pennsylvania, to begin construction of the first nuclear power
plant. Explain the role of neutrons in a nuclear power plant.

▲ President Dwight David Eisenhower with
neutron wand.

20.90 Control rods in a nuclear reactor are often made of boron
RAN because it absorbs neutrons. Write the nuclear equation in
which boron-10 absorbs a neutron to produce lithium-7 and
an alpha particle.

20.91 What is the difference between uranium fuel rods in a nuclear
power plant and uranium fuel for an atomic weapon?

20.92 Can fuel rods in a power plant be used to make an atomic
weapon without further treatment? Explain.

20.93 What are the benefits of using fusion over fission as a source of nuclear energy? Why have fusion reactors not been developed yet?

20.94 How much energy is released (in kJ) in the fusion reaction of ^{2}H to yield 1 mol of ^{3}He? The atomic mass of ^{2}H is 2.0141, and the atomic mass of ^{3}He is 3.0160.

$$2\,^2_1\text{H} \rightarrow \,^3_2\text{He} + \,^1_0\text{n}$$

20.95 How much energy (in kJ/mol) is produced in the following
RAN fission reaction of plutonium-239?

$$^{239}_{94}\text{Pu} + \,^1_0\text{n} \rightarrow \,^{146}_{56}\text{Ba} + \,^{91}_{38}\text{Sr} + 3\,^1_0\text{n}$$

The masses are ^{239}Pu (239.052 16), ^{146}Ba (145.930 22), ^{91}Sr (90.910 20), $^1_0$n (1.008 67).

20.96 How much energy (in kJ/mol) is released in the fusion reaction of ^{2}H with ^{3}He?

$$^2_1\text{H} + \,^3_2\text{He} \rightarrow \,^4_2\text{He} + \,^1_1\text{H}$$

The relevant masses are ^{2}H (2.0141), ^{3}He (3.0160), ^{4}He (4.0026), and ^{1}H (1.0078).

20.97 How much energy (in kJ/mol) is released in the fusion reac-
RAN tion of ^{1}H and ^{2}H atoms?

$$^1_1\text{H} + \,^2_1\text{H} \rightarrow \,^3_2\text{He}$$

The relevant masses are ^{1}H (1.007 83), ^{2}H (2.014 10) and ^{3}He (3.016 03).

Nuclear Transmutation (Section 20.8)

20.98 Give the products of the following nuclear reactions.
RAN (a) $^{109}_{47}\text{Ag} + \,^4_2\text{He} \rightarrow$?
  (b) $^{10}_5\text{B} + \,^4_2\text{He} \rightarrow$? $+ \,^1_0\text{n}$

20.99 Balance the following equations for the nuclear fission of ^{235}U.
  (a) $^{235}_{92}\text{U} \rightarrow \,^{160}_{62}\text{Sm} + \,^{72}_{30}\text{Zn} + ?\,^1_0\text{n}$
  (b) $^{235}_{92}\text{U} \rightarrow \,^{87}_{35}\text{Br} + ? + 2\,^1_0\text{n}$

20.100 Element 109 ($^{266}_{109}\text{Mt}$) was prepared in 1982 by bombardment of ^{209}Bi atoms with ^{58}Fe atoms. Identify the other product that must have formed, and write a balanced nuclear equation assuming no other products were formed.

20.101 Molybdenum-99 is formed by neutron bombardment of a naturally occurring isotope of Mo. If one neutron is absorbed and no by-products are formed, what is the starting isotope?

20.102 Write a balanced nuclear equation for the reaction of uranium-238 with a deuteron (^2_1H).

$$^{238}_{92}\text{U} + \,^2_1\text{H} \rightarrow ? + 2\,^1_0\text{n}$$

20.103 Balance the following transmutation reactions.
  (a) $^{246}_{96}\text{Cm} + \,^{12}_6\text{C} \rightarrow$? $+ 4\,^1_0\text{n}$
  (b) $^{253}_{99}\text{Es} + ? \rightarrow \,^{256}_{101}\text{Md} + \,^1_0\text{n}$
  (c) $^{250}_{98}\text{Cf} + \,^{11}_5\text{B} \rightarrow$? $+ 4\,^1_0\text{n}$

20.104 One of the new superheavy elements added to the periodic table in 2016 was synthesized when a beam of ^{48}Ca ions was directed at a target of ^{249}Cf.
  (a) Write a balanced nuclear equation for the formation of this element.

$$^{249}_{98}\text{Cf} + \,^{48}_{20}\text{Ca} \rightarrow ? + 3\,^1_0\text{n}$$

  (b) What isotope is formed after the nuclide formed in the nuclear transmutation reaction in part (a) emits four alpha particles?

20.105 One of the new superheavy elements added to the periodic table in 2016 was synthesized when a beam of ^{48}Ca ions was directed at a target of ^{243}Am.
  (a) Write a balanced nuclear equation for the formation of this element.

$$^{243}_{95}\text{Am} + \,^{48}_{20}\text{Ca} \rightarrow ? + 3\,^1_0\text{n}$$

  (b) What isotope is formed after the nuclide formed in the nuclear transmutation reaction in part (a) emits one alpha particle?
  (c) How many alpha particles were emitted to reach the isotope ^{268}Db, the final decay product?

20.106 Fraud in science is rare but does happen occasionally. In 1999, the creation of three superheavy elements (one new) was claimed when ^{208}Pb was bombarded with ^{86}Kr. The claim was subsequently found to be fraudulent and was withdrawn. Identify the isotopes X, Y, and Z that were claimed.

$$^{208}_{82}\text{Pb} + \,^{86}_{36}\text{Kr} \xrightarrow{-\text{n}} \text{X} \xrightarrow{-\alpha} \text{Y} \xrightarrow{-\alpha} \text{Z}$$

20.107 The most abundant isotope of uranium, ^{238}U, does not undergo fission. In a *breeder reactor*, however, a ^{238}U atom captures a neutron and emits two β particles to make a fissionable isotope of plutonium, which can then be used as fuel in a nuclear reactor. Write a balanced nuclear equation.

Detecting and Measuring Radioactivity (Section 20.9)

20.108 A gastrointestinal tract X ray exposes a patient to 5000 μSv
RAN of radiation. For X rays (1 Sv = 1 Gy). How many joules are absorbed by a 60-kg person?

20.109 A 255-gram laboratory rat is exposed to 23.2 rads. How
RAN many grays did the rat receive? How many joules were absorbed by the rat?

20.110 The maximum level of radon in drinking water is 4.0 pCi
RAN (4.0×10^{-12}Ci) per milliliter.
  (a) How many disintegrations occur per minute in 1 mL of water at the maximum radon level?
  (b) If the radioactive isotope is ^{222}Rn ($t_{1/2} = 3.8$ days), how many ^{222}Rn atoms are present in 1 mL of the water?

20.111 Determine the activity of 10.0 mg of ^{226}Ra in units of Bq
RAN and Ci. The half-life of ^{226}Ra is 1600 years.

20.112 Nitrogen-16 is formed in the cooling water flowing through a hot reactor core in a nuclear power plant. It is formed when oxygen captures a neutron and then emits a β particle. Determine the activity of 50.0 mg of ^{16}N in units of Bq and Ci.

20.113 The maximum allowable radiation dose for recovery workers at the Fukushima nuclear plant in Japan was set to 100 mSv. A gastrointestinal (GI) tract X ray is 5,000 μSv. The radiation exposure of a Fukushima worker is equivalent to how many GI tract X rays?

MULTICONCEPT PROBLEMS

20.114 A proposed nuclear theory suggests that the relative abundances of the uranium isotopes ^{235}U and ^{238}U were approximately equal at the time they were formed. Today, the observed ratio of these isotopes $^{235}U/^{238}U$ is 7.25×10^{-3}. Given that the half-lives for radioactive decay are 7.04×10^8 y for ^{235}U and 4.47×10^9 y for ^{238}U, calculate the age of the elements.

20.115 A positron has the same mass as an electron (9.109×10^{-31} kg) but an opposite charge. When the two particles encounter each other, annihilation occurs and only γ rays are produced. How much energy (in kJ/mol) is produced?

20.116 What is the wavelength (in nm) of γ rays whose energy is
RAN 1.50 MeV?

20.117 What is the frequency (in Hz) of X rays whose energy is
RAN 6.82 keV?

20.118 A small sample of wood from an archaeological site in Clovis, New Mexico, was burned in O_2 and the CO_2 produced
RAN was bubbled through a solution of $Ba(OH)_2$ to produce a precipitate of $BaCO_3$. When the $BaCO_3$ was collected by filtration, a 1.000 g sample was found to have a radioactivity of 4.0×10^{-3} Bq. The half-life of ^{14}C is 5715 y, and living organisms have a radioactivity due to ^{14}C of 15.3 disintegrations/min per gram of carbon. If the analysis was carried out in 1960, what is the date of the Clovis site?

20.119 Polonium-210, a naturally occurring radioisotope, is an α
RAN emitter, with $t_{1/2} = 138$ d. Assume that a sample of ^{210}Po with a mass of 0.700 mg was placed in a 250.0-mL flask, which was evacuated, sealed, and allowed to sit undisturbed. What would the pressure be inside the flask (in mm Hg) at 20 °C after 365 days if all the α particles emitted had become helium atoms?

20.120 Imagine that you have a 0.007 50 M aqueous $MgCl_2$ solu-
RAN tion, prepared so that it contains a small amount of radioactive ^{28}Mg. The half-life of ^{28}Mg is 20.91 h, and the initial activity of the $MgCl_2$ solution is 0.112 μCi/mL. Assume that 20.00 mL of this $MgCl_2$ solution is added to 15.00 mL of 0.012 50 M aqueous Na_2CO_3 solution and that the resultant precipitate is then removed by filtration to yield a clear filtrate. After a long break to go for a run, you find that the activity of the filtrate measured 2.40 h after beginning the experiment is 0.029 μCi/mL. What are the molar concentrations of Mg^{2+} and CO_3^{2-} in the filtrate, and what is the solubility product constant of $MgCO_3$?

chapter 21

Transition Elements and Coordination Chemistry

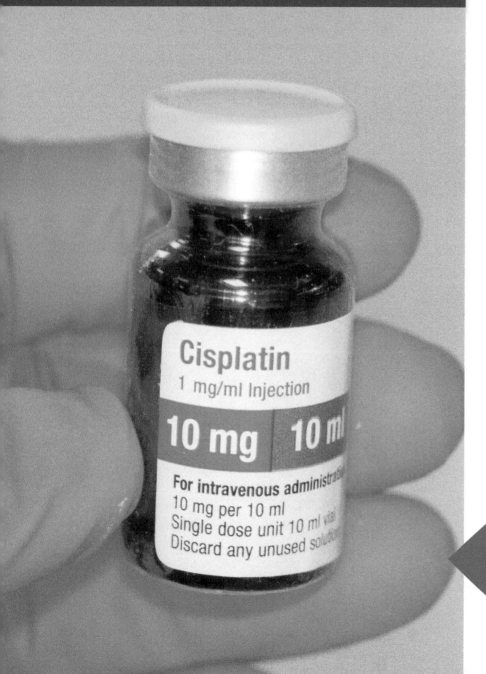

Contents

Cisplatin, $Pt(NH_3)_2Cl_2$, is a coordination compound and an antitumor drug that has been successful in treating testicular, ovarian, bladder, lung, and stomach cancer.

How does cisplatin kill cancer cells?

The answer to this question can be found on page 940 in the INQUIRY ?

The *transition elements* occupy the central part of the periodic table, bridging the gap between the active *s*-block metals of groups 1A and 2A on the left and the *p*-block metals, semimetals, and nonmetals of groups 3A–8A on the right (**FIGURE 21.1**). Because the *d* subshells are being filled in this region of the periodic table, the transition elements are also called the *d-block elements.*

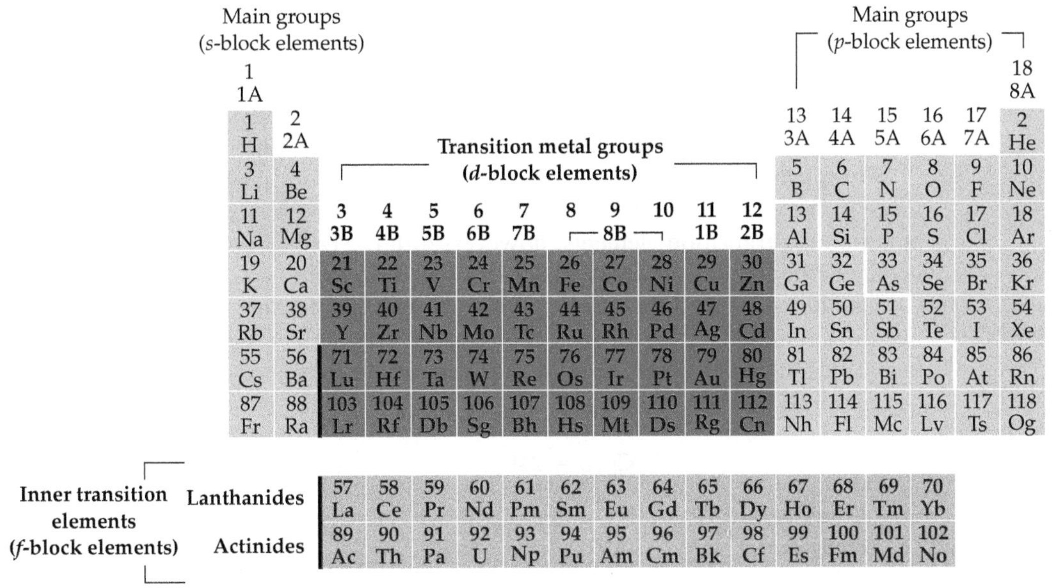

Main groups
(*s*-block elements)

Transition metal groups
(*d*-block elements)

Main groups
(*p*-block elements)

◀ **FIGURE 21.1**

The transition elements are located in the central region of the periodic table.

Inner transition elements
(*f*-block elements)

Lanthanides

Actinides

Each *d* subshell consists of five orbitals and can accommodate 10 electrons, so each transition series consists of 10 elements. The first series extends from scandium through zinc and includes many familiar metals, such as chromium, iron, and copper. The second series runs from yttrium through cadmium, and the third series runs from lutecium through mercury. In addition, there is a fourth transition series made up of the unstable, radioactive elements lawrencium through the recently discovered element 112, copernicium (Cn).

Tucked into the periodic table between barium (atomic number 56) and lutecium (atomic number 71) are the *lanthanides.* In this series of 14 metallic elements, the seven 4*f* orbitals are progressively filled. Following radium (atomic number 88) is a second series of 14 elements, the *actinides,* in which the 5*f* subshell is progressively filled. The lanthanides and actinides together comprise the *f-block elements,* or *inner transition elements.*

The transition metals iron and copper have been known since antiquity and have played an important role in the development of civilization. Iron, the main constituent of steel, is an important structural material. Worldwide production of steel was 1.6 billion metric tons in 2017. Other transition elements are used in newer technologies. For example, the strong, lightweight metal, titanium, is a major component in modern jet aircraft and is used in manufacturing artificial joints and dental implants. Transition metals are also used as heterogeneous catalysts in automobile catalytic converters and in the industrial synthesis of essential chemicals such as sulfuric acid, nitric acid, and ammonia.

The role of the transition elements in living systems is equally important. Iron is present in biomolecules such as hemoglobin, which transports oxygen from our lungs to other parts of the body. Cobalt is an essential component of vitamin B_{12}. Nickel, copper, and zinc are vital constituents of many enzymes, the large protein molecules that catalyze biochemical reactions.

In this chapter, we'll look at the properties and chemical behavior of transition metal compounds, paying special attention to *coordination compounds,* in which a central metal ion or atom—usually a transition metal—is attached to a group of surrounding molecules or ions by coordinate covalent bonds (Section 7.5).

21.1 ELECTRON CONFIGURATIONS

Look at the electron configurations of potassium and calcium, the *s*-block elements immediately preceding the first transition series. These atoms have 4*s* valence electrons, but no *d* electrons:

$$\textbf{K } (Z = 19)\text{: } [\text{Ar}]\, 3d^0\, 4s^1 \qquad \textbf{Ca } (Z = 20)\text{: } [\text{Ar}]\, 3d^0\, 4s^2$$

The filling of the 3*d* subshell begins at atomic number 21 (scandium) and continues until the subshell is completely filled at atomic number 30 (zinc):

$$\textbf{Sc } (Z = 21)\text{: } [\text{Ar}]\, 3d^1\, 4s^2 \longrightarrow \textbf{Zn } (Z = 30)\text{: } [\text{Ar}]\, 3d^{10}\, 4s^2$$

(It's convenient to list the valence electrons in order of principal quantum number rather than in order of orbital filling because the 4*s* electrons are lost first in forming ions.)

For main-group elements, the valence electrons are generally considered to be those in the outermost shell because they are the ones that are involved in chemical bonding. For transition elements, however, both the $(n - 1)d$ and the ns electrons are involved in bonding and are considered valence electrons.

The filling of the 3*d* subshell generally proceeds according to **Hund's rule** with one electron adding to each of the five 3*d* orbitals before a second electron adds to any one of them (Section 5.10). There are just two exceptions to the expected regular filling pattern, chromium and copper:

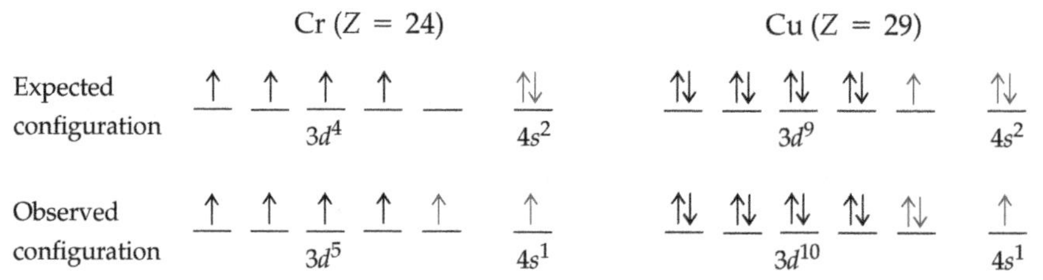

REMEMBER . . .
According to **Hund's rule**, if two or more orbitals with the same energy are available, one electron goes into each until all are half-filled and the electrons in the half-filled orbitals all have the same spin (Section 5.10).

	Cr (Z = 24)		Cu (Z = 29)	
Expected configuration	↑ ↑ ↑ ↑ __ ↑↓ 3d^4 4s^2		↑↓ ↑↓ ↑↓ ↑↓ ↑ ↑↓ 3d^9 4s^2	
Observed configuration	↑ ↑ ↑ ↑ ↑ ↑ 3d^5 4s^1		↑↓ ↑↓ ↑↓ ↑↓ ↑↓ ↑ 3d^{10} 4s^1	

Electron configurations depend on both orbital energies and electron–electron repulsions. Consequently, it's not always possible to predict configurations when two valence subshells have similar energies. It's often found, however, that exceptions from the expected orbital filling pattern result in either half-filled or completely filled subshells. In the case of chromium, for example, the 3*d* and 4*s* subshells have similar energies. It's evidently advantageous to shift one electron from the 4*s* to the 3*d* subshell, which decreases electron–electron repulsions and gives two half-filled subshells. Because each valence electron is in a separate orbital, the electron–electron repulsion that would otherwise occur between the two 4*s* electrons in the expected configuration is eliminated. A similar shift of one electron from 4*s* to 3*d* in copper gives a completely filled 3*d* subshell and a half-filled 4*s* subshell.

When a neutral atom loses one or more electrons, the remaining electrons are less shielded and the effective nuclear charge (Z_{eff}) increases. Consequently, the remaining electrons are more strongly attracted to the nucleus and their orbital energies decrease. It turns out that the valence *d* orbitals experience a steeper drop in energy with increasing Z_{eff} than does the valence *s* orbital, making the *d* orbitals in cations lower in energy than the *s* orbital. Lower energy configurations are more stable; therefore, the valence *s* orbital is vacant and all the valence electrons occupy the *d* orbitals in transition metal cations.

It is important to remember that electrons are lost from valence *s* orbitals before they are lost from valence *d* orbitals in the formation of cations from neutral transition metal elements. Iron, for example, which forms 2+ and 3+ cations, has the following valence electron configurations:

Remove two electrons from the 4*s* orbital to write an electron configuration for the Fe^{2+} cation.

If all electrons from the valence 4*s* orbital have been removed, then remove one 3*d* electron to write the electron configuration for the Fe^{3+} cation.

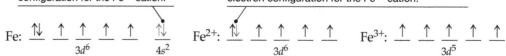

Fe: ↑↓ ↑ ↑ ↑ ↑   ↑↓ Fe^{2+}: ↑↓ ↑ ↑ ↑ ↑ Fe^{3+}: ↑ ↑ ↑ ↑ ↑
      3d^6 4s^2 3d^6 3d^5

In neutral molecules and complex anions, the metal atom usually has a positive oxidation state. It therefore has a partial positive charge and a higher Z_{eff} than that of the neutral atom. As a result, the valence d orbitals are again lower in energy than the valence s orbital, and so all the metal's valence electrons occupy the d orbitals. The metal atom in both VCl_4 and $MnO_4{}^{2-}$, for example, has the valence configuration $3d^1$ rather than $4s^1$. Electron configurations and other properties for atoms and common ions of first-series transition elements are summarized in **TABLE 21.1**.

TABLE 21.1   Selected Properties of First-Series Transition Elements

Group:	3B	4B	5B	6B	7B	8B			1B	2B
Element:	Sc	Ti	V	Cr	Mn	Fe	Co	Ni	Cu	Zn
Valence electron configuration										
M atom	$3d^1\,4s^2$	$3d^2\,4s^2$	$3d^3\,4s^2$	$3d^5\,4s^1$	$3d^5\,4s^2$	$3d^6\,4s^2$	$3d^7\,4s^2$	$3d^8\,4s^2$	$3d^{10}\,4s^1$	$3d^{10}\,4s^2$
M^{2+} ion	N/A	$3d^2$	$3d^3$	$3d^4$	$3d^5$	$3d^6$	$3d^7$	$3d^8$	$3d^9$	$3d^{10}$
M^{3+} ion	$3d^0$	$3d^1$	$3d^2$	$3d^3$	$3d^4$	$3d^5$	$3d^6$	N/A	N/A	N/A
Elec. conductivity*	3	4	8	11	1	17	24	24	96	27
Melting point (°C)	1541	1668	1910	1907	1246	1538	1495	1455	1085	420
Boiling point (°C)	2836	3287	3407	2671	2061	2861	2927	2913	2562	907
Density (g/cm³)	2.99	4.51	6.0	7.15	7.3	7.87	8.86	8.90	8.96	7.13
Atomic radius (pm)	162	147	134	128	127	126	125	124	128	134
E_i (kJ/mol)†										
First	633	659	651	653	717	762	760	737	745	906
Second	1235	1310	1410	1591	1509	1562	1648	1753	1958	1733
Third	2389	2653	2828	2987	3248	2957	3232	3395	3555	3833

*Electrical conductivity relative to an arbitrary value of 100 for silver

†Ionization energy

N/A means the ion does not exist.

Historically, transition elements are placed in groups of the periodic table designated 1B–8B (Figure 21.1) because their valence electron configurations are similar to those of analogous elements in the main groups 1A–8A. Thus, copper in group 1B ($[Ar]\,3d^{10}4s^1$) and zinc in group 2B ($[Ar]\,3d^{10}4s^2$) have valence electron configurations similar to those of potassium in group 1A ($[Ar]\,4s^1$) and calcium in group 2A ($[Ar]\,4s^2$). Similarly, scandium in group 3B ($[Ar]\,3d^14s^2$) through iron in group 8B ($[Ar]\,3d^64s^2$) have the same number of valence electrons as the p-block elements aluminum in group 3A ($[Ne]\,3s^23p^1$) through argon in group 8A ($[Ne]\,3s^23p^6$). Cobalt ($[Ar]\,3d^74s^2$) and nickel ($[Ar]\,3d^84s^2$) are also assigned to group 8B although there are no main-group elements with 9 or 10 valence electrons.

WORKED EXAMPLE 21.1

Writing Electron Configurations of Transition Metals

Write the electron configuration of the metal in each of the following atoms or ions:

(a) Ni   (b) Cr^{3+} (c) $FeO_4{}^{2-}$ (ferrate ion)

STRATEGY

In neutral atoms of the first transition series, the $4s$ orbital is usually filled with 2 electrons and the remaining electrons occupy the $3d$ orbitals. In writing electron configurations for transition metal ions, remove electrons from valence ns orbitals before valence $(n-1)\,d$ orbitals. For polyatomic ions, first determine the oxidation number of the transition metal.

continued on next page

SOLUTION

(a) Nickel ($Z = 28$) has a total of 28 electrons, including the argon core of 18. Two of the 10 valence electrons occupy the $4s$ orbital, and the remaining eight are assigned to the $3d$ orbitals. The electron configuration is therefore $[Ar]\,3d^8 4s^2$.

$$\text{Ni:} \quad \underset{3d}{\underline{\uparrow\downarrow}\;\underline{\uparrow\downarrow}\;\underline{\uparrow\downarrow}\;\underline{\uparrow}\;\underline{\uparrow}} \quad \underset{4s}{\underline{\uparrow\downarrow}}$$

(b) A neutral Cr atom ($Z = 24$) has a 6 valence electrons and the electron configuration is $[Ar]\,3d^5 4s^1$. A Cr^{3+} ion has lost 3 valence electrons and has the configuration $[Ar]\,3d^3$.

$$\text{Cr:}\quad \underset{3d}{\underline{\uparrow}\;\underline{\uparrow}\;\underline{\uparrow}\;\underline{\uparrow}\;\underline{\uparrow}}\quad \underset{4s}{\underline{\uparrow}} \qquad Cr^{3+}:\quad \underset{3d}{\underline{\uparrow}\;\underline{\uparrow}\;\underline{\uparrow}\;\underline{\quad}\;\underline{\quad}}\quad \underset{4s}{\underline{\quad}}$$

(c) The oxidation number of each of the four oxygens in FeO_4^{2-} is -2 and the overall charge on the oxoanion is -2, so the oxidation number of the iron must be $+6$. An iron atom has 8 valence electrons and the electron configuration $[Ar]\,3d^6 4s^2$. An iron(VI) atom has lost 6 valence electrons and the electron configuration is $[Ar]\,3d^2$.

$$\text{Fe:}\quad \underset{3d}{\underline{\uparrow\downarrow}\;\underline{\uparrow}\;\underline{\uparrow}\;\underline{\uparrow}\;\underline{\uparrow}}\quad \underset{4s}{\underline{\uparrow\downarrow}} \qquad Fe^{6+}:\quad \underset{3d}{\underline{\uparrow}\;\underline{\uparrow}\;\underline{\quad}\;\underline{\quad}\;\underline{\quad}}\quad \underset{4s}{\underline{\quad}}$$

▶ **PRACTICE 21.1** What is the electron configuration of Mn in MnO_2? How many unpaired electrons are in MnO_2?

▶ **CONCEPTUAL APPLY 21.2** On the periodic table below, locate the transition metal atom or ion with the following electron configurations. Identify each atom or ion.

(a) An atom: $[Ar]\,3d^5\,4s^2$ (b) A 2+ ion: $[Ar]\,3d^8$
(c) An atom: $[Kr]\,4d^{10}\,5s^1$ (d) A 3+ ion: $[Kr]\,4d^3$

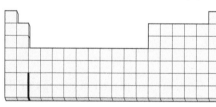

All **PRACTICE** and **APPLY** problems are interactive in the eText.

21.2 PROPERTIES OF TRANSITION ELEMENTS

Let's look at some trends in the properties of the transition elements shown in Table 21.1 and try to understand them in terms of electron configurations.

Metallic Properties

All the transition elements are metals. Like the metals of groups 1A and 2A, the transition metals are malleable, ductile, lustrous, and good conductors of heat and electricity. Silver has the highest electrical conductivity of any element at room temperature, with copper a close second. The transition metals are harder, have higher melting and boiling points, and are denser than the group 1A and 2A metals, largely because the sharing of d, as well as s, electrons results in stronger bonding.

From left to right across the first transition series, melting points increase from 1541 °C for Sc to a maximum of 1910 °C for V in group 5B, then decrease to 420 °C for Zn (Table 21.1, **FIGURE 21.2**). The second- and third-series elements exhibit a similar maximum in melting point, but at group 6B rather than 5B: 2623 °C for Mo and 3422 °C for W, the metal with the highest melting point. The melting points increase as

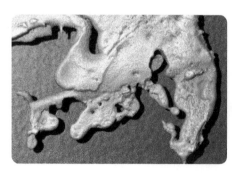

▲ The color of the Carmen Lucia Ruby is due to transitions of d electrons in Cr^{3+} ions.

▲ Native copper.

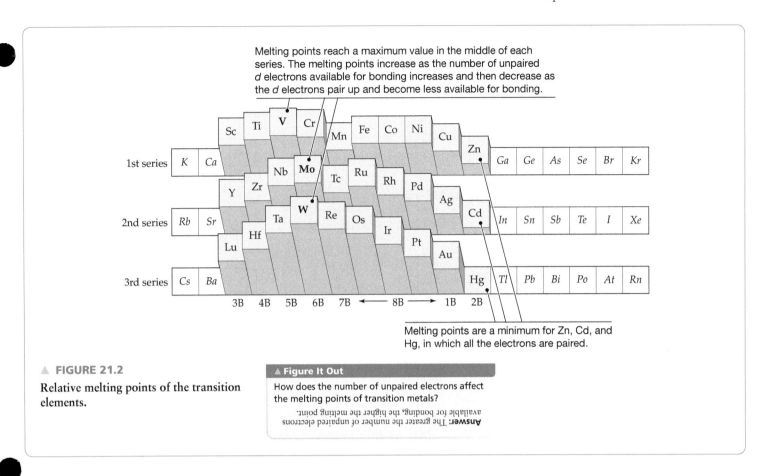

Melting points reach a maximum value in the middle of each series. The melting points increase as the number of unpaired d electrons available for bonding increases and then decrease as the d electrons pair up and become less available for bonding.

Melting points are a minimum for Zn, Cd, and Hg, in which all the electrons are paired.

▲ **FIGURE 21.2**

Relative melting points of the transition elements.

▲ **Figure It Out**

How does the number of unpaired electrons affect the melting points of transition metals?

Answer: The greater the number of unpaired electrons available for bonding, the higher the melting point.

the number of unpaired d electrons available for bonding increases and then decrease as the d electrons pair up and become less available for bonding. Zinc ($3d^{10}4s^2$), in which all the d and s electrons are paired, has a relatively low melting point (420 °C) and mercury ($4f^{14}\,5d^{10}\,6s^2$) is a liquid at room temperature (mp -39 °C).

Atomic Radii and Densities

Atomic radii are given in **FIGURE 21.3**. From left to right across a transition series, the atomic radii decrease, at first markedly and then more gradually after group 6B. Toward the end of each series, the radii increase again. The decrease in radii with increasing atomic number occurs because the added d electrons only partially **shield** the added nuclear charge (Section 5.9). As a result, the **effective nuclear charge Z_{eff}** increases. With increasing Z_{eff}, the electrons are more strongly attracted to the nucleus and atomic radii decrease. The upturn in radii toward the end of each series is probably due to more effective shielding and increasing electron–electron repulsion as double occupation of the d orbitals is completed. In contrast to the large variation in radii for main-group elements, all transition metal atoms have quite similar radii, which accounts for their ability to blend together in forming alloys such as brass (mostly copper and zinc).

The atomic radii of the second- and third-series transition elements from group 4B on are nearly identical, though we would expect an increase in size on adding an entire principal quantum shell of electrons. The smaller-than-expected sizes of the third-series atoms are associated with what is called the **lanthanide contraction**, the general decrease in atomic radii of the f-block lanthanide elements between the second and third transition series (**FIGURE 21.4**).

The lanthanide contraction is due to the increase in effective nuclear charge with increasing atomic number as the $4f$ subshell is filled. By the end of the lanthanides, the size *decrease* due to a larger Z_{eff} almost exactly compensates for the expected size *increase* due to an added quantum shell of electrons. Consequently, atoms of the third transition

REMEMBER . . .

The valence electrons are only weakly **shielded** by electrons in the same subshell, which are at the same distance from the nucleus (Section 5.9).

REMEMBER . . .

The nuclear charge actually felt by the electron is called **the effective nuclear charge, Z_{eff}** (Section 5.9).

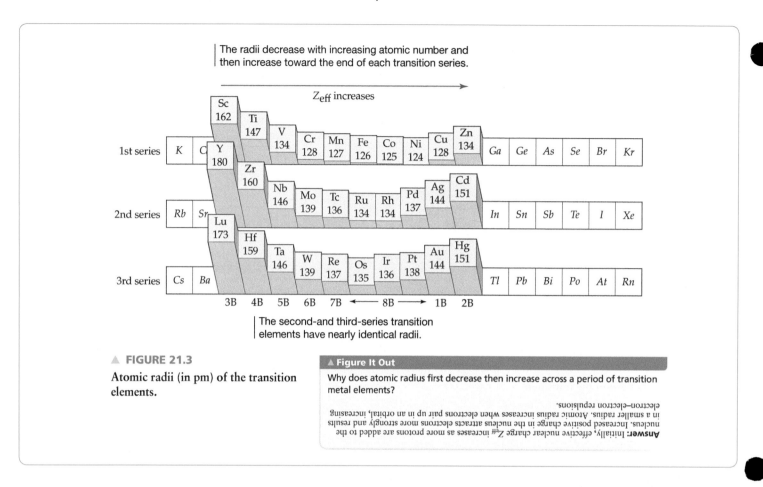

The radii decrease with increasing atomic number and then increase toward the end of each transition series.

Z_{eff} increases

The second-and third-series transition elements have nearly identical radii.

▲ **FIGURE 21.3**

Atomic radii (in pm) of the transition elements.

▲ **Figure It Out**

Why does atomic radius first decrease then increase across a period of transition metal elements?

Answer: Initially, effective nuclear charge Z_{eff} increases as more protons are added to the nucleus. Increased positive charge in the nucleus attracts electrons more strongly and results in a smaller radius. Atomic radius increases when electrons pair up in an orbital, increasing electron–electron repulsions.

▶ **FIGURE 21.4**

Atomic radii (in pm) of the lanthanide elements.

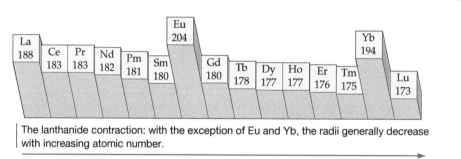

The lanthanide contraction: with the exception of Eu and Yb, the radii generally decrease with increasing atomic number.

$4f$ subshell fills; Z_{eff} increases

series have radii very similar to those of the second transition series (Figure 21.3). The irregularly large atomic radii in Figure 21.4 for Eu and Yb are related to their half-filled and completely filled $4f$ subshells, $[Xe]\,4f^7 6s^2$ for Eu and $[Xe]\,4f^{14} 6s^2$ for Yb, but a full explanation of their anomalous radii is beyond the scope of this discussion.

The densities of the transition metals are inversely related to their atomic radii (**FIGURE 21.5**). The densities initially increase from left to right across each transition series and then decrease toward the end of each series. Because the second- and third-series elements have nearly the same atomic volume, the much heavier third-series elements have unusually high densities: 22.6 g/cm^3 for osmium and iridium, the densest elements.

Ionization Energies and Oxidation Potentials

Ionization energies generally increase from left to right across each transition series, though there are some irregularities, as indicated in Table 21.1 for the atoms of the

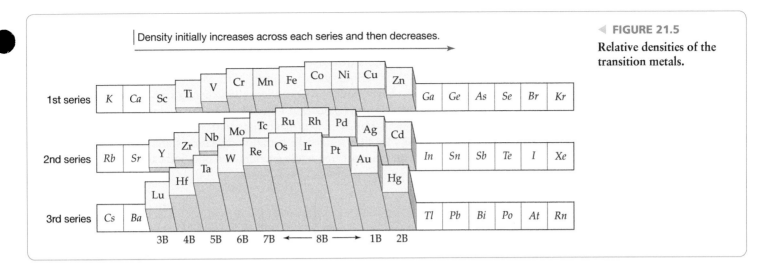

◀ FIGURE 21.5

Relative densities of the transition metals.

first transition series. The general trend correlates with an increase in effective nuclear charge and a decrease in atomic radius.

TABLE 21.2 lists standard potentials ($E°$) for oxidation of first-series transition metals. Note that these potentials are the negative of the corresponding standard reduction potentials that were given in Table 19.1. Except for copper, all the $E°$ values are positive, which means that the solid metal is oxidized to its aqueous cation more readily than H_2 gas is oxidized to $H^+(aq)$.

$$M(s) \longrightarrow M^{2+}(aq) + 2\ e^- \qquad E° > 0\ V \qquad \text{Product is } M^{3+} \text{ for } M = Sc$$
$$H_2(g) \longrightarrow 2\ H^+(aq) + 2\ e^- \qquad E° = 0\ V$$

In other words, the first-series metals, except for copper, are stronger reducing agents than H_2 gas and can therefore be oxidized by the H^+ ion in acids like HCl that lack an oxidizing anion:

$$M(s) + 2\ H^+(aq) \longrightarrow M^{2+}(aq) + H_2(g) \quad E° > 0\ V \text{ (except for } M = Cu)$$

Oxidation of copper requires a stronger oxidizing agent, such as HNO_3.

TABLE 21.2 Standard Potentials for Oxidation of First-Series Transition Metals

Oxidation Half-Reaction	$E°$ (V)	Oxidation Half-Reaction	$E°$ (V)
$Sc(s) \rightarrow Sc^{3+}(aq) + 3\ e^-$	2.08	$Fe(s) \rightarrow Fe^{2+}(aq) + 2\ e^-$	0.45
$Ti(s) \rightarrow Ti^{2+}(aq) + 2\ e^-$	1.63	$Co(s) \rightarrow Co^{2+}(aq) + 2e^-$	0.28
$V(s) \rightarrow V^{2+}(aq) + 2\ e^-$	1.18	$Ni(s) \rightarrow Ni^{2+}(aq) + 2\ e^-$	0.26
$Cr(s) \rightarrow Cr^{2+}(aq) + 2\ e^-$	0.91	$Cu(s) \rightarrow Cu^{2+}(aq) + 2\ e^-$	−0.34
$Mn(s) \rightarrow Mn^{2+}(aq) + 2\ e^-$	1.18	$Zn(s) \rightarrow Zn^{2+}(aq) + 2\ e^-$	0.76

The standard potential for the oxidation of a metal is a composite property that depends on $\Delta G°$ for sublimation of the metal, the ionization energies of the metal atom, and $\Delta G°$ for hydration of the metal ion:

$$M(s) \xrightarrow{\Delta G°_{subl}} M(g) \xrightarrow{E_i(-2e^-)} M^{2+}(g) \xrightarrow{\Delta G°_{hydr}} M^{2+}(aq)$$

Nevertheless, the general trend in the $E°$ values shown in Table 21.2 correlates with the general trend in the ionization energies in Table 21.1. The ease of oxidation of the metal decreases as the ionization energies increase across the transition series from Sc to Zn. (Only Mn and Zn deviate from the trend of decreasing $E°$ values.) Thus, the so-called *early transition metals,* those on the left side of the *d* block (Sc through Mn), are oxidized most easily and are the strongest reducing agents.

Go to eText

BIG IDEA Question 1

Which element in the second-series transition elements do you expect to be the easiest to oxidize: Mo, Y, or Zr?

21.3 OXIDATION STATES OF TRANSITION ELEMENTS

The transition elements differ from most main-group metals in that they exhibit a variety of oxidation states. Sodium, magnesium, and aluminum, for example, have a single oxidation state equal to their periodic group number (Na^+, Mg^{2+}, and Al^{3+}), but the transition elements frequently have oxidation states less than their group number. For example, manganese in group 7B shows oxidation states of +2 in $Mn^{2+}(aq)$, +3 in $Mn(OH)_3 (s)$, +4 in $MnO_2(s)$, +6 in $MnO_4^{2-}(aq)$ (manganate ion), and +7 in $MnO_4^-(aq)$ (permanganate ion). **FIGURE 21.6** summarizes the common oxidation states for elements of the first transition series, with the most frequently encountered ones indicated in red.

▶ Manganese has different oxidation states and different colors in these ions and solid compounds.

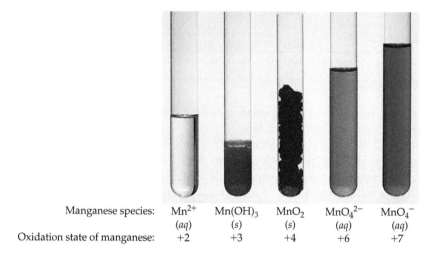

Manganese species:	Mn^{2+} (aq)	$Mn(OH)_3$ (s)	MnO_2 (s)	MnO_4^{2-} (aq)	MnO_4^- (aq)
Oxidation state of manganese:	+2	+3	+4	+6	+7

▶ **FIGURE 21.6**

Common oxidation states for first-series transition elements. The states encountered most frequently are shown in red. Most transition elements have more than one common oxidation state.

▶ **Figure It Out**

What is the highest oxidation state for Cr? Explain.

Answer: The highest oxidation state for Cr is +6 because there are 6 valence electrons in chromium.

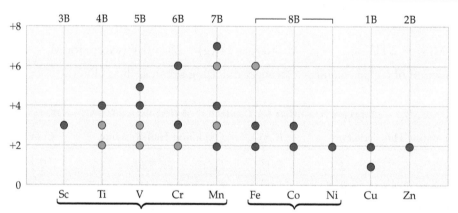

The highest oxidation state for the group 3B–7B metals is their periodic group number ...

... but the group 8B transition metals have a maximum oxidation state less than their group number.

All the first-series transition elements except scandium form a 2+ cation, corresponding to loss of the two $4s$ valence electrons. Because the $3d$ and $4s$ orbitals have similar energies, loss of a $3d$ electron is also possible, yielding 3+ cations such as $V^{3+}(aq)$, $Cr^{3+}(aq)$, and $Fe^{3+}(aq)$. Additional energy is required to remove the third electron, but this is more than compensated for by the larger (more negative) $\Delta G°$ of hydration of the more highly charged 3+ cation. Still higher oxidation states result from loss or sharing of additional d electrons. In their highest oxidation states, the transition elements are combined with the most electronegative elements, F and O. For example, vanadium forms $VF_5(l)$ and $V_2O_5(s)$, chromium forms CrO_4^{2-} (chromate ion) and $Cr_2O_7^{2-}$ (dichromate ion), and manganese forms MnO_4^-.

Note in Figure 21.6 that the highest oxidation state for the group 3B–7B metals is the group number, corresponding to loss or sharing of all the valence s and d electrons. For the group 8B transition metals, though, loss or sharing of all the valence electrons is energetically prohibitive because of the increasing value of Z_{eff}. Consequently, only lower oxidation states are accessible for these transition metals—for example, +6 in FeO_4^{2-} and +3 in Co^{3+}. Even these species have a great tendency to be reduced to still lower oxidation states. For example, the aqueous Co^{3+} ion is unstable because it oxidizes water to O_2 gas and is thereby reduced to Co^{2+}:

$$4\,Co^{3+}(aq) + 2\,H_2O(l) \longrightarrow 4\,Co^{2+}(aq) + O_2(g) + 4\,H^+(aq) \qquad E° = +0.58\ V$$

In general, ions that have the transition metal in a high oxidation state tend to be good oxidizing agents—for example, $Cr_2O_7^{2-}$, MnO_4^-, and FeO_4^{2-}. Conversely, early transition metal ions with the metal in a low oxidation state are good reducing agents—for example, V^{2+} and Cr^{2+}. Divalent ions of the later transition metals on the right side of the d block, such as Co^{2+}, Ni^{2+}, Cu^{2+}, and Zn^{2+}, are poor reducing agents because of the larger value of Z_{eff}. In fact, zinc has only one stable oxidation state (+2).

21.4 COORDINATION COMPOUNDS

A **coordination compound** is a compound in which a central metal ion or atom is attached to a group of surrounding molecules or ions by coordinate covalent bonds, where one atom contributes both electrons to the shared electron pair. A good example is the green hexaaquanickel(II) ion, $Ni(H_2O)_6^{2+}$, in which six H_2O molecules use lone pairs of electrons on oxygen atoms to bond to the nickel(II) ion:

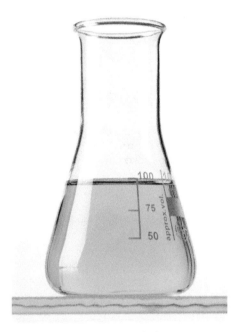

$$[Ni(H_2O)_6]^{2+}$$

The molecules or ions that surround the central metal ion in a coordination compound are called **ligands,** and the atoms that are attached directly to the metal are called **ligand donor atoms.** In $[Ni(H_2O)_6]^{2+}$ for example, the ligands are H_2O molecules and the ligand donor atoms are oxygens. The formation of a coordination compound is a **Lewis acid–base reaction** in which the ligands act as Lewis bases, or electron-pair donors, and the central metal ion behaves as a Lewis acid, an electron-pair acceptor (Section 16.15).

Many coordination compounds are salts, such as $[Ni(H_2O)_6]SO_4$ and $K_3[Fe(CN)_6]$, which contain a complex cation or anion along with enough ions of opposite charge to give a compound that is electrically neutral overall. To emphasize that the complex ion is a discrete structural unit, it is always enclosed in brackets in the formula of a salt. Thus, $[Ni(H_2O)_6]SO_4$ contains $[Ni(H_2O)_6]^{2+}$ cations and SO_4^{2-} anions. The term **metal complex,** or simply complex, refers to both complex ions, such as $[Ni(H_2O)_6]^{2+}$ and $[Fe(CN)_6]^{3-}$, and neutral molecules such as the anticancer drug cisplatin, $Pt(NH_3)_2Cl_2$, described in the *Inquiry.*

The number of ligand donor atoms that surround a central metal ion in a complex is called the **coordination number** of the metal. Thus, nickel(II) has a coordination number of 6 in $[Ni(H_2O)_6]^{2+}$ and platinum(II) has a coordination number of 4 in

Go to eText

BIG IDEA Question 2

Consider how effective nuclear charge, Z_{eff}, varies from left to right across the first transition series.

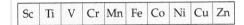

| Sc | Ti | V | Cr | Mn | Fe | Co | Ni | Cu | Zn |

Based on Z_{eff}, which M^{2+} ion should be the strongest reducing agent: Ti^{2+}, Cr^{2+}, Fe^{2+}, or Co^{2+}?

Go to eText

BIG IDEA Question 3

Based on Z_{eff}, which oxoanion should be the strongest oxidizing agent: CrO_4^{2-}, MnO_4^{2-}, or FeO_4^{2-}?

▲ The green hexaaquanickel(II) ion, $Ni(H_2O)_6^{2+}$ forms when nickel(II) sulfate is dissolved in water.

REMEMBER . . .
In a **Lewis acid–base reaction,** a coordinate covalent bond is formed when a Lewis acid accepts a share in a lone pair of electrons from a Lewis base, which acts as an electron-pair donor (Section 16.15).

$Pt(NH_3)_2Cl_2$. The most common coordination numbers are 4 and 6, but others are well known (**TABLE 21.3**). The coordination number of a metal ion in a particular complex depends on the metal ion's size, charge, and electron configuration, and on the size and shape of the ligands.

TABLE 21.3   Examples of Complexes with Various Coordination Numbers

Coordination Number	Complex
2	$[Ag(NH_3)_2]^+$, $[CuCl_2]^-$
3	$[HgI_3]^-$
4	$[Zn(NH_3)_4]^{2+}$, $[Ni(CN)_4]^{2-}$
5	$[Ni(CN)_5]^{3-}$, $Fe(CO)_5$
6	$[Cr(H_2O)_6]^{3+}$, $[Fe(CN)_6]^{3-}$
7	$[ZrF_7]^{3-}$
8	$[Mo(CN)_8]^{4-}$

Metal complexes have characteristic shapes that depend on the metal ion's coordination number. Two-coordinate complexes, such as $[Ag(NH_3)_2]^+$, are linear. Four-coordinate complexes are either tetrahedral or square planar; for example, $[Zn(NH_3)_4]^{2+}$ is tetrahedral, and $[Ni(CN)_4]^{2-}$ is square planar. Nearly all six-coordinate complexes are octahedral. The more common coordination geometries are illustrated in **FIGURE 21.7**. Coordination geometries were first deduced by the Swiss chemist Alfred Werner, who was awarded the 1913 Nobel Prize in Chemistry for his pioneering studies.

The charge on a metal complex equals the charge on the metal ion (its oxidation state) plus the sum of the charges on the ligands. Thus, if we know the charge on each ligand and the charge on the complex, we can easily find the oxidation state of the metal as shown in Worked Example 21.2.

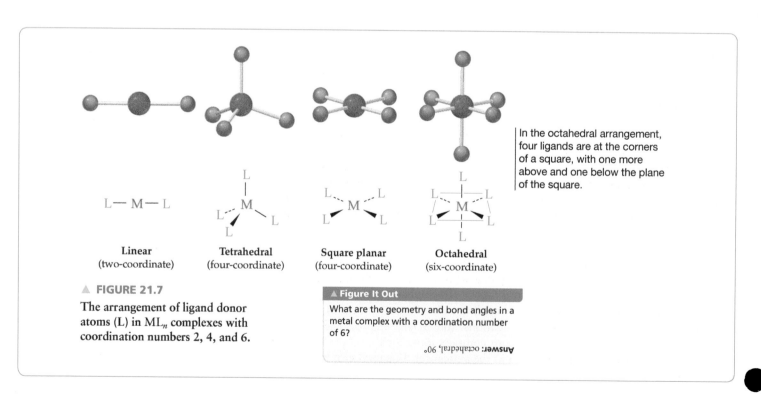

In the octahedral arrangement, four ligands are at the corners of a square, with one more above and one below the plane of the square.

L—M—L

Linear
(two-coordinate)

Tetrahedral
(four-coordinate)

Square planar
(four-coordinate)

Octahedral
(six-coordinate)

▲ **FIGURE 21.7**

The arrangement of ligand donor atoms (L) in ML_n complexes with coordination numbers 2, 4, and 6.

▲ **Figure It Out**

What are the geometry and bond angles in a metal complex with a coordination number of 6?

Answer: octahedral, 90°

WORKED EXAMPLE 21.2

Determining the Oxidation State of the Metal in a Coordination Compound

What is the oxidation state of platinum in the coordination compound $K[Pt(NH_3)Cl_5]$?

STRATEGY

Because the compound is electrically neutral overall and contains one K^+ cation per complex anion, the anion must be $[Pt(NH_3)Cl_5]^-$. Since ammonia is neutral and chloride has a charge of -1, the sum of the oxidation numbers is $+1 + n + 0 + (5)(-1) = 0$, where n is the oxidation number of Pt:

$$K^+ \quad Pt^{n+} \quad NH_3 \quad 5\ Cl^-$$

$$+1 + n + 0 + (5)(-1) = 0; \quad \text{so, } n = +4$$

SOLUTION

The oxidation state of platinum is $+4$.

▶ **PRACTICE 21.3** What is the oxidation state of iron in $Na_4[Fe(CN)_6]$?

▶ **APPLY 21.4** A chromium coordination complex with a charge of -1 contains two ammonia (NH_3) and four thiocyanate (SCN^-) ligands. What is the oxidation state of chromium?

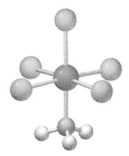

$[Pt(NH_3)Cl_5]^-$

21.5 LIGANDS

The structures of some typical ligands are shown in **FIGURE 21.8**. They can be classified as *monodentate* or *polydentate*, depending on the number of ligand donor atoms that bond to the metal. Ligands such as H_2O, NH_3, or Cl^- that bond using an electron pair of a single donor atom are called **monodentate ligands** (literally, "one-toothed" ligands). Those that bond through electron pairs on more than one donor atom are termed **polydentate ligands** ("many-toothed" ligands). For example, ethylenediamine ($NH_2CH_2CH_2NH_2$, abbreviated en) is a **bidentate ligand** because it bonds to a metal using the electron pair on each of its two nitrogen atoms. The **hexadentate ligand** ethylenediaminetetraacetate ion ($EDTA^{4-}$) bonds to a metal ion through electron pairs on six donor atoms (two N atoms and four O atoms).

Polydentate ligands are known as **chelating agents** (pronounced key-late-ing, from the Greek word *chele*, meaning "claw") because their multipoint attachment to a metal ion resembles the grasping of an object by the claws of a crab. For example, ethylenediamine holds a cobalt(III) ion with two claws, its two nitrogen donor atoms (**FIGURE 21.9**). The resulting five-membered ring consisting of the Co(III) ion, two N atoms, and two C atoms of the ligand is called a **chelate ring**. A complex such as $[Co(en)_3]^{3+}$ or $[Co(EDTA)]^-$ that contains one or more chelate rings is known as a metal **chelate**.

Because $EDTA^{4-}$ bonds to a metal ion through six donor atoms, it forms especially stable complexes and is often used to hold metal ions in solution. For example, in the treatment of lead poisoning, $EDTA^{4-}$ bonds to Pb^{2+}, which is then excreted by the kidneys as the soluble chelate $[Pb(EDTA)]^{2-}$. $EDTA^{4-}$ is commonly added to food products such as commercial salad dressings to complex any metal cations that might be present in trace amounts. The free metal ions might otherwise catalyze the oxidation of oils, thus causing the dressing to become rancid.

Naturally occurring chelating ligands are essential constituents of many important biomolecules. For example, the heme group in hemoglobin contains a planar,

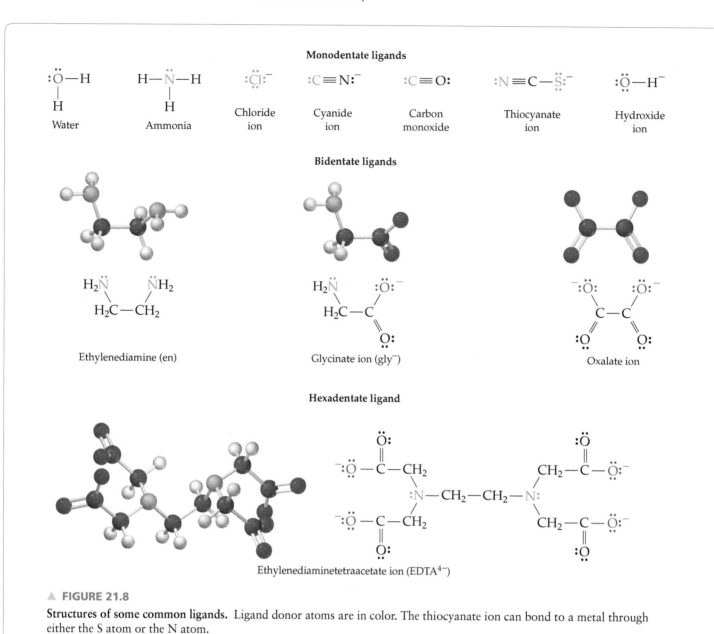

Monodentate ligands

Water

Ammonia

Chloride ion

Cyanide ion

Carbon monoxide

Thiocyanate ion

Hydroxide ion

Bidentate ligands

Ethylenediamine (en)

Glycinate ion (gly⁻)

Oxalate ion

Hexadentate ligand

Ethylenediaminetetraacetate ion (EDTA⁴⁻)

▲ **FIGURE 21.8**

Structures of some common ligands. Ligand donor atoms are in color. The thiocyanate ion can bond to a metal through either the S atom or the N atom.

BIG IDEA Question 4

Draw the structure of the coordination compound Cr(gly)₃, where gly⁻ is the glycinate ligand shown in Figure 21.8. (Hint: Use the abbreviation N⁻O for the glycinate ligand in your structure). What is the coordination number, geometry, ligand donor atom(s), and oxidation state of chromium?

tetradentate ligand that uses the lone pair of electrons on each of its four N atoms to bond to an iron(II) ion. The ligand in heme is a *porphyrin*, a derivative of the porphine molecule (**FIGURE 21.10a**) in which the porphine's peripheral H atoms are replaced by various substituent groups ($-CH_3$, $-CH=CH_2$, and $-CH_2CH_2CO_2^-$ in heme). Bonding of the porphyrin to the Fe(II) ion involves the prior loss of the two NH protons, which makes room for the Fe(II) to occupy the cavity between the four N atoms.

In hemoglobin, the heme is linked to a protein (globin) through an additional Fe–N bond, as shown in **FIGURE 21.10b**. In addition, the Fe(II) can bond to an O_2 molecule to give the six-coordinate, octahedral complex present in oxyhemoglobin. The three-dimensional shape of the protein part of the molecule makes possible the reversible binding of O_2.

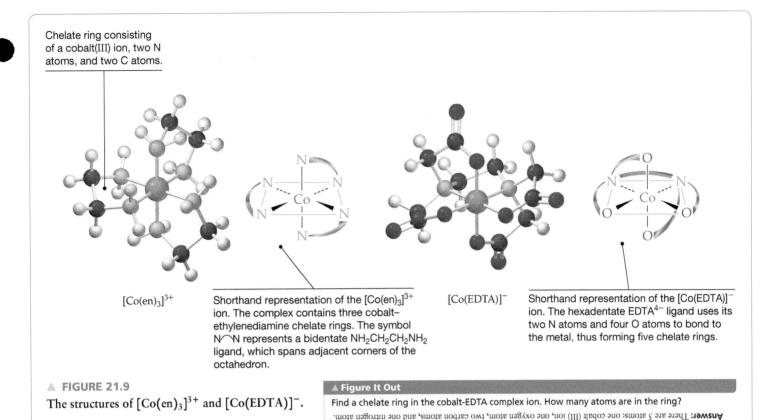

Chelate ring consisting of a cobalt(III) ion, two N atoms, and two C atoms.

$[Co(en)_3]^{3+}$

Shorthand representation of the $[Co(en)_3]^{3+}$ ion. The complex contains three cobalt–ethylenediamine chelate rings. The symbol N⌒N represents a bidentate $NH_2CH_2CH_2NH_2$ ligand, which spans adjacent corners of the octahedron.

$[Co(EDTA)]^-$

Shorthand representation of the $[Co(EDTA)]^-$ ion. The hexadentate $EDTA^{4-}$ ligand uses its two N atoms and four O atoms to bond to the metal, thus forming five chelate rings.

▲ **FIGURE 21.9**

The structures of $[Co(en)_3]^{3+}$ and $[Co(EDTA)]^-$.

▲ **Figure It Out**

Find a chelate ring in the cobalt-EDTA complex ion. How many atoms are in the ring?

Answer: There are 5 atoms: one cobalt (III) ion, one oxygen atom, two carbon atoms, and one nitrogen atom.

(a) Loss of the two NH protons gives a planar, tetradentate 2− ligand that can bond to a metal cation.

(b) The Fe(II) ion has a six-coordinate, octahedral environment, and the O_2 acts as a monodentate ligand.

Heme

Protein (globin)

▲ **FIGURE 21.10**

Chelating ligands in biomolecules. (a) The structure of the porphine molecule. The porphyrins are derivatives of porphine in which some of the peripheral H atoms are replaced by various substituent groups. (b) Schematic of the planar heme group, the attached protein chain, and the bound O_2 molecule in oxyhemoglobin.

▲ **Figure It Out**

What is the coordination number of iron in the heme complex shown here? What are the ligand donor atoms?

Answer: The coordination number of iron is 6. Fe forms coordinate covalent bonds with the ligand donor atoms nitrogen and oxygen.

21.6 NAMING COORDINATION COMPOUNDS

In the early days of coordination chemistry, coordination compounds were named after their discoverer or according to their color. Now, we use systematic names that specify the number of ligands of each particular type, the metal, and its oxidation state. Before listing the rules used to name coordination compounds, let's consider a few examples that will illustrate how to apply the rules:

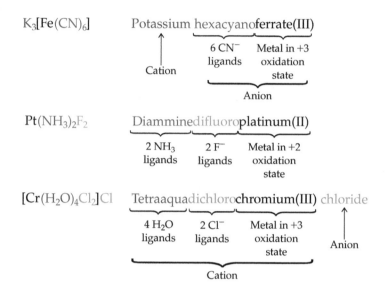

The following list summarizes the nomenclature rules recommended by the International Union of Pure and Applied Chemistry:

1. If the compound is a salt, name the cation first and then the anion, just as in naming simple salts. For example, $K_3[Fe(CN)_6]$ is potassium hexacyanoferrate(III).

2. In naming a complex ion or a neutral complex, name the ligands first and then the metal. The names of anionic ligands end in -o. As shown in **TABLE 21.4,** they are usually obtained by changing the anion endings -*ide* to -o and -*ate* to -*ato*. Neutral ligands are specified by their usual names, except for H_2O, NH_3, and CO, which are called aqua, ammine (note spelling), and carbonyl, respectively. The name of a complex is one word, with no space between the various ligand names and no space between the names of the last ligand and the metal.

TABLE 21.4   Names of Some Common Ligands

Anionic Ligand	Ligand Name	Neutral Ligand	Ligand Name
Bromide, Br^-	Bromo	Ammonia, NH_3	Ammine
Carbonate, CO_3^{2-}	Carbonato	Water, H_2O	Aqua
Chloride, Cl^-	Chloro	Carbon monoxide, CO	Carbonyl
Cyanide, CN^-	Cyano	Ethylenediamine, en	Ethylenediamine
Fluoride, F^-	Fluoro		
Glycinate, gly^-	Glycinato		
Hydroxide, OH^-	Hydroxo		
Oxalate, $C_2O_4^{2-}$	Oxalato		
Thiocyanate, SCN^-	Thiocyanato[*]		
	Isothiocyanato[†]		

[*]Ligand donor atom is S
[†]Ligand donor atom is N

3. If the complex contains more than one ligand of a particular type, indicate the number with the appropriate Greek prefix: *di-*, *tri-*, *tetra-*, *penta-*, *hexa-*, and so forth. The ligands are listed in alphabetical order, and the prefixes are ignored in determining the order. Thus, tetra*a*qua precedes di*c*hloro in the name for $[Cr(H_2O)_4Cl_2]Cl$: tetraaquadichlorochromium(III) chloride.

4. If the name of a ligand itself contains a Greek prefix—for example, ethylene*di*amine—put the ligand name in parentheses and use one of the following alternative prefixes to specify the number of ligands: *bis-* (2), *tris-* (3), *tetrakis-* (4), and so forth. Thus, the name of $[Co(en)_3]Cl_3$ is tris(ethylenediamine)cobalt(III) chloride.

5. Use a Roman numeral in parentheses immediately following the name of the metal to indicate the metal's oxidation state. As shown by the preceding examples, there is no space between the name of the metal and the parenthesis.

6. If a solid compound contains water molecules it is called a **hydrate**. $K_3[Fe(C_2O_4)_3] \cdot 3 H_2O$ is an example. Because the structures of hydrates are sometimes complex or unknown, a dot is used in the formula of a hydrate to specify the composition without indicating how the water is bound. Indicate the number of water molecules in a hydrate with the appropriate Greek prefix: *di-*, *tri-*, *tetra-*, *penta-*, *hexa-*, and so forth. The name of $K_3[Fe(C_2O_4)_3] \cdot 3 H_2O$ is potassium trioxalatoferrate(III) trihydrate.

7. In naming the metal, use the ending *-ate* if the metal is in an anionic complex. Thus, $[Fe(CN)_6]^{3-}$ is the hexacyanoferrate(III) anion. There are no simple rules for going from the name of the metal to the name of the metallate anion, partly because some of the anions have Latin names. Some common examples are given in **TABLE 21.5**.

▲ A sample of tris(ethylenediamine)cobalt(III) chloride, $[Co(en)_3]Cl_3$.

TABLE 21.5 Names of Some Common Metallate Anions

Metal	Anion Name	Metal	Anion Name
Aluminum	Aluminate	Iron	Ferrate
Chromium	Chromate	Manganese	Manganate
Cobalt	Cobaltate	Nickel	Nickelate
Copper	Cuprate	Platinum	Platinate
Gold	Aurate	Zinc	Zincate

▲ A sample of potassium trioxalatoferrate(III) trihydrate, $K_3[Fe(C_2O_4)_3] \cdot 3 H_2O$. $[Fe(C_2O_4)_3]^{3-}$, is the ion formed when Fe_2O_3 rust stains are dissolved in oxalic acid.

The rules for naming coordination compounds make it possible to go from a formula to the systematic name or from a systematic name to the appropriate formula. Worked Examples 21.3 and 21.4 provide some practice.

WORKED EXAMPLE 21.3

Determining the Name of a Coordination Compound from Its Formula

Name each of the following:

(a) $[Co(NH_3)_6]Cl_3$, prepared in 1798 by B. M. Tassaert and generally considered to be the first coordination compound

(b) $[Rh(NH_3)_5I]I_2$, a yellow compound obtained by heating $[Rh(NH_3)_5(H_2O)]I_3$ at 100 °C

(c) $Fe(CO)_5$, a highly toxic, volatile liquid

(d) $[PtCl_4]^{2-}$, a reddish orange salt used as a reagent in the preparation of other coordination complexes of platinum.

STRATEGY

First determine the oxidation state of the metal, as in Worked Example 21.2. Then apply the seven rules above to name the compound or ion.

SOLUTION

(a) Because the chloride ion has a charge of −1 and ammonia is neutral, the oxidation state of cobalt is +3. Use the prefix *hexa-* to indicate that the cation contains six NH_3 ligands,

continued on next page

and use a Roman numeral III in parentheses to indicate the oxidation state of cobalt. The name of $[Co(NH_3)_6]Cl_3$ is hexaamminecobalt(III) chloride.

(b) Because the iodide ion has a charge of -1, the complex cation is $[Rh(NH_3)_5I]^{2+}$ and rhodium has an oxidation state of $+3$. List the *ammine* ligands before the *iodo* ligand because ligands are listed in alphabetical order, and use the prefix *penta-* to indicate the presence of five NH_3 ligands. The name of $[Rh(NH_3)_5I]I_2$ is pentaammineiodorhodium(III) iodide.

(c) Because the carbonyl ligand is neutral, the oxidation state of iron is zero. The systematic name of $Fe(CO)_5$ is pentacarbonyliron(0), but the common name iron pentacarbonyl is often used.

(d) Because each chloride ligand (Cl^-) has a charge of -1 and because $[PtCl_4]^{2-}$ has an overall charge of -2, platinum must have an oxidation state of $+2$. Use the name platinate(II) for the metal because the complex is an anion. The name of $[PtCl_4]^{2-}$ is the tetrachloroplatinate(II) ion.

▶ **PRACTICE 21.5**   What is the name of $[Cu(NH_3)_2(H_2O)_2]Cl_2$?

▶ **CONCEPTUAL APPLY 21.6**   Name the coordination compound from its structural formula (brown = Co, green = Cl, blue = N, and white = H).

WORKED EXAMPLE 21.4

Determining the Formula of a Coordination Compound from Its Name

Write the formula for each of the following:

(a) Potassium tetracyanonickelate(II)
(b) Aquachlorobis(ethylenediamine)cobalt(III) chloride
(c) Diamminesilver(I) ion

STRATEGY

To find the formula and charge of the complex cation or anion, note the number of ligands of each type, their charge, and the oxidation state of the metal in the name of the compound or ion. If the name refers to an electrically neutral compound, balance the charge of the complex cation or anion with the appropriate number of ions of opposite charge.

SOLUTION

(a) The bonding of four CN^- ligands to Ni^{2+} gives an $[Ni(CN)_4]^{2-}$ anion, which must be balanced by two K^+ cations. The compound's formula is therefore $K_2[Ni(CN)_4]$. This compound is obtained when excess KCN is added to a solution of a nickel(II) salt.

(b) Because the complex cation contains one H_2O, one Cl^-, and two neutral ethylenediammine (en) ligands, and because the metal is Co^{3+}, the cation is $[Co(en)_2(H_2O)Cl]^{2+}$. The $2+$ charge of the cation must be balanced by two Cl^- anions, so the formula of the compound is $[Co(en)_2(H_2O)Cl]Cl_2$. The cation is the first product formed when $[Co(en)_2Cl_2]^+$ reacts with water.

(c) The diamminesilver(I) ion has the formula $[Ag(NH_3)_2]^+$. The Greek prefix *di* represents the number 2 and specifies

that there are two ammonia ligands. Ag has an oxidation state of $+1$ and ammonia is neutral. Therefore, the charge on the complex ion is $+1$.

▶ **PRACTICE 21.7**   What is the formula for diamminedihydroxoplatinum(IV) chloride?

▶ **CONCEPTUAL APPLY 21.8**   Write the formula and name for the coordination compound given the structural formula (brown = Pt, blue = N, green = Cl, white = H).

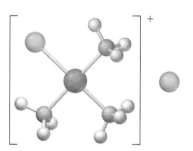

21.7 ISOMERS

One of the more interesting aspects of coordination chemistry is the existence of **isomers,** compounds that have the same formula but a different arrangement of their constituent atoms. Because their atoms are arranged differently, isomers are different compounds with different chemical reactivity and different physical properties such as color, solubility, and melting point. **FIGURE 21.11** shows a scheme for classifying some of the kinds of isomers in coordination chemistry. As we'll see in Chapter 23, isomers are also important in organic chemistry.

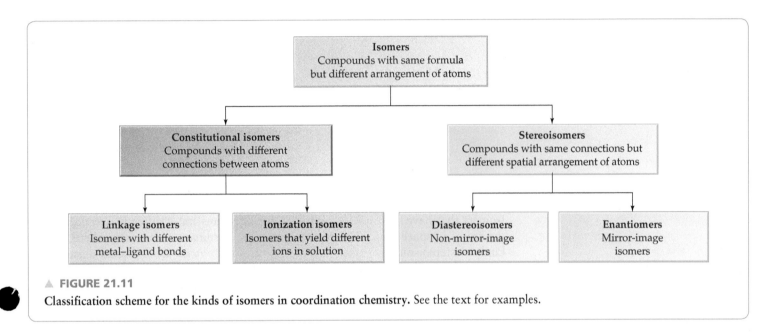

▲ **FIGURE 21.11**

Classification scheme for the kinds of isomers in coordination chemistry. See the text for examples.

Constitutional Isomers

Isomers that have different connections among their constituent atoms are called **constitutional isomers.** Of the various types of constitutional isomers, we'll discuss just two: *linkage isomers* and *ionization isomers.*

Linkage isomers arise when a ligand can bond to a metal through either of two different donor atoms. For example, the nitrite (NO_2^-) ion forms two different pentaamminecobalt(III) complexes: a yellow *nitro* complex that contains a Co—N bond, $[Co(NH_3)_5(NO_2)]^{2+}$, and a red *nitrito* complex that contains a Co—O bond, $[Co(NH_3)_5(ONO)]^{2+}$ (**FIGURE 21.12**). The ligand in the nitrito complex is written

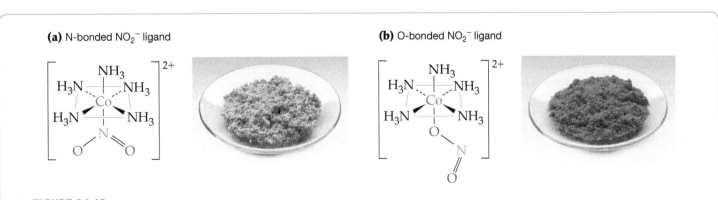

▲ **FIGURE 21.12**

Linkage isomers. Structures and samples of (a) the nitro complex $[Co(NH_3)_5(NO_2)]^{2+}$ and (b) the nitrito complex $[Co(NH_3)_5(ONO)]^{2+}$.

as ONO to emphasize that it's linked to the cobalt through an oxygen atom. The thiocyanate (SCN^-) ion is another ligand that gives linkage isomers because it can bond to a metal through either the sulfur atom to give a thiocyanato complex or the nitrogen atom to give an isothiocyanato complex.

Ionization isomers differ in the anion that is bonded to the metal ion. An example is the pair $[Co(NH_3)_5Br]SO_4$, a violet compound that has a Co—Br bond and a free sulfate anion, and $[Co(NH_3)_5SO_4]Br$, a red compound that has a Co—sulfate bond and a free bromide ion. Ionization isomers get their name because they yield different ions in solution.

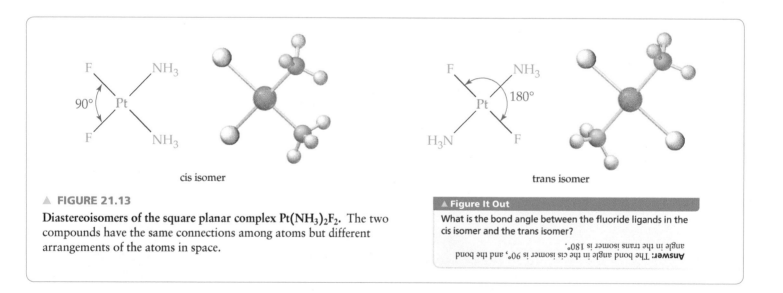

Stereoisomers

Isomers that have the same connections among atoms but a different arrangement of the atoms in space are called **stereoisomers**. In coordination chemistry, there are two kinds of stereoisomers: *diastereoisomers* and *enantiomers*. We'll discuss diastereoisomers in this section and enantiomers in Section 21.8.

Diastereoisomers, also called *geometric isomers,* are non-mirror-image stereoisomers. For example, in the square planar complex $Pt(NH_3)_2F_2$, the two Pt—F bonds can be oriented either adjacent at a 90° angle or opposite at a 180° angle, as shown in **FIGURE 21.13.** The isomer in which identical ligands occupy adjacent corners of the square is called the **cis isomer,** and the one in which identical ligands are across from each other is called the **trans isomer.** (The Latin word *cis* means "next to"; *trans* means "across.")

cis isomer

trans isomer

▲ **FIGURE 21.13**

Diastereoisomers of the square planar complex $Pt(NH_3)_2F_2$. The two compounds have the same connections among atoms but different arrangements of the atoms in space.

▲ **Figure It Out**

What is the bond angle between the fluoride ligands in the cis isomer and the trans isomer?

Answer: The bond angle in the cis isomer is 90°, and the bond angle in the trans isomer is 180°.

Cis and trans isomers are different compounds with different properties. Thus, *cis*-$Pt(NH_3)_2F_2$ is a polar molecule and is more soluble in water than *trans*-$Pt(NH_3)_2F_2$. The trans isomer is nonpolar because the two Pt—F and the two Pt—NH_3 bond dipoles point in opposite directions and therefore cancel. We'll see in the *Inquiry* at the end of the chapter that cis and trans isomers of $Pt(NH_3)_2Cl_2$ behave very differently in the body.

In general, square planar complexes of the type MA_2B_2 and MA_2BC—where M is a metal ion and A, B, and C are ligands—can exist as cis–trans isomers. No cis–trans isomers are possible, however, for four-coordinate tetrahedral complexes because all four corners of a tetrahedron are adjacent to one another.

Octahedral complexes of the type MA_4B_2 can also exist as diastereoisomers because the two B ligands can be on either adjacent or opposite corners of the octahedron. Examples are the violet compound *cis*-$[Co(NH_3)_4Cl_2]Cl$ and the green compound *trans*-$[Co(NH_3)_4Cl_2]Cl$. As **FIGURE 21.14** shows, there are several ways of drawing the cis and trans isomers because each complex can be rotated in space, changing the perspective but not the identity of the isomer.

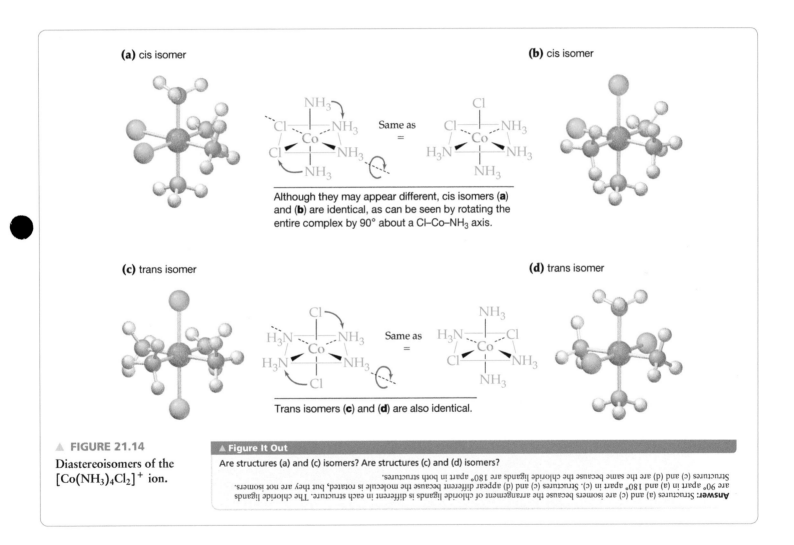

(a) cis isomer   **(b)** cis isomer

Although they may appear different, cis isomers **(a)** and **(b)** are identical, as can be seen by rotating the entire complex by 90° about a Cl–Co–NH₃ axis.

(c) trans isomer   **(d)** trans isomer

Trans isomers **(c)** and **(d)** are also identical.

▲ **FIGURE 21.14**
Diastereoisomers of the $[Co(NH_3)_4Cl_2]^+$ ion.

▲ **Figure It Out**
Are structures (a) and (c) isomers? Are structures (c) and (d) isomers?

Answer: Structures (a) and (c) are isomers because the arrangement of chloride ligands is different in each structure. The chloride ligands are 90° apart in (a) and 180° apart in (c). Structures (c) and (d) appear different because the molecule is rotated, but they are not isomers. Structures (c) and (d) are the same because the chloride ligands are 180° apart in both structures.

How can we be sure that there are only two diastereoisomers of the $[Co(NH_3)_4Cl_2]^+$ ion? The first Cl^- ligand can be located at any one of the six corners of the octahedron. Once one Cl^- is present, however, the five corners remaining are no longer equivalent. The second Cl^- can be located either on one of the four corners adjacent to the first Cl^-, which gives the cis isomer, or on the unique corner opposite to the first Cl^-, which gives the trans isomer. Thus, only two diastereoisomers are possible for complexes of the type MA_4B_2 (and MA_4BC).

CONCEPTUAL WORKED EXAMPLE 21.5

Identifying Diastereomers

Consider the following structural representations of the $[\text{Co}(\text{NH}_3)_4(\text{H}_2\text{O})\text{Cl}]^{2+}$ ion:

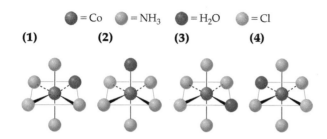

● = Co   ● = NH₃   ● = H₂O   ● = Cl

(1)   **(2)**   **(3)**   **(4)**

(a) Label the isomers as cis or trans.
(b) Which isomers are identical, and which are different?

STRATEGY

Diastereomers are compounds with the same chemical formula and connectivity of atoms, but different spatial arrangements. To identify diastereiomers evaluate structures to determine if the atoms have different spatial arrangements. Two structures may appear different, but they are not isomers if one can be rotated to turn it into to the other one. In octahedral complexes, cis and trans isomerism occurs when two ligands can be arranged with a bond angle of 90° (cis) or 180° (trans).

SOLUTION

(a) In the complex ion $[\text{Co}(\text{NH}_3)_4(\text{H}_2\text{O})\text{Cl}]^{2+}$ look at the location of the H_2O and Cl^- ligands. Isomers that have H_2O and Cl^- on adjacent corners of the octahedron with a bond angle of 90° are cis, whereas those that have H_2O and Cl^- on opposite corners with a bond angle of 180° are trans. Isomers (1) and (4) have H_2O and Cl^- on opposite corners of the octahedron and are therefore trans isomers. Isomers (2) and (3) have H_2O and Cl^- on adjacent corners of the octahedron and are therefore cis isomers.

(b) Because only two isomers are possible for complexes of the type MA_4BC, all the cis isomers are identical and all the trans isomers are identical. Structures (1) and (4) are identical, as can be seen by rotating (1) counterclockwise by 90° about the vertical (z) axis:

(1)   **(4)**

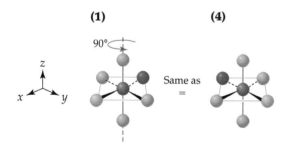

Same as
=

(2)   **(3)**

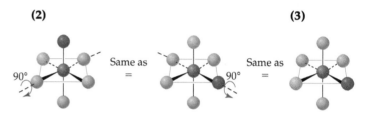

Same as   Same as
=   =

90°   90°

Structures (2) and (3) are identical, as can be seen by rotating (2) clockwise by 90° about the x axis and then counterclockwise by 90° about the y axis:

▶ **CONCEPTUAL PRACTICE 21.9** Which of the following $[\text{Pt}(\text{H}_2\text{O})_2\text{Cl}_3\text{Br}]$ structures are identical, and which are different?

● = Pt   ● = H₂O   ● = Cl   ● = Br

(1)   **(2)**   **(3)**   **(4)**

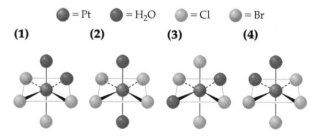

▶ **CONCEPTUAL APPLY 21.10** Which of the following $[\text{Co}(\text{en})(\text{NH}_3)_3\text{Cl}]^{2+}$ structures are identical, and which are different?

● = Co   ◠ = NH₂CH₂CH₂NH₂   ● = NH₃   ● = Cl

(1)   **(2)**   **(3)**   **(4)**

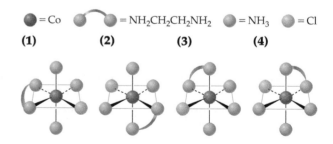

WORKED EXAMPLE 21.6

Drawing Diastereomers for Square Planar and Octahedral Complexes

Platinum(II) forms square planar complexes, and platinum(IV) gives octahedral complexes. How many diastereoisomers are possible for each of the following complexes? Draw their structures and identify cis and trans isomers.

(a) $[Pt(NH_3)_3Cl]^+$ (b) $[Pt(NH_3)Cl_5]^-$ (c) $Pt(NH_3)_2Cl(NO_2)$
(d) $[Pt(NH_3)_4ClBr]^{2+}$ (e) $Co(NH_3)_3Cl_3$

STRATEGY

Diastereomers are isomers with the same connectivity of atoms, but different spatial arrangements. To identify diastereomers, draw a representation of the molecule and change positions of atoms to create a different spatial arrangement. Two structures may appear different, but they are not isomers if one can be merely rotated to obtain the conformation of the other one. In square planar and octahedral complexes, cis and trans isomerism occurs when two ligands can be arranged with a bond angle of 90° (cis) or 180° (trans). Cis and trans isomers are not possible when only one ligand differs from the others, as in complexes of the type MA_3B and MA_5B.

SOLUTION

(a) No isomers are possible for $[Pt(NH_3)_3Cl]^+$, a square planar complex of the type MA_3B.

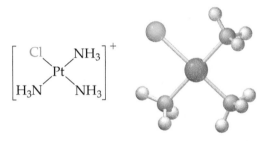

(b) No isomers are possible for $[Pt(NH_3)Cl_5]^-$, an octahedral complex of the type MA_5B.

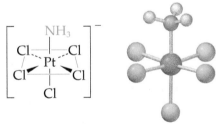

(c) Cis and trans isomers are possible for $Pt(NH_3)_2Cl(NO_2)$, a square planar complex of the type MA_2BC. The Cl^- and NO_2^- ligands can be on either adjacent (90° bond angle) or opposite corners of the square (180° bond angle)

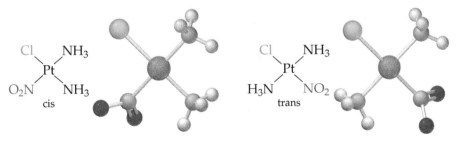

(d) Cis and trans isomers are possible for $[Pt(NH_3)_4ClBr]^{2+}$, an octahedral complex of the type MA_4BC. The Cl^- and Br^- ligands can be on either adjacent (90° bond angle) or opposite corners of the octahedron (180° bond angle).

continued on next page

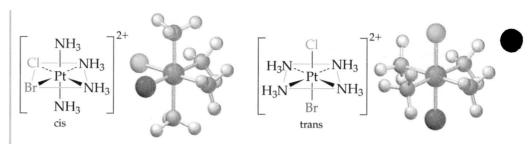

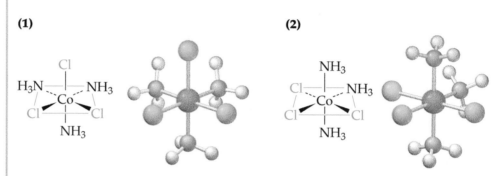

(e) Two diastereoisomers are possible for an octahedral complex of the type MA_3B_3. Isomer (1) shown below has the three Cl^- ligands in adjacent positions on one triangular face of the octahedron; isomer (2) has all three Cl^- ligands in a plane that contains the Co(III) ion. In isomer (1), all three Cl—Co—Cl bond angles are 90°, whereas in isomer (2) two Cl—Co—Cl angles are 90° and the third is 180°.

(1)   (2)

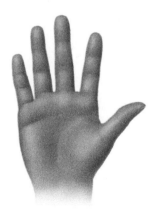

▶ **PRACTICE 21.11** How many diastereoisomers are possible for the complex ion $[Cr(en)_2Cl_2]^+$? Draw the structure of each diastereoisomer.

▶ **APPLY 21.12** How many diastereoisomers are possible for the following complexes? Draw the structure of each diastereoisomer.

(a) $[CoCl_2Br_2]^{2-}$ (tetrahedral)   (b) $Pt(en)Cl_2$ (square planar)
(c) $Pt(NH_3)_2(SCN)_2$ (square planar)   (d) $Co(NH_3)_3(NO_2)_3$

Symmetry plane

An achiral object like the coffee mug has a symmetry plane passing through it, making the two halves mirror images.

A chiral object like the hand has no symmetry plane because the two halves of the hand are not mirror images.

▲ **FIGURE 21.15**

The meaning of a symmetry plane.

▲ **Figure It Out**

Which of the following objects are chiral? chair, foot, corkscrew, football

Answer: Objects such as a corkscrew and a foot do not have a symmetry plane and are chiral. Objects such as a chair and football have a symmetry plane and are achiral.

21.8 ENANTIOMERS AND MOLECULAR HANDEDNESS

Diastereoisomers are relatively easy to distinguish because the various bonds in the cis and trans isomers point in different directions. *Enantiomers,* however, are stereoisomers that differ in a more subtle way. **Enantiomers** are molecules or ions that are nonidentical mirror images of each other and differ because of their handedness.

Handedness affects almost everything we do. We all know that a softball glove for the left hand will not fit on the right hand. The reason for these difficulties is that our hands aren't identical—they're mirror images. When you hold your left hand up to a mirror, the image you see looks like your right hand. (Try it.)

Not all objects are handed. There's no such thing as a right-handed tennis ball or a left-handed coffee mug. When a tennis ball or a coffee mug is held up to a mirror, the image reflected is identical to the ball or mug itself. Objects that do have a handedness to them are said to be **chiral** (pronounced ky-ral, from the Greek *cheir,* meaning "hand"), and objects that lack handedness are said to be nonchiral, or **achiral.**

Why is it that some objects are chiral but others aren't? In general, an object is not chiral if, like the coffee mug, it has a **symmetry plane** cutting through its middle so that one half of the object is a mirror image of the other half. If you were to cut the mug in half, one half of the mug would be the mirror image of the other half. A hand, however, has no symmetry plane and is therefore chiral. If you were to cut a hand in two, one "half" of the hand would not be a mirror image of the other half (**FIGURE 21.15**).

Just as certain objects like a hand are chiral, certain molecules and ions are also chiral. Consider, for example, the tris(ethylenediamine)cobalt(III) ion, $[Co(en)_3]^{3+}$, an ion with the shape of a three-bladed propeller (**FIGURE 21.16**). The $[Co(en)_3]^{3+}$ cation has no symmetry plane because its two halves aren't mirror images. Thus, $[Co(en)_3]^{3+}$ is chiral and can exist in two nonidentical mirror-image forms. When viewed as drawn in Figure 21.16, one enantiomer is "right-handed" because the three ethylenediamine ligands appear to spiral in a clockwise direction and the other enantiomer is "left-handed" because the ethylenediamine ligands spiral in a counterclockwise direction. By contrast, the analogous ammonia complex $[Co(NH_3)_6]^{3+}$ is achiral because, like a coffee mug, it has a symmetry plane. (Actually, $[Co(NH_3)_6]^{3+}$ has several symmetry planes, though only one is shown in Figure 21.16.) Thus, $[Co(NH_3)_6]^{3+}$ exists in a single form and does not have enantiomers.

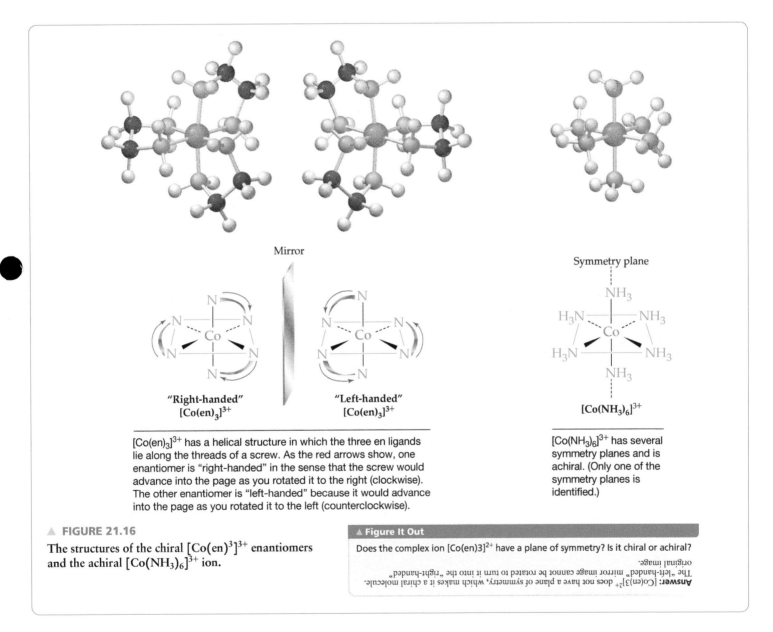

Mirror

"Right-handed"
$[Co(en)_3]^{3+}$

"Left-handed"
$[Co(en)_3]^{3+}$

Symmetry plane

$[Co(NH_3)_6]^{3+}$

$[Co(en)_3]^{3+}$ has a helical structure in which the three en ligands lie along the threads of a screw. As the red arrows show, one enantiomer is "right-handed" in the sense that the screw would advance into the page as you rotated it to the right (clockwise). The other enantiomer is "left-handed" because it would advance into the page as you rotated it to the left (counterclockwise).

$[Co(NH_3)_6]^{3+}$ has several symmetry planes and is achiral. (Only one of the symmetry planes is identified.)

▲ **FIGURE 21.16**

The structures of the chiral $[Co(en)^3]^{3+}$ enantiomers and the achiral $[Co(NH_3)_6]^{3+}$ ion.

▲ **Figure It Out**

Does the complex ion [Co(en)3]²⁺ have a plane of symmetry? Is it chiral or achiral?

Answer: [Co(en)3]²⁺ does not have a plane of symmetry, which makes it a chiral molecule. The "left-handed" mirror image cannot be rotated to turn it into the "right-handed" original image.

Enantiomers have identical properties except for their reactions with other chiral substances and their effect on *plane-polarized light*. In ordinary light, the electric oscillations of the light wave occur in all planes parallel to the direction in which the light is traveling, but in plane-polarized light, the electric oscillations are restricted to a single plane, as shown in **FIGURE 21.17**. Plane-polarized light is obtained by passing ordinary light through a polarizing filter, like that found in certain kinds of sunglasses. If the

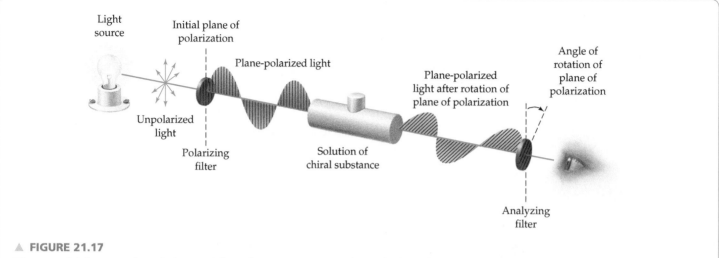

▲ **FIGURE 21.17**

The essential features of a polarimeter. The polarimeter measures the angle through which the plane of plane-polarized light is rotated when the light is passed through a solution of a chiral substance.

plane-polarized light is then passed through a solution of one enantiomer, the plane of polarization is rotated, either to the right (clockwise) or to the left (counterclockwise). If the light is passed through a solution of the other enantiomer, its plane of polarization is rotated through an equal angle, but in the opposite direction. Enantiomers are sometimes called *optical isomers* because of their effect on plane-polarized light.

Enantiomers are labeled $(+)$ or $(-)$, depending on the direction of rotation of the plane of polarization. For example, the isomer of $[Co(en)_3]^{3+}$ that rotates the plane of polarization to the right is labeled $(+) - [Co(en)_3]^{3+}$, and the isomer that rotates the plane of polarization to the left is designated $(-) - [Co(en)_3]^{3+}$. A 1:1 mixture of the $(+)$ and $(-)$ isomers, called a **racemic mixture**, produces no net optical rotation because the rotations produced by the individual enantiomers exactly cancel.

WORKED EXAMPLE 21.7   Go to eText

Drawing Diastereoisomers and Enantiomers

Draw the structures of all possible diastereoisomers and enantiomers of $[Co(en)_2Cl_2]^+$.

STRATEGY

Because of the relatively short distance between the two N atoms, ethylenediamine always spans adjacent corners of an octahedron (see Figure 21.8). The Cl^- ligands, however, can be on either adjacent or opposite corners. Therefore, there are two diastereoisomers, cis and trans. Because the trans isomer has several symmetry planes—one cuts through the Co and the en ligands—it is achiral and has no enantiomers. The cis isomer, however, is chiral and exists as a pair of enantiomers that are nonidentical mirror images.

SOLUTION

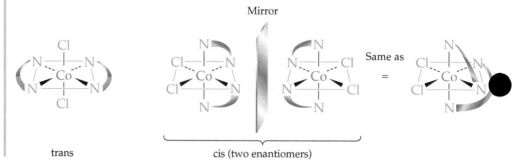

You can see that the cis enantiomer on the right is not the same as the one on the left if you rotate it by 180° about the vertical N—Co—N axis.

▶ **PRACTICE 21.13**   Is the complex ion $[Rh(en)Cl_2Br_2]^-$ chiral? If so, draw the structure of both enantiomers.

▶ **CONCEPTUAL APPLY 21.14**   Consider the following ethylenediamine complexes of rhodium:

(1)   **(2)**   **(3)**

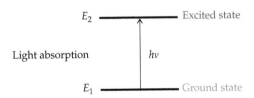

(a) Which complexes are chiral, and which are achiral?
(b) Draw the enantiomer of each chiral complex.

All **WORKED EXAMPLES** with this icon [Go to eText] have an interactive video in the eText.

21.9 COLOR OF TRANSITION METAL COMPLEXES

Most transition metal complexes have beautiful colors that depend on the identity of the metal and the ligands. The color of an aqua complex, for example, depends on the metal: $[Co(H_2O)_6]^{2+}$ is pink, $[Ni(H_2O)_6]^{2+}$ is green, $[Cu(H_2O)_6]^{2+}$ is blue, but $[Zn(H_2O)_6]^{2+}$ is colorless (**FIGURE 21.18**). If we keep the metal constant but vary the ligand, the color changes. For example, $[Ni(H_2O)_6]^{2+}$ is green, $[Ni(NH_3)_6]^{2+}$ is blue, and $[Ni(en)_3]^{2+}$ is violet.

How can we account for the color of transition metal complexes? Let's begin by recalling that white light consists of a continuous spectrum of wavelengths corresponding to different colors (Section 5.1). When white light strikes a colored substance, some wavelengths are transmitted while others are absorbed. Just as atoms can absorb light by undergoing electronic transitions between atomic energy levels, thereby giving rise to **atomic spectra**, a metal complex can absorb light by undergoing an electronic transition from its lowest (ground) energy state to a higher (excited) energy state.

$[Ni(H_2O)_6]^{2+}$ $[Cu(H_2O)_6]^{2+}$

$[Co(H_2O)_6]^{2+}$ $[Zn(H_2O)_6]^{2+}$

Different metals, but the same ligand

▲ **FIGURE 21.18**

Aqueous solutions of the nitrate salts of cobalt(II), nickel(II), copper(II), and zinc(II).

E_2 ———————— Excited state

Light absorption $h\nu$

E_1 ———————— Ground state

REMEMBER . . .
Atomic line spectra result when an electron moves between two states having different energy (Section 5.3).

REMEMBER . . .
To cause an electronic transition between two energy states, the energy of the photon must match the energy difference between the two states (Section 5.4).

The wavelength λ of the light absorbed by a metal complex depends on the **energy separation $\Delta E = E_2 - E_1$ between the two states**, as given by the equation

Go to
eText

BIG IDEA Question 5

When $NiSO_4$ dissolves in water, the solution appears green due to the formation of the complex ion $[Ni(H_2O)_6]^{2+}$. Which wavelength is absorbed most strongly in the absorption spectrum for this complex ion: 450 nm, 520 nm, 570 nm, or 660 nm?

$\Delta E = h\nu = hc/\lambda$ (Section 5.4), where h is Planck's constant, ν is the frequency of the light, and c is the speed of light:

$$\Delta E = E_2 - E_1 = h\nu = \frac{hc}{\lambda} \quad \text{or} \quad \lambda = \frac{hc}{\Delta E}$$

The measure of the amount of light absorbed by a substance is called the *absorbance,* and a plot of absorbance versus wavelength is called an **absorption spectrum.** For example, the absorption spectrum of the red-violet $[Ti(H_2O)_6]^{3+}$ ion has a broad absorption band at about 500 nm, a wavelength in the blue-green part of the visible spectrum (**FIGURE 21.19**). In general, the color that we see is complementary to the color absorbed (**FIGURE 21.20**).

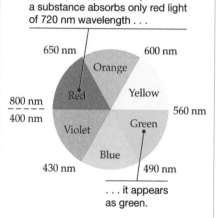

Observed and absorbed colors are generally complementary. Thus, if a substance absorbs only red light of 720 nm wavelength . . .

650 nm · Orange · 600 nm

800 nm / 400 nm · Red · Yellow · 560 nm

Violet · Green

Blue

430 nm · 490 nm

. . . it appears as green.

▲ **FIGURE 21.20**

An artist's color wheel. The observed color of a substance is complementary to the color of the light absorbed and appears on the opposite side of the wheel.

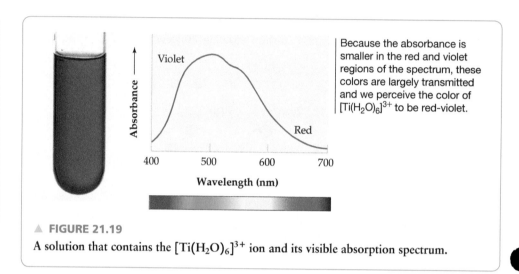

Violet

Because the absorbance is smaller in the red and violet regions of the spectrum, these colors are largely transmitted and we perceive the color of $[Ti(H_2O)_6]^{3+}$ to be red-violet.

Red

400 · 500 · 600 · 700

Wavelength (nm)

▲ **FIGURE 21.19**

A solution that contains the $[Ti(H_2O)_6]^{3+}$ ion and its visible absorption spectrum.

21.10 CRYSTAL FIELD THEORY

We saw in Section 21.9 that transition metal complexes are often colored solids, which produce colored solutions when dissolved in water. These complexes also have interesting and varied magnetic properties. To explain these properties, we turn to the **crystal field theory,** a model that views the bonding in complexes as arising from electrostatic interactions and considers the effect of the ligand charges on the energies of the metal ion d orbitals. This model was first applied to transition metal ions in ionic crystals—hence the name crystal field theory—but it also applies to metal complexes where the "crystal field" is the electric field due to the charges or dipoles of the ligands.

Octahedral Complexes

Let's first consider an octahedral complex such as $[TiF_6]^{3-}$ (**FIGURE 21.21**). According to the crystal field theory, the bonding is ionic and involves electrostatic attraction between the positively charged Ti^{3+} ion and the negatively charged F^- ligands. Of course, the F^- ligands repel one another, which is why they adopt the geometry (octahedral) that locates them as far apart from one another as possible. Because the metal–ligand attractions are greater than the ligand–ligand repulsions, the complex is more stable than the separated ions, thus accounting for the bonding.

Crystal field theory is different from other bonding models such as valence bond theory (Section 8.2) and molecular orbital theory (Section 8.7). The valence bond model describes a bond as the overlap of atomic or hybrid orbitals and will be applied to transition metal complexes in Section 21.11. The molecular orbital model describes the probability of finding an electron within a given region of space in a molecule, called a molecular orbital, which was formed by the mathematical mixing of atomic

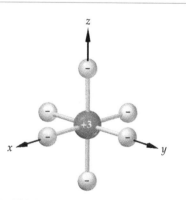

▲ **FIGURE 21.21**

Crystal field model of the octahedral $[TiF_6]^{3-}$ **complex.** The metal ion and ligands are regarded as charged particles held together by electrostatic attraction. The ligands lie along the $\pm x$, $\pm y$, and $\pm z$ directions.

orbitals. In crystal field theory, there are no covalent bonds, no shared electrons, and no hybrid orbitals—just electrostatic interactions within an array of ions. In complexes that contain neutral dipolar ligands, such as H_2O or NH_3, the electrostatic interactions are of the ion–dipole type (Section 8.6). For example, in $[Ti(H_2O)_6]^{3+}$, the Ti^{3+} ion attracts the negative end of the water dipoles.

To explain why complexes are colored, we need to look at the effect of the ligand charges on the energies of the d orbitals. Recall that four of the *d* orbitals are shaped like a cloverleaf, while the fifth (d_{z^2}) is shaped like a dumbbell inside a donut (Section 5.7). **FIGURE 21.22** shows the spatial orientation of the d orbitals with respect to an octahedral array of charged ligands located along the x, y, and z coordinate axes.

REMEMBER . . .
Although the *d* orbitals have different shapes, all five *d* orbitals have the same energy in a free metal ion (Section 5.7).

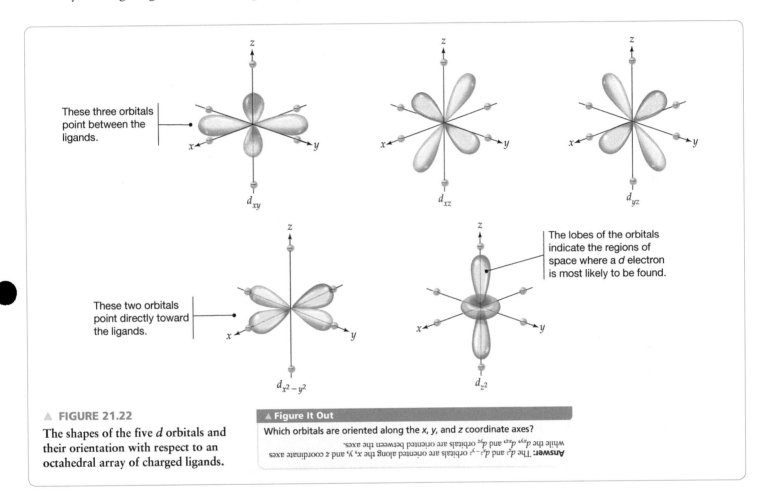

These three orbitals point between the ligands.

d_{xy} d_{xz} d_{yz}

These two orbitals point directly toward the ligands.

The lobes of the orbitals indicate the regions of space where a *d* electron is most likely to be found.

$d_{x^2-y^2}$ d_{z^2}

▲ **FIGURE 21.22**

The shapes of the five *d* orbitals and their orientation with respect to an octahedral array of charged ligands.

▲ **Figure It Out**

Which orbitals are oriented along the x, y, and z coordinate axes?

Answer: The d_{z^2} and $d_{x^2-y^2}$ orbitals are oriented along the x, y, and z coordinate axes while the d_{xy}, d_{xz}, and d_{yz} orbitals are oriented between the axes.

Because the d electrons are negatively charged, they are repelled by the negatively charged ligands. Thus, their orbital energies are higher in the complex than in the free metal ion. But not all the d orbitals are raised in energy by the same amount. As shown in **FIGURE 21.23**, the d_{z^2} and $d_{x^2-y^2}$ orbitals, which point directly at the ligands, are raised in energy to a greater extent than the d_{xy}, d_{xz}, and d_{yz} orbitals, which point between the ligands. This energy splitting between the two sets of d orbitals is called the **crystal field splitting** and is represented by the Greek letter Δ.

In general, the crystal field splitting energy Δ corresponds to wavelengths of light in the visible region of the spectrum, and the colors of complexes can therefore be attributed to electronic transitions between the lower- and higher-energy sets of d orbitals. Consider, for example, $[Ti(H_2O)_6]^{3+}$, a complex that contains a single d electron (Ti^{3+} has the electron configuration $[Ar]\ 3d^1$). In the ground-energy state, the d electron occupies one of the lower-energy orbitals—xy, xz, or yz (from now on we'll denote the d orbitals by their subscripts). When $[Ti(H_2O)_6]^{3+}$ absorbs blue-green light

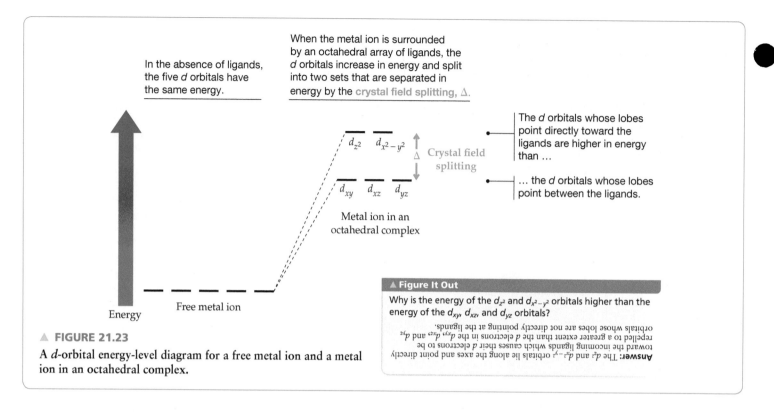

In the absence of ligands, the five *d* orbitals have the same energy.

When the metal ion is surrounded by an octahedral array of ligands, the *d* orbitals increase in energy and split into two sets that are separated in energy by the crystal field splitting, Δ.

d_{z^2} $d_{x^2-y^2}$ Δ Crystal field splitting

d_{xy} d_{xz} d_{yz}

Metal ion in an octahedral complex

The *d* orbitals whose lobes point directly toward the ligands are higher in energy than …

… the *d* orbitals whose lobes point between the ligands.

Free metal ion

Energy

▲ **FIGURE 21.23**

A *d*-orbital energy-level diagram for a free metal ion and a metal ion in an octahedral complex.

▲ **Figure It Out**

Why is the energy of the d_{z^2} and $d_{x^2-y^2}$ orbitals higher than the energy of the d_{xy}, d_{xz}, and d_{yz} orbitals?

Answer: The d_{z^2} and $d_{x^2-y^2}$ orbitals lie along the axes and point directly toward the incoming ligands which causes their *d* electrons to be repelled to a greater extent than the *d* electrons in the d_{xy}, d_{xz}, and d_{yz} orbitals whose lobes are not directly pointing at the ligands.

with a wavelength of about 500 nm, the absorbed energy promotes the *d* electron to one of the higher-energy orbitals, z^2 or $x^2 - y^2$:

z^2 x^2-y^2 Δ $\xrightarrow[\text{Light absorption}]{hv}$ z^2 x^2-y^2 Δ

Energy xy xz yz xy xz yz

We can calculate the value of Δ from the wavelength of the absorbed light.

> **Crystal field splitting energy** $\Delta = hv = \dfrac{hc}{\lambda}$

For $[\text{Ti(H}_2\text{O})_6]^{3+}$, which absorbs at about 500 nm, the value of Δ is given by

$$\Delta = \frac{hc}{\lambda} = \frac{(6.626 \times 10^{-34}\,\text{J}\cdot\text{s})(3.00 \times 10^{8}\,\text{m/s})}{500 \times 10^{-9}\,\text{m}} = 4.0 \times 10^{-19}\,\text{J}$$

This is the energy needed to excite a single $[\text{Ti(H}_2\text{O})_6]^{3+}$ ion. To express Δ on a per-mole basis, we multiply by Avogadro's number:

$$\Delta = \left(\frac{4.0 \times 10^{-19}\,\text{J}}{\text{ion}}\right)\left(\frac{6.02 \times 10^{23}\,\text{ions}}{\text{mol}}\right) = 2.4 \times 10^{5}\,\text{J/mol} = 240\,\text{kJ/mol}$$

The absorption spectra of different complexes indicate that the size of the crystal field splitting energy depends on the nature of the ligands. For example, Δ for Ni^{2+} ($[\text{Ar}]\,3d^8$) complexes increases as the ligand varies from H_2O to NH_3 to ethylenediamine (en). Accordingly, the electronic transitions shift to higher energy (shorter wavelength) as the ligand varies from H_2O to NH_3 to en, thus accounting for the observed variation in color. **FIGURE 21.24** shows the colors of the different nickel complexes in aqueous solution. $[Ni(H_2O)_6]^{2+}$ is a green solution because Δ is small and low-energy red light (~ 675 nm) is absorbed. $[Ni(NH_3)_6]^{2+}$ is a blue solution because Δ is larger and higher-energy orange light (~ 625 nm) is absorbed. $[Ni(en)_3]^{2+}$ is a violet solution because Δ is even larger and higher-energy yellow light (~ 580 nm) is absorbed.

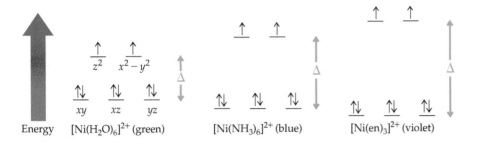

In general, the crystal field splitting increases as the ligand varies in the following order, known as the **spectrochemical series.**

Spectrochemical Series

Weak-field ligands $I^- < Br^- < Cl^- < F^- < H_2O < NH_3 < en < CN^-$ Strong-field ligands

Increasing Δ

Ligands such as halides and H_2O, which give a relatively small value of Δ, are called **weak-field ligands**. Ligands such as NH_3, en, and CN^-, which produce a relatively large value of Δ, are known as **strong-field ligands**. Different metal ions have different values of Δ, which explains why their complexes with the same ligand have different colors (Figure 21.18). Because d^0 ions such as Ti^{4+}, d^{10} ions such as Zn^{2+}, and main-group ions don't have partially filled d subshells, they can't undergo d–d electronic transitions and most of their compounds are therefore colorless.

Crystal field theory accounts for the magnetic properties of complexes as well as their color. Complexes with weak field ligands, such as $[CoF_6]^{3-}$, have small crystal field splitting energies and electrons occupy d orbitals according to Hund's rule, which states that electrons fill orbitals one at a time before pairing up. In $[CoF_6]^{3-}$, the six d electrons of Co^{3+} ($[Ar]3d^6$) occupy both the higher- and lower-energy d orbitals. Weak field ligands result in **high-spin complexes** to give the maximum number of unpaired electrons. Complexes with strong field ligands, such as $[Co(CN)_6]^{3-}$, have large crystal field splitting energies and electrons pair and fill the lower-energy d orbitals (d_{xy}, d_{xz}, d_{yz}) before occupying the higher-energy d orbitals (d_{z^2} and $d_{x^2-y^2}$). In $[Co(CN)_6]^{3-}$, all six d electrons are spin-paired in the lower-energy orbitals. Strong field ligands result in **low-spin complexes** in which d electrons are paired up in lower-energy d orbitals before occupying higher-energy d orbitals.

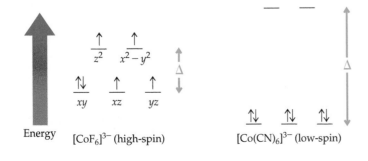

$[Ni(H_2O)_6]^{2+}$ $[Ni(NH_3)_6]^{2+}$ $[Ni(en)_3]^{2+}$

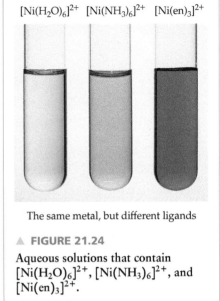

The same metal, but different ligands

▲ **FIGURE 21.24**

Aqueous solutions that contain $[Ni(H_2O)_6]^{2+}$, $[Ni(NH_3)_6]^{2+}$, **and** $[Ni(en)_3]^{2+}$.

Go to eText

BIG IDEA Question 6

Which complex ion will have a maximum in its absorption spectrum at the shortest wavelength of light: $[Co(NH_3)_5Cl]^{2+}$, $[Co(NH_3)_6]^{2+}$, or $[Co(NH_3)_5H_2O]^{3+}$?

What determines which of the two spin states (high-spin or low-spin) has the lower energy? In general, when an electron moves from a z^2 or $x^2 - y^2$ orbital to one of the lower-energy orbitals, the energy decreases by Δ. Due to electron–electron repulsion, it costs energy to put the electron into an orbital that already contains another electron. The energy required is called the spin-pairing energy P. If Δ is greater than P, as it is for $[Co(CN)_6]^{3-}$, then the low-spin arrangement has lower energy. If Δ is less than P, as it is for $[CoF_6]^{3-}$, the high-spin arrangement has lower energy. Thus, the observed spin state depends on the relative values of Δ and P. In general, strong-field ligands give low-spin complexes, and weak-field ligands give high-spin complexes.

A choice between the high-spin and low-spin electron configurations arises only for complexes of metal ions with four to seven d electrons, so-called $d^4 - d^7$ complexes. For $d^1 - d^3$ and $d^8 - d^{10}$ complexes, only one ground-state electron configuration is possible. In $d^1 - d^3$ complexes, all the electrons occupy the lower-energy d orbitals, independent of the value of Δ. In $d^8 - d^{10}$ complexes, the lower-energy set of d orbitals is filled with three pairs of electrons, while the higher-energy set contains two, three, or four electrons, again independent of the value of Δ.

WORKED EXAMPLE 21.8

Go to eText

Drawing Crystal Field Energy-Level Diagrams for Octahedral Complexes

Draw a crystal field orbital energy-level diagram, and predict the number of unpaired electrons for each of the following complexes:

(a) $[Cr(en)_3]^{3+}$ (b) $[Mn(CN)_6]^{3-}$ (c) $[Co(H_2O)_6]^{2+}$

STRATEGY

All three complexes are octahedral, so the energy-level diagrams will show three lower-energy and two higher-energy d orbitals. For $d^1 - d^3$ and $d^8 - d^{10}$ complexes, the electrons occupy the orbitals in accord with Hund's rule so as to give the maximum number of unpaired electrons. For $d^4 - d^7$ complexes, the orbital occupancy and number of unpaired electrons depend on the position of the ligand in the spectrochemical series.

SOLUTION

(a) Cr^{3+} ($[Ar]\, 3d^3$) has three unpaired electrons. In the complex, they occupy the lower-energy set of d orbitals as shown below.
(b) Mn^{3+} ($[Ar]\, 3d^4$) can have a high-spin or a low-spin configuration. Because CN^- is a strong-field ligand, all four d electrons go into the lower-energy d orbitals. The complex is low-spin, with two unpaired electrons.
(c) Co^{2+} ($[Ar]\, 3d^7$) has a high-spin configuration with three unpaired electrons because H_2O is a weak-field ligand.

In the following orbital energy-level diagrams, the relative values of the crystal field splitting Δ agree with the positions of the ligands in the spectrochemical series ($H_2O < en < CN^-$):

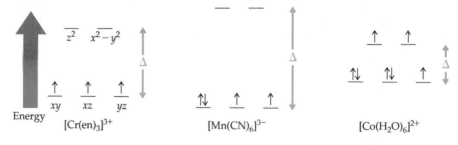

PRACTICE 21.15 Draw a crystal field d-orbital energy-level diagram, and predict the number of unpaired electrons in the complex $[Fe(CN)_6]^{4-}$.

APPLY 21.16 Which complex will be more paramagnetic? $[FeCl_6]^{3-}$ or $[Fe(CN)_6]^{3-}$?

Tetrahedral and Square Planar Complexes

Different geometric arrangements of the ligands give different energy splittings for the *d* orbitals. **FIGURE 21.25** shows *d*-orbital energy-level diagrams for tetrahedral and square planar complexes.

The splitting pattern in tetrahedral complexes is just the opposite of that in octahedral complexes; that is, the *xy, xz,* and *yz* orbitals have higher energy than the $x^2 - y^2$ and z^2 orbitals. (As with octahedral complexes, the energy ordering follows from the relative orientation of the orbital lobes and the ligands, but we won't derive the result.) Because none of the orbitals points directly at the ligands in tetrahedral geometry and because there are only four ligands instead of six, the crystal field splitting in tetrahedral complexes is only about half of that in octahedral complexes. Consequently, Δ is almost always smaller than the spin-pairing energy *P*, and nearly all tetrahedral complexes are high spin.

Square planar complexes look like octahedral ones (Figure 21.22) except that the two trans ligands along the *z* axis are missing. In square planar complexes, the $x^2 - y^2$ orbital is high in energy (Figure 21.25) because it points directly at all four ligands, which lie along the *x* and *y* axes. The splitting pattern is more complicated than for octahedral and tetrahedral complexes, but the main point to remember is that a large energy gap exists between the $x^2 - y^2$ orbital and the four lower-energy orbitals. Square planar geometry is most common for metal ions with electron configuration d^8 because this configuration favors low-spin complexes in which all four lower-energy orbitals are filled and the higher-energy $x^2 - y^2$ orbital is vacant. Common examples are $[Ni(CN)_4]^{2-}$, $[PdCl_4]^{2-}$, and $Pt(NH_3)_2Cl_2$.

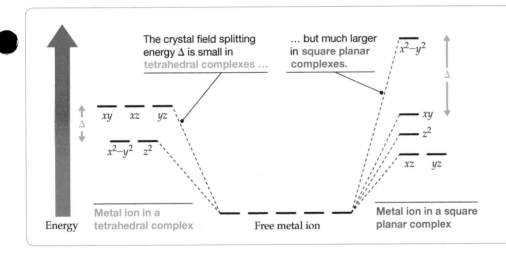

The crystal field splitting energy Δ is small in tetrahedral complexes ...

... but much larger in square planar complexes.

xy xz yz

x^2-y^2 z^2

Metal ion in a tetrahedral complex

Free metal ion

x^2-y^2

xy

z^2

xz yz

Metal ion in a square planar complex

Energy

◀ **FIGURE 21.25**

Energies of the *d* orbitals in tetrahedral and square planar complexes relative to their energy in the free metal ion.

WORKED EXAMPLE 21.9

Drawing Crystal Field Energy-Level Diagrams for Tetrahedral and Square Planar Complexes

Draw crystal field energy-level diagrams, and predict the number of unpaired electron for the following complexes:

(a) $[FeCl_4]^-$ (tetrahedral)

(b) $[PtCl_4]^{2-}$ (square planar)

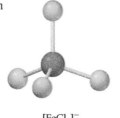

$[FeCl_4]^-$ $[PtCl_4]^{2-}$

continued on next page

STRATEGY

Begin with the energy-level diagrams in Figure 21.25, and remember that nearly all tetrahedral complexes are high-spin (small Δ) while square planar complexes have a large energy gap (large Δ) between the $x^2 - y^2$ orbital and the four lower-energy orbitals.

SOLUTION

(a) The five d electrons of Fe^{3+} ($[Ar]\,3d^5$) are distributed between the higher- and lower-energy orbitals as shown below. Because Δ is small in tetrahedral complexes, $[FeCl_4]^-$ is high-spin with five unpaired electrons.

(b) $Pt^{2+}([Xe]\,4f^{14}\,5d^8)$ has eight d electrons. Because Δ is large in square planar complexes, all the electrons occupy the four lower-energy d orbitals. There are no unpaired electrons, and the complex is diamagnetic.

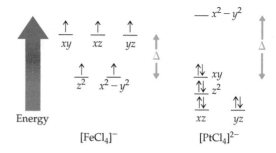

▶ **PRACTICE 21.17** Draw a crystal field energy-level diagram and predict the number of unpaired electrons for the complex ion $[MnCl_4]^{2-}$ with tetrahedral geometry.

▶ **APPLY 21.18** The complex $[Ni(CN)_4]^{2-}$ is found to be diamagnetic. What is the most likely geometry of this complex, tetrahedral, or square planar?

21.11 BONDING IN COMPLEXES: VALENCE BOND THEORY

Although crystal field theory is a useful model for explaining the color and magnetism in transition metal complexes, the valence bond model is needed to rationalize the geometry of coordination compounds. According to the **valence bond theory** (Section 8.2), the bonding in metal complexes arises when a filled ligand orbital containing a pair of electrons overlaps a vacant hybrid orbital on the metal ion to give a coordinate covalent bond:

Vacant metal   Occupied ligand   Coordinate covalent
hybrid atomic orbital   atomic orbital   bond

The **hybrid orbitals** that a metal ion uses to accept a share in the ligand electrons are those that point in the directions of the ligands. In linear complexes the metal uses sp hybrids, while in tetrahedral complexes the metal uses sp^3 hybrids. Remember that geometry and hybridization go together. Once you know the geometry of a complex, you automatically know which hybrid orbitals the metal ion uses.

In octahedral complexes, the metal ion uses valence d orbitals in addition to s and p orbitals to form six bonds to the ligands. Hybridization occurs by a combination of two d orbitals, one s orbital, and three p orbitals, resulting in either six d^2sp^3 hybrids or six sp^3d^2 hybrids. As we'll see later, the difference between d^2sp^3 and sp^3d^2 hybrids lies in the principal quantum number of the d orbitals. Both sets of hybrids, however, point toward the six corners of a regular octahedron, as shown in **FIGURE 21.26**. All six orbitals within each set are equivalent, and the angle between any two adjacent orbitals is 90°.

In square planar complexes, the metal ion uses a set of four equivalent dsp^2 hybrid orbitals formed by the hybridization of one d orbital, one s orbital, and two p orbitals. These hybrids point toward the four corners of a square, as shown in **FIGURE 21.27**.

The relationship between the geometry of a complex and the hybrid orbitals used by the metal ion is summarized in **TABLE 21.6**.

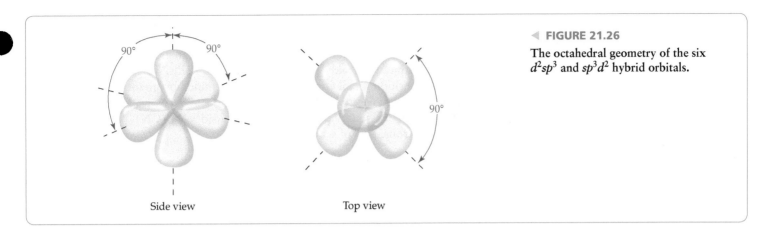

◀ **FIGURE 21.26**

The octahedral geometry of the six d^2sp^3 and sp^3d^2 hybrid orbitals.

Side view   Top view

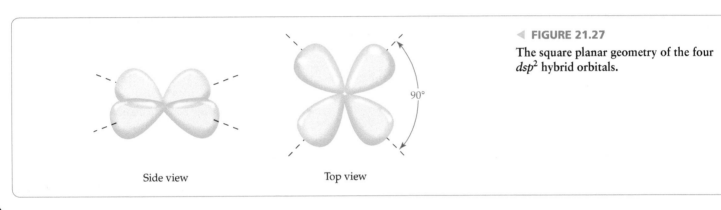

◀ **FIGURE 21.27**

The square planar geometry of the four dsp^2 hybrid orbitals.

Side view   Top view

TABLE 21.6 **Hybrid Orbitals for Common Coordination Geometries**			
Coordination Number	Geometry	Hybrid Orbitals	Example
2	Linear	sp	$[Ag(NH_3)_2]^+$
4	Tetrahedral	sp^3	$[CoCl_4]^{2-}$
4	Square planar	dsp^2	$[Ni(CN)_4]^{2-}$
6	Octahedral	d^2sp^3 or sp^3d^2	$[Cr(H_2O)_6]^{3+}, [Co(H_2O)_6]^{2+}$

To illustrate the relationship between geometry and hybridization, let's consider the tetrahedral complex $[CoCl_4]^{2-}$. A free Co^{2+} ion has the electron configuration [Ar] $3d^7$ and the following orbital diagram:

$$Co^{2+}: [Ar] \quad \underset{3d}{\uparrow\downarrow \; \uparrow\downarrow \; \uparrow \; \uparrow \; \uparrow} \quad \underset{4s}{\_\_} \quad \underset{4p}{\_\_ \; \_\_ \; \_\_}$$

Because the geometry of $[CoCl_4]^{2-}$ is tetrahedral, the hybrid orbitals that Co^{2+} uses to share the four pairs of ligand electrons must be sp^3 hybrids formed from the vacant $4s$ and $4p$ orbitals. The following orbital diagram represents the bonding in the complex, showing the hybridization of the metal orbitals and the four pairs of ligand electrons, now shared in the bonds between the metal and the ligands:

$$[CoCl_4]^{2-}: [Ar] \quad \underset{3d}{\uparrow\downarrow \; \uparrow\downarrow \; \uparrow \; \uparrow \; \uparrow} \quad \boxed{\underset{4s}{\uparrow\downarrow} \quad \underset{4p}{\uparrow\downarrow \; \uparrow\downarrow \; \uparrow\downarrow}}$$

Four sp^3 bonds to the ligands

REMEMBER...
The O_2 molecule has two unpaired π^*_{2p} electrons and is **paramagnetic**, whereas N_2 and F_2 have no unpaired electrons and are diamagnetic (Section 8.8).

In accord with this description, $[CoCl_4]^{2-}$ has three unpaired electrons and is paramagnetic. Recall from Section 8.8 that **paramagnetic** substances contain unpaired electrons and are attracted by magnetic fields, whereas diamagnetic substances contain only paired electrons and are weakly repelled by magnetic fields. The number of unpaired electrons in a transition metal complex can be determined by quantitatively measuring the force exerted on the complex by a magnetic field.

As an example of a square planar complex, consider $[Ni(CN)_4]^{2-}$. A free Ni^{2+} ion has eight $3d$ electrons, two of which are unpaired:

$$Ni^{2+}: [Ar] \quad \underset{3d}{\uparrow\downarrow \; \uparrow\downarrow \; \uparrow\downarrow \; \uparrow \; \uparrow} \qquad \underset{4s}{\_\_} \qquad \underset{4p}{\_\_ \; \_\_ \; \_\_}$$

Because $[Ni(CN)_4]^{2-}$ is square planar, the metal ion must use dsp^2 hybrids. By pairing up the two unpaired d electrons in one d orbital, we obtain a vacant $3d$ orbital that can be hybridized with the $4s$ orbital and two of the $4p$ orbitals to give the square planar dsp^2 hybrids. These hybrids form bonds to the ligands by accepting a share in the four pairs of ligand electrons:

$$[Ni(CN)_4]^{2-}: [Ar] \quad \underset{3d}{\uparrow\downarrow \; \uparrow\downarrow \; \uparrow\downarrow \; \uparrow\downarrow} \boxed{\uparrow\downarrow \qquad \underset{4s}{\uparrow\downarrow} \qquad \underset{4p}{\uparrow\downarrow \; \uparrow\downarrow}} \_\_$$

Four dsp^2 bonds to the ligands

In agreement with this description, $[Ni(CN)_4]^{2-}$ is diamagnetic.

In octahedral complexes, the metal ion uses either sp^3d^2 or d^2sp^3 hybrid orbitals. To see the difference between these two kinds of hybrids, let's consider the cobalt(III) complexes $[CoF_6]^{3-}$ and $[Co(CN)_6]^{3-}$. A free Co^{3+} ion has six $3d$ electrons, four of which are unpaired:

$$Co^{3+}: [Ar] \quad \underset{3d}{\uparrow\downarrow \; \uparrow \; \uparrow \; \uparrow \; \uparrow} \qquad \underset{4s}{\_\_} \qquad \underset{4p}{\_\_ \; \_\_ \; \_\_} \qquad \underset{4d}{\_\_ \; \_\_ \; \_\_ \; \_\_ \; \_\_}$$

Magnetic measurements indicate that $[CoF_6]^{3-}$ is paramagnetic and contains four unpaired electrons. Evidently, none of the $3d$ orbitals is available to accept a share in the ligand electrons because each is already at least partially occupied. Consequently, the octahedral hybrids that Co^{3+} uses are formed from the vacant $4s$, $4p$, and $4d$ orbitals. These orbitals, called sp^3d^2 hybrids, share in the six pairs of ligand electrons, as shown in the following orbital diagram:

$$[CoF_6]^{3-}: [Ar] \quad \underset{3d}{\uparrow\downarrow \; \uparrow \; \uparrow \; \uparrow \; \uparrow} \boxed{\underset{4s}{\uparrow\downarrow} \quad \underset{4p}{\uparrow\downarrow \; \uparrow\downarrow \; \uparrow\downarrow} \quad \uparrow\downarrow \; \uparrow\downarrow} \underset{4d}{\_\_ \; \_\_ \; \_\_}$$

Six sp^3d^2 bonds to the ligands

In contrast to $[CoF_6]^{3-}$, magnetic measurements indicate that $[Co(CN)_6]^{3-}$ is diamagnetic. All six $3d$ electrons are therefore paired and occupy just three of the five $3d$ orbitals. That leaves two vacant $3d$ orbitals, which combine with the vacant $4s$ and $4p$ orbitals to give a set of six octahedral hybrid orbitals called d^2sp^3 hybrids. The d^2sp^3 hybrids form bonds to the ligands by accepting a share in the six pairs of ligand electrons:

$$[Co(CN)_6]^{3-}: [Ar] \quad \underset{3d}{\uparrow\downarrow \; \uparrow\downarrow \; \uparrow\downarrow} \boxed{\uparrow\downarrow \; \uparrow\downarrow \qquad \underset{4s}{\uparrow\downarrow} \qquad \underset{4p}{\uparrow\downarrow \; \uparrow\downarrow \; \uparrow\downarrow}} \underset{4d}{\_\_ \; \_\_ \; \_\_ \; \_\_ \; \_\_}$$

Six d^2sp^3 bonds to the ligands

The difference between d^2sp^3 and sp^3d^2 hybrids lies in the principal quantum number of the d orbitals. In d^2sp^3 hybrids, the principal quantum number of the d orbitals is one less than the principal quantum number of the s and p orbitals. In sp^3d^2 hybrids, the s, p, and d orbitals have the same principal quantum number. To determine which set of hybrids is used in any given complex, we must know the magnetic properties of the complex.

Complexes of metals like Co^{3+} that exhibit more than one spin state are classified as *high-spin* or *low-spin*. A **high-spin complex,** such as $[CoF_6]^{3-}$, is one in which the d electrons are arranged according to Hund's rule to give the maximum number of unpaired electrons. A **low-spin complex,** such as $[Co(CN)_6]^{3-}$, is one in which the d electrons are paired up to give a maximum number of doubly occupied d orbitals and a minimum number of unpaired electrons.

WORKED EXAMPLE 21.10

Describing the Bonding in Complexes in Terms of Valence Bond Hybrid Orbitals

Give a valence bond description of the bonding in $[V(NH_3)_6]^{3+}$. Include orbital diagrams for the free metal ion and the metal ion in the complex. Tell which hybrid orbitals the metal ion uses and the number of unpaired electrons present.

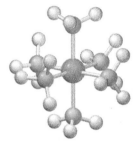

$[V(NH_3)_6]^{3+}$

STRATEGY AND SOLUTION

The free V^{3+} ion has the electron configuration $[Ar]\ 3d^2$ and the orbital diagram:

$$V^{3+}:\ [Ar]\quad \underset{3d}{\uparrow\ \uparrow\ \_\ \_\ \_}\quad \underset{4s}{\_}\quad \underset{4p}{\_\ \_\ \_}$$

Because $[V(NH_3)_6]^{3+}$ is octahedral, the V^{3+} ion must use either d^2sp^3 or sp^3d^2 hybrid orbitals in accepting a share in six pairs of electrons from the six NH_3 ligands. The preferred hybrids are d^2sp^3 because several $3d$ orbitals are vacant and d^2sp^3 hybrids have lower energy than sp^3d^2 hybrids (because the $3d$ orbitals have lower energy than the $4d$ orbitals). Thus, $[V(NH_3)_6]^{3+}$ has the following orbital diagram:

$$[V(NH_3)_6]^{3+}:\ [Ar]\quad \underset{3d}{\uparrow\ \uparrow\ \_}\quad \boxed{\underset{\quad}{\uparrow\downarrow\ \uparrow\downarrow}\quad \underset{4s}{\uparrow\downarrow}\quad \underset{4p}{\uparrow\downarrow\ \uparrow\downarrow\ \uparrow\downarrow}}$$

Six d^2sp^3 bonds to the ligands

The complex has two unpaired electrons and is therefore paramagnetic.

▶ **PRACTICE 21.19**   Give a valence bond description of the bonding in the square planar complex $[PtCl_4]^{2-}$. Draw orbital diagrams for the free metal ion and the metal ion in the complex. What hybrid orbitals are used by the metal ion, and how many unpaired electrons are in the complex?

▶ **APPLY 21.20**   Give a valence bond description of the bonding in each of the following complexes. Include orbital diagrams for the free metal ion and the metal ion in the complex. Tell which hybrid orbitals the metal ion uses and the number of unpaired electrons in each complex.

(a) $[Fe(CN)_6]^{3-}$ (low-spin)   **(b)** $[Co(H_2O)_6]^{2+}$ (high-spin)

In 1965, at Michigan State University, a biophysicist named Barnett Rosenberg made an astonishing discovery while investigating whether electrical currents played a role in cell division. He used platinum electrodes to run a current through an ammonium chloride buffered solution of *Escherichia coli* (*E. coli*) cells. After a period of time, he observed that the cells were about 300 times longer than normal. The change in shape was attributed to inhibition of cellular division. Rosenberg initially proposed that electricity prevented the cells from dividing, but he soon determined that the platinum electrode had reacted with the ammonium chloride salt to generate the transition metal complex $Pt(NH_3)_2Cl_2$. The cis isomer reacted with DNA, inhibiting replication, but the trans isomer was inactive (**FIGURE 21.28**).

As a result of these experiments, it was reasoned that the cis isomer of $Pt(NH_3)_2Cl_2$, named cisplatin, might be a useful antitumor drug due to its ability to inhibit replication in rapidly dividing cancer cells. Cisplatin was first used to treat John Cleland, a 22-year-old veterinary student with testicular cancer in the cancer ward at Indiana University in 1974. At that time, the survival rate of metastatic testes cancer, cancer that had migrated into the lymph nodes and lungs, was less than 5 percent. Larry Einhorn, a young oncologist, suggested a last-ditch effort to save Cleland: treatment with cisplatin in combination with two other drugs. Astonishingly, when Cleland had routine scans taken 10 days after treatment, the tumors in his lungs had vanished. Over the years, cisplatin has shown remarkable clinical success and is responsible for the cure of over 90% of testicular cancer cases including the well-known case of cyclist Lance Armstrong. It is also widely prescribed for ovarian, bladder, lung, and stomach cancers. The success of cisplatin was overwhelming, but patients suffered severe side effects: joint pain, weakness, and unrelenting nausea. On average patients vomited 12 times a day, but today newer drugs can somewhat ease the side effects of cisplatin treatment.

How does cisplatin inhibit DNA replication? Although the mechanism has not yet been fully elucidated, cisplatin is believed to kill cancer cells by binding to DNA, which eventually leads to cell death. Cisplatin and many chemotherapy agents kill all cells, but the effect is more pronounced on rapidly dividing cancer cells. Unfortunately other cells that divide rapidly such as those in hair and the stomach lining are also greatly affected. This leads to common side effects of treatment such as hair loss and nausea. Cisplatin is injected into the body where the chloride concentration in blood plasma is high enough (100 mM) that replacement of chloride ligands by water is limited. Once cisplatin diffuses through a cell membrane into the interior of a cell where the chloride concentration is lower (4-20 mM), cisplatin undergoes a reaction with water, producing the highly reactive charged platinum complex $[Pt(NH_3)_2(H_2O)Cl]^+$. This complex coordinates to DNA through a nitrogen atom of either a guanine or adenine base (**FIGURE 21.29**). The remaining chloride ligand can also be replaced by water enabling the platinum to bind to a second nucleotide base, significantly distorting the shape of the DNA strand (**FIGURE 21.30**). The bend in the DNA strand caused by the cisplatin-DNA adduct is recognized by a protein involved in a DNA "proofreading" process that is designed to detect defects in DNA. Detection of the DNA defect induced by cisplatin will initiate a signaling pathway that leads to cell death.

Although cisplatin is an effective drug, researchers have explored other platinum compounds that are more effective and have fewer side effects. The most common is carboplatin, which entered the U.S. market as paraplatin in 1989 for treatment of advanced ovarian cancer and now outranks cisplatin in sales. Carboplatin, or *cis*-diammine(1,1-cyclobutanedicarboxylato)

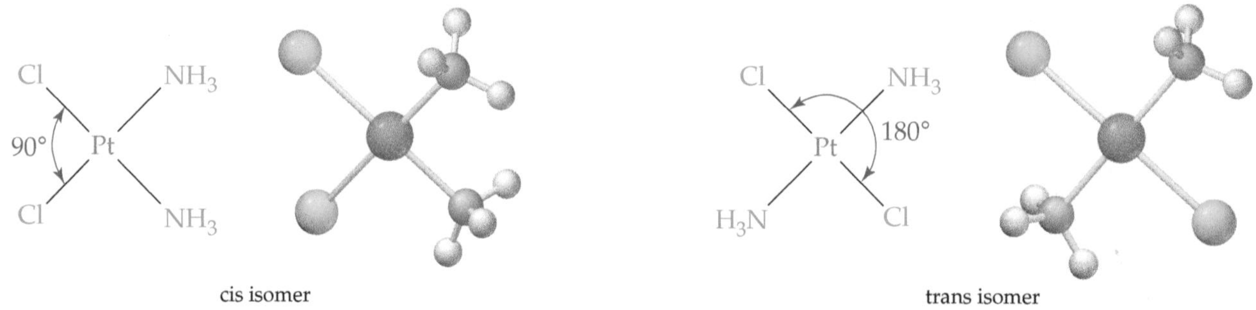

▲ FIGURE 21.28

Diastereomers of the square planar complex $Pt(NH_3)_2Cl_2$. The two compounds have the same connections among atoms but different arrangements of the atoms in space.

platinum(II), is less toxic because the dicarboxylate ligand slows down the degradation of the complex into toxic derivatives. In addition, compounds that possess more than one platinum center have also been studied. Compound BBR 3464 is more potent than cisplatin and shows promise against platinum-resistant cell lines.

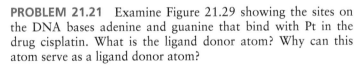

Carboplatin   Compound BBR 3464

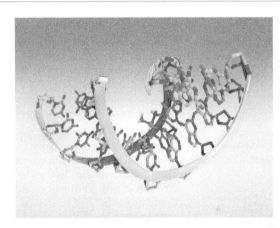

FIGURE 21.29

A schematic representation of a portion of a DNA strand showing sites available for Pt binding on the bases guanine and adenine. The nitrogen on guanine is the preferred site for initial binding of $[Pt(NH_3)_2(H_2O)Cl]^+$.

▲ **FIGURE 21.30**

Cisplatin coordinated to a section of DNA containing two guanine bases. When cisplatin binds it induces a a bend in DNA, which prevents the cells from dividing. The cells try to repair the damage but are unable to do so, and cell death is initiated.

PROBLEM 21.21 Examine Figure 21.29 showing the sites on the DNA bases adenine and guanine that bind with Pt in the drug cisplatin. What is the ligand donor atom? Why can this atom serve as a ligand donor atom?

PROBLEM 21.22 Answer the following questions about the antitumor drug cisplatin, $Pt(NH_3)_2Cl_2$.
(a) Cisplatin is a common name. What is the chemical name for the compound?
(b) What is the oxidation state and coordination number of Pt?
(c) What is the Lewis acid and Lewis base when ammonia and chloride react with Pt^{2+} to form cisplatin, $Pt(NH_3)_2Cl_2$?
(d) What is the electron configuration of Pt in cisplatin?

PROBLEM 21.23 Draw the crystal field d-orbital-energy-level diagram and predict the number of unpaired electrons in the square planar cisplatin molecule.

PROBLEM 21.24 The following equilibrium exists for cisplatin in aqueous solution.

$$Pt(NH_3)_2Cl_2(aq) + H_2O(l) \rightleftharpoons$$
$$[Pt(NH_3)_2(H_2O)Cl]^+(aq) + Cl^-(aq)$$

(a) Explain why the forward reaction is limited in blood plasma but occurs to a greater extent inside the cell.
(b) What is the name of the complex ion $[Pt(NH_3)_2(H_2O)Cl]^+$?
(c) Predict how the wavelength of maximum absorption would shift upon replacing a Cl^- ligand with a H_2O ligand.

PROBLEM 21.25 Another platinum compound effective in inhibiting DNA replication is $PtCl_4(NH_3)_2$.
(a) What is the oxidation state of platinum?
(b) What is the name of this compound?
(c) Draw the possible diastereomers of this compound.
(d) Can either of the diastereomers exist as enantiomers?

STUDY GUIDE

Section	Conceptual Summary	Learning Objectives	Test Your Understanding
21.1 Electron Configurations	*Transition elements,* or *d-block elements,* are the metallic elements in the central part of the periodic table. Most of the neutral atoms have the valence electron configuration $(n - 1)d^{1-10} \, ns^2$, and the cations have a configuration $(n - 1)d^{0-10}$.	**21.1** Write electron configurations for transition metals and predict the number of unpaired electrons in transition metal ions.	Worked Example 21.1; Problems 21.16–21.17, 21.37–21.38
21.2 Properties of Transition Elements	Transition metal elements are malleable and good conductors of heat and electricity. Trends in atomic radius, ionization energy, and oxidation potential can be explained by examining effective nuclear charge and electron–electron repulsions.	**20.2** Describe the trends in physical and chemical properties of transition metal elements across a period. Explain the reasons for trends.	Problems 21.28, 21.40, 21.42, 21.45, 21.48
21.3 Oxidation States of Transition Elements	Transition metals exhibit a variety of oxidation states. Ions with the metal in a high oxidation state tend to be good oxidizing agents ($Cr_2O_7^{2-}$, MnO_4^-), while ions with an early transition metal in a low oxidation state are good reducing agents (V^{2+}, Cr^{2+}).	**21.3** Relate the oxidizing and reducing strength of a transition metal ion to its oxidation state and location in the periodic table.	Problems 21.56, 21.58, 21.60
21.4 Coordination Compounds	**Coordination compounds** are compounds in which a central metal ion is attached to a group of surrounding **ligands** by coordinate covalent bonds. The number of ligand donor atoms bonded to a metal is called the **coordination number** of the metal. Common coordination numbers and geometries are 2 (linear), 4 (tetrahedral or square planar), and 6 (octahedral). Systematic names for complexes specify the number of ligands of each particular type, the metal, and its oxidation state.	**21.4** Identify Lewis acids, Lewis bases, ligands, and donor atoms in a coordination complex. **21.5** Determine the formula of a coordination complex, identify the oxidation state and coordination number of the metal atom, and draw the structure of the complex.	Problems 21.74–21.75 Worked Example 21.2; Problems 21.62, 21.66, 21.68, 21.70
21.5 Ligands	Ligands can be **monodentate** or **polydentate**, depending on the number of **ligand donor atoms** attached to the metal. Polydentate ligands are also called **chelating agents.** They form complexes known as metal **chelates** that contain rings of atoms known as **chelate rings.**	**21.6** Classify a ligand as mono, bi, tri, tetra, or hexadentate based on its chemical structure.	Problems 21.30–21.31
21.6 Naming Coordination Compounds	Coordination complexes are named using rules established by the International Union of Pure and Applied Chemistry. The nomenclature rules are similar to those for ionic compounds with the cation named first and the anion second. Ligands are named in alphabetical order and the number of ligands is specified using Greek prefixes. The oxidation state of the transition metal ion is designated using a Roman numeral in parentheses.	**21.7** Write the systematic name for a coordination complex.	Worked Examples 21.3–21.4; Problems 21.32, 21.78, 21.82

Section	Conceptual Summary	Learning Objectives	Test Your Understanding
21.7 Isomers	Many complexes exist as **isomers,** compounds that have the same formula but a different arrangement of the constituent atoms. **Constitutional isomers,** such as **linkage isomers** and **ionization isomers,** have different connections between their constituent atoms. **Stereoisomers** have the same connections but a different spatial arrangement of the atoms. **Diastereoisomers** are non-mirror-image stereoisomers, such as the **cis** and **trans isomers** of square planar and octahedral complexes.	**21.8** Classify isomers according to the scheme given in Figure 21.11.	Worked Example 21.5; Problems 21.84–21.85, 21.94
		21.9 Draw diastereomers for octahedral and square planar complexes.	Worked Example 21.6; Problems 21.86–21.87
21.8 Enantiomers and Molecular Handedness	Molecules that have handedness are said to be **chiral.** All other molecules are **achiral.** **Enantiomers** are mirror-image stereoisomers, such as "right-handed" and "left-handed" $[\text{Co(en)}_3]^{3+}$. One enantiomer rotates the plane of plane-polarized light to the right, and the other rotates the plane through an equal angle but in the opposite direction. A 1:1 mixture of the two enantiomers is called a **racemic mixture.**	**21.10** Classify a coordination complex as chiral or achiral and draw enantiomers for chiral compounds.	Worked Example 21.7; Problems 21.88, 21.90, 21.96
21.9 Color of Transition Metal Complexes	Many transition metal complexes have beautiful colors because the energy difference between the ground and excited state is in the visible region of the electromagnetic spectrum. The wavelength (λ) of the light absorbed by a metal complex depends on the **energy separation $\Delta E = E_2 - E_1$ between the two states,** as given by the equation $\Delta E = h\nu = hc/\lambda$, where h is Planck's constant, ν is the frequency of the light, and c is the speed of light.	**21.11** Relate the color of a metal complex, the wavelength of light it absorbs, and the energy difference between its ground and excited state.	Problems 21.100–21.101
21.10 Crystal Field Theory	**Crystal field theory** assumes that the metal–ligand bonding is entirely ionic. Because of electrostatic repulsions between the d electrons and the ligands, the d orbitals are raised in energy and are differentiated by an energy separation called the **crystal field splitting, Δ.** In octahedral complexes, the d_{z^2} and $d_{x^2-y^2}$ orbitals have higher energy than the d_{xy}, d_{xz}, and d_{yz} orbitals. Tetrahedral and square planar complexes exhibit different splitting patterns. The colors of complexes are due to electronic transitions from one set of d orbitals to another, and the transition energies depend on the position of the ligand in the **spectrochemical series. Weak-field ligands** give small Δ values, and **strong-field ligands** give large Δ values. Crystal field theory accounts for the magnetic properties of complexes in terms of the relative values of Δ and the spin-pairing energy P. Small Δ values favor **high-spin complexes,** and large Δ values favor **low-spin complexes.**	**21.12** Draw crystal field energy-level diagrams for octahedral, tetrahedral, and square complexes. Classify ligands as strong- or weak-field and predict (or account for) color and magnetic properties of complexes in terms of crystal field theory.	Worked Examples 21.8–21.9; Problems 21.104, 21.106, 21.108, 21.114, 21.116, 21.124
21.11 Bonding in Complexes: Valence Bond Theory	Valence bond theory describes the bonding in complexes in terms of two-electron, coordinate covalent bonds resulting from the overlap of filled ligand orbitals with vacant metal hybrid orbitals that point in the direction of the ligands: sp (linear), sp^3 (tetrahedral), dsp^2 (square planar), and d^2sp^3 or sp^3d^2 (octahedral).	**21.13** Describe the bonding in a metal complex in terms of valence bond theory, including orbital diagrams, the hybrid orbitals used by the metal atom, and the number of unpaired electrons.	Worked Example 21.10; Problems 21.16, 21.128

KEY TERMS

KEY EQUATION

- Relationship between the crystal field splitting energy **Δ** and the wavelength of light **λ** absorbed in a $d-d$ electronic transition (Section 21.9)

$$\Delta = h\nu = \frac{hc}{\lambda}$$

where h is Planck's constant, ν is the frequency of the light, and c is the speed of light.

PRACTICE TEST

After studying this chapter, you can assess your understanding with these practice test questions, which are correlated with chapter learning objectives. If you answer a question incorrectly, refer to the learning objectives in the end-of-chapter Study Guide for assistance. The Study Guide provides a conceptual summary, references a Worked Example to model how to solve the problem, and gives additional problems for more practice.

1. What is the electron configuration of Co^{2+} and how many unpaired electrons are in the free transition metal ion? (**LO 21.1**)
   (a) $[Ar]3d^5\,4s^2$; 5 unpaired electrons
   (b) $[Ar]3d^5\,4s^2$; 1 unpaired electron
   (c) $[Ar]3d^7$; 3 unpaired electrons
   (d) $[Ar]3d^7$; 1 unpaired electron

2. Based on effective nuclear charge (Z_{eff}), which ion is the strongest oxidizing agent? (**LO 21.3**)
   (a) Cu^{2+} (b) Ni^{2+}
   (c) Fe^{2+} (d) Mn^{2+}

3. What is the Lewis base in the reaction of oxalate with the manganese ion to form $[Mn(C_2O_4)_3]^{2-}$? What is the oxidation state of Mn and the coordination number of the complex? (**LO 21.4, 21.5**)
   (a) Lewis base is $C_2O_4{}^{2-}$; Mn oxidation number is +3; coordination number is 3.
   (b) Lewis base is $C_2O_4{}^{2-}$; Mn oxidation number is +4; coordination number is 6.
   (c) Lewis base is Mn^{2+}; Mn oxidation number is +2; coordination number is 3.
   (d) Lewis base is Mn^{4+}; Mn oxidation number is +4; coordination number is 6.

Refer to the figure showing the structure of various ligands to answer questions 4 and 5.

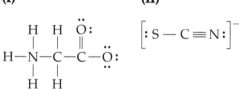

(I)   (II)

(III)   (IV)

4. Which ligand(s) are monodentate? (**LO 21.6**)
   (a) III   (b) I and II
   (c) IV   (d) I, II, and IV

5. Which ligand(s) can participate in linkage isomerism? (**LO 21.8**)
   (a) All of the ligands can participate in linkage isomerism.
   (b) I, II, and III
   (c) I and IV
   (d) II and IV

6. What is the name of the compound $[Fe(H_2O)_5(SCN)]Cl_2$? (**LO 21.7**)
   (a) pentaaquathiocyanatoiron(III) chloride
   (b) pentaaquachlorothiocyanato iron(III)
   (c) pentaaquathiocyanatoiron(III) dichloride
   (d) pentaaquathiocyanatoiron(II) chloride

7. Identify the *false* statement about the structures of the complex ion [Fe(en)₂Cl₂]⁺ shown below. (**LO 21.9, 21.10**)

(I)   **(II)**

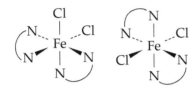

(III)   **(IV)**

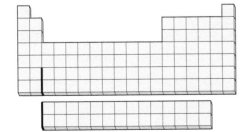

(a) Structures I and II are cis-trans isomers.
(b) Structures I and IV are cis-trans isomers.
(c) Structures I and III are enantiomers.
(d) Structures II and IV are enantiomers.

8. Based on the wavelength of maximum absorption of the cobalt complexes, arrange the following ligands in a spectrochemical series from weakest-field to strongest-field ligand. (**LO 21.11, 21.12**)

Coordination complex	Wavelength of maximum absorption
$[Co(NH_3)_6]^{2+}$	430 nm
$[Co(NH_3)_5Cl]^{2+}$	522 nm
$[Co(NH_3)_5NCS]^{3+}$	470 nm
$[Co(NH_3)_5H_2O]^{3+}$	500 nm

(a) $Cl^- < NCS^- < H_2O < NH_3$
(b) $NH_3 < NCS^- < H_2O < Cl^-$
(c) $H_2O < Cl^- < NH_3 < NCS^-$
(d) $Cl^- < H_2O < NCS^- < NH_3$

9. What is the crystal field energy level diagram for the complex $[Fe(NH_3)_6]^{3+}$? (**LO 21.12**)

(a)   **(b)**

$$\overline{d_{z^2}}\ \overline{d_{x^2-y^2}}$$

$$\underset{d_{xy}}{\uparrow\downarrow}\ \underset{d_{xz}}{\uparrow\downarrow}\ \underset{d_{yz}}{\uparrow\downarrow}$$

$$\overset{\uparrow}{d_{z^2}}\ \overset{\uparrow}{d_{x^2-y^2}}$$

$$\underset{d_{xy}}{\uparrow}\ \underset{d_{xz}}{\uparrow}\ \underset{d_{yz}}{\uparrow}$$

(c)   **(d)**

$$\overline{d_{z^2}}\ \overline{d_{x^2-y^2}}$$

$$\underset{d_{xy}}{\uparrow\downarrow}\ \underset{d_{xz}}{\uparrow\downarrow}\ \underset{d_{yz}}{\uparrow}$$

$$\overset{\uparrow}{d_{z^2}}\ \overset{\uparrow}{d_{x^2-y^2}}$$

$$\underset{d_{xy}}{\uparrow\downarrow}\ \underset{d_{xz}}{\uparrow}\ \underset{d_{yz}}{\uparrow}$$

10. What hybrid orbitals are used by the metal ion and how many unpaired electrons are present the complex ion $[VCl_4]^-$ with tetrahedral geometry? (**LO 21.13**)

(a) sp^3; 2 unpaired electrons
(b) sp^3; 3 unpaired electrons
(c) sp^3d^2; 3 unpaired electrons
(d) sp^3d^2; 4 unpaired electrons

Answers:
1. c, 2. a, 3. b, 4. d, 5. d, 6. a, 7. d, 8. d, 9. c, 10. a

Mastering Chemistry provides end-of-chapter exercises, feedback-enriched tutorial problems, animations, and interactive activities to encourage problem-solving practice and deeper understanding of key concepts and topics.

RAN   *Randomized in Mastering Chemistry*

CONCEPTUAL PROBLEMS

Problems 21.1–21.25 appear within the chapter.

21.26 Locate on the periodic table the transition elements with the following electron configurations. Identify each element.
(a) [Ar] $3d^7\ 4s^2$ (b) [Ar] $3d^5 4s^1$
(c) [Kr] $4d^2\ 5s^2$ (d) [Xe] $4f^3\ 6s^2$

21.27 Look at the location in the periodic table of elements A, B, C, and D.

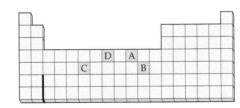

What is the electron configuration of the transition metal in each of the following ions?
(a) A^{2+} (b) B^+
(c) C^{3+} (d) DO_4^{2-}

21.28 What is the general trend in the following properties from left to right across the first-transition series (Sc to Zn)? Explain each trend.

Sc	Ti	V	Cr	Mn	Fe	Co	Ni	Cu	Zn

(a) Atomic radius
(b) Density
(c) Ionization energy
(d) Standard oxidation potential

21.29 The oxalate ion is a bidentate ligand as indicated in Figure 21.8. Would you expect the carbonate ion to be a monodentate or bidentate ligand? Explain your reasoning.

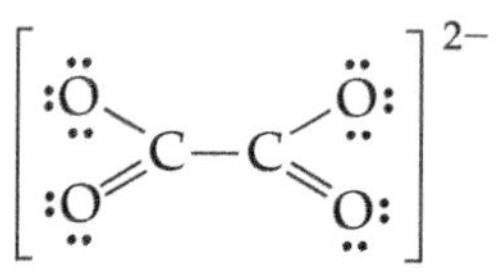

Oxalate ion 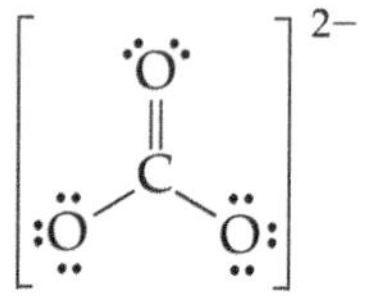   Carbonate ion

21.30 Classify the following ligands as monodentate, bidentate, tridentate, or tetradentate. Which can form chelate rings?

(a)

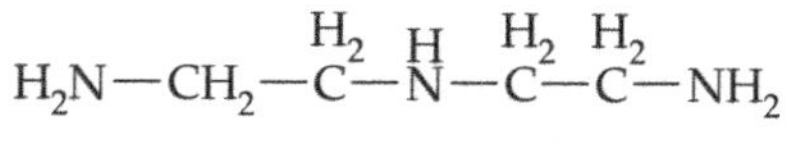

$$H_2N-CH_2-\overset{H_2}{\underset{}{C}}-\overset{H}{\underset{}{N}}-\overset{H_2}{\underset{}{C}}-\overset{H_2}{\underset{}{C}}-NH_2$$

(b)

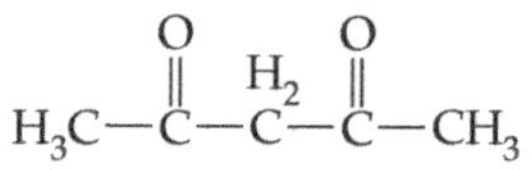

$$H_3C-\overset{O}{\underset{}{C}}-\overset{H_2}{\underset{}{C}}-\overset{O}{\underset{}{C}}-CH_3$$

(c)

HC=CH
HC—N   C—CH
HC   C—C   CH
HC=N   N=CH

(c)

$$\overset{H_2}{C}-CH_2$$
C=N   N=CH
HC=C   C—CH
HC   C—O⁻   ⁻O—C   CH
HC—CH   HC=CH

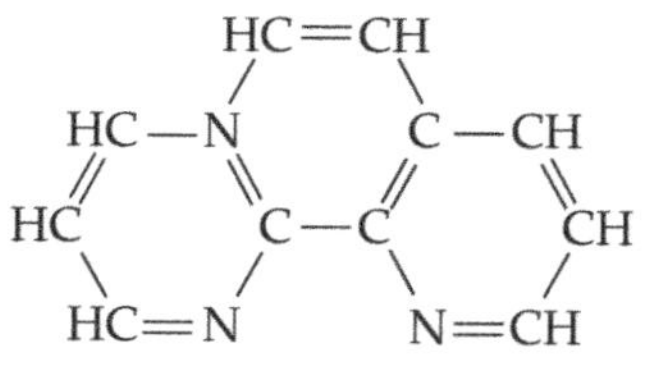

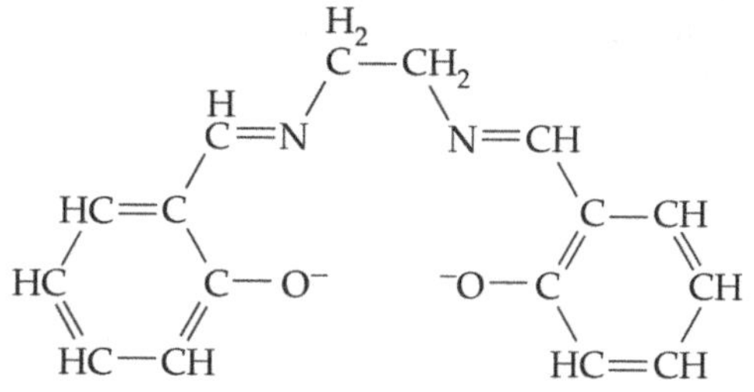

21.31 Classify the following ligands as monodentate, bidentate, or tridentate. Which can form chelate rings?

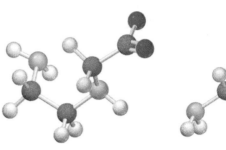

(a) $NH_2CH_2CH_2NH_2$ **(b)** $CH_3CH_2CH_2NH_2$

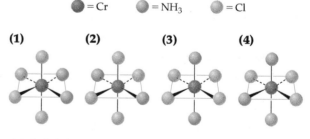

(c) $NH_2CH_2CH_2NHCH_2CO_2^-$ 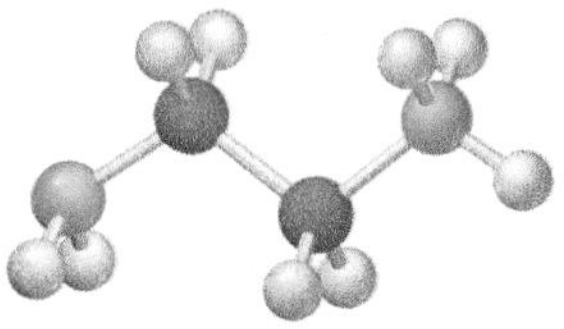   **(d)** $NH_2CH_2CH_2NH_3^+$

21.32 What is the systematic name for each of the following molecules or ions? Include *cis* or *trans* prefixes for diastereoisomers. Platinum is Pt(II) in square planar complexes and Pt(IV) in octahedral complexes.

● = Pt   ⬭ = $NH_2CH_2CH_2NH_2$ ○ = NH_3 ◔ = Cl   ● = H_2O

(1)   **(2)**   **(3)**   **(4)**

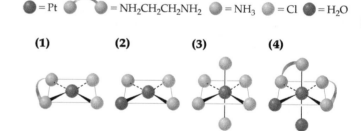

21.33 Consider the following isomers of $[Cr(NH_3)_2Cl_4]^-$.

● = Cr   ○ = NH_3 ○ = Cl

(1)   **(2)**   **(3)**   **(4)**

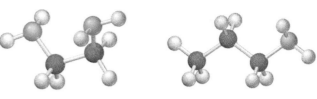

(a) Label the isomers as *cis* or *trans*.
(b) Which isomers are identical, and which are different?
(c) Do any of these isomers exist as enantiomers? Explain.

21.34 Consider the following ethylenediamine complexes.

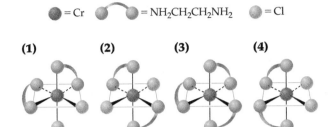

= Cr   = NH$_2$CH$_2$CH$_2$NH$_2$ = Cl

(1)   **(2)**   **(3)**   **(4)**

(a) Which complexes are chiral, and which are achiral?

(b) Draw the enantiomer of each chiral complex.

(c) Which, if any, of the chiral complexes are enantiomers of one another?

21.35 Imagine two complexes, one tetrahedral and one square planar, in which the central atom is bonded to four different ligands (shown here in four different colors). Is either complex chiral? Explain.

SECTION PROBLEMS

Electron Configurations (Section 21.1)

21.36 Use the periodic table to give the electron configuration for each of the following atoms and ions.

(a) Co   (b) Co^{2+}

(c) Co(V) in CoO$_4^{3-}$ (d) Co(IV) in CoF$_6^{2-}$

21.37 Specify the electron configuration for each of the following atoms and ions.

(a) Cr   (b) Cr^{3+}

(c) Cr(III) in Cr$_2$O$_4^{2-}$ (d) Cr(VI) in Cr$_2$O$_7^{2-}$

21.38 Predict the number of unpaired electrons for each of the following.

(a) Cu^{2+} (b) Ti^{2+}

(c) Zn^{2+} (d) Cr^{3+}

21.39 Predict the number of unpaired electrons for each of the following.

(a) Sc^{3+} (b) Co^{2+}

(c) Mn^{3+} (d) Cr^{2+}

Properties of Transition Elements (Section 21.2)

21.40 Titanium, used to make jet aircraft engines, is much harder than potassium or calcium. Explain.

21.41 Molybdenum (mp 2623 °C) has a higher melting point than yttrium (mp 1522 °C) or cadmium (mp 321 °C). Explain.

21.42 Briefly account for each of the following observations:

(a) Atomic radii decrease in the order Sc > Ti > V.

(b) Densities increase in the order Ti < V < Cr.

21.43 Arrange the following atoms in order of decreasing atomic radius, and account for the trend.

(a) Cr   (b) Ti   (c) Mn   (d) V

21.44 What is the lanthanide contraction, and why does it occur?

21.45 The atomic radii of zirconium (160 pm) and hafnium (159 pm) are nearly identical. Explain.

21.46 Calculate the sum of the first two ionization energies for the first-series transition elements, and account for the general trend.

21.47 What is the general trend in standard potentials for the oxidation of first-series transition metals from Sc to Zn? What is the reason for the trend?

21.48 Which element is more easily oxidized, Ti or Zn?
RAN

21.49 Which element is more easily oxidized, Sc or Co?
RAN

21.50 Write a balanced net ionic equation for the reaction of each of the following metals with hydrochloric acid in the absence of air. If no reaction occurs, indicate N.R.

(a) Cr   (b) Zn   (c) Cu   (d) Fe

21.51 Write a balanced net ionic equation for the reaction of each of the following metals with dilute sulfuric acid in the absence of air. If no reaction occurs, indicate N.R.

(a) Mn   (b) Ag   (c) Sc   (d) Ni

Oxidation States (Section 21.3)

21.52 Which of the following transition metals have more than one oxidation state?

(a) Ti   (b) V   (c) Cr   (d) Zn

21.53 Which of the following metals have only one oxidation state?

(a) Cu   (b) Mn   (c) Ni   (d) Sc

21.54 What is the highest oxidation state for each of the elements from Sc to Zn?

21.55 The highest oxidation state for the early transition metals Sc, Ti, V, Cr, and Mn is the periodic group number. The highest oxidation state for the later transition elements Fe, Co, and Ni is less than the periodic group number. Explain.

21.56 Which is the stronger oxidizing agent, Cr^{2+} or Cu^{2+}? Explain.

21.57 Which is the stronger reducing agent, Ti^{2+} or Ni^{2+}? Explain.

21.58 Do you expect a compound with vanadium in the +2 oxidation state to be an oxidizing or a reducing agent? Explain.

21.59 Will a compound that contains a Fe^{6+} ion be an oxidizing agent or a reducing agent? Explain.

21.60 Arrange the following substances in order of increasing strength as an oxidizing agent, and account for the trend.

(a) Mn^{2+} (b) MnO$_2$ (c) MnO$_4^-$

21.61 Arrange the following ions in order of increasing strength as a reducing agent, and account for the trend.

(a) Cr^{2+} (b) Cr^{3+} (c) $Cr_2O_7^{2-}$

Coordination Compounds (Section 21.4)

21.62 What is the coordination number of the metal in each of the
RAN following complexes?

(a) $AgCl_2^-$ (b) $[Cr(H_2O)_5Cl]^{2+}$
(c) $[Co(NCS)_4]^{2-}$ (d) $[ZrF_8]^{4-}$
(e) $[Fe(EDTA)(H_2O)]^-$

21.63 What is the coordination number of the metal in each of the
RAN following complexes?

(a) $[Ni(CN)_5]^{3-}$ (b) $Ni(CO)_4$
(c) $[Co(en)_2(H_2O)Br]^{2+}$ (d) $[Cu(H_2O)_2(C_2O_4)_2]^{2-}$
(e) $Co(NH_3)_3(NO_2)_3$

21.64 What is the oxidation state of the metal in each of the com-
RAN plexes in Problem 21.62?

21.65 What is the oxidation state of the metal in each of the com-
RAN plexes in Problem 21.63?

21.66 Identify the oxidation state of the metal in each of the following compounds.

(a) $Co(NH_3)_3(NO_2)_3$ (b) $[Ag(NH_3)_2]NO_3$
(c) $K_3[Cr(C_2O_4)_2Cl_2]$ (d) $Cs[CuCl_2]$

21.67 What is the oxidation state of the metal in each of the following compounds?

(a) $(NH_4)_3[RhCl_6]$ (b) $[Cr(NH_3)_4(SCN)_2]Br$
(c) $[Cu(en)_2]SO_4$ (d) $Na_2[Mn(EDTA)]$

21.68 What is the formula, including the charge, for each of the following complexes?

(a) An iridium(III) complex with three ammonia and three chloride ligands
(b) A chromium(III) complex with two water and two oxalate ligands
(c) A platinum(IV) complex with two ethylenediamine and two thiocyanate ligands

21.69 What is the formula, including the charge, for each of the
RAN following complexes?

(a) An iron(III) complex with six water ligands
(b) A nickel(II) complex with two ethylenediamine and two bromide ligands
(c) A platinum(II) complex with two chloride and two ammonia ligands

21.70 Draw the structure of the iron oxalate complex $[Fe(C_2O_4)_3]^{3-}$. Describe the coordination geometry, and identify any chelate rings. What are the coordination number and the oxidation number of the iron?

21.71 Draw the structure of the platinum ethylenediamine complex $[Pt(en)_2]^{2+}$. Describe the coordination geometry, and identify any chelate rings. What are the coordination number and the oxidation number of the platinum?

21.72 Draw the structure of the following complexes. What are the oxidation state, coordination number, and coordination geometry of the metal in each?

(a) $Na[Au(CN)_2]$ (b) $[Cr(NH_3)_2(C_2O_4)_2]NO_2$

21.73 Draw the structure of the following complexes. What are the oxidation state, coordination number, and coordination geometry of the metal in each?

(a) $Pt(en)_2$ (b) $[Co(H_2O)_5Cl]SO_4$

Ligands (Section 21.5)

21.74 (a) Identify the Lewis acid and the Lewis base in the reac-
RAN   tion of ethylenediamine ($H_2NCH_2CH_2NH_2$) with Ni^{2+} to form $[Ni(en)_3]^{2+}$.
(b) Identify the ligands and donor atoms.
(c) Give the coordination number and geometry of the metal in the complex.

21.75 Identify the Lewis acid and the Lewis base in the reaction of
RAN oxalate ions ($C_2O_4^{2-}$) with Fe^{3+} to form $[Fe(C_2O_4)_3]^{3-}$.

(a) Identify the ligands and donor atoms.
(b) Give the coordination number and geometry of the metal in the complex.

21.76 What role does $EDTA^{4-}$ play as a trace additive to mayonnaise? Would the glycinate ion ($H_2NCH_2CO_2^-$) be an effective substitute for $EDTA^{4-}$?

21.77 What role does $EDTA^{4-}$ play as a trace additive to shampoo? Would ethylene diammine ($H_2NCH_2CH_2NH_2$) be an effective substitute for $EDTA^{4-}$?

Naming Coordination Compounds (Section 21.6)

21.78 What is the systematic name for each of the following ions?

(a) $[MnCl_4]^{2-}$ (b) $[Ni(NH_3)_6]^{2+}$
(c) $[Co(CO_3)_3]^{3-}$ (d) $[Pt(en)_2(SCN)_2]^{2+}$

21.79 Assign a systematic name to each of the following ions.

(a) $[AuCl_4]^-$ (b) $[Fe(CN)_6]^{4-}$
(c) $[Fe(H_2O)_5NCS]^{2+}$ (d) $[Cr(NH_3)_2(C_2O_4)_2]^-$

21.80 What is the systematic name for each of the following coordination compounds?

(a) $Cs[FeCl_4]$ (b) $[V(H_2O)_6](NO_3)_3$
(c) $[Co(NH_3)_4Br_2]Br$ (d) $Cu(gly)_2$

21.81 What is the systematic name for each of the following compounds?

(a) $[Cu(NH_3)_4]SO_4$ (b) $Cr(CO)_6$
(c) $K_3[Fe(C_2O_4)_3]$ (d) $[Co(en)_2(NH_3)CN]Cl_2$

21.82 Write the formula for each of the following compounds.

(a) Tetraammineplatinum(II) chloride
(b) Sodium hexacyanoferrate(III)
(c) Tris(ethylenediamine)platinum(IV) sulfate
(d) Triamminetrithiocyanatorhodium(III)

21.83 Write the formula for each of the following compounds:

(a) Diamminesilver(I) nitrate
(b) Potassium diaquadioxalatocobaltate(III)
(c) Hexacarbonylmolybdenum(0)
(d) Diamminebis(ethylenediamine)chromium(III) chloride

Isomers (Sections 21.7–21.8)

21.84 Constitutional isomers of a ruthenium(II) coordination compound are shown below.
RAN

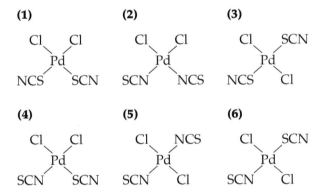

(1)

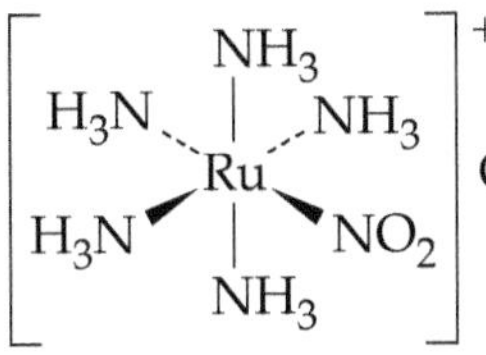

(2)

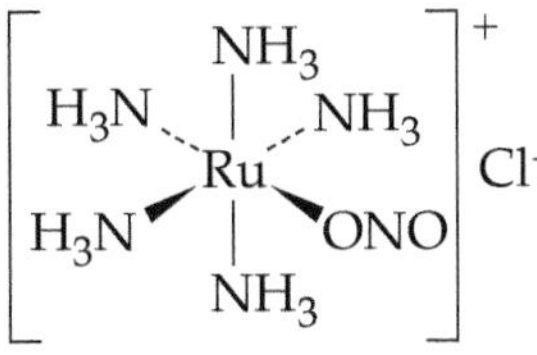

(3)

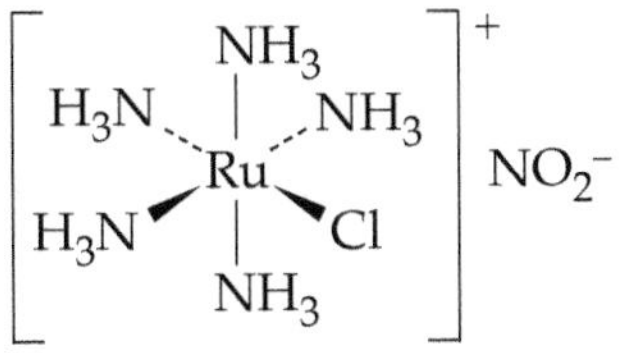

(a) Give the formula and name for structures 1–3.

(b) Which structures are linkage isomers?

(c) Which structures are ionization isomers?

21.85 Six isomers for a square planar palladium(II) complex that contains two Cl^- and two SCN^- ligands are shown below.

(1)   **(2)**   **(3)**

(4)   **(5)**   **(6)**

(a) Which structures are cis-trans isomers?

(b) Which structures are linkage isomers?

21.86 Which of the following complexes can exist as diastereoisomers?

(a) $[Cr(NH_3)_2Cl_4]^-$

(b) $[Co(NH_3)_5Br]^{2+}$

(c) $[MnCl_2Br_2]^{2-}$ (tetrahedral)

(d) $[Pt(NH_3)_2Br_2]^{2-}$ (square planar)

21.87 Tell how many diastereoisomers are possible for each of the following complexes, and draw their structures.

(a) $Pt(NH_3)_3Cl$ (square planar)

(b) $[FeBr_2Cl_2(en)]^-$

(c) $[Cr(H_2O)_4Cl_2]^+$

(d) $Ru(NH_3)_3I_3$

21.88 Which of the following complexes are chiral?

(a) $Pt(en)Cl_2$ (b) cis-$[Co(NH_3)_4Br_2]^+$

(c) cis-$[Cr(en)_2(H_2O)_2]^{3+}$ (d) $[Cr(C_2O_4)_3]^{3-}$

21.89 Which of the following complexes can exist as enantiomers? Draw their structures.

(a) $[Cr(en)_3]^{3+}$

(b) cis-$[Co(en)_2(NH_3)Cl]^{2+}$

(c) $trans$-$[Co(en)_2(NH_3)Cl]^{2+}$

(d) $[Pt(NH_3)_3Cl_3]^+$

21.90 Draw all possible diastereoisomers of $[Cr(C_2O_4)_2(H_2O)_2]^-$. Which can exist as a pair of enantiomers?

21.91 Draw the three possible diastereoisomers of the triethylenetetramine complex $[Co(trien)Cl_2]^+$. Abbreviate the flexible tetradentate trien ligand $H_2NCH_2CH_2NHCH_2CH_2NHCH_2CH_2NH_2$ as N⌢N⌢N⌢N. Which of the isomers can exist as a pair of enantiomers?

21.92 How does plane-polarized light differ from ordinary light? Draw the structure of a chromium complex that rotates the plane of plane-polarized light.

21.93 What is a racemic mixture? Does it affect plane-polarized light? Explain.

21.94 Draw the structure of all isomers of the octahedral complex $[NbX_2Cl_4]^-$ ($X^- = NCS^-$), and identify those that are linkage isomers.

21.95 The glycinate anion, $gly^- = NH_2CH_2CO_2^-$, bonds to metal ions through the N atom and one of the O atoms. Using N⌢O to represent gly^-, sketch the structures of the four stereoisomers of $Co(gly)_3$.

21.96 Draw the structures of all possible diastereoisomers of an octahedral complex with the formula $MA_2B_2C_2$. Which of the diastereoisomers, if any, can exist as enantiomers?

21.97 Tris(2-aminoethyl)amine, abbreviated tren, is the tetradentate ligand $N(CH_2CH_2NH_2)_3$. Using N⌢N to represent each of the three $NCH_2CH_2NH_2$ segments of the ligand, sketch all possible isomers of the octahedral complex $[Co(tren)BrCl]^+$.

21.98 Consider the octahedral complex $[Co(en)(dien)Cl]^{2+}$, where dien $= H_2NCH_2CH_2NHCH_2CH_2NH_2$, which can be abbreviated N⌢N⌢N.

(a) The dien (diethylenetriamine) ligand is a tridentate ligand. Explain what is meant by "tridentate" and why dien can act as a tridentate ligand.

(b) Draw all possible stereoisomers of $[Co(en)(dien)Cl]^{2+}$ (dien is a flexible ligand). Which stereoisomers are chiral, and which are achiral?

21.99 The reaction of the octahedral complex $Co(NH_3)_3(NO_2)_3$ with HCl yields a complex $[Co(NH_3)_3(H_2O)Cl_2]^+$ in which the two chloride ligands are trans to one another.

(a) Draw the two possible stereoisomers of the starting material $[Co(NH_3)_3(NO_2)_3]$. (All three NO_2^- ligands are bonded to Co through the N atom.)

(b) Assuming that the NH_3 groups remain in place, which of the two starting isomers could give rise to the observed product?

Color of Complexes and Crystal Field Theory (Sections 21.9–21.10)

21.100 What is an absorption spectrum? If the absorption spectrum of a complex has just one band at 455 nm, what is the color of the complex?

21.101 A red complex has just one absorption band in the visible region of the spectrum. Predict the approximate wavelength of this band.

21.102 Draw a crystal field energy-level diagram for the $3d$ orbitals of titanium in $[Ti(H_2O)_6]^{3+}$. Indicate the crystal field splitting, and explain why $[Ti(H_2O)_6]^{3+}$ is colored.

21.103 Use a sketch to explain why the d_{xy} and $d_{x^2-y^2}$ orbitals have different energies in an octahedral complex. Which of the two orbitals has higher energy?

21.104 The $[Ti(NCS)_6]^{3-}$ ion exhibits a single absorption band at 544 nm. Calculate the crystal field splitting energy Δ in kJ/mol. Is NCS^- a stronger or weaker field ligand than water? Predict the color of $[Ti(NCS)_6]^{3-}$.

21.105 The $[Cr(H_2O)_6]^{3+}$ ion is violet, and $[Cr(CN)_6]^{3-}$ is yellow. Explain this difference using crystal field theory. Use the colors to order H_2O and CN^- in the spectrochemical series.

21.106 For each of the following complexes, draw a crystal field energy-level diagram, assign the electrons to orbitals, and predict the number of unpaired electrons.
(a) $[CrF_6]^{3-}$
(b) $[V(H_2O)_6]^{3+}$
(c) $[Fe(CN)_6]^{3-}$

21.107 Draw a crystal field energy-level diagram, assign the electrons to orbitals, and predict the number of unpaired electrons for each of the following.
(a) $[Cu(en)_3]^{2+}$
(b) $[FeF_6]^{3-}$
(c) $[Co(en)_3]^{3+}$ (low-spin)

21.108 The $Ni^{2+}(aq)$ cation is green, but $Zn^{2+}(aq)$ is colorless. Explain.

21.109 The $Cr^{3+}(aq)$ cation is violet, but $Y^{3+}(aq)$ is colorless. Explain.

21.110 Weak-field ligands tend to give high-spin complexes, but strong-field ligands tend to give low-spin complexes. Explain.

21.111 Explain why nearly all tetrahedral complexes are high-spin.

21.112 Draw a crystal field energy-level diagram for a square planar complex, and explain why square planar geometry is especially common for d^8 complexes.

21.113 For each of the following complexes, draw a crystal field energy-level diagram, assign the electrons to orbitals, and predict the number of unpaired electrons.
(a) $[Pt(NH_3)_4]^{2+}$ (square planar)
(b) $[MnCl_4]^{2-}$ (tetrahedral)
(c) $[Co(NCS)_4]^{2-}$ (tetrahedral)
(d) $[Cu(en)_2]^{2+}$ (square planar)

21.114 Which of the following complexes are paramagnetic?
(a) $[Mn(CN)_6]^{3-}$
(b) $[Zn(NH_3)_4]^{2+}$ (tetrahedral)
(c) $[Fe(CN)_6]^{4-}$
(d) $[FeF_6]^{4-}$

21.115 Which of the following complexes are diamagnetic?
(a) $[Ni(H_2O)_6]^{2+}$
(b) $[Co(CN)_6]^{3-}$
(c) $[HgI_4]^{2-}$ (tetrahedral)
(d) $[Cu(NH_3)_4]^{2+}$ (square planar)

21.116 Although Cl^- is a weak-field ligand and CN^- is a strong-field ligand, $[CrCl_6]^{3-}$ and $[Cr(CN)_6]^{3-}$ exhibit approximately the same amount of paramagnetism. Explain.

21.117 In octahedral complexes, the choice between high-spin and low-spin electron configurations arises only for $d^4 - d^7$ complexes. Explain.

21.118 Draw a crystal field energy-level diagram, and predict the number of unpaired electrons for each of the following:
(a) $[Mn(H_2O)_6]^{2+}$
(b) $Pt(NH_3)_2Cl_2$
(c) $[FeO_4]^{2-}$
(d) $[Ru(NH_3)_6]^{2+}$ (low-spin)

21.119 Explain why $[CoCl_4]^{2-}$ (blue) and $[Co(H_2O)_6]^{2+}$ (pink) have different colors. Which complex has its absorption bands at longer wavelengths?

21.120 Look at the colors of the isomeric complexes in Figure 21.12, and predict which is the stronger field ligand, nitro ($-NO_2$) or nitrito ($-ONO$). Explain.

21.121 Predict the crystal field energy-level diagram for a linear ML_2 complex that has two ligands along the $\pm z$ axis:

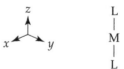

21.122 Predict the crystal field energy-level diagram for a square pyramidal ML_5 complex that has two ligands along the $\pm x$ and $\pm y$ axes but only one ligand along the z axis. Your diagram should be intermediate between those for an octahedral ML_6 complex and a square planar ML_4 complex.

21.123 For each of the following, (i) give the systematic name of the compound and specify the oxidation state of the transition metal, (ii) draw a crystal field energy-level diagram and assign the d electrons to orbitals, (iii) indicate whether the complex is high-spin or low-spin (for $d^4 - d^7$ complexes), and (iv) specify the number of unpaired electrons.

(a) $(NH_4)[Cr(H_2O)_6](SO_4)_2$

(b) $Mo(CO)_6$ (CO is a strong-field ligand)

(c) $[Ni(NH_3)_4(H_2O)_2](NO_3)_2$

(d) $K_4[Os(CN)_6]$

(e) $[Pt(NH_3)_4](ClO_4)_2$

(f) $Na_2[Fe(CO)_4]$

21.124 The drug Nipride, $Na_2[Fe(CN)_5NO]$, is an inorganic complex used as a source of NO to lower blood pressure during surgery.

(a) The nitrosyl ligand in this complex is believed to be NO^+ rather than neutral NO. What is the oxidation state of iron, and what is the systematic name for $Na_2[Fe(CN)_5NO]$?

(b) Draw a crystal field energy-level diagram for $[Fe(CN)_5NO]^{2-}$, assign the electrons to orbitals, and predict the number of unpaired electrons.

21.125 Based on the colors of their Cr(III) complexes, arrange the following ligands in a spectrochemical series in order of increasing value of the crystal field splitting Δ: acac⁻ (a bidentate ligand), $CH_3CO_2^-$ (acetate), Cl⁻, H_2O, NH_3, and urea. The colors of the Cr(III) complexes are red for $Cr(acac)_3$, violet for $[Cr(H_2O)_6]^{3+}$, green for $[CrCl_2(H_2O)_4]^+$, green for $[Cr(urea)_6]^{3+}$, yellow for $[Cr(NH_3)_6]^{3+}$, and blue-violet for $Cr(CH_3CO_2)_3(H_2O)_3$.

Valence Bond Theory (Section 21.11)

21.126 Give a valence bond description of the bonding in each of the following complexes. Include orbital diagrams for the

free metal ion and the metal ion in the complex. Indicate which hybrid orbitals the metal ion uses for bonding, and specify the number of unpaired electrons.

(a) $[Ti(H_2O)_6]^{3+}$

(b) $[NiBr_4]^{2-}$ (tetrahedral)

(c) $[Fe(CN)_6]^{3-}$ (low-spin)

(d) $[MnCl_6]^{32}$ (high-spin)

21.127 For each of the following complexes, describe the bonding using valence bond theory. Include orbital diagrams for the free metal ion and the metal ion in the complex. Indicate which hybrid orbitals the metal ion uses for bonding, and specify the number of unpaired electrons.

(a) $[AuCl_4]^2$ (square planar)

(b) $[Ag(NH_3)_2]^+$

(c) $[Fe(H_2O)_6]^{2+}$ (high-spin)

(d) $[Fe(CN)_6]^{4-}$ (low-spin)

21.128 There are two possible $[M(OH)_4]^-$ complexes of first-series transition metals that have three unpaired electrons.

(a) What are the oxidation state and the identity of M in these complexes?

(b) Using orbital diagrams, give a valence bond description of the bonding in each complex.

(c) Based on common oxidation states of first-series transition metals (Figure 21.6), which $[M(OH)_4]^-$ complex is more likely to exist?

21.129 Two first-series transition metals have three unpaired electrons in complex ions of the type $[MCl_4]^{2-}$.

(a) What are the oxidation state and the identity of M in these ions?

(b) Draw valence bond orbital diagrams for the two possible ions.

(c) Based on common oxidation states of first-series transition metals (Figure 21.6), which ion is more likely to exist?

MULTICONCEPT PROBLEMS

21.130 There are three coordination compounds with the empirical formula $Co(NH_3)_3(NO_2)_3$ in which all the nitrite ions are bonded through the N atom. Two isomers have the same molar mass but different nonzero dipole moments. The third compound is a salt with singly charged ions and a molar mass twice that of the other compounds.

(a) Draw the structures of the isomeric compounds.

(b) What is the chemical formula of the third compound?

21.131 Nickel(II) complexes with the formula NiX_2L_2, where X⁻ is Cl⁻ or N-bonded NCS⁻ and L is the monodentate triphenylphosphine ligand $P(C_6H_5)_3$, can be square planar or tetrahedral.

(a) Draw crystal field energy-level diagrams for a square planar and a tetrahedral nickel(II) complex, and show the population of the orbitals.

(b) If $NiCl_2L_2$ is paramagnetic and $Ni(NCS)_2L_2$ is diamagnetic, which of the two complexes is tetrahedral and which is square planar?

(c) Draw possible structures for each of the NiX_2L_2 complexes, and tell which ones have a dipole moment.

21.132 Describe the bonding in $[Mn(CN)_6]^{3-}$, using both crystal field theory and valence bond theory. Include the appropriate crystal field d-orbital energy-level diagram and the valence bond orbital diagram. Which model allows you to predict the number of unpaired electrons? How many do you expect?

21.133 The amount of paramagnetism for a first-series transition metal complex is related approximately to its spin-only magnetic moment. The spin-only value of the magnetic moment in units of Bohr magnetons (BM) is given by $\sqrt{n(n + 2)}$, where n is the number of unpaired electrons. Calculate the spin-only value of the magnetic moment for the 2+ ions of the first-series transition metals (except Sc) in octahedral complexes with (a) weak-field ligands and (b) strong-field ligands. For which electron configurations can the magnetic moment distinguish between high-spin and low-spin electron configurations?

21.134 Spinach contains a lot of iron but is not a good source of RAN dietary iron because nearly all the iron is tied up in the oxalate complex $[Fe(C_2O_4)_3]^{3-}$.

(a) The formation constant K_f for $[Fe(C_2O_4)_3]^{3-}$ is 3.3×10^{20}. Calculate the equilibrium concentration of free Fe^{3+} in a 0.100 M solution of $[Fe(C_2O_4)_3]^{3-}$. (Ignore any acid–base reactions.)

(b) Under the acidic conditions in the stomach, the Fe^{3+} concentration should be greater because of the reaction

$$[Fe(C_2O_4)_3]^{3-}(aq) + 6\,H_3O^+(aq) \rightleftharpoons$$
$$Fe^{3+}(aq) + 3\,H_2C_2O_4\,(aq) + 6\,H_2O(l)$$

Show, however, that this reaction is nonspontaneous under standard-state conditions. (For $H_2C_2O_4$, $K_{a1} = 5.9 \times 10^{-2}$ and $K_{a2} = 6.4 \times 10^{-5}$.)

(c) Draw a crystal field energy-level diagram for $[Fe(C_2O_4)_3]^{3-}$, and predict the number of unpaired electrons. ($C_2O_4^{2-}$ is a weak-field bidentate ligand.)

(d) Draw the structure of $[Fe(C_2O_4)_3]^{3-}$. Is the complex chiral or achiral?

21.135 Formation constants for the ammonia and ethylenediamine complexes of nickel(II) indicate that $Ni(en)_3^{2+}$ is much more stable than $Ni(NH_3)_6^{2+}$:

(1) $Ni(H_2O)_6^{2+}(aq) + 6\,NH_3(aq) \rightleftharpoons$
$\quad Ni(NH_3)_6^{2+}(aq) + 6\,H_2O(l) \qquad K_f = 2.0 \times 10^8$

(2) $Ni(H_2O)_6^{2+}(aq) + 3\,en(aq) \rightleftharpoons$
$\quad Ni(en)_3^{2+}(aq) + 6\,H_2O(l) \qquad K_f = 4 \times 10^{17}$

The enthalpy changes for the two reactions, $\Delta H°_1$ and $\Delta H°_2$, should be about the same because both complexes have six Ni − N bonds.

(a) Which of the two reactions should have the larger entropy change, $\Delta S°$? Explain.

(b) Account for the greater stability of $Ni(en)_3^{2+}$ in terms of the relative values of $\Delta S°$ for the two reactions.

(c) Assuming that $\Delta H°_2 - \Delta H°_1$ is zero, calculate the value of $\Delta S°_2 - \Delta S°_1$.

21.136 The percent iron in iron ore can be determined by dissolving
RAN the ore in acid, then reducing the iron to Fe^{2+}, and finally titrating the Fe^{2+} with aqueous $KMnO_4$. The reaction products are Fe^{2+} and Mn^{2+}.

(a) Write a balanced net ionic equation for the titration reaction.

(b) Use the $E°$ values in Appendix D to calculate $\Delta G°$ (in kilojoules) and the equilibrium constant for the reaction.

(c) Draw a crystal field energy-level diagram for the reactants and products, MnO_4^-, $[Fe(H_2O)_6]^{2+}$, $[Fe(H_2O)_6]^{3+}$, and $[Mn(H_2O)_6]^{2+}$, and predict the number of unpaired electrons for each.

(d) Does the paramagnetism of the solution increase or decrease as the reaction proceeds? Explain.

(e) What is the mass % Fe in the iron ore if titration of the Fe^{2+} from a 1.265-g sample of ore requires 34.83 mL of 0.051 32 M $KMnO_4$ to reach the equivalence point?

21.137 The complete reaction of 2.60 g of chromium metal with
RAN 50.00 mL of 1.200 M H_2SO_4 in the absence of air gave a blue solution and a colorless gas that was collected at 25 °C and a pressure of 735 mm Hg.

(a) Write a balanced net ionic equation for the reaction.

(b) How many liters of gas were produced?

(c) What is the pH of the solution?

(d) Describe the bonding in the blue-colored ion, using both the crystal field theory and the valence bond theory. Include the appropriate crystal field d-orbital energy-level diagram and the valence bond orbital diagram. Identify the hybrid orbitals used in the valence bond description.

(e) When an excess of KCN is added to the solution, the color changes, and the paramagnetism of the solution decreases. Explain.

21.138 In acidic aqueous solution, the complex $trans$-$[Co(en)_2Cl_2]^+$
RAN undergoes the following substitution reaction:

$$trans\text{-}[Co(en)_2Cl_2]^+(aq) + H_2O(l) \rightarrow$$
$$trans\text{-}[Co(en)_2(H_2O)Cl]^{2+}(aq) + Cl^-(aq)$$

The reaction is first order in $trans$-$[Co(en)_2Cl_2]^+$, and the rate constant at 25 °C is $3.2 \times 10^{-5}\ s^{-1}$.

(a) What is the half-life of the reaction in hours?

(b) If the initial concentration of $trans$-$[Co(en)_2Cl_2]^+$ is 0.138 M, what is its molarity after a reaction time of 16.5 h?

(c) Devise a possible reaction mechanism with a unimolecular rate-determining step.

(d) Is the reaction product chiral or achiral? Explain.

(e) Draw a crystal field energy-level diagram for $trans$-$[Co(en)_2Cl_2]^+$ that takes account of the fact that Cl^- is a weaker-field ligand than ethylenediamine.

21.139 Chromium forms three isomeric compounds A, B, and C
RAN with percent composition 19.52% Cr, 39.91% Cl, and 40.57% H_2O. When a sample of each compound was dissolved in water and aqueous $AgNO_3$ was added, a precipitate of AgCl formed immediately. A 0.225-g sample of compound A gave 0.363 g of AgCl, 0.263 g of B gave 0.283 g of AgCl, and 0.358 g of C gave 0.193 g of AgCl. One of the three compounds is violet, while the other two are green. In all three, chromium has coordination number 6.

(a) What are the empirical formulas of A, B, and C?

(b) What are the probable structural formulas of A, B, and C? Draw the structure of the cation in each compound. Which cation can exist as diastereoisomers?

(c) Which of the three compounds is likely to be the violet one? Explain.

(d) What are the approximate freezing points of 0.25 m solutions of A, B, and C, assuming complete dissociation?

21.140 Cobalt(III) trifluoroacetylacetonate, $Co(tfac)_3$, is a six-coordinate, octahedral metal chelate in which three planar, bidentate tfac ligands are attached to a central Co atom:

(a) Draw all possible diastereoisomers and enantiomers of $Co(tfac)_3$.

(b) Diastereoisomers A and B have dipole moments of 6.5 D and 3.8 D, respectively. Which of your diastereoisomers is A and which is B?

(c) The isomerization reaction $A \rightarrow B$ in chloroform solution has first-order rate constants of $0.0889 \ h^{-1}$ at 66.1 °C and $0.0870 \ min^{-1}$ at 99.2 °C. What is the activation energy for the reaction?

(d) Draw a crystal field energy-level diagram for $Co(tfac)_3$, and predict its magnetic properties. (In this complex, tfac is a strong-field ligand.)

21.141 Consider the following reaction, and assume that its equilibrium constant is 1.00×10^{14}:

$$2 \ CrO_4^{2-}(aq) + 2 \ H^+(aq) \rightleftharpoons Cr_2O_7^{2-}(aq) + H_2O(l)$$

(a) Write the equilibrium equation for the reaction, and explain why CrO_4^{2-} ions predominate in basic solutions and $Cr_2O_7^{2-}$ ions predominate in acidic solutions.

(b) Calculate the CrO_4^{2-} and $Cr_2O_7^{2-}$ concentrations in a solution that has a total chromium concentration of 0.100 M and a pH of 4.000.

(c) What are the CrO_4^{2-} and $Cr_2O_7^{2-}$ concentrations if the pH is 2.000?

21.142 An alternative to cyanide leaching of gold ores is leaching with thiocyanate ion, which forms a square planar gold(III) complex, $[Au(SCN)_4]^-$.

(a) If the formation constant for $[Au(SCN)_4]^-$ is $K_f = 10^{37}$, what is the equilibrium concentration of Au^{3+} in a 0.050 M solution of $[Au(SCN)_4]^-$?

(b) Draw a crystal field energy-level diagram for $[Au(SCN)_4]^-$, and predict the number of unpaired electrons.

chapter 22

The Main-Group Elements

Contents

Hydrogen is an enormously attractive fuel because it's environmentally clean, giving only water as a combustion product. As a result, some people envision what they call a "hydrogen economy" in which our energy needs are met by gaseous, liquid, and solid hydrogen.

What are the barriers to a hydrogen economy?

The answer to this question can be found on page 990 in the INQUIRY

The main-group elements are the 50 elements that occupy groups 1A–8A of the periodic table. They are subdivided into the *s*-block elements of groups 1A and 2A, with valence electron configuration ns^1 or ns^2, and the *p*-block elements of groups 3A–8A, with valence configurations $ns^2\,np^{1-6}$. Main-group elements are important because of their high natural abundance and their presence in commercially valuable chemicals. Eight of the ten most abundant elements in the Earth's crust and all ten of the most abundant elements in the human body are main-group elements (**FIGURE 22.1**). Many of the most widely used industrial chemicals contain only main-group elements. Examples include sulfuric acid (H_2SO_4), used in fertilizers, oil refining, and chemical manufacturing; ethylene (C_2H_4), used in plastics and antifreeze; lime (CaO), used in steel making and water treatment; ammonia (NH_3), used in fertilizers; and chlorine (Cl_2), used in water treatment and plastics.

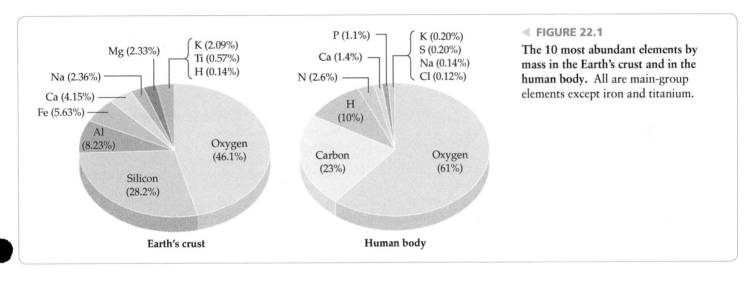

◄ **FIGURE 22.1**

The 10 most abundant elements by mass in the Earth's crust and in the human body. All are main-group elements except iron and titanium.

22.1 A REVIEW OF GENERAL PROPERTIES AND PERIODIC TRENDS

Let's begin our survey of the main-group elements by reviewing the periodic trends that make it possible to classify these elements as metals, nonmetals, or semimetals. **FIGURE 22.2** shows the main-group regions of the periodic table, with metals to the left of the heavy stairstep line, nonmetals to the right of the line, and semimetals—elements with intermediate properties—along the line. The elements usually classified as semimetals are boron (group 3A), silicon and germanium (group 4A), arsenic and antimony (group 5A), tellurium (group 6A), and astatine (group 7A).

From left to right across the periodic table, the **effective nuclear charge** Z_{eff} increases because each additional valence electron does not completely shield the additional nuclear charge (Section 5.10). As a result, the atom's electrons are more strongly attracted to the nucleus, ionization energy generally increases, atomic radius decreases, and electronegativity increases, as shown in Figure 22.2. The elements on the left side of the table tend to form cations by losing electrons, and those on the right tend to form anions by gaining electrons. Thus, metallic character decreases and nonmetallic character increases across the table from left to right. In the third row, for example, sodium, magnesium, and aluminum are metals, silicon is a semimetal, and phosphorus, sulfur, and chlorine are nonmetals.

From the top to the bottom of a group in the periodic table, additional shells of electrons are occupied and atomic radius therefore increases. Because the valence electrons are farther from the nucleus, though, ionization energy and electronegativity generally decrease. As a result, metallic character increases and nonmetallic character decreases down a group from top to bottom. In group 4A, for example, carbon is a nonmetal, silicon and germanium are semimetals, and tin and lead are metals. The horizontal and vertical periodic trends combine to locate the element with the most

REMEMBER . . .
The nuclear charge actually felt by an electron, called the **effective nuclear charge** Z_{eff}, is less than the actual nuclear charge because of electron shielding. The valence electrons are less well shielded by other valence electrons than by the inner-shell electrons (Section 5.10).

► **FIGURE 22.2**

Periodic trends in the properties of the main-group elements.

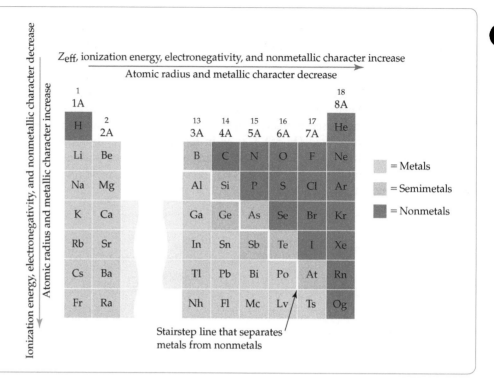

► **FIGURE 22.2**

Periodic trends in the properties of the main-group elements.

► **Figure It Out**

Where are nonmetals located in the periodic table? Which element has a higher value of Z_{eff} and higher electronegativity: S or Al?

Answer: Nonmetals are located on the upper-right side of the periodic table. S has a higher value of Z_{eff} and higher electronegativity because these properties increase across a period.

metallic character (francium) in the lower left of the periodic table, the element with the most nonmetallic character (fluorine) in the upper right, and the semimetals along the diagonal stairstep that stretches across the middle. **TABLE 22.1** summarizes some of the properties that distinguish metallic and nonmetallic elements.

TABLE 22.1   Properties of Metallic and Nonmetallic Elements

Metals	Nonmetals
All are solids at 25 °C except Hg, which is a liquid	Eleven are gases at 25 °C, one is a liquid (Br), and five are solids (C, P, S, Se, and I)
Most have a silvery shine	Most lack a metallic luster
Malleable and ductile	Nonmalleable and brittle
Good conductors of heat and electricity	Poor conductors of heat and electricity, except graphite
Relatively low ionization energies	Relatively high ionization energies
Relatively low electronegativities	Relatively high electronegativities
Lose electrons to form cations	Gain electrons to form anions; share electrons to form oxoanions

— **WORKED EXAMPLE 22.1**

Using the Periodic Table to Predict Metallic Character

Use the periodic table to predict which element in each of the following pairs has more metallic character.

(a) Ga or As   (b) P or Bi   (c) Sb or S   (d) Sn or Ba

STRATEGY

Because metallic character increases from right to left and from top to bottom in the periodic table, look to see which of the pair of elements lies farther toward the lower left of the table.

SOLUTION

(a) Ga and As are in the same row of the periodic table, but Ga (group 3A) lies to the left of As (group 5A). Therefore, Ga is more metallic.

(b) Bi lies below P in group 5A and is therefore more metallic.
(c) Sb (group 5A) has more metallic character because it lies below and to the left of S (group 6A).
(d) Ba (group 2A) is more metallic because it lies below and to the left of Sn (group 4A).

▶ **PRACTICE 22.1** Which element has more nonmetallic character: Cl or Sb?

▶ **APPLY 22.2** Element A has a lower ionization energy, lower electronegativity, and larger size than element B. Which element has more metallic character?

All **PRACTICE** and **APPLY** problems are interactive in the eText.

22.2 DISTINCTIVE PROPERTIES OF THE SECOND-ROW ELEMENTS

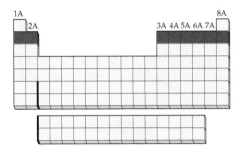

The properties of elements in the second row of the periodic table differ markedly from those of heavier elements in the same periodic group. The second-row atoms have especially small sizes and especially high electronegativities. In group 5A, for example, the electronegativity of N is 3.0, whereas the electronegativities of P, As, Sb, and Bi are all in the range 2.1–1.9. Figure 7.4 shows the discontinuity in electronegativity that distinguishes the second-row elements from other elements of the same periodic group.

The small sizes and high electronegativities of the second-row elements accentuate their nonmetallic behavior. Boron differs from the metallic elements of group 3A in forming mainly covalent, molecular compounds. For example, BF_3 (bp = -100 °C) is a gaseous, molecular halide, but AIF_3 (mp 1290°C) is a typical high-melting, ionic solid. Furthermore, **hydrogen bonding** interactions are generally restricted to compounds of the highly electronegative second-row elements N, O, and F (Section 8.6). Recall also that HF contrasts with HCl, HBr, and HI in being the only **weak hydrohalic acid** (Section 16.3).

Because of the small sizes of their atoms, the second-row elements generally form a maximum of four covalent bonds. By contrast, the larger atoms of the third-row elements can accommodate more than four nearest neighbors and can therefore form more than four bonds. Thus, nitrogen forms only NCl_3, but phosphorus forms both PCl_3 and PCl_5.

REMEMBER . . .
A **hydrogen bond** is an attractive interaction between an H atom bonded to a very electronegative atom (O, N, or F) and an unshared electron pair on another electronegative atom. (Section 8.6)

REMEMBER . . .
A **weak acid** HA is only partially dissociated in water to give H_3O^+ and A^- ions, whereas a strong acid is essentially 100% dissociated. (Section 15.3)

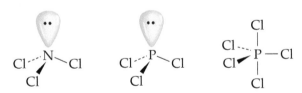

A further consequence of the small size of the second-row atoms C, N, and O is their ability to form multiple bonds by π overlap of the $2p$ orbitals. By contrast, the $3p$ orbitals of the corresponding third-row atoms Si, P, and S are more diffuse, and the longer bond distances for these larger atoms result in poor π overlap.

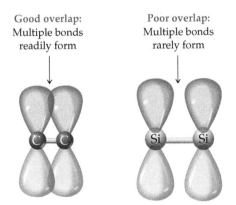

Good overlap:
Multiple bonds
readily form

Poor overlap:
Multiple bonds
rarely form

As a result of this poor overlap, π bonds involving p orbitals are rare for elements of the third and higher rows. Although compounds with C=C and C≡C multiple bonds are common, molecules with Si=Si and Si≡Si multiple bonds are uncommon and have been synthesized only recently. In group 5A, elemental nitrogen contains triply bonded N_2 molecules, whereas white phosphorus contains tetrahedral P_4 molecules in which each P atom forms three single bonds rather than one triple bond. Similarly, O_2 contains an $O{=}O$ double bond, whereas elemental sulfur contains crown-shaped S_8 rings in which each S atom forms two single bonds rather than one double bond.

$N{\equiv}N$

Group 5A

Group 6A

WORKED EXAMPLE 22.2

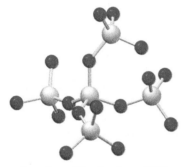

Carbon dioxide

A portion of the structure of SiO_2

Exploring Differences between Second- and Third-Row Elements

Account for the following observations:

(a) CO_2 is a gaseous molecular substance, whereas SiO_2 is a covalent network solid in which SiO_4 tetrahedra are linked to four neighboring SiO_4 tetrahedra by shared oxygen atoms.

$$O{=}C{=}O$$

(b) Glass made of SiO_2 is attacked by hydrofluoric acid with the formation of $SiF_6{}^{2-}$ anions. The analogous $CF_6{}^{2-}$ anion does not exist.

STRATEGY

To account for these differences, remember that:

(a) Second-row atoms are smaller and form stronger multiple bonds than third-row atoms.

(b) Third-row atoms can form more than four bonds because of their larger size.

SOLUTION

(a) Because of its small size and good π overlap with other small atoms, carbon forms strong double bonds with two oxygens to give discrete CO_2 molecules. Because the larger Si atom does not have good π overlap with other atoms, it uses its four valence electrons to form four single bonds rather than two double bonds.

(b) The larger silicon atom can bond to six F^- ions, whereas the smaller carbon atom can form a maximum of only four bonds.

▶ **PRACTICE 22.3**   The organic solvent acetone has the molecular formula $(CH_3)_2CO$. The silicon analogue, a thermally stable lubricant, is a polymer, $[(CH_3)_2SiO]_n$. What is the reason for the difference in structure?

Acetone

A portion of a dimethylsiloxane polymer chain

▶ **APPLY 22.4**   Phosphorus forms PCl_5, but nitrogen bonds to a maximum of three Cl atoms, yielding NCl_3. Explain.

22.3 GROUP 1A: HYDROGEN

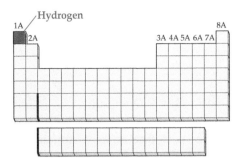

As described in Section 22.1, metallic character increases down a group because the valence electrons are farther from the nucleus and ionization energy and electronegativity decrease. Hydrogen, at the top of group 1A, is a nonmetal with different chemical and physical properties than the other alkali metal elements of group 1A.

Elemental Hydrogen

At ordinary temperatures and pressures, hydrogen is a colorless, odorless gas comprised of diatomic H_2 molecules. Because H_2 is a nonpolar molecule that contains only two electrons, intermolecular forces are extremely weak (Section 8.6). As a result, hydrogen has a very low melting point ($-259.2 \,°C$) and a very low boiling point ($-252.8 \,°C$). The bonding forces within the H_2 molecule are exceptionally strong, however, the H—H bond dissociation energy is 436 kJ/mol, greater than that for any other single bond between two atoms of the same element (Section 7.2):

$$H_2(g) \longrightarrow 2\,H(g) \qquad D = 436 \text{ kJ/mol}$$

Even at 2000 K, only 1 of every 2500 H_2 molecules is dissociated into H atoms at 1 atm pressure.

Hydrogen is thought to account for approximately 75% of the mass of the universe. Our Sun and other stars, for instance, are composed mainly of hydrogen, which serves as their nuclear fuel. On Earth, though, hydrogen is rarely found in uncombined form because the Earth's gravity is too weak to hold such a light molecule. The Earth's atmosphere contains only 0.53 ppm of H_2 by volume. In the Earth's crust and oceans, hydrogen is the ninth most abundant element on a mass basis (0.9 mass %) and the third most abundant on an atom basis (15.4 atom %). Hydrogen is found in water, petroleum, proteins, carbohydrates, fats, and many millions of other compounds.

We'll discuss large-scale production of hydrogen in the *Inquiry* at the end of the chapter, but small amounts of hydrogen are conveniently prepared in the laboratory by the reaction of dilute acid with an active metal such as zinc:

$$Zn(s) + 2\,H^+(aq) \longrightarrow H_2(g) + Zn^{2+}(aq)$$

A typical apparatus for generating and collecting hydrogen is shown in **FIGURE 22.3**.

▲ The Lagoon Nebula in the constellation Sagittarius. These interstellar gas clouds consist largely of atomic hydrogen, the most abundant element in the universe. The gas is heated by radiation from nearby stars. The characteristic red glow is due to the emission spectrum of hydrogen.

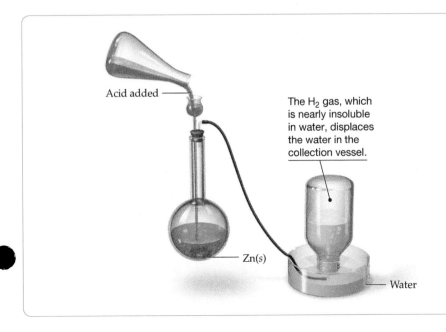

Acid added

The H_2 gas, which is nearly insoluble in water, displaces the water in the collection vessel.

Zn(s)

Water

◀ **FIGURE 22.3**

Preparation of hydrogen by the reaction of zinc metal with dilute acid.

Hydrogen Compounds

The hydrogen atom is the simplest of all atoms, containing only a single $1s$ electron and a single proton. In most versions of the periodic table, hydrogen is located in group 1A above the alkali metals because they too have just one valence electron. Alternatively, hydrogen could be placed in group 7A above the halogens because, like the halogens, it is just one electron short of a noble gas configuration (He). Thus, hydrogen has properties similar to those of both alkali metals and halogens. A hydrogen atom can lose an electron to form a hydrogen cation, H^+, or it can gain an electron to yield a hydride anion, H^-:

Alkali-metal-like reaction: $H(g) \longrightarrow H^+(g) + e^-$ $E_i = +1312 \text{ kJ/mol}$

Halogen-like reaction: $H(g) + e^- \longrightarrow H^-(g)$ $E_{ea} = -73 \text{ kJ/mol}$

Hydrogen doesn't completely transfer its valence electron in chemical reactions, because the amount of energy needed to ionize a hydrogen atom is so large ($E_i = 1312 \text{ kJ/mol}$). Instead, it shares this electron with a nonmetallic element to give a covalent compound such as CH_4, NH_3, H_2O, or HF. In this regard, hydrogen differs markedly from the alkali metals, which have much smaller ionization energies, ranging from 520 kJ/mol for Li to 376 kJ/mol for Cs, and form ionic compounds with nonmetals.

Although adding an electron to hydrogen ($E_{ea} = -73 \text{ kJ/mol}$) releases less energy than adding an electron to the halogens ($E_{ea} = -295$ to -349 kJ/mol), hydrogen will accept an electron from an active metal to give an ionic hydride, such as NaH or CaH_2. In this regard, the behavior of hydrogen parallels that of the halogens, which form ionic halides such as NaCl and $CaCl_2$.

Binary Hydrides

The **binary hydrides** are compounds that contain hydrogen and just one other element. Formulas and melting points of the simplest hydrides of the main-group elements are listed in **FIGURE 22.4.** Binary hydrides can be classified as ionic, covalent, or metallic. **Ionic hydrides** are saltlike, high-melting, white, crystalline compounds formed by the alkali metals and the heavier alkaline-earth metals Ca, Sr, and Ba. They can be prepared by direct reaction of the elements at about 400 °C:

$$2 \text{ Na}(l) + H_2(g) \longrightarrow 2 \text{ NaH}(s) \quad \Delta H° = -112.6 \text{ kJ}$$
$$\text{Ca}(s) + H_2(g) \longrightarrow \text{CaH}_2(s) \quad \Delta H° = -181.5 \text{ kJ}$$

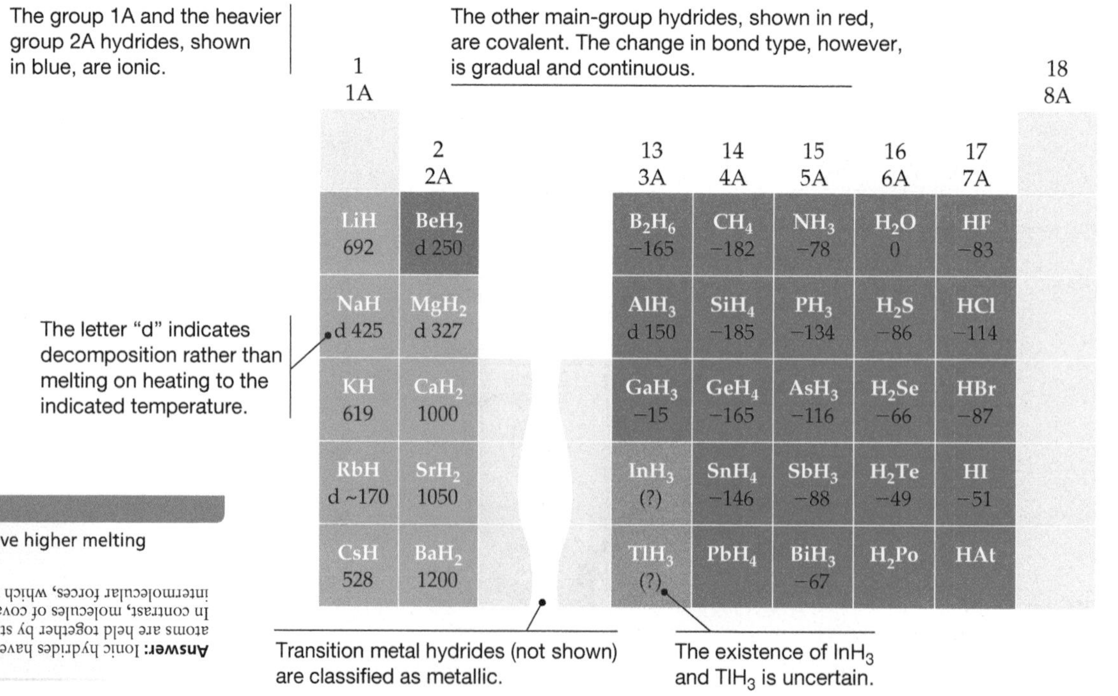

BIG IDEA Question 1   Go to eText

What is the chemical formula of the compound that contains strontium and hydrogen? Does the compound have ionic or covalent bonding?

The group 1A and the heavier group 2A hydrides, shown in blue, are ionic.

The other main-group hydrides, shown in red, are covalent. The change in bond type, however, is gradual and continuous.

1 1A	2 2A		13 3A	14 4A	15 5A	16 6A	17 7A	18 8A
LiH 692	BeH$_2$ d 250		B$_2$H$_6$ −165	CH$_4$ −182	NH$_3$ −78	H$_2$O 0	HF −83	
NaH d 425	MgH$_2$ d 327		AlH$_3$ d 150	SiH$_4$ −185	PH$_3$ −134	H$_2$S −86	HCl −114	
KH 619	CaH$_2$ 1000		GaH$_3$ −15	GeH$_4$ −165	AsH$_3$ −116	H$_2$Se −66	HBr −87	
RbH d ~170	SrH$_2$ 1050		InH$_3$ (?)	SnH$_4$ −146	SbH$_3$ −88	H$_2$Te −49	HI −51	
CsH 528	BaH$_2$ 1200		TlH$_3$ (?)	PbH$_4$	BiH$_3$ −67	H$_2$Po	HAt	

▶ **FIGURE 22.4**

Formulas and melting points (°C) of the simplest hydrides of the main-group elements.

The letter "d" indicates decomposition rather than melting on heating to the indicated temperature.

Transition metal hydrides (not shown) are classified as metallic.

The existence of InH$_3$ and TlH$_3$ is uncertain.

▶ **Figure It Out**

Do ionic or covalent hydrides have higher melting points? Explain.

Answer: Ionic hydrides have higher melting points because atoms are held together by strong ionic bonds in a crystal lattice. In contrast, molecules of covalent hydrides are held together by intermolecular forces, which are weaker than ionic bonds.

The alkali metal hydrides contain alkali metal cations and H^- anions in a face-centered cubic crystal structure like that of sodium chloride (Section 11.7).

The H^- anion is a good proton acceptor (Brønsted–Lowry base). Consequently, ionic hydrides react with water to give H_2 gas and OH^- ions:

$$CaH_2(s) + 2\ H_2O(l) \longrightarrow 2\ H_2\ (g) + Ca^{2+}(aq) + 2\ OH^-(aq)$$

Oxidation numbers: -1 (for CaH_2 H), $+1$ (for H_2O H), 0 (for H_2)

This reaction of an ionic hydride with water is a redox reaction, as well as an acid–base reaction, because the hydride reduces the water ($+1$ oxidation state for H) to H_2 (0 oxidation state). In turn, the hydride (-1 oxidation state for H) is oxidized to H_2. In general, ionic hydrides are good reducing agents.

Covalent hydrides, as their name implies, are compounds in which hydrogen is attached to another element by a covalent bond. The most common examples are hydrides of non-metallic elements, such as diborane (B_2H_6), methane (CH_4), ammonia (NH_3), water (H_2O), and the hydrogen halides (HX; X = F, Cl, Br, or I). Only the simplest covalent hydrides are listed in Figure 22.4, though more complex examples, such as hydrogen peroxide (H_2O_2) and hydrazine (N_2H_4), are also known. Because most covalent hydrides consist of discrete, small molecules that have relatively weak intermolecular forces, they are gases or volatile liquids at ordinary temperatures.

Metallic hydrides are formed by reaction of the lanthanide and actinide metals and certain d-block transition metals with variable amounts of hydrogen. These hydrides have the general formula MH_x, where the x subscript represents the number of H atoms in the simplest formula. They are often called **interstitial hydrides** because they are thought to consist of a crystal lattice of metal atoms with the smaller hydrogen atoms occupying holes, or *interstices,* between the larger metal atoms (**FIGURE 22.5**).

The nature of the bonding in metallic hydrides is not well understood, and it's not known whether the hydrogens are present as neutral H atoms, H^+ cations, or H^- anions. Because the hydrogen atoms can fill a variable number of interstices, many metallic hydrides are **nonstoichiometric compounds,** meaning that their atomic composition can't be expressed as a ratio of small whole numbers. Examples are $TiH_{1.7}$, $ZrH_{1.9}$, and PdH_x ($x < 1$). Other metallic hydrides, however, such as TiH_2 and UH_3, are stoichiometric compounds. The properties of metallic hydrides depend on their composition, which is a function of the partial pressure of H_2 gas in the surroundings. For example, PdH_x behaves as a metallic conductor for small values of x but becomes a **semiconductor** when x reaches about 0.5.

▲ Calcium hydride reacts with water to give bubbles of H_2 gas and OH^- ions. The pink color is due to added phenolphthalein, which turns from colorless to pink in the presence of a base.

> **REMEMBER . . .**
> A **semiconductor** is a material that has an electrical conductivity intermediate between that of a metal and that of an insulator (Section 12.7).

CONCEPTUAL WORKED EXAMPLE 22.3

Determining Formulas and Properties of Binary Hydrides

The following pictures represent binary hydrides AH_x, where A = K, Ti, C, or F. Ivory spheres represent H atoms or ions, and burgundy spheres represent atoms or ions of the element A.

(1)   **(2)**   **(3)**   **(4)**

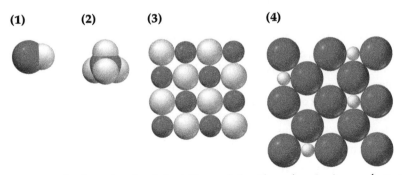

(a) Write the formula of each hydride, and classify each as ionic, covalent, or interstitial.
(b) Which hydride has the lowest melting point? Explain.
(c) Which hydride reacts with water at 25 °C to give H_2 gas?

The metal atoms (larger spheres) have a face-centered cubic structure.

The hydrogen atoms (smaller spheres) occupy interstices (holes) between the metal atoms.

▲ **FIGURE 22.5**

One plane of the structure of an interstitial metallic hydride.

continued on next page

STRATEGY

(a) The location of each element in the periodic table tells us the formula and the type of hydride it forms. Alkali metals form ionic hydrides and have the formula AH. Some transition elements form interstitial hydrides and often have nonstoichiometric formulas AH_x. Nonmetals form covalent hydrides and have formulas that depend on their periodic group.

(b) Covalent hydrides have low melting points because they have only relatively weak intermolecular forces between molecules.

(c) Ionic hydrides react with water to give H_2 gas.

SOLUTION

(a) Pictures (1) and (2) represent discrete AH and AH_4 molecules, respectively, and are therefore covalent hydrides HF and CH_4 of the nonmetallic elements F and C. Picture (3) shows a slice of the face-centered-cubic crystal structure of a solid ionic hydride with formula AH. Thus, hydride (3) is the ionic hydride KH. Picture (4) represents a slice of the structure of a nonstoichiometric interstitial hydride, which must be TiH_x.

(b) The lowest melting point belongs to one of the covalent hydrides, HF or CH_4. Polar HF molecules have dipole–dipole forces, hydrogen-bonding forces, and dispersion forces, while nonpolar CH_4 molecules have only dispersion forces. Thus, CH_4 has the lowest melting point.

(c) KH, the only ionic hydride, reacts with water at 25 °C to give H_2 gas:

$$KH(s) + H_2O(l) \longrightarrow H_2(g) + K^+(aq) + OH^-(aq).$$

▶ **CONCEPTUAL PRACTICE 22.5**   The following pictures represent binary hydrides AH_x, where A = Br, Li, P, or Zr. Ivory spheres represent H atoms or ions, and burgundy spheres represent atoms or ions of the element A.

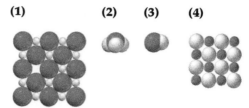

(1)   **(2)**   **(3)**   **(4)**

(a) Write the formula of each hydride, and classify each as ionic, covalent, or interstitial.

(b) Which hydrides are likely to be solids at 25 °C, and which are likely to be gases? Explain.

(c) Which hydride reacts with water at 25 °C to give a basic solution?

▶ **APPLY 22.6**   Look at the location of elements A, B, C, and D in the following periodic table:

(a) Write the formula of the simplest binary hydride of each element.

(b) Classify each binary hydride as ionic, covalent, or interstitial.

(c) Which of these hydrides are molecular? Which are solids with an infinitely extended three-dimensional crystal structure?

(d) What are the oxidation states of hydrogen and the other element in the hydrides of A, C, and D?

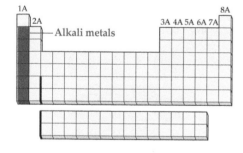

REMEMBER . . .

A reducing agent is a substance that loses one or more electrons and is itself oxidized. An oxidizing agent is a substance that gains one or more electrons and is itself reduced (Section 4.11).

22.4 GROUP 1A: ALKALI METALS AND GROUP 2A: ALKALINE EARTH METALS

The alkali metals of group 1A—Li, Na, K, Rb, Cs, and Fr—have the smallest ionization energies of all the elements because of their valence-shell ns^1 electron configurations. They therefore lose this ns^1 electron easily in chemical reactions to yield 1+ ions and are thus among the most powerful reducing agents in the periodic table. In fact, the chemistry of the alkali metals is dominated by their ability to donate an electron to another element or compound.

As their group name implies, the alkali metals are *metallic.* They have a bright, silvery appearance, are malleable, and are good conductors of electricity. Unlike the more common metals such as iron, though, the alkali metals are all soft enough to cut with a dull knife, have low melting points and densities, and are so reactive that they must be stored under oil to prevent their instantaneous reaction with oxygen and moisture. None are found in the elemental state in nature; they occur only in salts. Their properties are summarized in **TABLE 22.2.**

Alkali metals are produced commercially by reduction of their chloride salts, although the exact procedure differs for each element. Both lithium metal and sodium metal are produced by *electrolysis,* a process in which an electric current is passed through the molten salt (Section 19.13). A high reaction temperature is necessary to keep the salt liquid.

$$2\ LiCl(l) \xrightarrow[\text{450 °C}]{\text{Electrolysis in KCl}} 2\ Li(l) + Cl_2(g)$$

$$2\ NaCl(l) \xrightarrow[\text{580 °C}]{\text{Electrolysis in CaCl}_2} 2\ Na(l) + Cl_2(g)$$

TABLE 22.2   Properties of Alkali Metals

Name	Melting Point (°C)	Boiling Point (°C)	Density (g/cm^3)	First Ionization Energy (kJ/mol)	Abundance on Earth (%)	Atomic Radius (pm)	Ionic (M$^+$) Radius (pm)
Lithium	180.5	1342	0.534	520.2	0.002 0	152	68
Sodium	97.7	883	0.971	495.8	2.36	186	102
Potassium	63.3	759	0.862	418.8	2.09	227	138
Rubidium	39.3	688	1.532	403.0	0.009 0	248	152
Cesium	28.4	671	1.873	375.7	0.000 10	265	167
Francium	—	—	—	~400	Trace	—	—

Potassium, rubidium, and cesium metals are produced by chemical reduction rather than by electrolysis. Sodium is the reducing agent used in potassium production, and calcium is the reducing agent used for preparing rubidium and cesium.

Alkali Metal Compounds

Alkali metals react rapidly with group 7A elements (halogens) to yield colorless, crystalline ionic salts called *halides*. The reactivity of an alkali metal increases as its ionization energy decreases, giving a reactivity order Cs > Rb > K > Na > Li. Cesium is the most reactive, combining almost explosively with the halogens.

$$2\ M(s) + X_2 \longrightarrow 2\ MX(s)$$
A metal halide

where   M = Alkali metal (Li, Na, K, Rb, or Cs)

X = Halogen (F, Cl, Br, or I)

▲ A sample of sodium metal.

All the alkali metals also react rapidly with oxygen, but different metals give different kinds of products. Lithium reacts with O_2 to yield the *oxide*, Li_2O; sodium reacts to yield the *peroxide*, Na_2O_2; and the remaining alkali metals, K, Rb, and Cs, form either peroxides or *superoxides*, MO_2, depending on the reaction conditions and on how much oxygen is present. The reasons for the differences have to do largely with the differences in stability of the various products and with how the ions pack together in crystals. The alkali metal cations have a +1 charge in all cases, but the oxygen anions might be either O^{2-}, O_2^{2-}, or O_2^-.

$$4\ Li(s) + O_2(g) \longrightarrow 2\ Li_2O(s) \quad \text{An } oxide;\text{ the anion is } O^{2-}$$

$$2\ Na(s) + O_2(g) \longrightarrow Na_2O_2(s) \quad \text{A } peroxide;\text{ the anion is } O_2^{2-}$$

$$K(s) + O_2(g) \longrightarrow KO_2(s) \quad \text{A } superoxide;\text{ the anion is } O_2^-$$

Ionic solids can adopt a variety of different unit cells, depending on the size and charge of the ions (Section 12.4). **FIGURE 22.6** shows the unit cell for cesium chloride and sodium oxide. Cesium chloride has a body-centered cubic arrangement and sodium oxide has a face-centered cubic arrangement of the O^{2-} anions with Na^+ cations surrounded tetrahedrally.

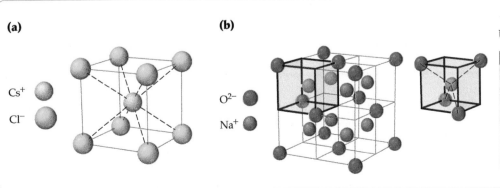

(a)

Cs$^+$

Cl$^-$

(b)

O^{2-}

Na$^+$

◀ **FIGURE 22.6**

Unit cells of (a) CsCl and (b) Na$_2$O.

◀ **Figure It Out**

How many cations and anions are in each unit cell?

Answer: CsCl has 1 Cs$^+$ cation because Cs$^+$ is contained in the center of the cell and 1 Cl$^-$ anion because each of the 8 Cl$^-$ corner ions contributes 1/8 of an ion to the unit cell. Na$_2$O has 8 Na$^+$ cations because 8 Na$^+$ ions are contained in the center of the cell and 4 O^{2-} anions because each corner O^{2-} ion contributes 1/8 of an ion to the unit cell and each face ion contributes 1/2 of an ion to the unit cell.

Go to
eText

BIG IDEA Question 2

What is the balanced reaction that occurs when solid Cs is placed in liquid water?

▶ All the alkali metals react with water to generate H_2 gas.

Perhaps the most well-known and dramatic reaction of the alkali metals is with water to yield hydrogen gas and an alkali metal hydroxide, MOH. In fact, it's this reaction that gives the elements their group name because the solution of metal hydroxide that results from adding an alkali metal to water is *alkaline*, or basic. Lithium undergoes the reaction with vigorous bubbling as hydrogen is released, sodium reacts rapidly with evolution of heat, and potassium reacts so violently that the hydrogen produced bursts instantly into flame. Rubidium and cesium react almost explosively.

$$2\,M(s) + 2\,H_2O(l) \longrightarrow 2\,M^+(aq) + 2\,OH^-(aq) + H_2(g)$$
$$\text{where}\quad M = \text{Li, Na, K, Rb, or Cs}$$

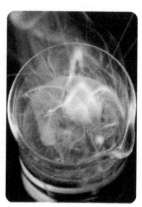

Lithium reacts vigorously, with bubbling.

Sodium reacts rapidly and releases heat.

Potassium reacts violently.

Group 2A: Alkaline Earth Metals

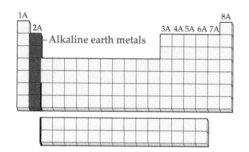

The alkaline earth elements of group 2A—Be, Mg, Ca, Sr, Ba, and Ra—are similar to the alkali metals in many respects. They differ, however, in that they have ns^2 valence-shell electron configurations and can therefore lose two electrons in forming doubly positive ions, M^{2+}. Because their first ionization energy is larger than that of alkali metals, the group 2A metals are somewhat less reactive than alkali metals. The general reactivity trend is Ba > Sr > Ca > Mg > Be.

Although harder than their neighbors in group 1A, the alkaline earth elements are still relatively soft, silvery metals. They tend, however, to have higher melting points and densities than alkali metals, as listed in **TABLE 22.3**. Alkaline earth elements are less reactive toward oxygen and water than alkali metals but are nevertheless found in nature only in salts, not in the elemental state.

TABLE 22.3 **Properties of Alkaline Earth Metals**

Name	Melting Point (°C)	Boiling Point (°C)	Density (g/cm³)	First Ionization Energy (kJ/mol)	Abundance on Earth (%)	Atomic Radius (pm)	Ionic (M²⁺) Radius (pm)
Beryllium	1287	2471	1.848	899.4	0.00028	112	44
Magnesium	650	1090	1.738	737.7	2.33	160	66
Calcium	842	1484	1.55	589.8	4.15	197	99
Strontium	777	1382	2.54	549.5	0.038	215	112
Barium	727	1897	3.62	502.9	0.042	222	134
Radium	700	1140	~5.0	509.3	Trace	223	143

Pure alkaline earth metals, like alkali metals, are produced commercially by reduction of their salts, either chemically or through electrolysis. Beryllium is prepared by reduction of BeF_2 with magnesium, and magnesium is prepared by electrolysis of its molten chloride.

$$BeF_2(l) + Mg(l) \xrightarrow{1300°C} Be(l) + MgF_2(l)$$

$$MgCl_2(l) \xrightarrow[750\,°C]{\text{Electrolysis}} Mg(l) + Cl_2(g)$$

Calcium, strontium, and barium are all made by high-temperature reduction of their oxides with aluminum metal.

$$3 \, MO(l) + 2 \, Al(l) \xrightarrow{\text{High temp}} 3 \, M(l) + Al_2O_3(s)$$
$$\text{where} \quad M = Ca, Sr, \text{ or } Ba$$

Alkaline Earth Metal Compounds

Alkaline earth metals react with halogens to yield ionic halide salts, MX_2, and with oxygen to form oxides, MO:

$$M + X_2 \longrightarrow MX_2 \quad \text{where} \quad M = Be, Mg, Ca, Sr, \text{ or } Ba$$
$$2 \, M + O_2 \longrightarrow 2 \, MO \quad \quad X = F, Cl, Br, \text{ or } I$$

Beryllium and magnesium are relatively unreactive toward oxygen at room temperature, but both burn with a brilliant white glare when ignited by a flame. Calcium, strontium, and barium are so reactive that they are best stored under oil to keep them from contact with air. Like the heavier alkali metals, strontium and barium form peroxides, MO_2.

With the exception of beryllium, the alkaline-earth elements react with water to yield metal hydroxides, $M(OH)_2$. Magnesium undergoes reaction only at temperatures above 100 °C, while calcium and strontium react slowly with liquid water at room temperature. Only barium reacts vigorously at room temperature.

$$M(s) + 2 \, H_2O(l) \longrightarrow M(OH)_2 + H_2(g)$$
$$\text{where} \quad M = Mg, Ca, Sr, \text{ or } Ba$$

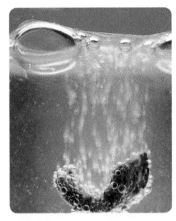

▲ Calcium metal reacts very slowly with water at room temperature.

Go to eText

BIG IDEA Question 3

How do the chemical and physical properties of the alkaline earth metals differ from those of the alkali metals?

22.5 GROUP 3A ELEMENTS

The elements of group 3A—boron, aluminum, gallium, indium, and thallium—are the first of the *p*-block elements (Section 5.13) and have the valence electron configuration $ns^2 \, np^1$. (The electron configuration and chemical properties of element 113, and also elements 114–118, are as yet unknown because only a few atoms of these recently discovered, radioactive elements have been synthesized.) With a valence electron configuration of ns^2np^1, the group 3A elements have a primary oxidation state of +3. In addition, the heavier elements exhibit a +1 state, which is uncommon for gallium and indium but is the most stable oxidation state for thallium.

Despite some irregularities, the properties of the group 3A elements are generally consistent with increasing metallic character down the group (**TABLE 22.4**). Boron has a smaller atomic radius and a higher electronegativity than the other elements, which leads to different properties. The group 3A elements are silvery metals and good conductors of electricity, but boron is a semimetal. Boron also shares its valence electrons in covalent bonds rather than transferring them to another element. Aluminum is the most important element of the group because it is used to make aircraft, electrical transmission lines, cooking utensils, soda cans, and numerous other commercial products. Gallium is remarkable for its unusually low melting point (29.8 °C) and unusually large liquid range (29.8–2204 °C). Its most important use is in making gallium arsenide (GaAs), a semiconductor material employed in the manufacture of diode lasers for laser printers, compact disc players, and fiber optic communication devices. Indium is also used in making semiconductor devices, such as transistors and electrical resistance thermometers called *thermistors*. Thallium is toxic, but thallium halides are used in specialized sample holders for infrared spectroscopy.

REMEMBER . . .
The periodic table can be divided into regions according to the subshell being filled. The group 1A and 2A elements are the *s*-block elements, and the group 3A–8A elements are the *p*-block elements (Section 5.13).

Elemental Boron

Boron is a relatively rare element, accounting for only 0.0010% of the Earth's crust by mass. Nevertheless, boron is readily available because it occurs in concentrated deposits of borate minerals such as borax, $Na_2B_4O_7 \cdot 10H_2O$.

High-purity, crystalline boron is best obtained by the reaction of boron tribromide and hydrogen on a heated tantalum filament at high temperatures:

$$2 \, BBr_3(g) + 3 \, H_2(g) \xrightarrow[\text{1200 °C}]{\text{Ta wire}} 2 \, B(s) + 6 \, HBr(g)$$

▲ Gallium metal (mp 29.8 °C) melts at body temperature.

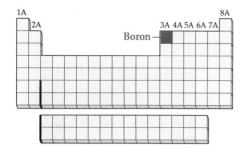

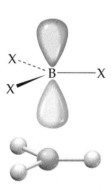

▲ The borate minerals in this open-pit borax mine near Boron, California, are believed to have been formed by the evaporation of water from hot springs that were once present.

REMEMBER . . .
A **composite material** is a ceramic material that is mixed with another component such as a metal, organic polymer, or another ceramic (Section 12.10).

TABLE 22.4 Properties of the Group 3A Elements

Property	Boron	Aluminum	Gallium	Indium	Thallium
Valence electron configuration	$2s^2 2p^1$	$3s^2 3p^1$	$4s^2 4p^1$	$5s^2 5p^1$	$6s^2 6p^1$
Melting point (°C)	2075	660	29.8	157	304
Boiling point (°C)	4000	2519	2204	2072	1473
Density (g/cm³)	2.34	2.70	5.91	7.31	11.8
Abundance in Earth's crust (mass %)	0.0010	8.23	0.0019	0.000 02	0.000 08
Common oxidation states	+3	+3	+3	+3	+3, +1
Atomic radius (pm)	83	143	135	167	170
M^{3+} ionic radius (pm)	—	54	62	80	89
First ionization energy (kJ/mol)	801	578	579	558	589
Electronegativity	2.0	1.5	1.6	1.7	1.8
Redox potential, $E°$ (V) for $M^{3+}(aq) + 3 e^- \longrightarrow M(s)$	−0.87*	−1.66	−0.55	−0.34	−0.34†

*$E°$ for the reaction $B(OH)_3(aq) + 3 H^+(aq) + 3 e^- \longrightarrow B(s) + 3 H_2O(l)$
†$E°$ for the reaction $Tl^+(aq) + e^- \longrightarrow Tl(s)$

Crystalline boron is a strong, hard, high-melting substance (mp 2075 °C) that is chemically inert at room temperature except for reaction with fluorine. These properties make boron fibers a desirable component in high-strength **composite materials** used in making sports equipment and military aircraft. Unlike Al, Ga, In, and Tl, which are metallic conductors, boron is a semiconductor.

Boron Compounds

Boron Halides The boron halides are highly reactive, volatile, covalent compounds that consist of trigonal planar BX_3 molecules. In their most important reactions, the boron halides behave as Lewis acids. For example, BF_3 reacts with ammonia to give the Lewis acid–base adduct F_3B-NH_3 (Section 16.15); it reacts with metal fluorides, yielding salts that contain the tetrahedral BF_4^- anion; and it acts as a Lewis acid catalyst in many industrially important organic reactions. In all these reactions, the boron atom uses its vacant $2p$ orbital to accept a pair of electrons from a Lewis base.

Boron Hydrides The boron hydrides, or **boranes**, are volatile, molecular compounds with formulas B_nH_m. The simplest is diborane (B_2H_6), the dimer of the unstable BH_3.

Boranes are of interest because of their unusual bonding and structures. The diborane molecule has a structure in which two BH_2 groups are connected by two bridging H atoms. The geometry about the B atoms is roughly tetrahedral, and the bridging $B-H$ bonds are significantly longer than the terminal $B-H$ bonds, 133 pm versus 119 pm. The structure differs from that of ethane (C_2H_6) and is unusual because hydrogen normally forms only one bond.

▲ The boron halides are Lewis acids because the boron atom has a vacant $2p$ orbital.

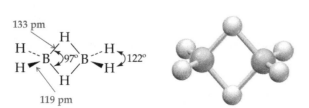

Diborane

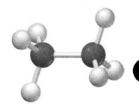

Ethane

If each line in the structural formula of diborane represented an ordinary 2-electron covalent bond between 2 atoms (a *two-center, two-electron bond,* or 2c—2e bond), there would be 8 pairs, or 16 valence electrons. Diborane, however, has a total of only 12 valence electrons—3 electrons from each boron and 1 from each hydrogen. Thus, diborane is said to be *electron-deficient:* It doesn't have enough electrons to form a 2c—2e bond between each pair of bonded atoms.

Because the geometry about the B atoms is roughly tetrahedral, we can assume that each boron uses sp^3 hybrid orbitals to bond to the four neighboring H atoms. The four terminal B—H bonds are assumed to be ordinary 2c—2e bonds, formed by the overlap of a boron sp^3 hybrid orbital and a hydrogen $1s$ orbital, thereby using 4 of the 6 pairs of electrons. Each of the 2 remaining pairs of electrons forms a **three-center, two-electron bond** (3c—2e bond), which joins each bridging H atom to *both* B atoms. Each electron pair occupies a three-center orbital formed by the interaction of three atomic orbitals—one sp^3 hybrid orbital from each B atom and the $1s$ orbital on one bridging H atom (**FIGURE 22.7**). Because the 2 electrons in the B—H—B bridge are spread out over 3 atoms, the electron density between adjacent atoms is less than in an ordinary 2c—2e bond. The bridging B—H bonds are therefore weaker and correspondingly longer than the terminal B—H bonds.

Each of the two 3-center bonds (one shown darker than the other) is formed by the interaction of an sp^3 hybrid orbital on each B atom and the 1s orbital on one bridging H atom.

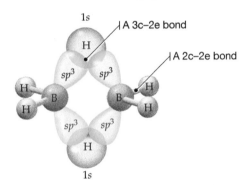

Each 3-center bond contains one pair of electrons and accounts for the bonding in one B–H–B bridge.

◄ **FIGURE 22.7**

Three-center bonding orbitals in diborane.

◄ **Figure It Out**

How is a 3c—2e bond different than a 2c—2e bond?

Answer: In 3c—2e bond, an electron pair is shared between 3 atoms, and therefore the electron density between atoms in the bond is less than in a 2c—2e bond. The 3c—2e bond is longer and weaker than a 2c—2e bond.

22.6 GROUP 4A ELEMENTS

The group 4A elements—carbon, silicon, germanium, tin, and lead—are especially important both in industry and in living organisms. Carbon is present in all plants and animals, accounts for 23% of the mass of the human body, and is an essential constituent of the molecules on which life is based. Silicon is equally important in the mineral world: It is present in numerous silicate minerals and is the second most abundant element in the Earth's crust. Both silicon and germanium are used in making modern solid-state electronic devices. Tin and lead have been known and used since ancient times.

The group 4A elements exemplify the increase in metallic character down a group in the periodic table: carbon is a nonmetal, silicon and germanium are semimetals, and tin and lead are metals. The usual periodic trends in atomic size, ionization energy, and electronegativity are evident in the data of **TABLE 22.5.**

The physical properties of the heavier group 4A elements nicely illustrate the gradual transition from semimetallic to metallic character. Germanium is a relatively high-melting, brittle semiconductor that has the same crystal structure as diamond and silicon. Tin exists in two allotropic forms: the usual silvery white metallic form called *white tin* and a brittle,

Go to
eText

BIG IDEA Question 4

Why doesn't diborane (B_2H_6) have a molecular structure similar to ethane (C_2H_6)?

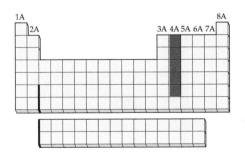

Germanium

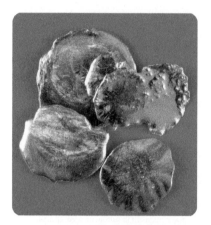

Tin

Lead

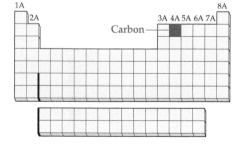

TABLE 22.5   Properties of the Group 4A Elements

Property	Carbon	Silicon	Germanium	Tin	Lead
Valence electron configuration	$2s^2\,2p^2$	$3s^2\,3p^2$	$4s^2\,4p^2$	$5s^2\,5p^2$	$6s^2\,6p^2$
Melting point (°C)	~8700*	1414	938	232†	327
Boiling point (°C)	—	3265	2833	2602	1749
Density (g/cm³)	3.51*	2.33	5.32	7.29†	11.3
Abundance in Earth's crust (mass %)	0.020	28.2	0.0005	0.0002	0.0013
Common oxidation states	+2, +4	+4	+4	+2, +4	+2, +4
Atomic radius (pm)	77	118	122	140	175
First ionization energy (kJ/mol)	1086	787	762	709	716
Electronegativity	2.5	1.8	1.8	1.8	1.9
Redox potential, $E°$ (V) for $M^{2+}(aq) + 2\,e^- \longrightarrow M(s)$	—	—	—	−0.14	−0.13

*Diamond; mp at 6—10 million atm.
†White Sn

semiconducting form with the diamond structure called *gray tin*. White tin is the stable form at room temperature, but when kept for long periods of time below the transition temperature of 13 °C it slowly crumbles to gray tin, a phenomenon known as tin disease:

$$\text{White tin} \underset{}{\overset{13\,°C}{\rightleftharpoons}} \text{Gray tin}$$

Both white tin and lead are soft, malleable, low-melting metals. Only the metallic form occurs for lead.

Because the group 4A elements have the valence electron configuration $ns^2\,np^2$, they adopt an oxidation state of +4, as in CCl_4, $SiCl_4$, $GeCl_4$, $SnCl_4$, and $PbCl_4$. These compounds are volatile, molecular liquids in which the group 4A atom uses tetrahedral sp^3 hybrid orbitals to form covalent bonds to the Cl atoms. The +2 oxidation state occurs for tin and lead and is the most stable oxidation state for lead. Both $Sn^{2+}(aq)$ and $Pb^{2+}(aq)$ are common solution species, but there are no simple $M^{4+}(aq)$ ions for any of the group 4A elements. Instead, M(IV) species exist in solution as covalently bonded complex ions—for example, SiF_6^{2-}, $GeCl_6^{2-}$, $Sn(OH)_6^{2-}$, and $Pb(OH)_6^{2-}$. In general, the +4 oxidation-state compounds are covalent, and the compounds with tin and lead in the +2 oxidation state are largely ionic.

Elemental Carbon

Carbon, although the second most abundant element in living organisms, accounts for only 0.02% of the mass of the Earth's crust. It is present in carbonate minerals, such as limestone $CaCO_3$, and in fossil fuels, such as coal, petroleum, and natural gas. Recall from Section 12.5 that carbon has more than 40 allotropes, or different elemental forms, including diamond, graphite, and fullerenes. Figure 12.13 shows the different covalent network structures of diamond and graphite. Figure 12.14 shows the structures of two fullerenes, which are molecules of carbon in the form of hollow spheres or tubes. Closely related to graphite is *graphene,* a remarkable two-dimensional array of hexagonally arranged carbon atoms that is just one atom thick. First prepared in 2004 by pulling apart layers of graphite with sticky tape, graphene is extremely strong and flexible and is a superb conductor of electricity. Each C atom in graphene uses sp^2 hybrid orbitals to form trigonal planar σ bonds to three neighboring C atoms. In addition, each C atom uses its remaining p orbital, which is perpendicular to the plane of the sheet, to form a π bond. The electrons in the π bond are delocalized and free to move in the plane of the sheet, resulting in a highly conducting material. Possible future applications of graphene include use in new, smaller, and faster computer chips, thin coatings for solar cells and LCD displays, and components of strong composite materials. For their work on the preparation and properties of graphene, Andre Geim and Konstantin Novoselov were awarded the 2010 Nobel Prize in Physics.

Carbon Compounds

Carbon forms more than 40 million known compounds, most of which are classified as *organic*; only CO, CO_2, $CaCO_3$, HCN, CaC_2, and a handful of others are considered to be *inorganic*. The distinction is a historical one rather than a scientific one, though, as discussed in more detail in Chapter 23. For the present, we'll look only at some simple inorganic compounds of carbon.

Oxides of Carbon The most important oxides of carbon are carbon monoxide (CO) and carbon dioxide (CO_2). Carbon monoxide is a colorless, odorless, toxic gas that forms when carbon or hydrocarbon fuels are burned in a limited supply of oxygen. In an excess of oxygen, CO burns to give CO_2:

$$2\ C(s) + O_2(g) \longrightarrow 2\ CO(g) \qquad \Delta H° = -221\ kJ$$
$$2\ CO(g) + O_2(g) \longrightarrow 2\ CO_2(g) \qquad \Delta H° = -566\ kJ$$

The high toxicity of CO results from its ability to bond strongly to the iron(II) atom of hemoglobin, the oxygen-carrying protein in red blood cells. Because hemoglobin has a greater affinity for CO than for O_2 by a factor of 200, even small concentrations of CO in the blood can convert a substantial fraction of the O_2-bonded hemoglobin, called *oxyhemoglobin*, to the CO-bonded form, called *carboxyhemoglobin*, thus impairing the ability of hemoglobin to carry O_2 to the tissues:

$$Hb{-}O_2 + CO \rightleftharpoons Hb{-}CO + O_2$$
$$\text{Oxyhemoglobin} \qquad \text{Carboxyhemoglobin}$$

A CO concentration in air of only 200 ppm can produce symptoms such as headache, dizziness, and nausea, and a concentration of 1000 ppm can cause death within 4 hours. One hazard of cigarette smoking is chronic exposure to low levels of CO. Because CO reduces the blood's ability to carry O_2, the heart must work harder to supply O_2 to the tissues, thus increasing the risk of heart attack.

Carbon dioxide, a colorless, odorless, nonpoisonous gas, is produced when fuels are burned in an excess of oxygen and is an end product of food metabolism in humans and animals. The concentration of carbon dioxide in the atmosphere has increased dramatically since the start of the Industrial Revolution and is of environmental concern because of its impact on climate change (Section 10.11) and the pH of oceans (Chapter 17, *Inquiry*).

Carbonates Carbonic acid, H_2CO_3, forms two series of salts: carbonates, which contain the trigonal planar $CO_3{}^{2-}$ ion, and hydrogen carbonates (bicarbonates), which contain the $HCO_3{}^{-}$ ion. The carbonate ion removes cations such as Ca^{2+} and Mg^{2+} from hard water, and it acts as a base to give OH^- ions, which help remove grease from fabrics:

$$Ca^{2+}(aq) + CO_3{}^{2-}(aq) \longrightarrow CaCO_3(s)$$
$$CO_3{}^{2-}(aq) + H_2O(l) \rightleftharpoons HCO_3{}^{-}(aq) + OH^-(aq)$$

Sodium hydrogen carbonate, $NaHCO_3$, is called baking soda because it reacts with acidic substances in food to yield bubbles of CO_2 gas that cause dough to rise:

$$NaHCO_3(s) + H^+(aq) \longrightarrow Na^+(aq) + CO_2(g) + H_2O(l)$$

Hydrogen Cyanide and Cyanides Hydrogen cyanide is a highly toxic, volatile substance (bp 26 °C) produced when metal cyanide solutions are acidified:

$$CN^-(aq) + H^+(aq) \longrightarrow HCN(aq)$$

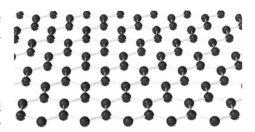

▲ A ball-and-stick model of the crystalline structure of graphene shows the two-dimensional array of hexagonally arranged carbon atoms.

▲ Known as "the silent killer," carbon monoxide is the cause of 500 accidental poisoning deaths in the United States each year. Household carbon monoxide detectors are designed to monitor CO concentrations over time and sound an alarm before life-threatening levels are reached.

Go to
eText

BIG IDEA Question 5

A treatment for carbon monoxide poisoning is to administer pure oxygen at a pressure of 3 atm to the patient. Why is this an effective treatment?

Aqueous solutions of HCN, known as hydrocyanic acid, are very weakly acidic ($K_a = 4.9 \times 10^{-10}$).

The cyanide ion, CN^-, acts as a Lewis base, bonding to transition metals through the lone pair of electrons on its carbon atom. In fact, the toxicity of HCN and other cyanides is due to the strong bonding of CN^- to iron(III) in cytochrome oxidase, an enzyme involved in the oxidation of food molecules. With CN^- attached to the iron, the enzyme is unable to function. Cellular energy production thus comes to a halt, and rapid death follows.

The bonding of CN^- to gold and silver is exploited in the extraction of these metals from their ores. The crushed rock containing small amounts of the precious metals is treated with an aerated cyanide solution, and the metals are then recovered from their $M(CN)_2^-$ complex ions by reduction with zinc. For gold, the reactions are

$$4\,Au(s) + 8\,CN^-(aq) + O_2(g) + 2\,H_2O(l) \longrightarrow 4\,Au(CN)_2^-(aq) + 4\,OH^-(aq)$$

$$2\,Au(CN)_2^-(aq) + Zn(s) \longrightarrow 2\,Au(s) + Zn(CN)_4^{2-}(aq)$$

Because of the toxicity of cyanides, this method is now banned in several U.S. states and foreign countries.

Carbides Carbon forms a number of binary inorganic compounds called **carbides,** in which the carbon atom has a negative oxidation state. Examples include ionic carbides of active metals such as CaC_2 and Al_4C_3, interstitial carbides of transition metals such as Fe_3C, and covalent network carbides such as SiC, the industrial abrasive called carborundum. Calcium carbide is a high-melting, colorless solid that has an NaCl type of structure with Ca^{2+} ions in place of Na^+ and C_2^{2-} ions in place of Cl^-. It is prepared by heating lime (CaO) and coke (C) at high temperatures and is used to prepare acetylene (C_2H_2) for oxyacetylene welding:

$$CaO(s) + 3\,C(s) \xrightarrow{2200\,°C} CaC_2(s) + CO(g)$$

$$CaC_2(s) + 2\,H_2O(l) \longrightarrow C_2H_2(g) + Ca(OH)_2(s)$$

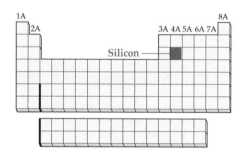

▲ A cyanide leaching pool for extracting gold operated by the Zolotaya Zvezda mining company in Khakassia, Russia.

Elemental Silicon

Silicon is a hard, gray, semiconducting solid that melts at 1414 °C. It crystallizes in a diamondlike structure but does not form a graphitelike allotrope because of the relatively poor overlap of silicon π orbitals. In nature, silicon is generally found combined with oxygen in SiO_2 and in various silicate minerals. It is obtained in elemental form by the reduction of silica sand (SiO_2) with coke (C) in an electric furnace:

$$SiO_2(l) + 2\,C(s) \xrightarrow{Heat} Si(l) + 2\,CO(g)$$

The silicon used for making solid-state semiconductor devices such as transistors, computer chips, and solar cells must be ultrapure, with impurities at a level of less than $10^{-7}\%$ (1 ppb). For electronic applications, silicon is purified by converting it to $SiCl_4$, a volatile liquid (bp 58 °C) that can be separated from impurities by fractional distillation (Section 12.9) and then converted back to elemental silicon by reduction with hydrogen:

$$Si(s) + 2\,Cl_2(g) \longrightarrow SiCl_4(l)$$

$$SiCl_4(g) + 2\,H_2(g) \xrightarrow{Heat} Si(s) + 4\,HCl(g)$$

The silicon is purified further by a process called **zone refining** (**FIGURE 22.8a**), in which a heater melts a narrow zone of a silicon rod. Because the impurities are more soluble in the liquid phase than in the solid, they concentrate in the molten zone. As the heater sweeps slowly down the rod, ultrapure silicon crystallizes at the trailing edge of the molten zone and the impurities are dragged to the rod's lower end. **FIGURE 22.8b** shows some samples of ultrapure silicon.

Silicon Compounds

Approximately 90% of the Earth's crust consists of **silicates,** ionic compounds that contain silicon oxoanions along with cations such as Na^+, K^+, Mg^{2+}, or Ca^{2+} to balance the negative charge of the anions. As shown in **FIGURE 22.9**, the basic structural building

(a)

The heater coil sweeps the molten zone and the impurities to the lower end of the rod. After the rod has cooled, the impurities are removed by cutting off the rod's lower end.

(b)

Silicon wafers are used to produce the integrated-circuit chips found in solid-state electronic devices.

◀ **FIGURE 22.8**

Ultrapure silicon for solid-state semiconductor devices. (a) Purification of silicon by zone refining. (b) A rod of ultrapure silicon and silicon wafers cut from the rod.

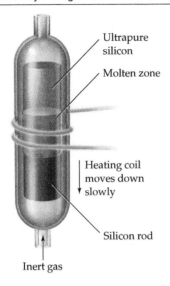

Ultrapure silicon

Molten zone

Heating coil moves down slowly

Silicon rod

Inert gas

(a) A view of the SiO_4^{4-} anion showing the tetrahedral SiO_4 structural unit.

(b) A view of the $Si_2O_7^{6-}$ anion.

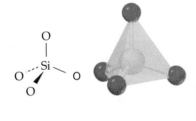

A tetrahedron is used as a shorthand representation of the SiO_4^{4-} anion. An O atom is located at each corner of the tetrahedron, and the Si atom is at the center.

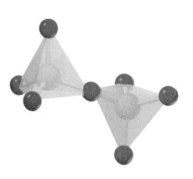

A shorthand representation of the $Si_2O_7^{6-}$ anion. The corner shared by the two tetrahedra represents a shared O atom.

▲ **FIGURE 22.9**

The SiO_4^{4-} and $Si_2O_7^{6-}$ anions and their shorthand representations.

block in silicates is the SiO_4 tetrahedron, a unit that occurs as the simple orthosilicate ion (SiO_4^{4-}) in the mineral zircon, $ZrSiO_4$. (Zircon is not to be confused with cubic zirconia, the cubic form of ZrO_2.) If two SiO_4 tetrahedra share a common O atom, the disilicate anion $Si_2O_7^{6-}$, found in $Sc_2Si_2O_7$, results.

Simple anions such as SiO_4^{4-} and $Si_2O_7^{6-}$ are relatively rare in silicate minerals. More common are larger anions in which two or more O atoms bridge between Si atoms to give rings, chains, layers, and extended three-dimensional structures. The sharing of two O atoms per SiO_4 tetrahedron gives either cyclic anions, such as $Si_6O_{18}^{12-}$, or infinitely extended chain anions with repeating $Si_2O_6^{4-}$ units (**FIGURE 22.10**). The $Si_6O_{18}^{12-}$ cyclic anion is present in the mineral beryl ($Be_3Al_2Si_6O_{18}$) and in the gemstone emerald, a beryl in which about 2% of the Al^{3+} is replaced by green Cr^{3+} cations. Chain anions are found in minerals such as diopside, $CaMgSi_2O_6$.

As shown in **FIGURE 22.11**, additional sharing of O atoms gives the double-stranded chain anions $(Si_4O_{11}^{6-})_n$ found in asbestos minerals such as tremolite,

▲ The mineral zircon ($ZrSiO_4$) is a relatively inexpensive gemstone.

▶ **FIGURE 22.10**

Samples of the silicate minerals emerald and diopside and the structures of their anions.

(a) Emerald, a green beryl ($Be_3Al_2Si_6O_{18}$) with about 2% Cr^{3+} ions substituting for Al^{3+}

(b) Diopside ($CaMgSi_2O_6$)

(c) The cyclic anion $Si_6O_{18}^{12-}$ in beryl

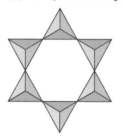

(d) The infinitely extended chain anion $(Si_2O_6^{4-})_n$ in diopside

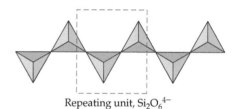

Repeating unit, $Si_2O_6^{4-}$

The number of negative charges on the $Si_2O_6^{4-}$ repeating unit equals the number of terminal (unshared) O atoms in that unit (four).

▶ **FIGURE 22.11**

Samples of asbestos and mica, and the structures of their anions.

(a) Asbestos is a fibrous material because of its chain structure.

(b) Mica cleaves into thin sheets because of its two-dimensional layer structure.

(c) The double-stranded chain anion $(Si_4O_{11}^{6-})_n$ in asbestos minerals.

Two of the single-stranded chains of Figure 22.10d are laid side by side, and half of the SiO_4 tetrahedra share an additional O atom.

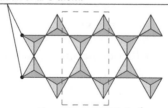

Repeating unit, $Si_4O_{11}^{6-}$

(d) The layer anion $(Si_4O_{10}^{4-})_n$ in mica is formed by the sharing of three O atoms per SiO_4 tetrahedron.

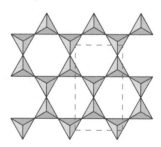

Repeating unit, $Si_4O_{10}^{4-}$

The number of negative charges on each repeating unit equals the number of terminal O atoms in that unit.

$Ca_2Mg_5(Si_4O_{11})_2(OH)_2$, and the infinitely extended two-dimensional layer anions $(Si_4O_{10}^{4-})_n$ found in clay minerals, micas, and talc, $Mg_3(OH)_2(Si_4O_{10})$. Asbestos is fibrous, as shown in the figure, because the ionic bonds between the silicate chain anions and the Ca^{2+} and Mg^{2+} cations that lie between the chains and hold them together are relatively weak and easily broken. Mica is sheetlike because the ionic bonds between the two-dimensional layer anions and the interposed metal cations are much weaker than the Si–O covalent bonds within the layer anions.

If the layer anions of Figure 22.11d are stacked on top of one another and the terminal O atoms are shared, an infinitely extended three-dimensional structure is obtained in which all four O atoms of each SiO_4 tetrahedron are shared between two Si atoms, resulting in *silica* (SiO_2). The mineral quartz is one of many crystalline forms of SiO_2.

▲ Pure crystalline quartz, one form of SiO_2, is colorless.

CONCEPTUAL WORKED EXAMPLE 22.4   Go to eText

Interpreting Representations of Silicate Anions

The following pictures represent silicate anions. What are the formula and charge of each anion?

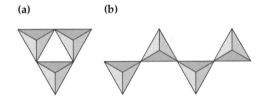

(a)   (b)

STRATEGY

Each tetrahedron represents an SiO_4 structural unit, with an Si atom at the center of the tetrahedron and an O atom at each corner. Each terminal (unshared) O atom has a formal charge of -1. Each bridging (shared) O atom completes its octet by forming bonds to two Si atoms and therefore has a formal charge of 0. To find the number of Si atoms in the formula, count the number of tetrahedra. To find the number of O atoms, count the number of corners (shared and unshared). To find the charge on the anion, count the number of unshared corners.

SOLUTION

(a) The picture contains three tetrahedra, with three shared corners and six unshared corners (nine in all). Therefore, the anion is $Si_3O_9^{6-}$.

(b) The picture contains four tetrahedra, with three shared corners and 10 unshared corners (13 in all). Therefore, the anion is $Si_4O_{13}^{10-}$.

CHECK

We can check the charge on a silicate anion by assigning the usual oxidation states of $+4$ to silicon and -2 to oxygen. Thus, the charge on anion **(a)** must be $(3)(4) + (9)(-2) = -6$, and the charge on anion **(b)** must be $(4)(4) + (13)(-2) = -10$.

▶ **CONCEPTUAL PRACTICE 22.7** What are the formula and charge of the silicate anion shown?

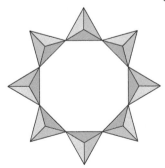

▶ **CONCEPTUAL APPLY 22.8** Suggest a plausible structure for the silicate anion in the mineral, wollastonite, $Ca_3Si_3O_9$. Use a tetrahedron to represent an SiO_4 structural unit.

All **WORKED EXAMPLES** with this icon have an interactive video in the eText.

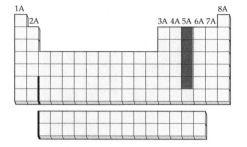

22.7 GROUP 5A ELEMENTS

The group 5A elements are nitrogen, phosphorus, arsenic, antimony, and bismuth. As shown in **TABLE 22.6**, these elements exhibit the expected trends of increasing atomic size, decreasing ionization energy, and decreasing electronegativity down the group from N to Bi. Accordingly, metallic character increases in the same order: N and P are typical nonmetals, As and Sb are semimetals, and Bi is a metal. Thus, nitrogen is a gaseous substance comprised of N_2 molecules, but bismuth is a silvery solid having an extended three-dimensional structure.

TABLE 22.6 Properties of the Group 5A Elements

Property	Nitrogen	Phosphorus	Arsenic	Antimony	Bismuth
Valence electron configuration	$2s^2 2p^3$	$3s^2 3p^3$	$4s^2 4p^3$	$5s^2 5p^3$	$6s^2 6p^3$
Melting point (°C)	−210	44*	614†	631	271
Boiling point (°C)	−196	280		1587	1564
Atomic radius (pm)	75	110	120	140	150
First ionization energy (kJ/mol)	1402	1012	944	831	703
Electronegativity	3.0	2.1	2.0	1.9	1.9

*White phosphorus

†Sublimes

Arsenic

Antimony

Bismuth

The valence electron configuration of the group 5A elements is $ns^2 np^3$. They exhibit a maximum oxidation state of +5 in compounds such as HNO_3 and PF_5, in which they share all 5 valence electrons with a more electronegative element. They show a minimum oxidation state of −3 in compounds such as NH_3 and PH_3, where they share 3 valence electrons with a less electronegative element. The −3 state also occurs in ionic compounds such as Li_3N and Mg_3N_2, which contain the N^{3-} anion.

Nitrogen and phosphorus are unusual in that they exhibit all oxidation states between −3 and +5. For arsenic and antimony, the most important oxidation states are +3, as in $AsCl_3$, As_2O_3, and H_3AsO_3, and +5, as in AsF_5, As_2O_5, and H_3AsO_4. The

+5 state becomes increasingly less stable from As to Sb to Bi. Another indication of increasing metallic character down the group is the existence of Sb^{3+} and Bi^{3+} cations in salts such as $Sb_2(SO_4)_3$ and $Bi(NO_3)_3$. By contrast, no simple cations are found in compounds of N or P.

As, Sb, and Bi are found in sulfide ores and are used to make various metal alloys. Arsenic is also used to make pesticides and semiconductors, such as GaAs. Bismuth compounds are present in some pharmaceuticals, such as Pepto-Bismol.

Elemental Nitrogen

Elemental nitrogen is a colorless, odorless, tasteless gas that makes up 78% of the Earth's atmosphere by volume. Because nitrogen (bp $-196\ °C$) is the most volatile component of liquid air, it is readily separated from the less volatile oxygen (bp $-183\ °C$) and argon (bp $-186\ °C$) by fractional distillation. Nitrogen gas is used as a protective inert atmosphere in manufacturing processes, and the liquid is used as a refrigerant. By far the most important use of nitrogen, however, is in the Haber process for the manufacture of ammonia, used in nitrogen fertilizers (Section 15.7).

Under most conditions, the N_2 molecule is unreactive because a large amount of energy is required to break its strong nitrogen–nitrogen triple bond:

$$:N\equiv N: \longrightarrow 2\ :\dot{N}\cdot \qquad \Delta H° = 945\ kJ$$

As a result, reactions involving N_2 often have a high activation energy and/or an unfavorable equilibrium constant. For example, the equilibrium constant for formation of nitric oxide from N_2 and O_2 is 2.0×10^{-31} at 25 °C:

$$N_2(g) + O_2(g) \rightleftharpoons 2\ NO(g) \qquad \Delta H° = 182.6\ kJ; \qquad K_c = 2.0 \times 10^{-31}\ at\ 25\ °C$$

At higher temperatures, however, this reaction does occur because it is endothermic and the **equilibrium constant increases as the temperature increases** (Section 15.9). Indeed, the high-temperature formation of NO from air in automobile engines is a major source of air pollution. Atmospheric N_2 and O_2 also react to form NO during electrical storms, where lightning discharges provide the energy required for the highly endothermic reaction.

Nitrogen Compounds

Among the more important compounds of nitrogen are ammonia, hydrazine, oxides, and oxoacids.

Ammonia Ammonia and its synthesis by the Haber process serve as the gateway to nitrogen chemistry because ammonia is the starting material for the industrial synthesis of other important nitrogen compounds, such as nitric acid. Used in agriculture as a fertilizer, ammonia is the most commercially important compound of nitrogen.

Ammonia is a colorless, pungent-smelling gas, consisting of polar, trigonal pyramidal NH_3 molecules that have a lone pair of electrons on the N atom. Because of hydrogen bonding (Section 8.6), gaseous NH_3 is extremely soluble in water and is easily condensed to liquid NH_3, which boils at $-33\ °C$. Neutralization of aqueous ammonia with acids yields ammonium salts, which resemble alkali metal salts in their solubility.

Hydrazine Hydrazine (H_2NNH_2) can be regarded as a derivative of NH_3 in which one H atom is replaced by an *amino* (NH_2) group. It can be prepared by the reaction of ammonia with a basic solution of sodium hypochlorite (NaOCl):

$$2\ NH_3(aq) + OCl^-(aq) \longrightarrow N_2H_4(aq) + H_2O(l) + Cl^-(aq)$$

You have perhaps heard that household cleaners should never be mixed because exothermic reactions may occur or dangerous products may form. The formation of hydrazine on mixing household ammonia and hypochlorite-containing chlorine bleaches is a case in point.

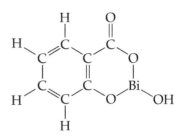

▲ Pepto-Bismol (bismuth subsalicylate, $C_7H_5BiO_4$)

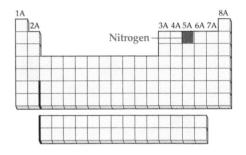

REMEMBER...
The temperature dependence of the equilibrium constant K depends on the sign of $\Delta H°$ for the reaction. For an endothermic reaction ($\Delta H° > 0$), K increases as T increases (Section 15.9).

Ammonia

Hydrazine

Pure hydrazine is a poisonous, colorless liquid that smells like ammonia, freezes at 2 °C, and boils at 114 °C. It is explosive when mixed with air or other oxidizing agents and is used as a rocket fuel. Hydrazine can be handled safely in aqueous solutions, where it behaves as a weak base ($K_b = 8.9 \times 10^{-7}$) and a versatile reducing agent. It reduces Fe^{3+} to Fe^{2+}, I_2 to I^-, and Ag^+ to metallic Ag, for example.

Oxides of Nitrogen Nitrogen forms a large number of oxides, but we'll discuss only three: nitrous oxide (dinitrogen monoxide, N_2O), nitric oxide (nitrogen monoxide, NO), and nitrogen dioxide (NO_2).

Nitrous oxide (N_2O) is a colorless, sweet-smelling gas obtained when molten ammonium nitrate is heated gently at about 270 °C. Strong heating can cause an explosion. Known as "laughing gas" because small doses are mildly intoxicating, nitrous oxide is used as a dental anesthetic and as a propellant for dispensing whipped cream and cooking sprays.

Nitric oxide (NO) is a colorless gas, produced in the laboratory when copper metal is treated with dilute nitric acid:

$$3\ Cu(s) + 2\ NO_3^-(aq) + 8\ H^+(aq) \longrightarrow 3\ Cu^{2+}(aq) + 2\ NO(g) + 4\ H_2O(l)$$

Nitrous oxide

Nitric oxide

Nitrogen dioxide

Nitric oxide is important in many biological processes, where it helps transmit messages between nerve cells and kills harmful bacteria. It also helps to protect the heart from insufficient oxygen levels by dilating blood vessels. The drug nitroglycerin, long used to relieve the pain of angina, is a source of NO, which in turn dilates constricted arteries.

Nitrogen dioxide (NO_2) is the highly toxic, reddish brown gas that forms rapidly when nitric oxide is exposed to air (**FIGURE 22.12**):

$$2\ NO(g) + O_2(g) \longrightarrow 2\ NO_2(g)$$

NO_2 is also produced when copper reacts with concentrated nitric acid:

$$Cu(s) + 2\ NO_3^-(aq) + 4\ H^+(aq) \longrightarrow Cu^{2+}(aq) + 2\ NO_2(g) + 2\ H_2O(l)$$

Go to
eText

BIG IDEA Question 6

In which compound does nitrogen have the lowest oxidation state: N_2O, NO_2, N_2H_4, or NH_3?

▶ **FIGURE 22.12**

Nitric oxide (NO) and nitrogen dioxide (NO_2).

| Nitric oxide is a colorless gas. | NO turns brown on contact with air because it is rapidly oxidized to NO_2. | Copper reacts with concentrated HNO_3, yielding noxious, red-brown fumes of NO_2. The blue color of the solution is due to Cu^{2+} ions. |

Because NO_2 has an odd number of valence electrons (17), it tends to dimerize, forming colorless, N_2O_4, in which the unpaired electrons of two NO_2 molecules pair up to give an N—N bond. In the gas phase, NO_2 and N_2O_4 are present in equilibrium:

$$O_2N\cdot + \cdot NO_2 \rightleftharpoons O_2N\!-\!NO_2 \qquad \Delta H° = -55.3\ kJ$$

Brown　　　　　　　　Colorless

Nitric Acid Nitric acid, one of the most important inorganic acids, is used mainly to make ammonium nitrate for fertilizers, but it is also used to manufacture explosives, plastics, and dyes. Annual U.S. production of HNO_3 is approximately 6.3 million metric tons.

In the laboratory, concentrated nitric acid often has a yellow-brown color due to the presence of a small amount of NO_2 produced by a slight amount of decomposition:

$$4 \, HNO_3(aq) \longrightarrow 4 \, NO_2(aq) + O_2(g) + 2 \, H_2O(l)$$

Nitric acid is a strong acid and is essentially 100% dissociated in water. It's also a strong oxidizing agent, as indicated by large, positive $E\,°$ values for reduction to lower oxidation states:

$$NO_3^-(aq) + 2 \, H^+(aq) + e^- \longrightarrow NO_2(g) + H_2O(l) \qquad E\,° = 0.79 \, V$$

$$NO_3^-(aq) + 4 \, H^+(aq) + 3 \, e^- \longrightarrow NO(g) + 2 \, H_2O(l) \quad E\,° = 0.96 \, V$$

Thus, nitric acid is a stronger oxidizing agent than $H^+(aq)$ and can oxidize relatively inactive metals like copper and silver that are not oxidized by aqueous HCl.

An even more potent oxidizing agent than HNO_3 is *aqua regia*, a mixture of concentrated HCl and concentrated HNO_3 in a 3:1 ratio by volume. Aqua regia can oxidize even inactive metals like gold, which do not react with either HCl or HNO_3 separately:

$$Au(s) + 3 \, NO_3^-(aq) + 6 \, H^+(aq) + 4 \, Cl^-(aq) \longrightarrow AuCl_4^-(aq) + 3 \, NO_2(g) + 3 \, H_2O(l)$$

The NO_3^- ion serves as the oxidizing agent, and Cl^- facilitates the reaction by converting the Au(III) oxidation product to the $AuCl_4^-$ complex ion.

Elemental Phosphorus

Phosphorus is the most abundant element of group 5A, accounting for 0.10% of the mass of the Earth's crust. It is found in phosphate rock, which is mostly calcium phosphate, $Ca_3(PO_4)_2$, and in fluorapatite, $Ca_5(PO_4)_3 \, F$. The apatites are phosphate minerals with the formula $3 \, Ca_3(PO_4)_2 \cdot CaX_2$, where X^- is usually F^- or OH^-. Phosphorus is also important in living systems and is the sixth most abundant element in the human body (Figure 22.1). Our bones are mostly hydroxyapatite, $Ca_5(PO_4)_3OH$, along with the fibrous protein collagen, and tooth enamel is almost pure hydroxyapatite. Phosphate groups are also an integral part of the nucleic acids DNA and RNA, the molecules that pass genetic information from generation to generation, and phospholipids, the phosphate-containing molecules that are major components of cell membranes.

Elemental phosphorus is produced industrially by heating phosphate rock, coke, and silica sand at about 1500 °C in an electric furnace. The reaction can be represented by the simplified equation

$$2 \, Ca_3(PO_4)_2(s) + 10 \, C(s) + 6 \, SiO_2(s) \longrightarrow P_4(s) + 10 \, CO(g) + 6 \, CaSiO_3(l)$$

Phosphorus is used to make phosphoric acid, one of the top 10 industrial chemicals.

Phosphorus exists as two common allotropes: white phosphorus and red phosphorus (**FIGURE 22.13**). White phosphorus, the form produced in the industrial synthesis, is a toxic, waxy, white solid that contains discrete tetrahedral P_4 molecules. Red phosphorus, by contrast, is essentially nontoxic and has a polymeric structure.

As expected for a molecular solid that contains small, nonpolar molecules, white phosphorus has a low melting point (44 °C) and is soluble in nonpolar solvents. It is highly reactive, bursting into flames when exposed to air, and is thus stored under water. When white phosphorus is heated in the absence of air at about 300 °C, it is converted to the more stable red form. Consistent with its polymeric structure, red phosphorus is higher melting (mp 579 °C), less soluble, and less reactive than white phosphorus, and it does not ignite on contact with air (Figure 22.13).

The high reactivity of white phosphorus is due to an unusual bonding that produces considerable strain in the tetrahedral P_4 molecules. If each P atom uses three $3p$ orbitals

 Go to eText

BIG IDEA Question 7

The dimerization of NO_2 to form N_2O_4 is an exothermic reaction. Which substance does the equilibrium favor at high temperatures?

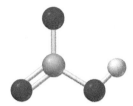

Nitric acid

▲ Freshly prepared concentrated nitric acid (left) turns yellow-brown on standing (right) because nitric acid decomposes to brown-colored NO_2.

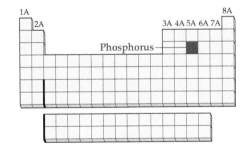

White phosphorus

A portion of the polymeric
structure of red phosphorus

▲ **FIGURE 22.13**

Red phosphorus (left) and white
phosphorus stored under water (right)
because it reacts with oxygen in air.

▲ **Figure It Out**

Based on the chemical structure of red and
white phosphorus, which form has a lower
melting point?

Answer: White phosphorus has a lower melting point because it contains P_4 molecules, which are held together by relatively weak intermolecular forces. Red phosphorus has strong covalent bonds between all P atoms giving it a higher melting point.

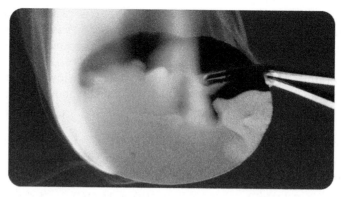

▲ A piece of filter paper coated with white phosphorus bursts
into flame in air.

to form its three P—P bonds, all the bond angles should be 90°. The geometry of P_4, however, requires that all the bonds have 60° angles, which means that the *p* orbitals can't overlap in a head-on fashion. As a result, the P—P bonds are "bent," relatively weak, and highly reactive (**FIGURE 22.14**).

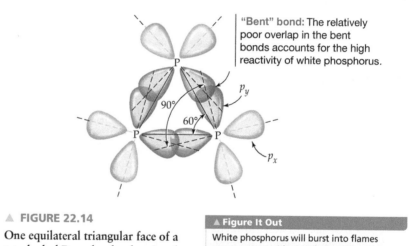

"Bent" bond: The relatively
poor overlap in the bent
bonds accounts for the high
reactivity of white phosphorus.

p_y

90°

60°

p_x

▲ **FIGURE 22.14**

One equilateral triangular face of a
tetrahedral P_4 molecule, showing the
60° bond angles and the 90° angles
between the *p* orbitals.

▲ **Figure It Out**

White phosphorus will burst into flames
upon contact with air. Why is white
phosphorus highly reactive?

Answer: The 60° bond angles in a P_4 molecule lead to poor overlap of the *p* orbitals and weak chemical bonds.

Phosphorus Compounds

Like nitrogen, phosphorus forms compounds in all oxidation states between -3 and $+5$, but the $+3$ state, as in PCl_3, P_4O_6, and H_3PO_3, and the $+5$ state, as in PCl_5, P_4O_{10}, and H_3PO_4, are the most common. Compared to nitrogen, phosphorus is more likely to be found in a positive oxidation state because of its lower electronegativity.

Phosphine Phosphine (PH_3), a colorless, extremely poisonous gas, is the simplest hydride of phosphorus. Like NH_3, phosphine has a trigonal pyramidal structure and has the group 5A atom in the -3 oxidation state. Unlike NH_3, however, its aqueous solutions are neutral, indicating that PH_3 is a poor proton acceptor. In accord with the low electronegativity of phosphorus, phosphine is easily oxidized, burning in air to form phosphoric acid:

Phosphine

$$PH_3(g) + 2 O_2(g) \longrightarrow H_3PO_4(l)$$

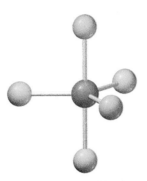

Phosphorus trichloride

Phosphorus pentachloride

Phosphorus Halides Phosphorus reacts with all the halogens, forming phosphorus(III) halides, PX_3, or phosphorus(V) halides, PX_5 (X = F, Cl, Br, or I), depending on the relative amounts of the reactants:

$$\text{Limited amount of } X_2: \quad P_4 + 6\,X_2 \longrightarrow 4\,PX_3$$
$$\text{Excess amount of } X_2: \quad P_4 + 10\,X_2 \longrightarrow 4\,PX_5$$

At room temperature, all of these halides are gases, volatile liquids, or low-melting solids. For example, phosphorus trichloride is a colorless liquid that boils at 76 °C, and phosphorus pentachloride is an off-white solid that melts at 167 °C.

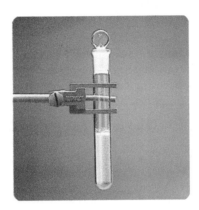

▲ Samples of PCl_3 (left) and PCl_5 (right).

Oxides and Oxoacids of Phosphorus When phosphorus burns in air or oxygen, it yields either tetraphosphorus hexoxide (P_4O_6, mp 24 °C) or tetraphosphorus decoxide (P_4O_{10}, mp 420 °C), depending on the amount of oxygen present:

$$\text{Limited amount of } O_2: \quad P_4(s) + 3\,O_2(g) \longrightarrow P_4O_6(s)$$
$$\text{Excess amount of } O_2: \quad P_4(s) + 5\,O_2(g) \longrightarrow P_4O_{10}(s)$$

Both oxides are molecular compounds and have structures with a tetrahedral array of P atoms, as in white phosphorus. One O atom bridges each of the six edges of the P_4 tetrahedron, and an additional, terminal O atom is bonded to each P atom in P_4O_{10}.

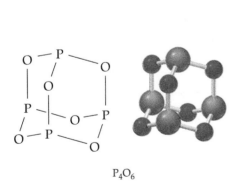

P_4O_6

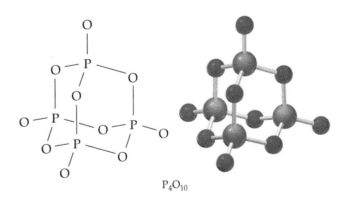

P_4O_{10}

Phosphorous acid (H_3PO_3) is a weak diprotic acid because only two of its three H atoms are bonded to oxygen. The H atom bonded directly to phosphorus is not acidic because phosphorus and hydrogen have the same electronegativity and so the P—H bond is nonpolar. In phosphoric acid, however, all three hydrogens are attached to oxygen, and thus phosphoric acid is a weak triprotic acid. The geometry about the P atom in both molecules is tetrahedral, as expected. Note that the **successive dissociation constants decrease** by a factor of about 10^5 (Section 16.11).

Pure phosphoric acid is a low-melting, colorless, crystalline solid (mp 42 °C), but the commercially available phosphoric acid used in the laboratory is a syrupy, aqueous

REMEMBER . . .
Successive dissociation constants decrease in the order $K_{a1} > K_{a2} > K_{a3}$ because it's easier to remove H^+ from a neutral molecule (H_3PO_4) than from a negatively charged anion ($H_2PO_4^-$). Similarly, it's easier to remove H^+ from $H_2PO_4^-$ than from a doubly charged anion (HPO_4^{2-}) (Section 16.11).

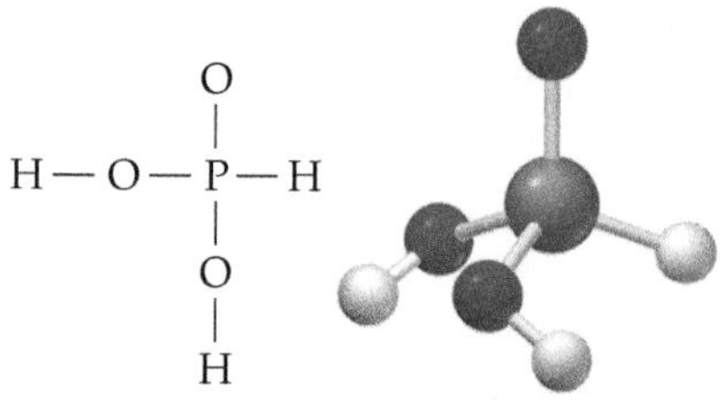

Phosphorous acid, H_3PO_3
$K_{a1} = 1.0 \times 10^{-2}$
$K_{a2} = 2.6 \times 10^{-7}$

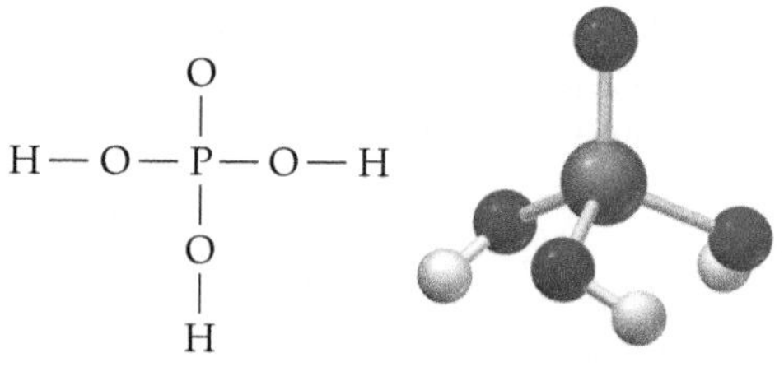

Phosphoric acid, H_3PO_4
$K_{a1} = 7.5 \times 10^{-3}$
$K_{a2} = 6.2 \times 10^{-8}$
$K_{a3} = 4.8 \times 10^{-13}$

solution containing about 82% H_3PO_4 by mass. For use as a food additive—for example, as the tart ingredient in soft drinks—pure phosphoric acid is made by burning molten phosphorus in a mixture of air and steam. For use in making fertilizers, an impure form of phosphoric acid is produced by treating phosphate rock with sulfuric acid:

$$Ca_3(PO_4)_2(s) + 3\ H_2SO_4(aq) \longrightarrow 2\ H_3PO_4(aq) + 3\ CaSO_4(aq)$$

22.8 GROUP 6A ELEMENTS

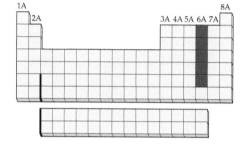

The group 6A elements are oxygen, sulfur, selenium, tellurium, and polonium. As shown in **TABLE 22.7**, their properties exhibit the usual periodic trends. Both oxygen and sulfur are typical nonmetals. Selenium and tellurium are primarily nonmetallic in character, though the most stable allotrope of selenium, gray selenium, is a lustrous semiconducting solid. Tellurium is also a semiconductor and is usually classified as a semimetal. Polonium, a radioactive element that occurs in trace amounts in uranium ores, is a silvery white metal.

Elemental selenium　　　　　　　Elemental tellurium

TABLE 22.7　Properties of the Group 6A Elements

Property	Oxygen	Sulfur	Selenium	Tellurium	Polonium
Valence electron configuration	$2s^2\ 2p^4$	$3s^2\ 3p^4$	$4s^2\ 4p^4$	$5s^2\ 5p^4$	$6s^2\ 6p^4$
Melting point (°C)	−219	113*	221†	450	254
Boiling point (°C)	−183	445	685	988	962
Atomic radius (pm)	66	104	116	143	167
X^{2-} ionic radius (pm)	140	184	198	221	—
First ionization energy (kJ/mol)	1314	1000	941	869	812
Electron affinity (kJ/mol)	−141	−200	−195	−190	−183
Electronegativity	3.5	2.5	2.4	2.1	2.0
Redox potential, $E°$ (V) for $X + 2H^+\ 2\ e^- \longrightarrow H_2X$	1.23	0.14	−0.40	−0.79	—

*Rhombic S
†Gray Se

With valence electron configuration $ns^2 np^4$, the group 6A elements are just two electrons short of an octet configuration and the −2 oxidation state is therefore common. The stability of the −2 state decreases, however, with increasing metallic character, as indicated by the $E°$ values in Appendix D. Thus, oxygen is a powerful oxidizing agent, but $E°$ values for reduction of Se and Te are negative, which means that H_2Se and H_2Te are reducing agents. Because S, Se, and Te are much less electronegative than oxygen, they are commonly found in positive oxidation states, especially +4, as in SF_4, SO_2, and H_2SO_3, and +6, as in SF_6, SO_3, and H_2SO_4.

Commercial uses of Se, Te, and Po are limited, though selenium is used to make red-colored glass and is used in photocopiers and laser printers. Tellurium is used in alloys to improve their machinability, and polonium (^{210}Po) has been used as a heat source in space equipment and as a source of alpha particles in scientific research.

▲ The color of the red glass in this traffic signal is due to cadmium selenide, CdSe.

Elemental Oxygen

Oxygen is the most abundant element on the surface of our planet and is crucial to human life. It's in the air we breathe, the water we drink, and the food we eat. It's the oxidizing agent in the metabolic "burning" of foods, and it's an important component of biological molecules: Approximately one-fourth of the atoms in living organisms are oxygen.

On a mass basis, oxygen constitutes 23% of the atmosphere (21% by volume), 46% of the lithosphere (the Earth's crust), and more than 85% of the hydrosphere. In the atmosphere, oxygen is found primarily as O_2, sometimes called *dioxygen*. The oxygen in the hydrosphere is in the form of H_2O, but enough dissolved O_2 is typically present within the water to maintain aquatic life. In the lithosphere, oxygen is combined with other elements in crustal rocks composed of silicates, carbonates, oxides, and other oxygen-containing minerals.

Gaseous O_2 condenses at −183 °C to form a pale blue liquid and freezes at −219 °C to give a pale blue solid. In all three phases—gas, liquid, and solid—O_2 is paramagnetic, as illustrated previously in Figure 8.20. The bond length in O_2 is 121 pm, appreciably shorter than the O—O single bond in H_2O_2 (148 pm), and the bond dissociation energy of O_2 (498 kJ/mol) is intermediate between that for the single bond in F_2 (159 kJ/mol) and the triple bond in N_2 (945 kJ/mol). These properties are consistent with the presence of a **double bond** in O_2 (Section 8.8).

Oxygen is produced on an industrial scale, along with nitrogen and argon, by the fractional distillation of liquefied air. When liquid air warms in a suitable distilling column, the more volatile components—nitrogen (bp −196 °C) and argon (bp −186 °C)—can be removed as gases from the top of the column. The less volatile oxygen (bp −183 °C) remains as a liquid at the bottom. Annual production of oxygen in the United States is approximately 29 million metric tons; only sulfuric acid and nitrogen are produced in greater quantities.

More than two-thirds of the oxygen produced industrially is used in making steel. Among its other uses, oxygen is employed in sewage treatment to destroy malodorous compounds and in paper bleaching to oxidize compounds that impart unwanted colors. In all its applications, O_2 serves as an inexpensive and readily available oxidizing agent.

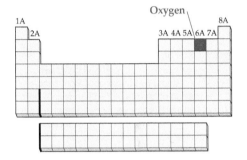

> **REMEMBER . . .**
> Molecular orbital theory predicts the presence of both paramagnetism and a **double bond** in the O_2 molecule (Section 8.8).

▲ Liquid oxygen has a pale blue color.

Oxygen Compounds

We can anticipate the reactivity of oxygen from the electron configuration of an oxygen atom ($1s^2 2s^2 2p^4$) and its high electronegativity. With six valence electrons, oxygen is just two electrons short of the octet configuration of neon, the next noble gas. Oxygen can therefore achieve an octet configuration either by accepting two electrons from an active metal or by gaining a share in two additional electrons through covalent bonding. Thus, oxygen reacts with active metals, such as lithium and magnesium, to give *ionic oxides:*

$$4 \text{ Li}(s) + O_2(g) \longrightarrow 2 \text{ Li}_2O(s)$$

$$2 \text{ Mg}(s) + O_2(g) \longrightarrow 2 \text{ MgO}(s)$$

▲ Crude iron is converted to steel by oxidizing impurities with O_2 gas.

On the other hand, with nonmetals, such as hydrogen, carbon, sulfur, and phosphorus, oxygen forms *covalent oxides:*

$$2\,H_2(g) + O_2(g) \longrightarrow 2\,H_2O(l)$$

$$C(s) + O_2(g) \longrightarrow CO_2(g)$$

$$S_8(s) + 8\,O_2(g) \longrightarrow 8\,SO_2(g)$$

$$P_4(s) + 5\,O_2(g) \longrightarrow P_4O_{10}(s)$$

Water   Carbon dioxide   Sulfur dioxide   Tetraphosphorus decoxide

REMEMBER . . .
Sigma (σ) bonds arise from head-on, σ overlap of orbitals. Pi (π) bonds arise from parallel, sideways, π **overlap** of orbitals (Section 8.4).

In covalent compounds, oxygen generally achieves an octet configuration either by forming two single bonds, as in H_2O, or one double bond, as in CO_2. Oxygen often forms a double bond to small atoms such as carbon and nitrogen because there is good π **overlap** between the relatively compact p orbitals of second-row atoms (Section 8.4). With larger atoms such as silicon, however, there is less efficient π overlap, and double bond formation is therefore less common (**FIGURE 22.15**).

▶ **FIGURE 22.15**

Pi overlap between the p orbitals of oxygen and other atoms.

▶ **Figure It Out**

Why does oxygen form a double bond with a small atom like carbon but a single bond with a larger atom like silicon?

Answer: A π bond requires significant overlap of p orbitals. Oxygen and silicon do not form a π bond because the $3p$ orbital in silicon is large which results in insufficient overlap with the $2p$ orbital in oxygen.

Good overlap: With a second-row atom, such as carbon, oxygen forms a strong π bond.

Poor overlap: With a larger atom, such as silicon, oxygen tends not to form π bonds because the longer Si–O distance and the larger, more diffuse silicon orbital result in poor π overlap.

 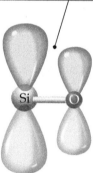

Acid–Base Properties of Oxides

Binary compounds with oxygen in the -2 oxidation state are called **oxides.** Oxides can be categorized as basic, acidic, or amphoteric (both basic and acidic), as shown in **FIGURE 22.16.** *Basic oxides,* also called base anhydrides, are ionic and are formed by metals on the left side of the periodic table. Water-soluble basic oxides, such as sodium oxide, Na_2O, dissolve by reacting with water to produce OH^- ions:

$$Na_2O(s) + H_2O(l) \longrightarrow 2\,Na^+(aq) + 2\,OH^-(aq)$$

Water-insoluble basic oxides, such as MgO, can dissolve in strong acids because H^+ ions from the acid combine with the O^{2-} ion to produce water:

$$MgO(s) + 2\,H^+(aq) \longrightarrow Mg^{2+}(aq) + H_2O(l)$$

Acidic oxides, also called acid anhydrides, are covalent and are formed by the nonmetals on the right side of the periodic table. Water-soluble acidic oxides, such as dinitrogen pentoxide, N_2O_5, dissolve by reacting with water to produce aqueous H^+ ions:

$$N_2O_5(s) + H_2O(l) \longrightarrow 2\,H^+(aq) + 2\,NO_3^-(aq)$$

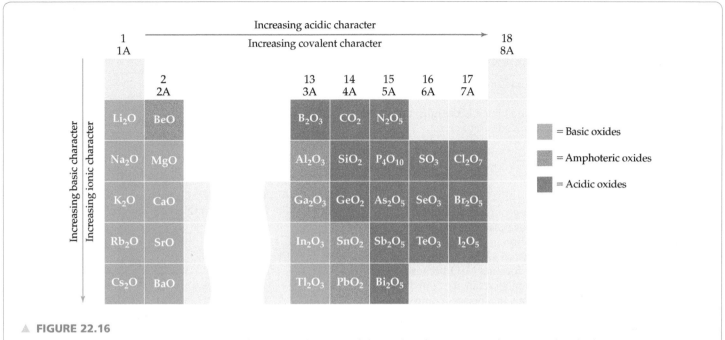

▲ **FIGURE 22.16**

Formulas, acid–base properties, and the covalent–ionic character of the oxides of main-group elements in their highest oxidation states.

Water-insoluble acidic oxides, such as silicon dioxide, SiO_2, can dissolve in strong bases:

$$SiO_2(s) + 2\ OH^-(aq) \longrightarrow SiO_3^{2-}(aq) + H_2O(l)$$

Amphoteric oxides, like the corresponding amphoteric hydroxides described in Section 17.12, exhibit both acidic and basic properties. For example, Al_2O_3 is insoluble in water, but it dissolves both in strong acids and in strong bases. Al_2O_3 behaves as a base when it reacts with acids, giving the Al^{3+} ion, but it behaves as an acid when it reacts with bases, yielding the aluminate ion, $Al(OH)_4^-$.

Basic behavior: $Al_2O_3(s) + 6\ H^+(aq) \longrightarrow 2\ Al^{3+}(aq) + 3\ H_2O(l)$

Acidic behavior: $Al_2O_3(s) + 2\ OH^-(aq) + 3\ H_2O(l) \longrightarrow 2\ Al(OH)_4^-(aq)$

The elements that form amphoteric oxides have intermediate electronegativities, and the bonds in their oxides have intermediate ionic–covalent character.

The acid–base properties and the ionic–covalent character of an element's oxide depend on the element's position in the periodic table. As Figure 22.16 shows, both the acidic character and the covalent character of an oxide increase across the periodic table, from the active metals on the left to the electronegative nonmetals on the right. In the third row, for example, Na_2O and MgO are basic, Al_2O_3 is amphoteric, and SiO_2, P_4O_{10}, SO_3, and Cl_2O_7 are acidic. Within a group in the periodic table, both the basic character and the ionic character of an oxide increase going down the table, from the more electronegative elements at the top to the less electronegative ones at the bottom. In group 3A, for example, B_2O_3 is acidic, Al_2O_3 and Ga_2O_3 are amphoteric, and In_2O_3 and Tl_2O_3 are basic. Combining the horizontal and vertical trends in acidity, we find the most acidic oxides in the upper right of the periodic table, the most basic oxides in the lower left, and the amphoteric oxides in a roughly diagonal band stretching across the middle.

Both the acidic character and the covalent character of different oxides of the same element increase with increasing oxidation number of the element. Thus, sulfur(VI) oxide (sulfur trioxide; SO_3) is more acidic than sulfur(IV) oxide (sulfur dioxide; SO_2). The reaction of SO_3 with water gives sulfuric acid (H_2SO_4), a strong acid, whereas the reaction of SO_2 with water yields sulfurous acid (H_2SO_3), a weak acid. The oxides of chromium exhibit the same trend. Chromium(VI) oxide (CrO_3) is acidic, chromium(III) oxide (Cr_2O_3) is amphoteric, and chromium(II) oxide (CrO) is basic.

REMEMBER . . .

Some hydroxides, such as $Al(OH)_3$, $Cr(OH)_3$, $Zn(OH)_2$, $Sn(OH)_2$, and $Pb(OH)_2$, are said to be **amphoteric** because they are soluble in both strongly acidic and strongly basic solutions (Section 17.12).

CONCEPTUAL WORKED EXAMPLE 22.5

Determining Formulas and Properties of Oxides

Look at the location of elements A, B, and C in the periodic table.

(a) Write the formula of the oxide that has each of these elements in its highest oxidation state.
(b) Which oxide is the most ionic, and which is the most covalent?
(c) Classify each oxide as basic, acidic, or amphoteric.

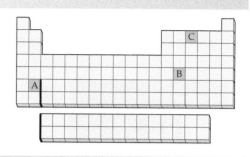

STRATEGY

(a) The oxidation number of a main-group element in its highest oxidation state and the formula of its oxide are determined by the element's location in the periodic table. For example, an element M in group 1A has an oxidation number of +1, and its oxide has formula M_2O because the oxidation number of oxygen in an oxide is −2 and the oxidation numbers of all the atoms in a neutral compound must sum to zero.
(b) Recall that ionic character decreases and covalent character increases in the periodic table from left to right and from bottom to top.
(c) Recall that basic character decreases and acidic character increases in the periodic table from left to right and from bottom to top.

SOLUTION

(a) Elements A, B, and C can be identified as Ba, Sn, and N, respectively. Because Ba, Sn, and N are in groups 2A, 4A, and 5A, respectively, the formulas of their highest-oxidation-state oxides are BaO, SnO_2, and N_2O_5.
(b) BaO is the most ionic because Ba is in the lower left region of the periodic table, and N_2O_5 is the most covalent because N is in the upper right region.
(c) BaO is basic, N_2O_5 is acidic, and SnO_2 is amphoteric. Note that Sn is one of five main-group elements that form an amphoteric oxide (Figure 22.16).

▶ **CONCEPTUAL PRACTICE 22.9** Look at the location of elements A, B, and C in the periodic table.

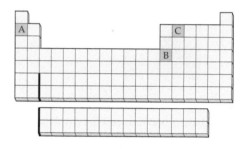

(a) Write the formula of the oxide that has each of these elements in its highest oxidation state.
(b) Which oxide is the most ionic, and which is the most covalent?
(c) Which oxide is the most acidic, and which is the most basic?
(d) Which oxide can react with both $H^+(aq)$ and $OH^-(aq)$?

▶ **APPLY 22.10** Write balanced net ionic equations for the following reactions:

(a) Dissolution of solid Li_2O in water
(b) Dissolution of SO_3 in water
(c) Dissolution of the amphoteric oxide Cr_2O_3 in strong acid
(d) Dissolution of Cr_2O_3 in strong base to give $Cr(OH)_4^-$ ions

Elemental Sulfur

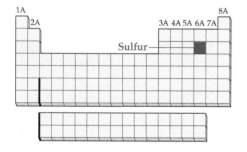

Sulfur occurs in elemental form in large underground deposits and is present in numerous minerals such as pyrite (FeS_2, which contains the S_2^{2-} ion), galena (PbS), cinnabar (HgS), and gypsum ($CaSO_4 \cdot 2\,H_2O$). Sulfur is also present in natural gas as H_2S and in crude oil as organic sulfur compounds. In plants and animals, sulfur occurs in various proteins, and it is one of the 10 most abundant elements in the human body (Figure 22.1).

Elemental sulfur is obtained from underground deposits and is recovered from natural gas and crude oil. In the United States, approximately 90% of the sulfur produced is used to manufacture sulfuric acid.

Sulfur exists in many allotropic forms, but the most stable at 25 °C is rhombic sulfur, a yellow crystalline solid (mp 113 °C) that contains crown-shaped S_8 rings.

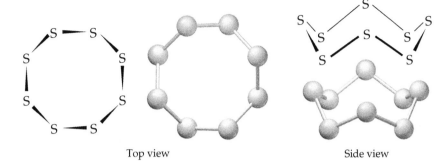

Top view   Side view

▲ Pyrite (FeS_2) is often called fool's gold because of its golden yellow color. It contains the disulfide ion (S_2^{2-}).

Above 95 °C, rhombic sulfur is less stable than monoclinic sulfur (mp ~ 119 °C), an allotrope in which the cyclic S_8 molecules pack differently in the crystal. The phase transition from rhombic to monoclinic sulfur is very slow, however, and rhombic sulfur simply melts at 113 °C when heated at an ordinary rate.

As shown in **FIGURE 22.17,** molten sulfur exhibits some striking changes when its temperature is increased. Just above its melting point, sulfur is a fluid, straw-colored liquid, but between 160 °C and 195 °C its color becomes dark reddish brown and its viscosity increases by a factor of more than 10,000. At still higher temperatures, the liquid becomes more fluid again and then boils at 445 °C. If the hot liquid is cooled rapidly by pouring it into water, the sulfur forms an amorphous, rubbery material called *plastic sulfur.*

The dramatic increase in the viscosity of molten sulfur at 160–195 °C is due to the opening of the S_8 rings, yielding S_8 chains that subsequently form long polymers with more than 200,000 S atoms in the chain:

▲ A sample of rhombic sulfur, the most stable allotrope of sulfur.

$$S_8 \text{ rings} \xrightarrow{\text{Heat}} \cdot S\!-\!S_6\!-\!S \cdot \text{chains}$$

$$\cdot S\!-\!S_6\!-\!S \cdot \text{ chains } + \text{ } S_8 \text{ rings} \longrightarrow \cdot S\!-\!S_{14}\!-\!S \cdot \text{chains}$$

$$\longrightarrow S_n \text{ chains}$$

$$n > 200{,}000$$

Fluid, straw-colored liquid sulfur at about 120 °C.

Viscous, reddish-brown liquid sulfur at about 180 °C.

Plastic sulfur, obtained by pouring liquid sulfur into water. Plastic sulfur is unstable and reverts to rhombic sulfur on standing at room temperature.

◀ **FIGURE 22.17**
Effect of temperature on the properties of sulfur.

Hydrogen sulfide

Whereas the small S_8 rings easily slide over one another in the liquid, the long polymer chains become entangled, thus accounting for the increase in viscosity. Above 200 °C, the polymer chains begin to fragment into smaller pieces, and the viscosity therefore decreases. On rapid cooling, the chains are temporarily frozen in a disordered, tangled arrangement, which accounts for the elastic properties of *plastic sulfur.*

Sulfur Compounds

Hydrogen Sulfide Hydrogen sulfide is a colorless gas (bp = −60 °C) with the strong, foul odor we associate with rotten eggs, in which it occurs because of the bacterial decomposition of sulfur-containing proteins. Hydrogen sulfide is extremely toxic, causing headaches and nausea at concentrations of 10 ppm and sudden paralysis and death at 100 ppm. On initial exposure, the odor of H_2S can be detected at about 0.02 ppm, but unfortunately the gas tends to dull the sense of smell. It is thus an extremely insidious poison, even more dangerous than HCN.

In the laboratory, H_2S can be prepared by treating iron(II) sulfide with dilute sulfuric acid:

$$FeS(s) + 2 H^+(aq) \longrightarrow H_2S(g) + Fe^{2+}(aq)$$

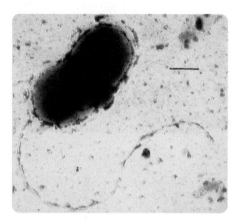

▲ *Desulfovibrio vulgaris* is in a class of sulfate-reducing bacteria that obtain energy by oxidizing organic compounds while reducing sulfate (SO_4^{2-}) to hydrogen sulfide (H_2S). These microbes are said to "breathe" sulfur because in low-oxygen environments sulfur instead of oxygen is reduced in chemical reactions that provide energy for the organism.

Hydrogen sulfide is a very weak diprotic acid ($K_{a1} = 1.0 \times 10^{-7}$; $K_{a2} \approx 10^{-19}$) and a mild reducing agent. In reactions with mild oxidizing agents, it is oxidized to a milky white suspension of elemental sulfur:

$$H_2S(aq) + 2 Fe^{3+}(aq) \longrightarrow S(s) + 2 Fe^{2+}(aq) + 2 H^+(aq)$$

Oxides and Oxoacids of Sulfur Sulfur dioxide (SO_2) and sulfur trioxide (SO_3) are the most important of the various oxides of sulfur. Because SO_2 is toxic to microorganisms, it is used for sterilizing wine and dried fruit. Sulfur dioxide, a colorless, toxic gas (bp −10 °C) with a pungent, choking odor, is formed when sulfur burns in air:

$$S(s) + O_2(g) \longrightarrow SO_2(g)$$

Sulfur dioxide

Sulfur dioxide is slowly oxidized in the atmosphere to SO_3, which dissolves in rainwater to give sulfuric acid. The burning of sulfur-containing fuels is thus a major cause of acid rain (Chapter 16, *Inquiry*).

The United States produces nearly 40 million metric tons of sulfuric acid annually. It is used mostly to manufacture soluble phosphate fertilizers but is essential in many other industries. **FIGURE 22.18** shows the major uses of sulfuric acid in the United States. The use of sulfuric acid is so widespread in industrial countries that it is sometimes regarded

Sulfur trioxide

$$\overset{\displaystyle :\ddot{O}:}{\underset{\displaystyle :\ddot{O}:}{H-\ddot{O}-S-\ddot{O}-H}}$$

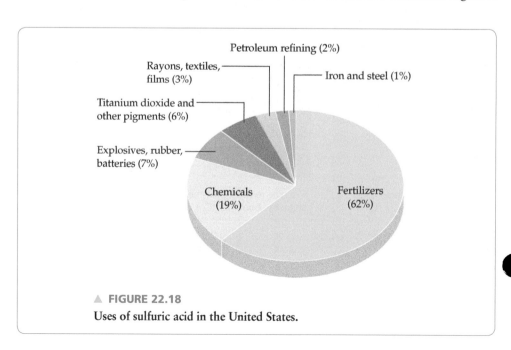

▲ **FIGURE 22.18**
Uses of sulfuric acid in the United States.

Sulfuric acid

as an indicator of economic activity. Sulfuric acid (H_2SO_4), the world's most important industrial chemical, is manufactured by the **contact process,** a three-step reaction sequence in which (1) sulfur burns in air to give SO_2, (2) SO_2 is oxidized to SO_3 in the presence of a vanadium(V) oxide catalyst, and (3) SO_3 reacts with water to give H_2SO_4:

1. $S(s) + O_2(g) \longrightarrow SO_2(g)$
2. $2 SO_2(g) + O_2(g) \xrightarrow[\text{V}_2\text{O}_5 \text{ catalyst}]{\text{Heat}} 2 SO_3(g)$
3. $SO_3(g) + H_2O$ (in conc H_2SO_4) $\longrightarrow H_2SO_4(aq)$

In the third step, the SO_3 is absorbed in concentrated sulfuric acid rather than in water because the dissolution of SO_3 in water is slow.

Sulfuric acid is a strong acid for the dissociation of its first proton and has $K_{a2} = 1.2 \times 10^{-2}$ for the dissociation of its second proton. As a diprotic acid, it forms two series of salts: hydrogen sulfates, such as $NaHSO_4$, and sulfates, such as Na_2SO_4.

The oxidizing properties of sulfuric acid depend on its concentration and temperature. In dilute solutions at room temperature, H_2SO_4 behaves like HCl, oxidizing metals that stand above hydrogen in the activity series (Table 4.5):

$$Fe(s) + 2 H^+(aq) \longrightarrow Fe^{2+}(aq) + H_2(g)$$

Hot, concentrated H_2SO_4 is a stronger oxidizing agent than the dilute, cold acid and can oxidize metals like copper, which are not oxidized by $H^+(aq)$. In the process, H_2SO_4 is reduced to SO_2:

$$Cu(s) + 2 H_2SO_4(l) \longrightarrow Cu^{2+}(aq) + SO_4^{2-}(aq) + SO_2(g) + 2 H_2O(l)$$

22.9 GROUP 7A: THE HALOGENS

The halogens of group 7A (F, Cl, Br, I, and At) are too reactive to occur in nature as free elements. Instead, they are found only as their anions in various salts and minerals. Even the name *halogen* implies reactivity, since it comes from the Greek words *hals* (salt) and *gennan* (to form). Thus, a halogen is literally a salt former. The halogens are nonmetals, and they exist as diatomic molecules rather than as individual atoms. They tend to gain rather than lose electrons when they enter into reactions because of their $ns^2 np^5$ electron configurations. In other words, the halogens are characterized by large negative electron affinities and large positive ionization energies. Some of their properties are listed in **TABLE 22.8**.

Go to eText

BIG IDEA Question 8

Consider the following sulfur-containing oxoanions:

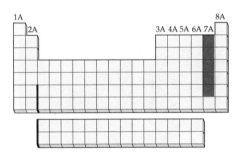

Which oxoanion is the strongest acid? Which oxoanion the strongest base?

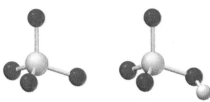

TABLE 22.8 Properties of Halogens

Name	Melting Point (°C)	Boiling Point (°C)	Density (g/cm³)	Electron Affinity (kJ/mol)	Abundance on Earth (%)	Atomic Radius (pm)	Ionic (X^-) Radius (pm)
Fluorine	−220	−188	1.50 (*l*)	−328	0.062	72	133
Chlorine	−101	−34	2.03 (*l*)	−349	0.013	99	181
Bromine	−7	59	3.12 (*l*)	−325	0.000 3	114	196
Iodine	114	184	4.930 (*s*)	−295	0.000 05	133	220
Astatine	—	—	—	−270	Trace	—	—

All the free halogens are produced commercially by oxidation of their anions. Fluorine and chlorine are both produced by electrolysis: fluorine from a molten 1:2 molar mixture of KF and HF, and chlorine from molten NaCl.

$$2 HF(l) \xrightarrow[\text{100 °C}]{\text{Electrolysis}} H_2(g) + F_2(g)$$

$$2 NaCl(l) \xrightarrow[\text{580 °C}]{\text{Electrolysis}} 2 Na(l) + Cl_2(g)$$

Bromine and iodine are both prepared by oxidation of the corresponding halide ion with chlorine. Naturally occurring aqueous solutions of bromide ion with concentrations of up to 5000 ppm are found in Arkansas and in the Dead Sea in Israel. Iodide ion solutions of up to 100 ppm concentration are found in Oklahoma and Michigan.

$$2\,Br^-(aq) + Cl_2(g) \longrightarrow Br_2(l) + 2\,Cl^-(aq)$$
$$2\,I^-(aq) + Cl_2(g) \longrightarrow I_2(s) + 2\,Cl^-(aq)$$

Halogen Compounds

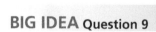 Go to eText

BIG IDEA Question 9

Which halogen is the most reactive, and what is the reason for the high reactivity?

Halogens are among the most reactive elements in the periodic table. Fluorine, in fact, forms compounds with every other element except the three noble gases He, Ne, and Ar. As noted previously, their large negative electron affinities allow halogens to accept electrons from other atoms to yield halide anions, X^-.

Halogens react with every metal in the periodic table to yield metal halides. With alkali and alkaline-earth metals, the formula of the halide product is easily predictable. With transition metals, though, more than one product can sometimes form depending on the reaction conditions and the amounts of reactants present. Iron, for instance, can react with Cl_2 to form either $FeCl_2$ or $FeCl_3$. The reaction can be generalized as

$$2\,M + n\,X_2 \longrightarrow 2\,MX_n \quad \text{where} \quad M = \text{Metal}$$
$$X = \text{F, Cl, Br, or I}$$

Unlike the metallic elements, halogens become less rather than more reactive going down the periodic table because of their generally decreasing electron affinity. Thus, their reactivity order is $F_2 > Cl_2 > Br_2 > I_2$. Fluorine often reacts violently, chlorine and bromine somewhat less so, and iodine often sluggishly.

In addition to their reaction with metals, halogens also react with hydrogen gas to yield hydrogen halides, HX. Fluorine reacts explosively with hydrogen as soon as the two gases come in contact. Chlorine also reacts explosively once the reaction is initiated by a spark or by ultraviolet light, but the mixture of gases is stable in the dark. Bromine and iodine react more slowly.

$$H_2(g) + X_2 \longrightarrow 2\,HX(g) \quad \text{where} \quad X = \text{F, Cl, Br, or I}$$

Hydrogen halides are useful because they are acids—that is, they produce H^+ ions when dissolved in water. An aqueous solution of HCl, for instance, is used throughout the chemical industry in a vast number of processes, from pickling steel (removing its iron oxide coating) to dissolving animal bones for producing gelatin.

$$HX \xrightarrow[\text{in } H_2O]{\text{Dissolve}} H^+(aq) + X^-(aq)$$

Hydrogen fluoride (HF) is used frequently for etching glass because it is one of the few substances that reacts with glass.

They also share one electron with nonmetals to give molecular compounds such as HCl, BCl_3, PF_5, and SF_6. In all these compounds, the halogen is in the -1 oxidation state. Among the most important compounds of halogens in positive oxidation states are the oxoacids of Cl, Br, and I (**TABLE 22.9**) and the corresponding oxoacid salts. In

▲ This beautiful piece of glass was etched with gaseous HF, one of the few substances that reacts with and etches glass (SiO_2), according to the equation $SiO_2(s) + 4\,HF(g) \longrightarrow SiF_4(g) + 2\,H_2O(l)$.

TABLE 22.9 Oxoacids of the Halogens

Oxidation State	Generic Name (formula)	Chlorine	Bromine	Iodine
+1	Hypohalous acid (HXO)	HClO	HBrO	HIO
+3	Halous acid (HXO_2)	$HClO_2$	—	—
+5	Halic acid (HXO_3)	$HClO_3$	$HBrO_3$	HIO_3
+7	Perhalic acid (HXO_4)	$HClO_4$	$HBrO_4$	HIO_4, H_5IO_6

these compounds, the halogen shares its valence electrons with oxygen, a more electro-negative element. (Electronegativities are O, 3.5; Cl, 3.0; Br, 2.8; I, 2.5.) The general formula for a halogen oxoacid is HXO_n, and the oxidation state of the halogen is +1, +3, +5, or +7, depending on the value of n.

Only four of the acids listed in Table 22.9 have been isolated in pure form: per-chloric acid ($HClO_4$), iodic acid (HIO_3), and the two periodic acids, metaperiodic acid (HIO_4) and paraperiodic acid (H_5IO_6). The others are stable only in aqueous solution or in the form of their salts. Bromous acid ($HBrO_2$) is known only as an unstable reaction intermediate, and iodous acid (HIO_2) is unknown.

The acid strength of the halogen oxoacids increases with the **increasing oxidation state of the halogen** (Section 16.3). For example, acid strength increases from HClO, a weak acid ($K_a = 3.5 \times 10^{-8}$), to $HClO_4$, a very strong acid ($K_a \gg 1$). The acidic proton is bonded to oxygen, not to the halogen, even though we usually write the molecular formula of these acids as HXO_n. All the halogen oxoacids and their salts are strong oxidizing agents.

> **REMEMBER . . .**
> The **higher the oxidation state of the halogen** in HXO_n, the greater the shift of electron density from the O—H bond toward the halogen, thus weakening the O—H bond, increasing its polarity, and facilitating proton transfer from HXO_n to a solvent water molecule (Section 16.3).

22.10 GROUP 8A: NOBLE GASES

The noble gases of group 8A—He, Ne, Ar, Kr, Xe, and Rn—are neither metals nor reactive nonmetals. Rather, they are colorless, odorless, unreactive gases. The $1s^2$ valence-shell electron configuration for He and $ns^2 np^6$ for the others already contain octets and thus make it difficult for the noble gases to either gain or lose electrons.

Although sometimes referred to as rare gases or inert gases, these older names are not really accurate because the group 8A elements are neither rare nor completely inert. Argon, for instance, makes up nearly 1% by volume of dry air, and there are sev-eral dozen known compounds of krypton and xenon, although none occur naturally. Some properties of the noble gases are listed in **TABLE 22.10**.

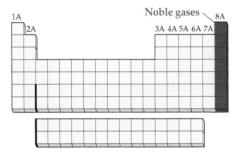

TABLE 22.10 Properties of Noble Gases

Name	Melting Point (°C)	Boiling Point (°C)	First Ionization Energy (kJ/mol)	Abundance Dry Air (vol %)
Helium	−272.2	−268.9	2372.3	5.2×10^{-4}
Neon	−248.6	−246.1	2080.6	1.8×10^{-3}
Argon	−189.3	−185.9	1520.4	0.93
Krypton	−157.4	−153.2	1350.7	1.1×10^{-4}
Xenon	−111.8	−108.0	1170.4	9×10^{-6}
Radon	−71	−61.7	1037	Trace

Helium and neon undergo no chemical reactions and form no known compounds. Argon forms only HArF, and krypton and xenon react only with fluorine. Depending on the reaction conditions and on the amounts of reactants present, xenon can form three different fluorides: XeF_2, XeF_4, and XeF_6.

XeF$_2$

XeF$_4$

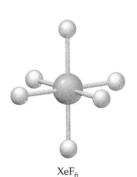

XeF$_6$

$$Xe(g) + F_2(g) \longrightarrow XeF_2(s)$$
$$Xe(g) + 2\,F_2(g) \longrightarrow XeF_4(s)$$
$$Xe(g) + 3\,F_2(g) \longrightarrow XeF_6(s)$$

The lack of reactivity of the noble gases is a consequence of their unusually large ionization energies and their unusually small electron affinities, which result from their valence-shell electron configurations.

ydrogen is an enormously attractive fuel because it's environmentally clean, giving only water as a combustion product. In addition, the amount of heat liberated when hydrogen burns is 242 kJ/mol (121 kJ/g), more than twice that of gasoline, oil, or natural gas on a mass basis.

$$H_2(g) + 1/2\ O_2(g) \longrightarrow H_2O(g) \quad \Delta H° = -242\ kJ$$

As a result, some people envision what they call a "hydrogen economy" in which our energy needs are met by gaseous, liquid, and solid hydrogen. The benefits of hydrogen are numerous, so what is keeping us from reaching a hydrogen economy?

Hydrogen Production

One barrier to a hydrogen economy is the availability of an inexpensive and nonpolluting method for producing hydrogen. Hydrogen is not available on Earth in convenient natural reservoirs like fossil fuels because most hydrogen is bonded to oxygen in water. Energy must be expended to extract hydrogen in a useful form for fuels. Most methods for the production of hydrogen involve water because it is the cheapest and most readily available source of hydrogen.

At present, the most important industrial method for producing hydrogen is the **steam–hydrocarbon reforming process**. The first step in the process is the conversion of steam and methane (CH_4) to a mixture of carbon monoxide and hydrogen known as *synthesis gas,* so called because it can be used as the starting material for the synthesis of liquid fuels. The reaction requires high temperature, moderately high pressure, and a nickel catalyst:

$$H_2O(g) + CH_4(g) \xrightarrow[\text{Ni catalyst}]{1100\ °C} CO(g) + 3\ H_2(g) \quad \Delta H° = +206\ kJ$$

In the second step, the synthesis gas and additional steam are passed over a metal oxide catalyst at about 400 °C. Under these conditions, the carbon monoxide component of the synthesis gas and the steam are converted to carbon dioxide and more hydrogen. This reaction of CO with H_2O is called the **water–gas shift reaction** because it shifts the composition of synthesis gas by removing the toxic carbon monoxide and producing more of the economically important hydrogen:

$$CO(g) + H_2O(g) \xrightarrow[\text{Catalyst}]{400\ °C} CO_2(g) + H_2(g) \quad \Delta H° = -41\ kJ$$

Although production of hydrogen from methane is economically feasible, it can contribute to climate change because it produces CO_2 as a byproduct. It may be possible, however, to capture the CO_2 and sequester it in depleted gas wells or deep saline aquifers, thus avoiding addition of CO_2 to the atmosphere.

A cleaner method of generating hydrogen is electrolysis (shown in the reactions below) because only high purity hydrogen and oxygen are produced (**FIGURE 22.19**).

The source of electricity used in the electrolysis reaction determines whether hydrogen is really a "clean" fuel or whether it is just as polluting as fossil fuels. Electricity produced by photovoltaic systems, nuclear power, or other alternative energy sources such as wind or hydroelectric will not generate undesired pollutants.

Another approach is to use solar energy to "split" water into H_2 and O_2. Professor Daniel Nocera at Massachusetts Institute of Technology has created a device referred to as the *artificial leaf* because just like photosynthesis it can turn the energy of sunlight directly into a chemical fuel. In photosynthesis, carbon dioxide and oxygen are converted to glucose ($C_6H_{12}O_6$) which is used as an energy source in the plant. The artificial leaf consists of a thin, flat, three-layered silicon solar cell with inexpensive catalysts made of cobalt and nickel bonded to both faces of the silicon. When placed in a beaker of water and exposed to sunlight, silicon absorbs photons, and water is split into hydrogen and oxygen (**FIGURE 22.20**).

▲ **FIGURE 22.19**

Electrolysis of water using a battery as a power source gives H_2 gas at one electrode and O_2 gas at the other electrode.

> ▲ **Figure It Out**
>
> Which gas is produced at which electrode?
>
> **Answer:** According to the balanced equation for the hydrolysis of water, two moles of H_2 are produced for every mole of O_2. The electrode on the right in the photo has twice the volume of gas indicating hydrogen is produced at that electrode.

Anode (oxidation):	$2\ H_2O(l) \longrightarrow O_2(g) + 4\ H^+(aq) + 4\ e^-$		$E° = -1.23\ V$
Cathode (reduction):	$4\ H_2O(l) + 4\ e^- \longrightarrow 2\ H_2(g) + 4\ OH^-(aq)$		$E° = -0.83\ V$
Overall cell reaction:	$6\ H_2O(l) \longrightarrow 2\ H_2(g) + O_2(g) + 4\ H^+(aq) + 4\ OH^-(aq)$		$E° = -2.06\ V$

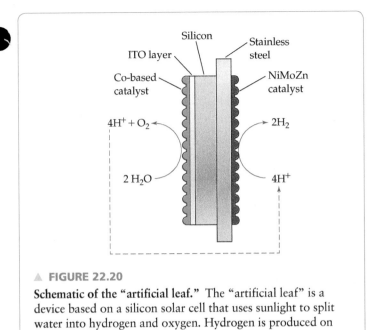

▲ **FIGURE 22.20**
Schematic of the "artificial leaf." The "artificial leaf" is a device based on a silicon solar cell that uses sunlight to split water into hydrogen and oxygen. Hydrogen is produced on one side, and oxygen is produced on the other.

Hydrogen Storage and Transport

Another barrier to the hydrogen economy is the availability of safe and widespread methods for storing and transporting hydrogen. For heating homes, gaseous hydrogen could be conveyed through underground pipes much like natural gas, while liquid hydrogen could be shipped by truck or by rail in large vacuum-insulated tanks. Automobiles, such as the one shown in the opening photo of Chapter 19, might be powered by compressed hydrogen gas, liquid hydrogen, or "solid hydrogen"—hydrogen stored in the form of solid metal hydrides, porous solids with high surface areas, or solid hydrogen storage compounds such as NH_3BH_3. Hydrogen gas is reacted with other materials to produce the hydrogen storage material, which can be transported relatively easily. At the point of use, the hydrogen storage material can be made to decompose, yielding hydrogen gas.

Interstitial transition metal hydrides (Section 22.3) are of interest as potential hydrogen-storage devices because they can contain a remarkably large amount of hydrogen. Palladium, for example, absorbs up to 935 times its own volume of H_2, an amount that corresponds to a density of hydrogen comparable to that in liquid hydrogen. For use as a fuel, hydrogen could be stored as PdH_x and then liberated when needed simply by heating the PdH_x.

$$\text{Pd}(s) + \frac{x}{2}\,H_2(g) \rightleftharpoons PdH_x(s)$$

Favored at   Favored at
higher temperature   lower temperature

This approach, however, may not be economical compared to other methods of storing and transporting hydrogen because of the high cost and weight of palladium. Another new chemical approach may be hydrogen generation from ammonia-borane materials by the following reactions:

$$NH_3BH_3(s) \xrightarrow{(\sim120\ °C)} NH_2BH_2(s) + H_2(g) \xrightarrow{(\sim160\ °C)}$$
$$NHBH(s) + H_2(g)$$

Ammonia borane (NH_3BH_3) releases two moles of hydrogen upon heating to modest temperatures that are easily achieved in a fuel cell.

PROBLEM 22.11   Liquid hydrogen has been used as a fuel in the U.S. space program for many years. The space shuttle fuel tank contains 1.45×10^6 L of liquid hydrogen, which has a density of 0.088 kg/L. How much heat (in kilojoules) is liberated when the hydrogen burns in an excess of oxygen? How many kilograms of oxygen are needed to oxidize the hydrogen?

PROBLEM 22.12

(a) Write balanced equations for the steam-hydrocarbon reforming process.

(b) Why does the steam-hydrocarbon reforming process contribute to climate change?

PROBLEM 22.13   Write a balanced equation for the production of synthesis gas, a mixture of H_2 and CO, from water and each of the following:

(a) graphite

(b) propane, $C_3H_8(g)$

PROBLEM 22.14   Hydrogen is a gas at ordinary temperatures. Explain how it can be stored as a solid.

PROBLEM 22.15   If palladium metal (density 12.0 g/cm³) dissolves 935 times its own volume of H_2 at STP, what is the value of x in the formula PdH_x? What is the density of hydrogen in PdH_x in units of g/cm³? Assume that the volume of palladium is unchanged when the H atoms go into the interstices.

PROBLEM 22.16   Titanium hydride, TiH_2, has a density of 3.75 g/cm³.

(a) Calculate the density of hydrogen in TiH_2, and compare it with that in liquid H_2 (0.070 g/cm³).

(b) How many cubic liters of H_2 at STP are absorbed in making 1.00 cm³ of TiH_2?

PROBLEM 22.17 (a) Identify the Lewis acid and the Lewis base when ammonia (NH_3) reacts with borane (BH_3) to form the hydrogen storage material ammonia borane (NH_3BH_3). (b) Draw the electron-dot structure for ammonia borane. Give the hybridization of N and B and the bond angles predicted by VSEPR theory.

STUDY GUIDE

Section	Concept Summary	Learning Objectives	Test Your Understanding
22.1 A Review of General Properties and Periodic Trends	The main-group elements are the *s*-block elements of groups 1A and 2A and the *p*-block elements of groups 3A–8A. From left to right across the periodic table, ionization energy, electronegativity, and nonmetallic character generally increase, while atomic radius and metallic character decrease. From top to bottom of a group in the periodic table, ionization energy, electronegativity, and nonmetallic character generally decrease, while atomic radius and metallic character increase.	**22.1** Use the periodic table to predict the chemical and physical properties of the main-group elements.	Worked Example 22.1; Problems 22.18–22.19, 22.32, 22.38, 22.40
22.2 Distinctive Properties of the Second-Row Elements	The second-row elements form strong multiple bonds but are generally unable to form more than four bonds because of the small size of their atoms.	**22.2** Compare the number and types of bonds in compounds with second-row elements and compounds with third- and higher-row elements.	Worked Example 22.2; Problems 22.29, 22.46, 22.48, 22.50, 22.51
22.3 Group 1A: Hydrogen	Hydrogen exists at normal temperatures as a stable, diatomic H_2 molecule. Hydrogen forms three types of **binary hydrides**. Active metals give **ionic hydrides,** such as LiH and CaH_2; nonmetals give **covalent hydrides,** such as NH_3, H_2O, and HF; and transition metals give **metallic** or **interstitial hydrides,** such as PdH_x. Interstitial hydrides are often **nonstoichiometric compounds.**	**22.3** Write and balance a chemical equation for the reaction of ionic hydrides with water.	Problems 22.66–22.67
		22.4 Classify binary hydrides and describe their bonding and properties.	Worked Example 22.3; Problems 22.21–22.23, 22.58, 22.60
22.4 Group 1A: Alkali Metals and Group 2A: Alkaline Earth Metals	Alkali elements are soft, silvery metals that are highly reactive. Alkali metals react rapidly with halogens to form ionic compounds called halides with the general formula MX and with oxygen to form oxides, peroxides, and superoxides. Alkali metals react with water to yield hydrogen gas and an alkali-metal hydroxide, MOH. Alkaline-earth elements are relatively soft, silvery metals that are slightly less reactive than alkali metals. They react with halogens to yield ionic halides with the formula MX_2. They also react with oxygen to form oxides, peroxides, and superoxides and with water to yield hydrogen gas and a metal hydroxide, M $(OH)_2$.	**22.5** Use periodic trends to predict chemical and physical properties of Group 1A and 2A elements.	Problem 22.68
		22.6 Write and balance reactions of alkali and alkaline-earth metals with halogens, oxygen, and water.	Problems 22.70, 22.72
		22.7 Write electrode reactions and perform calculations in the electrolysis of molten salts to produce alkali and alkaline-earth metals.	Problems 22.76–22.77
22.5 Group 3A Elements	The group 3A elements—B, Al, Ga, In, and Tl—are metals except for boron, which is a semimetal. Boron is a semiconductor and forms molecular compounds. **Boranes,** such as diborane (B_2H_6), are electron-deficient molecules that contain **three-center, two-electron bonds** (B–H–B).	**22.8** Use periodic trends to predict chemical and physical properties of Group 3A elements.	Problems 22.80–22.82
		22.9 Describe the structure and bonding in boron halides and diborane.	Problems 22.84–22.87
22.6 Group 4A Elements	The group 4A elements—C, Si, Ge, Sn, and Pb—exhibit the usual increase in metallic character down the group. They often adopt an oxidation state of +4, but the +2 state becomes increasingly more stable from Ge to Sn to Pb. In elemental form, carbon exists as diamond, graphite, graphene, and fullerene. Silicon, the second most abundant element in the Earth's crust, is obtained by reducing silica sand (SiO_2) with coke. It is purified for use in the semiconductor industry by **zone refining.** In the **silicates,** SiO_4 tetrahedra share common O atoms to give silicon oxoanions with ring, chain, layer, and extended three-dimensional structures.	**22.10** Draw electron-dot structures and predict hybrid orbitals and geometry for molecules and ions in main-group compounds containing elements from Group 4A.	Problems 22.90, 22.92–22.93
		22.11 Use the shorthand notation for a silicate anion to represent its structure and interpret the notation to find the formula and charge of a silicate anion.	Worked Example 22.5; Problems 22.31, 22.100–22.101

Section	Concept Summary	Learning Objectives	Test Your Understanding
22.7 Group 5A Elements	Molecular nitrogen (N_2) is unreactive because of its strong NN triple bond. Nitrogen exhibits all oxidation states between -3 and $+5$. Phosphorus, the most abundant group 5A element, exists in two common allotropic forms—white phosphorus, which contains highly reactive tetrahedral P_4 molecules, and red phosphorus, which is polymeric. The most common oxidation states of P are -3, as in phosphine (PH_3); $+3$ as in PCl_3; and $+5$, as in PCl_5.	**22.12** Write electron-dot structures and describe the structure, bonding, and properties of molecules and ions in main-group compounds containing nitrogen and phosphorus.	Problems 22.28, 22.108, 22.110, 22.114
		22.13 Determine oxidation states of nitrogen and phosphorus in acids and write acid–base reactions.	Problems 22.106–22.107, 22.112
22.8 Group 6A Elements	Oxygen exists as a diatomic molecule at normal temperatures and forms ionic oxides with metals and covalent oxides with nonmetals. Ionic oxides are generally basic and covalent oxides are generally acidic. Amphoteric oxides are formed when oxygen combines with elements that have intermediated electronegativities. Sulfur is obtained from underground deposits. The properties of sulfur change dramatically on heating as the S_8 rings of rhombic sulfur open and polymerize to give long chains, which then fragment at higher temperatures. The most common oxidation states of S are -2, as in H_2S; $+4$ as in SO_2; and $+6$, as in SO_3.	**22.14** Classify oxides as ionic or covalent and as acidic, basic, or amphoteric and write balanced equations for the reaction that occurs when oxides dissolve in water.	Worked Example 22.6; Problems 22.24, 22.26, 22.130
		22.15 Write electron-dot structures and describe the structure, bonding, and properties of compounds containing oxygen and sulfur.	Problems 22.140–22.143
22.9 Group 7A: The Halogens	The halogens are nonmetals that exist as diatomic molecules. Halogens react with every metal in the periodic table to form metal halides, MX_n. Chlorine, bromine, and iodine form a series of oxoacids: hypohalous acid (HXO), halous acid (HXO_2 for X = Cl), halic acid (HXO_3), and perhalic acid (HXO_4). Acid strength increases as the oxidation state of the halogen increases from $+1$ to $+7$. Halogen oxoacids and their salts are strong oxidizing agents.	**22.16** Write balanced equations for the formation of halogens and reactions of their compounds.	Problems 22.146, 22.148–22.149
		22.17 Determine oxidation states of halogens in oxoacids and predict the relative acid strengths of oxoacids.	Problems 22.150, 22.154
22.10 Group 8A: Noble Gases	The noble gases are colorless, odorless, unreactive gases. Helium and neon do not react, but argon, krypton, and xenon do react with fluorine to form molecular compounds.	**22.18** Draw electron-dot structures and predict geometry for molecules and ions containing noble gases.	Problems 22.157–22.158

KEY TERMS

binary hydride *960*	interstitial hydride *961*	oxide *982*	three-center, two-electron bond *967*
boranes *966*	ionic hydride *960*	silicate *970*	water–gas shift reaction *990*
carbide *970*	metallic hydride *961*	steam–hydrocarbon reforming process *990*	zone refining *970*
contact process *987*	nonstoichiometric compound *961*		
covalent hydride *961*			

PRACTICE TEST

After studying this chapter, you can assess your understanding with these practice test questions, which are correlated with chapter learning objectives. If you answer a question incorrectly, refer to the learning objectives in the end-of-chapter Study Guide for assistance. The Study Guide provides a conceptual summary, references a Worked Example to model how to solve the problem, and gives additional problems for more practice.

1. Which of the following elements is the best conductor of electricity? (**LO 22.1**)
    (a) N
    (b) As
    (c) P
    (d) Bi

2. Consider the distinctive properties of second-row elements, and select the best electron-dot structures for HNO_3 and H_3PO_4. (**LO 22.2**)

(a)

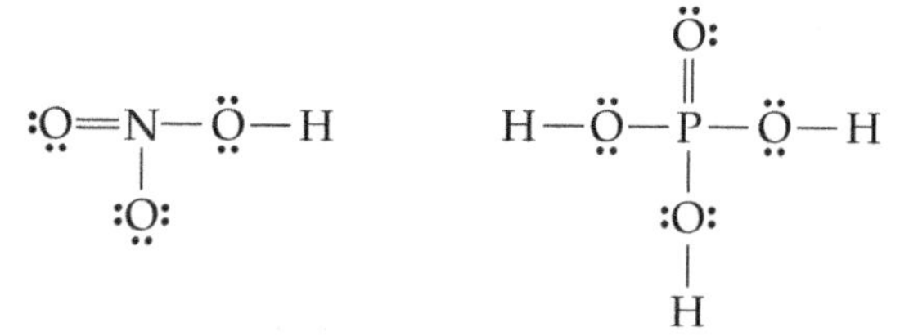

:Ö:
||
$:O=N-\ddot{O}-H$ $H-\ddot{O}-P-\ddot{O}-H$
| |
:O: :O:
                                |
                                H

(b)

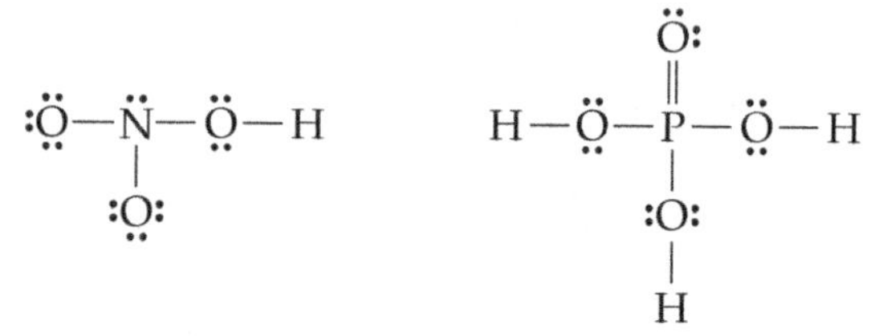

:Ö:
||
$:\ddot{O}-\ddot{N}-\ddot{O}-H$ $H-\ddot{O}-P-\ddot{O}-H$
| |
:O: :O:
                                           |
                                           H

(c)

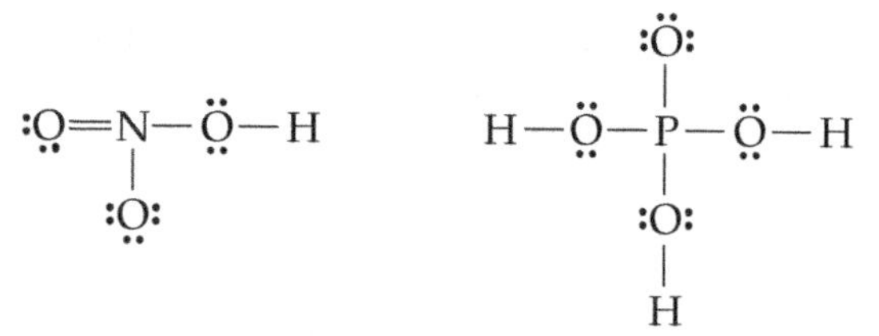

:Ö:
|
$:O=N-\ddot{O}-H$ $H-\ddot{O}-P-\ddot{O}-H$
| |
:O: :O:
                                |
                                H

(d)

:Ö:
|
$:\ddot{O}-\ddot{N}-\ddot{O}-H$ $H-\ddot{O}-P-\ddot{O}-H$
| |
:O: :O:
                                           |
                                           H

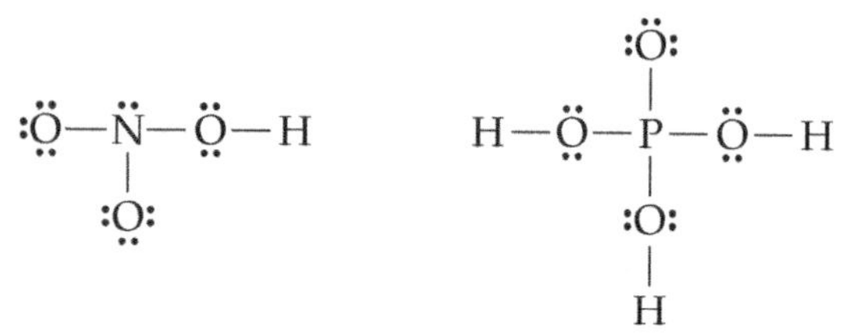

3. Which of the following elements (X) will form a covalent hydride with the formula (XH$_3$) that is a gas at room temperature? (**LO 22.4**)
   (a) Al   (b) As   (c) Ba   (d) Se

4. Which element will react most vigorously with water, and what is the reaction that occurs? (**LO 22.5, LO 22.6**)
   (a) Lithium; $2 Li(s) + 2 H_2O(l) \rightarrow 2 Li^+(aq) + 2 H^-(aq) + H_2O_2(aq)$
   (b) Potassium; $2 K(s) + 2 H_2O(l) \rightarrow 2 K^+(aq) + 2 OH^-(aq) + H_2(g)$
   (c) Magnesium; $Mg(s) + 2 H_2O(l) \rightarrow Mg^{2+}(aq) + 2 OH^-(aq) + H_2(g)$
   (d) Barium; $Ba(s) + 2 H_2O(l) \rightarrow Ba^{2+}(aq) + 2 H^-(aq) + H_2O_2(aq)$

5. Draw the electron-dot structure for BH_3. What are the hybrid orbitals used by boron to form bonds with hydrogen, and will BH_3 act as a Lewis acid or a Lewis base? (**LO 22.9**)
   (a) sp^3; Lewis acid   (b) sp^3; Lewis base
   (c) sp^2; Lewis acid   (d) sp^2; Lewis base

6. The silicate anion in the mineral kinoite is represented by the following structure. The mineral also contains Ca^{2+} ions, Cu^{2+} ions, and water molecules in a 1:1:1 ratio. What are the formula and charge of the silicate anion and the complete formula for the mineral? (**LO 22.11**)

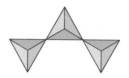

   (a) $Si_3O_{10}^{8-}$; $Ca_2Cu_2Si_3O_{10} \cdot 2 H_2O$
   (b) $Si_3O_8^{10-}$; $Ca_3Cu_2Si_3O_8 \cdot 2 H_2O$
   (c) $Si_3O_4^{4-}$; $CaCuSi_3O_8 \cdot H_2O$
   (d) $Si_3O_{12}^{12-}$; $Ca_3Cu_3Si_3O_{12} \cdot 3 H_2O$

7. Which of the following group 5A elements (X) cannot form a compound with the formula XCl_5? (**LO 22.12**)
   (a) N   (b) P   (c) As   (d) Sb

8. What is the balanced equation for the reaction between phosphorous acid (H_3PO_3) and a large excess of NaOH? (**LO 22.13**)

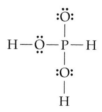

   (a) $H_3PO_3(aq) + NaOH(aq) \rightarrow NaH_2PO_3(aq) + H_2O(l)$
   (b) $H_3PO_3(aq) + 2 NaOH(aq) \rightarrow Na_2HPO_3(aq) + 2 H_2O(l)$
   (c) $H_3PO_3(aq) + 3 NaOH(aq) \rightarrow Na_3PO_3(aq) + 3 H_2O(l)$
   (d) No reaction occurs because phosphorous acid is a weak acid.

9. Which of the following oxides will be more soluble in acidic solution? (**LO 22.14**)
   (a) SiO_2 (b) NO_2
   (c) BaO   (d) B_2O_3

10. Consider the following oxoacids: HClO, $HClO_2$, $HClO_3$, and $HClO_4$. In which oxoacid does chlorine have an oxidation state of +5? Which oxoacid is the strongest? (**LO 22.17**)
    (a) HClO has a Cl oxidation state of +5, and $HClO_4$ is the strongest acid.
    (b) $HClO_2$ has a Cl oxidation state of +5, and HClO is the strongest acid.
    (c) $HClO_3$ has a Cl oxidation state of +5, and $HClO_4$ is the strongest acid.
    (d) $HClO_4$ has a Cl oxidation state of +5, and HClO is the strongest acid.

Answers:

1. d, 2. c, 3. b, 4. b, 5. c, 6. a, 7. a, 8. b, 9. c, 10. c

Mastering Chemistry   provides end-of-chapter exercises, feedback-enriched tutorial problems, animations, and interactive activities to encourage problem-solving practice and deeper understanding of key concepts and topics.

RAN   *Randomized in Mastering Chemistry*

CONCEPTUAL PROBLEMS

Problems 22.1–22.17 appear within the chapter.

22.18 Locate each of the following groups of elements on the periodic table.
  (a) Main-group elements
  (b) *s*-Block elements
  (c) *p*-Block elements
  (d) Main-group metals
  (e) Nonmetals
  (f) Semimetals

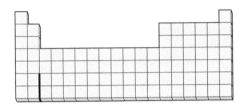

22.19 Locate each of the following elements on the periodic table.
  (a) Element with the lowest ionization energy
  (b) Most electronegative element
  (c) Group 4A element with the largest atomic radius
  (d) Group 6A element with the smallest atomic radius
  (e) Group 3A element that is a semiconductor
  (f) Group 5A element that forms the strongest π bonds

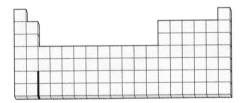

22.20 Look at the location of elements A, B, C, and D in the following periodic table:

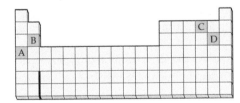

  (a) Write the formula of the simplest binary hydride of each element.
  (b) Which hydride has the lowest boiling point?
  (c) Which hydrides react with water to give H_2 gas? Write a balanced net ionic equation for each reaction.
  (d) Which hydrides react with water to give an acidic solution, and which give a basic solution?

22.21 In the following pictures of binary hydrides, ivory spheres represent H atoms or ions, and burgundy spheres represent atoms or ions of the other element.

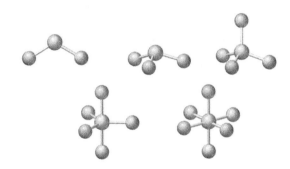

  (a) Identify each binary hydride as ionic, covalent, or interstitial.
  (b) What is the oxidation state of hydrogen in compounds (1), (2), and (3)? What is the oxidation state of the other element?

22.22 The following models represent the structures of binary hydrides of second-row elements:

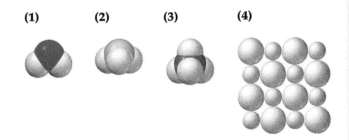

  (a) Identify the nonhydrogen atom in each case, and write the molecular formula for each hydride.
  (b) Draw an electron-dot structure for each hydride. For which hydride is there a problem in drawing the structure? Explain.

22.23 The following pictures represent structures of the hydrides of four second-row elements:

(1)   (2)   (3)   (4)

  (a) Which compound has the highest melting point?
  (b) Which compound has the lowest boiling point?
  (c) Which compounds yield H_2 gas when they are mixed together?

22.24 In the following pictures of oxides, red spheres represent O atoms or ions, and green spheres represent atoms or ions of a second- or third-row element in its highest oxidation state.

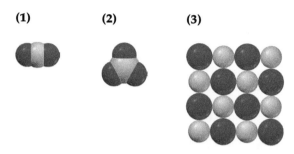

(1)　　**(2)**　　**(3)**

(a) What is the oxidation state of oxygen in each oxide? What is the oxidation state of the other element?

(b) Identify each oxide as ionic or covalent.

(c) Identify each oxide as acidic or basic.

(d) What is the identity of the other element in (1) and (2)?

22.25 In the following pictures of oxides, red spheres represent O atoms or ions, and green spheres represent atoms or ions of a first- or second-row element in its highest oxidation state.

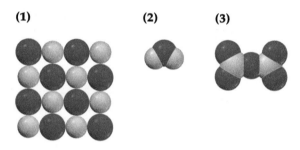

(1)　　**(2)**　　**(3)**

(a) What is the oxidation state of oxygen in each oxide? What is the oxidation state of the other element?

(b) Which of these oxides is (are) molecular, and which has (have) an infinitely extended three-dimensional structure?

(c) Which of these oxides is (are) likely to be a gas or a liquid, and which is (are) likely to be a high-melting solid?

(d) Identify the other element in (2) and (3).

22.26 Look at the location of elements A, B, C, and D in the following periodic table:

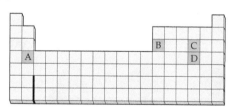

(a) Write the formula of the oxide that has each of these elements in its highest oxidation state.

(b) Classify each oxide as basic, acidic, or amphoteric.

(c) Which oxide is the most ionic? Which is the most covalent?

(d) Which of these oxides are molecular? Which are solids with an infinitely extended three-dimensional crystal structure?

(e) Which of these oxides has the highest melting point? Which has the lowest melting point?

22.27 Locate the following elements on the periodic table, and
RAN　write the formula of a compound that justifies each of your answers.

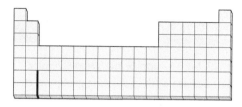

(a) Two nonmetals that can form more than four bonds

(b) Two nonmetals that form a maximum of four bonds

(c) Two nonmetals that form oxides that are gases at 25 °C

(d) A nonmetal that forms an oxide that is a solid at 25 °C

22.28 Consider the six second- and third-row elements in groups 5A–7A of the periodic table:

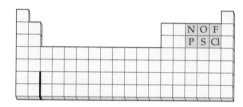

Possible molecular structures for common allotropes of these elements are shown below.

(a) What is the molecular structure of each of the six elements?

(b) Using electron-dot structures, explain why each element has its particular molecular structure.

(c) Explain why nitrogen and phosphorus have different molecular structures and why oxygen and sulfur have different molecular structures but fluorine and chlorine have the same molecular structure.

22.29 Consider the six second- and third-row elements in groups 4A–6A of the periodic table:

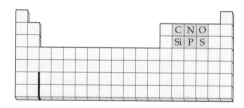

Possible structures for the binary fluorides of each of these elements in its highest oxidation state are shown below.

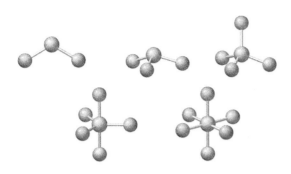

(a) Identify the nonfluorine atom in each case, and write the molecular formula of each fluoride.

(b) Explain why the fluorides of nitrogen and phosphorus have different molecular structures but the fluorides of carbon and silicon have the same molecular structure.

22.30 The following models represent the structures of binary oxides of second- and third-row elements in their highest oxidation states:

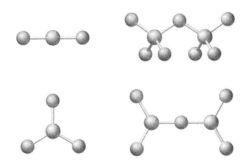

(a) Identify the non-oxygen atom in each case, and write the molecular formula for each oxide.

(b) Draw an electron-dot structure for each oxide. For which oxides are resonance structures needed?

22.31 The following pictures represent various silicate anions. Write the formula and charge of each anion.

(a)   **(b)**

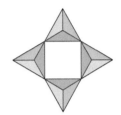

SECTION PROBLEMS

Periodic Trends (Section 22.1)

22.32 Which element in each of the following pairs has the higher ionization energy?

    (a) As or Cl   (b) Mg or Ba

22.33 Arrange the following elements in order of increasing ionization energy.

    (a) N   (b) Ca   (c) Rb   (d) Se

22.34 Which element in each of the following pairs has the larger atomic radius?

    (a) Sn or P   (b) Ge or Br

22.35 Arrange the following elements in order of increasing atomic radius.

    (a) Rb   (b) F   (c) Sb   (d) S

22.36 Which element in each of the following pairs has the higher electronegativity?

    (a) Sb or I   (b) P or Sb

22.37 Arrange the following elements in order of increasing electronegativity.

    (a) Sr   (b) Cl   (c) Sn   (d) Ge

22.38 Which element in each of the following pairs has more metallic character?

    (a) Bi or As   (b) Ge or Si

22.39 Which element in each of the following pairs has more non-metallic character?

    (a) Se or Te   (b) Br or As

22.40 Which compound in each of the following pairs is more ionic?

    (a) CaH_2 or NH_3 (b) P_4O_6 or Ga_2O_3

    (c) $SiCl_4$ or KCl (d) BCl_3 or $AlCl_3$

22.41 Which compound in each of the following pairs is more covalent?

    (a) PCl_3 or AlF_3 (b) CaO or NO

    (c) NH_3 or KH (d) SnO_2 or SiO_2

22.42 Which of the following compounds are molecular, and which have an extended three-dimensional structure?

    (a) B_2H_6 (b) $KAlSi_3O_8$

    (c) SO_3 (d) $GeCl_4$

22.43 Which of the following compounds are molecular, and which have an extended three-dimensional structure?
(a) KF
(b) P_4O_{10}
(c) $SiCl_4$
(d) $CaMgSi_2O_6$

22.44 Consider the elements C, Se, B, Sn, and Cl. Identify which of
RAN these elements:
(a) Has the largest atomic radius
(b) Is the most electronegative
(c) Is the best electrical conductor
(d) Has a maximum oxidation state of +6
(e) Forms a hydride with the empirical formula XH_3

22.45 Consider the elements N, Si, Al, S, and F. Identify which of these elements:
(a) Has the highest ionization energy
(b) Has the most metallic character
(c) Is a semiconductor
(d) Forms a 2− anion

Distinctive Properties of the Second-Row Elements (Section 22.2)

22.46 BF_3 reacts with F^- to give BF_4^-, but AlF_3 reacts with F^- to give AlF_6^{3-}. Explain.

22.47 $GeCl_4$ reacts with Cl^- to give $GeCl_6^{2-}$, but CCl_4 does not react with excess Cl^-. Explain.

22.48 At ordinary temperatures, sulfur exists as S_8 but oxygen exists as O_2. Explain.

22.49 Carbon, nitrogen, and oxygen form π bonds, but fluorine does not. Explain.

22.50 Elemental nitrogen exists as N_2, but white phosphorus exists as P_4. Explain.

22.51 Consider the elements Mn, Al, C, S, and Si. Which element forms the strongest π bonds?

Group 1A: Hydrogen (Section 22.3)

22.52 Write a balanced equation for the synthesis of hydrogen using each of the following starting materials.
(a) Zn
(b) C
(c) CH_4
(d) H_2O

22.53 Complete and balance the equation for each of the following reactions.
(a) $Fe(s) + H^+(aq) \rightarrow$
(b) $Ca(s) + H_2O(l) \rightarrow$
(c) $Al(s) + H^+(aq) \rightarrow$
(d) $C_2H_6(g) + H_2O(g) \xrightarrow[\text{Catalyst}]{\text{Heat}}$

22.54 Calcium hydride is a convenient, portable source of hydro-
RAN gen that is used, among other things, to inflate weather balloons. If the reaction of CaH_2 with water is used to inflate a balloon with 2.0×10^5 L of H_2 gas at 25 °C and 1.00 atm pressure, how many kilograms of CaH_2 is needed?

22.55 The hydrogen-filled dirigible Hindenburg had a volume of
RAN 1.99×10^8 L. If the hydrogen used was produced by the reaction of carbon with steam, how many kilograms of carbon would have been needed to produce enough hydrogen to fill the dirigible at 20 °C and 740 mm pressure?

$$C(s) + H_2O(g) \rightarrow CO(g) + H_2(g)$$

22.56 Write the chemical formula of a compound that contains hydrogen and each of the following elements. Which compounds are ionic, and which are covalent?
(a) Na
(b) C

22.57 Write the chemical formula of a compound that contains hydrogen and each of the following elements. Which compounds are ionic, and which are covalent?
(a) K
(b) S

22.58 In the following compounds, is hydrogen present as H^+, H^-, or a covalently bound H atom?
(a) MgH_2
(b) PH_3
(c) KH
(d) HBr

22.59 In the following compounds, is hydrogen present as H^+, H^-, or a covalently bound H atom?
(a) H_2Se
(b) RbH
(c) CaH_2
(d) GeH_4

22.60 Compare some of the physical properties of H_2S, NaH, and
RAN PdH_x.

22.61 Compare some of the physical properties of $TiH_{1.7}$, HCl, and CaH_2.

22.62 Describe the molecular geometry of:
(a) H_2Se
(b) AsH_3
(c) SiH_4

22.63 Describe the molecular geometry of:
(a) GeH_4
(b) H_2S
(c) NH_3

22.64 What is a nonstoichiometric compound? Give an example, and account for its lack of stoichiometry in terms of structure.

22.65 Explain why the hydrogen atoms in interstitial hydrides are mobile.

22.66 Write a balanced net ionic equation for the reaction of each of the following hydrides with water.
(a) SrH_2
(b) NH_3

22.67 Write a balanced net ionic equation for the reaction of each of the following hydrides with water.
(a) HI
(b) KH

Group 1A and 2A: Alkali and Alkaline Earth Metals (Section 22.4)

22.68 Look at the properties of the alkali metals summarized in Table 22.2, and predict reasonable values for the melting point, boiling point, density, and atomic radius of francium.

22.69 Why does chemical reactivity increase from top to bottom in groups 1A and 2A?

22.70 Write chemical equations for the reaction of potassium with
RAN the following substances, making sure that the numbers and kinds of atoms are the same on both sides of the equations. If no reaction occurs, write N.R.
(a) H_2O
(b) Br_2
(c) O_2

22.71 Write chemical equations for the reaction of calcium with the following substances, making sure that the numbers and kinds of atoms are the same on both sides of the equations. If no reaction occurs, write N.R.
(a) H_2O
(b) He
(c) Br_2
(d) O_2

22.72 Complete the following equations so that the same numbers
RAN and kinds of atoms appear on both sides of the reaction
arrow. If no reaction takes place, write N.R.
(a) $Cs(s) + H_2O(l) \rightarrow$?   (b) $Rb(s) + O_2(g) \rightarrow$?

22.73 Predict the products of the following reactions, and balance
the equations so that the numbers and kinds of atoms are
the same on both sides of the reaction arrows.
(a) $Be(s) + Br_2(l) \rightarrow$?
(b) $Sr(s) + H_2O(l) \rightarrow$?
(c) $Mg(s) + O_2(g) \rightarrow$?

22.74 Milk of magnesia, a widely used antacid, is an aqueous suspension of $Mg(OH)_2$. How would you prepare $Mg(OH)_2$
from magnesium metal?

22.75 Barium metal can be prepared from its oxide by heating with
RAN aluminum at 1200 °C. (Aluminum oxide is also formed.) At
the reaction temperature, the oxides are solids, aluminum is
a liquid, and barium is a gas. Write a balanced equation for
the reaction.

22.76 Magnesium metal is produced by electrolysis of molten magnesium chloride using inert electrodes. Predict the anode,
cathode, and overall cell reactions.

22.77 How many hours are required to produce 10.0 kg of mag-
RAN nesium by electrolysis of molten $MgCl_2$ with a constant current of 1.00×10^4 A? How many liters of Cl_2 at STP will be
obtained?

22.78 Assign charges to the oxygen-containing anions in the following compounds:
(a) Na_2O (b) K_2O_2

22.79 Assign charges to the oxygen-containing anions in the following compounds:
(a) CsO_2 (b) BaO_2

Group 3A Elements (Section 22.5)

22.80 Identify the group 3A element that best fits each of the fol-
RAN lowing descriptions.
(a) Is the most abundant element of the group
(b) Is stable in the +1 oxidation state
(c) Is a semiconductor

22.81 Identify the group 3A element that best fits each of the fol-
RAN lowing descriptions.
(a) Has an unusually low melting point
(b) Is the most electronegative
(c) Is extremely toxic

22.82 What is the most common oxidation state for each of the
group 3A elements?

22.83 What is the oxidation state of the group 3A element in each
of the following compounds?
(a) $NaBF_4$ (b) $GaCl_3$
(c) $TlCl$ (d) B_2H_6

22.84 List three ways in which the properties of boron differ from
those of the other group 3A elements.

22.85 Explain why the properties of boron differ so markedly
from the properties of the other group 3A elements.

22.86 (a) Describe what is meant by an electron-deficient molecule.
(b) Describe what is meant by a three-center, two-electron bond.
(c) Describe the structure of diborane (B_2H_6) and explain
why the bridging B—H bonds are longer than the terminal B—H bonds.

22.87 Suggest a structure for the mixed aluminum–boron hydride
$AlBH_6$.

Group 4A Elements (Section 22.6)

22.88 Identify the group 4A element that best fits each of the following descriptions.
(a) Prefers the +2 oxidation state
(b) Forms the strongest π bonds
(c) Is the second most abundant element in the Earth's crust
(d) Forms the most acidic oxide

22.89 Select the group 4A element that best fits each of the following descriptions.
(a) Forms the most basic oxide
(b) Is the least dense semimetal
(c) Is the second most abundant element in the human body
(d) Is the most electronegative

22.90 Describe the shape of each of the following molecules or
RAN ions, and tell which hybrid orbitals are used by the central
atom.
(a) $GeBr_4$ (b) CO_2
(c) CO_3^{2-} (d) $SnCl_3^-$

22.91 What is the shape of each of the following molecules or ions,
and which hybrid orbitals are used by the central atom?
(a) SiO_4^{4-} (b) CCl_4
(c) $SnCl_2$ (d) HCN

22.92 Draw the electron-dot structure for CO, CO_2, and CO_3^{2-},
and predict which substance will have the strongest carbon–
oxygen bond.

22.93 What is the hybridization and geometry around carbon
atoms in graphene? Explain why graphene is an excellent
conductor of electricity.

22.94 Which of the group 4A elements have allotropes with the diamond structure? Which have metallic allotropes? How does
the variation in the structure of the group 4A elements illustrate how metallic character varies down a periodic group?

22.95 Give an example of an ionic carbide. What is the oxidation
state of carbon in this substance?

22.96 Why are CO and CN^- so toxic to humans?

22.97 Describe the preparation of silicon from silica sand, and
tell how silicon is purified for use in semiconductor devices.
Write balanced equations for all reactions.

22.98 How do the structures and properties of elemental silicon
and germanium differ from those of tin and lead?

22.99 Using the shorthand notation of Figure 22.9, draw the structure of the silicate anion in:
(a) K_4SiO_4 (b) $Ag_{10}Si_4O_{13}$
What is the relationship between the charge on the anion
and the number of terminal O atoms?

22.100 Using the shorthand notation of Figure 22.9, draw the structure of the cyclic silicate anion in which four SiO_4 tetrahedra share O atoms to form an eight-membered ring of alternating Si and O atoms. Give the formula and charge of the anion.

22.101 Suggest a plausible structure for the silicate anion in the mineral thortveitite, $Sc_2Si_2O_7$.

22.102 Carbon is an essential element in the molecules on which life is based. Would silicon be equally satisfactory? Explain.

22.103 The organ pipes in unheated northern European churches are often observed to be pitted and crumbling to powder in places. Suggest an explanation, given that the pipes are made from tin.

Group 5A Elements (Section 22.7)

22.104 Identify the group 5A element(s) that best fits each of the following descriptions.
(a) Makes up part of bones and teeth
(b) Forms stable salts containing M^{3+} ions
(c) Is the most abundant element in the atmosphere
(d) Forms a basic oxide

22.105 Identify the group 5A element that best fits each of the following descriptions.
(a) Forms strong π bonds
(b) Is a metal
(c) Is the most abundant group 5A element in the Earth's crust
(d) Forms oxides with the group 5A element in the +1, +2 and +4 oxidation states

22.106 Give the chemical formula for each of the following compounds, and indicate the oxidation state of the group 5A element.
RAN
(a) Nitrous oxide   (b) Hydrazine
(c) Calcium phosphide   (d) Phosphorous acid
(e) Arsenic acid

22.107 Give the chemical formula for each of the following compounds, and indicate the oxidation state of the group 5A element.
(a) Nitric oxide
(b) Nitrous acid
(c) Phosphine
(d) Tetraphosphorus decoxide
(e) Phosphoric acid

22.108 Draw an electron-dot structure for N_2, and explain why this molecule is so unreactive.

22.109 Describe the structures of the white and red allotropes of phosphorus, and explain why white phosphorus is so reactive.

22.110 Predict the geometrical structure of each of the following molecules or ions.
RAN
(a) NO_2^- (b) PH_3
(c) PF_5 (d) PCl_4^+

22.111 Draw the structure of each of the following molecules.
(a) Tetraphosphorus hexoxide
(b) Tetraphosphorus decoxide
(c) Phosphorous acid
(d) Phosphoric acid

22.112 Account for each of the following observations.
(a) Nitric acid is a strong oxidizing agent, but phosphoric acid is not.
(b) Phosphorous acid is a diprotic acid.

22.113 Compare and contrast the properties of ammonia and phosphine.

22.114 Draw all the possible resonance structure for N_2O and assign formal charges. Which resonance structure makes the greatest contribution to the resonance hybrid?

22.115 Could the strain in the P_4 molecule be reduced by using sp^3 hybrid orbitals in bonding instead of pure p orbitals? Explain.

Group 6A Elements (Section 22.8)

22.116 Identify the group 6A element that best fits each of the following descriptions.
(a) Is the most electronegative
(b) Is a semimetal
(c) Is radioactive
(d) Is the most abundant element in the Earth's crust

22.117 Identify the group 6A element that best fits each of the following descriptions.
(a) Is a metal
(b) Is the most abundant element in the human body
(c) Is the strongest oxidizing agent
(d) Has the most negative electron affinity

22.118 In industry O_2 is prepared by fractional distillation of liquefied air. The main components of air are O_2, N_2, and Ar. Order these substances from lowest to highest boiling point and explain your reasoning.

22.119 In what forms is oxygen commonly found in nature?

22.120 Write a balanced equation for the reaction of an excess of
RAN O_2 with each of the following elements.
(a) Li   (b) P   (c) Al   (d) Si

22.121 Write a balanced equation for the reaction of an excess of O_2 with each of the following elements.
(a) Ca   (b) C   (c) As   (d) B

22.122 If an element forms an acidic oxide, is it more likely to form an ionic or covalent hydride?

22.123 If an element forms an ionic hydride, is it more likely to form an acidic or basic oxide?

22.124 Arrange the following oxides in order of increasing covalent character: B_2O_3, BeO, CO_2, Li_2O, N_2O_5.

22.125 Arrange the following oxides in order of increasing ionic character: SiO_2, K_2O, P_4O_{10}, Ga_2O_3, GeO_2.

22.126 Arrange the following oxides in order of increasing basic character: Al_2O_3, Cs_2O, K_2O, N_2O_5.

22.127 Arrange the following oxides in order of increasing acidic character: BaO, Cl_2O_7, SO_3, SnO_2.

22.128 Which is more acidic?
(a) Cr_2O_3 or CrO_3 (b) N_2O_5 or N_2O_3 (c) SO_2 or SO_3

22.129 Which is more basic?
(a) CrO or Cr_2O_3
(b) SnO_2 or SnO
(c) As_2O_3 or As_2O_5

22.130 Write a balanced net ionic equation for the reaction of each of the following oxides with water.
(a) Cl_2O_7 (b) K_2O (c) SO_3

22.131 Write a balanced net ionic equation for the reaction of each of the following oxides with water.
(a) BaO (b) Cs_2O (c) N_2O_5

22.132 Write a balanced net ionic equation for the reaction of the amphoteric oxide ZnO with:
(a) Hydrochloric acid
(b) Aqueous sodium hydroxide [The product is $Zn(OH)_4^{2-}$.]

22.133 Write a balanced net ionic equation for the reaction of the amphoteric oxide Ga_2O_3 with:
(a) Aqueous sulfuric acid
(b) Aqueous potassium hydroxide [The product is $Ga(OH)_4^-$.]

22.134 Describe the structure of the sulfur molecules in:
(a) Rhombic sulfur
(b) Monoclinic sulfur
(c) Plastic sulfur
(d) Liquid sulfur above 160 °C

22.135 The viscosity of liquid sulfur increases sharply at about 160 °C and then decreases again above 200 °C. Explain.

22.136 Write a balanced net ionic equation for each of the following reactions.
(a) $Zn(s) +$ dilute $H_2SO_4(aq) \rightarrow ?$
(b) $BaSO_3(s) + HCl(aq) \rightarrow ?$
(c) $Cu(s) +$ hot, concentrated $H_2SO_4(l) \rightarrow ?$
(d) $H_2S(aq) + I_2(aq) \rightarrow ?$

22.137 Write a balanced net ionic equation for each of the following reactions.
(a) $ZnS(s) + HCl(aq) \rightarrow ?$
(b) $H_2S(aq) + Fe(NO_3)_3(aq) \rightarrow ?$
(c) $Fe(s) +$ dilute $H_2SO_4(aq) \rightarrow ?$
(d) $BaO(s) + H_2SO_4(aq) \rightarrow ?$

22.138 Account for each of the following observations.
(a) H_2SO_4 is a stronger acid than H_2SO_3.
(b) SF_4 exists, but OF_4 does not.

22.139 Account for each of the following observations.
(a) Oxygen is more electronegative than sulfur.
(b) Sulfur forms long S_n chains, but oxygen does not.

22.140 Write electron-dot structures for each of the following molecules, and use VSEPR theory to predict the structure of each.
(a) H_2S (b) SO_2 (c) SO_3

22.141 (a) Why is the SO_3 molecule trigonal planar but the SO_3^{2-} ion is trigonal pyramidal?
(b) Why is the S_8 ring nonplanar?

22.142 The following pictures represent the structures of oxides of carbon and sulfur. Which has the stronger bonds? Explain.

(1)   **(2)**

22.143 Fuming sulfuric acid is formed when sulfur trioxide dissolves in anhydrous sulfuric acid to form $H_2S_2O_7$. Propose a structure for $H_2S_2O_7$, which contains an S – O – S linkage.

Group 7A and 8A: The Halogens and Noble Gases (Sections 22.9–22.10)

22.144 Little is known about the chemistry of astatine (At) from
RAN direct observation, but reasonable predictions can be made.
(a) Is astatine likely to be a gas, a liquid, or a solid?
(b) Is astatine likely to react with sodium? If so, what is the formula of the product?

22.145 Why does chemical reactivity decrease from top to bottom in group 7A?

22.146 The element bromine was first prepared by oxidation of aqueous potassium bromide with solid manganese(IV) oxide. Write a balanced net ionic equation for the reaction in aqueous acidic solution. (Mn^{2+} is also formed.)

22.147 The first preparation of chlorine was similar to the synthe-
RAN sis of elemental bromine (Problem 22.146). A later method entailed the reaction of hydrogen chloride and oxygen at 400 °C. Write a balanced equation for the reaction. (Water is also formed.)

22.148 Reaction of titanium and chlorine at 300 °C yields a metal halide that is 25.25% Ti by mass. The melting point (−24 °C) and boiling point (136 °C) of the halide suggest it is a molecular compound rather than an ionic one.
(a) What are the formula and name of the compound, assuming the molecular formula is the same as the empirical formula?
(b) Write the balanced equation for the reaction.
(c) When treated with magnesium, the compound yields high-purity titanium metal. Write a balanced equation for the reaction.

22.149 Niobium reacts with fluorine at room temperature to give a
RAN solid binary compound that is 49.44% Nb by mass.
(a) What is the empirical formula of the compound?
(b) Write a balanced equation for the reaction.
(c) The compound reacts with hydrogen to regenerate metallic niobium. Write a balanced equation for the reaction.

22.150 Write the formula for each of the following compounds,
RAN and indicate the oxidation state of the halogen.
(a) Bromic acid (b) Hypoiodous acid

22.151 Write the formula for each of the following compounds,
RAN and indicate the oxidation state of the halogen.
(a) Paraperiodic acid (b) Chlorous acid

22.152 Write an electron-dot structure for each of the following molecules or ions, and predict the geometrical structure.
(a) HIO_3 (b) ClO_2^-
(c) $HOCl$ (d) IO_6^{5-}

22.153 Write an electron-dot structure for each of the following molecules or ions, and predict the geometrical structure.
(a) BrO_4^- (b) ClO_3^-
(c) HIO_4 (d) $HOBr$

22.154 Explain why acid strength increases in the order $HClO < HClO_2 < HClO_3 < HClO_4$.

22.155 Explain why acid strength increases in the order $HIO < HBrO < HClO$.

22.156 Iodine forms the acid anhydride I_2O_5. Write a balanced equation for the reaction of this anhydride with water, and name the acid that is formed.

22.157 Chlorine reacts with molten sulfur to yield disulfur dichloride, a yellowish-red liquid. Propose a structure for disulfur dichloride.

22.158 In the early 1960s krypton was found to react with fluorine gas in an electrical discharge tube at $-183\,°C$. The compound formed was KrF_2. What is the oxidation state of Kr and F in this compound? What is the electron-dot structure and geometry?

22.159 Fluorine is the only element that directly reacts with xenon. However, other xenon compounds can be formed by reacting xenon halides with other compounds such as H_2O.

$$XeF_2(s) + H_2O(l) \rightarrow XeO_3(s) + 6\,HF(g)$$

What is the electron-dot structure and geometry of XeO_3?

MULTICONCEPT PROBLEMS

22.160 Give one example from main-group chemistry that illustrates each of the following descriptions.

 (a) Covalent network solid

 (b) Disproportionation reaction

 (c) Paramagnetic oxide

 (d) Polar molecule that violates the octet rule

 (e) Lewis acid

 (f) Amphoteric oxide

 (g) Semiconductor

 (h) Strong oxidizing agent

 (i) Allotropes

22.161 Of ammonia, hydrazine, and hydroxylamine, which reacts to the greatest extent with the weak acid HNO_2? Consult Appendix C for equilibrium constants.

22.162 Write balanced equations for the reactions of (a) H_3PO_4 and (b) $B(OH)_3$ with water. Classify each acid as a Brønsted–Lowry acid or a Lewis acid.

22.163 An important physiological reaction of nitric oxide (NO) is its interaction with the superoxide ion (O_2^-) to form the peroxynitrite ion $(ONOO^-)$.

 (a) Write electron-dot structures for NO, O_2^-, and $ONOO^-$, and predict the $O - N - O$ bond angle in $ONOO^-$.

 (b) The bond length in NO (115 pm) is intermediate between the length of an NO triple bond and an NO double bond. Account for the bond length and the paramagnetism of NO using molecular orbital theory.

22.164 Consider phosphorous acid, a polyprotic acid with formula H$_3$PO$_3$.
RAN

 (a) Draw two plausible structures for H_3PO_3. For each one, predict the shape of the pH titration curve for the titration of the H_3PO_3 ($K_{a1} = 1.0 \times 10^{-2}$) with aqueous NaOH.

 (b) For the structure with the H atoms in two different environments, calculate the pH at the first and second equivalence points, assuming that 30.00 mL of 0.1240 M H_3PO_3 ($K_{a2} = 2.6 \times 10^{-7}$) is titrated with 0.1000 M NaOH.

22.165 We've said that the +1 oxidation state is uncommon for indium but is the most stable state for thallium. Verify this statement by calculating $E\,°$ and $\Delta G\,°$ (in kilojoules) for the disproportionation reaction

$$3\,M^+(aq) \rightarrow M^{3+}(aq) + 2\,M(s) \qquad M = In\ or\ Tl$$

Is disproportionation a spontaneous reaction for In^+ and/or Tl^+? Standard reduction potentials for the relevant half-reactions are

$$In^{3+}(aq) + 2\,e^- \rightarrow In^+(aq) \qquad E° = -0.44\ V$$
$$In^+(aq) + e^- \rightarrow In(s) \qquad E° = -0.14\ V$$
$$Tl^{3+}(aq) + 2\,e^- \rightarrow Tl^+(aq) \qquad E° = +1.25\ V$$
$$Tl^+(aq) + e^- \rightarrow Tl(s) \qquad E° = -0.34\ V$$

22.166 A 5.00 g quantity of white phosphorus was burned in an
RAN excess of oxygen, and the product was dissolved in enough water to make 250.0 mL of solution.

 (a) Write balanced equations for the reactions.

 (b) What is the pH of the solution?

 (c) When the solution was treated with an excess of aqueous $Ca(NO_3)_2$, a white precipitate was obtained. Write a balanced equation for the reaction, and calculate the mass of the precipitate in grams.

 (d) The precipitate in part (c) was removed, and the solution that remained was treated with an excess of zinc, yielding a colorless gas that was collected at 20 °C and 742 mm Hg. Identify the gas, and determine its volume.

22.167 A 500.0 mL sample of an equilibrium mixture of gaseous
RAN N_2O_4 and NO_2 at 25 °C and 753 mm Hg pressure was allowed to react with enough water to make 250.0 mL of solution at 25 °C. You may assume that all the dissolved N_2O_4 is converted to NO_2, which disproportionates in water, yielding a solution of nitrous acid and nitric acid. Assume further that the disproportionation reaction goes to completion and that none of the nitrous acid disproportionates. The equilibrium constant K_p for the reaction $N_2O_4(g) \rightleftharpoons 2\,NO_2(g)$ is 0.113 at 25 °C. K_a for HNO_2 is 4.5×10^{-4} at 25 °C.

 (a) Write a balanced equation for the disproportionation reaction.

 (b) What is the molar concentration of NO_2^-, and what is the pH of the solution?

 (c) What is the osmotic pressure of the solution in atmospheres?

 (d) How many grams of lime (CaO) would be needed to neutralize the solution?

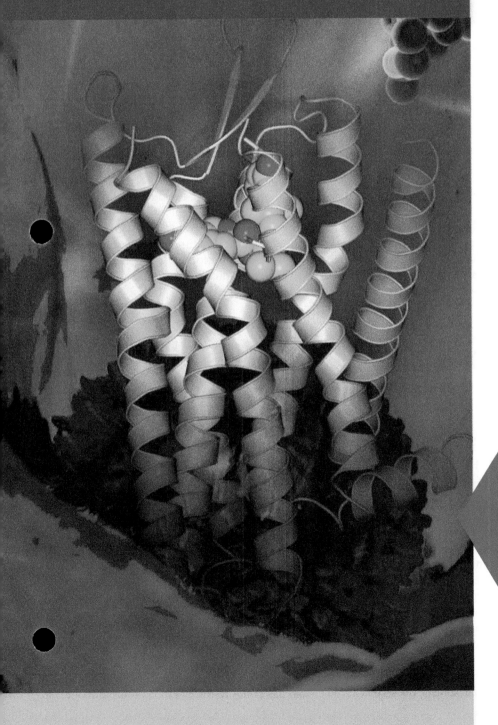

chapter 23

Organic and Biological Chemistry

Contents

Artist's rendition of the human opioid receptor, a molecule on the surface of brain cells that binds opioids and is involved in pleasure, pain, and addiction. The binding of molecules to biological receptors is governed by molecular shape. Enantiomers are molecules that are nonidentical mirror images with subtle differences in shape and very different biological effects.

Why do enantiomers have different biological responses?

The answer to this question can be found on page 1045 in the **INQUIRY**

If the ultimate goal of chemistry is to understand the world around us on a molecular level, then a knowledge of **biochemistry**—the chemistry of living organisms—is a central part of that goal. Biochemistry, in turn, is a branch of *organic chemistry*, a term originally used to mean the study of compounds from living organisms while *inorganic chemistry* was used for the study of compounds from nonliving sources. Today, however, we know that there are no fundamental differences between organic and inorganic compounds; the same principles apply to both. The only common characteristic of compounds from living sources is that all contain the element carbon. Thus, **organic chemistry** is now defined as the study of carbon compounds.

But why is carbon special, and why do chemists still treat organic chemistry as a separate branch of science? The answers to these questions involve the ability of carbon atoms to bond together, forming long chains and rings. Of all the elements, only carbon is able to form such an immense array of compounds, from methane, with one carbon atom, to deoxyribonucleic acid (DNA), with tens of billions of carbon atoms. In fact, more than 67 million organic compounds have been made in laboratories around the world, and living organisms contain additional millions.

In this chapter, we will review key concepts of structure and bonding from Chapters 7 and 8 and apply them to organic compounds. We'll show how key features of molecular structure play an important role in the function of the major classes of biological molecules: carbohydrates, lipids, proteins, and nucleic acids.

23.1 ORGANIC MOLECULES AND THEIR STRUCTURES: CONSITUTIONAL ISOMERS

Why are there so many organic compounds? The answer is that a relatively small number of atoms can bond together in a great many ways. Take molecules that contain only carbon and hydrogen—**hydrocarbons**—and have only single bonds. Such compounds belong to the family of organic molecules called **alkanes.** Because carbon atoms have four outer-shell electrons and form four covalent bonds (Section 7.5), the only possible one-carbon alkane is methane, CH_4. Similarly, the only possible two-carbon alkane is ethane, C_2H_6, and the only possible three-carbon alkane is propane, C_3H_8.

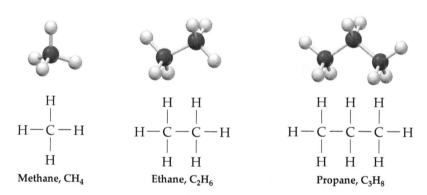

Methane, CH_4 Ethane, C_2H_6 Propane, C_3H_8

When larger numbers of carbons combine with hydrogen, however, more than one structure can result. There are two four-carbon alkanes with the formula C_4H_{10}, for instance. In one compound, the four carbons are in a row, while in the other they have a branched arrangement. Similarly, there are three alkanes with the formula C_5H_{12} and even more possibilities for larger alkanes. Compounds with all their carbons connected in a row are called **straight-chain alkanes,** and those with a branching connection of carbons are called **branched-chain alkanes.**

Compounds like the two different C_4H_{10} molecules and the three different C_5H_{12} molecules, which have the same molecular formula but different connections between atoms, are called **consitutional isomers.** For example, in butane, the second carbon from the left has bonds to two carbon atoms and two hydrogen atoms. In 2-methylpropane (a consitutional isomer), the second carbon atom has bonds to three carbon atoms and one

C_4H_{10}

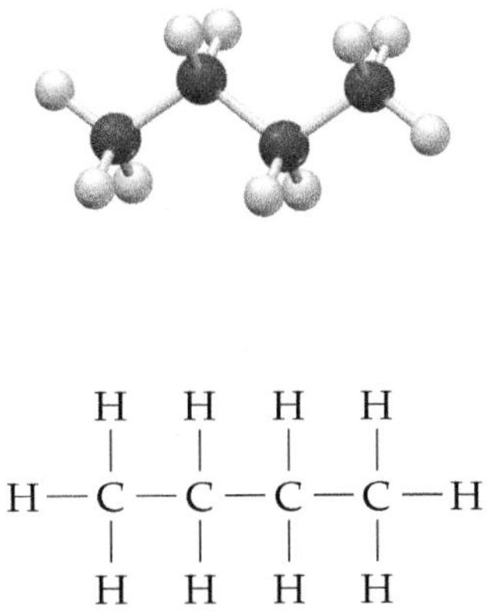

Butane (straight chain)

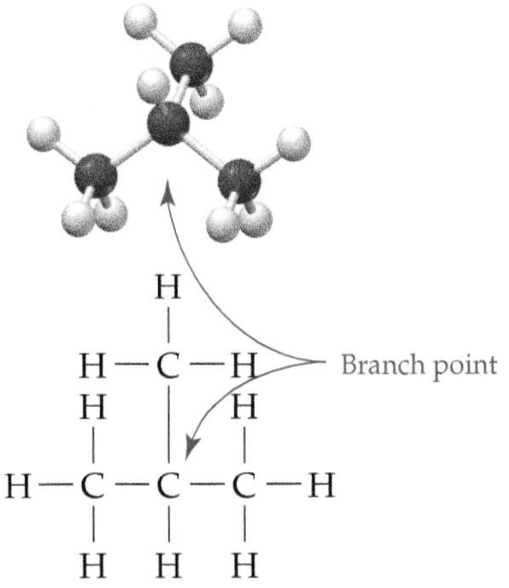

H—C—H → Branch point

2-Methylpropane (branched chain)

C_5H_{12}

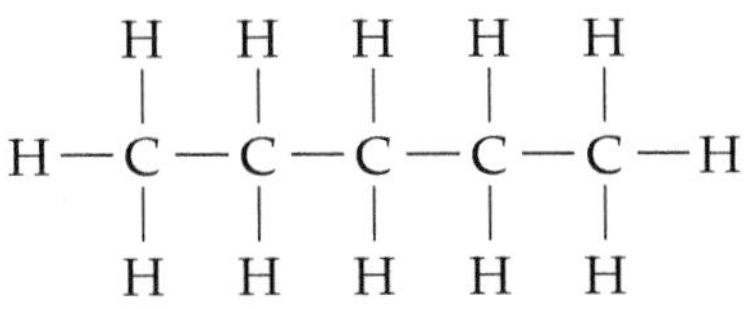

Pentane
(straight chain)

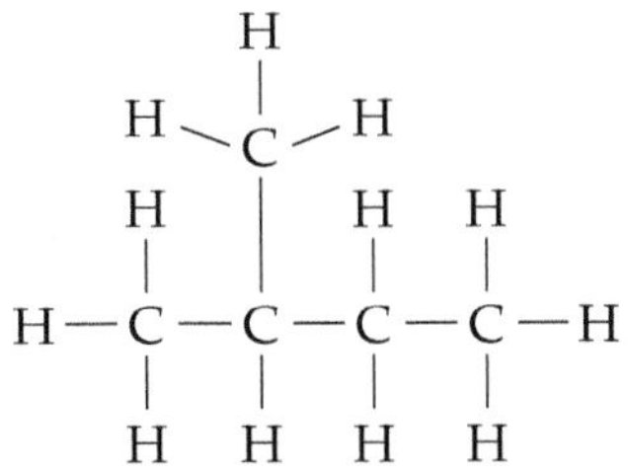

2-Methylbutane
(branched chain)

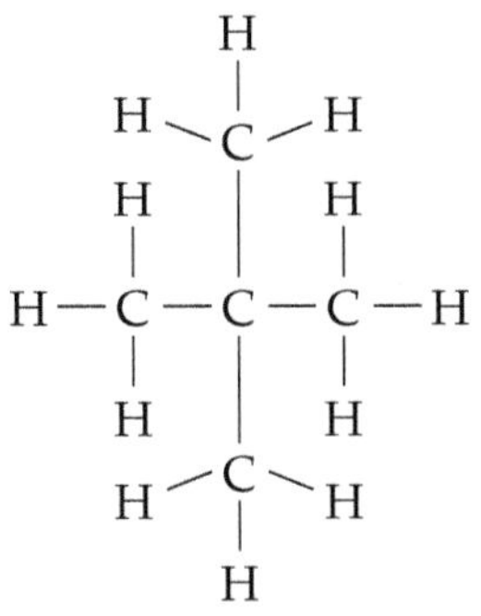

2,2-Dimethylpropane
(branched chain)

hydrogen atom. The number of possible alkane isomers grows rapidly as the number of carbon atoms increases, from five isomers for C_6H_{14} to more than 60 *trillion* possible isomers for $C_{40}H_{82}$! As noted in Section 21.7, different isomers are different chemical compounds. They have different structures, different chemical properties, and different physical properties, such as melting point and boiling point.

Because it's both awkward and time-consuming to draw all the bonds and all the atoms in organic molecules, chemists often use a shorthand way of drawing **condensed structures**. In condensed structures, carbon–hydrogen and most carbon–carbon single bonds aren't shown; rather, they're "understood." If a carbon atom has three hydrogens bonded to it, we write CH_3; if the carbon has two hydrogens bonded to it, we write CH_2; and so on. For example, the four-carbon, straight-chain alkane (called butane) and its branched-chain isomer (called 2-methylpropane, or isobutane) can be written in the following way:

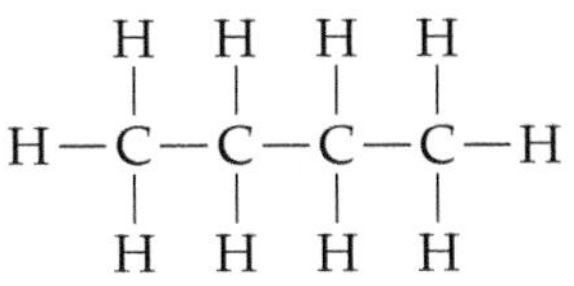

H—C—C—C—C—H = $CH_3CH_2CH_2CH_3$ Butane

In condensed
structures, most
single bonds are
omitted.

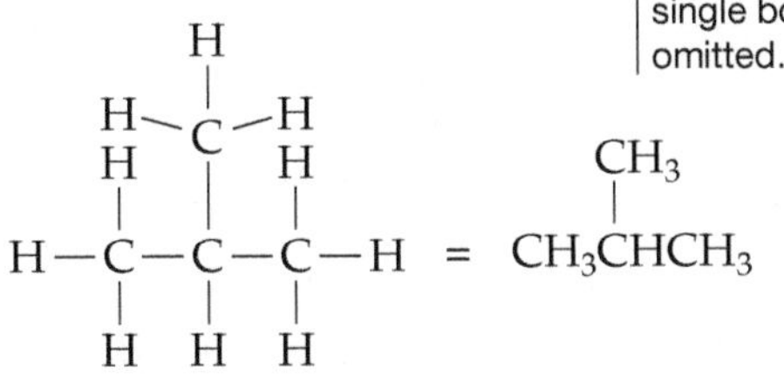

H—C—C—C—H = CH_3CHCH_3 2-Methylpropane
(Isobutane)

Note that the horizontal bonds between carbons aren't shown—the CH_3 and CH_2 units are simply placed next to each other—but the vertical bond in 2-methylpropane is shown for clarity.

The condensed structure of an organic molecule indicates the connections among its atoms but implies nothing about its three-dimensional shape, which can be predicted by the VSEPR model. Thus, a molecule can be arbitrarily drawn in many different ways. The branched-chain alkane called 2-methylbutane, for instance, might be represented by any of the following structures. All have four carbons connected in a row, with a —CH_3 branch on the second carbon from the end.

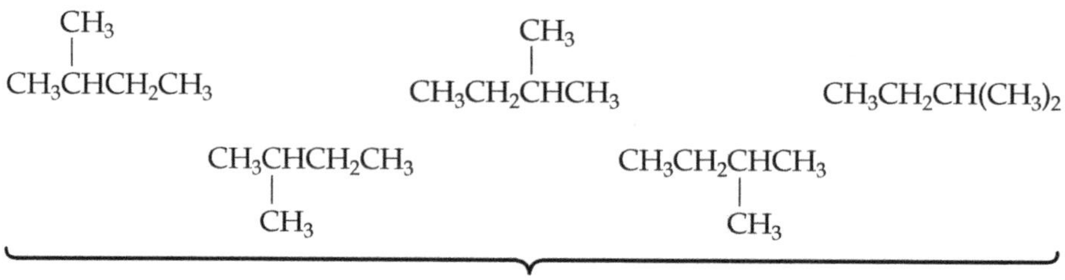

$$CH_3CHCH_2CH_3 \quad CH_3CH_2CHCH_3 \quad CH_3CH_2CH(CH_3)_2$$

with CH_3 branches:

$$CH_3CHCH_2CH_3 \quad\quad CH_3CH_2CHCH_3$$
$$\;\;\;\;\;\;\;\;|CH_3 \quad\quad\quad\quad\quad\quad |CH_3$$

Some representations of 2-Methylbutane

In fact, 2-methylbutane has no one single shape because rotation occurs around carbon–carbon single bonds. The two parts of a molecule joined by a C—C single bond are free to spin around the bond, giving rise to an infinite number of possible three-dimensional structures, or *conformations*. Thus, a large sample of 2-methylbutane contains a great many molecules that are constantly changing their shape. At any given instant, though, most of the molecules have an extended, zigzag shape, which is slightly more stable than other possibilities. The same is true for other alkanes. Remember when drawing three-dimensional structures that normal lines represent bonds in the plane of the page, wedges represent bonds coming out of the page toward the viewer, and dashed lines represent bonds receding into the page away from the viewer (Section 8.1).

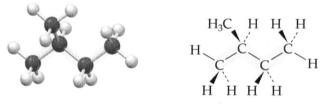

The carbon chain in 2-Methylbutane has a zigzag shape.

Line drawings show only bonds in the carbon chain and are another convenient shorthand representation of structure. A line drawing shows the carbon chain as a zigzag shape because a chain of carbon atoms assumes this shape as shown in the structure of 2-methylbutane above. Each vertex or endpoint represents a carbon atom. Bonds to hydrogen atoms are not shown, but the number of hydrogen atoms can be determined because *carbon atoms form four bonds*. Therefore, a carbon atom with one bond to another carbon will also have three bonds to hydrogen. A carbon atom with two bonds to other carbon atoms will have two bonds to hydrogen. The line drawing for 2-methylbutane is:

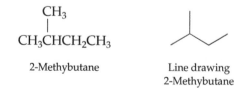

$$CH_3$$
$$|$$
$$CH_3CHCH_2CH_3$$

2-Methybutane

Line drawing
2-Methybutane

The alkanes we've seen thus far have been open-chain, or *acyclic*, compounds. *Cyclic* compounds, which contain *rings* of carbon atoms, are well known and widespread throughout nature. Compounds of all ring sizes from 3 through 30 carbons and beyond have been prepared.

The simplest cyclic compounds are the **cycloalkanes** which, like their open-chain counterparts, contain only C—C and C—H single bonds. The compounds having three carbons (cyclopropane), four carbons (cyclobutane), five carbons (cyclopentane), and six carbons (cyclohexane) are shown in **FIGURE 23.1**. Condensed structures are difficult to draw for cyclic molecules, so line drawings are typically used to represent the cyclo-alkanes. A carbon atom is at the junction of each line, and the number of hydrogen atoms can be determined because carbon forms *four* bonds. In a cyclic structure, each carbon atom has two bonds to other carbon atoms and therefore has two bonds to hydrogen.

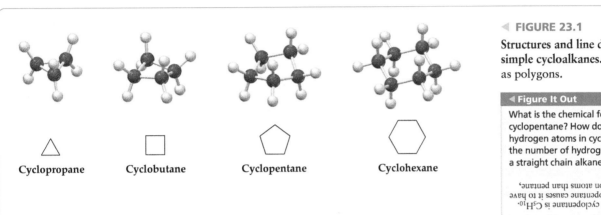

Cyclopropane   Cyclobutane   Cyclopentane   Cyclohexane

◀ **FIGURE 23.1**

Structures and line drawings of some simple cycloalkanes. The rings are shown as polygons.

◀ **Figure It Out**

What is the chemical formula for cyclopentane? How does the number of hydrogen atoms in cyclopentane compare to the number of hydrogen atoms in pentane, a straight chain alkane with 5 carbon atoms?

Answer: The formula for cyclopentane is C_5H_{10}. The ring structure of cyclopentane causes it to have two fewer hydrogen carbon atoms than pentane, C_5H_{12}.

As you might imagine, the C—C—C bond angles in cyclopropane and cyclobutane are considerably distorted from the ideal 109.5° tetrahedral value. Cyclopropane, for example, has the shape of an equilateral triangle, with C—C—C angles of 60°. As a result, the bonds in three- and four-membered rings are weaker than normal, and the molecules are more reactive than other alkanes. Cyclopentane, cyclohexane, and larger cycloalkanes, however, pucker into shapes that allow bond angles to be near their normal tetrahedral value, as shown by the models in Figure 23.1.

CONCEPTUAL WORKED EXAMPLE 23.1

Converting Molecular Models to Line Drawings

Convert the molecular models of the following alkanes into line drawings.

(a)

2,2-Dimethylpentane

(b)

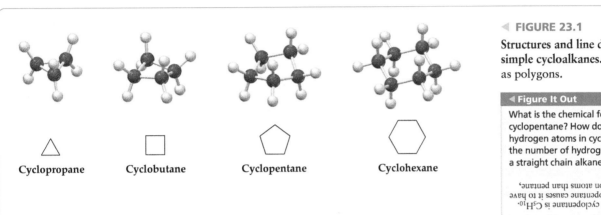

Methylcyclohexane

STRATEGY

The first step is to draw the carbon chain structure using each vertex or endpoint to represent a carbon atom. Connect the carbon atoms with single bonds.

SOLUTION

(a)

(b)

continued on next page

▶ **CONCEPTUAL PRACTICE 23.1**   Convert the molecular models of the following alkanes into line drawings.

(a) Ethylcyclopentane   **(b)** 3,5-Dimethylheptane

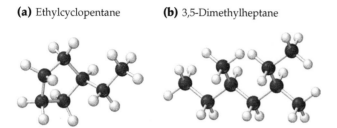

▶ **APPLY 23.2**   Determine the molecular formula of the following alkanes from the line drawings.

(a)   **(b)**

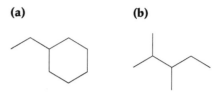

All **PRACTICE** and **APPLY** problems are interactive in the eText.

WORKED EXAMPLE 23.2

Identifying Constitutional Isomers

The following condensed structures have the same formula, C_8H_{18}. Which of them represent the same molecule and which are isomers?

(a)   **(b)**   **(c)**

$$\underset{|}{CH_3} \quad \underset{|}{CH_3}$$

$CH_3CHCH_2CHCH_2CH_3$ $CH_3CH_2CHCH_2CHCH_3$ $CH_3CHCH_2CH_2CHCH_3$

STRATEGY

Pay attention to the connections between atoms. Don't get confused by the apparent differences caused by writing a structure right to left versus left to right.

SOLUTION

Structure (a) has a straight chain of six carbons with $-CH_3$ branches on the second and fourth carbons from the end. Structure (b) also has a straight chain of six carbons with $-CH_3$ branches on the second and fourth carbons from the end and is therefore identical to (a). The only difference between (a) and (b) is that one is written "forward" and one is written "backward." Structure (c) has a straight chain of six carbons with $-CH_3$ branches on the second and *fifth* carbons from the end, so it is an isomer of (a) and (b).

▶ **PRACTICE 23.3**   Which of the following structures are identical?

(a)   **(b)**   **(c)**

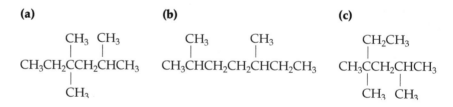

▲ A right hand fits only into a right-handed glove with a complementary shape, not into a left-handed glove.

▶ **APPLY 23.4**   Draw the five isomers of the alkane with the formula C_6H_{14} as condensed structures and as line drawings.

23.2 STEREOISOMERS: CHIRAL MOLECULES

REMEMBER . . .
Molecules that do not have a plane of symmetry are **chiral** molecules (Section 21.8).

One of the most important properties of organic and biological molecules is chirality. As we discussed in Chapter 21, a molecule or object is **chiral** if its mirror image cannot be superimposed on the original. To illustrate chirality, hold up your hands with the palms facing toward one another. The two hands are mirror images, but no matter how you turn them they cannot be superimposed on one another. (Try it!) In everyday life we observe the chiral nature of our hands because we know that a left-handed glove does not fit the right hand. Chiral objects are said to have a handedness, and they can be identified because they do not have a plane of symmetry. In contrast, achiral objects have a plane of symmetry, and the mirror image can be superimposed on the original object. **FIGURE 23.2** shows examples of achiral and chiral objects. The baseball is achiral because it has a plane of symmetry. A vertical line is drawn through the baseball, and the two halves are mirror images. No symmetry plane exists for the scissors, which indicates that they are chiral.

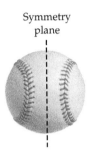

Symmetry plane

◀ **FIGURE 23.2**

Achiral and chiral objects.

◀ **Figure It Out**

Is a coffee mug a chiral or an achiral object?

Answer: A coffee mug is an achiral object because it has a symmetry plane in line with the handle.

A baseball is an achiral object because it has a symmetry plane passing through it. The symmetry plane divides the ball into two halves that are mirror images. (Only one symmetry plane is shown here, but there are others.)

Scissors are a chiral object because there is no symmetry plane. There are right-handed scissors and left-handed scissors.

How does chirality relate to molecular structure? Molecules, too, can have shapes that give them a handedness and can thus exist in mirror-image forms, one right-handed and one left-handed. The left-handed and right-handed isomers of chiral molecules are non-identical mirror images of one another and are called **enantiomers**. Enantiomers are in a class of isomers called stereoisomers. Unlike constitutional isomers, **stereoisomers** have the same connections between atoms, but stereoisomers differ in the arrangement of the atoms in space. Enantiomers have the same physical and chemical properties except for their reactions with other chiral substances and their effect on plane-polarized light. Enantiomers are often called **optical isomers** because of their ability to rotate plane-polarized light. We'll see in the Inquiry section of this chapter how enantiomers have very different biological responses such as taste, smell, and drug interactions.

The simplest example of a chiral molecule is a molecule with a tetrahedral central atom and four different substituent groups, designated as compound 1 below. This molecule is chiral because in the mirror image the substituent groups have a different geometric arrangement in space. Notice in compound 1 that the groups A, B, C, and D are in clockwise arrangement, while in the mirror image the same groups are arranged counterclockwise. The mirror image cannot be rotated to position its substituent groups in the same locations as in Compound 1. As a result, compound 1 and its mirror image are non-superimposable.

REMEMBER . . .
Stereoisomers are isomers that have the same connections between atoms but a different arrangement of the atoms in space (Section 21.7).

REMEMBER . . .
The chiral molecule and its non-superimposable mirror image are **enantiomers**. Enantiomers are often referred to as **optical isomers** due to their ability to rotate plane-polarized light (Section 21.8).

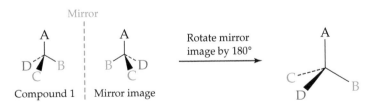

Mirror

Compound 1   Mirror image   Rotate mirror image by 180°

The rotated mirror image is not superimposable on compound 1. The positions of A and B are the same as in compound 1, but C and D are different.

Now let's examine a molecule with a tetrahedral central atom and two of the same substituent groups, designated as compound 2. To determine if compound 2 is chiral or achiral, we draw the mirror image and then rotate it to see if it can be superimposed on the original structure. Compound 2 is achiral because the mirror image can be rotated to superimpose it on the original molecule as shown. Compound 2 also has a symmetry plane that exists in in the plane of atoms A and B, which is the same as the flat plane of the page. Any molecule with a symmetry plane is achiral.

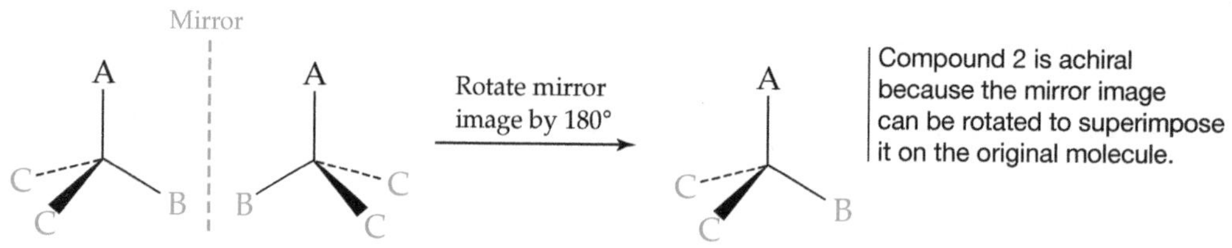

Mirror

Rotate mirror image by 180°

Compound 2 is achiral because the mirror image can be rotated to superimpose it on the original molecule.

Compound 2   Mirror image

Large organic molecules may have a number of chiral centers within the same molecule. Chiral centers are atoms with tetrahedral molecular geometry and four different substituent groups. At each chiral center, two different arrangements of the molecule are possible, the left-handed and the right-handed isomer. The maximum number of isomers increases with the number of different chiral centers; with n different chiral centers there are 2^n possible isomers.

Go to eText

WORKED EXAMPLE 23.3

Identifying Chiral Molecules

Amphetamine is a potent stimulant of the central nervous system and has been used in the treatment of attention deficit hyperactivity disorder (ADHD). Is amphetamine chiral? If so, identify any chiral centers.

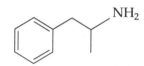

Amphetamine

STRATEGY

Chiral centers are atoms with tetrahedral molecular geometry and four different substituent groups.

SOLUTION

It may be useful to translate the line drawing into an electron-dot structure until you are more familiar with visualizing bonds to hydrogen atoms in line drawings. The electron-dot structure shows that there is one tetrahedral carbon atom with 4 different substituent groups. Therefore, amphetamine is a chiral molecule.

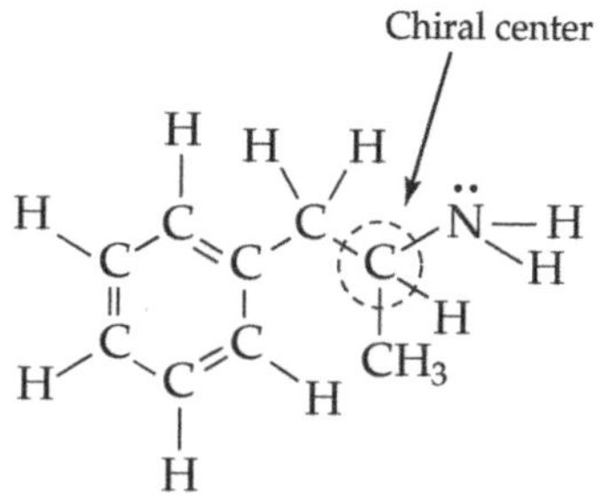

Chiral center

▶ **PRACTICE 23.5**   Ascorbic acid, also known as vitamin C, is found in citrus fruits and is used a dietary supplement. The disease

scurvy is prevented by having a sufficient amount of vitamin C in the diet. Is ascorbic acid chiral? If so, identify any chiral centers.

OH

HO

HO   OH

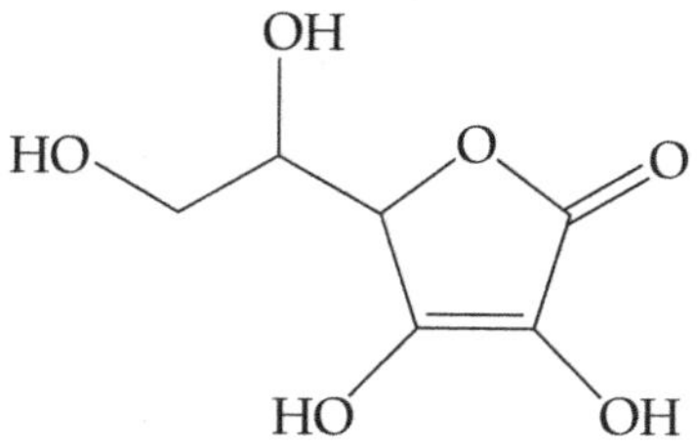

Ascorbic acid

▶ **APPLY 23.6**   The structure of the straight-chain form of glucose is shown below. Identify the chiral centers in glucose and give the number of isomers that exist as a result of the chiral centers.

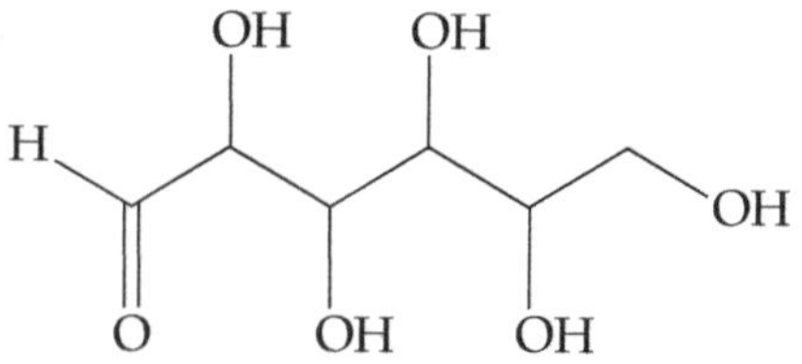

OH   OH

H

OH

O   OH   OH

All **WORKED EXAMPLES** with this icon have an interactive video in the eText.

23.3 FAMILIES OF ORGANIC COMPOUNDS: FUNCTIONAL GROUPS

Chemists have learned through experience that organic compounds can be classified into families according to their structural features and that the members of a given family often have similar chemical reactivity. Instead of 45 million compounds with random chemical behavior, there are a small number of families of compounds whose behavior is reasonably predictable.

The structural features that make it possible to classify compounds into families are called *functional groups*. A **functional group** is an atom or group of atoms within a molecule that has a characteristic chemical behavior and that undergoes the same kinds of reactions in every molecule where it occurs. Look at the carbon–carbon double-bond functional group, for instance. Ethylene ($H_2C=CH_2$), the simplest compound with a double bond, undergoes reactions that are remarkably similar to those of menthene ($C_{10}H_{18}$), a much larger and more complex molecule derived from peppermint oil. Both, for example, react with Br_2 to give products in which a Br atom has added to each of the double-bond carbons (**FIGURE 23.3**).

The example shown in Figure 23.3 is typical: the chemistry of an organic molecule, regardless of its size and complexity, is largely determined by the functional groups it contains.

TABLE 23.1 lists some of the most common functional groups and gives examples of their occurrence. Functional groups can be hydrocarbons with multiple bonds such as **alkenes** with a carbon-carbon double bond or **alkynes** with a carbon-carbon triple bond. Other functional groups contain single bonds to elements such as oxygen, nitrogen, or halogens; and still others have carbon-oxygen double bonds. Most biological molecules, in particular, contain a $C=O$, called a **carbonyl group**, which is present in several different functional groups.

REMEMBER . . .
A **double bond** is formed by sharing two pairs, or four electrons, between atoms (Section 7.5).

Go to eText

BIG IDEA Question 1

Locate and identify the functional groups in the following molecules.

(a) Lactic acid, from sour milk

$$CH_3-\overset{\overset{\displaystyle H}{|}}{C}-\overset{\overset{\displaystyle O}{\|}}{C}-OH$$
$$\underset{\displaystyle OH}{|}$$

(b) Styrene, used to make polystyrene

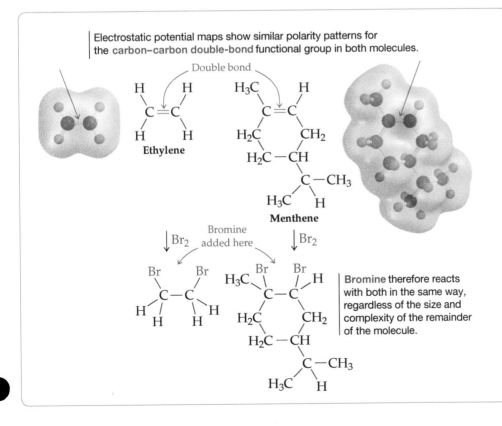

◀ **FIGURE 23.3**

Reactions of the carbon–carbon double bond in ethylene and menthene with bromine.

TABLE 23.1   Some Families of Organic Compounds

Family Name	Functional Group Structure	Simple Example	Name	Name Ending
Alkane	(contains only C—H and C—C single bonds)	CH_3CH_3	Ethane	*-ane*
Alkene	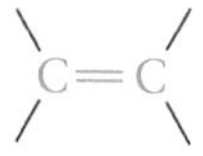 $\diagdown C = C \diagdown$	$H_2C = CH_2$	Ethene (Ethylene)	*-ene*
Alkyne	$- C \equiv C -$	$H - C \equiv C - H$	Ethyne (Acetylene)	*-yne*
Arene (aromatic)	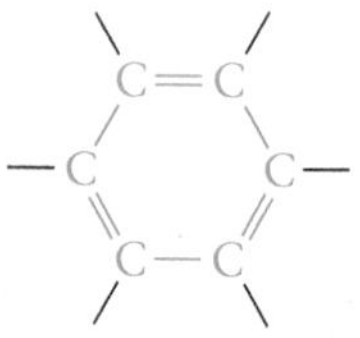	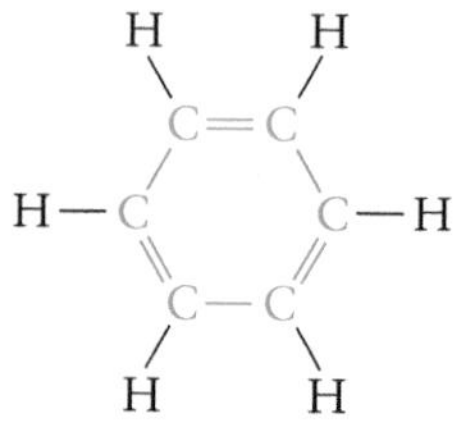	Benzene	None
Alcohol	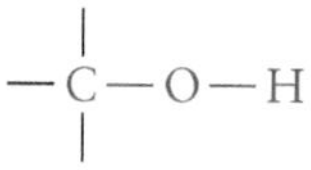 $- \overset{\vert}{\underset{\vert}{C}} - O - H$	CH_3OH	Methanol	*-ol*
Ether	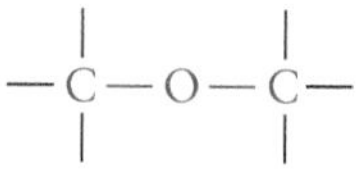 $- \overset{\vert}{\underset{\vert}{C}} - O - \overset{\vert}{\underset{\vert}{C}} -$	CH_3OCH_3	Dimethyl ether	*ether*
Amine	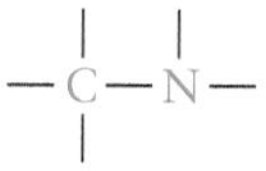 $- \overset{\vert}{\underset{\vert}{C}} - \overset{\vert}{N} -$	CH_3NH_2	Methylamine	*-amine*
Aldehyde	$- \overset{\vert}{\underset{\vert}{C}} - \overset{O}{\overset{\|}{C}} - H$	$CH_3 \overset{O}{\overset{\|}{C}}H$	Ethanal (Acetaldehyde)	*-al*
Ketone	$- \overset{\vert}{\underset{\vert}{C}} - \overset{O}{\overset{\|}{C}} - \overset{\vert}{\underset{\vert}{C}} -$	$CH_3 \overset{O}{\overset{\|}{C}}CH_3$	Propanone (Acetone)	*-one*
Carboxylic acid	$- \overset{\vert}{\underset{\vert}{C}} - \overset{O}{\overset{\|}{C}} - O - H$	$CH_3 \overset{O}{\overset{\|}{C}}OH$	Ethanoic acid (Acetic acid)	*-oic acid*
Ester	$- \overset{\vert}{\underset{\vert}{C}} - \overset{O}{\overset{\|}{C}} - O - \overset{\vert}{C} -$	$CH_3 \overset{O}{\overset{\|}{C}}OCH_3$	Methyl ethanoate (Methyl acetate)	*-oate*
Amide	$- \overset{\vert}{\underset{\vert}{C}} - \overset{O}{\overset{\|}{C}} - \overset{\vert}{N} -$	$CH_3 \overset{O}{\overset{\|}{C}}NH_2$	Ethanamide (Acetamide)	*-amide*

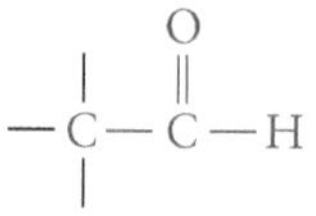

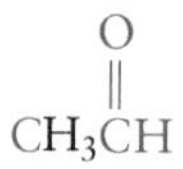

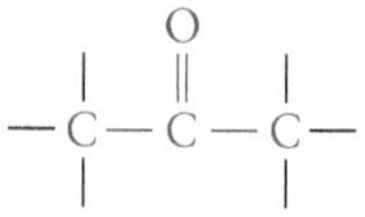

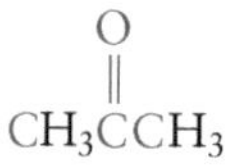

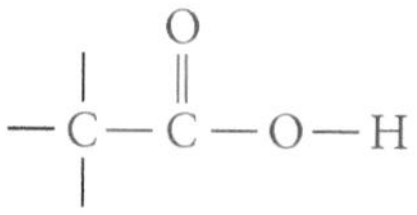

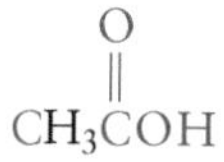

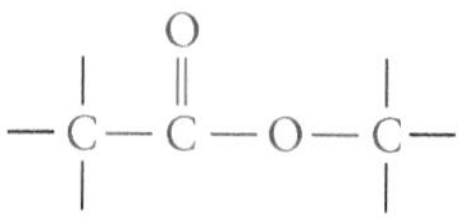

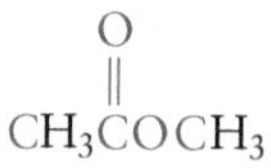

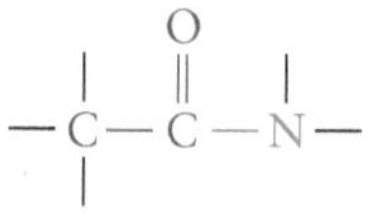

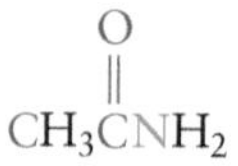

The bonds whose connections aren't specified are assumed to be attached to carbon or hydrogen atoms in the rest of the molecule.

— WORKED EXAMPLE 23.4

Interpreting Line Drawings of Molecules with Functional Groups

Draw a complete electron-dot structure for melatonin, a hormone produced in the body to regulate sleep patterns.

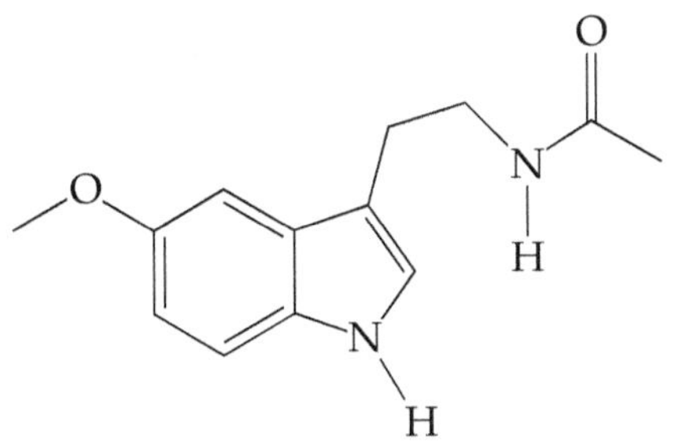

Melatonin

STRATEGY AND SOLUTION

Step 1. Draw the skeletal structure showing connecting atoms. Add a carbon at each endpoint or vertex in the line drawing as shown.

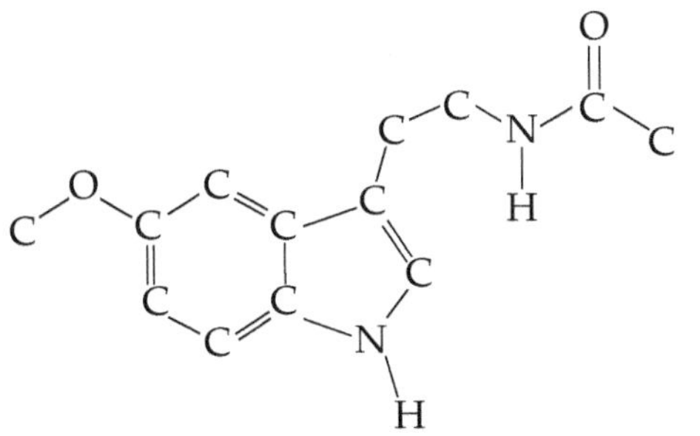

Step 2. Complete the octet for each atom. Line drawings do not show hydrogen atoms bonded to carbon but do show hydrogen atoms bonded to other elements such as nitrogen or oxygen. Line drawings also do not show lone pairs of electrons.

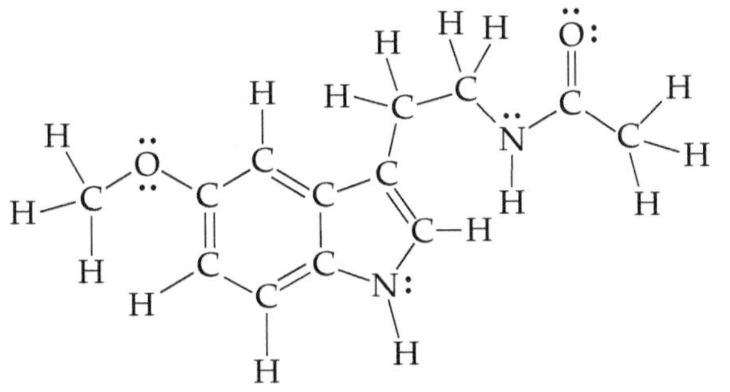

For each carbon atom, complete the octet by drawing bonds to hydrogen so the total number of bonds around the carbon atom is four. For atoms other than carbon, complete the octet by adding lone pairs of electrons.

▶ **PRACTICE 23.7** Draw a complete electron-dot structure for acetaminophen, a medicine used to treat pain and fever.

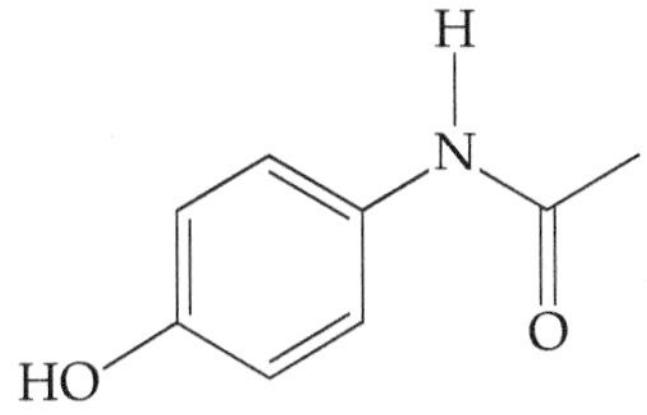

Acetaminophen

▶ **APPLY 23.8** Limonene is the major component in the oil of citrus fruit rind and has the fragrance of oranges. Convert the electron-dot structure into a line drawing.

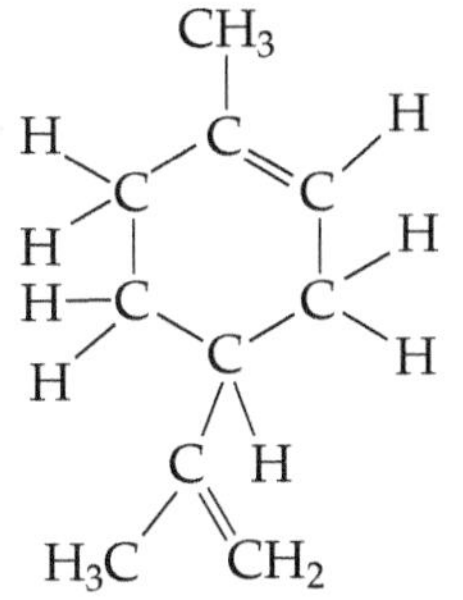

Limonene

23.4 CARBOHYDRATES: A BIOLOGICAL EXAMPLE OF ISOMERS

We have seen how isomers result when the atoms of a molecule can be connected in different ways, as in variations in the branching of alkanes and in the positions of functional groups, such as the double bond in alkenes. *Carbohydrates,* which are molecules found in every living organism, exhibit several different types of isomerism. Familiar carbohydrates are starch in food and cellulose in grass. Modified carbohydrates form part of the coating around all living cells, and other carbohydrates are found in the DNA that carries genetic information from one generation to the next.

The word *carbohydrate* was used originally to describe glucose, which has the formula $C_6H_{12}O_6$ and was once thought to be a "hydrate of carbon," with a general formula of $(C·H_2O)_n$ (where $n \geq 3$) such as $C_6(H_2O)_6$ for glucose. This view was soon abandoned, but the word persisted; the term **carbohydrate** is now used to refer to the large class of polyhydroxylated aldehydes and ketones that we commonly call *sugars.* Glucose, for example, is a six-carbon aldehyde with five hydroxyl (OH) groups.

Aldehyde

$$HOCH_2CHCHCHCHCH$$

Glucose—a pentahydroxy aldehyde   Hydroxyl group

Carbohydrates are classified as either *simple* or *complex.* Simple sugars, or **monosaccharides,** are carbohydrates such as glucose and fructose that can't be broken down into smaller molecules by hydrolysis with aqueous acid. Complex carbohydrates, or **polysaccharides,** are compounds such as cellulose and starch that are made of many simple sugars linked together and *can* be broken down by hydrolysis.

Monosaccharides

Monosaccharides are classified according to three different characteristics: the placement of its carbonyl group, the number of carbon atoms it contains, and its three-dimensional arrangement of atoms. An *aldose* contains a carbonyl group $(C=O)$ in an aldehyde functional group; a *ketose* contains a carbonyl group in a ketone functional group. The *-ose* suffix indicates a sugar, and the number of carbon atoms in the sugar is specified by using the appropriate numerical prefix *tri-, tetr-, pent-,* or *hex-.* Glucose, for example, is an aldohexose (a six-carbon aldehyde sugar), fructose is a ketohexose (a six-carbon ketone sugar), and ribose is an aldopentose (a five-carbon aldehyde sugar). Most commonly occurring sugars are either aldopentoses or aldohexoses.

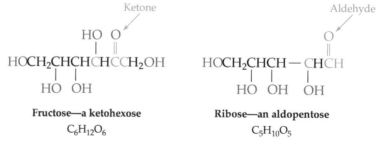

Ketone

$$HOCH_2CHCHCHCCH_2OH$$

Fructose—a ketohexose
$C_6H_{12}O_6$

Aldehyde

$$HOCH_2CHCH-CHCH$$

Ribose—an aldopentose
$C_5H_{10}O_5$

BIG IDEA Question 2   Go to eText

What is the classification of the carbohydrate shown?

$$HOCH_2CCH_2OH$$

Carbohydrates can exist in a variety of isomeric forms, but we'll just consider three types—*constitutional isomers, enantiomers,* and *anomers.* The smallest monosaccharides consist of three carbon atoms. Two of these, glyceraldehyde and dihydroxacetone, are examples of constitutional isomers because the atoms are connected in different ways.

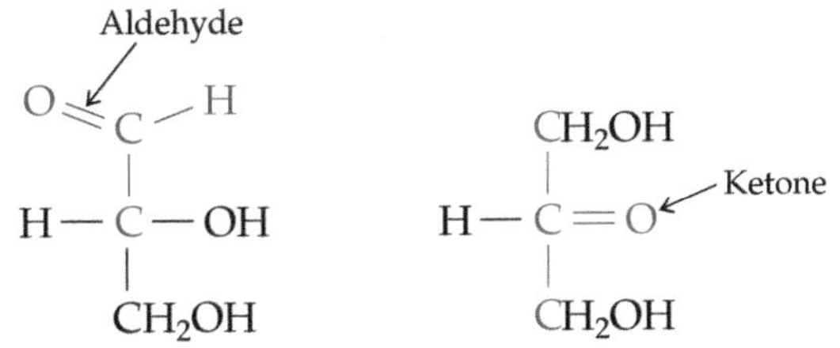

Glyceraldehyde
($C_3H_6O_3$)

Dihydroxyacetone
($C_3H_6O_3$)

Carbohydrates can also exist as enantiomers, nonidentical mirror images. Chiral molecules have enantiomers and are said to exhibit a "handedness" because just like your hands they are non-superimposable mirr or images. The Latin prefixes *dextro* (D) and *levo* (L) mean "right" and "left," respectively. Glyceraldehyde is a chiral molecule, and its enantiomers are:

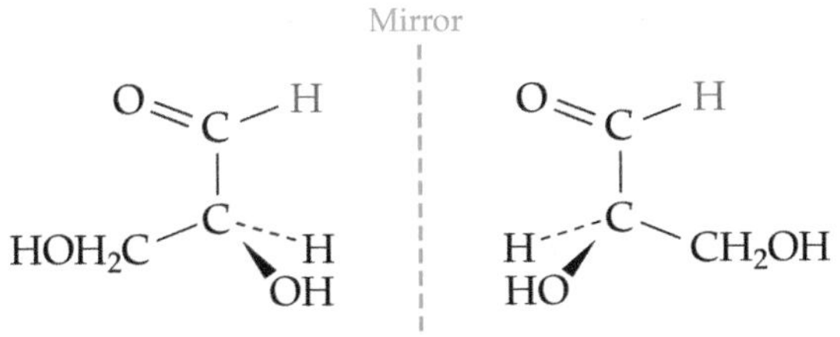

Mirror

D-Glyceraldehyde

L-Glyceraldehyde

Glucose and other monosaccharides are often shown for convenience as having open-chain structures. They actually exist, however, primarily as cyclic molecules in which an —OH group near one end of the chain adds to the carbonyl group at or near the other end of the chain to form a ring. In glucose, ring formation occurs between the —OH group on C5 and the C=O group at C1 (**FIGURE 23.4**).

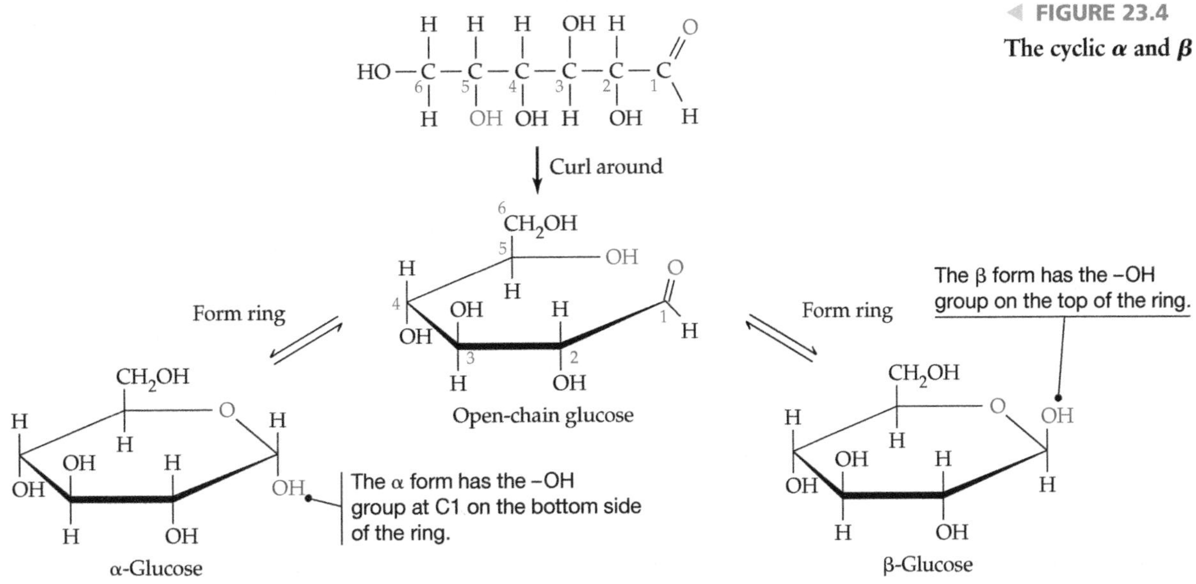

◀ **FIGURE 23.4**

The cyclic α and β forms of glucose.

Curl around

Open-chain glucose

Form ring

The β form has the –OH group on the top of the ring.

Form ring

The α form has the –OH group at C1 on the bottom side of the ring.

α-Glucose

β-Glucose

Two cyclic forms of glucose, called **anomers,** can result from ring formation, depending on whether the newly formed —OH group at C1 is on the bottom or top side of the ring. The ordinary crystalline glucose you might take from a bottle is entirely the cyclic α form, in which the C1—OH group is on the bottom side of the ring. At equilibrium in water solution, however, all three forms are present in the proportion 0.02% open-chain form, 36% α form, and 64% β form.

Polysaccharides

Sucrose, or plain table sugar, is probably the most common pure organic chemical in the world. Sucrose is found in many plants; sugar beets (20% by mass) and sugar cane (15% by mass) are the most common sources. Chemically, sucrose is a *disaccharide,* meaning a sugar composed of two monosaccharides—in this case, one molecule of glucose and one molecule of fructose join together with a glycosidic linkage. The 1:1 mixture of glucose and fructose that results from the hydrolysis of sucrose, often called *invert sugar,* is commonly used as a food additive.

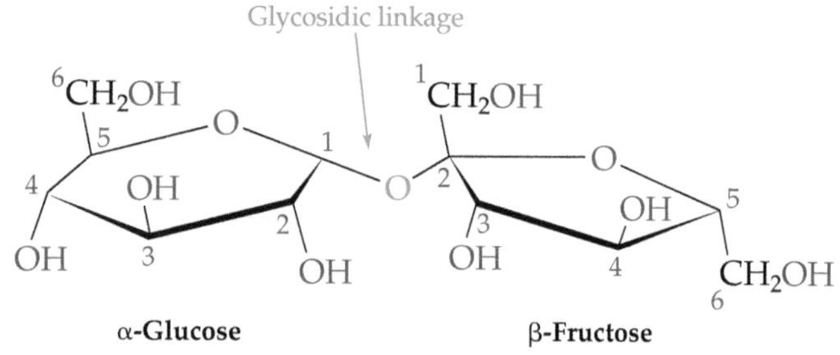

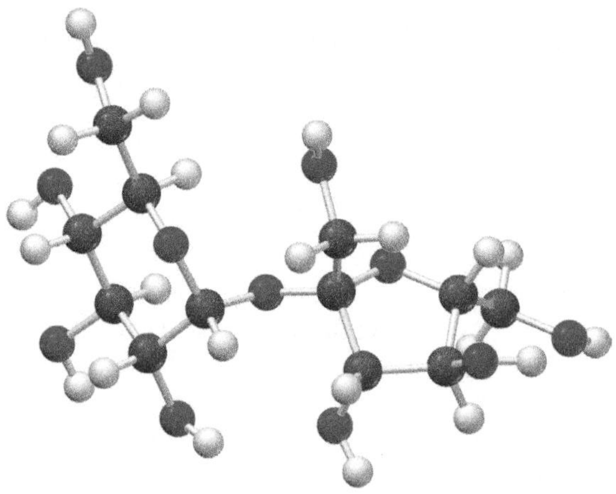

Sucrose

Cellulose, the fibrous substance that forms the structural material in grasses, leaves, and stems, is a polysaccharide composed of several thousand β-glucose molecules joined together to form an immense chain.

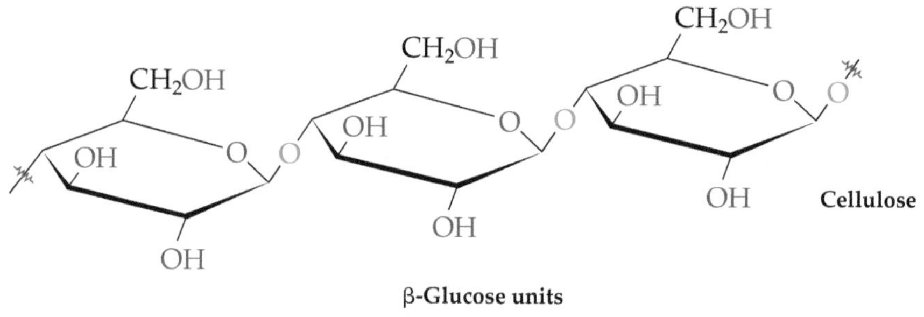

β-Glucose units

Starch is also made of several thousand glucose units but, unlike cellulose, is digestible for humans. Indeed, the starch in such vegetables as beans, rice, and potatoes is an essential part of the human diet. The difference between the two polysaccharides is that cellulose contains β-glucose units, while starch contains α-glucose units. The human stomach contains enzymes that are so specific in their action they are able to digest starch molecules while leaving cellulose untouched.

23.5 VALENCE BOND THEORY AND ORBITAL OVERLAP PICTURES

Electron-dot structures and line drawings are relatively simple models that indicate the connectivity of atoms and locations of multiple bonds and lone pairs of electrons in a molecule. However, these simplistic models have limitations. Consider cytosine, a component in deoxyribonucleic acid (DNA):

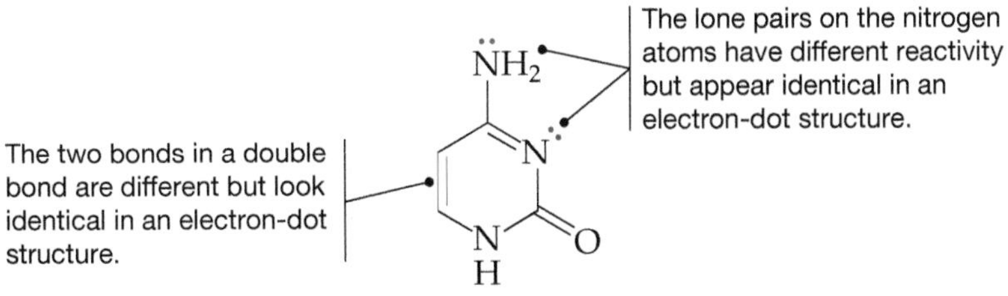

Cytosine

The lone pairs on the nitrogen atoms have different reactivity but appear identical in an electron-dot structure.

The two bonds in a double bond are different but look identical in an electron-dot structure.

In an electron-dot structure, the two bonds of the double bond (highlighted in blue) look identical. In reality, however, they are two kinds of bonds with different strengths and reactivity. In a similar argument, we would have no reason to expect the lone pairs on the two nitrogen atoms (in pink) to have differences in reactivity, but they are very different. If we want to understand these differences, we need a more complete picture of bonding and electron distribution in molecules.

Valence bond theory (Section 8.2) uses an orbital overlap picture to show how electron pairs are shared in a chemical bond. A covalent bond results when singly filled valence orbitals overlap spatially and electrons are attracted to both nuclei. Applying valence bond theory to organic molecules is not fundamentally different than what you have learned previously and visualizing orbital overlap is important in interpreting structure and reactivity. At this point, you should review valence bond theory and hybridization in Sections 8.2–8.4. You should be able to:

1. Determine the hybridization of a central atom in an electron-dot structure.
2. Describe the difference between a sigma (σ) and a pi (π) bond.
3. Sketch orbitals that overlap to form sigma (σ) and pi (π) bonds.

Worked Example 23.5 describes orbital overlap in an organic molecule with single bonds and Worked Example 23.6 describes orbital overlap in an organic molecule with double bonds.

WORKED EXAMPLE 23.5

Visualizing Orbital Overlap in Organic Molecules with Single Bonds

Draw an orbital overlap picture for methanol (commonly known as wood alcohol), CH_3OH.

STRATEGY

Draw the electron-dot structure for CH_3OH and determine the hybridization on each central atom. Overlap atomic or hybrid orbitals to make single bonds and fill in non-bonding orbitals with lone pairs as indicated by the electron-dot structure.

SOLUTION

The electron-dot structure for methanol is:

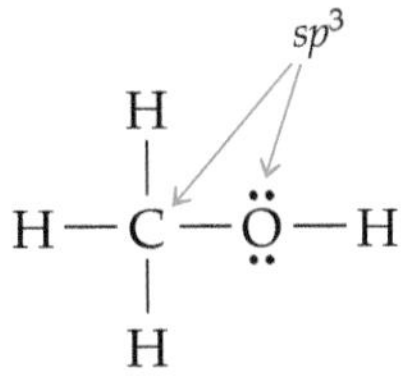

continued on next page

The carbon atom forms four bonds; three with hydrogen and one with oxygen and is therefore sp^3 hybridized. Oxygen has two bonds and two lone pairs and is also sp^3 hybridized. Overlap three carbon sp^3 orbitals with $1s$ orbitals of hydrogen to form C—H bonds. Overlap the remaining carbon sp^3 orbital with an oxygen sp^3 orbital to form the C—O bond. Account for the O—H bond by overlapping another oxygen sp^3 orbital with the $1s$ orbital of hydrogen. The two lone pairs on oxygen are placed in the two nonbonding sp^3 orbitals of oxygen.

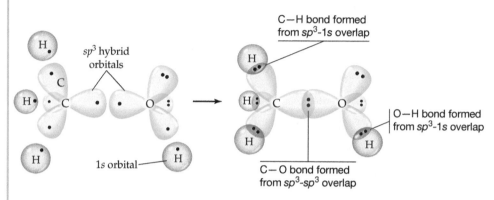

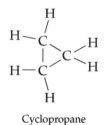

Cyclopropane

▶ **PRACTICE 23.9** Draw an orbital overlap picture for methyl amine CH_3NH_2, a compound with a strong odor similar to fish. Which orbitals overlap to form the C—H bonds, the C—N bond, and the N—H bonds?

▶ **APPLY 23.10** Cyclopropane is a cyclic organic compound that is relatively unstable. Draw an orbital overlap picture, and use it to explain this instability.

WORKED EXAMPLE 23.6

Visualizing Orbital Overlap in Organic Molecules with Multiple Bonds

Draw an orbital overlap picture for propene, the monomer used to make plastics for garbage bags. The structure of propene is $CH_2{=}CHCH_3$.

STRATEGY

Draw an electron-dot structure for propene and identify the hybridization of the central atoms. Overlap atomic or hybridized orbitals to make sigma (σ) bonds and overlap unhybridized p orbitals to make pi (π) bonds.

SOLUTION

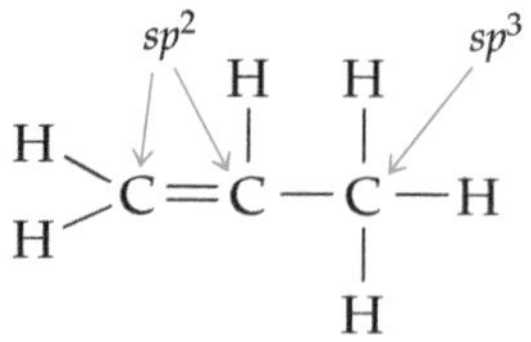

To make the C=C double bond, overlap one sp^2 hybrid orbital from each of the sp^2 hybridized carbon atoms to make a σ bond, and then line up the p orbitals for sideways overlap to make a π bond. The remaining sp^2 orbitals should all be in the same plane. Overlap three of them with the $1s$ orbitals of three hydrogen atoms to make C—H bonds. Overlap the third sp^2 orbital of the central carbon atom with an sp^3 orbital from the carbon on the right to make a C—C single bond. Make the remaining C—H bonds by overlapping the carbon sp^3 orbitals with the $1s$ orbitals of three hydrogen atoms.

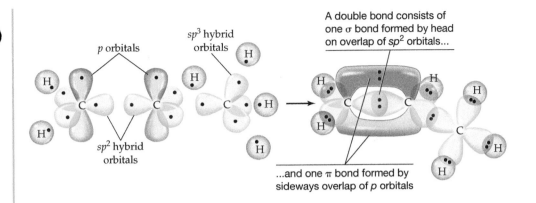

PRACTICE 23.11   Draw an orbital overlap picture for formaldehyde, $O{=}CH_2$.

(a) Are the lone pair electrons in the same plane as the π bond?
(b) Are the lone pairs in the same plane as the hydrogen atoms?

APPLY 23.12

(a) Draw an electron-dot structure for the compound C_2H_2O. (The carbon atoms are connected.)
(b) Draw an orbital overlap picture.

Although sp-hybridized atoms do occur in organic molecules, they are relatively rare. However, sp^2-hybridized atoms are common and very important. The orbital overlap creating a π bond restricts rotation around the bond and has important implications in the structure of organic molecules. For instance, there are two geometrical isomers, or **cis–trans isomers,** of 2-butene, which differ in their geometry about the double bond. The cis isomer has its two —CH_3 groups on the same side of the double bond, and the trans isomer has its two —CH_3 groups on opposite sides.

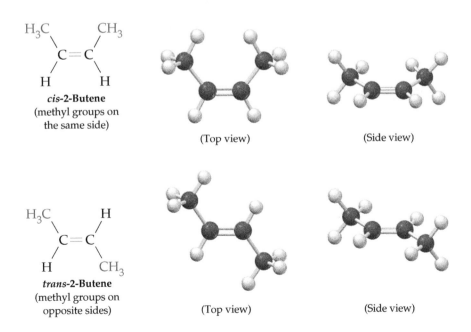

Cis–trans isomerism in alkenes arises because of the electronic structure of the double bond, which makes bond rotation energetically unfavorable. Were rotation to occur, it would break the π part of the double bond by disrupting the sideways overlap of two parallel p orbitals (**FIGURE 23.5**). An energy input of 240 kJ/mol is needed to cause bond rotation in 2-butene.

▶ **FIGURE 23.5**
Rotation around a carbon–carbon double bond.

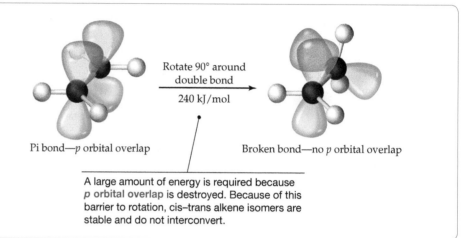

Rotate 90° around double bond
240 kJ/mol

Pi bond—p orbital overlap   Broken bond—no p orbital overlap

A large amount of energy is required because p orbital overlap is destroyed. Because of this barrier to rotation, cis–trans alkene isomers are stable and do not interconvert.

— **WORKED EXAMPLE 23.7**

Identifying and Drawing Cis–Trans Isomers

Which of the molecules (a–c) can exhibit cis–trans isomerism? For those that have this type of isomer, show the line drawing of the corresponding cis or trans isomer.

(a)   **(b)**   **(c)**

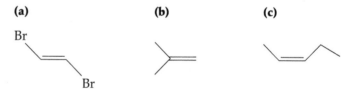

STRATEGY

It is often easier to visualize organic line drawings by converting them back into structural formulas. The presence of cis–trans isomerism exists when each carbon atom in a double bond has two *different* substituents attached.

SOLUTION

(a) Cis–trans isomers exist because the Br atoms can be placed on the same side or on opposite sides of the double bond.

$$
\begin{array}{ccc}
\underset{H}{\overset{Br}{\diagdown}}C=C\underset{Br}{\overset{H}{\diagup}} & \underset{Br}{\overset{H}{\diagdown}}C=C\underset{Br}{\overset{H}{\diagup}} & \\
\text{trans} & \text{cis} & \text{Line drawing of cis isomer}
\end{array}
$$

Br   Br

(b) This structure has two —CH$_3$ (methyl) groups bonded to the same C atom, therefore cis–trans isomers do not exist.

$$
\underset{H_3C}{\overset{H_3C}{\diagdown}}C=C\underset{H}{\overset{H}{\diagup}}
$$

(c) Cis–trans isomers exist because each carbon has two different substituents attached.

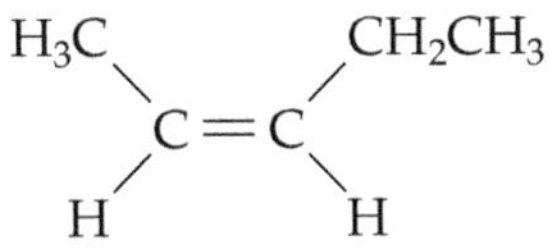

cis 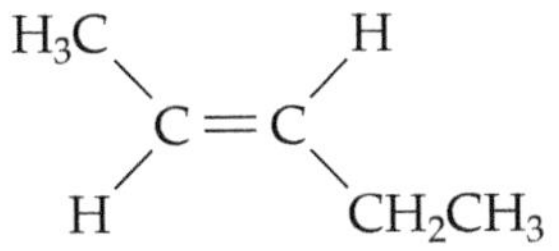   trans 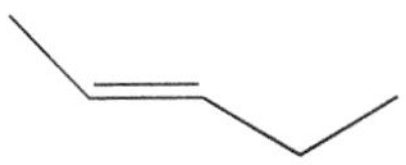   Line drawing of trans isomer

▶ **PRACTICE 23.13** Which of the following molecules can exhibit cis–trans isomerism? If these isomers exist, show the line drawing of the corresponding cis or trans isomer.

(a)   **(b)** 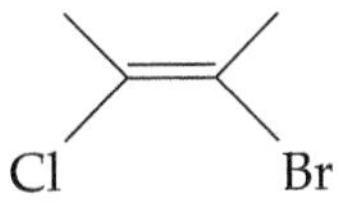   **(c)**

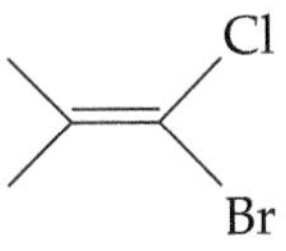

▶ **APPLY 23.14** The Krebs cycle (citric acid cycle) is a metabolic pathway in which energy is harvested from carbohydrates. In the process, succinic acid is converted to fumaric acid.

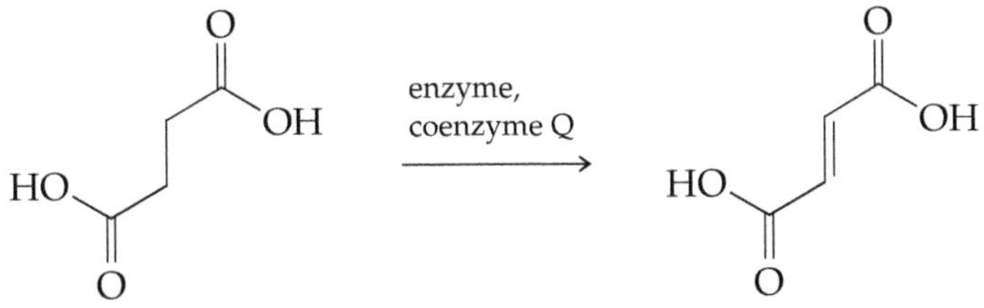

Succinic acid   Fumaric acid

(a) Is fumaric acid cis or trans?
(b) Draw the structure of maleic acid, the geometric isomer of fumaric acid.
(c) Does succinic acid have cis and trans isomers? Explain.

23.6 LIPIDS: A BIOLOGICAL EXAMPLE OF CIS–TRANS ISOMERISM

Lipids have many important biological functions, serving as sources of fuel, as protective coatings around many plants and insects, and as components of the membranes that enclose every living cell. Chemically, a **lipid** is a naturally occurring organic molecule that dissolves in a nonpolar organic solvent when a sample of plant or animal tissue is crushed or ground. Because they're defined by solubility, a physical property, rather than by chemical structure, it's not surprising that there are a great many different kinds of lipids (**FIGURE 23.6**). Note that all the lipids in Figure 23.6 contain large hydrocarbon portions, which accounts for their solubility behavior.

Animal fats and vegetable oils are the most abundant lipids in nature. Although they appear physically different—animal fats like butter and lard are usually solid at room temperature while vegetable oils like corn and peanut oil are liquid—their structures are similar. All fats and oils are **triacylglycerols**, or *triglycerides,* esters of glycerol (1,2,3-propanetriol) with three long-chain carboxylic acids called **fatty acids.** The fatty acids are usually unbranched and have an even number of carbon atoms in the range 12–22.

As shown by the triacylglycerol structure in Figure 23.6, the three fatty acids of a given molecule need not be the same. Furthermore, the fat or oil from a given source is a complex mixture of many different triacylglycerols. **TABLE 23.2** shows the structures of some commonly occurring fatty acids.

▶ **FIGURE 23.6**

Structures of some representative lipids. All are isolated from plant and animal tissue by extraction with nonpolar organic solvents, and all have large hydrocarbon portions.

▶ **Figure It Out**

Why are lipids soluble in nonpolar organic solvents?

Answer: Lipids have a high solubility in nonpolar organic solvents because a large portion of the molecule contains only carbon and hydrogen atoms and is nonpolar. Nonpolar solutes are soluble in nonpolar solvents.

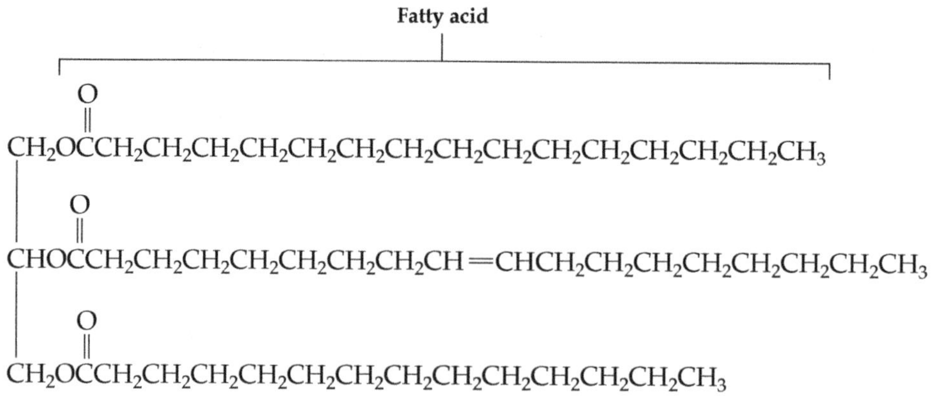

Fatty acid

CH₂OCCH₂CH₂CH₂CH₂CH₂CH₂CH₂CH₂CH₂CH₂CH₂CH₂CH₂CH₂CH₃

CHOCCH₂CH₂CH₂CH₂CH₂CH₂CH₂CH=CHCH₂CH₂CH₂CH₂CH₂CH₂CH₂CH₃

CH₂OCCH₂CH₂CH₂CH₂CH₂CH₂CH₂CH₂CH₂CH₂CH₂CH₂CH₃

A triglyceride (animal fat or vegetable oil)

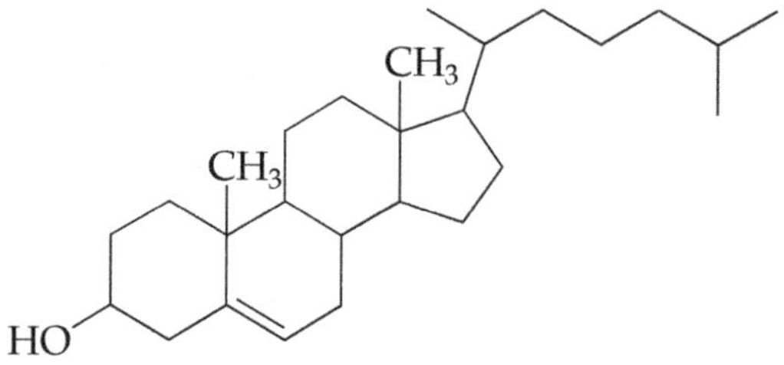

Cholesterol—a steroid

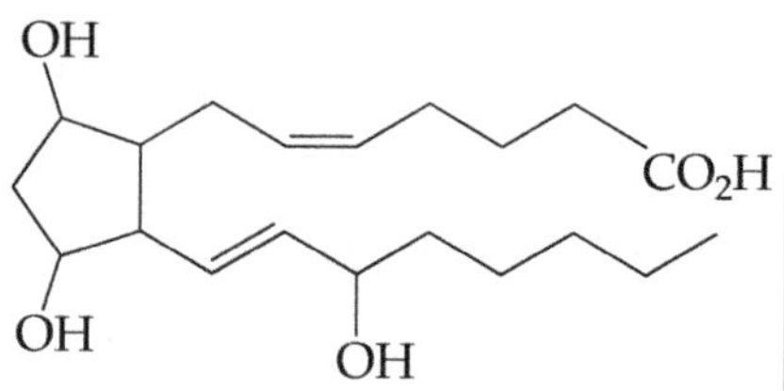

PGF₂α—a prostaglandin

TABLE 23.2	Structures of Some Common Fatty Acids		
Name	No. of Carbons	No. of Double Bonds	Structure
Saturated			
Myristic	14	0	$CH_3(CH_2)_{12}CO_2H$
Palmitic	16	0	$CH_3(CH_2)_{14}CO_2H$
Stearic	18	0	$CH_3(CH_2)_{16}CO_2H$
Unsaturated			
Oleic	18	1	$CH_3(CH_2)_7CH=CH(CH_2)_7CO_2H$ (cis)
Linoleic	18	2	$CH_3(CH_2)_4CH=CHCH_2CH=CH(CH_2)_7CO_2H$ (all cis)
Linolenic	18	3	$CH_3CH_2CH=CHCH_2CH=CHCH_2CH=CH(CH_2)_7CO_2H$ (all cis)

Fatty acids that contain double bonds are classified as **unsaturated** because they have fewer hydrogen atoms per carbon than fatty acids with an alkane chain that have the maximum possible number of hydrogens and are thus **saturated**. About 40 different fatty acids occur naturally. Palmitic acid (C_{16}) and stearic acid (C_{18}) are the most abundant saturated acids; oleic and linoleic acids (both C_{18}) are the most abundant unsaturated ones. Oleic acid is monounsaturated because it has only one double bond, but linoleic and linolenic acids are *polyunsaturated fatty acids* because they have more than one carbon–carbon double bond. A diet rich in saturated fats leads to higher levels of total cholesterol and low-density lipoprotein (LDL) cholesterol, which promotes the formation of blockages in the arteries and increases the risk of heart attack.

The main difference between animal fats and vegetable oils is that vegetable oils generally have a higher proportion of unsaturated fatty acids than do animal fats. This is useful in explaining their macroscopic properties. Why is animal lard a solid at room temperature while vegetable oil is a liquid? This can be explained by considering the number and geometry of the unsaturated fatty acids present in a fat. **FIGURE 23.7** shows that in a fat composed of mainly saturated fatty acids, the hydrocarbon tails are able to line up with one another to maximize **London dispersion forces**. Even though each individual London dispersion interaction is relatively weak, many interactions add up to strongly hold these fatty acid molecules together, making them solids. On

REMEMBER . . .
London dispersion forces are intermolecular forces that occur when the distribution of electrons in a molecule temporarily shifts causing a temporary dipole (Section 8.6).

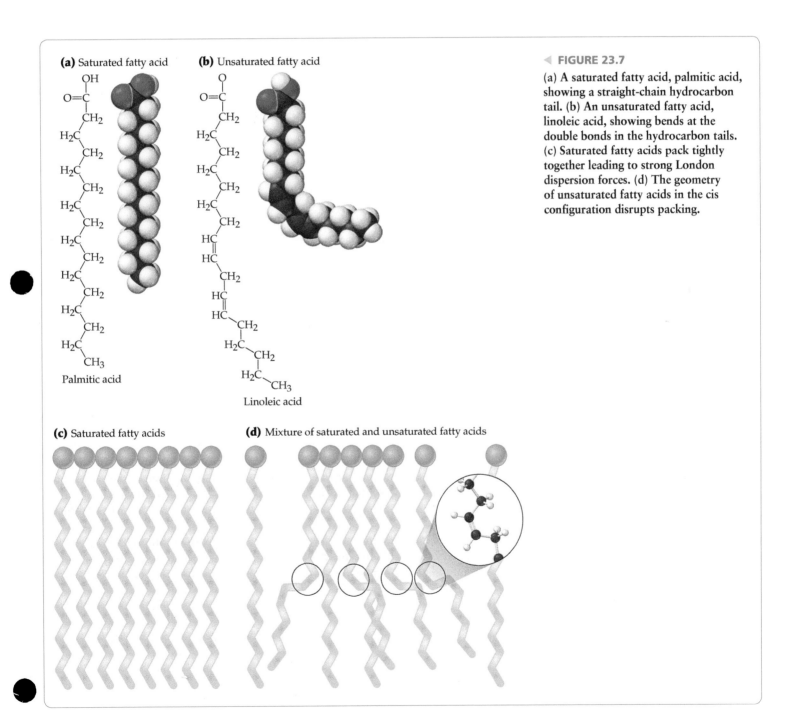

(a) Saturated fatty acid

Palmitic acid

(b) Unsaturated fatty acid

Linoleic acid

(c) Saturated fatty acids

(d) Mixture of saturated and unsaturated fatty acids

◀ **FIGURE 23.7**
(a) A saturated fatty acid, palmitic acid, showing a straight-chain hydrocarbon tail. (b) An unsaturated fatty acid, linoleic acid, showing bends at the double bonds in the hydrocarbon tails. (c) Saturated fatty acids pack tightly together leading to strong London dispersion forces. (d) The geometry of unsaturated fatty acids in the cis configuration disrupts packing.

the other hand, vegetable oils have a significant number of unsaturated fatty acids in the cis configuration. This geometry disrupts intermolecular forces substantially, making these oils liquid at room temperature.

It is possible to chemically process naturally occurring oils to give them different properties. Addition of hydrogen in the presence of a metal catalyst, called **hydrogenation,** is used commercially to convert unsaturated vegetable oils to the saturated fats used in margarine and cooking fats (**FIGURE 23.8**). By carefully controlling the extent of hydrogenation, the final product can have any desired consistency. Spreads with more cis double bonds have lower melting points and spreads with more saturated fats have higher melting points and are more solid at room temperature. Margarine, for example, is prepared so that only about two-thirds of the double bonds present in the starting vegetable oil are hydrogenated.

▶ **FIGURE 23.8**

Hydrogenation of a fatty acid in the presence of a metal catalyst.

$$CH_3CH_2CH_2CH_2CH_2CH_2CH_2 \qquad CH_2CH_2CH_2CH_2CH_2CH_2\overset{\overset{\displaystyle O}{\|}}{C} \rightleftharpoons$$

$$C=C$$

$$H \qquad H$$

Partial structure of a vegetable oil

$$\downarrow H_2, Pd\ catalyst$$

$$CH_3CH_2CH_2CH_2CH_2CH_2CH_2 \qquad CH_2CH_2CH_2CH_2CH_2CH_2\overset{\overset{\displaystyle O}{\|}}{C} \rightleftharpoons$$

$$C-C$$

$$H \qquad H$$

$$H \qquad H$$

Partial structure of a saturated fat

▲ Vegetable oils have more unsaturated fats and lower melting points than butter, which consists of saturated fats.

Partially hydrogenated fats take longer to spoil because fats that are unsaturated are highly susceptible to reacting with oxygen. In the days before refrigeration, this was even more important, because precious dairy products could be stored for longer periods. There is a down side, though. The hydrogenation process has, as a side reaction, the ability to isomerize double bonds. Therefore, a percentage of the remaining cis double bonds in margarine are transformed into trans double bonds. Diets with these so-called "trans fats" are correlated to higher incidence of heart disease because they raise levels of low-density lipoprotein (LDL) or "bad" cholesterol, which contributes to the buildup of fatty plaque in arteries. Therefore, chemists are coming up with new hydration processes that do not lead to trans fat formation.

Unsaturated fats are often considered "healthy" fats because diets higher in unsaturated fats correlate to better heart health. **Polyunsaturated fats** are fats with more than one π bond and appear to have additional health benefits. An example of a highly unsaturated fatty acid is arachidonic acid:

Arachidonic acid

Go to
eText

BIG IDEA Question 3

Label the configuration around the double bond in the fatty acids as cis or trans.

   (a) Elaidic acid ($C_{18}H_{34}O_2$)

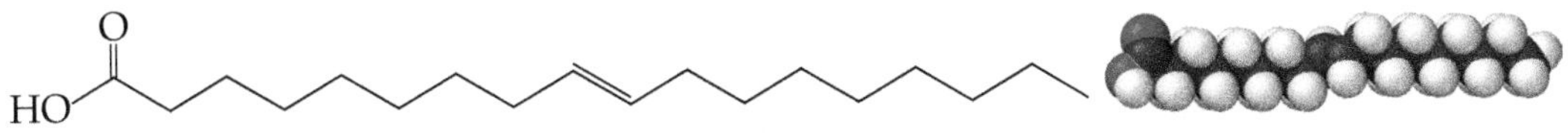

   (b) Oleic acid ($C_{18}H_{34}O_2$)

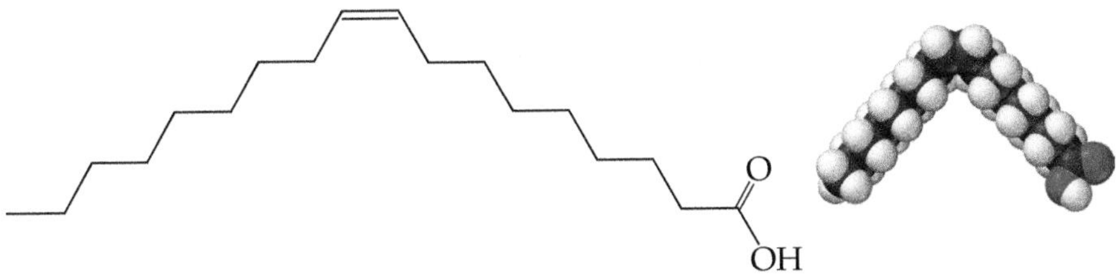

23.7 FORMAL CHARGE AND RESONANCE IN ORGANIC COMPOUNDS

Formal charge and **resonance** in organic compounds play an important role in understanding both structure and reactivity. In organic line drawings, formal charges must be indicated for correct interpretation. If formal charge is not indicated, the drawing is incorrect! Most organic reactions occur between regions of high and low electron density which is often shown by formal charge. Worked Example 23.8 shows how to calculate and represent formal charges in line drawings of organic molecules.

> **REMEMBER . . .**
> **Formal charge** is the number of valence electrons in an isolated atom minus the number of valence electrons assigned to an atom in an electron-dot structure (Section 7.10).

> **REMEMBER . . .**
> Different **resonance** forms of a substance differ only in the placement of bonding and nonbonding electrons. The connections between atoms and the relative positions of the atoms remain the same (Section 7.9).

WORKED EXAMPLE 23.8

Calculating Formal Charges in Organic Molecules

For the following electron-dot structures, give the line drawing and indicate any non-zero formal charges on atoms. (Note: It is possible to have resonance structures with electron-deficient carbon atoms because orbitals can be empty. However, carbon can never expand its octet.)

(a)

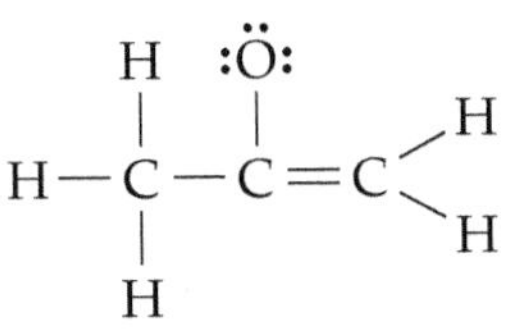

(b)

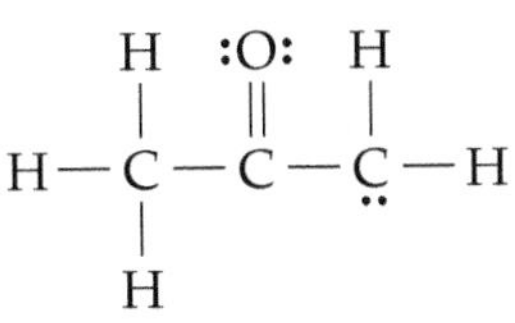

(c)

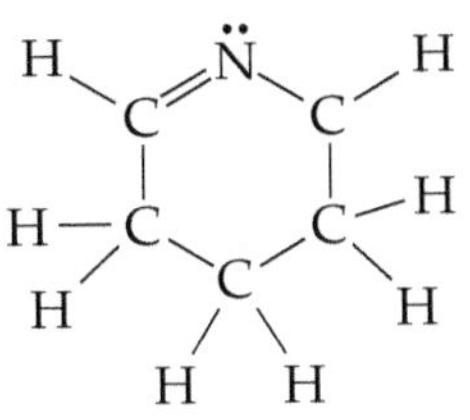

STRATEGY

Step 1. Apply the formula for calculating formal charge given in Section 7.10 to each carbon and oxygen atom in each structure as shown.

continued on next page

$$\text{Formal charge} = \begin{pmatrix} \text{Number of} \\ \text{valence electrons} \\ \text{in free atom} \end{pmatrix} - \frac{1}{2}\begin{pmatrix} \text{Number of} \\ \text{bonding} \\ \text{electrons} \end{pmatrix} - \begin{pmatrix} \text{Number of} \\ \text{nonbonding} \\ \text{electrons} \end{pmatrix}$$

(a)   **(b)**

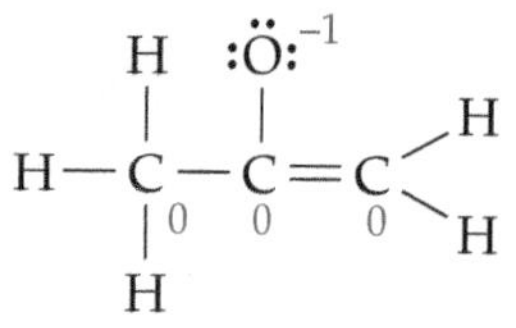

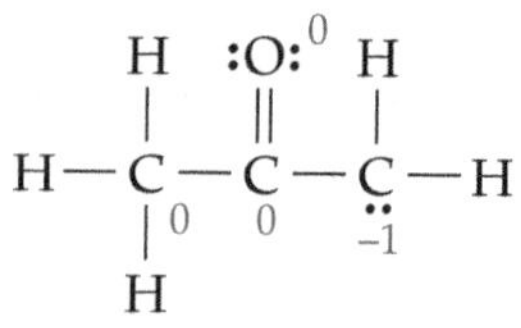

(c)

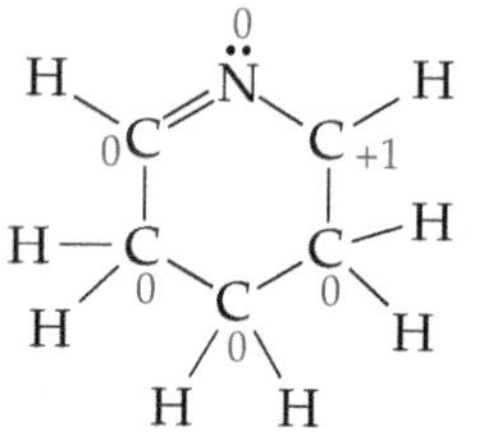

Step 2. Convert the electron-dot structure into a line drawing by indicating carbon–carbon bonds with a line. Do not show hydrogen atoms or lone pairs of electrons. Be sure to indicate all nonzero formal charges on atoms in the line drawing.

SOLUTION

(a)   **(b)** 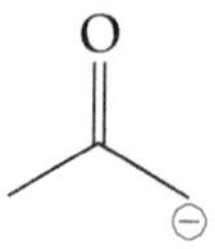   **(c)**

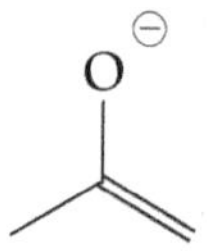

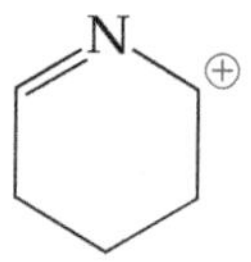

▶ **PRACTICE 23.15** For the following electron-dot structures, give the line drawing and indicate any nonzero formal charges on atoms.

(a)   **(b)** 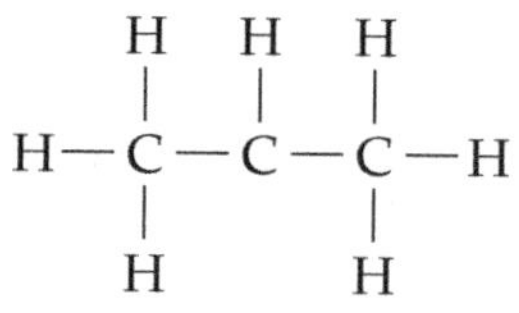   **(c)**

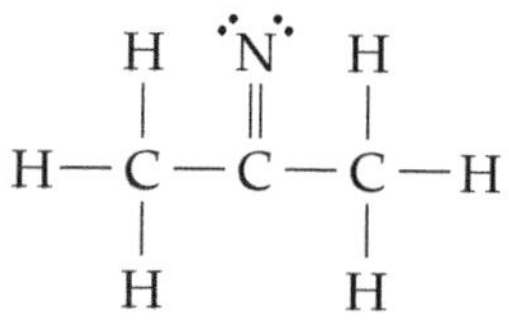

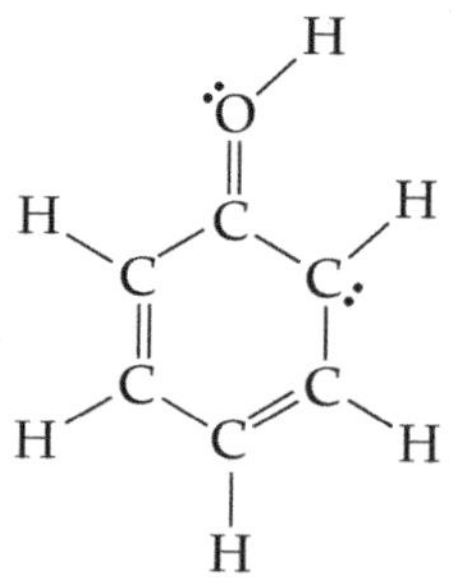

▶ **APPLY 23.16** Draw complete electron-dot structures from the line drawing by adding lone pairs of electrons and hydrogen atoms.

(a)   **(b)**   **(c)**

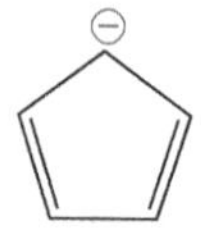

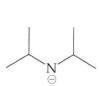
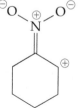

Resonance theory describes experimental observations about bond lengths and strengths in large organic molecules just as it does with small inorganic molecules like ozone (Section 7.9). It also helps explain how compounds react by predicting stability and sites of high and low electron density. The molecule benzene, C_6H_6, is a classic example of resonance in organic compounds. Benzene was first isolated from the oily residue in London street lamps in 1825, but is now an important industrial chemical used to manufacture pharmaceuticals, plastics, detergents, and pesticides. Benzene, C_6H_6, is a ring of six carbon atoms consisting of alternating single and double bonds:

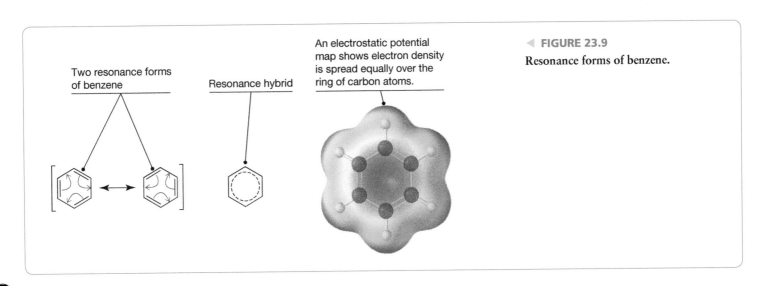

Structural formula   Line drawing

FIGURE 23.9 shows that two resonance structures can be drawn for benzene and curved arrows show the interconversion of one form into another. The resonance hybrid for benzene is typically represented as a hexagon with a circle in the center, to convey it is a blend of the two contributing resonance structures. The circle represents the spreading or delocalization of electrons in the double bonds over all six carbon atoms in the ring. Electron delocalization diffuses electron density over a greater volume, which reduces electron–electron repulsions, stabilizing the molecule. The resonance hybrid can be thought of as an *average* of the two resonance structures, thus resulting in carbon–carbon bonds that are between a single and a double bond. Experimental determination of the structure of benzene verifies that there are six equivalent carbon–carbon bonds with a bond length of 139 pm, which is between a $C-C$ single bond (154 pm) and a $C=C$ double bond (134 pm). The electrostatic potential map shows the electron density is spread equally over the ring of carbon atoms.

Two resonance forms of benzene

Resonance hybrid

An electrostatic potential map shows electron density is spread equally over the ring of carbon atoms.

◀ FIGURE 23.9

Resonance forms of benzene.

It is very helpful to recognize common patterns that indicate a resonance structure can be drawn for an organic molecule. Three important patterns are summarized here.

1. The first pattern occurs when there is a lone pair of electrons on an atom that is one σ bond away from a π bond. The ozone molecule (Section 7.9) is an example of this pattern and the resonance structures are:

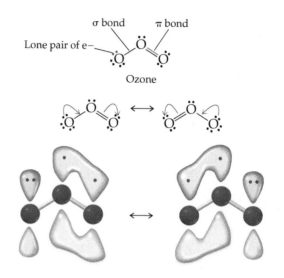

Ozone

Examples of organic molecules that fit the same resonance pattern as ozone are shown:

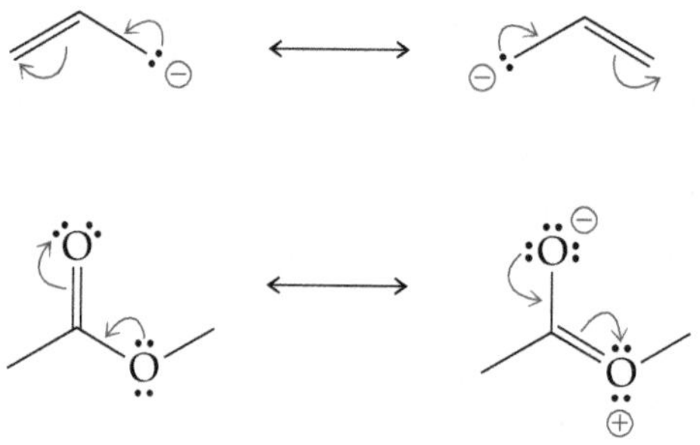

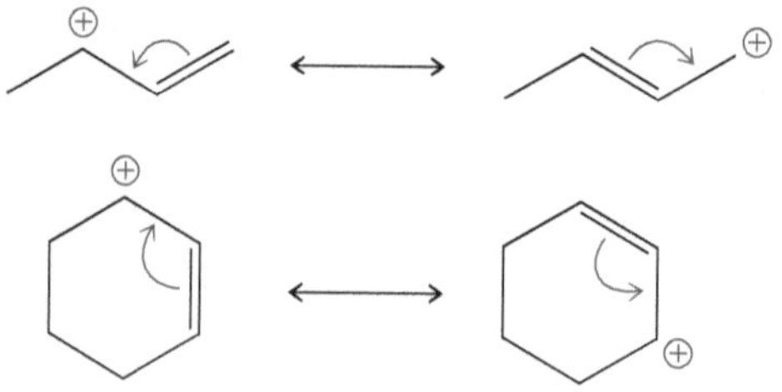

Go to eText

BIG IDEA Question 4

Can resonance structures be drawn for the following compound? If so, draw a resonance structure.

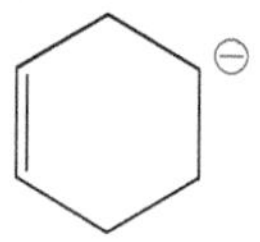

2. A second resonance pattern occurs with a polar π bond, such as C=O. In this pattern, the π bond can be rewritten as a lone pair on the electronegative oxygen atom.

3. A third resonance pattern occurs when a positive formal charge resides on a carbon atom one σ bond away from a π bond as shown:

WORKED EXAMPLE 23.9

Drawing Resonance Structures for Organic Molecules

Draw the electron-dot structure that results from the curved arrows shown below. If an incorrect structure results, explain why. For valid structures give the line drawing and indicate which resonance structure is preferred based on formal charge.

(a)

(b)

(c)

(d)

STRATEGY

Draw the complete electron-dot structure by adding hydrogen atoms and lone pairs. Then rearrange electrons as indicated by the curved arrows and check to make sure that the octet rule is not exceeded. Assign formal charges and evaluate the best structure based upon criteria given in Section 7.10 and summarized below.

- Smaller formal charges (either positive or negative) are preferable to larger ones. Zero is preferred over −1, but −1 is preferred over −2.
- Negative formal charges should reside on more electronegative atoms.
- Like charges should not be on adjacent atoms.

SOLUTION

(a) The resonance structure on the left is preferred because the negative formal charge is on the more electronegative oxygen atom.

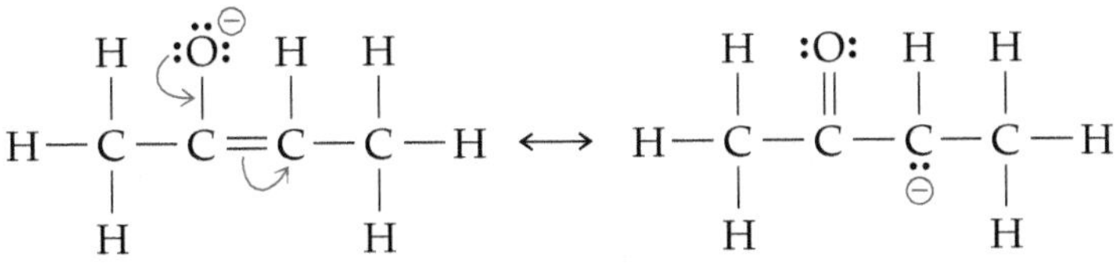

(b) The electron rearrangement indicated by the arrows results in an incorrect electron-dot structure because carbon cannot have five bonds.

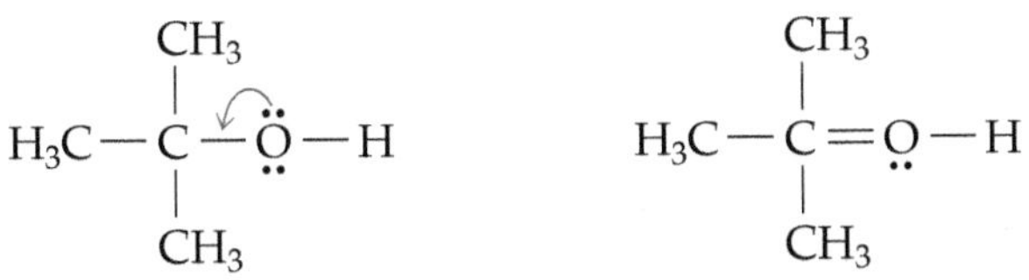

Incorrect electron-dot structure because carbon cannot have an expanded octet.

(c) The resonance structure on the left is preferred since the formal charge on all atoms is zero. Energy is required to separate + and − charges, so the structure without formal charges is probably lower in energy than the structure with formal charges.

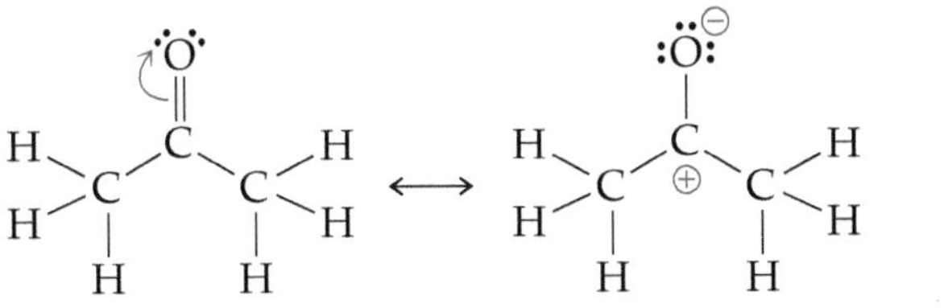

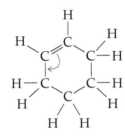

(d) The electron rearrangement indicated by the arrows results in an incorrect electron-dot structure because carbon cannot have five bonds.

This structure is incorrect because carbon cannot have an expanded octet.

continued on next page

▶ **PRACTICE 23.17** Determine if the arrow depicting electron rearrangement to form a new electron-dot structure is valid. If an incorrect structure results, explain why. If it is valid, show the line drawing of the resonance structure. Indicate which resonance structure is preferred based on formal charge.

(a)

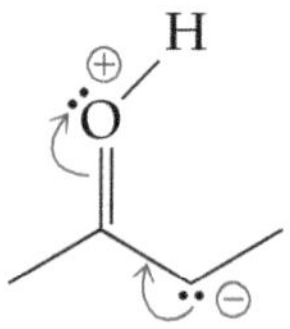

(b)

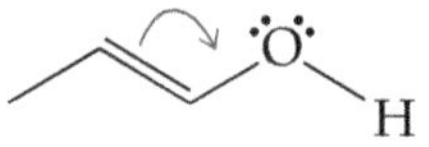

(c)

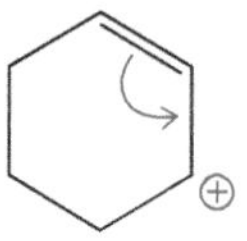

(d)

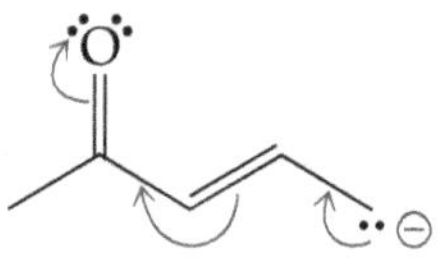

▶ **APPLY 23.18**

(a) Draw a complete electron-dot structure for cytosine, a constituent of DNA, by adding lone pairs and hydrogen atoms to the line drawing.

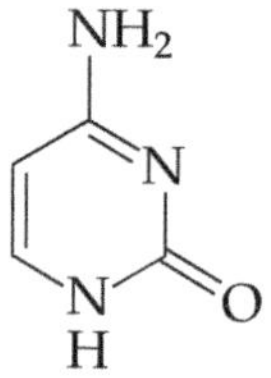

(b) Draw curved arrows in the electron-dot structure you drew in part (a) to generate the resonance structure shown.

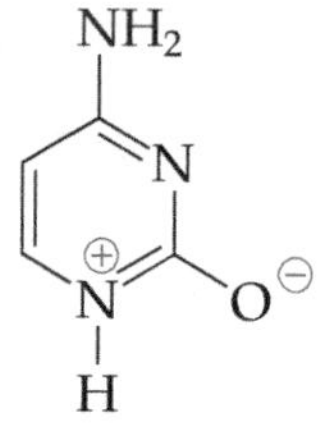

23.8 CONJUGATED SYSTEMS

Using both line drawings and orbital overlap pictures, we can get a pretty good picture of the bonding of organic molecules. The line drawing has the advantage of being simple while orbital overlap pictures help us to see the locations of electrons within molecules. When considering molecules with π bonds, it is sometimes beneficial to blend both approaches together. Because the electrons in π bonds are not directly between two nuclei, these electrons tend to be more reactive than the electrons in σ bonds. To make a picture that is both useful and as simple as possible, structures are sometimes drawn with σ bonds in a line drawing with the p orbitals superimposed on sp^2 hybridized atoms. For instance, geraniol is a major component of rose and citronella oil. It would be difficult to draw a full orbital overlap picture for this molecule, but we can draw a picture that emphasizes the p-orbital overlap as shown in **FIGURE 23.10**.

▶ **FIGURE 23.10**

A simplified orbital overlap picture of the π bonds in geraniol.

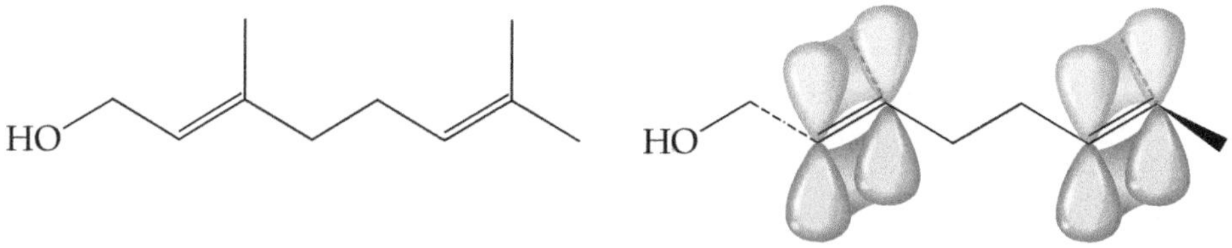

Pictures like this are even more useful in understanding a molecule like geranial, a molecule closely related to geraniol with a strong lemon smell (**FIGURE 23.11**). Geranial contains a **pi (π) system,** in which there are more than two p orbitals adjacent to each other. In geranial, there are four sp^2 hybridized atoms in a row. The two π bonds of this π system are not independent and allow electron delocalization over all four atoms.

The π system can be explained by examining the following picture of the p orbitals. Although the line drawing shows two distinct π bonds, one a C=O π bond, and one a C=C π bond, the orbital picture makes it clear that there is some overlap between the four p orbitals. All four of the atoms have been "joined together" in terms of their orbitals. **Conjugated systems** occur when p orbitals are connected in compounds with alternating single and multiple bonds. We therefore say that the two π bonds separated by a σ bond are conjugated.

Isolated π bond

◀ **FIGURE 23.11**
Conjugated **π** bonds and isolated **π** bonds in the molecule geranial.

Conjugated π bonds share adjacent p orbital overlap.

— **WORKED EXAMPLE 23.10**

Drawing Simplified Orbital Overlap Pictures for π Systems

Draw a simplified orbital overlap picture of vitamin A and identify the atoms in the conjugated system.

STRATEGY

Superimpose *p* orbitals for the π bonds on the line drawing for vitamin A. Identify conjugation by finding all atoms with *p* orbitals that overlap each other.

SOLUTION

Vitamin A has 10 carbon atoms in one π system. We can also say that all 10 carbon atoms are conjugated.

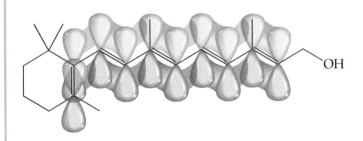

▶ **PRACTICE 23.19** Draw a simplified orbital overlap picture of beta-vetivone, an ingredient in perfumes. Label the numbered π bonds as conjugated or nonconjugated.

▶ **APPLY 23.20** In the molecule below, the triple bond has one π bond that is in conjugation with the neighboring double bond, and it has one π bond that is not conjugated. Visualize a simplified orbital overlap picture and explain why one π bond of the triple bond has delocalized electrons, but the other does not. (The hybridization of each atom is indicated.)

sp^2
sp^2 sp sp

Conjugation and Resonance

In line drawings, electrons appear to be localized in bonds between two atoms. However, in a conjugated system, the electrons are better described as **delocalized,** or associated with more than one or two atoms in a molecule. You have already been introduced to molecular orbitals (MOs) in Sections 8.7 and 8.8, which provide a good picture of

> **REMEMBER . . .**
> Molecular orbitals (MOs) are wave functions whose square gives the probability of finding an electron in a region of space spread over the entire molecule (Section 8.7).

delocalization and we'll explore this concept further in Section 23.10. For now, we can expand the concept of valence bond theory to explain delocalization.

One of the most useful aspects of conjugation is its ability to give a structural basis to the concept of resonance. Consider the molecule formamide:

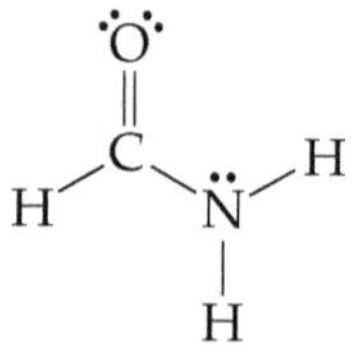

You learned to recognize the pattern of a lone pair of electrons on an atom that is one σ bond away from a π bond and to draw resonance structures in Section 23.7. Both structures are blended in a resonance hybrid, which is a better picture of electron distribution than either resonance structure by itself.

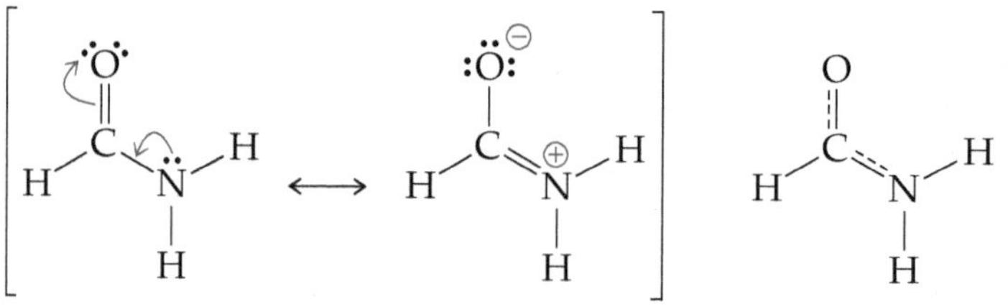

Resonance structures   Resonance hybrid

But what is the actual structure of the molecule? According to resonance theory, the double bond is not fluctuating between the C=O and C=N, but is in both places all the time. How is this possible? It is possible because the carbon, nitrogen, and oxygen atoms are all sp^2 hybridized, and the p orbitals are conjugated. A simplified orbital picture of formamide shows localized electrons in the σ bonds and delocalized electrons in the π bonds.

We see that bonding is possible across all three p orbitals, since all three overlap. A simple electron-dot structure implies that the electrons exist as a discrete π bond and a discrete lone pair, but a conjugated picture shows all four of these electrons as spread out over all three atoms, supporting the resonance concept. According to resonance theory, there is not a lone one pair on nitrogen. Rather, this lone pair is *resonance stabilized* by being delocalized over three atoms.

Conjugated system

The lone pairs on oxygen are in sp^2 hybrid orbitals and are not part of the conjugated system.

Conjugated system in formamide

Go to
eText

BIG IDEA Question 5

How many electrons are delocalized in the molecule shown?

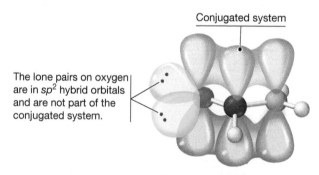

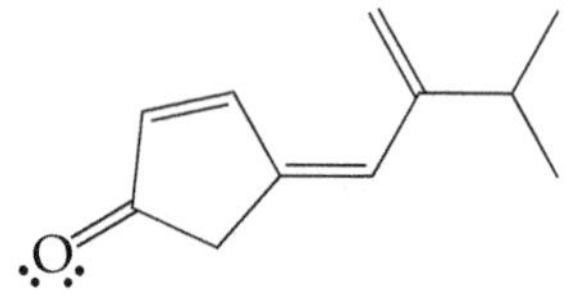

The simplified orbital overlap picture also provides a structural explanation of the double bond in the resonance hybrid. The hybrid model suggests double bond character between the carbon and nitrogen atoms as well as between the carbon and oxygen atoms. The lone pairs on oxygen, in contrast, are localized on one atom because they are in sp^2 hybrid orbitals.

Conjugation and Hybridization

The major principle of valence bond theory is that *atoms will adopt the hybridization that makes the molecule most stable by lowering the overall energy.* To make the most stable structural picture of a molecule, keep in mind the following two principles:

1. *Conjugation is immensely stabilizing; therefore, always look for conjugated systems within a molecule first.* You can recognize conjugated systems by the same traits that characterize resonance structures: a lone pair of electrons on an atom that is one σ bond away from a π bond, two π bonds separated by a σ bond, or a positive formal charge on an atom one σ bond away from a π bond. Once you have identified the atoms in a conjugated system, remember that they must be sp or sp^2 hybridized to have a p orbital for conjugation.
2. *If an atom is not conjugated, it will adopt a stable hybridization according to the VSEPR model.* Electrons that are not in conjugated systems tend to repel each other as far as possible.

> **REMEMBER . . .**
> The VSEPR (valence shell electron pair repulsion) model considers electrons in bonds and lone pairs as charge clouds that repel one another causing molecules to assume specific shapes (Section 8.1).

WORKED EXAMPLE 23.11

Determining Hybridization of Atoms in Organic Molecules with π Bonds

Formate is the compound that causes the sting of ant bites. Determine the hybridization of each atom in formate, then draw a simplified orbital overlap picture of formate, including lone pairs in appropriate orbitals.

STRATEGY

First, determine the hybridization of each atom by looking for conjugation, then moving to VSEPR if necessary. Next, draw an orbital overlap picture.

SOLUTION

The electron-dot structure shows a π bond between the carbon atom and one of the oxygen atoms. The other oxygen atom has a lone pair of electrons one σ bond away from a π bond and is therefore in the conjugated system. Thus, the carbon atom and both oxygen atoms are sp^2 hybridized. One lone pair on the oxygen atom is delocalized in the π system. All other lone pairs on oxygen are in sp^2 hybridized orbitals.

These four lone pairs are localized in sp^2 hybrid orbitals.

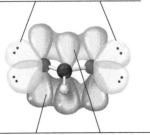

The π system is a series of three overlapped p orbitals with four delocalized electrons.

▶ **PRACTICE 23.21** Pyridoxine, the precursor to vitamin B$_6$, is shown below. What is the hybridization of each numbered oxygen atom?

▶ **APPLY 23.22** Niacinamide, the precursor for coenzyme NADH, is shown below. One of the nitrogen lone pairs is localized, and the other is delocalized. Label each lone pair on nitrogen as localized or delocalized.

23.9 PROTEINS: A BIOLOGICAL EXAMPLE OF CONJUGATION

A concept like "delocalized electrons" might seem a bit abstract. How could this apply to life? It turns out that conjugation and delocalization of electrons play a key role in the structure and function of an important class of biological molecules, proteins.

Taken from the Greek *proteios*, meaning "primary," the name *protein* aptly describes a group of biological molecules that are of primary importance to all living organisms. Approximately 50% of the human body's dry weight is protein, and almost all the reactions that occur in the body are catalyzed by proteins. In fact, a human body is thought to contain more than *150,000* different kinds of proteins.

Proteins have many different biological functions. Some, such as the keratin in skin, hair, and fingernails, serve a structural purpose. Others, such as the insulin that controls carbohydrate metabolism, act as hormones—chemical messengers that coordinate the activities of different cells in an organism. And still other proteins, such as DNA polymerase, are **enzymes,** the biological catalysts that carry out body chemistry, as discussed in the Ch 14 *Inquiry*.

Chemically, **proteins** are made up of many *amino acid* molecules linked together to form a long chain. As their name implies, amino acids contain two functional groups, a basic amino group ($-NH_2$) and an acidic ($-CO_2H$) group. Alanine is one of the simplest examples.

▲ Bird feathers are made largely of the protein keratin.

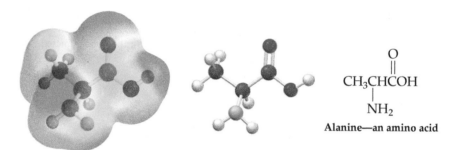

$$CH_3CHCOH$$
$$| \atop NH_2$$
with $O \atop \|$ above C

Alanine—an amino acid

Two or more amino acids can link together by forming amide bonds, usually called **peptide bonds,** between the $-NH_2$ group of one and the $-CO_2H$ group of the other. A *dipeptide* results when two amino acids link together by one amide bond, a *tripeptide* results when three amino acids link together with two peptide bonds, and so on. Short chains of up to 50 amino acids are usually called **peptides,** while the terms *polypeptide* and *protein* are generally used for longer chains.

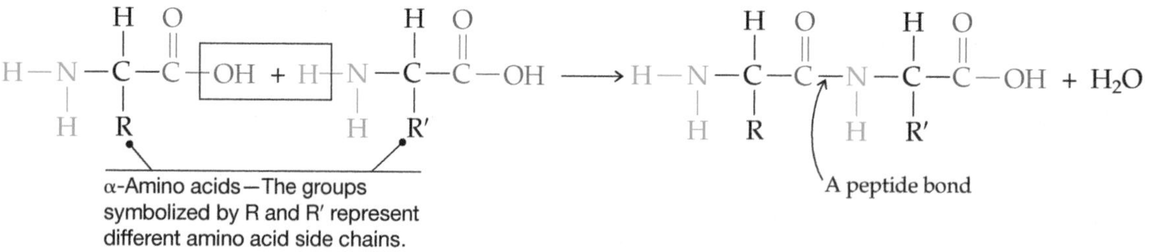

α-Amino acids—The groups symbolized by R and R′ represent different amino acid side chains.

A peptide bond

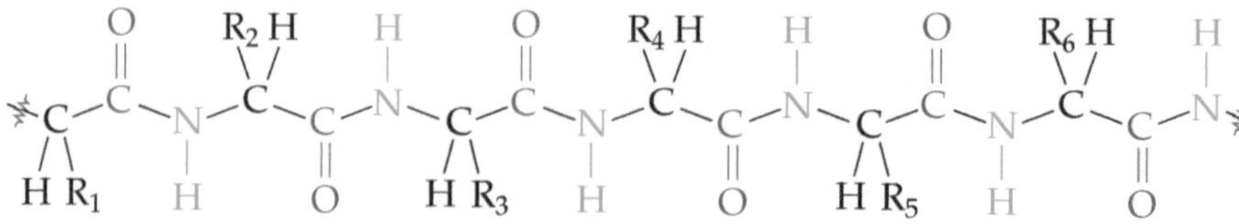

A segment of a protein backbone. The side-chain R groups of the individual amino acids are substituents on the backbone.

A polypeptide

The concepts of conjugation and hybridization discussed in the previous section are essential to understanding the structure of the peptide bond. The nitrogen atom has a lone pair of electrons that is one σ bond away from a π bond, and two resonance structures are shown:

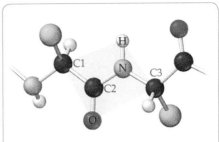

▲ **FIGURE 23.12**

Structural representation of a peptide bond. The peptide bond is planar because C2 and N are sp^2 hybridized. In a pair of linked amino acids, C1, C2, O, N, H, and C3 lie in a plane. Green balls represent amino acid side chains.

Resonance shows us that the nitrogen atom in the peptide bond is sp^2 hybridized, and its lone pair is in a conjugated system. Therefore, the carbon–nitrogen bond has significant p orbital overlap, giving the peptide bond significant double-bond character.

Several features of the experimentally determined structure show that the (C—N) bond in a peptide has a significant degree of double-bond character. The (C—N) bond length is 132 pm, which is between an average single (C—N) bond length of 147 pm and double (C=N) bond length of 127 pm. In addition, the atoms in the peptide bond all lie in one plane, as seen in the **FIGURE 23.12**.

Twenty different amino acids are commonly found in proteins, as shown in **FIGURE 23.13**. Our bodies can synthesize only 11 of the 20 amino acids. The remaining 9, highlighted in Figure 23.13, are called *essential amino acids* because they must be obtained from the diet. For convenience, there is a three-letter shorthand code for each amino acid, such as Ala (alanine), Gly (glycine), Pro (proline), and so on. All 20 are called **alpha-(α-) amino acids** because the amine nitrogen atom in each is connected to the carbon atom *alpha to* (next to) the carboxylic acid group. Nineteen of the 20 have an —NH$_2$ amino group, and one (proline) has an —NH— amino group as part of a ring.

The 20 amino acids differ in the nature of the group attached to the α carbon. Called the *side chain*, this group can be symbolized in a general way by the letter **R**.

▲ **Figure It Out**

What is the effect of resonance on the strength and rotation of the C—N bond?

Answer: Resonance gives the C—N bond some partial double bond character, which makes it stronger than a single bond and decreases rotation. Due to the restricted rotation, the peptide bond can have cis and trans isomers. Most peptide bonds have the trans configuration, with carbon atoms on opposite sides of the bond.

α carbon

R—CH—COH

Side chain   NH$_2$

O

Generalized structure of an α-amino acid

R

The 20 common amino acids are classified as *neutral, basic,* or *acidic,* depending on the structure of their side chains. Fifteen of the 20 have neutral side chains. Two (aspartic acid and glutamic acid) have an additional carboxylic acid group in their side chains and are classified as acidic amino acids. Three (lysine, arginine, and histidine) have an additional amine function in their side chains and are classified as basic amino acids. The 15 neutral amino acids can be further divided into those with nonpolar side chains and those with polar functional groups such as amide or hydroxyl groups. Nonpolar side chains are often described as *hydrophobic* (water fearing) because they are not attracted to water, while polar side chains are described as *hydrophilic* (water loving) because they *are* attracted to water. When a protein folds into a complex structure, amino acids with hydrophobic side chains are often in the nonpolar interior of the protein, while amino acids with hydrophilic side chains are exposed to polar water molecules on the exterior.

Because amino acids can be assembled in any order, depending on which —CO$_2$H group forms an amide bond with which —NH$_2$ group, the number of possible isomeric peptides increases rapidly as the number of amino acids increases. There are six ways in which three different amino acids can be joined, more than 40,000 ways in

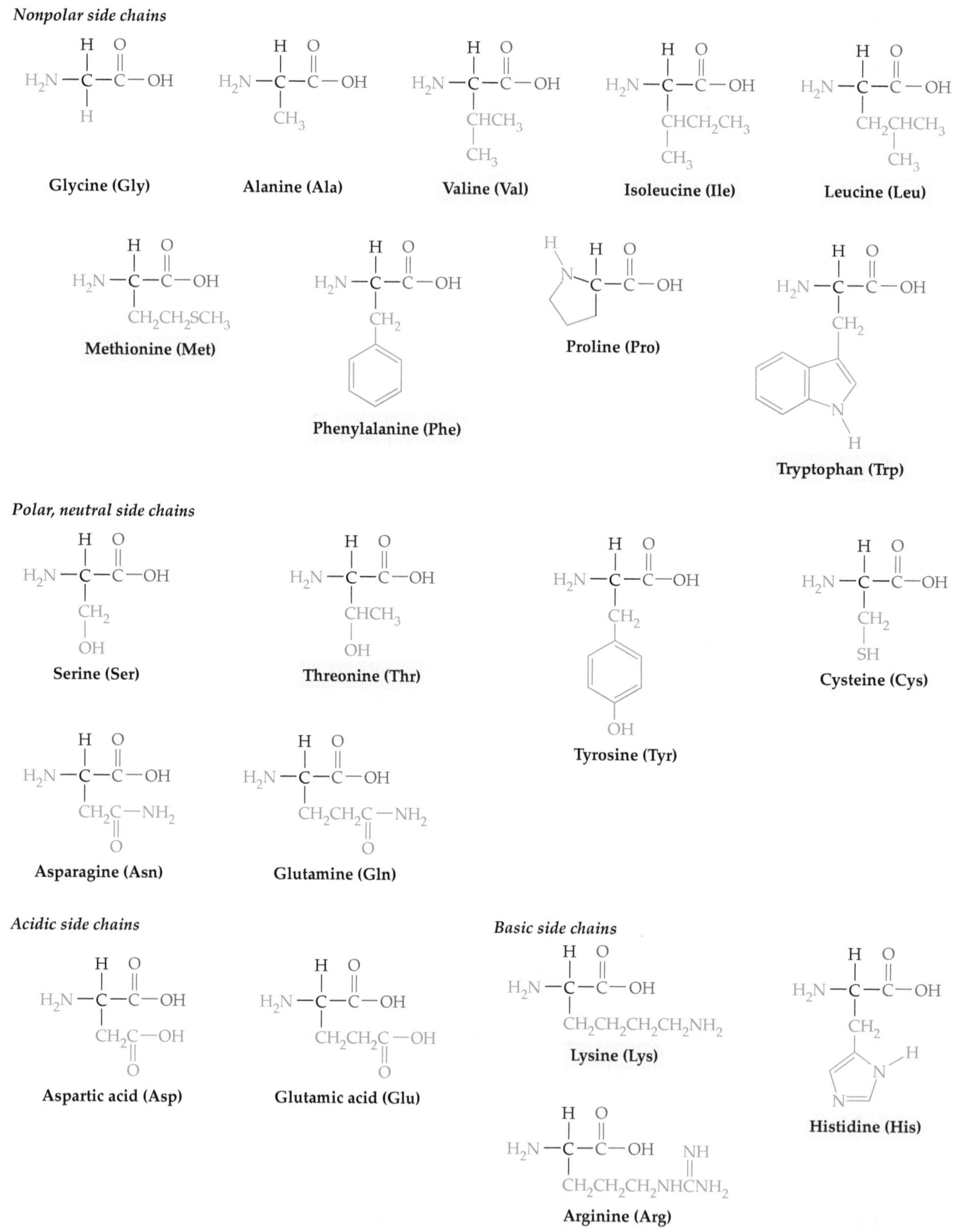

Nonpolar side chains

Glycine (Gly)   Alanine (Ala)   Valine (Val)   Isoleucine (Ile)   Leucine (Leu)

Methionine (Met)   Phenylalanine (Phe)   Proline (Pro)   Tryptophan (Trp)

Polar, neutral side chains

Serine (Ser)   Threonine (Thr)   Tyrosine (Tyr)   Cysteine (Cys)

Asparagine (Asn)   Glutamine (Gln)

Acidic side chains   *Basic side chains*

Aspartic acid (Asp)   Glutamic acid (Glu)   Lysine (Lys)   Histidine (His)

Arginine (Arg)

▲ **FIGURE 23.13**

Structures of the 20 α-amino acids found in proteins. Fifteen of the 20 have neutral side chains, 2 have acidic side chains, and 3 have basic side chains. The names of the 9 essential amino acids are highlighted.

which the eight amino acids present in the blood pressure–regulating hormone angio-tensin II can be joined (**FIGURE 23.14**), and a staggering number of ways in which the *1800* amino acids in myosin, the major component of muscle filaments, can be arranged.

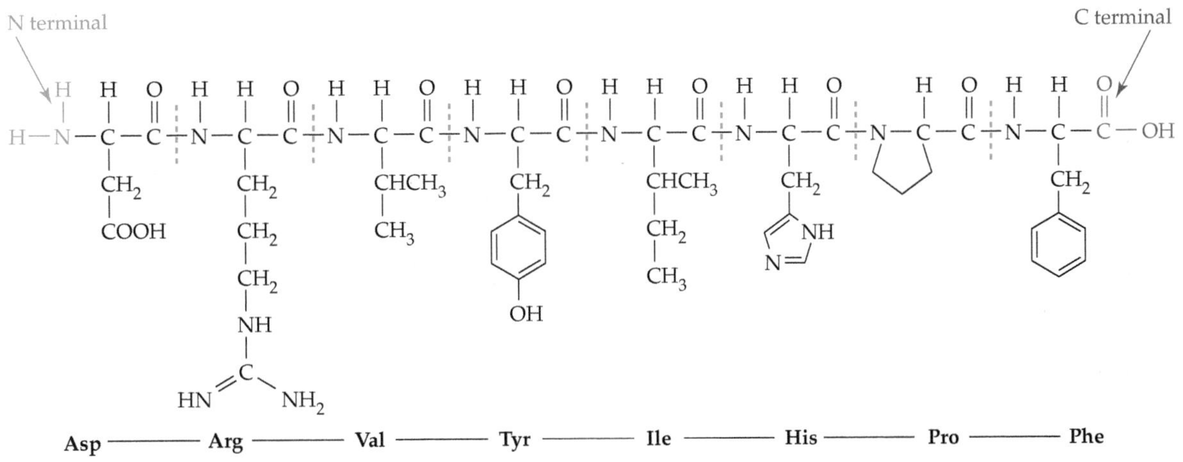

▲ **FIGURE 23.14**
The structure of angiotensin II, an octapeptide present in blood plasma.

No matter how long the chain, all noncyclic proteins have an *N-terminal amino acid* with a free $-NH_2$ on one end and a *C-terminal amino acid* with a free $-CO_2H$ on the other end. By convention, a protein is written with the free $-NH_2$ on the left and the free $-CO_2H$ on the right, and its name is indicated using the three-letter abbreviations listed in Figure 23.13.

WORKED EXAMPLE 23.12

Drawing a Peptide Structure

Draw the structure of the dipeptide Ala-Ser.

STRATEGY

First, look up the names and structures of the two amino acids, Ala (alanine) and Ser (serine). Since alanine is N-terminal and serine is C-terminal, Ala-Ser must have an amide bond between the alanine $-CO_2H$ and the serine $-NH_2$.

SOLUTION

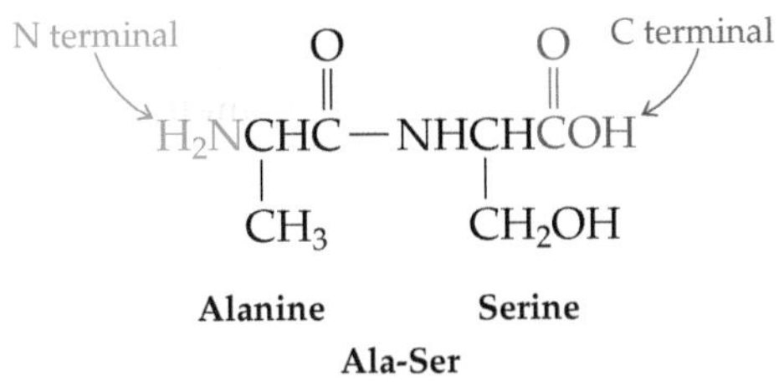

▶ **APPLY 23.24**

(a) How many amino acids are in the following peptide?
(b) Identify the amino acids and classify their side chains as polar or nonpolar.

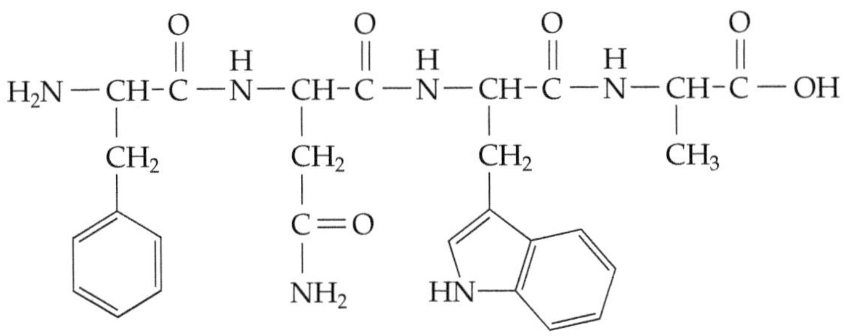

▶ **PRACTICE 23.23** Draw the structure of the tripeptide Val-Cys-Phe.

23.10 AROMATIC COMPOUNDS AND MOLECULAR ORBITAL THEORY

Although we have developed a fairly sophisticated view of molecular structure and bonding using electron-dot structures, hybridization, and valence bond theory, there are observations of molecular reactivity that cannot be explained with these models. Consider a group of highly unsaturated compounds, referred to as *aromatic compounds* in the early days of organic chemistry because they exist in fragrant substances found in fruits and other natural sources. Chemists soon realized, however, that substances grouped as aromatic behaved in a chemically different manner from most other organic compounds. Today, the term **aromatic** refers to a class of compounds that are cyclic, planar, and have adjacent *p* orbitals around a ring. Benzene, which has a six-membered ring with alternating single and double bonds, is the most common example of an aromatic compound. Aspirin, the steroid sex hormone estradiol, and many important biological molecules and pharmaceutical agents also contain aromatic rings.

▲ Benzaldehyde, a close structural relative of benzene, is an aromatic compound responsible for the odor of cherries.

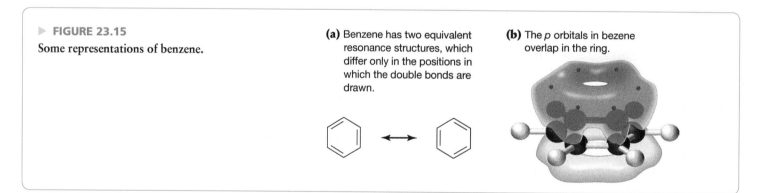

Benzene   **Aspirin**   **Estradiol**

Benzene has two equivalent resonance structures; each of the six carbons in benzene is sp^2-hybridized and has a *p* orbital perpendicular to the ring (**FIGURE 23.15**). The sideways overlap of the *p* orbitals results in six electrons in a conjugated π system.

▶ **FIGURE 23.15**

Some representations of benzene.

(a) Benzene has two equivalent resonance structures, which differ only in the positions in which the double bonds are drawn.

(b) The *p* orbitals in bezene overlap in the ring.

Although conjugation and resonance help us to understand the structure of benzene, valence bond theory does not explain its unusually high stability. If a molecule is more stable, it is less reactive. Molecular orbital (MO) theory is a more useful model in explaining electronic structure and reactivity of benzene. According to MO theory, electrons exist in molecular orbitals that cover the entire molecule. **FIGURE 23.16** shows six new MOs formed from the mathematical combination of six atomic *p* orbitals in benzene. Three of the molecular orbitals are bonding orbitals and are lower in energy than the atomic *p* orbitals from which they were formed. Bonding MOs result from the interaction of lobes with the same phase, and concentrate electron density between the nuclei. The other three molecular orbitals are antibonding orbitals and are higher in energy than the *p* orbitals from which they were formed. Antibonding MOs result from interaction of lobes with opposite phases and create nodes (or regions of zero electron probability) between the nuclei.

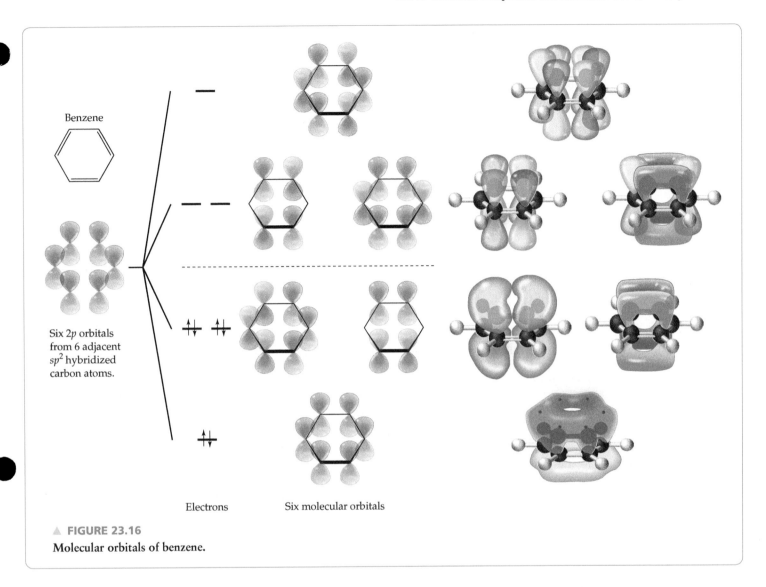

Benzene

Six 2p orbitals from 6 adjacent sp^2 hybridized carbon atoms.

Electrons   Six molecular orbitals

▲ **FIGURE 23.16**

Molecular orbitals of benzene.

MO theory is mathematically complicated, but we can understand its implications qualitatively by focusing on the energy levels of the MOs. In benzene, there are six p orbitals with one electron in each orbital, giving a total of six electrons. When the p orbitals combine to form molecular orbitals, the six electrons fill the MOs starting with the lowest energy. Therefore, the six electrons are paired in bonding MOs with lower energy than the atomic p orbitals, thus stabilizing the molecule.

The number of electrons in the π system is another very important characteristic of aromatic compounds. If there were only four electrons in the π system of benzene, two of the electrons would be unpaired in the MO diagram—a destabilizing configuration. If there were eight electrons in the π system, the last two electrons would be unpaired in antibonding MOs, which has a destabilizing effect as well. German physicist Erich Hückel articulated a requirement for the number of electron pairs in aromatic compounds. **Hückel's rule** states that a compound can only be aromatic if the number of π electrons in the ring is equal to $4n + 2$, where n equals zero or any positive integer. Therefore, the number of π electrons could be 2, 6, 10, 14, 18, and so on. Biological molecules that are aromatic most commonly have 6 or 10 electrons.

To summarize, aromatic compounds must have the following characteristics:

1. They must be cyclic.
2. They must be planar so p orbitals overlap and mathematically mix to form MOs.
3. They must have p orbitals on adjacent atoms around their ring.
4. They must have $(4n + 2)\pi$ electrons in the ring.

WORKED EXAMPLE 23.13

Predicting If an Organic Compound Is Aromatic

A number of amino acids can be categorized as aromatic amino acids because they have aromatic groups in their side chains. Can the amino acid histidine be categorized as an aromatic amino acid?

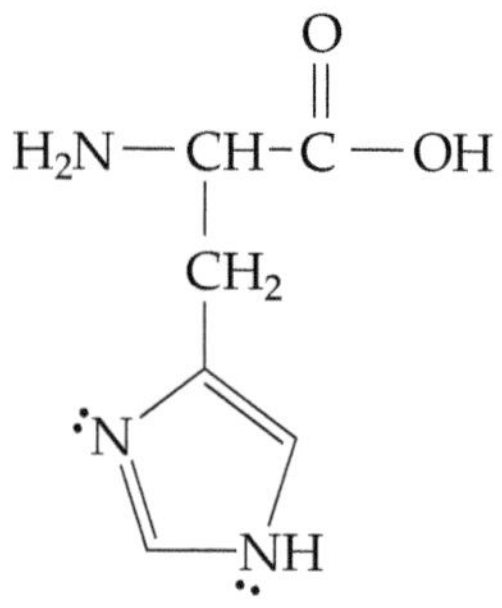

Histadine

STRATEGY

Determine if there is a ring system in which all the atoms are sp^2 hybridized and the π electrons part of a conjugated system. Next count the number of π electrons in the ring to evaluate if the $4n + 2$ rule is satisfied.

SOLUTION

The side chain in histidine has a five-member ring with three carbon and two nitrogen atoms, all of which are sp^2 hybridized. Two double bonds contribute four π electrons to the π system. Now evaluate lone pairs of electrons. Include lone pairs that are in p orbitals (these are the ones that are conjugated); *don't* include lone pairs in sp^2 hybrid orbitals because they are not in the π system!

The electrons on the NH group are adjacent to a double bond and conjugated. Therefore, they are in a p orbital and part of the π system. The other lone pair on the N atom in the ring is in a sp^2 hybrid orbital, and not part of the π system. Counting the four electrons in the π bonds and the two electrons in the lone pair of the NH group, there are a total of six electrons, making this amino acid aromatic.

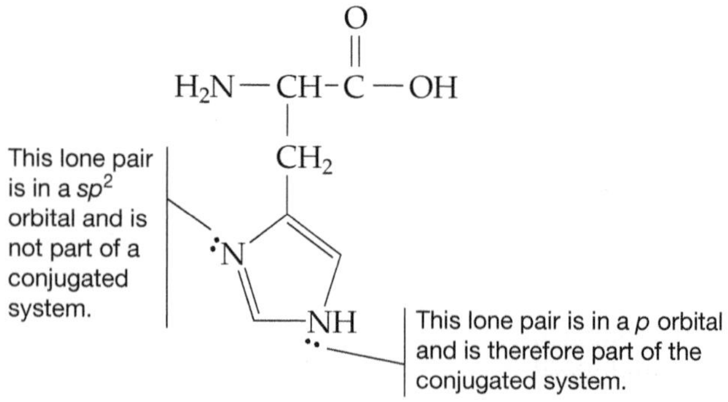

This lone pair is in a sp^2 orbital and is not part of a conjugated system.

This lone pair is in a p orbital and is therefore part of the conjugated system.

▶ **PRACTICE 23.25**   Can either proline or tyrosine be considered aromatic amino acids? Explain.

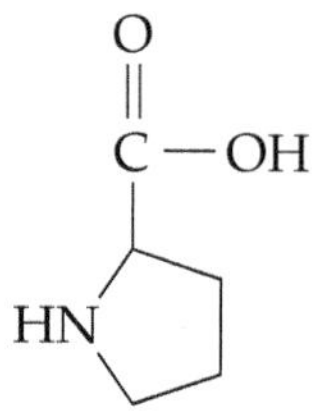

Proline

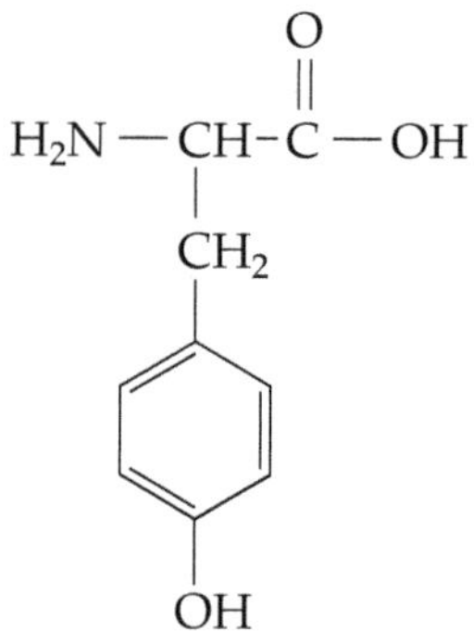

Tyrosine

▶ **APPLY 23.26**   The structure of tryptophan is drawn below.

(a) Draw a simplified orbital overlap picture of the π system of tryptophan.
(b) Use this picture and the concepts from this section to explain why tryptophan is an aromatic amino acid.

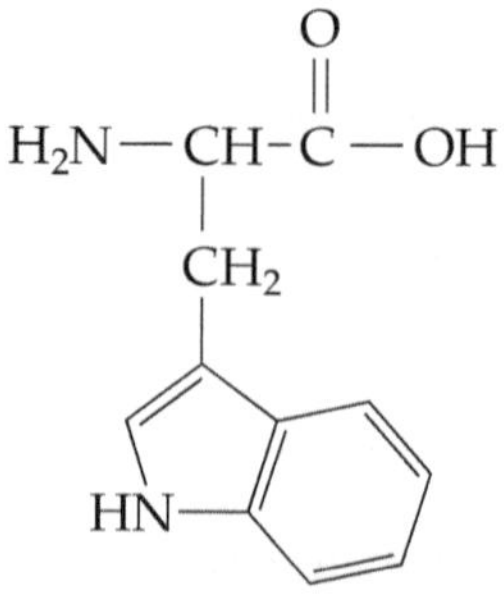

Tryptophan

23.11 NUCLEIC ACIDS: A BIOLOGICAL EXAMPLE OF AROMATICITY

How does a seed "know" what kind of plant to become? How does a fertilized ovum know how to grow into a human being? How does a cell know what part of the body it's in? The answers to such questions involve the biological molecules called **nucleic acids,** the information carrying molecules of the cell. We cannot understand the way nucleic acids function until we understand more about their structure and the key role of aromaticity.

Deoxyribonucleic acid (DNA) and **ribonucleic acid (RNA)** are the chemical carriers of an organism's genetic information. Coded in an organism's DNA is all the information that determines the nature of the organism and all the directions that are needed for producing the many thousands of different proteins required by the organism.

Just as proteins are made of amino acid units linked together, nucleic acids are made of **nucleotide** units linked together in a long chain. Each nucleotide is composed of a **nucleoside** plus phosphoric acid, H_3PO_4, and each nucleoside is composed of an aldopentose sugar plus an amine base.

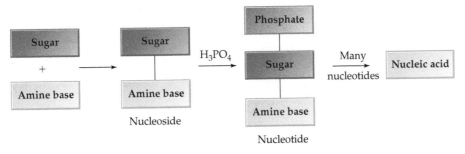

The sugar component in RNA is ribose, and the sugar in DNA is 2-deoxyribose, where "2-deoxy" means that oxygen is missing from C2 of ribose.

Ribose   **2-Deoxyribose**

Four different cyclic amine bases occur in DNA: adenine, guanine, cytosine, and thymine. Adenine, guanine, and cytosine also occur in RNA, but thymine is replaced in RNA by a related base called uracil.

Adenine (A)	**Guanine (G)**	**Cytosine (C)**	**Thymine (T)**	**Uracil (U)**
DNA	DNA	DNA	DNA	RNA
RNA	RNA	RNA		

In both DNA and RNA, the cyclic amine base is bonded to C1′ of the sugar, and the phosphoric acid is bonded to the C5′ sugar position. (Numbers with a prime superscript refer to the positions on the sugar component of the nucleotide.)

Nucleotides join together in nucleic acids by forming a bond between the phosphate group at the 5′ position of one nucleotide and the hydroxyl group on the sugar component at the 3′ position of another nucleotide (**FIGURE 23.17**).

▶ **FIGURE 23.17**

Generalized structure of a nucleic acid.

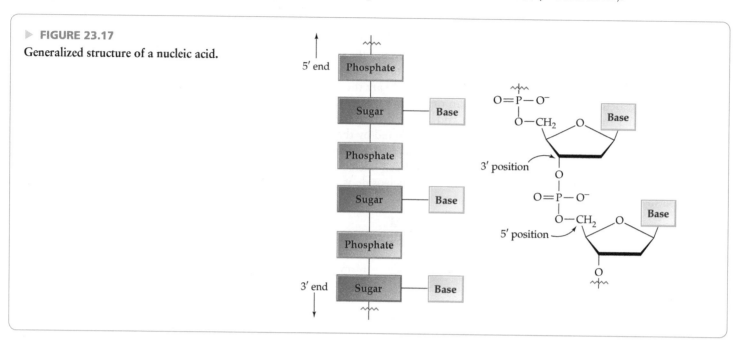

Just as the structure of a protein depends on the sequence of its individual amino acids, the structure of a nucleic acid depends on the sequence of its individual nucleotides. That sequence is described by starting at the 5′ phosphate end of the chain and identifying the bases in order. Abbreviations are used for each nucleotide: A for adenosine, G for guanosine, C for cytidine, T for thymidine, and U for uracil. Thus, a typical DNA sequence might be written as -T-A-G-G-C-T-.

Interestingly, molecules of DNA isolated from different tissues of the same species have the same proportions of nucleotides, but molecules from different species can have quite different proportions. For example, human DNA contains about 30% each of A and T and about 20% each of G and C, but the bacterium *Clostridium perfringens* contains about 37% each of A and T and only 13% each of G and C. Note that in both cases, the bases occur in pairs. Adenine and thymine are usually present in equal amounts, as are guanine and cytosine. Why should this be?

According to the **Watson–Crick model**, DNA consists of two polynucleotide strands coiled around each other in a *double helix* like the handrails on a spiral staircase. The sugar–phosphate backbone is on the outside of the helix, and the amine bases are on the inside, so that a base on one strand points directly in toward a base on the second strand. The two strands run in opposite directions and are held together by hydrogen bonds between pairs of bases. Adenine and thymine form two strong hydrogen bonds to each other, but not to G or C; G and C form three strong hydrogen bonds to each other, but not to A or T (**FIGURE 23.18**).

The two strands of the DNA double helix aren't identical; rather, they're complementary. Whenever a G base occurs in one strand, a C base occurs opposite it in the other strand because of hydrogen bonding. When an A base occurs in one strand, a T base occurs in the other strand. This complementary pairing of bases explains why A and T are always found in equal amounts, as are G and C. **FIGURE 23.19** shows how the two complementary strands coil into the double helix.

In order to fit into this complex structure, however, the bases must be planar so that they can stack on one another. This **base stacking** is the key to the stability of the DNA double helix, and the stacking interactions arise from the fact that all the bases are aromatic. These planar, aromatic bases stack exceptionally tightly for increased intermolecular interactions between bases.

Electrostatic potential maps show that the faces of the bases are relatively neutral (green), while the edges have positive (blue) and negative (red) regions.

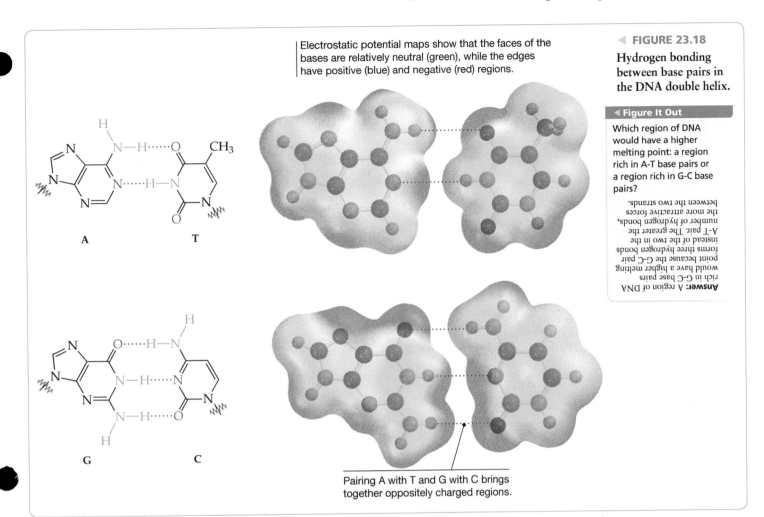

A   T

G   C

Pairing A with T and G with C brings together oppositely charged regions.

◀ **FIGURE 23.19**

The DNA double helix. The coil of the sugar–phosphate backbone is visible on the outside of the DNA double helix, while the hydrogen-bonded pairs of amine bases lie flat on the inside.

A
T
G
C

Thymine is incorporated into nucleosides to make thymidine, a building block of DNA. At first glance, this base might not appear to be aromatic due to the limitations of electron-dot structures. The nitrogen atoms appear to be sp^3 hybridized if we do not consider resonance structures. Remember the requirement of aromaticity is a planar, cyclic compound with sp^2 hybridized atoms.

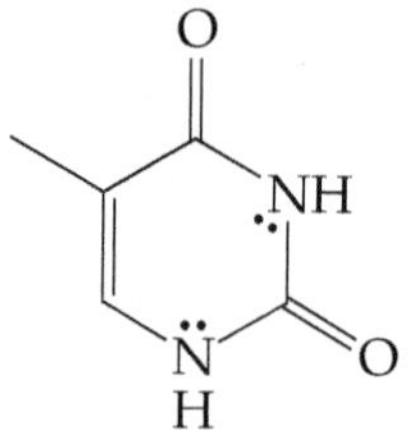

Thymine

Each nitrogen atom that is adjacent to a double bond has a lone pair and therefore a resonance structure for thymine can be drawn as shown:

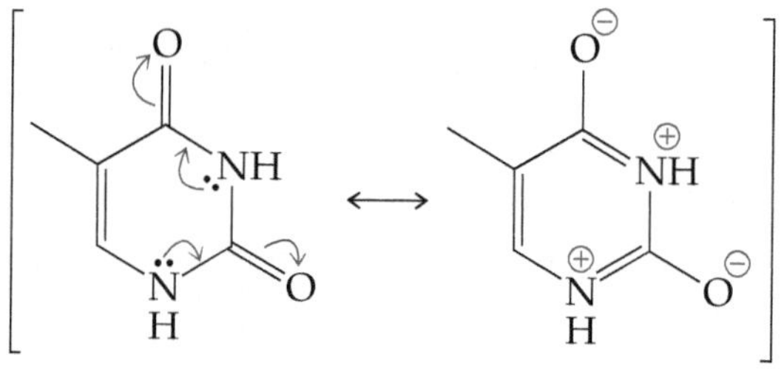

The resonance structure on the right has a six-member ring in which each atom is sp^2 hybridized. Therefore, each atom has an atomic p orbital that will overlap with p orbitals on adjacent atoms, enabling the delocalization of electrons in the π system. The three π bonds drawn contribute a total of six π electrons and therefore thymine is aromatic and planar. The other amine bases in DNA can also be shown to be aromatic by drawing resonance structures.

BIG IDEA Question 6

Go to
eText

Draw a resonance structure that shows cytosine is aromatic.

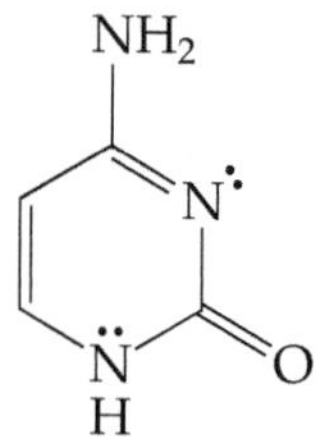

Cytosine

INQUIRY ? Why do enantiomers have different biological responses?

Have you ever wondered why medicines have specific effects in your body? Acetaminophen and morphine relieve pain, ephedrine and caffeine stimulate the central nervous system, and penicillin fights bacterial infections. The answer is that a small molecule (or a part of a large molecule) binds to a very specific receptor site on a biological molecule and either promotes or inhibits a response. Nearly every process that happens in your body, from your sense of taste and smell to nerve impulses to responses to medication occur because of molecular recognition between a biological receptor and specific target molecule. This rec-ognition depends on the molecule's shape and is often described as a lock-and-key fit. In some interactions the molecule is the same shape as the active site on the receptor, and in other cases the molecule triggers a change in the shape of the active site, allowing binding to occur. Although shape is one impor-tant criteria in binding, molecules must also be held in place in the receptor site. Some molecules bind irreversibly to a receptor by forming covalent bonds, but in many cases molecules bind reversibly to the target through weaker intermolecular forces such as electrostatic attraction, hydrogen bonds, dipole–dipole interactions, and dispersion forces.

A specific binding site in a large biological molecule such as an enzyme can distinguish between enantiomers, the right- and left-handed forms of a chiral molecule (Section 23.2). Let's consider a simple chiral molecule with a tetrahedral center and four different substituent groups (Compound 1). Compound 1 and its mirror image are enantiomers.

Examine the binding sites on the receptor shown in **FIGURE 23.20,** which match up with and therefore bind to spe-cific substituent groups in Compound 1 (Figure 23.20b). The substituent groups in the mirror image of Compound 1, how-ever, do not line up with the binding sites on the receptor. If the mirror image molecule is rotated so that substituent B aligns with its receptor site, the groups C and D do not align and binding does not occur (Figure 23.20c).

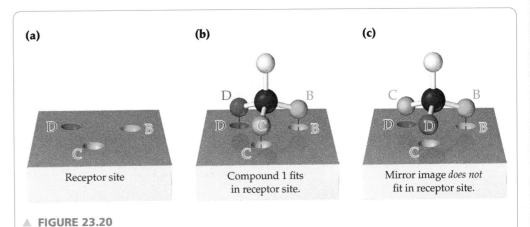

(a)   (b)   (c)

Receptor site   Compound 1 fits in receptor site.   Mirror image *does not* fit in receptor site.

▲ **FIGURE 23.20**

A schematic diagram of a receptor site capable of interacting with Compound 1 but not its mirror image.

Figure 23.20 explains why enantiomers have different bio-logical properties. Just like a right-handed glove only fits the right hand, chiral molecules have specific shapes that match only complementary-shaped receptor sites. The mirror-image forms of the molecules can't fit into the receptor sites and thus don't elicit the same biological response. The biological consequences of molecular shape can be dramatic. Look at the structures of dextromethorphan and levomethorphan, for instance. Both of these molecules bind to opioid receptors like the one depicted in the chapter-opening image. (The Latin pre-fixes *dextro-* and *levo-* mean "right" and "left," respectively.) Dextromethorphan is a common cough suppressant found in many over-the-counter cold medicines, but its mirror image, levomethorphan, is a powerful narcotic pain reliever with effects similar to morphine. The enantiomers have the same chemical formula but different "handedness," which leads to different biological responses. A substance called carvone is another example of the effect of chirality. Left-handed carvone occurs in mint plants and has the characteristic odor of spear-mint, while right-handed carvone occurs in several herbs and has the odor of caraway seeds. Again, the two structures are the same except for their handedness, yet they have entirely different odors.

Precise molecular shape is critically important to every liv-ing organism. Many biological molecules such as carbohydrates and amino acids are chiral, and usually only one of the enantio-mers occurs naturally in a given organism. The enantiomer can often be made in the laboratory but does not occur naturally. Numerous chemical interactions in living systems are governed by complementarity between chiral molecules and their chiral receptors. Scientists can experimentally determine the three-dimensional atomic structure of a molecule such as the opi-oid receptor (shown in the chapter-opening photo) that binds

▶ Chiral molecules have enantiomers, which are nonidentical mirror images. Enantiomers have different "handedness" and often have dramatically different biological responses.

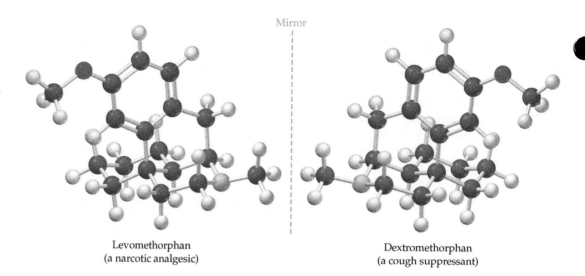

Levomethorphan
(a narcotic analgesic)

Dextromethorphan
(a cough suppressant)

dextromethorphan and levomethorphan along with dozens of other legal and illegal drugs. The detailed atomic shape paves the way for the design of safer and more effective drugs.

PROBLEM 23.27   Asparagine is a naturally occurring amino acid that exists in two mirror-image forms. One form tastes bitter, while the other form tastes sweet.

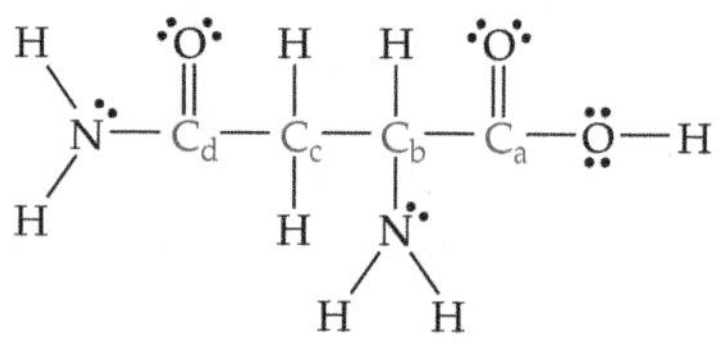

(a) Which of the carbon atoms (labeled a–d) is a chiral center?
(b) How many σ bonds and how many π bonds does the asparagine molecule contain?
(c) Indicate the hybridization of carbon atoms (labeled a–d).
(d) Circle the atoms that can participate in hydrogen bonding with the asparagine receptor site.

PROBLEM 23.28   The left- and right-handed carvone molecules fit different smell receptors due to differences in shape.

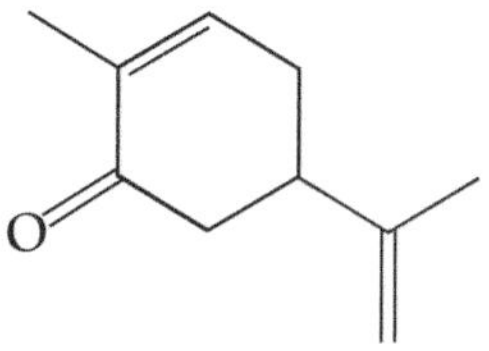

Carvone

(a) What is the chemical formula for carvone?
(b) Which of the carbon atoms is a chiral center?
(c) Is carvone aromatic?

PROBLEM 23.29   Naproxen, the active ingredient in Aleve, is a nonsteroidal anti-inflammatory drug.

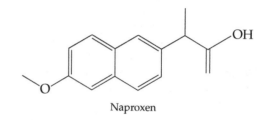

Naproxen

(a) Which of the carbon atoms is a chiral center?
(b) Is Naproxen aromatic?
(c) Draw a resonance structure for Naproxen.
(d) Circle the atoms that can participate in hydrogen bonding with the receptor site.

STUDY GUIDE

Section	Concept Summary	Learning Objectives	Test Your Understanding
23.1 Organic Molecules and Their Structures: Constitutional isomers	**Organic chemistry** is the study of carbon compounds. The simplest compounds are the **alkanes**, which contain only carbon and hydrogen (**hydrocarbons**) and have only single bonds. **Straight-chain alkanes** have all their carbons connected in a row, **branched-chain alkanes** have a branched connection of atoms in their chain, and **cycloalkanes** have a ring of carbon atoms. **Constitutional isomers** are compounds that have the same chemical formula but different connections of atoms.	**23.1** Represent the chemical structure of alkanes as condensed structures or line drawings. **23.2** Identify and draw constitutional isomers of alkanes.	Worked Example 23.1; Problems 23.42–23.43 Worked Example 23.2; Problems 23.44–23.45
23.2 Stereoisomers: Chiral Molecules	**Stereoisomers** are molecules that have the same chemical formula and same connections of atoms but different arrangements of the atoms in space. **Chiral molecules** do not have a plane of symmetry, and the mirror image of a chiral molecule cannot be superimposed on the original molecule.	**23.3** Identify chiral centers in a molecule and classify molecules as chiral or achiral.	Worked Example 23.3; Problems 23.49–23.52
23.3 Families of Organic Compounds: Functional Groups	The more than 45 million known organic compounds can be organized into families according to the functional groups they contain. A **functional group** is an atom or group of atoms that has characteristic chemical behavior.	**23.4** Identify and name functional groups in organic molecules. **23.5** Represent molecules with functional groups using condensed structures and line drawings.	Problems 23.33–23.34 Worked Example 23.4; Problems 23.58, 23.61, 23.64, 23.66
23.4 Carbohydrates: A Biological Example of Isomers	**Carbohydrates** are polyhydroxy aldehydes and ketones. Simple carbohydrates, or **monosaccharides**, can't be hydrolyzed to smaller molecules; complex carbohydrates such as starch and cellulose contain many simple sugars linked together. Monosaccharides exist in different isomeric forms including constitutional isomers, **enantiomers**, and **anomers**.	**23.6** Classify a monosaccharide as an aldose or ketose and draw a typical open chain structure. **23.7** Classify the type of isomerism in monosaccharides.	Problems 23.71–23.72 Problems 23.74–23.75
23.5 Valence Bond Theory and Cis–Trans Isomerism	**Cis–trans isomers** are possible for substituted alkenes because of the lack of rotation about the carbon–carbon double bond. The cis isomer has two substituents on the same side of the double bond, and the trans isomer has two substituents on opposite sides.	**23.8** Draw orbital overlap pictures to describe bonding in organic compounds. **23.9** Identify and draw cis–trans isomers.	Worked Example 23.6; Problems 23.76, 23.78–23.79 Worked Example 23.7; Problems 23.80–23.85
23.6 Lipids: A Biological Example of Cis–Trans Isomerism	**Lipids** are the naturally occurring organic molecules that dissolve in a nonpolar solvent. Animal fats and vegetable oils are **triacylglycerols**—esters of glycerol with three long-chain **fatty acids**. The fatty acids are unbranched, have an even number of carbon atoms, and may be either saturated or unsaturated.	**23.10** Recognize and draw the structure of triacyl glycerols. **23.11** Relate lipid structure to physical properties.	Problems 23.88–23.89 Problems 23.92–23.93

Section	Concept Summary	Learning Objectives	Test Your Understanding
23.7 Formal Charge and Resonance in Organic Compounds	Resonance structures and formal charge are important concepts in understanding the structure and reactivity of organic and biological compounds. Formal charge must be indicated in an organic line drawing for the structure to be correct. Formal charges are used to evaluate relative contributions of different resonance structures to the resonance hybrid.	**23.12** Use curved arrows to draw resonance structures and specify formal charges in organic line drawings.	Worked Examples 23.8–23.9; Problems 23.96, 23.98, 23.100, 23.102
23.8 Conjugated Systems	**Conjugated systems** consist of more than two p orbitals on adjacent atoms. The p orbitals overlap and delocalize electrons over all the atoms in the π system. Resonance structures can be drawn for conjugated systems and atoms adopt the hybridization that lowers the overall energy. Electrons in p orbitals that are part of conjugated systems are delocalized and electrons that are in hybrid orbitals are localized.	**23.13** Draw orbital overlap pictures for conjugated systems and identify localized and delocalized electrons.	Worked Example 23.10; Problems 23.112–23.113
		23.14 Identify conjugated systems and the hybridization of atoms included in the system.	Worked Example 23.11; Problems 23.108–23.109
23.9 Proteins: A Biological Example of Conjugation	Proteins are large biomolecules consisting of α-amino acids linked together by amide, or peptide bonds. Twenty amino acids are commonly found in proteins.	**23.15** Identify amino acids and draw structures for peptides.	Worked Example 23.12; Problems 23.115–23.117
23.10 Aromatic Compounds and Molecular Orbital Theory	**Aromatic** compounds are a class of compounds that are cyclic, planar, have adjacent p orbitals around a ring, and have $(4n + 2)\pi$ electrons. Aromatic compounds have extra stability due to paired electrons in bonding molecular orbitals.	**23.16** Predict if an organic compound is aromatic.	Worked Example 23.13; Problems 23.120–23.122, 23.124
23.11 Nucleic Acids: A Biological Example of Aromaticity	**Deoxyribonucleic acid (DNA)** and **ribonucleic acid (RNA)** are the chemical carriers of an organism's genetic information. They are made up of **nucleotides**, linked together to form a long chain. Each nucleotide consists of a cyclic amine base linked to C1 of a sugar, with the sugar in turn linked to phosphoric acid. The sugar component in RNA is ribose; the sugar in DNA is 2-deoxyribose. The bases in DNA are adenine (A), guanine (G), cytosine (C), and thymine (T); the bases in RNA are adenine, guanine, cytosine, and uracil (U). Molecules of DNA consist of two complementary nucleotide strands held together by hydrogen bonds and coiled into a double helix.	**23.17** Draw structures showing how different components of the nucleic acid are joined, and write the DNA sequence that is complementary to a specified strand.	Problems 23.128–23.130
		23.18 Draw resonance structures to show that DNA bases are aromatic.	Problems 23.132–23.133

KEY TERMS

PRACTICE TEST

After studying this chapter, you can assess your understanding with these practice test questions, which are correlated with chapter learning objectives. If you answer a question incorrectly, refer to the learning objectives in the end-of-chapter Study Guide for assistance. The Study Guide provides a conceptual summary, references a Worked Example to model how to solve the problem, and gives additional problems for more practice.

1. Which of the following pairs are constitutional isomers? (LO 23.1, 23.2)

(a)

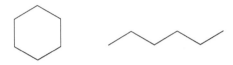

(b)

(c)

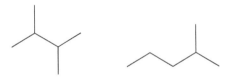

(d)

2. Ibuprofen is a chiral drug. Which circled atom represents a chiral center? (LO 23.3)

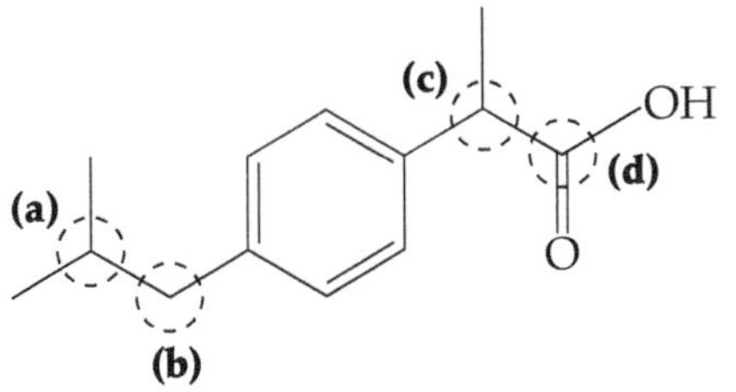

Ibuprofen

3. What is the molecular formula for the following compound, and what functional groups are present? (LO 23.4, 23.5)

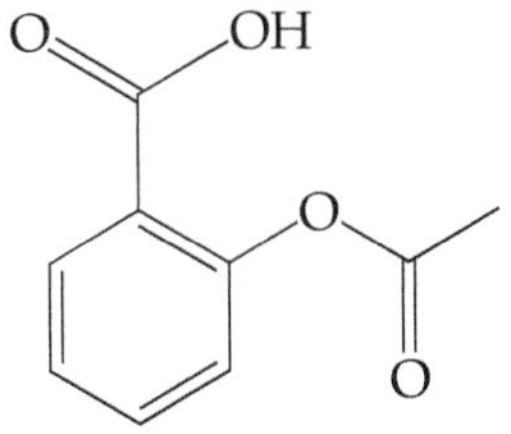

(a) $C_9H_8O_4$; aromatic ring, carboxylic acid, ketone, ether
(b) $C_8H_{10}O_4$; aromatic ring, ketone, alcohol, ester
(c) $C_8H_{10}O_4$; aromatic ring, ester, ketone, alcohol
(d) $C_9H_8O_4$; aromatic ring, ester, carboxylic acid

4. Which of the following structures is an aldopentose? (LO 23.6)

(a)

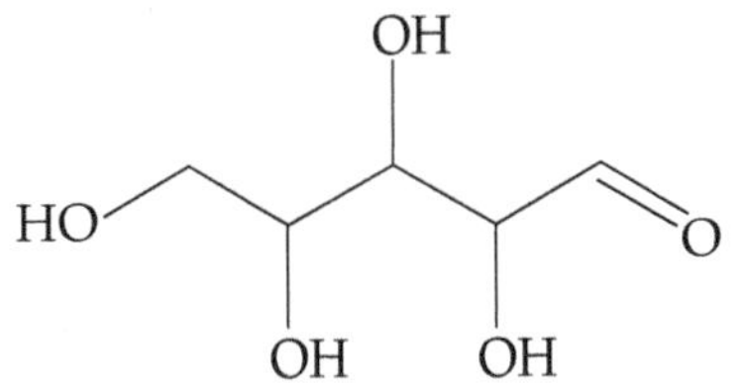

(b)

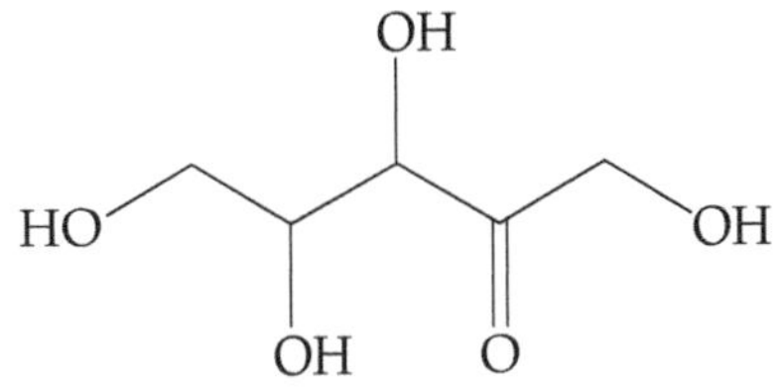

(c)

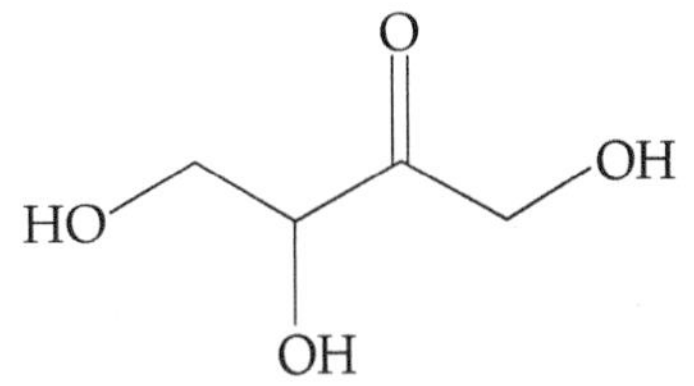

(d)

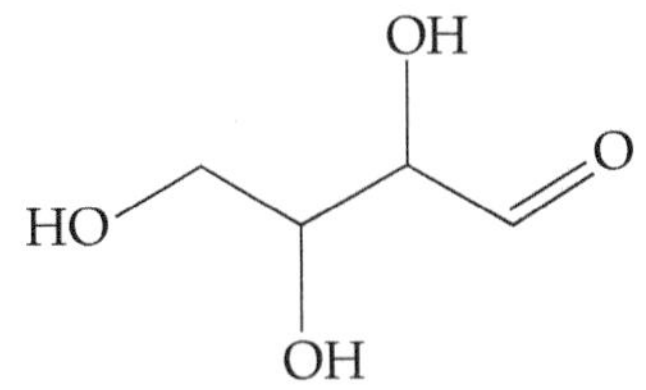

5. Which orbitals overlap to make the carbon–carbon bond in ascorbic acid indicated by the arrow? (LO 23.8)

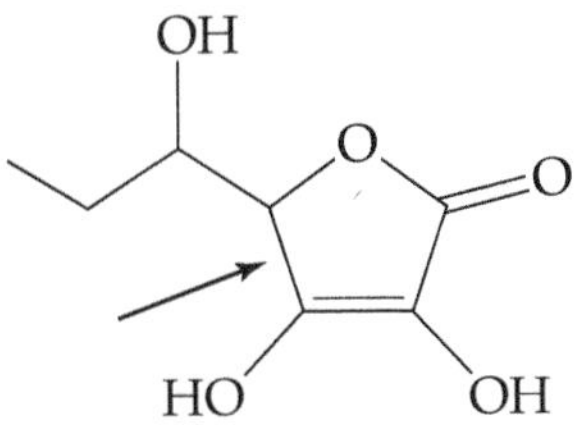

Ascorbic acid

(a) sp^2 with sp^2
(b) sp^3 with sp^3
(c) sp^3 with sp^2
(d) sp^2 with $2p$

6. Which of the following pairs represent cis–trans isomers?
   (LO 23.9)

   (a)

   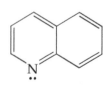

   (b)

   (c)

   (d)

7. Which of the following structures has a lone pair of electrons
   on nitrogen that is delocalized in a π system? **(LO 23.13, 23.14)**

   (a)   **(b)**

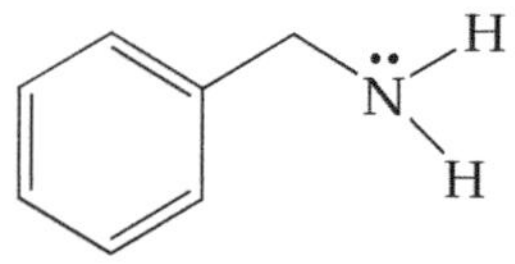

   (c)

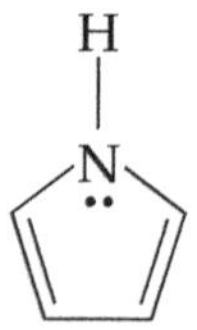

   (d) The lone pair on the nitrogen atom in all these structures
   is delocalized.

8. Which structure shows the dipeptide (Cys-Tyr)? **(LO 23.15)**

   (a)

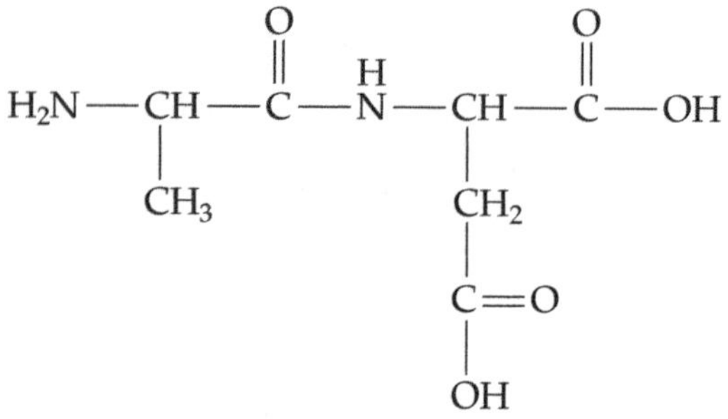

(b)

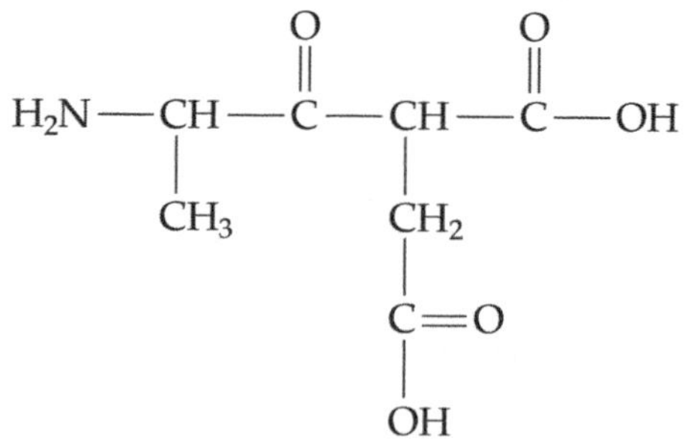

$$H_2N-CH-\overset{\overset{\displaystyle O}{\|}}{C}-CH-\overset{\overset{\displaystyle O}{\|}}{C}-OH$$

with CH₃ below first CH and CH₂—C(=O)—OH below second CH

(c)

$$H_2N-CH-\overset{\overset{\displaystyle O}{\|}}{C}-CH-\overset{\overset{\displaystyle O}{\|}}{C}-OH$$

with CH₂—SH and CH₂—(aromatic ring)—OH

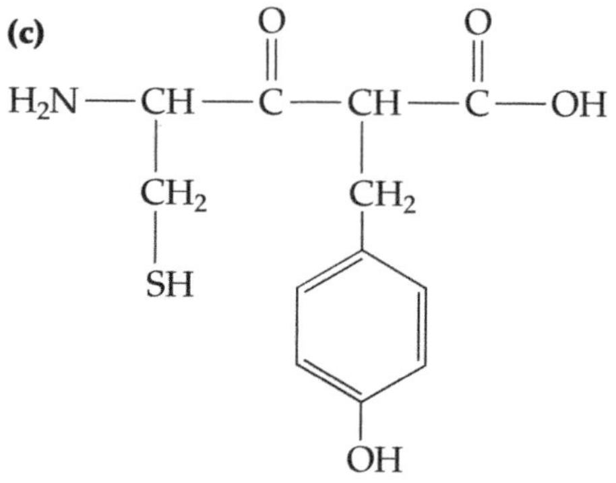

(d)

$$H_2N-CH-\overset{\overset{\displaystyle O}{\|}}{C}-\overset{H}{N}-CH-\overset{\overset{\displaystyle O}{\|}}{C}-OH$$

with CH₂—SH and CH₂—(aromatic ring)—OH

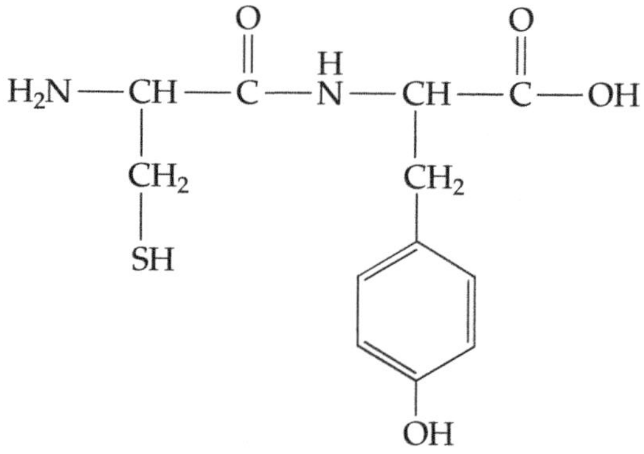

9. Which of the following molecules is aromatic? **(LO 23.16)**

   (a)   **(b)**

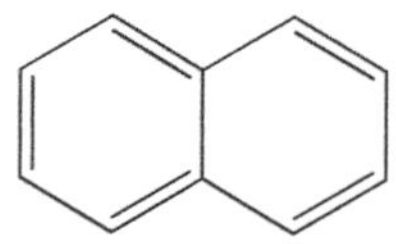

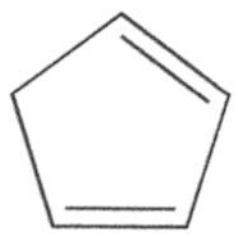

   (c) 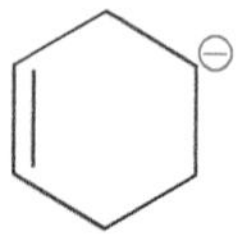   **(d)**

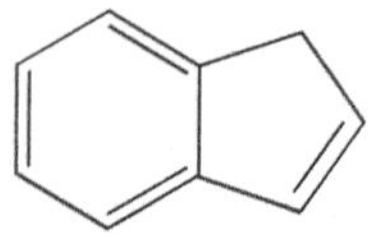

10. Resonance arrows are shown for the DNA base guanine. Which of the following is the resonance structure that results from the arrows indicating electron rearrangement? Be sure to label formal charges on atoms. **(LO 23.12, 23.18)**

(a)   **(b)**   **(c)**

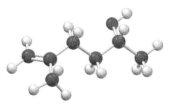

(d) These arrows do not result in a valid resonance structure.

Answers:
1. c, 2. c, 3. d, 4. a, 5. c, 6. b, 7. c, 8. d, 9. a, 10. b

Mastering Chemistry   provides end-of-chapter exercises, feedback-enriched tutorial problems, animations, and interactive activities to encourage problem-solving practice and deeper understanding of key concepts and topics.

RAN   *Randomized in Mastering Chemistry*

CONCEPTUAL PROBLEMS

Problems 23.1–23.29 appear within the chapter.

23.30 Convert the following model into a condensed structure and line drawing. Draw the structures of two isomeric compounds.

23.31 Convert the following models into a condensed structures and line drawings.

(a)   **(b)**

23.32 One of the following two molecules is chiral and can exist in two mirror-image forms; the other does not. Which is which? Why?

(a)   **(b)**

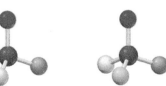

23.33 Identify the functional groups in each of the following compounds.

(a)   **(b)**

23.34 Identify the functional groups in cocaine.

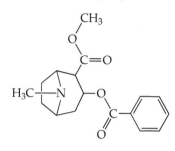

Cocaine

23.35 The following structure is a representation of tryptophan, an amino acid constituent of proteins. Complete the structure by showing where lone pairs are located. (Red = O, gray = C, blue = N, ivory = H.) Give the line drawing and molecular formula.

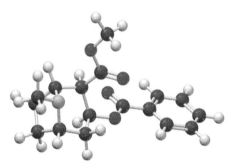

Tryptophan

23.36 The following structure is a representation of thymine, a constituent of DNA. Complete the structure by showing where lone pairs are located. (Red = O, gray = C, blue = N, ivory = H.) Give the line drawing and molecular formula.

Thymine

23.37 Draw three resonance forms for naphthalene, showing the positions of the double bonds.

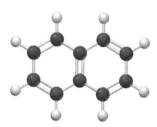

23.38 Identify the following amino acids. (Yellow = S.)

(a)

(b)

23.39 Identify the following dipeptide.

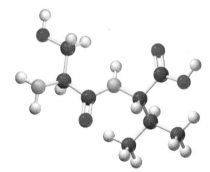

SECTION PROBLEMS

Organic Molecules and Constitutional Isomers (Section 23.1)

23.40 What is the difference between a straight-chain alkane and a branched-chain alkane?

23.41 What is the difference in the chemical formula between a straight-chain alkane and cycloalkane, each with the same number of carbon atoms?

23.42 Draw a straight-chain alkane with 6 carbon atoms as a condensed structure and as a line drawing.

23.43 Draw the straight-chain alkane with 7 carbon atoms as condensed structure and as a line drawing.

23.44 Decide whether the following pairs of alkanes are constitu-
RAN   tional isomers. If not, explain why.

(a)

(b)

(c)

23.45 Decide whether the following pairs of alkanes are constitu-
RAN   tional isomers. If not, explain why.

(a)

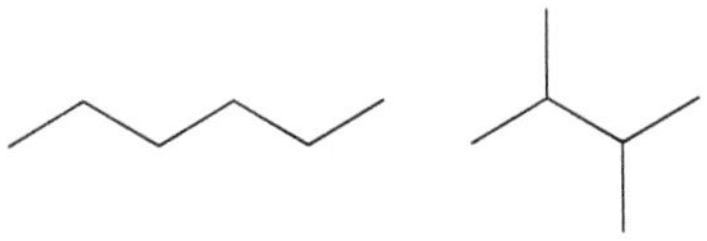

(b)

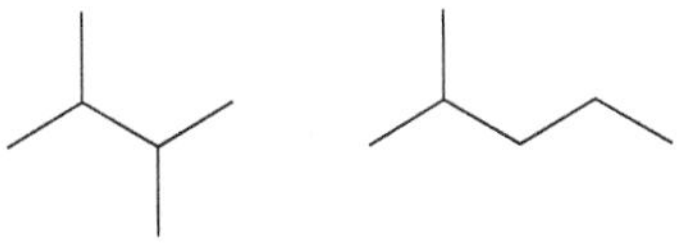

(c)

23.46 The following line drawings represent two amino acids, leucine and isoleucine. Are these two compounds constitutional isomers? Explain your reasoning.

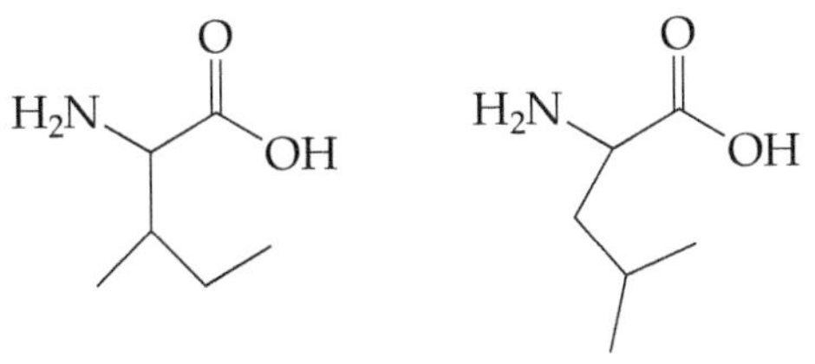

Isoleucine   Leucine

23.47 Are these two compounds constitutional isomers? Explain your reasoning.

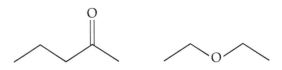

Stereoisomers: Chiral Molecules (Section 23.2)

23.48 Which of these objects are chiral (exhibit handedness)?
    (a) Basketball   (b) Foot
    (c) Golf club   (d) Coffee mug
    (e) Seashell with a helical twist

23.49 Identify which of the following molecules are chiral and can exist in two mirror-image forms.
    (a) Lactic acid: formed during muscle contraction

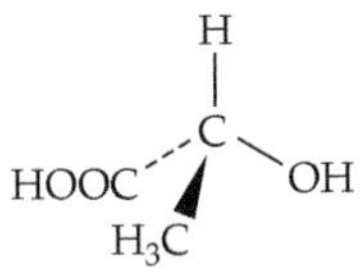

    (b) Acetic acid: the main component of vinegar

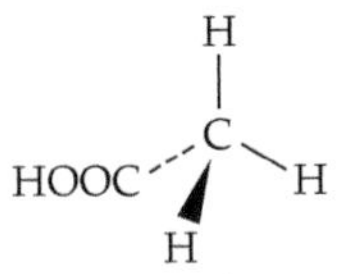

    (c) Propanoic acid: a preservative that inhibits the growth of mold

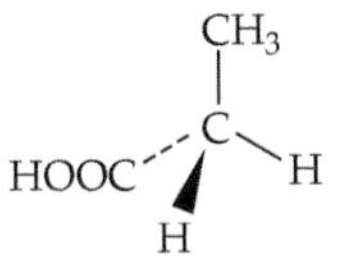

23.50 Are the following molecules chiral? If so, identify the chiral center(s).
    (a)   **(b)**

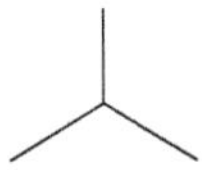

    (c)   **(d)**

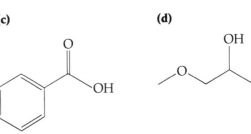

23.51 Are the following molecules chiral? If so, identify the chiral center(s).

(a)

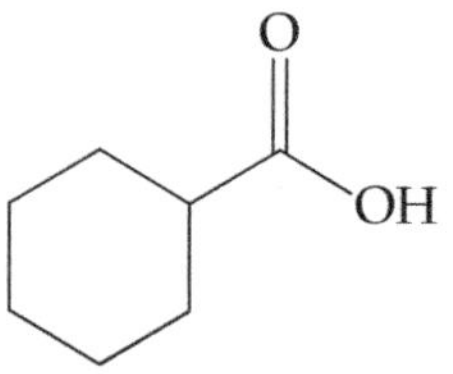

(b)

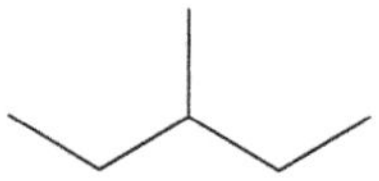

(c)

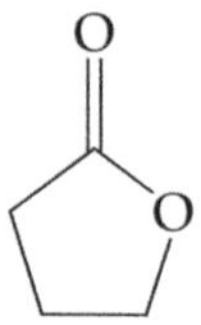

(d)

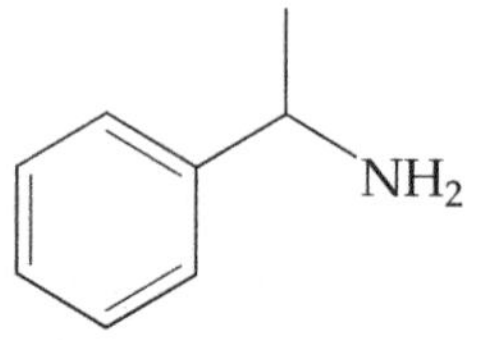

23.52 The structure for penicillamine, a chiral drug, is shown. One enantiomer is an effective treatment for chronic arthritis, and the other enantiomer is highly toxic. Identify the chiral center(s) in penicillamine.

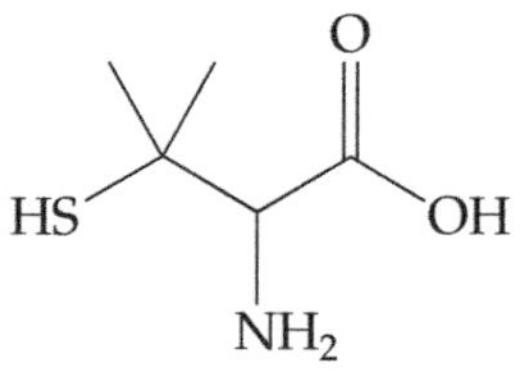

23.53 The structure for naproxen, a chiral drug, is shown. One enantiomer is an effective treatment for inflammation, and the other is a liver toxin. Identify the chiral center(s) in naproxen.

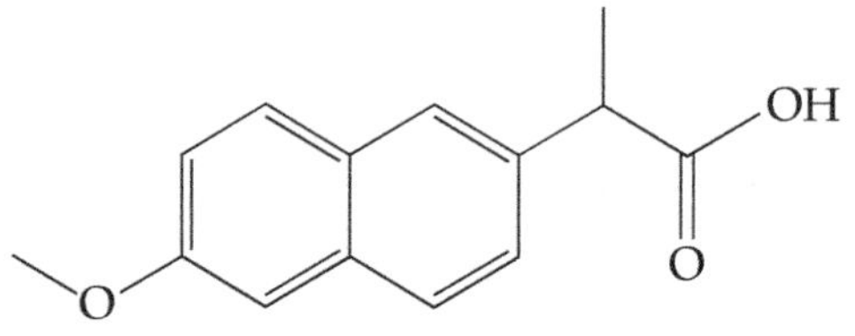

Functional Groups (Section 23.3)

23.54 What are functional groups, and why are they important?

23.55 Describe the structure of the functional group in each of the following families.
   (a) Alkene   (b) Alcohol
   (c) Ester   (d) Amine

23.56 Propose structures and draw condensed formulas for molecules that meet the following descriptions.
   (a) A ketone with the formula $C_5H_{10}O$
   (b) An ester with the formula $C_6H_{12}O_2$
   (c) A compound with formula $C_2H_5NO_2$ that is both an amine and a carboxylic acid

23.57 Give line drawings for each of the following molecular formulas. You may have to use rings and/or multiple bonds in some instances.
   (a) C_2H_7N (b) C_4H_8
   (c) C_2H_4O (d) CH_2O_2

23.58 Give a line drawing and molecular formula for the following.
   (a) A linear alkyne with three carbon atoms and one triple bond
   (b) A linear alkene with four carbon atoms and one double bond
   (c) An alkene with five carbons in a ring and two double bonds

23.59 Give a line drawing and molecular formula for the following.
   (a) A linear alkane with four carbon atoms
   (b) An alkene with six carbon atoms in a ring and one double bond
   (c) A linear alkyne with six carbon atoms and two triple bonds

23.60 How many dienes (compounds with two double bonds) are there with the formula C_5H_8? Draw as many structures as you can.

23.61 Propose structures and draw condensed formulas of the three isomers with the formula C_3H_8O.

23.62 If someone reported the preparation of a compound with the formula C_3H_9, most chemists would be skeptical. Why?

23.63 What is wrong with each of the following structures?
   (a) $CH_3 = HCH_2CH_2OH$

   (b)

$$CH_3CH_2CH{=}\overset{\overset{\textstyle O}{\|}}{C}CH_3$$

   (c) $CH_3CH_2C{\equiv}CH_2CH_3$

23.64 Draw the complete electron-dot structure and determine the molecular formulas for the following alkanes with functional groups.

(a) Alcohol   (d) Ether

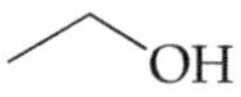

(b) Amine   (e) Halogen

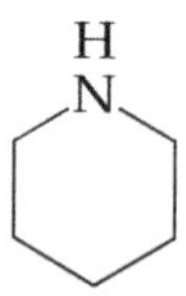

(c) Cyclic amine   (f) Ketone

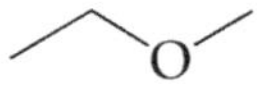

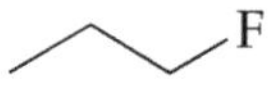

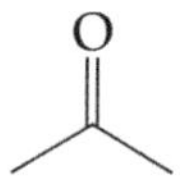

23.65 Draw the complete electron-dot structure and determine the molecular formulas for the following alkanes with functional groups.

(a)

Valine, an amino acid

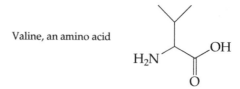

(b)

Methyl tert-butyl ether
(MTBE), a gasoline additive

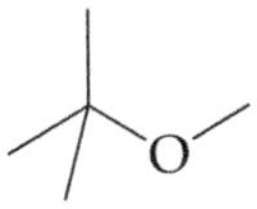

(c)

Barbituric acid, parent
compound of barbituate
drugs

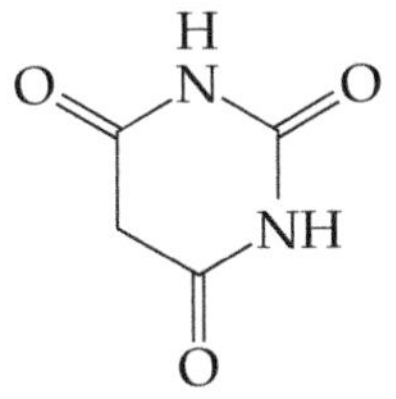

23.66 Draw the complete electron-dot structure and determine the
molecular formula for the following organic molecules.

(a)   **(b)** 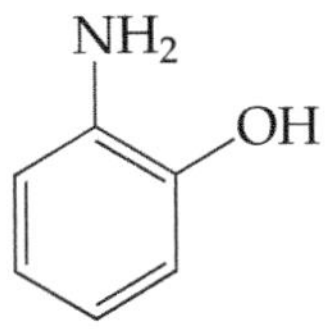   **(c)**

NH₂

OH

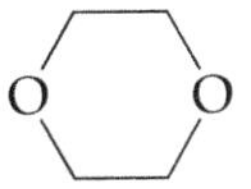

23.67 Draw the complete electron-dot structure and determine the
molecular formula for the following organic molecules.

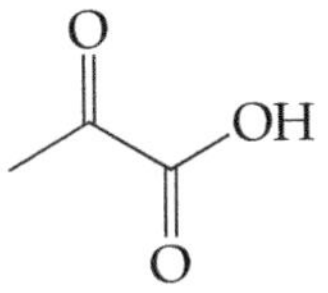

(a)   **(b)** 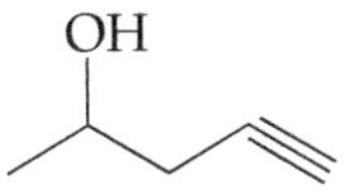   **(c)**

Cl   Cl   OH

H₂N   NH₂

O

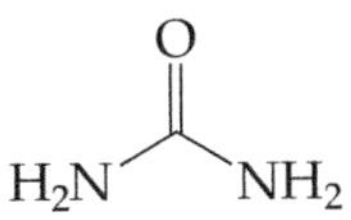

23.68 Methylphenidate is the active ingredient in Ritalin, a medica-
RAN   tion prescribed for attention deficit disorder.

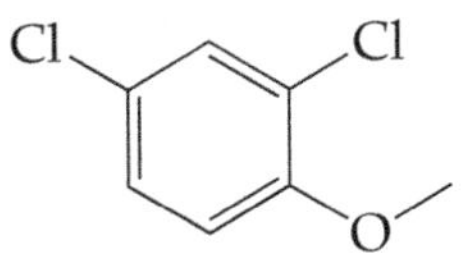

OCH₃

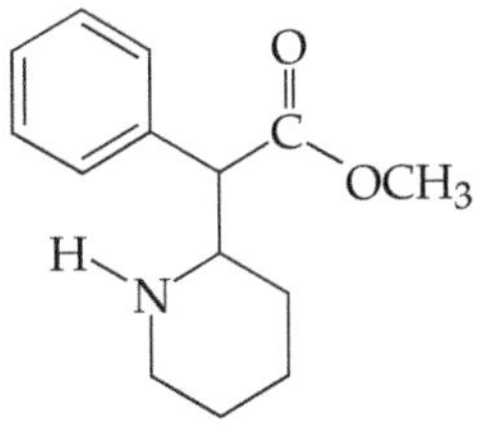

Methylphenidate

H–N

What functional groups are present in methylphenidate?

23.69 Benzyne, C_6H_4, is a highly energetic and reactive molecule.
The carbon atoms are connected in six-membered ring.
Give a line drawing for benzyne. What functional group is
present?

Carbohydrates (Section 23.4)

23.70 What is the structural difference between an aldose and a
ketose?

23.71 Classify each of the following carbohydrates by indicating
the nature of its carbonyl group and the number of carbon
atoms present. For example, glucose is an aldohexose.

(a)

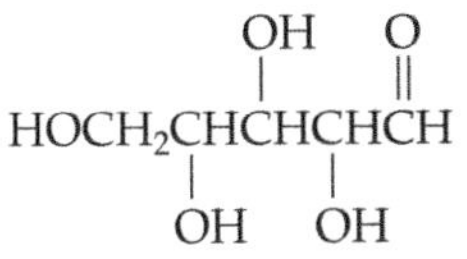

$$\underset{\text{OH OH}}{\underset{|\quad\;\; |}{\overset{\overset{\text{OH}}{|}}{\text{HOCH}_2\text{CHCHCHCH}}}}\overset{\overset{\text{O}}{\|}}{}$$

(b)

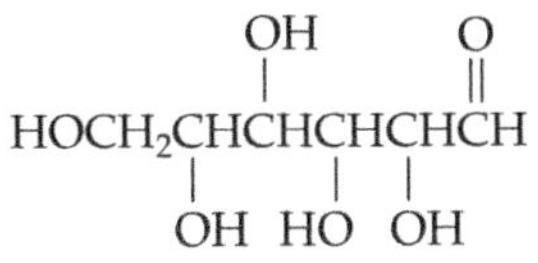

$$\underset{\text{OH HO OH}}{\underset{|\;\;\; |\;\; |}{\overset{\overset{\text{OH}}{|}}{\text{HOCH}_2\text{CHCHCHCHCH}}}}\overset{\overset{\text{O}}{\|}}{}$$

(c)

$$\underset{\text{HO OH}}{\underset{|\;\; |}{\overset{\overset{\text{OH}}{|}}{\text{HOCH}_2\text{CHCHCHCCH}_2\text{OH}}}}\overset{\overset{\text{O}}{\|}}{}$$

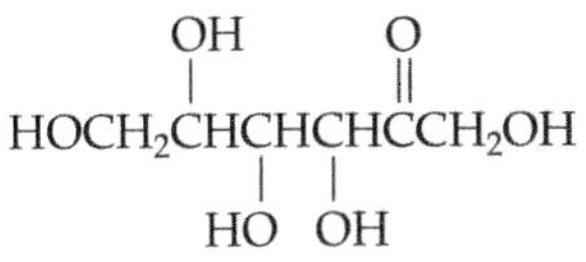

23.72 Write the open-chain structure of a ketotetrose.

23.73 Write the open-chain structure of a four-carbon deoxy sugar.

23.74 Are the following carbohydrates isomers? If so, classify the
type of isomer.

(a)

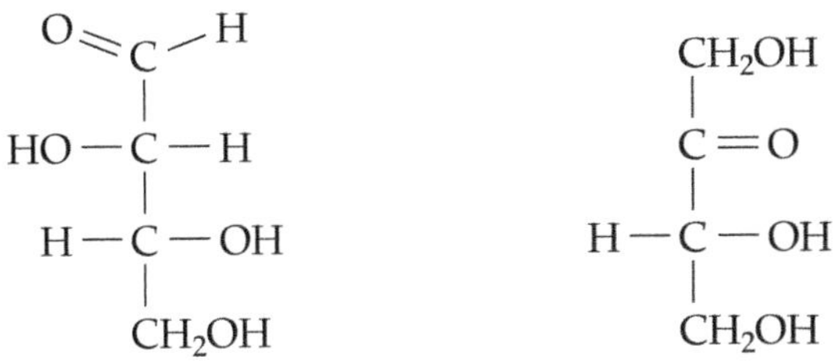

$$O=\underset{|}{\overset{}{C}}-H$$
$$HO-\underset{|}{\overset{}{C}}-H$$
$$H-\underset{|}{\overset{}{C}}-OH$$
$$CH_2OH$$

$$CH_2OH$$
$$C=O$$
$$H-\underset{|}{\overset{}{C}}-OH$$
$$CH_2OH$$

(b)

$$\underset{\text{OH}\quad\text{OH}}{CH_2OH\;\;\overset{O}{\diagup}\quad H}$$

$$\underset{\text{OH}\quad\text{OH}}{CH_2OH\;\;\overset{O}{\diagup}\quad OH}$$

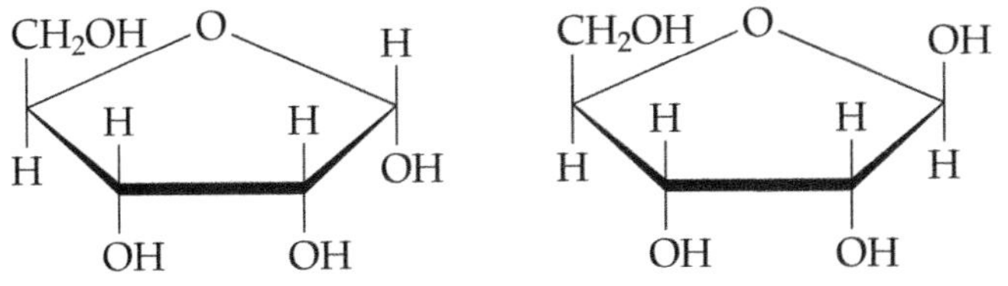

23.75 Are the following carbohydrates isomers? If so, classify the
type of isomer.

(a)

$$O=\underset{|}{\overset{}{C}}-H$$
$$HO-\underset{|}{\overset{}{C}}-H$$
$$H-\underset{|}{\overset{}{C}}-OH$$
$$CH_2OH$$

$$O=\underset{|}{\overset{}{C}}-H$$
$$H-\underset{|}{\overset{}{C}}-OH$$
$$CH_2OH$$

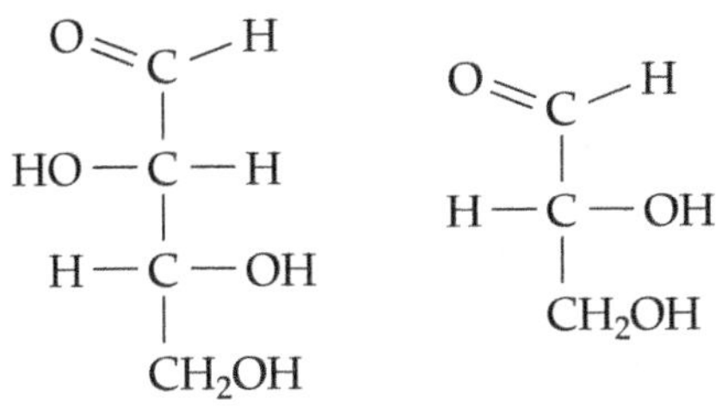

(b)

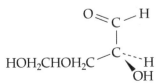

Valence Bond Theory, Cis–Trans Isomers (Section 23.5)

23.76 (a) Draw an orbital overlap picture of ethane (CH_3CH_3), assuming both carbon atoms are sp^3 hybridized. What are the C—H bond angles in this drawing?

(b) Draw an orbital overlap picture of ethane, assuming both carbon atoms are sp^2 hybridized. What are the C—H bond angles in this drawing?

(c) The real structure of ethane is like the picture you drew in part (a). Use VSEPR theory to explain why your picture from part (a) makes a more stable molecule than your picture in part (b).

23.77 Draw an orbital overlap picture of methane (CH_4) with the carbon in sp^3 hybridization and an orbital overlap picture of methane with carbon in sp^2 hybridization. Explain why the carbon atom of methane is more stable in sp^3 hybridization than in sp^2 hybridization.

23.78 In order to convert a double bond in the cis configuration into a double bond in the trans configuration, the π bond must be broken. This would involve a transition state in which the two p orbitals do not overlap at all. Draw a simplified orbital overlap picture of the π bond broken by filling in the p orbitals in the transition state:

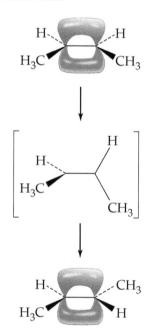

23.79 Draw a simplified orbital overlap picture of this molecule, and use it to explain why all the carbon and hydrogen atoms are in the same plane.

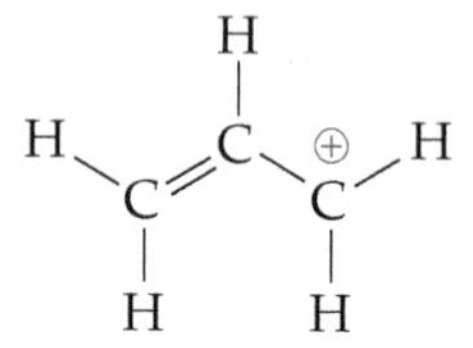

23.80 Palmerolide A is a chemical which has been isolated from marine organisms and shown to be able to kill melanoma cancer cells. A structure closely related to palmerolide A is shown. For each of its carbon–carbon double bonds, indicate whether they are cis, trans, or neither.

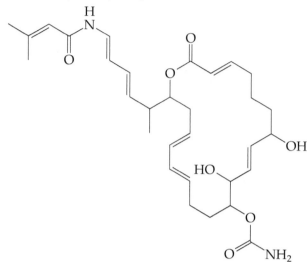

23.81 Unlike palmerolide A (Problem 23.80), which has a large ring, a cyclic molecule with only six atoms can only have a cis double bond, but not a trans double bond. Give a line drawing of a cyclic molecule with six carbon atoms and a cis double bond. Explain why it cannot have a trans double bond.

23.82 Which of the following compounds exhibit cis–trans isomerism? If isomers exist, show the line drawing for the corresponding cis or trans isomer to the compound shown.

(a) (b) (c)

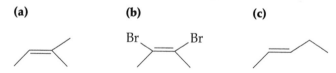

23.83 Which of the following compounds exhibit cis–trans isomerism? If isomers exist, show the line drawing for the corresponding cis or trans isomer to the compound shown.

(a) (b) (c)

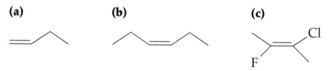

23.84 Which of the following compounds are capable of cis–trans isomerism?

(a) 1-Hexene (b) 2-Hexene (c) 3-Hexene

23.85 Which of the following compounds are capable of cis–trans isomerism?

(a)

$$CH_3CHCH=CHCH_3$$
(with CH_3 substituent)

(b)

$$CH_3CH_2CHCH_3$$
(with $CH=CH_2$ substituent)

(c)

$$CH_3CH=CHCHCH_2CH_3$$
(with Cl substituent)

Lipids (Section 23.6)

23.86 What is a fatty acid?

23.87 What does it mean to say that fats and oils are triacylglycerols?

23.88 Draw the structure of glycerol myristate, a fat made from glycerol and three myristic acid molecules (see Table 23.3).

23.89 Show the structure of glyceryl trioleate, a fat molecule whose components are glycerol and three oleic acid units.

23.90 Spermaceti, a fragrant substance isolated from sperm whales, was a common ingredient in cosmetics until its use was banned in 1976 to protect the whales from extinction. Chemically, spermaceti is cetyl palmitate, the ester of palmitic acid (see Table 23.3) with cetyl alcohol (the straight-chain C_{16} alcohol). Show the structure of spermaceti.

23.91 There are two isomeric fat molecules whose components are glycerol, one palmitic acid, and two stearic acids (see Table 23.3). Draw the structures of both, and explain how they differ.

23.92 Which of the fatty acids would most contribute to a triacylg-
RAN lyceride being an oil, rather than a fat, at room temperature?

(a)

(b)

(c)

23.93 Which of the fatty acids would most contribute to a triacylg-
RAN lyceride being a fat, rather than an oil, at room temperature?

(a)

(b)

(c)

23.94 Jojoba wax, used in candles and cosmetics, is partially composed of the ester of stearic acid and a straight-chain C_{22} alcohol. Draw the structure of this ester.

23.95 One of the constituents of the carnauba wax used in floor and furniture polish is an ester of a C_{32} straight-chain alcohol with a C_{20} straight-chain carboxylic acid. Draw the structure of this ester.

Formal Charge and Resonance (Section 23.7)

23.96 Draw the electron-dot structure resulting from the curved arrows shown. If the arrow is valid, give the line drawing of the resonance structure. For incorrect structures, explain why the structure does not exist.

(a)   **(b)**   **(c)**

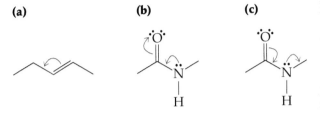

23.97 Draw the electron-dot structure resulting from the curved arrows shown. If the arrow is valid, give the line drawing of the resonance structure. For incorrect structures, explain why the structure does not exist.

(a)   **(b)**   **(c)**

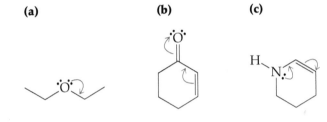

23.98 Draw the electron-dot structure resulting from the curved arrows shown. If the arrow is valid, give the line drawing of the resonance structure. For incorrect structures, explain why the structure does not exist.

(a)   **(b)**   **(c)**

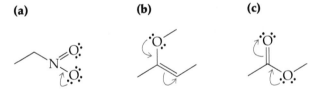

23.99 Draw the electron-dot structure resulting from the curved arrows shown. If the arrow is valid, give the line drawing of the resonance structure. For incorrect structures, explain why the structure does not exist.

(a)   **(b)**   **(c)**

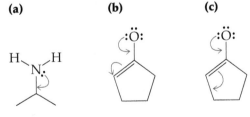

23.100 Draw the resonance structure that results from indicated arrows. Show formal charge and evaluate the relative contribution of each structure to the resonance hybrid.

(a)   **(b)**

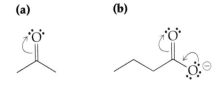

23.101 Draw the resonance structure that results from indicated arrows. Show formal charge and evaluate the relative contribution of each structure to the resonance hybrid.

(a)   **(b)**

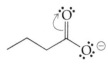

23.102 Draw the resonance structure that results from indicated arrows. Show formal charge and evaluate the relative contribution of each structure to the resonance hybrid.

(a)   **(b)**

23.103 Draw the resonance structure that results from indicated arrows. Show formal charge and evaluate the relative contribution of each structure to the resonance hybrid.

(a)   **(b)**

23.104 Are the following two structures constitutional isomers or
RAN   resonance structures?

23.105 Are the following two structures constitutional isomers or
RAN   resonance structures?

23.106 Draw curved arrows showing how to convert the first structure into the resonance structure shown.

(a)

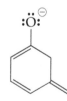

(b)

23.107 Naproxen is a nonsteroidal anti-inflammatory drug (NSAID) commonly used to relieve pain and fever.

(a) Draw a complete electron-dot structure for naproxen by adding lone pairs and hydrogen atoms to the line drawing.

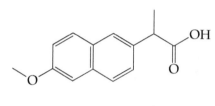

(b) Draw curved arrows in the electron-dot structure you drew in part (a) to generate the resonance structures shown.

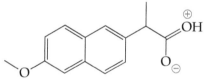

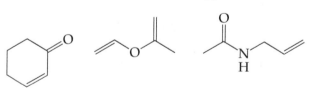

Conjugation Systems (Section 23.8)

23.108 In the following molecules, indicate which atoms are part of a conjugated system.

(a)   **(b)**   **(c)**

23.109 In the following molecules, indicate which atoms are part of a conjugated system.

(a)   **(b)**   **(c)**

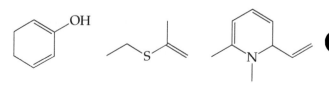

23.110 Label the hybridization of each N, O, and S atom in vitamin B₇.

23.111 Label the hybridization of each O atom in aspirin.

23.112 Draw simplified orbital overlap picture for this molecule, and indicate which lone pairs are localized and which are delocalized.

23.113 Draw simplified orbital overlap picture for this molecule, and indicate which lone pairs are localized and which are delocalized.

Proteins (Section 23.9)

23.114 What amino acids do the following abbreviations stand for?
(a) Ser   (b) Thr   (c) Pro
(d) Phe   (e) Cys

23.115 Name and draw the structures of amino acids that fit the following descriptions.
(a) Contains an isopropyl group
(b) Contains an alcohol group
(c) Contains a thiol (—SH) group
(d) Contains an aromatic ring

23.116 Identify the amino acids present in the following tripeptide.

23.117 Aspartame, marketed for use as a nonnutritive sweetener under such trade names as Equal, NutraSweet, and Canderel, is the methyl ester of a simple dipeptide. Identify the two amino acids present in aspartame.

Aspartame

23.118 Use the three-letter abbreviations to name all tripeptides containing methionine, isoleucine, and lysine.

23.119 How many tetrapeptides containing alanine, serine, leucine, and glutamic acid do you think there are? Use the three-letter abbreviations to name three.

Aromaticity and Molecular Orbital Theory (Section 23.10)

23.120 Which of the following molecules are aromatic?
RAN
(a)   (b)   (c)

23.121 Which of the following molecules are aromatic?
RAN
(a)   (b)   (c)

23.122 How many electrons are in the aromatic system of Tagamet®, a drug used to treat peptic ulcers and heartburn?
RAN

23.123 How many electrons are in the aromatic system of quinine, a drug used against malaria?
RAN

23.124 Caffeine is the most widely used stimulant psychoactive drug in the United States. It is found naturally in the seeds and leaves of some plants such as tea and coffee and can be synthesized in the laboratory.

    (a) Draw the resonance structure that results from the following arrows. Include any relevant formal charges.

    (b) What is the hybridization of each C and N atom in the two rings in the resonance structure?

    (c) Is caffeine an aromatic compound?

Caffeine

23.125 An abbreviated chemical structure of Viagra® is shown. Is the portion shown aromatic or not? Explain, using Hückel's rule. (Hint: Remember to consider conjugation and resonance!)

Nucleic Acids (Section 23.11)

23.126 What is a nucleotide, and what three kinds of components does it contain?

23.127 What are the names of the sugars in DNA and RNA, and how do they differ?

23.128 Show by drawing structures how the phosphate and sugar components of a nucleic acid are joined.

23.129 Show by drawing structures how the sugar and amine base components of a nucleic acid are joined.

23.130 If the sequence T-A-C-C-G-A appeared on one strand of DNA, what sequence would appear opposite it on the other strand?

23.131 The DNA from sea urchins contains about 32% A and about 18% G. What percentages of T and C would you expect in sea urchin DNA? Explain.

23.132 Draw a resonance structure for the RNA base uracil showing its aromatic conjugated double bonds. Give the hybridization for each atom in the ring in the resonance structure.

23.133 Draw a resonance structure for the DNA base guanine that shows its aromatic conjugated double bonds. Give the hybridization for each atom in the rings in the resonance structure.

MULTICONCEPT PROBLEMS

23.134 For the following molecule:

    (a) Draw simplified orbital overlap pictures to indicate the π system and localized lone pairs.

    (b) Indicate which lone pairs are localized and which are delocalized.

    (c) Draw a resonance structure for this compound.

23.135 Fumaric acid is an organic substance widely used as a food additive. Its elemental composition is 41.4% C, 3.5% H, and 55.1% O. A solution made by dissolving 0.1500 g of fumaric acid in water and diluting to a volume of 100.0 mL gave rise to an osmotic pressure of 240.3 mm Hg at 298 K. On titration of a sample weighing 0.573 g, 94.1 mL of 0.105 M NaOH was required to reach an equivalence point. Fumaric acid reacts with 1 mol of H_2 to give a hydrogenation product.

    (a) What is the empirical formula of fumaric acid?

    (b) What is the molecular weight of fumaric acid?

    (c) Draw three possible structures for fumaric acid.

    (d) If fumaric acid contains a trans double bond, which of your structures is correct?

23.136 Cytochrome *c* is an important enzyme found in the cells of all aerobic organisms. Elemental analysis of cytochrome *c* shows that it contains 0.43% iron. What is the minimum molecular weight of this enzyme?

23.137 The protonated form of a neutral amino acid such as alanine is a diprotic acid, H_2A^+, with two acid dissociation constants, one for the $-NH_3^+$ group and one for the $-CO_2H$ group.

$pK_a = 9.69$ $pK_a = 2.34$

$$H_3\overset{+}{N}-CH-\underset{\underset{CH_3}{|}}{\overset{\overset{O}{\|}}{C}}-OH$$

Alanine
(protonated)

(a) Which group is more acidic, the $-NH_3^+$ or the $-CO_2H$?

(b) What percentage of each form—protonated (H_2A^+), neutral (HA), and deprotonated (A^-)—is present in aqueous solution at $pH = 4.00$?

(c) What percentage of each form—protonated (H_2A^+), neutral (HA), and deprotonated (A^-)—is present in aqueous solution at $pH = 8.50$?

(d) At what pH is the neutral form present in maximum amount?

23.138 The relative amount of unsaturation in a fat or oil is expressed as an *iodine number*. Olive oil, for instance, is highly unsaturated and has an iodine number of 172, while butter is much less unsaturated and has an iodine number of 37. Defined as the number of grams of I_2 absorbed per 100 grams of fat, the iodine number is based on the fact that the carbon–carbon double bonds in fats and oils undergo an addition reaction with I_2. The larger the number of double bonds, the larger the amount of I_2 that reacts.

$$\underset{}{\overset{}{>}}C=C\overset{}{\underset{}{<}} \quad \xrightarrow{I_2} \quad \underset{}{\overset{I}{>}}C-C\overset{}{\underset{I}{<}}$$

To determine an iodine number, a known amount of fat is treated with a known amount of I_2. When the addition reaction is complete, the amount of excess I_2 remaining is determined by titration with $Na_2S_2O_3$ according to the equation

$$2\,Na_2S_2O_3(aq) + I_2(aq) \rightarrow Na_2S_4O_6(aq) + 2\,NaI(aq)$$

Knowing both the amount of I_2 originally added and the amount remaining after reaction, the iodine number can be calculated. Assume that 0.500 g of human milk fat is allowed to react with 25.0 mL of 0.200 M I_2 solution and that 81.99 mL of 0.100 M $Na_2S_2O_3$ is required for complete reaction with the excess I_2.

(a) What amount (in grams) of I_2 was added initially?

(b) How many grams of I_2 reacted with the milk fat, and how many grams were in excess?

(c) What is the iodine number of human milk fat?

(d) Assuming a molecular weight of 800, how many double bonds does an average molecule of milk fat contain?

Appendix A

Mathematical Operations

A.1 ▶ SCIENTIFIC NOTATION

The numbers that you encounter in chemistry are often either very large or very small. For example, there are about 33,000,000,000,000,000,000,000 H_2O molecules in 1.0 mL of water, and the distance between the H and O atoms in an H_2O molecule is 0.000 000 000 095 7 m. These quantities are more conveniently written in scientific notation as 3.3×10^{22} molecules and 9.57×10^{-11} m, respectively. In scientific notation, numbers are written in the exponential format $A \times 10^n$, where A is a number between 1 and 10, and the exponent n is a positive or negative integer.

How do you convert a number from ordinary notation to scientific notation? If the number is greater than or equal to 10, shift the decimal point to the *left* by n places until you obtain a number between 1 and 10. Then, multiply the result by 10^n. For example, the number 8137.6 is written in scientific notation as 8.1376×10^3:

$$8137.6 = 8.1376 \times 10^3$$

Shift decimal point to the left by 3 places to get a number between 1 and 10

Number of places decimal point was shifted to the left

When you shift the decimal point to the left by three places, you are in effect dividing the number by $10 \times 10 \times 10 = 1000 = 10^3$. Therefore, you must multiply the result by 10^3 so that the value of the number is unchanged.

To convert a number less than 1 to scientific notation, shift the decimal point to the *right* by n places until you obtain a number between 1 and 10. Then, multiply the result by 10^{-n}. For example, the number 0.012 is written in scientific notation as 1.2×10^{-2}:

$$0.012 = 1.2 \times 10^{-2}$$

Shift decimal point to the right by 2 places to get a number between 1 and 10

Number of places decimal point was shifted to the right

When you shift the decimal point to the right by two places, you are in effect multiplying the number by $10 \times 10 = 100 = 10^2$. Therefore, you must multiply the result by 10^{-2} so that the value of the number is unchanged. ($10^2 \times 10^{-2} = 10^0 = 1$.)

The following table gives some additional examples. To convert from scientific notation to ordinary notation, simply reverse the preceding process. Thus, to write the number 5.84×10^4 in ordinary notation, drop the factor of 10^4 and move the decimal point by 4 places to the *right* ($5.84 \times 10^4 = 58,400$). To write the number 3.5×10^{-1} in ordinary notation, drop the factor of 10^{-1} and move the decimal point by 1 place to the *left* ($3.5 \times 10^{-1} = 0.35$). Note that you don't need scientific notation for numbers between 1 and 10 because $10^0 = 1$.

Number	Scientific Notation
58,400	5.84×10^4
0.35	3.5×10^{-1}
7.296	7.296×10^0

Addition and Subtraction

To add or subtract two numbers expressed in scientific notation, both numbers must have the same exponent. Thus, to add 7.16×10^3 and 1.32×10^2, first write the latter number as 0.132×10^3 and then add:

$$\begin{array}{r} 7.16 \times 10^3 \\ +0.132 \times 10^3 \\ \hline 7.29 \times 10^3 \end{array}$$

The answer has three significant figures. (Significant figures are discussed in Section 1.9.) Alternatively, you can write the first number as 71.6×10^2 and then add:

$$\begin{array}{r} 71.6 \times 10^2 \\ +1.32 \times 10^2 \\ \hline 72.9 \times 10^2 = 7.29 \times 10^3 \end{array}$$

Multiplication and Division

To multiply two numbers expressed in scientific notation, multiply the factors in front of the powers of 10 and then add the exponents:

$$(A \times 10^n)(B \times 10^m) = AB \times 10^{n+m}$$

For example,

$$(2.5 \times 10^4)(4.7 \times 10^7) = (2.5)(4.7) \times 10^{4+7} = 12 \times 10^{11} = 1.2 \times 10^{12}$$

$$(3.46 \times 10^5)(2.2 \times 10^{-2}) = (3.46)(2.2) \times 10^{5+(-2)} = 7.6 \times 10^3$$

Both answers have two significant figures.

To divide two numbers expressed in scientific notation, divide the factors in front of the powers of 10 and then subtract the exponent in the denominator from the exponent in the numerator:

$$\frac{A \times 10^n}{B \times 10^m} = \frac{A}{B} \times 10^{n-m}$$

For example,

$$\frac{3 \times 10^6}{7.2 \times 10^2} = \frac{3}{7.2} \times 10^{6-2} = 0.4 \times 10^4 = 4 \times 10^3 \quad \text{(1 significant figure)}$$

$$\frac{7.50 \times 10^{-5}}{2.5 \times 10^{-7}} = \frac{7.50}{2.5} \times 10^{-5-(-7)} = 3.0 \times 10^2 \quad \text{(2 significant figures)}$$

Powers and Roots

To raise a number $A \times 10^n$ to a power m, raise the factor A to the power m and then multiply the exponent n by the power m:

$$(A \times 10^n)^m = A^m \times 10^{n \times m}$$

For example, 3.6×10^2 raised to the 3rd power is 4.7×10^7:

$$(3.6 \times 10^2)^3 = (3.6)^3 \times 10^{2 \times 3} = 47 \times 10^6 = 4.7 \times 10^7 \quad \text{(2 significant figures)}$$

To take the mth root of a number $A \times 10^n$, raise the number to the power $1/m$. That is, raise factor A to the power $1/m$ and then divide the exponent n by the root m:

$$\sqrt[m]{A \times 10^n} = (A \times 10^n)^{1/m} = A^{1/m} \times 10^{n/m}$$

For example, the square root of 9.0×10^8 is 3.0×10^4:

$$\sqrt{9.0 \times 10^8} = (9.0 \times 10^8)^{1/2} = (9.0)^{1/2} \times 10^{8/2} = 3.0 \times 10^4 \quad \text{(2 significant figures)}$$

Because the exponent in the answer (n/m) is an integer, we must sometimes rewrite the original number by shifting the decimal point so that the exponent n is an integral multiple of the root m. For example, to take the cube root of 6.4×10^{10}, we first rewrite this number as 64×10^9 so that the exponent (9) is an integral multiple of the root 3:

$$\sqrt[3]{6.4 \times 10^{10}} = \sqrt[3]{64 \times 10^9} = (64)^{1/3} \times 10^{9/3} = 4.0 \times 10^3$$

Scientific Notation and Electronic Calculators

With a scientific calculator you can carry out calculations in scientific notation. You should consult the instruction manual for your particular calculator to learn how to enter and manipulate numbers expressed in an exponential format. On most calculators, you enter the number $A \times 10^n$ by (i) entering the number A, (ii) pressing a key labeled EXP or EE, and (iii) entering the exponent n. If the exponent is negative, you press a key labeled $+/-$ before entering the value of n. (Note that you do not enter the number 10.) The calculator displays the number $A \times 10^n$ with the number A on the left followed by some space and then the exponent n. For example,

$$4.625 \times 10^2 \quad \text{is displayed as} \quad 4.625 \quad 02$$

To add, subtract, multiply, or divide exponential numbers, use the same sequence of keystrokes as you would in working with ordinary numbers. When you add or subtract on a calculator, the numbers need not have the same exponent; the calculator automatically takes account of the different exponents. Remember, though, that the calculator often gives more digits in the answer than the allowed number of significant figures. It's sometimes helpful to outline the calculation on paper, as in the preceding examples, in order to keep track of the number of significant figures.

Most calculators have x^2 and $\sqrt{x}$ keys for squaring a number and finding its square root. Just enter the number and press the appropriate key. You probably have a y^x (or a^x) key for raising a number to a power. To raise 4.625×10^2 to the 3rd power, for example, use the following keystrokes: (i) enter the number 4.625×10^2 in the usual way, (ii) press the y^x key, (iii) enter the power 3, and (iv) press the $=$ key. The result is displayed as 9.8931641 07, but it must be rounded to 4 significant figures. Therefore, $(4.625 \times 10^2)^3 = 9.893 \times 10^7$.

To take the mth root of a number, raise the number to the power $1/m$. For example, to take the 5th root of 4.52×10^{11}, use the following keystrokes: (i) enter the number 4.52×10^{11}, (ii) press the y^x key, (iii) enter the number 5 (for the 5th root), (iv) press the $1/x$ key (to convert the 5th root to the power 1/5), and (v) press the $=$ key. The result is

$$\sqrt[5]{4.52 \times 10^{11}} = (4.52 \times 10^{11})^{1/5} = 2.14 \times 10^2$$

The calculator is able to handle the nonintegral exponent 11/5, and there is therefore no need to enter the number as 45.2×10^{10} so that the exponent is an integral multiple of the root 5.

PROBLEM A.1 Perform the following calculations, expressing the result in scientific notation with the correct number of significant figures. (You don't need a calculator for these.)

(a) $(1.50 \times 10^4) + (5.04 \times 10^3)$ (b) $(2.5 \times 10^{-2}) - (5.0 \times 10^{-3})$

(c) $(4.0 \times 10^4)^2$ (d) $\sqrt[3]{8 \times 10^{12}}$ (e) $\sqrt{2.5 \times 10^5}$

ANSWERS:
(a) 2.00×10^4 (b) 2.0×10^{-2} (c) 1.6×10^9 (d) 2×10^4 (e) $\pm 5.0 \times 10^2$

PROBLEM A.2 Perform the following calculations, expressing the result in scientific notation with the correct number of significant figures. (Use a calculator for these.)

(a) $(9.72 \times 10^{-1}) + (3.4823 \times 10^2)$ (b) $(3.772 \times 10^3) - (2.891 \times 10^4)$

(c) $(7.62 \times 10^{-3})^4$ (d) $\sqrt[3]{8.2 \times 10^7}$ (e) $\sqrt[5]{3.47 \times 10^{-12}}$

ANSWERS:
(a) 3.4920×10^2 (b) -2.514×10^4 (c) 3.37×10^{-9}
(d) 4.3×10^2 (e) 5.11×10^{-3}

A.2 ▶ LOGARITHMS

Common Logarithms

Any positive number x can be written as 10 raised to some power z—that is, $x = 10^z$. The exponent z is called the *common*, or *base 10*, *logarithm* of the number x and is denoted $\log_{10} x$, or simply $\log x$:

$$x = 10^z \qquad \log x = z$$

For example, 100 can be written as 10^2, and log 100 is therefore equal to 2:

$$100 = 10^2 \qquad \log 100 = 2$$

Similarly,

$$10 = 10^1 \qquad \log 10 = 1$$
$$1 = 10^0 \qquad \log 1 = 0$$
$$0.1 = 10^{-1} \qquad \log 0.1 = -1$$

In general, the logarithm of a number x is the power z to which 10 must be raised to equal the number x.

As **FIGURE A.1** shows, the logarithm of a number greater than 1 is positive, the logarithm of 1 is zero, and the logarithm of a positive number less than 1 is negative. The logarithm of a *negative* number is undefined because 10 raised to any power is always positive ($x = 10^z > 0$).

You can use a calculator to find the logarithm of a number that is not an integral power of 10. For example, to find the logarithm of 61.2, simply enter 61.2, and press the LOG key. The logarithm should be between 1 and 2 because 61.2 is between 10^1 and 10^2. The calculator gives a value of 1.786751422, which must be rounded to 1.787 because 61.2 has three significant figures.

Significant Figures and Common Logarithms

The only significant figures in a logarithm are the digits to the right of the decimal point; the number to the left of the decimal point is an exact number related to the integral power of 10 in the exponential expression for the number whose logarithm is to be found. Thus, the logarithm of 61.2, which has three significant figures, can be written as follows:

$$\log \underbrace{61.2}_{\text{3 SF's}} = \log (\underbrace{6.12}_{\text{3 SF's}} \times \underbrace{10^1}_{\substack{\text{Exact} \\ \text{number}}}) = \log 6.12 + \log 10^1 = \underbrace{0.787}_{\text{3 SF's}} + \underbrace{1}_{\substack{\text{Exact} \\ \text{number}}} = 1.\underbrace{787}_{\substack{\text{Exact} \\ \text{number}}} \underbrace{}_{\text{3 SF's}}$$

The digit (1) to the left of the decimal point in the logarithm (1.787) is an exact number and is not a significant figure; it merely indicates the location of the decimal point in the number 61.2. There are only three significant figures in the logarithm (7, 8, 7) because 61.2 has only three significant figures. Similarly, log 61 = 1.79 (2 significant figures), and $\log (6 \times 10^1) = 1.8$ (1 significant figure).

Antilogarithms

The antilogarithm, denoted antilog, is the inverse of the common logarithm. If z is the logarithm of x, then x is the antilogarithm of z. But since x can be written as 10^z, the antilogarithm of z is 10^z:

$$\text{If} \quad z = \log x \qquad \text{then} \qquad x = \text{antilog } z = 10^z$$

In other words, the antilog of a number is 10 raised to a power equal to that number. For example, the antilog of 2 is $10^2 = 100$, and the antilog of 3.71 is $10^{3.71}$.

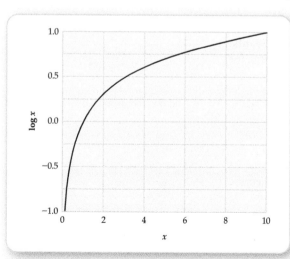

▲ **FIGURE A.1**

Values of log x for values of x in the range 0.1 to 10.

To find the value of antilog 3.71, use your calculator. If you have a 10^x key, enter 3.71 and press the 10^x key. If you have a y^x key, use the following keystrokes: (i) enter 10, (ii) press the y^x key, (iii) enter the exponent 3.71, and (iv) press the = key. If you have an INV (inverse) key, enter 3.71, press the INV key, and then press the LOG key. The calculator gives antilog $3.71 = 5.12861384 \times 10^3$, which must be rounded to 5.1×10^3 (2 significant figures) because the logarithm (3.71) has just two significant figures, the two digits to the right of the decimal point.

Natural Logarithms

The number $e = 2.718\ 28\ldots$, like $\pi = 3.141\ 59\ldots$, turns up in many scientific problems. It is therefore convenient to define a logarithm based on e, just as we defined a logarithm based on 10. Just as a number x can be written as 10^z, it can also be written as e^u. The exponent u is called the *natural*, or *base e*, *logarithm* of the number x and is denoted $\log_e x$, or more commonly, $\ln x$:

$$x = e^u \qquad \ln x = u$$

The natural logarithm of a number x is the power u to which e must be raised to equal the number x. For example, the number 10.0 can be written as $e^{2.303}$, and therefore the natural logarithm of 10.0 equals 2.303:

$$10.0 = e^{2.303} = (2.718\ 28\ldots)^{2.303} \quad \ln 10.0 = 2.303 \quad \text{(3 significant figures)}$$

To find the natural logarithm of a number on your calculator, simply enter the number, and press the ln key.

The natural antilogarithm, denoted antiln, is the inverse of the natural logarithm. If u is the natural logarithm of x, then $x\ (= e^u)$ is the natural antilogarithm of u:

$$\text{If} \quad u = \ln x \qquad \text{then} \qquad x = \text{antiln } u = e^u$$

In other words, the natural antilogarithm of a number is e raised to a power equal to that number. For example, the natural antilogarithm of 3.71 is $e^{3.71}$, which equals 41:

$$\text{antiln } 3.71 = e^{3.71} = 41 \qquad \text{(2 significant figures)}$$

Your calculator probably has an INV (inverse) key or an e^x key. To find the natural antilogarithm of a number—say, 3.71—enter 3.71, press the INV key, and then press the ln key. Alternatively, you can enter 3.71, and press the e^x key.

Some Mathematical Properties of Logarithms

Because logarithms are exponents, the algebraic properties of exponents can be used to derive the following useful relationships involving logarithms:

1. The logarithm (either common or natural) of a product xy equals the sum of the logarithm of x and the logarithm of y:

$$\log xy = \log x + \log y \quad \ln xy = \ln x + \ln y$$

2. The logarithm of a quotient x/y equals the difference between the logarithm of x and the logarithm of y:

$$\log \frac{x}{y} = \log x - \log y \quad \ln \frac{x}{y} = \ln x - \ln y$$

It follows from these relationships that

$$\log \frac{y}{x} = -\log \frac{x}{y} \quad \ln \frac{y}{x} = -\ln \frac{x}{y}$$

Because $\log 1 = \ln 1 = 0$, it also follows that

$$\log \frac{1}{x} = -\log x \quad \ln \frac{1}{x} = -\ln x$$

3. The logarithm of x raised to a power a equals a times the logarithm of x:

$$\log x^a = a \log x \qquad \ln x^a = a \ln x$$

Similarly,

$$\log x^{1/a} = \frac{1}{a} \log x \qquad \ln x^{1/a} = \frac{1}{a} \ln x$$

where

$$x^{1/a} = \sqrt[a]{x}.$$

What is the numerical relationship between the common logarithm and the natural logarithm? To derive it, we begin with the definitions of $\log x$ and $\ln x$:

$$\log x = z \qquad \text{where} \qquad x = 10^z$$

$$\log x = u \qquad \text{where} \qquad x = e^u$$

We then write $\ln x$ in terms of 10^z and make use of the property that $\ln x^a = a \ln x$:

$$\ln x = \ln 10^z = z \ln 10$$

Because $z = \log x$ and $\ln 10.0 = 2.303$, we find that the natural logarithm is 2.303 times the common logarithm:

$$\ln x = 2.303 \log x$$

Since the natural and common logarithms differ by a factor of only 2.303, the same rule can be used to find the number of significant figures in both: The only digits that are significant figures in both natural and common logarithms are those to the right of the decimal point.

PROBLEM A.3 Use a calculator to evaluate the following expressions, and round each result to the correct number of significant figures:
(a) $\log 705$ (b) $\ln (3.4 \times 10^{-6})$ (c) antilog (-2.56) (d) antiln 8.1

ANSWERS:
(a) 2.848   (b) −12.59   (c) 2.8×10^{-3} (d) 3×10^3

A.3 ▶ STRAIGHT-LINE GRAPHS AND LINEAR EQUATIONS

The results of a scientific experiment are often summarized in the form of a graph. Consider an experiment in which some property y is measured as a function of some variable x. (A real example would be measurement of the volume of a gas as a function of its temperature, but we'll use y and x to keep the discussion general.) Suppose that we obtain the following experimental data:

x	y
−1	−5
1	1
3	7
5	13

The graph in **FIGURE A.2** shows values of x, called the independent variable, along the horizontal axis and values of y, the dependent variable, along the vertical axis. Each pair of experimental values of x and y is represented by a point on the graph. For this particular experiment, the four data points lie on a straight line.

The equation of a straight line can be written as

$$y = mx + b$$

where m is the slope of the line and b is the intercept, the value of y at the point where the line crosses the y axis—that is, the value of y when $x = 0$. The slope of the line is the change in y (Δy) for a given change in x (Δx):

$$m = \text{slope} = \frac{\Delta y}{\Delta x}$$

The right-triangle in Figure A.2 shows that y changes from 4 to 13 when x changes from 2 to 5. Therefore, the slope of the line is 3:

$$m = \text{slope} = \frac{\Delta y}{\Delta x} = \frac{13 - 4}{5 - 2} = \frac{9}{3} = 3$$

The graph shows a y intercept of -2 ($b = -2$), and the equation of the line is therefore

$$y = 3x - 2$$

An equation of the form $y = mx + b$ is called a *linear equation* because values of x and y that satisfy such an equation are the coordinates of points that lie on a straight line. We also say that y is a *linear function* of x, or that y is *directly proportional* to x. In our example, the rate of change of y is 3 times that of x.

A.4 ▶ QUADRATIC EQUATIONS

A quadratic equation is an equation that can be written in the form

$$ax^2 + bx + c = 0$$

where a, b, and c are constants. The equation contains only powers of x and is called quadratic because the highest power of x is 2. The solutions to a quadratic equation (values of x that satisfy the equation) are given by the *quadratic formula*:

$$x = \frac{-b \pm \sqrt{b^2 - 4ac}}{2a}$$

The $\pm$ indicates that there are two solutions, one given by the $+$ sign and the other given by the $-$ sign.

As an example, let's solve the equation

$$x^2 = \frac{2 - 6x}{3}$$

First, we put the equation into the form $ax^2 + bx + c = 0$ by multiplying it by 3 and moving $2 - 6x$ to the left side. The result is

$$3x^2 + 6x - 2 = 0$$

Then we apply the quadratic formula with $a = 3$, $b = 6$, and $c = -2$:

$$x = \frac{-6 \pm \sqrt{(6)^2 - 4(3)(-2)}}{2(3)}$$

$$= \frac{-6 \pm \sqrt{36 + 24}}{6} = \frac{-6 \pm \sqrt{60}}{6} = \frac{-6 \pm 7.746}{6}$$

The two solutions are

$$x = \frac{-6 + 7.746}{6} = \frac{1.746}{6} = 0.291 \quad \text{and} \quad x = \frac{-6 - 7.746}{6} = \frac{-13.746}{6} = -2.291$$

A.5 ▶ CALCULUS DERIVATIONS OF INTEGRATED RATE LAWS

Zeroth-Order Reactions

The rate law for a zeroth-order reaction is

$$\text{Rate} = -\frac{\Delta[A]}{\Delta t} = k$$

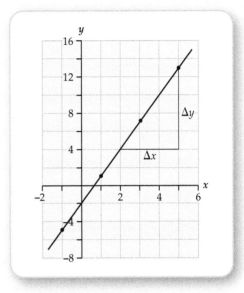

▲ **FIGURE A.2**

A straight-line y versus x plot of the data in the table.

In differential form, we can write this equation as

$$-\frac{d[A]}{dt} = k$$

$$\text{or} \quad d[A] = -kdt$$

Integrating from $t = 0$ to $t = t$ gives

$$\int_{[A]_0}^{[A]_t} d[A] = -k\int_0^t dt$$

$$[A]_t - [A]_0 = -kt$$

$$\text{or} \quad [A]_t = -kt + [A]_0$$

First-Order Reactions

The rate law for a first-order reaction is

$$\text{Rate} = -\frac{\Delta[A]}{\Delta t} = k[A]$$

In differential form, we can write this equation as

$$-\frac{d[A]}{dt} = k[A]$$

$$\text{or} \quad \frac{d[A]}{[A]} = -kdt$$

Integrating from $t = 0$ to $t = t$ gives

$$\int_{[A]_0}^{[A_t]} \frac{d[A]}{[A]} = -k\int_0^t dt$$

$$\ln[A]_t - \ln[A]_0 = -kt$$

$$\text{or} \quad \ln\frac{[A]_t}{[A]_0} = -kt$$

Second-Order Reactions

The rate law for a second-order reaction is

$$\text{Rate} = -\frac{\Delta[A]}{\Delta t} = k[A]^2$$

In differential form, we can write this equation as

$$-\frac{d[A]}{dt} = k[A]^2$$

$$\text{or} \quad \frac{d[A]}{[A]^2} = -kdt$$

Integrating from $t = 0$ to $t = t$ gives

$$\int_{[A]_0}^{[A]_t} \frac{d[A]}{[A]^2} = -k\int_0^t dt$$

$$-\frac{1}{[A]_t} + \frac{1}{[A]_0} = -kt$$

$$\text{or} \quad \frac{1}{[A]_t} = kt + \frac{1}{[A]_0}$$

Appendix B

Thermodynamic Properties at 25 °C

TABLE B.1 Inorganic Substances

Substance and State	$\Delta H°_f$ (kJ/mol)	$\Delta G°_f$ (kJ/mol)	$S°$ [J/(K·mol)]	Substance and State	$\Delta H°_f$ (kJ/mol)	$\Delta G°_f$ (kJ/mol)	$S°$ [J/(K·mol)]
Aluminum				*Calcium*			
Al(s)	0	0	28.3	Ca(s)	0	0	41.4
Al(g)	330.0	289.4	164.5	Ca(g)	177.8	144.0	154.8
$AlCl_3(s)$	−704.2	−628.8	109.3	$Ca^{2+}(aq)$	−542.8	−553.6	−53.1
$Al_2O_3(s)$	−1676	−1582	50.9	$CaF_2(s)$	−1228.0	−1175.6	68.5
				$CaCl_2(s)$	−795.4	−748.8	108.4
Barium				$CaH_2(s)$	−181.5	−142.5	41.4
Ba(s)	0	0	62.5	$CaC_2(s)$	−59.8	−64.8	70.0
Ba(g)	180.0	146.0	170.1	CaO(s)	−634.9	−603.3	38.1
$Ba^{2+}(aq)$	−537.6	−560.8	9.6	$Ca(OH)_2(s)$	−985.2	−897.5	83.4
$BaCl_2(s)$	−855.0	−806.7	123.7	$CaCO_3(s)$	−1207.6	−1129.1	91.7
BaO(s)	−548.0	−520.3	72.1	$CaSO_4(s)$	−1434.1	−1321.9	107
$BaCO_3(s)$	−1213.0	−1134.4	112.1	$Ca_3(PO_4)_2(s)$	−4120.8	−3884.7	236.0
$BaSO_4(s)$	−1473.2	−1362.2	132.2				
				Carbon			
Beryllium				C(s, graphite)	0	0	5.7
Be(s)	0	0	9.5	C(s, diamond)	1.9	2.9	2.4
BeO(s)	−609.4	−580.1	13.8	C(g)	716.7	671.3	158.0
$Be(OH)_2(s)$	−902.5	−815.0	45.5	CO(g)	−110.5	−137.2	197.6
				$CO_2(g)$	−393.5	−394.4	213.6
Boron				$CO_2(aq)$	−413.8	−386.0	117.6
B(s)	0	0	5.9	$CO_3^{2-}(aq)$	−677.1	−527.8	−56.9
$BF_3(g)$	−1136.0	−1119.4	254.3	$HCO_3^-(aq)$	−692.0	−586.8	91.2
$BCl_3(g)$	−403.8	−388.7	290.0	$H_2CO_3(aq)$	−699.7	−623.2	187.4
$B_2H_6(g)$	36.4	87.6	232.0	HCN(l)	108.9	125.0	112.8
$B_2O_3(s)$	−1273.5	−1194.3	54.0	HCN(g)	135.1	124.7	201.7
$H_3BO_3(s)$	−1094.3	−968.9	90.0	$CS_2(l)$	89.0	64.6	151.3
				$CS_2(g)$	116.7	67.1	237.7
Bromine				$COCl_2(g)$	−219.1	−204.9	283.4
Br(g)	111.9	82.4	174.9				
$Br^-(aq)$	−121.5	−104.0	82.4	*Cesium*			
$Br_2(l)$	0	0	152.2	Cs(s)	0	0	85.2
$Br_2(g)$	30.9	3.14	245.4	Cs(g)	76.5	49.6	175.6
HBr(g)	−36.3	−53.4	198.6	$Cs^+(aq)$	−258.3	−292.0	133.1
				CsF(s)	−553.5	−525.5	92.8
Cadmium				CsCl(s)	−443.0	−414.5	101.2
Cd(s)	0	0	51.8	CsBr(s)	−405.8	−391.4	113.1
Cd(g)	111.8	77.3	167.6	CsI(s)	−346.6	−340.6	123.1
$Cd^{2+}(aq)$	−75.9	−77.6	−73.2				
$CdCl_2(s)$	−391.5	−343.9	115.3				
CdO(s)	−258.4	−228.7	54.8				
CdS(s)	−161.9	−156.5	64.9				
$CdSO_4(s)$	−933.3	−822.7	123.0				

continued on next page

TABLE B.1   Inorganic Substances (continued)

Substance and State	$\Delta H°_f$ (kJ/mol)	$\Delta G°_f$ (kJ/mol)	$S°$ [J/(K·mol)]	Substance and State	$\Delta H°_f$ (kJ/mol)	$\Delta G°_f$ (kJ/mol)	$S°$ [J/(K·mol)]
Chlorine				**Iron**			
$Cl(g)$	121.3	105.3	165.1	$Fe(s)$	0	0	27.3
$Cl^-(aq)$	−167.2	−131.3	56.5	$Fe(g)$	416.3	370.3	180.5
$Cl_2(g)$	0	0	223.0	$FeCl_2(s)$	−341.8	−302.3	118.0
$HCl(g)$	−92.3	−95.3	186.8	$FeCl_3(s)$	−399.5	−334.0	142.3
$HCl(aq)$	−167.2	−131.2	56.5	$FeO(s)$	−272	−255	61
$ClO_2(g)$	102.5	120.5	256.7	$Fe_2O_3(s)$	−824.2	−742.2	87.4
$Cl_2O(g)$	80.3	97.9	266.1	$Fe_3O_4(s)$	−1118	−1015	146
Chromium				$FeS_2(s)$	−178.2	−166.9	52.9
$Cr(s)$	0	0	23.8	**Lead**			
$Cr(g)$	396.6	351.8	174.4	$Pb(s)$	0	0	64.8
$Cr_2O_3(s)$	−1140	−1058	81.2	$Pb(g)$	195.2	162.2	175.3
Cobalt				$PbCl_2(s)$	−359.4	−314.1	136.0
$Co(s)$	0	0	30.0	$PbBr_2(s)$	−278.7	−261.9	161.5
$CO(g)$	424.7	380.3	179.4	$PbO(s)$	−217.3	−187.9	68.7
$CoO(s)$	−237.9	−214.2	53.0	$PbO_2(s)$	−277	−217.4	68.6
Copper				$PbS(s)$	−100	−98.7	91.2
$Cu(s)$	0	0	33.1	$PbCO_3(s)$	−699.1	−625.5	131.0
$Cu(g)$	337.4	297.7	166.3	$PbSO_4(s)$	−919.9	−813.2	148.6
$Cu^{2+}(aq)$	64.8	65.5	−99.6	**Lithium**			
$CuCl(s)$	−137.2	−119.9	86.2	$Li(s)$	0	0	29.1
$CuCl_2(s)$	−220.1	−175.7	108.1	$Li(g)$	159.3	126.6	138.7
$CuO(s)$	−157.3	−129.7	42.6	$Li^+(aq)$	−278.5	−293.3	13
$Cu_2O(s)$	−168.6	−146.0	93.1	$LiF(s)$	−616.0	−587.7	35.7
$CuS(s)$	−53.1	−53.6	66.5	$LiCl(s)$	−408.6	−384.4	59.3
$Cu_2S(s)$	−79.5	−86.2	120.9	$LiBr(s)$	−351.2	−342.0	74.3
$CuSO_4(s)$	−771.4	−662.2	109.2	$LiI(s)$	−270.4	−270.3	86.8
Fluorine				$Li_2O(s)$	−597.9	−561.2	37.6
$F(g)$	79.4	62.3	158.7	$LiOH(s)$	−487.5	−441.5	42.8
$F^-(aq)$	−332.6	−278.8	−13.8	**Magnesium**			
$F_2(g)$	0	0	202.7	$Mg(s)$	0	0	32.7
$HF(g)$	−273.3	−275.4	173.7	$Mg(g)$	147.1	112.5	148.6
Hydrogen				$MgCl_2(s)$	−641.6	−591.8	89.6
$H(g)$	218.0	203.3	114.6	$MgO(s)$	−601.7	−569.4	26.9
$H^+(aq)$	0	0	0	$MgCO_3(s)$	−1096	−1012	65.7
$H_2(g)$	0	0	130.6	$MgSO_4(s)$	−1284.9	−1170.6	91.6
$OH^-(aq)$	−230.0	−157.3	−10.8	**Manganese**			
$H_2O(l)$	−285.8	−237.2	69.9	$Mn(s)$	0	0	32.0
$H_2O(g)$	−241.8	−228.6	188.7	$Mn(g)$	280.7	238.5	173.6
$H_2O_2(l)$	−187.8	−120.4	110	$MnO(s)$	−385.2	−362.9	59.7
$H_2O_2(g)$	−136.3	−105.6	232.6	$MnO_2(s)$	−520.0	−465.1	53.1
$H_2O_2(aq)$	−191.2	−134.1	144	**Mercury**			
Iodine				$Hg(l)$	0	0	76.0
$I(g)$	106.8	70.3	180.7	$Hg(g)$	61.32	31.85	174.8
$I^-(aq)$	−55.2	−51.6	111	$Hg^{2+}(aq)$	171.1	164.4	−32.2
$I_2(s)$	0	0	116.1	$Hg_2^{2+}(aq)$	172.4	153.5	84.5
$I_2(g)$	62.4	19.4	260.6	$HgCl_2(s)$	−224.3	−178.6	146.0
$HI(g)$	26.5	1.7	206.5	$Hg_2Cl_2(s)$	−265.4	−210.7	191.6

TABLE B.1   Inorganic Substances (continued)

Substance and State	$\Delta H°_f$ (kJ/mol)	$\Delta G°_f$ (kJ/mol)	$S°$ [J/(K·mol)]	Substance and State	$\Delta H°_f$ (kJ/mol)	$\Delta G°_f$ (kJ/mol)	$S°$ [J/(K·mol)]
HgO(s)	−90.8	−58.6	70.3	*Potassium*			
HgS(s)	−58.2	−50.6	82.4	K(s)	0	0	64.7
Nickel				K(g)	89.2	60.6	160.2
Ni(s)	0	0	29.9	K⁺(aq)	−252.4	−283.3	102.5
Ni(g)	429.7	384.5	182.1	KF(s)	−567.3	−537.8	66.6
NiCl₂(s)	−305.3	−259.1	97.7	KCl(s)	−436.5	−408.5	82.6
NiO(s)	−240	−212	38.0	KBr(s)	−393.8	−380.7	95.9
NiS(s)	−82.0	−79.5	53.0	KI(s)	−327.9	−324.9	106.3
Nitrogen				K₂O(s)	−361.5		
N(g)	472.7	455.6	153.2	K₂O₂(s)	−494.1	−425.1	102.1
N₂(g)	0	0	191.5	KO₂(s)	−284.9	−239.4	116.7
NH₃(g)	−46.1	−16.5	192.3	KOH(s)	−424.6	−379.4	81.2
NH₃(aq)	−80.3	−26.6	111	KOH(aq)	−482.4	−440.5	91.6
NH₄⁺(aq)	−132.5	−79.4	113	KClO₃(s)	−397.7	−296.3	143.1
N₂H₄(l)	50.6	149.2	121.2	KClO₄(s)	−432.8	−303.1	151.0
N₂H₄(g)	95.4	159.3	238.4	KNO₃(s)	−494.6	−394.9	133.1
NO(g)	91.3	87.6	210.7	*Rubidium*			
NO₂(g)	33.2	51.3	240.0	Rb(s)	0	0	76.8
N₂O(g)	82.0	104.2	219.7	Rb(g)	80.9	53.1	170.0
N₂O₄(g)	11.1	99.8	304.3	Rb⁺(aq)	−251.2	−284.0	121.5
N₂O₅(g)	13.3	117.1	355.6	RbF(s)	−557.7		
NOCl(g)	51.7	66.1	261.6	RbCl(s)	−435.4	−407.8	95.9
NO₂Cl(g)	12.6	54.4	272.1	RbBr(s)	−394.6	−381.8	110.0
HNO₃(l)	−174.1	−80.8	155.6	RbI(s)	−333.8	−328.9	118.4
HNO₃(g)	−133.9	−73.5	266.8	*Selenium*			
HNO₂(aq)	−119	−50.6	136	Se(s, black)	0	0	42.44
HNO₃(aq)	−207.4	−111.3	146.4	H₂Se(g)	29.7	15.9	219.0
NO₃⁻(aq)	−207.4	−111.3	146.4	*Silicon*			
NH₄Cl(s)	−314.4	−202.9	94.6	Si(s)	0	0	18.8
NH₄NO₃(s)	−365.6	−184.0	151.1	Si(g)	450.0	405.5	167.9
Oxygen				SiF₄(g)	−1615.0	−1572.8	282.7
O(g)	249.2	231.7	160.9	SiCl₄(l)	−687.0	−619.8	239.7
O₂(g)	0	0	205.0	SiO₂(s, quartz)	−910.7	−856.3	41.5
O₃(g)	143	163	238.8	*Silver*			
Phosphorus				Ag(s)	0	0	42.6
P(s, white)	0	0	41.1	Ag(g)	284.9	246.0	173.0
P(s, red)	−18	−12	22.8	Ag⁺(aq)	105.6	77.1	72.7
P₄(g)	58.9	24.5	279.9	AgF(s)	−204.6		
PH₃(g)	5.4	13.5	210.1	AgCl(s)	−127.1	−109.8	96.2
PCl₃(l)	−319.7	−272.3	217.1	AgBr(s)	−100.4	−96.9	107.1
PCl₃(g)	−287.0	−267.8	311.7	AgI(s)	−61.8	−66.2	115.5
PCl₅(s)	−443.5			Ag₂O(s)	−31.1	−11.2	121.3
PCl₅(g)	−374.9	−305.0	364.5	Ag₂S(s)	−32.6	−40.7	144.0
P₄O₁₀(s)	−2984	−2698	228.9	AgNO₃(s)	−124.4	−33.4	140.9
PO₄³⁻(aq)	−1277.4	−1018.7	−220.5	*Sodium*			
HPO₄²⁻(aq)	−1292.1	−1089.2	−33.5	Na(s)	0	0	51.2
H₂PO₄⁻(aq)	−1296.3	−1130.2	90.4	Na(g)	107.3	76.8	153.6
H₃PO₄(s)	−1284.4	−1124.3	110.5				

continued on next page

TABLE B.1 Inorganic Substances (continued)

Substance and State	$\Delta H°_f$ (kJ/mol)	$\Delta G°_f$ (kJ/mol)	$S°$ [J/(K · mol)]	Substance and State	$\Delta H°_f$ (kJ/mol)	$\Delta G°_f$ (kJ/mol)	$S°$ [J/(K · mol)]
$Na^+(aq)$	−240.1	−261.9	59.0	*Tin*			
$NaF(s)$	−576.6	−546.3	51.1	$Sn(s, white)$	0	0	51.2
$NaCl(s)$	−411.2	−384.2	72.1	$Sn(s, gray)$	−2.1	0.1	44.1
$NaBr(s)$	−361.1	−349.0	86.8	$Sn(g)$	301.2	266.2	168.4
$NaI(s)$	−287.8	−286.1	98.5	$SnCl_4(l)$	−511.3	−440.1	258.6
$NaH(s)$	−56.3	−33.5	40.0	$SnCl_4(g)$	−471.5	−432.2	365.8
$Na_2O(s)$	−414.2	−375.5	75.1	$SnO(s)$	−280.7	−251.9	57.4
$Na_2O_2(s)$	−510.9	−447.7	95.0	$SnO_2(s)$	−577.6	−515.8	49.0
$NaO_2(s)$	−260.2	−218.4	115.9	*Titanium*			
$NaOH(s)$	−425.6	−379.5	64.5	$Ti(s)$	0	0	30.6
$NaOH(aq)$	−470.1	−419.2	48.2	$Ti(g)$	473.0	428.4	180.2
$Na_2CO_3(s)$	−1130.7	−1044.5	135.0	$TiCl_4(l)$	−804.2	−737.2	252.3
$NaHCO_3(s)$	−950.8	−851.0	102	$TiCl_4(g)$	−763.2	−726.3	353.2
$NaNO_3(s)$	−467.9	−367.0	116.5	$TiO_2(s)$	−944.0	−888.8	50.6
$NaNO_3(aq)$	−447.5	−373.2	205.4	*Tungsten*			
$Na_2SO_4(s)$	−1387.1	−1270.2	149.6	$W(s)$	0	0	32.6
Sulfur				$W(g)$	849.4	807.1	174.0
$S(s, rhombic)$	0	0	31.8	$WO_3(s)$	−842.9	−764.0	75.9
$S(s, monoclinic)$	0.3			*Zinc*			
$S(g)$	277.2	236.7	167.7	$Zn(s)$	0	0	41.6
$S_2(g)$	128.6	79.7	228.2	$Zn(g)$	130.4	94.8	160.9
$H_2S(g)$	−20.6	−33.6	205.7	$Zn^{2+}(aq)$	−153.9	−147.1	−112.1
$H_2S(aq)$	−39.7	−27.9	121	$ZnCl_2(s)$	−415.1	−369.4	111.5
$HS^-(aq)$	−17.6	12.1	62.8	$ZnO(s)$	−350.5	−320.5	43.7
$SO_2(g)$	−296.8	−300.2	248.1	$ZnS(s)$	−206.0	−201.3	57.7
$SO_3(g)$	−395.7	−371.1	256.6	$ZnSO_4(s)$	−982.8	−871.5	110.5
$H_2SO_4(l)$	−814.0	−690.1	156.9				
$H_2SO_4(aq)$	−909.3	−744.6	20				
$HSO_4^-(aq)$	−887.3	−756.0	132				
$SO_4^{2-}(aq)$	−909.3	−744.6	20				

TABLE B.2 Organic Substances

Substance and State	Formula	$\Delta H°_f$ (kJ/mol)	$\Delta G°_f$ (kJ/mol)	$S°$ [J/(K · mol)]
Acetaldehyde(g)	CH_3CHO	−166.2	−133.0	263.8
Acetic acid(l)	CH_3CO_2H	−484.5	−390	160
Acetylene(g)	C_2H_2	227.4	209.9	200.8
Benzene(l)	C_6H_6	49.1	124.5	173.4
Butane(g)	C_4H_{10}	−126	−17	310
Carbon tetrachloride(l)	CCl_4	−135.4	−65.3	216.4
Dichloroethane(l)	CH_2ClCH_2Cl	−165.2	−79.6	208.5
Ethane(g)	C_2H_6	−84.0	−32.0	229.1
Ethanol(l)	C_2H_5OH	−277.7	−174.9	161
Ethanol(g)	C_2H_5OH	−234.8	−167.9	281.5
Ethylene(g)	C_2H_4	52.3	68.1	219.5
Ethylene oxide(g)	C_2H_4O	−52.6	−13.1	242.4
Formaldehyde(g)	$HCHO$	−108.6	−102.5	218.8
Formic acid(l)	HCO_2H	−424.7	−361.4	129.0
Glucose(s)	$C_6H_{12}O_6$	−1273.3	−910	209.2
Methane(g)	CH_4	−74.8	−50.8	186.2
Methanol(l)	CH_3OH	−239.2	−166.6	127
Methanol(g)	CH_3OH	−201.0	−162.3	239.8
Propane(g)	C_3H_8	−103.8	−23.4	270.2
Vinyl chloride(g)	$CH_2{=}CHCl$	35	51.9	263.9

Equilibrium Constants at 25 °C

TABLE C.1   Acid-Dissociation Constants at 25 °C

Acid	Formula	K_{a1}	K_{a2}	K_{a3}
Acetic	CH_3CO_2H	1.8×10^{-5}		
Acetylsalicylic	$C_9H_8O_4$	3.0×10^{-4}		
Arsenic	H_3AsO_4	5.6×10^{-3}	1.7×10^{-7}	4.0×10^{-12}
Arsenious	H_3AsO_3	6×10^{-10}		
Ascorbic	$C_6H_8O_6$	8.0×10^{-5}		
Benzoic	$C_6H_5CO_2H$	6.5×10^{-5}		
Boric	H_3BO_3	5.8×10^{-10}		
Carbonic	H_2CO_3	4.3×10^{-7}	5.6×10^{-11}	
Chloroacetic	CH_2ClCO_2H	1.4×10^{-3}		
Citric	$C_6H_8O_7$	7.1×10^{-4}	1.7×10^{-5}	4.1×10^{-7}
Formic	HCO_2H	1.8×10^{-4}		
Hydrazoic	HN_3	1.9×10^{-5}		
Hydrocyanic	HCN	4.9×10^{-10}		
Hydrofluoric	HF	3.5×10^{-4}		
Hydrogen peroxide	H_2O_2	2.4×10^{-12}		
Hydrosulfuric	H_2S	1.0×10^{-7}	$\sim 10^{-19}$	
Hypobromous	$HOBr$	2.0×10^{-9}		
Hypochlorous	$HOCl$	3.5×10^{-8}		
Hypoiodous	HOI	2.3×10^{-11}		
Iodic	HIO_3	1.7×10^{-1}		
Lactic	$HC_3H_5O_3$	1.4×10^{-4}		
Nitrous	HNO_2	4.5×10^{-4}		
Oxalic	$H_2C_2O_4$	5.9×10^{-2}	6.4×10^{-5}	
Phenol	C_6H_5OH	1.3×10^{-10}		
Phosphoric	H_3PO_4	7.5×10^{-3}	6.2×10^{-8}	4.8×10^{-13}
Phosphorous	H_3PO_3	1.0×10^{-2}	2.6×10^{-7}	
Saccharin	$C_7H_5NO_3S$	2.1×10^{-12}		
Selenic	H_2SeO_4	Very large	1.2×10^{-2}	
Selenious	H_2SeO_3	3.5×10^{-2}	5×10^{-8}	
Sulfuric	H_2SO_4	Very large	1.2×10^{-2}	
Sulfurous	H_2SO_3	1.5×10^{-2}	6.3×10^{-8}	
Tartaric	$C_4H_6O_6$	1.0×10^{-3}	4.6×10^{-5}	
Water	H_2O	1.8×10^{-16}		

TABLE C.2   Acid-Dissociation Constants at 25 °C for Hydrated Metal Cations

Cation	K_a
$Fe^{2+}(aq)$	3.2×10^{-10}
$Co^{2+}(aq)$	1.3×10^{-9}
$Ni^{2+}(aq)$	2.5×10^{-11}
$Zn^{2+}(aq)$	2.5×10^{-10}
$Be^{2+}(aq)$	3×10^{-7}
$Al^{3+}(aq)$	1.4×10^{-5}
$Cr^{3+}(aq)$	1.6×10^{-4}
$Fe^{3+}(aq)$	6.3×10^{-3}

Note: As an example, K_a for $Fe^{2+}(aq)$ is the equilibrium constant for the reaction

$$Fe(H_2O)_6{}^{2+}(aq) + H_2O(l) \rightleftharpoons H_3O^+(aq) + Fe(H_2O)_5(OH)^+(aq)$$

TABLE C.3   Base-Dissociation Constants at 25 °C

Base	Formula	K_b
Ammonia	NH_3	1.8×10^{-5}
Aniline	$C_6H_5NH_2$	4.3×10^{-10}
Codeine	$C_{18}H_{21}NO_3$	1.6×10^{-6}
Dimethylamine	$(CH_3)_2NH$	5.4×10^{-4}
Ethylamine	$C_2H_5NH_2$	6.4×10^{-4}
Hydrazine	N_2H_4	8.9×10^{-7}
Hydroxylamine	NH_2OH	9.1×10^{-9}
Methylamine	CH_3NH_2	3.7×10^{-4}
Morphine	$C_{17}H_{19}NO_3$	1.6×10^{-6}
Piperidine	$C_5H_{11}N$	1.3×10^{-3}
Propylamine	$C_3H_7NH_2$	5.1×10^{-4}
Pyridine	C_5H_5N	1.8×10^{-9}
Strychnine	$C_{21}H_{22}N_2O_2$	1.8×10^{-6}
Trimethylamine	$(CH_3)_3N$	6.5×10^{-5}

TABLE C.4   Solubility Product Constants at 25 °C

Compound	Formula	K_{sp}
Aluminum hydroxide	$Al(OH)_3$	1.9×10^{-33}
Barium carbonate	$BaCO_3$	2.6×10^{-9}
Barium chromate	$BaCrO_4$	1.2×10^{-10}
Barium fluoride	BaF_2	1.8×10^{-7}
Barium hydroxide	$Ba(OH)_2$	5.0×10^{-3}
Barium sulfate	$BaSO_4$	1.1×10^{-10}
Cadmium carbonate	$CdCO_3$	1.0×10^{-12}
Cadmium hydroxide	$Cd(OH)_2$	5.3×10^{-15}
Calcium carbonate	$CaCO_3$	5.0×10^{-9}
Calcium fluoride	CaF_2	3.5×10^{-11}
Calcium hydroxide	$Ca(OH)_2$	4.7×10^{-6}
Calcium phosphate	$Ca_3(PO_4)_2$	2.1×10^{-33}
Calcium sulfate	$CaSO_4$	7.1×10^{-5}
Chromium(III) hydroxide	$Cr(OH)_3$	6.7×10^{-31}
Cobalt(II) hydroxide	$Co(OH)_2$	5.9×10^{-15}
Copper(I) bromide	$CuBr$	6.3×10^{-9}
Copper(I) chloride	$CuCl$	1.7×10^{-7}
Copper(II) carbonate	$CuCO_3$	2.5×10^{-10}

continued on next page

TABLE C.4   Solubility Product Constants at 25 °C (continued)

Compound	Formula	K_{sp}
Copper(II) hydroxide	$Cu(OH)_2$	1.6×10^{-19}
Copper(II) phosphate	$Cu_3(PO_4)_2$	1.4×10^{-37}
Iron(II) hydroxide	$Fe(OH)_2$	4.9×10^{-17}
Iron(III) hydroxide	$Fe(OH)_3$	2.6×10^{-39}
Lead(II) bromide	$PbBr_2$	6.6×10^{-6}
Lead(II) chloride	$PbCl_2$	1.2×10^{-5}
Lead(II) chromate	$PbCrO_4$	2.8×10^{-13}
Lead(II) iodide	PbI_2	8.5×10^{-9}
Lead(II) sulfate	$PbSO_4$	1.8×10^{-8}
Magnesium carbonate	$MgCO_3$	6.8×10^{-6}
Magnesium fluoride	MgF_2	7.4×10^{-11}
Magnesium hydroxide	$Mg(OH)_2$	5.6×10^{-12}
Manganese(II) carbonate	$MnCO_3$	2.2×10^{-11}
Manganese(II) hydroxide	$Mn(OH)_2$	2.1×10^{-13}
Mercury(I) bromide	Hg_2Br_2	6.4×10^{-23}
Mercury(I) chloride	Hg_2Cl_2	1.4×10^{-18}
Mercury(I) iodide	Hg_2I_2	5.3×10^{-29}
Mercury(II) hydroxide	$Hg(OH)_2$	3.1×10^{-26}
Nickel(II) hydroxide	$Ni(OH)_2$	5.5×10^{-16}
Silver bromide	$AgBr$	5.4×10^{-13}
Silver carbonate	Ag_2CO_3	8.4×10^{-12}
Silver chloride	$AgCl$	1.8×10^{-10}
Silver chromate	Ag_2CrO_4	1.1×10^{-12}
Silver cyanide	$AgCN$	6.0×10^{-17}
Silver iodide	AgI	8.5×10^{-17}
Silver sulfate	Ag_2SO_4	1.2×10^{-5}
Silver sulfite	Ag_2SO_3	1.5×10^{-14}
Strontium carbonate	$SrCO_3$	5.6×10^{-10}
Tin(II) hydroxide	$Sn(OH)_2$	5.4×10^{-27}
Zinc carbonate	$ZnCO_3$	1.2×10^{-10}
Zinc hydroxide	$Zn(OH)_2$	4.1×10^{-17}

TABLE C.5   Solubility Products in Acid (K_{spa}) at 25 °C

Compound	Formula	K_{spa}
Cadmium sulfide	CdS	8×10^{-7}
Cobalt(II) sulfide	CoS	3
Copper(II) sulfide	CuS	6×10^{-16}
Iron(II) sulfide	FeS	6×10^{2}
Lead(II) sulfide	PbS	3×10^{-7}
Manganese(II) sulfide	MnS	3×10^{7}
Mercury(II) sulfide	HgS	2×10^{-32}
Nickel(II) sulfide	NiS	8×10^{-1}
Silver sulfide	Ag_2S	6×10^{-30}
Tin(II) sulfide	SnS	1×10^{-5}
Zinc sulfide	ZnS	3×10^{-2}

Note: K_{spa} for MS is the equilibrium constant for the reaction

$$MS(s) + 2 H_3O^+(aq) \rightleftharpoons M^{2+}(aq) + H_2S(aq) + 2 H_2O(l)$$

We use K_{spa} for metal sulfides rather than K_{sp} because the traditional values of K_{sp} are now known to be incorrect since they are based on a K_{a2} value for H_2S that is greatly in error (see R. J. Myers, *J. Chem. Educ.*, **1986**, *63*, 687–690).

TABLE C.6 Formation Constants for Complex Ions at 25 °C

Complex Ion	K_f
$Ag(CN)_2^-$	3.0×10^{20}
$Ag(NH_3)_2^+$	1.7×10^7
$Ag(S_2O_3)_2^{3-}$	4.7×10^{13}
$Al(OH)_4^-$	3×10^{33}
$Be(OH)_4^{2-}$	4×10^{18}
$Cr(OH)_4^-$	8×10^{29}
$Cu(NH_3)_4^{2+}$	5.6×10^{11}
$Fe(CN)_6^{4-}$	3×10^{35}
$Fe(CN)_6^{3-}$	4×10^{43}
$Ga(OH)_4^-$	3×10^{39}
$Ni(CN)_4^{2-}$	1.7×10^{30}
$Ni(NH_3)_6^{2+}$	2.0×10^8
$Ni(en)_3^{2+}$	4×10^{17}
$Pb(OH)_3^-$	8×10^{13}
$Sn(OH)_3^-$	3×10^{25}
$Zn(CN)_4^{2-}$	4.7×10^{19}
$Zn(NH_3)_4^{2+}$	7.8×10^8
$Zn(OH)_4^{2-}$	3×10^{15}

Appendix D

Standard Reduction Potentials at 25 °C

Half-Reaction	E° (V)
$F_2(g) + 2\,e^- \longrightarrow 2\,F^-(aq)$	2.87
$O_3(g) + 2\,H^+(aq) + 2\,e^- \longrightarrow O_2(g) + H_2O(l)$	2.08
$S_2O_8{}^{2-}(aq) + 2\,e^- \longrightarrow 2\,SO_4{}^{2-}(aq)$	2.01
$Co^{3+}(aq) + e^- \longrightarrow Co^{2+}(aq)$	1.81
$H_2O_2(aq) + 2\,H^+(aq) + 2\,e^- \longrightarrow 2\,H_2O(l)$	1.78
$Ce^{4+}(aq) + e^- \longrightarrow Ce^{3+}(aq)$	1.72
$MnO_4{}^-(aq) + 4\,H^+(aq) + 3\,e^- \longrightarrow MnO_2(s) + 2\,H_2O(l)$	1.68
$PbO_2(s) + 3\,H^+(aq) + HSO_4{}^-(aq) + 2\,e^- \longrightarrow PbSO_4(s) + 2\,H_2O(l)$	1.628
$2\,HClO(aq) + 2\,H^+(aq) + 2\,e^- \longrightarrow Cl_2(g) + 2\,H_2O(l)$	1.61
$Mn^{3+}(aq) + e^- \longrightarrow Mn^{2+}(aq)$	1.54
$MnO_4{}^-(aq) + 8\,H^+(aq) + 5\,e^- \longrightarrow Mn^{2+}(aq) + 4\,H_2O(l)$	1.51
$2\,BrO_3{}^-(aq) + 12\,H^+(aq) + 10\,e^- \longrightarrow Br_2(l) + 6\,H_2O(l)$	1.48
$ClO_3{}^-(aq) + 6\,H^+(aq) + 5\,e^- \longrightarrow 1/2\,Cl_2(g) + 3\,H_2O(l)$	1.47
$Au^{3+}(aq) + 2\,e^- \longrightarrow Au^+(aq)$	1.40
$Cl_2(g) + 2\,e^- \longrightarrow 2\,Cl^-(aq)$	1.36
$Cr_2O_7{}^{2-}(aq) + 14\,H^+(aq) + 6\,e^- \longrightarrow 2\,Cr^{3+}(aq) + 7\,H_2O(l)$	1.36
$O_2(g) + 4\,H^+(aq) + 4\,e^- \longrightarrow 2\,H_2O(l)$	1.23
$MnO_2(s) + 4\,H^+(aq) + 2\,e^- \longrightarrow Mn^{2+}(aq) + 2\,H_2O(l)$	1.22
$2\,IO_3{}^-(aq) + 12\,H^+(aq) + 10\,e^- \longrightarrow I_2(s) + 6\,H_2O(l)$	1.20
$Br_2(l) + 2\,e^- \longrightarrow 2\,Br^-(aq)$	1.09
$HNO_2(aq) + H^+(aq) + e^- \longrightarrow NO(g) + H_2O(l)$	0.98
$NO_3{}^-(aq) + 4\,Hl^+(aq) + 3\,e^- \longrightarrow NO(g) + 2\,H_2O(l)$	0.96
$2\,Hg^{2+}(aq) + 2\,e^- \longrightarrow Hg_2{}^{2+}(aq)$	0.92
$HO_2{}^-(aq) + H_2O(l) + 2\,e^- \longrightarrow 3\,OH^-(aq)$	0.88
$Hg^{2+}(aq) + 2\,e^- \longrightarrow Hg(l)$	0.85
$ClO^-(aq) + H_2O(l) + 2\,e^- \longrightarrow Cl^-(aq) + 2\,OH^-(aq)$	0.84
$Ag^+(aq) + e^- \longrightarrow Ag(s)$	0.80
$Hg_2{}^{2+}(aq) + 2\,e^- \longrightarrow 2\,Hg(l)$	0.80
$NO_3{}^-(aq) + 2\,H^+(aq) + e^- \longrightarrow NO_2(g) + H_2O(l)$	0.79
$Fe^{3+}(aq) + e^- \longrightarrow Fe^{2+}(aq)$	0.77
$O_2(g) + 2\,H^+(aq) + 2\,e^- \longrightarrow H_2O_2(aq)$	0.70
$MnO_4{}^-(aq) + e^- \longrightarrow MnO_4{}^{2-}(aq)$	0.56
$H_3AsO_4(aq) + 2\,H^+(aq) + 2\,e^- \longrightarrow H_3AsO_3(aq) + H_2O(l)$	0.56
$I_2(s) + 2\,e^- \longrightarrow 2\,I^-(aq)$	0.54

Half-Reaction	$E°$ (V)
$Cu^+(aq) + e^- \longrightarrow Cu(s)$	0.52
$H_2SO_3(aq) + 4 H^+(aq) + 4 e^- \longrightarrow S(s) + 3 H_2O(l)$	0.45
$O_2(g) + 2 H_2O(l) + 4 e^- \longrightarrow 4 OH^-(aq)$	0.40
$Cu^{2+}(aq) + 2 e^- \longrightarrow Cu(s)$	0.34
$BiO^+(aq) + 2 H^+(aq) + 3 e^- \longrightarrow Bi(s) + H_2O(l)$	0.32
$Hg_2Cl_2(s) + 2 e^- \longrightarrow 2 Hg(l) + 2 Cl^-(aq)$	0.28
$AgCl(s) + e^- \longrightarrow Ag(s) + Cl^-(aq)$	0.22
$SO_4{}^{2-}(aq) + 4 H^+(aq) + 2 e^- \longrightarrow H_2SO_3(aq) + H_2O(l)$	0.17
$Cu^{2+}(aq) + e^- \longrightarrow Cu^+(aq)$	0.15
$Sn^{4+}(aq) + 2 e^- \longrightarrow Sn^{2+}(aq)$	0.15
$S(s) + 2 H^+(aq) + 2 e^- \longrightarrow H_2S(aq)$	0.14
$AgBr(s) + e^- \longrightarrow Ag(s) + Br^-(aq)$	0.07
$2 H^+(aq) + 2 e^- \longrightarrow H_2(g)$	0
$Fe^{3+}(aq) + 3 e^- \longrightarrow Fe(s)$	-0.04
$Pb^{2+}(aq) + 2 e^- \longrightarrow Pb(s)$	-0.13
$CrO_4{}^{2-}(aq) + 4 H_2O(l) + 3 e^- \longrightarrow Cr(OH)_3(s) + 5 OH^-(aq)$	-0.13
$Sn^{2+}(aq) + 2 e^- \longrightarrow Sn(s)$	-0.14
$AgI(s) + e^- \longrightarrow Ag(s) + I^-(aq)$	-0.15
$Ni^{2+}(aq) + 2 e^- \longrightarrow Ni(s)$	-0.26
$Co^{2+}(aq) + 2 e^- \longrightarrow Co(s)$	-0.28
$PbSO_4(s) + H^+(aq) + 2 e^- \longrightarrow Pb(s) + HSO_4{}^-(aq)$	-0.296
$Tl^+(aq) + e^- \longrightarrow Tl(s)$	-0.34
$Se(s) + 2 H^+(aq) + 2 e^- \longrightarrow H_2Se(aq)$	-0.40
$Cd^{2+}(aq) + 2 e^- \longrightarrow Cd(s)$	-0.40
$Cr^{3+}(aq) + e^- \longrightarrow Cr^{2+}(aq)$	-0.41
$Fe^{2+}(aq) + 2 e^- \longrightarrow Fe(s)$	-0.45
$2 CO_2(g) + 2 H^+(aq) + 2 e^- \longrightarrow H_2C_2O_4(aq)$	-0.49
$Ga^{3+}(aq) + 3 e^- \longrightarrow Ga(s)$	-0.55
$Cr^{3+}(aq) + 3 e^- \longrightarrow Cr(s)$	-0.74
$Zn^{2+}(aq) + 2 e^- \longrightarrow Zn(s)$	-0.76
$2 H_2O(l) + 2 e^- \longrightarrow H_2(g) + 2 OH^-(aq)$	-0.83
$Cr^{2+}(aq) + 2 e^- \longrightarrow Cr(s)$	-0.91
$Mn^{2+}(aq) + 2 e^- \longrightarrow Mn(s)$	-1.18
$Al^{3+}(aq) + 3 e^- \longrightarrow Al(s)$	-1.66
$Mg^{2+}(aq) + 2 e^- \longrightarrow Mg(s)$	-2.37
$Na^+(aq) + e^- \longrightarrow Na(s)$	-2.71
$Ca^{2+}(aq) + 2 e^- \longrightarrow Ca(s)$	-2.87
$Ba^{2+}(aq) + 2 e^- \longrightarrow Ba(s)$	-2.91
$K^+(aq) + e^- \longrightarrow K(s)$	-2.93
$Li^+(aq) + e^- \longrightarrow Li(s)$	-3.04

Appendix E

Properties of Water

Normal melting point	0 °C = 273.15 K
Normal boiling point	100 °C = 373.15 K
Heat of fusion	6.01 kJ/mol at 0 °C
Heat of vaporization	44.94 kJ/mol at 0 °C
	44.02 kJ/mol at 25 °C
	40.67 kJ/mol at 100 °C
Specific heat	4.181 J/(g · °C) at 25 °C
Ion-product constant, K_w	1.15×10^{-15} at 0 °C
	1.01×10^{-14} at 25 °C
	5.43×10^{-13} at 100 °C

TABLE E.1   Vapor Pressure of Water at Various Temperatures

Temp (°C)	P_{vap} (mm Hg)	Temp (°C)	P_{vap} (mm Hg)
0	4.58	60	149.4
10	9.21	70	233.7
20	17.5	80	355.1
30	31.8	90	525.9
40	55.3	100	760.0
50	92.5	105	906.0

CHAPTER 1

1.2. (a) 7×10^{-5} m; (b) 2×10^{13} kg **1.3.** 801 °C, 1074 K
1.4. liquid state **1.5.** 6.32 mL **1.6.** the bracelet is made of pure silver.
1.8. 42.5 m/s **1.10.** 4.55 mL **1.12.** high precision accuracy is low
1.14. 2.23×10^{-2} g/mL **1.16.** 2.52 cm^3 **1.18.** 6×10^{-11} cm^3,
0.06 pL **1.21.** 1,000 times; (b) 10,000 times **1.26.** (a) 30%; (b) 15%

Conceptual Problems
1.28. 32.2°C
1.30.

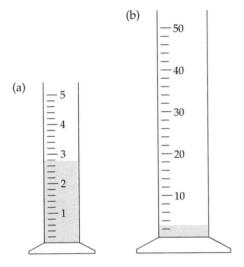

(b)

The 5 mL graduated cylinder can be read to ± 0.02 mL. The 50 mL graduated cylinder can only be read to ± 0.2 mL. The 5 mL graduated cylinder will give more accurate measurements.

Section Problems
Scientific Method (Section 1.1)

1.32. (a) experiment; (b) hypothesis; (c) observation **1.34.** (a), (b) and (d) are quantitative. (c) and (e) are qualitative.

SI Units and Scientific Notation (Section 1.2)

1.38. (a) kilogram, kg; (b) meter, m; (c) kelvin, K; (d) cubic meter, m^3; (e) joule, $(kg \cdot m^2)/s^2$; (f) kg/m^3 or g/cm^3 **1.42.** cL is centiliter $(10^{-2}$ L) **1.48.** (a) 4.5332×10^2 mg; (b) 4.21×10^{-5} mL; (c) 6.67×10^5 g.

Measurement of Mass, Length, and Temperature (Sections 1.3–1.5)

1.54. 103.8 °F (goat), 72.0 °F (Australian spiny anteater) **1.56.** 347 °F
1.58. (a) 1.021 °E/°C; (b) 0.5675 °E/°F; (c) H$_2$O melting point, 119.8 °E, H$_2$O boiling point, 222.0 °E; (d) 157.6 °E; (e) sweater or light jacket. **1.60.** melting point 801°C; 1474 °F boiling point 1413 °C; 2575 °F

Derived Units: Volume and Density (Sections 1.6–1.7)

1.66. 2.212 g/cm^3 **1.70.** H$_2$: 11.2 L, Cl$_2$: 11.03 L **1.72.** 11 $\frac{8}{cm^3}$ **1.74.** not pure silver. **1.76.** 75.85 mL

Energy (Section 1.8)

1.78. car **1.80.** 169 kcal **1.82.** (a) 2300 kJ; (b) 6.3 h

Accuracy, Precision, and Significant Figures (Sections 1.9–1.10)

1.84. (a) and (b) **1.86.** (a) 6; (b) 6; (c) 4; (d) 3; (e) 2, 3, 4, or 5; (f) 5 **1.88.** 3.666×10^6 m^3, 3.7×10^6 m^3 **1.90.** (a) 3.567×10^4 or 35,670 m; 35,670.1 m; (b) 69 g, 68.5 g; (c) 4.99×10^3 cm; (d) 2.3098×10^{-4} kg **1.92.** (a) 10.0; (b) 26; (c) 0.039; (d) 5526; (e) 87.6; (f) 13

Unit Conversions (Section 1.11)

1.94. (a) 110 g; (b) 443.2 m; (c) 76181×10^{12} m^2 **1.96.** 11.394 mi/h **1.98.** (a) 43,560 ft^3; (b) 3.92×10^8 acre–ft **1.100.** 121 lb **1.102.** 72.6 kg; 1.5 mg

Multiconcept Problems
1.112. (a) 200 kisses; (b) 3.3 mL; (c) 26 Cal/kiss; (d) 51 calories **1.114.** −40 °C (−40 °F).

CHAPTER 2

2.2. (a) K, potassium, metal; (b) shiny, metallic solid; (c) It would deform and not crack; (d) It would conduct electricity, but it is not a good choice for wiring because it reacts when exposed to oxygen and humidity (water) in the atmosphere. Potassium is also a soft metal which makes it unsuitable for use as a wire. **2.4.** SO_3 **2.5.** 2×10^4 Au atoms **2.6.** 40 times **2.7.** 34 protons, 34 electrons, 41 neutrons. **2.8.** Cr; $^{52}_{24}$Cr. **2.9.** 63.55 **2.10.** (a) gallium-69; (b) ^{69}Ga 60.11%; ^{71}Ga; 39.89% **2.11.** 0.0487 mol Pt 2.93×10^{22} atoms Pt **2.12.** Ca **2.15.** (b) **2.16.** (a) Figures (b) and (d); (b) Figures (a) and (c); (c) Figures (b) and (d). **2. 17.**

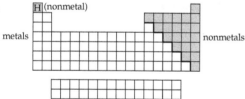

2.18. $C_9H_{13}NO_3$ **2.20.** Figure (a) ionic Figure; (b) molecular **2.21.** (a) MgF_2; (b) SnO_2; (c) Fe_2S_3 **2.22.** red – potassium sulfide, K_2S; green – strontium iodide, SrI_2; blue – gallium oxide, Ga_2O_3 **2.24.** Drawing 1 Only (c) $CaCl_2$. Drawing 2 (a) LiBr and (b) $NaNO_2$ **2.26.** (a) phosphorus pentachloride; (b) dinitrogen monoxide

Conceptual Problems

2.36. americium (Am) 95. actinide **2.38.** (a) **2.40.** (b) and (c) are isotopes (a) is a different element.

Section Problems
Elements and the Periodic Table (Sections 2.1–2.3)

2.46. 118; 90 naturally. **2.48.** (a) Gd; (b) Ge; (c) Tc; (d) As. **2.50.** (a) tellurium; (b) rhenium; (c) beryllium; (d) argon; (e) plutonium **2.52.** (a) Tin is Sn, Ti is titanium; (b) Manganese is Mn, Mg is magnesium; (c) Potassium is K, Po is Polonium; (d) The symbol for helium is He. **2.56.** periods groups. **2.58.** similar chemical properties. **2.60.**

```
H (nonmetal)
         [periodic table grid]
metals                          nonmetals
```

2.62. (a) Ti, metal; (b) Te, semimetal; (c) Se, nonmetal; (d) Sc, metal; (e) Si, semimetal **2.64.** (a) The alkali metals are shiny, soft, low-melting metals that react rapidly with water to form products that are alkaline; (b) The noble gases are gases of very low reactivity; (c) The halogens are nonmetallic and corrosive. They are found in nature only in combination with other elements. **2.66.** F, Cl, Br, and I. **2.68.** metal. **2.70.** nonmetal. **2.72.** 2A and 7A.

Atomic Theory (Sections 2.4–2.5)

2.76. The law of mass conservation in terms of Dalton's atomic theory states that chemical reactions only rearrange the way that atoms are combined; the atoms themselves are not changed. The law of definite proportions in terms of Dalton's atomic theory states that the chemical combination of elements to make different substances occurs when atoms join together in small, whole-number ratios. **2.78.** 3.7 g **2.80.** not methane.

$$\frac{\text{C:H mass ratio in "other"}}{\text{C:H mass ratio in methane}} = \frac{4}{3}$$

2.82. $\dfrac{\text{C:H mass ratio in benzene}}{\text{C:H mass ratio in ethane}} = \dfrac{3}{1}$,

$\dfrac{\text{C:H mass ratio in benzene}}{\text{C:H mass ratio in ethylene}} = \dfrac{2}{1}$,

$\dfrac{\text{C:H mass ratio in ethylene}}{\text{C:H mass ratio ethane}} = \dfrac{3}{2}$

2.84. $\dfrac{\text{O:C mass ratio in compound 2}}{\text{O:C mass ratio in compound 1}} = \dfrac{2}{1}$, CO_2.

Elements and Atoms (Sections 2.6–2.8)

2.86. electron **2.90.** (a) -1.010×10^{-18} C **2.92.** (a) The alpha particles would pass right through the gold foil with little to no deflection. **2.94.** 1.8×10^7 Pb atoms thick **2.96.** The atomic number is equal to the number of protons. The mass number is equal to the sum of the number of protons and the number of neutrons. **2.98.** one can readily look up the atomic number in the periodic table. **2.100.** (a) carbon, C; (b) argon, Ar; (c) vanadium, V. **2.102.** (a) $^{220}_{86}$Rn; (b) $^{210}_{84}$Po; (c) $^{197}_{79}$Au **2.104.** (a) 7 protons, 7 electrons, 8 neutrons; (b) 27 protons, 27 electrons, 33 neutrons; (c) 53 protons, 53 electrons, 78 neutrons, 58 protons, 58 electrons, 84 neutrons **2.106.** (a) magnesium; (b) nickel; (c) palladium; (d) tungsten **2.108.** $^{12}_{5}$C, $^{33}_{35}$Br, $^{11}_{5}$Bo

Atomic Weight, Moles, and Mass Spectrometry (Sections 2.9–2.10)

2.116. ^{63}Cu **2.118.** 10.8 **2.120.** 25.982 **2.122.** 63.55 **2.124.** (a) 72.04 g Ti; (b) 7.75 g Na; (c) 614.8 g U **2.126.** X grams. **2.128.** Kr.

Chemical Compounds (Sections 2.11–2.12)

2.134. A covalent bond results when two atoms share several (usually two) of their electrons. An ionic bond results from a complete transfer of one or more electrons from one atom to another.

2.136. Element symbols are composed of one or two letters. If the element symbol is two letters, the first letter is uppercase and the second is lowercase. CO stands for carbon and oxygen in carbon monoxide. **2.138.** (a) 4 protons and 2 electrons; (b) 37 protons and 36 electrons; (c) 34 protons and 36 electrons; (d) 79 protons and 76 electrons **2.140.** C_3H_8O **2.142.**

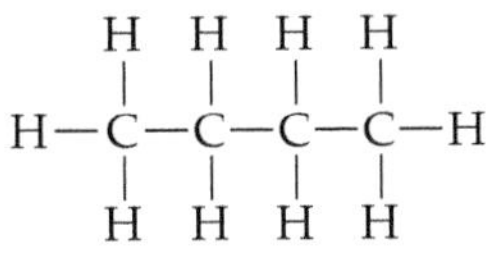

$$H-\underset{\underset{H}{|}}{\overset{\overset{H}{|}}{C}}-\underset{\underset{H}{|}}{\overset{\overset{H}{|}}{C}}-\underset{\underset{H}{|}}{\overset{\overset{H}{|}}{C}}-\underset{\underset{H}{|}}{\overset{\overset{H}{|}}{C}}-H$$

2.144.

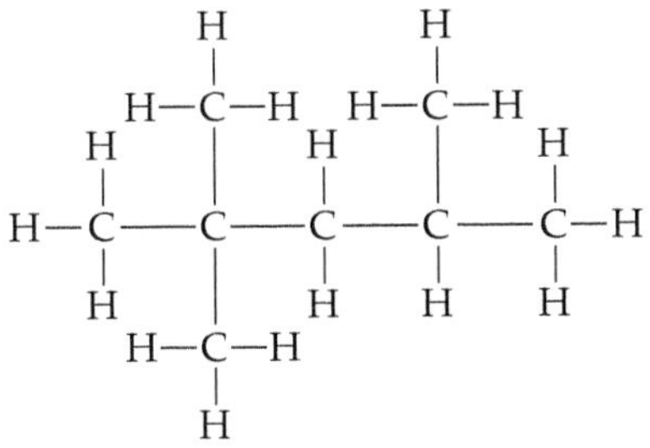

2.146. (a) cesium fluoride; (b) potassium oxide; (c) copper(II) oxide.
2.147. (a) barium sulfide; (b) beryllium bromide; (c) iron(III) chloride
2.148. (a) KCl; (b) $SnBr_2$; (c) CaO; (d) $BaCl_2$; (e) AlH_3
2.150. (a) $Ca(CH_3CO_2)_2$; (b) $Fe(CN)_2$; (c) $Na_2Cr_2O_7$; (d) $Cr_2(SO_4)_3$; (e) $Hg(ClO_4)_2$. **2.152.** (a) calcium hypochlorite; (b) silver(I) thiosulfate or silver thiosulfate; (c) sodium dihydrogen phosphate; (d) tin(II) nitrate; (e) lead(IV) acetate; (f) ammonium sulfate
2.154. (a) $CaBr_2$; (b) $CaSO_4$; (c) $Al_2(SO_4)_3$ **2.156.** (a) $CaCl_2$; (b) CaO; (c) CaS **2.158.** (a) SO_3^{2-}; (b) PO_4^{3-}; (c) Zr^{4+}; (d) CrO_4^{2-}; (e) $CH_3CO_2^-$; (f) $S_2O_3^{2-}$ **2.160.** (a) carbon tetrachloride; (b) chlorine dioxide; (c) dinitrogen monoxide; (d) dinitrogen trioxide.
2.162. (a) nitrogen monoxide; (b) dinitrogen monoxide; (c) nitrogen dioxide; (d) dinitrogen tetroxide; (e) dinitrogen pentoxide
2.164. (a) Na_2SO_4; (b) $Ba_3(PO_4)_2$; (c) $Ga_2(SO_4)_3$

Multiconcept Problems
2.166. 0.505 g H; 0.337 g H.

CHAPTER 3

3.1. $3 A_2 + 2 B \rightarrow 2 BA_3$. **3.2.** reactants, box (d), and products, box (c)
3.4. $8 KClO_3 + C_{12}H_{22}O_{11} \rightarrow 8 KCl + 12 CO_2 + 11 H_2O$
3.6. 342.3, 342.3 g/mol **3.7.** 0.0626 mol $NaHCO_3$ **3.8.** 15.0 g
4 tablets; 5.02×10^{22} molecules; **3.9.** 3.33 g **3.10.** 7.66 g; (b) 3.34 g
3.11. 63% **3.12.** 27.9 g; (b) 143 g **3.13.** (a) $A + B_2 \rightarrow AB_2$; (b) Limiting reactant is A, excess reactant is B_2; (c) 6 molecules of AB_2
3.14.

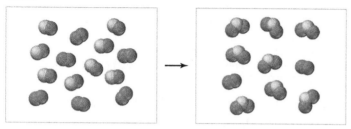

B_2 is in excess, AB is the limiting reactant. **3.16.** Yes, 411.7 g CO_2
3.17. $MgCO_3$. **3.18.** CH_2O, %C = 40.0%; %H = 6.7%,
%O = 53.3% **3.19.** $C_{10}H_{20}O$. **3.20.** Empirical formula is C_5H_{11} and
Molecular formula is $C_{10}H_{22}$ **3.21.** $C_{12}H_{10}$ **3.22.** C_5H_5N, empirical
formula and molecular formula are the same. **3.25.** (a) 56 million
tons of monoethanolamine; (b) 11 million tons of CO_2

Conceptual Problems

3.32. (a) $A_2 + 3 B_2 \rightarrow 2 AB_3$; B_2 is the limiting reactant; (b)
2/3 mol AB_3

Section Problems
Balancing Equations (Section 3.2)

3.36. (b) **3.38.** (a) $Mg + 2 HNO_3 \rightarrow H_2 + Mg(NO_3)_2$;
(b) $CaC_2 + 2 H_2O \rightarrow Ca(OH)_2 + C_2H_2$; (c) $2 S + 3 O_2 \rightarrow 2 SO_3$;
(d) $UO_2 + 4 HF \rightarrow UF_4 + 2 H_2O$ **3.40.**
(a) $SiCl_4 + 2 H_2O \rightarrow SiO_2 + 4 HCl$; (b) $P_4O_{10} + 6 H_2O \rightarrow 4 H_3PO_4$;
(c) $CaCN_2 + 3 H_2O \rightarrow CaCO_3 + 2 NH_3$;
(d) $3 NO_2 + H_2O \rightarrow 2 HNO_3 + NO$

Molecular Weights and Molar Mass (Section 3.3)

3.44. (a) 472.1; (b) 88.1; (c) 120.9 **3.46.** (a) 558.7; (b) 444.5;
(c) 321.8 **3.48.** (a) 47.87 g; (b) 159.81 g; (c) 200.59 g; (d) 18.02 g
3.50. 0.867 mol **3.52.** 119. **3.54.** 1.97×10^{-3} mol,
1.19×10^4 Fe(II)atoms **3.56.** 6.44×10^{-4} mol,
3.88×10^{20} molecules **3.58.** neon (Ne). **3.60.** 166.8 kg

Stoichiometry (Section 3.4)

3.62. (a) $2 Fe_2O_3 + 3 C \rightarrow 4 Fe + 3 CO_2$; (b) 4.93 mol C; (c) 59.2 g C
3.64. (a) $2 Mg + O_2 \rightarrow 2 MgO$; (b) 16.5 g O_2 41.5 g MgO; (c) 38.0 g
Mg, 63.0 g MgO **3.66.** (a) $2 HgO \rightarrow 2 Hg + O_2$; (b) 42.1 g Hg,
3.36 g O_2, 451 g HgO **3.68.** AgCl. **3.70.** (a) 581 g; (b) 1847 g

Reaction Yield and Limiting Reactants (Sections 3.5–3.6)

3.72. (a) 12.0 g; (b) 14.9 g **3.74.** 15.8 g NH_3 83.3 g N_2 left over
3.76. 5.22 g **3.78.** $CaCO_3$ is the limiting. 0.526 L CO_2 **3.80.** 3.2 g.
3.82. 86.8%

Percent Composition and Empirical Formulas (Section 3.7)

3.86. %C = 20.0% %H = 6.72% %N = 46.6% %O = 26.6%
3.88. (a) $C_9H_8O_4$.; (b) $FeTiO_3$; (c) $Na_2S_2O_3$. **3.90.** SnF_2.

Formulas and Elemental Analysis (Section 3.8)

3.92. C_7H_7Cl, **3.94.** C_7H_8. **3.96.** 13,000 u **3.98.** 6 **3.100.** $C_{12}Br_{10}O$.

Mass Spectrometry (Section 3.9)

3.102. A neutral molecule will travel in a straight, undeflected, path
in a mass spectrometer. Ionization is necessary as electric and magnetic
fields will only exert a force on a charged species, not a neutral
molecule. Ions of different masses are then accelerated by an electric
field and passed between the poles of a strong magnet, which deflects
them through a curved, evacuated pipe. The radius of deflection of a
charged ion, M^+, as it passes between the magnet poles depends on
its mass, with lighter ions deflected more strongly than heavier ones.
3.104. C_4H_6O. **3.106.** CH, C_6H_6

Multiconcept Problems

3.108. High resolution mass spectrometry is capable of measuring
the mass of molecules with a particular isotopic composition.
3.110. 18.1 lb **3.112.** the mass %'s for the pulverized rock are different
from the mass %'s for pure $CaCO_3$ **3.114.** 4.4 g Fe_2O_3 5.6 g FeO
3.116. $C_6H_{12}O_6 + 6 O_2 \rightarrow 6 CO_2 + 6 H_2O$; 97.2 g 56.1 L
3.118. 31.1, P **3.120.** mass ratio of NH_4NO_3 to $(NH_4)_2HPO_4$ in the
mixture is 2 to 1. **3.122.** 80; Br. 64; Cu **3.124.** (a) $C_{12}H_5Cl_5$;
(b) $C_{12}H_5Cl_5$; (c) No, combustion analysis only directly gives the
mass of carbon and hydrogen. The mass of *one* extra element can be
found by subtracting the mass of carbon and hydrogen from the original
mass of the compound. In this case, the mass of two elements is
unknown.

CHAPTER 4

4.5. 0.656 M **4.6.** 6.94 mL **4.7.** 0.675 M. **4.8.** (a) A_2Y is the strongest A_2X is the weakest; (b) = 0.700 M, Y ions = 0.350 M.; (c) 20% **4.10.** $Ag^+(aq) + Cl^-(aq) \rightarrow AgCl(s)$, $2\,AgNO_3(aq) + CaCl_2(aq) \rightarrow 2\,AgCl(s) + Ca(NO_3)_2(aq)$
4.12. $3\,Ca^{2+}(aq) + 2\,PO_4^{3-}(aq) \rightarrow Ca_3(PO_4)_2(s)$ **4.14.** $PbBr_2$.

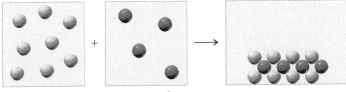

Anion = Br^- Cation = Pb^{2+}

4.16. (a) H_3PO_3; (b) H_2Se
4.18. $Mg^{2+}(aq) + 2\,OH^-(aq) + 2\,H^+(aq) + 2\,Cl^-(aq) \rightarrow Mg^{2+}(aq) + 2\,Cl^-(aq) + H_2O(l)$ $H^+(aq) + OH^-(aq) \rightarrow H_2O(l)$
4.19. 10.0 mL **4.20.** 0.0547 M **4.21.** 0.758 M **4.22.** 66.7 mL
4.23. (a) Cl -1, Sn $+4$; (b) O -2, Cr $+6$; (c) O -2, Cl -1, V $+5$; (d) O -2, V $+3$; (e) O -2, H $+1$, N $+5$; (f) O -2, S $+6$, Fe $+2$
4.24. (a) $+2$, ClO, chlorine monoxide; $+3$, Cl_2O_3, dichlorine trioxide; $+6$, ClO_3, chlorine trioxide; $+7$, Cl_2O_7, dichlorine heptoxide; (b) Cl_2O_7 **4.26.** (a) C in C_2H_5OH oxidized and Cr in $K_2Cr_2O_7$ gets reduced; (b) $K_2Cr_2O_7$ is the oxidizing agent; C_2H_5OH is the reducing agent. **4.29.** 1.98 M **4.30.** 50.0 mg **4.31.** Sodium chloride, sodium citrate, and potassium dihydrogen phosphate are strong electrolytes citric acid and vitamin B3 are weak electrolytes fructose is a nonelectrolyte; (b) Sodium chloride, potassium dihydrogen phosphate, and sodium citrate **4.32.** 0.018 M; 0.0025 M
4.33. 0.528 g **4.36.** 0.0190 M

Conceptual Problems
4.38. (a) box (1); (b) box (2); (c) box (3). **4.40.** HY is the strongest HX is the weakest **4.42.** 0.040 M; 67% **4.44.** (a) $Sr^+ + At \rightarrow Sr + At^+$ No reaction; (b) $Si + At^+ \rightarrow Si^+ + At$ Reaction would occur; (c) $Sr + Si^+ \rightarrow Sr^+ + Si$ Reaction would occur.

Section Problems
Molarity (Section 4.1)
4.46. (a) 0.0420 mol; (b) 0.12 mol **4.48.** 160 mL **4.50.** 0.0685 M
4.52. 0.958 M **4.54.** (a) 25.0 g; (b) 67.6 g

Dilutions (Section 4.2)
4.58. 1.71 M

Electrolytes (Section 4.3)
4.62. (a) bright; (b) dark; (c) dim **4.64.** $Ba(OH)_2$ dissociates into $Ba^{2+}(aq)$ and $2\,OH^-(aq)$, and conducts electricity. H_2SO_4 dissociates into $H^+(aq)$ and $HSO_4^-(aq)$. H_2SO_4 solutions conduct electricity. When equal molar solutions of $Ba(OH)_2$ and H_2SO_4 are mixed, the insoluble $BaSO_4$ is formed $BaSO_4$ does not produce any appreciable amount of ions and the mixture does not conduct electricity. **4.66.** (a) strong electrolyte; (b) weak electrolyte; (c) strong electrolyte; (d) strong electrolyte; (e) weak electrolyte; (f) nonelectrolyte **4.68.** (a) 2.25 M; (b) = 1.42 M
4.70. Na^+ 0.147 M Ca^{2+} 0.002 98 M K^+ 0.004 02 M Cl^- 0.157 M

Net Ionic Equations and Aqueous Reactions (Sections 4.4–4.5)
4.72. (a) precipitation; (b) redox; (c) acid-base neutralization

Precipitation Reactions and Solubility Guidelines (Section 4.6)
4.76. (a) insoluble; (b) soluble; (c) insoluble; (d) soluble **4.78.** (a) No; (b) $Fe^{2+}(aq) + 2\,OH^-(aq) \rightarrow Fe(OH)_2(s)$; (c) No (d) No
4.80. (a) not form; (b) $BaSO_4(s)$ will precipitate; (c) $AgCl(s)$ will precipitate
4.82. (a) $Pb(NO_3)_2(aq) + Na_2SO_4(aq) \rightarrow PbSO_4(s) + 2\,NaNO_3(aq)$; (b) $3\,MgCl_2(aq) + 2\,K_3PO_4(aq) \rightarrow Mg_3(PO_4)_2(s) + 6\,KCl(aq)$; (c) $ZnSO_4(aq) + Na_2CrO_4(aq) \rightarrow ZnCrO_4(s) + Na_2SO_4(aq)$
4.84. 0.645 g AgCl **4.86.** Add HCl(aq); it will selectively precipitate AgCl(s). **4.88.** Cs^+ and/or NH_4^+.
4.90. (a) Add HCl $Hg_2^{2+}(aq) + 2\,Cl^-(aq) \rightarrow Hg_2Cl_2(s)$; (b) Add H_2SO_4 $Pb^{2+}(aq) + SO_4^{2-}(aq) \rightarrow PbSO_4(s)$; (c) Add Na_2CO_3 $Ca^{2+}(aq) + CO_3^{2-}(aq) \rightarrow CaCO_3(s)$; (d) Add Na_2SO_4 $Ba^{2+}(aq) + SO_4^{2-}(aq) \rightarrow BaSO_4(s)$

Acids, Bases, and Neutralization Reactions (Section 4.7)
4.94. Add the solution to an active metal, such as magnesium. Bubbles of H_2 gas indicate the presence of an acid.
4.96. (a) $2\,H^+(aq) + 2\,ClO_4^-(aq) + Ca^{2+}(aq) + 2\,OH^-(aq) \rightarrow Ca^{2+}(aq) + 2\,ClO_4^-(aq) + 2\,H_2O(l)$; (b) $CH_3CO_2H(aq) + Na^+(aq) + OH^-(aq) \rightarrow CH_3CO_2^-(aq) + Na^+(aq) + H_2O(l)$ $H^+(aq) + OH^-(aq) \rightarrow H_2O(l)$ $H^+(aq) + OH^-(aq) \rightarrow H_2O(l)$

Solution Stoichiometry and Titration (Sections 4.8–4.9)

4.100. 15.5 g **4.102.** 57.2 mL **4.104.** (a) 9.0 mL; (b) 11.0 mL
4.106. (a) basic; (b) basic

Oxidation Numbers (Section 4.10)

4.108. (a) O -2, N $+4$; (b) O -2, S $+6$; (c) O -2, Cl -1, C $+4$;
(d) Cl -1, H $+1$, C 0; (e) O -2, K $+1$, Cl $+5$;
(f) O -2, H $+1$, N $+5$ **4.110.** (a) O -2, Cl $+5$; (b) O -2, S $+4$;
(c) O -2, C $+3$; (d) O -2, N $+3$; (e) O -2, Br $+1$;
(f) O -2, As $+5$ **4.112.** (a) $+1$, N_2O, nitrous oxide; $+2$, NO,
nitric oxide; $+4$, N_2O_4, dinitrogen tetroxide; $+5$, N_2O_5, dinitrogen
pentoxide; (b) N_2O_5

Redox Reactions (Section 4.11)

4.114. The best reducing agents are at the bottom left of the periodic
table. The best oxidizing agents are at the top right of the periodic
table (excluding the noble gases). **4.116.** (a) gains; (b) loses; (c) loses;
(d) gains **4.118.** (a) Ca(s) is oxidized Sn^{2+}(aq) is reduced; (b) No
oxidation numbers change.

Activity Series (Section 4.12)

4.120. (a) no reaction; (b) no reaction; (c) no reaction;
(d) Au^{3+}(aq) $+$ 3 Ag(s) $\rightarrow$ 3 Ag^+(aq) $+$ Au(s).
4.122. (a) A > B > C > D; (b) 1. no reaction. 2. no reaction.

Redox Titrations (Section 4.13)

4.124. 0.670 g I_2 **4.126.** 1.130 M **4.128.** 0.134 M As (III)
4.130. 80.32% **4.132.** 0.101%

Multiconcept Problems

4.138. 0.4450 % Cu **4.144.** Cu **4.148.** empirical formula is $C_7H_6O_3$.
empirical formula is the molecular formula **4.150.** (a) Ba;
(b) 0.1571 L CO_2 **4.152.** 6.5 g $Ba(OH)_2$ 3.5 g NaOH **4.154.**
(a) 14 H^+(aq) $+ Cr_2O_7^{2-}$(aq) $+$ 6 Cr^{2+}(aq) $\rightarrow$ 8 Cr^{3+}(aq) $+$ 7 H_2O(l);
(b) K^+ 0.0833 M NO_3^- 0.617 M H^+ 0.183 M $Cr_2O_7^{2-} = 0.0250$ M
$Cr^{3+} = 0.133$ M
4.156. (a) 1. 3 Cu(s) $+$ 8 H^+(aq) $+$ 2 NO_3^-(aq) $\rightarrow$
3 Cu^{2+}(aq) $+$ 2 NO(g) $+$ 4 H_2O(l)
2. 2 Cu^{2+}(aq) $+$ 2 SCN^-(aq) $+$ H_2O(l) $+$ HSO_3^-(aq) $\rightarrow$
2 CuSCN(s) $+$ HSO_4^-(aq) $+$ 2 H^+(aq),
3. 10 Cu^+(aq) $+$ 12 H^+(aq) $+$ 2 IO_3^-(aq) $\rightarrow$
10 Cu^{2+}(aq) $+$ I_2(aq) $+$ 6 H_2O(l)
4. I_2(aq) $+$ 2 $S_2O_3^{2-}$(aq) $\rightarrow$ 2 I^-(aq) $+$ $S_4O_6^{2-}$(aq),
5. 2 $ZnNH_4PO_4$ $\rightarrow$ $Zn_2P_2O_7$ $+$ H_2O $+$ 2 NH_3; (b) 77.1% Cu;
(c) 19.5% Zn **4.158.** (a) 5 H_3MO_3(aq) $+$ 2 MnO_4^-(aq) $+$
6 H^+(*aq*) $\rightarrow$ 5 H_3MO_4(*aq*) $+$ 2 Mn^{2+}(aq) $+$ 3 H_2O(l);
(b) 1.34 $\times$ 10^{-3} mol M_2O_3; 2.68 $\times$ 10^{-3} mol M; (c) As.

CHAPTER 5

5.2. (b) higher frequency; (b) more intense (b) represents blue light; (a) represents red light. **5.3.** IR, 77.2 kJ/mol; UV, 479 kJ/mol, X ray, 2.18×10^4 kJ/mol **5.4.** 1.4×10^3 mol photons **5.5.** Electrons will not be ejected. **5.6.** (a) Ag; (b) Rb **5.8.** (a) 1875 nm; (b) 820.4 nm **5.10.** 7×10^6 m/s
5.12.

n	l	m_l	Orbital	No. of Orbitals
5	0	0	5s	1
	1	$-1, 0, +1$	5p	3
	2	$-2, -1, 0, +1, +2$	5d	5
	3	$-3, -2, -1, 0, +1, +2, +3$	5f	7
	4	$-4, -3, -2, -1, 0, +1, +2, +3, +4$	5g	9
				25

5.13. $n = 4, l = 0, 4s$ **5.14.** four **5.16.** (a) Tc; (b) Ni **5.18.** C–I **5.21.** (a) The fluorescent bulb does not emit all the wavelengths of light that would be emitted from a white light source; (b) Fluorescent light does appear as "white light" because its line spectrum has contributions from all the colors **5.22.** (a) $[Xe]\ 6s^2 4f^{14} 5d^{10}$; (b) $[Xe]$

↑ ↑↓ ↑↓ ↑↓ ↑↓ ↑↓ ↑↓ ↑↓ ↑↓ ↑↓ ↑↓ ↑↓ ↑↓ ;
‾6s‾ ‾‾‾‾‾‾‾‾4f‾‾‾‾‾‾‾‾ ‾‾‾‾‾5d‾‾‾‾‾

(c) no unpaired electrons. **5.23.** (a) $[Xe]\ 6s^1 4f^{14} 5d^{10} 6p^1$;

(b) $[Xe]$ ↑ ↑↓ ↑↓ ↑↓ ↑↓ ↑↓ ↑↓ ↑↓
        ‾6s‾ ‾‾‾‾‾‾‾4f‾‾‾‾‾‾‾

↑↓ ↑↓ ↑↓ ↑↓ ↑↓ ↑ ___ ; (c) 2
‾‾‾‾‾5d‾‾‾‾‾ ‾‾‾‾‾6p‾‾‾‾‾

5.24. (a) $n = 7, l = 2, m_l = -2, -1, 0, 1, 2$; (b) $n = 6, l = 1, m_l = -1, 0, 1$; (c) 275 kJ/mol **5.25.** 126.8 nm corresponds to 8p → 6s; 140.2 nm corresponds to 7p → 6s; and 185.0 nm corresponds to 6p → 6s.

Conceptual Problems

5.26. (a) has the greater intensity. (a) has the higher energy radiation. (a) represents yellow light. (b) represents infrared radiation. **5.28.** (a) $3p_y\ n = 3, l = 1$; (b) $4d_{z^2}\ n = 4, l = 2$ **5.30.** molybdenum, Its anomalous electron configuration is $[Ar]\ 5s^1 4d^5$ because of the resulting half-filled d-orbitals. **5.32.** selenium

Se,   [Ar]   ↑↓   ↑↓ ↑↓ ↑↓ ↑↓ ↑↓   ↑↓ ↑ ↑
            ‾4s‾   ‾‾‾‾‾‾‾3d‾‾‾‾‾‾‾   ‾‾‾4p‾‾‾

Section Problems
Wave Properties of Radiant Energy (Section 5.1)

5.34. Violet has the higher frequency and energy. Red has the higher wavelength. **5.36.** visible completely within this range ultraviolet and infrared regions are partially in this range. **5.38.** 5.5×10^{-8} m **5.40.** (a) 36.4 cm; (b) 34.3 cm

Particlelike Properties of Radiant Energy (Section 5.2)

5.42. (a) 99.5 MHz 3.97×10^{-5} kJ/mol 1150 kHz 4.589×10^{-7} kJ/mol FM radio wave (99.5 MHz) has the higher energy; (b) 3.44×10^{-9} m 3.48×10^4 kJ/mol 6.71×10^{-2} m 1.78×10^{-3} kJ/mol X ray ($\lambda = 3.44 \times 10^{-9}$ m) has the higher energy. **5.44.** (a) 1320 nm, near IR; (b) 0.149 m, radio wave; (c) 65.4 nm, UV **5.46.** (a) 1.0×10^{-10} m; (b) 1.2×10^6 kJ/mol; (c) X rays **5.50.** (a) 779 nm 2.55×10^{-19} J; (b) 649 nm 3.06×10^{-19} J; (c) 405 nm 4.91×10^{-19} J **5.52.** Both (c) & (d) no electrons would be ejected; (b) would eject the least number **5.54.** 1.09×10^{15} Hz

Atomic Line Spectra and Quantized Energy (Section 5.3)

5.56. continuous **5.58.** n = 3; 182.3kJ/mol n = 4; 246.1kJ/mol n = 5; 275.6kJ/mol **5.60.** 91.16 nm 1312kJ/mol 1312 kJ/mol 2625 nm 45.60 kJ/mol, IR **5.62.** 4051 nm **5.64.** 410.2 nm 291.6 kJ/mol

Wavelike Properties of Particles (Section 5.4)

5.66. 2.45×10^{-12} m, γ ray **5.68.** 1.06×10^{-34} m too small **5.70.** 9.14×10^{-24} m/s

Orbitals and Quantum Mechanics (Sections 5.5–5.8)

5.72. 8×10^{-31} m **5.74.** The Heisenberg uncertainty principle states that one can never know both the position and the velocity of an electron beyond a certain level of precision. This means we cannot think of electrons circling the nucleus in specific orbital paths, but we can think of electrons as being found in certain three-dimensional regions of space around the nucleus, called orbitals. **5.76.** n is the principal quantum number. The size and energy level of an orbital depends on n. l is the angular-momentum quantum number. l defines the three-dimensional shape of an orbital. m_l is the magnetic quantum number. m_l defines the spatial orientation of an orbital. m_s is the spin quantum number. m_s indicates the spin of the electron and can have either of two values, $+^1/_2$ or $-^1/_2$. **5.78.** (a) $n = 4; l = 0; m_l = 0; m_s = \pm^1/_2$; (b) $l = 1$; $m_l = -1, 0, +1; m_s = \pm^1/_2$; (c) $l = 3$; $m_l = -3, -2, -1, 0, +1, +2, +3; m_s = \pm^1/_2$; (d) $n = 5; l = 2$; $m_l = -2, -1, 0, +1, +2; m_s = \pm^1/_2$ **5.80.** (c) **5.82.** 138 electrons **5.84.** 363kJ/mol
5.86.

n	l	m_l	m_s
1	0	0	$+^1/_2$
1	0	0	$-^1/_2$
2	0	0	$+^1/_2$
2	0	0	$-^1/_2$
2	1	-1	$+^1/_2$
2	1	0	$+^1/_2$

5.88.

n	l	m_l	m_s
5	0	0	$+^1/_2$
5	0	0	$-^1/_2$

5.90. three

4s orbital
nodes are white
regions of maximum electron probability are black

Orbital Energy Levels in Multielectron Atoms (Section 5.9)

5.94. Part of the electron-nucleus attraction is canceled by the electron-electron repulsion, an effect we describe by saying that the electrons are shielded from the nucleus by the other electrons. The net nuclear charge actually felt by an electron is called the effective nuclear charge, Z_{eff}, and is often substantially lower than the actual nuclear charge, Z_{actual}. **5.96.** 4s > 4d > 4f **5.98.** The number of elements in successive periods of the periodic table increases by the progression 2, 8, 18, 32 because the principal quantum number n increases by 1 from one period to the next. As the principal quantum number increases, the number of orbitals in a shell increases. The progression of elements parallels the number of electrons in a particular shell. **5.100.** (a) 5d; (b) 4s; (c) 6s **5.102.** (a) 3d; (b) 4p; (c) 6d; (d) 6s

Electron Configurations (Sections 5.10–5.12)

5.104. (a) $1s^2\ 2s^2\ 2p^6\ 3s^2\ 3p^6\ 4s^2\ 3d^2$;
(b) $1s^2\ 2s^2\ 2p^6\ 3s^2\ 3p^6\ 4s^2\ 3d^{10}\ 4p^6\ 5s^2\ 4d^6$;
(c) $1s^2\ 2s^2\ 2p^6\ 3s^2\ 3p^6\ 4s^2\ 3d^{10}\ 4p^6\ 5s^2\ 4d^{10}\ 5p^2$;
(d) $1s^2\ 2s^2\ 2p^6\ 3s^2\ 3p^6\ 4s^2\ 3d^{10}\ 4p^6\ 5s^2$;
(e) $1s^2\ 2s^2\ 2p^6\ 3s^2\ 3p^6\ 4s^2\ 3d^{10}\ 4p^4$ **5.106.** (a) $[Kr]\ \uparrow\ (5s)$;
(b) $[Xe]$ $\uparrow\downarrow$ (6s) $\uparrow\downarrow\ \uparrow\downarrow\ \uparrow\downarrow\ \uparrow\downarrow\ \uparrow\downarrow\ \uparrow\downarrow\ \uparrow\downarrow$ (4f) $\uparrow\ \uparrow\ \uparrow\ \uparrow$ (5d);
(c) $\uparrow\downarrow$ (4s) $\uparrow\downarrow\ \uparrow\downarrow\ \uparrow\downarrow\ \uparrow\downarrow\ \uparrow\downarrow\ \uparrow$ (3d) $\uparrow$ (4p);
(d) $\uparrow\downarrow$ (5s) $\uparrow\ \uparrow\ \_\ \_\ \_$ (4d)
5.108. (a) 2; (b) 2; (c) 1; (d) 3

5.110. (a) Ra $[Rn]\ 7s^2$ $[Rn]\ \uparrow\downarrow$ (7s);
(b) Sc $[Ar]\ 4s^2\ 3d^1$ $[Ar]\ \uparrow\downarrow$ (4s) $\uparrow\ \_\ \_\ \_\ \_$ (3d);

(c) Lr $[Rn]\ 7s^2\ 5f^{14}\ 6d^1$ $[Rn]\ \uparrow\downarrow$ (7s) $\uparrow\downarrow\ \uparrow\downarrow\ \uparrow\downarrow\ \uparrow\downarrow\ \uparrow\downarrow\ \uparrow\downarrow\ \uparrow\downarrow$ (5f) $\uparrow\ \_\ \_\ \_\ \_$ (6d);
(d) B $[He]\ 2s^2\ 2p^1$ $[He]\ \uparrow\downarrow$ (2s) $\uparrow\ \_\ \_$ (2p);
(e) Te $[Kr]\ 5s^2\ 4d^{10}\ 5p^4$ $[Kr]\ \uparrow\downarrow$ (5s) $\uparrow\downarrow\ \uparrow\downarrow\ \uparrow\downarrow\ \uparrow\downarrow\ \uparrow\downarrow$ (4d) $\uparrow\downarrow\ \uparrow\ \uparrow$ (5p)
5.112. $Z = 121$ **5.114.** $1s^2\ 2s^2\ 2p^6$ **5.116.** $[Rn]\ 7s^2\ 5f^{14}\ 6d^{10}\ 7p^4$

Electron Configurations and Periodic Properties (Section 5.13)

5.118. Atomic radii increase down a group because the electron shells are farther away from the nucleus. **5.120.** $F < O < S$ **5.122.** (a) K; (b) Ta; (c) V; (d) Ba

Multiconcept Problems

5.124. 164 nm **5.132.** 940 kJ/mol **5.134.** $\Delta E = \dfrac{Z^2 e^2}{2a_o}\left[\dfrac{1}{n_1^2} - \dfrac{1}{n_2^2}\right]$
This is similar to the Balmer-Rydberg equation where $1/\lambda$ or v for the emission spectra of atoms is proportional to $\left[\dfrac{1}{m^2} - \dfrac{1}{n^2}\right]$ where m and n are integers with $n > m$.
5.136. (a) $1.09 \times 10^{15}\ s^{-1}$; (b) 1.8 nm **5.138.** (a) 5f subshell: $n = 5, l = 3, m_l = -3, -2, -1, 0, +1, +2, +3$ 3d subshell: $n = 3, l = 2, m_l = -2, -1, 0, +1, +2$; (b) 1282 nm; (c) 146 kJ/mol

CHAPTER 6

6.1. (a) Ra^{2+} [Rn]; (b) Ni^{2+} [Ar] $3d^8$; (c) N^{3-} [Ne]. **6.2.** (b)(c).
6.4. Cl^- is yellow, K^+ is green, and Ca^{2+} is red. **6.6.** (c) < (b) <
(a) < (d). **6.8.** Al. **6.10.** The least favorable E_{ea} is for Kr (red). The
most favorable E_{ea} is for Ge (blue). **6.12.** gain 2 electrons. 2−.
6.13. −1119 kJ/mol. **6.14.** −863 kJ/mol. **6.16.** (a) is NaCl and
(b) is MgO. MgO has the larger lattice energy. **6.18.** The cation has
an irregular shape and one or both of the ions are large and bulky.
6.19. (a) Tetraheptylammonium bromide corresponds to picture
(ii) and tetraheptylammonium iodide corresponds to picture (i);
(b) Tetraheptylammonium bromide; (c) Tetraheptylammonium bromide has melting point of 88°C and tetraheptylammonium iodide has
a melting point of 39°C. **6.20.** (a) F^-, $1s^2\,2s^2\,2p^6$
Se^{2-}, $1s^2\,2s^2\,2p^6\,3s^2\,3p^6\,4s^2\,3d^{10}\,4p^6$
O^{2-}, $1s^2\,2s^2\,2p^6$
Br^-, $1s^2\,2s^2\,2p^6\,3s^2\,3p^6\,4s^2\,3d^{10}\,4p^6$;
(b) F^- and O^{2-}; Se^{2-} and Br^-. (c) Br^-.

Conceptual Problems
6.26. (a) ionic; (b) covalent

Section Problems
Electron Configuration of Ions (Section 6.1)
6.34. A covalent bond results when two atoms share several (usually
two) of their electrons. An ionic bond results from a complete transfer of one or more electrons from one atom to another.
6.36. A molecule is the unit of matter that results when two or more
atoms are joined by covalent bonds. An ion results when an atom
gains or loses electrons. **6.38.** (a) 4 protons and 2 electrons;
(b) 37 protons and 36 electrons; (c) 34 protons and 36 electrons;
(d) 79 protons and 76 electrons **6.40.** (a) [Xe]; (b) [Kr] $4d^{10}$;
(c) [Kr] $5s^2\,4d^{10}$. **6.42.** [Ar]; [Ar] $3d^2$ **6.44.** Mg^{2+}. **6.46.** $Cr^{2+}\,Fe^{2+}$

Ionic Radii (Section 6.2)
6.48. (a) S^{2-}; (b) Ca; (c) O^{2-}. **6.50.** $Sr^{2+} < Rb^+ < Br^- < Se^{2-}$.

Ionization Energy (Section 6.3)
6.56. largest 8A smallest 1A.

6.58.

	Lowest E_{i1}	Highest E_{i1}
(a)	K	Li;
(b)	B	Cl;
(c)	Ca	Cl

Higher Ionization Energies (Section 6.4)
6.60. (a) Ca; (b) Ca. **6.62.** (a) P; (b) Ar; (c) Ca Ar has the highest E_{i2}.
Ar has the lowest E_{i7}. **6.64.** boron.

Electron Affinity (Section 6.5)
6.66. same magnitude but opposite signs. **6.68.** Na^+ more than
either. **6.70.** Energy is usually released when an electron is added to
a neutral atom but absorbed when an electron is removed from a
neutral atom because of the positive Z_{eff}. **6.72.** The electron-electron
repulsion is large and Z_{eff} is low.

Octet Rule (Sections 6.6)
6.74. (a) [Ne], N^{3-}; (b) [Ar], Ca^{2+}; (c) [Ar], S^{2-}; (d) [Kr], Br^-
6.78. (a) metal; (b) nonmetal; (c) X_2Y_3; (d) X 3A Y 6A.

Formation of Ionic Compounds (Section 6.7)
6.82. −325 kJ/mol. **6.86.** −537 **6.88.** −176 kJ/mol
6.90. −294 kJ/mol. MgF(s) −1114 MgF_2(s) MgF_2 will form
6.94. +640 kJ/mol for Cl^+Na^-(s) formation of Cl^+Na^- is not favored
because the net energy change is positive whereas it is negative for
the formation of Na^+Cl^-.

Lattice Energy (Section 6.8)
6.98. $MgCl_2$ > LiCl > KCl > KBr.

Multiconcept Problems
6.100. When moving diagonally down and right on the periodic
table, the increase in atomic radius caused by going to a larger shell
is offset by a decrease caused by a higher Z_{eff}. Thus, there is little net
change in the charge density.

CHAPTER 7

7.2. H is positively polarized (blue). O is negatively polarized (red). This is consistent. **7.6.** (a) OF_2; (b) $SiCl_4$ **7.8.** (a)(b)(c) **7.12.** O_2^-

7.14.

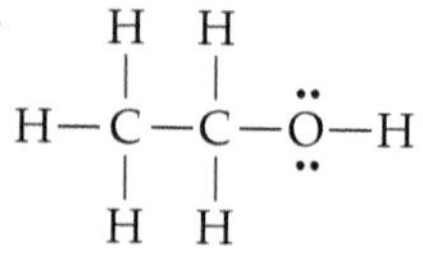

H—C—C—Ö—H   and   H—C—Ö—C—H

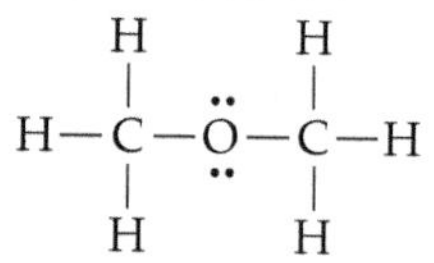

7.15. Molecular formula: $C_4H_5N_3O$

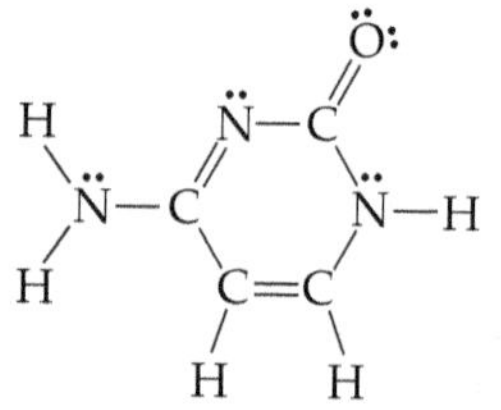

7.17. (a)

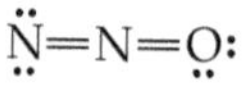

N=N=O:

(b)

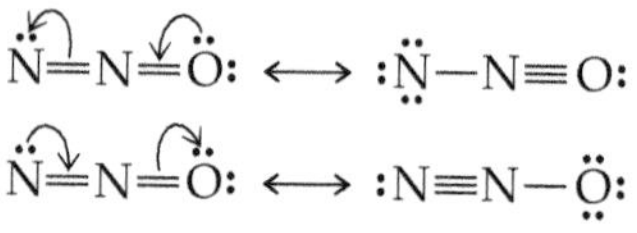

N=N=O: ⟷ :N—N≡O:

N=N=O: ⟷ :N≡N—O:

7.18.

(a)

:O—S=O: ⟷ :O=S—O:

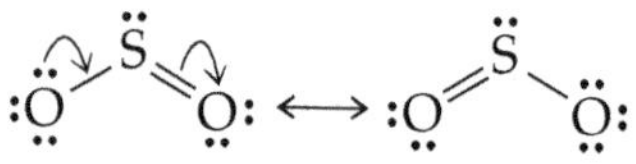

(b)

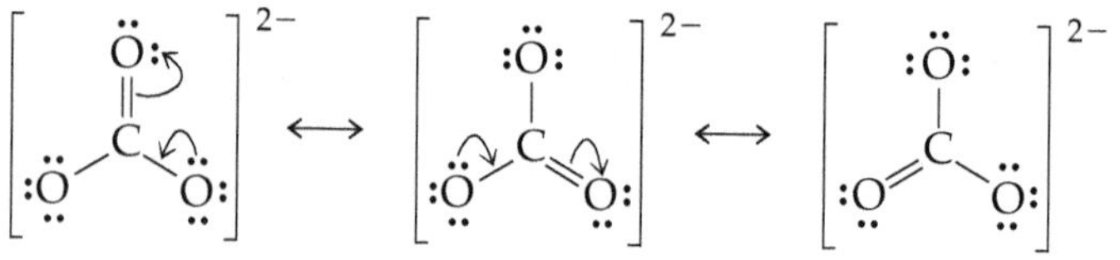

(c)

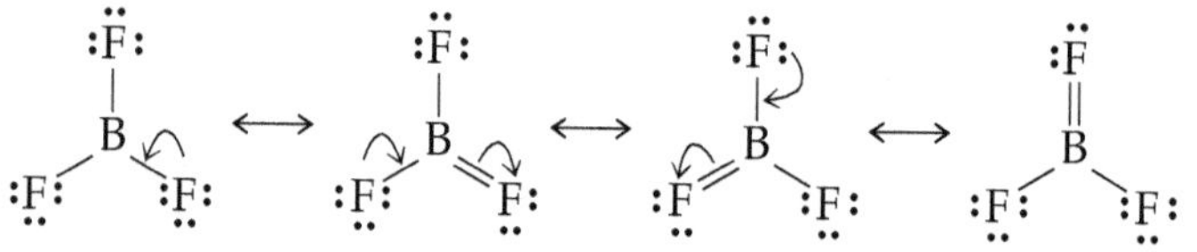

7.19. (a) This is a valid resonance structure; (b) not a valid resonance structure. **7.20.**

(a)

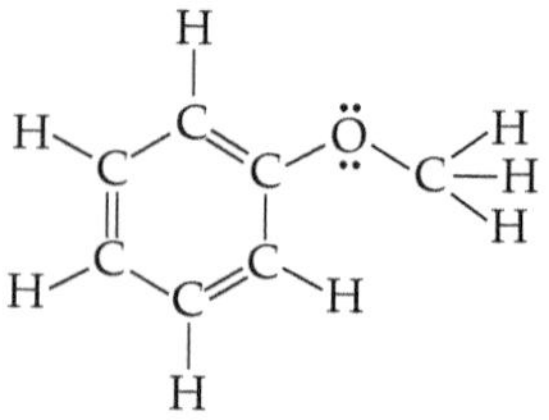

(b)

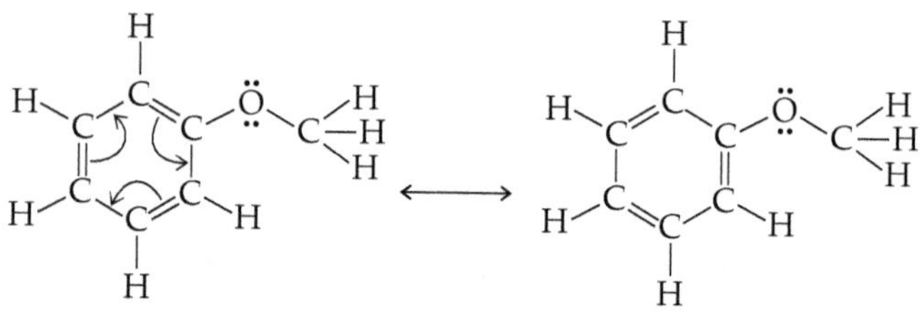

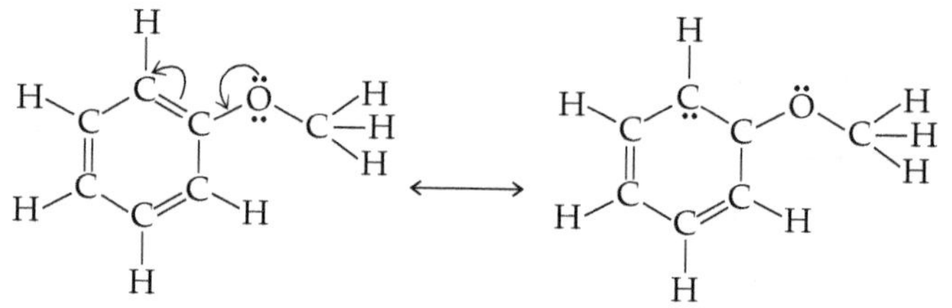

7.22.

(a)

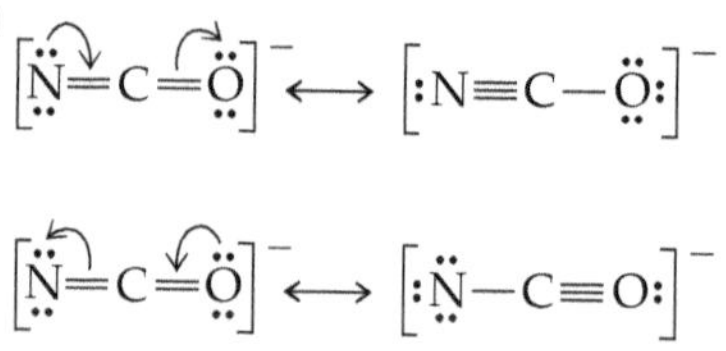

$$\left[N=C=O: \right]^- \longleftrightarrow \left[:N≡C-O: \right]^-$$

$$\left[N=C=O: \right]^- \longleftrightarrow \left[:N-C≡O: \right]^-$$

(b)

$$\left[N=C=O: \right]^-$$

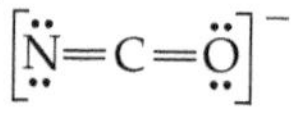

nitrogen: −1
carbon: 0
oxygen: 0
nitrogen: 0
carbon: 0
oxygen: −1
nitrogen: −2
carbon: 0
oxygen: +1 first two structures; (c) Carbon–nitrogen
7.23. All atoms 0
top oxygen: −1
right oxygen: +1
other atoms 0
structure without formal charges makes a larger contribution
7.24 All atoms in structures 1 & 2 have 0 structure 3: carbon: −1
oxygen: +1 other atoms 0 formal 1 and 2 larger contribution
7.26. (a) Cl; (b) CF_3 **7.27.** Phosphorus can utilize d orbitals and form more than four bonds. **7.28.** $C_{11}H_{19}N_2PSO_3$

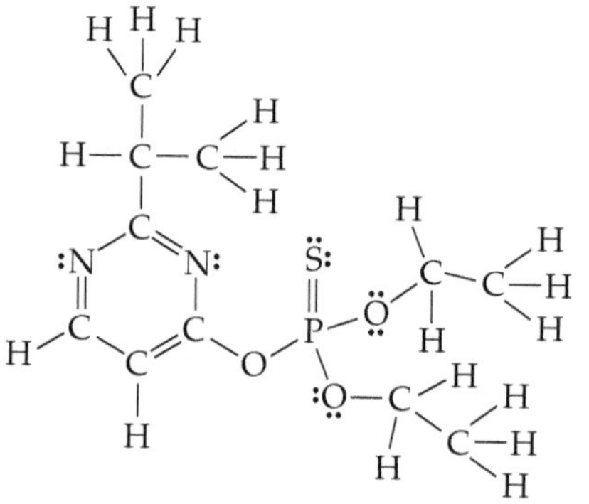

7.29. phosphorus: 0
top oxygen: 0
right oxygen: 0
fluorine: 0
carbon: 0

Conceptual Problems

7.32. C–D A–B **7.34.** (a) fluoroethane; (b) ethane; (c) ethanol;
(d) acetaldehyde
7.36. (a) $C_8H_9NO_2$

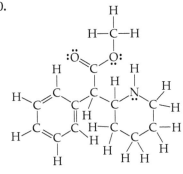

7.38.

Section Problems

Covalent Bonding (Section 7.1)

7.40 (a) ionic; (b) nonpolar covalent; (c) covalent

Strengths of Covalent Bonds (Section 7.2)

7.42. increased bond polarity leads to increased bond strength.
7.44. O—H

Polar Covalent Bonds: Electronegativity (Section 7.3)

7.46. Electronegativity increases from left to right across a period
and decreases down a group. **7.48.** K < Li < Mg < Pb < C < Br
7.56. (a) $CCl_4 \sim ClO_2$ < $TiCl_3$ < $BaCl_2$ **7.58.** (a) $MgBr_2$; (b) PBr_3
7.60. 5.05%

A Comparison of Ionic and Covalent Compounds (Section 7.4)

7.62. (a)

Electron-Dot Structures and Resonance (Sections 7.5–7.7)

7.64. The octet rule states that main-group elements tend to react
so that they attain a noble gas electron configuration. The transition
metals are characterized by partially filled d orbitals that can be used
to expand their valence shell beyond the normal octet of electrons.
7.66.

(a) (b) (c) (d) (e) (f)

7.68. (c) **7.70.**

7.72. (a) Al; (b) P.
7.74. (a) (b)

7.76. N_2H_2
7.78. (a) (b)

Electron-Dot Structures for Molecules with Second-Row Elements (Section 7.8)

7.80.

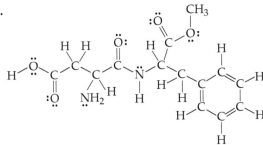

7.82.

Resonance (Section 7.9)

7.84.

(a)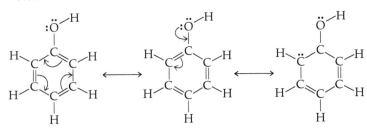

(b)

(c)

7.86. (a) yes; (b) no; (c) yes; (d) yes
7.88.

7.90. (a)

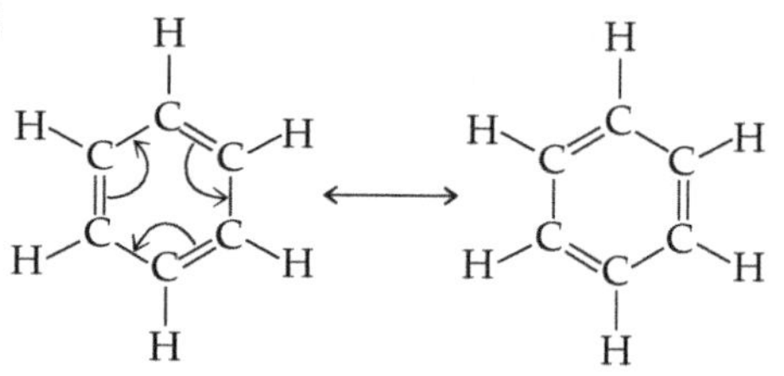

7.92. (a)

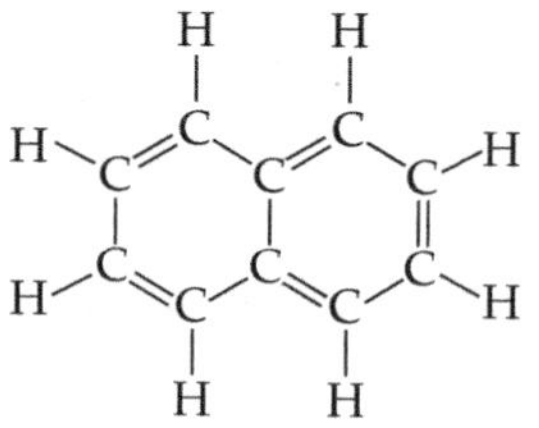

(b)

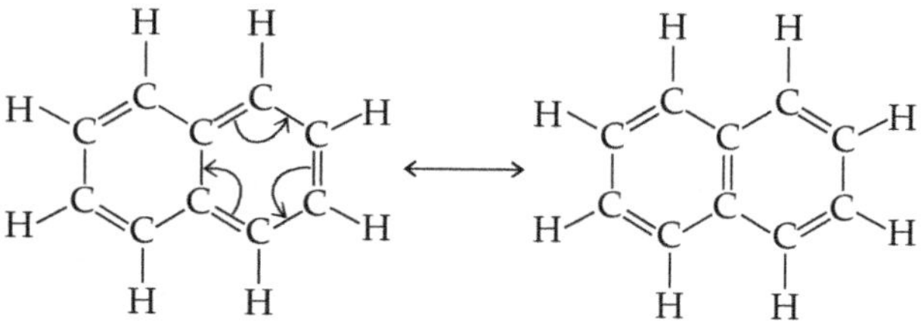

Formal Charges (Section 7.10)

7.94. carbon: −1
       oxygen: +1

7.96.

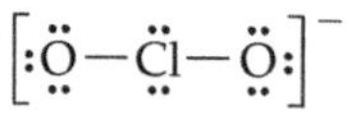

For both oxygens: −1
chlorine: +1

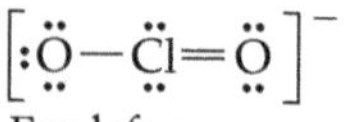

For left oxygen: −1
right oxygen: 0
chlorine: 0

7.98. (a)

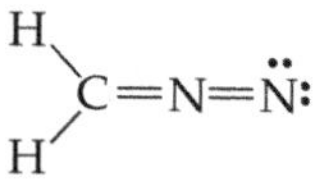

hydrogen: 0
nitrogen (central): +1
nitrogen (terminal): −1
carbon: 0; (b)

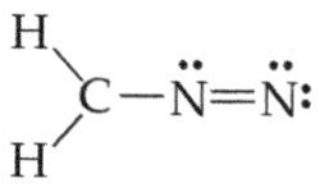

hydrogen: 0
nitrogen (central): 0
nitrogen (terminal): −1
carbon: +1 (a)

7.100.

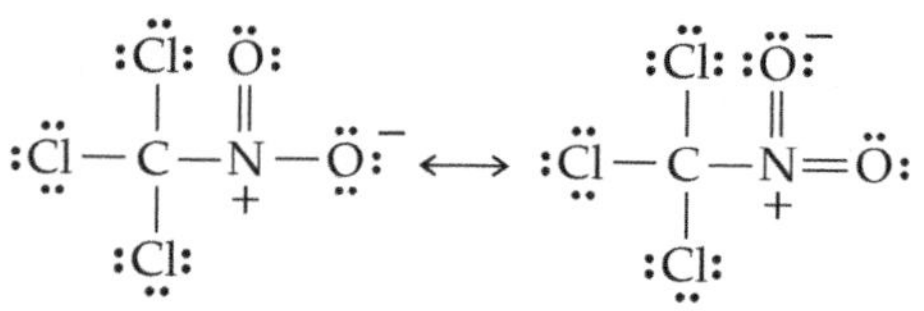

chlorine: 0
carbon: 0 nitrogen: +1 oxygen (double bonded): 0
oxygen (single bonded): −1
7.102. All atoms 0

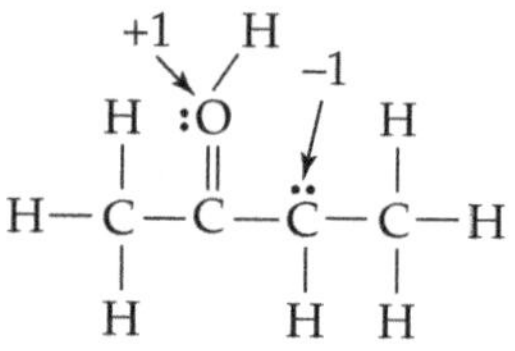

oxygen: +1
carbon: −1
other atoms 0
original structure is the larger contributor.

Multiconcept Problems
7.108.

(a)

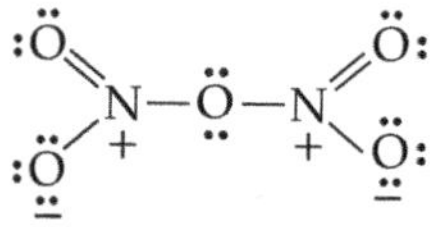

(b)

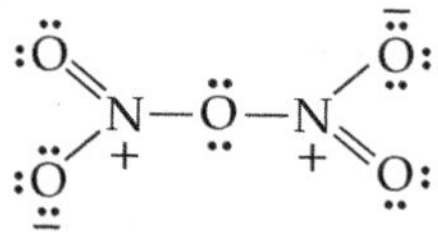

(d)

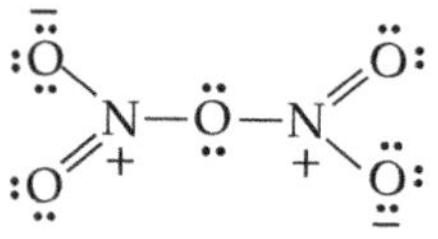

(e)

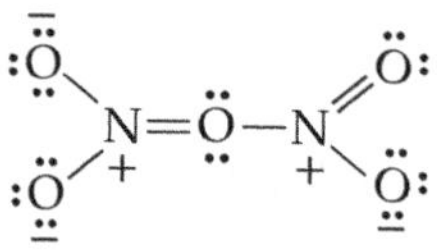

(g)

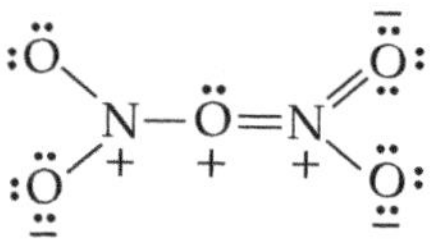

(h)

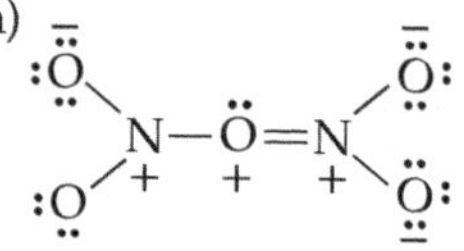

(c)

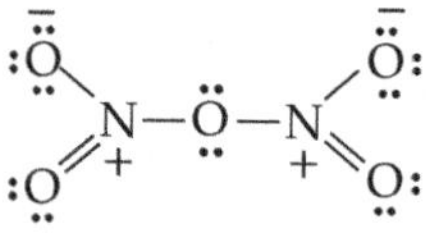

(f)

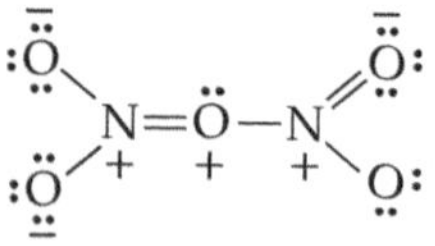

(a)–(d) make more important

CHAPTER 8

8.2. (a) 4 tetrahedral tetrahedral molecular geometry; (b) 5 trigonal bipyramidal seesaw molecular geometry.

8.3.

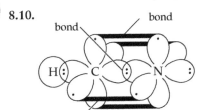

Tetrahedral (109.5° bond angles)   Trigonal planar (120° bond angles)

8.4. The bond angle around every carbon is 120° (trigonal planar).

8.5. The C—H bonds are formed by the overlap of one singly occupied sp³ orbital on C with a singly occupied H 1s orbital. The C—Cl bonds are formed by the overlap of one singly occupied sp³ orbital on C with a singly occupied Cl 2p orbital.

8.6. The C—C bonds are formed by the overlap of one singly occupied sp³ hybrid orbital from each C. The C—H bonds are formed by the overlap of one singly occupied sp³ orbital on C with a singly occupied H 1s orbital. The carbon in formaldehyde is sp² hybridized.

8.10. In HCN the carbon is sp hybridized.

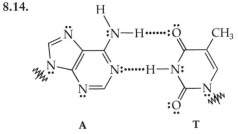

bond   bond   bond

8.12. (a) CF_4; (b) CH_2F_2; (c) CHF_3; (d) CH_3F

8.14.

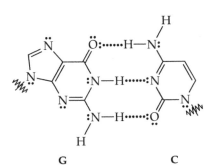

A   T

G   C

high in G–C pairs

8.16. H_2S dipole-dipole, dispersion
CH_3OH hydrogen bonding, dipole-dipole, dispersion
C_2H_6 dispersion
Ar   dispersion
Ar $< C_2H_6 < H_2S < CH_3OH$

8.17. For He_2^+ σ^*_{1s} ↑
σ_{1s} ↑↓
1/2 stable

8.18. He_2^{2+} stronger bond

8.20. The bond orders are: $O_2^{2-} = 1$, $O_2^- = 1.5$, $O_2 = 2$, $O_2^+ = 2.5$, $O_2^{2+} = 3$. The order from weakest to strongest bond is: $O_2^{2-} < O_2^- < O_2 < O_2^+ < O_2^{2+}$. The order from shortest to longest bond is: $O_2^{2+} < O_2^+ < O_2 < O_2^- < O_2^{2-}$.

8.21.

8.22.

Conceptual Problems

8.28. (a) trigonal bipyramidal; (b) tetrahedral; (c) (4 ligands one hidden) **8.30.** (a) sp²; (b) sp; (c) sp³ **8.32.** (a) $C_8H_9NO_2$ (b), (c) & (d)

all C's in ring, sp², trigonal planar
sp³, tetrahedral
sp², trigonal planar

8.34. The electronegative O atoms are electron rich (red), while the rest of the molecule is electron poor (blue). **8.36.** The N atom is electron rich (red). The C and H atoms are electron poor (blue)

Section Problems
The VSEPR Model (Section 8.1)

8.38. (a) trigonal planar; (b) trigonal bipyramidal; (c) linear; (d) octahedral **8.40.** (a) tetrahedral, 4; (b) octahedral, 6; (c) bent, 3 or 4; (d) linear, 2 or 5; (e) square pyramidal, 6; (f) trigonal pyramidal, 4.
8.42. (a) bent; (b) tetrahedral; (c) bent; (d) trigonal planar.
8.44. (a) trigonal bipyramidal; (b) see saw; (c) trigonal pyramidal; (d) tetrahedral. **8.46.** (a) tetrahedral; (b) tetrahedral; (c) tetrahedral; (d) trigonal pyramidal; (e) tetrahedral; (f) linear **8.48.** (a) 109°; (b) 120°; (c) 90°; (d) 120°.
8.50. H-C$_a$-H ~120°   C$_b$-C$_c$-N ~180°
H-C$_a$-C$_b$ ~120°   C$_a$-C$_b$-H ~120°
C$_a$-C$_b$-C$_c$ ~120°   H-C$_b$-C$_c$ ~120°

8.52.

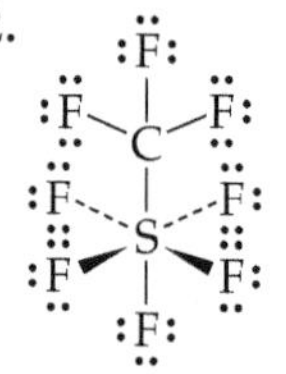

:F:
:F—C—F:
:F—S—F:
:F—S—F:
:F:

109.5° around carbon and 90° around S. **8.54.** 109°. Because the geometry about each carbon is tetrahedral, the cyclohexane ring cannot be flat.

Valence Bond Theory and Hybridization (Sections 8.2–8.4)

8.58. In a π bond, the shared electrons occupy a region above and below a line connecting the two nuclei. A σ bond has its shared electrons located along the axis between the two nuclei. **8.60.** (a) sp; (b) sp^2; (c) sp^3 **8.62.** (a) sp^2; (b) sp^2; (c) sp^3; (d) sp^2 **8.64.** Carbons a, b, and d are sp^2 hybridized and carbon c is sp^3 hybridized. carbons a, b, and d are ~120°. carbon c are ~109°. H—O—C bond angles are ~109°.

8.66.

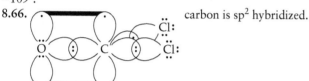 carbon is sp^2 hybridized.

8.68.

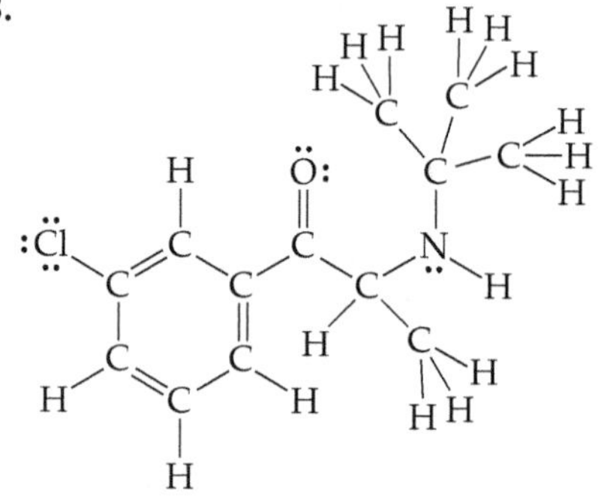

(a) 34 σ and 4 π; (b) and (c) Each C with four single bonds is sp^3 hybridized with bond angles of 109.5°. Each C with a double bond is sp^2 hybridized with bond angles of 120°; (d) The nitrogen is sp^3 hybridized. **8.70.** Both B and N are sp^2 hybridized. 120°. planar.

Dipole Moments (Sections 8.5)

8.78.

(a)   (b)

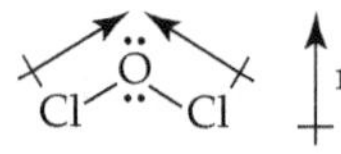

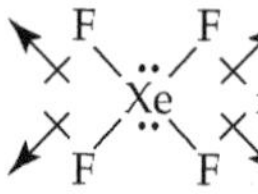

(c) 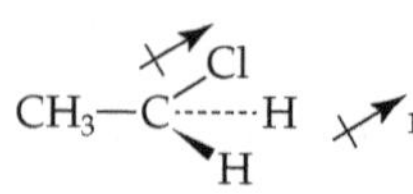   (d)

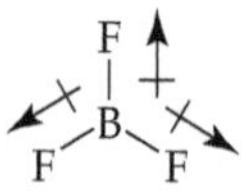

8.80. SO_2 is bent and the individual bond dipole moments add CO_2 is linear bond dipole moments can

8.82. (a)   (b)

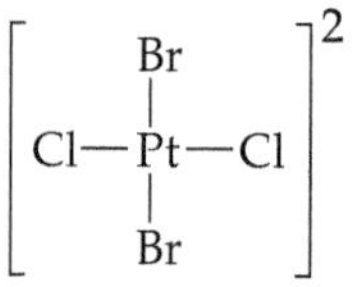

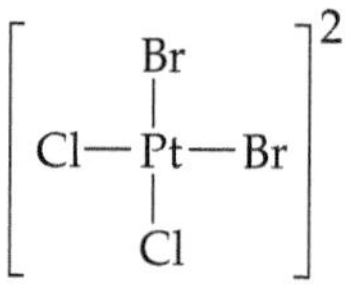

8.84. If a molecule has polar covalent bonds, the molecular shape (and location of lone pairs of electrons) determines whether the bond dipoles cancel and thus whether the molecule has a dipole moment.

Intermolecular Forces (Section 8.6)

8.88. (a) Dipole-dipole; (b) London dispersion; (c) London dispersion; (d) Dipole-dipole and hydrogen bonding **8.90.** CH_3OH can hydrogen bond; CH_4 cannot. For 1-decanol and decane, dispersion forces are comparable and relatively large. 1-decanol can hydrogen bond; decane cannot.

8.92.

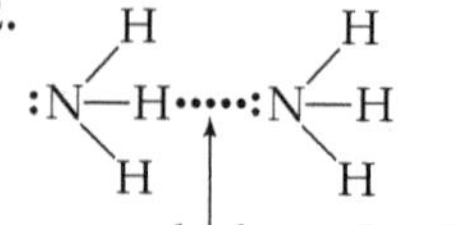

H   H
:N—H·····:N—H
  H   H
hydrogen bond

8.94. (ii) a nd (iii)

Molecular Orbital Theory (Sections 8.7–8.9)

8.96. Electrons in a bonding molecular orbital spend most of their time in the region between the two nuclei, helping to bond the atoms together. Electrons in an antibonding molecular orbital cannot occupy the central region between the nuclei and cannot contribute to bonding. **8.98.** O_2^+ 2.5

O_2 2

O_2^- 1.5. All are stable. All have unpaired electrons.

8.100. (a) 2; (b) Add one electron; (c) 2.5 **8.102.** (a) diamagnetic; (b) paramagnetic; (c) paramagnetic; (d) diamagnetic; (e) paramagnetic

Multiconcept Problems

8.110.

(a)   (b)   (c)

H—C≡C—H   H—N̈=N̈—H   :Ö:
                                            :Cl—S—Cl:

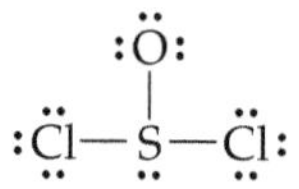

8.112.

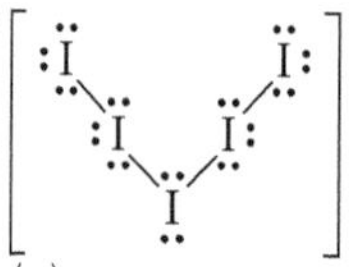

[:I: I:]
[:I: I:]
[:I:]

8.114. (a)

H   Cl   Cl   Cl   Cl   H
 C=C   C=C   C=C
H   Cl   H   H   H   Cl

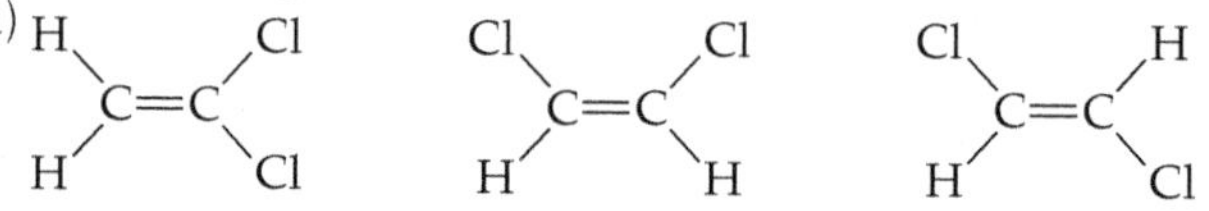

polar   polar   nonpolar

(b) All three molecules are planar. The first two structures are polar because they both have an unsymmetrical distribution of atoms about the center of the molecule and bond polarities do not cancel. Structure 3 is nonpolar because the H's and Cl's, respectively, are symmetrically distributed about the center of the molecule, both being opposite each other. In this arrangement, bond polarities cancel; (c) 599 kJ/mol; (d)

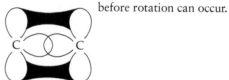 The π bond must be broken before rotation can occur.

CHAPTER 9

9.2. −0.25 kJ system does work on the surroundings. **9.4.** (a) PΔV is negative because the system volume is decreased at constant pressure; (b) ΔH is slightly more negative than ΔE; **9.6.** 39 kg C_3H_8
9.9. 26.9 J/(mol · C) **9.10.** 825 J/°C **9.15.** +131.3 kJ
9.16. (a) A + 2 B → D; ΔH° = −100 kJ + (−50 kJ) = −150 kJ; (b) red arrow step 1: A + B → C green arrow step 2: C + B → D blue arrow overall reaction; (c) top level A + 2 B. middle level C + B. bottom level D. **9.18.** −250 kJ/mol **9.21.** negative because the reaction decreases the number of moles of gaseous molecules. **9.22.** reaction proceeds from a solid and a gas (reactants) to all gas (product) ΔS° is positive. **9.23.** ΔH is negative. ΔS is negative. ΔG is negative. **9.24.** (a) 2 A_2 + B_2 → 2 A_2B; (b) ΔH is negative. ΔS is negative; (c) favored at low temperatures **9.25.** spontaneous. 190 °C **9.26.** The temperature should be increased to make the reaction spontaneous. **9.31.** (a) +2.47 kJ Work energy is transferred to the system; (b) −1365 kJ/mol

Conceptual Problems
9.34. ΔH > 0 and w < 0
9.36. (a)

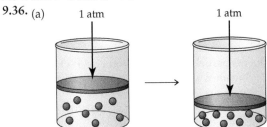

(b)

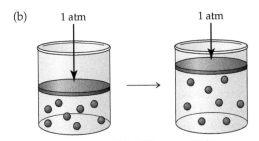

9.38. (a) Diagram 1; (b) ΔH° = + 30 kJ
9.40. (a) 2 AB_2 → A_2 + 2 B_2; (b) ΔH is negative. ΔS is positive; (c) at all temperatures. **9.42.** ΔG is negative ΔS is positive. ΔS is positive

Section Problems
Heat, Work, and Energy (Section 9.1–9.3)
9.44. Heat is the energy transferred from one object to another as the result of a temperature difference between them. Temperature is a measure of the kinetic energy of molecular motion. Energy is the capacity to do work or supply heat. Work is defined as the distance moved times the force that opposes the motion. Kinetic energy is the energy of motion. Potential energy is stored energy. **9.46.** car **9.48.** (a) and (b) (c) is not. **9.50.** −70 J; The energy change is negative. **9.52.** (a) 35 min; (b) 150 min. **9.54.** (a) −593 J; (b) 26.4 cm
9.56. = −15.1 L · atm

Energy and Enthalpy (Section 9.4)
9.58. ΔE = q_v is the heat change associated with a reaction at constant volume. Since ΔV = 0, no PV work is done. ΔH = q_p is the heat change associated with a reaction at constant pressure. Since ΔV ≠ 0, PV work can also be done. **9.60.** −45.2 kJ **9.62.** −0.30 kJ **9.64.** = 45.4 kJ

Thermochemical Equations for Chemical and Physical Changes (Section 9.5–9.6)
9.66. (a) to the system, endothermic, ΔH° > 0; (b) to the surroundings, exothermic, ΔH° < 0; (c) to the system, endothermic, ΔH° > 0 **9.68.** 25.5 kJ **9.70.** 131 kJ is released. **9.72.** (a) 780. kJ evolved; (b) 1.24 kJ absorbed. **9.74.** 0.388 kJ is evolved. exothermic.

Calorimetry and Heat Capacity (Section 9.7)
9.76. Heat capacity is the amount of heat required to raise the temperature of a substance a given amount. Specific heat is the amount of heat necessary to raise the temperature of exactly 1 g of a substance by exactly 1 °C. **9.78.** 1.23 J/(g · °C) **9.80.** −32 kJ **9.82.** 3.0°C **9.84.** −83.7 kJ **9.90.** −120 kJ/mol citric acid

Hess's Law (Section 9.8)
9.92. The standard state of an element is its most stable form at 1 atm and the specified temperature, usually 25 °C. The sum of the enthalpy changes for the individual steps in a reaction is equal to the overall enthalpy change for the entire reaction, a relationship known as Hess's law. **9.94.** −202 kJ

Heats of Formation (Section 9.9)
9.98. A compound's standard heat of formation is the amount of heat associated with the formation of 1 mole of a compound from its elements (in their standard states). **9.100.** (a) gas; (b) liquid; (c) gas; (d) solid **9.102.** (a) 2 Fe(s) + 3/2 O_2(g) → Fe_2O_3(s); (b) 12 C(s) + 11 H_2(g) + 11/2 O_2(g) → $C_{12}H_{22}O_{11}$(s); (c) U(s) + 3 F_2(g) → UF_6(s); **9.104.** −395.7 kJ/mol **9.106.** −909.3 kJ **9.108.** +104 kJ/mol **9.110.** −16.9 kJ/mol **9.112.** +179.2 kJ **9.114.** +179.2 kJ **9.116.** (a) +172.5 kJ; (b) −189.2 kJ; (c) −24.8 kJ

Free Energy and Entropy (Section 9.11–9.12)
9.122. Entropy is a measure of molecular randomness.
9.124. if ΔS is positive the TΔS term is larger than ΔH.
9.126. (a) positive; (b) negative **9.128.** (a) zero; (b) zero; (c) negative **9.130.** ΔS is positive. The reaction increases the total number of molecules. **9.132.** (a) spontaneous; exothermic; (b) nonspontaneous; exothermic; (c) spontaneous; endothermic; (d) nonspontaneous; endothermic. **9.134.** 570 K
9.136. (a) spontaneous at all temperatures; (b) crossover temperature; (c) crossover temperature; (d) nonspontaneous at all temperatures. **9.138.** 31.6 J/(K · mol) **9.140.** 59 °C **9.142.** −238.6 kJ spontaneous at <u>all</u> temperatures.

Multiconcept Problems
9.144. 7.22 kJ −224 kJ **9.146.** (a) ΔG = −TΔS$_{total}$; (b) −9399 J/(K · mol) **9.148.** 3.72 g $NaNO_3$ 6.3 g KF

CHAPTER 10

10.3. 0.650 atm
10.4. (a)

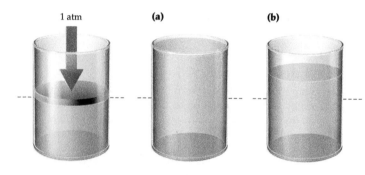

746 mm Hg   (b) 732 mm Hg

Gas

237 mm

10.6.

1 atm   **(a)**   **(b)**

10.8. 5.0 atm **10.9.** 28°C. **10.10.** 142 L,
10.12. 750.0 L H_2 250.0 L N_2 **10.13.** 34.1 H_2S, hydrogen sulfide.
10.14. (a) 1.80 g/L; (b) higher molar mass **10.16.** 0.025
10.18. 32.00; O_2.

Conceptual Problems

10.24. The gas pressure in the bulb in mm Hg is equal to the difference in the height of the Hg in the two arms of the manometer.
10.32. The picture on the right will be the same as that on the left

Section Problems
Gases and Gas Pressure (Section 10.1)

10.34. 14.7 psi 1.93×10^{-2} psi **10.38.** 0.632 atm; 6.40×10^4 Pa
10.40. 930 mm Hg **10.42.** 1.046×10^5 Pa **10.44.** 1155 mm Hg
10.46. 28.96

The Gas Laws (Sections 10.2–10.3)

10.48. (a) pressure would triple; (b) pressure would be $\frac{1}{3}$ the initial

pressure; (c) pressure would increase by 1.8 times; (d) The pressure would be 0.17 times the initial pressure. **10.50.** They all contain the same number of gas molecules. **10.52.** 7210 L 51.5 L

10.54. 2.1×10^4 mm Hg **10.56.** 1×10^{-17} mm Hg **10.60.** ice
10.62. Weigh the containers. heavier contains O_2.

Gas Stoichiometry (Section 10.4)

10.64. 1.5×10^4 g O_2 **10.66.** (a) 0.716 g/L; (b) 1.96 g/L; (c) 1.43 g/L
10.68. 34.0 **10.70.** 0.5469 L **10.72.** (a) 9.44 L; (b) 6.05 g Zn
10.74. (a) 380 g CO_2; (b) 5.4 days **10.76.** 0.464 g He **10.78.** 15 L

Dalton's Law and Mole Fraction (Section 10.5)

10.80. P_{N_2} 0.7808 atm; P_{O_2} 0.2095 atm; P_{Ar} 0.0093 atm; P_{CO_2} 0.000 38 atm the rest are negligible. **10.82.** P_{O_2} 0.970 atm; P_{CO_2} 0.007 11 atm; **10.84.** $X_{HCl} = 0.026$; $X_{H_2} = 0.094$; $X_{Ne} = 0.88$;
10.86. (a) 1.68 atm; (b) 0.219 atm; **10.88.** 723 mm Hg 3.36 g Mg

Kinetic-Molecular Theory (Section 10.6)

10.92. The kinetic-molecular theory is based on the following assumptions:
1. A gas consists of tiny particles, either atoms or molecules, moving about at random.
2. The volume of the particles themselves is negligible compared with the total volume of the gas; most of the volume of a gas is empty space.
3. The gas particles act independently; there are no attractive or repulsive forces between particles.
4. Collisions of the gas particles, either with other particles or with the walls of the container, are elastic; that is, the total kinetic energy of the gas particles is constant at constant T.
5. The average kinetic energy of the gas particles is proportional to the Kelvin temperature of the sample.
10.94. 443 m/s **10.96.** −32 °C **10.98.** He **10.100.** −272.83 °C

Graham's Law (Section 10.7)

10.106. HCl > F_2 > Ar

Real Gases (Section 10.8)

10.110. (a) high; (b) larger; **10.112.** 20.5 atm 20.3 atm

The Earth's Atmosphere, Greenhouse Gases, and Climate Change (Sections 10.9–10.11)

10.116. Troposphere, stratosphere, mesosphere, and thermosphere. Temperature changes **10.118.** The force of gravity is strongest at the earth's surface and becomes weaker at higher altitude.
10.120. (a) visible and UV; (b) UV; (c) infrared; (d) infrared
10.122. Because of nuclear fusion, the Sun emits all forms of electromagnetic radiation. The Earth absorbs visible radiation, warms up, and radiates Infrared radiation back to space. **10.124.** Nitrogen and oxygen are diatomic molecules that do not have a dipole moment. As the bond stretches in a vibration, the molecule still does not have a dipole moment. Since no change in dipole moment occurs, IR radiation will not be absorbed. **10.126.** Symmetric stretch does not absorb the asymmetric stretch absorbs **10.128.** The atmosphere has a much higher concentration of CO_2. **10.130.** CO_2, N_2O, and CH_4
10.132. The Earth's average temperature has risen about 1°C since 1900.

Multiconcept Problems

10.134. 1.68 mi **10.138.** (a) 0.007 06 mol; (b) 0.004 71 moles;
(c) 0.872 g/L; (d) 46.3;

(e) $Hg_2CO_3(s) + 6 HNO_3(aq) \rightarrow$

$$2 Hg(NO_3)_2(aq) + 3 H_2O(l) + CO_2(g) + 2 NO_2(g)$$

Daily Peak AQI (Combined $PM_{2.5}$ and O_3) August 8, 2012

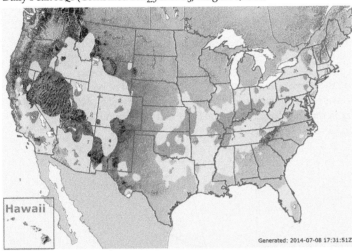

Daily Peak AQI (Combined $PM_{2.5}$ and O_3) December 5, 2012

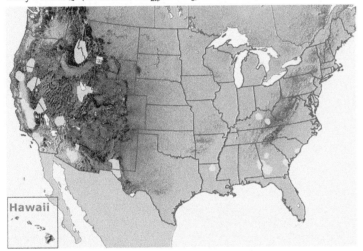

10.140. $P_{CO_2} = 0.600$ atm $P_{SO_2} = 1.20$ atm $P_{O_2} = 0.600$ atm
10.146. (a) C_2H_6O $C_2H_6O + 3 O_2 \rightarrow 2 CO_2 + 3 H_2O$; (b) It is an
empirical formula because it is the smallest whole number ratio of atoms. It is also a molecular formula because any higher multiple such
as $C_4H_{12}O_2$ does not correspond to a stable electron-dot structure;
(c)

$$H-\overset{\overset{\displaystyle H}{|}}{C}-\overset{..}{\underset{..}{O}}-\overset{\overset{\displaystyle H}{|}}{\underset{\displaystyle H}{C}}-H \qquad H-\overset{\overset{\displaystyle H}{|}}{\underset{\displaystyle H}{C}}-\overset{\overset{\displaystyle H}{|}}{\underset{\displaystyle H}{C}}-\overset{..}{\underset{..}{O}}-H$$

(d) -183.8 kJ/mol **10.148.** (a) 492 °R.; (b) $0.0456 \dfrac{L \cdot atm}{°R \cdot mol}$;

(c) 116 atm **10.150.** (a) $C_5H_{12}O$; (b) 88.1 g/mol the molecular
formula and empirical formula are the same. $C_5H_{12}O$;
(c) $C_5H_{12}O(l) + 15/2 O_2(g) \rightarrow 5 CO_2(g) + 6 H_2O(l)$; (d)
-313.6 kJ/mol

CHAPTER 11

11.1. 31.4 kJ/mol **11.2.** 83.0 °C **11.3.** 109.7 J/(K·mol) **11.4.** 334 K
11.5. 7.00 kJ **11.6.** −4.59 kJ
11.8. (a)

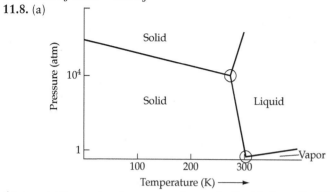

(b) two; (c) Increasing the pressure favors the liquid phase, liquid phase is more dense

Conceptual Problems

11.16. (a) normal boiling point ≈ 300 K; normal melting point ≈ 180 K; (b) (i) solid (ii) gas (iii)   supercritical fluid

Section Problems
Properties of Liquids (Section 11.1)

11.20. surface tension **11.22.** (a) CH_2Br_2; (b) Ethylene glycol
11.24. Oleic acid is a much larger molecule than H_2O with larger dispersion forces. It can also hydrogen bond.

Vapor Pressure and Boiling Point (Section 11.2)

11.26. 28.0 kJ/mol
11.30.

ln P_{vap}	1/T
4.383	0.003 802
4.8949	0.003 663
5.3627	0.003 534
5.7979	0.003 413
6.2054	0.003 300
6.5853	0.003 195

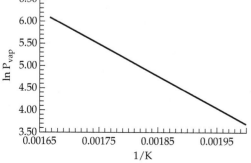

11.36. −30.7 °C

Phase Changes (Sections 11.3–11.4)

11.38. ΔH_{vap} is the heat required to overcome all intermolecular forces. **11.40.** 88.2 J/(K·mol) **11.44.** (a) Hg(l) → Hg(g); (b) no change of state; (c) Hg(g) → Hg(l) → Hg(s) **11.46.** 2.40 kJ **11.48.** 3.73 kJ

Phase Diagrams (Section 11.5)

11.52. As the pressure over the liquid H_2O is lowered, H_2O vapor is removed by the pump. As H_2O vapor is removed, more of the liquid H_2O is converted to H_2O vapor. This conversion is an endothermic process and the temperature decreases. The combination of both a decrease in pressure and temperature takes the system across the liquid/solid boundary in the phase diagram so the H_2O that remains turns to ice.
11.56.

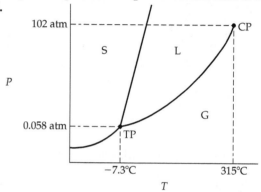

11.62.

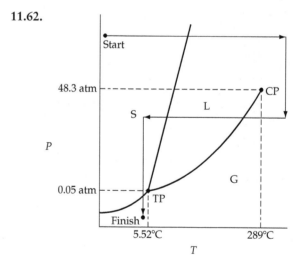

11.66. cannot be liquified

Multiconcept Problems

11.74. Al_2O_3, ionic (greater lattice energy than NaCl) F_2, dispersion; H_2O, H–bonding, dipole-dipole; Br_2, dispersion (larger and more polarizable than F_2), ICl, dipole-dipole, NaCl, ionic $F_2 < Br_2 < ICl < H_2O < NaCl < Al_2O_3$

CHAPTER 12

12.1. 557 pm **12.2.** 167 pm **12.3.** 9.31 g/cm³ **12.4.** 192.2 g/mol Ir.
12.5. (a) 1/8 S²⁻ at 8 corners = 1 S²⁻;
1/2 S²⁻ at 6 faces = 3 S²⁻; 4 Zn²⁺; (b) ZnS; (c) 2+; (d) tetrahedral
12.6. (a) $CaTiO_3$; (b) +4; (c) all octahedral. **12.7.** The electron configuration for Hg is [Xe] $4f^{14}$ $5d^{10}$ $6s^2$. Assuming the 5d and 6s bands overlap, the composite band can accomodate 12 valence electrons per metal atom. Weak bonding and a low melting point are expected for Hg because both the bonding and antibonding MOs are occupied.
12.8. (a) The s-d band is 1/4 full, so Hf is picture (1). The s-d band is 5/6 full, so Pt is picture (2). The s-d band is 7/12 full, so Re is picture (3); (b) Re has an excess of 5 bonding electrons and it has the highest melting point and is the hardest of the three; (c) Pt has an excess of only 2 bonding electrons and it has the lowest melting point and is the softest of the three. **12.9.** Ge doped with As is an n-type semiconductor because As has an additional valence electron. The extra electrons are in the conduction band. The number of electrons in the conduction band of the doped Ge is much higher than for pure Ge, and the conductivity of the doped semiconductor is higher. **12.10.** (a) (1), silicon; (2), white tin; (3), diamond; (4), silicon doped with aluminum; (b) (3) < (1) < (4) < (2), Diamond (3) is an insulator with a large band gap. Silicon (1) is a semiconductor with a band gap smaller than diamond. Silicon doped with aluminum (4) is a p-type semiconductor that has fewer electrons than needed for bonding and has vacancies (positive holes) in the valence band. White tin (2) has a partially filled s-p composite band and is a metallic conductor.
12.11. $Si(OCH_3)_4 + 4\ H_2O \longrightarrow Si(OH)_4 + 4\ HOCH_3$
12.12. $Ba[OCH(CH_3)_2]_2 + Ti[OCH(CH_3)_2]_4 + 6\ H_2O \longrightarrow$
$$BaTi(OH)_6(s) + 6\ HOCH(CH_3)_2$$
$BaTi(OH)_6(s) \xrightarrow{\text{Heat}} BaTiO_3(s) + 3\ H_2O(g)$
12.13. The color of quantum dots depends on the wavelength of light that they absorb, which is determined by band-gap energy. Different sizes have different band-gap energies. **12.14.** (a) 5.0 nm; (b) 2.2 nm; (c) 3.5nm **12.15.** 25000 **12.16.** (a) size (a) is green, size (b) is blue, size (c) is purple, and size (d) is red. (b) sizes (d), (c), (b), (a)
12.17. The smaller the particle, the larger the band gap and the greater the shift in the color of the emitted light from the red to the violet. The yellow quantum dot is larger because yellow is closer to the red than is the blue.

Conceptual Problems

12.22. (a) (1) and (4) are semiconductors; (2) is a metal; (3) is an insulator; (b) (3) < (1) < (4) < (2). The conductivity increases with decreasing band gap; (c) (1) and (4) increases; (2) decreases; (3) not much change.

12.24.

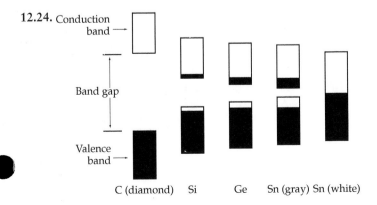

Section Problems
Types of Solids (Section 12.1)

12.26. molecular solid, CO_2, I_2; metallic solid, any metallic element; covalent network solid, diamond; ionic solid, NaCl **12.28.** (a) rubber; (b) Na_3PO_4; (c) CBr_4; (d) quartz; (e) Au **12.30.** covalent network solid.

X-Ray Crystallography (Section 12.2)
12.32. 201 pm

The Packing of Spheres in Crystalline Solids: Unit Cells (Section 12.3)
12.34. Hexagonal and cubic closest packing are the most efficient; Simple cubic packing is the least efficient **12.36.** 8.90 g/cm³ **12.38.** 404.9 pm **12.40.** 137 pm **12.42.** face-centered cubic **12.44.** face-centered cube **12.46.** 60% **12.48.** 6.01×10^{23} atoms/mol

Structures of Ionic Solids (Section 12.4)
12.50. Six Na⁺ six H⁻ **12.52.** 244 pm

Structures of Covalent Network Solids (Section 12.5)
12.54.

Bonding in Metals (Section 12.6)
12.58.

Each K has a single valence electron and has eight nearest neighbor K atoms. The valence electrons can't be localized in an electron-pair bond between any particular pair of K atoms. **12.60.** Malleability and ductility of metals follow from the fact that the delocalized bonding extends in all directions. When a metallic crystal is deformed, no localized bonds are broken. Instead, the electron sea simply adjusts to the new distribution of cations, and the energy of the deformed structure is similar to that of the original. Thus, the energy required to deform a metal is relatively small. **12.62.** Ionic bonding is stronger than metallic bonding **12.64.** The energy required to deform a transition metal like W is greater than that for Cs because W has more valence electrons and hence more electrostatic "glue." **12.66.** The difference in energy between successive MOs in a metal decreases as the number of metal atoms increases so that the MOs merge into an almost continuous band of energy levels. Consequently, MO theory for metals is often called band theory. **12.68.** The energy levels within a band occur in degenerate pairs; one set of energy levels applies to electrons moving to the right, and the other set applies to electrons moving to the left. In the absence of an electrical potential, the two sets of levels are equally populated. As a result there is no net electric current. In the presence of an electrical potential those electrons moving to the right are accelerated, those moving to the left are slowed down, and some change direction. The number of electrons moving to the right is now greater than the number moving to the left, and so there is a net electric current.

12.70. (a)   (b)

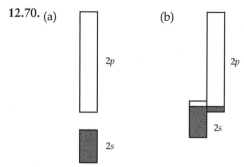

Diagram (b) shows the 2s and 2p bands overlapping in energy and the resulting composite band is only partially filled. Thus, Be is a good electrical conductor. **12.72.** Transition metals have a d band that can overlap the s band to give a composite band consisting of six MOs per metal atom. Half of the MOs are bonding and half are antibonding, and thus one expects maximum bonding for metals that have six valence electrons per metal atom. Accordingly, the melting points of the transition metals go through a maximum at or near group 6B.

Semiconductors (Section 12.7–12.8)

12.76. A semiconductor is a material that has an electrical conductivity intermediate between that of a metal and that of an insulator. Si, Ge, and Sn (gray) are semiconductors.

12.78. (a)   (b)

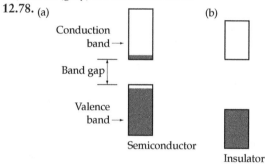

The MOs of a semiconductor are similar to those of an insulator, but the band gap in a semiconductor is smaller. As a result, a few electrons have enough energy to jump the gap and occupy the higher-energy, conduction band. When an electrical potential is applied to a semiconductor, it conducts a small amount of current because the potential can accelerate the electrons in the partially filled bands. **12.80.** As the band gap increases, the number of electrons able to jump the gap and occupy the higher-energy conduction band decreases, and thus the conductivity decreases. **12.82.** An n-type semiconductor is a semiconductor doped with a substance with more valence electrons than the semiconductor itself. Si doped with P is an example.

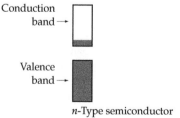

12.84. In the MO picture, the extra electrons occupy the conduction band. The number of electrons in the conduction band of the doped Ge is much greater than for pure Ge, and the conductivity of the doped semiconductor is correspondingly higher. **12.86.** (a) p-type; (b) n-type; (c) n-type

12.88. $Cd(CH_3)_2(g) + H_2Se(g) \longrightarrow CdSe(s) + 2\ CH_4(g)$

12.90. Al_2O_3 < Ge < Ge doped with In < Fe < Cu

12.92. In a diode, current flows only when the junction is under a forward bias (negative battery terminal on the n-type side). A p-n junction that is part of a circuit and subjected to an alternating potential acts as a rectifier, allowing current to flow in only one direction, thereby converting alternating current to direct current.

12.94. Both an LED and a photovoltaic cell contain p-n junctions, but the two devices involve opposite processes. An LED converts electrical energy to light; a photovoltaic, or solar, cell converts light to electricity. **12.98.** (a) GaN; (b) green GaP, ultraviolet GaN

12.100. GaP0.50As0.50 < GaP0.80As0.20 < GaP1.00As0.00

12.102. (a) 1118 nm; (b) no, at 1118 nm the intensity is about half of the maximum **12.104.** GaAs is a semiconductor. GaAs is isoelectronic with Ge.

12.108. (a) (b) The electrical conductivity of a semiconductor increases with increasing temperature because the number of electrons with sufficient energy to occupy the conduction band increases as the temperature rises; (c) (i) The conductivity of GaAs would increase when doped with Zn because it would produce positive holes in the valence band. Zn doped GaAs would be a p-type semiconductor. (ii) The conductivity of GaAs would increase when doped with S because it would put extra electrons in the conduction band. S doped GaAs would be an n-type semiconductor.

Superconductors (Section 12.9)

12.110. A superconductor is able to levitate a magnet. A superconductor has no electrical resistance. **12.112.** Some K^+ ions are surrounded octahedrally by six C_{60}^{3-} ions; others are surrounded tetrahedrally by four ions.

Ceramics and Composites (Section 12.10)

12.116. Ceramics are inorganic, nonmetallic, nonmolecular solids, including both crystalline and amorphous materials. Ceramics have higher melting points, and they are stiffer, harder, and more resistant to wear and corrosion than are metals. **12.118.** Ceramics are stiffer, harder, and more wear resistant than metals because they have stronger bonding. **12.120.** The brittleness of ceramics is due to strong chemical bonding. The strong, highly directional covalent bonds prevent the planes of atoms from sliding over one another when the solid is subjected to a stress. As a result, the solid can't deform to relieve the stress. It maintains its shape up to a point, but then the bonds give way suddenly and the material fails catastrophically when the stress exceeds a certain threshold value. By contrast, metals are able to deform under stress because their planes of metal cations can slide easily in the electron sea. **12.122.** Ceramic processing is the series of steps that leads from raw material to the finished ceramic object.

12.124. $Zr[OCH(CH_3)_2]_4 + 4 H_2O \longrightarrow Zr(OH)_4 + 4 HOCH(CH_3)_2$

12.126. $(HO)_3Si-O-H + H-O-Si(OH)_3 \longrightarrow (HO)_3$
$Si-O-Si(OH)_3 + H_2O$ On heating, SiO_2 is obtained.

12.128. $2 Ti(BH_4)_3(soln) \longrightarrow 2 TiB_2(s) + B_2H_6(g) + 9 H_2(g)$

12.130. $3 SiCl_4(g) + 4 NH_3(g) \longrightarrow Si_3N_4(s) + 12 HCl(g)$

12.132. because of their high strength-to-weight ratios.

Multiconcept Problems

12.134. (a) 1.926 g Fe; (b) Fe_3O_4; (c) 24 Fe 32 O

12.140. 1.0×10^{16} photons/s, 1.6 mA

12.142. (a)

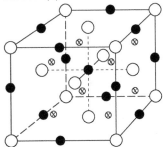

$\bigcirc$ = C_{60}^{3-}

● = M$^+$ in an octahedral hole

⊗ = M$^+$ in a tetrahedral hole

(b) There are 4 C_{60}^{3-} ions, 4 octahedral holes, and 8 tetrahedral holes per unit cell; (c) Octahedral holes:
(1/2, 1/2, 1/2), (1/2, 0, 0), (0, 1/2, 0), (0, 0, 1/2)
Tetrahedral holes: (1/4, 1/4, 1/4), (3/4, 1/4, 1/4),
(1/4, 3/4, 1/4), (3/4, 3/4, 1/4),(1/4, 1/4, 3/4),
(3/4, 1/4, 3/4), (1/4, 3/4, 3/4), (3/4, 3/4, 3/4);
(d) R(octahedral hole) = 207 pm,
R(tetrahedral hole) = 112 pm; (e) Na$^+$ will fit into the octahedral and tetrahedral holes without expanding the C_{60}^{3-} framework. K$^+$ and Rb$^+$ will fit into the octahedral holes without expanding the C_{60}^{3-} framework but will fit into the tetrahedral holes only if the C_{60}^{3-} framework is expanded.

CHAPTER 13

13.1. KBr < 1,5 pentanediol < toluene **13.2.** Vitamin E
13.3. 5.52 mass % **13.4.** 185 g H_2O **13.5.** 25.0 ppb exceeds limit
13.6. 4.6×10^{-5} g CO_2 **13.7.** 312 g **13.8.** 0.0249 m 2.96×10^{-3}
13.10. 0.251 M **13.13.** 223.8 mm Hg **13.14.** 2.6 **13.15.** 46.6 mm Hg
13.16. red and blue curves pure liquids green mixture. **13.18.** (a)
62 °C; (b) 2 m **13.19.** 7.06 atm **13.20.** (a) 0.17 M; (b) net transfer of
water from outside the cell to the inside of the cell and the cell would
swell. **13.21.** 180.5 g/mol **13.22.** powder is sucrose.
13.23. Both solvent molecules and small solute particles can pass
through a semipermeable dialysis membrane. Only solvent molecules
can pass through a semipermeable membrane. **13.24.** Solvent–solvent
is hydrogen bonding, solvent–solute is hydrogen bonding, and
solute–solute is hydrogen bonding. **13.25.** (a) 7.23 atm; (b) dialysis
solution to blood. **13.26.** 24 mg

Conceptual Problems
13.28. (a)<(b)<(c) **13.30.** (a) 37 °C (b) 1.5 m **13.32.** (a) red curve
solution of a volatile solute green curve represents the solution of a
nonvolatile solute. (b) & (d)

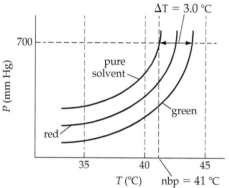

(c) 1.5 m **13.34.**

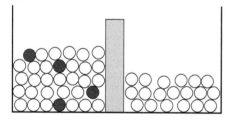

Section Problems
Solutions, Energy Changes, and Solubility (Sections 13.1–13.3)
13.36. The larger the surface area, the more rapidly the solid will
dissolve. Powdered NaCl has a much larger surface area **13.38.** (a)
Na^+; (b) Ba^{2+} **13.40.** Solvent-solvent is hydrogen bonding, solvent–
solute is dispersion and solute–solute is dispersion. I_2 is not soluble
in water. **13.42.** Both Br_2 and CCl_4 are nonpolar, and intermolecular

forces for both are dispersion forces. H_2O is a polar molecule with
dipole-dipole forces and hydrogen bonding. **13.44.** greater in water.
13.46. toluene < Br_2 < KBr
13.48. three

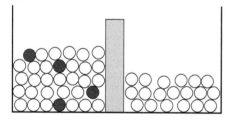

13.50. Ethyl alcohol and water are both polar with small dispersion
forces. They both can hydrogen bond. Pentyl alcohol is slightly polar
and can hydrogen bond. It has, however, a relatively large solute-
solute dispersion force because of its size, which limits its water
solubility. **13.52.** 42.5 °C

Units of Concentration (Section 13.4)
13.54. 0.614 M **13.56.** (a) 0.049 ppb; (b) 1.9×1^{-1} mol/L **13.58.** (a)
Dissolve 0.150 mol of glucose in water; dilute to 1.00 L; (b) Dissolve
1.135 mol of KBr in 1.00 kg of H_2O; (c) Mix together 0.15 mol of
CH_3OH with 0.85 mol of H_2O. **13.60.** Dissolve 4.42×10^{-3} mol
(0.540 g) of $C_7H_6O_2$ in enough $CHCl_3$ to make 165 mL of solution. **13.62.** (a) 0.500 M KCl; (b) 1.75 M **13.64.** (a) 11.2 mass %;
(b) 0.002 70 mass % KBr; (c) 3.65 mass % aspirin **13.66.** 0.20 ppm
13.68. (a) 0.196 m; (b) $X_{C_{10}H_{14}N_2}$ 0.0145; $X_{CH_2Cl_2}$ = 0.985 **13.70.**
1.81 M **13.74.** 3.7 g **13.76.** (a) 0.0187; (b) 16.0%; (c) 1.06 m

Solubility and Henry's Law (Section 13.5)
13.80. 0.068 M **13.82.** 0.06 atm **13.84.** 4.8×10^{-4} mol/L **13.86.**
$CaCl_2$ < $FeCl_3$ < glucose

Colligative Properties (Sections 13.6–13.9)
13.88. When solid $CaCl_2$ is added to liquid water, the temperature
rises because ΔH_{soln} for $CaCl_2$ is exothermic.
When solid $CaCl_2$ is added to ice at 0 °C, some of the ice will melt
(an endothermic process) and the temperature will fall because the
$CaCl_2$ lowers the freezing point of an ice/water mixture. **13.90.** 98.6
mm Hg **13.92.** −3.55 °C **13.94.** 1.9 **13.96.** NaCl is a nonvolatile solute. Methyl alcohol is a volatile solute. When NaCl is added to water, the vapor pressure of the solution is decreased, which means that
the boiling point of the solution will increase. When methyl alcohol
is added to water, the vapor pressure of the solution is increased
which means that the boiling point of the solution will decrease.

13.98.

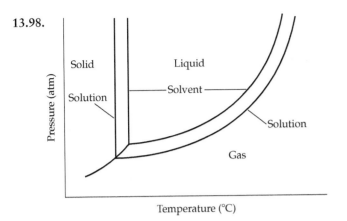

13.100. (a) 70.5 mm Hg; (b) 68.0 mm Hg **13.104.** 2.3 **13.106.** 0.514
13.108. 219 mm Hg **13.112.** 3.6 $\dfrac{°C \cdot kg}{mol}$ **13.114.** 0.573 m

13.116. (a) 13.0 atm; (b) 65.2 atm **13.118.** 0.197 M **13.120.** $\dfrac{°C \cdot kg}{mol}$

Reasonable amounts of salt are capable of lowering the freezing
point (ΔT_f) of the snow below an air temperature of -2 °C. Reasonable amounts of salt, however, are not capable of causing a ΔT_f of
more than 30 °C **13.122.** 342.5 **13.124.** HCl completely dissociates
into two solute particles per each HCl. Only a few percent of the HF
molecules dissociates into ions. **13.126.** 0.442 **13.128.** 1.3 g $C_{10}H_8$
13.130. freezing point -2.3 °C boiling point 100.63 °C
13.132. 2.45% **13.134.** 0.51 mm Hg **13.138.** 334 g/mol **13.142.**
538 $C_{40}H_{56}$

Multiconcept Problems
13.144. 271 $C_{18}H_{24}O_2$ **13.146.** (a) 24.07 g $BaSO_4$; (b) More precipitate will form **13.152.** 89.8 g of $C_{12}H_{22}O_{11}$ 10.2 g of NaCl

CHAPTER 14

14.1. (a) 1.6×10^{-4} M/s; (b) -3.2×10^{-4} M/s **14.2.** (a) -6.0×10^{-5} M/s, -1.2×10^{-4} M/s, 6.0×10^{-5} M/s; (b) $A + 2B \rightarrow C$ **14.4.** order for NO is 1, reaction order for Cl_2 is 2, 3rd order overall **14.6.** (a) Rate = $k[C_2H_4Br_2][I^-]$; (b) $4.98 \times 10^{-3}/(M \cdot s)$; (c) 1.12×10^{-4} M/s; (d) -3.36×10^{-4} M/s **14.8.** vessel 2, 0.02 M/s, vessel 3, 0.005 M/s **14.9.** 44.4 min **14.10.** (a) graph of $[C_2H_5OH]$ vs time is linear, zeroth order in $[C_2H_5OH]$; (b) 4.0×10^{-5} M/s; (c) 0.014 M **14.11.** (a) 0.080 M; (b) 61 h **14.12.** (a) 0.0562 h^{-1}; (b) 33.8 h **14.13.** A plot of ln[cyclopropane] versus time is linear, first-order reaction, $k = 6.6 \times 10^{-4}/s$ (0.040/min) **14.16.** $t_{1/2} = 5$ min, After 15 min, one A molecule would remain **14.18.** A graph of In Abs vs time is linear, first order in dye $t_{1/2} = 22.6$ s

14.20. Unsuccessful   Successful

Unsuccessful   Unsuccessful

14.22. 76.5 kJ/mol **14.23.** (a) $2NO_2(g) + F_2(g) \rightarrow 2NO_2F(g)$, F(g), reaction intermediate; (b) each elementary reaction is bimolecular **14.24.** (a) $2 H_2O_2(aq) \rightarrow 2 H_2O(l) + O_2(g)$, OH(aq) and HO_2(aq), reaction intermediates; (b) Step 1 is uni-molecular, Steps 2 and 3 are bimolecular **14.26.** (a) ii); (b) iii); (c) i) **14.27.** $NO(g) + Cl_2(g) \rightarrow NOCl(g) + Cl(g)$ (slow), $NO(g) + Cl(g) \rightarrow NOCl(g)$ (fast) **14.28.** $Co(CN)_5(H_2O)^{2-}(aq) \rightarrow Co(CN)_5^{2-}(aq) + H_2O(l)$ (slow), $Co(CN)_5^{2-}(aq) + I^-(aq) \rightarrow Co(CN)_5I^{3-}(aq)$ (fast), The pre-dicted rate law is the rate law for the first (slow) elementary reaction: Rate = $k[Co(CN)_5(H_2O)^{2-}]$ **14.30.** (a) $H_2(g) + I_2(g) \rightarrow 2 HI(g)$; (b) Rate = $k_3 \dfrac{k_1 \; k_2}{k_{-1}k_{-2}}[H_2][I_2]$; (c) $k = \dfrac{k_1k_2k_3}{k_{-1}k_{-2}}$

Conceptual Problems
14.40. (a)–(d) are 2 : 1 : 4 : 2; (b) k's are all the same
14.42. (a)

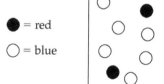

● = red

○ = blue

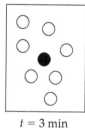

$t = 2$ min $t = 3$ min

(b) 1 minute. **14.44.** (a) second-order in A; (b) Rate = $k[A]^2$; (c) (For fourth box, $t = 7$ min)

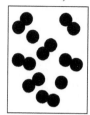

$t = 3$ min + 4 min = 7 min

14.46. (a) zeroth-order in B, second-order in A; (b) Rate = $k[A]^2$; (c) $2 A \rightarrow A_2$ (slow), $A_2 + B \rightarrow AB + A$; (d) A_2 is an intermedi-ate **14.48.** (a) $BC + D \rightarrow B + CD$; (b) 1. B—C + D (reactants), A (catalyst); 2. B—C—A (transition state), D (reactant), 3. A—C (intermediate), B (product), D (reactant), 4. A—C—D (transition state), B (product), 5. A (catalyst), C—D + B (products); (c) first step, Rate = $k[A][BC]$; (d) Endothermic

Section Problems
Reaction Rates (Section 14.1)
14.56. (a) The rate of consumption of H_2 is 3 times faster; (b) The rate of formation of NH_3 is 2 times faster. **14.58.** (a) -2.4×10^{-6} M/s; (b) 9.6×10^{-6} M/s

Rate Laws (Sections 14.2–14.3)
14.60. Rate = $k[H_2][ICl]$; units for k are $\dfrac{L}{mol \cdot s}$ or $1/(M \cdot s)$

14.62. (a) Rate = $k[CH_3Br][OH^-]$; (b) decrease by a factor of 5; (c) increase by a factor 4 **14.64.** (a) Rate = $k[Cu(C_{10}H_8N_2)_2^+]^2[O_2]$; (b) 3; (c) rate will decrease by 1/16 **14.66.** (a) Rate = $k[NH_4^+][NO_2^-]$; (b) $3.0 \times 10^{-4}(M \cdot s)$; (c) 6.1×10^{-6} M/s

Integrated Rate Law; Half-Life (Sections 14.4–14.6)
14.70. (a) 0.015 M; (b) 40 min; (c) 7.2 min **14.74.** (a) 1, 5.2×10^{-3} M; (b) 3.1 h **14.78.** $2.79 \times 10^{-3}/s$

The Arrhenius Equation (Sections 14.7–14.8)
14.86. Very few collisions involve a collision energy greater than or equal to the activation energy, and only a fraction of those have the proper orientation for reaction. **14.88.** there is an exponential increase in the fraction of the collisions that leads to products.

14.92.

(a)

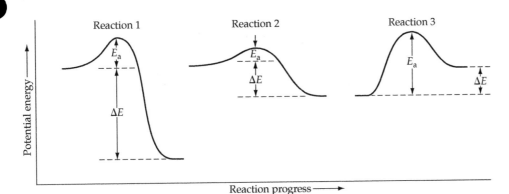

(b) Reaction 2 is the fastest (smallest E_a), and reaction 3 is the slowest (largest E_a); (c) Reaction 3 is the most endothermic, reaction 1 is the most exothermic **14.94.** (a) 134 kJ/mol; (b) 6.0/(M·s) **14.96.** 87 kJ/mol **14.98.** 68 kJ/mol, 1.7 **14.100.** (a) Measure the change in the concentration of A as a function of time at several different temperatures; (b) Plot ln [A] versus time, for each temperature. Straight line graphs will result and k at each temperature equals –slope. Graph ln k versus 1/K, Determine the slope of the line. $E_a = -R(slope)$

Reaction Mechanisms (Sections 14.9–14.11)

14.104. There is no relationship unless the overall reaction occurs in a single elementary step **14.106.** (a) $H_2(g) + 2ICl(g) \rightarrow I_2(g) + 2HCl(g)$; (b) $HI(g)$; (c) bimolecular **14.108.** (a) bimolecular, Rate = $k[O_3][Cl]$; (b) unimolecular, Rate = $k[NO_2]$; (c) bimolecular, Rate = $k[ClO][O]$; (d) termolecular, Rate = $k[Cl]^2[N_2]$ **14.110.** (a) $2\,NO_2Cl(g) \rightarrow 2\,NO_2(g) + Cl_2(g)$; (b) 1. unimolecular; 2. bimolecular; (c) Rate = $k[NO_2Cl]$ **14.112.** $NO_2(g) + F_2(g) \rightarrow NO_2F(g) + F(g)$ (slow), $F(g) + NO_2(g) \rightarrow NO_2F(g)$ (fast) **14.114.** (a)

$2\,NO(g) + O_2(g) \rightarrow 2\,NO_2(g)$; (b) $2k_2\dfrac{k_1}{k_{-1}}[NO]^2[O_2]$;

(c) $k = \dfrac{2k_2k_1}{k_{-1}}$

Catalysis (Sections 14.13–14.15)

14.116. A catalyst does participate in the reaction, but it is not consumed because it reacts in one step of the reaction and is regenerated in a subsequent step. **14.118.** (a) $O_3(g) + O_{(g)} \rightarrow 2\,O_2(g)$;

(b) Cl acts as a catalyst; (c) CIO is a reaction intermediate; (d) A catalyst reacts in one step and is regenerated in a subsequent step. A reaction intermediate is produced in one step and consumed in another. **14.120.** (a) $NH_2NO_2(aq) \rightarrow N_2O(g) + H_2O(l)$; (b) OH^- catalyst $NHNO_2^-$, intermediate; (c) decrease

Multiconcept Problems

14.126. (a) $I^-(aq) + OCl^-(aq) \rightarrow Cl^-(aq) + OI^-(aq)$;

(b) Rate = $k\dfrac{[I^-][OCl^-]}{[OH^-]}$, 60/s; (c) does not occur by a single-step mechanism because OH^- appears in the rate law but not in the overall reaction; (d) $OCl^-(aq) + H_2O(l) \rightleftharpoons$
$HOCl(aq) + OH^-(aq)$ (fast)

$HOCl(aq) + I^-(aq) \rightarrow HOI(aq) + Cl^-(aq)$ (slow)

$HOI(aq) + OH^-(aq) \rightarrow H_2O(l) + OI^-(aq)$ (fast)

Rate = $k\dfrac{[OCl^-][I^-]}{[OH^-]}$ where $k = \dfrac{k_1k_2[H_2O]}{k_{-1}}$; **14.128.** 49 °C

14.130. (a) 7.7×10^{-7} s; (b) speed of mixing **14.132.** 2.7×10^2 s **14.136.** 15 s

CHAPTER 15

15.2. (a) K_c(overall) $= \dfrac{[NO_2]^2}{[N_2][O_2]^2}$; (b) 2.8×10^{-15} **15.4.** (a)

$K_c = 1.38 \times 10^{-4}$; (b) 0.992 M **15.5.** For a mixture to be at equilibrium, [AB][B] must be equal to 1, (2), = 1 at equilibrium. (3), = 0.125 not at equilibrium. (4), = 1 at equilibrium.

15.6.

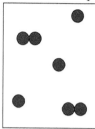

15.7. 9.48 **15.8.** 1.43 atm **15.10.** 0.335 **15.12.** 1.80×10^{-11}
15.16. (a) (2); (b) (1), reverse; (3), forward **15.17.**
$[CO_2] = [H_2] = 0.101$ M, $[CO] = [H_2O] = 0.049$ M
15.18. $[CO_2] = [H_2] = 0.135$ M $[CO] = [H_2O] = 0.065$ M
15.19. $[H_2] = 0.0032$ M, $[I_2] = 0.2032$, $[HI] = 0.1936$ M
15.20. $[N_2O_4] = 0.476$ M, $[NO_2] = 0.0470$ M
15.24. $P_{H_2O} = 0.90$ atm, $P_{CO} = 1.30$ atm, $P_{H_2} = 1.70$ atm
15.26. (a) The concentration of both Ca^{2+} and $C_2O_4^{2-}$ will increase as fluid is lost and the reaction will proceed towards products; (b) Increasing the concentration of Ca^{2+} will cause the reaction to proceed toward products; (c) Decreasing the concentration of $C_2O_4^{2-}$ will cause the reaction to proceed towards reactants and the kidney stone will dissolve; (d) The concentration of both Ca^{2+} and $C_2O_4^{2-}$ will decrease causing the reaction proceed towards reactants and the kidney stone will dissolve **15.28.** (a) 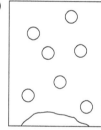 (b) the same

15.29. more NO, the higher the temperature.
15.30. decreases, increase in K_c. **15.31.** There are more AB(g) molecules at the higher temperature. The equilibrium shifted to the right at the higher temperature, which means the reaction is endothermic.
15.32. (a)   (b)

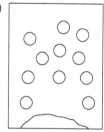

15.34. (a) k_r is larger; (b) < 1; (c) forward and reverse reactions both decrease. K_c decreases **15.35.** (a) 75%; (b) 20%; (c) An efficient unloading of oxygen occurs in working muscle tissue **15.36.** towards reactants, releasing oxygen from hemoglobin and making it available

for use by the muscles. **15.37.** (a) become larger; (b) more product favored; (c) greatly increased **15.38.** (a) 5.3×10^{19}; (b) 2.9×10^4
15.39. If CO binds to Hb, Hb is removed from the reaction and the reaction will shift to the left **15.40.** (a) 1.6; (b) increase in oxygen will shift the equilibrium toward reactants

Conceptual Problems
15.42. (a) $A_2 + C_2 \rightleftharpoons 2\ AC$; (b) $A_2 + B_2 \rightleftharpoons 2\ AB$ **15.44.** (a) $A_2 + 2\ B \rightleftharpoons 2\ AB$; (b) fewer molecules **15.46.** exothermic
15.48. (a)   (b)   (c)

15.50. (a) $A \rightarrow 2$ B; (b) (1) A increases, (2) A increases. (3) the same. (4) the same.

Section Problems
The Equilibrium State (Section 15.1)
15.52. (d)

Equilibrium Constant Expressions and Equilibrium Constants (Section 15.2–15.4)
15.54. $K_c = \dfrac{[C_2H_5OC_2H_5][H_2O]}{[C_2H_5OH]^2}$ **15.56.** (a) $K_c = \dfrac{[CO][H_2]^3}{[CH_4][H_2O]}$;
(b) $K_c = \dfrac{[ClF_3]^2}{[F_2]^3[Cl_2]}$; (c) $K_c = \dfrac{[HF]^2}{[H_2][F_2]}$ **15.60.** (a) 1.3×10^8;
(b) 1.2×10^4; (c) 5.6×10^{-17} **15.62.** (a) 1.4×10^{-8}; (b) 5.2×10^{15};
(c) 2.7×10^{-24} **15.64.** 4×10^3 **15.66.** 0.058 **15.68.** 9.0×10^3
15.70. 0.0313 atm, 1, 1.28×10^{-3} **15.72.** 3.48×10^{-9} M
15.74. (a) $K_c = \dfrac{[CO_2]^3}{[CO]^3}$, $K_p = \dfrac{(P_{CO_2})^3}{(P_{CO})^3}$; (b) $K_c = \dfrac{1}{[O_2]^3}$,
$K_p = \dfrac{1}{(P_{O_2})^3}$; (c) $K_c = [SO_3]$, $K_p = P_{SO_3}$; (d) $K_c = [Ba^{2+}][SO_4^{2-}]$

Using the Equilibrium Constant (Section 15.5)
15.76. (a) product; (b) reactants **15.78.** much SO_3 **15.80.** not at equilibrium reverse direction **15.82.** (a) 178; forward; (b) 0.0240 atm **15.84.** $[N_2O_4] = 0.0429$ M, $[NO_2] = 0.0141$ M
15.86. 29.0 **15.88.** $CH_3CO_2C_2H_5$(soln) $+ H_2O$(soln) $\rightleftharpoons$ CH_3CO_2H(soln) $+ C_2H_5OH$(soln), 0.29 **15.90.** (a) 0.49; reverse; (b) $[BrCl] = 0.068$ M, $[Br_2] = [Cl_2] = 0.026$ M
15.92. 3.7×10^{-3} M **15.94.** $[N_2] = 2.22$ M,
$[O_2] = 0.54$ M, $[NO] = 0.045$ M **15.96.** (a) 6.8 moles;
(b) mol $CH_3CO_2H = 0.03$ mol, mol $C_2H_5OH = 9.03$ mol,
mol $CH_3CO_2C_2H_5 = 0.97$ mol **15.98.** $P_{ClF} = P_{F_2} = 0.389$ atm,
$P_{ClF_3} = 1.08$ atm **15.100.** $[H_2] = [I_2] = 0.045$ M; $[HI] = 0.31$ M
15.102. $[CO] = [H_2] = 0.18$ M; $[H_2O] = 1.02$ M

Le Châtelier's Principle (Sections 15.6–15.9)
15.118. (a) increases; (b) increases; (c) decreases; (d) decreases decreasing $[Cl^-]$ increases Q_c reaction must go from right to left

15.120. (a) decreases; (b) same; (c) increases **15.122.** K_c increases
15.124. (a) increases; (b) increases; (c) decreases; (d) increases
15.126. (a) decreases; (b) increases; (c) no change; (d) increases;
(e) no change **15.128.** (a) no shift; (b) toward reactants; (c) toward
products; (d) toward reactants

Chemical Equilibrium and Chemical Kinetics (Section 15.10)

15.132. (a) same; (b) increases; (c) decreases; (d) same; (e) decreases

15.134. forward **15.136.** $\dfrac{k_f}{k_r} = \dfrac{[C]}{[A][B]} = K_c$ **15.138.** 210 **15.140.** E_a

(reverse) is greater than E_a (forward), exothermic

Multiconcept Problems

15.146. (a) $K_c = \dfrac{[C_2H_6][C_2H_2]}{[C_4H_{10}]}$ $K_p \dfrac{(P_{C_2H_6})(P_{C_2H_4})}{P_{C_4H_{10}}}$; (b) 0.19;

(c) 69 atm; (d) decrease **15.148.** (a) 1.6×10^{-3}; (b) 0.039

15.152. (a) 49.6; (b)

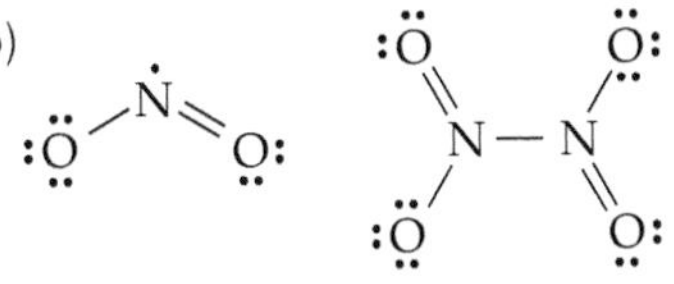

15.154. (a) 2.5; (b) 0.0034; (c) H_2O can hydrogen bond with acetic acid, thus preventing acetic acid dimer formation. Benzene cannot hydrogen bond with acetic acid. **15.156.** (a) 1.47; (b) 10.3; (c) endothermic **15.158.** (a) 1.52; (b) 6.65 kJ **15.160.** 6.0×10^2 O_3 molecules **15.162.** (a) endothermic; (b) 99.97%;

(c)   no dipole moment,

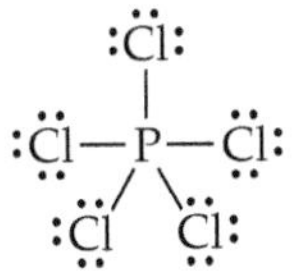

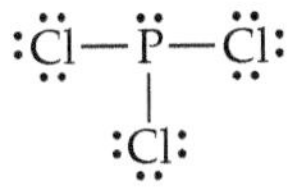

a dipole moment

CHAPTER 16

16.3. $HCl(aq) + NH_3(aq) \rightleftharpoons NH_4^+(aq) + Cl^-(aq)$
     Acid   Base   Acid   Base
              Conjugate acid–base pairs

16.4. (a) $CH_2NH_2(aq) + H_2O(l) \rightleftharpoons OH^-(aq) + CH_3NH_2^+(aq)$
          Base   Acid   Base   Acid
                   conjugate acid-base pairs

(b) $HNO_3(aq) + H_2O(l) \rightleftharpoons H_3O^+(aq) + NO_3^-(aq)$
     Acid   Base   Acid   Base
              conjugate acid-base pairs

16.6. (a) HY; (b) (X^-); (c) to the left. **16.9.** 7.1×10^{-11} M, acidic. **16.10.** $[H_3O^+] = [OH^-] = 2.3 \times 10^{-7}$ M **16.12.** -0.80 5×10^{-12} M **16.14.** 40 times **16.16.** 11.77 **16.18.** (a) $K_a(HY) < K_a(HX) < K_a(HZ)$; (b) HZ; (c) HY has the highest pH; HX has the lowest pH **16.20.** 0.57 M
16.23. $[OH^-] = 2.7 \times 10^{-3}$ M, $[NH_4^+] = 2.7 \times 10^{-3}$ M; $[NH_3] = 0.40$ M, $[H_3O^+] = 3.7 \times 10^{-12}$ M, pH = 11.43 **16.24.** 7.0×10^{-11} **16.26.** (a) HX; (b) Y^-; (c) X^- **16.28.** (a) 5.00, 2.5×10^{-3}%; (b) $Fe(H_2O)_6^{3+}$
16.30.

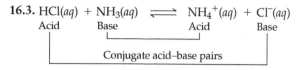

16.34. (a) 100; (b) $[H_3O^+] = 7.9 \times 10^{-6}$ M, $[OH^-] = 1.3 \times 10^{-9}$ M **16.35.** 3.921 **16.36.** Lewis acids include, cations and neutral molecules having vacant valence orbitals that can accept a share in a pair of electrons. The O^{2-} from CaO is the Lewis base and SO_2 is the Lewis acid. **16.37.** NH_4NO_3 is an acidic salt, $(NH_4)_2SO_4$ is an acidic salt.

Conceptual Problems
16.38. (a) acids, HCO_3^- and H_3O^+; bases, H_2O and CO_3^{2-}; (b) acids, HF and H_2CO_3; bases HCO_3^- and F^- **16.40.** (a) H_2S, weakest; HBr, strongest; (b) H_2SeO_3, weakest; $HClO_3$, strongest. **16.46.** (a) $H_3BO_3(aq) + H_2O(l) \rightleftharpoons H_3O^+(aq) + H_2BO_3^-(aq)$; (b) $H_3BO_3(aq) + 2 H_2O(l) \rightleftharpoons H_3O^+(aq) + B(OH)_4^-(aq)$

Section Problems
Acid–Base Concepts (Section 16.1)
16.48. (b) NH_3; (c) HCO_3^- **16.50.** (a) SO_4^{2-}; (b) HSO_3^-; (c) HPO_4^{2-}; (d) NH_3; (e) OH^-; (f) NH_2^-
16.52. (a) $CH_3CO_2H(aq) + NH_3(aq) \rightleftharpoons NH_4^+(aq) + CH_3CO_2^-(aq)$
             Acid   Base   Acid   Base

(b) $CO_3^{2-}(aq) + H_3O^+(aq) \rightleftharpoons H_2O(l) + HCO_3^-(aq)$
     Base   Acid   Base   Acid

(c) $HSO_3^-(aq) + H_2O(l) \rightleftharpoons H_3O^+(aq) + SO_3^{2-}(aq)$
     Acid   Base   Acid   Base

(d) $HSO_3^-(aq) + H_2O(l) \rightleftharpoons H_2SO_3(aq) + OH^-(aq)$
     Base   Acid   Acid   Base

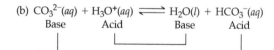

16.54. H_2S acid only, HS^- an acid and a base, S^{2-} base only, H_2O an acid and a base, H_3O^+ acid only, OH^- base only.

Acid and Base Strength (Section 16.2)
16.56. HSO_4^-, SO_4^{2-}

Factors That Affect Acid Strength (Section 16.3)
16.60. (a) $PH_3 < H_2S < HCl$; electronegativity increases from P to Cl; (b) $NH_3 < PH_3 < AsH_3$; X—H bond strength decreases from N to As (down a group); (c) $HBrO < HBrO_2 < HBrO_3$; acid strength increases with the number of O atoms **16.62.** (a) HCl; The strength of a binary acid H_nA increases as A moves from left to right and from top to bottom in the periodic table; (b) $HClO_3$; The strength of an oxoacid increases with increasing electronegativity and increasing oxidation state of the central atom; (c) HBr; The strength of a binary acid H_nA increases as A moves from left to right and from top to bottom in the periodic table. **16.64.** (a) H_2Te, weaker X—H bond; (b) H_3PO_4, P has higher electronegativity; (c) $H_2PO_4^-$, lower negative charge; (d) NH_4^+, higher positive charge and N is more electronegative than C **16.66.** (b)

Dissociation of Water; pH (Sections 16.4–16.6)
16.68. 5.0×10^{-9} M **16.70.** (a) 2.9×10^{-6} M, basic; (b) 1.0×10^{-12} M, basic; (c) 1.0×10^{-4} M, acidic; (d) 1.0×10^{-7} M, neutral; (e) 1.2×10^{-10} M, acidic **16.72.** $[H_3O^+] = [OH^-] = 3.9 \times 10^{-6}$ M, neutral. **16.74.** (a) 4.70; (b) 11.6; (c) 8.449; (d) 3; (e) 15.08 **16.76.** (a) 8×10^{-5} M; (b) 1.5×10^{-11} M; (c) 1.0 M; (d) 5.6×10^{-15} M; (e) 10 M; (f) 5.78×10^{-6} M **16.78.** (a) chlorphenol red; (b) thymol blue; (c) methyl orange

Strong Acids and Strong Bases (Section 16.7)
16.82. (a) 11.30; (b) 1.82; (c) 12.54 **16.84.** (a) 13.90; (b) 1.19; (c) 2.92 **16.86.** 0.0089 g CaO

Weak Acids (Sections 16.8–16.10)
16.88. (a) $C_6H_5OH < HOCl < CH_3CO_2H < HNO_3$; (b) $HNO_3 > CH_3CO_2H > HOCl > C_6H_5OH$
1 M HNO_3, $[H_3O^+] = 1$ M; (c)
1 M CH_3CO_2H, $[H_3O^+] = 4 \times 10^{-3}$ M,
1 M HOCl, $[H_3O^+] = 2 \times 10^{-4}$ M
1 M C_6H_5OH, $[H_3O^+] = [HA] \times K_a = 1 \times 10^{-5}$ M **16.90.** 2.0×10^9 **16.92.** 3.92×10^{-5}, 4.41 **16.94.** pH = 3.20 **16.96.** (a) 0.0029 M $= [H_3O^+] = [C_3H_3O_2^-]$, $[HC_3H_3O_2] = 0.147$ M pH = 2.54, $[OH^-] = 3.4 \times 10^{-12}$ M; (b) 3.3% **16.98.** 1.59, 1.7% **16.100.** (a) 8.0%; (b) 2.6%

Polyprotic Acids (Section 16.11)
16.102. $H_2SeO_4(aq) + H_2O(l) \rightleftharpoons H_3O^+(aq) + HSeO_4^-(aq)$;
$$K_{a1} = \frac{[H_3O^+][HSeOS_4^-]}{[H_2SeO_4]}$$
$HSeO_4^-(aq) + H_2O(l) \rightleftharpoons H_3O^+(aq) + SeO_4^{2-}(aq)$;
$$K_{a2} = \frac{[H_3O^+][SeO_4^{2-}]}{[HSeO_4^-]}$$
16.104. $[H_3O^+] = [HCO_3^-] = 6.6 \times 10^{-5}$ M; $[H_2CO_3] = 0.010$ M, $[CO_3^{2-}] = 5.6 \times 10^{-11}$ M, $[OH^-] = 1.5 \times 10^{-10}$ M, pH = 4.18 **16.106.** 1.08, 6.4×10^{-5} M **16.108.** $[H_2SeO_4] = 0$ M; $[HSeO_4^-] = 0.49$ M; $[SeO_4^{2-}] = 0.011$ M, $[H_3O^+] = 0.51$ M, pH = 0.29, $[OH^-] = 2.0 \times 10^{-14}$ M

Weak Bases (Section 16.12)

16.112.

(a) $(CH_3)_2NH(aq) + H_2O(l) \rightleftharpoons (CH_3)_2NH_2^+(aq) + OH^-(aq)$

$$K_b = \frac{[(CH_3)_2NH_2^+][OH^-]}{[(CH_3)_2NH]}$$

(b) $C_6H_5NH_2(aq) + H_2O(l) \rightleftharpoons C_6H_5NH_3^+(aq) + OH^-(aq)$

$$K_b = \frac{[C_6H_5NH_3^+][OH^-]}{[C_6H_5NH_2]}$$

(c) $CN^-(aq) + H_2O(l) \rightleftharpoons HCN(aq) + OH^-(aq)$;

$$K_b = \frac{[HCN][OH^-]}{[CN^-]}$$

16.114. 9.45 **16.116.** $K_b = 1 \times 10^{-6}$, $pK_b = 5.8$

16.118. 9.0×10^{-5} M $= [OH^-] = [HC_{18}H_{21}NO_4^+]$,

$[C_{18}H_{21}NO_4] = 0.0024$ M, $[H_3O^+] = 1.1 \times 10^{-10}$ M, pH $= 9.96$

Relation between K_a and K_b (Section 16.13)

16.120. (a) 2.0×10^{-11}; (b) 1.1×10^{-6}; (c) 2.3×10^{-5}; (d) 5.6×10^{-6} **16.122.** 1.0×10^{-8}, 7.7×10^{-4}

Acid–Base Properties of Salts (Section 16.14)

16.124. (a) $CH_3NH_3^+(aq) + H_2O(l) \rightleftharpoons H_3O^+(aq) + CH_3NH_2(aq)$
        Acid   Base   Acid   Base

(b) $Cr(H_2O)_6^{3+}(aq) + H_2O(l) \rightleftharpoons H_3O^+(aq) + Cr(H_2O)_5OH^{2+}(aq)$
        Acid   Base   Acid   Base

(c) $CH_3CO_2^-(aq) + H_2O(l) \rightleftharpoons CH_3CO_2H(aq) + OH^-(aq)$
        Base   Acid   Acid   Base

(d) $PO_4^{3-}(aq) + H_2O(l) \rightleftharpoons HPO_4^{2-}(aq) + OH^-(aq)$
        Base   Acid   Acid   Base

16.126. (a) basic; (b) neutral; (c) acidic; (d) neutral; (e) basic; (f) acidic **16.128.** (a) 1.2×10^{-6} M $= [H_3O^+] = [C_2H_5NH_2]$, pH $= 5.90$ $[C_2H_5NH_3^+] = 0.10$ M; $[NO_3^-] = 0.10$ M, $[OH^-] = 8.0 \times 10^{-9}$; (b) 7.5×10^{-6} M $= [CH_3CO_2H] = [OH^-]$, $[CH_3CO_2^-] = 0.10$ M; $[Na^+] = 0.10$ M, $[H_3O^+] = 1.3 \times 10^{-9}$ M, pH $= 8.89$; (c) $[Na^+] = [NO_3^-] = 0.10$ M, $[H_3O^+] = [OH^-] = 1.0 \times 10^{-7}$ M; pH $= 7.00$ **16.130.** For NH_4^+, 5.6×10^{-10}, For CN^-, 2.0×10^{-5}, basic. **16.132.** (a) $A^-(aq) + H_2O(l) \rightleftharpoons HA(aq) + OH^-(aq)$; basic; (b) $M(H_2O)_6^{3+}(aq) + H_2O(l) \rightleftharpoons H_3O^+(aq) + M(H_2O)_5(OH)^{2+}(aq)$; acidic; (c) $2 H_2O(l) \rightleftharpoons H_3O^+(aq) + OH^-(aq)$; neutral; (d) $M(H_2O)_6^{3+}(aq) + A^-(aq) \rightleftharpoons HA(aq) + M(H_2O)_5(OH)^{2+}(aq)$; acidic **16.136.** $HCO_3^-(aq) + Al(H_2O)_6^{3+}(aq) \rightarrow H_2O(l) + CO_2(g) + Al(H_2O)_5(OH)^{2+}(aq)$

Lewis Acids and Bases (Section 16.15)

16.138. (a) Lewis acid, SiF_4; Lewis base, F^-; (b) Lewis acid, Zn^{2+}; Lewis base, NH_3; (c) Lewis acid, $HgCl_2$; Lewis base, Cl^-; (d) Lewis acid, CO_2; Lewis base, H_2O **16.142.** (a) base; (b) acid; (c) base; (d) acid; (e) base; (f) acid; (g) base; (h) Lewis acid

Multiconcept Problems

16.144. pH $= 7.472$, fraction dissociated $= 6.09 \times 10^{-10}$, % dissociation $= 6.09 \times 10^{-8}$ % **16.148.** (a) 2.4×10^{-5} M; (b) 5.52; (c) will increase but only slightly. **16.150.** 2.54 **16.152.** 6.2×10^{-8} M $= [HPO_4^{2-}]$, 4.2×10^{-19} M $= [PO_4^{3-}]$, $[H_3PO_4] = 0.67$ M, $[H_2PO_4^-] = [H_3O^+] = 0.0708$ M, $[OH^-] = 1.4 \times 10^{-13}$ M, pH $= 1.15$ **16.154.** 0.25 m, -0.93 °C **16.156.** $H_3O^+(aq) + PO_4^{3-}(aq) \rightleftharpoons HPO_4^{2-}(aq) + H_2O(l)$, $H_3O^+(aq) + HPO_4^{2-}(aq) \rightleftharpoons H_2PO_4^-(aq) + H_2O(l)$, $H_3O^+(aq) + H_2PO_4^-(aq) \rightleftharpoons H_3PO_4(aq) + H_2O(l)$, pH $= 2.02$ **16.158.** (a) 1.88; (b) 0.48 L, because the concentration of HF that dissociates can't be neglected compared with the initial HF concentration. **16.160.** (a) Rate $= k \dfrac{[OCl^-][NH_3]}{[OH^-]}$, $17 s^{-1}$;

(b) 5.9×10^7 $M^{-1}s^{-1}$

CHAPTER 17

17.2. (a) $HF(aq) + OH^-(aq) \rightleftharpoons H_2O(l) + F^-(aq)$, 3.5×10^{10};
(b) $H_3O^+(aq) + OH^-(aq) \rightleftharpoons 2\,H_2O(l)$, 1.0×10^{14};
(c) $HF(aq) + NH_3(aq) \rightleftharpoons NH_4^+(aq) + F^-(aq)$, 6.3×10^5, reaction (c) < reaction (a) < reaction (b) **17.4.** pH = 10.08 **17.5.** solution 2 also five F^- ions. solution 2 has higher pH. solution 1 larger percent dissociation. **17.6.** Solution (2) largest percent dissociation. Solution (1) lowest pH. **17.8.** (a) pH = 3.76, pH = 3.5; (b) pH does not change. **17.11.** make the Na_2CO_3 concentration 1.4 times the concentration of $NaHCO_3$. **17.12.** (a) HOCl, $K_a = 3.5 \times 10^{-8}$, pK_a HOCl − NaOCl; (b) $\dfrac{[OCl^-]}{[HOCl]}$ 0.35 **17.14.** (a) 12.00; (b) 4.08;
(c) 2.15 **17.16.** (a) (3), only HA present; (b) (1), HA and A^- present; (c) (4), only A^- present; (d) (2), A^- and OH^- present **17.18.** (a) 5.97; (b) 9.62; (c) 10.91 **17.21.** 2.1×10^{-33} **17.22.** 0.077 M **17.24.** 1.10×10^{-10} **17.25.** 1.4×10^{-5} M **17.26.** 4.1×10^{-11} M **17.27.** 1.1×10^{-12} M **17.28.** $= 2 \times 10^{-40}$ M **17.29.** 25.4 0.045 mol/L **17.30.** **Step 1.** $Cu^{2+}(aq) + 2\,OH^-(aq) \rightleftharpoons Cu(OH)_2(s)$, **Step 2.**
$Cu(OH)_2(s) + 4\,NH_3(aq) \rightleftharpoons Cu(NH_3)_4^{2+}(aq) + 2\,OH^-(aq)$
17.32. For $Mn(OH)_2$, no precipitate For $Fe(OH)_2$, precipitate will form.

Conceptual Problems
17.38. (4); **17.40.** (a) (1), (3), and (4); (b) (4)
17.42. (a) i. (1), ii. (4), iii. (3), iv. (2); (b) less than 7

Section Problems
Buffer Solutions (Sections 17.3–17.4)
17.80. $[NH_4^+]$ 0.50 M; $[NH_3]$ 1.0 M 9.55

Weak Acid–Strong Base Titrations (Section 17.7)
17.88. (a)

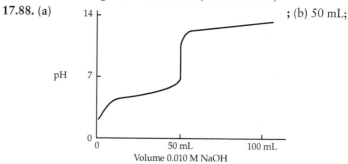

; (b) 50 mL;

(c) > 7.00; (d) pH = pK_a = 4.00

Precipitation; Qualitative Analysis (Sections 17.13–17.15)
17.132. Fe^{2+} and Sn^{2+} can be separated
17.134. (i) not precipitate, (ii) will precipitate
17.136. (a) add Cl^- to precipitate AgCl; (b) add CO_3^{2-} to precipitate $CaCO_3$; (c) add H_2S to precipitate MnS; (d) add NH_3 and NH_4Cl to precipitate $Cr(OH)_3$, (Need buffer to control $[OH^-]$;
17.138. Prepare aqueous solutions of the three salts. Add a solution of $(NH_4)_2HPO_4$. If a white precipitate forms, the solution contains Mg^{2+}. Perform flame test on the other two solutions. A yellow flame test indicates Na^+. A violet flame test indicates K^+.

CHAPTER 18

18.2. (a) negative; (b) positive; (c) positive; (d) negative; (e) negative;
18.4. (a) (larger volume, more randomness); (b) $\Delta S = -5.76$ J/K
18.6. (a) $C_3H_8(g) + 5\,O_2(g) \rightarrow 3\,CO_2(g) + 4\,H_2O(l)$; 6 mol of gas in the reactants are converted to 3 mol; (b) 272.0 J/(K · mol)
18.8. 58.4 °C **18.10.** $\Delta H < 0$, $\Delta S < 0$, $\Delta G < 0$ **18.12.** (a) positive; (b) high temperature **18.14.** (a) -2.9 kJ; (b) spontaneous; (c) reaction rate is extremely low. **18.16.** (a) Reaction (3) largest ΔG Reaction (2) smallest ΔG; (b) 15 kJ **18.17.** 9×10^{-24}
18.18. 80 kJ/mol **18.19.** 0.03 atm **18.20.** 6.27 kJ **18.22.** (a)
$C_6H_{12}O_6(s) + 6\,O_2(g) + 32\,ADP^{3-}(aq) + 32\,H_2PO_4{}^-(aq) \rightleftharpoons$
$6\,CO_2(g) + 32\,ATP^{4-}(aq) + 38\,H_2O(l)$; (b) spontaneous.
18.23. (a) not spontaneous; (b) spontaneous; (c) 652 **18.24.** (a)
(b) The amount of free-energy released increases at the concentrations present in muscle cells. **18.25.** -12.6 kJ

Conceptual Problems

18.26. (a) $A_2 + AB_3 \rightarrow 3\,AB$; (b) positive **18.28.** $\Delta H > 0$ (heat is absorbed), $\Delta S > 0$ (gas has more randomness), $\Delta G < 0$ spontaneous
18.30. $\Delta H = 0$ (ideal gas at constant temperature),
$\Delta S < 0$ less randomness,
$\Delta G > 0$ not spontaneous
18.32. (a) initial state 1, $Q_p < K_p$,
initial state 2, $Q_p > K_p$; (b) $\Delta H > 0$ involves bond breaking,
$\Delta S > 0$ more randomness,
$\Delta G < 0$ spontaneously; (c) $\Delta H < 0$ bond making,
$\Delta S < 0$ less randomness,
$\Delta G < 0$ spontaneously;
(d) State 1 lies to the left of the minimum State 2 lies to the right
18.34. (a) free energy curve must go through a minimum somewhere. At the minimum point, $\Delta G = 0$ and the system is at equilibrium;
(b) left side of the equilibrium composition is rich in reactants.
18.36. ($\Delta H < 0$). $\Delta S°$ is negative.

Section Problems
Spontaneous Processes (Section 18.1)

18.38. (a) and (d) nonspontaneous; (b) and (c) spontaneous
18.40. (b) and (d) spontaneous

Enthalpy, Entropy, and Spontaneous Processes (Section 18.2)

18.42. Molecular randomness is called entropy. For the following reaction, the entropy increases: $H_2O(s) \rightarrow H_2O(l)$ at 25 °C.;
18.44. (a) +; (b) −; (c) −; (d) +

Entropy (Sections 18.3–18.4)

18.52. 1.17 1.06 **18.56.** (a) 50 L; (b) 1 atm; (c) 100 °C; (d) 100 °C, 0.1 atm **18.58.** 0.14 J/K

18.60.

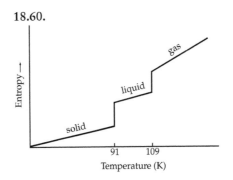

Temperature (K)

Standard Molar Entropies and Standard Entropies of Reaction (Section 18.5)

18.62. (a) $C_2H_6(g)$; more atoms/molecule; (b) $CO_2(g)$; more atoms/molecule; (c) $I_2(g)$; gas has more randomness; (d) $CH_3OH(g)$; gas has more randomness **18.64.** -173.0 J/K **18.66.** (a) +125 J/K (+, because moles of gas increase); (b) -181.2 J/K (−, because moles of gas decrease); (c) +137.4 J/K (+, because moles of gas increase); (d) -626.4 J/K (−, because moles of gas decrease)

Entropy and the Second Law of Thermodynamics (Section 18.6)

18.68. In any spontaneous process, the total entropy of a system and its surroundings always increases. **18.70.** $\Delta S_{surr} = \dfrac{-\Delta H}{T}$

(a) positive; (b) negative **18.72.** (a) +879.3 J/K; (b) no temperature
18.74. $\Delta S_{total} < 0$, not spontaneous **18.76.** +216.4 J/K,
$\Delta S_{surr} - 609.4$ J/K, $\Delta S_{total} - 393.0$ J/K, not spontaneous;
(b) 839.2 K **18.78.** (a) -89.5 J/(K · mol), -2.5 J/(K · mol);
(b) -87.0 J/(K · mol), 0; (c) -84.6 J/(K · mol),

+2.4 J/(K · mol), not boil at 70 °C ,normal boiling point is 80 °C

Free Energy (Section 18.7)

18.80. ΔH ΔS $\Delta G = \Delta H - T\Delta S$ Reaction Spontaneity,

ΔH	ΔS	$\Delta G = \Delta H - T\Delta S$	Reaction Spontaneity
−	+	−	Spontaneous at all temperatures,
−	−	− or +	Spontaneous at low temperatures where $|\Delta H| > |T\Delta S|$,
			Nonspontaneous at high temperatures where $|\Delta H| < |T\Delta S|$,
+	−	+	Nonspontaneous at all temperatures,
+	+	− or +	Spontaneous at high temperatures where $T\Delta S > \Delta H$,
			Nonspontaneous at low temperature where $T\Delta S < \Delta H$

18.82. (e) a negative free-energy change. **18.84.** (a) 0 °C $\Delta H > 0$, $\Delta S > 0$, $\Delta G > 0$; (b) 15 °C $\Delta H > 0$, $\Delta S > 0$, $\Delta G < 0$ **18.86.** (a) +0.9 kJ/mol, not boil; (b) 0 boiling point; (c) −0.9 kJ/mol; boils **18.88.** 122.4 °C **18.90.** 78 °C

Standard Free-Energy Changes and Standard Free Energies of Formation (Sections 18.8–18.9)

18.92. (a) $\Delta G°$ is the change in free energy that occurs when reactants in their standard states are converted to products in their standard states; (b) $\Delta G°_f$ is the free-energy change for formation of one mole of a substance in its standard state from the most stable form of the constituent elements in their standard states. **18.98.** Stable, (a) yes; (b) yes; (c) no; (d) no **18.100.** −147.7 kJ, can be synthesized **18.102.** −6.1 kJ reaction becomes nonspontaneous at high temperatures because $\Delta S°$ is negative, nonspontaneous at 72 °C. **18.104.** reaction is possible. Look for a catalyst, benzene from graphite and gaseous H_2 at 25 °C and 1 atm pressure is not possible.

Free Energy for Reactions under Nonstandard-State Conditions (Section 18.10)

18.106. $\Delta G = \Delta G° + RT \ln Q$ **18.108.** +11.8 kJ/mol, spontaneous in the reverse direction. **18.110.** (a) −176.0 kJ/mol; (b) −133.8 kJ/mol; (c) −141.8 kJ/mol

Free Energy and Chemical Equilibrium (Section 18.11)

18.112. $\Delta G° = -RT \ln K$, (a) negative; (b) 0; (c) positive. **18.114.** 0.059 atm **18.116.** 0.28 atm **18.118.** −162.4 kJ 2.9×10^{28} **18.122.** (a) 3×10^{-9} atm; (b) 0.39 atm; (c) 0.39 atm **18.124.** (a) $\Delta H° = -92.2$ kJ, $\Delta S° = -199$ J/K; (b) negative mol of gas molecules decreases; (c) spontaneous; (d) K_p 2.3×10^3, $K_c = 1.9 \times 10^6$

Multiconcept Problems

18.126. (a)

	$\Delta H_{vap}/T_{bp}$,
ammonia	98 J/K
benzene	87 J/K
carbon tetrachloride	85 J/K
chloroform	87 J/K
mercury	94 J/K

(b) All processes are the conversion of a liquid to a gas at the boiling point; (c) NH_3 deviates from Trouton's rule because of hydrogen bonding. Hg has metallic bonding

CHAPTER 19

19.1. $2 NO_3^-(aq) + 8 H^+(aq) + 3 Cu(s) \rightarrow 3 Cu^{2+}(aq) + 2 NO(g) + 4 H_2O(l)$

19.5.

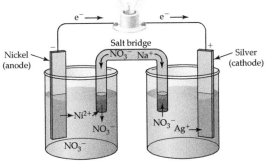

$$Ni(s) \rightarrow Ni^{2+}(aq) + 2 e^- \qquad Ag^+(aq) + e^- \rightarrow Ag(s)$$

19.6. $Al(s) \rightarrow Al^{3+}(aq) + e^-$ anode, $Ag^+(aq) + e^- \rightarrow Ag(s)$ cathode, $Al(s) + 3 Ag^+(aq) \rightarrow Al^{3+}(aq) + 3Ag(s)$, Anode Al(s) minus, cathode Ag(s) positive. **19.10.** nonspontaneous -3.10 V
19.12. (a) D is the strongest reducing agent. A^{3+} is the strongest oxidizing agent; (b) B^{2+} can oxidize C and D. C can reduce A^{3+} and B^{2+}; (c) $A^{3+} + 2D \rightarrow A^+ + 2D^+$ 2.85 V **19.15.**
$2Ag^+(aq) + Ni(s) \rightarrow 2 Ag(s) + Ni^{2+}(aq)$ 1.06 Ni(s)/Ni^{2+}
$(1M)//Ag^+ (1M)/Ag(s)$, Ni anode, Ag cathode **19.16.** -0.74 V
19.17. 0.25 V
19.18. (a)

anode: $4[Al(s) \rightarrow Al^{3+}(aq) + 3 e^-]$	$E° = 1.66$ V
cathode: $3[O_2(g) + 4 H^+(aq) + 4 e^- \rightarrow 2 H_2O(l)]$	$E° = 1.23$ V
overall: $4 Al(s) + 3 O_2(g) + 12 H^+(aq) \rightarrow 4 Al^{3+}(aq) + 6 H_2O(l)$	$E° = 2.89$ V

(b) &(c) $E = E° - \dfrac{2.303 \, RT}{nF} \log \dfrac{[Al^{3+}]^4}{(P_{O_2})^3 [H^+]^{12}}$, 2.63 V **19.21.** 6.9

19.22. 0.14 V **19.24.** -0.140 V **19.27.** A fuel cell and a battery are both galvanic cells that convert chemical energy into electrical energy utilizing a spontaneous redox reaction. A fuel cell differs from an ordinary battery in that the reactants are not contained within the cell but instead are continuously supplied from an external reservoir. **19.31.** cell potential = 0.80 V, $K = 10^{162}$ **19.32.** (a) H is reduced and is the oxidizing agent, carbon is oxidized and is the reducing agent; (b) Carbon is oxidized and is the reducing agent. H is reduced and is the oxidizing agent; (c) Production of CO_2 and pollution caused in production of high temperatures and use of fuels to generate high temperatures. **19.33.** (a) 4.7 g; (b) 5.36 hr

Conceptual Problems
19.38. (a)-(b)

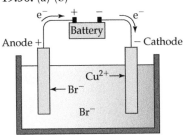

(c) anode reaction　$2 Br^-(aq) \rightarrow Br_2(aq) + 2 e^-$
　　cathode reaction $Cu^{2+}(aq) + 2 e^- \rightarrow Cu(s)$
　　overvall reaction $Cu^{2+}(aq) + 2 Br^-(aq) \rightarrow Cu(s) + Br_2(aq)$

19.40. (a) & (b)

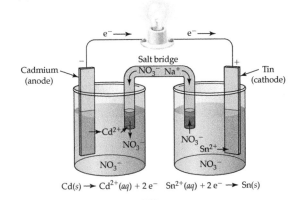

(c) anode reaction　$2 O^{2-} \rightarrow O_2(g) + 4 e^-$
　　cathode reaction $TiO_2(s) + 4 e^- \rightarrow Ti(s) + 2 O^{2-}$
　　overall reaction　$TiO_2(s) \rightarrow Ti(s) + O_2(g)$

Section Problems
Balacing Redox Reactions (Section 19.1)
19.44. (a) $3 e^- + 4 H^+(aq) + NO_3^-(aq) \rightarrow NO(g) + 2 H_2O(l)$;
(b) $Zn(s) \rightarrow Zn^{2+}(aq) + 2 e^-$; (c) $Ti^{3+}(aq) + 2 H_2O(l) \rightarrow TiO_2(s) + 4 H^+(aq) + e^-$; (d) $Sn^{4+}(aq) + 2 e^- \rightarrow Sn^{2+}(aq)$
19.46. (a) oxidation: $Te(s) \rightarrow TeO_2(s)$, reduction: $NO_3^-(aq) \rightarrow NO(g)$; (b) oxidation: $Fe^{2+}(aq) \rightarrow Fe^{3+}(aq)$, reduction: $H_2O_2(aq) \rightarrow H_2O(l)$ **19.48.**
(a) $14 H^+(aq) + Cr_2O_7^{2-}(aq) + 6 e^- \rightarrow 2 Cr^{3+}(aq) + 7 H_2O(l)$;
(b) $4 H_2O(l) + CrO_4^{2-}(aq) + 3 e^- \rightarrow Cr(OH)_4^-(aq) + 4 OH^-(aq)$;
(c) $Bi^{3+}(aq) + 6 OH^-(aq) \rightarrow BiO_3^-(aq) + 3 H_2O(l) + 2 e^-$;
(d) $H_2O(l) + ClO^-(aq) + 2 e^- \rightarrow Cl^-(aq) + 2 OH^-(aq)$

Galvanic Cells (Sections 19.2–19.3)
19.54. because electrons flow through the external circuit toward the positive electrode (the cathode).
19.56. (a)

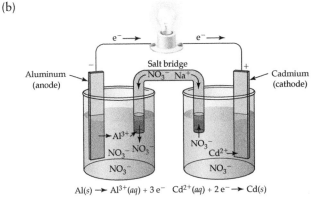

$$Cd(s) \rightarrow Cd^{2+}(aq) + 2 e^- \qquad Sn^{2+}(aq) + 2 e^- \rightarrow Sn(s)$$

(b)

$$Al(s) \rightarrow Al^{3+}(aq) + 3 e^- \qquad Cd^{2+}(aq) + 2 e^- \rightarrow Cd(s)$$

(c)

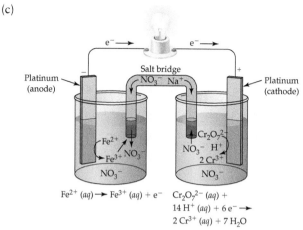

$$Fe^{2+}(aq) \rightarrow Fe^{3+}(aq) + e^-$$
$$Cr_2O_7^{2-}(aq) + 14\,H^+(aq) + 6\,e^- \rightarrow 2\,Cr^{3+}(aq) + 7\,H_2O$$

19.58. Inert electrodes are required because none of the reactants or products is an electrical conductor.

19.60. $Al(s)\,|\,Al^{3+}(aq)\,\|\,Cd^{2+}\,|\,Cd(s)$

19.62. (a)

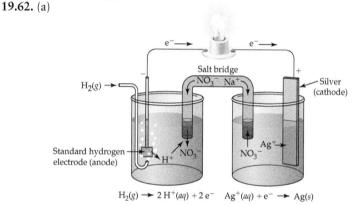

$$H_2(g) \rightarrow 2\,H^+(aq) + 2\,e^- \qquad Ag^+(aq) + e^- \rightarrow Ag(s)$$

(b) anode reaction $H_2(g) \rightarrow 2\,H^+(aq) + 2\,e^-$

   cathode reaction $2\,Ag^+(aq) + 2\,e^- \rightarrow 2\,Ag(s)$

   overall reaction $H_2(g) + 2\,Ag^+(aq) \rightarrow 2\,H^+(aq) + 2\,Ag(s)$

(c) $Pt(s)\,|\,H_2(g)\,|\,H^+(ag)\,\|\,Ag^+(aq)\,|\,Ag(s)$

19.64. (a) anode reaction $Co(s) \rightarrow Co^{2+}(aq) + 2\,e^-$

   cathode reaction $Cu^{2+}(aq) + 2\,e^- \rightarrow Cu(s)$

   overall reaction $Co(s) + Cu^{2+}(aq) \rightarrow Co^{2+}(aq) + Cu(s)$

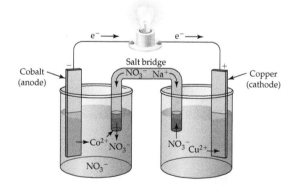

(b) anode reactoin $2\,Fe(s) \rightarrow 2\,Fe^{2+}(aq) + 4\,e^-$

cathode reaction $O_2(g) + 4\,H^+(aq) + 4\,e^- \rightarrow 2\,H_2O(l)$

overall reaction

$$2\,Fe(s) + O_2(g) + 4H^+(aq) \rightarrow 2\,Fe^{2+}(aq) + 2H_2O(l)$$

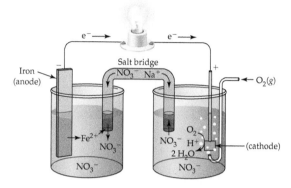

Cell Potentials and Free-Energy Changes (Section 19.4)

19.66. E is the standard cell potential (E°) when all reactants and products are in their standard states—solutes at 1 M concentrations, gases at a partial pressure of 1 atm, solids and liquids in pure form, all at 25°C. **19.68.** −309 kJ **19.70.** 3 mol e⁻ **19.72.** +1.23 V

Standard Reduction Potentials (Sections 19.5–19.6)

19.74. $Sn^{4+}(aq) < Br_2\,(aq) < MnO_4^-$ **19.76.** $Cr_2O_7^{2-}$ is the strongest oxidizing agent. $Fe^{2+}(aq)$ is the weakest oxidizing agent. **19.80.** (a) $Sn^{2+}(aq)$ can be oxidized by $Br_2(aq)$; (b) $Ni^{2+}(aq)$ cannot be reduced by $Sn^{2+}(aq)$; (c) Ag (s) cannot be oxidized by $Pb^{2+}(aq)$; (d) $I_2(s)$ can be reduced by H_2SO_3. **19.82.** (a) no reaction; (b) $Pb(s) + 2\,Ag^+(aq) \rightarrow Pb^{2+}(aq) + 2\,Ag(s)$; (c) $Cl_2(g) + H_2C_2O_4(aq) \rightarrow 2\,Cl^-(aq) + 2\,CO_2(g) + 2\,H^+(aq)$; (d) $Ni(s) + 2\,HClO(aq) + 2\,H^+(aq) \rightarrow Ni^{2+}(aq) + Cl_2(g) + H_2O(l)$ **19.84.** −0.36 V **19.86.** overall $I_2(s) + Co(s) \rightarrow Co^{2+}(aq) + 2\,I^-(aq)$ E° = 0.82 V, −1.6 × 10² kJ **19.88.** E° = 1.26 V, −730 kJ **19.90.** (a) −0.90 V, nonspontaneous; (b) 2.11 V, spontaneous.

The Nernst Equation (Sections 19.7–19.8)

19.96. 1 × 10⁻¹⁶ M **19.98.** $\dfrac{[Zn^{2+}]}{[Cu^{2+}]} = 10$ **19.100.** (a) 0.64 V; (b) 0.77 V; (c) −0.23 V; (d) −1.08 V **19.104.** $\dfrac{[Cd^{2+}]}{[Pb^{2+}]} = 1 \times 10^9$

Standard Cell Potetials and Equilibrium Constants (Section 19.9)

19.108. $K = 10^{\frac{nE°}{0.0592}}$ If E° is positive, the exponent is positive (because n is positive), and K is greater than 1. **19.110.** 6 × 10³⁵ **19.112.** 6.3 × 10⁸ **19.114.** 9 × 10⁻³

Batteries (Section 19.10)

19.122. (a)

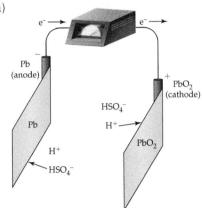

(b) anode:
$Pb(s) + HSO_4^-(aq) \rightarrow PbSO_4(s) + H^+(aq) + 2\ e^-$ $\quad E° = 0.296\ V$
cathode:
$PbO_2(s) + 3\ H^+(aq) + HSO_4^-(aq) + 2\ e^- \rightarrow PbSO_4(s) + 2\ H_2O(l)$
$\quad E° = 1.628\ V$

overall
$Pb(s) + PbO_2(s) + 2\ H^+(aq) + 2\ HSO_4^-(aq) \rightarrow 2\ PbSO_4(s) +$
$\quad 2\ H_2O(l) E° = 1.924\ V$

(c) $K = 1 \times 10^{65}$; (d) 0.

Corrosion (Section 19.11)

19.124. Rust is a hydrated form of iron(III) oxide ($Fe_2O_3 \cdot H_2O$). Rust forms from the oxidation of Fe in the pressure of O_2 an H_2O. Rust can be prevented by coating Fe with Zn (galvanizing). **19.126.** Cr forms a protective oxide coating similar to Al. **19.128.** (d) A strip of magnesium is attached to steel because the magnesium is more easily oxidized than iron. **19.130.** Mn and Al

Electrolysis (Sections 19.12–19.14)

19.132. (a)

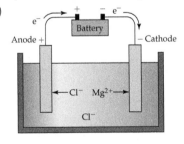

(b) anode: $\quad 2\ Cl^-(l) \rightarrow Cl_2(g) + 2\ e^-$
cathode: $Mg^{2+}(l) + 2\ e^- \rightarrow Mg(l)$
overall: $\overline{Mg^{2+}(l) + 2\ Cl^-(l) \rightarrow Mg(l) + Cl_2(g)}$

19.134. Possible anode reactions:
$2\ Cl^-(aq) \rightarrow Cl_2(g) + 2\ e^-$
$2\ H_2O(l) \rightarrow O_2(g) + 4\ H^+(aq) + 4\ e^-$
Possible cathode reactions:
$2\ H_2O(l) + 2\ e^- \rightarrow H_2(g) + 2\ OH^-(aq)$
$Mg^{2+}(aq) + 2\ e^- \rightarrow Mg(s)$
Actual reactions:
anode: $\quad 2\ Cl^-(aq) \rightarrow Cl_2(g) + 2\ e^-$
cathode: $2\ H_2O(l) + 2\ e^- \rightarrow H_2(g) + 2\ OH^-(aq)$
This anode reaction takes place because of a high overvoltage for formation of gaseous O_2. This cathode reaction takes place because H_2O is easier to reduce than Mg^{2+}.

19.136. (a) NaBr
anode: $\quad 2\ Br^-(aq) \rightarrow Br_2(l) + 2\ e^-$
cathode: $\underline{2\ H_2O(l) + 2\ e^- \rightarrow H_2(g) + 2\ OH^-(aq)}$
overall: $\quad 2\ H_2O(l) + 2\ Br^-(aq) \rightarrow Br_2(l) + H_2(g) + 2\ OH^-(aq)$
(b) $CuCl_2$
anode: $\quad 2\ Cl^-(aq) \rightarrow Cl_2(g) + 2\ e^-$
cathode: $\underline{Cu^{2+}(aq) + 2\ e^- \rightarrow Cu(s)}$
overall: $\quad Cu^{2+}(aq) + 2\ Cl^-(aq) \rightarrow Cu(s) + Cl_2(g)$
(c) LiOH
anode: $\quad 4\ OH^-(aq) \rightarrow O_2(g) + 2\ H_2O(l) + 4\ e^-$
cathode: $\underline{4\ H_2O(l) + 4\ e^- \rightarrow 2\ H_2(g) + 4\ OH^-(aq)}$
overall: $\quad 2\ H_2O\ (l) \rightarrow O_2(g) + 2\ H_2(g)$
19.138. 3.22 g **19.140.** 38.9 h, 4.87×10^5 L Cl_2 **19.144.** 62.6 min

Multiconcept Problems

19.152. (a) $E°_3 = \dfrac{n_1E°_1 + n_2E°_2}{n_3}$ (b) 0.78 V;
(c) $E°$ values would be additive ($E°_3 = E°_1 + E°_2$) if reaction (3) is an overall cell reaction because the electrons in the two half reactions, (1) and (2), cancel. **19.154.** $15\ H^+(aq) +$
$5\ H_2MoO_4(aq) + 3\ As(s) \rightarrow 5\ Mo^{3+}(aq) + 3\ H_3AsO_4(aq) +$
$8\ H_2O(l), +0.060\ V$ **19.156.** $E° = -2.91\ V$
19.158. (a) $Cr_2O_7^{2-}(aq) + 6\ Fe^{2+}(aq) + 14\ H^+(aq) \rightarrow$
$2\ Cr^{3+}(aq) + 6\ Fe^{3+}(aq) + 7\ H_2O(l)$; (b) 1.05 V

CHAPTER 20

20.1. (a) $^{106}_{44}\text{Ru} \rightarrow ^{0}_{-1}\text{e} + ^{106}_{45}\text{Rh}$; (b) $^{189}_{83}\text{Bi} \rightarrow ^{4}_{2}\text{He} + ^{185}_{81}\text{Tl}$; (c) $^{204}_{84}\text{Po} + ^{0}_{-1}\text{e} \rightarrow ^{204}_{83}\text{Bi}$ **20.2.** $^{148}_{69}\text{Tm}$ decays to $^{148}_{68}\text{Er}$ by either positron emission or electron capture. **20.3.** 64.2 h **20.4.** 0.181 d^{-1} **20.5.** 14.0% **20.6.** 30.0 y **20.7.** 44.5 d **20.8.** 141 d^{-1} **20.10.** 3.1817×10^{-24}, 1.91611 g/mol, 7.62 MeV/nucleon **20.12.** 6.01677 u, 5.35 Mev/nucleon **20.13.** -9.11×10^{-9} g **20.14.** 4.333×10^{-4} g **20.18.** $^{238}_{92}\text{U} + ^{12}_{6}\text{C} \rightarrow ^{246}_{98}\text{Cf} + 4\,^{1}_{0}\text{n}$ **20.24.** 1.8 s

Section Problems
Nuclear Reactions and Radioactivity (Sections 20.1–20.2)

20.26. Positron emission is the conversion of a proton in the nucleus into a neutron plus an ejected positron. Electron capture is the process in which a proton in the nucleus captures an inner-shell electron and is thereby converted into a neutron. **20.28.** In beta emission a neutron is converted to a proton and the atomic number increases. In positron emission a proton is converted to a neutron and the atomic number decreases. **20.30.** (a) $^{126}_{50}\text{Sn} \rightarrow ^{0}_{-1}\text{e} + ^{126}_{51}\text{Sb}$; (b) $^{210}_{88}\text{Ra} \rightarrow ^{4}_{2}\text{He} + ^{206}_{86}\text{Rn}$; (c) $^{77}_{37}\text{Rb} \rightarrow ^{0}_{1}\text{e} + ^{77}_{36}\text{Kr}$; (d) $^{76}_{36}\text{Kr} + ^{0}_{-1}\text{e} \rightarrow ^{76}_{35}\text{Br}$ **20.32.** alpha **20.34.** (a) $^{0}_{1}\text{e}$; (b) $^{4}_{2}\text{He}$; (c) $^{0}_{-1}\text{e}$ **20.36.** (a) $^{162}_{75}\text{Re} \rightarrow ^{158}_{73}\text{Ta} + ^{4}_{2}\text{He}$; (b) $^{138}_{62}\text{Sm} + ^{0}_{-1}\text{e} \rightarrow ^{138}_{61}\text{Pm}$; (c) $^{188}_{74}\text{W} \rightarrow ^{188}_{75}\text{Re} + ^{0}_{-1}\text{e}$; (d) $^{165}_{73}\text{Ta} \rightarrow ^{165}_{72}\text{Hf} + ^{0}_{1}\text{e}$ **20.38.** $^{100}_{43}\text{Tc} \rightarrow ^{0}_{1}\text{e} + ^{100}_{42}\text{Mo}$ (positron emission)
$^{100}_{43}\text{Tc} + ^{0}_{-1}\text{e} \rightarrow ^{100}_{42}\text{Mo}$ (electron capture)

Nuclear Stability (Section 20.3)

20.40. ^{160}W is neutron poor and decays by alpha emission. ^{185}W is neutron rich and decays by beta emission. **20.44.** $^{237}_{93}\text{Np}$, $^{233}_{91}\text{Pa}$, $^{233}_{92}\text{U}$, $^{229}_{90}\text{Th}$, $^{225}_{88}\text{Ra}$, $^{225}_{89}\text{Ac}$, $^{221}_{87}\text{Fr}$, $^{217}_{85}\text{At}$, $^{213}_{83}\text{Bi}$, $^{213}_{84}\text{Po}$, $^{209}_{82}\text{Pb}$, $^{209}_{83}\text{Bi}$ **20.46.** 6 alpha 4 β **20.48.** 7 alpha, 4 β

Radioactive Decay Rates (Section 20.4)

20.50. 0.247 d^{-1} **20.52.** N = 23.0% **20.54.** 621 **20.56.** 1.1×10^{6} y **20.58.** 34.6 d **20.60.** (a) 1/8; (b) 0.024 y^{-1}; (c) 193 y

Radiocarbon Dating (Section 20.5)

20.64. 1.6×10^{4} y **20.70.** 1.76×10^{9} y

Energy Changes during Nuclear Reactions (Section 20.6)

20.74. The lost mass is converted into the binding energy that is used to hold the nucleons together. **20.84.** 1.02×10^{-9} g

Fission and Fusion (Section 20.7)

20.88. (a) **20.90.** B-10 + n $\rightarrow$ Li-7 + He-4 **20.92.** Fuel rods contain too low a concentration of U-235.

Nuclear Transmutation (Section 20.8)

20.98. (a) $^{113}_{49}\text{In}$; (b) $^{13}_{7}\text{N} + ^{1}_{0}\text{n}$ **20.100.** $^{209}_{83}\text{Bi} + ^{58}_{26}\text{Fe} \rightarrow ^{266}_{109}\text{Mt} + ^{1}_{0}\text{n}$ **20.102.** $^{238}_{92}\text{U} + ^{2}_{1}\text{H} \rightarrow ^{238}_{93}\text{Np} + 2\,^{1}_{0}\text{n}$

Detecting and Measuring Radioactivity (Section 20.9)

20.108. 0.30 J **20.108.** $^{238}_{92}\text{U} + ^{1}_{0}\text{n} \rightarrow ^{239}_{94}\text{Pu} + 2\,^{0}_{-1}\text{e}$ **20.110.** (a) 8.9 dis/min; (b) 7.02×10^{4} atoms

Multiconcept Problems

20.114. 5.9×10^{9} y **20.116.** 0.000 828 nm

CHAPTER 21

21.2.

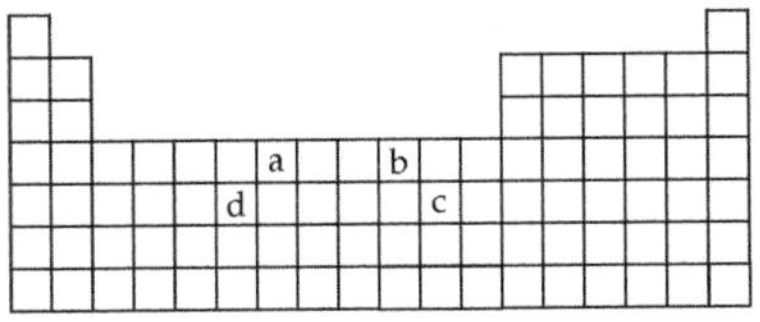

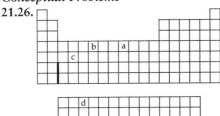

(a) Mn; (b) Ni^{2+}; (c) Ag; (d) Mo^{3+} **21.3.** +2. **21.6.** triamminetrichlorocobalt (III) **21.8.** triamminechloroplatinum(I) **21.16.** $[FeCl6]^{3-}$ **21.18.** square planar **21.20.**

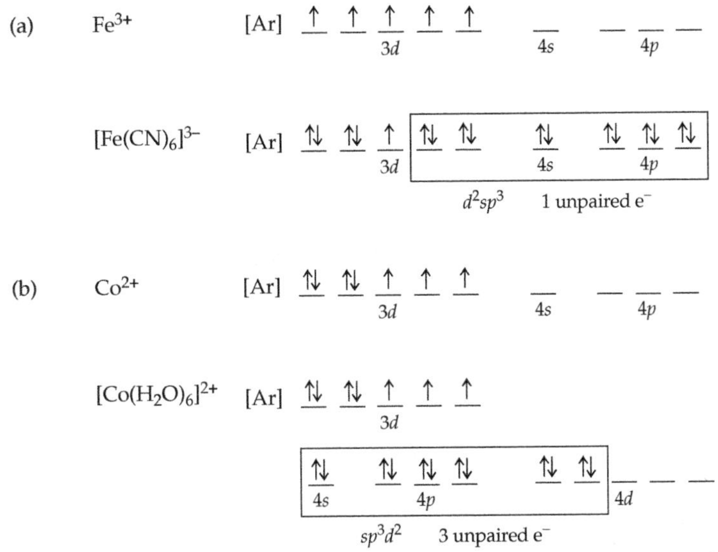

(a) Fe^{3+} [Ar] ↑ ↑ ↑ ↑ ↑   _ _ _ _
                    3d   4s   4p

(b) $[Fe(CN)_6]^{3-}$ [Ar] ↑↓ ↑↓ ↑ | ↑↓ ↑↓ | ↑↓ | ↑↓ ↑↓ ↑↓ |
                               3d   4s   4p
                        d^2sp^3 1 unpaired e^-

(b) Co^{2+} [Ar] ↑↓ ↑↓ ↑ ↑ ↑   _ _ _ _
                      3d   4s   4p

     $[Co(H_2O)_6]^{2+}$ [Ar] ↑↓ ↑↓ ↑ ↑ ↑
                                  3d

     | ↑↓ | ↑↓ ↑↓ ↑↓ | ↑↓ ↑↓ | _ _ _
       4s   4p   4d
        sp^3d^2 3 unpaired e^-

21.22. (a) diamminedichloroplatinum (II); (b) +2, 4; (c) Chloride and ammonia are bases, platinum is the acid; (d) Pt-[Xe] $5d^8$

21.23.

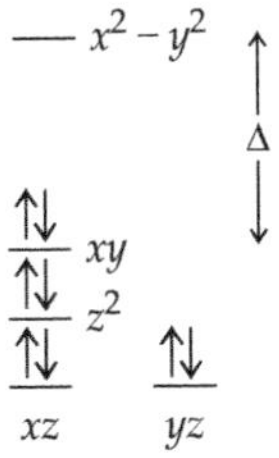

— $x^2 - y^2$ ↑

↑ Δ

↑↓ xy
↑↓ z^2 ↑↓
↑↓
xz   yz

no unpaired electrons. **21.24.** (a) The chloride concentration is higher in blood plasma than inside the cell; (b) diammineaquachloroplatinum (II); (c) reduce the wavelength.

21.25. (a) (IV); (b) diamminetetrachloroPlatinum(IV); (c) The diastereomers are like the cis and trans isomers on page 869 except these isomers have ammonia where those have chloride and these have chloride where those have ammonia.; (d) Both isomers are achiral.

Conceptual Problems
21.26.

(a) Co ; (b) Cr; (c) Zr ; (d) Pr **21.28.** (a) The atomic radii decrease, at first markedly and then more gradually. Toward the end of the series, the radii increase again. The decrease in atomic radii is a result of an increase in Z_{eff}. The increase is due to electron-electron repulsions in doubly occupied d orbitals; (b) The densities of the transition metals are inversely related to their atomic radii. The densities initially increase from left to right and then decrease toward the end of the series; (c) Ionization energies generally increase from left to right across the series. The general trend correlates with an increase in Z_{eff} and a decrease in atomic radii; (d) The standard oxidation potentials generally decrease from left to right across the first transition series. This correlates with the general trend in ionization energies.

21.34. (a) (1) chiral; (2) achiral; (3) chiral; (4) chiral;

(b)

(1)   enantiomer of (1)

(3)   enantiomer of (3)

(4)   enantiomer of (4)

(c) (1) and (4) are enantiomers.

Section Problems
Electron Configurations (Section 21.1)

21.38. (a) 1; (b) 2; (c) 0; (d) 3

Properties of Transition Elements (Section 21.2)

21.40. Ti is harder because the sharing of d, as well as s, electrons results in stronger metallic bonding. **21.42.** (a) The decrease in radii with increasing atomic number is expected because the added d electrons only partially shield the added nuclear charge; (b) The densities of the transition metals are inversely related to their atomic radii. **21.44.** The smaller than expected sizes of the third-transition series atoms are associated with what is called the lanthanide contraction, the general decrease in atomic radii of the f-block lanthanide elements. The lanthanide contraction is due to the increase in Z_{eff} as the 4f subshell is filled. **21.46.** Sc 1866 kJ/mol, Ti 1969 kJ/mol, V 2061 kJ/mol, Cr 2224 kJ/mol, Mn 2226 kJ/mol, Fe 2324 kJ/mol, Co 2408 kJ/mol, Ni 2490 kJ/mol, Cu 2703 kJ/mol, Zn 2639 kJ/mol, Across the first transition element series, Z_{eff} increases and there is an almost linear increase in the sum of the first two ionization energies. This is what is expected if the two electrons are removed from the 4s orbital. Higher than expected values for the sum of the first two ionization energies are observed for Cr and Cu because of their anomalous electron configurations (Cr $3d^5\ 4s^1$; Cu $3d^{10}\ 4s^1$). An increasing Z_{eff} affects 3d orbitals more than the 4s orbital and the second ionization energy for an electron from the 3d orbital is higher than expected. **21.48.** Ti
21.50. (a) $Cr(s) + 2\ H^+(aq) \longrightarrow Cr^{2+}(aq) + H_2(g)$;
(b) $Zn(s) + 2\ H^+(aq) \longrightarrow Zn^{2+}(aq) + H_2(g)$; (c) N.R.;
(d) $Fe(s) + 2\ H^+(aq) \longrightarrow Fe^{2+}(aq) + H_2(g)$

Oxidation States (Section 21.3)

21.54. Sc(III), Ti(IV), V(V), Cr(VI), Mn(VII), Fe(VI), Co(III), Ni(II), Cu(II), Zn(II) **21.56.** Cu^{2+} is a stronger oxidizing agent than Cr^{2+} because of a higher Z_{eff}. **21.58.** A compound with vanadium in the +2 oxidation state is expected to be a reducing agent, because early transition metal atoms have a relatively low effective nuclear charge and are easily oxidized to higher oxidation states.
21.60. $Mn^{2+} < MnO_2 < MnO_4^-$ because of increasing oxidation state of Mn.

Coordination Compounds (Section 21.4)

21.62. (a) 2; (b) 6; (c) 4; (d) 8; (e) 7 **21.66.** (a) +3; (b) +1; (c) +3; (d) +1 **21.68.** (a) $Ir(NH_3)_3Cl_3$; (b) $[Cr(H_2O)_2(C_2O_4)_2]^-$; (c) $[Pt(en)_2(SCN)_2]^{2+}$

21.70.

The iron is in the +3 oxidation state, and the coordination number is six. The geometry about the Fe is octahedral. The oxalate ligand is behaving as a bidentate chelating ligand. There are three chelate rings, one formed by each oxalate ligand.

Ligands (Section 21.5)

21.74. (a) base ethylenediamine, acid Ni^{2+}; (b) ligands ethylenediamine, donor atoms N; (c) 6 octahedral **21.76.** $EDTA^{4-}$ in mayonnaise will complex any metal cations that are present in trace amounts. Free metal ions can catalyze the oxidation of oils, causing the mayonnaise to become rancid. The bidentate ligand $H_2NCH_2CO_2^-$ will not bind to metal ions as strongly as does the hexadentate $EDTA^{4-}$ and so would not be an effective substitute for $EDTA^{4-}$.

Naming Coordination Compounds (Section 21.6)

21.78. (a) tetrachloromanganate(II); (b) hexaamminenickel(II); (c) tricarbonatocobaltate(III); (d) bis(ethylenediamine) dithiocyanatoplatinum(IV) **21.80.** (a) cesium tetrachloroferrate(III); (b) hexaaquavanadium(III) nitrate; (c) tetraamminedibromocobalt(III) bromide; (d) diglycinatocopper(II) **21.82.** (a) $[Pt(NH_3)_4]Cl_2$; (b) $Na_3[Fe(CN)_6]$; (c) $[Pt(en)_3](SO_4)_2$; (d) $Rh(NH_3)_3(SCN)_3$

Isomers (Sections 21.7–21.8)

21.84. (a) (1) $[(NH_3)_5(NO_2)Ru]^+Cl^-$ pentamminenitroruthenium(II) chloride (2) $[(NH_3)_5(ONO)Ru]^+Cl^-$ pentamminenitritoruthenium(II) chloride (3) $[(NH_3)_5ClRu]^+NO_2^-$ pentamminecholoruthenium(II) nitrite; (b) (1) and (2); (c) (1) and (3) **21.88.** (c) cis-$[Cr(en)_2(H_2O)_2]^{3+}$; (d) $[Cr(C_2O_4)_3]^{3-}$
21.90.

enantiomers

diastereoisomers

21.92. Plane-polarized light is light in which the electric vibrations of the light wave are restricted to a single plane. The following chromium complex can rotate the plane of plane-polarized light.

$[Cr(en)_3]^{3+}$

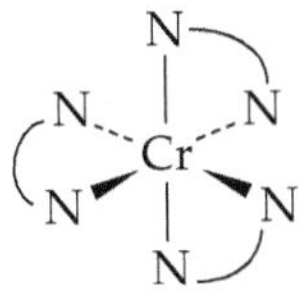

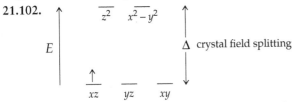

21.94.

Linkage isomers:

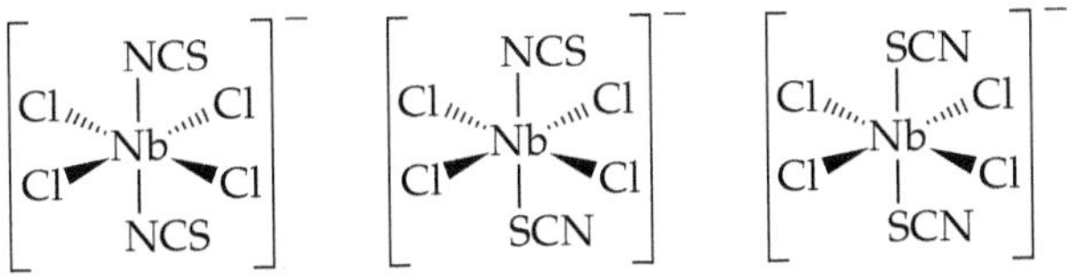

Linkage isomers:

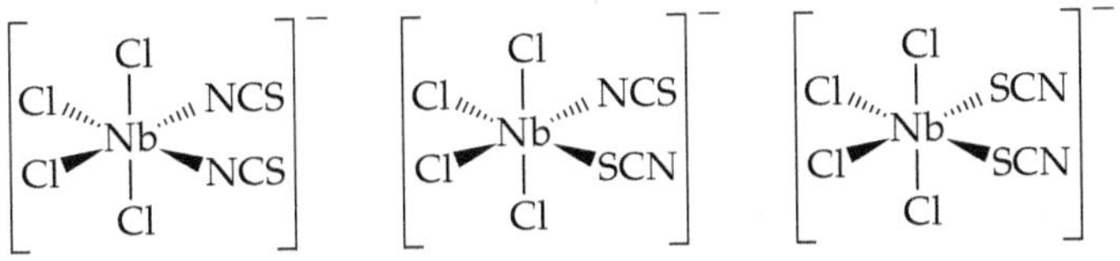

Color of Complexes and Crystal Field Theory (Sections 21.9–21.10)

21.100. The measure of the amount of light absorbed by a substance is called the absorbance, and a graph of absorbance versus wavelength is called an absorption spectrum orange

21.102.

E z^2 $x^2 - y^2$ Δ crystal field splitting

    xz yz xy

$[Ti(H_2O)_6]^{3+}$ is colored because it can absorb light in the visible region, exciting the electron to the higher-energy set of orbitals.
21.104. 220 kJ/mol, NCS^- is a weaker-field ligand than H_2O red
21.106. (a) $[CrF_6]^{3-}$ (b) $[V(H_2O)_6]^{3+}$ (c) $[Fe(CN)_6]^{3-}$

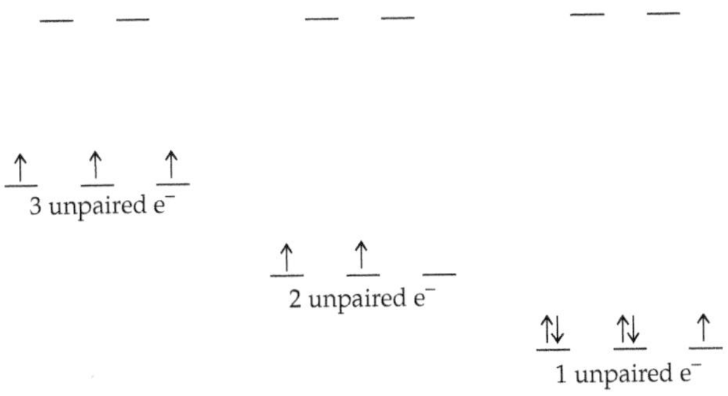

3 unpaired e^-

2 unpaired e^-

1 unpaired e^-

21.108. $Ni^{2+}(aq)$ is green because the Ni^{2+} ion can absorb light, which promotes electrons from the filled d orbitals to the higher energy half-filled d orbitals. $Zn^{2+}(aq)$ is colorless because the d orbitals are completely filled and no electrons can be promoted, so no light is absorbed. **21.110.** Weak-field ligands produce a small Δ. Strong-field ligands produce a large Δ. For a metal complex with weak-field ligands, $\Delta < P$, where P is the pairing energy, and it is easier to place an electron in either d_{z^2} or $d_{x^2-y^2}$ than to pair up electrons; high-spin complexed result. For a metal complex with strong-field ligands, $\Delta > P$ and it is easier to pair up electrons than to place them in either d; low-spin complexes result.
21.112. — $x^2 - y^2$

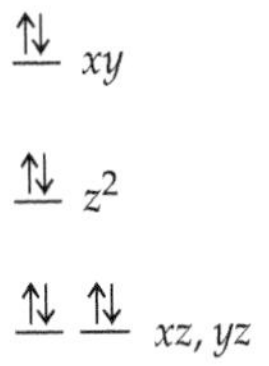

⇅ xy

⇅ z^2

⇅ ⇅ xz, yz

Square planar geometry is most common for metal ions with d^8 configurations because this configuration favors low-spin complexes in which all four lower energy d orbitals are filled, and the higher energy $d_{x^2-y^2}$ orbital is vacant. **21.116.** Cr^{3+} is a $3d^3$ ion. Regardless of the crystal field splitting energy, the three electrons singly occupy the three lower energy d orbitals.
21.118. (a) $[Mn(H_2O)_6]^{2+}$ high-spin Mn^{2+}, $3d^5$

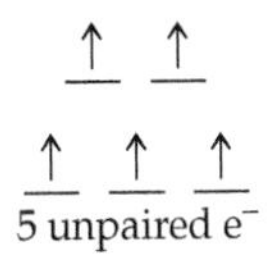

↑ ↑

↑ ↑ ↑
5 unpaired e^-

(b) $Pt(NH_3)_2Cl_2$ square-planar Pt^{2+}, $5d^8$

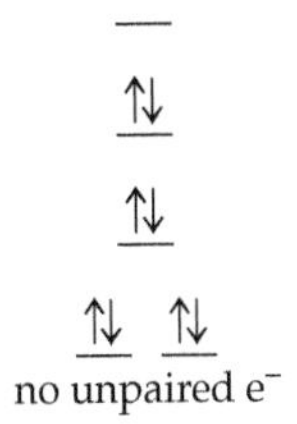

—

⇅

⇅

⇅ ⇅
no unpaired e^-

(c) $[FeO_4]^{2-}$ tetrahedral Fe(VI), $3d^2$

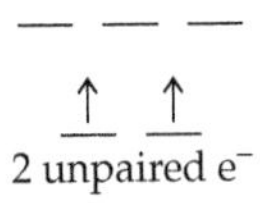

— — — —

↑ ↑
2 unpaired e^-

(d) $[Ru(NH_3)_6]^{2+}$ low-spin Ru^{2+}, $4d^6$

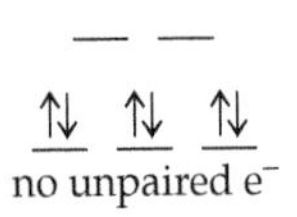

— —

⇅ ⇅ ⇅
no unpaired e^-

21.120. The nitro ($-NO_2$) complex is orange which means it absorbs in the blue region (see color wheel) of the visible spectrum. The nitrito ($-ONO$) is red which means it absorbs in the green region. The energy of the absorbed light is related to ligand field strength. Blue is higher energy than green, therefore nitro ($-NO_2$) is the stronger field ligand.

Valence Bond Theory (Section 21.11)

21.126. (a) $[Ti(H_2O)_6]^{3+}$

Ti^{3+}

[Ar] $\underset{3d}{\uparrow\ \_\ \_\ \_\ \_}$ $\underset{4s}{\_}$ $\underset{4p}{\_\ \_\ \_}$

$[Ti(H_2O)_6]^{3+}$

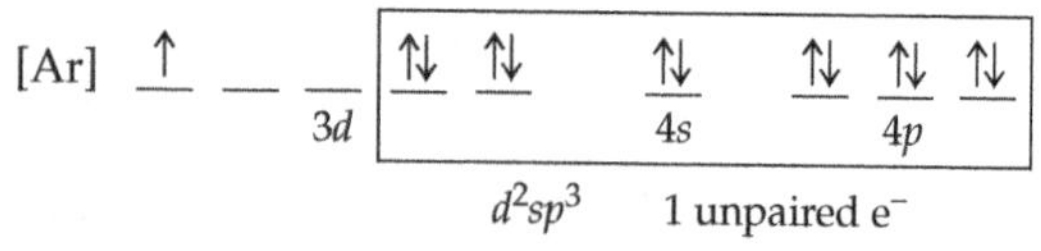

[Ar] $\underset{3d}{\uparrow\ \_\ \_}$ $\boxed{\underset{4s}{\uparrow\downarrow\ \uparrow\downarrow}\ \underset{4p}{\uparrow\downarrow}\ \uparrow\downarrow\ \uparrow\downarrow\ \uparrow\downarrow}$

d^2sp^3 1 unpaired e^-

(b) $[NiBr_4]^{2-}$

Ni^{2+}

[Ar] $\underset{3d}{\uparrow\downarrow\ \uparrow\downarrow\ \uparrow\downarrow\ \uparrow\ \uparrow}$ $\underset{4s}{\_}$ $\underset{4p}{\_\ \_\ \_}$

$[NiBr_4]^{2-}$

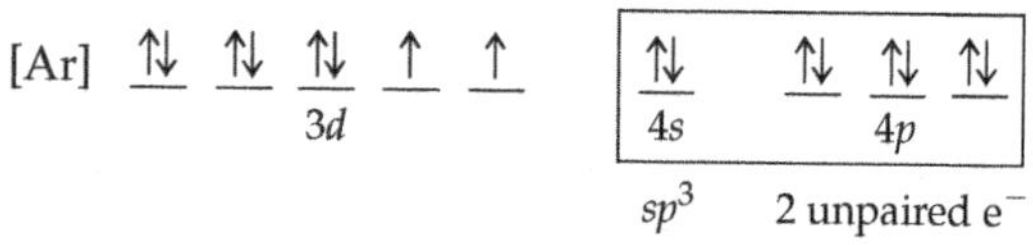

[Ar] $\underset{3d}{\uparrow\downarrow\ \uparrow\downarrow\ \uparrow\downarrow\ \uparrow\ \uparrow}$ $\boxed{\underset{4s}{\uparrow\downarrow}\ \underset{4p}{\uparrow\downarrow\ \uparrow\downarrow\ \uparrow\downarrow}}$

sp^3 2 unpaired e^-

(c) $[Fe(CN)_6]^{3-}$ (low-spin)

Fe^{3+}

[Ar] $\underset{3d}{\uparrow\ \uparrow\ \uparrow\ \uparrow\ \uparrow}$ $\underset{4s}{\_}$ $\underset{4p}{\_\ \_\ \_}$

$[Fe(CN)_6]^{3-}$

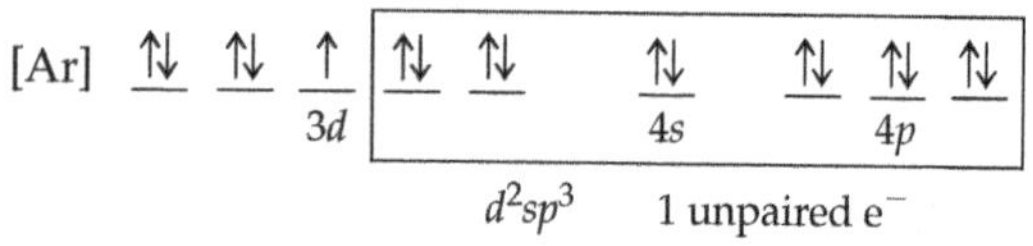

[Ar] $\underset{3d}{\uparrow\downarrow\ \uparrow\downarrow\ \uparrow}$ $\boxed{\uparrow\downarrow\ \uparrow\downarrow\ \underset{4s}{\uparrow\downarrow}\ \underset{4p}{\uparrow\downarrow\ \uparrow\downarrow\ \uparrow\downarrow}}$

d^2sp^3 1 unpaired e^-

(d) $[MnCl_6]^{3-}$ (high-spin)

Mn^{3+}

[Ar] $\underset{3d}{\uparrow\ \uparrow\ \uparrow\ \uparrow\ \_}$ $\underset{4s}{\_}$ $\underset{4p}{\_\ \_\ \_}$

$[MnCl_6]^{3-}$

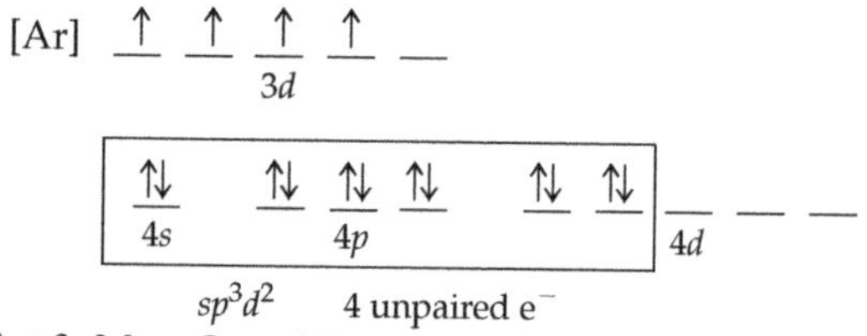

[Ar] $\underset{3d}{\uparrow\ \uparrow\ \uparrow\ \uparrow\ \_}$

$\boxed{\underset{4s}{\uparrow\downarrow}\ \underset{4p}{\uparrow\downarrow\ \uparrow\downarrow\ \uparrow\downarrow}\ \underset{4d}{\uparrow\downarrow\ \uparrow\downarrow}}\ \_\ \_\ \_$

sp^3d^2 4 unpaired e^-

21.128. (a) +3, M = Cr or Ni;

(b)

$[Cr(OH)_4]^-$: [Ar] $\underset{3d}{\uparrow\ \uparrow\ \uparrow\ \_\ \_}$ $\boxed{\underset{4s}{\uparrow\downarrow}\ \underset{4p}{\uparrow\downarrow\ \uparrow\downarrow\ \uparrow\downarrow}}$

Four sp^3 bonds to the ligands

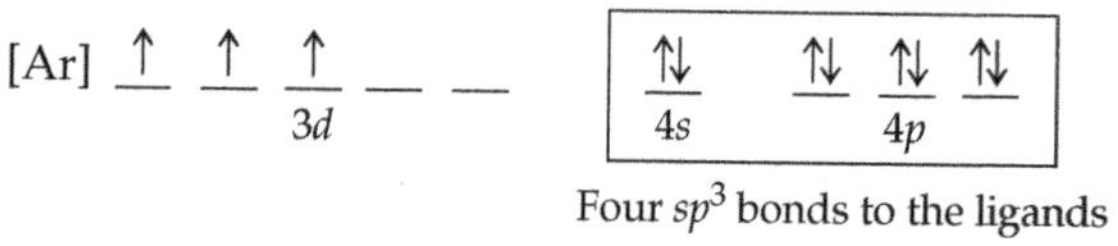

$[Ni(OH)_4]^-$: [Ar] $\underset{3d}{\uparrow\downarrow\ \uparrow\downarrow\ \uparrow\ \uparrow\ \uparrow}$ $\boxed{\underset{4s}{\uparrow\downarrow}\ \underset{4p}{\uparrow\downarrow\ \uparrow\downarrow\ \uparrow\downarrow}}$

Four sp^3 bonds to the ligands

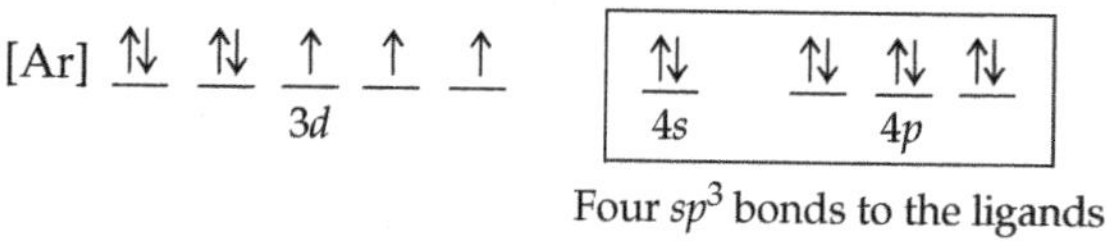

(c) $[Cr(OH)_4]^-$

CHAPTER 22

22.2. A **22.8.** The silicate ion is like (a) in worked example 22.5
22.9. (a) Li_2O, Ga_2O_3, CO_2; (b) Li_2O is the most ionic. CO_2 is the most covalent; (c) CO_2 is the most acidic. Li_2O is the most basic; (d) Ga_2O_3
22.10. (a) $Li_2O(s) + H_2O(l) \longrightarrow 2\ Li^+(aq) + 2\ OH^-(aq)$;
(b) $SO_3(l) + H_2O(l) \longrightarrow H^+(aq) + HSO_4^-(aq)$;
(c) $Cr_2O_3(s) + 6\ H^+(aq) \longrightarrow 2\ Cr^{3+}(aq) + 3\ H_2O(l)$;
(d) $Cr_2O_3(s) + 2\ OH^-(aq) + 3\ H_2O(l) \longrightarrow 2\ Cr(OH)_4^-(aq)$
22.11. 1.5×10^{10} kJ, 1.0×10^6 kg **22.13.** (a) 1000 °C
$H_2O(g) + C(s) \longrightarrow CO(g) + H_2(g)$;
(b) $C_3H_8(g) + 3\ H_2O(g) \longrightarrow 7\ H_2(g) + 3\ CO(g)$ **22.14.** Hydrogen can be stored as a solid in the form of solid interstitial hydrides or in the recently discovered tube-shaped molecules called carbon nanotubes. **22.15.** $PdH_{0.74}$, 0.0841 g/cm^3; **22.16.** (a) 0.15 g/cm^3; 2.1 times the density of liquid H_2; (b) 1.7×10^3 cm^3 **22.17.** (a) N is the Lewis base, B the acid (b) N and B have tetrahedral hybridization and bond angles

Conceptual Problems

22.28. (a) N_2, O_2, F_2, P_4 (tetrahedral), S_8 (crown-shaped ring), Cl_2;
(b) :N≡N: :O=O: :F̈—F̈:   :C̈l—C̈l:

(c) The smaller N and O can form strong π bonds, whereas P and S cannot. In both F_2 and Cl_2, the atoms are joined by a single bond.
22.30. (a) CO_2, Cl_2O_7, SO_3, N_2O_5;

(b)   :Ö=C=Ö:

(resonance structures are needed)

(resonance structures are needed)

Section problems
Periodic Trends (Sections 22.1)

22.40. (a) CaH_2; (b) Ga_2O_3; (c) KCl; (d) $AlCl_3$ **22.42.** Molecular, (a) B_2H_6; (b) $KAlSi_3O_8$; (c) SO_3; (d) $GeCl_4$, Extended three-dimensional structure **22.44.** (a) Sn; (b) Cl; (c) Sn; (d) Se; (e) B

Distinctive Properties of the Second-Row Elements (Section 22.2)

22.46. The smaller B atom can bond to a maximum of four nearest neighbors, whereas the larger Al atom can accommodate more than four nearest neighbors. **22.48.** In O_2 a π bond is formed by 2p orbitals on each O. S does not form strong π bonds with its 3p orbitals, which leads to the S_8 ring structure with single bonds.

Group 1A: Hydrogen (Section 22.3)

22.52. (a) $Zn(s) + 2\ H^+(aq) \longrightarrow H_2(g) + Zn^{2+}(aq)$;
(b) at 1000 °C, $H_2O(g) + C(s) \longrightarrow CO(g) + H_2(g)$; (c) at 1100 °C with a Ni catalyst, $H_2O(g) + CH_4(g) \longrightarrow CO(g) + 3\ H_2(g)$;
(d) There are a number of possibilities. (b) and (c) above are two; electrolysis is another: $2\ H_2O(l) \longrightarrow 2\ H_2(g) + O_2(g)$
22.54. $= 1.7 \times 10^2$ kg CaH_2 **22.58.** (a) MgH_2, H^-; (b) PH_3, covalent; (c) KH, H^-; (d) HBr, covalent
22.60. H_2S – covalent hydride, gas, weak acid in H_2O, NaH – ionic hydride, solid (salt like), reacts with H_2O to produce H_2, PdH_x – metallic (interstitial) hydride, solid, stores hydrogen
22.62. (a) bent; (b) trigonal pyramidal; (c) tetrahedral **22.64.** A non-stoichiometric compound is a compound whose atomic composition cannot be expressed as a ratio of small whole numbers. An example is PdH_x. The lack of stoichiometry results from the hydrogen occupying holes in the solid state structure.

Group 1A and 2A: Alkali and Alkaline Earth Metals (Section 22.4)

22.68. Predicted for Fr: melting point $\approx$ 23 °C boiling point $\approx$ 650 °C, density $\approx$ 2 g/cm^3, atomic radius $\approx$ 275 pm
22.70. (a) $2\ K(s) + 2\ H_2O(l) \longrightarrow 2\ K^+(aq) + 2\ OH^-(aq) + H_2(g)$;
(b) $2\ K(s) + Br_2(l) \longrightarrow 2\ KBr(s)$; (c) $K(s) + O_2(g) \longrightarrow KO_2(s)$
22.72. (a) $2\ Cs(s) + 2\ H_2O(l) \longrightarrow$
$2\ Cs^+(aq) + 2\ OH^-(aq) + H_2(g)$; (b) $Rb(s) + O_2(g) \longrightarrow RbO_2(s)$
22.74. $2\ Mg(s) + O_2(g) \longrightarrow 2\ MgO(s)$, $MgO(s) + H_2O(l) \longrightarrow$
$Mg(OH)_2(aq)$ **22.76.** anode $2\ Cl^- \longrightarrow Cl_2\ (g) + 2\ e^-$ cathode
$Mg^{2+} + 2\ e^- \longrightarrow Mg(l)$ overall $MgCl_2 \longrightarrow Mg(l) + Cl_2\ (g)$
22.78. (a) O^{2-}, (b) O_2^{2-}, (c) O_2^-

Group 3A Elements (Section 22.5)

22.80. (a) Al; (b) Tl; (c) B **22.82.** +3 for B, Al, Ga and In; +1 for Tl
22.84. Boron is a hard semiconductor with a high melting point. Boron forms only molecular compounds and does not form an aqueous B^{3+} ion. $B(OH)_3$ is an acid.

Group 4A Elements (Section 22.6)

22.88. (a) Pb; (b) C; (c) Si; (d) C **22.90.** (a) $GeBr_4$, tetrahedral; Ge is sp^3 hybridized; (b) CO_2, linear; C is sp hybridized; (c) CO_3^{2-}, trigonal planar; C is sp^2 hybridized; (d) $SnCl_3^-$, trigonal pyramidal; Sn is sp^3 hybridized. **22.92.** CO has the strongest bond **22.94.** C, Si, Ge and Sn have allotropes with the diamond structure. Sn and Pb have metallic allotropes. C (nonmetal), Si (semimetal), Ge (semimetal), Sn (semimetal and metal), Pb (metal) **22.96.** CO bonds to hemoglobin and prevents it from carrying O_2. CN^- bonds to cytochrome oxidase and interferes with the electron transfer associated with oxidative phosphorylation. **22.98.** Silicon and germanium are semimetals, and tin and lead are metals. Silicon is a hard, gray, semiconducting solid that melts at 1414 °C. It crystallizes in a diamond-like structure but does not form a graphite-like allotrope because of the relatively poor overlap of silicon p orbitals. Germanium is a relatively high-melting, brittle semiconductor that has the same crystal structure as diamond and silicon. Tin exists in two allotropic forms: the usual silvery white metallic form called white tin and a brittle, semiconducting form with the diamond structure called gray tin. Both white tin and lead are soft, malleable, low-melting metals. Only the metallic form occurs for lead.

22.100.

$Si_4O_{12}{}^{8-}$

22.102. Carbon is a versatile element that can form millions of very stable compounds with elements such as N, O, and H. Biomolecules contain chains and rings with many C—C bonds. Si—Si bonds are much less stable and chains of Si atoms are uncommon. In addition, carbon can form very stable $p\pi$-$p\pi$ multiple bonds. On the other hand, the chemistry of silicon (which cannot form stable $p\pi$-$p\pi$ bonds) is dominated by structures based on the $SiO_4{}^{4-}$ anion.

Group 5A Elements (Section 22.7)

22.104. (a) P; (b) Sb and Bi; (c) N; (d) Bi **22.106.** (a) N_2O, +1; (b) N_2H_4, −2; (c) Ca_3P_2, −3; (d) H_3PO_3, +3; (e) H_3AsO_4, +5 **22.108.** :N≡N: N_2 is unreactive because of the large amount of energy necessary to break the N≡N triple bond.

Group 6A Elements (Section 22.8)

22.116. (a) O; (b) Te; (c) Po; (d) O
22.120. (a) $4 Li(s) + O_2(g) \longrightarrow 2 Li_2O(s)$;
(b) $P_4(s) + 5 O_2(g) \longrightarrow P_4O_{10}(s)$;
(c) $4 Al(s) + 3 O_2(g) \longrightarrow 2 Al_2O_3(s)$;
(d) $Si(s) + O_2(g) \longrightarrow SiO_2(s)$ **22.122.** An element that forms an acidic oxide is more likely to form a covalent hydride. C and N are examples. **22.124.** $Li_2O < BeO < B_2O_3 < CO_2 < N_2O_5$

22.126. $N_2O_5 < Al_2O_3 < K_2O < Cs_2O$
22.128. (a) CrO_3; (b) N_2O_5; (c) SO_3
22.130. (a) $Cl_2O_7(l) + H_2O(l) \longrightarrow 2 H^+(aq) + 2 ClO_4{}^-(aq)$;
(b) $K_2O(s) + H_2O(l) \longrightarrow 2 K^+(aq) + 2 OH^-(aq)$;
(c) $SO_3(l) + H_2O(l) \longrightarrow H^+(aq) + HSO_4{}^-(aq)$
22.132. (a) $ZnO(s) + 2 H^+(aq) \longrightarrow Zn^{2+}(aq) + H_2O(l)$;
(b) $ZnO(s) + 2 OH^-(aq) + H_2O(l) \longrightarrow Zn(OH)_4{}^{2-}(aq)$
22.134. (a) rhombic sulfur; (b) an allotrope of sulfur in which the S_8 rings pack differently in the crystal; (c) when sulfur is cooled rapidly, the sulfur forms disordered, tangled chains, yielding an amorphous, rubbery material called plastic sulfur; (d) dark reddish-brown and very viscous forming long polymer chains (S_n, n > 200,000).
22.136. (a) $Zn(s) + 2 H_3O^+(aq) \longrightarrow$
$Zn^{2+}(aq) + H_2(g) + 2 H_2O(l)$; (b)
$BaSO_3(s) + 2 H_3O^+(aq) \longrightarrow H_2SO_3(aq) + Ba^{2+}(aq) + 2 H_2O(l)$;
(c) $Cu(s) + 2 H_2SO_4(l) \longrightarrow Cu^{2+}(aq) + SO_4{}^{2-}(aq) + SO_2(g) + 2 H_2O(l)$; (d) $H_2S(aq) + I_2(aq) \longrightarrow S(s) + 2 H^+(aq) + 2 I^-(aq)$
22.138. (a) Acid strength increases as the number of O atoms increases; (b) In comparison with S, O is much too electronegative to form compounds of O in the +4 oxidation state. Also, an S atom is large enough to accommodate four bond pairs and a lone pair in its valence shell, but an O atom is too small to do so; (c) Each S is sp^3 hybridized with two lone pairs of electrons. The bond angles are therefore 109.5°. A planar ring would require bond angles of 135°.
22.140.

(a)   H—S̈—H, bent;

(b)   :Ö—S̈=Ö: ⟷ :Ö=S̈—Ö:,   bent, S is sp^2 hybridized;

(c)   trigonal planar, S is sp^2 hybridized

22.142. (1) is CO_2. There are two C=O double bonds. (2) is SO_2. There is one S—O single bond and one S=O double bond. CO_2 has the stronger bonds.

Group 7A and 8A: The Halogens and Noble Gases (Sections 22.9–22.10)

22.144. (a) At, should be a solid; (b) At is likely to react with Na just like the other halogens, yielding NaAt. **22.146.** $MnO_2(s) + 2\,Br^-(aq) + 4\,H^+(aq) \longrightarrow Mn^{2+}(aq) + 2\,H_2O(l) + Br_2(aq)$ **22.148.** (a) $TiCl_4$; (b) $Ti(s) + 2\,Cl_2(g) \longrightarrow TiCl_4(g)$; (c) $TiCl_4(l) + 2\,Mg(s) \longrightarrow Ti(s) + 2\,MgCl_2(s)$ **22.150.** (a) $HBrO_3$, +5; (b) HIO, +1

22.152. (a) 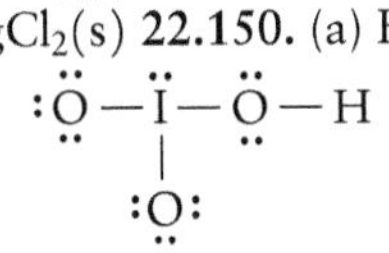 :Ö—Ï—Ö—H   trigonal pyramidal

(b) 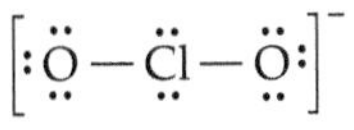 $\left[\ddot{O}—\ddot{C}l—\ddot{O}\right]^-$ bent

(c) 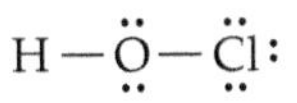 H—Ö—Ċl: bent

(d) 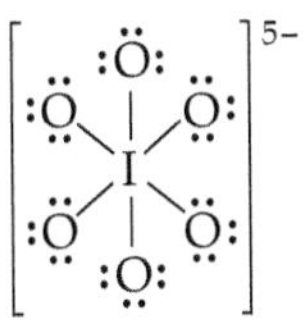   octahedral

22.154. Oxygen atoms are highly electronegative. Increasing the number of oxygen atoms increases the polarity of the O—H bond and increases the acid strength.

22.156. $I_2O_5(aq) + H_2O(l) \longrightarrow 2\,HIO_3(aq)$; HIO_3 is iodic acid.

Multiconcept Problems

22.160. (a) C as diamond; (b) $Cl_2(g) + H_2O(l) \longrightarrow HOCl(aq) + H^+(aq) + Cl^-(aq)$; (c) NO; (d) NO_2; (e) BF_3; (f) Al_2O_3; (g) Si; (h) HNO_3; (i) C as diamond, graphite, and fullerene.

22.164.

(a)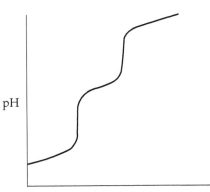

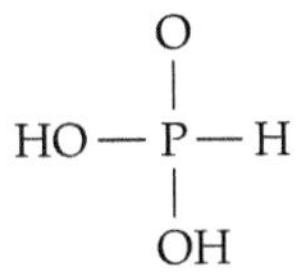

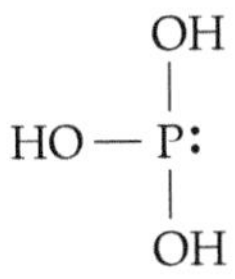

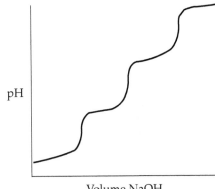

(b) 4.29, 9.57 **22.166.** (a) $P_4(s) + 5\,O_2(g) \longrightarrow P_4O_{10}(s)$ $P_4O_{10}(s) + 6\,H_2O(l) \longrightarrow 4\,H_3PO_4(aq)$; (b) 1.18; (c) $3\,Ca^{2+}(aq) + 2\,H_3PO_4(aq) \longrightarrow Ca_3(PO_4)_2(s) + 6\,H^+(aq)$, 25.0 g (d) H_2, 5.96 L

CHAPTER 23

23.1. (a) 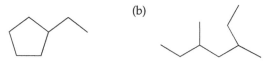   (b)

23.2. (a) C_8H_{16}; (b) C_7H_{16} **23.3.** Structures (a) and (c)

23.4.

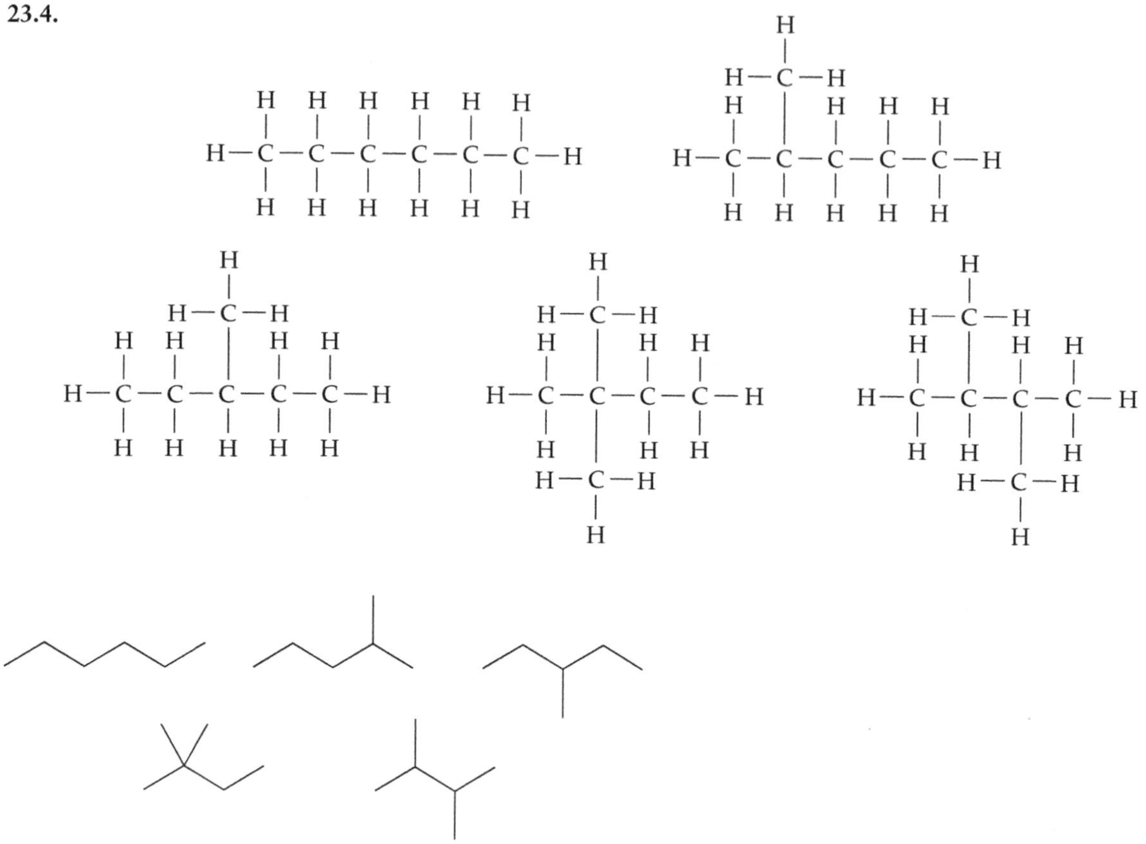

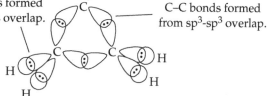

23.10. The molecule is unstable because the C—C bonds are relatively weak due to the less effective orbital overlap in a molecule with 60° bond angles.

C–H bonds formed from sp³-1s overlap.

C–C bonds formed from sp³-sp³ overlap.

23.12.

(a)

$H \\ H / C = C = O$

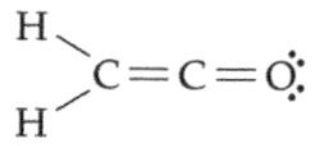

(b)

Carbon–Carbon double bond consists of one σ bond formed by head on overlap of sp² and sp orbitals...

Carbon–Oxygen double bond consists of one σ bond formed by head on overlap of sp and sp² orbitals...

...and one π bond formed by sideways overlap of p orbitals

...and one π bond formed by sideways overlap of p orbitals

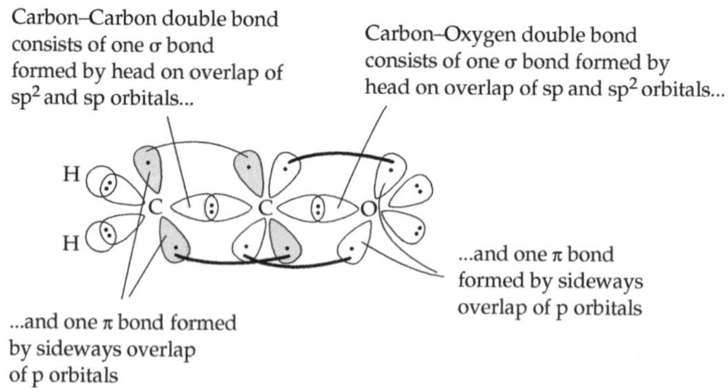

23.13. (a) Does not exhibit cis-trans isomerism; (b) The cis isomer was given in the problem. The structure for the trans isomer is:

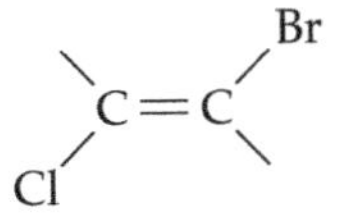

$$\text{Cl}\diagdown\text{C}=\text{C}\diagup\text{Br}$$

(c) Does not exhibit cis-trans isomerism.

23.14. (a) Fumaric acid is a trans isomer;
(b)   O   (c) No, because succinic acid does not have a carbon–carbon double bond.

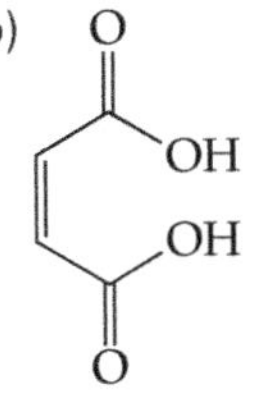

Maleic acid

23.15.

(a)   (b) 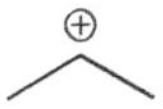   (c)

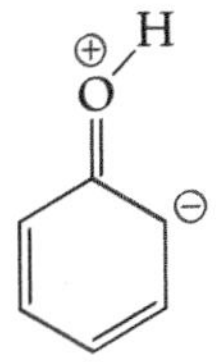

23.16.

(a)   CH₃   CH₃   (b)

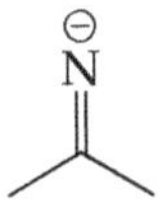

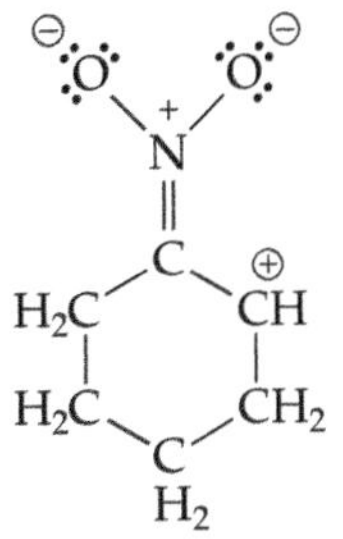

$$H_3C-\underset{\underset{H}{|}}{\overset{\overset{CH_3}{|}}{C}}-\overset{\ominus}{\underset{\cdot\cdot}{N}}-\underset{\underset{H}{|}}{\overset{\overset{CH_3}{|}}{C}}-CH_3$$

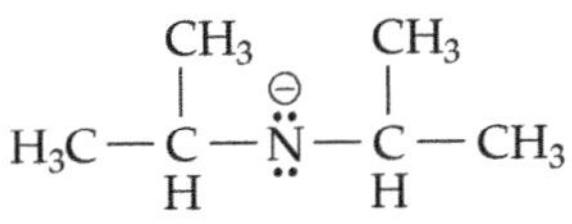

(c)   HC   CH

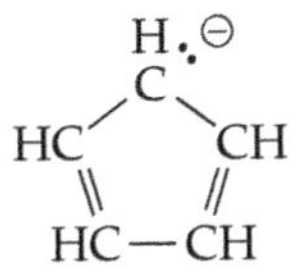

23.17.

(a)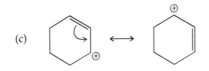

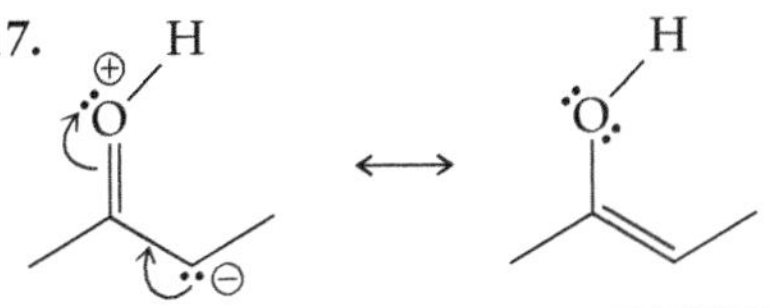

Preferred structure because formal charges are zero.

(b)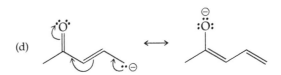

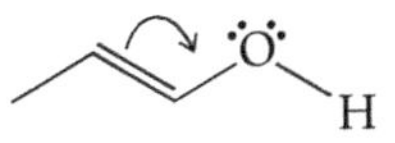

Moving electrons as indicated would result in an incorrect electron-dot structure because oxygen would have an expanded octet.

(c)

Both resonance structures are equivalent.

(d)

Preferred resonance structure because negative formal charge is on more electronegative oxygen.

23.18. (a)   (b)

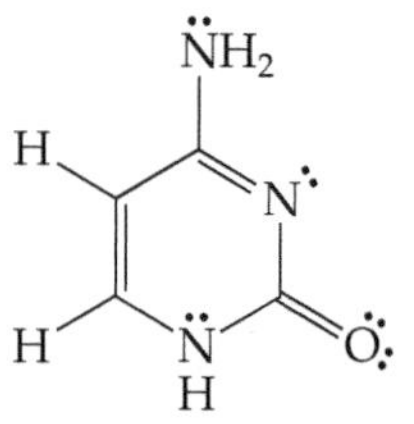

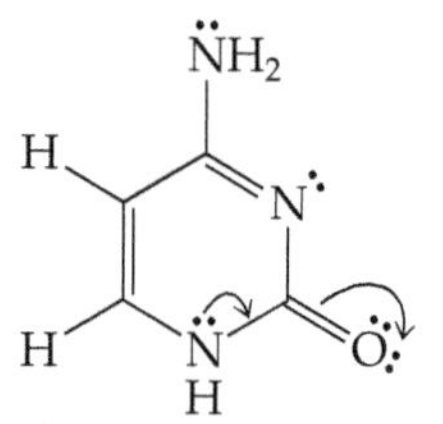

23.19.

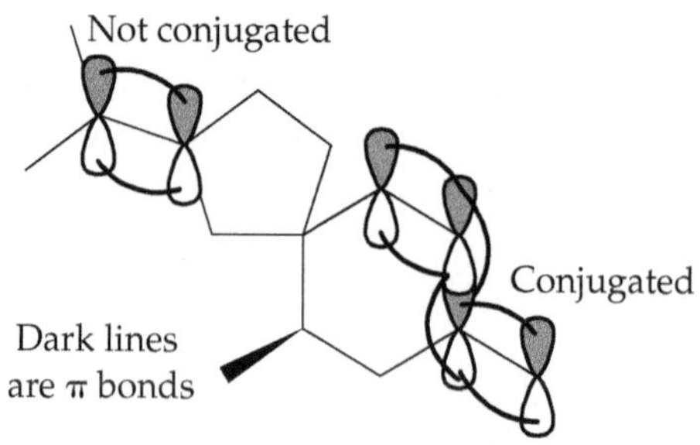

Not conjugated

Conjugated

Dark lines
are π bonds

23.20. In the triple bond one of the π bonds is perpendicular to the other π bonds in the molecule. Therefore, it cannot have sideways overlap with the p orbitals that make up the other π bonds

23.21.

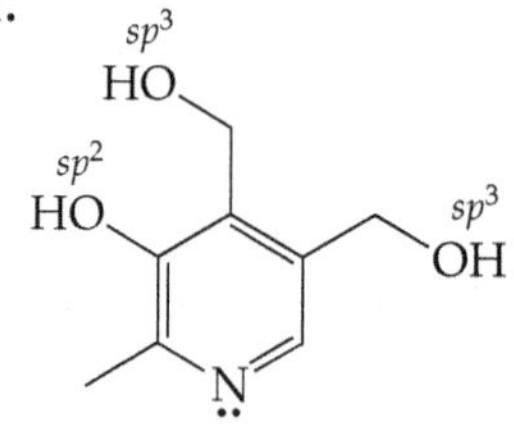

23.22.

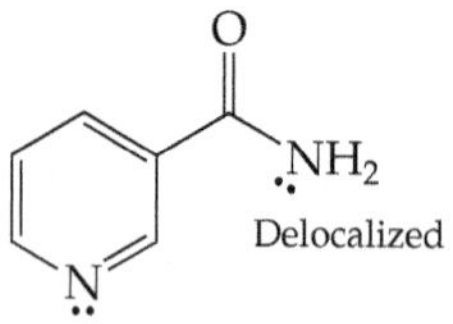

Delocalized

Localized because the electrons used
to make π bonds are all in the same plane
and delocalized. The lone pair of electrons
is in an sp^2 hybrid orbital in a different plane.

23.24. (a) 4; (b) Phe-nonpolar, Asn-polar, Trp-nonpolar, Ala-nonpolar

23.25. Proline is not aromatic because it does not have a ring of sp^2 hybridized atoms. Tyrosine is aromatic because it has a six-member ring of sp^2 hybridized carbon atoms that are part of a conjugated system. There are 6π electrons and satisfying the $4n + 2$ rule for aromaticity.

23.26. (a)

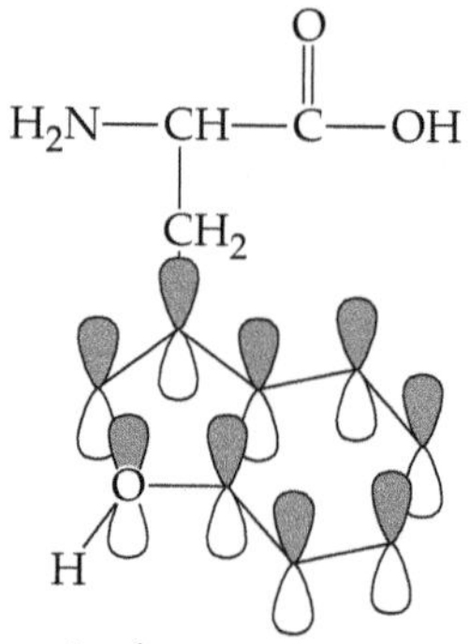

(b) All atoms in the two-ring system of the side chain of tryptophan are sp^2 hybridized and conjugated. Four double bonds contribute $8\ \pi$ electrons and one lone pair contributes $2\ \pi$ electrons for a total of $10\ \pi$ electrons, satisfying the $4n + 2$ rule.

Conceptual Problems

23.32. (a) no handedness; (b) is a handedness **23.34.** ester, aromatic ring, and amine **23.38.** (a) serine; (b) methionine

Section Problems

Organic Molecules and Constitutional Isomers (Section 23.1)

23.40. In a straight-chain alkane, all the carbons are connected in a row. In a branched-chain alkane, there are branching connections of carbons along the carbon chain.

Stereioisomers: Chiral Molecules (Section 23.2)

23.48. (b), (c) and (e)

Functional Groups (Section 23.3)

23.54. A functional group is a part of a larger molecule and is composed of an atom or group of atoms that has a characteristic chemical behavior. They are important because their chemistry controls the chemistry in molecules that contain them.

23.56.

(a)

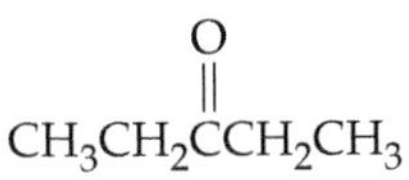

$$CH_3CH_2\overset{\overset{\displaystyle O}{\|}}{C}CH_2CH_3$$

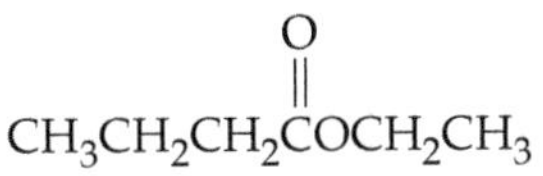

(b)

$$CH_3CH_2CH_2\overset{\overset{\displaystyle O}{\|}}{C}OCH_2CH_3$$

(c)

$$NH_2CH_2\overset{\overset{\displaystyle O}{\|}}{C}OH$$

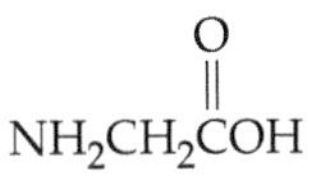

23.58. (a) 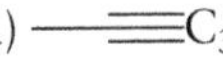 ──≡C₃H₄; (b) 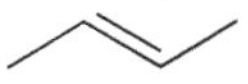 /=\ C₄H₈; (c) C₅H₆;

23.60.

$CH_2=C=CHCH_2CH_3$ $CH_2=CHCH=CHCH_3$ $CH_2=CHCH_2CH=CH_2$

$$CH_3CH=C=CHCH_3 \qquad CH_2=\overset{\overset{\displaystyle CH_3}{|}}{C}CH=CH_2 \qquad CH_2=C=\overset{\overset{\displaystyle CH_3}{|}}{C}CH_3$$

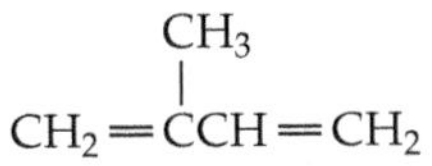

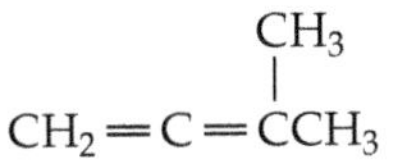

23.64. (a) 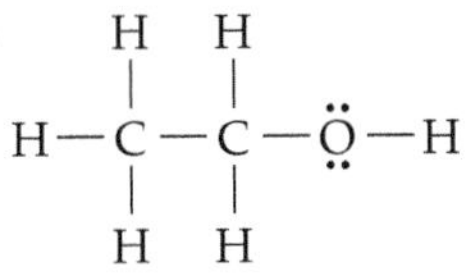 C_2H_6O;

$$H-\underset{\underset{\displaystyle H}{|}}{\overset{\overset{\displaystyle H}{|}}{C}}-\underset{\underset{\displaystyle H}{|}}{\overset{\overset{\displaystyle H}{|}}{C}}-\overset{\cdot\cdot}{\underset{\cdot\cdot}{O}}-H$$

(b) 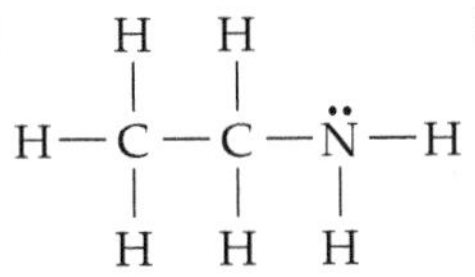 C_2H_7N; (c)

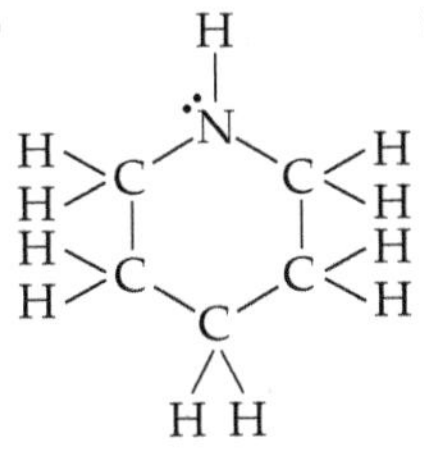

$$H-\underset{\underset{\displaystyle H}{|}}{\overset{\overset{\displaystyle H}{|}}{C}}-\underset{\underset{\displaystyle H}{|}}{\overset{\overset{\displaystyle H}{|}}{C}}-\underset{\underset{\displaystyle H}{|}}{N}-H$$

$C_5H_{11}N$;

(d) 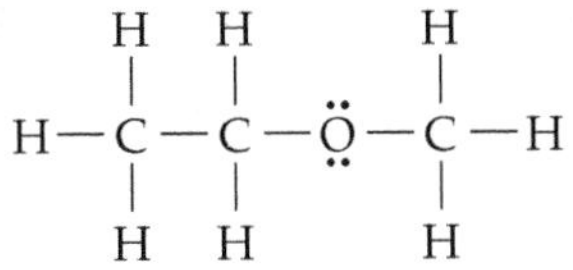 C_3H_8O;

$$H-\underset{\underset{\displaystyle H}{|}}{\overset{\overset{\displaystyle H}{|}}{C}}-\underset{\underset{\displaystyle H}{|}}{\overset{\overset{\displaystyle H}{|}}{C}}-\overset{\cdot\cdot}{\underset{\cdot\cdot}{O}}-\underset{\underset{\displaystyle H}{|}}{\overset{\overset{\displaystyle H}{|}}{C}}-H$$

(e) 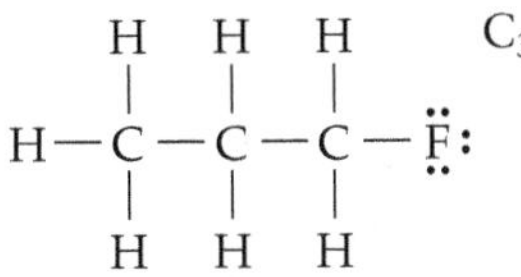 C_3H_7F; (f)

$$H-\underset{\underset{\displaystyle H}{|}}{\overset{\overset{\displaystyle H}{|}}{C}}-\underset{\underset{\displaystyle H}{|}}{\overset{\overset{\displaystyle H}{|}}{C}}-\underset{\underset{\displaystyle H}{|}}{\overset{\overset{\displaystyle H}{|}}{C}}-\overset{\cdot\cdot}{\underset{\cdot\cdot}{F}}:$$

$$H-\underset{\underset{\displaystyle H}{|}}{\overset{\overset{\displaystyle H}{|}}{C}}-\overset{\overset{\displaystyle :O:}{\|}}{C}-\underset{\underset{\displaystyle H}{|}}{\overset{\overset{\displaystyle H}{|}}{C}}-H$$

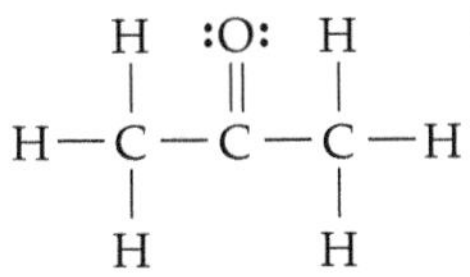

C_3H_6O

23.66. (a) 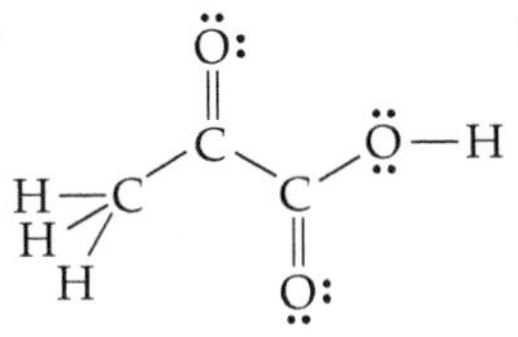 $C_3H_4O_3$;

(b) 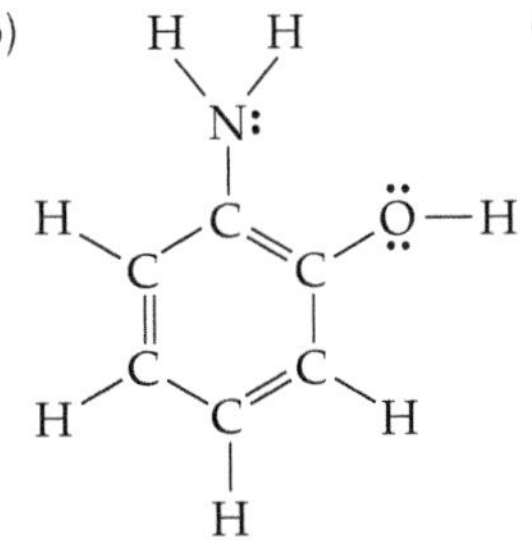 C_6H_7NO; (c) 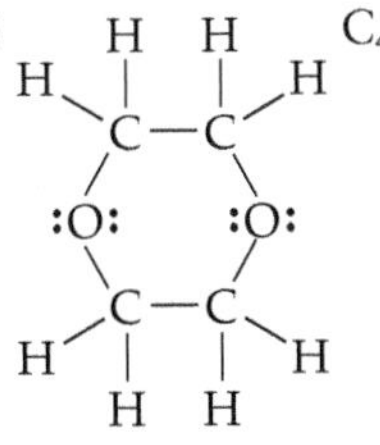 $C_4H_8O_2$

Carbohydrates (Section 23.4)

23.70. An aldose contains the aldehyde functional group while a ketose contains the ketone functional group.

23.72.

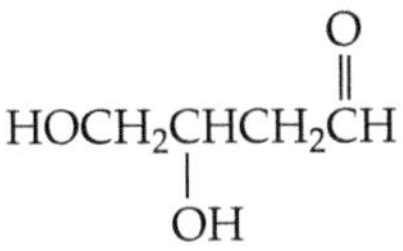

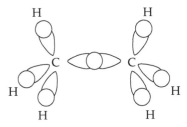

$$HOCH_2\overset{\overset{\displaystyle}{|}}{CH}CH_2\overset{\overset{\displaystyle O}{\|}}{CH}$$
$$\underset{\displaystyle OH}{|}$$

23.74. (a) constitutional isomers; (b) anomers

Valence Bond Theory, Cis–Trans Isomers (Section 23.5)

23.76. (a) All C—H bond angles are ~109.5°;

(b) Two C—H bond angles are ~120° and one is 90°;

(c) In structure (b), the 90° bond angle introduces a larger repulsion and lower stability.

23.78.

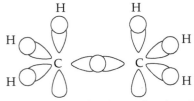

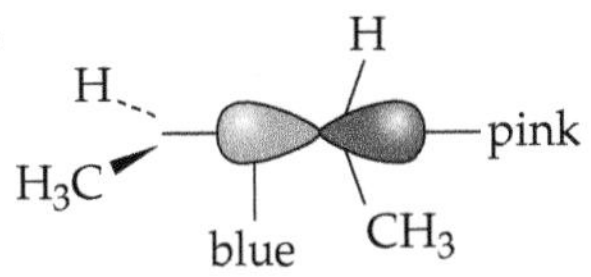

23.80.

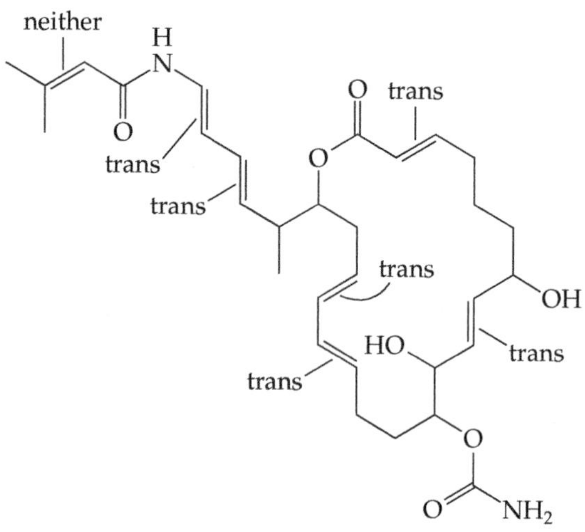

neither ... trans ... trans ... trans ... trans ... trans ... trans ... OH ... HO ... trans ... O ... NH₂

(complex macrolide structure with labeled cis/trans double bonds)

23.82. (b) 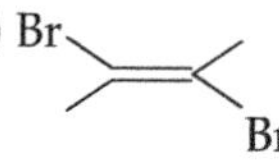 Br₂C=C(CH₃)... Br   (c) *(structure)*

23.84. (b) $CH_3CH{=}CHCH_2CH_2CH_3$;
(c) $CH_3CH_2CH{=}CHCH_2CH_3$

Lipids (Section 23.6)

23.86. Long-chain carboxylic acids are called fatty acids. Fatty acids are usually unbranched and have an even number of carbon atoms in the range of 12–22.

23.88.

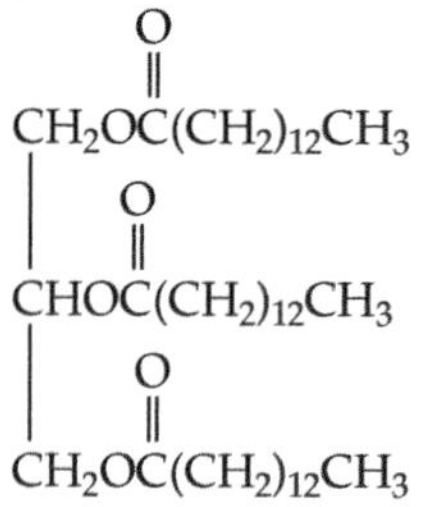

$CH_2OC(CH_2)_{12}CH_3$
|
$CHOC(CH_2)_{12}CH_3$
|
$CH_2OC(CH_2)_{12}CH_3$

(each ester carbon bearing a C=O)

23.90.

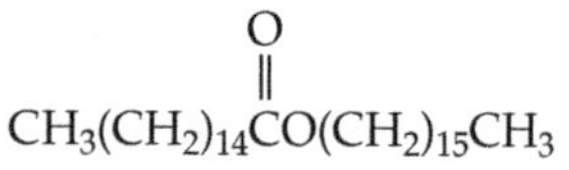

$CH_3(CH_2)_{14}CO(CH_2)_{15}CH_3$

23.92. (c) **23.94.**

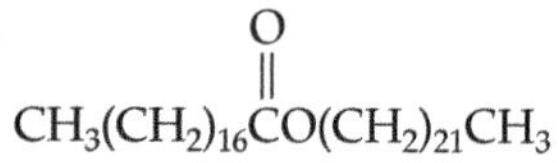

$CH_3(CH_2)_{16}CO(CH_2)_{21}CH_3$

Formal Charge and Resonance (Section 23.7)

23.96. (a)

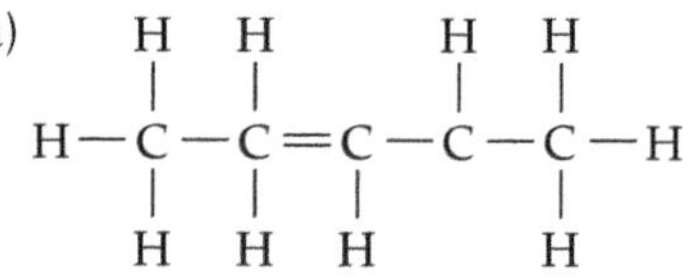

H−C−C=C−C−C−H *(with H substituents shown)*

This structure is not valid because there are five bonds on the second carbon.

(b) 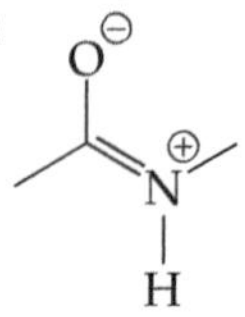 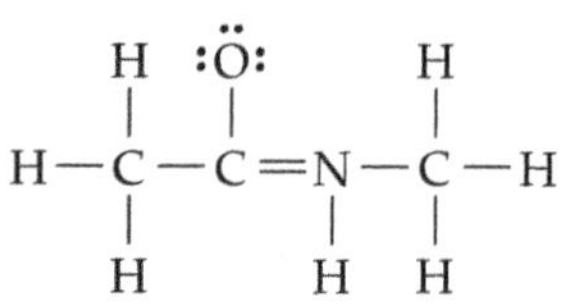 *(structures: acetamide-type with N⁺ and O⁻; and H−C−C=N−C−H structure)*

(c) *(structure: H−C−C=N=C−H)*

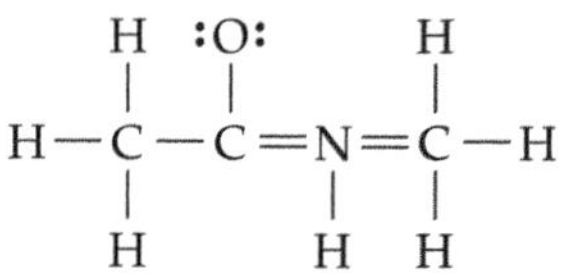

This structure is not valid because the nitrogen and one of the carbons bonded to it both have five bonds. The oxygen does not have a complete octet.

23.98. (a)

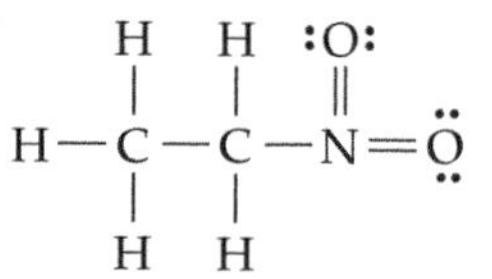

H−C−C−N=O *(with H substituents and :O:)*

This structure is not valid because there are five bonds on the nitrogen.

(b) 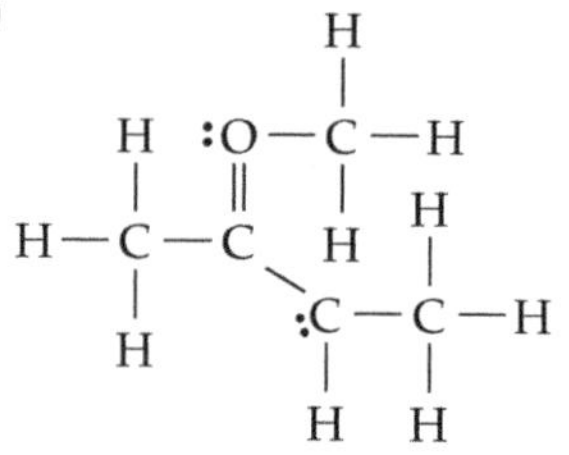 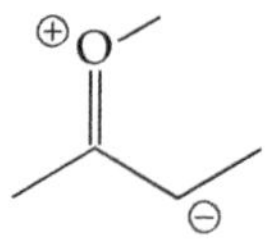 *(three resonance structures shown)*

(c) 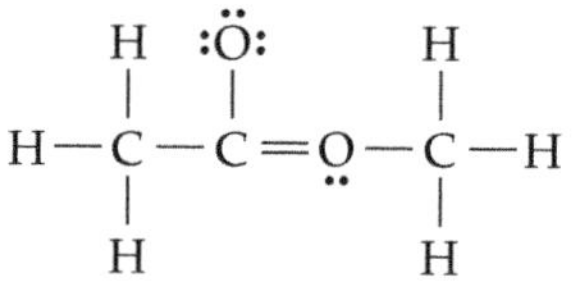 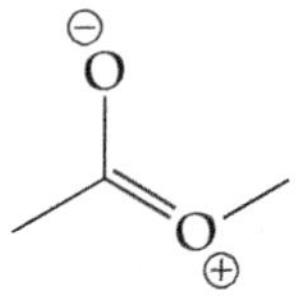 *(resonance structures with :Ö: and ⊖ O and ⊕ O)*

23.100. (a) 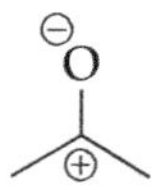 ⊖O / ⊕ The original structure contributes more to the resonance hybrid because all formal charges are zero.

(b) 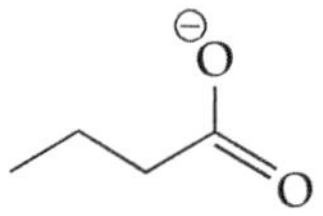 ⊖O *(butanoate-type structure)* The two structures are identical and are equal contributors to the resonance hybrid.

23.102. (a) 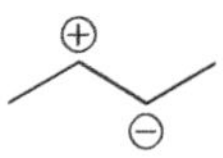 ⊕ ... ⊖ The original structure contributes more to the resonance hybrid because all formal charges are zero.

(b) 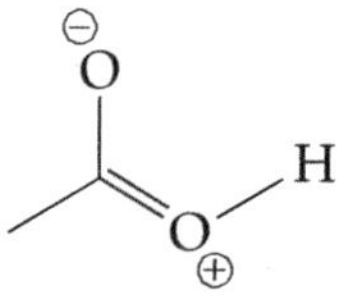 ⊖O ... H ... ⊕O The original structure contributes more to the resonance hybrid because all formal charges are zero.

23.104. They are resonance structures.

23.106. (a)  *(curved arrow structure with N and ⊖)* (b) 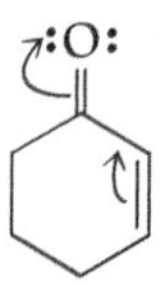 *(curved arrow structure with :O:)*

Conjugation Systems (Section 23.8)

23.108.

23.110.

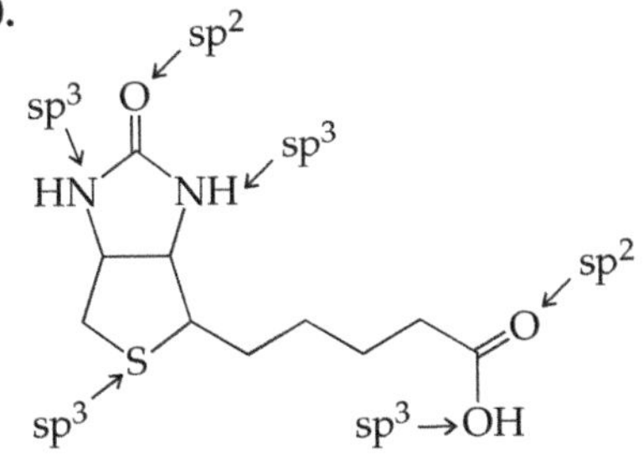

23.112. delocalized

localized

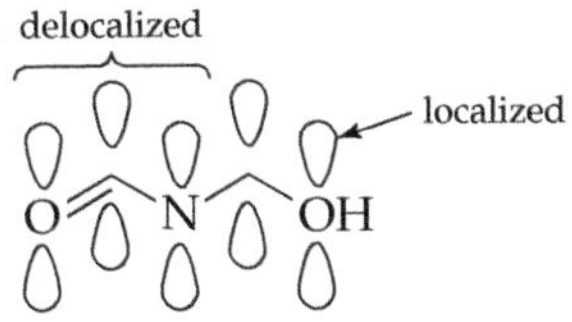

Proteins (Section 23.9)

23.114. (a) serine; (b) threonine; (c) proline; (d) phenylalanine; (e) cysteine **23.118.** Met-Ile-Lys, Met-Lys-Ile, Ile-Met-Lys, Ile-Lys-Met, Lys-Met-Ile, Lys-Ile-Met

Aromaticity and Molecular Orbital Theory (Section 23.10)

23.120. Only (b) is aromatic. **23.122.** There are 6 π-electrons, 2 each in the 2 π bonds and 2 in the p-orbital of the NH nitrogen.

Nucleic Acids (Section 23.11)

23.126. nucleic acids are polymers made up of nucleotide units linked together to form a long chain. Each nucleotide contains a phosphate group, an aldopentose sugar, and an amine base.

23.128.

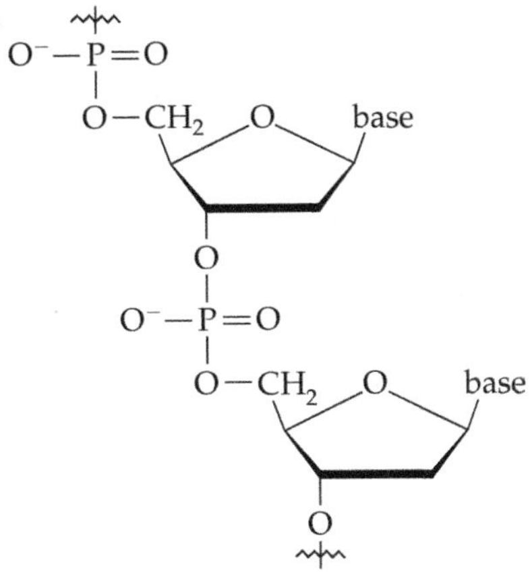

23.130. A—T—G—G—C—T **23.132.**

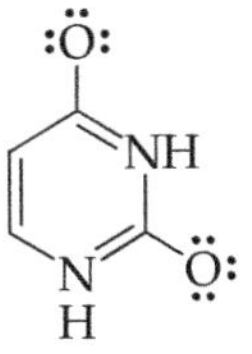

Multiconcept Problems

23.134. (a) CHO; (b) 116

(c)

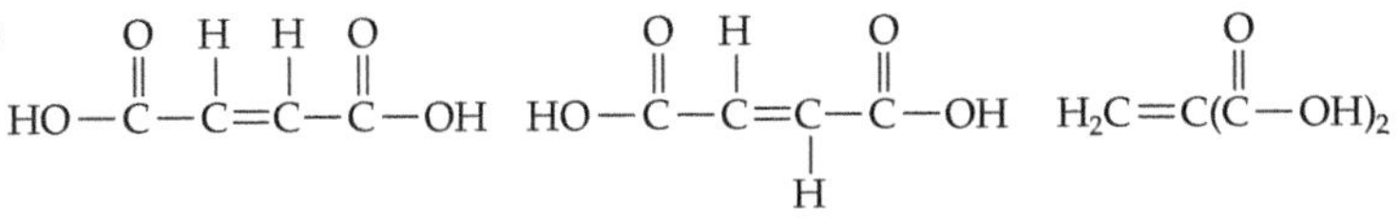

(d)

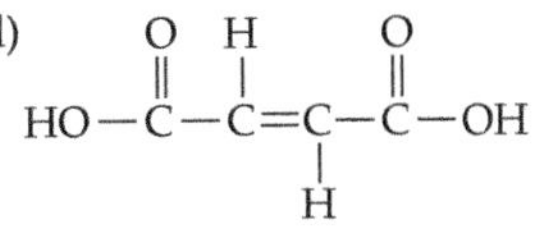

23.136. (a) —CO_2H; (b) HA = 98%, H_2A^+ = 2.1%; (c) HA = 94%, A^- = 6.1%; (d) 6.01.

Glossary

Absorption spectrum a plot of the amount of light absorbed versus wavelength (*Section 21.9*)

Accuracy how close to the true value a given measurement is (*Section 1.9*)

Achiral lacking handedness (*Section 21.8*)

Acid a substance that provides H$^+$ ions when dissolved in water (*Section 4.7*)

Acid–base indicator a substance that changes color in a specific pH range (*Section 16.6*)

Acid–base neutralization reaction a process in which an acid reacts with a base to yield water plus an ionic compound called a salt (*Section 4.4*)

Acid-dissociation constant (K_a) the equilibrium constant for the dissociation of an acid in water (*Section 16.8*)

Acidic oxides are formed by the nonmetals on the right side of the periodic table (*Section 22.8*)

Acid rain is a matter of serious concern because most species of fish die in waters having a pH lower than 4.5–5.0 (*Section 10.9*)

Actinide one of the 14 inner-transition metals starting with actinium in the periodic table (*Section 2.2*)

Activation energy (E_a) the height of the energy barrier between reactants and products (*Section 14.7 and Section 20.6*)

Active site a small three-dimensional region of an enzyme with the specific shape necessary to bind the substrate and catalyze the appropriate reaction (*Chapter 14 Inquiry*)

Activity series a list of elements in order of their reducing ability in aqueous solution (*Section 4.12*)

Advanced ceramic a ceramic material that has high-tech engineering, electronic, or biomedical applications (*Section 12.10*)

Alcohol an organic molecule that contains an —OH group (*Section 23.3*)

Aldehyde an organic molecule that contains one alkyl group and one hydrogen bonded to a C=O carbon (*Section 23.3*)

Alkane a compound that contains only carbon and hydrogen and has only single bonds (*Section 23.1*)

Alkene a hydrocarbon that has a carbon–carbon double bond (*Section 23.3*)

Alkyne a hydrocarbon that has a carbon–carbon triple bond (*Section 23.3*)

Allotropes different structural forms of an element (*Section 12.5*)

Alloy a solid solution of two or more metals (*Section 1.1*)

Alpha (α) radiation a type of radioactive emission; a helium nucleus (*Section 2.8 and Section 20.2*)

alpha-(∞-) amino acids acids such as Ala (alanine), Gly (glycine), Pro (proline), and so on are called alpha-(∞-) amino acids because the amine nitrogen atom in each is connected to the carbon atom *alpha* to the carboxylic acid group (*Section 23.9*)

Amide an organic molecule that contains one alkyl group and one nitrogen bonded to a C=O carbon (*Section 23.8*)

Amine an organic derivative of ammonia (*Section 23.7*)

Amino acids a molecule that contains both a basic amine group (—NH$_2$) and an acidic carboxyl group (—CO$_2$H); the building block from which proteins are made (*Section 23.9*)

Amorphous solid a solid whose constituent particles are randomly arranged and have no ordered, long-range structure (*Section 12.1*)

Ampere (1 A = 1 C/s) is the SI unit of electric current (*Section 19.14*)

Amphoteric exhibiting both acidic and basic properties (*Section 17.12*)

Amplitude a wave's height measured from the midpoint between peak and trough (*Section 5.1*)

Angular-momentum quantum number (l) a variable in the solutions to the Schrödinger wave equation that gives the three-dimensional shape of an orbital (*Section 5.6*)

Anion a negatively charged atom or group of atoms (*Section 2.12*)

Anisotropic properties properties that depend on the directional orientation of the molecules (*Section 11.6*)

Anode the electrode at which oxidation takes place (*Section 19.2*)

Anomers are two cyclic forms of glucose, and can result from ring formation, depending on whether the newly formed —OH group at C1 is on the bottom or top side of the ring (*Section 23.4*)

Antibonding molecular orbital a molecular orbital that is higher in energy than the atomic orbitals it is derived from (*Section 8.7*)

Aqueous solution a solution with water as solvent (*Chapter 4 Introduction*)

Aromatic compound the class of compounds related to benzene (*Section 23.10*)

Arrhenius acid a substance that provides H$^+$ ions when dissolved in water (*Section 16.1*)

Arrhenius base a substance that provides OH$^-$ ions when dissolved in water (*Section 16.1*)

Arrhenius equation an equation relating reaction rate constant, temperature, and activation energy; $k = Ae^{-E_a/RT}$ (*Section 14.7 and Section 15.10*)

Arrhenius plot an equation gives a straight line with slope $m = -E_a/R$ and intercept $b = \ln$ (*Section 14.9*)

Atmosphere (atm) a common unit of pressure measurement; standard atmospheric pressure at sea level is defined as exactly 760 mm Hg (*Section 10.1*)

Atom the smallest particle that retains the chemical properties of an element (*Section 2.5*)

Atom economy a concept which states that it is best to have all or most starting atoms end up in the desired product rather than in waste by-products (*Chapter 3 Inquiry*)

Atomic mass the weighted average mass of an element's naturally occurring atoms (*Section 2.9*)

Atomic number (Z) the number of protons in an atom's nucleus (*Section 2.8*)

Atomic radii half the distance between nuclei of two identical bonded atoms, decreases across a period and increases down a group (*Section 5.13 and Section 5.14*)

Atomic spectra result when an electron moves between two states having different energy (*Section 5.3*)

Atomic weight an element's atomic weight is the weighted average of the isotopic masses of the element's naturally occurring isotopes (*Section 2.9*)

Aufbau principle a set of rules that guides the electron filling order of orbitals in atoms (*Section 5.10 and Section 6.1*)

Avogadro's law the volume of a gas at a fixed pressure and temperature is proportional to its molar amount (*Section 10.2*)

Avogadro's number (N_A) the number of units in a mole; 6.022×10^{23} (*Section 2.9 and Section 3.3*)

Balanced equation   a chemical equation in which the numbers and kinds of atoms are the same on both sides of the reaction arrow (*Section 3.2*)

Balmer–Rydberg equation   an equation that accounts for the lines in the hydrogen spectrum (*Section 5.3*)

$$\frac{1}{\lambda} = R_\infty \left[\frac{1}{m^2} - \frac{1}{n^2} \right] \text{ or } \nu = R_\infty \cdot c \left[\frac{1}{m^2} - \frac{1}{n^2} \right]$$

Band   a set of MOs that are very closely spaced in energy (*Section 12.6*)

Band   the molecular orbital theory for metals (*Section 12.6*)

Band gap   the energy difference between the bonding MOs in the valence band and the antibonding MOs in the conduction band of a semiconductor (*Section 12.7*)

Band theory   the molecular orbital theory for metals (*Section 12.6*)

Bar   a unit of pressure (*Section 10.1*)

Base   a substance that provides OH⁻ ions when dissolved in water (*Section 4.7*)

Base stacking   is the key to the stability of the DNA double helix, and the stacking interactions arise from the fact that all the bases are aromatic (*Section 23.11*)

Base-dissociation constant (K_b)   the equilibrium constant for the reaction of a base with water (*Section 16.12*)

Basic oxides   are formed by metals on the left side of the periodic table (*Section 22.8*)

Battery   *see* Galvanic cell (*Section 19.4 and Section 19.10*)

Beta (β) radiation   a type of radioactive emission consisting of electrons (*Section 20.2*)

Bidentate ligand   a ligand that bonds to a metal using electron pairs on two donor atoms (*Section 21.5*)

Bimolecular reaction   an elementary reaction that results from a collision between two reactant molecules (*Section 14.9*)

Binary hydride   a compound that contains hydrogen and one other element (*Section 22.3*)

Binding energy   the energy that holds nucleons together in the nucleus of an atom (*Section 20.5 and Section 22.1*)

Biochemistry   the chemistry of living organisms (*Chapter 23 Introduction*)

Biofuel   a fuel such as biodiesel, made from a renewable plant source (*Chapter 9 Inquiry*)

Body-centered cubic packing   a packing arrangement of spheres into a body-centered cubic unit cell (*Section 12.3*)

Body-centered cubic unit cell   a cubic unit cell with an atom at each of its eight corners and an additional atom in the center of the cube (*Section 12.3*)

Bond angle   the angle at which two adjacent bonds intersect (*Section 8.1*)

Bond dissociation energy (D)   the amount of energy necessary to break a chemical bond in an isolated molecule in the gaseous state (*Section 7.2, Section 9.10, and Section 16.3*)

Bond length   the minimum-energy distance between nuclei in a covalent bond (*Section 7.1*)

Bond order   the number of electron pairs shared between two bonded atoms (*Section 7.2 and Section 8.7*)

Bonding molecular orbital   a molecular orbital that is lower in energy than the atomic orbitals it is derived from (*Section 8.7*)

Bonding pair   the shared electrons (*Section 7.5*)

Borane   any compound of boron and hydrogen (*Section 22.5*)

Born–Haber cycle   a pictorial way of viewing the energy changes in the various steps during formation of an ionic solid from its elements (*Section 6.7*)

Boyle's law   the volume of a fixed amount of gas at a constant temperature varies inversely with its pressure (*Section 10.2*)

Bragg equation   an equation used in X-ray crystallography for calculating the distance between atoms in a crystal (*Section 12.2*)

Branched-chain alkane   an alkane with a branching connection of carbons (*Section 23.1*)

Brønsted–Lowry acid   a substance that can transfer H⁺ to a base in an acid–base reaction (*Section 16.1*)

Brønsted–Lowry base   a substance that can accept H⁺ from an acid in an acid–base reaction (*Section 16.1*)

Brønsted–Lowry theory   an acid is any substance (molecule or ion) that can transfer a proton (H⁺ ion) to another substance, and a base is any substance that can accept a proton (*Section 16.1*)

Buffer capacity   a measure of the amount of acid or base that a buffer can absorb without a significant change in pH (*Section 17.3*)

Buffer solution   a solution of a weak acid and its conjugate base that resists drastic changes in pH (*Section 17.3*)

Calories   energy content of foods is typically reported in units of Calories (Cal) per gram (*Section 1.8*)

Carbide   a carbon compound in which the carbon atom has a negative oxidation state (*Section 22.6*)

Carbohydrate   a large class of organic molecules commonly called sugars and related to glucose (*Section 23.4*)

Carbonyl group   the C=O group (*Section 23.3*)

Carboxylic acid   an organic molecule that contains the —CO₂H group (*Section 23.3*)

Catalyst   a substance that increases the rate of a reaction without itself being consumed (*Section 11.1, Section 14.12, and Section 15.10*)

Cathode   the electrode at which reduction takes place (*Section 19.2*)

Cathode ray   the visible glow emitted when an electric potential is applied across two electrodes in an evacuated chamber (*Section 2.3*)

Cathodic protection   a technique for protecting a metal from corrosion by connecting it to a second metal that is more easily oxidized (*Section 19.11*)

Cation   a positively charged atom or group of atoms (*Section 2.12*)

Cell potential (E)   *see* Electromotive force (*Section 19.4*)

Cell voltage   *see* Electromotive force (*Section 19.4*)

Celsius degree (°C)   a common unit of temperature; 0°C = 273.15 K (*Section 1.5*)

Centimeter (cm)   a common unit of length; 1 cm = 0.01 m (*Section 1.4*)

Ceramic   an inorganic, nonmetallic, nonmolecular solid (*Section 12.10*)

Ceramic composite   a hybrid material made of two ceramics (*Section 12.10*)

Ceramic–metal composite (cermet)   a hybrid material made of a metal reinforced with a ceramic (*Section 12.10*)

Ceramic–polymer composite   a hybrid material made of a polymer reinforced with a ceramic (*Section 12.10*)

Chain reaction   a self-sustaining reaction whose product initiates further reaction (*Section 20.7*)

Charles's law   the volume of a fixed amount of gas at a constant pressure varies directly with its absolute temperature (*Section 10.2*)

Chelate   the cyclic complex formed by a metal atom and a polydentate ligand (*Section 21.5*)

Chelate ring   rings of atoms (*Section 21.5*)

Chelating agent   a polydentate ligand (*Section 21.5*)

Chemical bond the force that holds atoms together in chemical compounds (*Section 2.11*)

Chemical compound a chemical substance composed of atoms of more than one element (*Section 2.11*)

Chemical equation a format for writing a chemical reaction, listing reactants on the left, products on the right, and an arrow between them (*Section 2.4*)

Chemical equilibrium the state reached when the concentrations of reactants and products remain constant in time (*Chapter 14 and Chapter 15 Introduction*)

Chemical formula a format for listing the number and kind of constituent elements in a compound (*Section 2.4*)

Chemical kinetics the area of chemistry concerned with reaction rates and the sequence of steps by which reactions occur (*Chapter 14 Introduction*)

Chemical property a characteristic that results in a change in the chemical makeup of a sample (*Section 2.3*)

Chemical reaction the transformation of one substance into another (*Chapter 3 Introduction*)

Chemistry the study of the composition, properties, and transformations of matter (*Chapter 1*)

Chiral having handedness (*Section 21.8 and Section 23.2*)

Chlor-alkali industry the commercial method of production for Cl_2 and NaOH by electrolysis of aqueous sodium chloride (*Section 19.13*)

Cholesteric liquid crystal phase an orientation of the liquid crystal phase in which the molecules are arranged such that one layer is placed on top of another, but the long axes of the molecules are rotated at a slight angle to layer above and below it (*Section 11.6*)

Cis trans-isomer the isomer of a metal complex or alkene in which identical ligands or groups are opposite rather than adjacent (*Section 21.7 and Section 23.5*)

Clausius–Clapeyron equation a mathematical relationship between vapor pressure and heat of vaporization for a substance (*Section 11.2*)

Climate change occurs as a result of rising levels of greenhouse gases from human activities (*Section 10.11*)

Coefficient a number placed before a formula in a chemical equation to indicate how many formula units are required to balance the equation (*Section 3.2*)

Colligative property a property that depends only on the amount of dissolved solute rather than on the chemical identity of the solute (*Section 13.6*)

Collision theory a model by which bimolecular reactions occur when two properly oriented reactant molecules come together in a sufficiently energetic collision (*Section 14.7*)

Colloid a homogeneous mixture containing particles with diameters in the range 2–500 nm (*Section 13.1*)

Common-ion effect the shift in the position of an equilibrium on addition of a substance that provides an ion in common with one of the ions already involved in the equilibrium (*Section 17.2*)

Complex ion an ion that contains a metal cation bonded to one or more small molecules or ions (*Section 17.12*)

Composite material is a ceramic material that is mixed with another component such as a metal, organic polymer, or another ceramic (*Section 12.10*)

Concentration amount of substance in a defined space and is expressed in terms of mass per unit volume (*Section 16.4*)

Condensed structure a shorthand method for drawing organic structures in which C—H and C—C single bonds are "understood" rather than shown (*Section 23.1*)

Conduction band the antibonding molecular orbitals in a semiconductor (*Section 12.7*)

Conjugate acid the species HA formed by addition of H^+ to a base A^- (*Section 16.1*)

Conjugate acid–base pair chemical species whose formulas differ only by one proton (*Section 16.1*)

Conjugate base the species A^- formed by removal of H^+ from the acid HA (*Section 16.1*)

Conjugated systems systems occur when p orbitals are connected in compounds with alternating single and multiple bonds. It is therefore said that the two (π) bonds separated by a σ bond are conjugated. (*Section 23.8*)

Conservation of energy law energy cannot be created or destroyed; it can only be converted from one form into another (*Section 9.1*)

Constitutional isomers isomers that have different connections among their constituent atoms (*Section 21.7 and Section 23.4*)

Contact process the commercial process for making sulfuric acid from sulfur (*Section 22.8*)

Conversion factor an expression that describes the relationship between different units (*Section 1.11*)

Coordinate covalent bond a bond formed when one atom donates two electrons to another atom that has a vacant valence orbital (*Section 7.5*)

Coordination compound a compound in which a central metal ion is attached to a group of surrounding molecules or ions by coordinate covalent bonds (*Section 21.4*)

Coordination number the number of nearest-neighbor atoms in a crystal (*Section 10.8*) or the number of ligand donor atoms that surround a central metal ion in a complex (*Section 21.4*)

Core electrons inner-shell electrons (*Section 6.3*)

Corrosion the oxidative deterioration of a metal, such as the conversion of iron to rust (*Sections 19.11*)

Coulomb's law the force resulting from the interaction of two electric charges is equal to a constant k times the magnitude of the charges divided by the square of the distance between them (*Section 6.8*)

Covalent bond a bond that occurs when two atoms share several (usually two) electrons (*Section 2.11*)

Covalent hydride a compound in which hydrogen is attached to another element by a covalent bond (*Section 22.3*)

Covalent network solid a solid whose atoms are linked together by covalent bonds into a giant three-dimensional array (*Section 12.1*)

Critical mass the amount of material necessary for a nuclear chain reaction to become self-sustaining (*Section 20.7*)

Critical point a combination of temperature and pressure beyond which a gas cannot be liquefied (*Section 11.5*)

Crystal field splitting the energy splitting between two sets of d orbitals in a metal complex (*Section 21.10*)

Crystal field theory a model that views the bonding in metal complexes as arising from electrostatic interactions and considers the effect of the ligand charges on the energies of the metal ion d orbitals (*Section 21.10*)

Crystalline solid a solid whose atoms, ions, or molecules have an ordered arrangement extending over a long range (*Section 12.1*)

Cubic centimeter (cm^3) a common unit of volume, equal in size to the milliliter; $1 \ cm^3 = 10^{-6} \ m^3$ (*Section 1.6*)

Cubic closest-packed   a packing arrangement of spheres into a face-centered cubic unit cell with three alternating layers (*Section 12.3*)

Cubic closest-packing   a packing arrangement with three alternating layers, a-b-c-a-b-c. The a-b layers are identical to those in the hexagonal closest-packed arrangement, but the third layer is offset from both a and b layers. (*Section 12.3*)

Cubic decimeter (dm³)   a common unit of volume, equal in size to the metric liter (L); (1 dm³ = 0.001 m³) (*Section 1.6*)

Cubic meter (m³)   the SI unit of volume (*Section 1.6*)

Cycloalkane   an alkane that contains a ring of carbon atoms (*Section 23.1*)

***d*-Block element**   a transition metal element in which *d* orbitals are filled (*Section 5.12; Chapter 20*)

Dalton (Da)   an alternative name for the unified atomic mass unit (*Section 2.9*)

Dalton's law of partial pressures   The total pressure exerted by a mixture of gases in a container at constant V and T is equal to the sum of the pressures exerted by each individual gas in the container (*Section 10.5*)

de Broglie equation   an equation that relates mass, wavelength, and velocity, $m = h/\lambda v$ (*Section 5.4*)

Decay constant   the first-order rate constant for radioactive decay (*Section 20.4*)

Degenerate   having the same energy level (*Section 5.10, Section 8.8, and Section 21.4*)

Delocalized   spread out over the two oxygen–oxygen bonds (*Section 23.8*)

Density   an intensive physical property that relates the mass of an object to its volume (*Section 1.7*)

Deoxyribonucleic acid (DNA)   an immense biological molecule, made up of deoxyribonucleotide units and containing an organism's genetic information (*Section 8.6 and Section 23.11*)

Diamagnetic   a substance that has no unpaired electrons and is weakly repelled by a magnetic field (*Section 8.8*)

Diastereoisomers   non-mirror-image stereoisomers (*Section 21.7*)

Diffraction   scattering of a light beam by an object containing regularly spaced lines or points (*Section 12.2*)

Diffusion   the mixing of different gases by random molecular motion with frequent collisions (*Section 10.7*)

Dimensional-analysis method   a method of problem solving whereby problems are set up so that unwanted units cancel (*Section 1.11*)

Diode   a semiconductor device that permits electric current to flow in one direction but is highly resistant to current flow in the opposite direction (*Section 12.8*)

Dipole   a pair of separated electrical charges (*Section 8.5*)

Dipole–dipole force   an intermolecular force resulting from electrical interactions among dipoles on neighboring molecules (*Section 8.6*)

Dipole moment (μ)   the measure of net molecular polarity; $\mu = Q \times r$ (*Section 7.3, Section 8.5, and Section 10.10*)

Diprotic acid   an acid that has two dissociable protons (*Section 4.7*)

Dissociate   splitting apart to give ions when dissolved in water (*Section 4.3*)

Dissociation of water   the ability of water to act as an acid and as a base (*Section 16.4*)

Donor atom   the atom attached directly to a metal in a coordination compound (*Section 21.4*)

Doping   the addition of a small amount of an impurity to increase the conductivity of a semiconductor (*Section 12.7*)

Double bond   a covalent bond formed by sharing four electrons between atoms (*Section 7.5, Section 11.8, and Section 23.2*)

Effective nuclear charge (Z_{eff})   the net nuclear charge actually felt by an electron (*Section 5.9 and Section 6.2*)

Effusion   the escape of gas molecules through a tiny hole in a membrane without molecular collisions (*Section 10.7*)

Electrochemical cell   a device for interconverting chemical and electrical energy (*Section 19.2*)

Electrochemistry   the area of chemistry concerned with the interconversion of chemical and electrical energy (*Chapter 19*)

Electrode   a conductor through which electrical current enters or leaves a cell (*Section 19.2*)

Electrolysis   the process of using an electric current to bring about chemical change (*Section 19.12*)

Electrolyte   a substance that dissolves in water to produce ions (*Section 4.3*)

Electrolytic cell   an electrochemical cell in which an electric current drives a nonspontaneous reaction (*Section 19.12*)

Electromagnetic spectrum   the range of different kinds of electromagnetic radiation (*Section 5.1 and Section 11.5*)

Electromotive force (emf)   the electrical potential that pushes electrons away from the anode and pulls them toward the cathode (*Section 19.4*)

Electron   a negatively charged, fundamental atomic particle (*Section 2.6*)

Electron affinity (E_{ea})   the energy change that occurs when an electron is added to an isolated atom in the gaseous state (*Section 6.5*)

Electron capture   a nuclear reaction in which a proton in the nucleus captures an inner-shell electron and is thereby converted into a neutron (*Section 2.8 and Section 20.2*)

Electron configuration   a description of which orbitals in an atom are occupied by electrons (*Section 5.10*)

Electron deficient   surrounded by less than eight electrons (*Section 7.6*)

Electron-dot structure   a representation of a molecule that shows valence electrons as dots; also called a Lewis structure (*Section 7.5*)

Electron-sea model   a model that visualizes metals as a three-dimensional array of metal cations immersed in a sea of delocalized electrons that are free to move about (*Section 12.6*)

Electronegativity (EN)   the ability of an atom in a molecule to attract the shared electrons in a covalent bond (*Section 7.3, Section 8.5, and Section 12.8*)

Electroplating   the coating of one metal on the surface of another using electrolysis (*Section 19.13*)

Electrorefining   the purification of a metal by means of electrolysis (*Section 19.12*)

Electrostatic potential map   portrays the calculated electron distribution in a molecule (*Section 7.3 and Section 16.3*)

Element   a fundamental substance that can't be chemically changed or broken down into anything simpler (*Section 2.1*)

Elementary reaction   a single chemical step in a reaction mechanism (*Section 14.9*)

Elementary step   *see* Elementary reaction (*Section 14.9*)

Empirical formula   a formula that gives the ratios of atoms in a chemical compound but not necessarily the exact values (*Section 3.6*)

Enantiomers   stereoisomers that are nonidentical mirror images of each other (*Section 21.8 and Section 23.4*)

End point   the point in a titration at which stoichiometrically equivalent quantities of reactants have been mixed together (*Section 4.13*)

Endothermic   a reaction in which heat is absorbed and the temperature of the surroundings falls (*Section 9.6*)

Energy   the capacity to do work or supply heat (*Section 1.8 and Section 9.1*)

Energy separation $\Delta E = E_2 - E_1$ between the two states   given by the equation $\Delta E = hn = hc > 1$, where h is Planck's constant, ν is the frequency of the light, and c is the speed of light (*Section 21.9*)

Energy enthalpy (H)   the quantity $E + PV$ (*Section 9.4*)

Energy separation $\Delta E = E_2 - E_1$ between the two states   To cause an electronic transition between two energy states, the energy of the photon must match the energy difference between the two states (*Section 21.9*)

Enthalpy change (ΔH)   the heat change in a reaction or process at constant pressure; $\Delta H = \Delta E + P\Delta V$ (*Section 9.4*)

Enthalpy of solution (ΔH_{soln})   the enthalpy change during formation of a solution (*Section 13.2*)

Entropy (S)   the amount of molecular randomness in a system (*Section 18.2*)

Entropy of solution (ΔS_{soln})   the entropy change during formation of a solution (*Section 13.2*)

Enzyme   a large protein that acts as a catalyst for a biological reaction (*Section 14.14 and Section 23.9*)

Equilibrium constant (K_c)   the constant in the equilibrium equation (*Section 15.2*)

Equilibrium constant (K_p)   the equilibrium constant for reaction of gases, defined using partial pressures (*Section 15.3*)

Equilibrium equation   an equation that relates the concentrations in an equilibrium mixture (*Section 15.2*)

Equilibrium mixture   a mixture of reactants and products at equilibrium (*Chapter 15 Introduction*)

Equivalence point   the point in a titration at which stoichiometrically equivalent quantities of reactants have been mixed together (*Section 17.5*)

Ester   an organic molecule that contains the $-CO_2R$ group (*Section 23.3*)

Ether   an organic molecule that contains two alkyl groups bonded to the same oxygen atom (*Section 23.3*)

Exothermic   a reaction in which heat is evolved and the temperature of the surroundings rises (*Section 9.6*)

Expanded octet   form more bonds than predicted by the octet rule (*Section 7.6*)

Experiment   a procedure for testing the hypothesis (*Section 1.1*)

Extensive property   a property whose value depends on the sample size (*Section 2.3 and Section 9.7*)

f-Block element   a lanthanide or actinide element, in which f orbitals are filled (*Section 5.12*)

Face-centered cubic unit cell   a cubic unit cell with an atom at each of its eight corners and an additional atom on each of its six faces (*Section 12.5*)

Fahrenheit (°F2)   is the most common unit for measuring temperature (*Section 1.5*)

Faraday   the electrical charge on 1 mol of electrons (96,485 C/mol e⁻) (*Section 19.4*)

Fatty acid   a long-chain carboxylic acid found as a constituent of fats and oils (*Section 23.6*)

First law of thermodynamics   the total internal energy of an isolated system is constant (*Section 18.6*)

First-order reaction   a reaction whose rate depends on the concentration of a single reactant raised to the first power (*Section 14.5*)

Fission   fragmenting of heavy nuclei (*Section 20.7*)

Formal charge   an electron bookkeeping device that tells whether an atom in a molecule has gained or lost electrons compared to an isolated atom (*Section 7.10 and Section 23.7*)

Formation constant (K_f)   the equilibrium constant for formation of a complex ion (*Section 17.12*)

Formula unit   one unit (atom, ion, or molecule) corresponding to a given formula (*Section 3.2*)

Formula weight   the sum of atomic weights of all atoms in a formula unit of any compound, molecular or ionic (*Section 3.3*)

Free energy   a state function that indicates whether a reaction is spontaneous or nonspontaneous (*Section 18.7*)

Free-energy change (ΔG) $\Delta G = \Delta H - T\Delta S$ (*Section 9.12*)

Frequency (ν)   the number of wave maxima that pass by a fixed point per unit time (*Section 5.1*)

Frequency factor   the parameter A $(=pZ)$ in the Arrhenius equation (*Section 14.7*)

Fuel cell   a galvanic cell in which one of the reactants is a traditional fuel such as methane or hydrogen (*Section 1.1 and Chapter 19 Inquiry*)

Fullerene   the allotrope of carbon that contains C_{60} molecules (*Section 12.5*)

Functional group   a part of a larger molecule; composed of an atom or group of atoms that has characteristic chemical behavior (*Section 23.3*)

Fusion   melting, or the change of a solid to a liquid (*Section 10.4*); also, the joining

together of two nuclei in a nuclear reaction, accompanied by release of an enormous amount of energy (*Section 20.7*)

Galvanic cell   an electrochemical cell in which a spontaneous chemical reaction generates an electric current (*Section 19.2*)

Galvanizing   a process for protecting steel from corrosion by coating it with zinc (*Section 19.11*)

Gamma (γ) radiation   a type of radioactive emission consisting of a stream of high-energy photons (*Section 2.8 and Section 20.2*)

Gas constant (R)   the constant in the ideal gas law $PV = nRT$ (*Section 10.3*)

Gas laws   relationships among the variables P, V, n, and T for a gas sample (*Section 10.2*)

Geometric isomers   *see* Diastereoisomers (*Section 21.8*)

Gibbs free-energy change (ΔG) $\Delta G = \Delta H - T\Delta S$ (*Section 9.12*)

Global warming   refers to the idea that increasing concentrations of greenhouse gases will upset the delicate thermal balance of incoming and outgoing radiation on Earth (*Section 10.11*)

Graham's law   The rate of effusion of a gas is inversely proportional to the square root of its molar mass (*Section 10.7*)

Gram (g)   a common unit of mass; $1 g = 0.001$ kg (*Section 1.3*)

Green chemistry   a set of guidelines describing environmentally benign chemical reactions and practices (*Chapter 3 Inquiry*)

Greenhouse effect   the trapping of heat emitted from the Earth by gases that absorb infrared radiation (greenhouse gases) (*Chapter 3 Inquiry, Section 10.10*)

Greenhouse gases   gases that absorb infrared (IR) radiation (*Section 10.9*)

Ground-state electron configuration   the lowest-energy electron configuration of an atom (*Section 5.10 and Section 6.1*)

Group   a column of elements in the periodic table (*Section 2.2*)

Half-life ($t_{1/2}$)   the time required for a reactant concentration to drop to one-half its initial value (*Section 14.5*)

Half-reaction   the oxidation or reduction part of a redox reaction (*Section 4.11 and Section 19.1*)

Half-reaction method   a method for balancing redox equations (*Section 19.1*)

Hall–Héroult process   the commercial method for producing aluminum by electrolysis of a molten mixture of aluminum oxide and cryolite (*Section 19.13*)

Heat   the energy transferred from one object to another as the result of a temperature difference between them (*Section 9.1*)

Heat capacity (C)   the amount of heat required to raise the temperature of an object or substance a given amount (*Section 9.7*)

Heat of combustion   the amount of energy released on burning a substance (*Chapter 9 Inquiry*)

Heat of fusion (ΔH_{fusion})   the amount of heat required for melting a solid to a liquid (*Section 11.4*)

Heat of reaction (ΔH)   the enthalpy change for a reaction (*Section 9.4*)

Heat of vaporization (ΔH_{vap})   the amount of heat required for vaporization of a liquid to a gas (*Section 11.4*)

Heisenberg uncertainty principle   the position and the velocity of an electron can never both be known beyond a certain level of precision (*Section 5.5*)

Henderson–Hasselbalch equation   an equation relating the pH of a solution to the pK_a of the weak acid; pH = pK_a + log ([base]/[acid]) (*Section 17.4*)

Henry's law   the solubility of a gas in a liquid at a given temperature is directly proportional to the partial pressure of the gas over the solution (*Section 13.5*)

Hertz (Hz)   a unit of frequency; 1 Hz = 1 s^{-1} (*Section 5.1*)

Hess's law   the overall enthalpy change for a reaction is equal to the sum of the enthalpy changes for the individual steps in the reaction (*Section 9.8*)

Heterogeneous catalyst   a catalyst that exists in a different phase than the reactants (*Section 14.13*)

Heterogeneous equilibria   equilibria in which reactants and products are present in more than one phase (*Section 15.4*)

Hexadentate ligand   a ligand with six donor atoms that bonds to a metal (*Section 21.5*)

Hexagonal closest-packing   a packing arrangement two alternating layers, *a-b-a-b*. Each layer has a hexagonal arrangement of touching spheres, which are offset so that spheres in a *b* layer fit into the small triangular depressions between spheres in an *a* layer. (*Section 12.5*)

High-spin complex   a metal complex in which the *d* electrons are arranged to give the maximum number of unpaired electrons (*Section 21.11*)

Homogeneous catalyst   a catalyst that exists in the same phase as the reactants (*Section 14.13*)

Homogeneous equilibria   equilibria in which all reactants and products are in a single phase, usually either gaseous or solution (*Section 15.4*)

Hückel's rule   a compound can only be aromatic if the number of p electrons in the ring is equal to $4n + 2$, where *n* equals zero or any positive integer (*Section 23.10*)

Hund's rule   if two or more degenerate orbitals are available, one electron goes in each until all are half full (*Section 5.10*)

Hybrid atomic orbital   a wave function derived by combination of atomic wave functions (*Section 8.3*)

Hydrate   a solid compound that contains water molecules (*Section 21.6*)

Hydrocarbon   a compound that contains only carbon and hydrogen (*Section 23.1*)

Hydrogen bond   an attractive intermolecular force between a hydrogen atom bonded to an electronegative O, N, or F atom and an unshared electron pair on a nearby electronegative atom (*Section 8.6*)

Hydrogenation   the addition of H_2 to an alkene to yield an alkane (*Section 23.6*)

Hydronium ion   the protonated water molecule, H_3O^+ (*Section 16.1*)

Hypothesis   a possible explanation for the observation developed based upon facts collected from previous experiments as well as scientific knowledge and intuition (*Section 1.1*)

Ideal gas   a gas whose behavior exactly follows the ideal gas law (*Section 10.2*)

Ideal gas law   a description of how the volume of a gas is affected by changes in pressure, temperature, and amount; $PV = nRT$ (*Section 10.3*)

Initial rate   the instantaneous rate at the beginning of a reaction (*Section 14.1*)

Inner transition metal element   an element in the 14 groups shown separately at the bottom of the periodic table (*Section 2.2*)

Inner transition metal group   the 14 groups shown separately at the bottom of the table (*Section 2.2*)

Instantaneous rate   the rate of a reaction at a particular time (*Section 14.1*)

Integrated rate law   the integrated form of a rate law (*Section 14.4*)

Intensive property   a property whose value does not depend on the sample size (*Section 2.3*)

Intermolecular force   an attractive interaction between molecules (*Section 8.6 and Chapter 11*)

Internal energy (E)   the sum of kinetic and potential energies for each particle in a system (*Section 9.2*)

Interstitial hydride   a metallic hydride that consists of a crystal lattice of metal atoms with the smaller hydrogen atoms occupying holes between the larger metal atoms (*Section 22.3*)

Ion   a charged atom or group of atoms (*Section 2.11*)

Ion–dipole force   an intermolecular force resulting from electrical interactions between an ion and the partial charges on a polar molecule (*Section 21.10*)

Ion pair   a dissolved pair of closely spaced positive and negative ions that behave as one particle instead of two (*Section 13.7*)

Ion product (IP)   a number defined in the same way as K_{sp}, except that the concentrations in the expression for IP are not necessarily equilibrium values (*Section 17.13*)

Ion-product constant for water (K_w) $[H_3O^+][OH^-] = 1.0 \times 10^{-14}$ (*Section 16.4*)

Ionic bond   a bond that results from a transfer of one or more electrons between atoms (*Section 2.12 and Section 6.7*)

Ionic equation   a chemical equation written so that ions are explicitly shown (*Section 4.5*)

Ionic hydride   a saltlike, high-melting, white, crystalline compound formed by the alkali metals and the heavier alkaline earth metals (*Section 22.3*)

Ionic liquid   a liquid whose constituent particles are ions rather than molecules (*Chapter 6 Inquiry*)

Ionic solid   a solid whose constituent particles are ions ordered into a regular three–dimensional arrangement held together by ionic bonds (*Section 2.11, Section 6.7, and Section 12.1*)

Ionization energy (E_i)   the amount of energy necessary to remove the outermost electron from an isolated neutral atom in the gaseous state (*Section 6.3*)

Ionization isomers   isomers that differ in the anion bonded to the metal ion (*Section 21.7*)

Isoelectronic   atoms or ions with same electron configuration (*Section 6.1*)

Isomers   compounds that have the same formula but a different bonding arrangement of their constituent atoms (*Section 21.7 and Section 23.1*)

Isotopes   atoms with identical atomic numbers but different mass numbers (*Section 2.8 and Section 3.6*)

Joule (J)   the SI unit of energy, equal to 1 (kg · m^2)/s^2 (*Section 1.8 and Section 9.1*)

K_f   formation constant; the equilibrium constant for formation of a complex ion (*Section 17.12*)

K_{sp} solubility-product constant; the equilibrium constant for a dissolution reaction (*Section 17.10*)

K_{spa} solubility-product constant in acid; the equilibrium constant for a dissolution reaction in acid (*Section 17.14*)

Kelvin (K) the SI unit of temperature; $0\ K$ = absolute zero (*Section 1.5*)

Ketone an organic molecule that contains two alkyl groups bonded to a $C=O$ carbon (*Section 23.3*)

Kilogram (kg) the SI unit of mass; 1 kg = 2.205 U.S. lb (*Section 1.3*)

Kinetic energy (E_K) the energy of motion; $E_K = (1/2)mv^2$ (*Section 1.8 and Section 9.1*)

Kinetic–molecular theory a theory describing the quantitative behavior of gases (*Section 10.6 and Section 11.2*)

Lanthanide contraction the decrease in atomic radii across the *f*-block lanthanide elements (*Section 21.2*)

Lattice energy (U) the sum of the electrostatic interactions between ions in a solid that must be overcome to break a crystal into individual ions (*Section 6.8 and Section 7.4*)

Law of definite proportions Different samples of a pure chemical substance always contain the same proportion of elements by mass (*Section 2.4*)

Law of mass conservation Mass is neither created nor destroyed in chemical reactions (*Section 2.4 and Section 3.2*)

Law of multiple proportions When two elements combine in different ways to form different substances, the mass ratios are small, whole-number multiples of one another (*Section 2.5*)

Le Châtelier's principle If a stress is applied to a reaction mixture at equilibrium, reaction occurs in the direction that relieves the stress (*Section 15.6*)

Lewis acid an electron-pair acceptor (*Section 16.15*)

Lewis base an electron-pair donor (*Section 16.15*)

Ligand a molecule or ion that bonds to the central metal ion in a complex (*Section 20.5 and Section 21.4*)

Ligand donor atom an atom attached directly to the metal ion in a metal complex (*Section 21.4*)

Light-emitting diode (LED) a semiconductor device that converts electrical energy into light (*Section 12.8*)

Limiting reactant the reactant present in limiting amount that controls the extent to which a reaction occurs (*Section 3.5*)

Line drawing show only bonds in the carbon backbone and are another convenient shorthand representation of structure (*Section 23.1*)

Line spectrum the wavelengths of light emitted by an energetically excited atom (*Section 5.3*)

Lines between atoms term used to indicate covalent bonds (*Section 7.1*)

Linkage isomers isomers that arise when a ligand bonds to a metal through either of two different donor atoms (*Section 21.7*)

Lipid a naturally occurring organic molecule that dissolves in nonpolar organic solvents when a sample of plant or animal tissue is crushed or ground (*Section 23.6*)

Lipid number indicates the number of carbons, numbers of unsaturation and where the last double bond in the chain is located (*Section 23.6*)

Liquid crystal an intermediate phase that exhibits properties of both liquids and solids. Liquid crystals are ordered like solids, but flow like liquids. (*Section 11.6*)

Liter (L) a common unit of volume; $1\ L = 10^{-3}\ m^3$ (*Section 1.6*)

Lock-and-key model a model that pictures an enzyme as a large, irregularly shaped molecule with a cleft into which substrate can fit (*Chapter 13 Inquiry*)

London dispersion force an intermolecular force resulting from the presence of temporary dipoles in atoms or molecules (*Section 8.6 and Section 23.6*)

Lone pair non bonding electrons (*Section 7.5*)

Low-spin complex a metal complex in which the *d* electrons are paired up to give a maximum number of doubly occupied *d* orbitals and a minimum number of unpaired electrons (*Section 21.11*)

Macroscale items are large enough to be observed with the human eye and are measured with instruments such as rulers and calipers (*Chapter 1 Inquiry*)

Magnetic quantum number (m_l) a variable in the solutions to the Schrödinger wave equation that defines the spatial orientation of an orbital (*Section 5.6*)

Main group the two larger groups on the left and the six larger groups on the right of the periodic table (*Section 2.2*)

Manometer a simple instrument for measuring gas pressure; similar in principle to mercury barometer (*Section 10.1 and Section 11.2*)

Mass the amount of matter in an object (*Section 1.3*)

Mass defect the loss in mass that occurs when protons and neutrons combine to form a nucleus (*Section 20.6*)

Mass number (A) the total number of protons and neutrons in an atom (*Section 2.8*)

Mass percent (Mass %) a unit of concentration; the mass of one component divided by the total mass of the solution times 100% (*Section 13.4*)

Mass spectrometer an instrument that uses a magnetic field to separate ions of different mass-to-charge ratios (*Section 2.10*)

Mass spectrum a graph of ion intensity versus mass-to-charge ratio (*Section 2.10*)

Matter a term used to describe anything that has mass (*Section 1.3*)

Metal complex *see* Coordination compound (*Section 21.4*)

Metallic hydride a compound formed by reaction of lanthanide, actinide, or some *d*-block transition metals with variable amounts of hydrogen (*Section 22.3*)

Metallic solid a solid consisting of metal atoms, whose crystals have metallic properties such as electrical conductivity (*Section 12.1*)

Meter (m) the SI unit of length (*Section 1.4*)

Microgram (μg) a common unit of mass; $1\mu g = 0.001\ mg = 10^{-6}\ g$ (*Section 1.3*)

Micrometer (μm) a common unit of length; $1\ \mu m = 0.001\ mm = 10^{-6}\ m$ (*Section 1.4*)

Microscale is a smaller size regime and named so because dimensions of materials are in the micrometer range ($1\ \mu m = 1 \times 10^{-6}\ m$) (*Chapter 1 Inquiry*)

Milligram (mg) a common unit of mass; $1\ mg = 0.001\ g = 10^{-6}\ kg$ (*Section 1.3*)

Milliliter (mL) a common unit of volume; $1\ mL = 1\ cm^3$ (*Section 1.6*)

Millimeter (mm) a common unit of length; $1\ mm = 0.001\ m$ (*Section 1.4*)

Millimeter of mercury (mm Hg) a common unit of pressure; the millimeter of mercury, also called a *torr*, is based on atmospheric pressure measurements using a mercury barometer (*Section 10.1*)

Miscible mutually soluble in all proportions (*Section 13.5*)

Mixture a blend of two or more substances in some arbitrary proportion (*Section 2.11 and Chapter 13 Introduction*)

Molal boiling-point-elevation constant (K_b) the amount by which the boiling point of a solvent is raised by dissolved substances (*Section 13.8*)

Molal freezing-point-depression constant (K_f) the amount by which the melting point of a solvent is lowered by dissolved substances (*Section 13.8*)

Molality (m)   a unit of concentration; the number of moles of solute per kilogram of solvent (mol/kg) (*Section 4.1 and Section 13.4*)

Molar heat capacity (C_m)   the amount of heat necessary to raise the temperature of 1 mol of a substance 1°C (*Section 9.7 and Section 11.2*)

Molar mass   the mass of 1 mol of substance; equal to the molecular or formula mass of the substance in grams (*Section 2.9*)

Molarity (M)   a common unit of concentration; the number of moles of solute per liter of solution (*Section 4.1 and Section 13.4*)

Mole (mol)   the SI unit for amount of substance; the quantity of a substance that contains as many molecules or formula units as there are atoms in exactly 12 g of carbon-12 (*Section 2.9 and Section 3.3*)

Mole fraction (X)   a unit of concentration; the number of moles of a component divided by the total number of moles in the mixture (*Section 10.5 and Section 13.4*)

Molecular equation   a chemical equation written using the complete formulas of reactants and products (*Section 4.5*)

Molecular formula   a formula that tells the identity and numbers of atoms in a molecule (*Section 3.6*)

Molecular orbital   a wave function whose square gives the probability of finding an electron within a given region of space in a molecule (*Section 8.7 and Section 10.10*)

Molecular orbital (MO) theory   a quantum mechanical description of bonding in which electrons occupy molecular orbitals that belong to the entire molecule rather than to an individual atom (*Section 8.7*)

Molecular solid   a solid whose constituent particles are molecules held together by intermolecular forces (*Section 12.1*)

Molecular weight   sum of atomic weights of all atoms in a molecule (*Section 3.3*)

Molecularity   the number of molecules on the reactant side of the chemical equation for an elementary reaction (*Section 14.9*)

Molecule   the unit of matter that results when two or more atoms are joined by covalent bonds (*Section 2.11*)

Monodentate ligand   a ligand that bonds to a metal using the electron pair of a single donor atom (*Section 21.5*)

Monoprotic acid   an acid that has a single dissociable proton (*Section 4.7*)

Monosaccharide   a carbohydrate such as glucose that can't be broken down into smaller molecules by hydrolysis (*Section 23.4*)

N-Terminal amino acid   the amino acid with a free —NH_2 group on the end of a protein chain (*Section 23.9*)

n-Type semiconductor   a semiconductor doped with an impurity that has more electrons than necessary for bonding (*Section 12.8*)

Nanometer (nm)   a common unit of length; 1 nm = 10^{-9} m (*Section 1.4*)

Nanoscale   one thousand times smaller than the microscale and represents particles with nanometer sized (1 nm = 1×10^{-9} m) dimensions (*Chapter 1 Inquiry*)

Nanoscience   the study and production of materials and structures that have at least one dimension between 1 nm and 100 nm (*Section 1.1*)

Nematic liquid crystal display   an orientation of the liquid crystal phase in which the long axes of molecules are approximately parallel but their ends do not line up (*Section 11.6*)

Nernst equation   an equation for calculating cell potentials under non-standard-state conditions; $E = E° - (RT \ln Q)/(nF)$ (*Section 19.7*)

Net ionic equation   a chemical equation written so that spectator ions are removed (*Section 4.5*)

Neutralization reaction   *see* Acid–base neutralization reaction (*Sections 4.7 and Section 17.1*)

Neutron   a neutral, fundamental atomic particle in the nucleus of atoms (*Section 2.7*)

Newton (N)   the SI unit for force (*Section 10.1*)

Node   a region where a wave has zero amplitude (*Section 5.7*)

Nonelectrolyte   a substance that does not produce ions when dissolved in water (*Section 4.3*)

Nonstoichiometric compound   a compound whose atomic composition can't be expressed as a ratio of small whole numbers (*Section 22.3*)

Normal boiling point   the temperature at which boiling occurs when there is exactly 1 atm of external pressure (*Section 11.2*)

Normal melting point   the temperature at which melting occurs when there is exactly 1 atm of external pressure (*Section 11.5*)

Nucleon   is the general term for a nuclear particle, either proton or neutron (*Section 20.1*)

Nucleus   composed of two kinds of particles, called *protons* and *neutrons* (*Section 2.7*)

Nuclear chemistry   the study of the properties and reactions of atomic nuclei (*Section 2.7 and Chapter 20 Introduction*)

Nuclear equation   an equation for a nuclear reaction in which the sums of the nucleons are the same on both sides and the sums of the charges on the nuclei and any elementary particles are the same on both sides (*Section 2.7 and Section 20.1*)

Nuclear fission   the fragmenting of heavy nuclei (*Section 20.7*)

Nuclear fusion   the joining together of light nuclei (*Section 20.7*)

Nuclear reaction   a reaction that changes an atomic nucleus (*Section 20.1*)

Nuclear transmutation   the change of one element into another by a nuclear reaction (*Section 20.8*)

Nucleic acid   a biological molecule made up of nucleotide units linked together to form a long chain (*Section 23.11*)

Nucleon   a general term for nuclear particles, both protons and neutrons (*Section 20.1*)

Nucleoside   a constituent of nucleotides; composed of an aldopentose sugar plus an amine base (*Section 23.11*)

Nucleotide   a building block from which nucleic acids are made; composed of a nucleoside plus phosphoric acid (*Section 23.11*)

Nucleus   the central core of an atom consisting of protons and neutrons (*Section 2.7*)

Nuclide   the nucleus of a given isotope (*Section 20.1*)

Observations   a systematic recording of natural phenomena and may be qualitative, descriptive in nature, or quantitative, involving measurements (*Section 1.1*)

Octet rule   the statement that main-group elements tend to undergo reactions that leave them with eight valence electrons (*Section 6.6 and Section 7.5*)

Orbital   a solution to the Schrödinger wave equation, describing a region of space where an electron is likely to be found (*Section 5.6*)

Organic chemistry   the study of carbon compounds (*Chapter 23 Introduction*)

Organic compounds   molecules that contain carbon atoms bonded together in a chain (*Section 7.8*)

Osmosis   the passage of solvent through a membrane from the less concentrated side to the more concentrated side (*Section 13.9*)

Osmotic pressure the amount of pressure necessary to cause osmosis to stop (*Section 13.9*)

Overpotential the required additional voltage (*Section 19.12*)

Overvoltage the additional voltage required above that calculated for an electrolysis reaction (*Section 19.12*)

Oxidation the loss of one or more electrons by a substance (*Section 4.10*)

Oxidation number a value that measures whether an atom in a compound is neutral, electron-rich, or electron-poor compared to an isolated atom (*Section 4.10*)

Oxidation–reduction (redox) reaction a process in which one or more electrons are transferred between reaction partners (*Section 4.4*)

Oxide a binary compound with oxygen in the −2 oxidation state (*Section 22.8*)

Oxidizing agent a substance that causes an oxidation by accepting an electron (*Section 4.11*)

Oxoacid an acid that contains oxygen in addition to hydrogen and another element (*Section 4.7*)

Oxoanion an anion of an oxoacid (*Section 2.12*)

***p*-Block element** an element in groups 3A–8A, in which *p* orbitals are filled (*Section 5.12*)

***p*-Type semiconductor** a semiconductor doped with an impurity that has fewer electrons than necessary for bonding (*Section 12.8*)

Paramagnetic a substance that contains unpaired electrons and is attracted by a magnetic field (*Section 8.8 and Section 21.11*)

Parts per billion (ppb) a concentration unit for very dilute solutions; a concentration of 1 ppb means that each kilogram of solution contains 1 μg of solute (*Section 13.4*)

Parts per million (ppm) a concentration unit for very dilute solutions; a concentration of 1 ppm means that each kilogram of solution contains 1 mg of solute (*Section 13.4*)

Pascal (Pa) the SI unit for pressure (*Section 10.1*)

Pauli exclusion principle no two electrons in an atom can have the same four quantum numbers (*Section 5.8*)

Peptide consists of molecules called amino acids linked together in a chain (*Section 1.1 and Section 23.9*)

Peptide bond the amide bond linking two amino acids in a protein (*Section 23.9*)

Percent composition a list of elements present in a compound and the mass percent of each (*Section 3.6*)

Percent dissociation the concentration of the acid that dissociates divided by the initial concentration of the acid times 100% (*Section 16.10*)

Percent ionic character the extent of electron transfer from one element to another in a bond (*Section 7.3*)

Percent yield the amount of product actually formed in a reaction divided by the amount theoretically possible and multiplied by 100% (*Section 3.4*)

Period a row of elements in the periodic table (*Section 2.2*)

Periodic table a chart of the elements arranged by increasing atomic number so that elements in a given group have similar chemical properties (*Section 2.1*)

pH the negative base-10 logarithm of the molar hydronium ion concentration (*Section 16.5*)

pH titration curve a plot of the pH of a solution as a function of the volume of added base or acid (*Section 17.5*)

Phase a state of matter (*Sections 5.7, Section 8.2, and Section 11.3*)

Phase change a process in which the physical form but not the chemical identity of a substance changes (*Section 11.3*)

Phase diagram a plot showing the effects of pressure and temperature on the physical state of a substance (*Section 11.5 and Section 13.8*)

Photoelectric effect the ejection of electrons from a metal on exposure to radiant energy (*Section 5.2*)

Photon the smallest possible amount of radiant energy; a quantum (*Section 5.2*)

Photovoltaic cell a semiconductor device that converts light into electrical energy (*Section 12.8*)

Physical property a characteristic that can be determined without changing the chemical makeup of a sample (*Section 2.3*)

Pi (π) bond or pi system a covalent bond formed by sideways overlap of orbitals in which shared electrons occupy a region above and below a line connecting the two nuclei (*Section 8.4 and Section 23.8*)

Picometer (pm) a common unit of length; 1 pm = 10^{-12} m (*Section 1.4*)

Planck's postulate the energy of one quantum is related to the frequency of an electromagnetic wave (*Section 5.2*)

Polar covalent bond a bond in which the bonding electrons are attracted somewhat more strongly by one atom than by the other (*Section 7.3 and Section 8.5*)

Polarizability the ease with which a molecule's electron cloud can be distorted by a nearby electric field (*Section 8.6*)

Polyatomic ion a charged, covalently bonded group of atoms (*Section 2.12*)

Polydentate ligand a ligand that bonds to a metal through electron pairs on more than one donor atom (*Section 21.5*)

Polyprotic acid an acid that contains more than one dissociable proton (*Section 16.11*)

Polysaccharide a compound such as cellulose that is made of many simple sugars linked together (*Section 23.4*)

Positron emission a nuclear reaction that converts a proton into a neutron plus an ejected positron (*Section 2.8 and Section 20.2*)

Potential energy (E_P) energy that is stored, either in an object because of its position or in a molecule because of its chemical composition (*Section 1.8 and Section 9.1*)

Precipitation reaction a reaction in which an insoluble solid precipitate forms and drops out of solution (*Section 4.4*)

Precision how well a number of independent measurements agree with one another (*Section 1.9*)

Pressure (P) defined as a force (*F*) exerted per unit area (*A*) (*Section 10.1*)

Primitive-cubic unit cell a cubic unit cell with an atom at each of its eight corners (*Section 12.3*)

Principal quantum number (n) a variable in the solutions to the Schrödinger wave equation on which the size and energy level of an orbital primarily depends (*Section 5.6*)

Principal reaction the proton-transfer reaction that proceeds farther to the right when calculating equilibrium concentrations in solutions of weak acids (*Section 16.9*)

Property any characteristic that can be used to describe or identify matter (*Section 2.3*)

Protein a biological molecule made up of many amino acids linked together to form a long chain (*Section 23.9*)

Proton a positively charged, fundamental atomic particle in the nucleus of atoms (*Section 2.7*)

Qualitative observations that are descriptive in nature (*Section 1.1*)

Qualitative analysis a procedure for identifying the ions present in an unknown solution (*Section 17.15*)

Quantitative observations that involve measurements (*Section 1.1*)

Quantized changing only in discrete amounts (*Section 5.2*)

Quantum   the smallest possible amount of radiant energy (*Section 5.2*)

Quantum mechanical model   a model of atomic structure that concentrates on an electron's wavelike properties (*Section 5.5*)

Quantum number   a variable in solutions to the Schrödinger wave equation that describes the energy level and position in space where an electron is most likely to be found (*Section 5.6*)

Racemic mixture   a 1:1 mixture of enantiomers (*Section 21.8*)

Radial distribution plot   the radial distribution is the probability of finding the electron as a given point in space times the surface area of a sphere with radius (r) (*Section 5.8*)

Radiant energy   electromagnetic radiation (*Section 5.1*)

Radicals   elements having unpaired elements in a half-filled orbital (*Section 7.7*)

Radioactive   they undergo a spontaneous decay and emit some form of radiation (*Section 20.2*)

Radioactivity   the spontaneous emission of radiation accompanying a nuclear reaction (*Section 2.1 and Section 20.2*)

Radioactive decay chain   a series of nuclear decays (*Section 20.3*)

Radiocarbon dating   a technique for dating archaeological artifacts by measuring the amount of ^{14}C in the sample (*Section 20.5*)

Radioisotope   a radioactive isotope (*Section 2.8 and Section 20.2*)

Raoult's law   the vapor pressure of a solution containing a nonvolatile solute is equal to the vapor pressure of pure solvent times the mole fraction of the solvent (*Section 13.7*)

Rate constant   the proportionality constant in a rate law (*Section 14.2*)

Rate-determining step   the slowest step in a reaction mechanism (*Section 14.11*)

Rate law   an equation that tells how reaction rate depends on the concentration of each reactant (*Section 14.2 and Section 15.1*)

Reaction intermediate   a species that is formed in one step of a reaction mechanism and consumed in a subsequent step (*Section 14.9*)

Reaction mechanism   the sequence of molecular events that defines the pathway from reactants to products (*Section 14.9*)

Reaction order   the value of the exponents of concentration terms in the rate law (*Section 14.2*)

Reaction quotient (Q_c)   similar to the equilibrium constant K_c except that the concentrations in the equilibrium constant expression are not necessarily equilibrium values (*Section 15.5*)

Reaction rate   the increase in the concentration of a product per unit time or the decrease in the concentration of a reactant per unit time (*Section 14.1*)

Redox reaction   an oxidation–reduction reaction (*Section 1.1 and Section 4.4*)

Reducing agent   a substance that causes a reduction by donating an electron (*Section 4.11*)

Reduction   the gain of one or more electrons by a substance (*Section 4.10*)

Replication   the process by which identical copies of DNA are made (*Section 23.13*)

Resonance   in organic compounds play an important role in understanding both structure and reactivity (*Section 23.7*)

Resonance hybrid   an average of several valid electron-dot structures for a molecule (*Section 7.9*)

Resonance theory   describes many molecules by drawing two or more electron-dot structures and considering the actual molecule to be a composite of these two structures (*Section 7.9*)

Ribonucleic acid (RNA)   a biological polymer of ribonucleotide units that serves to transcribe the genetic information in DNA and uses that information to direct the synthesis of proteins (*Section 23.11*)

Rounding off   deleting digits to keep only the correct number of significant figures in a calculation (*Section 1.10*)

Rutherford's nuclear model   an atom must consist mostly of empty space, having a dense, positively charged nucleus at its center (*Section 5.4*)

s-Block element   an element in groups 1A or 2A, in which s orbitals are filled (*Section 5.12*)

Sacrificial anode   an easily oxidized metal that corrodes instead of a less reactive metal to which it is connected (*Section 19.11*)

Salt   an ionic compound formed in an acid–base neutralization reaction (*Section 4.7*)

Salt bridge   a tube that contains a gel permeated with a solution of an inert electrolyte connecting the two sides of an electrochemical cell (*Section 19.2*)

Saturated   when equilibrium is reached and no further solute dissolves in a given amount of solvent (*Section 13.5 and Section 23.3*)

Scientific method   a general approach to research (*Section 1.1*)

Scientific notation   a system in which a large or small number is written as a number between 1 and 10 times a power of 10 (*Section 1.2*)

Second law of thermodynamics   in any spontaneous process, the total entropy of a system and its surroundings always increases (*Section 18.6*)

Second-order reaction   a reaction whose rate depends on the concentration of a single reactant raised to the second power or on the concentrations of two different reactants, each raised to the first power (*Section 14.6*)

Semiconductor   a material that has an electrical conductivity intermediate between that of a metal and that of an insulator (*Section 12.7*)

Shell   a grouping of orbitals according to principal quantum number (*Section 5.6*)

Shield   conceal or guard (*Section 5.9 and Section 6.3*)

SI unit   units of measure established by the Système Internationale (*Section 1.2*)

Side chain   the group attached to the α carbon of an amino acid (*Section 23.10*)

Sigma (σ) bond   a covalent bond formed by head-on overlap of orbitals in which the shared electrons are centered about the axis between the two nuclei (*Section 8.2*)

Significant figure   the total number of digits in a measurement (*Section 1.9*)

Silicate   an ionic compound that contains silicon oxoanions along with cations, such as Na^+, K^+, Mg^{2+}, or Ca^{2+} (*Section 22.6*)

Simple cubic packing   a packing arrangement of spheres into a primitive-cubic unit cell (*Section 12.3*)

Single bond   a covalent bond formed by sharing two electrons between atoms (*Section 7.5*)

Sintering   a process in which the particles of a powder are "welded" together without completely melting (*Section 12.10*)

Smectic liquid crystal phase   an orientation of the liquid crystal phase in which the long axes of molecules are parallel, but the ends are also aligned, creating rows of molecules (*Section 11.6*)

Solubility   the amount of a substance that dissolves in a given volume of solvent at a given temperature (*Section 4.6*)

Solubility product (K_{sp})   the equilibrium constant for a dissolution reaction (*Section 17.10*)

Solute the dissolved substance in a solution (*Section 4.1 and Section 13.1*)

Solution a homogeneous mixture containing particles the size of a typical ion or covalent molecule (*Section 13.1*)

Solvent the major component in a solution (*Section 13.1*)

Sol–gel method a method of preparing ceramics, involving synthesis of a metal oxide powder from a metal alkoxide (*Section 12.10*)

sp Hybrid orbital a hybrid orbital formed by combination of one atomic s orbital with one p orbital (*Section 8.4*)

sp^2 Hybrid orbital a hybrid orbital formed by combination of one s and two p atomic orbitals (*Section 8.4*)

sp^3 Hybrid orbital a hybrid orbital formed by combination of one s and three p atomic orbitals (*Section 8.3*)

sp^3d^2 Hybrid orbital a hybrid orbital formed by combination of one s, three p, and two d atomic orbitals (*Section 21.11*)

Specific heat the amount of heat necessary to raise the temperature of 1 gram of a substance 1°C (*Section 9.7*)

Spectator ion an ion that appears on both sides of the reaction arrow (*Section 4.5*)

Spectrochemical series an ordered list of ligands in which crystal field splitting increases (*Section 21.10*)

Spin quantum number (m_s) a variable that describes the spin of an electron, either $+1/2$ or $-1/2$ (*Section 5.8*)

Spontaneous process one that proceeds on its own without any continuous external influence (*Section 9.11 and Section 18.1*)

Standard cell potential (E°) the cell potential when both reactants and products are in their standard states (*Section 19.4*)

Standard electrode potential *see* Standard reduction potential (*Section 19.5*)

Standard enthalpy of reaction ($\Delta H\delta$) enthalpy change under standard-state conditions (*Section 9.5*)

Standard entropy of reaction ($\Delta S°$) the entropy change for a chemical reaction under standard-state conditions (*Section 18.5*)

Standard free-energy change ($\Delta G°$) the free-energy change that occurs when reactants in their standard states are converted to products in their standard states (*Section 18.8*)

Standard free energy of formation ($\Delta G°_f$) the free-energy change for formation of 1 mol of a substance in its standard state from the

most stable form of the constituent elements in their standard states (*Section 18.9*)

Standard heat of formation ($\Delta H°_f$) the enthalpy change $\Delta H°_f$ for the hypothetical formation of 1 mol of a substance in its standard state from the most stable forms of its constituent elements in their standard states (*Section 9.9*)

Standard hydrogen electrode (S.H.E.) a reference half-cell consisting of a platinum electrode in contact with H_2 gas and aqueous H^+ ions at standard-state conditions (*Section 19.5*)

Standard molar entropy (S°) the entropy of 1 mol of a pure substance at 1 atm pressure and a specified temperature, usually 25 °C (*Section 18.5*)

Standard molar volume the volume of 1 mol of a gas at 0°C and 1 atm pressure; 22.414 L (*Section 10.2*)

Standard reduction potential the standard potential for a reduction half-cell (*Section 19.5*)

Standard temperature and pressure (STP) $T = 273.15$ K; $P = 1$ atm (*Section 10.3*)

State function a function or property whose value depends only on the present condition of the system, not on the path used to arrive at that condition (*Section 9.2*)

Steam–hydrocarbon re-forming process an important industrial method for producing hydrogen from methane (*Chapter 22 Inquiry*)

Stereoisomers isomers that have the same connections among atoms but have a different arrangement of the atoms in space (*Section 21.7*)

Steric factor the fraction of collisions with the proper orientation for converting reactants to products (*Section 14.7*)

Stoichiometry mole/mass relationships between reactants and products (*Section 3.3*)

Straight-chain alkane an alkane that has all its carbons connected in a row (*Section 23.1*)

Strong acid an acid that dissociates completely in water to give H^+ ions and is a strong electrolyte (*Section 16.2*)

Strong base a base that dissociates or reacts completely with water to give OH^- ions and is a strong electrolyte (*Section 4.7*)

Strong electrolyte a compound that dissociates completely into ions when dissolved in water (*Section 4.3*)

Strong-field ligand a ligand that has a large crystal field splitting (*Section 21.10*)

Structural formula a representation that shows the specific connections between atoms in a molecule (*Section 2.10*)

Sublimation the direct conversion of a solid to a vapor without going through a liquid state (*Section 9.6*)

Subshell a grouping of orbitals by angular-momentum quantum number (*Section 5.6*)

Subsidiary reaction any proton-transfer process other than the principal one in an acid–base reaction (*Section 16.9*)

Substrate the compound acted on by an enzyme (*Chapter 14 Inquiry*)

Superconducting transition temperature (T_c) the temperature below which a superconductor loses all electrical resistance (*Section 12.9*)

Superconductor a material that loses all electrical resistance below a certain temperature (*Section 12.9*)

Supercritical fluid a state of matter beyond the critical point that is neither liquid nor gas (*Section 11.5*)

Supersaturated a solution which contains a greater-than-equilibrium amount of solute (*Section 13.5*)

Surface tension the resistance of a liquid to spreading out and increasing its surface area (*Section 11.1*)

Surroundings everything found around the system (*Section 9.2*)

System substances in an experiment, the starting reactants and the final products (*Section 9.2*)

Symmetry plane a plane that cuts through an object so that one half of the object is a mirror image of the other half (*Section 21.8*)

Temperature a measure of the kinetic energy of molecular motion (*Section 9.1*)

Termolecular reaction an elementary reaction that results from collisions between three reactant molecules (*Section 14.9*)

Theory a consistent explanation of known observations (*Section 1.1*)

Thermochemical equation gives a balanced chemical equation along with the value of the enthalpy change, the amount of heat released or absorbed when reactants are converted to products (*Section 9.5*)

Thermochemistry a study of the heat changes that take place during reactions (*Chapter 9*)

Thermodynamic standard state conditions under which thermodynamic measurements are reported; 298.15 K (25°C), 1 atm pressure for each gas, 1 M concentration for solutions (*Section 9.5*)

Thermodynamics   the study of the interconversion of heat and other forms of energy (*Chapter 18*)

Third law of thermodynamics   the entropy of a perfectly ordered crystalline substance at 0 K is zero (*Section 18.4*)

Three-center, two-electron bond   a covalent bond in which three atoms share two electrons (*Section 22.5*)

Titration   a procedure for determining the concentration of a solution (*Section 4.9 and Section 4.13*)

Trans isomer   the isomer of a metal complex or alkene in which identical ligands or groups are opposite one another rather than adjacent (*Section 21.7 and Section 23.4*)

Transistor   a semiconductor device that controls and amplifies electrical signals (*Section 12.8*)

Transitional metal group   the 10 smaller groups in the middle of the table (*Section 2.2*)

Transition state   the configuration of atoms at the maximum in the potential energy profile for a reaction (*Section 14.7*)

Transuranium elements   the 26 artificially produced elements beyond uranium in the periodic table (*Section 20.3*)

Triacylglycerol   a triester of glycerol (1,2,3-propanetriol) with three long-chain carboxylic acids (*Section 23.6*)

Triple bond   a covalent bond formed by sharing six electrons between atoms (*Section 7.5*)

Triple point   a unique combination of pressure and temperature at which gas, liquid, and solid phases coexist in equilibrium (*Section 11.5*)

Triprotic acid   an acid that can provide three H^+ ions (*Section 4.7*)

Unified atomic mass unit (u)   the mass in grams of a single atom is much too small a number for convenience so chemists use a unit called an atomic mass unit (amu) (*Section 2.9 and Section 20.6*)

Unimolecular reaction   an elementary reaction that involves a single reactant molecule (*Section 14.9*)

Unit cell   a small repeating unit that makes up a crystal (*Section 12.3*)

Unsaturated   an organic molecule that contains a double or triple bond (*Section 23.3*)

Valence band   the bonding molecular orbitals in a semiconductor (*Section 12.7*)

Valence bond theory   a quantum mechanical description of bonding that pictures covalent bond formation as the overlap of two singly occupied atomic orbitals (*Section 8.2 and Section 21.11*)

Valence shell   the outermost electron shell (*Section 5.12*)

Valence-shell electron-pair repulsion (VSEPR) model   a model for predicting the approximate shape of a molecule (*Section 8.1 and Section 23.1*)

van der Waals equation   a modification of the ideal gas law that introduces correction factors to account for the behavior of real gases (*Section 10.8*)

van der Waals forces   an alternative name for intermolecular forces (*Section 8.6*)

van't Hoff factor (*i*)   a measure of the extent of dissociation of a substance, used in interpreting colligative property measurements (*Section 13.7*)

Vapor pressure (P_{vap})   the partial pressure of a gas in equilibrium with liquid (*Section 11.2*)

Viscosity   the measure of a liquid's resistance to flow (*Section 11.1*)

Voltaic cell   *see* Galvanic cell (*Section 19.2*)

Water-gas shift reaction   a method for the industrial preparation of H_2 by reaction of CO with H_2O (*Chapter 22 Inquiry*)

Watson–Crick model   a model of DNA structure, consisting of two polynucleotide strands coiled around each other in a double helix (*Section 23.11*)

Wave function   a solution to the Schrödinger wave equation (*Section 5.6*)

Wavelength (λ)   the length of a wave from one maximum to the next (*Section 5.1*)

Weak acid   an acid that dissociates incompletely in water and is a weak electrolyte (*Section 16.2*)

Weak base   a base that dissociates or reacts incompletely with water and is a weak electrolyte (*Section 4.7*)

Weak electrolyte   a compound that dissociates incompletely when dissolved in water (*Section 4.3*)

Weak-field ligand   a ligand that has a small crystal field splitting (*Section 21.10*)

Work (*w*)   the distance (*d*) moved times the force (*F*) that opposes the motion (*Section 9.3*)

Yield   the amount of product formed in a reaction (*Section 3.4*)

Zeroth-order reaction   one whose rate remains constant, independent of reactant concentrations (*Section 14.4*)

Zone refining   a purification technique in which a heater melts a narrow zone at the top of a rod of some material and then sweeps slowly down the rod bringing impurities with it (*Section 22.6*)

Index

Note: The page numbers set in bold show where the term or phrase is defined in the text.

Credits

Photo Credits

Chapter 1: **1**, Andrew Brookes, National Physical Laboratory/Science Source; **2**, U.S. Department of Energy; **3**, Skrabalak Research Group At Indiana University; **5**, Skrabalak Research Group At Indiana University; **8**, Mattesimages/Fotolia; **8**, Mettler-Toledo International Inc.; **8**, OHAUS Adventurer Pro www.ohaus.com; **9**, Tony Brian/Science Source; **11**, Richard Megna/Fundamental Photographs, NYC; **12**, Richard Megna/Fundamental Photographs, NYC; **13**, Christian Wilkinson/Shutterstock; **13**, Richard Megna/Fundamental Photographs, NYC; **14**, Chris Leachman/Fotolia; **15**, choneschones/123RF; **15**, ThomasLENNE/Shutterstock; **16**, Jill K. Robinson; **18**, Image used with permission by Texas Instruments INC.; **18**, Richard Megna/Fundamental Photographs, NYC; **20**, David W. Cerny/Reuters; **21**, Susumu Nishinaga/Science Source; **22**, Byelikova Oksana/Shutterstock; **22**, John R. Foster/Science Source; **23**, Anthony Bradshaw/Photographer's Choice RF/Getty Images; **23**, Kandela, Irawati; **24**, Daniel Mortell/Fotolia

Chapter 2: **33**, KRT/Newscom; **34**, Brandon E. Hirsch and Steve L. Tait, Indiana University; **34**, DariaRen/Shutterstock; **34**, John E. McMurry; **34**, MarcelClemens/Shutterstock; **34**, Margaret M Stewart/Shutterstock; **34**, Ventin/Shutterstock; **36**, sciencephotos/Alamy Stock Photo; **37**, Richard Megna/Fundamental Photographs, NYC; **38**, Richard Megna/Fundamental Photographs, NYC; **39**, Richard Megna/Fundamental Photographs, NYC; **40**, Elnavegante/Fotolia; **40**, Richard Megna/Fundamental Photographs, NYC; **40**, sciencephotos/Alamy Stock Photo; **42**, Richard Megna/Fundamental Photographs, NYC; **43**, Richard Megna/Fundamental Photographs, NYC; **44**, Richard Megna/Fundamental Photographs, NYC; **45**, Richard Megna/Fundamental Photographs, NYC; **46**, Richard Megna/Fundamental Photographs, NYC; **48**, Kiyoshi Ota/Reuters; **51**, Colin Monteath/AGE Fotostock; **53**, Richard Megna/Fundamental Photographs, NYC; **54**, NASA; **54**, Dmitry Nikolaev/Fotolia; **54**, gwimages/Fotolia; **54**, Nikolai Sorokin/Fotolia; **54**, Sebastien Burel/Fotolia; **57**, Tom Till/Alamy Stock Photo; **58**, Zurijeta/Shutterstock; **61**, Richard Megna/Fundamental Photographs, NYC; **62**, Peter Hermes Furian/Shutterstock; **63**, Menzl Guenter/Shutterstock; **64**, Richard Megna/Fundamental Photographs, NYC

Chapter 3: **83**, Sherry Yates Young/Shutterstock; **87**, LoloStock/Fotolia; **88**, Charles D. Winters/Science Source; **88**, Jeff Morgan 13/Alamy Stock Photo; **90**, Ted Foxx/Alamy Stock Photo; **92**, moodboard/Alamy Stock Photo

Chapter 4: **116**, wundervisuals/E+/Getty Images; **118**, Richard Megna/Fundamental Photographs, NYC; **119**, Paul Silverman/Fundamental Photographs, NYC; **120**, Richard Megna/Fundamental Photographs, NYC; **121**, Richard Megna/Fundamental Photographs, NYC; **123**, sciencephotos/Alamy Stock Photo; **125**, Richard Megna/Fundamental Photographs, NYC; **126**, Turtle Rock Scientific/Science Source; **129**, ggw/Shutterstock; **132**, Amy Etra/Photoedit; **133**, Paul Silverman/Fundamental Photographs, NYC; **134**, Richard Megna/Fundamental Photographs, NYC; **135**, Chip Clark/Fundamental Photographs, NYC; **135**, Richard Megna/Fundamental Photographs, NYC; **139**, bszef/Alamy Stock Photo; **141**, GIPhotoStock/Science Source; **141**, Michael Donne/Science Source; **142**, Trevor Clifford/Pearson Education Ltd.; **143**, Peticolas/Megna/Fundamental Photographs, NYC; **145**, Richard Megna/Fundamental Photographs, NYC; **146**, Paul Silverman/Fundamental Photographs, NYC; **147**, Pavel Ryabushkin/Shutterstock; **148**, Jill K. Robinson

Chapter 5: **164**, Turtle Rock Scientific/Science Source; **165**, Philip Long/The Image Bank/Getty Images; **167**, Natalia Bratslavsky/Shutterstock; **169**, Karin Hildebrand Lau/Shutterstock; **169**, Ted Kinsman/Science Source; **170**, Archive Image/Alamy Stock Photo; **170**, David Taylor/Science Source; **170**, Richard Megna/Fundamental Photographs, NYC; **170**, Richard Treptow/Science Source; **171**, Jay Bray/Alamy Stock Photo; **171**, Nestor Noci/Shutterstock; **172**, GIPhotoStock/Science Source; **177**, Dr. David Morgan; **180**, Akira Tonomura, "Direct observation of thitherto unobservable quantum phenomena by using electrons," Proceedings of the National Academy of Sciences (Oct 2005), 102 (42) 14952-14959. Copyright (2005) National Academy of Sciences, U.S.A.; **180**, Colorful High Speed Photographs/Moment Select/Getty Images; **185**, Richard Megna/Fundamental Photographs, NYC; **195**, Cultura Creative (RF)/Alamy Stock Photo

Chapter 6: **208**, BASF SE; **226**, Richard Megna/Fundamental Photographs, NYC; **232**, press photo BASF, ©BASF

Chapter 7: **242**, Carolina K. Smith MD; **246**, GoodWin777/Shutterstock; **246**, jamakosy/Shutterstock; **247**, Olga Galushko/Fotolia; **247**, Zastolskiy Victor/Shutterstock; **254**, Allik Camazine/Alamy Stock Photo; **254**, Chronicle/Alamy Stock Photo; **269**, Exotica.im 2/Alamy Stock Photo

Chapter 8: **282**, dream79/Fotolia; **303**, David Taylor/Science Source; **307**, Bronwyn Photo/Shutterstock; **313**, Richard Megna/Fundamental Photographs, NYC; **318**, kitharaa/Fotolia

Chapter 9: **327**, Javier Larrea/age fotostock/Getty Images; **331**, Lucidio Studio, Inc./Moment/Getty Images; **332**, lzf/Shutterstock; **335**, PhotoSky/Shutterstock; **338**, Charles D. Winters/Science Source; **339**, Richard Megna/Fundamental Photographs, NYC; **340**, Eiichi Onodera/Dex Image/Alamy Stock Photo; **340**, science photo/Shutterstock; **340**, Alias Studiot Oy/Shutterstock; **340**, Matt Meadows/Photolibrary/Getty Images; **342**, John E. McMurry; **344**, Eiichi Onodera/Dex Image/Alamy Stock Photo; **344**, Richard Megna/Fundamental Photographs, NYC; **352**, Avalon/World Pictures/Alamy Stock Photo; **353**, Alias Studiot Oy/Shutterstock; **353**, science photo/Shutterstock; **353**, tvwilks/Stockimo/Alamy Stock Photo; **356**, vlad_g/Fotolia; **359**, Goran Jakus/Shutterstock; **359**, Kyodo/Newscom; **360**, Pavel Cheiko/Fotolia

Chapter 10: **374**, herjua/Shutterstock; **376**, NASA; **377**, Andrew Lambert Photography/Science Source; **379**, Dmitry Pichugin/Shutterstock; **386**, Paulo Resende/Shutterstock; **387**, Caspar Benson/fStop Images GmbH/Alamy Stock Photo; **388**, GIPhotoStock/Science Source; **394**, nikkytok/Fotolia; **396**, Rich Carey/Shutterstock;

889, Philip Lange/Shutterstock; 891, TASS/ITAR-TASS News Agency/Alamy Stock Photo; 892, Ablestock.com/Getty Images; 894, Mark Kostich/Shutterstock; 895, Phillip Beron/Alamy Stock Photo; 895, CAVALLINI JAMES/BSIP SA/Alamy Stock Photo; 901, Walter Sanders/The LIFE Picture Collection/Getty Images

Chapter 21: 904, Dr P. Marazzi/Science Source; 908, AlessandroZocc/Shutterstock; 908, Alex Wong/Getty Images News/Getty Images; 912, Richard Megna/Fundamental Photographs, NYC; 913, Martyn F. Chillmaid/Science Source; 919, Richard Megna/Fundamental Photographs, NYC; 921, Richard Megna/Fundamental Photographs, NYC; 929, Richard Megna/Fundamental Photographs, NYC; 930, Richard Megna/Fundamental Photographs, NYC; 933, Richard Megna/Fundamental Photographs, NYC; 941, Ramón Andrade, 3Dciencia/Science Source; 950, Richard Megna/Fundamental Photographs, NYC; 950, Richard Megna/Fundamental Photographs, NYC

Chapter 22: 954, Petr Malinak/123RF; 959, NOAO/AURA/NSF; 961, Richard Megna/Fundamental Photographs, NYC; 963, Richard Megna/Fundamental Photographs, NYC; 964, Richard Megna/Fundamental Photographs, NYC; 965, Bjoern Wylezich/123RF; 965, Richard Megna/Fundamental Photographs, NYC; 966, U.S. Borax, Inc.; 968, Charles D. Winters/Science Source; 968, Richard Megna/Fundamental Photographs, NYC; 968, Russell Lappa/Science Source; 969, Martyn F. Chillmaid/Science Source; 970, Boyd Norton/Alamy Stock Photo; 971, Courtesy Texas Instruments; 971, Jewellery specialist/Alamy Stock Photo; 972, Charles D. Winters/Science Source; 972, Sergey Lavrentev/Shutterstock; 972, The Natural History Museum/Alamy Stock Photo; 973, gontar/Shutterstock; 974, Paul Silverman/Fundamental Photographs, NYC; 974, Russell Lappa/Science Source; 974, Vladimir Nenezic/123RF; 976, Chip Clark/Fundamental Photographs, NYC; 976, Richard Megna/Fundamental Photographs, NYC; 977, Kristen Brochmann/Fundamental Photographs, NYC; 978, Charles D. Winters/Science Source; 978, Richard Megna/Fundamental Photographs, NYC; 979, Richard Megna/Fundamental Photographs, NYC; 980, Charles D. Winters/Science Source; 980, Russ Lappa/Science Source; 981, Oleg - F/Shutterstock; 981, Rawf8/Alamy Stock Photo; 981, Richard Megna/Fundamental Photographs, NYC; 985, Mark A Schneider/Dembinsky Photo Associates/Alamy Stock Photo; 985, Richard Megna/Fundamental Photographs, NYC; 986, Graham Bradley; 988, forty40 photography/Alamy Stock Photo; 990, Charles D. Winters/Science Source

Chapter 23: 1003, Image by Katya Kadyshevskaya of GPCR Network at TSRI; 1008, Jaimie Duplass/iStock/Getty Images Plus; 1009, Dan Thornberg/Shutterstock; 1009, Sunisa Chukly/123RF; 1024, Tetra Images/Alamy Stock Photo; 1034, witthaya/Fotolia; 1038, Samo Trebizan/Shutterstock

Text Credits

Chapter 2: 37, Based from Holli Riebeek, "Paleoclimatology: the Oxygen Balance," NASA (May 6, 2005); 37, Data from NOAA National Centers for Environmental information, Climate at a Glance: Global Time Series; 38, Data from Jouzel, J., et al. 2007. EPICA Dome C Ice Core 800KYr Deuterium Data and Temperature Estimates. IGBP PAGES/World Data Center for Paleoclimatology Data Contribution Series # 2007-091. NOAA/NCDC Paleoclimatology Program, Boulder CO, USA; 47, Source: Nobel Laureates in Chemistry, 1901–1992, American Chemical Society, Washington D.C. edited by Laylin K. James (1993), p. 57

Chapter 3: 84, Dr. N. K. Verma, S. K. Khanna, Dr. B. Kapila, "Comprehensive Chemistry XI," Laxmi Publications, 2010–1124 pages; 106, Data from NIST Chemistry Webbook (http://webbook.nist.gov/chemistry); 106, Data from rom NIST Chemistry Webbook (http://webbook.nist.gov/chemistry); 107, Data from rom NIST Chemistry Webbook (http://webbook.nist.gov/chemistry); 112, Data from NIST Chemistry Webbook for Decane (C10H22) http://webbook.nist.gov/cgi/cbook.cgi?ID=C124185&Units=SI&Mask=200#Mass-Spec; 117, Data from NIST Chemistry Webbook (http://webbook.nist.gov/chemistry)

Chapter 4: 148, April Frawley Lacey, "UF celebrates 50 years of Gatorade," University of Florida (September 28, 2015)

Chapter 10: 406, U.S. Secretary of Commerce; 407, Intergovernmental Panel on Climate Change—Climate Change 2013: The Physical Science Basis; 408, National Aeronautics and Space Administration Goddard Institute for Space Studies; 408, U.S. Environmental Protection Agency, Climate Change, Greenhouse Gases; 409, National Research Council. Advancing the Science of Climate Change. The National Academies Press, 2010; 410, National Aeronautics and Space Administration, GISS Surface Temperature Analysis

Chapter 16: 695, National Atmospheric Deposition Program (NRSP-3). 2018. NADP Program Office, Wisconsin State Laboratory of Hygiene, 465 Henry Mall, Madison, WI 53706

Chapter 17: 754, NOAA Pacific Marine Environmental Laboratory

Useful Conversion Factors and Relationships

Length

SI unit: meter (m)

$1 \text{ km} = 10^3 \text{ m} = 0.621\ 37 \text{ mi}$

$1 \text{ mi} = 5280 \text{ ft} = 1760 \text{ yd} = 1.6093 \text{ km}$

$1 \text{ m} = 10^2 \text{ cm} = 1.0936 \text{ yd}$

$1 \text{ in.} = 2.54 \text{ cm (exactly)}$

$1 \text{ cm} = 0.393\ 70 \text{ in.}$

$1 \text{ Å} = 10^{-10} \text{ m} = 100 \text{ pm}$

Mass

SI unit: kilogram (kg)

$1 \text{ kg} = 10^3 \text{ g} = 2.2046 \text{ lb}$

$1 \text{ lb} = 16 \text{ oz} = 453.59 \text{ g}$

$1 \text{ oz} = 28.35 \text{ g}$

$1 \text{ ton} = 2000 \text{ lb} = 907.185 \text{ kg}$

$1 \text{ metric ton} = 10^3 \text{ kg} = 1.102 \text{ tons}$

$1 \text{ u} = 1.660\ 54 \times 10^{-27} \text{ kg}$

Temperature

SI unit: kelvin (K)

$0 \text{ K} = -273.15 \text{ °C} = -459.67 \text{ °F}$

$\text{K} = \text{°C} + 273.15$

$\text{°C} = \frac{5}{9} (\text{°F} - 32)$

$\text{°F} = \frac{9}{5} (\text{°C}) + 32$

Energy (derived)

SI unit: joule (J)

$1 \text{ J} = 1 \text{ (kg} \cdot \text{m}^2)/\text{s}^2 = 0.239\ 01 \text{ cal}$

$\quad = 1 \text{ C} \times 1 \text{ V}$

$1 \text{ cal} = 4.184 \text{ J (exactly)}$

$1 \text{ eV} = 1.602\ 176 \times 10^{-19} \text{ J}$

$1 \text{ MeV} = 1.602\ 176 \times 10^{-13} \text{ J}$

$1 \text{ kWh} = 3.600 \times 10^6 \text{ J}$

$1 \text{ Btu} = 1055 \text{ J}$

Pressure (derived)

SI unit: pascal (Pa)

$1 \text{ Pa} = 1 \text{ N/m}^2 = 1 \text{ kg}/(\text{m} \cdot \text{s}^2)$

$1 \text{ atm} = 101{,}325 \text{ Pa} = 1.013\ 25 \text{ bar}$

$\quad = 760 \text{ mm Hg (torr)}$

$\quad = 14.70 \text{ lb/in.}^2$

$1 \text{ bar} = 10^5 \text{ Pa}$

Volume (derived)

SI unit: cubic meter (m^3)

$1 \text{ L} = 10^{-3} \text{ m}^3 = 1 \text{ dm}^3 = 10^3 \text{ cm}^3$

$\quad = 1.0567 \text{ qt}$

$1 \text{ gal} = 4 \text{ qt} = 3.7854 \text{ L}$

$1 \text{ cm}^3 = 1 \text{ mL}$

$1 \text{ in.}^3 = 16.4 \text{ cm}^3$

Fundamental Constants

Atomic mass unit	1 u	$= 1.660\ 539 \times 10^{-27}$ kg
	1 g	$= 6.022\ 141 \times 10^{23}$ u
Avogadro's number	N_A	$= 6.022\ 141 \times 10^{23}$/mol
Boltzmann's constant	k	$= 1.380\ 649 \times 10^{-23}$ J/K
Electron charge	$-e$	$= -1.602\ 177 \times 10^{-19}$ C
Electron charge-to-mass ratio	$-e/m_e$	$= -1.758\ 820 \times 10^{11}$ C/kg
Electron mass	m_e	$= 5.485\ 798 \times 10^{-4}$ u
		$= 9.109\ 383 \times 10^{-31}$ kg
Elementary charge	e	$= 1.602\ 177 \times 10^{-19}$ C
Faraday's constant	F	$= 9.648\ 534 \times 10^4$ C/mol
Gas constant	R	$= 8.314\ 462$ J/(mol $\cdot$ K)
		$= 0.082\ 058\ 1$ (L $\cdot$ atm)/(mol $\cdot$ K)
Neutron mass	m_n	$= 1.008\ 665$ u
		$= 1.674\ 927 \times 10^{-27}$ kg
Pi	π	$= 3.141\ 592\ 6536$
Planck's constant	h	$= 6.626\ 069\ 57 \times 10^{-34}$ J $\cdot$ s
Proton mass	m_p	$= 1.007\ 276$ u
		$= 1.672\ 622 \times 10^{-27}$ kg
Rydberg constant	R_∞	$= 1.097\ 373 \times 10^7$/m
Speed of light	c	$= 2.997\ 924\ 58 \times 10^8$ m/s

Index of Important Information